Complete Solutions Manual

for

SINGLE VARIABLE CALCULUS EARLY TRANSCENDENTALS

EIGHTH EDITION

DANIEL ANDERSON
University of Iowa

JEFFERY A. COLE
Anoka-Ramsey Community College

DANIEL DRUCKER
Wayne State University

Australia · Brazil · Mexico · Singapore · United Kingdom · United States

ISBN: 978-1-305-27239-2

Cengage Learning
20 Channel Center Street
Boston, MA 02210
USA

Cengage Learning is a leading provider of customized learning solutions with employees residing in nearly 40 different countries and sales in more than 125 countries around the world. Find your local representative at: **www.cengage.com**.

Cengage Learning products are represented in Canada by Nelson Education, Ltd.

To learn more about Cengage Learning Solutions, visit **www.cengage.com**.

Purchase any of our products at your local college store or at our preferred online store **www.cengagebrain.com**.

Printed in the United States of America
Print Number: 01 Print Year: 2015

PREFACE

This *Complete Solutions Manual* contains solutions to all exercises in the text *Single Variable Calculus, Early Transcendentals,* Eighth Edition, by James Stewart. A student version of this manual is also available; it contains solutions to the odd-numbered exercises in each section, the review sections, the True-False Quizzes, and the Problem Solving sections. No solutions to the projects appear in the student version. It is our hope that by browsing through the solutions, professors will save time in determining appropriate assignments for their particular class.

We use some nonstandard notation in order to save space. If you see a symbol that you don't recognize, refer to the Table of Abbreviations and Symbols on page iv.

We appreciate feedback concerning errors, solution correctness or style, and manual style. Any comments may be sent directly to jeff-cole@comcast.net, or in care of the publisher: Cengage Learning, 20 Channel Center Street, Boston MA 02210.

We would like to thank Kira Abdallah, Kristina Elliott, Stephanie Kuhns, and Kathi Townes, of TECHarts, for their production services; and Samantha Lugtu, of Cengage Learning, for her patience and support. All of these people have provided invaluable help in creating this manual.

Jeffery A. Cole
Anoka-Ramsey Community College

James Stewart
McMaster University
and University of Toronto

Daniel Drucker
Wayne State University

Daniel Anderson
University of Iowa

ABBREVIATIONS AND SYMBOLS

CD	concave downward
CU	concave upward
D	the domain of f
FDT	First Derivative Test
HA	horizontal asymptote(s)
I	interval of convergence
IP	inflection point(s)
R	radius of convergence
VA	vertical asymptote(s)
$\overset{\text{CAS}}{=}$	indicates the use of a computer algebra system.
$\overset{\text{PR}}{\Rightarrow}$	indicates the use of the Product Rule.
$\overset{\text{QR}}{\Rightarrow}$	indicates the use of the Quotient Rule.
$\overset{\text{CR}}{\Rightarrow}$	indicates the use of the Chain Rule.
$\overset{\text{H}}{=}$	indicates the use of l'Hospital's Rule.
$\overset{j}{=}$	indicates the use of Formula j in the Table of Integrals in the back endpapers.
$\overset{\text{s}}{=}$	indicates the use of the substitution $\{u = \sin x,\ du = \cos x\, dx\}$.
$\overset{\text{c}}{=}$	indicates the use of the substitution $\{u = \cos x,\ du = -\sin x\, dx\}$.

CONTENTS

11 □ INFINITE SEQUENCES AND SERIES 957

□ APPENDIXES 1097

☐ DIAGNOSTIC TESTS

Test A Algebra

1. (a) $(-3)^4 = (-3)(-3)(-3)(-3) = 81$

(b) $-3^4 = -(3)(3)(3)(3) = -81$

(c) $3^{-4} = \dfrac{1}{3^4} = \dfrac{1}{81}$

(d) $\dfrac{5^{23}}{5^{21}} = 5^{23-21} = 5^2 = 25$

(e) $\left(\frac{2}{3}\right)^{-2} = \left(\frac{3}{2}\right)^2 = \frac{9}{4}$

(f) $16^{-3/4} = \dfrac{1}{16^{3/4}} = \dfrac{1}{\left(\sqrt[4]{16}\right)^3} = \dfrac{1}{2^3} = \dfrac{1}{8}$

2. (a) Note that $\sqrt{200} = \sqrt{100 \cdot 2} = 10\sqrt{2}$ and $\sqrt{32} = \sqrt{16 \cdot 2} = 4\sqrt{2}$. Thus $\sqrt{200} - \sqrt{32} = 10\sqrt{2} - 4\sqrt{2} = 6\sqrt{2}$.

(b) $(3a^3b^3)(4ab^2)^2 = 3a^3b^3 16a^2b^4 = 48a^5b^7$

(c) $\left(\dfrac{3x^{3/2}y^3}{x^2y^{-1/2}}\right)^{-2} = \left(\dfrac{x^2y^{-1/2}}{3x^{3/2}y^3}\right)^2 = \dfrac{(x^2y^{-1/2})^2}{(3x^{3/2}y^3)^2} = \dfrac{x^4y^{-1}}{9x^3y^6} = \dfrac{x^4}{9x^3y^6y} = \dfrac{x}{9y^7}$

3. (a) $3(x+6) + 4(2x-5) = 3x + 18 + 8x - 20 = 11x - 2$

(b) $(x+3)(4x-5) = 4x^2 - 5x + 12x - 15 = 4x^2 + 7x - 15$

(c) $\left(\sqrt{a}+\sqrt{b}\right)\left(\sqrt{a}-\sqrt{b}\right) = \left(\sqrt{a}\right)^2 - \sqrt{a}\sqrt{b} + \sqrt{a}\sqrt{b} - \left(\sqrt{b}\right)^2 = a - b$

Or: Use the formula for the difference of two squares to see that $\left(\sqrt{a}+\sqrt{b}\right)\left(\sqrt{a}-\sqrt{b}\right) = \left(\sqrt{a}\right)^2 - \left(\sqrt{b}\right)^2 = a - b$.

(d) $(2x+3)^2 = (2x+3)(2x+3) = 4x^2 + 6x + 6x + 9 = 4x^2 + 12x + 9$.

Note: A quicker way to expand this binomial is to use the formula $(a+b)^2 = a^2 + 2ab + b^2$ with $a = 2x$ and $b = 3$:

$(2x+3)^2 = (2x)^2 + 2(2x)(3) + 3^2 = 4x^2 + 12x + 9$

(e) See Reference Page 1 for the binomial formula $(a+b)^3 = a^3 + 3a^2b + 3ab^2 + b^3$. Using it, we get

$(x+2)^3 = x^3 + 3x^2(2) + 3x(2^2) + 2^3 = x^3 + 6x^2 + 12x + 8$.

4. (a) Using the difference of two squares formula, $a^2 - b^2 = (a+b)(a-b)$, we have

$4x^2 - 25 = (2x)^2 - 5^2 = (2x+5)(2x-5)$.

(b) Factoring by trial and error, we get $2x^2 + 5x - 12 = (2x-3)(x+4)$.

(c) Using factoring by grouping and the difference of two squares formula, we have

$x^3 - 3x^2 - 4x + 12 = x^2(x-3) - 4(x-3) = (x^2-4)(x-3) = (x-2)(x+2)(x-3)$.

(d) $x^4 + 27x = x(x^3 + 27) = x(x+3)(x^2 - 3x + 9)$

This last expression was obtained using the sum of two cubes formula, $a^3 + b^3 = (a+b)(a^2 - ab + b^2)$ with $a = x$ and $b = 3$. [See Reference Page 1 in the textbook.]

(e) The smallest exponent on x is $-\frac{1}{2}$, so we will factor out $x^{-1/2}$.

$3x^{3/2} - 9x^{1/2} + 6x^{-1/2} = 3x^{-1/2}(x^2 - 3x + 2) = 3x^{-1/2}(x-1)(x-2)$

(f) $x^3y - 4xy = xy(x^2 - 4) = xy(x-2)(x+2)$

5. (a) $\dfrac{x^2+3x+2}{x^2-x-2}=\dfrac{(x+1)(x+2)}{(x+1)(x-2)}=\dfrac{x+2}{x-2}$

(b) $\dfrac{2x^2-x-1}{x^2-9}\cdot\dfrac{x+3}{2x+1}=\dfrac{(2x+1)(x-1)}{(x-3)(x+3)}\cdot\dfrac{x+3}{2x+1}=\dfrac{x-1}{x-3}$

(c) $\dfrac{x^2}{x^2-4}-\dfrac{x+1}{x+2}=\dfrac{x^2}{(x-2)(x+2)}-\dfrac{x+1}{x+2}=\dfrac{x^2}{(x-2)(x+2)}-\dfrac{x+1}{x+2}\cdot\dfrac{x-2}{x-2}=\dfrac{x^2-(x+1)(x-2)}{(x-2)(x+2)}$

$$=\frac{x^2-(x^2-x-2)}{(x+2)(x-2)}=\frac{x+2}{(x+2)(x-2)}=\frac{1}{x-2}$$

(d) $\dfrac{\dfrac{y}{x}-\dfrac{x}{y}}{\dfrac{1}{y}-\dfrac{1}{x}}=\dfrac{\dfrac{y}{x}-\dfrac{x}{y}}{\dfrac{1}{y}-\dfrac{1}{x}}\cdot\dfrac{xy}{xy}=\dfrac{y^2-x^2}{x-y}=\dfrac{(y-x)(y+x)}{-(y-x)}=\dfrac{y+x}{-1}=-(x+y)$

6. (a) $\dfrac{\sqrt{10}}{\sqrt{5}-2}=\dfrac{\sqrt{10}}{\sqrt{5}-2}\cdot\dfrac{\sqrt{5}+2}{\sqrt{5}+2}=\dfrac{\sqrt{50}+2\sqrt{10}}{\left(\sqrt{5}\right)^2-2^2}=\dfrac{5\sqrt{2}+2\sqrt{10}}{5-4}=5\sqrt{2}+2\sqrt{10}$

(b) $\dfrac{\sqrt{4+h}-2}{h}=\dfrac{\sqrt{4+h}-2}{h}\cdot\dfrac{\sqrt{4+h}+2}{\sqrt{4+h}+2}=\dfrac{4+h-4}{h\left(\sqrt{4+h}+2\right)}=\dfrac{h}{h\left(\sqrt{4+h}+2\right)}=\dfrac{1}{\sqrt{4+h}+2}$

7. (a) $x^2+x+1=\left(x^2+x+\frac{1}{4}\right)+1-\frac{1}{4}=\left(x+\frac{1}{2}\right)^2+\frac{3}{4}$

(b) $2x^2-12x+11=2(x^2-6x)+11=2(x^2-6x+9-9)+11=2(x^2-6x+9)-18+11=2(x-3)^2-7$

8. (a) $x+5=14-\frac{1}{2}x \;\Leftrightarrow\; x+\frac{1}{2}x=14-5 \;\Leftrightarrow\; \frac{3}{2}x=9 \;\Leftrightarrow\; x=\frac{2}{3}\cdot 9 \;\Leftrightarrow\; x=6$

(b) $\dfrac{2x}{x+1}=\dfrac{2x-1}{x} \;\Rightarrow\; 2x^2=(2x-1)(x+1) \;\Leftrightarrow\; 2x^2=2x^2+x-1 \;\Leftrightarrow\; x=1$

(c) $x^2-x-12=0 \;\Leftrightarrow\; (x+3)(x-4)=0 \;\Leftrightarrow\; x+3=0$ or $x-4=0 \;\Leftrightarrow\; x=-3$ or $x=4$

(d) By the quadratic formula, $2x^2+4x+1=0 \;\Leftrightarrow$

$$x=\frac{-4\pm\sqrt{4^2-4(2)(1)}}{2(2)}=\frac{-4\pm\sqrt{8}}{4}=\frac{-4\pm 2\sqrt{2}}{4}=\frac{2\left(-2\pm\sqrt{2}\right)}{4}=\frac{-2\pm\sqrt{2}}{2}=-1\pm\tfrac{1}{2}\sqrt{2}.$$

(e) $x^4-3x^2+2=0 \;\Leftrightarrow\; (x^2-1)(x^2-2)=0 \;\Leftrightarrow\; x^2-1=0$ or $x^2-2=0 \;\Leftrightarrow\; x^2=1$ or $x^2=2 \;\Leftrightarrow$
$x=\pm 1$ or $x=\pm\sqrt{2}$

(f) $3\,|x-4|=10 \;\Leftrightarrow\; |x-4|=\frac{10}{3} \;\Leftrightarrow\; x-4=-\frac{10}{3}$ or $x-4=\frac{10}{3} \;\Leftrightarrow\; x=\frac{2}{3}$ or $x=\frac{22}{3}$

(g) Multiplying through $2x(4-x)^{-1/2}-3\sqrt{4-x}=0$ by $(4-x)^{1/2}$ gives $2x-3(4-x)=0 \;\Leftrightarrow$
$2x-12+3x=0 \;\Leftrightarrow\; 5x-12=0 \;\Leftrightarrow\; 5x=12 \;\Leftrightarrow\; x=\frac{12}{5}$.

9. (a) $-4<5-3x\leq 17 \;\Leftrightarrow\; -9<-3x\leq 12 \;\Leftrightarrow\; 3>x\geq -4$ or $-4\leq x<3$.
In interval notation, the answer is $[-4,3)$.

(b) $x^2<2x+8 \;\Leftrightarrow\; x^2-2x-8<0 \;\Leftrightarrow\; (x+2)(x-4)<0$. Now, $(x+2)(x-4)$ will change sign at the critical values $x=-2$ and $x=4$. Thus the possible intervals of solution are $(-\infty,-2)$, $(-2,4)$, and $(4,\infty)$. By choosing a single test value from each interval, we see that $(-2,4)$ is the only interval that satisfies the inequality.

(c) The inequality $x(x-1)(x+2) > 0$ has critical values of $-2, 0$, and 1. The corresponding possible intervals of solution are $(-\infty, -2)$, $(-2, 0)$, $(0, 1)$ and $(1, \infty)$. By choosing a single test value from each interval, we see that both intervals $(-2, 0)$ and $(1, \infty)$ satisfy the inequality. Thus, the solution is the union of these two intervals: $(-2, 0) \cup (1, \infty)$.

(d) $|x-4| < 3 \quad \Leftrightarrow \quad -3 < x - 4 < 3 \quad \Leftrightarrow \quad 1 < x < 7$. In interval notation, the answer is $(1, 7)$.

(e) $\dfrac{2x-3}{x+1} \leq 1 \quad \Leftrightarrow \quad \dfrac{2x-3}{x+1} - 1 \leq 0 \quad \Leftrightarrow \quad \dfrac{2x-3}{x+1} - \dfrac{x+1}{x+1} \leq 0 \quad \Leftrightarrow \quad \dfrac{2x-3-x-1}{x+1} \leq 0 \quad \Leftrightarrow \quad \dfrac{x-4}{x+1} \leq 0.$

Now, the expression $\dfrac{x-4}{x+1}$ may change signs at the critical values $x = -1$ and $x = 4$, so the possible intervals of solution are $(-\infty, -1)$, $(-1, 4]$, and $[4, \infty)$. By choosing a single test value from each interval, we see that $(-1, 4]$ is the only interval that satisfies the inequality.

10. (a) False. In order for the statement to be true, it must hold for all real numbers, so, to show that the statement is false, pick $p = 1$ and $q = 2$ and observe that $(1+2)^2 \neq 1^2 + 2^2$. In general, $(p+q)^2 = p^2 + 2pq + q^2$.

(b) True as long as a and b are nonnegative real numbers. To see this, think in terms of the laws of exponents: $\sqrt{ab} = (ab)^{1/2} = a^{1/2}b^{1/2} = \sqrt{a}\,\sqrt{b}$.

(c) False. To see this, let $p = 1$ and $q = 2$, then $\sqrt{1^2 + 2^2} \neq 1 + 2$.

(d) False. To see this, let $T = 1$ and $C = 2$, then $\dfrac{1 + 1(2)}{2} \neq 1 + 1$.

(e) False. To see this, let $x = 2$ and $y = 3$, then $\dfrac{1}{2-3} \neq \dfrac{1}{2} - \dfrac{1}{3}$.

(f) True since $\dfrac{1/x}{a/x - b/x} \cdot \dfrac{x}{x} = \dfrac{1}{a-b}$, as long as $x \neq 0$ and $a - b \neq 0$.

Test B Analytic Geometry

1. (a) Using the point $(2, -5)$ and $m = -3$ in the point-slope equation of a line, $y - y_1 = m(x - x_1)$, we get $y - (-5) = -3(x-2) \quad \Rightarrow \quad y + 5 = -3x + 6 \quad \Rightarrow \quad y = -3x + 1$.

(b) A line parallel to the x-axis must be horizontal and thus have a slope of 0. Since the line passes through the point $(2, -5)$, the y-coordinate of every point on the line is -5, so the equation is $y = -5$.

(c) A line parallel to the y-axis is vertical with undefined slope. So the x-coordinate of every point on the line is 2 and so the equation is $x = 2$.

(d) Note that $2x - 4y = 3 \quad \Rightarrow \quad -4y = -2x + 3 \quad \Rightarrow \quad y = \frac{1}{2}x - \frac{3}{4}$. Thus the slope of the given line is $m = \frac{1}{2}$. Hence, the slope of the line we're looking for is also $\frac{1}{2}$ (since the line we're looking for is required to be parallel to the given line). So the equation of the line is $y - (-5) = \frac{1}{2}(x-2) \quad \Rightarrow \quad y + 5 = \frac{1}{2}x - 1 \quad \Rightarrow \quad y = \frac{1}{2}x - 6$.

2. First we'll find the distance between the two given points in order to obtain the radius, r, of the circle: $r = \sqrt{[3-(-1)]^2 + (-2-4)^2} = \sqrt{4^2 + (-6)^2} = \sqrt{52}$. Next use the standard equation of a circle, $(x-h)^2 + (y-k)^2 = r^2$, where (h, k) is the center, to get $(x+1)^2 + (y-4)^2 = 52$.

3. We must rewrite the equation in standard form in order to identify the center and radius. Note that $x^2 + y^2 - 6x + 10y + 9 = 0 \;\Rightarrow\; x^2 - 6x + 9 + y^2 + 10y = 0$. For the left-hand side of the latter equation, we factor the first three terms and complete the square on the last two terms as follows: $x^2 - 6x + 9 + y^2 + 10y = 0 \;\Rightarrow$ $(x-3)^2 + y^2 + 10y + 25 = 25 \;\Rightarrow\; (x-3)^2 + (y+5)^2 = 25$. Thus, the center of the circle is $(3, -5)$ and the radius is 5.

4. (a) $A(-7, 4)$ and $B(5, -12) \;\Rightarrow\; m_{AB} = \dfrac{-12-4}{5-(-7)} = \dfrac{-16}{12} = -\dfrac{4}{3}$

(b) $y - 4 = -\frac{4}{3}[x - (-7)] \;\Rightarrow\; y - 4 = -\frac{4}{3}x - \frac{28}{3} \;\Rightarrow\; 3y - 12 = -4x - 28 \;\Rightarrow\; 4x + 3y + 16 = 0$. Putting $y = 0$, we get $4x + 16 = 0$, so the x-intercept is -4, and substituting 0 for x results in a y-intercept of $-\frac{16}{3}$.

(c) The midpoint is obtained by averaging the corresponding coordinates of both points: $\left(\frac{-7+5}{2}, \frac{4+(-12)}{2}\right) = (-1, -4)$.

(d) $d = \sqrt{[5-(-7)]^2 + (-12-4)^2} = \sqrt{12^2 + (-16)^2} = \sqrt{144 + 256} = \sqrt{400} = 20$

(e) The perpendicular bisector is the line that intersects the line segment $\overline{AB}$ at a right angle through its midpoint. Thus the perpendicular bisector passes through $(-1, -4)$ and has slope $\frac{3}{4}$ [the slope is obtained by taking the negative reciprocal of the answer from part (a)]. So the perpendicular bisector is given by $y + 4 = \frac{3}{4}[x - (-1)]$ or $3x - 4y = 13$.

(f) The center of the required circle is the midpoint of $\overline{AB}$, and the radius is half the length of $\overline{AB}$, which is 10. Thus, the equation is $(x+1)^2 + (y+4)^2 = 100$.

5. (a) Graph the corresponding horizontal lines (given by the equations $y = -1$ and $y = 3$) as solid lines. The inequality $y \geq -1$ describes the points (x, y) that lie on or *above* the line $y = -1$. The inequality $y \leq 3$ describes the points (x, y) that lie on or *below* the line $y = 3$. So the pair of inequalities $-1 \leq y \leq 3$ describes the points that lie on or *between* the lines $y = -1$ and $y = 3$.

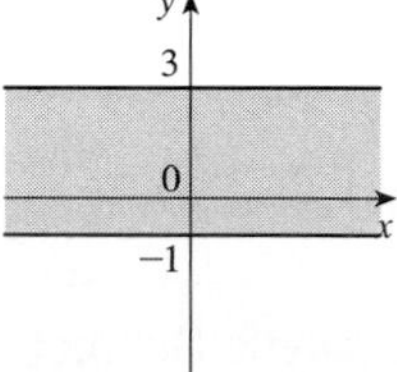

(b) Note that the given inequalities can be written as $-4 < x < 4$ and $-2 < y < 2$, respectively. So the region lies between the vertical lines $x = -4$ and $x = 4$ and between the horizontal lines $y = -2$ and $y = 2$. As shown in the graph, the region common to both graphs is a rectangle (minus its edges) centered at the origin.

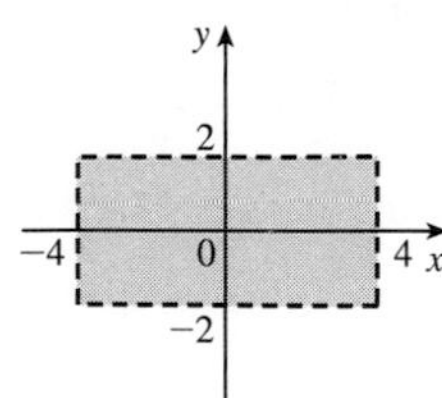

(c) We first graph $y = 1 - \frac{1}{2}x$ as a dotted line. Since $y < 1 - \frac{1}{2}x$, the points in the region lie *below* this line.

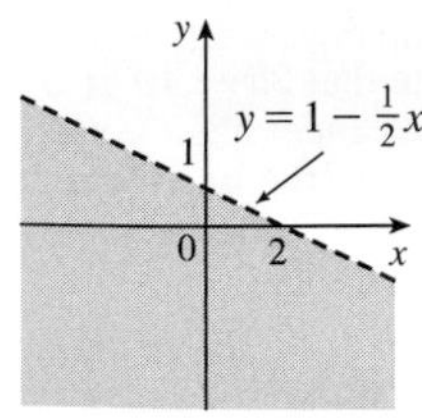

(d) We first graph the parabola $y = x^2 - 1$ using a solid curve. Since $y \geq x^2 - 1$, the points in the region lie on or *above* the parabola.

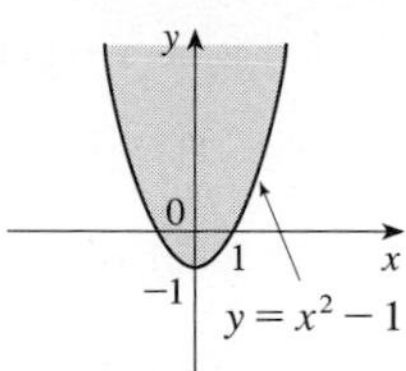

(e) We graph the circle $x^2 + y^2 = 4$ using a dotted curve. Since $\sqrt{x^2 + y^2} < 2$, the region consists of points whose distance from the origin is less than 2, that is, the points that lie *inside* the circle.

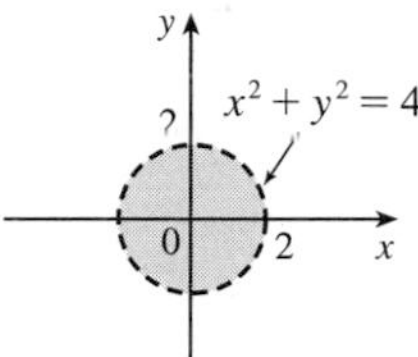

(f) The equation $9x^2 + 16y^2 = 144$ is an ellipse centered at $(0, 0)$. We put it in standard form by dividing by 144 and get $\dfrac{x^2}{16} + \dfrac{y^2}{9} = 1$. The x-intercepts are located at a distance of $\sqrt{16} = 4$ from the center while the y-intercepts are a distance of $\sqrt{9} = 3$ from the center (see the graph).

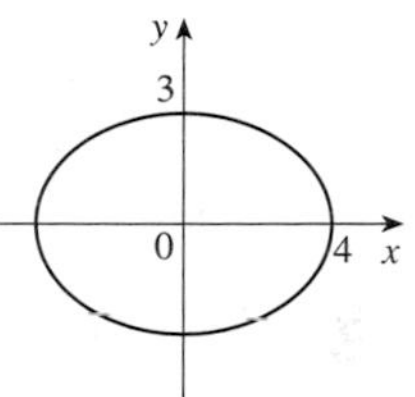

Test C Functions

1. (a) Locate -1 on the x-axis and then go down to the point on the graph with an x-coordinate of -1. The corresponding y-coordinate is the value of the function at $x = -1$, which is -2. So, $f(-1) = -2$.

(b) Using the same technique as in part (a), we get $f(2) \approx 2.8$.

(c) Locate 2 on the y-axis and then go left and right to find all points on the graph with a y-coordinate of 2. The corresponding x-coordinates are the x-values we are searching for. So $x = -3$ and $x = 1$.

(d) Using the same technique as in part (c), we get $x \approx -2.5$ and $x \approx 0.3$.

(e) The domain is all the x-values for which the graph exists, and the range is all the y-values for which the graph exists. Thus, the domain is $[-3, 3]$, and the range is $[-2, 3]$.

2. Note that $f(2 + h) = (2 + h)^3$ and $f(2) = 2^3 = 8$. So the difference quotient becomes

$$\frac{f(2+h) - f(2)}{h} = \frac{(2+h)^3 - 8}{h} = \frac{8 + 12h + 6h^2 + h^3 - 8}{h} = \frac{12h + 6h^2 + h^3}{h} = \frac{h(12 + 6h + h^2)}{h} = 12 + 6h + h^2.$$

3. (a) Set the denominator equal to 0 and solve to find restrictions on the domain: $x^2 + x - 2 = 0 \;\Rightarrow$ $(x - 1)(x + 2) = 0 \;\Rightarrow\; x = 1$ or $x = -2$. Thus, the domain is all real numbers except 1 or -2 or, in interval notation, $(-\infty, -2) \cup (-2, 1) \cup (1, \infty)$.

(b) Note that the denominator is always greater than or equal to 1, and the numerator is defined for all real numbers. Thus, the domain is $(-\infty, \infty)$.

(c) Note that the function h is the sum of two root functions. So h is defined on the intersection of the domains of these two root functions. The domain of a square root function is found by setting its radicand greater than or equal to 0. Now,

$4 - x \geq 0 \;\Rightarrow\; x \leq 4$ and $x^2 - 1 \geq 0 \;\Rightarrow\; (x-1)(x+1) \geq 0 \;\Rightarrow\; x \leq -1$ or $x \geq 1$. Thus, the domain of h is $(-\infty, -1] \cup [1, 4]$.

4. (a) Reflect the graph of f about the x-axis.

(b) Stretch the graph of f vertically by a factor of 2, then shift 1 unit downward.

(c) Shift the graph of f right 3 units, then up 2 units.

5. (a) Make a table and then connect the points with a smooth curve:

x	-2	-1	0	1	2
y	-8	-1	0	1	8

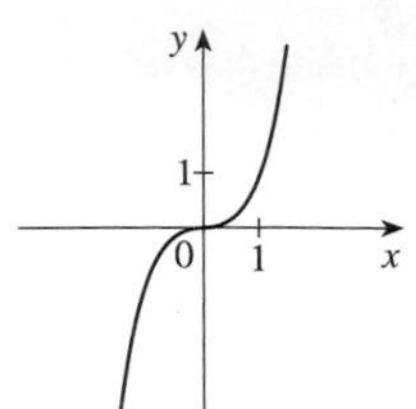

(b) Shift the graph from part (a) left 1 unit.

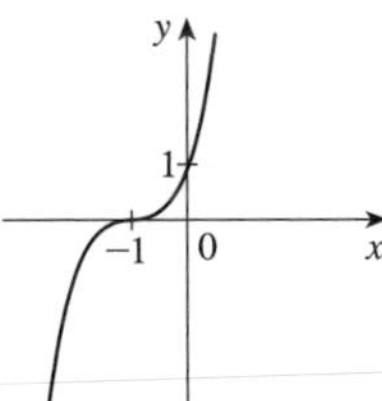

(c) Shift the graph from part (a) right 2 units and up 3 units.

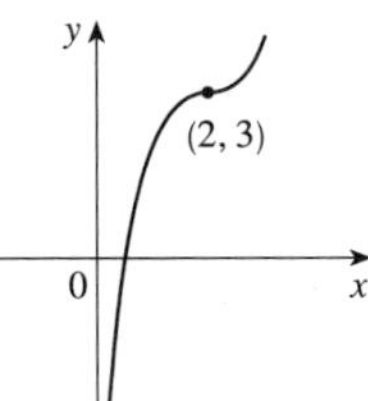

(d) First plot $y = x^2$. Next, to get the graph of $f(x) = 4 - x^2$, reflect f about the x-axis and then shift it upward 4 units.

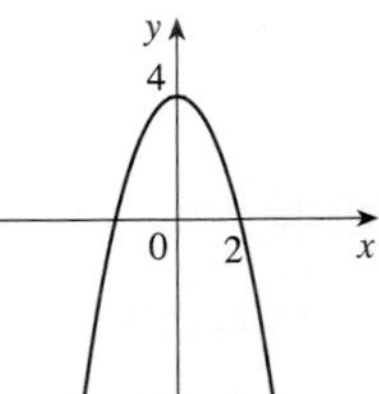

(e) Make a table and then connect the points with a smooth curve:

x	0	1	4	9
y	0	1	2	3

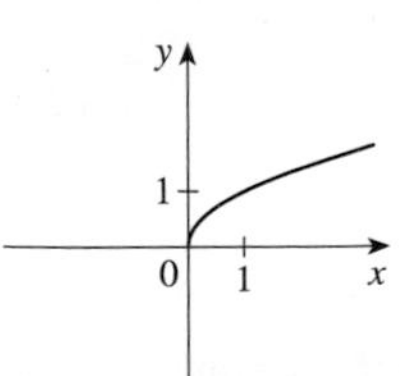

(f) Stretch the graph from part (e) vertically by a factor of two.

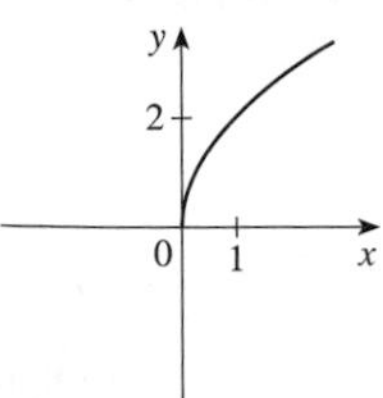

(g) First plot $y = 2^x$. Next, get the graph of $y = -2^x$ by reflecting the graph of $y = 2^x$ about the x-axis.

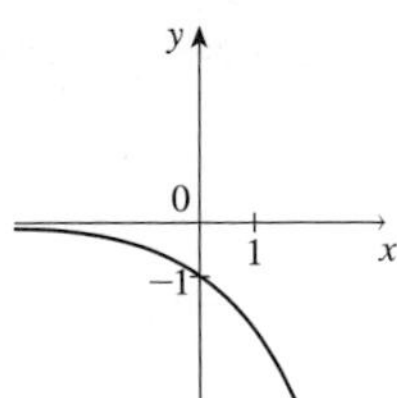

(h) Note that $y = 1 + x^{-1} = 1 + 1/x$. So first plot $y = 1/x$ and then shift it upward 1 unit.

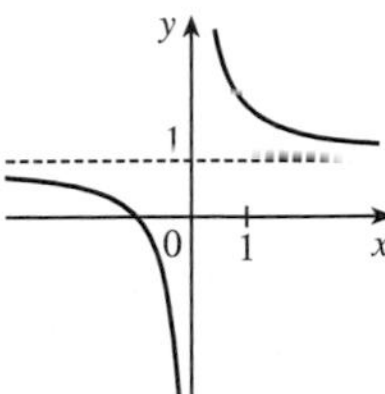

6. (a) $f(-2) = 1 - (-2)^2 = -3$ and $f(1) = 2(1) + 1 = 3$

(b) For $x \le 0$ plot $f(x) = 1 - x^2$ and, on the same plane, for $x > 0$ plot the graph of $f(x) = 2x + 1$.

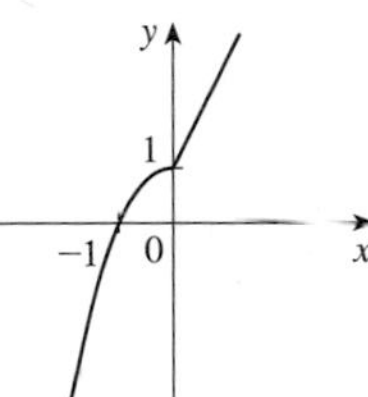

7. (a) $(f \circ g)(x) = f(g(x)) = f(2x - 3) = (2x - 3)^2 + 2(2x - 3) - 1 = 4x^2 - 12x + 9 + 4x - 6 - 1 = 4x^2 - 8x + 2$

(b) $(g \circ f)(x) = g(f(x)) = g(x^2 + 2x - 1) = 2(x^2 + 2x - 1) - 3 = 2x^2 + 4x - 2 - 3 = 2x^2 + 4x - 5$

(c) $(g \circ g \circ g)(x) = g(g(g(x))) = g(g(2x - 3)) = g(2(2x - 3) - 3) = g(4x - 9) = 2(4x - 9) - 3$
$= 8x - 18 - 3 = 8x - 21$

Test D Trigonometry

1. (a) $300° = 300°\left(\dfrac{\pi}{180°}\right) = \dfrac{300\pi}{180} = \dfrac{5\pi}{3}$ (b) $-18° = -18°\left(\dfrac{\pi}{180°}\right) = -\dfrac{18\pi}{180} = -\dfrac{\pi}{10}$

2. (a) $\dfrac{5\pi}{6} = \dfrac{5\pi}{6}\left(\dfrac{180}{\pi}\right)^\circ = 150°$ (b) $2 = 2\left(\dfrac{180}{\pi}\right)^\circ = \left(\dfrac{360}{\pi}\right)^\circ \approx 114.6°$

3. We will use the arc length formula, $s = r\theta$, where s is arc length, r is the radius of the circle, and θ is the measure of the central angle in radians. First, note that $30° = 30°\left(\dfrac{\pi}{180°}\right) = \dfrac{\pi}{6}$. So $s = (12)\left(\dfrac{\pi}{6}\right) = 2\pi$ cm.

4. (a) $\tan(\pi/3) = \sqrt{3}$ [You can read the value from a right triangle with sides 1, 2, and $\sqrt{3}$.]

(b) Note that $7\pi/6$ can be thought of as an angle in the third quadrant with reference angle $\pi/6$. Thus, $\sin(7\pi/6) = -\frac{1}{2}$, since the sine function is negative in the third quadrant.

(c) Note that $5\pi/3$ can be thought of as an angle in the fourth quadrant with reference angle $\pi/3$. Thus, $\sec(5\pi/3) = \dfrac{1}{\cos(5\pi/3)} = \dfrac{1}{1/2} = 2$, since the cosine function is positive in the fourth quadrant.

5. $\sin\theta = a/24 \;\Rightarrow\; a = 24\sin\theta$ and $\cos\theta = b/24 \;\Rightarrow\; b = 24\cos\theta$

6. $\sin x = \frac{1}{3}$ and $\sin^2 x + \cos^2 x = 1 \;\Rightarrow\; \cos x = \sqrt{1-\frac{1}{9}} = \dfrac{2\sqrt{2}}{3}$. Also, $\cos y = \frac{4}{5} \;\Rightarrow\; \sin y = \sqrt{1-\frac{16}{25}} = \frac{3}{5}$.

So, using the sum identity for the sine, we have

$$\sin(x+y) = \sin x\,\cos y + \cos x\,\sin y = \frac{1}{3}\cdot\frac{4}{5} + \frac{2\sqrt{2}}{3}\cdot\frac{3}{5} = \frac{4+6\sqrt{2}}{15} = \frac{1}{15}\left(4+6\sqrt{2}\right)$$

7. (a) $\tan\theta\,\sin\theta + \cos\theta = \dfrac{\sin\theta}{\cos\theta}\sin\theta + \cos\theta = \dfrac{\sin^2\theta}{\cos\theta} + \dfrac{\cos^2\theta}{\cos\theta} = \dfrac{1}{\cos\theta} = \sec\theta$

(b) $\dfrac{2\tan x}{1+\tan^2 x} = \dfrac{2\sin x/(\cos x)}{\sec^2 x} = 2\,\dfrac{\sin x}{\cos x}\cos^2 x = 2\sin x\,\cos x = \sin 2x$

8. $\sin 2x = \sin x \;\Leftrightarrow\; 2\sin x\,\cos x = \sin x \;\Leftrightarrow\; 2\sin x\,\cos x - \sin x = 0 \;\Leftrightarrow\; \sin x\,(2\cos x - 1) = 0 \;\Leftrightarrow$

$\sin x = 0$ or $\cos x = \frac{1}{2} \;\Rightarrow\; x = 0, \frac{\pi}{3}, \pi, \frac{5\pi}{3}, 2\pi$.

9. We first graph $y = \sin 2x$ (by compressing the graph of $\sin x$ by a factor of 2) and then shift it upward 1 unit.

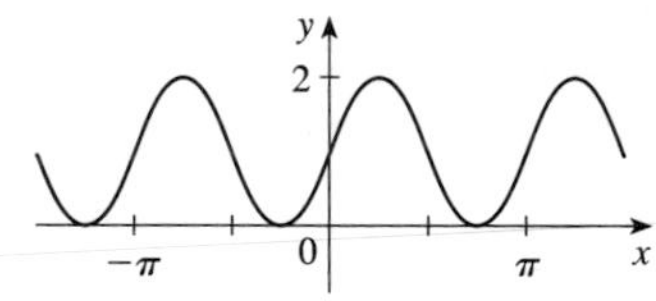

1 ☐ FUNCTIONS AND MODELS

1.1 Four Ways to Represent a Function

1. The functions $f(x) = x + \sqrt{2 - x}$ and $g(u) = u + \sqrt{2 - u}$ give exactly the same output values for every input value, so f and g are equal.

2. $f(x) = \dfrac{x^2 - x}{x - 1} = \dfrac{x(x-1)}{x - 1} = x$ for $x - 1 \neq 0$, so f and g [where $g(x) = x$] are not equal because $f(1)$ is undefined and $g(1) = 1$.

3. (a) The point $(1, 3)$ is on the graph of f, so $f(1) = 3$.

(b) When $x = -1$, y is about -0.2, so $f(-1) \approx -0.2$.

(c) $f(x) = 1$ is equivalent to $y = 1$. When $y = 1$, we have $x = 0$ and $x = 3$.

(d) A reasonable estimate for x when $y = 0$ is $x = -0.8$.

(e) The domain of f consists of all x-values on the graph of f. For this function, the domain is $-2 \leq x \leq 4$, or $[-2, 4]$. The range of f consists of all y-values on the graph of f. For this function, the range is $-1 \leq y \leq 3$, or $[-1, 3]$.

(f) As x increases from -2 to 1, y increases from -1 to 3. Thus, f is increasing on the interval $[-2, 1]$.

4. (a) The point $(-4, -2)$ is on the graph of f, so $f(-4) = -2$. The point $(3, 4)$ is on the graph of g, so $g(3) = 4$.

(b) We are looking for the values of x for which the y-values are equal. The y-values for f and g are equal at the points $(-2, 1)$ and $(2, 2)$, so the desired values of x are -2 and 2.

(c) $f(x) = -1$ is equivalent to $y = -1$. When $y = -1$, we have $x = -3$ and $x = 4$.

(d) As x increases from 0 to 4, y decreases from 3 to -1. Thus, f is decreasing on the interval $[0, 4]$.

(e) The domain of f consists of all x-values on the graph of f. For this function, the domain is $-4 \leq x \leq 4$, or $[-4, 4]$. The range of f consists of all y-values on the graph of f. For this function, the range is $-2 \leq y \leq 3$, or $[-2, 3]$.

(f) The domain of g is $[-4, 3]$ and the range is $[0.5, 4]$.

5. From Figure 1 in the text, the lowest point occurs at about $(t, a) = (12, -85)$. The highest point occurs at about $(17, 115)$. Thus, the range of the vertical ground acceleration is $-85 \leq a \leq 115$. Written in interval notation, we get $[-85, 115]$.

6. *Example 1:* A car is driven at 60 mi/h for 2 hours. The distance d traveled by the car is a function of the time t. The domain of the function is $\{t \mid 0 \leq t \leq 2\}$, where t is measured in hours. The range of the function is $\{d \mid 0 \leq d \leq 120\}$, where d is measured in miles.

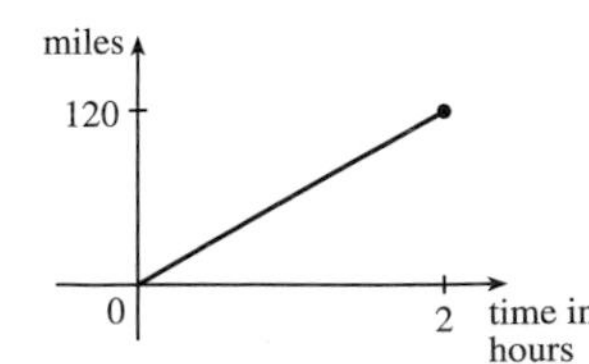

Example 2: At a certain university, the number of students N on campus at any time on a particular day is a function of the time t after midnight. The domain of the function is $\{t \mid 0 \leq t \leq 24\}$, where t is measured in hours. The range of the function is $\{N \mid 0 \leq N \leq k\}$, where N is an integer and k is the largest number of students on campus at once.

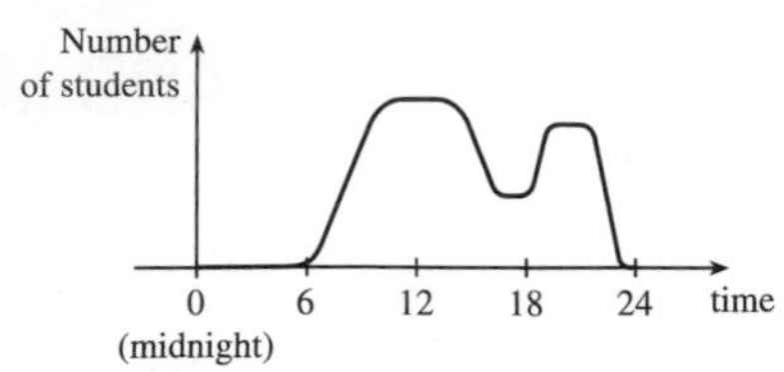

Example 3: A certain employee is paid \$8.00 per hour and works a maximum of 30 hours per week. The number of hours worked is rounded down to the nearest quarter of an hour. This employee's gross weekly pay P is a function of the number of hours worked h. The domain of the function is $[0, 30]$ and the range of the function is $\{0, 2.00, 4.00, \ldots, 238.00, 240.00\}$.

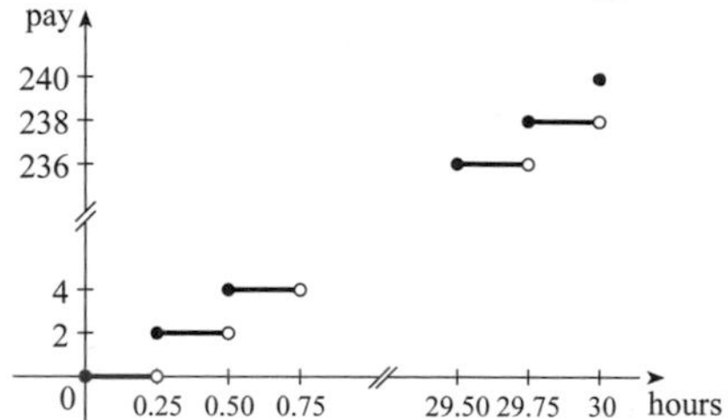

7. No, the curve is not the graph of a function because a vertical line intersects the curve more than once. Hence, the curve fails the Vertical Line Test.

8. Yes, the curve is the graph of a function because it passes the Vertical Line Test. The domain is $[-2, 2]$ and the range is $[-1, 2]$.

9. Yes, the curve is the graph of a function because it passes the Vertical Line Test. The domain is $[-3, 2]$ and the range is $[-3, -2) \cup [-1, 3]$.

10. No, the curve is not the graph of a function since for $x = 0, \pm 1$, and ± 2, there are infinitely many points on the curve.

11. (a) When $t = 1950$, $T \approx 13.8°\text{C}$, so the global average temperature in 1950 was about 13.8°C.

(b) When $T = 14.2°\text{C}$, $t \approx 1990$.

(c) The global average temperature was smallest in 1910 (the year corresponding to the lowest point on the graph) and largest in 2005 (the year corresponding to the highest point on the graph).

(d) When $t = 1910$, $T \approx 13.5°\text{C}$, and when $t = 2005$, $T \approx 14.5°\text{C}$. Thus, the range of T is about $[13.5, 14.5]$.

12. (a) The ring width varies from near 0 mm to about 1.6 mm, so the range of the ring width function is approximately $[0, 1.6]$.

(b) According to the graph, the earth gradually cooled from 1550 to 1700, warmed into the late 1700s, cooled again into the late 1800s, and has been steadily warming since then. In the mid-19th century, there was variation that could have been associated with volcanic eruptions.

13. The water will cool down almost to freezing as the ice melts. Then, when the ice has melted, the water will slowly warm up to room temperature.

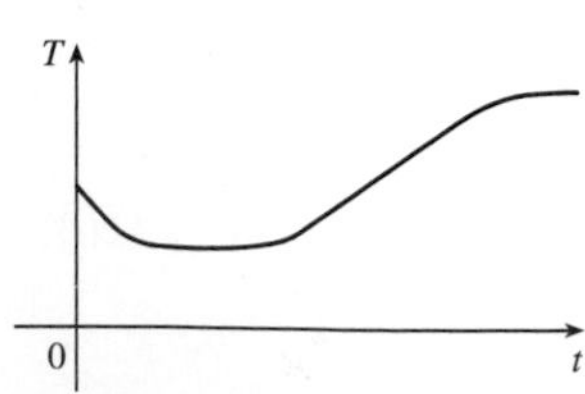

14. Runner A won the race, reaching the finish line at 100 meters in about 15 seconds, followed by runner B with a time of about 19 seconds, and then by runner C who finished in around 23 seconds. B initially led the race, followed by C, and then A. C then passed B to lead for a while. Then A passed first B, and then passed C to take the lead and finish first. Finally, B passed C to finish in second place. All three runners completed the race.

15. (a) The power consumption at 6 AM is 500 MW, which is obtained by reading the value of power P when $t = 6$ from the graph. At 6 PM we read the value of P when $t = 18$, obtaining approximately 730 MW.

(b) The minimum power consumption is determined by finding the time for the lowest point on the graph, $t = 4$, or 4 AM. The maximum power consumption corresponds to the highest point on the graph, which occurs just before $t = 12$, or right before noon. These times are reasonable, considering the power consumption schedules of most individuals and businesses.

16. The summer solstice (the longest day of the year) is around June 21, and the winter solstice (the shortest day) is around December 22. (Exchange the dates for the southern hemisphere.)

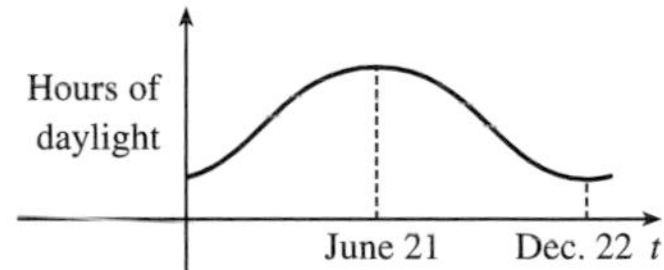

17. Of course, this graph depends strongly on the geographical location!

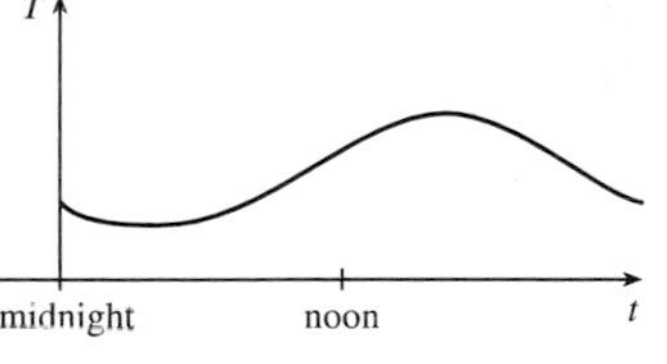

18. The value of the car decreases fairly rapidly initially, then somewhat less rapidly.

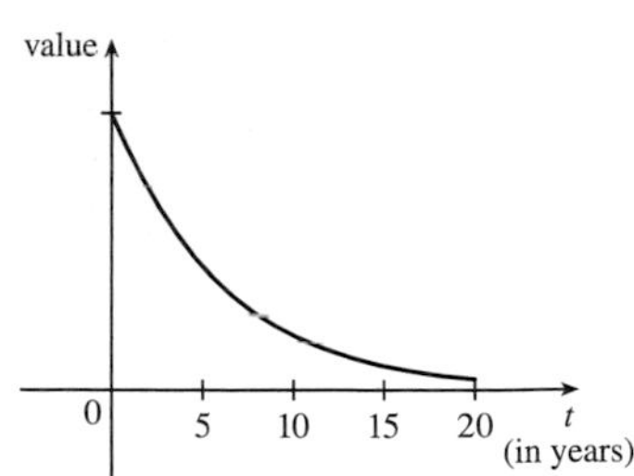

19. As the price increases, the amount sold decreases.

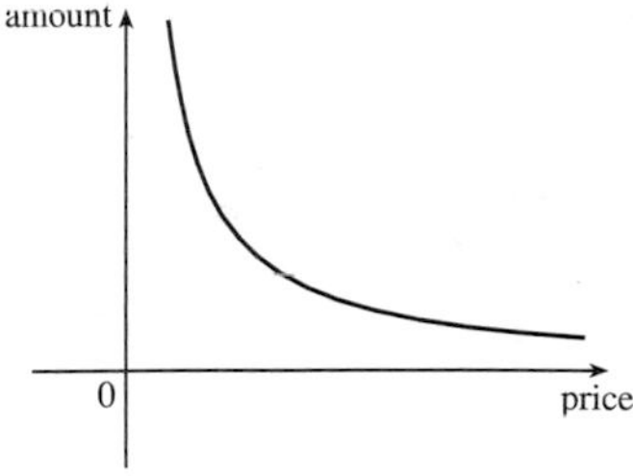

20. The temperature of the pie would increase rapidly, level off to oven temperature, decrease rapidly, and then level off to room temperature.

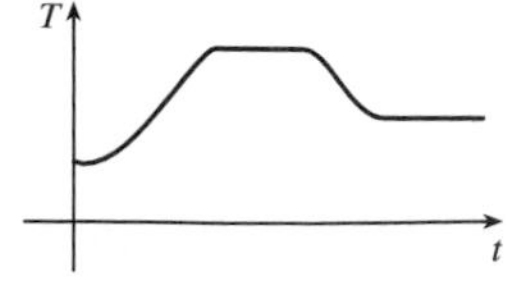

21.

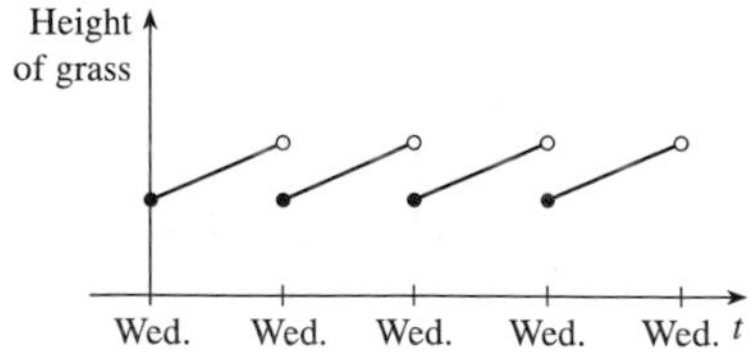

22. (a)

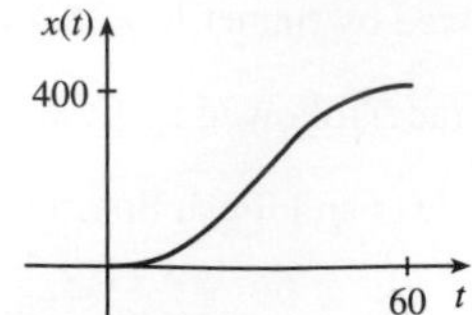

(b)

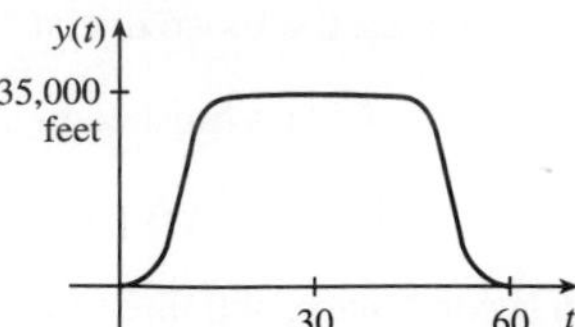

(c)

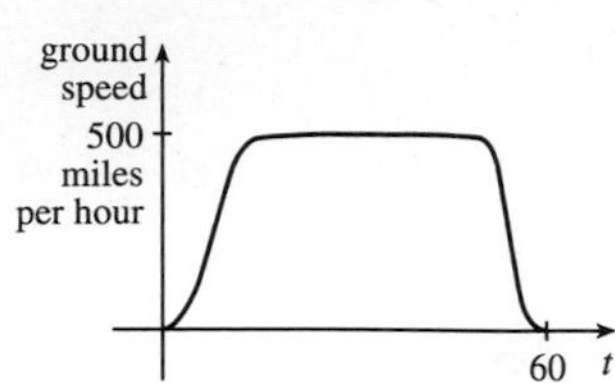

(d)

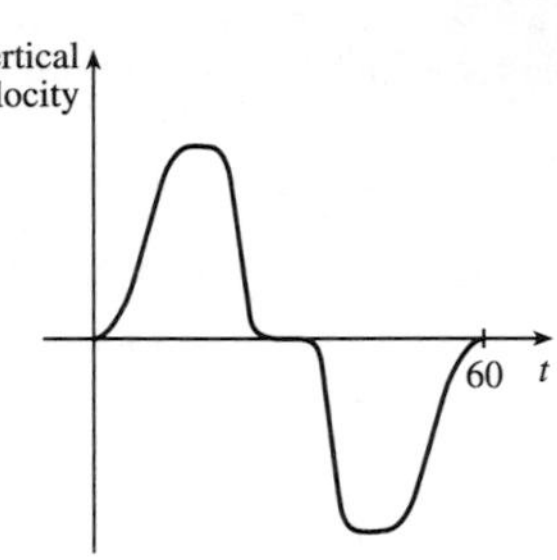

23. (a)

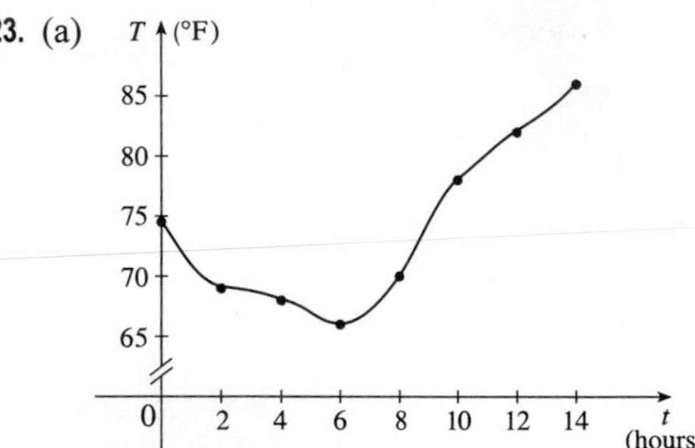

(b) 9:00 AM corresponds to $t = 9$. When $t = 9$, the temperature T is about 74°F.

24. (a)

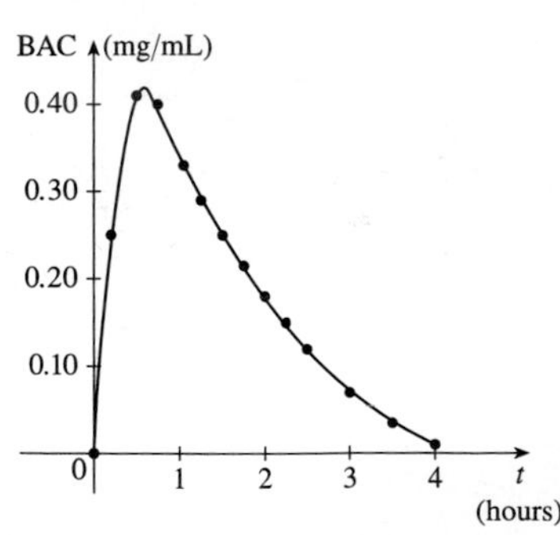

(b) The blood alcohol concentration rises rapidly, then slowly decreases to near zero. Note that the BAC in this exercise is measured in mg/mL, not percent.

25. $f(x) = 3x^2 - x + 2.$

$f(2) = 3(2)^2 - 2 + 2 = 12 - 2 + 2 = 12.$

$f(-2) = 3(-2)^2 - (-2) + 2 = 12 + 2 + 2 = 16.$

$f(a) = 3a^2 - a + 2.$

$f(-a) = 3(-a)^2 - (-a) + 2 = 3a^2 + a + 2.$

$f(a + 1) = 3(a + 1)^2 - (a + 1) + 2 = 3(a^2 + 2a + 1) - a - 1 + 2 = 3a^2 + 6a + 3 - a + 1 = 3a^2 + 5a + 4.$

$2f(a) = 2 \cdot f(a) = 2(3a^2 - a + 2) = 6a^2 - 2a + 4.$

$f(2a) = 3(2a)^2 - (2a) + 2 = 3(4a^2) - 2a + 2 = 12a^2 - 2a + 2.$

$f(a^2) = 3(a^2)^2 - (a^2) + 2 = 3(a^4) - a^2 + 2 = 3a^4 - a^2 + 2.$

$$[f(a)]^2 = \left[3a^2 - a + 2\right]^2 = \left(3a^2 - a + 2\right)\left(3a^2 - a + 2\right)$$
$$= 9a^4 - 3a^3 + 6a^2 - 3a^3 + a^2 - 2a + 6a^2 - 2a + 4 = 9a^4 - 6a^3 + 13a^2 - 4a + 4.$$

$f(a+h) = 3(a+h)^2 - (a+h) + 2 = 3(a^2 + 2ah + h^2) - a - h + 2 = 3a^2 + 6ah + 3h^2 - a - h + 2.$

26. A spherical balloon with radius $r+1$ has volume $V(r+1) = \frac{4}{3}\pi(r+1)^3 = \frac{4}{3}\pi(r^3 + 3r^2 + 3r + 1)$. We wish to find the amount of air needed to inflate the balloon from a radius of r to $r+1$. Hence, we need to find the difference $V(r+1) - V(r) = \frac{4}{3}\pi(r^3 + 3r^2 + 3r + 1) - \frac{4}{3}\pi r^3 = \frac{4}{3}\pi(3r^2 + 3r + 1)$.

27. $f(x) = 4 + 3x - x^2$, so $f(3+h) = 4 + 3(3+h) - (3+h)^2 = 4 + 9 + 3h - (9 + 6h + h^2) = 4 - 3h - h^2$,

and $\dfrac{f(3+h) - f(3)}{h} = \dfrac{(4 - 3h - h^2) - 4}{h} = \dfrac{h(-3-h)}{h} = -3 - h.$

28. $f(x) = x^3$, so $f(a+h) = (a+h)^3 = a^3 + 3a^2h + 3ah^2 + h^3$,

and $\dfrac{f(a+h) - f(a)}{h} = \dfrac{(a^3 + 3a^2h + 3ah^2 + h^3) - a^3}{h} = \dfrac{h(3a^2 + 3ah + h^2)}{h} = 3a^2 + 3ah + h^2.$

29. $$\frac{f(x) - f(a)}{x - a} = \frac{\frac{1}{x} - \frac{1}{a}}{x - a} = \frac{\frac{a - x}{xa}}{x - a} = \frac{a - x}{xa(x-a)} = \frac{-1(x-a)}{xa(x-a)} = -\frac{1}{ax}$$

30. $$\frac{f(x) - f(1)}{x - 1} = \frac{\frac{x+3}{x+1} - 2}{x - 1} = \frac{\frac{x + 3 - 2(x+1)}{x+1}}{x - 1} = \frac{x + 3 - 2x - 2}{(x+1)(x-1)}$$
$$= \frac{-x + 1}{(x+1)(x-1)} = \frac{-(x-1)}{(x+1)(x-1)} = -\frac{1}{x+1}$$

31. $f(x) = (x+4)/(x^2 - 9)$ is defined for all x except when $0 = x^2 - 9 \iff 0 = (x+3)(x-3) \iff x = -3$ or 3, so the domain is $\{x \in \mathbb{R} \mid x \neq -3, 3\} = (-\infty, -3) \cup (-3, 3) \cup (3, \infty)$.

32. $f(x) = (2x^3 - 5)/(x^2 + x - 6)$ is defined for all x except when $0 = x^2 + x - 6 \iff 0 = (x+3)(x-2) \iff x = -3$ or 2, so the domain is $\{x \in \mathbb{R} \mid x \neq -3, 2\} = (-\infty, -3) \cup (-3, 2) \cup (2, \infty)$.

33. $f(t) = \sqrt[3]{2t - 1}$ is defined for all real numbers. In fact $\sqrt[3]{p(t)}$, where $p(t)$ is a polynomial, is defined for all real numbers. Thus, the domain is $\mathbb{R}$, or $(-\infty, \infty)$.

34. $g(t) = \sqrt{3 - t} - \sqrt{2 + t}$ is defined when $3 - t \geq 0 \iff t \leq 3$ and $2 + t \geq 0 \iff t \geq -2$. Thus, the domain is $-2 \leq t \leq 3$, or $[-2, 3]$.

35. $h(x) = 1 \big/ \sqrt[4]{x^2 - 5x}$ is defined when $x^2 - 5x > 0 \iff x(x-5) > 0$. Note that $x^2 - 5x \neq 0$ since that would result in division by zero. The expression $x(x-5)$ is positive if $x < 0$ or $x > 5$. (See Appendix A for methods for solving inequalities.) Thus, the domain is $(-\infty, 0) \cup (5, \infty)$.

36. $f(u) = \dfrac{u+1}{1+\dfrac{1}{u+1}}$ is defined when $u+1 \neq 0$ $[u \neq -1]$ and $1 + \dfrac{1}{u+1} \neq 0$. Since $1 + \dfrac{1}{u+1} = 0 \;\Leftrightarrow$

$\dfrac{1}{u+1} = -1 \;\Leftrightarrow\; 1 = -u - 1 \;\Leftrightarrow\; u = -2$, the domain is $\{u \mid u \neq -2,\, u \neq -1\} = (-\infty, -2) \cup (-2, -1) \cup (-1, \infty)$.

37. $F(p) = \sqrt{2 - \sqrt{p}}$ is defined when $p \geq 0$ and $2 - \sqrt{p} \geq 0$. Since $2 - \sqrt{p} \geq 0 \;\Leftrightarrow\; 2 \geq \sqrt{p} \;\Leftrightarrow\; \sqrt{p} \leq 2 \;\Leftrightarrow$ $0 \leq p \leq 4$, the domain is $[0, 4]$.

38. $h(x) = \sqrt{4 - x^2}$. Now $y = \sqrt{4 - x^2} \;\Rightarrow\; y^2 = 4 - x^2 \;\Leftrightarrow\; x^2 + y^2 = 4$, so the graph is the top half of a circle of radius 2 with center at the origin. The domain is $\{x \mid 4 - x^2 \geq 0\} = \{x \mid 4 \geq x^2\} = \{x \mid 2 \geq |x|\} = [-2, 2]$. From the graph, the range is $0 \leq y \leq 2$, or $[0, 2]$.

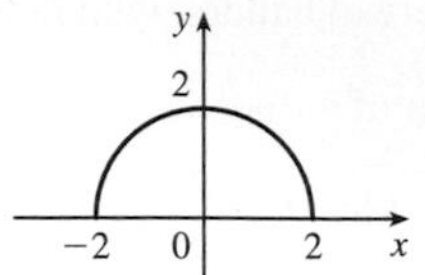

39. The domain of $f(x) = 1.6x - 2.4$ is the set of all real numbers, denoted by $\mathbb{R}$ or $(-\infty, \infty)$. The graph of f is a line with slope 1.6 and y-intercept -2.4.

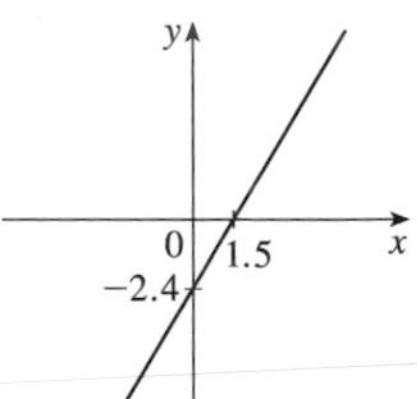

40. Note that $g(t) = \dfrac{t^2 - 1}{t + 1} = \dfrac{(t+1)(t-1)}{t+1} = t - 1$ for $t + 1 \neq 0$, i.e., $t \neq -1$. The domain of g is the set of all real numbers except -1. In interval notation, we have $(-\infty, -1) \cup (-1, \infty)$. The graph of g is a line with slope 1, y-intercept -1, and a hole at $t = -1$.

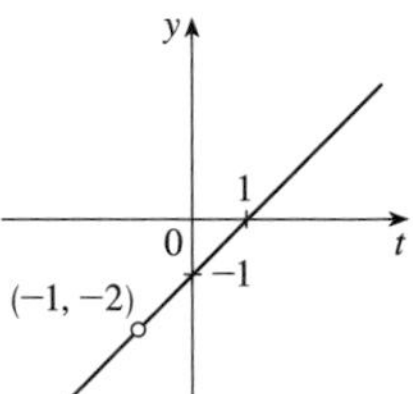

41. $f(x) = \begin{cases} x + 2 & \text{if } x < 0 \\ 1 - x & \text{if } x \geq 0 \end{cases}$

$f(-3) = -3 + 2 = -1$, $f(0) = 1 - 0 = 1$, and $f(2) = 1 - 2 = -1$.

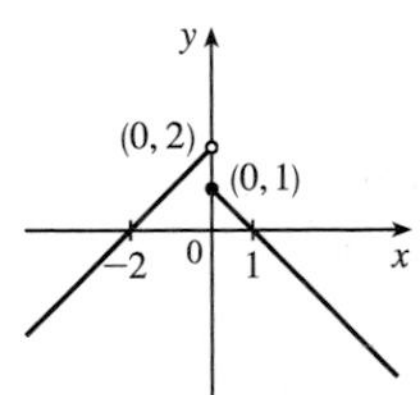

42. $f(x) = \begin{cases} 3 - \frac{1}{2}x & \text{if } x < 2 \\ 2x - 5 & \text{if } x \geq 2 \end{cases}$

$f(-3) = 3 - \frac{1}{2}(-3) = \frac{9}{2}$, $f(0) = 3 - \frac{1}{2}(0) = 3$, and $f(2) = 2(2) - 5 = -1$.

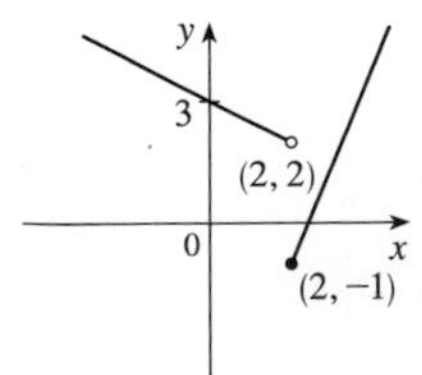

43. $f(x) = \begin{cases} x + 1 & \text{if } x \leq -1 \\ x^2 & \text{if } x > -1 \end{cases}$

$f(-3) = -3 + 1 = -2$, $f(0) = 0^2 = 0$, and $f(2) = 2^2 = 4$.

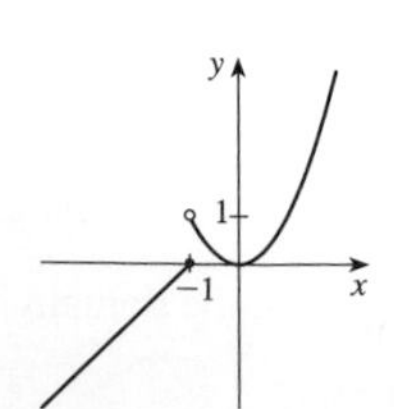

44. $f(x) = \begin{cases} -1 & \text{if } x \le 1 \\ 7 - 2x & \text{if } x > 1 \end{cases}$

$f(-3) = -1$, $f(0) = -1$, and $f(2) = 7 - 2(2) = 3$.

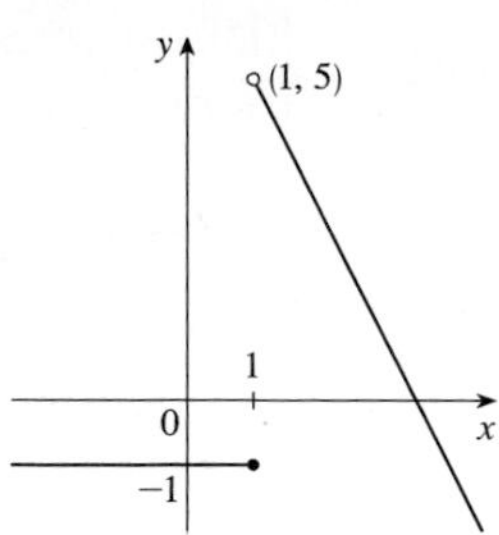

45. $|x| = \begin{cases} x & \text{if } x \ge 0 \\ -x & \text{if } x < 0 \end{cases}$

so $f(x) = x + |x| = \begin{cases} 2x & \text{if } x \ge 0 \\ 0 & \text{if } x < 0 \end{cases}$

Graph the line $y = 2x$ for $x \ge 0$ and graph $y = 0$ (the x-axis) for $x < 0$.

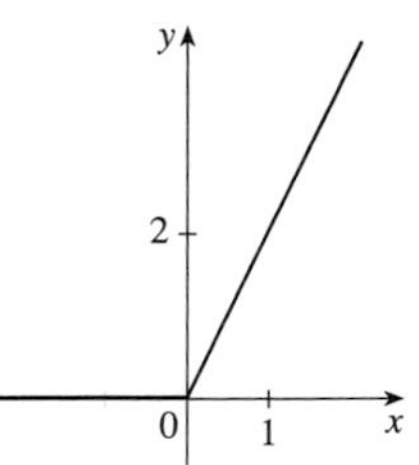

46. $f(x) = |x + 2| = \begin{cases} x + 2 & \text{if } x + 2 \ge 0 \\ -(x + 2) & \text{if } x + 2 < 0 \end{cases}$

$= \begin{cases} x + 2 & \text{if } x \ge -2 \\ -x - 2 & \text{if } x < -2 \end{cases}$

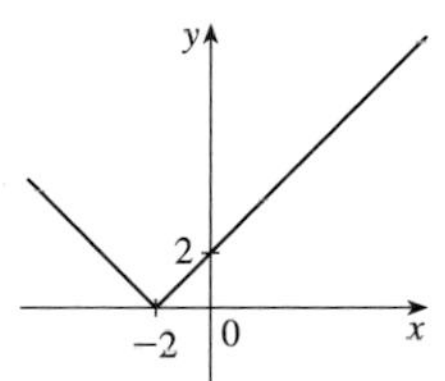

47. $g(t) = |1 - 3t| = \begin{cases} 1 - 3t & \text{if } 1 - 3t \ge 0 \\ -(1 - 3t) & \text{if } 1 - 3t < 0 \end{cases}$

$= \begin{cases} 1 - 3t & \text{if } t \le \frac{1}{3} \\ 3t - 1 & \text{if } t > \frac{1}{3} \end{cases}$

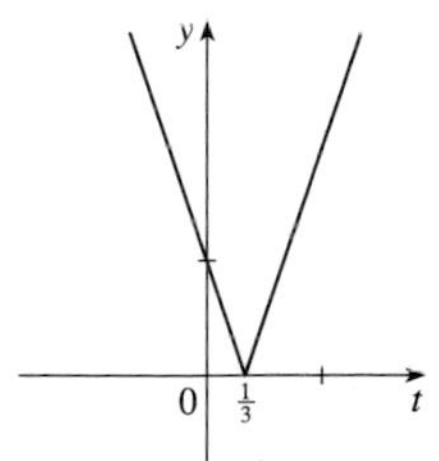

48. $|t| = \begin{cases} t & \text{if } t \ge 0 \\ -t & \text{if } t < 0 \end{cases}$ and

$|t + 1| = \begin{cases} t + 1 & \text{if } t + 1 \ge 0 \\ -(t + 1) & \text{if } t + 1 < 0 \end{cases} = \begin{cases} t + 1 & \text{if } t \ge -1 \\ -t - 1 & \text{if } t < -1 \end{cases}$

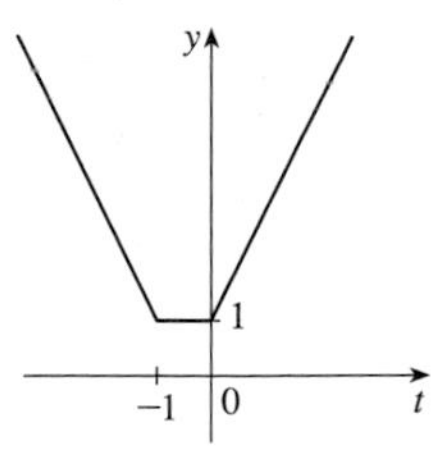

so $h(t) = |t| + |t + 1| = \begin{cases} t + (t + 1) & \text{if } t \ge 0 \\ -t + (t + 1) & \text{if } -1 \le t < 0 \\ -t + (-t - 1) & \text{if } t < -1 \end{cases} = \begin{cases} 2t + 1 & \text{if } t \ge 0 \\ 1 & \text{if } -1 \le t < 0 \\ -2t - 1 & \text{if } t < -1 \end{cases}$

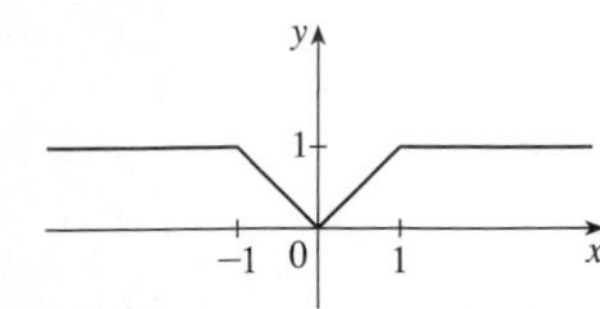

49. To graph $f(x) = \begin{cases} |x| & \text{if } |x| \le 1 \\ 1 & \text{if } |x| > 1 \end{cases}$, graph $y = |x|$ (Figure 16) for $-1 \le x \le 1$ and graph $y = 1$ for $x > 1$ and for $x < -1$.

We could rewrite f as $f(x) = \begin{cases} 1 & \text{if } x < -1 \\ -x & \text{if } -1 \le x < 0 \\ x & \text{if } 0 \le x \le 1 \\ 1 & \text{if } x > 1 \end{cases}$.

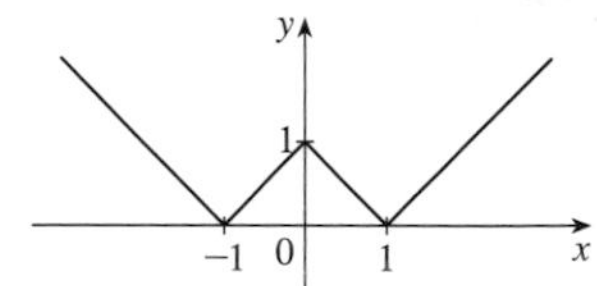

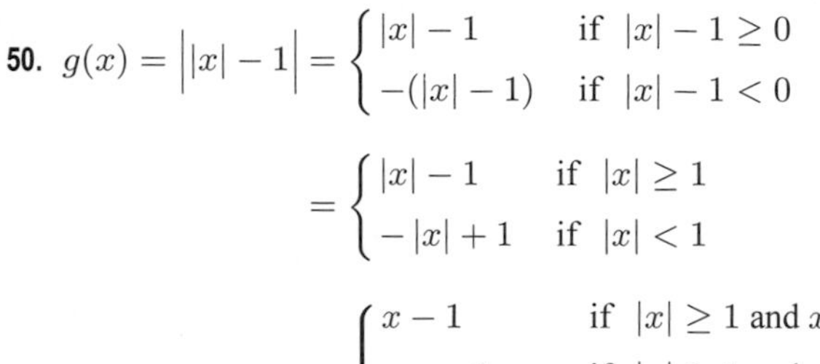

50. $g(x) = \big||x| - 1\big| = \begin{cases} |x| - 1 & \text{if } |x| - 1 \ge 0 \\ -(|x| - 1) & \text{if } |x| - 1 < 0 \end{cases}$

$= \begin{cases} |x| - 1 & \text{if } |x| \ge 1 \\ -|x| + 1 & \text{if } |x| < 1 \end{cases}$

$= \begin{cases} x - 1 & \text{if } |x| \ge 1 \text{ and } x \ge 0 \\ -x - 1 & \text{if } |x| \ge 1 \text{ and } x < 0 \\ -x + 1 & \text{if } |x| < 1 \text{ and } x \ge 0 \\ -(-x) + 1 & \text{if } |x| < 1 \text{ and } x < 0 \end{cases} = \begin{cases} x - 1 & \text{if } x \ge 1 \\ -x - 1 & \text{if } x \le -1 \\ -x + 1 & \text{if } 0 \le x < 1 \\ x + 1 & \text{if } -1 < x < 0 \end{cases}$

51. Recall that the slope m of a line between the two points (x_1, y_1) and (x_2, y_2) is $m = \dfrac{y_2 - y_1}{x_2 - x_1}$ and an equation of the line connecting those two points is $y - y_1 = m(x - x_1)$. The slope of the line segment joining the points $(1, -3)$ and $(5, 7)$ is $\dfrac{7 - (-3)}{5 - 1} = \dfrac{5}{2}$, so an equation is $y - (-3) = \frac{5}{2}(x - 1)$. The function is $f(x) = \frac{5}{2}x - \frac{11}{2}$, $1 \le x \le 5$.

52. The slope of the line segment joining the points $(-5, 10)$ and $(7, -10)$ is $\dfrac{-10 - 10}{7 - (-5)} = -\dfrac{5}{3}$, so an equation is $y - 10 = -\frac{5}{3}[x - (-5)]$. The function is $f(x) = -\frac{5}{3}x + \frac{5}{3}$, $-5 \le x \le 7$.

53. We need to solve the given equation for y. $x + (y - 1)^2 = 0 \;\Leftrightarrow\; (y - 1)^2 = -x \;\Leftrightarrow\; y - 1 = \pm\sqrt{-x} \;\Leftrightarrow\; y = 1 \pm \sqrt{-x}$. The expression with the positive radical represents the top half of the parabola, and the one with the negative radical represents the bottom half. Hence, we want $f(x) = 1 - \sqrt{-x}$. Note that the domain is $x \le 0$.

54. $x^2 + (y - 2)^2 = 4 \;\Leftrightarrow\; (y - 2)^2 = 4 - x^2 \;\Leftrightarrow\; y - 2 = \pm\sqrt{4 - x^2} \;\Leftrightarrow\; y = 2 \pm \sqrt{4 - x^2}$. The top half is given by the function $f(x) = 2 + \sqrt{4 - x^2}$, $-2 \le x \le 2$.

55. For $0 \le x \le 3$, the graph is the line with slope -1 and y-intercept 3, that is, $y = -x + 3$. For $3 < x \le 5$, the graph is the line with slope 2 passing through $(3, 0)$; that is, $y - 0 = 2(x - 3)$, or $y = 2x - 6$. So the function is

$$f(x) = \begin{cases} -x + 3 & \text{if } 0 \le x \le 3 \\ 2x - 6 & \text{if } 3 < x \le 5 \end{cases}$$

56. For $-4 \le x \le -2$, the graph is the line with slope $-\frac{3}{2}$ passing through $(-2, 0)$; that is, $y - 0 = -\frac{3}{2}[x - (-2)]$, or $y = -\frac{3}{2}x - 3$. For $-2 < x < 2$, the graph is the top half of the circle with center $(0, 0)$ and radius 2. An equation of the circle is $x^2 + y^2 = 4$, so an equation of the top half is $y = \sqrt{4 - x^2}$. For $2 \le x \le 4$, the graph is the line with slope $\frac{3}{2}$ passing through $(2, 0)$; that is, $y - 0 = \frac{3}{2}(x - 2)$, or $y = \frac{3}{2}x - 3$. So the function is

$$f(x) = \begin{cases} -\frac{3}{2}x - 3 & \text{if } -4 \le x \le -2 \\ \sqrt{4 - x^2} & \text{if } -2 < x < 2 \\ \frac{3}{2}x - 3 & \text{if } \ 2 \le x \le 4 \end{cases}$$

57. Let the length and width of the rectangle be L and W. Then the perimeter is $2L + 2W = 20$ and the area is $A = LW$. Solving the first equation for W in terms of L gives $W = \dfrac{20 - 2L}{2} = 10 - L$. Thus, $A(L) = L(10 - L) = 10L - L^2$. Since lengths are positive, the domain of A is $0 < L < 10$. If we further restrict L to be larger than W, then $5 < L < 10$ would be the domain.

58. Let the length and width of the rectangle be L and W. Then the area is $LW = 16$, so that $W = 16/L$. The perimeter is $P = 2L + 2W$, so $P(L) = 2L + 2(16/L) = 2L + 32/L$, and the domain of P is $L > 0$, since lengths must be positive quantities. If we further restrict L to be larger than W, then $L > 4$ would be the domain.

59. Let the length of a side of the equilateral triangle be x. Then by the Pythagorean Theorem, the height y of the triangle satisfies $y^2 + \left(\frac{1}{2}x\right)^2 = x^2$, so that $y^2 = x^2 - \frac{1}{4}x^2 = \frac{3}{4}x^2$ and $y = \frac{\sqrt{3}}{2}x$. Using the formula for the area A of a triangle, $A = \frac{1}{2}(\text{base})(\text{height})$, we obtain $A(x) = \frac{1}{2}(x)\left(\frac{\sqrt{3}}{2}x\right) = \frac{\sqrt{3}}{4}x^2$, with domain $x > 0$.

60. Let the length, width, and height of the closed rectangular box be denoted by L, W, and H, respectively. The length is twice the width, so $L = 2W$. The volume V of the box is given by $V = LWH$. Since $V = 8$, we have $8 = (2W)WH \ \Rightarrow$ $8 = 2W^2H \ \Rightarrow \ H = \dfrac{8}{2W^2} = \dfrac{4}{W^2}$, and so $H = f(W) = \dfrac{4}{W^2}$.

61. Let each side of the base of the box have length x, and let the height of the box be h. Since the volume is 2, we know that $2 = hx^2$, so that $h = 2/x^2$, and the surface area is $S = x^2 + 4xh$. Thus, $S(x) = x^2 + 4x(2/x^2) = x^2 + (8/x)$, with domain $x > 0$.

62. The area of the window is $A = xh + \frac{1}{2}\pi\left(\frac{1}{2}x\right)^2 = xh + \dfrac{\pi x^2}{8}$, where h is the height of the rectangular portion of the window. The perimeter is $P = 2h + x + \frac{1}{2}\pi x = 30 \ \Leftrightarrow \ 2h = 30 - x - \frac{1}{2}\pi x \ \Leftrightarrow \ h = \frac{1}{4}(60 - 2x - \pi x)$. Thus,

$$A(x) = x\,\frac{60 - 2x - \pi x}{4} + \frac{\pi x^2}{8} = 15x - \tfrac{1}{2}x^2 - \tfrac{\pi}{4}x^2 + \tfrac{\pi}{8}x^2 = 15x - \tfrac{4}{8}x^2 - \tfrac{\pi}{8}x^2 = 15x - x^2\left(\frac{\pi + 4}{8}\right).$$

Since the lengths x and h must be positive quantities, we have $x > 0$ and $h > 0$. For $h > 0$, we have $2h > 0 \ \Leftrightarrow$ $30 - x - \frac{1}{2}\pi x > 0 \ \Leftrightarrow \ 60 > 2x + \pi x \ \Leftrightarrow \ x < \dfrac{60}{2 + \pi}$. Hence, the domain of A is $0 < x < \dfrac{60}{2 + \pi}$.

63. The height of the box is x and the length and width are $L = 20 - 2x$, $W = 12 - 2x$. Then $V = LWx$ and so $V(x) = (20 - 2x)(12 - 2x)(x) = 4(10 - x)(6 - x)(x) = 4x(60 - 16x + x^2) = 4x^3 - 64x^2 + 240x$.
The sides L, W, and x must be positive. Thus, $L > 0 \ \Leftrightarrow \ 20 - 2x > 0 \ \Leftrightarrow \ x < 10$;
$W > 0 \ \Leftrightarrow \ 12 - 2x > 0 \ \Leftrightarrow \ x < 6$; and $x > 0$. Combining these restrictions gives us the domain $0 < x < 6$.

64. We can summarize the monthly cost with a piecewise defined function.

$$C(x) = \begin{cases} 35 & \text{if } 0 \le x \le 400 \\ 35 + 0.10(x - 400) & \text{if } x > 400 \end{cases}$$

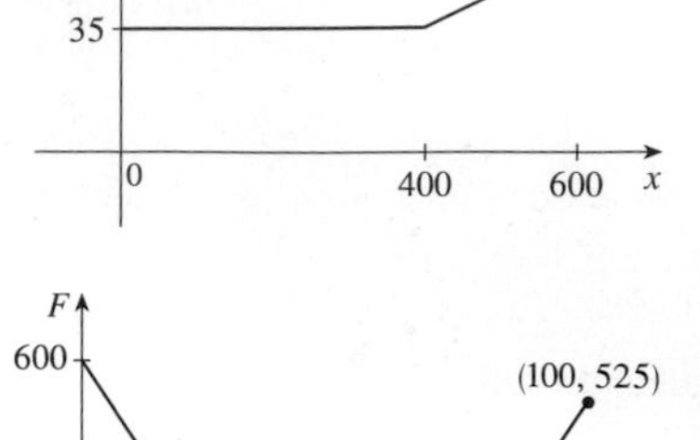

65. We can summarize the amount of the fine with a piecewise defined function.

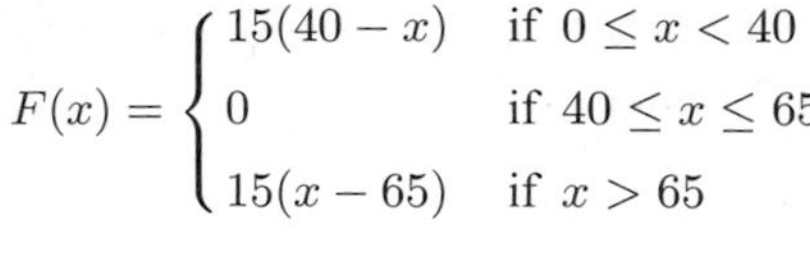

$$F(x) = \begin{cases} 15(40 - x) & \text{if } 0 \le x < 40 \\ 0 & \text{if } 40 \le x \le 65 \\ 15(x - 65) & \text{if } x > 65 \end{cases}$$

F
600
(100, 525)
0 40 65 100 x

66. For the first 1200 kWh, $E(x) = 10 + 0.06x$.

For usage over 1200 kWh, the cost is

$E(x) = 10 + 0.06(1200) + 0.07(x - 1200) = 82 + 0.07(x - 1200)$.

Thus,

$$E(x) = \begin{cases} 10 + 0.06x & \text{if } 0 \le x \le 1200 \\ 82 + 0.07(x - 1200) & \text{if } x > 1200 \end{cases}$$

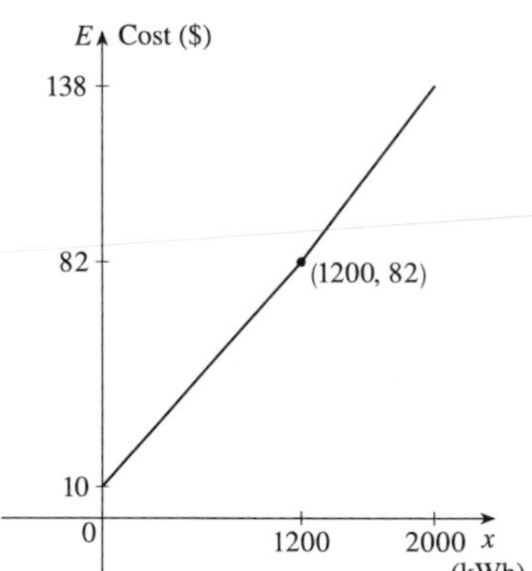

67. (a)

R(%)
15
10
0 10,000 20,000 I (in dollars)

(b) On \$14,000, tax is assessed on \$4000, and $10\%(\$4000) = \$400$.

On \$26,000, tax is assessed on \$16,000, and

$10\%(\$10{,}000) + 15\%(\$6000) = \$1000 + \$900 = \$1900$.

(c) As in part (b), there is \$1000 tax assessed on \$20,000 of income, so the graph of T is a line segment from $(10{,}000, 0)$ to $(20{,}000, 1000)$. The tax on \$30,000 is \$2500, so the graph of T for $x > 20{,}000$ is the ray with initial point $(20{,}000, 1000)$ that passes through $(30{,}000, 2500)$.

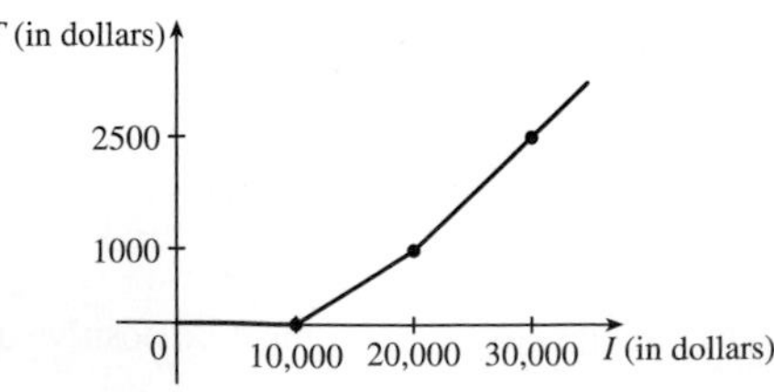

68. One example is the amount paid for cable or telephone system repair in the home, usually measured to the nearest quarter hour. Another example is the amount paid by a student in tuition fees, if the fees vary according to the number of credits for which the student has registered.

69. f is an odd function because its graph is symmetric about the origin. g is an even function because its graph is symmetric with respect to the y-axis.

70. f is not an even function since it is not symmetric with respect to the y-axis. f is not an odd function since it is not symmetric about the origin. Hence, f is *neither* even nor odd. g is an even function because its graph is symmetric with respect to the y-axis.

71. (a) Because an even function is symmetric with respect to the y-axis, and the point $(5, 3)$ is on the graph of this even function, the point $(-5, 3)$ must also be on its graph.

(b) Because an odd function is symmetric with respect to the origin, and the point $(5, 3)$ is on the graph of this odd function, the point $(-5, -3)$ must also be on its graph.

72. (a) If f is even, we get the rest of the graph by reflecting about the y-axis.

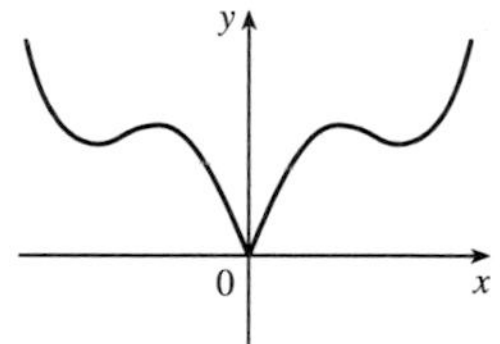

(b) If f is odd, we get the rest of the graph by rotating 180° about the origin.

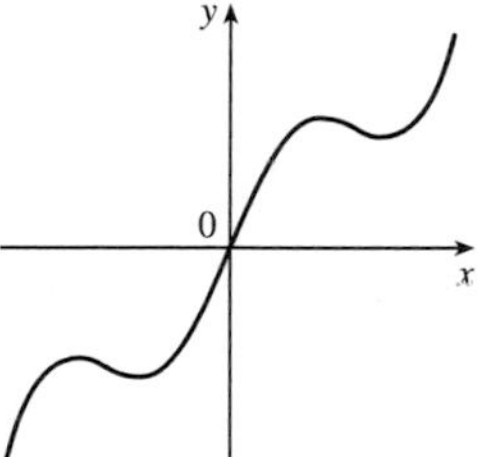

73. $f(x) = \dfrac{x}{x^2 + 1}$.

$$f(-x) = \frac{-x}{(-x)^2 + 1} = \frac{-x}{x^2 + 1} = -\frac{x}{x^2 + 1} = -f(x).$$

Since $f(-x) = -f(x)$, f is an odd function.

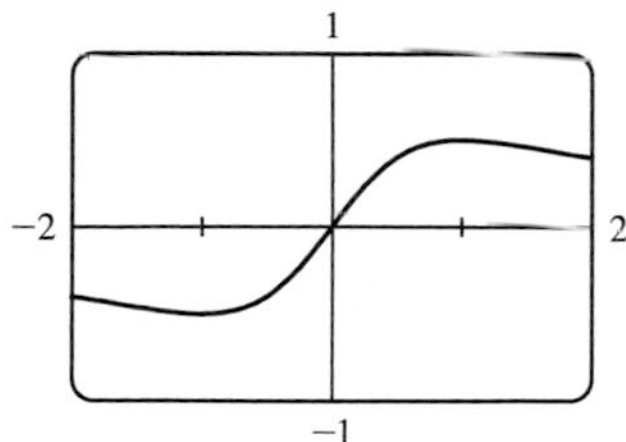

74. $f(x) = \dfrac{x^2}{x^4 + 1}$.

$$f(-x) = \frac{(-x)^2}{(-x)^4 + 1} = \frac{x^2}{x^4 + 1} = f(x).$$

Since $f(-x) = f(x)$, f is an even function.

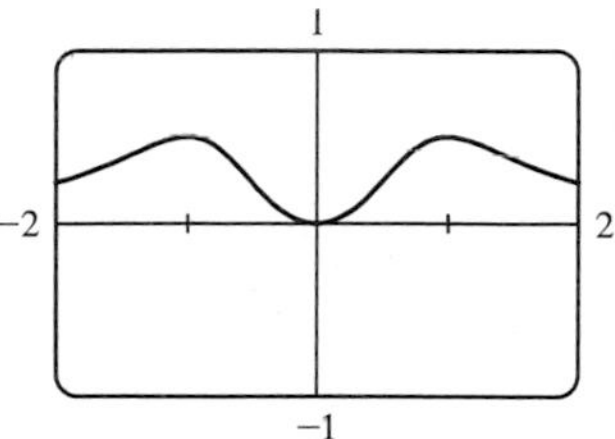

75. $f(x) = \dfrac{x}{x + 1}$, so $f(-x) = \dfrac{-x}{-x + 1} = \dfrac{x}{x - 1}$.

Since this is neither $f(x)$ nor $-f(x)$, the function f is neither even nor odd.

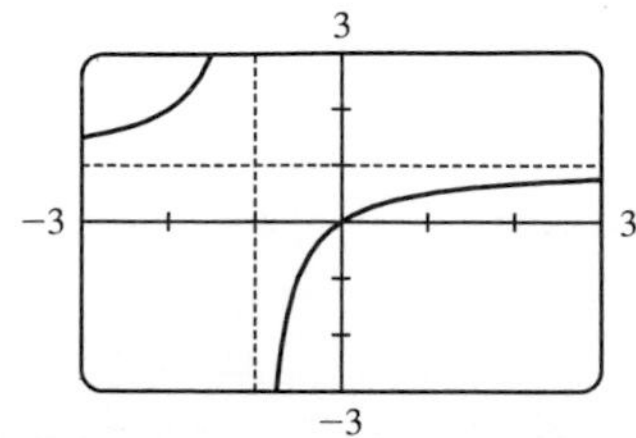

76. $f(x) = x\,|x|$.

$$f(-x) = (-x)\,|-x| = (-x)\,|x| = -(x\,|x|) = -f(x)$$

Since $f(-x) = -f(x)$, f is an odd function.

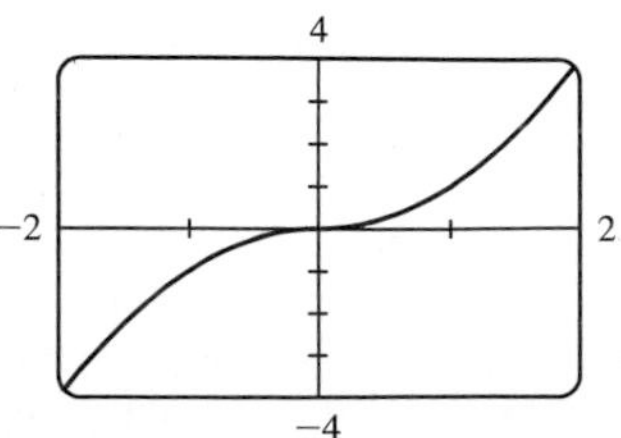

77. $f(x) = 1 + 3x^2 - x^4$.

$f(-x) = 1+3(-x)^2-(-x)^4 = 1+3x^2-x^4 = f(x)$.

Since $f(-x) = f(x)$, f is an even function.

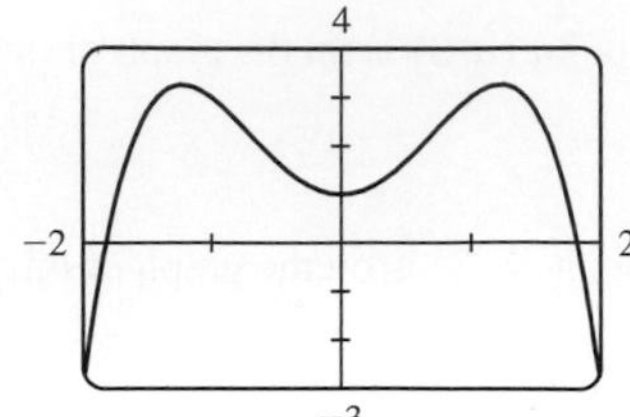

78. $f(x) = 1 + 3x^3 - x^5$, so

$$f(-x) = 1 + 3(-x)^3 - (-x)^5 = 1 + 3(-x^3) - (-x^5)$$
$$= 1 - 3x^3 + x^5$$

Since this is neither $f(x)$ nor $-f(x)$, the function f is neither even nor odd.

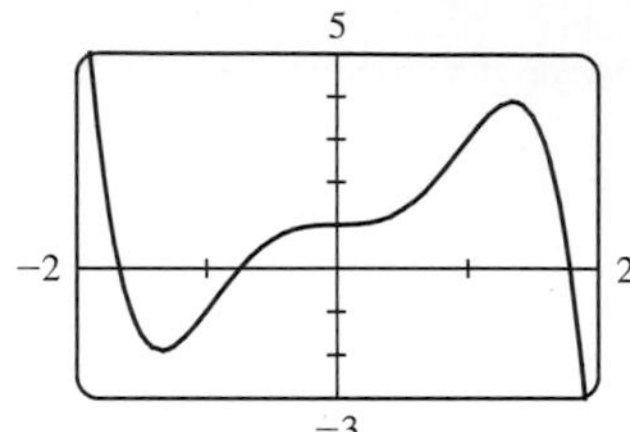

79. (i) If f and g are both even functions, then $f(-x) = f(x)$ and $g(-x) = g(x)$. Now $(f + g)(-x) = f(-x) + g(-x) = f(x) + g(x) = (f + g)(x)$, so $f + g$ is an *even* function.

(ii) If f and g are both odd functions, then $f(-x) = -f(x)$ and $g(-x) = -g(x)$. Now $(f + g)(-x) = f(-x) + g(-x) = -f(x) + [-g(x)] = -[f(x) + g(x)] = -(f + g)(x)$, so $f + g$ is an *odd* function.

(iii) If f is an even function and g is an odd function, then $(f+g)(-x) = f(-x)+g(-x) = f(x)+[-g(x)] = f(x)-g(x)$, which is not $(f + g)(x)$ nor $-(f + g)(x)$, so $f + g$ is *neither* even nor odd. (Exception: if f is the zero function, then $f + g$ will be *odd*. If g is the zero function, then $f + g$ will be *even*.)

80. (i) If f and g are both even functions, then $f(-x) = f(x)$ and $g(-x) = g(x)$. Now $(fg)(-x) = f(-x)g(-x) = f(x)g(x) = (fg)(x)$, so fg is an *even* function.

(ii) If f and g are both odd functions, then $f(-x) = -f(x)$ and $g(-x) = -g(x)$. Now $(fg)(-x) = f(-x)g(-x) = [-f(x)][-g(x)] = f(x)g(x) = (fg)(x)$, so fg is an *even* function.

(iii) If f is an even function and g is an odd function, then $(fg)(-x) = f(-x)g(-x) = f(x)[-g(x)] = -[f(x)g(x)] = -(fg)(x)$, so fg is an *odd* function.

1.2 Mathematical Models: A Catalog of Essential Functions

1. (a) $f(x) = \log_2 x$ is a logarithmic function.

(b) $g(x) = \sqrt[4]{x}$ is a root function with $n = 4$.

(c) $h(x) = \dfrac{2x^3}{1 - x^2}$ is a rational function because it is a ratio of polynomials.

(d) $u(t) = 1 - 1.1t + 2.54t^2$ is a polynomial of degree 2 (also called a *quadratic function*).

(e) $v(t) = 5^t$ is an exponential function.

(f) $w(\theta) = \sin\theta\, \cos^2\theta$ is a trigonometric function.

2. (a) $y = \pi^x$ is an exponential function (notice that x is the *exponent*).

(b) $y = x^\pi$ is a power function (notice that x is the *base*).

(c) $y = x^2(2 - x^3) = 2x^2 - x^5$ is a polynomial of degree 5.

(d) $y = \tan t - \cos t$ is a trigonometric function.

(e) $y = s/(1 + s)$ is a rational function because it is a ratio of polynomials.

(f) $y = \sqrt{x^3 - 1}/(1 + \sqrt[3]{x})$ is an algebraic function because it involves polynomials and roots of polynomials.

3. We notice from the figure that g and h are even functions (symmetric with respect to the y-axis) and that f is an odd function (symmetric with respect to the origin). So (b) $[y = x^5]$ must be f. Since g is flatter than h near the origin, we must have (c) $[y = x^8]$ matched with g and (a) $[y = x^2]$ matched with h.

4. (a) The graph of $y = 3x$ is a line (choice G).

(b) $y = 3^x$ is an exponential function (choice f).

(c) $y = x^3$ is an odd polynomial function or power function (choice F).

(d) $y = \sqrt[3]{x} = x^{1/3}$ is a root function (choice g).

5. The denominator cannot equal 0, so $1 - \sin x \neq 0 \iff \sin x \neq 1 \iff x \neq \frac{\pi}{2} + 2n\pi$. Thus, the domain of $f(x) = \dfrac{\cos x}{1 - \sin x}$ is $\left\{x \mid x \neq \frac{\pi}{2} + 2n\pi, n \text{ an integer}\right\}$.

6. The denominator cannot equal 0, so $1 - \tan x \neq 0 \iff \tan x \neq 1 \iff x \neq \frac{\pi}{4} + n\pi$. The tangent function is not defined if $x \neq \frac{\pi}{2} + n\pi$. Thus, the domain of $g(x) = \dfrac{1}{1 - \tan x}$ is $\left\{x \mid x \neq \frac{\pi}{4} + n\pi, x \neq \frac{\pi}{2} + n\pi, n \text{ an integer}\right\}$.

7. (a) An equation for the family of linear functions with slope 2 is $y = f(x) = 2x + b$, where b is the y-intercept.

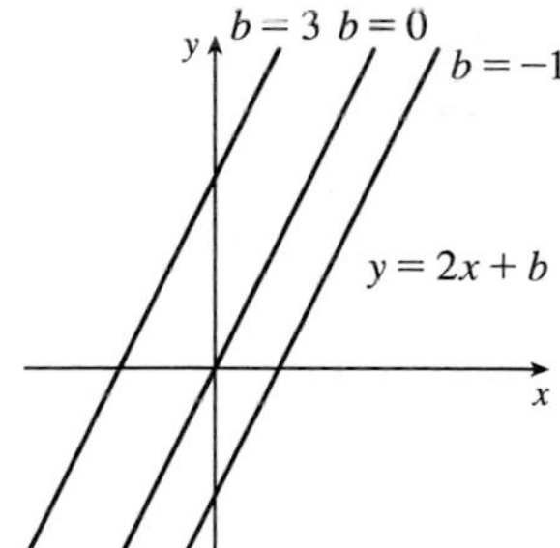

(b) $f(2) = 1$ means that the point $(2, 1)$ is on the graph of f. We can use the point-slope form of a line to obtain an equation for the family of linear functions through the point $(2, 1)$. $y - 1 = m(x - 2)$, which is equivalent to $y = mx + (1 - 2m)$ in slope-intercept form.

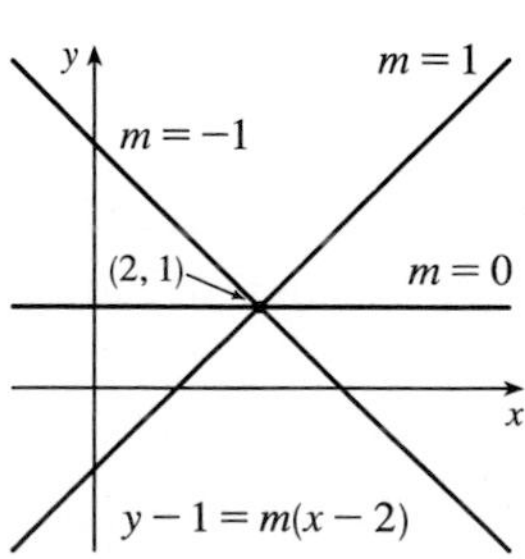

(c) To belong to both families, an equation must have slope $m = 2$, so the equation in part (b), $y = mx + (1 - 2m)$, becomes $y = 2x - 3$. It is the *only* function that belongs to both families.

8. All members of the family of linear functions $f(x) = 1 + m(x + 3)$ have graphs that are lines passing through the point $(-3, 1)$.

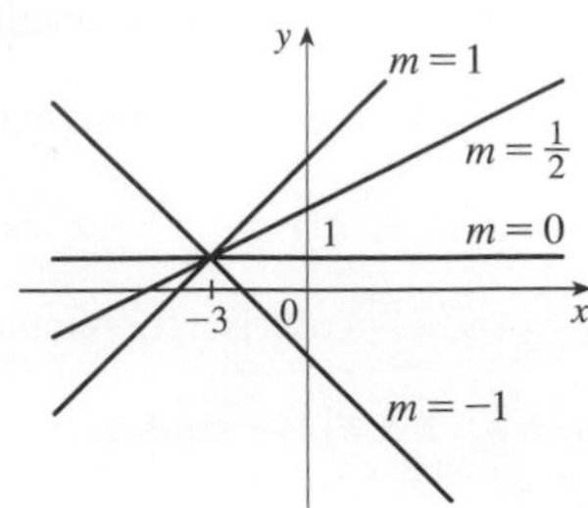

9. All members of the family of linear functions $f(x) = c - x$ have graphs that are lines with slope -1. The y-intercept is c.

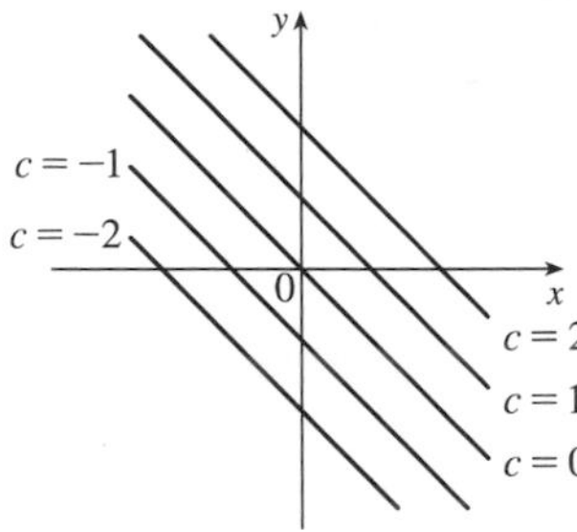

10. The vertex of the parabola on the left is $(3, 0)$, so an equation is $y = a(x - 3)^2 + 0$. Since the point $(4, 2)$ is on the parabola, we'll substitute 4 for x and 2 for y to find a. $2 = a(4 - 3)^2 \quad\Rightarrow\quad a = 2$, so an equation is $f(x) = 2(x - 3)^2$.

The y-intercept of the parabola on the right is $(0, 1)$, so an equation is $y = ax^2 + bx + 1$. Since the points $(-2, 2)$ and $(1, -2.5)$ are on the parabola, we'll substitute -2 for x and 2 for y as well as 1 for x and -2.5 for y to obtain two equations with the unknowns a and b.

$$(-2, 2): \qquad 2 = 4a - 2b + 1 \quad\Rightarrow\quad 4a - 2b = 1 \qquad \textbf{(1)}$$

$$(1, -2.5): \qquad -2.5 = a + b + 1 \quad\Rightarrow\quad a + b = -3.5 \qquad \textbf{(2)}$$

$2 \cdot$ **(2)** + **(1)** gives us $6a = -6 \quad\Rightarrow\quad a = -1$. From **(2)**, $-1 + b = -3.5 \quad\Rightarrow\quad b = -2.5$, so an equation is $g(x) = -x^2 - 2.5x + 1$.

11. Since $f(-1) = f(0) = f(2) = 0$, f has zeros of -1, 0, and 2, so an equation for f is $f(x) = a[x - (-1)](x - 0)(x - 2)$, or $f(x) = ax(x + 1)(x - 2)$. Because $f(1) = 6$, we'll substitute 1 for x and 6 for $f(x)$.
$6 = a(1)(2)(-1) \quad\Rightarrow\quad -2a = 6 \quad\Rightarrow\quad a = -3$, so an equation for f is $f(x) = -3x(x + 1)(x - 2)$.

12. (a) For $T = 0.02t + 8.50$, the slope is 0.02, which means that the average surface temperature of the world is increasing at a rate of $0.02\,^\circ\text{C}$ per year. The T-intercept is 8.50, which represents the average surface temperature in $^\circ\text{C}$ in the year 1900.

(b) $t = 2100 - 1900 = 200 \quad\Rightarrow\quad T = 0.02(200) + 8.50 = 12.50\,^\circ\text{C}$

13. (a) $D = 200$, so $c = 0.0417D(a + 1) = 0.0417(200)(a + 1) = 8.34a + 8.34$. The slope is 8.34, which represents the change in mg of the dosage for a child for each change of 1 year in age.

(b) For a newborn, $a = 0$, so $c = 8.34$ mg.

14. (a)

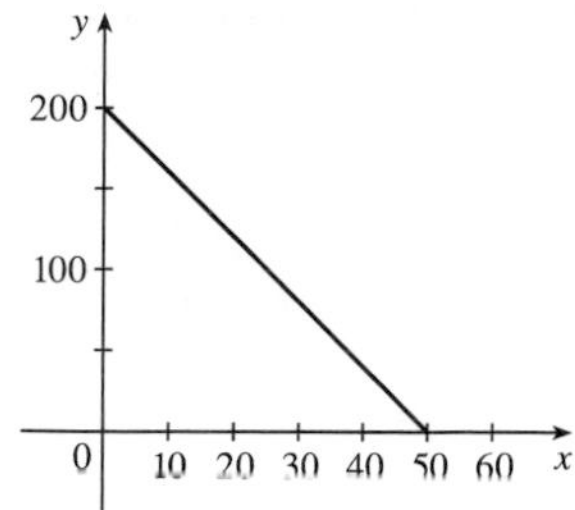

(b) The slope of -4 means that for each increase of 1 dollar for a rental space, the number of spaces rented *decreases* by 4. The y-intercept of 200 is the number of spaces that would be occupied if there were no charge for each space. The x-intercept of 50 is the smallest rental fee that results in no spaces rented.

15. (a)

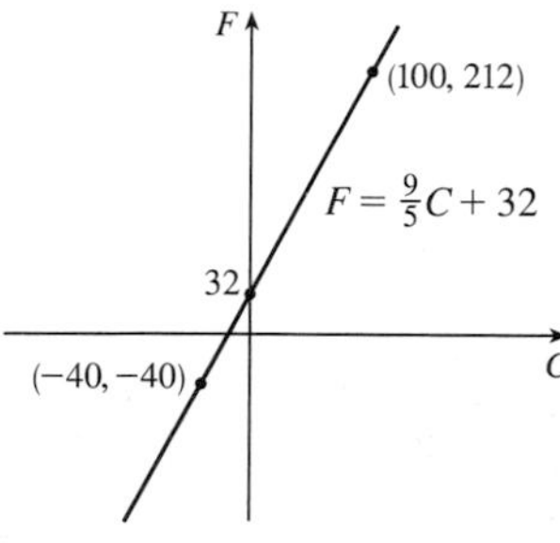

(b) The slope of $\frac{9}{5}$ means that F increases $\frac{9}{5}$ degrees for each increase of $1°$C. (Equivalently, F increases by 9 when C increases by 5 and F decreases by 9 when C decreases by 5.) The F-intercept of 32 is the Fahrenheit temperature corresponding to a Celsius temperature of 0.

16. (a) Let $d =$ distance traveled (in miles) and $t =$ time elapsed (in hours). At $t = 0$, $d = 0$ and at $t = 50$ minutes $= 50 \cdot \frac{1}{60} = \frac{5}{6}$ h, $d = 40$. Thus we have two points: $(0, 0)$ and $\left(\frac{5}{6}, 40\right)$, so $m = \dfrac{40 - 0}{\frac{5}{6} - 0} = 48$ and so $d = 48t$.

(b)

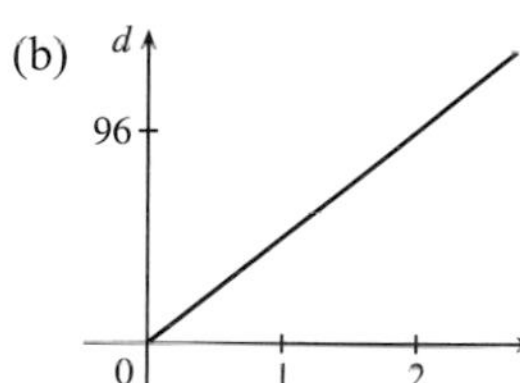

(c) The slope is 48 and represents the car's speed in mi/h.

17. (a) Using N in place of x and T in place of y, we find the slope to be $\dfrac{T_2 - T_1}{N_2 - N_1} = \dfrac{80 - 70}{173 - 113} = \dfrac{10}{60} = \dfrac{1}{6}$. So a linear equation is $T - 80 = \frac{1}{6}(N - 173) \Leftrightarrow T - 80 = \frac{1}{6}N - \frac{173}{6} \Leftrightarrow T = \frac{1}{6}N + \frac{307}{6}$ $\left[\frac{307}{6} = 51.1\overline{6}\right]$.

(b) The slope of $\frac{1}{6}$ means that the temperature in Fahrenheit degrees increases one-sixth as rapidly as the number of cricket chirps per minute. Said differently, each increase of 6 cricket chirps per minute corresponds to an increase of $1°$F.

(c) When $N = 150$, the temperature is given approximately by $T = \frac{1}{6}(150) + \frac{307}{6} = 76.1\overline{6}\,°\text{F} \approx 76\,°\text{F}$.

18. (a) Let x denote the number of chairs produced in one day and y the associated cost. Using the points $(100, 2200)$ and $(300, 4800)$, we get the slope $\frac{4800-2200}{300-100} = \frac{2600}{200} = 13$. So $y - 2200 = 13(x - 100) \Leftrightarrow$ $y = 13x + 900$.

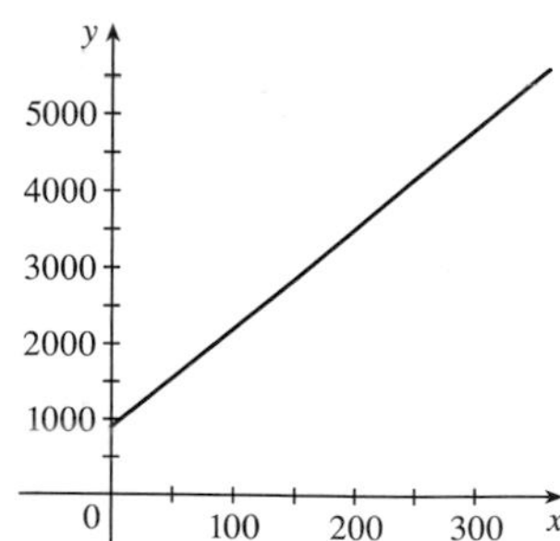

(b) The slope of the line in part (a) is 13 and it represents the cost (in dollars) of producing each additional chair.

(c) The y-intercept is 900 and it represents the fixed daily costs of operating the factory.

19. (a) We are given $\dfrac{\text{change in pressure}}{10 \text{ feet change in depth}} = \dfrac{4.34}{10} = 0.434$. Using P for pressure and d for depth with the point $(d, P) = (0, 15)$, we have the slope-intercept form of the line, $P = 0.434d + 15$.

(b) When $P = 100$, then $100 = 0.434d + 15 \;\Leftrightarrow\; 0.434d = 85 \;\Leftrightarrow\; d = \frac{85}{0.434} \approx 195.85$ feet. Thus, the pressure is $100\ \text{lb/in}^2$ at a depth of approximately 196 feet.

20. (a) Using d in place of x and C in place of y, we find the slope to be $\dfrac{C_2 - C_1}{d_2 - d_1} = \dfrac{460 - 380}{800 - 480} = \dfrac{80}{320} = \dfrac{1}{4}$.

So a linear equation is $C - 460 = \frac{1}{4}(d - 800) \;\Leftrightarrow\; C - 460 = \frac{1}{4}d - 200 \;\Leftrightarrow\; C = \frac{1}{4}d + 260$.

(b) Letting $d = 1500$ we get $C = \frac{1}{4}(1500) + 260 = 635$.

The cost of driving 1500 miles is \$635.

(c)

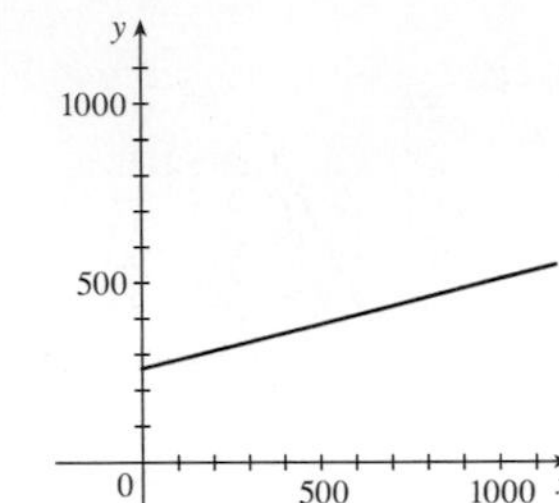

The slope of the line represents the cost per mile, \$0.25.

(d) The y-intercept represents the fixed cost, \$260.

(e) A linear function gives a suitable model in this situation because you have fixed monthly costs such as insurance and car payments, as well as costs that increase as you drive, such as gasoline, oil, and tires, and the cost of these for each additional mile driven is a constant.

21. (a) The data appear to be periodic and a sine or cosine function would make the best model. A model of the form $f(x) = a\cos(bx) + c$ seems appropriate.

(b) The data appear to be decreasing in a linear fashion. A model of the form $f(x) = mx + b$ seems appropriate.

22. (a) The data appear to be increasing exponentially. A model of the form $f(x) = a \cdot b^x$ or $f(x) = a \cdot b^x + c$ seems appropriate.

(b) The data appear to be decreasing similarly to the values of the reciprocal function. A model of the form $f(x) = a/x$ seems appropriate.

Exercises 23–28: Some values are given to many decimal places. These are the results given by several computer algebra systems — rounding is left to the reader.

23. (a)

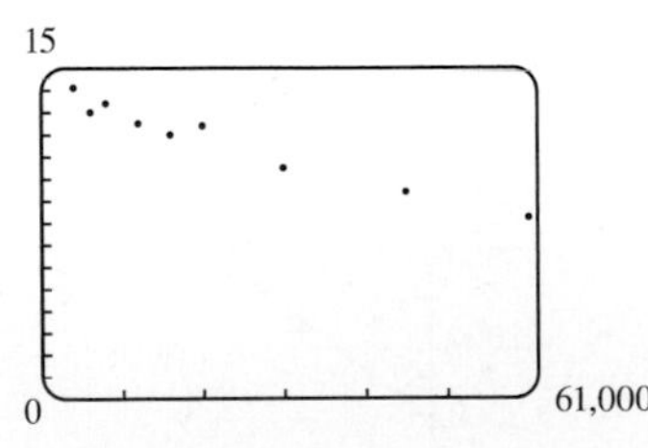

A linear model does seem appropriate.

(b) Using the points (4000, 14.1) and (60,000, 8.2), we obtain

$y - 14.1 = \dfrac{8.2 - 14.1}{60{,}000 - 4000}(x - 4000)$ or, equivalently,

$y \approx -0.000105357x + 14.521429$.

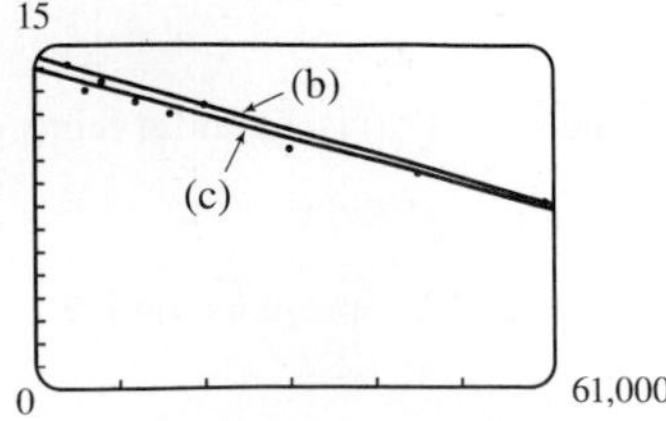

(c) Using a computing device, we obtain the least squares regression line $y = -0.0000997855x + 13.950764$.

The following commands and screens illustrate how to find the least squares regression line on a TI-84 Plus.

Enter the data into list one (L1) and list two (L2). Press STAT 1 to enter the editor.

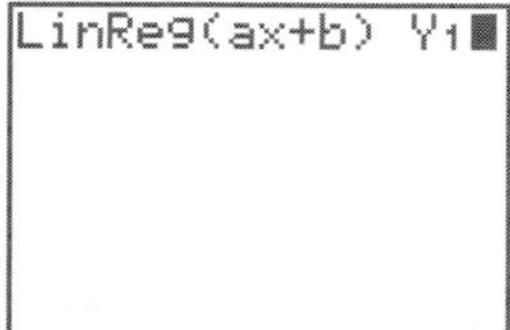

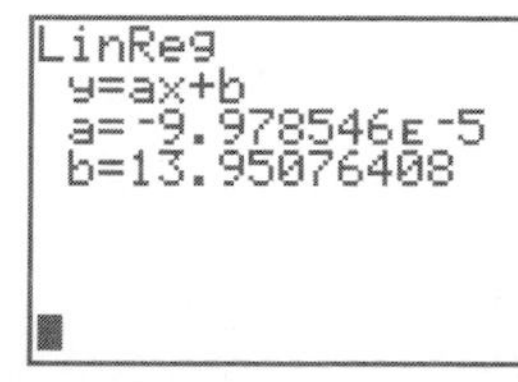

Find the regession line and store it in Y_1. Press 2nd QUIT STAT ▸ 4 VARS ▸ 1 1 ENTER .

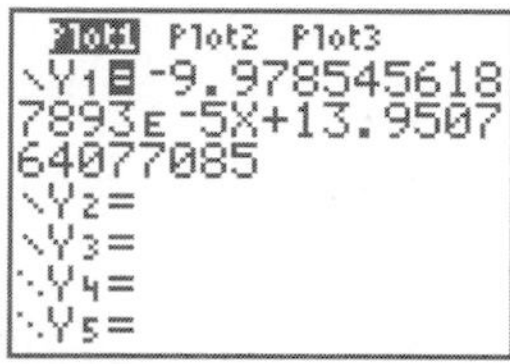

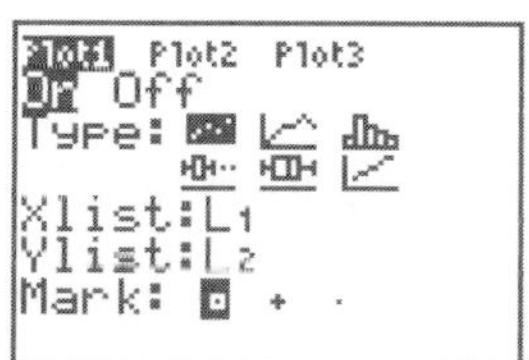

Note from the last figure that the regression line has been stored in Y_1 and that Plot1 has been turned on (Plot1 is highlighted). You can turn on Plot1 from the Y= menu by placing the cursor on Plot1 and pressing ENTER or by pressing 2nd STAT PLOT 1 ENTER .

Now press ZOOM 9 to produce a graph of the data and the regression line. Note that choice 9 of the ZOOM menu automatically selects a window that displays all of the data.

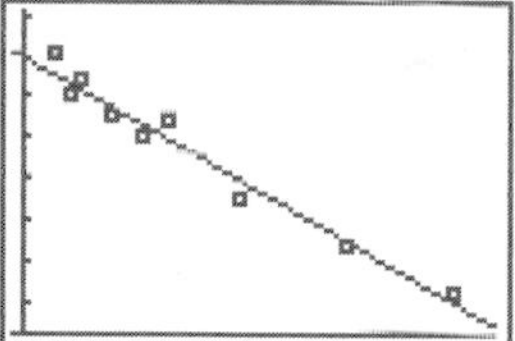

(d) When $x = 25{,}000$, $y \approx 11.456$; or about 11.5 per 100 population.

(e) When $x = 80{,}000$, $y \approx 5.968$; or about a 6% chance.

(f) When $x = 200{,}000$, y is negative, so the model does not apply.

24. (a)

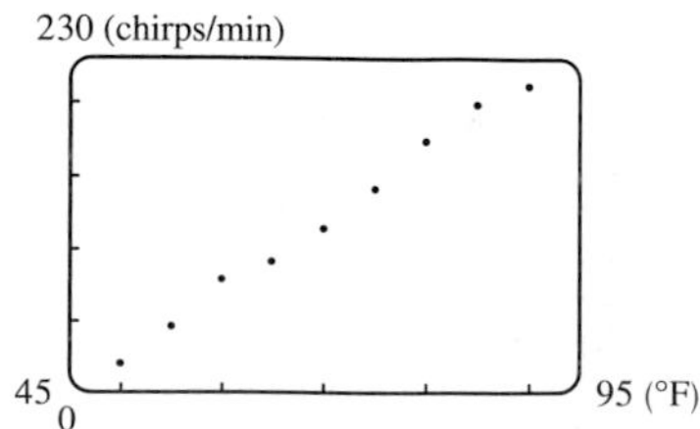

(b)

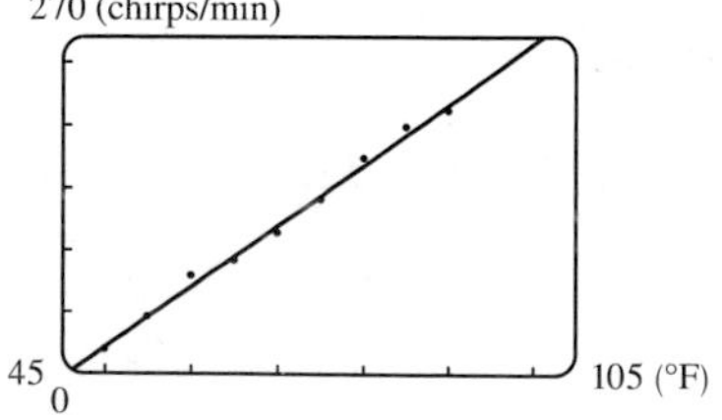

Using a computing device, we obtain the least squares regression line $y = 4.85\overline{6}x - 220.9\overline{6}$.

(c) When $x = 100°\text{F}$, $y = 264.7 \approx 265$ chirps/min.

25. (a)

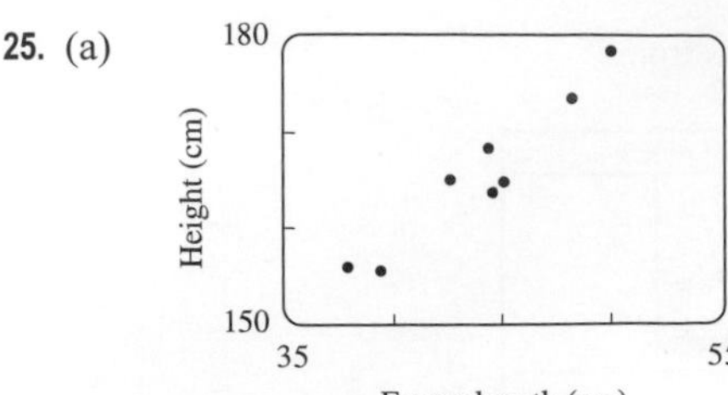

(b) Using a computing device, we obtain the regression line $y = 1.88074x + 82.64974$.

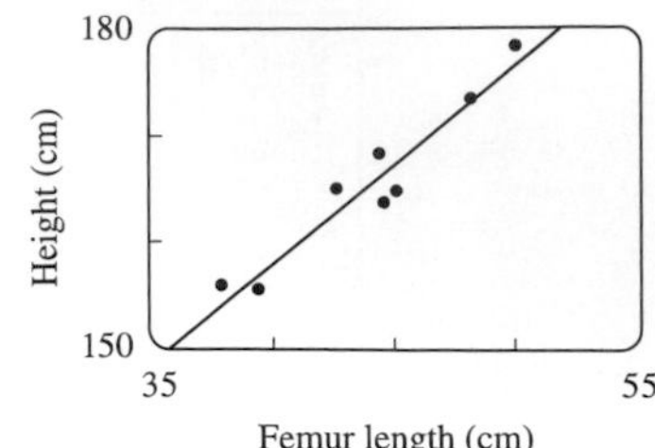

(c) When $x = 53$ cm, $y \approx 182.3$ cm.

26. (a) Using a computing device, we obtain the regression line $y = 0.01879x + 0.30480$.

(b) The regression line appears to be a suitable model for the data.

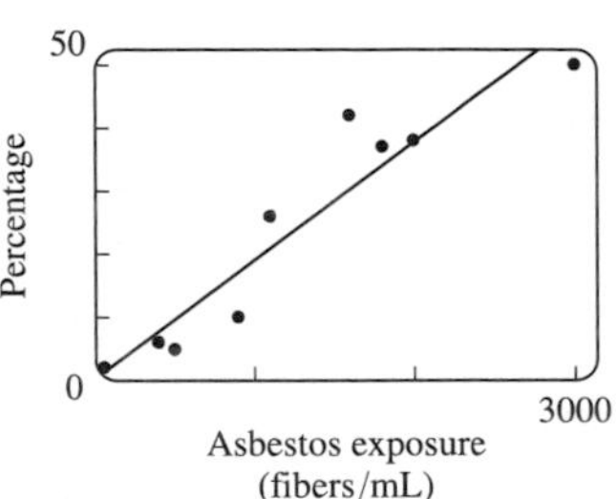

(c) The y-intercept represents the percentage of laboratory rats that develop lung tumors when *not* exposed to asbestos fibers.

27. (a) See the scatter plot in part (b). A linear model seems appropriate.

(b) Using a computing device, we obtain the regression line $y = 1116.64x + 60{,}188.33$.

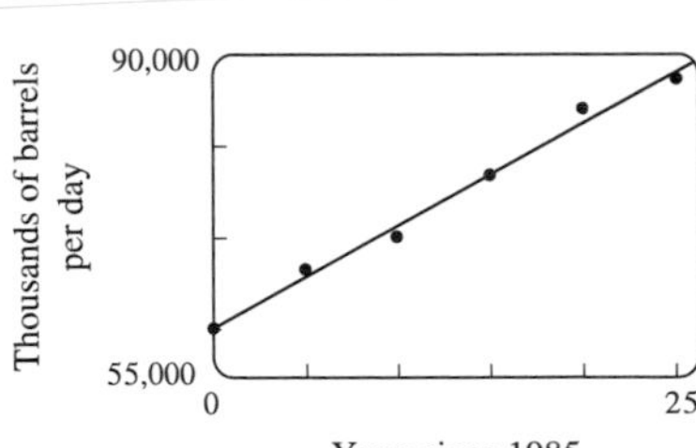

(c) For 2002, $x = 17$ and $y \approx 79{,}171$ thousands of barrels per day. For 2012, $x = 27$ and $y \approx 90{,}338$ thousands of barrels per day.

28. (a) See the scatter plot in part (b). A linear model seems appropriate.

(b) Using a computing device, we obtain the regression line $y = 0.33089x + 8.07321$.

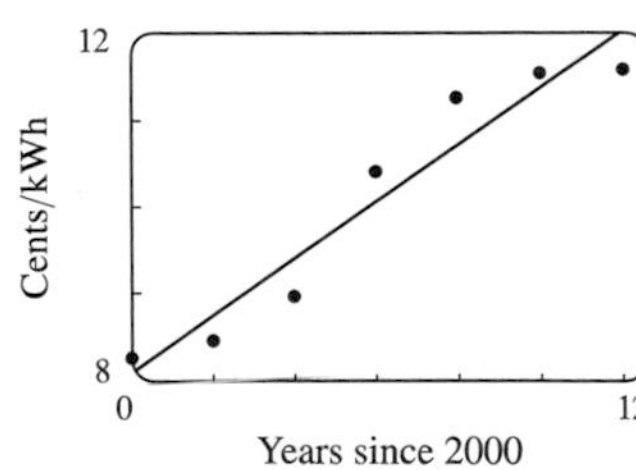

(c) For 2005, $x = 5$ and $y \approx 9.73$ cents/kWh. For 2013, $x = 13$ and $y \approx 12.37$ cents/kWh.

29. If x is the original distance from the source, then the illumination is $f(x) = kx^{-2} = k/x^2$. Moving halfway to the lamp gives us an illumination of $f\left(\frac{1}{2}x\right) = k\left(\frac{1}{2}x\right)^{-2} = k(2/x)^2 = 4(k/x^2)$, so the light is 4 times as bright.

30. (a) If $A = 60$, then $S = 0.7A^{0.3} \approx 2.39$, so you would expect to find 2 species of bats in that cave.

(b) $S = 4 \;\Rightarrow\; 4 = 0.7A^{0.3} \;\Rightarrow\; \frac{40}{7} = A^{3/10} \;\Rightarrow\; A = \left(\frac{40}{7}\right)^{10/3} \approx 333.6$, so we estimate the surface area of the cave to be 334 m^2.

31. (a) Using a computing device, we obtain a power function $N = cA^b$, where $c \approx 3.1046$ and $b \approx 0.308$.

(b) If $A = 291$, then $N = cA^b \approx 17.8$, so you would expect to find 18 species of reptiles and amphibians on Dominica.

32. (a) $T = 1.000\,431\,227d^{\,1.499\,528\,750}$

(b) The power model in part (a) is approximately $T = d^{1.5}$. Squaring both sides gives us $T^2 = d^3$, so the model matches Kepler's Third Law, $T^2 = kd^3$.

1.3 New Functions from Old Functions

1. (a) If the graph of f is shifted 3 units upward, its equation becomes $y = f(x) + 3$.

(b) If the graph of f is shifted 3 units downward, its equation becomes $y = f(x) - 3$.

(c) If the graph of f is shifted 3 units to the right, its equation becomes $y = f(x - 3)$.

(d) If the graph of f is shifted 3 units to the left, its equation becomes $y = f(x + 3)$.

(e) If the graph of f is reflected about the x-axis, its equation becomes $y = -f(x)$.

(f) If the graph of f is reflected about the y-axis, its equation becomes $y = f(-x)$.

(g) If the graph of f is stretched vertically by a factor of 3, its equation becomes $y = 3f(x)$.

(h) If the graph of f is shrunk vertically by a factor of 3, its equation becomes $y = \frac{1}{3}f(x)$.

2. (a) To obtain the graph of $y = f(x) + 8$ from the graph of $y = f(x)$, shift the graph 8 units upward.

(b) To obtain the graph of $y = f(x + 8)$ from the graph of $y = f(x)$, shift the graph 8 units to the left.

(c) To obtain the graph of $y = 8f(x)$ from the graph of $y = f(x)$, stretch the graph vertically by a factor of 8.

(d) To obtain the graph of $y = f(8x)$ from the graph of $y = f(x)$, shrink the graph horizontally by a factor of 8.

(e) To obtain the graph of $y = -f(x) - 1$ from the graph of $y = f(x)$, first reflect the graph about the x-axis, and then shift it 1 unit downward.

(f) To obtain the graph of $y = 8f(\frac{1}{8}x)$ from the graph of $y = f(x)$, stretch the graph horizontally and vertically by a factor of 8.

3. (a) (graph 3) The graph of f is shifted 4 units to the right and has equation $y = f(x - 4)$.

(b) (graph 1) The graph of f is shifted 3 units upward and has equation $y = f(x) + 3$.

(c) (graph 4) The graph of f is shrunk vertically by a factor of 3 and has equation $y = \frac{1}{3}f(x)$.

(d) (graph 5) The graph of f is shifted 4 units to the left and reflected about the x-axis. Its equation is $y = -f(x + 4)$.

(e) (graph 2) The graph of f is shifted 6 units to the left and stretched vertically by a factor of 2. Its equation is $y = 2f(x + 6)$.

4. (a) $y = f(x) - 3$: Shift the graph of f 3 units down.

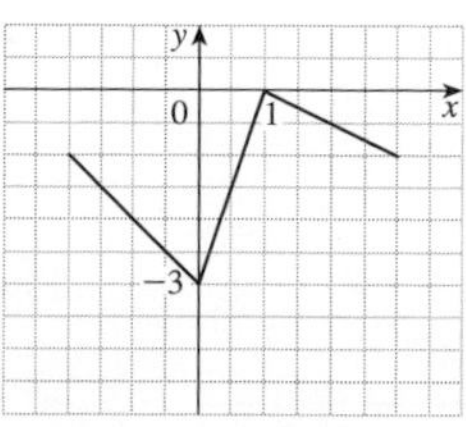

(b) $y = f(x + 1)$: Shift the graph of f 1 unit to the left.

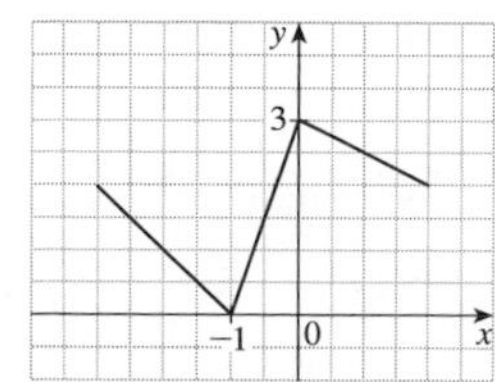

(c) $y = \frac{1}{2}f(x)$: Shrink the graph of f vertically by a factor of 2.

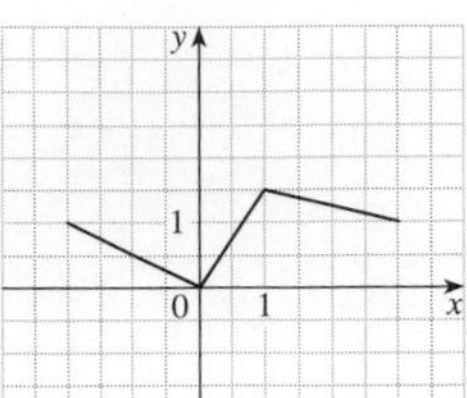

(d) $y = -f(x)$: Reflect the graph of f about the x-axis.

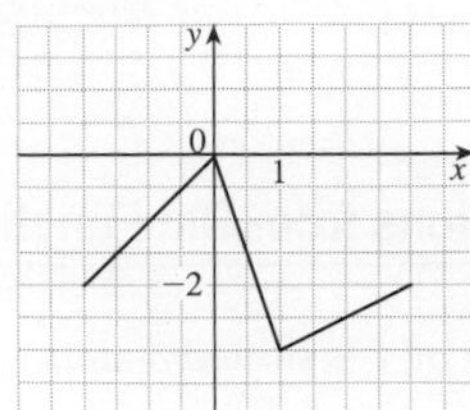

5. (a) To graph $y = f(2x)$ we shrink the graph of f horizontally by a factor of 2.

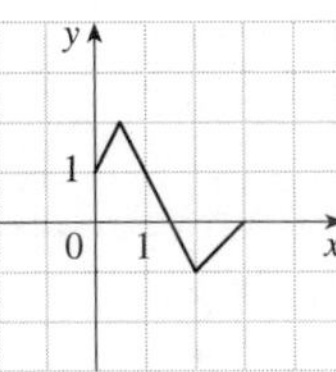

The point $(4, -1)$ on the graph of f corresponds to the point $\left(\frac{1}{2} \cdot 4, -1\right) = (2, -1)$.

(b) To graph $y = f\left(\frac{1}{2}x\right)$ we stretch the graph of f horizontally by a factor of 2.

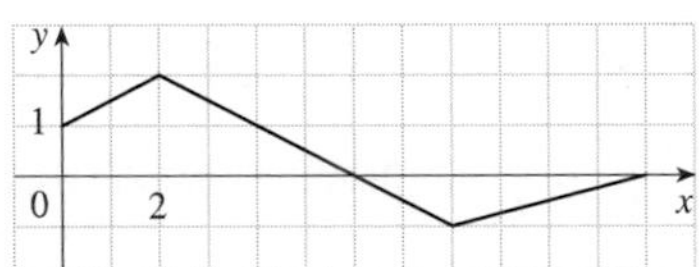

The point $(4, -1)$ on the graph of f corresponds to the point $(2 \cdot 4, -1) = (8, -1)$.

(c) To graph $y = f(-x)$ we reflect the graph of f about the y-axis.

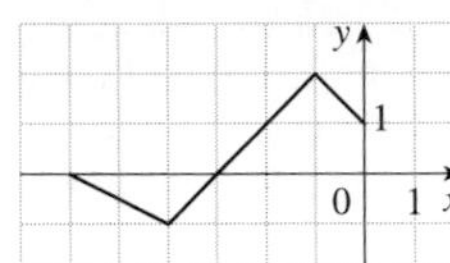

The point $(4, -1)$ on the graph of f corresponds to the point $(-1 \cdot 4, -1) = (-4, -1)$.

(d) To graph $y = -f(-x)$ we reflect the graph of f about the y-axis, then about the x-axis.

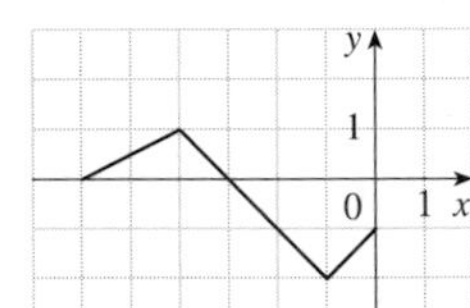

The point $(4, -1)$ on the graph of f corresponds to the point $(-1 \cdot 4, -1 \cdot -1) = (-4, 1)$.

6. The graph of $y = f(x) = \sqrt{3x - x^2}$ has been shifted 2 units to the right and stretched vertically by a factor of 2. Thus, a function describing the graph is

$$y = 2f(x-2) = 2\sqrt{3(x-2) - (x-2)^2} = 2\sqrt{3x - 6 - (x^2 - 4x + 4)} = 2\sqrt{-x^2 + 7x - 10}$$

7. The graph of $y = f(x) = \sqrt{3x - x^2}$ has been shifted 4 units to the left, reflected about the x-axis, and shifted downward 1 unit. Thus, a function describing the graph is

$$y = \underbrace{-1 \cdot}_{\substack{\text{reflect} \\ \text{about } x\text{-axis}}} \quad \underbrace{f\ (x+4)}_{\substack{\text{shift} \\ \text{4 units left}}} \quad \underbrace{-\ 1}_{\substack{\text{shift} \\ \text{1 unit left}}}$$

This function can be written as

$$\begin{aligned} y &= -f(x+4) - 1 = -\sqrt{3(x+4) - (x+4)^2} - 1 \\ &= -\sqrt{3x + 12 - (x^2 + 8x + 16)} - 1 = -\sqrt{-x^2 - 5x - 4} - 1 \end{aligned}$$

8. (a) The graph of $y = 2\sin x$ can be obtained from the graph of $y = \sin x$ by stretching it vertically by a factor of 2.

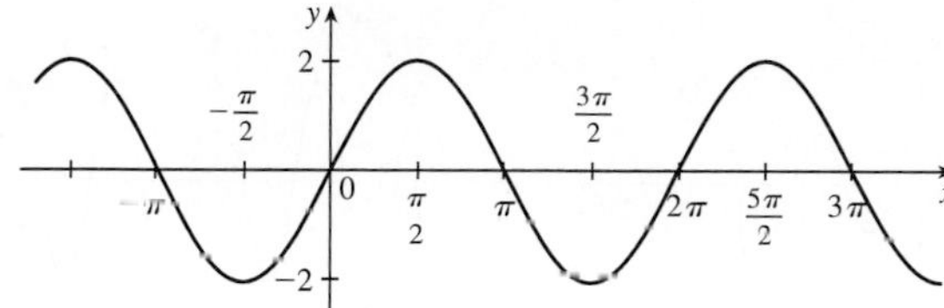

(b) The graph of $y = 1 + \sqrt{x}$ can be obtained from the graph of $y = \sqrt{x}$ by shifting it upward 1 unit.

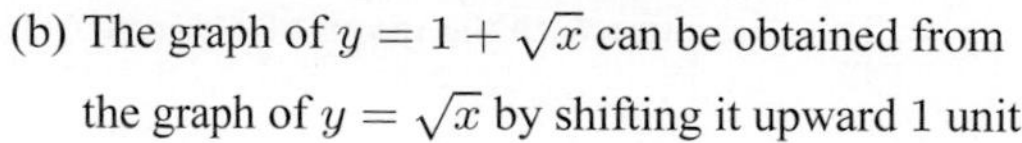

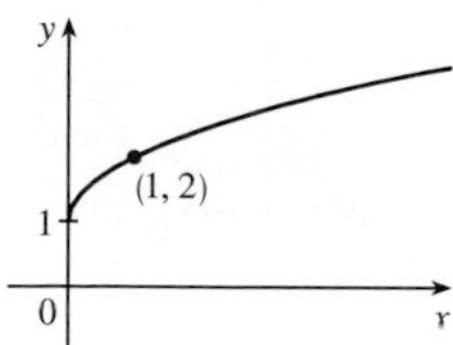

9. $y = -x^2$: Start with the graph of $y = x^2$ and reflect about the x-axis.

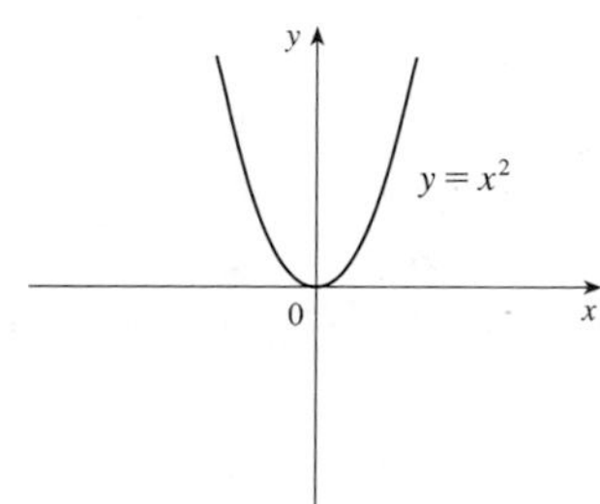

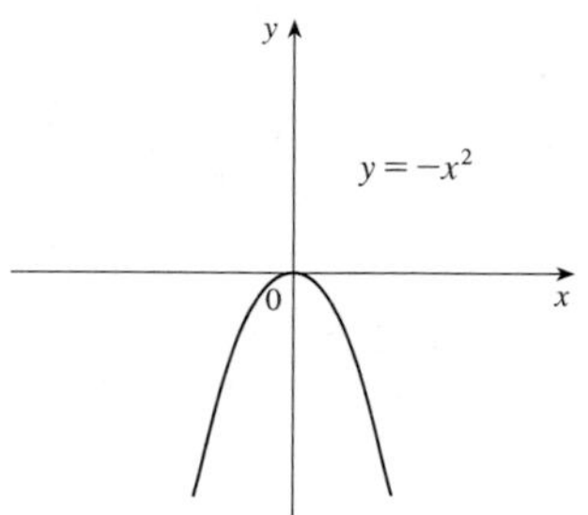

10. $y = (x - 3)^2$: Start with the graph of $y = x^2$ and shift 3 units to the right.

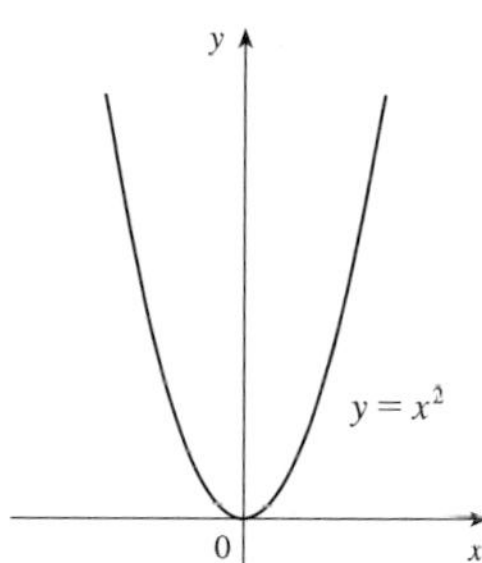

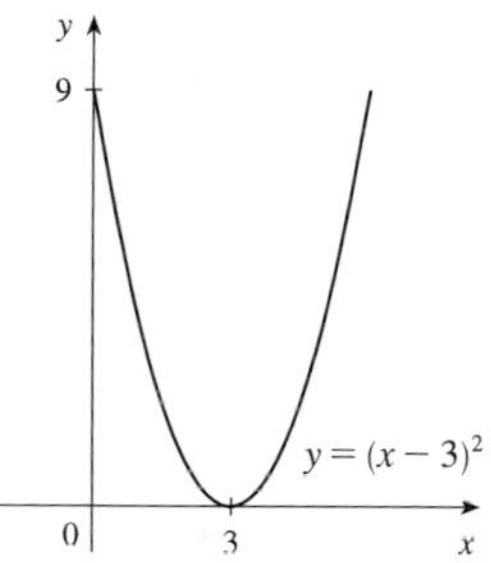

11. $y = x^3 + 1$: Start with the graph of $y = x^3$ and shift upward 1 unit.

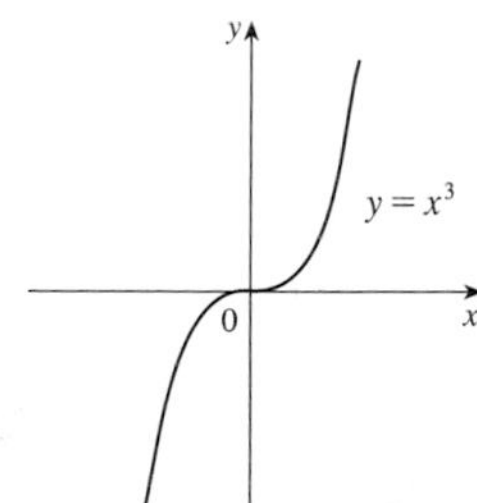

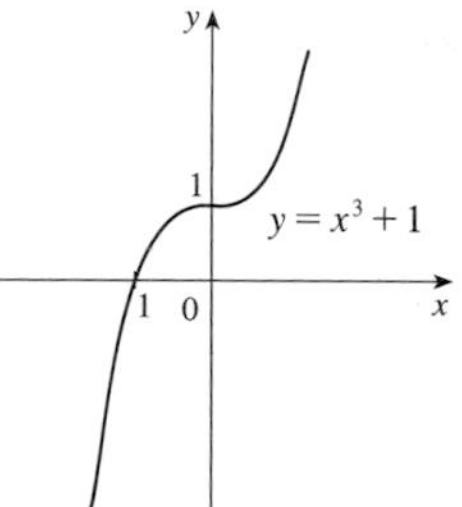

12. $y = 1 - \dfrac{1}{x} = -\dfrac{1}{x} + 1$: Start with the graph of $y = \dfrac{1}{x}$, reflect about the x-axis, and shift upward 1 unit.

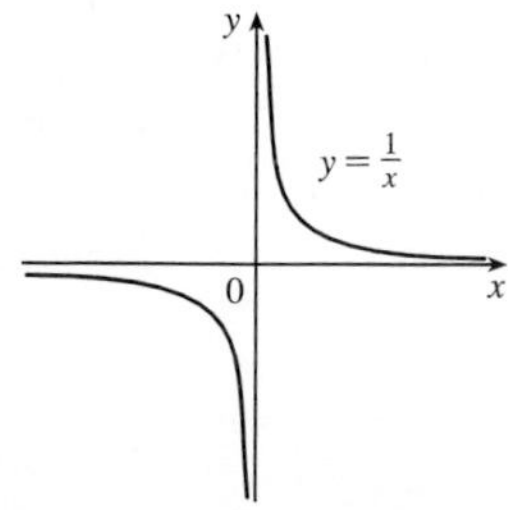

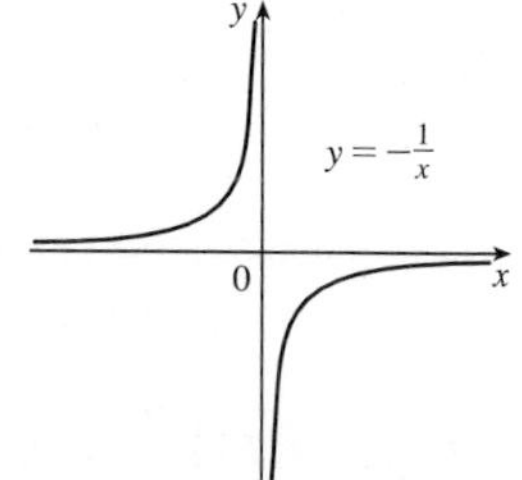

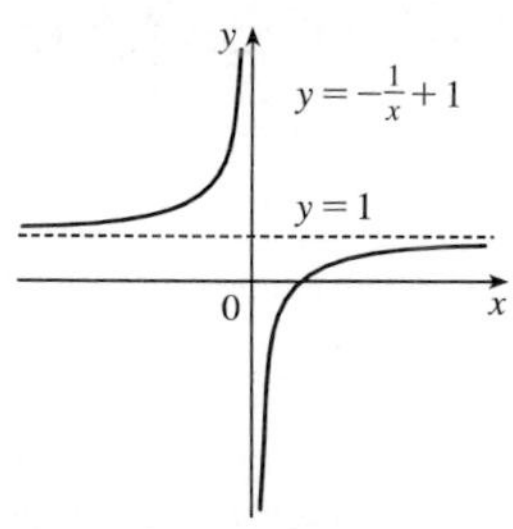

13. $y = 2\cos 3x$: Start with the graph of $y = \cos x$, compress horizontally by a factor of 3, and then stretch vertically by a factor of 2.

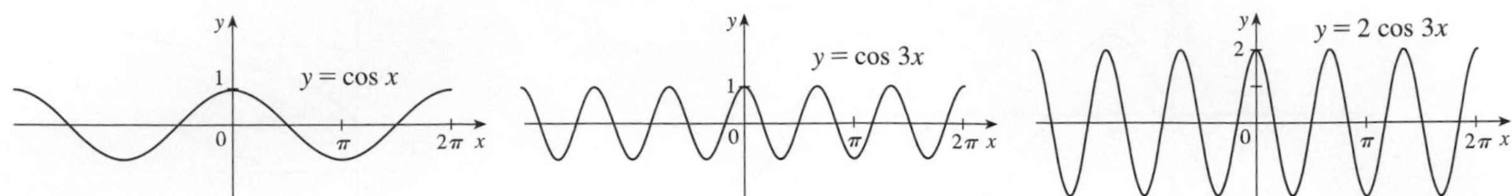

14. $y = 2\sqrt{x+1}$: Start with the graph of $y = \sqrt{x}$, shift 1 unit to the left, and then stretch vertically by a factor of 2.

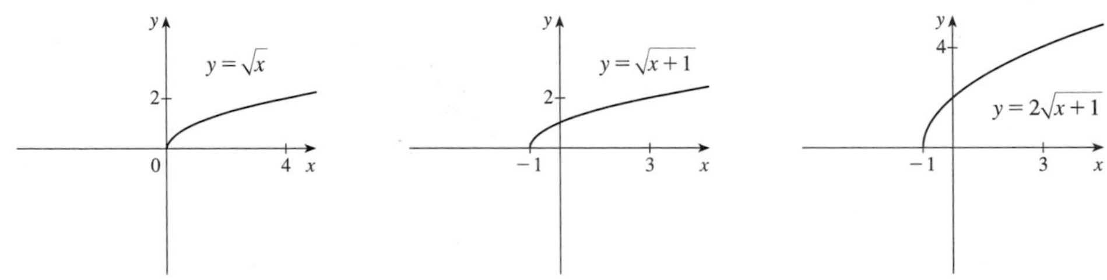

15. $y = x^2 - 4x + 5 = (x^2 - 4x + 4) + 1 = (x - 2)^2 + 1$: Start with the graph of $y = x^2$, shift 2 units to the right, and then shift upward 1 unit.

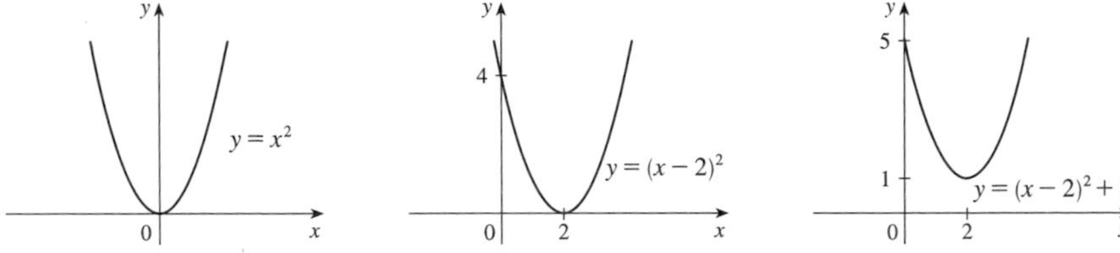

16. $y = 1 + \sin \pi x$: Start with the graph of $y = \sin x$, compress horizontally by a factor of π, and then shift upward 1 unit.

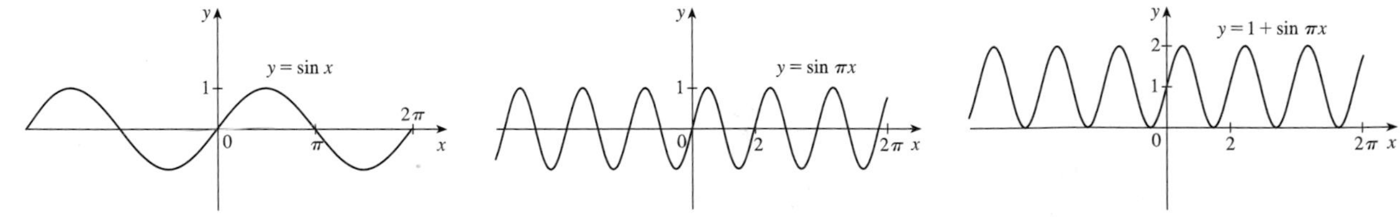

17. $y = 2 - \sqrt{x}$: Start with the graph of $y = \sqrt{x}$, reflect about the x-axis, and then shift 2 units upward.

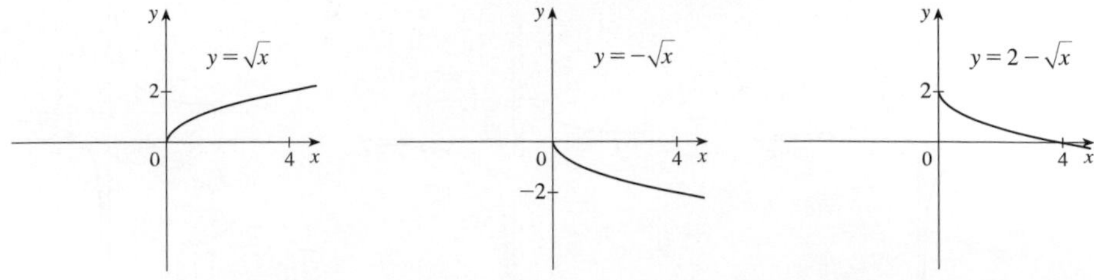

18. $y = 3 - 2\cos x$: Start with the graph of $y = \cos x$, stretch vertically by a factor of 2, reflect about the x-axis, and then shift 3 units upward.

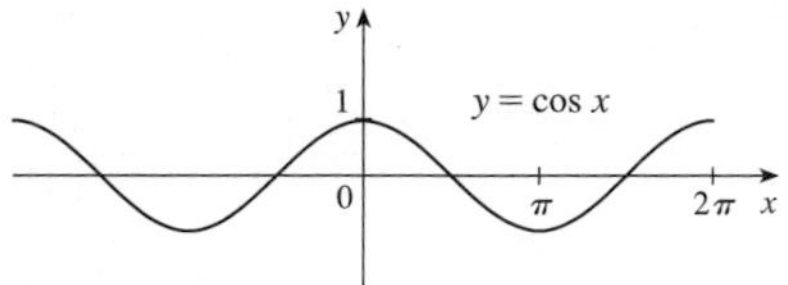

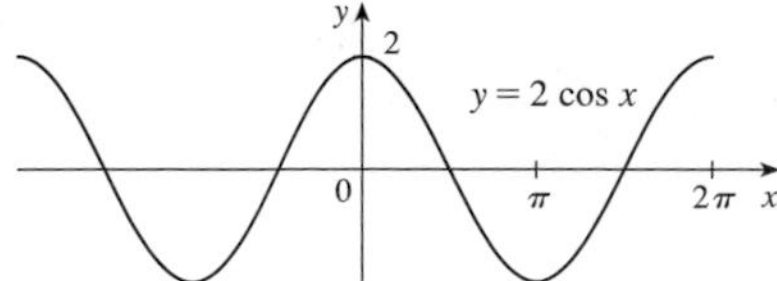

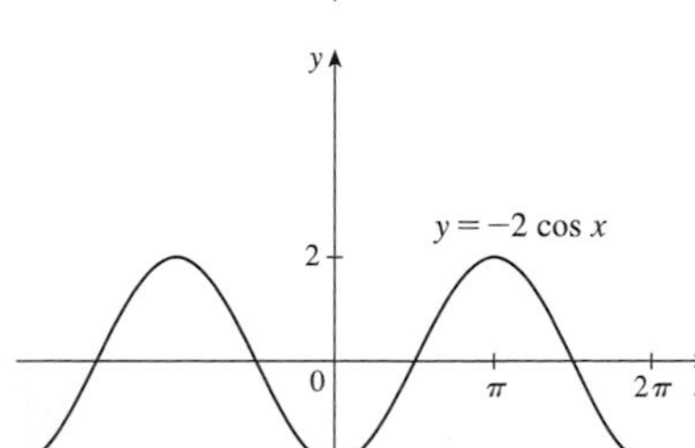

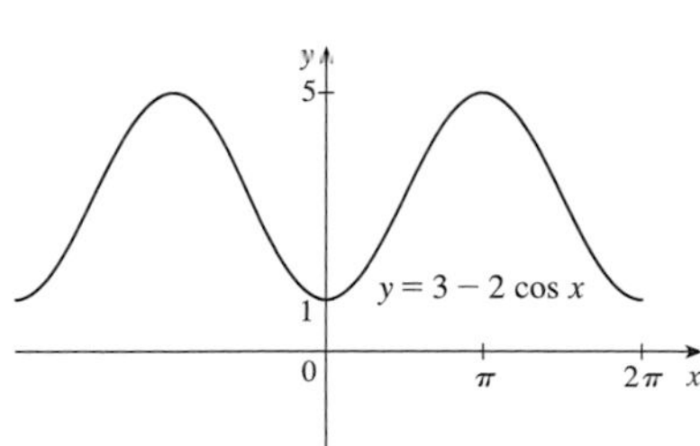

19. $y = \sin(x/2)$: Start with the graph of $y = \sin x$ and stretch horizontally by a factor of 2.

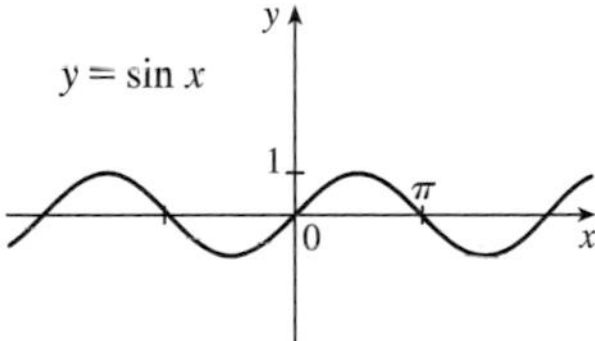

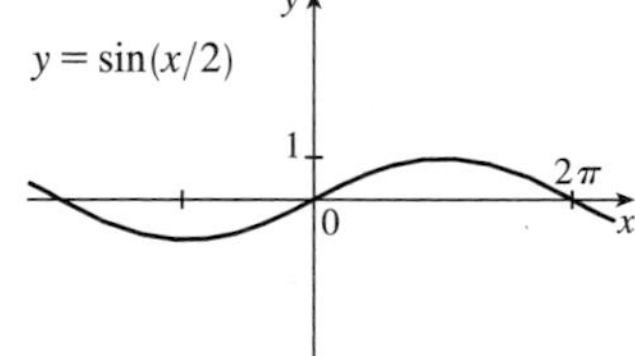

20. $y = |x| - 2$: Start with the graph of $y = |x|$ and shift 2 units downward.

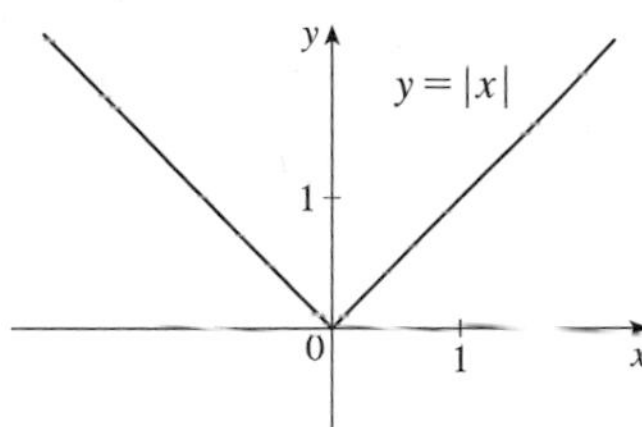

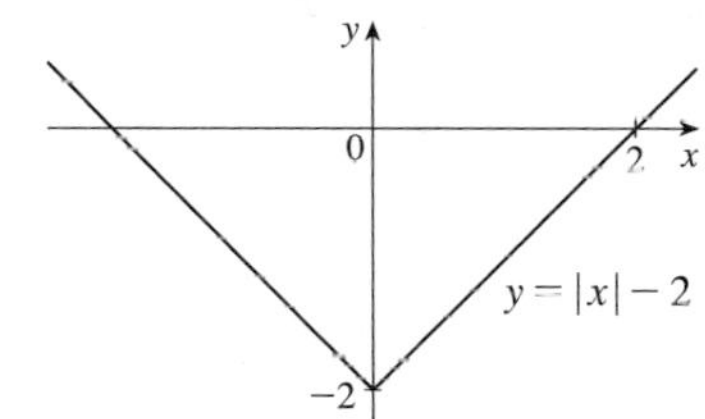

21. $y = |x - 2|$: Start with the graph of $y = |x|$ and shift 2 units to the right.

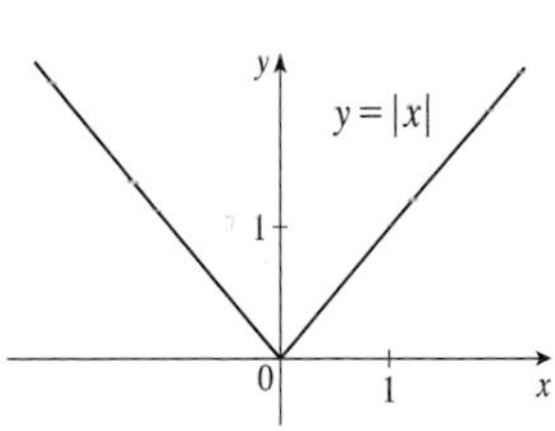

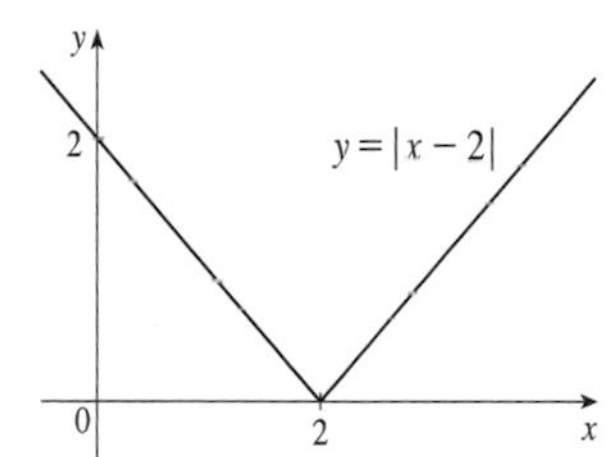

22. $y = \frac{1}{4}\tan(x - \frac{\pi}{4})$: Start with the graph of $y = \tan x$, shift $\frac{\pi}{4}$ units to the right, and then compress vertically by a factor of 4.

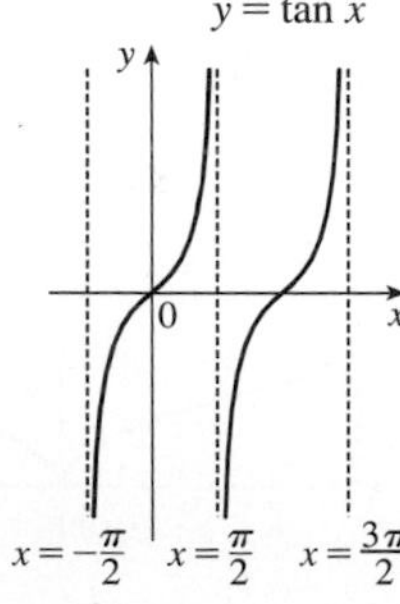

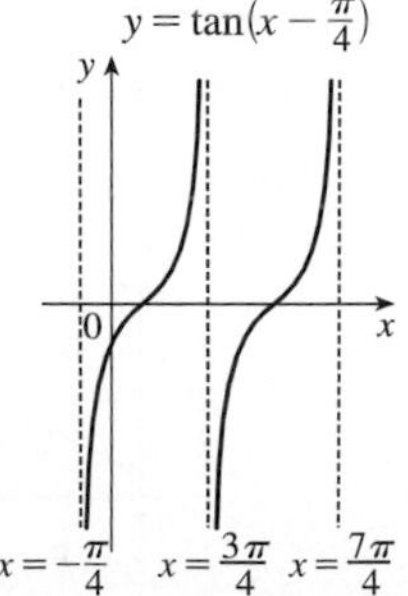

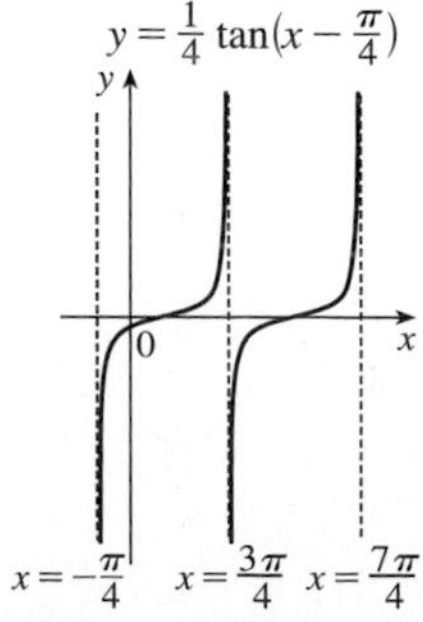

23. $y = |\sqrt{x} - 1|$: Start with the graph of $y = \sqrt{x}$, shift it 1 unit downward, and then reflect the portion of the graph below the x-axis about the x-axis.

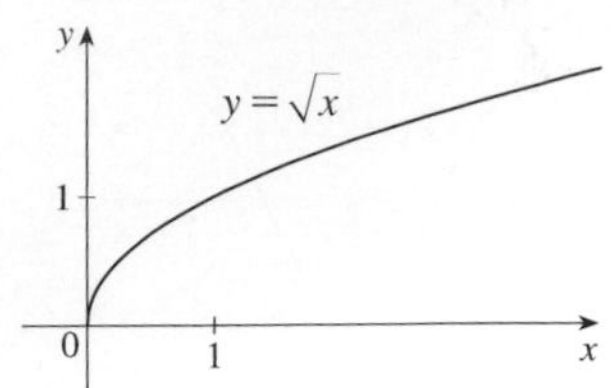

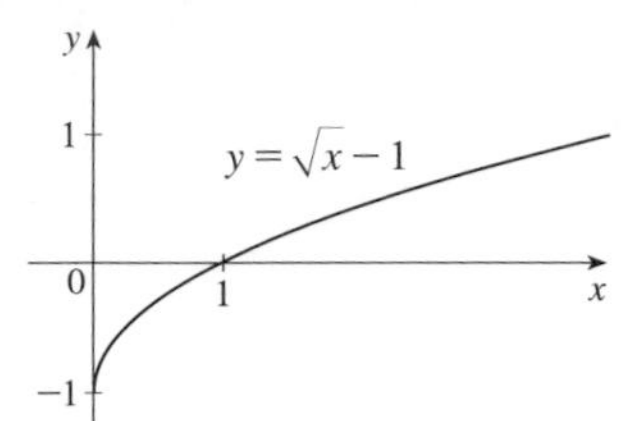

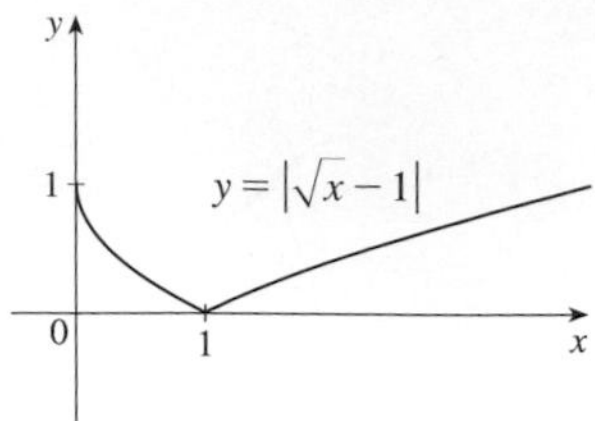

24. $y = |\cos \pi x|$: Start with the graph of $y = \cos x$, shrink it horizontally by a factor of π, and reflect all the parts of the graph below the x-axis about the x-axis.

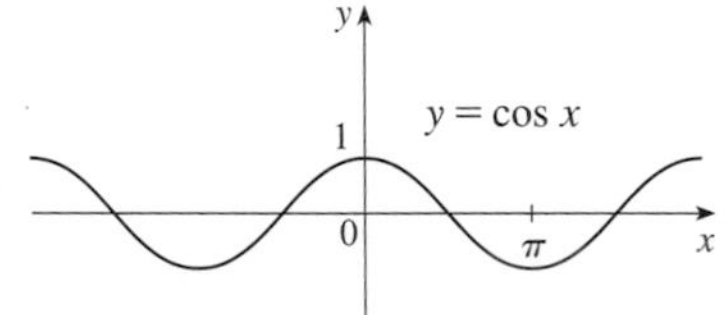

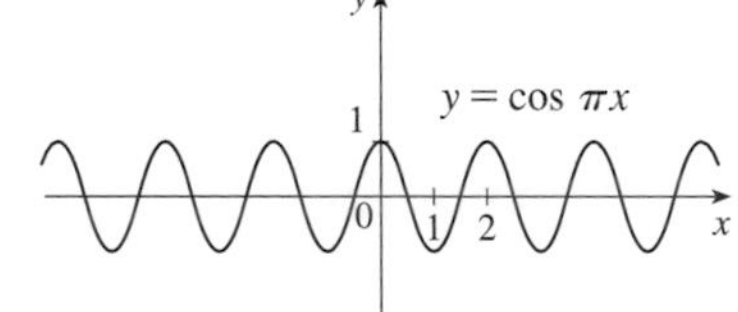

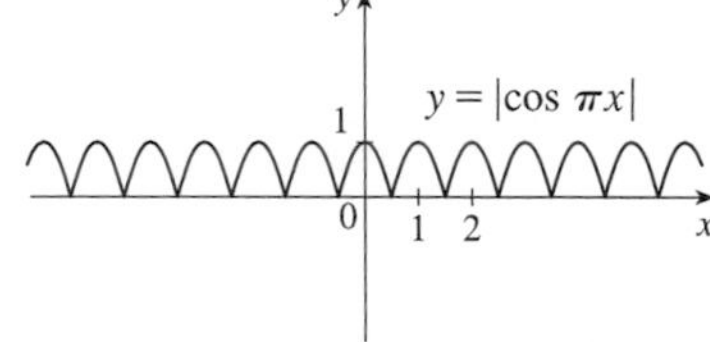

25. This is just like the solution to Example 4 except the amplitude of the curve (the 30°N curve in Figure 9 on June 21) is $14 - 12 = 2$. So the function is $L(t) = 12 + 2\sin\left[\frac{2\pi}{365}(t - 80)\right]$. March 31 is the 90th day of the year, so the model gives $L(90) \approx 12.34$ h. The daylight time (5:51 AM to 6:18 PM) is 12 hours and 27 minutes, or 12.45 h. The model value differs from the actual value by $\frac{12.45-12.34}{12.45} \approx 0.009$, less than 1%.

26. Using a sine function to model the brightness of Delta Cephei as a function of time, we take its period to be 5.4 days, its amplitude to be 0.35 (on the scale of magnitude), and its average magnitude to be 4.0. If we take $t = 0$ at a time of average brightness, then the magnitude (brightness) as a function of time t in days can be modeled by the formula $M(t) = 4.0 + 0.35\sin\left(\frac{2\pi}{5.4}t\right)$.

27. The water depth $D(t)$ can be modeled by a cosine function with amplitude $\dfrac{12 - 2}{2} = 5$ m, average magnitude $\dfrac{12 + 2}{2} = 7$ m, and period 12 hours. High tide occurred at time 6:45 AM ($t = 6.75$ h), so the curve begins a cycle at time $t = 6.75$ h (shift 6.75 units to the right). Thus, $D(t) = 5\cos\left[\frac{2\pi}{12}(t - 6.75)\right] + 7 = 5\cos\left[\frac{\pi}{6}(t - 6.75)\right] + 7$, where D is in meters and t is the number of hours after midnight.

28. The total volume of air $V(t)$ in the lungs can be modeled by a sine function with amplitude $\dfrac{2500 - 2000}{2} = 250$ mL, average volume $\dfrac{2500 + 2000}{2} = 2250$ mL, and period 4 seconds. Thus, $V(t) = 250\sin\frac{2\pi}{4}t + 2250 = 250\sin\frac{\pi}{2}t + 2250$, where V is in mL and t is in seconds.

29. (a) To obtain $y = f(|x|)$, the portion of the graph of $y = f(x)$ to the right of the y-axis is reflected about the y-axis.

(b) $y = \sin|x|$

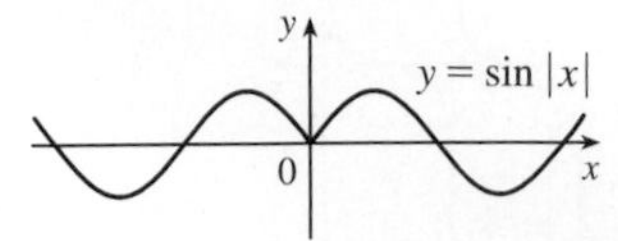

(c) 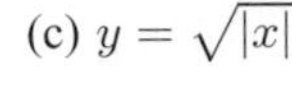 $y = \sqrt{|x|}$

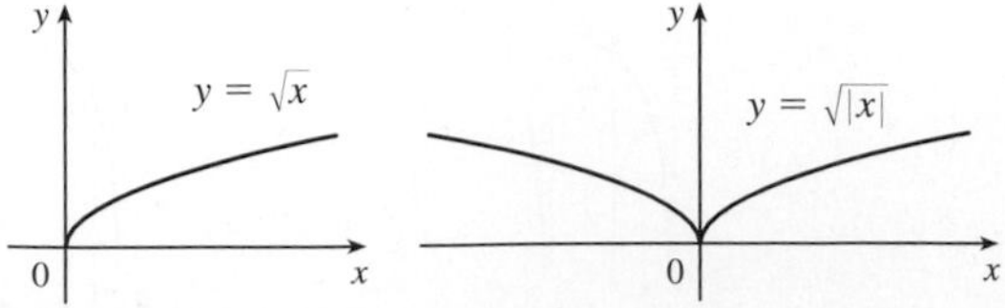

30. The most important features of the given graph are the x-intercepts and the maximum and minimum points. The graph of $y = 1/f(x)$ has vertical asymptotes at the x-values where there are x-intercepts on the graph of $y = f(x)$. The maximum of 1 on the graph of $y = f(x)$ corresponds to a minimum of $1/1 = 1$ on $y = 1/f(x)$. Similarly, the minimum on the graph of $y = f(x)$ corresponds to a maximum on the graph of $y = 1/f(x)$. As the values of y get large (positively or negatively) on the graph of $y = f(x)$, the values of y get close to zero on the graph of $y = 1/f(x)$.

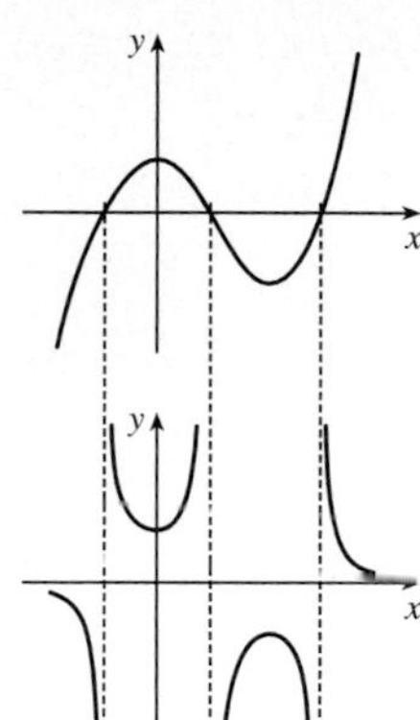

31. $f(x) = x^3 + 2x^2$; $g(x) = 3x^2 - 1$. $D = \mathbb{R}$ for both f and g.

(a) $(f + g)(x) = (x^3 + 2x^2) + (3x^2 - 1) = x^3 + 5x^2 - 1$, $D = (-\infty, \infty)$, or $\mathbb{R}$.

(b) $(f - g)(x) = (x^3 + 2x^2) - (3x^2 - 1) = x^3 - x^2 + 1$, $D = \mathbb{R}$.

(c) $(fg)(x) = (x^3 + 2x^2)(3x^2 - 1) = 3x^5 + 6x^4 - x^3 - 2x^2$, $D = \mathbb{R}$.

(d) $\left(\dfrac{f}{g}\right)(x) = \dfrac{x^3 + 2x^2}{3x^2 - 1}$, $D = \left\{x \mid x \neq \pm\dfrac{1}{\sqrt{3}}\right\}$ since $3x^2 - 1 \neq 0$.

32. $f(x) = \sqrt{3 - x}$, $D = (-\infty, 3]$; $g(x) = \sqrt{x^2 - 1}$, $D = (-\infty, -1] \cup [1, \infty)$.

(a) $(f + g)(x) = \sqrt{3 - x} + \sqrt{x^2 - 1}$, $D = (-\infty, -1] \cup [1, 3]$, which is the intersection of the domains of f and g.

(b) $(f - g)(x) = \sqrt{3 - x} - \sqrt{x^2 - 1}$, $D = (-\infty, -1] \cup [1, 3]$.

(c) $(fg)(x) = \sqrt{3 - x} \cdot \sqrt{x^2 - 1}$, $D = (-\infty, -1] \cup [1, 3]$.

(d) $\left(\dfrac{f}{g}\right)(x) = \dfrac{\sqrt{3 - x}}{\sqrt{x^2 - 1}}$, $D = (-\infty, -1) \cup (1, 3]$. We must exclude $x = \pm 1$ since these values would make $\dfrac{f}{g}$ undefined.

33. $f(x) = 3x + 5$; $g(x) = x^2 + x$. $D = \mathbb{R}$ for both f and g, and hence for their composites.

(a) $(f \circ g)(x) = f(g(x)) = f(x^2 + x) = 3(x^2 + x) + 5 = 3x^2 + 3x + 5$, $D = \mathbb{R}$.

(b) $(g \circ f)(x) = g(f(x)) = g(3x + 5) = (3x + 5)^2 + (3x + 5)$

$$= 9x^2 + 30x + 25 + 3x + 5 = 9x^2 + 33x + 30, \ D = \mathbb{R}.$$

(c) $(f \circ f) = f(f(x)) = f(3x + 5) = 3(3x + 5) + 5 = 9x + 15 + 5 = 9x + 20$, $D = \mathbb{R}$.

(d) $(g \circ g)(x) = g(g(x)) = g(x^2 + x) = (x^2 + x)^2 + (x^2 + x)$

$$= x^4 + 2x^3 + x^2 + x^2 + x = x^4 + 2x^3 + 2x^2 + x, \ D = \mathbb{R}.$$

34. $f(x) = x^3 - 2$; $g(x) = 1 - 4x$. $D = \mathbb{R}$ for both f and g, and hence for their composites.

(a) $(f \circ g)(x) = f(g(x)) = f(1 - 4x) = (1 - 4x)^3 - 2$

$$= (1)^3 - 3(1)^2(4x) + 3(1)(4x)^2 - (4x)^3 - 2 = 1 - 12x + 48x^2 - 64x^3 - 2$$

$$= -1 - 12x + 48x^2 - 64x^3, \ D = \mathbb{R}.$$

(b) $(g \circ f)(x) = g(f(x)) = g(x^3 - 2) = 1 - 4(x^3 - 2) = 1 - 4x^3 + 8 = 9 - 4x^3$, $D = \mathbb{R}$.

(c) $(f \circ f)(x) = f(f(x)) = f(x^3 - 2) = (x^3 - 2)^3 - 2$

$$= (x^3)^3 - 3(x^3)^2(2) + 3(x^3)(2)^2 - (2)^3 - 2 = x^9 - 6x^6 + 12x^3 - 10, \;\; D = \mathbb{R}.$$

(d) $(g \circ g)(x) = g(g(x)) = g(1 - 4x) = 1 - 4(1 - 4x) = 1 - 4 + 16x = -3 + 16x, \;\; D = \mathbb{R}.$

35. $f(x) = \sqrt{x+1}, D = \{x \mid x \geq -1\}; \;\; g(x) = 4x - 3, \;\; D = \mathbb{R}.$

(a) $(f \circ g)(x) = f(g(x)) = f(4x - 3) = \sqrt{(4x-3)+1} = \sqrt{4x - 2}$

The domain of $f \circ g$ is $\{x \mid 4x - 3 \geq -1\} = \{x \mid 4x \geq 2\} = \{x \mid x \geq \frac{1}{2}\} = [\frac{1}{2}, \infty)$.

(b) $(g \circ f)(x) = g(f(x)) = g(\sqrt{x+1}) = 4\sqrt{x+1} - 3$

The domain of $g \circ f$ is $\{x \mid x$ is in the domain of f and $f(x)$ is in the domain of $g\}$. This is the domain of f, that is, $\{x \mid x + 1 \geq 0\} = \{x \mid x \geq -1\} = [-1, \infty)$.

(c) $(f \circ f)(x) = f(f(x)) = f(\sqrt{x+1}) = \sqrt{\sqrt{x+1}+1}$

For the domain, we need $x + 1 \geq 0$, which is equivalent to $x \geq -1$, and $\sqrt{x+1} \geq -1$, which is true for all real values of x. Thus, the domain of $f \circ f$ is $[-1, \infty)$.

(d) $(g \circ g)(x) = g(g(x)) = g(4x - 3) = 4(4x - 3) - 3 = 16x - 12 - 3 = 16x - 15, \;\; D = \mathbb{R}.$

36. $f(x) = \sin x; \;\; g(x) = x^2 + 1.$ $\;\; D = \mathbb{R}$ for both f and g, and hence for their composites.

(a) $(f \circ g)(x) = f(g(x)) = f(x^2 + 1) = \sin(x^2 + 1), \;\; D = \mathbb{R}.$

(b) $(g \circ f) = g(f(x)) = g(\sin x) = (\sin x)^2 + 1 = \sin^2 x + 1, \;\; D = \mathbb{R}.$

(c) $(f \circ f)(x) = f(f(x)) = f(\sin x) = \sin(\sin x), \;\; D = \mathbb{R}.$

(d) $(g \circ g)(x) = g(g(x)) = g(x^2 + 1) = (x^2 + 1)^2 + 1 = x^4 + 2x^2 + 1 + 1 = x^4 + 2x^2 + 2, \;\; D = \mathbb{R}.$

37. $f(x) = x + \dfrac{1}{x}, \;\; D = \{x \mid x \neq 0\}; \;\; g(x) = \dfrac{x+1}{x+2}, \;\; D = \{x \mid x \neq -2\}$

(a) $(f \circ g)(x) = f(g(x)) = f\left(\dfrac{x+1}{x+2}\right) = \dfrac{x+1}{x+2} + \dfrac{1}{\dfrac{x+1}{x+2}} = \dfrac{x+1}{x+2} + \dfrac{x+2}{x+1}$

$$= \frac{(x+1)(x+1) + (x+2)(x+2)}{(x+2)(x+1)} = \frac{(x^2 + 2x + 1) + (x^2 + 4x + 4)}{(x+2)(x+1)} = \frac{2x^2 + 6x + 5}{(x+2)(x+1)}$$

Since $g(x)$ is not defined for $x = -2$ and $f(g(x))$ is not defined for $x = -2$ and $x = -1$, the domain of $(f \circ g)(x)$ is $D = \{x \mid x \neq -2, -1\}$.

(b) $(g \circ f)(x) = g(f(x)) = g\left(x + \dfrac{1}{x}\right) = \dfrac{\left(x + \dfrac{1}{x}\right) + 1}{\left(x + \dfrac{1}{x}\right) + 2} = \dfrac{\dfrac{x^2 + 1 + x}{x}}{\dfrac{x^2 + 1 + 2x}{x}} = \dfrac{x^2 + x + 1}{x^2 + 2x + 1} = \dfrac{x^2 + x + 1}{(x+1)^2}$

Since $f(x)$ is not defined for $x = 0$ and $g(f(x))$ is not defined for $x = -1$, the domain of $(g \circ f)(x)$ is $D = \{x \mid x \neq -1, 0\}$.

(c) $(f\circ f)(x)=f(f(x))=f\left(x+\frac{1}{x}\right)=\left(x+\frac{1}{x}\right)+\frac{1}{x+\frac{1}{x}}=x+\frac{1}{x}+\frac{1}{\frac{x^2+1}{x}}=x+\frac{1}{x}+\frac{x}{x^2+1}$

$$=\frac{x(x)(x^2+1)+1\,(x^2+1)+x(x)}{x(x^2+1)}=\frac{x^4+x^2+x^2+1+x^2}{x(x^2+1)}$$

$$=\frac{x^4+3x^2+1}{x(x^2+1)},\quad D=\{x\mid x\neq 0\}$$

(d) $(g\circ g)(x)=g(g(x))=g\left(\frac{x+1}{x+2}\right)=\frac{\frac{x+1}{x+2}+1}{\frac{x+1}{x+2}+2}=\frac{\frac{x+1+1(x+2)}{x+2}}{\frac{x+1+2(x+2)}{x+2}}=\frac{x+1+x+2}{x+1+2x+4}=\frac{2x+3}{3x+5}$

Since $g(x)$ is not defined for $x=-2$ and $g(g(x))$ is not defined for $x=-\frac{5}{3}$,

the domain of $(g\circ g)(x)$ is $D=\left\{x\mid x\neq -2,-\frac{5}{3}\right\}$.

38. $f(x)=\frac{x}{1+x}$, $D=\{x\mid x\neq -1\}$; $g(x)=\sin 2x$, $D=\mathbb{R}$.

(a) $(f\circ g)(x)=f(g(x))=f(\sin 2x)=\frac{\sin 2x}{1+\sin 2x}$

Domain: $1+\sin 2x\neq 0 \;\Rightarrow\; \sin 2x\neq -1 \;\Rightarrow\; 2x\neq\frac{3\pi}{2}+2\pi n \;\Rightarrow\; x\neq\frac{3\pi}{4}+\pi n$ [n an integer].

(b) $(g\circ f)(x)=g(f(x))=g\left(\frac{x}{1+x}\right)=\sin\left(\frac{2x}{1+x}\right)$.

Domain: $\{x\mid x\neq -1\}$

(c) $(f\circ f)(x)=f(f(x))=f\left(\frac{x}{1+x}\right)=\frac{\frac{x}{1+x}}{1+\frac{x}{1+x}}=\frac{\left(\frac{x}{1+x}\right)\cdot(1+x)}{\left(1+\frac{x}{1+x}\right)\cdot(1+x)}=\frac{x}{1+x+x}=\frac{x}{2x+1}$

Since $f(x)$ is not defined for $x=-1$, and $f(f(x))$ is not defined for $x=-\frac{1}{2}$,

the domain of $(f\circ f)(x)$ is $D=\{x\mid x\neq -1,-\frac{1}{2}\}$.

(d) $(g\circ g)(g)=g(g(x))=g(\sin 2x)=\sin(2\sin 2x)$.

Domain: $\mathbb{R}$

39. $(f\circ g\circ h)(x)=f(g(h(x)))=f(g(x^2))=f(\sin(x^2))=3\sin(x^2)-2$

40. $(f\circ g\circ h)(x)=f(g(h(x)))=f(g(\sqrt{x}))=f(2^{\sqrt{x}})=\left|2^{\sqrt{x}}-4\right|$

41. $(f\circ g\circ h)(x)=f(g(h(x)))=f(g(x^3+2))=f[(x^3+2)^2]$

$$=f(x^6+4x^3+4)=\sqrt{(x^6+4x^3+4)-3}=\sqrt{x^6+4x^3+1}$$

42. $(f\circ g\circ h)(x)=f(g(h(x)))=f(g(\sqrt[3]{x}))=f\left(\frac{\sqrt[3]{x}}{\sqrt[3]{x}-1}\right)=\tan\left(\frac{\sqrt[3]{x}}{\sqrt[3]{x}-1}\right)$

43. Let $g(x)=2x+x^2$ and $f(x)=x^4$. Then $(f\circ g)(x)=f(g(x))=f(2x+x^2)=(2x+x^2)^4=F(x)$.

44. Let $g(x)=\cos x$ and $f(x)=x^2$. Then $(f\circ g)(x)=f(g(x))=f(\cos x)=(\cos x)^2=\cos^2 x=F(x)$.

45. Let $g(x) = \sqrt[3]{x}$ and $f(x) = \dfrac{x}{1+x}$. Then $(f \circ g)(x) = f(g(x)) = f(\sqrt[3]{x}) = \dfrac{\sqrt[3]{x}}{1+\sqrt[3]{x}} = F(x)$.

46. Let $g(x) = \dfrac{x}{1+x}$ and $f(x) = \sqrt[3]{x}$. Then $(f \circ g)(x) = f(g(x)) = f\left(\dfrac{x}{1+x}\right) = \sqrt[3]{\dfrac{x}{1+x}} = G(x)$.

47. Let $g(t) = t^2$ and $f(t) = \sec t \tan t$. Then $(f \circ g)(t) = f(g(t)) = f(t^2) = \sec(t^2)\tan(t^2) = v(t)$.

48. Let $g(t) = \tan t$ and $f(t) = \dfrac{t}{1+t}$. Then $(f \circ g)(t) = f(g(t)) = f(\tan t) = \dfrac{\tan t}{1+\tan t} = u(t)$.

49. Let $h(x) = \sqrt{x}$, $g(x) = x - 1$, and $f(x) = \sqrt{x}$. Then

$(f \circ g \circ h)(x) = f(g(h(x))) = f(g(\sqrt{x})) = f(\sqrt{x} - 1) = \sqrt{\sqrt{x} - 1} = R(x)$.

50. Let $h(x) = |x|$, $g(x) = 2 + x$, and $f(x) = \sqrt[8]{x}$. Then

$(f \circ g \circ h)(x) = f(g(h(x))) = f(g(|x|)) = f\,(2 + |x|) = \sqrt[8]{2 + |x|} = H(x)$.

51. Let $h(t) = \cos t$, $g(t) = \sin t$, and $f(t) = t^2$. Then

$(f \circ g \circ h)(t) = f(g(h(t))) = f(g(\cos t)) = f(\sin(\cos t)) = [\sin(\cos t)]^2 = \sin^2(\cos t) = S(t)$.

52. (a) $f(g(1)) = f(6) = 5$

(b) $g(f(1)) = g(3) = 2$

(c) $f(f(1)) = f(3) = 4$

(d) $g(g(1)) = g(6) = 3$

(e) $(g \circ f)(3) = g(f(3)) = g(4) = 1$

(f) $(f \circ g)(6) = f(g(6)) = f(3) = 4$

53. (a) $g(2) = 5$, because the point $(2, 5)$ is on the graph of g. Thus, $f(g(2)) = f(5) = 4$, because the point $(5, 4)$ is on the graph of f.

(b) $g(f(0)) = g(0) = 3$

(c) $(f \circ g)(0) = f(g(0)) = f(3) = 0$

(d) $(g \circ f)(6) = g(f(6)) = g(6)$. This value is not defined, because there is no point on the graph of g that has x-coordinate 6.

(e) $(g \circ g)(-2) = g(g(-2)) = g(1) = 4$

(f) $(f \circ f)(4) = f(f(4)) = f(2) = -2$

54. To find a particular value of $f(g(x))$, say for $x = 0$, we note from the graph that $g(0) \approx 2.8$ and $f(2.8) \approx -0.5$. Thus, $f(g\,(0)) \approx f(2.8) \approx -0.5$. The other values listed in the table were obtained in a similar fashion.

x	$g(x)$	$f(g(x))$
-5	-0.2	-4
-4	1.2	-3.3
-3	2.2	-1.7
-2	2.8	-0.5
-1	3	-0.2

x	$g(x)$	$f(g(x))$
0	2.8	-0.5
1	2.2	-1.7
2	1.2	-3.3
3	-0.2	-4
4	-1.9	-2.2
5	-4.1	1.9

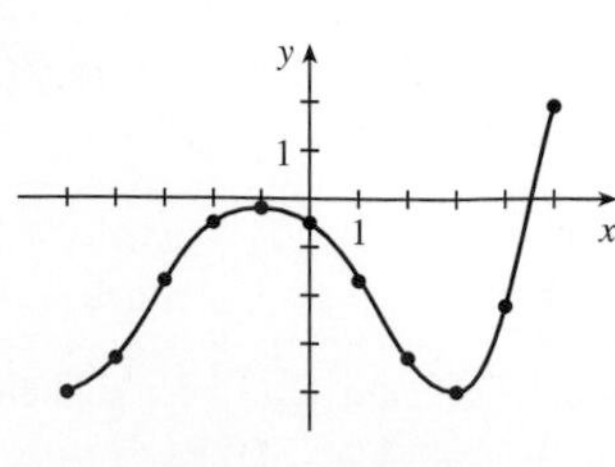

55. (a) Using the relationship *distance* = *rate* · *time* with the radius r as the distance, we have $r(t) = 60t$.

(b) $A = \pi r^2 \quad \Rightarrow \quad (A \circ r)(t) = A(r(t)) = \pi(60t)^2 = 3600\pi t^2$. This formula gives us the extent of the rippled area (in cm^2) at any time t.

56. (a) The radius r of the balloon is increasing at a rate of 2 cm/s, so $r(t) = (2\text{ cm/s})(t\text{ s}) = 2t$ (in cm).

(b) Using $V = \frac{4}{3}\pi r^3$, we get $(V \circ r)(t) = V(r(t)) = V(2t) = \frac{4}{3}\pi(2t)^3 = \frac{32}{3}\pi t^3$.

The result, $V = \frac{32}{3}\pi t^3$, gives the volume of the balloon (in cm^3) as a function of time (in s).

57. (a) From the figure, we have a right triangle with legs 6 and d, and hypotenuse s.

By the Pythagorean Theorem, $d^2 + 6^2 = s^2 \quad \Rightarrow \quad s = f(d) = \sqrt{d^2 + 36}$.

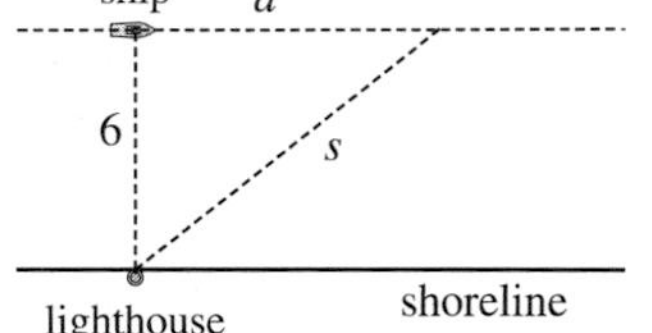

(b) Using $d = rt$, we get $d = (30\text{ km/h})(t\text{ hours}) = 30t$ (in km). Thus,

$d = g(t) = 30t$.

(c) $(f \circ g)(t) = f(g(t)) = f(30t) = \sqrt{(30t)^2 + 36} = \sqrt{900t^2 + 36}$. This function represents the distance between the lighthouse and the ship as a function of the time elapsed since noon.

58. (a) $d = rt \quad \Rightarrow \quad d(t) = 350t$

(b) There is a Pythagorean relationship involving the legs with lengths d and 1 and the hypotenuse with length s:

$d^2 + 1^2 = s^2$. Thus, $s(d) = \sqrt{d^2 + 1}$.

(c) $(s \circ d)(t) = s(d(t)) = s(350t) = \sqrt{(350t)^2 + 1}$

59. (a)

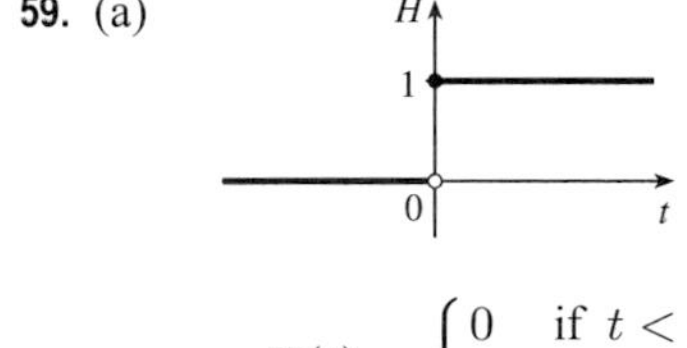

$$H(t) = \begin{cases} 0 & \text{if } t < 0 \\ 1 & \text{if } t \geq 0 \end{cases}$$

(b)

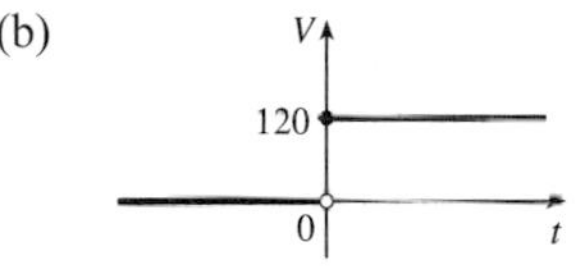

$$V(t) = \begin{cases} 0 & \text{if } t < 0 \\ 120 & \text{if } t \geq 0 \end{cases} \quad \text{so } V(t) = 120H(t).$$

(c)

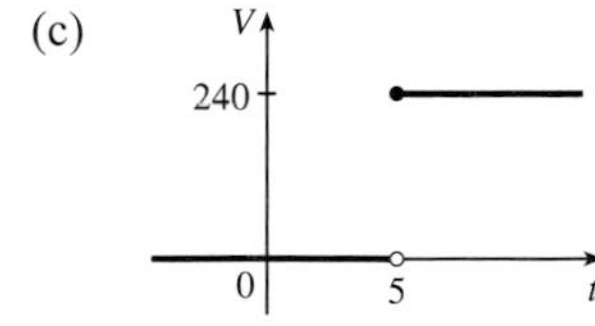

Starting with the formula in part (b), we replace 120 with 240 to reflect the different voltage. Also, because we are starting 5 units to the right of $t = 0$, we replace t with $t - 5$. Thus, the formula is $V(t) = 240H(t - 5)$.

60. (a) $R(t) = tH(t)$

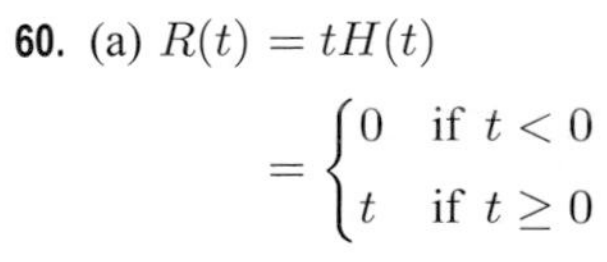

$$= \begin{cases} 0 & \text{if } t < 0 \\ t & \text{if } t \geq 0 \end{cases}$$

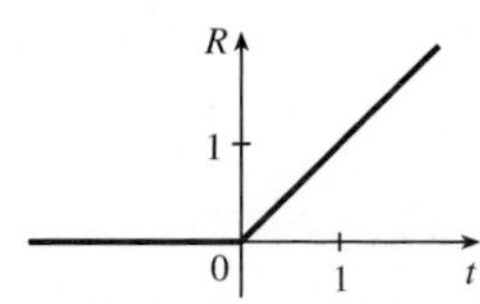

(b) $V(t) = \begin{cases} 0 & \text{if } t < 0 \\ 2t & \text{if } 0 \leq t \leq 60 \end{cases}$

so $V(t) = 2tH(t),\ t \leq 60$.

(c) $V(t) = \begin{cases} 0 & \text{if } t < 7 \\ 4(t - 7) & \text{if } 7 \leq t \leq 32 \end{cases}$

so $V(t) = 4(t - 7)H(t - 7),\ t \leq 32$.

61. If $f(x) = m_1x + b_1$ and $g(x) = m_2x + b_2$, then

$(f \circ g)(x) = f(g(x)) = f(m_2x + b_2) = m_1(m_2x + b_2) + b_1 = m_1m_2x + m_1b_2 + b_1.$

So $f \circ g$ is a linear function with slope m_1m_2.

62. If $A(x) = 1.04x$, then

$(A \circ A)(x) = A(A(x)) = A(1.04x) = 1.04(1.04x) = (1.04)^2x,$

$(A \circ A \circ A)(x) = A((A \circ A)(x)) = A((1.04)^2x) = 1.04(1.04)^2x = (1.04)^3x$, and

$(A \circ A \circ A \circ A)(x) = A((A \circ A \circ A)(x)) = A((1.04)^3x) = 1.04(1.04)^3x, = (1.04)^4x.$

These compositions represent the amount of the investment after 2, 3, and 4 years.

Based on this pattern, when we compose n copies of A, we get the formula $\underbrace{(A \circ A \circ \cdots \circ A)}_{n\ A\text{'s}}(x) = (1.04)^nx$.

63. (a) By examining the variable terms in g and h, we deduce that we must square g to get the terms $4x^2$ and $4x$ in h. If we let $f(x) = x^2 + c$, then $(f \circ g)(x) = f(g(x)) = f(2x + 1) = (2x + 1)^2 + c = 4x^2 + 4x + (1 + c)$. Since $h(x) = 4x^2 + 4x + 7$, we must have $1 + c = 7$. So $c = 6$ and $f(x) = x^2 + 6$.

(b) We need a function g so that $f(g(x)) = 3(g(x)) + 5 = h(x)$. But
$h(x) = 3x^2 + 3x + 2 = 3(x^2 + x) + 2 = 3(x^2 + x - 1) + 5$, so we see that $g(x) = x^2 + x - 1$.

64. We need a function g so that $g(f(x)) = g(x + 4) = h(x) = 4x - 1 = 4(x + 4) - 17$. So we see that the function g must be $g(x) = 4x - 17$.

65. We need to examine $h(-x)$.

$$h(-x) = (f \circ g)(-x) = f(g(-x)) = f(g(x)) \quad \text{[because } g \text{ is even]} \quad = h(x)$$

Because $h(-x) = h(x)$, h is an even function.

66. $h(-x) = f(g(-x)) = f(-g(x))$. At this point, we can't simplify the expression, so we might try to find a counterexample to show that h is not an odd function. Let $g(x) = x$, an odd function, and $f(x) = x^2 + x$. Then $h(x) = x^2 + x$, which is neither even nor odd.

Now suppose f is an odd function. Then $f(-g(x)) = -f(g(x)) = -h(x)$. Hence, $h(-x) = -h(x)$, and so h is odd if both f and g are odd.

Now suppose f is an even function. Then $f(-g(x)) = f(g(x)) = h(x)$. Hence, $h(-x) = h(x)$, and so h is even if g is odd and f is even.

1.4 Exponential Functions

1. (a) $\dfrac{4^{-3}}{2^{-8}} = \dfrac{2^8}{4^3} = \dfrac{2^8}{(2^2)^3} = \dfrac{2^8}{2^6} = 2^{8-6} = 2^2 = 4$

(b) $\dfrac{1}{\sqrt[3]{x^4}} = \dfrac{1}{x^{4/3}} = x^{-4/3}$

2. (a) $8^{4/3} = (8^{1/3})^4 = 2^4 = 16$

(b) $x(3x^2)^3 = x \cdot 3^3(x^2)^3 = 27x \cdot x^6 = 27x^7$

3. (a) $b^8(2b)^4 = b^8 \cdot 2^4b^4 = 16b^{12}$

(b) $\dfrac{(6y^3)^4}{2y^5} = \dfrac{6^4(y^3)^4}{2y^5} = \dfrac{1296y^{12}}{2y^5} = 648y^7$

4. (a) $\dfrac{x^{2n}\cdot x^{3n-1}}{x^{n+2}}=\dfrac{x^{2n+3n-1}}{x^{n+2}}=\dfrac{x^{5n-1}}{x^{n+2}}=x^{4n-3}$

(b) $\dfrac{\sqrt{a\sqrt{b}}}{\sqrt[3]{ab}}=\dfrac{\sqrt{a}\sqrt{\sqrt{b}}}{\sqrt[3]{a}\sqrt[3]{b}}=\dfrac{a^{1/2}b^{1/4}}{a^{1/3}b^{1/3}}=a^{(1/2-1/3)}b^{(1/4-1/3)}=a^{1/6}b^{-1/12}$

5. (a) $f(x)=b^x,\ b>0$ (b) $\mathbb{R}$ (c) $(0,\infty)$ (d) See Figures 4(c), 4(b), and 4(a), respectively.

6. (a) The number e is the value of a such that the slope of the tangent line at $x=0$ on the graph of $y=a^x$ is exactly 1.

(b) $e\approx 2.71828$ (c) $f(x)=e^x$

7. All of these graphs approach 0 as $x\to-\infty$, all of them pass through the point $(0,1)$, and all of them are increasing and approach ∞ as $x\to\infty$. The larger the base, the faster the function increases for $x>0$, and the faster it approaches 0 as $x\to-\infty$.

Note: The notation "$x\to\infty$" can be thought of as "x becomes large" at this point. More details on this notation are given in Chapter 2.

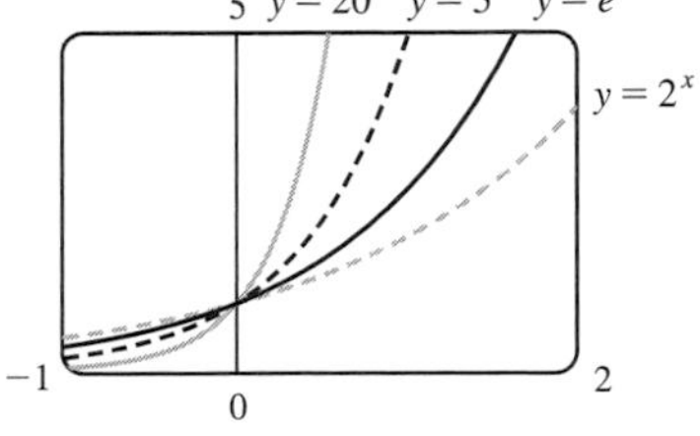

8. The graph of e^{-x} is the reflection of the graph of e^x about the y-axis, and the graph of 8^{-x} is the reflection of that of 8^x about the y-axis. The graph of 8^x increases more quickly than that of e^x for $x>0$, and approaches 0 faster as $x\to-\infty$.

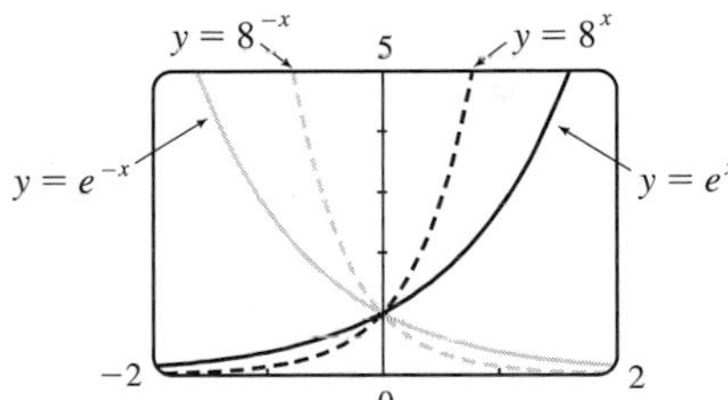

9. The functions with bases greater than 1 (3^x and 10^x) are increasing, while those with bases less than 1 $\left[\left(\frac{1}{3}\right)^x \text{ and } \left(\frac{1}{10}\right)^x\right]$ are decreasing. The graph of $\left(\frac{1}{3}\right)^x$ is the reflection of that of 3^x about the y-axis, and the graph of $\left(\frac{1}{10}\right)^x$ is the reflection of that of 10^x about the y-axis. The graph of 10^x increases more quickly than that of 3^x for $x>0$, and approaches 0 faster as $x\to-\infty$.

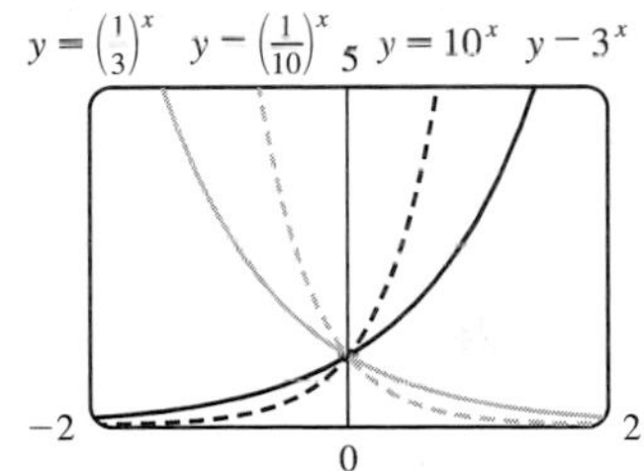

10. Each of the graphs approaches ∞ as $x\to-\infty$, and each approaches 0 as $x\to\infty$. The smaller the base, the faster the function grows as $x\to-\infty$, and the faster it approaches 0 as $x\to\infty$.

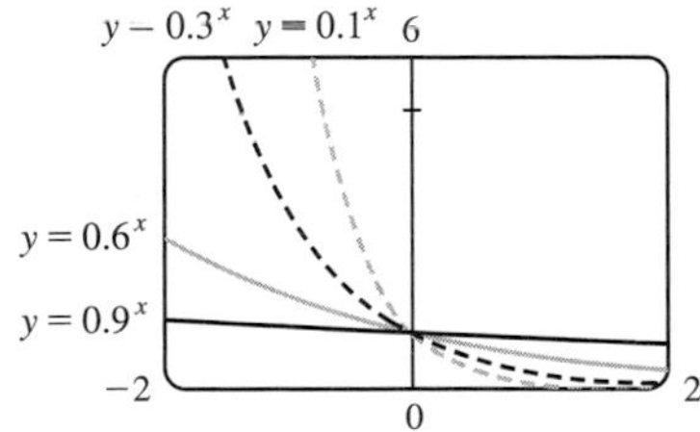

11. We start with the graph of $y=4^x$ (Figure 3) and shift it 1 unit down to obtain the graph of $y=4^x-1$.

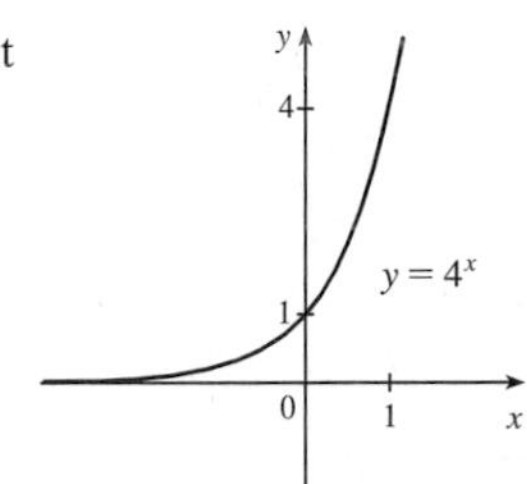

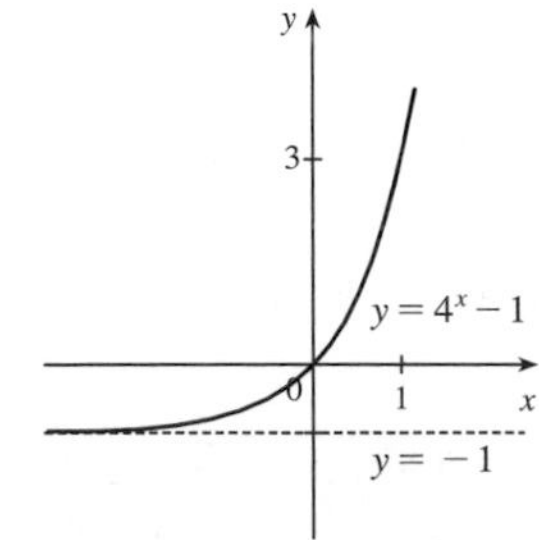

12. We start with the graph of $y = (0.5)^x$ (Figure 3) and shift it 1 unit to the right to obtain the graph of $y = (0.5)^{x-1}$.

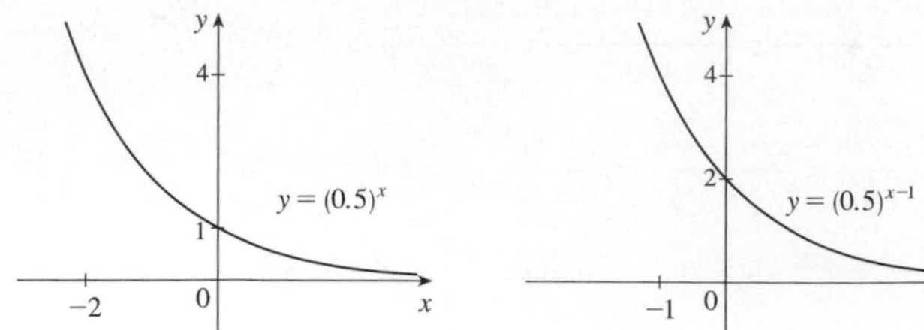

13. We start with the graph of $y = 2^x$ (Figure 16), reflect it about the y-axis, and then about the x-axis (or just rotate $180°$ to handle both reflections) to obtain the graph of $y = -2^{-x}$. In each graph, $y = 0$ is the horizontal asymptote.

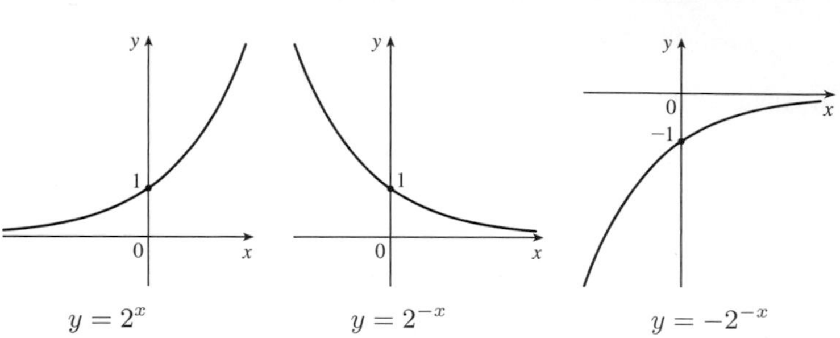

14. We start with the graph of $y = e^x$ (Figure 16) and reflect the portion of the graph in the first quadrant about the y-axis to obtain the graph of $y = e^{|x|}$.

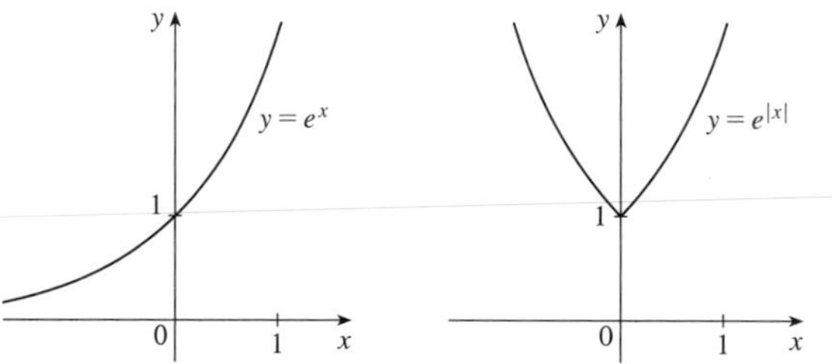

15. We start with the graph of $y = e^x$ (Figure 16) and reflect about the y-axis to get the graph of $y = e^{-x}$. Then we compress the graph vertically by a factor of 2 to obtain the graph of $y = \frac{1}{2}e^{-x}$ and then reflect about the x-axis to get the graph of $y = -\frac{1}{2}e^{-x}$. Finally, we shift the graph upward one unit to get the graph of $y = 1 - \frac{1}{2}e^{-x}$.

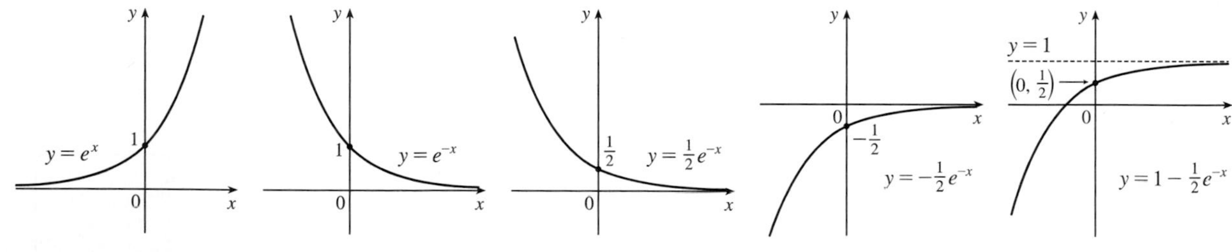

16. We start with the graph of $y = e^x$ (Figure 13) and reflect about the x-axis to get the graph of $y = -e^x$. Then shift the graph upward one unit to get the graph of $y = 1 - e^x$. Finally, we stretch the graph vertically by a factor of 2 to obtain the graph of $y = 2(1 - e^x)$.

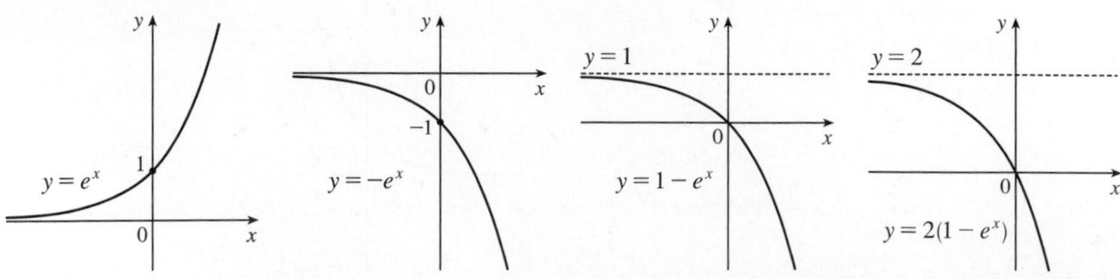

17. (a) To find the equation of the graph that results from shifting the graph of $y = e^x$ 2 units downward, we subtract 2 from the original function to get $y = e^x - 2$.

(b) To find the equation of the graph that results from shifting the graph of $y = e^x$ 2 units to the right, we replace x with $x - 2$ in the original function to get $y = e^{(x-2)}$.

(c) To find the equation of the graph that results from reflecting the graph of $y = e^x$ about the x-axis, we multiply the original function by -1 to get $y = -e^x$.

(d) To find the equation of the graph that results from reflecting the graph of $y = e^x$ about the y-axis, we replace x with $-x$ in the original function to get $y = e^{-x}$.

(e) To find the equation of the graph that results from reflecting the graph of $y = e^x$ about the x-axis and then about the y-axis, we first multiply the original function by -1 (to get $y = -e^x$) and then replace x with $-x$ in this equation to get $y = -e^{-x}$.

18. (a) This reflection consists of first reflecting the graph about the x-axis (giving the graph with equation $y = -e^x$) and then shifting this graph $2 \cdot 4 = 8$ units upward. So the equation is $y = -e^x + 8$.

(b) This reflection consists of first reflecting the graph about the y-axis (giving the graph with equation $y = e^{-x}$) and then shifting this graph $2 \cdot 2 = 4$ units to the right. So the equation is $y = e^{-(x-4)}$.

19. (a) The denominator is zero when $1 - e^{1-x^2} = 0 \quad \Leftrightarrow \quad e^{1-x^2} = 1 \quad \Leftrightarrow \quad 1 - x^2 = 0 \quad \Leftrightarrow \quad x = \pm 1$. Thus, the function $f(x) = \dfrac{1 - e^{x^2}}{1 - e^{1-x^2}}$ has domain $\{x \mid x \neq \pm 1\} = (-\infty, -1) \cup (-1, 1) \cup (1, \infty)$.

(b) The denominator is never equal to zero, so the function $f(x) = \dfrac{1 + x}{e^{\cos x}}$ has domain $\mathbb{R}$, or $(-\infty, \infty)$.

20. (a) The function $g(t) = \sqrt{10^t - 100}$ has domain $\{t \mid 10^t - 100 \geq 0\} = \{t \mid 10^t \geq 10^2\} = \{t \mid t \geq 2\} = [2, \infty)$.

(b) The sine and exponential functions have domain $\mathbb{R}$, so $g(t) = \sin(e^t - 1)$ also has domain $\mathbb{R}$.

21. Use $y = Cb^x$ with the points $(1, 6)$ and $(3, 24)$. $6 = Cb^1 \quad \left[C = \frac{6}{b}\right]$ and $24 = Cb^3 \quad \Rightarrow \quad 24 = \left(\dfrac{6}{b}\right) b^3 \quad \Rightarrow$ $4 = b^2 \quad \Rightarrow \quad b = 2 \quad [\text{since } b > 0]$ and $C = \frac{6}{2} = 3$. The function is $f(x) = 3 \cdot 2^x$.

22. Use $y = Cb^x$ with the points $(-1, 3)$ and $\left(1, \frac{4}{3}\right)$. From the point $(-1, 3)$, we have $3 = Cb^{-1}$, hence $C = 3b$. Using this and the point $\left(1, \frac{4}{3}\right)$, we get $\frac{4}{3} = Cb^1 \quad \Rightarrow \quad \frac{4}{3} = (3b)b \quad \Rightarrow \quad \frac{4}{9} = b^2 \quad \Rightarrow \quad b = \frac{2}{3} \quad [\text{since } b > 0]$ and $C = 3\left(\frac{2}{3}\right) = 2$. The function is $f(x) = 2\left(\frac{2}{3}\right)^x$.

23. If $f(x) = 5^x$, then $\dfrac{f(x + h) - f(x)}{h} = \dfrac{5^{x+h} - 5^x}{h} = \dfrac{5^x 5^h - 5^x}{h} = \dfrac{5^x\left(5^h - 1\right)}{h} = 5^x\left(\dfrac{5^h - 1}{h}\right)$.

24. Suppose the month is February. Your payment on the 28th day would be $2^{28-1} = 2^{27} = 134{,}217{,}728$ cents, or \$1,342,177.28. Clearly, the second method of payment results in a larger amount for any month.

25. $2 \text{ ft} = 24 \text{ in}$, $f(24) = 24^2 \text{ in} = 576 \text{ in} = 48 \text{ ft}$. $g(24) = 2^{24} \text{ in} = 2^{24}/(12 \cdot 5280) \text{ mi} \approx 265 \text{ mi}$

26. We see from the graphs that for x less than about 1.8, $g(x) = 5^x > f(x) = x^5$, and then near the point $(1.8, 17.1)$ the curves intersect. Then $f(x) > g(x)$ from $x \approx 1.8$ until $x = 5$. At $(5, 3125)$ there is another point of intersection, and for $x > 5$ we see that $g(x) > f(x)$. In fact, g increases much more rapidly than f beyond that point.

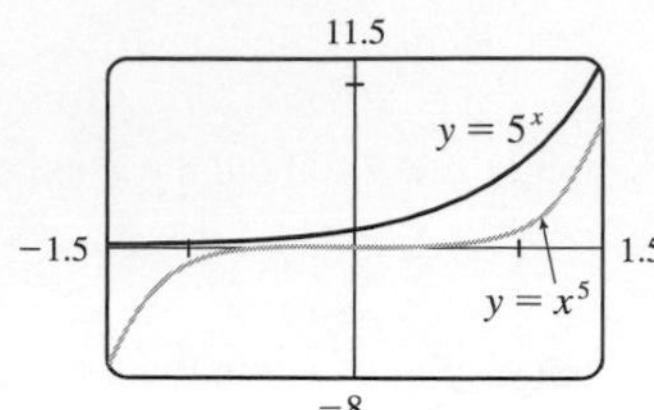

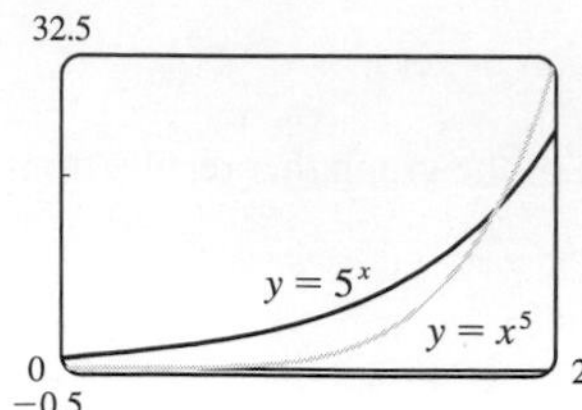

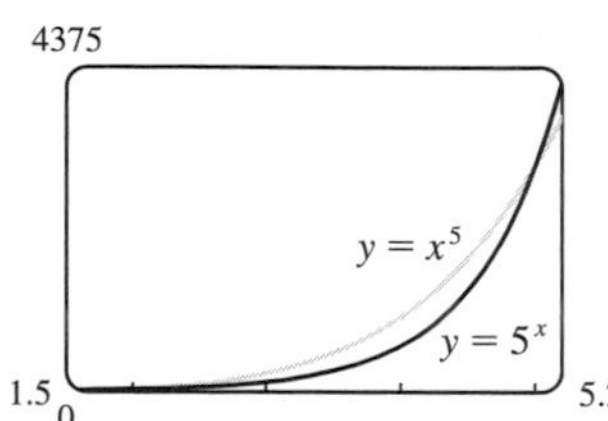

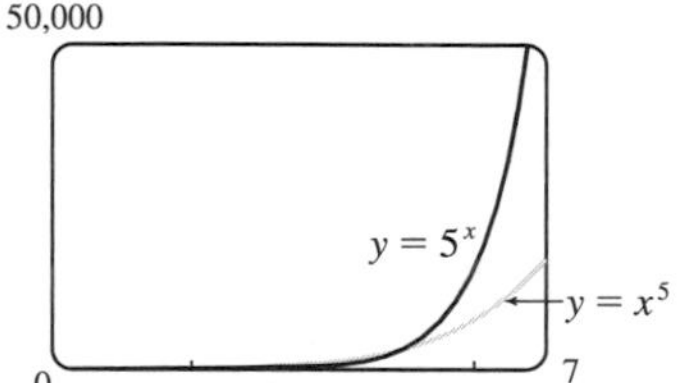

27. The graph of g finally surpasses that of f at $x \approx 35.8$.

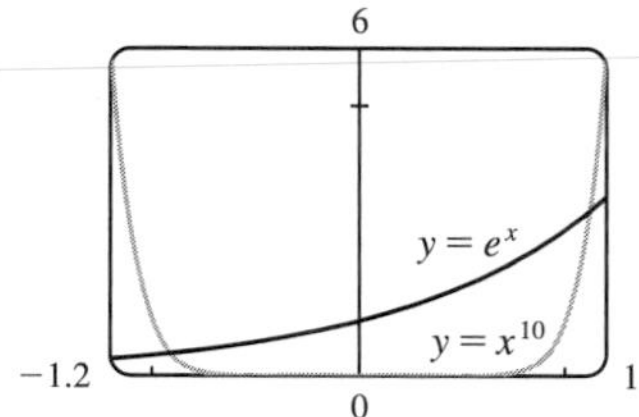

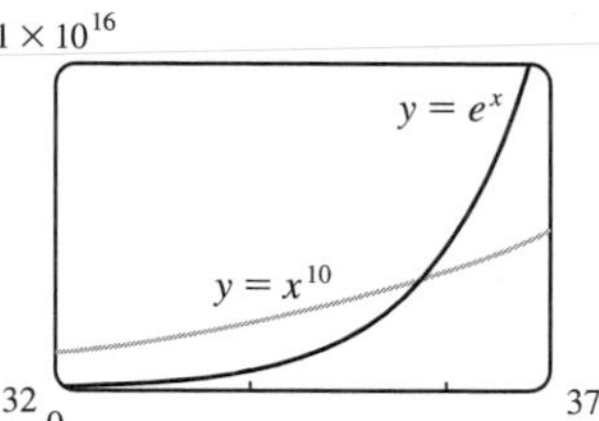

28. We graph $y = e^x$ and $y = 1{,}000{,}000{,}000$ and determine where $e^x = 1 \times 10^9$. This seems to be true at $x \approx 20.723$, so $e^x > 1 \times 10^9$ for $x > 20.723$.

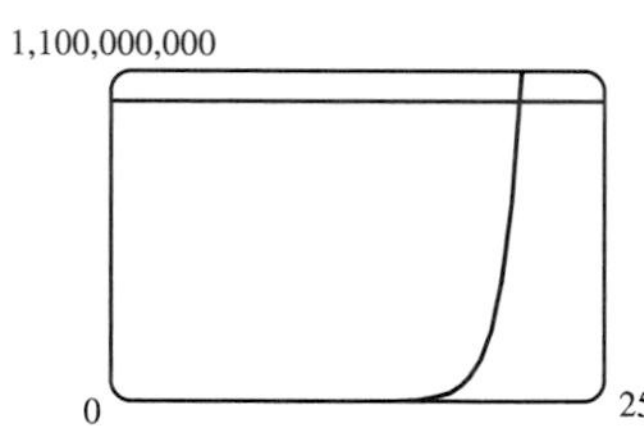

29. (a)

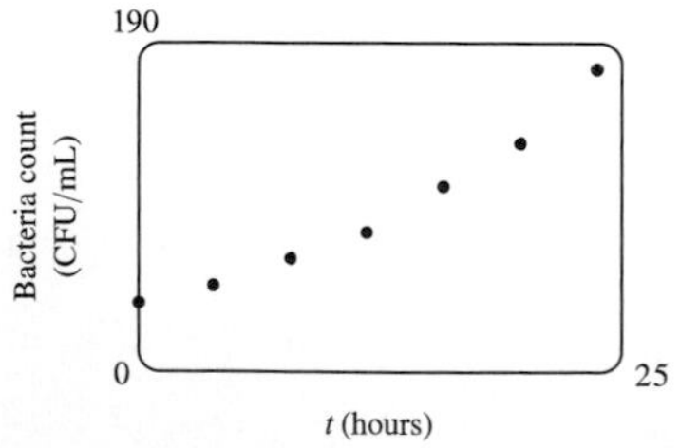

190

Bacteria count (CFU/mL)

0

25

t (hours)

(b) Using a graphing calculator, we obtain the exponential curve $f(t) = 36.89301(1.06614)^t$.

(c) Using the TRACE and zooming in, we find that the bacteria count doubles from 37 to 74 in about 10.87 hours.

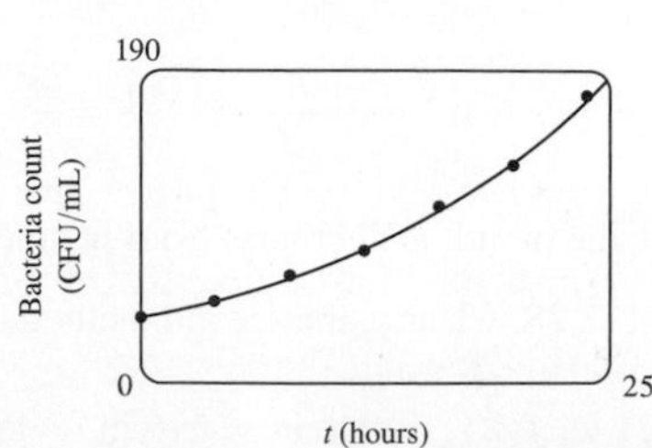

30. (a) Three hours represents 6 doubling periods (one doubling period is 30 minutes). $500 \cdot 2^6 = 32{,}000$

(b) In t hours, there will be $2t$ doubling periods. The initial population is 500, so the population y at time t is $y = 500 \cdot 2^{2t}$.

(c) $t = \frac{40}{60} = \frac{2}{3} \quad \Rightarrow \quad y = 500 \cdot 2^{2(2/3)} \approx 1260$

(d) We graph $y_1 = 500 \cdot 2^{2t}$ and $y_2 = 100{,}000$. The two curves intersect at $t \approx 3.82$, so the population reaches 100,000 in about 3.82 hours.

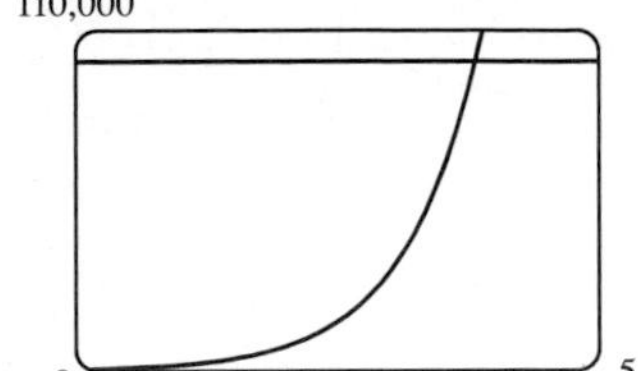

31. (a) Fifteen days represents 3 half-life periods (one half-life period is 5 days). $200\left(\frac{1}{2}\right)^3 = 25$ mg

(b) In t hours, there will be $t/5$ half-life periods. The initial amount is 200 mg, so the amount remaining after t days is $y = 200\left(\frac{1}{2}\right)^{t/5}$ mg, or equivalently, $y = 200 \cdot 2^{-t/5}$ mg.

(c) $t = 3$ weeks $= 21$ days $\quad \Rightarrow \quad y = 200 \cdot 2^{-21/5} \approx 10.9$ mg

(d) We graph $y_1 = 200 \cdot 2^{-t/5}$ and $y_2 = 1$. The two curves intersect at $t \approx 38.2$, so the mass will be reduced to 1 mg in about 38.2 days.

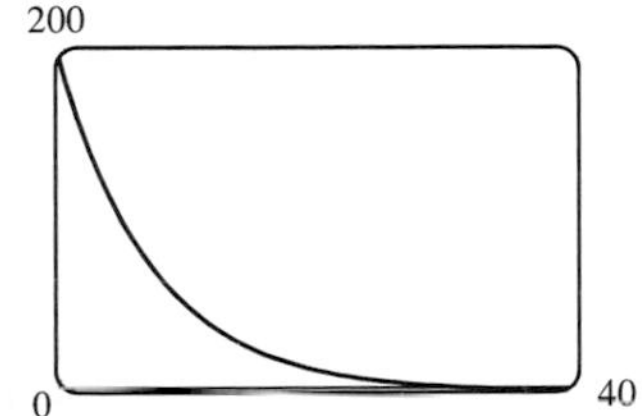

32. (a) Sixty hours represents 4 half-life periods. $2 \cdot \left(\frac{1}{2}\right)^4 = \frac{1}{8}$ g

(b) In t hours, there will be $t/15$ half-life periods. The initial mass is 2 g, so the mass y at time t is $y = 2 \cdot \left(\frac{1}{2}\right)^{t/15}$.

(c) 4 days $= 4 \cdot 24 = 96$ hours. $t = 96 \quad \Rightarrow \quad y = 2 \cdot \left(\frac{1}{2}\right)^{96/15} \approx 0.024$ g

(d) $y = 0.01 \quad \Rightarrow \quad t \approx 114.7$ hours

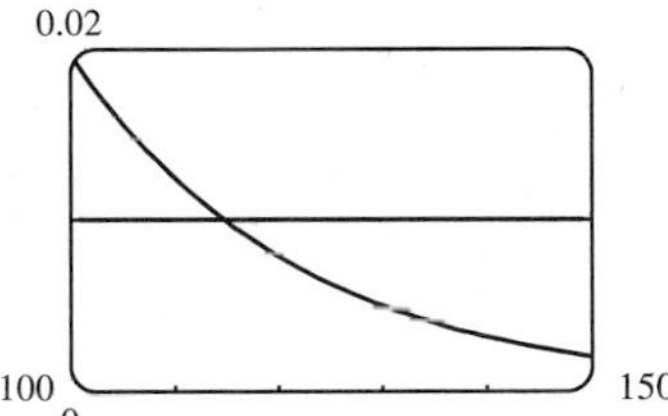

33. From the table, we see that $V(1) = 76$. In Figure 11, we estimate that $V = 38$ (half of 76) when $t \approx 4.5$. This gives us a half-life of $4.5 - 1 = 3.5$ days.

34. (a) The exponential decay model has the form $C(t) = a\left(\frac{1}{2}\right)^{t/1.5}$, where t is the number of hours after midnight and $C(t)$ is the BAC. When $t = 0$, $C(t) = 0.6$, so $0.6 = a\left(\frac{1}{2}\right)^0 \quad \Leftrightarrow \quad a = 0.6$. Thus, the model is $C(t) = 0.6\left(\frac{1}{2}\right)^{t/1.5}$.

(b) From the graph, we estimate that the BAC is 0.08 mg/mL when $t \approx 4.4$ hours. (Note that the legal limit is often 0.08%, which is not 0.08 mg/mL.)

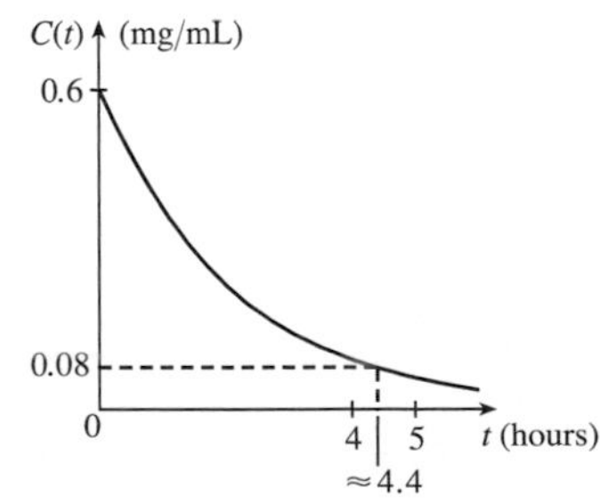

35. Let $t = 0$ correspond to 1950 to get the model $P = ab^t$, where $a \approx 2614.086$ and $b \approx 1.01693$. To estimate the population in 1993, let $t = 43$ to obtain $P \approx 5381$ million. To predict the population in 2020, let $t = 70$ to obtain $P \approx 8466$ million.

36. Let $t = 0$ correspond to 1900 to get the model $P = ab^t$, where $a \approx 80.8498$ and $b \approx 1.01269$. To estimate the population in 1925, let $t = 25$ to obtain $P \approx 111$ million. To predict the population in 2020, let $t = 120$ to obtain $P \approx 367$ million.

37. 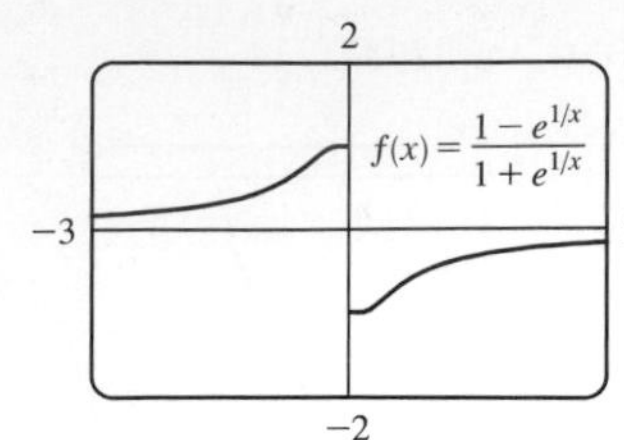

From the graph, it appears that f is an odd function (f is undefined for $x = 0$). To prove this, we must show that $f(-x) = -f(x)$.

$$f(-x) = \frac{1 - e^{1/(-x)}}{1 + e^{1/(-x)}} = \frac{1 - e^{(-1/x)}}{1 + e^{(-1/x)}} = \frac{1 - \dfrac{1}{e^{1/x}}}{1 + \dfrac{1}{e^{1/x}}} \cdot \frac{e^{1/x}}{e^{1/x}} = \frac{e^{1/x} - 1}{e^{1/x} + 1}$$

$$= -\frac{1 - e^{1/x}}{1 + e^{1/x}} = -f(x)$$

so f is an odd function.

38. We'll start with $b = -1$ and graph $f(x) = \dfrac{1}{1 + ae^{bx}}$ for $a = 0.1$, 1, and 5.

From the graph, we see that there is a horizontal asymptote $y = 0$ as $x \to -\infty$ and a horizontal asymptote $y = 1$ as $x \to \infty$. If $a = 1$, the y-intercept is $\left(0, \frac{1}{2}\right)$. As a gets smaller (close to 0), the graph of f moves left. As a gets larger, the graph of f moves right.

2
$a = 1$
$a = 0.1$
$a = 5$
−7
7
$b = -1$
−1

As b changes from -1 to 0, the graph of f is stretched horizontally. As b changes through large negative values, the graph of f is compressed horizontally. (This takes care of negatives values of b.)

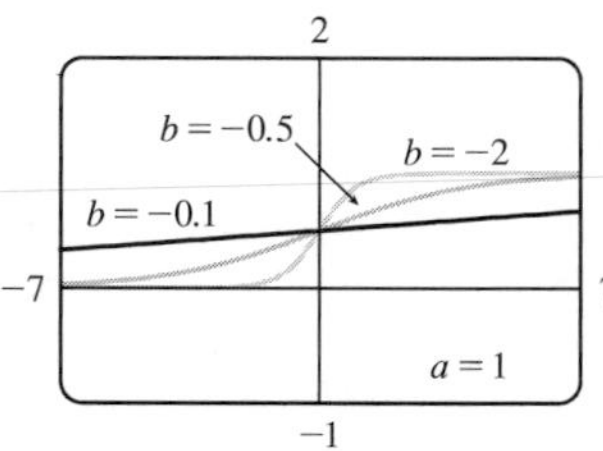

If b is positive, the graph of f is reflected through the y-axis.

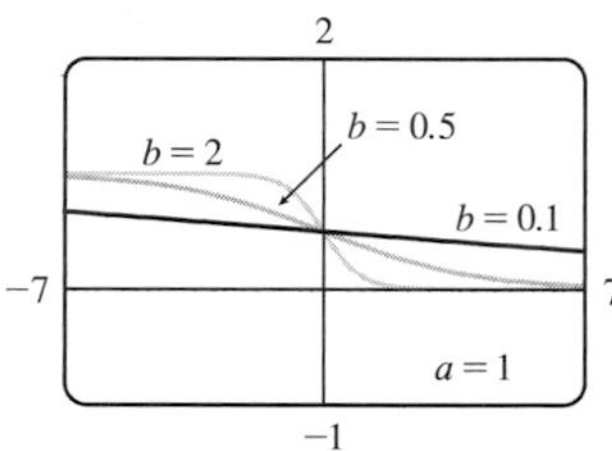

Last, if $b = 0$, the graph of f is the horizontal line $y = 1/(1 + a)$.

1.5 Inverse Functions and Logarithms

1. (a) See Definition 1.

(b) It must pass the Horizontal Line Test.

2. (a) $f^{-1}(y) = x \iff f(x) = y$ for any y in B. The domain of f^{-1} is B and the range of f^{-1} is A.

(b) See the steps in (5).

(c) Reflect the graph of f about the line $y = x$.

3. f is not one-to-one because $2 \neq 6$, but $f(2) = 2.0 = f(6)$.

4. f is one-to-one because it never takes on the same value twice.

5. We could draw a horizontal line that intersects the graph in more than one point. Thus, by the Horizontal Line Test, the function is not one-to-one.

6. No horizontal line intersects the graph more than once. Thus, by the Horizontal Line Test, the function is one-to-one.

7. No horizontal line intersects the graph more than once. Thus, by the Horizontal Line Test, the function is one-to-one.

8. We could draw a horizontal line that intersects the graph in more than one point. Thus, by the Horizontal Line Test, the function is not one-to-one.

9. The graph of $f(x) = 2x - 3$ is a line with slope 2. It passes the Horizontal Line Test, so f is one-to-one.

Algebraic solution: If $x_1 \neq x_2$, then $2x_1 \neq 2x_2 \Rightarrow 2x_1 - 3 \neq 2x_2 - 3 \Rightarrow f(x_1) \neq f(x_2)$, so f is one-to-one.

10. The graph of $f(x) = x^4 - 16$ is symmetric with respect to the y-axis. Pick any x-values equidistant from 0 to find two equal function values. For example, $f(-1) = -15$ and $f(1) = -15$, so f is not one-to-one.

11. $g(x) = 1 - \sin x$. $g(0) = 1$ and $g(\pi) = 1$, so g is not one-to-one.

12. The graph of $g(x) = \sqrt[3]{x}$ passes the Horizontal Line Test, so g is one-to-one.

13. A football will attain every height h up to its maximum height twice: once on the way up, and again on the way down. Thus, even if t_1 does not equal t_2, $f(t_1)$ may equal $f(t_2)$, so f is not 1-1.

14. f is not 1-1 because eventually we all stop growing and therefore, there are two times at which we have the same height.

15. (a) Since f is 1-1, $f(6) = 17 \Leftrightarrow f^{-1}(17) = 6$.

(b) Since f is 1-1, $f^{-1}(3) = 2 \Leftrightarrow f(2) = 3$.

16. First, we must determine x such that $f(x) = 3$. By inspection, we see that if $x = 1$, then $f(1) = 3$. Since f is 1-1 (f is an increasing function), it has an inverse, and $f^{-1}(3) = 1$. If f is a 1-1 function, then $f(f^{-1}(a)) = a$, so $f(f^{-1}(2)) = 2$.

17. First, we must determine x such that $g(x) = 4$. By inspection, we see that if $x = 0$, then $g(x) = 4$. Since g is 1-1 (g is an increasing function), it has an inverse, and $g^{-1}(4) = 0$.

18. (a) f is 1-1 because it passes the Horizontal Line Test.

(b) Domain of $f = [-3, 3]$ = Range of f^{-1}. Range of $f = [-1, 3]$ = Domain of f^{-1}.

(c) Since $f(0) = 2$, $f^{-1}(2) = 0$.

(d) Since $f(-1.7) \approx 0$, $f^{-1}(0) \approx -1.7$.

19. We solve $C = \frac{5}{9}(F - 32)$ for F: $\frac{9}{5}C = F - 32 \Rightarrow F = \frac{9}{5}C + 32$. This gives us a formula for the inverse function, that is, the Fahrenheit temperature F as a function of the Celsius temperature C. $F \geq -459.67 \Rightarrow \frac{9}{5}C + 32 \geq -459.67 \Rightarrow \frac{9}{5}C \geq -491.67 \Rightarrow C \geq -273.15$, the domain of the inverse function.

20. $m = \dfrac{m_0}{\sqrt{1 - v^2/c^2}} \Rightarrow 1 - \dfrac{v^2}{c^2} = \dfrac{m_0^2}{m^2} \Rightarrow \dfrac{v^2}{c^2} = 1 - \dfrac{m_0^2}{m^2} \Rightarrow v^2 = c^2\left(1 - \dfrac{m_0^2}{m^2}\right) \Rightarrow v = c\sqrt{1 - \dfrac{m_0^2}{m^2}}$.

This formula gives us the speed v of the particle in terms of its mass m, that is, $v = f^{-1}(m)$.

21. $y = f(x) = 1 + \sqrt{2 + 3x}$ $(y \geq 1) \Rightarrow y - 1 = \sqrt{2 + 3x} \Rightarrow (y - 1)^2 = 2 + 3x \Rightarrow (y - 1)^2 - 2 = 3x \Rightarrow x = \frac{1}{3}(y - 1)^2 - \frac{2}{3}$. Interchange x and y: $y = \frac{1}{3}(x - 1)^2 - \frac{2}{3}$. So $f^{-1}(x) = \frac{1}{3}(x - 1)^2 - \frac{2}{3}$. Note that the domain of f^{-1} is $x \geq 1$.

22. $y = f(x) = \dfrac{4x-1}{2x+3} \Rightarrow y(2x+3) = 4x-1 \Rightarrow 2xy+3y = 4x-1 \Rightarrow 3y+1 = 4x-2xy \Rightarrow$

$3y+1 = (4-2y)x \Rightarrow x = \dfrac{3y+1}{4-2y}$. Interchange x and y: $y = \dfrac{3x+1}{4-2x}$. So $f^{-1}(x) = \dfrac{3x+1}{4-2x}$.

23. $y = f(x) = e^{2x-1} \Rightarrow \ln y = 2x-1 \Rightarrow 1+\ln y = 2x \Rightarrow x = \frac{1}{2}(1+\ln y)$.

Interchange x and y: $y = \frac{1}{2}(1+\ln x)$. So $f^{-1}(x) = \frac{1}{2}(1+\ln x)$.

24. $y = f(x) = x^2 - x \quad (x \geq \frac{1}{2}) \Rightarrow y = x^2 - x + \frac{1}{4} - \frac{1}{4} \Rightarrow y = (x-\frac{1}{2})^2 - \frac{1}{4} \Rightarrow$

$y + \frac{1}{4} = (x-\frac{1}{2})^2 \Rightarrow x - \frac{1}{2} = \sqrt{y+\frac{1}{4}} \Rightarrow x = \frac{1}{2} + \sqrt{y+\frac{1}{4}}$. Interchange x and y: $y = \frac{1}{2} + \sqrt{x+\frac{1}{4}}$. So

$f^{-1}(x) = \frac{1}{2} + \sqrt{x+\frac{1}{4}}$.

25. $y = f(x) = \ln(x+3) \Rightarrow x+3 = e^y \Rightarrow x = e^y - 3$. Interchange x and y: $y = e^x - 3$. So $f^{-1}(x) = e^x - 3$.

26. $y = f(x) = \dfrac{1-e^{-x}}{1+e^{-x}} \Rightarrow y(1+e^{-x}) = 1-e^{-x} \Rightarrow y + ye^{-x} = 1 - e^{-x} \Rightarrow ye^x + y = e^x - 1$ [multiply

each term by e^x] $\Rightarrow ye^x - e^x = -y-1 \Rightarrow e^x(y-1) = -y-1 \Rightarrow e^x = \dfrac{1+y}{1-y} \Rightarrow x = \ln\left(\dfrac{1+y}{1-y}\right)$.

Interchange x and y: $y = \ln\left(\dfrac{1+x}{1-x}\right)$. So $f^{-1}(x) = \ln\left(\dfrac{1+x}{1-x}\right)$.

27. $y = f(x) = \sqrt{4x+3} \; (y \geq 0) \Rightarrow y^2 = 4x+3 \Rightarrow x = \dfrac{y^2-3}{4}$.

Interchange x and y: $y = \dfrac{x^2-3}{4}$. So $f^{-1}(x) = \dfrac{x^2-3}{4} \; (x \geq 0)$. From the graph, we see that f and f^{-1} are reflections about the line $y = x$.

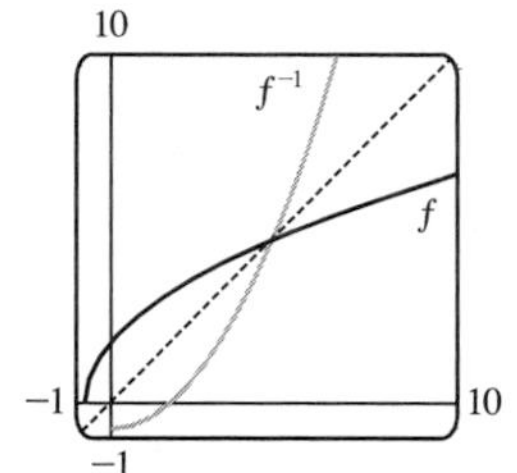

28. $y = f(x) = 1 + e^{-x} \Rightarrow e^{-x} = y-1 \Rightarrow -x = \ln(y-1) \Rightarrow$

$x = -\ln(y-1)$. Interchange x and y: $y = -\ln(x-1)$.

So $f^{-1}(x) = -\ln(x-1)$. From the graph, we see that f and f^{-1} are reflections about the line $y = x$.

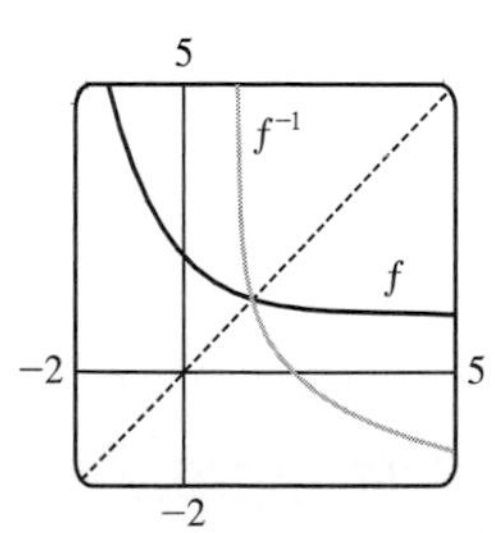

29. Reflect the graph of f about the line $y = x$. The points $(-1,-2)$, $(1,-1)$, $(2,2)$, and $(3,3)$ on f are reflected to $(-2,-1)$, $(-1,1)$, $(2,2)$, and $(3,3)$ on f^{-1}.

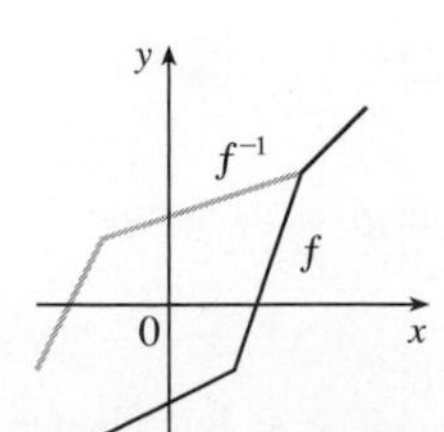

30. Reflect the graph of f about the line $y = x$.

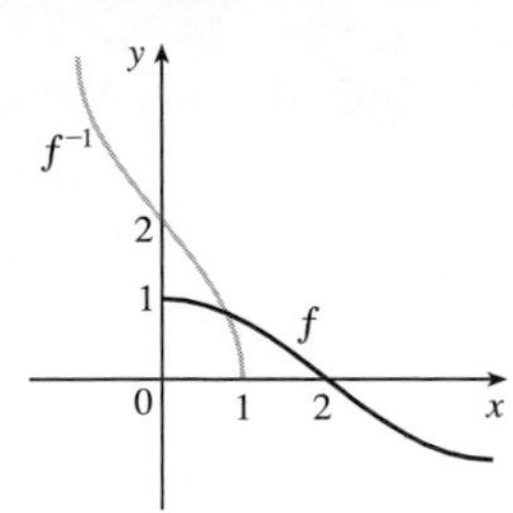

31. (a) $y = f(x) = \sqrt{1 - x^2}$ $\quad (0 \le x \le 1$ and note that $y \ge 0) \quad \Rightarrow$

$y^2 = 1 - x^2 \quad \Rightarrow \quad x^2 = 1 - y^2 \quad \Rightarrow \quad x = \sqrt{1 - y^2}$. So

$f^{-1}(x) = \sqrt{1 - x^2},\ 0 \le x \le 1$. We see that $f^{-1}$and f are the same function.

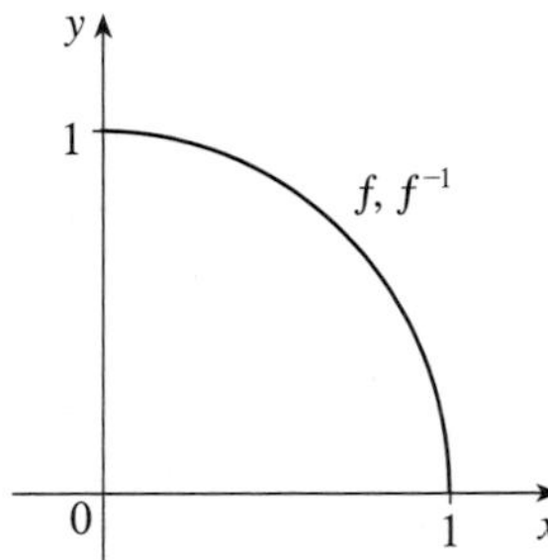

(b) The graph of f is the portion of the circle $x^2 + y^2 = 1$ with $0 \le x \le 1$ and $0 \le y \le 1$ (quarter-circle in the first quadrant). The graph of f is symmetric with respect to the line $y = x$, so its reflection about $y = x$ is itself, that is, $f^{-1} = f$.

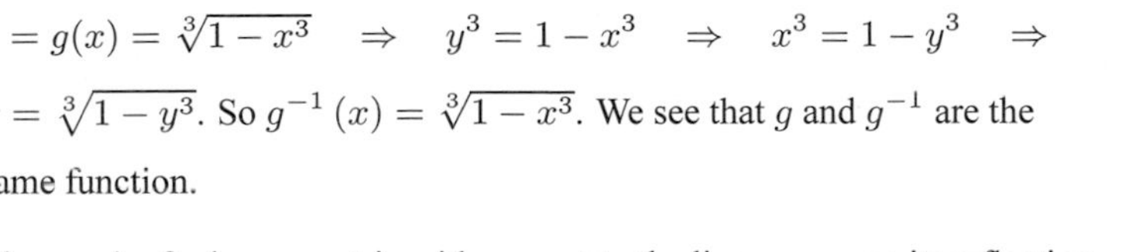

32. (a) $y = g(x) = \sqrt[3]{1 - x^3} \quad \Rightarrow \quad y^3 = 1 - x^3 \quad \Rightarrow \quad x^3 = 1 - y^3 \quad \Rightarrow$

$x = \sqrt[3]{1 - y^3}$. So $g^{-1}(x) = \sqrt[3]{1 - x^3}$. We see that g and g^{-1} are the same function.

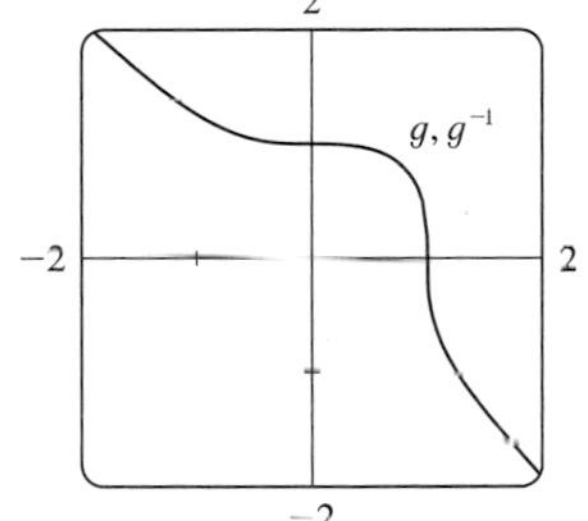

(b) The graph of g is symmetric with respect to the line $y = x$, so its reflection about $y = x$ is itself, that is, $g^{-1} = g$.

33. (a) It is defined as the inverse of the exponential function with base b, that is, $\log_b x = y \quad \Leftrightarrow \quad b^y = x$.

(b) $(0, \infty)$ $\qquad$ (c) $\mathbb{R}$ $\qquad$ (d) See Figure 11.

34. (a) The natural logarithm is the logarithm with base e, denoted $\ln x$.

(b) The common logarithm is the logarithm with base 10, denoted $\log x$.

(c) See Figure 13.

35. (a) $\log_2 32 = \log_2 2^5 = 5$ by (7).

(b) $\log_8 2 = \log_8 8^{1/3} = \frac{1}{3}$ by (7).

Another method: Set the logarithm equal to x and change to an exponential equation.

$\log_8 2 = x \quad \Leftrightarrow \quad 8^x = 2 \quad \Leftrightarrow \quad (2^3)^x = 2 \quad \Leftrightarrow \quad 2^{3x} = 2^1 \quad \Leftrightarrow \quad 3x = 1 \quad \Leftrightarrow \quad x = \frac{1}{3}$.

36. (a) $\log_5 \dfrac{1}{125} = \log_5 \dfrac{1}{5^3} = \log_5 5^{-3} = -3$ by (7).

(b) $\ln(1/e^2) = \ln e^{-2} = -2$ by (9).

37. (a)
$$\begin{aligned} \log_{10} 40 + \log_{10} 2.5 &= \log_{10}[(40)(2.5)] && \text{[by Law 1]} \\ &= \log_{10} 100 \\ &= \log_{10} 10^2 = 2 && \text{[by (7)]} \end{aligned}$$

(b) $\log_8 60 - \log_8 3 - \log_8 5 = \log_8 \frac{60}{3} - \log_8 5$ [by Law 2]

$= \log_8 20 - \log_8 5$

$= \log_8 \frac{20}{5}$ [by Law 2]

$= \log_8 4 = \log_8 8^{2/3} = \frac{2}{3}$ [by (7)]

38. (a) $e^{-\ln 2} = \dfrac{1}{e^{\ln 2}} = \dfrac{1}{2}$ by (9). *Or:* $e^{-\ln 2} = \left(e^{\ln 2}\right)^{-1} = 2^{-1} = \dfrac{1}{2}$

(b) $e^{\ln(\ln e^3)} = e^{\ln 3}$ [by (9)] $= 3$ by (9).

39. $\ln 10 + 2\ln 5 = \ln 10 + \ln 5^2$ [by Law 3]

$= \ln[(10)(25)]$ [by Law 1]

$= \ln 250$

40. $\ln b + 2\ln c - 3\ln d = \ln b + \ln c^2 - \ln d^3$ [by Law 3]

$= \ln bc^2 - \ln d^3$ [by Law 1]

$= \ln \dfrac{bc^2}{d^3}$ [by Law 2]

41. $\frac{1}{3}\ln(x+2)^3 + \frac{1}{2}\left[\ln x - \ln(x^2+3x+2)^2\right] = \ln[(x+2)^3]^{1/3} + \frac{1}{2}\ln \dfrac{x}{(x^2+3x+2)^2}$ [by Laws 3, 2]

$= \ln(x+2) + \ln \dfrac{\sqrt{x}}{x^2+3x+2}$ [by Law 3]

$= \ln \dfrac{(x+2)\sqrt{x}}{(x+1)(x+2)}$ [by Law 1]

$= \ln \dfrac{\sqrt{x}}{x+1}$

Note that since $\ln x$ is defined for $x > 0$, we have $x+1$, $x+2$, and x^2+3x+2 all positive, and hence their logarithms are defined.

42. (a) $\log_5 10 = \dfrac{\ln 10}{\ln 5}$ [by (10)] ≈ 1.430677 (b) $\log_3 57 = \dfrac{\ln 57}{\ln 3}$ [by (10)] ≈ 3.680144

43. To graph these functions, we use $\log_{1.5} x = \dfrac{\ln x}{\ln 1.5}$ and $\log_{50} x = \dfrac{\ln x}{\ln 50}$. These graphs all approach $-\infty$ as $x \to 0^+$, and they all pass through the point $(1, 0)$. Also, they are all increasing, and all approach ∞ as $x \to \infty$. The functions with larger bases increase extremely slowly, and the ones with smaller bases do so somewhat more quickly. The functions with large bases approach the y-axis more closely as $x \to 0^+$.

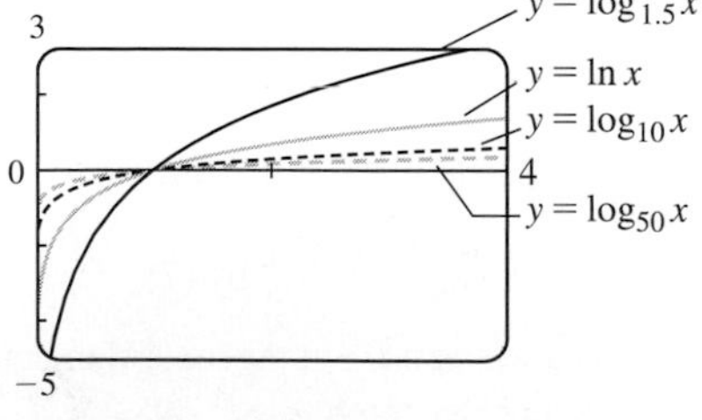

44. We see that the graph of $\ln x$ is the reflection of the graph of e^x about the line $y = x$, and that the graph of $\log_{10} x$ is the reflection of the graph of 10^x about the same line. The graph of 10^x increases more quickly than that of e^x. Also note that $\log_{10} x \to \infty$ as $x \to \infty$ more slowly than $\ln x$.

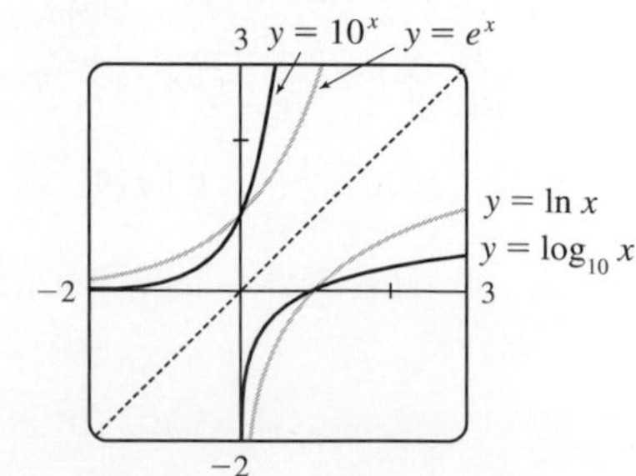

45. 3 ft = 36 in, so we need x such that $\log_2 x = 36 \Leftrightarrow x = 2^{36} = 68{,}719{,}476{,}736$. In miles, this is

$$68{,}719{,}476{,}736 \text{ in} \cdot \frac{1 \text{ ft}}{12 \text{ in}} \cdot \frac{1 \text{ mi}}{5280 \text{ ft}} \approx 1{,}084{,}587.7 \text{ mi.}$$

46.

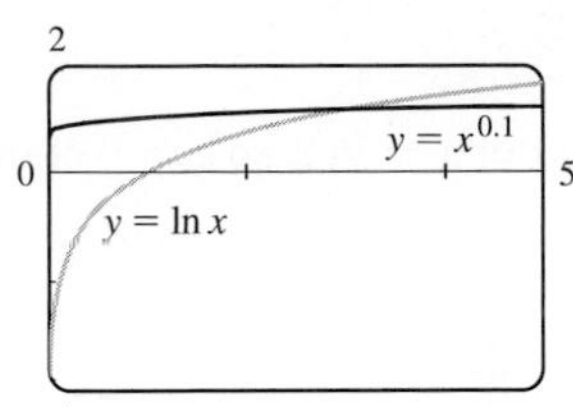

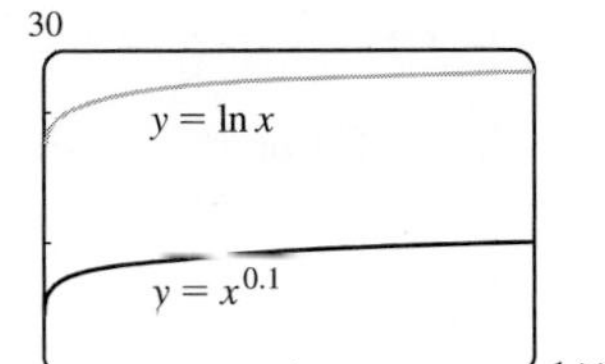

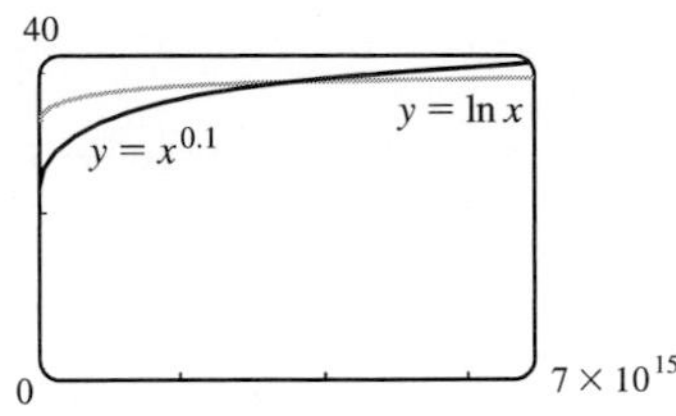

From the graphs, we see that $f(x) = x^{0.1} > g(x) = \ln x$ for approximately $0 < x < 3.06$, and then $g(x) > f(x)$ for $3.06 < x < 3.43 \times 10^{15}$ (approximately). At that point, the graph of f finally surpasses the graph of g for good.

47. (a) Shift the graph of $y = \log_{10} x$ five units to the left to obtain the graph of $y = \log_{10}(x+5)$. Note the vertical asymptote of $x = -5$.

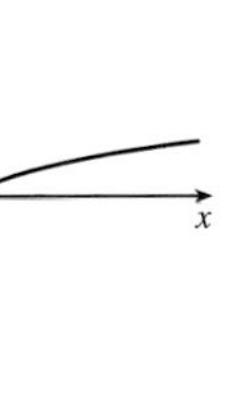

$y = \log_{10} x$

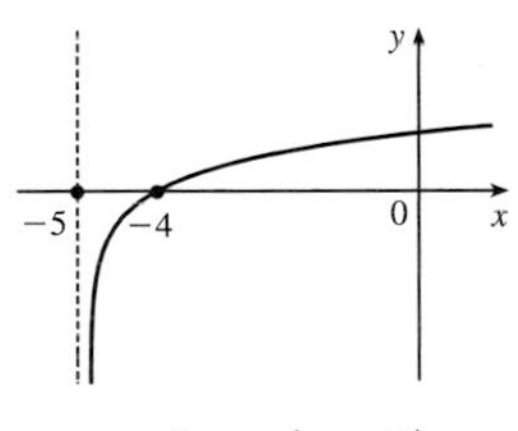

$y = \log_{10}(x+5)$

(b) Reflect the graph of $y = \ln x$ about the x-axis to obtain the graph of $y = -\ln x$.

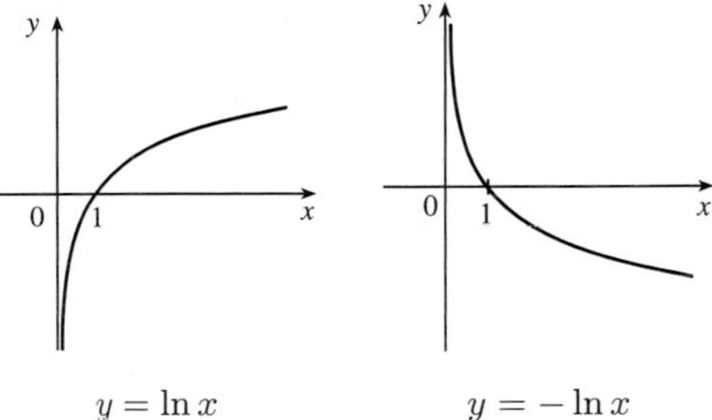

$y = \ln x$ $y = -\ln x$

48. (a) Reflect the graph of $y = \ln x$ about the y-axis to obtain the graph of $y = \ln(-x)$.

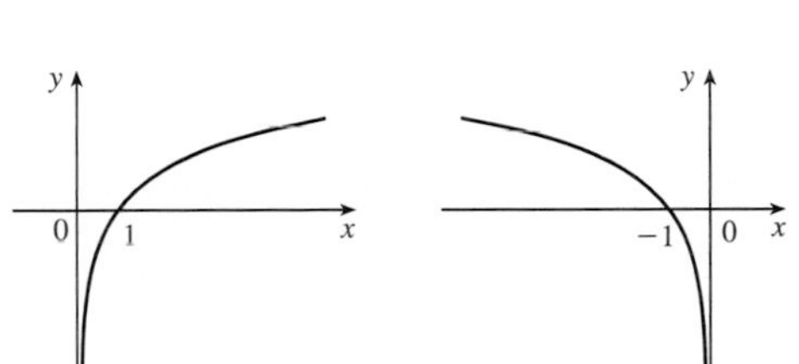

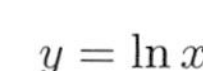

$y = \ln x$

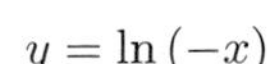

$y = \ln(-x)$

(b) Reflect the portion of the graph of $y = \ln x$ to the right of the y-axis about the y-axis. The graph of $y = \ln|x|$ is that reflection in addition to the original portion.

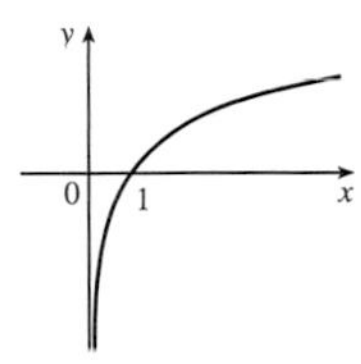

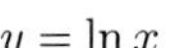

$y = \ln x$

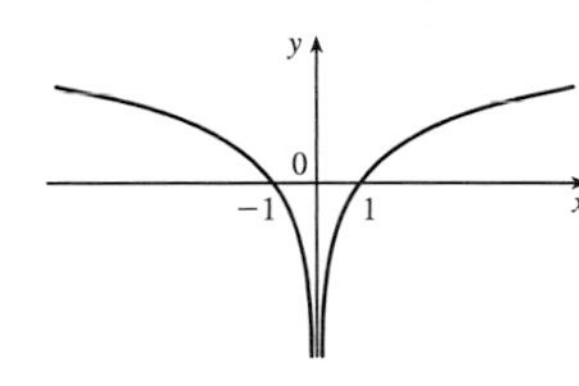

$y = \ln|x|$

49. (a) The domain of $f(x) = \ln x + 2$ is $x > 0$ and the range is $\mathbb{R}$.

(b) $y = 0 \Rightarrow \ln x + 2 = 0 \Rightarrow \ln x = -2 \Rightarrow x = e^{-2}$

(c) We shift the graph of $y = \ln x$ two units upward.

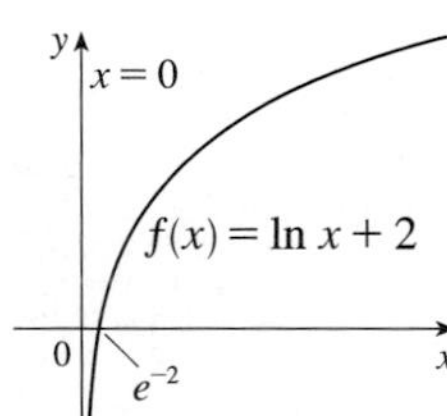

50. (a) The domain of $f(x) = \ln(x-1) - 1$ is $x > 1$ and the range is $\mathbb{R}$.

(b) $y = 0 \Rightarrow \ln(x-1) - 1 = 0 \Rightarrow \ln(x-1) = 1 \Rightarrow$
$x - 1 = e^1 \Rightarrow x = e + 1$

(c) We shift the graph of $y = \ln x$ one unit to the right and one unit downward.

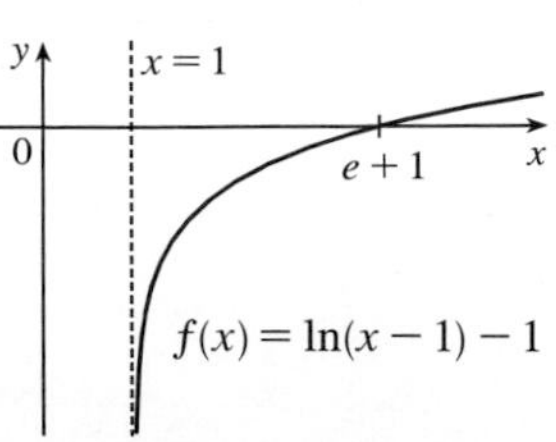

51. (a) $e^{7-4x} = 6 \quad \Leftrightarrow \quad 7 - 4x = \ln 6 \quad \Leftrightarrow \quad 7 - \ln 6 = 4x \quad \Leftrightarrow \quad x = \frac{1}{4}(7 - \ln 6)$

(b) $\ln(3x - 10) = 2 \quad \Leftrightarrow \quad 3x - 10 = e^2 \quad \Leftrightarrow \quad 3x = e^2 + 10 \quad \Leftrightarrow \quad x = \frac{1}{3}(e^2 + 10)$

52. (a) $\ln(x^2 - 1) = 3 \quad \Leftrightarrow \quad x^2 - 1 = e^3 \quad \Leftrightarrow \quad x^2 = 1 + e^3 \quad \Leftrightarrow \quad x = \pm\sqrt{1 + e^3}$.

(b) $e^{2x} - 3e^x + 2 = 0 \Leftrightarrow (e^x - 1)(e^x - 2) = 0 \Leftrightarrow e^x = 1$ or $e^x = 2 \Leftrightarrow x = \ln 1$ or $x = \ln 2$, so $x = 0$ or $\ln 2$.

53. (a) $2^{x-5} = 3 \quad \Leftrightarrow \quad \log_2 3 = x - 5 \quad \Leftrightarrow \quad x = 5 + \log_2 3$.

Or: $2^{x-5} = 3 \quad \Leftrightarrow \quad \ln(2^{x-5}) = \ln 3 \quad \Leftrightarrow \quad (x - 5)\ln 2 = \ln 3 \quad \Leftrightarrow \quad x - 5 = \dfrac{\ln 3}{\ln 2} \quad \Leftrightarrow \quad x = 5 + \dfrac{\ln 3}{\ln 2}$

(b) $\ln x + \ln(x - 1) = \ln(x(x - 1)) = 1 \quad \Leftrightarrow \quad x(x - 1) = e^1 \quad \Leftrightarrow \quad x^2 - x - e = 0$. The quadratic formula (with $a = 1$, $b = -1$, and $c = -e$) gives $x = \frac{1}{2}\left(1 \pm \sqrt{1 + 4e}\right)$, but we reject the negative root since the natural logarithm is not defined for $x < 0$. So $x = \frac{1}{2}\left(1 + \sqrt{1 + 4e}\right)$.

54. (a) $\ln(\ln x) = 1 \quad \Leftrightarrow \quad e^{\ln(\ln x)} = e^1 \quad \Leftrightarrow \quad \ln x = e^1 = e \quad \Leftrightarrow \quad e^{\ln x} = e^e \quad \Leftrightarrow \quad x = e^e$

(b) $e^{ax} = Ce^{bx} \quad \Leftrightarrow \quad \ln e^{ax} = \ln[C(e^{bx})] \quad \Leftrightarrow \quad ax = \ln C + \ln e^{bx} \quad \Leftrightarrow \quad ax = \ln C + bx \quad \Leftrightarrow$

$ax - bx = \ln C \quad \Leftrightarrow \quad (a - b)x = \ln C \quad \Leftrightarrow \quad x = \dfrac{\ln C}{a - b}$

55. (a) $\ln x < 0 \quad \Rightarrow \quad x < e^0 \quad \Rightarrow \quad x < 1$. Since the domain of $f(x) = \ln x$ is $x > 0$, the solution of the original inequality is $0 < x < 1$.

(b) $e^x > 5 \quad \Rightarrow \quad \ln e^x > \ln 5 \quad \Rightarrow \quad x > \ln 5$

56. (a) $1 < e^{3x-1} < 2 \quad \Rightarrow \quad \ln 1 < 3x - 1 < \ln 2 \quad \Rightarrow \quad 0 < 3x - 1 < \ln 2 \quad \Rightarrow \quad 1 < 3x < 1 + \ln 2 \quad \Rightarrow$
$\frac{1}{3} < x < \frac{1}{3}(1 + \ln 2)$

(b) $1 - 2\ln x < 3 \quad \Rightarrow \quad -2\ln x < 2 \quad \Rightarrow \quad \ln x > -1 \quad \Rightarrow \quad x > e^{-1}$

57. (a) We must have $e^x - 3 > 0 \quad \Leftrightarrow \quad e^x > 3 \quad \Leftrightarrow \quad x > \ln 3$. Thus, the domain of $f(x) = \ln(e^x - 3)$ is $(\ln 3, \infty)$.

(b) $y = \ln(e^x - 3) \quad \Rightarrow \quad e^y = e^x - 3 \quad \Rightarrow \quad e^x = e^y + 3 \quad \Rightarrow \quad x = \ln(e^y + 3)$, so $f^{-1}(x) = \ln(e^x + 3)$.

Now $e^x + 3 > 0 \quad \Rightarrow \quad e^x > -3$, which is true for any real x, so the domain of f^{-1} is $\mathbb{R}$.

58. (a) By (9), $e^{\ln 300} = 300$ and $\ln(e^{300}) = 300$.

(b) A calculator gives $e^{\ln 300} = 300$ and an error message for $\ln(e^{300})$ since e^{300} is larger than most calculators can evaluate.

59. We see that the graph of $y = f(x) = \sqrt{x^3 + x^2 + x + 1}$ is increasing, so f is 1-1.

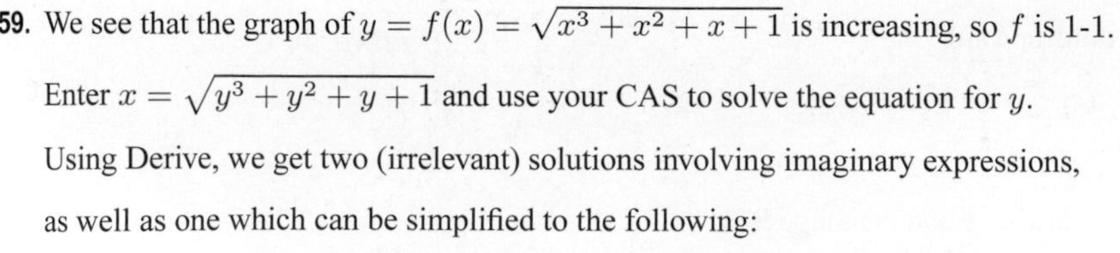

Enter $x = \sqrt{y^3 + y^2 + y + 1}$ and use your CAS to solve the equation for y.

Using Derive, we get two (irrelevant) solutions involving imaginary expressions, as well as one which can be simplified to the following:

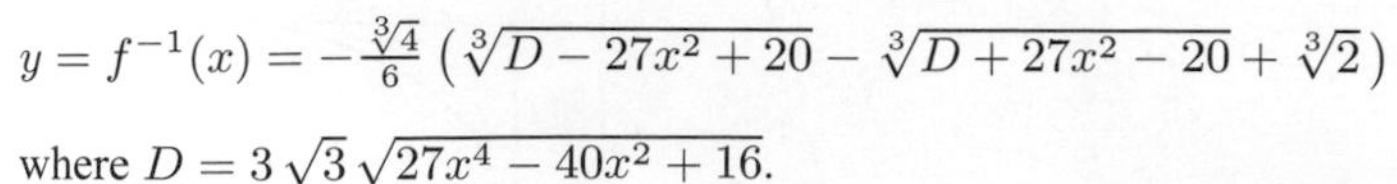

$$y = f^{-1}(x) = -\frac{\sqrt[3]{4}}{6}\left(\sqrt[3]{D - 27x^2 + 20} - \sqrt[3]{D + 27x^2 - 20} + \sqrt[3]{2}\right)$$

where $D = 3\sqrt{3}\sqrt{27x^4 - 40x^2 + 16}$.

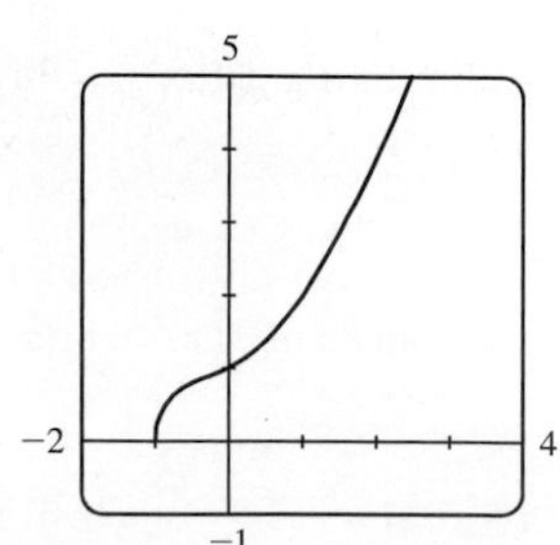

[continued]

Maple and Mathematica each give two complex expressions and one real expression, and the real expression is equivalent to that given by Derive. For example, Maple's expression simplifies to $\frac{1}{6}\frac{M^{2/3}-8-2M^{1/3}}{2M^{1/3}}$, where $M = 108x^2 + 12\sqrt{48 - 120x^2 + 81x^4} - 80$.

60. (a) If we use Derive, then solving $x = y^6 + y^4$ for y gives us six solutions of the form $y = \pm\frac{\sqrt{3}}{3}\sqrt{B-1}$, where $B \in \left\{-2\sin\frac{A}{3}, 2\sin\left(\frac{A}{3}+\frac{\pi}{3}\right), -2\cos\left(\frac{A}{3}+\frac{\pi}{6}\right)\right\}$ and $A = \sin^{-1}\left(\frac{27x-2}{2}\right)$. The inverse for $y = x^6 + x^4$ $(x \geq 0)$ is $y = \frac{\sqrt{3}}{3}\sqrt{B-1}$ with $B = 2\sin\left(\frac{A}{3}+\frac{\pi}{3}\right)$, but because the domain of A is $\left[0, \frac{4}{27}\right]$, this expression is only valid for $x \in \left[0, \frac{4}{27}\right]$.

Happily, Maple gives us the rest of the solution! We solve $x = y^6 + y^4$ for y to get the two real solutions $\pm\frac{\sqrt{6}}{6}\frac{\sqrt{C^{1/3}(C^{2/3}-2C^{1/3}+4)}}{C^{1/3}}$, where $C = 108x + 12\sqrt{3}\sqrt{x(27x-4)}$, and the inverse for $y = x^6 + x^4$ $(x \geq 0)$ is the positive solution, whose domain is $\left[\frac{4}{27}, \infty\right)$.

Mathematica also gives two real solutions, equivalent to those of Maple. The positive one is $\frac{\sqrt{6}}{6}\left(\sqrt[3]{4}D^{1/3} + 2\sqrt[3]{2}D^{-1/3} - 2\right)$, where $D = -2 + 27x + 3\sqrt{3}\sqrt{x}\sqrt{27x-4}$. Although this expression also has domain $\left[\frac{4}{27}, \infty\right)$, Mathematica is mysteriously able to plot the solution for all $x \geq 0$.

(b)

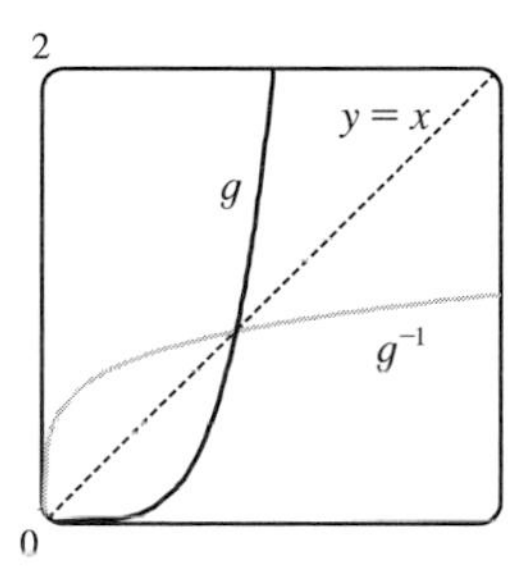

61. (a) $n = f(t) = 100 \cdot 2^{t/3} \Rightarrow \frac{n}{100} = 2^{t/3} \Rightarrow \log_2\left(\frac{n}{100}\right) = \frac{t}{3} \Rightarrow t = 3\log_2\left(\frac{n}{100}\right)$. Using formula (10), we can write this as $t = f^{-1}(n) = 3 \cdot \frac{\ln(n/100)}{\ln 2}$. This function tells us how long it will take to obtain n bacteria (given the number n).

(b) $n = 50{,}000 \Rightarrow t = f^{-1}(50{,}000) = 3 \cdot \frac{\ln\left(\frac{50{,}000}{100}\right)}{\ln 2} = 3\left(\frac{\ln 500}{\ln 2}\right) \approx 26.9$ hours

62. (a) $Q = Q_0(1 - e^{-t/a}) \Rightarrow \frac{Q}{Q_0} = 1 - e^{-t/a} \Rightarrow e^{-t/a} = 1 - \frac{Q}{Q_0} \Rightarrow -\frac{t}{a} = \ln\left(1 - \frac{Q}{Q_0}\right) \Rightarrow$ $t = -a\ln(1 - Q/Q_0)$. This gives us the time t necessary to obtain a given charge Q.

(b) $Q = 0.9Q_0$ and $a = 2 \Rightarrow t = -2\ln(1 - 0.9Q_0/Q_0) = -2\ln 0.1 \approx 4.6$ seconds.

63. (a) $\cos^{-1}(-1) = \pi$ because $\cos\pi = -1$ and π is in the interval $[0, \pi]$ (the range of $\cos^{-1}$).

(b) $\sin^{-1}(0.5) = \frac{\pi}{6}$ because $\sin\frac{\pi}{6} = 0.5$ and $\frac{\pi}{6}$ is in the interval $\left[-\frac{\pi}{2}, \frac{\pi}{2}\right]$ (the range of $\sin^{-1}$).

64. (a) $\tan^{-1}\sqrt{3} = \frac{\pi}{3}$ because $\tan\frac{\pi}{3} = \sqrt{3}$ and $\frac{\pi}{3}$ is in the interval $\left(-\frac{\pi}{2}, \frac{\pi}{2}\right)$ (the range of $\tan^{-1}$).

(b) $\arctan(-1) = -\frac{\pi}{4}$ because $\tan\left(-\frac{\pi}{4}\right) = -1$ and $-\frac{\pi}{4}$ is in the interval $\left(-\frac{\pi}{2}, \frac{\pi}{2}\right)$ (the range of arctan).

65. (a) $\csc^{-1}\sqrt{2} = \frac{\pi}{4}$ because $\csc\frac{\pi}{4} = \sqrt{2}$ and $\frac{\pi}{4}$ is in $\left(0, \frac{\pi}{2}\right] \cup \left(\pi, \frac{3\pi}{2}\right]$ (the range of $\csc^{-1}$).

(b) $\arcsin 1 = \frac{\pi}{2}$ because $\sin\frac{\pi}{2} = 1$ and $\frac{\pi}{2}$ is in $\left[-\frac{\pi}{2}, \frac{\pi}{2}\right]$ (the range of arcsin).

66. (a) $\sin^{-1}(-1/\sqrt{2}) = -\frac{\pi}{4}$ because $\sin\left(-\frac{\pi}{4}\right) = -1/\sqrt{2}$ and $-\frac{\pi}{4}$ is in $\left[-\frac{\pi}{2}, \frac{\pi}{2}\right]$.

(b) $\cos^{-1}\left(\sqrt{3}/2\right) = \frac{\pi}{6}$ because $\cos\frac{\pi}{6} = \sqrt{3}/2$ and $\frac{\pi}{6}$ is in $[0, \pi]$.

67. (a) $\cot^{-1}\left(-\sqrt{3}\right) = \frac{5\pi}{6}$ because $\cot\frac{5\pi}{6} = -\sqrt{3}$ and $\frac{5\pi}{6}$ is in $(0, \pi)$ (the range of $\cot^{-1}$).

(b) $\sec^{-1} 2 = \frac{\pi}{3}$ because $\sec\frac{\pi}{3} = 2$ and $\frac{\pi}{3}$ is in $\left[0, \frac{\pi}{2}\right) \cup \left[\pi, \frac{3\pi}{2}\right)$ (the range of $\sec^{-1}$).

68. (a) $\arcsin(\sin(5\pi/4)) = \arcsin\left(-1/\sqrt{2}\right) = -\frac{\pi}{4}$ because $\sin\left(-\frac{\pi}{4}\right) = -1/\sqrt{2}$ and $-\frac{\pi}{4}$ is in $\left[-\frac{\pi}{2}, \frac{\pi}{2}\right]$.

(b) Let $\theta = \sin^{-1}\left(\frac{5}{13}\right)$ [see the figure].

$$\cos\left(2\sin^{-1}\left(\tfrac{5}{13}\right)\right) = \cos 2\theta = \cos^2\theta - \sin^2\theta$$
$$= \left(\tfrac{12}{13}\right)^2 - \left(\tfrac{5}{13}\right)^2 = \tfrac{144}{169} - \tfrac{25}{169} = \tfrac{119}{169}$$

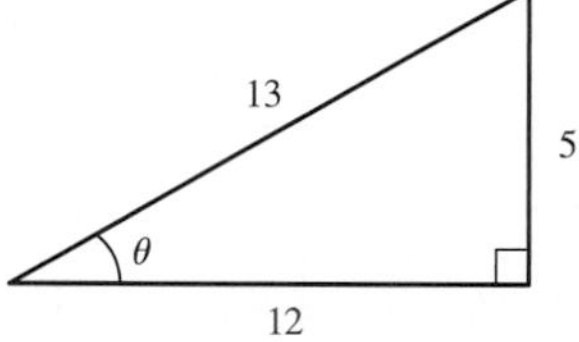

69. Let $y = \sin^{-1} x$. Then $-\frac{\pi}{2} \le y \le \frac{\pi}{2} \Rightarrow \cos y \ge 0$, so $\cos(\sin^{-1} x) = \cos y = \sqrt{1 - \sin^2 y} = \sqrt{1 - x^2}$.

70. Let $y = \sin^{-1} x$. Then $\sin y = x$, so from the triangle (which illustrates the case $y > 0$), we see that

$$\tan(\sin^{-1} x) = \tan y = \frac{x}{\sqrt{1 - x^2}}.$$

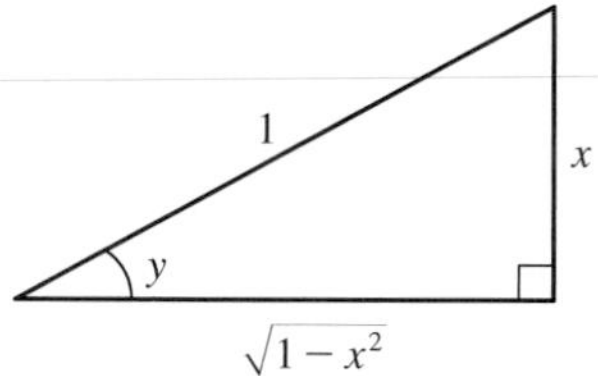

71. Let $y = \tan^{-1} x$. Then $\tan y = x$, so from the triangle (which illustrates the case $y > 0$), we see that

$$\sin(\tan^{-1} x) = \sin y = \frac{x}{\sqrt{1 + x^2}}.$$

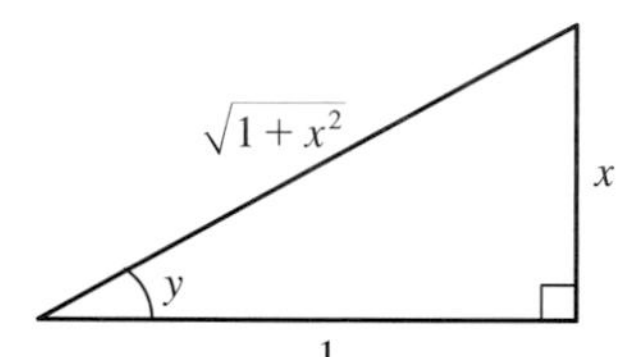

72. Let $y = \arccos x$. Then $\cos y = x$, so from the triangle (which illustrates the case $y > 0$), we see that

$$\sin(2\arccos x) = \sin 2y = 2\sin y\,\cos y$$
$$= 2\left(\sqrt{1 - x^2}\right)(x) = 2x\sqrt{1 - x^2}$$

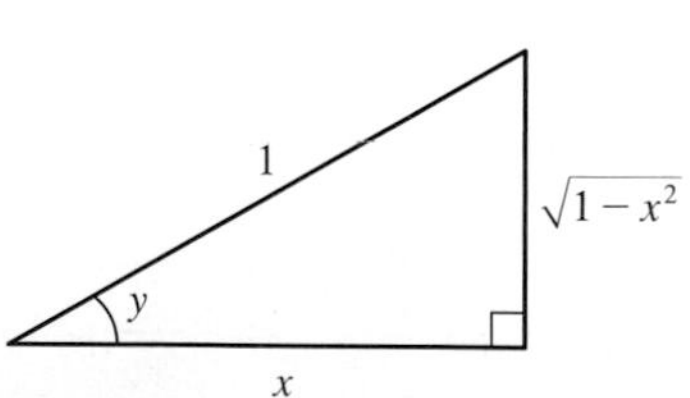

73.

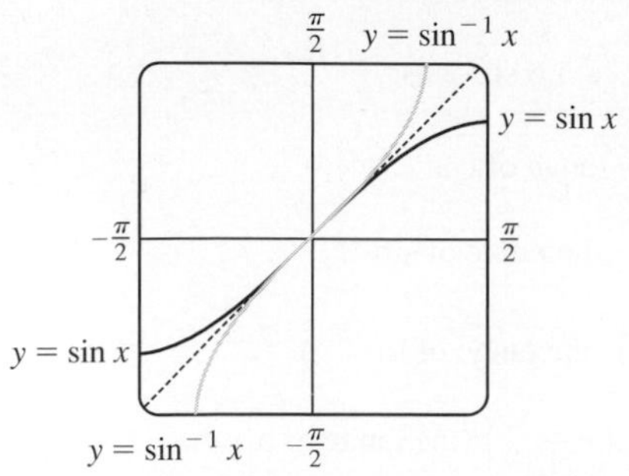

The graph of $\sin^{-1} x$ is the reflection of the graph of $\sin x$ about the line $y = x$.

74.

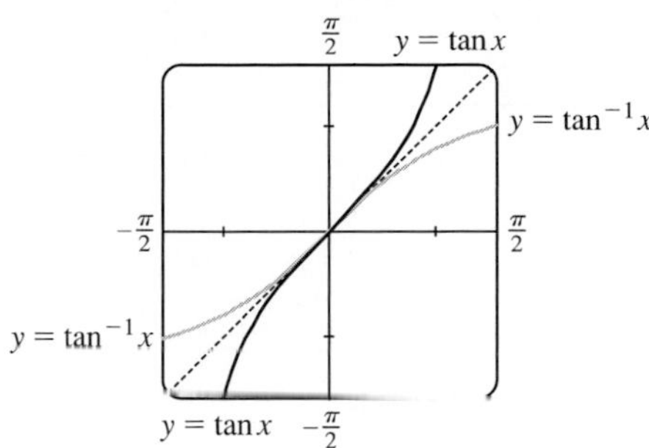

The graph of $\tan^{-1} x$ is the reflection of the graph of $\tan x$ about the line $y = x$.

75. $g(x) = \sin^{-1}(3x + 1)$.

Domain $(g) = \{x \mid -1 \le 3x + 1 \le 1\} = \{x \mid -2 \le 3x \le 0\} = \{x \mid -\frac{2}{3} \le x \le 0\} = [-\frac{2}{3}, 0]$.

Range $(g) = \{y \mid -\frac{\pi}{2} \le y \le \frac{\pi}{2}\} = [-\frac{\pi}{2}, \frac{\pi}{2}]$.

76. (a) $f(x) = \sin(\sin^{-1} x)$

Since one function undoes what the other one does, we get the identity function, $y = x$, on the restricted domain $-1 \le x \le 1$.

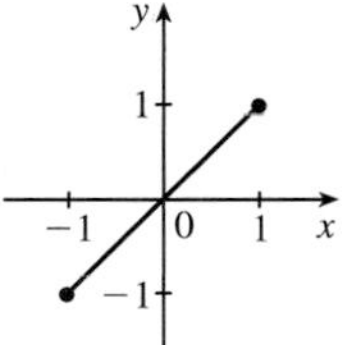

(b) $g(x) = \sin^{-1}(\sin x)$

This is similar to part (a), but with domain $\mathbb{R}$.

Equations for g on intervals of the form $(-\frac{\pi}{2} + \pi n, \frac{\pi}{2} + \pi n)$, for any integer n, can be found using $g(x) = (-1)^n x + (-1)^{n+1} n\pi$.

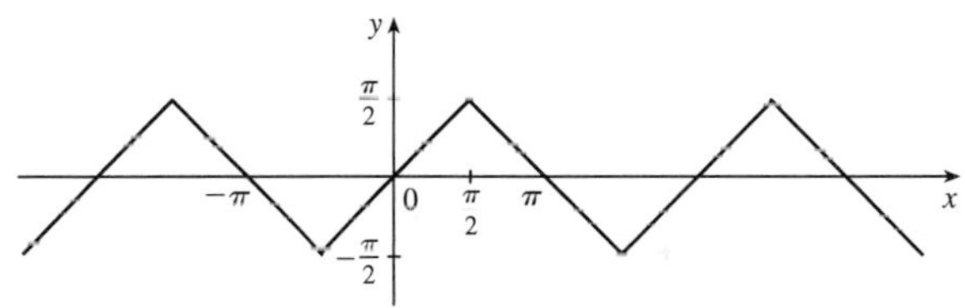

The sine function is monotonic on each of these intervals, and hence, so is g (but in a linear fashion).

77. (a) If the point (x, y) is on the graph of $y = f(x)$, then the point $(x - c, y)$ is that point shifted c units to the left. Since f is 1-1, the point (y, x) is on the graph of $y = f^{-1}(x)$ and the point corresponding to $(x - c, y)$ on the graph of f is $(y, x - c)$ on the graph of f^{-1}. Thus, the curve's reflection is shifted *down* the same number of units as the curve itself is shifted to the left. So an expression for the inverse function is $g^{-1}(x) = f^{-1}(x) - c$.

(b) If we compress (or stretch) a curve horizontally, the curve's reflection in the line $y = x$ is compressed (or stretched) *vertically* by the same factor. Using this geometric principle, we see that the inverse of $h(x) = f(cx)$ can be expressed as $h^{-1}(x) = (1/c)\, f^{-1}(x)$.

1 Review

TRUE-FALSE QUIZ

1. False. Let $f(x) = x^2$, $s = -1$, and $t = 1$. Then $f(s+t) = (-1+1)^2 = 0^2 = 0$, but $f(s) + f(t) = (-1)^2 + 1^2 = 2 \neq 0 = f(s+t)$.

2. False. Let $f(x) = x^2$. Then $f(-2) = 4 = f(2)$, but $-2 \neq 2$.

3. False. Let $f(x) = x^2$. Then $f(3x) = (3x)^2 = 9x^2$ and $3f(x) = 3x^2$. So $f(3x) \neq 3f(x)$.

4. True. If $x_1 < x_2$ and f is a decreasing function, then the y-values get smaller as we move from left to right. Thus, $f(x_1) > f(x_2)$.

5. True. See the Vertical Line Test.

6. False. Let $f(x) = x^2$ and $g(x) = 2x$. Then $(f \circ g)(x) = f(g(x)) = f(2x) = (2x)^2 = 4x^2$ and $(g \circ f)(x) = g(f(x)) = g(x^2) = 2x^2$. So $f \circ g \neq g \circ f$.

7. False. Let $f(x) = x^3$. Then f is one-to-one and $f^{-1}(x) = \sqrt[3]{x}$. But $1/f(x) = 1/x^3$, which is not equal to $f^{-1}(x)$.

8. True. We can divide by e^x since $e^x \neq 0$ for every x.

9. True. The function $\ln x$ is an increasing function on $(0, \infty)$.

10. False. Let $x = e$. Then $(\ln x)^6 = (\ln e)^6 = 1^6 = 1$, but $6 \ln x = 6 \ln e = 6 \cdot 1 = 6 \neq 1 = (\ln x)^6$. What *is* true, however, is that $\ln(x^6) = 6 \ln x$ for $x > 0$.

11. False. Let $x = e^2$ and $a = e$. Then $\dfrac{\ln x}{\ln a} = \dfrac{\ln e^2}{\ln e} = \dfrac{2 \ln e}{\ln e} = 2$ and $\ln \dfrac{x}{a} = \ln \dfrac{e^2}{e} = \ln e = 1$, so in general the statement is false. What *is* true, however, is that $\ln \dfrac{x}{a} = \ln x - \ln a$.

12. False. It is true that $\tan \frac{3\pi}{4} = -1$, but since the range of $\tan^{-1}$ is $\left(-\frac{\pi}{2}, \frac{\pi}{2}\right)$, we must have $\tan^{-1}(-1) = -\frac{\pi}{4}$.

13. False. For example, $\tan^{-1} 20$ is defined; $\sin^{-1} 20$ and $\cos^{-1} 20$ are not.

14. False. For example, if $x = -3$, then $\sqrt{(-3)^2} = \sqrt{9} = 3$, not -3.

EXERCISES

1. (a) When $x = 2$, $y \approx 2.7$. Thus, $f(2) \approx 2.7$.

(b) $f(x) = 3 \;\Rightarrow\; x \approx 2.3, 5.6$

(c) The domain of f is $-6 \le x \le 6$, or $[-6, 6]$.

(d) The range of f is $-4 \le y \le 4$, or $[-4, 4]$.

(e) f is increasing on $[-4, 4]$, that is, on $-4 \le x \le 4$.

(f) f is not one-to-one since it fails the Horizontal Line Test.

(g) f is odd since its graph is symmetric about the origin.

2. (a) When $x = 2$, $y = 3$. Thus, $g(2) = 3$.

(b) g is one-to-one because it passes the Horizontal Line Test.

(c) When $y = 2$, $x \approx 0.2$. So $g^{-1}(2) \approx 0.2$.

(d) The range of g is $[-1, 3.5]$, which is the same as the domain of g^{-1}.

(e) We reflect the graph of g through the line $y = x$ to obtain the graph of g^{-1}.

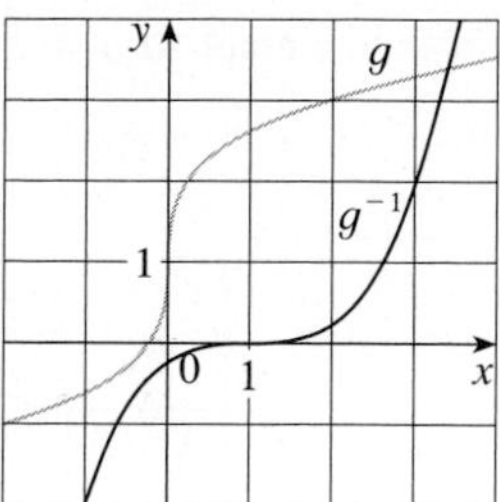

3. $f(x) = x^2 - 2x + 3$, so $f(a+h) = (a+h)^2 - 2(a+h) + 3 = a^2 + 2ah + h^2 - 2a - 2h + 3$, and

$$\frac{f(a+h) - f(a)}{h} = \frac{(a^2 + 2ah + h^2 - 2a - 2h + 3) - (a^2 - 2a + 3)}{h} = \frac{h(2a + h - 2)}{h} = 2a + h - 2.$$

4. There will be some yield with no fertilizer, increasing yields with increasing fertilizer use, a leveling-off of yields at some point, and disaster with too much fertilizer use.

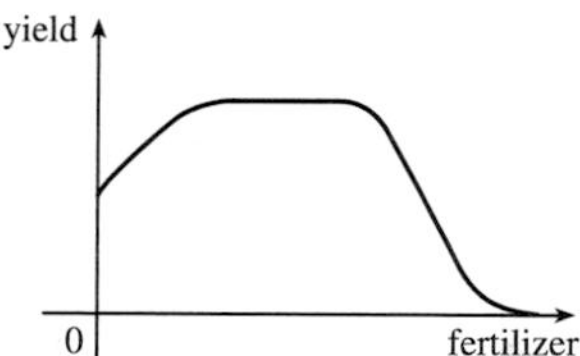

5. $f(x) = 2/(3x - 1)$. Domain: $3x - 1 \neq 0 \Rightarrow 3x \neq 1 \Rightarrow x \neq \frac{1}{3}$. $D = \left(-\infty, \frac{1}{3}\right) \cup \left(\frac{1}{3}, \infty\right)$

Range: all reals except 0 ($y = 0$ is the horizontal asymptote for f.)

$R = (-\infty, 0) \cup (0, \infty)$

6. $g(x) = \sqrt{16 - x^4}$. Domain: $16 - x^4 \geq 0 \Rightarrow x^4 \leq 16 \Rightarrow |x| \leq \sqrt[4]{16} \Rightarrow |x| \leq 2$. $D = [-2, 2]$

Range: $y \geq 0$ and $y \leq \sqrt{16} \Rightarrow 0 \leq y \leq 4$.

$R = [0, 4]$

7. $h(x) = \ln(x + 6)$. Domain: $x + 6 > 0 \Rightarrow x > -6$. $D = (-6, \infty)$

Range: $x + 6 > 0$, so $\ln(x + 6)$ takes on all real numbers and, hence, the range is $\mathbb{R}$.

$R = (-\infty, \infty)$

8. $y = F(t) = 3 + \cos 2t$. Domain: $\mathbb{R}$. $D = (-\infty, \infty)$

Range: $-1 \leq \cos 2t \leq 1 \Rightarrow 2 \leq 3 + \cos 2t \leq 4 \Rightarrow 2 \leq y \leq 4$.

$R = [2, 4]$

9. (a) To obtain the graph of $y = f(x) + 8$, we shift the graph of $y = f(x)$ up 8 units.

(b) To obtain the graph of $y = f(x + 8)$, we shift the graph of $y = f(x)$ left 8 units.

(c) To obtain the graph of $y = 1 + 2f(x)$, we stretch the graph of $y = f(x)$ vertically by a factor of 2, and then shift the resulting graph 1 unit upward.

(d) To obtain the graph of $y = f(x - 2) - 2$, we shift the graph of $y = f(x)$ right 2 units (for the "-2" inside the parentheses), and then shift the resulting graph 2 units downward.

(e) To obtain the graph of $y = -f(x)$, we reflect the graph of $y = f(x)$ about the x-axis.

(f) To obtain the graph of $y = f^{-1}(x)$, we reflect the graph of $y = f(x)$ about the line $y = x$ (assuming f is one–to-one).

10. (a) To obtain the graph of $y = f(x - 8)$, we shift the graph of $y = f(x)$ right 8 units.

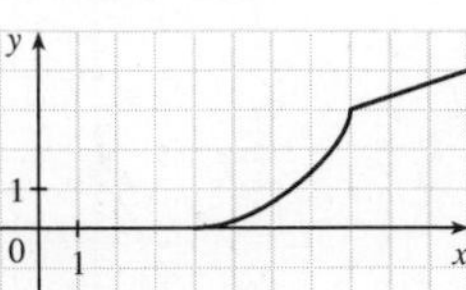

(b) To obtain the graph of $y = -f(x)$, we reflect the graph of $y = f(x)$ about the x-axis.

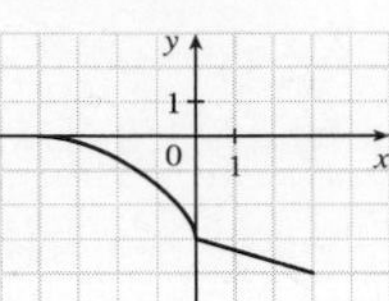

(c) To obtain the graph of $y = 2 - f(x)$, we reflect the graph of $y = f(x)$ about the x-axis, and then shift the resulting graph 2 units upward.

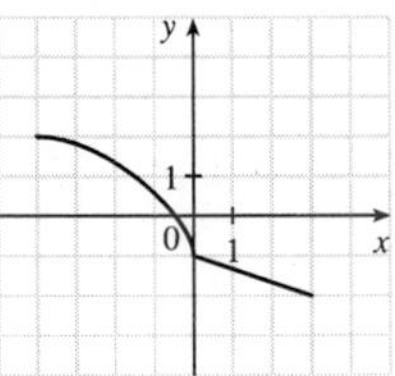

(d) To obtain the graph of $y = \frac{1}{2}f(x) - 1$, we shrink the graph of $y = f(x)$ by a factor of 2, and then shift the resulting graph 1 unit downward.

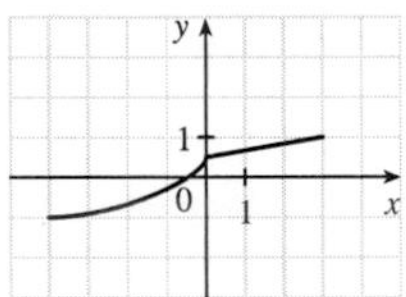

(e) To obtain the graph of $y = f^{-1}(x)$, we reflect the graph of $y = f(x)$ about the line $y = x$.

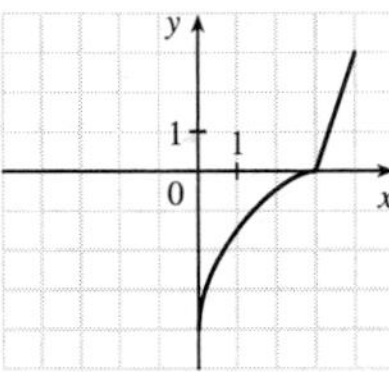

(f) To obtain the graph of $y = f^{-1}(x + 3)$, we reflect the graph of $y = f(x)$ about the line $y = x$ [see part (e)], and then shift the resulting graph left 3 units.

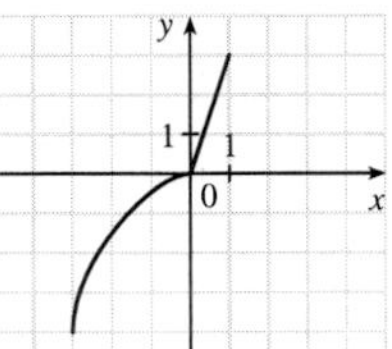

11. $y = (x - 2)^3$: Start with the graph of $y = x^3$ and shift 2 units to the right.

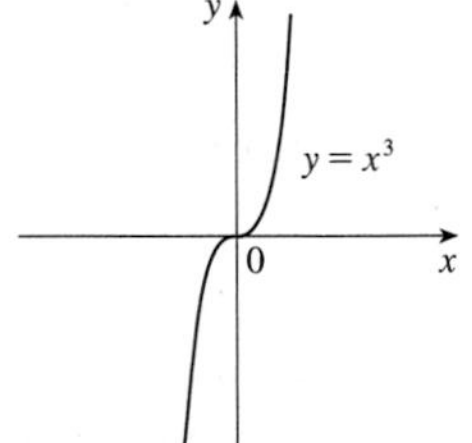

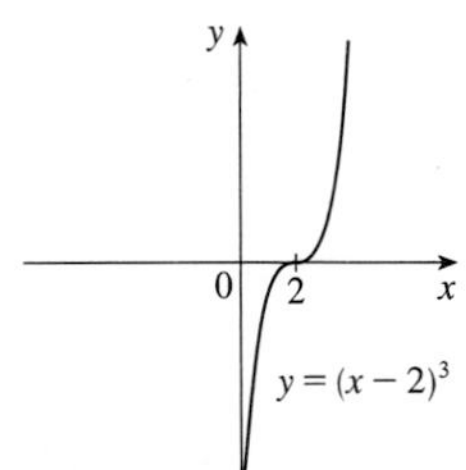

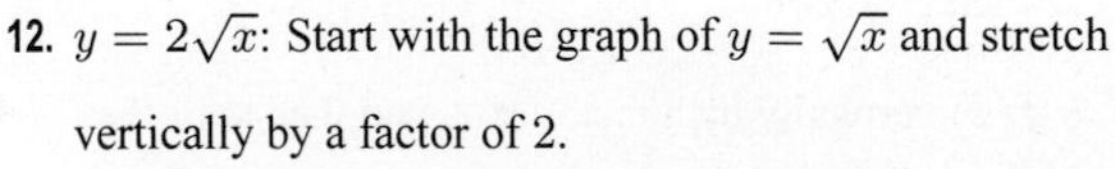

12. $y = 2\sqrt{x}$: Start with the graph of $y = \sqrt{x}$ and stretch vertically by a factor of 2.

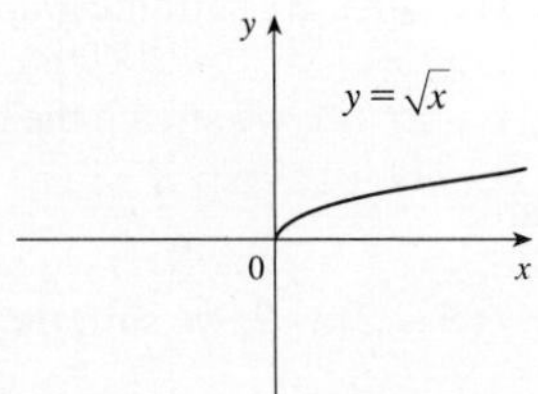

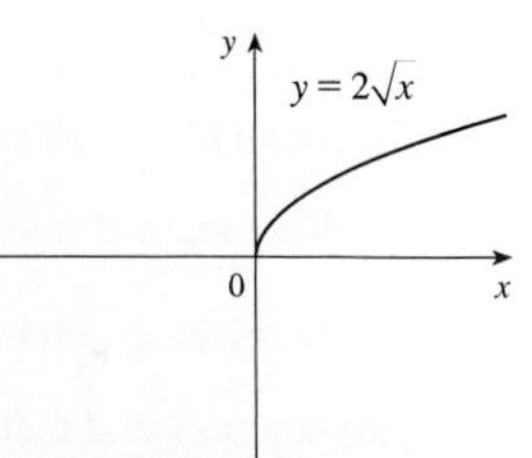

13. $y = x^2 - 2x + 2 = (x^2 - 2x + 1) + 1 = (x - 1)^2 + 1$: Start with the graph of $y = x^2$, shift 1 unit to the right, and shift 1 unit upward.

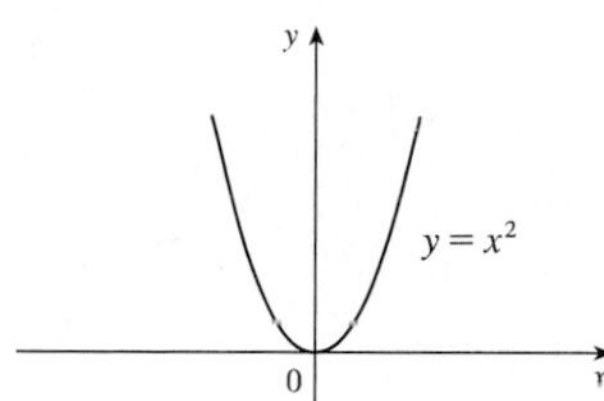

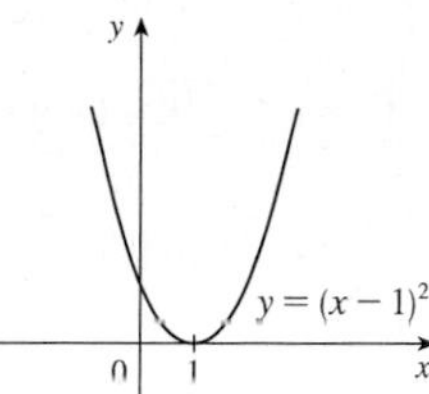

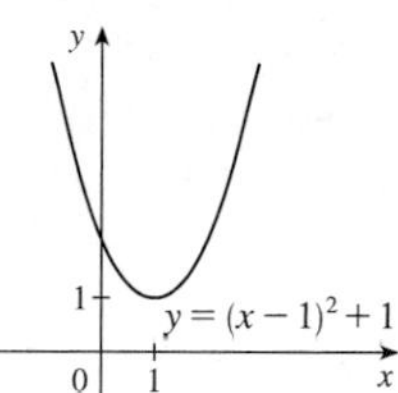

14. $y = \ln(x + 1)$: Start with the graph of $y = \ln x$ and shift left 1 unit.

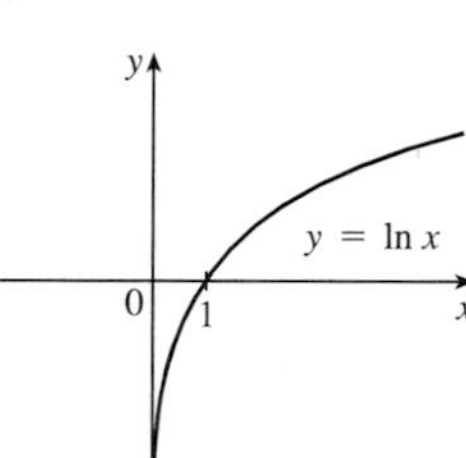

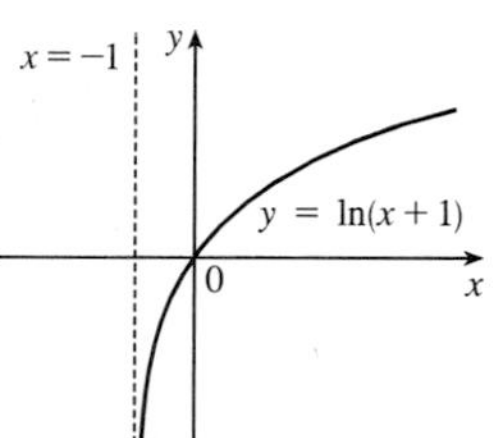

15. $f(x) = -\cos 2x$: Start with the graph of $y = \cos x$, shrink horizontally by a factor of 2, and reflect about the x-axis.

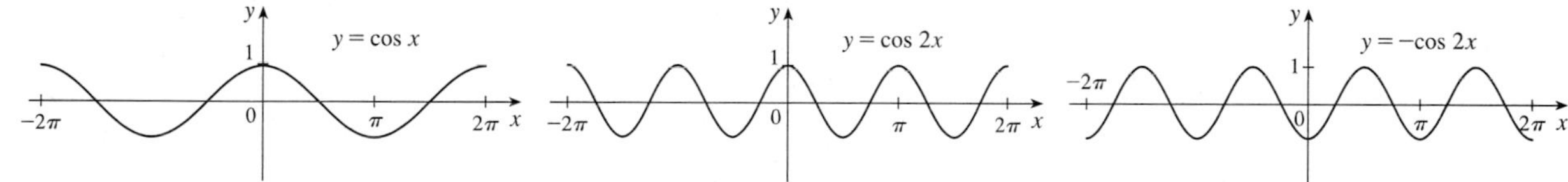

16. $f(x) = \begin{cases} -x & \text{if } x < 0 \\ e^x - 1 & \text{if } x \geq 0 \end{cases}$

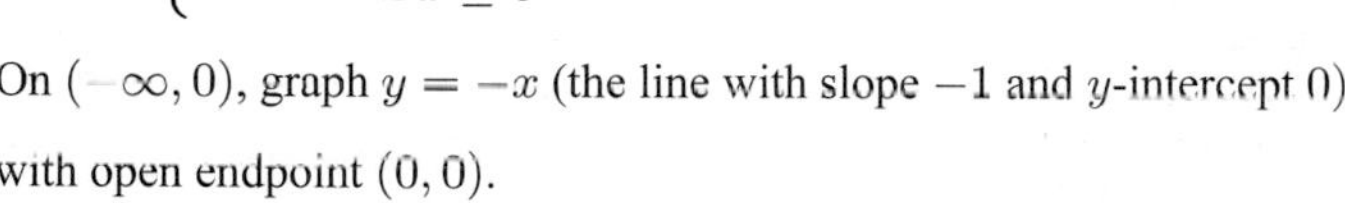

On $(-\infty, 0)$, graph $y = -x$ (the line with slope -1 and y-intercept 0) with open endpoint $(0, 0)$.

On $[0, \infty)$, graph $y = e^x - 1$ (the graph of $y = e^x$ shifted 1 unit downward) with closed endpoint $(0, 0)$.

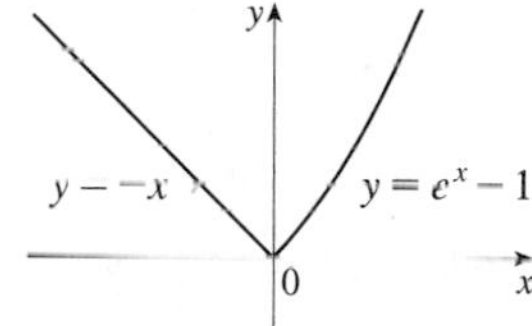

17. (a) The terms of f are a mixture of odd and even powers of x, so f is neither even nor odd.

(b) The terms of f are all odd powers of x, so f is odd.

(c) $f(-x) = e^{-(-x)^2} = e^{-x^2} = f(x)$, so f is even.

(d) $f(-x) = 1 + \sin(-x) = 1 - \sin x$. Now $f(-x) \neq f(x)$ and $f(-x) \neq -f(x)$, so f is neither even nor odd.

18. For the line segment from $(-2, 2)$ to $(-1, 0)$, the slope is $\dfrac{0 - 2}{-1 + 2} = -2$, and an equation is $y - 0 = -2(x + 1)$ or, equivalently, $y = -2x - 2$. The circle has equation $x^2 + y^2 = 1$; the top half has equation $y = \sqrt{1 - x^2}$ (we have solved for positive y). Thus, $f(x) = \begin{cases} -2x - 2 & \text{if } -2 \leq x \leq -1 \\ \sqrt{1 - x^2} & \text{if } -1 < x \leq 1 \end{cases}$

19. $f(x) = \ln x$, $D = (0, \infty)$; $g(x) = x^2 - 9$, $D = \mathbb{R}$.

(a) $(f \circ g)(x) = f(g(x)) = f(x^2 - 9) = \ln(x^2 - 9)$.

Domain: $x^2 - 9 > 0 \;\Rightarrow\; x^2 > 9 \;\Rightarrow\; |x| > 3 \;\Rightarrow\; x \in (-\infty, -3) \cup (3, \infty)$

(b) $(g \circ f)(x) = g(f(x)) = g(\ln x) = (\ln x)^2 - 9$. Domain: $x > 0$, or $(0, \infty)$

(c) $(f \circ f)(x) = f(f(x)) = f(\ln x) = \ln(\ln x)$. Domain: $\ln x > 0 \;\Rightarrow\; x > e^0 = 1$, or $(1, \infty)$

(d) $(g \circ g)(x) = g(g(x)) = g(x^2 - 9) = (x^2 - 9)^2 - 9$. Domain: $x \in \mathbb{R}$, or $(-\infty, \infty)$

20. Let $h(x) = x + \sqrt{x}$, $g(x) = \sqrt{x}$, and $f(x) = 1/x$. Then $(f \circ g \circ h)(x) = \dfrac{1}{\sqrt{x + \sqrt{x}}} = F(x)$.

21.

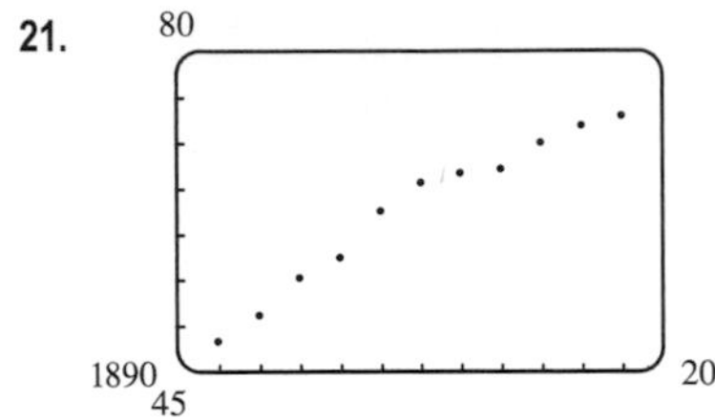

Many models appear to be plausible. Your choice depends on whether you think medical advances will keep increasing life expectancy, or if there is bound to be a natural leveling-off of life expectancy. A linear model, $y = 0.2493x - 423.4818$, gives us an estimate of 77.6 years for the year 2010.

22. (a) Let x denote the number of toaster ovens produced in one week and

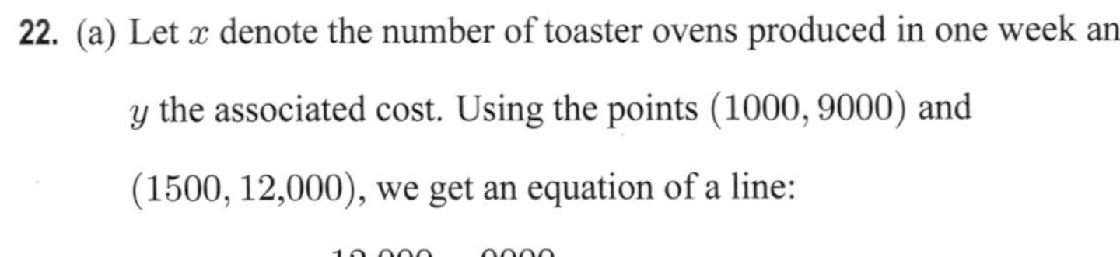

y the associated cost. Using the points $(1000, 9000)$ and $(1500, 12{,}000)$, we get an equation of a line:

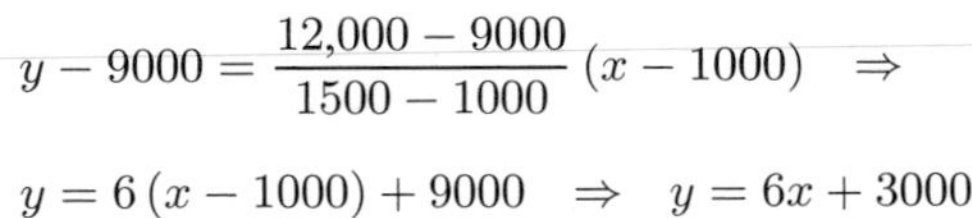

$$y - 9000 = \frac{12{,}000 - 9000}{1500 - 1000}\,(x - 1000) \;\Rightarrow$$

$$y = 6\,(x - 1000) + 9000 \;\Rightarrow\; y = 6x + 3000.$$

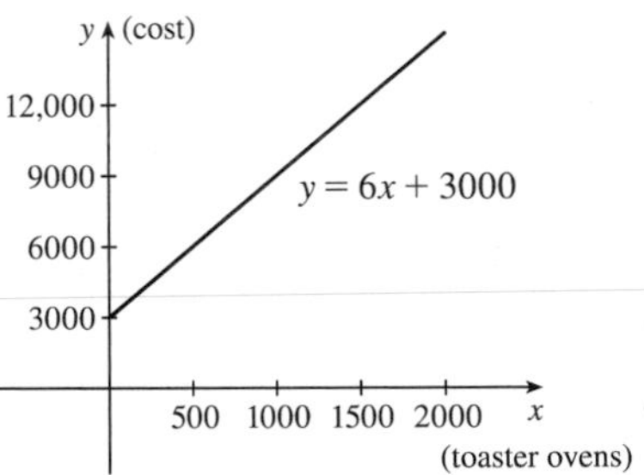

(b) The slope of 6 means that each additional toaster oven produced adds \$6 to the weekly production cost.

(c) The y-intercept of 3000 represents the overhead cost—the cost incurred without producing anything.

23. We need to know the value of x such that $f(x) = 2x + \ln x = 2$. Since $x = 1$ gives us $y = 2$, $f^{-1}(2) = 1$.

24. $y = \dfrac{x+1}{2x+1}$. Interchanging x and y gives us $x = \dfrac{y+1}{2y+1} \;\Rightarrow\; 2xy + x = y + 1 \;\Rightarrow\; 2xy - y = 1 - x \;\Rightarrow$

$y(2x - 1) = 1 - x \;\Rightarrow\; y = \dfrac{1-x}{2x-1} = f^{-1}(x)$.

25. (a) $e^{2\ln 3} = (e^{\ln 3})^2 = 3^2 = 9$

(b) $\log_{10} 25 + \log_{10} 4 = \log_{10}(25 \cdot 4) = \log_{10} 100 = \log_{10} 10^2 = 2$

(c) $\tan\left(\arcsin \frac{1}{2}\right) = \tan \frac{\pi}{6} = \frac{1}{\sqrt{3}}$

(d) Let $\theta = \cos^{-1} \frac{4}{5}$, so $\cos\theta = \frac{4}{5}$. Then $\sin\left(\cos^{-1}\left(\frac{4}{5}\right)\right) = \sin\theta = \sqrt{1 - \cos^2\theta} = \sqrt{1 - \left(\frac{4}{5}\right)^2} = \sqrt{\frac{9}{25}} = \frac{3}{5}$.

26. (a) $e^x = 5 \;\Rightarrow\; x = \ln 5$

(b) $\ln x = 2 \;\Rightarrow\; x = e^2$

(c) $e^{e^x} = 2 \;\Rightarrow\; e^x = \ln 2 \;\Rightarrow\; x = \ln(\ln 2)$

(d) $\tan^{-1} x = 1 \;\Rightarrow\; \tan\tan^{-1} x = \tan 1 \;\Rightarrow\; x = \tan 1 \;(\approx 1.5574)$

27. (a) After 4 days, $\frac{1}{2}$ gram remains; after 8 days, $\frac{1}{4}$ g; after 12 days, $\frac{1}{8}$ g; after 16 days, $\frac{1}{16}$ g.

(b) $m(4) = \dfrac{1}{2}$, $m(8) = \dfrac{1}{2^2}$, $m(12) = \dfrac{1}{2^3}$, $m(16) = \dfrac{1}{2^4}$. From the pattern, we see that $m(t) = \dfrac{1}{2^{t/4}}$, or $2^{-t/4}$.

(c) $m = 2^{-t/4} \quad \Rightarrow \quad \log_2 m = -t/4 \quad \Rightarrow \quad t = -4 \log_2 m$; this is the time elapsed when there are m grams of ^{100}Pd.

(d) $m = 0.01 \quad \Rightarrow \quad t = -4 \log_2 0.01 = -4\left(\dfrac{\ln 0.01}{\ln 2}\right) \approx 26.6$ days

28. (a)

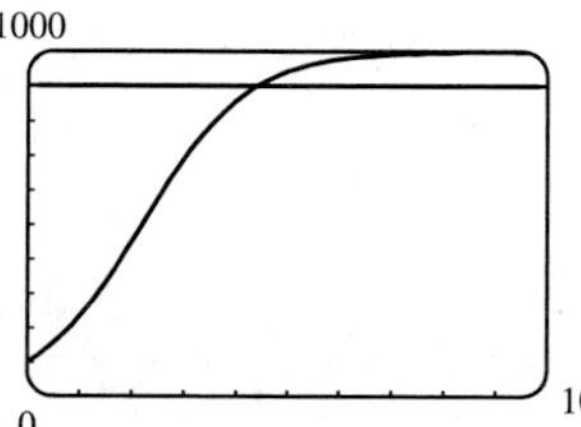

The population would reach 900 in about 4.4 years.

(b) $P = \dfrac{100{,}000}{100 + 900e^{-t}} \quad \Rightarrow \quad 100P + 900Pe^{-t} = 100{,}000 \Rightarrow \quad 900Pe^{-t} = 100{,}000 - 100P \quad \Rightarrow$

$e^{-t} = \dfrac{100{,}000 - 100P}{900P} \quad \Rightarrow \quad -t = \ln\left(\dfrac{1000 - P}{9P}\right) \quad \Rightarrow \quad t = -\ln\left(\dfrac{1000 - P}{9P}\right)$, or $\ln\left(\dfrac{9P}{1000 - P}\right)$;

this is the time required for the population to reach a given number P.

(c) $P = 900 \quad \Rightarrow \quad t = \ln\left(\dfrac{9 \cdot 900}{1000 - 900}\right) = \ln 81 \approx 4.4$ years, as in part (a).

☐ PRINCIPLES OF PROBLEM SOLVING

1.

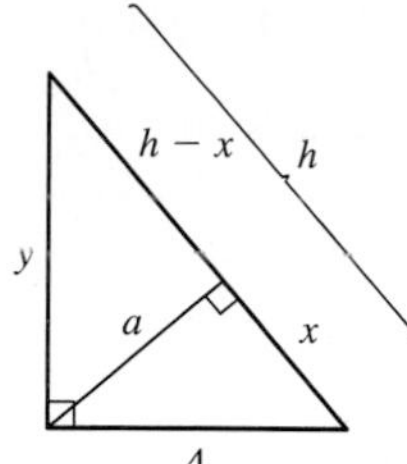

By using the area formula for a triangle, $\frac{1}{2}$ (base) (height), in two ways, we see that $\frac{1}{2}(4)(y) = \frac{1}{2}(h)(a)$, so $a = \dfrac{4y}{h}$. Since $4^2 + y^2 = h^2$, $y = \sqrt{h^2 - 16}$, and $a = \dfrac{4\sqrt{h^2 - 16}}{h}$.

2.

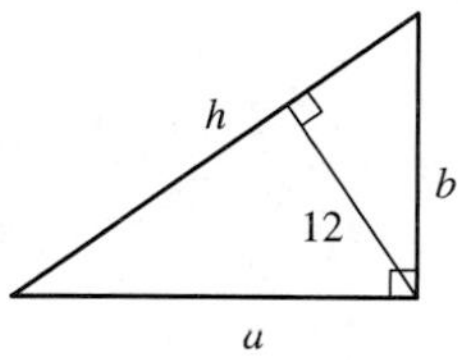

Refer to Example 1, where we obtained $h = \dfrac{P^2 - 100}{2P}$. The 100 came from 4 times the area of the triangle. In this case, the area of the triangle is $\frac{1}{2}(h)(12) = 6h$. Thus, $h = \dfrac{P^2 - 4\,(6h)}{2P} \;\Rightarrow\; 2Ph = P^2 - 24h \;\Rightarrow$

$2Ph + 24h = P^2 \;\Rightarrow\; h\,(2P + 24) = P^2 \;\Rightarrow\; h = \dfrac{P^2}{2P + 24}$.

3. $|2x - 1| = \begin{cases} 2x - 1 & \text{if } x \geq \frac{1}{2} \\ 1 - 2x & \text{if } x < \frac{1}{2} \end{cases}$ and $|x + 5| = \begin{cases} x + 5 & \text{if } x \geq -5 \\ -x - 5 & \text{if } x < -5 \end{cases}$

Therefore, we consider the three cases $x < -5$, $-5 \leq x < \frac{1}{2}$, and $x \geq \frac{1}{2}$.

If $x < -5$, we must have $1 - 2x - (-x - 5) = 3 \;\Leftrightarrow\; x = 3$, which is false, since we are considering $x < -5$.

If $-5 \leq x < \frac{1}{2}$, we must have $1 - 2x - (x + 5) = 3 \;\Leftrightarrow\; x = -\frac{7}{3}$.

If $x \geq \frac{1}{2}$, we must have $2x - 1 - (x + 5) = 3 \;\Leftrightarrow\; x = 9$.

So the two solutions of the equation are $x = -\frac{7}{3}$ and $x = 9$.

4. $|x - 1| = \begin{cases} x - 1 & \text{if } x \geq 1 \\ 1 - x & \text{if } x < 1 \end{cases}$ and $|x - 3| = \begin{cases} x - 3 & \text{if } x \geq 3 \\ 3 - x & \text{if } x < 3 \end{cases}$

Therefore, we consider the three cases $x < 1$, $1 \leq x < 3$, and $x \geq 3$.

If $x < 1$, we must have $1 - x - (3 - x) \geq 5 \;\Leftrightarrow\; 0 > 7$, which is false.

If $1 \leq x < 3$, we must have $x - 1 - (3 - x) \geq 5 \;\Leftrightarrow\; x \geq \frac{9}{2}$, which is false because $x < 3$.

If $x \geq 3$, we must have $x - 1 - (x - 3) \geq 5 \;\Leftrightarrow\; 2 \geq 5$, which is false.

All three cases lead to falsehoods, so the inequality has no solution.

5. $f(x) = \left|x^2 - 4\,|x| + 3\right|$. If $x \geq 0$, then $f(x) = \left|x^2 - 4x + 3\right| = |(x - 1)(x - 3)|$.

Case (i): If $0 < x \leq 1$, then $f(x) = x^2 - 4x + 3$.

Case (ii): If $1 < x \leq 3$, then $f(x) = -(x^2 - 4x + 3) = -x^2 + 4x - 3$.

Case (iii): If $x > 3$, then $f(x) = x^2 - 4x + 3$.

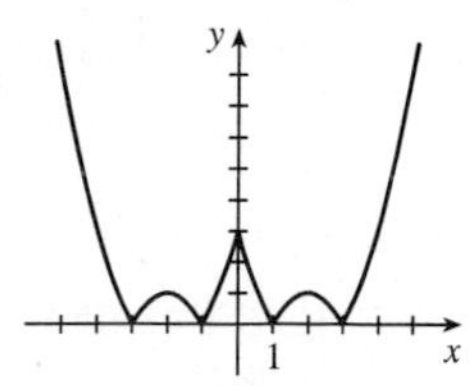

This enables us to sketch the graph for $x \geq 0$. Then we use the fact that f is an even function to reflect this part of the graph about the y-axis to obtain the entire graph. Or, we could consider also the cases $x < -3$, $-3 \leq x < -1$, and $-1 \leq x < 0$.

6. $g(x) = |x^2 - 1| - |x^2 - 4|$.

$$|x^2 - 1| = \begin{cases} x^2 - 1 & \text{if } |x| \geq 1 \\ 1 - x^2 & \text{if } |x| < 1 \end{cases} \quad \text{and} \quad |x^2 - 4| = \begin{cases} x^2 - 4 & \text{if } |x| \geq 2 \\ 4 - x^2 & \text{if } |x| < 2 \end{cases}$$

So for $0 \leq |x| < 1, g(x) = 1 - x^2 - (4 - x^2) = -3$, for $1 \leq |x| < 2, g(x) = x^2 - 1 - (4 - x^2) = 2x^2 - 5$, and for $|x| \geq 2, g(x) = x^2 - 1 - (x^2 - 4) = 3$.

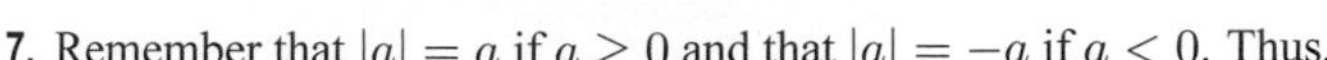

7. Remember that $|a| = a$ if $a \geq 0$ and that $|a| = -a$ if $a < 0$. Thus,

$$x + |x| = \begin{cases} 2x & \text{if } x \geq 0 \\ 0 & \text{if } x < 0 \end{cases} \quad \text{and} \quad y + |y| = \begin{cases} 2y & \text{if } y \geq 0 \\ 0 & \text{if } y < 0 \end{cases}$$

We will consider the equation $x + |x| = y + |y|$ in four cases.

(1) $x \geq 0, y \geq 0$	(2) $x \geq 0, y < 0$	(3) $x < 0, y \geq 0$	(4) $x < 0, y < 0$
$2x = 2y$	$2x = 0$	$0 = 2y$	$0 = 0$
$x = y$	$x = 0$	$0 = y$	

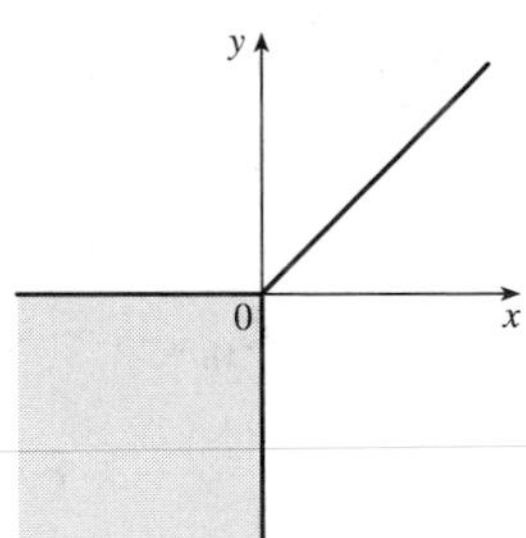

Case 1 gives us the line $y = x$ with nonnegative x and y.

Case 2 gives us the portion of the y-axis with y negative.

Case 3 gives us the portion of the x-axis with x negative.

Case 4 gives us the entire third quadrant.

8. $|x - y| + |x| - |y| \leq 2$ [call this inequality $(\star)$]

Case (i):	$x \geq y \geq 0$.	Then $(\star)$	$\Leftrightarrow$	$x - y + x - y \leq 2$	$\Leftrightarrow$	$x - y \leq 1$	$\Leftrightarrow$	$y \geq x - 1$.
Case (ii):	$y \geq x \geq 0$.	Then $(\star)$	$\Leftrightarrow$	$y - x + x - y \leq 2$	$\Leftrightarrow$	$0 \leq 2$ (true).		
Case (iii):	$x \geq 0$ and $y \leq 0$.	Then $(\star)$	$\Leftrightarrow$	$x - y + x + y \leq 2$	$\Leftrightarrow$	$2x \leq 2$	$\Leftrightarrow$	$x \leq 1$.
Case (iv):	$x \leq 0$ and $y \geq 0$.	Then $(\star)$	$\Leftrightarrow$	$y - x - x - y \leq 2$	$\Leftrightarrow$	$-2x \leq 2$	$\Leftrightarrow$	$x \geq -1$.
Case (v):	$y \leq x \leq 0$.	Then $(\star)$	$\Leftrightarrow$	$x - y - x + y \leq 2$	$\Leftrightarrow$	$0 \leq 2$ (true).		
Case (vi):	$x \leq y \leq 0$.	Then $(\star)$	$\Leftrightarrow$	$y - x - x + y \leq 2$	$\Leftrightarrow$	$y - x \leq 1$	$\Leftrightarrow$	$y \leq x + 1$.

Note: Instead of considering cases (iv), (v), and (vi), we could have noted that the region is unchanged if x and y are replaced by $-x$ and $-y$, so the region is symmetric about the origin. Therefore, we need only draw cases (i), (ii), and (iii), and rotate through $180°$ about the origin.

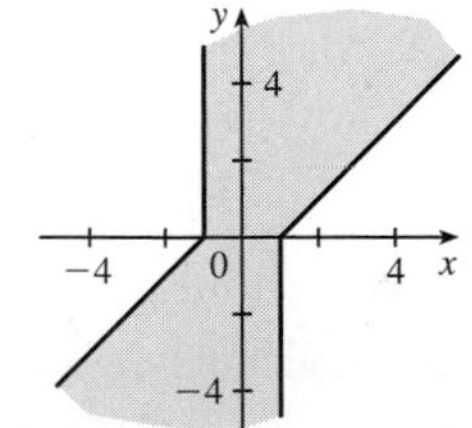

9. (a) To sketch the graph of $f(x) = \max\{x, 1/x\}$, we first graph $g(x) = x$ and $h(x) = 1/x$ on the same coordinate axes. Then create the graph of f by plotting the largest y-value of g and h for every value of x.

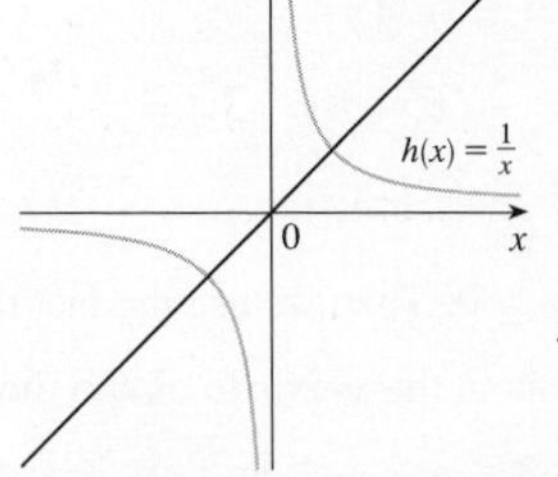

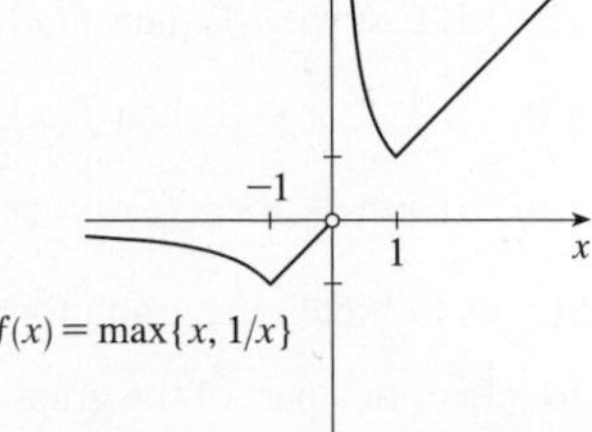

(b)

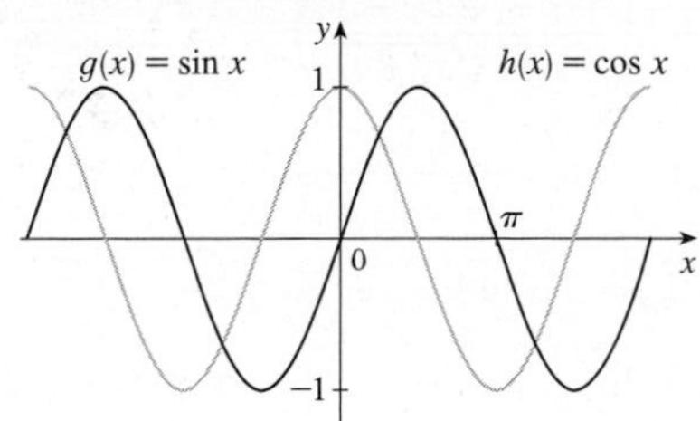

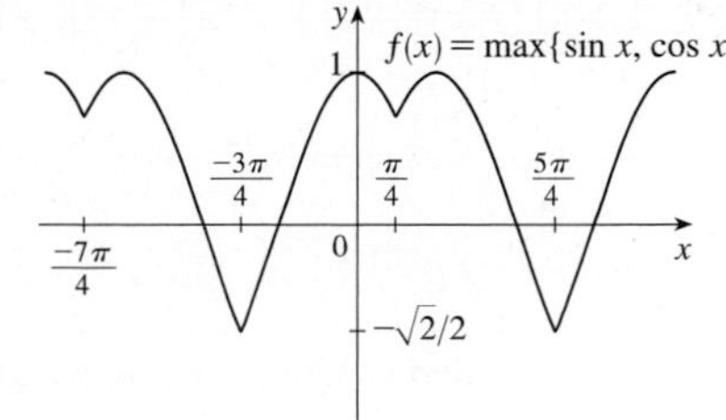

(c)

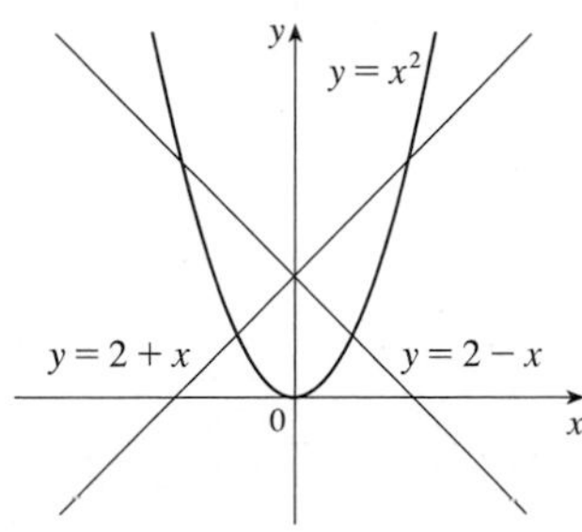

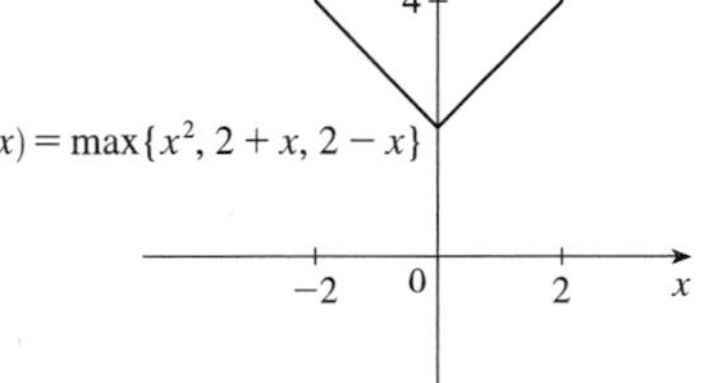

On the TI-84 Plus, max is found under `LIST`, then under `MATH`. To graph $f(x) = \max\left\{x^2, 2 + x, 2 - x\right\}$, use `Y` $= \max(x^2, \max(2 + x, 2 - x))$.

10. (a) If $\max\{x, 2y\} = 1$, then either $x = 1$ *and* $2y \le 1$ ***or*** $x \le 1$ *and* $2y = 1$. Thus, we obtain the set of points such that $x = 1$ *and* $y \le \frac{1}{2}$ [a vertical line with highest point $(1, \frac{1}{2})$] ***or*** $x \le 1$ *and* $y = \frac{1}{2}$ [a horizontal line with rightmost point $(1, \frac{1}{2})$].

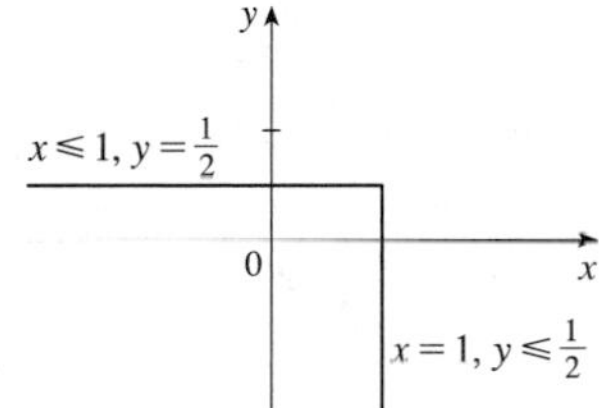

(b) The graph of $\max\{x, 2y\} = 1$ is shown in part (a), and the graph of $\max\{x, 2y\} = -1$ can be found in a similar manner. The inequalities in $-1 \le \max\{x, 2y\} \le 1$ give us all the points on or inside the boundaries.

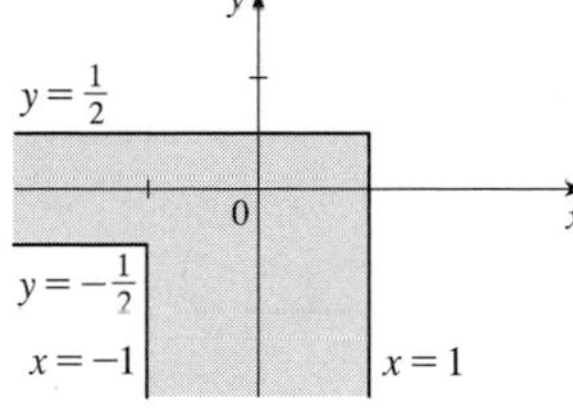

(c) $\max\{x, y^2\} = 1 \quad \Leftrightarrow$

$x = 1$ *and* $y^2 \le 1$ $[-1 \le y \le 1]$

or $x \le 1$ *and* $y^2 = 1$ $[y = \pm 1]$.

y
$y = 1, x \le 1$
$x = 1,$
$-1 \le y \le 1$
0
x
$y = -1, x \le 1$

11. $(\log_2 3)(\log_3 4)(\log_4 5) \cdots (\log_{31} 32) = \left(\dfrac{\ln 3}{\ln 2}\right)\left(\dfrac{\ln 4}{\ln 3}\right)\left(\dfrac{\ln 5}{\ln 4}\right) \cdots \left(\dfrac{\ln 32}{\ln 31}\right) = \dfrac{\ln 32}{\ln 2} = \dfrac{\ln 2^5}{\ln 2} = \dfrac{5 \ln 2}{\ln 2} = 5$

12. (a) $f(-x) = \ln\left(-x + \sqrt{(-x)^2 + 1}\right) = \ln\left(-x + \sqrt{x^2+1} \cdot \dfrac{-x - \sqrt{x^2+1}}{-x - \sqrt{x^2+1}}\right)$

$= \ln\left(\dfrac{x^2 - (x^2+1)}{-x - \sqrt{x^2+1}}\right) = \ln\left(\dfrac{-1}{-x - \sqrt{x^2+1}}\right) = \ln\left(\dfrac{1}{x + \sqrt{x^2+1}}\right)$

$= \ln 1 - \ln\left(x + \sqrt{x^2+1}\right) = -\ln\left(x + \sqrt{x^2-1}\right) = -f(x)$

(b) $y = \ln\left(x + \sqrt{x^2+1}\right)$. Interchanging x and y, we get $x = \ln\left(y + \sqrt{y^2+1}\right) \;\Rightarrow\; e^x = y + \sqrt{y^2+1} \;\Rightarrow$

$e^x - y = \sqrt{y^2+1} \;\Rightarrow\; e^{2x} - 2ye^x + y^2 = y^2 + 1 \;\Rightarrow\; e^{2x} - 1 = 2ye^x \;\Rightarrow\; y = \dfrac{e^{2x}-1}{2e^x} = f^{-1}(x)$

13. $\ln(x^2 - 2x - 2) \le 0 \;\Rightarrow\; x^2 - 2x - 2 \le e^0 = 1 \;\Rightarrow\; x^2 - 2x - 3 \le 0 \;\Rightarrow\; (x-3)(x+1) \le 0 \;\Rightarrow\; x \in [-1, 3]$.

Since the argument must be positive, $x^2 - 2x - 2 > 0 \;\Rightarrow\; \left[x - \left(1 - \sqrt{3}\right)\right]\left[x - \left(1 + \sqrt{3}\right)\right] > 0 \;\Rightarrow$

$x \in \left(-\infty, 1 - \sqrt{3}\right) \cup \left(1 + \sqrt{3}, \infty\right)$. The intersection of these intervals is $\left[-1, 1 - \sqrt{3}\right) \cup \left(1 + \sqrt{3}, 3\right]$.

14. Assume that $\log_2 5$ is rational. Then $\log_2 5 = m/n$ for natural numbers m and n. Changing to exponential form gives us $2^{m/n} = 5$ and then raising both sides to the nth power gives $2^m = 5^n$. But 2^m is even and 5^n is odd. We have arrived at a contradiction, so we conclude that our hypothesis, that $\log_2 5$ is rational, is false. Thus, $\log_2 5$ is irrational.

15. Let d be the distance traveled on each half of the trip. Let t_1 and t_2 be the times taken for the first and second halves of the trip. For the first half of the trip we have $t_1 = d/30$ and for the second half we have $t_2 = d/60$. Thus, the average speed for the entire trip is $\dfrac{\text{total distance}}{\text{total time}} = \dfrac{2d}{t_1 + t_2} = \dfrac{2d}{\frac{d}{30} + \frac{d}{60}} \cdot \dfrac{60}{60} = \dfrac{120d}{2d + d} = \dfrac{120d}{3d} = 40$. The average speed for the entire trip is 40 mi/h.

16. Let $f(x) = \sin x$, $g(x) = x$, and $h(x) = x$. Then the left-hand side of the equation is $[f \circ (g + h)](x) = \sin(x + x) = \sin 2x = 2 \sin x \cos x$; and the right-hand side is $(f \circ g)(x) + (f \circ h)(x) = \sin x + \sin x = 2 \sin x$. The two sides are not equal, so the given statement is false.

17. Let S_n be the statement that $7^n - 1$ is divisible by 6.

- S_1 is true because $7^1 - 1 = 6$ is divisible by 6.
- Assume S_k is true, that is, $7^k - 1$ is divisible by 6. In other words, $7^k - 1 = 6m$ for some positive integer m. Then $7^{k+1} - 1 = 7^k \cdot 7 - 1 = (6m + 1) \cdot 7 - 1 = 42m + 6 = 6(7m + 1)$, which is divisible by 6, so S_{k+1} is true.
- Therefore, by mathematical induction, $7^n - 1$ is divisible by 6 for every positive integer n.

18. Let S_n be the statement that $1 + 3 + 5 + \cdots + (2n - 1) = n^2$.

- S_1 is true because $[2(1) - 1] = 1 = 1^2$.
- Assume S_k is true, that is, $1 + 3 + 5 + \cdots + (2k - 1) = k^2$. Then

  $1 + 3 + 5 + \cdots + (2k-1) + [2(k+1) - 1] = 1 + 3 + 5 + \cdots + (2k-1) + (2k+1) = k^2 + (2k+1) = (k+1)^2$

  which shows that S_{k+1} is true.
- Therefore, by mathematical induction, $1 + 3 + 5 + \cdots + (2n - 1) = n^2$ for every positive integer n.

19. $f_0(x) = x^2$ and $f_{n+1}(x) = f_0(f_n(x))$ for $n = 0, 1, 2, \ldots$.

$f_1(x) = f_0(f_0(x)) = f_0(x^2) = (x^2)^2 = x^4$, $f_2(x) = f_0(f_1(x)) = f_0(x^4) = (x^4)^2 = x^8$,

$f_3(x) = f_0(f_2(x)) = f_0(x^8) = (x^8)^2 = x^{16}, \ldots$. Thus, a general formula is $f_n(x) = x^{2^{n+1}}$.

20. (a) $f_0(x) = 1/(2 - x)$ and $f_{n+1} = f_0 \circ f_n$ for $n = 0, 1, 2, \ldots$.

$$f_1(x) = f_0\left(\frac{1}{2-x}\right) = \frac{1}{2 - \dfrac{1}{2-x}} = \frac{2-x}{2(2-x)-1} = \frac{2-x}{3-2x},$$

$$f_2(x) = f_0\left(\frac{2-x}{3-2x}\right) = \frac{1}{2 - \dfrac{2-x}{3-2x}} = \frac{3-2x}{2(3-2x)-(2-x)} = \frac{3-2x}{4-3x},$$

$$f_3(x) = f_0\left(\frac{3-2x}{4-3x}\right) = \frac{1}{2 - \dfrac{3-2x}{4-3x}} = \frac{4-3x}{2(4-3x)-(3-2x)} = \frac{4-3x}{5-4x}, \ldots$$

Thus, we conjecture that the general formula is $f_n(x) = \dfrac{n+1-nx}{n+2-(n+1)x}$.

To prove this, we use the Principle of Mathematical Induction. We have already verified that f_n is true for $n = 1$. Assume that the formula is true for $n = k$; that is, $f_k(x) = \dfrac{k+1-kx}{k+2-(k+1)x}$. Then

$$f_{k+1}(x) = (f_0 \circ f_k)(x) = f_0(f_k(x)) = f_0\left(\frac{k+1-kx}{k+2-(k+1)x}\right) = \frac{1}{2 - \dfrac{k+1-kx}{k+2-(k+1)x}}$$

$$= \frac{k+2-(k+1)x}{2\,[k+2-(k+1)x] - (k+1-kx)} = \frac{k+2-(k+1)x}{k+3-(k+2)x}$$

This shows that the formula for f_n is true for $n = k+1$. Therefore, by mathematical induction, the formula is true for all positive integers n.

(b) From the graph, we can make several observations:

- The values at each fixed $x = a$ keep increasing as n increases.
- The vertical asymptote gets closer to $x = 1$ as n increases.
- The horizontal asymptote gets closer to $y = 1$ as n increases.
- The x-intercept for f_{n+1} is the value of the vertical asymptote for f_n.
- The y-intercept for f_n is the value of the horizontal asymptote for f_{n+1}.

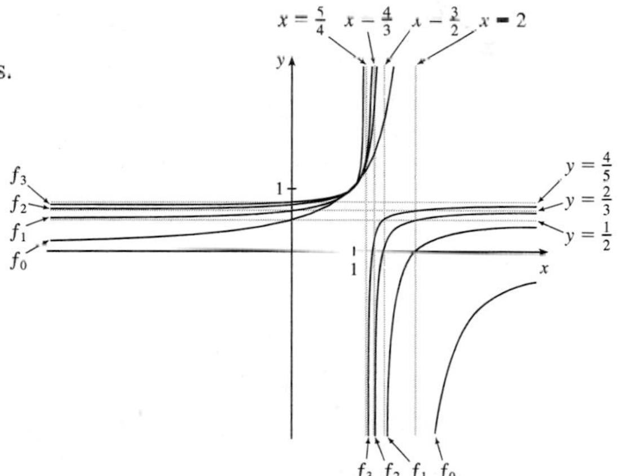

2 ☐ LIMITS AND DERIVATIVES

2.1 The Tangent and Velocity Problems

1. (a) Using $P(15, 250)$, we construct the following table:

t	Q	slope $= m_{PQ}$
5	$(5, 694)$	$\frac{694-250}{5-15} = -\frac{444}{10} = -44.4$
10	$(10, 444)$	$\frac{444-250}{10-15} = -\frac{194}{5} = -38.8$
20	$(20, 111)$	$\frac{111-250}{20-15} = -\frac{139}{5} = -27.8$
25	$(25, 28)$	$\frac{28-250}{25-15} = -\frac{222}{10} = -22.2$
30	$(30, 0)$	$\frac{0-250}{30-15} = -\frac{250}{15} = -16.\overline{6}$

(b) Using the values of t that correspond to the points closest to P ($t = 10$ and $t = 20$), we have

$$\frac{-38.8 + (-27.8)}{2} = -33.3$$

(c) From the graph, we can estimate the slope of the tangent line at P to be $\frac{-300}{9} = -33.\overline{3}$.

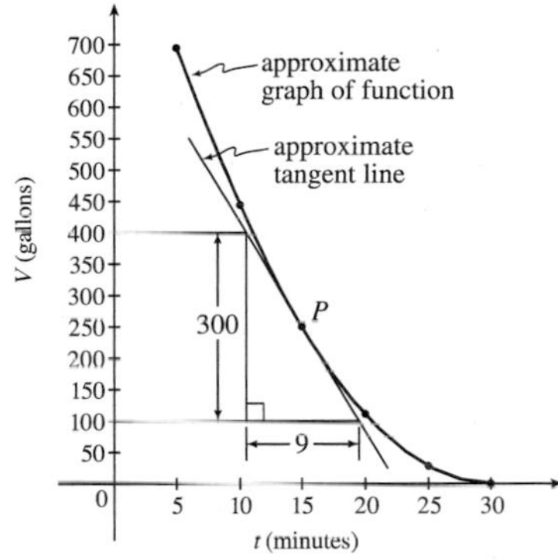

2. (a) Slope $= \frac{2948 - 2530}{42 - 36} = \frac{418}{6} \approx 69.67$

(b) Slope $= \frac{2948 - 2661}{42 - 38} = \frac{287}{4} = 71.75$

(c) Slope $= \frac{2948 - 2806}{42 - 40} = \frac{142}{2} = 71$

(d) Slope $= \frac{3080 - 2948}{44 - 42} = \frac{132}{2} = 66$

From the data, we see that the patient's heart rate is decreasing from 71 to 66 heartbeats/minute after 42 minutes. After being stable for a while, the patient's heart rate is dropping.

3. (a) $y = \dfrac{1}{1 - x}$, $P(2, -1)$

	x	$Q(x, 1/(1 - x))$	m_{PQ}
(i)	1.5	$(1.5, -2)$	2
(ii)	1.9	$(1.9, -1.111\,111)$	1.111 111
(iii)	1.99	$(1.99, -1.010\,101)$	1.010 101
(iv)	1.999	$(1.999, -1.001\,001)$	1.001 001
(v)	2.5	$(2.5, -0.666\,667)$	0.666 667
(vi)	2.1	$(2.1, -0.909\,091)$	0.909 091
(vii)	2.01	$(2.01, -0.990\,099)$	0.990 099
(viii)	2.001	$(2.001, -0.999\,001)$	0.999 001

(b) The slope appears to be 1.

(c) Using $m = 1$, an equation of the tangent line to the curve at $P(2, -1)$ is $y - (-1) = 1(x - 2)$, or $y = x - 3$.

4. (a) $y = \cos \pi x$, $P(0.5, 0)$

	x	Q	m_{PQ}
(i)	0	$(0, 1)$	-2
(ii)	0.4	$(0.4, 0.309017)$	-3.090170
(iii)	0.49	$(0.49, 0.031411)$	-3.141076
(iv)	0.499	$(0.499, 0.003142)$	-3.141587
(v)	1	$(1, -1)$	-2
(vi)	0.6	$(0.6, -0.309017)$	-3.090170
(vii)	0.51	$(0.51, -0.031411)$	-3.141076
(viii)	0.501	$(0.501, -0.003142)$	-3.141587

(b) The slope appears to be $-\pi$.

(c) $y - 0 = -\pi(x - 0.5)$ or $y = -\pi x + \frac{1}{2}\pi$.

(d)

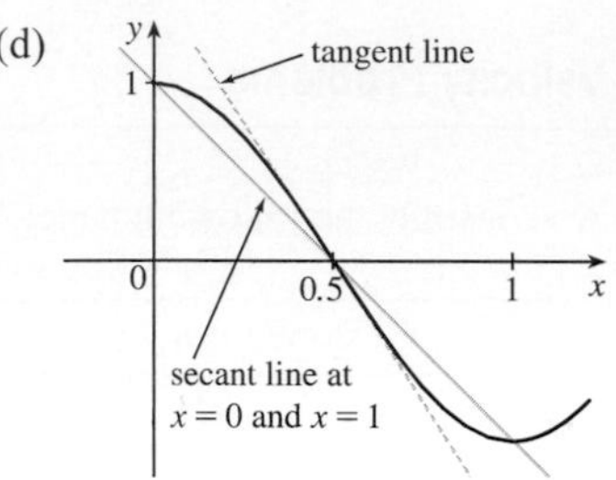

5. (a) $y = y(t) = 40t - 16t^2$. At $t = 2$, $y = 40(2) - 16(2)^2 = 16$. The average velocity between times 2 and $2 + h$ is

$$v_{\text{ave}} = \frac{y(2+h) - y(2)}{(2+h) - 2} = \frac{\left[40(2+h) - 16(2+h)^2\right] - 16}{h} = \frac{-24h - 16h^2}{h} = -24 - 16h, \text{ if } h \neq 0.$$

(i) $[2, 2.5]$: $h = 0.5$, $v_{\text{ave}} = -32$ ft/s

(ii) $[2, 2.1]$: $h = 0.1$, $v_{\text{ave}} = -25.6$ ft/s

(iii) $[2, 2.05]$: $h = 0.05$, $v_{\text{ave}} = -24.8$ ft/s

(iv) $[2, 2.01]$: $h = 0.01$, $v_{\text{ave}} = -24.16$ ft/s

(b) The instantaneous velocity when $t = 2$ (h approaches 0) is -24 ft/s.

6. (a) $y = y(t) = 10t - 1.86t^2$. At $t = 1$, $y = 10(1) - 1.86(1)^2 = 8.14$. The average velocity between times 1 and $1 + h$ is

$$v_{\text{ave}} = \frac{y(1+h) - y(1)}{(1+h) - 1} = \frac{\left[10(1+h) - 1.86(1+h)^2\right] - 8.14}{h} = \frac{6.28h - 1.86h^2}{h} = 6.28 - 1.86h, \text{ if } h \neq 0.$$

(i) $[1, 2]$: $h = 1$, $v_{\text{ave}} = 4.42$ m/s

(ii) $[1, 1.5]$: $h = 0.5$, $v_{\text{ave}} = 5.35$ m/s

(iii) $[1, 1.1]$: $h = 0.1$, $v_{\text{ave}} = 6.094$ m/s

(iv) $[1, 1.01]$: $h = 0.01$, $v_{\text{ave}} = 6.2614$ m/s

(v) $[1, 1.001]$: $h = 0.001$, $v_{\text{ave}} = 6.27814$ m/s

(b) The instantaneous velocity when $t = 1$ (h approaches 0) is 6.28 m/s.

7. (a) (i) On the interval $[2, 4]$, $v_{\text{ave}} = \dfrac{s(4) - s(2)}{4 - 2} = \dfrac{79.2 - 20.6}{2} = 29.3$ ft/s.

(ii) On the interval $[3, 4]$, $v_{\text{ave}} = \dfrac{s(4) - s(3)}{4 - 3} = \dfrac{79.2 - 46.5}{1} = 32.7$ ft/s.

(iii) On the interval $[4, 5]$, $v_{\text{ave}} = \dfrac{s(5) - s(4)}{5 - 4} = \dfrac{124.8 - 79.2}{1} = 45.6$ ft/s.

(iv) On the interval $[4, 6]$, $v_{\text{ave}} = \dfrac{s(6) - s(4)}{6 - 4} = \dfrac{176.7 - 79.2}{2} = 48.75$ ft/s.

(b) Using the points $(2, 16)$ and $(5, 105)$ from the approximate tangent line, the instantaneous velocity at $t = 3$ is about $\dfrac{105 - 16}{5 - 2} = \dfrac{89}{3} \approx 29.7$ ft/s.

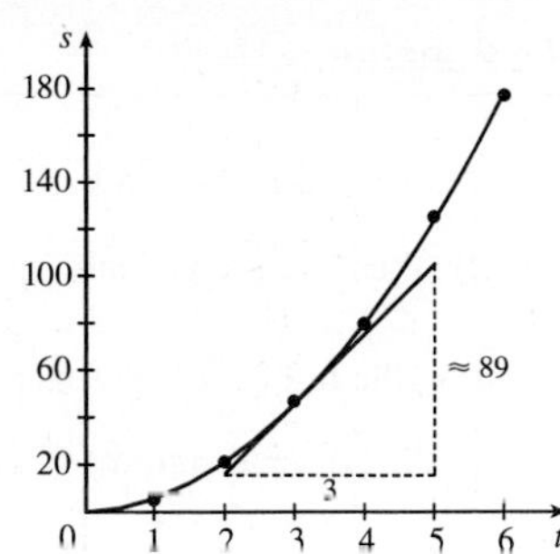

8. (a) (i) $s = s(t) = 2\sin \pi t + 3\cos \pi t$. On the interval $[1, 2]$, $v_{\text{ave}} = \dfrac{s(2) - s(1)}{2 - 1} = \dfrac{3 - (-3)}{1} = 6$ cm/s.

(ii) On the interval $[1, 1.1]$, $v_{\text{ave}} = \dfrac{s(1.1) - s(1)}{1.1 - 1} \approx \dfrac{-3.471 - (-3)}{0.1} = -4.71$ cm/s.

(iii) On the interval $[1, 1.01]$, $v_{\text{ave}} = \dfrac{s(1.01) - s(1)}{1.01 - 1} \approx \dfrac{-3.0613 - (-3)}{0.01} = -6.13$ cm/s.

(iv) On the interval $[1, 1.001]$, $v_{\text{ave}} = \dfrac{s(1.001) - s(1)}{1.001 - 1} \approx \dfrac{-3.00627 - (-3)}{0.001} = -6.27$ cm/s.

(b) The instantaneous velocity of the particle when $t = 1$ appears to be about -6.3 cm/s.

9. (a) For the curve $y = \sin(10\pi/x)$ and the point $P(1, 0)$:

x	Q	m_{PQ}
2	$(2, 0)$	0
1.5	$(1.5, 0.8660)$	1.7321
1.4	$(1.4, -0.4339)$	-1.0847
1.3	$(1.3, -0.8230)$	-2.7433
1.2	$(1.2, 0.8660)$	4.3301
1.1	$(1.1, -0.2817)$	-2.8173

x	Q	m_{PQ}
0.5	$(0.5, 0)$	0
0.6	$(0.6, 0.8660)$	-2.1651
0.7	$(0.7, 0.7818)$	-2.6061
0.8	$(0.8, 1)$	-5
0.9	$(0.9, -0.3420)$	3.4202

As x approaches 1, the slopes do not appear to be approaching any particular value.

(b)

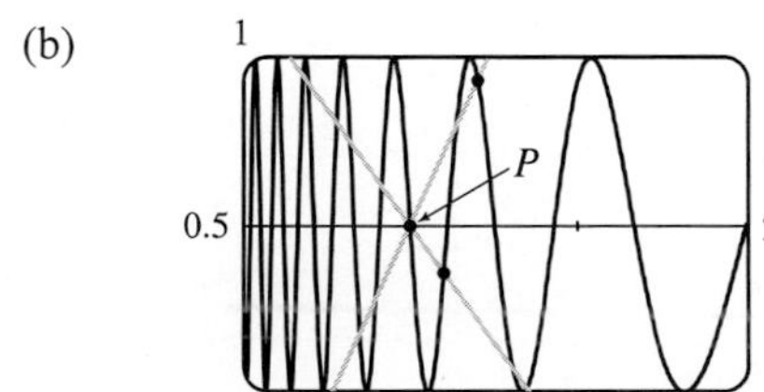

We see that problems with estimation are caused by the frequent oscillations of the graph. The tangent is so steep at P that we need to take x-values much closer to 1 in order to get accurate estimates of its slope.

(c) If we choose $x = 1.001$, then the point Q is $(1.001, -0.0314)$ and $m_{PQ} \approx -31.3794$. If $x = 0.999$, then Q is $(0.999, 0.0314)$ and $m_{PQ} = -31.4422$. The average of these slopes is -31.4108. So we estimate that the slope of the tangent line at P is about -31.4.

2.2 The Limit of a Function

1. As x approaches 2, $f(x)$ approaches 5. [Or, the values of $f(x)$ can be made as close to 5 as we like by taking x sufficiently close to 2 (but $x \neq 2$).] Yes, the graph could have a hole at $(2, 5)$ and be defined such that $f(2) = 3$.

2. As x approaches 1 from the left, $f(x)$ approaches 3; and as x approaches 1 from the right, $f(x)$ approaches 7. No, the limit does not exist because the left- and right-hand limits are different.

3. (a) $\lim_{x \to -3} f(x) = \infty$ means that the values of $f(x)$ can be made arbitrarily large (as large as we please) by taking x sufficiently close to -3 (but not equal to -3).

(b) $\lim_{x \to 4^+} f(x) = -\infty$ means that the values of $f(x)$ can be made arbitrarily large negative by taking x sufficiently close to 4 through values larger than 4.

4. (a) As x approaches 2 from the left, the values of $f(x)$ approach 3, so $\lim_{x \to 2^-} f(x) = 3$.

(b) As x approaches 2 from the right, the values of $f(x)$ approach 1, so $\lim_{x \to 2^+} f(x) = 1$.

(c) $\lim_{x \to 2} f(x)$ does not exist since the left-hand limit does not equal the right-hand limit.

(d) When $x = 2$, $y = 3$, so $f(2) = 3$.

(e) As x approaches 4, the values of $f(x)$ approach 4, so $\lim_{x \to 4} f(x) = 4$.

(f) There is no value of $f(x)$ when $x = 4$, so $f(4)$ does not exist.

5. (a) As x approaches 1, the values of $f(x)$ approach 2, so $\lim_{x \to 1} f(x) = 2$.

(b) As x approaches 3 from the left, the values of $f(x)$ approach 1, so $\lim_{x \to 3^-} f(x) = 1$.

(c) As x approaches 3 from the right, the values of $f(x)$ approach 4, so $\lim_{x \to 3^+} f(x) = 4$.

(d) $\lim_{x \to 3} f(x)$ does not exist since the left-hand limit does not equal the right-hand limit.

(e) When $x = 3$, $y = 3$, so $f(3) = 3$.

6. (a) $h(x)$ approaches 4 as x approaches -3 from the left, so $\lim_{x \to -3^-} h(x) = 4$.

(b) $h(x)$ approaches 4 as x approaches -3 from the right, so $\lim_{x \to -3^+} h(x) = 4$.

(c) $\lim_{x \to -3} h(x) = 4$ because the limits in part (a) and part (b) are equal.

(d) $h(-3)$ is not defined, so it doesn't exist.

(e) $h(x)$ approaches 1 as x approaches 0 from the left, so $\lim_{x \to 0^-} h(x) = 1$.

(f) $h(x)$ approaches -1 as x approaches 0 from the right, so $\lim_{x \to 0^+} h(x) = -1$.

(g) $\lim_{x \to 0} h(x)$ does not exist because the limits in part (e) and part (f) are not equal.

(h) $h(0) = 1$ since the point $(0, 1)$ is on the graph of h.

(i) Since $\lim_{x \to 2^-} h(x) = 2$ and $\lim_{x \to 2^+} h(x) = 2$, we have $\lim_{x \to 2} h(x) = 2$.

(j) $h(2)$ is not defined, so it doesn't exist.

(k) $h(x)$ approaches 3 as x approaches 5 from the right, so $\lim\limits_{x \to 5^+} h(x) = 3$.

(l) $h(x)$ does not approach any one number as x approaches 5 from the left, so $\lim\limits_{x \to 5^-} h(x)$ does not exist.

7. (a) $\lim\limits_{t \to 0^-} g(t) = -1$ (b) $\lim\limits_{t \to 0^+} g(t) = -2$

(c) $\lim\limits_{t \to 0} g(t)$ does not exist because the limits in part (a) and part (b) are not equal.

(d) $\lim\limits_{t \to 2^-} g(t) = 2$ (e) $\lim\limits_{t \to 2^+} g(t) = 0$

(f) $\lim\limits_{t \to 2} g(t)$ does not exist because the limits in part (d) and part (e) are not equal.

(g) $g(2) = 1$ (h) $\lim\limits_{t \to 4} g(t) = 3$

8. (a) $\lim\limits_{x \to -3} A(x) = \infty$ (b) $\lim\limits_{x \to 2} A(x)$ does not exist. (c) $\lim\limits_{x \to 2^-} A(x) = -\infty$

(d) $\lim\limits_{x \to 2^+} A(x) = \infty$ (e) $\lim\limits_{x \to -1} A(x) = -\infty$

(f) The equations of the vertical asymptotes are $x = -3$, $x = -1$ and $x = 2$.

9. (a) $\lim\limits_{x \to -7} f(x) = -\infty$ (b) $\lim\limits_{x \to -3} f(x) = \infty$ (c) $\lim\limits_{x \to 0} f(x) = \infty$

(d) $\lim\limits_{x \to 6^-} f(x) = -\infty$ (e) $\lim\limits_{x \to 6^+} f(x) = \infty$

(f) The equations of the vertical asymptotes are $x = -7$, $x = -3$, $x = 0$, and $x = 6$.

10. $\lim\limits_{t \to 12^-} f(t) = 150$ mg and $\lim\limits_{t \to 12^+} f(t) = 300$ mg. These limits show that there is an abrupt change in the amount of drug in the patient's bloodstream at $t = 12$ h. The left-hand limit represents the amount of the drug just before the fourth injection. The right-hand limit represents the amount of the drug just after the fourth injection.

11. From the graph of

$$f(x) = \begin{cases} 1 + x & \text{if } x < -1 \\ x^2 & \text{if } -1 \le x < 1, \\ 2 - x & \text{if } x \ge 1 \end{cases}$$

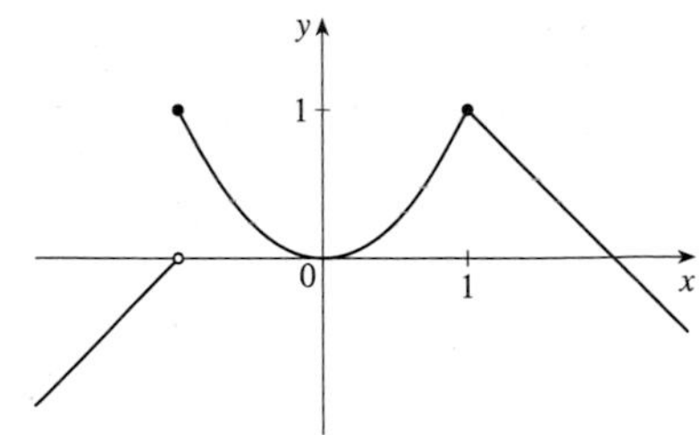

we see that $\lim\limits_{x \to a} f(x)$ exists for all a except $a = -1$. Notice that the right and left limits are different at $a = -1$.

12. From the graph of

$$f(x) = \begin{cases} 1 + \sin x & \text{if } x < 0 \\ \cos x & \text{if } 0 \le x \le \pi, \\ \sin x & \text{if } x > \pi \end{cases}$$

we see that $\lim\limits_{x \to a} f(x)$ exists for all a except $a = \pi$. Notice that the right and left limits are different at $a = \pi$.

13. (a) $\lim_{x \to 0^-} f(x) = 1$

(b) $\lim_{x \to 0^+} f(x) = 0$

(c) $\lim_{x \to 0} f(x)$ does not exist because the limits in part (a) and part (b) are not equal.

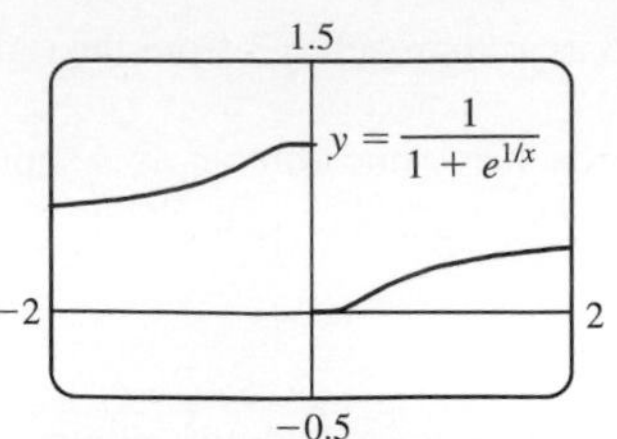

14. (a) $\lim_{x \to 0^-} f(x) = -1$

(b) $\lim_{x \to 0^+} f(x) = 1$

(c) $\lim_{x \to 0} f(x)$ does not exist because the limits in part (a) and part (b) are not equal.

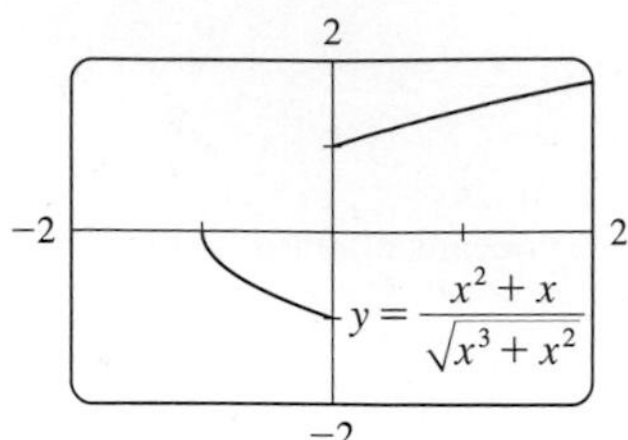

15. $\lim_{x \to 0^-} f(x) = -1$, $\lim_{x \to 0^+} f(x) = 2$, $f(0) = 1$

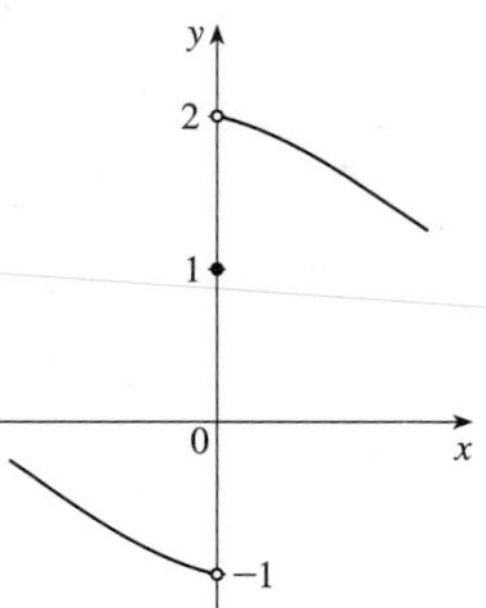

16. $\lim_{x \to 0} f(x) = 1$, $\lim_{x \to 3^-} f(x) = -2$, $\lim_{x \to 3^+} f(x) = 2$, $f(0) = -1$, $f(3) = 1$

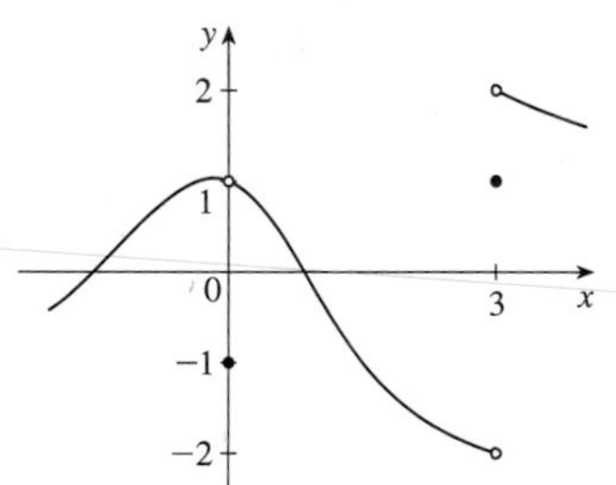

17. $\lim_{x \to 3^+} f(x) = 4$, $\lim_{x \to 3^-} f(x) = 2$, $\lim_{x \to -2} f(x) = 2$, $f(3) = 3$, $f(-2) = 1$

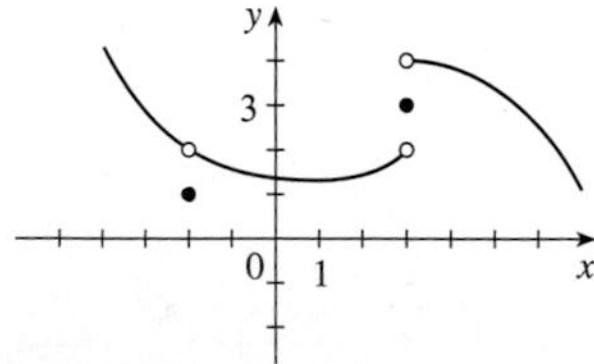

18. $\lim_{x \to 0^-} f(x) = 2$, $\lim_{x \to 0^+} f(x) = 0$, $\lim_{x \to 4^-} f(x) = 3$, $\lim_{x \to 4^+} f(x) = 0$, $f(0) = 2$, $f(4) = 1$

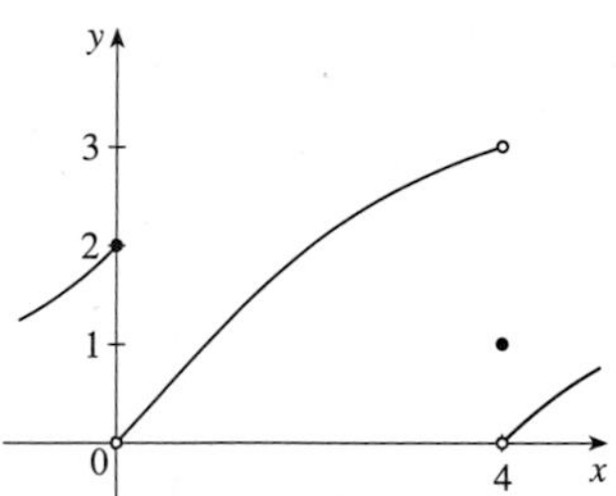

19. For $f(x) = \frac{x^2 - 3x}{x^2 - 9}$:

x	$f(x)$
3.1	0.508 197
3.05	0.504 132
3.01	0.500 832
3.001	0.500 083
3.0001	0.500 008

x	$f(x)$
2.9	0.491 525
2.95	0.495 798
2.99	0.499 165
2.999	0.499 917
2.9999	0.499 992

It appears that $\lim_{x \to 3} \frac{x^2 - 3x}{x^2 - 9} = \frac{1}{2}$.

20. For $f(x) = \dfrac{x^2 - 3x}{x^2 - 9}$:

x	$f(x)$
-2.5	-5
-2.9	-29
-2.95	-59
-2.99	-299
-2.999	-2999
-2.9999	$-29{,}999$

x	$f(x)$
-3.5	7
-3.1	31
-3.05	61
-3.01	301
-3.001	3001
-3.0001	$30{,}001$

It appears that $\lim\limits_{x \to -3^+} f(x) = -\infty$ and that $\lim\limits_{x \to -3^-} f(x) = \infty$, so $\lim\limits_{x \to -3} \dfrac{x^2 - 3x}{x^2 - 9}$ does not exist.

21. For $f(t) = \dfrac{e^{5t} - 1}{t}$:

t	$f(t)$
0.5	22.364 988
0.1	6.487 213
0.01	5.127 110
0.001	5.012 521
0.0001	5.001 250

t	$f(t)$
-0.5	1.835 830
-0.1	3.934 693
-0.01	4.877 058
-0.001	4.987 521
-0.0001	4.998 750

It appears that $\lim\limits_{t \to 0} \dfrac{e^{5t} - 1}{t} = 5$.

22. For $f(h) = \dfrac{(2+h)^5 - 32}{h}$:

h	$f(h)$
0.5	131.312 500
0.1	88.410 100
0.01	80.804 010
0.001	80.080 040
0.0001	80.008 000

h	$f(h)$
-0.5	48.812 500
-0.1	72.390 100
-0.01	79.203 990
-0.001	79.920 040
-0.0001	79.992 000

It appears that $\lim\limits_{h \to 0} \dfrac{(2+h)^5 - 32}{h} = 80$.

23. For $f(x) = \dfrac{\ln x - \ln 4}{x - 4}$:

x	$f(x)$
3.9	0.253 178
3.99	0.250 313
3.999	0.250 031
3.9999	0.250 003

x	$f(x)$
4.1	0.246 926
4.01	0.249 688
4.001	0.249 969
4.0001	0.249 997

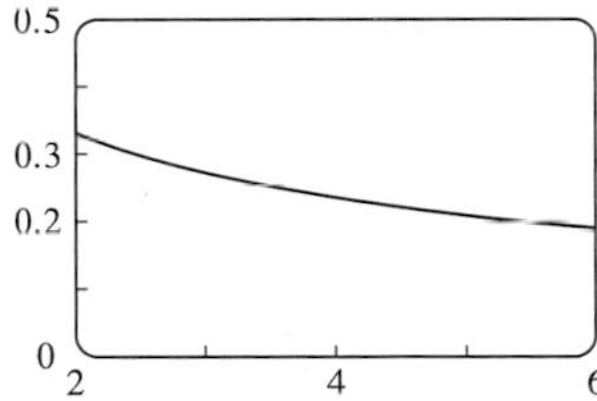

It appears that $\lim\limits_{x \to 4} f(x) = 0.25$. The graph confirms that result.

24. For $f(p) = \dfrac{1 + p^9}{1 + p^{15}}$:

p	$f(p)$
-1.1	0.427 397
-1.01	0.582 008
-1.001	0.598 200
-1.0001	0.599 820

p	$f(p)$
-0.9	0.771 405
-0.99	0.617 992
-0.999	0.601 800
-0.9999	0.600 180

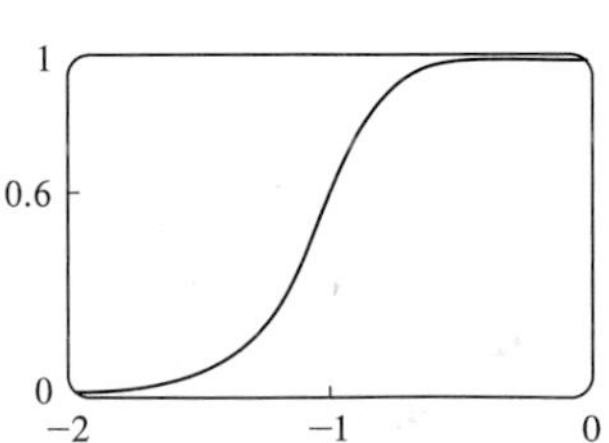

It appears that $\lim\limits_{p \to -1} f(p) = 0.6$. The graph confirms that result.

25. For $f(\theta) = \dfrac{\sin 3\theta}{\tan 2\theta}$:

θ	$f(\theta)$
± 0.1	1.457 847
± 0.01	1.499 575
± 0.001	1.499 996
± 0.0001	1.500 000

It appears that $\lim\limits_{\theta \to 0} \dfrac{\sin 3\theta}{\tan 2\theta} = 1.5$.

The graph confirms that result.

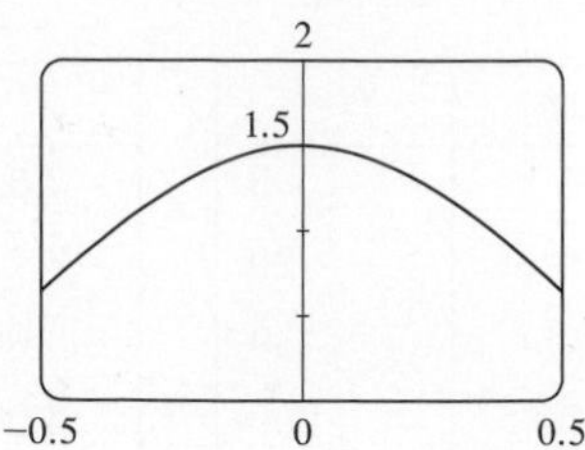

26. For $f(t) = \dfrac{5^t - 1}{t}$:

t	$f(t)$
0.1	1.746 189
0.01	1.622 459
0.001	1.610 734
0.0001	1.609 567

t	$f(t)$
-0.1	1.486 601
-0.01	1.596 556
-0.001	1.608 143
-0.0001	1.609 308

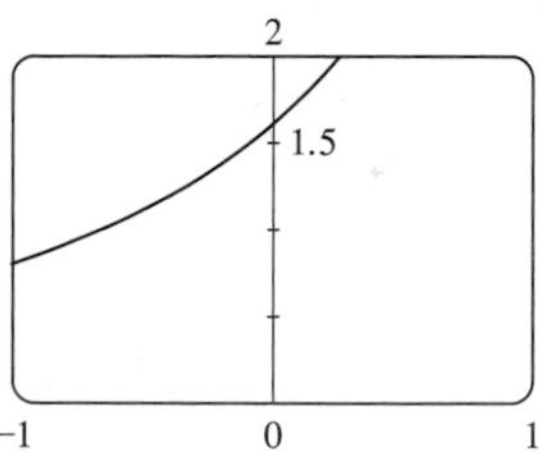

It appears that $\lim\limits_{t \to 0} f(t) \approx 1.6094$. The graph confirms that result.

27. For $f(x) = x^x$:

x	$f(x)$
0.1	0.794 328
0.01	0.954 993
0.001	0.993 116
0.0001	0.999 079

It appears that $\lim\limits_{x \to 0^+} f(x) = 1$.

The graph confirms that result.

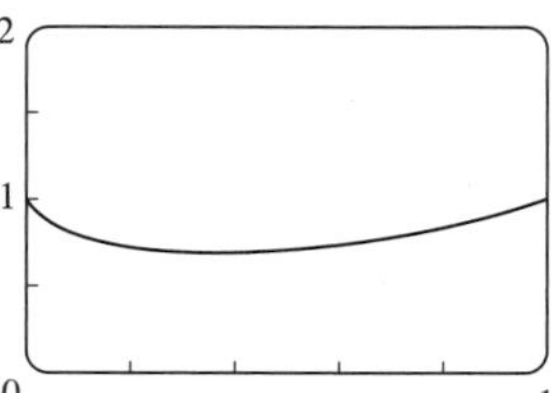

28. For $f(x) = x^2 \ln x$:

x	$f(x)$
0.1	$-0.023\,026$
0.01	$-0.000\,461$
0.001	$-0.000\,007$
0.0001	$-0.000\,000$

It appears that $\lim\limits_{x \to 0^+} f(x) = 0$.

The graph confirms that result.

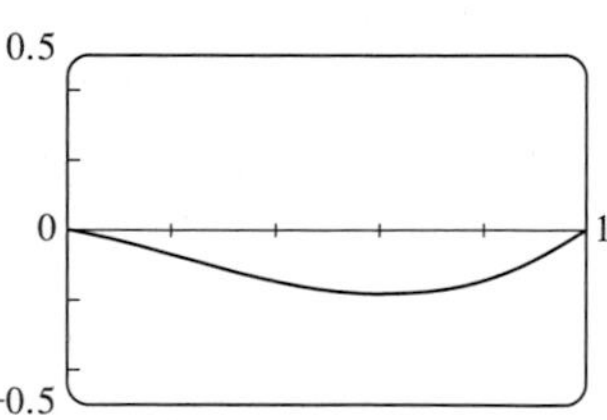

29. (a) From the graphs, it seems that $\lim\limits_{x \to 0} \dfrac{\cos 2x - \cos x}{x^2} = -1.5$.

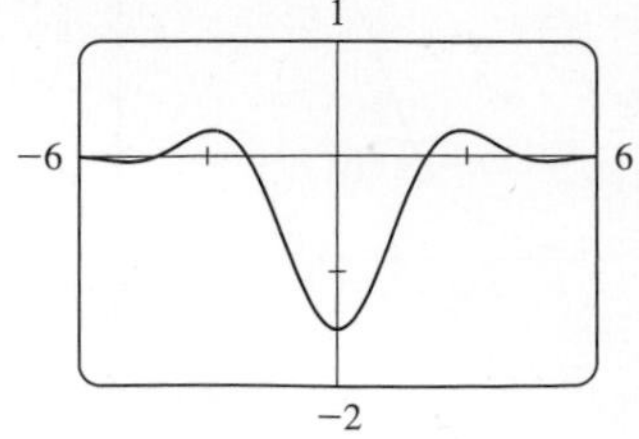

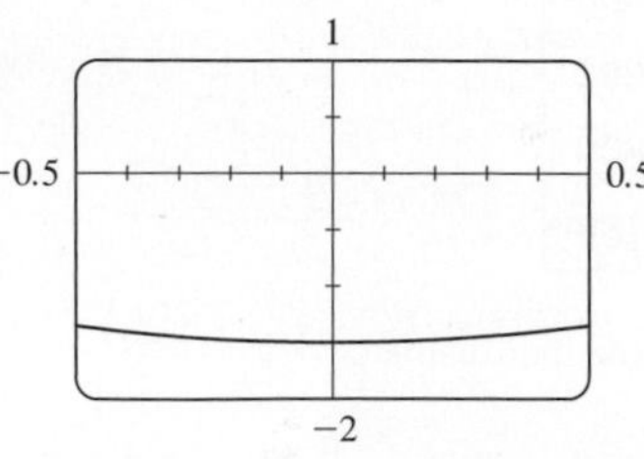

(b)

x	$f(x)$
± 0.1	$-1.493\,759$
± 0.01	$-1.499\,938$
± 0.001	$-1.499\,999$
± 0.0001	$-1.500\,000$

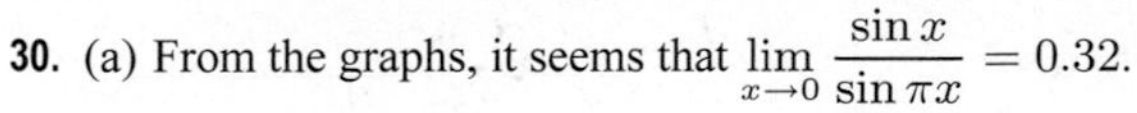

30. (a) From the graphs, it seems that $\lim_{x\to 0} \dfrac{\sin x}{\sin \pi x} = 0.32$.

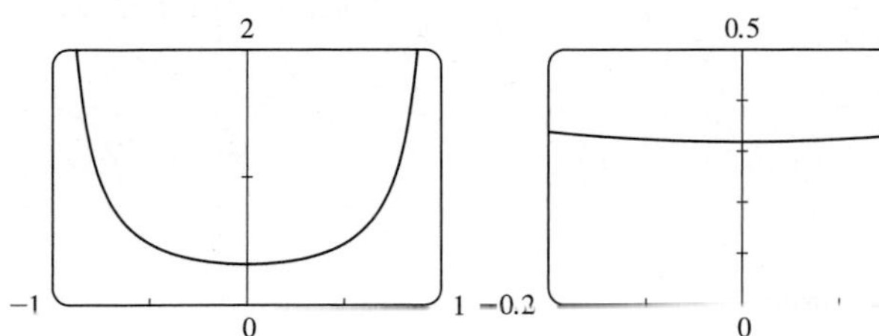

(b)

x	$f(x)$
± 0.1	0.323 068
± 0.01	0.318 357
± 0.001	0.318 310
± 0.0001	0.318 310

Later we will be able to show that the exact value is $\dfrac{1}{\pi}$.

31. $\lim_{x\to 5^+} \dfrac{x+1}{x-5} = \infty$ since the numerator is positive and the denominator approaches 0 from the positive side as $x \to 5^+$.

32. $\lim_{x\to 5^-} \dfrac{x+1}{x-5} = -\infty$ since the numerator is positive and the denominator approaches 0 from the negative side as $x \to 5^-$.

33. $\lim_{x\to 1} \dfrac{2-x}{(x-1)^2} = \infty$ since the numerator is positive and the denominator approaches 0 through positive values as $x \to 1$.

34. $\lim_{x\to 3^-} \dfrac{\sqrt{x}}{(x-3)^5} = -\infty$ since the numerator is positive and the denominator approaches 0 from the negative side as $x \to 3^-$.

35. Let $t = x^2 - 9$. Then as $x \to 3^+$, $t \to 0^+$, and $\lim_{x\to 3^+} \ln(x^2 - 9) = \lim_{t\to 0^+} \ln t = -\infty$ by (5).

36. $\lim_{x\to 0^+} \ln(\sin x) = -\infty$ since $\sin x \to 0^+$ as $x \to 0^+$.

37. $\lim_{x\to(\pi/2)^+} \dfrac{1}{x} \sec x = -\infty$ since $\dfrac{1}{x}$ is positive and $\sec x \to -\infty$ as $x \to (\pi/2)^+$.

38. $\lim_{x\to \pi^-} \cot x = \lim_{x\to \pi^-} \dfrac{\cos x}{\sin x} = -\infty$ since the numerator is negative and the denominator approaches 0 through positive values as $x \to \pi^-$.

39. $\lim_{x\to 2\pi^-} x \csc x = \lim_{x\to 2\pi^-} \dfrac{x}{\sin x} = -\infty$ since the numerator is positive and the denominator approaches 0 through negative values as $x \to 2\pi^-$.

40. $\lim_{x\to 2^-} \dfrac{x^2 - 2x}{x^2 - 4x + 4} = \lim_{x\to 2^-} \dfrac{x(x-2)}{(x-2)^2} = \lim_{x\to 2^-} \dfrac{x}{x-2} = -\infty$ since the numerator is positive and the denominator approaches 0 through negative values as $x \to 2^-$.

41. $\lim_{x\to 2^+} \dfrac{x^2 - 2x - 8}{x^2 - 5x + 6} = \lim_{x\to 2^+} \dfrac{(x-4)(x+2)}{(x-3)(x-2)} = \infty$ since the numerator is negative and the denominator approaches 0 through negative values as $x \to 2^+$.

42. $\lim_{x\to 0^+} \left(\dfrac{1}{x} - \ln x\right) = \infty$ since $\dfrac{1}{x} \to \infty$ and $\ln x \to -\infty$ as $x \to 0^+$.

43. $\lim_{x\to 0} (\ln x^2 - x^{-2}) = -\infty$ since $\ln x^2 \to -\infty$ and $x^{-2} \to \infty$ as $x \to 0$.

44. (a) The denominator of $y = \dfrac{x^2+1}{3x-2x^2} = \dfrac{x^2+1}{x(3-2x)}$ is equal to zero when $x=0$ and $x=\frac{3}{2}$ (and the numerator is not), so $x=0$ and $x=1.5$ are vertical asymptotes of the function.

(b)

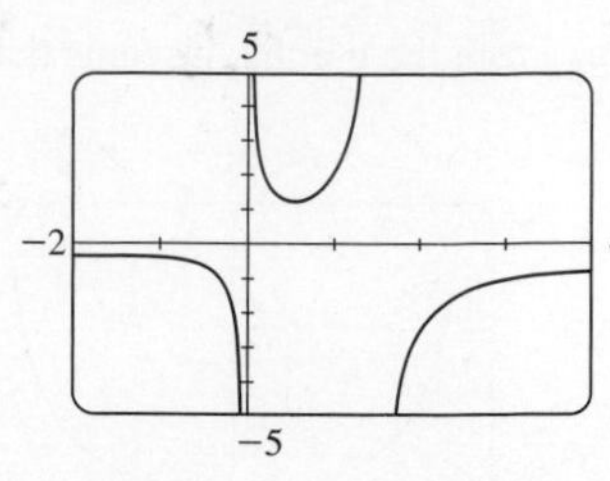

45. (a) $f(x) = \dfrac{1}{x^3-1}$.

From these calculations, it seems that

$\lim\limits_{x\to 1^-} f(x) = -\infty$ and $\lim\limits_{x\to 1^+} f(x) = \infty$.

x	$f(x)$
0.5	−1.14
0.9	−3.69
0.99	−33.7
0.999	−333.7
0.9999	−3333.7
0.99999	−33,333.7

x	$f(x)$
1.5	0.42
1.1	3.02
1.01	33.0
1.001	333.0
1.0001	3333.0
1.00001	33,333.3

(b) If x is slightly smaller than 1, then x^3-1 will be a negative number close to 0, and the reciprocal of x^3-1, that is, $f(x)$, will be a negative number with large absolute value. So $\lim\limits_{x\to 1^-} f(x) = -\infty$.

If x is slightly larger than 1, then x^3-1 will be a small positive number, and its reciprocal, $f(x)$, will be a large positive number. So $\lim\limits_{x\to 1^+} f(x) = \infty$.

(c) It appears from the graph of f that

$\lim\limits_{x\to 1^-} f(x) = -\infty$ and $\lim\limits_{x\to 1^+} f(x) = \infty$.

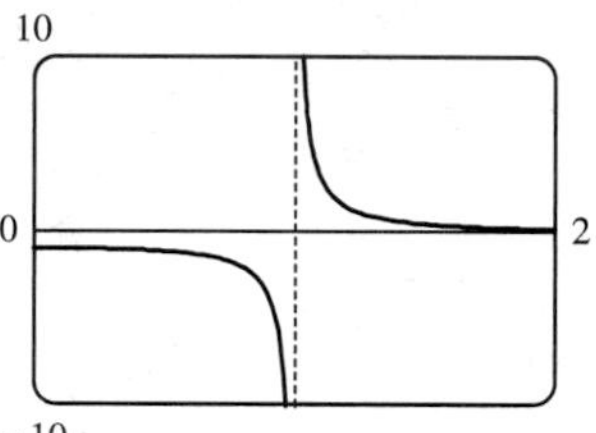

46. (a) From the graphs, it seems that $\lim\limits_{x\to 0} \dfrac{\tan 4x}{x} = 4$.

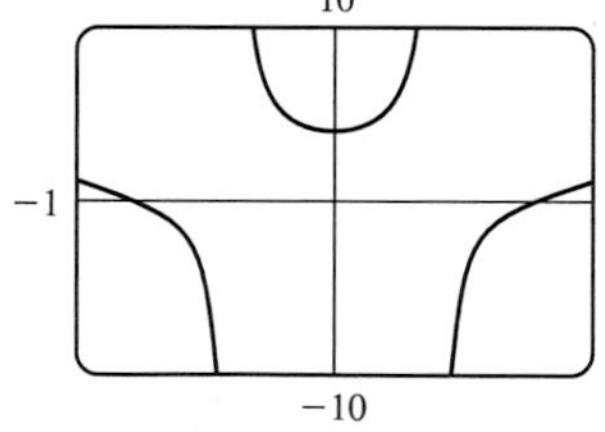

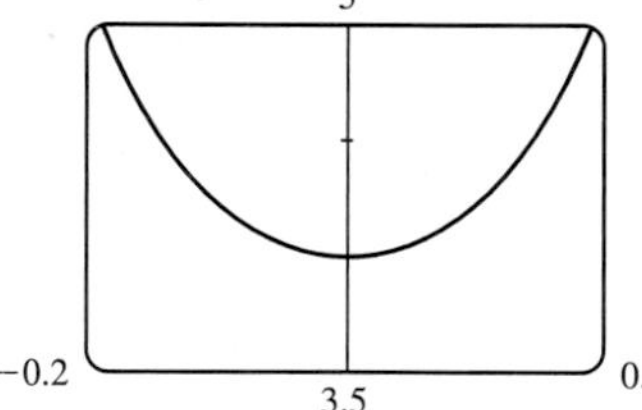

(b)

x	$f(x)$
±0.1	4.227 932
±0.01	4.002 135
±0.001	4.000 021
±0.0001	4.000 000

47. (a) Let $h(x) = (1+x)^{1/x}$.

x	$h(x)$
−0.001	2.71964
−0.0001	2.71842
−0.00001	2.71830
−0.000001	2.71828
0.000001	2.71828
0.00001	2.71827
0.0001	2.71815
0.001	2.71692

(b)

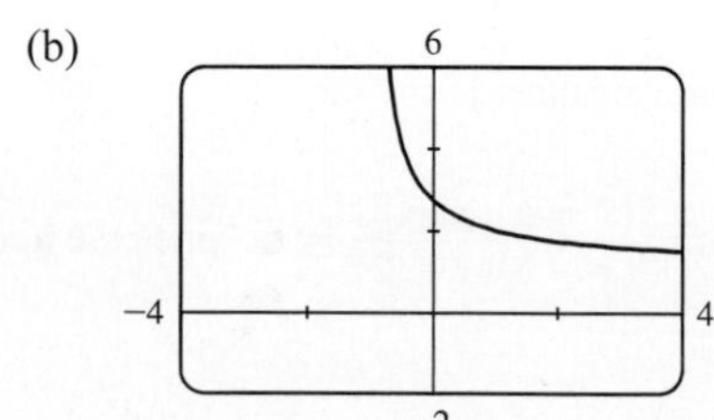

It appears that $\lim\limits_{x\to 0} (1+x)^{1/x} \approx 2.71828$, which is approximately e.

In Section 3.6 we will see that the value of the limit is exactly e.

48. (a)

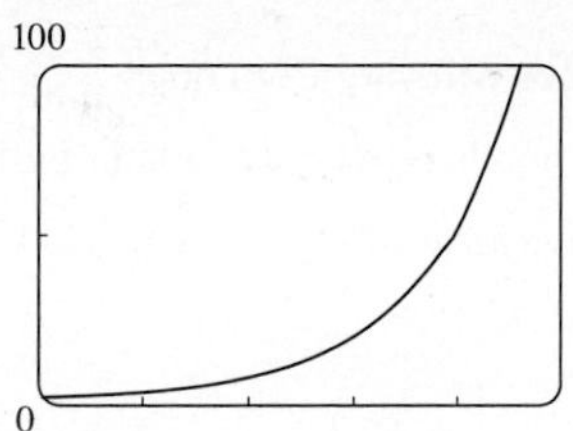

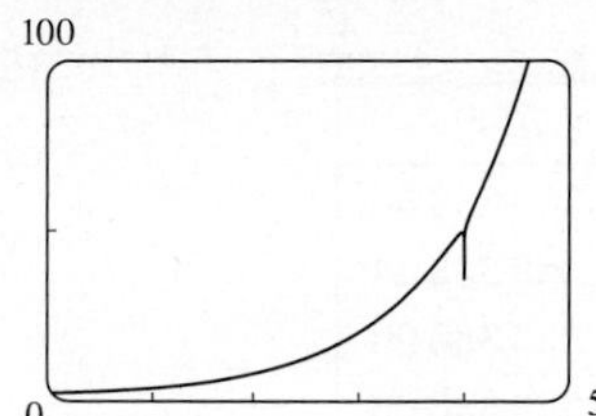

No, because the calculator-produced graph of $f(x) = e^x + \ln|x - 4|$ looks like an exponential function, but the graph of f has an infinite discontinuity at $x = 4$. A second graph, obtained by increasing the `numpoints` option in Maple, begins to reveal the discontinuity at $x = 4$.

(b) There isn't a single graph that shows all the features of f. Several graphs are needed since f looks like $\ln|x - 4|$ for large negative values of x and like e^x for $x > 5$, but yet has the infinite discontiuity at $x = 4$.

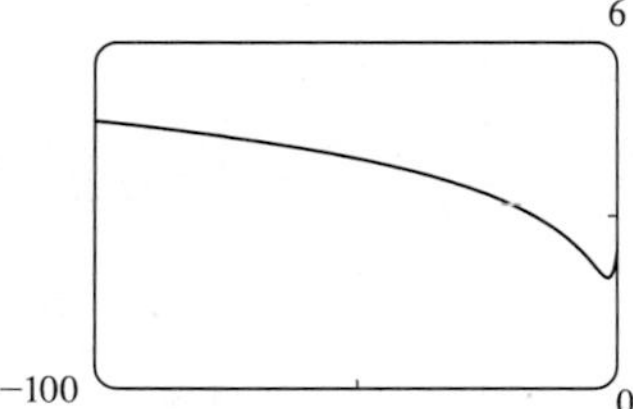

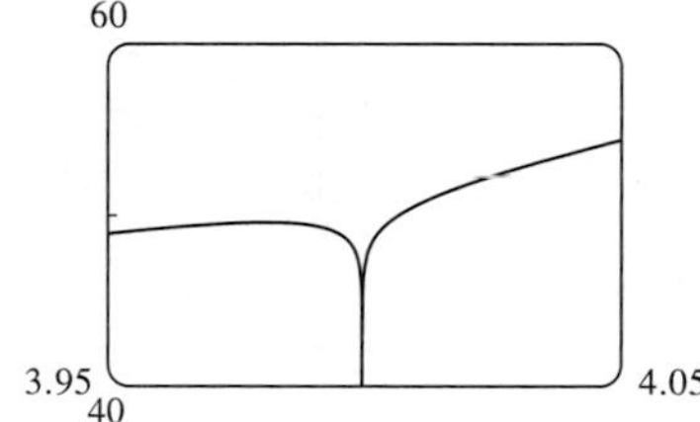

A hand-drawn graph, though distorted, might be better at revealing the main features of this function.

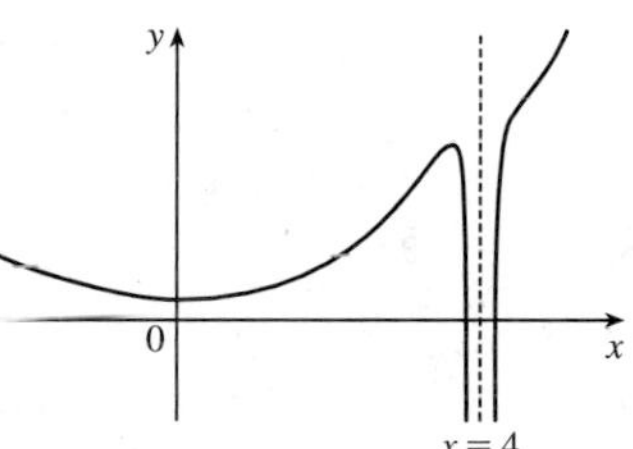

49. For $f(x) = x^2 - (2^x/1000)$:

(a)

x	$f(x)$
1	0.998 000
0.8	0.638 259
0.6	0.358 484
0.4	0.158 680
0.2	0.038 851
0.1	0.008 928
0.05	0.001 465

It appears that $\lim_{x\to 0} f(x) = 0$.

(b)

x	$f(x)$
0.04	0.000 572
0.02	−0.000 614
0.01	−0.000 907
0.005	−0.000 978
0.003	−0.000 993
0.001	−0.001 000

It appears that $\lim_{x\to 0} f(x) = -0.001$.

50. For $h(x) = \dfrac{\tan x - x}{x^3}$:

(a)

x	$h(x)$
1.0	0.557 407 73
0.5	0.370 419 92
0.1	0.334 672 09
0.05	0.333 667 00
0.01	0.333 346 67
0.005	0.333 336 67

(b) It seems that $\lim_{x\to 0} h(x) = \frac{1}{3}$.

(c)

x	$h(x)$
0.001	0.333 333 50
0.0005	0.333 333 44
0.0001	0.333 330 00
0.00005	0.333 336 00
0.00001	0.333 000 00
0.000001	0.000 000 00

Here the values will vary from one calculator to another. Every calculator will eventually give *false values*.

(d) As in part (c), when we take a small enough viewing rectangle we get incorrect output.

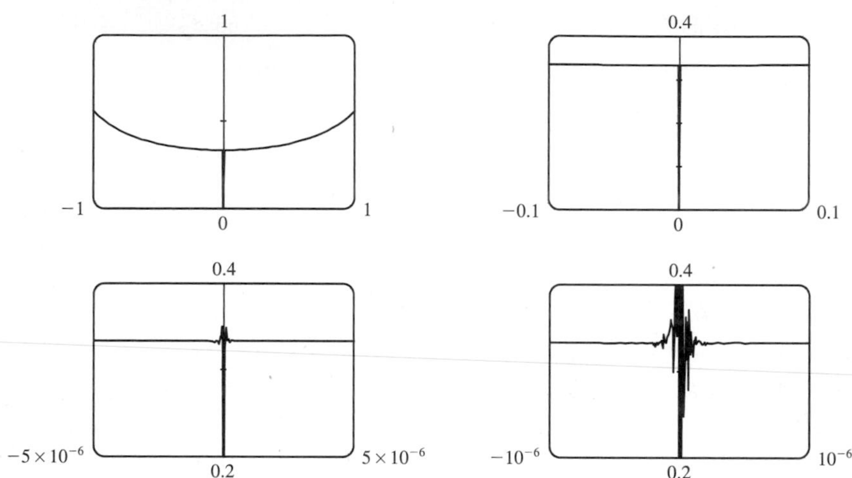

51. No matter how many times we zoom in toward the origin, the graphs of $f(x) = \sin(\pi/x)$ appear to consist of almost-vertical lines. This indicates more and more frequent oscillations as $x \to 0$.

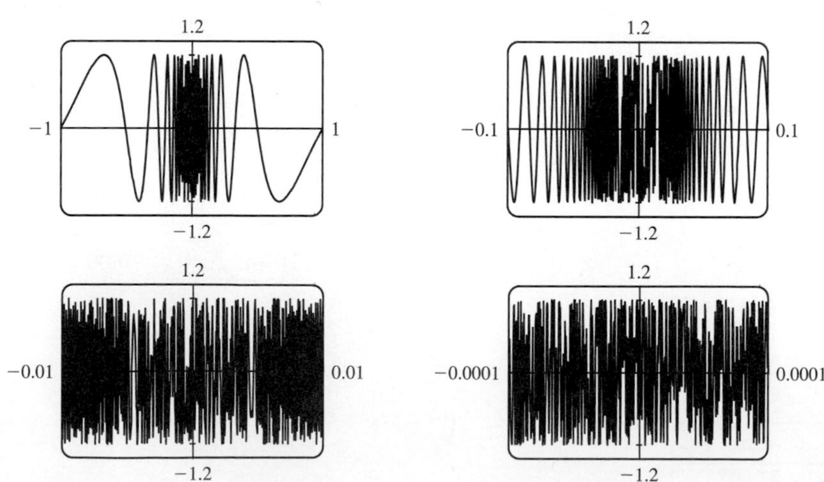

52. (a) For any positive integer n, if $x = \dfrac{1}{n\pi}$, then $f(x) = \tan\dfrac{1}{x} = \tan(n\pi) = 0$. (Remember that the tangent function has period π.)

(b) For any nonnegative number n, if $x = \dfrac{4}{(4n+1)\pi}$, then

$$f(x) = \tan\frac{1}{x} = \tan\frac{(4n+1)\pi}{4} = \tan\left(\frac{4n\pi}{4} + \frac{\pi}{4}\right) = \tan\left(n\pi + \frac{\pi}{4}\right) = \tan\frac{\pi}{4} = 1$$

(c) From part (a), $f(x) = 0$ infinitely often as $x \to 0$. From part (b), $f(x) = 1$ infinitely often as $x \to 0$. Thus, $\lim\limits_{x\to 0} \tan\dfrac{1}{x}$ does not exist since $f(x)$ does not get close to a fixed number as $x \to 0$.

53. 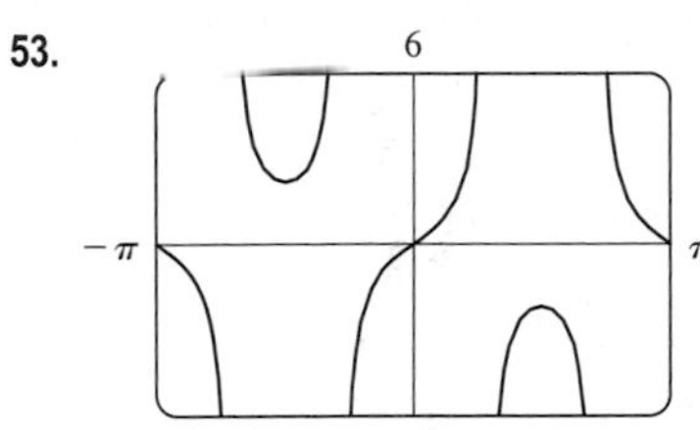

There appear to be vertical asymptotes of the curve $y = \tan(2\sin x)$ at $x \approx \pm 0.90$ and $x \approx \pm 2.24$. To find the exact equations of these asymptotes, we note that the graph of the tangent function has vertical asymptotes at $x = \frac{\pi}{2} + \pi n$. Thus, we must have $2\sin x = \frac{\pi}{2} + \pi n$, or equivalently, $\sin x = \frac{\pi}{4} + \frac{\pi}{2}n$. Since $-1 \le \sin x \le 1$, we must have $\sin x = \pm\frac{\pi}{4}$ and so $x = \pm\sin^{-1}\frac{\pi}{4}$ (corresponding to $x \approx \pm 0.90$). Just as 150° is the reference angle for 30°, $\pi - \sin^{-1}\frac{\pi}{4}$ is the reference angle for $\sin^{-1}\frac{\pi}{4}$. So $x = \pm\left(\pi - \sin^{-1}\frac{\pi}{4}\right)$ are also equations of vertical asymptotes (corresponding to $x \approx \pm 2.24$).

54. $\lim\limits_{v\to c^-} m = \lim\limits_{v\to c^-} \dfrac{m_0}{\sqrt{1 - v^2/c^2}}$. As $v \to c^-$, $\sqrt{1 - v^2/c^2} \to 0^+$, and $m \to \infty$.

55. (a) Let $y = \dfrac{x^3 - 1}{\sqrt{x} - 1}$.

From the table and the graph, we guess that the limit of y as x approaches 1 is 6.

x	y
0.99	5.925 31
0.999	5.992 50
0.9999	5.999 25
1.01	6.075 31
1.001	6.007 50
1.0001	6.000 75

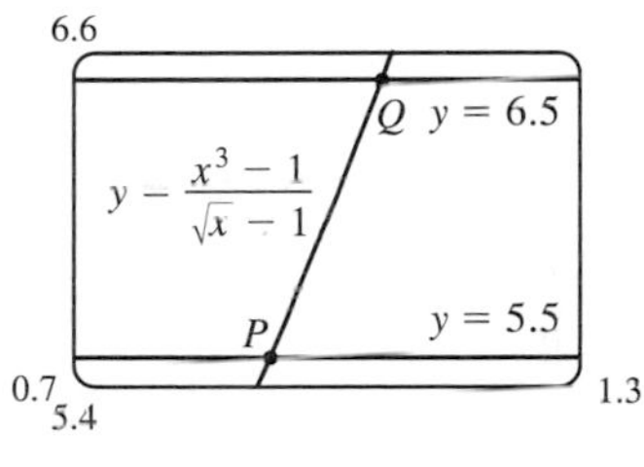

(b) We need to have $5.5 < \dfrac{x^3 - 1}{\sqrt{x} - 1} < 6.5$. From the graph we obtain the approximate points of intersection $P(0.9314, 5.5)$ and $Q(1.0649, 6.5)$. Now $1 - 0.9314 = 0.0686$ and $1.0649 - 1 = 0.0649$, so by requiring that x be within 0.0649 of 1, we ensure that y is within 0.5 of 6.

2.3 Calculating Limits Using the Limit Laws

1. (a) $\lim\limits_{x\to 2}[f(x) + 5g(x)] = \lim\limits_{x\to 2} f(x) + \lim\limits_{x\to 2}[5g(x)]$ [Limit Law 1]

$= \lim\limits_{x\to 2} f(x) + 5\lim\limits_{x\to 2} g(x)$ [Limit Law 3]

$= 4 + 5(-2) = -6$

(b) $\lim\limits_{x\to 2}[g(x)]^3 = \left[\lim\limits_{x\to 2} g(x)\right]^3$ [Limit Law 6]

$= (-2)^3 = -8$

(c) $\lim_{x\to 2} \sqrt{f(x)} = \sqrt{\lim_{x\to 2} f(x)}$ [Limit Law 11]

$= \sqrt{4} = 2$

(d) $\lim_{x\to 2} \dfrac{3f(x)}{g(x)} = \dfrac{\lim_{x\to 2} [3f(x)]}{\lim_{x\to 2} g(x)}$ [Limit Law 5]

$= \dfrac{3 \lim_{x\to 2} f(x)}{\lim_{x\to 2} g(x)}$ [Limit Law 3]

$= \dfrac{3(4)}{-2} = -6$

(e) Because the limit of the denominator is 0, we can't use Limit Law 5. The given limit, $\lim_{x\to 2} \dfrac{g(x)}{h(x)}$, does not exist because the denominator approaches 0 while the numerator approaches a nonzero number.

(f) $\lim_{x\to 2} \dfrac{g(x)\,h(x)}{f(x)} = \dfrac{\lim_{x\to 2} [g(x)\,h(x)]}{\lim_{x\to 2} f(x)}$ [Limit Law 5]

$= \dfrac{\lim_{x\to 2} g(x) \cdot \lim_{x\to 2} h(x)}{\lim_{x\to 2} f(x)}$ [Limit Law 4]

$= \dfrac{-2 \cdot 0}{4} = 0$

2. (a) $\lim_{x\to 2} [f(x) + g(x)] = \lim_{x\to 2} f(x) + \lim_{x\to 2} g(x)$ [Limit Law 1]

$= -1 + 2$

$= 1$

(b) $\lim_{x\to 0} f(x)$ exists, but $\lim_{x\to 0} g(x)$ does not exist, so we cannot apply Limit Law 2 to $\lim_{x\to 0} [f(x) - g(x)]$. The limit does not exist.

(c) $\lim_{x\to -1} [f(x)\,g(x)] = \lim_{x\to -1} f(x) \cdot \lim_{x\to -1} g(x)$ [Limit Law 4]

$= 1 \cdot 2$

$= 2$

(d) $\lim_{x\to 3} f(x) = 1$, but $\lim_{x\to 3} g(x) = 0$, so we cannot apply Limit Law 5 to $\lim_{x\to 3} \dfrac{f(x)}{g(x)}$. The limit does not exist.

Note: $\lim_{x\to 3^-} \dfrac{f(x)}{g(x)} = \infty$ since $g(x) \to 0^+$ as $x \to 3^-$ and $\lim_{x\to 3^+} \dfrac{f(x)}{g(x)} = -\infty$ since $g(x) \to 0^-$ as $x \to 3^+$. Therefore, the limit does not exist, even as an infinite limit.

(e) $\lim_{x\to 2} [x^2 f(x)] = \lim_{x\to 2} x^2 \cdot \lim_{x\to 2} f(x)$ [Limit Law 4]

$= 2^2 \cdot (-1)$

$= -4$

(f) $f(-1) + \lim_{x\to -1} g(x)$ is undefined since $f(-1)$ is not defined.

3. $\lim_{x\to 3} (5x^3 - 3x^2 + x - 6) = \lim_{x\to 3} (5x^3) - \lim_{x\to 3} (3x^2) + \lim_{x\to 3} x - \lim_{x\to 3} 6$ [Limit Laws 2 and 1]

$= 5 \lim_{x\to 3} x^3 - 3 \lim_{x\to 3} x^2 + \lim_{x\to 3} x - \lim_{x\to 3} 6$ [3]

$= 5(3^3) - 3(3^2) + 3 - 6$ [9, 8, and 7]

$= 105$

4. $\lim\limits_{x\to -1}(x^4-3x)(x^2+5x+3) = \lim\limits_{x\to -1}(x^4-3x)\lim\limits_{x\to -1}(x^2+5x+3)$ [Limit Law 4]

$= \left(\lim\limits_{x\to -1}x^4 - \lim\limits_{x\to -1}3x\right)\left(\lim\limits_{x\to -1}x^2+\lim\limits_{x\to -1}5x+\lim\limits_{x\to -1}3\right)$ [2, 1]

$= \left(\lim\limits_{x\to -1}x^4 - 3\lim\limits_{x\to -1}x\right)\left(\lim\limits_{x\to -1}x^2+5\lim\limits_{x\to -1}x+\lim\limits_{x\to -1}3\right)$ [3]

$= (1+3)(1-5+3)$ [9, 8, and 7]

$= 4(-1) = -4$

5. $\lim\limits_{t\to -2}\dfrac{t^4-2}{2t^2-3t+2} = \dfrac{\lim\limits_{t\to -2}(t^4-2)}{\lim\limits_{t\to -2}(2t^2-3t+2)}$ [Limit Law 5]

$= \dfrac{\lim\limits_{t\to -2}t^4 - \lim\limits_{t\to -2}2}{2\lim\limits_{t\to -2}t^2 - 3\lim\limits_{t\to -2}t + \lim\limits_{t\to -2}2}$ [1, 2, and 3]

$= \dfrac{16-2}{2(4)-3(-2)+2}$ [9, 7, and 8]

$= \dfrac{14}{16} = \dfrac{7}{8}$

6. $\lim\limits_{u\to -2}\sqrt{u^4+3u+6} = \sqrt{\lim\limits_{u\to -2}(u^4+3u+6)}$ [11]

$= \sqrt{\lim\limits_{u\to -2}u^4 + 3\lim\limits_{u\to -2}u + \lim\limits_{u\to -2}6}$ [1, 2, and 3]

$= \sqrt{(-2)^4+3(-2)+6}$ [9, 8, and 7]

$= \sqrt{16-6+6} = \sqrt{16} = 4$

7. $\lim\limits_{x\to 8}(1+\sqrt[3]{x})(2-6x^2+x^3) = \lim\limits_{x\to 8}(1+\sqrt[3]{x})\cdot\lim\limits_{x\to 8}(2-6x^2+x^3)$ [Limit Law 4]

$= \left(\lim\limits_{x\to 8}1+\lim\limits_{x\to 8}\sqrt[3]{x}\right)\cdot\left(\lim\limits_{x\to 8}2-6\lim\limits_{x\to 8}x^2+\lim\limits_{x\to 8}x^3\right)$ [1, 2, and 3]

$= (1+\sqrt[3]{8})\cdot(2-6\cdot 8^2+8^3)$ [7, 10, 9]

$= (3)(130) = 390$

8. $\lim\limits_{t\to 2}\left(\dfrac{t^2-2}{t^3-3t+5}\right)^2 = \left(\lim\limits_{t\to 2}\dfrac{t^2-2}{t^3-3t+5}\right)^2$ [Limit Law 6]

$= \left(\dfrac{\lim\limits_{t\to 2}(t^2-2)}{\lim\limits_{t\to 2}(t^3-3t+5)}\right)^2$ [5]

$= \left(\dfrac{\lim\limits_{t\to 2}t^2-\lim\limits_{t\to 2}2}{\lim\limits_{t\to 2}t^3-3\lim\limits_{t\to 2}t+\lim\limits_{t\to 2}5}\right)^2$ [1, 2, and 3]

$= \left(\dfrac{4-2}{8-3(2)+5}\right)^2$ [9, 7, and 8]

$= \left(\dfrac{2}{7}\right)^2 = \dfrac{4}{49}$

9. $\lim\limits_{x\to 2}\sqrt{\dfrac{2x^2+1}{3x-2}} = \sqrt{\lim\limits_{x\to 2}\dfrac{2x^2+1}{3x-2}}$ [Limit Law 11]

$= \sqrt{\dfrac{\lim\limits_{x\to 2}(2x^2+1)}{\lim\limits_{x\to 2}(3x-2)}}$ [5]

$= \sqrt{\dfrac{2\lim\limits_{x\to 2}x^2+\lim\limits_{x\to 2}1}{3\lim\limits_{x\to 2}x-\lim\limits_{x\to 2}2}}$ [1, 2, and 3]

$= \sqrt{\dfrac{2(2)^2+1}{3(2)-2}} = \sqrt{\dfrac{9}{4}} = \dfrac{3}{2}$ [9, 8, and 7]

10. (a) The left-hand side of the equation is not defined for $x = 2$, but the right-hand side is.

(b) Since the equation holds for all $x \neq 2$, it follows that both sides of the equation approach the same limit as $x \to 2$, just as in Example 3. Remember that in finding $\lim\limits_{x\to a} f(x)$, we never consider $x = a$.

11. $\lim\limits_{x\to 5}\dfrac{x^2-6x+5}{x-5} = \lim\limits_{x\to 5}\dfrac{(x-5)(x-1)}{x-5} = \lim\limits_{x\to 5}(x-1) = 5-1 = 4$

12. $\lim\limits_{x\to -3}\dfrac{x^2+3x}{x^2-x-12} = \lim\limits_{x\to -3}\dfrac{x(x+3)}{(x-4)(x+3)} = \lim\limits_{x\to -3}\dfrac{x}{x-4} = \dfrac{-3}{-3-4} = \dfrac{3}{7}$

13. $\lim\limits_{x\to 5}\dfrac{x^2-5x+6}{x-5}$ does not exist since $x-5\to 0$, but $x^2-5x+6\to 6$ as $x\to 5$.

14. $\lim\limits_{x\to 4}\dfrac{x^2+3x}{x^2-x-12} = \lim\limits_{x\to 4}\dfrac{x(x+3)}{(x-4)(x+3)} = \lim\limits_{x\to 4}\dfrac{x}{x-4}$. The last limit does not exist since $\lim\limits_{x\to 4^-}\dfrac{x}{x-4} = -\infty$ and $\lim\limits_{x\to 4^+}\dfrac{x}{x-4} = \infty$.

15. $\lim\limits_{t\to -3}\dfrac{t^2-9}{2t^2+7t+3} = \lim\limits_{t\to -3}\dfrac{(t+3)(t-3)}{(2t+1)(t+3)} = \lim\limits_{t\to -3}\dfrac{t-3}{2t+1} = \dfrac{-3-3}{2(-3)+1} = \dfrac{-6}{-5} = \dfrac{6}{5}$

16. $\lim\limits_{x\to -1}\dfrac{2x^2+3x+1}{x^2-2x-3} = \lim\limits_{x\to -1}\dfrac{(2x+1)(x+1)}{(x-3)(x+1)} = \lim\limits_{x\to -1}\dfrac{2x+1}{x-3} = \dfrac{2(-1)+1}{-1-3} = \dfrac{-1}{-4} = \dfrac{1}{4}$

17. $\lim\limits_{h\to 0}\dfrac{(-5+h)^2-25}{h} = \lim\limits_{h\to 0}\dfrac{(25-10h+h^2)-25}{h} = \lim\limits_{h\to 0}\dfrac{-10h+h^2}{h} = \lim\limits_{h\to 0}\dfrac{h(-10+h)}{h} = \lim\limits_{h\to 0}(-10+h) = -10$

18. $\lim\limits_{h\to 0}\dfrac{(2+h)^3-8}{h} = \lim\limits_{h\to 0}\dfrac{(8+12h+6h^2+h^3)-8}{h} = \lim\limits_{h\to 0}\dfrac{12h+6h^2+h^3}{h}$

$= \lim\limits_{h\to 0}(12+6h+h^2) = 12+0+0 = 12$

19. By the formula for the sum of cubes, we have

$$\lim_{x\to -2}\frac{x+2}{x^3+8} = \lim_{x\to -2}\frac{x+2}{(x+2)(x^2-2x+4)} = \lim_{x\to -2}\frac{1}{x^2-2x+4} = \frac{1}{4+4+4} = \frac{1}{12}.$$

20. We use the difference of squares in the numerator and the difference of cubes in the denominator.

$$\lim_{t\to 1}\frac{t^4-1}{t^3-1}=\lim_{t\to 1}\frac{(t^2-1)(t^2+1)}{(t-1)(t^2+t+1)}=\lim_{t\to 1}\frac{(t-1)(t+1)(t^2+1)}{(t-1)(t^2+t+1)}=\lim_{t\to 1}\frac{(t+1)(t^2+1)}{t^2+t+1}=\frac{2(2)}{3}=\frac{4}{3}$$

21. $$\lim_{h\to 0}\frac{\sqrt{9+h}-3}{h}=\lim_{h\to 0}\frac{\sqrt{9+h}-3}{h}\cdot\frac{\sqrt{9+h}+3}{\sqrt{9+h}+3}=\lim_{h\to 0}\frac{\left(\sqrt{9+h}\right)^2-3^2}{h\left(\sqrt{9+h}+3\right)}=\lim_{h\to 0}\frac{(9+h)-9}{h\left(\sqrt{9+h}+3\right)}$$

$$=\lim_{h\to 0}\frac{h}{h\left(\sqrt{9+h}+3\right)}=\lim_{h\to 0}\frac{1}{\sqrt{9+h}+3}=\frac{1}{3+3}=\frac{1}{6}$$

22. $$\lim_{u\to 2}\frac{\sqrt{4u+1}-3}{u-2}=\lim_{u\to 2}\frac{\sqrt{4u+1}-3}{u-2}\cdot\frac{\sqrt{4u+1}+3}{\sqrt{4u+1}+3}=\lim_{u\to 2}\frac{\left(\sqrt{4u+1}\right)^2-3^2}{(u-2)\left(\sqrt{4u+1}+3\right)}$$

$$=\lim_{u\to 2}\frac{4u+1-9}{(u-2)\left(\sqrt{4u+1}+3\right)}=\lim_{u\to 2}\frac{4(u-2)}{(u-2)\left(\sqrt{4u+1}+3\right)}$$

$$=\lim_{u\to 2}\frac{4}{\sqrt{4u+1}+3}=\frac{4}{\sqrt{9}+3}=\frac{2}{3}$$

23. $$\lim_{x\to 3}\frac{\frac{1}{x}-\frac{1}{3}}{x-3}=\lim_{x\to 3}\frac{\frac{1}{x}-\frac{1}{3}}{x-3}\cdot\frac{3x}{3x}=\lim_{x\to 3}\frac{3-x}{3x(x-3)}=\lim_{x\to 3}\frac{-1}{3x}=-\frac{1}{9}$$

24. $$\lim_{h\to 0}\frac{(3+h)^{-1}-3^{-1}}{h}=\lim_{h\to 0}\frac{\frac{1}{3+h}-\frac{1}{3}}{h}=\lim_{h\to 0}\frac{3-(3+h)}{h(3+h)3}=\lim_{h\to 0}\frac{-h}{h(3+h)3}$$

$$=\lim_{h\to 0}\left[-\frac{1}{3(3+h)}\right]=-\frac{1}{\lim\limits_{h\to 0}[3(3+h)]}=-\frac{1}{3(3+0)}=-\frac{1}{9}$$

25. $$\lim_{t\to 0}\frac{\sqrt{1+t}-\sqrt{1-t}}{t}=\lim_{t\to 0}\frac{\sqrt{1+t}-\sqrt{1-t}}{t}\cdot\frac{\sqrt{1+t}+\sqrt{1-t}}{\sqrt{1+t}+\sqrt{1-t}}=\lim_{t\to 0}\frac{\left(\sqrt{1+t}\right)^2-\left(\sqrt{1-t}\right)^2}{t\left(\sqrt{1+t}+\sqrt{1-t}\right)}$$

$$=\lim_{t\to 0}\frac{(1+t)-(1-t)}{t\left(\sqrt{1+t}+\sqrt{1-t}\right)}=\lim_{t\to 0}\frac{2t}{t\left(\sqrt{1+t}+\sqrt{1-t}\right)}=\lim_{t\to 0}\frac{2}{\sqrt{1+t}+\sqrt{1-t}}$$

$$=\frac{2}{\sqrt{1}+\sqrt{1}}=\frac{2}{2}=1$$

26. $$\lim_{t\to 0}\left(\frac{1}{t}-\frac{1}{t^2+t}\right)=\lim_{t\to 0}\left(\frac{1}{t}-\frac{1}{t(t+1)}\right)=\lim_{t\to 0}\frac{t+1-1}{t(t+1)}=\lim_{t\to 0}\frac{1}{t+1}=\frac{1}{0+1}=1$$

27. $$\lim_{x\to 16}\frac{4-\sqrt{x}}{16x-x^2}=\lim_{x\to 16}\frac{(4-\sqrt{x})(4+\sqrt{x})}{(16x-x^2)(4+\sqrt{x})}=\lim_{x\to 16}\frac{16-x}{x(16-x)(4+\sqrt{x})}$$

$$=\lim_{x\to 16}\frac{1}{x(4+\sqrt{x})}=\frac{1}{16\left(4+\sqrt{16}\right)}=\frac{1}{16(8)}=\frac{1}{128}$$

28. $$\lim_{x\to 2}\frac{x^2-4x+4}{x^4-3x^2-4}=\lim_{x\to 2}\frac{(x-2)^2}{(x^2-4)(x^2+1)}=\lim_{x\to 2}\frac{(x-2)^2}{(x+2)(x-2)(x^2+1)}$$

$$=\lim_{x\to 2}\frac{x-2}{(x+2)(x^2+1)}=\frac{0}{4\cdot 5}=0$$

29. $\lim_{t\to 0}\left(\frac{1}{t\sqrt{1+t}}-\frac{1}{t}\right)=\lim_{t\to 0}\frac{1-\sqrt{1+t}}{t\sqrt{1+t}}=\lim_{t\to 0}\frac{\left(1-\sqrt{1+t}\right)\left(1+\sqrt{1+t}\right)}{t\sqrt{t+1}\left(1+\sqrt{1+t}\right)}=\lim_{t\to 0}\frac{-t}{t\sqrt{1+t}\left(1+\sqrt{1+t}\right)}$

$$=\lim_{t\to 0}\frac{-1}{\sqrt{1+t}\left(1+\sqrt{1+t}\right)}=\frac{-1}{\sqrt{1+0}\left(1+\sqrt{1+0}\right)}=-\frac{1}{2}$$

30. $\lim_{x\to -4}\frac{\sqrt{x^2+9}-5}{x+4}=\lim_{x\to -4}\frac{\left(\sqrt{x^2+9}-5\right)\left(\sqrt{x^2+9}+5\right)}{(x+4)\left(\sqrt{x^2+9}+5\right)}=\lim_{x\to -4}\frac{(x^2+9)-25}{(x+4)\left(\sqrt{x^2+9}+5\right)}$

$$=\lim_{x\to -4}\frac{x^2-16}{(x+4)\left(\sqrt{x^2+9}+5\right)}=\lim_{x\to -4}\frac{(x+4)(x-4)}{(x+4)\left(\sqrt{x^2+9}+5\right)}$$

$$=\lim_{x\to -4}\frac{x-4}{\sqrt{x^2+9}+5}=\frac{-4-4}{\sqrt{16+9}+5}=\frac{-8}{5+5}=-\frac{4}{5}$$

31. $\lim_{h\to 0}\frac{(x+h)^3-x^3}{h}=\lim_{h\to 0}\frac{(x^3+3x^2h+3xh^2+h^3)-x^3}{h}=\lim_{h\to 0}\frac{3x^2h+3xh^2+h^3}{h}$

$$=\lim_{h\to 0}\frac{h(3x^2+3xh+h^2)}{h}=\lim_{h\to 0}(3x^2+3xh+h^2)=3x^2$$

32. $\lim_{h\to 0}\frac{\frac{1}{(x+h)^2}-\frac{1}{x^2}}{h}=\lim_{h\to 0}\frac{\frac{x^2-(x+h)^2}{(x+h)^2x^2}}{h}=\lim_{h\to 0}\frac{x^2-(x^2+2xh+h^2)}{hx^2(x+h)^2}=\lim_{h\to 0}\frac{-h(2x+h)}{hx^2(x+h)^2}$

$$=\lim_{h\to 0}\frac{-(2x+h)}{x^2(x+h)^2}=\frac{-2x}{x^2\cdot x^2}=-\frac{2}{x^3}$$

33. (a)

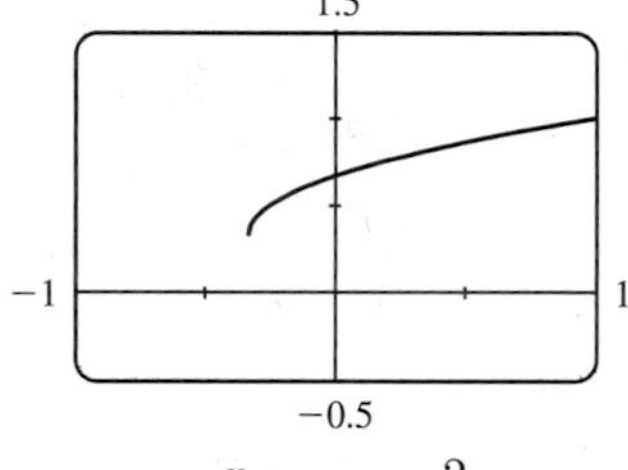

$\lim_{x\to 0}\frac{x}{\sqrt{1+3x}-1}\approx\frac{2}{3}$

(b)

x	$f(x)$
−0.001	0.666 166 3
−0.000 1	0.666 616 7
−0.000 01	0.666 661 7
−0.000 001	0.666 666 2
0.000 001	0.666 667 2
0.000 01	0.666 671 7
0.000 1	0.666 716 7
0.001	0.667 166 3

The limit appears to be $\frac{2}{3}$.

(c) $\lim_{x\to 0}\left(\frac{x}{\sqrt{1+3x}-1}\cdot\frac{\sqrt{1+3x}+1}{\sqrt{1+3x}+1}\right)=\lim_{x\to 0}\frac{x\left(\sqrt{1+3x}+1\right)}{(1+3x)-1}=\lim_{x\to 0}\frac{x\left(\sqrt{1+3x}+1\right)}{3x}$

$=\frac{1}{3}\lim_{x\to 0}\left(\sqrt{1+3x}+1\right)$ [Limit Law 3]

$=\frac{1}{3}\left[\sqrt{\lim_{x\to 0}(1+3x)}+\lim_{x\to 0}1\right]$ [1 and 11]

$=\frac{1}{3}\left(\sqrt{\lim_{x\to 0}1+3\lim_{x\to 0}x}+1\right)$ [1, 3, and 7]

$=\frac{1}{3}\left(\sqrt{1+3\cdot 0}+1\right)$ [7 and 8]

$=\frac{1}{3}(1+1)=\frac{2}{3}$

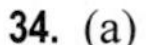

34. (a)

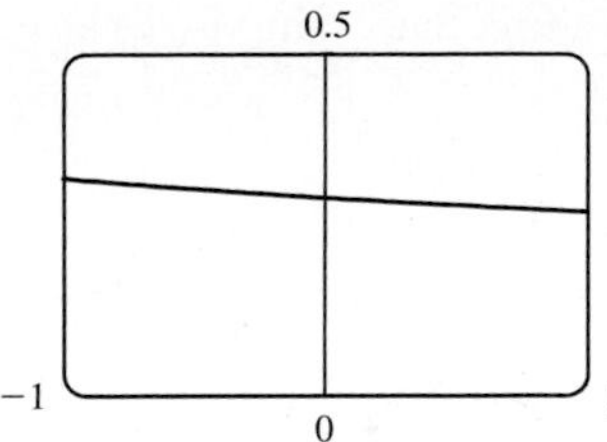

$\lim\limits_{x\to 0} \dfrac{\sqrt{3+x}-\sqrt{3}}{x} \approx 0.29$

(b)

x	$f(x)$
-0.001	$0.288\,699\,2$
$-0.000\,1$	$0.288\,677\,5$
$-0.000\,01$	$0.288\,675\,4$
$-0.000\,001$	$0.288\,675\,2$
$0.000\,001$	$0.288\,675\,1$
$0.000\,01$	$0.288\,674\,9$
$0.000\,1$	$0.288\,672\,7$
0.001	$0.288\,651\,1$

The limit appears to be approximately 0.2887.

(c)
$$\lim_{x\to 0}\left(\frac{\sqrt{3+x}-\sqrt{3}}{x}\cdot\frac{\sqrt{3+x}+\sqrt{3}}{\sqrt{3+x}+\sqrt{3}}\right) = \lim_{x\to 0}\frac{(3+x)-3}{x\left(\sqrt{3+x}+\sqrt{3}\right)} = \lim_{x\to 0}\frac{1}{\sqrt{3+x}+\sqrt{3}}$$

$$= \frac{\lim\limits_{x\to 0} 1}{\lim\limits_{x\to 0}\sqrt{3+x}+\lim\limits_{x\to 0}\sqrt{3}} \qquad \text{[Limit Laws 5 and 1]}$$

$$= \frac{1}{\sqrt{\lim\limits_{x\to 0}(3+x)}+\sqrt{3}} \qquad \text{[7 and 11]}$$

$$= \frac{1}{\sqrt{3+0}+\sqrt{3}} \qquad \text{[1, 7, and 8]}$$

$$= \frac{1}{2\sqrt{3}}$$

35. Let $f(x) = -x^2$, $g(x) = x^2\cos 20\pi x$ and $h(x) = x^2$. Then

$-1 \le \cos 20\pi x \le 1 \;\Rightarrow\; -x^2 \le x^2\cos 20\pi x \le x^2 \;\Rightarrow\; f(x) \le g(x) \le h(x)$.

So since $\lim\limits_{x\to 0} f(x) = \lim\limits_{x\to 0} h(x) = 0$, by the Squeeze Theorem we have

$\lim\limits_{x\to 0} g(x) = 0$.

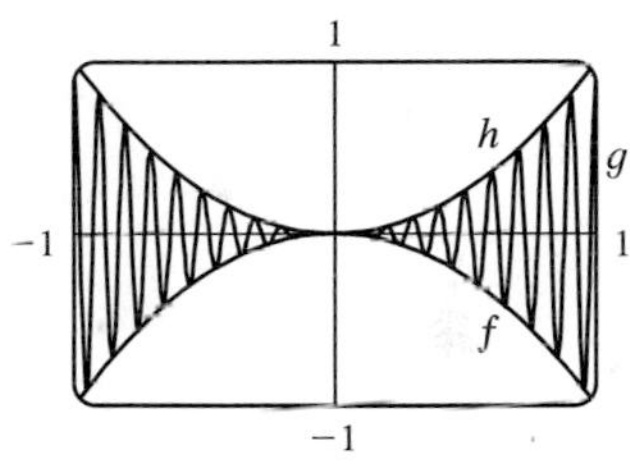

36. Let $f(x) = -\sqrt{x^3+x^2}$, $g(x) = \sqrt{x^3+x^2}\sin(\pi/x)$, and $h(x) = \sqrt{x^3+x^2}$. Then

$-1 \le \sin(\pi/x) \le 1 \;\Rightarrow\; -\sqrt{x^3+x^2} \le \sqrt{x^3+x^2}\sin(\pi/x) \le \sqrt{x^3+x^2} \;\Rightarrow$

$f(x) \le g(x) \le h(x)$. So since $\lim\limits_{x\to 0} f(x) = \lim\limits_{x\to 0} h(x) = 0$, by the Squeeze Theorem

we have $\lim\limits_{x\to 0} g(x) = 0$.

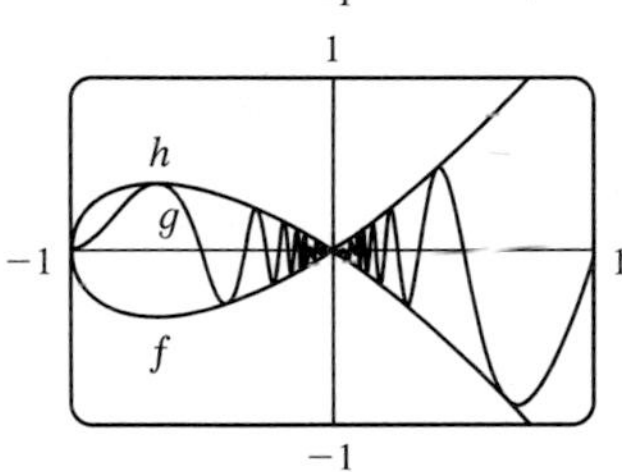

37. We have $\lim\limits_{x\to 4}(4x-9) = 4(4)-9 = 7$ and $\lim\limits_{x\to 4}\left(x^2-4x+7\right) = 4^2-4(4)+7 = 7$. Since $4x-9 \le f(x) \le x^2-4x+7$ for $x \ge 0$, $\lim\limits_{x\to 4} f(x) = 7$ by the Squeeze Theorem.

38. We have $\lim\limits_{x\to 1}(2x) = 2(1) = 2$ and $\lim\limits_{x\to 1}(x^4-x^2+2) = 1^4-1^2+2 = 2$. Since $2x \le g(x) \le x^4-x^2+2$ for all x, $\lim\limits_{x\to 1} g(x) = 2$ by the Squeeze Theorem.

39. $-1 \le \cos(2/x) \le 1 \;\Rightarrow\; -x^4 \le x^4\cos(2/x) \le x^4$. Since $\lim\limits_{x\to 0}\left(-x^4\right) = 0$ and $\lim\limits_{x\to 0} x^4 = 0$, we have $\lim\limits_{x\to 0}\left[x^4\cos(2/x)\right] = 0$ by the Squeeze Theorem.

40. $-1 \le \sin(\pi/x) \le 1 \;\Rightarrow\; e^{-1} \le e^{\sin(\pi/x)} \le e^{1} \;\Rightarrow\; \sqrt{x}/e \le \sqrt{x}\,e^{\sin(\pi/x)} \le \sqrt{x}\,e$. Since $\lim\limits_{x \to 0^+} (\sqrt{x}/e) = 0$ and $\lim\limits_{x \to 0^+} (\sqrt{x}\,e) = 0$, we have $\lim\limits_{x \to 0^+} \left[\sqrt{x}\,e^{\sin(\pi/x)}\right] = 0$ by the Squeeze Theorem.

41. $|x-3| = \begin{cases} x-3 & \text{if } x-3 \ge 0 \\ -(x-3) & \text{if } x-3 < 0 \end{cases} = \begin{cases} x-3 & \text{if } x \ge 3 \\ 3-x & \text{if } x < 3 \end{cases}$

Thus, $\lim\limits_{x \to 3^+} (2x + |x-3|) = \lim\limits_{x \to 3^+} (2x + x - 3) = \lim\limits_{x \to 3^+} (3x - 3) = 3(3) - 3 = 6$ and

$\lim\limits_{x \to 3^-} (2x + |x-3|) = \lim\limits_{x \to 3^-} (2x + 3 - x) = \lim\limits_{x \to 3^-} (x + 3) = 3 + 3 = 6$. Since the left and right limits are equal,

$\lim\limits_{x \to 3} (2x + |x-3|) = 6$.

42. $|x+6| = \begin{cases} x+6 & \text{if } x+6 \ge 0 \\ -(x+6) & \text{if } x+6 < 0 \end{cases} = \begin{cases} x+6 & \text{if } x \ge -6 \\ -(x+6) & \text{if } x < -6 \end{cases}$

We'll look at the one-sided limits.

$$\lim_{x \to -6^+} \frac{2x+12}{|x+6|} = \lim_{x \to -6^+} \frac{2(x+6)}{x+6} = 2 \quad \text{and} \quad \lim_{x \to -6^-} \frac{2x+12}{|x+6|} = \lim_{x \to -6^-} \frac{2(x+6)}{-(x+6)} = -2$$

The left and right limits are different, so $\lim\limits_{x \to -6} \dfrac{2x+12}{|x+6|}$ does not exist.

43. $\left|2x^3 - x^2\right| = \left|x^2(2x-1)\right| = \left|x^2\right| \cdot |2x-1| = x^2\,|2x-1|$

$|2x-1| = \begin{cases} 2x-1 & \text{if } 2x-1 \ge 0 \\ -(2x-1) & \text{if } 2x-1 < 0 \end{cases} = \begin{cases} 2x-1 & \text{if } x \ge 0.5 \\ -(2x-1) & \text{if } x < 0.5 \end{cases}$

So $\left|2x^3 - x^2\right| = x^2[-(2x-1)]$ for $x < 0.5$.

Thus, $\lim\limits_{x \to 0.5^-} \dfrac{2x-1}{\left|2x^3 - x^2\right|} = \lim\limits_{x \to 0.5^-} \dfrac{2x-1}{x^2[-(2x-1)]} = \lim\limits_{x \to 0.5^-} \dfrac{-1}{x^2} = \dfrac{-1}{(0.5)^2} = \dfrac{-1}{0.25} = -4$.

44. Since $|x| = -x$ for $x < 0$, we have $\lim\limits_{x \to -2} \dfrac{2-|x|}{2+x} = \lim\limits_{x \to -2} \dfrac{2-(-x)}{2+x} = \lim\limits_{x \to -2} \dfrac{2+x}{2+x} = \lim\limits_{x \to -2} 1 = 1$.

45. Since $|x| = -x$ for $x < 0$, we have $\lim\limits_{x \to 0^-} \left(\dfrac{1}{x} - \dfrac{1}{|x|}\right) = \lim\limits_{x \to 0^-} \left(\dfrac{1}{x} - \dfrac{1}{-x}\right) = \lim\limits_{x \to 0^-} \dfrac{2}{x}$, which does not exist since the denominator approaches 0 and the numerator does not.

46. Since $|x| = x$ for $x > 0$, we have $\lim\limits_{x \to 0^+} \left(\dfrac{1}{x} - \dfrac{1}{|x|}\right) = \lim\limits_{x \to 0^+} \left(\dfrac{1}{x} - \dfrac{1}{x}\right) = \lim\limits_{x \to 0^+} 0 = 0$.

47. (a)

y
1
0
x

(b) (i) Since $\operatorname{sgn} x = 1$ for $x > 0$, $\lim\limits_{x \to 0^+} \operatorname{sgn} x = \lim\limits_{x \to 0^+} 1 = 1$.

(ii) Since $\operatorname{sgn} x = -1$ for $x < 0$, $\lim\limits_{x \to 0^-} \operatorname{sgn} x = \lim\limits_{x \to 0^-} -1 = -1$.

(iii) Since $\lim\limits_{x \to 0^-} \operatorname{sgn} x \ne \lim\limits_{x \to 0^+} \operatorname{sgn} x$, $\lim\limits_{x \to 0} \operatorname{sgn} x$ does not exist.

(iv) Since $|\operatorname{sgn} x| = 1$ for $x \ne 0$, $\lim\limits_{x \to 0} |\operatorname{sgn} x| = \lim\limits_{x \to 0} 1 = 1$.

48. (a) $g(x) = \operatorname{sgn}(\sin x) = \begin{cases} -1 & \text{if } \sin x < 0 \\ 0 & \text{if } \sin x = 0 \\ 1 & \text{if } \sin x > 0 \end{cases}$

(i) $\lim\limits_{x \to 0^+} g(x) = \lim\limits_{x \to 0^+} \operatorname{sgn}(\sin x) = 1$ since $\sin x$ is positive for small positive values of x.

(ii) $\lim\limits_{x \to 0^-} g(x) = \lim\limits_{x \to 0^-} \operatorname{sgn}(\sin x) = -1$ since $\sin x$ is negative for small negative values of x.

(iii) $\lim\limits_{x \to 0} g(x)$ does not exist since $\lim\limits_{x \to 0^+} g(x) \neq \lim\limits_{x \to 0^-} g(x)$.

(iv) $\lim\limits_{x \to \pi^+} g(x) = \lim\limits_{x \to \pi^+} \operatorname{sgn}(\sin x) = -1$ since $\sin x$ is negative for values of x slightly greater than π.

(v) $\lim\limits_{x \to \pi^-} g(x) = \lim\limits_{x \to \pi^-} \operatorname{sgn}(\sin x) = 1$ since $\sin x$ is positive for values of x slightly less than π.

(vi) $\lim\limits_{x \to \pi} g(x)$ does not exist since $\lim\limits_{x \to \pi^+} g(x) \neq \lim\limits_{x \to \pi^-} g(x)$.

(b) The sine function changes sign at every integer multiple of π, so the signum function equals 1 on one side and -1 on the other side of $n\pi$, n an integer. Thus, $\lim\limits_{x \to a} g(x)$ does not exist for $a = n\pi$, n an integer.

(c)

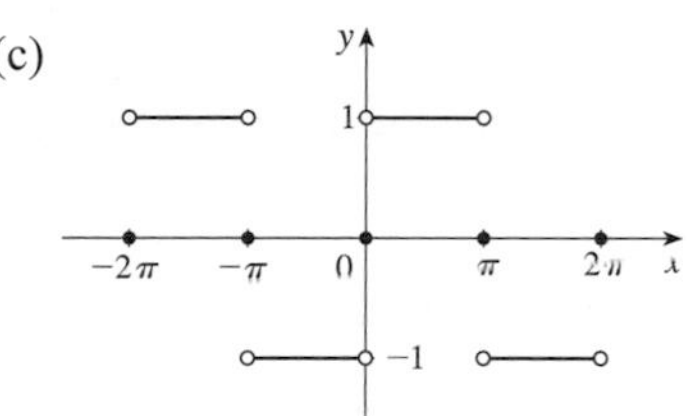

49. (a) (i) $\lim\limits_{x \to 2^+} g(x) = \lim\limits_{x \to 2^+} \dfrac{x^2 + x - 6}{|x - 2|} = \lim\limits_{x \to 2^+} \dfrac{(x + 3)(x - 2)}{|x - 2|}$

$$= \lim_{x \to 2^+} \frac{(x + 3)(x - 2)}{x - 2} \quad [\text{since } x - 2 > 0 \text{ if } x \to 2^+]$$

$$= \lim_{x \to 2^+} (x + 3) = 5$$

(ii) The solution is similar to the solution in part (i), but now $|x - 2| = 2 - x$ since $x - 2 < 0$ if $x \to 2^-$.

Thus, $\lim\limits_{x \to 2^-} g(x) = \lim\limits_{x \to 2^-} -(x + 3) = -5$.

(b) Since the right-hand and left-hand limits of g at $x = 2$ are not equal, $\lim\limits_{x \to 2} g(x)$ does not exist.

(c)

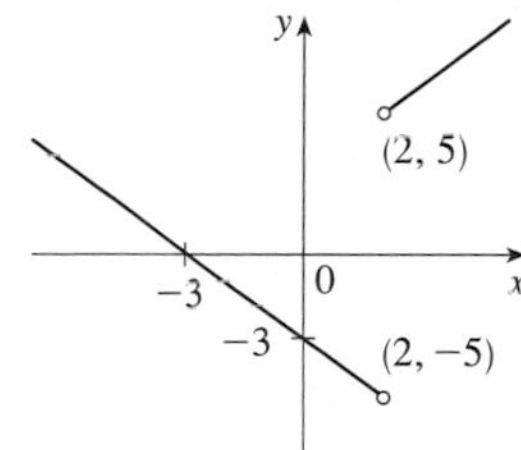

50. (a) $f(x) = \begin{cases} x^2 + 1 & \text{if } x < 1 \\ (x - 2)^2 & \text{if } x \geq 1 \end{cases}$

$\lim\limits_{x \to 1^-} f(x) = \lim\limits_{x \to 1^-} (x^2 + 1) = 1^2 + 1 = 2, \quad \lim\limits_{x \to 1^+} f(x) = \lim\limits_{x \to 1^+} (x - 2)^2 = (-1)^2 = 1$

(b) Since the right-hand and left-hand limits of f at $x = 1$ are not equal, $\lim\limits_{x \to 1} f(x)$ does not exist.

(c)

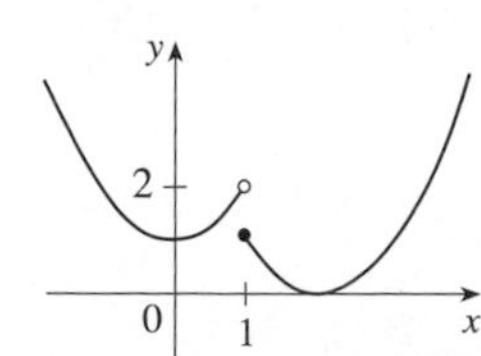

51. For the $\lim\limits_{t \to 2} B(t)$ to exist, the one-sided limits at $t = 2$ must be equal. $\lim\limits_{t \to 2^-} B(t) = \lim\limits_{t \to 2^-} \left(4 - \frac{1}{2}t\right) = 4 - 1 = 3$ and

$\lim\limits_{t \to 2^+} B(t) = \lim\limits_{t \to 2^+} \sqrt{t + c} = \sqrt{2 + c}$. Now $3 = \sqrt{2 + c} \;\Rightarrow\; 9 = 2 + c \;\Leftrightarrow\; c = 7$.

52. (a) (i) $\lim\limits_{x \to 1^-} g(x) = \lim\limits_{x \to 1^-} x = 1$

(ii) $\lim\limits_{x \to 1^+} g(x) = \lim\limits_{x \to 1^+} (2 - x^2) = 2 - 1^2 = 1$. Since $\lim\limits_{x \to 1^-} g(x) = 1$ and $\lim\limits_{x \to 1^+} g(x) = 1$, we have $\lim\limits_{x \to 1} g(x) = 1$.

Note that the fact $g(1) = 3$ does not affect the value of the limit.

(iii) When $x = 1$, $g(x) = 3$, so $g(1) = 3$.

(iv) $\lim\limits_{x \to 2^-} g(x) = \lim\limits_{x \to 2^-} (2 - x^2) = 2 - 2^2 = 2 - 4 = -2$

(v) $\lim\limits_{x \to 2^+} g(x) = \lim\limits_{x \to 2^+} (x - 3) = 2 - 3 = -1$

(vi) $\lim\limits_{x \to 2} g(x)$ does not exist since $\lim\limits_{x \to 2^-} g(x) \neq \lim\limits_{x \to 2^+} g(x)$.

(b)

$$g(x) = \begin{cases} x & \text{if } x < 1 \\ 3 & \text{if } x = 1 \\ 2 - x^2 & \text{if } 1 < x \le 2 \\ x - 3 & \text{if } x > 2 \end{cases}$$

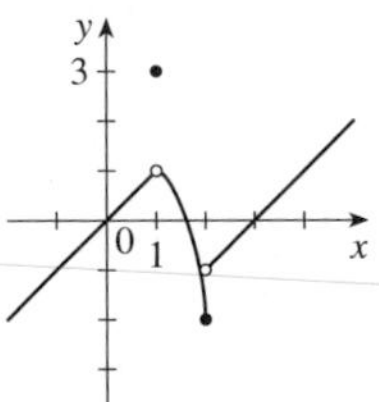

53. (a) (i) $[\![x]\!] = -2$ for $-2 \le x < -1$, so $\lim\limits_{x \to -2^+} [\![x]\!] = \lim\limits_{x \to -2^+} (-2) = -2$

(ii) $[\![x]\!] = -3$ for $-3 \le x < -2$, so $\lim\limits_{x \to -2^-} [\![x]\!] = \lim\limits_{x \to -2^-} (-3) = -3$.

The right and left limits are different, so $\lim\limits_{x \to -2} [\![x]\!]$ does not exist.

(iii) $[\![x]\!] = -3$ for $-3 \le x < -2$, so $\lim\limits_{x \to -2.4} [\![x]\!] = \lim\limits_{x \to -2.4} (-3) = -3$.

(b) (i) $[\![x]\!] = n - 1$ for $n - 1 \le x < n$, so $\lim\limits_{x \to n^-} [\![x]\!] = \lim\limits_{x \to n^-} (n - 1) = n - 1$.

(ii) $[\![x]\!] = n$ for $n \le x < n + 1$, so $\lim\limits_{x \to n^+} [\![x]\!] = \lim\limits_{x \to n^+} n = n$.

(c) $\lim\limits_{x \to a} [\![x]\!]$ exists $\;\Leftrightarrow\;$ a is not an integer.

54. (a) See the graph of $y = \cos x$.

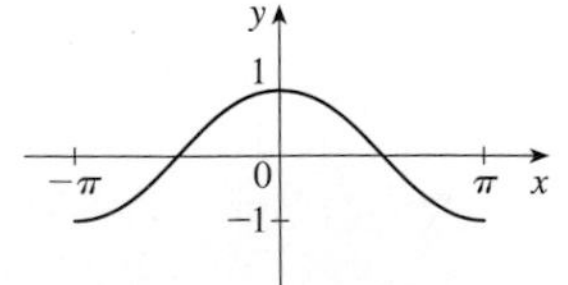

Since $-1 \le \cos x < 0$ on $[-\pi, -\pi/2)$, we have $y = f(x) = [\![\cos x]\!] = -1$ on $[-\pi, -\pi/2)$.

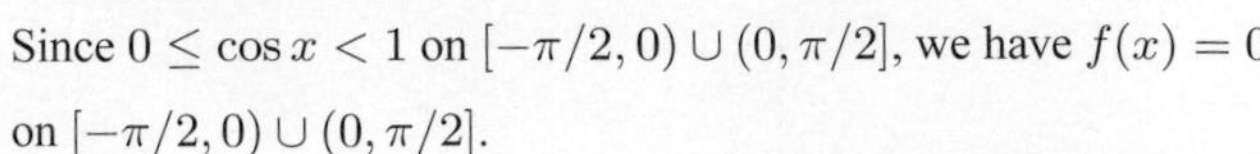

Since $0 \le \cos x < 1$ on $[-\pi/2, 0) \cup (0, \pi/2]$, we have $f(x) = 0$ on $[-\pi/2, 0) \cup (0, \pi/2]$.

Since $-1 \le \cos x < 0$ on $(\pi/2, \pi]$, we have $f(x) = -1$ on $(\pi/2, \pi]$.

Note that $f(0) = 1$.

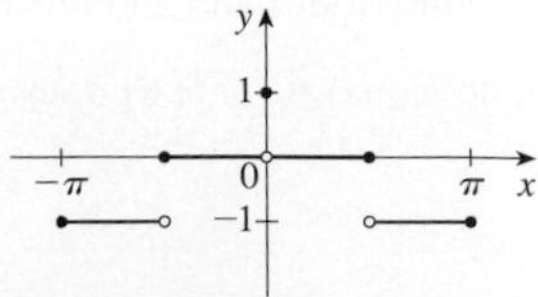

(b) (i) $\lim_{x\to 0^-} f(x) = 0$ and $\lim_{x\to 0^+} f(x) = 0$, so $\lim_{x\to 0} f(x) = 0$.

(ii) As $x \to (\pi/2)^-$, $f(x) \to 0$, so $\lim_{x\to(\pi/2)^-} f(x) = 0$.

(iii) As $x \to (\pi/2)^+$, $f(x) \to -1$, so $\lim_{x\to(\pi/2)^+} f(x) = -1$.

(iv) Since the answers in parts (ii) and (iii) are not equal, $\lim_{x\to\pi/2} f(x)$ does not exist.

(c) $\lim_{x\to a} f(x)$ exists for all a in the open interval $(-\pi, \pi)$ except $a = -\pi/2$ and $a = \pi/2$.

55. The graph of $f(x) = [\![x]\!] + [\![-x]\!]$ is the same as the graph of $g(x) = -1$ with holes at each integer, since $f(a) = 0$ for any integer a. Thus, $\lim_{x\to 2^-} f(x) = -1$ and $\lim_{x\to 2^+} f(x) = -1$, so $\lim_{x\to 2} f(x) = -1$. However, $f(2) = [\![2]\!] + [\![-2]\!] = 2 + (-2) = 0$, so $\lim_{x\to 2} f(x) \neq f(2)$.

56. $\lim_{v\to c^-} \left(L_0 \sqrt{1 - \frac{v^2}{c^2}} \right) = L_0\sqrt{1-1} = 0$. As the velocity approaches the speed of light, the length approaches 0.

A left-hand limit is necessary since L is not defined for $v > c$.

57. Since $p(x)$ is a polynomial, $p(x) = a_0 + a_1x + a_2x^2 + \cdots + a_nx^n$. Thus, by the Limit Laws,

$$\begin{aligned}\lim_{x\to a} p(x) &= \lim_{x\to a} \left(a_0 + a_1x + a_2x^2 + \cdots + a_nx^n\right) = a_0 + a_1 \lim_{x\to a} x + a_2 \lim_{x\to a} x^2 + \cdots + a_n \lim_{x\to a} x^n \\ &= a_0 + a_1a + a_2a^2 + \cdots + a_na^n = p(a)\end{aligned}$$

Thus, for any polynomial p, $\lim_{x\to a} p(x) = p(a)$.

58. Let $r(x) = \frac{p(x)}{q(x)}$ where $p(x)$ and $q(x)$ are any polynomials, and suppose that $q(a) \neq 0$. Then

$$\lim_{x\to a} r(x) = \lim_{x\to a} \frac{p(x)}{q(x)} = \frac{\lim_{x\to a} p(x)}{\lim_{x\to a} q(x)} \quad \text{[Limit Law 5]} \quad = \frac{p(a)}{q(a)} \quad \text{[Exercise 57]} \quad = r(a).$$

59. $\lim_{x\to 1} [f(x) - 8] = \lim_{x\to 1} \left[\frac{f(x) - 8}{x - 1} \cdot (x - 1) \right] = \lim_{x\to 1} \frac{f(x) - 8}{x - 1} \cdot \lim_{x\to 1} (x - 1) = 10 \cdot 0 = 0$.

Thus, $\lim_{x\to 1} f(x) = \lim_{x\to 1} \{[f(x) - 8] + 8\} = \lim_{x\to 1} [f(x) - 8] + \lim_{x\to 1} 8 = 0 + 8 = 8$.

Note: The value of $\lim_{x\to 1} \frac{f(x) - 8}{x - 1}$ does not affect the answer since it's multiplied by 0. What's important is that $\lim_{x\to 1} \frac{f(x) - 8}{x - 1}$ exists.

60. (a) $\lim_{x\to 0} f(x) = \lim_{x\to 0} \left[\frac{f(x)}{x^2} \cdot x^2 \right] = \lim_{x\to 0} \frac{f(x)}{x^2} \cdot \lim_{x\to 0} x^2 = 5 \cdot 0 = 0$

(b) $\lim_{x\to 0} \frac{f(x)}{x} = \lim_{x\to 0} \left[\frac{f(x)}{x^2} \cdot x \right] = \lim_{x\to 0} \frac{f(x)}{x^2} \cdot \lim_{x\to 0} x = 5 \cdot 0 = 0$

61. Observe that $0 \le f(x) \le x^2$ for all x, and $\lim_{x\to 0} 0 = 0 = \lim_{x\to 0} x^2$. So, by the Squeeze Theorem, $\lim_{x\to 0} f(x) = 0$.

62. Let $f(x) = [\![x]\!]$ and $g(x) = -[\![x]\!]$. Then $\lim_{x\to 3} f(x)$ and $\lim_{x\to 3} g(x)$ do not exist [Example 10]

but $\lim_{x\to 3}[f(x)+g(x)] = \lim_{x\to 3}([\![x]\!] - [\![x]\!]) = \lim_{x\to 3} 0 = 0$.

63. Let $f(x) = H(x)$ and $g(x) = 1 - H(x)$, where H is the Heaviside function defined in Exercise 1.3.59.

Thus, either f or g is 0 for any value of x. Then $\lim_{x\to 0} f(x)$ and $\lim_{x\to 0} g(x)$ do not exist, but $\lim_{x\to 0}[f(x)g(x)] = \lim_{x\to 0} 0 = 0$.

64. $$\lim_{x\to 2}\frac{\sqrt{6-x}-2}{\sqrt{3-x}-1} = \lim_{x\to 2}\left(\frac{\sqrt{6-x}-2}{\sqrt{3-x}-1}\cdot\frac{\sqrt{6-x}+2}{\sqrt{6-x}+2}\cdot\frac{\sqrt{3-x}+1}{\sqrt{3-x}+1}\right)$$

$$= \lim_{x\to 2}\left[\frac{\left(\sqrt{6-x}\right)^2-2^2}{\left(\sqrt{3-x}\right)^2-1^2}\cdot\frac{\sqrt{3-x}+1}{\sqrt{6-x}+2}\right] = \lim_{x\to 2}\left(\frac{6-x-4}{3-x-1}\cdot\frac{\sqrt{3-x}+1}{\sqrt{6-x}+2}\right)$$

$$= \lim_{x\to 2}\frac{(2-x)\left(\sqrt{3-x}+1\right)}{(2-x)\left(\sqrt{6-x}+2\right)} = \lim_{x\to 2}\frac{\sqrt{3-x}+1}{\sqrt{6-x}+2} = \frac{1}{2}$$

65. Since the denominator approaches 0 as $x \to -2$, the limit will exist only if the numerator also approaches 0 as $x \to -2$. In order for this to happen, we need $\lim_{x\to -2}\left(3x^2+ax+a+3\right) = 0 \Leftrightarrow$

$3(-2)^2 + a(-2) + a + 3 = 0 \Leftrightarrow 12 - 2a + a + 3 = 0 \Leftrightarrow a = 15$. With $a = 15$, the limit becomes

$$\lim_{x\to -2}\frac{3x^2+15x+18}{x^2+x-2} = \lim_{x\to -2}\frac{3(x+2)(x+3)}{(x-1)(x+2)} = \lim_{x\to -2}\frac{3(x+3)}{x-1} = \frac{3(-2+3)}{-2-1} = \frac{3}{-3} = -1.$$

66. *Solution 1:* First, we find the coordinates of P and Q as functions of r. Then we can find the equation of the line determined by these two points, and thus find the x-intercept (the point R), and take the limit as $r \to 0$. The coordinates of P are $(0, r)$. The point Q is the point of intersection of the two circles $x^2 + y^2 = r^2$ and $(x-1)^2 + y^2 = 1$. Eliminating y from these equations, we get $r^2 - x^2 = 1 - (x-1)^2 \Leftrightarrow r^2 = 1 + 2x - 1 \Leftrightarrow x = \frac{1}{2}r^2$. Substituting back into the equation of the shrinking circle to find the y-coordinate, we get $\left(\frac{1}{2}r^2\right)^2 + y^2 = r^2 \Leftrightarrow y^2 = r^2\left(1 - \frac{1}{4}r^2\right) \Leftrightarrow y = r\sqrt{1-\frac{1}{4}r^2}$ (the positive y-value). So the coordinates of Q are $\left(\frac{1}{2}r^2, r\sqrt{1-\frac{1}{4}r^2}\right)$. The equation of the line joining P and Q is thus

$y - r = \dfrac{r\sqrt{1-\frac{1}{4}r^2} - r}{\frac{1}{2}r^2 - 0}(x - 0)$. We set $y = 0$ in order to find the x-intercept, and get

$$x = -r\frac{\frac{1}{2}r^2}{r\left(\sqrt{1-\frac{1}{4}r^2}-1\right)} = \frac{-\frac{1}{2}r^2\left(\sqrt{1-\frac{1}{4}r^2}+1\right)}{1-\frac{1}{4}r^2-1} = 2\left(\sqrt{1-\frac{1}{4}r^2}+1\right)$$

Now we take the limit as $r \to 0^+$: $\lim_{r\to 0^+} x = \lim_{r\to 0^+} 2\left(\sqrt{1-\frac{1}{4}r^2}+1\right) = \lim_{r\to 0^+} 2\left(\sqrt{1}+1\right) = 4$.

So the limiting position of R is the point $(4, 0)$.

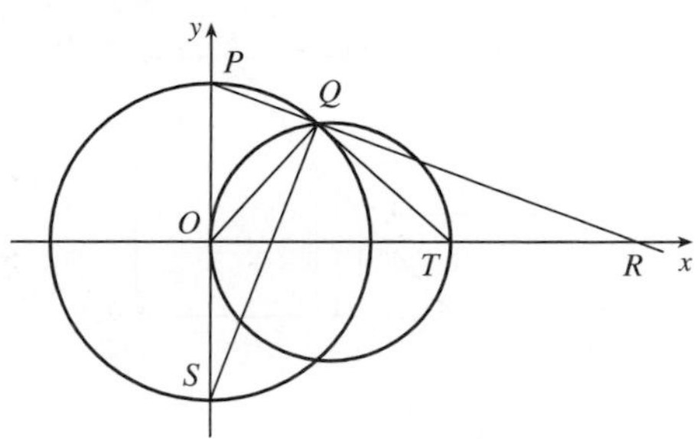

Solution 2: We add a few lines to the diagram, as shown. Note that $\angle PQS = 90°$ (subtended by diameter PS). So $\angle SQR = 90° = \angle OQT$ (subtended by diameter OT). It follows that $\angle OQS = \angle TQR$. Also $\angle PSQ = 90° - \angle SPQ = \angle ORP$. Since $\triangle QOS$ is isosceles, so is $\triangle QTR$, implying that $QT = TR$. As the circle C_2 shrinks, the point Q plainly approaches the origin, so the point R must approach a point twice as far from the origin as T, that is, the point $(4, 0)$, as above.

2.4 The Precise Definition of a Limit

1. If $|f(x) - 1| < 0.2$, then $-0.2 < f(x) - 1 < 0.2 \quad \Rightarrow \quad 0.8 < f(x) < 1.2$. From the graph, we see that the last inequality is true if $0.7 < x < 1.1$, so we can choose $\delta = \min\{1 - 0.7, 1.1 - 1\} = \min\{0.3, 0.1\} = 0.1$ (or any smaller positive number).

2. If $|f(x) - 2| < 0.5$, then $-0.5 < f(x) - 2 < 0.5 \quad \Rightarrow \quad 1.5 < f(x) < 2.5$. From the graph, we see that the last inequality is true if $2.6 < x < 3.8$, so we can take $\delta = \min\{3 - 2.6, 3.8 - 3\} = \min\{0.4, 0.8\} = 0.4$ (or any smaller positive number). Note that $x \neq 3$.

3. The leftmost question mark is the solution of $\sqrt{x} = 1.6$ and the rightmost, $\sqrt{x} = 2.4$. So the values are $1.6^2 = 2.56$ and $2.4^2 = 5.76$. On the left side, we need $|x - 4| < |2.56 - 4| = 1.44$. On the right side, we need $|x - 4| < |5.76 - 4| = 1.76$. To satisfy both conditions, we need the more restrictive condition to hold—namely, $|x - 4| < 1.44$. Thus, we can choose $\delta = 1.44$, or any smaller positive number.

4. The leftmost question mark is the positive solution of $x^2 = \frac{1}{2}$, that is, $x = \frac{1}{\sqrt{2}}$, and the rightmost question mark is the positive solution of $x^2 = \frac{3}{2}$, that is, $x = \sqrt{\frac{3}{2}}$. On the left side, we need $|x - 1| < \left|\frac{1}{\sqrt{2}} - 1\right| \approx 0.292$ (rounding down to be safe). On the right side, we need $|x - 1| < \left|\sqrt{\frac{3}{2}} - 1\right| \approx 0.224$. The more restrictive of these two conditions must apply, so we choose $\delta = 0.224$ (or any smaller positive number).

5.

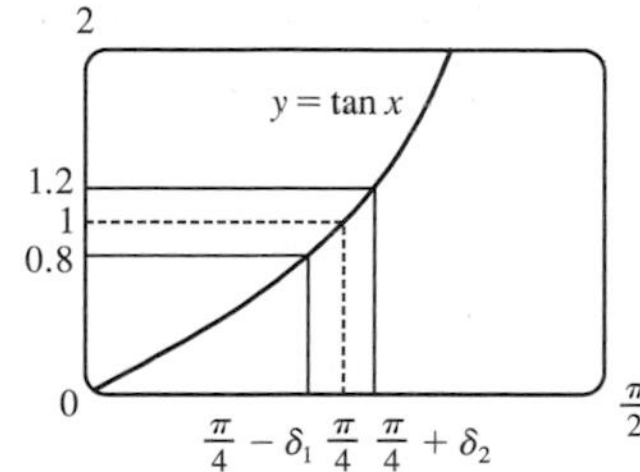

From the graph, we find that $y = \tan x = 0.8$ when $x \approx 0.675$, so $\frac{\pi}{4} - \delta_1 \approx 0.675 \quad \Rightarrow \quad \delta_1 \approx \frac{\pi}{4} - 0.675 \approx 0.1106$. Also, $y = \tan x = 1.2$ when $x \approx 0.876$, so $\frac{\pi}{4} + \delta_2 \approx 0.876 \quad \Rightarrow \quad \delta_2 = 0.876 - \frac{\pi}{4} \approx 0.0906$. Thus, we choose $\delta = 0.0906$ (or any smaller positive number) since this is the smaller of δ_1 and δ_2.

6.

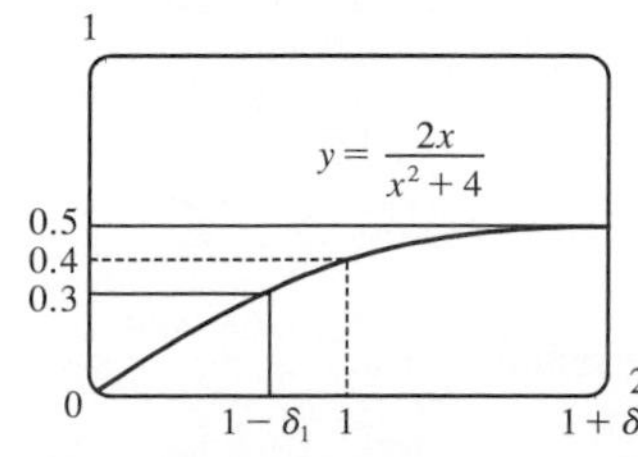

From the graph, we find that $y = 2x/(x^2 + 4) = 0.3$ when $x = \frac{2}{3}$, so $1 - \delta_1 = \frac{2}{3} \quad \Rightarrow \quad \delta_1 = \frac{1}{3}$. Also, $y = 2x/(x^2 + 4) = 0.5$ when $x = 2$, so $1 + \delta_2 = 2 \quad \Rightarrow \quad \delta_2 = 1$. Thus, we choose $\delta = \frac{1}{3}$ (or any smaller positive number) since this is the smaller of δ_1 and δ_2.

7.

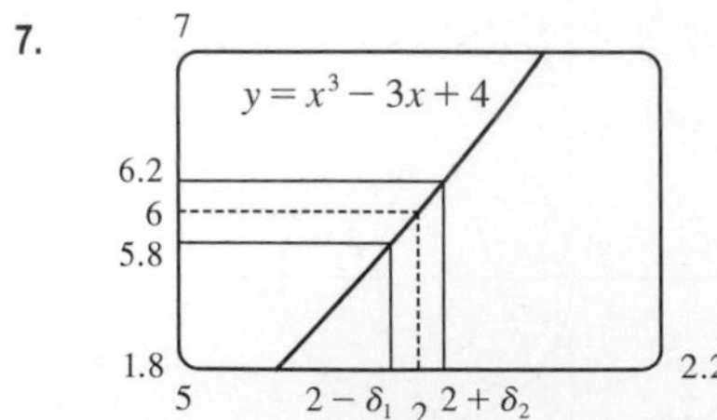

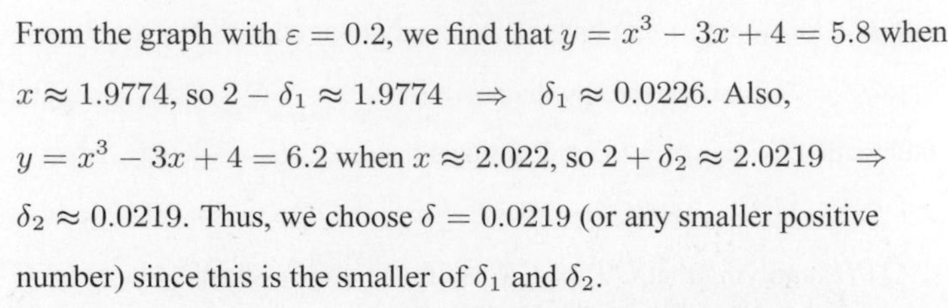

From the graph with $\varepsilon = 0.2$, we find that $y = x^3 - 3x + 4 = 5.8$ when $x \approx 1.9774$, so $2 - \delta_1 \approx 1.9774 \;\Rightarrow\; \delta_1 \approx 0.0226$. Also, $y = x^3 - 3x + 4 = 6.2$ when $x \approx 2.022$, so $2 + \delta_2 \approx 2.0219 \;\Rightarrow\; \delta_2 \approx 0.0219$. Thus, we choose $\delta = 0.0219$ (or any smaller positive number) since this is the smaller of δ_1 and δ_2.

For $\varepsilon = 0.1$, we get $\delta_1 \approx 0.0112$ and $\delta_2 \approx 0.0110$, so we choose

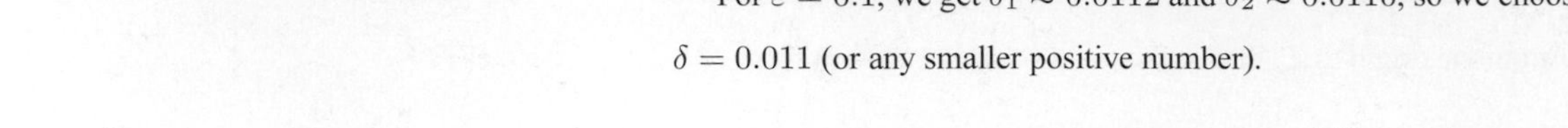

$\delta = 0.011$ (or any smaller positive number).

8.

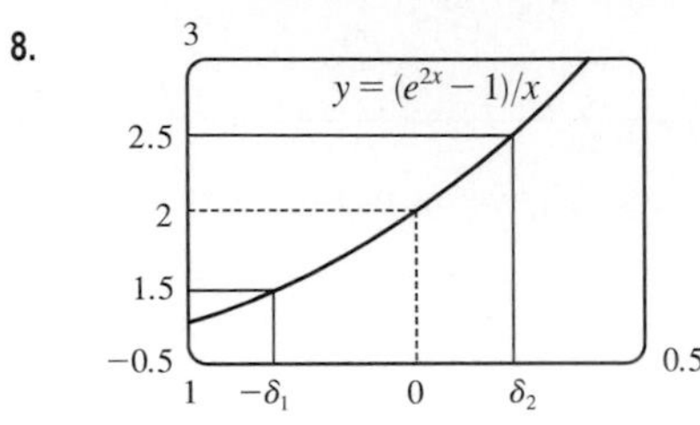

From the graph with $\varepsilon = 0.5$, we find that $y = (e^{2x} - 1)/x = 1.5$ when $x \approx -0.303$, so $\delta_1 \approx 0.303$. Also, $y = (e^{2x} - 1)/x = 2.5$ when $x \approx 0.215$, so $\delta_2 \approx 0.215$. Thus, we choose $\delta = 0.215$ (or any smaller positive number) since this is the smaller of δ_1 and δ_2.

For $\varepsilon = 0.1$, we get $\delta_1 \approx 0.052$ and $\delta_2 \approx 0.048$, so we choose $\delta = 0.048$ (or any smaller positive number).

9. (a)

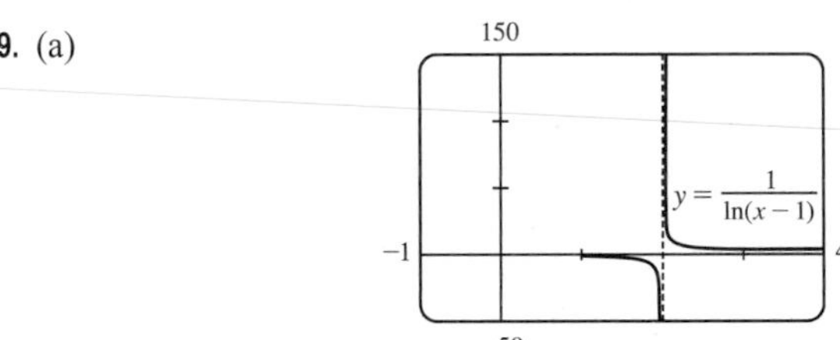

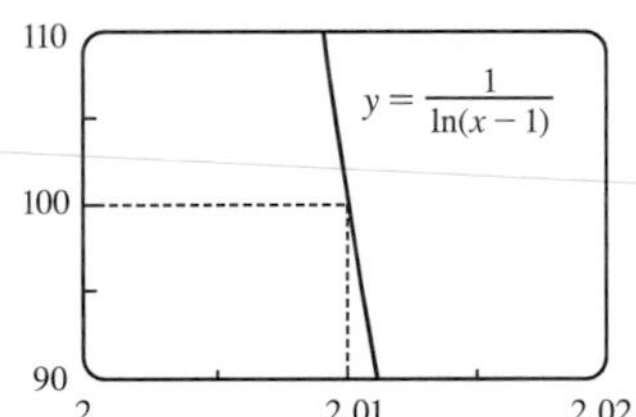

The first graph of $y = \dfrac{1}{\ln(x - 1)}$ shows a vertical asymptote at $x = 2$. The second graph shows that $y = 100$ when $x \approx 2.01$ (more accurately, 2.01005). Thus, we choose $\delta = 0.01$ (or any smaller positive number).

(b) From part (a), we see that as x gets closer to 2 from the right, y increases without bound. In symbols,

$$\lim_{x \to 2^+} \frac{1}{\ln(x - 1)} = \infty.$$

10. We graph $y = \csc^2 x$ and $y = 500$. The graphs intersect at $x \approx 3.186$, so we choose $\delta = 3.186 - \pi \approx 0.044$. Thus, if $0 < |x - \pi| < 0.044$, then $\csc^2 x > 500$. Similarly, for $M = 1000$, we get $\delta = 3.173 - \pi \approx 0.031$.

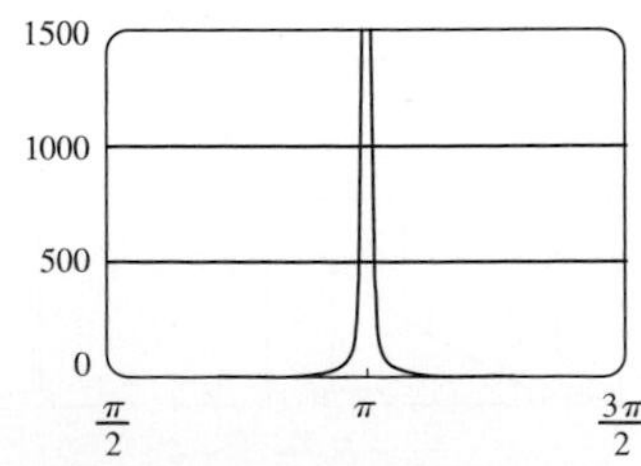

11. (a) $A = \pi r^2$ and $A = 1000 \text{ cm}^2 \;\Rightarrow\; \pi r^2 = 1000 \;\Rightarrow\; r^2 = \frac{1000}{\pi} \;\Rightarrow\; r = \sqrt{\frac{1000}{\pi}} \quad (r > 0) \quad \approx 17.8412$ cm.

(b) $|A - 1000| \le 5 \;\Rightarrow\; -5 \le \pi r^2 - 1000 \le 5 \;\Rightarrow\; 1000 - 5 \le \pi r^2 \le 1000 + 5 \;\Rightarrow$

$\sqrt{\frac{995}{\pi}} \le r \le \sqrt{\frac{1005}{\pi}} \;\Rightarrow\; 17.7966 \le r \le 17.8858$. $\sqrt{\frac{1000}{\pi}} - \sqrt{\frac{995}{\pi}} \approx 0.04466$ and $\sqrt{\frac{1005}{\pi}} - \sqrt{\frac{1000}{\pi}} \approx 0.04455$. So if the machinist gets the radius within 0.0445 cm of 17.8412, the area will be within 5 cm^2 of 1000.

(c) x is the radius, $f(x)$ is the area, a is the target radius given in part (a), L is the target area (1000), ε is the tolerance in the area (5), and δ is the tolerance in the radius given in part (b).

12. (a) $T = 0.1w^2 + 2.155w + 20$ and $T = 200 \;\Rightarrow$

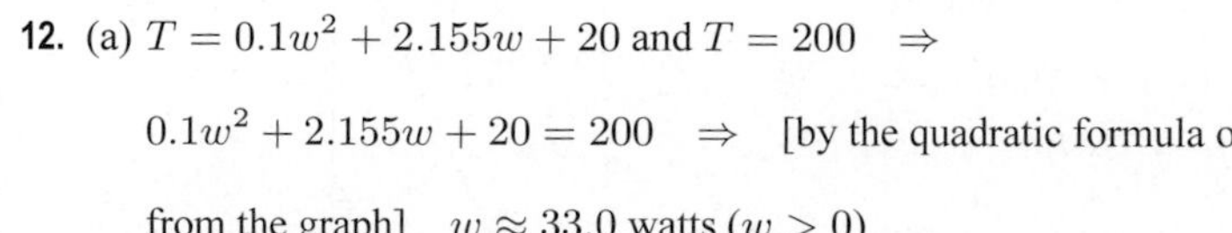

$0.1w^2 + 2.155w + 20 = 200 \;\Rightarrow$ [by the quadratic formula or from the graph] $w \approx 33.0$ watts $(w > 0)$

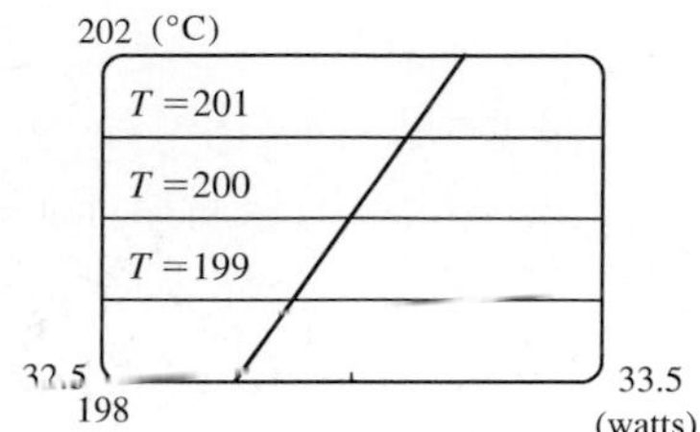

(b) From the graph, $199 \leq T \leq 201 \;\Rightarrow\; 32.89 < w < 33.11$.

(c) x is the input power, $f(x)$ is the temperature, a is the target input power given in part (a), L is the target temperature (200), ε is the tolerance in the temperature (1), and δ is the tolerance in the power input in watts indicated in part (b) (0.11 watts).

13. (a) $|4x - 8| = 4\,|x - 2| < 0.1 \;\Leftrightarrow\; |x - 2| < \dfrac{0.1}{4}$, so $\delta = \dfrac{0.1}{4} = 0.025$.

(b) $|4x - 8| = 4\,|x - 2| < 0.01 \;\Leftrightarrow\; |x - 2| < \dfrac{0.01}{4}$, so $\delta = \dfrac{0.01}{4} = 0.0025$.

14. $|(5x - 7) - 3| = |5x - 10| = |5(x - 2)| = 5\,|x - 2|$. We must have $|f(x) - L| < \varepsilon$, so $5\,|x - 2| < \varepsilon \;\Leftrightarrow\; |x - 2| < \varepsilon/5$. Thus, choose $\delta = \varepsilon/5$. For $\varepsilon = 0.1$, $\delta = 0.02$; for $\varepsilon = 0.05$, $\delta = 0.01$; for $\varepsilon = 0.01$, $\delta = 0.002$.

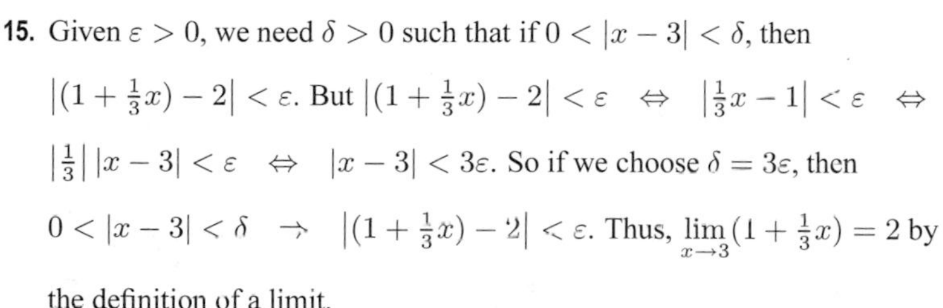

15. Given $\varepsilon > 0$, we need $\delta > 0$ such that if $0 < |x - 3| < \delta$, then $\left|\left(1 + \frac{1}{3}x\right) - 2\right| < \varepsilon$. But $\left|\left(1 + \frac{1}{3}x\right) - 2\right| < \varepsilon \;\Leftrightarrow\; \left|\frac{1}{3}x - 1\right| < \varepsilon \;\Leftrightarrow\; \left|\frac{1}{3}\right|\,|x - 3| < \varepsilon \;\Leftrightarrow\; |x - 3| < 3\varepsilon$. So if we choose $\delta = 3\varepsilon$, then $0 < |x - 3| < \delta \;\Rightarrow\; \left|\left(1 + \frac{1}{3}x\right) - 2\right| < \varepsilon$. Thus, $\lim\limits_{x \to 3}\left(1 + \frac{1}{3}x\right) = 2$ by the definition of a limit.

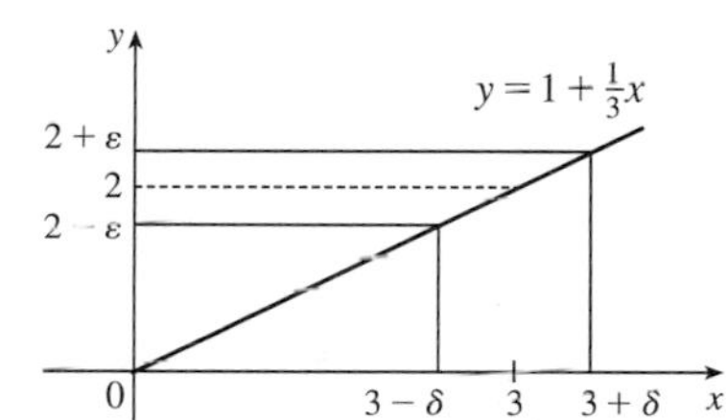

16. Given $\varepsilon > 0$, we need $\delta > 0$ such that if $0 < |x - 4| < \delta$, then $|(2x - 5) - 3| < \varepsilon$. But $|(2x - 5) - 3| < \varepsilon \;\Leftrightarrow\; |2x - 8| < \varepsilon \;\Leftrightarrow\; |2|\,|x - 4| < \varepsilon \;\Leftrightarrow\; |x - 4| < \varepsilon/2$. So if we choose $\delta = \varepsilon/2$, then $0 < |x - 4| < \delta \;\Rightarrow\; |(2x - 5) - 3| < \varepsilon$. Thus, $\lim\limits_{x \to 4}(2x - 5) = 3$ by the definition of a limit.

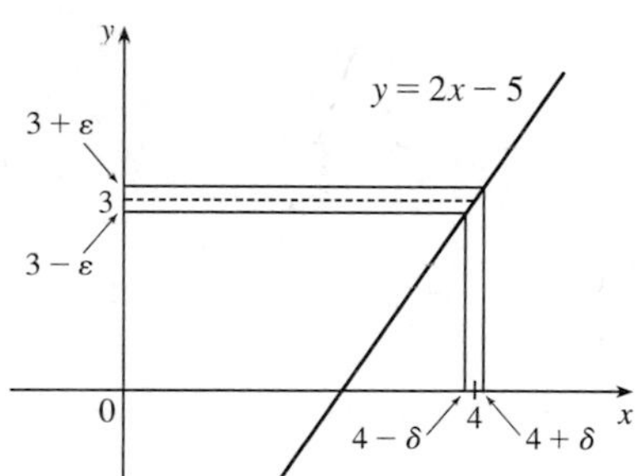

17. Given $\varepsilon > 0$, we need $\delta > 0$ such that if $0 < |x - (-3)| < \delta$, then $|(1 - 4x) - 13| < \varepsilon$. But $|(1 - 4x) - 13| < \varepsilon \;\Leftrightarrow\; |-4x - 12| < \varepsilon \;\Leftrightarrow\; |-4|\,|x + 3| < \varepsilon \;\Leftrightarrow\; |x - (-3)| < \varepsilon/4$. So if we choose $\delta = \varepsilon/4$, then $0 < |x - (-3)| < \delta \;\Rightarrow\; |(1 - 4x) - 13| < \varepsilon$. Thus, $\lim\limits_{x \to -3}(1 - 4x) = 13$ by the definition of a limit.

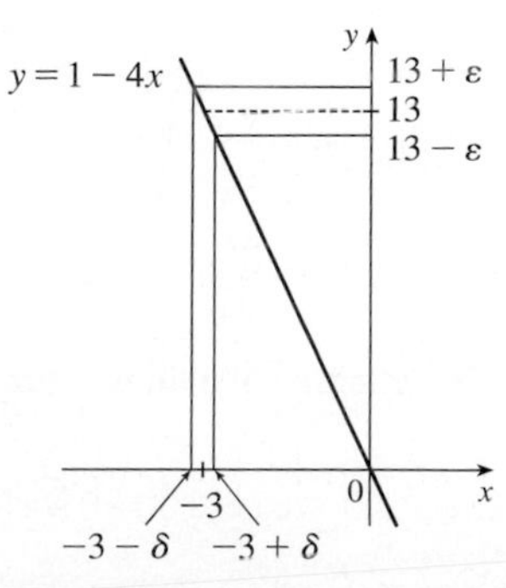

18. Given $\varepsilon > 0$, we need $\delta > 0$ such that if $0 < |x - (-2)| < \delta$, then $|(3x + 5) - (-1)| < \varepsilon$. But $|(3x + 5) - (-1)| < \varepsilon \;\Leftrightarrow\; |3x + 6| < \varepsilon \;\Leftrightarrow\; |3|\,|x + 2| < \varepsilon \;\Leftrightarrow\; |x + 2| < \varepsilon/3$. So if we choose $\delta = \varepsilon/3$, then $0 < |x + 2| < \delta \;\Rightarrow\; |(3x + 5) - (-1)| < \varepsilon$. Thus, $\lim\limits_{x \to -2}(3x + 5) = -1$ by the definition of a limit.

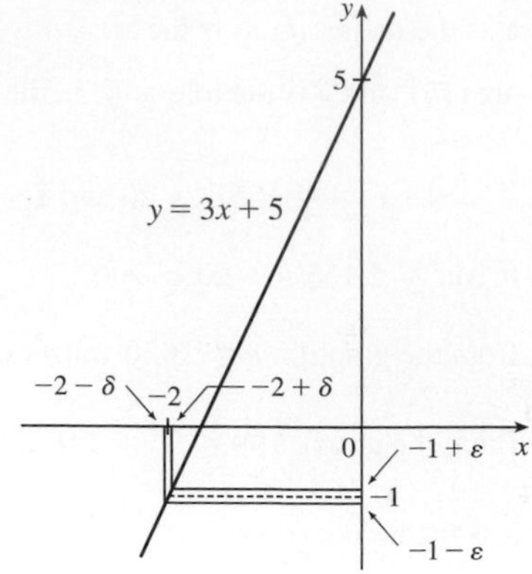

19. Given $\varepsilon > 0$, we need $\delta > 0$ such that if $0 < |x - 1| < \delta$, then $\left|\dfrac{2 + 4x}{3} - 2\right| < \varepsilon$. But $\left|\dfrac{2 + 4x}{3} - 2\right| < \varepsilon \;\Leftrightarrow$ $\left|\dfrac{4x - 4}{3}\right| < \varepsilon \;\Leftrightarrow\; \left|\frac{4}{3}\right|\,|x - 1| < \varepsilon \;\Leftrightarrow\; |x - 1| < \frac{3}{4}\varepsilon$. So if we choose $\delta = \frac{3}{4}\varepsilon$, then $0 < |x - 1| < \delta \;\Rightarrow$ $\left|\dfrac{2 + 4x}{3} - 2\right| < \varepsilon$. Thus, $\lim\limits_{x \to 1} \dfrac{2 + 4x}{3} = 2$ by the definition of a limit.

20. Given $\varepsilon > 0$, we need $\delta > 0$ such that if $0 < |x - 10| < \delta$, then $\left|3 - \frac{4}{5}x - (-5)\right| < \varepsilon$. But $\left|3 - \frac{4}{5}x - (-5)\right| < \varepsilon \;\Leftrightarrow$ $\left|8 - \frac{4}{5}x\right| < \varepsilon \;\Leftrightarrow\; \left|-\frac{4}{5}\right|\,|x - 10| < \varepsilon \;\Leftrightarrow\; |x - 10| < \frac{5}{4}\varepsilon$. So if we choose $\delta = \frac{5}{4}\varepsilon$, then $0 < |x - 10| < \delta \;\Rightarrow$ $\left|3 - \frac{4}{5}x - (-5)\right| < \varepsilon$. Thus, $\lim\limits_{x \to 10}\left(3 - \frac{4}{5}x\right) = -5$ by the definition of a limit.

21. Given $\varepsilon > 0$, we need $\delta > 0$ such that if $0 < |x - 4| < \delta$, then $\left|\dfrac{x^2 - 2x - 8}{x - 4} - 6\right| < \varepsilon \;\Leftrightarrow$ $\left|\dfrac{(x - 4)(x + 2)}{x - 4} - 6\right| < \varepsilon \;\Leftrightarrow\; |x + 2 - 6| < \varepsilon \;\; [x \neq 4] \;\Leftrightarrow\; |x - 4| < \varepsilon$. So choose $\delta = \varepsilon$. Then $0 < |x - 4| < \delta \;\Rightarrow\; |x - 4| < \varepsilon \;\Rightarrow\; |x + 2 - 6| < \varepsilon \;\Rightarrow\; \left|\dfrac{(x - 4)(x + 2)}{x - 4} - 6\right| < \varepsilon \;\; [x \neq 4] \;\Rightarrow$ $\left|\dfrac{x^2 - 2x - 8}{x - 4} - 6\right| < \varepsilon$. By the definition of a limit, $\lim\limits_{x \to 4} \dfrac{x^2 - 2x - 8}{x - 4} = 6$.

22. Given $\varepsilon > 0$, we need $\delta > 0$ such that if $0 < |x + 1.5| < \delta$, then $\left|\dfrac{9 - 4x^2}{3 + 2x} - 6\right| < \varepsilon \;\Leftrightarrow$ $\left|\dfrac{(3 + 2x)(3 - 2x)}{3 + 2x} - 6\right| < \varepsilon \;\Leftrightarrow\; |3 - 2x - 6| < \varepsilon \;\; [x \neq -1.5] \;\Leftrightarrow\; |-2x - 3| < \varepsilon \;\Leftrightarrow\; |-2|\,|x + 1.5| < \varepsilon \;\Leftrightarrow$ $|x + 1.5| < \varepsilon/2$. So choose $\delta = \varepsilon/2$. Then $0 < |x + 1.5| < \delta \;\Rightarrow\; |x + 1.5| < \varepsilon/2 \;\Rightarrow\; |-2|\,|x + 1.5| < \varepsilon \;\Rightarrow$ $|-2x - 3| < \varepsilon \;\Rightarrow\; |3 - 2x - 6| < \varepsilon \;\Rightarrow\; \left|\dfrac{(3 + 2x)(3 - 2x)}{3 + 2x} - 6\right| < \varepsilon \;\; [x \neq -1.5] \;\Rightarrow\; \left|\dfrac{9 - 4x^2}{3 + 2x} - 6\right| < \varepsilon$.

By the definition of a limit, $\lim\limits_{x \to -1.5} \dfrac{9 - 4x^2}{3 + 2x} = 6$.

23. Given $\varepsilon > 0$, we need $\delta > 0$ such that if $0 < |x - a| < \delta$, then $|x - a| < \varepsilon$. So $\delta = \varepsilon$ will work.

24. Given $\varepsilon > 0$, we need $\delta > 0$ such that if $0 < |x - a| < \delta$, then $|c - c| < \varepsilon$. But $|c - c| = 0$, so this will be true no matter what δ we pick.

25. Given $\varepsilon > 0$, we need $\delta > 0$ such that if $0 < |x - 0| < \delta$, then $|x^2 - 0| < \varepsilon \Leftrightarrow x^2 < \varepsilon \Leftrightarrow |x| < \sqrt{\varepsilon}$. Take $\delta = \sqrt{\varepsilon}$. Then $0 < |x - 0| < \delta \Rightarrow |x^2 - 0| < \varepsilon$. Thus, $\lim_{x \to 0} x^2 = 0$ by the definition of a limit.

26. Given $\varepsilon > 0$, we need $\delta > 0$ such that if $0 < |x - 0| < \delta$, then $|x^3 - 0| < \varepsilon \Leftrightarrow |x|^3 < \varepsilon \Leftrightarrow |x| < \sqrt[3]{\varepsilon}$. Take $\delta = \sqrt[3]{\varepsilon}$. Then $0 < |x - 0| < \delta \Rightarrow |x^3 - 0| < \delta^3 = \varepsilon$. Thus, $\lim_{x \to 0} x^3 = 0$ by the definition of a limit.

27. Given $\varepsilon > 0$, we need $\delta > 0$ such that if $0 < |x - 0| < \delta$, then $\big||x| - 0\big| < \varepsilon$. But $\big||x|\big| = |x|$. So this is true if we pick $\delta = \varepsilon$. Thus, $\lim_{x \to 0} |x| = 0$ by the definition of a limit.

28. Given $\varepsilon > 0$, we need $\delta > 0$ such that if $0 < x - (-6) < \delta$, then $\left|\sqrt[8]{6 + x} - 0\right| < \varepsilon$. But $\left|\sqrt[8]{6 + x} - 0\right| < \varepsilon \Leftrightarrow \sqrt[8]{6 + x} < \varepsilon \Leftrightarrow 6 + x < \varepsilon^8 \Leftrightarrow x - (-6) < \varepsilon^8$. So if we choose $\delta = \varepsilon^8$, then $0 < x - (-6) < \delta \Rightarrow \left|\sqrt[8]{6 + x} - 0\right| < \varepsilon$. Thus, $\lim_{x \to -6^+} \sqrt[8]{6 + x} = 0$ by the definition of a right-hand limit.

29. Given $\varepsilon > 0$, we need $\delta > 0$ such that if $0 < |x - 2| < \delta$, then $\left|(x^2 - 4x + 5) - 1\right| < \varepsilon \Leftrightarrow \left|x^2 - 4x + 4\right| < \varepsilon \Leftrightarrow \left|(x - 2)^2\right| < \varepsilon$. So take $\delta = \sqrt{\varepsilon}$. Then $0 < |x - 2| < \delta \Leftrightarrow |x - 2| < \sqrt{\varepsilon} \Leftrightarrow \left|(x - 2)^2\right| < \varepsilon$. Thus, $\lim_{x \to 2} (x^2 - 4x + 5) = 1$ by the definition of a limit.

30. Given $\varepsilon > 0$, we need $\delta > 0$ such that if $0 < |x - 2| < \delta$, then $\left|(x^2 + 2x - 7) - 1\right| < \varepsilon$. But $\left|(x^2 + 2x - 7) - 1\right| < \varepsilon \Leftrightarrow \left|x^2 + 2x - 8\right| < \varepsilon \Leftrightarrow |x + 4|\,|x - 2| < \varepsilon$. Thus our goal is to make $|x - 2|$ small enough so that its product with $|x + 4|$ is less than ε. Suppose we first require that $|x - 2| < 1$. Then $-1 < x - 2 < 1 \Rightarrow 1 < x < 3 \Rightarrow 5 < x + 4 < 7 \Rightarrow |x + 4| < 7$, and this gives us $7\,|x - 2| < \varepsilon \Rightarrow |x - 2| < \varepsilon/7$. Choose $\delta = \min\{1, \varepsilon/7\}$. Then if $0 < |x - 2| < \delta$, we have $|x - 2| < \varepsilon/7$ and $|x + 4| < 7$, so $\left|(x^2 + 2x - 7) - 1\right| = |(x + 4)(x - 2)| = |x + 4|\,|x - 2| < 7(\varepsilon/7) = \varepsilon$, as desired. Thus, $\lim_{x \to 2} (x^2 + 2x - 7) = 1$ by the definition of a limit.

31. Given $\varepsilon > 0$, we need $\delta > 0$ such that if $0 < |x - (-2)| < \delta$, then $\left|(x^2 - 1) - 3\right| < \varepsilon$ or upon simplifying we need $\left|x^2 - 4\right| < \varepsilon$ whenever $0 < |x + 2| < \delta$. Notice that if $|x + 2| < 1$, then $-1 < x + 2 < 1 \Rightarrow -5 < x - 2 < -3 \Rightarrow |x - 2| < 5$. So take $\delta = \min\{\varepsilon/5, 1\}$. Then $0 < |x + 2| < \delta \Rightarrow |x - 2| < 5$ and $|x + 2| < \varepsilon/5$, so $\left|(x^2 - 1) - 3\right| = |(x + 2)(x - 2)| = |x + 2|\,|x - 2| < (\varepsilon/5)(5) = \varepsilon$. Thus, by the definition of a limit, $\lim_{x \to -2} (x^2 - 1) = 3$.

32. Given $\varepsilon > 0$, we need $\delta > 0$ such that if $0 < |x - 2| < \delta$, then $\left|x^3 - 8\right| < \varepsilon$. Now $\left|x^3 - 8\right| = \left|(x - 2)(x^2 + 2x + 4)\right|$. If $|x - 2| < 1$, that is, $1 < x < 3$, then $x^2 + 2x + 4 < 3^2 + 2(3) + 4 = 19$ and so $\left|x^3 - 8\right| = |x - 2|\,(x^2 + 2x + 4) < 19\,|x - 2|$. So if we take $\delta = \min\left\{1, \frac{\varepsilon}{19}\right\}$, then $0 < |x - 2| < \delta \Rightarrow \left|x^3 - 8\right| = |x - 2|\,(x^2 + 2x + 4) < \frac{\varepsilon}{19} \cdot 19 = \varepsilon$. Thus, by the definition of a limit, $\lim_{x \to 2} x^3 = 8$.

33. Given $\varepsilon > 0$, we let $\delta = \min\left\{2, \frac{\varepsilon}{8}\right\}$. If $0 < |x - 3| < \delta$, then $|x - 3| < 2 \Rightarrow -2 < x - 3 < 2 \Rightarrow$ $4 < x + 3 < 8 \Rightarrow |x + 3| < 8$. Also $|x - 3| < \frac{\varepsilon}{8}$, so $\left|x^2 - 9\right| = |x + 3|\,|x - 3| < 8 \cdot \frac{\varepsilon}{8} = \varepsilon$. Thus, $\lim\limits_{x \to 3} x^2 = 9$.

34. From the figure, our choices for δ are $\delta_1 = 3 - \sqrt{9 - \varepsilon}$ and $\delta_2 = \sqrt{9 + \varepsilon} - 3$. The *largest* possible choice for δ is the minimum value of $\{\delta_1, \delta_2\}$; that is, $\delta = \min\{\delta_1, \delta_2\} = \delta_2 = \sqrt{9 + \varepsilon} - 3$.

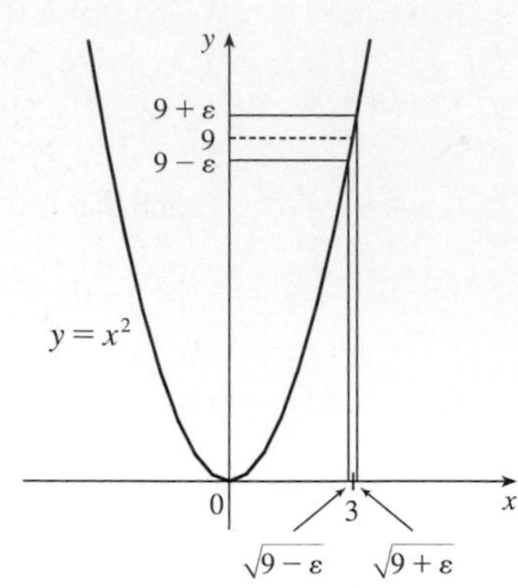

35. (a) The points of intersection in the graph are $(x_1, 2.6)$ and $(x_2, 3.4)$ with $x_1 \approx 0.891$ and $x_2 \approx 1.093$. Thus, we can take δ to be the smaller of $1 - x_1$ and $x_2 - 1$. So $\delta = x_2 - 1 \approx 0.093$.

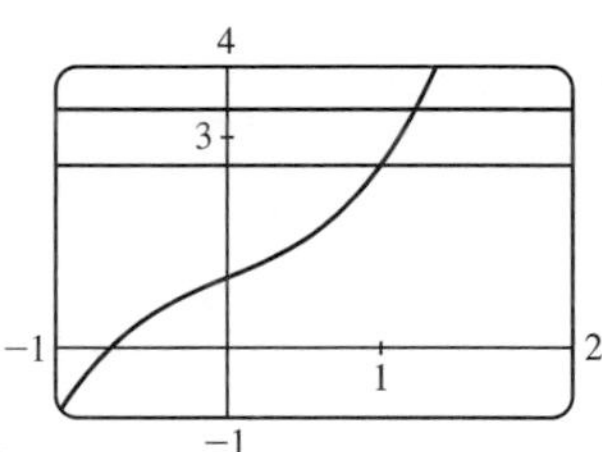

(b) Solving $x^3 + x + 1 = 3 + \varepsilon$ gives us two nonreal complex roots and one real root, which is

$$x(\varepsilon) = \frac{\left(216 + 108\varepsilon + 12\sqrt{336 + 324\varepsilon + 81\varepsilon^2}\,\right)^{2/3} - 12}{6\left(216 + 108\varepsilon + 12\sqrt{336 + 324\varepsilon + 81\varepsilon^2}\,\right)^{1/3}}.\ \text{Thus, } \delta = x(\varepsilon) - 1.$$

(c) If $\varepsilon = 0.4$, then $x(\varepsilon) \approx 1.093\,272\,342$ and $\delta = x(\varepsilon) - 1 \approx 0.093$, which agrees with our answer in part (a).

36. *1. Guessing a value for* δ Let $\varepsilon > 0$ be given. We have to find a number $\delta > 0$ such that $\left|\frac{1}{x} - \frac{1}{2}\right| < \varepsilon$ whenever $0 < |x - 2| < \delta$. But $\left|\frac{1}{x} - \frac{1}{2}\right| = \left|\frac{2 - x}{2x}\right| = \frac{|x - 2|}{|2x|} < \varepsilon$. We find a positive constant C such that $\frac{1}{|2x|} < C \Rightarrow$ $\frac{|x - 2|}{|2x|} < C\,|x - 2|$ and we can make $C\,|x - 2| < \varepsilon$ by taking $|x - 2| < \frac{\varepsilon}{C} = \delta$. We restrict x to lie in the interval $|x - 2| < 1 \Rightarrow 1 < x < 3$ so $1 > \frac{1}{x} > \frac{1}{3} \Rightarrow \frac{1}{6} < \frac{1}{2x} < \frac{1}{2} \Rightarrow \frac{1}{|2x|} < \frac{1}{2}$. So $C = \frac{1}{2}$ is suitable. Thus, we should choose $\delta = \min\{1, 2\varepsilon\}$.

2. Showing that δ *works* Given $\varepsilon > 0$ we let $\delta = \min\{1, 2\varepsilon\}$. If $0 < |x - 2| < \delta$, then $|x - 2| < 1 \Rightarrow 1 < x < 3 \Rightarrow$ $\frac{1}{|2x|} < \frac{1}{2}$ (as in part 1). Also $|x - 2| < 2\varepsilon$, so $\left|\frac{1}{x} - \frac{1}{2}\right| = \frac{|x - 2|}{|2x|} < \frac{1}{2} \cdot 2\varepsilon = \varepsilon$. This shows that $\lim\limits_{x \to 2}(1/x) = \frac{1}{2}$.

37. *1. Guessing a value for* δ Given $\varepsilon > 0$, we must find $\delta > 0$ such that $\left|\sqrt{x} - \sqrt{a}\right| < \varepsilon$ whenever $0 < |x - a| < \delta$. But $\left|\sqrt{x} - \sqrt{a}\right| = \frac{|x - a|}{\sqrt{x} + \sqrt{a}} < \varepsilon$ (from the hint). Now if we can find a positive constant C such that $\sqrt{x} + \sqrt{a} > C$ then

$\dfrac{|x-a|}{\sqrt{x}+\sqrt{a}} < \dfrac{|x-a|}{C} < \varepsilon$, and we take $|x-a| < C\varepsilon$. We can find this number by restricting x to lie in some interval centered at a. If $|x-a| < \frac{1}{2}a$, then $-\frac{1}{2}a < x-a < \frac{1}{2}a \Rightarrow \frac{1}{2}a < x < \frac{3}{2}a \Rightarrow \sqrt{x}+\sqrt{a} > \sqrt{\frac{1}{2}a}+\sqrt{a}$, and so $C = \sqrt{\frac{1}{2}a}+\sqrt{a}$ is a suitable choice for the constant. So $|x-a| < \left(\sqrt{\frac{1}{2}a}+\sqrt{a}\right)\varepsilon$. This suggests that we let $\delta = \min\left\{\frac{1}{2}a, \left(\sqrt{\frac{1}{2}a}+\sqrt{a}\right)\varepsilon\right\}$.

2. *Showing that δ works* Given $\varepsilon > 0$, we let $\delta = \min\left\{\frac{1}{2}a, \left(\sqrt{\frac{1}{2}a}+\sqrt{a}\right)\varepsilon\right\}$. If $0 < |x-a| < \delta$, then $|x-a| < \frac{1}{2}a \Rightarrow \sqrt{x}+\sqrt{a} > \sqrt{\frac{1}{2}a}+\sqrt{a}$ (as in part 1). Also $|x-a| < \left(\sqrt{\frac{1}{2}a}+\sqrt{a}\right)\varepsilon$, so $\left|\sqrt{x}-\sqrt{a}\right| = \dfrac{|x-a|}{\sqrt{x}+\sqrt{a}} < \dfrac{\left(\sqrt{a/2}+\sqrt{a}\right)\varepsilon}{\left(\sqrt{a/2}+\sqrt{a}\right)} = \varepsilon$. Therefore, $\lim\limits_{x\to a}\sqrt{x} = \sqrt{a}$ by the definition of a limit.

38. Suppose that $\lim\limits_{t\to 0} H(t) = L$. Given $\varepsilon = \frac{1}{2}$, there exists $\delta > 0$ such that $0 < |t| < \delta \Rightarrow |H(t) - L| < \frac{1}{2} \Leftrightarrow L - \frac{1}{2} < H(t) < L + \frac{1}{2}$. For $0 < t < \delta$, $H(t) = 1$, so $1 < L + \frac{1}{2} \Rightarrow L > \frac{1}{2}$. For $-\delta < t < 0$, $H(t) = 0$, so $L - \frac{1}{2} < 0 \Rightarrow L < \frac{1}{2}$. This contradicts $L > \frac{1}{2}$. Therefore, $\lim\limits_{t\to 0} H(t)$ does not exist.

39. Suppose that $\lim\limits_{x\to 0} f(x) = L$. Given $\varepsilon = \frac{1}{2}$, there exists $\delta > 0$ such that $0 < |x| < \delta \Rightarrow |f(x) - L| < \frac{1}{2}$. Take any rational number r with $0 < |r| < \delta$. Then $f(r) = 0$, so $|0 - L| < \frac{1}{2}$, so $L \le |L| < \frac{1}{2}$. Now take any irrational number s with $0 < |s| < \delta$. Then $f(s) = 1$, so $|1 - L| < \frac{1}{2}$. Hence, $1 - L < \frac{1}{2}$, so $L > \frac{1}{2}$. This contradicts $L < \frac{1}{2}$, so $\lim\limits_{x\to 0} f(x)$ does not exist.

40. First suppose that $\lim\limits_{x\to a} f(x) = L$. Then, given $\varepsilon > 0$ there exists $\delta > 0$ so that $0 < |x-a| < \delta \Rightarrow |f(x) - L| < \varepsilon$. Then $a - \delta < x < a \Rightarrow 0 < |x-a| < \delta$ so $|f(x) - L| < \varepsilon$. Thus, $\lim\limits_{x\to a^-} f(x) = L$. Also $a < x < a + \delta \Rightarrow 0 < |x-a| < \delta$ so $|f(x) - L| < \varepsilon$. Hence, $\lim\limits_{x\to a^+} f(x) = L$.

Now suppose $\lim\limits_{x\to a^-} f(x) = L = \lim\limits_{x\to a^+} f(x)$. Let $\varepsilon > 0$ be given. Since $\lim\limits_{x\to a^-} f(x) = L$, there exists $\delta_1 > 0$ so that $a - \delta_1 < x < a \Rightarrow |f(x) - L| < \varepsilon$. Since $\lim\limits_{x\to a^+} f(x) = L$, there exists $\delta_2 > 0$ so that $a < x < a + \delta_2 \Rightarrow |f(x) - L| < \varepsilon$. Let δ be the smaller of δ_1 and δ_2. Then $0 < |x-a| < \delta \Rightarrow a - \delta_1 < x < a$ or $a < x < a + \delta_2$ so $|f(x) - L| < \varepsilon$. Hence, $\lim\limits_{x\to a} f(x) = L$. So we have proved that $\lim\limits_{x\to a} f(x) = L \Leftrightarrow \lim\limits_{x\to a^-} f(x) = L = \lim\limits_{x\to a^+} f(x)$.

41. $\dfrac{1}{(x+3)^4} > 10{,}000 \Leftrightarrow (x+3)^4 < \dfrac{1}{10{,}000} \Leftrightarrow |x+3| < \dfrac{1}{\sqrt[4]{10{,}000}} \Leftrightarrow |x-(-3)| < \dfrac{1}{10}$

42. Given $M > 0$, we need $\delta > 0$ such that $0 < |x+3| < \delta \Rightarrow 1/(x+3)^4 > M$. Now $\dfrac{1}{(x+3)^4} > M \Leftrightarrow (x+3)^4 < \dfrac{1}{M} \Leftrightarrow |x+3| < \dfrac{1}{\sqrt[4]{M}}$. So take $\delta = \dfrac{1}{\sqrt[4]{M}}$. Then $0 < |x+3| < \delta = \dfrac{1}{\sqrt[4]{M}} \Rightarrow \dfrac{1}{(x+3)^4} > M$, so $\lim\limits_{x\to -3}\dfrac{1}{(x+3)^4} = \infty$.

43. Given $M < 0$ we need $\delta > 0$ so that $\ln x < M$ whenever $0 < x < \delta$; that is, $x = e^{\ln x} < e^M$ whenever $0 < x < \delta$. This suggests that we take $\delta = e^M$. If $0 < x < e^M$, then $\ln x < \ln e^M = M$. By the definition of a limit, $\lim\limits_{x \to 0^+} \ln x = -\infty$.

44. (a) Let M be given. Since $\lim\limits_{x \to a} f(x) = \infty$, there exists $\delta_1 > 0$ such that $0 < |x - a| < \delta_1 \Rightarrow f(x) > M + 1 - c$. Since $\lim\limits_{x \to a} g(x) = c$, there exists $\delta_2 > 0$ such that $0 < |x - a| < \delta_2 \Rightarrow |g(x) - c| < 1 \Rightarrow g(x) > c - 1$. Let δ be the smaller of δ_1 and δ_2. Then $0 < |x - a| < \delta \Rightarrow f(x) + g(x) > (M + 1 - c) + (c - 1) = M$. Thus, $\lim\limits_{x \to a} [f(x) + g(x)] = \infty$.

(b) Let $M > 0$ be given. Since $\lim\limits_{x \to a} g(x) = c > 0$, there exists $\delta_1 > 0$ such that $0 < |x - a| < \delta_1 \Rightarrow |g(x) - c| < c/2 \Rightarrow g(x) > c/2$. Since $\lim\limits_{x \to a} f(x) = \infty$, there exists $\delta_2 > 0$ such that $0 < |x - a| < \delta_2 \Rightarrow f(x) > 2M/c$. Let $\delta = \min\{\delta_1, \delta_2\}$. Then $0 < |x - a| < \delta \Rightarrow f(x)\,g(x) > \dfrac{2M}{c}\dfrac{c}{2} = M$, so $\lim\limits_{x \to a} f(x)\,g(x) = \infty$.

(c) Let $N < 0$ be given. Since $\lim\limits_{x \to a} g(x) = c < 0$, there exists $\delta_1 > 0$ such that $0 < |x - a| < \delta_1 \Rightarrow |g(x) - c| < -c/2 \Rightarrow g(x) < c/2$. Since $\lim\limits_{x \to a} f(x) = \infty$, there exists $\delta_2 > 0$ such that $0 < |x - a| < \delta_2 \Rightarrow f(x) > 2N/c$. (Note that $c < 0$ and $N < 0 \Rightarrow 2N/c > 0$.) Let $\delta = \min\{\delta_1, \delta_2\}$. Then $0 < |x - a| < \delta \Rightarrow f(x) > 2N/c \Rightarrow f(x)\,g(x) < \dfrac{2N}{c} \cdot \dfrac{c}{2} = N$, so $\lim\limits_{x \to a} f(x)\,g(x) = -\infty$.

2.5 Continuity

1. From Definition 1, $\lim\limits_{x \to 4} f(x) = f(4)$.

2. The graph of f has no hole, jump, or vertical asymptote.

3. (a) f is discontinuous at -4 since $f(-4)$ is not defined and at -2, 2, and 4 since the limit does not exist (the left and right limits are not the same).

(b) f is continuous from the left at -2 since $\lim\limits_{x \to -2^-} f(x) = f(-2)$. f is continuous from the right at 2 and 4 since $\lim\limits_{x \to 2^+} f(x) = f(2)$ and $\lim\limits_{x \to 4^+} f(x) = f(4)$. It is continuous from neither side at -4 since $f(-4)$ is undefined.

4. From the graph of g, we see that g is continuous on the intervals $[-3, -2)$, $(-2, -1)$, $(-1, 0]$, $(0, 1)$, and $(1, 3]$.

5. The graph of $y = f(x)$ must have a discontinuity at $x = 2$ and must show that $\lim\limits_{x \to 2^+} f(x) = f(2)$.

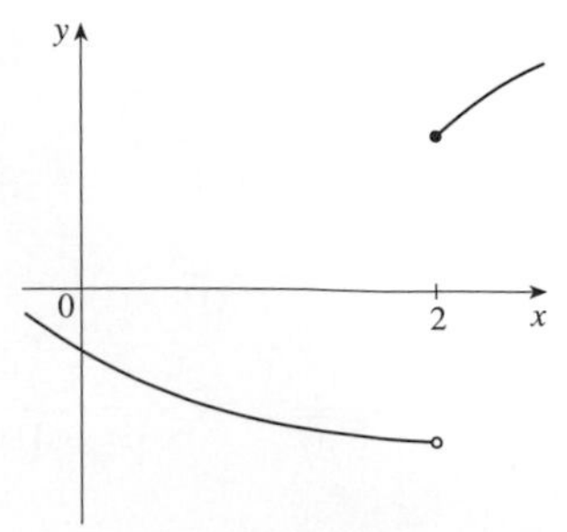

6. The graph of $y = f(x)$ must have discontinuities at $x = -1$ and $x = 4$. It must show that $\lim\limits_{x \to -1^-} f(x) = f(-1)$ and $\lim\limits_{x \to 4^+} f(x) = f(4)$.

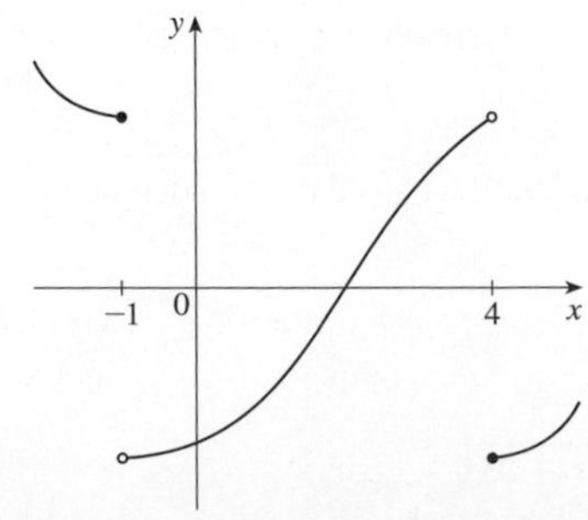

7. The graph of $y = f(x)$ must have a removable discontinuity (a hole) at $x = 3$ and a jump discontinuity at $x = 5$.

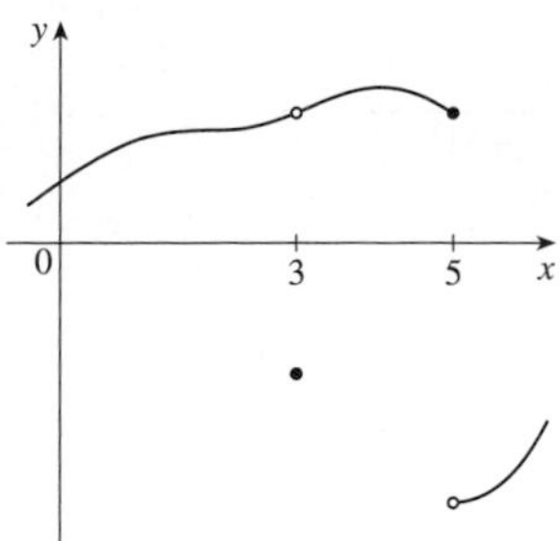

8. The graph of $y = f(x)$ must have a discontinuity at $x = -2$ with $\lim_{x \to -2^-} f(x) \neq f(-2)$ and $\lim_{x \to -2^+} f(x) \neq f(-2)$. It must also show that $\lim_{x \to 2^-} f(x) = f(2)$ and $\lim_{x \to 2^+} f(x) \neq f(2)$.

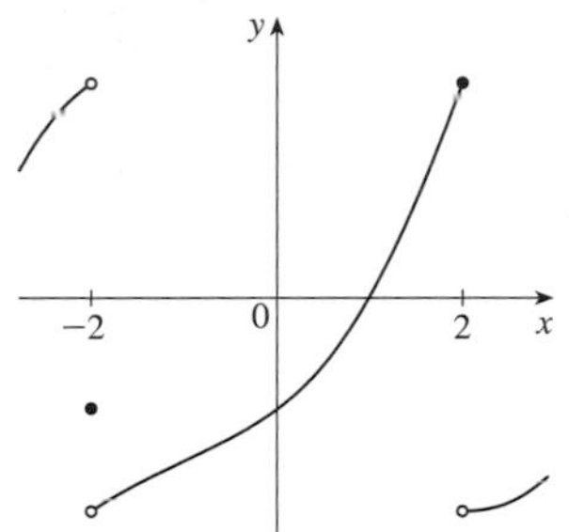

9. (a) The toll is \$7 between 7:00 AM and 10:00 AM and between 4:00 PM and 7:00 PM.

(b) The function T has jump discontinuities at $t = 7$, 10, 16, and 19. Their significance to someone who uses the road is that, because of the sudden jumps in the toll, they may want to avoid the higher rates between $t = 7$ and $t = 10$ and between $t = 16$ and $t = 19$ if feasible.

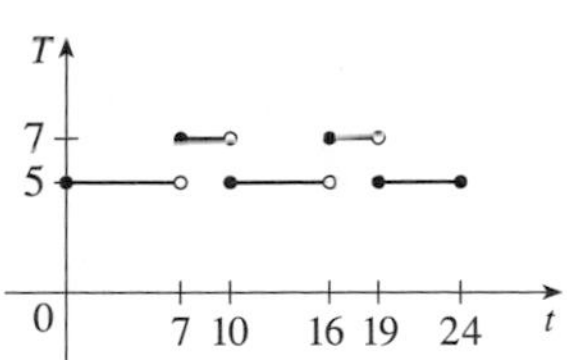

10. (a) Continuous; at the location in question, the temperature changes smoothly as time passes, without any instantaneous jumps from one temperature to another.

(b) Continuous; the temperature at a specific time changes smoothly as the distance due west from New York City increases, without any instantaneous jumps.

(c) Discontinuous; as the distance due west from New York City increases, the altitude above sea level may jump from one height to another without going through all of the intermediate values — at a cliff, for example.

(d) Discontinuous; as the distance traveled increases, the cost of the ride jumps in small increments.

(e) Discontinuous; when the lights are switched on (or off), the current suddenly changes between 0 and some nonzero value, without passing through all of the intermediate values. This is debatable, though, depending on your definition of current.

11. $\lim_{x \to -1} f(x) = \lim_{x \to -1} (x + 2x^3)^4 = \left(\lim_{x \to -1} x + 2 \lim_{x \to -1} x^3 \right)^4 = \left[-1 + 2(-1)^3\right]^4 = (-3)^4 = 81 = f(-1)$.

By the definition of continuity, f is continuous at $a = -1$.

12. $\lim_{t \to 2} g(t) = \lim_{t \to 2} \dfrac{t^2 + 5t}{2t + 1} = \dfrac{\lim_{t \to 2}(t^2 + 5t)}{\lim_{t \to 2}(2t + 1)} = \dfrac{\lim_{t \to 2} t^2 + 5 \lim_{t \to 2} t}{2 \lim_{t \to 2} t + \lim_{t \to 2} 1} = \dfrac{2^2 + 5(2)}{2(2) + 1} = \dfrac{14}{5} = g(2)$.

By the definition of continuity, g is continuous at $a = 2$.

13. $\lim\limits_{v\to 1} p(v) = \lim\limits_{v\to 1} 2\sqrt{3v^2+1} = 2\lim\limits_{v\to 1}\sqrt{3v^2+1} = 2\sqrt{\lim\limits_{v\to 1}(3v^2+1)} = 2\sqrt{3\lim\limits_{v\to 1} v^2 + \lim\limits_{v\to 1} 1}$

$= 2\sqrt{3(1)^2+1} = 2\sqrt{4} = 4 = p(1)$

By the definition of continuity, p is continuous at $a = 1$.

14. $\lim\limits_{x\to 2} f(x) = \lim\limits_{x\to 2}\left(3x^4 - 5x + \sqrt[3]{x^2+4}\right) = 3\lim\limits_{x\to 2} x^4 - 5\lim\limits_{x\to 2} x + \sqrt[3]{\lim\limits_{x\to 2}(x^2+4)}$

$= 3(2)^4 - 5(2) + \sqrt[3]{2^2+4} = 48 - 10 + 2 = 40 = f(2)$

By the definition of continuity, f is continuous at $a = 2$.

15. For $a > 4$, we have

$$\begin{aligned} \lim_{x\to a} f(x) &= \lim_{x\to a}(x + \sqrt{x-4}) = \lim_{x\to a} x + \lim_{x\to a}\sqrt{x-4} && \text{[Limit Law 1]} \\ &= a + \sqrt{\lim_{x\to a} x - \lim_{x\to a} 4} && \text{[8, 11, and 2]} \\ &= a + \sqrt{a-4} && \text{[8 and 7]} \\ &= f(a) \end{aligned}$$

So f is continuous at $x = a$ for every a in $(4,\infty)$. Also, $\lim\limits_{x\to 4^+} f(x) = 4 = f(4)$, so f is continuous from the right at 4.

Thus, f is continuous on $[4,\infty)$.

16. For $a < -2$, we have

$$\begin{aligned} \lim_{x\to a} g(x) &= \lim_{x\to a}\frac{x-1}{3x+6} = \frac{\lim\limits_{x\to a}(x-1)}{\lim\limits_{x\to a}(3x+6)} && \text{[Limit Law 5]} \\ &= \frac{\lim\limits_{x\to a} x - \lim\limits_{x\to a} 1}{3\lim\limits_{x\to a} x + \lim\limits_{x\to a} 6} && \text{[2, 1, and 3]} \\ &= \frac{a-1}{3a+6} && \text{[8 and 7]} \end{aligned}$$

Thus, g is continuous at $x = a$ for every a in $(-\infty,-2)$; that is, g is continuous on $(-\infty,-2)$.

17. $f(x) = \dfrac{1}{x+2}$ is discontinuous at $a = -2$ because $f(-2)$ is undefined.

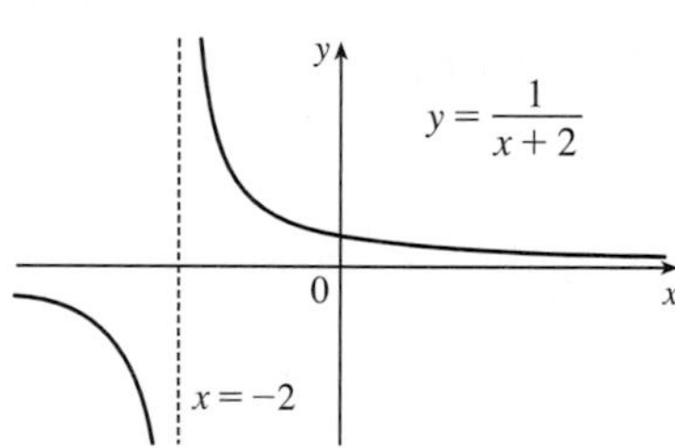

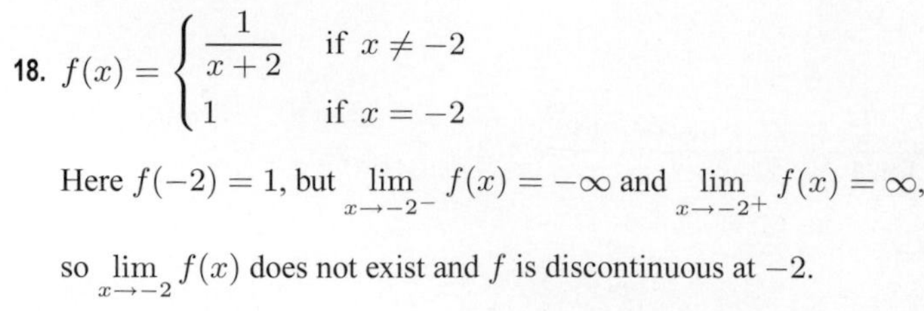

18. $f(x) = \begin{cases} \dfrac{1}{x+2} & \text{if } x \neq -2 \\ 1 & \text{if } x = -2 \end{cases}$

Here $f(-2) = 1$, but $\lim\limits_{x\to -2^-} f(x) = -\infty$ and $\lim\limits_{x\to -2^+} f(x) = \infty$,

so $\lim\limits_{x\to -2} f(x)$ does not exist and f is discontinuous at -2.

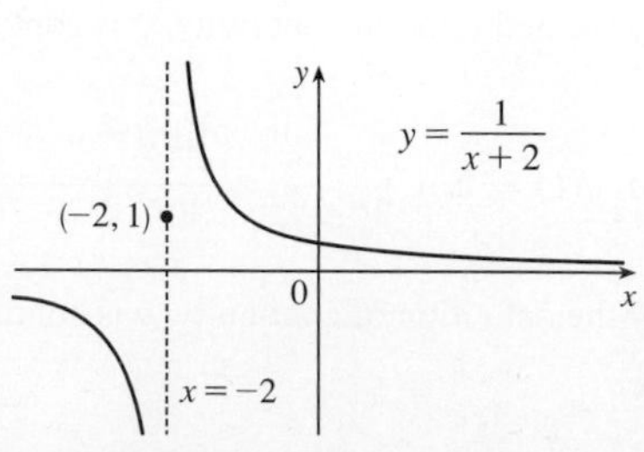

19. $f(x) = \begin{cases} x+3 & \text{if } x \le -1 \\ 2^x & \text{if } x > -1 \end{cases}$

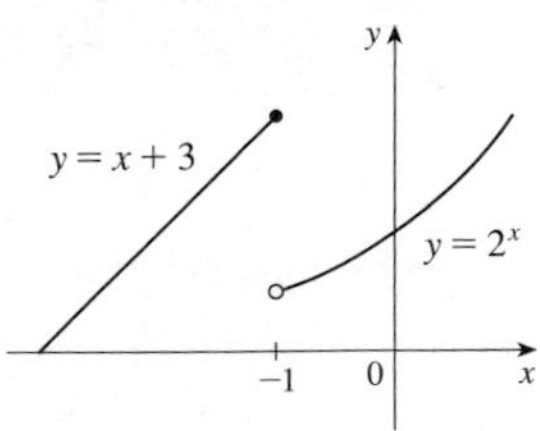

$\lim\limits_{x\to -1^-} f(x) = \lim\limits_{x\to -1^-} (x+3) = -1+3 = 2$ and

$\lim\limits_{x\to -1^+} f(x) = \lim\limits_{x\to -1^+} 2^x = 2^{-1} = \frac{1}{2}$. Since the left-hand and the right-hand limits of f at -1 are not equal, $\lim\limits_{x\to -1} f(x)$ does not exist, and f is discontinuous at -1.

20. $f(x) = \begin{cases} \dfrac{x^2-x}{x^2-1} & \text{if } x \ne 1 \\ 1 & \text{if } x = 1 \end{cases}$

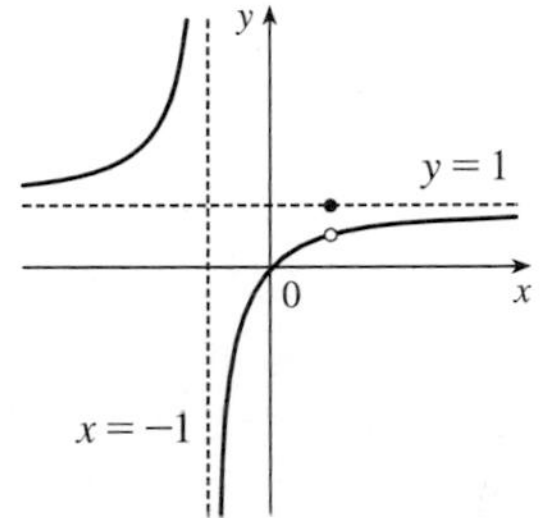

$\lim\limits_{x\to 1} f(x) = \lim\limits_{x\to 1} \dfrac{x^2-x}{x^2-1} = \lim\limits_{x\to 1} \dfrac{x(x-1)}{(x+1)(x-1)} = \lim\limits_{x\to 1} \dfrac{x}{x+1} = \dfrac{1}{2}$,

but $f(1) = 1$, so f is discontinous at 1.

21. $f(x) = \begin{cases} \cos x & \text{if } x < 0 \\ 0 & \text{if } x = 0 \\ 1 - x^2 & \text{if } x > 0 \end{cases}$

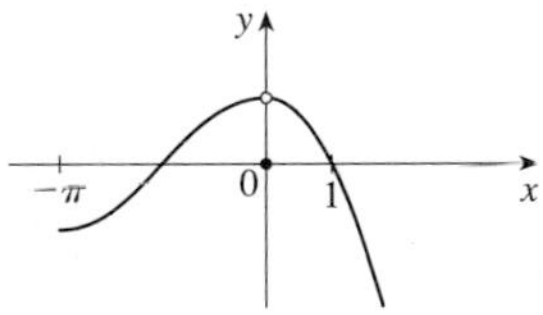

$\lim\limits_{x\to 0} f(x) = 1$, but $f(0) = 0 \ne 1$, so f is discontinuous at 0.

22. $f(x) = \begin{cases} \dfrac{2x^2-5x-3}{x-3} & \text{if } x \ne 3 \\ 6 & \text{if } x = 3 \end{cases}$

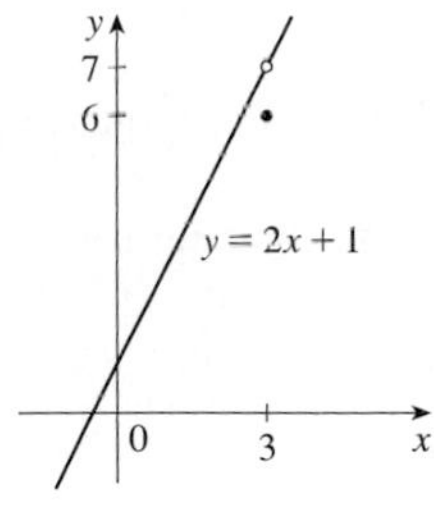

$\lim\limits_{x\to 3} f(x) = \lim\limits_{x\to 3} \dfrac{2x^2-5x-3}{x-3} = \lim\limits_{x\to 3} \dfrac{(2x+1)(x-3)}{x-3} = \lim\limits_{x\to 3}(2x+1) = 7$,

but $f(3) = 6$, so f is discontinuous at 3.

23. $f(x) = \dfrac{x^2-x-2}{x-2} = \dfrac{(x-2)(x+1)}{x-2} = x+1$ for $x \ne 2$. Since $\lim\limits_{x\to 2} f(x) = 2+1 = 3$, define $f(2) = 3$. Then f is continuous at 2.

24. $f(x) = \dfrac{x^3-8}{x^2-4} = \dfrac{(x-2)(x^2+2x+4)}{(x-2)(x+2)} = \dfrac{x^2+2x+4}{x+2}$ for $x \ne 2$. Since $\lim\limits_{x\to 2} f(x) = \dfrac{4+4+4}{2+2} = 3$, define $f(2) = 3$. Then f is continuous at 2.

25. $F(x) = \dfrac{2x^2-x-1}{x^2+1}$ is a rational function, so it is continuous on its domain, $(-\infty, \infty)$, by Theorem 5(b).

26. $G(x) = \dfrac{x^2+1}{2x^2-x-1} = \dfrac{x^2+1}{(2x+1)(x-1)}$ is a rational function, so it is continuous on its domain, $\left(-\infty, -\frac{1}{2}\right) \cup \left(-\frac{1}{2}, 1\right) \cup (1, \infty)$, by Theorem 5(b).

27. $x^3 - 2 = 0 \Rightarrow x^3 = 2 \Rightarrow x = \sqrt[3]{2}$, so $Q(x) = \dfrac{\sqrt[3]{x-2}}{x^3-2}$ has domain $(-\infty, \sqrt[3]{2}) \cup (\sqrt[3]{2}, \infty)$. Now $x^3 - 2$ is continuous everywhere by Theorem 5(a) and $\sqrt[3]{x-2}$ is continuous everywhere by Theorems 5(a), 7, and 9. Thus, Q is continuous on its domain by part 5 of Theorem 4.

28. The domain of $R(t) = \dfrac{e^{\sin t}}{2 + \cos \pi t}$ is $(-\infty, \infty)$ since the denominator is never 0 $[\cos \pi t \geq -1 \Rightarrow 2 + \cos \pi t \geq 1]$. By Theorems 7 and 9, $e^{\sin t}$ and $\cos \pi t$ are continuous on $\mathbb{R}$. By part 1 of Theorem 4, $2 + \cos \pi t$ is continuous on $\mathbb{R}$ and by part 5 of Theorem 4, R is continuous on $\mathbb{R}$.

29. By Theorem 5(a), the polynomial $1 + 2t$ is continuous on $\mathbb{R}$. By Theorem 7, the inverse trigonometric function $\arcsin x$ is continuous on its domain, $[-1, 1]$. By Theorem 9, $A(t) = \arcsin(1 + 2t)$ is continuous on its domain, which is $\{t \mid -1 \leq 1 + 2t \leq 1\} = \{t \mid -2 \leq 2t \leq 0\} = \{t \mid -1 \leq t \leq 0\} = [-1, 0]$.

30. By Theorem 7, the trigonometric function $\tan x$ is continuous on its domain, $\left\{x \mid x \neq \frac{\pi}{2} + \pi n\right\}$. By Theorems 5(a), 7, and 9, the composite function $\sqrt{4 - x^2}$ is continuous on its domain $[-2, 2]$. By part 5 of Theorem 4, $B(x) = \dfrac{\tan x}{\sqrt{4 - x^2}}$ is continuous on its domain, $(-2, -\pi/2) \cup (-\pi/2, \pi/2) \cup (\pi/2, 2)$.

31. $M(x) = \sqrt{1 + \dfrac{1}{x}} = \sqrt{\dfrac{x+1}{x}}$ is defined when $\dfrac{x+1}{x} \geq 0 \Rightarrow x + 1 \geq 0$ and $x > 0$ or $x + 1 \leq 0$ and $x < 0 \Rightarrow x > 0$ or $x \leq -1$, so M has domain $(-\infty, -1] \cup (0, \infty)$. M is the composite of a root function and a rational function, so it is continuous at every number in its domain by Theorems 7 and 9.

32. By Theorems 7 and 9, the composite function e^{-r^2} is continuous on $\mathbb{R}$. By part 1 of Theorem 4, $1 + e^{-r^2}$ is continuous on $\mathbb{R}$. By Theorem 7, the inverse trigonometric function $\tan^{-1}$ is continuous on its domain, $\mathbb{R}$. By Theorem 9, the composite function $N(r) = \tan^{-1}\left(1 + e^{-r^2}\right)$ is continuous on its domain, $\mathbb{R}$.

33. The function $y = \dfrac{1}{1 + e^{1/x}}$ is discontinuous at $x = 0$ because the left- and right-hand limits at $x = 0$ are different.

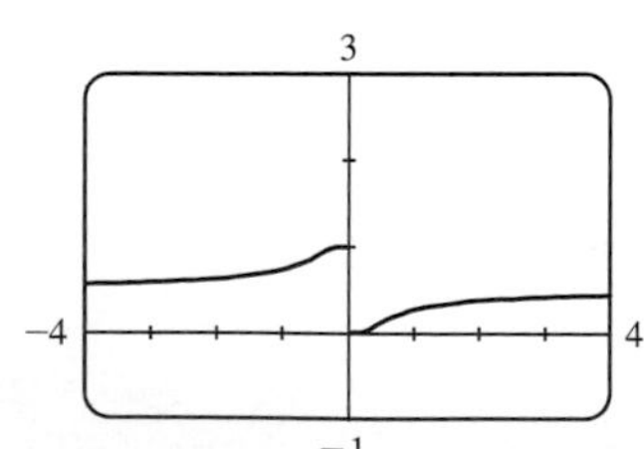

34. The function $y = \tan^2 x$ is discontinuous at $x = \frac{\pi}{2} + \pi k$, where k is any integer. The function $y = \ln(\tan^2 x)$ is also discontinuous where $\tan^2 x$ is 0, that is, at $x = \pi k$. So $y = \ln(\tan^2 x)$ is discontinuous at $x = \frac{\pi}{2} n$, n any integer.

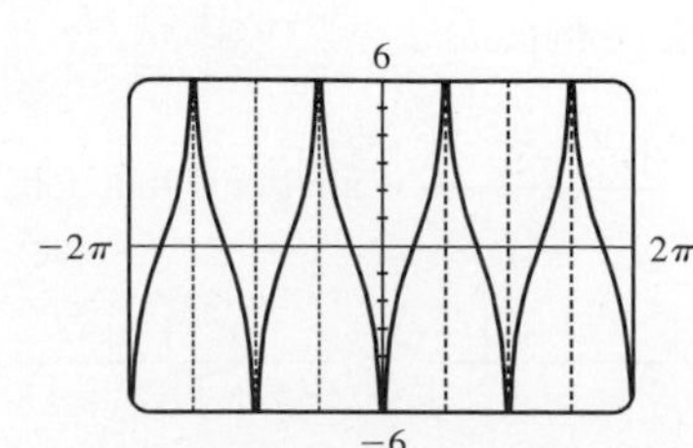

35. Because x is continuous on $\mathbb{R}$ and $\sqrt{20 - x^2}$ is continuous on its domain, $-\sqrt{20} \leq x \leq \sqrt{20}$, the product $f(x) = x\sqrt{20 - x^2}$ is continuous on $-\sqrt{20} \leq x \leq \sqrt{20}$. The number 2 is in that domain, so f is continuous at 2, and $\lim\limits_{x \to 2} f(x) = f(2) = 2\sqrt{16} = 8$.

36. Because x is continuous on $\mathbb{R}$, $\sin x$ is continuous on $\mathbb{R}$, and $x + \sin x$ is continuous on $\mathbb{R}$, the composite function $f(x) = \sin(x + \sin x)$ is continuous on $\mathbb{R}$, so $\lim\limits_{x \to \pi} f(x) = f(\pi) = \sin(\pi + \sin \pi) = \sin \pi = 0$.

37. The function $f(x) = \ln\left(\dfrac{5 - x^2}{1 + x}\right)$ is continuous throughout its domain because it is the composite of a logarithm function and a rational function. For the domain of f, we must have $\dfrac{5 - x^2}{1 + x} > 0$, so the numerator and denominator must have the same sign, that is, the domain is $(-\infty, -\sqrt{5}\,] \cup (-1, \sqrt{5}\,]$. The number 1 is in that domain, so f is continuous at 1, and $\lim\limits_{x \to 1} f(x) = f(1) = \ln \dfrac{5 - 1}{1 + 1} = \ln 2$.

38. The function $f(x) = 3^{\sqrt{x^2 - 2x - 4}}$ is continuous throughout its domain because it is the composite of an exponential function, a root function, and a polynomial. Its domain is

$$\begin{aligned}\{x \mid x^2 - 2x - 4 \geq 0\} &= \{x \mid x^2 - 2x + 1 \geq 5\} = \{x \mid (x - 1)^2 \geq 5\}\\ &= \left\{x \,\middle|\, |x - 1| \geq \sqrt{5}\,\right\} = (-\infty, 1 - \sqrt{5}\,] \cup [1 + \sqrt{5}, \infty)\end{aligned}$$

The number 4 is in that domain, so f is continuous at 4, and $\lim\limits_{x \to 4} f(x) = f(4) = 3^{\sqrt{16-8-4}} = 3^2 - 9$.

39. $f(x) = \begin{cases} 1 - x^2 & \text{if } x \leq 1 \\ \ln x & \text{if } x > 1 \end{cases}$

By Theorem 5, since $f(x)$ equals the polynomial $1 - x^2$ on $(-\infty, 1]$, f is continuous on $(-\infty, 1]$.

By Theorem 7, since $f(x)$ equals the logarithm function $\ln x$ on $(1, \infty)$, f is continuous on $(1, \infty)$.

At $x = 1$, $\lim\limits_{x \to 1^-} f(x) = \lim\limits_{x \to 1^-} (1 - x^2) = 1 - 1^2 = 0$ and $\lim\limits_{x \to 1^+} f(x) = \lim\limits_{x \to 1^+} \ln x = \ln 1 = 0$. Thus, $\lim\limits_{x \to 1} f(x)$ exists and equals 0. Also, $f(1) = 1 - 1^2 = 0$. Thus, f is continuous at $x = 1$. We conclude that f is continuous on $(-\infty, \infty)$.

40. $f(x) = \begin{cases} \sin x & \text{if } x < \pi/4 \\ \cos x & \text{if } x \geq \pi/4 \end{cases}$

By Theorem 7, the trigonometric functions are continuous. Since $f(x) = \sin x$ on $(-\infty, \pi/4)$ and $f(x) = \cos x$ on $(\pi/4, \infty)$, f is continuous on $(-\infty, \pi/4) \cup (\pi/4, \infty)$. $\lim\limits_{x \to (\pi/4)^-} f(x) = \lim\limits_{x \to (\pi/4)^-} \sin x = \sin \frac{\pi}{4} = 1/\sqrt{2}$ since the sine function is continuous at $\pi/4$. Similarly, $\lim\limits_{x \to (\pi/4)^+} f(x) = \lim\limits_{x \to (\pi/4)^+} \cos x = 1/\sqrt{2}$ by continuity of the cosine function at $\pi/4$. Thus, $\lim\limits_{x \to (\pi/4)} f(x)$ exists and equals $1/\sqrt{2}$, which agrees with the value $f(\pi/4)$. Therefore, f is continuous at $\pi/4$, so f is continuous on $(-\infty, \infty)$.

41. $f(x) = \begin{cases} x^2 & \text{if } x < -1 \\ x & \text{if } -1 \le x < 1 \\ 1/x & \text{if } x \ge 1 \end{cases}$

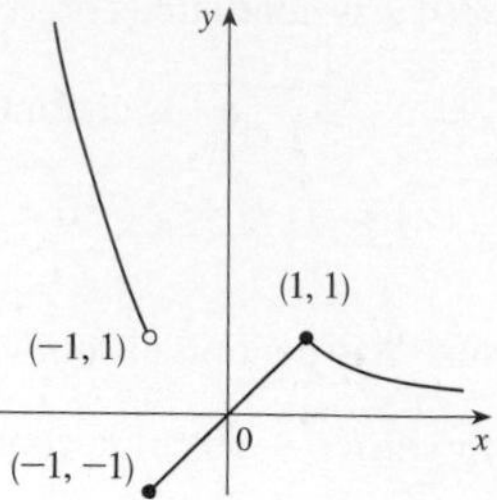

f is continuous on $(-\infty, -1)$, $(-1, 1)$, and $(1, \infty)$, where it is a polynomial, a polynomial, and a rational function, respectively.

Now $\lim\limits_{x \to -1^-} f(x) = \lim\limits_{x \to -1^-} x^2 = 1$ and $\lim\limits_{x \to -1^+} f(x) = \lim\limits_{x \to -1^+} x = -1$, so f is discontinuous at -1. Since $f(-1) = -1$, f is continuous from the right at -1. Also, $\lim\limits_{x \to 1^-} f(x) = \lim\limits_{x \to 1^-} x = 1$ and $\lim\limits_{x \to 1^+} f(x) = \lim\limits_{x \to 1^+} \dfrac{1}{x} = 1 = f(1)$, so f is continuous at 1.

42. $f(x) = \begin{cases} 2^x & \text{if } x \le 1 \\ 3 - x & \text{if } 1 < x \le 4 \\ \sqrt{x} & \text{if } x > 4 \end{cases}$

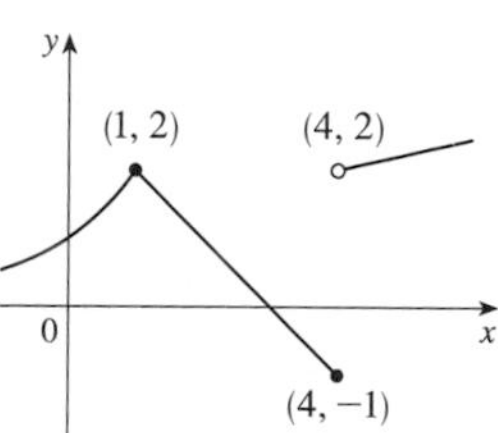

f is continuous on $(-\infty, 1)$, $(1, 4)$, and $(4, \infty)$, where it is an exponential, a polynomial, and a root function, respectively.

Now $\lim\limits_{x \to 1^-} f(x) = \lim\limits_{x \to 1^-} 2^x = 2$ and $\lim\limits_{x \to 1^+} f(x) = \lim\limits_{x \to 1^+} (3 - x) = 2$. Since $f(1) = 2$ we have continuity at 1. Also, $\lim\limits_{x \to 4^-} f(x) = \lim\limits_{x \to 4^-} (3 - x) = -1 = f(4)$ and $\lim\limits_{x \to 4^+} f(x) = \lim\limits_{x \to 4^+} \sqrt{x} = 2$, so f is discontinuous at 4, but it is continuous from the left at 4.

43. $f(x) = \begin{cases} x + 2 & \text{if } x < 0 \\ e^x & \text{if } 0 \le x \le 1 \\ 2 - x & \text{if } x > 1 \end{cases}$

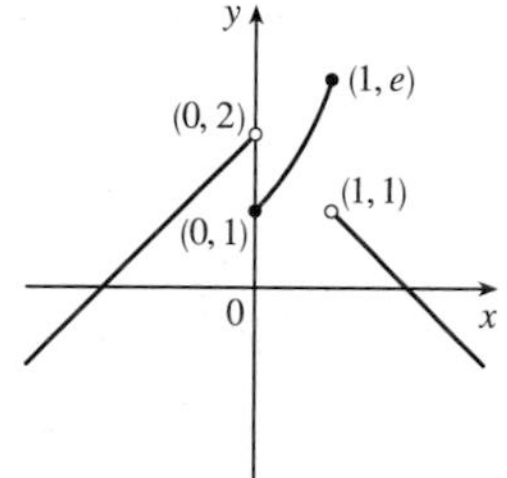

f is continuous on $(-\infty, 0)$ and $(1, \infty)$ since on each of these intervals it is a polynomial; it is continuous on $(0, 1)$ since it is an exponential.

Now $\lim\limits_{x \to 0^-} f(x) = \lim\limits_{x \to 0^-} (x + 2) = 2$ and $\lim\limits_{x \to 0^+} f(x) = \lim\limits_{x \to 0^+} e^x = 1$, so f is discontinuous at 0. Since $f(0) = 1$, f is continuous from the right at 0. Also $\lim\limits_{x \to 1^-} f(x) = \lim\limits_{x \to 1^-} e^x = e$ and $\lim\limits_{x \to 1^+} f(x) = \lim\limits_{x \to 1^+} (2 - x) = 1$, so f is discontinuous at 1. Since $f(1) = e$, f is continuous from the left at 1.

44. By Theorem 5, each piece of F is continuous on its domain. We need to check for continuity at $r = R$.

$\lim\limits_{r \to R^-} F(r) = \lim\limits_{r \to R^-} \dfrac{GMr}{R^3} = \dfrac{GM}{R^2}$ and $\lim\limits_{r \to R^+} F(r) = \lim\limits_{r \to R^+} \dfrac{GM}{r^2} = \dfrac{GM}{R^2}$, so $\lim\limits_{r \to R} F(r) = \dfrac{GM}{R^2}$. Since $F(R) = \dfrac{GM}{R^2}$, F is continuous at R. Therefore, F is a continuous function of r.

45. $f(x) = \begin{cases} cx^2 + 2x & \text{if } x < 2 \\ x^3 - cx & \text{if } x \ge 2 \end{cases}$

f is continuous on $(-\infty, 2)$ and $(2, \infty)$. Now $\lim\limits_{x \to 2^-} f(x) = \lim\limits_{x \to 2^-} (cx^2 + 2x) = 4c + 4$ and

$\lim_{x\to 2^+} f(x) = \lim_{x\to 2^+}(x^3 - cx) = 8 - 2c$. So f is continuous $\Leftrightarrow$ $4c + 4 = 8 - 2c$ $\Leftrightarrow$ $6c = 4$ $\Leftrightarrow$ $c = \frac{2}{3}$. Thus, for f to be continuous on $(-\infty, \infty)$, $c = \frac{2}{3}$.

46. $f(x) = \begin{cases} \dfrac{x^2 - 4}{x - 2} & \text{if } x < 2 \\ ax^2 - bx + 3 & \text{if } 2 \le x < 3 \\ 2x - a + b & \text{if } x \ge 3 \end{cases}$

At $x = 2$: $\lim_{x\to 2^-} f(x) = \lim_{x\to 2^-} \dfrac{x^2 - 4}{x - 2} = \lim_{x\to 2^-} \dfrac{(x+2)(x-2)}{x - 2} = \lim_{x\to 2^-}(x + 2) = 2 + 2 = 4$

$\lim_{x\to 2^+} f(x) = \lim_{x\to 2^+}(ax^2 - bx + 3) = 4a - 2b + 3$

We must have $4a - 2b + 3 = 4$, or $\mathbf{4a - 2b = 1}$ **(1)**.

At $x = 3$: $\lim_{x\to 3^-} f(x) = \lim_{x\to 3^-}(ax^2 - bx + 3) = 9a - 3b + 3$

$\lim_{x\to 3^+} f(x) = \lim_{x\to 3^+}(2x - a + b) = 6 - a + b$

We must have $9a - 3b + 3 = 6 - a + b$, or $\mathbf{10a - 4b = 3}$ **(2)**.

Now solve the system of equations by adding -2 times equation **(1)** to equation **(2)**.

$$\begin{aligned} -8a + 4b &= -2 \\ 10a - 4b &= 3 \\ \hline 2a \phantom{{}-4b} &= 1 \end{aligned}$$

So $a = \frac{1}{2}$. Substituting $\frac{1}{2}$ for a in **(1)** gives us $-2b = -1$, so $b = \frac{1}{2}$ as well. Thus, for f to be continuous on $(-\infty, \infty)$, $a = b = \frac{1}{2}$.

47. If f and g are continuous and $g(2) = 6$, then $\lim_{x\to 2}[3f(x) + f(x)\,g(x)] = 36$ $\Rightarrow$

$3\lim_{x\to 2} f(x) + \lim_{x\to 2} f(x) \cdot \lim_{x\to 2} g(x) = 36$ $\Rightarrow$ $3f(2) + f(2) \cdot 6 = 36$ $\Rightarrow$ $9f(2) = 36$ $\Rightarrow$ $f(2) = 4$.

48. (a) $f(x) = \dfrac{1}{x}$ and $g(x) = \dfrac{1}{x^2}$, so $(f \circ g)(x) = f(g(x)) = f(1/x^2) = 1/(1/x^2) = x^2$.

(b) The domain of $f \circ g$ is the set of numbers x in the domain of g (all nonzero reals) such that $g(x)$ is in the domain of f (also all nonzero reals). Thus, the domain is $\left\{ x \;\middle|\; x \ne 0 \text{ and } \dfrac{1}{x^2} \ne 0 \right\} = \{x \mid x \ne 0\}$ or $(-\infty, 0) \cup (0, \infty)$. Since $f \circ g$ is the composite of two rational functions, it is continuous throughout its domain; that is, everywhere except $x = 0$.

49. (a) $f(x) = \dfrac{x^4 - 1}{x - 1} = \dfrac{(x^2+1)(x^2-1)}{x - 1} = \dfrac{(x^2+1)(x+1)(x-1)}{x - 1} = (x^2+1)(x+1)$ [or $x^3 + x^2 + x + 1$]

for $x \ne 1$. The discontinuity is removable and $g(x) = x^3 + x^2 + x + 1$ agrees with f for $x \ne 1$ and is continuous on $\mathbb{R}$.

(b) $f(x) = \dfrac{x^3 - x^2 - 2x}{x - 2} = \dfrac{x(x^2 - x - 2)}{x - 2} = \dfrac{x(x-2)(x+1)}{x - 2} = x(x+1)$ [or $x^2 + x$] for $x \ne 2$. The discontinuity is removable and $g(x) = x^2 + x$ agrees with f for $x \ne 2$ and is continuous on $\mathbb{R}$.

(c) $\lim_{x\to \pi^-} f(x) = \lim_{x\to \pi^-} [\![\sin x]\!] = \lim_{x\to \pi^-} 0 = 0$ and $\lim_{x\to \pi^+} f(x) = \lim_{x\to \pi^+} [\![\sin x]\!] = \lim_{x\to \pi^+}(-1) = -1$, so $\lim_{x\to \pi} f(x)$ does not exist. The discontinuity at $x = \pi$ is a jump discontinuity.

50.

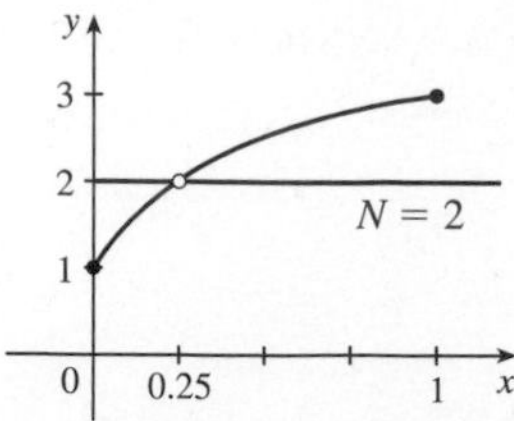

f does not satisfy the conclusion of the Intermediate Value Theorem.

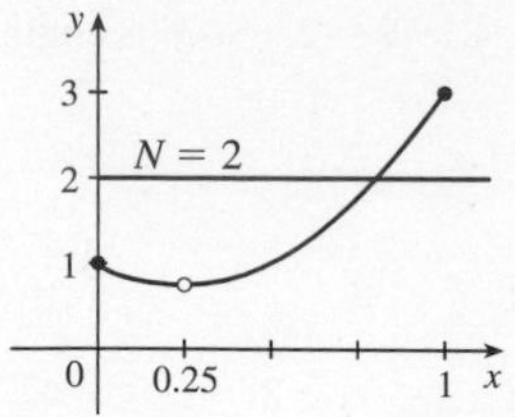

f does satisfy the conclusion of the Intermediate Value Theorem.

51. $f(x) = x^2 + 10\sin x$ is continuous on the interval $[31, 32]$, $f(31) \approx 957$, and $f(32) \approx 1030$. Since $957 < 1000 < 1030$, there is a number c in $(31, 32)$ such that $f(c) = 1000$ by the Intermediate Value Theorem. *Note:* There is also a number c in $(-32, -31)$ such that $f(c) = 1000$.

52. Suppose that $f(3) < 6$. By the Intermediate Value Theorem applied to the continuous function f on the closed interval $[2, 3]$, the fact that $f(2) = 8 > 6$ and $f(3) < 6$ implies that there is a number c in $(2, 3)$ such that $f(c) = 6$. This contradicts the fact that the only solutions of the equation $f(x) = 6$ are $x = 1$ and $x = 4$. Hence, our supposition that $f(3) < 6$ was incorrect. It follows that $f(3) \geq 6$. But $f(3) \neq 6$ because the only solutions of $f(x) = 6$ are $x = 1$ and $x = 4$. Therefore, $f(3) > 6$.

53. $f(x) = x^4 + x - 3$ is continuous on the interval $[1, 2]$, $f(1) = -1$, and $f(2) = 15$. Since $-1 < 0 < 15$, there is a number c in $(1, 2)$ such that $f(c) = 0$ by the Intermediate Value Theorem. Thus, there is a root of the equation $x^4 + x - 3 = 0$ in the interval $(1, 2)$.

54. The equation $\ln x = x - \sqrt{x}$ is equivalent to the equation $\ln x - x + \sqrt{x} = 0$. $f(x) = \ln x - x + \sqrt{x}$ is continuous on the interval $[2, 3]$, $f(2) = \ln 2 - 2 + \sqrt{2} \approx 0.107$, and $f(3) = \ln 3 - 3 + \sqrt{3} \approx -0.169$. Since $f(2) > 0 > f(3)$, there is a number c in $(2, 3)$ such that $f(c) = 0$ by the Intermediate Value Theorem. Thus, there is a root of the equation $\ln x - x + \sqrt{x} = 0$, or $\ln x = x - \sqrt{x}$, in the interval $(2, 3)$.

55. The equation $e^x = 3 - 2x$ is equivalent to the equation $e^x + 2x - 3 = 0$. $f(x) = e^x + 2x - 3$ is continuous on the interval $[0, 1]$, $f(0) = -2$, and $f(1) = e - 1 \approx 1.72$. Since $-2 < 0 < e - 1$, there is a number c in $(0, 1)$ such that $f(c) = 0$ by the Intermediate Value Theorem. Thus, there is a root of the equation $e^x + 2x - 3 = 0$, or $e^x = 3 - 2x$, in the interval $(0, 1)$.

56. The equation $\sin x = x^2 - x$ is equivalent to the equation $\sin x - x^2 + x = 0$. $f(x) = \sin x - x^2 + x$ is continuous on the interval $[1, 2]$, $f(1) = \sin 1 \approx 0.84$, and $f(2) = \sin 2 - 2 \approx -1.09$. Since $\sin 1 > 0 > \sin 2 - 2$, there is a number c in $(1, 2)$ such that $f(c) = 0$ by the Intermediate Value Theorem. Thus, there is a root of the equation $\sin x - x^2 + x = 0$, or $\sin x = x^2 - x$, in the interval $(1, 2)$.

57. (a) $f(x) = \cos x - x^3$ is continuous on the interval $[0, 1]$, $f(0) = 1 > 0$, and $f(1) = \cos 1 - 1 \approx -0.46 < 0$. Since $1 > 0 > -0.46$, there is a number c in $(0, 1)$ such that $f(c) = 0$ by the Intermediate Value Theorem. Thus, there is a root of the equation $\cos x - x^3 = 0$, or $\cos x = x^3$, in the interval $(0, 1)$.

(b) $f(0.86) \approx 0.016 > 0$ and $f(0.87) \approx -0.014 < 0$, so there is a root between 0.86 and 0.87, that is, in the interval $(0.86, 0.87)$.

58. (a) $f(x) = \ln x - 3 + 2x$ is continuous on the interval $[1, 2]$, $f(1) = -1 < 0$, and $f(2) = \ln 2 + 1 \approx 1.7 > 0$. Since $-1 < 0 < 1.7$, there is a number c in $(1, 2)$ such that $f(c) = 0$ by the Intermediate Value Theorem. Thus, there is a root of the equation $\ln x - 3 + 2x = 0$, or $\ln x = 3 - 2x$, in the interval $(1, 2)$.

(b) $f(1.34) \approx -0.03 < 0$ and $f(1.35) \approx 0.0001 > 0$, so there is a root between 1.34 and 1.35, that is, in the interval $(1.34, 1.35)$.

59. (a) Let $f(x) = 100e^{-x/100} - 0.01x^2$. Then $f(0) = 100 > 0$ and $f(100) = 100e^{-1} - 100 \approx -63.2 < 0$. So by the Intermediate Value Theorem, there is a number c in $(0, 100)$ such that $f(c) = 0$. This implies that $100e^{-c/100} = 0.01c^2$.

(b) Using the intersect feature of the graphing device, we find that the root of the equation is $x = 70.347$, correct to three decimal places.

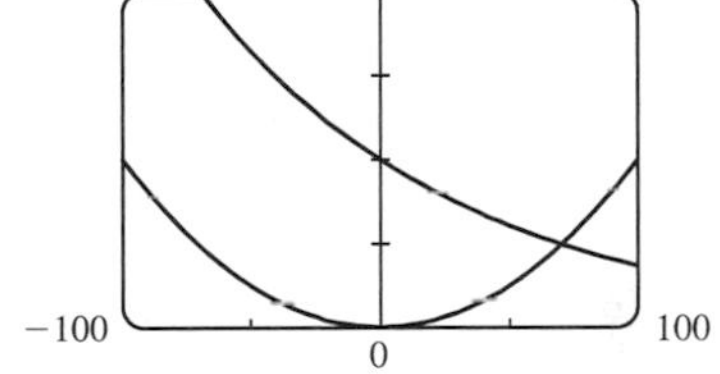

60. (a) Let $f(x) = \arctan x + x - 1$. Then $f(0) = -1 < 0$ and $f(1) = \frac{\pi}{4} > 0$. So by the Intermediate Value Theorem, there is a number c in $(0, 1)$ such that $f(c) = 0$. This implies that $\arctan c = 1 - c$.

(b) Using the intersect feature of the graphing device, we find that the root of the equation is $x = 0.520$, correct to three decimal places.

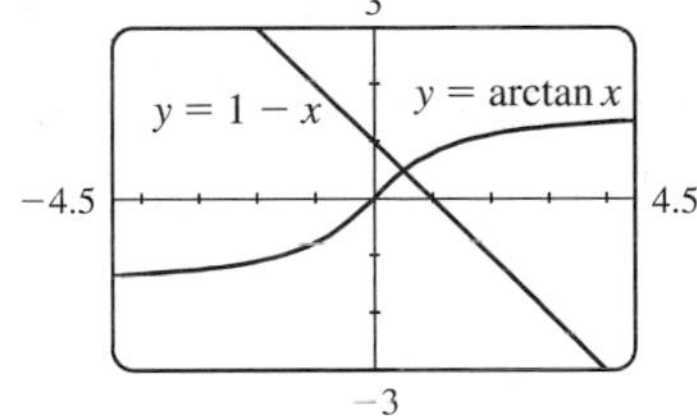

61. Let $f(x) = \sin x^3$. Then f is continuous on $[1, 2]$ since f is the composite of the sine function and the cubing function, both of which are continuous on $\mathbb{R}$. The zeros of the sine are at $n\pi$, so we note that $0 < 1 < \pi < \frac{3}{2}\pi < 2\pi < 8 < 3\pi$, and that the pertinent cube roots are related by $1 < \sqrt[3]{\frac{3}{2}\pi}$ [call this value A] < 2. [By observation, we might notice that $x = \sqrt[3]{\pi}$ and $x = \sqrt[3]{2\pi}$ are zeros of f.]

Now $f(1) = \sin 1 > 0$, $f(A) = \sin \frac{3}{2}\pi = -1 < 0$, and $f(2) = \sin 8 > 0$. Applying the Intermediate Value Theorem on $[1, A]$ and then on $[A, 2]$, we see there are numbers c and d in $(1, A)$ and $(A, 2)$ such that $f(c) = f(d) = 0$. Thus, f has at least two x-intercepts in $(1, 2)$.

62. Let $f(x) = x^2 - 3 + 1/x$. Then f is continuous on $(0, 2]$ since f is a rational function whose domain is $(0, \infty)$. By inspection, we see that $f\left(\frac{1}{4}\right) = \frac{17}{16} > 0$, $f(1) = -1 < 0$, and $f(2) = \frac{3}{2} > 0$. Appling the Intermediate Value Theorem on $\left[\frac{1}{4}, 1\right]$ and then on $[1, 2]$, we see there are numbers c and d in $\left(\frac{1}{4}, 1\right)$ and $(1, 2)$ such that $f(c) = f(d) = 0$. Thus, f has at least two x-intercepts in $(0, 2)$.

63. ($\Rightarrow$) If f is continuous at a, then by Theorem 8 with $g(h) = a + h$, we have

$$\lim_{h \to 0} f(a+h) = f\left(\lim_{h \to 0}(a+h)\right) = f(a).$$

($\Leftarrow$) Let $\varepsilon > 0$. Since $\lim\limits_{h \to 0} f(a+h) = f(a)$, there exists $\delta > 0$ such that $0 < |h| < \delta \;\Rightarrow$

$|f(a+h) - f(a)| < \varepsilon$. So if $0 < |x - a| < \delta$, then $|f(x) - f(a)| = |f(a + (x-a)) - f(a)| < \varepsilon$.

Thus, $\lim\limits_{x \to a} f(x) = f(a)$ and so f is continuous at a.

64.
$$\begin{aligned}\lim_{h \to 0} \sin(a+h) &= \lim_{h \to 0}(\sin a \cos h + \cos a \sin h) = \lim_{h \to 0}(\sin a \cos h) + \lim_{h \to 0}(\cos a \sin h)\\ &= \left(\lim_{h \to 0} \sin a\right)\left(\lim_{h \to 0} \cos h\right) + \left(\lim_{h \to 0} \cos a\right)\left(\lim_{h \to 0} \sin h\right) = (\sin a)(1) + (\cos a)(0) = \sin a\end{aligned}$$

65. As in the previous exercise, we must show that $\lim\limits_{h \to 0} \cos(a+h) = \cos a$ to prove that the cosine function is continuous.

$$\begin{aligned}\lim_{h \to 0} \cos(a+h) &= \lim_{h \to 0}(\cos a \cos h - \sin a \sin h) = \lim_{h \to 0}(\cos a \cos h) - \lim_{h \to 0}(\sin a \sin h)\\ &= \left(\lim_{h \to 0} \cos a\right)\left(\lim_{h \to 0} \cos h\right) - \left(\lim_{h \to 0} \sin a\right)\left(\lim_{h \to 0} \sin h\right) = (\cos a)(1) - (\sin a)(0) = \cos a\end{aligned}$$

66. (a) Since f is continuous at a, $\lim\limits_{x \to a} f(x) = f(a)$. Thus, using the Constant Multiple Law of Limits, we have

$\lim\limits_{x \to a}(cf)(x) = \lim\limits_{x \to a} cf(x) = c \lim\limits_{x \to a} f(x) = cf(a) = (cf)(a)$. Therefore, cf is continuous at a.

(b) Since f and g are continuous at a, $\lim\limits_{x \to a} f(x) = f(a)$ and $\lim\limits_{x \to a} g(x) = g(a)$. Since $g(a) \neq 0$, we can use the Quotient Law

of Limits: $\lim\limits_{x \to a}\left(\dfrac{f}{g}\right)(x) = \lim\limits_{x \to a}\dfrac{f(x)}{g(x)} = \dfrac{\lim\limits_{x \to a} f(x)}{\lim\limits_{x \to a} g(x)} = \dfrac{f(a)}{g(a)} = \left(\dfrac{f}{g}\right)(a)$. Thus, $\dfrac{f}{g}$ is continuous at a.

67. $f(x) = \begin{cases} 0 & \text{if } x \text{ is rational} \\ 1 & \text{if } x \text{ is irrational} \end{cases}$ is continuous nowhere. For, given any number a and any $\delta > 0$, the interval $(a - \delta, a + \delta)$

contains both infinitely many rational and infinitely many irrational numbers. Since $f(a) = 0$ or 1, there are infinitely many

numbers x with $0 < |x - a| < \delta$ and $|f(x) - f(a)| = 1$. Thus, $\lim\limits_{x \to a} f(x) \neq f(a)$. [In fact, $\lim\limits_{x \to a} f(x)$ does not even exist.]

68. $g(x) = \begin{cases} 0 & \text{if } x \text{ is rational} \\ x & \text{if } x \text{ is irrational} \end{cases}$ is continuous at 0. To see why, note that $-|x| \le g(x) \le |x|$, so by the Squeeze Theorem

$\lim\limits_{x \to 0} g(x) = 0 = g(0)$. But g is continuous nowhere else. For if $a \neq 0$ and $\delta > 0$, the interval $(a - \delta, a + \delta)$ contains both

infinitely many rational and infinitely many irrational numbers. Since $g(a) = 0$ or a, there are infinitely many numbers x with

$0 < |x - a| < \delta$ and $|g(x) - g(a)| > |a|/2$. Thus, $\lim\limits_{x \to a} g(x) \neq g(a)$.

69. If there is such a number, it satisfies the equation $x^3 + 1 = x \;\Leftrightarrow\; x^3 - x + 1 = 0$. Let the left-hand side of this equation be called $f(x)$. Now $f(-2) = -5 < 0$, and $f(-1) = 1 > 0$. Note also that $f(x)$ is a polynomial, and thus continuous. So by the Intermediate Value Theorem, there is a number c between -2 and -1 such that $f(c) = 0$, so that $c = c^3 + 1$.

70. $\dfrac{a}{x^3 + 2x^2 - 1} + \dfrac{b}{x^3 + x - 2} = 0 \;\Rightarrow\; a(x^3 + x - 2) + b(x^3 + 2x^2 - 1) = 0$. Let $p(x)$ denote the left side of the last

equation. Since p is continuous on $[-1, 1]$, $p(-1) = -4a < 0$, and $p(1) = 2b > 0$, there exists a c in $(-1, 1)$ such that

$p(c) = 0$ by the Intermediate Value Theorem. Note that the only root of either denominator that is in $(-1, 1)$ is $(-1 + \sqrt{5})/2 = r$, but $p(r) = (3\sqrt{5} - 9)a/2 \neq 0$. Thus, c is not a root of either denominator, so $p(c) = 0 \Rightarrow x = c$ is a root of the given equation.

71. $f(x) = x^4 \sin(1/x)$ is continuous on $(-\infty, 0) \cup (0, \infty)$ since it is the product of a polynomial and a composite of a trigonometric function and a rational function. Now since $-1 \leq \sin(1/x) \leq 1$, we have $-x^4 \leq x^4 \sin(1/x) \leq x^4$. Because $\lim_{x\to 0}(-x^4) = 0$ and $\lim_{x\to 0} x^4 = 0$, the Squeeze Theorem gives us $\lim_{x\to 0}(x^4 \sin(1/x)) = 0$, which equals $f(0)$. Thus, f is continuous at 0 and, hence, on $(-\infty, \infty)$.

72. (a) $\lim_{x\to 0^+} F(x) = 0$ and $\lim_{x\to 0^-} F(x) = 0$, so $\lim_{x\to 0} F(x) = 0$, which is $F(0)$, and hence F is continuous at $x = a$ if $a = 0$. For $a > 0$, $\lim_{x\to a} F(x) = \lim_{x\to a} x = a = F(a)$. For $a < 0$, $\lim_{x\to a} F(x) = \lim_{x\to a}(-x) = -a = F(a)$. Thus, F is continuous at $x = a$; that is, continuous everywhere.

(b) Assume that f is continuous on the interval I. Then for $a \in I$, $\lim_{x\to a} |f(x)| = \left|\lim_{x\to a} f(x)\right| = |f(a)|$ by Theorem 8. (If a is an endpoint of I, use the appropriate one-sided limit.) So $|f|$ is continuous on I.

(c) No, the converse is false. For example, the function $f(x) = \begin{cases} 1 & \text{if } x \geq 0 \\ -1 & \text{if } x < 0 \end{cases}$ is not continuous at $x = 0$, but $|f(x)| = 1$ is continuous on $\mathbb{R}$.

73. Define $u(t)$ to be the monk's distance from the monastery, as a function of time t (in hours), on the first day, and define $d(t)$ to be his distance from the monastery, as a function of time, on the second day. Let D be the distance from the monastery to the top of the mountain. From the given information we know that $u(0) = 0$, $u(12) = D$, $d(0) = D$ and $d(12) = 0$. Now consider the function $u - d$, which is clearly continuous. We calculate that $(u - d)(0) = -D$ and $(u - d)(12) = D$. So by the Intermediate Value Theorem, there must be some time t_0 between 0 and 12 such that $(u - d)(t_0) = 0 \Leftrightarrow u(t_0) = d(t_0)$. So at time t_0 after 7:00 AM, the monk will be at the same place on both days.

2.6 Limits at Infinity; Horizontal Asymptotes

1. (a) As x becomes large, the values of $f(x)$ approach 5.

(b) As x becomes large negative, the values of $f(x)$ approach 3.

2. (a) The graph of a function can intersect a vertical asymptote in the sense that it can meet but not cross it.

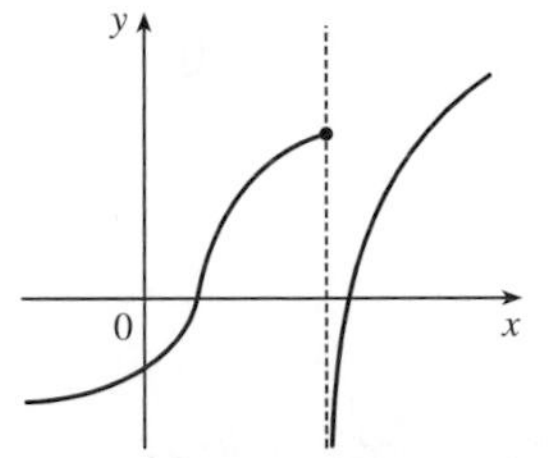

The graph of a function can intersect a horizontal asymptote. It can even intersect its horizontal asymptote an infinite number of times.

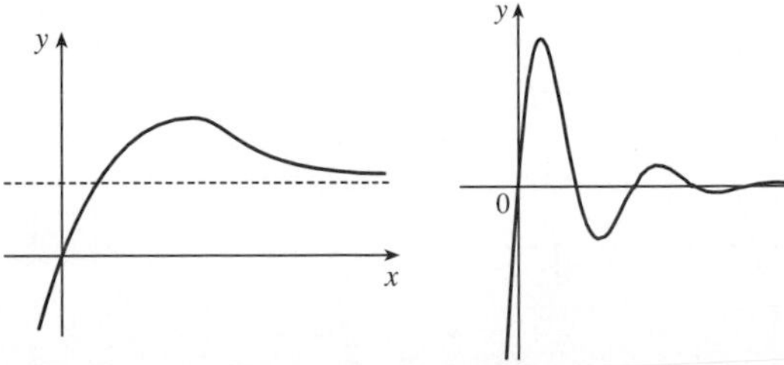

(b) The graph of a function can have 0, 1, or 2 horizontal asymptotes. Representative examples are shown.

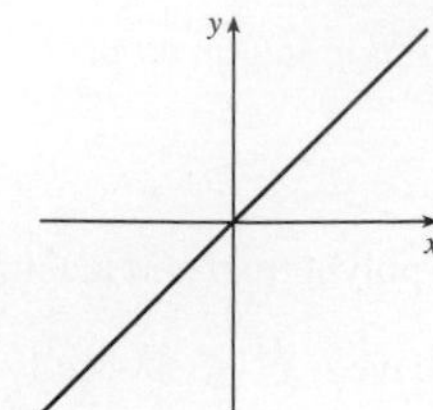

No horizontal asymptote

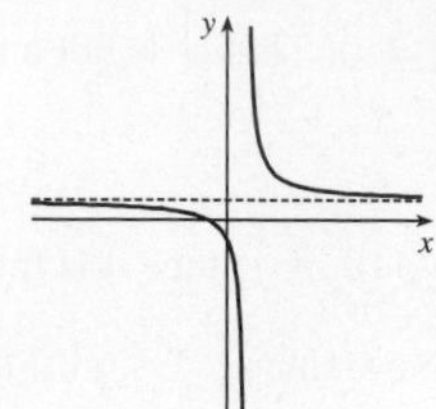

One horizontal asymptote

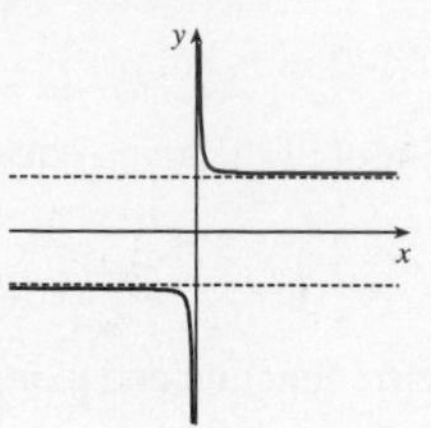

Two horizontal asymptotes

3. (a) $\lim_{x\to\infty} f(x) = -2$ (b) $\lim_{x\to-\infty} f(x) = 2$ (c) $\lim_{x\to1} f(x) = \infty$

(d) $\lim_{x\to3} f(x) = -\infty$ (e) Vertical: $x = 1$, $x = 3$; horizontal: $y = -2$, $y = 2$

4. (a) $\lim_{x\to\infty} g(x) = 2$ (b) $\lim_{x\to-\infty} g(x) = -1$ (c) $\lim_{x\to0} g(x) = -\infty$

(d) $\lim_{x\to2^-} g(x) = -\infty$ (e) $\lim_{x\to2^+} g(x) = \infty$ (f) Vertical: $x = 0$, $x = 2$; horizontal: $y = -1$, $y = 2$

5. $\lim_{x\to0} f(x) = -\infty$,

$\lim_{x\to-\infty} f(x) = 5$,

$\lim_{x\to\infty} f(x) = -5$

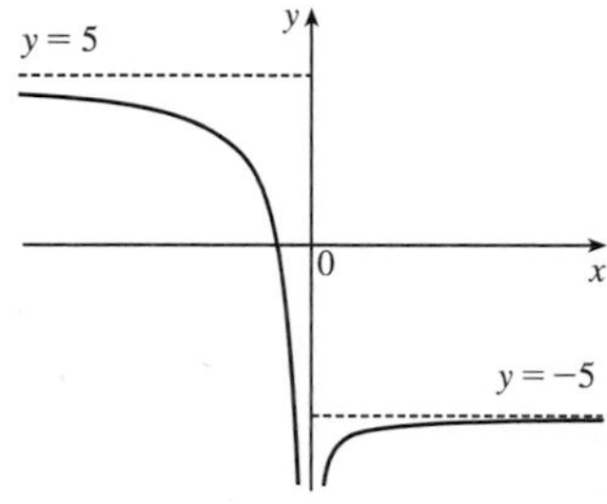

6. $\lim_{x\to2} f(x) = \infty$, $\lim_{x\to-2^+} f(x) = \infty$,

$\lim_{x\to-2^-} f(x) = -\infty$, $\lim_{x\to-\infty} f(x) = 0$,

$\lim_{x\to\infty} f(x) = 0$, $f(0) = 0$

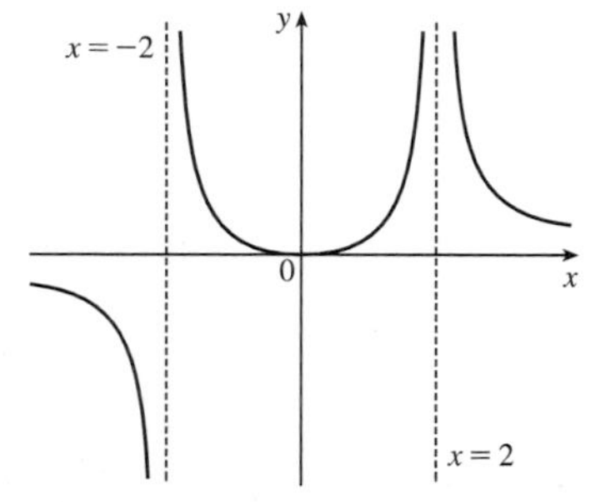

7. $\lim_{x\to2} f(x) = -\infty$, $\lim_{x\to\infty} f(x) = \infty$,

$\lim_{x\to-\infty} f(x) = 0$, $\lim_{x\to0^+} f(x) = \infty$,

$\lim_{x\to0^-} f(x) = -\infty$

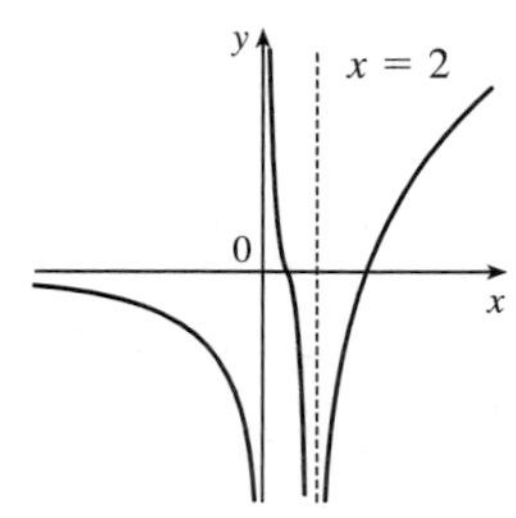

8. $\lim_{x\to\infty} f(x) = 3$,

$\lim_{x\to2^-} f(x) = \infty$,

$\lim_{x\to2^+} f(x) = -\infty$,

f is odd

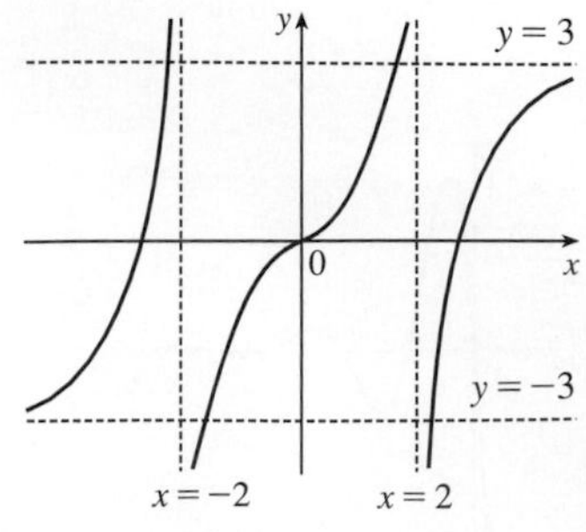

9. $f(0) = 3$, $\lim_{x\to0^-} f(x) = 4$,

$\lim_{x\to0^+} f(x) = 2$,

$\lim_{x\to-\infty} f(x) = -\infty$, $\lim_{x\to4^-} f(x) = -\infty$,

$\lim_{x\to4^+} f(x) = \infty$, $\lim_{x\to\infty} f(x) = 3$

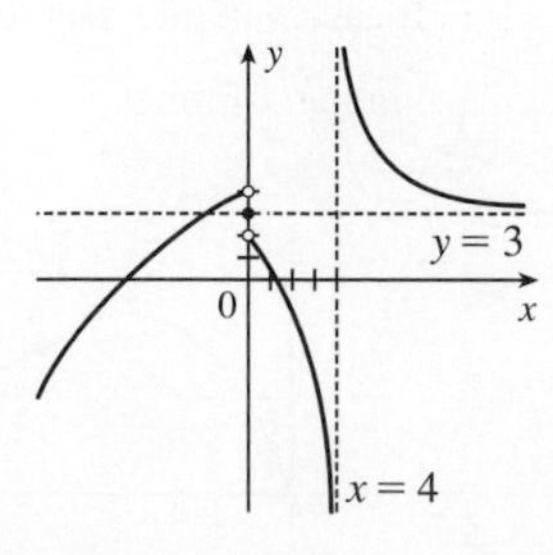

10. $\lim_{x\to3} f(x) = -\infty$, $\lim_{x\to\infty} f(x) = 2$,

$f(0) = 0$, f is even

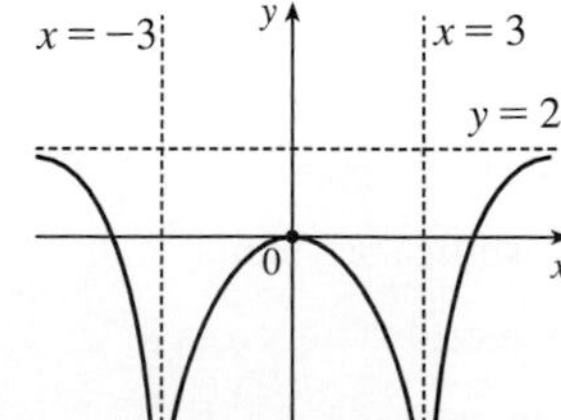

11. If $f(x) = x^2/2^x$, then a calculator gives $f(0) = 0$, $f(1) = 0.5$, $f(2) = 1$, $f(3) = 1.125$, $f(4) = 1$, $f(5) = 0.78125$, $f(6) = 0.5625$, $f(7) = 0.3828125$, $f(8) = 0.25$, $f(9) = 0.158203125$, $f(10) = 0.09765625$, $f(20) \approx 0.00038147$, $f(50) \approx 2.2204 \times 10^{-12}$, $f(100) \approx 7.8886 \times 10^{-27}$. It appears that $\lim\limits_{x\to\infty} (x^2/2^x) = 0$.

12. (a) From a graph of $f(x) = (1 - 2/x)^x$ in a window of $[0, 10{,}000]$ by $[0, 0.2]$, we estimate that $\lim\limits_{x\to\infty} f(x) = 0.14$ (to two decimal places.)

(b)

x	$f(x)$
10,000	0.135 308
100,000	0.135 333
1,000,000	0.135 335

From the table, we estimate that $\lim\limits_{x\to\infty} f(x) = 0.1353$ (to four decimal places.)

13.
$$\lim_{x\to\infty} \frac{2x^2 - 7}{5x^2 + x - 3} = \lim_{x\to\infty} \frac{(2x^2 - 7)/x^2}{(5x^2 + x - 3)/x^2}$$
[Divide both the numerator and denominator by x^2 (the highest power of x that appears in the denominator)]

$$= \frac{\lim\limits_{x\to\infty} (2 - 7/x^2)}{\lim\limits_{x\to\infty} (5 + 1/x - 3/x^2)}$$
[Limit Law 5]

$$= \frac{\lim\limits_{x\to\infty} 2 - \lim\limits_{x\to\infty} (7/x^2)}{\lim\limits_{x\to\infty} 5 + \lim\limits_{x\to\infty} (1/x) - \lim\limits_{x\to\infty} (3/x^2)}$$
[Limit Laws 1 and 2]

$$= \frac{2 - 7\lim\limits_{x\to\infty} (1/x^2)}{5 + \lim\limits_{x\to\infty} (1/x) - 3\lim\limits_{x\to\infty} (1/x^2)}$$
[Limit Laws 7 and 3]

$$= \frac{2 - 7(0)}{5 + 0 + 3(0)}$$
[Theorem 2.6.5]

$$= \frac{2}{5}$$

14.
$$\lim_{x\to\infty} \sqrt{\frac{9x^3 + 8x - 4}{3 - 5x + x^3}} = \sqrt{\lim_{x\to\infty} \frac{9x^3 + 8x - 4}{3 - 5x + x^3}}$$
[Limit Law 11]

$$= \sqrt{\lim_{x\to\infty} \frac{9 + 8/x^2 - 4/x^3}{3/x^3 - 5/x^2 + 1}}$$
[Divide by x^3]

$$= \sqrt{\frac{\lim\limits_{x\to\infty} (9 + 8/x^2 - 4/x^3)}{\lim\limits_{x\to\infty} (3/x^3 - 5/x^2 + 1)}}$$
[Limit Law 5]

$$= \sqrt{\frac{\lim\limits_{x\to\infty} 9 + \lim\limits_{x\to\infty} (8/x^2) - \lim\limits_{x\to\infty} (4/x^3)}{\lim\limits_{x\to\infty} (3/x^3) - \lim\limits_{x\to\infty} (5/x^2) + \lim\limits_{x\to\infty} 1}}$$
[Limit Laws 1 and 2]

$$= \sqrt{\frac{9 + 8\lim\limits_{x\to\infty} (1/x^2) - 4\lim\limits_{x\to\infty} (1/x^3)}{3\lim\limits_{x\to\infty} (1/x^3) - 5\lim\limits_{x\to\infty} (1/x^2) + 1}}$$
[Limit Laws 7 and 3]

$$= \sqrt{\frac{9 + 8(0) - 4(0)}{3(0) - 5(0) + 1}}$$
[Theorem 2.6.5]

$$= \sqrt{\frac{9}{1}} = \sqrt{9} = 3$$

15. $\displaystyle\lim_{x\to\infty}\frac{3x-2}{2x+1}=\lim_{x\to\infty}\frac{(3x-2)/x}{(2x+1)/x}=\lim_{x\to\infty}\frac{3-2/x}{2+1/x}=\frac{\lim\limits_{x\to\infty}3-2\lim\limits_{x\to\infty}1/x}{\lim\limits_{x\to\infty}2+\lim\limits_{x\to\infty}1/x}=\frac{3-2(0)}{2+0}=\frac{3}{2}$

16. $\displaystyle\lim_{x\to\infty}\frac{1-x^2}{x^3-x+1}=\lim_{x\to\infty}\frac{(1-x^2)/x^3}{(x^3-x+1)/x^3}=\lim_{x\to\infty}\frac{1/x^3-1/x}{1-1/x^2+1/x^3}$

$$=\frac{\lim\limits_{x\to\infty}1/x^3-\lim\limits_{x\to\infty}1/x}{\lim\limits_{x\to\infty}1-\lim\limits_{x\to\infty}1/x^2+\lim\limits_{x\to\infty}1/x^3}=\frac{0-0}{1-0+0}=0$$

17. $\displaystyle\lim_{x\to-\infty}\frac{x-2}{x^2+1}=\lim_{x\to-\infty}\frac{(x-2)/x^2}{(x^2+1)/x^2}=\lim_{x\to-\infty}\frac{1/x-2/x^2}{1+1/x^2}=\frac{\lim\limits_{x\to-\infty}1/x-2\lim\limits_{x\to-\infty}1/x^2}{\lim\limits_{x\to-\infty}1+\lim\limits_{x\to-\infty}1/x^2}=\frac{0-2(0)}{1+0}=0$

18. $\displaystyle\lim_{x\to-\infty}\frac{4x^3+6x^2-2}{2x^3-4x+5}=\lim_{x\to-\infty}\frac{(4x^3+6x^2-2)/x^3}{(2x^3-4x+5)/x^3}=\lim_{x\to-\infty}\frac{4+6/x-2/x^3}{2-4/x^2+5/x^3}=\frac{4+0-0}{2-0+0}=2$

19. $\displaystyle\lim_{t\to\infty}\frac{\sqrt{t}+t^2}{2t-t^2}=\lim_{t\to\infty}\frac{(\sqrt{t}+t^2)/t^2}{(2t-t^2)/t^2}=\lim_{t\to\infty}\frac{1/t^{3/2}+1}{2/t-1}=\frac{0+1}{0-1}=-1$

20. $\displaystyle\lim_{t\to\infty}\frac{t-t\sqrt{t}}{2t^{3/2}+3t-5}=\lim_{t\to\infty}\frac{\left(t-t\sqrt{t}\right)/t^{3/2}}{\left(2t^{3/2}+3t-5\right)/t^{3/2}}=\lim_{t\to\infty}\frac{1/t^{1/2}-1}{2+3/t^{1/2}-5/t^{3/2}}=\frac{0-1}{2+0-0}=-\frac{1}{2}$

21. $\displaystyle\lim_{x\to\infty}\frac{(2x^2+1)^2}{(x-1)^2(x^2+x)}=\lim_{x\to\infty}\frac{(2x^2+1)^2/x^4}{[(x-1)^2(x^2+x)]/x^4}=\lim_{x\to\infty}\frac{[(2x^2+1)/x^2]^2}{[(x^2-2x+1)/x^2][(x^2+x)/x^2]}$

$$=\lim_{x\to\infty}\frac{(2+1/x^2)^2}{(1-2/x+1/x^2)(1+1/x)}=\frac{(2+0)^2}{(1-0+0)(1+0)}=4$$

22. $\displaystyle\lim_{x\to\infty}\frac{x^2}{\sqrt{x^4+1}}=\lim_{x\to\infty}\frac{x^2/x^2}{\sqrt{x^4+1}/x^2}=\lim_{x\to\infty}\frac{1}{\sqrt{(x^4+1)/x^4}}$ [since $x^2=\sqrt{x^4}$ for $x>0$]

$$=\lim_{x\to\infty}\frac{1}{\sqrt{1+1/x^4}}=\frac{1}{\sqrt{1+0}}=1$$

23. $\displaystyle\lim_{x\to\infty}\frac{\sqrt{1+4x^6}}{2-x^3}=\lim_{x\to\infty}\frac{\sqrt{1+4x^6}/x^3}{(2-x^3)/x^3}=\frac{\lim\limits_{x\to\infty}\sqrt{(1+4x^6)/x^6}}{\lim\limits_{x\to\infty}(2/x^3-1)}$ [since $x^3=\sqrt{x^6}$ for $x>0$]

$$=\frac{\lim\limits_{x\to\infty}\sqrt{1/x^6+4}}{\lim\limits_{x\to\infty}(2/x^3)-\lim\limits_{x\to\infty}1}=\frac{\sqrt{\lim\limits_{x\to\infty}(1/x^6)+\lim\limits_{x\to\infty}4}}{0-1}$$

$$=\frac{\sqrt{0+4}}{-1}=\frac{2}{-1}=-2$$

24. $\displaystyle\lim_{x\to-\infty}\frac{\sqrt{1+4x^6}}{2-x^3}=\lim_{x\to-\infty}\frac{\sqrt{1+4x^6}/x^3}{(2-x^3)/x^3}=\frac{\lim\limits_{x\to-\infty}-\sqrt{(1+4x^6)/x^6}}{\lim\limits_{x\to-\infty}(2/x^3-1)}$ [since $x^3=-\sqrt{x^6}$ for $x<0$]

$$=\frac{\lim\limits_{x\to-\infty}-\sqrt{1/x^6+4}}{2\lim\limits_{x\to-\infty}(1/x^3)-\lim\limits_{x\to-\infty}1}=\frac{-\sqrt{\lim\limits_{x\to-\infty}(1/x^6)+\lim\limits_{x\to-\infty}4}}{2(0)-1}$$

$$=\frac{-\sqrt{0+4}}{-1}=\frac{-2}{-1}=2$$

25. $\lim\limits_{x\to\infty}\dfrac{\sqrt{x+3x^2}}{4x-1}=\lim\limits_{x\to\infty}\dfrac{\sqrt{x+3x^2}/x}{(4x-1)/x}=\dfrac{\lim\limits_{x\to\infty}\sqrt{(x+3x^2)/x^2}}{\lim\limits_{x\to\infty}(4-1/x)}$ [since $x=\sqrt{x^2}$ for $x>0$]

$$=\frac{\lim\limits_{x\to\infty}\sqrt{1/x+3}}{\lim\limits_{x\to\infty}4-\lim\limits_{x\to\infty}(1/x)}=\frac{\sqrt{\lim\limits_{x\to\infty}(1/x)+\lim\limits_{x\to\infty}3}}{4-0}=\frac{\sqrt{0+3}}{4}=\frac{\sqrt{3}}{4}$$

26. $\lim\limits_{x\to\infty}\dfrac{x+3x^2}{4x-1}=\lim\limits_{x\to\infty}\dfrac{(x+3x^2)/x}{(4x-1)/x}=\lim\limits_{x\to\infty}\dfrac{1+3x}{4-1/x}$

$=\infty$ since $1+3x\to\infty$ and $4-1/x\to 4$ as $x\to\infty$.

27. $\lim\limits_{x\to\infty}\left(\sqrt{9x^2+x}-3x\right)=\lim\limits_{x\to\infty}\dfrac{\left(\sqrt{9x^2+x}-3x\right)\left(\sqrt{9x^2+x}+3x\right)}{\sqrt{9x^2+x}+3x}=\lim\limits_{x\to\infty}\dfrac{\left(\sqrt{9x^2+x}\right)^2-(3x)^2}{\sqrt{9x^2+x}+3x}$

$$=\lim_{x\to\infty}\frac{(9x^2+x)-9x^2}{\sqrt{9x^2+x}+3x}=\lim_{x\to\infty}\frac{x}{\sqrt{9x^2+x}+3x}\cdot\frac{1/x}{1/x}$$

$$=\lim_{x\to\infty}\frac{x/x}{\sqrt{9x^2/x^2+x/x^2}+3x/x}=\lim_{x\to\infty}\frac{1}{\sqrt{9+1/x}+3}=\frac{1}{\sqrt{9}+3}=\frac{1}{3+3}=\frac{1}{6}$$

28. $\lim\limits_{x\to-\infty}\left(\sqrt{4x^2+3x}+2x\right)=\lim\limits_{x\to-\infty}\left(\sqrt{4x^2+3x}+2x\right)\left[\dfrac{\sqrt{4x^2+3x}-2x}{\sqrt{4x^2+3x}-2x}\right]$

$$=\lim_{x\to-\infty}\frac{(4x^2+3x)-(2x)^2}{\sqrt{4x^2+3x}-2x}=\lim_{x\to-\infty}\frac{3x}{\sqrt{4x^2+3x}-2x}$$

$=\lim\limits_{x\to-\infty}\dfrac{3x/x}{\left(\sqrt{4x^2+3x}-2x\right)/x}=\lim\limits_{x\to-\infty}\dfrac{3}{-\sqrt{4+3/x}-2}$ [since $x=-\sqrt{x^2}$ for $x<0$]

$$=\frac{3}{-\sqrt{4+0}-2}=-\frac{3}{4}$$

29. $\lim\limits_{x\to\infty}\left(\sqrt{x^2+ax}-\sqrt{x^2+bx}\right)=\lim\limits_{x\to\infty}\dfrac{\left(\sqrt{x^2+ax}-\sqrt{x^2+bx}\right)\left(\sqrt{x^2+ax}+\sqrt{x^2+bx}\right)}{\sqrt{x^2+ax}+\sqrt{x^2+bx}}$

$$=\lim_{x\to\infty}\frac{(x^2+ax)-(x^2+bx)}{\sqrt{x^2+ax}+\sqrt{x^2+bx}}=\lim_{x\to\infty}\frac{[(a-b)x]/x}{\left(\sqrt{x^2+ax}+\sqrt{x^2+bx}\right)/\sqrt{x^2}}$$

$$=\lim_{x\to\infty}\frac{a-b}{\sqrt{1+a/x}+\sqrt{1+b/x}}=\frac{a-b}{\sqrt{1+0}+\sqrt{1+0}}=\frac{a-b}{2}$$

30. For $x>0$, $\sqrt{x^2+1}>\sqrt{x^2}=x$. So as $x\to\infty$, we have $\sqrt{x^2+1}\to\infty$, that is, $\lim\limits_{x\to\infty}\sqrt{x^2+1}=\infty$.

31. $\lim\limits_{x\to\infty}\dfrac{x^4-3x^2+x}{x^3-x+2}=\lim\limits_{x\to\infty}\dfrac{(x^4-3x^2+x)/x^3}{(x^3-x+2)/x^3}$ [divide by the highest power of x in the denominator] $=\lim\limits_{x\to\infty}\dfrac{x-3/x+1/x^2}{1-1/x^2+2/x^3}=\infty$

since the numerator increases without bound and the denominator approaches 1 as $x\to\infty$.

32. $\lim\limits_{x\to\infty}(e^{-x}+2\cos 3x)$ does not exist. $\lim\limits_{x\to\infty}e^{-x}=0$, but $\lim\limits_{x\to\infty}(2\cos 3x)$ does not exist because the values of $2\cos 3x$ oscillate between the values of -2 and 2 infinitely often, so the given limit does not exist.

33. $\lim\limits_{x \to -\infty}(x^2 + 2x^7) = \lim\limits_{x \to -\infty} x^7\left(\dfrac{1}{x^5} + 2\right)$ [factor out the largest power of x] $= -\infty$ because $x^7 \to -\infty$ and $1/x^5 + 2 \to 2$ as $x \to -\infty$.

Or: $\lim\limits_{x \to -\infty}(x^2 + 2x^7) = \lim\limits_{x \to -\infty} x^2(1 + 2x^5) = -\infty$.

34. $\lim\limits_{x \to -\infty} \dfrac{1 + x^6}{x^4 + 1} = \lim\limits_{x \to -\infty} \dfrac{(1 + x^6)/x^4}{(x^4 + 1)/x^4}$ $\left[\begin{array}{c}\text{divide by the highest power}\\ \text{of } x \text{ in the denominator}\end{array}\right]$ $= \lim\limits_{x \to -\infty} \dfrac{1/x^4 + x^2}{1 + 1/x^4} = \infty$ since the numerator increases without bound and the denominator approaches 1 as $x \to -\infty$.

35. Let $t = e^x$. As $x \to \infty$, $t \to \infty$. $\lim\limits_{x \to \infty} \arctan(e^x) = \lim\limits_{t \to \infty} \arctan t = \frac{\pi}{2}$ by (3).

36. Divide numerator and denominator by e^{3x}: $\lim\limits_{x \to \infty} \dfrac{e^{3x} - e^{-3x}}{e^{3x} + e^{-3x}} = \lim\limits_{x \to \infty} \dfrac{1 - e^{-6x}}{1 + e^{-6x}} = \dfrac{1 - 0}{1 + 0} = 1$

37. $\lim\limits_{x \to \infty} \dfrac{1 - e^x}{1 + 2e^x} = \lim\limits_{x \to \infty} \dfrac{(1 - e^x)/e^x}{(1 + 2e^x)/e^x} = \lim\limits_{x \to \infty} \dfrac{1/e^x - 1}{1/e^x + 2} = \dfrac{0 - 1}{0 + 2} = -\dfrac{1}{2}$

38. Since $0 \le \sin^2 x \le 1$, we have $0 \le \dfrac{\sin^2 x}{x^2 + 1} \le \dfrac{1}{x^2 + 1}$. We know that $\lim\limits_{x \to \infty} 0 = 0$ and $\lim\limits_{x \to \infty} \dfrac{1}{x^2 + 1} = 0$, so by the Squeeze Theorem, $\lim\limits_{x \to \infty} \dfrac{\sin^2 x}{x^2 + 1} = 0$.

39. Since $-1 \le \cos x \le 1$ and $e^{-2x} > 0$, we have $-e^{-2x} \le e^{-2x} \cos x \le e^{-2x}$. We know that $\lim\limits_{x \to \infty}(-e^{-2x}) = 0$ and $\lim\limits_{x \to \infty}(e^{-2x}) = 0$, so by the Squeeze Theorem, $\lim\limits_{x \to \infty}(e^{-2x} \cos x) = 0$.

40. Let $t = \ln x$. As $x \to 0^+$, $t \to -\infty$. $\lim\limits_{x \to 0^+} \tan^{-1}(\ln x) = \lim\limits_{t \to -\infty} \tan^{-1} t = -\frac{\pi}{2}$ by (4).

41. $\lim\limits_{x \to \infty}[\ln(1 + x^2) - \ln(1 + x)] = \lim\limits_{x \to \infty} \ln \dfrac{1 + x^2}{1 + x} = \ln\left(\lim\limits_{x \to \infty} \dfrac{1 + x^2}{1 + x}\right) = \ln\left(\lim\limits_{x \to \infty} \dfrac{\frac{1}{x} + x}{\frac{1}{x} + 1}\right) = \infty$, since the limit in parentheses is ∞.

42. $\lim\limits_{x \to \infty}[\ln(2 + x) - \ln(1 + x)] = \lim\limits_{x \to \infty} \ln\left(\dfrac{2 + x}{1 + x}\right) = \lim\limits_{x \to \infty} \ln\left(\dfrac{2/x + 1}{1/x + 1}\right) = \ln \dfrac{1}{1} = \ln 1 = 0$

43. (a) (i) $\lim\limits_{x \to 0^+} f(x) = \lim\limits_{x \to 0^+} \dfrac{x}{\ln x} = 0$ since $x \to 0^+$ and $\ln x \to -\infty$ as $x \to 0^+$.

(ii) $\lim\limits_{x \to 1^-} f(x) = \lim\limits_{x \to 1^-} \dfrac{x}{\ln x} = -\infty$ since $x \to 1$ and $\ln x \to 0^-$ as $x \to 1^-$.

(iii) $\lim\limits_{x \to 1^+} f(x) = \lim\limits_{x \to 1^+} \dfrac{x}{\ln x} = \infty$ since $x \to 1$ and $\ln x \to 0^+$ as $x \to 1^+$.

(b)

x	$f(x)$
10,000	1085.7
100,000	8685.9
1,000,000	72,382.4

It appears that $\lim\limits_{x \to \infty} f(x) = \infty$.

(c)

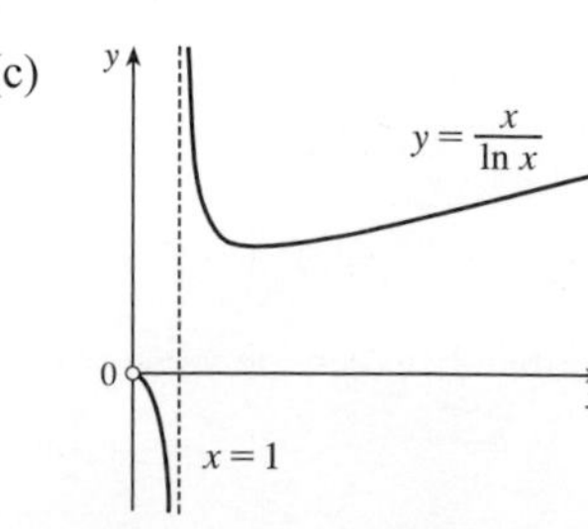

44. (a) $\lim\limits_{x\to\infty} f(x) = \lim\limits_{x\to\infty}\left(\dfrac{2}{x} - \dfrac{1}{\ln x}\right) = 0$

since $\dfrac{2}{x} \to 0$ and $\dfrac{1}{\ln x} \to 0$ as $x \to \infty$.

(b) $\lim\limits_{x\to 0^+} f(x) = \lim\limits_{x\to 0^+}\left(\dfrac{2}{x} - \dfrac{1}{\ln x}\right) = \infty$

since $\dfrac{2}{x} \to \infty$ and $\dfrac{1}{\ln x} \to 0$ as $x \to 0^+$.

(c) $\lim\limits_{x\to 1^-} f(x) = \lim\limits_{x\to 1^-}\left(\dfrac{2}{x} - \dfrac{1}{\ln x}\right) = \infty$ since $\dfrac{2}{x} \to 2$ and $\dfrac{1}{\ln x} \to -\infty$ as $x \to 1^-$.

(d) $\lim\limits_{x\to 1^+} f(x) = \lim\limits_{x\to 1^+}\left(\dfrac{2}{x} - \dfrac{1}{\ln x}\right) = -\infty$ since $\dfrac{2}{x} \to 2$ and $\dfrac{1}{\ln x} \to \infty$ as $x \to 1^+$.

(e)

45. (a)

From the graph of $f(x) = \sqrt{x^2+x+1} + x$, we estimate the value of $\lim\limits_{x\to-\infty} f(x)$ to be -0.5.

(b)

x	$f(x)$
$-10{,}000$	$-0.499\,962\,5$
$-100{,}000$	$-0.499\,996\,2$
$-1{,}000{,}000$	$-0.499\,999\,6$

From the table, we estimate the limit to be -0.5.

(c) $\lim\limits_{x\to-\infty}\left(\sqrt{x^2+x+1}+x\right) = \lim\limits_{x\to-\infty}\left(\sqrt{x^2+x+1}+x\right)\left[\dfrac{\sqrt{x^2+x+1}-x}{\sqrt{x^2+x+1}-x}\right] = \lim\limits_{x\to-\infty}\dfrac{(x^2+x+1)-x^2}{\sqrt{x^2+x+1}-x}$

$$= \lim_{x\to-\infty}\frac{(x+1)(1/x)}{\left(\sqrt{x^2+x+1}-x\right)(1/x)} = \lim_{x\to-\infty}\frac{1+(1/x)}{-\sqrt{1+(1/x)+(1/x^2)}-1}$$

$$= \frac{1+0}{-\sqrt{1+0+0}-1} = -\frac{1}{2}$$

Note that for $x < 0$, we have $\sqrt{x^2} = |x| = -x$, so when we divide the radical by x, with $x < 0$, we get $\dfrac{1}{x}\sqrt{x^2+x+1} = -\dfrac{1}{\sqrt{x^2}}\sqrt{x^2+x+1} = -\sqrt{1+(1/x)+(1/x^2)}$.

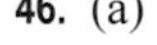
46. (a)

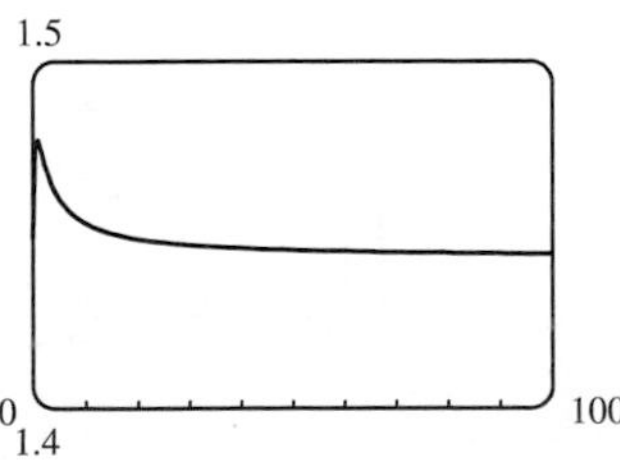

From the graph of $f(x) = \sqrt{3x^2+8x+6} - \sqrt{3x^2+3x+1}$, we estimate (to one decimal place) the value of $\lim\limits_{x\to\infty} f(x)$ to be 1.4.

(b)

x	$f(x)$
10,000	1.443 39
100,000	1.443 38
1,000,000	1.443 38

From the table, we estimate (to four decimal places) the limit to be 1.4434.

(c) $\lim\limits_{x\to\infty} f(x) = \lim\limits_{x\to\infty} \dfrac{\left(\sqrt{3x^2+8x+6} - \sqrt{3x^2+3x+1}\right)\left(\sqrt{3x^2+8x+6} + \sqrt{3x^2+3x+1}\right)}{\sqrt{3x^2+8x+6} + \sqrt{3x^2+3x+1}}$

$= \lim\limits_{x\to\infty} \dfrac{(3x^2+8x+6) - (3x^2+3x+1)}{\sqrt{3x^2+8x+6} + \sqrt{3x^2+3x+1}} = \lim\limits_{x\to\infty} \dfrac{(5x+5)(1/x)}{\left(\sqrt{3x^2+8x+6} + \sqrt{3x^2+3x+1}\right)(1/x)}$

$= \lim\limits_{x\to\infty} \dfrac{5+5/x}{\sqrt{3+8/x+6/x^2} + \sqrt{3+3/x+1/x^2}} = \dfrac{5}{\sqrt{3}+\sqrt{3}} = \dfrac{5}{2\sqrt{3}} = \dfrac{5\sqrt{3}}{6} \approx 1.443376$

47. $\lim\limits_{x\to\pm\infty} \dfrac{5+4x}{x+3} = \lim\limits_{x\to\pm\infty} \dfrac{(5+4x)/x}{(x+3)/x} = \lim\limits_{x\to\pm\infty} \dfrac{5/x+4}{1+3/x} = \dfrac{0+4}{1+0} = 4$, so

$y = 4$ is a horizontal asymptote. $y = f(x) = \dfrac{5+4x}{x+3}$, so $\lim\limits_{x\to-3^+} f(x) = -\infty$

since $5+4x \to -7$ and $x+3 \to 0^+$ as $x \to -3^+$. Thus, $x = -3$ is a vertical asymptote. The graph confirms our work.

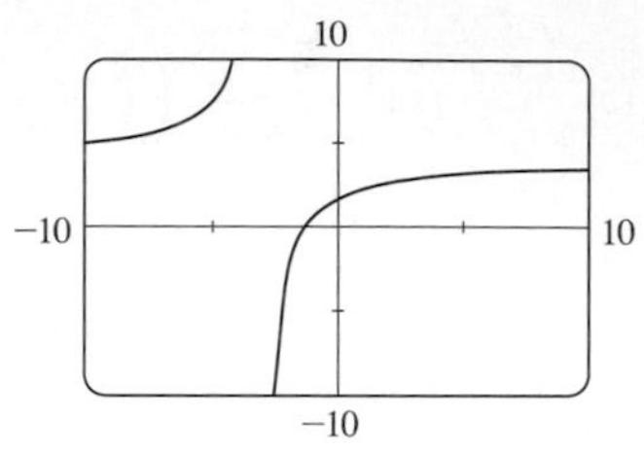

48. $\lim\limits_{x\to\pm\infty} \dfrac{2x^2+1}{3x^2+2x-1} = \lim\limits_{x\to\pm\infty} \dfrac{(2x^2+1)/x^2}{(3x^2+2x-1)/x^2}$

$= \lim\limits_{x\to\pm\infty} \dfrac{2+1/x^2}{3+2/x-1/x^2} = \dfrac{2}{3}$

so $y = \dfrac{2}{3}$ is a horizontal asymptote. $y = f(x) = \dfrac{2x^2+1}{3x^2+2x-1} = \dfrac{2x^2+1}{(3x-1)(x+1)}$.

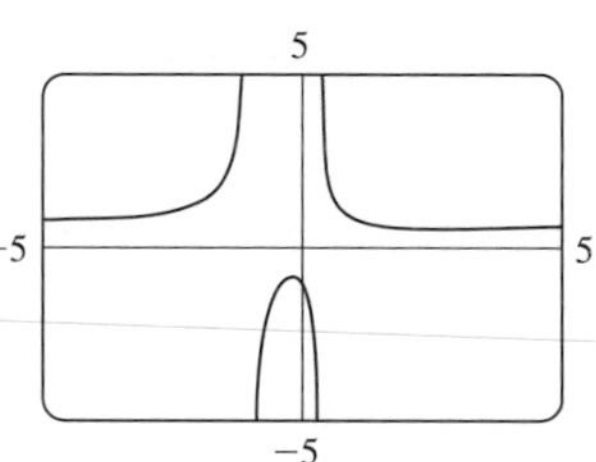

The denominator is zero when $x = \frac{1}{3}$ and -1, but the numerator is nonzero, so $x = \frac{1}{3}$ and $x = -1$ are vertical asymptotes. The graph confirms our work.

49. $\lim\limits_{x\to\pm\infty} \dfrac{2x^2+x-1}{x^2+x-2} = \lim\limits_{x\to\pm\infty} \dfrac{\dfrac{2x^2+x-1}{x^2}}{\dfrac{x^2+x-2}{x^2}} = \lim\limits_{x\to\pm\infty} \dfrac{2+\dfrac{1}{x}-\dfrac{1}{x^2}}{1+\dfrac{1}{x}-\dfrac{2}{x^2}} = \dfrac{\lim\limits_{x\to\pm\infty}\left(2+\dfrac{1}{x}-\dfrac{1}{x^2}\right)}{\lim\limits_{x\to\pm\infty}\left(1+\dfrac{1}{x}-\dfrac{2}{x^2}\right)}$

$= \dfrac{\lim\limits_{x\to\pm\infty} 2 + \lim\limits_{x\to\pm\infty} \dfrac{1}{x} - \lim\limits_{x\to\pm\infty} \dfrac{1}{x^2}}{\lim\limits_{x\to\pm\infty} 1 + \lim\limits_{x\to\pm\infty} \dfrac{1}{x} - 2\lim\limits_{x\to\pm\infty} \dfrac{1}{x^2}} = \dfrac{2+0-0}{1+0-2(0)} = 2$, so $y = 2$ is a horizontal asymptote.

$y = f(x) = \dfrac{2x^2+x-1}{x^2+x-2} = \dfrac{(2x-1)(x+1)}{(x+2)(x-1)}$, so $\lim\limits_{x\to-2^-} f(x) = \infty$,

$\lim\limits_{x\to-2^+} f(x) = -\infty$, $\lim\limits_{x\to1^-} f(x) = -\infty$, and $\lim\limits_{x\to1^+} f(x) = \infty$. Thus, $x = -2$ and $x = 1$ are vertical asymptotes. The graph confirms our work.

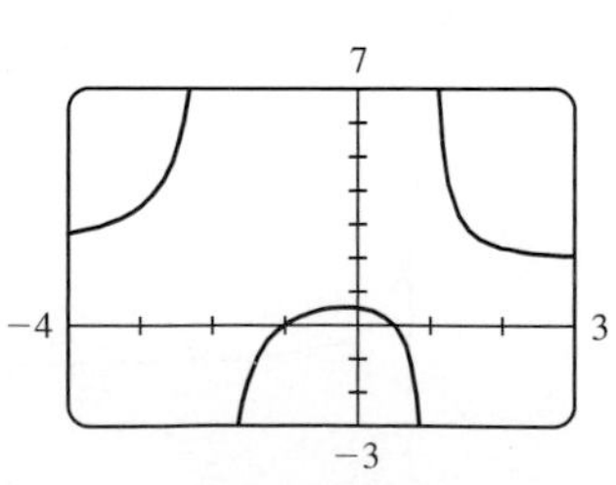

50. $\lim\limits_{x\to\pm\infty} \dfrac{1+x^4}{x^2-x^4} = \lim\limits_{x\to\pm\infty} \dfrac{\dfrac{1+x^4}{x^4}}{\dfrac{x^2-x^4}{x^4}} = \lim\limits_{x\to\pm\infty} \dfrac{\dfrac{1}{x^4}+1}{\dfrac{1}{x^2}-1} = \dfrac{\lim\limits_{x\to\pm\infty}\left(\dfrac{1}{x^4}+1\right)}{\lim\limits_{x\to\pm\infty}\left(\dfrac{1}{x^2}-1\right)} = \dfrac{\lim\limits_{x\to\pm\infty}\dfrac{1}{x^4} + \lim\limits_{x\to\pm\infty} 1}{\lim\limits_{x\to\pm\infty}\dfrac{1}{x^2} - \lim\limits_{x\to\pm\infty} 1}$

$= \dfrac{0+1}{0-1} = -1$, so $y = -1$ is a horizontal asymptote.

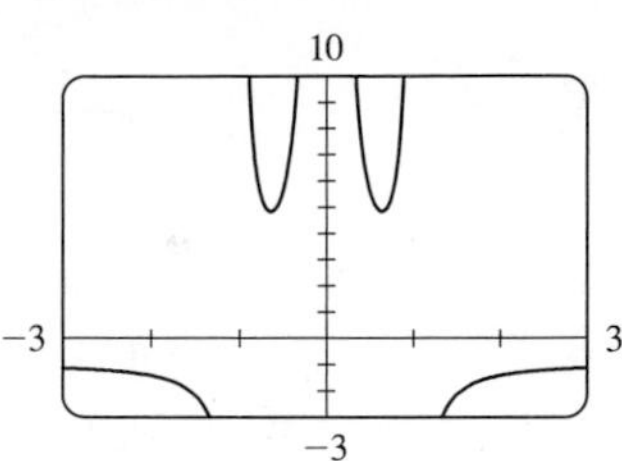

$y = f(x) = \dfrac{1+x^4}{x^2 - x^4} = \dfrac{1+x^4}{x^2(1-x^2)} = \dfrac{1+x^4}{x^2(1+x)(1-x)}$. The denominator is zero when $x = 0, -1$, and 1, but the numerator is nonzero, so $x = 0$, $x = -1$, and $x = 1$ are vertical asymptotes. Notice that as $x \to 0$, the numerator and denominator are both positive, so $\lim\limits_{x\to 0} f(x) = \infty$. The graph confirms our work.

51. $y = f(x) = \dfrac{x^3 - x}{x^2 - 6x + 5} = \dfrac{x(x^2-1)}{(x-1)(x-5)} = \dfrac{x(x+1)(x-1)}{(x-1)(x-5)} = \dfrac{x(x+1)}{x-5} = g(x)$ for $x \neq 1$.

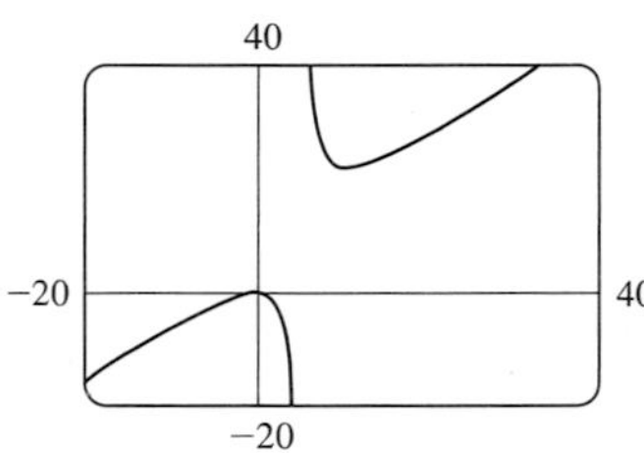

The graph of g is the same as the graph of f with the exception of a hole in the graph of f at $x = 1$. By long division, $g(x) = \dfrac{x^2+x}{x-5} = x + 6 + \dfrac{30}{x-5}$.

As $x \to \pm\infty$, $g(x) \to \pm\infty$, so there is no horizontal asymptote. The denominator of g is zero when $x = 5$. $\lim\limits_{x\to 5^-} g(x) = -\infty$ and $\lim\limits_{x\to 5^+} g(x) = \infty$, so $x = 5$ is a vertical asymptote. The graph confirms our work.

52. $\lim\limits_{x\to\infty} \dfrac{2e^x}{e^x - 5} = \lim\limits_{x\to\infty} \dfrac{2e^x}{e^x-5} \cdot \dfrac{1/e^x}{1/e^x} = \lim\limits_{x\to\infty} \dfrac{2}{1-(5/e^x)} = \dfrac{2}{1-0} = 2$, so $y = 2$ is a horizontal asymptote.

$\lim\limits_{x\to-\infty} \dfrac{2e^x}{e^x-5} = \dfrac{2(0)}{0-5} = 0$, so $y = 0$ is a horizontal asymptote. The denominator is zero (and the numerator isn't) when $e^x - 5 = 0 \quad\Rightarrow\quad e^x = 5 \quad\Rightarrow\quad x = \ln 5$.

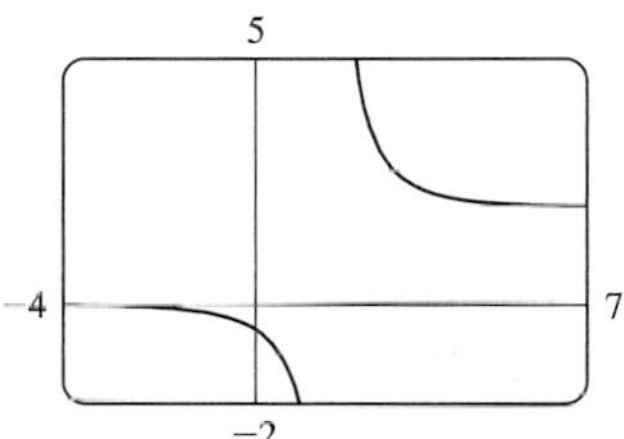

$\lim\limits_{x\to(\ln 5)^+} \dfrac{2e^x}{e^x-5} = \infty$ since the numerator approaches 10 and the denominator approaches 0 through positive values as $x \to (\ln 5)^+$. Similarly, $\lim\limits_{x\to(\ln 5)^-} \dfrac{2e^x}{e^x-5} = -\infty$. Thus, $x = \ln 5$ is a vertical asymptote. The graph confirms our work.

53. From the graph, it appears $y = 1$ is a horizontal asymptote.

$$\begin{aligned}
\lim_{x\to\pm\infty} \frac{3x^3 + 500x^2}{x^3 + 500x^2 + 100x + 2000} &= \lim_{x\to\pm\infty} \frac{\dfrac{3x^3+500x^2}{x^3}}{\dfrac{x^3+500x^2+100x+2000}{x^3}} \\
&= \lim_{x\to\pm\infty} \frac{3 + (500/x)}{1 + (500/x) + (100/x^2) + (2000/x^3)} \\
&= \frac{3+0}{1+0+0+0} = 3, \quad \text{so } y = 3 \text{ is a horizontal asymptote.}
\end{aligned}$$

The discrepancy can be explained by the choice of the viewing window. Try $[-100{,}000, 100{,}000]$ by $[-1, 4]$ to get a graph that lends credibility to our calculation that $y = 3$ is a horizontal asymptote.

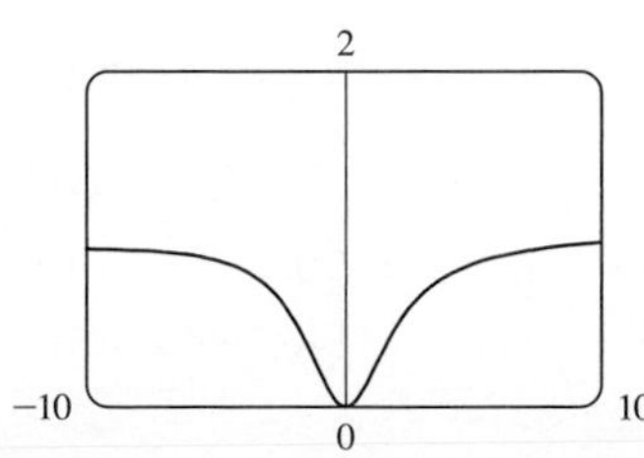

54. (a)

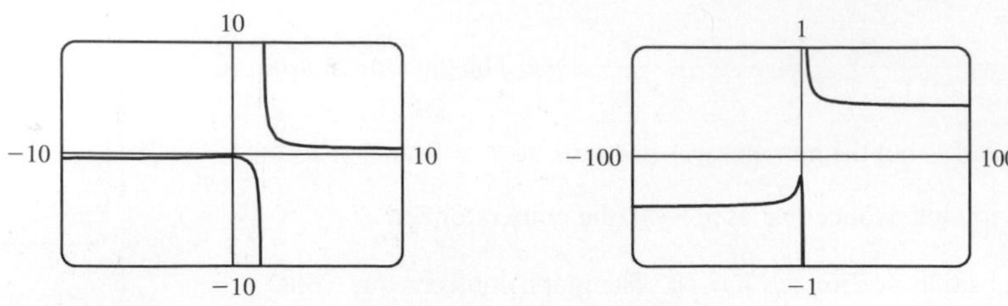

From the graph, it appears at first that there is only one horizontal asymptote, at $y \approx 0$, and a vertical asymptote at $x \approx 1.7$. However, if we graph the function with a wider and shorter viewing rectangle, we see that in fact there seem to be two horizontal asymptotes: one at $y \approx 0.5$ and one at $y \approx -0.5$. So we estimate that

$$\lim_{x\to\infty} \frac{\sqrt{2x^2+1}}{3x-5} \approx 0.5 \qquad \text{and} \qquad \lim_{x\to-\infty} \frac{\sqrt{2x^2+1}}{3x-5} \approx -0.5$$

(b) $f(1000) \approx 0.4722$ and $f(10{,}000) \approx 0.4715$, so we estimate that $\lim\limits_{x\to\infty} \dfrac{\sqrt{2x^2+1}}{3x-5} \approx 0.47$.

$f(-1000) \approx -0.4706$ and $f(-10{,}000) \approx -0.4713$, so we estimate that $\lim\limits_{x\to-\infty} \dfrac{\sqrt{2x^2+1}}{3x-5} \approx -0.47$.

(c) $\lim\limits_{x\to\infty} \dfrac{\sqrt{2x^2+1}}{3x-5} = \lim\limits_{x\to\infty} \dfrac{\sqrt{2+1/x^2}}{3-5/x}$ [since $\sqrt{x^2} = x$ for $x > 0$] $= \dfrac{\sqrt{2}}{3} \approx 0.471404$.

For $x < 0$, we have $\sqrt{x^2} = |x| = -x$, so when we divide the numerator by x, with $x < 0$, we get $\dfrac{1}{x}\sqrt{2x^2+1} = -\dfrac{1}{\sqrt{x^2}}\sqrt{2x^2+1} = -\sqrt{2+1/x^2}$. Therefore,

$$\lim_{x\to-\infty} \frac{\sqrt{2x^2+1}}{3x-5} = \lim_{x\to-\infty} \frac{-\sqrt{2+1/x^2}}{3-5/x} = -\frac{\sqrt{2}}{3} \approx -0.471404.$$

55. Divide the numerator and the denominator by the highest power of x in $Q(x)$.

(a) If $\deg P < \deg Q$, then the numerator $\to 0$ but the denominator doesn't. So $\lim\limits_{x\to\infty} [P(x)/Q(x)] = 0$.

(b) If $\deg P > \deg Q$, then the numerator $\to \pm\infty$ but the denominator doesn't, so $\lim\limits_{x\to\infty} [P(x)/Q(x)] = \pm\infty$ (depending on the ratio of the leading coefficients of P and Q).

56.

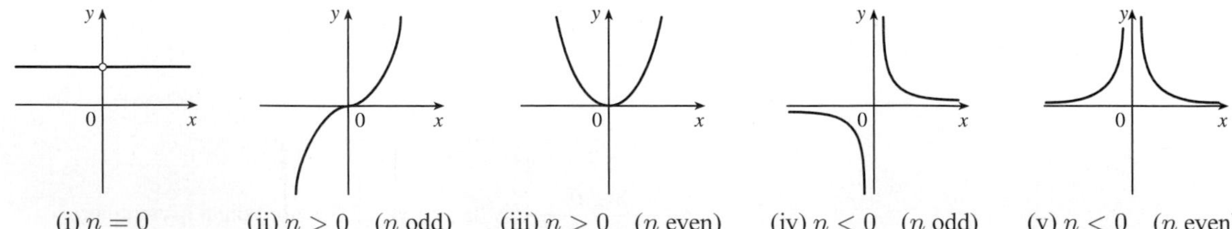

From these sketches we see that

(a) $\lim\limits_{x\to0^+} x^n = \begin{cases} 1 & \text{if } n = 0 \\ 0 & \text{if } n > 0 \\ \infty & \text{if } n < 0 \end{cases}$ (b) $\lim\limits_{x\to0^-} x^n = \begin{cases} 1 & \text{if } n = 0 \\ 0 & \text{if } n > 0 \\ -\infty & \text{if } n < 0,\ n \text{ odd} \\ \infty & \text{if } n < 0,\ n \text{ even} \end{cases}$

(c) $\lim_{x\to\infty} x^n = \begin{cases} 1 & \text{if } n = 0 \\ \infty & \text{if } n > 0 \\ 0 & \text{if } n < 0 \end{cases}$

(d) $\lim_{x\to-\infty} x^n = \begin{cases} 1 & \text{if } n = 0 \\ -\infty & \text{if } n > 0,\ n \text{ odd} \\ \infty & \text{if } n > 0,\ n \text{ even} \\ 0 & \text{if } n < 0 \end{cases}$

57. Let's look for a rational function.

(1) $\lim_{x\to\pm\infty} f(x) = 0 \;\Rightarrow\;$ degree of numerator < degree of denominator

(2) $\lim_{x\to 0} f(x) = -\infty \;\Rightarrow\;$ there is a factor of x^2 in the denominator (not just x, since that would produce a sign change at $x = 0$), and the function is negative near $x = 0$.

(3) $\lim_{x\to 3^-} f(x) = \infty$ and $\lim_{x\to 3^+} f(x) = -\infty \;\Rightarrow\;$ vertical asymptote at $x = 3$; there is a factor of $(x-3)$ in the denominator.

(4) $f(2) = 0 \;\Rightarrow\;$ 2 is an x-intercept; there is at least one factor of $(x-2)$ in the numerator.

Combining all of this information and putting in a negative sign to give us the desired left- and right-hand limits gives us $f(x) = \dfrac{2-x}{x^2(x-3)}$ as one possibility.

58. Since the function has vertical asymptotes $x = 1$ and $x = 3$, the denominator of the rational function we are looking for must have factors $(x-1)$ and $(x-3)$. Because the horizontal asymptote is $y = 1$, the degree of the numerator must equal the degree of the denominator, and the ratio of the leading coefficients must be 1. One possibility is $f(x) = \dfrac{x^2}{(x-1)(x-3)}$.

59. (a) We must first find the function f. Since f has a vertical asymptote $x = 4$ and x-intercept $x = 1$, $x - 4$ is a factor of the denominator and $x - 1$ is a factor of the numerator. There is a removable discontinuity at $x = -1$, so $x - (-1) = x + 1$ is a factor of both the numerator and denominator. Thus, f now looks like this: $f(x) = \dfrac{a(x-1)(x+1)}{(x-4)(x+1)}$, where a is still to be determined. Then $\lim_{x\to -1} f(x) = \lim_{x\to -1} \dfrac{a(x-1)(x+1)}{(x-4)(x+1)} = \lim_{x\to -1} \dfrac{a(x-1)}{x-4} = \dfrac{a(-1-1)}{(-1-4)} = \dfrac{2}{5}a$, so $\dfrac{2}{5}a = 2$, and $a = 5$. Thus $f(x) = \dfrac{5(x-1)(x+1)}{(x-4)(x+1)}$ is a ratio of quadratic functions satisfying all the given conditions and $f(0) = \dfrac{5(-1)(1)}{(-4)(1)} = \dfrac{5}{4}$.

(b) $\lim_{x\to\infty} f(x) = 5 \lim_{x\to\infty} \dfrac{x^2-1}{x^2-3x-4} = 5 \lim_{x\to\infty} \dfrac{(x^2/x^2) - (1/x^2)}{(x^2/x^2) - (3x/x^2) - (4/x^2)} = 5\dfrac{1-0}{1-0-0} = 5(1) = 5$

60. $y = f(x) = 2x^3 - x^4 = x^3(2-x)$. The y-intercept is $f(0) = 0$. The x-intercepts are 0 and 2. There are sign changes at 0 and 2 (odd exponents on x and $2 - x$). As $x \to \infty$, $f(x) \to -\infty$ because $x^3 \to \infty$ and $2 - x \to -\infty$. As $x \to -\infty$, $f(x) \to -\infty$ because $x^3 \to -\infty$ and $2 - x \to \infty$. Note that the graph of f near $x = 0$ flattens out (looks like $y = x^3$).

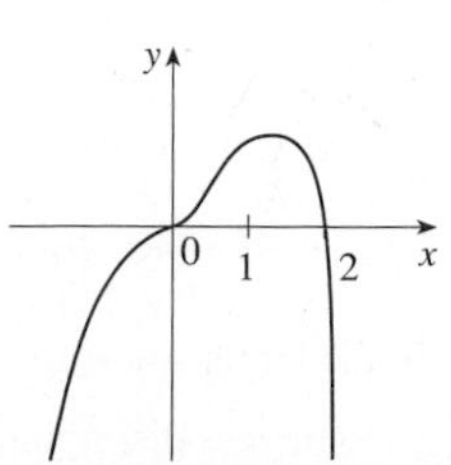

61. $y = f(x) = x^4 - x^6 = x^4(1 - x^2) = x^4(1 + x)(1 - x)$. The y-intercept is $f(0) = 0$. The x-intercepts are 0, -1, and 1 [found by solving $f(x) = 0$ for x]. Since $x^4 > 0$ for $x \neq 0$, f doesn't change sign at $x = 0$. The function does change sign at $x = -1$ and $x = 1$. As $x \to \pm\infty$, $f(x) = x^4(1 - x^2)$ approaches $-\infty$ because $x^4 \to \infty$ and $(1 - x^2) \to -\infty$.

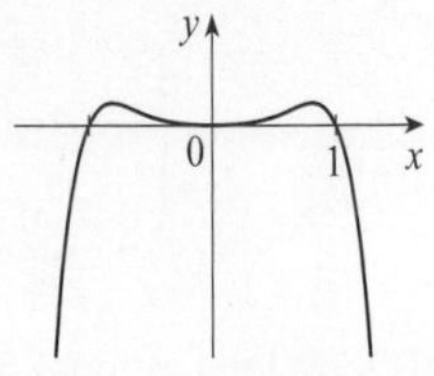

62. $y = f(x) = x^3(x + 2)^2(x - 1)$. The y-intercept is $f(0) = 0$. The x-intercepts are 0, -2, and 1. There are sign changes at 0 and 1 (odd exponents on x and $x - 1$). There is no sign change at -2. Also, $f(x) \to \infty$ as $x \to \infty$ because all three factors are large. And $f(x) \to \infty$ as $x \to -\infty$ because $x^3 \to -\infty$, $(x + 2)^2 \to \infty$, and $(x - 1) \to -\infty$. Note that the graph of f at $x = 0$ flattens out (looks like $y = -x^3$).

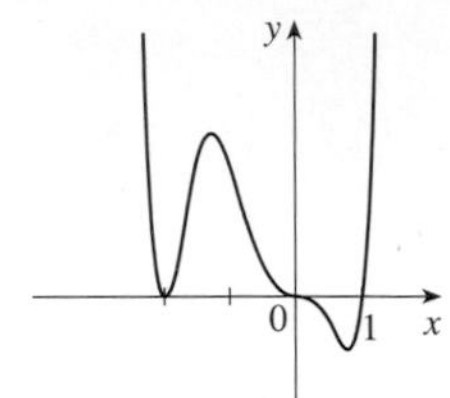

63. $y = f(x) = (3 - x)(1 + x)^2(1 - x)^4$. The y-intercept is $f(0) = 3(1)^2(1)^4 = 3$. The x-intercepts are 3, -1, and 1. There is a sign change at 3, but not at -1 and 1. When x is large positive, $3 - x$ is negative and the other factors are positive, so $\lim\limits_{x \to \infty} f(x) = -\infty$. When x is large negative, $3 - x$ is positive, so $\lim\limits_{x \to -\infty} f(x) = \infty$.

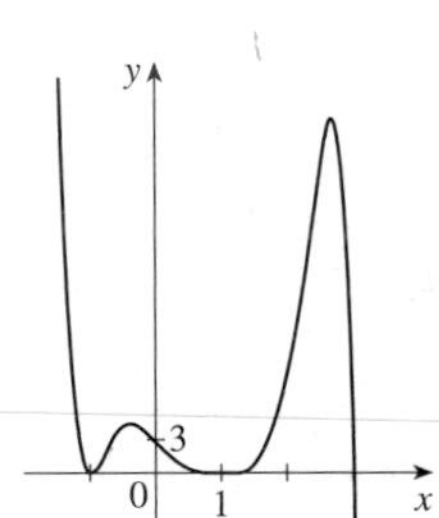

64. $y = f(x) = x^2(x^2 - 1)^2(x + 2) = x^2(x + 1)^2(x - 1)^2(x + 2)$. The y-intercept is $f(0) = 0$. The x-intercepts are 0, -1, 1, and -2. There is a sign change at -2, but not at 0, -1, and 1. When x is large positive, all the factors are positive, so $\lim\limits_{x \to \infty} f(x) = \infty$. When x is large negative, only $x + 2$ is negative, so $\lim\limits_{x \to -\infty} f(x) = -\infty$.

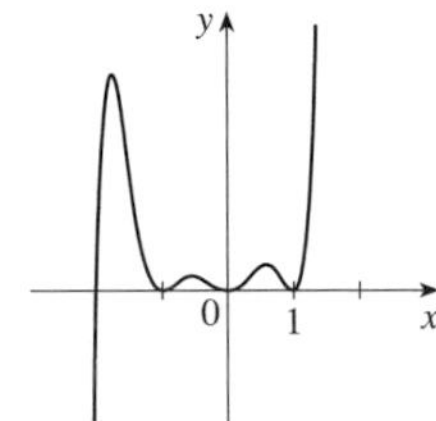

65. (a) Since $-1 \leq \sin x \leq 1$ for all x, $-\dfrac{1}{x} \leq \dfrac{\sin x}{x} \leq \dfrac{1}{x}$ for $x > 0$. As $x \to \infty$, $-1/x \to 0$ and $1/x \to 0$, so by the Squeeze Theorem, $(\sin x)/x \to 0$. Thus, $\lim\limits_{x \to \infty} \dfrac{\sin x}{x} = 0$.

(b) From part (a), the horizontal asymptote is $y = 0$. The function $y = (\sin x)/x$ crosses the horizontal asymptote whenever $\sin x = 0$; that is, at $x = \pi n$ for every integer n. Thus, the graph crosses the asymptote *an infinite number of times.*

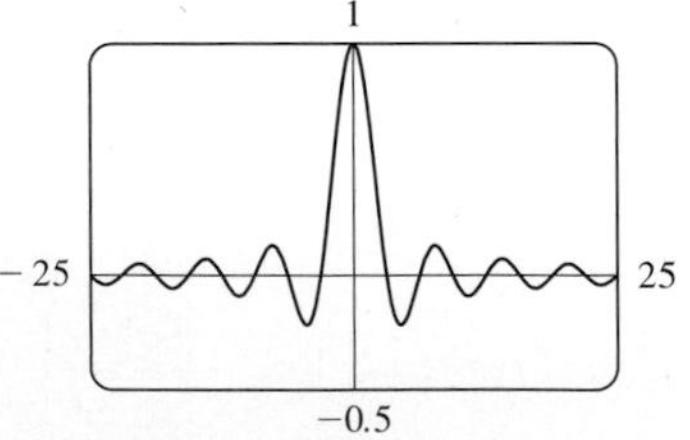

66. (a) In both viewing rectangles,

$\lim\limits_{x \to \infty} P(x) = \lim\limits_{x \to \infty} Q(x) = \infty$ and

$\lim\limits_{x \to -\infty} P(x) = \lim\limits_{x \to -\infty} Q(x) = -\infty$.

In the larger viewing rectangle, P and Q become less distinguishable.

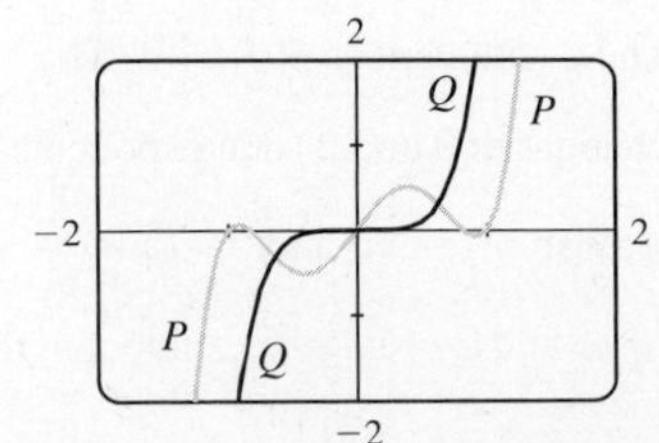

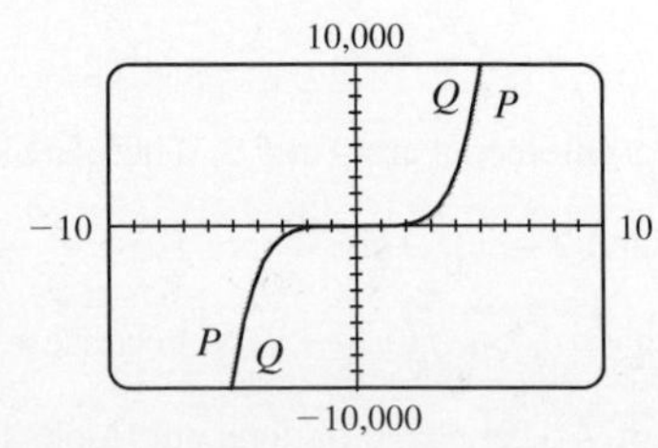

(b) $\lim_{x\to\infty} \frac{P(x)}{Q(x)} = \lim_{x\to\infty} \frac{3x^5 - 5x^3 + 2x}{3x^5} = \lim_{x\to\infty} \left(1 - \frac{5}{3}\cdot\frac{1}{x^2} + \frac{2}{3}\cdot\frac{1}{x^4}\right) = 1 - \frac{5}{3}(0) + \frac{2}{3}(0) = 1 \Rightarrow$

P and Q have the same end behavior.

67. $\lim_{x\to\infty} \frac{5\sqrt{x}}{\sqrt{x-1}} \cdot \frac{1/\sqrt{x}}{1/\sqrt{x}} = \lim_{x\to\infty} \frac{5}{\sqrt{1-(1/x)}} = \frac{5}{\sqrt{1-0}} = 5$ and

$\lim_{x\to\infty} \frac{10e^x - 21}{2e^x} \cdot \frac{1/e^x}{1/e^x} = \lim_{x\to\infty} \frac{10 - (21/e^x)}{2} = \frac{10-0}{2} = 5$. Since $\frac{10e^x - 21}{2e^x} < f(x) < \frac{5\sqrt{x}}{\sqrt{x-1}}$,

we have $\lim_{x\to\infty} f(x) = 5$ by the Squeeze Theorem.

68. (a) After t minutes, $25t$ liters of brine with 30 g of salt per liter has been pumped into the tank, so it contains

$(5000 + 25t)$ liters of water and $25t \cdot 30 = 750t$ grams of salt. Therefore, the salt concentration at time t will be

$C(t) = \frac{750t}{5000 + 25t} = \frac{30t}{200 + t} \frac{\text{g}}{\text{L}}$.

(b) $\lim_{t\to\infty} C(t) = \lim_{t\to\infty} \frac{30t}{200+t} = \lim_{t\to\infty} \frac{30t/t}{200/t + t/t} = \frac{30}{0+1} = 30$. So the salt concentration approaches that of the brine being pumped into the tank.

69. (a) $\lim_{t\to\infty} v(t) = \lim_{t\to\infty} v^*\left(1 - e^{-gt/v^*}\right) = v^*(1-0) = v^*$

(b) We graph $v(t) = 1 - e^{-9.8t}$ and $v(t) = 0.99v^*$, or in this case, $v(t) = 0.99$. Using an intersect feature or zooming in on the point of intersection, we find that $t \approx 0.47$ s.

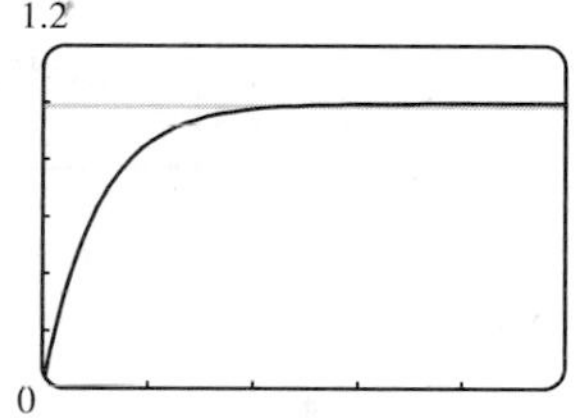

70. (a) $y = e^{-x/10}$ and $y = 0.1$ intersect at $x_1 \approx 23.03$.

If $x > x_1$, then $e^{-x/10} < 0.1$.

(b) $e^{-x/10} < 0.1 \Rightarrow -x/10 < \ln 0.1 \Rightarrow$

$x > -10 \ln \frac{1}{10} = -10 \ln 10^{-1} = 10 \ln 10 \approx 23.03$

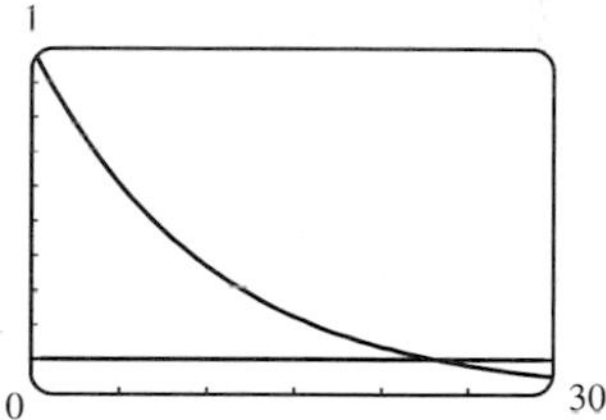

71. Let $g(x) = \frac{3x^2 + 1}{2x^2 + x + 1}$ and $f(x) = |g(x) - 1.5|$. Note that

$\lim_{x\to\infty} g(x) = \frac{3}{2}$ and $\lim_{x\to\infty} f(x) = 0$. We are interested in finding the x-value at which $f(x) < 0.05$. From the graph, we find that $x \approx 14.804$, so we choose $N = 15$ (or any larger number).

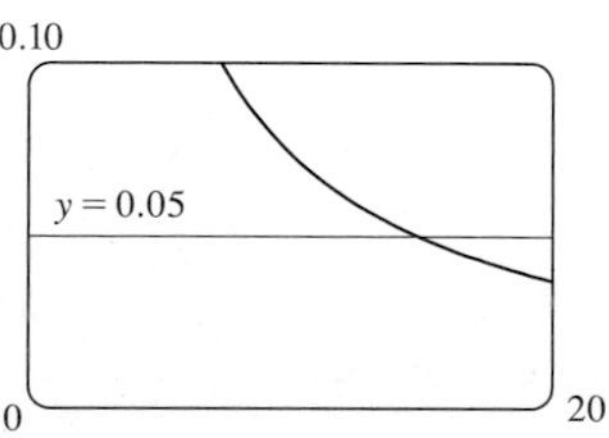

72. We want to find a value of N such that $x > N \Rightarrow \left|\frac{1-3x}{\sqrt{x^2+1}} - (-3)\right| < \varepsilon$, or equivalently,

$-3 - \varepsilon < \frac{1-3x}{\sqrt{x^2+1}} < -3 + \varepsilon$. When $\varepsilon = 0.1$, we graph $y = f(x) = \frac{1-3x}{\sqrt{x^2+1}}$, $y = -3.1$, and $y = -2.9$. From the graph, we find that $f(x) = -2.9$ at about $x = 11.283$, so we choose $N = 12$ (or any larger number). Similarly for $\varepsilon = 0.05$, we find that $f(x) = -2.95$ at about $x = 21.379$, so we choose $N = 22$ (or any larger number).

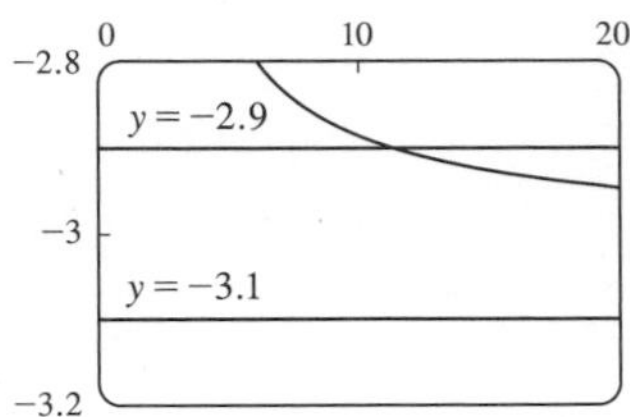

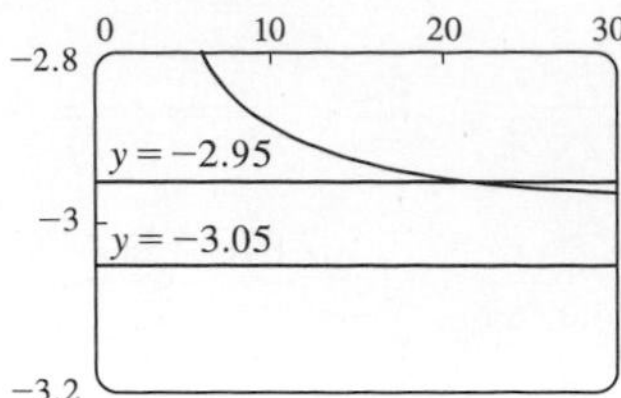

73. We want a value of N such that $x < N \Rightarrow \left| \dfrac{1-3x}{\sqrt{x^2+1}} - 3 \right| < \varepsilon$, or equivalently, $3 - \varepsilon < \dfrac{1-3x}{\sqrt{x^2+1}} < 3 + \varepsilon$. When $\varepsilon = 0.1$, we graph $y = f(x) = \dfrac{1-3x}{\sqrt{x^2+1}}$, $y = 3.1$, and $y = 2.9$. From the graph, we find that $f(x) = 3.1$ at about $x = -8.092$, so we choose $N = -9$ (or any lesser number). Similarly for $\varepsilon = 0.05$, we find that $f(x) = 3.05$ at about $x = -18.338$, so we choose $N = -19$ (or any lesser number).

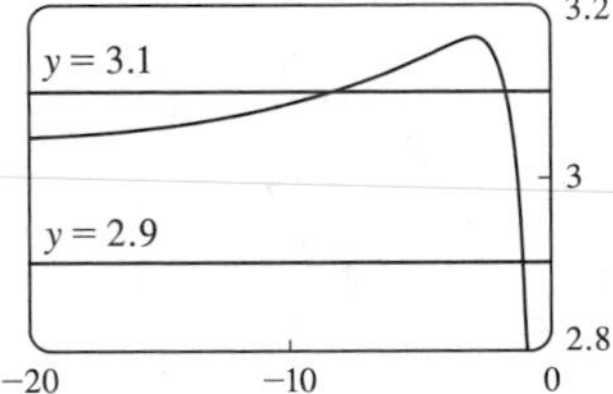

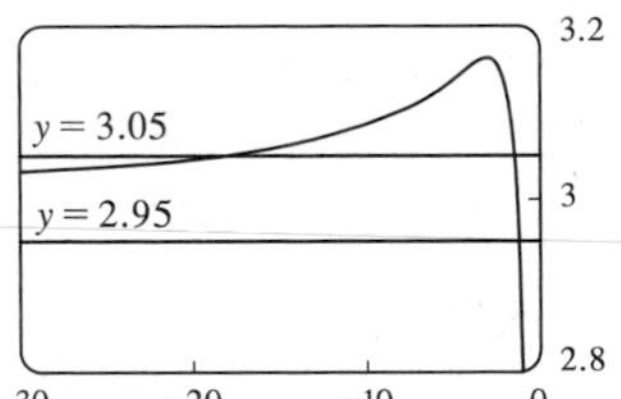

74. We want to find a value of N such that $x > N \quad \Rightarrow \quad \sqrt{x \ln x} > 100$. We graph $y = f(x) = \sqrt{x \ln x}$ and $y = 100$. From the graph, we find that $f(x) = 100$ at about $x = 1382.773$, so we choose $N = 1383$ (or any larger number).

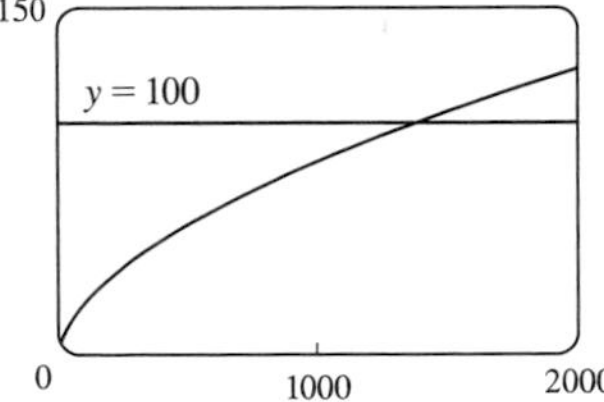

75. (a) $1/x^2 < 0.0001 \quad \Leftrightarrow \quad x^2 > 1/0.0001 = 10\,000 \quad \Leftrightarrow \quad x > 100 \quad (x > 0)$

(b) If $\varepsilon > 0$ is given, then $1/x^2 < \varepsilon \quad \Leftrightarrow \quad x^2 > 1/\varepsilon \quad \Leftrightarrow \quad x > 1/\sqrt{\varepsilon}$. Let $N = 1/\sqrt{\varepsilon}$.

Then $x > N \quad \Rightarrow \quad x > \dfrac{1}{\sqrt{\varepsilon}} \quad \Rightarrow \quad \left| \dfrac{1}{x^2} - 0 \right| = \dfrac{1}{x^2} < \varepsilon$, so $\displaystyle\lim_{x\to\infty} \frac{1}{x^2} = 0$.

76. (a) $1/\sqrt{x} < 0.0001 \quad \Leftrightarrow \quad \sqrt{x} > 1/0.0001 = 10^4 \quad \Leftrightarrow \quad x > 10^8$

(b) If $\varepsilon > 0$ is given, then $1/\sqrt{x} < \varepsilon \quad \Leftrightarrow \quad \sqrt{x} > 1/\varepsilon \quad \Leftrightarrow \quad x > 1/\varepsilon^2$. Let $N = 1/\varepsilon^2$.

Then $x > N \quad \Rightarrow \quad x > \dfrac{1}{\varepsilon^2} \quad \Rightarrow \quad \left| \dfrac{1}{\sqrt{x}} - 0 \right| = \dfrac{1}{\sqrt{x}} < \varepsilon$, so $\displaystyle\lim_{x\to\infty} \frac{1}{\sqrt{x}} = 0$.

77. For $x < 0$, $|1/x - 0| = -1/x$. If $\varepsilon > 0$ is given, then $-1/x < \varepsilon \quad \Leftrightarrow \quad x < -1/\varepsilon$.
Take $N = -1/\varepsilon$. Then $x < N \quad \Rightarrow \quad x < -1/\varepsilon \quad \Rightarrow \quad |(1/x) - 0| = -1/x < \varepsilon$, so $\displaystyle\lim_{x\to-\infty} (1/x) = 0$.

78. Given $M > 0$, we need $N > 0$ such that $x > N \quad \Rightarrow \quad x^3 > M$. Now $x^3 > M \quad \Leftrightarrow \quad x > \sqrt[3]{M}$, so take $N = \sqrt[3]{M}$. Then $x > N = \sqrt[3]{M} \quad \Rightarrow \quad x^3 > M$, so $\displaystyle\lim_{x\to\infty} x^3 = \infty$.

79. Given $M > 0$, we need $N > 0$ such that $x > N \Rightarrow e^x > M$. Now $e^x > M \Leftrightarrow x > \ln M$, so take $N = \max(1, \ln M)$. (This ensures that $N > 0$.) Then $x > N = \max(1, \ln M) \Rightarrow e^x > \max(e, M) \geq M$, so $\lim\limits_{x \to \infty} e^x = \infty$.

80. Definition Let f be a function defined on some interval $(-\infty, a)$. Then $\lim\limits_{x \to -\infty} f(x) = -\infty$ means that for every negative number M there is a corresponding negative number N such that $f(x) < M$ whenever $x < N$. Now we use the definition to prove that $\lim\limits_{x \to -\infty} (1 + x^3) = -\infty$. Given a negative number M, we need a negative number N such that $x < N \Rightarrow 1 + x^3 < M$. Now $1 + x^3 < M \Leftrightarrow x^3 < M - 1 \Leftrightarrow x < \sqrt[3]{M - 1}$. Thus, we take $N = \sqrt[3]{M - 1}$ and find that $x < N \Rightarrow 1 + x^3 < M$. This proves that $\lim\limits_{x \to -\infty} (1 + x^3) = -\infty$.

81. (a) Suppose that $\lim\limits_{x \to \infty} f(x) = L$. Then for every $\varepsilon > 0$ there is a corresponding positive number N such that $|f(x) - L| < \varepsilon$ whenever $x > N$. If $t = 1/x$, then $x > N \Leftrightarrow 0 < 1/x < 1/N \Leftrightarrow 0 < t < 1/N$. Thus, for every $\varepsilon > 0$ there is a corresponding $\delta > 0$ (namely $1/N$) such that $|f(1/t) - L| < \varepsilon$ whenever $0 < t < \delta$. This proves that $\lim\limits_{t \to 0^+} f(1/t) = L = \lim\limits_{x \to \infty} f(x)$.

Now suppose that $\lim\limits_{x \to -\infty} f(x) = L$. Then for every $\varepsilon > 0$ there is a corresponding negative number N such that $|f(x) - L| < \varepsilon$ whenever $x < N$. If $t = 1/x$, then $x < N \Leftrightarrow 1/N < 1/x < 0 \Leftrightarrow 1/N < t < 0$. Thus, for every $\varepsilon > 0$ there is a corresponding $\delta > 0$ (namely $-1/N$) such that $|f(1/t) - L| < \varepsilon$ whenever $-\delta < t < 0$. This proves that $\lim\limits_{t \to 0^-} f(1/t) = L = \lim\limits_{x \to -\infty} f(x)$.

(b)
$$\begin{aligned}
\lim_{x \to 0^+} x \sin \frac{1}{x} &= \lim_{t \to 0^+} t \sin \frac{1}{t} && [\text{let } x = t] \\
&= \lim_{y \to \infty} \frac{1}{y} \sin y && [\text{part (a) with } y = 1/t] \\
&= \lim_{x \to \infty} \frac{\sin x}{x} && [\text{let } y = x] \\
&= 0 && [\text{by Exercise 65}]
\end{aligned}$$

2.7 Derivatives and Rates of Change

1. (a) This is just the slope of the line through two points: $m_{PQ} = \dfrac{\Delta y}{\Delta x} = \dfrac{f(x) - f(3)}{x - 3}$.

(b) This is the limit of the slope of the secant line PQ as Q approaches P: $m = \lim\limits_{x \to 3} \dfrac{f(x) - f(3)}{x - 3}$.

2. The curve looks more like a line as the viewing rectangle gets smaller.

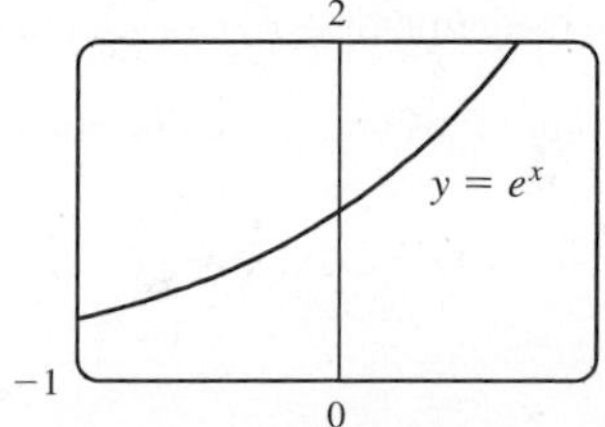

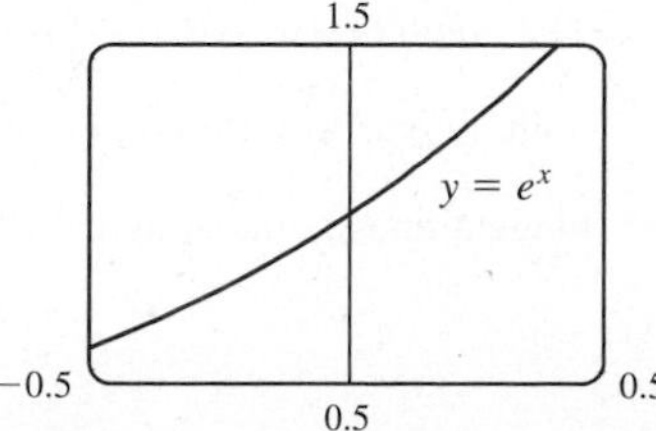

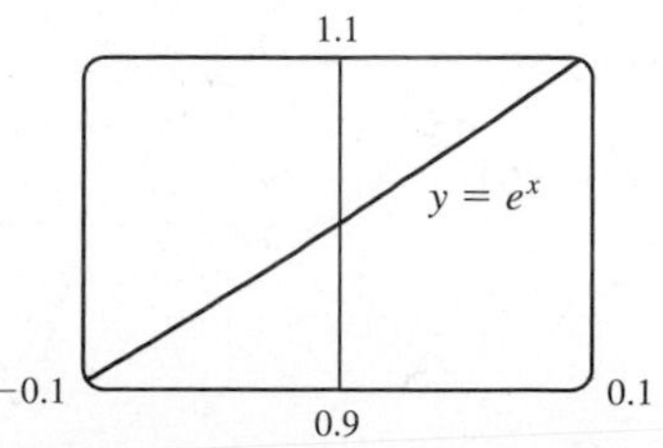

3. (a) (i) Using Definition 1 with $f(x) = 4x - x^2$ and $P(1, 3)$,

$$m = \lim_{x \to a} \frac{f(x) - f(a)}{x - a} = \lim_{x \to 1} \frac{(4x - x^2) - 3}{x - 1} = \lim_{x \to 1} \frac{-(x^2 - 4x + 3)}{x - 1} = \lim_{x \to 1} \frac{-(x - 1)(x - 3)}{x - 1}$$

$$= \lim_{x \to 1} (3 - x) = 3 - 1 = 2$$

(ii) Using Equation 2 with $f(x) = 4x - x^2$ and $P(1, 3)$,

$$m = \lim_{h \to 0} \frac{f(a + h) - f(a)}{h} = \lim_{h \to 0} \frac{f(1 + h) - f(1)}{h} = \lim_{h \to 0} \frac{\left[4(1 + h) - (1 + h)^2\right] - 3}{h}$$

$$= \lim_{h \to 0} \frac{4 + 4h - 1 - 2h - h^2 - 3}{h} = \lim_{h \to 0} \frac{-h^2 + 2h}{h} = \lim_{h \to 0} \frac{h(-h + 2)}{h} = \lim_{h \to 0} (-h + 2) = 2$$

(b) An equation of the tangent line is $y - f(a) = f'(a)(x - a) \;\Rightarrow\; y - f(1) = f'(1)(x - 1) \;\Rightarrow\; y - 3 = 2(x - 1)$, or $y = 2x + 1$.

(c)

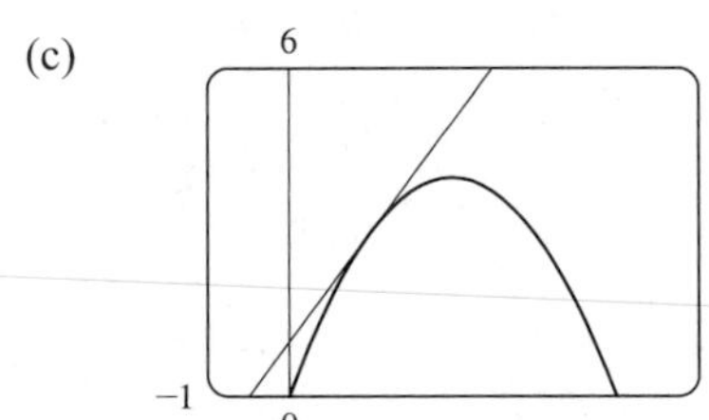

The graph of $y = 2x + 1$ is tangent to the graph of $y = 4x - x^2$ at the point $(1, 3)$. Now zoom in toward the point $(1, 3)$ until the parabola and the tangent line are indistiguishable.

4. (a) (i) Using Definition 1 with $f(x) = x - x^3$ and $P(1, 0)$,

$$m = \lim_{x \to 1} \frac{f(x) - 0}{x - 1} = \lim_{x \to 1} \frac{x - x^3}{x - 1} = \lim_{x \to 1} \frac{x(1 - x^2)}{x - 1} = \lim_{x \to 1} \frac{x(1 + x)(1 - x)}{x - 1}$$

$$= \lim_{x \to 1} \left[-x(1 + x)\right] = -1(2) = -2$$

(ii) Using Equation 2 with $f(x) = x - x^3$ and $P(1, 0)$,

$$m = \lim_{h \to 0} \frac{f(a + h) - f(a)}{h} = \lim_{h \to 0} \frac{f(1 + h) - f(1)}{h} = \lim_{h \to 0} \frac{\left[(1 + h) - (1 + h)^3\right] - 0}{h}$$

$$= \lim_{h \to 0} \frac{1 + h - (1 + 3h + 3h^2 + h^3)}{h} = \lim_{h \to 0} \frac{-h^3 - 3h^2 - 2h}{h} = \lim_{h \to 0} \frac{h(-h^2 - 3h - 2)}{h}$$

$$= \lim_{h \to 0} (-h^2 - 3h - 2) = -2$$

(b) An equation of the tangent line is $y - f(a) = f'(a)(x - a) \;\Rightarrow\; y - f(1) = f'(1)(x - 1) \;\Rightarrow\; y - 0 = -2(x - 1)$, or $y = -2x + 2$.

(c)

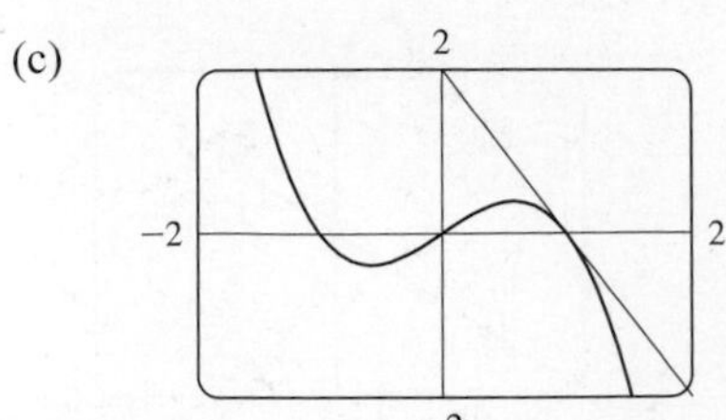

The graph of $y = -2x + 2$ is tangent to the graph of $y = x - x^3$ at the point $(1, 0)$. Now zoom in toward the point $(1, 0)$ until the cubic and the tangent line are indistinguishable.

5. Using (1) with $f(x) = 4x - 3x^2$ and $P(2, -4)$ [we could also use (2)],

$$m = \lim_{x \to a} \frac{f(x) - f(a)}{x - a} = \lim_{x \to 2} \frac{(4x - 3x^2) - (-4)}{x - 2} = \lim_{x \to 2} \frac{-3x^2 + 4x + 4}{x - 2}$$

$$= \lim_{x \to 2} \frac{(-3x - 2)(x - 2)}{x - 2} = \lim_{x \to 2} (-3x - 2) = -3(2) - 2 = -8$$

Tangent line: $y - (-4) = -8(x - 2) \Leftrightarrow y + 4 = -8x + 16 \Leftrightarrow y = -8x + 12.$

6. Using (2) with $f(x) = x^3 - 3x + 1$ and $P(2, 3)$,

$$m = \lim_{h \to 0} \frac{f(a + h) - f(a)}{h} = \lim_{h \to 0} \frac{f(2 + h) - f(2)}{h} = \lim_{h \to 0} \frac{(2 + h)^3 - 3(2 + h) + 1 - 3}{h}$$

$$= \lim_{h \to 0} \frac{8 + 12h + 6h^2 + h^3 - 6 - 3h - 2}{h} = \lim_{h \to 0} \frac{9h + 6h^2 + h^3}{h} = \lim_{h \to 0} \frac{h(9 + 6h + h^2)}{h}$$

$$= \lim_{h \to 0} (9 + 6h + h^2) = 9$$

Tangent line: $y - 3 = 9(x - 2) \Leftrightarrow y - 3 = 9x - 18 \Leftrightarrow y = 9x - 15$

7. Using (1), $m = \lim\limits_{x \to 1} \dfrac{\sqrt{x} - \sqrt{1}}{x - 1} = \lim\limits_{x \to 1} \dfrac{(\sqrt{x} - 1)(\sqrt{x} + 1)}{(x - 1)(\sqrt{x} + 1)} = \lim\limits_{x \to 1} \dfrac{x - 1}{(x - 1)(\sqrt{x} + 1)} = \lim\limits_{x \to 1} \dfrac{1}{\sqrt{x} + 1} = \dfrac{1}{2}.$

Tangent line: $y - 1 = \frac{1}{2}(x - 1) \Leftrightarrow y = \frac{1}{2}x + \frac{1}{2}$

8. Using (1) with $f(x) = \dfrac{2x + 1}{x + 2}$ and $P(1, 1)$,

$$m = \lim_{x \to a} \frac{f(x) - f(a)}{x - a} = \lim_{x \to 1} \frac{\dfrac{2x + 1}{x + 2} - 1}{x - 1} = \lim_{x \to 1} \frac{\dfrac{2x + 1 - (x + 2)}{x + 2}}{x - 1} = \lim_{x \to 1} \frac{x - 1}{(x - 1)(x + 2)}$$

$$= \lim_{x \to 1} \frac{1}{x + 2} = \frac{1}{1 + 2} = \frac{1}{3}$$

Tangent line: $y - 1 = \frac{1}{3}(x - 1) \Leftrightarrow y - 1 = \frac{1}{3}x - \frac{1}{3} \Leftrightarrow y = \frac{1}{3}x + \frac{2}{3}$

9. (a) Using (2) with $y = f(x) = 3 + 4x^2 - 2x^3$,

$$m = \lim_{h \to 0} \frac{f(a + h) - f(a)}{h} = \lim_{h \to 0} \frac{3 + 4(a + h)^2 - 2(a + h)^3 - (3 + 4a^2 - 2a^3)}{h}$$

$$= \lim_{h \to 0} \frac{3 + 4(a^2 + 2ah + h^2) - 2(a^3 + 3a^2h + 3ah^2 + h^3) - 3 - 4a^2 + 2a^3}{h}$$

$$= \lim_{h \to 0} \frac{3 + 4a^2 + 8ah + 4h^2 - 2a^3 - 6a^2h - 6ah^2 - 2h^3 - 3 - 4a^2 + 2a^3}{h}$$

$$= \lim_{h \to 0} \frac{8ah + 4h^2 - 6a^2h - 6ah^2 - 2h^3}{h} = \lim_{h \to 0} \frac{h(8a + 4h - 6a^2 - 6ah - 2h^2)}{h}$$

$$= \lim_{h \to 0} (8a + 4h - 6a^2 - 6ah - 2h^2) = 8a - 6a^2$$

(b) At $(1, 5)$: $m = 8(1) - 6(1)^2 = 2$, so an equation of the tangent line is $y - 5 = 2(x - 1) \Leftrightarrow y = 2x + 3$.

At $(2, 3)$: $m = 8(2) - 6(2)^2 = -8$, so an equation of the tangent line is $y - 3 = -8(x - 2) \Leftrightarrow y = -8x + 19$.

(c)

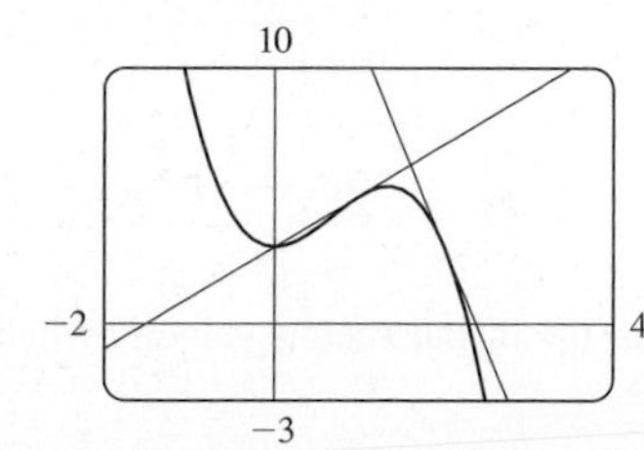

10. (a) Using (1),

$$m = \lim_{x \to a} \frac{\dfrac{1}{\sqrt{x}} - \dfrac{1}{\sqrt{a}}}{x - a} = \lim_{x \to a} \frac{\dfrac{\sqrt{a} - \sqrt{x}}{\sqrt{ax}}}{x - a} = \lim_{x \to a} \frac{(\sqrt{a} - \sqrt{x})(\sqrt{a} + \sqrt{x})}{\sqrt{ax}\,(x - a)(\sqrt{a} + \sqrt{x})} = \lim_{x \to a} \frac{a - x}{\sqrt{ax}\,(x - a)(\sqrt{a} + \sqrt{x})}$$

$$= \lim_{x \to a} \frac{-1}{\sqrt{ax}\,(\sqrt{a} + \sqrt{x})} = \frac{-1}{\sqrt{a^2}\,(2\sqrt{a})} = -\frac{1}{2a^{3/2}} \text{ or } -\frac{1}{2}a^{-3/2} \quad [a > 0]$$

(b) At $(1, 1)$: $m = -\frac{1}{2}$, so an equation of the tangent line is $y - 1 = -\frac{1}{2}(x - 1) \Leftrightarrow y = -\frac{1}{2}x + \frac{3}{2}$.

At $\left(4, \frac{1}{2}\right)$: $m = -\frac{1}{16}$, so an equation of the tangent line is $y - \frac{1}{2} = -\frac{1}{16}(x - 4) \Leftrightarrow y = -\frac{1}{16}x + \frac{3}{4}$.

(c)

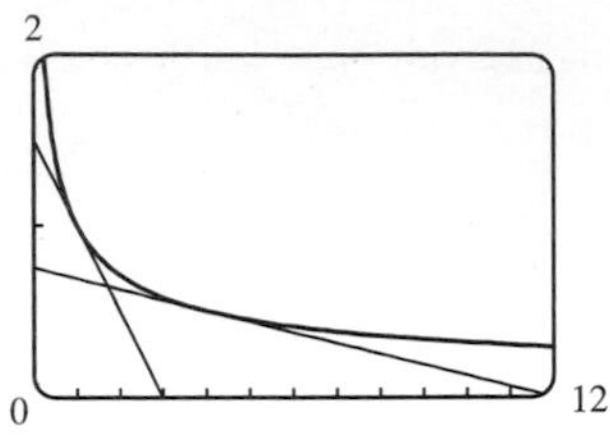

11. (a) The particle is moving to the right when s is increasing; that is, on the intervals $(0, 1)$ and $(4, 6)$. The particle is moving to the left when s is decreasing; that is, on the interval $(2, 3)$. The particle is standing still when s is constant; that is, on the intervals $(1, 2)$ and $(3, 4)$.

(b) The velocity of the particle is equal to the slope of the tangent line of the graph. Note that there is no slope at the corner points on the graph. On the interval $(0, 1)$, the slope is $\dfrac{3 - 0}{1 - 0} = 3$. On the interval $(2, 3)$, the slope is $\dfrac{1 - 3}{3 - 2} = -2$. On the interval $(4, 6)$, the slope is $\dfrac{3 - 1}{6 - 4} = 1$.

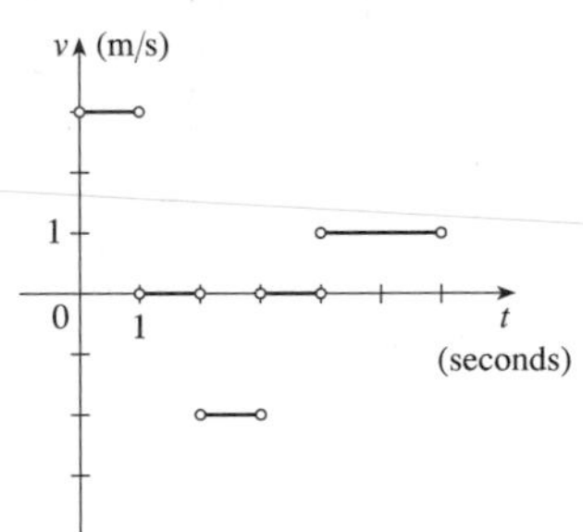

12. (a) **Runner A** runs the entire 100-meter race at the same velocity since the slope of the position function is constant. **Runner B** starts the race at a slower velocity than runner A, but finishes the race at a faster velocity.

(b) The distance between the runners is the greatest at the time when the largest vertical line segment fits between the two graphs—this appears to be somewhere between 9 and 10 seconds.

(c) The runners had the same velocity when the slopes of their respective position functions are equal—this also appears to be at about 9.5 s. Note that the answers for parts (b) and (c) must be the same for these graphs because as soon as the velocity for runner B overtakes the velocity for runner A, the distance between the runners starts to decrease.

13. Let $s(t) = 40t - 16t^2$.

$$v(2) = \lim_{t \to 2} \frac{s(t) - s(2)}{t - 2} = \lim_{t \to 2} \frac{(40t - 16t^2) - 16}{t - 2} = \lim_{t \to 2} \frac{-16t^2 + 40t - 16}{t - 2} = \lim_{t \to 2} \frac{-8(2t^2 - 5t + 2)}{t - 2}$$

$$= \lim_{t \to 2} \frac{-8(t - 2)(2t - 1)}{t - 2} = -8 \lim_{t \to 2}(2t - 1) = -8(3) = -24$$

Thus, the instantaneous velocity when $t = 2$ is -24 ft/s.

14. (a) Let $H(t) = 10t - 1.86t^2$.

$$v(1) = \lim_{h\to 0} \frac{H(1+h) - H(1)}{h} = \lim_{h\to 0} \frac{\left[10(1+h) - 1.86(1+h)^2\right] - (10 - 1.86)}{h}$$

$$= \lim_{h\to 0} \frac{10 + 10h - 1.86(1 + 2h + h^2) - 10 + 1.86}{h}$$

$$= \lim_{h\to 0} \frac{10 + 10h - 1.86 - 3.72h - 1.86h^2 - 10 + 1.86}{h}$$

$$= \lim_{h\to 0} \frac{6.28h - 1.86h^2}{h} = \lim_{h\to 0} (6.28 - 1.86h) = 6.28$$

The velocity of the rock after one second is 6.28 m/s.

(b) $$v(a) = \lim_{h\to 0} \frac{H(a+h) - H(a)}{h} = \lim_{h\to 0} \frac{\left[10(a+h) - 1.86(a+h)^2\right] - (10a - 1.86a^2)}{h}$$

$$= \lim_{h\to 0} \frac{10a + 10h - 1.86(a^2 + 2ah + h^2) - 10a + 1.86a^2}{h}$$

$$= \lim_{h\to 0} \frac{10a + 10h - 1.86a^2 - 3.72ah - 1.86h^2 - 10a + 1.86a^2}{h} = \lim_{h\to 0} \frac{10h - 3.72ah - 1.86h^2}{h}$$

$$= \lim_{h\to 0} \frac{h(10 - 3.72a - 1.86h)}{h} = \lim_{h\to 0} (10 - 3.72a - 1.86h) = 10 - 3.72a$$

The velocity of the rock when $t = a$ is $(10 - 3.72a)$ m/s.

(c) The rock will hit the surface when $H = 0 \;\Leftrightarrow\; 10t - 1.86t^2 = 0 \;\Leftrightarrow\; t(10 - 1.86t) = 0 \;\Leftrightarrow\; t = 0$ or $1.86t = 10$.

The rock hits the surface when $t = 10/1.86 \approx 5.4$ s.

(d) The velocity of the rock when it hits the surface is $v\left(\frac{10}{1.86}\right) = 10 - 3.72\left(\frac{10}{1.86}\right) = 10 - 20 = -10$ m/s.

15. $$v(a) = \lim_{h\to 0} \frac{s(a+h) - s(a)}{h} = \lim_{h\to 0} \frac{\dfrac{1}{(a+h)^2} - \dfrac{1}{a^2}}{h} = \lim_{h\to 0} \frac{\dfrac{a^2 - (a+h)^2}{a^2(a+h)^2}}{h} = \lim_{h\to 0} \frac{a^2 - (a^2 + 2ah + h^2)}{ha^2(a+h)^2}$$

$$= \lim_{h\to 0} \frac{-(2ah + h^2)}{ha^2(a+h)^2} = \lim_{h\to 0} \frac{-h(2a+h)}{ha^2(a+h)^2} = \lim_{h\to 0} \frac{-(2a+h)}{a^2(a+h)^2} = \frac{-2a}{a^2 \cdot a^2} = \frac{-2}{a^3} \text{ m/s}$$

So $v(1) = \dfrac{-2}{1^3} = -2$ m/s, $v(2) = \dfrac{-2}{2^3} = -\dfrac{1}{4}$ m/s, and $v(3) = \dfrac{-2}{3^3} = -\dfrac{2}{27}$ m/s.

16. (a) The average velocity between times t and $t + h$ is

$$\frac{s(t+h) - s(t)}{(t+h) - t} = \frac{\frac{1}{2}(t+h)^2 - 6(t+h) + 23 - \left(\frac{1}{2}t^2 - 6t + 23\right)}{h}$$

$$= \frac{\frac{1}{2}t^2 + th + \frac{1}{2}h^2 - 6t - 6h + 23 - \frac{1}{2}t^2 + 6t - 23}{h}$$

$$= \frac{th + \frac{1}{2}h^2 - 6h}{h} = \frac{h\left(t + \frac{1}{2}h - 6\right)}{h} = \left(t + \tfrac{1}{2}h - 6\right) \text{ ft/s}$$

(i) $[4, 8]$: $t = 4$, $h = 8 - 4 = 4$, so the average velocity is $4 + \frac{1}{2}(4) - 6 = 0$ ft/s.

(ii) $[6, 8]$: $t = 6$, $h = 8 - 6 = 2$, so the average velocity is $6 + \frac{1}{2}(2) - 6 = 1$ ft/s.

(iii) $[8, 10]$: $t = 8$, $h = 10 - 8 = 2$, so the average velocity is $8 + \frac{1}{2}(2) - 6 = 3$ ft/s.

(iv) $[8, 12]$: $t = 8$, $h = 12 - 8 = 4$, so the average velocity is $8 + \frac{1}{2}(4) - 6 = 4$ ft/s.

(b) $v(t) = \lim\limits_{h\to 0} \dfrac{s(t+h) - s(t)}{h} = \lim\limits_{h\to 0} \left(t + \frac{1}{2}h - 6\right)$
$= t - 6$, so $v(8) = 2$ ft/s.

(c)

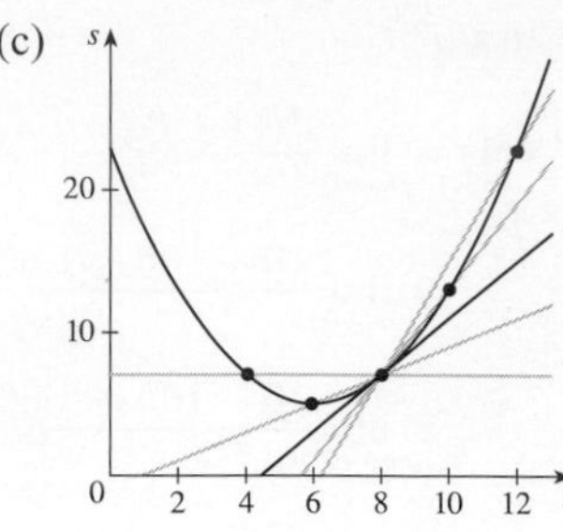

17. $g'(0)$ is the only negative value. The slope at $x = 4$ is smaller than the slope at $x = 2$ and both are smaller than the slope at $x = -2$. Thus, $g'(0) < 0 < g'(4) < g'(2) < g'(-2)$.

18. (a) On $[20, 60]$: $\dfrac{f(60) - f(20)}{60 - 20} = \dfrac{700 - 300}{40} = \dfrac{400}{40} = 10$

(b) Pick any interval that has the same y-value at its endpoints. $[0, 57]$ is such an interval since $f(0) = 600$ and $f(57) = 600$.

(c) On $[40, 60]$: $\dfrac{f(60) - f(40)}{60 - 40} = \dfrac{700 - 200}{20} = \dfrac{500}{20} = 25$

On $[40, 70]$: $\dfrac{f(70) - f(40)}{70 - 40} = \dfrac{900 - 200}{30} = \dfrac{700}{30} = 23\frac{1}{3}$

Since $25 > 23\frac{1}{3}$, the average rate of change on $[40, 60]$ is larger.

(d) $\dfrac{f(40) - f(10)}{40 - 10} = \dfrac{200 - 400}{30} = \dfrac{-200}{30} = -6\frac{2}{3}$

This value represents the slope of the line segment from $(10, f(10))$ to $(40, f(40))$.

19. (a) The tangent line at $x = 50$ appears to pass through the points $(43, 200)$ and $(60, 640)$, so

$f'(50) \approx \dfrac{640 - 200}{60 - 43} = \dfrac{440}{17} \approx 26$.

(b) The tangent line at $x = 10$ is steeper than the tangent line at $x = 30$, so it is larger in magnitude, but less in numerical value, that is, $f'(10) < f'(30)$.

(c) The slope of the tangent line at $x = 60$, $f'(60)$, is greater than the slope of the line through $(40, f(40))$ and $(80, f(80))$.

So yes, $f'(60) > \dfrac{f(80) - f(40)}{80 - 40}$.

20. Since $g(5) = -3$, the point $(5, -3)$ is on the graph of g. Since $g'(5) = 4$, the slope of the tangent line at $x = 5$ is 4. Using the point-slope form of a line gives us $y - (-3) = 4(x - 5)$, or $y = 4x - 23$.

21. For the tangent line $y = 4x - 5$: when $x = 2$, $y = 4(2) - 5 = 3$ and its slope is 4 (the coefficient of x). At the point of tangency, these values are shared with the curve $y = f(x)$; that is, $f(2) = 3$ and $f'(2) = 4$.

22. Since $(4, 3)$ is on $y = f(x)$, $f(4) = 3$. The slope of the tangent line between $(0, 2)$ and $(4, 3)$ is $\frac{1}{4}$, so $f'(4) = \frac{1}{4}$.

23. We begin by drawing a curve through the origin with a slope of 3 to satisfy $f(0) = 0$ and $f'(0) = 3$. Since $f'(1) = 0$, we will round off our figure so that there is a horizontal tangent directly over $x = 1$. Last, we make sure that the curve has a slope of -1 as we pass over $x = 2$. Two of the many possibilities are shown.

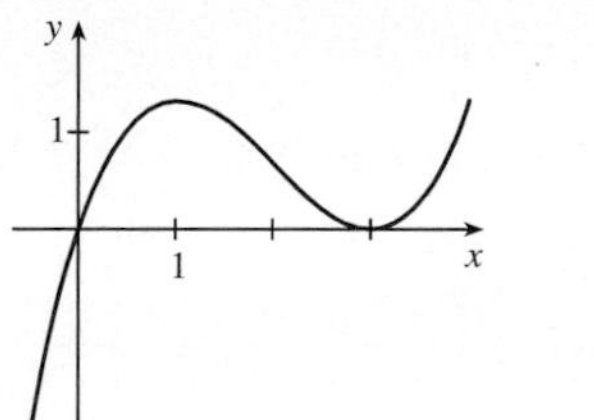

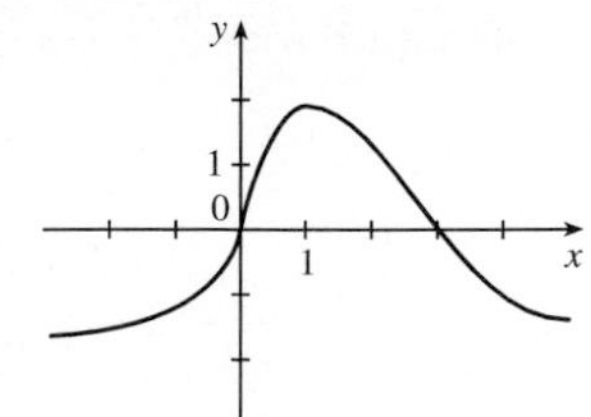

24. We begin by drawing a curve through the origin with a slope of 1 to satisfy $g(0) = 0$ and $g'(0) = 1$. We round off our figure at $x = 1$ to satisfy $g'(1) = 0$, and then pass through $(2, 0)$ with slope -1 to satisfy $g(2) = 0$ and $g'(2) = -1$. We round the figure at $x = 3$ to satisfy $g'(3) = 0$, and then pass through $(4, 0)$ with slope 1 to satisfy $g(4) = 0$ and $g'(4) = 1$. Finally we extend the curve on both ends to satisfy $\lim_{x \to \infty} g(x) = \infty$ and $\lim_{x \to -\infty} g(x) = -\infty$.

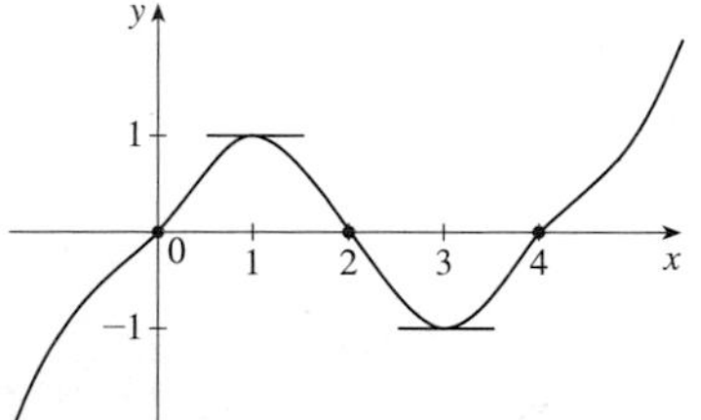

25. We begin by drawing a curve through $(0, 1)$ with a slope of 1 to satisfy $g(0) = 1$ and $g'(0) = 1$. We round off our figure at $x = -2$ to satisfy $g'(-2) = 0$. As $x \to -5^+$, $y \to \infty$, so we draw a vertical asymptote at $x = -5$. As $x \to 5^-$, $y \to 3$, so we draw a dot at $(5, 3)$ [the dot could be open or closed].

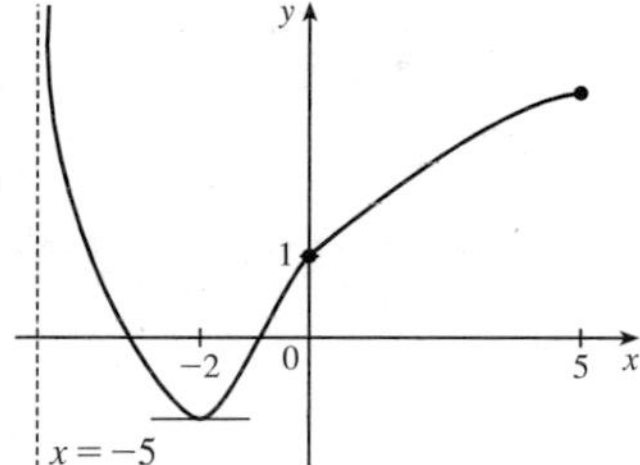

26. We begin by drawing an odd function (symmetric with respect to the origin) through the origin with slope -2 to satisfy $f'(0) = -2$. Now draw a curve starting at $x = 1$ and increasing without bound as $x \to 2^-$ since $\lim_{x \to 2^-} f(x) = \infty$. Lastly, reflect the last curve through the origin (rotate $180°$) since f is an odd function.

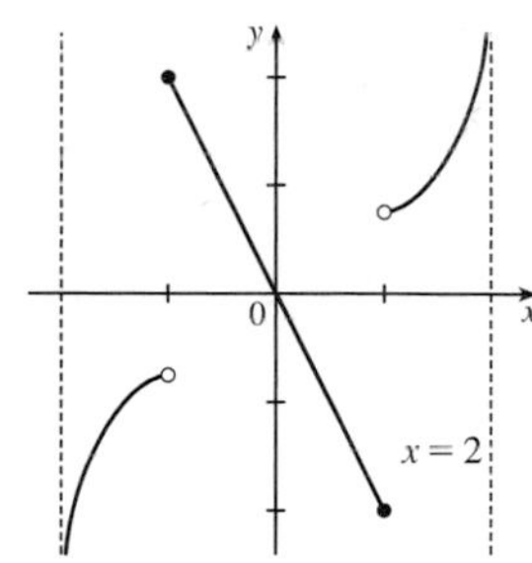

27. Using (4) with $f(x) = 3x^2 - x^3$ and $a = 1$,

$$\begin{aligned} f'(1) &= \lim_{h \to 0} \frac{f(1+h) - f(1)}{h} = \lim_{h \to 0} \frac{[3(1+h)^2 - (1+h)^3] - 2}{h} \\ &= \lim_{h \to 0} \frac{(3 + 6h + 3h^2) - (1 + 3h + 3h^2 + h^3) - 2}{h} = \lim_{h \to 0} \frac{3h - h^3}{h} = \lim_{h \to 0} \frac{h(3 - h^2)}{h} \\ &= \lim_{h \to 0} (3 - h^2) = 3 - 0 = 3 \end{aligned}$$

Tangent line: $y - 2 = 3(x - 1) \;\Leftrightarrow\; y - 2 = 3x - 3 \;\Leftrightarrow\; y = 3x - 1$

28. Using (5) with $g(x) = x^4 - 2$ and $a = 1$,

$$g'(1) = \lim_{x \to 1} \frac{g(x) - g(1)}{x - 1} = \lim_{x \to 1} \frac{(x^4 - 2) - (-1)}{x - 1} = \lim_{x \to 1} \frac{x^4 - 1}{x - 1} = \lim_{x \to 1} \frac{(x^2 + 1)(x^2 - 1)}{x - 1}$$

$$= \lim_{x \to 1} \frac{(x^2 + 1)(x + 1)(x - 1)}{x - 1} = \lim_{x \to 1} [(x^2 + 1)(x + 1)] = 2(2) = 4$$

Tangent line: $y - (-1) = 4(x - 1) \Leftrightarrow y + 1 = 4x - 4 \Leftrightarrow y = 4x - 5$

29. (a) Using (4) with $F(x) = 5x/(1 + x^2)$ and the point $(2, 2)$, we have

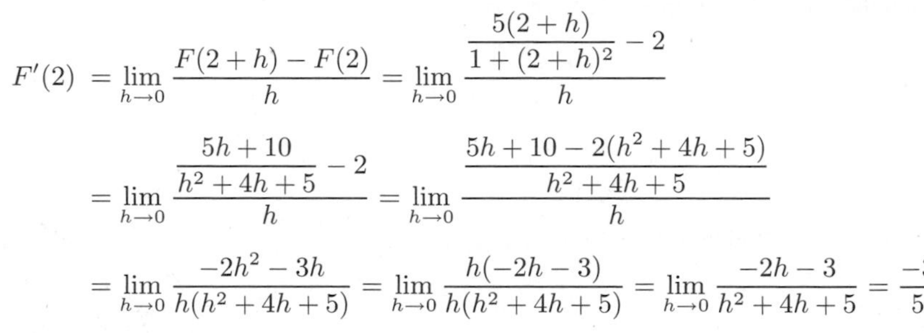

$$F'(2) = \lim_{h \to 0} \frac{F(2 + h) - F(2)}{h} = \lim_{h \to 0} \frac{\dfrac{5(2 + h)}{1 + (2 + h)^2} - 2}{h}$$

$$= \lim_{h \to 0} \frac{\dfrac{5h + 10}{h^2 + 4h + 5} - 2}{h} = \lim_{h \to 0} \frac{\dfrac{5h + 10 - 2(h^2 + 4h + 5)}{h^2 + 4h + 5}}{h}$$

$$= \lim_{h \to 0} \frac{-2h^2 - 3h}{h(h^2 + 4h + 5)} = \lim_{h \to 0} \frac{h(-2h - 3)}{h(h^2 + 4h + 5)} = \lim_{h \to 0} \frac{-2h - 3}{h^2 + 4h + 5} = \frac{-3}{5}$$

So an equation of the tangent line at $(2, 2)$ is $y - 2 = -\frac{3}{5}(x - 2)$ or $y = -\frac{3}{5}x + \frac{16}{5}$.

(b)

4

−1 6

−2

30. (a) Using (4) with $G(x) = 4x^2 - x^3$, we have

$$G'(a) = \lim_{h \to 0} \frac{G(a + h) - G(a)}{h} = \lim_{h \to 0} \frac{[4(a + h)^2 - (a + h)^3] - (4a^2 - a^3)}{h}$$

$$= \lim_{h \to 0} \frac{4a^2 + 8ah + 4h^2 - (a^3 + 3a^2h + 3ah^2 + h^3) - 4a^2 + a^3}{h}$$

$$= \lim_{h \to 0} \frac{8ah + 4h^2 - 3a^2h - 3ah^2 - h^3}{h} = \lim_{h \to 0} \frac{h(8a + 4h - 3a^2 - 3ah - h^2)}{h}$$

$$= \lim_{h \to 0} (8a + 4h - 3a^2 - 3ah - h^2) = 8a - 3a^2$$

At the point $(2, 8)$, $G'(2) = 16 - 12 = 4$, and an equation of the tangent line is $y - 8 = 4(x - 2)$, or $y = 4x$. At the point $(3, 9)$, $G'(3) = 24 - 27 = -3$, and an equation of the tangent line is $y - 9 = -3(x - 3)$, or $y = -3x + 18$.

(b)

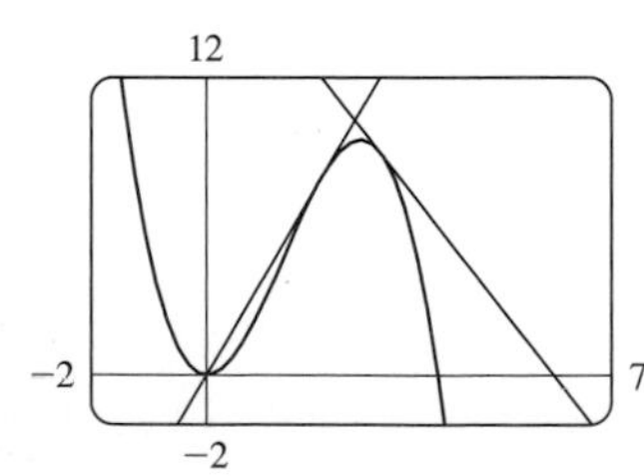

31. Use (4) with $f(x) = 3x^2 - 4x + 1$.

$$f'(a) = \lim_{h \to 0} \frac{f(a + h) - f(a)}{h} = \lim_{h \to 0} \frac{[3(a + h)^2 - 4(a + h) + 1] - (3a^2 - 4a + 1)}{h}$$

$$= \lim_{h \to 0} \frac{3a^2 + 6ah + 3h^2 - 4a - 4h + 1 - 3a^2 + 4a - 1}{h} = \lim_{h \to 0} \frac{6ah + 3h^2 - 4h}{h}$$

$$= \lim_{h \to 0} \frac{h(6a + 3h - 4)}{h} = \lim_{h \to 0} (6a + 3h - 4) = 6a - 4$$

32. Use (4) with $f(t) = 2t^3 + t$.

$$f'(a) = \lim_{h \to 0} \frac{f(a+h) - f(a)}{h} = \lim_{h \to 0} \frac{[2(a+h)^3 + (a+h)] - (2a^3 + a)}{h}$$

$$= \lim_{h \to 0} \frac{2a^3 + 6a^2h + 6ah^2 + 2h^3 + a + h - 2a^3 - a}{h} = \lim_{h \to 0} \frac{6a^2h + 6ah^2 + 2h^3 + h}{h}$$

$$= \lim_{h \to 0} \frac{h(6a^2 + 6ah + 2h^2 + 1)}{h} = \lim_{h \to 0} (6a^2 + 6ah + 2h^2 + 1) = 6a^2 + 1$$

33. Use (4) with $f(t) = (2t + 1)/(t + 3)$.

$$f'(a) = \lim_{h \to 0} \frac{f(a+h) - f(a)}{h} = \lim_{h \to 0} \frac{\dfrac{2(a+h)+1}{(a+h)+3} - \dfrac{2a+1}{a+3}}{h}$$

$$= \lim_{h \to 0} \frac{(2a + 2h + 1)(a + 3) - (2a + 1)(a + h + 3)}{h(a + h + 3)(a + 3)}$$

$$= \lim_{h \to 0} \frac{(2a^2 + 6a + 2ah + 6h + a + 3) - (2a^2 + 2ah + 6a + a + h + 3)}{h(a + h + 3)(a + 3)}$$

$$= \lim_{h \to 0} \frac{5h}{h(a + h + 3)(a + 3)} = \lim_{h \to 0} \frac{5}{(a + h + 3)(a + 3)} = \frac{5}{(a + 3)^2}$$

34. Use (4) with $f(x) = x^{-2} = 1/x^2$.

$$f'(a) = \lim_{h \to 0} \frac{f(a+h) - f(a)}{h} = \lim_{h \to 0} \frac{\dfrac{1}{(a+h)^2} - \dfrac{1}{a^2}}{h} = \lim_{h \to 0} \frac{\dfrac{a^2 - (a+h)^2}{a^2(a+h)^2}}{h}$$

$$= \lim_{h \to 0} \frac{a^2 - (a^2 + 2ah + h^2)}{ha^2(a+h)^2} = \lim_{h \to 0} \frac{-2ah - h^2}{ha^2(a+h)^2} = \lim_{h \to 0} \frac{h(-2a - h)}{ha^2(a+h)^2}$$

$$= \lim_{h \to 0} \frac{-2a - h}{a^2(a+h)^2} = \frac{-2a}{a^2(a^2)} = \frac{-2}{a^3}$$

35. Use (4) with $f(x) = \sqrt{1 - 2x}$.

$$f'(a) = \lim_{h \to 0} \frac{f(a+h) - f(a)}{h} = \lim_{h \to 0} \frac{\sqrt{1 - 2(a+h)} - \sqrt{1 - 2a}}{h}$$

$$= \lim_{h \to 0} \frac{\sqrt{1 - 2(a+h)} - \sqrt{1 - 2a}}{h} \cdot \frac{\sqrt{1 - 2(a+h)} + \sqrt{1 - 2a}}{\sqrt{1 - 2(a+h)} + \sqrt{1 - 2a}}$$

$$= \lim_{h \to 0} \frac{\left(\sqrt{1 - 2(a+h)}\right)^2 - \left(\sqrt{1 - 2a}\right)^2}{h\left(\sqrt{1 - 2(a+h)} + \sqrt{1 - 2a}\right)} = \lim_{h \to 0} \frac{(1 - 2a - 2h) - (1 - 2a)}{h\left(\sqrt{1 - 2(a+h)} + \sqrt{1 - 2a}\right)}$$

$$= \lim_{h \to 0} \frac{-2h}{h\left(\sqrt{1 - 2(a+h)} + \sqrt{1 - 2a}\right)} = \lim_{h \to 0} \frac{-2}{\sqrt{1 - 2(a+h)} + \sqrt{1 - 2a}}$$

$$= \frac{-2}{\sqrt{1 - 2a} + \sqrt{1 - 2a}} = \frac{-2}{2\sqrt{1 - 2a}} = \frac{-1}{\sqrt{1 - 2a}}$$

36. Use (4) with $f(x) = \dfrac{4}{\sqrt{1-x}}$.

$$f'(a) = \lim_{h\to 0} \frac{f(a+h)-f(a)}{h} = \lim_{h\to 0} \frac{\dfrac{4}{\sqrt{1-(a+h)}} - \dfrac{4}{\sqrt{1-a}}}{h}$$

$$= 4\lim_{h\to 0} \frac{\dfrac{\sqrt{1-a}-\sqrt{1-a-h}}{\sqrt{1-a-h}\,\sqrt{1-a}}}{h} = 4\lim_{h\to 0} \frac{\sqrt{1-a}-\sqrt{1-a-h}}{h\sqrt{1-a-h}\,\sqrt{1-a}}$$

$$= 4\lim_{h\to 0} \frac{\sqrt{1-a}-\sqrt{1-a-h}}{h\sqrt{1-a-h}\,\sqrt{1-a}} \cdot \frac{\sqrt{1-a}+\sqrt{1-a-h}}{\sqrt{1-a}+\sqrt{1-a-h}} = 4\lim_{h\to 0} \frac{(\sqrt{1-a})^2-(\sqrt{1-a-h})^2}{h\sqrt{1-a-h}\,\sqrt{1-a}(\sqrt{1-a}+\sqrt{1-a-h})}$$

$$= 4\lim_{h\to 0} \frac{(1-a)-(1-a-h)}{h\sqrt{1-a-h}\,\sqrt{1-a}(\sqrt{1-a}+\sqrt{1-a-h})} = 4\lim_{h\to 0} \frac{h}{h\sqrt{1-a-h}\,\sqrt{1-a}(\sqrt{1-a}+\sqrt{1-a-h})}$$

$$= 4\lim_{h\to 0} \frac{1}{\sqrt{1-a-h}\,\sqrt{1-a}(\sqrt{1-a}+\sqrt{1-a-h})} = 4\cdot\frac{1}{\sqrt{1-a}\,\sqrt{1-a}(\sqrt{1-a}+\sqrt{1-a})}$$

$$= \frac{4}{(1-a)(2\sqrt{1-a})} = \frac{2}{(1-a)^1(1-a)^{1/2}} = \frac{2}{(1-a)^{3/2}}$$

37. By (4), $\displaystyle\lim_{h\to 0} \frac{\sqrt{9+h}-3}{h} = f'(9)$, where $f(x) = \sqrt{x}$ and $a = 9$.

38. By (4), $\displaystyle\lim_{h\to 0} \frac{e^{-2+h}-e^{-2}}{h} = f'(-2)$, where $f(x) = e^x$ and $a = -2$.

39. By Equation 5, $\displaystyle\lim_{x\to 2} \frac{x^6-64}{x-2} = f'(2)$, where $f(x) = x^6$ and $a = 2$.

40. By Equation 5, $\displaystyle\lim_{x\to 1/4} \frac{\dfrac{1}{x}-4}{x-\dfrac{1}{4}} = f'(4)$, where $f(x) = \dfrac{1}{x}$ and $a = \dfrac{1}{4}$.

41. By (4), $\displaystyle\lim_{h\to 0} \frac{\cos(\pi+h)+1}{h} = f'(\pi)$, where $f(x) = \cos x$ and $a = \pi$.

Or: By (4), $\displaystyle\lim_{h\to 0} \frac{\cos(\pi+h)+1}{h} = f'(0)$, where $f(x) = \cos(\pi+x)$ and $a = 0$.

42. By Equation 5, $\displaystyle\lim_{\theta\to \pi/6} \frac{\sin\theta-\frac{1}{2}}{\theta-\dfrac{\pi}{6}} = f'\left(\frac{\pi}{6}\right)$, where $f(\theta) = \sin\theta$ and $a = \dfrac{\pi}{6}$.

43. $\displaystyle v(4) = f'(4) = \lim_{h\to 0} \frac{f(4+h)-f(4)}{h} = \lim_{h\to 0} \frac{\left[80(4+h)-6(4+h)^2\right]-\left[80(4)-6(4)^2\right]}{h}$

$$= \lim_{h\to 0} \frac{(320+80h-96-48h-6h^2)-(320-96)}{h} = \lim_{h\to 0} \frac{32h-6h^2}{h}$$

$$= \lim_{h\to 0} \frac{h(32-6h)}{h} = \lim_{h\to 0}(32-6h) = 32 \text{ m/s}$$

The speed when $t = 4$ is $|32| = 32$ m/s.

44. $v(4) = f'(4) = \lim\limits_{h\to 0} \dfrac{f(4+h) - f(4)}{h} = \lim\limits_{h\to 0} \dfrac{\left(10 + \dfrac{45}{4+h+1}\right) - \left(10 + \dfrac{45}{4+1}\right)}{h} = \lim\limits_{h\to 0} \dfrac{\dfrac{45}{5+h} - 9}{h}$

$$= \lim_{h\to 0} \frac{45 - 9(5+h)}{h(5+h)} = \lim_{h\to 0} \frac{-9h}{h(5+h)} = \lim_{h\to 0} \frac{-9}{5+h} = -\frac{9}{5} \text{ m/s.}$$

The speed when $t = 4$ is $\left|-\frac{9}{5}\right| = \frac{9}{5}$ m/s.

45. The sketch shows the graph for a room temperature of $72°$ and a refrigerator temperature of $38°$. The initial rate of change is greater in magnitude than the rate of change after an hour.

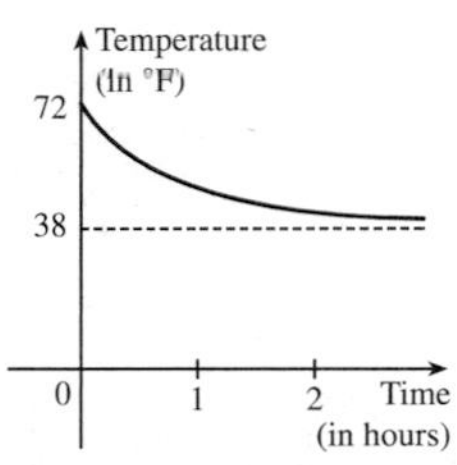

46. The slope of the tangent (that is, the rate of change of temperature with respect to time) at $t = 1$ h seems to be about $\dfrac{75 - 168}{132 - 0} \approx -0.7\,°\text{F/min}$.

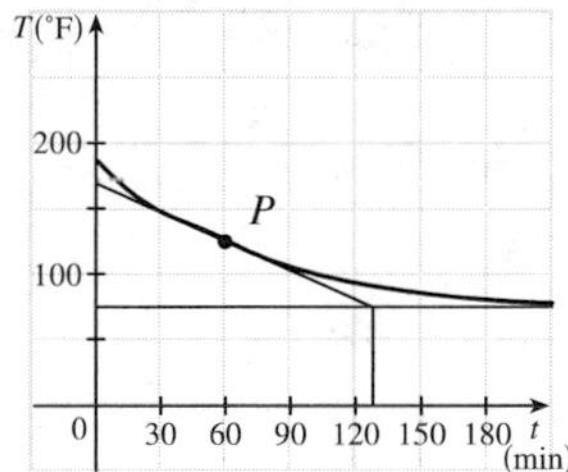

47. (a) (i) $[1.0, 2.0]$: $\dfrac{C(2) - C(1)}{2 - 1} = \dfrac{0.18 - 0.33}{1} = -0.15 \dfrac{\text{mg/mL}}{\text{h}}$

(ii) $[1.5, 2.0]$: $\dfrac{C(2) - C(1.5)}{2 - 1.5} = \dfrac{0.18 - 0.24}{0.5} = \dfrac{-0.06}{0.5} = -0.12 \dfrac{\text{mg/mL}}{\text{h}}$

(iii) $[2.0, 2.5]$: $\dfrac{C(2.5) - C(2)}{2.5 - 2} = \dfrac{0.12 - 0.18}{0.5} = \dfrac{-0.06}{0.5} = -0.12 \dfrac{\text{mg/mL}}{\text{h}}$

(iv) $[2.0, 3.0]$: $\dfrac{C(3) - C(2)}{3 - 2} = \dfrac{0.07 - 0.18}{1} = -0.11 \dfrac{\text{mg/mL}}{\text{h}}$

(b) We estimate the instantaneous rate of change at $t = 2$ by averaging the average rates of change for $[1.5, 2.0]$ and $[2.0, 2.5]$:

$\dfrac{-0.12 + (-0.12)}{2} = -0.12 \dfrac{\text{mg/mL}}{\text{h}}$. After 2 hours, the BAC is decreasing at a rate of 0.12 (mg/mL)/h.

48. (a) (i) $[2006, 2008]$: $\dfrac{N(2008) - N(2006)}{2008 - 2006} = \dfrac{16{,}680 - 12{,}440}{2} = \dfrac{4240}{2} = 2120$ locations/year

(ii) $[2008, 2010]$: $\dfrac{N(2010) - N(2008)}{2010 - 2008} = \dfrac{16{,}858 - 16{,}680}{2} = \dfrac{178}{2} = 89$ locations/year.

The rate of growth decreased over the period from 2006 to 2010.

(b) $[2010, 2012]$: $\dfrac{N(2012) - N(2010)}{2012 - 2010} = \dfrac{18{,}066 - 16{,}858}{2} = \dfrac{1208}{2} = 604$ locations/year.

Using that value and the value from part (a)(ii), we have $\dfrac{89 + 604}{2} = \dfrac{693}{2} = 346.5$ locations/year.

(c) The tangent segment has endpoints (2008, 16,250) and (2012, 17,500).

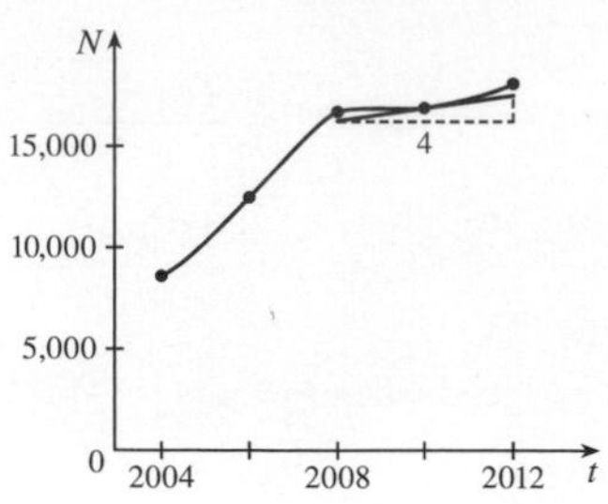

An estimate of the instantaneous rate of growth in 2010 is

$$\frac{17{,}500 - 16{,}250}{2012 - 2008} = \frac{1250}{4} = 312.5 \text{ locations/year.}$$

49. (a) $[1990, 2005]$: $\dfrac{84{,}077 - 66{,}533}{2005 - 1990} = \dfrac{17{,}544}{15} = 1169.6$ thousands of barrels per day per year. This means that oil consumption rose by an average of 1169.6 thousands of barrels per day each year from 1990 to 2005.

(b) $[1995, 2000]$: $\dfrac{76{,}784 - 70{,}099}{2000 - 1995} = \dfrac{6685}{5} = 1337$

$[2000, 2005]$: $\dfrac{84{,}077 - 76{,}784}{2005 - 2000} = \dfrac{7293}{5} = 1458.6$

An estimate of the instantaneous rate of change in 2000 is $\frac{1}{2}\,(1337 + 1458.6) = 1397.8$ thousands of barrels per day per year.

50. (a) (i) $[4, 11]$: $\dfrac{V(11) - V(4)}{11 - 4} = \dfrac{9.4 - 53}{7} = \dfrac{-43.6}{7} \approx -6.23\ \dfrac{\text{RNA copies/mL}}{\text{day}}$

(ii) $[8, 11]$: $\dfrac{V(11) - V(8)}{11 - 8} = \dfrac{9.4 - 18}{3} = \dfrac{-8.6}{3} \approx -2.87\ \dfrac{\text{RNA copies/mL}}{\text{day}}$

(iii) $[11, 15]$: $\dfrac{V(15) - V(11)}{15 - 11} = \dfrac{5.2 - 9.4}{4} = \dfrac{-4.2}{4} = -1.05\ \dfrac{\text{RNA copies/mL}}{\text{day}}$

(iv) $[11, 22]$: $\dfrac{V(22) - V(11)}{22 - 11} = \dfrac{3.6 - 9.4}{11} = \dfrac{-5.8}{11} \approx -0.53\ \dfrac{\text{RNA copies/mL}}{\text{day}}$

(b) An estimate of $V'(11)$ is the average of the answers from part (a)(ii) and (iii).

$$V'(11) \approx \tfrac{1}{2}\,[-2.87 + (-1.05)] = -1.96\ \frac{\text{RNA copies/mL}}{\text{day}}.$$

$V'(11)$ measures the instantaneous rate of change of patient 303's viral load 11 days after ABT-538 treatment began.

51. (a) (i) $\dfrac{\Delta C}{\Delta x} = \dfrac{C(105) - C(100)}{105 - 100} = \dfrac{6601.25 - 6500}{5} = \$20.25/\text{unit.}$

(ii) $\dfrac{\Delta C}{\Delta x} = \dfrac{C(101) - C(100)}{101 - 100} = \dfrac{6520.05 - 6500}{1} = \$20.05/\text{unit.}$

(b) $$\frac{C(100 + h) - C(100)}{h} = \frac{\left[5000 + 10(100 + h) + 0.05(100 + h)^2\right] - 6500}{h} = \frac{20h + 0.05h^2}{h}$$
$$= 20 + 0.05h,\ h \neq 0$$

So the instantaneous rate of change is $\displaystyle\lim_{h \to 0} \frac{C(100 + h) - C(100)}{h} = \lim_{h \to 0}\,(20 + 0.05h) = \$20/\text{unit.}$

52. $\Delta V = V(t+h) - V(t) = 100{,}000\left(1 - \frac{t+h}{60}\right)^2 - 100{,}000\left(1 - \frac{t}{60}\right)^2$

$$= 100{,}000\left[\left(1 - \frac{t+h}{30} + \frac{(t+h)^2}{3600}\right) - \left(1 - \frac{t}{30} + \frac{t^2}{3600}\right)\right] = 100{,}000\left(-\frac{h}{30} + \frac{2th}{3600} + \frac{h^2}{3600}\right)$$

$$= \frac{100{,}000}{3600}h\,(-120 + 2t + h) = \frac{250}{9}h\,(-120 + 2t + h)$$

Dividing ΔV by h and then letting $h \to 0$, we see that the instantaneous rate of change is $\frac{500}{9}(t - 60)$ gal/min.

t	Flow rate (gal/min)	Water remaining $V(t)$ (gal)
0	$-3333.\overline{3}$	100,000
10	$-2777.\overline{7}$	$69{,}444.\overline{4}$
20	$-2222.\overline{2}$	$44{,}444.\overline{4}$
30	$-1666.\overline{6}$	25,000
40	$-1111.\overline{1}$	$11{,}111.\overline{1}$
50	$-555.\overline{5}$	$2{,}777.\overline{7}$
60	0	0

The magnitude of the flow rate is greatest at the beginning and gradually decreases to 0.

53. (a) $f'(x)$ is the rate of change of the production cost with respect to the number of ounces of gold produced. Its units are dollars per ounce.

(b) After 800 ounces of gold have been produced, the rate at which the production cost is increasing is \$17/ounce. So the cost of producing the 800th (or 801st) ounce is about \$17.

(c) In the short term, the values of $f'(x)$ will decrease because more efficient use is made of start-up costs as x increases. But eventually $f'(x)$ might increase due to large-scale operations.

54. (a) $f'(5)$ is the rate of growth of the bacteria population when $t = 5$ hours. Its units are bacteria per hour.

(b) With unlimited space and nutrients, f' should increase as t increases; so $f'(5) < f'(10)$. If the supply of nutrients is limited, the growth rate slows down at some point in time, and the opposite may be true.

55. (a) $H'(58)$ is the rate at which the daily heating cost changes with respect to temperature when the outside temperature is 58 °F. The units are dollars/ °F.

(b) If the outside temperature increases, the building should require less heating, so we would expect $H'(58)$ to be negative.

56. (a) $f'(8)$ is the rate of change of the quantity of coffee sold with respect to the price per pound when the price is \$8 per pound. The units for $f'(8)$ are pounds/(dollars/pound).

(b) $f'(8)$ is negative since the quantity of coffee sold will decrease as the price charged for it increases. People are generally less willing to buy a product when its price increases.

57. (a) $S'(T)$ is the rate at which the oxygen solubility changes with respect to the water temperature. Its units are (mg/L)/°C.

(b) For $T = 16°$C, it appears that the tangent line to the curve goes through the points $(0, 14)$ and $(32, 6)$. So

$S'(16) \approx \frac{6 - 14}{32 - 0} = -\frac{8}{32} = -0.25$ (mg/L)/°C. This means that as the temperature increases past 16°C, the oxygen solubility is decreasing at a rate of 0.25 (mg/L)/°C.

58. (a) $S'(T)$ is the rate of change of the maximum sustainable speed of Coho salmon with respect to the temperature. Its units are (cm/s)/°C.

(b) For $T = 15°\text{C}$, it appears the tangent line to the curve goes through the points $(10, 25)$ and $(20, 32)$. So $S'(15) \approx \dfrac{32-25}{20-10} = 0.7$ (cm/s)/°C. This tells us that at $T = 15°\text{C}$, the maximum sustainable speed of Coho salmon is changing at a rate of 0.7 (cm/s)/°C. In a similar fashion for $T = 25°\text{C}$, we can use the points $(20, 35)$ and $(25, 25)$ to obtain $S'(25) \approx \dfrac{25-35}{25-20} = -2$ (cm/s)/°C. As it gets warmer than 20°C, the maximum sustainable speed decreases rapidly.

59. Since $f(x) = x\sin(1/x)$ when $x \neq 0$ and $f(0) = 0$, we have

$f'(0) = \lim\limits_{h\to 0} \dfrac{f(0+h)-f(0)}{h} = \lim\limits_{h\to 0} \dfrac{h\sin(1/h)-0}{h} = \lim\limits_{h\to 0} \sin(1/h)$. This limit does not exist since $\sin(1/h)$ takes the values -1 and 1 on any interval containing 0. (Compare with Example 2.2.4.)

60. Since $f(x) = x^2\sin(1/x)$ when $x \neq 0$ and $f(0) = 0$, we have

$f'(0) = \lim\limits_{h\to 0} \dfrac{f(0+h)-f(0)}{h} = \lim\limits_{h\to 0} \dfrac{h^2\sin(1/h)-0}{h} = \lim\limits_{h\to 0} h\sin(1/h)$. Since $-1 \le \sin\dfrac{1}{h} \le 1$, we have

$-|h| \le |h|\sin\dfrac{1}{h} \le |h| \;\Rightarrow\; -|h| \le h\sin\dfrac{1}{h} \le |h|$. Because $\lim\limits_{h\to 0}(-|h|) = 0$ and $\lim\limits_{h\to 0}|h| = 0$, we know that

$\lim\limits_{h\to 0}\left(h\sin\dfrac{1}{h}\right) = 0$ by the Squeeze Theorem. Thus, $f'(0) = 0$.

61. (a) The slope at the origin appears to be 1.

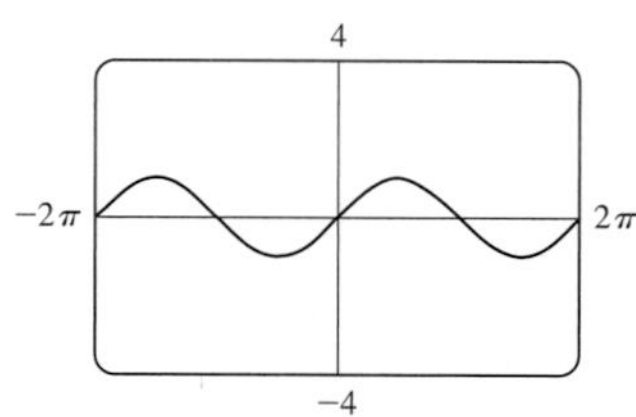

(b) The slope at the origin still appears to be 1.

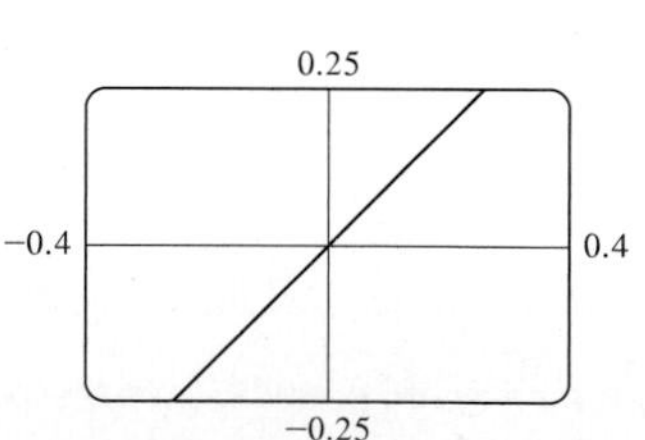

(c) Yes, the slope at the origin now appears to be 0.

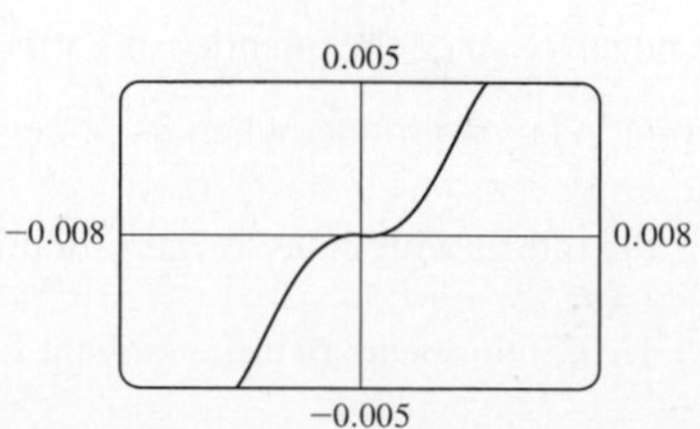

2.8 The Derivative as a Function

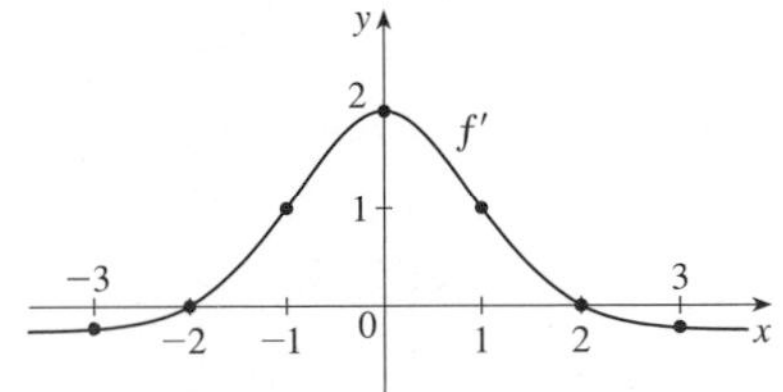

1. It appears that f is an odd function, so f' will be an even function—that is, $f'(-a) = f'(a)$.

(a) $f'(-3) \approx -0.2$

(b) $f'(-2) \approx 0$ (c) $f'(-1) \approx 1$ (d) $f'(0) \approx 2$

(e) $f'(1) \approx 1$ (f) $f'(2) \approx 0$ (g) $f'(3) \approx -0.2$

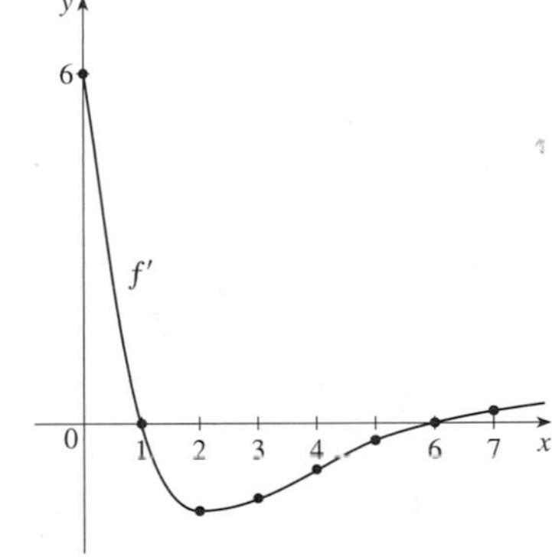

2. Your answers may vary depending on your estimates.

(a) *Note:* By estimating the slopes of tangent lines on the graph of f, it appears that $f'(0) \approx 6$.

(b) $f'(1) \approx 0$

(c) $f'(2) \approx -1.5$ (d) $f'(3) \approx -1.3$ (e) $f'(4) \approx -0.8$

(f) $f'(5) \approx -0.3$ (g) $f'(6) \approx 0$ (h) $f'(7) \approx 0.2$

3. (a)$' =$ II, since from left to right, the slopes of the tangents to graph (a) start out negative, become 0, then positive, then 0, then negative again. The actual function values in graph II follow the same pattern.

(b)$' =$ IV, since from left to right, the slopes of the tangents to graph (b) start out at a fixed positive quantity, then suddenly become negative, then positive again. The discontinuities in graph IV indicate sudden changes in the slopes of the tangents.

(c)$' =$ I, since the slopes of the tangents to graph (c) are negative for $x < 0$ and positive for $x > 0$, as are the function values of graph I.

(d)$' =$ III, since from left to right, the slopes of the tangents to graph (d) are positive, then 0, then negative, then 0, then positive, then 0, then negative again, and the function values in graph III follow the same pattern.

Hints for Exercises 4–11: First plot x-intercepts on the graph of f' for any horizontal tangents on the graph of f. Look for any corners on the graph of f—there will be a discontinuity on the graph of f'. On any interval where f has a tangent with positive (or negative) slope, the graph of f' will be positive (or negative). If the graph of the function is linear, the graph of f' will be a horizontal line.

4.

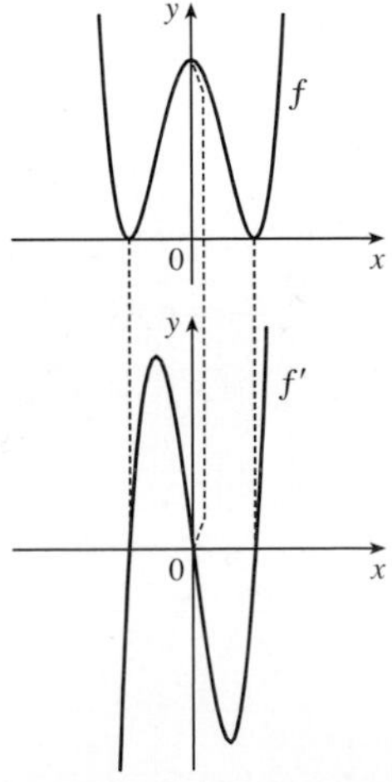

5.

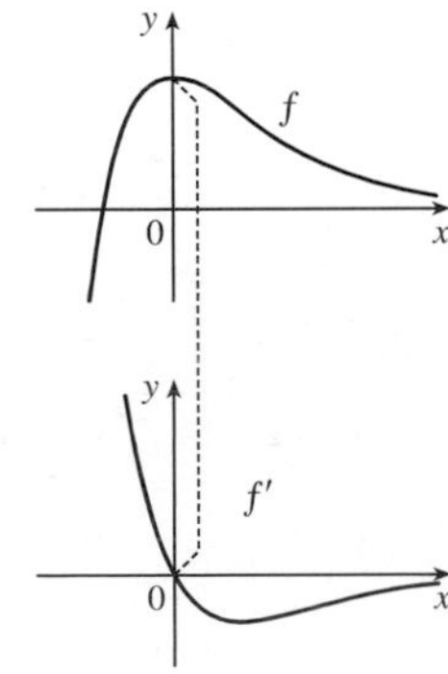

6.

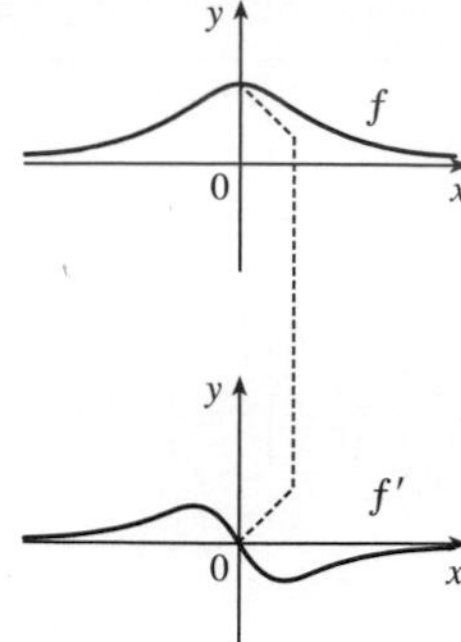

7.

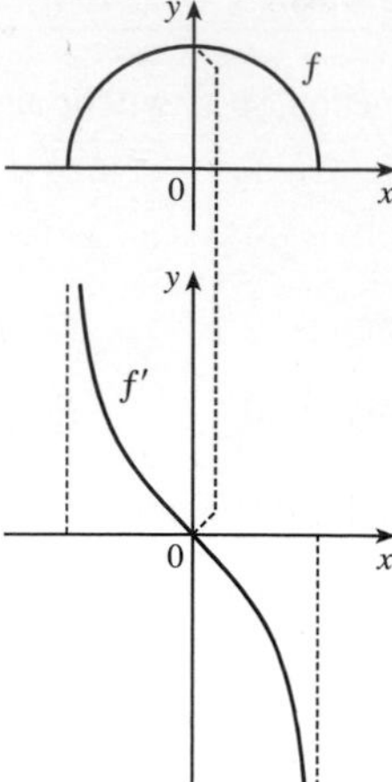

8.

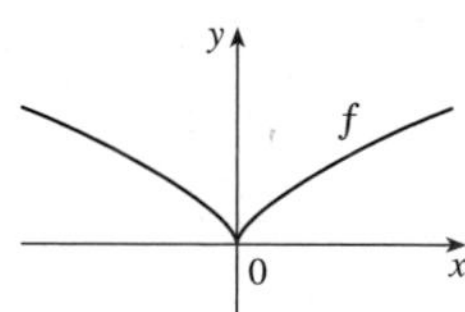

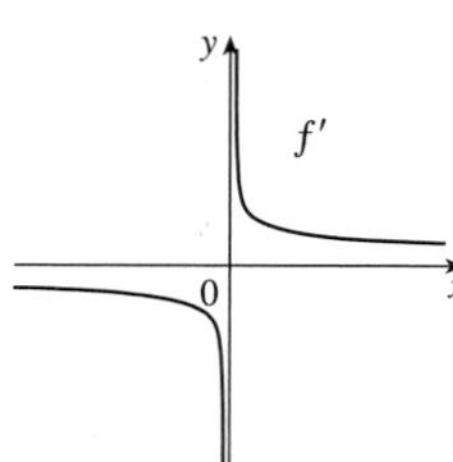

9.

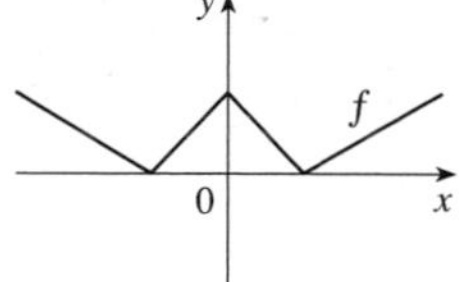

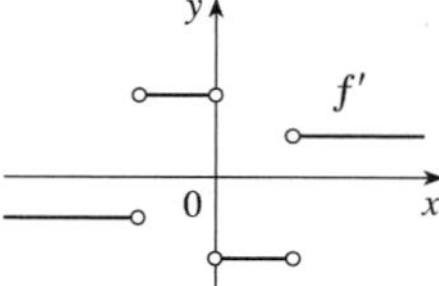

10.

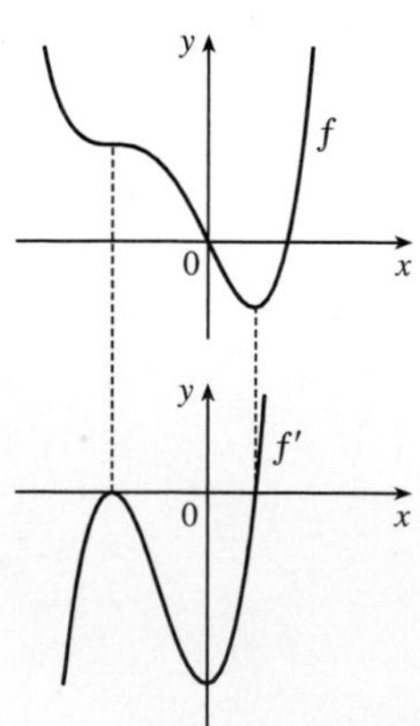

11.

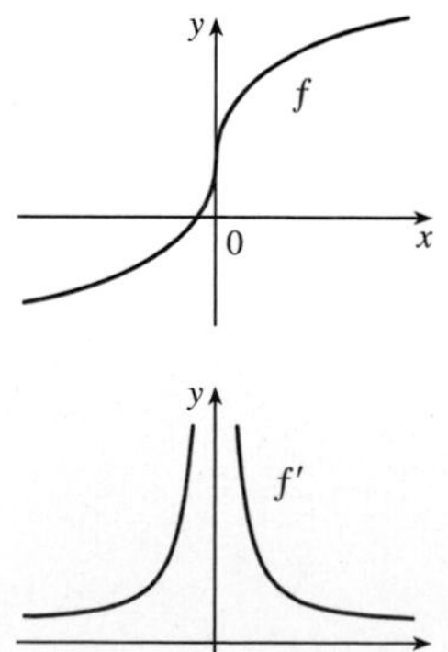

12. The slopes of the tangent lines on the graph of $y = P(t)$ are always positive, so the y-values of $y = P'(t)$ are always positive. These values start out relatively small and keep increasing, reaching a maximum at about $t = 6$. Then the y-values of $y = P'(t)$ decrease and get close to zero. The graph of P' tells us that the yeast culture grows most rapidly after 6 hours and then the growth rate declines.

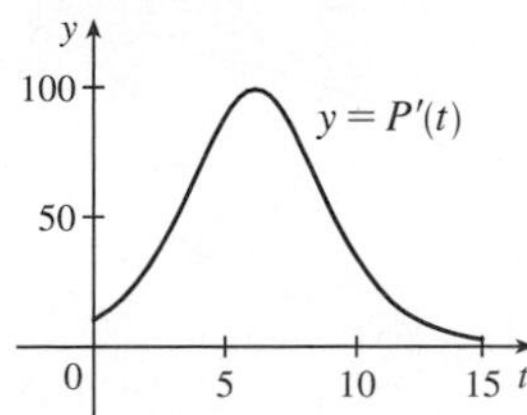

13. (a) $C'(t)$ is the instantaneous rate of change of percentage of full capacity with respect to elapsed time in hours.

(b) The graph of $C'(t)$ tells us that the rate of change of percentage of full capacity is decreasing and approaching 0.

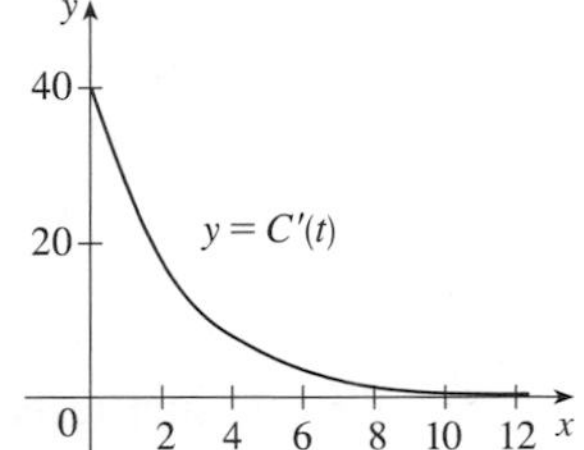

14. (a) $F'(v)$ is the instantaneous rate of change of fuel economy with respect to speed.

(b) Graphs will vary depending on estimates of F', but will change from positive to negative at about $v = 50$.

(c) To save on gas, drive at the speed where F is a maximum and F' is 0, which is about 50 mi/h.

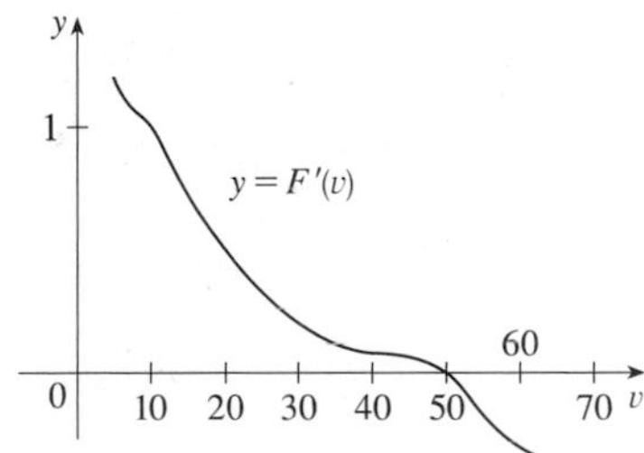

15. It appears that there are horizontal tangents on the graph of M for $t = 1963$ and $t = 1971$. Thus, there are zeros for those values of t on the graph of M'. The derivative is negative for the years 1963 to 1971.

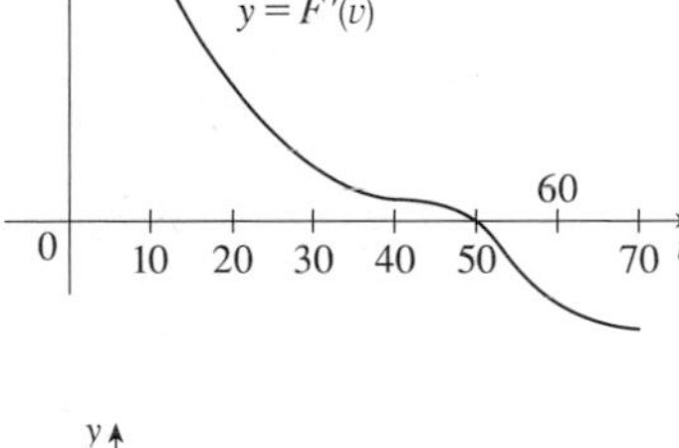

16. See Figure 3.3.1.

17.

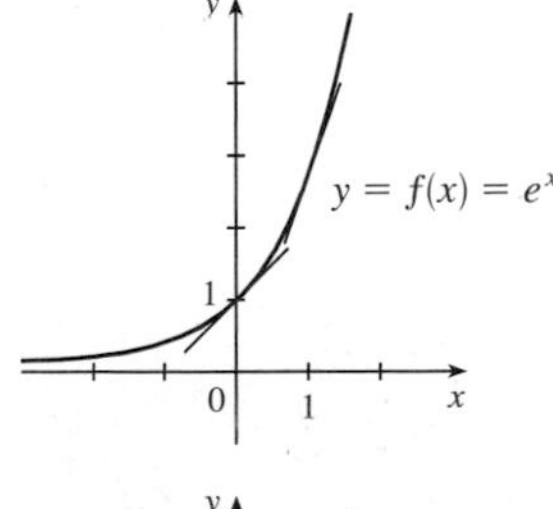

y = f'(x)

The slope at 0 appears to be 1 and the slope at 1 appears to be 2.7. As x decreases, the slope gets closer to 0. Since the graphs are so similar, we might guess that $f'(x) = e^x$.

18.

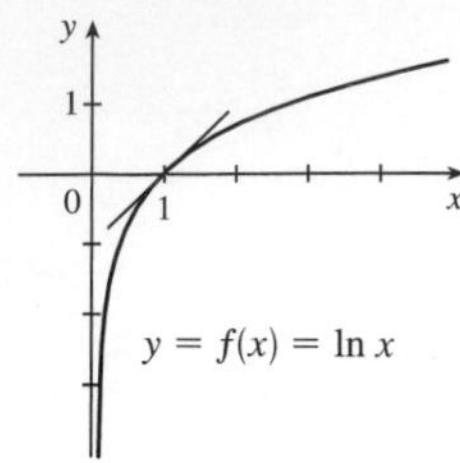

y

y = f′(x)

1

0 1 x

As x increases toward 1, $f'(x)$ decreases from very large numbers to 1. As x becomes large, $f'(x)$ gets closer to 0. As a guess, $f'(x) = 1/x^2$ or $f'(x) = 1/x$ makes sense.

19. (a) By zooming in, we estimate that $f'(0) = 0$, $f'\left(\frac{1}{2}\right) = 1$, $f'(1) = 2$, and $f'(2) = 4$.

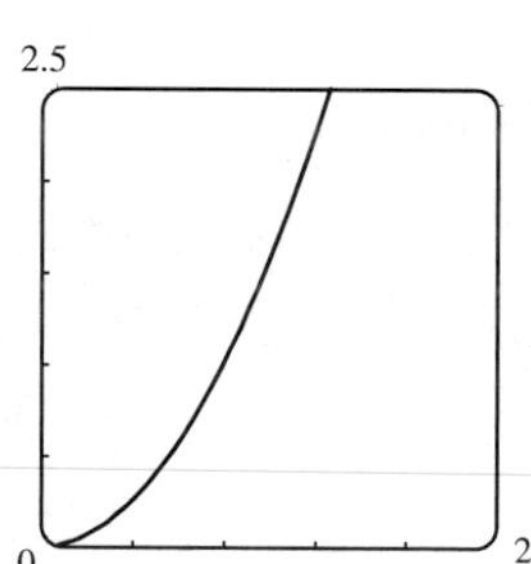

(b) By symmetry, $f'(-x) = -f'(x)$. So $f'\left(-\frac{1}{2}\right) = -1$, $f'(-1) = -2$, and $f'(-2) = -4$.

(c) It appears that $f'(x)$ is twice the value of x, so we guess that $f'(x) = 2x$.

(d) $$f'(x) = \lim_{h\to 0}\frac{f(x+h)-f(x)}{h} = \lim_{h\to 0}\frac{(x+h)^2 - x^2}{h}$$

$$= \lim_{h\to 0}\frac{\left(x^2+2hx+h^2\right)-x^2}{h} = \lim_{h\to 0}\frac{2hx+h^2}{h} = \lim_{h\to 0}\frac{h(2x+h)}{h} = \lim_{h\to 0}(2x+h) = 2x$$

20. (a) By zooming in, we estimate that $f'(0) = 0$, $f'\left(\frac{1}{2}\right) \approx 0.75$, $f'(1) \approx 3$, $f'(2) \approx 12$, and $f'(3) \approx 27$.

(b) By symmetry, $f'(-x) = f'(x)$. So $f'\left(-\frac{1}{2}\right) \approx 0.75$, $f'(-1) \approx 3$, $f'(-2) \approx 12$, and $f'(-3) \approx 27$.

(c)

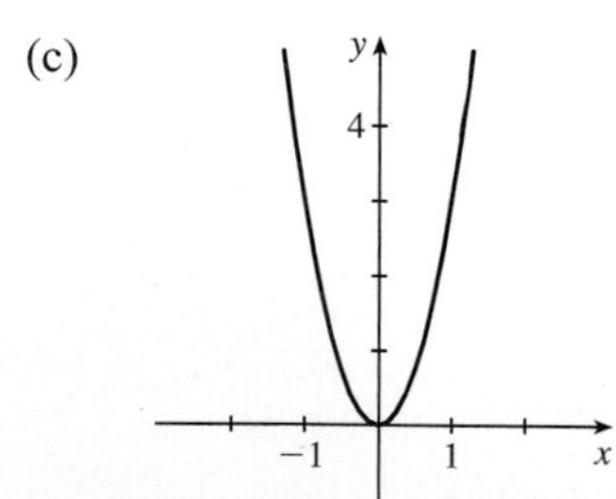

(d) Since $f'(0) = 0$, it appears that f' may have the form $f'(x) = ax^2$. Using $f'(1) = 3$, we have $a = 3$, so $f'(x) = 3x^2$.

(e) $$f'(x) = \lim_{h\to 0}\frac{f(x+h)-f(x)}{h} = \lim_{h\to 0}\frac{(x+h)^3 - x^3}{h} = \lim_{h\to 0}\frac{\left(x^3+3x^2h+3xh^2+h^3\right)-x^3}{h}$$

$$= \lim_{h\to 0}\frac{3x^2h+3xh^2+h^3}{h} = \lim_{h\to 0}\frac{h\left(3x^2+3xh+h^2\right)}{h} = \lim_{h\to 0}\left(3x^2+3xh+h^2\right) = 3x^2$$

21. $f'(x) = \lim\limits_{h \to 0} \dfrac{f(x+h) - f(x)}{h} = \lim\limits_{h \to 0} \dfrac{[3(x+h) - 8] - (3x - 8)}{h} = \lim\limits_{h \to 0} \dfrac{3x + 3h - 8 - 3x + 8}{h}$

$= \lim\limits_{h \to 0} \dfrac{3h}{h} = \lim\limits_{h \to 0} 3 = 3$

Domain of f = domain of $f' = \mathbb{R}$.

22. $f'(x) = \lim\limits_{h \to 0} \dfrac{f(x+h) - f(x)}{h} = \lim\limits_{h \to 0} \dfrac{[m(x+h) + b] - (mx + b)}{h} = \lim\limits_{h \to 0} \dfrac{mx + mh + b - mx - b}{h}$

$= \lim\limits_{h \to 0} \dfrac{mh}{h} = \lim\limits_{h \to 0} m = m$

Domain of f = domain of $f' = \mathbb{R}$.

23. $f'(t) = \lim\limits_{h \to 0} \dfrac{f(t+h) - f(t)}{h} = \lim\limits_{h \to 0} \dfrac{\left[2.5(t+h)^2 + 6(t+h)\right] - \left(2.5t^2 + 6t\right)}{h}$

$= \lim\limits_{h \to 0} \dfrac{2.5(t^2 + 2th + h^2) + 6t + 6h - 2.5t^2 - 6t}{h} = \lim\limits_{h \to 0} \dfrac{2.5t^2 + 5th + 2.5h^2 + 6h - 2.5t^2}{h}$

$= \lim\limits_{h \to 0} \dfrac{5th + 2.5h^2 + 6h}{h} = \lim\limits_{h \to 0} \dfrac{h\,(5t + 2.5h + 6)}{h} = \lim\limits_{h \to 0} (5t + 2.5h + 6)$

$= 5t + 6$

Domain of f = domain of $f' = \mathbb{R}$.

24. $f'(x) = \lim\limits_{h \to 0} \dfrac{f(x+h) - f(x)}{h} = \lim\limits_{h \to 0} \dfrac{\left[4 + 8(x+h) - 5(x+h)^2\right] - (4 + 8x - 5x^2)}{h}$

$= \lim\limits_{h \to 0} \dfrac{4 + 8x + 8h - 5(x^2 + 2xh + h^2) - 4 - 8x + 5x^2}{h} = \lim\limits_{h \to 0} \dfrac{8h - 5x^2 - 10xh - 5h^2 + 5x^2}{h}$

$= \lim\limits_{h \to 0} \dfrac{8h - 10xh - 5h^2}{h} = \lim\limits_{h \to 0} \dfrac{h(8 - 10x - 5h)}{h} = \lim\limits_{h \to 0} (8 - 10x - 5h)$

$= 8 - 10x$

Domain of f = domain of $f' = \mathbb{R}$.

25. $f'(x) = \lim\limits_{h \to 0} \dfrac{f(x+h) - f(x)}{h} = \lim\limits_{h \to 0} \dfrac{[(x+h)^2 - 2(x+h)^3] - (x^2 - 2x^3)}{h}$

$= \lim\limits_{h \to 0} \dfrac{x^2 + 2xh + h^2 - 2x^3 - 6x^2h - 6xh^2 - 2h^3 - x^2 + 2x^3}{h}$

$= \lim\limits_{h \to 0} \dfrac{2xh + h^2 - 6x^2h - 6xh^2 - 2h^3}{h} = \lim\limits_{h \to 0} \dfrac{h(2x + h - 6x^2 - 6xh - 2h^2)}{h}$

$= \lim\limits_{h \to 0} (2x + h - 6x^2 - 6xh - 2h^2) = 2x - 6x^2$

Domain of f = domain of $f' = \mathbb{R}$.

26. $g'(t) = \lim\limits_{h \to 0} \dfrac{g(t+h) - g(t)}{h} = \lim\limits_{h \to 0} \dfrac{\dfrac{1}{\sqrt{t+h}} - \dfrac{1}{\sqrt{t}}}{h} = \lim\limits_{h \to 0} \dfrac{\dfrac{\sqrt{t} - \sqrt{t+h}}{\sqrt{t+h}\,\sqrt{t}}}{h} = \lim\limits_{h \to 0} \left(\dfrac{\sqrt{t} - \sqrt{t+h}}{h\,\sqrt{t+h}\,\sqrt{t}} \cdot \dfrac{\sqrt{t} + \sqrt{t+h}}{\sqrt{t} + \sqrt{t+h}}\right)$

$= \lim\limits_{h \to 0} \dfrac{t - (t+h)}{h\,\sqrt{t+h}\,\sqrt{t}\,(\sqrt{t} + \sqrt{t+h})} = \lim\limits_{h \to 0} \dfrac{-h}{h\,\sqrt{t+h}\,\sqrt{t}\,(\sqrt{t} + \sqrt{t+h})} = \lim\limits_{h \to 0} \dfrac{-1}{\sqrt{t+h}\,\sqrt{t}\,(\sqrt{t} + \sqrt{t+h})}$

$= \dfrac{-1}{\sqrt{t}\,\sqrt{t}\,(\sqrt{t} + \sqrt{t})} = \dfrac{-1}{t\,(2\sqrt{t})} = -\dfrac{1}{2t^{3/2}}$

Domain of g = domain of $g' = (0, \infty)$.

27. $g'(x) = \lim\limits_{h \to 0} \dfrac{g(x+h) - g(x)}{h} = \lim\limits_{h \to 0} \dfrac{\sqrt{9-(x+h)} - \sqrt{9-x}}{h} \left[\dfrac{\sqrt{9-(x+h)} + \sqrt{9-x}}{\sqrt{9-(x+h)} + \sqrt{9-x}} \right]$

$$= \lim_{h \to 0} \frac{[9-(x+h)] - (9-x)}{h\left[\sqrt{9-(x+h)} + \sqrt{9-x}\right]} = \lim_{h \to 0} \frac{-h}{h\left[\sqrt{9-(x+h)} + \sqrt{9-x}\right]}$$

$$= \lim_{h \to 0} \frac{-1}{\sqrt{9-(x+h)} + \sqrt{9-x}} = \frac{-1}{2\sqrt{9-x}}$$

Domain of $g = (-\infty, 9]$, domain of $g' = (-\infty, 9)$.

28. $f'(x) = \lim\limits_{h \to 0} \dfrac{f(x+h) - f(x)}{h} = \lim\limits_{h \to 0} \dfrac{\dfrac{(x+h)^2 - 1}{2(x+h) - 3} - \dfrac{x^2 - 1}{2x - 3}}{h}$

$$= \lim_{h \to 0} \frac{\dfrac{[(x+h)^2 - 1](2x-3) - [2(x+h) - 3](x^2 - 1)}{[2(x+h) - 3](2x - 3)}}{h}$$

$$= \lim_{h \to 0} \frac{(x^2 + 2xh + h^2 - 1)(2x - 3) - (2x + 2h - 3)(x^2 - 1)}{h[2(x+h) - 3](2x - 3)}$$

$$= \lim_{h \to 0} \frac{(2x^3 + 4x^2h + 2xh^2 - 2x - 3x^2 - 6xh - 3h^2 + 3) - (2x^3 + 2x^2h - 3x^2 - 2x - 2h + 3)}{h(2x + 2h - 3)(2x - 3)}$$

$$= \lim_{h \to 0} \frac{4x^2h + 2xh^2 - 6xh - 3h^2 - 2x^2h + 2h}{h(2x + 2h - 3)(2x - 3)} = \lim_{h \to 0} \frac{h(2x^2 + 2xh - 6x - 3h + 2)}{h(2x + 2h - 3)(2x - 3)}$$

$$= \lim_{h \to 0} \frac{2x^2 + 2xh - 6x - 3h + 2}{(2x + 2h - 3)(2x - 3)} = \frac{2x^2 - 6x + 2}{(2x - 3)^2}$$

Domain of f = domain of $f' = (-\infty, \frac{3}{2}) \cup (\frac{3}{2}, \infty)$.

29. $G'(t) = \lim\limits_{h \to 0} \dfrac{G(t+h) - G(t)}{h} = \lim\limits_{h \to 0} \dfrac{\dfrac{1 - 2(t+h)}{3 + (t+h)} - \dfrac{1 - 2t}{3 + t}}{h}$

$$= \lim_{h \to 0} \frac{\dfrac{[1 - 2(t+h)](3 + t) - [3 + (t+h)](1 - 2t)}{[3 + (t+h)](3 + t)}}{h}$$

$$= \lim_{h \to 0} \frac{3 + t - 6t - 2t^2 - 6h - 2ht - (3 - 6t + t - 2t^2 + h - 2ht)}{h[3 + (t+h)](3 + t)} = \lim_{h \to 0} \frac{-6h - h}{h(3 + t + h)(3 + t)}$$

$$= \lim_{h \to 0} \frac{-7h}{h(3 + t + h)(3 + t)} = \lim_{h \to 0} \frac{-7}{(3 + t + h)(3 + t)} = \frac{-7}{(3 + t)^2}$$

Domain of G = domain of $G' = (-\infty, -3) \cup (-3, \infty)$.

30. $f'(x) = \lim\limits_{h \to 0} \dfrac{f(x+h) - f(x)}{h} = \lim\limits_{h \to 0} \dfrac{(x+h)^{3/2} - x^{3/2}}{h} = \lim\limits_{h \to 0} \dfrac{[(x+h)^{3/2} - x^{3/2}][(x+h)^{3/2} + x^{3/2}]}{h\,[(x+h)^{3/2} + x^{3/2}]}$

$$= \lim_{h \to 0} \frac{(x+h)^3 - x^3}{h[(x+h)^{3/2} + x^{3/2}]} = \lim_{h \to 0} \frac{x^3 + 3x^2h + 3xh^2 + h^3 - x^3}{h[(x+h)^{3/2} + x^{3/2}]} = \lim_{h \to 0} \frac{h\,(3x^2 + 3xh + h^2)}{h[(x+h)^{3/2} + x^{3/2}]}$$

$$= \lim_{h \to 0} \frac{3x^2 + 3xh + h^2}{(x+h)^{3/2} + x^{3/2}} = \frac{3x^2}{2x^{3/2}} = \tfrac{3}{2}x^{1/2}$$

Domain of f = domain of $f' = [0, \infty)$. Strictly speaking, the domain of f' is $(0, \infty)$ because the limit that defines $f'(0)$ does not exist (as a two-sided limit). But the right-hand derivative (in the sense of Exercise 64) does exist at 0, so in that sense one could regard the domain of f' to be $[0, \infty)$.

31. $f'(x) = \lim_{h\to 0} \dfrac{f(x+h)-f(x)}{h} = \lim_{h\to 0} \dfrac{(x+h)^4 - x^4}{h} = \lim_{h\to 0} \dfrac{(x^4 + 4x^3h + 6x^2h^2 + 4xh^3 + h^4) - x^4}{h}$

$= \lim_{h\to 0} \dfrac{4x^3h + 6x^2h^2 + 4xh^3 + h^4}{h} = \lim_{h\to 0} (4x^3 + 6x^2h + 4xh^2 + h^3) = 4x^3$

Domain of f = domain of f' = $\mathbb{R}$.

32. (a)

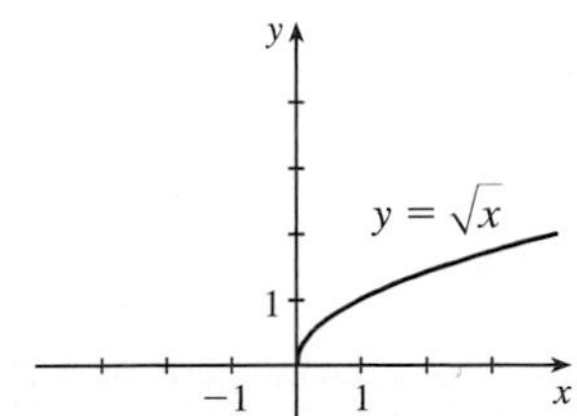

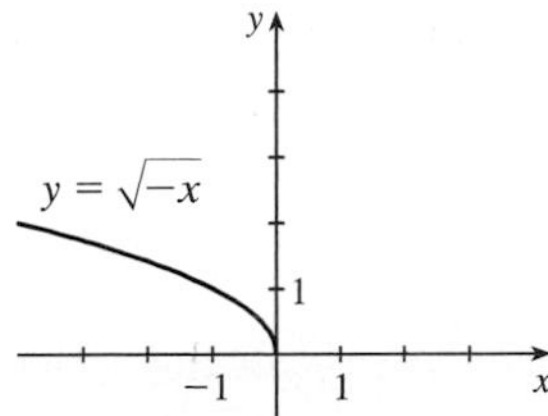

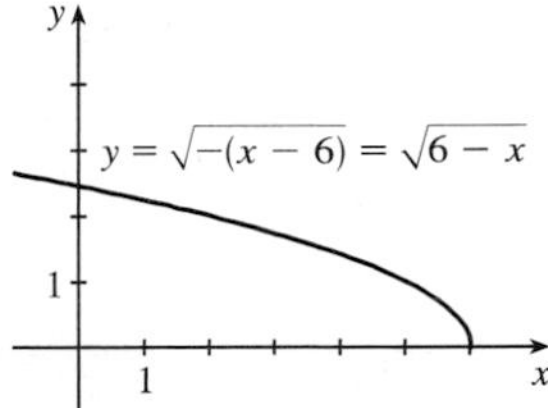

(b) Note that the third graph in part (a) has small negative values for its slope, f'; but as $x \to 6^-$, $f' \to -\infty$.

See the graph in part (d).

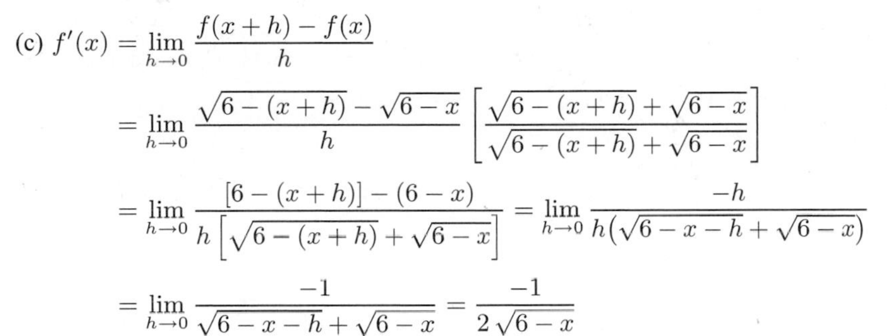

(c) $f'(x) = \lim_{h\to 0} \dfrac{f(x+h)-f(x)}{h}$

$= \lim_{h\to 0} \dfrac{\sqrt{6-(x+h)} - \sqrt{6-x}}{h}\left[\dfrac{\sqrt{6-(x+h)} + \sqrt{6-x}}{\sqrt{6-(x+h)} + \sqrt{6-x}}\right]$

$= \lim_{h\to 0} \dfrac{[6-(x+h)] - (6-x)}{h\left[\sqrt{6-(x+h)} + \sqrt{6-x}\right]} = \lim_{h\to 0} \dfrac{-h}{h\left(\sqrt{6-x-h} + \sqrt{6-x}\right)}$

$= \lim_{h\to 0} \dfrac{-1}{\sqrt{6-x-h} + \sqrt{6-x}} = \dfrac{-1}{2\sqrt{6-x}}$

(d)

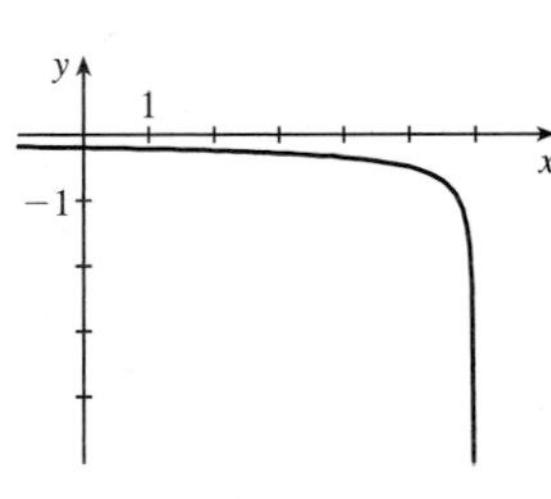

Domain of $f = (-\infty, 6]$, domain of $f' = (-\infty, 6)$.

33. (a) $f'(x) = \lim_{h\to 0} \dfrac{f(x+h)-f(x)}{h} = \lim_{h\to 0} \dfrac{[(x+h)^4 + 2(x+h)] - (x^4 + 2x)}{h}$

$= \lim_{h\to 0} \dfrac{x^4 + 4x^3h + 6x^2h^2 + 4xh^3 + h^4 + 2x + 2h - x^4 - 2x}{h}$

$= \lim_{h\to 0} \dfrac{4x^3h + 6x^2h^2 + 4xh^3 + h^4 + 2h}{h} = \lim_{h\to 0} \dfrac{h(4x^3 + 6x^2h + 4xh^2 + h^3 + 2)}{h}$

$= \lim_{h\to 0} (4x^3 + 6x^2h + 4xh^2 + h^3 + 2) = 4x^3 + 2$

(b) Notice that $f'(x) = 0$ when f has a horizontal tangent, $f'(x)$ is positive when the tangents have positive slope, and $f'(x)$ is negative when the tangents have negative slope.

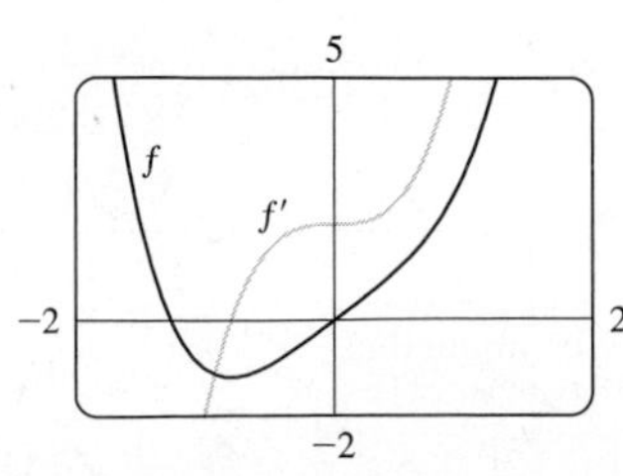

34. (a) $f'(x) = \lim_{h\to 0} \dfrac{f(x+h)-f(x)}{h} = \lim_{h\to 0} \dfrac{[(x+h)+1/(x+h)]-(x+1/x)}{h} = \lim_{h\to 0} \dfrac{\dfrac{(x+h)^2+1}{x+h} - \dfrac{x^2+1}{x}}{h}$

$$= \lim_{h\to 0} \frac{x[(x+h)^2+1]-(x+h)(x^2+1)}{h(x+h)x} = \lim_{h\to 0} \frac{(x^3+2hx^2+xh^2+x)-(x^3+x+hx^2+h)}{h(x+h)x}$$

$$= \lim_{h\to 0} \frac{hx^2+xh^2-h}{h(x+h)x} = \lim_{h\to 0} \frac{h(x^2+xh-1)}{h(x+h)x} = \lim_{h\to 0} \frac{x^2+xh-1}{(x+h)x} = \frac{x^2-1}{x^2}, \text{ or } 1-\frac{1}{x^2}$$

(b) Notice that $f'(x) = 0$ when f has a horizontal tangent, $f'(x)$ is positive when the tangents have positive slope, and $f'(x)$ is negative when the tangents have negative slope. Both functions are discontinuous at $x = 0$.

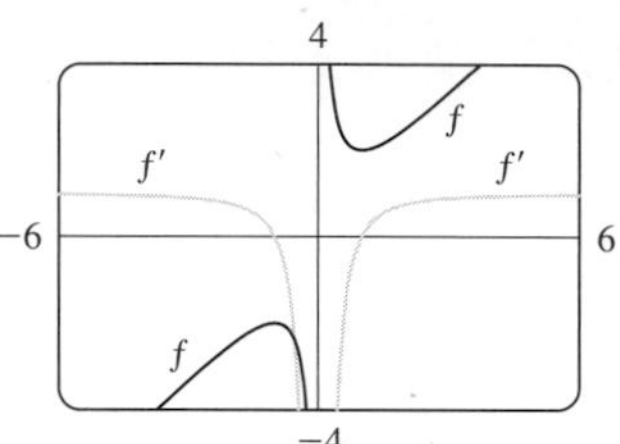

35. (a) $U'(t)$ is the rate at which the unemployment rate is changing with respect to time. Its units are percent unemployed per year.

(b) To find $U'(t)$, we use $\lim_{h\to 0} \dfrac{U(t+h)-U(t)}{h} \approx \dfrac{U(t+h)-U(t)}{h}$ for small values of h.

For 2003: $U'(2003) \approx \dfrac{U(2004)-U(2003)}{2004-2003} = \dfrac{5.5-6.0}{1} = -0.5$

For 2004: We estimate $U'(2004)$ by using $h=-1$ and $h=1$, and then average the two results to obtain a final estimate.

$h=-1 \;\Rightarrow\; U'(2004) \approx \dfrac{U(2003)-U(2004)}{2003-2004} = \dfrac{6.0-5.5}{-1} = -0.5;$

$h=1 \;\Rightarrow\; U'(2004) \approx \dfrac{U(2005)-U(2004)}{2005-2004} = \dfrac{5.1-5.5}{1} = -0.4.$

So we estimate that $U'(2004) \approx \frac{1}{2}[-0.5+(-0.4)] = -0.45$.

t	2003	2004	2005	2006	2007	2008	2009	2010	2011	2012
$U'(t)$	−0.50	−0.45	−0.45	−0.25	0.60	2.35	1.90	−0.20	−0.75	−0.80

36. (a) $N'(t)$ is the rate at which the number of minimally invasive cosmetic surgery procedures performed in the United States is changing with respect to time. Its units are thousands of surgeries per year.

(b) To find $N'(t)$, we use $\lim_{h\to 0} \dfrac{N(t+h)-N(t)}{h} \approx \dfrac{N(t+h)-N(t)}{h}$ for small values of h.

For 2000: $N'(2000) \approx \dfrac{N(2002)-N(2000)}{2002-2000} = \dfrac{4897-5500}{2} = -301.5$

For 2002: We estimate $N'(2002)$ by using $h=-2$ and $h=2$, and then average the two results to obtain a final estimate.

$h=-2 \;\Rightarrow\; N'(2002) \approx \dfrac{N(2000)-N(2002)}{2000-2002} = \dfrac{5500-4897}{-2} = -301.5$

$h=2 \;\Rightarrow\; N'(2002) \approx \dfrac{N(2004)-N(2002)}{2004-2002} = \dfrac{7470-4897}{2} = 1286.5$

So we estimate that $N'(2002) \approx \frac{1}{2}[-301.5+1286.5] = 492.5$.

t	2000	2002	2004	2006	2008	2010	2012
$N'(t)$	−301.5	492.5	1060.25	856.75	605.75	534.5	737

(c)

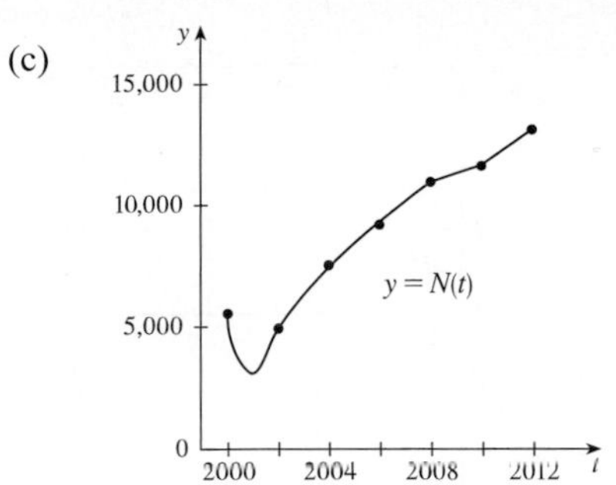

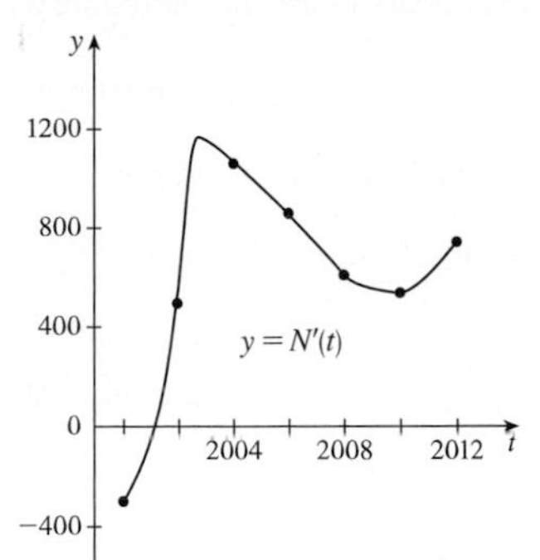

(d) We could get more accurate values for $N'(t)$ by obtaining data for more values of t.

37. As in Exercise 35, we use one-sided difference quotients for the first and last values, and average two difference quotients for all other values.

t	14	21	28	35	42	49
$H(t)$	41	54	64	72	78	83
$H'(t)$	$\frac{13}{7}$	$\frac{23}{14}$	$\frac{18}{14}$	$\frac{14}{14}$	$\frac{11}{14}$	$\frac{5}{7}$

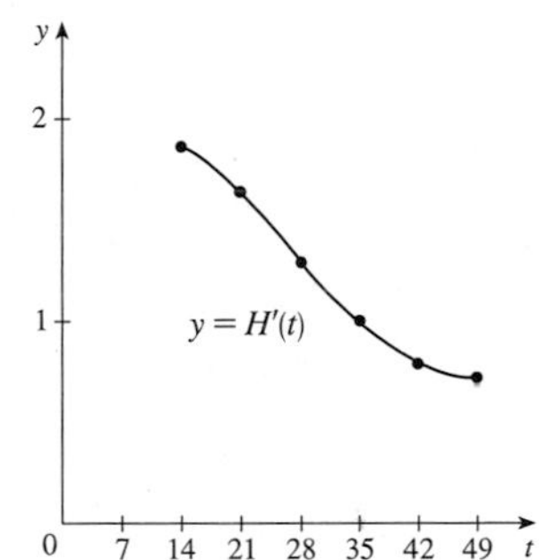

38. As in Exercise 35, we use one-sided difference quotients for the first and last values, and average two difference quotients for all other values. The units for $W'(x)$ are grams per degree (g/°C).

x	15.5	17.7	20.0	22.4	24.4
$W(x)$	37.2	31.0	19.8	9.7	−9.8
$W'(x)$	−2.82	−3.87	−4.53	−6.73	−9.75

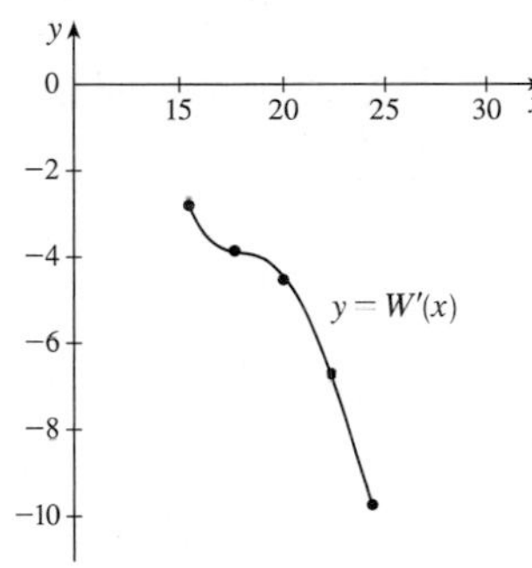

39. (a) dP/dt is the rate at which the percentage of the city's electrical power produced by solar panels changes with respect to time t, measured in percentage points per year.

(b) 2 years after January 1, 2000 (January 1, 2002), the percentage of electrical power produced by solar panels was increasing at a rate of 3.5 percentage points per year.

40. dN/dp is the rate at which the number of people who travel by car to another state for a vacation changes with respect to the price of gasoline. If the price of gasoline goes up, we would expect fewer people to travel, so we would expect dN/dp to be negative.

41. f is not differentiable at $x = -4$, because the graph has a corner there, and at $x = 0$, because there is a discontinuity there.

42. f is not differentiable at $x = -1$, because there is a discontinuity there, and at $x = 2$, because the graph has a corner there.

43. f is not differentiable at $x = 1$, because f is not defined there, and at $x = 5$, because the graph has a vertical tangent there.

44. f is not differentiable at $x = -2$ and $x = 3$, because the graph has corners there, and at $x = 1$, because there is a discontinuity there.

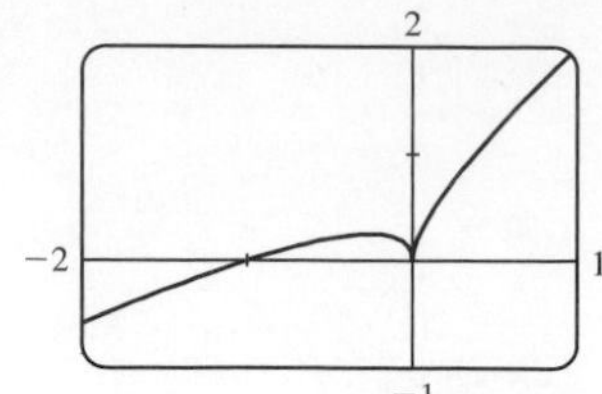

45. As we zoom in toward $(-1, 0)$, the curve appears more and more like a straight line, so $f(x) = x + \sqrt{|x|}$ is differentiable at $x = -1$. But no matter how much we zoom in toward the origin, the curve doesn't straighten out—we can't eliminate the sharp point (a cusp). So f is not differentiable at $x = 0$.

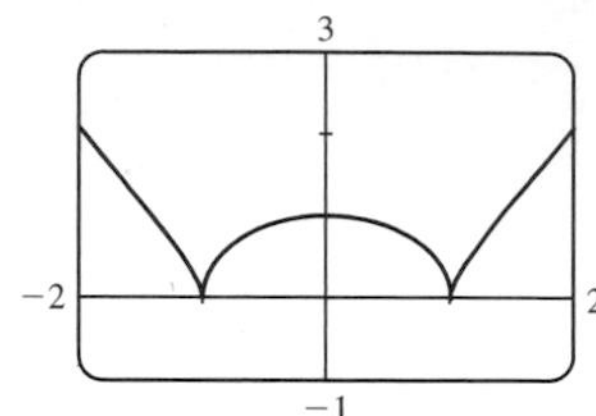

46. As we zoom in toward $(0, 1)$, the curve appears more and more like a straight line, so f is differentiable at $x = 0$. But no matter how much we zoom in toward $(1, 0)$ or $(-1, 0)$, the curve doesn't straighten out—we can't eliminate the sharp point (a cusp). So f is not differentiable at $x = \pm 1$.

47. Call the curve with the positive y-intercept g and the other curve h. Notice that g has a maximum (horizontal tangent) at $x = 0$, but $h \neq 0$, so h cannot be the derivative of g. Also notice that where g is positive, h is increasing. Thus, $h = f$ and $g = f'$. Now $f'(-1)$ is negative since f' is below the x-axis there and $f''(1)$ is positive since f is concave upward at $x = 1$. Therefore, $f''(1)$ is greater than $f'(-1)$.

48. Call the curve with the smallest positive x-intercept g and the other curve h. Notice that where g is positive in the first quadrant, h is increasing. Thus, $h = f$ and $g = f'$. Now $f'(-1)$ is positive since f' is above the x-axis there and $f''(1)$ appears to be zero since f has an inflection point at $x = 1$. Therefore, $f'(1)$ is greater than $f''(-1)$.

49. $a = f, b = f', c = f''$. We can see this because where a has a horizontal tangent, $b = 0$, and where b has a horizontal tangent, $c = 0$. We can immediately see that c can be neither f nor f', since at the points where c has a horizontal tangent, neither a nor b is equal to 0.

50. Where d has horizontal tangents, only c is 0, so $d' = c$. c has negative tangents for $x < 0$ and b is the only graph that is negative for $x < 0$, so $c' = b$. b has positive tangents on $\mathbb{R}$ (except at $x = 0$), and the only graph that is positive on the same domain is a, so $b' = a$. We conclude that $d = f$, $c = f'$, $b = f''$, and $a = f'''$.

51. We can immediately see that a is the graph of the acceleration function, since at the points where a has a horizontal tangent, neither c nor b is equal to 0. Next, we note that $a = 0$ at the point where b has a horizontal tangent, so b must be the graph of the velocity function, and hence, $b' = a$. We conclude that c is the graph of the position function.

52. a must be the jerk since none of the graphs are 0 at its high and low points. a is 0 where b has a maximum, so $b' = a$. b is 0 where c has a maximum, so $c' = b$. We conclude that d is the position function, c is the velocity, b is the acceleration, and a is the jerk.

53. $$f'(x) = \lim_{h \to 0} \frac{f(x+h) - f(x)}{h} = \lim_{h \to 0} \frac{[3(x+h)^2 + 2(x+h) + 1] - (3x^2 + 2x + 1)}{h}$$

$$= \lim_{h \to 0} \frac{(3x^2 + 6xh + 3h^2 + 2x + 2h + 1) - (3x^2 + 2x + 1)}{h} = \lim_{h \to 0} \frac{6xh + 3h^2 + 2h}{h}$$

$$= \lim_{h \to 0} \frac{h(6x + 3h + 2)}{h} = \lim_{h \to 0} (6x + 3h + 2) = 6x + 2$$

$$f''(x) = \lim_{h \to 0} \frac{f'(x+h) - f'(x)}{h} = \lim_{h \to 0} \frac{[6(x+h)+2] - (6x+2)}{h} = \lim_{h \to 0} \frac{(6x+6h+2) - (6x+2)}{h}$$

$$= \lim_{h \to 0} \frac{6h}{h} = \lim_{h \to 0} 6 = 6$$

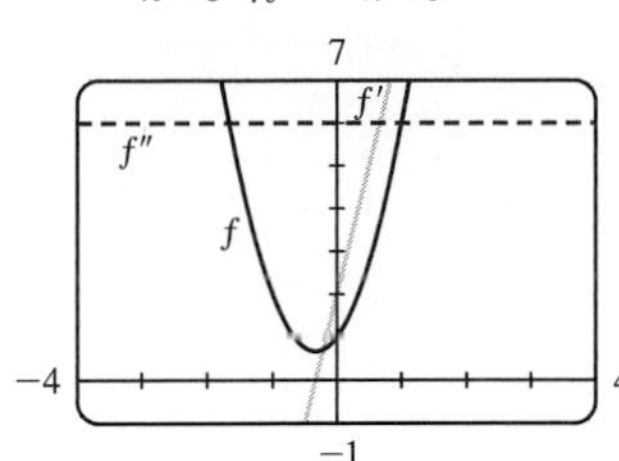

We see from the graph that our answers are reasonable because the graph of f' is that of a linear function and the graph of f'' is that of a constant function.

54. $f'(x) = \lim_{h \to 0} \frac{f(x+h) - f(x)}{h} = \lim_{h \to 0} \frac{[(x+h)^3 - 3(x+h)] - (x^3 - 3x)}{h}$

$$= \lim_{h \to 0} \frac{(x^3 + 3x^2h + 3xh^2 + h^3 - 3x - 3h) - (x^3 - 3x)}{h} = \lim_{h \to 0} \frac{3x^2h + 3xh^2 + h^3 - 3h}{h}$$

$$= \lim_{h \to 0} \frac{h(3x^2 + 3xh + h^2 - 3)}{h} = \lim_{h \to 0} (3x^2 + 3xh + h^2 - 3) = 3x^2 - 3$$

$$f''(x) = \lim_{h \to 0} \frac{f'(x+h) - f'(x)}{h} = \lim_{h \to 0} \frac{[3(x+h)^2 - 3] - (3x^2 - 3)}{h} = \lim_{h \to 0} \frac{(3x^2 + 6xh + 3h^2 - 3) - (3x^2 - 3)}{h}$$

$$= \lim_{h \to 0} \frac{6xh + 3h^2}{h} = \lim_{h \to 0} \frac{h(6x + 3h)}{h} = \lim_{h \to 0} (6x + 3h) = 6x$$

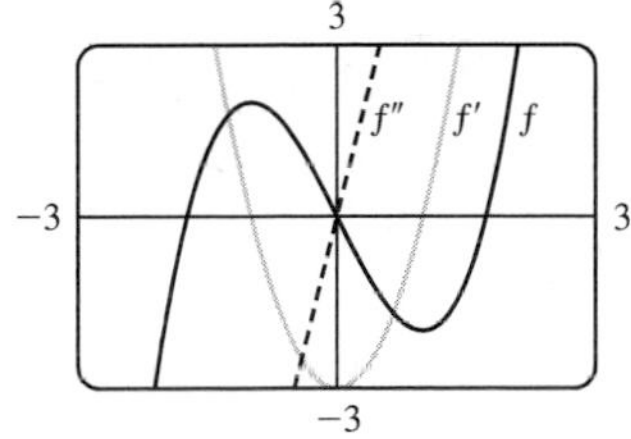

We see from the graph that our answers are reasonable because the graph of f' is that of an even function (f is an odd function) and the graph of f'' is that of an odd function. Furthermore, $f' = 0$ when f has a horizontal tangent and $f'' = 0$ when f' has a horizontal tangent.

55. $f'(x) = \lim_{h \to 0} \frac{f(x+h) - f(x)}{h} = \lim_{h \to 0} \frac{\left[2(x+h)^2 - (x+h)^3\right] - (2x^2 - x^3)}{h}$

$$= \lim_{h \to 0} \frac{h(4x + 2h - 3x^2 - 3xh - h^2)}{h} = \lim_{h \to 0} (4x + 2h - 3x^2 - 3xh - h^2) = 4x - 3x^2$$

$$f''(x) = \lim_{h \to 0} \frac{f'(x+h) - f'(x)}{h} = \lim_{h \to 0} \frac{\left[4(x+h) - 3(x+h)^2\right] - (4x - 3x^2)}{h} = \lim_{h \to 0} \frac{h(4 - 6x - 3h)}{h}$$

$$= \lim_{h \to 0} (4 - 6x - 3h) = 4 - 6x$$

$$f'''(x) = \lim_{h \to 0} \frac{f''(x+h) - f''(x)}{h} = \lim_{h \to 0} \frac{[4 - 6(x+h)] - (4 - 6x)}{h} = \lim_{h \to 0} \frac{-6h}{h} = \lim_{h \to 0} (-6) = -6$$

$$f^{(4)}(x) = \lim_{h \to 0} \frac{f'''(x+h) - f'''(x)}{h} = \lim_{h \to 0} \frac{-6 - (-6)}{h} = \lim_{h \to 0} \frac{0}{h} = \lim_{h \to 0} (0) = 0$$

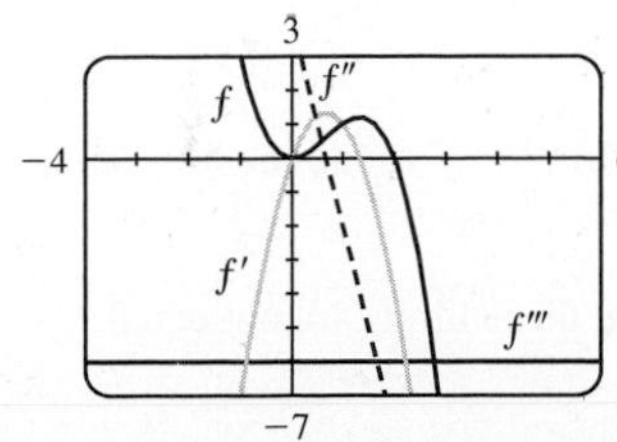

The graphs are consistent with the geometric interpretations of the derivatives because f' has zeros where f has a local minimum and a local maximum, f'' has a zero where f' has a local maximum, and f''' is a constant function equal to the slope of f''.

56. (a) Since we estimate the velocity to be a maximum at $t = 10$, the acceleration is 0 at $t = 10$.

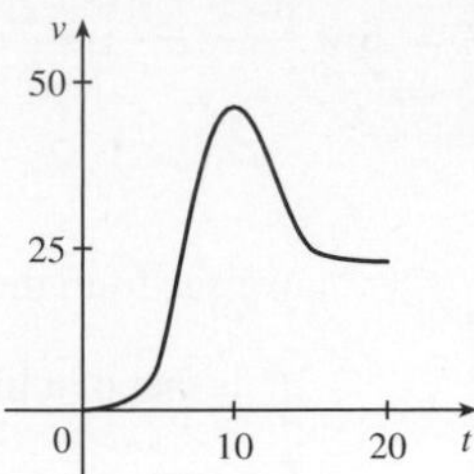

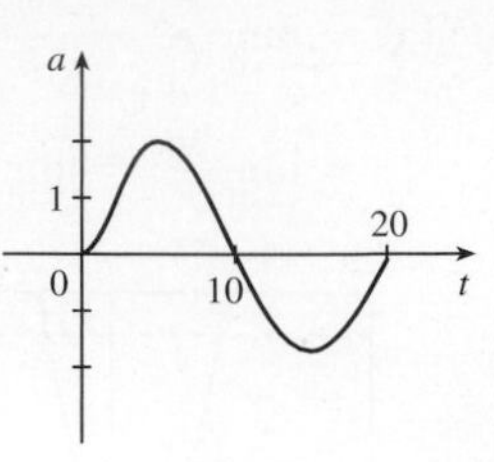

(b) Drawing a tangent line at $t = 10$ on the graph of a, a appears to decrease by 10 ft/s^2 over a period of 20 s. So at $t = 10$ s, the jerk is approximately $-10/20 = -0.5$ (ft/s^2)/s or ft/s^3.

57. (a) Note that we have factored $x - a$ as the difference of two cubes in the third step.

$$f'(a) = \lim_{x \to a} \frac{f(x) - f(a)}{x - a} = \lim_{x \to a} \frac{x^{1/3} - a^{1/3}}{x - a} = \lim_{x \to a} \frac{x^{1/3} - a^{1/3}}{(x^{1/3} - a^{1/3})(x^{2/3} + x^{1/3}a^{1/3} + a^{2/3})}$$

$$= \lim_{x \to a} \frac{1}{x^{2/3} + x^{1/3}a^{1/3} + a^{2/3}} = \frac{1}{3a^{2/3}} \text{ or } \tfrac{1}{3}a^{-2/3}$$

(b) $f'(0) = \lim\limits_{h \to 0} \dfrac{f(0 + h) - f(0)}{h} = \lim\limits_{h \to 0} \dfrac{\sqrt[3]{h} - 0}{h} = \lim\limits_{h \to 0} \dfrac{1}{h^{2/3}}$. This function increases without bound, so the limit does not exist, and therefore $f'(0)$ does not exist.

(c) $\lim\limits_{x \to 0} |f'(x)| = \lim\limits_{x \to 0} \dfrac{1}{3x^{2/3}} = \infty$ and f is continuous at $x = 0$ (root function), so f has a vertical tangent at $x = 0$.

58. (a) $g'(0) = \lim\limits_{x \to 0} \dfrac{g(x) - g(0)}{x - 0} = \lim\limits_{x \to 0} \dfrac{x^{2/3} - 0}{x} = \lim\limits_{x \to 0} \dfrac{1}{x^{1/3}}$, which does not exist.

(b) $$g'(a) = \lim_{x \to a} \frac{g(x) - g(a)}{x - a} = \lim_{x \to a} \frac{x^{2/3} - a^{2/3}}{x - a} = \lim_{x \to a} \frac{(x^{1/3} - a^{1/3})(x^{1/3} + a^{1/3})}{(x^{1/3} - a^{1/3})(x^{2/3} + x^{1/3}a^{1/3} + a^{2/3})}$$

$$= \lim_{x \to a} \frac{x^{1/3} + a^{1/3}}{x^{2/3} + x^{1/3}a^{1/3} + a^{2/3}} = \frac{2a^{1/3}}{3a^{2/3}} = \frac{2}{3a^{1/3}} \text{ or } \tfrac{2}{3}a^{-1/3}$$

(c) $g(x) = x^{2/3}$ is continuous at $x = 0$ and $\lim\limits_{x \to 0} |g'(x)| = \lim\limits_{x \to 0} \dfrac{2}{3\,|x|^{1/3}} = \infty$. This shows that g has a vertical tangent line at $x = 0$.

(d)

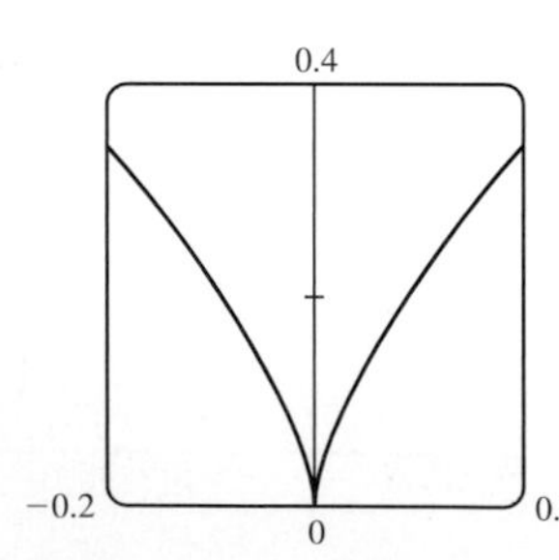

59. $f(x) = |x - 6| = \begin{cases} x - 6 & \text{if } x - 6 \geq 6 \\ -(x - 6) & \text{if } x - 6 < 0 \end{cases} = \begin{cases} x - 6 & \text{if } x \geq 6 \\ 6 - x & \text{if } x < 6 \end{cases}$

So the right-hand limit is $\lim\limits_{x \to 6^+} \dfrac{f(x) - f(6)}{x - 6} = \lim\limits_{x \to 6^+} \dfrac{|x - 6| - 0}{x - 6} = \lim\limits_{x \to 6^+} \dfrac{x - 6}{x - 6} = \lim\limits_{x \to 6^+} 1 = 1$, and the left-hand limit is $\lim\limits_{x \to 6^-} \dfrac{f(x) - f(6)}{x - 6} = \lim\limits_{x \to 6^-} \dfrac{|x - 6| - 0}{x - 6} = \lim\limits_{x \to 6^-} \dfrac{6 - x}{x - 6} = \lim\limits_{x \to 6^-} (-1) = -1$. Since these limits are not equal,

$f'(6) = \lim\limits_{x \to 6} \dfrac{f(x) - f(6)}{x - 6}$ does not exist and f is not differentiable at 6.

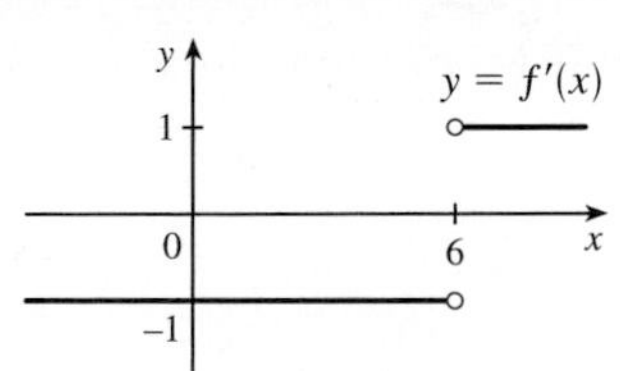

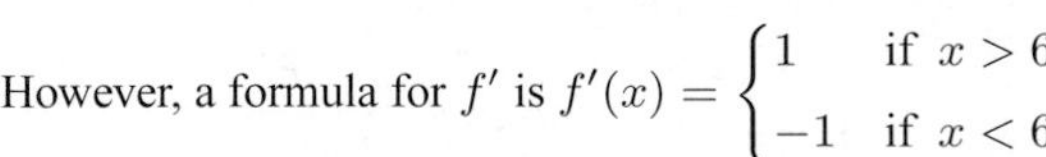

However, a formula for f' is $f'(x) = \begin{cases} 1 & \text{if } x > 6 \\ -1 & \text{if } x < 6 \end{cases}$

Another way of writing the formula is $f'(x) = \dfrac{x - 6}{|x - 6|}$.

60. $f(x) = [\![x]\!]$ is not continuous at any integer n, so f is not differentiable at n by the contrapositive of Theorem 4. If a is not an integer, then f is constant on an open interval containing a, so $f'(a) = 0$. Thus, $f'(x) = 0$, x not an integer.

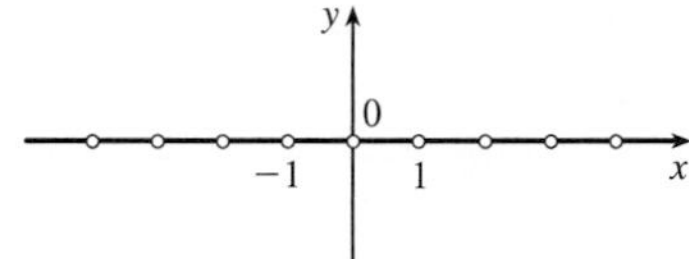

61. (a) $f(x) = x\,|x| = \begin{cases} x^2 & \text{if } x \geq 0 \\ -x^2 & \text{if } x < 0 \end{cases}$

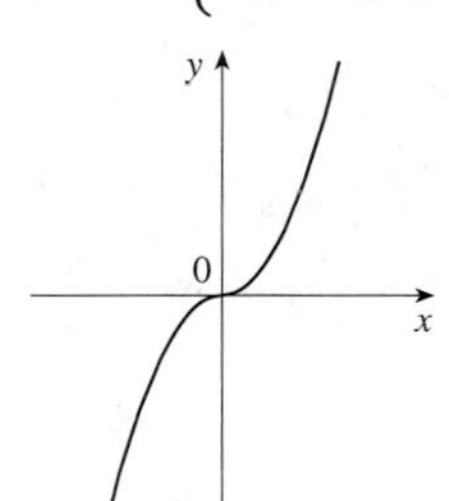

(b) Since $f(x) = x^2$ for $x \geq 0$, we have $f'(x) = 2x$ for $x > 0$. [See Exercise 19(d).] Similarly, since $f(x) = -x^2$ for $x < 0$, we have $f'(x) = -2x$ for $x < 0$. At $x = 0$, we have

$$f'(0) = \lim_{x \to 0} \frac{f(x) - f(0)}{x - 0} = \lim_{x \to 0} \frac{x\,|x|}{x} = \lim_{x \to 0} |x| = 0.$$

So f is differentiable at 0. Thus, f is differentiable for all x.

(c) From part (b), we have $f'(x) = \left\{ \begin{array}{ll} 2x & \text{if } x \geq 0 \\ 2x & \text{if } x < 0 \end{array} \right\} = 2\,|x|$.

62. (a) $|x| = \begin{cases} x & \text{if } x \geq 0 \\ -x & \text{if } x < 0 \end{cases}$

so $f(x) = x + |x| = \begin{cases} 2x & \text{if } x \geq 0 \\ 0 & \text{if } x < 0 \end{cases}$.

Graph the line $y = 2x$ for $x \geq 0$ and graph $y = 0$ (the x-axis) for $x < 0$.

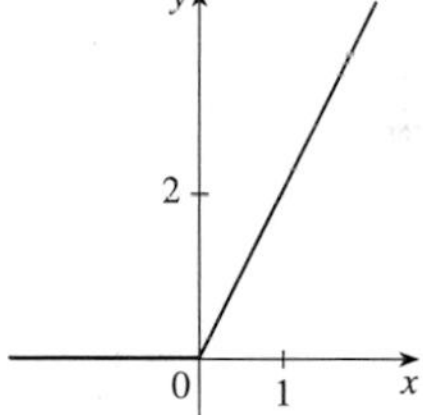

(b) g is not differentiable at $x = 0$ because the graph has a corner there, but is differentiable at all other values; that is, g is differentiable on $(-\infty, 0) \cup (0, \infty)$.

(c) $g(x) = \begin{cases} 2x & \text{if } x \geq 0 \\ 0 & \text{if } x < 0 \end{cases} \quad \Rightarrow \quad g'(x) = \begin{cases} 2 & \text{if } x > 0 \\ 0 & \text{if } x < 0 \end{cases}$

Another way of writing the formula is $g'(x) = 1 + \operatorname{sgn} x$ for $x \neq 0$.

63. (a) If f is even, then

$$\begin{aligned} f'(-x) &= \lim_{h \to 0} \frac{f(-x+h) - f(-x)}{h} = \lim_{h \to 0} \frac{f[-(x-h)] - f(-x)}{h} \\ &= \lim_{h \to 0} \frac{f(x-h) - f(x)}{h} = -\lim_{h \to 0} \frac{f(x-h) - f(x)}{-h} \qquad [\text{let } \Delta x = -h] \\ &= -\lim_{\Delta x \to 0} \frac{f(x + \Delta x) - f(x)}{\Delta x} = -f'(x) \end{aligned}$$

Therefore, f' is odd.

(b) If f is odd, then

$$\begin{aligned} f'(-x) &= \lim_{h\to 0}\frac{f(-x+h)-f(-x)}{h} = \lim_{h\to 0}\frac{f[-(x-h)]-f(-x)}{h} \\ &= \lim_{h\to 0}\frac{-f(x-h)+f(x)}{h} = \lim_{h\to 0}\frac{f(x-h)-f(x)}{-h} \qquad [\text{let } \Delta x = -h] \\ &= \lim_{\Delta x\to 0}\frac{f(x+\Delta x)-f(x)}{\Delta x} = f'(x) \end{aligned}$$

Therefore, f' is even.

64. (a) $$\begin{aligned} f'_-(4) &= \lim_{h\to 0^-}\frac{f(4+h)-f(4)}{h} = \lim_{h\to 0^-}\frac{5-(4+h)-1}{h} \\ &= \lim_{h\to 0^-}\frac{-h}{h} = -1 \end{aligned}$$

and

$$\begin{aligned} f'_+(4) &= \lim_{h\to 0^+}\frac{f(4+h)-f(4)}{h} = \lim_{h\to 0^+}\frac{\dfrac{1}{5-(4+h)}-1}{h} \\ &= \lim_{h\to 0^+}\frac{1-(1-h)}{h(1-h)} = \lim_{h\to 0^+}\frac{1}{1-h} = 1 \end{aligned}$$

(b)

(c) $$f(x) = \begin{cases} 0 & \text{if } x \le 0 \\ 5-x & \text{if } 0 < x < 4 \\ 1/(5-x) & \text{if } x \ge 4 \end{cases}$$

At 4 we have $\lim_{x\to 4^-} f(x) = \lim_{x\to 4^-}(5-x) = 1$ and $\lim_{x\to 4^+} f(x) = \lim_{x\to 4^+}\frac{1}{5-x} = 1$, so $\lim_{x\to 4} f(x) = 1 = f(4)$ and f is continuous at 4. Since $f(5)$ is not defined, f is discontinuous at 5. These expressions show that f is continuous on the intervals $(-\infty, 0)$, $(0, 4)$, $(4, 5)$ and $(5, \infty)$. Since $\lim_{x\to 0^+} f(x) = \lim_{x\to 0^+}(5-x) = 5 \ne 0 = \lim_{x\to 0^-} f(x)$, $\lim_{x\to 0} f(x)$ does not exist, so f is discontinuous (and therefore not differentiable) at 0.

(d) From (a), f is not differentiable at 4 since $f'_-(4) \ne f'_+(4)$, and from (c), f is not differentiable at 0 or 5.

65. These graphs are idealizations conveying the spirit of the problem. In reality, changes in speed are not instantaneous, so the graph in (a) would not have corners and the graph in (b) would be continuous.

(a)

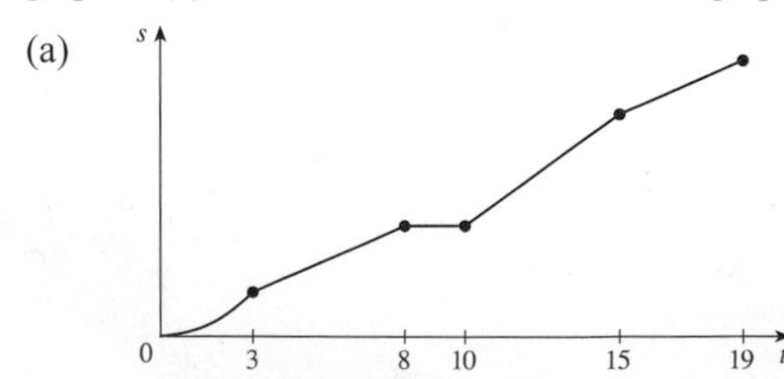

(b)

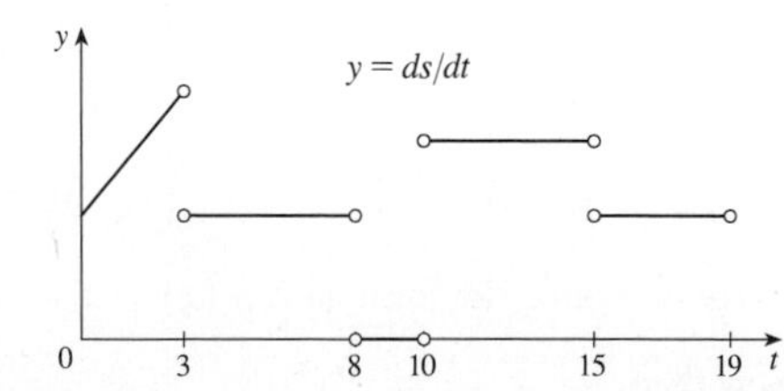

66. (a)

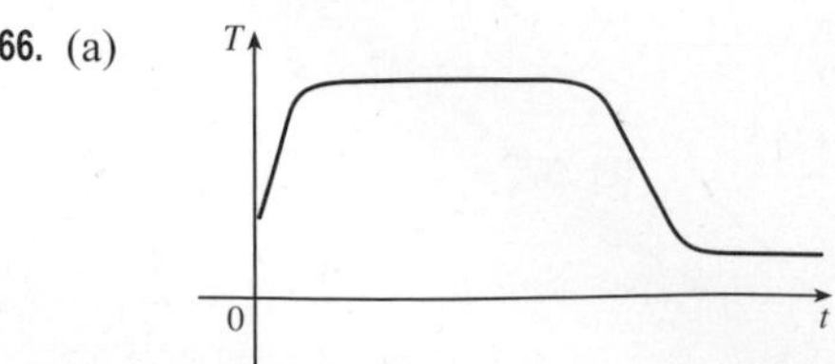

(b) The initial temperature of the water is close to room temperature because of the water that was in the pipes. When the water from the hot water tank starts coming out, dT/dt is large and positive as T increases to the temperature of the water in the tank. In the next phase, $dT/dt = 0$ as the water comes out at a constant, high temperature. After some time, dT/dt becomes small and negative as the contents of the hot water tank are exhausted. Finally, when the hot water has run out, dT/dt is once again 0 as the water maintains its (cold) temperature.

(c)

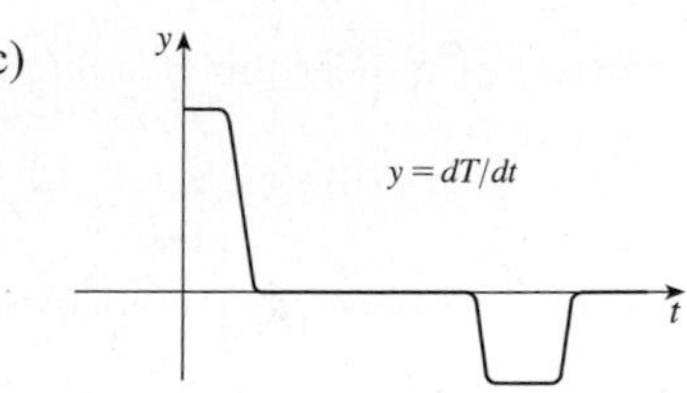

67.

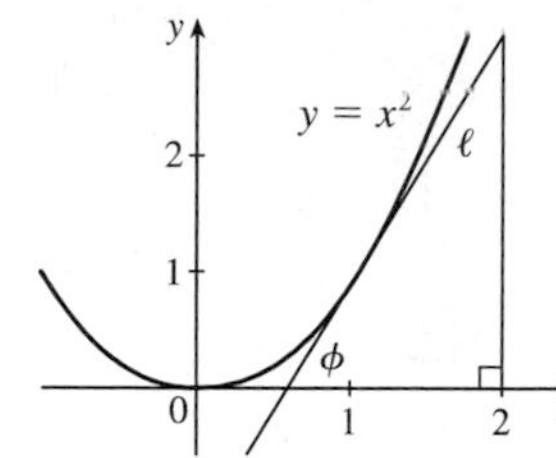

In the right triangle in the diagram, let Δy be the side opposite angle ϕ and Δx the side adjacent to angle ϕ. Then the slope of the tangent line ℓ is $m = \Delta y/\Delta x = \tan\phi$. Note that $0 < \phi < \frac{\pi}{2}$. We know (see Exercise 19) that the derivative of $f(x) = x^2$ is $f'(x) = 2x$. So the slope of the tangent to the curve at the point $(1, 1)$ is 2. Thus, ϕ is the angle between 0 and $\frac{\pi}{2}$ whose tangent is 2; that is, $\phi = \tan^{-1} 2 \approx 63°$.

2 Review

TRUE-FALSE QUIZ

1. False. Limit Law 2 applies only if the individual limits exist (these don't).

2. False. Limit Law 5 cannot be applied if the limit of the denominator is 0 (it is).

3. True. Limit Law 5 applies.

4. False. $\dfrac{x^2 - 9}{x - 3}$ is not defined when $x = 3$, but $x + 3$ is.

5. True. $\displaystyle\lim_{x\to 3} \frac{x^2 - 9}{x - 3} = \lim_{x\to 3} \frac{(x+3)(x-3)}{(x-3)} = \lim_{x\to 3} (x + 3)$

6. True. The limit doesn't exist since $f(x)/g(x)$ doesn't approach any real number as x approaches 5. (The denominator approaches 0 and the numerator doesn't.)

7. False. Consider $\displaystyle\lim_{x\to 5} \frac{x(x-5)}{x-5}$ or $\displaystyle\lim_{x\to 5} \frac{\sin(x-5)}{x-5}$. The first limit exists and is equal to 5. By Example 2.2.3, we know that the latter limit exists (and it is equal to 1).

8. False. If $f(x) = 1/x$, $g(x) = -1/x$, and $a = 0$, then $\lim_{x\to 0} f(x)$ does not exist, $\lim_{x\to 0} g(x)$ does not exist, but $\lim_{x\to 0} [f(x) + g(x)] = \lim_{x\to 0} 0 = 0$ exists.

9. True. Suppose that $\lim_{x\to a} [f(x) + g(x)]$ exists. Now $\lim_{x\to a} f(x)$ exists and $\lim_{x\to a} g(x)$ does not exist, but $\lim_{x\to a} g(x) = \lim_{x\to a} \{[f(x) + g(x)] - f(x)\} = \lim_{x\to a} [f(x) + g(x)] - \lim_{x\to a} f(x)$ [by Limit Law 2], which exists, and we have a contradiction. Thus, $\lim_{x\to a} [f(x) + g(x)]$ does not exist.

10. False. Consider $\lim_{x \to 6} [f(x)g(x)] = \lim_{x \to 6} \left[(x-6)\dfrac{1}{x-6}\right]$. It exists (its value is 1) but $f(6) = 0$ and $g(6)$ does not exist, so $f(6)g(6) \neq 1$.

11. True. A polynomial is continuous everywhere, so $\lim_{x \to b} p(x)$ exists and is equal to $p(b)$.

12. False. Consider $\lim_{x \to 0} [f(x) - g(x)] = \lim_{x \to 0} \left(\dfrac{1}{x^2} - \dfrac{1}{x^4}\right)$. This limit is $-\infty$ (not 0), but each of the individual functions approaches ∞.

13. True. See Figure 2.6.8.

14. False. Consider $f(x) = \sin x$ for $x \geq 0$. $\lim_{x \to \infty} f(x) \neq \pm\infty$ and f has no horizontal asymptote.

15. False. Consider $f(x) = \begin{cases} 1/(x-1) & \text{if } x \neq 1 \\ 2 & \text{if } x = 1 \end{cases}$

16. False. The function f must be *continuous* in order to use the Intermediate Value Theorem. For example, let $f(x) = \begin{cases} 1 & \text{if } 0 \leq x < 3 \\ -1 & \text{if } x = 3 \end{cases}$ There is no number $c \in [0, 3]$ with $f(c) = 0$.

17. True. Use Theorem 2.5.8 with $a = 2$, $b = 5$, and $g(x) = 4x^2 - 11$. Note that $f(4) = 3$ is not needed.

18. True. Use the Intermediate Value Theorem with $a = -1$, $b = 1$, and $N = \pi$, since $3 < \pi < 4$.

19. True, by the definition of a limit with $\varepsilon = 1$.

20. False. For example, let $f(x) = \begin{cases} x^2 + 1 & \text{if } x \neq 0 \\ 2 & \text{if } x = 0 \end{cases}$

Then $f(x) > 1$ for all x, but $\lim_{x \to 0} f(x) = \lim_{x \to 0} (x^2 + 1) = 1$.

21. False. See the note after Theorem 2.8.4.

22. True. $f'(r)$ exists $\Rightarrow$ f is differentiable at r $\Rightarrow$ f is continuous at r $\Rightarrow$ $\lim_{x \to r} f(x) = f(r)$.

23. False. $\dfrac{d^2y}{dx^2}$ is the second derivative while $\left(\dfrac{dy}{dx}\right)^2$ is the first derivative squared. For example, if $y = x$, then $\dfrac{d^2y}{dx^2} = 0$, but $\left(\dfrac{dy}{dx}\right)^2 = 1$.

24. True. $f(x) = x^{10} - 10x^2 + 5$ is continuous on the interval $[0, 2]$, $f(0) = 5$, $f(1) = -4$, and $f(2) = 989$. Since $-4 < 0 < 5$, there is a number c in $(0, 1)$ such that $f(c) = 0$ by the Intermediate Value Theorem. Thus, there is a root of the equation $x^{10} - 10x^2 + 5 = 0$ in the interval $(0, 1)$. Similarly, there is a root in $(1, 2)$.

25. True. See Exercise 2.5.72(b).

26. False See Exercise 2.5.72(b).

EXERCISES

1. (a) (i) $\lim_{x \to 2^+} f(x) = 3$ (ii) $\lim_{x \to -3^+} f(x) = 0$

(iii) $\lim_{x \to -3} f(x)$ does not exist since the left and right limits are not equal. (The left limit is -2.)

(iv) $\lim_{x \to 4} f(x) = 2$

(v) $\lim_{x \to 0} f(x) = \infty$ (vi) $\lim_{x \to 2^-} f(x) = -\infty$

(vii) $\lim_{x \to \infty} f(x) = 4$ (viii) $\lim_{x \to -\infty} f(x) = -1$

(b) The equations of the horizontal asymptotes are $y = -1$ and $y = 4$.

(c) The equations of the vertical asymptotes are $x = 0$ and $x = 2$.

(d) f is discontinuous at $x = -3$, 0, 2, and 4. The discontinuities are jump, infinite, infinite, and removable, respectively.

2. $\lim_{x \to -\infty} f(x) = -2$, $\lim_{x \to \infty} f(x) = 0$, $\lim_{x \to -3} f(x) = \infty$,

$\lim_{x \to 3^-} f(x) = -\infty$, $\lim_{x \to 3^+} f(x) = 2$,

f is continuous from the right at 3

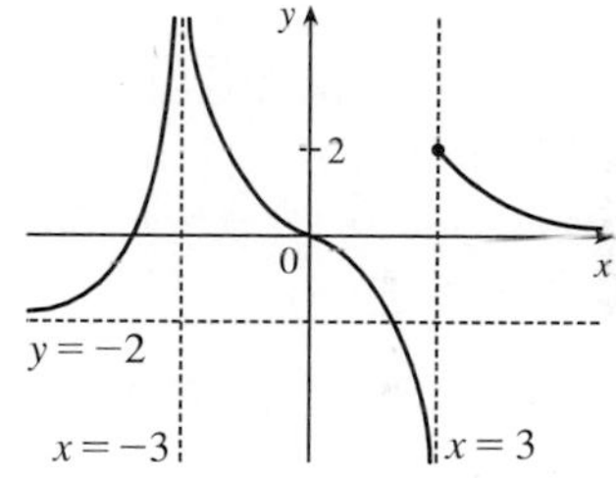

3. Since the exponential function is continuous, $\lim_{x \to 1} e^{x^3 - x} = e^{1-1} = e^0 = 1$.

4. Since rational functions are continuous, $\lim_{x \to 3} \dfrac{x^2 - 9}{x^2 + 2x - 3} = \dfrac{3^2 - 9}{3^2 + 2(3) - 3} = \dfrac{0}{12} = 0$.

5. $$\lim_{x \to -3} \frac{x^2 - 9}{x^2 + 2x - 3} = \lim_{x \to -3} \frac{(x+3)(x-3)}{(x+3)(x-1)} = \lim_{x \to -3} \frac{x-3}{x-1} = \frac{-3-3}{-3-1} = \frac{-6}{-4} = \frac{3}{2}$$

6. $\lim_{x \to 1^+} \dfrac{x^2 - 9}{x^2 + 2x - 3} = -\infty$ since $x^2 + 2x - 3 \to 0^+$ as $x \to 1^+$ and $\dfrac{x^2 - 9}{x^2 + 2x - 3} < 0$ for $1 < x < 3$.

7. $$\lim_{h \to 0} \frac{(h-1)^3 + 1}{h} = \lim_{h \to 0} \frac{(h^3 - 3h^2 + 3h - 1) + 1}{h} = \lim_{h \to 0} \frac{h^3 - 3h^2 + 3h}{h} = \lim_{h \to 0} \left(h^2 - 3h + 3\right) = 3$$

Another solution: Factor the numerator as a sum of two cubes and then simplify.

$$\begin{aligned} \lim_{h \to 0} \frac{(h-1)^3 + 1}{h} &= \lim_{h \to 0} \frac{(h-1)^3 + 1^3}{h} = \lim_{h \to 0} \frac{[(h-1) + 1]\left[(h-1)^2 - 1(h-1) + 1^2\right]}{h} \\ &= \lim_{h \to 0} \left[(h-1)^2 - h + 2\right] = 1 - 0 + 2 = 3 \end{aligned}$$

8. $$\lim_{t \to 2} \frac{t^2 - 4}{t^3 - 8} = \lim_{t \to 2} \frac{(t+2)(t-2)}{(t-2)(t^2 + 2t + 4)} = \lim_{t \to 2} \frac{t+2}{t^2 + 2t + 4} = \frac{2+2}{4+4+4} = \frac{4}{12} = \frac{1}{3}$$

9. $\lim_{r \to 9} \dfrac{\sqrt{r}}{(r-9)^4} = \infty$ since $(r-9)^4 \to 0^+$ as $r \to 9$ and $\dfrac{\sqrt{r}}{(r-9)^4} > 0$ for $r \neq 9$.

10. $\lim\limits_{v\to 4^+}\dfrac{4-v}{|4-v|}=\lim\limits_{v\to 4^+}\dfrac{4-v}{-(4-v)}=\lim\limits_{v\to 4^+}\dfrac{1}{-1}=-1$

11. $\lim\limits_{u\to 1}\dfrac{u^4-1}{u^3+5u^2-6u}=\lim\limits_{u\to 1}\dfrac{(u^2+1)(u^2-1)}{u(u^2+5u-6)}=\lim\limits_{u\to 1}\dfrac{(u^2+1)(u+1)(u-1)}{u(u+6)(u-1)}=\lim\limits_{u\to 1}\dfrac{(u^2+1)(u+1)}{u(u+6)}=\dfrac{2(2)}{1(7)}=\dfrac{4}{7}$

12. $\lim\limits_{x\to 3}\dfrac{\sqrt{x+6}-x}{x^3-3x^2}=\lim\limits_{x\to 3}\left[\dfrac{\sqrt{x+6}-x}{x^2(x-3)}\cdot\dfrac{\sqrt{x+6}+x}{\sqrt{x+6}+x}\right]=\lim\limits_{x\to 3}\dfrac{(\sqrt{x+6}\,)^2-x^2}{x^2(x-3)(\sqrt{x+6}+x)}$

$$=\lim_{x\to 3}\frac{x+6-x^2}{x^2(x-3)\left(\sqrt{x+6}+x\right)}=\lim_{x\to 3}\frac{-(x^2-x-6)}{x^2(x-3)\left(\sqrt{x+6}+x\right)}=\lim_{x\to 3}\frac{-(x-3)(x+2)}{x^2(x-3)\left(\sqrt{x+6}+x\right)}$$

$$=\lim_{x\to 3}\frac{-(x+2)}{x^2\left(\sqrt{x+6}+x\right)}=-\frac{5}{9(3+3)}=-\frac{5}{54}$$

13. Since x is positive, $\sqrt{x^2}=|x|=x$. Thus,

$$\lim_{x\to\infty}\frac{\sqrt{x^2-9}}{2x-6}=\lim_{x\to\infty}\frac{\sqrt{x^2-9}/\sqrt{x^2}}{(2x-6)/x}=\lim_{x\to\infty}\frac{\sqrt{1-9/x^2}}{2-6/x}=\frac{\sqrt{1-0}}{2-0}=\frac{1}{2}$$

14. Since x is negative, $\sqrt{x^2}=|x|=-x$. Thus,

$$\lim_{x\to-\infty}\frac{\sqrt{x^2-9}}{2x-6}=\lim_{x\to-\infty}\frac{\sqrt{x^2-9}/\sqrt{x^2}}{(2x-6)/(-x)}=\lim_{x\to-\infty}\frac{\sqrt{1-9/x^2}}{-2+6/x}=\frac{\sqrt{1-0}}{-2+0}=-\frac{1}{2}$$

15. Let $t=\sin x$. Then as $x\to\pi^-$, $\sin x\to 0^+$, so $t\to 0^+$. Thus, $\lim\limits_{x\to\pi^-}\ln(\sin x)=\lim\limits_{t\to 0^+}\ln t=-\infty$.

16. $\lim\limits_{x\to-\infty}\dfrac{1-2x^2-x^4}{5+x-3x^4}=\lim\limits_{x\to-\infty}\dfrac{(1-2x^2-x^4)/x^4}{(5+x-3x^4)/x^4}=\lim\limits_{x\to-\infty}\dfrac{1/x^4-2/x^2-1}{5/x^4+1/x^3-3}=\dfrac{0-0-1}{0+0-3}=\dfrac{-1}{-3}=\dfrac{1}{3}$

17. $\lim\limits_{x\to\infty}\left(\sqrt{x^2+4x+1}-x\right)=\lim\limits_{x\to\infty}\left[\dfrac{\sqrt{x^2+4x+1}-x}{1}\cdot\dfrac{\sqrt{x^2+4x+1}+x}{\sqrt{x^2+4x+1}+x}\right]=\lim\limits_{x\to\infty}\dfrac{(x^2+4x+1)-x^2}{\sqrt{x^2+4x+1}+x}$

$$=\lim_{x\to\infty}\frac{(4x+1)/x}{\left(\sqrt{x^2+4x+1}+x\right)/x}\qquad\left[\text{divide by } x=\sqrt{x^2}\text{ for } x>0\right]$$

$$=\lim_{x\to\infty}\frac{4+1/x}{\sqrt{1+4/x+1/x^2}+1}=\frac{4+0}{\sqrt{1+0+0}+1}=\frac{4}{2}=2$$

18. Let $t=x-x^2=x(1-x)$. Then as $x\to\infty$, $t\to-\infty$, and $\lim\limits_{x\to\infty}e^{x-x^2}=\lim\limits_{t\to-\infty}e^t=0$.

19. Let $t=1/x$. Then as $x\to 0^+$, $t\to\infty$, and $\lim\limits_{x\to 0^+}\tan^{-1}(1/x)=\lim\limits_{t\to\infty}\tan^{-1}t=\dfrac{\pi}{2}$.

20. $\lim\limits_{x\to 1}\left(\dfrac{1}{x-1}+\dfrac{1}{x^2-3x+2}\right)=\lim\limits_{x\to 1}\left[\dfrac{1}{x-1}+\dfrac{1}{(x-1)(x-2)}\right]=\lim\limits_{x\to 1}\left[\dfrac{x-2}{(x-1)(x-2)}+\dfrac{1}{(x-1)(x-2)}\right]$

$$=\lim_{x\to 1}\left[\frac{x-1}{(x-1)(x-2)}\right]=\lim_{x\to 1}\frac{1}{x-2}=\frac{1}{1-2}=-1$$

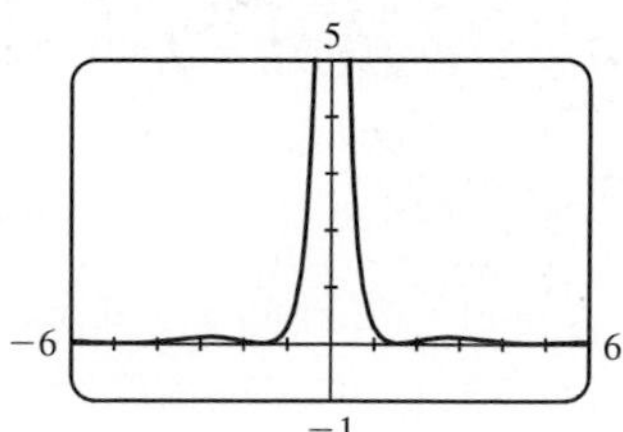

21. From the graph of $y = (\cos^2 x)/x^2$, it appears that $y = 0$ is the horizontal asymptote and $x = 0$ is the vertical asymptote. Now $0 \le (\cos x)^2 \le 1 \Rightarrow \frac{0}{x^2} \le \frac{\cos^2 x}{x^2} \le \frac{1}{x^2} \Rightarrow 0 \le \frac{\cos^2 x}{x^2} \le \frac{1}{x^2}$. But $\lim\limits_{x\to\pm\infty} 0 = 0$ and $\lim\limits_{x\to\pm\infty} \frac{1}{x^2} = 0$, so by the Squeeze Theorem, $\lim\limits_{x\to\pm\infty} \frac{\cos^2 x}{x^2} = 0$.

Thus, $y = 0$ is the horizontal asymptote. $\lim\limits_{x\to 0} \frac{\cos^2 x}{x^2} = \infty$ because $\cos^2 x \to 1$ and $x^2 \to 0^+$ as $x \to 0$, so $x = 0$ is the vertical asymptote.

22. From the graph of $y = f(x) = \sqrt{x^2+x+1} - \sqrt{x^2-x}$, it appears that there are 2 horizontal asymptotes and possibly 2 vertical asymptotes. To obtain a different form for f, let's multiply and divide it by its conjugate.

$$f_1(x) = \left(\sqrt{x^2+x+1} - \sqrt{x^2-x}\right)\frac{\sqrt{x^2+x+1} + \sqrt{x^2-x}}{\sqrt{x^2+x+1} + \sqrt{x^2-x}} = \frac{(x^2+x+1)-(x^2-x)}{\sqrt{x^2+x+1} + \sqrt{x^2-x}}$$
$$= \frac{2x+1}{\sqrt{x^2+x+1} + \sqrt{x^2-x}}$$

Now

$$\lim_{x\to\infty} f_1(x) = \lim_{x\to\infty} \frac{2x+1}{\sqrt{x^2+x+1} + \sqrt{x^2-x}}$$
$$= \lim_{x\to\infty} \frac{2+(1/x)}{\sqrt{1+(1/x)+(1/x^2)} + \sqrt{1-(1/x)}} \qquad \left[\text{since } \sqrt{x^2} = x \text{ for } x > 0\right]$$
$$= \frac{2}{1+1} = 1,$$

so $y = 1$ is a horizontal asymptote. For $x < 0$, we have $\sqrt{x^2} = |x| = -x$, so when we divide the denominator by x, with $x < 0$, we get

$$\frac{\sqrt{x^2+x+1} + \sqrt{x^2-x}}{x} = -\frac{\sqrt{x^2+x+1} + \sqrt{x^2-x}}{\sqrt{x^2}} = -\left[\sqrt{1+\frac{1}{x}+\frac{1}{x^2}} + \sqrt{1-\frac{1}{x}}\right]$$

Therefore,

$$\lim_{x\to-\infty} f_1(x) = \lim_{x\to-\infty} \frac{2x+1}{\sqrt{x^2+x+1} + \sqrt{x^2-x}} = \lim_{x\to\infty} \frac{2+(1/x)}{-\left[\sqrt{1+(1/x)+(1/x^2)} + \sqrt{1-(1/x)}\right]}$$
$$= \frac{2}{-(1+1)} = -1,$$

so $y = -1$ is a horizontal asymptote.

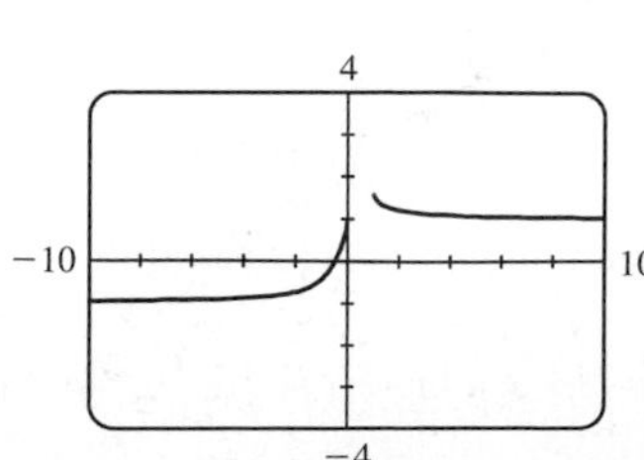

The domain of f is $(-\infty, 0] \cup [1, \infty)$. As $x \to 0^-$, $f(x) \to 1$, so $x = 0$ is *not* a vertical asymptote. As $x \to 1^+$, $f(x) \to \sqrt{3}$, so $x = 1$ is *not* a vertical asymptote and hence there are no vertical asymptotes.

23. Since $2x - 1 \le f(x) \le x^2$ for $0 < x < 3$ and $\lim_{x \to 1}(2x - 1) = 1 = \lim_{x \to 1} x^2$, we have $\lim_{x \to 1} f(x) = 1$ by the Squeeze Theorem.

24. Let $f(x) = -x^2$, $g(x) = x^2 \cos(1/x^2)$ and $h(x) = x^2$. Then since $\left|\cos(1/x^2)\right| \le 1$ for $x \ne 0$, we have $f(x) \le g(x) \le h(x)$ for $x \ne 0$, and so $\lim_{x \to 0} f(x) = \lim_{x \to 0} h(x) = 0 \quad \Rightarrow \quad \lim_{x \to 0} g(x) = 0$ by the Squeeze Theorem.

25. Given $\varepsilon > 0$, we need $\delta > 0$ such that if $0 < |x - 2| < \delta$, then $|(14 - 5x) - 4| < \varepsilon$. But $|(14 - 5x) - 4| < \varepsilon \quad \Leftrightarrow$ $|-5x + 10| < \varepsilon \quad \Leftrightarrow \quad |-5|\,|x - 2| < \varepsilon \quad \Leftrightarrow \quad |x - 2| < \varepsilon/5$. So if we choose $\delta = \varepsilon/5$, then $0 < |x - 2| < \delta \quad \Rightarrow$ $|(14 - 5x) - 4| < \varepsilon$. Thus, $\lim_{x \to 2}(14 - 5x) = 4$ by the definition of a limit.

26. Given $\varepsilon > 0$ we must find $\delta > 0$ so that if $0 < |x - 0| < \delta$, then $|\sqrt[3]{x} - 0| < \varepsilon$. Now $|\sqrt[3]{x} - 0| = |\sqrt[3]{x}| < \varepsilon \quad \Rightarrow$ $|x| = |\sqrt[3]{x}|^3 < \varepsilon^3$. So take $\delta = \varepsilon^3$. Then $0 < |x - 0| = |x| < \varepsilon^3 \quad \Rightarrow \quad |\sqrt[3]{x} - 0| = |\sqrt[3]{x}| = \sqrt[3]{|x|} < \sqrt[3]{\varepsilon^3} = \varepsilon$.
Therefore, by the definition of a limit, $\lim_{x \to 0} \sqrt[3]{x} = 0$.

27. Given $\varepsilon > 0$, we need $\delta > 0$ so that if $0 < |x - 2| < \delta$, then $\left|x^2 - 3x - (-2)\right| < \varepsilon$. First, note that if $|x - 2| < 1$, then $-1 < x - 2 < 1$, so $0 < x - 1 < 2 \quad \Rightarrow \quad |x - 1| < 2$. Now let $\delta = \min\{\varepsilon/2, 1\}$. Then $0 < |x - 2| < \delta \quad \Rightarrow$ $\left|x^2 - 3x - (-2)\right| = |(x - 2)(x - 1)| = |x - 2|\,|x - 1| < (\varepsilon/2)(2) = \varepsilon$.
Thus, $\lim_{x \to 2}(x^2 - 3x) = -2$ by the definition of a limit.

28. Given $M > 0$, we need $\delta > 0$ such that if $0 < x - 4 < \delta$, then $2/\sqrt{x - 4} > M$. This is true $\quad \Leftrightarrow \quad \sqrt{x - 4} < 2/M \quad \Leftrightarrow$ $x - 4 < 4/M^2$. So if we choose $\delta = 4/M^2$, then $0 < x - 4 < \delta \quad \Rightarrow \quad 2/\sqrt{x - 4} > M$. So by the definition of a limit, $\lim_{x \to 4^+}\left(2/\sqrt{x - 4}\right) = \infty$.

29. (a) $f(x) = \sqrt{-x}$ if $x < 0$, $f(x) = 3 - x$ if $0 \le x < 3$, $f(x) = (x - 3)^2$ if $x > 3$.

(i) $\lim_{x \to 0^+} f(x) = \lim_{x \to 0^+}(3 - x) = 3$

(ii) $\lim_{x \to 0^-} f(x) = \lim_{x \to 0^-}\sqrt{-x} = 0$

(iii) Because of (i) and (ii), $\lim_{x \to 0} f(x)$ does not exist.

(iv) $\lim_{x \to 3^-} f(x) = \lim_{x \to 3^-}(3 - x) = 0$

(v) $\lim_{x \to 3^+} f(x) = \lim_{x \to 3^+}(x - 3)^2 = 0$

(vi) Because of (iv) and (v), $\lim_{x \to 3} f(x) = 0$.

(b) f is discontinuous at 0 since $\lim_{x \to 0} f(x)$ does not exist.

f is discontinuous at 3 since $f(3)$ does not exist.

(c)

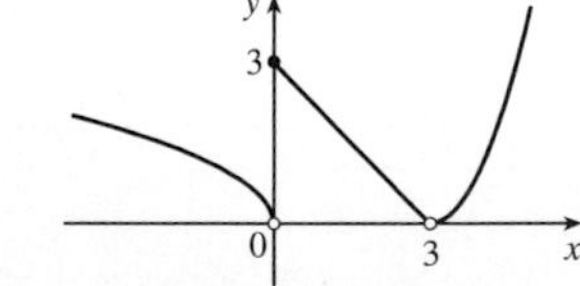

30. (a) $g(x) = 2x - x^2$ if $0 \le x \le 2$, $g(x) = 2 - x$ if $2 < x \le 3$, $g(x) = x - 4$ if $3 < x < 4$, $g(x) = \pi$ if $x \ge 4$.

Therefore, $\lim_{x \to 2^-} g(x) = \lim_{x \to 2^-}\left(2x - x^2\right) = 0$ and $\lim_{x \to 2^+} g(x) = \lim_{x \to 2^+}(2 - x) = 0$. Thus, $\lim_{x \to 2} g(x) = 0 = g(2)$, so g is continuous at 2. $\lim_{x \to 3^-} g(x) = \lim_{x \to 3^-}(2 - x) = -1$ and $\lim_{x \to 3^+} g(x) = \lim_{x \to 3^+}(x - 4) = -1$. Thus,

$\lim_{x\to 3} g(x) = -1 = g(3)$, so g is continuous at 3.

(b)

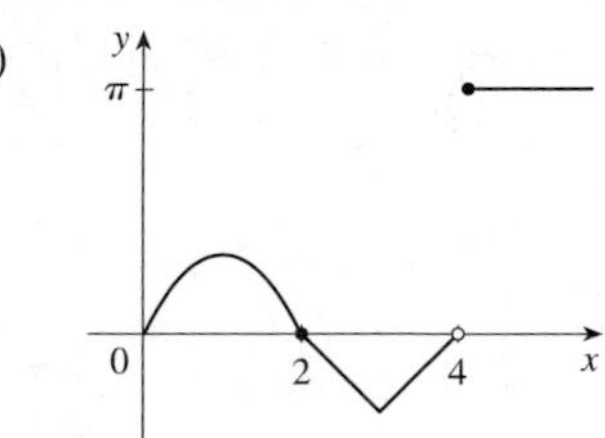

$\lim_{x\to 4^-} g(x) = \lim_{x\to 4^-} (x-4) = 0$ and $\lim_{x\to 4^+} g(x) = \lim_{x\to 4^+} \pi = \pi$.

Thus, $\lim_{x\to 4} g(x)$ does not exist, so g is discontinuous at 4. But

$\lim_{x\to 4^+} g(x) = \pi = g(4)$, so g is continuous from the right at 4.

31. $\sin x$ and e^x are continuous on $\mathbb{R}$ by Theorem 2.5.7. Since e^x is continuous on $\mathbb{R}$, $e^{\sin x}$ is continuous on $\mathbb{R}$ by Theorem 2.5.9. Lastly, x is continuous on $\mathbb{R}$ since it's a polynomial and the product $xe^{\sin x}$ is continuous on its domain $\mathbb{R}$ by Theorem 2.5.4.

32. $x^2 - 9$ is continuous on $\mathbb{R}$ since it is a polynomial and $\sqrt{x}$ is continuous on $[0,\infty)$ by Theorem 2.5.7, so the composition $\sqrt{x^2-9}$ is continuous on $\{x \mid x^2 - 9 \geq 0\} = (-\infty,-3] \cup [3,\infty)$ by Theorem 2.5.9. Note that $x^2 - 2 \neq 0$ on this set and so the quotient function $g(x) = \dfrac{\sqrt{x^2-9}}{x^2-2}$ is continuous on its domain, $(-\infty,-3] \cup [3,\infty)$ by Theorem 2.5.4.

33. $f(x) = x^5 - x^3 + 3x - 5$ is continuous on the interval $[1,2]$, $f(1) = -2$, and $f(2) = 25$. Since $-2 < 0 < 25$, there is a number c in $(1,2)$ such that $f(c) = 0$ by the Intermediate Value Theorem. Thus, there is a root of the equation $x^5 - x^3 + 3x - 5 = 0$ in the interval $(1,2)$.

34. $f(x) = \cos\sqrt{x} - e^x + 2$ is continuous on the interval $[0,1]$, $f(0) = 2$, and $f(1) \approx -0.2$. Since $-0.2 < 0 < 2$, there is a number c in $(0,1)$ such that $f(c) = 0$ by the Intermediate Value Theorem. Thus, there is a root of the equation $\cos\sqrt{x} - e^x + 2 = 0$, or $\cos\sqrt{x} = e^x - 2$, in the interval $(0,1)$.

35. (a) The slope of the tangent line at $(2,1)$ is

$$\begin{aligned}\lim_{x\to 2}\frac{f(x)-f(2)}{x-2} &= \lim_{x\to 2}\frac{9-2x^2-1}{x-2} = \lim_{x\to 2}\frac{8-2x^2}{x-2} = \lim_{x\to 2}\frac{-2(x^2-4)}{x-2} = \lim_{x\to 2}\frac{-2(x-2)(x+2)}{x-2}\\ &= \lim_{x\to 2}\left[-2(x+2)\right] = -2\cdot 4 = -8\end{aligned}$$

(b) An equation of this tangent line is $y - 1 = -8(x-2)$ or $y = -8x + 17$.

36. For a general point with x-coordinate a, we have

$$\begin{aligned}m &= \lim_{x\to a}\frac{2/(1-3x) - 2/(1-3a)}{x-a} = \lim_{x\to a}\frac{2(1-3a)-2(1-3x)}{(1-3a)(1-3x)(x-a)} = \lim_{x\to a}\frac{6(x-a)}{(1-3a)(1-3x)(x-a)}\\ &= \lim_{x\to a}\frac{6}{(1-3a)(1-3x)} = \frac{6}{(1-3a)^2}\end{aligned}$$

For $a = 0$, $m = 6$ and $f(0) = 2$, so an equation of the tangent line is $y - 2 = 6(x-0)$ or $y = 6x + 2$. For $a = -1$, $m = \frac{3}{8}$ and $f(-1) = \frac{1}{2}$, so an equation of the tangent line is $y - \frac{1}{2} = \frac{3}{8}(x+1)$ or $y = \frac{3}{8}x + \frac{7}{8}$.

37. (a) $s = s(t) = 1 + 2t + t^2/4$. The average velocity over the time interval $[1, 1+h]$ is

$$v_{\text{ave}} = \frac{s(1+h)-s(1)}{(1+h)-1} = \frac{1+2(1+h)+(1+h)^2/4 - 13/4}{h} = \frac{10h+h^2}{4h} = \frac{10+h}{4}$$

[continued]

So for the following intervals the average velocities are:

(i) $[1, 3]$: $h = 2$, $v_{\text{ave}} = (10 + 2)/4 = 3$ m/s (ii) $[1, 2]$: $h = 1$, $v_{\text{ave}} = (10 + 1)/4 = 2.75$ m/s

(iii) $[1, 1.5]$: $h = 0.5$, $v_{\text{ave}} = (10 + 0.5)/4 = 2.625$ m/s (iv) $[1, 1.1]$: $h = 0.1$, $v_{\text{ave}} = (10 + 0.1)/4 = 2.525$ m/s

(b) When $t = 1$, the instantaneous velocity is $\lim\limits_{h \to 0} \dfrac{s(1 + h) - s(1)}{h} = \lim\limits_{h \to 0} \dfrac{10 + h}{4} = \dfrac{10}{4} = 2.5$ m/s.

38. (a) When V increases from 200 in^3 to 250 in^3, we have $\Delta V = 250 - 200 = 50$ in^3, and since $P = 800/V$,

$\Delta P = P(250) - P(200) = \dfrac{800}{250} - \dfrac{800}{200} = 3.2 - 4 = -0.8$ lb/in^2. So the average rate of change

is $\dfrac{\Delta P}{\Delta V} = \dfrac{-0.8}{50} = -0.016 \dfrac{\text{lb/in}^2}{\text{in}^3}$.

(b) Since $V = 800/P$, the instantaneous rate of change of V with respect to P is

$$\lim_{h \to 0} \frac{\Delta V}{\Delta P} = \lim_{h \to 0} \frac{V(P + h) - V(P)}{h} = \lim_{h \to 0} \frac{800/(P + h) - 800/P}{h} = \lim_{h \to 0} \frac{800\,[P - (P + h)]}{h(P + h)P}$$

$$= \lim_{h \to 0} \frac{-800}{(P + h)P} = -\frac{800}{P^2}$$

which is inversely proportional to the square of P.

39. (a) $f'(2) = \lim\limits_{x \to 2} \dfrac{f(x) - f(2)}{x - 2} = \lim\limits_{x \to 2} \dfrac{x^3 - 2x - 4}{x - 2}$

$= \lim\limits_{x \to 2} \dfrac{(x - 2)(x^2 + 2x + 2)}{x - 2} = \lim\limits_{x \to 2} (x^2 + 2x + 2) = 10$

(b) $y - 4 = 10(x - 2)$ or $y = 10x - 16$

(c)

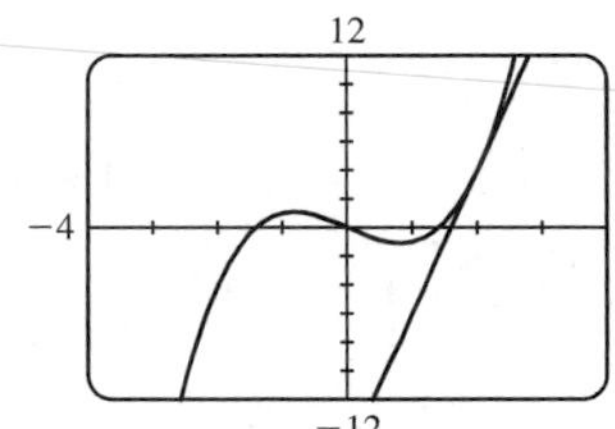

40. $2^6 = 64$, so $f(x) = x^6$ and $a = 2$.

41. (a) $f'(r)$ is the rate at which the total cost changes with respect to the interest rate. Its units are dollars/(percent per year).

(b) The total cost of paying off the loan is increasing by \$1200/(percent per year) as the interest rate reaches 10%. So if the interest rate goes up from 10% to 11%, the cost goes up approximately \$1200.

(c) As r increases, C increases. So $f'(r)$ will always be positive.

42.

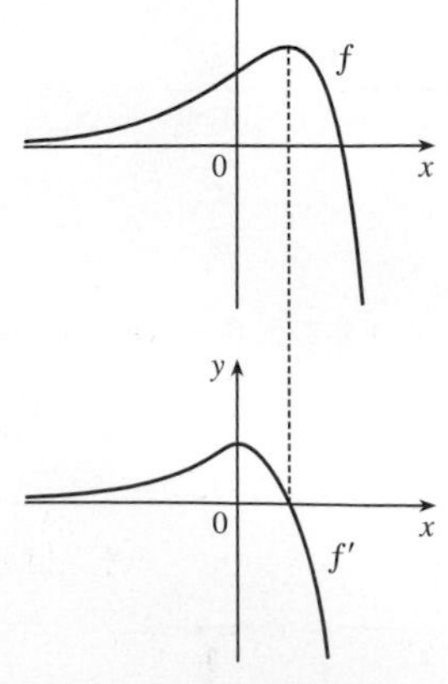

43.

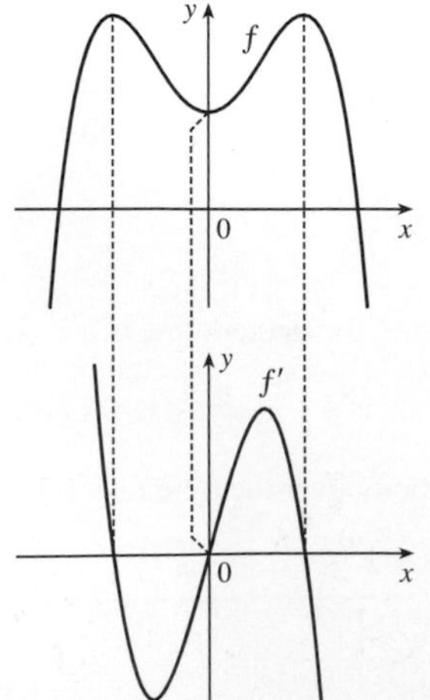

44.

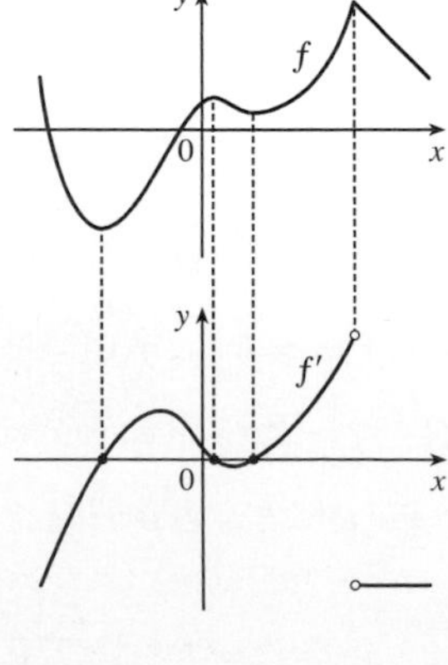

45. (a) $f'(x) = \lim\limits_{h\to 0} \dfrac{f(x+h)-f(x)}{h} = \lim\limits_{h\to 0} \dfrac{\sqrt{3-5(x+h)}-\sqrt{3-5x}}{h} \dfrac{\sqrt{3-5(x+h)}+\sqrt{3-5x}}{\sqrt{3-5(x+h)}+\sqrt{3-5x}}$

$$= \lim_{h\to 0} \frac{[3-5(x+h)]-(3-5x)}{h\left(\sqrt{3-5(x+h)}+\sqrt{3-5x}\right)} = \lim_{h\to 0} \frac{-5}{\sqrt{3-5(x+h)}+\sqrt{3-5x}} = \frac{-5}{2\sqrt{3-5x}}$$

(b) Domain of f: (the radicand must be nonnegative) $3-5x \geq 0 \Rightarrow$

$5x \leq 3 \Rightarrow x \in \left(-\infty, \frac{3}{5}\right]$

Domain of f': exclude $\frac{3}{5}$ because it makes the denominator zero;

$x \in \left(-\infty, \frac{3}{5}\right)$

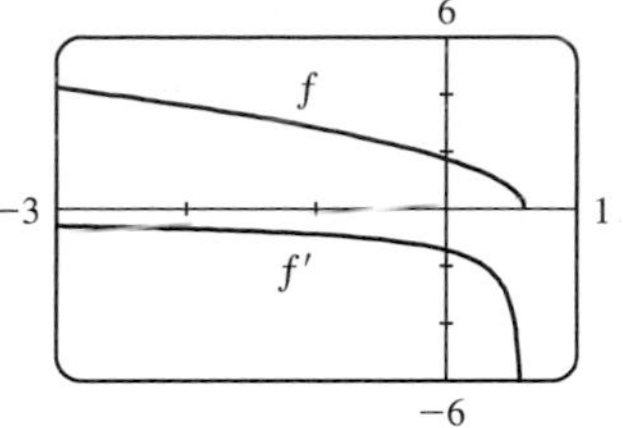

(c) Our answer to part (a) is reasonable because $f'(x)$ is always negative and f is always decreasing.

46. (a) As $x \to \pm\infty$, $f(x) = (4-x)/(3+x) \to -1$, so there is a horizontal asymptote at $y=-1$. As $x \to -3^+$, $f(x) \to \infty$, and as $x \to -3^-$, $f(x) \to -\infty$. Thus, there is a vertical asymptote at $x=-3$.

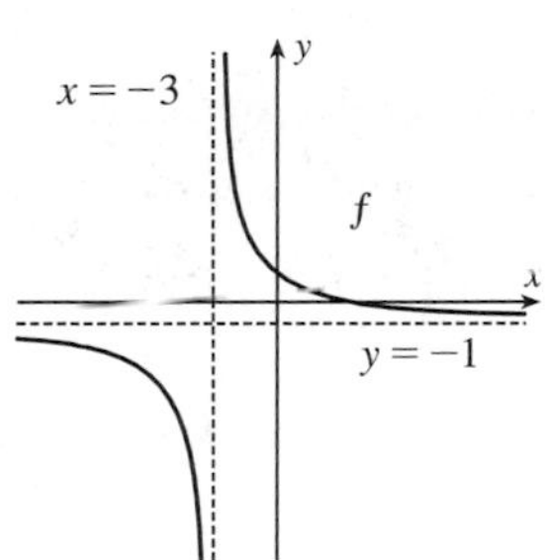

(b) Note that f is decreasing on $(-\infty,-3)$ and $(-3,\infty)$, so f' is negative on those intervals. As $x \to \pm\infty$, $f' \to 0$. As $x \to -3^-$ and as $x \to -3^+$, $f' \to -\infty$.

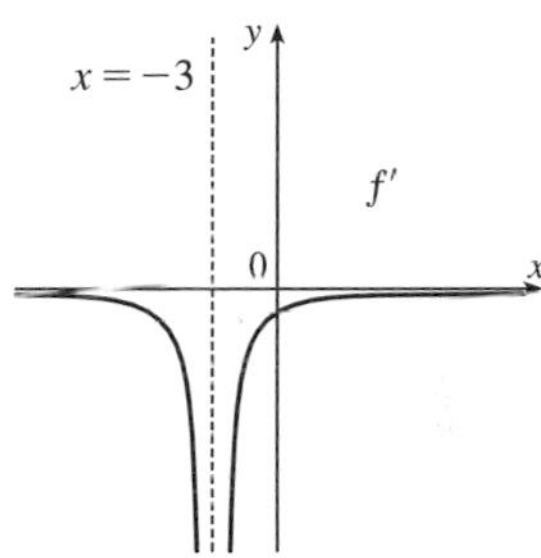

(c) $f'(x) = \lim\limits_{h\to 0} \dfrac{f(x+h)-f(x)}{h} = \lim\limits_{h\to 0} \dfrac{\dfrac{4-(x+h)}{3+(x+h)} - \dfrac{4-x}{3+x}}{h} = \lim\limits_{h\to 0} \dfrac{(3+x)\,[4-(x+h)]-(4-x)\,[3+(x+h)]}{h\,[3+(x+h)]\,(3+x)}$

$$= \lim_{h\to 0} \frac{(12-3x-3h+4x-x^2-hx)-(12+4x+4h-3x-x^2-hx)}{h[3+(x+h)](3+x)}$$

$$= \lim_{h\to 0} \frac{-7h}{h\,[3+(x+h)]\,(3+x)} = \lim_{h\to 0} \frac{-7}{[3+(x+h)]\,(3+x)} = -\frac{7}{(3+x)^2}$$

(d) The graphing device confirms our graph in part (b).

47. f is not differentiable: at $x=-4$ because f is not continuous, at $x=-1$ because f has a corner, at $x=2$ because f is not continuous, and at $x=5$ because f has a vertical tangent.

48. The graph of a has tangent lines with positive slope for $x<0$ and negative slope for $x>0$, and the values of c fit this pattern, so c must be the graph of the derivative of the function for a. The graph of c has horizontal tangent lines to the left and right of the x-axis and b has zeros at these points. Hence, b is the graph of the derivative of the function for c. Therefore, a is the graph of f, c is the graph of f', and b is the graph of f''.

49. Domain: $(-\infty, 0) \cup (0, \infty)$; $\lim_{x \to 0^-} f(x) = 1$; $\lim_{x \to 0^+} f(x) = 0$;

$f'(x) > 0$ for all x in the domain; $\lim_{x \to -\infty} f'(x) = 0$; $\lim_{x \to \infty} f'(x) = 1$

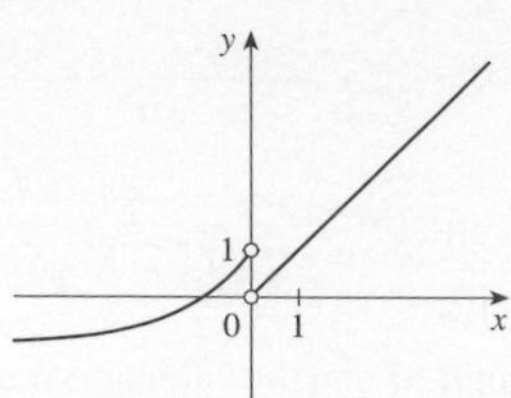

50. (a) $P'(t)$ is the rate at which the percentage of Americans under the age of 18 is changing with respect to time. Its units are percent per year (%/yr).

(b) To find $P'(t)$, we use $\lim_{h \to 0} \dfrac{P(t+h) - P(t)}{h} \approx \dfrac{P(t+h) - P(t)}{h}$ for small values of h.

For 1950: $P'(1950) \approx \dfrac{P(1960) - P(1950)}{1960 - 1950} = \dfrac{35.7 - 31.1}{10} = 0.46$

For 1960: We estimate $P'(1960)$ by using $h = -10$ and $h = 10$, and then average the two results to obtain a final estimate.

$h = -10 \;\Rightarrow\; P'(1960) \approx \dfrac{P(1950) - P(1960)}{1950 - 1960} = \dfrac{31.1 - 35.7}{-10} = 0.46$

$h = 10 \;\Rightarrow\; P'(1960) \approx \dfrac{P(1970) - P(1960)}{1970 - 1960} = \dfrac{34.0 - 35.7}{10} = -0.17$

So we estimate that $P'(1960) \approx \frac{1}{2}[0.46 + (-0.17)] = 0.145$.

t	1950	1960	1970	1980	1990	2000	2010
$P'(t)$	0.460	0.145	−0.385	−0.415	−0.115	−0.085	−0.170

(c)

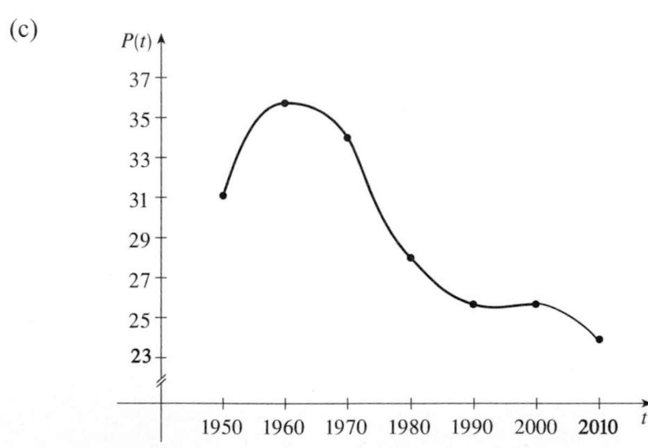

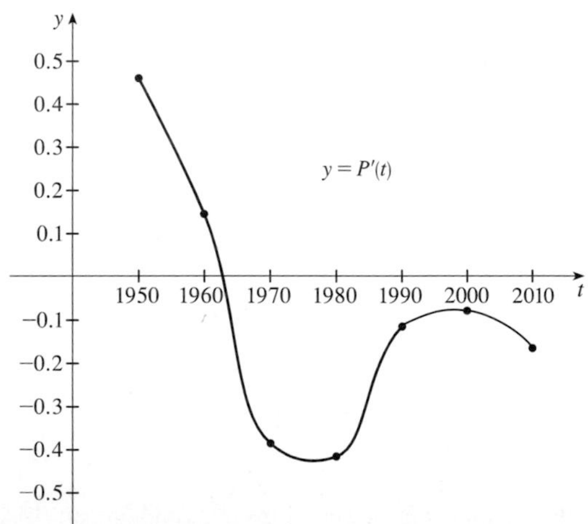

(d) We could get more accurate values for $P'(t)$ by obtaining data for the mid-decade years 1955, 1965, 1975, 1985, 1995, and 2005.

51. $B'(t)$ is the rate at which the number of US \$20 bills in circulation is changing with respect to time. Its units are billions of bills per year. We use a symmetric difference quotient to estimate $B'(2000)$.

$B'(2000) \approx \dfrac{B(2005) - B(1995)}{2005 - 1995} = \dfrac{5.77 - 4.21}{10} = 0.156$ billions of bills per year (or 156 million bills per year).

52. (a) Drawing slope triangles, we obtain the following estimates: $F'(1950) \approx \frac{1.1}{10} = 0.11$, $F'(1965) \approx \frac{-1.6}{10} = -0.16$, and $F'(1987) \approx \frac{0.2}{10} = 0.02$.

(b) The rate of change of the average number of children born to each woman was increasing by 0.11 in 1950, decreasing by 0.16 in 1965, and increasing by 0.02 in 1987.

(c) There are many possible reasons:

- In the baby-boom era (post-WWII), there was optimism about the economy and family size was rising.
- In the baby-bust era, there was less economic optimism, and it was considered less socially responsible to have a large family.
- In the baby-boomlet era, there was increased economic optimism and a return to more conservative attitudes.

53. $|f(x)| \leq g(x) \quad \Leftrightarrow \quad -g(x) \leq f(x) \leq g(x)$ and $\lim\limits_{x \to a} g(x) = 0 = \lim\limits_{x \to a} -g(x)$.

Thus, by the Squeeze Theorem, $\lim\limits_{x \to a} f(x) = 0$.

54. (a) Note that f is an even function since $f(x) = f(-x)$. Now for any integer n, $[\![n]\!] + [\![-n]\!] = n - n = 0$, and for any real number k which is not an integer, $[\![k]\!] + [\![-k]\!] = [\![k]\!] + (-[\![k]\!] - 1) = -1$. So $\lim\limits_{x \to a} f(x)$ exists (and is equal to -1) for all values of a.

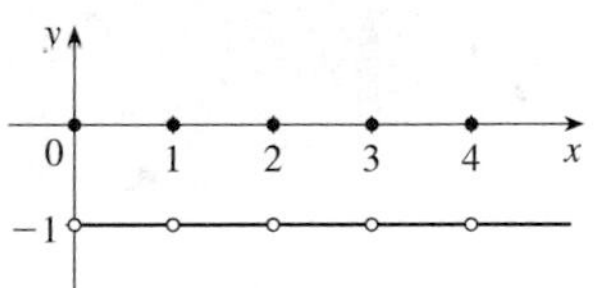

(b) f is discontinuous at all integers.

PROBLEMS PLUS

1. Let $t = \sqrt[6]{x}$, so $x = t^6$. Then $t \to 1$ as $x \to 1$, so

$$\lim_{x\to 1}\frac{\sqrt[3]{x}-1}{\sqrt{x}-1} = \lim_{t\to 1}\frac{t^2-1}{t^3-1} = \lim_{t\to 1}\frac{(t-1)(t+1)}{(t-1)\left(t^2+t+1\right)} = \lim_{t\to 1}\frac{t+1}{t^2+t+1} = \frac{1+1}{1^2+1+1} = \frac{2}{3}.$$

Another method: Multiply both the numerator and the denominator by $(\sqrt{x}+1)\left(\sqrt[3]{x^2}+\sqrt[3]{x}+1\right)$.

2. First rationalize the numerator: $\lim\limits_{x\to 0}\dfrac{\sqrt{ax+b}-2}{x}\cdot\dfrac{\sqrt{ax+b}+2}{\sqrt{ax+b}+2} = \lim\limits_{x\to 0}\dfrac{ax+b-4}{x\left(\sqrt{ax+b}+2\right)}$. Now since the denominator approaches 0 as $x \to 0$, the limit will exist only if the numerator also approaches 0 as $x \to 0$. So we require that $a(0)+b-4=0 \;\Rightarrow\; b=4$. So the equation becomes $\lim\limits_{x\to 0}\dfrac{a}{\sqrt{ax+4}+2} = 1 \;\Rightarrow\; \dfrac{a}{\sqrt{4}+2} = 1 \;\Rightarrow\; a = 4$.

Therefore, $a = b = 4$.

3. For $-\frac{1}{2} < x < \frac{1}{2}$, we have $2x-1<0$ and $2x+1>0$, so $|2x-1| = -(2x-1)$ and $|2x+1| = 2x+1$.

Therefore, $\lim\limits_{x\to 0}\dfrac{|2x-1|-|2x+1|}{x} = \lim\limits_{x\to 0}\dfrac{-(2x-1)-(2x+1)}{x} = \lim\limits_{x\to 0}\dfrac{-4x}{x} = \lim\limits_{x\to 0}(-4) = -4.$

4. Let R be the midpoint of OP, so the coordinates of R are $\left(\frac{1}{2}x, \frac{1}{2}x^2\right)$ since the coordinates of P are $\left(x, x^2\right)$. Let $Q = (0, a)$.

Since the slope $m_{OP} = \dfrac{x^2}{x} = x$, $m_{QR} = -\dfrac{1}{x}$ (negative reciprocal). But $m_{QR} = \dfrac{\frac{1}{2}x^2 - a}{\frac{1}{2}x - 0} = \dfrac{x^2-2a}{x}$, so we conclude that $-1 = x^2 - 2a \;\Rightarrow\; 2a = x^2+1 \;\Rightarrow\; a = \frac{1}{2}x^2 + \frac{1}{2}$. As $x \to 0$, $a \to \frac{1}{2}$, and the limiting position of Q is $\left(0, \frac{1}{2}\right)$.

5. (a) For $0 < x < 1$, $[\![x]\!] = 0$, so $\dfrac{[\![x]\!]}{x} = 0$, and $\lim\limits_{x\to 0^+}\dfrac{[\![x]\!]}{x} = 0$. For $-1 < x < 0$, $[\![x]\!] = -1$, so $\dfrac{[\![x]\!]}{x} = \dfrac{-1}{x}$, and

$\lim\limits_{x\to 0^-}\dfrac{[\![x]\!]}{x} = \lim\limits_{x\to 0^-}\left(\dfrac{-1}{x}\right) = \infty$. Since the one-sided limits are not equal, $\lim\limits_{x\to 0}\dfrac{[\![x]\!]}{x}$ does not exist.

(b) For $x > 0$, $1/x - 1 \le [\![1/x]\!] \le 1/x \;\Rightarrow\; x(1/x-1) \le x[\![1/x]\!] \le x(1/x) \;\Rightarrow\; 1 - x \le x[\![1/x]\!] \le 1$.

As $x \to 0^+$, $1 - x \to 1$, so by the Squeeze Theorem, $\lim\limits_{x\to 0^+} x[\![1/x]\!] = 1$.

For $x < 0$, $1/x - 1 \le [\![1/x]\!] \le 1/x \;\Rightarrow\; x(1/x-1) \ge x[\![1/x]\!] \ge x(1/x) \;\Rightarrow\; 1 - x \ge x[\![1/x]\!] \ge 1$.

As $x \to 0^-$, $1 - x \to 1$, so by the Squeeze Theorem, $\lim\limits_{x\to 0^-} x[\![1/x]\!] = 1$.

Since the one-sided limits are equal, $\lim\limits_{x\to 0} x[\![1/x]\!] = 1$.

6. (a) $[\![x]\!]^2 + [\![y]\!]^2 = 1$. Since $[\![x]\!]^2$ and $[\![y]\!]^2$ are positive integers or 0, there are only 4 cases:

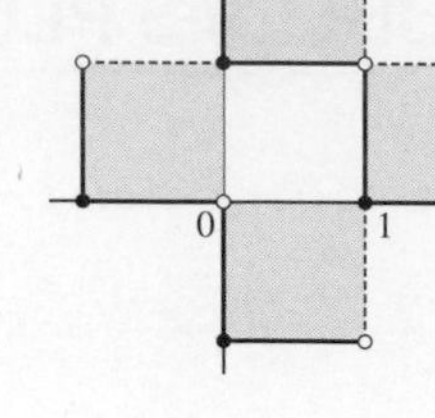

Case (i): $[\![x]\!] = 1,\ [\![y]\!] = 0 \ \Rightarrow 1 \le x < 2$ and $0 \le y < 1$

Case (ii): $[\![x]\!] = -1,\ [\![y]\!] = 0 \Rightarrow -1 \le x < 0$ and $0 \le y < 1$

Case (iii): $[\![x]\!] = 0,\ [\![y]\!] = 1 \ \Rightarrow 0 \le x < 1$ and $1 \le y < 2$

Case (iv): $[\![x]\!] = 0,\ [\![y]\!] = -1 \Rightarrow 0 \le x < 1$ and $-1 \le y < 0$

(b) $[\![x]\!]^2 - [\![y]\!]^2 = 3$. The only integral solution of $n^2 - m^2 = 3$ is $n = \pm 2$ and $m = \pm 1$. So the graph is

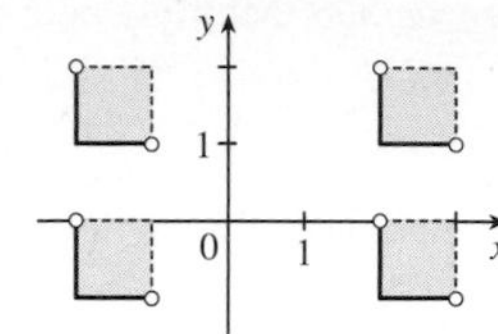

$$\{(x, y) \mid [\![x]\!] = \pm 2,\ [\![y]\!] = \pm 1\} = \left\{ (x, y) \ \middle| \ \begin{array}{l} 2 \le x \le 3 \ \text{ or } \ -2 \le x < 1, \\ 1 \le y < 2 \ \text{ or } \ -1 \le y < 0 \end{array} \right\}.$$

(c) $[\![x + y]\!]^2 = 1 \ \Rightarrow \ [\![x + y]\!] = \pm 1 \ \Rightarrow \ 1 \le x + y < 2$ or $-1 \le x + y < 0$

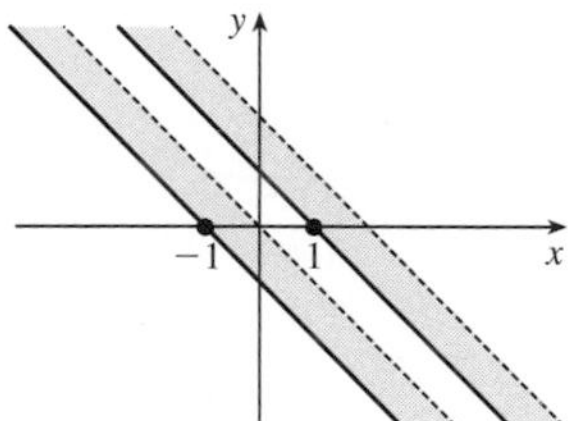

(d) For $n \le x < n + 1$, $[\![x]\!] = n$. Then $[\![x]\!] + [\![y]\!] = 1 \ \Rightarrow \ [\![y]\!] = 1 - n \ \Rightarrow$ $1 - n \le y < 2 - n$. Choosing integer values for n produces the graph.

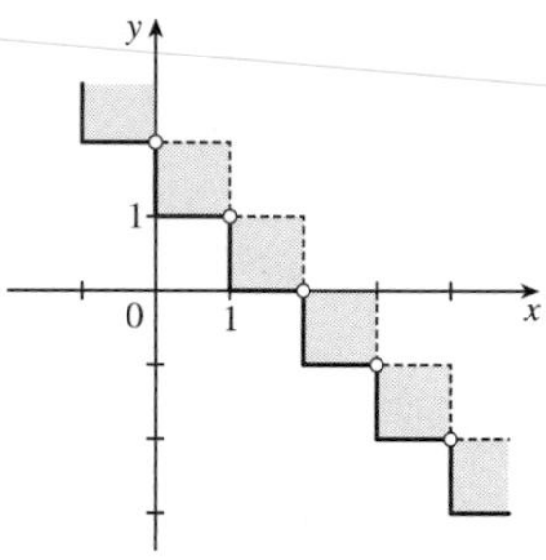

7. f is continuous on $(-\infty, a)$ and (a, ∞). To make f continuous on $\mathbb{R}$, we must have continuity at a. Thus,

$$\lim_{x \to a^+} f(x) = \lim_{x \to a^-} f(x) \ \Rightarrow \ \lim_{x \to a^+} x^2 = \lim_{x \to a^-} (x + 1) \ \Rightarrow \ a^2 = a + 1 \ \Rightarrow \ a^2 - a - 1 = 0 \ \Rightarrow$$

[by the quadratic formula] $a = \left(1 \pm \sqrt{5}\,\right)/2 \approx 1.618$ or -0.618.

8. (a) Here are a few possibilities:

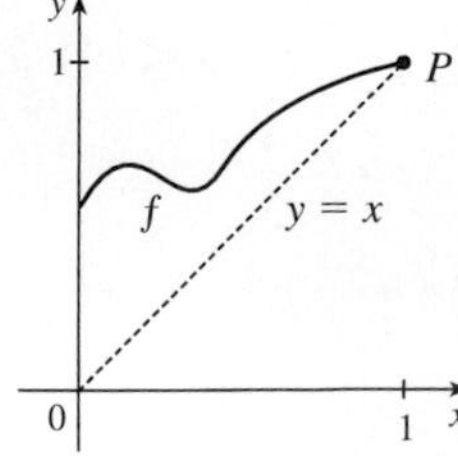

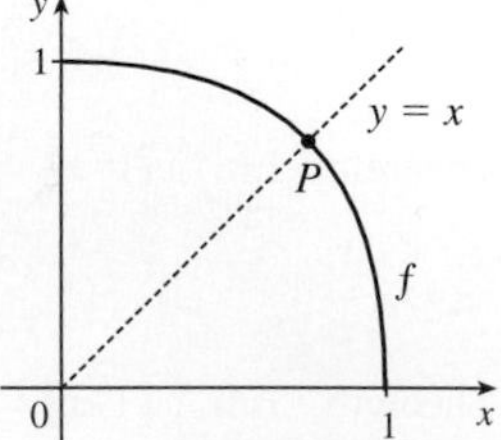

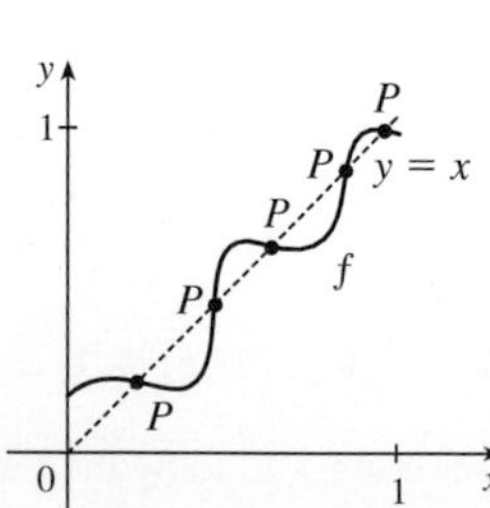

(b) The "obstacle" is the line $x = y$ (see diagram). Any intersection of the graph of f with the line $y = x$ constitutes a fixed point, and if the graph of the function does not cross the line somewhere in $(0, 1)$, then it must either start at $(0, 0)$ (in which case 0 is a fixed point) or finish at $(1, 1)$ (in which case 1 is a fixed point).

(c) Consider the function $F(x) = f(x) - x$, where f is any continuous function with domain $[0, 1]$ and range in $[0, 1]$. We shall prove that f has a fixed point. Now if $f(0) = 0$ then we are done: f has a fixed point (the number 0), which is what we are trying to prove. So assume $f(0) \neq 0$. For the same reason we can assume that $f(1) \neq 1$. Then $F(0) = f(0) > 0$ and $F(1) = f(1) - 1 < 0$. So by the Intermediate Value Theorem, there exists some number c in the interval $(0, 1)$ such that $F(c) = f(c) - c = 0$. So $f(c) = c$, and therefore f has a fixed point.

9. $\begin{cases} \lim\limits_{x \to a} [f(x) + g(x)] = 2 \\ \lim\limits_{x \to a} [f(x) - g(x)] = 1 \end{cases} \Rightarrow \begin{cases} \lim\limits_{x \to a} f(x) + \lim\limits_{x \to a} g(x) = 2 & \textbf{(1)} \\ \lim\limits_{x \to a} f(x) - \lim\limits_{x \to a} g(x) = 1 & \textbf{(2)} \end{cases}$

Adding equations **(1)** and **(2)** gives us $2 \lim\limits_{x \to a} f(x) = 3 \Rightarrow \lim\limits_{x \to a} f(x) = \frac{3}{2}$. From equation **(1)**, $\lim\limits_{x \to a} g(x) = \frac{1}{2}$. Thus,

$\lim\limits_{x \to a} [f(x)\, g(x)] = \lim\limits_{x \to a} f(x) \cdot \lim\limits_{x \to a} g(x) = \frac{3}{2} \cdot \frac{1}{2} = \frac{3}{4}$.

10. (a) *Solution 1:* We introduce a coordinate system and drop a perpendicular from P, as shown. We see from $\angle NCP$ that $\tan 2\theta = \dfrac{y}{1 - x}$, and from $\angle NBP$ that $\tan \theta = y/x$. Using the double-angle formula for tangents, we get $\dfrac{y}{1 - x} = \tan 2\theta = \dfrac{2 \tan \theta}{1 - \tan^2 \theta} = \dfrac{2(y/x)}{1 - (y/x)^2}$. After a bit of simplification, this becomes $\dfrac{1}{1 - x} = \dfrac{2x}{x^2 - y^2} \Leftrightarrow y^2 = x\,(3x - 2)$.

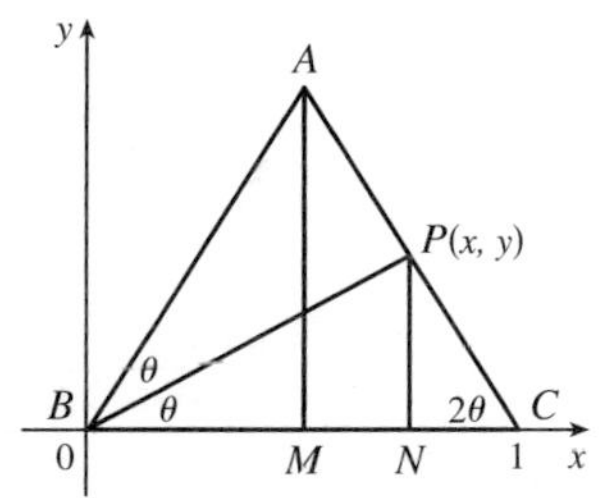

As the altitude AM decreases in length, the point P will approach the x-axis, that is, $y \to 0$, so the limiting location of P must be one of the roots of the equation $x(3x - 2) = 0$. Obviously it is not $x = 0$ (the point P can never be to the left of the altitude AM, which it would have to be in order to approach 0) so it must be $3x - 2 = 0$, that is, $x = \frac{2}{3}$.

Solution 2: We add a few lines to the original diagram, as shown. Now note that $\angle BPQ = \angle PBC$ (alternate angles; $QP \parallel BC$ by symmetry) and similarly $\angle CQP = \angle QCB$. So ΔBPQ and ΔCQP are isosceles, and the line segments BQ, QP and PC are all of equal length. As $|AM| \to 0$, P and Q approach points on the base, and the point P is seen to approach a position two-thirds of the way between B and C, as above.

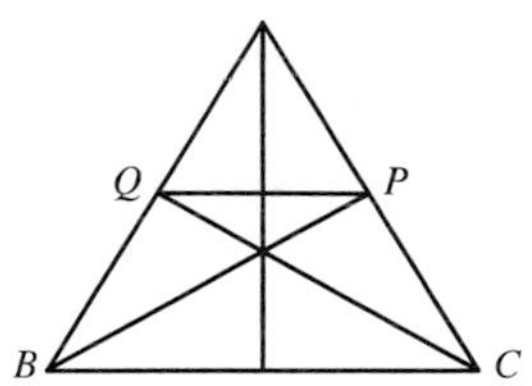

(b) The equation $y^2 = x(3x - 2)$ calculated in part (a) is the equation of the curve traced out by P. Now as $|AM| \to \infty$, $2\theta \to \frac{\pi}{2}$, $\theta \to \frac{\pi}{4}$, $x \to 1$, and since $\tan \theta = y/x$, $y \to 1$. Thus, P only traces out the part of the curve with $0 \leq y < 1$.

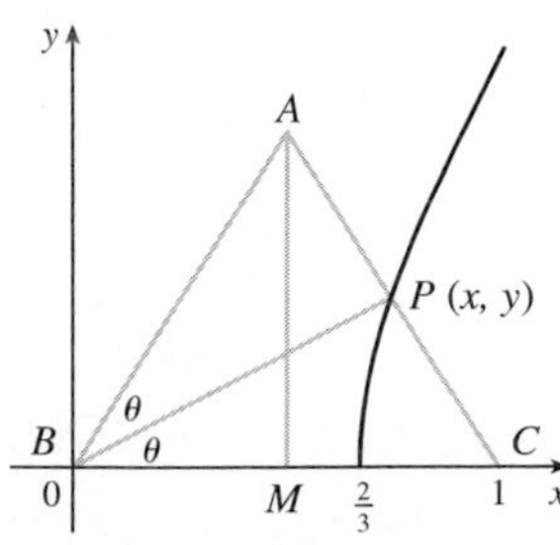

11. (a) Consider $G(x) = T(x + 180°) - T(x)$. Fix any number a. If $G(a) = 0$, we are done: Temperature at $a =$ Temperature at $a + 180°$. If $G(a) > 0$, then $G(a + 180°) = T(a + 360°) - T(a + 180°) = T(a) - T(a + 180°) = -G(a) < 0$. Also, G is continuous since temperature varies continuously. So, by the Intermediate Value Theorem, G has a zero on the interval $[a, a + 180°]$. If $G(a) < 0$, then a similar argument applies.

(b) Yes. The same argument applies.

(c) The same argument applies for quantities that vary continuously, such as barometric pressure. But one could argue that altitude above sea level is sometimes discontinuous, so the result might not always hold for that quantity.

12. $$g'(x) = \lim_{h\to 0} \frac{g(x+h) - g(x)}{h} = \lim_{h\to 0} \frac{(x+h)f(x+h) - xf(x)}{h} = \lim_{h\to 0} \left[\frac{xf(x+h) - xf(x)}{h} + \frac{hf(x+h)}{h}\right]$$
$$= x \lim_{h\to 0} \frac{f(x+h) - f(x)}{h} + \lim_{h\to 0} f(x+h) = xf'(x) + f(x)$$

because f is differentiable and therefore continuous.

13. (a) Put $x = 0$ and $y = 0$ in the equation: $f(0 + 0) = f(0) + f(0) + 0^2 \cdot 0 + 0 \cdot 0^2 \;\Rightarrow\; f(0) = 2f(0)$.

Subtracting $f(0)$ from each side of this equation gives $f(0) = 0$.

(b) $$f'(0) = \lim_{h\to 0} \frac{f(0+h) - f(0)}{h} = \lim_{h\to 0} \frac{\left[f(0) + f(h) + 0^2h + 0h^2\right] - f(0)}{h} = \lim_{h\to 0} \frac{f(h)}{h} = \lim_{x\to 0} \frac{f(x)}{x} = 1$$

(c) $$f'(x) = \lim_{h\to 0} \frac{f(x+h) - f(x)}{h} = \lim_{h\to 0} \frac{\left[f(x) + f(h) + x^2h + xh^2\right] - f(x)}{h} = \lim_{h\to 0} \frac{f(h) + x^2h + xh^2}{h}$$
$$= \lim_{h\to 0} \left[\frac{f(h)}{h} + x^2 + xh\right] = 1 + x^2$$

14. We are given that $|f(x)| \le x^2$ for all x. In particular, $|f(0)| \le 0$, but $|a| \ge 0$ for all a. The only conclusion is that $f(0) = 0$. Now $\left|\dfrac{f(x) - f(0)}{x - 0}\right| = \left|\dfrac{f(x)}{x}\right| = \dfrac{|f(x)|}{|x|} \le \dfrac{x^2}{|x|} = \dfrac{|x^2|}{|x|} = |x| \;\Rightarrow\; -|x| \le \dfrac{f(x) - f(0)}{x - 0} \le |x|$.

But $\lim_{x\to 0}(-|x|) = 0 = \lim_{x\to 0}|x|$, so by the Squeeze Theorem, $\lim_{x\to 0} \dfrac{f(x) - f(0)}{x - 0} = 0$. So by the definition of a derivative, f is differentiable at 0 and, furthermore, $f'(0) = 0$.

3 ☐ DIFFERENTIATION RULES

3.1 Derivatives of Polynomials and Exponential Functions

1. (a) e is the number such that $\lim\limits_{h\to 0}\dfrac{e^h-1}{h}=1$.

(b)

x	$\dfrac{2.7^x-1}{x}$
-0.001	0.9928
-0.0001	0.9932
0.001	0.9937
0.0001	0.9933

x	$\dfrac{2.8^x-1}{x}$
-0.001	1.0291
-0.0001	1.0296
0.001	1.0301
0.0001	1.0297

From the tables (to two decimal places),

$\lim\limits_{h\to 0}\dfrac{2.7^h-1}{h}=0.99$ and $\lim\limits_{h\to 0}\dfrac{2.8^h-1}{h}=1.03$.

Since $0.99<1<1.03$, $2.7<e<2.8$.

2. (a) The function value at $x=0$ is 1 and the slope at $x=0$ is 1.

(b) $f(x)=e^x$ is an exponential function and $g(x)=x^e$ is a power function. $\dfrac{d}{dx}(e^x)=e^x$ and $\dfrac{d}{dx}(x^e)=cx^{e-1}$.

(c) $f(x)=e^x$ grows more rapidly than $g(x)=x^e$ when x is large.

3. $f(x)=2^{40}$ is a constant function, so its derivative is 0, that is, $f'(x)=0$.

4. $f(x)=e^5$ is a constant function, so its derivative is 0, that is, $f'(x)=0$.

5. $f(x)=5.2x+2.3 \quad\Rightarrow\quad f'(x)=5.2(1)+0=5.2$

6. $g(x)=\frac{7}{4}x^2-3x+12 \quad\Rightarrow\quad g'(x)=\frac{7}{4}(2x)-3(1)+0=\frac{7}{2}x-3$

7. $f(t)=2t^3-3t^2-4t \quad\Rightarrow\quad f'(t)=2(3t^2)-3(2t)-4(1)=6t^2-6t-4$

8. $f(t)=1.4t^5-2.5t^2+6.7 \quad\Rightarrow\quad f'(t)=1.4(5t^4)-2.5(2t)+0=7t^4-5t$

9. $g(x)=x^2(1-2x)=x^2-2x^3 \quad\Rightarrow\quad g'(x)=2x-2(3x^2)=2x-6x^2$

10. $H(u)=(3u-1)(u+2)=3u^2+5u-2 \quad\Rightarrow\quad H'(u)=3(2u)+5(1)-0=6u+5$

11. $g(t)=2t^{-3/4} \quad\Rightarrow\quad g'(t)=2\left(-\frac{3}{4}t^{-7/4}\right)=-\frac{3}{2}t^{-7/4}$

12. $B(y)=cy^{-6} \quad\Rightarrow\quad B'(y)=c(-6y^{-7})=-6cy^{-7}$

13. $F(r)=\dfrac{5}{r^3}=5r^{-3} \quad\Rightarrow\quad F'(r)=5(-3r^{-4})=-15r^{-4}=-\dfrac{15}{r^4}$

14. $y=x^{5/3}-x^{2/3} \quad\Rightarrow\quad y'=\frac{5}{3}x^{2/3}-\frac{2}{3}x^{-1/3}$

15. $R(a) = (3a+1)^2 = 9a^2 + 6a + 1 \quad \Rightarrow \quad R'(a) = 9(2a) + 6(1) + 0 = 18a + 6$

16. $h(t) = \sqrt[4]{t} - 4e^t = t^{1/4} - 4e^t \quad \Rightarrow \quad h'(t) = \frac{1}{4}t^{-3/4} - 4(e^t) = \frac{1}{4}t^{-3/4} - 4e^t$

17. $S(p) = \sqrt{p} - p = p^{1/2} - p \quad \Rightarrow \quad S'(p) = \frac{1}{2}p^{-1/2} - 1$ or $\dfrac{1}{2\sqrt{p}} - 1$

18. $y = \sqrt[3]{x}\,(2+x) = 2x^{1/3} + x^{4/3} \quad \Rightarrow \quad y' = 2\left(\frac{1}{3}x^{-2/3}\right) + \frac{4}{3}x^{1/3} = \frac{2}{3}x^{-2/3} + \frac{4}{3}x^{1/3}$ or $\dfrac{2}{3\sqrt[3]{x^2}} + \dfrac{4}{3}\sqrt[3]{x}$

19. $y = 3e^x + \dfrac{4}{\sqrt[3]{x}} = 3e^x + 4x^{-1/3} \quad \Rightarrow \quad y' = 3(e^x) + 4(-\frac{1}{3})x^{-4/3} = 3e^x - \frac{4}{3}x^{-4/3}$

20. $S(R) = 4\pi R^2 \quad \Rightarrow \quad S'(R) = 4\pi(2R) = 8\pi R$

21. $h(u) = Au^3 + Bu^2 + Cu \quad \Rightarrow \quad h'(u) = A(3u^2) + B(2u) + C(1) = 3Au^2 + 2Bu + C$

22. $y = \dfrac{\sqrt{x} + x}{x^2} = \dfrac{\sqrt{x}}{x^2} + \dfrac{x}{x^2} = x^{1/2-2} + x^{1-2} = x^{-3/2} + x^{-1} \quad \Rightarrow \quad y' = -\frac{3}{2}x^{-5/2} + (-1x^{-2}) = -\frac{3}{2}x^{-5/2} - x^{-2}$

23. $y = \dfrac{x^2 + 4x + 3}{\sqrt{x}} = x^{3/2} + 4x^{1/2} + 3x^{-1/2} \quad \Rightarrow$

$y' = \frac{3}{2}x^{1/2} + 4(\frac{1}{2})x^{-1/2} + 3(-\frac{1}{2})x^{-3/2} = \frac{3}{2}\sqrt{x} + \dfrac{2}{\sqrt{x}} - \dfrac{3}{2x\sqrt{x}}$ [note that $x^{3/2} = x^{2/2} \cdot x^{1/2} = x\sqrt{x}$]

The last expression can be written as $\dfrac{3x^2}{2x\sqrt{x}} + \dfrac{4x}{2x\sqrt{x}} - \dfrac{3}{2x\sqrt{x}} = \dfrac{3x^2 + 4x - 3}{2x\sqrt{x}}$.

24. $G(t) = \sqrt{5t} + \dfrac{\sqrt{7}}{t} = \sqrt{5}\,t^{1/2} + \sqrt{7}\,t^{-1} \quad \Rightarrow \quad G'(t) = \sqrt{5}\left(\frac{1}{2}t^{-1/2}\right) + \sqrt{7}\left(-1t^{-2}\right) = \dfrac{\sqrt{5}}{2\sqrt{t}} - \dfrac{\sqrt{7}}{t^2}$

25. $j(x) = x^{2.4} + e^{2.4} \quad \Rightarrow \quad j'(x) = 2.4x^{1.4} + 0 = 2.4x^{1.4}$

26. $k(r) = e^r + r^e \quad \Rightarrow \quad k'(r) = e^r + er^{e-1}$

27. $G(q) = \left(1 + q^{-1}\right)^2 = 1 + 2q^{-1} + q^{-2} \quad \Rightarrow \quad G'(q) = 0 + 2(-1q^{-2}) + (-2q^{-3}) = -2q^{-2} - 2q^{-3}$

28. $F(z) = \dfrac{A + Bz + Cz^2}{z^2} = \dfrac{A}{z^2} + \dfrac{Bz}{z^2} + \dfrac{Cz^2}{z^2} = Az^{-2} + Bz^{-1} + C \quad \Rightarrow$

$F'(z) = A(-2z^{-3}) + B(-1z^{-2}) + 0 = -2Az^{-3} - Bz^{-2} = -\dfrac{2A}{z^3} - \dfrac{B}{z^2}$ or $-\dfrac{2A + Bz}{z^3}$

29. $f(v) = \dfrac{\sqrt[3]{v} - 2ve^v}{v} = \dfrac{\sqrt[3]{v}}{v} - \dfrac{2ve^v}{v} = v^{-2/3} - 2e^v \quad \Rightarrow \quad f'(v) = -\frac{2}{3}v^{-5/3} - 2e^v$

30. $D(t) = \dfrac{1 + 16t^2}{(4t)^3} = \dfrac{1 + 16t^2}{64t^3} = \frac{1}{64}t^{-3} + \frac{1}{4}t^{-1} \quad \Rightarrow$

$D'(t) = \frac{1}{64}(-3t^{-4}) + \frac{1}{4}(-1t^{-2}) = -\frac{3}{64}t^{-4} - \frac{1}{4}t^{-2}$ or $-\dfrac{3}{64t^4} - \dfrac{1}{4t^2}$

31. $z = \dfrac{A}{y^{10}} + Be^y = Ay^{-10} + Be^y \quad \Rightarrow \quad z' = -10Ay^{-11} + Be^y = -\dfrac{10A}{y^{11}} + Be^y$

32. $y = e^{x+1} + 1 = e^x e^1 + 1 = e \cdot e^x + 1 \quad \Rightarrow \quad y' = e \cdot e^x = e^{x+1}$

33. $y = 2x^3 - x^2 + 2 \;\Rightarrow\; y' = 6x^2 - 2x$. At $(1, 3)$, $y' = 6(1)^2 - 2(1) = 4$ and an equation of the tangent line is $y - 3 = 4(x - 1)$ or $y = 4x - 1$.

34. $y = 2e^x + x \;\Rightarrow\; y' = 2e^x + 1$. At $(0, 2)$, $y' = 2e^0 + 1 = 3$ and an equation of the tangent line is $y - 2 = 3(x - 0)$ or $y = 3x + 2$.

35. $y = x + \dfrac{2}{x} = x + 2x^{-1} \;\Rightarrow\; y' = 1 - 2x^{-2}$. At $(2, 3)$, $y' = 1 - 2(2)^{-2} = \frac{1}{2}$ and an equation of the tangent line is $y - 3 = \frac{1}{2}(x - 2)$ or $y = \frac{1}{2}x + 2$

36. $y = \sqrt[4]{x} - x = x^{1/4} - x \;\Rightarrow\; y' = \frac{1}{4}x^{-3/4} - 1 = \dfrac{1}{4\sqrt[4]{x^3}} - 1$. At $(1, 0)$, $y' = \frac{1}{4} - 1 = -\frac{3}{4}$ and an equation of the tangent line is $y - 0 = -\frac{3}{4}(x - 1)$ or $y = -\frac{3}{4}x + \frac{3}{4}$.

37. $y = x^4 + 2e^x \;\Rightarrow\; y' = 4x^3 + 2e^x$. At $(0, 2)$, $y' = 2$ and an equation of the tangent line is $y - 2 = 2(x - 0)$ or $y = 2x + 2$. The slope of the normal line is $-\frac{1}{2}$ (the negative reciprocal of 2) and an equation of the normal line is $y - 2 = -\frac{1}{2}(x - 0)$ or $y = -\frac{1}{2}x + 2$.

38. $y^2 = x^3 \;\Rightarrow\; y = x^{3/2}$ [since x and y are positive at $(1, 1)$] $\Rightarrow\; y' = \frac{3}{2}x^{1/2}$. At $(1, 1)$, $y' = \frac{3}{2}$ and an equation of the tangent line is $y - 1 = \frac{3}{2}(x - 1)$ or $y = \frac{3}{2}x - \frac{1}{2}$. The slope of the normal line is $-\frac{2}{3}$ (the negative reciprocal of $\frac{3}{2}$) and an equation of the normal line is $y - 1 = -\frac{2}{3}(x - 1)$ or $y = -\frac{2}{3}x + \frac{5}{3}$.

39. $y = 3x^2 - x^3 \;\Rightarrow\; y' = 6x - 3x^2$.

At $(1, 2)$, $y' = 6 - 3 = 3$, so an equation of the tangent line is $y - 2 = 3(x - 1)$ or $y = 3x - 1$.

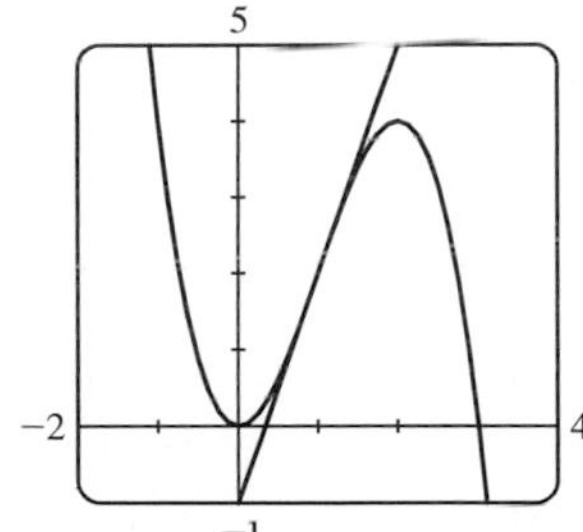

40. $y = x - \sqrt{x} \;\Rightarrow\; y' = 1 - \frac{1}{2}x^{-1/2} = 1 - \dfrac{1}{2\sqrt{x}}$.

At $(1, 0)$, $y' = \frac{1}{2}$, so an equation of the tangent line is $y - 0 = \frac{1}{2}(x - 1)$ or $y = \frac{1}{2}x - \frac{1}{2}$.

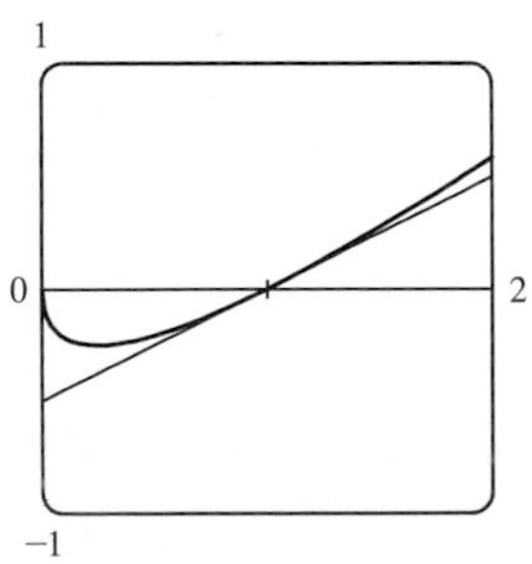

41. $f(x) = x^4 - 2x^3 + x^2 \;\Rightarrow\; f'(x) = 4x^3 - 6x^2 + 2x$

Note that $f'(x) = 0$ when f has a horizontal tangent, f' is positive when f is increasing, and f' is negative when f is decreasing.

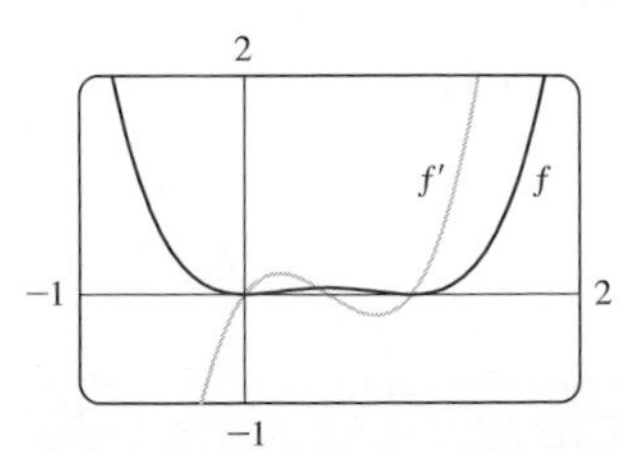

42. $f(x) = x^5 - 2x^3 + x - 1 \quad \Rightarrow \quad f'(x) = 5x^4 - 6x^2 + 1$

Note that $f'(x) = 0$ when f has a horizontal tangent, f' is positive when f is increasing, and f' is negative when f is decreasing.

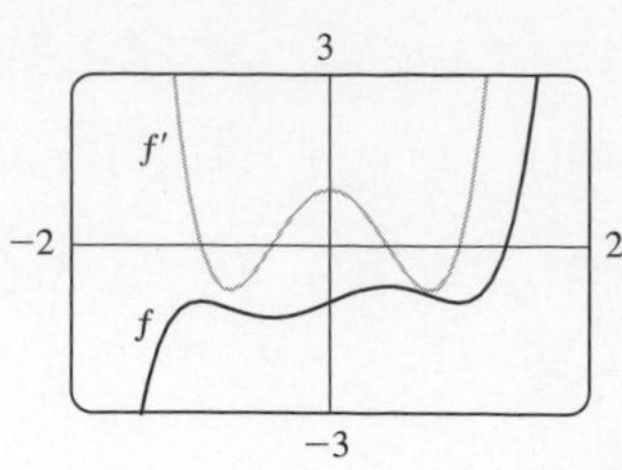

43. (a)

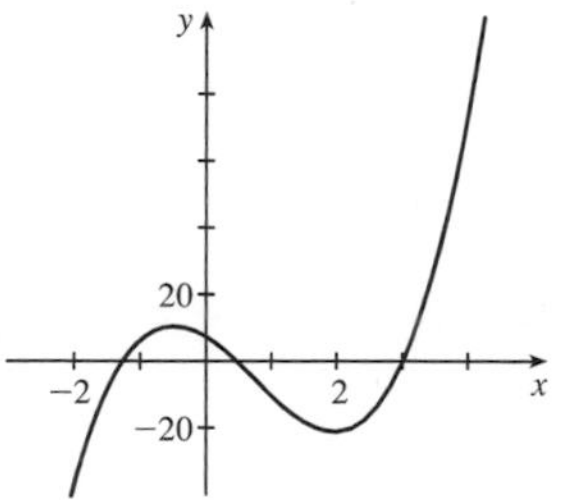

(b) From the graph in part (a), it appears that f' is zero at $x_1 \approx -1.25$, $x_2 \approx 0.5$, and $x_3 \approx 3$. The slopes are negative (so f' is negative) on $(-\infty, x_1)$ and (x_2, x_3). The slopes are positive (so f' is positive) on (x_1, x_2) and (x_3, ∞).

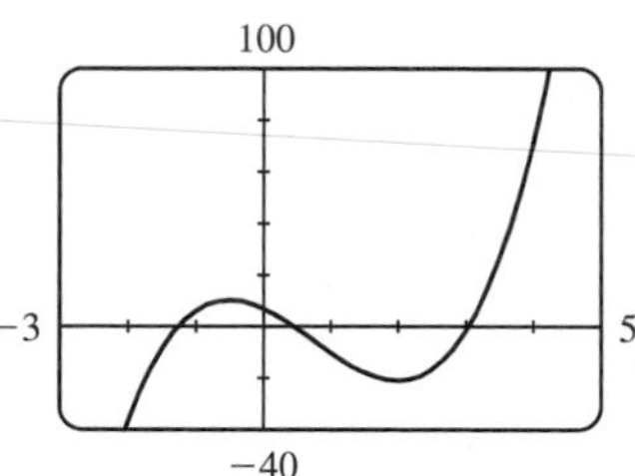

(c) $f(x) = x^4 - 3x^3 - 6x^2 + 7x + 30 \quad \Rightarrow$

$f'(x) = 4x^3 - 9x^2 - 12x + 7$

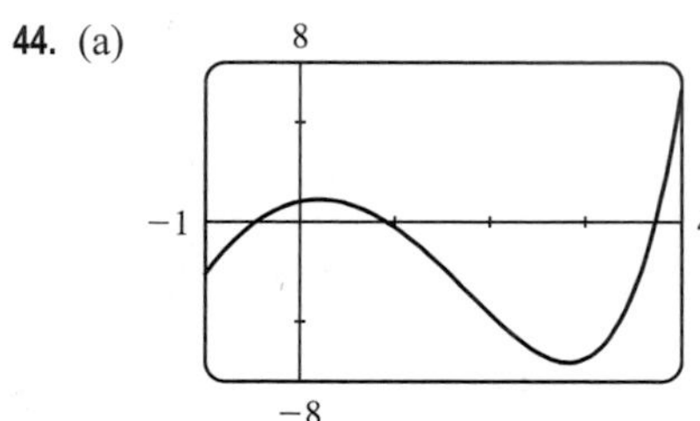

44. (a)

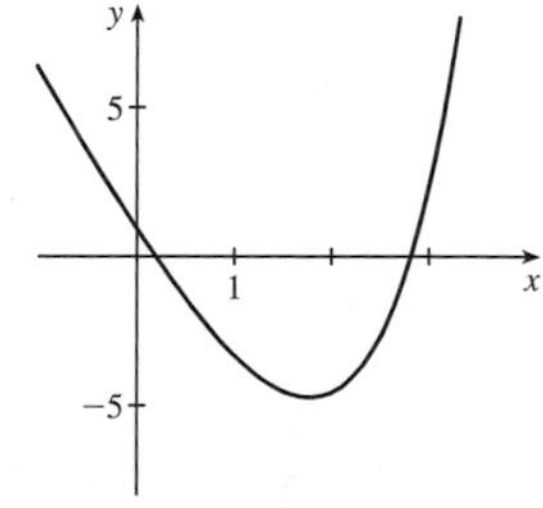

(b) From the graph in part (a), it appears that f' is zero at $x_1 \approx 0.2$ and $x_2 \approx 2.8$. The slopes are positive (so f' is positive) on $(-\infty, x_1)$ and (x_2, ∞). The slopes are negative (so f' is negative) on (x_1, x_2).

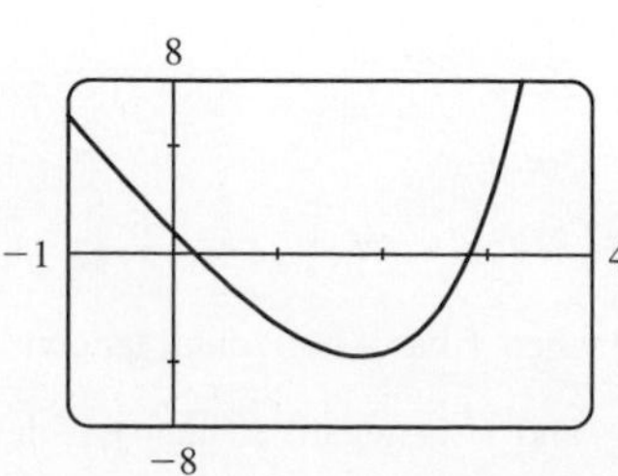

(c) $g(x) = e^x - 3x^2 \quad \Rightarrow \quad g'(x) = e^x - 6x$

45. $f(x) = 0.001x^5 - 0.02x^3 \quad \Rightarrow \quad f'(x) = 0.005x^4 - 0.06x^2 \quad \Rightarrow \quad f''(x) = 0.02x^3 - 0.12x$

46. $G(r) = \sqrt{r} + \sqrt[3]{r} \Rightarrow G'(r) = \frac{1}{2}r^{-1/2} + \frac{1}{3}r^{-2/3} \Rightarrow G''(r) = -\frac{1}{4}r^{-3/2} - \frac{2}{9}r^{-5/3}$

47. $f(x) = 2x - 5x^{3/4} \Rightarrow f'(x) = 2 - \frac{15}{4}x^{-1/4} \Rightarrow f''(x) = \frac{15}{16}x^{-5/4}$

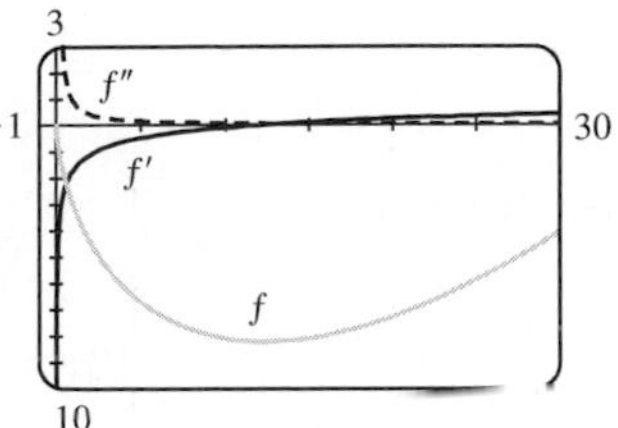

Note that f' is negative when f is decreasing and positive when f is increasing. f'' is always positive since f' is always increasing.

48. $f(x) = e^x - x^3 \Rightarrow f'(x) = e^x - 3x^2 \Rightarrow f''(x) = e^x - 6x$

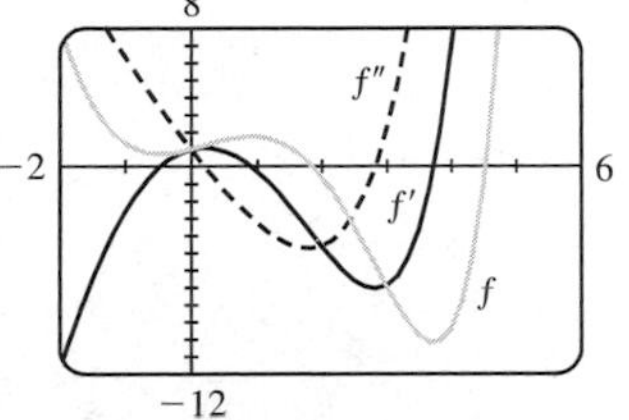

Note that $f'(x) = 0$ when f has a horizontal tangent and that $f''(x) = 0$ when f' has a horizontal tangent.

49. (a) $s = t^3 - 3t \Rightarrow v(t) = s'(t) = 3t^2 - 3 \Rightarrow a(t) = v'(t) = 6t$

(b) $a(2) = 6(2) = 12 \text{ m/s}^2$

(c) $v(t) = 3t^2 - 3 = 0$ when $t^2 = 1$, that is, $t = 1$ $[t \geq 0]$ and $a(1) = 6 \text{ m/s}^2$.

50. (a) $s = t^4 - 2t^3 + t^2 - t \Rightarrow$

$v(t) = s'(t) = 4t^3 - 6t^2 + 2t - 1 \Rightarrow$

$a(t) = v'(t) = 12t^2 - 12t + 2$

(b) $a(1) = 12(1)^2 - 12(1) + 2 = 2 \text{ m/s}^2$

(c)

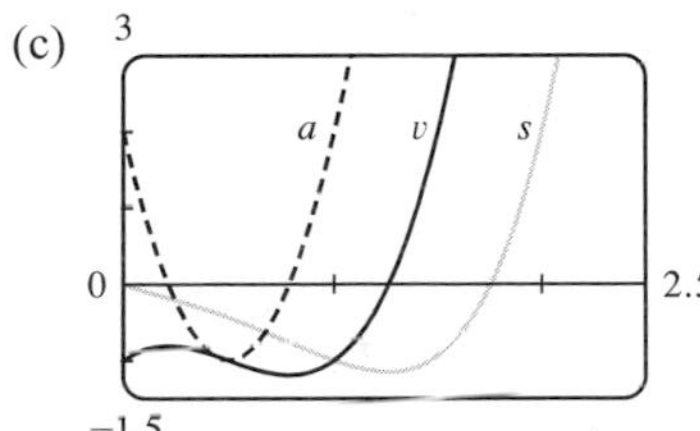

51. $L = 0.0155A^3 - 0.372A^2 + 3.95A + 1.21 \Rightarrow \dfrac{dL}{dA} = 0.0465A^2 - 0.744A + 3.95$, so

$\left.\dfrac{dL}{dA}\right|_{A=12} = 0.0465(12)^2 - 0.744(12) + 3.95 = 1.718$. The derivative is the instantaneous rate of change of the length of an Alaskan rockfish with respect to its age when its age is 12 years.

52. $S(A) = 0.882A^{0.842} \Rightarrow S'(A) = 0.882(0.842A^{-0.158}) = 0.742644A^{-0.158}$, so

$S'(100) = 0.742644(100)^{-0.158} \approx 0.36$. The derivative is the instantaneous rate of change of the number of tree species with respect to area. Its units are number of species per square meter.

53. (a) $P = \dfrac{k}{V}$ and $P = 50$ when $V = 0.106$, so $k = PV = 50(0.106) = 5.3$. Thus, $P = \dfrac{5.3}{V}$ and $V = \dfrac{5.3}{P}$.

(b) $V = 5.3P^{-1} \Rightarrow \dfrac{dV}{dP} = 5.3(-1P^{-2}) = -\dfrac{5.3}{P^2}$. When $P = 50$, $\dfrac{dV}{dP} = -\dfrac{5.3}{50^2} = -0.00212$. The derivative is the instantaneous rate of change of the volume with respect to the pressure at 25 °C. Its units are m^3/kPa.

54. (a) $L = aP^2 + bP + c$, where $a \approx -0.275428$, $b \approx 19.74853$, and $c \approx -273.55234$.

(b) $\dfrac{dL}{dP} = 2aP + b$. When $P = 30$, $\dfrac{dL}{dP} \approx 3.2$, and when $P = 40$, $\dfrac{dL}{dP} \approx -2.3$. The derivative is the instantaneous rate of change of tire life with respect to pressure. Its units are (thousands of miles)/(lb/in^2). When $\dfrac{dL}{dP}$ is positive, tire life is increasing, and when $\dfrac{dL}{dP} < 0$, tire life is decreasing.

55. The curve $y = 2x^3 + 3x^2 - 12x + 1$ has a horizontal tangent when $y' = 6x^2 + 6x - 12 = 0 \iff 6(x^2 + x - 2) = 0 \iff 6(x + 2)(x - 1) = 0 \iff x = -2$ or $x = 1$. The points on the curve are $(-2, 21)$ and $(1, -6)$.

56. $f(x) = e^x - 2x \Rightarrow f'(x) = e^x - 2$. $f'(x) = 0 \Rightarrow e^x = 2 \Rightarrow x = \ln 2$, so f has a horizontal tangent when $x = \ln 2$.

57. $y = 2e^x + 3x + 5x^3 \Rightarrow y' = 2e^x + 3 + 15x^2$. Since $2e^x > 0$ and $15x^2 \geq 0$, we must have $y' > 0 + 3 + 0 = 3$, so no tangent line can have slope 2.

58. $y = x^4 + 1 \Rightarrow y' = 4x^3$. The slope of the line $32x - y = 15$ (or $y = 32x - 15$) is 32, so the slope of any line parallel to it is also 32. Thus, $y' = 32 \iff 4x^3 = 32 \iff x^3 = 8 \iff x = 2$, which is the x-coordinate of the point on the curve at which the slope is 32. The y-coordinate is $2^4 + 1 = 17$, so an equation of the tangent line is $y - 17 = 32(x - 2)$ or $y = 32x - 47$.

59. The slope of the line $3x - y = 15$ (or $y = 3x - 15$) is 3, so the slope of both tangent lines to the curve is 3.
$y = x^3 - 3x^2 + 3x - 3 \Rightarrow y' = 3x^2 - 6x + 3 = 3(x^2 - 2x + 1) = 3(x - 1)^2$. Thus, $3(x - 1)^2 = 3 \Rightarrow (x - 1)^2 = 1 \Rightarrow x - 1 = \pm 1 \Rightarrow x = 0$ or 2, which are the x-coordinates at which the tangent lines have slope 3. The points on the curve are $(0, -3)$ and $(2, -1)$, so the tangent line equations are $y - (-3) = 3(x - 0)$ or $y = 3x - 3$ and $y - (-1) = 3(x - 2)$ or $y = 3x - 7$.

60. The slope of $y = 1 + 2e^x - 3x$ is given by $m = y' = 2e^x - 3$.
The slope of $3x - y = 5 \iff y = 3x - 5$ is 3.
$m = 3 \Rightarrow 2e^x - 3 = 3 \Rightarrow e^x = 3 \Rightarrow x = \ln 3$.
This occurs at the point $(\ln 3, 7 - 3\ln 3) \approx (1.1, 3.7)$.

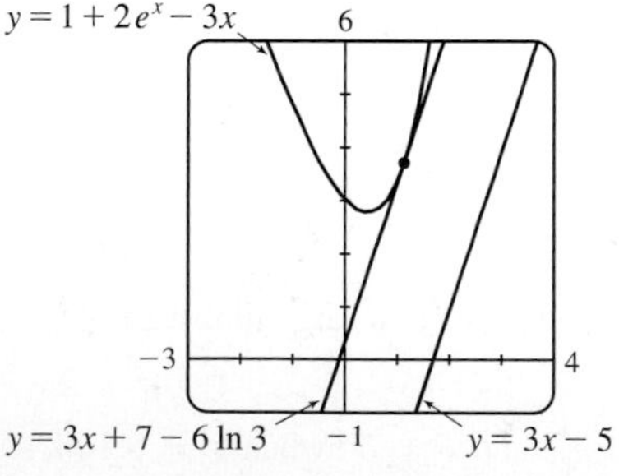

61. The slope of $y = \sqrt{x}$ is given by $y = \frac{1}{2}x^{-1/2} = \dfrac{1}{2\sqrt{x}}$. The slope of $2x + y = 1$ (or $y = -2x + 1$) is -2, so the desired normal line must have slope -2, and hence, the tangent line to the curve must have slope $\frac{1}{2}$. This occurs if $\dfrac{1}{2\sqrt{x}} = \dfrac{1}{2} \Rightarrow \sqrt{x} = 1 \Rightarrow x = 1$. When $x = 1$, $y = \sqrt{1} = 1$, and an equation of the normal line is $y - 1 = -2(x - 1)$ or $y = -2x + 3$.

62. $y = f(x) = x^2 - 1 \;\Rightarrow\; f'(x) = 2x$. So $f'(-1) = -2$, and the slope of the normal line is $\frac{1}{2}$. The equation of the normal line at $(-1, 0)$ is $y - 0 = \frac{1}{2}[x - (-1)]$ or $y = \frac{1}{2}x + \frac{1}{2}$. Substituting this into the equation of the parabola, we obtain $\frac{1}{2}x + \frac{1}{2} = x^2 - 1 \;\Leftrightarrow\; x + 1 = 2x^2 - 2 \;\Leftrightarrow$ $2x^2 - x - 3 = 0 \;\Leftrightarrow\; (2x - 3)(x + 1) = 0 \;\Leftrightarrow\; x = \frac{3}{2}$ or -1. Substituting $\frac{3}{2}$ into the equation of the normal line gives us $y = \frac{5}{4}$. Thus, the second point of intersection is $\left(\frac{3}{2}, \frac{5}{4}\right)$, as shown in the sketch.

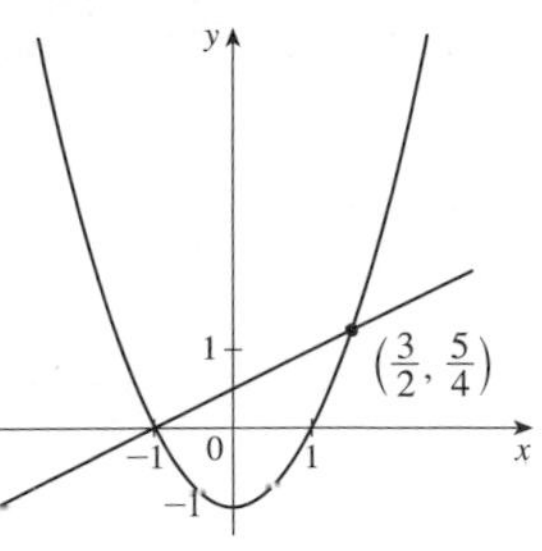

63.

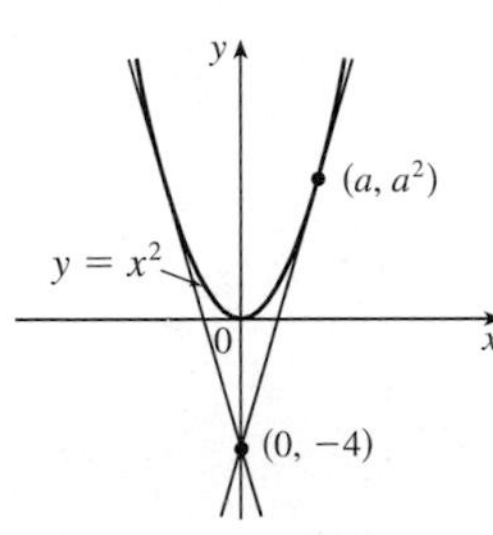

Let (a, a^2) be a point on the parabola at which the tangent line passes through the point $(0, -4)$. The tangent line has slope $2a$ and equation $y - (-4) = 2a(x - 0) \;\Leftrightarrow\; y = 2ax - 4$. Since (a, a^2) also lies on the line, $a^2 = 2a(a) - 4$, or $a^2 = 4$. So $a = \pm 2$ and the points are $(2, 4)$ and $(-2, 4)$.

64. (a) If $y = x^2 + x$, then $y' = 2x + 1$. If the point at which a tangent meets the parabola is $(a, a^2 + a)$, then the slope of the tangent is $2a + 1$. But since it passes through $(2, -3)$, the slope must also be $\dfrac{\Delta y}{\Delta x} = \dfrac{a^2 + a + 3}{a - 2}$.

Therefore, $2a + 1 = \dfrac{a^2 + a + 3}{a - 2}$. Solving this equation for a we get $a^2 + a + 3 = 2a^2 - 3a - 2 \;\Leftrightarrow$ $a^2 - 4a - 5 = (a - 5)(a + 1) = 0 \;\Leftrightarrow\; a = 5$ or -1. If $a = -1$, the point is $(-1, 0)$ and the slope is -1, so the equation is $y - 0 = (-1)(x + 1)$ or $y = -x - 1$. If $a = 5$, the point is $(5, 30)$ and the slope is 11, so the equation is $y - 30 = 11(x - 5)$ or $y = 11x - 25$.

(b) As in part (a), but using the point $(2, 7)$, we get the equation

$$2a + 1 = \frac{a^2 + a - 7}{a - 2} \;\Rightarrow\; 2a^2 - 3a - 2 = a^2 + a - 7 \;\Leftrightarrow\; a^2 - 4a + 5 = 0.$$

The last equation has no real solution (discriminant $= -16 < 0$), so there is no line through the point $(2, 7)$ that is tangent to the parabola. The diagram shows that the point $(2, 7)$ is "inside" the parabola, but tangent lines to the parabola do not pass through points inside the parabola.

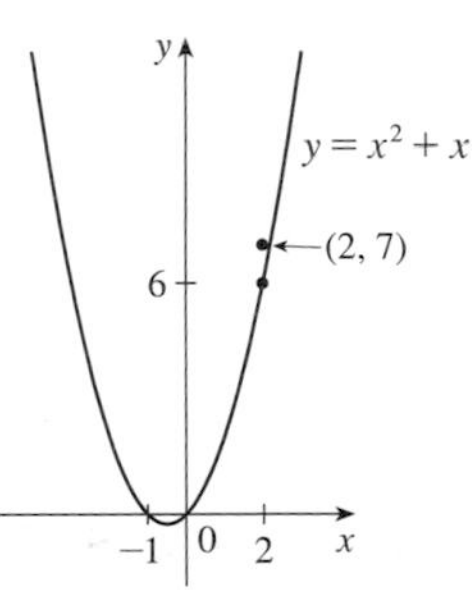

65. $f'(x) = \displaystyle\lim_{h \to 0} \frac{f(x + h) - f(x)}{h} = \lim_{h \to 0} \frac{\dfrac{1}{x + h} - \dfrac{1}{x}}{h} = \lim_{h \to 0} \frac{x - (x + h)}{hx(x + h)} = \lim_{h \to 0} \frac{-h}{hx(x + h)} = \lim_{h \to 0} \frac{-1}{x(x + h)} = -\frac{1}{x^2}$

66. (a) $f(x) = x^n \;\Rightarrow\; f'(x) = nx^{n-1} \;\Rightarrow\; f''(x) = n(n - 1)x^{n-2} \;\Rightarrow\; \cdots \;\Rightarrow$

$f^{(n)}(x) = n(n - 1)(n - 2) \cdots 2 \cdot 1x^{n-n} = n!$

(b) $f(x) = x^{-1} \;\Rightarrow\; f'(x) = (-1)x^{-2} \;\Rightarrow\; f''(x) = (-1)(-2)x^{-3} \;\Rightarrow\; \cdots \;\Rightarrow$

$f^{(n)}(x) = (-1)(-2)(-3) \cdots (-n)x^{-(n+1)} = (-1)^n n! x^{-(n+1)}$ or $\dfrac{(-1)^n\, n!}{x^{n+1}}$

67. Let $P(x) = ax^2 + bx + c$. Then $P'(x) = 2ax + b$ and $P''(x) = 2a$. $P''(2) = 2 \;\Rightarrow\; 2a = 2 \;\Rightarrow\; a = 1$.

$P'(2) = 3 \;\Rightarrow\; 2(1)(2) + b = 3 \;\Rightarrow\; 4 + b = 3 \;\Rightarrow\; b = -1$.

$P(2) = 5 \;\Rightarrow\; 1(2)^2 + (-1)(2) + c = 5 \;\Rightarrow\; 2 + c = 5 \;\Rightarrow\; c = 3$. So $P(x) = x^2 - x + 3$.

68. $y = Ax^2 + Bx + C \;\Rightarrow\; y' = 2Ax + B \;\Rightarrow\; y'' = 2A$. We substitute these expressions into the equation $y'' + y' - 2y = x^2$ to get

$$(2A) + (2Ax + B) - 2(Ax^2 + Bx + C) = x^2$$
$$2A + 2Ax + B - 2Ax^2 - 2Bx - 2C = x^2$$
$$(-2A)x^2 + (2A - 2B)x + (2A + B - 2C) = (1)x^2 + (0)x + (0)$$

The coefficients of x^2 on each side must be equal, so $-2A = 1 \;\Rightarrow\; A = -\frac{1}{2}$. Similarly, $2A - 2B = 0 \;\Rightarrow\; A = B = -\frac{1}{2}$ and $2A + B - 2C = 0 \;\Rightarrow\; -1 - \frac{1}{2} - 2C = 0 \;\Rightarrow\; C = -\frac{3}{4}$.

69. $y = f(x) = ax^3 + bx^2 + cx + d \;\Rightarrow\; f'(x) = 3ax^2 + 2bx + c$. The point $(-2, 6)$ is on f, so $f(-2) = 6 \;\Rightarrow\; -8a + 4b - 2c + d = 6$ **(1)**. The point $(2, 0)$ is on f, so $f(2) = 0 \;\Rightarrow\; 8a + 4b + 2c + d = 0$ **(2)**. Since there are horizontal tangents at $(-2, 6)$ and $(2, 0)$, $f'(\pm 2) = 0$. $f'(-2) = 0 \;\Rightarrow\; 12a - 4b + c = 0$ **(3)** and $f'(2) = 0 \;\Rightarrow\; 12a + 4b + c = 0$ **(4)**. Subtracting equation **(3)** from **(4)** gives $8b = 0 \;\Rightarrow\; b = 0$. Adding **(1)** and **(2)** gives $8b + 2d = 6$, so $d = 3$ since $b = 0$. From **(3)** we have $c = -12a$, so **(2)** becomes $8a + 4(0) + 2(-12a) + 3 = 0 \;\Rightarrow\; 3 = 16a \;\Rightarrow\; a = \frac{3}{16}$. Now $c = -12a = -12\left(\frac{3}{16}\right) = -\frac{9}{4}$ and the desired cubic function is $y = \frac{3}{16}x^3 - \frac{9}{4}x + 3$.

70. $y = ax^2 + bx + c \;\Rightarrow\; y'(x) = 2ax + b$. The parabola has slope 4 at $x = 1$ and slope -8 at $x = -1$, so $y'(1) = 4 \;\Rightarrow\; 2a + b = 4$ **(1)** and $y'(-1) = -8 \;\Rightarrow\; -2a + b = -8$ **(2)**. Adding **(1)** and **(2)** gives us $2b = -4 \;\Leftrightarrow\; b = -2$. From **(1)**, $2a - 2 = 4 \;\Leftrightarrow\; a = 3$. Thus, the equation of the parabola is $y = 3x^2 - 2x + c$. Since it passes through the point $(2, 15)$, we have $15 = 3(2)^2 - 2(2) + c \;\Rightarrow\; c = 7$, so the equation is $y = 3x^2 - 2x + 7$.

71. $f(x) = \begin{cases} x^2 + 1 & \text{if } x < 1 \\ x + 1 & \text{if } x \geq 1 \end{cases}$

Calculate the left- and right-hand derivatives as defined in Exercise 2.8.64:

$$f'_-(1) = \lim_{h \to 0^-} \frac{f(1+h) - f(1)}{h} = \lim_{h \to 0^-} \frac{[(1+h)^2 + 1] - (1+1)}{h} = \lim_{h \to 0^-} \frac{h^2 + 2h}{h} = \lim_{h \to 0^-} (h + 2) = 2 \text{ and}$$

$$f'_+(1) = \lim_{h \to 0^+} \frac{f(1+h) - f(1)}{h} = \lim_{h \to 0^+} \frac{[(1+h) + 1] - (1+1)}{h} = \lim_{h \to 0^+} \frac{h}{h} = \lim_{h \to 0^+} 1 = 1.$$

Since the left and right limits are different, $\displaystyle\lim_{h \to 0} \frac{f(1+h) - f(1)}{h}$ does not exist, that is, $f'(1)$ does not exist. Therefore, f is not differentiable at 1.

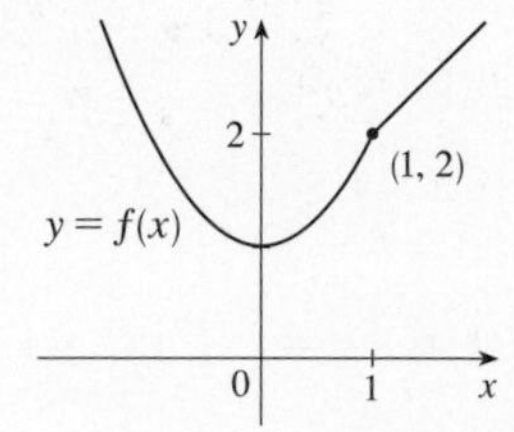

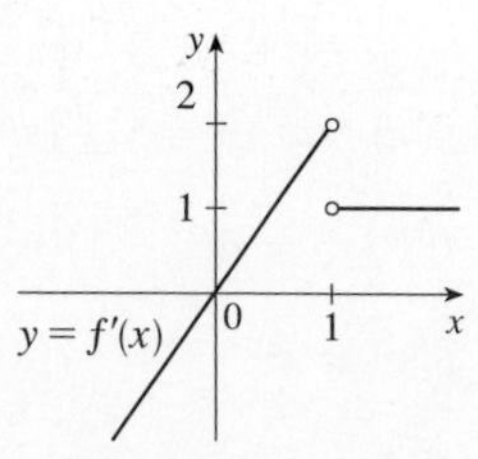

72. $g(x) = \begin{cases} 2x & \text{if } x \le 0 \\ 2x - x^2 & \text{if } 0 < x < 2 \\ 2 - x & \text{if } x \ge 2 \end{cases}$

Investigate the left- and right-hand derivatives at $x = 0$ and $x = 2$:

$$g'_-(0) = \lim_{h \to 0^-} \frac{g(0+h) - g(0)}{h} = \lim_{h \to 0^-} \frac{2h - 2(0)}{h} = 2 \text{ and}$$

$$g'_+(0) = \lim_{h \to 0^+} \frac{g(0+h) - g(0)}{h} = \lim_{h \to 0^+} \frac{(2h - h^2) - 2(0)}{h} = \lim_{h \to 0^+} (2 - h) = 2, \text{ so } g \text{ is differentiable at } x = 0.$$

$$g'_-(2) = \lim_{h \to 0^-} \frac{g(2+h) - g(2)}{h} = \lim_{h \to 0^-} \frac{2(2+h) - (2+h)^2 - (2-2)}{h} = \lim_{h \to 0^-} \frac{-2h - h^2}{h} = \lim_{h \to 0^-} (-2 - h) = -2$$

and

$$g'_+(2) = \lim_{h \to 0^+} \frac{g(2+h) - g(2)}{h} = \lim_{h \to 0^+} \frac{[2 - (2+h)] - (2-2)}{h} = \lim_{h \to 0^+} \frac{-h}{h} = \lim_{h \to 0^+} (-1) = -1,$$

so g is not differentiable at $x = 2$. Thus, a formula for g' is

$$g'(x) = \begin{cases} 2 & \text{if } x \le 0 \\ 2 - 2x & \text{if } 0 < x < 2 \\ -1 & \text{if } x > 2 \end{cases}$$

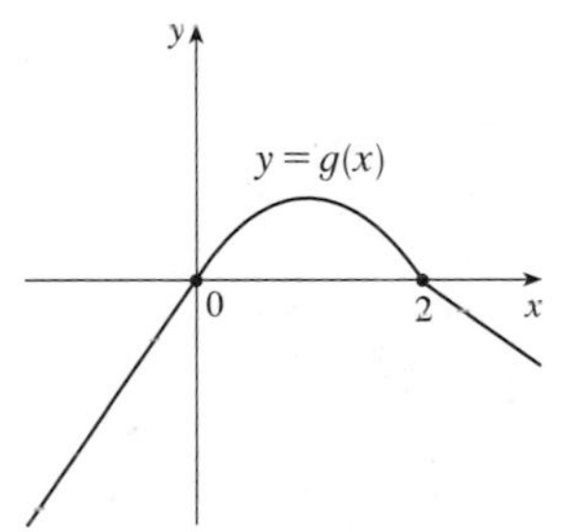

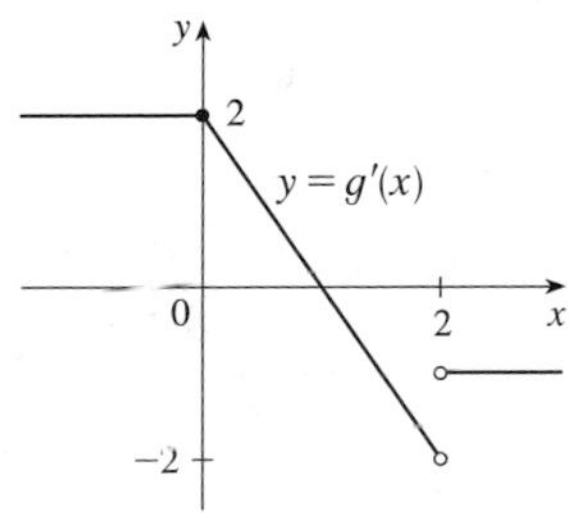

73. (a) Note that $x^2 - 9 < 0$ for $x^2 < 9 \iff |x| < 3 \iff -3 < x < 3$. So

$$f(x) = \begin{cases} x^2 - 9 & \text{if } x \le -3 \\ -x^2 + 9 & \text{if } -3 < x < 3 \\ x^2 - 9 & \text{if } x \ge 3 \end{cases} \Rightarrow f'(x) = \begin{cases} 2x & \text{if } x < -3 \\ -2x & \text{if } -3 < x < 3 \\ 2x & \text{if } x > 3 \end{cases} = \begin{cases} 2x & \text{if } |x| > 3 \\ -2x & \text{if } |x| < 3 \end{cases}$$

To show that $f'(3)$ does not exist we investigate $\lim\limits_{h \to 0} \dfrac{f(3+h) - f(3)}{h}$ by computing the left- and right-hand derivatives defined in Exercise 2.8.64.

$$f'_-(3) = \lim_{h \to 0^-} \frac{f(3+h) - f(3)}{h} = \lim_{h \to 0^-} \frac{[-(3+h)^2 + 9] - 0}{h} = \lim_{h \to 0^-} (-6 - h) = -6 \quad \text{and}$$

$$f'_+(3) = \lim_{h \to 0^+} \frac{f(3+h) - f(3)}{h} = \lim_{h \to 0^+} \frac{[(3+h)^2 - 9] - 0}{h} = \lim_{h \to 0^+} \frac{6h + h^2}{h} = \lim_{h \to 0^+} (6 + h) = 6.$$

Since the left and right limits are different, $\lim\limits_{h \to 0} \dfrac{f(3+h) - f(3)}{h}$ does not exist, that is, $f'(3)$ does not exist. Similarly, $f'(-3)$ does not exist. Therefore, f is not differentiable at 3 or at -3.

(b)

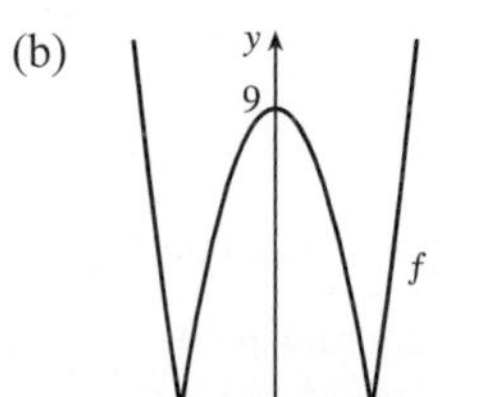

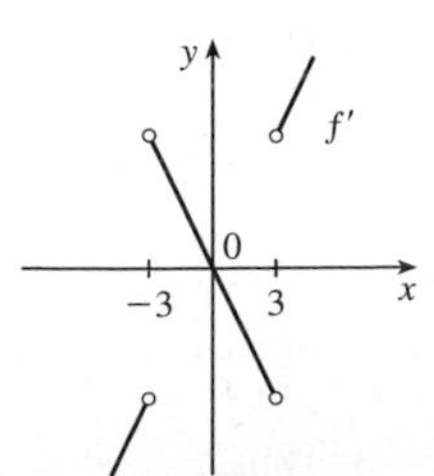

74. If $x \geq 1$, then $h(x) = |x-1| + |x+2| = x - 1 + x + 2 = 2x + 1$.

If $-2 < x < 1$, then $h(x) = -(x-1) + x + 2 = 3$.

If $x \leq -2$, then $h(x) = -(x-1) - (x+2) = -2x - 1$. Therefore,

$$h(x) = \begin{cases} -2x - 1 & \text{if } x \leq -2 \\ 3 & \text{if } -2 < x < 1 \\ 2x + 1 & \text{if } x \geq 1 \end{cases} \quad \Rightarrow \quad h'(x) = \begin{cases} -2 & \text{if } x < -2 \\ 0 & \text{if } -2 < x < 1 \\ 2 & \text{if } x > 1 \end{cases}$$

To see that $h'(1) = \lim\limits_{x \to 1} \dfrac{h(x) - h(1)}{x - 1}$ does not exist,

observe that $\lim\limits_{x \to 1^-} \dfrac{h(x) - h(1)}{x - 1} = \lim\limits_{x \to 1^-} \dfrac{3 - 3}{3 - 1} = 0$ but

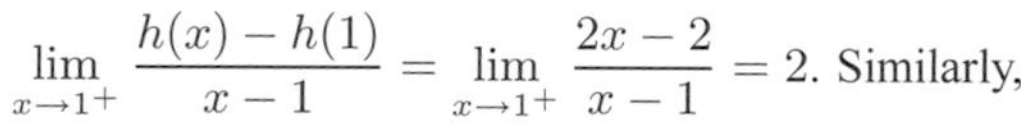

$\lim\limits_{x \to 1^+} \dfrac{h(x) - h(1)}{x - 1} = \lim\limits_{x \to 1^+} \dfrac{2x - 2}{x - 1} = 2$. Similarly,

$h'(-2)$ does not exist.

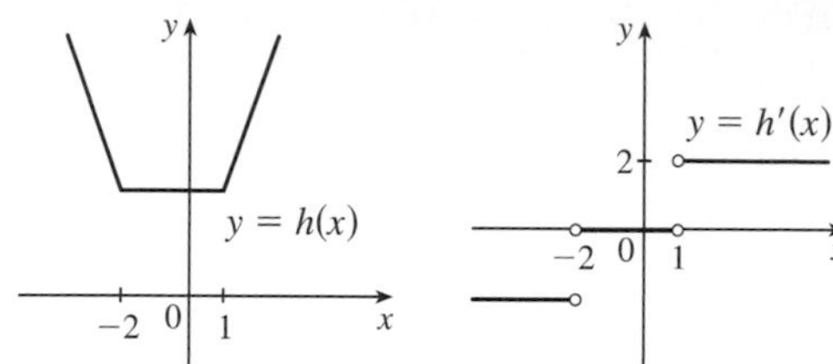

75. Substituting $x = 1$ and $y = 1$ into $y = ax^2 + bx$ gives us $a + b = 1$ **(1)**. The slope of the tangent line $y = 3x - 2$ is 3 and the slope of the tangent to the parabola at (x, y) is $y' = 2ax + b$. At $x = 1$, $y' = 3 \;\Rightarrow\; 3 = 2a + b$ **(2)**. Subtracting **(1)** from **(2)** gives us $2 = a$ and it follows that $b = -1$. The parabola has equation $y = 2x^2 - x$.

76. $y = x^4 + ax^3 + bx^2 + cx + d \;\Rightarrow\; y(0) = d$. Since the tangent line $y = 2x + 1$ is equal to 1 at $x = 0$, we must have $d = 1$. $y' = 4x^3 + 3ax^2 + 2bx + c \;\Rightarrow\; y'(0) = c$. Since the slope of the tangent line $y = 2x + 1$ at $x = 0$ is 2, we must have $c = 2$. Now $y(1) = 1 + a + b + c + d = a + b + 4$ and the tangent line $y = 2 - 3x$ at $x = 1$ has y-coordinate -1, so $a + b + 4 = -1$ or $a + b = -5$ **(1)**. Also, $y'(1) = 4 + 3a + 2b + c = 3a + 2b + 6$ and the slope of the tangent line $y = 2 - 3x$ at $x = 1$ is -3, so $3a + 2b + 6 = -3$ or $3a + 2b = -9$ **(2)**. Adding -2 times **(1)** to **(2)** gives us $a = 1$ and hence, $b = -6$. The curve has equation $y = x^4 + x^3 - 6x^2 + 2x + 1$.

77. $y = f(x) = ax^2 \;\Rightarrow\; f'(x) = 2ax$. So the slope of the tangent to the parabola at $x = 2$ is $m = 2a(2) = 4a$. The slope of the given line, $2x + y = b \;\Leftrightarrow\; y = -2x + b$, is seen to be -2, so we must have $4a = -2 \;\Leftrightarrow\; a = -\frac{1}{2}$. So when $x = 2$, the point in question has y-coordinate $-\frac{1}{2} \cdot 2^2 = -2$. Now we simply require that the given line, whose equation is $2x + y = b$, pass through the point $(2, -2)$: $2(2) + (-2) = b \;\Leftrightarrow\; b = 2$. So we must have $a = -\frac{1}{2}$ and $b = 2$.

78. The slope of the curve $y = c\sqrt{x}$ is $y' = \dfrac{c}{2\sqrt{x}}$ and the slope of the tangent line $y = \frac{3}{2}x + 6$ is $\frac{3}{2}$. These must be equal at the point of tangency $\left(a, c\sqrt{a}\right)$, so $\dfrac{c}{2\sqrt{a}} = \dfrac{3}{2} \;\Rightarrow\; c = 3\sqrt{a}$. The y-coordinates must be equal at $x = a$, so

$c\sqrt{a} = \frac{3}{2}a + 6 \;\Rightarrow\; \left(3\sqrt{a}\right)\sqrt{a} = \frac{3}{2}a + 6 \;\Rightarrow\; 3a = \frac{3}{2}a + 6 \;\Rightarrow\; \frac{3}{2}a = 6 \;\Rightarrow\; a = 4$. Since $c = 3\sqrt{a}$, we have $c = 3\sqrt{4} = 6$.

79. The line $y = 2x + 3$ has slope 2. The parabola $y = cx^2 \;\Rightarrow\; y' = 2cx$ has slope $2ca$ at $x = a$. Equating slopes gives us $2ca = 2$, or $ca = 1$. Equating y-coordinates at $x = a$ gives us $ca^2 = 2a + 3 \;\Leftrightarrow\; (ca)a = 2a + 3 \;\Leftrightarrow\; 1a = 2a + 3 \;\Leftrightarrow\; a = -3$. Thus, $c = \dfrac{1}{a} = -\dfrac{1}{3}$.

80. $f(x) = ax^2 + bx + c \Rightarrow f'(x) = 2ax + b$. The slope of the tangent line at $x = p$ is $2ap + b$, the slope of the tangent line at $x = q$ is $2aq + b$, and the average of those slopes is $\dfrac{(2ap + b) + (2aq + b)}{2} = ap + aq + b$. The midpoint of the interval $[p, q]$ is $\dfrac{p + q}{2}$ and the slope of the tangent line at the midpoint is $2a\left(\dfrac{p + q}{2}\right) + b = a(p + q) + b$. This is equal to $ap + aq + b$, as required.

81. f is clearly differentiable for $x < 2$ and for $x > 2$. For $x < 2$, $f'(x) = 2x$, so $f'_{-}(2) = 4$. For $x > 2$, $f'(x) = m$, so $f'_{+}(2) = m$. For f to be differentiable at $x = 2$, we need $4 = f'_{-}(2) = f'_{+}(2) = m$. So $f(x) = 4x + b$. We must also have continuity at $x = 2$, so $4 = f(2) = \lim\limits_{x \to 2^+} f(x) = \lim\limits_{x \to 2^+} (4x + b) = 8 + b$. Hence, $b = -4$.

82. (a) $xy = c \Rightarrow y = \dfrac{c}{x}$. Let $P = \left(a, \dfrac{c}{a}\right)$. The slope of the tangent line at $x = a$ is $y'(a) = -\dfrac{c}{a^2}$. Its equation is $y - \dfrac{c}{a} = -\dfrac{c}{a^2}(x - a)$ or $y = -\dfrac{c}{a^2}x + \dfrac{2c}{a}$, so its y-intercept is $\dfrac{2c}{a}$. Setting $y = 0$ gives $x = 2a$, so the x-intercept is $2a$. The midpoint of the line segment joining $\left(0, \dfrac{2c}{a}\right)$ and $(2a, 0)$ is $\left(a, \dfrac{c}{a}\right) = P$.

(b) We know the x- and y-intercepts of the tangent line from part (a), so the area of the triangle bounded by the axes and the tangent is $\frac{1}{2}(\text{base})(\text{height}) = \frac{1}{2}xy = \frac{1}{2}(2a)(2c/a) = 2c$, a constant.

83. *Solution 1:* Let $f(x) = x^{1000}$. Then, by the definition of a derivative, $f'(1) = \lim\limits_{x \to 1} \dfrac{f(x) - f(1)}{x - 1} = \lim\limits_{x \to 1} \dfrac{x^{1000} - 1}{x - 1}$. But this is just the limit we want to find, and we know (from the Power Rule) that $f'(x) = 1000x^{999}$, so $f'(1) = 1000(1)^{999} = 1000$. So $\lim\limits_{x \to 1} \dfrac{x^{1000} - 1}{x - 1} = 1000$.

Solution 2: Note that $(x^{1000} - 1) = (x - 1)(x^{999} + x^{998} + x^{997} + \cdots + x^2 + x + 1)$. So

$$\lim_{x \to 1} \frac{x^{1000} - 1}{x - 1} = \lim_{x \to 1} \frac{(x - 1)(x^{999} + x^{998} + x^{997} + \cdots + x^2 + x + 1)}{x - 1} = \lim_{x \to 1} (x^{999} + x^{998} + x^{997} + \cdots + x^2 + x + 1)$$

$$= \underbrace{1 + 1 + 1 + \cdots + 1 + 1 + 1}_{1000 \text{ ones}} = 1000, \text{ as above.}$$

84. In order for the two tangents to intersect on the y-axis, the points of tangency must be at equal distances from the y-axis, since the parabola $y = x^2$ is symmetric about the y-axis. Say the points of tangency are (a, a^2) and $(-a, a^2)$, for some $a > 0$. Then since the derivative of $y = x^2$ is $dy/dx = 2x$, the left-hand tangent has slope $-2a$ and equation $y - a^2 = -2a(x + a)$, or $y = -2ax - a^2$, and similarly the right-hand tangent line has equation $y - a^2 = 2a(x - a)$, or $y = 2ax - a^2$. So the two lines intersect at $(0, -a^2)$. Now if the lines are perpendicular, then the product of their slopes is -1, so $(-2a)(2a) = -1 \Leftrightarrow a^2 = \frac{1}{4} \Leftrightarrow a = \frac{1}{2}$. So the lines intersect at $\left(0, -\frac{1}{4}\right)$.

85. $y = x^2 \;\Rightarrow\; y' = 2x$, so the slope of a tangent line at the point (a, a^2) is $y' = 2a$ and the slope of a normal line is $-1/(2a)$, for $a \neq 0$. The slope of the normal line through the points (a, a^2) and $(0, c)$ is $\dfrac{a^2 - c}{a - 0}$, so $\dfrac{a^2 - c}{a} = -\dfrac{1}{2a} \;\Rightarrow$ $a^2 - c = -\frac{1}{2} \;\Rightarrow\; a^2 = c - \frac{1}{2}$. The last equation has two solutions if $c > \frac{1}{2}$, one solution if $c = \frac{1}{2}$, and no solution if $c < \frac{1}{2}$. Since the y-axis is normal to $y = x^2$ regardless of the value of c (this is the case for $a = 0$), we have three normal lines if $c > \frac{1}{2}$ and one normal line if $c \leq \frac{1}{2}$.

86. From the sketch, it appears that there may be a line that is tangent to both curves. The slope of the line through the points $P(a, a^2)$ and $Q(b, b^2 - 2b + 2)$ is $\dfrac{b^2 - 2b + 2 - a^2}{b - a}$. The slope of the tangent line at P is $2a$ $\;[y' = 2x]\;$ and at Q is $2b - 2$ $\;[y' = 2x - 2]$. All three slopes are equal, so $2a = 2b - 2 \;\Leftrightarrow\; a = b - 1$.

Also, $2b - 2 = \dfrac{b^2 - 2b + 2 - a^2}{b - a} \;\Rightarrow\; 2b - 2 = \dfrac{b^2 - 2b + 2 - (b-1)^2}{b - (b - 1)} \;\Rightarrow\; 2b - 2 = b^2 - 2b + 2 - b^2 + 2b - 1 \;\Rightarrow$ $2b = 3 \;\Rightarrow\; b = \frac{3}{2}$ and $a = \frac{3}{2} - 1 = \frac{1}{2}$. Thus, an equation of the tangent line at P is $y - \left(\frac{1}{2}\right)^2 = 2\left(\frac{1}{2}\right)\left(x - \frac{1}{2}\right)$ or $y = x - \frac{1}{4}$.

APPLIED PROJECT Building a Better Roller Coaster

1. (a) $f(x) = ax^2 + bx + c \;\Rightarrow\; f'(x) = 2ax + b$.

The origin is at P:	$f(0) = 0$	$\Rightarrow$	$c = 0$
The slope of the ascent is 0.8:	$f'(0) = 0.8$	$\Rightarrow$	$b = 0.8$
The slope of the drop is -1.6:	$f'(100) = -1.6$	$\Rightarrow$	$200a + b = -1.6$

(b) $b = 0.8$, so $200a + b = -1.6 \;\Rightarrow\; 200a + 0.8 = -1.6 \;\Rightarrow\; 200a = -2.4 \;\Rightarrow\; a = -\dfrac{2.4}{200} = -0.012$.

Thus, $f(x) = -0.012x^2 + 0.8x$.

(c) Since L_1 passes through the origin with slope 0.8, it has equation $y = 0.8x$. The horizontal distance between P and Q is 100, so the y-coordinate at Q is $f(100) = -0.012(100)^2 + 0.8(100) = -40$. Since L_2 passes through the point $(100, -40)$ and has slope -1.6, it has equation $y + 40 = -1.6(x - 100)$ or $y = -1.6x + 120$.

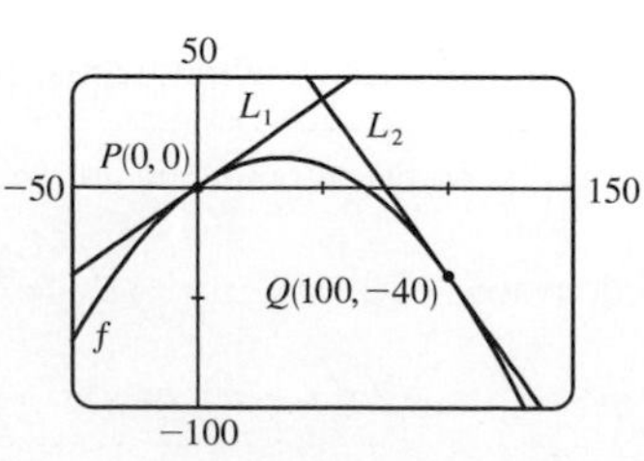

(d) The difference in elevation between $P(0, 0)$ and $Q(100, -40)$ is $0 - (-40) = 40$ feet.

2. (a)

Interval	Function	First Derivative	Second Derivative
$(-\infty, 0)$	$L_1(x) = 0.8x$	$L_1'(x) = 0.8$	$L_1''(x) = 0$
$[0, 10)$	$g(x) = kx^3 + lx^2 + mx + n$	$g'(x) = 3kx^2 + 2lx + m$	$g''(x) = 6kx + 2l$
$[10, 90]$	$q(x) = ax^2 + bx + c$	$q'(x) = 2ax + b$	$q''(x) = 2a$
$(90, 100]$	$h(x) = px^3 + qx^2 + rx + s$	$h'(x) = 3px^2 + 2qx + r$	$h''(x) = 6px + 2q$
$(100, \infty)$	$L_2(x) = -1.6x + 120$	$L_2'(x) = -1.6$	$L_2''(x) = 0$

There are 4 values of x (0, 10, 90, and 100) for which we must make sure the function values are equal, the first derivative values are equal, and the second derivative values are equal. The third column in the following table contains the value of each side of the condition—these are found after solving the system in part (b).

At $x =$	Condition	Value	Resulting Equation
0	$g(0) = L_1(0)$	0	$n = 0$
	$g'(0) = L_1'(0)$	$\frac{4}{5}$	$m = 0.8$
	$g''(0) = L_1''(0)$	0	$2l = 0$
10	$g(10) = q(10)$	$\frac{68}{9}$	$1000k + 100l + 10m + n = 100a + 10b + c$
	$g'(10) = q'(10)$	$\frac{2}{3}$	$300k + 20l + m = 20a + b$
	$g''(10) = q''(10)$	$-\frac{2}{75}$	$60k + 2l = 2a$
90	$h(90) = q(90)$	$-\frac{220}{9}$	$729{,}000p + 8100q + 90r + s = 8100a + 90b + c$
	$h'(90) = q'(90)$	$-\frac{22}{15}$	$24{,}300p + 180q + r = 180a + b$
	$h''(90) = q''(90)$	$-\frac{2}{75}$	$540p + 2q = 2a$
100	$h(100) = L_2(100)$	-40	$1{,}000{,}000p + 10{,}000q + 100r + s = -40$
	$h'(100) = L_2'(100)$	$-\frac{8}{5}$	$30{,}000p + 200q + r = -1.6$
	$h''(100) = L_2''(100)$	0	$600p + 2q = 0$

(b) We can arrange our work in a 12×12 matrix as follows.

a	b	c	k	l	m	n	p	q	r	s	constant
0	0	0	0	0	0	1	0	0	0	0	0
0	0	0	0	0	1	0	0	0	0	0	0.8
0	0	0	0	2	0	0	0	0	0	0	0
−100	−10	−1	1000	100	10	1	0	0	0	0	0
−20	−1	0	300	20	1	0	0	0	0	0	0
−2	0	0	60	2	0	0	0	0	0	0	0
−8100	−90	−1	0	0	0	0	729,000	8100	90	1	0
−180	−1	0	0	0	0	0	24,300	180	1	0	0
−2	0	0	0	0	0	0	540	2	0	0	0
0	0	0	0	0	0	0	1,000,000	10,000	100	1	−40
0	0	0	0	0	0	0	30,000	200	1	0	−1.6
0	0	0	0	0	0	0	600	2	0	0	0

Solving the system gives us the formulas for q, g, and h.

$$\left.\begin{aligned} a &= -0.01\overline{3} = -\tfrac{1}{75} \\ b &= 0.9\overline{3} = \tfrac{14}{15} \\ c &= -0.\overline{4} = -\tfrac{4}{9} \end{aligned}\right\} q(x) = -\tfrac{1}{75}x^2 + \tfrac{14}{15}x - \tfrac{4}{9}$$

$$\left.\begin{aligned} k &= -0.000\overline{4} = -\tfrac{1}{2250} \\ l &= 0 \\ m &= 0.8 = \tfrac{4}{5} \\ n &= 0 \end{aligned}\right\} g(x) = -\tfrac{1}{2250}x^3 + \tfrac{4}{5}x$$

$$\left.\begin{aligned} p &= 0.000\overline{4} = \tfrac{1}{2250} \\ q &= -0.1\overline{3} = -\tfrac{2}{15} \\ r &= 11.7\overline{3} = \tfrac{176}{15} \\ s &= -324.\overline{4} = -\tfrac{2920}{9} \end{aligned}\right\} h(x) = \tfrac{1}{2250}x^3 - \tfrac{2}{15}x^2 + \tfrac{176}{15}x - \tfrac{2920}{9}$$

(c) Graph of L_1, q, g, h, and L_2:

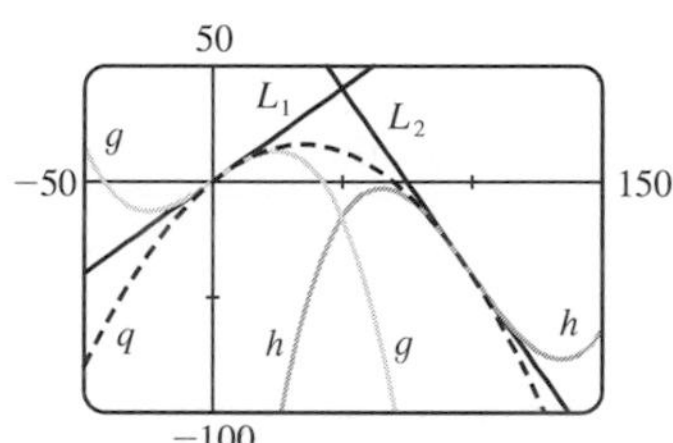

The graph of the five functions as a piecewise-defined function:

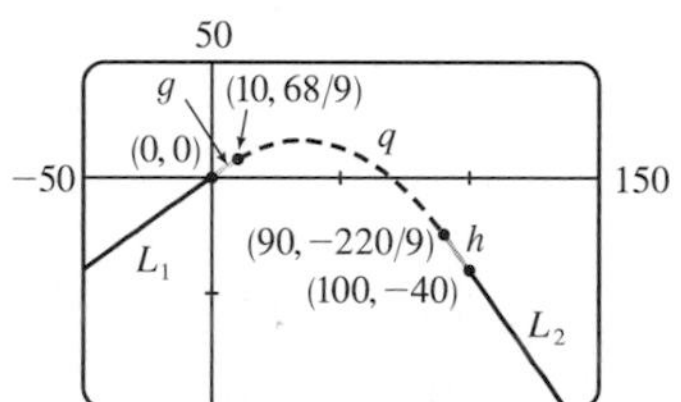

This is the piecewise-defined function assignment on a TI-83/4 Plus calculator, where $Y_2 = L_1$, $Y_6 = g$, $Y_5 = q$, $Y_7 = h$, and $Y_3 = L_2$.

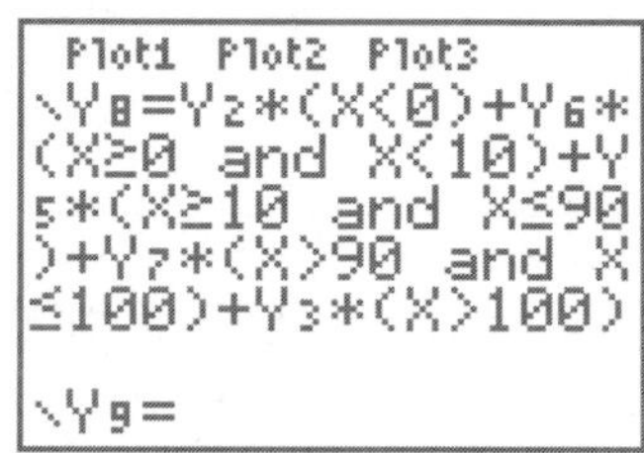

```
Plot1 Plot2 Plot3
\Y8=Y2*(X<0)+Y6*
(X≥0 and X<10)+Y
5*(X≥10 and X≤90
)+Y7*(X>90 and X
≤100)+Y3*(X>100)

\Y9=
```

A comparison of the graphs in part 1(c) and part 2(c):

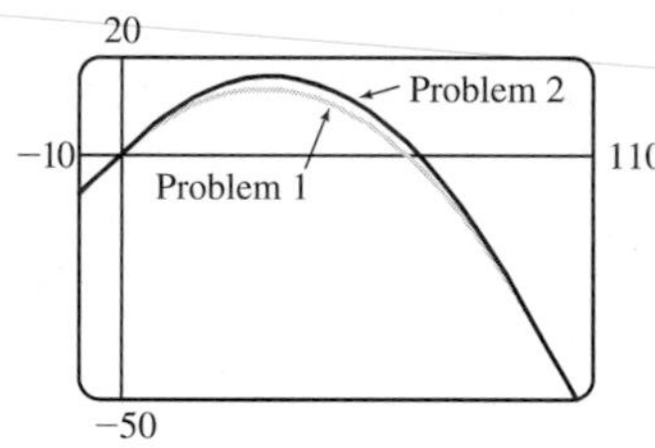

3.2 The Product and Quotient Rules

1. Product Rule: $f(x) = (1 + 2x^2)(x - x^2) \quad \Rightarrow$

$$f'(x) = (1 + 2x^2)(1 - 2x) + (x - x^2)(4x) = 1 - 2x + 2x^2 - 4x^3 + 4x^2 - 4x^3 = 1 - 2x + 6x^2 - 8x^3.$$

Multiplying first: $f(x) = (1 + 2x^2)(x - x^2) = x - x^2 + 2x^3 - 2x^4 \quad \Rightarrow \quad f'(x) = 1 - 2x + 6x^2 - 8x^3$ (equivalent).

2. Quotient Rule: $F(x) = \dfrac{x^4 - 5x^3 + \sqrt{x}}{x^2} = \dfrac{x^4 - 5x^3 + x^{1/2}}{x^2} \quad \Rightarrow$

$$F'(x) = \frac{x^2(4x^3 - 15x^2 + \frac{1}{2}x^{-1/2}) - (x^4 - 5x^3 + x^{1/2})(2x)}{(x^2)^2} = \frac{4x^5 - 15x^4 + \frac{1}{2}x^{3/2} - 2x^5 + 10x^4 - 2x^{3/2}}{x^4}$$

$$= \frac{2x^5 - 5x^4 - \frac{3}{2}x^{3/2}}{x^4} = 2x - 5 - \tfrac{3}{2}x^{-5/2}$$

Simplifying first: $F(x) = \dfrac{x^4 - 5x^3 + \sqrt{x}}{x^2} = x^2 - 5x + x^{-3/2} \quad \Rightarrow \quad F'(x) = 2x - 5 - \frac{3}{2}x^{-5/2}$ (equivalent).

For this problem, simplifying first seems to be the better method.

3. By the Product Rule, $f(x) = (3x^2 - 5x)e^x \Rightarrow$

$$f'(x) = (3x^2 - 5x)(e^x)' + e^x(3x^2 - 5x)' = (3x^2 - 5x)e^x + e^x(6x - 5)$$
$$= e^x[(3x^2 - 5x) + (6x - 5)] = e^x(3x^2 + x - 5)$$

4. By the Product Rule, $g(x) = (x + 2\sqrt{x})\, e^x \Rightarrow$

$$g'(x) = (x + 2\sqrt{x})(e^x)' + e^x(x + 2\sqrt{x})' = (x + 2\sqrt{x})e^x + e^x\left(1 + 2 \cdot \tfrac{1}{2}x^{-1/2}\right)$$
$$= e^x\left[(x + 2\sqrt{x}) + \left(1 + 1/\sqrt{x}\right)\right] = e^x\left(x + 2\sqrt{x} + 1 + 1/\sqrt{x}\right)$$

5. By the Quotient Rule, $y = \dfrac{x}{e^x} \quad\Rightarrow\quad y' = \dfrac{e^x(1) - x(e^x)}{(e^x)^2} = \dfrac{e^x(1 - x)}{(e^x)^2} = \dfrac{1 - x}{e^x}$.

6. By the Quotient Rule, $y = \dfrac{e^x}{1 - e^x} \quad\Rightarrow\quad y' = \dfrac{(1 - e^x)e^x - e^x(-e^x)}{(1 - e^x)^2} = \dfrac{e^x - e^{2x} + e^{2x}}{(1 - e^x)^2} = \dfrac{e^x}{(1 - e^x)^2}$.

The notations $\overset{\text{PR}}{\Rightarrow}$ and $\overset{\text{QR}}{\Rightarrow}$ indicate the use of the Product and Quotient Rules, respectively.

7. $g(x) = \dfrac{1 + 2x}{3 - 4x} \quad\overset{\text{QR}}{\Rightarrow}\quad g'(x) = \dfrac{(3 - 4x)(2) - (1 + 2x)(-4)}{(3 - 4x)^2} = \dfrac{6 - 8x + 4 + 8x}{(3 - 4x)^2} = \dfrac{10}{(3 - 4x)^2}$

8. $G(x) = \dfrac{x^2 - 2}{2x + 1} \quad\overset{\text{QR}}{\Rightarrow}\quad G'(x) = \dfrac{(2x + 1)(2x) - (x^2 - 2)(2)}{(2x + 1)^2} = \dfrac{4x^2 + 2x - 2x^2 + 4}{(2x + 1)^2} = \dfrac{2x^2 + 2x + 4}{(2x + 1)^2}$

9. $H(u) = (u - \sqrt{u})(u + \sqrt{u}) \quad\overset{\text{PR}}{\Rightarrow}$

$$H'(u) = (u - \sqrt{u})\left(1 + \frac{1}{2\sqrt{u}}\right) + (u + \sqrt{u})\left(1 - \frac{1}{2\sqrt{u}}\right) = u + \tfrac{1}{2}\sqrt{u} - \sqrt{u} - \tfrac{1}{2} + u - \tfrac{1}{2}\sqrt{u} + \sqrt{u} - \tfrac{1}{2} = 2u - 1.$$

An easier method is to simplify first and then differentiate as follows:

$$H(u) = (u - \sqrt{u})(u + \sqrt{u}) = u^2 - (\sqrt{u})^2 = u^2 - u \quad\Rightarrow\quad H'(u) = 2u - 1$$

10. $J(v) = (v^3 - 2v)(v^{-4} + v^{-2}) \quad\overset{\text{PR}}{\Rightarrow}$

$$J'(v) = (v^3 - 2v)(-4v^{-5} - 2v^{-3}) + (v^{-4} + v^{-2})(3v^2 - 2)$$
$$= -4v^{-2} - 2v^0 + 8v^{-4} + 4v^{-2} + 3v^{-2} - 2v^{-4} + 3v^0 - 2v^{-2} = 1 + v^{-2} + 6v^{-4}$$

11. $F(y) = \left(\dfrac{1}{y^2} - \dfrac{3}{y^4}\right)(y + 5y^3) = (y^{-2} - 3y^{-4})(y + 5y^3) \quad\overset{\text{PR}}{\Rightarrow}$

$$F'(y) = (y^{-2} - 3y^{-4})(1 + 15y^2) + (y + 5y^3)(-2y^{-3} + 12y^{-5})$$
$$= (y^{-2} + 15 - 3y^{-4} - 45y^{-2}) + (-2y^{-2} + 12y^{-4} - 10 + 60y^{-2})$$
$$= 5 + 14y^{-2} + 9y^{-4} \text{ or } 5 + 14/y^2 + 9/y^4$$

12. $f(z) = (1 - e^z)(z + e^z) \quad\overset{\text{PR}}{\Rightarrow}$

$$f'(z) = (1 - e^z)(1 + e^z) + (z + e^z)(-e^z) = 1^2 - (e^z)^2 - ze^z - (e^z)^2 = 1 - ze^z - 2e^{2z}$$

13. $y = \dfrac{x^2+1}{x^3-1} \overset{\text{QR}}{\Rightarrow}$

$$y' = \frac{(x^3-1)(2x)-(x^2+1)(3x^2)}{(x^3-1)^2} = \frac{x\left[(x^3-1)(2)-(x^2+1)(3x)\right]}{(x^3-1)^2} = \frac{x(2x^3-2-3x^3-3x)}{(x^3-1)^2} = \frac{x(-x^3-3x-2)}{(x^3-1)^2}$$

14. $y = \dfrac{\sqrt{x}}{2+x} \overset{\text{QR}}{\Rightarrow}$

$$y' = \frac{(2+x)\left(\dfrac{1}{2\sqrt{x}}\right)-\sqrt{x}\,(1)}{(2+x)^2} = \frac{\dfrac{1}{\sqrt{x}}+\dfrac{\sqrt{x}}{2}-\sqrt{x}}{(2+x)^2} = \frac{\dfrac{2+x-2x}{2\sqrt{x}}}{(2+x)^2} = \frac{2-x}{2\sqrt{x}(2+x)^2}$$

15. $y = \dfrac{t^3+3t}{t^2-4t+3} \overset{\text{QR}}{\Rightarrow}$

$$y' = \frac{(t^2-4t+3)(3t^2+3)-(t^3+3t)(2t-4)}{(t^2-4t+3)^2}$$

$$= \frac{3t^4+3t^2-12t^3-12t+9t^2+9-(2t^4-4t^3+6t^2-12t)}{(t^2-4t+3)^2} = \frac{t^4-8t^3+6t^2+9}{(t^2-4t+3)^2}$$

16. $y = \dfrac{1}{t^3+2t^2-1} \overset{\text{QR}}{\Rightarrow} y' = \dfrac{(t^3+2t^2-1)(0)-1(3t^2+4t)}{(t^3+2t^2-1)^2} = -\dfrac{3t^2+4t}{(t^3+2t^2-1)^2}$

17. $y = e^p(p+p\sqrt{p}) = e^p(p+p^{3/2}) \overset{\text{PR}}{\Rightarrow} y' = e^p\left(1+\frac{3}{2}p^{1/2}\right)+(p+p^{3/2})e^p = e^p\left(1+\frac{3}{2}\sqrt{p}+p+p\sqrt{p}\right)$

18. $h(r) = \dfrac{ae^r}{b+e^r} \overset{\text{QR}}{\Rightarrow} h'(r) = \dfrac{(b+e^r)(ae^r)-(ae^r)(e^r)}{(b+e^r)^2} = \dfrac{abe^r+ae^{2r}-ae^{2r}}{(b+e^r)^2} = \dfrac{abe^r}{(b+e^r)^2}$

19. $y = \dfrac{s-\sqrt{s}}{s^2} = \dfrac{s}{s^2}-\dfrac{\sqrt{s}}{s^2} = s^{-1}-s^{-3/2} \Rightarrow y' = -s^{-2}+\frac{3}{2}s^{-5/2} = \dfrac{-1}{s^2}+\dfrac{3}{2s^{5/2}} = \dfrac{3-2\sqrt{s}}{2s^{5/2}}$

20. $y = (z^2+e^z)\sqrt{z} \overset{\text{PR}}{\Rightarrow}$

$$y' = (z^2+e^z)\left(\frac{1}{2\sqrt{z}}\right)+\sqrt{z}\,(2z+e^z) = \frac{z^2}{2\sqrt{z}}+\frac{e^z}{2\sqrt{z}}+2z\sqrt{z}+\sqrt{z}\,e^z$$

$$= \frac{z^2+e^z+4z^2+2ze^z}{2\sqrt{z}} = \frac{5z^2+e^z+2ze^z}{2\sqrt{z}}$$

21. $f(t) = \dfrac{\sqrt[3]{t}}{t-3} \overset{\text{QR}}{\Rightarrow}$

$$f'(t) = \frac{(t-3)\left(\frac{1}{3}t^{-2/3}\right)-t^{1/3}(1)}{(t-3)^2} = \frac{\frac{1}{3}t^{1/3}-t^{-2/3}-t^{1/3}}{(t-3)^2} = \frac{-\frac{2}{3}t^{1/3}-t^{-2/3}}{(t-3)^2} = \frac{\dfrac{-2t}{3t^{2/3}}-\dfrac{3}{3t^{2/3}}}{(t-3)^2} = \frac{-2t-3}{3t^{2/3}(t-3)^2}$$

22. $V(t) = \dfrac{4+t}{te^t} \overset{\text{QR}}{\Rightarrow}$

$$V'(t) = \frac{te^t(1)-(4+t)(te^t+e^t(1))}{(te^t)^2} = \frac{te^t-4te^t-4e^t-t^2e^t-te^t}{t^2e^{2t}}$$

$$= \frac{-4te^t-4e^t-t^2e^t}{t^2e^{2t}} = \frac{-e^t(t^2+4t+4)}{t^2e^{2t}} = -\frac{(t+2)^2}{t^2e^t}$$

23. $f(x) = \dfrac{x^2 e^x}{x^2 + e^x} \overset{\text{QR}}{\Rightarrow}$

$$f'(x) = \frac{(x^2 + e^x)\left[x^2 e^x + e^x(2x)\right] - x^2 e^x(2x + e^x)}{(x^2 + e^x)^2} = \frac{x^4 e^x + 2x^3 e^x + x^2 e^{2x} + 2xe^{2x} - 2x^3 e^x - x^2 e^{2x}}{(x^2 + e^x)^2}$$

$$= \frac{x^4 e^x + 2xe^{2x}}{(x^2 + e^x)^2} = \frac{xe^x(x^3 + 2e^x)}{(x^2 + e^x)^2}$$

24. $F(t) = \dfrac{At}{Bt^2 + Ct^3} = \dfrac{A}{Bt + Ct^2} \overset{\text{QR}}{\Rightarrow}$

$$F'(t) = \frac{(Bt + Ct^2)(0) - A(B + 2Ct)}{(Bt + Ct^2)^2} = \frac{-A(B + 2Ct)}{(t)^2(B + Ct)^2} = -\frac{A(B + 2Ct)}{t^2(B + Ct)^2}$$

25. $f(x) = \dfrac{x}{x + c/x} \Rightarrow f'(x) = \dfrac{(x + c/x)(1) - x(1 - c/x^2)}{\left(x + \frac{c}{x}\right)^2} = \dfrac{x + c/x - x + c/x}{\left(\frac{x^2 + c}{x}\right)^2} = \dfrac{2c/x}{\frac{(x^2 + c)^2}{x^2}} \cdot \dfrac{x^2}{x^2} = \dfrac{2cx}{(x^2 + c)^2}$

26. $f(x) = \dfrac{ax + b}{cx + d} \Rightarrow f'(x) = \dfrac{(cx + d)(a) - (ax + b)(c)}{(cx + d)^2} = \dfrac{acx + ad - acx - bc}{(cx + d)^2} = \dfrac{ad - bc}{(cx + d)^2}$

27. $f(x) = (x^3 + 1)e^x \overset{\text{PR}}{\Rightarrow}$

$f'(x) = (x^3 + 1)e^x + e^x(3x^2) = e^x\left[(x^3 + 1) + 3x^2\right] = e^x(x^3 + 3x^2 + 1) \overset{\text{PR}}{\Rightarrow}$

$f''(x) = e^x(3x^2 + 6x) + (x^3 + 3x^2 + 1)e^x = e^x\left[(3x^2 + 6x) + (x^3 + 3x^2 + 1)\right] = e^x(x^3 + 6x^2 + 6x + 1)$

28. $f(x) = \sqrt{x}\, e^x \overset{\text{PR}}{\Rightarrow} f'(x) = \sqrt{x}\, e^x + e^x\left(\dfrac{1}{2\sqrt{x}}\right) = \left(\sqrt{x} + \dfrac{1}{2\sqrt{x}}\right)e^x = \dfrac{2x + 1}{2\sqrt{x}}\, e^x.$

Using the Product Rule and $f'(x) = \left(x^{1/2} + \frac{1}{2}x^{-1/2}\right)e^x$, we get

$$f''(x) = \left(x^{1/2} + \tfrac{1}{2}x^{-1/2}\right)e^x + e^x\left(\tfrac{1}{2}x^{-1/2} - \tfrac{1}{4}x^{-3/2}\right) = \left(x^{1/2} + x^{-1/2} - \tfrac{1}{4}x^{-3/2}\right)e^x = \frac{4x^2 + 4x - 1}{4x^{3/2}}\, e^x$$

29. $f(x) = \dfrac{x^2}{1 + e^x} \overset{\text{QR}}{\Rightarrow} f'(x) = \dfrac{(1 + e^x)(2x) - x^2(e^x)}{(1 + e^x)^2} = \dfrac{x[(1 + e^x)2 - xe^x]}{(1 + e^x)^2} = \dfrac{x(2 + 2e^x - xe^x)}{(1 + e^x)^2}.$

Using the Quotient and Product Rules and $f'(x) = \dfrac{2x + 2xe^x - x^2 e^x}{(1 + e^x)^2}$, we get

$$f''(x) = \frac{(1 + e^x)^2\left[2 + 2(xe^x + e^x) - (x^2 e^x + 2xe^x)\right] - (2x + 2xe^x - x^2 e^x)\left[(1 + e^x)e^x + (1 + e^x)e^x\right]}{[(1 + e^x)^2]^2}$$

$$= \frac{(1 + e^x)\left\{\left[(1 + e^x)(2 + 2xe^x + 2e^x - x^2 e^x - 2xe^x)\right] - (2x + 2xe^x - x^2 e^x)(2e^x)\right\}}{(1 + e^x)^4}$$

$$= \frac{(1 + e^x)(2 + 2e^x - x^2 e^x) - 4xe^x - 4xe^{2x} + 2x^2 e^{2x}}{(1 + e^x)^3}$$

$$= \frac{2 + 2e^x - x^2 e^x + 2e^x + 2e^{2x} - x^2 e^{2x} - 4xe^x - 4xe^{2x} + 2x^2 e^{2x}}{(1 + e^x)^3}$$

$$= \frac{2 + 4e^x - x^2 e^x - 4xe^x + 2e^{2x} + x^2 e^{2x} - 4xe^{2x}}{(1 + e^x)^3}$$

30. $f(x) = \dfrac{x}{x^2 - 1} \quad\Rightarrow\quad f'(x) = \dfrac{(x^2-1)(1) - x(2x)}{(x^2-1)^2} = \dfrac{x^2 - 1 - 2x^2}{(x^2-1)^2} = \dfrac{-x^2-1}{(x^2-1)^2} \quad\Rightarrow$

$$f''(x) = \frac{(x^2-1)^2(-2x) - (-x^2-1)(x^4 - 2x^2 + 1)'}{[(x^2-1)^2]^2} = \frac{(x^2-1)^2(-2x) + (x^2+1)(4x^3 - 4x)}{(x^2-1)^4}$$

$$= \frac{(x^2-1)^2(-2x) + (x^2+1)(4x)(x^2-1)}{(x^2-1)^4} = \frac{(x^2-1)[(x^2-1)(-2x) + (x^2+1)(4x)]}{(x^2-1)^4}$$

$$= \frac{-2x^3 + 2x + 4x^3 + 4x}{(x^2-1)^3} = \frac{2x^3 + 6x}{(x^2-1)^3}$$

31. $y = \dfrac{x^2 - 1}{x^2 + x + 1} \quad\Rightarrow$

$$y' = \frac{(x^2+x+1)(2x) - (x^2-1)(2x+1)}{(x^2+x+1)^2} = \frac{2x^3 + 2x^2 + 2x - 2x^3 - x^2 + 2x + 1}{(x^2+x+1)^2} = \frac{x^2 + 4x + 1}{(x^2+x+1)^2}.$$

At $(1, 0)$, $y' = \dfrac{6}{3^2} = \dfrac{2}{3}$, and an equation of the tangent line is $y - 0 = \frac{2}{3}(x - 1)$, or $y = \frac{2}{3}x - \frac{2}{3}$.

32. $y = \dfrac{1+x}{1+e^x} \quad\Rightarrow\quad y' = \dfrac{(1+e^x)(1) - (1+x)e^x}{(1+e^x)^2} = \dfrac{1 + e^x - e^x - xe^x}{(1+e^x)^2} = \dfrac{1 - xe^x}{(1+e^x)^2}.$

At $\left(0, \frac{1}{2}\right)$, $y' = \dfrac{1}{(1+1)^2} = \dfrac{1}{4}$, and an equation of the tangent line is $y - \frac{1}{2} = \frac{1}{4}(x - 0)$ or $y = \frac{1}{4}x + \frac{1}{2}$.

33. $y = 2xe^x \quad\Rightarrow\quad y' = 2(x \cdot e^x + e^x \cdot 1) = 2e^x(x+1)$.

At $(0, 0)$, $y' = 2e^0(0+1) = 2 \cdot 1 \cdot 1 = 2$, and an equation of the tangent line is $y - 0 = 2(x - 0)$, or $y = 2x$. The slope of the normal line is $-\frac{1}{2}$, so an equation of the normal line is $y - 0 = -\frac{1}{2}(x - 0)$, or $y = -\frac{1}{2}x$.

34. $y = \dfrac{2x}{x^2+1} \quad\Rightarrow\quad y' = \dfrac{(x^2+1)(2) - 2x(2x)}{(x^2+1)^2} = \dfrac{2 - 2x^2}{(x^2+1)^2}$. At $(1, 1)$, $y' = 0$, and an equation of the tangent line is $y - 1 = 0(x - 1)$, or $y = 1$. The slope of the normal line is undefined, so an equation of the normal line is $x = 1$.

35. (a) $y = f(x) = \dfrac{1}{1+x^2} \quad\Rightarrow$

$f'(x) = \dfrac{(1+x^2)(0) - 1(2x)}{(1+x^2)^2} = \dfrac{-2x}{(1+x^2)^2}$. So the slope of the tangent line at the point $\left(-1, \frac{1}{2}\right)$ is $f'(-1) = \dfrac{2}{2^2} = \frac{1}{2}$ and its equation is $y - \frac{1}{2} = \frac{1}{2}(x+1)$ or $y = \frac{1}{2}x + 1$.

(b)

1.5
(−1, 0.5)
−4
4
−0.5

36. (a) $y = f(x) = \dfrac{x}{1+x^2} \quad\Rightarrow$

$f'(x) = \dfrac{(1+x^2)1 - x(2x)}{(1+x^2)^2} = \dfrac{1 - x^2}{(1+x^2)^2}$. So the slope of the tangent line at the point $(3, 0.3)$ is $f'(3) = \frac{-8}{100}$ and its equation is $y - 0.3 = -0.08(x - 3)$ or $y = -0.08x + 0.54$.

(b)

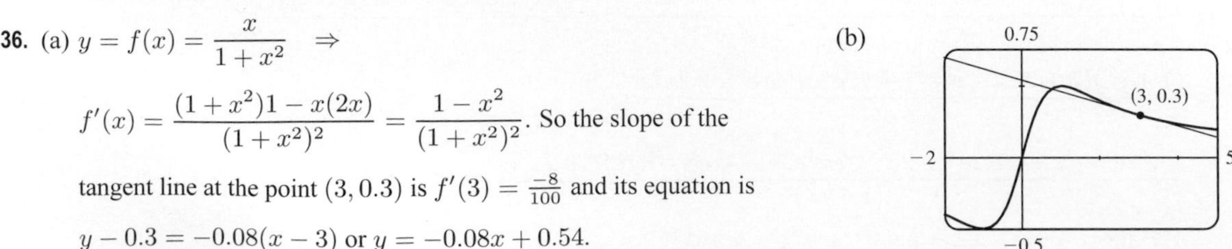

37. (a) $f(x) = (x^3 - x)e^x \quad \Rightarrow \quad f'(x) = (x^3 - x)e^x + e^x(3x^2 - 1) = e^x(x^3 + 3x^2 - x - 1)$

(b)

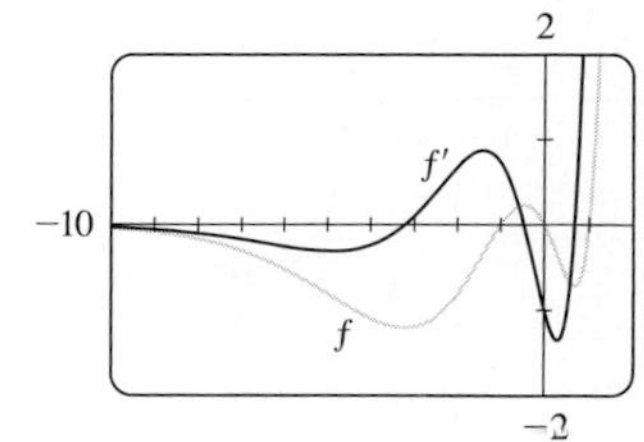

$f' = 0$ when f has a horizontal tangent line, f' is negative when f is decreasing, and f' is positive when f is increasing.

38. (a) $f(x) = \dfrac{e^x}{2x^2 + x + 1} \quad \Rightarrow$

$$f'(x) = \frac{(2x^2 + x + 1)e^x - e^x(4x + 1)}{(2x^2 + x + 1)^2} = \frac{e^x(2x^2 + x + 1 - 4x - 1)}{(2x^2 + x + 1)^2} = \frac{e^x(2x^2 - 3x)}{(2x^2 + x + 1)^2}$$

(b)

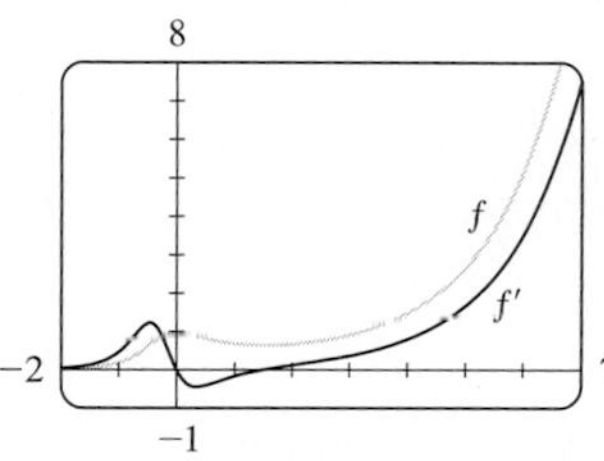

$f' = 0$ when f has a horizontal tangent line, f' is negative when f is decreasing, and f' is positive when f is increasing.

39. (a) $f(x) = \dfrac{x^2 - 1}{x^2 + 1} \quad \Rightarrow$

$$f'(x) = \frac{(x^2 + 1)(2x) - (x^2 - 1)(2x)}{(x^2 + 1)^2} = \frac{(2x)[(x^2 + 1) - (x^2 - 1)]}{(x^2 + 1)^2} = \frac{(2x)(2)}{(x^2 + 1)^2} = \frac{4x}{(x^2 + 1)^2} \quad \Rightarrow$$

$$f''(x) = \frac{(x^2 + 1)^2(4) - 4x(x^4 + 2x^2 + 1)'}{[(x^2 + 1)^2]^2} = \frac{4(x^2 + 1)^2 - 4x(4x^3 + 4x)}{(x^2 + 1)^4}$$

$$= \frac{4(x^2 + 1)^2 - 16x^2(x^2 + 1)}{(x^2 + 1)^4} = \frac{4(x^2 + 1)[(x^2 + 1) - 4x^2]}{(x^2 + 1)^4} = \frac{4(1 - 3x^2)}{(x^2 + 1)^3}$$

(b)

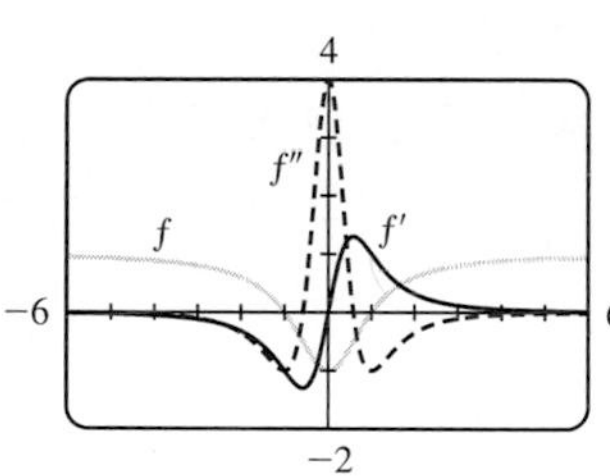

$f' = 0$ when f has a horizontal tangent and $f'' = 0$ when f' has a horizontal tangent. f' is negative when f is decreasing and positive when f is increasing. f'' is negative when f' is decreasing and positive when f' is increasing. f'' is negative when f is concave down and positive when f is concave up.

40. (a) $f(x) = (x^2 - 1)e^x \quad \Rightarrow \quad f'(x) = (x^2 - 1)e^x + e^x(2x) = e^x(x^2 + 2x - 1) \quad \Rightarrow$

$f''(x) = e^x(2x + 2) + (x^2 + 2x - 1)e^x = e^x(x^2 + 4x + 1)$

(b)

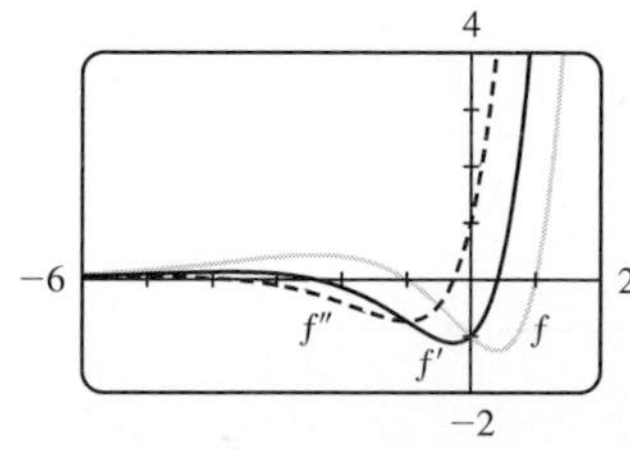

We can see that our answers are plausible, since f has horizontal tangents where $f'(x) = 0$, and f' has horizontal tangents where $f''(x) = 0$.

41. $f(x) = \dfrac{x^2}{1+x} \quad \Rightarrow \quad f'(x) = \dfrac{(1+x)(2x) - x^2(1)}{(1+x)^2} = \dfrac{2x + 2x^2 - x^2}{(1+x)^2} = \dfrac{x^2 + 2x}{x^2 + 2x + 1} \quad \Rightarrow$

$$f''(x) = \frac{(x^2+2x+1)(2x+2) - (x^2+2x)(2x+2)}{(x^2+2x+1)^2} = \frac{(2x+2)(x^2+2x+1-x^2-2x)}{[(x+1)^2]^2}$$

$$= \frac{2(x+1)(1)}{(x+1)^4} = \frac{2}{(x+1)^3},$$

so $f''(1) = \dfrac{2}{(1+1)^3} = \dfrac{2}{8} = \dfrac{1}{4}$.

42. $g(x) = \dfrac{x}{e^x} \quad \Rightarrow \quad g'(x) = \dfrac{e^x \cdot 1 - x \cdot e^x}{(e^x)^2} = \dfrac{e^x(1-x)}{(e^x)^2} = \dfrac{1-x}{e^x} \quad \Rightarrow$

$$g''(x) = \frac{e^x \cdot (-1) - (1-x)e^x}{(e^x)^2} = \frac{e^x[-1-(1-x)]}{(e^x)^2} = \frac{x-2}{e^x} \quad \Rightarrow$$

$$g'''(x) = \frac{e^x \cdot 1 - (x-2)e^x}{(e^x)^2} = \frac{e^x[1-(x-2)]}{(e^x)^2} = \frac{3-x}{e^x} \quad \Rightarrow$$

$$g^{(4)}(x) = \frac{e^x \cdot (-1) - (3-x)e^x}{(e^x)^2} = \frac{e^x[-1-(3-x)]}{(e^x)^2} = \frac{x-4}{e^x}.$$

The pattern suggests that $g^{(n)}(x) = \dfrac{(x-n)(-1)^n}{e^x}$. (We could use mathematical induction to prove this formula.)

43. We are given that $f(5) = 1$, $f'(5) = 6$, $g(5) = -3$, and $g'(5) = 2$.

(a) $(fg)'(5) = f(5)g'(5) + g(5)f'(5) = (1)(2) + (-3)(6) = 2 - 18 = -16$

(b) $\left(\dfrac{f}{g}\right)'(5) = \dfrac{g(5)f'(5) - f(5)g'(5)}{[g(5)]^2} = \dfrac{(-3)(6) - (1)(2)}{(-3)^2} = -\dfrac{20}{9}$

(c) $\left(\dfrac{g}{f}\right)'(5) = \dfrac{f(5)g'(5) - g(5)f'(5)}{[f(5)]^2} = \dfrac{(1)(2) - (-3)(6)}{(1)^2} = 20$

44. We are given that $f(4) = 2$, $g(4) = 5$, $f'(4) = 6$, and $g'(4) = -3$.

(a) $h(x) = 3f(x) + 8g(x) \quad \Rightarrow \quad h'(x) = 3f'(x) + 8g'(x)$, so

$h'(4) = 3f'(4) + 8g'(4) = 3(6) + 8(-3) = 18 - 24 = -6$.

(b) $h(x) = f(x)\,g(x) \quad \Rightarrow \quad h'(x) = f(x)\,g'(x) + g(x)\,f'(x)$, so

$h'(4) = f(4)\,g'(4) + g(4)\,f'(4) = 2(-3) + 5(6) = -6 + 30 = 24$.

(c) $h(x) = \dfrac{f(x)}{g(x)} \quad \Rightarrow \quad h'(x) = \dfrac{g(x)\,f'(x) - f(x)\,g'(x)}{[g(x)]^2}$, so

$$h'(4) = \frac{g(4)\,f'(4) - f(4)\,g'(4)}{[g(4)]^2} = \frac{5(6) - 2(-3)}{5^2} = \frac{30+6}{25} = \frac{36}{25}.$$

(d) $h(x) = \dfrac{g(x)}{f(x) + g(x)} \quad \Rightarrow$

$$h'(4) = \frac{[f(4) + g(4)]\,g'(4) - g(4)\,[f'(4) + g'(4)]}{[f(4) + g(4)]^2} = \frac{(2+5)\,(-3) - 5\,[6 + (-3)]}{(2+5)^2} = \frac{-21-15}{7^2} = -\frac{36}{49}$$

45. $f(x) = e^x g(x) \quad \Rightarrow \quad f'(x) = e^x g'(x) + g(x)e^x = e^x[g'(x) + g(x)]$. $f'(0) = e^0[g'(0) + g(0)] = 1(5 + 2) = 7$

46. $\dfrac{d}{dx}\left[\dfrac{h(x)}{x}\right] = \dfrac{xh'(x) - h(x) \cdot 1}{x^2} \quad \Rightarrow \quad \dfrac{d}{dx}\left[\dfrac{h(x)}{x}\right]_{x=2} = \dfrac{2h'(2) - h(2)}{2^2} = \dfrac{2(-3) - (4)}{4} = \dfrac{-10}{4} = -2.5$

47. $g(x) = xf(x) \quad \Rightarrow \quad g'(x) = xf'(x) + f(x) \cdot 1$. Now $g(3) = 3f(3) = 3 \cdot 4 = 12$ and $g'(3) = 3f'(3) + f(3) = 3(-2) + 4 = -2$. Thus, an equation of the tangent line to the graph of g at the point where $x = 3$ is $y - 12 = -2(x - 3)$, or $y = -2x + 18$.

48. $f'(x) = x^2 f(x) \quad \Rightarrow \quad f''(x) = x^2 f'(x) + f(x) \cdot 2x$. Now $f'(2) = 2^2 f(2) = 4(10) = 40$, so $f''(2) = 2^2(40) + 10(4) = 200$.

49. (a) From the graphs of f and g, we obtain the following values: $f(1) = 2$ since the point $(1, 2)$ is on the graph of f; $g(1) = 1$ since the point $(1, 1)$ is on the graph of g; $f'(1) = 2$ since the slope of the line segment between $(0, 0)$ and $(2, 4)$ is $\dfrac{4 - 0}{2 - 0} = 2$; $g'(1) = -1$ since the slope of the line segment between $(-2, 4)$ and $(2, 0)$ is $\dfrac{0 - 4}{2 - (-2)} = -1$.

Now $u(x) = f(x)g(x)$, so $u'(1) = f(1)g'(1) + g(1)\,f'(1) = 2 \cdot (-1) + 1 \cdot 2 = 0$.

(b) $v(x) = f(x)/g(x)$, so $v'(5) = \dfrac{g(5)f'(5) - f(5)g'(5)}{[g(5)]^2} = \dfrac{2\left(-\frac{1}{3}\right) - 3 \cdot \frac{2}{3}}{2^2} = \dfrac{-\frac{8}{3}}{4} = -\dfrac{2}{3}$

50. (a) $P(x) = F(x)\,G(x)$, so $P'(2) = F(2)\,G'(2) + G(2)\,F'(2) = 3 \cdot \frac{2}{4} + 2 \cdot 0 = \frac{3}{2}$.

(b) $Q(x) = F(x)/G(x)$, so $Q'(7) = \dfrac{G(7)\,F'(7) - F(7)\,G'(7)}{[G(7)]^2} = \dfrac{1 \cdot \frac{1}{4} - 5 \cdot \left(-\frac{2}{3}\right)}{1^2} = \dfrac{1}{4} + \dfrac{10}{3} = \dfrac{43}{12}$

51. (a) $y = xg(x) \quad \Rightarrow \quad y' = xg'(x) + g(x) \cdot 1 = xg'(x) + g(x)$

(b) $y = \dfrac{x}{g(x)} \quad \Rightarrow \quad y' = \dfrac{g(x) \cdot 1 - xg'(x)}{[g(x)]^2} = \dfrac{g(x) - xg'(x)}{[g(x)]^2}$

(c) $y = \dfrac{g(x)}{x} \quad \Rightarrow \quad y' = \dfrac{xg'(x) - g(x) \cdot 1}{(x)^2} = \dfrac{xg'(x) - g(x)}{x^2}$

52. (a) $y = x^2 f(x) \quad \Rightarrow \quad y' = x^2 f'(x) + f(x)(2x)$

(b) $y = \dfrac{f(x)}{x^2} \quad \Rightarrow \quad y' = \dfrac{x^2 f'(x) - f(x)(2x)}{(x^2)^2} = \dfrac{xf'(x) - 2f(x)}{x^3}$

(c) $y = \dfrac{x^2}{f(x)} \quad \Rightarrow \quad y' = \dfrac{f(x)(2x) - x^2 f'(x)}{[f(x)]^2}$

(d) $y = \dfrac{1 + xf(x)}{\sqrt{x}} \quad \Rightarrow$

$$y' = \frac{\sqrt{x}\,[xf'(x) + f(x)] - [1 + xf(x)]\,\dfrac{1}{2\sqrt{x}}}{\left(\sqrt{x}\,\right)^2}$$

$$= \frac{x^{3/2} f'(x) + x^{1/2} f(x) - \frac{1}{2}x^{-1/2} - \frac{1}{2}x^{1/2} f(x)}{x} \cdot \frac{2x^{1/2}}{2x^{1/2}} = \frac{xf(x) + 2x^2 f'(x) - 1}{2x^{3/2}}$$

53. If $y = f(x) = \dfrac{x}{x+1}$, then $f'(x) = \dfrac{(x+1)(1) - x(1)}{(x+1)^2} = \dfrac{1}{(x+1)^2}$. When $x = a$, the equation of the tangent line is

$y - \dfrac{a}{a+1} = \dfrac{1}{(a+1)^2}(x-a)$. This line passes through $(1,2)$ when $2 - \dfrac{a}{a+1} = \dfrac{1}{(a+1)^2}(1-a) \quad \Leftrightarrow$

$2(a+1)^2 - a(a+1) = 1 - a \quad \Leftrightarrow \quad 2a^2 + 4a + 2 - a^2 - a - 1 + a = 0 \quad \Leftrightarrow \quad a^2 + 4a + 1 = 0.$

The quadratic formula gives the roots of this equation as $a = \dfrac{-4 \pm \sqrt{4^2 - 4(1)(1)}}{2(1)} = \dfrac{-4 \pm \sqrt{12}}{2} = -2 \pm \sqrt{3}$,

so there are two such tangent lines. Since

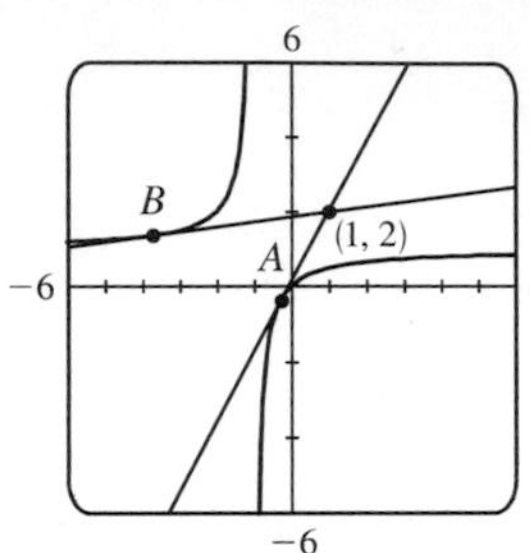

$$f(-2 \pm \sqrt{3}) = \frac{-2 \pm \sqrt{3}}{-2 \pm \sqrt{3} + 1} = \frac{-2 \pm \sqrt{3}}{-1 \pm \sqrt{3}} \cdot \frac{-1 \mp \sqrt{3}}{-1 \mp \sqrt{3}}$$
$$= \frac{2 \pm 2\sqrt{3} \mp \sqrt{3} - 3}{1 - 3} = \frac{-1 \pm \sqrt{3}}{-2} = \frac{1 \mp \sqrt{3}}{2},$$

the lines touch the curve at $A\left(-2 + \sqrt{3}, \frac{1 - \sqrt{3}}{2}\right) \approx (-0.27, -0.37)$

and $B\left(-2 - \sqrt{3}, \frac{1 + \sqrt{3}}{2}\right) \approx (-3.73, 1.37)$.

54. $y = \dfrac{x-1}{x+1} \quad \Rightarrow \quad y' = \dfrac{(x+1)(1) - (x-1)(1)}{(x+1)^2} = \dfrac{2}{(x+1)^2}$. If the tangent intersects

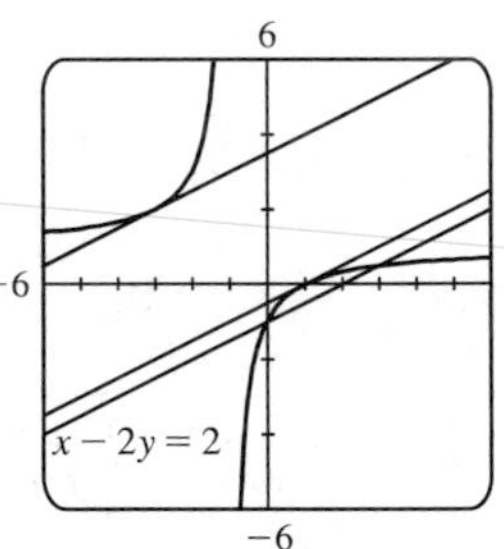

the curve when $x = a$, then its slope is $2/(a+1)^2$. But if the tangent is parallel to

$x - 2y = 2$, that is, $y = \frac{1}{2}x - 1$, then its slope is $\frac{1}{2}$. Thus, $\dfrac{2}{(a+1)^2} = \dfrac{1}{2} \quad \Rightarrow$

$(a+1)^2 = 4 \quad \Rightarrow \quad a + 1 = \pm 2 \quad \Rightarrow \quad a = 1$ or -3. When $a = 1$, $y = 0$ and the

equation of the tangent is $y - 0 = \frac{1}{2}(x-1)$ or $y = \frac{1}{2}x - \frac{1}{2}$. When $a = -3$, $y = 2$ and

the equation of the tangent is $y - 2 = \frac{1}{2}(x+3)$ or $y = \frac{1}{2}x + \frac{7}{2}$.

55. $R = \dfrac{f}{g} \quad \Rightarrow \quad R' = \dfrac{gf' - fg'}{g^2}$. For $f(x) = x - 3x^3 + 5x^5$, $f'(x) = 1 - 9x^2 + 25x^4$,

and for $g(x) = 1 + 3x^3 + 6x^6 + 9x^9$, $g'(x) = 9x^2 + 36x^5 + 81x^8$.

Thus, $R'(0) = \dfrac{g(0)f'(0) - f(0)g'(0)}{[g(0)]^2} = \dfrac{1 \cdot 1 - 0 \cdot 0}{1^2} = \dfrac{1}{1} = 1$.

56. $Q = \dfrac{f}{g} \quad \Rightarrow \quad Q' = \dfrac{gf' - fg'}{g^2}$. For $f(x) = 1 + x + x^2 + xe^x$, $f'(x) = 1 + 2x + xe^x + e^x$,

and for $g(x) = 1 - x + x^2 - xe^x$, $g'(x) = -1 + 2x - xe^x - e^x$.

Thus, $Q'(0) = \dfrac{g(0)f'(0) - f(0)g'(0)}{[g(0)]^2} = \dfrac{1 \cdot 2 - 1 \cdot (-2)}{1^2} = \dfrac{4}{1} = 4$.

57. If $P(t)$ denotes the population at time t and $A(t)$ the average annual income, then $T(t) = P(t)A(t)$ is the total personal income. The rate at which $T(t)$ is rising is given by $T'(t) = P(t)A'(t) + A(t)P'(t) \quad \Rightarrow$

$$T'(1999) = P(1999)A'(1999) + A(1999)P'(1999) = (961{,}400)(\$1400/\text{yr}) + (\$30{,}593)(9200/\text{yr})$$
$$= \$1{,}345{,}960{,}000/\text{yr} + \$281{,}455{,}600/\text{yr} = \$1{,}627{,}415{,}600/\text{yr}$$

So the total personal income was rising by about $\$1.627$ billion per year in 1999.

The term $P(t)A'(t) \approx \$1.346$ billion represents the portion of the rate of change of total income due to the existing population's increasing income. The term $A(t)P'(t) \approx \$281$ million represents the portion of the rate of change of total income due to increasing population.

58. (a) $f(20) = 10{,}000$ means that when the price of the fabric is $\$20/\text{yard}$, 10,000 yards will be sold.

$f'(20) = -350$ means that as the price of the fabric increases past $\$20/\text{yard}$, the amount of fabric which will be sold is decreasing at a rate of 350 yards per (dollar per yard).

(b) $R(p) = pf(p) \quad \Rightarrow \quad R'(p) = pf'(p) + f(p) \cdot 1 \quad \Rightarrow \quad R'(20) = 20f'(20) + f(20) \cdot 1 = 20(-350) + 10{,}000 = 3000.$

This means that as the price of the fabric increases past $\$20/\text{yard}$, the total revenue is increasing at $\$3000/(\$/\text{yard})$. Note that the Product Rule indicates that we will lose $\$7000/(\$/\text{yard})$ due to selling less fabric, but this loss is more than made up for by the additional revenue due to the increase in price.

59. $v = \dfrac{0.14[\text{S}]}{0.015 + [\text{S}]} \quad \Rightarrow \quad \dfrac{dv}{d[\text{S}]} = \dfrac{(0.015 + [\text{S}])(0.14) - (0.14[\text{S}])(1)}{(0.015 + [\text{S}])^2} = \dfrac{0.0021}{(0.015 + [\text{S}])^2}.$

$dv/d[\text{S}]$ represents the rate of change of the rate of an enzymatic reaction with respect to the concentration of a substrate S.

60. $B(t) = N(t)\,M(t) \quad \Rightarrow \quad B'(t) = N(t)\,M'(t) + M(t)\,N'(t)$, so

$B'(4) = N(4)\,M'(4) + M(4)\,N'(4) = 820(0.14) + 1.2(50) = 174.8\text{ g/week}.$

61. (a) $(fgh)' = [(fg)h]' = (fg)'h + (fg)h' = (f'g + fg')h + (fg)h' = f'gh + fg'h + fgh'$

(b) Putting $f = g = h$ in part (a), we have $\dfrac{d}{dx}[f(x)]^3 = (fff)' = f'ff + ff'f + fff' = 3fff' = 3[f(x)]^2 f'(x)$.

(c) $\dfrac{d}{dx}\left(e^{3x}\right) = \dfrac{d}{dx}\left(e^x\right)^3 = 3(e^x)^2 e^x = 3e^{2x}e^x = 3e^{3x}$

62. (a) We use the Product Rule repeatedly: $F = fg \quad \Rightarrow \quad F' = f'g + fg' \quad \Rightarrow$

$F'' = (f''g + f'g') + (f'g' + fg'') = f''g + 2f'g' + fg''.$

(b) $F''' = f'''g + f''g' + 2\left(f''g' + f'g''\right) + f'g'' + fg''' = f'''g + 3f''g' + 3f'g'' + fg''' \quad \Rightarrow$

$$\begin{aligned} F^{(4)} &= f^{(4)}g + f'''g' + 3\left(f'''g' + f''g''\right) + 3\left(f''g'' + f'g'''\right) + f'g''' + fg^{(4)} \\ &= f^{(4)}g + 4f'''g' + 6f''g'' + 4f'g''' + fg^{(4)} \end{aligned}$$

(c) By analogy with the Binomial Theorem, we make the guess:

$$F^{(n)} = f^{(n)}g + nf^{(n-1)}g' + \binom{n}{2} f^{(n-2)}g'' + \cdots + \binom{n}{k} f^{(n-k)}g^{(k)} + \cdots + nf'g^{(n-1)} + fg^{(n)},$$

where $\dbinom{n}{k} = \dfrac{n!}{k!\,(n-k)!} = \dfrac{n(n-1)(n-2)\cdots(n-k+1)}{k!}$.

63. For $f(x) = x^2 e^x$, $f'(x) = x^2 e^x + e^x(2x) = e^x(x^2 + 2x)$. Similarly, we have

$$\begin{aligned} f''(x) &= e^x(x^2 + 4x + 2) \\ f'''(x) &= e^x(x^2 + 6x + 6) \\ f^{(4)}(x) &= e^x(x^2 + 8x + 12) \\ f^{(5)}(x) &= e^x(x^2 + 10x + 20) \end{aligned}$$

It appears that the coefficient of x in the quadratic term increases by 2 with each differentiation. The pattern for the constant terms seems to be $0 = 1 \cdot 0$, $2 = 2 \cdot 1$, $6 = 3 \cdot 2$, $12 = 4 \cdot 3$, $20 = 5 \cdot 4$. So a reasonable guess is that $f^{(n)}(x) = e^x[x^2 + 2nx + n(n-1)]$.

Proof: Let S_n be the statement that $f^{(n)}(x) = e^x[x^2 + 2nx + n(n-1)]$.

1. S_1 is true because $f'(x) = e^x(x^2 + 2x)$.
2. Assume that S_k is true; that is, $f^{(k)}(x) = e^x[x^2 + 2kx + k(k-1)]$. Then

$$\begin{aligned} f^{(k+1)}(x) &= \frac{d}{dx}\left[f^{(k)}(x)\right] = e^x(2x + 2k) + [x^2 + 2kx + k(k-1)]e^x \\ &= e^x[x^2 + (2k+2)x + (k^2 + k)] = e^x[x^2 + 2(k+1)x + (k+1)k] \end{aligned}$$

   This shows that S_{k+1} is true.

3. Therefore, by mathematical induction, S_n is true for all n; that is, $f^{(n)}(x) = e^x[x^2 + 2nx + n(n-1)]$ for every positive integer n.

64. (a) $\displaystyle \frac{d}{dx}\left(\frac{1}{g(x)}\right) = \frac{g(x) \cdot \frac{d}{dx}(1) - 1 \cdot \frac{d}{dx}[g(x)]}{[g(x)]^2}$ [Quotient Rule] $\displaystyle = \frac{g(x) \cdot 0 - 1 \cdot g'(x)}{[g(x)]^2} = \frac{0 - g'(x)}{[g(x)]^2} = -\frac{g'(x)}{[g(x)]^2}$

(b) $\displaystyle \frac{d}{dt}\left(\frac{1}{t^3 + 2t^2 - 1}\right) = -\frac{(t^3 + 2t^2 - 1)'}{(t^3 + 2t^2 - 1)^2} = -\frac{3t^2 + 4t}{(t^3 + 2t^2 - 1)^2}$

(c) $\displaystyle \frac{d}{dx}(x^{-n}) = \frac{d}{dx}\left(\frac{1}{x^n}\right) = -\frac{(x^n)'}{(x^n)^2}$ [by the Reciprocal Rule] $\displaystyle = -\frac{nx^{n-1}}{x^{2n}} = -nx^{n-1-2n} = -nx^{-n-1}$

3.3 Derivatives of Trigonometric Functions

1. $f(x) = x^2 \sin x \overset{\text{PR}}{\Rightarrow} f'(x) = x^2 \cos x + (\sin x)(2x) = x^2 \cos x + 2x \sin x$

2. $f(x) = x \cos x + 2 \tan x \Rightarrow f'(x) = x(-\sin x) + (\cos x)(1) + 2\sec^2 x = \cos x - x \sin x + 2\sec^2 x$

3. $f(x) = e^x \cos x \Rightarrow f'(x) = e^x(-\sin x) + (\cos x)e^x = e^x(\cos x - \sin x)$

4. $y = 2\sec x - \csc x \Rightarrow y' = 2(\sec x \tan x) - (-\csc x \cot x) = 2\sec x \tan x + \csc x \cot x$

5. $y = \sec\theta \tan\theta \Rightarrow y' = \sec\theta(\sec^2\theta) + \tan\theta(\sec\theta\tan\theta) = \sec\theta(\sec^2\theta + \tan^2\theta)$. Using the identity $1 + \tan^2\theta = \sec^2\theta$, we can write alternative forms of the answer as $\sec\theta(1 + 2\tan^2\theta)$ or $\sec\theta(2\sec^2\theta - 1)$.

6. $g(\theta) = e^\theta(\tan\theta - \theta) \Rightarrow g'(\theta) = e^\theta(\sec^2\theta - 1) + (\tan\theta - \theta)e^\theta = e^\theta(\sec^2\theta - 1 + \tan\theta - \theta)$

7. $y = c\cos t + t^2 \sin t \quad \Rightarrow \quad y' = c(-\sin t) + t^2(\cos t) + \sin t\,(2t) = -c\sin t + t(t\cos t + 2\sin t)$

8. $f(t) = \dfrac{\cot t}{e^t} \quad \Rightarrow \quad f'(t) = \dfrac{e^t(-\csc^2 t) - (\cot t)e^t}{(e^t)^2} = \dfrac{e^t(-\csc^2 t - \cot t)}{(e^t)^2} = -\dfrac{\csc^2 t + \cot t}{e^t}$

9. $y = \dfrac{x}{2 - \tan x} \quad \Rightarrow \quad y' = \dfrac{(2 - \tan x)(1) - x(-\sec^2 x)}{(2 - \tan x)^2} = \dfrac{2 - \tan x + x\sec^2 x}{(2 - \tan x)^2}$

10. $y = \sin\theta\,\cos\theta \quad \Rightarrow \quad y' = \sin\theta(-\sin\theta) + \cos\theta(\cos\theta) = \cos^2\theta - \sin^2\theta \quad [\text{or } \cos 2\theta]$

11. $f(\theta) = \dfrac{\sin\theta}{1 + \cos\theta} \quad \Rightarrow$

$$f'(\theta) = \frac{(1 + \cos\theta)\cos\theta - (\sin\theta)(-\sin\theta)}{(1 + \cos\theta)^2} = \frac{\cos\theta + \cos^2\theta + \sin^2\theta}{(1 + \cos\theta)^2} = \frac{\cos\theta + 1}{(1 + \cos\theta)^2} = \frac{1}{1 + \cos\theta}$$

12. $y = \dfrac{\cos x}{1 - \sin x} \quad \Rightarrow$

$$y' = \frac{(1 - \sin x)(-\sin x) - \cos x(-\cos x)}{(1 - \sin x)^2} = \frac{-\sin x + \sin^2 x + \cos^2 x}{(1 - \sin x)^2} = \frac{-\sin x + 1}{(1 - \sin x)^2} = \frac{1}{1 - \sin x}$$

13. $y = \dfrac{t\sin t}{1 + t} \quad \Rightarrow$

$$y' = \frac{(1 + t)(t\cos t + \sin t) - t\sin t(1)}{(1 + t)^2} = \frac{t\cos t + \sin t + t^2\cos t + t\sin t - t\sin t}{(1 + t)^2} = \frac{(t^2 + t)\cos t + \sin t}{(1 + t)^2}$$

14. $y = \dfrac{\sin t}{1 + \tan t} \quad \Rightarrow$

$$y' = \frac{(1 + \tan t)\cos t - (\sin t)\sec^2 t}{(1 + \tan t)^2} = \frac{\cos t + \sin t - \dfrac{\sin t}{\cos^2 t}}{(1 + \tan t)^2} = \frac{\cos t + \sin t - \tan t\sec t}{(1 + \tan t)^2}$$

15. Using Exercise 3.2.61(a), $f(\theta) = \theta\cos\theta\,\sin\theta \quad \Rightarrow$

$$\begin{aligned} f'(\theta) &= 1\cos\theta\,\sin\theta + \theta(-\sin\theta)\sin\theta + \theta\cos\theta(\cos\theta) = \cos\theta\,\sin\theta - \theta\sin^2\theta + \theta\cos^2\theta \\ &= \sin\theta\,\cos\theta + \theta(\cos^2\theta - \sin^2\theta) = \tfrac{1}{2}\sin 2\theta + \theta\cos 2\theta \quad [\text{using double-angle formulas}] \end{aligned}$$

16. Using Exercise 3.2.61(a), $f(t) = te^t\cot t \quad \Rightarrow$

$$f'(t) = 1e^t\cot t + te^t\cot t + te^t(-\csc^2 t) = e^t(\cot t + t\cot t - t\csc^2 t)$$

17. $\dfrac{d}{dx}(\csc x) = \dfrac{d}{dx}\left(\dfrac{1}{\sin x}\right) = \dfrac{(\sin x)(0) - 1(\cos x)}{\sin^2 x} = \dfrac{-\cos x}{\sin^2 x} = -\dfrac{1}{\sin x} \cdot \dfrac{\cos x}{\sin x} = -\csc x\,\cot x$

18. $\dfrac{d}{dx}(\sec x) = \dfrac{d}{dx}\left(\dfrac{1}{\cos x}\right) = \dfrac{(\cos x)(0) - 1(-\sin x)}{\cos^2 x} = \dfrac{\sin x}{\cos^2 x} = \dfrac{1}{\cos x} \cdot \dfrac{\sin x}{\cos x} = \sec x\,\tan x$

19. $\dfrac{d}{dx}(\cot x) = \dfrac{d}{dx}\left(\dfrac{\cos x}{\sin x}\right) = \dfrac{(\sin x)(-\sin x) - (\cos x)(\cos x)}{\sin^2 x} = -\dfrac{\sin^2 x + \cos^2 x}{\sin^2 x} = -\dfrac{1}{\sin^2 x} = -\csc^2 x$

20. $f(x) = \cos x \quad \Rightarrow$

$$f'(x) = \lim_{h \to 0} \frac{f(x+h) - f(x)}{h} = \lim_{h \to 0} \frac{\cos(x+h) - \cos x}{h} = \lim_{h \to 0} \frac{\cos x \cos h - \sin x \sin h - \cos x}{h}$$

$$= \lim_{h \to 0} \left(\cos x \frac{\cos h - 1}{h} - \sin x \frac{\sin h}{h} \right) = \cos x \lim_{h \to 0} \frac{\cos h - 1}{h} - \sin x \lim_{h \to 0} \frac{\sin h}{h}$$

$$= (\cos x)(0) - (\sin x)(1) = -\sin x$$

21. $y = \sin x + \cos x \quad \Rightarrow \quad y' = \cos x - \sin x$, so $y'(0) = \cos 0 - \sin 0 = 1 - 0 = 1$. An equation of the tangent line to the curve $y = \sin x + \cos x$ at the point $(0, 1)$ is $y - 1 = 1(x - 0)$ or $y = x + 1$.

22. $y = e^x \cos x \quad \Rightarrow \quad y' = e^x(-\sin x) + (\cos x)e^x = e^x(\cos x - \sin x) \quad \Rightarrow$ the slope of the tangent line at $(0, 1)$ is $e^0(\cos 0 - \sin 0) = 1(1 - 0) = 1$ and an equation is $y - 1 = 1(x - 0)$ or $y = x + 1$.

23. $y = \cos x - \sin x \quad \Rightarrow \quad y' = -\sin x - \cos x$, so $y'(\pi) = -\sin \pi - \cos \pi = 0 - (-1) = 1$. An equation of the tangent line to the curve $y = \cos x - \sin x$ at the point $(\pi, -1)$ is $y - (-1) = 1(x - \pi)$ or $y = x - \pi - 1$.

24. $y = x + \tan x \quad \Rightarrow \quad y' = 1 + \sec^2 x$, so $y'(\pi) = 1 + (-1)^2 = 2$. An equation of the tangent line to the curve $y = x + \tan x$ at the point (π, π) is $y - \pi = 2(x - \pi)$ or $y = 2x - \pi$.

25. (a) $y = 2x \sin x \quad \Rightarrow \quad y' = 2(x \cos x + \sin x \cdot 1)$. At $\left(\frac{\pi}{2}, \pi\right)$, $y' = 2\left(\frac{\pi}{2} \cos \frac{\pi}{2} + \sin \frac{\pi}{2}\right) = 2(0 + 1) = 2$, and an equation of the tangent line is $y - \pi = 2\left(x - \frac{\pi}{2}\right)$, or $y = 2x$.

(b)

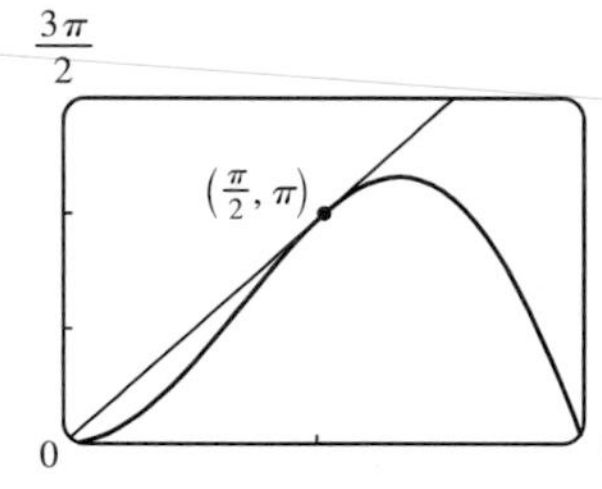

26. (a) $y = 3x + 6 \cos x \quad \Rightarrow \quad y' = 3 - 6 \sin x$. At $\left(\frac{\pi}{3}, \pi + 3\right)$, $y' = 3 - 6 \sin \frac{\pi}{3} = 3 - 6 \frac{\sqrt{3}}{2} = 3 - 3\sqrt{3}$, and an equation of the tangent line is $y - (\pi + 3) = \left(3 - 3\sqrt{3}\right)\left(x - \frac{\pi}{3}\right)$, or $y = \left(3 - 3\sqrt{3}\right)x + 3 + \pi\sqrt{3}$.

(b)

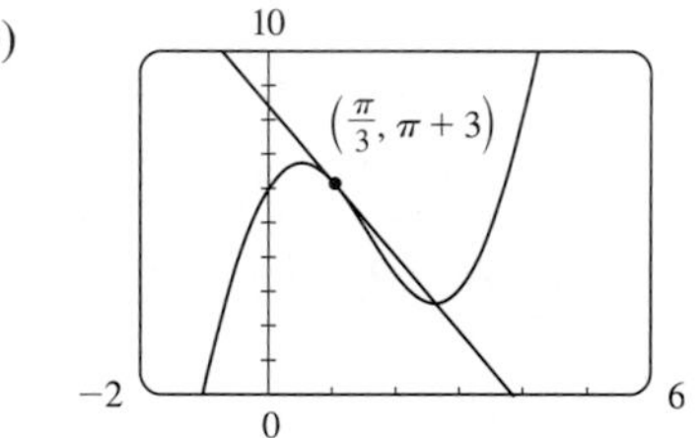

27. (a) $f(x) = \sec x - x \quad \Rightarrow \quad f'(x) = \sec x \tan x - 1$

(b)

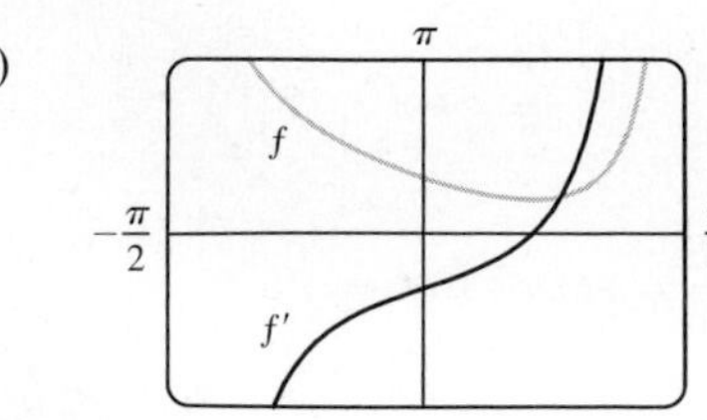

Note that $f' = 0$ where f has a minimum. Also note that f' is negative when f is decreasing and f' is positive when f is increasing.

28. (a) $f(x) = e^x \cos x \quad \Rightarrow \quad f'(x) = e^x(-\sin x) + (\cos x)\,e^x = e^x(\cos x - \sin x) \quad \Rightarrow$

$f''(x) = e^x(-\sin x - \cos x) + (\cos x - \sin x)\,e^x = e^x(-\sin x - \cos x + \cos x - \sin x) = -2e^x \sin x$

(b) 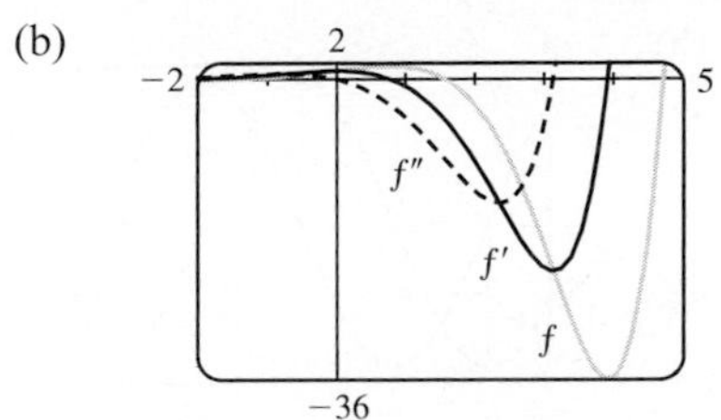

Note that $f' = 0$ where f has a minimum and $f'' = 0$ where f' has a minimum. Also note that f' is negative when f is decreasing and f'' is negative when f' is decreasing.

29. $H(\theta) = \theta\sin\theta \;\Rightarrow\; H'(\theta) = \theta(\cos\theta) + (\sin\theta)\cdot 1 = \theta\cos\theta + \sin\theta \;\Rightarrow$

$H''(\theta) = \theta(-\sin\theta) + (\cos\theta)\cdot 1 + \cos\theta = -\theta\sin\theta + 2\cos\theta$

30. $f(t) = \sec t \;\Rightarrow\; f'(t) = \sec t\,\tan t \;\Rightarrow\; f''(t) = (\sec t)\sec^2 t + (\tan t)\sec t\,\tan t = \sec^3 t + \sec t\,\tan^2 t$, so

$f''(\frac{\pi}{4}) = (\sqrt{2})^3 + \sqrt{2}(1)^2 = 2\sqrt{2} + \sqrt{2} = 3\sqrt{2}.$

31. (a) $f(x) = \dfrac{\tan x - 1}{\sec x} \;\Rightarrow$

$$f'(x) = \frac{\sec x(\sec^2 x) - (\tan x - 1)(\sec x\tan x)}{(\sec x)^2} = \frac{\sec x(\sec^2 x - \tan^2 x + \tan x)}{\sec^2 x} = \frac{1 + \tan x}{\sec x}$$

(b) $f(x) = \dfrac{\tan x - 1}{\sec x} = \dfrac{\dfrac{\sin x}{\cos x} - 1}{\dfrac{1}{\cos x}} = \dfrac{\dfrac{\sin x - \cos x}{\cos x}}{\dfrac{1}{\cos x}} = \sin x - \cos x \;\Rightarrow\; f'(x) = \cos x - (-\sin x) = \cos x + \sin x$

(c) From part (a), $f'(x) = \dfrac{1 + \tan x}{\sec x} = \dfrac{1}{\sec x} + \dfrac{\tan x}{\sec x} = \cos x + \sin x$, which is the expression for $f'(x)$ in part (b).

32. (a) $g(x) = f(x)\sin x \;\Rightarrow\; g'(x) = f(x)\cos x + \sin x\cdot f'(x)$, so

$g'(\frac{\pi}{3}) = f(\frac{\pi}{3})\cos\frac{\pi}{3} + \sin\frac{\pi}{3}\cdot f'(\frac{\pi}{3}) = 4\cdot\frac{1}{2} + \frac{\sqrt{3}}{2}\cdot(-2) = 2 - \sqrt{3}$

(b) $h(x) = \dfrac{\cos x}{f(x)} \;\Rightarrow\; h'(x) = \dfrac{f(x)\cdot(-\sin x) - \cos x\cdot f'(x)}{[f(x)]^2}$, so

$$h'(\tfrac{\pi}{3}) = \frac{f(\frac{\pi}{3})\cdot(-\sin\frac{\pi}{3}) - \cos\frac{\pi}{3}\cdot f'(\frac{\pi}{3})}{[f(\frac{\pi}{3})]^2} = \frac{4\left(-\frac{\sqrt{3}}{2}\right) - (\frac{1}{2})(-2)}{4^2} = \frac{-2\sqrt{3} + 1}{16} = \frac{1 - 2\sqrt{3}}{16}$$

33. $f(x) = x + 2\sin x$ has a horizontal tangent when $f'(x) = 0 \;\Leftrightarrow\; 1 + 2\cos x = 0 \;\Leftrightarrow\; \cos x = -\frac{1}{2} \;\Leftrightarrow$ $x = \frac{2\pi}{3} + 2\pi n$ or $\frac{4\pi}{3} + 2\pi n$, where n is an integer. Note that $\frac{4\pi}{3}$ and $\frac{2\pi}{3}$ are $\pm\frac{\pi}{3}$ units from π. This allows us to write the solutions in the more compact equivalent form $(2n + 1)\pi \pm \frac{\pi}{3}$, n an integer.

34. $f(x) = e^x\cos x$ has a horizontal tangent when $f'(x) = 0$. $f'(x) = e^x(-\sin x) + (\cos x)e^x = e^x(\cos x - \sin x)$.

$f'(x) = 0 \;\Leftrightarrow\; \cos x - \sin x = 0 \;\Leftrightarrow\; \cos x = \sin x \;\Leftrightarrow\; \tan x = 1 \;\Leftrightarrow\; x = \frac{\pi}{4} + n\pi$, n an integer.

35. (a) $x(t) = 8\sin t \;\Rightarrow\; v(t) = x'(t) = 8\cos t \;\Rightarrow\; a(t) = x''(t) = -8\sin t$

(b) The mass at time $t = \frac{2\pi}{3}$ has position $x(\frac{2\pi}{3}) = 8\sin\frac{2\pi}{3} = 8\left(\frac{\sqrt{3}}{2}\right) = 4\sqrt{3}$, velocity $v(\frac{2\pi}{3}) = 8\cos\frac{2\pi}{3} = 8(-\frac{1}{2}) = -4$, and acceleration $a(\frac{2\pi}{3}) = -8\sin\frac{2\pi}{3} = -8\left(\frac{\sqrt{3}}{2}\right) = -4\sqrt{3}$. Since $v(\frac{2\pi}{3}) < 0$, the particle is moving to the left.

36. (a) $s(t) = 2\cos t + 3\sin t \;\Rightarrow\; v(t) = -2\sin t + 3\cos t \;\Rightarrow$

$a(t) = -2\cos t - 3\sin t$

(b)

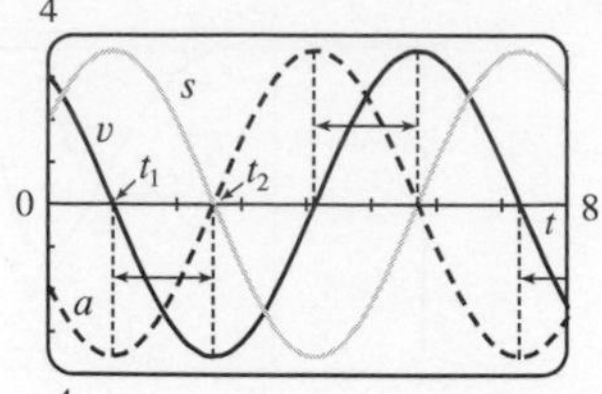

(c) $s = 0 \;\Rightarrow\; t_2 \approx 2.55.$ So the mass passes through the equilibrium position for the first time when $t \approx 2.55$ s.

(d) $v = 0 \;\Rightarrow\; t_1 \approx 0.98,\; s(t_1) \approx 3.61$ cm. So the mass travels a maximum of about 3.6 cm (upward and downward) from its equilibrium position.

(e) The speed $|v|$ is greatest when $s = 0$, that is, when $t = t_2 + n\pi$, n a positive integer.

37. From the diagram we can see that $\sin\theta = x/10 \;\Leftrightarrow\; x = 10\sin\theta$. We want to find the rate of change of x with respect to θ, that is, $dx/d\theta$. Taking the derivative of $x = 10\sin\theta$, we get $dx/d\theta = 10(\cos\theta)$. So when $\theta = \frac{\pi}{3}$, $\frac{dx}{d\theta} = 10\cos\frac{\pi}{3} = 10\left(\frac{1}{2}\right) = 5$ ft/rad.

38. (a) $$F = \frac{\mu W}{\mu\sin\theta + \cos\theta} \;\Rightarrow\; \frac{dF}{d\theta} = \frac{(\mu\sin\theta + \cos\theta)(0) - \mu W(\mu\cos\theta - \sin\theta)}{(\mu\sin\theta + \cos\theta)^2} = \frac{\mu W(\sin\theta - \mu\cos\theta)}{(\mu\sin\theta + \cos\theta)^2}$$

(b) $$\frac{dF}{d\theta} = 0 \;\Leftrightarrow\; \mu W(\sin\theta - \mu\cos\theta) = 0 \;\Leftrightarrow\; \sin\theta = \mu\cos\theta \;\Leftrightarrow\; \tan\theta = \mu \;\Leftrightarrow\; \theta = \tan^{-1}\mu$$

(c) From the graph of $F = \dfrac{0.6(50)}{0.6\sin\theta + \cos\theta}$ for $0 \le \theta \le 1$, we see that $\dfrac{dF}{d\theta} = 0 \;\Rightarrow\; \theta \approx 0.54$. Checking this with part (b) and $\mu = 0.6$, we calculate $\theta = \tan^{-1} 0.6 \approx 0.54$. So the value from the graph is consistent with the value in part (b).

39. $$\lim_{x\to 0}\frac{\sin 5x}{3x} = \lim_{x\to 0}\frac{5}{3}\left(\frac{\sin 5x}{5x}\right) = \frac{5}{3}\lim_{x\to 0}\frac{\sin 5x}{5x} = \frac{5}{3}\lim_{\theta\to 0}\frac{\sin\theta}{\theta} \quad [\theta = 5x] \quad = \frac{5}{3}\cdot 1 = \frac{5}{3}$$

40. $$\lim_{x\to 0}\frac{\sin x}{\sin \pi x} = \lim_{x\to 0}\frac{\sin x}{x}\cdot\frac{\pi x}{\sin \pi x}\cdot\frac{1}{\pi} = \lim_{x\to 0}\frac{\sin x}{x}\cdot\lim_{\theta\to 0}\frac{\theta}{\sin\theta}\cdot\frac{1}{\pi} \quad [\theta = \pi x]$$
$$= 1\cdot\lim_{\theta\to 0}\frac{1}{\dfrac{\sin\theta}{\theta}}\cdot\frac{1}{\pi} = 1\cdot 1\cdot\frac{1}{\pi} = \frac{1}{\pi}$$

41. $$\lim_{t\to 0}\frac{\tan 6t}{\sin 2t} = \lim_{t\to 0}\left(\frac{\sin 6t}{t}\cdot\frac{1}{\cos 6t}\cdot\frac{t}{\sin 2t}\right) = \lim_{t\to 0}\frac{6\sin 6t}{6t}\cdot\lim_{t\to 0}\frac{1}{\cos 6t}\cdot\lim_{t\to 0}\frac{2t}{2\sin 2t}$$
$$= 6\lim_{t\to 0}\frac{\sin 6t}{6t}\cdot\lim_{t\to 0}\frac{1}{\cos 6t}\cdot\frac{1}{2}\lim_{t\to 0}\frac{2t}{\sin 2t} = 6(1)\cdot\frac{1}{1}\cdot\frac{1}{2}(1) = 3$$

42. $$\lim_{\theta\to 0}\frac{\cos\theta - 1}{\sin\theta} = \lim_{\theta\to 0}\frac{\dfrac{\cos\theta - 1}{\theta}}{\dfrac{\sin\theta}{\theta}} = \frac{\lim\limits_{\theta\to 0}\dfrac{\cos\theta - 1}{\theta}}{\lim\limits_{\theta\to 0}\dfrac{\sin\theta}{\theta}} = \frac{0}{1} = 0$$

43. $\displaystyle\lim_{x\to 0}\frac{\sin 3x}{5x^3-4x}=\lim_{x\to 0}\left(\frac{\sin 3x}{3x}\cdot\frac{3}{5x^2-4}\right)=\lim_{x\to 0}\frac{\sin 3x}{3x}\cdot\lim_{x\to 0}\frac{3}{5x^2-4}=1\cdot\left(\frac{3}{-4}\right)=-\frac{3}{4}$

44. $\displaystyle\lim_{x\to 0}\frac{\sin 3x\sin 5x}{x^2}=\lim_{x\to 0}\left(\frac{3\sin 3x}{3x}\cdot\frac{5\sin 5x}{5x}\right)=\lim_{x\to 0}\frac{3\sin 3x}{3x}\cdot\lim_{x\to 0}\frac{5\sin 5x}{5x}$

$$=3\lim_{x\to 0}\frac{\sin 3x}{3x}\cdot 5\lim_{x\to 0}\frac{\sin 5x}{5x}=3(1)\cdot 5(1)=15$$

45. Divide numerator and denominator by θ. ($\sin\theta$ also works.)

$$\lim_{\theta\to 0}\frac{\sin\theta}{\theta+\tan\theta}=\lim_{\theta\to 0}\frac{\dfrac{\sin\theta}{\theta}}{1+\dfrac{\sin\theta}{\theta}\cdot\dfrac{1}{\cos\theta}}=\frac{\lim\limits_{\theta\to 0}\dfrac{\sin\theta}{\theta}}{1+\lim\limits_{\theta\to 0}\dfrac{\sin\theta}{\theta}\lim\limits_{\theta\to 0}\dfrac{1}{\cos\theta}}=\frac{1}{1+1\cdot 1}=\frac{1}{2}$$

46. $\displaystyle\lim_{x\to 0}\csc x\,\sin(\sin x)=\lim_{x\to 0}\frac{\sin(\sin x)}{\sin x}=\lim_{\theta\to 0}\frac{\sin\theta}{\theta}$ [As $x\to 0$, $\theta=\sin x\to 0$.] $=1$

47. $\displaystyle\lim_{\theta\to 0}\frac{\cos\theta-1}{2\theta^2}=\lim_{\theta\to 0}\frac{\cos\theta-1}{2\theta^2}\cdot\frac{\cos\theta+1}{\cos\theta+1}=\lim_{\theta\to 0}\frac{\cos^2\theta-1}{2\theta^2(\cos\theta+1)}=\lim_{\theta\to 0}\frac{-\sin^2\theta}{2\theta^2(\cos\theta+1)}$

$$=-\frac{1}{2}\lim_{\theta\to 0}\frac{\sin\theta}{\theta}\cdot\frac{\sin\theta}{\theta}\cdot\frac{1}{\cos\theta+1}=-\frac{1}{2}\lim_{\theta\to 0}\frac{\sin\theta}{\theta}\cdot\lim_{\theta\to 0}\frac{\sin\theta}{\theta}\cdot\lim_{\theta\to 0}\frac{1}{\cos\theta+1}$$

$$=-\frac{1}{2}\cdot 1\cdot 1\cdot\frac{1}{1+1}=-\frac{1}{4}$$

48. $\displaystyle\lim_{x\to 0}\frac{\sin(x^2)}{x}=\lim_{x\to 0}\left[x\cdot\frac{\sin(x^2)}{x\cdot x}\right]=\lim_{x\to 0}x\cdot\lim_{x\to 0}\frac{\sin(x^2)}{x^2}=0\cdot\lim_{y\to 0^+}\frac{\sin y}{y}$ [where $y=x^2$] $=0\cdot 1=0$

49. $\displaystyle\lim_{x\to\pi/4}\frac{1-\tan x}{\sin x-\cos x}=\lim_{x\to\pi/4}\frac{\left(1-\dfrac{\sin x}{\cos x}\right)\cdot\cos x}{(\sin x-\cos x)\cdot\cos x}=\lim_{x\to\pi/4}\frac{\cos x-\sin x}{(\sin x-\cos x)\cos x}=\lim_{x\to\pi/4}\frac{-1}{\cos x}=\frac{-1}{1/\sqrt{2}}=-\sqrt{2}$

50. $\displaystyle\lim_{x\to 1}\frac{\sin(x-1)}{x^2+x-2}=\lim_{x\to 1}\frac{\sin(x-1)}{(x+2)(x-1)}=\lim_{x\to 1}\frac{1}{x+2}\lim_{x\to 1}\frac{\sin(x-1)}{x-1}=\frac{1}{3}\cdot 1=\frac{1}{3}$

51. $\displaystyle\frac{d}{dx}(\sin x)=\cos x\;\Rightarrow\;\frac{d^2}{dx^2}(\sin x)=-\sin x\;\Rightarrow\;\frac{d^3}{dx^3}(\sin x)=-\cos x\;\Rightarrow\;\frac{d^4}{dx^4}(\sin x)=\sin x.$

The derivatives of $\sin x$ occur in a cycle of four. Since $99=4(24)+3$, we have $\displaystyle\frac{d^{99}}{dx^{99}}(\sin x)=\frac{d^3}{dx^3}(\sin x)=-\cos x.$

52. Let $f(x)=x\sin x$ and $h(x)=\sin x$, so $f(x)=xh(x)$. Then $f'(x)=h(x)+xh'(x)$,

$f''(x)=h'(x)+h'(x)+xh''(x)=2h'(x)+xh''(x)$,

$f'''(x)=2h''(x)+h''(x)+xh'''(x)=3h''(x)+xh'''(x),\cdots,f^{(n)}(x)=nh^{(n-1)}(x)+xh^{(n)}(x)$.

Since $34=4(8)+2$, we have $h^{(34)}(x)=h^{(2)}(x)=\dfrac{d^2}{dx^2}(\sin x)=-\sin x$ and $h^{(35)}(x)=-\cos x$.

Thus, $\dfrac{d^{35}}{dx^{35}}(x\sin x)=35h^{(34)}(x)+xh^{(35)}(x)=-35\sin x-x\cos x$.

53. $y = A\sin x + B\cos x \;\Rightarrow\; y' = A\cos x - B\sin x \;\Rightarrow\; y'' = -A\sin x - B\cos x$. Substituting these expressions for y, y', and y'' into the given differential equation $y'' + y' - 2y = \sin x$ gives us $(-A\sin x - B\cos x) + (A\cos x - B\sin x) - 2(A\sin x + B\cos x) = \sin x \;\Leftrightarrow$ $-3A\sin x - B\sin x + A\cos x - 3B\cos x = \sin x \;\Leftrightarrow\; (-3A - B)\sin x + (A - 3B)\cos x = 1\sin x$, so we must have $-3A - B = 1$ and $A - 3B = 0$ (since 0 is the coefficient of $\cos x$ on the right side). Solving for A and B, we add the first equation to three times the second to get $B = -\frac{1}{10}$ and $A = -\frac{3}{10}$.

54. (a) Let $\theta = \dfrac{1}{x}$. Then as $x \to \infty$, $\theta \to 0^+$, and $\displaystyle\lim_{x\to\infty} x\sin\frac{1}{x} = \lim_{\theta\to 0^+}\frac{1}{\theta}\sin\theta = \lim_{\theta\to 0}\frac{\sin\theta}{\theta} = 1$.

(b) Since $-1 \le \sin(1/x) \le 1$, we have (as illustrated in the figure) $-|x| \le x\sin(1/x) \le |x|$. We know that $\displaystyle\lim_{x\to 0}(|x|) = 0$ and $\displaystyle\lim_{x\to 0}(-|x|) = 0$; so by the Squeeze Theorem, $\displaystyle\lim_{x\to 0} x\sin(1/x) = 0$.

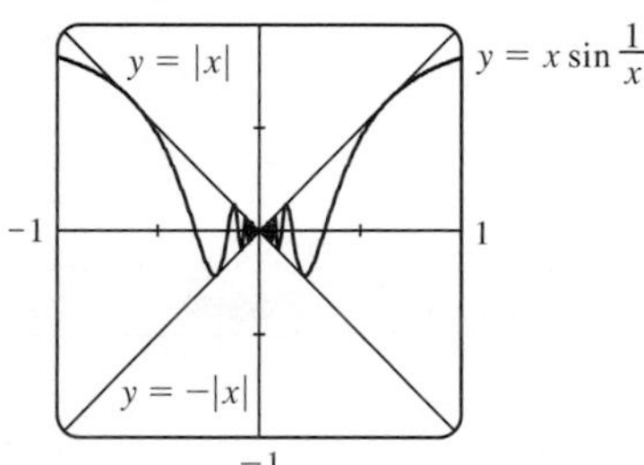

(c)

55. (a) $\dfrac{d}{dx}\tan x = \dfrac{d}{dx}\dfrac{\sin x}{\cos x} \;\Rightarrow\; \sec^2 x = \dfrac{\cos x\,\cos x - \sin x\,(-\sin x)}{\cos^2 x} = \dfrac{\cos^2 x + \sin^2 x}{\cos^2 x}$. So $\sec^2 x = \dfrac{1}{\cos^2 x}$.

(b) $\dfrac{d}{dx}\sec x = \dfrac{d}{dx}\dfrac{1}{\cos x} \;\Rightarrow\; \sec x\,\tan x = \dfrac{(\cos x)(0) - 1(-\sin x)}{\cos^2 x}$. So $\sec x\,\tan x = \dfrac{\sin x}{\cos^2 x}$.

(c) $\dfrac{d}{dx}(\sin x + \cos x) = \dfrac{d}{dx}\dfrac{1 + \cot x}{\csc x} \;\Rightarrow$

$$\cos x - \sin x = \frac{\csc x\,(-\csc^2 x) - (1 + \cot x)(-\csc x\,\cot x)}{\csc^2 x} = \frac{\csc x\,[-\csc^2 x + (1 + \cot x)\,\cot x]}{\csc^2 x}$$

$$= \frac{-\csc^2 x + \cot^2 x + \cot x}{\csc x} = \frac{-1 + \cot x}{\csc x}$$

So $\cos x - \sin x = \dfrac{\cot x - 1}{\csc x}$.

56. We get the following formulas for r and h in terms of θ:

$$\sin\frac{\theta}{2} = \frac{r}{10} \;\Rightarrow\; r = 10\sin\frac{\theta}{2} \quad\text{and}\quad \cos\frac{\theta}{2} = \frac{h}{10} \;\Rightarrow\; h = 10\cos\frac{\theta}{2}$$

Now $A(\theta) = \frac{1}{2}\pi r^2$ and $B(\theta) = \frac{1}{2}(2r)h = rh$. So

$$\lim_{\theta\to 0^+}\frac{A(\theta)}{B(\theta)} = \lim_{\theta\to 0^+}\frac{\frac{1}{2}\pi r^2}{rh} = \tfrac{1}{2}\pi\lim_{\theta\to 0^+}\frac{r}{h} = \tfrac{1}{2}\pi\lim_{\theta\to 0^+}\frac{10\sin(\theta/2)}{10\cos(\theta/2)}$$

$$= \tfrac{1}{2}\pi\lim_{\theta\to 0^+}\tan(\theta/2) = 0$$

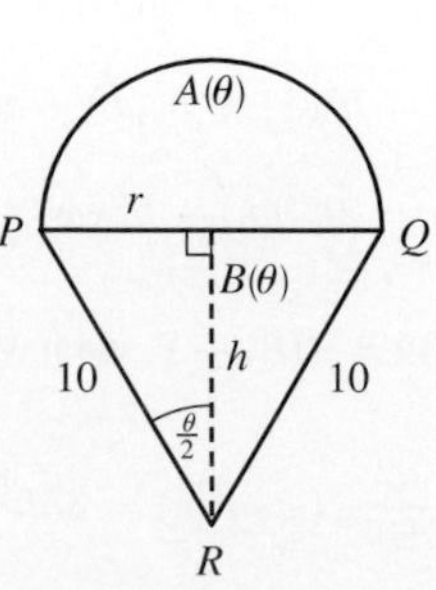

57. By the definition of radian measure, $s = r\theta$, where r is the radius of the circle. By drawing the bisector of the angle θ, we can see that $\sin\frac{\theta}{2} = \frac{d/2}{r} \Rightarrow d = 2r\sin\frac{\theta}{2}$. So $\lim\limits_{\theta\to 0^+}\frac{s}{d} = \lim\limits_{\theta\to 0^+}\frac{r\theta}{2r\sin(\theta/2)} = \lim\limits_{\theta\to 0^+}\frac{2\cdot(\theta/2)}{2\sin(\theta/2)} = \lim\limits_{\theta\to 0}\frac{\theta/2}{\sin(\theta/2)} = 1$.

[This is just the reciprocal of the limit $\lim\limits_{x\to 0}\frac{\sin x}{x} = 1$ combined with the fact that as $\theta\to 0$, $\frac{\theta}{2}\to 0$ also.]

58. (a)

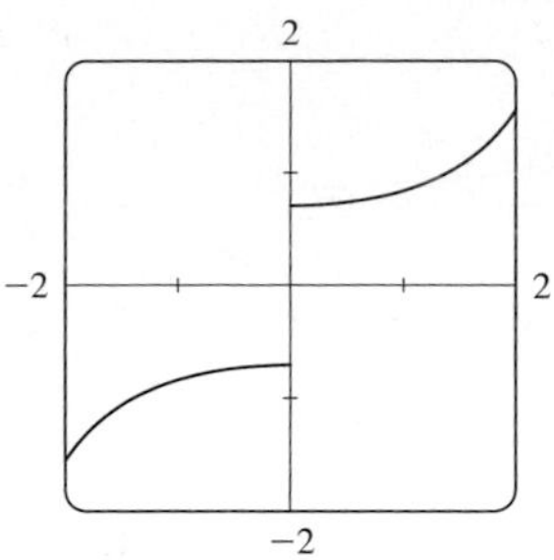

It appears that $f(x) = \dfrac{x}{\sqrt{1-\cos 2x}}$ has a jump discontinuity at $x = 0$.

(b) Using the identity $\cos 2x = 1 - \sin^2 x$, we have $\dfrac{x}{\sqrt{1-\cos 2x}} = \dfrac{x}{\sqrt{1-(1-2\sin^2 x)}} = \dfrac{x}{\sqrt{2\sin^2 x}} = \dfrac{x}{\sqrt{2}\,|\sin x|}$.

Thus,

$$\lim_{x\to 0^-}\frac{x}{\sqrt{1-\cos 2x}} = \lim_{x\to 0^-}\frac{x}{\sqrt{2}\,|\sin x|} = \frac{1}{\sqrt{2}}\lim_{x\to 0^-}\frac{x}{-(\sin x)}$$

$$= -\frac{1}{\sqrt{2}}\lim_{x\to 0^-}\frac{1}{\sin x/x} = -\frac{1}{\sqrt{2}}\cdot\frac{1}{1} = -\frac{\sqrt{2}}{2}$$

Evaluating $\lim\limits_{x\to 0^+} f(x)$ is similar, but $|\sin x| = +\sin x$, so we get $\frac{1}{2}\sqrt{2}$. These values appear to be reasonable values for the graph, so they confirm our answer to part (a).

Another method: Multiply numerator and denominator by $\sqrt{1+\cos 2x}$.

3.4 The Chain Rule

1. Let $u = g(x) = 1 + 4x$ and $y = f(u) = \sqrt[3]{u}$. Then $\dfrac{dy}{dx} = \dfrac{dy}{du}\dfrac{du}{dx} = \left(\frac{1}{3}u^{-2/3}\right)(4) = \dfrac{4}{3\sqrt[3]{(1+4x)^2}}$.

2. Let $u = g(x) = 2x^3 + 5$ and $y = f(u) = u^4$. Then $\dfrac{dy}{dx} = \dfrac{dy}{du}\dfrac{du}{dx} = (4u^3)(6x^2) = 24x^2(2x^3+5)^3$.

3. Let $u = g(x) = \pi x$ and $y = f(u) = \tan u$. Then $\dfrac{dy}{dx} = \dfrac{dy}{du}\dfrac{du}{dx} = (\sec^2 u)(\pi) = \pi\sec^2 \pi x$.

4. Let $u = g(x) = \cot x$ and $y = f(u) = \sin u$. Then $\dfrac{dy}{dx} = \dfrac{dy}{du}\dfrac{du}{dx} = (\cos u)(-\csc^2 x) = -\cos(\cot x)\csc^2 x$.

5. Let $u = g(x) = \sqrt{x}$ and $y = f(u) = e^u$. Then $\dfrac{dy}{dx} = \dfrac{dy}{du}\dfrac{du}{dx} = (e^u)\left(\frac{1}{2}x^{-1/2}\right) = e^{\sqrt{x}}\cdot\dfrac{1}{2\sqrt{x}} = \dfrac{e^{\sqrt{x}}}{2\sqrt{x}}$.

6. Let $u = g(x) = 2 - e^x$ and $y = f(u) = \sqrt{u}$. Then $\dfrac{dy}{dx} = \dfrac{dy}{du}\dfrac{du}{dx} = \left(\frac{1}{2}u^{-1/2}\right)(-e^x) = -\dfrac{e^x}{2\sqrt{2-e^x}}$.

7. $F(x) = (5x^6 + 2x^3)^4 \Rightarrow F'(x) = 4(5x^6+2x^3)^3\cdot\dfrac{d}{dx}(5x^6+2x^3) = 4(5x^6+2x^3)^3(30x^5+6x^2)$.

We can factor as follows: $4(x^3)^3(5x^3+2)^3 6x^2(5x^3+1) = 24x^{11}(5x^3+2)^3(5x^3+1)$

8. $F(x) = (1 + x + x^2)^{99} \quad \Rightarrow \quad F'(x) = 99(1 + x + x^2)^{98} \cdot \dfrac{d}{dx}\,(1 + x + x^2) = 99(1 + x + x^2)^{98}(1 + 2x)$

9. $f(x) = \sqrt{5x + 1} = (5x + 1)^{1/2} \quad \Rightarrow \quad f'(x) = \frac{1}{2}(5x + 1)^{-1/2}(5) = \dfrac{5}{2\sqrt{5x + 1}}$

10. $f(x) = \dfrac{1}{\sqrt[3]{x^2 - 1}} = (x^2 - 1)^{-1/3} \quad \Rightarrow \quad f'(x) = -\dfrac{1}{3}(x^2 - 1)^{-4/3}(2x) = \dfrac{-2x}{3(x^2 - 1)^{4/3}}$

11. $f(\theta) = \cos(\theta^2) \quad \Rightarrow \quad f'(\theta) = -\sin(\theta^2) \cdot \dfrac{d}{d\theta}\,(\theta^2) = -\sin(\theta^2) \cdot (2\theta) = -2\theta \sin(\theta^2)$

12. $g(\theta) = \cos^2 \theta = (\cos \theta)^2 \quad \Rightarrow \quad g'(\theta) = 2\,(\cos \theta)^1\,(-\sin \theta) = -2 \sin \theta\, \cos \theta = -\sin 2\theta$

13. $y = x^2 e^{-3x} \quad \Rightarrow \quad y' = x^2 e^{-3x}(-3) + e^{-3x}(2x) = e^{-3x}(-3x^2 + 2x) = xe^{-3x}(2 - 3x)$

14. $f(t) = t \sin \pi t \quad \Rightarrow \quad f'(t) = t(\cos \pi t) \cdot \pi + (\sin \pi t) \cdot 1 = \pi t \cos \pi t + \sin \pi t$

15. $f(t) = e^{at} \sin bt \quad \Rightarrow \quad f'(t) = e^{at}(\cos bt) \cdot b + (\sin bt)e^{at} \cdot a = e^{at}(b \cos bt + a \sin bt)$

16. $g(x) = e^{x^2 - x} \quad \Rightarrow \quad g'(x) = e^{x^2 - x}(2x - 1)$

17. $f(x) = (2x - 3)^4(x^2 + x + 1)^5 \quad \Rightarrow$

$$\begin{aligned} f'(x) &= (2x - 3)^4 \cdot 5(x^2 + x + 1)^4(2x + 1) + (x^2 + x + 1)^5 \cdot 4(2x - 3)^3 \cdot 2 \\ &= (2x - 3)^3(x^2 + x + 1)^4[(2x - 3) \cdot 5(2x + 1) + (x^2 + x + 1) \cdot 8] \\ &= (2x - 3)^3(x^2 + x + 1)^4(20x^2 - 20x - 15 + 8x^2 + 8x + 8) = (2x - 3)^3(x^2 + x + 1)^4(28x^2 - 12x - 7) \end{aligned}$$

18. $g(x) = (x^2 + 1)^3(x^2 + 2)^6 \quad \Rightarrow$

$$\begin{aligned} g'(x) &= (x^2 + 1)^3 \cdot 6(x^2 + 2)^5 \cdot 2x + (x^2 + 2)^6 \cdot 3(x^2 + 1)^2 \cdot 2x \\ &= 6x(x^2 + 1)^2(x^2 + 2)^5[2(x^2 + 1) + (x^2 + 2)] = 6x(x^2 + 1)^2(x^2 + 2)^5(3x^2 + 4) \end{aligned}$$

19. $h(t) = (t + 1)^{2/3}(2t^2 - 1)^3 \quad \Rightarrow$

$$\begin{aligned} h'(t) &= (t + 1)^{2/3} \cdot 3(2t^2 - 1)^2 \cdot 4t + (2t^2 - 1)^3 \cdot \tfrac{2}{3}(t + 1)^{-1/3} = \tfrac{2}{3}(t + 1)^{-1/3}(2t^2 - 1)^2[18t(t + 1) + (2t^2 - 1)] \\ &= \tfrac{2}{3}(t + 1)^{-1/3}(2t^2 - 1)^2(20t^2 + 18t - 1) \end{aligned}$$

20. $F(t) = (3t - 1)^4(2t + 1)^{-3} \quad \Rightarrow$

$$\begin{aligned} F'(t) &= (3t - 1)^4(-3)(2t + 1)^{-4}(2) + (2t + 1)^{-3} \cdot 4(3t - 1)^3(3) \\ &= 6(3t - 1)^3(2t + 1)^{-4}[-(3t - 1) + 2(2t + 1)] = 6(3t - 1)^3(2t + 1)^{-4}(t + 3) \end{aligned}$$

21. $y = \sqrt{\dfrac{x}{x + 1}} = \left(\dfrac{x}{x + 1}\right)^{1/2} \quad \Rightarrow$

$$\begin{aligned} y' &= \frac{1}{2}\left(\frac{x}{x + 1}\right)^{-1/2} \frac{d}{dx}\left(\frac{x}{x + 1}\right) = \frac{1}{2}\,\frac{x^{-1/2}}{(x + 1)^{-1/2}}\,\frac{(x + 1)(1) - x(1)}{(x + 1)^2} \\ &= \frac{1}{2}\,\frac{(x + 1)^{1/2}}{x^{1/2}}\,\frac{1}{(x + 1)^2} = \frac{1}{2\sqrt{x}\,(x + 1)^{3/2}} \end{aligned}$$

22. $y = \left(x + \frac{1}{x}\right)^5 \quad\Rightarrow\quad y' = 5\left(x + \frac{1}{x}\right)^4 \frac{d}{dx}\left(x + \frac{1}{x}\right) = 5\left(x + \frac{1}{x}\right)^4 \left(1 - \frac{1}{x^2}\right).$

Another form of the answer is $\dfrac{5(x^2+1)^4(x^2-1)}{x^6}$.

23. $y = e^{\tan\theta} \quad\Rightarrow\quad y' = e^{\tan\theta} \frac{d}{d\theta}(\tan\theta) = (\sec^2\theta)e^{\tan\theta}$

24. Using Formula 5 and the Chain Rule, $f(t) = 2^{t^3} \quad\Rightarrow\quad f'(t) = 2^{t^3} \ln 2 \frac{d}{dt}(t^3) = 3(\ln 2)t^2 2^{t^3}$.

25. $g(u) = \left(\dfrac{u^3-1}{u^3+1}\right)^8 \quad\Rightarrow$

$$g'(u) = 8\left(\frac{u^3-1}{u^3+1}\right)^7 \frac{d}{du}\frac{u^3-1}{u^3+1} = 8\frac{(u^3-1)^7}{(u^3+1)^7}\frac{(u^3+1)(3u^2)-(u^3-1)(3u^2)}{(u^3+1)^2}$$

$$= 8\frac{(u^3-1)^7}{(u^3+1)^7}\frac{3u^2[(u^3+1)-(u^3-1)]}{(u^3+1)^2} = 8\frac{(u^3-1)^7}{(u^3+1)^7}\frac{3u^2(2)}{(u^3+1)^2} = \frac{48u^2(u^3-1)^7}{(u^3+1)^9}$$

26. $s(t) = \sqrt{\dfrac{1+\sin t}{1+\cos t}} = \left(\dfrac{1+\sin t}{1+\cos t}\right)^{1/2} \quad\Rightarrow$

$$s'(t) = \frac{1}{2}\left(\frac{1+\sin t}{1+\cos t}\right)^{-1/2} \frac{(1+\cos t)\cos t - (1+\sin t)(-\sin t)}{(1+\cos t)^2}$$

$$= \frac{1}{2}\frac{(1+\sin t)^{-1/2}}{(1+\cos t)^{-1/2}}\frac{\cos t + \cos^2 t + \sin t + \sin^2 t}{(1+\cos t)^2} = \frac{\cos t + \sin t + 1}{2\sqrt{1+\sin t}\,(1+\cos t)^{3/2}}$$

27. Using Formula 5 and the Chain Rule, $r(t) = 10^{2\sqrt{t}} \quad\Rightarrow$

$$r'(t) = 10^{2\sqrt{t}} \ln 10 \frac{d}{dt}\left(2\sqrt{t}\right) = 10^{2\sqrt{t}} \ln 10 \left(2 \cdot \tfrac{1}{2}t^{-1/2}\right) = \frac{(\ln 10)\,10^{2\sqrt{t}}}{\sqrt{t}}$$

28. $f(z) = e^{z/(z-1)} \quad\Rightarrow\quad f'(z) = e^{z/(z-1)} \dfrac{d}{dz}\dfrac{z}{z-1} = e^{z/(z-1)}\dfrac{(z-1)(1)-z(1)}{(z-1)^2} = -\dfrac{e^{z/(z-1)}}{(z-1)^2}$

29. $H(r) = \dfrac{(r^2-1)^3}{(2r+1)^5} \quad\Rightarrow$

$$H'(r) = \frac{(2r+1)^5 \cdot 3(r^2-1)^2(2r) - (r^2-1)^3 \cdot 5(2r+1)^4(2)}{[(2r+1)^5]^2} = \frac{2(2r+1)^4(r^2-1)^2[3r(2r+1)-5(r^2-1)]}{(2r+1)^{10}}$$

$$= \frac{2(r^2-1)^2(6r^2+3r-5r^2+5)}{(2r+1)^6} = \frac{2(r^2-1)^2(r^2+3r+5)}{(2r+1)^6}$$

30. $J(\theta) = \tan^2(n\theta) = [\tan(n\theta)]^2 \quad\Rightarrow$

$$J'(\theta) = 2\,[\tan(n\theta)]^1 \frac{d}{d\theta}\tan(n\theta) = 2\tan(n\theta)\,\sec^2(n\theta)\cdot n = 2n\tan(n\theta)\,\sec^2(n\theta)$$

31. By (9), $F(t) = e^{t\sin 2t} \quad\Rightarrow$

$$F'(t) = e^{t\sin 2t}(t\sin 2t)' = e^{t\sin 2t}(t \cdot 2\cos 2t + \sin 2t \cdot 1) = e^{t\sin 2t}(2t\cos 2t + \sin 2t)$$

32. $F(t)=\dfrac{t^2}{\sqrt{t^3+1}} \quad \Rightarrow$

$$F'(t)=\frac{(t^3+1)^{1/2}(2t)-t^2\cdot\frac{1}{2}(t^3+1)^{-1/2}(3t^2)}{\left(\sqrt{t^3+1}\right)^2}=\frac{t(t^3+1)^{-1/2}\left[2(t^3+1)-\frac{3}{2}t^3\right]}{(t^3+1)^1}$$

$$=\frac{t\left(\frac{1}{2}t^3+2\right)}{(t^3+1)^{3/2}}=\frac{t(t^3+4)}{2(t^3+1)^{3/2}}$$

33. Using Formula 5 and the Chain Rule, $G(x)=4^{C/x} \quad \Rightarrow$

$$G'(x)=4^{C/x}\,(\ln 4)\,\frac{d}{dx}\frac{C}{x}\quad\left[\frac{C}{x}=Cx^{-1}\right]\quad=4^{C/x}\,(\ln 4)\left(-Cx^{-2}\right)=-C\,(\ln 4)\,\frac{4^{C/x}}{x^2}$$

34. $U(y)=\left(\dfrac{y^4+1}{y^2+1}\right)^5 \quad \Rightarrow$

$$U'(y)=5\left(\frac{y^4+1}{y^2+1}\right)^4\frac{(y^2+1)(4y^3)-(y^4+1)(2y)}{(y^2+1)^2}=\frac{5(y^4+1)^4 2y[2y^2(y^2+1)-(y^4+1)]}{(y^2+1)^4(y^2+1)^2}$$

$$=\frac{10y(y^4+1)^4(y^4+2y^2-1)}{(y^2+1)^6}$$

35. $y=\cos\left(\dfrac{1-e^{2x}}{1+e^{2x}}\right) \quad \Rightarrow$

$$y'=-\sin\left(\frac{1-e^{2x}}{1+e^{2x}}\right)\cdot\frac{d}{dx}\left(\frac{1-e^{2x}}{1+e^{2x}}\right)=-\sin\left(\frac{1-e^{2x}}{1+e^{2x}}\right)\cdot\frac{(1+e^{2x})(-2e^{2x})-(1-e^{2x})(2e^{2x})}{(1+e^{2x})^2}$$

$$=-\sin\left(\frac{1-e^{2x}}{1+e^{2x}}\right)\cdot\frac{-2e^{2x}\left[(1+e^{2x})+(1-e^{2x})\right]}{(1+e^{2x})^2}=-\sin\left(\frac{1-e^{2x}}{1+e^{2x}}\right)\cdot\frac{-2e^{2x}(2)}{(1+e^{2x})^2}=\frac{4e^{2x}}{(1+e^{2x})^2}\cdot\sin\left(\frac{1-e^{2x}}{1+e^{2x}}\right)$$

36. $y=x^2e^{-1/x} \quad \Rightarrow \quad y'=x^2e^{-1/x}\left(\dfrac{1}{x^2}\right)+e^{-1/x}(2x)=e^{-1/x}+2xe^{-1/x}=e^{-1/x}(1+2x)$

37. $y=\cot^2(\sin\theta)=[\cot(\sin\theta)]^2 \quad \Rightarrow$

$$y'=2[\cot(\sin\theta)]\cdot\frac{d}{d\theta}[\cot(\sin\theta)]=2\cot(\sin\theta)\cdot[-\csc^2(\sin\theta)\cdot\cos\theta]=-2\cos\theta\,\cot(\sin\theta)\,\csc^2(\sin\theta)$$

38. $y=\sqrt{1+xe^{-2x}} \quad \Rightarrow \quad y'=\frac{1}{2}\left(1+xe^{-2x}\right)^{-1/2}\left[x(-2e^{-2x})+e^{-2x}\right]=\dfrac{e^{-2x}(-2x+1)}{2\sqrt{1+xe^{-2x}}}$

39. $f(t)=\tan(\sec(\cos t)) \quad \Rightarrow$

$$f'(t)=\sec^2(\sec(\cos t))\,\frac{d}{dt}\,\sec(\cos t)=\sec^2(\sec(\cos t))[\sec(\cos t)\,\tan(\cos t)]\,\frac{d}{dt}\,\cos t$$

$$=-\sec^2(\sec(\cos t))\,\sec(\cos t)\,\tan(\cos t)\,\sin t$$

40. $y=e^{\sin 2x}+\sin(e^{2x}) \quad \Rightarrow$

$$y'=e^{\sin 2x}\,\frac{d}{dx}\,\sin 2x+\cos(e^{2x})\frac{d}{dx}e^{2x}=e^{\sin 2x}(\cos 2x)\cdot 2+\cos(e^{2x})\,e^{2x}\cdot 2$$

$$=2\cos 2x\,e^{\sin 2x}+2e^{2x}\cos(e^{2x})$$

41. $f(t) = \sin^2\left(e^{\sin^2 t}\right) = \left[\sin\left(e^{\sin^2 t}\right)\right]^2 \Rightarrow$

$$f'(t) = 2\left[\sin\left(e^{\sin^2 t}\right)\right] \cdot \frac{d}{dt}\sin\left(e^{\sin^2 t}\right) = 2\sin\left(e^{\sin^2 t}\right) \cdot \cos\left(e^{\sin^2 t}\right) \cdot \frac{d}{dt}e^{\sin^2 t}$$

$$= 2\sin\left(e^{\sin^2 t}\right)\cos\left(e^{\sin^2 t}\right) \cdot e^{\sin^2 t} \cdot \frac{d}{dt}\sin^2 t = 2\sin\left(e^{\sin^2 t}\right)\cos\left(e^{\sin^2 t}\right)e^{\sin^2 t} \cdot 2\sin t\,\cos t$$

$$= 4\sin\left(e^{\sin^2 t}\right)\cos\left(e^{\sin^2 t}\right)e^{\sin^2 t}\sin t\,\cos t$$

42. $y = \sqrt{x + \sqrt{x + \sqrt{x}}} \Rightarrow y' = \frac{1}{2}\left(x + \sqrt{x + \sqrt{x}}\right)^{-1/2}\left[1 + \frac{1}{2}\left(x + \sqrt{x}\right)^{-1/2}\left(1 + \frac{1}{2}x^{-1/2}\right)\right]$

43. $g(x) = (2ra^{rx} + n)^p \Rightarrow$

$$g'(x) = p(2ra^{rx} + n)^{p-1} \cdot \frac{d}{dx}(2ra^{rx} + n) = p(2ra^{rx} + n)^{p-1} \cdot 2ra^{rx}(\ln a) \cdot r = 2r^2 p(\ln a)(2ra^{rx} + n)^{p-1}a^{rx}$$

44. $y = 2^{3^{4^x}} \Rightarrow$

$$y' = 2^{3^{4^x}}(\ln 2)\frac{d}{dx}3^{4^x} = 2^{3^{4^x}}(\ln 2)\,3^{4^x}(\ln 3)\frac{d}{dx}4^x = 2^{3^{4^x}}(\ln 2)\,3^{4^x}(\ln 3)\,4^x(\ln 4) = (\ln 2)(\ln 3)(\ln 4)4^x 3^{4^x} 2^{3^{4^x}}$$

45. $y = \cos\sqrt{\sin(\tan \pi x)} = \cos(\sin(\tan \pi x))^{1/2} \Rightarrow$

$$y' = -\sin(\sin(\tan \pi x))^{1/2} \cdot \frac{d}{dx}(\sin(\tan \pi x))^{1/2} = -\sin(\sin(\tan \pi x))^{1/2} \cdot \tfrac{1}{2}(\sin(\tan \pi x))^{-1/2} \cdot \frac{d}{dx}(\sin(\tan \pi x))$$

$$= \frac{-\sin\sqrt{\sin(\tan \pi x)}}{2\sqrt{\sin(\tan \pi x)}} \cdot \cos(\tan \pi x) \cdot \frac{d}{dx}\tan \pi x = \frac{-\sin\sqrt{\sin(\tan \pi x)}}{2\sqrt{\sin(\tan \pi x)}} \cdot \cos(\tan \pi x) \cdot \sec^2(\pi x) \cdot \pi$$

$$= \frac{-\pi\cos(\tan \pi x)\sec^2(\pi x)\sin\sqrt{\sin(\tan \pi x)}}{2\sqrt{\sin(\tan \pi x)}}$$

46. $y = \left[x + (x + \sin^2 x)^3\right]^4 \Rightarrow y' = 4\left[x + (x + \sin^2 x)^3\right]^3 \cdot \left[1 + 3(x + \sin^2 x)^2 \cdot (1 + 2\sin x\,\cos x)\right]$

47. $y = \cos(\sin 3\theta) \Rightarrow y' = -\sin(\sin 3\theta) \cdot (\cos 3\theta) \cdot 3 = -3\cos 3\theta\,\sin(\sin 3\theta) \Rightarrow$

$$y'' = -3\left[(\cos 3\theta)\cos(\sin 3\theta)(\cos 3\theta) \cdot 3 + \sin(\sin 3\theta)(-\sin 3\theta) \cdot 3\right] = -9\cos^2(3\theta)\cos(\sin 3\theta) + 9(\sin 3\theta)\sin(\sin 3\theta)$$

48. $y = \dfrac{1}{(1 + \tan x)^2} = (1 + \tan x)^{-2} \Rightarrow y' = -2(1 + \tan x)^{-3}\sec^2 x = \dfrac{-2\sec^2 x}{(1 + \tan x)^3}$.

Using the Product Rule with $y' = \left[-2(1 + \tan x)^{-3}\right](\sec x)^2$, we get

$$y'' = -2(1 + \tan x)^{-3} \cdot 2(\sec x)(\sec x\,\tan x) + (\sec x)^2 \cdot 6(1 + \tan x)^{-4}\sec^2 x$$

$$= 2\sec^2 x\,(1 + \tan x)^{-4}\left[-2(1 + \tan x)\tan x + 3\sec^2 x\right] \qquad \left[\begin{array}{c}\text{2 is the lesser exponent for } \sec x \\ \text{and } -4 \text{ for } (1 + \tan x)\end{array}\right]$$

$$= 2\sec^2 x\,(1 + \tan x)^{-4}\left[-2\tan x - 2\tan^2 x + 3(\tan^2 x + 1)\right]$$

$$= \frac{2\sec^2 x\,(\tan^2 x - 2\tan x + 3)}{(1 + \tan x)^4}$$

49. $y = \sqrt{1 - \sec t} \Rightarrow y' = \frac{1}{2}(1 - \sec t)^{-1/2}(-\sec t\,\tan t) = \dfrac{-\sec t\,\tan t}{2\sqrt{1 - \sec t}}$.

Using the Product Rule with $y' = \left(-\frac{1}{2}\sec t\,\tan t\right)(1 - \sec t)^{-1/2}$, we get

$$y'' = \left(-\tfrac{1}{2}\sec t\,\tan t\right)\left[-\tfrac{1}{2}(1-\sec t)^{-3/2}(-\sec t\,\tan t)\right] + (1-\sec t)^{-1/2}\left(-\tfrac{1}{2}\right)[\sec t\,\sec^2 t + \tan t\,\sec t\,\tan t].$$

Now factor out $-\frac{1}{2}\sec t(1-\sec t)^{-3/2}$. Note that $-\frac{3}{2}$ is the lesser exponent on $(1-\sec t)$. Continuing,

$$\begin{aligned} y'' &= -\tfrac{1}{2}\sec t\,(1-\sec t)^{-3/2}\left[\tfrac{1}{2}\sec t\,\tan^2 t + (1-\sec t)(\sec^2 t + \tan^2 t)\right] \\ &= -\tfrac{1}{2}\sec t\,(1-\sec t)^{-3/2}\left(\tfrac{1}{2}\sec t\,\tan^2 t + \sec^2 t + \tan^2 t - \sec^3 t - \sec t\,\tan^2 t\right) \\ &= -\tfrac{1}{2}\sec t\,(1-\sec t)^{-3/2}\left[-\tfrac{1}{2}\sec t\,(\sec^2 t - 1) + \sec^2 t + (\sec^2 t - 1) - \sec^3 t\right] \\ &= -\tfrac{1}{2}\sec t\,(1-\sec t)^{-3/2}\left(-\tfrac{3}{2}\sec^3 t + 2\sec^2 t + \tfrac{1}{2}\sec t - 1\right) \\ &= \sec t\,(1-\sec t)^{-3/2}\left(\tfrac{3}{4}\sec^3 t - \sec^2 t - \tfrac{1}{4}\sec t + \tfrac{1}{2}\right) \\ &= \frac{\sec t\,(3\sec^3 t - 4\sec^2 t - \sec t + 2)}{4(1-\sec t)^{3/2}} \end{aligned}$$

There are many other correct forms of y'', such as $y'' = \dfrac{\sec t\,(3\sec t + 2)\sqrt{1-\sec t}}{4}$. We chose to find a factored form with only secants in the final form.

50. $y = e^{e^x} \quad\Rightarrow\quad y' = e^{e^x}\cdot(e^x)' = e^{e^x}\cdot e^x \quad\Rightarrow$

$y'' = e^{e^x}\cdot(e^x)' + e^x\cdot\left(e^{e^x}\right)' = e^{e^x}\cdot e^x + e^x\cdot e^{e^x}\cdot e^x = e^{e^x}\cdot e^x(1+e^x) \quad\text{or}\quad e^{e^x+x}(1+e^x)$

51. $y = 2^x \quad\Rightarrow\quad y' = 2^x\ln 2$. At $(0,1)$, $y' = 2^0\ln 2 = \ln 2$, and an equation of the tangent line is $y - 1 = (\ln 2)(x-0)$ or $y = (\ln 2)x + 1$.

52. $y = \sqrt{1+x^3} = (1+x^3)^{1/2} \quad\Rightarrow\quad y' = \frac{1}{2}(1+x^3)^{-1/2}\cdot 3x^2 = \dfrac{3x^2}{2\sqrt{1+x^3}}$. At $(2,3)$, $y' = \dfrac{3\cdot 4}{2\sqrt{9}} = 2$, and an equation of the tangent line is $y - 3 = 2(x-2)$, or $y = 2x - 1$.

53. $y = \sin(\sin x) \quad\Rightarrow\quad y' = \cos(\sin x)\cdot\cos x$. At $(\pi, 0)$, $y' = \cos(\sin\pi)\cdot\cos\pi = \cos(0)\cdot(-1) = 1(-1) = -1$, and an equation of the tangent line is $y - 0 = -1(x-\pi)$, or $y = -x + \pi$.

54. $y = xe^{-x^2} \quad\Rightarrow\quad y' = xe^{-x^2}(-2x) + e^{-x^2}(1) = e^{-x^2}(-2x^2+1)$. At $(0,0)$, $y' = e^0(1) = 1$, and an equation of the tangent line is $y - 0 = 1(x-0)$ or $y = x$.

55. (a) $y = \dfrac{2}{1+e^{-x}} \quad\Rightarrow\quad y' = \dfrac{(1+e^{-x})(0) - 2(-e^{-x})}{(1+e^{-x})^2} = \dfrac{2e^{-x}}{(1+e^{-x})^2}$.

At $(0,1)$, $y' = \dfrac{2e^0}{(1+e^0)^2} = \dfrac{2(1)}{(1+1)^2} = \dfrac{2}{2^2} = \dfrac{1}{2}$. So an equation of the tangent line is $y - 1 = \frac{1}{2}(x-0)$ or $y = \frac{1}{2}x + 1$.

(b)

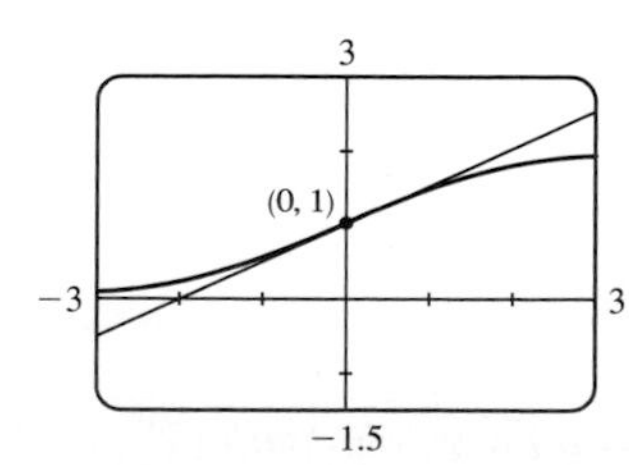

56. (a) For $x > 0$, $|x| = x$, and $y = f(x) = \dfrac{x}{\sqrt{2-x^2}} \quad\Rightarrow$

$$\begin{aligned} f'(x) &= \frac{\sqrt{2-x^2}\,(1) - x\left(\frac{1}{2}\right)(2-x^2)^{-1/2}(-2x)}{\left(\sqrt{2-x^2}\right)^2}\cdot\frac{(2-x^2)^{1/2}}{(2-x^2)^{1/2}} \\ &= \frac{(2-x^2)+x^2}{(2-x^2)^{3/2}} = \frac{2}{(2-x^2)^{3/2}} \end{aligned}$$

(b)

So at $(1,1)$, the slope of the tangent line is $f'(1) = 2$ and its equation is $y - 1 = 2(x-1)$ or $y = 2x - 1$.

57. (a) $f(x) = x\sqrt{2 - x^2} = x(2 - x^2)^{1/2} \quad \Rightarrow$

$$f'(x) = x \cdot \tfrac{1}{2}(2 - x^2)^{-1/2}(-2x) + (2 - x^2)^{1/2} \cdot 1 = (2 - x^2)^{-1/2}\left[-x^2 + (2 - x^2)\right] = \frac{2 - 2x^2}{\sqrt{2 - x^2}}$$

(b) $f' = 0$ when f has a horizontal tangent line, f' is negative when f is decreasing, and f' is positive when f is increasing.

58. (a)

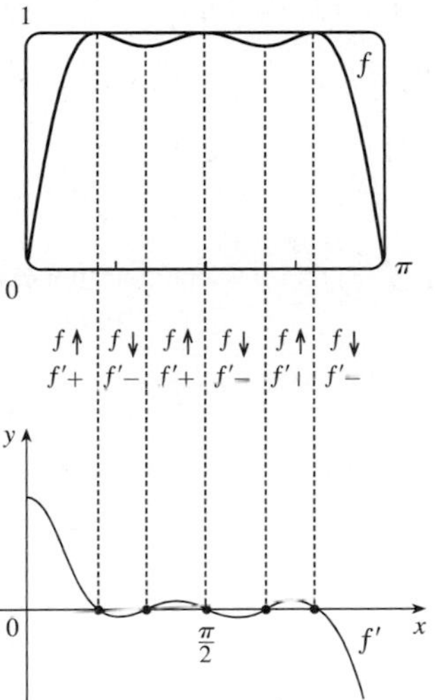

From the graph of f, we see that there are 5 horizontal tangents, so there must be 5 zeros on the graph of f'. From the symmetry of the graph of f, we must have the graph of f' as high at $x = 0$ as it is low at $x = \pi$. The intervals of increase and decrease as well as the signs of f' are indicated in the figure.

(b) $f(x) = \sin(x + \sin 2x) \quad \Rightarrow$

$$f'(x) = \cos(x + \sin 2x) \cdot \frac{d}{dx}(x + \sin 2x) = \cos(x + \sin 2x)(1 + 2\cos 2x)$$

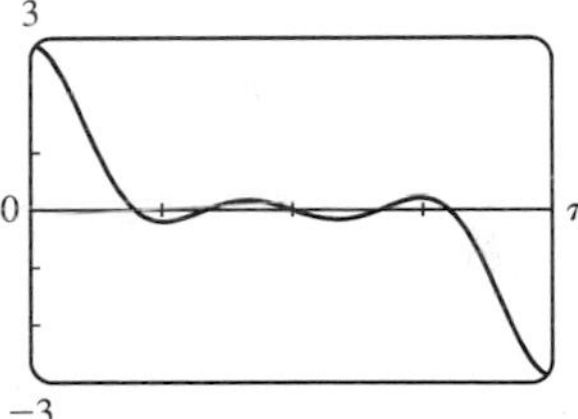

59. For the tangent line to be horizontal, $f'(x) = 0$. $f(x) = 2\sin x + \sin^2 x \quad \Rightarrow \quad f'(x) = 2\cos x + 2\sin x\,\cos x = 0 \quad \Leftrightarrow$ $2\cos x(1 + \sin x) = 0 \quad \Leftrightarrow \quad \cos x = 0$ or $\sin x = -1$, so $x = \frac{\pi}{2} + 2n\pi$ or $\frac{3\pi}{2} + 2n\pi$, where n is any integer. Now $f\left(\frac{\pi}{2}\right) = 3$ and $f\left(\frac{3\pi}{2}\right) = -1$, so the points on the curve with a horizontal tangent are $\left(\frac{\pi}{2} + 2n\pi, 3\right)$ and $\left(\frac{3\pi}{2} + 2n\pi, -1\right)$, where n is any integer.

60. $y = \sqrt{1 + 2x} \quad \Rightarrow \quad y' = \frac{1}{2}(1 + 2x)^{-1/2} \cdot 2 = \dfrac{1}{\sqrt{1 + 2x}}$. The line $6x + 2y = 1$ $\left(\text{or } y = -3x + \frac{1}{2}\right)$ has slope -3, so the tangent line perpendicular to it must have slope $\frac{1}{3}$. Thus, $\dfrac{1}{3} = \dfrac{1}{\sqrt{1 + 2x}} \quad \Leftrightarrow \quad \sqrt{1 + 2x} = 3 \quad \Rightarrow \quad 1 + 2x = 9 \quad \Leftrightarrow$ $2x = 8 \quad \Leftrightarrow \quad x = 4$. When $x = 4$, $y = \sqrt{1 + 2(4)} = 3$, so the point is $(4, 3)$.

61. $F(x) = f(g(x)) \quad \Rightarrow \quad F'(x) = f'(g(x)) \cdot g'(x)$, so $F'(5) = f'(g(5)) \cdot g'(5) = f'(-2) \cdot 6 = 4 \cdot 6 = 24$.

62. $h(x) = \sqrt{4 + 3f(x)} \;\Rightarrow\; h'(x) = \frac{1}{2}(4 + 3f(x))^{-1/2} \cdot 3f'(x)$, so

$h'(1) = \frac{1}{2}(4 + 3f(1))^{-1/2} \cdot 3f'(1) = \frac{1}{2}(4 + 3 \cdot 7)^{-1/2} \cdot 3 \cdot 4 = \frac{6}{\sqrt{25}} = \frac{6}{5}$.

63. (a) $h(x) = f(g(x)) \;\Rightarrow\; h'(x) = f'(g(x)) \cdot g'(x)$, so $h'(1) = f'(g(1)) \cdot g'(1) = f'(2) \cdot 6 = 5 \cdot 6 = 30$.

(b) $H(x) = g(f(x)) \;\Rightarrow\; H'(x) = g'(f(x)) \cdot f'(x)$, so $H'(1) = g'(f(1)) \cdot f'(1) = g'(3) \cdot 4 = 9 \cdot 4 = 36$.

64. (a) $F(x) = f(f(x)) \;\Rightarrow\; F'(x) = f'(f(x)) \cdot f'(x)$, so $F'(2) = f'(f(2)) \cdot f'(2) = f'(1) \cdot 5 = 4 \cdot 5 = 20$.

(b) $G(x) = g(g(x)) \;\Rightarrow\; G'(x) = g'(g(x)) \cdot g'(x)$, so $G'(3) = g'(g(3)) \cdot g'(3) = g'(2) \cdot 9 = 7 \cdot 9 = 63$.

65. (a) $u(x) = f(g(x)) \;\Rightarrow\; u'(x) = f'(g(x))g'(x)$. So $u'(1) = f'(g(1))g'(1) = f'(3)g'(1)$. To find $f'(3)$, note that f is linear from $(2, 4)$ to $(6, 3)$, so its slope is $\dfrac{3-4}{6-2} = -\dfrac{1}{4}$. To find $g'(1)$, note that g is linear from $(0, 6)$ to $(2, 0)$, so its slope is $\dfrac{0-6}{2-0} = -3$. Thus, $f'(3)g'(1) = \left(-\frac{1}{4}\right)(-3) = \frac{3}{4}$.

(b) $v(x) = g(f(x)) \;\Rightarrow\; v'(x) = g'(f(x))f'(x)$. So $v'(1) = g'(f(1))f'(1) = g'(2)f'(1)$, which does not exist since $g'(2)$ does not exist.

(c) $w(x) = g(g(x)) \;\Rightarrow\; w'(x) = g'(g(x))g'(x)$. So $w'(1) = g'(g(1))g'(1) = g'(3)g'(1)$. To find $g'(3)$, note that g is linear from $(2, 0)$ to $(5, 2)$, so its slope is $\dfrac{2-0}{5-2} = \dfrac{2}{3}$. Thus, $g'(3)g'(1) = \left(\frac{2}{3}\right)(-3) = -2$.

66. (a) $h(x) = f(f(x)) \;\Rightarrow\; h'(x) = f'(f(x))f'(x)$. So $h'(2) = f'(f(2))f'(2) = f'(1)f'(2) \approx (-1)(-1) = 1$.

(b) $g(x) = f(x^2) \;\Rightarrow\; g'(x) = f'(x^2) \cdot \dfrac{d}{dx}(x^2) = f'(x^2)(2x)$. So $g'(2) = f'(2^2)(2 \cdot 2) = 4f'(4) \approx 4(2) = 8$.

67. The point $(3, 2)$ is on the graph of f, so $f(3) = 2$. The tangent line at $(3, 2)$ has slope $\dfrac{\Delta y}{\Delta x} = \dfrac{-4}{6} = -\dfrac{2}{3}$.

$g(x) = \sqrt{f(x)} \;\Rightarrow\; g'(x) = \frac{1}{2}[f(x)]^{-1/2} \cdot f'(x) \;\Rightarrow$

$g'(3) = \frac{1}{2}[f(3)]^{-1/2} \cdot f'(3) = \frac{1}{2}(2)^{-1/2}\left(-\frac{2}{3}\right) = -\dfrac{1}{3\sqrt{2}}$ or $-\frac{1}{6}\sqrt{2}$.

68. (a) $F(x) = f(x^\alpha) \;\Rightarrow\; F'(x) = f'(x^\alpha)\dfrac{d}{dx}(x^\alpha) = f'(x^\alpha)\alpha x^{\alpha-1}$

(b) $G(x) = [f(x)]^\alpha \;\Rightarrow\; G'(x) = \alpha\,[f(x)]^{\alpha-1}\,f'(x)$

69. (a) $F(x) = f(e^x) \;\Rightarrow\; F'(x) = f'(e^x)\dfrac{d}{dx}(e^x) = f'(e^x)e^x$

(b) $G(x) = e^{f(x)} \;\Rightarrow\; G'(x) = e^{f(x)}\dfrac{d}{dx}f(x) = e^{f(x)}f'(x)$

70. (a) $g(x) = e^{cx} + f(x) \;\Rightarrow\; g'(x) = e^{cx} \cdot c + f'(x) \;\Rightarrow\; g'(0) = e^0 \cdot c + f'(0) = c + 5$.

$g'(x) = ce^{cx} + f'(x) \;\Rightarrow\; g''(x) = ce^{cx} \cdot c + f''(x) \;\Rightarrow\; g''(0) = c^2e^0 + f''(0) = c^2 - 2$.

(b) $h(x) = e^{kx}f(x) \;\Rightarrow\; h'(x) = e^{kx}f'(x) + f(x) \cdot ke^{kx} \;\Rightarrow\; h'(0) = e^0f'(0) + f(0) \cdot ke^0 = 5 + 3k$.

An equation of the tangent line to the graph of h at the point $(0, h(0)) = (0, f(0)) = (0, 3)$ is $y - 3 = (5 + 3k)(x - 0)$ or $y = (5 + 3k)x + 3$.

71. $r(x) = f(g(h(x))) \quad \Rightarrow \quad r'(x) = f'(g(h(x))) \cdot g'(h(x)) \cdot h'(x)$, so

$r'(1) = f'(g(h(1))) \cdot g'(h(1)) \cdot h'(1) = f'(g(2)) \cdot g'(2) \cdot 4 = f'(3) \cdot 5 \cdot 4 = 6 \cdot 5 \cdot 4 = 120$

72. $f(x) = xg(x^2) \quad \Rightarrow \quad f'(x) = xg'(x^2)\,2x + g(x^2) \cdot 1 = 2x^2g'(x^2) + g(x^2) \quad \Rightarrow$

$f''(x) = 2x^2g''(x^2)\,2x + g'(x^2)\,4x + g'(x^2)\,2x = 4x^3g''(x^2) + 4xg'(x^2) + 2xg'(x^2) = 6xg'(x^2) + 4x^3g''(x^2)$

73. $F(x) = f(3f(4f(x))) \quad \Rightarrow$

$$\begin{aligned} F'(x) &= f'(3f(4f(x))) \cdot \frac{d}{dx}(3f(4f(x))) = f'(3f(4f(x))) \cdot 3f'(4f(x)) \cdot \frac{d}{dx}(4f(x)) \\ &= f'(3f(4f(x))) \cdot 3f'(4f(x)) \cdot 4f'(x), \quad \text{so} \end{aligned}$$

$F'(0) = f'(3f(4f(0))) \cdot 3f'(4f(0)) \cdot 4f'(0) = f'(3f(4 \cdot 0)) \cdot 3f'(4 \cdot 0) \cdot 4 \cdot 2 = f'(3 \cdot 0) \cdot 3 \cdot 2 \cdot 4 \cdot 2 = 2 \cdot 3 \cdot 2 \cdot 4 \cdot 2 = 96.$

74. $F(x) = f(xf(xf(x))) \quad \Rightarrow$

$$\begin{aligned} F'(x) &= f'(xf(xf(x))) \cdot \frac{d}{dx}(xf(xf(x))) = f'(xf(xf(x))) \cdot \left[x \cdot f'(xf(x)) \cdot \frac{d}{dx}(xf(x)) + f(xf(x)) \cdot 1\right] \\ &= f'(xf(xf(x))) \cdot [xf'(xf(x)) \cdot (xf'(x) + f(x) \cdot 1) + f(xf(x))], \quad \text{so} \end{aligned}$$

$$\begin{aligned} F'(1) &= f'(f(f(1))) \cdot [f'(f(1)) \cdot (f'(1) + f(1)) + f(f(1))] = f'(f(2)) \cdot [f'(2) \cdot (4 + 2) + f(2)] \\ &= f'(3) \cdot [5 \cdot 6 + 3] = 6 \cdot 33 = 198. \end{aligned}$$

75. $y = e^{2x}(A\cos 3x + B\sin 3x) \quad \Rightarrow$

$$\begin{aligned} y' &= e^{2x}(-3A\sin 3x + 3B\cos 3x) + (A\cos 3x + B\sin 3x) \cdot 2e^{2x} \\ &= e^{2x}(-3A\sin 3x + 3B\cos 3x + 2A\cos 3x + 2B\sin 3x) \\ &= e^{2x}[(2A + 3B)\cos 3x + (2B - 3A)\sin 3x] \quad \Rightarrow \end{aligned}$$

$$\begin{aligned} y'' &= e^{2x}[-3(2A + 3B)\sin 3x + 3(2B - 3A)\cos 3x] + [(2A + 3B)\cos 3x + (2B - 3A)\sin 3x] \cdot 2e^{2x} \\ &= e^{2x}\{[-3(2A + 3B) + 2(2B - 3A)]\sin 3x + [3(2B - 3A) + 2(2A + 3B)]\cos 3x\} \\ &= e^{2x}[(-12A - 5B)\sin 3x + (-5A + 12B)\cos 3x] \end{aligned}$$

Substitute the expressions for y, y', and y'' in $y'' - 4y' + 13y$ to get

$$\begin{aligned} y'' - 4y' + 13y &= e^{2x}[(-12A - 5B)\sin 3x + (-5A + 12B)\cos 3x] \\ &\qquad - 4e^{2x}[(2A + 3B)\cos 3x + (2B - 3A)\sin 3x] + 13e^{2x}(A\cos 3x + B\sin 3x) \\ &= e^{2x}[(-12A - 5B - 8B + 12A + 13B)\sin 3x + (-5A + 12B - 8A - 12B + 13A)\cos 3x] \\ &= e^{2x}[(0)\sin 3x + (0)\cos 3x] = 0 \end{aligned}$$

Thus, the function y satisfies the differential equation $y'' - 4y' + 13y = 0$.

76. $y = e^{rx} \quad \Rightarrow \quad y' = re^{rx} \quad \Rightarrow \quad y'' = r^2e^{rx}$. Substituting y, y', and y'' into $y'' - 4y' + y = 0$ gives us

$r^2e^{rx} - 4re^{rx} + e^{rx} = 0 \quad \Rightarrow \quad e^{rx}(r^2 - 4r + 1) = 0$. Since $e^{rx} \neq 0$, we must have

$r^2 - 4r + 1 = 0 \quad \Rightarrow \quad r = \dfrac{4 \pm \sqrt{16 - 4}}{2} = 2 \pm \sqrt{3}.$

77. The use of $D, D^2, \ldots, D^n$ is just a derivative notation (see text page 159). In general, $Df(2x) = 2f'(2x)$, $D^2 f(2x) = 4f''(2x), \ldots, D^n f(2x) = 2^n f^{(n)}(2x)$. Since $f(x) = \cos x$ and $50 = 4(12) + 2$, we have $f^{(50)}(x) = f^{(2)}(x) = -\cos x$, so $D^{50} \cos 2x = -2^{50} \cos 2x$.

78. $f(x) = xe^{-x}$, $f'(x) = e^{-x} - xe^{-x} = (1-x)e^{-x}$, $f''(x) = -e^{-x} + (1-x)(-e^{-x}) = (x-2)e^{-x}$. Similarly, $f'''(x) = (3-x)e^{-x}$, $f^{(4)}(x) = (x-4)e^{-x}, \ldots, f^{(1000)}(x) = (x-1000)e^{-x}$.

79. $s(t) = 10 + \frac{1}{4}\sin(10\pi t) \Rightarrow$ the velocity after t seconds is $v(t) = s'(t) = \frac{1}{4}\cos(10\pi t)(10\pi) = \frac{5\pi}{2}\cos(10\pi t)$ cm/s.

80. (a) $s = A\cos(\omega t + \delta) \Rightarrow$ velocity $= s' = -\omega A \sin(\omega t + \delta)$.

(b) If $A \neq 0$ and $\omega \neq 0$, then $s' = 0 \Leftrightarrow \sin(\omega t + \delta) = 0 \Leftrightarrow \omega t + \delta = n\pi \Leftrightarrow t = \dfrac{n\pi - \delta}{\omega}$, n an integer.

81. (a) $B(t) = 4.0 + 0.35 \sin \dfrac{2\pi t}{5.4} \Rightarrow \dfrac{dB}{dt} = \left(0.35 \cos \dfrac{2\pi t}{5.4}\right)\left(\dfrac{2\pi}{5.4}\right) = \dfrac{0.7\pi}{5.4} \cos \dfrac{2\pi t}{5.4} = \dfrac{7\pi}{54} \cos \dfrac{2\pi t}{5.4}$

(b) At $t = 1$, $\dfrac{dB}{dt} = \dfrac{7\pi}{54} \cos \dfrac{2\pi}{5.4} \approx 0.16$.

82. $L(t) = 12 + 2.8 \sin\left(\frac{2\pi}{365}(t - 80)\right) \Rightarrow L'(t) = 2.8 \cos\left(\frac{2\pi}{365}(t-80)\right)\left(\frac{2\pi}{365}\right)$.

On March 21, $t = 80$, and $L'(80) \approx 0.0482$ hours per day. On May 21, $t = 141$, and $L'(141) \approx 0.02398$, which is approximately one-half of $L'(80)$.

83. $s(t) = 2e^{-1.5t} \sin 2\pi t \Rightarrow$

$v(t) = s'(t) = 2[e^{-1.5t}(\cos 2\pi t)(2\pi) + (\sin 2\pi t)e^{-1.5t}(-1.5)] = 2e^{-1.5t}(2\pi \cos 2\pi t - 1.5 \sin 2\pi t)$

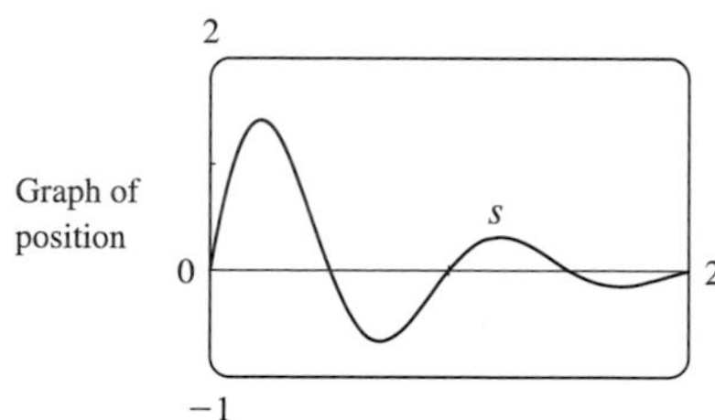

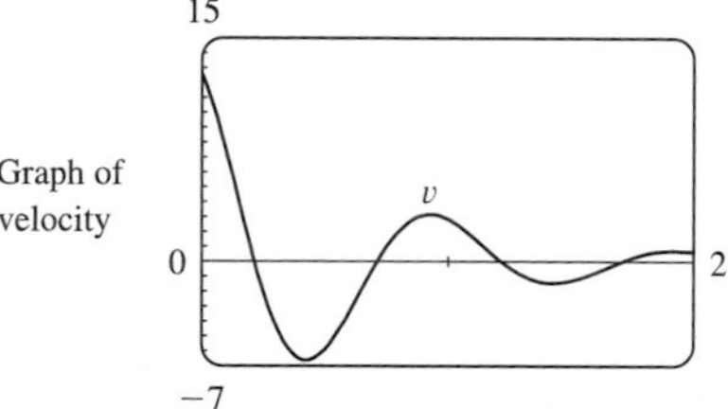

84. (a) $\lim\limits_{t\to\infty} p(t) = \lim\limits_{t\to\infty} \dfrac{1}{1 + ae^{-kt}} = \dfrac{1}{1 + a \cdot 0} = 1$, since $k > 0 \Rightarrow -kt \to -\infty \Rightarrow e^{-kt} \to 0$.

(b) $p(t) = (1 + ae^{-kt})^{-1} \Rightarrow \dfrac{dp}{dt} = -(1 + ae^{-kt})^{-2}(-kae^{-kt}) = \dfrac{kae^{-kt}}{(1 + ae^{-kt})^2}$

(c)

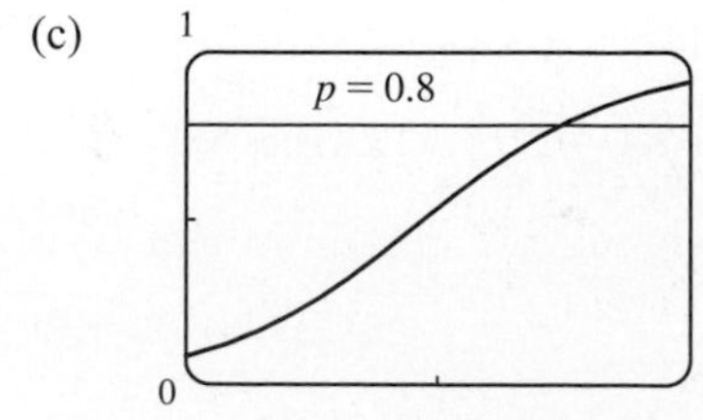

From the graph of $p(t) = (1 + 10e^{-0.5t})^{-1}$, it seems that $p(t) = 0.8$ (indicating that 80% of the population has heard the rumor) when $t \approx 7.4$ hours.

85. (a) Use $C(t) = ate^{bt}$ with $a = 0.0225$ and $b = -0.0467$ to get $C'(t) = a(te^{bt} \cdot b + e^{bt} \cdot 1) = a(bt + 1)e^{bt}$. $C'(10) = 0.0225(0.533)e^{-0.467} \approx 0.0075$, so the BAC was increasing at approximately 0.0075 (mg/mL)/min after 10 minutes.

(b) A half an hour later gives us $t = 10 + 30 = 40$. $C'(40) = 0.0225(-0.868)e^{-18.68} \approx -0.0030$, so the BAC was decreasing at approximately $0.0030\ (\text{mg/mL})/\text{min}$ after 40 minutes.

86. $P(t) = (1436.53) \cdot (1.01395)^t \quad \Rightarrow \quad P'(t) = (1436.53) \cdot (1.01395)^t (\ln 1.01395)$. The units for $P'(t)$ are millions of people per year. The rates of increase for 1920, 1950, and 2000 are $P'(20) \approx 26.25$, $P'(50) \approx 39.78$, and $P'(100) \approx 79.53$, respectively.

87. By the Chain Rule, $a(t) = \dfrac{dv}{dt} = \dfrac{dv}{ds}\dfrac{ds}{dt} = \dfrac{dv}{ds}v(t) = v(t)\dfrac{dv}{ds}$. The derivative dv/dt is the rate of change of the velocity with respect to time (in other words, the acceleration) whereas the derivative dv/ds is the rate of change of the velocity with respect to the displacement.

88. (a) The derivative dV/dr represents the rate of change of the volume with respect to the radius and the derivative dV/dt represents the rate of change of the volume with respect to time.

(b) Since $V = \frac{4}{3}\pi r^3$, $\dfrac{dV}{dt} = \dfrac{dV}{dr}\dfrac{dr}{dt} = 4\pi r^2 \dfrac{dr}{dt}$.

89. (a) Using a calculator or CAS, we obtain the model $Q = ab^t$ with $a \approx 100.0124369$ and $b \approx 0.000045145933$.

(b) Use $Q'(t) = ab^t \ln b$ (from Formula 5) with the values of a and b from part (a) to get $Q'(0.04) \approx -670.63\ \mu\text{A}$. The result of Example 2.1.2 was $-670\ \mu\text{A}$.

90. (a) $P = ab^t$ with $a = 4.502714 \times 10^{-20}$ and $b = 1.029953851$, where P is measured in thousands of people. The fit appears to be very good.

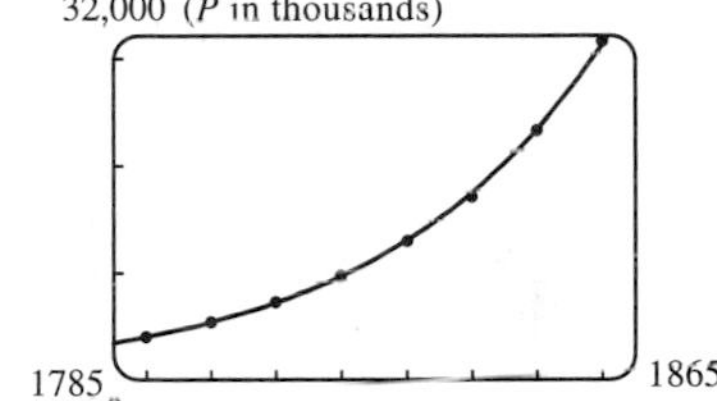

(b) **For 1800:** $m_1 = \dfrac{5308 - 3929}{1800 - 1790} = 137.9$, $m_2 = \dfrac{7240 - 5308}{1810 - 1800} = 193.2$.

So $P'(1800) \approx (m_1 + m_2)/2 = 165.55$ thousand people/year.

For 1850: $m_1 = \dfrac{23{,}192 - 17{,}063}{1850 - 1840} = 612.9$, $m_2 = \dfrac{31{,}443 - 23{,}192}{1860 - 1850} = 825.1$.

So $P'(1850) \approx (m_1 + m_2)/2 = 719$ thousand people/year.

(c) Using $P'(t) = ab^t \ln b$ (from Formula 7) with the values of a and b from part (a), we get $P'(1800) \approx 156.85$ and $P'(1850) \approx 686.07$. These estimates are somewhat less than the ones in part (b).

(d) $P(1870) \approx 41{,}946.56$. The difference of 3.4 million people is most likely due to the Civil War (1861–1865).

91. (a) Derive gives $g'(t) = \dfrac{45(t-2)^8}{(2t+1)^{10}}$ without simplifying. With either Maple or Mathematica, we first get

$g'(t) = 9\dfrac{(t-2)^8}{(2t+1)^9} - 18\dfrac{(t-2)^9}{(2t+1)^{10}}$, and the simplification command results in the expression given by Derive.

(b) Derive gives $y' = 2(x^3 - x + 1)^3(2x+1)^4(17x^3 + 6x^2 - 9x + 3)$ without simplifying. With either Maple or Mathematica, we first get $y' = 10(2x+1)^4(x^3 - x + 1)^4 + 4(2x+1)^5(x^3 - x + 1)^3(3x^2 - 1)$. If we use

Mathematica's `Factor` or `Simplify`, or Maple's `factor`, we get the above expression, but Maple's `simplify` gives the polynomial expansion instead. For locating horizontal tangents, the factored form is the most helpful.

92. (a) $f(x) = \left(\dfrac{x^4 - x + 1}{x^4 + x + 1}\right)^{1/2}$. Derive gives $f'(x) = \dfrac{(3x^4 - 1)\sqrt{\dfrac{x^4 - x + 1}{x^4 + x + 1}}}{(x^4 + x + 1)(x^4 - x + 1)}$ whereas either Maple or Mathematica give $f'(x) = \dfrac{3x^4 - 1}{\sqrt{\dfrac{x^4 - x + 1}{x^4 + x + 1}}\,(x^4 + x + 1)^2}$ after simplification.

(b) $f'(x) = 0 \;\Leftrightarrow\; 3x^4 - 1 = 0 \;\Leftrightarrow\; x = \pm\sqrt[4]{\tfrac{1}{3}} \approx \pm 0.7598.$

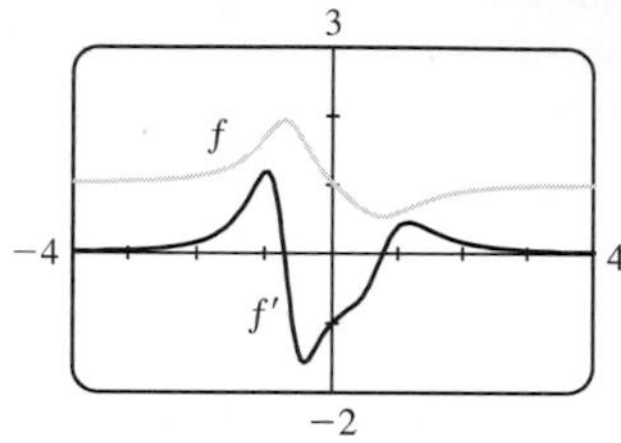

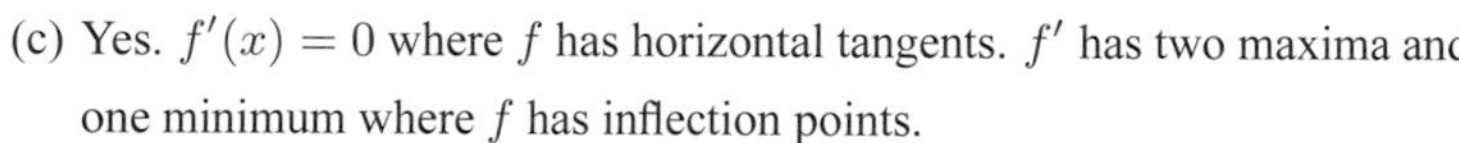

(c) Yes. $f'(x) = 0$ where f has horizontal tangents. f' has two maxima and one minimum where f has inflection points.

93. (a) If f is even, then $f(x) = f(-x)$. Using the Chain Rule to differentiate this equation, we get

$f'(x) = f'(-x)\dfrac{d}{dx}(-x) = -f'(-x)$. Thus, $f'(-x) = -f'(x)$, so f' is odd.

(b) If f is odd, then $f(x) = -f(-x)$. Differentiating this equation, we get $f'(x) = -f'(-x)(-1) = f'(-x)$, so f' is even.

94.
$$\left[\frac{f(x)}{g(x)}\right]' = \left\{f(x)\,[g(x)]^{-1}\right\}' = f'(x)\,[g(x)]^{-1} + (-1)\,[g(x)]^{-2}\,g'(x)f(x)$$
$$= \frac{f'(x)}{g(x)} - \frac{f(x)g'(x)}{[g(x)]^2} = \frac{f'(x)g(x) - f(x)g'(x)}{[g(x)]^2}$$

This is an alternative derivation of the *formula* in the Quotient Rule. But part of the purpose of the Quotient Rule is to show that if f and g are differentiable, so is f/g. The proof in Section 3.2 does that; this one doesn't.

95. (a)
$$\begin{aligned}
\frac{d}{dx}(\sin^n x\,\cos nx) &= n\sin^{n-1}x\,\cos x\,\cos nx + \sin^n x\,(-n\sin nx) && \text{[Product Rule]}\\
&= n\sin^{n-1}x\,(\cos nx\,\cos x - \sin nx\,\sin x) && \text{[factor out } n\sin^{n-1}x]\\
&= n\sin^{n-1}x\cos(nx + x) && \text{[Addition Formula for cosine]}\\
&= n\sin^{n-1}x\cos[(n+1)x] && \text{[factor out } x]
\end{aligned}$$

(b)
$$\begin{aligned}
\frac{d}{dx}(\cos^n x\,\cos nx) &= n\cos^{n-1}x\,(-\sin x)\cos nx + \cos^n x\,(-n\sin nx) && \text{[Product Rule]}\\
&= -n\cos^{n-1}x\,(\cos nx\,\sin x + \sin nx\,\cos x) && \text{[factor out } -n\cos^{n-1}x]\\
&= -n\cos^{n-1}x\sin(nx + x) && \text{[Addition Formula for sine]}\\
&= -n\cos^{n-1}x\sin[(n+1)x] && \text{[factor out } x]
\end{aligned}$$

96. "The rate of change of y^5 with respect to x is eighty times the rate of change of y with respect to x" $\Leftrightarrow$

$\dfrac{d}{dx}y^5 = 80\dfrac{dy}{dx} \;\Leftrightarrow\; 5y^4\dfrac{dy}{dx} = 80\dfrac{dy}{dx} \;\Leftrightarrow\; 5y^4 = 80$ (Note that $dy/dx \neq 0$ since the curve never has a horizontal tangent) $\Leftrightarrow\; y^4 = 16 \;\Leftrightarrow\; y = 2$ (since $y > 0$ for all x)

97. Since $\theta^\circ = \left(\frac{\pi}{180}\right)\theta$ rad, we have $\dfrac{d}{d\theta}(\sin\theta^\circ) = \dfrac{d}{d\theta}\left(\sin\frac{\pi}{180}\theta\right) = \frac{\pi}{180}\cos\frac{\pi}{180}\theta = \frac{\pi}{180}\cos\theta^\circ$.

98. (a) $f(x) = |x| = \sqrt{x^2} = (x^2)^{1/2} \quad\Rightarrow\quad f'(x) = \frac{1}{2}(x^2)^{-1/2}(2x) = x/\sqrt{x^2} = x/|x|$ for $x \neq 0$.

f is not differentiable at $x = 0$.

(b) $f(x) = |\sin x| = \sqrt{\sin^2 x} \quad\Rightarrow$

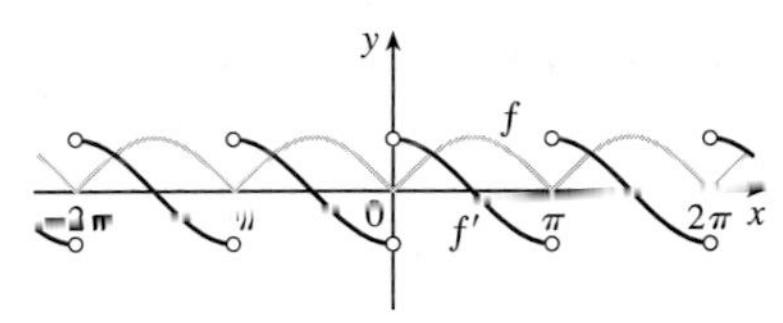

$$f'(x) = \tfrac{1}{2}(\sin^2 x)^{-1/2} 2\sin x\,\cos x = \frac{\sin x}{|\sin x|}\cos x$$

$$= \begin{cases} \cos x & \text{if } \sin x > 0 \\ -\cos x & \text{if } \sin x < 0 \end{cases}$$

f is not differentiable when $x = n\pi$, n an integer.

(c) $g(x) = \sin|x| = \sin\sqrt{x^2} \quad\Rightarrow$

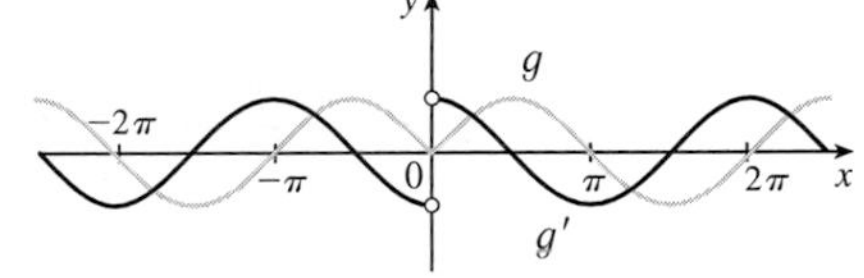

$$g'(x) = \cos|x| \cdot \frac{x}{|x|} = \frac{x}{|x|}\cos x = \begin{cases} \cos x & \text{if } x > 0 \\ -\cos x & \text{if } x < 0 \end{cases}$$

g is not differentiable at 0.

99. The Chain Rule says that $\dfrac{dy}{dx} = \dfrac{dy}{du}\dfrac{du}{dx}$, so

$$\frac{d^2y}{dx^2} = \frac{d}{dx}\left(\frac{dy}{dx}\right) = \frac{d}{dx}\left(\frac{dy}{du}\frac{du}{dx}\right) = \left[\frac{d}{dx}\left(\frac{dy}{du}\right)\right]\frac{du}{dx} + \frac{dy}{du}\frac{d}{dx}\left(\frac{du}{dx}\right) \quad \text{[Product Rule]}$$

$$= \left[\frac{d}{du}\left(\frac{dy}{du}\right)\frac{du}{dx}\right]\frac{du}{dx} + \frac{dy}{du}\frac{d^2u}{dx^2} = \frac{d^2y}{du^2}\left(\frac{du}{dx}\right)^2 + \frac{dy}{du}\frac{d^2u}{dx^2}$$

100. From Exercise 99, $\dfrac{d^2y}{dx^2} = \dfrac{d^2y}{du^2}\left(\dfrac{du}{dx}\right)^2 + \dfrac{dy}{du}\dfrac{d^2u}{dx^2} \quad\Rightarrow$

$$\frac{d^3y}{dx^3} = \frac{d}{dx}\frac{d^2y}{dx^2} = \frac{d}{dx}\left[\frac{d^2y}{du^2}\left(\frac{du}{dx}\right)^2\right] + \frac{d}{dx}\left[\frac{dy}{du}\frac{d^2u}{dx^2}\right]$$

$$= \left[\frac{d}{dx}\left(\frac{d^2y}{du^2}\right)\right]\left(\frac{du}{dx}\right)^2 + \left[\frac{d}{dx}\left(\frac{du}{dx}\right)^2\right]\frac{d^2y}{du^2} + \left[\frac{d}{dx}\left(\frac{dy}{du}\right)\right]\frac{d^2u}{dx^2} + \left[\frac{d}{dx}\left(\frac{d^2u}{dx^2}\right)\right]\frac{dy}{du}$$

$$= \left[\frac{d}{du}\left(\frac{d^2y}{du^2}\right)\frac{du}{dx}\right]\left(\frac{du}{dx}\right)^2 + 2\frac{du}{dx}\frac{d^2u}{dx^2}\frac{d^2y}{du^2} + \left[\frac{d}{du}\left(\frac{dy}{du}\right)\frac{du}{dx}\right]\left(\frac{d^2u}{dx^2}\right) + \frac{d^3u}{dx^3}\frac{dy}{du}$$

$$= \frac{d^3y}{du^3}\left(\frac{du}{dx}\right)^3 + 3\frac{du}{dx}\frac{d^2u}{dx^2}\frac{d^2y}{du^2} + \frac{dy}{du}\frac{d^3u}{dx^3}$$

APPLIED PROJECT Where Should a Pilot Start Descent?

1. Condition (i) will hold if and only if all of the following four conditions hold:

(α) $P(0) = 0$

(β) $P'(0) = 0$ (for a smooth landing)

(γ) $P'(\ell) = 0$ (since the plane is cruising horizontally when it begins its descent)

(δ) $P(\ell) = h$.

First of all, condition α implies that $P(0) = d = 0$, so $P(x) = ax^3 + bx^2 + cx \Rightarrow P'(x) = 3ax^2 + 2bx + c$. But $P'(0) = c = 0$ by condition β. So $P'(\ell) = 3a\ell^2 + 2b\ell = \ell(3a\ell + 2b)$. Now by condition γ, $3a\ell + 2b = 0 \Rightarrow a = -\frac{2b}{3\ell}$. Therefore, $P(x) = -\frac{2b}{3\ell}x^3 + bx^2$. Setting $P(\ell) = h$ for condition δ, we get $P(\ell) = -\frac{2b}{3\ell}\ell^3 + b\ell^2 = h \Rightarrow$ $-\frac{2}{3}b\ell^2 + b\ell^2 = h \Rightarrow \frac{1}{3}b\ell^2 = h \Rightarrow b = \frac{3h}{\ell^2} \Rightarrow a = -\frac{2h}{\ell^3}$. So $y = P(x) = -\frac{2h}{\ell^3}x^3 + \frac{3h}{\ell^2}x^2$.

2. By condition (ii), $\frac{dx}{dt} = -v$ for all t, so $x(t) = \ell - vt$. Condition (iii) states that $\left|\frac{d^2y}{dt^2}\right| \leq k$. By the Chain Rule,

we have $\frac{dy}{dt} = \frac{dy}{dx}\frac{dx}{dt} = -\frac{2h}{\ell^3}(3x^2)\frac{dx}{dt} + \frac{3h}{\ell^2}(2x)\frac{dx}{dt} = \frac{6hx^2v}{\ell^3} - \frac{6hxv}{\ell^2}$ (for $x \leq \ell$) $\Rightarrow$

$\frac{d^2y}{dt^2} = \frac{6hv}{\ell^3}(2x)\frac{dx}{dt} - \frac{6hv}{\ell^2}\frac{dx}{dt} = -\frac{12hv^2}{\ell^3}x + \frac{6hv^2}{\ell^2}$. In particular, when $t = 0$, $x = \ell$ and so

$\left.\frac{d^2y}{dt^2}\right|_{t=0} = -\frac{12hv^2}{\ell^3}\ell + \frac{6hv^2}{\ell^2} = -\frac{6hv^2}{\ell^2}$. Thus, $\left|\left.\frac{d^2y}{dt^2}\right|_{t=0}\right| = \frac{6hv^2}{\ell^2} \leq k$. (This condition also follows from taking $x = 0$.)

3. We substitute $k = 860\text{ mi/h}^2$, $h = 35{,}000\text{ ft} \times \frac{1\text{ mi}}{5280\text{ ft}}$, and $v = 300\text{ mi/h}$ into the result of part (b):

$$\frac{6\left(35{,}000 \cdot \frac{1}{5280}\right)(300)^2}{\ell^2} \leq 860 \Rightarrow \ell \geq 300\sqrt{6 \cdot \frac{35{,}000}{5280 \cdot 860}} \approx 64.5 \text{ miles.}$$

4. Substituting the values of h and ℓ in Problem 3 into

$P(x) = -\frac{2h}{\ell^3}x^3 + \frac{3h}{\ell^2}x^2$ gives us $P(x) = ax^3 + bx^2$,

where $a \approx -4.937 \times 10^{-5}$ and $b \approx 4.78 \times 10^{-3}$.

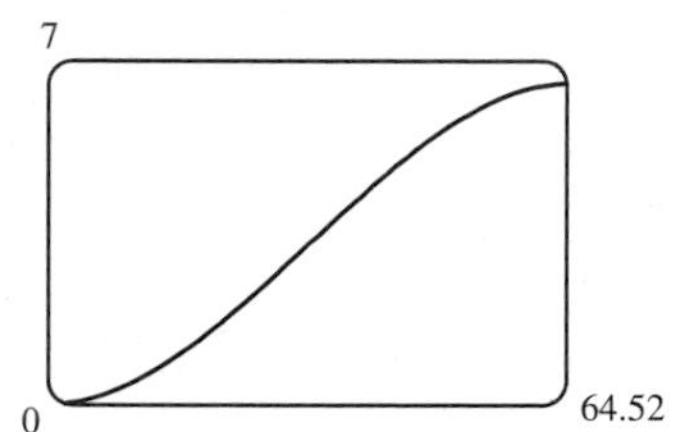

3.5 Implicit Differentiation

1. (a) $\frac{d}{dx}(9x^2 - y^2) = \frac{d}{dx}(1) \Rightarrow 18x - 2y\,y' = 0 \Rightarrow 2y\,y' = 18x \Rightarrow y' = \frac{9x}{y}$

(b) $9x^2 - y^2 = 1 \Rightarrow y^2 = 9x^2 - 1 \Rightarrow y = \pm\sqrt{9x^2 - 1}$, so $y' = \pm\frac{1}{2}(9x^2 - 1)^{-1/2}(18x) = \pm\frac{9x}{\sqrt{9x^2 - 1}}$.

(c) From part (a), $y' = \frac{9x}{y} = \frac{9x}{\pm\sqrt{9x^2 - 1}}$, which agrees with part (b).

2. (a) $\frac{d}{dx}(2x^2 + x + xy) = \frac{d}{dx}(1) \Rightarrow 4x + 1 + xy' + y \cdot 1 = 0 \Rightarrow xy' = -4x - y - 1 \Rightarrow y' = -\frac{4x + y + 1}{x}$

(b) $2x^2 + x + xy = 1 \Rightarrow xy = 1 - 2x^2 - x \Rightarrow y = \frac{1}{x} - 2x - 1$, so $y' = -\frac{1}{x^2} - 2$

(c) From part (a),

$$y' = -\frac{4x + y + 1}{x} = -4 - \frac{1}{x}y - \frac{1}{x} = -4 - \frac{1}{x}\left(\frac{1}{x} - 2x - 1 - \frac{1}{x}\right) = -4 - \frac{1}{x^2} + 2 + \frac{1}{x} - \frac{1}{x} = -\frac{1}{x^2} - 2, \text{ which}$$

agrees with part (b).

3. (a) $\dfrac{d}{dx}\left(\sqrt{x}+\sqrt{y}\right)=\dfrac{d}{dx}(1) \;\Rightarrow\; \dfrac{1}{2}x^{-1/2}+\dfrac{1}{2}y^{-1/2}y'=0 \;\Rightarrow\; \dfrac{1}{2\sqrt{y}}\,y'=-\dfrac{1}{2\sqrt{x}} \;\Rightarrow\; y'=-\dfrac{\sqrt{y}}{\sqrt{x}}$

(b) $\sqrt{x}+\sqrt{y}=1 \;\Rightarrow\; \sqrt{y}=1-\sqrt{x} \;\Rightarrow\; y=(1-\sqrt{x})^2 \;\Rightarrow\; y=1-2\sqrt{x}+x$, so

$y'=-2\cdot\dfrac{1}{2}x^{-1/2}+1=1-\dfrac{1}{\sqrt{x}}$.

(c) From part (a), $y'=-\dfrac{\sqrt{y}}{\sqrt{x}}=-\dfrac{1-\sqrt{x}}{\sqrt{x}}$ [from part (b)] $=-\dfrac{1}{\sqrt{x}}+1$, which agrees with part (b).

4. (a) $\dfrac{d}{dx}\left(\dfrac{2}{x}-\dfrac{1}{y}\right)=\dfrac{d}{dx}(4) \;\Rightarrow\; -2x^{-2}+y^{-2}y'=0 \;\Rightarrow\; \dfrac{1}{y^2}\,y'=\dfrac{2}{x^2} \;\Rightarrow\; y'=\dfrac{2y^2}{x^2}$

(b) $\dfrac{2}{x}-\dfrac{1}{y}=4 \;\Rightarrow\; \dfrac{1}{y}=\dfrac{2}{x}-4 \;\Rightarrow\; \dfrac{1}{y}=\dfrac{2-4x}{x} \;\Rightarrow\; y=\dfrac{x}{2-4x}$, so

$y'=\dfrac{(2-4x)(1)-x(-4)}{(2-4x)^2}=\dfrac{2}{(2-4x)^2}\left[\text{or } \dfrac{1}{2(1-2x)^2}\right]$.

(c) From part (a), $y'=\dfrac{2y^2}{x^2}=\dfrac{2\left(\dfrac{x}{2-4x}\right)^2}{x^2}$ [from part (b)] $=\dfrac{2x^2}{x^2(2-4x)^2}=\dfrac{2}{(2-4x)^2}$, which agrees with part (b).

5. $\dfrac{d}{dx}(x^2-4xy+y^2)=\dfrac{d}{dx}(4) \;\Rightarrow\; 2x-4[xy'+y(1)]+2y\,y'=0 \;\Rightarrow\; 2y\,y'-4xy'=4y-2x \;\Rightarrow$

$y'(y-2x)=2y-x \;\Rightarrow\; y'=\dfrac{2y-x}{y-2x}$

6. $\dfrac{d}{dx}(2x^2+xy-y^2)=\dfrac{d}{dx}(2) \;\Rightarrow\; 4x+xy'+y(1)-2y\,y'=0 \;\Rightarrow\; xy'-2y\,y'=-4x-y \;\Rightarrow$

$(x-2y)y'=-4x-y \;\Rightarrow\; y'=\dfrac{-4x-y}{x-2y}$

7. $\dfrac{d}{dx}(x^4+x^2y^2+y^3)=\dfrac{d}{dx}(5) \;\Rightarrow\; 4x^3+x^2\cdot 2y\,y'+y^2\cdot 2x+3y^2y'=0 \;\Rightarrow\; 2x^2y\,y'+3y^2y'=-4x^3-2xy^2 \;\Rightarrow$

$(2x^2y+3y^2)y'=-4x^3-2xy^2 \;\Rightarrow\; y'=\dfrac{-4x^3-2xy^2}{2x^2y+3y^2}=-\dfrac{2x(2x^2+y^2)}{y(2x^2+3y)}$

8. $\dfrac{d}{dx}(x^3-xy^2+y^3)=\dfrac{d}{dx}(1) \;\Rightarrow\; 3x^2-x\cdot 2y\,y'-y^2\cdot 1+3y^2y'=0 \;\Rightarrow\; 3y^2y'-2x\,y\,y'=y^2-3x^2 \;\Rightarrow$

$(3y^2-2xy)\,y'=y^2-3x^2 \;\Rightarrow\; y'=\dfrac{y^2-3x^2}{3y^2-2xy}=\dfrac{y^2-3x^2}{y(3y-2x)}$

9. $\dfrac{d}{dx}\left(\dfrac{x^2}{x+y}\right)=\dfrac{d}{dx}(y^2+1) \;\Rightarrow\; \dfrac{(x+y)(2x)-x^2(1+y')}{(x+y)^2}=2y\,y' \;\Rightarrow$

$2x^2+2xy-x^2-x^2\,y'=2y(x+y)^2\,y' \;\Rightarrow\; x^2+2xy=2y(x+y)^2\,y'+x^2\,y' \;\Rightarrow$

$x(x+2y)=[2y(x^2+2xy+y^2)+x^2]\,y' \;\Rightarrow\; y'=\dfrac{x(x+2y)}{2x^2y+4xy^2+2y^3+x^2}$

Or: Start by clearing fractions and then differentiate implicitly.

10. $\dfrac{d}{dx}(xe^y)=\dfrac{d}{dx}(x-y) \;\Rightarrow\; xe^y\,y'+e^y\cdot 1=1-y' \;\Rightarrow\; xe^y\,y'+y'=1-e^y \;\Rightarrow\; y'(xe^y+1)=1-e^y \;\Rightarrow$

$y'=\dfrac{1-e^y}{xe^y+1}$

11. $\dfrac{d}{dx}(y\cos x) = \dfrac{d}{dx}(x^2 + y^2) \;\Rightarrow\; y(-\sin x) + \cos x \cdot y' = 2x + 2y\,y' \;\Rightarrow\; \cos x \cdot y' - 2y\,y' = 2x + y\sin x \;\Rightarrow$

$y'(\cos x - 2y) = 2x + y\sin x \;\Rightarrow\; y' = \dfrac{2x + y\sin x}{\cos x - 2y}$

12. $\dfrac{d}{dx}\cos(xy) = \dfrac{d}{dx}(1 + \sin y) \;\Rightarrow\; -\sin(xy)(xy' + y \cdot 1) = \cos y \cdot y' \;\Rightarrow\; -xy'\sin(xy) - \cos y \cdot y' = y\sin(xy) \;\Rightarrow$

$y'[-x\sin(xy) - \cos y] = y\sin(xy) \;\Rightarrow\; y' = \dfrac{y\sin(xy)}{-x\sin(xy) - \cos y} = -\dfrac{y\sin(xy)}{x\sin(xy) + \cos y}$

13. $\dfrac{d}{dx}\sqrt{x+y} = \dfrac{d}{dx}\left(x^4 + y^4\right) \;\Rightarrow\; \frac{1}{2}(x+y)^{-1/2}(1 + y') = 4x^3 + 4y^3y' \;\Rightarrow$

$\dfrac{1}{2\sqrt{x+y}} + \dfrac{1}{2\sqrt{x+y}}\,y' = 4x^3 + 4y^3y' \;\Rightarrow\; \dfrac{1}{2\sqrt{x+y}} - 4x^3 = 4y^3y' - \dfrac{1}{2\sqrt{x+y}}\,y' \;\Rightarrow$

$\dfrac{1 - 8x^3\sqrt{x+y}}{2\sqrt{x+y}} = \dfrac{8y^3\sqrt{x+y} - 1}{2\sqrt{x+y}}\,y' \;\Rightarrow\; y' = \dfrac{1 - 8x^3\sqrt{x+y}}{8y^3\sqrt{x+y} - 1}$

14. $\dfrac{d}{dx}(e^y\sin x) = \dfrac{d}{dx}(x + xy) \;\Rightarrow\; e^y\cos x + \sin x \cdot e^y y' = 1 + xy' + y \cdot 1 \;\Rightarrow$

$e^y\sin x \cdot y' - xy' = 1 + y - e^y\cos x \;\Rightarrow\; y'(e^y\sin x - x) = 1 + y - e^y\cos x \;\Rightarrow\; y' = \dfrac{1 + y - e^y\cos x}{e^y\sin x - x}$

15. $\dfrac{d}{dx}(e^{x/y}) = \dfrac{d}{dx}(x - y) \;\Rightarrow\; e^{x/y} \cdot \dfrac{d}{dx}\left(\dfrac{x}{y}\right) = 1 - y' \;\Rightarrow$

$e^{x/y} \cdot \dfrac{y \cdot 1 - x \cdot y'}{y^2} = 1 - y' \;\Rightarrow\; e^{x/y} \cdot \dfrac{1}{y} - \dfrac{xe^{x/y}}{y^2} \cdot y' = 1 - y' \;\Rightarrow\; y' - \dfrac{xe^{x/y}}{y^2} \cdot y' = 1 - \dfrac{e^{x/y}}{y} \;\Rightarrow$

$y'\left(1 - \dfrac{xe^{x/y}}{y^2}\right) = \dfrac{y - e^{x/y}}{y} \;\Rightarrow\; y' = \dfrac{\dfrac{y - e^{x/y}}{y}}{\dfrac{y^2 - xe^{x/y}}{y^2}} = \dfrac{y(y - e^{x/y})}{y^2 - xe^{x/y}}$

16. $\dfrac{d}{dx}(xy) = \dfrac{d}{dx}\sqrt{x^2 + y^2} \;\Rightarrow\; xy' + y(1) = \frac{1}{2}\left(x^2 + y^2\right)^{-1/2}(2x + 2y\,y') \;\Rightarrow$

$xy' + y = \dfrac{x}{\sqrt{x^2+y^2}} + \dfrac{y}{\sqrt{x^2+y^2}}\,y' \;\Rightarrow\; xy' - \dfrac{y}{\sqrt{x^2+y^2}}\,y' = \dfrac{x}{\sqrt{x^2+y^2}} - y \;\Rightarrow$

$\dfrac{x\sqrt{x^2+y^2} - y}{\sqrt{x^2+y^2}}\,y' = \dfrac{x - y\sqrt{x^2+y^2}}{\sqrt{x^2+y^2}} \;\Rightarrow\; y' = \dfrac{x - y\sqrt{x^2+y^2}}{x\sqrt{x^2+y^2} - y}$

17. $\dfrac{d}{dx}\tan^{-1}(x^2y) = \dfrac{d}{dx}(x + xy^2) \;\Rightarrow\; \dfrac{1}{1 + (x^2y)^2}(x^2y' + y \cdot 2x) = 1 + x \cdot 2y\,y' + y^2 \cdot 1 \;\Rightarrow$

$\dfrac{x^2}{1 + x^4y^2}\,y' - 2xy\,y' = 1 + y^2 - \dfrac{2xy}{1 + x^4y^2} \;\Rightarrow\; y'\left(\dfrac{x^2}{1 + x^4y^2} - 2xy\right) = 1 + y^2 - \dfrac{2xy}{1 + x^4y^2} \;\Rightarrow$

$y' = \dfrac{1 + y^2 - \dfrac{2xy}{1 + x^4y^2}}{\dfrac{x^2}{1 + x^4y^2} - 2xy}$ or $y' = \dfrac{1 + x^4y^2 + y^2 + x^4y^4 - 2xy}{x^2 - 2xy - 2x^5y^3}$

18. $\dfrac{d}{dx}(x\sin y + y\sin x) = \dfrac{d}{dx}(1) \;\Rightarrow\; x\cos y\cdot y' + \sin y\cdot 1 + y\cos x + \sin x\cdot y' = 0 \;\Rightarrow$

$x\cos y\cdot y' + \sin x\cdot y' = -\sin y - y\cos x \;\Rightarrow\; y'(x\cos y + \sin x) = -\sin y - y\cos x \;\Rightarrow\; y' = \dfrac{-\sin y - y\cos x}{x\cos y + \sin x}$

19. $\dfrac{d}{dx}\sin(xy) = \dfrac{d}{dx}\cos(x+y) \;\Rightarrow\; \cos(xy)\cdot(xy' + y\cdot 1) = -\sin(x+y)\cdot(1+y') \;\Rightarrow$

$x\cos(xy)\,y' + y\cos(xy) = -\sin(x+y) - y'\sin(x+y) \;\Rightarrow$

$x\cos(xy)\,y' + y'\sin(x+y) = -y\cos(xy) - \sin(x+y) \;\Rightarrow$

$[x\cos(xy) + \sin(x+y)]\,y' = -1\,[y\cos(xy) + \sin(x+y)] \;\Rightarrow\; y' = -\dfrac{y\cos(xy) + \sin(x+y)}{x\cos(xy) + \sin(x+y)}$

20. $\tan(x-y) = \dfrac{y}{1+x^2} \;\Rightarrow\; (1+x^2)\tan(x-y) = y \;\Rightarrow\; (1+x^2)\sec^2(x-y)\cdot(1-y') + \tan(x-y)\cdot 2x = y' \;\Rightarrow$

$(1+x^2)\sec^2(x-y) - (1+x^2)\sec^2(x-y)\cdot y' + 2x\tan(x-y) = y' \;\Rightarrow$

$(1+x^2)\sec^2(x-y) + 2x\tan(x-y) = \left[1 + (1+x^2)\sec^2(x-y)\right]\cdot y' \;\Rightarrow$

$y' = \dfrac{(1+x^2)\sec^2(x-y) + 2x\tan(x-y)}{1 + (1+x^2)\sec^2(x-y)}$

21. $\dfrac{d}{dx}\left\{f(x) + x^2[f(x)]^3\right\} = \dfrac{d}{dx}(10) \;\Rightarrow\; f'(x) + x^2\cdot 3[f(x)]^2\cdot f'(x) + [f(x)]^3\cdot 2x = 0.$ If $x = 1$, we have

$f'(1) + 1^2\cdot 3[f(1)]^2\cdot f'(1) + [f(1)]^3\cdot 2(1) = 0 \;\Rightarrow\; f'(1) + 1\cdot 3\cdot 2^2\cdot f'(1) + 2^3\cdot 2 = 0 \;\Rightarrow$

$f'(1) + 12f'(1) = -16 \;\Rightarrow\; 13f'(1) = -16 \;\Rightarrow\; f'(1) = -\frac{16}{13}.$

22. $\dfrac{d}{dx}[g(x) + x\sin g(x)] = \dfrac{d}{dx}(x^2) \;\Rightarrow\; g'(x) + x\cos g(x)\cdot g'(x) + \sin g(x)\cdot 1 = 2x.$ If $x = 0$, we have

$g'(0) + 0 + \sin g(0) = 2(0) \;\Rightarrow\; g'(0) + \sin 0 = 0 \;\Rightarrow\; g'(0) + 0 = 0 \;\Rightarrow\; g'(0) = 0.$

23. $\dfrac{d}{dy}(x^4y^2 - x^3y + 2xy^3) = \dfrac{d}{dy}(0) \;\Rightarrow\; x^4\cdot 2y + y^2\cdot 4x^3\,x' - (x^3\cdot 1 + y\cdot 3x^2\,x') + 2(x\cdot 3y^2 + y^3\cdot x') = 0 \;\Rightarrow$

$4x^3y^2\,x' - 3x^2y\,x' + 2y^3\,x' = -2x^4y + x^3 - 6xy^2 \;\Rightarrow\; (4x^3y^2 - 3x^2y + 2y^3)\,x' = -2x^4y + x^3 - 6xy^2 \;\Rightarrow$

$x' = \dfrac{dx}{dy} = \dfrac{-2x^4y + x^3 - 6xy^2}{4x^3y^2 - 3x^2y + 2y^3}$

24. $\dfrac{d}{dy}(y\sec x) = \dfrac{d}{dy}(x\tan y) \;\Rightarrow\; y\cdot\sec x\,\tan x\cdot x' + \sec x\cdot 1 = x\cdot\sec^2 y + \tan y\cdot x' \;\Rightarrow$

$y\sec x\,\tan x\cdot x' - \tan y\cdot x' = x\sec^2 y - \sec x \;\Rightarrow\; (y\sec x\,\tan x - \tan y)\,x' = x\sec^2 y - \sec x \;\Rightarrow$

$x' = \dfrac{dx}{dy} = \dfrac{x\sec^2 y - \sec x}{y\sec x\,\tan x - \tan y}$

25. $y\sin 2x = x\cos 2y \;\Rightarrow\; y\cdot\cos 2x\cdot 2 + \sin 2x\cdot y' = x(-\sin 2y\cdot 2y') + \cos(2y)\cdot 1 \;\Rightarrow$

$\sin 2x\cdot y' + 2x\sin 2y\cdot y' = -2y\cos 2x + \cos 2y \;\Rightarrow\; y'(\sin 2x + 2x\sin 2y) = -2y\cos 2x + \cos 2y \;\Rightarrow$

$y' = \dfrac{-2y\cos 2x + \cos 2y}{\sin 2x + 2x\sin 2y}$. When $x = \frac{\pi}{2}$ and $y = \frac{\pi}{4}$, we have $y' = \dfrac{(-\pi/2)(-1) + 0}{0 + \pi \cdot 1} = \dfrac{\pi/2}{\pi} = \dfrac{1}{2}$, so an equation of the tangent line is $y - \frac{\pi}{4} = \frac{1}{2}(x - \frac{\pi}{2})$, or $y = \frac{1}{2}x$.

26. $\sin(x+y) = 2x - 2y \Rightarrow \cos(x+y)\cdot(1+y') = 2 - 2y' \Rightarrow \cos(x+y)\cdot y' + 2y' = 2 - \cos(x+y) \Rightarrow$ $y'[\cos(x+y) + 2] = 2 - \cos(x+y) \Rightarrow y' = \dfrac{2 - \cos(x+y)}{\cos(x+y) + 2}$. When $x = \pi$ and $y = \pi$, we have $y' = \dfrac{2-1}{1+2} = \dfrac{1}{3}$, so an equation of the tangent line is $y - \pi = \frac{1}{3}(x - \pi)$, or $y = \frac{1}{3}x + \frac{2\pi}{3}$.

27. $x^2 - xy - y^2 = 1 \Rightarrow 2x - (xy' + y\cdot 1) - 2y\,y' = 0 \Rightarrow 2x - xy' - y - 2y\,y' = 0 \Rightarrow 2x - y = xy' + 2y\,y' \Rightarrow$ $2x - y = (x + 2y)\,y' \Rightarrow y' = \dfrac{2x - y}{x + 2y}$. When $x = 2$ and $y = 1$, we have $y' = \dfrac{4-1}{2+2} = \dfrac{3}{4}$, so an equation of the tangent line is $y - 1 = \frac{3}{4}(x - 2)$, or $y = \frac{3}{4}x - \frac{1}{2}$.

28. $x^2 + 2xy + 4y^2 = 12 \Rightarrow 2x + 2x\,y' + 2y + 8y\,y' = 0 \Rightarrow 2x\,y' + 8y\,y' = -2x - 2y \Rightarrow$ $(x + 4y)\,y' = -x - y \Rightarrow y' = -\dfrac{x+y}{x+4y}$. When $x = 2$ and $y = 1$, we have $y' = -\dfrac{2+1}{2+4} = -\dfrac{1}{2}$, so an equation of the tangent line is $y - 1 = -\frac{1}{2}(x - 2)$ or $y = -\frac{1}{2}x + 2$.

29. $x^2 + y^2 = (2x^2 + 2y^2 - x)^2 \Rightarrow 2x + 2y\,y' = 2(2x^2 + 2y^2 - x)(4x + 4y\,y' - 1)$. When $x = 0$ and $y = \frac{1}{2}$, we have $0 + y' = 2(\frac{1}{2})(2y' - 1) \Rightarrow y' = 2y' - 1 \Rightarrow y' = 1$, so an equation of the tangent line is $y - \frac{1}{2} = 1(x - 0)$ or $y = x + \frac{1}{2}$.

30. $x^{2/3} + y^{2/3} = 4 \Rightarrow \frac{2}{3}x^{-1/3} + \frac{2}{3}y^{-1/3}y' = 0 \Rightarrow \dfrac{1}{\sqrt[3]{x}} + \dfrac{y'}{\sqrt[3]{y}} = 0 \Rightarrow y' = -\dfrac{\sqrt[3]{y}}{\sqrt[3]{x}}$. When $x = -3\sqrt{3}$ and $y = 1$, we have $y' = -\dfrac{1}{(-3\sqrt{3})^{1/3}} = -\dfrac{(-3\sqrt{3})^{2/3}}{-3\sqrt{3}} = \dfrac{3}{3\sqrt{3}} = \dfrac{1}{\sqrt{3}}$, so an equation of the tangent line is $y - 1 = \frac{1}{\sqrt{3}}(x + 3\sqrt{3})$ or $y = \frac{1}{\sqrt{3}}x + 4$.

31. $2(x^2 + y^2)^2 = 25(x^2 - y^2) \Rightarrow 4(x^2 + y^2)(2x + 2y\,y') = 25(2x - 2y\,y') \Rightarrow$ $4(x + y\,y')(x^2 + y^2) = 25(x - y\,y') \Rightarrow 4y\,y'(x^2 + y^2) + 25yy' = 25x - 4x(x^2 + y^2) \Rightarrow$ $y' = \dfrac{25x - 4x(x^2 + y^2)}{25y + 4y(x^2 + y^2)}$. When $x = 3$ and $y = 1$, we have $y' = \frac{75 - 120}{25 + 40} = -\frac{45}{65} = -\frac{9}{13}$, so an equation of the tangent line is $y - 1 = -\frac{9}{13}(x - 3)$ or $y = -\frac{9}{13}x + \frac{40}{13}$.

32. $y^2(y^2 - 4) = x^2(x^2 - 5) \Rightarrow y^4 - 4y^2 = x^4 - 5x^2 \Rightarrow 4y^3\,y' - 8y\,y' = 4x^3 - 10x$.
When $x = 0$ and $y = -2$, we have $-32y' + 16y' = 0 \Rightarrow -16y' = 0 \Rightarrow y' = 0$, so an equation of the tangent line is $y + 2 = 0(x - 0)$ or $y = -2$.

33. (a) $y^2 = 5x^4 - x^2 \Rightarrow 2y\,y' = 5(4x^3) - 2x \Rightarrow y' = \dfrac{10x^3 - x}{y}$.

So at the point $(1, 2)$ we have $y' = \dfrac{10(1)^3 - 1}{2} = \dfrac{9}{2}$, and an equation of the tangent line is $y - 2 = \frac{9}{2}(x - 1)$ or $y = \frac{9}{2}x - \frac{5}{2}$.

(b)

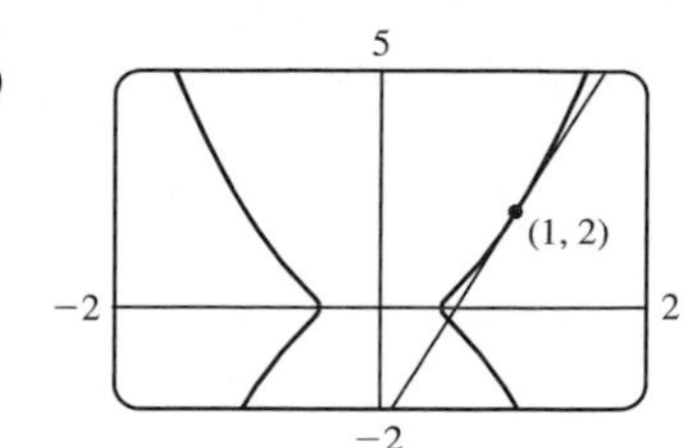

34. (a) $y^2 = x^3 + 3x^2 \Rightarrow 2y\,y' = 3x^2 + 3(2x) \Rightarrow y' = \dfrac{3x^2 + 6x}{2y}$. So at the point $(1, -2)$ we have

$y' = \dfrac{3(1)^2 + 6(1)}{2(-2)} = -\dfrac{9}{4}$, and an equation of the tangent line is $y + 2 = -\frac{9}{4}(x - 1)$ or $y = -\frac{9}{4}x + \frac{1}{4}$.

(b) The curve has a horizontal tangent where $y' = 0 \Leftrightarrow$

$3x^2 + 6x = 0 \Leftrightarrow 3x(x + 2) = 0 \Leftrightarrow x = 0$ or $x = -2$.

But note that at $x = 0$, $y = 0$ also, so the derivative does not exist.

At $x = -2$, $y^2 = (-2)^3 + 3(-2)^2 = -8 + 12 = 4$, so $y = \pm 2$.

So the two points at which the curve has a horizontal tangent are $(-2, -2)$ and $(-2, 2)$.

(c)

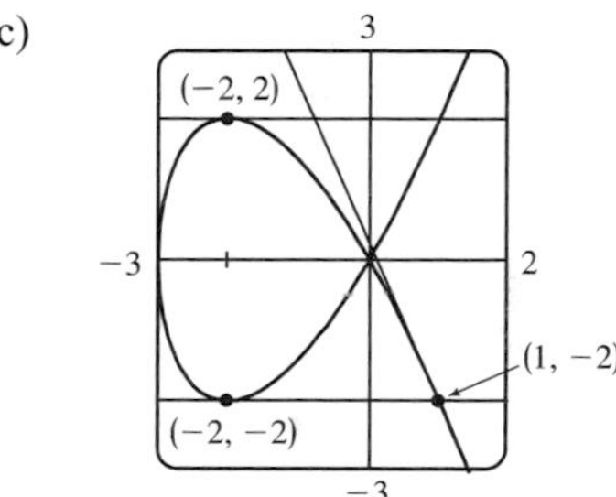

35. $x^2 + 4y^2 = 4 \Rightarrow 2x + 8y\,y' = 0 \Rightarrow y' = -x/(4y) \Rightarrow$

$$y'' = -\frac{1}{4}\frac{y \cdot 1 - x \cdot y'}{y^2} = -\frac{1}{4}\frac{y - x[-x/(4y)]}{y^2} = -\frac{1}{4}\frac{4y^2 + x^2}{4y^3} = -\frac{1}{4}\frac{4}{4y^3} \qquad \left[\begin{array}{c}\text{since } x \text{ and } y \text{ must satisfy the} \\ \text{original equation } x^2 + 4y^2 = 4\end{array}\right]$$

Thus, $y'' = -\dfrac{1}{4y^3}$.

36. $x^2 + xy + y^2 = 3 \Rightarrow 2x + xy' + y + 2y\,y' = 0 \Rightarrow (x + 2y)y' = -2x - y \Rightarrow y' = \dfrac{-2x - y}{x + 2y}$.

Differentiating $2x + xy' + y + 2y\,y' = 0$ to find y'' gives $2 + xy'' + y' + y' + 2y\,y'' + 2y'y' = 0 \Rightarrow$

$$(x + 2y)\,y'' = -2 - 2y' - 2(y')^2 = -2\left[1 - \frac{2x + y}{x + 2y} + \left(\frac{2x + y}{x + 2y}\right)^2\right] \Rightarrow$$

$$y'' = -\frac{2}{x + 2y}\left[\frac{(x + 2y)^2 - (2x + y)(x + 2y) + (2x + y)^2}{(x + 2y)^2}\right]$$

$$= -\frac{2}{(x + 2y)^3}\,(x^2 + 4xy + 4y^2 - 2x^2 - 4xy - xy - 2y^2 + 4x^2 + 4xy + y^2)$$

$$= -\frac{2}{(x + 2y)^3}\,(3x^2 + 3xy + 3y^2) = -\frac{2}{(x + 2y)^3}\,(9) \qquad \left[\begin{array}{c}\text{since } x \text{ and } y \text{ must satisfy the} \\ \text{original equation } x^2 + xy + y^2 = 3\end{array}\right]$$

Thus, $y'' = -\dfrac{18}{(x + 2y)^3}$.

37. $\sin y + \cos x = 1 \quad\Rightarrow\quad \cos y \cdot y' - \sin x = 0 \quad\Rightarrow\quad y' = \dfrac{\sin x}{\cos y} \quad\Rightarrow$

$$y'' = \frac{\cos y\,\cos x - \sin x(-\sin y)\,y'}{(\cos y)^2} = \frac{\cos y\,\cos x + \sin x\,\sin y(\sin x/\cos y)}{\cos^2 y}$$

$$= \frac{\cos^2 y\,\cos x + \sin^2 x\,\sin y}{\cos^2 y\,\cos y} = \frac{\cos^2 y\,\cos x + \sin^2 x\,\sin y}{\cos^3 y}$$

Using $\sin y + \cos x = 1$, the expression for y'' can be simplified to $y'' = (\cos^2 x + \sin y)/\cos^3 y$.

38. $x^3 - y^3 = 7 \quad\Rightarrow\quad 3x^2 - 3y^2y' = 0 \quad\Rightarrow\quad y' = \dfrac{x^2}{y^2} \quad\Rightarrow$

$$y'' = \frac{y^2(2x) - x^2(2y\,y')}{(y^2)^2} = \frac{2xy[y - x(x^2/y^2)]}{y^4} = \frac{2x(y - x^3/y^2)}{y^3} = \frac{2x(y^3 - x^3)}{y^3y^2} = \frac{2x(-7)}{y^5} = \frac{-14x}{y^5}$$

39. If $x = 0$ in $xy + e^y = e$, then we get $0 + e^y = e$, so $y = 1$ and the point where $x = 0$ is $(0, 1)$. Differentiating implicitly with respect to x gives us $xy' + y \cdot 1 + e^y y' = 0$. Substituting 0 for x and 1 for y gives us $0 + 1 + ey' = 0 \quad\Rightarrow\quad ey' = -1 \quad\Rightarrow\quad y' = -1/e$. Differentiating $xy' + y + e^y y' = 0$ implicitly with respect to x gives us $xy'' + y' \cdot 1 + y' + e^y y'' + y' \cdot e^y y' = 0$. Now substitute 0 for x, 1 for y, and $-1/e$ for y'.

$$0 + \left(-\frac{1}{e}\right) + \left(-\frac{1}{e}\right) + ey'' + \left(-\frac{1}{e}\right)(e)\left(-\frac{1}{e}\right) = 0 \quad\Rightarrow\quad -\frac{2}{e} + ey'' + \frac{1}{e} = 0 \quad\Rightarrow\quad ey'' = \frac{1}{e} \quad\Rightarrow\quad y'' = \frac{1}{e^2}.$$

40. If $x = 1$ in $x^2 + xy + y^3 = 1$, then we get $1 + y + y^3 = 1 \quad\Rightarrow\quad y^3 + y = 0 \quad\Rightarrow\quad y(y^2 + 1) \quad\Rightarrow\quad y = 0$, so the point where $x = 1$ is $(1, 0)$. Differentiating implicitly with respect to x gives us $2x + xy' + y \cdot 1 + 3y^2 \cdot y' = 0$. Substituting 1 for x and 0 for y gives us $2 + y' + 0 + 0 = 0 \quad\Rightarrow\quad y' = -2$. Differentiating $2x + xy' + y + 3y^2y' = 0$ implicitly with respect to x gives us $2 + xy'' + y' \cdot 1 + y' + 3(y^2y'' + y' \cdot 2yy') = 0$. Now substitute 1 for x, 0 for y, and -2 for y'. $2 + y'' + (-2) + (-2) + 3(0 + 0) = 0 \quad\Rightarrow\quad y'' = 2$. Differentiating $2 + xy'' + 2y' + 3y^2y'' + 6y(y')^2 = 0$ implicitly with respect to x gives us $xy''' + y'' \cdot 1 + 2y'' + 3(y^2y''' + y'' \cdot 2yy') + 6[y \cdot 2y'y'' + (y')^2y'] = 0$. Now substitute 1 for x, 0 for y, -2 for y', and 2 for y''. $y''' + 2 + 4 + 3(0 + 0) + 6[0 + (-8)] = 0 \quad\Rightarrow\quad y''' = -2 - 4 + 48 = 42$.

41. (a) There are eight points with horizontal tangents: four at $x \approx 1.57735$ and four at $x \approx 0.42265$.

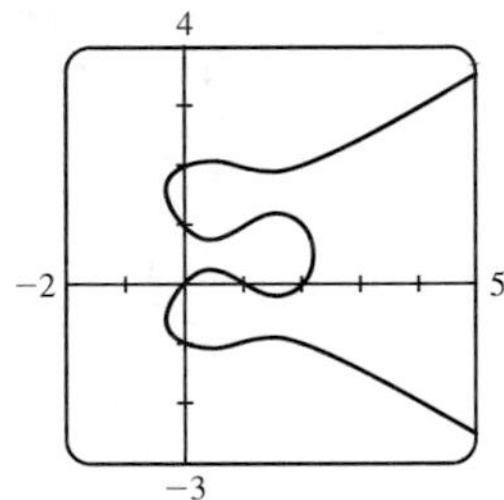

(b) 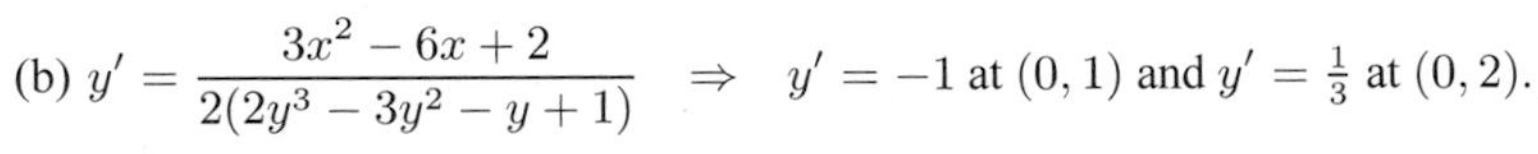

$y' = \dfrac{3x^2 - 6x + 2}{2(2y^3 - 3y^2 - y + 1)} \quad\Rightarrow\quad y' = -1$ at $(0, 1)$ and $y' = \frac{1}{3}$ at $(0, 2)$.

Equations of the tangent lines are $y = -x + 1$ and $y = \frac{1}{3}x + 2$.

(c) $y' = 0 \quad\Rightarrow\quad 3x^2 - 6x + 2 = 0 \quad\Rightarrow\quad x = 1 \pm \frac{1}{3}\sqrt{3}$

(d) By multiplying the right side of the equation by $x - 3$, we obtain the first graph. By modifying the equation in other ways, we can generate the other graphs.

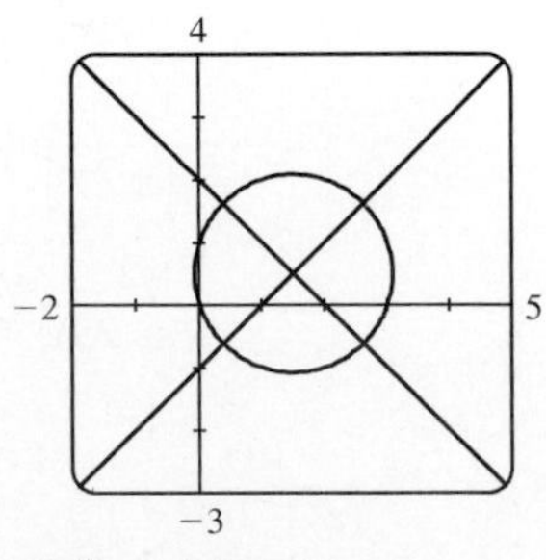

$y(y^2 - 1)(y - 2)$
$= x(x - 1)(x - 2)(x - 3)$

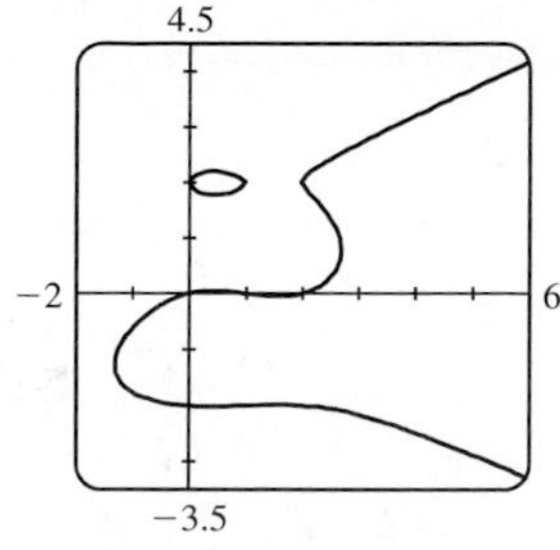

$y(y^2-4)(y-2)$
$= x(x-1)(x-2)$

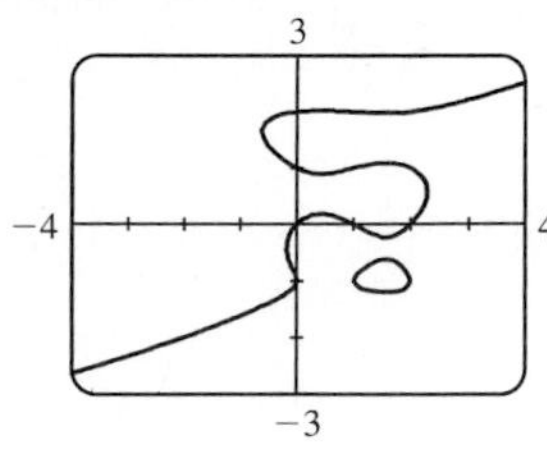

$y(y+1)(y^2-1)(y-2)$
$= x(x-1)(x-2)$

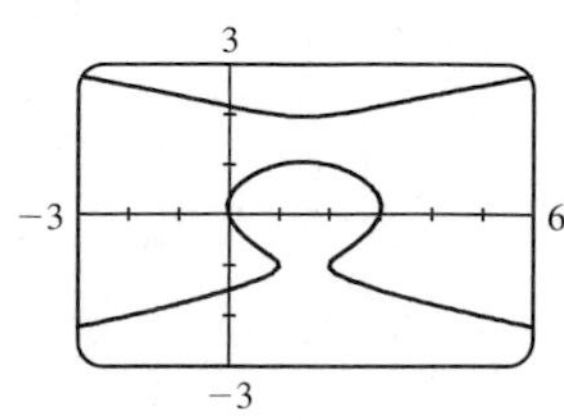

$(y+1)(y^2-1)(y-2)$
$= (x-1)(x-2)$

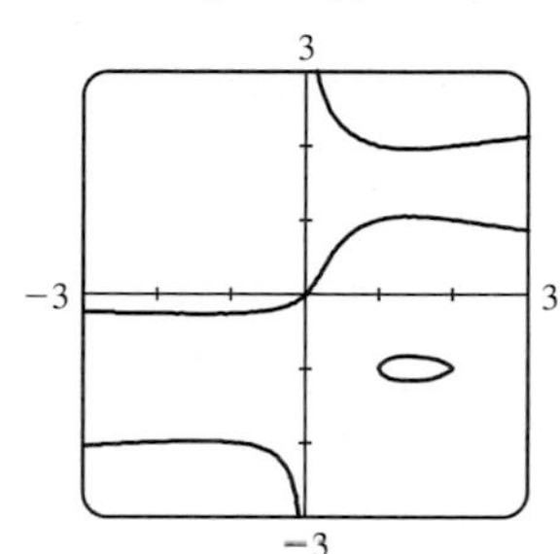

$x(y+1)(y^2-1)(y-2)$
$= y(x-1)(x-2)$

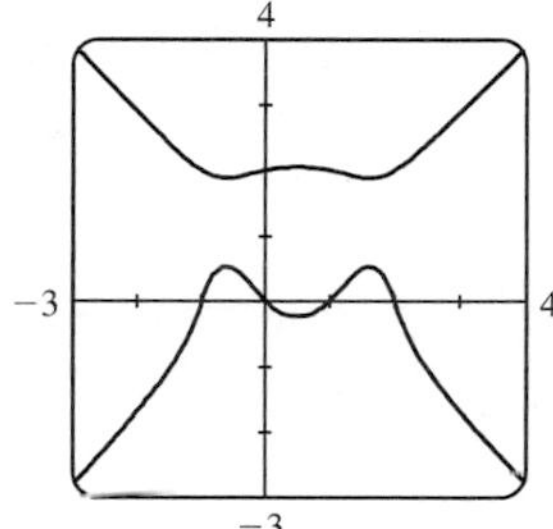

$y(y^2+1)(y-2)$
$= x(x^2-1)(x-2)$

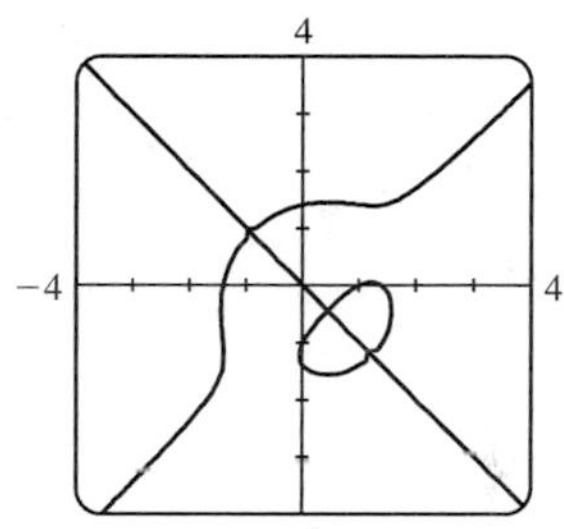

$y(y+1)(y^2-2)$
$= x(x-1)(x^2-2)$

42. (a)

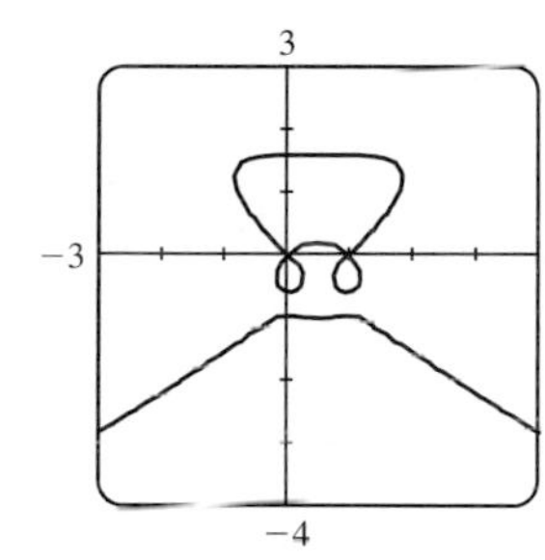

(b) $\dfrac{d}{dx}(2y^3+y^2-y^5) = \dfrac{d}{dx}(x^4-2x^3+x^2) \;\Rightarrow$

$6y^2y' + 2y\,y' - 5y^4y' = 4x^3 - 6x^2 + 2x \;\Rightarrow$

$y' = \dfrac{2x(2x^2-3x+1)}{6y^2+2y-5y^4} = \dfrac{2x(2x-1)(x-1)}{y(6y+2-5y^3)}$. From the graph and the values for which $y'=0$, we speculate that there are 9 points with horizontal tangents: 3 at $x=0$, 3 at $x=\frac{1}{2}$, and 3 at $x=1$. The three horizontal tangents along the top of the wagon are hard to find, but by limiting the y-range of the graph (to $[1.6, 1.7]$, for example) they are distinguishable.

43. From Exercise 31, a tangent to the lemniscate will be horizontal if $y'=0 \;\Rightarrow\; 25x - 4x(x^2+y^2) = 0 \;\Rightarrow$ $x[25-4(x^2+y^2)] = 0 \;\Rightarrow\; x^2+y^2 = \frac{25}{4}$ **(1)**. (Note that when x is 0, y is also 0, and there is no horizontal tangent at the origin.) Substituting $\frac{25}{4}$ for x^2+y^2 in the equation of the lemniscate, $2(x^2+y^2)^2 = 25(x^2-y^2)$, we get $x^2-y^2 = \frac{25}{8}$ **(2)**. Solving **(1)** and **(2)**, we have $x^2 = \frac{75}{16}$ and $y^2 = \frac{25}{16}$, so the four points are $\left(\pm\frac{5\sqrt{3}}{4}, \pm\frac{5}{4}\right)$.

44. $\dfrac{x^2}{a^2} + \dfrac{y^2}{b^2} = 1 \;\Rightarrow\; \dfrac{2x}{a^2} + \dfrac{2yy'}{b^2} = 0 \;\Rightarrow\; y' = -\dfrac{b^2x}{a^2y} \;\Rightarrow$ an equation of the tangent line at (x_0, y_0) is

$y - y_0 = \dfrac{-b^2x_0}{a^2y_0}(x-x_0)$. Multiplying both sides by $\dfrac{y_0}{b^2}$ gives $\dfrac{y_0y}{b^2} - \dfrac{y_0^2}{b^2} = -\dfrac{x_0x}{a^2} + \dfrac{x_0^2}{a^2}$. Since (x_0, y_0) lies on the ellipse,

we have $\dfrac{x_0x}{a^2} + \dfrac{y_0y}{b^2} = \dfrac{x_0^2}{a^2} + \dfrac{y_0^2}{b^2} = 1$.

45. $\dfrac{x^2}{a^2} - \dfrac{y^2}{b^2} = 1 \;\Rightarrow\; \dfrac{2x}{a^2} - \dfrac{2yy'}{b^2} = 0 \;\Rightarrow\; y' = \dfrac{b^2 x}{a^2 y} \;\Rightarrow$ an equation of the tangent line at (x_0, y_0) is

$y - y_0 = \dfrac{b^2 x_0}{a^2 y_0}(x - x_0)$. Multiplying both sides by $\dfrac{y_0}{b^2}$ gives $\dfrac{y_0 y}{b^2} - \dfrac{y_0^2}{b^2} = \dfrac{x_0 x}{a^2} - \dfrac{x_0^2}{a^2}$. Since (x_0, y_0) lies on the hyperbola,

we have $\dfrac{x_0 x}{a^2} - \dfrac{y_0 y}{b^2} = \dfrac{x_0^2}{a^2} - \dfrac{y_0^2}{b^2} = 1$.

46. $\sqrt{x} + \sqrt{y} = \sqrt{c} \;\Rightarrow\; \dfrac{1}{2\sqrt{x}} + \dfrac{y'}{2\sqrt{y}} = 0 \;\Rightarrow\; y' = -\dfrac{\sqrt{y}}{\sqrt{x}} \;\Rightarrow$ an equation of the tangent line at (x_0, y_0)

is $y - y_0 = -\dfrac{\sqrt{y_0}}{\sqrt{x_0}}(x - x_0)$. Now $x = 0 \;\Rightarrow\; y = y_0 - \dfrac{\sqrt{y_0}}{\sqrt{x_0}}(-x_0) = y_0 + \sqrt{x_0}\sqrt{y_0}$, so the y-intercept is

$y_0 + \sqrt{x_0}\sqrt{y_0}$. And $y = 0 \;\Rightarrow\; -y_0 = -\dfrac{\sqrt{y_0}}{\sqrt{x_0}}(x - x_0) \;\Rightarrow\; x - x_0 = \dfrac{y_0\sqrt{x_0}}{\sqrt{y_0}} \;\Rightarrow$

$x = x_0 + \sqrt{x_0}\sqrt{y_0}$, so the x-intercept is $x_0 + \sqrt{x_0}\sqrt{y_0}$. The sum of the intercepts is

$$\left(y_0 + \sqrt{x_0}\sqrt{y_0}\right) + \left(x_0 + \sqrt{x_0}\sqrt{y_0}\right) = x_0 + 2\sqrt{x_0}\sqrt{y_0} + y_0 = \left(\sqrt{x_0} + \sqrt{y_0}\right)^2 = \left(\sqrt{c}\right)^2 = c.$$

47. If the circle has radius r, its equation is $x^2 + y^2 = r^2 \;\Rightarrow\; 2x + 2yy' = 0 \;\Rightarrow\; y' = -\dfrac{x}{y}$, so the slope of the tangent line at $P(x_0, y_0)$ is $-\dfrac{x_0}{y_0}$. The negative reciprocal of that slope is $\dfrac{-1}{-x_0/y_0} = \dfrac{y_0}{x_0}$, which is the slope of OP, so the tangent line at P is perpendicular to the radius OP.

48. $y^q = x^p \;\Rightarrow\; qy^{q-1}y' = px^{p-1} \;\Rightarrow\; y' = \dfrac{px^{p-1}}{qy^{q-1}} = \dfrac{px^{p-1}y}{qy^q} = \dfrac{px^{p-1}x^{p/q}}{qx^p} = \dfrac{p}{q}x^{(p/q)-1}$

49. $y = (\tan^{-1}x)^2 \;\Rightarrow\; y' = 2(\tan^{-1}x)^1 \cdot \dfrac{d}{dx}(\tan^{-1}x) = 2\tan^{-1}x \cdot \dfrac{1}{1+x^2} = \dfrac{2\tan^{-1}x}{1+x^2}$

50. $y = \tan^{-1}(x^2) \;\Rightarrow\; y' = \dfrac{1}{1+(x^2)^2} \cdot \dfrac{d}{dx}(x^2) = \dfrac{1}{1+x^4} \cdot 2x = \dfrac{2x}{1+x^4}$

51. $y = \sin^{-1}(2x+1) \;\Rightarrow$

$$y' = \frac{1}{\sqrt{1-(2x+1)^2}} \cdot \frac{d}{dx}(2x+1) = \frac{1}{\sqrt{1-(4x^2+4x+1)}} \cdot 2 = \frac{2}{\sqrt{-4x^2-4x}} = \frac{1}{\sqrt{-x^2-x}}$$

52. $g(x) = \arccos\sqrt{x} \;\Rightarrow\; g'(x) = -\dfrac{1}{\sqrt{1-(\sqrt{x})^2}}\dfrac{d}{dx}\sqrt{x} = -\dfrac{1}{\sqrt{1-x}}\left(\dfrac{1}{2}x^{-1/2}\right) = -\dfrac{1}{2\sqrt{x}\sqrt{1-x}}$

53. $F(x) = x\sec^{-1}(x^3) \;\overset{\text{PR}}{\Rightarrow}$

$$F'(x) = x \cdot \frac{1}{x^3\sqrt{(x^3)^2-1}}\frac{d}{dx}(x^3) + \sec^{-1}(x^3) \cdot 1 = \frac{x(3x^2)}{x^3\sqrt{x^6-1}} + \sec^{-1}(x^3) = \frac{3}{\sqrt{x^6-1}} + \sec^{-1}(x^3)$$

54. $y = \tan^{-1}\left(x - \sqrt{x^2+1}\right) \Rightarrow$

$$y' = \frac{1}{1 + \left(x - \sqrt{x^2+1}\right)^2}\left(1 - \frac{x}{\sqrt{x^2+1}}\right) = \frac{1}{1 + x^2 - 2x\sqrt{x^2+1} + x^2 + 1}\left(\frac{\sqrt{x^2+1} - x}{\sqrt{x^2+1}}\right)$$

$$= \frac{\sqrt{x^2+1} - x}{2\left(1 + x^2 - x\sqrt{x^2+1}\right)\sqrt{x^2+1}} = \frac{\sqrt{x^2+1} - x}{2\left[\sqrt{x^2+1}\,(1+x^2) - x(x^2+1)\right]} = \frac{\sqrt{x^2+1} - x}{2\left[(1+x^2)\left(\sqrt{x^2+1} - x\right)\right]}$$

$$= \frac{1}{2(1+x^2)}$$

55. $h(t) = \cot^{-1}(t) + \cot^{-1}(1/t) \Rightarrow$

$$h'(t) = -\frac{1}{1+t^2} - \frac{1}{1+(1/t)^2}\cdot\frac{d}{dt}\frac{1}{t} = -\frac{1}{1+t^2} - \frac{t^2}{t^2+1}\cdot\left(-\frac{1}{t^2}\right) = -\frac{1}{1+t^2} + \frac{1}{t^2+1} = 0.$$

Note that this makes sense because $h(t) = \frac{\pi}{2}$ for $t > 0$ and $h(t) = \frac{3\pi}{2}$ for $t < 0$.

56. $R(t) = \arcsin(1/t) \Rightarrow$

$$R'(t) = \frac{1}{\sqrt{1-(1/t)^2}}\,\frac{d}{dt}\frac{1}{t} = \frac{1}{\sqrt{1-1/t^2}}\left(-\frac{1}{t^2}\right) = -\frac{1}{\sqrt{1-1/t^2}}\,\frac{1}{\sqrt{t^4}}$$

$$= -\frac{1}{\sqrt{t^4 - t^2}} = -\frac{1}{\sqrt{t^2(t^2-1)}} = -\frac{1}{|\,t\,|\sqrt{t^2-1}}$$

57. $y = x\sin^{-1}x + \sqrt{1-x^2} \Rightarrow$

$$y' = x\cdot\frac{1}{\sqrt{1-x^2}} + (\sin^{-1}x)(1) + \frac{1}{2}(1-x^2)^{-1/2}(-2x) = \frac{x}{\sqrt{1-x^2}} + \sin^{-1}x - \frac{x}{\sqrt{1-x^2}} = \sin^{-1}x$$

58. $y = \cos^{-1}(\sin^{-1}t) \Rightarrow y' = -\frac{1}{\sqrt{1-(\sin^{-1}t)^2}}\cdot\frac{d}{dt}\sin^{-1}t = -\frac{1}{\sqrt{1-(\sin^{-1}t)^2}}\cdot\frac{1}{\sqrt{1-t^2}}$

59. $y = \arccos\left(\frac{b + a\cos x}{a + b\cos x}\right) \Rightarrow$

$$y' = -\frac{1}{\sqrt{1 - \left(\frac{b+a\cos x}{a+b\cos x}\right)^2}}\,\frac{(a+b\cos x)(-a\sin x) - (b+a\cos x)(-b\sin x)}{(a+b\cos x)^2}$$

$$= \frac{1}{\sqrt{a^2 + b^2\cos^2 x - b^2 - a^2\cos^2 x}}\,\frac{(a^2-b^2)\sin x}{|a+b\cos x|}$$

$$= \frac{1}{\sqrt{a^2-b^2}\sqrt{1-\cos^2 x}}\,\frac{(a^2-b^2)\sin x}{|a+b\cos x|} = \frac{\sqrt{a^2-b^2}}{|a+b\cos x|}\,\frac{\sin x}{|\sin x|}$$

But $0 \le x \le \pi$, so $|\sin x| = \sin x$. Also $a > b > 0 \Rightarrow b\cos x \ge -b > -a$, so $a + b\cos x > 0$.

Thus $y' = \frac{\sqrt{a^2-b^2}}{a+b\cos x}$.

60. $y = \arctan\sqrt{\dfrac{1-x}{1+x}} = \arctan\left(\dfrac{1-x}{1+x}\right)^{1/2} \Rightarrow$

$$y' = \frac{1}{1+\left(\sqrt{\dfrac{1-x}{1+x}}\right)^2}\cdot\frac{d}{dx}\left(\frac{1-x}{1+x}\right)^{1/2} = \frac{1}{1+\dfrac{1-x}{1+x}}\cdot\frac{1}{2}\left(\frac{1-x}{1+x}\right)^{-1/2}\cdot\frac{(1+x)(-1)-(1-x)(1)}{(1+x)^2}$$

$$= \frac{1}{\dfrac{1+x}{1+x}+\dfrac{1-x}{1+x}}\cdot\frac{1}{2}\left(\frac{1+x}{1-x}\right)^{1/2}\cdot\frac{-2}{(1+x)^2} = \frac{1+x}{2}\cdot\frac{1}{2}\cdot\frac{(1+x)^{1/2}}{(1-x)^{1/2}}\cdot\frac{-2}{(1+x)^2}$$

$$= \frac{-1}{2(1-x)^{1/2}(1+x)^{1/2}} = \frac{-1}{2\sqrt{1-x^2}}$$

61. $f(x) = \sqrt{1-x^2}\arcsin x \Rightarrow f'(x) = \sqrt{1-x^2}\cdot\dfrac{1}{\sqrt{1-x^2}} + \arcsin x\cdot\dfrac{1}{2}\left(1-x^2\right)^{-1/2}(-2x) = 1 - \dfrac{x\arcsin x}{\sqrt{1-x^2}}$

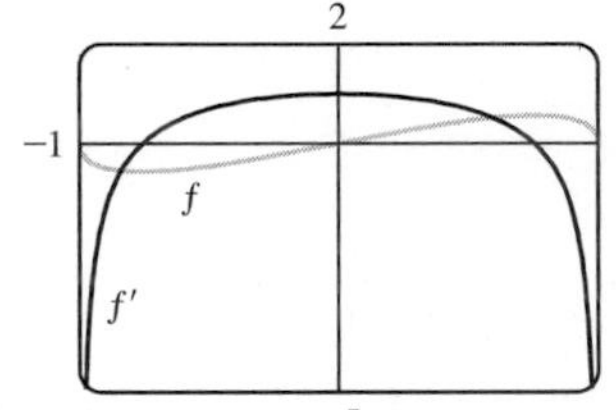

Note that $f' = 0$ where the graph of f has a horizontal tangent. Also note that f' is negative when f is decreasing and f' is positive when f is increasing.

62. $f(x) = \arctan(x^2 - x) \Rightarrow f'(x) = \dfrac{1}{1+(x^2-x)^2}\cdot\dfrac{d}{dx}(x^2-x) = \dfrac{2x-1}{1+(x^2-x)^2}$

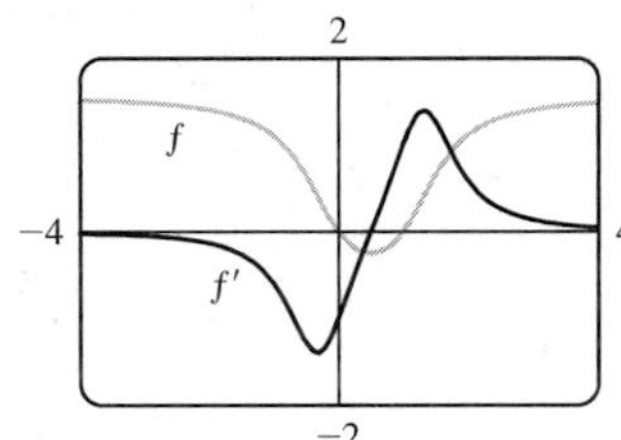

Note that $f' = 0$ where the graph of f has a horizontal tangent. Also note that f' is negative when f is decreasing and f' is positive when f is increasing.

63. Let $y = \cos^{-1}x$. Then $\cos y = x$ and $0 \le y \le \pi \Rightarrow -\sin y\,\dfrac{dy}{dx} = 1 \Rightarrow$

$\dfrac{dy}{dx} = -\dfrac{1}{\sin y} = -\dfrac{1}{\sqrt{1-\cos^2 y}} = -\dfrac{1}{\sqrt{1-x^2}}$. [Note that $\sin y \ge 0$ for $0 \le y \le \pi$.]

64. (a) Let $y = \sec^{-1}x$. Then $\sec y = x$ and $y \in \left(0, \frac{\pi}{2}\right] \cup \left(\pi, \frac{3\pi}{2}\right]$. Differentiate with respect to x: $\sec y\,\tan y\left(\dfrac{dy}{dx}\right) = 1 \Rightarrow$

$\dfrac{dy}{dx} = \dfrac{1}{\sec y\,\tan y} = \dfrac{1}{\sec y\,\sqrt{\sec^2 y - 1}} = \dfrac{1}{x\,\sqrt{x^2-1}}$. Note that $\tan^2 y = \sec^2 y - 1 \Rightarrow \tan y = \sqrt{\sec^2 y - 1}$

since $\tan y > 0$ when $0 < y < \frac{\pi}{2}$ or $\pi < y < \frac{3\pi}{2}$.

(b) $y = \sec^{-1}x \Rightarrow \sec y = x \Rightarrow \sec y\,\tan y\,\dfrac{dy}{dx} = 1 \Rightarrow \dfrac{dy}{dx} = \dfrac{1}{\sec y\,\tan y}$. Now $\tan^2 y = \sec^2 y - 1 = x^2 - 1$,

so $\tan y = \pm\sqrt{x^2-1}$. For $y \in \left[0, \frac{\pi}{2}\right)$, $x \ge 1$, so $\sec y = x = |x|$ and $\tan y \ge 0 \Rightarrow$

$\dfrac{dy}{dx} = \dfrac{1}{x\,\sqrt{x^2-1}} = \dfrac{1}{|x|\,\sqrt{x^2-1}}$. For $y \in \left(\frac{\pi}{2}, \pi\right]$, $x \le -1$, so $|x| = -x$ and $\tan y = -\sqrt{x^2-1} \Rightarrow$

$\dfrac{dy}{dx} = \dfrac{1}{\sec y\,\tan y} = \dfrac{1}{x\left(-\sqrt{x^2-1}\right)} = \dfrac{1}{(-x)\,\sqrt{x^2-1}} = \dfrac{1}{|x|\,\sqrt{x^2-1}}$.

65. $x^2 + y^2 = r^2$ is a circle with center O and $ax + by = 0$ is a line through O [assume a and b are not both zero]. $x^2 + y^2 = r^2 \Rightarrow 2x + 2yy' = 0 \Rightarrow y' = -x/y$, so the slope of the tangent line at $P_0\,(x_0, y_0)$ is $-x_0/y_0$. The slope of the line OP_0 is y_0/x_0, which is the negative reciprocal of $-x_0/y_0$. Hence, the curves are orthogonal, and the families of curves are orthogonal trajectories of each other.

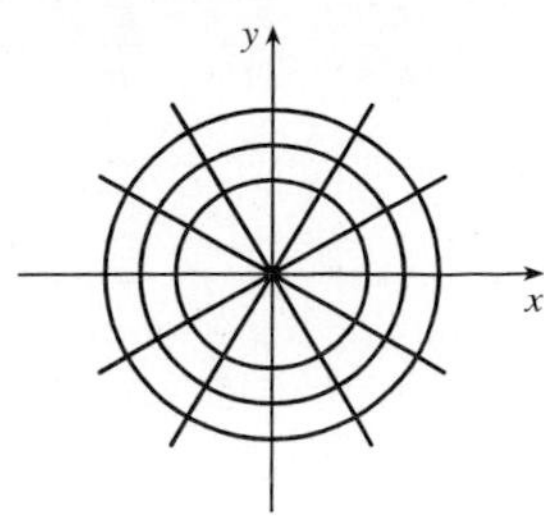

66. The circles $x^2 + y^2 = ax$ and $x^2 + y^2 = by$ intersect at the origin where the tangents are vertical and horizontal [assume a and b are both nonzero]. If (x_0, y_0) is the other point of intersection, then $x_0^2 + y_0^2 = ax_0$ **(1)** and $x_0^2 + y_0^2 = by_0$ **(2)**.

Now $x^2 + y^2 = ax \Rightarrow 2x + 2yy' = a \Rightarrow y' = \dfrac{a - 2x}{2y}$ and $x^2 + y^2 = by \Rightarrow$

$2x + 2yy' = by' \Rightarrow y' = \dfrac{2x}{b - 2y}$. Thus, the curves are orthogonal at $(x_0, y_0) \Leftrightarrow$

$\dfrac{a - 2x_0}{2y_0} = -\dfrac{b - 2y_0}{2x_0} \Leftrightarrow 2ax_0 - 4x_0^2 = 4y_0^2 - 2by_0 \Leftrightarrow ax_0 + by_0 = 2(x_0^2 + y_0^2)$,

which is true by **(1)** and **(2)**.

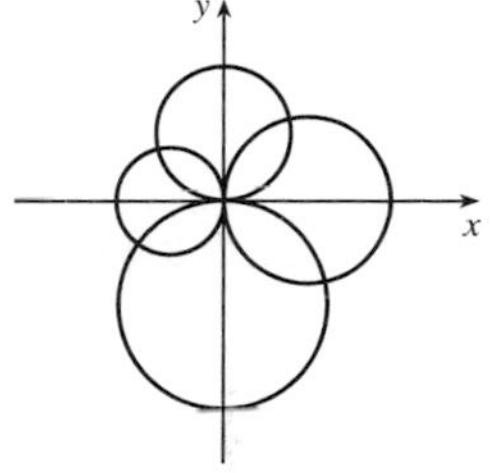

67. $y = cx^2 \Rightarrow y' = 2cx$ and $x^2 + 2y^2 = k$ [assume $k > 0$] $\Rightarrow 2x + 4yy' = 0 \Rightarrow$

$2yy' = -x \Rightarrow y' = -\dfrac{x}{2(y)} = -\dfrac{x}{2(cx^2)} = -\dfrac{1}{2cx}$, so the curves are orthogonal if

$c \neq 0$. If $c = 0$, then the horizontal line $y = cx^2 = 0$ intersects $x^2 + 2y^2 = k$ orthogonally

at $\left(\pm\sqrt{k}, 0\right)$, since the ellipse $x^2 + 2y^2 = k$ has vertical tangents at those two points.

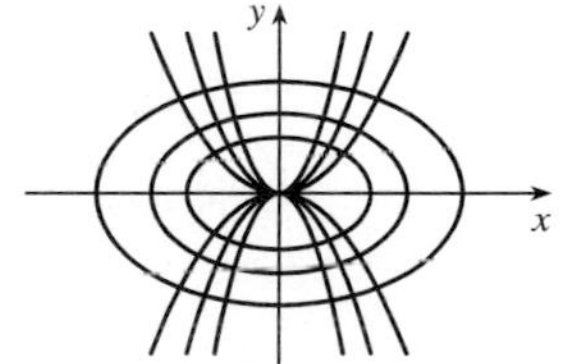

68. $y = ax^3 \Rightarrow y' = 3ax^2$ and $x^2 + 3y^2 = b$ [assume $b > 0$] $\Rightarrow 2x + 6yy' = 0 \Rightarrow$

$3yy' = -x \Rightarrow y' = -\dfrac{x}{3(y)} = -\dfrac{x}{3(ax^3)} = -\dfrac{1}{3ax^2}$, so the curves are orthogonal if

$a \neq 0$. If $a = 0$, then the horizontal line $y = ax^3 = 0$ intesects $x^2 + 3y^2 = b$ orthogonally

at $\left(\pm\sqrt{b}, 0\right)$, since the ellipse $x^2 + 3y^2 = b$ has vertical tangents at those two points.

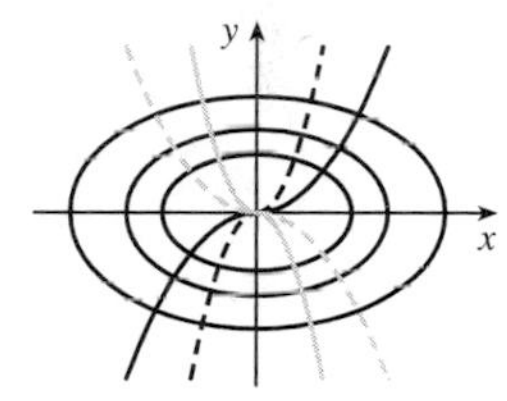

69. Since $A^2 < a^2$, we are assured that there are four points of intersection.

(1) $\dfrac{x^2}{a^2} + \dfrac{y^2}{b^2} = 1 \Rightarrow \dfrac{2x}{a^2} + \dfrac{2yy'}{b^2} = 0 \Rightarrow \dfrac{yy'}{b^2} = -\dfrac{x}{a^2} \Rightarrow$

$y' = m_1 = -\dfrac{xb^2}{ya^2}$.

(2) $\dfrac{x^2}{A^2} - \dfrac{y^2}{B^2} = 1 \Rightarrow \dfrac{2x}{A^2} - \dfrac{2yy'}{B^2} = 0 \Rightarrow \dfrac{yy'}{B^2} = \dfrac{x}{A^2} \Rightarrow$

$y' = m_2 = \dfrac{xB^2}{yA^2}$.

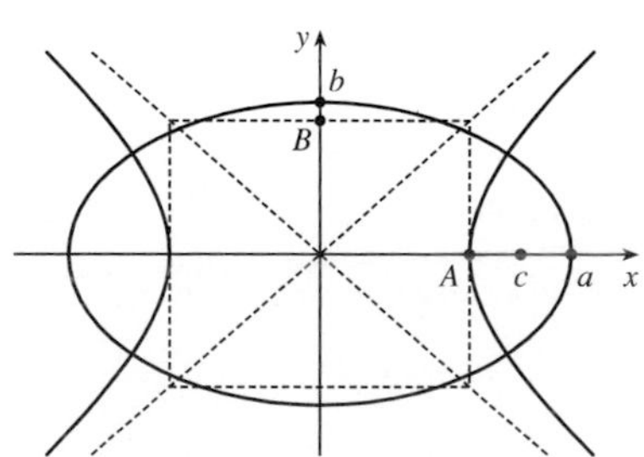

Now $m_1 m_2 = -\dfrac{xb^2}{ya^2} \cdot \dfrac{xB^2}{yA^2} = -\dfrac{b^2B^2}{a^2A^2} \cdot \dfrac{x^2}{y^2}$ **(3)**. Subtracting equations, **(1)** − **(2)**, gives us $\dfrac{x^2}{a^2} + \dfrac{y^2}{b^2} - \dfrac{x^2}{A^2} + \dfrac{y^2}{B^2} = 0 \Rightarrow$

$\dfrac{y^2}{b^2}+\dfrac{y^2}{B^2}=\dfrac{x^2}{A^2}-\dfrac{x^2}{a^2} \;\Rightarrow\; \dfrac{y^2B^2+y^2b^2}{b^2B^2}=\dfrac{x^2a^2-x^2A^2}{A^2a^2} \;\Rightarrow\; \dfrac{y^2(b^2+B^2)}{b^2B^2}=\dfrac{x^2(a^2-A^2)}{a^2A^2}$ **(4)**. Since $a^2-b^2=A^2+B^2$, we have $a^2-A^2=b^2+B^2$. Thus, equation **(4)** becomes $\dfrac{y^2}{b^2B^2}=\dfrac{x^2}{A^2a^2} \;\Rightarrow\; \dfrac{x^2}{y^2}=\dfrac{A^2a^2}{b^2B^2}$, and substituting for $\dfrac{x^2}{y^2}$ in equation **(3)** gives us $m_1m_2=-\dfrac{b^2B^2}{a^2A^2}\cdot\dfrac{a^2A^2}{b^2B^2}=-1$. Hence, the ellipse and hyperbola are orthogonal trajectories.

70. $y=(x+c)^{-1} \;\Rightarrow\; y'=-(x+c)^{-2}$ and $y=a(x+k)^{1/3} \;\Rightarrow\; y'=\frac{1}{3}a(x+k)^{-2/3}$, so the curves are othogonal if the product of the slopes is -1, that is, $\dfrac{-1}{(x+c)^2}\cdot\dfrac{a}{3(x+k)^{2/3}}=-1 \;\Rightarrow\; a=3(x+c)^2(x+k)^{2/3} \;\Rightarrow$

$a=3\left(\dfrac{1}{y}\right)^2\left(\dfrac{y}{a}\right)^2$ [since $y^2=(x+c)^{-2}$ and $y^2=a^2(x+k)^{2/3}$] $\;\Rightarrow\; a=3\left(\dfrac{1}{a^2}\right) \;\Rightarrow\; a^3=3 \;\Rightarrow\; a=\sqrt[3]{3}$.

71. (a) $\left(P+\dfrac{n^2a}{V^2}\right)(V-nb)=nRT \;\Rightarrow\; PV-Pnb+\dfrac{n^2a}{V}-\dfrac{n^3ab}{V^2}=nRT \;\Rightarrow$

$$\frac{d}{dP}(PV-Pnb+n^2aV^{-1}-n^3abV^{-2})=\frac{d}{dP}(nRT) \;\Rightarrow$$

$$PV'+V\cdot 1-nb-n^2aV^{-2}\cdot V'+2n^3abV^{-3}\cdot V'=0 \;\Rightarrow\; V'(P-n^2aV^{-2}+2n^3abV^{-3})=nb-V \;\Rightarrow$$

$$V'=\frac{nb-V}{P-n^2aV^{-2}+2n^3abV^{-3}} \text{ or } \frac{dV}{dP}=\frac{V^3(nb-V)}{PV^3-n^2aV+2n^3ab}$$

(b) Using the last expression for dV/dP from part (a), we get

$$\frac{dV}{dP}=\frac{(10\text{ L})^3[(1\text{ mole})(0.04267\text{ L/mole})-10\text{ L}]}{\left[\begin{array}{l}(2.5\text{ atm})(10\text{ L})^3-(1\text{ mole})^2(3.592\text{ L}^2\text{-atm/ mole}^2)(10\text{ L})\\ \qquad +2(1\text{ mole})^3(3.592\text{ L}^2\text{-atm/ mole}^2)(0.04267\text{ L/ mole})\end{array}\right]}$$

$$=\frac{-9957.33\text{ L}^4}{2464.386541\text{ L}^3\text{-atm}}\approx -4.04\text{ L/ atm.}$$

72. (a) $x^2+xy+y^2+1=0 \;\Rightarrow\; 2x+xy'+y\cdot 1+2yy'+0=0 \;\Rightarrow\; y'(x+2y)=-2x-y \;\Rightarrow\; y'=\dfrac{-2x-y}{x+2y}$

(b) Plotting the curve in part (a) gives us an empty graph, that is, there are no points that satisfy the equation. If there were any points that satisfied the equation, then x and y would have opposite signs; otherwise, all the terms are positive and their sum can not equal 0. $x^2+xy+y^2+1=0 \;\Rightarrow\; x^2+2xy+y^2-xy+1=0 \;\Rightarrow\; (x+y)^2=xy-1$. The left side of the last equation is nonnegative, but the right side is at most -1, so that proves there are no points that satisfy the equation.

Another solution: $x^2+xy+y^2+1=\frac{1}{2}x^2+xy+\frac{1}{2}y^2+\frac{1}{2}x^2+\frac{1}{2}y^2+1=\frac{1}{2}(x^2+2xy+y^2)+\frac{1}{2}(x^2+y^2)+1$

$$=\tfrac{1}{2}(x+y)^2+\tfrac{1}{2}(x^2+y^2)+1\geq 1$$

Another solution: Regarding $x^2+xy+y^2+1=0$ as a quadratic in x, the discriminant is $y^2-4(y^2+1)=-3y^2-4$. This is negative, so there are no real solutions.

(c) The expression for y' in part (a) is meaningless; that is, since the equation in part (a) has no solution, it does not implicitly define a function y of x, and therefore it is meaningless to consider y'.

73. To find the points at which the ellipse $x^2 - xy + y^2 = 3$ crosses the x-axis, let $y = 0$ and solve for x.

$y = 0 \Rightarrow x^2 - x(0) + 0^2 = 3 \Leftrightarrow x = \pm\sqrt{3}$. So the graph of the ellipse crosses the x-axis at the points $(\pm\sqrt{3}, 0)$.

Using implicit differentiation to find y', we get $2x - xy' - y + 2yy' = 0 \Rightarrow y'(2y - x) = y - 2x \Leftrightarrow y' = \dfrac{y - 2x}{2y - x}$.

So y' at $(\sqrt{3}, 0)$ is $\dfrac{0 - 2\sqrt{3}}{2(0) - \sqrt{3}} = 2$ and y' at $(-\sqrt{3}, 0)$ is $\dfrac{0 + 2\sqrt{3}}{2(0) + \sqrt{3}} = 2$. Thus, the tangent lines at these points are parallel.

74. (a) We use implicit differentiation to find $y' = \dfrac{y - 2x}{2y - x}$ as in Exercise 73. The slope of the tangent line at $(-1, 1)$ is $m = \dfrac{1 - 2(-1)}{2(1) - (-1)} = \dfrac{3}{3} = 1$, so the slope of the normal line is $-\dfrac{1}{m} = -1$, and its equation is $y - 1 = -1(x + 1) \Leftrightarrow$ $y = -x$. Substituting $-x$ for y in the equation of the ellipse, we get $x^2 - x(-x) + (-x)^2 = 3 \Rightarrow 3x^2 = 3 \Leftrightarrow x = \pm 1$. So the normal line must intersect the ellipse again at $x = 1$, and since the equation of the line is $y = -x$, the other point of intersection must be $(1, -1)$.

(b)

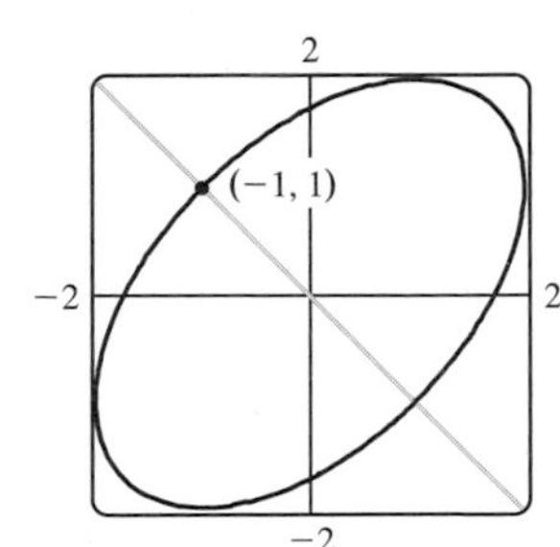

75. $x^2y^2 + xy = 2 \Rightarrow x^2 \cdot 2yy' + y^2 \cdot 2x + x \cdot y' + y \cdot 1 = 0 \Leftrightarrow y'(2x^2y + x) = -2xy^2 - y \Leftrightarrow$

$y' = -\dfrac{2xy^2 + y}{2x^2y + x}$. So $-\dfrac{2xy^2 + y}{2x^2y + x} = -1 \Leftrightarrow 2xy^2 + y = 2x^2y + x \Leftrightarrow y(2xy + 1) = x(2xy + 1) \Leftrightarrow$

$y(2xy + 1) - x(2xy + 1) = 0 \Leftrightarrow (2xy + 1)(y - x) = 0 \Leftrightarrow xy = -\frac{1}{2}$ or $y = x$. But $xy = -\frac{1}{2} \Rightarrow$

$x^2y^2 + xy = \frac{1}{4} - \frac{1}{2} \neq 2$, so we must have $x = y$. Then $x^2y^2 + xy = 2 \Rightarrow x^4 + x^2 = 2 \Leftrightarrow x^4 + x^2 - 2 = 0 \Leftrightarrow$

$(x^2 + 2)(x^2 - 1) = 0$. So $x^2 = -2$, which is impossible, or $x^2 = 1 \Leftrightarrow x = \pm 1$. Since $x = y$, the points on the curve where the tangent line has a slope of -1 are $(-1, -1)$ and $(1, 1)$.

76. $x^2 + 4y^2 = 36 \Rightarrow 2x + 8yy' = 0 \Rightarrow y' = -\dfrac{x}{4y}$. Let (a, b) be a point on $x^2 + 4y^2 = 36$ whose tangent line passes through $(12, 3)$. The tangent line is then $y - 3 = -\dfrac{a}{4b}(x - 12)$, so $b - 3 = -\dfrac{a}{4b}(a - 12)$. Multiplying both sides by $4b$ gives $4b^2 - 12b = -a^2 + 12a$, so $4b^2 + a^2 = 12(a + b)$. But $4b^2 + a^2 = 36$, so $36 = 12(a + b) \Rightarrow a + b = 3 \Rightarrow$ $b = 3 - a$. Substituting $3 - a$ for b into $a^2 + 4b^2 = 36$ gives $a^2 + 4(3 - a)^2 = 36 \Leftrightarrow a^2 + 36 - 24a + 4a^2 = 36 \Leftrightarrow$ $5a^2 - 24a = 0 \Leftrightarrow a(5a - 24) = 0$, so $a = 0$ or $a = \frac{24}{5}$. If $a = 0$, $b = 3 - 0 = 3$, and if $a = \frac{24}{5}$, $b = 3 - \frac{24}{5} = -\frac{9}{5}$.

So the two points on the ellipse are $(0, 3)$ and $\left(\frac{24}{5}, -\frac{9}{5}\right)$. Using

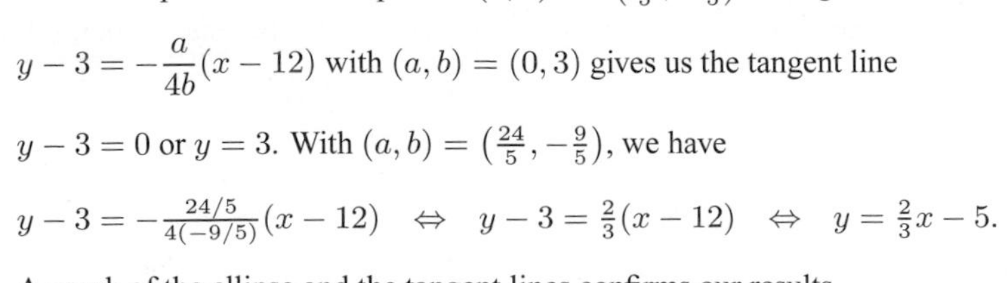

$y - 3 = -\dfrac{a}{4b}(x - 12)$ with $(a, b) = (0, 3)$ gives us the tangent line

$y - 3 = 0$ or $y = 3$. With $(a, b) = \left(\frac{24}{5}, -\frac{9}{5}\right)$, we have

$y - 3 = -\frac{24/5}{4(-9/5)}(x - 12) \Leftrightarrow y - 3 = \frac{2}{3}(x - 12) \Leftrightarrow y = \frac{2}{3}x - 5$.

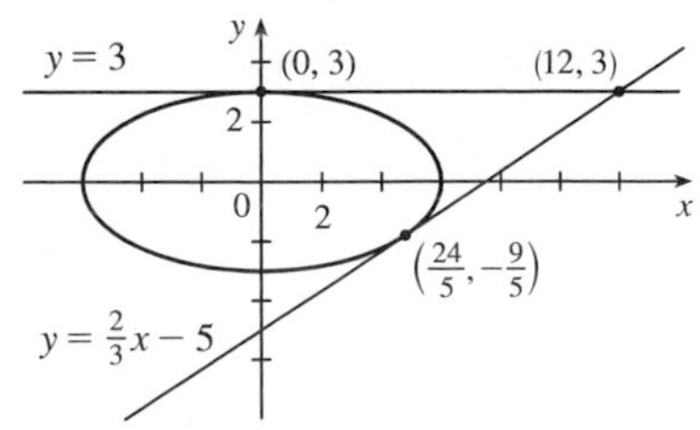

A graph of the ellipse and the tangent lines confirms our results.

77. (a) If $y = f^{-1}(x)$, then $f(y) = x$. Differentiating implicitly with respect to x and remembering that y is a function of x, we get $f'(y)\,\dfrac{dy}{dx} = 1$, so $\dfrac{dy}{dx} = \dfrac{1}{f'(y)} \quad\Rightarrow\quad (f^{-1})'(x) = \dfrac{1}{f'(f^{-1}(x))}$.

(b) $f(4) = 5 \;\Rightarrow\; f^{-1}(5) = 4$. By part (a), $(f^{-1})'(5) = \dfrac{1}{f'(f^{-1}(5))} = \dfrac{1}{f'(4)} = 1/\left(\frac{2}{3}\right) = \frac{3}{2}$.

78. (a) Assume $a < b$. Since e^x is an increasing function, $e^a < e^b$, and hence, $a + e^a < b + e^b$; that is, $f(a) < f(b)$. So $f(x) = x + e^x$ is an increasing function and therefore one-to-one.

(b) $f^{-1}(1) = a \;\Leftrightarrow\; f(a) = 1$, so we need to find a such that $f(a) = 1$. By inspection, we see that $f(0) = 0 + e^0 = 1$, so $a = 0$, and hence, $f^{-1}(1) = 0$.

(c) $(f^{-1})'(1) = \dfrac{1}{f'(f^{-1}(1))} = \dfrac{1}{f'(0)}$ [by part (b)]. Now $f(x) = x + e^x \;\Rightarrow\; f'(x) = 1 + e^x$, so $f'(0) = 1 + e^0 = 2$. Thus, $(f^{-1})'(1) = \frac{1}{2}$.

79. (a) $y = J(x)$ and $xy'' + y' + xy = 0 \;\Rightarrow\; xJ''(x) + J'(x) + xJ(x) = 0$. If $x = 0$, we have $0 + J'(0) + 0 = 0$, so $J'(0) = 0$.

(b) Differentiating $xy'' + y' + xy = 0$ implicitly, we get $xy''' + y'' \cdot 1 + y'' + xy' + y \cdot 1 = 0 \;\Rightarrow$ $xy''' + 2y'' + xy' + y = 0$, so $xJ'''(x) + 2J''(x) + xJ'(x) + J(x) = 0$. If $x = 0$, we have $0 + 2J''(0) + 0 + 1$ [$J(0) = 1$ is given] $= 0 \;\Rightarrow\; 2J''(0) = -1 \;\Rightarrow\; J''(0) = -\frac{1}{2}$.

80. $x^2 + 4y^2 = 5 \;\Rightarrow\; 2x + 4(2yy') = 0 \;\Rightarrow\; y' = -\dfrac{x}{4y}$. Now let h be the height of the lamp, and let (a, b) be the point of tangency of the line passing through the points $(3, h)$ and $(-5, 0)$. This line has slope $(h - 0)/[3 - (-5)] = \frac{1}{8}h$. But the slope of the tangent line through the point (a, b) can be expressed as $y' = -\dfrac{a}{4b}$, or as $\dfrac{b - 0}{a - (-5)} = \dfrac{b}{a + 5}$ [since the line passes through $(-5, 0)$ and (a, b)], so $-\dfrac{a}{4b} = \dfrac{b}{a + 5} \;\Leftrightarrow\; 4b^2 = -a^2 - 5a \;\Leftrightarrow\; a^2 + 4b^2 = -5a$. But $a^2 + 4b^2 = 5$ [since (a, b) is on the ellipse], so $5 = -5a \;\Leftrightarrow\; a = -1$. Then $4b^2 = -a^2 - 5a = -1 - 5(-1) = 4 \;\Rightarrow\; b = 1$, since the point is on the top half of the ellipse. So $\dfrac{h}{8} = \dfrac{b}{a + 5} = \dfrac{1}{-1 + 5} = \dfrac{1}{4} \;\Rightarrow\; h = 2$. So the lamp is located 2 units above the x-axis.

LABORATORY PROJECT Families of Implicit Curves

1. (a) There appear to be nine points of intersection. The "inner four" near the origin are about $(\pm 0.2, -0.9)$ and $(\pm 0.3, -1.1)$. The "outer five" are about $(2.0, -8.9)$, $(-2.8, -8.8)$, $(-7.5, -7.7)$, $(-7.8, -4.7)$, and $(-8.0, 1.5)$.

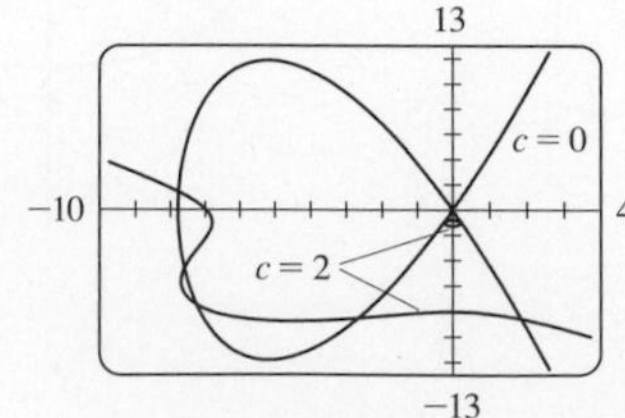

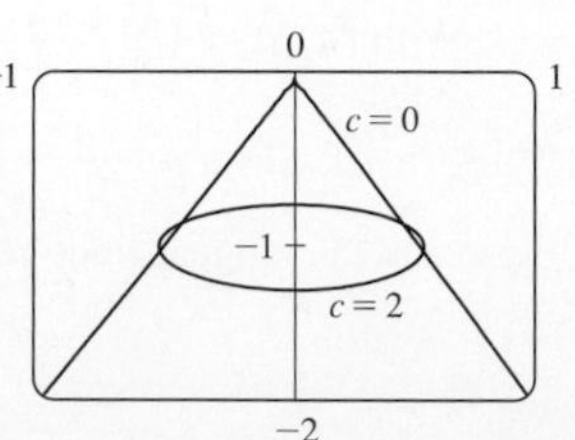

(b) We see from the graphs with $c = 5$ and $c = 10$, and for other values of c, that the curves change shape but the nine points of intersection are the same.

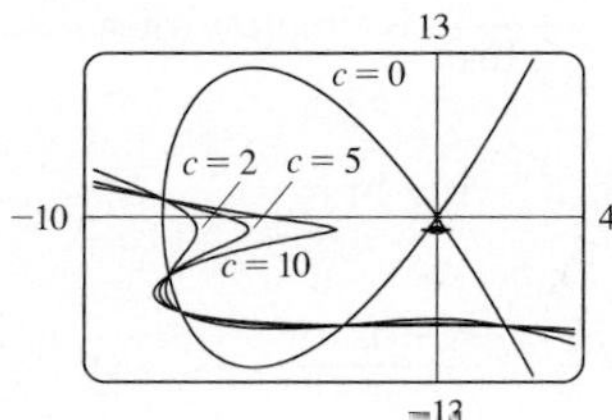

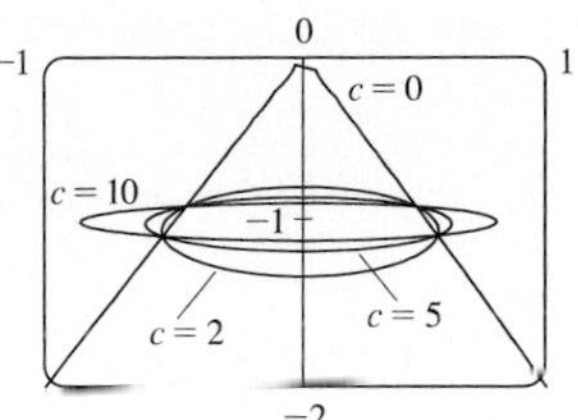

2. (a) If $c = 0$, the graph is the unit circle. As c increases, the graph looks more diamondlike and then more crosslike (see the graph for $c \geq 0$).

For $-1 < c < 0$ (see the graph), there are four hyperboliclike branches as well as an ellipticlike curve bounded by $|x| \leq 1$ and $|y| \leq 1$ for values of c close to 0. As c gets closer to -1, the branches and the curve become more rectangular, approaching the lines $|x| = 1$ and $|y| = 1$.

For $c = -1$, we get the lines $x = \pm 1$ and $y = \pm 1$. As c decreases, we get four test-tubelike curves (see the graph) that are bounded by $|x| = 1$ and $|y| = 1$, and get thinner as $|c|$ gets larger.

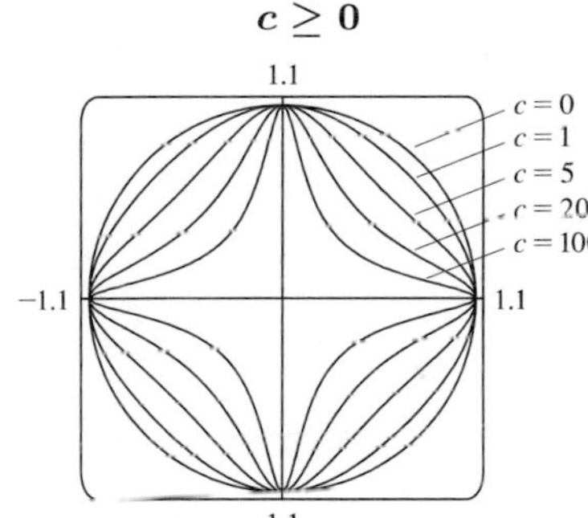

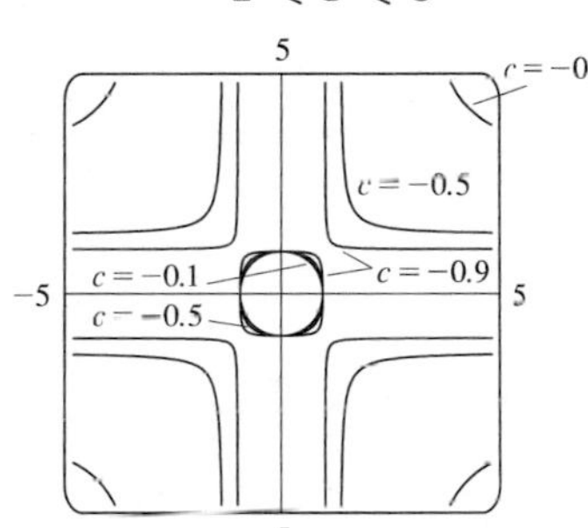

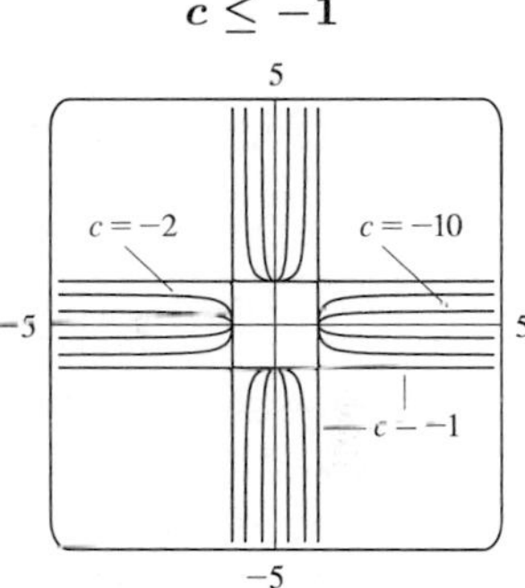

(b) The curve for $c = -1$ is described in part (a). When $c = -1$, we get $x^2 + y^2 - x^2y^2 = 1 \;\Leftrightarrow$ $0 = x^2y^2 - x^2 - y^2 + 1 \;\Leftrightarrow\; 0 = (x^2 - 1)(y^2 - 1) \;\Leftrightarrow\; x = \pm 1$ or $y = \pm 1$, which algebraically proves that the graph consists of the stated lines.

(c) $\dfrac{d}{dx}(x^2 + y^2 + cx^2y^2) = \dfrac{d}{dx}(1) \;\Rightarrow\; 2x + 2y\,y' + c(x^2 \cdot 2y\,y' + y^2 \cdot 2x) = 0 \;\Rightarrow$

$2y\,y' + 2cx^2y\,y' = -2x - 2cxy^2 \;\Rightarrow\; 2y(1 + cx^2)y' = -2x(1 + cy^2) \;\Rightarrow\; y' = -\dfrac{x(1 + cy^2)}{y(1 + cx^2)}$.

For $c = -1$, $y' = -\dfrac{x(1 - y^2)}{y(1 - x^2)} = -\dfrac{x(1 + y)(1 - y)}{y(1 + x)(1 - x)}$, so $y' = 0$ when $y = \pm 1$ or $x = 0$ (which leads to $y = \pm 1$) and y' is undefined when $x = \pm 1$ or $y = 0$ (which leads to $x = \pm 1$). Since the graph consists of the lines $x = \pm 1$ and $y = \pm 1$, the slope at any point on the graph is undefined or 0, which is consistent with the expression found for y'.

3.6 Derivatives of Logarithmic Functions

1. The differentiation formula for logarithmic functions, $\dfrac{d}{dx}(\log_a x) = \dfrac{1}{x \ln a}$, is simplest when $a = e$ because $\ln e = 1$.

2. $f(x) = x \ln x - x \quad \Rightarrow \quad f'(x) = x \cdot \dfrac{1}{x} + (\ln x) \cdot 1 - 1 = 1 + \ln x - 1 = \ln x$

3. $f(x) = \sin(\ln x) \quad \Rightarrow \quad f'(x) = \cos(\ln x) \cdot \dfrac{d}{dx} \ln x = \cos(\ln x) \cdot \dfrac{1}{x} = \dfrac{\cos(\ln x)}{x}$

4. $f(x) = \ln(\sin^2 x) = \ln(\sin x)^2 = 2 \ln |\sin x| \quad \Rightarrow \quad f'(x) = 2 \cdot \dfrac{1}{\sin x} \cdot \cos x = 2 \cot x$

5. $f(x) = \ln \dfrac{1}{x} \quad \Rightarrow \quad f'(x) = \dfrac{1}{1/x} \dfrac{d}{dx}\left(\dfrac{1}{x}\right) = x\left(-\dfrac{1}{x^2}\right) = -\dfrac{1}{x}.$

Another solution: $f(x) = \ln \dfrac{1}{x} = \ln 1 - \ln x = -\ln x \quad \Rightarrow \quad f'(x) = -\dfrac{1}{x}.$

6. $y = \dfrac{1}{\ln x} = (\ln x)^{-1} \quad \Rightarrow \quad y' = -1(\ln x)^{-2} \cdot \dfrac{1}{x} = \dfrac{-1}{x(\ln x)^2}$

7. $f(x) = \log_{10}(1 + \cos x) \quad \Rightarrow \quad f'(x) = \dfrac{1}{(1 + \cos x) \ln 10} \dfrac{d}{dx}(1 + \cos x) = \dfrac{-\sin x}{(1 + \cos x) \ln 10}$

8. $f(x) = \log_{10} \sqrt{x} \quad \Rightarrow \quad f'(x) = \dfrac{1}{\sqrt{x} \ln 10} \dfrac{d}{dx} \sqrt{x} = \dfrac{1}{\sqrt{x} \ln 10} \dfrac{1}{2\sqrt{x}} = \dfrac{1}{2(\ln 10)x}$

Or: $f(x) = \log_{10} \sqrt{x} = \log_{10} x^{1/2} = \frac{1}{2} \log_{10} x \quad \Rightarrow \quad f'(x) = \dfrac{1}{2} \dfrac{1}{x \ln 10} = \dfrac{1}{2\,(\ln 10)\,x}$

9. $g(x) = \ln(xe^{-2x}) = \ln x + \ln e^{-2x} = \ln x - 2x \quad \Rightarrow \quad g'(x) = \dfrac{1}{x} - 2$

10. $g(t) = \sqrt{1 + \ln t} \quad \Rightarrow \quad g'(t) = \frac{1}{2}(1 + \ln t)^{-1/2} \dfrac{d}{dt}(1 + \ln t) = \dfrac{1}{2\sqrt{1 + \ln t}} \cdot \dfrac{1}{t} = \dfrac{1}{2t\sqrt{1 + \ln t}}$

11. $F(t) = (\ln t)^2 \sin t \quad \Rightarrow \quad F'(t) = (\ln t)^2 \cos t + \sin t \cdot 2 \ln t \cdot \dfrac{1}{t} = \ln t \left(\ln t \, \cos t + \dfrac{2 \sin t}{t}\right)$

12. $h(x) = \ln\left(x + \sqrt{x^2 - 1}\right) \quad \Rightarrow \quad h'(x) = \dfrac{1}{x + \sqrt{x^2 - 1}}\left(1 + \dfrac{x}{\sqrt{x^2 - 1}}\right) = \dfrac{1}{x + \sqrt{x^2 - 1}} \cdot \dfrac{\sqrt{x^2 - 1} + x}{\sqrt{x^2 - 1}} = \dfrac{1}{\sqrt{x^2 - 1}}$

13. $G(y) = \ln \dfrac{(2y + 1)^5}{\sqrt{y^2 + 1}} = \ln(2y + 1)^5 - \ln(y^2 + 1)^{1/2} = 5 \ln(2y + 1) - \frac{1}{2} \ln(y^2 + 1) \quad \Rightarrow$

$$G'(y) = 5 \cdot \frac{1}{2y + 1} \cdot 2 - \frac{1}{2} \cdot \frac{1}{y^2 + 1} \cdot 2y = \frac{10}{2y + 1} - \frac{y}{y^2 + 1} \quad \left[\text{or } \frac{8y^2 - y + 10}{(2y + 1)(y^2 + 1)}\right]$$

14. $P(v) = \dfrac{\ln v}{1 - v} \quad \Rightarrow \quad P'(v) = \dfrac{(1 - v)(1/v) - (\ln v)(-1)}{(1 - v)^2} \cdot \dfrac{v}{v} = \dfrac{1 - v + v \ln v}{v(1 - v)^2}$

15. $F(s) = \ln \ln s \quad \Rightarrow \quad F'(s) = \dfrac{1}{\ln s} \dfrac{d}{ds} \ln s = \dfrac{1}{\ln s} \cdot \dfrac{1}{s} = \dfrac{1}{s \ln s}$

16. $y = \ln|1 + t - t^3| \Rightarrow y' = \dfrac{1}{1 + t - t^3}\dfrac{d}{dt}(1 + t - t^3) = \dfrac{1 - 3t^2}{1 + t - t^3}$

17. $T(z) = 2^z \log_2 z \Rightarrow T'(z) = 2^z \dfrac{1}{z \ln 2} + \log_2 z \cdot 2^z \ln 2 = 2^z \left(\dfrac{1}{z \ln 2} + \log_2 z\,(\ln 2)\right).$

Note that $\log_2 z\,(\ln 2) = \dfrac{\ln z}{\ln 2}(\ln 2) = \ln z$ by the change of base theorem. Thus, $T'(z) = 2^z\left(\dfrac{1}{z \ln 2} + \ln z\right).$

18. $y = \ln(\csc x - \cot x) \Rightarrow$

$$y' = \frac{1}{\csc x - \cot x}\frac{d}{dx}(\csc x - \cot x) = \frac{1}{\csc x - \cot x}(-\csc x\,\cot x + \csc^2 x) = \frac{\csc x(\csc x - \cot x)}{\csc x - \cot x} = \csc x$$

19. $y = \ln(e^{-x} + xe^{-x}) = \ln(e^{-x}(1 + x)) = \ln(e^{-x}) + \ln(1 + x) = -x + \ln(1 + x) \Rightarrow$

$$y' = -1 + \frac{1}{1 + x} = \frac{-1 - x + 1}{1 + x} = -\frac{x}{1 + x}$$

20. $H(z) = \ln\sqrt{\dfrac{a^2 - z^2}{a^2 + z^2}} = \ln\left(\dfrac{a^2 - z^2}{a^2 + z^2}\right)^{1/2} = \dfrac{1}{2}\ln\left(\dfrac{a^2 - z^2}{a^2 + z^2}\right) = \tfrac{1}{2}\ln(a^2 - z^2) - \tfrac{1}{2}\ln(a^2 + z^2) \Rightarrow$

$$H'(z) = \frac{1}{2}\cdot\frac{1}{a^2 - z^2}\cdot(-2z) - \frac{1}{2}\cdot\frac{1}{a^2 + z^2}\cdot(2z) = \frac{z}{z^2 - a^2} - \frac{z}{z^2 + a^2} = \frac{z(z^2 + a^2) - z(z^2 - a^2)}{(z^2 - a^2)(z^2 + a^2)}$$

$$= \frac{z^3 + za^2 - z^3 + za^2}{(z^2 - a^2)(z^2 + a^2)} = \frac{2a^2 z}{z^4 - a^4}$$

21. $y = \tan[\ln(ax + b)] \Rightarrow y' = \sec^2[\ln(ax + b)]\cdot\dfrac{1}{ax + b}\cdot a = \sec^2[\ln(ax + b)]\dfrac{a}{ax + b}$

22. $y = \log_2(x \log_5 x) \Rightarrow$

$$y' = \frac{1}{(x\log_5 x)(\ln 2)}\frac{d}{dx}(x\log_5 x) = \frac{1}{(x\log_5 x)(\ln 2)}\left(x\cdot\frac{1}{x\ln 5} + \log_5 x\right) = \frac{1}{(x\log_5 x)(\ln 5)(\ln 2)} + \frac{1}{x(\ln 2)}.$$

Note that $\log_5 x(\ln 5) = \dfrac{\ln x}{\ln 5}(\ln 5) = \ln x$ by the change of base theorem. Thus, $y' = \dfrac{1}{x\ln x\ln 2} + \dfrac{1}{x\ln 2} = \dfrac{1 + \ln x}{x\ln x\ln 2}.$

23. $y = \sqrt{x}\ln x \Rightarrow y' = \sqrt{x}\cdot\dfrac{1}{x} + (\ln x)\dfrac{1}{2\sqrt{x}} = \dfrac{2 + \ln x}{2\sqrt{x}} \Rightarrow$

$$y'' = \frac{2\sqrt{x}\,(1/x) - (2 + \ln x)(1/\sqrt{x}\,)}{(2\sqrt{x}\,)^2} = \frac{2/\sqrt{x} - (2 + \ln x)(1/\sqrt{x}\,)}{4x} = \frac{2 - (2 + \ln x)}{\sqrt{x}(4x)} = -\frac{\ln x}{4x\sqrt{x}}$$

24. $y = \dfrac{\ln x}{1 + \ln x} \Rightarrow y' = \dfrac{(1 + \ln x)(1/x) - (\ln x)(1/x)}{(1 + \ln x)^2} = \dfrac{1}{x(1 + \ln x)^2} \Rightarrow$

$$y'' = -\frac{\frac{d}{dx}[x(1 + \ln x)^2]}{[x(1 + \ln x)^2]^2} \quad \text{[Reciprocal Rule]} \quad = -\frac{x\cdot 2(1 + \ln x)\cdot(1/x) + (1 + \ln x)^2}{x^2(1 + \ln x)^4}$$

$$= -\frac{(1 + \ln x)[2 + (1 + \ln x)]}{x^2(1 + \ln x)^4} = -\frac{3 + \ln x}{x^2(1 + \ln x)^3}$$

25. $y = \ln|\sec x| \Rightarrow y' = \dfrac{1}{\sec x}\dfrac{d}{dx}\sec x = \dfrac{1}{\sec x}\sec x\,\tan x = \tan x \Rightarrow y'' = \sec^2 x$

26. $y = \ln(1 + \ln x) \quad \Rightarrow \quad y' = \dfrac{1}{1 + \ln x} \cdot \dfrac{1}{x} = \dfrac{1}{x(1 + \ln x)} \quad \Rightarrow$

$$y'' = -\frac{\frac{d}{dx}[x(1 + \ln x)]}{[x(1 + \ln x)]^2} \quad \text{[Reciprocal Rule]} \quad = -\frac{x(1/x) + (1 + \ln x)(1)}{x^2(1 + \ln x)^2} = -\frac{1 + 1 + \ln x}{x^2(1 + \ln x)^2} = -\frac{2 + \ln x}{x^2(1 + \ln x)^2}$$

27. $f(x) = \dfrac{x}{1 - \ln(x - 1)} \quad \Rightarrow$

$$f'(x) = \frac{[1 - \ln(x - 1)] \cdot 1 - x \cdot \dfrac{-1}{x - 1}}{[1 - \ln(x - 1)]^2} = \frac{\dfrac{(x - 1)[1 - \ln(x - 1)] + x}{x - 1}}{[1 - \ln(x - 1)]^2} = \frac{x - 1 - (x - 1)\ln(x - 1) + x}{(x - 1)[1 - \ln(x - 1)]^2}$$

$$= \frac{2x - 1 - (x - 1)\ln(x - 1)}{(x - 1)[1 - \ln(x - 1)]^2}$$

$\text{Dom}(f) = \{x \mid x - 1 > 0 \quad \text{and} \quad 1 - \ln(x - 1) \neq 0\} = \{x \mid x > 1 \quad \text{and} \quad \ln(x - 1) \neq 1\}$

$= \{x \mid x > 1 \quad \text{and} \quad x - 1 \neq e^1\} = \{x \mid x > 1 \quad \text{and} \quad x \neq 1 + e\} = (1, 1 + e) \cup (1 + e, \infty)$

28. $f(x) = \sqrt{2 + \ln x} = (2 + \ln x)^{1/2} \quad \Rightarrow \quad f'(x) = \dfrac{1}{2}(2 + \ln x)^{-1/2} \cdot \dfrac{1}{x} = \dfrac{1}{2x\sqrt{2 + \ln x}}$

$\text{Dom}(f) = \{x \mid 2 + \ln x \geq 0\} = \{x \mid \ln x \geq -2\} = \{x \mid x \geq e^{-2}\} = [e^{-2}, \infty)$.

29. $f(x) = \ln(x^2 - 2x) \quad \Rightarrow \quad f'(x) = \dfrac{1}{x^2 - 2x}(2x - 2) = \dfrac{2(x - 1)}{x(x - 2)}$.

$\text{Dom}(f) = \{x \mid x(x - 2) > 0\} = (-\infty, 0) \cup (2, \infty)$.

30. $f(x) = \ln \ln \ln x \quad \Rightarrow \quad f'(x) = \dfrac{1}{\ln \ln x} \cdot \dfrac{1}{\ln x} \cdot \dfrac{1}{x}$.

$\text{Dom}(f) = \{x \mid \ln \ln x > 0\} = \{x \mid \ln x > 1\} = \{x \mid x > e\} = (e, \infty)$.

31. $f(x) = \ln(x + \ln x) \quad \Rightarrow \quad f'(x) = \dfrac{1}{x + \ln x} \dfrac{d}{dx}(x + \ln x) = \dfrac{1}{x + \ln x}\left(1 + \dfrac{1}{x}\right)$.

Substitute 1 for x to get $f'(1) = \dfrac{1}{1 + \ln 1}\left(1 + \dfrac{1}{1}\right) = \dfrac{1}{1 + 0}(1 + 1) = 1 \cdot 2 = 2$.

32. $f(x) = \cos(\ln x^2) \quad \Rightarrow \quad f'(x) = -\sin(\ln x^2)\dfrac{d}{dx}\ln x^2 = -\sin(\ln x^2)\dfrac{1}{x^2}(2x) = -\dfrac{2\sin(\ln x^2)}{x}$.

Substitute 1 for x to get $f'(1) = -\dfrac{2\sin(\ln 1^2)}{1} = -2\sin 0 = 0$.

33. $y = \ln(x^2 - 3x + 1) \quad \Rightarrow \quad y' = \dfrac{1}{x^2 - 3x + 1} \cdot (2x - 3) \quad \Rightarrow \quad y'(3) = \frac{1}{1} \cdot 3 = 3$, so an equation of a tangent line at $(3, 0)$ is $y - 0 = 3(x - 3)$, or $y = 3x - 9$.

34. $y = x^2 \ln x \quad \Rightarrow \quad y' = x^2 \cdot \dfrac{1}{x} + (\ln x)(2x) \quad \Rightarrow \quad y'(1) = 1 + 0 = 1$, so an equation of a tangent line at $(1, 0)$ is $y - 0 = 1(x - 1)$, or $y = x - 1$.

35. $f(x) = \sin x + \ln x \quad \Rightarrow \quad f'(x) = \cos x + 1/x$.

This is reasonable, because the graph shows that f increases when f' is positive, and $f'(x) = 0$ when f has a horizontal tangent.

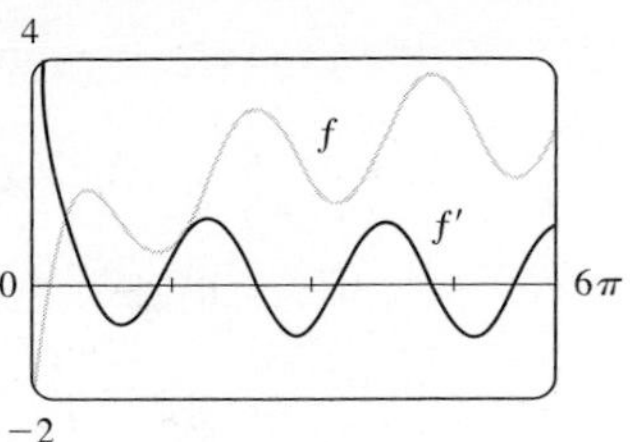

36. $y = \dfrac{\ln x}{x} \quad \Rightarrow \quad y' = \dfrac{x(1/x) - \ln x}{x^2} = \dfrac{1 - \ln x}{x^2}$.

$y'(1) = \dfrac{1 - 0}{1^2} = 1$ and $y'(e) = \dfrac{1 - 1}{e^2} = 0 \quad \Rightarrow \quad$ equations of tangent lines are $y - 0 = 1(x - 1)$ or $y = x - 1$ and $y - 1/e = 0(x - e)$ or $y = 1/e$.

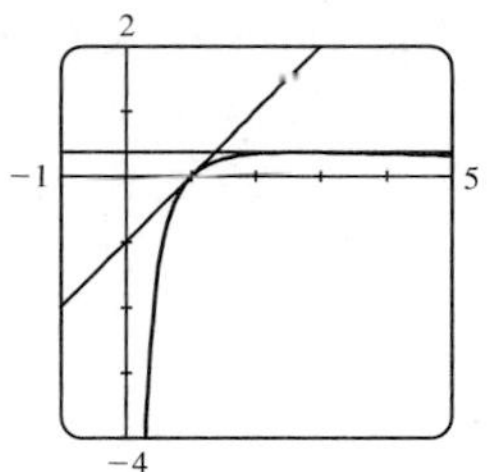

37. $f(x) = cx + \ln(\cos x) \quad \Rightarrow \quad f'(x) = c + \dfrac{1}{\cos x} \cdot (-\sin x) = c - \tan x$.

$f'(\frac{\pi}{4}) = 6 \quad \Rightarrow \quad c - \tan\frac{\pi}{4} = 6 \quad \Rightarrow \quad c - 1 = 6 \quad \Rightarrow \quad c = 7$.

38. $f(x) = \log_a(3x^2 - 2) \quad \Rightarrow \quad f'(x) = \dfrac{1}{(3x^2 - 2)\ln a} \cdot 6x$.

$f'(1) = 3 \quad \Rightarrow \quad \dfrac{1}{\ln a} \cdot 6 = 3 \quad \Rightarrow \quad 2 = \ln a \quad \Rightarrow \quad a = e^2$.

39. $y = (x^2 + 2)^2(x^4 + 4)^4 \quad \Rightarrow \quad \ln y = \ln[(x^2 + 2)^2(x^4 + 4)^4] \quad \Rightarrow \quad \ln y = 2\ln(x^2 + 2) + 4\ln(x^4 + 4) \quad \Rightarrow$

$$\frac{1}{y}y' = 2 \cdot \frac{1}{x^2 + 2} \cdot 2x + 4 \cdot \frac{1}{x^4 + 4} \cdot 4x^3 \quad \Rightarrow \quad y' = y\left(\frac{4x}{x^2 + 2} + \frac{16x^3}{x^4 + 4}\right) \quad \Rightarrow$$

$$y' = (x^2 + 2)^2(x^4 + 4)^4\left(\frac{4x}{x^2 + 2} + \frac{16x^3}{x^4 + 4}\right)$$

40. $y = \dfrac{e^{-x}\cos^2 x}{x^2 + x + 1} \quad \Rightarrow \quad \ln y = \ln\dfrac{e^{-x}\cos^2 x}{x^2 + x + 1} \quad \Rightarrow$

$$\ln y = \ln e^{-x} + \ln|\cos x|^2 - \ln(x^2 + x + 1) = -x + 2\ln|\cos x| - \ln(x^2 + x + 1) \quad \Rightarrow$$

$$\frac{1}{y}y' = -1 + 2 \cdot \frac{1}{\cos x}(-\sin x) - \frac{1}{x^2 + x + 1}(2x + 1) \quad \Rightarrow \quad y' = y\left(-1 - 2\tan x - \frac{2x + 1}{x^2 + x + 1}\right) \quad \Rightarrow$$

$$y' = -\frac{e^{-x}\cos^2 x}{x^2 + x + 1}\left(1 + 2\tan x + \frac{2x + 1}{x^2 + x + 1}\right)$$

41. $y = \sqrt{\dfrac{x - 1}{x^4 + 1}} \quad \Rightarrow \quad \ln y = \ln\left(\dfrac{x - 1}{x^4 + 1}\right)^{1/2} \quad \Rightarrow \quad \ln y = \dfrac{1}{2}\ln(x - 1) - \dfrac{1}{2}\ln(x^4 + 1) \quad \Rightarrow$

$$\frac{1}{y}y' = \frac{1}{2}\frac{1}{x - 1} - \frac{1}{2}\frac{1}{x^4 + 1} \cdot 4x^3 \quad \Rightarrow \quad y' = y\left(\frac{1}{2(x - 1)} - \frac{2x^3}{x^4 + 1}\right) \quad \Rightarrow \quad y' = \sqrt{\frac{x - 1}{x^4 + 1}}\left(\frac{1}{2x - 2} - \frac{2x^3}{x^4 + 1}\right)$$

42. $y = \sqrt{x}\, e^{x^2 - x}(x + 1)^{2/3} \quad \Rightarrow \quad \ln y = \ln\left[x^{1/2}e^{x^2 - x}(x + 1)^{2/3}\right] \quad \Rightarrow$

$$\ln y = \tfrac{1}{2}\ln x + (x^2 - x) + \tfrac{2}{3}\ln(x + 1) \quad \Rightarrow \quad \frac{1}{y}y' = \frac{1}{2} \cdot \frac{1}{x} + 2x - 1 + \frac{2}{3} \cdot \frac{1}{x + 1} \quad \Rightarrow$$

$$y' = y\left(\frac{1}{2x} + 2x - 1 + \frac{2}{3x + 3}\right) \quad \Rightarrow \quad y' = \sqrt{x}\, e^{x^2 - x}(x + 1)^{2/3}\left(\frac{1}{2x} + 2x - 1 + \frac{2}{3x + 3}\right)$$

43. $y = x^x \Rightarrow \ln y = \ln x^x \Rightarrow \ln y = x\ln x \Rightarrow y'/y = x(1/x) + (\ln x)\cdot 1 \Rightarrow y' = y(1 + \ln x) \Rightarrow$
$y' = x^x(1 + \ln x)$

44. $y = x^{\cos x} \Rightarrow \ln y = \ln x^{\cos x} \Rightarrow \ln y = \cos x \ln x \Rightarrow \frac{1}{y}\,y' = \cos x \cdot \frac{1}{x} + \ln x \cdot (-\sin x) \Rightarrow$
$y' = y\left(\frac{\cos x}{x} - \ln x \sin x\right) \Rightarrow y' = x^{\cos x}\left(\frac{\cos x}{x} - \ln x \sin x\right)$

45. $y = x^{\sin x} \Rightarrow \ln y = \ln x^{\sin x} \Rightarrow \ln y = \sin x \ln x \Rightarrow \frac{y'}{y} = (\sin x)\cdot\frac{1}{x} + (\ln x)(\cos x) \Rightarrow$
$y' = y\left(\frac{\sin x}{x} + \ln x \cos x\right) \Rightarrow y' = x^{\sin x}\left(\frac{\sin x}{x} + \ln x\,\cos x\right)$

46. $y = \sqrt{x}^{\,x} \Rightarrow \ln y = \ln \sqrt{x}^{\,x} \Rightarrow \ln y = x \ln x^{1/2} \Rightarrow \ln y = \frac{1}{2}x \ln x \Rightarrow \frac{1}{y}\,y' = \frac{1}{2}x \cdot \frac{1}{x} + \ln x \cdot \frac{1}{2} \Rightarrow$
$y' = y\left(\frac{1}{2} + \frac{1}{2}\ln x\right) \Rightarrow y' = \frac{1}{2}\sqrt{x}^{\,x}(1 + \ln x)$

47. $y = (\cos x)^x \Rightarrow \ln y = \ln(\cos x)^x \Rightarrow \ln y = x \ln \cos x \Rightarrow \frac{1}{y}\,y' = x \cdot \frac{1}{\cos x}\cdot(-\sin x) + \ln \cos x \cdot 1 \Rightarrow$
$y' = y\left(\ln \cos x - \frac{x \sin x}{\cos x}\right) \Rightarrow y' = (\cos x)^x(\ln \cos x - x \tan x)$

48. $y = (\sin x)^{\ln x} \Rightarrow \ln y = \ln(\sin x)^{\ln x} \Rightarrow \ln y = \ln x \cdot \ln \sin x \Rightarrow \frac{1}{y}\,y' = \ln x \cdot \frac{1}{\sin x}\cdot \cos x + \ln \sin x \cdot \frac{1}{x} \Rightarrow$
$y' = y\left(\ln x \cdot \frac{\cos x}{\sin x} + \frac{\ln \sin x}{x}\right) \Rightarrow y' = (\sin x)^{\ln x}\left(\ln x \cot x + \frac{\ln \sin x}{x}\right)$

49. $y = (\tan x)^{1/x} \Rightarrow \ln y = \ln(\tan x)^{1/x} \Rightarrow \ln y = \frac{1}{x}\ln \tan x \Rightarrow$
$\frac{1}{y}\,y' = \frac{1}{x}\cdot\frac{1}{\tan x}\cdot \sec^2 x + \ln \tan x \cdot\left(-\frac{1}{x^2}\right) \Rightarrow y' = y\left(\frac{\sec^2 x}{x \tan x} - \frac{\ln \tan x}{x^2}\right) \Rightarrow$
$y' = (\tan x)^{1/x}\left(\frac{\sec^2 x}{x \tan x} - \frac{\ln \tan x}{x^2}\right)$ or $y' = (\tan x)^{1/x}\cdot\frac{1}{x}\left(\csc x \sec x - \frac{\ln \tan x}{x}\right)$

50. $y = (\ln x)^{\cos x} \Rightarrow \ln y = \cos x \ln(\ln x) \Rightarrow \frac{y'}{y} = \cos x \cdot \frac{1}{\ln x}\cdot\frac{1}{x} + (\ln \ln x)(-\sin x) \Rightarrow$
$y' = (\ln x)^{\cos x}\left(\frac{\cos x}{x \ln x} - \sin x \ln \ln x\right)$

51. $y = \ln(x^2 + y^2) \Rightarrow y' = \frac{1}{x^2 + y^2}\frac{d}{dx}(x^2 + y^2) \Rightarrow y' = \frac{2x + 2yy'}{x^2 + y^2} \Rightarrow x^2y' + y^2y' = 2x + 2yy' \Rightarrow$
$x^2y' + y^2y' - 2yy' = 2x \Rightarrow (x^2 + y^2 - 2y)y' = 2x \Rightarrow y' = \frac{2x}{x^2 + y^2 - 2y}$

52. $x^y = y^x \Rightarrow y \ln x = x \ln y \Rightarrow y \cdot \frac{1}{x} + (\ln x)\cdot y' = x \cdot \frac{1}{y}\cdot y' + \ln y \Rightarrow y' \ln x - \frac{x}{y}\,y' = \ln y - \frac{y}{x} \Rightarrow$
$y' = \frac{\ln y - y/x}{\ln x - x/y}$

53. $f(x) = \ln(x-1) \Rightarrow f'(x) = \dfrac{1}{(x-1)} = (x-1)^{-1} \Rightarrow f''(x) = -(x-1)^{-2} \Rightarrow f'''(x) = 2(x-1)^{-3} \Rightarrow$

$f^{(4)}(x) = -2 \cdot 3(x-1)^{-4} \Rightarrow \cdots \Rightarrow f^{(n)}(x) = (-1)^{n-1} \cdot 2 \cdot 3 \cdot 4 \cdot \cdots \cdot (n-1)(x-1)^{-n} = (-1)^{n-1}\dfrac{(n-1)!}{(x-1)^n}$

54. $y = x^8 \ln x$, so $D^9 y = D^8 y' = D^8(8x^7 \ln x + x^7)$. But the eighth derivative of x^7 is 0, so we now have

$D^8(8x^7 \ln x) = D^7(8 \cdot 7x^6 \ln x + 8x^6) = D^7(8 \cdot 7x^6 \ln x) = D^6(8 \cdot 7 \cdot 6x^5 \ln x) = \cdots = D(8!\, x^0 \ln x) = 8!/x$.

55. If $f(x) = \ln(1+x)$, then $f'(x) = \dfrac{1}{1+x}$, so $f'(0) = 1$.

Thus, $\displaystyle\lim_{x\to 0} \frac{\ln(1+x)}{x} = \lim_{x\to 0} \frac{f(x)}{x} = \lim_{x\to 0} \frac{f(x) - f(0)}{x - 0} = f'(0) = 1$.

56. Let $m = n/x$. Then $n = xm$, and as $n \to \infty$, $m \to \infty$.

Therefore, $\displaystyle\lim_{n\to\infty} \left(1 + \frac{x}{n}\right)^n = \lim_{m\to\infty} \left(1 + \frac{1}{m}\right)^{mx} = \left[\lim_{m\to\infty} \left(1 + \frac{1}{m}\right)^m\right]^x = e^x$ by Equation 6.

3.7 Rates of Change in the Natural and Social Sciences

1. (a) $s = f(t) = t^3 - 8t^2 + 24t$ (in feet) $\Rightarrow$ $v(t) = f'(t) = 3t^2 - 16t + 24$ (in ft/s)

(b) $v(1) = 3(1)^2 - 16(1) + 24 = 11$ ft/s

(c) The particle is at rest when $v(t) = 0$. $3t^2 - 16t + 24 = 0 \Rightarrow \dfrac{-(-16) \pm \sqrt{(-16)^2 - 4(3)(24)}}{2(3)} = \dfrac{16 \pm \sqrt{-32}}{6}$.

The negative discriminant indicates that v is never 0 and that the particle never rests.

(d) From parts (b) and (c), we see that $v(t) > 0$ for all t, so the particle is always moving in the positive direction.

(e) The total distance traveled during the first 6 seconds (since the particle doesn't change direction) is $f(6) - f(0) = 72 - 0 = 72$ ft.

(f)

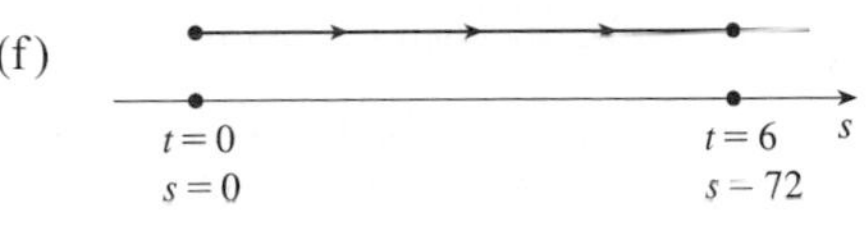

(g) $v(t) = 3t^2 - 16t + 24 \Rightarrow$

$a(t) = v'(t) = 6t - 16$ (in (ft/s)/s or ft/s^2).

$a(1) = 6(1) - 16 = -10$ ft/s^2

(h)

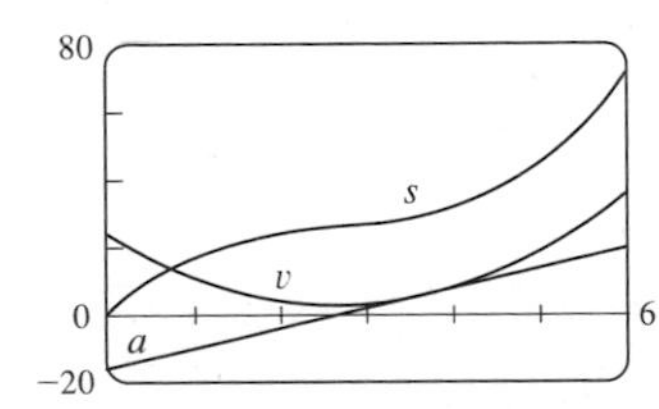

(i) The particle is speeding up when v and a have the same sign. v is always positive and a is positive when $6t - 16 > 0 \Rightarrow$ $t > \frac{8}{3}$, so the particle is speeding up when $t > \frac{8}{3}$. It is slowing down when v and a have opposite signs; that is, when $0 \le t < \frac{8}{3}$.

2. (a) $s = f(t) = \dfrac{9t}{t^2 + 9}$ (in feet) $\Rightarrow$ $v(t) = f'(t) = \dfrac{(t^2+9)(9) - 9t(2t)}{(t^2+9)^2} = \dfrac{-9t^2 + 81}{(t^2+9)^2} = \dfrac{-9(t^2 - 9)}{(t^2+9)^2}$ (in ft/s)

(b) $v(1) = \dfrac{-9(1-9)}{(1+9)^2} = \dfrac{72}{100} = 0.72$ ft/s

(c) The particle is at rest when $v(t) = 0$. $\dfrac{-9(t^2-9)}{(t^2+9)^2} = 0 \quad \Leftrightarrow \quad t^2 - 9 = 0 \quad \Rightarrow \quad t = 3 \text{ s}$ [since $t \geq 0$].

(d) The particle is moving in the positive direction when $v(t) > 0$.

$$\frac{-9(t^2-9)}{(t^2+9)^2} > 0 \quad \Rightarrow \quad -9(t^2-9) > 0 \quad \Rightarrow \quad t^2 - 9 < 0 \quad \Rightarrow \quad t^2 < 9 \quad \Rightarrow \quad 0 \leq t < 3.$$

(e) Since the particle is moving in the positve direction and in the negative direction, we need to calculate the distance traveled in the intervals $[0, 3]$ and$[3, 6]$, respectively.

$|f(3) - f(0)| = \left|\frac{27}{18} - 0\right| = \frac{3}{2}$

$|f(6) - f(3)| = \left|\frac{54}{45} - \frac{27}{18}\right| = \frac{3}{10}$

The total distance is $\frac{3}{2} + \frac{3}{10} = \frac{9}{5}$ or 1.8 ft.

(f)

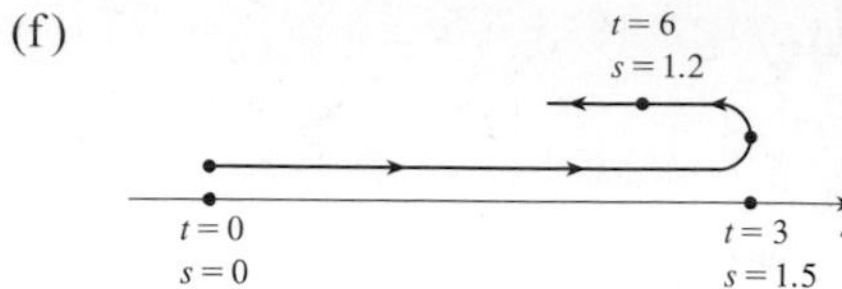

(g) $v(t) = -9\dfrac{t^2-9}{(t^2+9)^2} \quad \Rightarrow$

$$a(t) = v'(t) = -9\frac{(t^2+9)^2(2t) - (t^2-9)2(t^2+9)(2t)}{[(t^2+9)^2]^2} = -9\frac{2t(t^2+9)[(t^2+9) - 2(t^2-9)]}{(t^2+9)^4} = \frac{18t(t^2-27)}{(t^2+9)^3}.$$

$$a(1) = \frac{18(-26)}{10^3} = -0.468 \text{ ft/s}^2$$

(h)

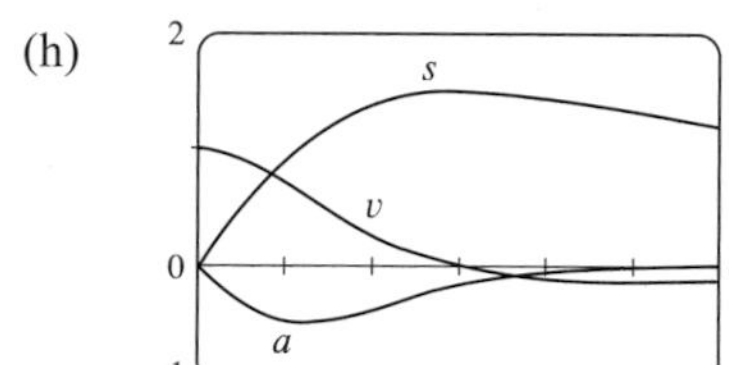

(i) The particle is speeding up when v and a have the same sign. a is negative for $0 < t < \sqrt{27}$ [≈ 5.2], so from the figure in part (h), we see that v and a are both negative for $3 < t < 3\sqrt{3}$. The particle is slowing down when v and a have opposite signs. This occurs when $0 < t < 3$ and when $t > 3\sqrt{3}$.

3. (a) $s = f(t) = \sin(\pi t/2)$ (in feet) $\Rightarrow \quad v(t) = f'(t) = \cos(\pi t/2) \cdot (\pi/2) = \frac{\pi}{2}\cos(\pi t/2)$ (in ft/s)

(b) $v(1) = \frac{\pi}{2}\cos\frac{\pi}{2} = \frac{\pi}{2}(0) = 0$ ft/s

(c) The particle is at rest when $v(t) = 0$. $\frac{\pi}{2}\cos\frac{\pi}{2}t = 0 \quad \Leftrightarrow \quad \cos\frac{\pi}{2}t = 0 \quad \Leftrightarrow \quad \frac{\pi}{2}t = \frac{\pi}{2} + n\pi \quad \Leftrightarrow \quad t = 1 + 2n$, where n is a nonnegative integer since $t \geq 0$.

(d) The particle is moving in the positive direction when $v(t) > 0$. From part (c), we see that v changes sign at every positive odd integer. v is positive when $0 < t < 1$, $3 < t < 5$, $7 < t < 9$, and so on.

(e) v changes sign at $t = 1$, 3, and 5 in the interval $[0, 6]$. The total distance traveled during the first 6 seconds is

$$\begin{aligned}|f(1) - f(0)| + |f(3) - f(1)| + |f(5) - f(3)| + |f(6) - f(5)| &= |1 - 0| + |-1 - 1| + |1 - (-1)| + |0 - 1| \\ &= 1 + 2 + 2 + 1 = 6 \text{ ft}\end{aligned}$$

(f)

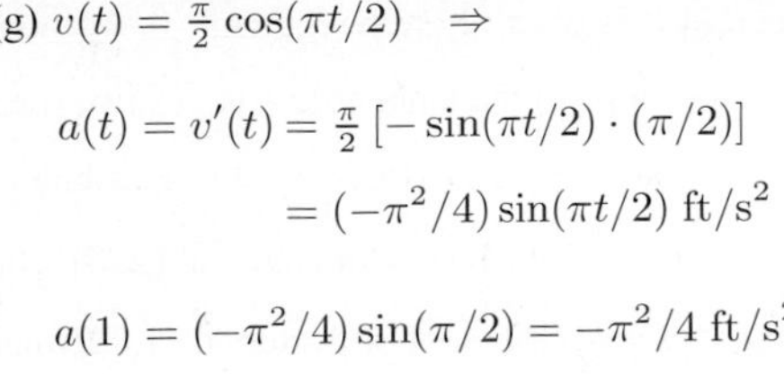

(g) $v(t) = \frac{\pi}{2}\cos(\pi t/2) \quad \Rightarrow$

$$a(t) = v'(t) = \tfrac{\pi}{2}\left[-\sin(\pi t/2)\cdot(\pi/2)\right]$$
$$= (-\pi^2/4)\sin(\pi t/2)\ \text{ft/s}^2$$

$$a(1) = (-\pi^2/4)\sin(\pi/2) = -\pi^2/4\ \text{ft/s}^2$$

(h)

(i) The particle is speeding up when v and a have the same sign. From the figure in part (h), we see that v and a are both positive when $3 < t < 4$ and both negative when $1 < t < 2$ and $5 < t < 6$. Thus, the particle is speeding up when $1 < t < 2$, $3 < t < 4$, and $5 < t < 6$. The particle is slowing down when v and a have opposite signs; that is, when $0 < t < 1$, $2 < t < 3$, and $4 < t < 5$.

4. (a) $s = f(t) = t^2e^{-t}$ (in feet) $\quad\Rightarrow\quad v(t) = f'(t) = t^2(-e^{-t}) + e^{-t}(2t) = te^{-t}(-t+2)$ (in ft/s)

(b) $v(1) = (1)e^{-1}(-1+2) = 1/e$ ft/s

(c) The particle is at rest when $v(t) = 0$. $v(t) = 0 \quad\Leftrightarrow\quad t = 0$ or 2 s.

(d) The particle is moving in the positive direction when $v(t) > 0 \quad\Leftrightarrow\quad te^{-t}(-t+2) > 0 \quad\Leftrightarrow\quad t(-t+2) > 0 \quad\Leftrightarrow$ $0 < t < 2$.

(e) v changes sign at $t = 2$ in the interval $[0, 6]$. The total distance traveled during the first 6 seconds is

$$|f(2) - f(0)| + |f(6) - f(2)| = \left|4e^{-2} - 0\right| + \left|36e^{-6} - 4e^{-2}\right| = 4e^{-2} + 4e^{-2} - 36e^{-6}$$
$$= 8e^{-2} - 36e^{-6} \approx 0.99\ \text{ft}$$

(f)

(g) $v(t) = (2t - t^2)e^{-t} \quad\Rightarrow$

$$a(t) = v'(t) = (2t - t^2)(-e^{-t}) + e^{-t}(2 - 2t)$$
$$= e^{-t}\left[-(2t - t^2) + (2 - 2t)\right]$$
$$= e^{-t}(t^2 - 4t + 2)\ \text{ft/s}^2$$

$$a(1) = e^{-1}(1 - 4 + 2) = -1/e\ \text{ft/s}^2$$

(h)

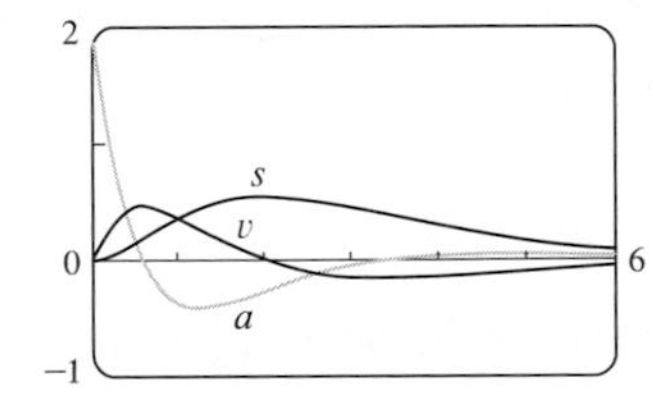

(i) $a(t) = 0 \quad\Leftrightarrow\quad t^2 - 4t + 2 = 0 \quad [e^{-t} \neq 0] \quad\Leftrightarrow\quad t = \dfrac{4 \pm \sqrt{8}}{2} = 2 \pm \sqrt{2} \quad [\approx 0.6$ and $3.4]$. The particle is speeding up when v and a have the same sign. Using the previous information and the figure in part (h), we see that v and a are both positive when $0 < t < 2 - \sqrt{2}$ and both negative when $2 < t < 2 + \sqrt{2}$. The particle is slowing down when v and a have opposite signs. This occurs when $2 - \sqrt{2} < t < 2$ and $t > 2 + \sqrt{2}$.

5. (a) From the figure, the velocity v is positive on the interval $(0, 2)$ and negative on the interval $(2, 3)$. The acceleration a is positive (negative) when the slope of the tangent line is positive (negative), so the acceleration is positive on the interval $(0, 1)$, and negative on the interval $(1, 3)$. The particle is speeding up when v and a have the same sign, that is, on the interval $(0, 1)$ when $v > 0$ and $a > 0$, and on the interval $(2, 3)$ when $v < 0$ and $a < 0$. The particle is slowing down when v and a have opposite signs, that is, on the interval $(1, 2)$ when $v > 0$ and $a < 0$.

(b) $v > 0$ on $(0, 3)$ and $v < 0$ on $(3, 4)$. $a > 0$ on $(1, 2)$ and $a < 0$ on $(0, 1)$ and $(2, 4)$. The particle is speeding up on $(1, 2)$ $[v > 0, a > 0]$ and on $(3, 4)$ $[v < 0, a < 0]$. The particle is slowing down on $(0, 1)$ and $(2, 3)$ $[v > 0, a < 0]$.

6. (a) The velocity v is positive when s is increasing, that is, on the intervals $(0, 1)$ and $(3, 4)$; and it is negative when s is decreasing, that is, on the interval $(1, 3)$. The acceleration a is positive when the graph of s is concave upward (v is increasing), that is, on the interval $(2, 4)$; and it is negative when the graph of s is concave downward (v is decreasing), that is, on the interval $(0, 2)$. The particle is speeding up on the interval $(1, 2)$ $[v < 0, a < 0]$ and on $(3, 4)$ $[v > 0, a > 0]$. The particle is slowing down on the interval $(0, 1)$ $[v > 0, a < 0]$ and on $(2, 3)$ $[v < 0, a > 0]$.

(b) The velocity v is positive on $(3, 4)$ and negative on $(0, 3)$. The acceleration a is positive on $(0, 1)$ and $(2, 4)$ and negative on $(1, 2)$. The particle is speeding up on the interval $(1, 2)$ $[v < 0, a < 0]$ and on $(3, 4)$ $[v > 0, a > 0]$. The particle is slowing down on the interval $(0, 1)$ $[v < 0, a > 0]$ and on $(2, 3)$ $[v < 0, a > 0]$.

7. (a) $h(t) = 2 + 24.5t - 4.9t^2 \;\Rightarrow\; v(t) = h'(t) = 24.5 - 9.8t$. The velocity after 2 s is $v(2) = 24.5 - 9.8(2) = 4.9$ m/s and after 4 s is $v(4) = 24.5 - 9.8(4) = -14.7$ m/s.

(b) The projectile reaches its maximum height when the velocity is zero. $v(t) = 0 \;\Leftrightarrow\; 24.5 - 9.8t = 0 \;\Leftrightarrow$

$$t = \frac{24.5}{9.8} = 2.5 \text{ s.}$$

(c) The maximum height occurs when $t = 2.5$. $h(2.5) = 2 + 24.5(2.5) - 4.9(2.5)^2 = 32.625$ m $\left[\text{or } 32\tfrac{5}{8} \text{ m}\right]$.

(d) The projectile hits the ground when $h = 0 \;\Leftrightarrow\; 2 + 24.5t - 4.9t^2 = 0 \;\Leftrightarrow$

$$t = \frac{-24.5 \pm \sqrt{24.5^2 - 4(-4.9)(2)}}{2(-4.9)} \;\Rightarrow\; t = t_f \approx 5.08 \text{ s} \;\; [\text{since } t \geq 0].$$

(e) The projectile hits the ground when $t = t_f$. Its velocity is $v(t_f) = 24.5 - 9.8t_f \approx -25.3$ m/s [downward].

8. (a) At maximum height the velocity of the ball is 0 ft/s. $v(t) = s'(t) = 80 - 32t = 0 \;\Leftrightarrow\; 32t = 80 \;\Leftrightarrow\; t = \frac{5}{2}$.

So the maximum height is $s\left(\frac{5}{2}\right) = 80\left(\frac{5}{2}\right) - 16\left(\frac{5}{2}\right)^2 = 200 - 100 = 100$ ft.

(b) $s(t) = 80t - 16t^2 = 96 \;\Leftrightarrow\; 16t^2 - 80t + 96 = 0 \;\Leftrightarrow\; 16(t^2 - 5t + 6) = 0 \;\Leftrightarrow\; 16(t - 3)(t - 2) = 0$.
So the ball has a height of 96 ft on the way up at $t = 2$ and on the way down at $t = 3$. At these times the velocities are $v(2) = 80 - 32(2) = 16$ ft/s and $v(3) = 80 - 32(3) = -16$ ft/s, respectively.

9. (a) $h(t) = 15t - 1.86t^2 \;\Rightarrow\; v(t) = h'(t) = 15 - 3.72t$. The velocity after 2 s is $v(2) = 15 - 3.72(2) = 7.56$ m/s.

(b) $25 = h \;\Leftrightarrow\; 1.86t^2 - 15t + 25 = 0 \;\Leftrightarrow\; t = \dfrac{15 \pm \sqrt{15^2 - 4(1.86)(25)}}{2(1.86)} \;\Leftrightarrow\; t = t_1 \approx 2.35$ or $t = t_2 \approx 5.71$.

The velocities are $v(t_1) = 15 - 3.72t_1 \approx 6.24$ m/s [upward] and $v(t_2) = 15 - 3.72t_2 \approx -6.24$ m/s [downward].

10. (a) $s(t) = t^4 - 4t^3 - 20t^2 + 20t \Rightarrow v(t) = s'(t) = 4t^3 - 12t^2 - 40t + 20.$ $v = 20 \Leftrightarrow$
$4t^3 - 12t^2 - 40t + 20 = 20 \Leftrightarrow 4t^3 - 12t^2 - 40t = 0 \Leftrightarrow 4t(t^2 - 3t - 10) = 0 \Leftrightarrow$
$4t(t-5)(t+2) = 0 \Leftrightarrow t = 0$ s or 5 s [for $t \geq 0$].

(b) $a(t) = v'(t) = 12t^2 - 24t - 40.$ $a = 0 \Leftrightarrow 12t^2 - 24t - 40 = 0 \Leftrightarrow 4(3t^2 - 6t - 10) = 0 \Leftrightarrow$
$t = \dfrac{6 \pm \sqrt{6^2 - 4(3)(-10)}}{2(3)} = 1 \pm \frac{1}{3}\sqrt{39} \approx 3.08$ s [for $t \geq 0$]. At this time, the acceleration changes from negative to positive and the velocity attains its minimum value.

11. (a) $A(x) = x^2 \Rightarrow A'(x) = 2x$. $A'(15) = 30\ \text{mm}^2/\text{mm}$ is the rate at which the area is increasing with respect to the side length as x reaches 15 mm.

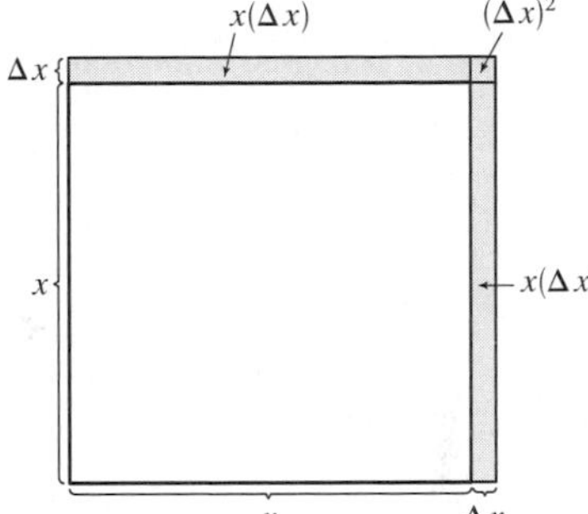

(b) The perimeter is $P(x) = 4x$, so $A'(x) = 2x = \frac{1}{2}(4x) = \frac{1}{2}P(x)$. The figure suggests that if Δx is small, then the change in the area of the square is approximately half of its perimeter (2 of the 4 sides) times Δx. From the figure, $\Delta A = 2x\,(\Delta x) + (\Delta x)^2$. If Δx is small, then $\Delta A \approx 2x\,(\Delta x)$ and so $\Delta A/\Delta x \approx 2x$.

12. (a) $V(x) = x^3 \Rightarrow \dfrac{dV}{dx} = 3x^2.$ $\left.\dfrac{dV}{dx}\right|_{x=3} = 3(3)^2 = 27\ \text{mm}^3/\text{mm}$ is the rate at which the volume is increasing as x increases past 3 mm.

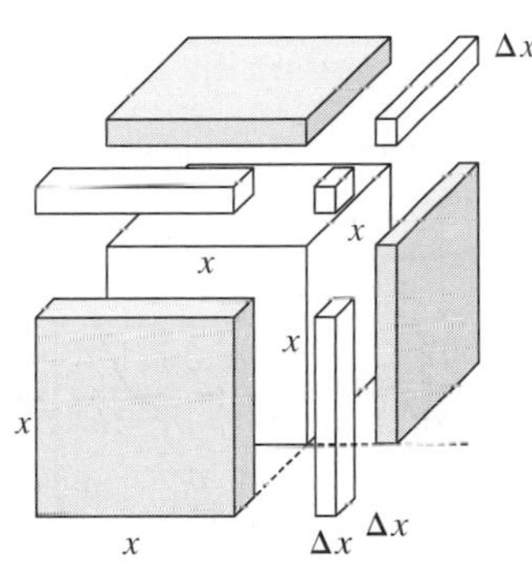

(b) The surface area is $S(x) = 6x^2$, so $V'(x) = 3x^2 = \frac{1}{2}(6x^2) = \frac{1}{2}S(x)$. The figure suggests that if Δx is small, then the change in the volume of the cube is approximately half of its surface area (the area of 3 of the 6 faces) times Δx. From the figure, $\Delta V = 3x^2(\Delta x) + 3x(\Delta x)^2 + (\Delta x)^3$. If Δx is small, then $\Delta V \approx 3x^2(\Delta x)$ and so $\Delta V/\Delta x \approx 3x^2$.

13. (a) Using $A(r) = \pi r^2$, we find that the average rate of change is:

(i) $\dfrac{A(3) - A(2)}{3 - 2} = \dfrac{9\pi - 4\pi}{1} = 5\pi$ (ii) $\dfrac{A(2.5) - A(2)}{2.5 - 2} = \dfrac{6.25\pi - 4\pi}{0.5} = 4.5\pi$

(iii) $\dfrac{A(2.1) - A(2)}{2.1 - 2} = \dfrac{4.41\pi - 4\pi}{0.1} = 4.1\pi$

(b) $A(r) = \pi r^2 \Rightarrow A'(r) = 2\pi r$, so $A'(2) = 4\pi$.

(c) The circumference is $C(r) = 2\pi r = A'(r)$. The figure suggests that if Δr is small, then the change in the area of the circle (a ring around the outside) is approximately equal to its circumference times Δr. Straightening out this ring gives us a shape that is approximately rectangular with length $2\pi r$ and width Δr, so $\Delta A \approx 2\pi r(\Delta r)$.

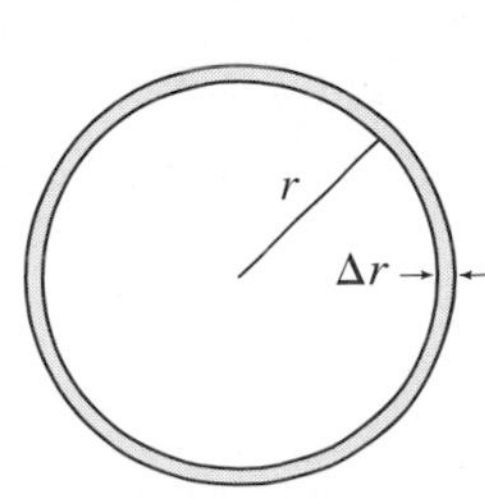

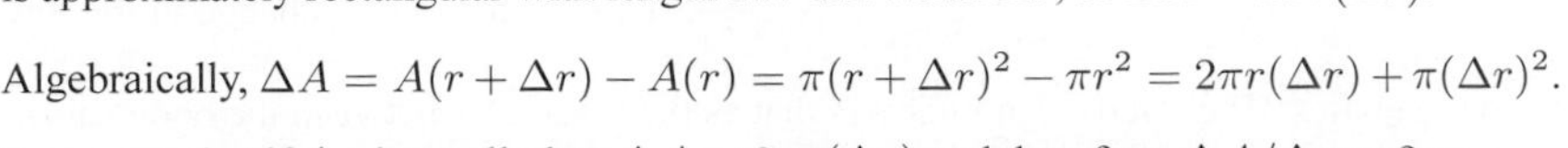
Algebraically, $\Delta A = A(r + \Delta r) - A(r) = \pi(r + \Delta r)^2 - \pi r^2 = 2\pi r(\Delta r) + \pi(\Delta r)^2$.
So we see that if Δr is small, then $\Delta A \approx 2\pi r(\Delta r)$ and therefore, $\Delta A/\Delta r \approx 2\pi r$.

14. After t seconds the radius is $r = 60t$, so the area is $A(t) = \pi(60t)^2 = 3600\pi t^2 \Rightarrow A'(t) = 7200\pi t \Rightarrow$

(a) $A'(1) = 7200\pi \text{ cm}^2/\text{s}$ (b) $A'(3) = 21,600\pi \text{ cm}^2/\text{s}$ (c) $A'(5) = 36{,}000\pi \text{ cm}^2/\text{s}$

As time goes by, the area grows at an increasing rate. In fact, the rate of change is linear with respect to time.

15. $S(r) = 4\pi r^2 \Rightarrow S'(r) = 8\pi r \Rightarrow$

(a) $S'(1) = 8\pi \text{ ft}^2/\text{ft}$ (b) $S'(2) = 16\pi \text{ ft}^2/\text{ft}$ (c) $S'(3) = 24\pi \text{ ft}^2/\text{ft}$

As the radius increases, the surface area grows at an increasing rate. In fact, the rate of change is linear with respect to the radius.

16. (a) Using $V(r) = \frac{4}{3}\pi r^3$, we find that the average rate of change is:

(i) $\dfrac{V(8) - V(5)}{8 - 5} = \dfrac{\frac{4}{3}\pi(512) - \frac{4}{3}\pi(125)}{3} = 172\pi \ \mu\text{m}^3/\mu\text{m}$

(ii) $\dfrac{V(6) - V(5)}{6 - 5} = \dfrac{\frac{4}{3}\pi(216) - \frac{4}{3}\pi(125)}{1} = 121.\overline{3}\pi \ \mu\text{m}^3/\mu\text{m}$

(iii) $\dfrac{V(5.1) - V(5)}{5.1 - 5} = \dfrac{\frac{4}{3}\pi(5.1)^3 - \frac{4}{3}\pi(5)^3}{0.1} = 102.01\overline{3}\pi \ \mu\text{m}^3/\mu\text{m}$

(b) $V'(r) = 4\pi r^2$, so $V'(5) = 100\pi \ \mu\text{m}^3/\mu\text{m}$.

(c) $V(r) = \frac{4}{3}\pi r^3 \Rightarrow V'(r) = 4\pi r^2 = S(r)$. By analogy with Exercise 13(c), we can say that the change in the volume of the spherical shell, ΔV, is approximately equal to its thickness, Δr, times the surface area of the inner sphere. Thus, $\Delta V \approx 4\pi r^2(\Delta r)$ and so $\Delta V/\Delta r \approx 4\pi r^2$.

17. The mass is $f(x) = 3x^2$, so the linear density at x is $\rho(x) = f'(x) = 6x$.

(a) $\rho(1) = 6 \text{ kg/m}$ (b) $\rho(2) = 12 \text{ kg/m}$ (c) $\rho(3) = 18 \text{ kg/m}$

Since ρ is an increasing function, the density will be the highest at the right end of the rod and lowest at the left end.

18. $V(t) = 5000\left(1 - \frac{1}{40}t\right)^2 \Rightarrow V'(t) = 5000 \cdot 2\left(1 - \frac{1}{40}t\right)\left(-\frac{1}{40}\right) = -250\left(1 - \frac{1}{40}t\right)$

(a) $V'(5) = -250\left(1 - \frac{5}{40}\right) = -218.75 \text{ gal/min}$ (b) $V'(10) = -250\left(1 - \frac{10}{40}\right) = -187.5 \text{ gal/min}$

(c) $V'(20) = -250\left(1 - \frac{20}{40}\right) = -125 \text{ gal/min}$ (d) $V'(40) = -250\left(1 - \frac{40}{40}\right) = 0 \text{ gal/min}$

The water is flowing out the fastest at the beginning—when $t = 0$, $V'(t) = -250$ gal/min. The water is flowing out the slowest at the end—when $t = 40$, $V'(t) = 0$. As the tank empties, the water flows out more slowly.

19. The quantity of charge is $Q(t) = t^3 - 2t^2 + 6t + 2$, so the current is $Q'(t) = 3t^2 - 4t + 6$.

(a) $Q'(0.5) = 3(0.5)^2 - 4(0.5) + 6 = 4.75 \text{ A}$ (b) $Q'(1) = 3(1)^2 - 4(1) + 6 = 5 \text{ A}$

The current is lowest when Q' has a minimum. $Q''(t) = 6t - 4 < 0$ when $t < \frac{2}{3}$. So the current decreases when $t < \frac{2}{3}$ and increases when $t > \frac{2}{3}$. Thus, the current is lowest at $t = \frac{2}{3}$ s.

20. (a) $F = \dfrac{GmM}{r^2} = (GmM)r^{-2} \Rightarrow \dfrac{dF}{dr} = -2(GmM)r^{-3} = -\dfrac{2GmM}{r^3}$, which is the rate of change of the force with respect to the distance between the bodies. The minus sign indicates that as the distance r between the bodies increases, the magnitude of the force F exerted by the body of mass m on the body of mass M is decreasing.

(b) Given $F'(20{,}000) = -2$, find $F'(10{,}000)$. $-2 = -\dfrac{2GmM}{20{,}000^3} \Rightarrow GmM = 20{,}000^3$.

$$F'(10,000) = -\frac{2(20{,}000^3)}{10{,}000^3} = -2 \cdot 2^3 = -16 \text{ N/km}$$

21. With $m = m_0\left(1 - \dfrac{v^2}{c^2}\right)^{-1/2}$,

$$F = \frac{d}{dt}(mv) = m\frac{d}{dt}(v) + v\frac{d}{dt}(m) = m_0\left(1 - \frac{v^2}{c^2}\right)^{-1/2} \cdot a + v \cdot m_0\left[-\frac{1}{2}\left(1 - \frac{v^2}{c^2}\right)^{-3/2}\right]\left(-\frac{2v}{c^2}\right)\frac{d}{dt}(v)$$

$$= m_0\left(1 - \frac{v^2}{c^2}\right)^{-3/2} \cdot a\left[\left(1 - \frac{v^2}{c^2}\right) + \frac{v^2}{c^2}\right] = \frac{m_0 a}{(1 - v^2/c^2)^{3/2}}$$

Note that we factored out $(1 - v^2/c^2)^{-3/2}$ since $-3/2$ was the lesser exponent. Also note that $\dfrac{d}{dt}(v) = a$.

22. (a) $D(t) = 7 + 5\cos[0.503(t - 6.75)] \Rightarrow D'(t) = -5\sin[0.503(t - 6.75)](0.503) = -2.515\sin[0.503(t - 6.75)]$.

At 3:00 AM, $t = 3$, and $D'(3) = -2.515\sin[0.503(-3.75)] \approx 2.39$ m/h (rising).

(b) At 6:00 AM, $t = 6$, and $D'(6) = -2.515\sin[0.503(-0.75)] \approx 0.93$ m/h (rising).

(c) At 9:00 AM, $t = 9$, and $D'(9) = -2.515\sin[0.503(2.25)] \approx -2.28$ m/h (falling).

(d) At noon, $t = 12$, and $D'(12) = -2.515\sin[0.503(5.25)] \approx -1.21$ m/h (falling).

23. (a) To find the rate of change of volume with respect to pressure, we first solve for V in terms of P.

$$PV = C \Rightarrow V = \frac{C}{P} \Rightarrow \frac{dV}{dP} = -\frac{C}{P^2}.$$

(b) From the formula for dV/dP in part (a), we see that as P increases, the absolute value of dV/dP decreases.

Thus, the volume is decreasing more rapidly at the beginning.

(c) $\beta = -\dfrac{1}{V}\dfrac{dV}{dP} = -\dfrac{1}{V}\left(-\dfrac{C}{P^2}\right) = \dfrac{C}{(PV)P} = \dfrac{C}{CP} = \dfrac{1}{P}$

24. (a) $[\text{C}] = \dfrac{a^2kt}{akt + 1} \Rightarrow$ rate of reaction $= \dfrac{d[\text{C}]}{dt} = \dfrac{(akt + 1)(a^2k) - (a^2kt)(ak)}{(akt + 1)^2} = \dfrac{a^2k(akt + 1 - akt)}{(akt + 1)^2} = \dfrac{a^2k}{(akt + 1)^2}$

(b) If $x = [\text{C}]$, then $a - x = a - \dfrac{a^2kt}{akt + 1} = \dfrac{a^2kt + a - a^2kt}{akt + 1} = \dfrac{a}{akt + 1}$.

So $k(a - x)^2 = k\left(\dfrac{a}{akt + 1}\right)^2 = \dfrac{a^2k}{(akt + 1)^2} = \dfrac{d[\text{C}]}{dt}$ [from part (a)] $= \dfrac{dx}{dt}$.

(c) As $t \to \infty$, $[\text{C}] = \dfrac{a^2kt}{akt + 1} = \dfrac{(a^2kt)/t}{(akt + 1)/t} = \dfrac{a^2k}{ak + (1/t)} \to \dfrac{a^2k}{ak} = a$ moles/L.

(d) As $t \to \infty$, $\dfrac{d[\text{C}]}{dt} = \dfrac{a^2k}{(akt + 1)^2} \to 0$.

(e) As t increases, nearly all of the reactants A and B are converted into product C. In practical terms, the reaction virtually stops.

25. In Example 6, the population function was $n = 2^t n_0$. Since we are tripling instead of doubling and the initial population is 400, the population function is $n(t) = 400 \cdot 3^t$. The rate of growth is $n'(t) = 400 \cdot 3^t \cdot \ln 3$, so the rate of growth after 2.5 hours is $n'(2.5) = 400 \cdot 3^{2.5} \cdot \ln 3 \approx 6850$ bacteria/hour.

26. $n = f(t) = \dfrac{a}{1 + be^{-0.7t}} \;\Rightarrow\; n' = -\dfrac{a \cdot be^{-0.7t}(-0.7)}{(1 + be^{-0.7t})^2}$ [Reciprocal Rule]. When $t = 0$, $n = 20$ and $n' = 12$.

$f(0) = 20 \;\Rightarrow\; 20 = \dfrac{a}{1+b} \;\Rightarrow\; a = 20(1+b)$. $f'(0) = 12 \;\Rightarrow\; 12 = \dfrac{0.7ab}{(1+b)^2} \;\Rightarrow\; 12 = \dfrac{0.7(20)(1+b)b}{(1+b)^2} \;\Rightarrow$

$\dfrac{12}{14} = \dfrac{b}{1+b} \;\Rightarrow\; 6(1+b) = 7b \;\Rightarrow\; 6 + 6b = 7b \;\Rightarrow\; b = 6$ and $a = 20(1+6) = 140$. For the long run, we let t increase without bound. $\lim\limits_{t\to\infty} f(t) = \lim\limits_{t\to\infty} \dfrac{140}{1+6e^{-0.7t}} = \dfrac{140}{1 + 6 \cdot 0} = 140$, indicating that the yeast population stabilizes at 140 cells.

27. (a) **1920:** $m_1 = \dfrac{1860 - 1750}{1920 - 1910} = \dfrac{110}{10} = 11$, $m_2 = \dfrac{2070 - 1860}{1930 - 1920} = \dfrac{210}{10} = 21$,

$(m_1 + m_2)/2 = (11 + 21)/2 = 16$ million/year

1980: $m_1 = \dfrac{4450 - 3710}{1980 - 1970} = \dfrac{740}{10} = 74$, $m_2 = \dfrac{5280 - 4450}{1990 - 1980} = \dfrac{830}{10} = 83$,

$(m_1 + m_2)/2 = (74 + 83)/2 = 78.5$ million/year

(b) $P(t) = at^3 + bt^2 + ct + d$ (in millions of people), where $a \approx -0.000\,284\,900\,3$, $b \approx 0.522\,433\,122\,43$, $c \approx -6.395\,641\,396$, and $d \approx 1720.586\,081$.

(c) $P(t) = at^3 + bt^2 + ct + d \;\Rightarrow\; P'(t) = 3at^2 + 2bt + c$ (in millions of people per year)

(d) 1920 corresponds to $t = 20$ and $P'(20) \approx 14.16$ million/year. 1980 corresponds to $t = 80$ and $P'(80) \approx 71.72$ million/year. These estimates are smaller than the estimates in part (a).

(e) $f(t) = pq^t$ (where $p = 1.43653 \times 10^9$ and $q = 1.01395$) $\;\Rightarrow\; f'(t) = pq^t \ln q$ (in millions of people per year)

(f) $f'(20) \approx 26.25$ million/year [much larger than the estimates in part (a) and (d)].

$f'(80) \approx 60.28$ million/year [much smaller than the estimates in parts (a) and (d)].

(g) $P'(85) \approx 76.24$ million/year and $f'(85) \approx 64.61$ million/year. The first estimate is probably more accurate.

28. (a) $A(t) = at^4 + bt^3 + ct^2 + dt + e$, where $a \approx -1.199\,781 \times 10^{-6}$, $b \approx 9.545\,853 \times 10^3$, $c \approx -28.478\,550$, $d \approx 37{,}757.105\,467$, and $e \approx -1.877\,031 \times 10^7$.

(b) $A(t) = at^4 + bt^3 + ct^2 + dt + e \;\Rightarrow\; A'(t) = 4at^3 + 3bt^2 + 2ct + d$.

(c) Part (b) gives $A'(1990) \approx 0.106$ years of age per year.

(d)

29. (a) Using $v = \dfrac{P}{4\eta l}(R^2 - r^2)$ with $R = 0.01$, $l = 3$, $P = 3000$, and $\eta = 0.027$, we have v as a function of r:

$$v(r) = \frac{3000}{4(0.027)3}(0.01^2 - r^2).\ v(0) = 0.\overline{925}\ \text{cm/s},\ v(0.005) = 0.69\overline{4}\ \text{cm/s},\ v(0.01) = 0.$$

(b) $v(r) = \dfrac{P}{4\eta l}(R^2 - r^2) \;\Rightarrow\; v'(r) = \dfrac{P}{4\eta l}(-2r) = -\dfrac{Pr}{2\eta l}$. When $l = 3$, $P = 3000$, and $\eta = 0.027$, we have

$$v'(r) = -\frac{3000r}{2(0.027)3}.\ v'(0) = 0,\ v'(0.005) = -92.\overline{592}\ \text{(cm/s)/cm, and}\ v'(0.01) = -185.\overline{185}\ \text{(cm/s)/cm}.$$

(c) The velocity is greatest where $r = 0$ (at the center) and the velocity is changing most where $r = R = 0.01$ cm (at the edge).

30. (a) (i) $f = \dfrac{1}{2L}\sqrt{\dfrac{T}{\rho}} = \left(\dfrac{1}{2}\sqrt{\dfrac{T}{\rho}}\right)L^{-1} \;\Rightarrow\; \dfrac{df}{dL} = -\left(\dfrac{1}{2}\sqrt{\dfrac{T}{\rho}}\right)L^{-2} = -\dfrac{1}{2L^2}\sqrt{\dfrac{T}{\rho}}$

(ii) $f = \dfrac{1}{2L}\sqrt{\dfrac{T}{\rho}} = \left(\dfrac{1}{2L\sqrt{\rho}}\right)T^{1/2} \;\Rightarrow\; \dfrac{df}{dT} = \dfrac{1}{2}\left(\dfrac{1}{2L\sqrt{\rho}}\right)T^{-1/2} = \dfrac{1}{4L\sqrt{T\rho}}$

(iii) $f = \dfrac{1}{2L}\sqrt{\dfrac{T}{\rho}} = \left(\dfrac{\sqrt{T}}{2L}\right)\rho^{-1/2} \;\Rightarrow\; \dfrac{df}{d\rho} = -\dfrac{1}{2}\left(\dfrac{\sqrt{T}}{2L}\right)\rho^{-3/2} = -\dfrac{\sqrt{T}}{4L\rho^{3/2}}$

(b) *Note:* Illustrating tangent lines on the generic figures may help to explain the results.

(i) $\dfrac{df}{dL} < 0$ and L is decreasing $\Rightarrow$ f is increasing $\Rightarrow$ higher note

(ii) $\dfrac{df}{dT} > 0$ and T is increasing $\Rightarrow$ f is increasing $\Rightarrow$ higher note

(iii) $\dfrac{df}{d\rho} < 0$ and ρ is increasing $\Rightarrow$ f is decreasing $\Rightarrow$ lower note

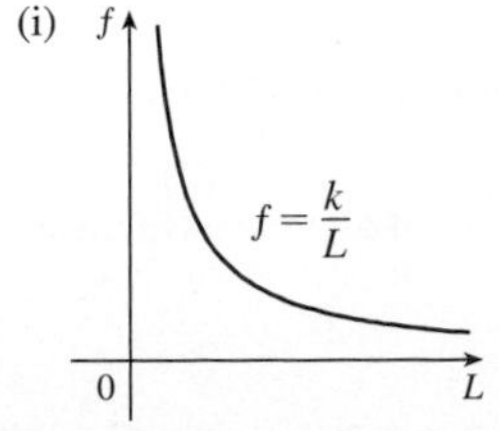

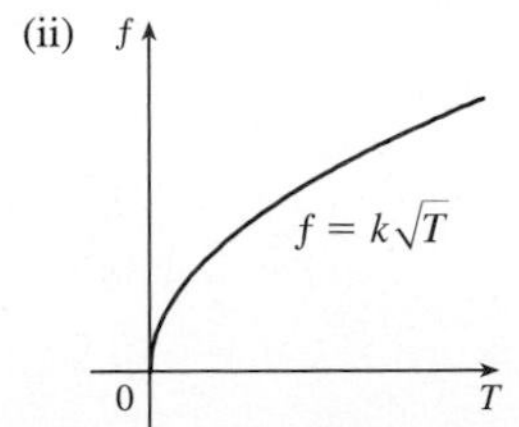

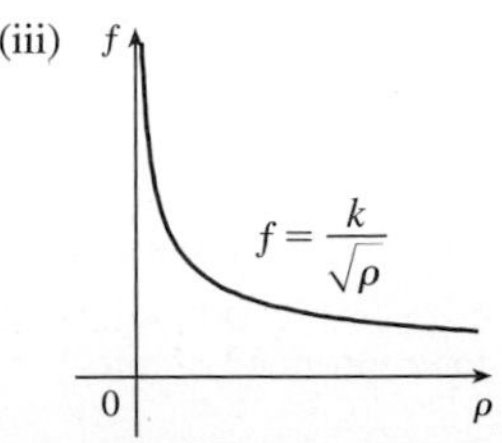

31. (a) $C(x) = 2000 + 3x + 0.01x^2 + 0.0002x^3 \quad \Rightarrow \quad C'(x) = 0 + 3(1) + 0.01(2x) + 0.0002(3x^2) = 3 + 0.02x + 0.0006x^2$

(b) $C'(100) = 3 + 0.02(100) + 0.0006(100)^2 = 3 + 2 + 6 = \$11/\text{pair}$. $C'(100)$ is the rate at which the cost is increasing as the 100th pair of jeans is produced. It predicts the (approximate) cost of the 101st pair.

(c) The cost of manufacturing the 101st pair of jeans is

$C(101) - C(100) = 2611.0702 - 2600 = 11.0702 \approx \$11.07$. This is close to the marginal cost from part (b).

32. (a) $C(q) = 84 + 0.16q - 0.0006q^2 + 0.000003q^3 \quad \Rightarrow \quad C'(q) = 0.16 - 0.0012q + 0.000009q^2$, and

$C'(100) = 0.16 - 0.0012(100) + 0.000009(100)^2 = 0.13$. This is the rate at which the cost is increasing as the 100th item is produced.

(b) The actual cost of producing the 101st item is $C(101) - C(100) = 97.13030299 - 97 \approx \$0.13$

33. (a) $A(x) = \dfrac{p(x)}{x} \quad \Rightarrow \quad A'(x) = \dfrac{xp'(x) - p(x) \cdot 1}{x^2} = \dfrac{xp'(x) - p(x)}{x^2}$.

$A'(x) > 0 \quad \Rightarrow \quad A(x)$ is increasing; that is, the average productivity increases as the size of the workforce increases.

(b) $p'(x)$ is greater than the average productivity $\Rightarrow \quad p'(x) > A(x) \quad \Rightarrow \quad p'(x) > \dfrac{p(x)}{x} \quad \Rightarrow \quad xp'(x) > p(x) \quad \Rightarrow$

$xp'(x) - p(x) > 0 \quad \Rightarrow \quad \dfrac{xp'(x) - p(x)}{x^2} > 0 \quad \Rightarrow \quad A'(x) > 0.$

34. (a) $$S = \frac{dR}{dx} = \frac{(1 + 4x^{0.4})(9.6x^{-0.6}) - (40 + 24x^{0.4})(1.6x^{-0.6})}{(1 + 4x^{0.4})^2}$$

$$= \frac{9.6x^{-0.6} + 38.4x^{-0.2} - 64x^{-0.6} - 38.4x^{-0.2}}{(1 + 4x^{0.4})^2} = -\frac{54.4x^{-0.6}}{(1 + 4x^{0.4})^2}$$

(b) At low levels of brightness, R is quite large $[R(0) = 40]$ and is quickly decreasing, that is, S is negative with large absolute value. This is to be expected: at low levels of brightness, the eye is more sensitive to slight changes than it is at higher levels of brightness.

35. $$t = \ln\left(\frac{3c + \sqrt{9c^2 - 8c}}{2}\right) = \ln\left(3c + \sqrt{9c^2 - 8c}\right) - \ln 2 \quad \Rightarrow$$

$$\frac{dt}{dc} = \frac{1}{3c + \sqrt{9c^2 - 8c}} \frac{d}{dc}\left(3c + \sqrt{9c^2 - 8c}\right) - 0 = \frac{3 + \frac{1}{2}(9c^2 - 8c)^{-1/2}(18c - 8)}{3c + \sqrt{9c^2 - 8c}}$$

$$= \frac{3 + \dfrac{9c - 4}{\sqrt{9c^2 - 8c}}}{3c + \sqrt{9c^2 - 8c}} = \frac{3\sqrt{9c^2 - 8c} + 9c - 4}{\sqrt{9c^2 - 8c}\left(3c + \sqrt{9c^2 - 8c}\right)}.$$

This derivative represents the rate of change of duration of dialysis required with respect to the initial urea concentration.

36. $f(r) = 2\sqrt{Dr} \quad \Rightarrow \quad f'(r) = 2 \cdot \frac{1}{2}(Dr)^{-1/2} \cdot D = \dfrac{D}{\sqrt{Dr}} = \sqrt{\dfrac{D}{r}}$. $f'(r)$ is the rate of change of the wave speed with respect to the reproductive rate.

37. $PV = nRT \Rightarrow T = \dfrac{PV}{nR} = \dfrac{PV}{(10)(0.0821)} = \dfrac{1}{0.821}(PV)$. Using the Product Rule, we have

$$\frac{dT}{dt} = \frac{1}{0.821}[P(t)V'(t) + V(t)P'(t)] = \frac{1}{0.821}[(8)(-0.15) + (10)(0.10)] \approx -0.2436 \text{ K/min}.$$

38. (a) If $dP/dt = 0$, the population is stable (it is constant).

(b) $\dfrac{dP}{dt} = 0 \Rightarrow \beta P = r_0\left(1 - \dfrac{P}{P_c}\right)P \Rightarrow \dfrac{\beta}{r_0} = 1 - \dfrac{P}{P_c} \Rightarrow \dfrac{P}{P_c} = 1 - \dfrac{\beta}{r_0} \Rightarrow P = P_c\left(1 - \dfrac{\beta}{r_0}\right)$.

If $P_c = 10{,}000$, $r_0 = 5\% = 0.05$, and $\beta = 4\% = 0.04$, then $P = 10{,}000\left(1 - \frac{4}{5}\right) = 2000$.

(c) If $\beta = 0.05$, then $P = 10{,}000\left(1 - \frac{5}{5}\right) = 0$. There is no stable population.

39. (a) If the populations are stable, then the growth rates are neither positive nor negative; that is, $\dfrac{dC}{dt} = 0$ and $\dfrac{dW}{dt} = 0$.

(b) "The caribou go extinct" means that the population is zero, or mathematically, $C = 0$.

(c) We have the equations $\dfrac{dC}{dt} = aC - bCW$ and $\dfrac{dW}{dt} = -cW + dCW$. Let $dC/dt = dW/dt = 0$, $a = 0.05$, $b = 0.001$, $c = 0.05$, and $d = 0.0001$ to obtain $0.05C - 0.001CW = 0$ **(1)** and $-0.05W + 0.0001CW = 0$ **(2)**. Adding 10 times **(2)** to **(1)** eliminates the CW-terms and gives us $0.05C - 0.5W = 0 \Rightarrow C = 10W$. Substituting $C = 10W$ into **(1)** results in $0.05(10W) - 0.001(10W)W = 0 \Leftrightarrow 0.5W - 0.01W^2 = 0 \Leftrightarrow 50W - W^2 = 0 \Leftrightarrow W(50 - W) = 0 \Leftrightarrow W = 0$ or 50. Since $C = 10W$, $C = 0$ or 500. Thus, the population pairs (C, W) that lead to stable populations are $(0, 0)$ and $(500, 50)$. So it is possible for the two species to live in harmony.

3.8 Exponential Growth and Decay

1. The relative growth rate is $\dfrac{1}{P}\dfrac{dP}{dt} = 0.7944$, so $\dfrac{dP}{dt} = 0.7944P$ and, by Theorem 2, $P(t) = P(0)e^{0.7944t} = 2e^{0.7944t}$.

Thus, $P(6) = 2e^{0.7944(6)} \approx 234.99$ or about 235 members.

2. (a) By Theorem 2, $P(t) = P(0)e^{kt} = 50e^{kt}$. In 20 minutes $\left(\frac{1}{3}\text{ hour}\right)$, there are 100 cells, so $P\left(\frac{1}{3}\right) = 50e^{k/3} = 100 \Rightarrow e^{k/3} = 2 \Rightarrow k/3 = \ln 2 \Rightarrow k = 3\ln 2 = \ln(2^3) = \ln 8$.

(b) $P(t) = 50e^{(\ln 8)t} = 50 \cdot 8^t$

(c) $P(6) = 50 \cdot 8^6 = 50 \cdot 2^{18} = 13{,}107{,}200$ cells

(d) $\dfrac{dP}{dt} = kP \Rightarrow P'(6) = kP(6) = (\ln 8)P(6) \approx 27{,}255{,}656$ cells/h

(e) $P(t) = 10^6 \Leftrightarrow 50 \cdot 8^t = 1{,}000{,}000 \Leftrightarrow 8^t = 20{,}000 \Leftrightarrow t\ln 8 = \ln 20{,}000 \Leftrightarrow t = \dfrac{\ln 20{,}000}{\ln 8} \approx 4.76$ h

3. (a) By Theorem 2, $P(t) = P(0)e^{kt} = 100e^{kt}$. Now $P(1) = 100e^{k(1)} = 420 \Rightarrow e^k = \frac{420}{100} \Rightarrow k = \ln 4.2$.

So $P(t) = 100e^{(\ln 4.2)t} = 100(4.2)^t$.

(b) $P(3) = 100(4.2)^3 = 7408.8 \approx 7409$ bacteria

(c) $dP/dt = kP \;\Rightarrow\; P'(3) = k \cdot P(3) = (\ln 4.2)\big(100(4.2)^3\big)$ [from part (a)] $\approx 10{,}632$ bacteria/h

(d) $P(t) = 100(4.2)^t = 10{,}000 \;\Rightarrow\; (4.2)^t = 100 \;\Rightarrow\; t = (\ln 100)/(\ln 4.2) \approx 3.2$ hours

4. (a) $y(t) = y(0)e^{kt} \;\Rightarrow\; y(2) = y(0)e^{2k} = 400$ and $y(6) = y(0)e^{6k} = 25{,}600$. Dividing these equations, we get $e^{6k}/e^{2k} = 25{,}600/400 \;\Rightarrow\; e^{4k} = 64 \;\Rightarrow\; 4k = \ln 2^6 = 6\ln 2 \;\Rightarrow\; k = \frac{3}{2}\ln 2 \approx 1.0397$, about 104% per hour.

(b) $400 = y(0)e^{2k} \;\Rightarrow\; y(0) = 400/e^{2k} \;\Rightarrow\; y(0) = 400/e^{3\ln 2} = 400/\left(e^{\ln 2}\right)^3 = 400/2^3 = 50.$

(c) $y(t) = y(0)e^{kt} = 50e^{(3/2)(\ln 2)t} = 50(e^{\ln 2})^{(3/2)t} \;\Rightarrow\; y(t) = 50(2)^{1.5t}$

(d) $y(4.5) = 50(2)^{1.5(4.5)} = 50(2)^{6.75} \approx 5382$ bacteria

(e) $\dfrac{dy}{dt} = ky = \left(\dfrac{3}{2}\ln 2\right)\left(50(2)^{6.75}\right) \approx 5596$ bacteria/h

(f) $y(t) = 50{,}000 \;\Rightarrow\; 50{,}000 = 50(2)^{1.5t} \;\Rightarrow\; 1000 = (2)^{1.5t} \;\Rightarrow\; \ln 1000 = 1.5t\ln 2 \;\Rightarrow$

$t = \dfrac{\ln 1000}{1.5\ln 2} \approx 6.64$ h

5. (a) Let the population (in millions) in the year t be $P(t)$. Since the initial time is the year 1750, we substitute $t - 1750$ for t in Theorem 2, so the exponential model gives $P(t) = P(1750)e^{k(t-1750)}$. Then $P(1800) = 980 = 790e^{k(1800-1750)} \;\Rightarrow$ $\frac{980}{790} = e^{k(50)} \;\Rightarrow\; \ln\frac{980}{790} = 50k \;\Rightarrow\; k = \frac{1}{50}\ln\frac{980}{790} \approx 0.0043104$. So with this model, we have $P(1900) = 790e^{k(1900-1750)} \approx 1508$ million, and $P(1950) = 790e^{k(1950-1750)} \approx 1871$ million. Both of these estimates are much too low.

(b) In this case, the exponential model gives $P(t) = P(1850)e^{k(t-1850)} \;\Rightarrow\; P(1900) = 1650 = 1260e^{k(1900-1850)} \;\Rightarrow$ $\ln\frac{1650}{1260} = k(50) \;\Rightarrow\; k = \frac{1}{50}\ln\frac{1650}{1260} \approx 0.005393$. So with this model, we estimate $P(1950) = 1260e^{k(1950-1850)} \approx 2161$ million. This is still too low, but closer than the estimate of $P(1950)$ in part (a).

(c) The exponential model gives $P(t) = P(1900)e^{k(t-1900)} \;\Rightarrow\; P(1950) = 2560 = 1650e^{k(1950-1900)} \;\Rightarrow$ $\ln\frac{2560}{1650} = k(50) \;\Rightarrow\; k = \frac{1}{50}\ln\frac{2560}{1650} \approx 0.008785$. With this model, we estimate $P(2000) = 1650e^{k(2000-1900)} \approx 3972$ million. This is much too low. The discrepancy is explained by the fact that the world birth rate (average yearly number of births per person) is about the same as always, whereas the mortality rate (especially the infant mortality rate) is much lower, owing mostly to advances in medical science and to the wars in the first part of the twentieth century. The exponential model assumes, among other things, that the birth and mortality rates will remain constant.

6. (a) Let $P(t)$ be the population (in millions) in the year t. Since the initial time is the year 1950, we substitute $t - 1950$ for t in Theorem 2, and find that the exponential model gives $P(t) = P(1950)e^{k(t-1950)} \;\Rightarrow$ $P(1960) = 100 = 83e^{k(1960-1950)} \;\Rightarrow\; \frac{100}{83} = e^{10k} \;\Rightarrow\; k = \frac{1}{10}\ln\frac{100}{83} \approx 0.0186$. With this model, we estimate $P(1980) = 83e^{k(1980-1950)} = 83e^{30k} \approx 145$ million, which is an underestimate of the actual population of 150 million.

(b) As in part (a), $P(t) = P(1960)e^{k(t-1960)} \Rightarrow P(1980) = 150 = 100e^{20k} \Rightarrow 20k = \ln\frac{150}{100} \Rightarrow$

$k = \frac{1}{20}\ln\frac{3}{2} \approx 0.0203$. Thus, $P(2000) = 100e^{40k} = 225$ million, which is an overestimate of the actual population of 214 million.

(c) As in part (a), $P(t) = P(1980)e^{k(t-1980)} \Rightarrow P(2000) = 214 = 150e^{20k} \Rightarrow 20k = \ln\frac{214}{150} \Rightarrow$

$k = \frac{1}{20}\ln\frac{214}{150} \approx 0.0178$. Thus, $P(2010) = 150e^{30k} \approx 256$, which is an overestimate of the actual population of 243 million.

(d) $P(2020) = 150e^{k(2020-1980)} \approx 305$ million. This estimate will probably be an overestimate since this model gave us an overestimate in part (c) — indicating that k is too large. Creating a model with more recent data would likely result in an improved estimate.

7. (a) If $y = [N_2O_5]$ then by Theorem 2, $\dfrac{dy}{dt} = -0.0005y \Rightarrow y(t) = y(0)e^{-0.0005t} = Ce^{-0.0005t}$.

(b) $y(t) = Ce^{-0.0005t} = 0.9C \Rightarrow e^{-0.0005t} = 0.9 \Rightarrow -0.0005t = \ln 0.9 \Rightarrow t = -2000\ln 0.9 \approx 211$ s

8. (a) The mass remaining after t days is $y(t) = y(0)\,e^{kt} = 50e^{kt}$. Since the half-life is 28 days, $y(28) = 50e^{28k} = 25 \Rightarrow$

$e^{28k} = \frac{1}{2} \Rightarrow 28k = \ln\frac{1}{2} \Rightarrow k = -(\ln 2)/28$, so $y(t) = 50e^{-(\ln 2)t/28} = 50 \cdot 2^{-t/28}$.

(b) $y(40) = 50 \cdot 2^{-40/28} \approx 18.6$ mg

(c) $y(t) = 2 \Rightarrow 2 = 50 \cdot 2^{-t/28} \Rightarrow \frac{2}{50} = 2^{-t/28} \Rightarrow$

$(-t/28)\ln 2 = \ln\frac{1}{25} \Rightarrow t = \left(-28\ln\frac{1}{25}\right)/\ln 2 \approx 130$ days

(d)

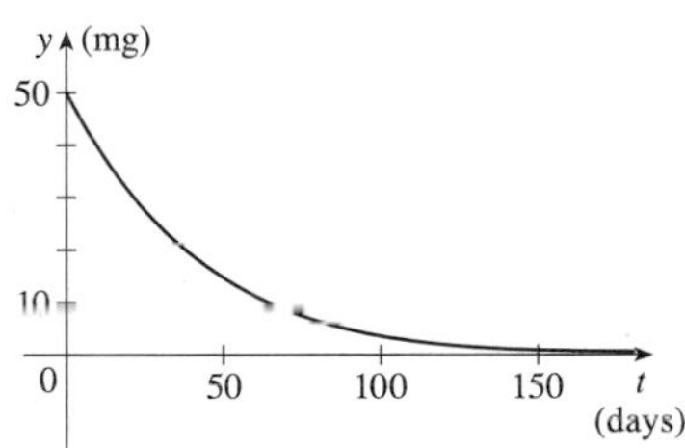

9. (a) If $y(t)$ is the mass (in mg) remaining after t years, then $y(t) = y(0)e^{kt} = 100e^{kt}$.

$y(30) = 100e^{30k} = \frac{1}{2}(100) \Rightarrow e^{30k} = \frac{1}{2} \Rightarrow k = -(\ln 2)/30 \Rightarrow y(t) = 100e^{-(\ln 2)t/30} = 100 \cdot 2^{-t/30}$

(b) $y(100) = 100 \cdot 2^{-100/30} \approx 9.92$ mg

(c) $100e^{-(\ln 2)t/30} = 1 \Rightarrow -(\ln 2)t/30 = \ln\frac{1}{100} \Rightarrow t = -30\,\frac{\ln 0.01}{\ln 2} \approx 199.3$ years

10. (a) If $y(t)$ is the mass after t days and $y(0) = A$, then $y(t) = Ae^{kt}$.

$y(1) = Ae^{k} = 0.945A \Rightarrow e^{k} = 0.945 \Rightarrow k = \ln 0.945$.

Then $Ae^{(\ln 0.945)t} = \frac{1}{2}A \Leftrightarrow \ln e^{(\ln 0.945)t} = \ln\frac{1}{2} \Leftrightarrow (\ln 0.945)t = \ln\frac{1}{2} \Leftrightarrow t = -\frac{\ln 2}{\ln 0.945} \approx 12.25$ years.

(b) $Ae^{(\ln 0.945)t} = 0.20A \Leftrightarrow (\ln 0.945)t = \ln\frac{1}{5} \Leftrightarrow t = -\frac{\ln 5}{\ln 0.945} \approx 28.45$ years

11. Let $y(t)$ be the level of radioactivity. Thus, $y(t) = y(0)e^{-kt}$ and k is determined by using the half-life:

$$y(5730) = \tfrac{1}{2}y(0) \Rightarrow y(0)e^{-k(5730)} = \tfrac{1}{2}y(0) \Rightarrow e^{-5730k} = \tfrac{1}{2} \Rightarrow -5730k = \ln\tfrac{1}{2} \Rightarrow k = -\frac{\ln\frac{1}{2}}{5730} = \frac{\ln 2}{5730}.$$

If 74% of the ^{14}C remains, then we know that $y(t) = 0.74y(0) \Rightarrow 0.74 = e^{-t(\ln 2)/5730} \Rightarrow \ln 0.74 = -\dfrac{t\ln 2}{5730} \Rightarrow$

$$t = -\frac{5730(\ln 0.74)}{\ln 2} \approx 2489 \approx 2500 \text{ years.}$$

12. From Exercise 11, we have the model $y(t) = y(0)e^{-kt}$ with $k = (\ln 2)/5730$. Thus,

$y(68{,}000{,}000) = y(0)e^{-68{,}000{,}000k} \approx y(0) \cdot 0 = 0$. There would be an undetectable amount of ^{14}C remaining for a 68-million-year-old dinosaur.

Now let $y(t) = 0.1\%\, y(0)$, so $0.001y(0) = y(0)e^{-kt} \;\Rightarrow\; 0.001 = e^{-kt} \;\Rightarrow\; \ln 0.001 = -kt \;\Rightarrow$

$t = \dfrac{\ln 0.001}{-k} = \dfrac{\ln 0.001}{-(\ln 2)/5730} \approx 57{,}104$, which is the maximum age of a fossil that we could date using ^{14}C.

13. Let t measure time since a dinosaur died in millions of years, and let $y(t)$ be the amount of ^{40}K in the dinosaur's bones at time t. Then $y(t) = y(0)e^{-kt}$ and k is determined by the half-life: $y(1250) = \frac{1}{2}y(0) \;\Rightarrow\; y(0)e^{-k(1250)} = \frac{1}{2}y(0) \;\Rightarrow$

$e^{-1250k} = \frac{1}{2} \;\Rightarrow\; -1250k = \ln\frac{1}{2} \;\Rightarrow\; k = -\dfrac{\ln\frac{1}{2}}{1250} = \dfrac{\ln 2}{1250}$. To determine if a dinosaur dating of 68 million years is possible, we find that $y(68) = y(0)e^{-k(68)} \approx 0.963y(0)$, indicating that about 96% of the ^{40}K is remaining, which is clearly detectable. To determine the maximum age of a fossil by using ^{40}K, we solve $y(t) = 0.1\%y(0)$ for t.

$y(0)e^{-kt} = 0.001y(0) \;\Leftrightarrow\; e^{-kt} = 0.001 \;\Leftrightarrow\; -kt = \ln 0.001 \;\Leftrightarrow\; t = \dfrac{\ln 0.001}{-(\ln 2)/1250} \approx 12{,}457$ million, or 12.457 billion years.

14. From the information given, we know that $\dfrac{dy}{dx} = 2y \;\Rightarrow\; y = Ce^{2x}$ by Theorem 2. To calculate C we use the point $(0, 5)$:

$5 = Ce^{2(0)} \;\Rightarrow\; C = 5$. Thus, the equation of the curve is $y = 5e^{2x}$.

15. (a) Using Newton's Law of Cooling, $\dfrac{dT}{dt} = k(T - T_s)$, we have $\dfrac{dT}{dt} = k(T - 75)$. Now let $y = T - 75$, so

$y(0) = T(0) - 75 = 185 - 75 = 110$, so y is a solution of the initial-value problem $dy/dt = ky$ with $y(0) = 110$ and by Theorem 2 we have $y(t) = y(0)e^{kt} = 110e^{kt}$.

$y(30) = 110e^{30k} = 150 - 75 \;\Rightarrow\; e^{30k} = \frac{75}{110} = \frac{15}{22} \;\Rightarrow\; k = \frac{1}{30}\ln\frac{15}{22}$, so $y(t) = 110e^{\frac{1}{30}t\ln\left(\frac{15}{22}\right)}$ and

$y(45) = 110e^{\frac{45}{30}\ln\left(\frac{15}{22}\right)} \approx 62°\text{F}$. Thus, $T(45) \approx 62 + 75 = 137°\text{F}$.

(b) $T(t) = 100 \;\Rightarrow\; y(t) = 25$. $y(t) = 110e^{\frac{1}{30}t\ln\left(\frac{15}{22}\right)} = 25 \;\Rightarrow\; e^{\frac{1}{30}t\ln\left(\frac{15}{22}\right)} = \frac{25}{110} \;\Rightarrow\; \frac{1}{30}t\ln\frac{15}{22} = \ln\frac{25}{110} \;\Rightarrow$

$t = \dfrac{30\ln\frac{25}{110}}{\ln\frac{15}{22}} \approx 116$ min.

16. Let $T(t)$ be the temperature of the body t hours after 1:30 PM. Then $T(0) = 32.5$ and $T(1) = 30.3$. Using Newton's Law of Cooling, $\dfrac{dT}{dt} = k(T - T_s)$, we have $\dfrac{dT}{dt} = k(T - 20)$. Now let $y = T - 20$, so $y(0) = T(0) - 20 = 32.5 - 20 = 12.5$, so y is a solution to the initial value problem $dy/dt = ky$ with $y(0) = 12.5$ and by Theorem 2 we have $y(t) = y(0)e^{kt} = 12.5e^{kt}$.

$y(1) = 30.3 - 20 \;\Rightarrow\; 10.3 = 12.5e^{k(1)} \;\Rightarrow\; e^k = \frac{10.3}{12.5} \;\Rightarrow\; k = \ln\frac{10.3}{12.5}$. The murder occurred when

$y(t) = 37 - 20 \Rightarrow 12.5e^{kt} = 17 \Rightarrow e^{kt} = \frac{17}{12.5} \Rightarrow kt = \ln\frac{17}{12.5} \Rightarrow t = \left(\ln\frac{17}{12.5}\right)/\ln\frac{10.3}{12.5} \approx -1.588$ h ≈ -95 minutes. Thus, the murder took place about 95 minutes before 1:30 PM, or 11:55 AM.

17. $\frac{dT}{dt} = k(T - 20)$. Letting $y = T - 20$, we get $\frac{dy}{dt} = ky$, so $y(t) = y(0)e^{kt}$. $y(0) = T(0) - 20 = 5 - 20 = -15$, so $y(25) = y(0)e^{25k} = -15e^{25k}$, and $y(25) = T(25) - 20 = 10 - 20 = -10$, so $-15e^{25k} = -10 \Rightarrow e^{25k} = \frac{2}{3}$. Thus, $25k = \ln\left(\frac{2}{3}\right)$ and $k = \frac{1}{25}\ln\left(\frac{2}{3}\right)$, so $y(t) = y(0)e^{kt} = -15e^{(1/25)\ln(2/3)t}$. More simply, $e^{25k} = \frac{2}{3} \Rightarrow e^k = \left(\frac{2}{3}\right)^{1/25} \Rightarrow e^{kt} = \left(\frac{2}{3}\right)^{t/25} \Rightarrow y(t) = -15\cdot\left(\frac{2}{3}\right)^{t/25}$.

(a) $T(50) = 20 + y(50) = 20 - 15\cdot\left(\frac{2}{3}\right)^{50/25} = 20 - 15\cdot\left(\frac{2}{3}\right)^2 = 20 - \frac{20}{3} = 13.\bar{3}\,°\text{C}$

(b) $15 = T(t) = 20 + y(t) = 20 - 15\cdot\left(\frac{2}{3}\right)^{t/25} \Rightarrow 15\cdot\left(\frac{2}{3}\right)^{t/25} = 5 \Rightarrow \left(\frac{2}{3}\right)^{t/25} = \frac{1}{3} \Rightarrow$

$(t/25)\ln\left(\frac{2}{3}\right) = \ln\left(\frac{1}{3}\right) \Rightarrow t = 25\ln\left(\frac{1}{3}\right)/\ln\left(\frac{2}{3}\right) \approx 67.74$ min.

18. $\frac{dT}{dt} = k(T - 20)$. Let $y = T - 20$. Then $\frac{dy}{dt} = ky$, so $y(t) = y(0)e^{kt}$. $y(0) = T(0) - 20 = 95 - 20 = 75$, so $y(t) = 75e^{kt}$. When $T(t) = 70$, $\frac{dT}{dt} = -1°\text{C/min}$. Equivalently, $\frac{dy}{dt} = -1$ when $y(t) = 50$. Thus, $-1 = \frac{dy}{dt} = ky(t) = 50k$ and $50 = y(t) = 75e^{kt}$. The first relation implies $k = -1/50$, so the second relation says $50 = 75e^{-t/50}$. Thus, $e^{-t/50} = \frac{2}{3} \Rightarrow -t/50 = \ln\left(\frac{2}{3}\right) \Rightarrow t = -50\ln\left(\frac{2}{3}\right) \approx 20.27$ min.

19. (a) Let $P(h)$ be the pressure at altitude h. Then $dP/dh = kP \Rightarrow P(h) = P(0)e^{kh} = 101.3e^{kh}$.

$P(1000) = 101.3e^{1000k} = 87.14 \Rightarrow 1000k = \ln\left(\frac{87.14}{101.3}\right) \Rightarrow k = \frac{1}{1000}\ln\left(\frac{87.14}{101.3}\right) \Rightarrow$

$P(h) = 101.3\,e^{\frac{1}{1000}h\ln\left(\frac{87.14}{101.3}\right)}$, so $P(3000) = 101.3e^{3\ln\left(\frac{87.14}{101.3}\right)} \approx 64.5$ kPa.

(b) $P(6187) = 101.3\,e^{\frac{6187}{1000}\ln\left(\frac{87.14}{101.3}\right)} \approx 39.9$ kPa

20. (a) Using $A = A_0\left(1 + \frac{r}{n}\right)^{nt}$ with $A_0 = 1000$, $r = 0.08$, and $t = 3$, we have:

(i) Annually: $n = 1$; $A = 1000\left(1 + \frac{0.08}{1}\right)^{1\cdot3} = \$1259.71$

(ii) Quarterly: $n = 4$; $A = 1000\left(1 + \frac{0.08}{4}\right)^{4\cdot3} = \$1268.24$

(iii) Monthly: $n = 12$; $A = 1000\left(1 + \frac{0.08}{12}\right)^{12\cdot3} = \$1270.24$

(iv) Weekly: $n = 52$ $A = 1000\left(1 + \frac{0.08}{52}\right)^{52\cdot3} = \$1271.01$

(v) Daily: $n = 365$; $A = 1000\left(1 + \frac{0.08}{365}\right)^{365\cdot3} = \$1271.22$

(vi) Hourly: $n = 365\cdot24$; $A = 1000\left(1 + \frac{0.08}{365\cdot24}\right)^{365\cdot24\cdot3} = \$1271.25$

(vii) Continuously: $A = 1000e^{(0.08)3} = \$1271.25$

(b)

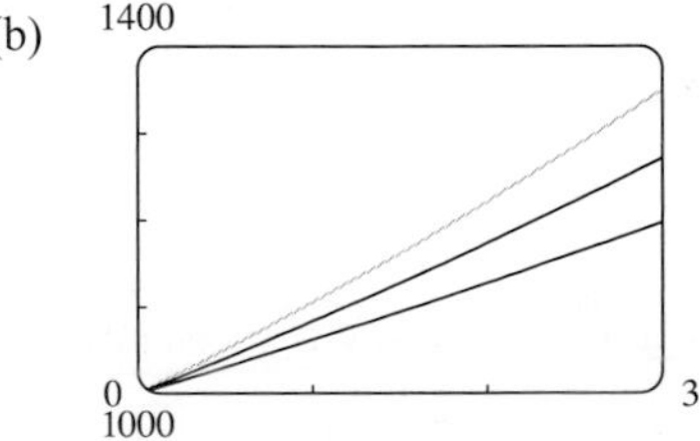

$A_{0.10}(3) = \$1349.86$, $A_{0.08}(3) = \$1271.25$, and $A_{0.06}(3) = \$1197.22$.

21. (a) Using $A = A_0\left(1 + \frac{r}{n}\right)^{nt}$ with $A_0 = 3000$, $r = 0.05$, and $t = 5$, we have:

(i) Annually: $n = 1$; $A = 3000\left(1 + \frac{0.05}{1}\right)^{1 \cdot 5} = \$3828.84$

(ii) Semiannually: $n = 2$; $A = 3000\left(1 + \frac{0.05}{2}\right)^{2 \cdot 5} = \$3840.25$

(iii) Monthly: $n = 12$; $A = 3000\left(1 + \frac{0.05}{12}\right)^{12 \cdot 5} = \$3850.08$

(iv) Weekly: $n = 52$; $A = 3000\left(1 + \frac{0.05}{52}\right)^{52 \cdot 5} = \$3851.61$

(v) Daily: $n = 365$; $A = 3000\left(1 + \frac{0.05}{365}\right)^{365 \cdot 5} = \$3852.01$

(vi) Continuously: $A = 3000e^{(0.05)5} = \$3852.08$

(b) $dA/dt = 0.05A$ and $A(0) = 3000$.

22. (a) $A_0 e^{0.06t} = 2A_0 \Leftrightarrow e^{0.06t} = 2 \Leftrightarrow 0.06t = \ln 2 \Leftrightarrow t = \frac{50}{3}\ln 2 \approx 11.55$, so the investment will double in about 11.55 years.

(b) The annual interest rate in $A = A_0(1 + r)^t$ is r. From part (a), we have $A = A_0 e^{0.06t}$. These amounts must be equal, so $(1 + r)^t = e^{0.06t} \Rightarrow 1 + r = e^{0.06} \Rightarrow r = e^{0.06} - 1 \approx 0.0618 = 6.18\%$, which is the equivalent annual interest rate.

APPLIED PROJECT Controlling Red Blood Cell Loss During Surgery

1. Let $R(t)$ be the volume of RBCs (in liters) at time t (in hours). Since the total volume of blood is 5 L, the concentration of RBCs is $R/5$. The patient bleeds 2 L of blood in 4 hours, so

$$\frac{dR}{dt} = -\frac{2L}{4h} \cdot \frac{R}{5} = -\frac{1}{10}R$$

From Section 3.8, we know that $dR/dt = kR$ has solution $R(t) = R(0)e^{kt}$. In this case, $R(0) = 45\%$ of $5 = \frac{9}{4}$ and $k = -\frac{1}{10}$, so $R(t) = \frac{9}{4}e^{-t/10}$. At the end of the operation, the volume of RBCs is $R(4) = \frac{9}{4}e^{-0.4} \approx 1.51$ L.

2. Let V be the volume of blood that is extracted and replaced with saline solution. Let $R_A(t)$ be the volume of RBCs with the ANH procedure. Then $R_A(0)$ is 45% of $(5 - V)$, or $\frac{9}{20}(5 - V)$, and hence $R_A(t) = \frac{9}{20}(5 - V)e^{-t/10}$. We want $R_A(4) \geq 25\%$ of $5 \Leftrightarrow \frac{9}{20}(5 - V)e^{-0.4} \geq \frac{5}{4} \Leftrightarrow 5 - V \geq \frac{25}{9}e^{0.4} \Leftrightarrow V \leq 5 - \frac{25}{9}e^{0.4} \approx 0.86$ L. To maximize the effect of the ANH procedure, the surgeon should remove 0.86 L of blood and replace it with saline solution.

3. The RBC loss *without* the ANH procedure is $R(0) - R(4) = \frac{9}{4} - \frac{9}{4}e^{-0.4} \approx 0.74$ L. The RBC loss *with* the ANH procedure is $R_A(0) - R_A(4) = \frac{9}{20}(5 - V) - \frac{9}{20}(5 - V)e^{-0.4} = \frac{9}{20}(5 - V)(1 - e^{-0.4})$. Now let $V = 5 - \frac{25}{9}e^{0.4}$ [from Problem 2] to get $R_A(0) - R_A(4) = \frac{9}{20}\left[5 - \left(5 - \frac{25}{9}e^{0.4}\right)\right](1 - e^{0.4}) = \frac{9}{20} \cdot \frac{25}{9}e^{0.4}(1 - e^{0.4}) = \frac{5}{4}(e^{0.4} - 1) \approx 0.61$ L. Thus, the ANH procedure reduces the RBC loss by about $0.74 - 0.61 = 0.13$ L (about 4.4 fluid ounces).

3.9 Related Rates

1. $V = x^3 \;\Rightarrow\; \dfrac{dV}{dt} = \dfrac{dV}{dx}\dfrac{dx}{dt} = 3x^2\dfrac{dx}{dt}$

2. (a) $A = \pi r^2 \;\Rightarrow\; \dfrac{dA}{dt} = \dfrac{dA}{dr}\dfrac{dr}{dt} = 2\pi r\dfrac{dr}{dt}$ (b) $\dfrac{dA}{dt} = 2\pi r\dfrac{dr}{dt} = 2\pi(30\text{ m})(1\text{ m/s}) = 60\pi\text{ m}^2\text{/s}$

3. Let s denote the side of a square. The square's area A is given by $A = s^2$. Differentiating with respect to t gives us $\dfrac{dA}{dt} = 2s\dfrac{ds}{dt}$. When $A = 16$, $s = 4$. Substitution 4 for s and 6 for $\dfrac{ds}{dt}$ gives us $\dfrac{dA}{dt} = 2(4)(6) = 48\text{ cm}^2\text{/s}$.

4. $A = \ell w \;\Rightarrow\; \dfrac{dA}{dt} = \ell\cdot\dfrac{dw}{dt} + w\cdot\dfrac{d\ell}{dt} = 20(3) + 10(8) = 140\text{ cm}^2\text{/s}$.

5. $V = \pi r^2 h = \pi(5)^2 h = 25\pi h \;\Rightarrow\; \dfrac{dV}{dt} = 25\pi\dfrac{dh}{dt} \;\Rightarrow\; 3 = 25\pi\dfrac{dh}{dt} \;\Rightarrow\; \dfrac{dh}{dt} = \dfrac{3}{25\pi}\text{ m/min}$.

6. $V = \frac{4}{3}\pi r^3 \;\Rightarrow\; \dfrac{dV}{dt} = \frac{4}{3}\pi\cdot 3r^2\dfrac{dr}{dt} \;\Rightarrow\; \dfrac{dV}{dt} = 4\pi\left(\frac{1}{2}\cdot 80\right)^2(4) = 25{,}600\pi\text{ mm}^3\text{/s}$.

7. $S = 4\pi r^2 \;\Rightarrow\; \dfrac{dS}{dt} = 4\pi\cdot 2r\dfrac{dr}{dt} \;\Rightarrow\; \dfrac{dS}{dt} = 4\pi\cdot 2\cdot 8\cdot 2 = 128\pi\text{ cm}^2\text{/min}$.

8. (a) $A = \frac{1}{2}ab\sin\theta \;\Rightarrow\; \dfrac{dA}{dt} = \frac{1}{2}ab\cos\theta\dfrac{d\theta}{dt} = \frac{1}{2}(2)(3)\left(\cos\frac{\pi}{3}\right)(0.2) = 3\left(\frac{1}{2}\right)(0.2) = 0.3\text{ cm}^2\text{/min}$.

(b) $A = \frac{1}{2}ab\sin\theta \;\Rightarrow$

$$\frac{dA}{dt} = \frac{1}{2}a\left(b\cos\theta\frac{d\theta}{dt} + \sin\theta\frac{db}{dt}\right) = \tfrac{1}{2}(2)\left[3\left(\cos\tfrac{\pi}{3}\right)(0.2) + \left(\sin\tfrac{\pi}{3}\right)(1.5)\right]$$
$$= 3\left(\tfrac{1}{2}\right)(0.2) + \tfrac{1}{2}\sqrt{3}\left(\tfrac{3}{2}\right) = 0.3 + \tfrac{3}{4}\sqrt{3}\text{ cm}^2\text{/min}\quad[\approx 1.6]$$

(c) $A = \frac{1}{2}ab\sin\theta \;\Rightarrow$

$$\frac{dA}{dt} = \frac{1}{2}\left(\frac{da}{dt}b\sin\theta + a\frac{db}{dt}\sin\theta + ab\cos\theta\frac{d\theta}{dt}\right)\quad\text{[by Exercise 3.2.61(a)]}$$
$$= \tfrac{1}{2}\left[(2.5)(3)\left(\tfrac{1}{2}\sqrt{3}\right) + (2)(1.5)\left(\tfrac{1}{2}\sqrt{3}\right) + (2)(3)\left(\tfrac{1}{2}\right)(0.2)\right]$$
$$= \left(\tfrac{15}{8}\sqrt{3} + \tfrac{3}{4}\sqrt{3} + 0.3\right) = \left(\tfrac{21}{8}\sqrt{3} + 0.3\right)\text{ cm}^2\text{/min}\quad[\approx 4.85]$$

Note how this answer relates to the answer in part (a) [θ changing] and part (b) [b and θ changing].

9. (a) $y = \sqrt{2x+1}$ and $\dfrac{dx}{dt} = 3 \;\Rightarrow\; \dfrac{dy}{dt} = \dfrac{dy}{dx}\dfrac{dx}{dt} = \dfrac{1}{2}(2x+1)^{-1/2}\cdot 2\cdot 3 = \dfrac{3}{\sqrt{2x+1}}$. When $x = 4$, $\dfrac{dy}{dt} = \dfrac{3}{\sqrt{9}} = 1$.

(b) $y = \sqrt{2x+1} \;\Rightarrow\; y^2 = 2x+1 \;\Rightarrow\; 2x = y^2 - 1 \;\Rightarrow\; x = \frac{1}{2}y^2 - \frac{1}{2}$ and $\dfrac{dy}{dt} = 5 \;\Rightarrow$

$\dfrac{dx}{dt} = \dfrac{dx}{dy}\dfrac{dy}{dt} = y\cdot 5 = 5y$. When $x = 12$, $y = \sqrt{25} = 5$, so $\dfrac{dx}{dt} = 5(5) = 25$.

10. (a) $\dfrac{d}{dt}(4x^2 + 9y^2) = \dfrac{d}{dt}(36) \;\Rightarrow\; 8x\dfrac{dx}{dt} + 18y\dfrac{dy}{dt} = 0 \;\Rightarrow\; 4x\dfrac{dx}{dt} + 9y\dfrac{dy}{dt} = 0 \;\Rightarrow$

$4(2)\dfrac{dx}{dt} + 9\left(\dfrac{2}{3}\sqrt{5}\right)\left(\dfrac{1}{3}\right) = 0 \;\Rightarrow\; 8\dfrac{dx}{dt} = -2\sqrt{5} \;\Rightarrow\; \dfrac{dx}{dt} = -\dfrac{1}{4}\sqrt{5}$

(b) $4x\,\dfrac{dx}{dt}+9y\,\dfrac{dy}{dt}=0 \;\Rightarrow\; 4(-2)(3)+9\left(\dfrac{2}{3}\sqrt{5}\right)\dfrac{dy}{dt}=0 \;\Rightarrow\; 6\sqrt{5}\,\dfrac{dy}{dt}=24 \;\Rightarrow\; \dfrac{dy}{dt}=\dfrac{4}{\sqrt{5}}$

11. $\dfrac{d}{dt}\left(x^2+y^2+z^2\right)=\dfrac{d}{dt}(9) \;\Rightarrow\; 2x\,\dfrac{dx}{dt}+2y\,\dfrac{dy}{dt}+2z\,\dfrac{dz}{dt}=0 \;\Rightarrow\; x\,\dfrac{dx}{dt}+y\,\dfrac{dy}{dt}+z\,\dfrac{dz}{dt}=0.$

If $\dfrac{dx}{dt}=5$, $\dfrac{dy}{dt}=4$ and $(x,y,z)=(2,2,1)$, then $2(5)+2(4)+1\,\dfrac{dz}{dt}=0 \;\Rightarrow\; \dfrac{dz}{dt}=-18.$

12. $\dfrac{d}{dt}(xy)=\dfrac{d}{dt}(8) \;\Rightarrow\; x\,\dfrac{dy}{dt}+y\,\dfrac{dx}{dt}=0.$ If $\dfrac{dy}{dt}=-3$ cm/s and $(x,y)=(4,2)$, then $4(-3)+2\,\dfrac{dx}{dt}=0 \;\Rightarrow$

$\dfrac{dx}{dt}=6$. Thus, the x-coordinate is increasing at a rate of 6 cm/s.

13. (a) Given: a plane flying horizontally at an altitude of 1 mi and a speed of 500 mi/h passes directly over a radar station. If we let t be time (in hours) and x be the horizontal distance traveled by the plane (in mi), then we are given that $dx/dt=500$ mi/h.

(b) Unknown: the rate at which the distance from the plane to the station is increasing when it is 2 mi from the station. If we let y be the distance from the plane to the station, then we want to find dy/dt when $y=2$ mi.

(c)

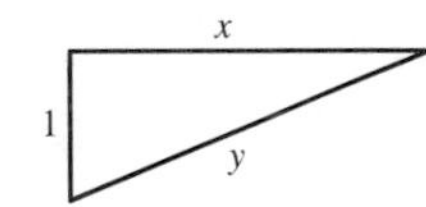

(d) By the Pythagorean Theorem, $y^2=x^2+1 \;\Rightarrow\; 2y\,(dy/dt)=2x\,(dx/dt)$.

(e) $\dfrac{dy}{dt}=\dfrac{x}{y}\dfrac{dx}{dt}=\dfrac{x}{y}(500)$. Since $y^2=x^2+1$, when $y=2$, $x=\sqrt{3}$, so $\dfrac{dy}{dt}=\dfrac{\sqrt{3}}{2}(500)=250\sqrt{3}\approx 433$ mi/h.

14. (a) Given: the rate of decrease of the surface area is 1 cm^2/min. If we let t be time (in minutes) and S be the surface area (in cm^2), then we are given that $dS/dt=-1$ cm^2/s.

(b) Unknown: the rate of decrease of the diameter when the diameter is 10 cm. If we let x be the diameter, then we want to find dx/dt when $x=10$ cm.

(c)

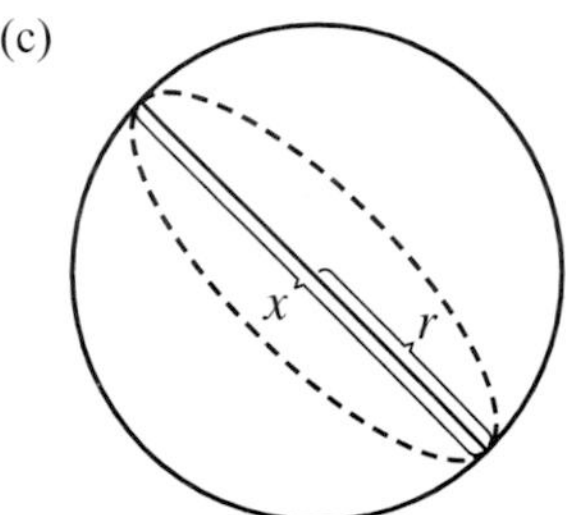

(d) If the radius is r and the diameter $x=2r$, then $r=\frac{1}{2}x$ and

$$S=4\pi r^2=4\pi\left(\tfrac{1}{2}x\right)^2=\pi x^2 \;\Rightarrow\; \frac{dS}{dt}=\frac{dS}{dx}\frac{dx}{dt}=2\pi x\,\frac{dx}{dt}.$$

(e) $-1=\dfrac{dS}{dt}=2\pi x\,\dfrac{dx}{dt} \;\Rightarrow\; \dfrac{dx}{dt}=-\dfrac{1}{2\pi x}$. When $x=10$, $\dfrac{dx}{dt}=-\dfrac{1}{20\pi}$. So the rate of decrease is $\dfrac{1}{20\pi}$ cm/min.

15. (a) Given: a man 6 ft tall walks away from a street light mounted on a 15-ft-tall pole at a rate of 5 ft/s. If we let t be time (in s) and x be the distance from the pole to the man (in ft), then we are given that $dx/dt=5$ ft/s.

(b) Unknown: the rate at which the tip of his shadow is moving when he is 40 ft from the pole. If we let y be the distance from the man to the tip of his shadow (in ft), then we want to find $\dfrac{d}{dt}(x+y)$ when $x=40$ ft.

(c)

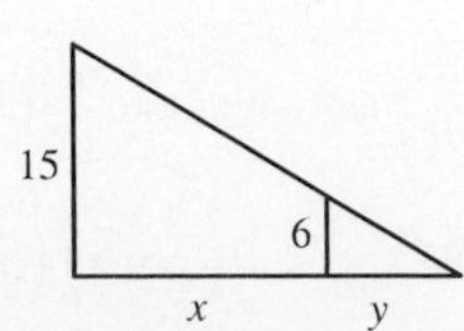

(d) By similar triangles, $\dfrac{15}{6} = \dfrac{x+y}{y} \quad\Rightarrow\quad 15y = 6x + 6y \quad\Rightarrow\quad 9y = 6x \quad\Rightarrow\quad y = \frac{2}{3}x.$

(e) The tip of the shadow moves at a rate of $\dfrac{d}{dt}(x+y) = \dfrac{d}{dt}\left(x + \dfrac{2}{3}x\right) = \dfrac{5}{3}\dfrac{dx}{dt} = \frac{5}{3}(5) = \frac{25}{3}$ ft/s.

16. (a) Given: at noon, ship A is 150 km west of ship B; ship A is sailing east at 35 km/h, and ship B is sailing north at 25 km/h. If we let t be time (in hours), x be the distance traveled by ship A (in km), and y be the distance traveled by ship B (in km), then we are given that $dx/dt = 35$ km/h and $dy/dt = 25$ km/h.

(b) Unknown: the rate at which the distance between the ships is changing at 4:00 PM. If we let z be the distance between the ships, then we want to find dz/dt when $t = 4$ h.

(c)

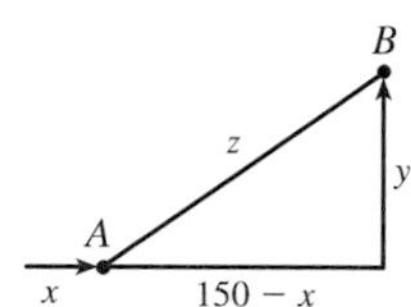

(d) $z^2 = (150 - x)^2 + y^2 \quad\Rightarrow\quad 2z\,\dfrac{dz}{dt} = 2(150 - x)\left(-\dfrac{dx}{dt}\right) + 2y\,\dfrac{dy}{dt}$

(e) At 4:00 PM, $x = 4(35) = 140$ and $y = 4(25) = 100 \quad\Rightarrow\quad z = \sqrt{(150 - 140)^2 + 100^2} = \sqrt{10{,}100}.$

So $\dfrac{dz}{dt} = \dfrac{1}{z}\left[(x - 150)\dfrac{dx}{dt} + y\,\dfrac{dy}{dt}\right] = \dfrac{-10(35) + 100(25)}{\sqrt{10{,}100}} = \dfrac{215}{\sqrt{101}} \approx 21.4$ km/h.

17.

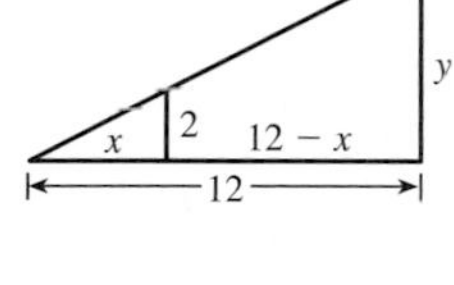

We are given that $\dfrac{dx}{dt} = 60$ mi/h and $\dfrac{dy}{dt} = 25$ mi/h. $z^2 = x^2 + y^2 \quad\Rightarrow$

$2z\,\dfrac{dz}{dt} = 2x\,\dfrac{dx}{dt} + 2y\,\dfrac{dy}{dt} \quad\Rightarrow\quad z\,\dfrac{dz}{dt} = x\,\dfrac{dx}{dt} + y\,\dfrac{dy}{dt} \quad\Rightarrow\quad \dfrac{dz}{dt} = \dfrac{1}{z}\left(x\,\dfrac{dx}{dt} + y\,\dfrac{dy}{dt}\right).$

After 2 hours, $x = 2\,(60) = 120$ and $y = 2\,(25) = 50 \quad\Rightarrow\quad z = \sqrt{120^2 + 50^2} = 130,$

so $\dfrac{dz}{dt} = \dfrac{1}{z}\left(x\,\dfrac{dx}{dt} + y\,\dfrac{dy}{dt}\right) = \dfrac{120(60) + 50(25)}{130} = 65$ mi/h.

18.

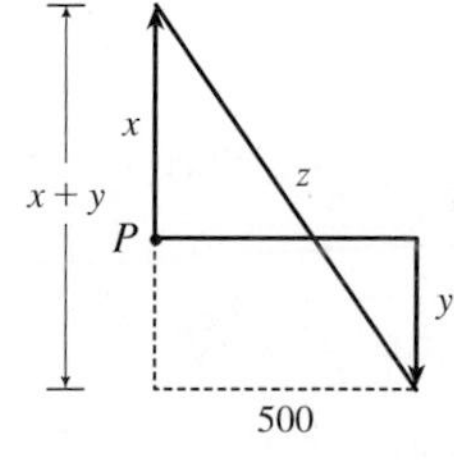

We are given that $\dfrac{dx}{dt} = 1.6$ m/s. By similar triangles, $\dfrac{y}{12} = \dfrac{2}{x} \quad\Rightarrow\quad y = \dfrac{24}{x} \quad\Rightarrow$

$\dfrac{dy}{dt} = -\dfrac{24}{x^2}\dfrac{dx}{dt} = -\dfrac{24}{x^2}(1.6)$. When $x = 8$, $\dfrac{dy}{dt} = -\dfrac{24(1.6)}{64} = -0.6$ m/s, so the shadow is decreasing at a rate of 0.6 m/s.

19.

We are given that $\dfrac{dx}{dt} = 4$ ft/s and $\dfrac{dy}{dt} = 5$ ft/s. $z^2 = (x+y)^2 + 500^2 \quad\Rightarrow$

$2z\,\dfrac{dz}{dt} = 2(x+y)\left(\dfrac{dx}{dt} + \dfrac{dy}{dt}\right)$. 15 minutes after the woman starts, we have

$x = (4 \text{ ft/s})(20 \text{ min})(60 \text{ s/min}) = 4800$ ft and $y = 5 \cdot 15 \cdot 60 = 4500 \quad\Rightarrow$

$z = \sqrt{(4800 + 4500)^2 + 500^2} = \sqrt{86{,}740{,}000}$, so

$\dfrac{dz}{dt} = \dfrac{x+y}{z}\left(\dfrac{dx}{dt} + \dfrac{dy}{dt}\right) = \dfrac{4800 + 4500}{\sqrt{86{,}740{,}000}}(4+5) = \dfrac{837}{\sqrt{8674}} \approx 8.99$ ft/s.

20. We are given that $\dfrac{dx}{dt} = 24$ ft/s.

(a)

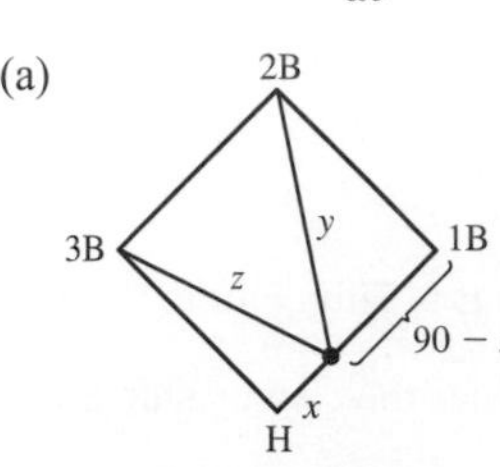

$y^2 = (90 - x)^2 + 90^2 \;\Rightarrow\; 2y\,\dfrac{dy}{dt} = 2(90 - x)\left(-\dfrac{dx}{dt}\right)$. When $x = 45$,

$y = \sqrt{45^2 + 90^2} = 45\sqrt{5}$, so $\dfrac{dy}{dt} = \dfrac{90 - x}{y}\left(-\dfrac{dx}{dt}\right) = \dfrac{45}{45\sqrt{5}}(-24) = -\dfrac{24}{\sqrt{5}}$,

so the distance from second base is decreasing at a rate of $\frac{24}{\sqrt{5}} \approx 10.7$ ft/s.

(b) Due to the symmetric nature of the problem in part (a), we expect to get the same answer—and we do.

$z^2 = x^2 + 90^2 \;\Rightarrow\; 2z\,\dfrac{dz}{dt} = 2x\,\dfrac{dx}{dt}$. When $x = 45$, $z = 45\sqrt{5}$, so $\dfrac{dz}{dt} = \dfrac{45}{45\sqrt{5}}(24) = \dfrac{24}{\sqrt{5}} \approx 10.7$ ft/s.

21. $A = \frac{1}{2}bh$, where b is the base and h is the altitude. We are given that $\dfrac{dh}{dt} = 1$ cm/min and $\dfrac{dA}{dt} = 2\text{ cm}^2/\text{min}$. Using the Product Rule, we have $\dfrac{dA}{dt} = \dfrac{1}{2}\left(b\,\dfrac{dh}{dt} + h\,\dfrac{db}{dt}\right)$. When $h = 10$ and $A = 100$, we have $100 = \frac{1}{2}b(10) \;\Rightarrow\; \frac{1}{2}b = 10 \;\Rightarrow$ $b = 20$, so $2 = \dfrac{1}{2}\left(20 \cdot 1 + 10\,\dfrac{db}{dt}\right) \;\Rightarrow\; 4 = 20 + 10\,\dfrac{db}{dt} \;\Rightarrow\; \dfrac{db}{dt} = \dfrac{4 - 20}{10} = -1.6$ cm/min.

22. Given $\dfrac{dy}{dt} = -1$ m/s, find $\dfrac{dx}{dt}$ when $x = 8$ m. $y^2 = x^2 + 1 \;\Rightarrow\; 2y\,\dfrac{dy}{dt} = 2x\,\dfrac{dx}{dt} \;\Rightarrow$ $\dfrac{dx}{dt} = \dfrac{y}{x}\dfrac{dy}{dt} = -\dfrac{y}{x}$. When $x = 8$, $y = \sqrt{65}$, so $\dfrac{dx}{dt} = -\dfrac{\sqrt{65}}{8}$. Thus, the boat approaches the dock at $\dfrac{\sqrt{65}}{8} \approx 1.01$ m/s.

23. We are given that $\dfrac{dx}{dt} = 35$ km/h and $\dfrac{dy}{dt} = 25$ km/h. $z^2 = (x + y)^2 + 100^2 \;\Rightarrow$

$2z\,\dfrac{dz}{dt} = 2(x + y)\left(\dfrac{dx}{dt} + \dfrac{dy}{dt}\right)$. At 4:00 PM, $x = 4(35) = 140$ and $y = 4(25) = 100 \;\Rightarrow$

$z = \sqrt{(140 + 100)^2 + 100^2} = \sqrt{67{,}600} = 260$, so

$$\frac{dz}{dt} = \frac{x + y}{z}\left(\frac{dx}{dt} + \frac{dy}{dt}\right) = \frac{140 + 100}{260}(35 + 25) = \frac{720}{13} \approx 55.4 \text{ km/h}.$$

24. The distance z of the particle to the origin is given by $z = \sqrt{x^2 + y^2}$, so $z^2 = x^2 + [2\sin(\pi x/2)]^2 \;\Rightarrow$

$2z\,\dfrac{dz}{dt} = 2x\,\dfrac{dx}{dt} + 4 \cdot 2\sin\left(\dfrac{\pi}{2}x\right)\cos\left(\dfrac{\pi}{2}x\right) \cdot \dfrac{\pi}{2}\dfrac{dx}{dt} \;\Rightarrow\; z\,\dfrac{dz}{dt} = x\,\dfrac{dx}{dt} + 2\pi\sin\left(\dfrac{\pi}{2}x\right)\cos\left(\dfrac{\pi}{2}x\right)\dfrac{dx}{dt}$. When

$(x, y) = \left(\dfrac{1}{3}, 1\right)$, $z = \sqrt{\left(\dfrac{1}{3}\right)^2 + 1^2} = \sqrt{\dfrac{10}{9}} = \dfrac{1}{3}\sqrt{10}$, so $\dfrac{1}{3}\sqrt{10}\,\dfrac{dz}{dt} = \dfrac{1}{3}\sqrt{10} + 2\pi\,\sin\dfrac{\pi}{6}\,\cos\dfrac{\pi}{6} \cdot \sqrt{10} \;\Rightarrow$

$\dfrac{1}{3}\dfrac{dz}{dt} = \dfrac{1}{3} + 2\pi\left(\dfrac{1}{2}\right)\left(\dfrac{1}{2}\sqrt{3}\right) \;\Rightarrow\; \dfrac{dz}{dt} = 1 + \dfrac{3\sqrt{3}\,\pi}{2}$ cm/s.

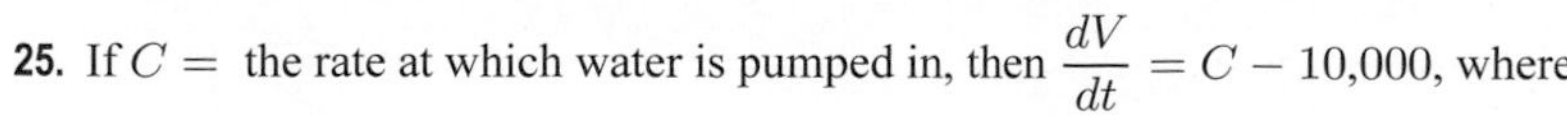

25. If $C =$ the rate at which water is pumped in, then $\dfrac{dV}{dt} = C - 10{,}000$, where

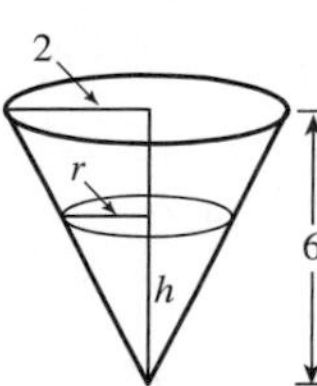

$V = \frac{1}{3}\pi r^2 h$ is the volume at time t. By similar triangles, $\dfrac{r}{2} = \dfrac{h}{6} \;\Rightarrow\; r = \dfrac{1}{3}h \;\Rightarrow$

$V = \frac{1}{3}\pi\left(\frac{1}{3}h\right)^2 h = \frac{\pi}{27}h^3 \;\Rightarrow\; \dfrac{dV}{dt} = \dfrac{\pi}{9}h^2\dfrac{dh}{dt}$. When $h = 200$ cm,

$\dfrac{dh}{dt} = 20$ cm/min, so $C - 10{,}000 = \dfrac{\pi}{9}(200)^2(20) \;\Rightarrow\; C = 10{,}000 + \dfrac{800{,}000}{9}\pi \approx 289{,}253\ \text{cm}^3/\text{min}$.

26. By similar triangles, $\dfrac{3}{1} = \dfrac{b}{h}$, so $b = 3h$. The trough has volume

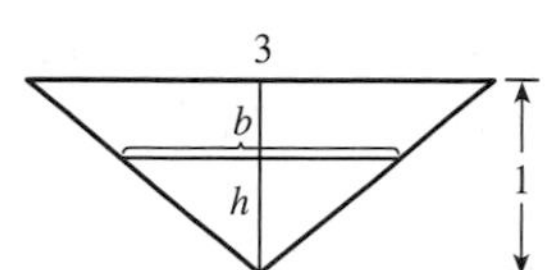

$V = \frac{1}{2}bh(10) = 5(3h)h = 15h^2 \;\Rightarrow\; 12 = \dfrac{dV}{dt} = 30h\dfrac{dh}{dt} \;\Rightarrow\; \dfrac{dh}{dt} = \dfrac{2}{5h}$.

When $h = \frac{1}{2}$, $\dfrac{dh}{dt} = \dfrac{2}{5 \cdot \frac{1}{2}} = \dfrac{4}{5}$ ft/min.

27.

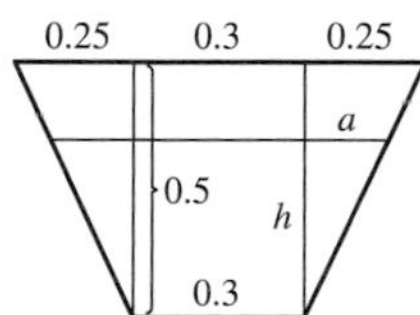

The figure is labeled in meters. The area A of a trapezoid is $\frac{1}{2}(\text{base}_1 + \text{base}_2)(\text{height})$, and the volume V of the 10-meter-long trough is $10A$. Thus, the volume of the trapezoid with height h is $V = (10)\frac{1}{2}[0.3 + (0.3 + 2a)]h$.

By similar triangles, $\dfrac{a}{h} = \dfrac{0.25}{0.5} = \dfrac{1}{2}$, so $2a = h \;\Rightarrow\; V = 5(0.6 + h)h = 3h + 5h^2$.

Now $\dfrac{dV}{dt} = \dfrac{dV}{dh}\dfrac{dh}{dt} \;\Rightarrow\; 0.2 = (3 + 10h)\dfrac{dh}{dt} \;\Rightarrow\; \dfrac{dh}{dt} = \dfrac{0.2}{3 + 10h}$. When $h = 0.3$,

$\dfrac{dh}{dt} = \dfrac{0.2}{3 + 10(0.3)} = \dfrac{0.2}{6}$ m/min $= \dfrac{1}{30}$ m/min or $\dfrac{10}{3}$ cm/min.

28. The figure is drawn without the top 3 feet.

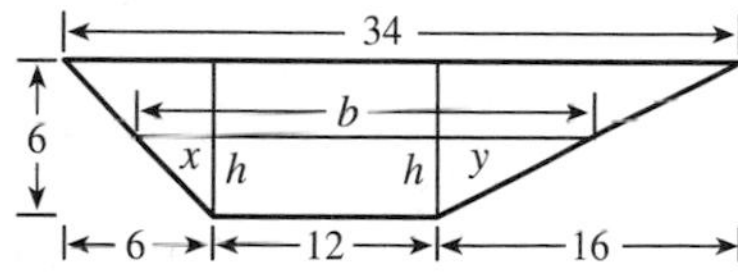

$V = \frac{1}{2}(b + 12)h(20) = 10(b + 12)h$ and, from similar triangles,

$\dfrac{x}{h} = \dfrac{6}{6}$ and $\dfrac{y}{h} = \dfrac{16}{6} = \dfrac{8}{3}$, so $b = x + 12 + y = h + 12 + \dfrac{8h}{3} = 12 + \dfrac{11h}{3}$.

Thus, $V = 10\left(24 + \dfrac{11h}{3}\right)h = 240h + \dfrac{110h^2}{3}$ and so $0.8 = \dfrac{dV}{dt} = \left(240 + \dfrac{220}{3}h\right)\dfrac{dh}{dt}$.

When $h = 5$, $\dfrac{dh}{dt} = \dfrac{0.8}{240 + 5(220/3)} = \dfrac{3}{2275} \approx 0.00132$ ft/min.

29. We are given that $\dfrac{dV}{dt} = 30\ \text{ft}^3/\text{min}$. $V = \dfrac{1}{3}\pi r^2 h = \dfrac{1}{3}\pi\left(\dfrac{h}{2}\right)^2 h = \dfrac{\pi h^3}{12} \;\Rightarrow$

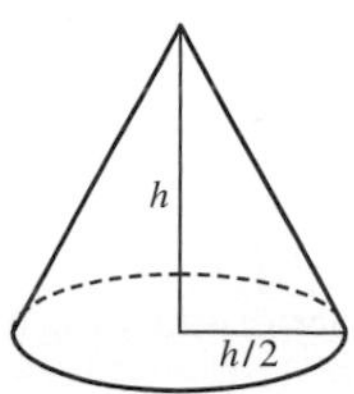

$\dfrac{dV}{dt} = \dfrac{dV}{dh}\dfrac{dh}{dt} \;\Rightarrow\; 30 = \dfrac{\pi h^2}{4}\dfrac{dh}{dt} \;\Rightarrow\; \dfrac{dh}{dt} = \dfrac{120}{\pi h^2}$.

When $h = 10$ ft, $\dfrac{dh}{dt} = \dfrac{120}{10^2\pi} = \dfrac{6}{5\pi} \approx 0.38$ ft/min.

30. We are given $dx/dt = 8$ ft/s. $\cot\theta = \dfrac{x}{100} \Rightarrow x = 100\cot\theta \Rightarrow$

$\dfrac{dx}{dt} = -100\csc^2\theta\,\dfrac{d\theta}{dt} \Rightarrow \dfrac{d\theta}{dt} = -\dfrac{\sin^2\theta}{100}\cdot 8$. When $y = 200$, $\sin\theta = \dfrac{100}{200} = \dfrac{1}{2} \Rightarrow$

$\dfrac{d\theta}{dt} = -\dfrac{(1/2)^2}{100}\cdot 8 = -\dfrac{1}{50}$ rad/s. The angle is decreasing at a rate of $\frac{1}{50}$ rad/s.

31. The area A of an equilateral triangle with side s is given by $A = \frac{1}{4}\sqrt{3}\,s^2$.

$\dfrac{dA}{dt} = \frac{1}{4}\sqrt{3}\cdot 2s\,\dfrac{ds}{dt} = \frac{1}{4}\sqrt{3}\cdot 2(30)(10) = 150\sqrt{3}\ \text{cm}^2/\text{min}$.

32. $\cos\theta = \dfrac{x}{10} \Rightarrow -\sin\theta\,\dfrac{d\theta}{dt} = \dfrac{1}{10}\dfrac{dx}{dt}$. From Example 2, $\dfrac{dx}{dt} = 1$ and

when $x = 6$, $y = 8$, so $\sin\theta = \dfrac{8}{10}$.

Thus, $-\dfrac{8}{10}\dfrac{d\theta}{dt} = \dfrac{1}{10}(1) \Rightarrow \dfrac{d\theta}{dt} = -\dfrac{1}{8}$ rad/s.

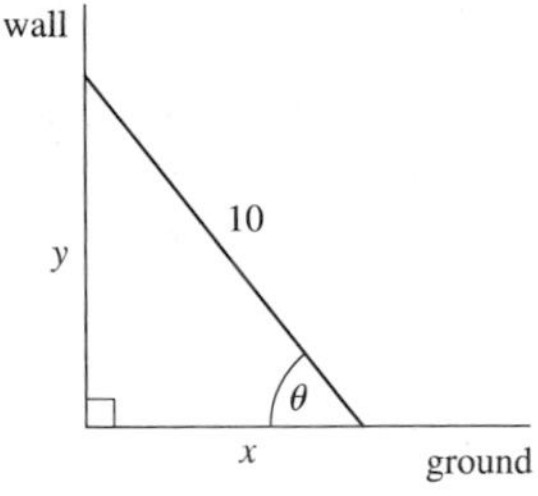

33. From the figure and given information, we have $x^2 + y^2 = L^2$, $\dfrac{dy}{dt} = -0.15$ m/ s, and

$\dfrac{dx}{dt} = 0.2$ m/ s when $x = 3$ m. Differentiating implicitly with respect to t, we get

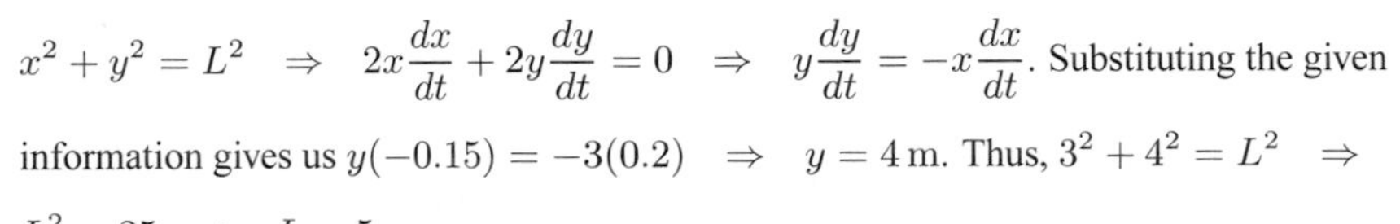

$x^2 + y^2 = L^2 \Rightarrow 2x\dfrac{dx}{dt} + 2y\dfrac{dy}{dt} = 0 \Rightarrow y\dfrac{dy}{dt} = -x\dfrac{dx}{dt}$. Substituting the given

information gives us $y(-0.15) = -3(0.2) \Rightarrow y = 4$ m. Thus, $3^2 + 4^2 = L^2 \Rightarrow$

$L^2 = 25 \Rightarrow L = 5$ m.

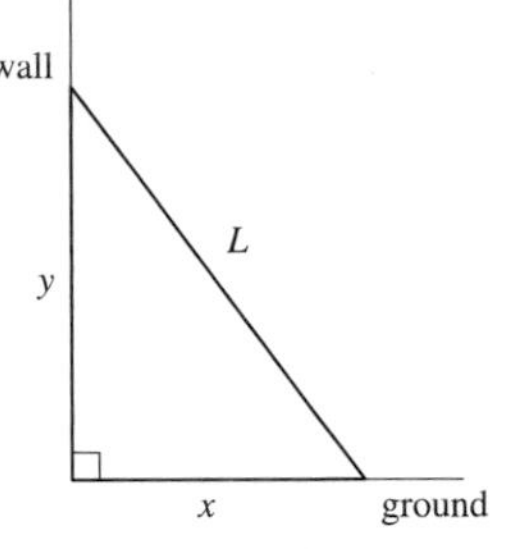

34. According to the model in Example 2, $\dfrac{dy}{dt} = -\dfrac{x}{y}\dfrac{dx}{dt} \to -\infty$ as $y \to 0$, which doesn't make physical sense. For example, the model predicts that for sufficiently small y, the tip of the ladder moves at a speed greater than the speed of light. Therefore the model is not appropriate for small values of y. What actually happens is that the tip of the ladder leaves the wall at some point in its descent. For a discussion of the true situation see the article "The Falling Ladder Paradox" by Paul Scholten and Andrew Simoson in *The College Mathematics Journal,* 27, (1), January 1996, pages 49–54. Also see "On Mathematical and Physical Ladders" by M. Freeman and P. Palffy-Muhoray in the *American Journal of Physics,* 53 (3), March 1985, pages 276–277.

35. The area A of a sector of a circle with radius r and angle θ is given by $A = \frac{1}{2}r^2\theta$. Here r is constant and θ varies, so

$\dfrac{dA}{dt} = \dfrac{1}{2}r^2\dfrac{d\theta}{dt}$. The minute hand rotates through $360^\circ = 2\pi$ radians each hour, so $\dfrac{dA}{dt} = \dfrac{1}{2}r^2(2\pi) = \pi r^2\ \text{cm}^2/\text{h}$. This answer makes sense because the minute hand sweeps through the full area of a circle, πr^2, each hour.

36. The volume of a hemisphere is $\frac{2}{3}\pi r^3$, so the volume of a hemispherical basin of radius 30 cm is $\frac{2}{3}\pi(30)^3 = 18{,}000\pi\ \text{cm}^3$. If the basin is half full, then $V = \pi\left(rh^2 - \frac{1}{3}h^3\right) \Rightarrow 9000\pi = \pi\left(30h^2 - \frac{1}{3}h^3\right) \Rightarrow \frac{1}{3}h^3 - 30h^2 + 9000 = 0 \Rightarrow$ $h = H \approx 19.58$ [from a graph or numerical rootfinder; the other two solutions are less than 0 and greater than 30].

$$V = \pi\left(30h^2 - \tfrac{1}{3}h^3\right) \Rightarrow \frac{dV}{dt} = \pi\left(60h\,\frac{dh}{dt} - h^2\,\frac{dh}{dt}\right) \Rightarrow \left(2\,\frac{\text{L}}{\text{min}}\right)\left(1000\,\frac{\text{cm}^3}{\text{L}}\right) = \pi(60h - h^2)\,\frac{dh}{dt} \Rightarrow$$

$$\frac{dh}{dt} = \frac{2000}{\pi(60H - H^2)} \approx 0.804\ \text{cm/min}.$$

37. Differentiating both sides of $PV = C$ with respect to t and using the Product Rule gives us $P\,\frac{dV}{dt} + V\,\frac{dP}{dt} = 0 \Rightarrow$ $\frac{dV}{dt} = -\frac{V}{P}\frac{dP}{dt}$. When $V = 600$, $P = 150$ and $\frac{dP}{dt} = 20$, so we have $\frac{dV}{dt} = -\frac{600}{150}(20) = -80$. Thus, the volume is decreasing at a rate of $80\ \text{cm}^3/\text{min}$.

38. $PV^{1.4} = C \Rightarrow P \cdot 1.4V^{0.4}\,\frac{dV}{dt} + V^{1.4}\,\frac{dP}{dt} = 0 \Rightarrow \frac{dV}{dt} = -\frac{V^{1.4}}{P \cdot 1.4V^{0.4}}\frac{dP}{dt} = -\frac{V}{1.4P}\frac{dP}{dt}$.

When $V = 400$, $P = 80$ and $\frac{dP}{dt} = -10$, so we have $\frac{dV}{dt} = -\frac{400}{1.4(80)}(-10) = \frac{250}{7}$. Thus, the volume is increasing at a rate of $\frac{250}{7} \approx 36\ \text{cm}^3/\text{min}$.

39. With $R_1 = 80$ and $R_2 = 100$, $\frac{1}{R} = \frac{1}{R_1} + \frac{1}{R_2} = \frac{1}{80} + \frac{1}{100} = \frac{180}{8000} = \frac{9}{400}$, so $R = \frac{400}{9}$. Differentiating $\frac{1}{R} = \frac{1}{R_1} + \frac{1}{R_2}$ with respect to t, we have $-\frac{1}{R^2}\frac{dR}{dt} = -\frac{1}{R_1^2}\frac{dR_1}{dt} - \frac{1}{R_2^2}\frac{dR_2}{dt} \Rightarrow \frac{dR}{dt} = R^2\left(\frac{1}{R_1^2}\frac{dR_1}{dt} + \frac{1}{R_2^2}\frac{dR_2}{dt}\right)$. When $R_1 = 80$ and $R_2 = 100$, $\frac{dR}{dt} = \frac{400^2}{9^2}\left[\frac{1}{80^2}(0.3) + \frac{1}{100^2}(0.2)\right] = \frac{107}{810} \approx 0.132\ \Omega/\text{s}$.

40. We want to find $\frac{dB}{dt}$ when $L = 18$ using $B = 0.007W^{2/3}$ and $W = 0.12L^{2.53}$.

$$\frac{dB}{dt} = \frac{dB}{dW}\frac{dW}{dL}\frac{dL}{dt} = \left(0.007 \cdot \tfrac{2}{3}W^{-1/3}\right)\left(0.12 \cdot 2.53 \cdot L^{1.53}\right)\left(\frac{20 - 15}{10{,}000{,}000}\right)$$

$$= \left[0.007 \cdot \tfrac{2}{3}\left(0.12 \cdot 18^{2.53}\right)^{-1/3}\right]\left(0.12 \cdot 2.53 \cdot 18^{1.53}\right)\left(\frac{5}{10^7}\right) \approx 1.045 \times 10^{-8}\ \text{g/yr}$$

41. We are given $d\theta/dt = 2^\circ/\text{min} = \frac{\pi}{90}$ rad/min. By the Law of Cosines,

$x^2 = 12^2 + 15^2 - 2(12)(15)\cos\theta = 369 - 360\cos\theta \Rightarrow$

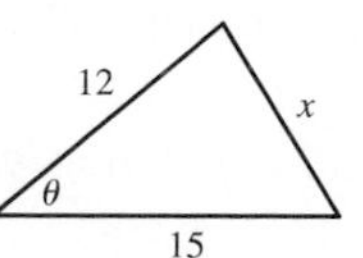

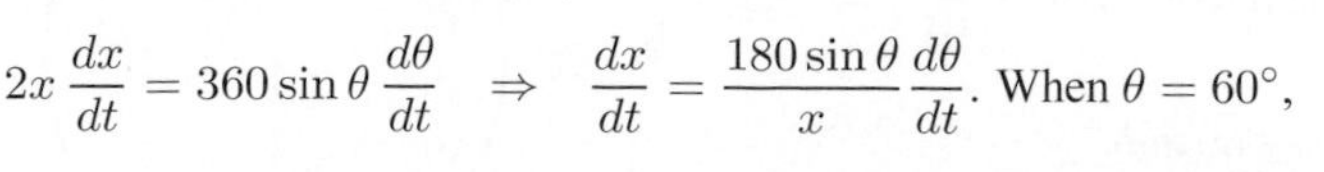

$2x\,\frac{dx}{dt} = 360\sin\theta\,\frac{d\theta}{dt} \Rightarrow \frac{dx}{dt} = \frac{180\sin\theta}{x}\frac{d\theta}{dt}$. When $\theta = 60^\circ$,

$x = \sqrt{369 - 360\cos 60^\circ} = \sqrt{189} = 3\sqrt{21}$, so $\frac{dx}{dt} = \frac{180\sin 60^\circ}{3\sqrt{21}}\frac{\pi}{90} = \frac{\pi\sqrt{3}}{3\sqrt{21}} = \frac{\sqrt{7}\,\pi}{21} \approx 0.396\ \text{m/min}$.

42. Using Q for the origin, we are given $\frac{dx}{dt} = -2$ ft/s and need to find $\frac{dy}{dt}$ when $x = -5$.

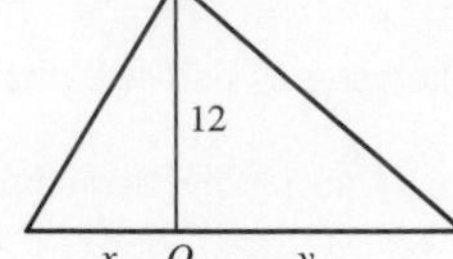

Using the Pythagorean Theorem twice, we have $\sqrt{x^2+12^2} + \sqrt{y^2+12^2} = 39$, the total length of the rope. Differentiating with respect to t, we get

$$\frac{x}{\sqrt{x^2+12^2}}\frac{dx}{dt} + \frac{y}{\sqrt{y^2+12^2}}\frac{dy}{dt} = 0, \text{ so } \frac{dy}{dt} = -\frac{x\sqrt{y^2+12^2}}{y\sqrt{x^2+12^2}}\frac{dx}{dt}.$$

Now when $x = -5$, $39 = \sqrt{(-5)^2+12^2} + \sqrt{y^2+12^2} = 13 + \sqrt{y^2+12^2} \Leftrightarrow \sqrt{y^2+12^2} = 26$, and

$y = \sqrt{26^2 - 12^2} = \sqrt{532}$. So when $x = -5$, $\frac{dy}{dt} = -\frac{(-5)(26)}{\sqrt{532}\,(13)}(-2) = -\frac{10}{\sqrt{133}} \approx -0.87$ ft/s.

So cart B is moving towards Q at about 0.87 ft/s.

43. (a) By the Pythagorean Theorem, $4000^2 + y^2 = \ell^2$. Differentiating with respect to t,

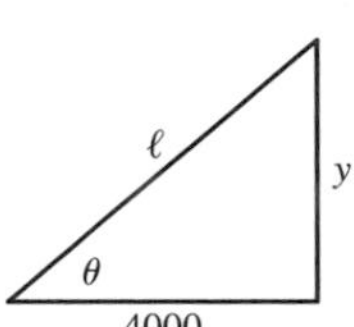

we obtain $2y\frac{dy}{dt} = 2\ell\frac{d\ell}{dt}$. We know that $\frac{dy}{dt} = 600$ ft/s, so when $y = 3000$ ft,

$\ell = \sqrt{4000^2 + 3000^2} = \sqrt{25{,}000{,}000} = 5000$ ft

and $\frac{d\ell}{dt} = \frac{y}{\ell}\frac{dy}{dt} = \frac{3000}{5000}(600) = \frac{1800}{5} = 360$ ft/s.

(b) Here $\tan\theta = \frac{y}{4000} \Rightarrow \frac{d}{dt}(\tan\theta) = \frac{d}{dt}\left(\frac{y}{4000}\right) \Rightarrow \sec^2\theta\,\frac{d\theta}{dt} = \frac{1}{4000}\frac{dy}{dt} \Rightarrow \frac{d\theta}{dt} = \frac{\cos^2\theta}{4000}\frac{dy}{dt}$. When $y = 3000$ ft, $\frac{dy}{dt} = 600$ ft/s, $\ell = 5000$ and $\cos\theta = \frac{4000}{\ell} = \frac{4000}{5000} = \frac{4}{5}$, so $\frac{d\theta}{dt} = \frac{(4/5)^2}{4000}(600) = 0.096$ rad/s.

44. We are given that $\frac{d\theta}{dt} = 4(2\pi) = 8\pi$ rad/min. $x = 3\tan\theta \Rightarrow$

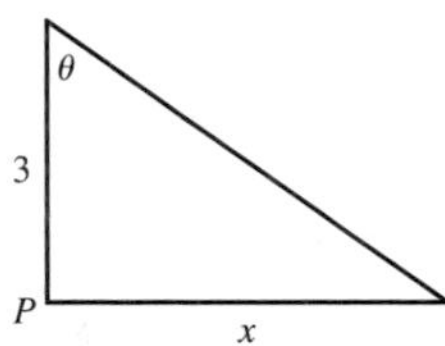

$\frac{dx}{dt} = 3\sec^2\theta\,\frac{d\theta}{dt}$. When $x = 1$, $\tan\theta = \frac{1}{3}$, so $\sec^2\theta = 1 + \left(\frac{1}{3}\right)^2 = \frac{10}{9}$

and $\frac{dx}{dt} = 3\left(\frac{10}{9}\right)(8\pi) = \frac{80}{3}\pi \approx 83.8$ km/min.

45. $\cot\theta = \frac{x}{5} \Rightarrow -\csc^2\theta\,\frac{d\theta}{dt} = \frac{1}{5}\frac{dx}{dt} \Rightarrow -\left(\csc\frac{\pi}{3}\right)^2\left(-\frac{\pi}{6}\right) = \frac{1}{5}\frac{dx}{dt} \Rightarrow$

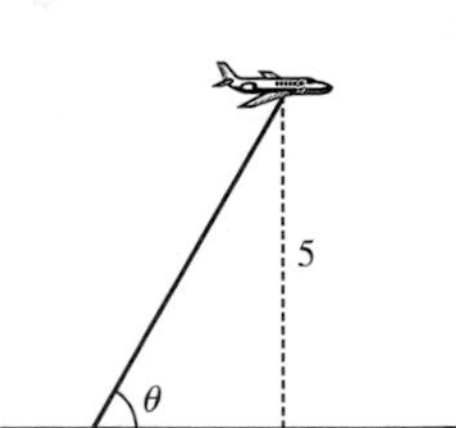

$\frac{dx}{dt} = \frac{5\pi}{6}\left(\frac{2}{\sqrt{3}}\right)^2 = \frac{10}{9}\pi$ km/min [≈ 130 mi/h]

46. We are given that $\frac{d\theta}{dt} = \frac{2\pi \text{ rad}}{2 \text{ min}} = \pi$ rad/min. By the Pythagorean Theorem, when

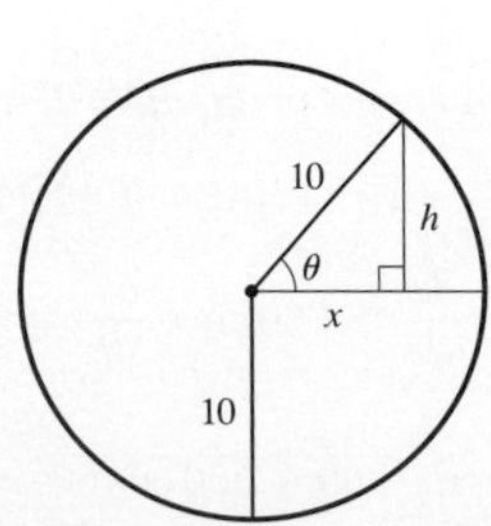

$h = 6$, $x = 8$, so $\sin\theta = \frac{6}{10}$ and $\cos\theta = \frac{8}{10}$. From the figure, $\sin\theta = \frac{h}{10} \Rightarrow$

$h = 10\sin\theta$, so $\frac{dh}{dt} = 10\cos\theta\,\frac{d\theta}{dt} = 10\left(\frac{8}{10}\right)\pi = 8\pi$ m/min.

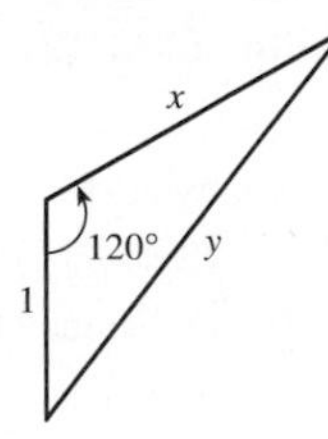

47. We are given that $\frac{dx}{dt} = 300$ km/h. By the Law of Cosines,

$y^2 = x^2 + 1^2 - 2(1)(x)\cos 120° = x^2 + 1 - 2x\left(-\frac{1}{2}\right) = x^2 + x + 1$, so

$2y\,\frac{dy}{dt} = 2x\,\frac{dx}{dt} + \frac{dx}{dt} \quad\Rightarrow\quad \frac{dy}{dt} = \frac{2x+1}{2y}\frac{dx}{dt}$. After 1 minute, $x = \frac{300}{60} = 5$ km $\Rightarrow$

$y = \sqrt{5^2 + 5 + 1} = \sqrt{31}$ km $\Rightarrow \quad \frac{dy}{dt} = \frac{2(5)+1}{2\sqrt{31}}(300) = \frac{1650}{\sqrt{31}} \approx 296$ km/h.

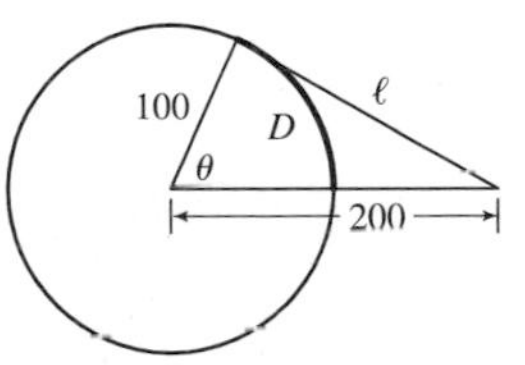

48. We are given that $\frac{dx}{dt} = 3$ mi/h and $\frac{dy}{dt} = 2$ mi/h. By the Law of Cosines,

$z^2 = x^2 + y^2 - 2xy\cos 45° = x^2 + y^2 - \sqrt{2}\,xy \quad\Rightarrow$

$2z\frac{dz}{dt} = 2x\frac{dx}{dt} + 2y\frac{dy}{dt} - \sqrt{2}\,x\frac{dy}{dt} - \sqrt{2}\,y\frac{dx}{dt}$. After 15 minutes $\left[= \frac{1}{4}\text{ h}\right]$,

we have $x = \frac{3}{4}$ and $y = \frac{2}{4} = \frac{1}{2} \quad\Rightarrow\quad z^2 = \left(\frac{3}{4}\right)^2 + \left(\frac{2}{4}\right)^2 - \sqrt{2}\left(\frac{3}{4}\right)\left(\frac{2}{4}\right) \quad\Rightarrow\quad z = \frac{\sqrt{13 - 6\sqrt{2}}}{4}$ and

$$\frac{dz}{dt} = \frac{2}{\sqrt{13-6\sqrt{2}}}\left[2\left(\tfrac{3}{4}\right)3 + 2\left(\tfrac{1}{2}\right)2 - \sqrt{2}\left(\tfrac{3}{4}\right)2 - \sqrt{2}\left(\tfrac{1}{2}\right)3\right] = \frac{2}{\sqrt{13-6\sqrt{2}}}\,\frac{13-6\sqrt{2}}{2} = \sqrt{13-6\sqrt{2}} \approx 2.125 \text{ mi/h}.$$

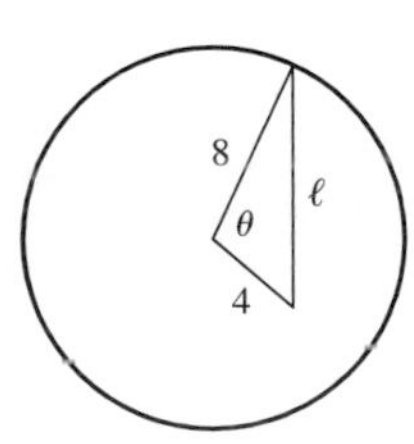

49. Let the distance between the runner and the friend be ℓ. Then by the Law of Cosines,

$\ell^2 = 200^2 + 100^2 - 2 \cdot 200 \cdot 100 \cdot \cos\theta = 50{,}000 - 40{,}000\cos\theta$ $(\star)$. Differentiating implicitly with respect to t, we obtain $2\ell\,\frac{d\ell}{dt} = -40{,}000(-\sin\theta)\,\frac{d\theta}{dt}$. Now if D is the distance run when the angle is θ radians, then by the formula for the length of an arc on a circle, $s = r\theta$, we have $D = 100\theta$, so $\theta = \frac{1}{100}D \quad\Rightarrow\quad \frac{d\theta}{dt} = \frac{1}{100}\frac{dD}{dt} = \frac{7}{100}$. To substitute into the expression for $\frac{d\ell}{dt}$, we must know $\sin\theta$ at the time when $\ell = 200$, which we find from $(\star)$: $200^2 = 50{,}000 - 40{,}000\cos\theta \quad\Leftrightarrow$

$\cos\theta = \frac{1}{4} \quad\Rightarrow\quad \sin\theta = \sqrt{1 - \left(\frac{1}{4}\right)^2} = \frac{\sqrt{15}}{4}$. Substituting, we get $2(200)\,\frac{d\ell}{dt} = 40{,}000\frac{\sqrt{15}}{4}\left(\frac{7}{100}\right) \quad\Rightarrow$

$d\ell/dt = \frac{7\sqrt{15}}{4} \approx 6.78$ m/s. Whether the distance between them is increasing or decreasing depends on the direction in which the runner is running.

50. The hour hand of a clock goes around once every 12 hours or, in radians per hour, $\frac{2\pi}{12} = \frac{\pi}{6}$ rad/h. The minute hand goes around once an hour, or at the rate of 2π rad/h. So the angle θ between them (measuring clockwise from the minute hand to the hour hand) is changing at the rate of $d\theta/dt = \frac{\pi}{6} - 2\pi = -\frac{11\pi}{6}$ rad/h. Now, to relate θ to ℓ, we use the Law of Cosines: $\ell^2 = 4^2 + 8^2 - 2 \cdot 4 \cdot 8 \cdot \cos\theta = 80 - 64\cos\theta$ $(\star)$.

Differentiating implicitly with respect to t, we get $2\ell\,\frac{d\ell}{dt} = -64(-\sin\theta)\frac{d\theta}{dt}$. At 1:00, the angle between the two hands is one-twelfth of the circle, that is, $\frac{2\pi}{12} = \frac{\pi}{6}$ radians. We use ($\star$) to find ℓ at 1:00: $\ell = \sqrt{80 - 64\cos\frac{\pi}{6}} = \sqrt{80 - 32\sqrt{3}}$.

Substituting, we get $2\ell\,\dfrac{d\ell}{dt} = 64\sin\dfrac{\pi}{6}\left(-\dfrac{11\pi}{6}\right) \Rightarrow \dfrac{d\ell}{dt} = \dfrac{64\left(\frac{1}{2}\right)\left(-\frac{11\pi}{6}\right)}{2\sqrt{80 - 32\sqrt{3}}} = -\dfrac{88\pi}{3\sqrt{80 - 32\sqrt{3}}} \approx -18.6.$

So at 1:00, the distance between the tips of the hands is decreasing at a rate of 18.6 mm/h ≈ 0.005 mm/s.

3.10 Linear Approximations and Differentials

1. $f(x) = x^3 - x^2 + 3 \Rightarrow f'(x) = 3x^2 - 2x$, so $f(-2) = -9$ and $f'(-2) = 16$. Thus,
$L(x) = f(-2) + f'(-2)(x - (-2)) = -9 + 16(x + 2) = 16x + 23.$

2. $f(x) = \sin x \Rightarrow f'(x) = \cos x$, so $f\left(\frac{\pi}{6}\right) = \frac{1}{2}$ and $f'\left(\frac{\pi}{6}\right) = \frac{1}{2}\sqrt{3}$. Thus,
$L(x) = f\left(\frac{\pi}{6}\right) + f'\left(\frac{\pi}{6}\right)\left(x - \frac{\pi}{6}\right) = \frac{1}{2} + \frac{1}{2}\sqrt{3}\left(x - \frac{\pi}{6}\right) = \frac{1}{2}\sqrt{3}\,x + \frac{1}{2} - \frac{1}{12}\sqrt{3}\,\pi.$

3. $f(x) = \sqrt{x} \Rightarrow f'(x) = \frac{1}{2}x^{-1/2} = 1/(2\sqrt{x})$, so $f(4) = 2$ and $f'(4) = \frac{1}{4}$. Thus,
$L(x) = f(4) + f'(4)(x - 4) = 2 + \frac{1}{4}(x - 4) = 2 + \frac{1}{4}x - 1 = \frac{1}{4}x + 1.$

4. $f(x) = 2^x \Rightarrow f'(x) = 2^x \ln 2$, so $f(0) = 1$ and $f'(0) = \ln 2$. Thus, $L(x) = f(0) + f'(0)(x - 0) = 1 + (\ln 2)x$.

5. $f(x) = \sqrt{1 - x} \Rightarrow f'(x) = \dfrac{-1}{2\sqrt{1 - x}}$, so $f(0) = 1$ and $f'(0) = -\frac{1}{2}$.

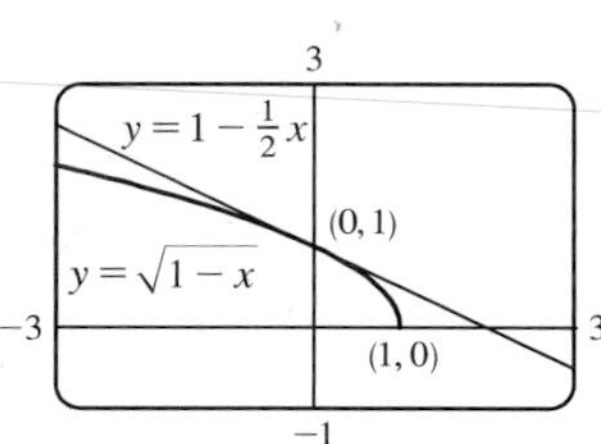

Therefore,

$\sqrt{1 - x} = f(x) \approx f(0) + f'(0)(x - 0) = 1 + \left(-\frac{1}{2}\right)(x - 0) = 1 - \frac{1}{2}x.$

So $\sqrt{0.9} = \sqrt{1 - 0.1} \approx 1 - \frac{1}{2}(0.1) = 0.95$

and $\sqrt{0.99} = \sqrt{1 - 0.01} \approx 1 - \frac{1}{2}(0.01) = 0.995.$

6. $g(x) = \sqrt[3]{1 + x} = (1 + x)^{1/3} \Rightarrow g'(x) = \frac{1}{3}(1 + x)^{-2/3}$, so $g(0) = 1$ and
$g'(0) = \frac{1}{3}$. Therefore, $\sqrt[3]{1 + x} = g(x) \approx g(0) + g'(0)(x - 0) = 1 + \frac{1}{3}x.$

So $\sqrt[3]{0.95} = \sqrt[3]{1 + (-0.05)} \approx 1 + \frac{1}{3}(-0.05) = 0.98\overline{3}$,

and $\sqrt[3]{1.1} = \sqrt[3]{1 + 0.1} \approx 1 + \frac{1}{3}(0.1) = 1.0\overline{3}$.

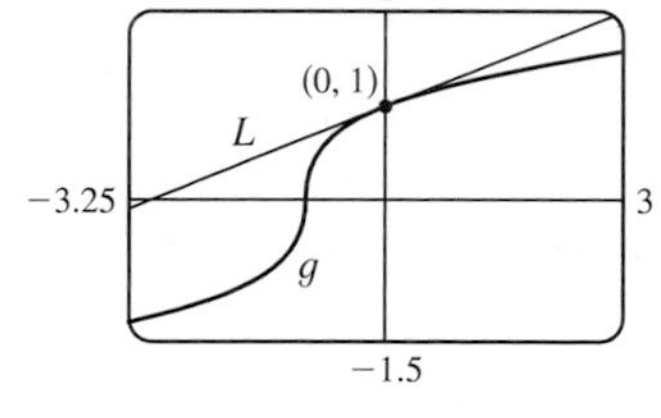

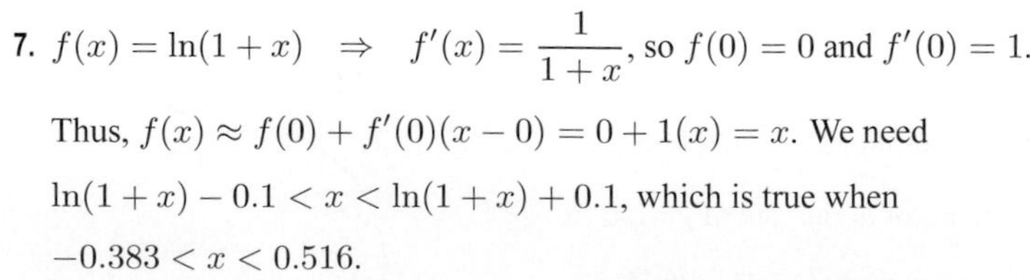

7. $f(x) = \ln(1 + x) \Rightarrow f'(x) = \dfrac{1}{1 + x}$, so $f(0) = 0$ and $f'(0) = 1$.

Thus, $f(x) \approx f(0) + f'(0)(x - 0) = 0 + 1(x) = x$. We need
$\ln(1 + x) - 0.1 < x < \ln(1 + x) + 0.1$, which is true when
$-0.383 < x < 0.516$.

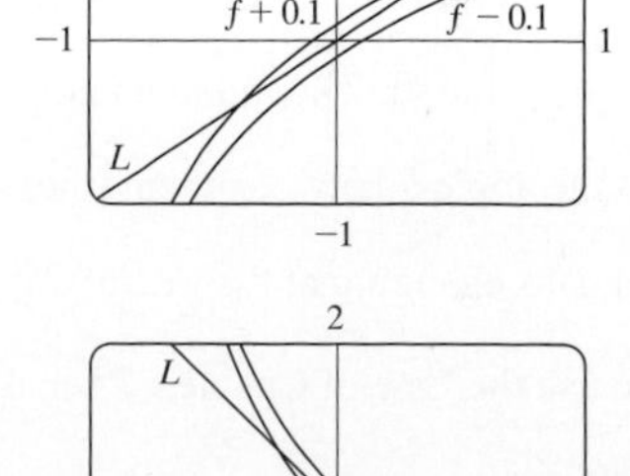

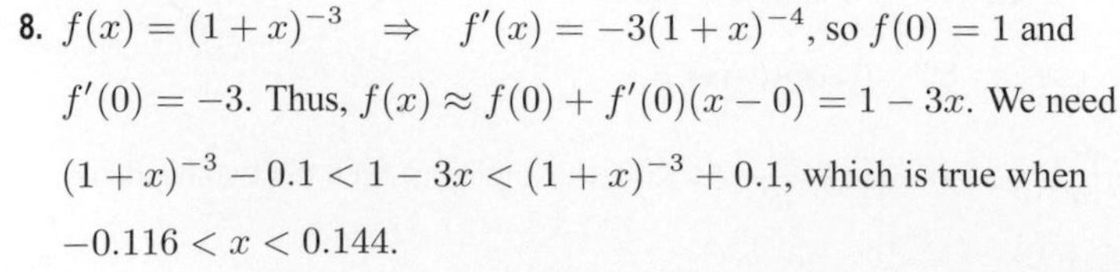

8. $f(x) = (1 + x)^{-3} \Rightarrow f'(x) = -3(1 + x)^{-4}$, so $f(0) = 1$ and
$f'(0) = -3$. Thus, $f(x) \approx f(0) + f'(0)(x - 0) = 1 - 3x$. We need
$(1 + x)^{-3} - 0.1 < 1 - 3x < (1 + x)^{-3} + 0.1$, which is true when
$-0.116 < x < 0.144$.

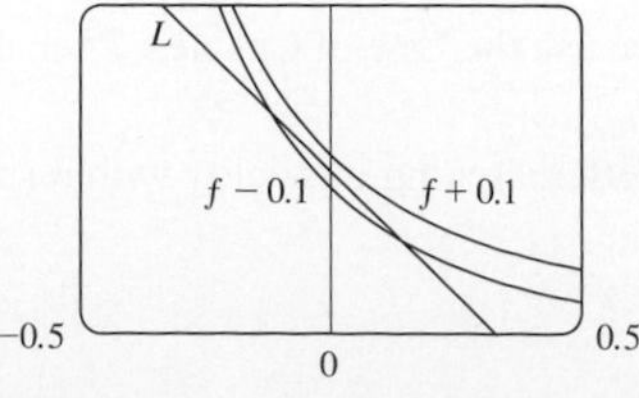

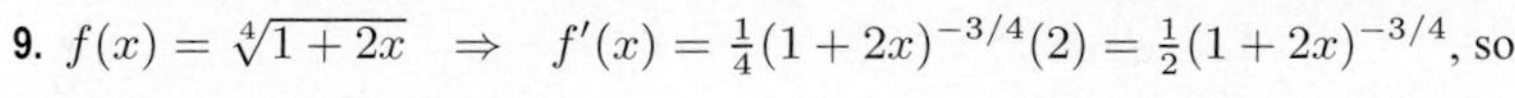

9. $f(x) = \sqrt[4]{1+2x} \;\Rightarrow\; f'(x) = \frac{1}{4}(1+2x)^{-3/4}(2) = \frac{1}{2}(1+2x)^{-3/4}$, so

$f(0) = 1$ and $f'(0) = \frac{1}{2}$. Thus, $f(x) \approx f(0) + f'(0)(x-0) = 1 + \frac{1}{2}x$.

We need $\sqrt[4]{1+2x} - 0.1 < 1 + \frac{1}{2}x < \sqrt[4]{1+2x} + 0.1$, which is true when

$-0.368 < x < 0.677$.

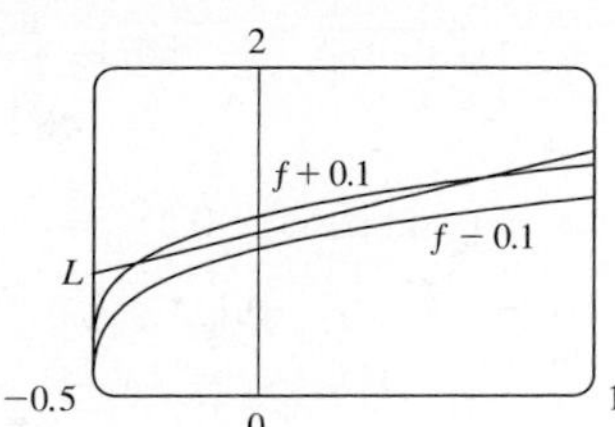

10. $f(x) = e^x \cos x \;\Rightarrow\; f'(x) = e^x(-\sin x) + (\cos x)e^x = e^x(\cos x - \sin x)$,

so $f(0) = 1$ and $f'(0) = 1$. Thus, $f(x) \approx f(0) + f'(0)(x-0) = 1 + x$.

We need $e^x \cos x - 0.1 < 1 + x < e^x \cos x + 0.1$, which is true when

$-0.762 < x < 0.607$.

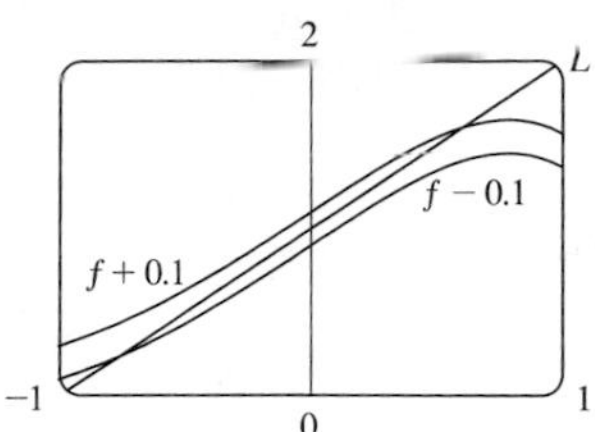

11. (a) The differential dy is defined in terms of dx by the equation $dy = f'(x)\,dx$. For $y = f(x) = xe^{-4x}$,

$f'(x) = xe^{-4x}(-4) + e^{-4x} \cdot 1 = e^{-4x}(-4x+1)$, so $dy = (1-4x)e^{-4x}dx$.

(b) For $y = f(t) = \sqrt{1-t^4}$, $f'(t) = \frac{1}{2}(1-t^4)^{-1/2}(-4t^3) = -\dfrac{2t^3}{\sqrt{1-t^4}}$, so $dy = -\dfrac{2t^3}{\sqrt{1-t^4}}\,dt$.

12. (a) For $y = f(u) = \dfrac{1+2u}{1+3u}$, $f'(u) = \dfrac{(1+3u)(2) - (1+2u)(3)}{(1+3u)^2} = \dfrac{-1}{(1+3u)^2}$, so $dy = \dfrac{-1}{(1+3u)^2}\,du$.

(b) For $y = f(\theta) = \theta^2 \sin 2\theta$, $f'(\theta) = \theta^2(\cos 2\theta)(2) + (\sin 2\theta)(2\theta)$, so $dy = 2\theta(\theta\cos 2\theta + \sin 2\theta)\,d\theta$.

13. (a) For $y = f(t) = \tan\sqrt{t}$, $f'(t) = \sec^2\sqrt{t} \cdot \dfrac{1}{2}t^{-1/2} = \dfrac{\sec^2\sqrt{t}}{2\sqrt{t}}$, so $dy = \dfrac{\sec^2\sqrt{t}}{2\sqrt{t}}\,dt$.

(b) For $y = f(v) = \dfrac{1-v^2}{1+v^2}$,

$$f'(v) = \frac{(1+v^2)(-2v) - (1-v^2)(2v)}{(1+v^2)^2} = \frac{-2v[(1+v^2)+(1-v^2)]}{(1+v^2)^2} = \frac{-2v(2)}{(1+v^2)^2} = \frac{-4v}{(1+v^2)^2},$$

so $dy = \dfrac{-4v}{(1+v^2)^2}\,dv$.

14. (a) For $y = f(\theta) = \ln(\sin\theta)$, $f'(\theta) = \dfrac{1}{\sin\theta}\cos\theta = \cot\theta$, so $dy = \cot\theta\,d\theta$.

(b) For $y = f(x) = \dfrac{e^x}{1-e^x}$, $f'(x) = \dfrac{(1-e^x)e^x - e^x(-e^x)}{(1-e^x)^2} = \dfrac{e^x[(1-e^x)-(-e^x)]}{(1-e^x)^2} = \dfrac{e^x}{(1-e^x)^2}$, so

$dy = \dfrac{e^x}{(1-e^x)^2}\,dx$.

15. (a) $y = e^{x/10} \;\Rightarrow\; dy = e^{x/10} \cdot \frac{1}{10}\,dx = \frac{1}{10}e^{x/10}dx$

(b) $x = 0$ and $dx = 0.1 \;\Rightarrow\; dy = \frac{1}{10}e^{0/10}(0.1) = 0.01$.

16. (a) $y = \cos\pi x \;\Rightarrow\; dy = -\sin\pi x \cdot \pi\,dx = -\pi\sin\pi x\,dx$

(b) $x = \frac{1}{3}$ and $dx = -0.02 \;\Rightarrow\; dy = -\pi\sin\frac{\pi}{3}(-0.02) = \pi\left(\sqrt{3}/2\right)(0.02) = 0.01\pi\sqrt{3} \approx 0.054$.

17. (a) $y = \sqrt{3 + x^2} \Rightarrow dy = \frac{1}{2}(3 + x^2)^{-1/2}(2x)\,dx = \frac{x}{\sqrt{3 + x^2}}\,dx$

(b) $x = 1$ and $dx = -0.1 \Rightarrow dy = \frac{1}{\sqrt{3 + 1^2}}(-0.1) = \frac{1}{2}(-0.1) = -0.05.$

18. (a) $y = \frac{x + 1}{x - 1} \Rightarrow dy = \frac{(x - 1)(1) - (x + 1)(1)}{(x - 1)^2}\,dx = \frac{-2}{(x - 1)^2}\,dx$

(b) $x = 2$ and $dx = 0.05 \Rightarrow dy = \frac{-2}{(2 - 1)^2}(0.05) = -2(0.05) = -0.1.$

19. $y = f(x) = x^2 - 4x,\ x = 3,\ \Delta x = 0.5 \Rightarrow$

$\Delta y = f(3.5) - f(3) = -1.75 - (-3) = 1.25$

$dy = f'(x)\,dx = (2x - 4)\,dx = (6 - 4)(0.5) = 1$

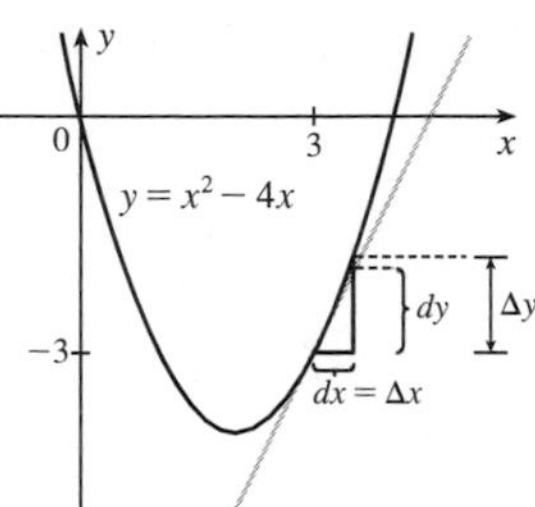

20. $y = f(x) = x - x^3,\ x = 0,\ \Delta x = -0.3 \Rightarrow$

$\Delta y = f(-0.3) - f(0) = -0.273 - 0 = -0.273$

$dy = f'(x)\,dx = (1 - 3x^2)\,dx = (1 - 0)(-0.3) = -0.3$

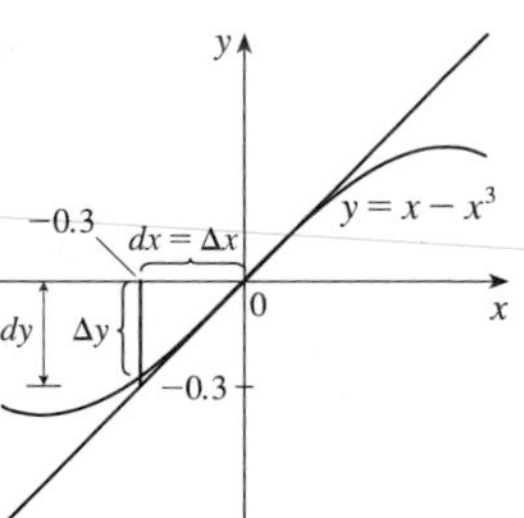

21. $y = f(x) = \sqrt{x - 2},\ x = 3,\ \Delta x = 0.8 \Rightarrow$

$\Delta y = f(3.8) - f(3) = \sqrt{1.8} - 1 \approx 0.34$

$dy = f'(x)\,dx = \frac{1}{2\sqrt{x - 2}}\,dx = \frac{1}{2(1)}(0.8) = 0.4$

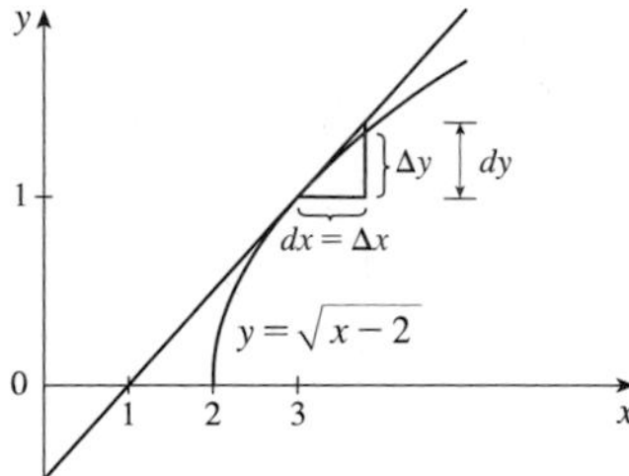

22. $y = f(x) = e^x,\ x = 0,\ \Delta x = 0.5 \Rightarrow$

$\Delta y = f(0.5) - f(0) = \sqrt{e} - 1\ [\approx 0.65]$

$dy = e^x\,dx = e^0(0.5) = 0.5$

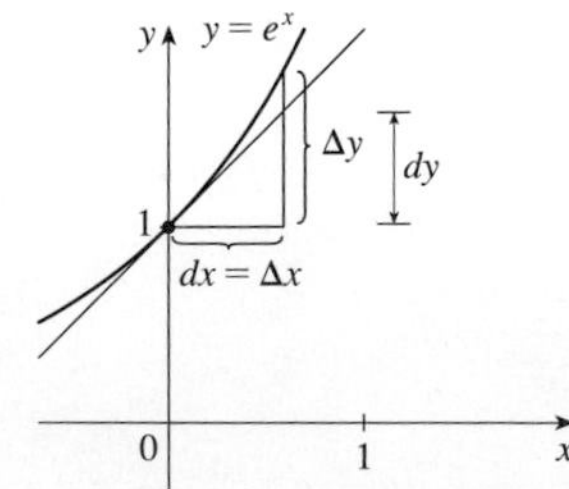

23. To estimate $(1.999)^4$, we'll find the linearization of $f(x) = x^4$ at $a = 2$. Since $f'(x) = 4x^3$, $f(2) = 16$, and $f'(2) = 32$, we have $L(x) = 16 + 32(x - 2)$. Thus, $x^4 \approx 16 + 32(x - 2)$ when x is near 2, so $(1.999)^4 \approx 16 + 32(1.999 - 2) = 16 - 0.032 = 15.968.$

24. $y = f(x) = 1/x \;\Rightarrow\; dy = -1/x^2\,dx$. When $x = 4$ and $dx = 0.002$, $dy = -\frac{1}{16}(0.002) = -\frac{1}{8000}$, so
$\frac{1}{4.002} \approx f(4) + dy = \frac{1}{4} - \frac{1}{8000} = \frac{1999}{8000} = 0.249875$.

25. $y = f(x) = \sqrt[3]{x} \;\Rightarrow\; dy = \frac{1}{3}x^{-2/3}\,dx$. When $x = 1000$ and $dx = 1$, $dy = \frac{1}{3}(1000)^{-2/3}(1) = \frac{1}{300}$, so
$\sqrt[3]{1001} = f(1001) \approx f(1000) + dy = 10 + \frac{1}{300} = 10.00\overline{3} \approx 10.003$.

26. $y = f(x) = \sqrt{x} \;\Rightarrow\; dy = \frac{1}{2}x^{-1/2}\,dx$. When $x = 100$ and $dx = 0.5$, $dy = \frac{1}{2}(100)^{-1/2}\left(\frac{1}{2}\right) = \frac{1}{40}$, so
$\sqrt{100.5} = f(100.5) \approx f(100) + dy = 10 + \frac{1}{40} = 10.025$.

27. $y = f(x) = e^x \;\Rightarrow\; dy = e^x\,dx$. When $x = 0$ and $dx = 0.1$, $dy = e^0(0.1) = 0.1$, so
$e^{0.1} = f(0.1) \approx f(0) + dy = 1 + 0.1 = 1.1$.

28. $y = f(x) = \cos x \;\Rightarrow\; dy = -\sin x\,dx$. When $x = 30^\circ$ $[\pi/6]$ and $dx = -1^\circ$ $[-\pi/180]$,
$dy = \left(-\sin\frac{\pi}{6}\right)\left(-\frac{\pi}{180}\right) = -\frac{1}{2}\left(-\frac{\pi}{180}\right) = \frac{\pi}{360}$, so $\cos 29^\circ = f(29^\circ) \approx f(30^\circ) + dy = \frac{1}{2}\sqrt{3} + \frac{\pi}{360} \approx 0.875$.

29. $y = f(x) = \sec x \;\Rightarrow\; f'(x) = \sec x\,\tan x$, so $f(0) = 1$ and $f'(0) = 1 \cdot 0 = 0$. The linear approximation of f at 0 is $f(0) + f'(0)(x - 0) = 1 + 0(x) = 1$. Since 0.08 is close to 0, approximating $\sec 0.08$ with 1 is reasonable.

30. $y = f(x) = \sqrt{x} \;\Rightarrow\; f'(x) = 1/(2\sqrt{x}\,)$, so $f(4) = 2$ and $f'(4) = \frac{1}{4}$. The linear approximation of f at 4 is $f(4) + f'(4)(x - 4) = 2 + \frac{1}{4}(x - 4)$. Now $f(4.02) = \sqrt{4.02} \approx 2 + \frac{1}{4}(0.02) = 2 + 0.005 = 2.005$, so the approximation is reasonable.

31. $y = f(x) = 1/x \;\Rightarrow\; f'(x) = -1/x^2$, so $f(10) = 0.1$ and $f'(10) = -0.01$. The linear approximation of f at 10 is $f(10) + f'(10)(x - 10) = 0.1 - 0.01(x - 10)$. Now $f(9.98) = 1/9.98 \approx 0.1 - 0.01(-0.02) = 0.1 + 0.0002 = 0.1002$, so the approximation is reasonable.

32. (a) $f(x) = (x - 1)^2 \;\Rightarrow\; f'(x) = 2(x - 1)$, so $f(0) = 1$ and $f'(0) = -2$.

Thus, $f(x) \approx L_f(x) = f(0) + f'(0)(x - 0) = 1 - 2x$.

$g(x) = e^{-2x} \;\Rightarrow\; g'(x) = -2e^{-2x}$, so $g(0) = 1$ and $g'(0) = -2$.

Thus, $g(x) \approx L_g(x) = g(0) + g'(0)(x - 0) = 1 - 2x$.

$h(x) = 1 + \ln(1 - 2x) \;\Rightarrow\; h'(x) = \dfrac{-2}{1 - 2x}$, so $h(0) = 1$ and $h'(0) = -2$.

Thus, $h(x) \approx L_h(x) = h(0) + h'(0)(x - 0) = 1 - 2x$.

Notice that $L_f = L_g = L_h$. This happens because f, g, and h have the same function values and the same derivative values at $a = 0$.

(b) The linear approximation appears to be the best for the function f since it is closer to f for a larger domain than it is to g and h. The approximation looks worst for h since h moves away from L faster than f and g do.

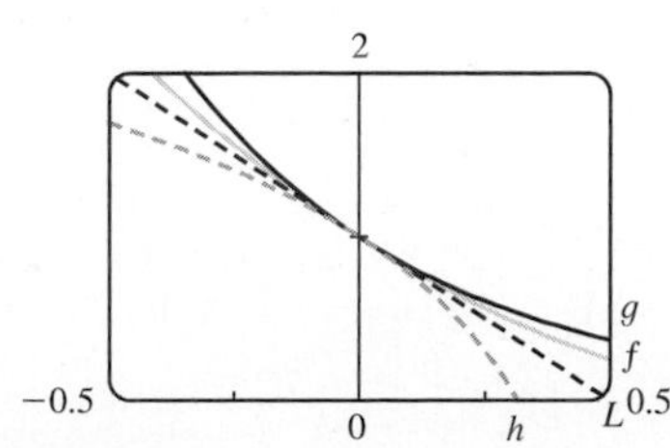

33. (a) If x is the edge length, then $V = x^3 \quad \Rightarrow \quad dV = 3x^2\,dx$. When $x = 30$ and $dx = 0.1$, $dV = 3(30)^2(0.1) = 270$, so the maximum possible error in computing the volume of the cube is about 270 cm^3. The relative error is calculated by dividing the change in V, ΔV, by V. We approximate ΔV with dV.

Relative error $= \dfrac{\Delta V}{V} \approx \dfrac{dV}{V} = \dfrac{3x^2\,dx}{x^3} = 3\,\dfrac{dx}{x} = 3\left(\dfrac{0.1}{30}\right) = 0.01$.

Percentage error $=$ relative error $\times\, 100\% = 0.01 \times 100\% = 1\%$.

(b) $S = 6x^2 \quad \Rightarrow \quad dS = 12x\,dx$. When $x = 30$ and $dx = 0.1$, $dS = 12(30)(0.1) = 36$, so the maximum possible error in computing the surface area of the cube is about 36 cm^2.

Relative error $= \dfrac{\Delta S}{S} \approx \dfrac{dS}{S} = \dfrac{12x\,dx}{6x^2} = 2\,\dfrac{dx}{x} = 2\left(\dfrac{0.1}{30}\right) = 0.00\overline{6}$.

Percentage error $=$ relative error $\times\, 100\% = 0.00\overline{6} \times 100\% = 0.\overline{6}\%$.

34. (a) $A = \pi r^2 \quad \Rightarrow \quad dA = 2\pi r\,dr$. When $r = 24$ and $dr = 0.2$, $dA = 2\pi(24)(0.2) = 9.6\pi$, so the maximum possible error in the calculated area of the disk is about $9.6\pi \approx 30\text{ cm}^2$.

(b) Relative error $= \dfrac{\Delta A}{A} \approx \dfrac{dA}{A} = \dfrac{2\pi r\,dr}{\pi r^2} = \dfrac{2\,dr}{r} = \dfrac{2(0.2)}{24} = \dfrac{0.2}{12} = \dfrac{1}{60} = 0.01\overline{6}$.

Percentage error $=$ relative error $\times 100\% = 0.01\overline{6} \times 100\% = 1.\overline{6}\%$.

35. (a) For a sphere of radius r, the circumference is $C = 2\pi r$ and the surface area is $S = 4\pi r^2$, so

$r = \dfrac{C}{2\pi} \quad \Rightarrow \quad S = 4\pi\left(\dfrac{C}{2\pi}\right)^2 = \dfrac{C^2}{\pi} \quad \Rightarrow \quad dS = \dfrac{2}{\pi}C\,dC$. When $C = 84$ and $dC = 0.5$, $dS = \dfrac{2}{\pi}(84)(0.5) = \dfrac{84}{\pi}$,

so the maximum error is about $\dfrac{84}{\pi} \approx 27\text{ cm}^2$. Relative error $\approx \dfrac{dS}{S} = \dfrac{84/\pi}{84^2/\pi} = \dfrac{1}{84} \approx 0.012 = 1.2\%$

(b) $V = \dfrac{4}{3}\pi r^3 = \dfrac{4}{3}\pi\left(\dfrac{C}{2\pi}\right)^3 = \dfrac{C^3}{6\pi^2} \quad \Rightarrow \quad dV = \dfrac{1}{2\pi^2}C^2\,dC$. When $C = 84$ and $dC = 0.5$,

$dV = \dfrac{1}{2\pi^2}(84)^2(0.5) = \dfrac{1764}{\pi^2}$, so the maximum error is about $\dfrac{1764}{\pi^2} \approx 179\text{ cm}^3$.

The relative error is approximately $\dfrac{dV}{V} = \dfrac{1764/\pi^2}{(84)^3/(6\pi^2)} = \dfrac{1}{56} \approx 0.018 = 1.8\%$.

36. For a hemispherical dome, $V = \frac{2}{3}\pi r^3 \quad \Rightarrow \quad dV = 2\pi r^2\,dr$. When $r = \frac{1}{2}(50) = 25$ m and $dr = 0.05$ cm $= 0.0005$ m, $dV = 2\pi(25)^2(0.0005) = \frac{5\pi}{8}$, so the amount of paint needed is about $\frac{5\pi}{8} \approx 2\text{ m}^3$.

37. (a) $V = \pi r^2 h \quad \Rightarrow \quad \Delta V \approx dV = 2\pi rh\,dr = 2\pi rh\,\Delta r$

(b) The error is

$$\Delta V - dV = [\pi(r + \Delta r)^2 h - \pi r^2 h] - 2\pi rh\,\Delta r = \pi r^2 h + 2\pi rh\,\Delta r + \pi(\Delta r)^2 h - \pi r^2 h - 2\pi rh\,\Delta r = \pi(\Delta r)^2 h.$$

38. (a) $\sin\theta = \dfrac{20}{x} \;\Rightarrow\; x = 20\csc\theta \;\Rightarrow$

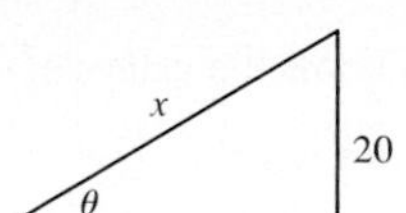

$$dx = 20(-\csc\theta\,\cot\theta)\,d\theta = -20\csc 30^\circ\,\cot 30^\circ\,(\pm 1^\circ)$$

$$= -20(2)\left(\sqrt{3}\right)\left(\pm\frac{\pi}{180}\right) = \pm\frac{2\sqrt{3}}{9}\pi$$

So the maximum error is about $\pm\frac{2}{9}\sqrt{3}\,\pi \approx \pm 1.21$ cm.

(b) The relative error is $\dfrac{\Delta x}{x} \approx \dfrac{dx}{x} = \dfrac{\pm\frac{2}{9}\sqrt{3}\,\pi}{20(2)} = \pm\dfrac{\sqrt{3}}{180}\pi \approx \pm 0.03$, so the percentage error is approximately $\pm 3\%$.

39. $V = RI \;\Rightarrow\; I = \dfrac{V}{R} \;\Rightarrow\; dI = -\dfrac{V}{R^2}\,dR$. The relative error in calculating I is $\dfrac{\Delta I}{I} \approx \dfrac{dI}{I} = \dfrac{-(V/R^2)\,dR}{V/R} = -\dfrac{dR}{R}$.

Hence, the relative error in calculating I is approximately the same (in magnitude) as the relative error in R.

40. $F = kR^4 \;\Rightarrow\; dF = 4kR^3\,dR \;\Rightarrow\; \dfrac{dF}{F} = \dfrac{4kR^3\,dR}{kR^4} = 4\left(\dfrac{dR}{R}\right)$. Thus, the relative change in F is about 4 times the relative change in R. So a 5% increase in the radius corresponds to a 20% increase in blood flow.

41. (a) $dc = \dfrac{dc}{dx}\,dx = 0\,dx = 0$

(b) $d(cu) = \dfrac{d}{dx}(cu)\,dx = c\,\dfrac{du}{dx}\,dx = c\,du$

(c) $d(u+v) = \dfrac{d}{dx}(u+v)\,dx = \left(\dfrac{du}{dx} + \dfrac{dv}{dx}\right)dx = \dfrac{du}{dx}\,dx + \dfrac{dv}{dx}\,dx = du + dv$

(d) $d(uv) = \dfrac{d}{dx}(uv)\,dx = \left(u\,\dfrac{dv}{dx} + v\,\dfrac{du}{dx}\right)dx = u\,\dfrac{dv}{dx}\,dx + v\,\dfrac{du}{dx}\,dx = u\,dv + v\,du$

(e) $d\left(\dfrac{u}{v}\right) = \dfrac{d}{dx}\left(\dfrac{u}{v}\right)dx = \dfrac{v\,\dfrac{du}{dx} - u\,\dfrac{dv}{dx}}{v^2}\,dx = \dfrac{v\,\dfrac{du}{dx}\,dx - u\,\dfrac{dv}{dx}\,dx}{v^2} = \dfrac{v\,du - u\,dv}{v^2}$

(f) $d\left(x^n\right) = \dfrac{d}{dx}\left(x^n\right)dx = nx^{n-1}\,dx$

42. (a) $f(x) = \sin x \;\Rightarrow\; f'(x) = \cos x$, so $f(0) = 0$ and $f'(0) = 1$. Thus, $f(x) \approx f(0) + f'(0)(x-0) = 0 + 1(x-0) = x$.

(b)

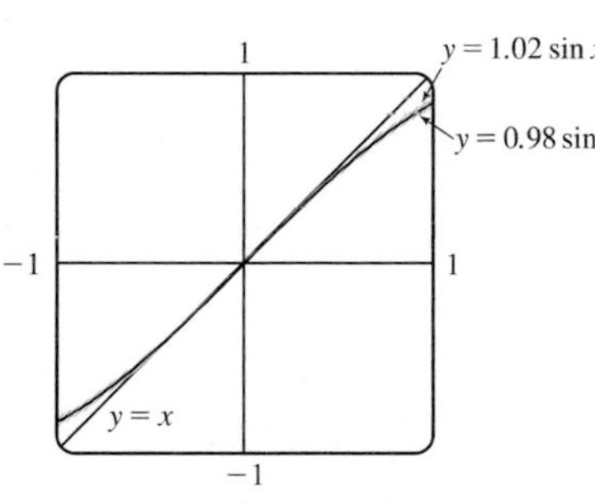

0.36
y = x
y = 1.02 sin x
y = 0.98 sin x
0.33
0.36

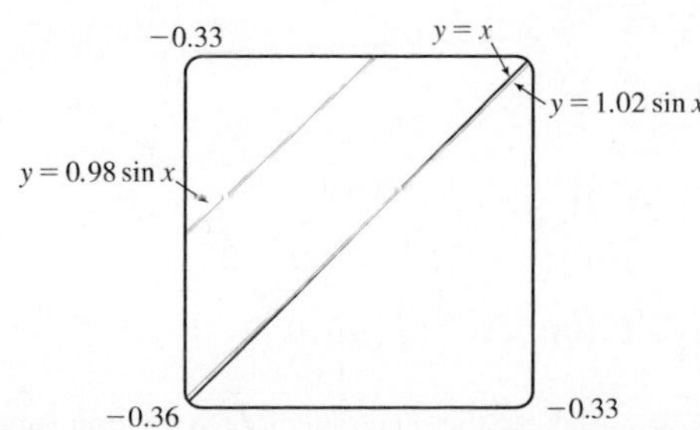

[continued]

We want to know the values of x for which $y = x$ approximates $y = \sin x$ with less than a 2% difference; that is, the values of x for which

$$\left|\frac{x - \sin x}{\sin x}\right| < 0.02 \quad \Leftrightarrow \quad -0.02 < \frac{x - \sin x}{\sin x} < 0.02 \quad \Leftrightarrow$$

$$\begin{cases} -0.02 \sin x < x - \sin x < 0.02 \sin x & \text{if } \sin x > 0 \\ -0.02 \sin x > x - \sin x > 0.02 \sin x & \text{if } \sin x < 0 \end{cases} \quad \Leftrightarrow \quad \begin{cases} 0.98 \sin x < x < 1.02 \sin x & \text{if } \sin x > 0 \\ 1.02 \sin x < x < 0.98 \sin x & \text{if } \sin x < 0 \end{cases}$$

In the first figure, we see that the graphs are very close to each other near $x = 0$. Changing the viewing rectangle and using an intersect feature (see the second figure) we find that $y = x$ intersects $y = 1.02 \sin x$ at $x \approx 0.344$. By symmetry, they also intersect at $x \approx -0.344$ (see the third figure). Converting 0.344 radians to degrees, we get $0.344\left(\frac{180^\circ}{\pi}\right) \approx 19.7^\circ \approx 20^\circ$, which verifies the statement.

43. (a) The graph shows that $f'(1) = 2$, so $L(x) = f(1) + f'(1)(x - 1) = 5 + 2(x - 1) = 2x + 3$. $f(0.9) \approx L(0.9) = 4.8$ and $f(1.1) \approx L(1.1) = 5.2$.

(b) From the graph, we see that $f'(x)$ is positive and decreasing. This means that the slopes of the tangent lines are positive, but the tangents are becoming less steep. So the tangent lines lie *above* the curve. Thus, the estimates in part (a) are too large.

44. (a) $g'(x) = \sqrt{x^2 + 5} \;\Rightarrow\; g'(2) = \sqrt{9} = 3$. $g(1.95) \approx g(2) + g'(2)(1.95 - 2) = -4 + 3(-0.05) = -4.15$. $g(2.05) \approx g(2) + g'(2)(2.05 - 2) = -4 + 3(0.05) = -3.85$.

(b) The formula $g'(x) = \sqrt{x^2 + 5}$ shows that $g'(x)$ is positive and increasing. This means that the slopes of the tangent lines are positive and the tangents are getting steeper. So the tangent lines lie *below* the graph of g. Hence, the estimates in part (a) are too small.

LABORATORY PROJECT Taylor Polynomials

1. We first write the functions described in conditions (i), (ii), and (iii):

$$\begin{aligned} P(x) &= A + Bx + Cx^2 & \qquad f(x) &= \cos x \\ P'(x) &= B + 2Cx & f'(x) &= -\sin x \\ P''(x) &= 2C & f''(x) &= -\cos x \end{aligned}$$

So, taking $a = 0$, our three conditions become

$$\begin{aligned} P(0) &= f(0): & A &= \cos 0 = 1 \\ P'(0) &= f'(0): & B &= -\sin 0 = 0 \\ P''(0) &= f''(0): & 2C &= -\cos 0 = -1 \;\Rightarrow\; C = -\tfrac{1}{2} \end{aligned}$$

The desired quadratic function is $P(x) = 1 - \frac{1}{2}x^2$, so the quadratic approximation is $\cos x \approx 1 - \frac{1}{2}x^2$.

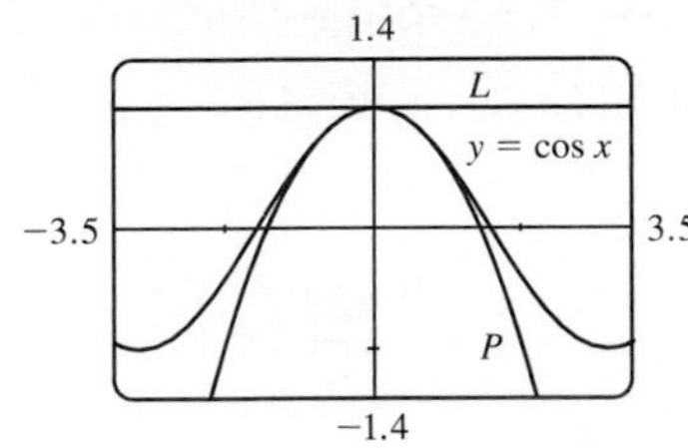

The figure shows a graph of the cosine function together with its linear approximation $L(x) = 1$ and quadratic approximation $P(x) = 1 - \frac{1}{2}x^2$ near 0. You can see that the quadratic approximation is much better than the linear one.

2. Accuracy to within 0.1 means that $\left|\cos x - \left(1 - \frac{1}{2}x^2\right)\right| < 0.1 \quad \Leftrightarrow \quad -0.1 < \cos x - \left(1 - \frac{1}{2}x^2\right) < 0.1 \quad \Leftrightarrow$

$0.1 > \left(1 - \frac{1}{2}x^2\right) - \cos x > -0.1 \quad \Leftrightarrow \quad \cos x + 0.1 > 1 - \frac{1}{2}x^2 > \cos x - 0.1 \quad \Leftrightarrow \quad \cos x - 0.1 < 1 - \frac{1}{2}x^2 < \cos x + 0.1.$

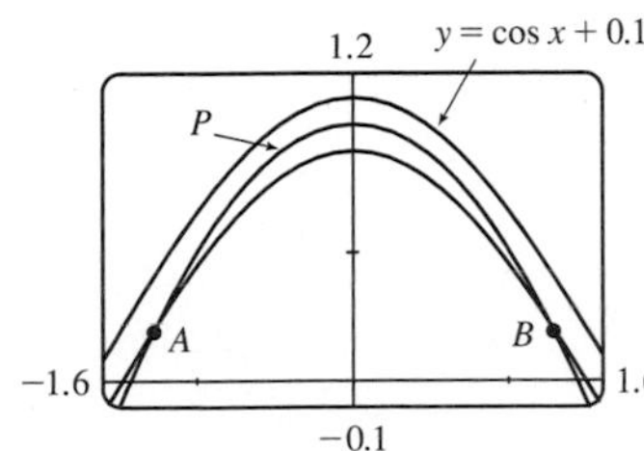

From the figure we see that this is true between A and B. Zooming in or using an intersect feature, we find that the x-coordinates of B and A are about ± 1.26. Thus, the approximation $\cos x \approx 1 - \frac{1}{2}x^2$ is accurate to within 0.1 when $-1.26 < x < 1.26$.

3. If $P(x) = A + B(x - a) + C(x - a)^2$, then $P'(x) = B + 2C(x - a)$ and $P''(x) = 2C$. Applying the conditions (i), (ii), and (iii), we get

$$\begin{aligned} P(a) = f(a){:} \quad & A = f(a) \\ P'(a) = f'(a){:} \quad & B = f'(a) \\ P''(a) = f''(a){:} \quad & 2C = f''(a) \quad \Rightarrow \quad C = \tfrac{1}{2}f''(a) \end{aligned}$$

Thus, $P(x) = A + B(x - a) + C(x - a)^2$ can be written in the form $P(x) = f(a) + f'(a)(x - a) + \frac{1}{2}f''(a)(x - a)^2$.

4. From Example 3.10.1, we have $f(1) = 2$, $f'(1) = \frac{1}{4}$, and $f'(x) = \frac{1}{2}(x + 3)^{-1/2}$.

So $f''(x) = -\frac{1}{4}(x + 3)^{-3/2} \quad \Rightarrow \quad f''(1) = -\frac{1}{32}$.

From Problem 3, the quadratic approximation $P(x)$ is

$\sqrt{x + 3} \approx f(1) + f'(1)(x - 1) + \frac{1}{2}f''(1)(x - 1)^2 = 2 + \frac{1}{4}(x - 1) - \frac{1}{64}(x - 1)^2$.

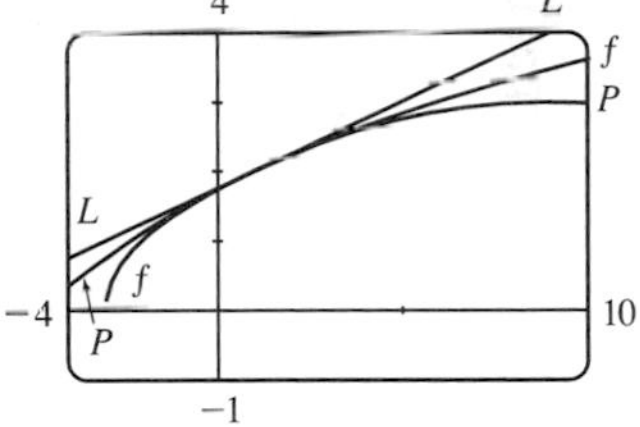

The figure shows the function $f(x) = \sqrt{x + 3}$ together with its linear approximation $L(x) = \frac{1}{4}x + \frac{7}{4}$ and its quadratic approximation $P(x)$. You can see that $P(x)$ is a better approximation than $L(x)$ and this is borne out by the numerical values in the following chart.

	from $L(x)$	actual value	from $P(x)$
$\sqrt{3.98}$	1.9950	1.99499373...	1.99499375
$\sqrt{4.05}$	2.0125	2.01246118...	2.01246094
$\sqrt{4.2}$	2.0500	2.04939015...	2.04937500

5. $T_n(x) = c_0 + c_1(x - a) + c_2(x - a)^2 + c_3(x - a)^3 + \cdots + c_n(x - a)^n$. If we put $x = a$ in this equation, then all terms after the first are 0 and we get $T_n(a) = c_0$. Now we differentiate $T_n(x)$ and obtain $T_n'(x) = c_1 + 2c_2(x - a) + 3c_3(x - a)^2 + 4c_4(x - a)^3 + \cdots + nc_n(x - a)^{n-1}$. Substituting $x = a$ gives $T_n'(a) = c_1$. Differentiating again, we have $T_n''(x) = 2c_2 + 2 \cdot 3c_3(x - a) + 3 \cdot 4c_4(x - a^2) + \cdots + (n - 1)nc_n(x - a)^{n-2}$ and so

$T_n''(a) = 2c_2$. Continuing in this manner, we get $T_n'''(x) = 2 \cdot 3c_3 + 2 \cdot 3 \cdot 4c_4(x-a) + \cdots + (n-2)(n-1)nc_n(x-a)^{n-3}$ and $T_n'''(a) = 2 \cdot 3c_3$. By now we see the pattern. If we continue to differentiate and substitute $x = a$, we obtain $T_n^{(4)}(a) = 2 \cdot 3 \cdot 4c_4$ and in general, for any integer k between 1 and n, $T_n^{(k)}(a) = 2 \cdot 3 \cdot 4 \cdot 5 \cdot \cdots \cdot kc_k = k!\, c_k \Rightarrow$ $c_k = \dfrac{T_n^{(k)}(a)}{k!}$. Because we want T_n and f to have the same derivatives at a, we require that $c_k = \dfrac{f^{(k)}(a)}{k!}$ for $k = 1, 2, \ldots, n$.

6. $T_n(x) = f(a) + f'(a)(x-a) + \dfrac{f''(a)}{2!}(x-a)^2 + \cdots + \dfrac{f^{(n)}(a)}{n!}(x-a)^n$. To compute the coefficients in this equation we need to calculate the derivatives of f at 0:

$$\begin{aligned} f(x) &= \cos x & f(0) &= \cos 0 = 1 \\ f'(x) &= -\sin x & f'(0) &= -\sin 0 = 0 \\ f''(x) &= -\cos x & f''(0) &= -1 \\ f'''(x) &= \sin x & f'''(0) &= 0 \\ f^{(4)}(x) &= \cos x & f^{(4)}(0) &= 1 \end{aligned}$$

We see that the derivatives repeat in a cycle of length 4, so $f^{(5)}(0) = 0$, $f^{(6)}(0) = -1$, $f^{(7)}(0) = 0$, and $f^{(8)}(0) = 1$.

From the original expression for $T_n(x)$, with $n = 8$ and $a = 0$, we have

$$\begin{aligned} T_8(x) &= f(0) + f'(0)(x-0) + \frac{f''(0)}{2!}(x-0)^2 + \frac{f'''(0)}{3!}(x-0)^3 + \cdots + \frac{f^{(8)}(0)}{8!}(x-0)^8 \\ &= 1 + 0 \cdot x + \frac{-1}{2!}x^2 + 0 \cdot x^3 + \frac{1}{4!}x^4 + 0 \cdot x^5 + \frac{-1}{6!}x^6 + 0 \cdot x^7 + \frac{1}{8!}x^8 = 1 - \frac{x^2}{2!} + \frac{x^4}{4!} - \frac{x^6}{6!} + \frac{x^8}{8!} \end{aligned}$$

and the desired approximation is $\cos x \approx 1 - \dfrac{x^2}{2!} + \dfrac{x^4}{4!} - \dfrac{x^6}{6!} + \dfrac{x^8}{8!}$. The Taylor polynomials T_2, T_4, and T_6 consist of the initial terms of T_8 up through degree 2, 4, and 6, respectively. Therefore, $T_2(x) = 1 - \dfrac{x^2}{2!}$, $T_4(x) = 1 - \dfrac{x^2}{2!} + \dfrac{x^4}{4!}$, and $T_6(x) = 1 - \dfrac{x^2}{2!} + \dfrac{x^4}{4!} - \dfrac{x^6}{6!}$. We graph T_2, T_4, T_6, T_8, and f:

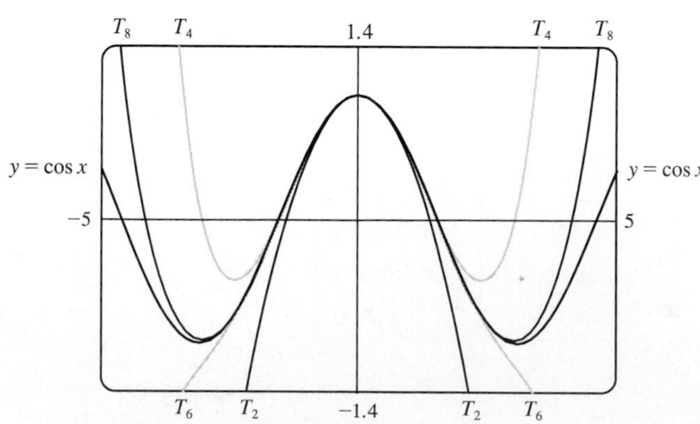

Notice that $T_2(x)$ is a good approximation to $\cos x$ near 0, $T_4(x)$ is a good approximation on a larger interval, $T_6(x)$ is a better approximation, and $T_8(x)$ is better still. Each successive Taylor polynomial is a good approximation on a larger interval than the previous one.

3.11 Hyperbolic Functions

1. (a) $\sinh 0 = \frac{1}{2}(e^0 - e^{-0}) = 0$ (b) $\cosh 0 = \frac{1}{2}(e^0 + e^{-0}) = \frac{1}{2}(1+1) = 1$

2. (a) $\tanh 0 = \dfrac{(e^0 - e^{-0})/2}{(e^0 + e^{-0})/2} = 0$ (b) $\tanh 1 = \dfrac{e^1 - e^{-1}}{e^1 + e^{-1}} = \dfrac{e^2 - 1}{e^2 + 1} \approx 0.76159$

3. (a) $\cosh(\ln 5) = \frac{1}{2}(e^{\ln 5} + e^{-\ln 5}) = \frac{1}{2}\left(5 + (e^{\ln 5})^{-1}\right) = \frac{1}{2}(5 + 5^{-1}) = \frac{1}{2}\left(5 + \frac{1}{5}\right) = \frac{13}{5}$

(b) $\cosh 5 = \frac{1}{2}(e^5 + e^{-5}) \approx 74.20995$

4. (a) $\sinh 4 = \frac{1}{2}(e^4 - e^{-4}) \approx 27.28992$

(b) $\sinh(\ln 4) = \frac{1}{2}(e^{\ln 4} - e^{-\ln 4}) = \frac{1}{2}\left(4 - (e^{\ln 4})^{-1}\right) = \frac{1}{2}(4 - 4^{-1}) = \frac{1}{2}\left(4 - \frac{1}{4}\right) = \frac{15}{8}$

5. (a) $\operatorname{sech} 0 = \dfrac{1}{\cosh 0} = \dfrac{1}{1} = 1$ (b) $\cosh^{-1} 1 = 0$ because $\cosh 0 = 1$.

6. (a) $\sinh 1 = \frac{1}{2}(e^1 - e^{-1}) \approx 1.17520$

(b) Using Equation 3, we have $\sinh^{-1} 1 = \ln\left(1 + \sqrt{1^2 + 1}\right) = \ln\left(1 + \sqrt{2}\right) \approx 0.88137$.

7. $\sinh(-x) = \frac{1}{2}[e^{-x} - e^{-(-x)}] = \frac{1}{2}(e^{-x} - e^x) = -\frac{1}{2}(e^{-x} - e^x) = -\sinh x$

8. $\cosh(-x) = \frac{1}{2}[e^{-x} + e^{-(-x)}] = \frac{1}{2}(e^{-x} + e^x) = \frac{1}{2}(e^x + e^{-x}) = \cosh x$

9. $\cosh x + \sinh x = \frac{1}{2}(e^x + e^{-x}) + \frac{1}{2}(e^x - e^{-x}) = \frac{1}{2}(2e^x) = e^x$

10. $\cosh x - \sinh x = \frac{1}{2}(e^x + e^{-x}) - \frac{1}{2}(e^x - e^{-x}) = \frac{1}{2}(2e^{-x}) = e^{-x}$

11.
$$\begin{aligned}\sinh x \cosh y + \cosh x \sinh y &= \left[\tfrac{1}{2}(e^x - e^{-x})\right]\left[\tfrac{1}{2}(e^y + e^{-y})\right] + \left[\tfrac{1}{2}(e^x + e^{-x})\right]\left[\tfrac{1}{2}(e^y - e^{-y})\right] \\ &= \tfrac{1}{4}[(e^{x+y} + e^{x-y} - e^{-x+y} - e^{-x-y}) + (e^{x+y} - e^{x-y} + e^{-x+y} - e^{-x-y})] \\ &= \tfrac{1}{4}(2e^{x+y} - 2e^{-x-y}) = \tfrac{1}{2}[e^{x+y} - e^{-(x+y)}] = \sinh(x+y)\end{aligned}$$

12.
$$\begin{aligned}\cosh x \cosh y + \sinh x \sinh y &= \left[\tfrac{1}{2}(e^x + e^{-x})\right]\left[\tfrac{1}{2}(e^y + e^{-y})\right] + \left[\tfrac{1}{2}(e^x - e^{-x})\right]\left[\tfrac{1}{2}(e^y - e^{-y})\right] \\ &= \tfrac{1}{4}\left[(e^{x+y} + e^{x-y} + e^{-x+y} + e^{-x-y}) + (e^{x+y} - e^{x-y} - e^{-x+y} + e^{-x-y})\right] \\ &= \tfrac{1}{4}(2e^{x+y} + 2e^{-x-y}) = \tfrac{1}{2}\left[e^{x+y} + e^{-(x+y)}\right] = \cosh(x+y)\end{aligned}$$

13. Divide both sides of the identity $\cosh^2 x - \sinh^2 x = 1$ by $\sinh^2 x$:

$$\frac{\cosh^2 x}{\sinh^2 x} - \frac{\sinh^2 x}{\sinh^2 x} = \frac{1}{\sinh^2 x} \quad\Leftrightarrow\quad \coth^2 x - 1 = \operatorname{csch}^2 x.$$

14.
$$\tanh(x+y) = \frac{\sinh(x+y)}{\cosh(x+y)} = \frac{\sinh x \cosh y + \cosh x \sinh y}{\cosh x \cosh y + \sinh x \sinh y} = \frac{\dfrac{\sinh x \cosh y}{\cosh x \cosh y} + \dfrac{\cosh x \sinh y}{\cosh x \cosh y}}{\dfrac{\cosh x \cosh y}{\cosh x \cosh y} + \dfrac{\sinh x \sinh y}{\cosh x \cosh y}}$$
$$= \frac{\tanh x + \tanh y}{1 + \tanh x \tanh y}$$

15. Putting $y = x$ in the result from Exercise 11, we have

$\sinh 2x = \sinh(x + x) = \sinh x \cosh x + \cosh x \sinh x = 2 \sinh x \cosh x$.

16. Putting $y = x$ in the result from Exercise 12, we have

$$\cosh 2x = \cosh(x + x) = \cosh x \cosh x + \sinh x \sinh x = \cosh^2 x + \sinh^2 x.$$

17. $\tanh(\ln x) = \dfrac{\sinh(\ln x)}{\cosh(\ln x)} = \dfrac{(e^{\ln x} - e^{-\ln x})/2}{(e^{\ln x} + e^{-\ln x})/2} = \dfrac{x - (e^{\ln x})^{-1}}{x + (e^{\ln x})^{-1}} = \dfrac{x - x^{-1}}{x + x^{-1}} = \dfrac{x - 1/x}{x + 1/x} = \dfrac{(x^2 - 1)/x}{(x^2 + 1)/x} = \dfrac{x^2 - 1}{x^2 + 1}$

18. $\dfrac{1 + \tanh x}{1 - \tanh x} = \dfrac{1 + (\sinh x)/\cosh x}{1 - (\sinh x)/\cosh x} = \dfrac{\cosh x + \sinh x}{\cosh x - \sinh x} = \dfrac{\frac{1}{2}(e^x + e^{-x}) + \frac{1}{2}(e^x - e^{-x})}{\frac{1}{2}(e^x + e^{-x}) - \frac{1}{2}(e^x - e^{-x})} = \dfrac{e^x}{e^{-x}} = e^{2x}$

Or: Using the results of Exercises 9 and 10, $\dfrac{\cosh x + \sinh x}{\cosh x - \sinh x} = \dfrac{e^x}{e^{-x}} = e^{2x}$

19. By Exercise 9, $(\cosh x + \sinh x)^n = (e^x)^n = e^{nx} = \cosh nx + \sinh nx$.

20. $\coth x = \dfrac{1}{\tanh x} \quad\Rightarrow\quad \coth x = \dfrac{1}{\tanh x} = \dfrac{1}{12/13} = \dfrac{13}{12}$.

$\operatorname{sech}^2 x = 1 - \tanh^2 x = 1 - \left(\frac{12}{13}\right)^2 = \frac{25}{169} \quad\Rightarrow\quad \operatorname{sech} x = \frac{5}{13}$ [sech, like cosh, is positive].

$\cosh x = \dfrac{1}{\operatorname{sech} x} \quad\Rightarrow\quad \cosh x = \dfrac{1}{5/13} = \dfrac{13}{5}$.

$\tanh x = \dfrac{\sinh x}{\cosh x} \quad\Rightarrow\quad \sinh x = \tanh x \cosh x \quad\Rightarrow\quad \sinh x = \dfrac{12}{13} \cdot \dfrac{13}{5} = \dfrac{12}{5}$.

$\operatorname{csch} x = \dfrac{1}{\sinh x} \quad\Rightarrow\quad \operatorname{csch} x = \dfrac{1}{12/5} = \dfrac{5}{12}$.

21. $\operatorname{sech} x = \dfrac{1}{\cosh x} \quad\Rightarrow\quad \operatorname{sech} x = \dfrac{1}{5/3} = \dfrac{3}{5}$.

$\cosh^2 x - \sinh^2 x = 1 \quad\Rightarrow\quad \sinh^2 x = \cosh^2 x - 1 = \left(\frac{5}{3}\right)^2 - 1 = \frac{16}{9} \quad\Rightarrow\quad \sinh x = \frac{4}{3}$ [because $x > 0$].

$\operatorname{csch} x = \dfrac{1}{\sinh x} \quad\Rightarrow\quad \operatorname{csch} x = \dfrac{1}{4/3} = \dfrac{3}{4}$.

$\tanh x = \dfrac{\sinh x}{\cosh x} \quad\Rightarrow\quad \tanh x = \dfrac{4/3}{5/3} = \dfrac{4}{5}$.

$\coth x = \dfrac{1}{\tanh x} \quad\Rightarrow\quad \coth x = \dfrac{1}{4/5} = \dfrac{5}{4}$.

22. (a)

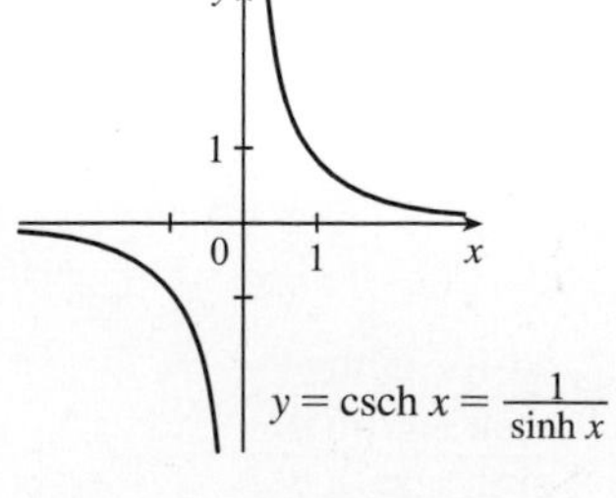

y
1
0
1
2
x
$y = \operatorname{sech} x = \dfrac{1}{\cosh x}$

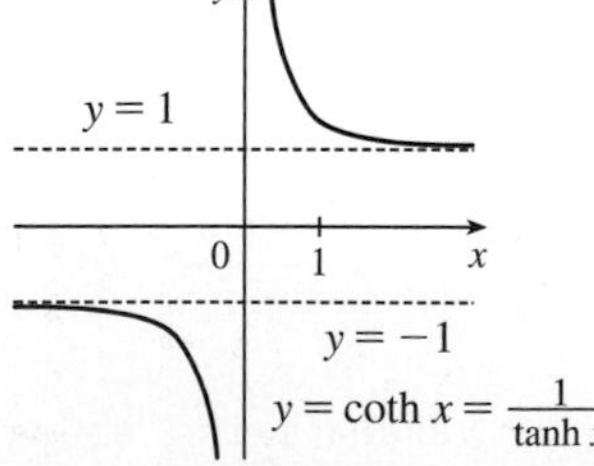

23. (a) $\displaystyle\lim_{x\to\infty} \tanh x = \lim_{x\to\infty} \frac{e^x - e^{-x}}{e^x + e^{-x}} \cdot \frac{e^{-x}}{e^{-x}} = \lim_{x\to\infty} \frac{1 - e^{-2x}}{1 + e^{-2x}} = \frac{1 - 0}{1 + 0} = 1$

(b) $\displaystyle\lim_{x\to-\infty} \tanh x = \lim_{x\to-\infty} \frac{e^x - e^{-x}}{e^x + e^{-x}} \cdot \frac{e^x}{e^x} = \lim_{x\to-\infty} \frac{e^{2x} - 1}{e^{2x} + 1} = \frac{0 - 1}{0 + 1} = -1$

(c) $\lim\limits_{x\to\infty} \sinh x = \lim\limits_{x\to\infty} \dfrac{e^x - e^{-x}}{2} = \infty$

(d) $\lim\limits_{x\to-\infty} \sinh x = \lim\limits_{x\to-\infty} \dfrac{e^x - e^{-x}}{2} = -\infty$

(e) $\lim\limits_{x\to\infty} \operatorname{sech} x = \lim\limits_{x\to\infty} \dfrac{2}{e^x + e^{-x}} = 0$

(f) $\lim\limits_{x\to\infty} \coth x = \lim\limits_{x\to\infty} \dfrac{e^x + e^{-x}}{e^x - e^{-x}} \cdot \dfrac{e^{-x}}{e^{-x}} = \lim\limits_{x\to\infty} \dfrac{1 + e^{-2x}}{1 - e^{-2x}} = \dfrac{1+0}{1-0} = 1$ [*Or:* Use part (a)]

(g) $\lim\limits_{x\to0^+} \coth x = \lim\limits_{x\to0^+} \dfrac{\cosh x}{\sinh x} = \infty$, since $\sinh x \to 0$ through positive values and $\cosh x \to 1$.

(h) $\lim\limits_{x\to0^-} \coth x = \lim\limits_{x\to0^-} \dfrac{\cosh x}{\sinh x} = -\infty$, since $\sinh x \to 0$ through negative values and $\cosh x \to 1$.

(i) $\lim\limits_{x\to-\infty} \operatorname{csch} x = \lim\limits_{x\to-\infty} \dfrac{2}{e^x - e^{-x}} = 0$

(j) $\lim\limits_{x\to\infty} \dfrac{\sinh x}{e^x} = \lim\limits_{x\to\infty} \dfrac{e^x - e^{-x}}{2e^x} = \lim\limits_{x\to\infty} \dfrac{1 - e^{-2x}}{2} = \dfrac{1-0}{2} = \dfrac{1}{2}$

24. (a) $\dfrac{d}{dx}(\cosh x) = \dfrac{d}{dx}\left[\frac{1}{2}(e^x + e^{-x})\right] = \frac{1}{2}(e^x - e^{-x}) = \sinh x$

(b) $\dfrac{d}{dx}(\tanh x) = \dfrac{d}{dx}\left(\dfrac{\sinh x}{\cosh x}\right) = \dfrac{\cosh x \cosh x - \sinh x \sinh x}{\cosh^2 x} = \dfrac{\cosh^2 x - \sinh^2 x}{\cosh^2 x} = \dfrac{1}{\cosh^2 x} = \operatorname{sech}^2 x$

(c) $\dfrac{d}{dx}(\operatorname{csch} x) = \dfrac{d}{dx}\left(\dfrac{1}{\sinh x}\right) = -\dfrac{\cosh x}{\sinh^2 x} = -\dfrac{1}{\sinh x} \cdot \dfrac{\cosh x}{\sinh x} = -\operatorname{csch} x \coth x$

(d) $\dfrac{d}{dx}(\operatorname{sech} x) = \dfrac{d}{dx}\left(\dfrac{1}{\cosh x}\right) = -\dfrac{\sinh x}{\cosh^2 x} = -\dfrac{1}{\cosh x} \cdot \dfrac{\sinh x}{\cosh x} = -\operatorname{sech} x \tanh x$

(e) $\dfrac{d}{dx}(\coth x) = \dfrac{d}{dx}\left(\dfrac{\cosh x}{\sinh x}\right) = \dfrac{\sinh x \sinh x - \cosh x \cosh x}{\sinh^2 x} = \dfrac{\sinh^2 x - \cosh^2 x}{\sinh^2 x} = -\dfrac{1}{\sinh^2 x} = -\operatorname{csch}^2 x$

25. Let $y = \sinh^{-1} x$. Then $\sinh y = x$ and, by Example 1(a), $\cosh^2 y - \sinh^2 y = 1 \;\Rightarrow\;$ [with $\cosh y > 0$]

$\cosh y = \sqrt{1 + \sinh^2 y} = \sqrt{1 + x^2}$. So by Exercise 9, $e^y = \sinh y + \cosh y = x + \sqrt{1 + x^2} \;\Rightarrow\; y = \ln\left(x + \sqrt{1 + x^2}\right)$.

26. Let $y = \cosh^{-1} x$. Then $\cosh y = x$ and $y \geq 0$, so $\sinh y = \sqrt{\cosh^2 y - 1} = \sqrt{x^2 - 1}$. So, by Exercise 9,

$e^y = \cosh y + \sinh y = x + \sqrt{x^2 - 1} \;\Rightarrow\; y = \ln\left(x + \sqrt{x^2 - 1}\right)$.

Another method: Write $x = \cosh y = \frac{1}{2}\left(e^y + e^{-y}\right)$ and solve a quadratic, as in Example 3.

27. (a) Let $y = \tanh^{-1} x$. Then $x = \tanh y = \dfrac{\sinh y}{\cosh y} = \dfrac{(e^y - e^{-y})/2}{(e^y + e^{-y})/2} \cdot \dfrac{e^y}{e^y} = \dfrac{e^{2y} - 1}{e^{2y} + 1} \;\Rightarrow\; xe^{2y} + x = e^{2y} - 1 \;\Rightarrow$

$1 + x = e^{2y} - xe^{2y} \;\Rightarrow\; 1 + x = e^{2y}(1 - x) \;\Rightarrow\; e^{2y} = \dfrac{1+x}{1-x} \;\Rightarrow\; 2y = \ln\left(\dfrac{1+x}{1-x}\right) \;\Rightarrow\; y = \frac{1}{2}\ln\left(\dfrac{1+x}{1-x}\right)$.

(b) Let $y = \tanh^{-1} x$. Then $x = \tanh y$, so from Exercise 18 we have

$$e^{2y} = \frac{1 + \tanh y}{1 - \tanh y} = \frac{1+x}{1-x} \quad \Rightarrow \quad 2y = \ln\left(\frac{1+x}{1-x}\right) \quad \Rightarrow \quad y = \tfrac{1}{2}\ln\left(\frac{1+x}{1-x}\right).$$

28. (a) (i) $y = \operatorname{csch}^{-1} x \quad \Leftrightarrow \quad \operatorname{csch} y = x \quad (x \neq 0)$

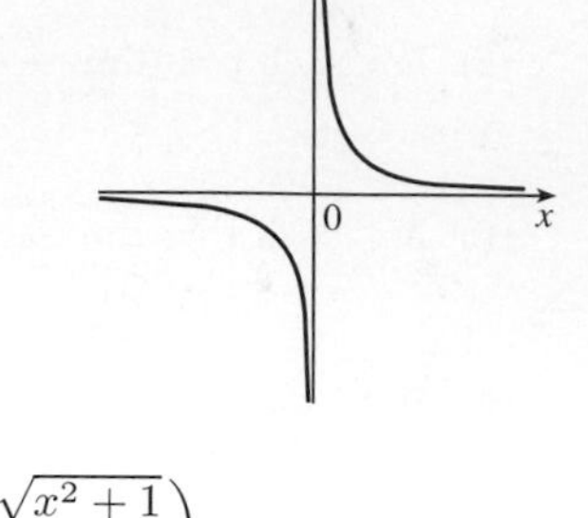

(ii) We sketch the graph of csch^{-1} by reflecting the graph of csch (see Exercise 22) about the line $y = x$.

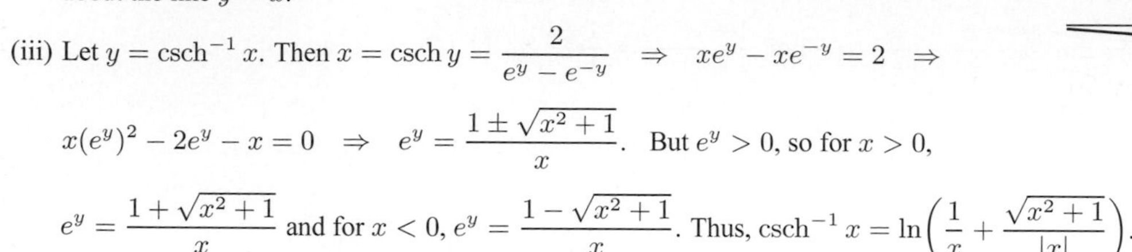

(iii) Let $y = \operatorname{csch}^{-1} x$. Then $x = \operatorname{csch} y = \dfrac{2}{e^y - e^{-y}} \quad \Rightarrow \quad xe^y - xe^{-y} = 2 \quad \Rightarrow$

$x(e^y)^2 - 2e^y - x = 0 \quad \Rightarrow \quad e^y = \dfrac{1 \pm \sqrt{x^2+1}}{x}$. But $e^y > 0$, so for $x > 0$,

$e^y = \dfrac{1 + \sqrt{x^2+1}}{x}$ and for $x < 0$, $e^y = \dfrac{1 - \sqrt{x^2+1}}{x}$. Thus, $\operatorname{csch}^{-1} x = \ln\left(\dfrac{1}{x} + \dfrac{\sqrt{x^2+1}}{|x|}\right)$.

(b) (i) $y = \operatorname{sech}^{-1} x \quad \Leftrightarrow \quad \operatorname{sech} y = x$ and $y > 0$.

(ii) We sketch the graph of sech^{-1} by reflecting the graph of sech (see Exercise 22) about the line $y = x$.

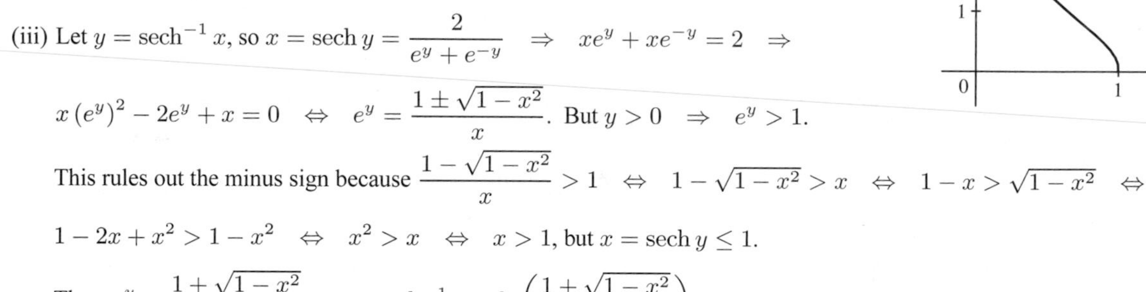

(iii) Let $y = \operatorname{sech}^{-1} x$, so $x = \operatorname{sech} y = \dfrac{2}{e^y + e^{-y}} \quad \Rightarrow \quad xe^y + xe^{-y} = 2 \quad \Rightarrow$

$x\,(e^y)^2 - 2e^y + x = 0 \quad \Leftrightarrow \quad e^y = \dfrac{1 \pm \sqrt{1-x^2}}{x}$. But $y > 0 \quad \Rightarrow \quad e^y > 1$.

This rules out the minus sign because $\dfrac{1 - \sqrt{1-x^2}}{x} > 1 \quad \Leftrightarrow \quad 1 - \sqrt{1-x^2} > x \quad \Leftrightarrow \quad 1 - x > \sqrt{1-x^2} \quad \Leftrightarrow$

$1 - 2x + x^2 > 1 - x^2 \quad \Leftrightarrow \quad x^2 > x \quad \Leftrightarrow \quad x > 1$, but $x = \operatorname{sech} y \leq 1$.

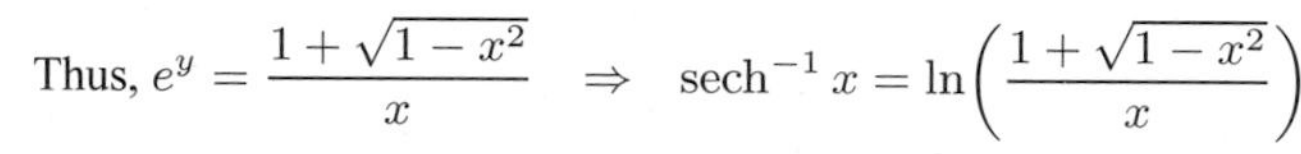

Thus, $e^y = \dfrac{1 + \sqrt{1-x^2}}{x} \quad \Rightarrow \quad \operatorname{sech}^{-1} x = \ln\left(\dfrac{1 + \sqrt{1-x^2}}{x}\right)$.

(c) (i) $y = \coth^{-1} x \quad \Leftrightarrow \quad \coth y = x$

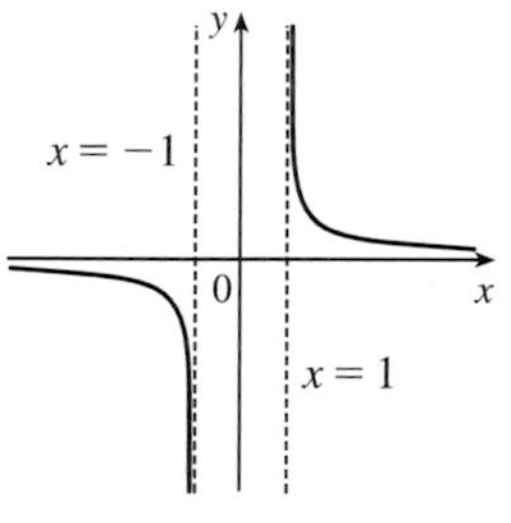

(ii) We sketch the graph of $\coth^{-1}$ by reflecting the graph of coth (see Exercise 22) about the line $y = x$.

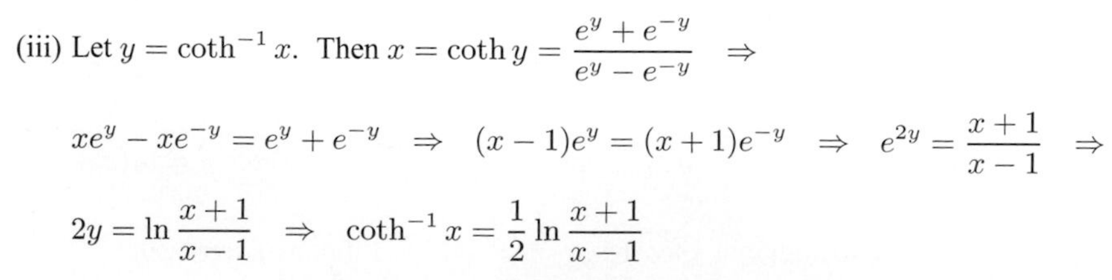

(iii) Let $y = \coth^{-1} x$. Then $x = \coth y = \dfrac{e^y + e^{-y}}{e^y - e^{-y}} \quad \Rightarrow$

$xe^y - xe^{-y} = e^y + e^{-y} \quad \Rightarrow \quad (x-1)e^y = (x+1)e^{-y} \quad \Rightarrow \quad e^{2y} = \dfrac{x+1}{x-1} \quad \Rightarrow$

$2y = \ln \dfrac{x+1}{x-1} \quad \Rightarrow \quad \coth^{-1} x = \dfrac{1}{2}\ln\dfrac{x+1}{x-1}$

29. (a) Let $y = \cosh^{-1} x$. Then $\cosh y = x$ and $y \geq 0 \quad \Rightarrow \quad \sinh y\,\dfrac{dy}{dx} = 1 \quad \Rightarrow$

$\dfrac{dy}{dx} = \dfrac{1}{\sinh y} = \dfrac{1}{\sqrt{\cosh^2 y - 1}} = \dfrac{1}{\sqrt{x^2-1}}$ [since $\sinh y \geq 0$ for $y \geq 0$]. *Or:* Use Formula 4.

(b) Let $y = \tanh^{-1} x$. Then $\tanh y = x \quad \Rightarrow \quad \operatorname{sech}^2 y\,\dfrac{dy}{dx} = 1 \quad \Rightarrow \quad \dfrac{dy}{dx} = \dfrac{1}{\operatorname{sech}^2 y} = \dfrac{1}{1 - \tanh^2 y} = \dfrac{1}{1-x^2}$.

Or: Use Formula 5.

(c) Let $y = \operatorname{csch}^{-1} x$. Then $\operatorname{csch} y = x \Rightarrow -\operatorname{csch} y \coth y \dfrac{dy}{dx} = 1 \Rightarrow \dfrac{dy}{dx} = -\dfrac{1}{\operatorname{csch} y \coth y}$. By Exercise 13,

$\coth y = \pm\sqrt{\operatorname{csch}^2 y + 1} = \pm\sqrt{x^2 + 1}$. If $x > 0$, then $\coth y > 0$, so $\coth y = \sqrt{x^2 + 1}$. If $x < 0$, then $\coth y < 0$,

so $\coth y = -\sqrt{x^2 + 1}$. In either case we have $\dfrac{dy}{dx} = -\dfrac{1}{\operatorname{csch} y \coth y} = -\dfrac{1}{|x|\sqrt{x^2 + 1}}$.

(d) Let $y = \operatorname{sech}^{-1} x$. Then $\operatorname{sech} y = x \Rightarrow -\operatorname{sech} y \tanh y \dfrac{dy}{dx} = 1 \Rightarrow$

$$\frac{dy}{dx} = -\frac{1}{\operatorname{sech} y \tanh y} = -\frac{1}{\operatorname{sech} y \sqrt{1 - \operatorname{sech}^2 y}} = -\frac{1}{x\sqrt{1 - x^2}}.$$ [Note that $y > 0$ and so $\tanh y > 0$.]

(e) Let $y = \coth^{-1} x$. Then $\coth y = x \Rightarrow -\operatorname{csch}^2 y \dfrac{dy}{dx} = 1 \Rightarrow \dfrac{dy}{dx} = -\dfrac{1}{\operatorname{csch}^2 y} = \dfrac{1}{1 - \coth^2 y} = \dfrac{1}{1 - x^2}$

by Exercise 13.

30. $f(x) = e^x \cosh x \overset{\text{PR}}{\Rightarrow} f'(x) = e^x \sinh x + (\cosh x)e^x = e^x(\sinh x + \cosh x)$, or, using Exercise 9, $e^x(e^x) = e^{2x}$.

31. $f(x) = \tanh\sqrt{x} \Rightarrow f'(x) = \operatorname{sech}^2\sqrt{x}\,\dfrac{d}{dx}\sqrt{x} = \operatorname{sech}^2\sqrt{x}\left(\dfrac{1}{2\sqrt{x}}\right) = \dfrac{\operatorname{sech}^2\sqrt{x}}{2\sqrt{x}}$

32. $g(x) = \sinh^2 x = (\sinh x)^2 \Rightarrow g'(x) = 2(\sinh x)^1 \dfrac{d}{dx}(\sinh x) = 2\sinh x \cosh x$, or, using Exercise 15, $\sinh 2x$.

33. $h(x) = \sinh(x^2) \Rightarrow h'(x) = \cosh(x^2)\dfrac{d}{dx}(x^2) = 2x\cosh(x^2)$

34. $F(t) = \ln(\sinh t) \Rightarrow F'(t) = \dfrac{1}{\sinh t}\dfrac{d}{dt}\sinh t = \dfrac{1}{\sinh t}\cosh t = \coth t$

35. $G(t) = \sinh(\ln t) \Rightarrow G'(t) = \cosh(\ln t)\dfrac{d}{dt}\ln t = \dfrac{1}{2}\left(e^{\ln t} + e^{-\ln t}\right)\left(\dfrac{1}{t}\right) = \dfrac{1}{2t}\left(t + \dfrac{1}{t}\right) = \dfrac{1}{2t}\left(\dfrac{t^2 + 1}{t}\right) = \dfrac{t^2 + 1}{2t^2}$

Or: $G(t) = \sinh(\ln t) = \dfrac{1}{2}(e^{\ln t} - e^{-\ln t}) = \dfrac{1}{2}\left(t - \dfrac{1}{t}\right) \Rightarrow G'(t) = \dfrac{1}{2}\left(1 + \dfrac{1}{t^2}\right) = \dfrac{t^2 + 1}{2t^2}$

36. $y = \operatorname{sech} x(1 + \ln\operatorname{sech} x) \overset{\text{PR}}{\Rightarrow}$

$$\begin{aligned} y' &= \operatorname{sech} x\,\frac{d}{dx}(1 + \ln\operatorname{sech} x) + (1 + \ln\operatorname{sech} x)\frac{d}{dx}\operatorname{sech} x \\ &= \operatorname{sech} x\left(\frac{-\operatorname{sech} x \tanh x}{\operatorname{sech} x}\right) + (1 + \ln\operatorname{sech} x)(-\operatorname{sech} x \tanh x) \\ &= -\operatorname{sech} x \tanh x\,[1 + (1 + \ln\operatorname{sech} x)] = -\operatorname{sech} x \tanh x\,(2 + \ln\operatorname{sech} x) \end{aligned}$$

37. $y = e^{\cosh 3x} \Rightarrow y' = e^{\cosh 3x} \cdot \sinh 3x \cdot 3 = 3e^{\cosh 3x}\sinh 3x$

38. $f(t) = \dfrac{1 + \sinh t}{1 - \sinh t} \overset{\text{QR}}{\Rightarrow}$

$$\begin{aligned} f'(t) &= \frac{(1 - \sinh t)\cosh t - (1 + \sinh t)(-\cosh t)}{(1 - \sinh t)^2} = \frac{\cosh t - \sinh t\cosh t + \cosh t + \sinh t\cosh t}{(1 - \sinh t)^2} \\ &= \frac{2\cosh t}{(1 - \sinh t)^2} \end{aligned}$$

39. $g(t) = t\coth\sqrt{t^2+1} \overset{\text{PR}}{\Rightarrow}$

$$g'(t) = t\left[-\operatorname{csch}^2\sqrt{t^2+1}\left(\tfrac{1}{2}(t^2+1)^{-1/2}\cdot 2t\right)\right] + \left(\coth\sqrt{t^2+1}\right)(1) = \coth\sqrt{t^2+1} - \frac{t^2}{\sqrt{t^2+1}}\operatorname{csch}^2\sqrt{t^2+1}$$

40. $y = \sinh^{-1}(\tan x) \Rightarrow y' = \dfrac{1}{\sqrt{1+(\tan x)^2}}\dfrac{d}{dx}(\tan x) = \dfrac{\sec^2 x}{\sqrt{\sec^2 x}} = \dfrac{|\sec^2 x|}{|\sec x|} = |\sec x|$

41. $y = \cosh^{-1}\sqrt{x} \Rightarrow y' = \dfrac{1}{\sqrt{(\sqrt{x})^2-1}}\dfrac{d}{dx}(\sqrt{x}) = \dfrac{1}{\sqrt{x-1}}\dfrac{1}{2\sqrt{x}} = \dfrac{1}{2\sqrt{x(x-1)}}$

42. $y = x\tanh^{-1}x + \ln\sqrt{1-x^2} = x\tanh^{-1}x + \frac{1}{2}\ln(1-x^2) \Rightarrow$

$$y' = \tanh^{-1}x + \frac{x}{1-x^2} + \frac{1}{2}\left(\frac{1}{1-x^2}\right)(-2x) = \tanh^{-1}x$$

43. $y = x\sinh^{-1}(x/3) - \sqrt{9+x^2} \Rightarrow$

$$y' = \sinh^{-1}\left(\frac{x}{3}\right) + x\frac{1/3}{\sqrt{1+(x/3)^2}} - \frac{2x}{2\sqrt{9+x^2}} = \sinh^{-1}\left(\frac{x}{3}\right) + \frac{x}{\sqrt{9+x^2}} - \frac{x}{\sqrt{9+x^2}} = \sinh^{-1}\left(\frac{x}{3}\right)$$

44. $y = \operatorname{sech}^{-1}(e^{-x}) \Rightarrow y' = -\dfrac{1}{e^{-x}\sqrt{1-(e^{-x})^2}}\dfrac{d}{dx}(e^{-x}) = -\dfrac{1}{e^{-x}\sqrt{1-e^{-2x}}}(-e^{-x}) = \dfrac{1}{\sqrt{1-e^{-2x}}}$

45. $y = \coth^{-1}(\sec x) \Rightarrow$

$$y' = \frac{1}{1-(\sec x)^2}\frac{d}{dx}(\sec x) = \frac{\sec x\,\tan x}{1-\sec^2 x} = \frac{\sec x\,\tan x}{1-(\tan^2 x+1)} = \frac{\sec x\,\tan x}{-\tan^2 x}$$

$$= -\frac{\sec x}{\tan x} = -\frac{1/\cos x}{\sin x/\cos x} = -\frac{1}{\sin x} = -\csc x$$

46. $\dfrac{1+\tanh x}{1-\tanh x} = \dfrac{1+(\sinh x)/\cosh x}{1-(\sinh x)/\cosh x} = \dfrac{\cosh x+\sinh x}{\cosh x-\sinh x} = \dfrac{e^x}{e^{-x}}$ [by Exercises 9 and 10] $= e^{2x}$, so

$\sqrt[4]{\dfrac{1+\tanh x}{1-\tanh x}} = \sqrt[4]{e^{2x}} = e^{x/2}$. Thus, $\dfrac{d}{dx}\sqrt[4]{\dfrac{1+\tanh x}{1-\tanh x}} = \dfrac{d}{dx}(e^{x/2}) = \dfrac{1}{2}e^{x/2}$.

47. $\dfrac{d}{dx}\arctan(\tanh x) = \dfrac{1}{1+(\tanh x)^2}\dfrac{d}{dx}(\tanh x) = \dfrac{\operatorname{sech}^2 x}{1+\tanh^2 x} = \dfrac{1/\cosh^2 x}{1+(\sinh^2 x)/\cosh^2 x}$

$= \dfrac{1}{\cosh^2 x+\sinh^2 x} = \dfrac{1}{\cosh 2x}$ [by Exercise 16] $= \operatorname{sech} 2x$

48. (a) Let $a = 0.03291765$. A graph of the central curve,

$y = f(x) = 211.49 - 20.96\cosh ax$, is shown.

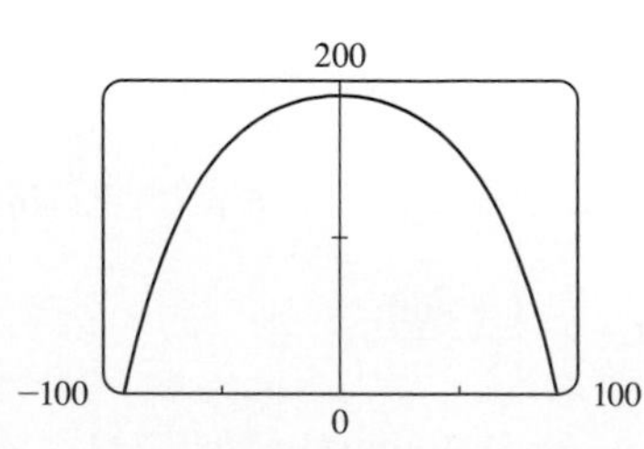

(b) $f(0) = 211.49 - 20.96\cosh 0 = 211.49 - 20.96(1) = 190.53$ m.

(c) $y = 100 \Rightarrow 100 = 211.49 - 20.96\cosh ax \Rightarrow$

$20.96\cosh ax = 111.49 \Rightarrow \cosh ax = \dfrac{111.49}{20.96} \Rightarrow$

$ax = \pm\cosh^{-1}\dfrac{111.49}{20.96} \Rightarrow x = \pm\dfrac{1}{a}\cosh^{-1}\dfrac{111.49}{20.96} \approx \pm 71.56$ m. The points are approximately $(\pm 71.56, 100)$.

(d) $f(x) = 211.49 - 20.96\cosh ax \quad\Rightarrow\quad f'(x) = -20.96\sinh ax \cdot a$.

$$f'\left(\pm\frac{1}{a}\cosh^{-1}\frac{111.49}{20.96}\right) = -20.96a\sinh\left[a\left(\pm\frac{1}{a}\cosh^{-1}\frac{111.49}{20.96}\right)\right] = -20.96a\sinh\left(\pm\cosh^{-1}\frac{111.49}{20.96}\right) \approx \mp 3.6.$$

So the slope at $(71.56, 100)$ is about -3.6 and the slope at $(-71.56, 100)$ is about 3.6.

49. As the depth d of the water gets large, the fraction $\dfrac{2\pi d}{L}$ gets large, and from Figure 3 or Exercise 23(a), $\tanh\left(\dfrac{2\pi d}{L}\right)$ approaches 1. Thus, $v = \sqrt{\dfrac{gL}{2\pi}\tanh\left(\dfrac{2\pi d}{L}\right)} \approx \sqrt{\dfrac{gL}{2\pi}(1)} = \sqrt{\dfrac{gL}{2\pi}}$.

50.

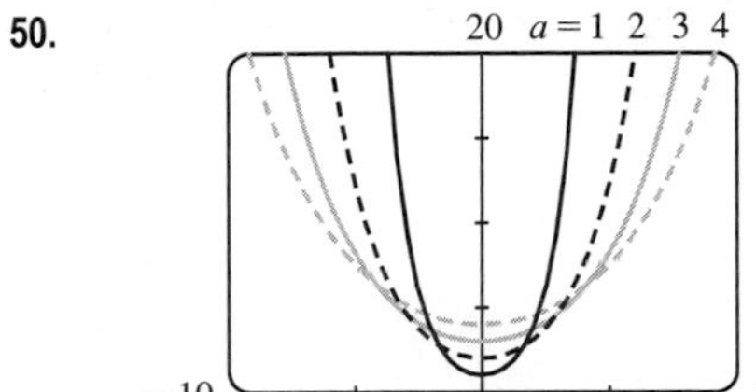

For $y = a\cosh(x/a)$ with $a > 0$, we have the y-intercept equal to a. As a increases, the graph flattens.

51. (a) $y = 20\cosh(x/20) - 15 \quad\Rightarrow\quad y' = 20\sinh(x/20)\cdot\frac{1}{20} = \sinh(x/20)$. Since the right pole is positioned at $x = 7$, we have $y'(7) = \sinh\frac{7}{20} \approx 0.3572$.

(b) If α is the angle between the tangent line and the x-axis, then $\tan\alpha =$ slope of the line $= \sinh\frac{7}{20}$, so $\alpha = \tan^{-1}\left(\sinh\frac{7}{20}\right) \approx 0.343$ rad $\approx 19.66^\circ$. Thus, the angle between the line and the pole is $\theta = 90^\circ - \alpha \approx 70.34^\circ$.

52. We differentiate the function twice, then substitute into the differential equation: $y = \dfrac{T}{\rho g}\cosh\dfrac{\rho g x}{T} \quad\Rightarrow$

$\dfrac{dy}{dx} = \dfrac{T}{\rho g}\sinh\left(\dfrac{\rho g x}{T}\right)\dfrac{\rho g}{T} = \sinh\dfrac{\rho g x}{T} \quad\Rightarrow\quad \dfrac{d^2y}{dx^2} = \cosh\left(\dfrac{\rho g x}{T}\right)\dfrac{\rho g}{T} = \dfrac{\rho g}{T}\cosh\dfrac{\rho g x}{T}$. We evaluate the two sides separately: LHS $= \dfrac{d^2y}{dx^2} = \dfrac{\rho g}{T}\cosh\dfrac{\rho g x}{T}$ and RHS $= \dfrac{\rho g}{T}\sqrt{1 + \left(\dfrac{dy}{dx}\right)^2} = \dfrac{\rho g}{T}\sqrt{1 + \sinh^2\dfrac{\rho g x}{T}} = \dfrac{\rho g}{T}\cosh\dfrac{\rho g x}{T}$, by the identity proved in Example 1(a).

53. (a) From Exercise 52, the shape of the cable is given by $y = f(x) = \dfrac{T}{\rho g}\cosh\left(\dfrac{\rho g x}{T}\right)$. The shape is symmetric about the y-axis, so the lowest point is $(0, f(0)) = \left(0, \dfrac{T}{\rho g}\right)$ and the poles are at $x = \pm 100$. We want to find T when the lowest point is 60 m, so $\dfrac{T}{\rho g} = 60 \quad\Rightarrow\quad T = 60\rho g = (60\text{ m})(2\text{ kg/m})(9.8\text{ m/s}^2) = 1176\,\dfrac{\text{kg-m}}{\text{s}^2}$, or 1176 N (newtons).

The height of each pole is $f(100) = \dfrac{T}{\rho g}\cosh\left(\dfrac{\rho g\cdot 100}{T}\right) = 60\cosh\left(\dfrac{100}{60}\right) \approx 164.50$ m.

(b) If the tension is doubled from T to $2T$, then the low point is doubled since $\dfrac{T}{\rho g} = 60 \quad\Rightarrow\quad \dfrac{2T}{\rho g} = 120$. The height of the poles is now $f(100) = \dfrac{2T}{\rho g}\cosh\left(\dfrac{\rho g\cdot 100}{2T}\right) = 120\cosh\left(\dfrac{100}{120}\right) \approx 164.13$ m, just a slight decrease.

54. (a) $\lim\limits_{t\to\infty} v(t) = \lim\limits_{t\to\infty} \sqrt{\dfrac{mg}{k}}\tanh\left(t\sqrt{\dfrac{gk}{m}}\right) = \sqrt{\dfrac{mg}{k}}\lim\limits_{t\to\infty}\tanh\left(t\sqrt{\dfrac{gk}{m}}\right) = \sqrt{\dfrac{mg}{k}}\cdot 1 \quad \left[\begin{matrix}\text{as } t\to\infty,\\ t\sqrt{gk/m}\to\infty\end{matrix}\right] \quad = \sqrt{\dfrac{mg}{k}}$

(b) Belly-to-earth: $g = 9.8$, $k = 0.515$, $m = 60$, so the terminal velocity is $\sqrt{\dfrac{60(9.8)}{0.515}} \approx 33.79$ m/s.

Feet-first: $g = 9.8$, $k = 0.067$, $m = 60$, so the terminal velocity is $\sqrt{\dfrac{60(9.8)}{0.067}} \approx 93.68$ m/s.

55. (a) $y = A\sinh mx + B\cosh mx \;\Rightarrow\; y' = mA\cosh mx + mB\sinh mx \;\Rightarrow$

$y'' = m^2A\sinh mx + m^2B\cosh mx = m^2(A\sinh mx + B\cosh mx) = m^2y$

(b) From part (a), a solution of $y'' = 9y$ is $y(x) = A\sinh 3x + B\cosh 3x$. So $-4 = y(0) = A\sinh 0 + B\cosh 0 = B$, so $B = -4$. Now $y'(x) = 3A\cosh 3x - 12\sinh 3x \;\Rightarrow\; 6 = y'(0) = 3A \;\Rightarrow\; A = 2$, so $y = 2\sinh 3x - 4\cosh 3x$.

56. $\cosh x = \cosh[\ln(\sec\theta + \tan\theta)] = \dfrac{1}{2}\left[e^{\ln(\sec\theta+\tan\theta)} + e^{-\ln(\sec\theta+\tan\theta)}\right] = \dfrac{1}{2}\left[\sec\theta + \tan\theta + \dfrac{1}{\sec\theta+\tan\theta}\right]$

$= \dfrac{1}{2}\left[\sec\theta + \tan\theta + \dfrac{\sec\theta - \tan\theta}{(\sec\theta+\tan\theta)(\sec\theta-\tan\theta)}\right] = \dfrac{1}{2}\left[\sec\theta + \tan\theta + \dfrac{\sec\theta-\tan\theta}{\sec^2\theta - \tan^2\theta}\right]$

$= \frac{1}{2}(\sec\theta + \tan\theta + \sec\theta - \tan\theta) = \sec\theta$

57. The tangent to $y = \cosh x$ has slope 1 when $y' = \sinh x = 1 \;\Rightarrow\; x = \sinh^{-1} 1 = \ln\left(1+\sqrt{2}\right)$, by Equation 3.

Since $\sinh x = 1$ and $y = \cosh x = \sqrt{1+\sinh^2 x}$, we have $\cosh x = \sqrt{2}$. The point is $\left(\ln\left(1+\sqrt{2}\right), \sqrt{2}\right)$.

58. $f_n(x) = \tanh(n\sin x)$, where n is a positive integer. Note that $f_n(x+2\pi) = f_n(x)$; that is, f_n is periodic with period 2π. Also, from Figure 3, $-1 < \tanh x < 1$, so we can choose a viewing rectangle of $[0, 2\pi]\times[-1,1]$. From the graph, we see that $f_n(x)$ becomes more rectangular looking as n increases. As n becomes large, the graph of f_n approaches the graph of $y = 1$ on the intervals $(2k\pi, (2k+1)\pi)$ and $y = -1$ on the intervals $((2k-1)\pi, 2k\pi)$.

59. If $ae^x + be^{-x} = \alpha\cosh(x+\beta)$ [or $\alpha\sinh(x+\beta)$], then

$ae^x + be^{-x} = \frac{\alpha}{2}\left(e^{x+\beta} \pm e^{-x-\beta}\right) = \frac{\alpha}{2}\left(e^x e^\beta \pm e^{-x}e^{-\beta}\right) = \left(\frac{\alpha}{2}e^\beta\right)e^x \pm \left(\frac{\alpha}{2}e^{-\beta}\right)e^{-x}$. Comparing coefficients of e^x and e^{-x}, we have $a = \frac{\alpha}{2}e^\beta$ **(1)** and $b = \pm\frac{\alpha}{2}e^{-\beta}$ **(2)**. We need to find α and β. Dividing equation **(1)** by equation **(2)** gives us $\frac{a}{b} = \pm e^{2\beta} \;\Rightarrow\; (\star)\;\; 2\beta = \ln\left(\pm\frac{a}{b}\right) \;\Rightarrow\; \beta = \frac{1}{2}\ln\left(\pm\frac{a}{b}\right)$. Solving equations **(1)** and **(2)** for e^β gives us

$e^\beta = \dfrac{2a}{\alpha}$ and $e^\beta = \pm\dfrac{\alpha}{2b}$, so $\dfrac{2a}{\alpha} = \pm\dfrac{\alpha}{2b} \;\Rightarrow\; \alpha^2 = \pm 4ab \;\Rightarrow\; \alpha = 2\sqrt{\pm ab}$.

$(\star)$ If $\frac{a}{b} > 0$, we use the + sign and obtain a cosh function, whereas if $\frac{a}{b} < 0$, we use the − sign and obtain a sinh function.

In summary, if a and b have the same sign, we have $ae^x + be^{-x} = 2\sqrt{ab}\cosh\left(x + \frac{1}{2}\ln\frac{a}{b}\right)$, whereas, if a and b have the opposite sign, then $ae^x + be^{-x} = 2\sqrt{-ab}\sinh\left(x + \frac{1}{2}\ln\left(-\frac{a}{b}\right)\right)$.

3 Review

TRUE-FALSE QUIZ

1. True. This is the Sum Rule.

2. False. See the warning before the Product Rule.

3. True. This is the Chain Rule.

4. True. $\dfrac{d}{dx}\sqrt{f(x)} = \dfrac{d}{dx}[f(x)]^{1/2} = \dfrac{1}{2}[f(x)]^{-1/2}\,f'(x) = \dfrac{f'(x)}{2\sqrt{f(x)}}$

5. False. $\dfrac{d}{dx}\,f(\sqrt{x}\,) = f'(\sqrt{x}\,)\cdot\frac{1}{2}x^{-1/2} = \dfrac{f'(\sqrt{x}\,)}{2\sqrt{x}}$, which is not $\dfrac{f'(x)}{2\sqrt{x}}$.

6. False. $y = e^2$ is a constant, so $y' = 0$, not $2e$.

7. False. $\dfrac{d}{dx}(10^x) = 10^x \ln 10$, which is not equal to $x10^{x-1}$.

8. False. $\ln 10$ is a constant, so its derivative, $\dfrac{d}{dx}(\ln 10)$, is 0, not $\frac{1}{10}$.

9. True. $\dfrac{d}{dx}(\tan^2 x) = 2\,\tan x\,\sec^2 x$, and $\dfrac{d}{dx}(\sec^2 x) = 2\,\sec x\,(\sec x\,\tan x) = 2\,\tan x\,\sec^2 x$.

Or: $\dfrac{d}{dx}(\sec^2 x) = \dfrac{d}{dx}(1+\tan^2 x) = \dfrac{d}{dx}(\tan^2 x)$.

10. False. $f(x) = \left|x^2 + x\right| = x^2 + x$ for $x \geq 0$ or $x \leq -1$ and $\left|x^2 + x\right| = -(x^2 + x)$ for $-1 < x < 0$.
So $f'(x) = 2x + 1$ for $x > 0$ or $x < -1$ and $f'(x) = -(2x+1)$ for $-1 < x < 0$. But $|2x+1| = 2x+1$ for $x \geq -\frac{1}{2}$ and $|2x+1| = -2x-1$ for $x < -\frac{1}{2}$.

11. True. If $p(x) = a_n x^n + a_{n-1}x^{n-1} + \cdots + a_1 x + a_0$, then $p'(x) = na_n x^{n-1} + (n-1)a_{n-1}x^{n-2} + \cdots + a_1$, which is a polynomial.

12. True. $f(x) = (x^6 - x^4)^5$ is a polynomial of degree 30, so its 31st derivative, $f^{(31)}(x)$, is 0.

13. True. If $r(x) = \dfrac{p(x)}{q(x)}$, then $r'(x) = \dfrac{q(x)p'(x) - p(x)q'(x)}{[q(x)]^2}$, which is a quotient of polynomials, that is, a rational function.

14. False. A tangent line to the parabola $y = x^2$ has slope $dy/dx = 2x$, so at $(-2, 4)$ the slope of the tangent is $2(-2) = -4$ and an equation of the tangent line is $y - 4 = -4(x+2)$. [The given equation, $y - 4 = 2x(x+2)$, is not even linear!]

15. True. $g(x) = x^5 \;\Rightarrow\; g'(x) = 5x^4 \;\Rightarrow\; g'(2) = 5(2)^4 = 80$, and by the definition of the derivative,

$$\lim_{x\to 2}\frac{g(x) - g(2)}{x - 2} = g'(2) = 5(2)^4 = 80.$$

EXERCISES

1. $y = (x^2 + x^3)^4 \quad \Rightarrow \quad y' = 4(x^2 + x^3)^3(2x + 3x^2) = 4(x^2)^3(1 + x)^3 x(2 + 3x) = 4x^7(x + 1)^3(3x + 2)$

2. $y = \dfrac{1}{\sqrt{x}} - \dfrac{1}{\sqrt[5]{x^3}} = x^{-1/2} - x^{-3/5} \quad \Rightarrow \quad y' = -\dfrac{1}{2}x^{-3/2} + \dfrac{3}{5}x^{-8/5}$ or $\dfrac{3}{5x\sqrt[5]{x^3}} - \dfrac{1}{2x\sqrt{x}}$ or $\dfrac{1}{10}x^{-8/5}(-5x^{1/10} + 6)$

3. $y = \dfrac{x^2 - x + 2}{\sqrt{x}} = x^{3/2} - x^{1/2} + 2x^{-1/2} \quad \Rightarrow \quad y' = \dfrac{3}{2}x^{1/2} - \dfrac{1}{2}x^{-1/2} - x^{-3/2} = \dfrac{3}{2}\sqrt{x} - \dfrac{1}{2\sqrt{x}} - \dfrac{1}{\sqrt{x^3}}$

4. $y = \dfrac{\tan x}{1 + \cos x} \quad \Rightarrow \quad y' = \dfrac{(1 + \cos x)\sec^2 x - \tan x(-\sin x)}{(1 + \cos x)^2} = \dfrac{(1 + \cos x)\sec^2 x + \tan x\,\sin x}{(1 + \cos x)^2}$

5. $y = x^2 \sin \pi x \quad \Rightarrow \quad y' = x^2(\cos \pi x)\pi + (\sin \pi x)(2x) = x(\pi x \cos \pi x + 2\sin \pi x)$

6. $y = x\cos^{-1} x \quad \Rightarrow \quad y' = x\left(-\dfrac{1}{\sqrt{1 - x^2}}\right) + (\cos^{-1} x)(1) = \cos^{-1} x - \dfrac{x}{\sqrt{1 - x^2}}$

7. $y = \dfrac{t^4 - 1}{t^4 + 1} \quad \Rightarrow \quad y' = \dfrac{(t^4 + 1)4t^3 - (t^4 - 1)4t^3}{(t^4 + 1)^2} = \dfrac{4t^3[(t^4 + 1) - (t^4 - 1)]}{(t^4 + 1)^2} = \dfrac{8t^3}{(t^4 + 1)^2}$

8. $\dfrac{d}{dx}(xe^y) = \dfrac{d}{dx}(y\sin x) \quad \Rightarrow \quad xe^y y' + e^y \cdot 1 = y\cos x + \sin x \cdot y' \quad \Rightarrow \quad xe^y y' - \sin x \cdot y' = y\cos x - e^y \quad \Rightarrow$

$(xe^y - \sin x)y' = y\cos x - e^y \quad \Rightarrow \quad y' = \dfrac{y\cos x - e^y}{xe^y - \sin x}$

9. $y = \ln(x\ln x) \quad \Rightarrow \quad y' = \dfrac{1}{x\ln x}(x\ln x)' = \dfrac{1}{x\ln x}\left(x \cdot \dfrac{1}{x} + \ln x \cdot 1\right) = \dfrac{1 + \ln x}{x\ln x}$

Another method: $y = \ln(x\ln x) = \ln x + \ln\ln x \quad \Rightarrow \quad y' = \dfrac{1}{x} + \dfrac{1}{\ln x} \cdot \dfrac{1}{x} = \dfrac{\ln x + 1}{x\ln x}$

10. $y = e^{mx}\cos nx \quad \Rightarrow$

$y' = e^{mx}(\cos nx)' + \cos nx\,(e^{mx})' = e^{mx}(-\sin nx \cdot n) + \cos nx\,(e^{mx} \cdot m) = e^{mx}(m\cos nx - n\sin nx)$

11. $y = \sqrt{x}\cos\sqrt{x} \quad \Rightarrow$

$y' = \sqrt{x}\left(\cos\sqrt{x}\right)' + \cos\sqrt{x}\left(\sqrt{x}\right)' = \sqrt{x}\left[-\sin\sqrt{x}\left(\tfrac{1}{2}x^{-1/2}\right)\right] + \cos\sqrt{x}\left(\tfrac{1}{2}x^{-1/2}\right)$

$= \tfrac{1}{2}x^{-1/2}\left(-\sqrt{x}\sin\sqrt{x} + \cos\sqrt{x}\right) = \dfrac{\cos\sqrt{x} - \sqrt{x}\sin\sqrt{x}}{2\sqrt{x}}$

12. $y = (\arcsin 2x)^2 \quad \Rightarrow \quad y' = 2(\arcsin 2x) \cdot (\arcsin 2x)' = 2\arcsin 2x \cdot \dfrac{1}{\sqrt{1 - (2x)^2}} \cdot 2 = \dfrac{4\arcsin 2x}{\sqrt{1 - 4x^2}}$

13. $y = \dfrac{e^{1/x}}{x^2} \quad \Rightarrow \quad y' = \dfrac{x^2(e^{1/x})' - e^{1/x}\,(x^2)'}{(x^2)^2} = \dfrac{x^2(e^{1/x})(-1/x^2) - e^{1/x}(2x)}{x^4} = \dfrac{-e^{1/x}(1 + 2x)}{x^4}$

14. $y = \ln\sec x \quad \Rightarrow \quad y' = \dfrac{1}{\sec x}\dfrac{d}{dx}(\sec x) = \dfrac{1}{\sec x}(\sec x\,\tan x) = \tan x$

15. $\dfrac{d}{dx}(y + x\cos y) = \dfrac{d}{dx}(x^2 y) \Rightarrow y' + x(-\sin y \cdot y') + \cos y \cdot 1 = x^2 y' + y \cdot 2x \Rightarrow$

$y' - x\sin y \cdot y' - x^2 y' = 2xy - \cos y \Rightarrow (1 - x\sin y - x^2)y' = 2xy - \cos y \Rightarrow y' = \dfrac{2xy - \cos y}{1 - x\sin y - x^2}$

16. $y = \left(\dfrac{u-1}{u^2+u+1}\right)^4 \Rightarrow$

$$y' = 4\left(\frac{u-1}{u^2+u+1}\right)^3 \frac{d}{du}\left(\frac{u-1}{u^2+u+1}\right) = 4\left(\frac{u-1}{u^2+u+1}\right)^3 \frac{(u^2+u+1)(1)-(u-1)(2u+1)}{(u^2+u+1)^2}$$

$$= \frac{4(u-1)^3}{(u^2+u+1)^3}\,\frac{u^2+u+1-2u^2+u+1}{(u^2+u+1)^2} = \frac{4(u-1)^3(-u^2+2u+2)}{(u^2+u+1)^5}$$

17. $y = \sqrt{\arctan x} \Rightarrow y' = \dfrac{1}{2}(\arctan x)^{-1/2}\dfrac{d}{dx}(\arctan x) = \dfrac{1}{2\sqrt{\arctan x}\,(1+x^2)}$

18. $y = \cot(\csc x) \Rightarrow y' = -\csc^2(\csc x)\dfrac{d}{dx}(\csc x) = -\csc^2(\csc x)\cdot(-\csc x\cot x) = \csc^2(\csc x)\csc x\cot x$

19. $y = \tan\left(\dfrac{t}{1+t^2}\right) \Rightarrow$

$$y' = \sec^2\left(\frac{t}{1+t^2}\right)\frac{d}{dt}\left(\frac{t}{1+t^2}\right) = \sec^2\left(\frac{t}{1+t^2}\right)\cdot\frac{(1+t^2)(1)-t(2t)}{(1+t^2)^2} = \frac{1-t^2}{(1+t^2)^2}\sec^2\left(\frac{t}{1+t^2}\right)$$

20. $y = e^{x\sec x} \Rightarrow y' = e^{x\sec x}\dfrac{d}{dx}(x\sec x) = e^{x\sec x}(x\sec x\tan x + \sec x \cdot 1) = \sec x\, e^{x\sec x}(x\tan x + 1)$

21. $y = 3^{x\ln x} \Rightarrow y' = 3^{x\ln x}(\ln 3)\dfrac{d}{dx}(x\ln x) = 3^{x\ln x}(\ln 3)\left(x\cdot\dfrac{1}{x} + \ln x \cdot 1\right) = 3^{x\ln x}(\ln 3)(1+\ln x)$

22. $y = \sec(1+x^2) \Rightarrow y' = 2x\sec(1+x^2)\tan(1+x^2)$

23. $y = (1-x^{-1})^{-1} \Rightarrow$

$y' = -1(1-x^{-1})^{-2}[-(-1x^{-2})] = -(1-1/x)^{-2}x^{-2} = -((x-1)/x)^{-2}x^{-2} = -(x-1)^{-2}$

24. $y = \dfrac{1}{\sqrt[3]{x+\sqrt{x}}} = \left(x+\sqrt{x}\right)^{-1/3} \Rightarrow y' = -\frac{1}{3}\left(x+\sqrt{x}\right)^{-4/3}\left(1+\dfrac{1}{2\sqrt{x}}\right)$

25. $\sin(xy) = x^2 - y \Rightarrow \cos(xy)(xy' + y\cdot 1) = 2x - y' \Rightarrow x\cos(xy)y' + y' = 2x - y\cos(xy) \Rightarrow$

$y'[x\cos(xy)+1] = 2x - y\cos(xy) \Rightarrow y' = \dfrac{2x - y\cos(xy)}{x\cos(xy)+1}$

26. $y = \sqrt{\sin\sqrt{x}} \Rightarrow y' = \frac{1}{2}\left(\sin\sqrt{x}\right)^{-1/2}\left(\cos\sqrt{x}\right)\left(\dfrac{1}{2\sqrt{x}}\right) = \dfrac{\cos\sqrt{x}}{4\sqrt{x\sin\sqrt{x}}}$

27. $y = \log_5(1+2x) \Rightarrow y' = \dfrac{1}{(1+2x)\ln 5}\dfrac{d}{dx}(1+2x) = \dfrac{2}{(1+2x)\ln 5}$

28. $y = (\cos x)^x \Rightarrow \ln y = \ln(\cos x)^x = x\ln\cos x \Rightarrow \dfrac{y'}{y} = x\cdot\dfrac{1}{\cos x}\cdot(-\sin x) + \ln\cos x \cdot 1 \Rightarrow$

$y' = (\cos x)^x(\ln\cos x - x\tan x)$

29. $y = \ln \sin x - \frac{1}{2}\sin^2 x \quad \Rightarrow \quad y' = \dfrac{1}{\sin x} \cdot \cos x - \frac{1}{2} \cdot 2\sin x \cdot \cos x = \cot x - \sin x\, \cos x$

30. $y = \dfrac{(x^2+1)^4}{(2x+1)^3(3x-1)^5} \quad \Rightarrow$

$$\ln y = \ln \frac{(x^2+1)^4}{(2x+1)^3(3x-1)^5} = \ln(x^2+1)^4 - \ln[(2x+1)^3(3x-1)^5] = 4\ln(x^2+1) - [\ln(2x+1)^3 + \ln(3x-1)^5]$$

$$= 4\ln(x^2+1) - 3\ln(2x+1) - 5\ln(3x-1) \quad \Rightarrow$$

$$\frac{y'}{y} = 4 \cdot \frac{1}{x^2+1} \cdot 2x - 3 \cdot \frac{1}{2x+1} \cdot 2 - 5 \cdot \frac{1}{3x-1} \cdot 3 \quad \Rightarrow \quad y' = \frac{(x^2+1)^4}{(2x+1)^3(3x-1)^5}\left(\frac{8x}{x^2+1} - \frac{6}{2x+1} - \frac{15}{3x-1}\right).$$

[The answer could be simplified to $y' = -\dfrac{(x^2+56x+9)(x^2+1)^3}{(2x+1)^4(3x-1)^6}$, but this is unnecessary.]

31. $y = x\tan^{-1}(4x) \quad \Rightarrow \quad y' = x \cdot \dfrac{1}{1+(4x)^2} \cdot 4 + \tan^{-1}(4x) \cdot 1 = \dfrac{4x}{1+16x^2} + \tan^{-1}(4x)$

32. $y = e^{\cos x} + \cos(e^x) \quad \Rightarrow \quad y' = e^{\cos x}(-\sin x) + [-\sin(e^x) \cdot e^x] = -\sin x\, e^{\cos x} - e^x \sin(e^x)$

33. $y = \ln|\sec 5x + \tan 5x| \quad \Rightarrow$

$$y' = \frac{1}{\sec 5x + \tan 5x}(\sec 5x\, \tan 5x \cdot 5 + \sec^2 5x \cdot 5) = \frac{5\sec 5x\,(\tan 5x + \sec 5x)}{\sec 5x + \tan 5x} = 5\sec 5x$$

34. $y = 10^{\tan \pi\theta} \quad \Rightarrow \quad y' = 10^{\tan \pi\theta} \cdot \ln 10 \cdot \sec^2 \pi\theta \cdot \pi = \pi(\ln 10)10^{\tan \pi\theta}\sec^2 \pi\theta$

35. $y = \cot(3x^2+5) \quad \Rightarrow \quad y' = -\csc^2(3x^2+5)(6x) = -6x\csc^2(3x^2+5)$

36. $y = \sqrt{t\ln(t^4)} \quad \Rightarrow$

$$y' = \frac{1}{2}[t\ln(t^4)]^{-1/2}\frac{d}{dt}[t\ln(t^4)] = \frac{1}{2\sqrt{t\ln(t^4)}} \cdot \left[1 \cdot \ln(t^4) + t \cdot \frac{1}{t^4} \cdot 4t^3\right] = \frac{1}{2\sqrt{t\ln(t^4)}} \cdot [\ln(t^4) + 4] = \frac{\ln(t^4)+4}{2\sqrt{t\ln(t^4)}}$$

Or: Since y is only defined for $t > 0$, we can write $y = \sqrt{t \cdot 4\ln t} = 2\sqrt{t\ln t}$. Then

$y' = 2 \cdot \dfrac{1}{2\sqrt{t\ln t}} \cdot \left(1 \cdot \ln t + t \cdot \dfrac{1}{t}\right) = \dfrac{\ln t + 1}{\sqrt{t\ln t}}$. This agrees with our first answer since

$$\frac{\ln(t^4)+4}{2\sqrt{t\ln(t^4)}} = \frac{4\ln t + 4}{2\sqrt{t \cdot 4\ln t}} = \frac{4(\ln t + 1)}{2 \cdot 2\sqrt{t\ln t}} = \frac{\ln t + 1}{\sqrt{t\ln t}}.$$

37. $y = \sin\left(\tan\sqrt{1+x^3}\right) \quad \Rightarrow \quad y' = \cos\left(\tan\sqrt{1+x^3}\right)\left(\sec^2\sqrt{1+x^3}\right)\left[3x^2/\left(2\sqrt{1+x^3}\right)\right]$

38. $y = \arctan\left(\arcsin\sqrt{x}\right) \quad \Rightarrow \quad y' = \dfrac{1}{1+\left(\arcsin\sqrt{x}\right)^2} \cdot \dfrac{1}{\sqrt{1-x}} \cdot \dfrac{1}{2\sqrt{x}}$

39. $y = \tan^2(\sin\theta) = [\tan(\sin\theta)]^2 \quad \Rightarrow \quad y' = 2[\tan(\sin\theta)] \cdot \sec^2(\sin\theta) \cdot \cos\theta$

40. $xe^y = y - 1 \quad \Rightarrow \quad xe^y y' + e^y = y' \quad \Rightarrow \quad e^y = y' - xe^y y' \quad \Rightarrow \quad y' = e^y/(1 - xe^y)$

41. $y = \dfrac{\sqrt{x+1}\,(2-x)^5}{(x+3)^7} \quad\Rightarrow\quad \ln y = \frac{1}{2}\ln(x+1) + 5\ln(2-x) - 7\ln(x+3) \quad\Rightarrow\quad \dfrac{y'}{y} = \dfrac{1}{2(x+1)} + \dfrac{-5}{2-x} - \dfrac{7}{x+3} \quad\Rightarrow$

$$y' = \frac{\sqrt{x+1}\,(2-x)^5}{(x+3)^7}\left[\frac{1}{2(x+1)} - \frac{5}{2-x} - \frac{7}{x+3}\right] \quad\text{or}\quad y' = \frac{(2-x)^4(3x^2-55x-52)}{2\sqrt{x+1}\,(x+3)^8}.$$

42. $y = \dfrac{(x+\lambda)^4}{x^4+\lambda^4} \quad\Rightarrow\quad y' = \dfrac{(x^4+\lambda^4)(4)(x+\lambda)^3 - (x+\lambda)^4(4x^3)}{(x^4+\lambda^4)^2} = \dfrac{4(x+\lambda)^3(\lambda^4 - \lambda x^3)}{(x^4+\lambda^4)^2}$

43. $y = x\sinh(x^2) \quad\Rightarrow\quad y' = x\cosh(x^2)\cdot 2x + \sinh(x^2)\cdot 1 = 2x^2\cosh(x^2) + \sinh(x^2)$

44. $y = (\sin mx)/x \quad\Rightarrow\quad y' = (mx\cos mx - \sin mx)/x^2$

45. $y = \ln(\cosh 3x) \quad\Rightarrow\quad y' = (1/\cosh 3x)(\sinh 3x)(3) = 3\tanh 3x$

46. $y = \ln\left|\dfrac{x^2-4}{2x+5}\right| = \ln|x^2-4| - \ln|2x+5| \quad\Rightarrow\quad y' = \dfrac{2x}{x^2-4} - \dfrac{2}{2x+5}$ or $\dfrac{2(x+1)(x+4)}{(x+2)(x-2)(2x+5)}$

47. $y = \cosh^{-1}(\sinh x) \quad\Rightarrow\quad y' = \dfrac{1}{\sqrt{(\sinh x)^2 - 1}}\cdot\cosh x = \dfrac{\cosh x}{\sqrt{\sinh^2 x - 1}}$

48. $y = x\tanh^{-1}\sqrt{x} \quad\Rightarrow\quad y' = \tanh^{-1}\sqrt{x} + x\,\dfrac{1}{1-\left(\sqrt{x}\right)^2}\,\dfrac{1}{2\sqrt{x}} = \tanh^{-1}\sqrt{x} + \dfrac{\sqrt{x}}{2(1-x)}$

49. $y = \cos\left(e^{\sqrt{\tan 3x}}\right) \quad\Rightarrow$

$$y' = -\sin\left(e^{\sqrt{\tan 3x}}\right)\cdot\left(e^{\sqrt{\tan 3x}}\right)' = -\sin\left(e^{\sqrt{\tan 3x}}\right)\,e^{\sqrt{\tan 3x}}\cdot\tfrac{1}{2}(\tan 3x)^{-1/2}\cdot\sec^2(3x)\cdot 3$$

$$= \frac{-3\sin\left(e^{\sqrt{\tan 3x}}\right)e^{\sqrt{\tan 3x}}\sec^2(3x)}{2\sqrt{\tan 3x}}$$

50. $y = \sin^2\left(\cos\sqrt{\sin \pi x}\right) = \left[\sin\left(\cos\sqrt{\sin \pi x}\right)\right]^2 \quad\Rightarrow$

$$y' = 2\left[\sin\left(\cos\sqrt{\sin\pi x}\right)\right]\left[\sin\left(\cos\sqrt{\sin\pi x}\right)\right]' = 2\sin\left(\cos\sqrt{\sin\pi x}\right)\cos\left(\cos\sqrt{\sin\pi x}\right)\left(\cos\sqrt{\sin\pi x}\right)'$$

$$= 2\sin\left(\cos\sqrt{\sin\pi x}\right)\cos\left(\cos\sqrt{\sin\pi x}\right)\left(-\sin\sqrt{\sin\pi x}\right)\left(\sqrt{\sin\pi x}\right)'$$

$$= -2\sin\left(\cos\sqrt{\sin\pi x}\right)\cos\left(\cos\sqrt{\sin\pi x}\right)\sin\sqrt{\sin\pi x}\cdot\tfrac{1}{2}(\sin\pi x)^{-1/2}(\sin\pi x)'$$

$$= \frac{-\sin\left(\cos\sqrt{\sin\pi x}\right)\cos\left(\cos\sqrt{\sin\pi x}\right)\sin\sqrt{\sin\pi x}}{\sqrt{\sin\pi x}}\cdot\cos\pi x\cdot\pi$$

$$= \frac{-\pi\sin\left(\cos\sqrt{\sin\pi x}\right)\cos\left(\cos\sqrt{\sin\pi x}\right)\sin\sqrt{\sin\pi x}\,\cos\pi x}{\sqrt{\sin\pi x}}$$

51. $f(t) = \sqrt{4t+1} \quad\Rightarrow\quad f'(t) = \frac{1}{2}(4t+1)^{-1/2}\cdot 4 = 2(4t+1)^{-1/2} \quad\Rightarrow$

$f''(t) = 2(-\frac{1}{2})(4t+1)^{-3/2}\cdot 4 = -4/(4t+1)^{3/2}$, so $f''(2) = -4/9^{3/2} = -\frac{4}{27}$.

52. $g(\theta) = \theta\sin\theta \;\Rightarrow\; g'(\theta) = \theta\cos\theta + \sin\theta\cdot 1 \;\Rightarrow\; g''(\theta) = \theta(-\sin\theta) + \cos\theta\cdot 1 + \cos\theta = 2\cos\theta - \theta\sin\theta$,

so $g''(\pi/6) = 2\cos(\pi/6) - (\pi/6)\sin(\pi/6) = 2\left(\sqrt{3}/2\right) - (\pi/6)(1/2) = \sqrt{3} - \pi/12$.

53. $x^6 + y^6 = 1 \;\Rightarrow\; 6x^5 + 6y^5y' = 0 \;\Rightarrow\; y' = -x^5/y^5 \;\Rightarrow$

$$y'' = -\frac{y^5(5x^4) - x^5(5y^4y')}{(y^5)^2} = -\frac{5x^4y^4\left[y - x(-x^5/y^5)\right]}{y^{10}} = -\frac{5x^4\left[(y^6 + x^6)/y^5\right]}{y^6} = -\frac{5x^4}{y^{11}}$$

54. $f(x) = (2-x)^{-1} \;\Rightarrow\; f'(x) = (2-x)^{-2} \;\Rightarrow\; f''(x) = 2(2-x)^{-3} \;\Rightarrow\; f'''(x) = 2\cdot 3(2-x)^{-4} \;\Rightarrow$

$f^{(4)}(x) = 2\cdot 3\cdot 4(2-x)^{-5}$. In general, $f^{(n)}(x) = 2\cdot 3\cdot 4\cdot\cdots\cdot n(2-x)^{-(n+1)} = \dfrac{n!}{(2-x)^{(n+1)}}$.

55. We first show it is true for $n = 1$: $f(x) = xe^x \;\Rightarrow\; f'(x) = xe^x + e^x = (x+1)e^x$. We now assume it is true for $n = k$: $f^{(k)}(x) = (x+k)e^x$. With this assumption, we must show it is true for $n = k+1$:

$$f^{(k+1)}(x) = \frac{d}{dx}\left[f^{(k)}(x)\right] = \frac{d}{dx}\left[(x+k)e^x\right] = (x+k)e^x + e^x = [(x+k)+1]e^x = [x+(k+1)]e^x.$$

Therefore, $f^{(n)}(x) = (x+n)e^x$ by mathematical induction.

56. $$\lim_{t\to 0}\frac{t^3}{\tan^3 2t} = \lim_{t\to 0}\frac{t^3\cos^3 2t}{\sin^3 2t} = \lim_{t\to 0}\cos^3 2t\cdot\frac{1}{8\dfrac{\sin^3 2t}{(2t)^3}} = \lim_{t\to 0}\frac{\cos^3 2t}{8\left(\lim\limits_{t\to 0}\dfrac{\sin 2t}{2t}\right)^3} = \frac{1}{8\cdot 1^3} = \frac{1}{8}$$

57. $y = 4\sin^2 x \;\Rightarrow\; y' = 4\cdot 2\sin x\cos x$. At $\left(\frac{\pi}{6}, 1\right)$, $y' = 8\cdot\frac{1}{2}\cdot\frac{\sqrt{3}}{2} = 2\sqrt{3}$, so an equation of the tangent line is $y - 1 = 2\sqrt{3}\left(x - \frac{\pi}{6}\right)$, or $y = 2\sqrt{3}\,x + 1 - \pi\sqrt{3}/3$.

58. $y = \dfrac{x^2-1}{x^2+1} \;\Rightarrow\; y' = \dfrac{(x^2+1)(2x) - (x^2-1)(2x)}{(x^2+1)^2} = \dfrac{4x}{(x^2+1)^2}$.

At $(0,-1)$, $y' = 0$, so an equation of the tangent line is $y + 1 = 0(x-0)$, or $y = -1$.

59. $y = \sqrt{1 + 4\sin x} \;\Rightarrow\; y' = \frac{1}{2}(1+4\sin x)^{-1/2}\cdot 4\cos x = \dfrac{2\cos x}{\sqrt{1+4\sin x}}$.

At $(0,1)$, $y' = \dfrac{2}{\sqrt{1}} = 2$, so an equation of the tangent line is $y - 1 = 2(x-0)$, or $y = 2x + 1$.

60. $x^2 + 4xy + y^2 = 13 \;\Rightarrow\; 2x + 4(xy' + y\cdot 1) + 2yy' = 0 \;\Rightarrow\; x + 2xy' + 2y + yy' = 0 \;\Rightarrow$

$2xy' + yy' = -x - 2y \;\Rightarrow\; y'(2x+y) = -x - 2y \;\Rightarrow\; y' = \dfrac{-x-2y}{2x+y}$.

At $(2,1)$, $y' = \dfrac{-2-2}{4+1} = -\dfrac{4}{5}$, so an equation of the tangent line is $y - 1 = -\frac{4}{5}(x-2)$, or $y = -\frac{4}{5}x + \frac{13}{5}$.

The slope of the normal line is $\frac{5}{4}$, so an equation of the normal line is $y - 1 = \frac{5}{4}(x-2)$, or $y = \frac{5}{4}x - \frac{3}{2}$.

61. $y = (2+x)e^{-x} \;\Rightarrow\; y' = (2+x)(-e^{-x}) + e^{-x}\cdot 1 = e^{-x}[-(2+x)+1] = e^{-x}(-x-1)$.

At $(0,2)$, $y' = 1(-1) = -1$, so an equation of the tangent line is $y - 2 = -1(x-0)$, or $y = -x + 2$.

The slope of the normal line is 1, so an equation of the normal line is $y - 2 = 1(x-0)$, or $y = x + 2$.

62. $f(x) = xe^{\sin x} \;\Rightarrow\; f'(x) = x[e^{\sin x}(\cos x)] + e^{\sin x}(1) = e^{\sin x}(x\cos x + 1)$. As a check on our work, we notice from the graphs that $f'(x) > 0$ when f is increasing. Also, we see in the larger viewing rectangle a certain similarity in the graphs of f and f': the sizes of the oscillations of f and f' are linked.

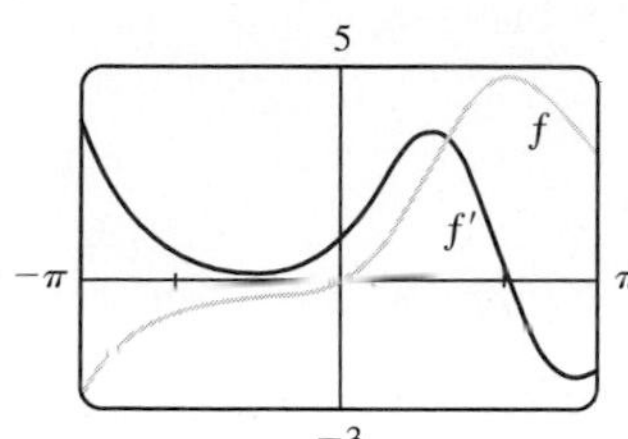

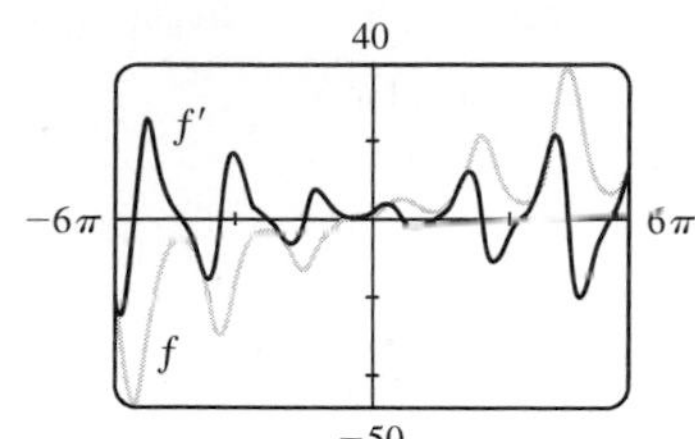

63. (a) $f(x) = x\sqrt{5-x} \;\Rightarrow$

$$f'(x) = x\left[\frac{1}{2}(5-x)^{-1/2}(-1)\right] + \sqrt{5-x} = \frac{-x}{2\sqrt{5-x}} + \sqrt{5-x}\cdot\frac{2\sqrt{5-x}}{2\sqrt{5-x}} = \frac{-x}{2\sqrt{5-x}} + \frac{2(5-x)}{2\sqrt{5-x}}$$

$$= \frac{-x+10-2x}{2\sqrt{5-x}} = \frac{10-3x}{2\sqrt{5-x}}$$

(b) At $(1,2)$: $f'(1) = \frac{7}{4}$.

So an equation of the tangent line is $y - 2 = \frac{7}{4}(x-1)$ or $y = \frac{7}{4}x + \frac{1}{4}$.

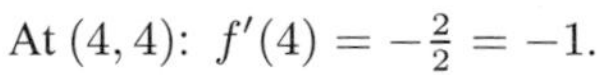

At $(4,4)$: $f'(4) = -\frac{2}{2} = -1$.

So an equation of the tangent line is $y - 4 = -1(x-4)$ or $y = -x + 8$.

(c)

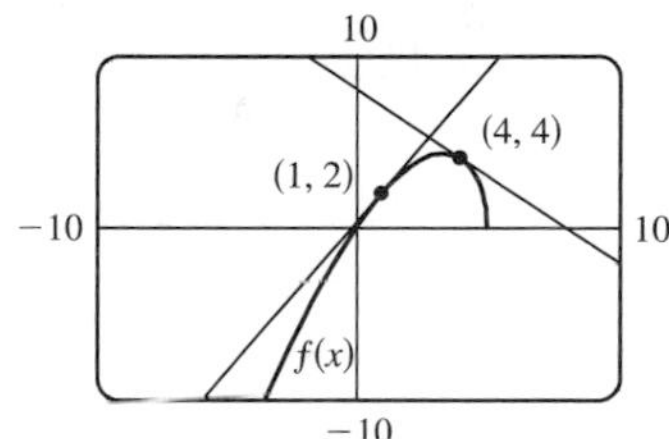

(d) The graphs look reasonable, since f' is positive where f has tangents with positive slope, and f' is negative where f has tangents with negative slope.

64. (a) $f(x) = 4x - \tan x \;\Rightarrow\; f'(x) = 4 - \sec^2 x \;\Rightarrow\; f''(x) = -2\sec x\,(\sec x\,\tan x) = -2\sec^2 x\,\tan x$.

(b)

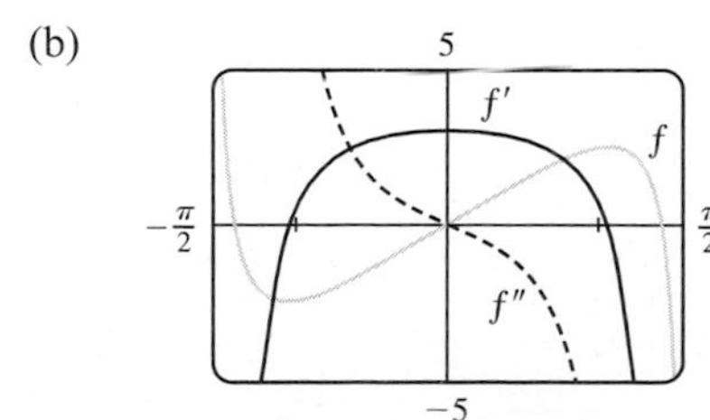

We can see that our answers are reasonable, since the graph of f' is 0 where f has a horizontal tangent, and the graph of f' is positive where f has tangents with positive slope and negative where f has tangents with negative slope. The same correspondence holds between the graphs of f' and f''.

65. $y = \sin x + \cos x \;\Rightarrow\; y' = \cos x - \sin x = 0 \;\Leftrightarrow\; \cos x = \sin x$ and $0 \le x \le 2\pi \;\Leftrightarrow\; x = \frac{\pi}{4}$ or $\frac{5\pi}{4}$, so the points are $\left(\frac{\pi}{4}, \sqrt{2}\right)$ and $\left(\frac{5\pi}{4}, -\sqrt{2}\right)$.

66. $x^2 + 2y^2 = 1 \;\Rightarrow\; 2x + 4yy' = 0 \;\Rightarrow\; y' = -x/(2y) = 1 \;\Leftrightarrow\; x = -2y$. Since the points lie on the ellipse, we have $(-2y)^2 + 2y^2 = 1 \;\Rightarrow\; 6y^2 = 1 \;\Rightarrow\; y = \pm\frac{1}{\sqrt{6}}$. The points are $\left(-\frac{2}{\sqrt{6}}, \frac{1}{\sqrt{6}}\right)$ and $\left(\frac{2}{\sqrt{6}}, -\frac{1}{\sqrt{6}}\right)$.

67. $f(x) = (x-a)(x-b)(x-c) \;\Rightarrow\; f'(x) = (x-b)(x-c) + (x-a)(x-c) + (x-a)(x-b)$.

So $\dfrac{f'(x)}{f(x)} = \dfrac{(x-b)(x-c)+(x-a)(x-c)+(x-a)(x-b)}{(x-a)(x-b)(x-c)} = \dfrac{1}{x-a} + \dfrac{1}{x-b} + \dfrac{1}{x-c}$.

Or: $f(x) = (x-a)(x-b)(x-c) \;\Rightarrow\; \ln|f(x)| = \ln|x-a| + \ln|x-b| + \ln|x-c| \;\Rightarrow$

$\dfrac{f'(x)}{f(x)} = \dfrac{1}{x-a} + \dfrac{1}{x-b} + \dfrac{1}{x-c}$

68. (a) $\cos 2x = \cos^2 x - \sin^2 x \;\Rightarrow\; -2\sin 2x = -2\cos x \sin x - 2\sin x \cos x \;\Leftrightarrow\; \sin 2x = 2\sin x \cos x$

(b) $\sin(x+a) = \sin x\,\cos a + \cos x\,\sin a \;\Rightarrow\; \cos(x+a) = \cos x\,\cos a - \sin x\,\sin a$.

69. (a) $S(x) = f(x) + g(x) \;\Rightarrow\; S'(x) = f'(x) + g'(x) \;\Rightarrow\; S'(1) = f'(1) + g'(1) = 3 + 1 = 4$

(b) $P(x) = f(x)\,g(x) \;\Rightarrow\; P'(x) = f(x)\,g'(x) + g(x)\,f'(x) \;\Rightarrow$

$P'(2) = f(2)\,g'(2) + g(2)f'(2) = 1(4) + 1(2) = 4 + 2 = 6$

(c) $Q(x) = \dfrac{f(x)}{g(x)} \;\Rightarrow\; Q'(x) = \dfrac{g(x)\,f'(x) - f(x)\,g'(x)}{[g(x)]^2} \;\Rightarrow$

$Q'(1) = \dfrac{g(1)\,f'(1) - f(1)\,g'(1)}{[g(1)]^2} = \dfrac{3(3) - 2(1)}{3^2} = \dfrac{9-2}{9} = \dfrac{7}{9}$

(d) $C(x) = f(g(x)) \;\Rightarrow\; C'(x) = f'(g(x))\,g'(x) \;\Rightarrow\; C'(2) = f'(g(2))\,g'(2) = f'(1)\cdot 4 = 3 \cdot 4 = 12$

70. (a) $P(x) = f(x)\,g(x) \;\Rightarrow\; P'(x) = f(x)\,g'(x) + g(x)\,f'(x) \;\Rightarrow$

$P'(2) = f(2)\,g'(2) + g(2)\,f'(2) = (1)\left(\frac{6-0}{3-0}\right) + (4)\left(\frac{0-3}{3-0}\right) = (1)(2) + (4)(-1) = 2 - 4 = -2$

(b) $Q(x) = \dfrac{f(x)}{g(x)} \;\Rightarrow\; Q'(x) = \dfrac{g(x)\,f'(x) - f(x)\,g'(x)}{[g(x)]^2} \;\Rightarrow$

$Q'(2) = \dfrac{g(2)\,f'(2) - f(2)\,g'(2)}{[g(2)]^2} = \dfrac{(4)(-1) - (1)(2)}{4^2} = \dfrac{-6}{16} = -\dfrac{3}{8}$

(c) $C(x) = f(g(x)) \;\Rightarrow\; C'(x) = f'(g(x))g'(x) \;\Rightarrow$

$C'(2) = f'(g(2))g'(2) = f'(4)g'(2) = \left(\frac{6-0}{5-3}\right)(2) = (3)(2) = 6$

71. $f(x) = x^2 g(x) \;\Rightarrow\; f'(x) = x^2 g'(x) + g(x)(2x) = x[xg'(x) + 2g(x)]$

72. $f(x) = g(x^2) \;\Rightarrow\; f'(x) = g'(x^2)(2x) = 2xg'(x^2)$

73. $f(x) = [g(x)]^2 \;\Rightarrow\; f'(x) = 2[g(x)] \cdot g'(x) = 2g(x)\,g'(x)$

74. $f(x) = g(g(x)) \;\Rightarrow\; f'(x) = g'(g(x))\,g'(x)$

75. $f(x) = g(e^x) \;\Rightarrow\; f'(x) = g'(e^x)\,e^x$

76. $f(x) = e^{g(x)} \;\Rightarrow\; f'(x) = e^{g(x)} g'(x)$

77. $f(x) = \ln|g(x)| \;\Rightarrow\; f'(x) = \dfrac{1}{g(x)} g'(x) = \dfrac{g'(x)}{g(x)}$

78. $f(x) = g(\ln x) \;\Rightarrow\; f'(x) = g'(\ln x) \cdot \dfrac{1}{x} = \dfrac{g'(\ln x)}{x}$

79. $h(x) = \dfrac{f(x)\,g(x)}{f(x) + g(x)} \;\Rightarrow$

$$h'(x) = \frac{[f(x) + g(x)]\,[f(x)\,g'(x) + g(x)\,f'(x)] - f(x)\,g(x)\,[f'(x) + g'(x)]}{[f(x) + g(x)]^2}$$

$$= \frac{[f(x)]^2\,g'(x) + f(x)\,g(x)\,f'(x) + f(x)\,g(x)\,g'(x) + [g(x)]^2\,f'(x) - f(x)\,g(x)\,f'(x) - f(x)\,g(x)\,g'(x)}{[f(x) + g(x)]^2}$$

$$= \frac{f'(x)\,[g(x)]^2 + g'(x)\,[f(x)]^2}{[f(x) + g(x)]^2}$$

80. $h(x) = \sqrt{\dfrac{f(x)}{g(x)}} \;\Rightarrow\; h'(x) = \dfrac{f'(x)\,g(x) - f(x)\,g'(x)}{2\,\sqrt{f(x)/g(x)}\,[g(x)]^2} = \dfrac{f'(x)\,g(x) - f(x)\,g'(x)}{2[g(x)]^{3/2}\sqrt{f(x)}}$

81. Using the Chain Rule repeatedly, $h(x) = f(g(\sin 4x)) \;\Rightarrow$

$h'(x) = f'(g(\sin 4x)) \cdot \dfrac{d}{dx}\,(g(\sin 4x)) = f'(g(\sin 4x)) \cdot g'(\sin 4x) \cdot \dfrac{d}{dx}\,(\sin 4x) = f'(g(\sin 4x))g'(\sin 4x)(\cos 4x)(4).$

82. (a)

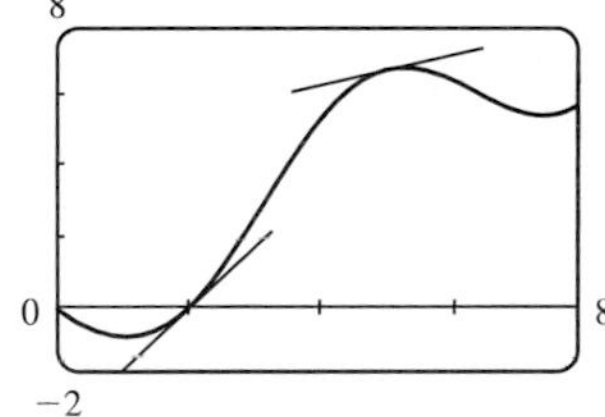

(b) The average rate of change is larger on $[2, 3]$.

(c) The instantaneous rate of change (the slope of the tangent) is larger at $x = 2$.

(d) $f(x) = x - 2\sin x \;\Rightarrow\; f'(x) = 1 - 2\cos x$,

so $f'(2) = 1 - 2\cos 2 \approx 1.8323$ and $f'(5) = 1 - 2\cos 5 \approx 0.4327$.

So $f'(2) > f'(5)$, as predicted in part (c).

83. $y = [\ln(x + 4)]^2 \;\Rightarrow\; y' = 2[\ln(x + 4)]^1 \cdot \dfrac{1}{x + 4} \cdot 1 = 2\,\dfrac{\ln(x + 4)}{x + 4}$ and $y' = 0 \;\Leftrightarrow\; \ln(x + 4) = 0 \;\Leftrightarrow$

$x + 4 = e^0 \;\Rightarrow\; x + 4 = 1 \;\Leftrightarrow\; x = -3$, so the tangent is horizontal at the point $(-3, 0)$.

84. (a) The line $x - 4y = 1$ has slope $\frac{1}{4}$. A tangent to $y = e^x$ has slope $\frac{1}{4}$ when $y' = e^x = \frac{1}{4} \;\Rightarrow\; x = \ln\frac{1}{4} = -\ln 4$.

Since $y = e^x$, the y-coordinate is $\frac{1}{4}$ and the point of tangency is $\left(-\ln 4, \frac{1}{4}\right)$. Thus, an equation of the tangent line

is $y - \frac{1}{4} = \frac{1}{4}(x + \ln 4)$ or $y = \frac{1}{4}x + \frac{1}{4}(\ln 4 + 1)$.

(b) The slope of the tangent at the point (a, e^a) is $\left.\dfrac{d}{dx}\,e^x\right|_{x=a} = e^a$. Thus, an equation of the tangent line is

$y - e^a = e^a(x - a)$. We substitute $x = 0$, $y = 0$ into this equation, since we want the line to pass through the origin:

$0 - e^a = e^a(0 - a) \;\Leftrightarrow\; -e^a = e^a(-a) \;\Leftrightarrow\; a = 1$. So an equation of the tangent line at the point $(a, e^a) = (1, e)$

is $y - e = e(x - 1)$ or $y = ex$.

85. $y = f(x) = ax^2 + bx + c \;\Rightarrow\; f'(x) = 2ax + b$. We know that $f'(-1) = 6$ and $f'(5) = -2$, so $-2a + b = 6$ and

$10a + b = -2$. Subtracting the first equation from the second gives $12a = -8 \;\Rightarrow\; a = -\frac{2}{3}$. Substituting $-\frac{2}{3}$ for a in the

first equation gives $b = \frac{14}{3}$. Now $f(1) = 4 \;\Rightarrow\; 4 = a + b + c$, so $c = 4 + \frac{2}{3} - \frac{14}{3} = 0$ and hence, $f(x) = -\frac{2}{3}x^2 + \frac{14}{3}x$.

86. (a) $\lim_{t\to\infty} C(t) = \lim_{t\to\infty} [K(e^{-at} - e^{-bt})] = K \lim_{t\to\infty} (e^{-at} - e^{-bt}) = K(0-0) = 0$ because $-at \to -\infty$ and $-bt \to -\infty$ as $t \to \infty$.

(b) $C(t) = K(e^{-at} - e^{-bt}) \Rightarrow C'(t) = K(e^{-at}(-a) - e^{-bt}(-b)) = K(-ae^{-at} + be^{-bt})$

(c) $C'(t) = 0 \Leftrightarrow be^{-bt} = ae^{-at} \Leftrightarrow \dfrac{b}{a} = e^{(-a+b)t} \Leftrightarrow \ln\dfrac{b}{a} = (b-a)t \Leftrightarrow t = \dfrac{\ln(b/a)}{b-a}$

87. $s(t) = Ae^{-ct}\cos(\omega t + \delta) \Rightarrow$

$v(t) = s'(t) = A\{e^{-ct}[-\omega\sin(\omega t+\delta)] + \cos(\omega t+\delta)(-ce^{-ct})\} = -Ae^{-ct}[\omega\sin(\omega t+\delta) + c\cos(\omega t+\delta)] \Rightarrow$

$$\begin{aligned} a(t) = v'(t) &= -A\{e^{-ct}[\omega^2\cos(\omega t+\delta) - c\omega\sin(\omega t+\delta)] + [\omega\sin(\omega t+\delta) + c\cos(\omega t+\delta)](-ce^{-ct})\} \\ &= -Ae^{-ct}[\omega^2\cos(\omega t+\delta) - c\omega\sin(\omega t+\delta) - c\omega\sin(\omega t+\delta) - c^2\cos(\omega t+\delta)] \\ &= -Ae^{-ct}[(\omega^2 - c^2)\cos(\omega t+\delta) - 2c\omega\sin(\omega t+\delta)] = Ae^{-ct}[(c^2-\omega^2)\cos(\omega t+\delta) + 2c\omega\sin(\omega t+\delta)] \end{aligned}$$

88. (a) $x = \sqrt{b^2 + c^2t^2} \Rightarrow v(t) = x' = \left[1/\left(2\sqrt{b^2+c^2t^2}\right)\right]2c^2t = c^2t/\sqrt{b^2+c^2t^2} \Rightarrow$

$$a(t) = v'(t) = \frac{c^2\sqrt{b^2+c^2t^2} - c^2t\left(c^2t/\sqrt{b^2+c^2t^2}\right)}{b^2+c^2t^2} = \frac{b^2c^2}{(b^2+c^2t^2)^{3/2}}$$

(b) $v(t) > 0$ for $t > 0$, so the particle always moves in the positive direction.

89. (a) $y = t^3 - 12t + 3 \Rightarrow v(t) = y' = 3t^2 - 12 \Rightarrow a(t) = v'(t) = 6t$

(b) $v(t) = 3(t^2 - 4) > 0$ when $t > 2$, so it moves upward when $t > 2$ and downward when $0 \le t < 2$.

(c) Distance upward $= y(3) - y(2) = -6 - (-13) = 7$,

Distance downward $= y(0) - y(2) = 3 - (-13) = 16$. Total distance $= 7 + 16 = 23$.

(d)

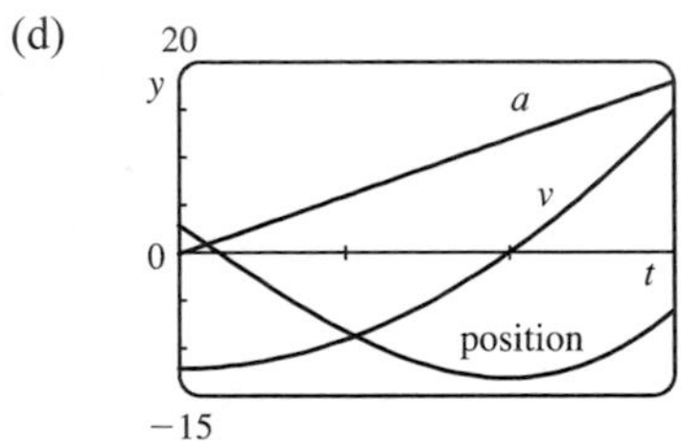

(e) The particle is speeding up when v and a have the same sign, that is, when $t > 2$. The particle is slowing down when v and a have opposite signs; that is, when $0 < t < 2$.

90. (a) $V = \frac{1}{3}\pi r^2 h \Rightarrow dV/dh = \frac{1}{3}\pi r^2$ [r constant]

(b) $V = \frac{1}{3}\pi r^2 h \Rightarrow dV/dr = \frac{2}{3}\pi rh$ [h constant]

91. The linear density ρ is the rate of change of mass m with respect to length x.

$m = x\left(1 + \sqrt{x}\right) = x + x^{3/2} \Rightarrow \rho = dm/dx = 1 + \frac{3}{2}\sqrt{x}$, so the linear density when $x = 4$ is $1 + \frac{3}{2}\sqrt{4} = 4$ kg/m.

92. (a) $C(x) = 920 + 2x - 0.02x^2 + 0.00007x^3 \Rightarrow C'(x) = 2 - 0.04x + 0.00021x^2$

(b) $C'(100) = 2 - 4 + 2.1 = \$0.10/\text{unit}$. This value represents the rate at which costs are increasing as the hundredth unit is produced, and is the approximate cost of producing the 101st unit.

(c) The cost of producing the 101st item is $C(101) - C(100) = 990.10107 - 990 = \$0.10107$, slightly larger than $C'(100)$.

93. (a) $y(t) = y(0)e^{kt} = 200e^{kt} \Rightarrow y(0.5) = 200e^{0.5k} = 360 \Rightarrow e^{0.5k} = 1.8 \Rightarrow 0.5k = \ln 1.8 \Rightarrow$
$k = 2\ln 1.8 = \ln(1.8)^2 = \ln 3.24 \Rightarrow y(t) = 200e^{(\ln 3.24)t} = 200(3.24)^t$

(b) $y(4) = 200(3.24)^4 \approx 22{,}040$ bacteria

(c) $y'(t) = 200(3.24)^t \cdot \ln 3.24$, so $y'(4) = 200(3.24)^4 \cdot \ln 3.24 \approx 25{,}910$ bacteria per hour

(d) $200(3.24)^t = 10{,}000 \Rightarrow (3.24)^t = 50 \Rightarrow t\ln 3.24 = \ln 50 \Rightarrow t = \ln 50/\ln 3.24 \approx 3.33$ hours

94. (a) If $y(t)$ is the mass remaining after t years, then $y(t) = y(0)e^{kt} = 100e^{kt}$. $y(5.24) = 100e^{5.24k} = \frac{1}{2}\cdot 100 \Rightarrow$
$e^{5.24k} = \frac{1}{2} \Rightarrow 5.24k = -\ln 2 \Rightarrow k = -\frac{1}{5.24}\ln 2 \Rightarrow y(t) = 100e^{-(\ln 2)t/5.24} = 100 \cdot 2^{-t/5.24}$. Thus,
$y(20) = 100 \cdot 2^{-20/5.24} \approx 7.1$ mg.

(b) $100 \cdot 2^{-t/5.24} = 1 \Rightarrow 2^{-t/5.24} = \dfrac{1}{100} \Rightarrow -\dfrac{t}{5.24}\ln 2 = \ln\dfrac{1}{100} \Rightarrow t = 5.24\dfrac{\ln 100}{\ln 2} \approx 34.8$ years

95. (a) $C'(t) = -kC(t) \Rightarrow C(t) = C(0)e^{-kt}$ by Theorem 3.8.2. But $C(0) = C_0$, so $C(t) = C_0e^{-kt}$.

(b) $C(30) = \frac{1}{2}C_0$ since the concentration is reduced by half. Thus, $\frac{1}{2}C_0 = C_0e^{-30k} \Rightarrow \ln\frac{1}{2} = -30k \Rightarrow$
$k = -\frac{1}{30}\ln\frac{1}{2} = \frac{1}{30}\ln 2$. Since 10% of the original concentration remains if 90% is eliminated, we want the value of t
such that $C(t) = \frac{1}{10}C_0$. Therefore, $\frac{1}{10}C_0 = C_0e^{-t(\ln 2)/30} \Rightarrow \ln 0.1 = -t(\ln 2)/30 \Rightarrow t = -\frac{30}{\ln 2}\ln 0.1 \approx 100$ h.

96. (a) If $y = u - 20$, $u(0) = 80 \Rightarrow y(0) = 80 - 20 = 60$, and the initial-value problem is $dy/dt = ky$ with $y(0) = 60$.
So the solution is $y(t) = 60e^{kt}$. Now $y(0.5) = 60e^{k(0.5)} = 60 - 20 \Rightarrow e^{0.5k} = \frac{40}{60} = \frac{2}{3} \Rightarrow k = 2\ln\frac{2}{3} = \ln\frac{4}{9}$,
so $y(t) = 60e^{(\ln 4/9)t} = 60(\frac{4}{9})^t$. Thus, $y(1) = 60(\frac{4}{9})^1 = \frac{80}{3} = 26\frac{2}{3}$ °C and $u(1) = 46\frac{2}{3}$ °C.

(b) $u(t) = 40 \Rightarrow y(t) = 20$. $y(t) = 60\left(\dfrac{4}{9}\right)^t = 20 \Rightarrow \left(\dfrac{4}{9}\right)^t = \dfrac{1}{3} \Rightarrow t\ln\dfrac{4}{9} = \ln\dfrac{1}{3} \Rightarrow t = \dfrac{\ln\frac{1}{3}}{\ln\frac{4}{9}} \approx 1.35$ h
or 81.3 min.

97. If x = edge length, then $V = x^3 \Rightarrow dV/dt = 3x^2\,dx/dt = 10 \Rightarrow dx/dt = 10/(3x^2)$ and $S = 6x^2 \Rightarrow$
$dS/dt = (12x)\,dx/dt = 12x[10/(3x^2)] = 40/x$. When $x = 30$, $dS/dt = \frac{40}{30} = \frac{4}{3}$ cm²/min.

98. Given $dV/dt = 2$, find dh/dt when $h = 5$. $V = \frac{1}{3}\pi r^2h$ and, from similar
triangles, $\dfrac{r}{h} = \dfrac{3}{10} \Rightarrow V = \dfrac{\pi}{3}\left(\dfrac{3h}{10}\right)^2 h = \dfrac{3\pi}{100}h^3$, so

$$2 = \frac{dV}{dt} = \frac{9\pi}{100}h^2\frac{dh}{dt} \Rightarrow \frac{dh}{dt} = \frac{200}{9\pi h^2} = \frac{200}{9\pi(5)^2} = \frac{8}{9\pi}\text{ cm/s}$$

when $h = 5$.

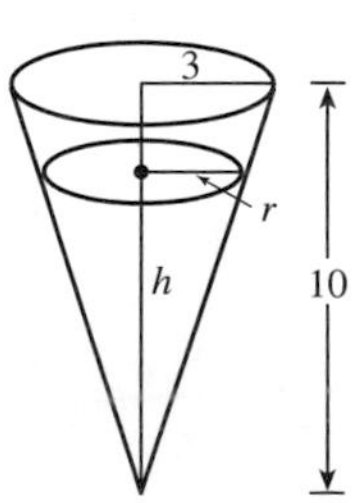

99. Given $dh/dt = 5$ and $dx/dt = 15$, find dz/dt. $z^2 = x^2 + h^2 \Rightarrow$

$$2z\frac{dz}{dt} = 2x\frac{dx}{dt} + 2h\frac{dh}{dt} \Rightarrow \frac{dz}{dt} = \frac{1}{z}(15x + 5h).$$ When $t = 3$,
$h = 45 + 3(5) = 60$ and $x = 15(3) = 45 \Rightarrow z = \sqrt{45^2 + 60^2} = 75$,
so $\dfrac{dz}{dt} = \dfrac{1}{75}[15(45) + 5(60)] = 13$ ft/s.

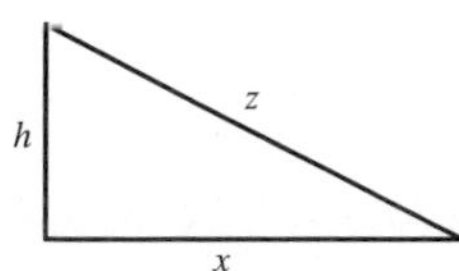

100. We are given $dz/dt = 30$ ft/s. By similar triangles, $\dfrac{y}{z} = \dfrac{4}{\sqrt{241}} \;\Rightarrow$

$y = \dfrac{4}{\sqrt{241}}z$, so $\dfrac{dy}{dt} = \dfrac{4}{\sqrt{241}}\dfrac{dz}{dt} = \dfrac{120}{\sqrt{241}} \approx 7.7$ ft/s.

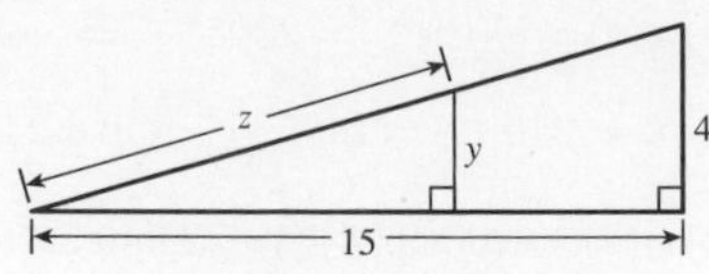

101. We are given $d\theta/dt = -0.25$ rad/h. $\tan\theta = 400/x \;\Rightarrow$

$x = 400\cot\theta \;\Rightarrow\; \dfrac{dx}{dt} = -400\csc^2\theta\,\dfrac{d\theta}{dt}$. When $\theta = \frac{\pi}{6}$,

$\dfrac{dx}{dt} = -400(2)^2(-0.25) = 400$ ft/h.

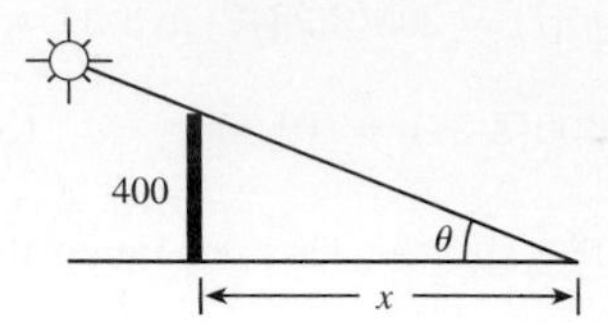

102. (a) $f(x) = \sqrt{25 - x^2} \;\Rightarrow\; f'(x) = \dfrac{-2x}{2\sqrt{25 - x^2}} = -x(25 - x^2)^{-1/2}$.

So the linear approximation to $f(x)$ near 3

is $f(x) \approx f(3) + f'(3)(x - 3) = 4 - \frac{3}{4}(x - 3)$.

(b)

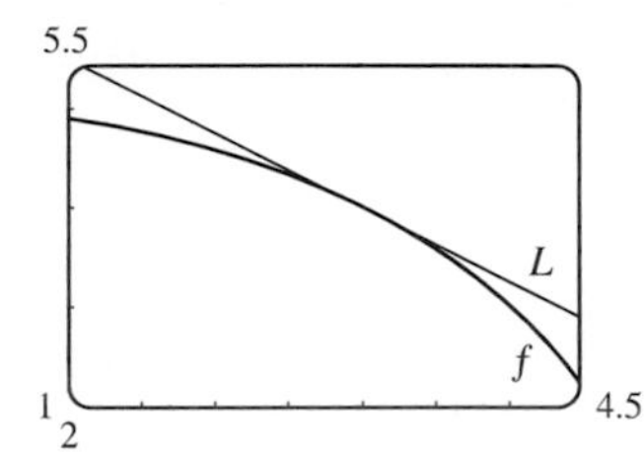

(c) For the required accuracy, we want $\sqrt{25 - x^2} - 0.1 < 4 - \frac{3}{4}(x - 3)$ and

$4 - \frac{3}{4}(x - 3) < \sqrt{25 - x^2} + 0.1$. From the graph, it appears that these both

hold for $2.24 < x < 3.66$.

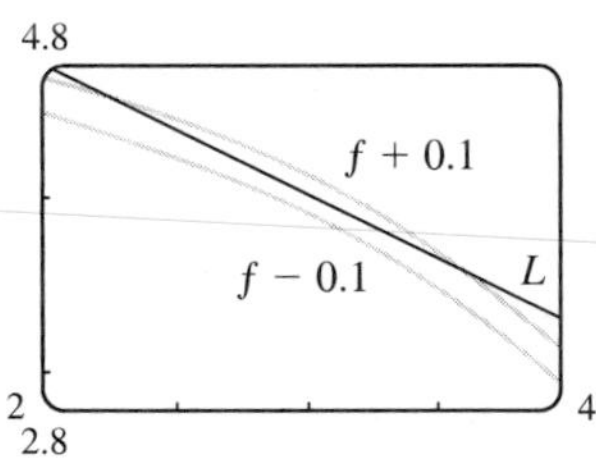

103. (a) $f(x) = \sqrt[3]{1 + 3x} = (1 + 3x)^{1/3} \;\Rightarrow\; f'(x) = (1 + 3x)^{-2/3}$, so the linearization of f at $a = 0$ is

$L(x) = f(0) + f'(0)(x - 0) = 1^{1/3} + 1^{-2/3}x = 1 + x$. Thus, $\sqrt[3]{1 + 3x} \approx 1 + x \;\Rightarrow$

$\sqrt[3]{1.03} = \sqrt[3]{1 + 3(0.01)} \approx 1 + (0.01) = 1.01$.

(b) The linear approximation is $\sqrt[3]{1 + 3x} \approx 1 + x$, so for the required accuracy

we want $\sqrt[3]{1 + 3x} - 0.1 < 1 + x < \sqrt[3]{1 + 3x} + 0.1$. From the graph,

it appears that this is true when $-0.235 < x < 0.401$.

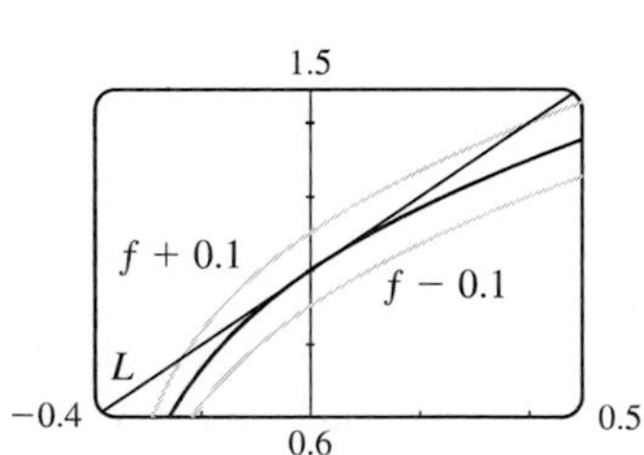

104. $y = x^3 - 2x^2 + 1 \;\Rightarrow\; dy = (3x^2 - 4x)\,dx$. When $x = 2$ and $dx = 0.2$, $dy = \left[3(2)^2 - 4(2)\right](0.2) = 0.8$.

105. $A = x^2 + \frac{1}{2}\pi\left(\frac{1}{2}x\right)^2 = \left(1 + \frac{\pi}{8}\right)x^2 \;\Rightarrow\; dA = \left(2 + \frac{\pi}{4}\right)x\,dx$. When $x = 60$

and $dx = 0.1$, $dA = \left(2 + \frac{\pi}{4}\right)60(0.1) = 12 + \frac{3\pi}{2}$, so the maximum error is

approximately $12 + \frac{3\pi}{2} \approx 16.7\ \text{cm}^2$.

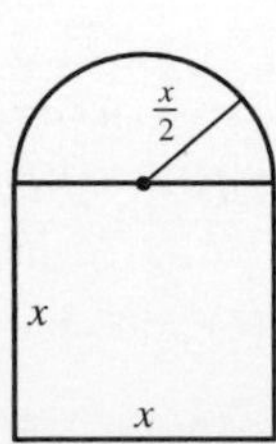

106. $\lim\limits_{x\to 1} \dfrac{x^{17}-1}{x-1} = \left[\dfrac{d}{dx}\, x^{17}\right]_{x=1} = 17(1)^{16} = 17$

107. $\lim\limits_{h\to 0} \dfrac{\sqrt[4]{16+h}-2}{h} = \left[\dfrac{d}{dx}\,\sqrt[4]{x}\right]_{x=16} = \dfrac{1}{4}x^{-3/4}\Big|_{x=16} = \dfrac{1}{4\left(\sqrt[4]{16}\right)^3} = \dfrac{1}{32}$

108. $\lim\limits_{\theta\to \pi/3} \dfrac{\cos\theta - 0.5}{\theta - \pi/3} = \left[\dfrac{d}{d\theta}\cos\theta\right]_{\theta=\pi/3} = -\sin\dfrac{\pi}{3} = -\dfrac{\sqrt{3}}{2}$

109.
$$\begin{aligned}
\lim_{x\to 0}\frac{\sqrt{1+\tan x}-\sqrt{1+\sin x}}{x^3} &= \lim_{x\to 0}\frac{\left(\sqrt{1+\tan x}-\sqrt{1+\sin x}\right)\left(\sqrt{1+\tan x}+\sqrt{1+\sin x}\right)}{x^3\left(\sqrt{1+\tan x}+\sqrt{1+\sin x}\right)} \\
&= \lim_{x\to 0}\frac{(1+\tan x)-(1+\sin x)}{x^3\left(\sqrt{1+\tan x}+\sqrt{1+\sin x}\right)} = \lim_{x\to 0}\frac{\sin x\,(1/\cos x - 1)}{x^3\left(\sqrt{1+\tan x}+\sqrt{1+\sin x}\right)}\cdot\frac{\cos x}{\cos x} \\
&= \lim_{x\to 0}\frac{\sin x\,(1-\cos x)}{x^3\left(\sqrt{1+\tan x}+\sqrt{1+\sin x}\right)\cos x}\cdot\frac{1+\cos x}{1+\cos x} \\
&= \lim_{x\to 0}\frac{\sin x\cdot\sin^2 x}{x^3\left(\sqrt{1+\tan x}+\sqrt{1+\sin x}\right)\cos x\,(1+\cos x)} \\
&= \left(\lim_{x\to 0}\frac{\sin x}{x}\right)^3 \lim_{x\to 0}\frac{1}{\left(\sqrt{1+\tan x}+\sqrt{1+\sin x}\right)\cos x\,(1+\cos x)} \\
&= 1^3\cdot\frac{1}{\left(\sqrt{1}+\sqrt{1}\right)\cdot 1\cdot(1+1)} = \frac{1}{4}
\end{aligned}$$

110. Differentiating the first given equation implicitly with respect to x and using the Chain Rule, we obtain $f(g(x)) = x \;\Rightarrow$ $f'(g(x))\,g'(x) = 1 \;\Rightarrow\; g'(x) = \dfrac{1}{f'(g(x))}$. Using the second given equation to expand the denominator of this expression gives $g'(x) = \dfrac{1}{1+[f(g(x))]^2}$. But the first given equation states that $f(g(x)) = x$, so $g'(x) = \dfrac{1}{1+x^2}$.

111. $\dfrac{d}{dx}\,[f(2x)] = x^2 \;\Rightarrow\; f'(2x)\cdot 2 = x^2 \;\Rightarrow\; f'(2x) = \frac{1}{2}x^2$. Let $t = 2x$. Then $f'(t) = \frac{1}{2}\left(\frac{1}{2}t\right)^2 = \frac{1}{8}t^2$, so $f'(x) = \frac{1}{8}x^2$.

112. Let (b, c) be on the curve, that is, $b^{2/3}+c^{2/3} = a^{2/3}$. Now $x^{2/3}+y^{2/3} = a^{2/3} \;\Rightarrow\; \frac{2}{3}x^{-1/3}+\frac{2}{3}y^{-1/3}\dfrac{dy}{dx} = 0$, so $\dfrac{dy}{dx} = -\dfrac{y^{1/3}}{x^{1/3}} = -\left(\dfrac{y}{x}\right)^{1/3}$, so at (b, c) the slope of the tangent line is $-(c/b)^{1/3}$ and an equation of the tangent line is $y - c = -(c/b)^{1/3}(x-b)$ or $y = -(c/b)^{1/3}x + (c + b^{2/3}c^{1/3})$. Setting $y = 0$, we find that the x-intercept is $b^{1/3}c^{2/3} + b = b^{1/3}(c^{2/3}+b^{2/3}) = b^{1/3}a^{2/3}$ and setting $x = 0$ we find that the y-intercept is $c + b^{2/3}c^{1/3} = c^{1/3}(c^{2/3}+b^{2/3}) = c^{1/3}a^{2/3}$. So the length of the tangent line between these two points is

$$\begin{aligned}
\sqrt{(b^{1/3}a^{2/3})^2+(c^{1/3}a^{2/3})^2} &= \sqrt{b^{2/3}a^{4/3}+c^{2/3}a^{4/3}} = \sqrt{(b^{2/3}+c^{2/3})a^{4/3}} \\
&= \sqrt{a^{2/3}a^{4/3}} = \sqrt{a^2} = a = \text{constant}
\end{aligned}$$

PROBLEMS PLUS

1. Let a be the x-coordinate of Q. Since the derivative of $y = 1 - x^2$ is $y' = -2x$, the slope at Q is $-2a$. But since the triangle is equilateral, $\overline{AO}/\overline{OC} = \sqrt{3}/1$, so the slope at Q is $-\sqrt{3}$. Therefore, we must have that $-2a = -\sqrt{3} \;\Rightarrow\; a = \frac{\sqrt{3}}{2}$.
Thus, the point Q has coordinates $\left(\frac{\sqrt{3}}{2}, 1 - \left(\frac{\sqrt{3}}{2}\right)^2\right) = \left(\frac{\sqrt{3}}{2}, \frac{1}{4}\right)$ and by symmetry, P has coordinates $\left(-\frac{\sqrt{3}}{2}, \frac{1}{4}\right)$.

2. $y = x^3 - 3x + 4 \;\Rightarrow\; y' = 3x^2 - 3$, and $y = 3(x^2 - x) \;\Rightarrow\; y' = 6x - 3$.

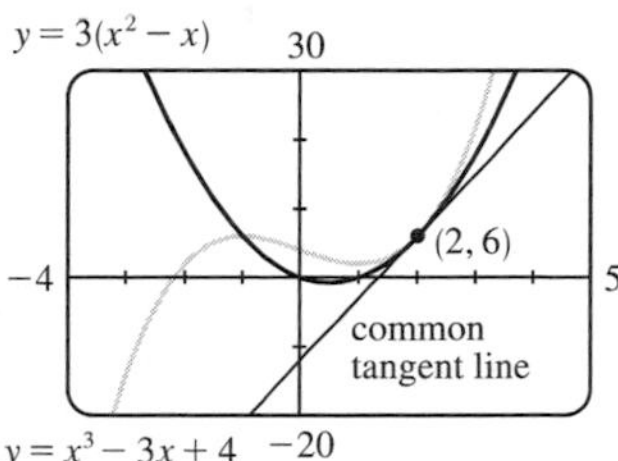

The slopes of the tangents of the two curves are equal when $3x^2 - 3 = 6x - 3$; that is, when $x = 0$ or 2. At $x = 0$, both tangents have slope -3, but the curves do not intersect. At $x = 2$, both tangents have slope 9 and the curves intersect at $(2, 6)$. So there is a common tangent line at $(2, 6)$, $y = 9x - 12$.

3.

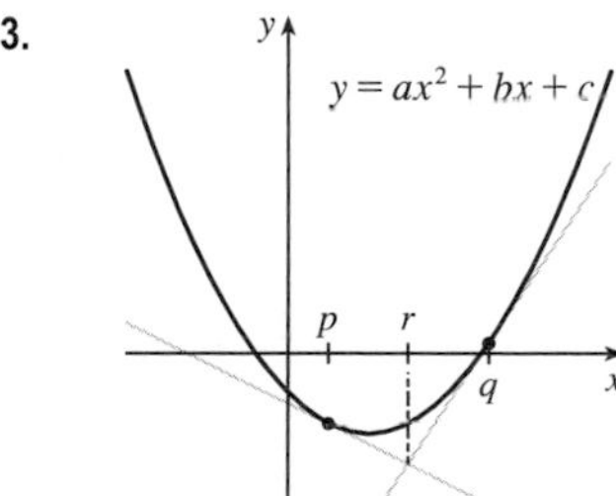

We must show that r (in the figure) is halfway between p and q, that is, $r = (p + q)/2$. For the parabola $y = ax^2 + bx + c$, the slope of the tangent line is given by $y' = 2ax + b$. An equation of the tangent line at $x = p$ is $y - (ap^2 + bp + c) = (2ap + b)(x - p)$. Solving for y gives us

$$y = (2ap + b)x - 2ap^2 - bp + (ap^2 + bp + c)$$

or

$$y = (2ap + b)x + c - ap^2 \qquad \textbf{(1)}$$

Similarly, an equation of the tangent line at $x = q$ is

$$y = (2aq + b)x + c - aq^2 \qquad \textbf{(2)}$$

We can eliminate y and solve for x by subtracting equation **(1)** from equation **(2)**.

$$\begin{aligned}
[(2aq + b) - (2ap + b)]x - aq^2 + ap^2 &= 0 \\
(2aq - 2ap)x &= aq^2 - ap^2 \\
2a(q - p)x &= a(q^2 - p^2) \\
x &= \frac{a(q + p)(q - p)}{2a(q - p)} = \frac{p + q}{2}
\end{aligned}$$

Thus, the x-coordinate of the point of intersection of the two tangent lines, namely r, is $(p + q)/2$.

4. We could differentiate and then simplify or we can simplify and then differentiate. The latter seems to be the simpler method.

$$\begin{aligned}
\frac{\sin^2 x}{1 + \cot x} + \frac{\cos^2 x}{1 + \tan x} &= \frac{\sin^2 x}{1 + \dfrac{\cos x}{\sin x}} \cdot \frac{\sin x}{\sin x} + \frac{\cos^2 x}{1 + \dfrac{\sin x}{\cos x}} \cdot \frac{\cos x}{\cos x} = \frac{\sin^3 x}{\sin x + \cos x} + \frac{\cos^3 x}{\cos x + \sin x} \\
&= \frac{\sin^3 x + \cos^3 x}{\sin x + \cos x} \quad \text{[factor sum of cubes]} \quad = \frac{(\sin x + \cos x)(\sin^2 x - \sin x\, \cos x + \cos^2 x)}{\sin x + \cos x} \\
&= \sin^2 x - \sin x\, \cos x + \cos^2 x = 1 - \sin x\, \cos x = 1 - \tfrac{1}{2}(2 \sin x\, \cos x) = 1 - \tfrac{1}{2} \sin 2x
\end{aligned}$$

Thus, $\dfrac{d}{dx}\left(\dfrac{\sin^2 x}{1 + \cot x} + \dfrac{\cos^2 x}{1 + \tan x}\right) = \dfrac{d}{dx}\left(1 - \tfrac{1}{2} \sin 2x\right) = -\tfrac{1}{2} \cos 2x \cdot 2 = -\cos 2x$.

5. Using $f'(a) = \lim_{x \to a} \dfrac{f(x) - f(a)}{x - a}$, we recognize the given expression, $f(x) = \lim_{t \to x} \dfrac{\sec t - \sec x}{t - x}$, as $g'(x)$ with $g(x) = \sec x$. Now $f'(\frac{\pi}{4}) = g''(\frac{\pi}{4})$, so we will find $g''(x)$. $g'(x) = \sec x \tan x \Rightarrow$

$g''(x) = \sec x \sec^2 x + \tan x \sec x \tan x = \sec x(\sec^2 x + \tan^2 x)$, so $g''(\frac{\pi}{4}) = \sqrt{2}(\sqrt{2}^2 + 1^2) = \sqrt{2}(2 + 1) = 3\sqrt{2}$.

6. Using $f'(0) = \lim_{x \to 0} \dfrac{f(x) - f(0)}{x - 0}$, we see that for the given equation, $\lim_{x \to 0} \dfrac{\sqrt[3]{ax + b} - 2}{x} = \dfrac{5}{12}$, we have $f(x) = \sqrt[3]{ax + b}$, $f(0) = 2$, and $f'(0) = \frac{5}{12}$. Now $f(0) = 2 \Leftrightarrow \sqrt[3]{b} = 2 \Leftrightarrow b = 8$. Also $f'(x) = \frac{1}{3}(ax + b)^{-2/3} \cdot a$, so

$f'(0) = \frac{5}{12} \Leftrightarrow \frac{1}{3}(8)^{-2/3} \cdot a = \frac{5}{12} \Leftrightarrow \frac{1}{3}(\frac{1}{4})a = \frac{5}{12} \Leftrightarrow a = 5$.

7. Let $y = \tan^{-1} x$. Then $\tan y = x$, so from the triangle we see that

$\sin(\tan^{-1} x) = \sin y = \dfrac{x}{\sqrt{1 + x^2}}$. Using this fact we have that

$\sin(\tan^{-1}(\sinh x)) = \dfrac{\sinh x}{\sqrt{1 + \sinh^2 x}} = \dfrac{\sinh x}{\cosh x} = \tanh x$.

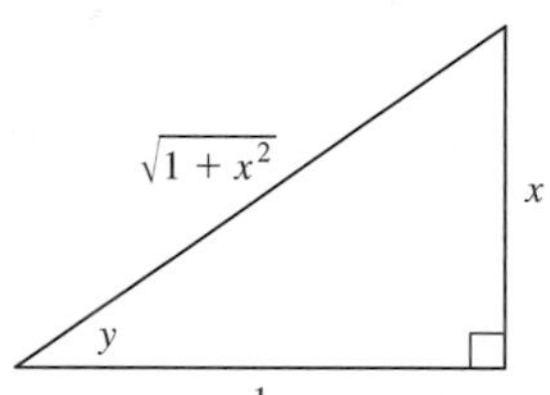

Hence, $\sin^{-1}(\tanh x) = \sin^{-1}(\sin(\tan^{-1}(\sinh x))) = \tan^{-1}(\sinh x)$.

8. We find the equation of the parabola by substituting the point $(-100, 100)$, at which the car is situated, into the general equation $y = ax^2$: $100 = a(-100)^2 \Rightarrow a = \frac{1}{100}$. Now we find the equation of a tangent to the parabola at the point (x_0, y_0). We can show that $y' = a(2x) = \frac{1}{100}(2x) = \frac{1}{50}x$, so an equation of the tangent is $y - y_0 = \frac{1}{50}x_0(x - x_0)$. Since the point (x_0, y_0) is on the parabola, we must have $y_0 = \frac{1}{100}x_0^2$, so our equation of the tangent can be simplified to $y = \frac{1}{100}x_0^2 + \frac{1}{50}x_0(x - x_0)$. We want the statue to be located on the tangent line, so we substitute its coordinates $(100, 50)$ into this equation: $50 = \frac{1}{100}x_0^2 + \frac{1}{50}x_0(100 - x_0) \Rightarrow x_0^2 - 200x_0 + 5000 = 0 \Rightarrow$

$x_0 = \frac{1}{2}\left[200 \pm \sqrt{200^2 - 4(5000)}\right] \Rightarrow x_0 = 100 \pm 50\sqrt{2}$. But $x_0 < 100$, so the car's headlights illuminate the statue when it is located at the point $(100 - 50\sqrt{2}, 150 - 100\sqrt{2}) \approx (29.3, 8.6)$, that is, about 29.3 m east and 8.6 m north of the origin.

9. We use mathematical induction. Let S_n be the statement that $\dfrac{d^n}{dx^n}(\sin^4 x + \cos^4 x) = 4^{n-1}\cos(4x + n\pi/2)$.

S_1 is true because

$$\begin{aligned}\frac{d}{dx}(\sin^4 x + \cos^4 x) &= 4\sin^3 x\,\cos x - 4\cos^3 x\,\sin x = 4\sin x\,\cos x\,(\sin^2 x - \cos^2 x)\,x \\ &= -4\sin x\,\cos x\,\cos 2x = -2\sin 2x\,\cos 2 = -\sin 4x = \sin(-4x) \\ &= \cos\left(\tfrac{\pi}{2} - (-4x)\right) = \cos\left(\tfrac{\pi}{2} + 4x\right) = 4^{n-1}\cos\left(4x + n\tfrac{\pi}{2}\right) \text{ when } n = 1\end{aligned}$$

[continued]

Now assume S_k is true, that is, $\dfrac{d^k}{dx^k}\left(\sin^4 x + \cos^4 x\right) = 4^{k-1}\cos\left(4x + k\frac{\pi}{2}\right)$. Then

$$\begin{aligned}\frac{d^{k+1}}{dx^{k+1}}\left(\sin^4 x + \cos^4 x\right) &= \frac{d}{dx}\left[\frac{d^k}{dx^k}\left(\sin^4 x + \cos^4 x\right)\right] = \frac{d}{dx}\left[4^{k-1}\cos\left(4x + k\tfrac{\pi}{2}\right)\right] \\ &= -4^{k-1}\sin\left(4x + k\tfrac{\pi}{2}\right)\cdot\frac{d}{dx}\left(4x + k\tfrac{\pi}{2}\right) = -4^k\sin\left(4x + k\tfrac{\pi}{2}\right) \\ &= 4^k\sin\left(-4x - k\tfrac{\pi}{2}\right) = 4^k\cos\left(\tfrac{\pi}{2} - \left(-4x - k\tfrac{\pi}{2}\right)\right) = 4^k\cos\left(4x + (k+1)\tfrac{\pi}{2}\right)\end{aligned}$$

which shows that S_{k+1} is true.

Therefore, $\dfrac{d^n}{dx^n}\left(\sin^4 x + \cos^4 x\right) = 4^{n-1}\cos\left(4x + n\frac{\pi}{2}\right)$ for every positive integer n, by mathematical induction.

Another proof: First write

$$\sin^4 x + \cos^4 x = (\sin^2 x + \cos^2 x)^2 - 2\sin^2 x\cos^2 x = 1 - \tfrac{1}{2}\sin^2 2x = 1 - \tfrac{1}{4}(1 - \cos 4x) = \tfrac{3}{4} + \tfrac{1}{4}\cos 4x$$

Then we have $\dfrac{d^n}{dx^n}\left(\sin^4 x + \cos^4 x\right) = \dfrac{d^n}{dx^n}\left(\dfrac{3}{4} + \dfrac{1}{4}\cos 4x\right) = \dfrac{1}{4}\cdot 4^n\cos\left(4x + n\dfrac{\pi}{2}\right) = 4^{n-1}\cos\left(4x + n\dfrac{\pi}{2}\right)$.

10. $$\begin{aligned}\lim_{x\to a}\frac{f(x) - f(a)}{\sqrt{x} - \sqrt{a}} &= \lim_{x\to a}\left[\frac{f(x) - f(a)}{\sqrt{x} - \sqrt{a}}\cdot\frac{\sqrt{x} + \sqrt{a}}{\sqrt{x} + \sqrt{a}}\right] = \lim_{x\to a}\left[\frac{f(x) - f(a)}{x - a}\cdot\left(\sqrt{x} + \sqrt{a}\right)\right] \\ &= \lim_{x\to a}\frac{f(x) - f(a)}{x - a}\cdot\lim_{x\to a}\left(\sqrt{x} + \sqrt{a}\right) = f'(a)\cdot\left(\sqrt{a} + \sqrt{a}\right) = 2\sqrt{a}\,f'(a)\end{aligned}$$

11. We must find a value x_0 such that the normal lines to the parabola $y = x^2$ at $x = \pm x_0$ intersect at a point one unit from the points $\left(\pm x_0, x_0^2\right)$. The normals to $y = x^2$ at $x = \pm x_0$ have slopes $-\dfrac{1}{\pm 2x_0}$ and pass through $\left(\pm x_0, x_0^2\right)$ respectively, so the normals have the equations $y - x_0^2 = -\dfrac{1}{2x_0}(x - x_0)$ and $y - x_0^2 = \dfrac{1}{2x_0}(x + x_0)$. The common y-intercept is $x_0^2 + \dfrac{1}{2}$.

We want to find the value of x_0 for which the distance from $\left(0, x_0^2 + \frac{1}{2}\right)$ to $\left(x_0, x_0^2\right)$ equals 1. The square of the distance is $(x_0 - 0)^2 + \left[x_0^2 - \left(x_0^2 + \frac{1}{2}\right)\right]^2 = x_0^2 + \frac{1}{4} = 1 \quad\Leftrightarrow\quad x_0 = \pm\frac{\sqrt{3}}{2}$. For these values of x_0, the y-intercept is $x_0^2 + \frac{1}{2} = \frac{5}{4}$, so the center of the circle is at $\left(0, \frac{5}{4}\right)$.

Another solution: Let the center of the circle be $(0, a)$. Then the equation of the circle is $x^2 + (y - a)^2 = 1$. Solving with the equation of the parabola, $y = x^2$, we get $x^2 + (x^2 - a)^2 = 1 \quad\Leftrightarrow\quad x^2 + x^4 - 2ax^2 + a^2 = 1 \quad\Leftrightarrow$ $x^4 + (1 - 2a)x^2 + a^2 - 1 = 0$. The parabola and the circle will be tangent to each other when this quadratic equation in x^2 has equal roots; that is, when the discriminant is 0. Thus, $(1 - 2a)^2 - 4(a^2 - 1) = 0 \quad\Leftrightarrow$ $1 - 4a + 4a^2 - 4a^2 + 4 = 0 \quad\Leftrightarrow\quad 4a = 5$, so $a = \frac{5}{4}$. The center of the circle is $\left(0, \frac{5}{4}\right)$.

12. See the figure. The parabolas $y = 4x^2$ and $x = c + 2y^2$ intersect each other at right angles at the point (a, b) if and only if (a, b) satisfies both equations and the tangent lines at (a, b) are perpendicular. $y = 4x^2 \Rightarrow y' = 8x$ and $x = c + 2y^2 \Rightarrow 1 = 4y\,y' \Rightarrow y' = \dfrac{1}{4y}$, so at (a, b) we must

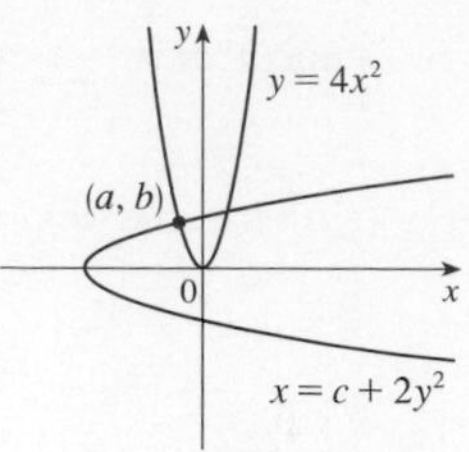

have $8a = -\dfrac{1}{1/(4b)} \Rightarrow 8a = -4b \Rightarrow b = -2a$. Since (a, b) is on both parabolas, we have **(1)** $b = 4a^2$ and **(2)** $a = c + 2b^2$. Substituting $-2a$ for b in **(1)** gives us $-2a = 4a^2 \Rightarrow 4a^2 + 2a = 0 \Rightarrow 2a(2a + 1) = 0 \Rightarrow a = 0$ or $a = -\frac{1}{2}$.

If $a = 0$, then $b = 0$ and $c = 0$, and the tangent lines at $(0, 0)$ are $y = 0$ and $x = 0$.

If $a = -\frac{1}{2}$, then $b = -2\left(-\frac{1}{2}\right) = 1$ and $-\frac{1}{2} = c + 2(1)^2 \Rightarrow c = -\frac{5}{2}$, and the tangent lines at $\left(-\frac{1}{2}, 1\right)$ are $y - 1 = -4\left(x + \frac{1}{2}\right)$ [or $y = -4x - 1$] and $y - 1 = \frac{1}{4}\left(x + \frac{1}{2}\right)$ $\left[\text{or } y = \frac{1}{4}x + \frac{9}{8}\right]$.

13. See the figure. Clearly, the line $y = 2$ is tangent to both circles at the point $(0, 2)$. We'll look for a tangent line L through the points (a, b) and (c, d), and if such a line exists, then its reflection through the y-axis is another such line. The slope of L is the same at (a, b) and (c, d). Find those slopes: $x^2 + y^2 = 4 \Rightarrow$ $2x + 2y\,y' = 0 \Rightarrow y' = -\dfrac{x}{y}$ $\left[= -\dfrac{a}{b}\right]$ and $x^2 + (y - 3)^2 = 1 \Rightarrow$

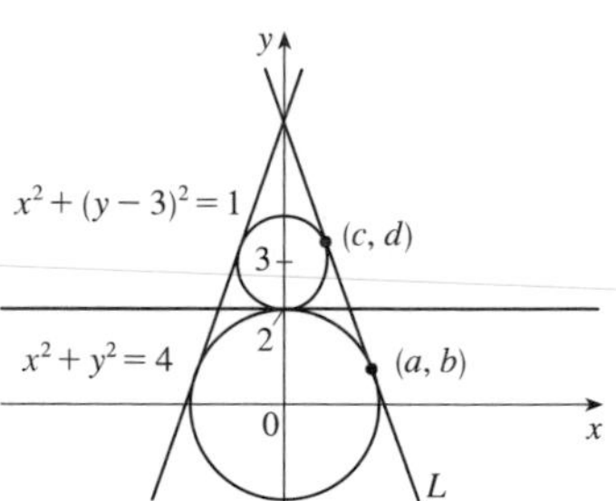

$$2x + 2(y - 3)y' = 0 \Rightarrow y' = -\frac{x}{y - 3} \quad \left[= -\frac{c}{d - 3}\right].$$

Now an equation for L can be written using either point-slope pair, so we get $y - b = -\dfrac{a}{b}(x - a)$ $\left[\text{or } y = -\dfrac{a}{b}x + \dfrac{a^2}{b} + b\right]$ and $y - d = -\dfrac{c}{d - 3}(x - c)$ $\left[\text{or } y = -\dfrac{c}{d - 3}x + \dfrac{c^2}{d - 3} + d\right]$. The slopes are equal, so $-\dfrac{a}{b} = -\dfrac{c}{d - 3} \Leftrightarrow$ $d - 3 = \dfrac{bc}{a}$. Since (c, d) is a solution of $x^2 + (y - 3)^2 = 1$, we have $c^2 + (d - 3)^2 = 1$, so $c^2 + \left(\dfrac{bc}{a}\right)^2 = 1 \Rightarrow$ $a^2c^2 + b^2c^2 = a^2 \Rightarrow c^2(a^2 + b^2) = a^2 \Rightarrow 4c^2 = a^2$ [since (a, b) is a solution of $x^2 + y^2 = 4$] $\Rightarrow a = 2c$.

Now $d - 3 = \dfrac{bc}{a} \Rightarrow d = 3 + \dfrac{bc}{2c}$, so $d = 3 + \dfrac{b}{2}$. The y-intercepts are equal, so $\dfrac{a^2}{b} + b = \dfrac{c^2}{d - 3} + d \Leftrightarrow$

$$\frac{a^2}{b} + b = \frac{(a/2)^2}{b/2} + \left(3 + \frac{b}{2}\right) \Leftrightarrow \left[\frac{a^2}{b} + b = \frac{a^2}{2b} + 3 + \frac{b}{2}\right](2b) \Leftrightarrow 2a^2 + 2b^2 = a^2 + 6b + b^2 \Leftrightarrow$$

$a^2 + b^2 = 6b \Leftrightarrow 4 = 6b \Leftrightarrow b = \frac{2}{3}$. It follows that $d = 3 + \dfrac{b}{2} = \dfrac{10}{3}$, $a^2 = 4 - b^2 = 4 - \frac{4}{9} = \frac{32}{9} \Rightarrow a = \frac{4}{3}\sqrt{2}$, and $c^2 = 1 - (d - 3)^2 = 1 - \left(\frac{1}{3}\right)^2 = \frac{8}{9} \Rightarrow c = \frac{2}{3}\sqrt{2}$. Thus, L has equation $y - \dfrac{2}{3} = -\dfrac{(4/3)\sqrt{2}}{2/3}\left(x - \dfrac{4}{3}\sqrt{2}\right) \Leftrightarrow$ $y - \frac{2}{3} = -2\sqrt{2}\left(x - \frac{4}{3}\sqrt{2}\right) \Leftrightarrow y = -2\sqrt{2}\,x + 6$. Its reflection has equation $y = 2\sqrt{2}\,x + 6$.

[continued]

In summary, there are three lines tangent to both circles: $y = 2$ touches at $(0, 2)$, L touches at $\left(\frac{4}{3}\sqrt{2}, \frac{2}{3}\right)$ and $\left(\frac{2}{3}\sqrt{2}, \frac{10}{3}\right)$, and its reflection through the y-axis touches at $\left(-\frac{4}{3}\sqrt{2}, \frac{2}{3}\right)$ and $\left(-\frac{2}{3}\sqrt{2}, \frac{10}{3}\right)$.

14. $f(x) = \dfrac{x^{46} + x^{45} + 2}{1 + x} = \dfrac{x^{45}(x+1) + 2}{x+1} = \dfrac{x^{45}(x+1)}{x+1} + \dfrac{2}{x+1} = x^{45} + 2(x+1)^{-1}$, so

$f^{(46)}(x) = (x^{45})^{(46)} + 2\left[(x+1)^{-1}\right]^{(46)}$. The forty-sixth derivative of any forty-fifth degree polynomial is 0, so $(x^{45})^{46} = 0$. Thus, $f^{(46)}(x) = 2\left[(-1)(-2)(-3)\cdots(-46)(x+1)^{-47}\right] = 2(46!)(x+1)^{-47}$ and $f^{(46)}(3) = 2(46!)(4)^{-47}$ or $(46!)2^{-93}$.

15. We can assume without loss of generality that $\theta = 0$ at time $t = 0$, so that $\theta = 12\pi t$ rad. [The angular velocity of the wheel is 360 rpm $= 360 \cdot (2\pi \text{ rad})/(60 \text{ s}) = 12\pi$ rad/s.] Then the position of A as a function of time is

$A = (40\cos\theta, 40\sin\theta) = (40\cos 12\pi t, 40\sin 12\pi t)$, so $\sin\alpha = \dfrac{y}{1.2 \text{ m}} = \dfrac{40\sin\theta}{120} = \dfrac{\sin\theta}{3} = \dfrac{1}{3}\sin 12\pi t$.

(a) Differentiating the expression for $\sin\alpha$, we get $\cos\alpha \cdot \dfrac{d\alpha}{dt} = \dfrac{1}{3} \cdot 12\pi \cdot \cos 12\pi t = 4\pi\cos\theta$. When $\theta = \dfrac{\pi}{3}$, we have

$$\sin\alpha = \frac{1}{3}\sin\theta = \frac{\sqrt{3}}{6}, \text{ so } \cos\alpha = \sqrt{1 - \left(\frac{\sqrt{3}}{6}\right)^2} = \sqrt{\frac{11}{12}} \text{ and } \frac{d\alpha}{dt} = \frac{4\pi\cos\frac{\pi}{3}}{\cos\alpha} = \frac{2\pi}{\sqrt{11/12}} = \frac{4\pi\sqrt{3}}{\sqrt{11}} \approx 6.56 \text{ rad/s.}$$

(b) By the Law of Cosines, $|AP|^2 = |OA|^2 + |OP|^2 - 2\,|OA|\,|OP|\cos\theta \;\Rightarrow$

$120^2 = 40^2 + |OP|^2 - 2 \cdot 40\,|OP|\cos\theta \;\Rightarrow\; |OP|^2 - (80\cos\theta)\,|OP| - 12{,}800 = 0 \;\Rightarrow$

$|OP| = \frac{1}{2}\left(80\cos\theta \pm \sqrt{6400\cos^2\theta + 51{,}200}\right) = 40\cos\theta \pm 40\sqrt{\cos^2\theta + 8} = 40\left(\cos\theta + \sqrt{8 + \cos^2\theta}\right)$ cm

[since $|OP| > 0$]. As a check, note that $|OP| = 160$ cm when $\theta = 0$ and $|OP| = 80\sqrt{2}$ cm when $\theta = \frac{\pi}{2}$.

(c) By part (b), the x-coordinate of P is given by $x = 40\left(\cos\theta + \sqrt{8 + \cos^2\theta}\right)$, so

$$\frac{dx}{dt} = \frac{dx}{d\theta}\frac{d\theta}{dt} = 40\left(-\sin\theta - \frac{2\cos\theta\sin\theta}{2\sqrt{8 + \cos^2\theta}}\right) \cdot 12\pi = -480\pi\sin\theta\left(1 + \frac{\cos\theta}{\sqrt{8 + \cos^2\theta}}\right) \text{ cm/s.}$$

In particular, $dx/dt = 0$ cm/s when $\theta = 0$ and $dx/dt = -480\pi$ cm/s when $\theta = \frac{\pi}{2}$.

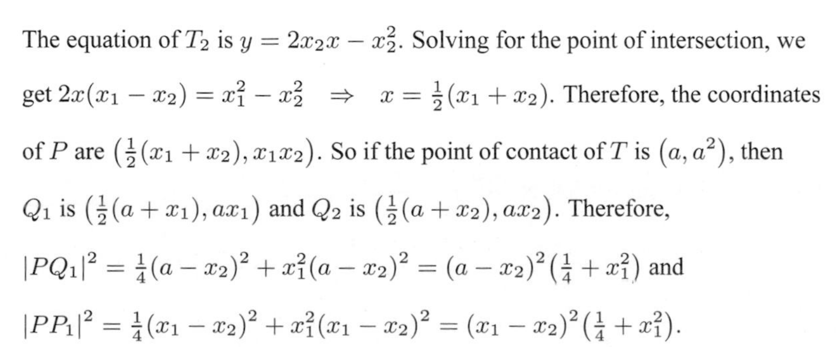

16. The equation of T_1 is $y - x_1^2 = 2x_1(x - x_1) = 2x_1 x - 2x_1^2$ or $y = 2x_1 x - x_1^2$. The equation of T_2 is $y = 2x_2 x - x_2^2$. Solving for the point of intersection, we get $2x(x_1 - x_2) = x_1^2 - x_2^2 \;\Rightarrow\; x = \frac{1}{2}(x_1 + x_2)$. Therefore, the coordinates of P are $\left(\frac{1}{2}(x_1 + x_2), x_1 x_2\right)$. So if the point of contact of T is (a, a^2), then Q_1 is $\left(\frac{1}{2}(a + x_1), a x_1\right)$ and Q_2 is $\left(\frac{1}{2}(a + x_2), a x_2\right)$. Therefore,

$|PQ_1|^2 = \frac{1}{4}(a - x_2)^2 + x_1^2(a - x_2)^2 = (a - x_2)^2\left(\frac{1}{4} + x_1^2\right)$ and

$|PP_1|^2 = \frac{1}{4}(x_1 - x_2)^2 + x_1^2(x_1 - x_2)^2 = (x_1 - x_2)^2\left(\frac{1}{4} + x_1^2\right)$.

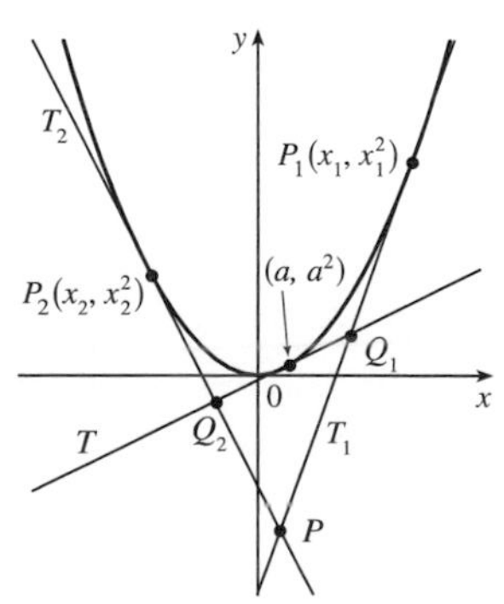

So $\dfrac{|PQ_1|^2}{|PP_1|^2} = \dfrac{(a - x_2)^2}{(x_1 - x_2)^2}$, and similarly $\dfrac{|PQ_2|^2}{|PP_2|^2} = \dfrac{(x_1 - a)^2}{(x_1 - x_2)^2}$. Finally, $\dfrac{|PQ_1|}{|PP_1|} + \dfrac{|PQ_2|}{|PP_2|} = \dfrac{a - x_2}{x_1 - x_2} + \dfrac{x_1 - a}{x_1 - x_2} = 1$.

17. Consider the statement that $\dfrac{d^n}{dx^n}(e^{ax}\sin bx) = r^n e^{ax}\sin(bx + n\theta)$. For $n = 1$,

$$\frac{d}{dx}(e^{ax}\sin bx) = ae^{ax}\sin bx + be^{ax}\cos bx, \text{ and}$$

$$re^{ax}\sin(bx+\theta) = re^{ax}[\sin bx\cos\theta + \cos bx\sin\theta] = re^{ax}\left(\frac{a}{r}\sin bx + \frac{b}{r}\cos bx\right) = ae^{ax}\sin bx + be^{ax}\cos bx$$

since $\tan\theta = \dfrac{b}{a} \Rightarrow \sin\theta = \dfrac{b}{r}$ and $\cos\theta = \dfrac{a}{r}$. So the statement is true for $n = 1$.

Assume it is true for $n = k$. Then

$$\begin{aligned}\frac{d^{k+1}}{dx^{k+1}}(e^{ax}\sin bx) &= \frac{d}{dx}\left[r^k e^{ax}\sin(bx+k\theta)\right] = r^k ae^{ax}\sin(bx+k\theta) + r^k e^{ax} b\cos(bx+k\theta) \\ &= r^k e^{ax}[a\sin(bx+k\theta) + b\cos(bx+k\theta)]\end{aligned}$$

But

$$\sin[bx+(k+1)\theta] = \sin[(bx+k\theta)+\theta] = \sin(bx+k\theta)\cos\theta + \sin\theta\cos(bx+k\theta) = \tfrac{a}{r}\sin(bx+k\theta) + \tfrac{b}{r}\cos(bx+k\theta).$$

Hence, $a\sin(bx+k\theta) + b\cos(bx+k\theta) = r\sin[bx+(k+1)\theta]$. So

$$\frac{d^{k+1}}{dx^{k+1}}(e^{ax}\sin bx) = r^k e^{ax}[a\sin(bx+k\theta)+b\cos(bx+k\theta)] = r^k e^{ax}[r\sin(bx+(k+1)\theta)] = r^{k+1}e^{ax}[\sin(bx+(k+1)\theta)].$$

Therefore, the statement is true for all n by mathematical induction.

18. We recognize this limit as the definition of the derivative of the function $f(x) = e^{\sin x}$ at $x = \pi$, since it is of the form $\displaystyle\lim_{x\to\pi}\frac{f(x)-f(\pi)}{x-\pi}$. Therefore, the limit is equal to $f'(\pi) = (\cos\pi)e^{\sin\pi} = -1\cdot e^0 = -1$.

19. It seems from the figure that as P approaches the point $(0, 2)$ from the right, $x_T \to \infty$ and $y_T \to 2^+$. As P approaches the point $(3, 0)$ from the left, it appears that $x_T \to 3^+$ and $y_T \to \infty$. So we guess that $x_T \in (3, \infty)$ and $y_T \in (2, \infty)$. It is more difficult to estimate the range of values for x_N and y_N. We might perhaps guess that $x_N \in (0, 3)$, and $y_N \in (-\infty, 0)$ or $(-2, 0)$.

In order to actually solve the problem, we implicitly differentiate the equation of the ellipse to find the equation of the tangent line: $\dfrac{x^2}{9} + \dfrac{y^2}{4} = 1 \Rightarrow \dfrac{2x}{9} + \dfrac{2y}{4}y' = 0$, so $y' = -\dfrac{4}{9}\dfrac{x}{y}$. So at the point (x_0, y_0) on the ellipse, an equation of the tangent line is $y - y_0 = -\dfrac{4}{9}\dfrac{x_0}{y_0}(x - x_0)$ or $4x_0x + 9y_0y = 4x_0^2 + 9y_0^2$. This can be written as $\dfrac{x_0x}{9} + \dfrac{y_0y}{4} = \dfrac{x_0^2}{9} + \dfrac{y_0^2}{4} = 1$, because (x_0, y_0) lies on the ellipse. So an equation of the tangent line is $\dfrac{x_0x}{9} + \dfrac{y_0y}{4} = 1$.

Therefore, the x-intercept x_T for the tangent line is given by $\dfrac{x_0x_T}{9} = 1 \Leftrightarrow x_T = \dfrac{9}{x_0}$, and the y-intercept y_T is given by $\dfrac{y_0y_T}{4} = 1 \Leftrightarrow y_T = \dfrac{4}{y_0}$.

So as x_0 takes on all values in $(0, 3)$, x_T takes on all values in $(3, \infty)$, and as y_0 takes on all values in $(0, 2)$, y_T takes on all values in $(2, \infty)$. At the point (x_0, y_0) on the ellipse, the slope of the normal line is $-\dfrac{1}{y'(x_0, y_0)} = \dfrac{9}{4}\dfrac{y_0}{x_0}$, and its equation is $y - y_0 = \dfrac{9}{4}\dfrac{y_0}{x_0}(x - x_0)$. So the x-intercept x_N for the normal line is given by $0 - y_0 = \dfrac{9}{4}\dfrac{y_0}{x_0}(x_N - x_0) \Rightarrow$ $x_N = -\dfrac{4x_0}{9} + x_0 = \dfrac{5x_0}{9}$, and the y-intercept y_N is given by $y_N - y_0 = \dfrac{9}{4}\dfrac{y_0}{x_0}(0 - x_0) \Rightarrow y_N = -\dfrac{9y_0}{4} + y_0 = -\dfrac{5y_0}{4}$.

So as x_0 takes on all values in $(0, 3)$, x_N takes on all values in $\left(0, \frac{5}{3}\right)$, and as y_0 takes on all values in $(0, 2)$, y_N takes on all values in $\left(-\frac{5}{2}, 0\right)$.

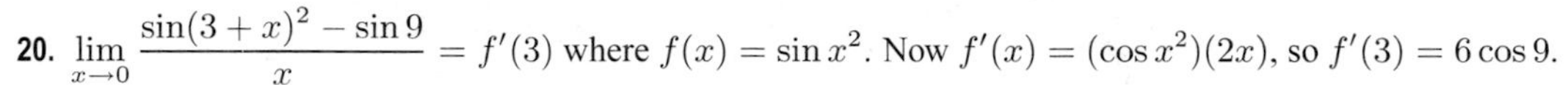

20. $\displaystyle\lim_{x \to 0} \frac{\sin(3 + x)^2 - \sin 9}{x} = f'(3)$ where $f(x) = \sin x^2$. Now $f'(x) = (\cos x^2)(2x)$, so $f'(3) = 6\cos 9$.

21. (a) If the two lines L_1 and L_2 have slopes m_1 and m_2 and angles of inclination ϕ_1 and ϕ_2, then $m_1 = \tan\phi_1$ and $m_2 = \tan\phi_2$. The triangle in the figure shows that $\phi_1 + \alpha + (180° - \phi_2) = 180°$ and so $\alpha = \phi_2 - \phi_1$. Therefore, using the identity for $\tan(x - y)$, we have $\tan\alpha = \tan(\phi_2 - \phi_1) = \dfrac{\tan\phi_2 - \tan\phi_1}{1 + \tan\phi_2 \tan\phi_1}$ and so $\tan\alpha = \dfrac{m_2 - m_1}{1 + m_1 m_2}$.

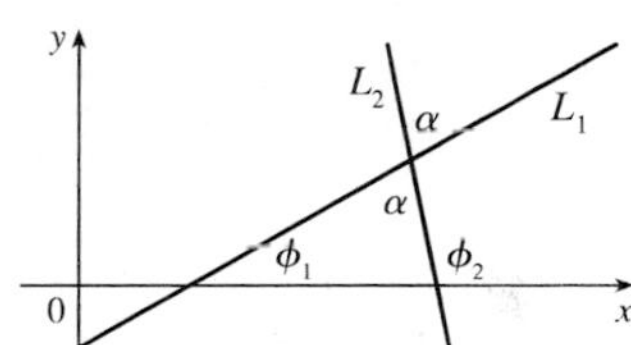

(b) (i) The parabolas intersect when $x^2 = (x - 2)^2 \Rightarrow x = 1$. If $y = x^2$, then $y' = 2x$, so the slope of the tangent to $y = x^2$ at $(1, 1)$ is $m_1 = 2(1) = 2$. If $y = (x - 2)^2$, then $y' = 2(x - 2)$, so the slope of the tangent to $y = (x - 2)^2$ at $(1, 1)$ is $m_2 = 2(1 - 2) = -2$. Therefore, $\tan\alpha = \dfrac{m_2 - m_1}{1 + m_1 m_2} = \dfrac{-2 - 2}{1 + 2(-2)} = \dfrac{4}{3}$ and so $\alpha = \tan^{-1}\left(\frac{4}{3}\right) \approx 53°$ [or $127°$].

(ii) $x^2 - y^2 = 3$ and $x^2 - 4x + y^2 + 3 = 0$ intersect when $x^2 - 4x + (x^2 - 3) + 3 = 0 \Leftrightarrow 2x(x - 2) = 0 \Rightarrow$ $x = 0$ or 2, but 0 is extraneous. If $x = 2$, then $y = \pm 1$. If $x^2 - y^2 = 3$ then $2x - 2yy' = 0 \Rightarrow y' = x/y$ and $x^2 - 4x + y^2 + 3 = 0 \Rightarrow 2x - 4 + 2yy' = 0 \Rightarrow y' = \dfrac{2 - x}{y}$. At $(2, 1)$ the slopes are $m_1 = 2$ and $m_2 = 0$, so $\tan\alpha = \frac{0 - 2}{1 + 2 \cdot 0} = -2 \Rightarrow \alpha \approx 117°$. At $(2, -1)$ the slopes are $m_1 = -2$ and $m_2 = 0$, so $\tan\alpha = \dfrac{0 - (-2)}{1 + (-2)(0)} = 2 \Rightarrow \alpha \approx 63°$ [or $117°$].

22. $y^2 = 4px \Rightarrow 2yy' = 4p \Rightarrow y' = 2p/y \Rightarrow$ slope of tangent at $P(x_1, y_1)$ is $m_1 = 2p/y_1$. The slope of FP is $m_2 = \dfrac{y_1}{x_1 - p}$, so by the formula from Problem 19(a),

$$\tan\alpha = \frac{\dfrac{y_1}{x_1 - p} - \dfrac{2p}{y_1}}{1 + \left(\dfrac{2p}{y_1}\right)\left(\dfrac{y_1}{x_1 - p}\right)} \cdot \frac{y_1(x_1 - p)}{y_1(x_1 - p)} = \frac{y_1^2 - 2p(x_1 - p)}{y_1(x_1 - p) + 2py_1} = \frac{4px_1 - 2px_1 + 2p^2}{x_1 y_1 - py_1 + 2py_1}$$

$$= \frac{2p(p + x_1)}{y_1(p + x_1)} = \frac{2p}{y_1} = \text{slope of tangent at } P = \tan\beta$$

Since $0 \le \alpha, \beta \le \frac{\pi}{2}$, this proves that $\alpha = \beta$.

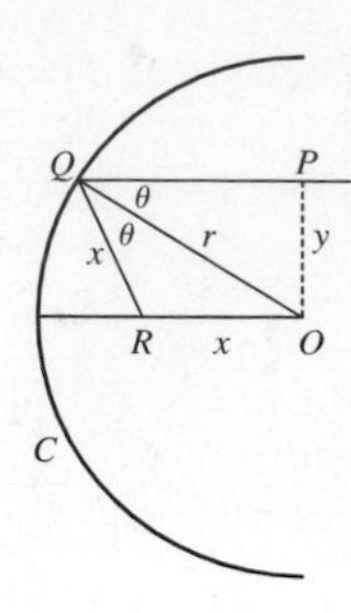

23. Since $\angle ROQ = \angle OQP = \theta$, the triangle QOR is isosceles, so $|QR| = |RO| = x$. By the Law of Cosines, $x^2 = x^2 + r^2 - 2rx\cos\theta$. Hence, $2rx\cos\theta = r^2$, so $x = \dfrac{r^2}{2r\cos\theta} = \dfrac{r}{2\cos\theta}$. Note that as $y \to 0^+$, $\theta \to 0^+$ (since $\sin\theta = y/r$), and hence $x \to \dfrac{r}{2\cos 0} = \dfrac{r}{2}$. Thus, as P is taken closer and closer to the x-axis, the point R approaches the midpoint of the radius AO.

24. $$\lim_{x\to 0}\frac{f(x)}{g(x)} = \lim_{x\to 0}\frac{f(x)-0}{g(x)-0} = \lim_{x\to 0}\frac{f(x)-f(0)}{g(x)-g(0)} = \lim_{x\to 0}\frac{\dfrac{f(x)-f(0)}{x-0}}{\dfrac{g(x)-g(0)}{x-0}} = \frac{\displaystyle\lim_{x\to 0}\frac{f(x)-f(0)}{x-0}}{\displaystyle\lim_{x\to 0}\frac{g(x)-g(0)}{x-0}} = \frac{f'(0)}{g'(0)}$$

25. $$\lim_{x\to 0}\frac{\sin(a+2x) - 2\sin(a+x) + \sin a}{x^2}$$

$$= \lim_{x\to 0}\frac{\sin a\,\cos 2x + \cos a\,\sin 2x - 2\sin a\,\cos x - 2\cos a\,\sin x + \sin a}{x^2}$$

$$= \lim_{x\to 0}\frac{\sin a\,(\cos 2x - 2\cos x + 1) + \cos a\,(\sin 2x - 2\sin x)}{x^2}$$

$$= \lim_{x\to 0}\frac{\sin a\,(2\cos^2 x - 1 - 2\cos x + 1) + \cos a\,(2\sin x\,\cos x - 2\sin x)}{x^2}$$

$$= \lim_{x\to 0}\frac{\sin a\,(2\cos x)(\cos x - 1) + \cos a\,(2\sin x)(\cos x - 1)}{x^2}$$

$$= \lim_{x\to 0}\frac{2(\cos x - 1)[\sin a\,\cos x + \cos a\,\sin x](\cos x + 1)}{x^2(\cos x + 1)}$$

$$= \lim_{x\to 0}\frac{-2\sin^2 x\,[\sin(a+x)]}{x^2(\cos x + 1)} = -2\lim_{x\to 0}\left(\frac{\sin x}{x}\right)^2 \cdot \frac{\sin(a+x)}{\cos x + 1} = -2(1)^2\,\frac{\sin(a+0)}{\cos 0 + 1} = -\sin a$$

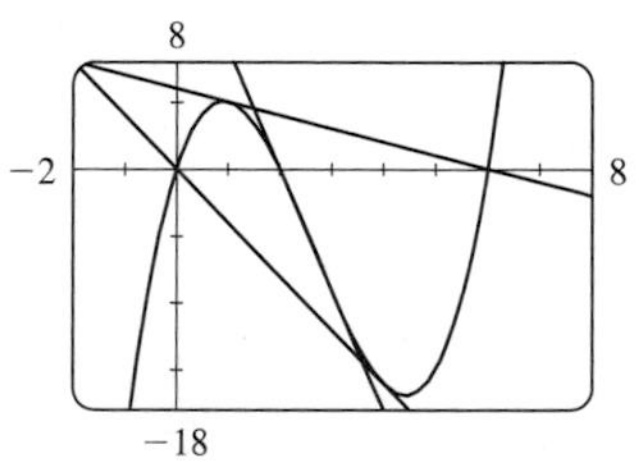

26. (a) $f(x) = x(x-2)(x-6) = x^3 - 8x^2 + 12x \;\Rightarrow$ $f'(x) = 3x^2 - 16x + 12$. The average of the first pair of zeros is $(0+2)/2 = 1$. At $x = 1$, the slope of the tangent line is $f'(1) = -1$, so an equation of the tangent line has the form $y = -1x + b$. Since $f(1) = 5$, we have $5 = -1 + b \;\Rightarrow\; b = 6$ and the tangent has equation $y = -x + 6$.

Similarly, at $x = \dfrac{0+6}{2} = 3$, $y = -9x + 18$; at $x = \dfrac{2+6}{2} = 4$, $y = -4x$. From the graph, we see that each tangent line drawn at the average of two zeros intersects the graph of f at the third zero.

(b) A CAS gives $f'(x) = (x-b)(x-c) + (x-a)(x-c) + (x-a)(x-b)$ or $f'(x) = 3x^2 - 2(a+b+c)x + ab + ac + bc$. Using the `Simplify` command, we get $f'\left(\dfrac{a+b}{2}\right) = -\dfrac{(a-b)^2}{4}$ and $f\left(\dfrac{a+b}{2}\right) = -\dfrac{(a-b)^2}{8}(a+b-2c)$, so an equation of the tangent line at $x = \dfrac{a+b}{2}$

is $y = -\dfrac{(a-b)^2}{4}\left(x - \dfrac{a+b}{2}\right) - \dfrac{(a-b)^2}{8}(a+b-2c)$. To find the x-intercept, let $y = 0$ and use the `Solve` command. The result is $x = c$.

Using Derive, we can begin by authoring the expression $(x-a)(x-b)(x-c)$. Now load the utility file `DifferentiationApplications`. Next we author tangent $(\#1, x, (a+b)/2)$—this is the command to find an equation of the tangent line of the function in #1 whose independent variable is x at the x-value $(a+b)/2$. We then simplify that expression and obtain the equation $y = \#4$. The form in expression #4 makes it easy to see that the x-intercept is the third zero, namely c. In a similar fashion we see that b is the x-intercept for the tangent line at $(a+c)/2$ and a is the x-intercept for the tangent line at $(b+c)/2$.

```
#1:  (x - a)·(x - b)·(x - c)

#2:  LOAD(C:\Program Files\TI Education\Derive 6\Math\DifferentiationApplications.mth

                                               a + b
#3:  TANGENT ((x - a)·(x - b)·(x - c), x, ─────)
                                                 2

#4:
```

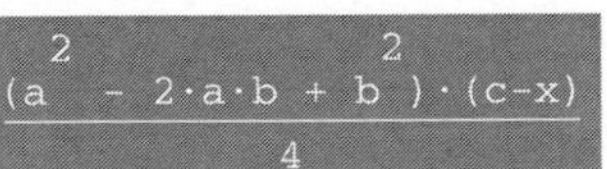

27.

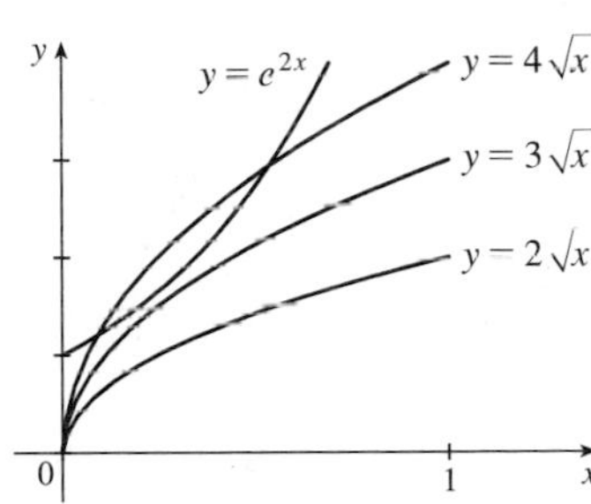

Let $f(x) = e^{2x}$ and $g(x) = k\sqrt{x}$ $[k > 0]$. From the graphs of f and g, we see that f will intersect g exactly once when f and g share a tangent line. Thus, we must have $f = g$ and $f' = g'$ at $x = a$.

$$f(a) = g(a) \quad\Rightarrow\quad e^{2a} = k\sqrt{a} \quad (\star)$$

and
$$f'(a) = g'(a) \quad\Rightarrow\quad 2e^{2a} = \frac{k}{2\sqrt{a}} \quad\Rightarrow\quad e^{2a} = \frac{k}{4\sqrt{a}}.$$

So we must have $k\sqrt{a} = \dfrac{k}{4\sqrt{a}} \quad\Rightarrow\quad \left(\sqrt{a}\right)^2 = \dfrac{k}{4k} \quad\Rightarrow\quad a = \frac{1}{4}$. From $(\star)$, $e^{2(1/4)} = k\sqrt{1/4} \quad\Rightarrow$

$k = 2e^{1/2} = 2\sqrt{e} \approx 3.297$.

28. We see that at $x = 0$, $f(x) = a^x = 1 + x = 1$, so if $y = a^x$ is to lie above $y = 1 + x$, the two curves must just touch at $(0, 1)$, that is, we must have $f'(0) = 1$. [To see this analytically, note that $a^x \geq 1 + x \quad\Rightarrow\quad a^x - 1 \geq x \quad\Rightarrow\quad \dfrac{a^x - 1}{x} \geq 1$ for $x > 0$, so

$f'(0) = \displaystyle\lim_{x\to 0^+} \frac{a^x - 1}{x} \geq 1$. Similarly, for $x < 0$, $a^x - 1 \geq x \quad\Rightarrow\quad \dfrac{a^x - 1}{x} \leq 1$, so

$f'(0) = \displaystyle\lim_{x\to 0^-} \frac{a^x - 1}{x} \leq 1$.

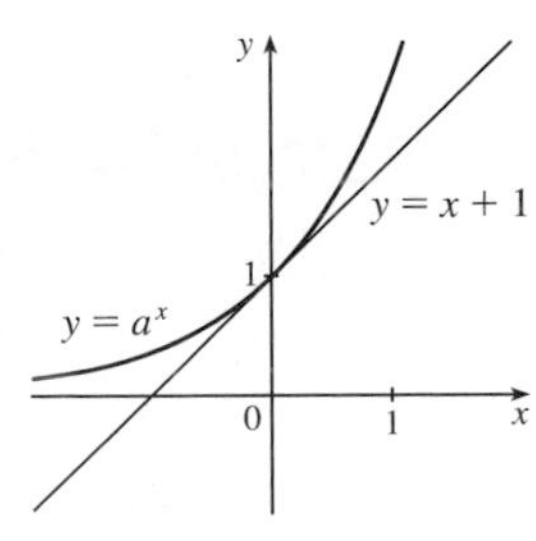

[continued]

Since $1 \le f'(0) \le 1$, we must have $f'(0) = 1$.] But $f'(x) = a^x \ln a \;\Rightarrow\; f'(0) = \ln a$, so we have $\ln a = 1 \;\Leftrightarrow\; a = e$.

Another method: The inequality certainly holds for $x \le -1$, so consider $x > -1$, $x \ne 0$. Then $a^x \ge 1 + x \;\Rightarrow$ $a \ge (1+x)^{1/x}$ for $x > 0 \;\Rightarrow\; a \ge \lim\limits_{x \to 0^+} (1+x)^{1/x} = e$, by Equation 3.6.5. Also, $a^x \ge 1 + x \;\Rightarrow\; a \le (1+x)^{1/x}$ for $x < 0 \;\Rightarrow\; a \le \lim\limits_{x \to 0^-} (1+x)^{1/x} = e$. So since $e \le a \le e$, we must have $a = e$.

29. $y = \dfrac{x}{\sqrt{a^2-1}} - \dfrac{2}{\sqrt{a^2-1}} \arctan \dfrac{\sin x}{a + \sqrt{a^2-1} + \cos x}$. Let $k = a + \sqrt{a^2-1}$. Then

$$y' = \frac{1}{\sqrt{a^2-1}} - \frac{2}{\sqrt{a^2-1}} \cdot \frac{1}{1 + \sin^2 x/(k+\cos x)^2} \cdot \frac{\cos x(k+\cos x) + \sin^2 x}{(k+\cos x)^2}$$

$$= \frac{1}{\sqrt{a^2-1}} - \frac{2}{\sqrt{a^2-1}} \cdot \frac{k\cos x + \cos^2 x + \sin^2 x}{(k+\cos x)^2 + \sin^2 x} = \frac{1}{\sqrt{a^2-1}} - \frac{2}{\sqrt{a^2-1}} \cdot \frac{k\cos x + 1}{k^2 + 2k\cos x + 1}$$

$$= \frac{k^2 + 2k\cos x + 1 - 2k\cos x - 2}{\sqrt{a^2-1}\,(k^2 + 2k\cos x + 1)} = \frac{k^2 - 1}{\sqrt{a^2-1}\,(k^2 + 2k\cos x + 1)}$$

But $k^2 = 2a^2 + 2a\sqrt{a^2-1} - 1 = 2a\left(a + \sqrt{a^2-1}\right) - 1 = 2ak - 1$, so $k^2 + 1 = 2ak$, and $k^2 - 1 = 2(ak - 1)$.

So $y' = \dfrac{2(ak-1)}{\sqrt{a^2-1}\,(2ak + 2k\cos x)} = \dfrac{ak-1}{\sqrt{a^2-1}\,k\,(a + \cos x)}$. But $ak - 1 = a^2 + a\sqrt{a^2-1} - 1 = k\sqrt{a^2-1}$,

so $y' = 1/(a + \cos x)$.

30. Suppose that $y = mx + c$ is a tangent line to the ellipse. Then it intersects the ellipse at only one point, so the discriminant of the equation $\dfrac{x^2}{a^2} + \dfrac{(mx+c)^2}{b^2} = 1 \;\Leftrightarrow\; (b^2 + a^2m^2)x^2 + 2mca^2x + a^2c^2 - a^2b^2 = 0$ must be 0; that is,

$$0 = (2mca^2)^2 - 4(b^2 + a^2m^2)(a^2c^2 - a^2b^2) = 4a^4c^2m^2 - 4a^2b^2c^2 + 4a^2b^4 - 4a^4c^2m^2 + 4a^4b^2m^2$$
$$= 4a^2b^2(a^2m^2 + b^2 - c^2)$$

Therefore, $a^2m^2 + b^2 - c^2 = 0$.

Now if a point (α, β) lies on the line $y = mx + c$, then $c = \beta - m\alpha$, so from above,

$$0 = a^2m^2 + b^2 - (\beta - m\alpha)^2 = (a^2 - \alpha^2)m^2 + 2\alpha\beta m + b^2 - \beta^2 \;\Leftrightarrow\; m^2 + \frac{2\alpha\beta}{a^2 - \alpha^2}m + \frac{b^2 - \beta^2}{a^2 - \alpha^2} = 0.$$

(a) Suppose that the two tangent lines from the point (α, β) to the ellipse have slopes m and $\dfrac{1}{m}$. Then m and $\dfrac{1}{m}$ are roots of the equation $z^2 + \dfrac{2\alpha\beta}{a^2 - \alpha^2}z + \dfrac{b^2 - \beta^2}{a^2 - \alpha^2} = 0$. This implies that $(z - m)\left(z - \dfrac{1}{m}\right) = 0 \;\Leftrightarrow$ $z^2 - \left(m + \dfrac{1}{m}\right)z + m\left(\dfrac{1}{m}\right) = 0$, so equating the constant terms in the two quadratic equations, we get $\dfrac{b^2 - \beta^2}{a^2 - \alpha^2} = m\left(\dfrac{1}{m}\right) = 1$, and hence $b^2 - \beta^2 = a^2 - \alpha^2$. So (α, β) lies on the hyperbola $x^2 - y^2 = a^2 - b^2$.

$\dfrac{x^2}{a^2} + \dfrac{y^2}{b^2} = 1$

$x^2 - y^2 = a^2 - b^2$

(b) If the two tangent lines from the point (α, β) to the ellipse have slopes m and $-\dfrac{1}{m}$, then m and $-\dfrac{1}{m}$ are roots of the quadratic equation, and so $(z - m)\left(z + \dfrac{1}{m}\right) = 0$, and equating the constant terms as in part (a), we get $\dfrac{b^2 - \beta^2}{a^2 - \alpha^2} = -1$, and hence $b^2 - \beta^2 = \alpha^2 - a^2$. So the point (α, β) lies on the circle $x^2 + y^2 = a^2 + b^2$.

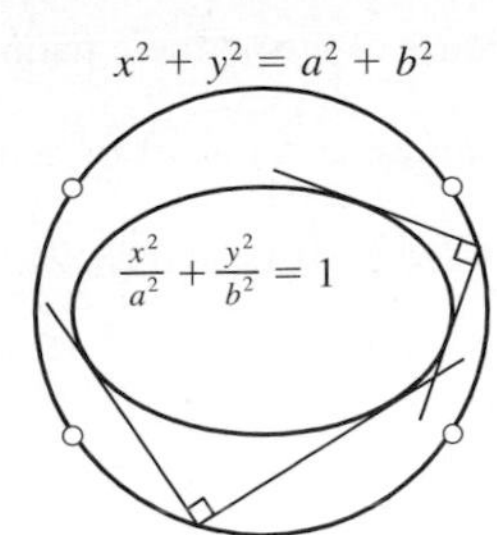

31. $y = x^4 - 2x^2 - x \Rightarrow y' = 4x^3 - 4x - 1$. The equation of the tangent line at $x = a$ is $y - (a^4 - 2a^2 - a) = (4a^3 - 4a - 1)(x - a)$ or $y = (4a^3 - 4a - 1)x + (-3a^4 + 2a^2)$ and similarly for $x = b$. So if at $x = a$ and $x = b$ we have the same tangent line, then $4a^3 - 4a - 1 = 4b^3 - 4b - 1$ and $-3a^4 + 2a^2 = -3b^4 + 2b^2$. The first equation gives $a^3 - b^3 = a - b \Rightarrow (a - b)(a^2 + ab + b^2) = (a - b)$. Assuming $a \neq b$, we have $1 = a^2 + ab + b^2$. The second equation gives $3(a^4 - b^4) = 2(a^2 - b^2) \Rightarrow 3(a^2 - b^2)(a^2 + b^2) = 2(a^2 - b^2)$ which is true if $a = -b$. Substituting into $1 = a^2 + ab + b^2$ gives $1 = a^2 - a^2 + a^2 \Rightarrow a = \pm 1$ so that $a = 1$ and $b = -1$ or vice versa. Thus, the points $(1, -2)$ and $(-1, 0)$ have a common tangent line.

As long as there are only two such points, we are done. So we show that these are in fact the only two such points. Suppose that $a^2 - b^2 \neq 0$. Then $3(a^2 - b^2)(a^2 + b^2) = 2(a^2 - b^2)$ gives $3(a^2 + b^2) = 2$ or $a^2 + b^2 = \frac{2}{3}$. Thus, $ab = (a^2 + ab + b^2) - (a^2 + b^2) = 1 - \dfrac{2}{3} = \dfrac{1}{3}$, so $b = \dfrac{1}{3a}$. Hence, $a^2 + \dfrac{1}{9a^2} = \dfrac{2}{3}$, so $9a^4 + 1 = 6a^2 \Rightarrow 0 = 9a^4 - 6a^2 + 1 = (3a^2 - 1)^2$. So $3a^2 - 1 = 0 \Rightarrow a^2 = \dfrac{1}{3} \Rightarrow b^2 = \dfrac{1}{9a^2} = \dfrac{1}{3} = a^2$, contradicting our assumption that $a^2 \neq b^2$.

32. Suppose that the normal lines at the three points (a_1, a_1^2), (a_2, a_2^2), and (a_3, a_3^2) intersect at a common point. Now if one of the a_i is 0 (suppose $a_1 = 0$) then by symmetry $a_2 = -a_3$, so $a_1 + a_2 + a_3 = 0$. So we can assume that none of the a_i is 0.

The slope of the tangent line at (a_i, a_i^2) is $2a_i$, so the slope of the normal line is $-\dfrac{1}{2a_i}$ and its equation is $y - a_i^2 = -\dfrac{1}{2a_i}(x - a_i)$. We solve for the x-coordinate of the intersection of the normal lines from (a_1, a_1^2) and (a_2, a_2^2):

$$y = a_1^2 - \frac{1}{2a_1}(x - a_1) = a_2^2 - \frac{1}{2a_2}(x - a_2) \Rightarrow x\left(\frac{1}{2a_2} - \frac{1}{2a_1}\right) = a_2^2 - a_1^2 \Rightarrow$$

$x\left(\dfrac{a_1 - a_2}{2a_1a_2}\right) = (-a_1 - a_2)(a_1 + a_2) \Leftrightarrow x = -2a_1a_2(a_1 + a_2)$ **(1)**. Similarly, solving for the x-coordinate of the intersections of the normal lines from (a_1, a_1^2) and (a_3, a_3^2) gives $x = -2a_1a_3(a_1 + a_3)$ **(2)**.

Equating **(1)** and **(2)** gives $a_2(a_1 + a_2) = a_3(a_1 + a_3) \Leftrightarrow a_1(a_2 - a_3) = a_3^2 - a_2^2 = -(a_2 + a_3)(a_2 - a_3) \Leftrightarrow a_1 = -(a_2 + a_3) \Leftrightarrow a_1 + a_2 + a_3 = 0$.

33. Because of the periodic nature of the lattice points, it suffices to consider the points in the 5×2 grid shown. We can see that the minimum value of r occurs when there is a line with slope $\frac{2}{5}$ which touches the circle centered at $(3, 1)$ and the circles centered at $(0, 0)$ and $(5, 2)$.

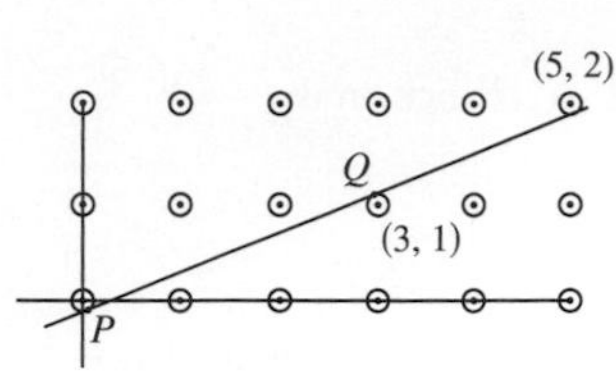

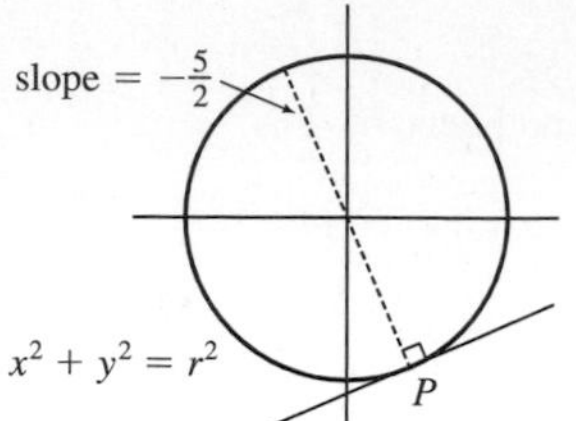

To find P, the point at which the line is tangent to the circle at $(0, 0)$, we simultaneously solve $x^2 + y^2 = r^2$ and $y = -\frac{5}{2}x \;\Rightarrow\; x^2 + \frac{25}{4}x^2 = r^2 \;\Rightarrow\; x^2 = \frac{4}{29}r^2 \;\Rightarrow\; x = \frac{2}{\sqrt{29}}r,\; y = -\frac{5}{\sqrt{29}}r$. To find Q, we either use symmetry or solve $(x-3)^2 + (y-1)^2 = r^2$ and $y - 1 = -\frac{5}{2}(x-3)$. As above, we get $x = 3 - \frac{2}{\sqrt{29}}r,\; y = 1 + \frac{5}{\sqrt{29}}r$. Now the slope of the line PQ is $\frac{2}{5}$, so $m_{PQ} = \dfrac{1 + \frac{5}{\sqrt{29}}r - \left(-\frac{5}{\sqrt{29}}r\right)}{3 - \frac{2}{\sqrt{29}}r - \frac{2}{\sqrt{29}}r} = \dfrac{1 + \frac{10}{\sqrt{29}}r}{3 - \frac{4}{\sqrt{29}}r} = \dfrac{\sqrt{29} + 10r}{3\sqrt{29} - 4r} = \dfrac{2}{5} \;\Rightarrow$

$5\sqrt{29} + 50r = 6\sqrt{29} - 8r \;\Leftrightarrow\; 58r = \sqrt{29} \;\Leftrightarrow\; r = \frac{\sqrt{29}}{58}$. So the minimum value of r for which any line with slope $\frac{2}{5}$ intersects circles with radius r centered at the lattice points on the plane is $r = \frac{\sqrt{29}}{58} \approx 0.093$.

34.

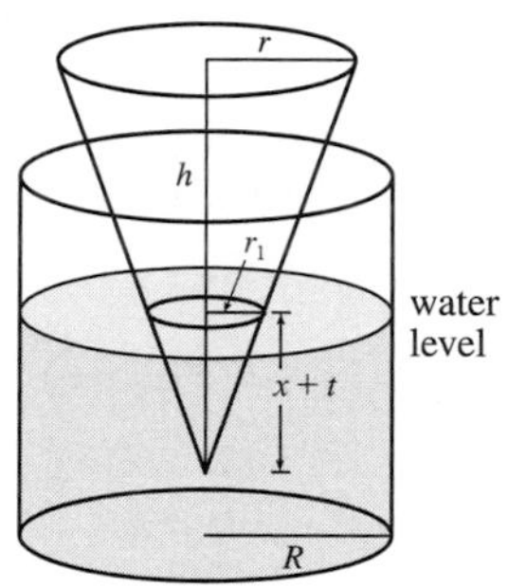

Assume the axes of the cone and the cylinder are parallel. Let H denote the initial height of the water. When the cone has been dropping for t seconds, the water level has risen x centimeters, so the tip of the cone is $x + 1t$ centimeters below the water line. We want to find dx/dt when $x + t = h$ (when the cone is completely submerged).

Using similar triangles, $\dfrac{r_1}{x+t} = \dfrac{r}{h} \;\Rightarrow\; r_1 = \dfrac{r}{h}(x+t)$.

$$\begin{array}{cccccc} \text{volume of water and cone at time } t & = & \text{original volume of water} & + & \text{volume of submerged part of cone} \\ \pi R^2(H + x) & = & \pi R^2 H & + & \frac{1}{3}\pi r_1^2(x+t) \\ \pi R^2 H + \pi R^2 x & = & \pi R^2 H & + & \frac{1}{3}\pi \dfrac{r^2}{h^2}(x+t)^3 \\ 3h^2 R^2 x & = & r^2(x+t)^3 & & \end{array}$$

Differentiating implicitly with respect to t gives us $3h^2R^2\dfrac{dx}{dt} = r^2\left[3(x+t)^2\dfrac{dx}{dt} + 3(x+t)^2\dfrac{dt}{dt}\right] \;\Rightarrow$

$\dfrac{dx}{dt} = \dfrac{r^2(x+t)^2}{h^2R^2 - r^2(x+t)^2} \;\Rightarrow\; \left.\dfrac{dx}{dt}\right|_{x+t=h} = \dfrac{r^2h^2}{h^2R^2 - r^2h^2} = \dfrac{r^2}{R^2 - r^2}$. Thus, the water level is rising at a rate of $\dfrac{r^2}{R^2 - r^2}$ cm/s at the instant the cone is completely submerged.

35. 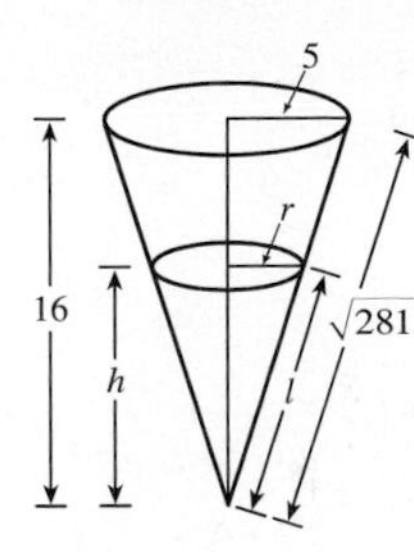

By similar triangles, $\frac{r}{5} = \frac{h}{16} \Rightarrow r = \frac{5h}{16}$. The volume of the cone is $V = \frac{1}{3}\pi r^2 h = \frac{1}{3}\pi\left(\frac{5h}{16}\right)^2 h = \frac{25\pi}{768}h^3$, so $\frac{dV}{dt} = \frac{25\pi}{256}h^2\,\frac{dh}{dt}$. Now the rate of change of the volume is also equal to the difference of what is being added ($2\text{ cm}^3/\text{min}$) and what is oozing out ($k\pi rl$, where πrl is the area of the cone and k is a proportionality constant). Thus, $\frac{dV}{dt} = 2 - k\pi rl$.

Equating the two expressions for $\frac{dV}{dt}$ and substituting $h = 10$, $\frac{dh}{dt} = -0.3$, $r = \frac{5(10)}{16} = \frac{25}{8}$, and $\frac{l}{\sqrt{281}} = \frac{10}{16} \Leftrightarrow$ $l = \frac{5}{8}\sqrt{281}$, we get $\frac{25\pi}{256}(10)^2(-0.3) = 2 - k\pi\frac{25}{8}\cdot\frac{5}{8}\sqrt{281} \Leftrightarrow \frac{125k\pi\sqrt{281}}{64} = 2 + \frac{750\pi}{256}$. Solving for k gives us $k = \frac{256 + 375\pi}{250\pi\sqrt{281}}$. To maintain a certain height, the rate of oozing, $k\pi rl$, must equal the rate of the liquid being poured in; that is, $\frac{dV}{dt} = 0$. Thus, the rate at which we should pour the liquid into the container is

$$k\pi rl = \frac{256 + 375\pi}{250\pi\sqrt{281}}\cdot\pi\cdot\frac{25}{8}\cdot\frac{5\sqrt{281}}{8} = \frac{256 + 375\pi}{128} \approx 11.204\text{ cm}^3/\text{min}$$

4 □ APPLICATIONS OF DIFFERENTIATION

4.1 Maximum and Minimum Values

1. A function f has an **absolute minimum** at $x = c$ if $f(c)$ is the smallest function value on the entire domain of f, whereas f has a **local minimum** at c if $f(c)$ is the smallest function value when x is near c.

2. (a) The Extreme Value Theorem

(b) See the Closed Interval Method.

3. Absolute maximum at s, absolute minimum at r, local maximum at c, local minima at b and r, neither a maximum nor a minimum at a and d.

4. Absolute maximum at r; absolute minimum at a; local maxima at b and r; local minimum at d; neither a maximum nor a minimum at c and s.

5. Absolute maximum value is $f(4) = 5$; there is no absolute minimum value; local maximum values are $f(4) = 5$ and $f(6) = 4$; local minimum values are $f(2) = 2$ and $f(1) = f(5) = 3$.

6. There is no absolute maximum value; absolute minimum value is $g(4) = 1$; local maximum values are $g(3) = 4$ and $g(6) = 3$; local minimum values are $g(2) = 2$ and $g(4) = 1$.

7. Absolute maximum at 5, absolute minimum at 2, local maximum at 3, local minima at 2 and 4

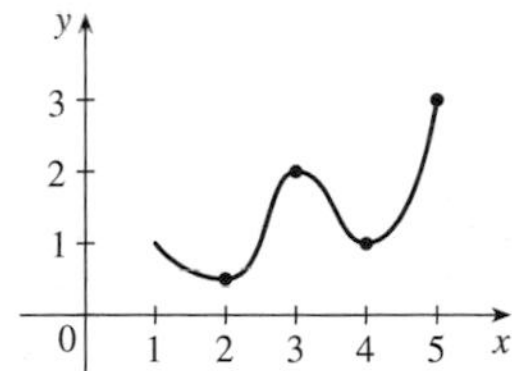

8. Absolute maximum at 4, absolute minimum at 5, local maximum at 2, local minimum at 3

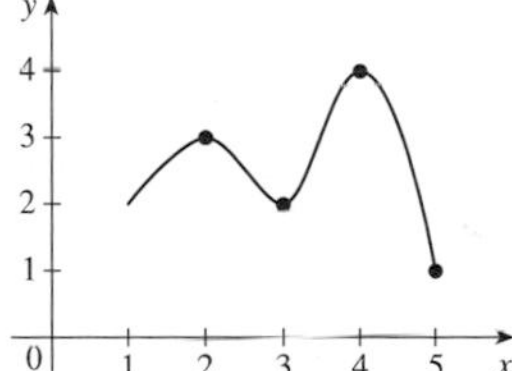

9. Absolute minimum at 3, absolute maximum at 4, local maximum at 2

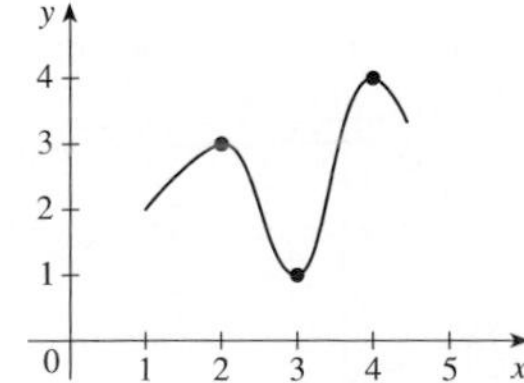

10. Absolute maximum at 2, absolute minimum at 5, 4 is a critial number but there is no local maximum or minimum there.

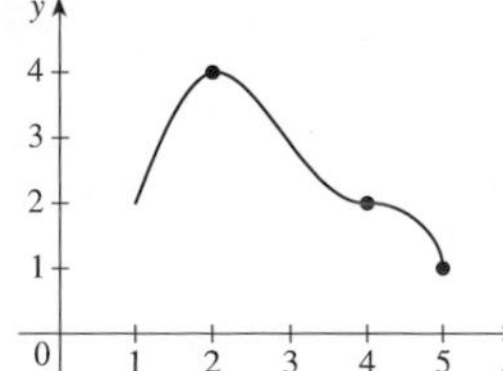

11. (a)

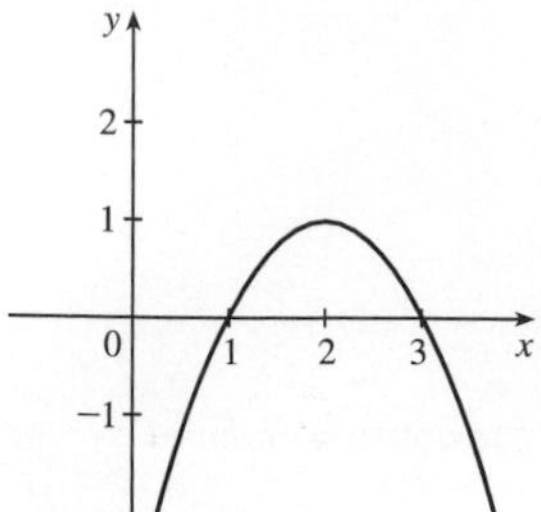

(b)

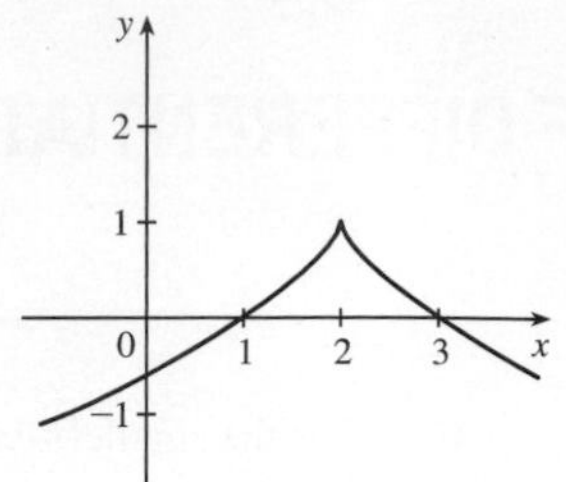

(c)

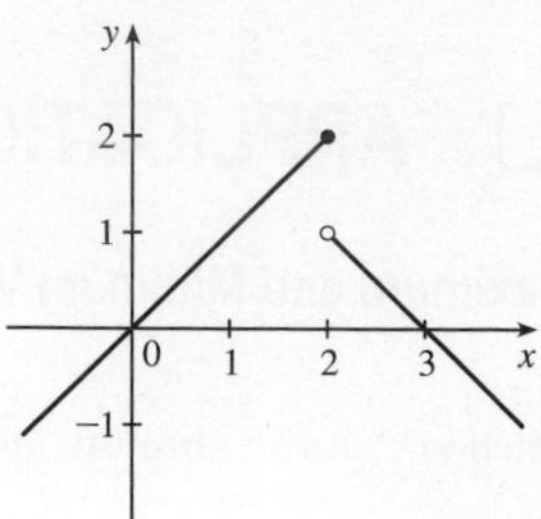

12. (a) Note that a local maximum cannot occur at an endpoint.

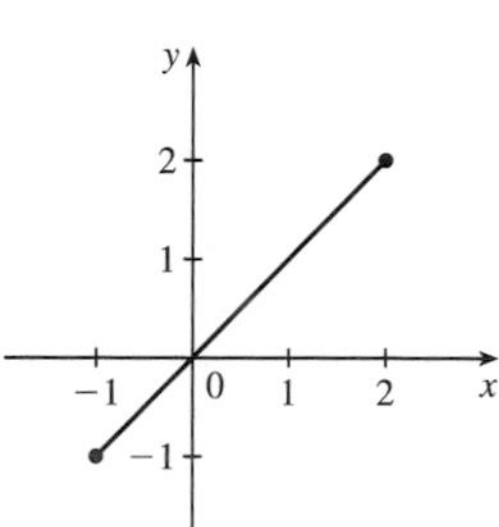

(b)

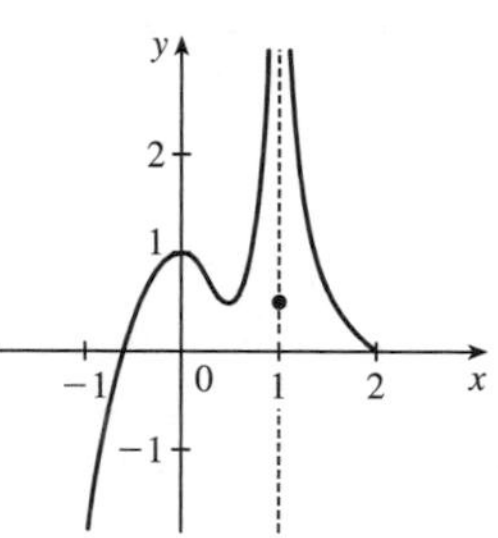

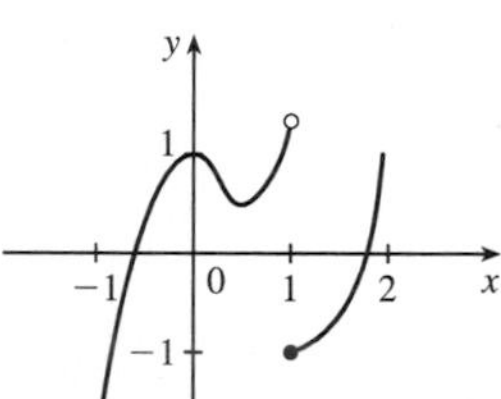

Note: By the Extreme Value Theorem, f must *not* be continuous.

13. (a) *Note:* By the Extreme Value Theorem, f must *not* be continuous; because if it were, it would attain an absolute minimum.

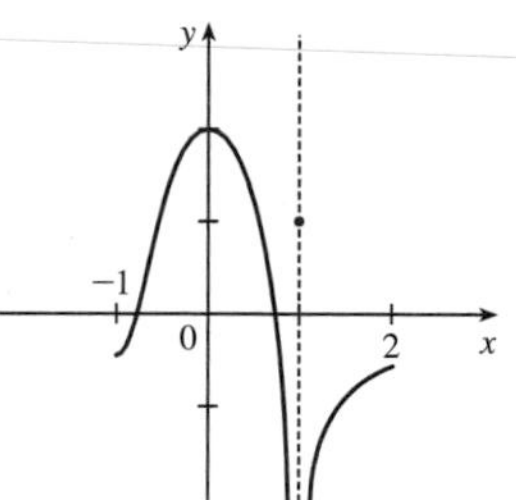

(b)

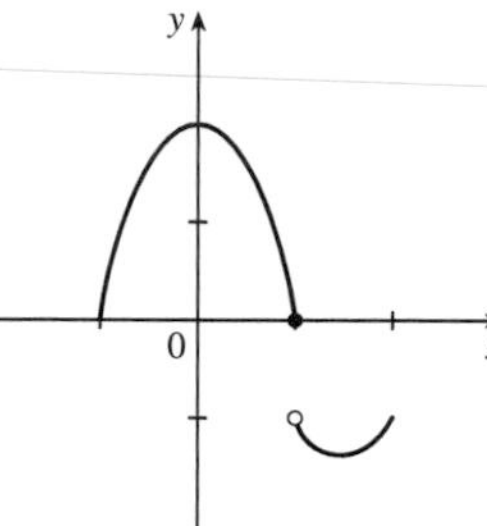

14. (a)

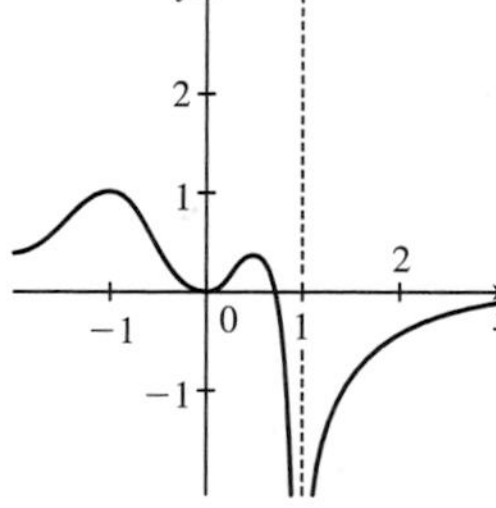

(b)

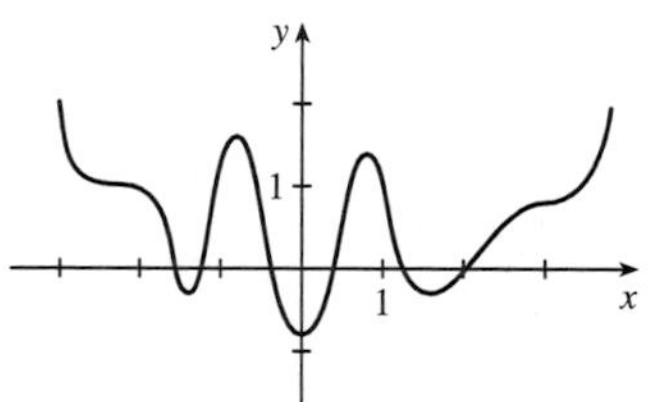

15. $f(x) = \frac{1}{2}(3x - 1)$, $x \le 3$. Absolute maximum $f(3) = 4$; no local maximum. No absolute or local minimum.

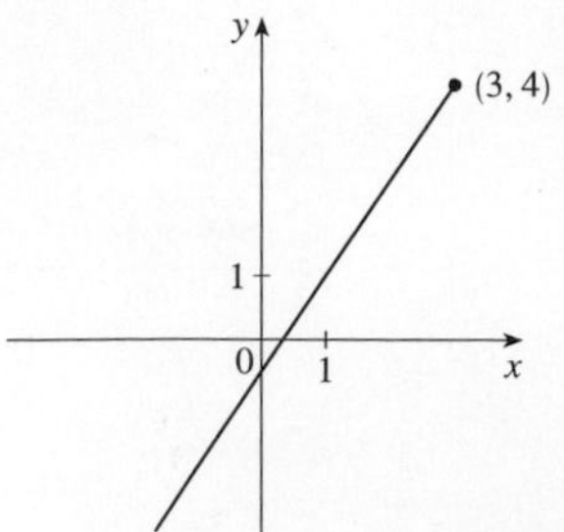

16. $f(x) = 2 - \frac{1}{3}x$, $x \ge -2$. Absolute maximum $f(-2) = \frac{8}{3}$; no local maximum. No absolute or local minimum.

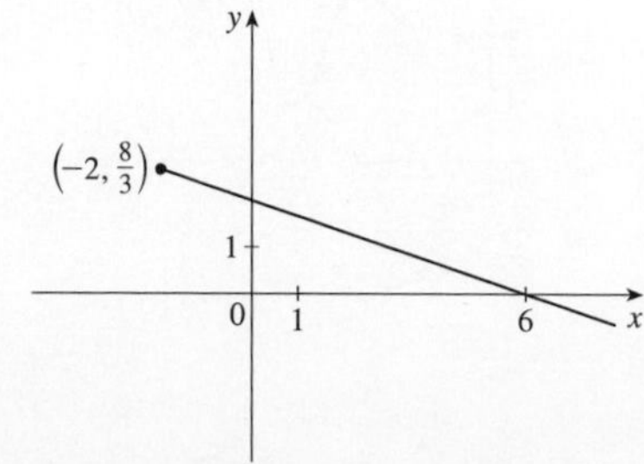

17. $f(x) = 1/x$, $x \geq 1$. Absolute maximum $f(1) = 1$; no local maximum. No absolute or local minimum.

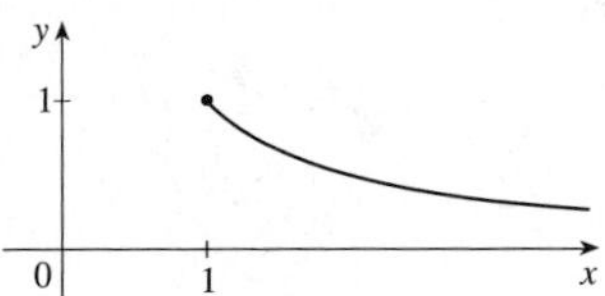

18. $f(x) = 1/x$, $1 < x < 3$. No absolute or local maximum. No absolute or local minimum.

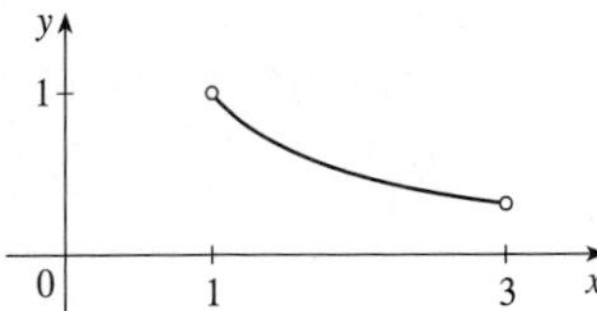

19. $f(x) = \sin x$, $0 \leq x < \pi/2$. No absolute or local maximum. Absolute minimum $f(0) = 0$; no local minimum.

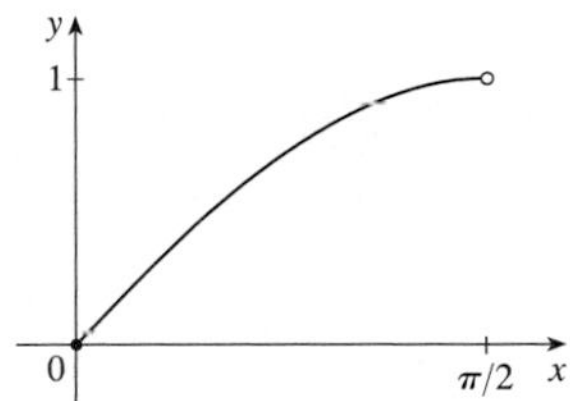

20. $f(x) = \sin x$, $0 < x \leq \pi/2$. Absolute maximum $f\left(\frac{\pi}{2}\right) = 1$; no local maximum. No absolute or local minimum.

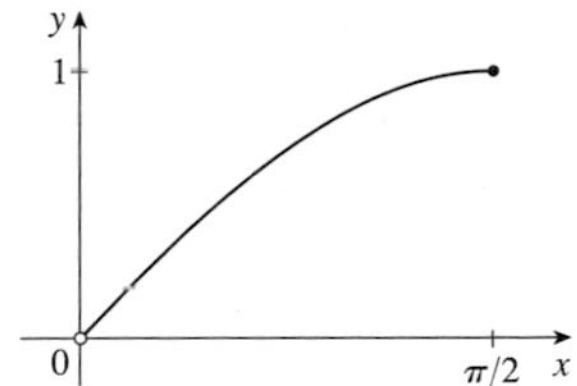

21. $f(x) = \sin x$, $-\pi/2 \leq x \leq \pi/2$. Absolute maximum $f\left(\frac{\pi}{2}\right) = 1$; no local maximum. Absolute minimum $f\left(-\frac{\pi}{2}\right) = -1$; no local minimum.

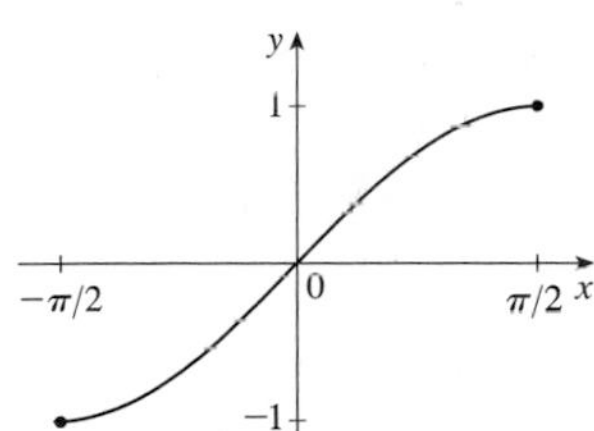

22. $f(t) = \cos t$, $-\frac{3\pi}{2} \leq t \leq \frac{3\pi}{2}$. Absolute and local maximum $f(0) = 1$; absolute and local minima $f(\pm\pi, -1)$.

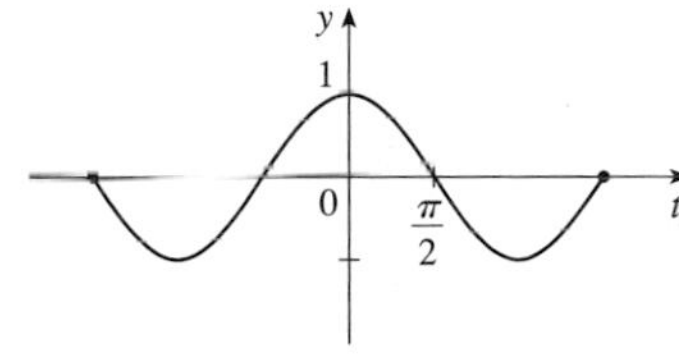

23. $f(x) = \ln x$, $0 < x \leq 2$. Absolute maximum $f(2) = \ln 2 \approx 0.69$; no local maximum. No absolute or local minimum.

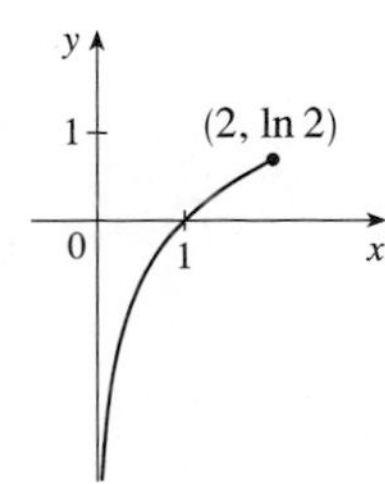

24. $f(x) = |\, x \,|$. No absolute or local maximum. Absolute and local minimum $f(0) = 0$.

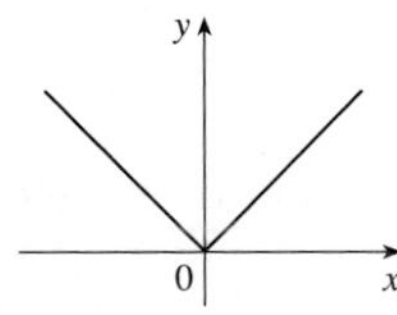

25. $f(x) = 1 - \sqrt{x}$. Absolute maximum $f(0) = 1$; no local maximum. No absolute or local minimum.

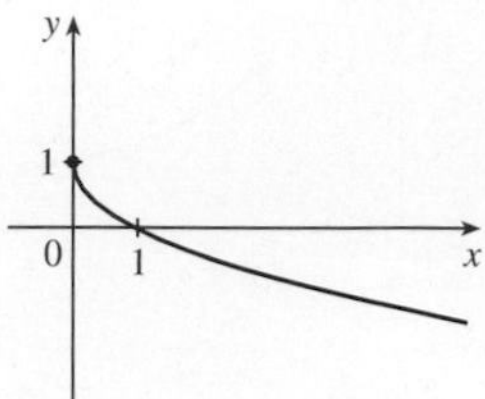

26. $f(x) = e^x$. No absolute or local maximum or minimum value.

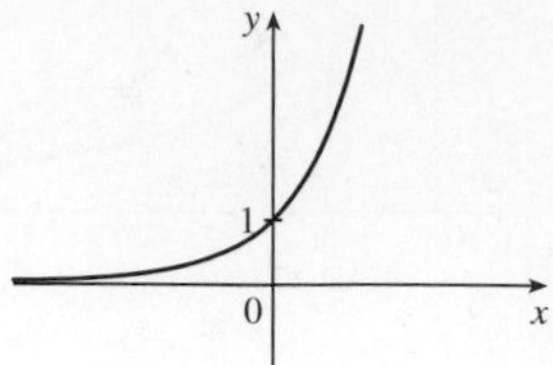

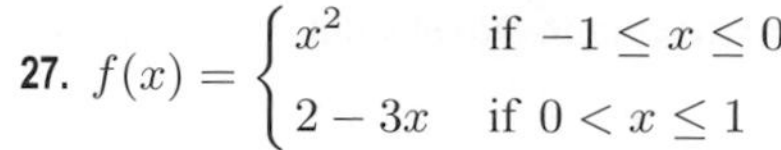

27. $f(x) = \begin{cases} x^2 & \text{if } -1 \le x \le 0 \\ 2 - 3x & \text{if } 0 < x \le 1 \end{cases}$

No absolute or local maximum.

Absolute minimum $f(1) = -1$.

Local minimum $f(0) = 0$.

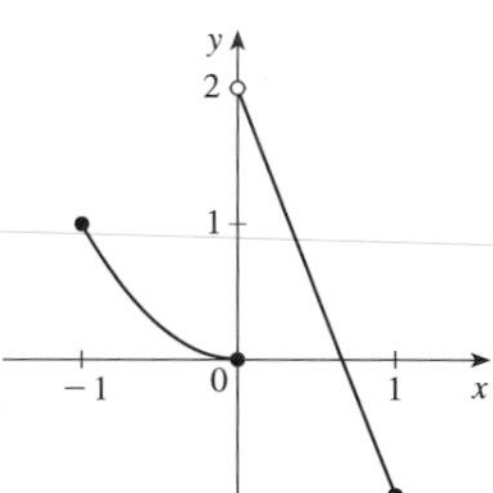

28. $f(x) = \begin{cases} 2x + 1 & \text{if } 0 \le x < 1 \\ 4 - 2x & \text{if } 1 \le x \le 3 \end{cases}$

No absolute or local maximum.

Absolute minimum $f(3) = -2$; no local minimum.

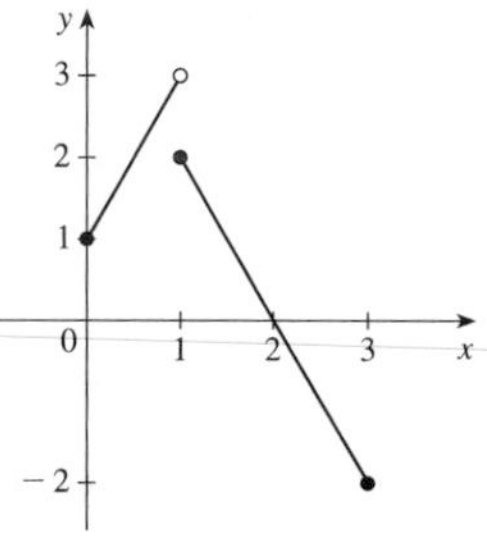

29. $f(x) = 4 + \frac{1}{3}x - \frac{1}{2}x^2 \;\Rightarrow\; f'(x) = \frac{1}{3} - x.\;\; f'(x) = 0 \;\Rightarrow\; x = \frac{1}{3}$. This is the only critical number.

30. $f(x) = x^3 + 6x^2 - 15x \;\Rightarrow\; f'(x) = 3x^2 + 12x - 15 = 3(x^2 + 4x - 5) = 3(x + 5)(x - 1)$.

$f'(x) = 0 \;\Rightarrow\; x = -5, 1$. These are the only critical numbers.

31. $f(x) = 2x^3 - 3x^2 - 36x \;\Rightarrow\; f'(x) = 6x^2 - 6x - 36 = 6(x^2 - x - 6) = 6(x + 2)(x - 3)$.

$f'(x) = 0 \;\Leftrightarrow\; x = -2, 3$. These are the only critical numbers.

32. $f(x) = 2x^3 + x^2 + 2x \;\Rightarrow\; f'(x) = 6x^2 + 2x + 2 = 2(3x^2 + x + 1)$. Using the quadratic formula, $f'(x) = 0 \;\Leftrightarrow$ $x = \dfrac{-1 \pm \sqrt{-11}}{6}$. Since the discrimininant, -11, is negative, there are no real soutions, and hence, there are no critical numbers.

33. $g(t) = t^4 + t^3 + t^2 + 1 \;\Rightarrow\; g'(t) = 4t^3 + 3t^2 + 2t = t(4t^2 + 3t + 2)$. Using the quadratic formula, we see that $4t^2 + 3t + 2 = 0$ has no real solution (its discriminant is negative), so $g'(t) = 0$ only if $t = 0$. Hence, the only critical number is 0.

34. $g(t) = |3t - 4| = \begin{cases} 3t - 4 & \text{if } 3t - 4 \ge 0 \\ -(3t - 4) & \text{if } 3t - 4 < 0 \end{cases} = \begin{cases} 3t - 4 & \text{if } t \ge \frac{4}{3} \\ 4 - 3t & \text{if } t < \frac{4}{3} \end{cases}$

$g'(t) = \begin{cases} 3 & \text{if } t > \frac{4}{3} \\ -3 & \text{if } t < \frac{4}{3} \end{cases}$ and $g'(t)$ does not exist at $t = \frac{4}{3}$, so $t = \frac{4}{3}$ is a critical number.

35. $g(y) = \dfrac{y-1}{y^2-y+1} \quad \Rightarrow$

$$g'(y) = \frac{(y^2-y+1)(1)-(y-1)(2y-1)}{(y^2-y+1)^2} = \frac{y^2-y+1-(2y^2-3y+1)}{(y^2-y+1)^2} = \frac{-y^2+2y}{(y^2-y+1)^2} = \frac{y(2-y)}{(y^2-y+1)^2}.$$

$g'(y) = 0 \;\Rightarrow\; y = 0, 2$. The expression $y^2 - y + 1$ is never equal to 0, so $g'(y)$ exists for all real numbers. The critical numbers are 0 and 2.

36. $h(p) = \dfrac{p-1}{p^2+4} \;\Rightarrow\; h'(p) = \dfrac{(p^2+4)(1)-(p-1)(2p)}{(p^2+4)^2} = \dfrac{p^2+4-2p^2+2p}{(p^2+4)^2} = \dfrac{-p^2+2p+4}{(p^2+4)^2}$.

$h'(p) = 0 \;\Rightarrow\; p = \dfrac{-2 \pm \sqrt{4+16}}{-2} = 1 \pm \sqrt{5}$. The critical numbers are $1 \pm \sqrt{5}$. [$h'(p)$ exists for all real numbers.]

37. $h(t) = t^{3/4} - 2t^{1/4} \;\Rightarrow\; h'(t) = \frac{3}{4}t^{-1/4} - \frac{2}{4}t^{-3/4} = \frac{1}{4}t^{-3/4}(3t^{1/2} - 2) = \dfrac{3\sqrt{t}-2}{4\sqrt[4]{t^3}}$.

$h'(t) = 0 \;\Rightarrow\; 3\sqrt{t} = 2 \;\Rightarrow\; \sqrt{t} = \frac{2}{3} \;\Rightarrow\; t = \frac{4}{9}$. $h'(t)$ does not exist at $t = 0$, so the critical numbers are 0 and $\frac{4}{9}$.

38. $g(x) = \sqrt[3]{4-x^2} = (4-x^2)^{1/3} \;\Rightarrow\; g'(x) = \frac{1}{3}(4-x^2)^{-2/3}(-2x) = \dfrac{-2x}{3(4-x^2)^{2/3}}$. $g'(x) = 0 \;\Rightarrow\; x = 0$.

$g'(\pm 2)$ do not exist. Thus, the three critical numbers are -2, 0, and 2.

39. $F(x) = x^{4/5}(x-4)^2 \;\Rightarrow$

$$F'(x) = x^{4/5} \cdot 2(x-4) + (x-4)^2 \cdot \tfrac{4}{5}x^{-1/5} = \tfrac{1}{5}x^{-1/5}(x-4)[5 \cdot x \cdot 2 + (x-4) \cdot 4]$$
$$= \frac{(x-4)(14x-16)}{5x^{1/5}} = \frac{2(x-4)(7x-8)}{5x^{1/5}}$$

$F'(x) = 0 \;\Rightarrow\; x = 4, \frac{8}{7}$. $F'(0)$ does not exist. Thus, the three critical numbers are 0, $\frac{8}{7}$, and 4.

40. $g(\theta) = 4\theta - \tan\theta \;\Rightarrow\; g'(\theta) = 4 - \sec^2\theta$. $g'(\theta) = 0 \;\Rightarrow\; \sec^2\theta = 4 \;\Rightarrow\; \sec\theta = \pm 2 \;\Rightarrow\; \cos\theta = \pm\frac{1}{2} \;\Rightarrow$

$\theta = \frac{\pi}{3} + 2n\pi$, $\frac{5\pi}{3} + 2n\pi$, $\frac{2\pi}{3} + 2n\pi$, and $\frac{4\pi}{3} + 2n\pi$ are critical numbers.

Note: The values of θ that make $g'(\theta)$ undefined are not in the domain of g.

41. $f(\theta) = 2\cos\theta + \sin^2\theta \;\Rightarrow\; f'(\theta) = -2\sin\theta + 2\sin\theta\cos\theta$. $f'(\theta) = 0 \;\Rightarrow\; 2\sin\theta(\cos\theta - 1) = 0 \;\Rightarrow\; \sin\theta = 0$ or $\cos\theta = 1 \;\Rightarrow\; \theta = n\pi$ [n an integer] or $\theta = 2n\pi$. The solutions $\theta = n\pi$ include the solutions $\theta = 2n\pi$, so the critical numbers are $\theta = n\pi$.

42. $h(t) = 3t - \arcsin t \;\Rightarrow\; h'(t) = 3 - \dfrac{1}{\sqrt{1-t^2}}$. $h'(t) = 0 \;\Rightarrow\; 3 = \dfrac{1}{\sqrt{1-t^2}} \;\Rightarrow\; \sqrt{1-t^2} = \frac{1}{3} \;\Rightarrow$

$1 - t^2 = \frac{1}{9} \;\Rightarrow\; t^2 = \frac{8}{9} \;\Rightarrow\; t = \pm\frac{2}{3}\sqrt{2} \approx \pm 0.94$, both in the domain of h, which is $[-1, 1]$.

43. $f(x) = x^2e^{-3x} \;\Rightarrow\; f'(x) = x^2(-3e^{-3x}) + e^{-3x}(2x) = xe^{-3x}(-3x+2)$. $f'(x) = 0 \;\Rightarrow\; x = 0, \frac{2}{3}$ [e^{-3x} is never equal to 0]. $f'(x)$ always exists, so the critical numbers are 0 and $\frac{2}{3}$.

44. $f(x) = x^{-2}\ln x \;\Rightarrow\; f'(x) = x^{-2}(1/x) + (\ln x)(-2x^{-3}) = x^{-3} - 2x^{-3}\ln x = x^{-3}(1 - 2\ln x) = \dfrac{1-2\ln x}{x^3}$.

$f'(x) = 0 \;\Rightarrow\; 1 - 2\ln x = 0 \;\Rightarrow\; \ln x = \frac{1}{2} \;\Rightarrow\; x = e^{1/2} \approx 1.65$. $f'(0)$ does not exist, but 0 is not in the domain of f, so the only critical number is $\sqrt{e}$.

45. The graph of $f'(x) = 5e^{-0.1|x|} \sin x - 1$ has 10 zeros and exists everywhere, so f has 10 critical numbers.

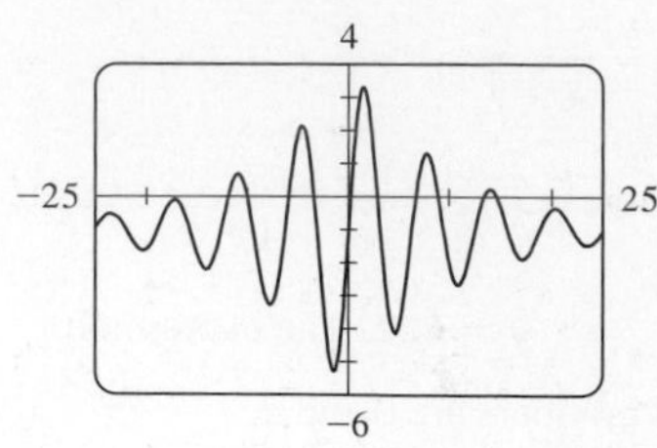

46. A graph of $f'(x) = \dfrac{100 \cos^2 x}{10 + x^2} - 1$ is shown. There are 7 zeros between 0 and 10, and 7 more zeros since f' is an even function. f' exists everywhere, so f has 14 critical numbers.

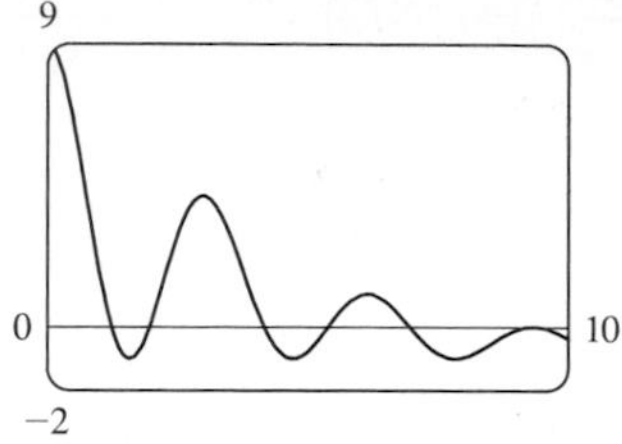

47. $f(x) = 12 + 4x - x^2$, $[0, 5]$. $f'(x) = 4 - 2x = 0 \Leftrightarrow x = 2$. $f(0) = 12$, $f(2) = 16$, and $f(5) = 7$. So $f(2) = 16$ is the absolute maximum value and $f(5) = 7$ is the absolute minimum value.

48. $f(x) = 5 + 54x - 2x^3$, $[0, 4]$. $f'(x) = 54 - 6x^2 = 6(9 - x^2) = 6(3 + x)(3 - x) = 0 \Leftrightarrow x = -3, 3$. $f(0) = 5$, $f(3) = 113$, and $f(4) = 93$. So $f(3) = 113$ is the absolute maximum value and $f(0) = 5$ is the absolute minimum value.

49. $f(x) = 2x^3 - 3x^2 - 12x + 1$, $[-2, 3]$. $f'(x) = 6x^2 - 6x - 12 = 6(x^2 - x - 2) = 6(x - 2)(x + 1) = 0 \Leftrightarrow$ $x = 2, -1$. $f(-2) = -3$, $f(-1) = 8$, $f(2) = -19$, and $f(3) = -8$. So $f(-1) = 8$ is the absolute maximum value and $f(2) = -19$ is the absolute minimum value.

50. $x^3 - 6x^2 + 5$, $[-3, 5]$. $f'(x) = 3x^2 - 12x = 3x(x - 4) = 0 \Leftrightarrow x = 0, 4$. $f(-3) = -76$, $f(0) = 5$, $f(4) = -27$, and $f(5) = -20$. So $f(0) = 5$ is the absolute maximum value and $f(-3) = -76$ is the absolute minimum value.

51. $f(x) = 3x^4 - 4x^3 - 12x^2 + 1$, $[-2, 3]$. $f'(x) = 12x^3 - 12x^2 - 24x = 12x(x^2 - x - 2) = 12x(x + 1)(x - 2) = 0 \Leftrightarrow$ $x = -1, 0, 2$. $f(-2) = 33$, $f(-1) = -4$, $f(0) = 1$, $f(2) = -31$, and $f(3) = 28$. So $f(-2) = 33$ is the absolute maximum value and $f(2) = -31$ is the absolute minimum value.

52. $f(t) = (t^2 - 4)^3$, $[-2, 3]$. $f'(t) = 3(t^2 - 4)^2(2t) = 6t(t + 2)^2(t - 2)^2 = 0 \Leftrightarrow t = -2, 0, 2$. $f(\pm 2) = 0$, $f(0) = -64$, and $f(3) = 5^3 = 125$. So $f(3) = 125$ is the absolute maximum value and $f(0) = -64$ is the absolute minimum value.

53. $f(x) = x + \dfrac{1}{x}$, $[0.2, 4]$. $f'(x) = 1 - \dfrac{1}{x^2} = \dfrac{x^2 - 1}{x^2} = \dfrac{(x + 1)(x - 1)}{x^2} = 0 \Leftrightarrow x = \pm 1$, but $x = -1$ is not in the given interval, $[0.2, 4]$. $f'(x)$ does not exist when $x = 0$, but 0 is not in the given interval, so 1 is the only critical nuumber. $f(0.2) = 5.2$, $f(1) = 2$, and $f(4) = 4.25$. So $f(0.2) = 5.2$ is the absolute maximum value and $f(1) = 2$ is the absolute minimum value.

54. $f(x) = \dfrac{x}{x^2 - x + 1}$, $[0, 3]$.

$$f'(x) = \frac{(x^2 - x + 1) - x(2x - 1)}{(x^2 - x + 1)^2} = \frac{x^2 - x + 1 - 2x^2 + x}{(x^2 - x + 1)^2} = \frac{1 - x^2}{(x^2 - x + 1)^2} = \frac{(1 + x)(1 - x)}{(x^2 - x + 1)^2} = 0 \Leftrightarrow$$

$x = \pm 1$, but $x = -1$ is not in the given interval, $[0, 3]$. $f(0) = 0$, $f(1) = 1$, and $f(3) = \frac{3}{7}$. So $f(1) = 1$ is the absolute maximum value and $f(0) = 0$ is the absolute minimum value.

55. $f(t) = t - \sqrt[3]{t}$, $[-1, 4]$. $f'(t) = 1 - \frac{1}{3}t^{-2/3} = 1 - \dfrac{1}{3t^{2/3}}$. $f'(t) = 0 \Leftrightarrow 1 = \dfrac{1}{3t^{2/3}} \Leftrightarrow t^{2/3} = \dfrac{1}{3} \Leftrightarrow$

$t = \pm\left(\dfrac{1}{3}\right)^{3/2} = \pm\sqrt{\dfrac{1}{27}} = \pm\dfrac{1}{3\sqrt{3}} = \pm\dfrac{\sqrt{3}}{9}$. $f'(t)$ does not exist when $t = 0$. $f(-1) = 0$, $f(0) = 0$,

$f\left(\dfrac{-1}{3\sqrt{3}}\right) = \dfrac{-1}{3\sqrt{3}} - \dfrac{-1}{\sqrt{3}} = \dfrac{-1+3}{3\sqrt{3}} = \dfrac{2\sqrt{3}}{9} \approx 0.3849$, $f\left(\dfrac{1}{3\sqrt{3}}\right) = \dfrac{1}{3\sqrt{3}} - \dfrac{1}{\sqrt{3}} = -\dfrac{2\sqrt{3}}{9}$, and

$f(4) = 4 - \sqrt[3]{4} \approx 2.413$. So $f(4) = 4 - \sqrt[3]{4}$ is the absolute maximum value and $f\left(\dfrac{\sqrt{3}}{9}\right) = -\dfrac{2\sqrt{3}}{9}$ is the absolute minimum value.

56. $f(t) = \dfrac{\sqrt{t}}{1+t^2}$, $[0, 2]$. $f'(t) = \dfrac{(1+t^2)(1/(2\sqrt{t})) - \sqrt{t}\,(2t)}{(1+t^2)^2} = \dfrac{(1+t^2) - 2\sqrt{t}\,\sqrt{t}\,(2t)}{2\sqrt{t}\,(1+t^2)^2} = \dfrac{1-3t^2}{2\sqrt{t}\,(1+t^2)^2}$.

$f'(t) = 0 \Leftrightarrow 1 - 3t^2 = 0 \Leftrightarrow t^2 = \dfrac{1}{3} \Leftrightarrow t = \pm\dfrac{1}{\sqrt{3}}$, but $t = -\dfrac{1}{\sqrt{3}}$ is not in the given interval, $[0, 2]$. $f'(t)$ does not exist when $t = 0$, which is an endpoint. $f(0) = 0$, $f\left(\dfrac{1}{\sqrt{3}}\right) = \dfrac{1/\sqrt[4]{3}}{1+1/3} = \dfrac{3^{-1/4}}{4/3} = \dfrac{3^{3/4}}{4} \approx 0.570$, and

$f(2) = \dfrac{\sqrt{2}}{5} \approx 0.283$. So $f\left(\dfrac{1}{\sqrt{3}}\right) = \dfrac{3^{3/4}}{4}$ is the absolute maximum value and $f(0) = 0$ is the absolute minimum value.

57. $f(t) = 2\cos t + \sin 2t$, $[0, \pi/2]$.

$f'(t) = -2\sin t + \cos 2t \cdot 2 = -2\sin t + 2(1 - 2\sin^2 t) = -2(2\sin^2 t + \sin t - 1) = -2(2\sin t - 1)(\sin t + 1)$.

$f'(t) = 0 \Rightarrow \sin t = \frac{1}{2}$ or $\sin t = -1 \Rightarrow t = \frac{\pi}{6}$. $f(0) = 2$, $f(\frac{\pi}{6}) = \sqrt{3} + \frac{1}{2}\sqrt{3} = \frac{3}{2}\sqrt{3} \approx 2.60$, and $f(\frac{\pi}{2}) = 0$.

So $f(\frac{\pi}{6}) = \frac{3}{2}\sqrt{3}$ is the absolute maximum value and $f(\frac{\pi}{2}) = 0$ is the absolute minimum value.

58. $f(t) = t + \cot(t/2)$, $[\pi/4, 7\pi/4]$. $f'(t) = 1 - \csc^2(t/2) \cdot \frac{1}{2}$.

$f'(t) = 0 \Rightarrow \frac{1}{2}\csc^2(t/2) = 1 \Rightarrow \csc^2(t/2) = 2 \Rightarrow \csc(t/2) = \pm\sqrt{2} \Rightarrow \frac{1}{2}t = \frac{\pi}{4}$ or $\frac{1}{2}t = \frac{3\pi}{4}$

$\left[\frac{\pi}{4} \le t \le \frac{7\pi}{4} \Rightarrow \frac{\pi}{8} \le \frac{1}{2}t \le \frac{7\pi}{8}\right.$ and $\csc(t/2) \ne -\sqrt{2}$ in the last interval$\left.\right] \Rightarrow t = \frac{\pi}{2}$ or $t = \frac{3\pi}{2}$.

$f\left(\frac{\pi}{4}\right) = \frac{\pi}{4} + \cot\frac{\pi}{8} \approx 3.20$, $f\left(\frac{\pi}{2}\right) = \frac{\pi}{2} + \cot\frac{\pi}{4} = \frac{\pi}{2} + 1 \approx 2.57$, $f\left(\frac{3\pi}{2}\right) = \frac{3\pi}{2} + \cot\frac{3\pi}{4} = \frac{3\pi}{2} - 1 \approx 3.71$, and

$f\left(\frac{7\pi}{4}\right) = \frac{7\pi}{4} + \cot\frac{7\pi}{8} \approx 3.08$. So $f\left(\frac{3\pi}{2}\right) = \frac{3\pi}{2} - 1$ is the absolute maximum value and $f\left(\frac{\pi}{2}\right) = \frac{\pi}{2} + 1$ is the absolute minimum value.

59. $f(x) = x^{-2}\ln x$, $\left[\frac{1}{2}, 4\right]$. $f'(x) = x^{-2} \cdot \dfrac{1}{x} + (\ln x)(-2x^{-3}) = x^{-3} - 2x^{-3}\ln x = x^{-3}(1 - 2\ln x) = \dfrac{1 - 2\ln x}{x^3}$.

$f'(x) = 0 \Leftrightarrow 1 - 2\ln x = 0 \Leftrightarrow 2\ln x = 1 \Leftrightarrow \ln x = \frac{1}{2} \Leftrightarrow x = e^{1/2} \approx 1.65$. $f'(x)$ does not exist when $x = 0$, which is not in the given interval, $\left[\frac{1}{2}, 4\right]$. $f(\frac{1}{2}) = \dfrac{\ln 1/2}{(1/2)^2} = \dfrac{\ln 1 - \ln 2}{1/4} = -4\ln 2 \approx -2.773$,

$f(e^{1/2}) = \dfrac{\ln e^{1/2}}{(e^{1/2})^2} = \dfrac{1/2}{e} = \dfrac{1}{2e} \approx 0.184$, and $f(4) = \dfrac{\ln 4}{4^2} = \dfrac{\ln 4}{16} \approx 0.087$. So $f(e^{1/2}) = \dfrac{1}{2e}$ is the absolute maximum value and $f(\frac{1}{2}) = -4\ln 2$ is the absolute minimum value.

60. $f(x) = xe^{x/2}$, $[-3,1]$. $f'(x) = xe^{x/2}\left(\frac{1}{2}\right) + e^{x/2}(1) = e^{x/2}\left(\frac{1}{2}x + 1\right)$. $f'(x) = 0 \;\Leftrightarrow\; \frac{1}{2}x + 1 = 0 \;\Leftrightarrow\; x = -2$. $f(-3) = -3e^{-3/2} \approx -0.669$, $f(-2) = -2e^{-1} \approx -0.736$, and $f(1) = e^{1/2} \approx 1.649$. So $f(1) = e^{1/2}$ is the absolute maximum value and $f(-2) = -2/e$ is the absolute minimum value.

61. $f(x) = \ln(x^2 + x + 1)$, $[-1,1]$. $f'(x) = \dfrac{1}{x^2 + x + 1} \cdot (2x + 1) = 0 \;\Leftrightarrow\; x = -\frac{1}{2}$. Since $x^2 + x + 1 > 0$ for all x, the domain of f and f' is $\mathbb{R}$. $f(-1) = \ln 1 = 0$, $f\left(-\frac{1}{2}\right) = \ln\frac{3}{4} \approx -0.29$, and $f(1) = \ln 3 \approx 1.10$. So $f(1) = \ln 3 \approx 1.10$ is the absolute maximum value and $f\left(-\frac{1}{2}\right) = \ln\frac{3}{4} \approx -0.29$ is the absolute minimum value.

62. $f(x) = x - 2\tan^{-1}x$, $[0,4]$. $f'(x) = 1 - 2\cdot\dfrac{1}{1+x^2} = 0 \;\Leftrightarrow\; 1 = \dfrac{2}{1+x^2} \;\Leftrightarrow\; 1 + x^2 = 2 \;\Leftrightarrow\; x^2 = 1 \;\Leftrightarrow\; x = \pm 1$. $f(0) = 0$, $f(1) = 1 - \frac{\pi}{2} \approx -0.57$, and $f(4) = 4 - 2\tan^{-1}4 \approx 1.35$. So $f(4) = 4 - 2\tan^{-1}4$ is the absolute maximum value and $f(1) = 1 - \frac{\pi}{2}$ is the absolute minimum value.

63. $f(x) = x^a(1-x)^b$, $0 \le x \le 1$, $a > 0$, $b > 0$.

$$\begin{aligned} f'(x) &= x^a \cdot b(1-x)^{b-1}(-1) + (1-x)^b \cdot ax^{a-1} = x^{a-1}(1-x)^{b-1}[x \cdot b(-1) + (1-x)\cdot a] \\ &= x^{a-1}(1-x)^{b-1}(a - ax - bx) \end{aligned}$$

At the endpoints, we have $f(0) = f(1) = 0$ [the minimum value of f]. In the interval $(0,1)$, $f'(x) = 0 \;\Leftrightarrow\; x = \dfrac{a}{a+b}$.

$$f\left(\frac{a}{a+b}\right) = \left(\frac{a}{a+b}\right)^a\left(1 - \frac{a}{a+b}\right)^b = \frac{a^a}{(a+b)^a}\left(\frac{a+b-a}{a+b}\right)^b = \frac{a^a}{(a+b)^a}\cdot\frac{b^b}{(a+b)^b} = \frac{a^ab^b}{(a+b)^{a+b}}.$$

So $f\left(\dfrac{a}{a+b}\right) = \dfrac{a^ab^b}{(a+b)^{a+b}}$ is the absolute maximum value.

64.

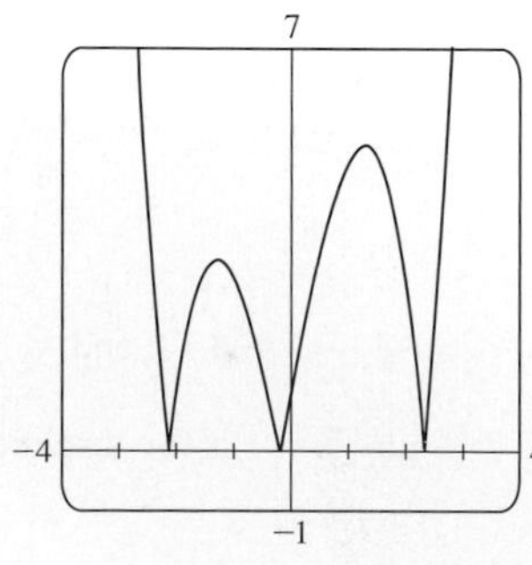

The graph of $f(x) = \left|1 + 5x - x^3\right|$ indicates that $f'(x) = 0$ at $x \approx \pm 1.3$ and that $f'(x)$ does not exist at $x \approx -2.1$, -0.2, and 2.3. Those five values of x are the critical numbers of f.

65. (a)

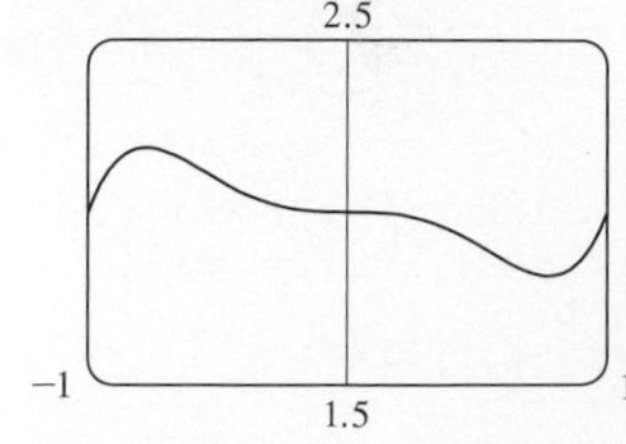

From the graph, it appears that the absolute maximum value is about $f(-0.77) = 2.19$, and the absolute minimum value is about $f(0.77) = 1.81$.

(b) $f(x) = x^5 - x^3 + 2 \quad\Rightarrow\quad f'(x) = 5x^4 - 3x^2 = x^2(5x^2 - 3)$. So $f'(x) = 0 \quad\Rightarrow\quad x = 0, \pm\sqrt{\frac{3}{5}}$.

$$f\left(-\sqrt{\tfrac{3}{5}}\right) = \left(-\sqrt{\tfrac{3}{5}}\right)^5 - \left(-\sqrt{\tfrac{3}{5}}\right)^3 + 2 = -\left(\tfrac{3}{5}\right)^2\sqrt{\tfrac{3}{5}} + \tfrac{3}{5}\sqrt{\tfrac{3}{5}} + 2$$
$$= \left(\tfrac{3}{5} - \tfrac{9}{25}\right)\sqrt{\tfrac{3}{5}} + 2 = \tfrac{6}{25}\sqrt{\tfrac{3}{5}} + 2 \quad \text{(maximum)}$$

and similarly, $f\left(\sqrt{\frac{3}{5}}\right) = -\frac{6}{25}\sqrt{\frac{3}{5}} + 2$ (minimum).

66. (a)

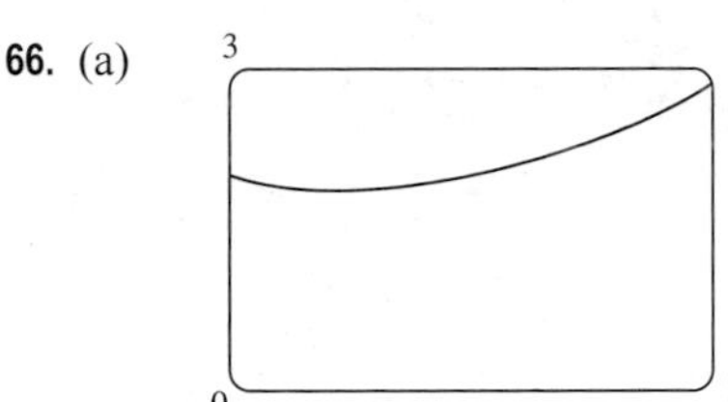

From the graph, it appears that the absolute maximum value is about $f(1) = 2.85$, and the absolute minimum value is about $f(0.23) = 1.89$.

(b) $f(x) = e^x + e^{-2x} \quad\Rightarrow\quad f'(x) = e^x - 2e^{-2x} = e^{-2x}(e^{3x} - 2)$. So $f'(x) = 0 \quad\Leftrightarrow\quad e^{3x} = 2 \quad\Leftrightarrow\quad 3x = \ln 2 \quad\Leftrightarrow$

$x = \frac{1}{3}\ln 2\ [\approx 0.23]$. $f\left(\frac{1}{3}\ln 2\right) = (e^{\ln 2})^{1/3} + (e^{\ln 2})^{-2/3} = 2^{1/3} + 2^{-2/3}\ [\approx 1.89]$, the minimum value.

$f(1) = e^1 + e^{-2}\ [\approx 2.85]$, the maximum.

67. (a)

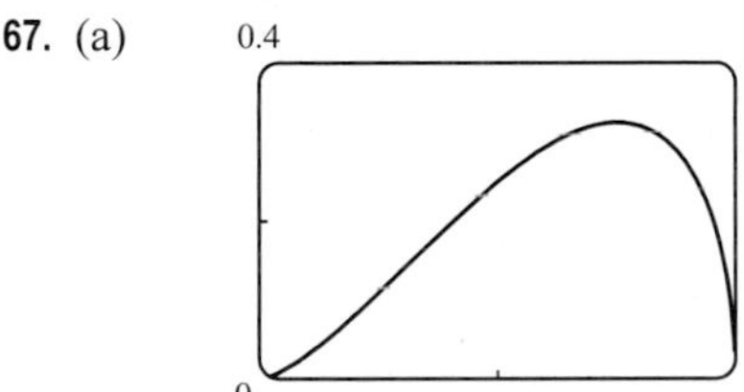

From the graph, it appears that the absolute maximum value is about $f(0.75) = 0.32$, and the absolute minimum value is $f(0) = f(1) = 0$; that is, at both endpoints.

(b) $f(x) = x\sqrt{x - x^2} \quad\Rightarrow\quad f'(x) = x \cdot \dfrac{1 - 2x}{2\sqrt{x - x^2}} + \sqrt{x - x^2} = \dfrac{(x - 2x^2) + (2x - 2x^2)}{2\sqrt{x - x^2}} = \dfrac{3x - 4x^2}{2\sqrt{x - x^2}}$.

So $f'(x) = 0 \quad\Rightarrow\quad 3x - 4x^2 = 0 \quad\Rightarrow\quad x(3 - 4x) = 0 \quad\Rightarrow\quad x = 0$ or $\frac{3}{4}$.

$f(0) = f(1) = 0$ (minimum), and $f\left(\frac{3}{4}\right) = \frac{3}{4}\sqrt{\frac{3}{4} - \left(\frac{3}{4}\right)^2} = \frac{3}{4}\sqrt{\frac{3}{16}} = \frac{3\sqrt{3}}{16}$ (maximum).

68. (a)

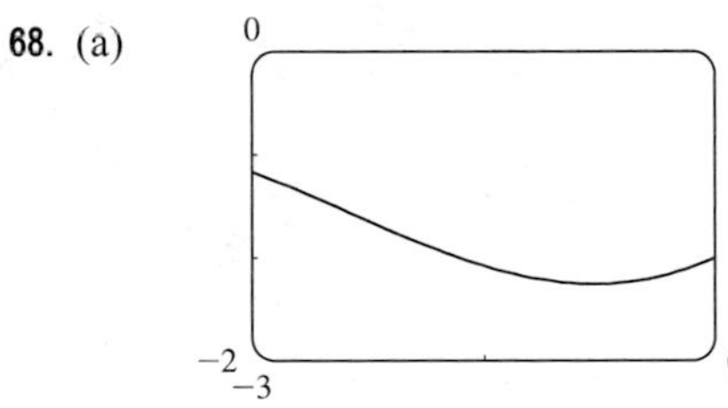

From the graph, it appears that the absolute maximum value is about $f(-2) = -1.17$, and the absolute minimum value is about $f(-0.52) = -2.26$.

(b) $f(x) = x - 2\cos x \quad\Rightarrow\quad f'(x) = 1 + 2\sin x$. So $f'(x) = 0 \quad\Rightarrow\quad \sin x = -\frac{1}{2} \quad\Rightarrow\quad x = -\frac{\pi}{6}$ on $[-2, 0]$.

$f(-2) = -2 - 2\cos(-2)$ (maximum) and $f\left(-\frac{\pi}{6}\right) = -\frac{\pi}{6} - 2\cos\left(-\frac{\pi}{6}\right) = -\frac{\pi}{6} - 2\left(\frac{\sqrt{3}}{2}\right) = -\frac{\pi}{6} - \sqrt{3}$ (minimum).

69. Let $a = 1.35$ and $b = -2.802$. Then $C(t) = ate^{bt} \quad\Rightarrow\quad C'(t) = a(te^{bt} \cdot b + e^{bt} \cdot 1) = ae^{bt}(bt + 1)$. $C'(t) = 0 \quad\Leftrightarrow$

$bt + 1 = 0 \quad\Leftrightarrow\quad t = -\dfrac{1}{b} \approx 0.36$ h. $C(0) = 0$, $C(-1/b) = -\dfrac{a}{b}e^{-1} = -\dfrac{a}{be} \approx 0.177$, and $C(3) = 3ae^{3b} \approx 0.0009$. The maximum average BAC during the first three hours is about 0.177 mg/mL and it occurs at approximately 0.36 h (21.4 min).

70. $C(t) = 8(e^{-0.4t} - e^{-0.6t}) \Rightarrow C'(t) = 8(-0.4e^{-0.4t} + 0.6e^{-0.6t})$. $C'(t) = 0 \Leftrightarrow 0.6e^{-0.6t} = 0.4e^{-0.4t} \Leftrightarrow$ $\frac{0.6}{0.4} = e^{-0.4t+0.6t} \Leftrightarrow \frac{3}{2} = e^{0.2t} \Leftrightarrow 0.2t = \ln\frac{3}{2} \Leftrightarrow t = 5\ln\frac{3}{2} \approx 2.027$ h. $C(0) = 8(1-1) = 0$, $C(5\ln\frac{3}{2}) = 8(e^{-2\ln 3/2} - e^{-3\ln 3/2}) = 8\left[\left(\frac{3}{2}\right)^{-2} - \left(\frac{3}{2}\right)^{-3}\right] = 8\left(\frac{4}{9} - \frac{8}{27}\right) = \frac{32}{27} \approx 1.185$, and $C(12) = 8(e^{-4.8} - e^{-7.2}) \approx 0.060$. The maximum concentration of the antibiotic during the first 12 hours is $\frac{32}{27}\ \mu\text{g/mL}$.

71. The density is defined as $\rho = \frac{\text{mass}}{\text{volume}} = \frac{1000}{V(T)}$ (in g/cm^3). But a critical point of ρ will also be a critical point of V $\left[\text{since } \frac{d\rho}{dT} = -1000V^{-2}\frac{dV}{dT} \text{ and } V \text{ is never } 0\right]$, and V is easier to differentiate than ρ.

$V(T) = 999.87 - 0.06426T + 0.0085043T^2 - 0.0000679T^3 \Rightarrow V'(T) = -0.06426 + 0.0170086T - 0.0002037T^2$.

Setting this equal to 0 and using the quadratic formula to find T, we get

$$T = \frac{-0.0170086 \pm \sqrt{0.0170086^2 - 4 \cdot 0.0002037 \cdot 0.06426}}{2(-0.0002037)} \approx 3.9665°\text{C} \text{ or } 79.5318°\text{C}.$$

Since we are only interested in the region $0°\text{C} \le T \le 30°\text{C}$, we check the density ρ at the endpoints and at 3.9665°C: $\rho(0) \approx \frac{1000}{999.87} \approx 1.00013$; $\rho(30) \approx \frac{1000}{1003.7628} \approx 0.99625$; $\rho(3.9665) \approx \frac{1000}{999.7447} \approx 1.000255$. So water has its maximum density at about 3.9665°C.

72. $F = \frac{\mu W}{\mu\sin\theta + \cos\theta} \Rightarrow \frac{dF}{d\theta} = \frac{(\mu\sin\theta + \cos\theta)(0) - \mu W(\mu\cos\theta - \sin\theta)}{(\mu\sin\theta + \cos\theta)^2} = \frac{-\mu W(\mu\cos\theta - \sin\theta)}{(\mu\sin\theta + \cos\theta)^2}$.

So $\frac{dF}{d\theta} = 0 \Rightarrow \mu\cos\theta - \sin\theta = 0 \Rightarrow \mu = \frac{\sin\theta}{\cos\theta} = \tan\theta$. Substituting $\tan\theta$ for μ in F gives us

$$F = \frac{(\tan\theta)W}{(\tan\theta)\sin\theta + \cos\theta} = \frac{W\tan\theta}{\frac{\sin^2\theta}{\cos\theta} + \cos\theta} = \frac{W\tan\theta\cos\theta}{\sin^2\theta + \cos^2\theta} = \frac{W\sin\theta}{1} = W\sin\theta.$$

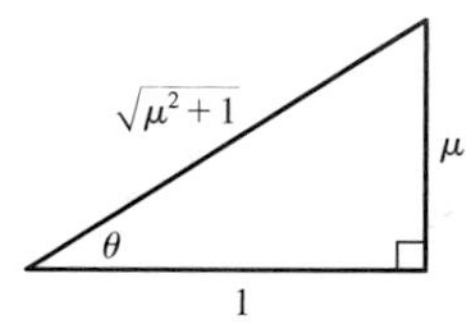

If $\tan\theta = \mu$, then $\sin\theta = \frac{\mu}{\sqrt{\mu^2+1}}$ (see the figure), so $F = \frac{\mu}{\sqrt{\mu^2+1}}W$.

We compare this with the value of F at the endpoints: $F(0) = \mu W$ and $F\left(\frac{\pi}{2}\right) = W$.

Now because $\frac{\mu}{\sqrt{\mu^2+1}} \le 1$ and $\frac{\mu}{\sqrt{\mu^2+1}} \le \mu$, we have that $\frac{\mu}{\sqrt{\mu^2+1}}W$ is less than or equal to each of $F(0)$ and $F\left(\frac{\pi}{2}\right)$.

Hence, $\frac{\mu}{\sqrt{\mu^2+1}}W$ is the absolute minimum value of $F(\theta)$, and it occurs when $\tan\theta = \mu$.

73. $L(t) = 0.01441t^3 - 0.4177t^2 + 2.703t + 1060.1 \Rightarrow L'(t) = 0.04323t^2 - 0.8354t + 2.703$. Use the quadratic formula to solve $L'(t) = 0$. $t = \frac{0.8354 \pm \sqrt{(0.8354)^2 - 4(0.04323)(2.703)}}{2(0.04323)} \approx 4.1$ or 15.2. For $0 \le t \le 12$, we have $L(0) = 1060.1$, $L(4.1) \approx 1065.2$, and $L(12) \approx 1057.3$. Thus, the water level was highest during 2012 about 4.1 months after January 1.

74. (a) The equation of the graph in the figure is

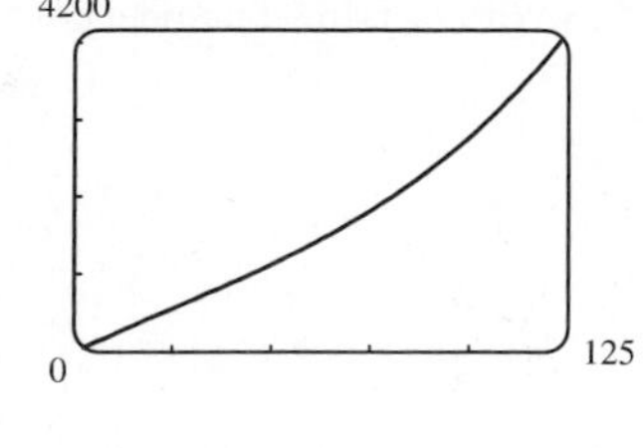

$v(t) = 0.00146t^3 - 0.11553t^2 + 24.98169t - 21.26872.$

(b) $a(t) = v'(t) = 0.00438t^2 - 0.23106t + 24.98169 \Rightarrow$

$a'(t) = 0.00876t - 0.23106.$

$a'(t) = 0 \Rightarrow t_1 = \frac{0.23106}{0.00876} \approx 26.4.$ $a(0) \approx 24.98$, $a(t_1) \approx 21.93$,

and $a(125) \approx 64.54$.

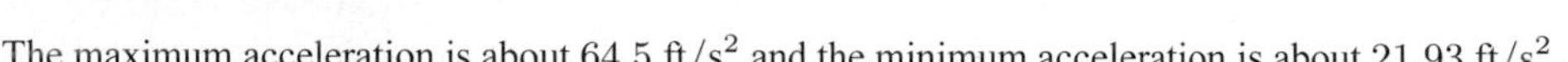

The maximum acceleration is about 64.5 ft/s^2 and the minimum acceleration is about 21.93 ft/s^2.

75. (a) $v(r) = k(r_0 - r)r^2 = kr_0r^2 - kr^3 \Rightarrow v'(r) = 2kr_0r - 3kr^2$. $v'(r) = 0 \Rightarrow kr(2r_0 - 3r) = 0 \Rightarrow$

$r = 0$ or $\frac{2}{3}r_0$ (but 0 is not in the interval). Evaluating v at $\frac{1}{2}r_0$, $\frac{2}{3}r_0$, and r_0, we get $v\left(\frac{1}{2}r_0\right) = \frac{1}{8}kr_0^3$, $v\left(\frac{2}{3}r_0\right) = \frac{4}{27}kr_0^3$,

and $v(r_0) = 0$. Since $\frac{4}{27} > \frac{1}{8}$, v attains its maximum value at $r = \frac{2}{3}r_0$. This supports the statement in the text.

(b) From part (a), the maximum value of v is $\frac{4}{27}kr_0^3$.

(c)

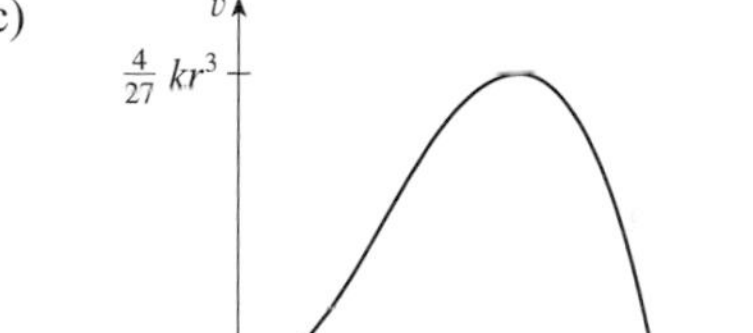

76. $g(x) = 2 + (x-5)^3 \Rightarrow g'(x) = 3(x-5)^2 \Rightarrow g'(5) = 0$, so 5 is a critical number. But $g(5) = 2$ and g takes on values > 2 and values < 2 in any open interval containing 5, so g does not have a local maximum or minimum at 5.

77. $f(x) = x^{101} + x^{51} + x + 1 \Rightarrow f'(x) = 101x^{100} + 51x^{50} + 1 \geq 1$ for all x, so $f'(x) = 0$ has no solution. Thus, $f(x)$ has no critical number, so $f(x)$ can have no local maximum or minimum.

78. Suppose that f has a minimum value at c, so $f(x) \geq f(c)$ for all x near c. Then $g(x) = -f(x) \leq -f(c) = g(c)$ for all x near c, so $g(x)$ has a maximum value at c.

79. If f has a local minimum at c, then $g(x) = -f(x)$ has a local maximum at c, so $g'(c) = 0$ by the case of Fermat's Theorem proved in the text. Thus, $f'(c) = -g'(c) = 0$.

80. (a) $f(x) = ax^3 + bx^2 + cx + d$, $a \neq 0$. So $f'(x) = 3ax^2 + 2bx + c$ is a quadratic and hence has either 2, 1, or 0 real roots, so $f(x)$ has either 2, 1 or 0 critical numbers.

Case (i) [2 critical numbers]: $f(x) = x^3 - 3x \Rightarrow$

$f'(x) = 3x^2 - 3$, so $x = -1, 1$

are critical numbers.

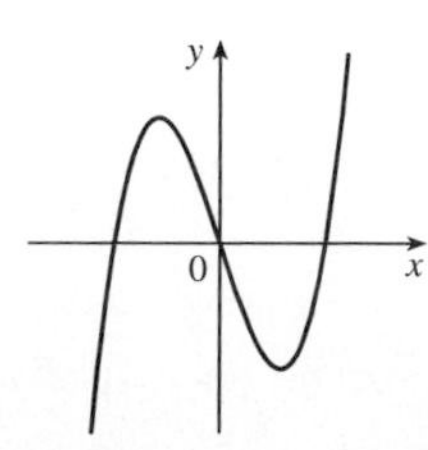

Case (ii) [1 critical number]: $f(x) = x^3 \Rightarrow$
$f'(x) = 3x^2$, so $x = 0$
is the only critical number.

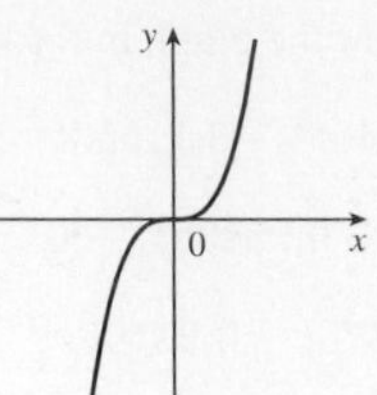

Case (iii) [no critical number]: $f(x) = x^3 + 3x \Rightarrow$
$f'(x) = 3x^2 + 3$,
so there is no critical number.

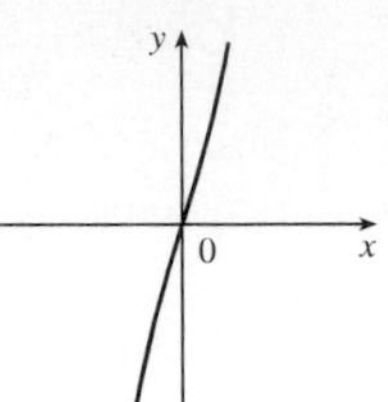

(b) Since there are at most two critical numbers, it can have at most two local extreme values and by (i) this can occur. By (iii) it can have no local extreme value. However, if there is only one critical number, then there is no local extreme value.

APPLIED PROJECT The Calculus of Rainbows

1. From Snell's Law, we have $\sin\alpha = k\sin\beta \approx \frac{4}{3}\sin\beta \Leftrightarrow \beta \approx \arcsin\left(\frac{3}{4}\sin\alpha\right)$. We substitute this into $D(\alpha) = \pi + 2\alpha - 4\beta = \pi + 2\alpha - 4\arcsin\left(\frac{3}{4}\sin\alpha\right)$, and then differentiate to find the minimum:

$D'(\alpha) = 2 - 4\left[1 - \left(\frac{3}{4}\sin\alpha\right)^2\right]^{-1/2}\left(\frac{3}{4}\cos\alpha\right) = 2 - \dfrac{3\cos\alpha}{\sqrt{1 - \frac{9}{16}\sin^2\alpha}}$. This is 0 when $\dfrac{3\cos\alpha}{\sqrt{1 - \frac{9}{16}\sin^2\alpha}} = 2 \Leftrightarrow$

$\frac{9}{4}\cos^2\alpha = 1 - \frac{9}{16}\sin^2\alpha \Leftrightarrow \frac{9}{4}\cos^2\alpha = 1 - \frac{9}{16}\left(1 - \cos^2\alpha\right) \Leftrightarrow \frac{27}{16}\cos^2\alpha = \frac{7}{16} \Leftrightarrow \cos\alpha = \sqrt{\frac{7}{27}} \Leftrightarrow$

$\alpha = \arccos\sqrt{\frac{7}{27}} \approx 59.4^\circ$, and so the local minimum is $D(59.4^\circ) \approx 2.4$ radians $\approx 138^\circ$.

To see that this is an absolute minimum, we check the endpoints, which in this case are $\alpha = 0$ and $\alpha = \frac{\pi}{2}$:

$D(0) = \pi$ radians $= 180^\circ$, and $D\left(\frac{\pi}{2}\right) \approx 166^\circ$.

Another method: We first calculate $\dfrac{d\beta}{d\alpha}$: $\sin\alpha = \frac{4}{3}\sin\beta \Leftrightarrow \cos\alpha = \frac{4}{3}\cos\beta\,\dfrac{d\beta}{d\alpha} \Leftrightarrow \dfrac{d\beta}{d\alpha} = \dfrac{3\cos\alpha}{4\cos\beta}$, so since

$D'(\alpha) = 2 - 4\,\dfrac{d\beta}{d\alpha} = 0 \Leftrightarrow \dfrac{d\beta}{d\alpha} = \dfrac{1}{2}$, the minimum occurs when $3\cos\alpha = 2\cos\beta$. Now we square both sides and substitute $\sin\alpha = \frac{4}{3}\sin\beta$, leading to the same result.

2. If we repeat Problem 1 with k in place of $\frac{4}{3}$, we get $D(\alpha) = \pi + 2\alpha - 4\arcsin\left(\dfrac{1}{k}\sin\alpha\right) \Rightarrow$

$D'(\alpha) = 2 - \dfrac{4\cos\alpha}{k\sqrt{1 - \left(\dfrac{\sin\alpha}{k}\right)^2}}$, which is 0 when $\dfrac{2\cos\alpha}{k} = \sqrt{1 - \left(\dfrac{\sin\alpha}{k}\right)^2} \Leftrightarrow \left(\dfrac{2\cos\alpha}{k}\right)^2 = 1 - \left(\dfrac{\sin\alpha}{k}\right)^2 \Leftrightarrow$

$4\cos^2\alpha = k^2 - \sin^2\alpha \Leftrightarrow 3\cos^2\alpha = k^2 - 1 \Leftrightarrow \alpha = \arccos\sqrt{\dfrac{k^2 - 1}{3}}$. So for $k \approx 1.3318$ (red light) the minimum occurs at $\alpha_1 \approx 1.038$ radians, and so the rainbow angle is about $\pi - D(\alpha_1) \approx 42.3^\circ$. For $k \approx 1.3435$ (violet light) the minimum occurs at $\alpha_2 \approx 1.026$ radians, and so the rainbow angle is about $\pi - D(\alpha_2) \approx 40.6^\circ$.

Another method: As in Problem 1, we can instead find $D'(\alpha)$ in terms of $\frac{d\beta}{d\alpha}$, and then substitute $\frac{d\beta}{d\alpha} = \frac{\cos\alpha}{k\cos\beta}$.

3. At each reflection or refraction, the light is bent in a counterclockwise direction: the bend at A is $\alpha - \beta$, the bend at B is $\pi - 2\beta$, the bend at C is again $\pi - 2\beta$, and the bend at D is $\alpha - \beta$. So the total bend is

$D(\alpha) = 2(\alpha - \beta) + 2(\pi - 2\beta) = 2\alpha - 6\beta + 2\pi$, as required. We substitute $\beta = \arcsin\left(\frac{\sin\alpha}{k}\right)$ and differentiate, to get

$D'(\alpha) = 2 - \dfrac{6\cos\alpha}{k\sqrt{1 - \left(\frac{\sin\alpha}{k}\right)^2}}$, which is 0 when $\dfrac{3\cos\alpha}{k} = \sqrt{1 - \left(\frac{\sin\alpha}{k}\right)^2} \Leftrightarrow 9\cos^2\alpha = k^2 - \sin^2\alpha \Leftrightarrow$

$8\cos^2\alpha = k^2 - 1 \Leftrightarrow \cos\alpha = \sqrt{\frac{1}{8}(k^2 - 1)}$. If $k = \frac{4}{3}$, then the minimum occurs at

$\alpha_1 = \arccos\sqrt{\dfrac{(4/3)^2 - 1}{8}} \approx 1.254$ radians. Thus, the minimum *counterclockwise* rotation is $D(\alpha_1) \approx 231^\circ$, which is equivalent to a *clockwise* rotation of $360^\circ - 231^\circ = 129^\circ$ (see the figure). So the rainbow angle for the secondary rainbow is about $180^\circ - 129^\circ = 51^\circ$, as required. In general, the rainbow angle for the secondary rainbow is $\pi - [2\pi - D(\alpha)] = D(\alpha) - \pi$.

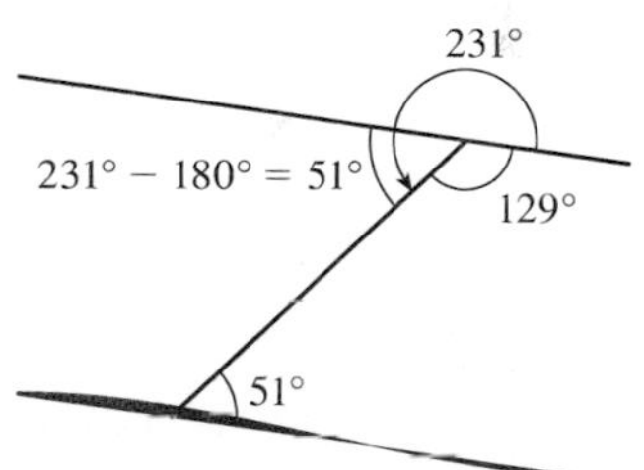

4. In the primary rainbow, the rainbow angle gets smaller as k gets larger, as we found in Problem 2, so the colors appear from top to bottom in order of increasing k. But in the secondary rainbow, the rainbow angle gets larger as k gets larger. To see this, we find the minimum deviations for red light and for violet light in the secondary rainbow. For $k \approx 1.3318$ (red light) the minimum occurs at $\alpha_1 \approx \arccos\sqrt{\dfrac{1.3318^2 - 1}{8}} \approx 1.255$ radians, and so the rainbow angle is $D(\alpha_1) - \pi \approx 50.6^\circ$. For $k \approx 1.3435$ (violet light) the minimum occurs at $\alpha_2 \approx \arccos\sqrt{\dfrac{1.3435^2 - 1}{8}} \approx 1.248$ radians, and so the rainbow angle is $D(\alpha_2) - \pi \approx 53.6^\circ$. Consequently, the rainbow angle is larger for colors with higher indices of refraction, and the colors appear from bottom to top in order of increasing k, the reverse of their order in the primary rainbow.

Note that our calculations above also explain why the secondary rainbow is more spread out than the primary rainbow: in the primary rainbow, the difference between rainbow angles for red and violet light is about 1.7°, whereas in the secondary rainbow it is about 3°.

4.2 The Mean Value Theorem

1. **(1)** f is continuous on the closed interval $[0, 8]$.

(2) f is differentiable on the open interval $(0, 8)$.

(3) $f(0) = 3$ and $f(8) = 3$

Thus, f satisfies the hypotheses of Rolle's Theorem. The numbers $c = 1$ and $c = 5$ satisfy the conclusion of Rolle's Theorem since $f'(1) = f'(5) = 0$.

2. The possible graphs fall into two general categories: **(1)** Not continuous and therefore not differentiable, **(2)** Continuous, but not differentiable.

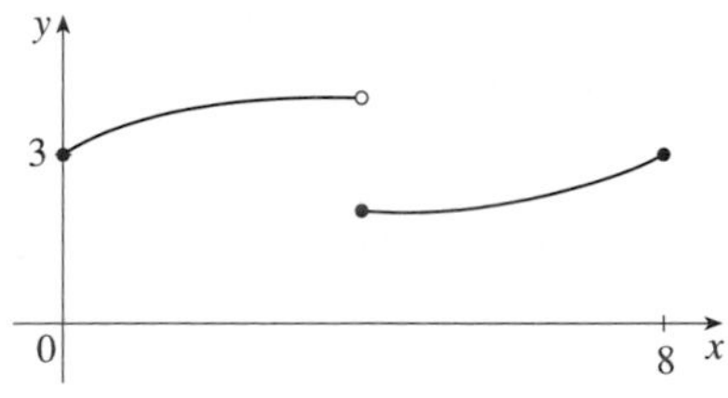

Not continuous

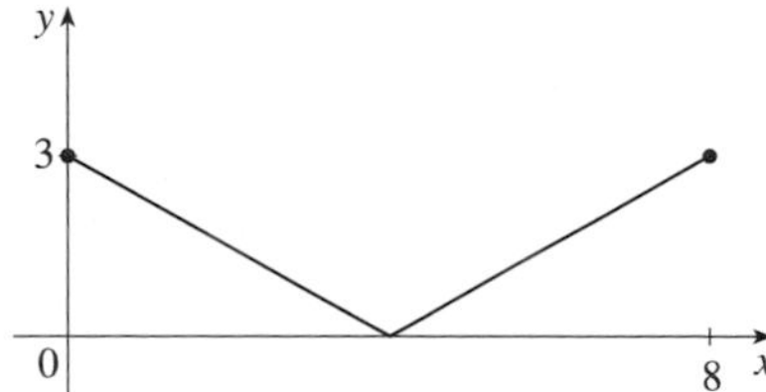

Not differentiable

In either case, there is no number c such that $f'(c) = 0$.

3. (a) **(1)** g is continuous on the closed interval $[0, 8]$.

(2) g is differentiable on the open interval $(0, 8)$.

(b) $g'(c) = \dfrac{g(8) - g(0)}{8 - 0} = \dfrac{4 - 1}{8} = \dfrac{3}{8}$.

It appears that $g'(c) = \frac{3}{8}$ when $c \approx 2.2$ and 6.4.

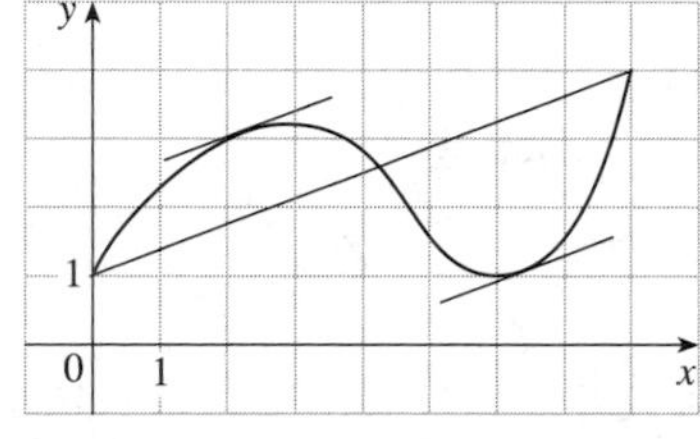

(c) $g'(c) = \dfrac{g(6) - g(2)}{6 - 2} = \dfrac{1 - 3}{4} = -\dfrac{1}{2}$.

It appears that $g'(c) = -\frac{1}{2}$ when $c \approx 3.7$ and 5.5.

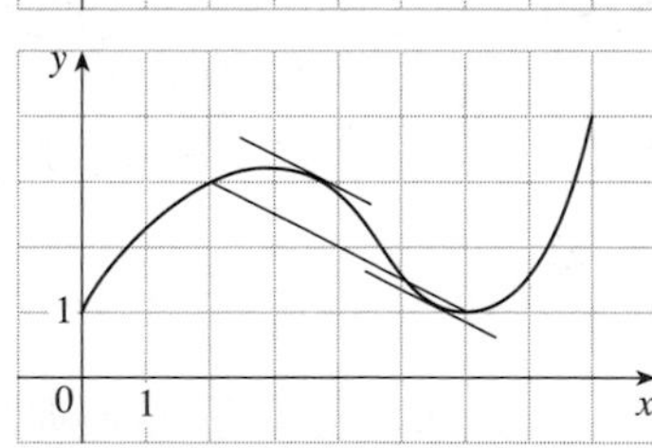

4.

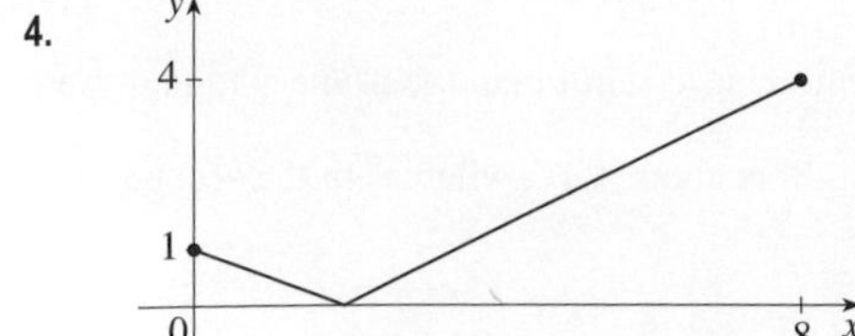

The function shown in the figure is continuous on $[0, 8]$ [but not differentiable on $(0, 8)$] with $f(0) = 1$ and $f(8) = 4$. The line passing through the two points has slope $\frac{3}{8}$. There is no number c in $(0, 8)$ such that $f'(c) = \frac{3}{8}$.

5. $f(x) = 2x^2 - 4x + 5$, $[-1, 3]$. f is a polynomial, so it's continuous and differentiable on $\mathbb{R}$, and hence, continuous on $[-1, 3]$ and differentiable on $(-1, 3)$. Since $f(-1) = 11$ and $f(3) = 11$, f satisfies all the hypotheses of Rolle's

Theorem. $f'(c) = 4c - 4$ and $f'(c) = 0 \Leftrightarrow 4c - 4 = 0 \Leftrightarrow c = 1$. $c = 1$ is in the interval $(-1, 3)$, so 1 satisfies the conclusion of Rolle's Theorem.

6. $f(x) = x^3 - 2x^2 - 4x + 2$, $[-2, 2]$. f is a polynomial, so it's continuous and differentiable on $\mathbb{R}$, and hence, continuous on $[-2, 2]$ and differentiable on $(-2, 2)$. Since $f(-2) = -6$ and $f(2) = -6$, f satisfies all the hypotheses of Rolle's Theorem. $f'(c) = 3c^2 - 4c - 4$ and $f'(c) = 0 \Leftrightarrow (3c + 2)(c - 2) = 0 \Leftrightarrow c = -\frac{2}{3}$ or 2. $c = -\frac{2}{3}$ is in the open interval $(-2, 2)$ (but 2 *isn't*), so only $-\frac{2}{3}$ satisfies the conclusion of Rolle's Theorem.

7. $f(x) = \sin(x/2)$, $[\pi/2, 3\pi/2]$. f, being the composite of the sine function and the polynomial $x/2$, is continuous and differentiable on $\mathbb{R}$, so it is continuous on $[\pi/2, 3\pi/2]$ and differentiable on $(\pi/2, 3\pi/2)$. Also, $f\left(\frac{\pi}{2}\right) = \frac{1}{2}\sqrt{2} = f\left(\frac{3\pi}{2}\right)$. $f'(c) = 0 \Leftrightarrow \frac{1}{2}\cos(c/2) = 0 \Leftrightarrow \cos(c/2) = 0 \Leftrightarrow c/2 = \frac{\pi}{2} + n\pi \Leftrightarrow c = \pi + 2n\pi$, n an integer.
Only $c = \pi$ is in $(\pi/2, 3\pi/2)$, so π satisfies the conclusion of Rolle's Theorem.

8. $f(x) = x + 1/x$, $\left[\frac{1}{2}, 2\right]$. $f'(x) = 1 - 1/x^2 = \dfrac{x^2 - 1}{x^2}$. f is a rational function that is continuous on its domain, $(-\infty, 0) \cup (0, \infty)$, so it is continuous on $\left[\frac{1}{2}, 2\right]$. f' has the same domain and is differentiable on $\left(\frac{1}{2}, 2\right)$. Also, $f\left(\frac{1}{2}\right) = \frac{5}{2} = f(2)$. $f'(c) = 0 \Leftrightarrow \dfrac{c^2 - 1}{c^2} = 0 \Leftrightarrow c^2 - 1 = 0 \Leftrightarrow c = \pm 1$. Only 1 is in $\left(\frac{1}{2}, 2\right)$, so 1 satisfies the conclusion of Rolle's Theorem.

9. $f(x) = 1 - x^{2/3}$. $f(-1) = 1 - (-1)^{2/3} = 1 - 1 = 0 = f(1)$. $f'(x) = -\frac{2}{3}x^{-1/3}$, so $f'(c) = 0$ has no solution. This does not contradict Rolle's Theorem, since $f'(0)$ does not exist, and so f is not differentiable on $(-1, 1)$.

10. $f(x) = \tan x$. $f(0) = \tan 0 = 0 = \tan \pi = f(\pi)$. $f'(x) = \sec^2 x \geq 1$, so $f'(c) = 0$ has no solution. This does not contradict Rolle's Theorem, since $f'\left(\frac{\pi}{2}\right)$ does not exist, and so f is not differentiable on $(0, \pi)$. (Also, $f(x)$ is not continuous on $[0, \pi]$.)

11. $f(x) = 2x^2 - 3x + 1$, $[0, 2]$. f is continuous on $[0, 2]$ and differentiable on $(0, 2)$ since polynomials are continuous and differentiable on $\mathbb{R}$. $f'(c) = \dfrac{f(b) - f(a)}{b - a} \Leftrightarrow 4c - 3 = \dfrac{f(2) - f(0)}{2 - 0} = \dfrac{3 - 1}{2} = 1 \Leftrightarrow 4c = 4 \Leftrightarrow c = 1$, which is in $(0, 2)$.

12. $f(x) = x^3 - 3x + 2$, $[-2, 2]$. f is continuous on $[-2, 2]$ and differentiable on $(-2, 2)$ since polynomials are continuous and differentiable on $\mathbb{R}$. $f'(c) = \dfrac{f(b) - f(a)}{b - a} \Leftrightarrow 3c^2 - 3 = \dfrac{f(2) - f(-2)}{2 - (-2)} = \dfrac{4 - 0}{4} = 1 \Leftrightarrow 3c^2 = 4 \Leftrightarrow$ $c^2 = \dfrac{4}{3} \Leftrightarrow c = \pm\dfrac{2}{\sqrt{3}}$, which are both in $(-2, 2)$.

13. $f(x) = \ln x$, $[1, 4]$. f is continuous and differentiable on $(0, \infty)$, so f is continuous on $[1, 4]$ and differentiable on $(1, 4)$. $f'(c) = \dfrac{f(b) - f(a)}{b - a} \Leftrightarrow \dfrac{1}{c} = \dfrac{f(4) - f(1)}{4 - 1} = \dfrac{\ln 4 - 0}{3} = \dfrac{\ln 4}{3} \Leftrightarrow c = \dfrac{3}{\ln 4} \approx 2.16$, which is in $(1, 4)$.

14. $f(x) = \dfrac{1}{x}$, $[1, 3]$. f is continuous and differentiable on $(-\infty, 0) \cup (0, \infty)$, so f is continous on $[1, 3]$ and differentiable on $(1, 3)$. $f'(c) = \dfrac{f(b) - f(a)}{b - a} \iff -\dfrac{1}{c^2} = \dfrac{f(3) - f(1)}{3 - 1} = \dfrac{\frac{1}{3} - 1}{2} = -\dfrac{1}{3} \iff c^2 = 3 \iff c = \pm\sqrt{3}$, but only $\sqrt{3}$ is in $(1, 3)$.

15. $f(x) = \sqrt{x}$, $[0, 4]$. $f'(c) = \dfrac{f(4) - f(0)}{4 - 0} \iff \dfrac{1}{2\sqrt{c}} = \dfrac{2 - 0}{4} \iff \dfrac{1}{2\sqrt{c}} = \dfrac{1}{2} \iff \sqrt{c} = 1 \iff c = 1$. The secant line and the tangent line are parallel.

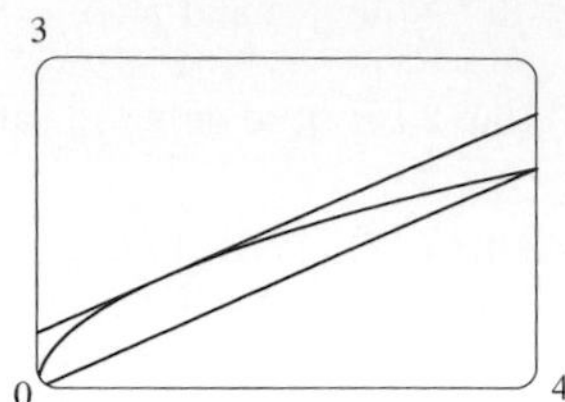

16. $f(x) = e^{-x}$, $[0, 2]$. $f'(c) = \dfrac{f(2) - f(0)}{2 - 0} \iff -e^{-c} = \dfrac{e^{-2} - 1}{2} \iff e^{-c} = \dfrac{1 - e^{-2}}{2} \iff -c = \ln\dfrac{1 - e^{-2}}{2} \iff c = -\ln\dfrac{1 - e^{-2}}{2} \approx 0.8386$. The secant line and the tangent line are parallel.

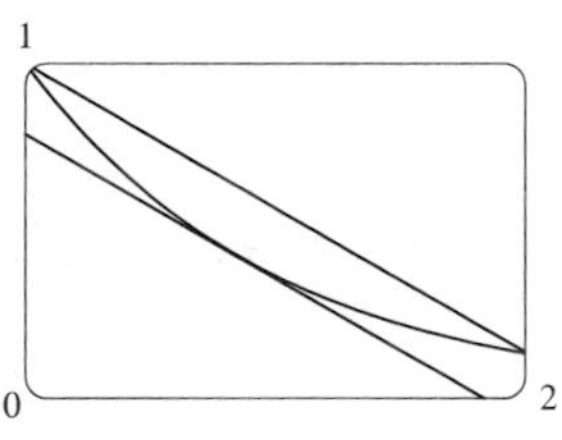

17. $f(x) = (x - 3)^{-2} \Rightarrow f'(x) = -2\,(x - 3)^{-3}$. $f(4) - f(1) = f'(c)(4 - 1) \Rightarrow \dfrac{1}{1^2} - \dfrac{1}{(-2)^2} = \dfrac{-2}{(c - 3)^3} \cdot 3 \Rightarrow \dfrac{3}{4} = \dfrac{-6}{(c - 3)^3} \Rightarrow (c - 3)^3 = -8 \Rightarrow c - 3 = -2 \Rightarrow c = 1$, which is not in the open interval $(1, 4)$. This does not contradict the Mean Value Theorem since f is not continuous at $x = 3$.

18. $f(x) = 2 - |2x - 1| = \begin{cases} 2 - (2x - 1) & \text{if } 2x - 1 \geq 0 \\ 2 - [-(2x - 1)] & \text{if } 2x - 1 < 0 \end{cases} = \begin{cases} 3 - 2x & \text{if } x \geq \frac{1}{2} \\ 1 + 2x & \text{if } x < \frac{1}{2} \end{cases} \Rightarrow f'(x) = \begin{cases} -2 & \text{if } x > \frac{1}{2} \\ 2 & \text{if } x < \frac{1}{2} \end{cases}$

$f(3) - f(0) = f'(c)(3 - 0) \Rightarrow -3 - 1 = f'(c) \cdot 3 \Rightarrow f'(c) = -\frac{4}{3}$ [not ± 2]. This does not contradict the Mean Value Theorem since f is not differentiable at $x = \frac{1}{2}$.

19. Let $f(x) = 2x + \cos x$. Then $f(-\pi) = -2\pi - 1 < 0$ and $f(0) = 1 > 0$. Since f is the sum of the polynomial $2x$ and the trignometric function $\cos x$, f is continuous and differentiable for all x. By the Intermediate Value Theorem, there is a number c in $(-\pi, 0)$ such that $f(c) = 0$. Thus, the given equation has at least one real root. If the equation has distinct real roots a and b with $a < b$, then $f(a) = f(b) = 0$. Since f is continuous on $[a, b]$ and differentiable on (a, b), Rolle's Theorem implies that there is a number r in (a, b) such that $f'(r) = 0$. But $f'(r) = 2 - \sin r > 0$ since $\sin r \leq 1$. This contradiction shows that the given equation can't have two distinct real roots, so it has exactly one root.

20. Let $f(x) = x^3 + e^x$. Then $f(-1) = -1 + 1/e < 0$ and $f(0) = 1 > 0$. Since f is the sum of a polynomial and the natural exponential function, f is continous and differentiable for all x. By the Intermediate Value Theorem, there is a number c in $(-1, 0)$ such that $f(c) = 0$. Thus, the given equation has at least one real root. If the equation has distinct real roots a and b

with $a < b$, then $f(a) = f(b) = 0$. Since f is continuous on $[a, b]$ and differentiable on (a, b), Rolle's Theorem implies that there is a number r in (a, b) such that $f'(r) = 0$. But $f'(r) = 3r^2 + e^r > 0$. This contradiction shows that the given equation can't have two distinct real roots, so it has exactly one root.

21. Let $f(x) = x^3 - 15x + c$ for x in $[-2, 2]$. If f has two real roots a and b in $[-2, 2]$, with $a < b$, then $f(a) = f(b) = 0$. Since the polynomial f is continuous on $[a, b]$ and differentiable on (a, b), Rolle's Theorem implies that there is a number r in (a, b) such that $f'(r) = 0$. Now $f'(r) = 3r^2 - 15$. Since r is in (a, b), which is contained in $[-2, 2]$, we have $|r| < 2$, so $r^2 < 4$. It follows that $3r^2 - 15 < 3 \cdot 4 - 15 = -3 < 0$. This contradicts $f'(r) = 0$, so the given equation can't have two real roots in $[-2, 2]$. Hence, it has at most one real root in $[-2, 2]$.

22. $f(x) = x^4 + 4x + c$. Suppose that $f(x) = 0$ has three distinct real roots a, b, d where $a < b < d$. Then $f(a) = f(b) = f(d) = 0$. By Rolle's Theorem there are numbers c_1 and c_2 with $a < c_1 < b$ and $b < c_2 < d$ and $0 = f'(c_1) = f'(c_2)$, so $f'(x) = 0$ must have at least two real solutions. However $0 = f'(x) = 4x^3 + 4 = 4(x^3 + 1) = 4(x + 1)(x^2 - x + 1)$ has as its only real solution $x = -1$. Thus, $f(x)$ can have at most two real roots.

23. (a) Suppose that a cubic polynomial $P(x)$ has roots $a_1 < a_2 < a_3 < a_4$, so $P(a_1) = P(a_2) = P(a_3) = P(a_4)$. By Rolle's Theorem there are numbers c_1, c_2, c_3 with $a_1 < c_1 < a_2$, $a_2 < c_2 < a_3$ and $a_3 < c_3 < a_4$ and $P'(c_1) = P'(c_2) = P'(c_3) = 0$. Thus, the second-degree polynomial $P'(x)$ has three distinct real roots, which is impossible.

(b) We prove by induction that a polynomial of degree n has at most n real roots. This is certainly true for $n = 1$. Suppose that the result is true for all polynomials of degree n and let $P(x)$ be a polynomial of degree $n + 1$. Suppose that $P(x)$ has more than $n + 1$ real roots, say $a_1 < a_2 < a_3 < \cdots < a_{n+1} < a_{n+2}$. Then $P(a_1) = P(a_2) = \cdots = P(a_{n+2}) = 0$. By Rolle's Theorem there are real numbers $c_1, \ldots, c_{n+1}$ with $a_1 < c_1 < a_2, \ldots, a_{n+1} < c_{n+1} < a_{n+2}$ and $P'(c_1) = \cdots = P'(c_{n+1}) = 0$. Thus, the nth degree polynomial $P'(x)$ has at least $n + 1$ roots. This contradiction shows that $P(x)$ has at most $n + 1$ real roots.

24. (a) Suppose that $f(a) = f(b) = 0$ where $a < b$. By Rolle's Theorem applied to f on $[a, b]$ there is a number c such that $a < c < b$ and $f'(c) = 0$.

(b) Suppose that $f(a) = f(b) = f(c) = 0$ where $a < b < c$. By Rolle's Theorem applied to $f(x)$ on $[a, b]$ and $[b, c]$ there are numbers $a < d < b$ and $b < e < c$ with $f'(d) = 0$ and $f'(e) = 0$. By Rolle's Theorem applied to $f'(x)$ on $[d, e]$ there is a number g with $d < g < e$ such that $f''(g) = 0$.

(c) Suppose that f is n times differentiable on $\mathbb{R}$ and has $n + 1$ distinct real roots. Then $f^{(n)}$ has at least one real root.

25. By the Mean Value Theorem, $f(4) - f(1) = f'(c)(4 - 1)$ for some $c \in (1, 4)$. But for every $c \in (1, 4)$ we have $f'(c) \geq 2$. Putting $f'(c) \geq 2$ into the above equation and substituting $f(1) = 10$, we get $f(4) = f(1) + f'(c)(4 - 1) = 10 + 3f'(c) \geq 10 + 3 \cdot 2 = 16$. So the smallest possible value of $f(4)$ is 16.

26. If $3 \leq f'(x) \leq 5$ for all x, then by the Mean Value Theorem, $f(8) - f(2) = f'(c) \cdot (8-2)$ for some c in $[2, 8]$. (f is differentiable for all x, so, in particular, f is differentiable on $(2, 8)$ and continuous on $[2, 8]$. Thus, the hypotheses of the Mean Value Theorem are satisfied.) Since $f(8) - f(2) = 6f'(c)$ and $3 \leq f'(c) \leq 5$, it follows that $6 \cdot 3 \leq 6f'(c) \leq 6 \cdot 5 \quad\Rightarrow\quad 18 \leq f(8) - f(2) \leq 30$.

27. Suppose that such a function f exists. By the Mean Value Theorem there is a number $0 < c < 2$ with $f'(c) = \dfrac{f(2) - f(0)}{2 - 0} = \dfrac{5}{2}$. But this is impossible since $f'(x) \leq 2 < \frac{5}{2}$ for all x, so no such function can exist.

28. Let $h = f - g$. Note that since $f(a) = g(a)$, $h(a) = f(a) - g(a) = 0$. Then since f and g are continuous on $[a, b]$ and differentiable on (a, b), so is h, and thus h satisfies the assumptions of the Mean Value Theorem. Therefore, there is a number c with $a < c < b$ such that $h(b) = h(b) - h(a) = h'(c)(b - a)$. Since $h'(c) < 0$, $h'(c)(b - a) < 0$, so $f(b) - g(b) = h(b) < 0$ and hence $f(b) < g(b)$.

29. Consider the function $f(x) = \sin x$, which is continuous and differentiable on $\mathbb{R}$. Let a be a number such that $0 < a < 2\pi$. Then f is continuous on $[0, a]$ and differentiable on $(0, a)$. By the Mean Value Theorem, there is a number c in $(0, a)$ such that $f(a) - f(0) = f'(c)(a - 0)$; that is, $\sin a - 0 = (\cos c)(a)$. Now $\cos c < 1$ for $0 < c < 2\pi$, so $\sin a < 1 \cdot a = a$. We took a to be an arbitrary number in $(0, 2\pi)$, so $\sin x < x$ for all x satisfying $0 < x < 2\pi$.

30. f satisfies the conditions for the Mean Value Theorem, so we use this theorem on the interval $[-b, b]$: $\dfrac{f(b) - f(-b)}{b - (-b)} = f'(c)$ for some $c \in (-b, b)$. But since f is odd, $f(-b) = -f(b)$. Substituting this into the above equation, we get $\dfrac{f(b) + f(b)}{2b} = f'(c) \quad\Rightarrow\quad \dfrac{f(b)}{b} = f'(c)$.

31. Let $f(x) = \sin x$ and let $b < a$. Then $f(x)$ is continuous on $[b, a]$ and differentiable on (b, a). By the Mean Value Theorem, there is a number $c \in (b, a)$ with $\sin a - \sin b = f(a) - f(b) = f'(c)(a - b) = (\cos c)(a - b)$. Thus, $|\sin a - \sin b| \leq |\cos c|\,|b - a| \leq |a - b|$. If $a < b$, then $|\sin a - \sin b| = |\sin b - \sin a| \leq |b - a| = |a - b|$. If $a = b$, both sides of the inequality are 0.

32. Suppose that $f'(x) = c$. Let $g(x) = cx$, so $g'(x) = c$. Then, by Corollary 7, $f(x) = g(x) + d$, where d is a constant, so $f(x) = cx + d$.

33. For $x > 0$, $f(x) = g(x)$, so $f'(x) = g'(x)$. For $x < 0$, $f'(x) = (1/x)' = -1/x^2$ and $g'(x) = (1 + 1/x)' = -1/x^2$, so again $f'(x) = g'(x)$. However, the domain of $g(x)$ is not an interval [it is $(-\infty, 0) \cup (0, \infty)$] so we cannot conclude that $f - g$ is constant (in fact it is not).

34. Let $f(x) = 2\sin^{-1} x - \cos^{-1}(1 - 2x^2)$. Then $f'(x) = \dfrac{2}{\sqrt{1 - x^2}} - \dfrac{4x}{\sqrt{1 - (1 - 2x^2)^2}} = \dfrac{2}{\sqrt{1 - x^2}} - \dfrac{4x}{2x\sqrt{1 - x^2}} = 0$ [since $x \geq 0$]. Thus, $f'(x) = 0$ for all $x \in (0, 1)$. Thus, $f(x) = C$ on $(0, 1)$. To find C, let $x = 0.5$. Thus, $2\sin^{-1}(0.5) - \cos^{-1}(0.5) = 2\left(\frac{\pi}{6}\right) - \frac{\pi}{3} = 0 = C$. We conclude that $f(x) = 0$ for x in $(0, 1)$. By continuity of f, $f(x) = 0$ on $[0, 1]$. Therefore, we see that $f(x) = 2\,\sin^{-1} x - \cos^{-1}(1 - 2x^2) = 0 \quad\Rightarrow\quad 2\,\sin^{-1} x = \cos^{-1}(1 - 2x^2)$.

35. Let $f(x) = \arcsin\left(\dfrac{x-1}{x+1}\right) - 2\arctan\sqrt{x} + \frac{\pi}{2}$. Note that the domain of f is $[0,\infty)$. Thus,

$$f'(x) = \frac{1}{\sqrt{1-\left(\dfrac{x-1}{x+1}\right)^2}}\,\frac{(x+1)-(x-1)}{(x+1)^2} - \frac{2}{1+x}\cdot\frac{1}{2\sqrt{x}} = \frac{1}{\sqrt{x}\,(x+1)} - \frac{1}{\sqrt{x}\,(x+1)} = 0.$$

Then $f(x) = C$ on $(0,\infty)$ by Theorem 5. By continuity of f, $f(x) = C$ on $[0,\infty)$. To find C, we let $x = 0 \;\Rightarrow$ $\arcsin(-1) - 2\arctan(0) + \frac{\pi}{2} = C \;\Rightarrow\; -\frac{\pi}{2} - 0 + \frac{\pi}{2} = 0 = C$. Thus, $f(x) = 0 \;\Rightarrow$

$$\arcsin\left(\frac{x-1}{x+1}\right) = 2\arctan\sqrt{x} - \frac{\pi}{2}.$$

36. Let $v(t)$ be the velocity of the car t hours after 2:00 PM. Then $\dfrac{v(1/6)-v(0)}{1/6-0} = \dfrac{50-30}{1/6} = 120$. By the Mean Value Theorem, there is a number c such that $0 < c < \frac{1}{6}$ with $v'(c) = 120$. Since $v'(t)$ is the acceleration at time t, the acceleration c hours after 2:00 PM is exactly 120 mi/h^2.

37. Let $g(t)$ and $h(t)$ be the position functions of the two runners and let $f(t) = g(t) - h(t)$. By hypothesis, $f(0) = g(0) - h(0) = 0$ and $f(b) = g(b) - h(b) = 0$, where b is the finishing time. Then by the Mean Value Theorem, there is a time c, with $0 < c < b$, such that $f'(c) = \dfrac{f(b)-f(0)}{b-0}$. But $f(b) = f(0) = 0$, so $f'(c) = 0$. Since $f'(c) = g'(c) - h'(c) = 0$, we have $g'(c) = h'(c)$. So at time c, both runners have the same speed $g'(c) = h'(c)$.

38. Assume that f is differentiable (and hence continuous) on $\mathbb{R}$ and that $f'(x) \neq 1$ for all x. Suppose f has more than one fixed point. Then there are numbers a and b such that $a < b$, $f(a) = a$, and $f(b) = b$. Applying the Mean Value Theorem to the function f on $[a,b]$, we find that there is a number c in (a,b) such that $f'(c) = \dfrac{f(b)-f(a)}{b-a}$. But then $f'(c) = \dfrac{b-a}{b-a} = 1$, contradicting our assumption that $f'(x) \neq 1$ for every real number x. This shows that our supposition was wrong, that is, that f cannot have more than one fixed point.

4.3 How Derivatives Affect the Shape of a Graph

1. (a) f is increasing on $(1,3)$ and $(4,6)$.

(b) f is decreasing on $(0,1)$ and $(3,4)$.

(c) f is concave upward on $(0,2)$.

(d) f is concave downward on $(2,4)$ and $(4,6)$.

(e) The point of inflection is $(2,3)$.

2. (a) f is increasing on $(0,1)$ and $(3,7)$.

(b) f is decreasing on $(1,3)$.

(c) f is concave upward on $(2,4)$ and $(5,7)$.

(d) f is concave downward on $(0,2)$ and $(4,5)$.

(e) The points of inflection are $(2,2)$, $(4,3)$, and $(5,4)$.

3. (a) Use the Increasing/Decreasing (I/D) Test.

(b) Use the Concavity Test.

(c) At any value of x where the concavity changes, we have an inflection point at $(x, f(x))$.

4. (a) See the First Derivative Test.

(b) See the Second Derivative Test and the note that precedes Example 7.

5. (a) Since $f'(x) > 0$ on $(1, 5)$, f is increasing on this interval. Since $f'(x) < 0$ on $(0, 1)$ and $(5, 6)$, f is decreasing on these intervals.

(b) Since $f'(x) = 0$ at $x = 1$ and f' changes from negative to positive there, f changes from decreasing to increasing and has a local minimum at $x = 1$. Since $f'(x) = 0$ at $x = 5$ and f' changes from positive to negative there, f changes from increasing to decreasing and has a local maximum at $x = 5$.

6. (a) $f'(x) > 0$ and f is increasing on $(0, 1)$ and $(5, 7)$. $f'(x) < 0$ and f is decreasing on $(1, 5)$ and $(7, 8)$.

(b) Since $f'(x) = 0$ at $x = 1$ and $x = 7$ and f' changes from positive to negative at both values, f changes from increasing to decreasing and has local maxima at $x = 1$ and $x = 7$. Since $f'(x) = 0$ at $x = 5$ and f' changes from negative to positive there, f changes from decreasing to increasing and has a local minimum at $x = 5$.

7. (a) There is an IP at $x = 3$ because the graph of f changes from CD to CU there. There is an IP at $x = 5$ because the graph of f changes from CU to CD there.

(b) There is an IP at $x = 2$ and at $x = 6$ because $f'(x)$ has a maximum value there, and so $f''(x)$ changes from positive to negative there. There is an IP at $x = 4$ because $f'(x)$ has a minimum value there and so $f''(x)$ changes from negative to positive there.

(c) There is an inflection point at $x = 1$ because $f''(x)$ changes from negative to positive there, and so the graph of f changes from concave downward to concave upward. There is an inflection point at $x = 7$ because $f''(x)$ changes from positive to negative there, and so the graph of f changes from concave upward to concave downward.

8. (a) f is increasing when f' is positive. This happens on the intervals $(0, 4)$ and $(6, 8)$.

(b) f has a local maximum where it changes from increasing to decreasing, that is, where f' changes from positive to negative (at $x = 4$ and $x = 8$). Similarly, f has a local minimum where f' changes from negative to positive (at $x = 6$).

(c) f is concave upward where f' is increasing (hence f'' is positive). This happens on $(0, 1)$, $(2, 3)$, and $(5, 7)$. Similarly, f is concave downward where f' is decreasing, that is, on $(1, 2)$, $(3, 5)$, and $(7, 9)$.

(d) f has an inflection point where the concavity changes. This happens at $x = 1$, 2, 3, 5, and 7.

9. (a) $f(x) = x^3 - 3x^2 - 9x + 4 \Rightarrow f'(x) = 3x^2 - 6x - 9 = 3(x^2 - 2x - 3) = 3(x + 1)(x - 3)$.

Interval	$x + 1$	$x - 3$	$f'(x)$	f
$x < -1$	$-$	$-$	$+$	increasing on $(-\infty, -1)$
$-1 < x < 3$	$+$	$-$	$-$	decreasing on $(-1, 3)$
$x > 3$	$+$	$+$	$+$	increasing on $(3, \infty)$

(b) f changes from increasing to decreasing at $x = -1$ and from decreasing to increasing at $x = 3$. Thus, $f(-1) = 9$ is a local maximum value and $f(3) = -23$ is a local minimum value.

(c) $f''(x) = 6x - 6 = 6(x - 1)$. $f''(x) > 0 \Leftrightarrow x > 1$ and $f''(x) < 0 \Leftrightarrow x < 1$. Thus, f is concave upward on $(1, \infty)$ and concave downward on $(-\infty, 1)$. There is an inflection point at $(1, -7)$.

10. (a) $f(x) = 2x^3 - 9x^2 + 12x - 3 \Rightarrow f'(x) = 6x^2 - 18x + 12 = 6(x^2 - 3x + 2) = 6(x-1)(x-2)$.

Interval	$x-1$	$x-2$	$f'(x)$	f
$x < 1$	$-$	$-$	$+$	increasing on $(-\infty, 1)$
$1 < x < 2$	$+$	$-$	$-$	decreasing on $(1, 2)$
$x > 2$	$+$	$+$	$+$	increasing on $(2, \infty)$

(b) f changes from increasing to decreasing at $x = 1$ and from decreasing to increasing at $x = 2$. Thus, $f(1) = 2$ is a local maximum value and $f(2) = 1$ is a local minimum value.

(c) $f''(x) = 12x - 18 = 12\left(x - \frac{3}{2}\right)$. $f''(x) > 0 \Leftrightarrow x > \frac{3}{2}$ and $f''(x) < 0 \Leftrightarrow x < \frac{3}{2}$. Thus, f is concave upward on $\left(\frac{3}{2}, \infty\right)$ and concave downward on $\left(-\infty, \frac{3}{2}\right)$. There is an inflection point at $\left(\frac{3}{2}, \frac{3}{2}\right)$.

11. (a) $f(x) = x^4 - 2x^2 + 3 \Rightarrow f'(x) = 4x^3 - 4x = 4x(x^2 - 1) = 4x(x+1)(x-1)$.

Interval	$x+1$	x	$x-1$	$f'(x)$	f
$x < -1$	$-$	$-$	$-$	$-$	decreasing on $(-\infty, -1)$
$-1 < x < 0$	$+$	$-$	$-$	$+$	increasing on $(-1, 0)$
$0 < x < 1$	$+$	$+$	$-$	$-$	decreasing on $(0, 1)$
$x > 1$	$+$	$+$	$+$	$+$	increasing on $(1, \infty)$

(b) f changes from increasing to decreasing at $x = 0$ and from decreasing to increasing at $x = -1$ and $x = 1$. Thus, $f(0) = 3$ is a local maximum value and $f(\pm 1) = 2$ are local minimum values.

(c) $f''(x) = 12x^2 - 4 = 12\left(x^2 - \frac{1}{3}\right) = 12\left(x + 1/\sqrt{3}\right)\left(x - 1/\sqrt{3}\right)$. $f''(x) > 0 \Leftrightarrow x < -1/\sqrt{3}$ or $x > 1/\sqrt{3}$ and $f''(x) < 0 \Leftrightarrow -1/\sqrt{3} < x < 1/\sqrt{3}$. Thus, f is concave upward on $\left(-\infty, -\sqrt{3}/3\right)$ and $\left(\sqrt{3}/3, \infty\right)$ and concave downward on $\left(-\sqrt{3}/3, \sqrt{3}/3\right)$. There are inflection points at $\left(\pm\sqrt{3}/3, \frac{22}{9}\right)$.

12. (a) $f(x) = \dfrac{x}{x^2+1} \Rightarrow f'(x) = \dfrac{(x^2+1)(1) - x(2x)}{(x^2+1)^2} = \dfrac{1-x^2}{(x^2+1)^2} = -\dfrac{(x+1)(x-1)}{(x^2+1)^2}$. Thus, $f'(x) > 0$ if $(x+1)(x-1) < 0 \Leftrightarrow -1 < x < 1$, and $f'(x) < 0$ if $x < -1$ or $x > 1$. So f is increasing on $(-1, 1)$ and f is decreasing on $(-\infty, -1)$ and $(1, \infty)$.

(b) f changes from decreasing to increasing at $x = -1$ and from increasing to decreasing at $x = 1$. Thus, $f(-1) = -\frac{1}{2}$ is a local minimum value and $f(1) = \frac{1}{2}$ is a local maximum value.

(c) $f''(x) = \dfrac{(x^2+1)^2(-2x) - (1-x^2)[2(x^2+1)(2x)]}{[(x^2+1)^2]^2} = \dfrac{(x^2+1)(-2x)[(x^2+1) + 2(1-x^2)]}{(x^2+1)^4} = \dfrac{2x(x^2-3)}{(x^2+1)^3}$.

$f''(x) > 0 \Leftrightarrow -\sqrt{3} < x < 0$ or $x > \sqrt{3}$, and $f''(x) < 0 \Leftrightarrow x < -\sqrt{3}$ or $0 < x < \sqrt{3}$. Thus, f is concave upward on $\left(-\sqrt{3}, 0\right)$ and $\left(\sqrt{3}, \infty\right)$ and concave downward on $\left(-\infty, -\sqrt{3}\right)$ and $\left(0, \sqrt{3}\right)$. There are inflection points at $\left(-\sqrt{3}, -\sqrt{3}/4\right)$, $(0, 0)$, and $\left(\sqrt{3}, \sqrt{3}/4\right)$.

13. (a) $f(x) = \sin x + \cos x,\ 0 \le x \le 2\pi$. $f'(x) = \cos x - \sin x = 0 \Rightarrow \cos x = \sin x \Rightarrow 1 = \dfrac{\sin x}{\cos x} \Rightarrow$ $\tan x = 1 \Rightarrow x = \frac{\pi}{4}$ or $\frac{5\pi}{4}$. Thus, $f'(x) > 0 \Leftrightarrow \cos x - \sin x > 0 \Leftrightarrow \cos x > \sin x \Leftrightarrow 0 < x < \frac{\pi}{4}$ or $\frac{5\pi}{4} < x < 2\pi$ and $f'(x) < 0 \Leftrightarrow \cos x < \sin x \Leftrightarrow \frac{\pi}{4} < x < \frac{5\pi}{4}$. So f is increasing on $\left(0, \frac{\pi}{4}\right)$ and $\left(\frac{5\pi}{4}, 2\pi\right)$ and f is decreasing on $\left(\frac{\pi}{4}, \frac{5\pi}{4}\right)$.

(b) f changes from increasing to decreasing at $x = \frac{\pi}{4}$ and from decreasing to increasing at $x = \frac{5\pi}{4}$. Thus, $f\left(\frac{\pi}{4}\right) = \sqrt{2}$ is a local maximum value and $f\left(\frac{5\pi}{4}\right) = -\sqrt{2}$ is a local minimum value.

(c) $f''(x) = -\sin x - \cos x = 0 \Rightarrow -\sin x = \cos x \Rightarrow \tan x = -1 \Rightarrow x = \frac{3\pi}{4}$ or $\frac{7\pi}{4}$. Divide the interval $(0, 2\pi)$ into subintervals with these numbers as endpoints and complete a second derivative chart.

Interval	$f''(x) = -\sin x - \cos x$	Concavity
$\left(0, \frac{3\pi}{4}\right)$	$f''\left(\frac{\pi}{2}\right) = -1 < 0$	downward
$\left(\frac{3\pi}{4}, \frac{7\pi}{4}\right)$	$f''(\pi) = 1 > 0$	upward
$\left(\frac{7\pi}{4}, 2\pi\right)$	$f''\left(\frac{11\pi}{6}\right) = \frac{1}{2} - \frac{1}{2}\sqrt{3} < 0$	downward

There are inflection points at $\left(\frac{3\pi}{4}, 0\right)$ and $\left(\frac{7\pi}{4}, 0\right)$.

14. (a) $f(x) = \cos^2 x - 2\sin x,\ 0 \le x \le 2\pi$. $f'(x) = -2\cos x\,\sin x - 2\cos x = -2\cos x\,(1 + \sin x)$. Note that $1 + \sin x \ge 0$ [since $\sin x \ge -1$], with equality $\Leftrightarrow \sin x = -1 \Leftrightarrow x = \frac{3\pi}{2}$ [since $0 \le x \le 2\pi$] $\Rightarrow$ $\cos x = 0$. Thus, $f'(x) > 0 \Leftrightarrow \cos x < 0 \Leftrightarrow \frac{\pi}{2} < x < \frac{3\pi}{2}$ and $f'(x) < 0 \Leftrightarrow \cos x > 0 \Leftrightarrow 0 < x < \frac{\pi}{2}$ or $\frac{3\pi}{2} < x < 2\pi$. Thus, f is increasing on $\left(\frac{\pi}{2}, \frac{3\pi}{2}\right)$ and f is decreasing on $\left(0, \frac{\pi}{2}\right)$ and $\left(\frac{3\pi}{2}, 2\pi\right)$.

(b) f changes from decreasing to increasing at $x = \frac{\pi}{2}$ and from increasing to decreasing at $x = \frac{3\pi}{2}$. Thus, $f\left(\frac{\pi}{2}\right) = -2$ is a local minimum value and $f\left(\frac{3\pi}{2}\right) = 2$ is a local maximum value.

(c)
$$\begin{aligned} f''(x) &= 2\sin x\,(1 + \sin x) - 2\cos^2 x = 2\sin x + 2\sin^2 x - 2(1 - \sin^2 x) \\ &= 4\sin^2 x + 2\sin x - 2 = 2(2\sin x - 1)(\sin x + 1) \end{aligned}$$

so $f''(x) > 0 \Leftrightarrow \sin x > \frac{1}{2} \Leftrightarrow \frac{\pi}{6} < x < \frac{5\pi}{6}$, and $f''(x) < 0 \Leftrightarrow \sin x < \frac{1}{2}$ and $\sin x \ne -1 \Leftrightarrow$ $0 < x < \frac{\pi}{6}$ or $\frac{5\pi}{6} < x < \frac{3\pi}{2}$ or $\frac{3\pi}{2} < x < 2\pi$. Thus, f is concave upward on $\left(\frac{\pi}{6}, \frac{5\pi}{6}\right)$ and concave downward on $\left(0, \frac{\pi}{6}\right)$, $\left(\frac{5\pi}{6}, \frac{3\pi}{2}\right)$, and $\left(\frac{3\pi}{2}, 2\pi\right)$. There are inflection points at $\left(\frac{\pi}{6}, -\frac{1}{4}\right)$ and $\left(\frac{5\pi}{6}, -\frac{1}{4}\right)$.

15. (a) $f(x) = e^{2x} + e^{-x} \Rightarrow f'(x) = 2e^{2x} - e^{-x}$. $f'(x) > 0 \Leftrightarrow 2e^{2x} > e^{-x} \Leftrightarrow e^{3x} > \frac{1}{2} \Leftrightarrow 3x > \ln\frac{1}{2} \Leftrightarrow$ $x > \frac{1}{3}(\ln 1 - \ln 2) \Leftrightarrow x > -\frac{1}{3}\ln 2$ [≈ -0.23] and $f'(x) < 0$ if $x < -\frac{1}{3}\ln 2$. So f is increasing on $\left(-\frac{1}{3}\ln 2, \infty\right)$ and f is decreasing on $\left(-\infty, -\frac{1}{3}\ln 2\right)$.

(b) f changes from decreasing to increasing at $x = -\frac{1}{3}\ln 2$. Thus,

$$f\left(-\tfrac{1}{3}\ln 2\right) = f\left(\ln\sqrt[3]{1/2}\right) = e^{2\ln\sqrt[3]{1/2}} + e^{-\ln\sqrt[3]{1/2}} = e^{\ln\sqrt[3]{1/4}} + e^{\ln\sqrt[3]{2}} = \sqrt[3]{1/4} + \sqrt[3]{2} = 2^{-2/3} + 2^{1/3}\ [\approx 1.89]$$

is a local minimum value.

(c) $f''(x) = 4e^{2x} + e^{-x} > 0$ [the sum of two positive terms]. Thus, f is concave upward on $(-\infty, \infty)$ and there is no point of inflection.

16. (a) $f(x) = x^2 \ln x \Rightarrow f'(x) = x^2(1/x) + (\ln x)(2x) = x + 2x \ln x = x(1 + 2\ln x)$. The domain of f is $(0, \infty)$, so the sign of f' is determined solely by the factor $1 + 2\ln x$. $f'(x) > 0 \Leftrightarrow \ln x > -\frac{1}{2} \Leftrightarrow x > e^{-1/2}$ $[\approx 0.61]$ and $f'(x) < 0 \Leftrightarrow 0 < x < e^{-1/2}$. So f is increasing on $(e^{-1/2}, \infty)$ and f is decreasing on $(0, e^{-1/2})$.

(b) f changes from decreasing to increasing at $x = e^{-1/2}$. Thus, $f(e^{-1/2}) = (e^{-1/2})^2 \ln(e^{-1/2}) = e^{-1}(-1/2) = -1/(2e)$ $[\approx -0.18]$ is a local minimum value.

(c) $f'(x) = x(1 + 2\ln x) \Rightarrow f''(x) = x(2/x) + (1 + 2\ln x) \cdot 1 = 2 + 1 + 2\ln x = 3 + 2\ln x$. $f''(x) > 0 \Leftrightarrow 3 + 2\ln x > 0 \Leftrightarrow \ln x > -3/2 \Leftrightarrow x > e^{-3/2}$ $[\approx 0.22]$. Thus, f is concave upward on $(e^{-3/2}, \infty)$ and f is concave downward on $(0, e^{-3/2})$. $f(e^{-3/2}) = (e^{-3/2})^2 \ln e^{-3/2} = e^{-3}(-3/2) = -3/(2e^3)$ $[\approx -0.07]$. There is a point of inflection at $\left(e^{-3/2}, f(e^{-3/2})\right) = \left(e^{-3/2}, -3/(2e^3)\right)$.

17. (a) $f(x) = x^2 - x - \ln x \Rightarrow f'(x) = 2x - 1 - \dfrac{1}{x} = \dfrac{2x^2 - x - 1}{x} = \dfrac{(2x+1)(x-1)}{x}$. Thus, $f'(x) > 0$ if $x > 1$ [note that $x > 0$] and $f'(x) < 0$ if $0 < x < 1$. So f is increasing on $(1, \infty)$ and f is decreasing on $(0, 1)$.

(b) f changes from decreasing to increasing at $x = 1$. Thus, $f(1) = 0$ is a local minimum value.

(c) $f''(x) = 2 + 1/x^2 > 0$ for all x, so f is concave upward on $(0, \infty)$. There is no inflection point.

18. (a) $f(x) = x^4 e^{-x} \Rightarrow f'(x) = x^4(-e^{-x}) + e^{-x}(4x^3) = x^3 e^{-x}(-x + 4)$. Thus, $f'(x) > 0$ if $0 < x < 4$ and $f'(x) < 0$ if $x < 0$ or $x > 4$. So f is increasing on $(0, 4)$ and decreasing on $(-\infty, 0)$ and $(4, \infty)$.

(b) f changes from decreasing to increasing at $x = 0$ and from increasing to decreasing at $x = 4$. Thus, $f(0) = 0$ is a local minimum value and $f(4) = 256/e^4$ is a local maximum value.

(c) $f'(x) = e^{-x}(-x^4 + 4x^3) \Rightarrow$

$$\begin{aligned} f''(x) &= e^{-x}(-4x^3 + 12x^2) + (-x^4 + 4x^3)(-e^{-x}) = e^{-x}[(-4x^3 + 12x^2) - (-x^4 + 4x^3)] \\ &= e^{-x}(x^4 - 8x^3 + 12x^2) = x^2 e^{-x}(x^2 - 8x + 12) = x^2 e^{-x}(x-2)(x-6) \end{aligned}$$

$f''(x) > 0 \Leftrightarrow x < 2$ [excluding 0] or $x > 6$ and $f''(x) < 0 \Leftrightarrow 2 < x < 6$. Thus, f is concave upward on $(-\infty, 2)$ and $(6, \infty)$ and f is concave downward on $(2, 6)$. There are inflection points at $(2, 16e^{-2})$ and $(6, 1296e^{-6})$.

19. $f(x) = 1 + 3x^2 - 2x^3 \Rightarrow f'(x) = 6x - 6x^2 = 6x(1 - x)$.

First Derivative Test: $f'(x) > 0 \Rightarrow 0 < x < 1$ and $f'(x) < 0 \Rightarrow x < 0$ or $x > 1$. Since f' changes from negative to positive at $x = 0$, $f(0) = 1$ is a local minimum value; and since f' changes from positive to negative at $x = 1$, $f(1) = 2$ is a local maximum value.

Second Derivative Test: $f''(x) = 6 - 12x$. $f'(x) = 0 \Leftrightarrow x = 0, 1$. $f''(0) = 6 > 0 \Rightarrow f(0) = 1$ is a local minimum value. $f''(1) = -6 < 0 \Rightarrow f(1) = 2$ is a local maximum value.

Preference: For this function, the two tests are equally easy.

20. $f(x) = \dfrac{x^2}{x-1} \quad\Rightarrow\quad f'(x) = \dfrac{(x-1)(2x) - x^2(1)}{(x-1)^2} = \dfrac{x^2 - 2x}{(x-1)^2} = \dfrac{x(x-2)}{(x-1)^2}$.

First Derivative Test: $f'(x) > 0 \;\Rightarrow\; x < 0$ or $x > 2$ and $f'(x) < 0 \;\Rightarrow\; 0 < x < 1$ or $1 < x < 2$. Since f' changes from positive to negative at $x = 0$, $f(0) = 0$ is a local maximum value; and since f' changes from negative to positive at $x = 2$, $f(2) = 4$ is a local minimum value.

Second Derivative Test:

$$f''(x) = \frac{(x-1)^2(2x-2) - (x^2-2x)2(x-1)}{[(x-1)^2]^2} = \frac{2(x-1)[(x-1)^2 - (x^2-2x)]}{(x-1)^4} = \frac{2}{(x-1)^3}.$$

$f'(x) = 0 \;\Leftrightarrow\; x = 0, 2$. $f''(0) = -2 < 0 \;\Rightarrow\; f(0) = 0$ is a local maximum value. $f''(2) = 2 > 0 \;\Rightarrow\; f(2) = 4$ is a local minimum value.

Preference: Since calculating the second derivative is fairly difficult, the First Derivative Test is easier to use for this function.

21. $f(x) = \sqrt{x} - \sqrt[4]{x} \;\Rightarrow\; f'(x) = \frac{1}{2}x^{-1/2} - \frac{1}{4}x^{-3/4} = \frac{1}{4}x^{-3/4}(2x^{1/4} - 1) = \dfrac{2\sqrt[4]{x} - 1}{4\sqrt[4]{x^3}}$

First Derivative Test: $2\sqrt[4]{x} - 1 > 0 \;\Rightarrow\; x > \frac{1}{16}$, so $f'(x) > 0 \;\Rightarrow\; x > \frac{1}{16}$ and $f'(x) < 0 \;\Rightarrow\; 0 < x < \frac{1}{16}$. Since f' changes from negative to positive at $x = \frac{1}{16}$, $f(\frac{1}{16}) = \frac{1}{4} - \frac{1}{2} = -\frac{1}{4}$ is a local minimum value.

Second Derivative Test: $f''(x) = -\frac{1}{4}x^{-3/2} + \frac{3}{16}x^{-7/4} = -\dfrac{1}{4\sqrt{x^3}} + \dfrac{3}{16\sqrt[4]{x^7}}$.

$f'(x) = 0 \;\Leftrightarrow\; x = \frac{1}{16}$. $f''(\frac{1}{16}) = -16 + 24 = 8 > 0 \;\Rightarrow\; f(\frac{1}{16}) = -\frac{1}{4}$ is a local minimum value.

Preference: The First Derivative Test may be slightly easier to apply in this case.

22. (a) $f(x) = x^4(x-1)^3 \;\Rightarrow\; f'(x) = x^4 \cdot 3(x-1)^2 + (x-1)^3 \cdot 4x^3 = x^3(x-1)^2\,[3x + 4(x-1)] = x^3(x-1)^2(7x-4)$

The critical numbers are 0, 1, and $\frac{4}{7}$.

(b)
$$\begin{aligned} f''(x) &= 3x^2(x-1)^2(7x-4) + x^3 \cdot 2(x-1)(7x-4) + x^3(x-1)^2 \cdot 7 \\ &= x^2(x-1)\,[3(x-1)(7x-4) + 2x(7x-4) + 7x(x-1)] \end{aligned}$$

Now $f''(0) = f''(1) = 0$, so the Second Derivative Test gives no information for $x = 0$ or $x = 1$.

$f''(\frac{4}{7}) = (\frac{4}{7})^2(\frac{4}{7} - 1)\,[0 + 0 + 7(\frac{4}{7})(\frac{4}{7} - 1)] = (\frac{4}{7})^2(-\frac{3}{7})(4)(-\frac{3}{7}) > 0$, so there is a local minimum at $x = \frac{4}{7}$.

(c) f' is positive on $(-\infty, 0)$, negative on $(0, \frac{4}{7})$, positive on $(\frac{4}{7}, 1)$, and positive on $(1, \infty)$. So f has a local maximum at $x = 0$, a local minimum at $x = \frac{4}{7}$, and no local maximum or minimum at $x = 1$.

23. (a) By the Second Derivative Test, if $f'(2) = 0$ and $f''(2) = -5 < 0$, f has a local maximum at $x = 2$.

(b) If $f'(6) = 0$, we know that f has a horizontal tangent at $x = 6$. Knowing that $f''(6) = 0$ does not provide any additional information since the Second Derivative Test fails. For example, the first and second derivatives of $y = (x-6)^4$, $y = -(x-6)^4$, and $y = (x-6)^3$ all equal zero for $x = 6$, but the first has a local minimum at $x = 6$, the second has a local maximum at $x = 6$, and the third has an inflection point at $x = 6$.

24. (a) $f'(x) < 0$ and $f''(x) < 0$ for all x

The function must be always decreasing (since the first derivative is always negative) and concave downward (since the second derivative is always negative).

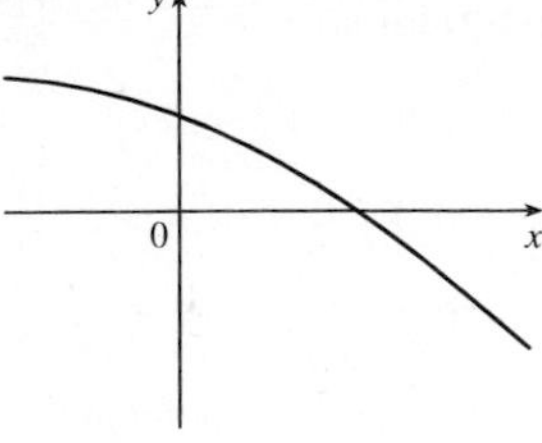

(b) $f'(x) > 0$ and $f''(x) > 0$ for all x

The function must be always increasing (since the first derivative is always positive) and concave upward (since the second derivative is always positive).

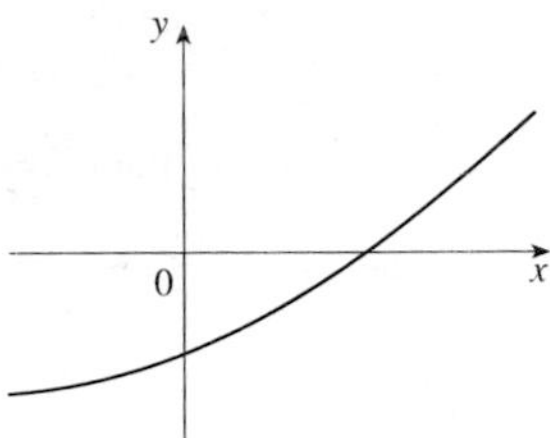

25. (a) $f'(x) > 0$ and $f''(x) < 0$ for all x

The function must be always increasing (since the first derivative is always positive) and concave downward (since the second derivative is always negative).

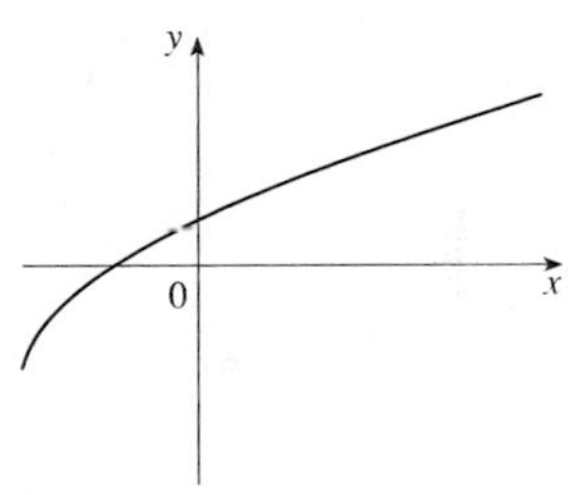

(b) $f'(x) < 0$ and $f''(x) > 0$ for all x

The function must be always decreasing (since the first derivative is always negative) and concave upward (since the second derivative is always positive).

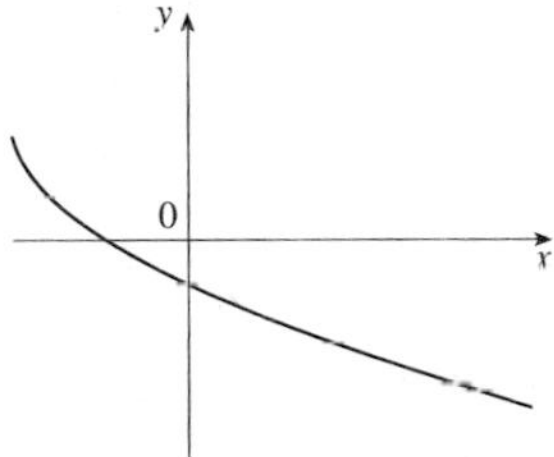

26. Vertical asymptote $x = 0$

$f'(x) > 0$ if $x < -2$ $\Rightarrow$ f is increasing on $(-\infty, -2)$.

$f'(x) < 0$ if $x > -2$ $(x \neq 0)$ $\Rightarrow$ f is decreasing on $(-2, 0)$ and $(0, \infty)$.

$f''(x) < 0$ if $x < 0$ $\Rightarrow$ f is concave downward on $(-\infty, 0)$.

$f''(x) > 0$ if $x > 0$ $\Rightarrow$ f is concave upward on $(0, \infty)$.

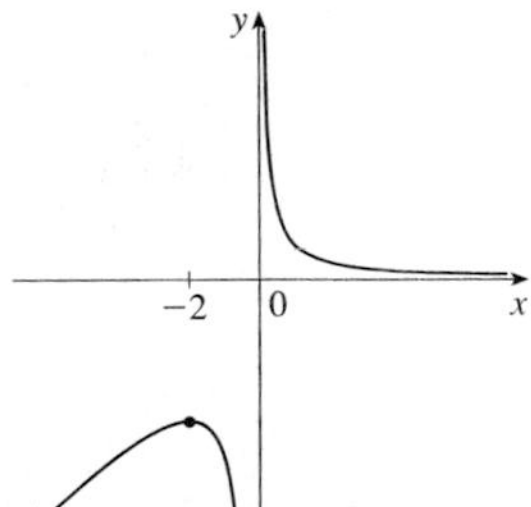

27. $f'(0) = f'(2) = f'(4) = 0$ $\Rightarrow$ horizontal tangents at $x = 0, 2, 4$.

$f'(x) > 0$ if $x < 0$ or $2 < x < 4$ $\Rightarrow$ f is increasing on $(-\infty, 0)$ and $(2, 4)$.

$f'(x) < 0$ if $0 < x < 2$ or $x > 4$ $\Rightarrow$ f is decreasing on $(0, 2)$ and $(4, \infty)$.

$f''(x) > 0$ if $1 < x < 3$ $\Rightarrow$ f is concave upward on $(1, 3)$.

$f''(x) < 0$ if $x < 1$ or $x > 3$ $\Rightarrow$ f is concave downward on $(-\infty, 1)$ and $(3, \infty)$. There are inflection points when $x = 1$ and 3.

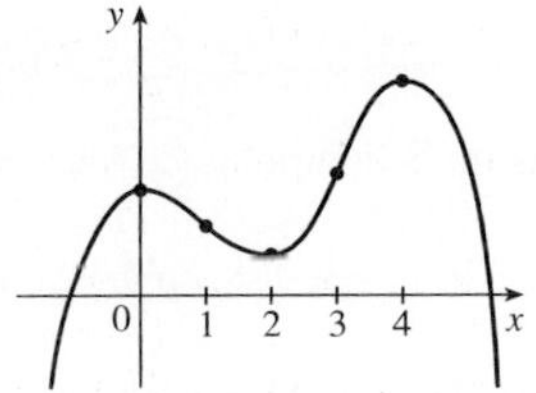

28. $f'(x) > 0$ for all $x \neq 1 \quad \Rightarrow \quad f$ is increasing on $(-\infty, 1)$ and $(1, \infty)$.

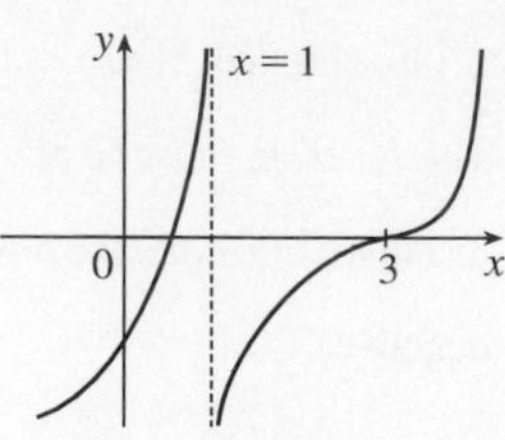

Vertical asymptote $x = 1$

$f''(x) > 0$ if $x < 1$ or $x > 3 \quad \Rightarrow \quad f$ is concave upward on $(-\infty, 1)$ and $(3, \infty)$.

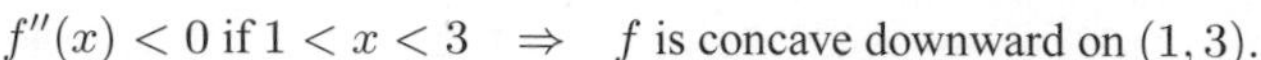

$f''(x) < 0$ if $1 < x < 3 \quad \Rightarrow \quad f$ is concave downward on $(1, 3)$.

There is an inflection point at $x = 3$.

29. $f'(5) = 0 \quad \Rightarrow \quad$ horizontal tangent at $x = 5$.

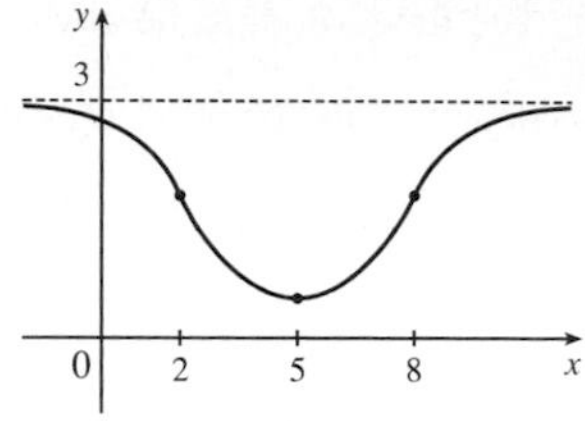

$f'(x) < 0$ when $x < 5 \quad \Rightarrow \quad f$ is decreasing on $(-\infty, 5)$.

$f'(x) > 0$ when $x > 5 \quad \Rightarrow \quad f$ is increasing on $(5, \infty)$.

$f''(2) = 0, \quad f''(8) = 0, \quad f''(x) < 0$ when $x < 2$ or $x > 8$,

$f''(x) > 0$ for $2 < x < 8 \quad \Rightarrow \quad f$ is concave upward on $(2, 8)$ and concave downward on $(-\infty, 2)$ and $(8, \infty)$.

There are inflection points at $x = 2$ and $x = 8$.

$\lim\limits_{x \to \infty} f(x) = 3, \quad \lim\limits_{x \to -\infty} f(x) = 3 \quad \Rightarrow \quad y = 3$ is a horizontal asymptote.

30. $f'(0) = f'(4) = 0 \quad \Rightarrow \quad$ horizontal tangents at $x = 0$ and 4.

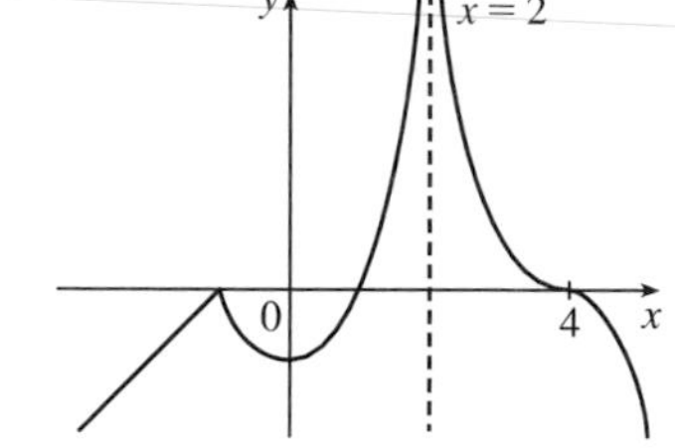

$f'(x) = 1$ if $x < -1 \quad \Rightarrow \quad f$ is a line with slope 1 on $(-\infty, -1)$.

$f'(x) > 0$ if $0 < x < 2 \quad \Rightarrow \quad f$ is increasing on $(0, 2)$.

$f'(x) < 0$ if $-1 < x < 0$ or $2 < x < 4$ or $x > 4 \quad \Rightarrow \quad f$ is decreasing on $(-1, 0)$, $(2, 4)$, and $(4, \infty)$.

$\lim\limits_{x \to 2^-} f'(x) = \infty \quad \Rightarrow \quad f'$ increases without bound as $x \to 2^-$.

$\lim\limits_{x \to 2^+} f'(x) = -\infty \quad \Rightarrow \quad f'$ decreases without bound as $x \to 2^+$.

$f''(x) > 0$ if $-1 < x < 2$ or $2 < x < 4 \quad \Rightarrow \quad f$ is concave upward on $(-1, 2)$ and $(2, 4)$.

$f''(x) < 0$ if $x > 4 \quad \Rightarrow \quad f$ is concave downward on $(4, \infty)$.

31. $f'(x) > 0$ if $x \neq 2 \quad \Rightarrow \quad f$ is increasing on $(-\infty, 2)$ and $(2, \infty)$.

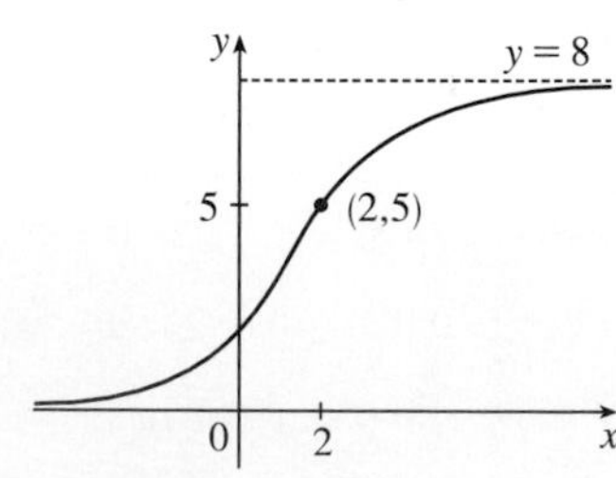

$f''(x) > 0$ if $x < 2 \quad \Rightarrow \quad f$ is concave upward on $(-\infty, 2)$.

$f''(x) < 0$ if $x > 2 \quad \Rightarrow \quad f$ is concave downward on $(2, \infty)$.

f has inflection point $(2, 5) \quad \Rightarrow \quad f$ changes concavity at the point $(2, 5)$.

$\lim\limits_{x \to \infty} f(x) = 8 \quad \Rightarrow \quad f$ has a horizontal asymptote of $y = 8$ as $x \to \infty$.

$\lim\limits_{x \to -\infty} f(x) = 0 \quad \Rightarrow \quad f$ has a horizontal asymptote of $y = 0$ as $x \to -\infty$.

32. (a) $f(3) = 2 \Rightarrow$ the point $(3, 2)$ is on the graph of f. $f'(3) = \frac{1}{2} \Rightarrow$ the slope of the tangent line at $(3, 2)$ is $\frac{1}{2}$. $f'(x) > 0$ for all $x \Rightarrow f$ is increasing on $\mathbb{R}$. $f''(x) < 0$ for all $x \Rightarrow f$ is concave downward on $\mathbb{R}$. A possible graph for f is shown.

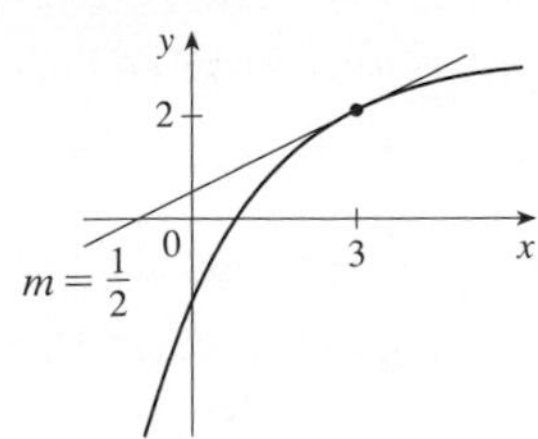

(b) The tangent line at $(3, 2)$ has equation $y - 2 = \frac{1}{2}(x - 3)$, or $y = \frac{1}{2}x + \frac{1}{2}$, and x-intercept -1. Since f is concave downward on $\mathbb{R}$, f is below the x-axis at $x = -1$, and hence changes sign at least once. Since f is increasing on $\mathbb{R}$, it changes sign at most once. Thus, it changes sign exactly once and there is one solution of the equation $f(x) = 0$.

(c) $f'' < 0 \Rightarrow f'$ is decreasing. Since $f'(3) = \frac{1}{2}$, $f'(2)$ must be greater than $\frac{1}{2}$, so no, it is not possible that $f'(2) = \frac{1}{3}$.

33. (a) Intuitively, since f is continuous, increasing, and concave upward for $x > 2$, it cannot have an absolute maximum. For a proof, we appeal to the MVT. Let $x = d > 2$. Then by the MVT, $f(d) - f(2) = f'(c)(d - 2)$ for some c such that $2 < c < d$. So $f(d) = f(2) + f'(c)(d - 2)$ where $f(2)$ is positive since $f(x) > 0$ for all x and $f'(c)$ is positive since $f'(x) > 0$ for $x > 2$. Thus, as $d \to \infty$, $f(d) \to \infty$, and no absolute maximum exists.

(b) Yes, the local minimum at $x = 2$ can be an absolute minimum.

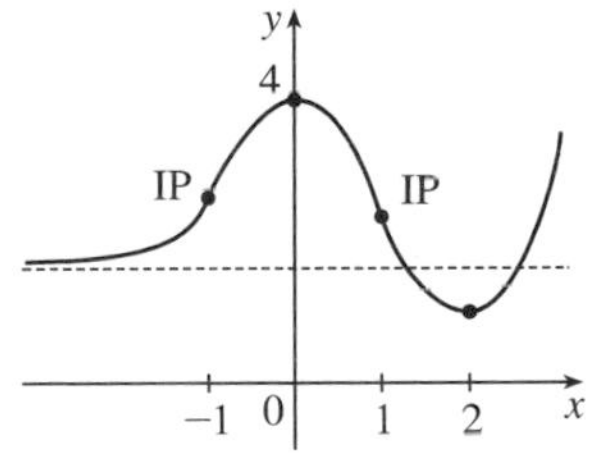

(c) Here $f(x) \to 0$ as $x \to -\infty$, but f does not achieve an absolute minimum.

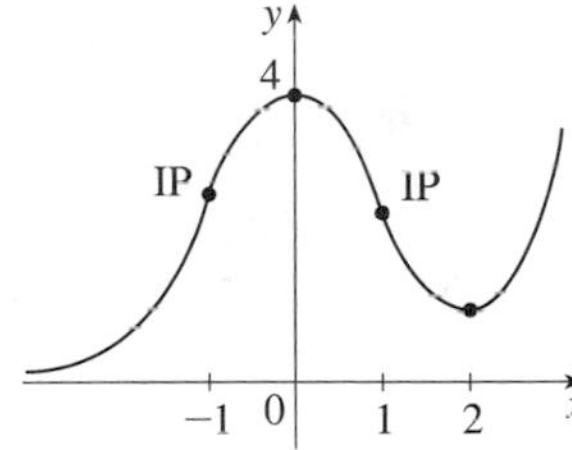

34. (a) $\dfrac{dy}{dx} > 0$ (f is increasing) and $\dfrac{d^2y}{dx^2} > 0$ (f is concave upward) at point B.

(b) $\dfrac{dy}{dx} < 0$ (f is decreasing) and $\dfrac{d^2y}{dx^2} < 0$ (f is concave downward) at point E.

(c) $\dfrac{dy}{dx} < 0$ (f is decreasing) and $\dfrac{d^2y}{dx^2} > 0$ (f is concave upward) at point A.

Note: At C, $\dfrac{dy}{dx} > 0$ and $\dfrac{d^2y}{dx^2} < 0$. At D, $\dfrac{dy}{dx} = 0$ and $\dfrac{d^2y}{dx^2} \le 0$.

35. (a) f is increasing where f' is positive, that is, on $(0, 2)$, $(4, 6)$, and $(8, \infty)$; and decreasing where f' is negative, that is, on $(2, 4)$ and $(6, 8)$.

(b) f has local maxima where f' changes from positive to negative, at $x = 2$ and at $x = 6$, and local minima where f' changes from negative to positive, at $x = 4$ and at $x = 8$.

(c) f is concave upward (CU) where f' is increasing, that is, on $(3, 6)$ and $(6, \infty)$, and concave downward (CD) where f' is decreasing, that is, on $(0, 3)$.

(d) There is a point of inflection where f changes from being CD to being CU, that is, at $x = 3$.

(e) 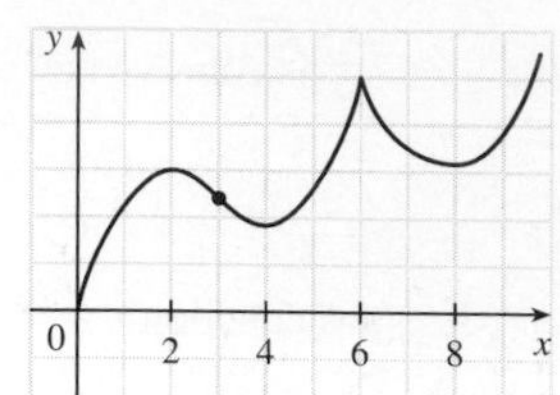

36. (a) f is increasing where f' is positive, on $(1, 6)$ and $(8, \infty)$, and decreasing where f' is negative, on $(0, 1)$ and $(6, 8)$.

(b) f has a local maximum where f' changes from positive to negative, at $x = 6$, and local minima where f' changes from negative to positive, at $x = 1$ and at $x = 8$.

(c) f is concave upward where f' is increasing, that is, on $(0, 2)$, $(3, 5)$, and $(7, \infty)$, and concave downward where f' is decreasing, that is, on $(2, 3)$ and $(5, 7)$.

(d) There are points of inflection where f changes its direction of concavity, at $x = 2$, $x = 3$, $x = 5$ and $x = 7$.

(e) 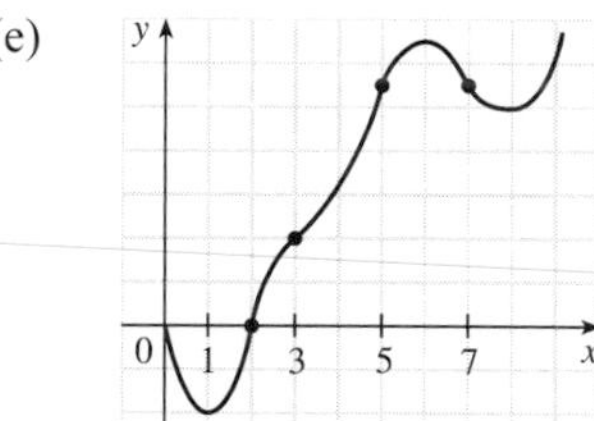

37. (a) $f(x) = x^3 - 12x + 2 \quad \Rightarrow \quad f'(x) = 3x^2 - 12 = 3(x^2 - 4) = 3(x + 2)(x - 2)$. $f'(x) > 0 \quad \Leftrightarrow \quad x < -2$ or $x > 2$ and $f'(x) < 0 \quad \Leftrightarrow \quad -2 < x < 2$. So f is increasing on $(-\infty, -2)$ and $(2, \infty)$ and f is decreasing on $(-2, 2)$.

(b) f changes from increasing to decreasing at $x = -2$, so $f(-2) = 18$ is a local maximum value. f changes from decreasing to increasing at $x = 2$, so $f(2) = -14$ is a local minimum value.

(c) $f''(x) = 6x$. $f''(x) = 0 \quad \Leftrightarrow \quad x = 0$. $f''(x) > 0$ on $(0, \infty)$ and $f''(x) < 0$ on $(-\infty, 0)$. So f is concave upward on $(0, \infty)$ and f is concave downward on $(-\infty, 0)$. There is an inflection point at $(0, 2)$.

(d) 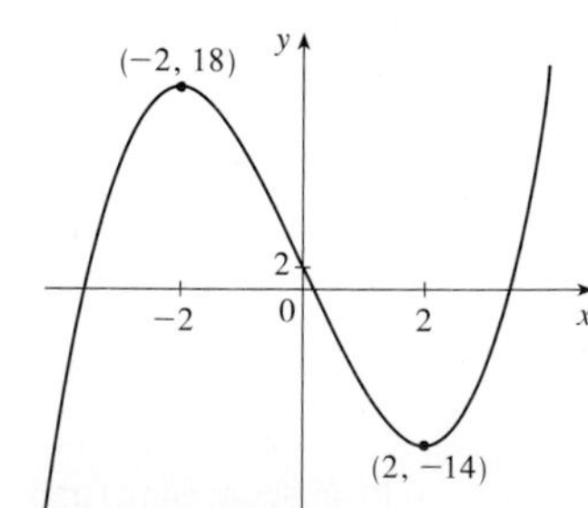

38. (a) $f(x) = 36x + 3x^2 - 2x^3 \quad \Rightarrow \quad f'(x) = 36 + 6x - 6x^2 = -6(x^2 - x - 6) = -6(x + 2)(x - 3)$. $f'(x) > 0 \quad \Leftrightarrow$ $-2 < x < 3$ and $f'(x) < 0 \quad \Leftrightarrow \quad x < -2$ or $x > 3$. So f is increasing on $(-2, 3)$ and f is decreasing on $(-\infty, -2)$ and $(3, \infty)$.

(b) f changes from increasing to decreasing at $x = 3$, so $f(3) = 81$ is a local maximum value. f changes from decreasing to increasing at $x = -2$, so $f(-2) = -44$ is a local minimum value.

(c) $f''(x) = 6 - 12x$. $f''(x) = 0 \Leftrightarrow x = \frac{1}{2}$. $f''(x) > 0$ on $\left(-\infty, \frac{1}{2}\right)$ and $f''(x) < 0$ on $\left(\frac{1}{2}, \infty\right)$. So f is CU on $\left(-\infty, \frac{1}{2}\right)$ and f is CD on $\left(\frac{1}{2}, \infty\right)$. There is an inflection point at $\left(\frac{1}{2}, \frac{37}{2}\right)$.

(d)

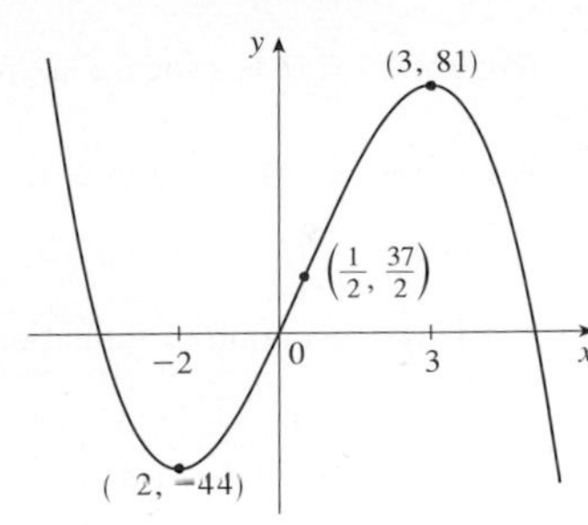

39. (a) $f(x) = \frac{1}{2}x^4 - 4x^2 + 3 \Rightarrow f'(x) = 2x^3 - 8x = 2x(x^2 - 4) = 2x(x+2)(x-2)$. $f'(x) > 0 \Leftrightarrow -2 < x < 0$ or $x > 2$, and $f'(x) < 0 \Leftrightarrow x < -2$ or $0 < x < 2$. So f is increasing on $(-2, 0)$ and $(2, \infty)$ and f is decreasing on $(-\infty, -2)$ and $(0, 2)$.

(b) f changes from increasing to decreasing at $x = 0$, so $f(0) = 3$ is a local maximum value.

f changes from decreasing to increasing at $x = \pm 2$, so $f(\pm 2) = -5$ is a local minimum value.

(c) $f''(x) = 6x^2 - 8 = 6\left(x^2 - \frac{4}{3}\right) = 6\left(x + \frac{2}{\sqrt{3}}\right)\left(x - \frac{2}{\sqrt{3}}\right)$.

$f''(x) = 0 \Leftrightarrow x = \pm\frac{2}{\sqrt{3}}$. $f''(x) > 0$ on $\left(-\infty, -\frac{2}{\sqrt{3}}\right)$ and $\left(\frac{2}{\sqrt{3}}, \infty\right)$ and $f''(x) < 0$ on $\left(-\frac{2}{\sqrt{3}}, \frac{2}{\sqrt{3}}\right)$. So f is CU on $\left(-\infty, -\frac{2}{\sqrt{3}}\right)$ and $\left(\frac{2}{\sqrt{3}}, \infty\right)$, and f is CD on $\left(-\frac{2}{\sqrt{3}}, \frac{2}{\sqrt{3}}\right)$. There are inflection points at $\left(\pm\frac{2}{\sqrt{3}}, -\frac{13}{9}\right)$.

(d)

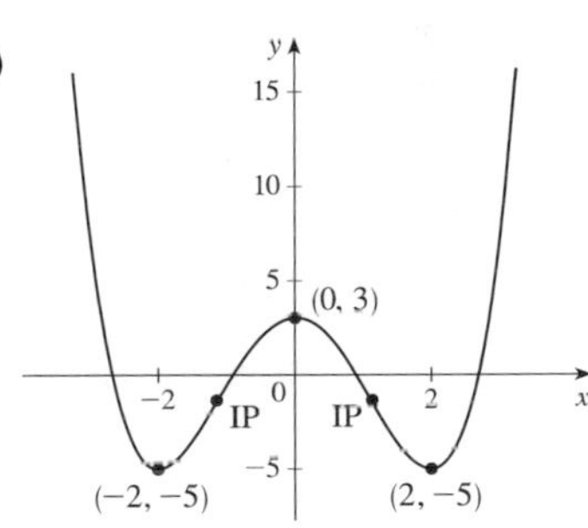

40. (a) $g(x) = 200 + 8x^3 + x^4 \Rightarrow g'(x) = 24x^2 + 4x^3 = 4x^2(6 + x) = 0$ when $x = -6$ and when $x = 0$. $g'(x) > 0 \Leftrightarrow x > -6$ $[x \neq 0]$ and $g'(x) < 0 \Leftrightarrow x < -6$, so g is decreasing on $(-\infty, -6)$ and g is increasing on $(-6, \infty)$, with a horizontal tangent at $x = 0$.

(b) $g(-6) = -232$ is a local minimum value. There is no local maximum value.

(c) $g''(x) = 48x + 12x^2 = 12x(4 + x) = 0$ when $x = -4$ and when $x = 0$. $g''(x) > 0 \Leftrightarrow x < -4$ or $x > 0$ and $g''(x) < 0 \Leftrightarrow -4 < x < 0$, so g is CU on $(-\infty, -4)$ and $(0, \infty)$, and g is CD on $(-4, 0)$. There are inflection points at $(-4, -56)$ and $(0, 200)$.

(d)

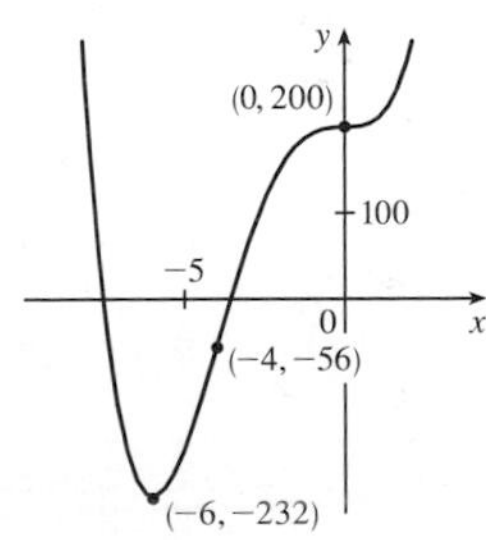

41. (a) $h(x) = (x+1)^5 - 5x - 2 \Rightarrow h'(x) = 5(x+1)^4 - 5$. $h'(x) = 0 \Leftrightarrow 5(x+1)^4 = 5 \Leftrightarrow (x+1)^4 = 1 \Rightarrow (x+1)^2 = 1 \Rightarrow x + 1 = 1$ or $x + 1 = -1 \Rightarrow x = 0$ or $x = -2$. $h'(x) > 0 \Leftrightarrow x < -2$ or $x > 0$ and $h'(x) < 0 \Leftrightarrow -2 < x < 0$. So h is increasing on $(-\infty, -2)$ and $(0, \infty)$ and h is decreasing on $(-2, 0)$.

(b) $h(-2) = 7$ is a local maximum value and $h(0) = -1$ is a local minimum value.

(c) $h''(x) = 20(x+1)^3 = 0 \quad \Leftrightarrow \quad x = -1.\quad h''(x) > 0 \quad \Leftrightarrow \quad x > -1$ and $h''(x) < 0 \quad \Leftrightarrow \quad x < -1$, so h is CU on $(-1, \infty)$ and h is CD on $(-\infty, -1)$. There is a point of inflection at $(-1, h(-1)) = (-1, 3)$.

(d)

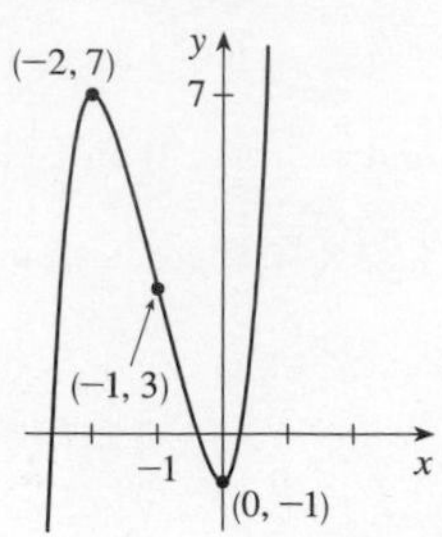

42. (a) $h(x) = 5x^3 - 3x^5 \quad \Rightarrow \quad h'(x) = 15x^2 - 15x^4 = 15x^2(1 - x^2) = 15x^2(1+x)(1-x).\quad h'(x) > 0 \quad \Leftrightarrow$ $-1 < x < 0$ and $0 < x < 1$ [note that $h'(0) = 0$] and $h'(x) < 0 \quad \Leftrightarrow \quad x < -1$ or $x > 1$. So h is increasing on $(-1, 1)$ and h is decreasing on $(-\infty, -1)$ and $(1, \infty)$.

(b) h changes from decreasing to increasing at $x = -1$, so $h(-1) = -2$ is a local minimum value. h changes from increasing to decreasing at $x = 1$, so $h(1) = 2$ is a local maximum value.

(c) $h''(x) = 30x - 60x^3 = 30x(1 - 2x^2).\quad h''(x) = 0 \quad \Leftrightarrow \quad x = 0$ or $1 - 2x^2 = 0 \quad \Leftrightarrow \quad x = 0$ or $x = \pm 1/\sqrt{2}$. $h''(x) > 0$ on $\left(-\infty, -1/\sqrt{2}\right)$ and $\left(0, 1/\sqrt{2}\right)$, and $h''(x) < 0$ on $\left(-1/\sqrt{2}, 0\right)$ and $\left(1/\sqrt{2}, \infty\right)$. So h is CU on $\left(-\infty, -1/\sqrt{2}\right)$ and $\left(0, 1/\sqrt{2}\right)$, and h is CD on $\left(-1/\sqrt{2}, 0\right)$ and $\left(1/\sqrt{2}, \infty\right)$. There are inflection points at $\left(-1/\sqrt{2}, -7/\left(4\sqrt{2}\right)\right)$, $(0, 0)$, and $\left(1/\sqrt{2}, 7/\left(4\sqrt{2}\right)\right)$.

(d)

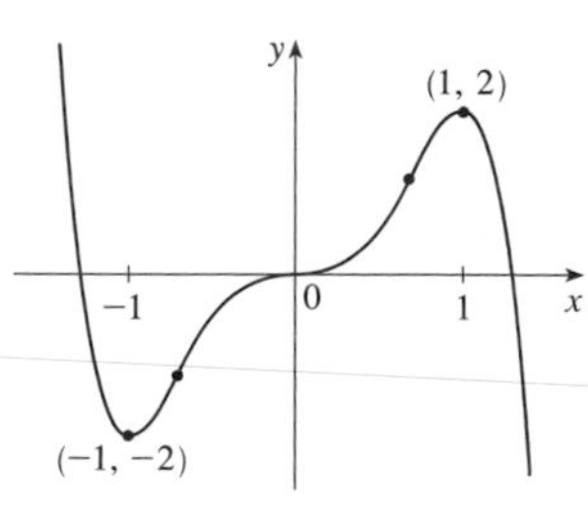

43. (a) $F(x) = x\sqrt{6 - x} \quad \Rightarrow$

$$F'(x) = x \cdot \tfrac{1}{2}(6-x)^{-1/2}(-1) + (6-x)^{1/2}(1) = \tfrac{1}{2}(6-x)^{-1/2}[-x + 2(6-x)] = \frac{-3x+12}{2\sqrt{6-x}}.$$

$F'(x) > 0 \quad \Leftrightarrow \quad -3x + 12 > 0 \quad \Leftrightarrow \quad x < 4$ and $F'(x) < 0 \quad \Leftrightarrow \quad 4 < x < 6$. So F is increasing on $(-\infty, 4)$ and F is decreasing on $(4, 6)$.

(b) F changes from increasing to decreasing at $x = 4$, so $F(4) = 4\sqrt{2}$ is a local maximum value. There is no local minimum value.

(c) $F'(x) = -\frac{3}{2}(x-4)(6-x)^{-1/2} \quad \Rightarrow$

$$F''(x) = -\tfrac{3}{2}\left[(x-4)\left(-\tfrac{1}{2}(6-x)^{-3/2}(-1)\right) + (6-x)^{-1/2}(1)\right]$$

$$= -\frac{3}{2} \cdot \frac{1}{2}(6-x)^{-3/2}[(x-4) + 2(6-x)] = \frac{3(x-8)}{4(6-x)^{3/2}}$$

$F''(x) < 0$ on $(-\infty, 6)$, so F is CD on $(-\infty, 6)$. There is no inflection point.

(d)

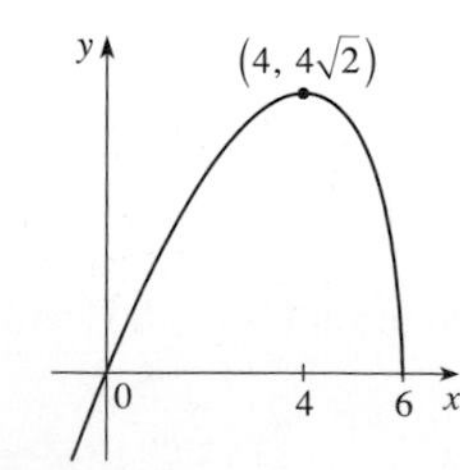

44. (a) $G(x) = 5x^{2/3} - 2x^{5/3} \quad \Rightarrow \quad G'(x) = \frac{10}{3}x^{-1/3} - \frac{10}{3}x^{2/3} = \frac{10}{3}x^{-1/3}(1 - x) = \dfrac{10(1-x)}{3x^{1/3}}.$

$G'(x) > 0 \quad \Leftrightarrow \quad 0 < x < 1$ and $G'(x) < 0 \quad \Leftrightarrow \quad x < 0$ or $x > 1$. So G is increasing on $(0, 1)$ and G is decreasing on $(-\infty, 0)$ and $(1, \infty)$.

(b) G changes from decreasing to increasing at $x = 0$, so $G(0) = 0$ is a local minimum value. G changes from increasing to decreasing at $x = 1$, so $G(1) = 3$ is a local maximum value. Note that the First Derivative Test applies at $x = 0$ even though G' is not defined at $x = 0$, since G is continuous at 0.

(c) $G''(x) = -\frac{10}{9}x^{-4/3} - \frac{20}{9}x^{-1/3} = -\frac{10}{9}x^{-4/3}(1 + 2x)$. $G''(x) > 0 \Leftrightarrow x < -\frac{1}{2}$ and $G''(x) < 0 \Leftrightarrow -\frac{1}{2} < x < 0$ or $x > 0$. So G is CU on $\left(-\infty, -\frac{1}{2}\right)$ and G is CD on $\left(-\frac{1}{2}, 0\right)$ and $(0, \infty)$. The only change in concavity occurs at $x = -\frac{1}{2}$, so there is an inflection point at $\left(-\frac{1}{2}, 6/\sqrt[3]{4}\right)$.

(d)

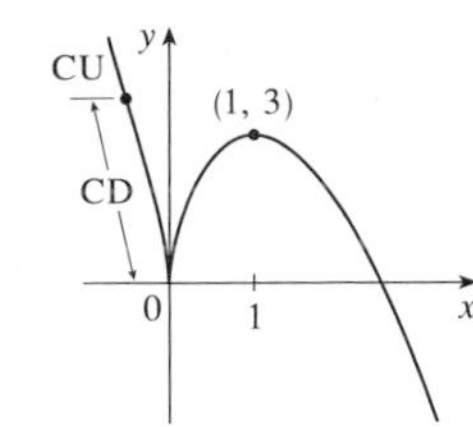

45. (a) $C(x) = x^{1/3}(x + 4) = x^{4/3} + 4x^{1/3} \Rightarrow C'(x) = \frac{4}{3}x^{1/3} + \frac{4}{3}x^{-2/3} = \frac{4}{3}x^{-2/3}(x + 1) = \dfrac{4(x + 1)}{3\sqrt[3]{x^2}}$. $C'(x) > 0$ if $-1 < x < 0$ or $x > 0$ and $C'(x) < 0$ for $x < -1$, so C is increasing on $(-1, \infty)$ and C is decreasing on $(-\infty, -1)$.

(b) $C(-1) = -3$ is a local minimum value.

(c) $C''(x) = \frac{4}{9}x^{-2/3} - \frac{8}{9}x^{-5/3} = \frac{4}{9}x^{-5/3}(x - 2) = \dfrac{4(x - 2)}{9\sqrt[3]{x^5}}$.

$C''(x) < 0$ for $0 < x < 2$ and $C''(x) > 0$ for $x < 0$ and $x > 2$, so C is concave downward on $(0, 2)$ and concave upward on $(-\infty, 0)$ and $(2, \infty)$. There are inflection points at $(0, 0)$ and $\left(2, 6\sqrt[3]{2}\right) \approx (2, 7.56)$.

(d)

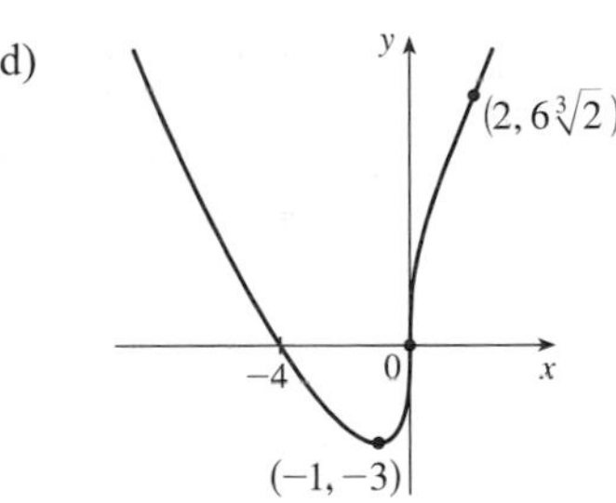

46. (a) $f(x) = \ln(x^2 + 9) \Rightarrow f'(x) = \dfrac{1}{x^2 + 9} \cdot 2x = \dfrac{2x}{x^2 + 9}$. $f'(x) > 0 \Leftrightarrow 2x > 0 \Leftrightarrow x > 0$ and $f'(x) < 0 \Leftrightarrow x < 0$. So f is increasing on $(0, \infty)$ and f is decreasing on $(-\infty, 0)$.

(b) f changes from decreasing to increasing at $x = 0$, so $f(0) = \ln 9$ is a local minimum value. There is no local maximum value.

(c) $f''(x) = \dfrac{(x^2 + 9) \cdot 2 - 2x(2x)}{(x^2 + 9)^2} = \dfrac{18 - 2x^2}{(x^2 + 9)^2} = \dfrac{-2(x + 3)(x - 3)}{(x^2 + 9)^2}$.

$f''(x) = 0 \Leftrightarrow x = \pm 3$. $f''(x) > 0$ on $(-3, 3)$ and $f''(x) < 0$ on $(-\infty, -3)$ and $(3, \infty)$. So f is CU on $(-3, 3)$, and f is CD on $(-\infty, -3)$ and $(3, \infty)$. There are inflection points at $(\pm 3, \ln 18)$.

(d)

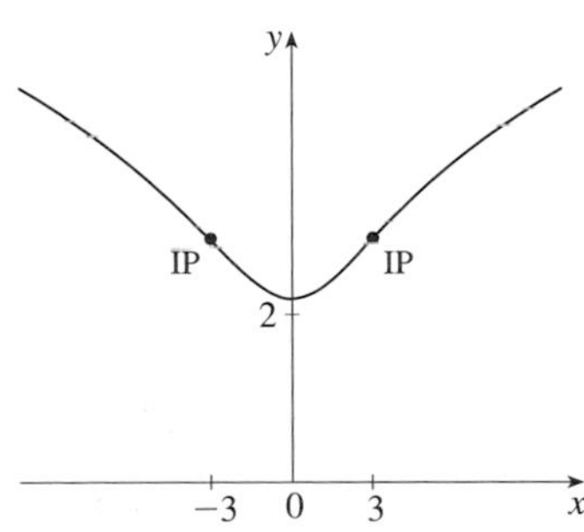

47. (a) $f(\theta) = 2\cos\theta + \cos^2\theta$, $0 \le \theta \le 2\pi \Rightarrow f'(\theta) = -2\sin\theta + 2\cos\theta\,(-\sin\theta) = -2\sin\theta\,(1 + \cos\theta)$. $f'(\theta) = 0 \Leftrightarrow \theta = 0, \pi$, and 2π. $f'(\theta) > 0 \Leftrightarrow \pi < \theta < 2\pi$ and $f'(\theta) < 0 \Leftrightarrow 0 < \theta < \pi$. So f is increasing on $(\pi, 2\pi)$ and f is decreasing on $(0, \pi)$.

(b) $f(\pi) = -1$ is a local minimum value.

(c) $f'(\theta) = -2\sin\theta\,(1+\cos\theta) \;\Rightarrow$

$$\begin{aligned} f''(\theta) &= -2\sin\theta\,(-\sin\theta) + (1+\cos\theta)(-2\cos\theta) = 2\sin^2\theta - 2\cos\theta - 2\cos^2\theta \\ &= 2(1-\cos^2\theta) - 2\cos\theta - 2\cos^2\theta = -4\cos^2\theta - 2\cos\theta + 2 \\ &= -2(2\cos^2\theta + \cos\theta - 1) = -2(2\cos\theta - 1)(\cos\theta + 1) \end{aligned}$$

Since $-2(\cos\theta+1) < 0$ [for $\theta \neq \pi$], $f''(\theta) > 0 \;\Rightarrow\; 2\cos\theta - 1 < 0 \;\Rightarrow\; \cos\theta < \frac{1}{2} \;\Rightarrow\; \frac{\pi}{3} < \theta < \frac{5\pi}{3}$ and $f''(\theta) < 0 \;\Rightarrow\; \cos\theta > \frac{1}{2} \;\Rightarrow\; 0 < \theta < \frac{\pi}{3}$ or $\frac{5\pi}{3} < \theta < 2\pi$. So f is CU on $\left(\frac{\pi}{3}, \frac{5\pi}{3}\right)$ and f is CD on $\left(0, \frac{\pi}{3}\right)$ and $\left(\frac{5\pi}{3}, 2\pi\right)$. There are points of inflection at $\left(\frac{\pi}{3}, f\left(\frac{\pi}{3}\right)\right) = \left(\frac{\pi}{3}, \frac{5}{4}\right)$ and $\left(\frac{5\pi}{3}, f\left(\frac{5\pi}{3}\right)\right) = \left(\frac{5\pi}{3}, \frac{5}{4}\right)$.

(d)

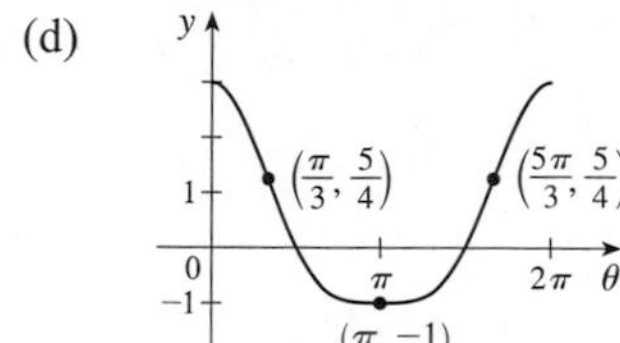

48. (a) $S(x) = x - \sin x,\ 0 \le x \le 4\pi \;\Rightarrow\; S'(x) = 1 - \cos x$. $S'(x) = 0 \;\Leftrightarrow\; \cos x = 1 \;\Leftrightarrow\; x = 0, 2\pi$, and 4π. $S'(x) > 0 \;\Leftrightarrow\; \cos x < 1$, which is true for all x except integer multiples of 2π, so S is increasing on $(0, 4\pi)$ since $S'(2\pi) = 0$.

(b) There is no local maximum or minimum.

(c)

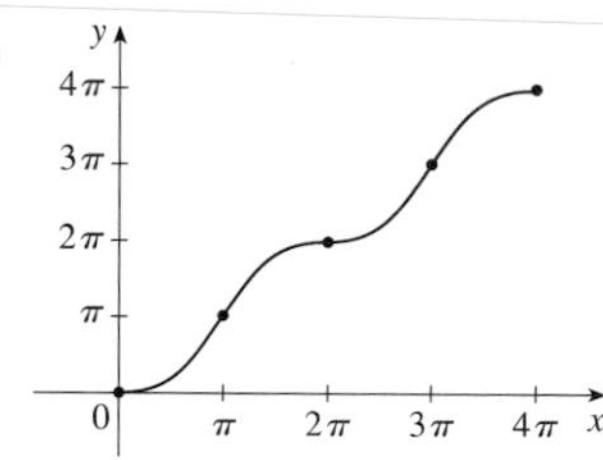

(d) $S''(x) = \sin x$. $S''(x) > 0$ if $0 < x < \pi$ or $2\pi < x < 3\pi$, and $S''(x) < 0$ if $\pi < x < 2\pi$ or $3\pi < x < 4\pi$. So S is CU on $(0, \pi)$ and $(2\pi, 3\pi)$, and S is CD on $(\pi, 2\pi)$ and $(3\pi, 4\pi)$. There are inflection points at (π, π), $(2\pi, 2\pi)$, and $(3\pi, 3\pi)$.

49. $f(x) = 1 + \dfrac{1}{x} - \dfrac{1}{x^2}$ has domain $(-\infty, 0) \cup (0, \infty)$.

(a) $\displaystyle\lim_{x\to\pm\infty}\left(1 + \frac{1}{x} - \frac{1}{x^2}\right) = 1$, so $y = 1$ is a HA. $\displaystyle\lim_{x\to 0^+}\left(1 + \frac{1}{x} - \frac{1}{x^2}\right) = \lim_{x\to 0^+}\left(\frac{x^2 + x - 1}{x^2}\right) = -\infty$ since $(x^2 + x - 1) \to -1$ and $x^2 \to 0$ as $x \to 0^+$ [a similar argument can be made for $x \to 0^-$], so $x = 0$ is a VA.

(b) $f'(x) = -\dfrac{1}{x^2} + \dfrac{2}{x^3} = -\dfrac{1}{x^3}(x - 2)$. $f'(x) = 0 \;\Leftrightarrow\; x = 2$. $f'(x) > 0 \;\Leftrightarrow\; 0 < x < 2$ and $f'(x) < 0 \;\Leftrightarrow\; x < 0$ or $x > 2$. So f is increasing on $(0, 2)$ and f is decreasing on $(-\infty, 0)$ and $(2, \infty)$.

(c) f changes from increasing to decreasing at $x = 2$, so $f(2) = \frac{5}{4}$ is a local maximum value. There is no local minimum value.

(d) $f''(x) = \dfrac{2}{x^3} - \dfrac{6}{x^4} = \dfrac{2}{x^4}(x - 3)$. $f''(x) = 0 \;\Leftrightarrow\; x = 3$. $f''(x) > 0 \;\Leftrightarrow\; x > 3$ and $f''(x) < 0 \;\Leftrightarrow\; x < 0$ or $0 < x < 3$. So f is CU on $(3, \infty)$ and f is CD on $(-\infty, 0)$ and $(0, 3)$. There is an inflection point at $\left(3, \frac{11}{9}\right)$.

(e)

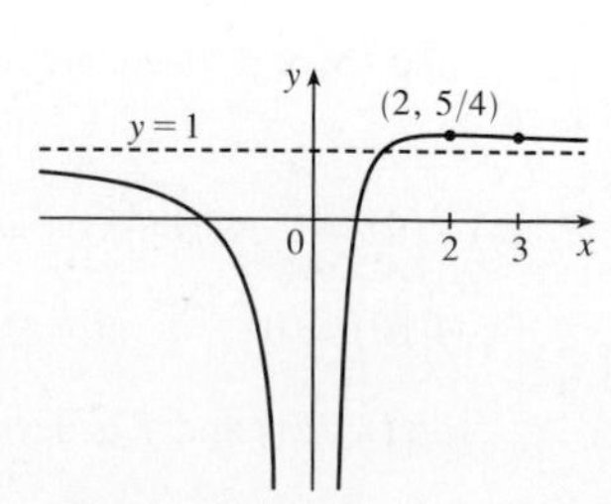

50. $f(x) = \dfrac{x^2 - 4}{x^2 + 4}$ has domain $\mathbb{R}$.

(a) $\displaystyle\lim_{x\to\pm\infty} \frac{x^2-4}{x^2+4} = \lim_{x\to\pm\infty} \frac{1-4/x^2}{1+4/x^2} = \frac{1}{1} = 1$, so $y = 1$ is a HA. There is no vertical asymptote.

(b) $f'(x) = \dfrac{(x^2+4)(2x) - (x^2-4)(2x)}{(x^2+4)^2} = \dfrac{2x[(x^2+4)-(x^2-4)]}{(x^2+4)^2} = \dfrac{16x}{(x^2+4)^2}$. $f'(x) > 0 \Leftrightarrow x > 0$ and $f'(x) < 0 \Leftrightarrow x < 0$. So f is increasing on $(0, \infty)$ and f is decreasing on $(-\infty, 0)$.

(c) f changes from decreasing to increasing at $x = 0$, so $f(0) = -1$ is a local minimum value.

(d) $f''(x) = \dfrac{(x^2+4)^2(16) - 16x \cdot 2(x^2+4)(2x)}{[(x^2+4)^2]^2} = \dfrac{16(x^2+4)[(x^2+4)-4x^2]}{(x^2+4)^4} = \dfrac{16(4-3x^2)}{(x^2+4)^3}$.

$f''(x) = 0 \Leftrightarrow x = \pm 2/\sqrt{3}$. $f''(x) > 0 \Leftrightarrow -2/\sqrt{3} < x < 2/\sqrt{3}$ and $f''(x) < 0 \Leftrightarrow x < -2/\sqrt{3}$ or $x > 2/\sqrt{3}$. So f is CU on $\left(-2/\sqrt{3}, 2/\sqrt{3}\right)$ and f is CD on $\left(-\infty, -2/\sqrt{3}\right)$ and $\left(2/\sqrt{3}, \infty\right)$. There are inflection points at $\left(\pm 2/\sqrt{3}, -\frac{1}{2}\right)$.

(e)

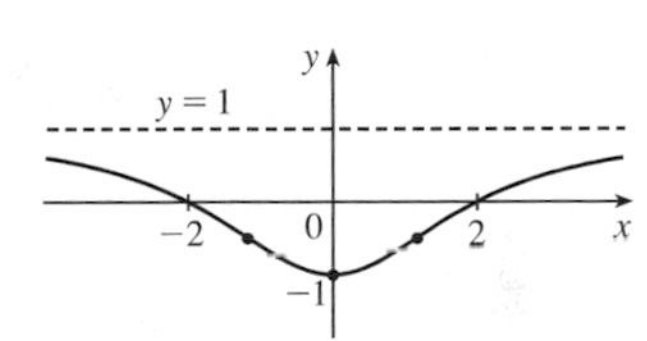

51. (a) $\displaystyle\lim_{x\to-\infty} \left(\sqrt{x^2+1} - x\right) = \infty$ and

$$\lim_{x\to\infty} \left(\sqrt{x^2+1} - x\right) = \lim_{x\to\infty} \left(\sqrt{x^2+1} - x\right) \frac{\sqrt{x^2+1}+x}{\sqrt{x^2+1}+x} = \lim_{x\to\infty} \frac{1}{\sqrt{x^2+1}+x} = 0\text{, so } y = 0 \text{ is a HA.}$$

(b) $f(x) = \sqrt{x^2+1} - x \Rightarrow f'(x) = \dfrac{x}{\sqrt{x^2+1}} - 1$. Since $\dfrac{x}{\sqrt{x^2+1}} < 1$ for all x, $f'(x) < 0$, so f is decreasing on $\mathbb{R}$.

(c) No minimum or maximum

(d) $$f''(x) = \frac{(x^2+1)^{1/2}(1) - x \cdot \frac{1}{2}(x^2+1)^{-1/2}(2x)}{\left(\sqrt{x^2+1}\right)^2}$$

$$= \frac{(x^2+1)^{1/2} - \dfrac{x^2}{(x^2+1)^{1/2}}}{x^2+1} = \frac{(x^2+1)-x^2}{(x^2+1)^{3/2}} = \frac{1}{(x^2+1)^{3/2}} > 0,$$

so f is CU on $\mathbb{R}$. No IP

(e)

52. $f(x) = \dfrac{e^x}{1-e^x}$ has domain $\{x \mid 1 - e^x \neq 0\} = \{x \mid e^x \neq 1\} = \{x \mid x \neq 0\}$.

(a) $\displaystyle\lim_{x\to\infty} \frac{e^x}{1-e^x} = \lim_{x\to\infty} \frac{e^x/e^x}{(1-e^x)/e^x} = \lim_{x\to\infty} \frac{1}{1/e^x - 1} = \frac{1}{0-1} = -1$, so $y = -1$ is a HA.

$\displaystyle\lim_{x\to-\infty} \frac{e^x}{1-e^x} = \frac{0}{1-0} = 0$, so $y = 0$ is a HA. $\displaystyle\lim_{x\to 0^+} \frac{e^x}{1-e^x} = -\infty$ and $\displaystyle\lim_{x\to 0^-} \frac{e^x}{1-e^x} = \infty$, so $x = 0$ is a VA.

(b) $f'(x) = \dfrac{(1-e^x)e^x - e^x(-e^x)}{(1-e^x)^2} = \dfrac{e^x[(1-e^x)+e^x]}{(1-e^x)^2} = \dfrac{e^x}{(1-e^x)^2}$. $f'(x) > 0$ for $x \neq 0$, so f is increasing on $(-\infty, 0)$ and $(0, \infty)$.

(c) There is no local maximum or minimum.

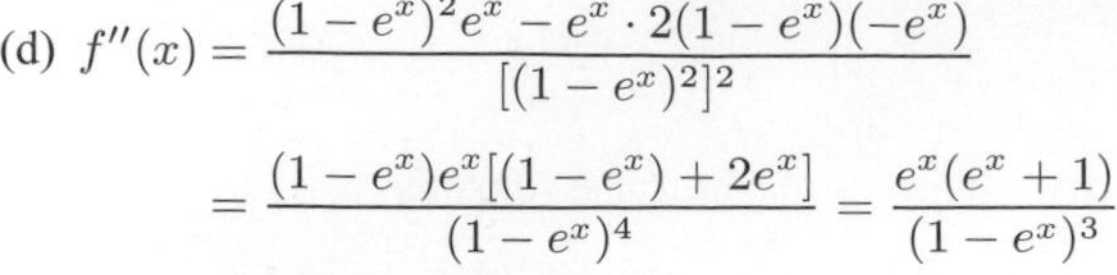

(d) $f''(x) = \dfrac{(1-e^x)^2 e^x - e^x \cdot 2(1-e^x)(-e^x)}{[(1-e^x)^2]^2}$

$= \dfrac{(1-e^x)e^x[(1-e^x)+2e^x]}{(1-e^x)^4} = \dfrac{e^x(e^x+1)}{(1-e^x)^3}$

$f''(x) > 0 \;\Leftrightarrow\; (1-e^x)^3 > 0 \;\Leftrightarrow\; e^x < 1 \;\Leftrightarrow\; x < 0$ and $f''(x) < 0 \;\Leftrightarrow\; x > 0$. So f is CU on $(-\infty, 0)$ and f is CD on $(0, \infty)$. There is no inflection point.

(e)

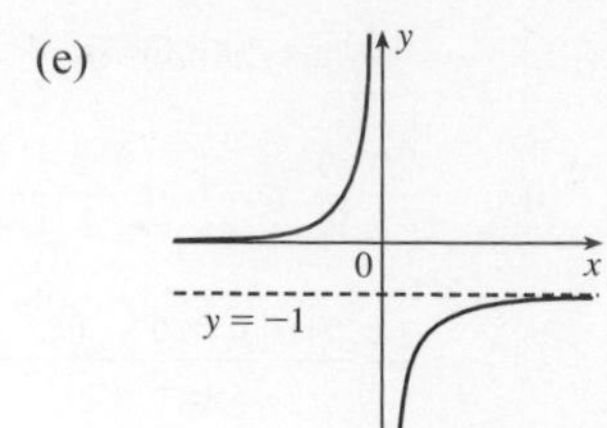

53. (a) $\lim\limits_{x\to\pm\infty} e^{-x^2} = \lim\limits_{x\to\pm\infty} \dfrac{1}{e^{x^2}} = 0$, so $y = 0$ is a HA. There is no VA.

(b) $f(x) = e^{-x^2} \;\Rightarrow\; f'(x) = e^{-x^2}(-2x)$. $f'(x) = 0 \;\Leftrightarrow\; x = 0$. $f'(x) > 0 \;\Leftrightarrow\; x < 0$ and $f'(x) < 0 \;\Leftrightarrow\; x > 0$. So f is increasing on $(-\infty, 0)$ and f is decreasing on $(0, \infty)$.

(c) f changes from increasing to decreasing at $x = 0$, so $f(0) = 1$ is a local maximum value. There is no local minimum value.

(d) $f''(x) = e^{-x^2}(-2) + (-2x)e^{-x^2}(-2x) = -2e^{-x^2}(1-2x^2)$. $f''(x) = 0 \;\Leftrightarrow\; x^2 = \frac{1}{2} \;\Leftrightarrow\; x = \pm 1/\sqrt{2}$. $f''(x) > 0 \;\Leftrightarrow\; x < -1/\sqrt{2}$ or $x > 1/\sqrt{2}$ and $f''(x) < 0 \;\Leftrightarrow\; -1/\sqrt{2} < x < 1/\sqrt{2}$. So f is CU on $(-\infty, -1/\sqrt{2})$ and $(1/\sqrt{2}, \infty)$, and f is CD on $(-1/\sqrt{2}, 1/\sqrt{2})$. There are inflection points at $\left(\pm 1/\sqrt{2}, e^{-1/2}\right)$.

(e)

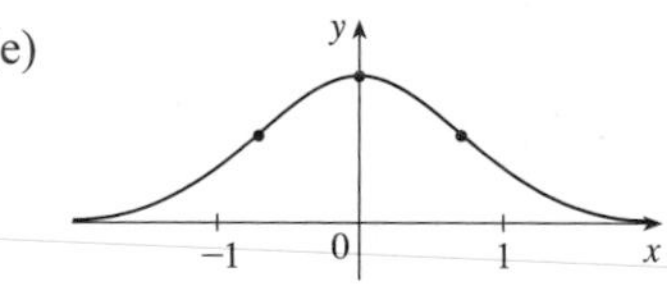

54. $f(x) = x - \frac{1}{6}x^2 - \frac{2}{3}\ln x$ has domain $(0, \infty)$.

(a) $\lim\limits_{x\to 0^+} \left(x - \frac{1}{6}x^2 - \frac{2}{3}\ln x\right) = \infty$ since $\ln x \to -\infty$ as $x \to 0^+$, so $x = 0$ is a VA. There is no HA.

(b) $f'(x) = 1 - \dfrac{1}{3}x - \dfrac{2}{3x} = \dfrac{3x - x^2 - 2}{3x} = \dfrac{-(x^2 - 3x + 2)}{3x} = \dfrac{-(x-1)(x-2)}{3x}$. $f'(x) > 0 \;\Leftrightarrow\; (x-1)(x-2) < 0 \;\Leftrightarrow\; 1 < x < 2$ and $f'(x) < 0 \;\Leftrightarrow\; 0 < x < 1$ or $x > 2$. So f is increasing on $(1, 2)$ and f is decreasing on $(0, 1)$ and $(2, \infty)$.

(c) f changes from decreasing to increasing at $x = 1$, so $f(1) = \frac{5}{6}$ is a local minimum value. f changes from increasing to decreasing at $x = 2$, so $f(2) = \frac{4}{3} - \frac{2}{3}\ln 2 \approx 0.87$ is a local maximum value.

(d) $f''(x) = -\dfrac{1}{3} + \dfrac{2}{3x^2} = \dfrac{2 - x^2}{3x^2}$. $f''(x) > 0 \;\Leftrightarrow\; 0 < x < \sqrt{2}$ and $f''(x) < 0 \;\Leftrightarrow\; x > \sqrt{2}$. So f is CU on $(0, \sqrt{2})$ and f is CD on $(\sqrt{2}, \infty)$. There is an inflection point at $\left(\sqrt{2}, \sqrt{2} - \frac{1}{3} - \frac{1}{3}\ln 2\right)$.

(e)

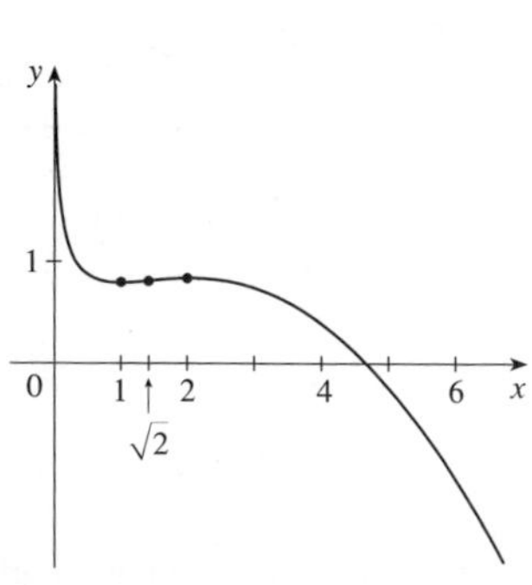

55. $f(x) = \ln(1 - \ln x)$ is defined when $x > 0$ (so that $\ln x$ is defined) and $1 - \ln x > 0$ [so that $\ln(1 - \ln x)$ is defined]. The second condition is equivalent to $1 > \ln x \quad \Leftrightarrow \quad x < e$, so f has domain $(0, e)$.

(a) As $x \to 0^+$, $\ln x \to -\infty$, so $1 - \ln x \to \infty$ and $f(x) \to \infty$. As $x \to e^-$, $\ln x \to 1^-$, so $1 - \ln x \to 0^+$ and $f(x) \to -\infty$. Thus, $x = 0$ and $x = e$ are vertical asymptotes. There is no horizontal asymptote.

(b) $f'(x) = \dfrac{1}{1 - \ln x}\left(-\dfrac{1}{x}\right) = -\dfrac{1}{x(1 - \ln x)} < 0$ on $(0, e)$. Thus, f is decreasing on its domain, $(0, e)$.

(c) $f'(x) \neq 0$ on $(0, e)$, so f has no local maximum or minimum value.

(e)

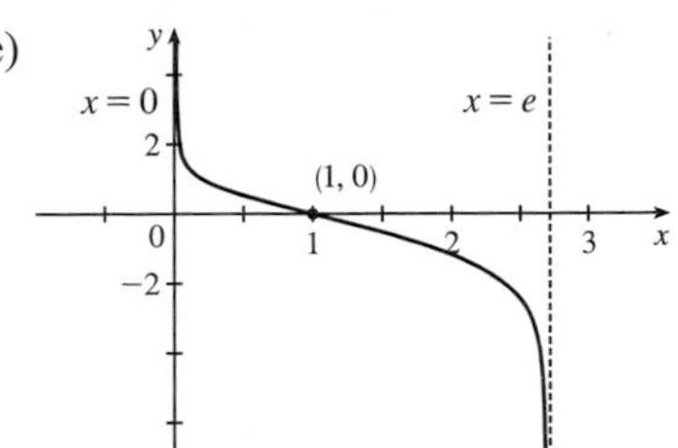

(d) $f''(x) = -\dfrac{-[x(1 - \ln x)]'}{[x(1 - \ln x)]^2} = \dfrac{x(-1/x) + (1 - \ln x)}{x^2(1 - \ln x)^2}$

$= -\dfrac{\ln x}{x^2(1 - \ln x)^2}$

so $f''(x) > 0 \quad \Leftrightarrow \quad \ln x < 0 \quad \Leftrightarrow \quad 0 < x < 1$. Thus, f is CU on $(0, 1)$ and CD on $(1, e)$. There is an inflection point at $(1, 0)$.

56. (a) $\lim\limits_{x\to\infty} \arctan x = \frac{\pi}{2}$, so $\lim\limits_{x\to\infty} e^{\arctan x} = e^{\pi/2}$ [≈ 4.81], so $y = e^{\pi/2}$ is a HA.

$\lim\limits_{x\to-\infty} e^{\arctan x} = e^{-\pi/2}$ [≈ 0.21], so $y = e^{-\pi/2}$ is a HA. No VA.

(b) $f(x) = e^{\arctan x} \quad \Rightarrow \quad f'(x) = e^{\arctan x} \cdot \dfrac{1}{1 + x^2} > 0$ for all x. Thus, f is increasing on $\mathbb{R}$.

(c) There is no local maximum or minimum.

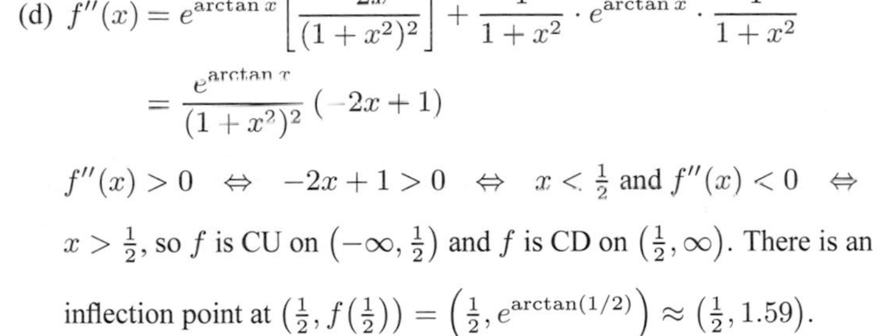

(d) $f''(x) = e^{\arctan x}\left[\dfrac{-2x}{(1 + x^2)^2}\right] + \dfrac{1}{1 + x^2} \cdot e^{\arctan x} \cdot \dfrac{1}{1 + x^2}$

$= \dfrac{e^{\arctan x}}{(1 + x^2)^2}(-2x + 1)$

$f''(x) > 0 \quad \Leftrightarrow \quad -2x + 1 > 0 \quad \Leftrightarrow \quad x < \frac{1}{2}$ and $f''(x) < 0 \quad \Leftrightarrow \quad x > \frac{1}{2}$, so f is CU on $\left(-\infty, \frac{1}{2}\right)$ and f is CD on $\left(\frac{1}{2}, \infty\right)$. There is an inflection point at $\left(\frac{1}{2}, f\left(\frac{1}{2}\right)\right) = \left(\frac{1}{2}, e^{\arctan(1/2)}\right) \approx \left(\frac{1}{2}, 1.59\right)$.

(e)

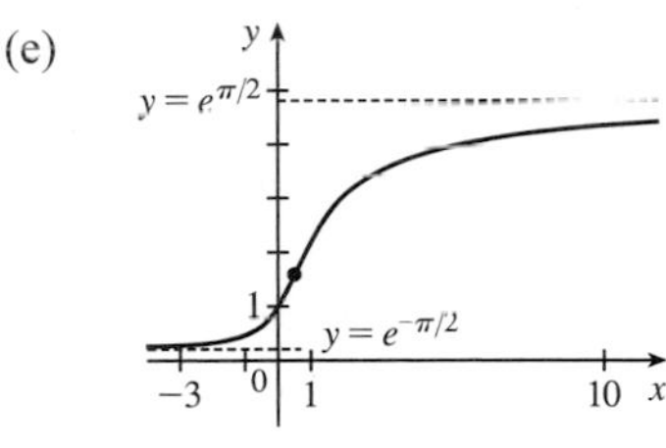

57. The nonnegative factors $(x + 1)^2$ and $(x - 6)^4$ do not affect the sign of $f'(x) = (x + 1)^2(x - 3)^5(x - 6)^4$. So $f'(x) > 0 \quad \Rightarrow \quad (x - 3)^5 > 0 \quad \Rightarrow \quad x - 3 > 0 \quad \Rightarrow \quad x > 3$. Thus, f is increasing on the interval $(3, \infty)$.

58. $y = f(x) = x^3 - 3a^2x + 2a^3$, $a > 0$. The y-intercept is $f(0) = 2a^3$. $y' = 3x^2 - 3a^2 = 3(x^2 - a^2) = 3(x + a)(x - a)$. The critical numbers are $-a$ and a. $f' < 0$ on $(-a, a)$, so f is decreasing on $(-a, a)$ and f is increasing on $(-\infty, -a)$ and (a, ∞). $f(-a) = 4a^3$ is a local maximum value and $f(a) = 0$ is a local minimum value. Since $f(a) = 0$, a is an x-intercept, and $x - a$ is a factor of f. Synthetically dividing $y = x^3 - 3a^2x + 2a^3$ by $x - a$ gives us the following result: $y = x^3 - 3a^2x + 2a^3 = (x - a)(x^2 + ax - 2a^2) = (x - a)(x - a)(x + 2a) = (x - a)^2(x + 2a)$, which tells us that the only x-intercepts are $-2a$ and a. $y' = 3x^2 - 3a^2 \Rightarrow y'' = 6x$, so $y'' > 0$ on $(0, \infty)$ and $y'' < 0$ on $(-\infty, 0)$. This tells us that f is CU on $(0, \infty)$ and CD on $(-\infty, 0)$. There is an inflection point at $(0, 2a^3)$. The graph illustrates these features.

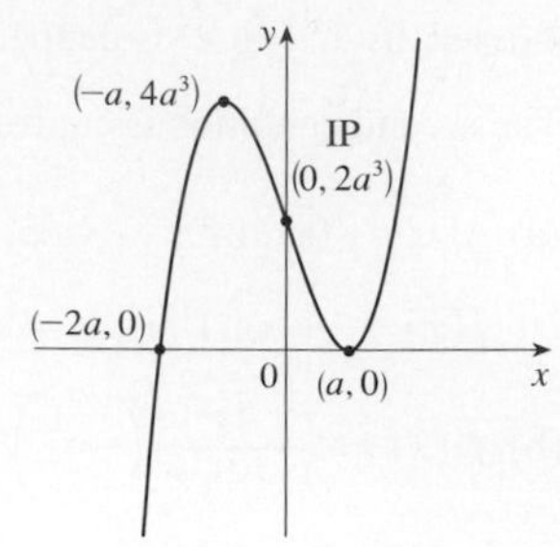

What the curves in the family have in common is that they are all CD on $(-\infty, 0)$, CU on $(0, \infty)$, and have the same basic shape. But as a increases, the four key points shown in the figure move further away from the origin.

59. (a)

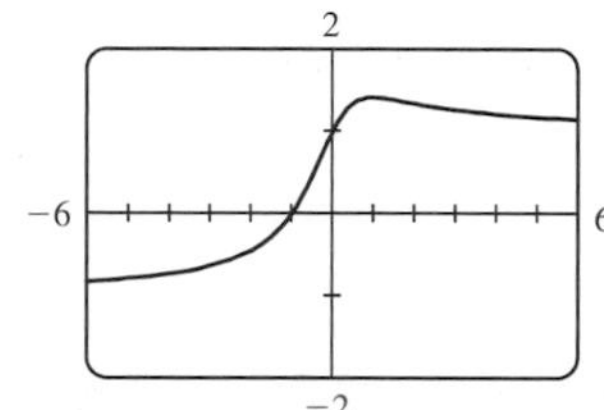

From the graph, we get an estimate of $f(1) \approx 1.41$ as a local maximum value, and no local minimum value.

$f(x) = \dfrac{x+1}{\sqrt{x^2+1}} \Rightarrow f'(x) = \dfrac{1-x}{(x^2+1)^{3/2}}$.

$f'(x) = 0 \Leftrightarrow x = 1$. $f(1) = \frac{2}{\sqrt{2}} = \sqrt{2}$ is the exact value.

(b) From the graph in part (a), f increases most rapidly somewhere between $x = -\frac{1}{2}$ and $x = -\frac{1}{4}$. To find the exact value, we need to find the maximum value of f', which we can do by finding the critical numbers of f'.

$f''(x) = \dfrac{2x^2 - 3x - 1}{(x^2+1)^{5/2}} = 0 \Leftrightarrow x = \dfrac{3 \pm \sqrt{17}}{4}$. $x = \dfrac{3+\sqrt{17}}{4}$ corresponds to the *minimum* value of f'.

The maximum value of f' occurs at $x = \frac{3-\sqrt{17}}{4} \approx -0.28$.

60. (a)

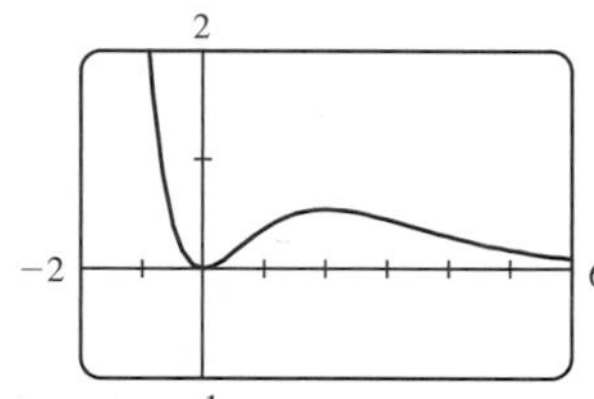

Tracing the graph gives us estimates of $f(0) = 0$ for a local minimum value and $f(2) = 0.54$ for a local maximum value.

$f(x) = x^2 e^{-x} \Rightarrow f'(x) = xe^{-x}(2-x)$. $f'(x) = 0 \Leftrightarrow x = 0$ or 2.

$f(0) = 0$ and $f(2) = 4e^{-2}$ are the exact values.

(b) From the graph in part (a), f increases most rapidly around $x = \frac{3}{4}$. To find the exact value, we need to find the maximum value of f', which we can do by finding the critical numbers of f'. $f''(x) = e^{-x}(x^2 - 4x + 2) = 0 \Rightarrow x = 2 \pm \sqrt{2}$. $x = 2 + \sqrt{2}$ corresponds to the *minimum* value of f'. The maximum value of f' is at $\left(2-\sqrt{2}, (2-\sqrt{2})^2 e^{-2+\sqrt{2}}\right) \approx (0.59, 0.19)$.

61. $f(x) = \sin 2x + \sin 4x \Rightarrow f'(x) = 2\cos 2x + 4\cos 4x \Rightarrow f''(x) = -4\sin 2x - 16\sin 4x$

(a) From the graph of f, it seems that f is CD on $(0, 0.8)$, CU on $(0.8, 1.6)$, CD on $(1.6, 2.3)$, and CU on $(2.3, \pi)$. The inflection points appear to be at $(0.8, 0.7)$, $(1.6, 0)$, and $(2.3, -0.7)$.

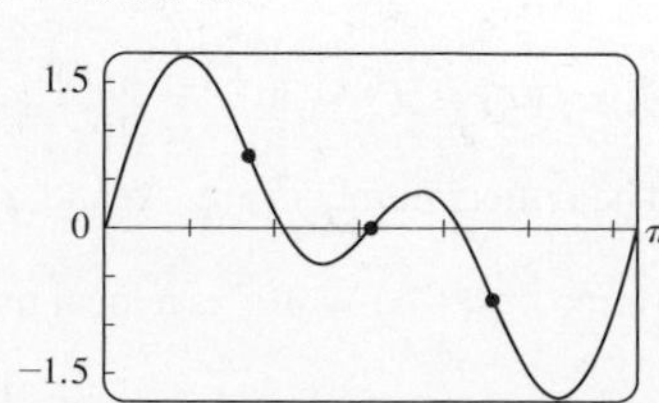

(b) From the graph of f'' (and zooming in near the zeros), it seems that f is CD on $(0, 0.85)$, CU on $(0.85, 1.57)$, CD on $(1.57, 2.29)$, and CU on $(2.29, \pi)$. Refined estimates of the inflection points are $(0.85, 0.74)$, $(1.57, 0)$, and $(2.29, -0.74)$.

62. $f(x) = (x-1)^2(x+1)^3 \quad \Rightarrow$

$$\begin{aligned} f'(x) &= (x-1)^2 3(x+1)^2 + (x+1)^3 2(x-1) \\ &= (x-1)(x+1)^2\,[3(x-1) + 2(x+1)] = (x-1)(x+1)^2(5x-1) \quad \Rightarrow \end{aligned}$$

$$\begin{aligned} f''(x) &= (1)(x+1)^2(5x-1) + (x-1)(2)(x+1)(5x-1) + (x-1)(x+1)^2(5) \\ &= (x+1)[(x+1)(5x-1) + 2(x-1)(5x-1) + 5(x-1)(x+1)] \\ &= (x+1)[5x^2 + 4x - 1 + 10x^2 - 12x + 2 + 5x^2 - 5] \\ &= (x+1)(20x^2 - 8x - 4) = 4(x+1)(5x^2 - 2x - 1) \end{aligned}$$

(a) From the graph of f, it seems that f is CD on $(-\infty, -1)$, CU on $(-1, -0.3)$, CD on $(-0.3, 0.7)$, and CU on $(0.7, \infty)$. The inflection points appear to be at $(-1, 0)$, $(-0.3, 0.6)$, and $(0.7, 0.5)$.

(b) From the graph of f'' (and zooming in near the zeros), it seems that f is CD on $(-1, 0)$, CU on $(-1, -0.29)$, CD on $(-0.29, 0.69)$, and CU on $(0.69, \infty)$. Refined estimates of the inflection points are $(-1, 0)$, $(-0.29, 0.60)$, and $(0.69, 0.46)$.

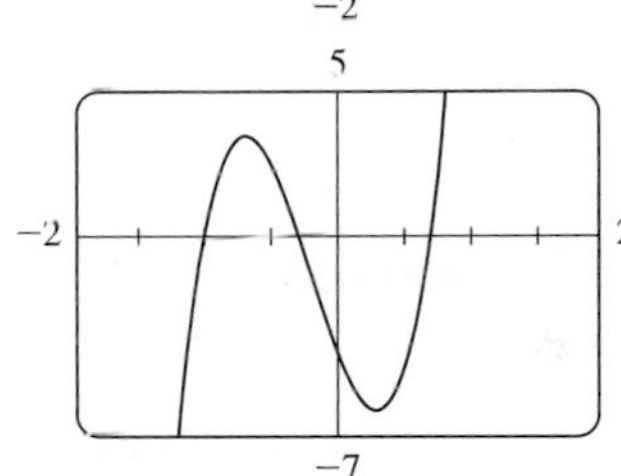

63. $f(x) = \dfrac{x^4 + x^3 + 1}{\sqrt{x^2 + x + 1}}$. In Maple, we define f and then use the command `plot(diff(diff(f,x),x),x=-2..2);`. In Mathematica, we define f and then use `Plot[Dt[Dt[f,x],x],{x,-2,2}]`. We see that $f'' > 0$ for $x < -0.6$ and $x > 0.0$ $[\approx 0.03]$ and $f'' < 0$ for $-0.6 < x < 0.0$. So f is CU on $(-\infty, -0.6)$ and $(0.0, \infty)$ and CD on $(-0.6, 0.0)$.

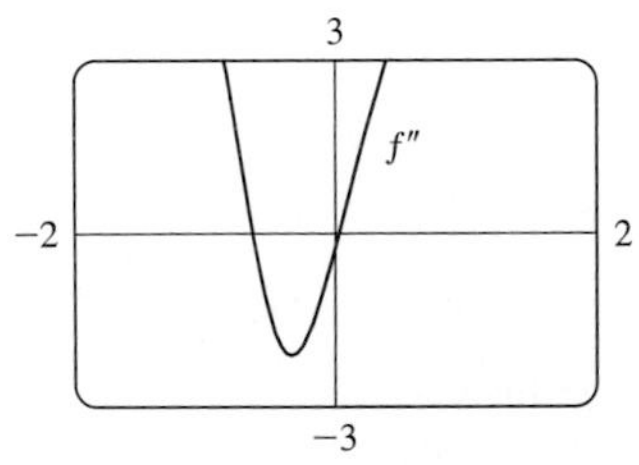

64. $f(x) = \dfrac{x^2 \tan^{-1} x}{1 + x^3}$. It appears that f'' is positive (and thus f is concave upward) on $(-\infty, -1)$, $(0, 0.7)$, and $(2.5, \infty)$; and f'' is negative (and thus f is concave downward) on $(-1, 0)$ and $(0.7, 2.5)$.

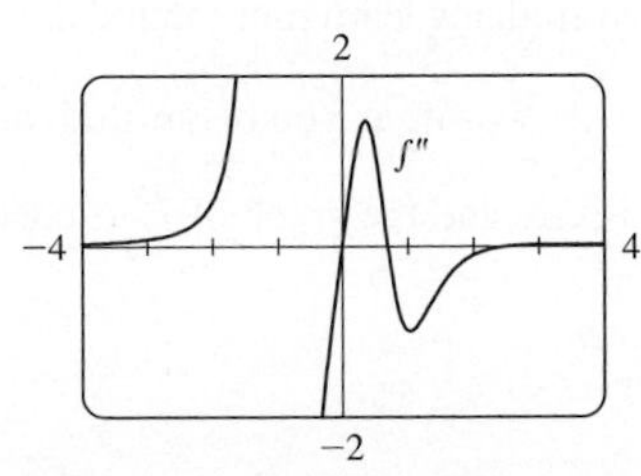

65. (a) The rate of increase of the population is initially very small, then gets larger until it reaches a maximum at about $t = 8$ hours, and decreases toward 0 as the population begins to level off.

(b) The rate of increase has its maximum value at $t = 8$ hours.

(c) The population function is concave upward on $(0, 8)$ and concave downward on $(8, 18)$.

(d) At $t = 8$, the population is about 350, so the inflection point is about $(8, 350)$.

66. If $S(t)$ is the average SAT score as a function of time t, then $S'(t) < 0$ (since the SAT scores are declining) and $S''(t) > 0$ (since the rate of decrease of the scores is increasing—becoming less negative).

67. If $D(t)$ is the size of the national deficit as a function of time t, then at the time of the speech $D'(t) > 0$ (since the deficit is increasing), and $D''(t) < 0$ (since the rate of increase of the deficit is decreasing).

68. (a) I'm very unhappy. It's uncomfortably hot and $f'(3) = 2$ indicates that the temperature is increasing, and $f''(3) = 4$ indicates that the rate of increase is increasing. (The temperature is rapidly getting warmer.)

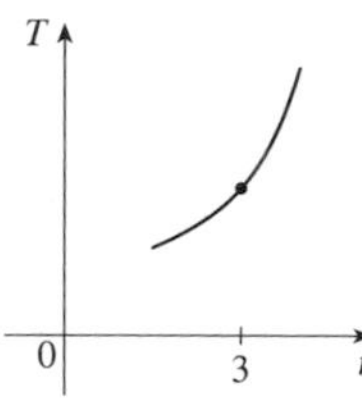

(b) I'm still unhappy, but not as unhappy as in part (a). It's uncomfortably hot and $f'(3) = 2$ indicates that the temperature is increasing, but $f''(3) = -4$ indicates that the rate of increase is decreasing. (The temperature is slowly getting warmer.)

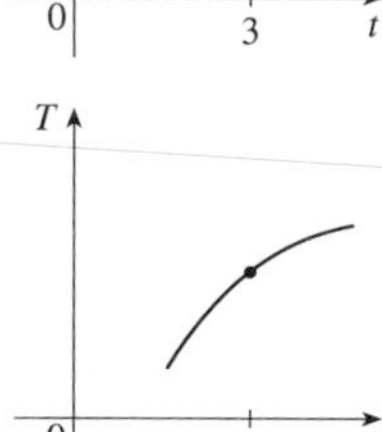

(c) I'm somewhat happy. It's uncomfortably hot and $f'(3) = -2$ indicates that the temperature is decreasing, but $f''(3) = 4$ indicates that the rate of change is increasing. (The rate of change is negative but it's becoming less negative. The temperature is slowly getting cooler.)

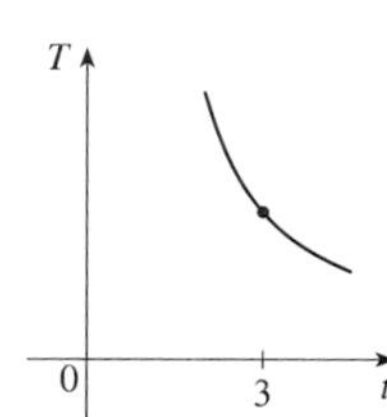

(d) I'm very happy. It's uncomfortably hot and $f'(3) = -2$ indicates that the temperature is decreasing, and $f''(3) = -4$ indicates that the rate of change is decreasing, that is, becoming more negative. (The temperature is rapidly getting cooler.)

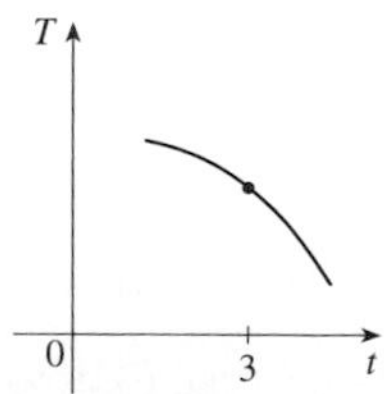

69. Most students learn more in the third hour of studying than in the eighth hour, so $K(3) - K(2)$ is larger than $K(8) - K(7)$. In other words, as you begin studying for a test, the rate of knowledge gain is large and then starts to taper off, so $K'(t)$ decreases and the graph of K is concave downward.

70. At first the depth increases slowly because the base of the mug is wide. But as the mug narrows, the coffee rises more quickly. Thus, the depth d increases at an increasing rate and its graph is concave upward. The rate of increase of d has a maximum where the mug is narrowest; that is, when the mug is half full. It is there that the inflection point (IP) occurs. Then the rate of increase of d starts to decrease as the mug widens and the graph becomes concave down.

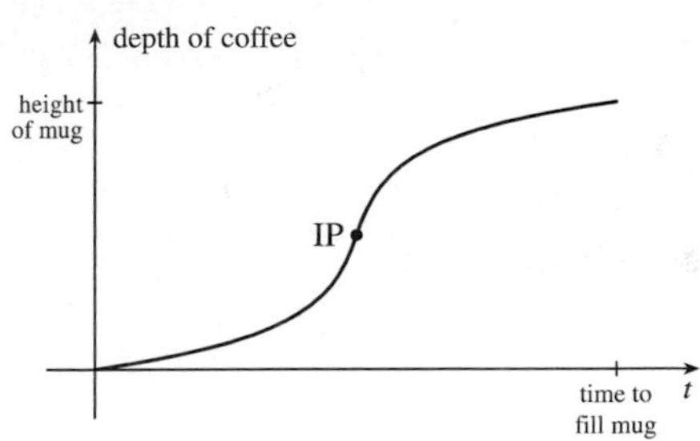

71. $S(t) = At^p e^{-kt}$ with $A = 0.01$, $p = 4$, and $k = 0.07$. We will find the zeros of f'' for $f(t) = t^p e^{-kt}$.

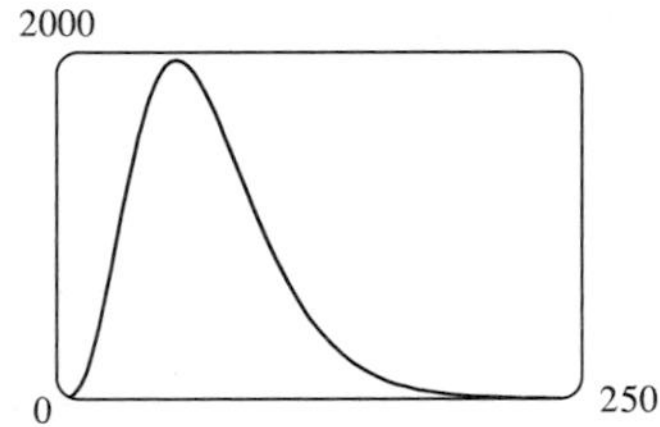

$f'(t) = t^p(-ke^{-kt}) + e^{-kt}(pt^{p-1}) = e^{-kt}(-kt^p + pt^{p-1})$

$$\begin{aligned} f''(t) &= e^{-kt}(-kpt^{p-1} + p(p-1)t^{p-2}) + (-kt^p + pt^{p-1})(-ke^{-kt}) \\ &= t^{p-2}e^{-kt}[-kpt + p(p-1) + k^2t^2 - kpt] \\ &= t^{p-2}e^{-kt}(k^2t^2 - 2kpt + p^2 - p) \end{aligned}$$

Using the given values of p and k gives us $f''(t) = t^2 e^{-0.07t}(0.0049t^2 - 0.56t + 12)$. So $S''(t) = 0.01f''(t)$ and its zeros are $t = 0$ and the solutions of $0.0049t^2 - 0.56t + 12 = 0$, which are $t_1 = \frac{200}{7} \approx 28.57$ and $t_2 = \frac{600}{7} \approx 85.71$.

At t_1 minutes, the rate of increase of the level of medication in the bloodstream is at its greatest and at t_2 minutes, the rate of decrease is the greatest.

72. (a) As $|x| \to \infty$, $t = -x^2/(2\sigma^2) \to -\infty$, and $e^t \to 0$. The HA is $y = 0$. Since t takes on its maximum value at $x = 0$, so does e^t. Showing this result using derivatives, we have $f(x) = e^{-x^2/(2\sigma^2)} \Rightarrow f'(x) = e^{-x^2/(2\sigma^2)}(-x/\sigma^2)$.

$f'(x) = 0 \Leftrightarrow x = 0$. Because f' changes from positive to negative at $x = 0$, $f(0) = 1$ is a local maximum. For inflection points, we find $f''(x) = -\dfrac{1}{\sigma^2}\left[e^{-x^2/(2\sigma^2)} \cdot 1 + xe^{-x^2/(2\sigma^2)}(-x/\sigma^2)\right] = \dfrac{-1}{\sigma^2}e^{-x^2/(2\sigma^2)}(1 - x^2/\sigma^2)$.

$f''(x) = 0 \Leftrightarrow x^2 = \sigma^2 \Leftrightarrow x = \pm\sigma$. $f''(x) < 0 \Leftrightarrow x^2 < \sigma^2 \Leftrightarrow -\sigma < x < \sigma$.

So f is CD on $(-\sigma, \sigma)$ and CU on $(-\infty, -\sigma)$ and (σ, ∞). IP at $(\pm\sigma, e^{-1/2})$.

(b) Since we have IP at $x = \pm\sigma$, the inflection points move away from the y-axis as σ increases.

(c)

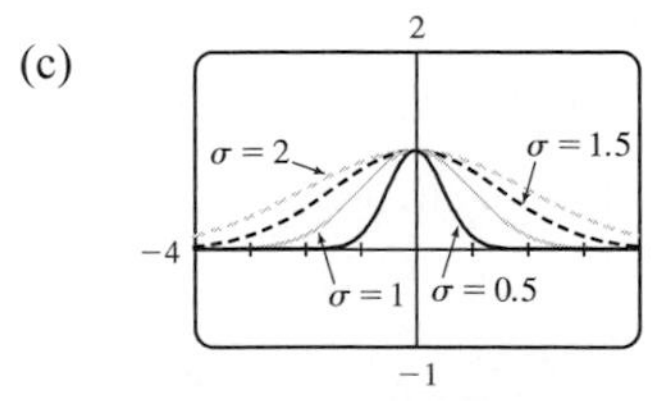

From the graph, we see that as σ increases, the graph tends to spread out and there is more area between the curve and the x-axis.

73. $f(x) = ax^3 + bx^2 + cx + d \Rightarrow f'(x) = 3ax^2 + 2bx + c$.

We are given that $f(1) = 0$ and $f(-2) = 3$, so $f(1) = a + b + c + d = 0$ and $f(-2) = -8a + 4b - 2c + d = 3$. Also $f'(1) = 3a + 2b + c = 0$ and $f'(-2) = 12a - 4b + c = 0$ by Fermat's Theorem. Solving these four equations, we get $a = \frac{2}{9}$, $b = \frac{1}{3}$, $c = -\frac{4}{3}$, $d = \frac{7}{9}$, so the function is $f(x) = \frac{1}{9}(2x^3 + 3x^2 - 12x + 7)$.

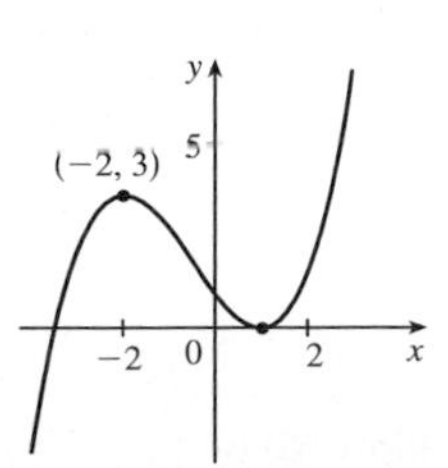

74. $f(x) = axe^{bx^2} \Rightarrow f'(x) = a\left[xe^{bx^2} \cdot 2bx + e^{bx^2} \cdot 1\right] = ae^{bx^2}(2bx^2 + 1)$. For $f(2) = 1$ to be a maximum value, we must have $f'(2) = 0$. $f(2) = 1 \Rightarrow 1 = 2ae^{4b}$ and $f'(2) = 0 \Rightarrow 0 = (8b + 1)ae^{4b}$. So $8b + 1 = 0$ $[a \neq 0] \Rightarrow$ $b = -\frac{1}{8}$ and now $1 = 2ae^{-1/2} \Rightarrow a = \sqrt{e}/2$.

75. (a) $f(x) = x^3 + ax^2 + bx \Rightarrow f'(x) = 3x^2 + 2ax + b$. f has the local minimum value $-\frac{2}{9}\sqrt{3}$ at $x = 1/\sqrt{3}$, so $f'(\frac{1}{\sqrt{3}}) = 0 \Rightarrow 1 + \frac{2}{\sqrt{3}}a + b = 0$ **(1)** and $f(\frac{1}{\sqrt{3}}) = -\frac{2}{9}\sqrt{3} \Rightarrow \frac{1}{9}\sqrt{3} + \frac{1}{3}a + \frac{1}{3}\sqrt{3}b = -\frac{2}{9}\sqrt{3}$ **(2)**.

Rewrite the system of equations as

$$\begin{aligned} \tfrac{2}{3}\sqrt{3}a &+ b &= -1 \quad &\textbf{(3)} \\ \tfrac{1}{3}a &+ \tfrac{1}{3}\sqrt{3}b &= -\tfrac{1}{3}\sqrt{3} \quad &\textbf{(4)} \end{aligned}$$

and then multiplying **(4)** by $-2\sqrt{3}$ gives us the system

$$\begin{aligned} \tfrac{2}{3}\sqrt{3}a &+ b &= -1 \\ -\tfrac{2}{3}\sqrt{3}a &- 2b &= 2 \end{aligned}$$

Adding the equations gives us $-b = 1 \Rightarrow b = -1$. Substituting -1 for b into **(3)** gives us $\frac{2}{3}\sqrt{3}a - 1 = -1 \Rightarrow \frac{2}{3}\sqrt{3}a = 0 \Rightarrow a = 0$. Thus, $f(x) = x^3 - x$.

(b) To find the smallest slope, we want to find the minimum of the slope function, f', so we'll find the critical numbers of f'. $f(x) = x^3 - x \Rightarrow f'(x) = 3x^2 - 1 \Rightarrow f''(x) = 6x$. $f''(x) = 0 \Leftrightarrow x = 0$. At $x = 0$, $y = 0$, $f'(x) = -1$, and f'' changes from negative to positive. Thus, we have a minimum for f' and $y - 0 = -1(x - 0)$, or $y = -x$, is the tangent line that has the smallest slope.

76. The original equation can be written as $(x^2 + b)y + ax = 0$. Call this **(1)**. Since $(2, 2.5)$ is on this curve, we have $(4 + b)(\frac{5}{2}) + 2a = 0$, or $20 + 5b + 4a = 0$. Let's rewrite that as $4a + 5b = -20$ and call it **(A)**. Differentiating **(1)** gives (after regrouping) $(x^2 + b)y' = -(2xy + a)$. Call this **(2)**. Differentiating again gives $(x^2 + b)y'' + (2x)y' = -2xy' - 2y$, or $(x^2 + b)y'' + 4xy' + 2y = 0$. Call this **(3)**. At $(2, 2.5)$, equations **(2)** and **(3)** say that $(4 + b)y' = -(10 + a)$ and $(4 + b)y'' + 8y' + 5 = 0$. If $(2, 2.5)$ is an inflection point, then $y'' = 0$ there, so the second condition becomes $8y' + 5 = 0$, or $y' = -\frac{5}{8}$. Substituting in the first condition, we get $-(4 + b)\frac{5}{8} = -(10 + a)$, or $20 + 5b = 80 + 8a$, which simplifies to $-8a + 5b = 60$. Call this **(B)**. Subtracting **(B)** from **(A)** yields $12a = -80$, so $a = -\frac{20}{3}$. Substituting that value in **(A)** gives $-\frac{80}{3} + 5b = -20 = -\frac{60}{3}$, so $5b = \frac{20}{3}$ and $b = \frac{4}{3}$. Thus far we've shown that IF the curve has an inflection point at $(2, 2.5)$, then $a = -\frac{20}{3}$ and $b = \frac{4}{3}$.

To prove the converse, suppose that $a = -\frac{20}{3}$ and $b = \frac{4}{3}$. Then by **(1)**, **(2)**, and **(3)**, our curve satisfies

$$(x^2 + \tfrac{4}{3})y = \tfrac{20}{3}x \qquad \textbf{(4)}$$

$$(x^2 + \tfrac{4}{3})y' = -2xy + \tfrac{20}{3} \qquad \textbf{(5)}$$

and

$$(x^2 + \tfrac{4}{3})y'' + 4xy' + 2y = 0. \qquad \textbf{(6)}$$

Multiply **(6)** by $(x^2 + \frac{4}{3})$ and substitute from **(4)** and **(5)** to obtain $(x^2 + \frac{4}{3})^2 y'' + 4x(-2xy + \frac{20}{3}) + 2(\frac{20}{3}x) = 0$, or

$\left(x^2+\frac{4}{3}\right)^2 y''-8x^2y+40x=0$. Now multiply by (x^2+b) again and substitute from the first equation to obtain $\left(x^2+\frac{4}{3}\right)^3 y''-8x^2\left(\frac{20}{3}x\right)+40x\left(x^2+\frac{4}{3}\right)=0$, or $\left(x^2+\frac{4}{3}\right)^3 y''-\frac{40}{3}(x^3-4x)=0$. The coefficient of y'' is positive, so the sign of y'' is the same as the sign of $\frac{40}{3}(x^3-4x)$, which is a positive multiple of $x(x+2)(x-2)$. It is clear from this expression that y'' changes sign at $x=0$, $x=-2$, and $x=2$, so the curve changes its direction of concavity at those values of x. By **(4)**, the corresponding y-values are 0, -2.5, and 2.5, respectively. Thus when $a=-\frac{20}{3}$ and $b=\frac{4}{3}$, the curve has inflection points, not only at $(2,2.5)$, but also at $(0,0)$ and $(-2,-2.5)$.

77. $y=\dfrac{1+x}{1+x^2} \quad\Rightarrow\quad y'=\dfrac{(1+x^2)(1)-(1+x)(2x)}{(1+x^2)^2}=\dfrac{1-2x-x^2}{(1+x^2)^2} \quad\Rightarrow$

$$y''=\frac{(1+x^2)^2(-2-2x)-(1-2x-x^2)\cdot 2(1+x^2)(2x)}{[(1+x^2)^2]^2}=\frac{2(1+x^2)[(1+x^2)(-1-x)-(1-2x-x^2)(2x)]}{(1+x^2)^4}$$

$$=\frac{2(-1-x-x^2-x^3-2x+4x^2+2x^3)}{(1+x^2)^3}=\frac{2(x^3+3x^2-3x-1)}{(1+x^2)^3}=\frac{2(x-1)(x^2+4x+1)}{(1+x^2)^3}$$

So $y''=0 \;\Rightarrow\; x=1,\,-2\pm\sqrt{3}$. Let $a=-2-\sqrt{3}$, $b=-2+\sqrt{3}$, and $c=1$. We can show that $f(a)=\frac{1}{4}\left(1-\sqrt{3}\right)$, $f(b)=\frac{1}{4}\left(1+\sqrt{3}\right)$, and $f(c)=1$. To show that these three points of inflection lie on one straight line, we'll show that the slopes m_{ac} and m_{bc} are equal.

$$m_{ac}=\frac{f(c)-f(a)}{c-a}=\frac{1-\frac{1}{4}\left(1-\sqrt{3}\right)}{1-\left(-2-\sqrt{3}\right)}=\frac{\frac{3}{4}+\frac{1}{4}\sqrt{3}}{3+\sqrt{3}}=\frac{1}{4}$$

$$m_{bc}=\frac{f(c)-f(b)}{c-b}=\frac{1-\frac{1}{4}\left(1+\sqrt{3}\right)}{1-\left(-2+\sqrt{3}\right)}=\frac{\frac{3}{4}-\frac{1}{4}\sqrt{3}}{3-\sqrt{3}}=\frac{1}{4}$$

78. $y=f(x)=e^{-x}\sin x \quad\Rightarrow\quad y'=e^{-x}\cos x+\sin x(-e^{-x})=e^{-x}(\cos x-\sin x) \quad\Rightarrow$

$y''=e^{-x}(-\sin x-\cos x)+(\cos x-\sin x)(-e^{-x})=e^{-x}(-\sin x-\cos x-\cos x+\sin x)=e^{-x}(-2\cos x)$.

So $y''=0 \;\Rightarrow\; \cos x=0 \;\Rightarrow\; x=\frac{\pi}{2}+n\pi$. At these values of x, f has points of inflection and since $\sin\left(\frac{\pi}{2}+n\pi\right)=\pm1$, we get $y=\pm e^{-x}$, so f intersects the other curves at its inflection points.

Let $g(x)=e^{-x}$ and $h(x)=-e^{-x}$, so that $g'(x)=-e^{-x}$ and $h'(x)=e^{-x}$. Now

$f'\left(\frac{\pi}{2}+n\pi\right)=e^{-(\pi/2+n\pi)}\left[\cos\left(\frac{\pi}{2}+n\pi\right)-\sin\left(\frac{\pi}{2}+n\pi\right)\right]=-e^{-(\pi/2+n\pi)}\sin\left(\frac{\pi}{2}+n\pi\right)$. If n is odd, then $f'\left(\frac{\pi}{2}+n\pi\right)=e^{-(\pi/2+n\pi)}=h'\left(\frac{\pi}{2}+n\pi\right)$. If n is even, then $f'\left(\frac{\pi}{2}+n\pi\right)=-e^{-(\pi/2+n\pi)}=g'\left(\frac{\pi}{2}+n\pi\right)$.

Thus, at $x=\frac{\pi}{2}+n\pi$, f has the same slope as either g or h, and hence, g and h touch f at its inflection points.

79. $y=x\sin x \;\Rightarrow\; y'=x\cos x+\sin x \;\Rightarrow\; y''=-x\sin x+2\cos x$. $\;y''=0 \;\Rightarrow\; 2\cos x=x\sin x$ [which is y] $\;\Rightarrow$ $(2\cos x)^2=(x\sin x)^2 \;\Rightarrow\; 4\cos^2 x=x^2\sin^2 x \;\Rightarrow\; 4\cos^2 x=x^2(1-\cos^2 x) \;\Rightarrow\; 4\cos^2 x+x^2\cos^2 x=x^2 \;\Rightarrow$ $\cos^2 x(4+x^2)=x^2 \;\Rightarrow\; 4\cos^2 x(x^2+4)=4x^2 \;\Rightarrow\; y^2(x^2+4)=4x^2$ since $y=2\cos x$ when $y''=0$.

80. (a) We will make use of the converse of the Concavity Test (along with the stated assumptions); that is, if f is concave upward on I, then $f''>0$ on I. If f and g are CU on I, then $f''>0$ and $g''>0$ on I, so $(f+g)''=f''+g''>0$ on $I \;\Rightarrow$ $f+g$ is CU on I.

(b) Since f is positive and CU on I, $f > 0$ and $f'' > 0$ on I. So $g(x) = [f(x)]^2 \Rightarrow g' = 2ff' \Rightarrow$ $g'' = 2f'f' + 2ff'' = 2(f')^2 + 2ff'' > 0 \Rightarrow g$ is CU on I.

81. (a) Since f and g are positive, increasing, and CU on I with f'' and g'' never equal to 0, we have $f > 0$, $f' \geq 0$, $f'' > 0$, $g > 0$, $g' \geq 0$, $g'' > 0$ on I. Then $(fg)' = f'g + fg' \Rightarrow (fg)'' = f''g + 2f'g' + fg'' \geq f''g + fg'' > 0$ on I $\Rightarrow$ fg is CU on I.

(b) In part (a), if f and g are both decreasing instead of increasing, then $f' \leq 0$ and $g' \leq 0$ on I, so we still have $2f'g' \geq 0$ on I. Thus, $(fg)'' = f''g + 2f'g' + fg'' \geq f''g + fg'' > 0$ on I $\Rightarrow$ fg is CU on I as in part (a).

(c) Suppose f is increasing and g is decreasing [with f and g positive and CU]. Then $f' \geq 0$ and $g' \leq 0$ on I, so $2f'g' \leq 0$ on I and the argument in parts (a) and (b) fails.

Example 1. $I = (0, \infty)$, $f(x) = x^3$, $g(x) = 1/x$. Then $(fg)(x) = x^2$, so $(fg)'(x) = 2x$ and $(fg)''(x) = 2 > 0$ on I. Thus, fg is CU on I.

Example 2. $I = (0, \infty)$, $f(x) = 4x\sqrt{x}$, $g(x) = 1/x$. Then $(fg)(x) = 4\sqrt{x}$, so $(fg)'(x) = 2/\sqrt{x}$ and $(fg)''(x) = -1/\sqrt{x^3} < 0$ on I. Thus, fg is CD on I.

Example 3. $I = (0, \infty)$, $f(x) = x^2$, $g(x) = 1/x$. Thus, $(fg)(x) = x$, so fg is linear on I.

82. Since f and g are CU on $(-\infty, \infty)$, $f'' > 0$ and $g'' > 0$ on $(-\infty, \infty)$. $h(x) = f(g(x)) \Rightarrow$ $h'(x) = f'(g(x))\,g'(x) \Rightarrow h''(x) = f''(g(x))\,g'(x)\,g'(x) + f'(g(x))\,g''(x) = f''(g(x))[g'(x)]^2 + f'(g(x))\,g''(x) > 0$ if $f' > 0$. So h is CU if f is increasing.

83. $f(x) = \tan x - x \Rightarrow f'(x) = \sec^2 x - 1 > 0$ for $0 < x < \frac{\pi}{2}$ since $\sec^2 x > 1$ for $0 < x < \frac{\pi}{2}$. So f is increasing on $\left(0, \frac{\pi}{2}\right)$. Thus, $f(x) > f(0) = 0$ for $0 < x < \frac{\pi}{2} \Rightarrow \tan x - x > 0 \Rightarrow \tan x > x$ for $0 < x < \frac{\pi}{2}$.

84. (a) Let $f(x) = e^x - 1 - x$. Now $f(0) = e^0 - 1 = 0$, and for $x \geq 0$, we have $f'(x) = e^x - 1 \geq 0$. Now, since $f(0) = 0$ and f is increasing on $[0, \infty)$, $f(x) \geq 0$ for $x \geq 0 \Rightarrow e^x - 1 - x \geq 0 \Rightarrow e^x \geq 1 + x$.

(b) Let $f(x) = e^x - 1 - x - \frac{1}{2}x^2$. Thus, $f'(x) = e^x - 1 - x$, which is positive for $x \geq 0$ by part (a). Thus, $f(x)$ is increasing on $(0, \infty)$, so on that interval, $0 = f(0) \leq f(x) = e^x - 1 - x - \frac{1}{2}x^2 \Rightarrow e^x \geq 1 + x + \frac{1}{2}x^2$.

(c) By part (a), the result holds for $n = 1$. Suppose that $e^x \geq 1 + x + \dfrac{x^2}{2!} + \cdots + \dfrac{x^k}{k!}$ for $x \geq 0$.

Let $f(x) = e^x - 1 - x - \dfrac{x^2}{2!} - \cdots - \dfrac{x^k}{k!} - \dfrac{x^{k+1}}{(k+1)!}$. Then $f'(x) = e^x - 1 - x - \cdots - \dfrac{x^k}{k!} \geq 0$ by assumption. Hence, $f(x)$ is increasing on $(0, \infty)$. So $0 \leq x$ implies that $0 = f(0) \leq f(x) = e^x - 1 - x - \cdots - \dfrac{x^k}{k!} - \dfrac{x^{k+1}}{(k+1)!}$, and hence $e^x \geq 1 + x + \cdots + \dfrac{x^k}{k!} + \dfrac{x^{k+1}}{(k+1)!}$ for $x \geq 0$. Therefore, for $x \geq 0$, $e^x \geq 1 + x + \dfrac{x^2}{2!} + \cdots + \dfrac{x^n}{n!}$ for every positive integer n, by mathematical induction.

85. Let the cubic function be $f(x) = ax^3 + bx^2 + cx + d \Rightarrow f'(x) = 3ax^2 + 2bx + c \Rightarrow f''(x) = 6ax + 2b$.

So f is CU when $6ax + 2b > 0 \Leftrightarrow x > -b/(3a)$, CD when $x < -b/(3a)$, and so the only point of inflection occurs when $x = -b/(3a)$. If the graph has three x-intercepts x_1, x_2 and x_3, then the expression for $f(x)$ must factor as $f(x) = a(x - x_1)(x - x_2)(x - x_3)$. Multiplying these factors together gives us

$$f(x) = a[x^3 - (x_1 + x_2 + x_3)x^2 + (x_1x_2 + x_1x_3 + x_2x_3)x - x_1x_2x_3]$$

Equating the coefficients of the x^2-terms for the two forms of f gives us $b = -a(x_1 + x_2 + x_3)$. Hence, the x-coordinate of the point of inflection is $-\dfrac{b}{3a} = -\dfrac{-a(x_1 + x_2 + x_3)}{3a} = \dfrac{x_1 + x_2 + x_3}{3}$.

86. $P(x) = x^4 + cx^3 + x^2 \Rightarrow P'(x) = 4x^3 + 3cx^2 + 2x \Rightarrow P''(x) = 12x^2 + 6cx + 2$. The graph of $P''(x)$ is a parabola. If $P''(x)$ has two roots, then it changes sign twice and so has two inflection points. This happens when the discriminant of $P''(x)$ is positive, that is, $(6c)^2 - 4 \cdot 12 \cdot 2 > 0 \Leftrightarrow 36c^2 - 96 > 0 \Leftrightarrow |c| > \frac{2\sqrt{6}}{3} \approx 1.63$. If $36c^2 - 96 = 0 \Leftrightarrow c = \pm\frac{2\sqrt{6}}{3}$, $P''(x)$ is 0 at one point, but there is still no inflection point since $P''(x)$ never changes sign, and if $36c^2 - 96 < 0 \Leftrightarrow |c| < \frac{2\sqrt{6}}{3}$, then $P''(x)$ never changes sign, and so there is no inflection point.

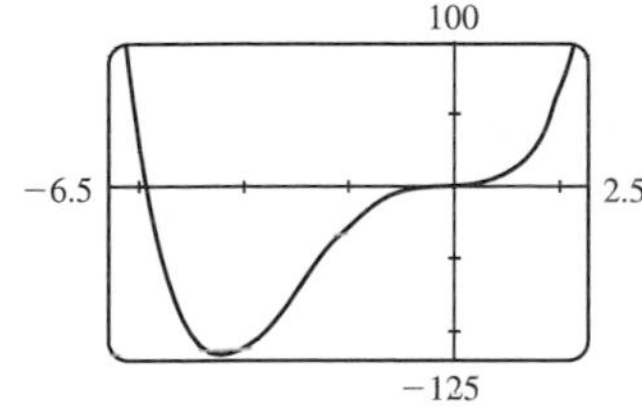

$c = 6$

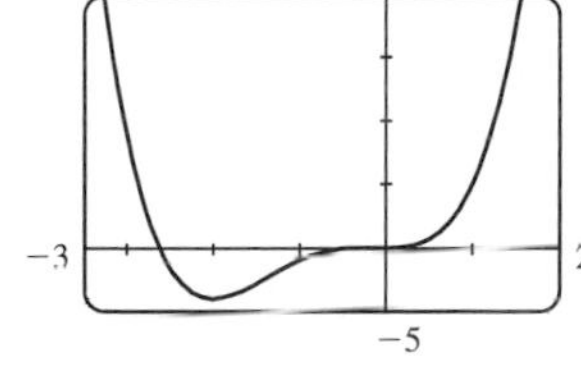

$c = 3$

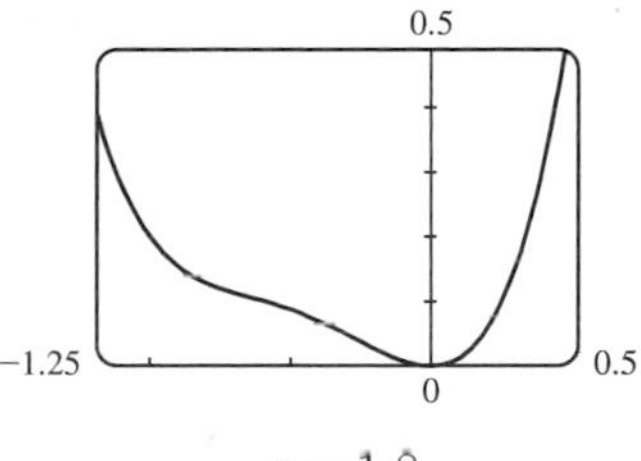

$c = 1.8$

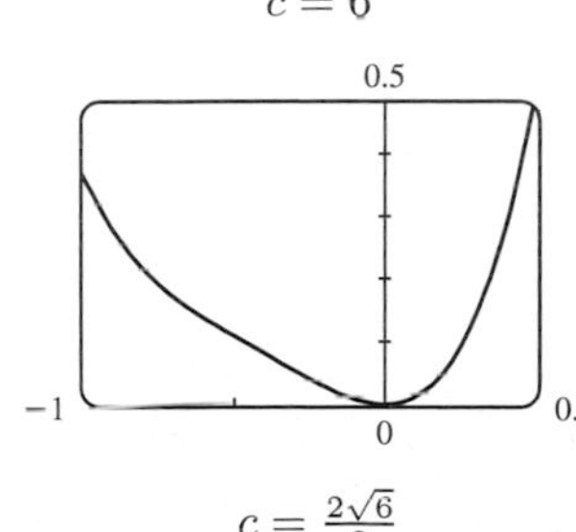

$c = \frac{2\sqrt{6}}{3}$

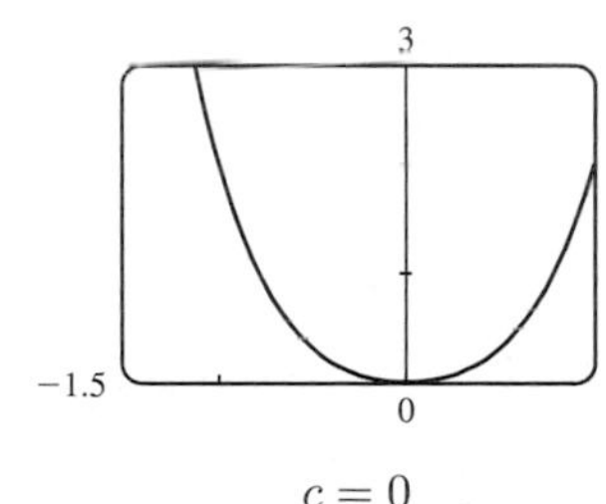

$c = 0$

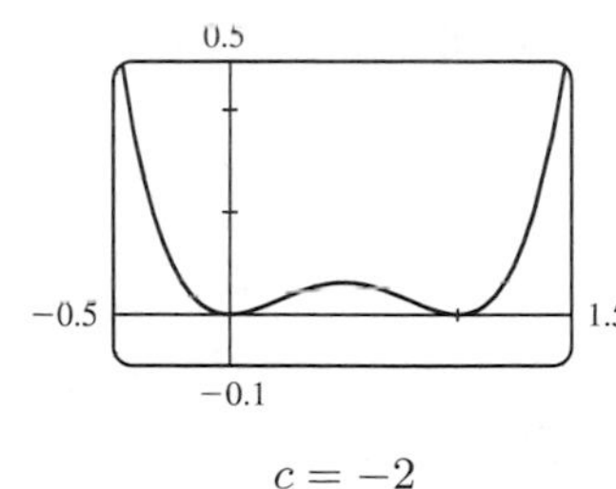

$c = -2$

For large positive c, the graph of f has two inflection points and a large dip to the left of the y-axis. As c decreases, the graph of f becomes flatter for $x < 0$, and eventually the dip rises above the x-axis, and then disappears entirely, along with the inflection points. As c continues to decrease, the dip and the inflection points reappear, to the right of the origin.

87. By hypothesis $g = f'$ is differentiable on an open interval containing c. Since $(c, f(c))$ is a point of inflection, the concavity changes at $x = c$, so $f''(x)$ changes signs at $x = c$. Hence, by the First Derivative Test, f' has a local extremum at $x = c$. Thus, by Fermat's Theorem $f''(c) = 0$.

88. $f(x) = x^4 \Rightarrow f'(x) = 4x^3 \Rightarrow f''(x) = 12x^2 \Rightarrow f''(0) = 0$. For $x < 0$, $f''(x) > 0$, so f is CU on $(-\infty, 0)$; for $x > 0$, $f''(x) > 0$, so f is also CU on $(0, \infty)$. Since f does not change concavity at 0, $(0, 0)$ is not an inflection point.

89. Using the fact that $|x| = \sqrt{x^2}$, we have that $g(x) = x\,|x| = x\,\sqrt{x^2} \;\Rightarrow\; g'(x) = \sqrt{x^2} + \sqrt{x^2} = 2\,\sqrt{x^2} = 2\,|x| \;\Rightarrow$

$g''(x) = 2x\left(x^2\right)^{-1/2} = \dfrac{2x}{|x|} < 0$ for $x < 0$ and $g''(x) > 0$ for $x > 0$, so $(0,0)$ is an inflection point. But $g''(0)$ does not exist.

90. There must exist some interval containing c on which f''' is positive, since $f'''(c)$ is positive and f''' is continuous. On this interval, f'' is increasing (since f''' is positive), so $f'' = (f')'$ changes from negative to positive at c. So by the First Derivative Test, f' has a local minimum at $x = c$ and thus cannot change sign there, so f has no maximum or minimum at c. But since f'' changes from negative to positive at c, f has a point of inflection at c (it changes from concave down to concave up).

91. Suppose that f is differentiable on an interval I and $f'(x) > 0$ for all x in I except $x = c$. To show that f is increasing on I, let x_1, x_2 be two numbers in I with $x_1 < x_2$.

Case 1 $x_1 < x_2 < c$. Let J be the interval $\{x \in I \mid x < c\}$. By applying the Increasing/Decreasing Test to f on J, we see that f is increasing on J, so $f(x_1) < f(x_2)$.

Case 2 $c < x_1 < x_2$. Apply the Increasing/Decreasing Test to f on $K = \{x \in I \mid x > c\}$.

Case 3 $x_1 < x_2 = c$. Apply the proof of the Increasing/Decreasing Test, using the Mean Value Theorem (MVT) on the interval $[x_1, x_2]$ and noting that the MVT does not require f to be differentiable at the endpoints of $[x_1, x_2]$.

Case 4 $c = x_1 < x_2$. Same proof as in Case 3.

Case 5 $x_1 < c < x_2$. By Cases 3 and 4, f is increasing on $[x_1, c]$ and on $[c, x_2]$, so $f(x_1) < f(c) < f(x_2)$.

In all cases, we have shown that $f(x_1) < f(x_2)$. Since x_1, x_2 were any numbers in I with $x_1 < x_2$, we have shown that f is increasing on I.

92. $f(x) = cx + \dfrac{1}{x^2+3} \;\Rightarrow\; f'(x) = c - \dfrac{2x}{(x^2+3)^2}$. $f'(x) > 0 \;\Leftrightarrow\; c > \dfrac{2x}{(x^2+3)^2}$ [call this $g(x)$].

Now f' is positive (and hence f increasing) if $c > g$, so we'll find the maximum value of g.

$$g'(x) = \frac{(x^2+3)^2 \cdot 2 - 2x \cdot 2(x^2+3) \cdot 2x}{[(x^2+3)^2]^2} = \frac{2(x^2+3)[(x^2+3) - 4x^2]}{(x^2+3)^4} = \frac{2(3-3x^2)}{(x^2+3)^3} = \frac{6(1+x)(1-x)}{(x^2+3)^3}.$$

$g'(x) = 0 \;\Leftrightarrow\; x = \pm 1$. $g'(x) > 0$ on $(0,1)$ and $g'(x) < 0$ on $(1,\infty)$, so g is increasing on $(0,1)$ and decreasing on $(1,\infty)$, and hence g has a maximum value on $(0,\infty)$ of $g(1) = \frac{2}{16} = \frac{1}{8}$. Also since $g(x) \le 0$ if $x \le 0$, the maximum value of g on $(-\infty,\infty)$ is $\frac{1}{8}$. Thus, when $c > \frac{1}{8}$, f is increasing. When $c = \frac{1}{8}$, $f'(x) > 0$ on $(-\infty,1)$ and $(1,\infty)$, and hence f is increasing on these intervals. Since f is continuous, we may conclude that f is also increasing on $(-\infty,\infty)$ if $c = \frac{1}{8}$. Therefore, f is increasing on $(-\infty,\infty)$ if $c \ge \frac{1}{8}$.

93. (a) $f(x) = x^4 \sin\frac{1}{x} \Rightarrow f'(x) = x^4 \cos\frac{1}{x}\left(-\frac{1}{x^2}\right) + \sin\frac{1}{x}\,(4x^3) = 4x^3 \sin\frac{1}{x} - x^2 \cos\frac{1}{x}$.

$g(x) = x^4\left(2 + \sin\frac{1}{x}\right) = 2x^4 + f(x) \Rightarrow g'(x) = 8x^3 + f'(x)$.

$h(x) = x^4\left(-2 + \sin\frac{1}{x}\right) = -2x^4 + f(x) \Rightarrow h'(x) = -8x^3 + f'(x)$.

It is given that $f(0) = 0$, so $f'(0) = \lim\limits_{x\to 0}\frac{f(x) - f(0)}{x - 0} = \lim\limits_{x\to 0}\frac{x^4 \sin\frac{1}{x} - 0}{x} = \lim\limits_{x\to 0} x^3 \sin\frac{1}{x}$. Since $-\left|x^3\right| \le x^3 \sin\frac{1}{x} \le \left|x^3\right|$ and $\lim\limits_{x\to 0}\left|x^3\right| = 0$, we see that $f'(0) = 0$ by the Squeeze Theorem. Also, $g'(0) = 8(0)^3 + f'(0) = 0$ and $h'(0) = -8(0)^3 + f'(0) = 0$, so 0 is a critical number of f, g, and h.

For $x_{2n} = \frac{1}{2n\pi}$ [n a nonzero integer], $\sin\frac{1}{x_{2n}} = \sin 2n\pi = 0$ and $\cos\frac{1}{x_{2n}} = \cos 2n\pi = 1$, so $f'(x_{2n}) = -x_{2n}^2 < 0$.

For $x_{2n+1} = \frac{1}{(2n+1)\pi}$, $\sin\frac{1}{x_{2n+1}} = \sin(2n+1)\pi = 0$ and $\cos\frac{1}{x_{2n+1}} = \cos(2n+1)\pi = -1$, so $f'(x_{2n+1}) = x_{2n+1}^2 > 0$. Thus, f' changes sign infinitely often on both sides of 0.

Next, $g'(x_{2n}) = 8x_{2n}^3 + f'(x_{2n}) = 8x_{2n}^3 - x_{2n}^2 = x_{2n}^2(8x_{2n} - 1) < 0$ for $x_{2n} < \frac{1}{8}$, but $g'(x_{2n+1}) = 8x_{2n+1}^3 + x_{2n+1}^2 = x_{2n+1}^2(8x_{2n+1} + 1) > 0$ for $x_{2n+1} > -\frac{1}{8}$, so g' changes sign infinitely often on both sides of 0.

Last, $h'(x_{2n}) = -8x_{2n}^3 + f'(x_{2n}) = -8x_{2n}^3 - x_{2n}^2 = -x_{2n}^2(8x_{2n} + 1) < 0$ for $x_{2n} > -\frac{1}{8}$ and $h'(x_{2n+1}) = -8x_{2n+1}^3 + x_{2n+1}^2 = x_{2n+1}^2(-8x_{2n+1} + 1) > 0$ for $x_{2n+1} < \frac{1}{8}$, so h' changes sign infinitely often on both sides of 0.

(b) $f(0) = 0$ and since $\sin\frac{1}{x}$ and hence $x^4 \sin\frac{1}{x}$ is both positive and negative inifinitely often on both sides of 0, and arbitrarily close to 0, f has neither a local maximum nor a local minimum at 0.

Since $2 + \sin\frac{1}{x} \ge 1$, $g(x) = x^4\left(2 + \sin\frac{1}{x}\right) > 0$ for $x \ne 0$, so $g(0) = 0$ is a local minimum.

Since $-2 + \sin\frac{1}{x} \le -1$, $h(x) = x^4\left(-2 + \sin\frac{1}{x}\right) < 0$ for $x \ne 0$, so $h(0) = 0$ is a local maximum.

4.4 Indeterminate Forms and l'Hospital's Rule

Note: The use of l'Hospital's Rule is indicated by an H above the equal sign: $\overset{\text{H}}{=}$

1. (a) $\lim\limits_{x\to a}\frac{f(x)}{g(x)}$ is an indeterminate form of type $\frac{0}{0}$.

(b) $\lim\limits_{x\to a}\frac{f(x)}{p(x)} = 0$ because the numerator approaches 0 while the denominator becomes large.

(c) $\lim\limits_{x\to a}\frac{h(x)}{p(x)} = 0$ because the numerator approaches a finite number while the denominator becomes large.

(d) If $\lim_{x\to a} p(x) = \infty$ and $f(x) \to 0$ through positive values, then $\lim_{x\to a} \dfrac{p(x)}{f(x)} = \infty$. [For example, take $a = 0$, $p(x) = 1/x^2$, and $f(x) = x^2$.] If $f(x) \to 0$ through negative values, then $\lim_{x\to a} \dfrac{p(x)}{f(x)} = -\infty$. [For example, take $a = 0$, $p(x) = 1/x^2$, and $f(x) = -x^2$.] If $f(x) \to 0$ through both positive and negative values, then the limit might not exist. [For example, take $a = 0$, $p(x) = 1/x^2$, and $f(x) = x$.]

(e) $\lim_{x\to a} \dfrac{p(x)}{q(x)}$ is an indeterminate form of type $\dfrac{\infty}{\infty}$.

2. (a) $\lim_{x\to a} [f(x)p(x)]$ is an indeterminate form of type $0 \cdot \infty$.

(b) When x is near a, $p(x)$ is large and $h(x)$ is near 1, so $h(x)p(x)$ is large. Thus, $\lim_{x\to a} [h(x)p(x)] = \infty$.

(c) When x is near a, $p(x)$ and $q(x)$ are both large, so $p(x)q(x)$ is large. Thus, $\lim_{x\to a} [p(x)q(x)] = \infty$.

3. (a) When x is near a, $f(x)$ is near 0 and $p(x)$ is large, so $f(x) - p(x)$ is large negative. Thus, $\lim_{x\to a} [f(x) - p(x)] = -\infty$.

(b) $\lim_{x\to a} [p(x) - q(x)]$ is an indeterminate form of type $\infty - \infty$.

(c) When x is near a, $p(x)$ and $q(x)$ are both large, so $p(x) + q(x)$ is large. Thus, $\lim_{x\to a} [p(x) + q(x)] = \infty$.

4. (a) $\lim_{x\to a} [f(x)]^{g(x)}$ is an indeterminate form of type 0^0.

(b) If $y = [f(x)]^{p(x)}$, then $\ln y = p(x) \ln f(x)$. When x is near a, $p(x) \to \infty$ and $\ln f(x) \to -\infty$, so $\ln y \to -\infty$. Therefore, $\lim_{x\to a} [f(x)]^{p(x)} = \lim_{x\to a} y = \lim_{x\to a} e^{\ln y} = 0$, provided f^p is defined.

(c) $\lim_{x\to a} [h(x)]^{p(x)}$ is an indeterminate form of type 1^∞.

(d) $\lim_{x\to a} [p(x)]^{f(x)}$ is an indeterminate form of type ∞^0.

(e) If $y = [p(x)]^{q(x)}$, then $\ln y = q(x) \ln p(x)$. When x is near a, $q(x) \to \infty$ and $\ln p(x) \to \infty$, so $\ln y \to \infty$. Therefore, $\lim_{x\to a} [p(x)]^{q(x)} = \lim_{x\to a} y = \lim_{x\to a} e^{\ln y} = \infty$.

(f) $\lim_{x\to a} \sqrt[q(x)]{p(x)} = \lim_{x\to a} [p(x)]^{1/q(x)}$ is an indeterminate form of type ∞^0.

5. From the graphs of f and g, we see that $\lim_{x\to 2} f(x) = 0$ and $\lim_{x\to 2} g(x) = 0$, so l'Hospital's Rule applies.

$$\lim_{x\to 2} \frac{f(x)}{g(x)} = \lim_{x\to 2} \frac{f'(x)}{g'(x)} = \frac{\lim_{x\to 2} f'(x)}{\lim_{x\to 2} g'(x)} = \frac{f'(2)}{g'(2)} = \frac{1.8}{\frac{4}{5}} = \frac{9}{4}$$

6. From the graphs of f and g, we see that $\lim_{x\to 2} f(x) = 0$ and $\lim_{x\to 2} g(x) = 0$, so l'Hospital's Rule applies.

$$\lim_{x\to 2} \frac{f(x)}{g(x)} = \lim_{x\to 2} \frac{f'(x)}{g'(x)} = \frac{\lim_{x\to 2} f'(x)}{\lim_{x\to 2} g'(x)} = \frac{f'(2)}{g'(2)} = \frac{1.5}{-1} = -\frac{3}{2}$$

7. f and $g = e^x - 1$ are differentiable and $g' = e^x \neq 0$ on an open interval that contains 0. $\lim\limits_{x \to 0} f(x) = 0$ and $\lim\limits_{x \to 0} g(x) = 0$, so we have the indeterminate form $\frac{0}{0}$ and can apply l'Hospital's Rule.

$$\lim_{x \to 0} \frac{f(x)}{e^x - 1} \overset{\text{H}}{=} \lim_{x \to 0} \frac{f'(x)}{e^x} = \frac{1}{1} = 1$$

Note that $\lim\limits_{x \to 0} f'(x) = 1$ since the graph of f has the same slope as the line $y = x$ at $x = 0$.

8. This limit has the form $\frac{0}{0}$. $\lim\limits_{x \to 3} \dfrac{x - 3}{x^2 - 9} = \lim\limits_{x \to 3} \dfrac{x - 3}{(x + 3)(x - 3)} = \lim\limits_{x \to 3} \dfrac{1}{x + 3} = \dfrac{1}{3 + 3} = \dfrac{1}{6}$

Note: Alternatively, we could apply l'Hospital's Rule.

9. This limit has the form $\frac{0}{0}$. $\lim\limits_{x \to 4} \dfrac{x^2 - 2x - 8}{x - 4} = \lim\limits_{x \to 4} \dfrac{(x - 4)(x + 2)}{x - 4} = \lim\limits_{x \to 4} (x + 2) = 4 + 2 = 6$

Note: Alternatively, we could apply l'Hospital's Rule.

10. This limit has the form $\frac{0}{0}$. $\lim\limits_{x \to -2} \dfrac{x^3 + 8}{x + 2} \overset{\text{H}}{=} \lim\limits_{x \to -2} \dfrac{3x^2}{1} = 3(-2)^2 = 12$

Note: Alternatively, we could factor and simplify.

11. This limit has the form $\frac{0}{0}$. $\lim\limits_{x \to 1} \dfrac{x^3 - 2x^2 + 1}{x^3 - 1} \overset{\text{H}}{=} \lim\limits_{x \to 1} \dfrac{3x^2 - 4x}{3x^2} = -\dfrac{1}{3}$

Note: Alternatively, we could factor and simplify.

12. This limit has the form $\frac{0}{0}$. $\lim\limits_{x \to 1/2} \dfrac{6x^2 + 5x - 4}{4x^2 + 16x - 9} \overset{\text{H}}{=} \lim\limits_{x \to 1/2} \dfrac{12x + 5}{8x + 16} = \dfrac{6 + 5}{4 + 16} = \dfrac{11}{20}$

Note: Alternatively, we could factor and simplify.

13. This limit has the form $\frac{0}{0}$. $\lim\limits_{x \to (\pi/2)^+} \dfrac{\cos x}{1 - \sin x} \overset{\text{H}}{=} \lim\limits_{x \to (\pi/2)^+} \dfrac{-\sin x}{-\cos x} = \lim\limits_{x \to (\pi/2)^+} \tan x = -\infty$.

14. This limit has the form $\frac{0}{0}$. $\lim\limits_{x \to 0} \dfrac{\tan 3x}{\sin 2x} \overset{\text{H}}{=} \lim\limits_{x \to 0} \dfrac{3 \sec^2 3x}{2 \cos 2x} = \dfrac{3(1)^2}{2(1)} = \dfrac{3}{2}$

15. This limit has the form $\frac{0}{0}$. $\lim\limits_{t \to 0} \dfrac{e^{2t} - 1}{\sin t} \overset{\text{H}}{=} \lim\limits_{t \to 0} \dfrac{2e^{2t}}{\cos t} = \dfrac{2(1)}{1} = 2$

16. This limit has the form $\frac{0}{0}$. $\lim\limits_{x \to 0} \dfrac{x^2}{1 - \cos x} \overset{\text{H}}{=} \lim\limits_{x \to 0} \dfrac{2x}{\sin x} = \lim\limits_{x \to 0} \dfrac{2}{(\sin x)/x} = \dfrac{2}{1} = 2$

17. This limit has the form $\frac{0}{0}$. $\lim\limits_{\theta \to \pi/2} \dfrac{1 - \sin \theta}{1 + \cos 2\theta} \overset{\text{H}}{=} \lim\limits_{\theta \to \pi/2} \dfrac{-\cos \theta}{-2 \sin 2\theta} \overset{\text{H}}{=} \lim\limits_{\theta \to \pi/2} \dfrac{\sin \theta}{-4 \cos 2\theta} = \dfrac{1}{4}$

18. The limit can be evaluated by substituting π for θ. $\lim\limits_{\theta \to \pi} \dfrac{1 + \cos \theta}{1 - \cos \theta} = \dfrac{1 + (-1)}{1 - (-1)} = \dfrac{0}{2} = 0$

19. This limit has the form $\frac{\infty}{\infty}$. $\lim\limits_{x \to \infty} \dfrac{\ln x}{\sqrt{x}} \overset{\text{H}}{=} \lim\limits_{x \to \infty} \dfrac{1/x}{\frac{1}{2} x^{-1/2}} = \lim\limits_{x \to \infty} \dfrac{2}{\sqrt{x}} = 0$

20. This limit has the form $\frac{\infty}{\infty}$. $\lim\limits_{x \to \infty} \dfrac{x + x^2}{1 - 2x^2} \overset{\text{H}}{=} \lim\limits_{x \to \infty} \dfrac{1 + 2x}{-4x} \overset{\text{H}}{=} \lim\limits_{x \to \infty} \dfrac{2}{-4} = -\dfrac{1}{2}$.

A better method is to divide the numerator and the denominator by x^2: $\lim\limits_{x \to \infty} \dfrac{x + x^2}{1 - 2x^2} = \lim\limits_{x \to \infty} \dfrac{\frac{1}{x} + 1}{\frac{1}{x^2} - 2} = \dfrac{0 + 1}{0 - 2} = -\dfrac{1}{2}$.

21. $\lim\limits_{x\to 0^+} [(\ln x)/x] = -\infty$ since $\ln x \to -\infty$ as $x \to 0^+$ and dividing by small values of x just increases the magnitude of the quotient $(\ln x)/x$. L'Hospital's Rule does not apply.

22. This limit has the form $\frac{\infty}{\infty}$. $\lim\limits_{x\to\infty} \dfrac{\ln\sqrt{x}}{x^2} = \lim\limits_{x\to\infty} \dfrac{\frac{1}{2}\ln x}{x^2} \overset{\text{H}}{=} \lim\limits_{x\to\infty} \dfrac{\frac{1}{2x}}{2x} = \lim\limits_{x\to\infty} \dfrac{1}{4x^2} = 0$

23. This limit has the form $\frac{0}{0}$. $\lim\limits_{t\to 1} \dfrac{t^8-1}{t^5-1} \overset{\text{H}}{=} \lim\limits_{t\to 1} \dfrac{8t^7}{5t^4} = \dfrac{8}{5}\lim\limits_{t\to 1} t^3 = \dfrac{8}{5}(1) = \dfrac{8}{5}$

24. This limit has the form $\frac{0}{0}$. $\lim\limits_{t\to 0} \dfrac{8^t-5^t}{t} \overset{\text{H}}{=} \lim\limits_{t\to 0} \dfrac{8^t\ln 8 - 5^t \ln 5}{1} = \ln 8 - \ln 5 = \ln\dfrac{8}{5}$

25. This limit has the form $\frac{0}{0}$.

$$\lim_{x\to 0} \frac{\sqrt{1+2x}-\sqrt{1-4x}}{x} \overset{\text{H}}{=} \lim_{x\to 0} \frac{\frac{1}{2}(1+2x)^{-1/2}\cdot 2 - \frac{1}{2}(1-4x)^{-1/2}(-4)}{1}$$

$$= \lim_{x\to 0}\left(\frac{1}{\sqrt{1+2x}} + \frac{2}{\sqrt{1-4x}}\right) = \frac{1}{\sqrt{1}} + \frac{2}{\sqrt{1}} = 3$$

26. This limit has the form $\frac{\infty}{\infty}$.

$$\lim_{u\to\infty} \frac{e^{u/10}}{u^3} \overset{\text{H}}{=} \lim_{u\to\infty} \frac{e^{u/10}\cdot\frac{1}{10}}{3u^2} \overset{\text{H}}{=} \frac{1}{30}\lim_{u\to\infty} \frac{e^{u/10}\cdot\frac{1}{10}}{2u} \overset{\text{H}}{=} \frac{1}{600}\lim_{u\to\infty} \frac{e^{u/10}\cdot\frac{1}{10}}{1} = \frac{1}{6000}\lim_{u\to\infty} e^{u/10} = \infty$$

27. This limit has the form $\frac{0}{0}$. $\lim\limits_{x\to 0} \dfrac{e^x-1-x}{x^2} \overset{\text{H}}{=} \lim\limits_{x\to 0} \dfrac{e^x-1}{2x} \overset{\text{H}}{=} \lim\limits_{x\to 0} \dfrac{e^x}{2} = \dfrac{1}{2}$

28. This limit has the form $\frac{0}{0}$. $\lim\limits_{x\to 0} \dfrac{\sinh x - x}{x^3} \overset{\text{H}}{=} \lim\limits_{x\to 0} \dfrac{\cosh x - 1}{3x^2} \overset{\text{H}}{=} \lim\limits_{x\to 0} \dfrac{\sinh x}{6x} \overset{\text{H}}{=} \lim\limits_{x\to 0} \dfrac{\cosh x}{6} = \dfrac{1}{6}$

29. This limit has the form $\frac{0}{0}$. $\lim\limits_{x\to 0} \dfrac{\tanh x}{\tan x} \overset{\text{H}}{=} \lim\limits_{x\to 0} \dfrac{\operatorname{sech}^2 x}{\sec^2 x} = \dfrac{\operatorname{sech}^2 0}{\sec^2 0} = \dfrac{1}{1} = 1$

30. This limit has the form $\frac{0}{0}$.

$$\lim_{x\to 0} \frac{x-\sin x}{x-\tan x} \overset{\text{H}}{=} \lim_{x\to 0} \frac{1-\cos x}{1-\sec^2 x} \overset{\text{H}}{=} \lim_{x\to 0} \frac{-(-\sin x)}{-2\sec x\,(\sec x\,\tan x)} = -\frac{1}{2}\lim_{x\to 0} \frac{\sin x\left(\dfrac{\cos x}{\sin x}\right)}{\sec^2 x}$$

$$= -\tfrac{1}{2}\lim_{x\to 0}\cos^3 x = -\tfrac{1}{2}(1)^3 = -\tfrac{1}{2}$$

Another method is to write the limit as $\lim\limits_{x\to 0} \dfrac{1-\dfrac{\sin x}{x}}{1-\dfrac{\tan x}{x}}$.

31. This limit has the form $\frac{0}{0}$. $\lim\limits_{x\to 0} \dfrac{\sin^{-1}x}{x} \overset{\text{H}}{=} \lim\limits_{x\to 0} \dfrac{1/\sqrt{1-x^2}}{1} = \lim\limits_{x\to 0} \dfrac{1}{\sqrt{1-x^2}} = \dfrac{1}{1} = 1$

32. This limit has the form $\frac{\infty}{\infty}$. $\lim\limits_{x\to\infty} \dfrac{(\ln x)^2}{x} \overset{\text{H}}{=} \lim\limits_{x\to\infty} \dfrac{2(\ln x)(1/x)}{1} = 2\lim\limits_{x\to\infty} \dfrac{\ln x}{x} \overset{\text{H}}{=} 2\lim\limits_{x\to\infty} \dfrac{1/x}{1} = 2(0) = 0$

33. This limit has the form $\frac{0}{0}$. $\lim\limits_{x\to 0} \dfrac{x3^x}{3^x-1} \overset{\text{H}}{=} \lim\limits_{x\to 0} \dfrac{x3^x\ln 3 + 3^x}{3^x \ln 3} = \lim\limits_{x\to 0} \dfrac{3^x(x\ln 3+1)}{3^x\ln 3} = \lim\limits_{x\to 0} \dfrac{x\ln 3+1}{\ln 3} = \dfrac{1}{\ln 3}$

34. This limit has the form $\frac{0}{0}$.

$$\lim_{x\to 0}\frac{\cos mx-\cos nx}{x^2}\overset{\text{H}}{=}\lim_{x\to 0}\frac{-m\sin mx+n\sin nx}{2x}\overset{\text{H}}{=}\lim_{x\to 0}\frac{-m^2\cos mx+n^2\cos nx}{2}=\tfrac{1}{2}(n^2-m^2)$$

35. This limit can be evaluated by substituting 0 for x. $\displaystyle\lim_{x\to 0}\frac{\ln(1+x)}{\cos x+e^x-1}=\frac{\ln 1}{1+1-1}=\frac{0}{1}=0$

36. This limit has the form $\frac{0}{0}$. $\displaystyle\lim_{x\to 1}\frac{x\sin(x-1)}{2x^2-x-1}\overset{\text{H}}{=}\lim_{x\to 1}\frac{x\cos(x-1)+\sin(x-1)}{4x-1}=\frac{\cos 0}{4-1}=\frac{1}{3}$

37. This limit has the form $\frac{0}{\infty}$, so l'Hospital's Rule doesn't apply. As $x\to 0^+$, $\arctan(2x)\to 0$ and $\ln x\to -\infty$, so

$$\lim_{x\to 0^+}\frac{\arctan(2x)}{\ln x}=0.$$

38. $\displaystyle\lim_{x\to 0^+}\frac{x^x-1}{\ln x+x-1}$. From Example 9, $\displaystyle\lim_{x\to 0^+}x^x=1$, so $\displaystyle\lim_{x\to 0^+}(x^x-1)=0$. As $x\to 0^+$, $\ln x\to -\infty$, so

$\ln x+x-1\to -\infty$ as $x\to 0^+$. Thus, $\displaystyle\lim_{x\to 0^+}\frac{x^x-1}{\ln x+x-1}=0$.

39. This limit has the form $\frac{0}{0}$. $\displaystyle\lim_{x\to 1}\frac{x^a-1}{x^b-1}$ [for $b\neq 0$] $\displaystyle\overset{\text{H}}{=}\lim_{x\to 1}\frac{ax^{a-1}}{bx^{b-1}}=\frac{a(1)}{b(1)}=\frac{a}{b}$

40. This limit has the form $\frac{0}{0}$. $\displaystyle\lim_{x\to 0}\frac{e^x-e^{-x}-2x}{x-\sin x}\overset{\text{H}}{=}\lim_{x\to 0}\frac{e^x+e^{-x}-2}{1-\cos x}\overset{\text{H}}{=}\lim_{x\to 0}\frac{e^x-e^{-x}}{\sin x}\overset{\text{H}}{=}\lim_{x\to 0}\frac{e^x+e^{-x}}{\cos x}=\frac{1+1}{1}=2$

41. This limit has the form $\frac{0}{0}$. $\displaystyle\lim_{x\to 0}\frac{\cos x-1+\frac{1}{2}x^2}{x^4}\overset{\text{H}}{=}\lim_{x\to 0}\frac{-\sin x+x}{4x^3}\overset{\text{H}}{=}\lim_{x\to 0}\frac{-\cos x+1}{12x^2}\overset{\text{H}}{=}\lim_{x\to 0}\frac{\sin x}{24x}\overset{\text{H}}{=}\lim_{x\to 0}\frac{\cos x}{24}=\frac{1}{24}$

42. This limit has the form $\frac{\infty}{\infty}$.

$$\lim_{x\to a^+}\frac{\cos x\ln(x-a)}{\ln(e^x-e^a)}=\lim_{x\to a^+}\cos x\lim_{x\to a^+}\frac{\ln(x-a)}{\ln(e^x-e^a)}\overset{\text{H}}{=}\cos a\lim_{x\to a^+}\frac{\dfrac{1}{x-a}}{\dfrac{1}{e^x-e^a}\cdot e^x}$$

$$=\cos a\lim_{x\to a^+}\frac{1}{e^x}\cdot\lim_{x\to a^+}\frac{e^x-e^a}{x-a}\overset{\text{H}}{=}\cos a\cdot\frac{1}{e^a}\lim_{x\to a^+}\frac{e^x}{1}=\cos a\cdot\frac{1}{e^a}\cdot e^a=\cos a$$

43. This limit has the form $\infty\cdot 0$. We'll change it to the form $\frac{0}{0}$.

$$\lim_{x\to\infty}x\sin(\pi/x)=\lim_{x\to\infty}\frac{\sin(\pi/x)}{1/x}\overset{\text{H}}{=}\lim_{x\to\infty}\frac{\cos(\pi/x)(-\pi/x^2)}{-1/x^2}=\pi\lim_{x\to\infty}\cos(\pi/x)=\pi(1)=\pi$$

44. This limit has the form $\infty\cdot 0$. We'll change it to the form $\frac{\infty}{\infty}$.

$$\lim_{x\to\infty}\sqrt{x}\,e^{-x/2}=\lim_{x\to\infty}\frac{\sqrt{x}}{e^{x/2}}\overset{\text{H}}{=}\lim_{x\to\infty}\frac{\frac{1}{2}x^{-1/2}}{\frac{1}{2}e^{x/2}}=\lim_{x\to\infty}\frac{1}{\sqrt{x}\,e^{x/2}}=0$$

45. This limit has the form $0\cdot\infty$. We'll change it to the form $\frac{0}{0}$. $\displaystyle\lim_{x\to 0}\sin 5x\csc 3x=\lim_{x\to 0}\frac{\sin 5x}{\sin 3x}\overset{\text{H}}{=}\lim_{x\to 0}\frac{5\cos 5x}{3\cos 3x}=\frac{5\cdot 1}{3\cdot 1}=\frac{5}{3}$

46. This limit has the form $(-\infty)\cdot 0$.

$$\lim_{x\to-\infty}x\ln\!\left(1-\frac{1}{x}\right)=\lim_{x\to-\infty}\frac{\ln\left(1-\dfrac{1}{x}\right)}{\dfrac{1}{x}}\overset{\text{H}}{=}\lim_{x\to-\infty}\frac{\dfrac{1}{1-1/x}\cdot\dfrac{1}{x^2}}{-\dfrac{1}{x^2}}=\lim_{x\to-\infty}\frac{-1}{1-\dfrac{1}{x}}=\frac{-1}{1}=-1$$

47. This limit has the form $\infty \cdot 0$. $\lim\limits_{x\to\infty} x^3e^{-x^2} = \lim\limits_{x\to\infty} \dfrac{x^3}{e^{x^2}} \overset{\text{H}}{=} \lim\limits_{x\to\infty} \dfrac{3x^2}{2xe^{x^2}} = \lim\limits_{x\to\infty} \dfrac{3x}{2e^{x^2}} \overset{\text{H}}{=} \lim\limits_{x\to\infty} \dfrac{3}{4xe^{x^2}} = 0$

48. This limit has the form $\infty \cdot 0$. $\lim\limits_{x\to\infty} x^{3/2}\sin(1/x) = \lim\limits_{x\to\infty} x^{1/2} \cdot \dfrac{\sin(1/x)}{1/x} = \lim\limits_{t\to 0^+} \dfrac{1}{\sqrt{t}}\,\dfrac{\sin t}{t}$ [where $t = 1/x$] $= \infty$

since as $t \to 0^+$, $\dfrac{1}{\sqrt{t}} \to \infty$ and $\dfrac{\sin t}{t} \to 1$.

49. This limit has the form $0 \cdot (-\infty)$.

$$\lim_{x\to 1^+} \ln x\,\tan(\pi x/2) = \lim_{x\to 1^+} \frac{\ln x}{\cot(\pi x/2)} \overset{\text{H}}{=} \lim_{x\to 1^+} \frac{1/x}{(-\pi/2)\csc^2(\pi x/2)} = \frac{1}{(-\pi/2)(1)^2} = -\frac{2}{\pi}$$

50. This limit has the form $0 \cdot \infty$. $\lim\limits_{x\to(\pi/2)^-} \cos x\,\sec 5x = \lim\limits_{x\to(\pi/2)^-} \dfrac{\cos x}{\cos 5x} \overset{\text{H}}{=} \lim\limits_{x\to(\pi/2)^-} \dfrac{-\sin x}{-5\sin 5x} = \dfrac{-1}{-5} = \dfrac{1}{5}$

51. This limit has the form $\infty - \infty$.

$$\lim_{x\to 1}\left(\frac{x}{x-1} - \frac{1}{\ln x}\right) = \lim_{x\to 1} \frac{x\ln x - (x-1)}{(x-1)\ln x} \overset{\text{H}}{=} \lim_{x\to 1} \frac{x(1/x) + \ln x - 1}{(x-1)(1/x) + \ln x} = \lim_{x\to 1} \frac{\ln x}{1 - (1/x) + \ln x}$$

$$\overset{\text{H}}{=} \lim_{x\to 1} \frac{1/x}{1/x^2 + 1/x} \cdot \frac{x^2}{x^2} = \lim_{x\to 1} \frac{x}{1+x} = \frac{1}{1+1} = \frac{1}{2}$$

52. This limit has the form $\infty - \infty$. $\lim\limits_{x\to 0}(\csc x - \cot x) = \lim\limits_{x\to 0}\left(\dfrac{1}{\sin x} - \dfrac{\cos x}{\sin x}\right) = \lim\limits_{x\to 0} \dfrac{1-\cos x}{\sin x} \overset{\text{H}}{=} \lim\limits_{x\to 0} \dfrac{\sin x}{\cos x} = 0$

53. This limit has the form $\infty - \infty$.

$$\lim_{x\to 0^+}\left(\frac{1}{x} - \frac{1}{e^x - 1}\right) = \lim_{x\to 0^+} \frac{e^x - 1 - x}{x(e^x - 1)} \overset{\text{H}}{=} \lim_{x\to 0^+} \frac{e^x - 1}{xe^x + e^x - 1} \overset{\text{H}}{=} \lim_{x\to 0^+} \frac{e^x}{xe^x + e^x + e^x} = \frac{1}{0+1+1} = \frac{1}{2}$$

54. This limit has the form $\infty - \infty$.

$$\lim_{x\to 0^+}\left(\frac{1}{x} - \frac{1}{\tan^{-1} x}\right) = \lim_{x\to 0^+} \frac{\tan^{-1} x - x}{x\tan^{-1} x} \overset{\text{H}}{=} \lim_{x\to 0^+} \frac{1/(1+x^2) - 1}{x/(1+x^2) + \tan^{-1} x} = \lim_{x\to 0^+} \frac{1 - (1+x^2)}{x + (1+x^2)\tan^{-1} x}$$

$$= \lim_{x\to 0^+} \frac{-x^2}{x + (1+x^2)\tan^{-1} x} \overset{\text{H}}{=} \lim_{x\to 0^+} \frac{-2x}{1 + (1+x^2)(1/(1+x^2)) + (\tan^{-1} x)(2x)}$$

$$= \lim_{x\to 0^+} \frac{-2x}{2 + 2x\tan^{-1} x} = \frac{0}{2+0} = 0$$

55. The limit has the form $\infty - \infty$ and we will change the form to a product by factoring out x.

$$\lim_{x\to\infty}(x - \ln x) = \lim_{x\to\infty} x\left(1 - \frac{\ln x}{x}\right) = \infty \text{ since } \lim_{x\to\infty} \frac{\ln x}{x} \overset{\text{H}}{=} \lim_{x\to\infty} \frac{1/x}{1} = 0.$$

56. This limit has the form $\infty - \infty$.

$$\lim_{x\to 1^+}[\ln(x^7 - 1) - \ln(x^5 - 1)] = \lim_{x\to 1^+} \ln \frac{x^7 - 1}{x^5 - 1} = \ln \lim_{x\to 1^+} \frac{x^7 - 1}{x^5 - 1} \overset{\text{H}}{=} \ln \lim_{x\to 1^+} \frac{7x^6}{5x^4} = \ln\frac{7}{5}$$

57. $y = x^{\sqrt{x}} \;\Rightarrow\; \ln y = \sqrt{x}\,\ln x$, so

$$\lim_{x\to 0^+} \ln y = \lim_{x\to 0^+} \sqrt{x}\,\ln x = \lim_{x\to 0^+} \frac{\ln x}{x^{-1/2}} \overset{\text{H}}{=} \lim_{x\to 0^+} \frac{1/x}{-\frac{1}{2}x^{-3/2}} = -2\lim_{x\to 0^+} \sqrt{x} = 0 \;\Rightarrow$$

$$\lim_{x\to 0^+} x^{\sqrt{x}} = \lim_{x\to 0^+} e^{\ln y} = e^0 = 1.$$

58. $y = (\tan 2x)^x \quad \Rightarrow \quad \ln y = x \cdot \ln \tan 2x$, so

$$\lim_{x\to 0^+} \ln y = \lim_{x\to 0^+} x \cdot \ln \tan 2x = \lim_{x\to 0^+} \frac{\ln \tan 2x}{1/x} \overset{\text{H}}{=} \lim_{x\to 0^+} \frac{(1/\tan 2x)(2\sec^2 2x)}{-1/x^2} = \lim_{x\to 0^+} \frac{-2x^2 \cos 2x}{\sin 2x \cos^2 2x}$$

$$= \lim_{x\to 0^+} \frac{2x}{\sin 2x} \cdot \lim_{x\to 0^+} \frac{-x}{\cos 2x} = 1 \cdot 0 = 0 \quad \Rightarrow$$

$$\lim_{x\to 0^+} (\tan 2x)^x = \lim_{x\to 0^+} e^{\ln y} = e^0 = 1.$$

59. $y = (1-2x)^{1/x} \quad \Rightarrow \quad \ln y = \dfrac{1}{x}\ln(1-2x)$, so $\displaystyle\lim_{x\to 0} \ln y = \lim_{x\to 0} \frac{\ln(1-2x)}{x} \overset{\text{H}}{=} \lim_{x\to 0} \frac{-2/(1-2x)}{1} = -2 \quad \Rightarrow$

$$\lim_{x\to 0} (1-2x)^{1/x} = \lim_{x\to 0} e^{\ln y} = e^{-2}.$$

60. $y = \left(1 + \dfrac{a}{x}\right)^{bx} \quad \Rightarrow \quad \ln y = bx \ln\left(1 + \dfrac{a}{x}\right)$, so

$$\lim_{x\to\infty} \ln y = \lim_{x\to\infty} \frac{b\ln(1+a/x)}{1/x} \overset{\text{H}}{=} \lim_{x\to\infty} \frac{b\left(\dfrac{1}{1+a/x}\right)\left(-\dfrac{a}{x^2}\right)}{-1/x^2} = \lim_{x\to\infty} \frac{ab}{1+a/x} = ab \quad \Rightarrow$$

$$\lim_{x\to\infty} \left(1 + \frac{a}{x}\right)^{bx} = \lim_{x\to\infty} e^{\ln y} = c^{ab}.$$

61. $y = x^{1/(1-x)} \quad \Rightarrow \quad \ln y = \dfrac{1}{1-x}\ln x$, so $\displaystyle\lim_{x\to 1^+} \ln y = \lim_{x\to 1^+} \frac{1}{1-x}\ln x = \lim_{x\to 1^+} \frac{\ln x}{1-x} \overset{\text{H}}{=} \lim_{x\to 1^+} \frac{1/x}{-1} = -1 \quad \Rightarrow$

$$\lim_{x\to 1^+} x^{1/(1-x)} = \lim_{x\to 1^+} e^{\ln y} = e^{-1} = \frac{1}{e}.$$

62. $y = x^{(\ln 2)/(1+\ln x)} \quad \Rightarrow \quad \ln y = \dfrac{\ln 2}{1+\ln x}\ln x \quad \Rightarrow$

$$\lim_{x\to\infty} \ln y = \lim_{x\to\infty} \frac{(\ln 2)(\ln x)}{1+\ln x} \overset{\text{H}}{=} \lim_{x\to\infty} \frac{(\ln 2)(1/x)}{1/x} = \lim_{x\to\infty} \ln 2 = \ln 2, \text{ so } \lim_{x\to\infty} x^{(\ln 2)/(1+\ln x)} = \lim_{x\to\infty} e^{\ln y} = e^{\ln 2} = 2.$$

63. $y = x^{1/x} \quad \Rightarrow \quad \ln y = (1/x)\ln x \quad \Rightarrow \quad \displaystyle\lim_{x\to\infty} \ln y = \lim_{x\to\infty} \frac{\ln x}{x} \overset{\text{H}}{=} \lim_{x\to\infty} \frac{1/x}{1} = 0 \quad \Rightarrow$

$$\lim_{x\to\infty} x^{1/x} = \lim_{x\to\infty} e^{\ln y} = e^0 = 1$$

64. $y = x^{e^{-x}} \quad \Rightarrow \quad \ln y = e^{-x}\ln x \quad \Rightarrow \quad \displaystyle\lim_{x\to\infty} \ln y = \lim_{x\to\infty} \frac{\ln x}{e^x} \overset{\text{H}}{=} \lim_{x\to\infty} \frac{1/x}{e^x} = \lim_{x\to\infty} \frac{1}{xe^x} = 0 \quad \Rightarrow$

$$\lim_{x\to\infty} x^{e^{-x}} = \lim_{x\to\infty} e^{\ln y} = e^0 = 1$$

65. $y = (4x+1)^{\cot x} \quad \Rightarrow \quad \ln y = \cot x \ln(4x+1)$, so $\displaystyle\lim_{x\to 0^+} \ln y = \lim_{x\to 0^+} \frac{\ln(4x+1)}{\tan x} \overset{\text{H}}{=} \lim_{x\to 0^+} \frac{\dfrac{4}{4x+1}}{\sec^2 x} = 4 \quad \Rightarrow$

$$\lim_{x\to 0^+} (4x+1)^{\cot x} = \lim_{x\to 0^+} e^{\ln y} = e^4.$$

66. $y = (2-x)^{\tan(\pi x/2)} \quad \Rightarrow \quad \ln y = \tan\left(\dfrac{\pi x}{2}\right)\ln(2-x) \quad \Rightarrow$

$$\lim_{x\to 1} \ln y = \lim_{x\to 1}\left[\tan\left(\frac{\pi x}{2}\right)\ln(2-x)\right] = \lim_{x\to 1} \frac{\ln(2-x)}{\cot\left(\dfrac{\pi x}{2}\right)} \overset{\text{H}}{=} \lim_{x\to 1} \frac{\dfrac{1}{2-x}(-1)}{-\csc^2\left(\dfrac{\pi x}{2}\right)\cdot\dfrac{\pi}{2}} = \frac{2}{\pi}\lim_{x\to 1} \frac{\sin^2\left(\dfrac{\pi x}{2}\right)}{2-x}$$

$$= \frac{2}{\pi}\cdot\frac{1^2}{1} = \frac{2}{\pi} \quad \Rightarrow \quad \lim_{x\to 1}(2-x)^{\tan(\pi x/2)} = \lim_{x\to 1} e^{\ln y} = e^{(2/\pi)}$$

67. $y = (1 + \sin 3x)^{1/x} \quad \Rightarrow \quad \ln y = \dfrac{1}{x}\ln(1 + \sin 3x) \quad \Rightarrow$

$$\lim_{x\to 0^+} \ln y = \lim_{x\to 0^+} \frac{\ln(1+\sin 3x)}{x} \overset{\text{H}}{=} \lim_{x\to 0^+} \frac{[1/(1+\sin 3x)]\cdot 3\cos 3x}{1} = \lim_{x\to 0^+} \frac{3\cos 3x}{1+\sin 3x} = \frac{3\cdot 1}{1+0} = 3 \quad \Rightarrow$$

$$\lim_{x\to 0^+} (1+\sin 3x)^{1/x} = \lim_{x\to 0^+} e^{\ln y} = e^3$$

68. $y = \left(\dfrac{2x-3}{2x+5}\right)^{2x+1} \quad \Rightarrow \quad \ln y = (2x+1)\ln\left(\dfrac{2x-3}{2x+5}\right) \quad \Rightarrow$

$$\lim_{x\to\infty} \ln y = \lim_{x\to\infty} \frac{\ln(2x-3) - \ln(2x+5)}{1/(2x+1)} \overset{\text{H}}{=} \lim_{x\to\infty} \frac{2/(2x-3) - 2/(2x+5)}{-2/(2x+1)^2} = \lim_{x\to\infty} \frac{-8(2x+1)^2}{(2x-3)(2x+5)}$$

$$= \lim_{x\to\infty} \frac{-8(2+1/x)^2}{(2-3/x)(2+5/x)} = -8 \quad \Rightarrow \quad \lim_{x\to\infty}\left(\frac{2x-3}{2x+5}\right)^{2x+1} = e^{-8}$$

69.

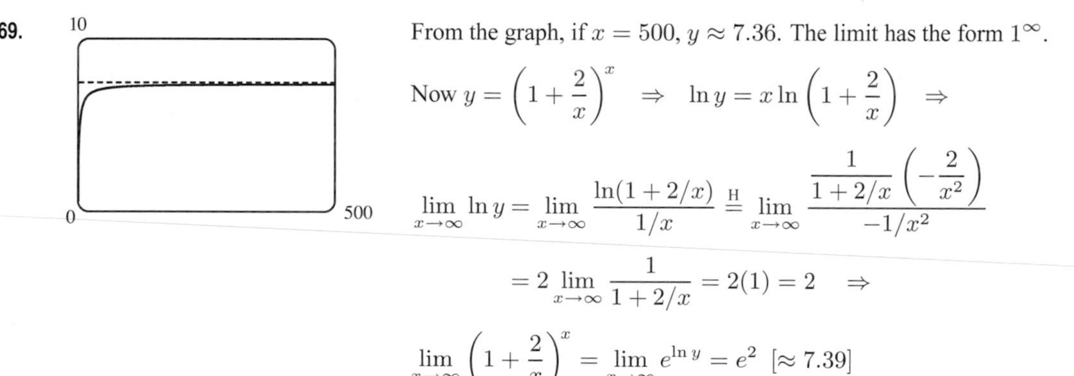

From the graph, if $x = 500$, $y \approx 7.36$. The limit has the form 1^∞.

Now $y = \left(1 + \dfrac{2}{x}\right)^x \quad \Rightarrow \quad \ln y = x\ln\left(1+\dfrac{2}{x}\right) \quad \Rightarrow$

$$\lim_{x\to\infty} \ln y = \lim_{x\to\infty} \frac{\ln(1+2/x)}{1/x} \overset{\text{H}}{=} \lim_{x\to\infty} \frac{\dfrac{1}{1+2/x}\left(-\dfrac{2}{x^2}\right)}{-1/x^2}$$

$$= 2\lim_{x\to\infty} \frac{1}{1+2/x} = 2(1) = 2 \quad \Rightarrow$$

$$\lim_{x\to\infty}\left(1+\frac{2}{x}\right)^x = \lim_{x\to\infty} e^{\ln y} = e^2 \quad [\approx 7.39]$$

70.

From the graph, as $x \to 0$, $y \approx 0.55$. The limit has the form $\frac{0}{0}$.

$$\lim_{x\to 0} \frac{5^x - 4^x}{3^x - 2^x} \overset{\text{H}}{=} \lim_{x\to 0} \frac{5^x \ln 5 - 4^x \ln 4}{3^x \ln 3 - 2^x \ln 2} = \frac{\ln 5 - \ln 4}{\ln 3 - \ln 2} = \frac{\ln\frac{5}{4}}{\ln\frac{3}{2}} \quad [\approx 0.55]$$

71.

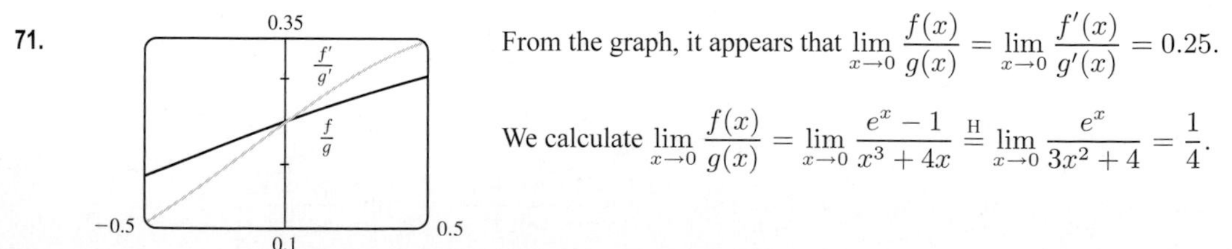

From the graph, it appears that $\displaystyle\lim_{x\to 0}\frac{f(x)}{g(x)} = \lim_{x\to 0}\frac{f'(x)}{g'(x)} = 0.25$.

We calculate $\displaystyle\lim_{x\to 0}\frac{f(x)}{g(x)} = \lim_{x\to 0}\frac{e^x - 1}{x^3 + 4x} \overset{\text{H}}{=} \lim_{x\to 0}\frac{e^x}{3x^2+4} = \frac{1}{4}$.

72.

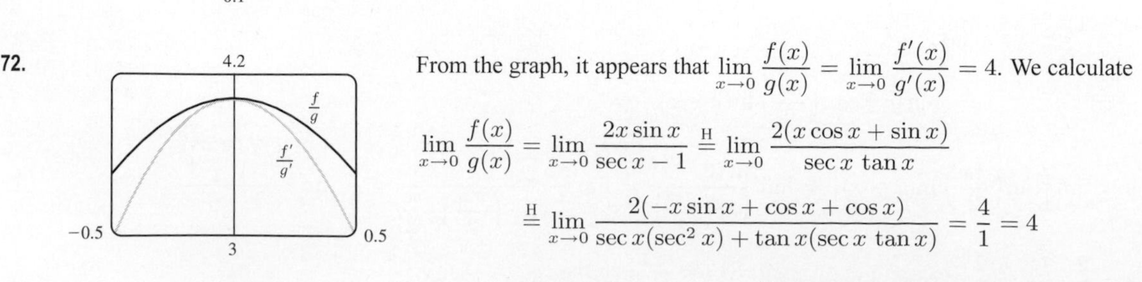

From the graph, it appears that $\displaystyle\lim_{x\to 0}\frac{f(x)}{g(x)} = \lim_{x\to 0}\frac{f'(x)}{g'(x)} = 4$. We calculate

$$\lim_{x\to 0}\frac{f(x)}{g(x)} = \lim_{x\to 0}\frac{2x\sin x}{\sec x - 1} \overset{\text{H}}{=} \lim_{x\to 0}\frac{2(x\cos x + \sin x)}{\sec x\,\tan x}$$

$$\overset{\text{H}}{=} \lim_{x\to 0}\frac{2(-x\sin x + \cos x + \cos x)}{\sec x(\sec^2 x) + \tan x(\sec x\,\tan x)} = \frac{4}{1} = 4$$

73. $\lim\limits_{x\to\infty}\dfrac{e^x}{x^n}\overset{\text{H}}{=}\lim\limits_{x\to\infty}\dfrac{e^x}{nx^{n-1}}\overset{\text{H}}{=}\lim\limits_{x\to\infty}\dfrac{e^x}{n(n-1)x^{n-2}}\overset{\text{H}}{=}\cdots\overset{\text{H}}{=}\lim\limits_{x\to\infty}\dfrac{e^x}{n!}=\infty$

74. This limit has the form $\frac{\infty}{\infty}$. $\lim\limits_{x\to\infty}\dfrac{\ln x}{x^p}\overset{\text{H}}{=}\lim\limits_{x\to\infty}\dfrac{1/x}{px^{p-1}}=\lim\limits_{x\to\infty}\dfrac{1}{px^p}=0$ since $p>0$.

75. $\lim\limits_{x\to\infty}\dfrac{x}{\sqrt{x^2+1}}\overset{\text{H}}{=}\lim\limits_{x\to\infty}\dfrac{1}{\frac{1}{2}(x^2+1)^{-1/2}(2x)}=\lim\limits_{x\to\infty}\dfrac{\sqrt{x^2+1}}{x}$. Repeated applications of l'Hospital's Rule result in the original limit or the limit of the reciprocal of the function. Another method is to try dividing the numerator and denominator by x: $\lim\limits_{x\to\infty}\dfrac{x}{\sqrt{x^2+1}}=\lim\limits_{x\to\infty}\dfrac{x/x}{\sqrt{x^2/x^2+1/x^2}}=\lim\limits_{x\to\infty}\dfrac{1}{\sqrt{1+1/x^2}}=\dfrac{1}{1}=1$

76. $\lim\limits_{x\to(\pi/2)^-}\dfrac{\sec x}{\tan x}\overset{\text{H}}{=}\lim\limits_{x\to(\pi/2)^-}\dfrac{\sec x\,\tan x}{\sec^2 x}=\lim\limits_{x\to(\pi/2)^-}\dfrac{\tan x}{\sec x}$. Repeated applications of l'Hospital's Rule result in the original limit or the limit of the reciprocal of the function. Another method is to simplify first:

$$\lim_{x\to(\pi/2)^-}\frac{\sec x}{\tan x}=\lim_{x\to(\pi/2)^-}\frac{1/\cos x}{\sin x/\cos x}=\lim_{x\to(\pi/2)^-}\frac{1}{\sin x}=\frac{1}{1}=1$$

77. $f(x)=e^x-cx \;\Rightarrow\; f'(x)=e^x-c=0 \;\Leftrightarrow\; e^x=c \;\Leftrightarrow\; x=\ln c,\ c>0$. $f''(x)=e^x>0$, so f is CU on $(-\infty,\infty)$. $\lim\limits_{x\to\infty}(e^x-cx)=\lim\limits_{x\to\infty}\left[x\left(\dfrac{e^x}{x}-c\right)\right]=L_1$. Now $\lim\limits_{x\to\infty}\dfrac{e^x}{x}\overset{\text{H}}{=}\lim\limits_{x\to\infty}\dfrac{e^x}{1}=\infty$, so $L_1=\infty$, regardless of the value of c. For $L=\lim\limits_{x\to-\infty}(e^x-cx)$, $e^x\to 0$, so L is determined by $-cx$. If $c>0$, $-cx\to\infty$, and $L=\infty$. If $c<0$, $-cx\to-\infty$, and $L=-\infty$. Thus, f has an absolute minimum for $c>0$. As c increases, the minimum points $(\ln c, c-c\ln c)$, get farther away from the origin.

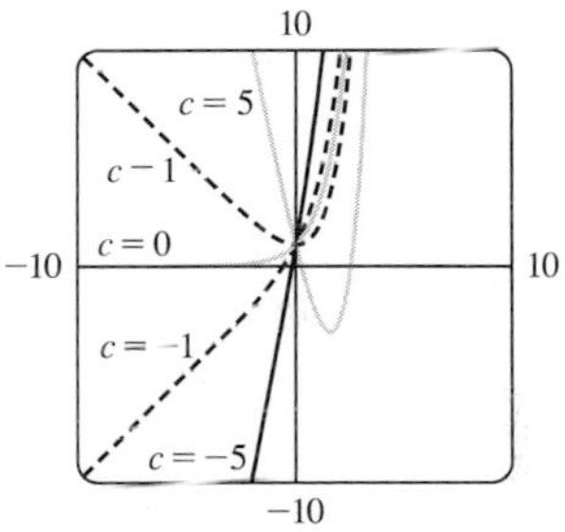

78. (a) $\lim\limits_{t\to\infty}v=\lim\limits_{t\to\infty}\dfrac{mg}{c}\left(1-e^{-ct/m}\right)=\dfrac{mg}{c}\lim\limits_{t\to\infty}\left(1-e^{-ct/m}\right)=\dfrac{mg}{c}(1-0)$ [because $-ct/m\to-\infty$ as $t\to\infty$]

$=\dfrac{mg}{c}$, which is the speed the object approaches as time goes on, the so-called limiting velocity.

(b) $\lim\limits_{c\to0^+}v=\lim\limits_{c\to0^+}\dfrac{mg}{c}(1-e^{-ct/m})=mg\lim\limits_{c\to0^+}\dfrac{1-e^{-ct/m}}{c}$ [form is $\frac{0}{0}$]

$\overset{\text{H}}{=}mg\lim\limits_{c\to0^+}\dfrac{(-e^{-ct/m})\cdot(-t/m)}{1}=\dfrac{mgt}{m}\lim\limits_{c\to0^+}e^{-ct/m}=gt(1)=gt$

The velocity of a falling object in a vacuum is directly proportional to the amount of time it falls.

79. First we will find $\lim\limits_{n\to\infty}\left(1+\dfrac{r}{n}\right)^{nt}$, which is of the form 1^∞. $y=\left(1+\dfrac{r}{n}\right)^{nt}\;\Rightarrow\;\ln y=nt\ln\left(1+\dfrac{r}{n}\right)$, so

$\lim\limits_{n\to\infty}\ln y=\lim\limits_{n\to\infty}nt\ln\left(1+\dfrac{r}{n}\right)=t\lim\limits_{n\to\infty}\dfrac{\ln(1+r/n)}{1/n}\overset{\text{H}}{=}t\lim\limits_{n\to\infty}\dfrac{(-r/n^2)}{(1+r/n)(-1/n^2)}=t\lim\limits_{n\to\infty}\dfrac{r}{1+i/n}=tr\;\Rightarrow$

$\lim\limits_{n\to\infty}y=e^{rt}$. Thus, as $n\to\infty$, $A=A_0\left(1+\dfrac{r}{n}\right)^{nt}\to A_0e^{rt}$.

80. (a) $r=3,\ \rho=0.05\;\Rightarrow\;P=\dfrac{1-10^{-\rho r^2}}{\rho r^2\ln 10}=\dfrac{1-10^{-0.45}}{0.45\ln 10}\approx 0.62$, or about 62%.

(b) $r = 2$, $\rho = 0.05 \Rightarrow P = \dfrac{1 - 10^{-0.2}}{0.2 \ln 10} \approx 0.80$, or about 80%.

Yes, it makes sense. Since measured brightness decreases with light entering farther from the center of the pupil, a smaller pupil radius means that the average brightness measurements are higher than when including light entering at larger radii.

(c) $\displaystyle\lim_{r \to 0^+} P = \lim_{r \to 0^+} \frac{1 - 10^{-\rho r^2}}{\rho r^2 \ln 10} \overset{\text{H}}{=} \lim_{r \to 0^+} \frac{-10^{-\rho r^2}(\ln 10)(-2\rho r)}{2\rho r(\ln 10)} = \lim_{r \to 0^+} \frac{1}{10^{\rho r^2}} = 1$, or 100%.

We might expect that 100% of the brightness is sensed at the very center of the pupil, so a limit of 1 would make sense in this context if the radius r could approach 0. This result isn't physically possible because there are limitations on how small the pupil can shrink.

81. (a) $\displaystyle\lim_{t \to \infty} P(t) = \lim_{t \to \infty} \frac{M}{1 + Ae^{-kt}} = \frac{M}{1 + A \cdot 0} = M$

It is to be expected that a population that is growing will eventually reach the maximum population size that can be supported.

(b) $\displaystyle\lim_{M \to \infty} P(t) = \lim_{M \to \infty} \frac{M}{1 + \dfrac{M - P_0}{P_0} e^{-kt}} = \lim_{M \to \infty} \frac{M}{1 + \left(\dfrac{M}{P_0} - 1\right) e^{-kt}} \overset{\text{H}}{=} \lim_{M \to \infty} \frac{1}{\dfrac{1}{P_0} e^{-kt}} = P_0 e^{kt}$

$P_0 e^{kt}$ is an exponential function.

82. (a) $\displaystyle\lim_{R \to r^+} v = \lim_{R \to r^+} \left[-c\left(\frac{r}{R}\right)^2 \ln\left(\frac{r}{R}\right)\right] = -cr^2 \lim_{R \to r^+} \left[\left(\frac{1}{R}\right)^2 \ln\left(\frac{r}{R}\right)\right] = -cr^2 \cdot \frac{1}{r^2} \cdot \ln 1 = -c \cdot 0 = 0$

As the insulation of a metal cable becomes thinner, the velocity of an electrical impulse in the cable approaches zero.

(b) $\displaystyle\lim_{r \to 0^+} v = \lim_{r \to 0^+} \left[-c\left(\frac{r}{R}\right)^2 \ln\left(\frac{r}{R}\right)\right] = -\frac{c}{R^2} \lim_{r \to 0^+} \left[r^2 \ln\left(\frac{r}{R}\right)\right]$ [form is $0 \cdot \infty$]

$$= -\frac{c}{R^2} \lim_{r \to 0^+} \frac{\ln\left(\frac{r}{R}\right)}{\frac{1}{r^2}} \quad [\text{form is } \infty/\infty] \quad \overset{\text{H}}{=} -\frac{c}{R^2} \lim_{r \to 0^+} \frac{\frac{R}{r} \cdot \frac{1}{R}}{\frac{-2}{r^3}} = -\frac{c}{R^2} \lim_{r \to 0^+} \left(-\frac{r^2}{2}\right) = 0$$

As the radius of the metal cable approaches zero, the velocity of an electrical impulse in the cable approaches zero.

83. We see that both numerator and denominator approach 0, so we can use l'Hospital's Rule:

$$\lim_{x \to a} \frac{\sqrt{2a^3x - x^4} - a\sqrt[3]{aax}}{a - \sqrt[4]{ax^3}} \overset{\text{H}}{=} \lim_{x \to a} \frac{\frac{1}{2}(2a^3x - x^4)^{-1/2}(2a^3 - 4x^3) - a\left(\frac{1}{3}\right)(aax)^{-2/3}a^2}{-\frac{1}{4}(ax^3)^{-3/4}(3ax^2)}$$

$$= \frac{\frac{1}{2}(2a^3a - a^4)^{-1/2}(2a^3 - 4a^3) - \frac{1}{3}a^3(a^2a)^{-2/3}}{-\frac{1}{4}(aa^3)^{-3/4}(3aa^2)}$$

$$= \frac{(a^4)^{-1/2}(-a^3) - \frac{1}{3}a^3(a^3)^{-2/3}}{-\frac{3}{4}a^3(a^4)^{-3/4}} = \frac{-a - \frac{1}{3}a}{-\frac{3}{4}} = \frac{4}{3}\left(\frac{4}{3}a\right) = \frac{16}{9}a$$

84. Let the radius of the circle be r. We see that $A(\theta)$ is the area of the whole figure (a sector of the circle with radius 1), minus the area of $\triangle OPR$. But the area of the sector of the circle is $\frac{1}{2}r^2\theta$ (see Reference Page 1), and the area of the triangle is $\frac{1}{2}r\,|PQ| = \frac{1}{2}r(r\sin\theta) = \frac{1}{2}r^2\sin\theta$. So we have $A(\theta) = \frac{1}{2}r^2\theta - \frac{1}{2}r^2\sin\theta = \frac{1}{2}r^2(\theta - \sin\theta)$. Now by elementary trigonometry, $B(\theta) = \frac{1}{2}|QR||PQ| = \frac{1}{2}(r - |OQ|)|PQ| = \frac{1}{2}(r - r\cos\theta)(r\sin\theta) = \frac{1}{2}r^2(1-\cos\theta)\sin\theta$.

So the limit we want is

$$\lim_{\theta\to 0^+}\frac{A(\theta)}{B(\theta)} = \lim_{\theta\to 0^+}\frac{\frac{1}{2}r^2(\theta-\sin\theta)}{\frac{1}{2}r^2(1-\cos\theta)\sin\theta} \overset{\text{H}}{=} \lim_{\theta\to 0^+}\frac{1-\cos\theta}{(1-\cos\theta)\cos\theta+\sin\theta\,(\sin\theta)}$$

$$= \lim_{\theta\to 0^+}\frac{1-\cos\theta}{\cos\theta-\cos^2\theta+\sin^2\theta} \overset{\text{H}}{=} \lim_{\theta\to 0^+}\frac{\sin\theta}{-\sin\theta-2\cos\theta\,(-\sin\theta)+2\sin\theta\,(\cos\theta)}$$

$$= \lim_{\theta\to 0^+}\frac{\sin\theta}{-\sin\theta+4\sin\theta\cos\theta} = \lim_{\theta\to 0^+}\frac{1}{-1+4\cos\theta} = \frac{1}{-1+4\cos 0} = \frac{1}{3}$$

85. The limit, $L = \lim_{x\to\infty}\left[x - x^2\ln\left(\frac{1+x}{x}\right)\right] = \lim_{x\to\infty}\left[x - x^2\ln\left(\frac{1}{x}+1\right)\right]$. Let $t = 1/x$, so as $x\to\infty$, $t\to 0^+$.

$$L = \lim_{t\to 0^+}\left[\frac{1}{t}-\frac{1}{t^2}\ln(t+1)\right] = \lim_{t\to 0^+}\frac{t-\ln(t+1)}{t^2} \overset{\text{H}}{=} \lim_{t\to 0^+}\frac{1-\dfrac{1}{t+1}}{2t} = \lim_{t\to 0^+}\frac{t/(t+1)}{2t} = \lim_{t\to 0^+}\frac{1}{2\,(t+1)} = \frac{1}{2}$$

Note: Starting the solution by factoring out x or x^2 leads to a more complicated solution.

86. $y = [f(x)]^{g(x)} \Rightarrow \ln y = g(x)\ln f(x)$. Since f is a positive function, $\ln f(x)$ is defined. Now $\lim_{x\to a}\ln y = \lim_{x\to a} g(x)\ln f(x) = -\infty$ since $\lim_{x\to a} g(x) = \infty$ and $\lim_{x\to a} f(x) = 0 \Rightarrow \lim_{x\to a}\ln f(x) = -\infty$. Thus, if $t = \ln y$, $\lim_{x\to a} y = \lim_{t\to -\infty} e^t = 0$. Note that the limit, $\lim_{x\to a} g(x)\ln f(x)$, is *not* of the form $\infty\cdot 0$.

87. Since $f(2) = 0$, the given limit has the form $\frac{0}{0}$.

$$\lim_{x\to 0}\frac{f(2+3x)+f(2+5x)}{x} \overset{\text{H}}{=} \lim_{x\to 0}\frac{f'(2+3x)\cdot 3+f'(2+5x)\cdot 5}{1} = f'(2)\cdot 3+f'(2)\cdot 5 = 8f'(2) = 8\cdot 7 = 56$$

88. $L = \lim_{x\to 0}\left(\frac{\sin 2x}{x^3}+a+\frac{b}{x^2}\right) = \lim_{x\to 0}\frac{\sin 2x+ax^3+bx}{x^3} \overset{\text{H}}{=} \lim_{x\to 0}\frac{2\cos 2x+3ax^2+b}{3x^2}$. As $x\to 0$, $3x^2\to 0$, and $(2\cos 2x+3ax^2+b)\to b+2$, so the last limit exists only if $b+2=0$, that is, $b=-2$. Thus,

$$\lim_{x\to 0}\frac{2\cos 2x+3ax^2-2}{3x^2} \overset{\text{H}}{=} \lim_{x\to 0}\frac{-4\sin 2x+6ax}{6x} \overset{\text{H}}{=} \lim_{x\to 0}\frac{-8\cos 2x+6a}{6} = \frac{6a-8}{6},$$ which is equal to 0 if and only if $a = \frac{4}{3}$. Hence, $L = 0$ if and only if $b = -2$ and $a = \frac{4}{3}$.

89. Since $\lim_{h\to 0}[f(x+h)-f(x-h)] = f(x)-f(x) = 0$ (f is differentiable and hence continuous) and $\lim_{h\to 0} 2h = 0$, we use l'Hospital's Rule:

$$\lim_{h\to 0}\frac{f(x+h)-f(x-h)}{2h} \overset{\text{H}}{=} \lim_{h\to 0}\frac{f'(x+h)(1)-f'(x-h)(-1)}{2} = \frac{f'(x)+f'(x)}{2} = \frac{2f'(x)}{2} = f'(x)$$

$\dfrac{f(x+h)-f(x-h)}{2h}$ is the slope of the secant line between $(x-h, f(x-h))$ and $(x+h, f(x+h))$. As $h\to 0$, this line gets closer to the tangent line and its slope approaches $f'(x)$.

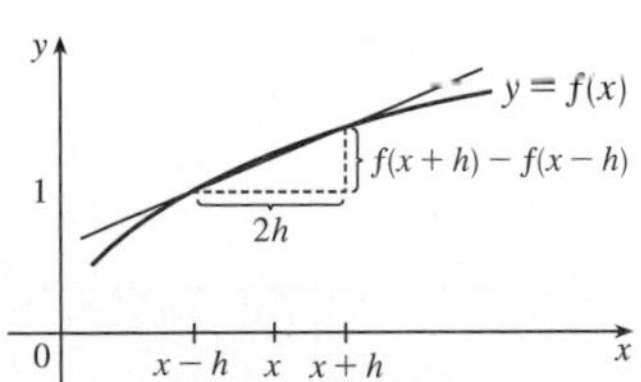

90. Since $\lim_{h\to 0}[f(x+h)-2f(x)+f(x-h)] = f(x)-2f(x)+f(x)=0$ [f is differentiable and hence continuous] and $\lim_{h\to 0} h^2 = 0$, we can apply l'Hospital's Rule:

$$\lim_{h\to 0}\frac{f(x+h)-2f(x)+f(x-h)}{h^2} \overset{\text{H}}{=} \lim_{h\to 0}\frac{f'(x+h)-f'(x-h)}{2h} = f''(x)$$

At the last step, we have applied the result of Exercise 89 to $f'(x)$.

91. (a) We show that $\lim_{x\to 0}\frac{f(x)}{x^n}=0$ for every integer $n\geq 0$. Let $y=\frac{1}{x^2}$. Then

$$\lim_{x\to 0}\frac{f(x)}{x^{2n}} = \lim_{x\to 0}\frac{e^{-1/x^2}}{(x^2)^n} = \lim_{y\to\infty}\frac{y^n}{e^y} \overset{\text{H}}{=} \lim_{y\to\infty}\frac{ny^{n-1}}{e^y} \overset{\text{H}}{=} \cdots \overset{\text{H}}{=} \lim_{y\to\infty}\frac{n!}{e^y}=0 \Rightarrow$$

$$\lim_{x\to 0}\frac{f(x)}{x^n} = \lim_{x\to 0}x^n\frac{f(x)}{x^{2n}} = \lim_{x\to 0}x^n\lim_{x\to 0}\frac{f(x)}{x^{2n}} = 0. \text{ Thus, } f'(0)=\lim_{x\to 0}\frac{f(x)-f(0)}{x-0} = \lim_{x\to 0}\frac{f(x)}{x}=0.$$

(b) Using the Chain Rule and the Quotient Rule we see that $f^{(n)}(x)$ exists for $x\neq 0$. In fact, we prove by induction that for each $n\geq 0$, there is a polynomial p_n and a non-negative integer k_n with $f^{(n)}(x)=p_n(x)f(x)/x^{k_n}$ for $x\neq 0$. This is true for $n=0$; suppose it is true for the nth derivative. Then $f'(x)=f(x)(2/x^3)$, so

$$\begin{aligned} f^{(n+1)}(x) &= \left[x^{k_n}\left[p_n'(x)\,f(x)+p_n(x)\,f'(x)\right]-k_nx^{k_n-1}p_n(x)\,f(x)\right]x^{-2k_n} \\ &= \left[x^{k_n}p_n'(x)+p_n(x)(2/x^3)-k_nx^{k_n-1}p_n(x)\right]f(x)x^{-2k_n} \\ &= \left[x^{k_n+3}p_n'(x)+2p_n(x)-k_nx^{k_n+2}p_n(x)\right]f(x)x^{-(2k_n+3)} \end{aligned}$$

which has the desired form.

Now we show by induction that $f^{(n)}(0)=0$ for all n. By part (a), $f'(0)=0$. Suppose that $f^{(n)}(0)=0$. Then

$$\begin{aligned} f^{(n+1)}(0) &= \lim_{x\to 0}\frac{f^{(n)}(x)-f^{(n)}(0)}{x-0} = \lim_{x\to 0}\frac{f^{(n)}(x)}{x} = \lim_{x\to 0}\frac{p_n(x)\,f(x)/x^{k_n}}{x} = \lim_{x\to 0}\frac{p_n(x)\,f(x)}{x^{k_n+1}} \\ &= \lim_{x\to 0}p_n(x)\lim_{x\to 0}\frac{f(x)}{x^{k_n+1}} = p_n(0)\cdot 0 = 0 \end{aligned}$$

92. (a) For f to be continuous, we need $\lim_{x\to 0}f(x)=f(0)=1$. We note that for $x\neq 0$, $\ln f(x)=\ln|x|^x = x\ln|x|$.

So $\lim_{x\to 0}\ln f(x) = \lim_{x\to 0}x\ln|x| = \lim_{x\to 0}\frac{\ln|x|}{1/x} \overset{\text{H}}{=} \lim_{x\to 0}\frac{1/x}{-1/x^2}=0$. Therefore, $\lim_{x\to 0}f(x)=\lim_{x\to 0}e^{\ln f(x)}=e^0=1$.

So f is continuous at 0.

(b) From the graphs, it appears that f is differentiable at 0.

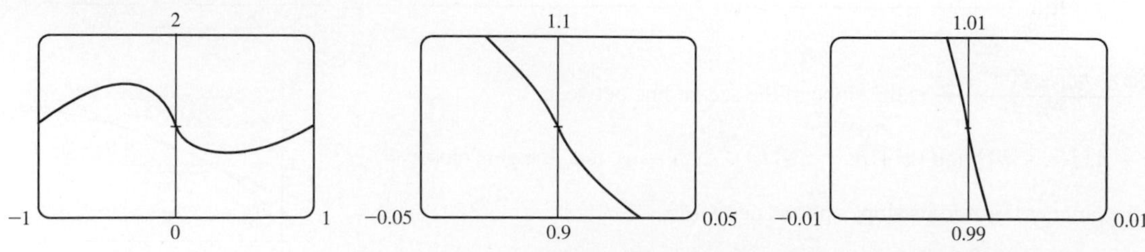

(c) To find f', we use logarithmic differentiation: $\ln f(x) = x \ln|x| \;\Rightarrow\; \dfrac{f'(x)}{f(x)} = x\left(\dfrac{1}{x}\right) + \ln|x| \;\Rightarrow$

$f'(x) = f(x)(1 + \ln|x|) = |x|^x(1 + \ln|x|)$, $x \neq 0$. Now $f'(x) \to -\infty$ as $x \to 0$ [since $|x|^x \to 1$ and $(1 + \ln|x|) \to -\infty$], so the curve has a vertical tangent at $(0, 1)$ and is therefore not differentiable there. The fact cannot be seen in the graphs in part (b) because $\ln|x| \to -\infty$ very slowly as $x \to 0$.

4.5 Summary of Curve Sketching

1. $y = f(x) = x^3 + 3x^2 = x^2(x + 3)$ **A.** f is a polynomial, so $D = \mathbb{R}$. **B.** y-intercept $= f(0) = 0$, x-intercepts are 0 and -3 **C.** No symmetry **D.** No asympote **E.** $f'(x) = 3x^2 + 6x = 3x(x + 2) > 0 \;\Leftrightarrow\; x < -2$ or $x > 0$, so f is increasing on $(-\infty, -2)$ and $(0, \infty)$, and decreasing on $(-2, 0)$. **F.** Local maximum value $f(-2) = 4$, local minimum value $f(0) = 0$ **G.** $f''(x) = 6x + 6 = 6(x + 1) > 0 \;\Leftrightarrow\; x > -1$, so f is CU on $(-1, \infty)$ and CD on $(-\infty, -1)$. IP at $(-1, 2)$

H.

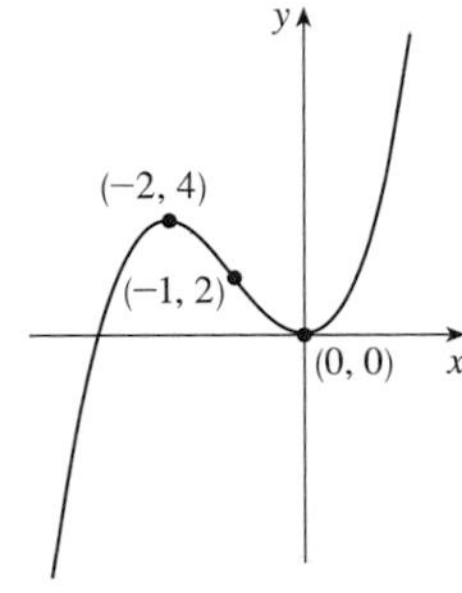

2. $y = f(x) = 2 + 3x^2 - x^3$ **A.** $D = \mathbb{R}$ **B.** y-intercept $= f(0) = 2$ **C.** No symmetry **D.** No asymptote **E.** $f'(x) = 6x - 3x^2 = 3x(2 - x) > 0 \;\Leftrightarrow\; 0 < x < 2$, so f is increasing on $(0, 2)$ and decreasing on $(-\infty, 0)$ and $(2, \infty)$. **F.** Local maximum value $f(2) = 6$, local minimum value $f(0) = 2$ **G.** $f''(x) = 6 - 6x = 6(1 - x) > 0 \;\Leftrightarrow\; x < 1$, so f is CU on $(-\infty, 1)$ and CD on $(1, \infty)$. IP at $(1, 4)$

H.

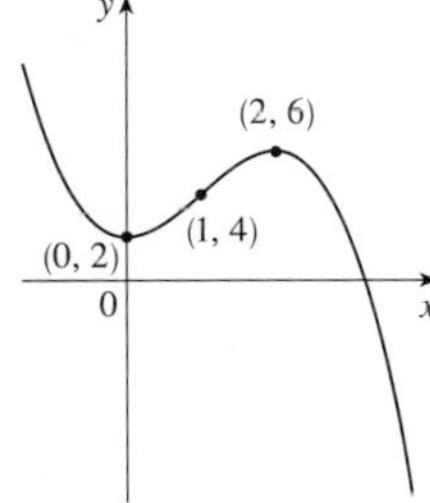

3. $y = f(x) = x^4 - 4x = x(x^3 - 4)$ **A.** $D = \mathbb{R}$ **B.** x-intercepts are 0 and $\sqrt[3]{4}$, y-intercept $= f(0) = 0$ **C.** No symmetry **D.** No asymptote **E.** $f'(x) = 4x^3 - 4 = 4(x^3 - 1) = 4(x - 1)(x^2 + x + 1) > 0 \;\Leftrightarrow\; x > 1$, so f is increasing on $(1, \infty)$ and decreasing on $(-\infty, 1)$. **F.** Local minimum value $f(1) = -3$, no local maximum **G.** $f''(x) = 12x^2 > 0$ for all x, so f is CU on $(-\infty, \infty)$. No IP

H.

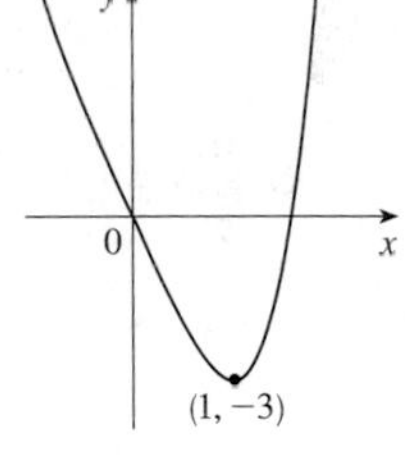

4. $y = f(x) = x^4 - 8x^2 + 8$ **A.** $D = \mathbb{R}$ **B.** y-intercept $f(0) = 8$; x-intercepts: $f(x) = 0 \;\Rightarrow$ [by the quadratic formula] $x = \pm\sqrt{4 \pm 2\sqrt{2}} \approx \pm 2.61, \pm 1.08$ **C.** $f(-x) = f(x)$, so f is even and symmetric about the y-axis **D.** No asymptote **E.** $f'(x) = 4x^3 - 16x = 4x(x^2 - 4) = 4x(x + 2)(x - 2) > 0 \;\Leftrightarrow\; -2 < x < 0$ or $x > 2$, so f is increasing on $(-2, 0)$ and $(2, \infty)$, and f is decreasing on $(-\infty, -2)$ and $(0, 2)$. **F.** Local maximum value $f(0) = 8$, local minimum values $f(\pm 2) = -8$ **G.** $f''(x) = 12x^2 - 16 = 4(3x^2 - 4) > 0 \;\Rightarrow\; |x| > 2/\sqrt{3}$ [≈ 1.15], so f is CU on $\left(-\infty, -2/\sqrt{3}\right)$ and $\left(2/\sqrt{3}, \infty\right)$, and f is CD on $\left(-2/\sqrt{3}, 2/\sqrt{3}\right)$. IP at $\left(\pm 2/\sqrt{3}, -\frac{8}{9}\right)$

H.

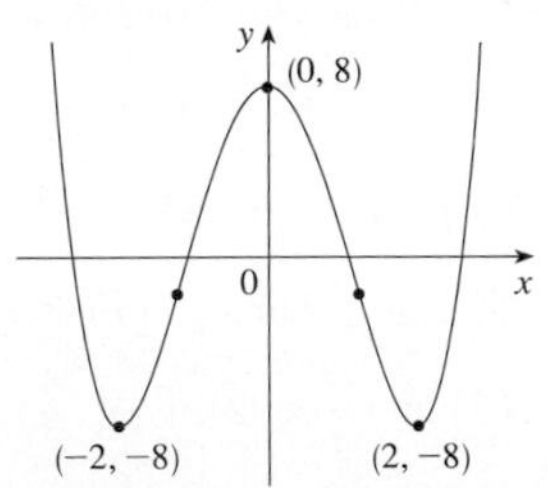

5. $y = f(x) = x(x-4)^3$ **A.** $D = \mathbb{R}$ **B.** x-intercepts are 0 and 4, y-intercept $f(0) = 0$ **C.** No symmetry

D. No asymptote

H.

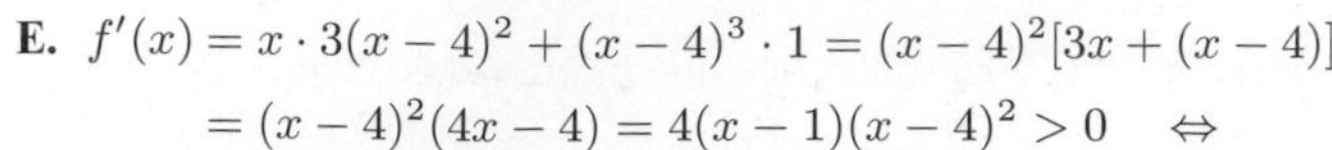

E. $f'(x) = x \cdot 3(x-4)^2 + (x-4)^3 \cdot 1 = (x-4)^2[3x + (x-4)]$
$= (x-4)^2(4x-4) = 4(x-1)(x-4)^2 > 0 \quad \Leftrightarrow$

$x > 1$, so f is increasing on $(1, \infty)$ and decreasing on $(-\infty, 1)$.

F. Local minimum value $f(1) = -27$, no local maximum value

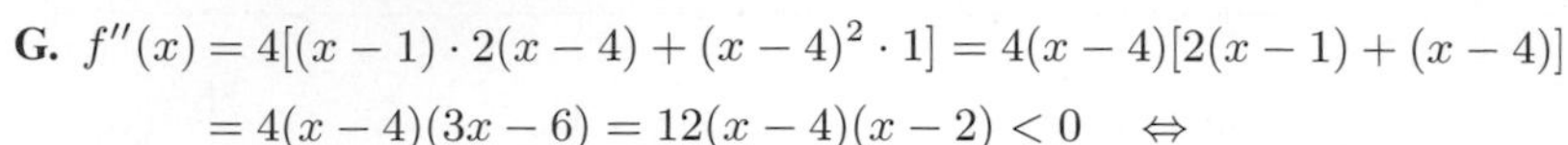

G. $f''(x) = 4[(x-1) \cdot 2(x-4) + (x-4)^2 \cdot 1] = 4(x-4)[2(x-1) + (x-4)]$
$= 4(x-4)(3x-6) = 12(x-4)(x-2) < 0 \quad \Leftrightarrow$

$2 < x < 4$, so f is CD on $(2, 4)$ and CU on $(-\infty, 2)$ and $(4, \infty)$. IPs at $(2, -16)$ and $(4, 0)$

6. $y = f(x) = x^5 - 5x = x(x^4 - 5)$ **A.** $D = \mathbb{R}$ **B.** x-intercepts $\pm\sqrt[4]{5}$ and 0, y-intercept $= f(0) = 0$

C. $f(-x) = -f(x)$, so f is odd; the curve is symmetric about the origin. **D.** No asymptote

E. $f'(x) = 5x^4 - 5 = 5(x^4 - 1) = 5(x^2 - 1)(x^2 + 1)$
$= 5(x+1)(x-1)(x^2+1) > 0 \quad \Leftrightarrow$

H.

$x < -1$ or $x > 1$, so f is increasing on $(-\infty, -1)$ and $(1, \infty)$, and f is decreasing on $(-1, 1)$. **F.** Local maximum value $f(-1) = 4$, local minimum value $f(1) = -4$ **G.** $f''(x) = 20x^3 > 0 \quad \Leftrightarrow \quad x > 0$, so f is CU on $(0, \infty)$ and CD on $(-\infty, 0)$. IP at $(0, 0)$

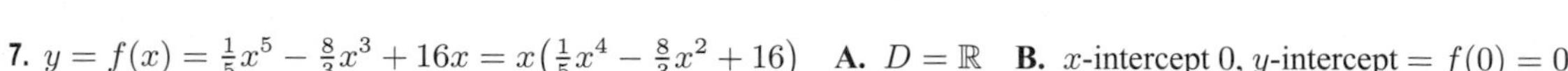

7. $y = f(x) = \frac{1}{5}x^5 - \frac{8}{3}x^3 + 16x = x\left(\frac{1}{5}x^4 - \frac{8}{3}x^2 + 16\right)$ **A.** $D = \mathbb{R}$ **B.** x-intercept 0, y-intercept $= f(0) = 0$

C. $f(-x) = -f(x)$, so f is odd; the curve is symmetric about the origin. **D.** No asymptote

E. $f'(x) = x^4 - 8x^2 + 16 = (x^2 - 4)^2 = (x+2)^2(x-2)^2 > 0$ for all x except ± 2, so f is increasing on $\mathbb{R}$. **F.** There is no local maximum or minimum value.

H.

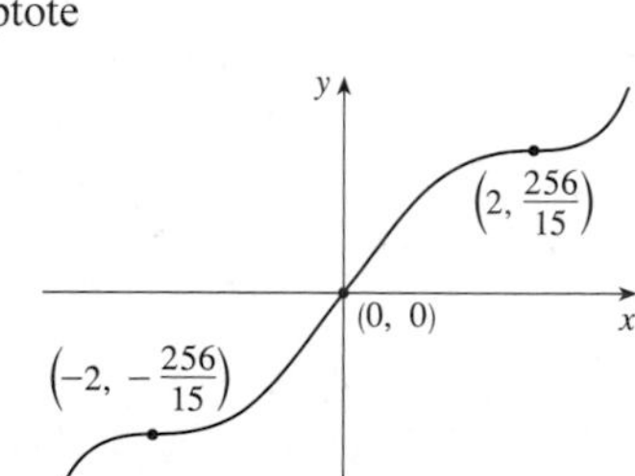

G. $f''(x) = 4x^3 - 16x = 4x(x^2 - 4) = 4x(x+2)(x-2) > 0 \quad \Leftrightarrow$
$-2 < x < 0$ or $x > 2$, so f is CU on $(-2, 0)$ and $(2, \infty)$, and f is CD on $(-\infty, -2)$ and $(0, 2)$. IP at $\left(-2, -\frac{256}{15}\right)$, $(0, 0)$, and $\left(2, \frac{256}{15}\right)$

8. $y = f(x) = (4 - x^2)^5$ **A.** $D = \mathbb{R}$ **B.** y-intercept: $f(0) = 4^5 = 1024$; x-intercepts: ± 2 **C.** $f(-x) = f(x) \Rightarrow$ f is even; the curve is symmetric about the y-axis. **D.** No asymptote **E.** $f'(x) = 5(4 - x^2)^4(-2x) = -10x(4 - x^2)^4$, so for $x \neq \pm 2$ we have $f'(x) > 0 \quad \Leftrightarrow \quad x < 0$ and $f'(x) < 0 \quad \Leftrightarrow \quad x > 0$. Thus, f is increasing on $(-\infty, 0)$ and decreasing on $(0, \infty)$. **F.** Local maximum value $f(0) = 1024$

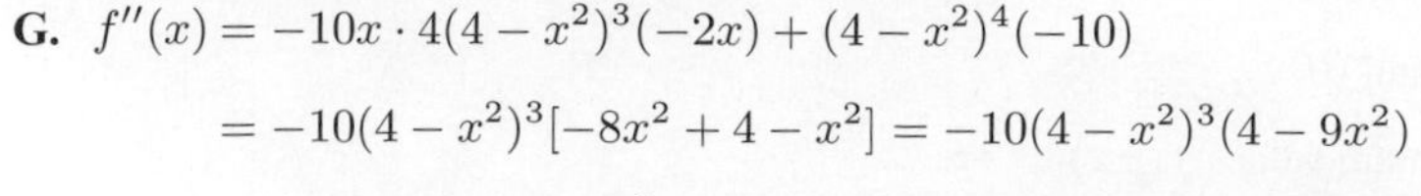

G. $f''(x) = -10x \cdot 4(4 - x^2)^3(-2x) + (4 - x^2)^4(-10)$
$= -10(4 - x^2)^3[-8x^2 + 4 - x^2] = -10(4 - x^2)^3(4 - 9x^2)$

H.

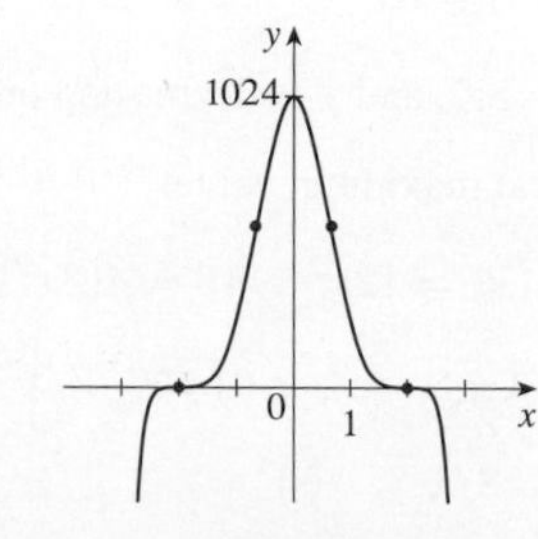

so $f''(x) = 0 \quad \Leftrightarrow \quad x = \pm 2, \pm\frac{2}{3}$. $f''(x) > 0 \quad \Leftrightarrow \quad -2 < x < -\frac{2}{3}$ and $\frac{2}{3} < x < 2$ and $f''(x) < 0 \quad \Leftrightarrow \quad x < -2, -\frac{2}{3} < x < \frac{2}{3}$, and $x > 2$, so f is CU on $(-\infty, 2)$, $\left(-\frac{2}{3}, \frac{2}{3}\right)$, and $(2, \infty)$, and CD on $\left(-2, -\frac{2}{3}\right)$ and $\left(\frac{2}{3}, 2\right)$.

IP at $(\pm 2, 0)$ and $\left(\pm\frac{2}{3}, \left(\frac{32}{9}\right)^5\right) \approx (\pm 0.67, 568.25)$

9. $y = f(x) = x/(x-1)$ **A.** $D = \{x \mid x \neq 1\} = (-\infty, 1) \cup (1, \infty)$ **B.** x-intercept $= 0$, y-intercept $= f(0) = 0$

C. No symmetry **D.** $\displaystyle\lim_{x\to\pm\infty} \frac{x}{x-1} = 1$, so $y = 1$ is a HA. $\displaystyle\lim_{x\to 1^-} \frac{x}{x-1} = -\infty$, $\displaystyle\lim_{x\to 1^+} \frac{x}{x-1} = \infty$, so $x = 1$ is a VA.

E. $f'(x) = \dfrac{(x-1)-x}{(x-1)^2} = \dfrac{-1}{(x-1)^2} < 0$ for $x \neq 1$, so f is decreasing on $(-\infty, 1)$ and $(1, \infty)$. **F.** No extreme values

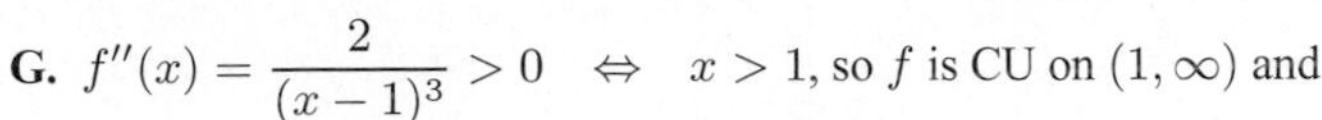

G. $f''(x) = \dfrac{2}{(x-1)^3} > 0 \;\Leftrightarrow\; x > 1$, so f is CU on $(1, \infty)$ and CD on $(-\infty, 1)$. No IP

H.

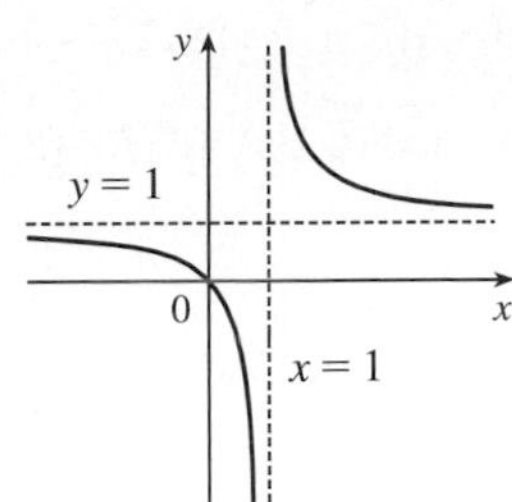

10. $y = f(x) = \dfrac{x^2+5x}{25-x^2} = \dfrac{x(x+5)}{(5+x)(5-x)} = \dfrac{x}{5-x}$ for $x \neq -5$. There is a hole in the graph at $\left(-5, -\frac{1}{2}\right)$.

A. $D = \{x \mid x \neq \pm 5\} = (-\infty, -5) \cup (-5, 5) \cup (5, \infty)$ **B.** x-intercept $= 0$, y-intercept $= f(0) = 0$ **C.** No symmetry

D. $\displaystyle\lim_{x\to\pm\infty} \frac{x}{5-x} = -1$, so $y = -1$ is a HA. $\displaystyle\lim_{x\to 5^-} \frac{x}{5-x} = \infty$, $\displaystyle\lim_{x\to 5^+} \frac{x}{5-x} = -\infty$, so $x = 5$ is a VA.

E. $f'(x) = \dfrac{(5-x)(1)-x(-1)}{(5-x)^2} = \dfrac{5}{(5-x)^2} > 0$ for all x in D, so f is increasing on $(-\infty, -5)$, $(-5, 5)$, and $(5, \infty)$. **F.** No extrema

G. $f'(x) = 5(5-x)^{-2} \;\Rightarrow$

$f''(x) = -10(5-x)^{-3}(-1) = \dfrac{10}{(5-x)^3} > 0 \;\Leftrightarrow\; x < 5$, so f is CU on $(-\infty, -5)$ and $(-5, 5)$, and f is CD on $(5, \infty)$. No IP

H.

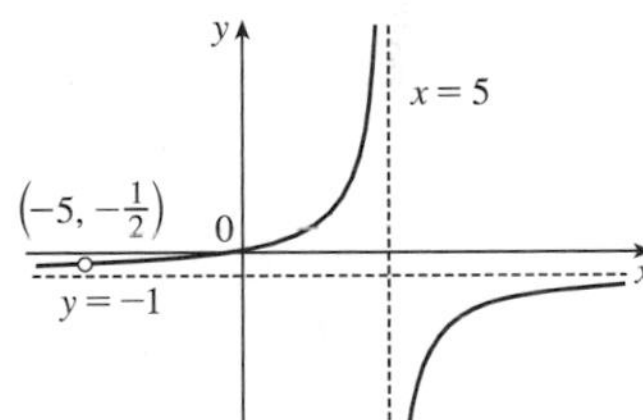

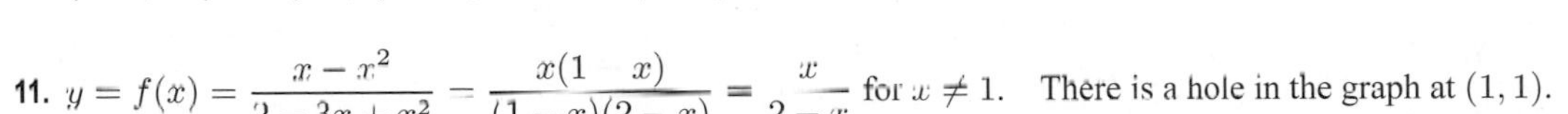

11. $y = f(x) = \dfrac{x-x^2}{2-3x+x^2} = \dfrac{x(1-x)}{(1-x)(2-x)} = \dfrac{x}{2-x}$ for $x \neq 1$. There is a hole in the graph at $(1, 1)$.

A. $D = \{x \mid x \neq 1, 2\} = (-\infty, 1) \cup (1, 2) \cup (2, \infty)$ **B.** x-intercept $= 0$, y-intercept $= f(0) = 0$ **C.** No symmetry

D. $\displaystyle\lim_{x\to\pm\infty} \frac{x}{2-x} = -1$, so $y = -1$ is a HA. $\displaystyle\lim_{x\to 2^-} \frac{x}{2-x} = \infty$, $\displaystyle\lim_{x\to 2^+} \frac{x}{2-x} = -\infty$, so $x = 2$ is a VA.

E. $f'(x) = \dfrac{(2-x)(1)-x(-1)}{(2-x)^2} = \dfrac{2}{(2-x)^2} > 0$ $[x \neq 1, 2]$, so f is increasing on $(-\infty, 1)$, $(1, 2)$, and $(2, \infty)$. **F.** No extrema

G. $f'(x) = 2(2-x)^{-2} \;\Rightarrow$

$f''(x) = -4(2-x)^{-3}(-1) = \dfrac{4}{(2-x)^3} > 0 \;\Leftrightarrow\; x < 2$, so f is CU on $(-\infty, 1)$ and $(1, 2)$, and f is CD on $(2, \infty)$. No IP

H.

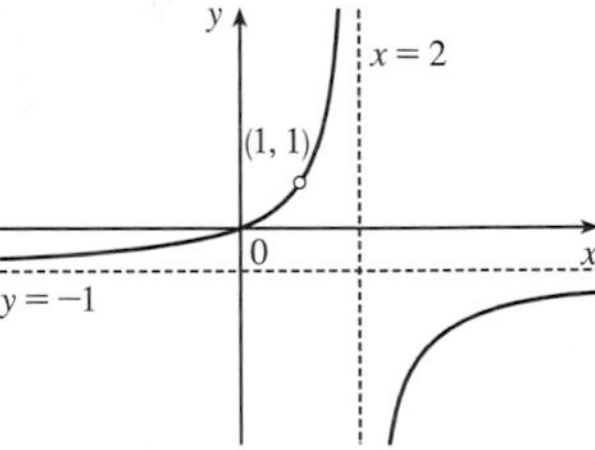

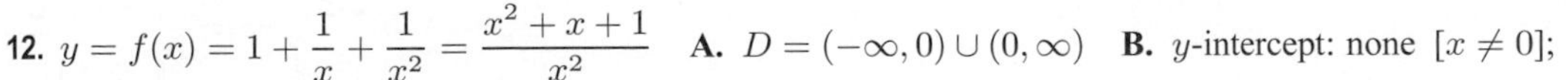

12. $y = f(x) = 1 + \dfrac{1}{x} + \dfrac{1}{x^2} = \dfrac{x^2+x+1}{x^2}$ **A.** $D = (-\infty, 0) \cup (0, \infty)$ **B.** y-intercept: none $[x \neq 0]$;

x-intercepts: $f(x) = 0 \;\Leftrightarrow\; x^2 + x + 1 = 0$, there is no real solution, and hence, no x-intercept **C.** No symmetry

D. $\displaystyle\lim_{x\to\pm\infty}\left(1 + \frac{1}{x} + \frac{1}{x^2}\right) = 1$, so $y = 1$ is a HA. $\displaystyle\lim_{x\to 0} f(x) = \infty$, so $x = 0$ is a VA. **E.** $f'(x) = -\dfrac{1}{x^2} - \dfrac{2}{x^3} = \dfrac{-x-2}{x^3}$.

$f'(x) > 0 \;\Leftrightarrow\; -2 < x < 0$ and $f'(x) < 0 \;\Leftrightarrow\; x < -2$ or $x > 0$, so f is increasing on $(-2, 0)$ and decreasing on $(-\infty, -2)$ and $(0, \infty)$. **F.** Local minimum value $f(-2) = \frac{3}{4}$; no local maximum **G.** $f''(x) = \dfrac{2}{x^3} + \dfrac{6}{x^4} = \dfrac{2x+6}{x^4}$. $f''(x) < 0 \Leftrightarrow x < -3$ and $f''(x) > 0 \Leftrightarrow -3 < x < 0$ and $x > 0$, so f is CD on $(-\infty, -3)$ and CU on $(-3, 0)$ and $(0, \infty)$. IP at $\left(-3, \frac{7}{9}\right)$

H.

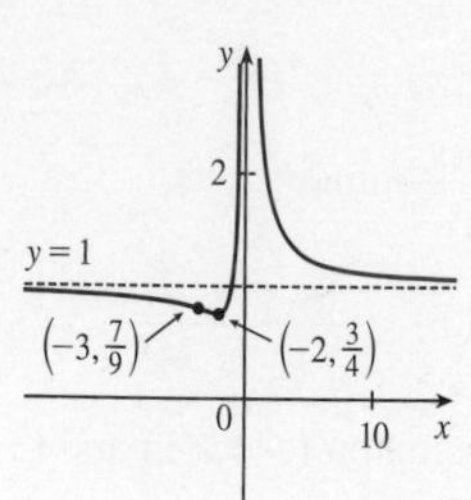

13. $y = f(x) = \dfrac{x}{x^2-4} = \dfrac{x}{(x+2)(x-2)}$ **A.** $D = (-\infty, -2) \cup (-2, 2) \cup (2, \infty)$ **B.** x-intercept $= 0$, y-intercept $= f(0) = 0$ **C.** $f(-x) = -f(x)$, so f is odd; the graph is symmetric about the origin.

D. $\displaystyle\lim_{x\to 2^+} \frac{x}{x^2-4} = \infty$, $\displaystyle\lim_{x\to 2^-} f(x) = -\infty$, $\displaystyle\lim_{x\to -2^+} f(x) = \infty$, $\displaystyle\lim_{x\to -2^-} f(x) = -\infty$, so $x = \pm 2$ are VAs.

$\displaystyle\lim_{x\to \pm\infty} \frac{x}{x^2-4} = 0$, so $y = 0$ is a HA. **E.** $f'(x) = \dfrac{(x^2-4)(1) - x(2x)}{(x^2-4)^2} = -\dfrac{x^2+4}{(x^2-4)^2} < 0$ for all x in D, so f is decreasing on $(-\infty, -2)$, $(-2, 2)$, and $(2, \infty)$.

F. No local extrema

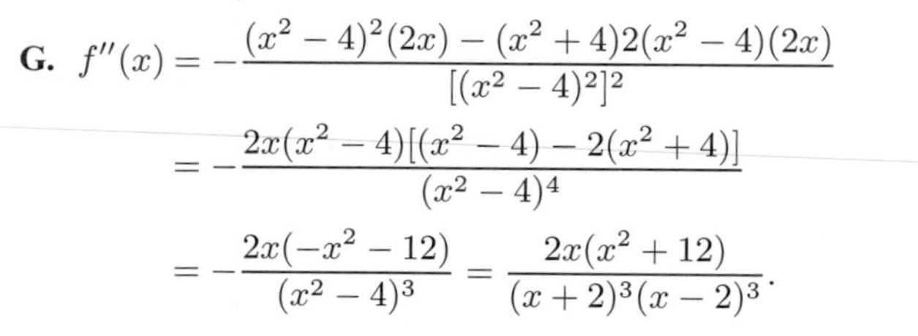

G.
$$\begin{aligned} f''(x) &= -\frac{(x^2-4)^2(2x) - (x^2+4)2(x^2-4)(2x)}{[(x^2-4)^2]^2} \\ &= -\frac{2x(x^2-4)[(x^2-4) - 2(x^2+4)]}{(x^2-4)^4} \\ &= -\frac{2x(-x^2-12)}{(x^2-4)^3} = \frac{2x(x^2+12)}{(x+2)^3(x-2)^3}. \end{aligned}$$

H.

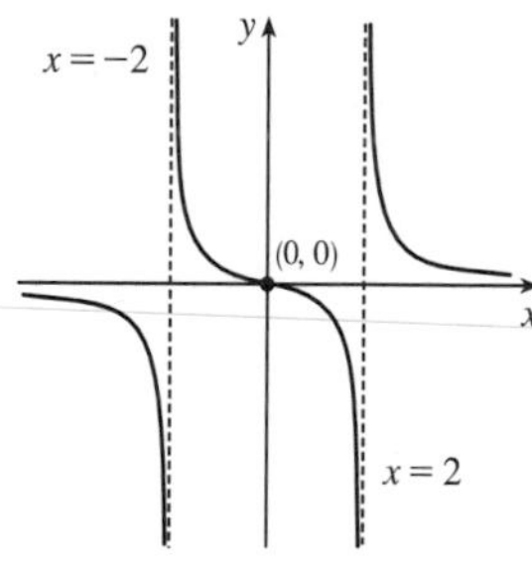

$f''(x) < 0$ if $x < -2$ or $0 < x < 2$, so f is CD on $(-\infty, -2)$ and $(0, 2)$, and CU on $(-2, 0)$ and $(2, \infty)$. IP at $(0, 0)$

14. $y = f(x) = \dfrac{1}{x^2-4} = \dfrac{1}{(x+2)(x-2)}$ **A.** $D = (-\infty, -2) \cup (-2, 2) \cup (2, \infty)$ **B.** No x-intercept, y-intercept $= f(0) = -\frac{1}{4}$ **C.** $f(-x) = f(x)$, so f is even; the graph is symmetric about the y-axis.

D. $\displaystyle\lim_{x\to 2^+} \frac{1}{x^2-4} = \infty$, $\displaystyle\lim_{x\to 2^-} f(x) = -\infty$, $\displaystyle\lim_{x\to -2^+} f(x) = -\infty$, $\displaystyle\lim_{x\to -2^-} f(x) = \infty$, so $x = \pm 2$ are VAs. $\displaystyle\lim_{x\to \pm\infty} f(x) = 0$, so $y = 0$ is a HA. **E.** $f'(x) = -\dfrac{2x}{(x^2-4)^2}$ [Reciprocal Rule] > 0 if $x < 0$ and x is in D, so f is increasing on $(-\infty, -2)$ and $(-2, 0)$. f is decreasing on $(0, 2)$ and $(2, \infty)$.

F. Local maximum value $f(0) = -\frac{1}{4}$, no local minimum value

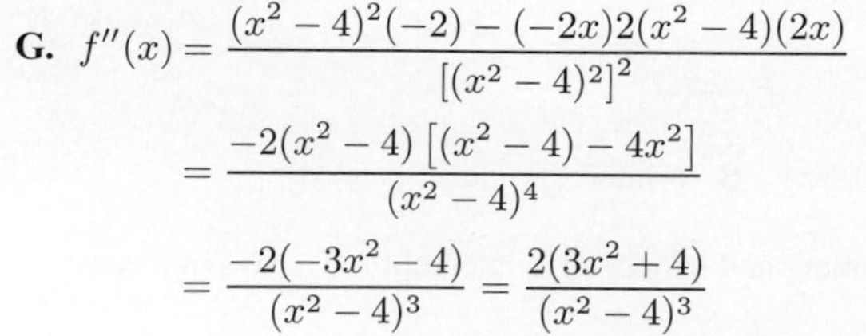

G.
$$\begin{aligned} f''(x) &= \frac{(x^2-4)^2(-2) - (-2x)2(x^2-4)(2x)}{[(x^2-4)^2]^2} \\ &= \frac{-2(x^2-4)\left[(x^2-4) - 4x^2\right]}{(x^2-4)^4} \\ &= \frac{-2(-3x^2-4)}{(x^2-4)^3} = \frac{2(3x^2+4)}{(x^2-4)^3} \end{aligned}$$

H.

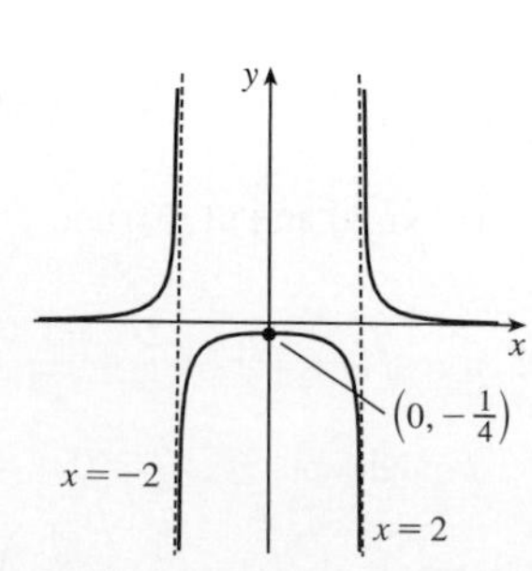

$f''(x) < 0 \Leftrightarrow -2 < x < 2$, so f is CD on $(-2, 2)$ and CU on $(-\infty, -2)$ and $(2, \infty)$. No IP

15. $y = f(x) = \dfrac{x^2}{x^2+3} = \dfrac{(x^2+3)-3}{x^2+3} = 1 - \dfrac{3}{x^2+3}$ **A.** $D = \mathbb{R}$ **B.** y-intercept: $f(0) = 0$;

x-intercepts: $f(x) = 0 \iff x = 0$ **C.** $f(-x) = f(x)$, so f is even; the graph is symmetric about the y-axis.

D. $\displaystyle\lim_{x\to\pm\infty} \frac{x^2}{x^2+3} = 1$, so $y = 1$ is a HA. No VA. **E.** Using the Reciprocal Rule, $f'(x) = -3 \cdot \dfrac{-2x}{(x^2+3)^2} = \dfrac{6x}{(x^2+3)^2}$.

$f'(x) > 0 \iff x > 0$ and $f'(x) < 0 \iff x < 0$, so f is decreasing on $(-\infty, 0)$ and increasing on $(0, \infty)$.

F. Local minimum value $f(0) = 0$, no local maximum.

G. $f''(x) = \dfrac{(x^2+3)^2 \cdot 6 - 6x \cdot 2(x^2+3) \cdot 2x}{[(x^2+3)^2]^2}$

$= \dfrac{6(x^2+3)[(x^2+3) - 4x^2]}{(x^2+3)^4} = \dfrac{6(3-3x^2)}{(x^2+3)^3} = \dfrac{-18(x+1)(x-1)}{(x^2+3)^3}$

$f''(x)$ is negative on $(-\infty, -1)$ and $(1, \infty)$ and positive on $(-1, 1)$,

so f is CD on $(-\infty, -1)$ and $(1, \infty)$ and CU on $(-1, 1)$. IP at $\left(\pm 1, \frac{1}{4}\right)$

H.

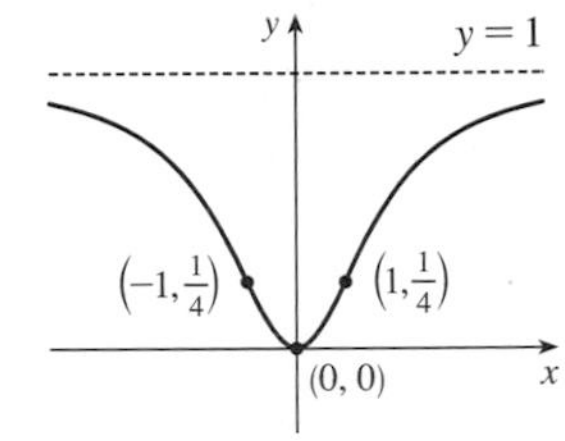

16. $y = f(x) = \dfrac{(x-1)^2}{x^2+1} \geq 0$ with equality $\iff x = 1$. **A.** $D = \mathbb{R}$ **B.** y-intercept $= f(0) = 1$; x-intercept 1 **C.** No symmetry **D.** $\displaystyle\lim_{x\to\pm\infty} f(x) = \lim_{x\to\pm\infty} \frac{x^2-2x+1}{x^2+1} = \lim_{x\to\pm\infty} \frac{1-2/x+1/x^2}{1+1/x^2} = 1$, so $y = 1$ is a HA. No VA

E. $f'(x) = \dfrac{(x^2+1)2(x-1) - (x-1)^2(2x)}{(x^2+1)^2} = \dfrac{2(x-1)\left[(x^2+1) - x(x-1)\right]}{(x^2+1)^2} = \dfrac{2(x-1)(x+1)}{(x^2+1)^2} < 0 \iff$

$-1 < x < 1$, so f is decreasing on $(-1, 1)$ and increasing on $(-\infty, -1)$ and $(1, \infty)$ **F.** Local maximum value $f(-1) = 2$, local minimum value $f(1) = 0$

G. $f''(x) = \dfrac{(x^2+1)^2(4x) - (2x^2-2)2(x^2+1)(2x)}{[(x^2+1)^2]^2} = \dfrac{4x(x^2+1)\left[(x^2+1) - (2x^2-2)\right]}{(x^2+1)^4} = \dfrac{4x(3-x^2)}{(x^2+1)^3}$.

$f''(x) > 0 \iff x < -\sqrt{3}$ or $0 < x < \sqrt{3}$, so f is CU on $\left(-\infty, -\sqrt{3}\right)$ and $\left(0, \sqrt{3}\right)$, and f is CD on $\left(-\sqrt{3}, 0\right)$ and $\left(\sqrt{3}, \infty\right)$.

$f\left(\pm\sqrt{3}\right) = \frac{1}{4}\left(\sqrt{3} \mp 1\right)^2 = \frac{1}{4}\left(4 \mp 2\sqrt{3}\right) = 1 \mp \frac{1}{2}\sqrt{3}$ $[\approx 0.13, 1.87]$, so there are IPs at $\left(-\sqrt{3}, 1 + \frac{1}{2}\sqrt{3}\right)$, $(0, 1)$, and $\left(\sqrt{3}, 1 - \frac{1}{2}\sqrt{3}\right)$. Note that the graph is symmetric about the point $(0, 1)$.

H.

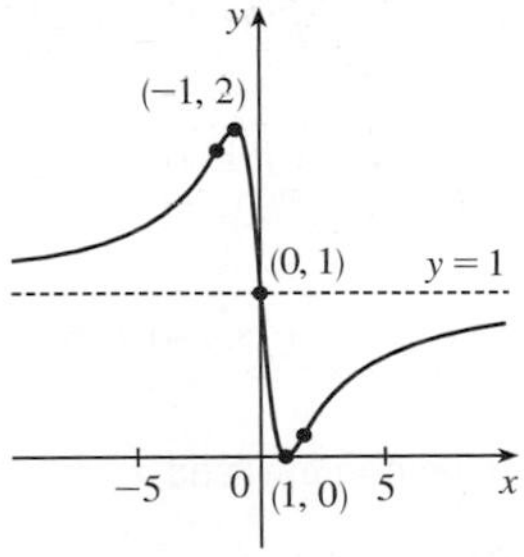

17. $y = f(x) = \dfrac{x-1}{x^2}$ **A.** $D = \{x \mid x \neq 0\} = (-\infty, 0) \cup (0, \infty)$ **B.** No y-intercept; x-intercept: $f(x) = 0 \iff x = 1$

C. No symmetry **D.** $\displaystyle\lim_{x\to\pm\infty} \frac{x-1}{x^2} = 0$, so $y = 0$ is a HA. $\displaystyle\lim_{x\to 0} \frac{x-1}{x^2} = -\infty$, so $x = 0$ is a VA.

E. $f'(x) = \dfrac{x^2 \cdot 1 - (x-1) \cdot 2x}{(x^2)^2} = \dfrac{-x^2+2x}{x^4} = \dfrac{-(x-2)}{x^3}$, so $f'(x) > 0 \iff 0 < x < 2$ and $f'(x) < 0 \iff$

$x < 0$ or $x > 2$. Thus, f is increasing on $(0, 2)$ and decreasing on $(-\infty, 0)$ and $(2, \infty)$. **F.** No local minimum, local maximum value $f(2) = \frac{1}{4}$.

G. $f''(x) = \dfrac{x^3 \cdot (-1) - [-(x-2)] \cdot 3x^2}{(x^3)^2} = \dfrac{2x^3 - 6x^2}{x^6} = \dfrac{2(x-3)}{x^4}$.

$f''(x)$ is negative on $(-\infty, 0)$ and $(0, 3)$ and positive on $(3, \infty)$, so f is CD on $(-\infty, 0)$ and $(0, 3)$ and CU on $(3, \infty)$. IP at $\left(3, \frac{2}{9}\right)$

H.

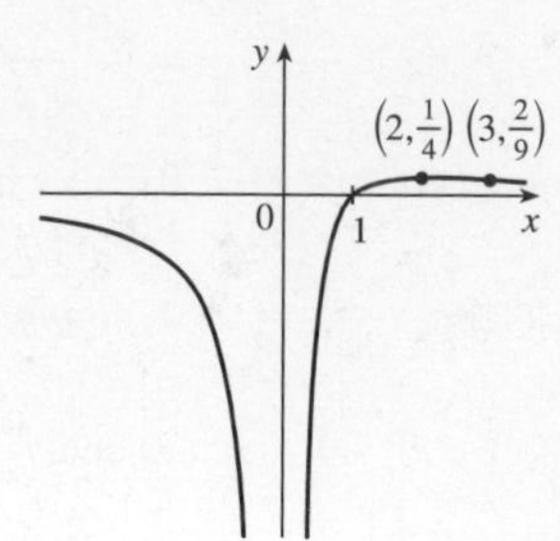

18. $y = f(x) = \dfrac{x}{x^3 - 1}$ **A.** $D = (-\infty, 1) \cup (1, \infty)$ **B.** y-intercept: $f(0) = 0$; x-intercept: $f(x) = 0 \Leftrightarrow x = 0$

C. No symmetry **D.** $\displaystyle\lim_{x \to \pm\infty} \frac{x}{x^3 - 1} = 0$, so $y = 0$ is a HA. $\displaystyle\lim_{x \to 1^-} f(x) = -\infty$ and $\displaystyle\lim_{x \to 1^+} f(x) = \infty$, so $x = 1$ is a VA.

E. $f'(x) = \dfrac{(x^3 - 1)(1) - x(3x^2)}{(x^3 - 1)^2} = \dfrac{-2x^3 - 1}{(x^3 - 1)^2}$. $f'(x) = 0 \Rightarrow x = -\sqrt[3]{1/2}$. $f'(x) > 0 \Leftrightarrow x < -\sqrt[3]{1/2}$ and

$f'(x) < 0 \Leftrightarrow -\sqrt[3]{1/2} < x < 1$ and $x > 1$, so f is increasing on $\left(-\infty, -\sqrt[3]{1/2}\right)$ and decreasing on $\left(-\sqrt[3]{1/2}, 1\right)$ and $(1, \infty)$. **F.** Local maximum value $f\left(-\sqrt[3]{1/2}\right) = \frac{2}{3}\sqrt[3]{1/2}$; no local minimum

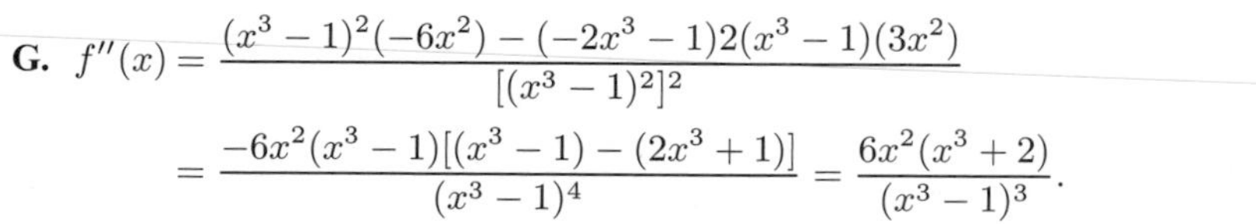

G. $$f''(x) = \frac{(x^3 - 1)^2(-6x^2) - (-2x^3 - 1)2(x^3 - 1)(3x^2)}{[(x^3 - 1)^2]^2} = \frac{-6x^2(x^3 - 1)[(x^3 - 1) - (2x^3 + 1)]}{(x^3 - 1)^4} = \frac{6x^2(x^3 + 2)}{(x^3 - 1)^3}.$$

$f''(x) > 0 \Leftrightarrow x < -\sqrt[3]{2}$ and $x > 1$, $f''(x) < 0 \Leftrightarrow -\sqrt[3]{2} < x < 0$ and $0 < x < 1$, so f is CU on $\left(-\infty, -\sqrt[3]{2}\right)$ and $(1, \infty)$ and CD on $\left(-\sqrt[3]{2}, 1\right)$. IP at $\left(-\sqrt[3]{2}, \frac{1}{3}\sqrt[3]{2}\right)$

H.

19. $y = f(x) = \dfrac{x^3}{x^3 + 1} = \dfrac{x^3}{(x + 1)(x^2 - x + 1)}$ **A.** $D = (-\infty, -1) \cup (-1, \infty)$ **B.** y-intercept: $f(0) = 0$; x-intercept:

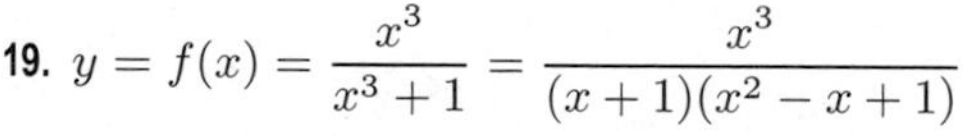

$f(x) = 0 \Leftrightarrow x = 0$ **C.** No symmetry **D.** $\displaystyle\lim_{x \to \pm\infty} \frac{x^3}{x^3 + 1} = \frac{1}{1 + 1/x^3} = 1$, so $y = 1$ is a HA. $\displaystyle\lim_{x \to -1^-} f(x) = \infty$ and $\displaystyle\lim_{x \to -1^+} f(x) = -\infty$, so $x = -1$ is a VA. **E.** $f'(x) = \dfrac{(x^3 + 1)(3x^2) - x^3(3x^2)}{(x^3 + 1)^2} = \dfrac{3x^2}{(x^3 + 1)^2}$. $f'(x) > 0$ for $x \neq -1$ (not in the domain) and $x \neq 0$ ($f' = 0$), so f is increasing on $(-\infty, -1)$, $(-1, 0)$, and $(0, \infty)$, and furthermore, by Exercise 4.3.91, f is increasing on $(-\infty, -1)$, and $(-1, \infty)$. **F.** No local extrema

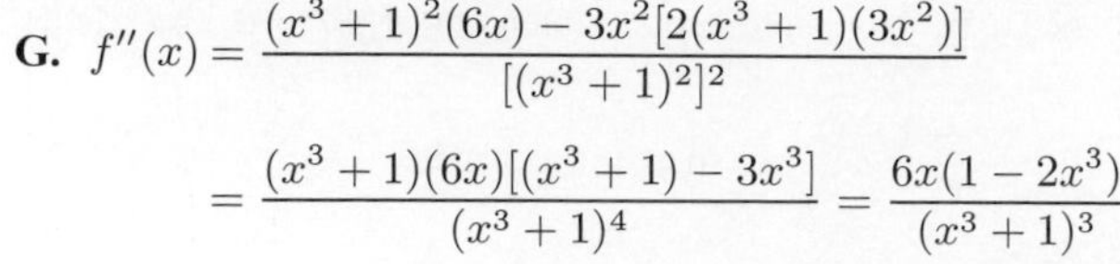

G. $$f''(x) = \frac{(x^3 + 1)^2(6x) - 3x^2[2(x^3 + 1)(3x^2)]}{[(x^3 + 1)^2]^2} = \frac{(x^3 + 1)(6x)[(x^3 + 1) - 3x^3]}{(x^3 + 1)^4} = \frac{6x(1 - 2x^3)}{(x^3 + 1)^3}$$

$f''(x) > 0 \Leftrightarrow x < -1$ or $0 < x < \sqrt[3]{\frac{1}{2}}$ [≈ 0.79], so f is CU on $(-\infty, -1)$ and $\left(0, \sqrt[3]{\frac{1}{2}}\right)$ and CD on $(-1, 0)$ and $\left(\sqrt[3]{\frac{1}{2}}, \infty\right)$. There are IPs at $(0, 0)$ and $\left(\sqrt[3]{\frac{1}{2}}, \frac{1}{3}\right)$.

H.

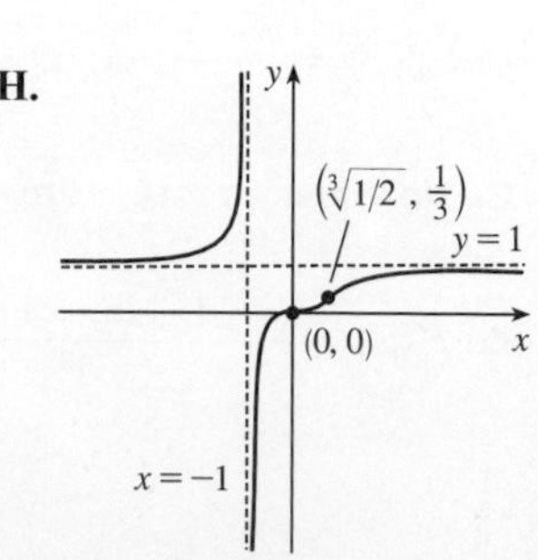

20. $y = f(x) = \dfrac{x^3}{x-2} = x^2 + 2x + 4 + \dfrac{8}{x-2}$ [by long division] **A.** $D = (-\infty, 2) \cup (2, \infty)$ **B.** x-intercept $= 0$, y-intercept $= f(0) = 0$ **C.** No symmetry **D.** $\displaystyle\lim_{x\to 2^-} \frac{x^3}{x-2} = -\infty$ and $\displaystyle\lim_{x\to 2^+} \frac{x^3}{x-2} = \infty$, so $x = 2$ is a VA.

There are no horizontal or slant asymptotes. *Note:* Since $\displaystyle\lim_{x\to\pm\infty} \frac{8}{x-2} = 0$, the parabola $y = x^2 + 2x + 4$ is approached asymptotically as $x \to \pm\infty$.

E. $f'(x) = \dfrac{(x-2)(3x^2) - x^3(1)}{(x-2)^2} = \dfrac{x^2[3(x-2) - x]}{(x-2)^2} = \dfrac{x^2(2x-6)}{(x-2)^2} = \dfrac{2x^2(x-3)}{(x-2)^2} > 0 \Leftrightarrow x > 3$ and

$f'(x) < 0 \Leftrightarrow x < 0$ or $0 < x < 2$ or $2 < x < 3$, so f is increasing on $(3, \infty)$ and f is decreasing on $(-\infty, 2)$ and $(2, 3)$.

F. Local minimum value $f(3) = 27$, no local maximum value **G.** $f'(x) = 2\dfrac{x^3 - 3x^2}{(x-2)^2} \Rightarrow$

$$f''(x) = 2\frac{(x-2)^2(3x^2-6x) - (x^3-3x^2)2(x-2)}{[(x-2)^2]^2}$$
$$= 2\frac{(x-2)x[(x-2)(3x-6) - (x^2-3x)2]}{(x-2)^4}$$
$$= \frac{2x(3x^2 - 12x + 12 - 2x^2 + 6x)}{(x-2)^3}$$
$$= \frac{2x(x^2 - 6x + 12)}{(x-2)^3} > 0 \Leftrightarrow$$

H.

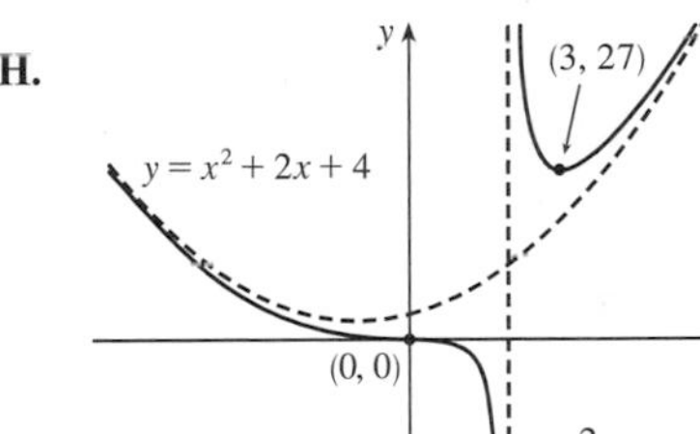

$x < 0$ or $x > 2$, so f is CU on $(-\infty, 0)$ and $(2, \infty)$, and f is CD on $(0, 2)$. IP at $(0, 0)$

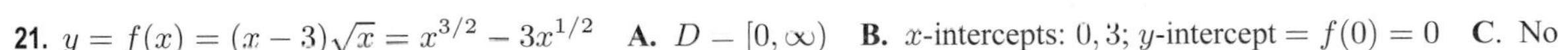

21. $y = f(x) = (x-3)\sqrt{x} = x^{3/2} - 3x^{1/2}$ **A.** $D = [0, \infty)$ **B.** x-intercepts: $0, 3$; y-intercept $= f(0) = 0$ **C.** No symmetry **D.** No asymptote **E.** $f'(x) = \frac{3}{2}x^{1/2} - \frac{3}{2}x^{-1/2} = \frac{3}{2}x^{-1/2}(x-1) = \dfrac{3(x-1)}{2\sqrt{x}} > 0 \Leftrightarrow x > 1$,

so f is increasing on $(1, \infty)$ and decreasing on $(0, 1)$.

F. Local minimum value $f(1) = -2$, no local maximum value

G. $f''(x) = \frac{3}{4}x^{-1/2} + \frac{3}{4}x^{-3/2} = \frac{3}{4}x^{-3/2}(x+1) = \dfrac{3(x+1)}{4x^{3/2}} > 0$ for $x > 0$,

so f is CU on $(0, \infty)$. No IP

H.

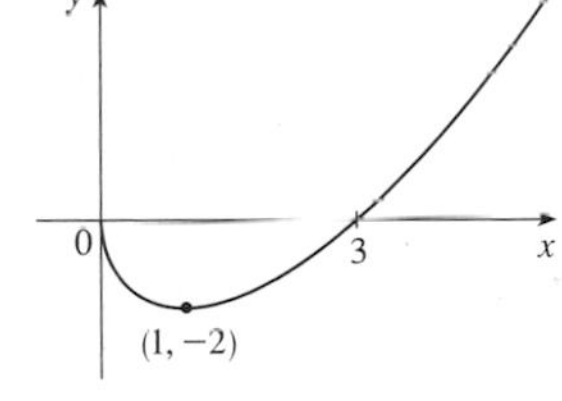

22. $y = f(x) = (x-4)\sqrt[3]{x} = x^{4/3} - 4x^{1/3}$ **A.** $D = \mathbb{R}$ **B.** y-intercept $= f(0) = 0$; x-intercepts: 0 and 4

C. No symmetry **D.** No asymptote

E. $f'(x) = \frac{4}{3}x^{1/3} - \frac{4}{3}x^{-2/3} = \frac{4}{3}x^{-2/3}(x-1) = \dfrac{4(x-1)}{3x^{2/3}}$. $f'(x) > 0 \Leftrightarrow$

$x > 1$, so f is increasing on $(1, \infty)$ and f is decreasing on $(-\infty, 1)$.

F. Local minimum value $f(1) = -3$

G. $f''(x) = \frac{4}{9}x^{-2/3} + \frac{8}{9}x^{-5/3} = \frac{4}{9}x^{-5/3}(x+2) = \dfrac{4(x+2)}{9x^{5/3}}$.

$f''(x) < 0 \Leftrightarrow -2 < x < 0$, so f is CD on $(-2, 0)$, and f is CU on $(-\infty, -2)$ and $(0, \infty)$. There are IPs at $\left(-2, 6\sqrt[3]{2}\right)$ and $(0, 0)$.

H.

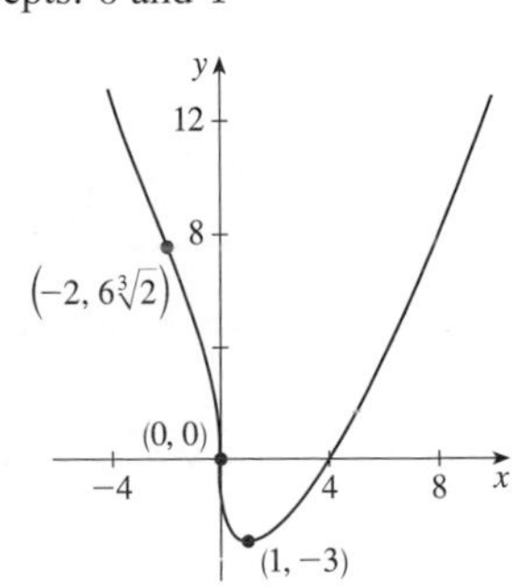

23. $y = f(x) = \sqrt{x^2 + x - 2} = \sqrt{(x+2)(x-1)}$ **A.** $D = \{x \mid (x+2)(x-1) \geq 0\} = (-\infty, -2] \cup [1, \infty)$

B. y-intercept: none; x-intercepts: -2 and 1 **C.** No symmetry **D.** No asymptote

E. $f'(x) = \frac{1}{2}(x^2 + x - 2)^{-1/2}(2x+1) = \dfrac{2x+1}{2\sqrt{x^2 + x - 2}}$, $f'(x) = 0$ if $x = -\frac{1}{2}$, but $-\frac{1}{2}$ is not in the domain.

$f'(x) > 0 \;\Rightarrow\; x > -\frac{1}{2}$ and $f'(x) < 0 \;\Rightarrow\; x < -\frac{1}{2}$, so (considering the domain) f is increasing on $(1, \infty)$ and f is decreasing on $(-\infty, -2)$. **F.** No local extrema

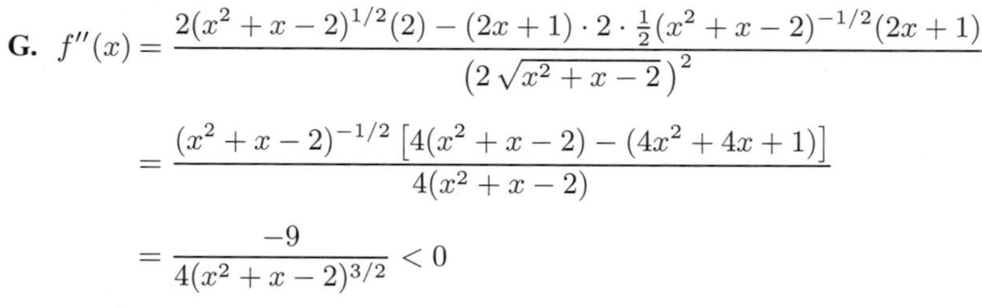

G. $$f''(x) = \frac{2(x^2 + x - 2)^{1/2}(2) - (2x+1) \cdot 2 \cdot \frac{1}{2}(x^2 + x - 2)^{-1/2}(2x+1)}{\left(2\sqrt{x^2 + x - 2}\right)^2}$$

$$= \frac{(x^2 + x - 2)^{-1/2}\left[4(x^2 + x - 2) - (4x^2 + 4x + 1)\right]}{4(x^2 + x - 2)}$$

$$= \frac{-9}{4(x^2 + x - 2)^{3/2}} < 0$$

so f is CD on $(-\infty, -2)$ and $(1, \infty)$. No IP

H.

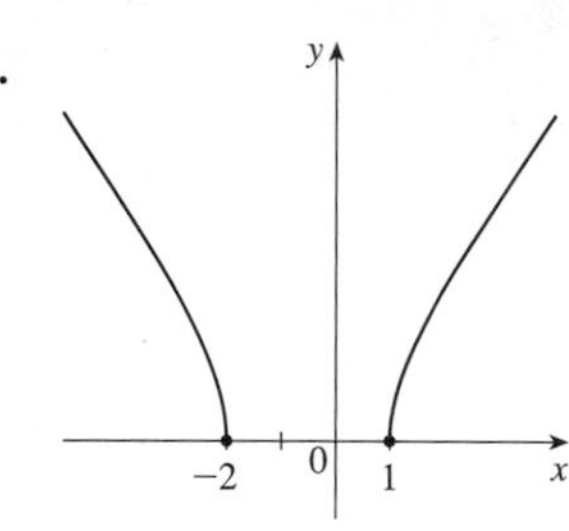

24. $y = f(x) = \sqrt{x^2 + x} - x = \sqrt{x(x+1)} - x$ **A.** $D = (-\infty, -1] \cup [0, \infty)$ **B.** y-intercept: $f(0) = 0$; x-intercepts: $f(x) = 0 \;\Rightarrow\; \sqrt{x^2 + x} = x \;\Rightarrow\; x^2 + x = x^2 \;\Rightarrow\; x = 0$ **C.** No symmetry

D. $$\lim_{x\to\infty} f(x) = \lim_{x\to\infty} \left(\sqrt{x^2 + x} - x\right) \frac{\sqrt{x^2 + x} + x}{\sqrt{x^2 + x} + x} = \lim_{x\to\infty} \frac{x^2 + x - x^2}{\sqrt{x^2 + x} + x}$$

$$= \lim_{x\to\infty} \frac{x/x}{\left(\sqrt{x^2 + x} + x\right)/x} = \lim_{x\to\infty} \frac{1}{\sqrt{1 + 1/x} + 1} = \frac{1}{2},$$ so $y = \frac{1}{2}$ is a HA. No VA

E. $f'(x) = \frac{1}{2}(x^2 + x)^{-1/2}(2x+1) - 1 = \dfrac{2x+1}{2\sqrt{x^2 + x}} - 1 > 0 \;\Leftrightarrow\; 2x + 1 > 2\sqrt{x^2 + x} \;\Leftrightarrow$

$x + \frac{1}{2} > \sqrt{\left(x + \frac{1}{2}\right)^2 - \frac{1}{4}}$. Keep in mind that the domain excludes the interval $(-1, 0)$. When $x + \frac{1}{2}$ is positive (for $x \geq 0$), the last inequality is *true* since the value of the radical is less than $x + \frac{1}{2}$. When $x + \frac{1}{2}$ is negative (for $x \leq -1$), the last inequality is *false* since the value of the radical is positive. So f is increasing on $(0, \infty)$ and decreasing on $(-\infty, -1)$.

F. No local extrema

G. $$f''(x) = \frac{2(x^2 + x)^{1/2}(2) - (2x+1) \cdot 2 \cdot \frac{1}{2}(x^2 + x)^{-1/2}(2x+1)}{\left(2\sqrt{x^2 + x}\right)^2}$$

$$= \frac{(x^2 + x)^{-1/2}[4(x^2 + x) - (2x+1)^2]}{4(x^2 + x)} = \frac{-1}{4(x^2 + x)^{3/2}}.$$

$f''(x) < 0$ when it is defined, so f is CD on $(-\infty, -1)$ and $(0, \infty)$. No IP

H.

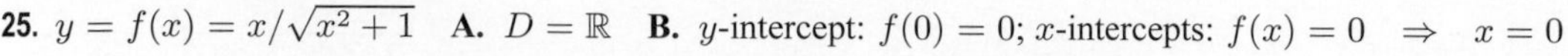

25. $y = f(x) = x/\sqrt{x^2 + 1}$ **A.** $D = \mathbb{R}$ **B.** y-intercept: $f(0) = 0$; x-intercepts: $f(x) = 0 \;\Rightarrow\; x = 0$

C. $f(-x) = -f(x)$, so f is odd; the graph is symmetric about the origin.

D. $$\lim_{x\to\infty} f(x) = \lim_{x\to\infty} \frac{x}{\sqrt{x^2 + 1}} = \lim_{x\to\infty} \frac{x/x}{\sqrt{x^2 + 1}/x} = \lim_{x\to\infty} \frac{x/x}{\sqrt{x^2 + 1}/\sqrt{x^2}} = \lim_{x\to\infty} \frac{1}{\sqrt{1 + 1/x^2}} = \frac{1}{\sqrt{1 + 0}} = 1$$

and

$$\lim_{x\to-\infty} f(x) = \lim_{x\to-\infty} \frac{x}{\sqrt{x^2+1}} = \lim_{x\to-\infty} \frac{x/x}{\sqrt{x^2+1}/x} = \lim_{x\to-\infty} \frac{x/x}{\sqrt{x^2+1}/\left(-\sqrt{x^2}\right)} = \lim_{x\to-\infty} \frac{1}{-\sqrt{1+1/x^2}}$$

$= \dfrac{1}{-\sqrt{1+0}} = -1$ so $y = \pm 1$ are HA. No VA

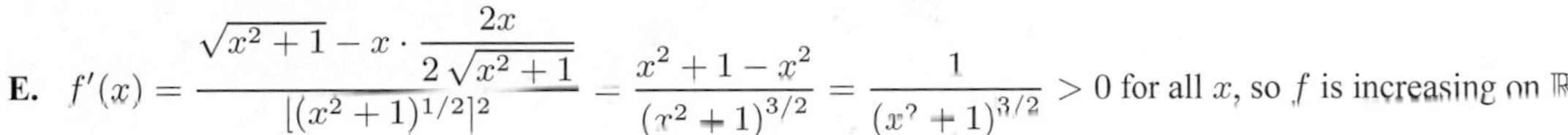

E. $f'(x) = \dfrac{\sqrt{x^2+1} - x \cdot \dfrac{2x}{2\sqrt{x^2+1}}}{[(x^2+1)^{1/2}]^2} = \dfrac{x^2+1-x^2}{(x^2+1)^{3/2}} = \dfrac{1}{(x^2+1)^{3/2}} > 0$ for all x, so f is increasing on $\mathbb{R}$.

F. No extreme values

G. $f''(x) = -\frac{3}{2}(x^2+1)^{-5/2} \cdot 2x = \dfrac{-3x}{(x^2+1)^{5/2}}$, so $f''(x) > 0$ for $x < 0$ and $f''(x) < 0$ for $x > 0$. Thus, f is CU on $(-\infty, 0)$ and CD on $(0, \infty)$. IP at $(0, 0)$

H.

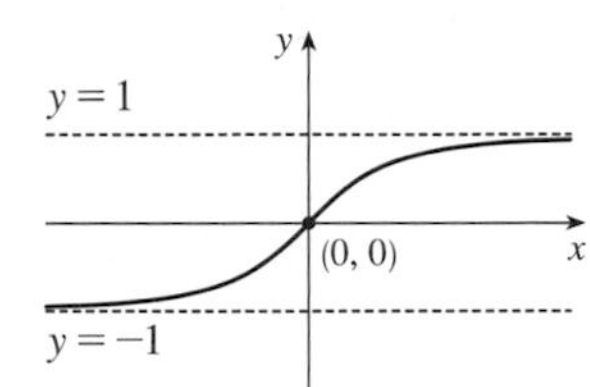

26. $y = f(x) = x\sqrt{2-x^2}$ **A.** $D = \left[-\sqrt{2}, \sqrt{2}\right]$ **B.** y-intercept: $f(0) = 0$; x-intercepts: $f(x) = 0 \Rightarrow x = 0, \pm\sqrt{2}$. **C.** $f(-x) = -f(x)$, so f is odd; the graph is symmetric about the origin. **D.** No asymptote

E. $f'(x) = x \cdot \dfrac{-x}{\sqrt{2-x^2}} + \sqrt{2-x^2} = \dfrac{-x^2+2-x^2}{\sqrt{2-x^2}} = \dfrac{2(1+x)(1-x)}{\sqrt{2-x^2}}$. $f'(x)$ is negative for $-\sqrt{2} < x < -1$ and $1 < x < \sqrt{2}$, and positive for $-1 < x < 1$, so f is decreasing on $\left(-\sqrt{2}, -1\right)$ and $\left(1, \sqrt{2}\right)$ and increasing on $(-1, 1)$.

F. Local minimum value $f(-1) = -1$, local maximum value $f(1) = 1$.

G. $$f''(x) = \frac{\sqrt{2-x^2}(-4x) - (2-2x^2)\dfrac{-x}{\sqrt{2-x^2}}}{[(2-x^2)^{1/2}]^2} = \frac{(2-x^2)(-4x)+(2-2x^2)x}{(2-x^2)^{3/2}} = \frac{2x^3-6x}{(2-x^2)^{3/2}} = \frac{2x(x^2-3)}{(2-x^2)^{3/2}}$$

Since $x^2 - 3 < 0$ for x in $\left[-\sqrt{2}, \sqrt{2}\right]$, $f''(x) > 0$ for $\sqrt{2} < x < 0$ and $f''(x) < 0$ for $0 < x < \sqrt{2}$. Thus, f is CU on $\left(-\sqrt{2}, 0\right)$ and CD on $\left(0, \sqrt{2}\right)$. The only IP is $(0, 0)$.

H.

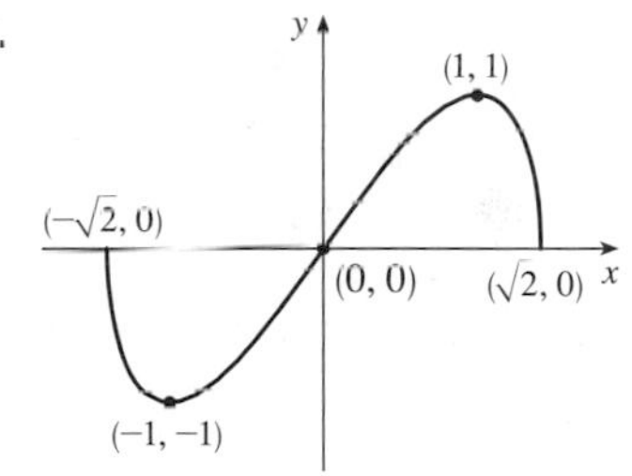

27. $y = f(x) = \sqrt{1-x^2}/x$ **A.** $D = \{x \mid |x| \leq 1, x \neq 0\} = [-1, 0) \cup (0, 1]$ **B.** x-intercepts ± 1, no y-intercept

C. $f(-x) = -f(x)$, so the curve is symmetric about $(0, 0)$. **D.** $\displaystyle\lim_{x\to 0^+} \frac{\sqrt{1-x^2}}{x} = \infty$, $\displaystyle\lim_{x\to 0^-} \frac{\sqrt{1-x^2}}{x} = -\infty$, so $x = 0$ is a VA. **E.** $f'(x) = \dfrac{\left(-x^2/\sqrt{1-x^2}\right) - \sqrt{1-x^2}}{x^2} = -\dfrac{1}{x^2\sqrt{1-x^2}} < 0$, so f is decreasing on $(-1, 0)$ and $(0, 1)$. **F.** No extreme values

G. $f''(x) = \dfrac{2-3x^2}{x^3(1-x^2)^{3/2}} > 0 \Leftrightarrow -1 < x < -\sqrt{\frac{2}{3}}$ or $0 < x < \sqrt{\frac{2}{3}}$, so f is CU on $\left(-1, -\sqrt{\frac{2}{3}}\right)$ and $\left(0, \sqrt{\frac{2}{3}}\right)$ and CD on $\left(-\sqrt{\frac{2}{3}}, 0\right)$ and $\left(\sqrt{\frac{2}{3}}, 1\right)$. IP at $\left(\pm\sqrt{\frac{2}{3}}, \pm\frac{1}{\sqrt{2}}\right)$

H.

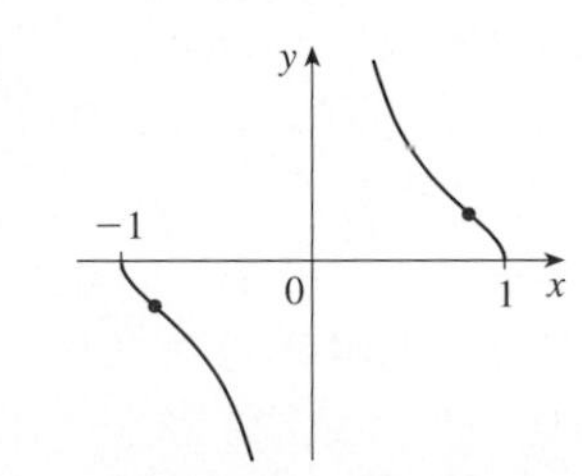

28. $y = f(x) = x/\sqrt{x^2 - 1}$ **A.** $D = (-\infty, -1) \cup (1, \infty)$ **B.** No intercepts **C.** $f(-x) = -f(x)$, so f is odd; the graph is symmetric about the origin. **D.** $\lim\limits_{x \to \infty} \dfrac{x}{\sqrt{x^2 - 1}} = 1$ and $\lim\limits_{x \to -\infty} \dfrac{x}{\sqrt{x^2 - 1}} = -1$, so $y = \pm 1$ are HA.

$\lim\limits_{x \to 1^+} f(x) = +\infty$ and $\lim\limits_{x \to -1^-} f(x) = -\infty$, so $x = \pm 1$ are VA.

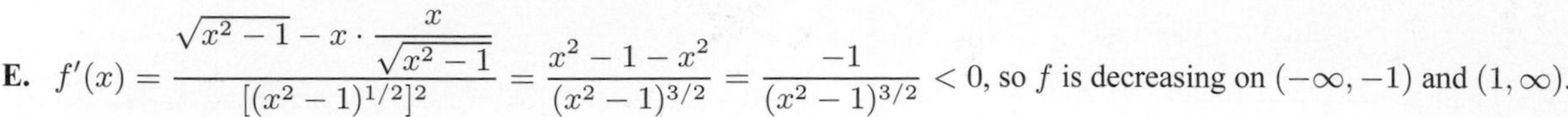

E. $f'(x) = \dfrac{\sqrt{x^2 - 1} - x \cdot \dfrac{x}{\sqrt{x^2 - 1}}}{[(x^2 - 1)^{1/2}]^2} = \dfrac{x^2 - 1 - x^2}{(x^2 - 1)^{3/2}} = \dfrac{-1}{(x^2 - 1)^{3/2}} < 0$, so f is decreasing on $(-\infty, -1)$ and $(1, \infty)$.

F. No extreme values

G. $f''(x) = (-1)\left(-\frac{3}{2}\right)(x^2 - 1)^{-5/2} \cdot 2x = \dfrac{3x}{(x^2 - 1)^{5/2}}$.

$f''(x) < 0$ on $(-\infty, -1)$ and $f''(x) > 0$ on $(1, \infty)$, so f is CD on $(-\infty, -1)$ and CU on $(1, \infty)$. No IP

H.

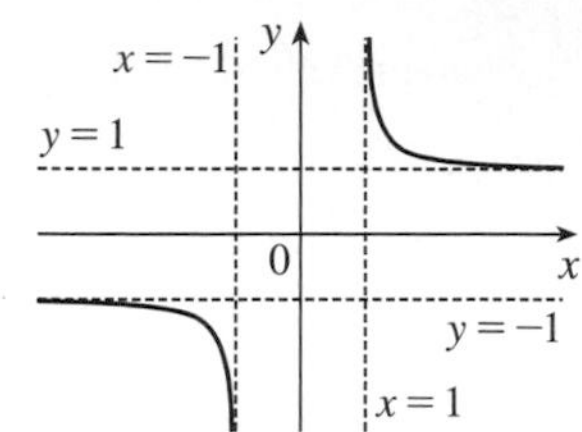

29. $y = f(x) = x - 3x^{1/3}$ **A.** $D = \mathbb{R}$ **B.** y-intercept: $f(0) = 0$; x-intercepts: $f(x) = 0 \;\Rightarrow\; x = 3x^{1/3} \;\Rightarrow\; x^3 = 27x \;\Rightarrow\; x^3 - 27x = 0 \;\Rightarrow\; x(x^2 - 27) = 0 \;\Rightarrow\; x = 0, \pm 3\sqrt{3}$ **C.** $f(-x) = -f(x)$, so f is odd; the graph is symmetric about the origin. **D.** No asymptote **E.** $f'(x) = 1 - x^{-2/3} = 1 - \dfrac{1}{x^{2/3}} = \dfrac{x^{2/3} - 1}{x^{2/3}}$.

$f'(x) > 0$ when $|x| > 1$ and $f'(x) < 0$ when $0 < |x| < 1$, so f is increasing on $(-\infty, -1)$ and $(1, \infty)$, and decreasing on $(-1, 0)$ and $(0, 1)$ [hence decreasing on $(-1, 1)$ since f is continuous on $(-1, 1)$]. **F.** Local maximum value $f(-1) = 2$, local minimum value $f(1) = -2$ **G.** $f''(x) = \frac{2}{9}x^{-5/3} < 0$ when $x < 0$ and $f''(x) > 0$ when $x > 0$, so f is CD on $(-\infty, 0)$ and CU on $(0, \infty)$. IP at $(0, 0)$

H.

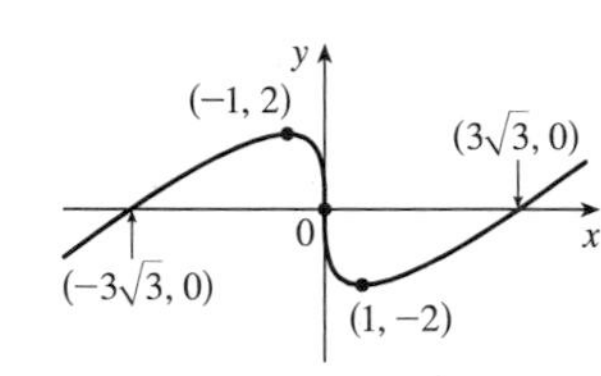

30. $y = f(x) = x^{5/3} - 5x^{2/3} = x^{2/3}(x - 5)$ **A.** $D = \mathbb{R}$ **B.** x-intercepts 0, 5; y-intercept 0 **C.** No symmetry **D.** $\lim\limits_{x \to \pm\infty} x^{2/3}(x - 5) = \pm\infty$, so there is no asymptote **E.** $f'(x) = \frac{5}{3}x^{2/3} - \frac{10}{3}x^{-1/3} = \frac{5}{3}x^{-1/3}(x - 2) > 0 \;\Leftrightarrow\; x < 0$ or $x > 2$, so f is increasing on $(-\infty, 0)$, $(2, \infty)$ and decreasing on $(0, 2)$.

F. Local maximum value $f(0) = 0$, local minimum value $f(2) = -3\sqrt[3]{4}$

G. $f''(x) = \frac{10}{9}x^{-1/3} + \frac{10}{9}x^{-4/3} = \frac{10}{9}x^{-4/3}(x + 1) > 0 \;\Leftrightarrow\; x > -1$, so f is CU on $(-1, 0)$ and $(0, \infty)$, CD on $(-\infty, -1)$. IP at $(-1, -6)$

H.

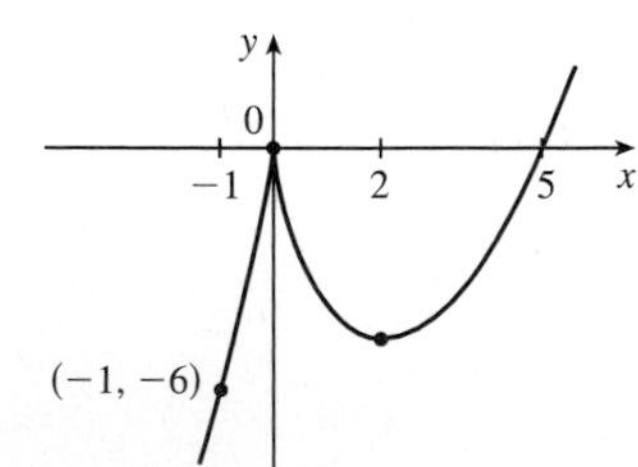

31. $y = f(x) = \sqrt[3]{x^2 - 1}$ **A.** $D = \mathbb{R}$ **B.** y-intercept: $f(0) = -1$; x-intercepts: $f(x) = 0 \;\Leftrightarrow\; x^2 - 1 = 0 \;\Leftrightarrow\; x = \pm 1$ **C.** $f(-x) = f(x)$, so the curve is symmetric about the y-axis. **D.** No asymptote

E. $f'(x) = \frac{1}{3}(x^2 - 1)^{-2/3}(2x) = \dfrac{2x}{3\sqrt[3]{(x^2 - 1)^2}}$. $f'(x) > 0 \;\Leftrightarrow\; x > 0$ and $f'(x) < 0 \;\Leftrightarrow\; x < 0$, so f is increasing on $(0, \infty)$ and decreasing on $(-\infty, 0)$. **F.** Local minimum value $f(0) = -1$

G. $f''(x) = \frac{2}{3} \cdot \frac{(x^2-1)^{2/3}(1) - x \cdot \frac{2}{3}(x^2-1)^{-1/3}(2x)}{[(x^2-1)^{2/3}]^2}$

$= \frac{2}{9} \cdot \frac{(x^2-1)^{-1/3}[3(x^2-1) - 4x^2]}{(x^2-1)^{4/3}} = -\frac{2(x^2+3)}{9(x^2-1)^{5/3}}$

$f''(x) > 0 \Leftrightarrow -1 < x < 1$ and $f''(x) < 0 \Leftrightarrow x < -1$ or $x > 1$, so f is CU on $(-1, 1)$ and f is CD on $(-\infty, -1)$ and $(1, \infty)$. IP at $(\pm 1, 0)$

H.

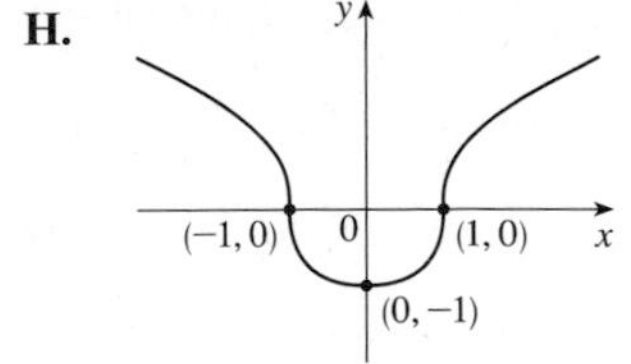

32. $y = f(x) = \sqrt[3]{x^3+1}$ **A.** $D = \mathbb{R}$ **B.** y-intercept: $f(0) = 1$; x-intercept: $f(x) = 0 \Leftrightarrow x^3 + 1 = 0 \Rightarrow x = -1$

C. No symmetry **D.** No asymptote **E.** $f'(x) = \frac{1}{3}(x^3+1)^{-2/3}(3x^2) = \frac{x^2}{\sqrt[3]{(x^3+1)^2}}$. $f'(x) > 0$ if $x < -1$, $-1 < x < 0$, and $x > 0$, so f is increasing on $\mathbb{R}$. **F.** No local extrema

G. $f''(x) = \frac{(x^3+1)^{2/3}(2x) - x^2 \cdot \frac{2}{3}(x^3+1)^{-1/3}(3x^2)}{[(x^3+1)^{2/3}]^2}$

$= \frac{x(x^3+1)^{-1/3}[2(x^3+1) - 2x^3]}{(x^3+1)^{4/3}} = \frac{2x}{(x^3+1)^{5/3}}$

$f''(x) > 0 \Leftrightarrow x < -1$ or $x > 0$ and $f''(x) < 0 \Leftrightarrow -1 < x < 0$, so f is CU on $(-\infty, -1)$ and $(0, \infty)$ and CD on $(-1, 0)$. IP at $(-1, 0)$ and $(0, 1)$

H.

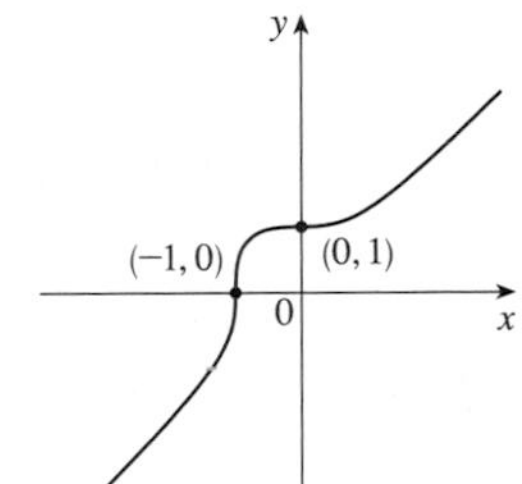

33. $y = f(x) = \sin^3 x$ **A.** $D = \mathbb{R}$ **B.** x-intercepts: $f(x) = 0 \Leftrightarrow x = n\pi$, n an integer; y-intercept $= f(0) = 0$

C. $f(-x) = -f(x)$, so f is odd and the curve is symmetric about the origin. Also, $f(x + 2\pi) = f(x)$, so f is periodic with period 2π, and we determine **E–G** for $0 \le x \le \pi$. Since f is odd, we can reflect the graph of f on $[0, \pi]$ about the origin to obtain the graph of f on $[-\pi, \pi]$, and then since f has period 2π, we can extend the graph of f for all real numbers.

D. No asymptote **E.** $f'(x) = 3\sin^2 x \cos x > 0 \Leftrightarrow \cos x > 0$ and $\sin x \ne 0 \Leftrightarrow 0 < x < \frac{\pi}{2}$, so f is increasing on $\left(0, \frac{\pi}{2}\right)$ and f is decreasing on $\left(\frac{\pi}{2}, \pi\right)$. **F.** Local maximum value $f\left(\frac{\pi}{2}\right) = 1$ $\left[\text{local minimum value } f\left(-\frac{\pi}{2}\right) = -1\right]$

G. $f''(x) = 3\sin^2 x\,(-\sin x) + 3\cos x\,(2\sin x \cos x) = 3\sin x\,(2\cos^2 x - \sin^2 x)$

$= 3\sin x\,[2(1 - \sin^2 x) - \sin^2 x] = 3\sin x(2 - 3\sin^2 x) > 0 \Leftrightarrow$

$\sin x > 0$ and $\sin^2 x < \frac{2}{3} \Leftrightarrow 0 < x < \pi$ and $0 < \sin x < \sqrt{\frac{2}{3}} \Leftrightarrow 0 < x < \sin^{-1}\sqrt{\frac{2}{3}}$ $\left[\text{let } \alpha = \sin^{-1}\sqrt{\frac{2}{3}}\right]$ or $\pi - \alpha < x < \pi$, so f is CU on $(0, \alpha)$ and $(\pi - \alpha, \pi)$, and f is CD on $(\alpha, \pi - \alpha)$. There are inflection points at $x = 0$, π, α, and $x = \pi - \alpha$.

H.

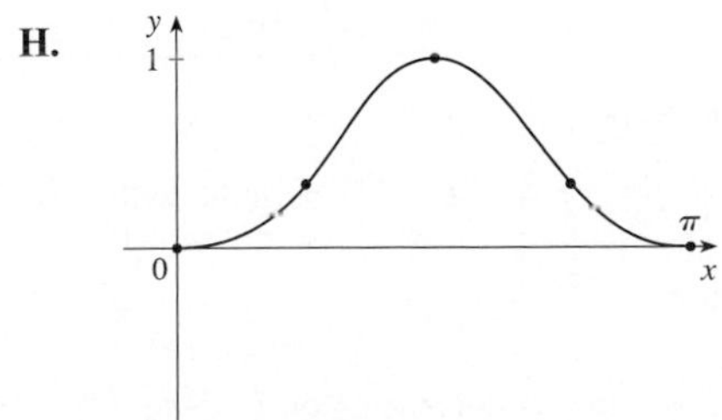

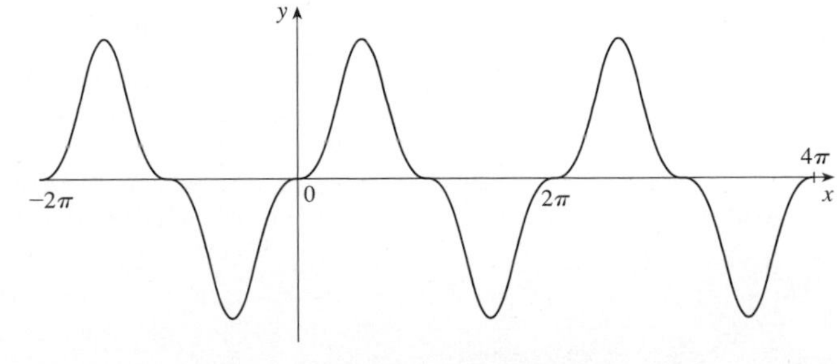

34. $y = f(x) = x + \cos x$ **A.** $D = \mathbb{R}$ **B.** y-intercept: $f(0) = 1$; the x-intercept is about -0.74 and can be found using Newton's method **C.** No symmetry **D.** No asymptote **E.** $f'(x) = 1 - \sin x > 0$ except for $x = \frac{\pi}{2} + 2n\pi$, so f is increasing on $\mathbb{R}$. **F.** No local extrema

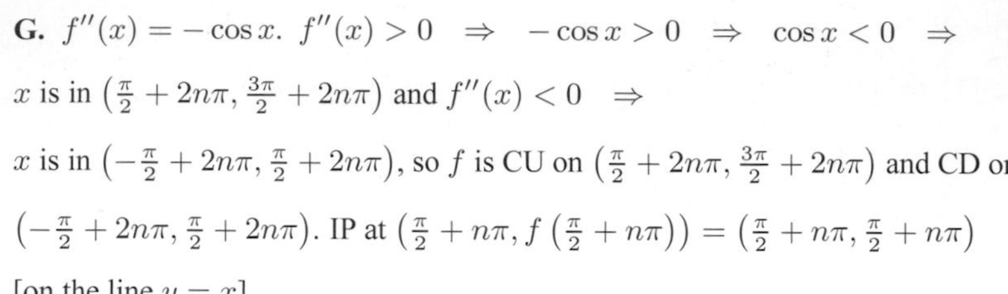

G. $f''(x) = -\cos x$. $f''(x) > 0 \Rightarrow -\cos x > 0 \Rightarrow \cos x < 0 \Rightarrow$ x is in $\left(\frac{\pi}{2} + 2n\pi, \frac{3\pi}{2} + 2n\pi\right)$ and $f''(x) < 0 \Rightarrow$ x is in $\left(-\frac{\pi}{2} + 2n\pi, \frac{\pi}{2} + 2n\pi\right)$, so f is CU on $\left(\frac{\pi}{2} + 2n\pi, \frac{3\pi}{2} + 2n\pi\right)$ and CD on $\left(-\frac{\pi}{2} + 2n\pi, \frac{\pi}{2} + 2n\pi\right)$. IP at $\left(\frac{\pi}{2} + n\pi, f\left(\frac{\pi}{2} + n\pi\right)\right) = \left(\frac{\pi}{2} + n\pi, \frac{\pi}{2} + n\pi\right)$ [on the line $y = x$]

H.

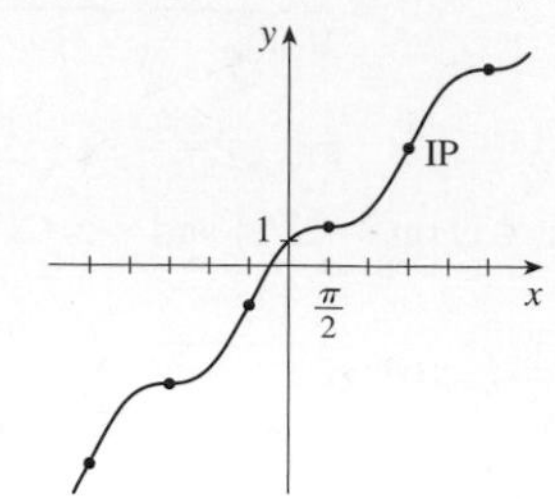

35. $y = f(x) = x \tan x$, $-\frac{\pi}{2} < x < \frac{\pi}{2}$ **A.** $D = \left(-\frac{\pi}{2}, \frac{\pi}{2}\right)$ **B.** Intercepts are 0 **C.** $f(-x) = f(x)$, so the curve is symmetric about the y-axis. **D.** $\lim\limits_{x \to (\pi/2)^-} x \tan x = \infty$ and $\lim\limits_{x \to -(\pi/2)^+} x \tan x = \infty$, so $x = \frac{\pi}{2}$ and $x = -\frac{\pi}{2}$ are VA.

E. $f'(x) = \tan x + x \sec^2 x > 0 \Leftrightarrow 0 < x < \frac{\pi}{2}$, so f increases on $\left(0, \frac{\pi}{2}\right)$ and decreases on $\left(-\frac{\pi}{2}, 0\right)$. **F.** Absolute and local minimum value $f(0) = 0$.

G. $y'' = 2\sec^2 x + 2x \tan x \sec^2 x > 0$ for $-\frac{\pi}{2} < x < \frac{\pi}{2}$, so f is CU on $\left(-\frac{\pi}{2}, \frac{\pi}{2}\right)$. No IP

H.

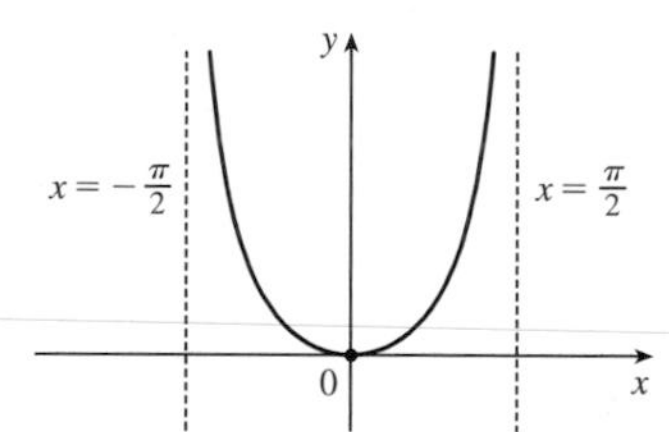

36. $y = f(x) = 2x - \tan x$, $-\frac{\pi}{2} < x < \frac{\pi}{2}$ **A.** $D = \left(-\frac{\pi}{2}, \frac{\pi}{2}\right)$ **B.** y-intercept: $f(0) = 0$; x-intercepts: $f(x) = 0 \Leftrightarrow$ $2x = \tan x \Leftrightarrow x = 0$ or $x \approx \pm 1.17$ **C.** $f(-x) = -f(x)$, so f is odd; the graph is symmetric about the origin.

D. $\lim\limits_{x \to (-\pi/2)^+} (2x - \tan x) = \infty$ and $\lim\limits_{x \to (\pi/2)^-} (2x - \tan x) = -\infty$, so $x = \pm\frac{\pi}{2}$ are VA. No HA.

E. $f'(x) = 2 - \sec^2 x < 0 \Leftrightarrow |\sec x| > \sqrt{2}$ and $f'(x) > 0 \Leftrightarrow |\sec x| < \sqrt{2}$, so f is decreasing on $\left(-\frac{\pi}{2}, -\frac{\pi}{4}\right)$, increasing on $\left(-\frac{\pi}{4}, \frac{\pi}{4}\right)$, and decreasing again on $\left(\frac{\pi}{4}, \frac{\pi}{2}\right)$ **F.** Local maximum value $f\left(\frac{\pi}{4}\right) = \frac{\pi}{2} - 1$, local minimum value $f\left(-\frac{\pi}{4}\right) = -\frac{\pi}{2} + 1$

G. $f''(x) = -2\sec x \cdot \sec x \tan x = -2\tan x \sec^2 x = -2\tan x(\tan^2 x + 1)$ so $f''(x) > 0 \Leftrightarrow \tan x < 0 \Leftrightarrow -\frac{\pi}{2} < x < 0$, and $f''(x) < 0 \Leftrightarrow$ $\tan x > 0 \Leftrightarrow 0 < x < \frac{\pi}{2}$. Thus, f is CU on $\left(-\frac{\pi}{2}, 0\right)$ and CD on $\left(0, \frac{\pi}{2}\right)$. IP at $(0, 0)$

H.

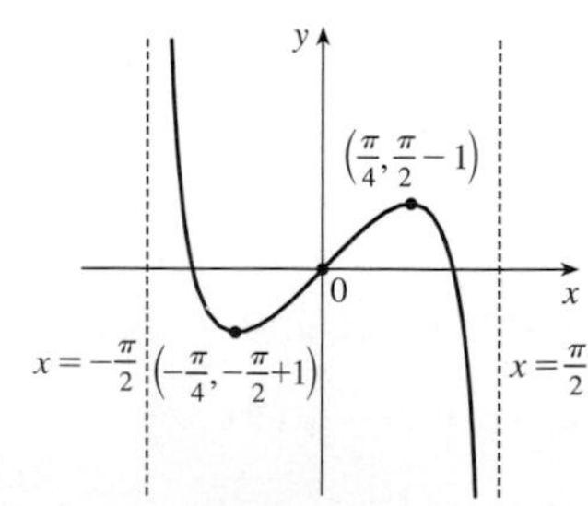

37. $y = f(x) = \sin x + \sqrt{3}\cos x$, $-2\pi \le x \le 2\pi$ **A.** $D = [-2\pi, 2\pi]$ **B.** y-intercept: $f(0) = \sqrt{3}$; x-intercepts: $f(x) = 0 \Leftrightarrow \sin x = -\sqrt{3}\cos x \Leftrightarrow \tan x = -\sqrt{3} \Leftrightarrow x = -\frac{4\pi}{3}, -\frac{\pi}{3}, \frac{2\pi}{3}$, or $\frac{5\pi}{3}$ **C.** f is periodic with period 2π. **D.** No asymptote **E.** $f'(x) = \cos x - \sqrt{3}\sin x$. $f'(x) = 0 \Leftrightarrow \cos x = \sqrt{3}\sin x \Leftrightarrow \tan x = \dfrac{1}{\sqrt{3}} \Leftrightarrow$ $x = -\frac{11\pi}{6}, -\frac{5\pi}{6}, \frac{\pi}{6}$, or $\frac{7\pi}{6}$. $f'(x) < 0 \Leftrightarrow -\frac{11\pi}{6} < x < -\frac{5\pi}{6}$ or $\frac{\pi}{6} < x < \frac{7\pi}{6}$, so f is decreasing on $\left(-\frac{11\pi}{6}, -\frac{5\pi}{6}\right)$

and $\left(\frac{\pi}{6}, \frac{7\pi}{6}\right)$, and f is increasing on $\left(-2\pi, -\frac{11\pi}{6}\right)$, $\left(-\frac{5\pi}{6}, \frac{\pi}{6}\right)$, and $\left(\frac{7\pi}{6}, 2\pi\right)$. **F.** Local maximum value $f\left(-\frac{11\pi}{6}\right) = f\left(\frac{\pi}{6}\right) = \frac{1}{2} + \sqrt{3}\left(\frac{1}{2}\sqrt{3}\right) = 2$, local minimum value $f\left(-\frac{5\pi}{6}\right) = f\left(\frac{7\pi}{6}\right) = -\frac{1}{2} + \sqrt{3}\left(-\frac{1}{2}\sqrt{3}\right) = -2$

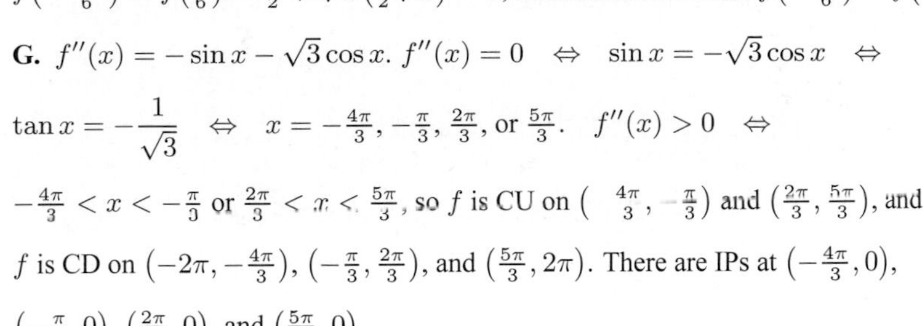

G. $f''(x) = -\sin x - \sqrt{3}\cos x$. $f''(x) = 0 \Leftrightarrow \sin x = -\sqrt{3}\cos x \Leftrightarrow$ $\tan x = -\dfrac{1}{\sqrt{3}} \Leftrightarrow x = -\frac{4\pi}{3}, -\frac{\pi}{3}, \frac{2\pi}{3}$, or $\frac{5\pi}{3}$. $f''(x) > 0 \Leftrightarrow$ $-\frac{4\pi}{3} < x < -\frac{\pi}{3}$ or $\frac{2\pi}{3} < x < \frac{5\pi}{3}$, so f is CU on $\left(-\frac{4\pi}{3}, -\frac{\pi}{3}\right)$ and $\left(\frac{2\pi}{3}, \frac{5\pi}{3}\right)$, and f is CD on $\left(-2\pi, -\frac{4\pi}{3}\right)$, $\left(-\frac{\pi}{3}, \frac{2\pi}{3}\right)$, and $\left(\frac{5\pi}{3}, 2\pi\right)$. There are IPs at $\left(-\frac{4\pi}{3}, 0\right)$, $\left(-\frac{\pi}{3}, 0\right)$, $\left(\frac{2\pi}{3}, 0\right)$, and $\left(\frac{5\pi}{3}, 0\right)$.

H.

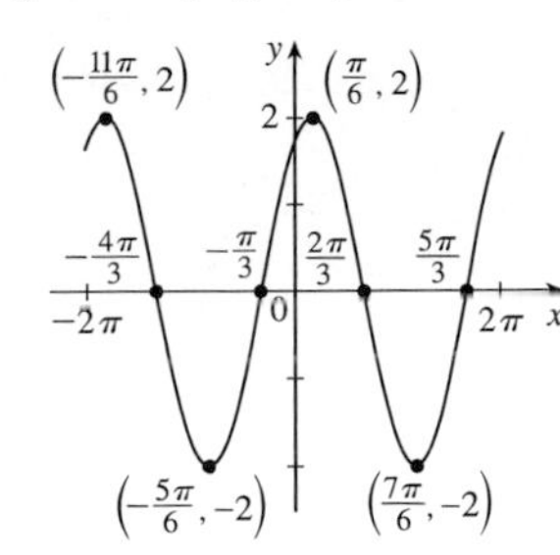

38. $y = f(x) = \csc x - 2\sin x$, $0 < x < \pi$ **A.** $D = (0, \pi)$ **B.** No y-intercept; x-intercept: $f(x) = 0 \Leftrightarrow$ $\csc x = 2\sin x \Leftrightarrow \frac{1}{2} = \sin^2 x \Leftrightarrow \sin x = \pm\frac{1}{2}\sqrt{2} \Leftrightarrow x = \frac{\pi}{4}$ or $\frac{3\pi}{4}$ **C.** No symmetry

D. $\lim\limits_{x\to 0^+} f(x) = \infty$ and $\lim\limits_{x\to\pi^-} f(x) = \infty$, so $x = 0$ and $x = \pi$ are VAs.

E. $f'(x) = -\csc x\cot x - 2\cos x = -\dfrac{\cos x}{\sin^2 x} - 2\cos x = -\cos x\left(\dfrac{1}{\sin^2 x} + 2\right)$. $f'(x) > 0$ when $-\cos x > 0 \Leftrightarrow$ $\cos x < 0 \Leftrightarrow \frac{\pi}{2} < x < \pi$, so f' is increasing on $\left(\frac{\pi}{2}, \pi\right)$, and f is decreasing on $\left(0, \frac{\pi}{2}\right)$. **F.** Local minimum value $f\left(\frac{\pi}{2}\right) = -1$

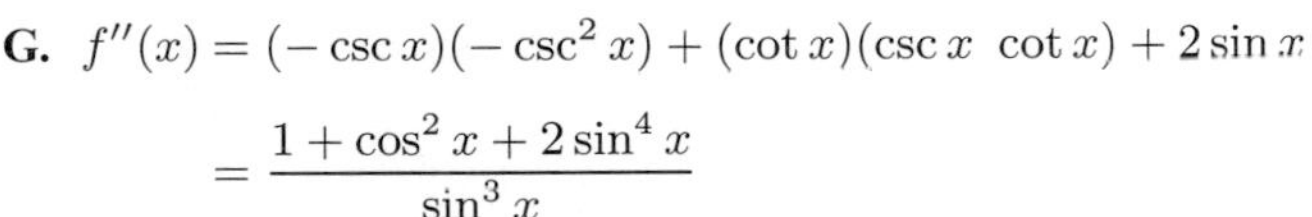

G. $f''(x) = (-\csc x)(-\csc^2 x) + (\cot x)(\csc x \cot x) + 2\sin x$

$$= \frac{1 + \cos^2 x + 2\sin^4 x}{\sin^3 x}$$

f'' has the same sign as $\sin x$, which is positive on $(0, \pi)$, so f is CU on $(0, \pi)$. No IP

H.

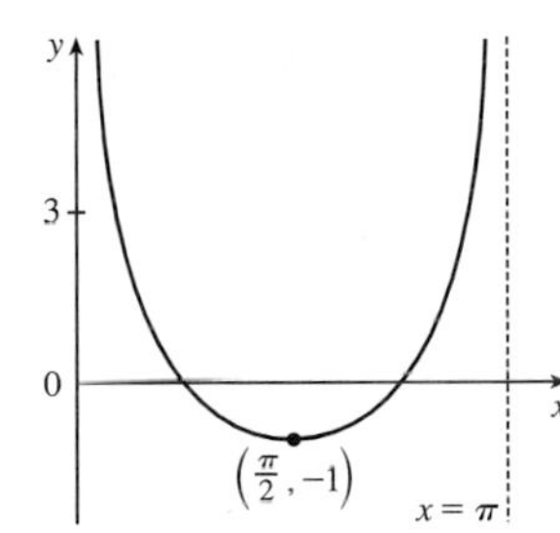

39. $y = f(x) = \dfrac{\sin x}{1 + \cos x}$ $\left[\overset{\text{when } \cos x \neq 1}{=} \dfrac{\sin x}{1 + \cos x} \cdot \dfrac{1 - \cos x}{1 - \cos x} = \dfrac{\sin x\,(1 - \cos x)}{\sin^2 x} = \dfrac{1 - \cos x}{\sin x} = \csc x - \cot x\right]$

A. The domain of f is the set of all real numbers except odd integer multiples of π; that is, all reals except $(2n + 1)\pi$, where n is an integer. **B.** y-intercept: $f(0) = 0$; x-intercepts: $x = 2n\pi$, n an integer. **C.** $f(-x) = -f(x)$, so f is an odd function; the graph is symmetric about the origin and has period 2π. **D.** When n is an odd integer, $\lim\limits_{x\to(n\pi)^-} f(x) = \infty$ and $\lim\limits_{x\to(n\pi)^+} f(x) = -\infty$, so $x = n\pi$ is a VA for each odd integer n. No HA.

E. $f'(x) = \dfrac{(1 + \cos x)\cdot\cos x - \sin x(-\sin x)}{(1 + \cos x)^2} = \dfrac{1 + \cos x}{(1 + \cos x)^2} = \dfrac{1}{1 + \cos x}$. $f'(x) > 0$ for all x except odd multiples of π, so f is increasing on $((2k - 1)\pi, (2k + 1)\pi)$ for each integer k. **F.** No extreme values

G. $f''(x) = \dfrac{\sin x}{(1 + \cos x)^2} > 0 \Rightarrow \sin x > 0 \Rightarrow$ $x \in (2k\pi, (2k + 1)\pi)$ and $f''(x) < 0$ on $((2k - 1)\pi, 2k\pi)$ for each integer k. f is CU on $(2k\pi, (2k + 1)\pi)$ and CD on $((2k - 1)\pi, 2k\pi)$ for each integer k. f has IPs at $(2k\pi, 0)$ for each integer k.

H.

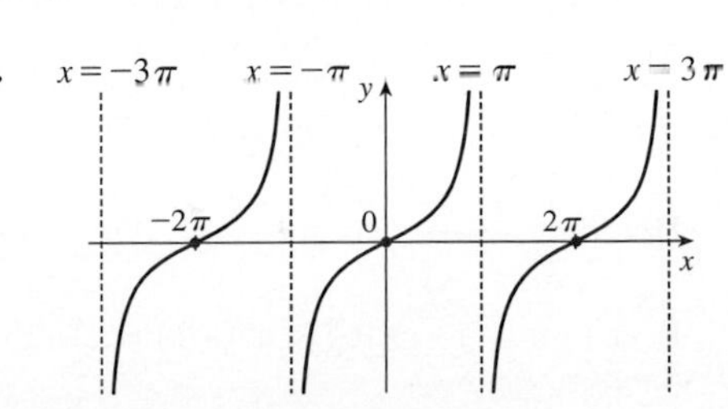

40. $y = f(x) = \dfrac{\sin x}{2 + \cos x}$ **A.** $D = \mathbb{R}$ **B.** y-intercept: $f(0) = 0$; x-intercepts: $f(x) = 0 \Leftrightarrow \sin x = 0 \Leftrightarrow x = n\pi$

C. $f(-x) = -f(x)$, so the curve is symmetric about the origin. f is periodic with period 2π, so we determine **E–G** for $0 \le x \le 2\pi$. **D.** No asymptote

E. $f'(x) = \dfrac{(2 + \cos x)\cos x - \sin x(-\sin x)}{(2 + \cos x)^2} = \dfrac{2\cos x + \cos^2 x + \sin^2 x}{(2 + \cos x)^2} = \dfrac{2\cos x + 1}{(2 + \cos x)^2}$.

$f'(x) > 0 \Leftrightarrow 2\cos x + 1 > 0 \Leftrightarrow \cos x > -\frac{1}{2} \Leftrightarrow x$ is in $\left(0, \frac{2\pi}{3}\right)$ or $\left(\frac{4\pi}{3}, 2\pi\right)$, so f is increasing on $\left(0, \frac{2\pi}{3}\right)$ and $\left(\frac{4\pi}{3}, 2\pi\right)$, and f is decreasing on $\left(\frac{2\pi}{3}, \frac{4\pi}{3}\right)$.

F. Local maximum value $f\left(\frac{2\pi}{3}\right) = \dfrac{\sqrt{3}/2}{2 - (1/2)} = \dfrac{\sqrt{3}}{3}$ and local minimum value $f\left(\frac{4\pi}{3}\right) = \dfrac{-\sqrt{3}/2}{2 - (1/2)} = -\dfrac{\sqrt{3}}{3}$

G. $$f''(x) = \frac{(2 + \cos x)^2(-2\sin x) - (2\cos x + 1)2(2 + \cos x)(-\sin x)}{[(2 + \cos x)^2]^2}$$
$$= \frac{-2\sin x\,(2 + \cos x)[(2 + \cos x) - (2\cos x + 1)]}{(2 + \cos x)^4} = \frac{-2\sin x\,(1 - \cos x)}{(2 + \cos x)^3}$$

$f''(x) > 0 \Leftrightarrow -2\sin x > 0 \Leftrightarrow \sin x < 0 \Leftrightarrow x$ is in $(\pi, 2\pi)$ [f is CU] and $f''(x) < 0 \Leftrightarrow x$ is in $(0, \pi)$ [f is CD]. The inflection points are $(0, 0)$, $(\pi, 0)$, and $(2\pi, 0)$.

H.

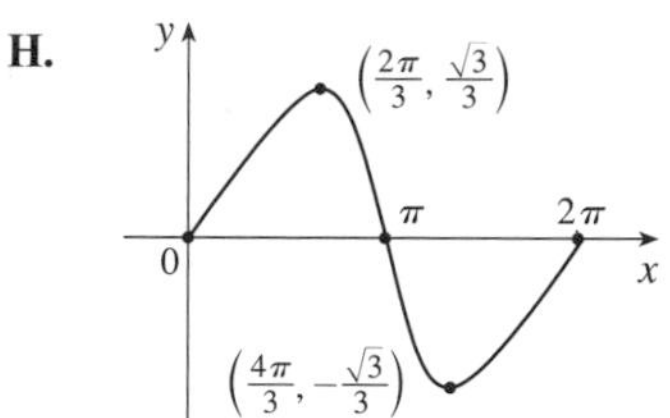

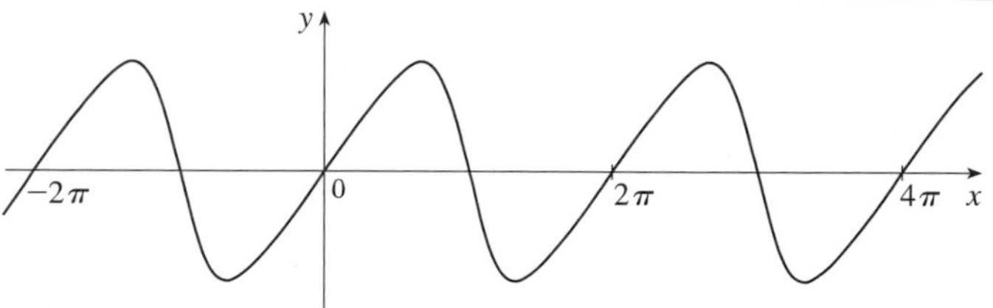

41. $y = f(x) = \arctan(e^x)$ **A.** $D = \mathbb{R}$ **B.** y-intercept $= f(0) = \arctan 1 = \frac{\pi}{4}$. $f(x) > 0$ so there are no x-intercepts.

C. No symmetry **D.** $\lim\limits_{x \to -\infty} \arctan(e^x) = 0$ and $\lim\limits_{x \to \infty} \arctan(e^x) = \frac{\pi}{2}$, so $y = 0$ and $y = \frac{\pi}{2}$ are HAs. No VA

E. $f'(x) = \dfrac{1}{1 + (e^x)^2}\dfrac{d}{dx}e^x = \dfrac{e^x}{1 + e^{2x}} > 0$, so f is increasing on $(-\infty, \infty)$. **F.** No local extrema

G. $$f''(x) = \frac{(1 + e^{2x})e^x - e^x(2e^{2x})}{(1 + e^{2x})^2} = \frac{e^x[(1 + e^{2x}) - 2e^{2x}]}{(1 + e^{2x})^2}$$
$$= \frac{e^x(1 - e^{2x})}{(1 + e^{2x})^2} > 0 \Leftrightarrow$$

$1 - e^{2x} > 0 \Leftrightarrow e^{2x} < 1 \Leftrightarrow 2x < 0 \Leftrightarrow x < 0$, so f is CU on $(-\infty, 0)$ and CD on $(0, \infty)$. IP at $\left(0, \frac{\pi}{4}\right)$

H.

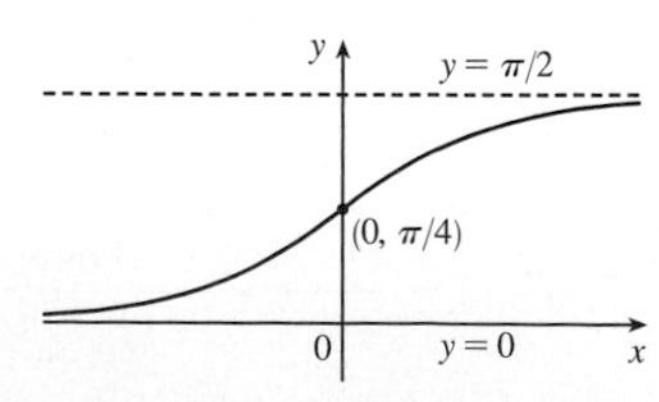

42. $y = f(x) = (1 - x)e^x$ **A.** $D = \mathbb{R}$ **B.** x-intercept 1, y-intercept $= f(0) = 1$ **C.** No symmetry

D. $\lim\limits_{x \to -\infty} \dfrac{1 - x}{e^{-x}}$ $\left[\text{form } \frac{\infty}{\infty}\right] \overset{\text{H}}{=} \lim\limits_{x \to -\infty} \dfrac{-1}{-e^{-x}} = 0$, so $y = 0$ is a HA. No VA

E. $f'(x) = (1 - x)e^x + e^x(-1) = e^x[(1 - x) + (-1)] = -xe^x > 0 \Leftrightarrow x < 0$, so f is increasing on $(-\infty, 0)$

and decreasing on $(0, \infty)$.

F. Local maximum value $f(0) = 1$, no local minimum value

G. $f''(x) = -xe^x + e^x(-1) = e^x(-x-1) = -(x+1)e^x > 0 \iff$ $x < -1$, so f is CU on $(-\infty, -1)$ and CD on $(-1, \infty)$. IP at $(-1, 2/e)$

H.

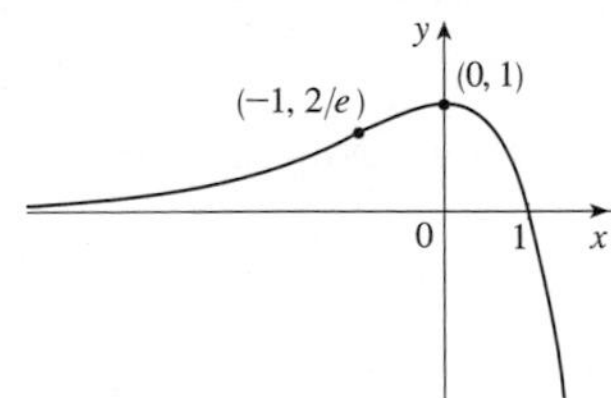

43. $y = 1/(1 + e^{-x})$ **A.** $D = \mathbb{R}$ **B.** No x-intercept; y-intercept $= f(0) = \frac{1}{2}$. **C.** No symmetry

D. $\lim\limits_{x\to\infty} 1/(1+e^{-x}) = \frac{1}{1+0} = 1$ and $\lim\limits_{x\to-\infty} 1/(1+e^{-x}) = 0$ since $\lim\limits_{x\to-\infty} e^{-x} = \infty$, so f has horizontal asymptotes $y = 0$ and $y = 1$. **E.** $f'(x) = -(1+e^{-x})^{-2}(-e^{-x}) = e^{-x}/(1+e^{-x})^2$. This is positive for all x, so f is increasing on $\mathbb{R}$.

F. No extreme values **G.** $f''(x) = \dfrac{(1+e^{-x})^2(-e^{-x}) - e^{-x}(2)(1+e^{-x})(-e^{-x})}{(1+e^{-x})^4} = \dfrac{e^{-x}(e^{-x}-1)}{(1+e^{-x})^3}$

The second factor in the numerator is negative for $x > 0$ and positive for $x < 0$, and the other factors are always positive, so f is CU on $(-\infty, 0)$ and CD on $(0, \infty)$. IP at $\left(0, \frac{1}{2}\right)$

H.

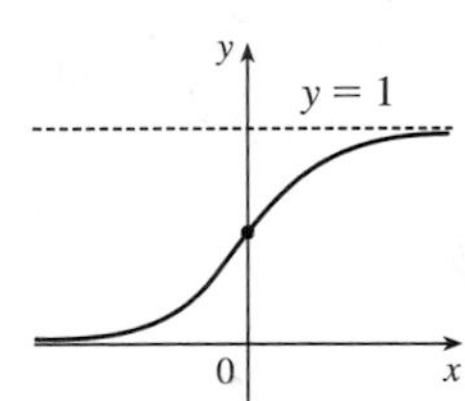

44. $y = f(x) = e^{-x}\sin x$, $0 \le x \le 2\pi$ **A.** $D = \mathbb{R}$ **B.** y-intercept: $f(0) = 0$; x-intercepts: $f(x) = 0 \iff \sin x = 0 \iff$ $x = 0, \pi$, and 2π. **C.** No symmetry **D.** No asymptote **E.** $f'(x) = e^{-x}\cos x + \sin x\,(-e^{-x}) = e^{-x}(\cos x - \sin x)$.

$f'(x) = 0 \iff \cos x = \sin x \iff x = \frac{\pi}{4}, \frac{5\pi}{4}$. $f'(x) > 0$ if x is in $\left(0, \frac{\pi}{4}\right)$ or $\left(\frac{5\pi}{4}, 2\pi\right)$ [f is increasing] and $f'(x) < 0$ if x is in $\left(\frac{\pi}{4}, \frac{5\pi}{4}\right)$ [f is decreasing]. **F.** Local maximum value $f\left(\frac{\pi}{4}\right)$ and local minimum value $f\left(\frac{5\pi}{4}\right)$

G. $f''(x) = e^{-x}(-\sin x - \cos x) + (\cos x - \sin x)(-e^{-x}) = e^{-x}(-2\cos x)$. $f''(x) > 0 \iff -2\cos x > 0 \iff$

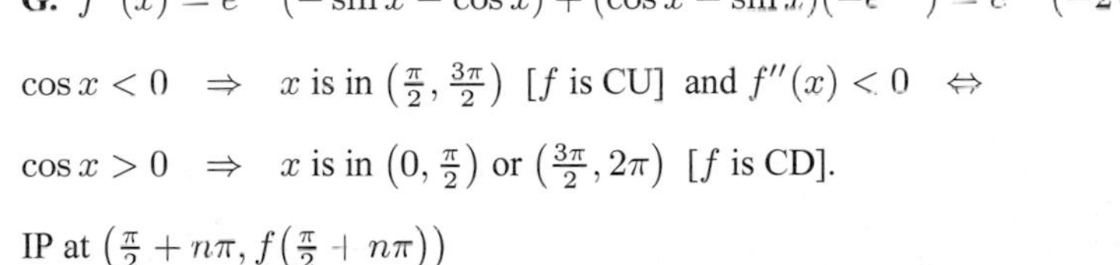

$\cos x < 0 \Rightarrow x$ is in $\left(\frac{\pi}{2}, \frac{3\pi}{2}\right)$ [f is CU] and $f''(x) < 0 \iff$ $\cos x > 0 \Rightarrow x$ is in $\left(0, \frac{\pi}{2}\right)$ or $\left(\frac{3\pi}{2}, 2\pi\right)$ [f is CD].

IP at $\left(\frac{\pi}{2} + n\pi, f\left(\frac{\pi}{2} + n\pi\right)\right)$

H.

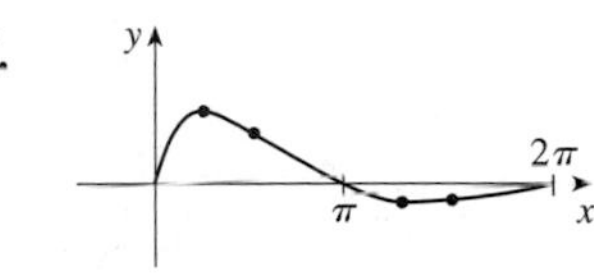

45. $y = f(x) = \dfrac{1}{x} + \ln x$ **A.** $D = (0, \infty)$ [same as $\ln x$] **B.** No y-intercept; no x-intercept $\left[\dfrac{1}{x} \text{ and } \ln x \text{ are both positive on } D\right]$ **C.** No symmetry. **D.** $\lim\limits_{x\to 0^+} f(x) = \infty$, so $x = 0$ is a VA.

E. $f'(x) = -\dfrac{1}{x^2} + \dfrac{1}{x} = \dfrac{x-1}{x^2}$. $f'(x) > 0$ for $x > 1$, so f is increasing on $(1, \infty)$ and f is decreasing on $(0, 1)$.

F. Local minimum value $f(1) = 1$ **G.** $f''(x) = \dfrac{2}{x^3} - \dfrac{1}{x^2} = \dfrac{2-x}{x^3}$.

$f''(x) > 0$ for $0 < x < 2$, so f is CU on $(0, 2)$, and f is CD on $(2, \infty)$.

IP at $\left(2, \frac{1}{2} + \ln 2\right)$

H.

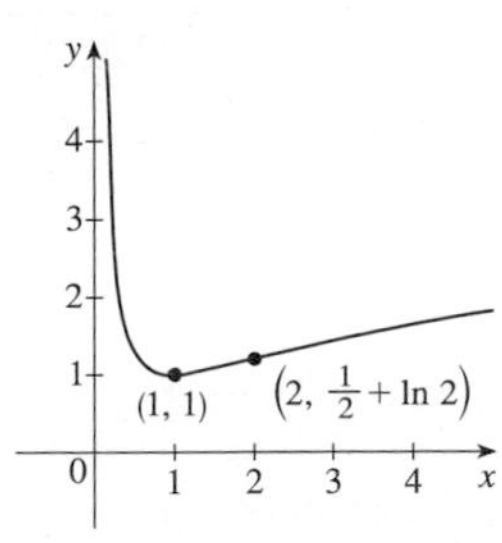

46. $y = f(x) = e^{2x} - e^x$ **A.** $D = \mathbb{R}$ **B.** y-intercept: $f(0) = 0$; x-intercepts: $f(x) = 0 \;\Rightarrow\; e^{2x} = e^x \;\Rightarrow\; e^x = 1 \;\Rightarrow$ $x = 0$. **C.** No symmetry **D.** $\lim\limits_{x \to -\infty} e^{2x} - e^x = 0$, so $y = 0$ is a HA. No VA. **E.** $f'(x) = 2e^{2x} - e^x = e^x(2e^x - 1)$, so $f'(x) > 0 \;\Leftrightarrow\; e^x > \frac{1}{2} \;\Leftrightarrow\; x > \ln\frac{1}{2} = -\ln 2$ and $f'(x) < 0 \;\Leftrightarrow$ $e^x < \frac{1}{2} \;\Leftrightarrow\; x < \ln\frac{1}{2}$, so f is decreasing on $\left(-\infty, \ln\frac{1}{2}\right)$ and increasing on $\left(\ln\frac{1}{2}, \infty\right)$.

F. Local minimum value $f\left(\ln\frac{1}{2}\right) = e^{2\ln(1/2)} - e^{\ln(1/2)} = \left(\frac{1}{2}\right)^2 - \frac{1}{2} = -\frac{1}{4}$

G. $f''(x) = 4e^{2x} - e^x = e^x(4e^x - 1)$, so $f''(x) > 0 \;\Leftrightarrow\; e^x > \frac{1}{4} \;\Leftrightarrow$ $x > \ln\frac{1}{4}$ and $f''(x) < 0 \;\Leftrightarrow\; x < \ln\frac{1}{4}$. Thus, f is CD on $\left(-\infty, \ln\frac{1}{4}\right)$ and CU on $\left(\ln\frac{1}{4}, \infty\right)$. IP at $\left(\ln\frac{1}{4}, \left(\frac{1}{4}\right)^2 - \frac{1}{4}\right) = \left(\ln\frac{1}{4}, -\frac{3}{16}\right)$

H.

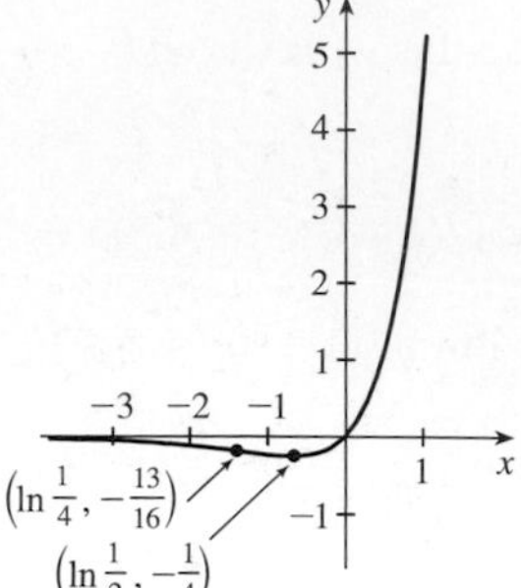

47. $y = f(x) = (1 + e^x)^{-2} = \dfrac{1}{(1 + e^x)^2}$ **A.** $D = \mathbb{R}$ **B.** y-intercept: $f(0) = \frac{1}{4}$. x-intercepts: none [since $f(x) > 0$]

C. No symmetry **D.** $\lim\limits_{x \to \infty} f(x) = 0$ and $\lim\limits_{x \to -\infty} f(x) = 1$, so $y = 0$ and $y = 1$ are HA; no VA

E. $f'(x) = -2(1 + e^x)^{-3}e^x = \dfrac{-2e^x}{(1 + e^x)^3} < 0$, so f is decreasing on $\mathbb{R}$ **F.** No local extrema

G. $f''(x) = (1 + e^x)^{-3}(-2e^x) + (-2e^x)(-3)(1 + e^x)^{-4}e^x$

$$= -2e^x(1 + e^x)^{-4}[(1 + e^x) - 3e^x] = \frac{-2e^x(1 - 2e^x)}{(1 + e^x)^4}.$$

$f''(x) > 0 \;\Leftrightarrow\; 1 - 2e^x < 0 \;\Leftrightarrow\; e^x > \frac{1}{2} \;\Leftrightarrow\; x > \ln\frac{1}{2}$ and $f''(x) < 0 \;\Leftrightarrow\; x < \ln\frac{1}{2}$, so f is CU on $\left(\ln\frac{1}{2}, \infty\right)$ and CD on $\left(-\infty, \ln\frac{1}{2}\right)$. IP at $\left(\ln\frac{1}{2}, \frac{4}{9}\right)$

H.

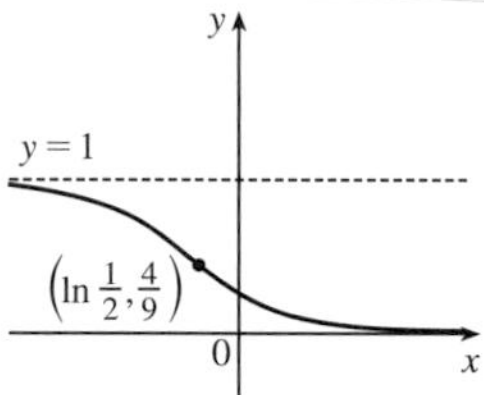

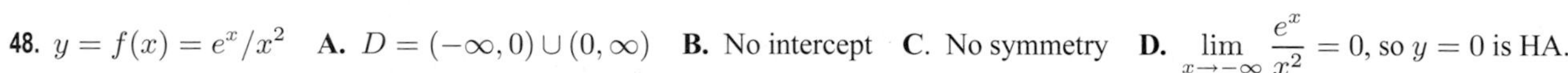

48. $y = f(x) = e^x/x^2$ **A.** $D = (-\infty, 0) \cup (0, \infty)$ **B.** No intercept **C.** No symmetry **D.** $\lim\limits_{x \to -\infty} \dfrac{e^x}{x^2} = 0$, so $y = 0$ is HA.

$\lim\limits_{x \to 0} \dfrac{e^x}{x^2} = \infty$, so $x = 0$ is a VA. **E.** $f'(x) = \dfrac{x^2e^x - e^x(2x)}{(x^2)^2} = \dfrac{xe^x(x - 2)}{x^4} = \dfrac{e^x(x - 2)}{x^3} > 0 \;\Leftrightarrow\; x < 0$ or $x > 2$, so f is increasing on $(-\infty, 0)$ and $(2, \infty)$, and f is decreasing on $(0, 2)$.

F. Local minimum value $f(2) = e^2/4 \approx 1.85$, no local maximum value

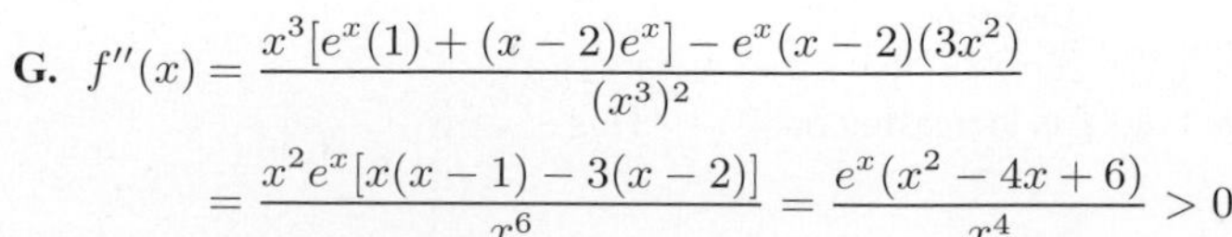

G. $f''(x) = \dfrac{x^3[e^x(1) + (x - 2)e^x] - e^x(x - 2)(3x^2)}{(x^3)^2}$

$$= \frac{x^2e^x[x(x - 1) - 3(x - 2)]}{x^6} = \frac{e^x(x^2 - 4x + 6)}{x^4} > 0$$

for all x in the domain of f; that is, f is CU on $(-\infty, 0)$ and $(0, \infty)$. No IP

H.

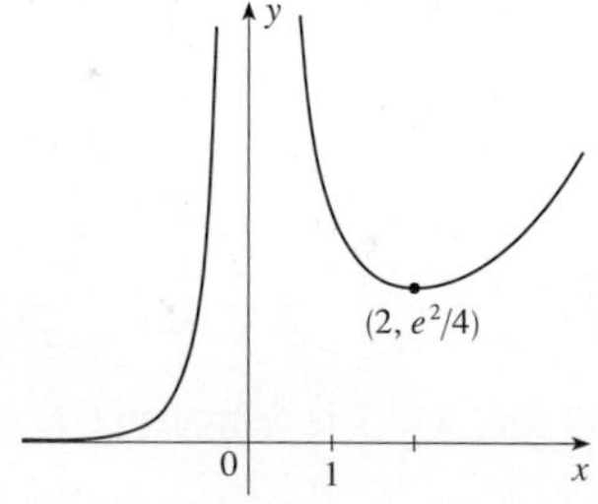

49. $y = f(x) = \ln(\sin x)$

A. $D = \{x \text{ in } \mathbb{R} \mid \sin x > 0\} = \bigcup\limits_{n=-\infty}^{\infty} (2n\pi, (2n + 1)\pi) = \cdots \cup (-4\pi, -3\pi) \cup (-2\pi, -\pi) \cup (0, \pi) \cup (2\pi, 3\pi) \cup \cdots$

B. No y-intercept; x-intercepts: $f(x) = 0 \Leftrightarrow \ln(\sin x) = 0 \Leftrightarrow \sin x = e^0 = 1 \Leftrightarrow x = 2n\pi + \frac{\pi}{2}$ for each integer n. **C.** f is periodic with period 2π. **D.** $\lim\limits_{x \to (2n\pi)^+} f(x) = -\infty$ and $\lim\limits_{x \to [(2n+1)\pi]^-} f(x) = -\infty$, so the lines $x = n\pi$ are VAs for all integers n. **E.** $f'(x) = \dfrac{\cos x}{\sin x} = \cot x$, so $f'(x) > 0$ when $2n\pi < x < 2n\pi + \frac{\pi}{2}$ for each integer n, and $f'(x) < 0$ when $2n\pi + \frac{\pi}{2} < x < (2n+1)\pi$. Thus, f is increasing on $\left(2n\pi, 2n\pi + \frac{\pi}{2}\right)$ and decreasing on $\left(2n\pi + \frac{\pi}{2}, (2n+1)\pi\right)$ for each integer n.

F. Local maximum values $f\left(2n\pi + \frac{\pi}{2}\right) = 0$, no local minimum.

G. $f''(x) = -\csc^2 x < 0$, so f is CD on $(2n\pi, (2n+1)\pi)$ for each integer n. No IP

H.

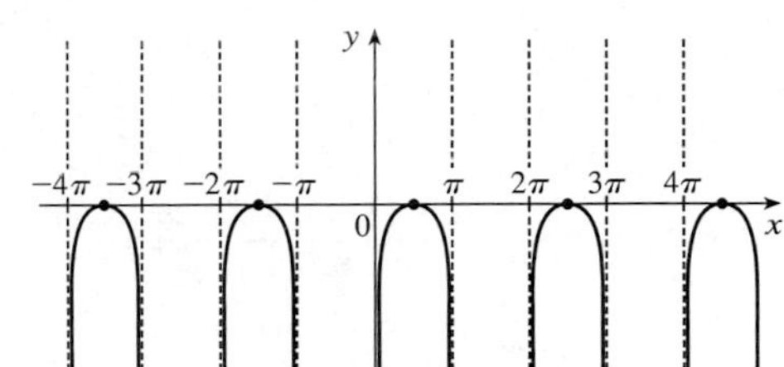

50. $y = f(x) = \ln(1 + x^3)$ **A.** $1 + x^3 > 0 \Leftrightarrow x^3 > -1 \Leftrightarrow x > -1$, so $D = (-1, \infty)$. **B.** y-intercept: $f(0) = \ln 1 = 0$; x-intercept: $f(x) = 0 \Leftrightarrow \ln(1 + x^3) = 0 \Leftrightarrow 1 + x^3 = e^0 \Leftrightarrow x^3 = 0 \Leftrightarrow x = 0$ **C.** No symmetry. **D.** $\lim\limits_{x \to -1^+} f(x) = -\infty$, so $x = -1$ is a VA **E.** $f'(x) = \dfrac{3x^2}{1 + x^3}$. $f'(x) > 0$ on $(-1, 0)$ and $(0, \infty)$ $[f'(x) = 0$ at $x = 0]$, so by Exercise 4.3.91, f is increasing on $(-1, \infty)$. **F.** No extreme values

G. $$f''(x) = \frac{(1 + x^3)(6x) - 3x^2(3x^2)}{(1 + x^3)^2} = \frac{3x[2(1 + x^3) - 3x^3]}{(1 + x^3)^2} = \frac{3x(2 - x^3)}{(1 + x^3)^2}$$

$f''(x) > 0 \Leftrightarrow 0 < x < \sqrt[3]{2}$, so f is CU on $\left(0, \sqrt[3]{2}\right)$ and f is CD on $(-1, 0)$ and $\left(\sqrt[3]{2}, \infty\right)$. IP at $(0, 0)$ and $\left(\sqrt[3]{2}, \ln 3\right)$

H.

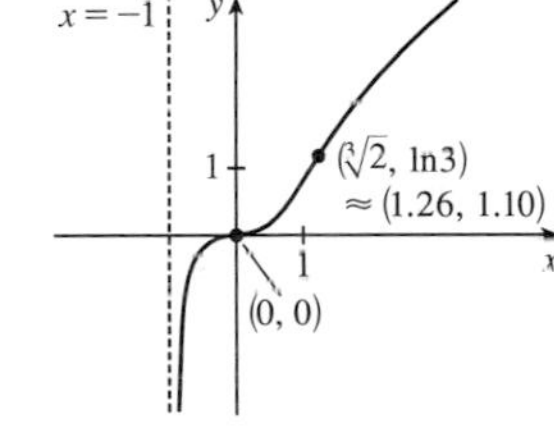

51. $y = f(x) = xe^{-1/x}$ **A.** $D = (-\infty, 0) \cup (0, \infty)$ **B.** No intercept **C.** No symmetry

D. $\lim\limits_{x \to 0^-} \dfrac{e^{-1/x}}{1/x} \overset{\text{H}}{=} \lim\limits_{x \to 0^-} \dfrac{e^{-1/x}(1/x^2)}{-1/x^2} = -\lim\limits_{x \to 0^-} e^{-1/x} = -\infty$, so $x = 0$ is a VA. Also, $\lim\limits_{x \to 0^+} xe^{-1/x} = 0$, so the graph approaches the origin as $x \to 0^+$. **E.** $f'(x) = xe^{-1/x}\left(\dfrac{1}{x^2}\right) + e^{-1/x}(1) = e^{-1/x}\left(\dfrac{1}{x} + 1\right) = \dfrac{x+1}{xe^{1/x}} > 0 \Leftrightarrow$ $x < -1$ or $x > 0$, so f is increasing on $(-\infty, -1)$ and $(0, \infty)$, and f is decreasing on $(-1, 0)$.

F. Local maximum value $f(-1) = -e$, no local minimum value

G. $f'(x) = e^{-1/x}\left(\dfrac{1}{x} + 1\right) \Rightarrow$

$$f''(x) = e^{-1/x}\left(-\frac{1}{x^2}\right) + \left(\frac{1}{x} + 1\right)e^{-1/x}\left(\frac{1}{x^2}\right) = \frac{1}{x^2}e^{-1/x}\left[-1 + \left(\frac{1}{x} + 1\right)\right] = \frac{1}{x^3 e^{1/x}} > 0 \Leftrightarrow$$

$x > 0$, so f is CU on $(0, \infty)$ and CD on $(-\infty, 0)$. No IP

H.

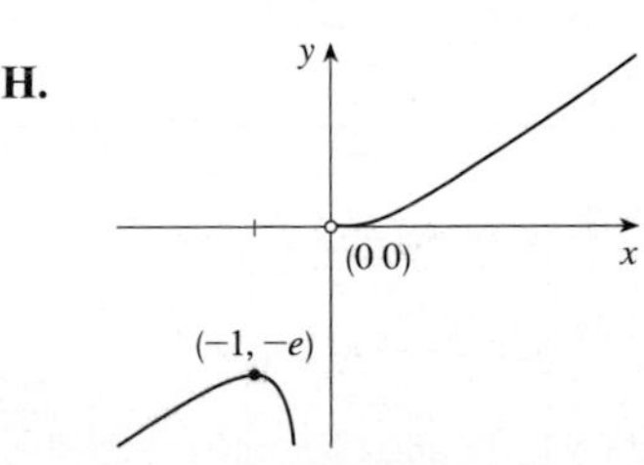

52. $y = f(x) = \dfrac{\ln x}{x^2}$ **A.** $D = (0, \infty)$ **B.** y-intercept: none; x-intercept: $f(x) = 0 \Leftrightarrow \ln x = 0 \Leftrightarrow x = 1$

C. No symmetry **D.** $\lim\limits_{x \to 0^+} f(x) = -\infty$, so $x = 0$ is a VA; $\lim\limits_{x \to \infty} \dfrac{\ln x}{x^2} \overset{\text{H}}{=} \lim\limits_{x \to \infty} \dfrac{1/x}{2x} = 0$, so $y = 0$ is a HA.

E. $f'(x) = \dfrac{x^2(1/x) - (\ln x)(2x)}{(x^2)^2} = \dfrac{x(1 - 2\ln x)}{x^4} = \dfrac{1 - 2\ln x}{x^3}$. $f'(x) > 0 \Leftrightarrow 1 - 2\ln x > 0 \Leftrightarrow \ln x < \frac{1}{2} \Rightarrow$

$0 < x < e^{1/2}$ and $f'(x) < 0 \Rightarrow x > e^{1/2}$, so f is increasing on $\left(0, \sqrt{e}\right)$ and decreasing on $\left(\sqrt{e}, \infty\right)$.

F. Local maximum value $f(e^{1/2}) = \dfrac{1/2}{e} = \dfrac{1}{2e}$

G. $f''(x) = \dfrac{x^3(-2/x) - (1 - 2\ln x)(3x^2)}{(x^3)^2}$

$= \dfrac{x^2[-2 - 3(1 - 2\ln x)]}{x^6} = \dfrac{-5 + 6\ln x}{x^4}$

$f''(x) > 0 \Leftrightarrow -5 + 6\ln x > 0 \Leftrightarrow \ln x > \frac{5}{6} \Rightarrow x > e^{5/6}$ [f is CU]

and $f''(x) < 0 \Leftrightarrow 0 < x < e^{5/6}$ [f is CD]. IP at $(e^{5/6}, 5/(6e^{5/3}))$

H.

53. $y = f(x) = e^{\arctan x}$ **A.** $D = \mathbb{R}$ **B.** y-intercept: $f(0) = e^0 = 1$; no x-intercept since $e^{\arctan x}$ is positive for all x.

C. No symmetry **D.** $\lim\limits_{x \to -\infty} f(x) = e^{-\pi/2}$ [≈ 0.21], so $y = e^{-\pi/2}$ is a HA. $\lim\limits_{x \to \infty} f(x) = e^{\pi/2}$ [≈ 4.81], so $y = e^{\pi/2}$ is a HA. **E.** $f'(x) = e^{\arctan x}\left(\dfrac{1}{1 + x^2}\right)$. $f'(x) > 0$ for all x, so f is increasing on $\mathbb{R}$. **F.** No local extrema

G. $f''(x) = \dfrac{(1 + x^2)e^{\arctan x}\left(\dfrac{1}{1 + x^2}\right) - e^{\arctan x}(2x)}{(1 + x^2)^2}$

$= \dfrac{e^{\arctan x}(1 - 2x)}{(1 + x^2)^2}$

$f''(x) > 0$ for $x < \frac{1}{2}$, so f is CU on $\left(-\infty, \frac{1}{2}\right)$ and f is CD on $\left(\frac{1}{2}, \infty\right)$.

IP at $\left(\frac{1}{2}, e^{\arctan 1/2}\right) \approx (0.5, 1.59)$

H.

54. $y = f(x) = \tan^{-1}\left(\dfrac{x - 1}{x + 1}\right)$ **A.** $D = \{x \mid x \neq -1\}$ **B.** x-intercept $= 1$, y-intercept $= f(0) = \tan^{-1}(-1) = -\frac{\pi}{4}$

C. No symmetry **D.** $\lim\limits_{x \to \pm\infty} \tan^{-1}\left(\dfrac{x - 1}{x + 1}\right) = \lim\limits_{x \to \pm\infty} \tan^{-1}\left(\dfrac{1 - 1/x}{1 + 1/x}\right) = \tan^{-1} 1 = \frac{\pi}{4}$, so $y = \frac{\pi}{4}$ is a HA.

Also $\lim\limits_{x \to -1^+} \tan^{-1}\left(\dfrac{x - 1}{x + 1}\right) = -\dfrac{\pi}{2}$ and $\lim\limits_{x \to -1^-} \tan^{-1}\left(\dfrac{x - 1}{x + 1}\right) = \dfrac{\pi}{2}$.

E. $f'(x) = \dfrac{1}{1 + [(x - 1)/(x + 1)]^2} \dfrac{(x + 1) - (x - 1)}{(x + 1)^2} = \dfrac{2}{(x + 1)^2 + (x - 1)^2} = \dfrac{1}{x^2 + 1} > 0$,

so f is increasing on $(-\infty, -1)$ and $(-1, \infty)$. **F.** No extreme values **H.**

G. $f''(x) = -2x/\left(x^2 + 1\right)^2 > 0 \Leftrightarrow x < 0$, so f is CU on $(-\infty, -1)$ and $(-1, 0)$, and CD on $(0, \infty)$. IP at $\left(0, -\frac{\pi}{4}\right)$

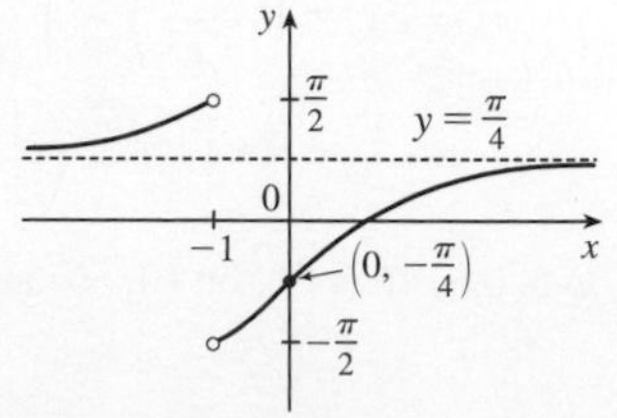

55. $m = f(v) = \dfrac{m_0}{\sqrt{1 - v^2/c^2}}$. The m-intercept is $f(0) = m_0$. There are no v-intercepts. $\lim\limits_{v \to c^-} f(v) = \infty$, so $v = c$ is a VA.

$$f'(v) = -\tfrac{1}{2}m_0(1 - v^2/c^2)^{-3/2}(-2v/c^2) = \frac{m_0 v}{c^2(1 - v^2/c^2)^{3/2}} = \frac{m_0 v}{c^2\dfrac{(c^2 - v^2)^{3/2}}{c^3}} = \frac{m_0 cv}{(c^2 - v^2)^{3/2}} > 0, \text{ so } f \text{ is}$$

increasing on $(0, c)$. There are no local extreme values.

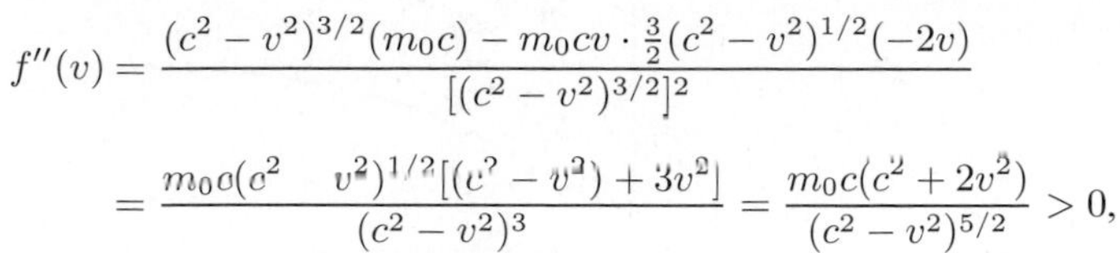

$$f''(v) = \frac{(c^2 - v^2)^{3/2}(m_0 c) - m_0 cv \cdot \frac{3}{2}(c^2 - v^2)^{1/2}(-2v)}{[(c^2 - v^2)^{3/2}]^2}$$

$$= \frac{m_0 c(c^2 - v^2)^{1/2}[(c^2 - v^2) + 3v^2]}{(c^2 - v^2)^3} = \frac{m_0 c(c^2 + 2v^2)}{(c^2 - v^2)^{5/2}} > 0,$$

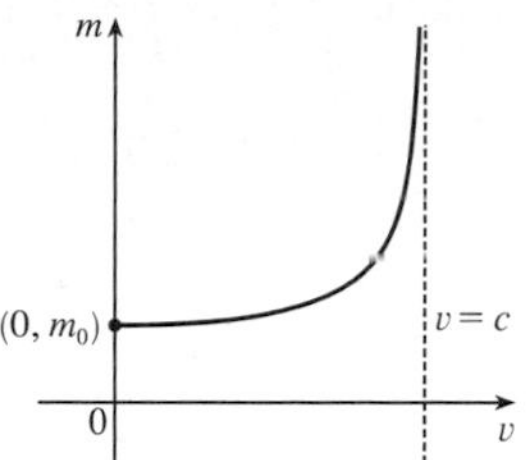

so f is CU on $(0, c)$. There are no inflection points.

56. Let $a = m_0^2 c^4$ and $b = h^2 c^2$, so the equation can be written as $E = f(\lambda) = \sqrt{a + b/\lambda^2} = \sqrt{\dfrac{a\lambda^2 + b}{\lambda^2}} = \dfrac{\sqrt{a\lambda^2 + b}}{\lambda}$.

$\lim\limits_{\lambda \to 0^+} \dfrac{\sqrt{a\lambda^2 + b}}{\lambda} = \infty$, so $\lambda = 0$ is a VA.

$\lim\limits_{\lambda \to \infty} \dfrac{\sqrt{a\lambda^2 + b}}{\lambda} = \lim\limits_{\lambda \to \infty} \dfrac{\sqrt{a\lambda^2 + b}/\lambda}{\lambda/\lambda} = \lim\limits_{\lambda \to \infty} \dfrac{\sqrt{a + b/\lambda^2}}{1} = \sqrt{a}$, so $E = \sqrt{a} = m_0 c^2$ is a HA.

$$f'(\lambda) = \frac{\lambda \cdot \frac{1}{2}(a\lambda^2 + b)^{-1/2}(2a\lambda) - (a\lambda^2 + b)^{1/2}(1)}{\lambda^2} = \frac{(a\lambda^2 + b)^{-1/2}[a\lambda^2 - (a\lambda^2 + b)]}{\lambda^2} = \frac{-b}{\lambda^2\sqrt{a\lambda^2 + b}} < 0,$$

so f is decreasing on $(0, \infty)$. Using the Reciprocal Rule,

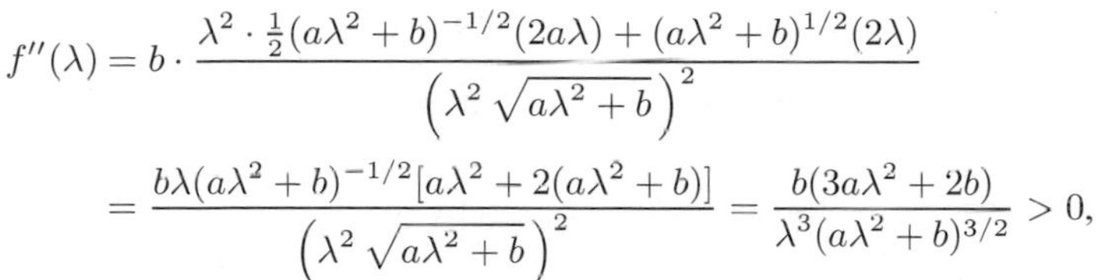

$$f''(\lambda) = b \cdot \frac{\lambda^2 \cdot \frac{1}{2}(a\lambda^2 + b)^{-1/2}(2a\lambda) + (a\lambda^2 + b)^{1/2}(2\lambda)}{\left(\lambda^2\sqrt{a\lambda^2 + b}\right)^2}$$

$$= \frac{b\lambda(a\lambda^2 + b)^{-1/2}[a\lambda^2 + 2(a\lambda^2 + b)]}{\left(\lambda^2\sqrt{a\lambda^2 + b}\right)^2} = \frac{b(3a\lambda^2 + 2b)}{\lambda^3(a\lambda^2 + b)^{3/2}} > 0,$$

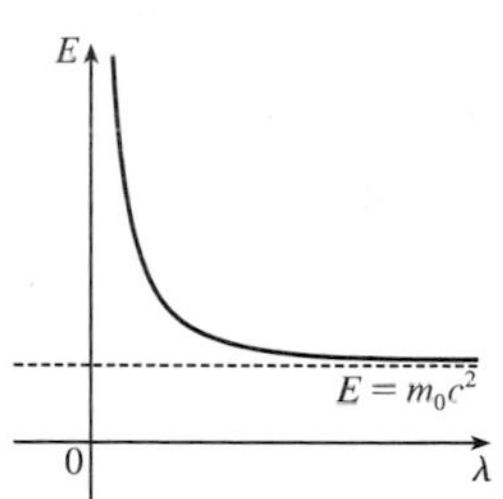

so f is CU on $(0, \infty)$. There are no extrema or inflection points. The graph shows that as λ decreases, the energy increases and as λ increases, the energy decreases. For large wavelengths, the energy is very close to the energy at rest.

57. (a) $p(t) = \dfrac{1}{2} \;\Rightarrow\; \dfrac{1}{2} = \dfrac{1}{1 + ae^{-kt}} \;\Leftrightarrow\; 1 + ae^{-kt} = 2 \;\Leftrightarrow\; ae^{-kt} = 1 \;\Leftrightarrow\; e^{-kt} = \dfrac{1}{a} \;\Leftrightarrow$

$\ln e^{-kt} = \ln a^{-1} \;\Leftrightarrow\; -kt = -\ln a \;\Leftrightarrow\; t = \dfrac{\ln a}{k}$, which is when half the population will have heard the rumor.

(b) The rate of spread is given by $p'(t) = \dfrac{ake^{-kt}}{(1 + ae^{-kt})^2}$. To find the greatest rate of spread, we'll apply the First Derivative Test to $p'(t)$ [not $p(t)$].

$$[p'(t)]' = p''(t) = \frac{(1 + ae^{-kt})^2(-ak^2e^{-kt}) - ake^{-kt} \cdot 2(1 + ae^{-kt})(-ake^{-kt})}{[(1 + ae^{-kt})^2]^2}$$

$$= \frac{(1 + ae^{-kt})(-ake^{-kt})[k(1 + ae^{-kt}) - 2ake^{-kt}]}{(1 + ae^{-kt})^4} = \frac{-ake^{-kt}(k)(1 - ae^{-kt})}{(1 + ae^{-kt})^3} = \frac{ak^2e^{-kt}(ae^{-kt} - 1)}{(1 + ae^{-kt})^3}$$

$p''(t) > 0 \Leftrightarrow ae^{-kt} > 1 \Leftrightarrow -kt > \ln a^{-1} \Leftrightarrow t < \dfrac{\ln a}{k}$, so $p'(t)$ is increasing for $t < \dfrac{\ln a}{k}$ and $p'(t)$ is decreasing for $t > \dfrac{\ln a}{k}$. Thus, $p'(t)$, the rate of spread of the rumor, is greatest at the same time, $\dfrac{\ln a}{k}$, as when half the population [by part (a)] has heard it.

(c) $p(0) = \dfrac{1}{1+a}$ and $\lim\limits_{t\to\infty} p(t) = 1$. The graph is shown with $a = 4$ and $k = \frac{1}{2}$.

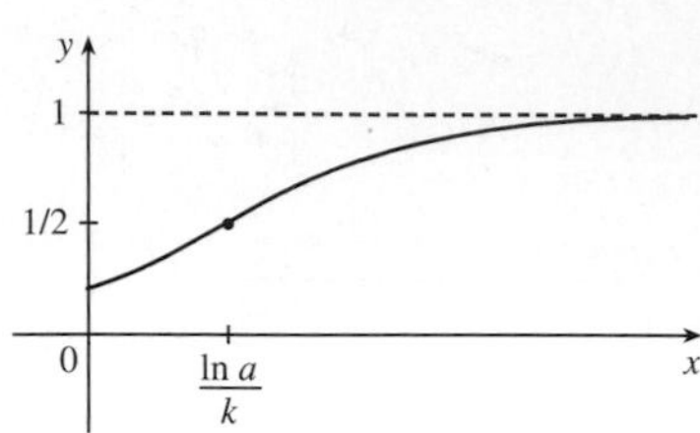

58. $C(t) = K(e^{-at} - e^{-bt})$, where $K > 0$ and $b > a > 0$. $C(0) = K(1-1) = 0$ is the only intercept. $\lim\limits_{t\to\infty} C(t) = 0$, so $C = 0$ is a HA. $C'(t) = K(-ae^{-at} + be^{-bt}) > 0 \Leftrightarrow be^{-bt} > ae^{-at} \Leftrightarrow e^{at}e^{-bt} > \dfrac{a}{b} \Leftrightarrow e^{(a-b)t} > \dfrac{a}{b} \Leftrightarrow$ $(a-b)t > \ln \dfrac{a}{b} \Leftrightarrow t > \dfrac{\ln(a/b)}{a-b}$ or $\dfrac{\ln(b/a)}{b-a}$ [call this value D]. C is increasing for $t < D$ and decreasing for $t > D$, so $C(D)$ is a local maximum [and absolute] value. $C''(t) = K(a^2e^{-at} - b^2e^{-bt}) > 0 \Leftrightarrow a^2e^{-at} > b^2e^{-bt} \Leftrightarrow$ $e^{bt}e^{-at} > \dfrac{b^2}{a^2} \Leftrightarrow e^{(b-a)t} > \left(\dfrac{b}{a}\right)^2 \Leftrightarrow (b-a)t > \ln\left(\dfrac{b}{a}\right)^2 \Leftrightarrow t > \dfrac{2\ln(b/a)}{b-a} = 2D$, so C is CU on $(2D, \infty)$ and CD on $(0, 2D)$. The inflection point is $(2D, C(2D))$. For the graph shown, $K = 1, a = 1, b = 2, D = \ln 2, C(D) = \frac{1}{4}$, and $C(2D) = \frac{3}{16}$. The graph tells us that when the drug is injected into the bloodstream, its concentration rises rapidly to a maximum at time D, then falls, reaching its maximum rate of decrease at time $2D$, then continues to decrease more and more slowly, approaching 0 as $t \to \infty$.

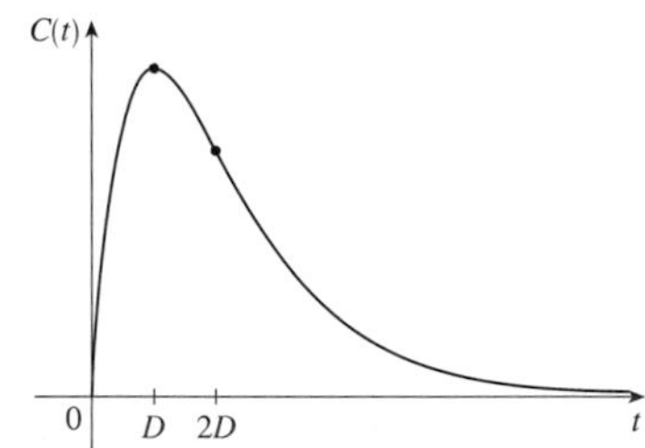

59. $y = -\dfrac{W}{24EI}x^4 + \dfrac{WL}{12EI}x^3 - \dfrac{WL^2}{24EI}x^2 = -\dfrac{W}{24EI}x^2(x^2 - 2Lx + L^2)$

$= \dfrac{-W}{24EI}x^2(x-L)^2 = cx^2(x-L)^2$

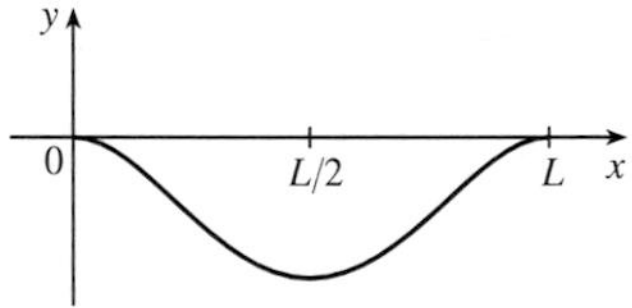

where $c = -\dfrac{W}{24EI}$ is a negative constant and $0 \le x \le L$. We sketch $f(x) = cx^2(x-L)^2$ for $c = -1$. $f(0) = f(L) = 0$.

$f'(x) = cx^2[2(x-L)] + (x-L)^2(2cx) = 2cx(x-L)[x + (x-L)] = 2cx(x-L)(2x-L)$. So for $0 < x < L$, $f'(x) > 0 \Leftrightarrow x(x-L)(2x-L) < 0$ [since $c < 0$] $\Leftrightarrow L/2 < x < L$ and $f'(x) < 0 \Leftrightarrow 0 < x < L/2$. Thus, f is increasing on $(L/2, L)$ and decreasing on $(0, L/2)$, and there is a local and absolute minimum at the point $(L/2, f(L/2)) = (L/2, cL^4/16)$. $f'(x) = 2c[x(x-L)(2x-L)] \Rightarrow$

$f''(x) = 2c[1(x-L)(2x-L) + x(1)(2x-L) + x(x-L)(2)] = 2c(6x^2 - 6Lx + L^2) = 0 \Leftrightarrow$

$x = \dfrac{6L \pm \sqrt{12L^2}}{12} = \frac{1}{2}L \pm \frac{\sqrt{3}}{6}L$, and these are the x-coordinates of the two inflection points.

60. $F(x) = -\dfrac{k}{x^2} + \dfrac{k}{(x-2)^2}$, where $k > 0$ and $0 < x < 2$. For $0 < x < 2$, $x - 2 < 0$, so

$F'(x) = \dfrac{2k}{x^3} - \dfrac{2k}{(x-2)^3} > 0$ and F is increasing. $\lim\limits_{x\to 0^+} F(x) = -\infty$ and

$\lim\limits_{x\to 2^-} F(x) = \infty$, so $x = 0$ and $x = 2$ are vertical asymptotes. Notice that when the middle particle is at $x = 1$, the net force acting on it is 0. When $x > 1$, the net force is positive, meaning that it acts to the right. And if the particle approaches $x = 2$, the force on it rapidly becomes very large. When $x < 1$, the net force is negative, so it acts to the left. If the particle approaches 0, the force becomes very large to the left.

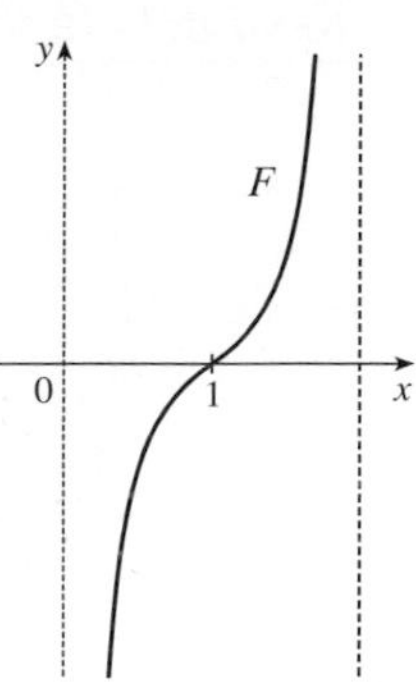

61. $y = \dfrac{x^2+1}{x+1}$. Long division gives us:

$$\begin{array}{r|l} & \quad x - 1 \\ \hline x+1 & x^2 \qquad +1 \\ & x^2 + x \\ \hline & \quad -x + 1 \\ & \quad -x - 1 \\ \hline & \qquad\quad 2 \end{array}$$

Thus, $y = f(x) = \dfrac{x^2+1}{x+1} = x - 1 + \dfrac{2}{x+1}$ and $f(x) - (x-1) = \dfrac{2}{x+1} = \dfrac{\frac{2}{x}}{1 + \frac{1}{x}}$ [for $x \neq 0$] $\to 0$ as $x \to \pm\infty$.

So the line $y = x - 1$ is a slant asymptote (SA).

62. $y = \dfrac{4x^3 - 10x^2 - 11x + 1}{x^2 - 3x}$. Long division gives us:

$$\begin{array}{r|l} & \qquad\qquad 4x + 2 \\ \hline x^2 - 3x & 4x^3 - 10x^2 - 11x + 1 \\ & 4x^3 - 12x^2 \\ \hline & \qquad 2x^2 - 11x \\ & \qquad 2x^2 - \ \ 6x \\ \hline & \qquad\qquad -5x + 1 \end{array}$$

Thus, $y = f(x) = \dfrac{4x^3 - 10x^2 - 11x + 1}{x^2 - 3x} = 4x + 2 + \dfrac{-5x+1}{x^2 - 3x}$ and $f(x) - (4x+2) = \dfrac{-5x+1}{x^2-3x} = \dfrac{-\frac{5}{x} + \frac{1}{x^2}}{1 - \frac{3}{x}}$

[for $x \neq 0$] $\to \frac{0}{1} = 0$ as $x \to \pm\infty$. So the line $y = 4x + 2$ is a slant asymptote (SA).

63. $y = \dfrac{2x^3 - 5x^2 + 3x}{x^2 - x - 2}$. Long division gives us:

$$\begin{array}{r|l} & \qquad\qquad 2x - 3 \\ \hline x^2 - x - 2 & 2x^3 - 5x^2 + 3x \\ & 2x^3 - 2x^2 - 4x \\ \hline & \quad -3x^2 + 7x \\ & \quad -3x^2 + 3x + 6 \\ \hline & \qquad\qquad 4x - 6 \end{array}$$

Thus, $y = f(x) = \dfrac{2x^3 - 5x^2 + 3x}{x^2 - x - 2} = 2x - 3 + \dfrac{4x-6}{x^2-x-2}$ and $f(x) - (2x-3) = \dfrac{4x-6}{x^2-x-2} = \dfrac{\frac{4}{x} - \frac{6}{x^2}}{1 - \frac{1}{x} - \frac{1}{x^2}}$

[for $x \neq 0$] $\to \frac{0}{1} = 0$ as $x \to \pm\infty$. So the line $y = 2x - 3$ is a slant asymptote (SA).

64. $y = \dfrac{-6x^4 + 2x^3 + 3}{2x^3 - x}$. Long division gives us:

$$\begin{array}{r|l} & -3x + 1 \\ \hline 2x^3 - x & -6x^4 + 2x^3 \qquad\qquad + 3 \\ & -6x^4 \qquad\quad + 3x^2 \\ \hline & \qquad 2x^3 - 3x^2 \\ & \qquad 2x^3 \qquad\quad - x \\ \hline & \qquad -3x^2 + x + 3 \end{array}$$

Thus, $y = f(x) = \dfrac{-6x^4 + 2x^3 + 3}{2x^3 - x} = -3x + 1 + \dfrac{-3x^2 + x + 3}{2x^3 - x}$ and

$f(x) - (-3x + 1) = \dfrac{-3x^2 + x + 3}{2x^3 - x} = \dfrac{-\dfrac{3}{x} + \dfrac{1}{x^2} + \dfrac{3}{x^3}}{2 - \dfrac{1}{x^2}}$ [for $x \neq 0$] $\to \frac{0}{2} = 0$ as $x \to \pm\infty$. So the line $y = -3x + 1$ is a slant asymptote (SA).

65. $y = f(x) = \dfrac{x^2}{x - 1} = x + 1 + \dfrac{1}{x - 1}$ **A.** $D = (-\infty, 1) \cup (1, \infty)$ **B.** x-intercept: $f(x) = 0 \Leftrightarrow x = 0$; y-intercept: $f(0) = 0$ **C.** No symmetry **D.** $\lim\limits_{x \to 1^-} f(x) = -\infty$ and $\lim\limits_{x \to 1^+} f(x) = \infty$, so $x = 1$ is a VA.

$\lim\limits_{x \to \pm\infty} [f(x) - (x + 1)] = \lim\limits_{x \to \pm\infty} \dfrac{1}{x - 1} = 0$, so the line $y = x + 1$ is a SA.

E. $f'(x) = 1 - \dfrac{1}{(x - 1)^2} = \dfrac{(x - 1)^2 - 1}{(x - 1)^2} = \dfrac{x^2 - 2x}{(x - 1)^2} = \dfrac{x(x - 2)}{(x - 1)^2} > 0$ for $x < 0$ or $x > 2$, so f is increasing on $(-\infty, 0)$ and $(2, \infty)$, and f is decreasing on $(0, 1)$ and $(1, 2)$. **F.** Local maximum value $f(0) = 0$, local minimum value $f(2) = 4$ **G.** $f''(x) = \dfrac{2}{(x - 1)^3} > 0$ for $x > 1$, so f is CU on $(1, \infty)$ and f is CD on $(-\infty, 1)$. No IP

H.

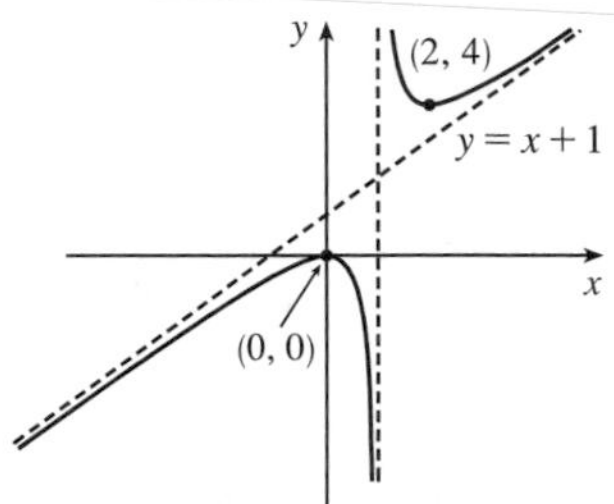

66. $y = f(x) = \dfrac{1 + 5x - 2x^2}{x - 2} = -2x + 1 + \dfrac{3}{x - 2}$ **A.** $D = (-\infty, 2) \cup (2, \infty)$ **B.** x-intercepts: $f(x) = 0 \Leftrightarrow$ $1 + 5x - 2x^2 = 0 \Rightarrow x = \dfrac{-5 \pm \sqrt{33}}{-4} \Rightarrow x \approx -0.19, 2.69$; y-intercept: $f(0) = -\frac{1}{2}$ **C.** No symmetry

D. $\lim\limits_{x \to 2^-} f(x) = -\infty$ and $\lim\limits_{x \to 2^+} f(x) = \infty$, so $x = 2$ is a VA. $\lim\limits_{x \to \pm\infty} [f(x) - (-2x + 1)] = \lim\limits_{x \to \pm\infty} \dfrac{3}{x - 2} = 0$, so $y = -2x + 1$ is a SA.

E. $f'(x) = -2 - \dfrac{3}{(x - 2)^2} = \dfrac{-2(x^2 - 4x + 4) - 3}{(x - 2)^2}$
$= \dfrac{-2x^2 + 8x - 11}{(x - 2)^2} < 0$

for $x \neq 2$, so f is decreasing on $(-\infty, 2)$ and $(2, \infty)$. **F.** No local extrema

G. $f''(x) = \dfrac{6}{(x - 2)^3} > 0$ for $x > 2$, so f is CU on $(2, \infty)$ and CD on $(-\infty, 2)$.

No IP

H.

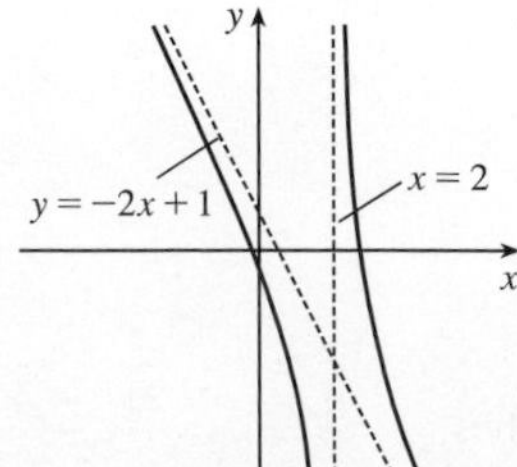

67. $y = f(x) = \dfrac{x^3+4}{x^2} = x + \dfrac{4}{x^2}$ **A.** $D = (-\infty, 0) \cup (0, \infty)$ **B.** x-intercept: $f(x) = 0 \;\Leftrightarrow\; x = -\sqrt[3]{4}$; no y-intercept

C. No symmetry **D.** $\lim\limits_{x\to 0} f(x) = \infty$, so $x = 0$ is a VA. $\lim\limits_{x\to\pm\infty} [f(x) - x] = \lim\limits_{x\to\pm\infty} \dfrac{4}{x^2} = 0$, so $y = x$ is a SA.

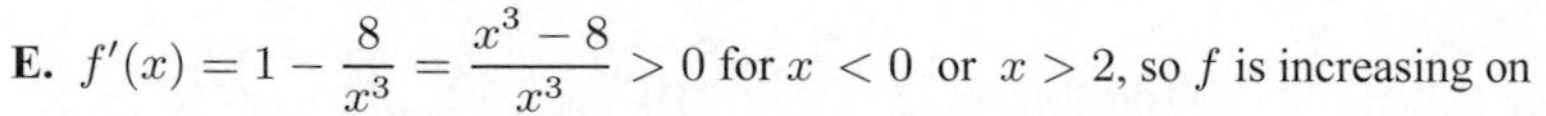

E. $f'(x) = 1 - \dfrac{8}{x^3} = \dfrac{x^3-8}{x^3} > 0$ for $x < 0$ or $x > 2$, so f is increasing on $(-\infty, 0)$ and $(2, \infty)$, and f is decreasing on $(0, 2)$. **F.** Local minimum value $f(2) = 3$, no local maximum value **G.** $f''(x) = \dfrac{24}{x^4} > 0$ for $x \neq 0$, so f is CU on $(-\infty, 0)$ and $(0, \infty)$. No IP

H.

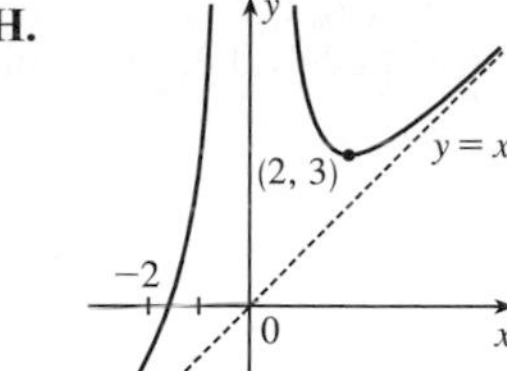

68. $y = f(x) = \dfrac{x^3}{(x+1)^2} = x - 2 + \dfrac{3x+2}{(x+1)^2}$ **A.** $D = (-\infty, -1) \cup (-1, \infty)$ **B.** x-intercept: 0; y-intercept: $f(0) = 0$

C. No symmetry **D.** $\lim\limits_{x\to -1^-} f(x) = -\infty$ and $\lim\limits_{x\to -1^+} f(x) = -\infty$, so $x = -1$ is a VA.

$\lim\limits_{x\to\pm\infty} [f(x) - (x-2)] = \lim\limits_{x\to\pm\infty} \dfrac{3x+2}{(x+1)^2} = 0$, so $y = x - 2$ is a SA.

E. $f'(x) = \dfrac{(x+1)^2(3x^2) - x^3 \cdot 2(x+1)}{[(x+1)^2]^2} = \dfrac{x^2(x+1)[3(x+1) - 2x]}{(x+1)^4} = \dfrac{x^2(x+3)}{(x+1)^3} > 0 \;\Leftrightarrow\; x < -3$ or

$x > -1$ $[x \neq 0]$, so f is increasing on $(-\infty, -3)$ and $(-1, \infty)$, and f is decreasing on $(-3, -1)$.

F. Local maximum value $f(-3) = -\frac{27}{4}$, no local minimum

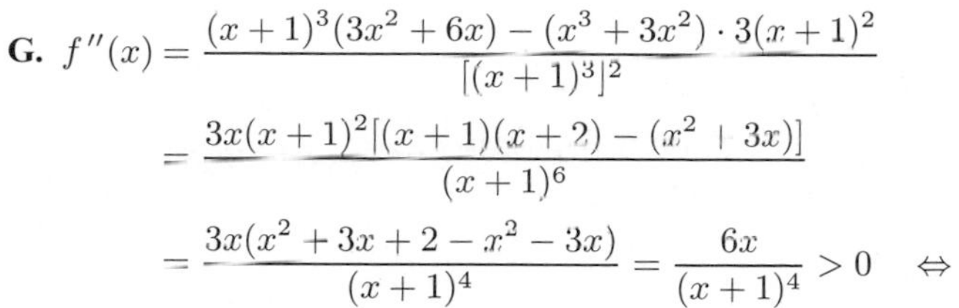

G. $$f''(x) = \frac{(x+1)^3(3x^2+6x) - (x^3+3x^2)\cdot 3(x+1)^2}{[(x+1)^3]^2}$$
$$= \frac{3x(x+1)^2[(x+1)(x+2) - (x^2+3x)]}{(x+1)^6}$$
$$= \frac{3x(x^2+3x+2-x^2-3x)}{(x+1)^4} = \frac{6x}{(x+1)^4} > 0 \;\Leftrightarrow$$

$x > 0$, so f is CU on $(0, \infty)$ and f is CD on $(-\infty, -1)$ and $(-1, 0)$. IP at $(0, 0)$

H.

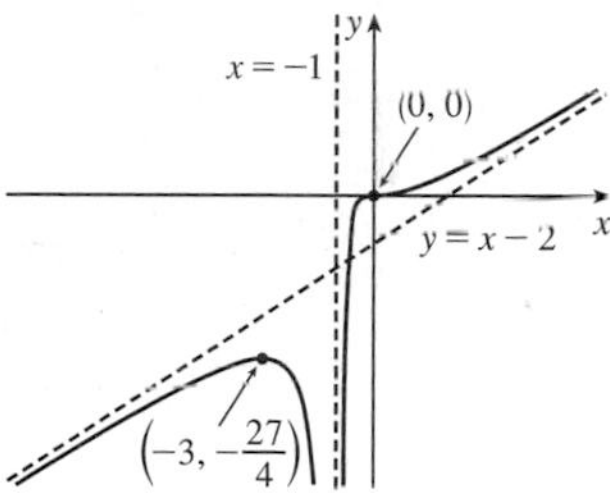

69. $y = f(x) = 1 + \frac{1}{2}x + e^{-x}$ **A.** $D = \mathbb{R}$ **B.** y-intercept $= f(0) = 2$, no x-intercept [see part **F**] **C.** No symmetry

D. No VA or HA. $\lim\limits_{x\to\infty} \left[f(x) - \left(1 + \frac{1}{2}x\right)\right] = \lim\limits_{x\to\infty} e^{-x} = 0$, so $y = 1 + \frac{1}{2}x$ is a SA. **E.** $f'(x) = \frac{1}{2} - e^{-x} > 0 \;\Leftrightarrow$

$\frac{1}{2} > e^{-x} \;\Leftrightarrow\; -x < \ln\frac{1}{2} \;\Leftrightarrow\; x > -\ln 2^{-1} \;\Leftrightarrow\; x > \ln 2$, so f is increasing on $(\ln 2, \infty)$ and decreasing

on $(-\infty, \ln 2)$. **F.** Local and absolute minimum value

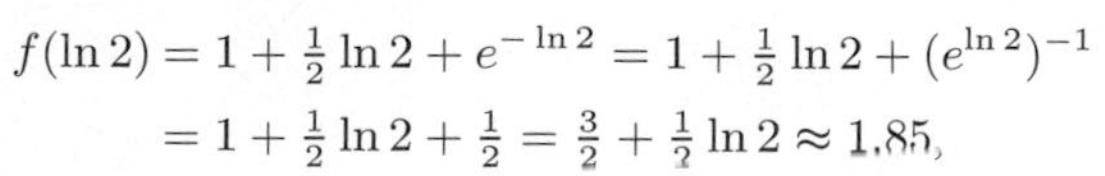

$$f(\ln 2) = 1 + \tfrac{1}{2}\ln 2 + e^{-\ln 2} = 1 + \tfrac{1}{2}\ln 2 + (e^{\ln 2})^{-1}$$
$$= 1 + \tfrac{1}{2}\ln 2 + \tfrac{1}{2} = \tfrac{3}{2} + \tfrac{1}{2}\ln 2 \approx 1.85,$$

no local maximum value **G.** $f''(x) = e^{-x} > 0$ for all x, so f is CU on $(-\infty, \infty)$. No IP

H.

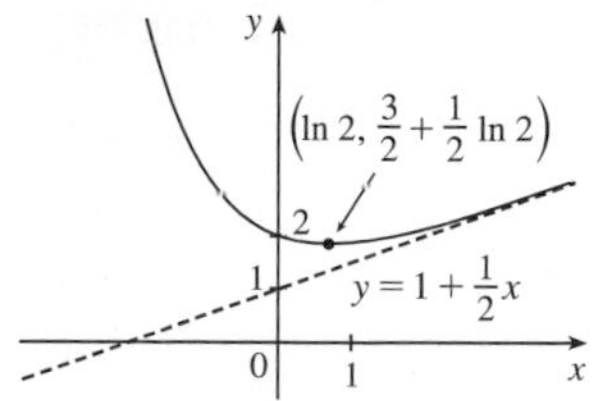

70. $y = f(x) = 1 - x + e^{1+x/3}$ **A.** $D = \mathbb{R}$ **B.** y-intercept $= f(0) = 1 + e$, no x-intercept [see part **F**]

C. No symmetry **D.** No VA or HA $\lim\limits_{x \to -\infty} [f(x) - (1 - x)] = \lim\limits_{x \to -\infty} e^{1+x/3} = 0$, so $y = 1 - x$ is a SA.

E. $f'(x) = -1 + \frac{1}{3}e^{1+x/3} > 0 \Leftrightarrow \frac{1}{3}e^{1+x/3} > 1 \Leftrightarrow e^{1+x/3} > 3 \Leftrightarrow 1 + \dfrac{x}{3} > \ln 3 \Leftrightarrow \dfrac{x}{3} > \ln 3 - 1 \Leftrightarrow$

$x > 3(\ln 3 - 1) \approx 0.3$, so f isincreasing on $(3\ln 3 - 3, \infty)$ and decreasing on $(-\infty, 3\ln 3 - 3)$. **F.** Local and absolute minimum value $f(3\ln 3 - 3) = 1 - (3\ln 3 - 3) + e^{1+\ln 3 - 1} = 4 - 3\ln 3 + 3 = 7 - 3\ln 3 \approx 3.7$, no local maximum value **G.** $f''(x) = \frac{1}{9}e^{1+x/3} > 0$ for all x, so f is CU on $(-\infty, \infty)$. No IP

H.

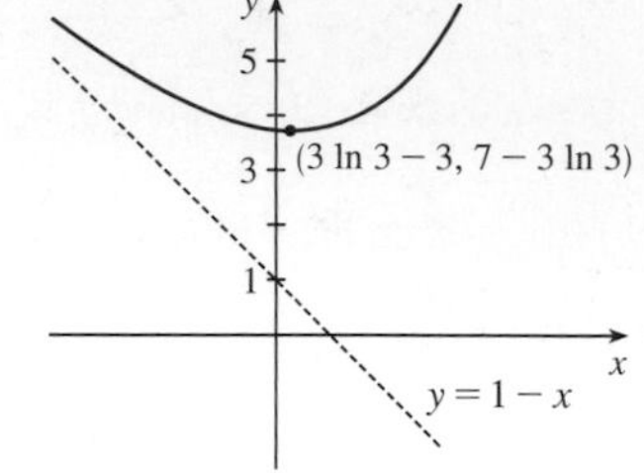

71. $y = f(x) = x - \tan^{-1} x$, $f'(x) = 1 - \dfrac{1}{1+x^2} = \dfrac{1 + x^2 - 1}{1+x^2} = \dfrac{x^2}{1+x^2}$,

$$f''(x) = \frac{(1+x^2)(2x) - x^2(2x)}{(1+x^2)^2} = \frac{2x(1+x^2-x^2)}{(1+x^2)^2} = \frac{2x}{(1+x^2)^2}.$$

$\lim\limits_{x \to \infty} \left[f(x) - \left(x - \frac{\pi}{2}\right)\right] = \lim\limits_{x \to \infty} \left(\frac{\pi}{2} - \tan^{-1} x\right) = \frac{\pi}{2} - \frac{\pi}{2} = 0$, so $y = x - \frac{\pi}{2}$ is a SA.

Also, $\lim\limits_{x \to -\infty} \left[f(x) - \left(x + \frac{\pi}{2}\right)\right] = \lim\limits_{x \to -\infty} \left(-\frac{\pi}{2} - \tan^{-1} x\right) = -\frac{\pi}{2} - \left(-\frac{\pi}{2}\right) = 0$,

so $y = x + \frac{\pi}{2}$ is also a SA. $f'(x) \geq 0$ for all x, with equality $\Leftrightarrow$ $x = 0$, so f is increasing on $\mathbb{R}$. $f''(x)$ has the same sign as x, so f is CD on $(-\infty, 0)$ and CU on $(0, \infty)$. $f(-x) = -f(x)$, so f is an odd function; its graph is symmetric about the origin. f has no local extreme values. Its only IP is at $(0, 0)$.

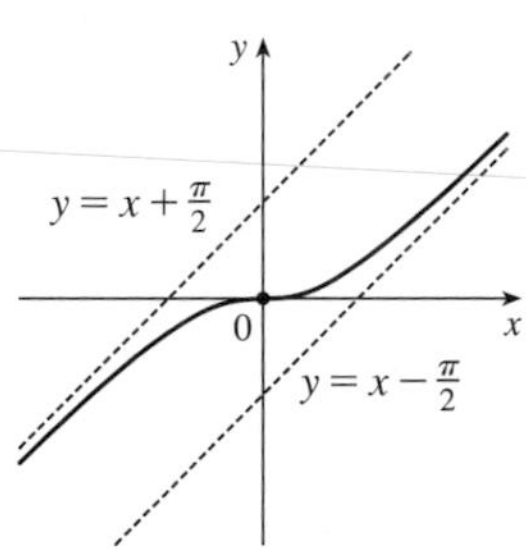

72. $y = f(x) = \sqrt{x^2 + 4x} = \sqrt{x(x+4)}$. $x(x+4) \geq 0 \Leftrightarrow x \leq -4$ or $x \geq 0$, so $D = (-\infty, -4] \cup [0, \infty)$.

y-intercept: $f(0) = 0$; x-intercepts: $f(x) = 0 \Rightarrow x = -4, 0$.

$$\sqrt{x^2+4x} \mp (x+2) = \frac{\sqrt{x^2+4x} \mp (x+2)}{1} \cdot \frac{\sqrt{x^2+4x} \pm (x+2)}{\sqrt{x^2+4x} \pm (x+2)} = \frac{(x^2+4x) - (x^2+4x+4)}{\sqrt{x^2+4x} \pm (x+2)}$$
$$= \frac{-4}{\sqrt{x^2+4x} \pm (x+2)}$$

so $\lim\limits_{x \to \pm\infty} [f(x) \mp (x+2)] = 0$. Thus, the graph of f approaches the slant asymptote $y = x + 2$ as $x \to \infty$ and it approaches the slant asymptote $y = -(x+2)$ as $x \to -\infty$. $f'(x) = \dfrac{x+2}{\sqrt{x^2+4x}}$, so $f'(x) < 0$ for $x < -4$ and $f'(x) > 0$ for $x > 0$; that is, f is decreasing on $(-\infty, -4)$ and increasing on $(0, \infty)$. There are no local extreme values. $f'(x) = (x+2)(x^2+4x)^{-1/2}$ $\Rightarrow$

$$\begin{aligned} f''(x) &= (x+2) \cdot \left(-\tfrac{1}{2}\right)(x^2+4x)^{-3/2} \cdot (2x+4) + (x^2+4x)^{-1/2} \\ &= (x^2+4x)^{-3/2}\left[-(x+2)^2 + (x^2+4x)\right] = -4(x^2+4x)^{-3/2} < 0 \text{ on } D \end{aligned}$$

so f is CD on $(-\infty, -4)$ and $(0, \infty)$. No IP

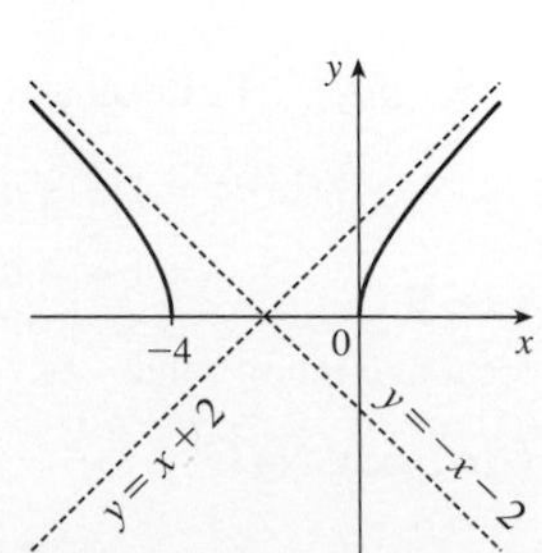

73. $\dfrac{x^2}{a^2} - \dfrac{y^2}{b^2} = 1 \;\Rightarrow\; y = \pm\dfrac{b}{a}\sqrt{x^2 - a^2}$. Now

$$\lim_{x\to\infty}\left[\frac{b}{a}\sqrt{x^2-a^2} - \frac{b}{a}x\right] = \frac{b}{a}\cdot\lim_{x\to\infty}\left(\sqrt{x^2-a^2} - x\right)\frac{\sqrt{x^2-a^2}+x}{\sqrt{x^2-a^2}+x} = \frac{b}{a}\cdot\lim_{x\to\infty}\frac{-a^2}{\sqrt{x^2-a^2}+x} = 0,$$

which shows that $y = \dfrac{b}{a}x$ is a slant asymptote. Similarly,

$$\lim_{x\to\infty}\left[-\frac{b}{a}\sqrt{x^2-a^2} - \left(-\frac{b}{a}x\right)\right] = -\frac{b}{a}\cdot\lim_{x\to\infty}\frac{-a^2}{\sqrt{x^2-a^2}+x} = 0, \text{ so } y = -\frac{b}{a}x \text{ is a slant asymptote.}$$

74. $f(x) - x^2 = \dfrac{x^3+1}{x} - x^2 = \dfrac{x^3+1-x^3}{x} = \dfrac{1}{x}$, and $\displaystyle\lim_{x\to\pm\infty}\frac{1}{x} = 0$. Therefore, $\displaystyle\lim_{x\to\pm\infty}\left[f(x) - x^2\right] = 0$,

and so the graph of f is asymptotic to that of $y = x^2$. For purposes of differentiation, we will use $f(x) = x^2 + 1/x$.

A. $D = \{x \mid x \neq 0\}$ **B.** No y-intercept; to find the x-intercept, we set $y = 0 \;\Leftrightarrow\; x = -1$.

C. No symmetry **D.** $\displaystyle\lim_{x\to 0^+}\frac{x^3+1}{x} = \infty$ and $\displaystyle\lim_{x\to 0^-}\frac{x^3+1}{x} = -\infty$, so $x = 0$ is a vertical asymptote. Also, the graph is asymptotic to the parabola $y = x^2$, as shown above. **E.** $f'(x) = 2x - 1/x^2 > 0 \;\Leftrightarrow\; x > \frac{1}{\sqrt[3]{2}}$, so f is increasing on $\left(\frac{1}{\sqrt[3]{2}}, \infty\right)$ and decreasing on $(-\infty, 0)$ and $\left(0, \frac{1}{\sqrt[3]{2}}\right)$.

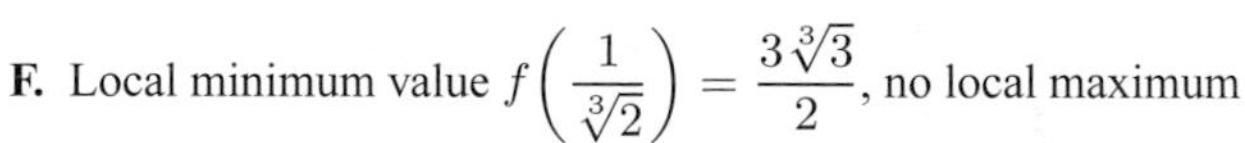

F. Local minimum value $f\left(\dfrac{1}{\sqrt[3]{2}}\right) = \dfrac{3\sqrt[3]{3}}{2}$, no local maximum

G. $f''(x) = 2 + 2/x^3 > 0 \;\Leftrightarrow\; x < -1$ or $x > 0$, so f is CU on $(-\infty, -1)$ and $(0, \infty)$, and CD on $(-1, 0)$. IP at $(-1, 0)$

H.

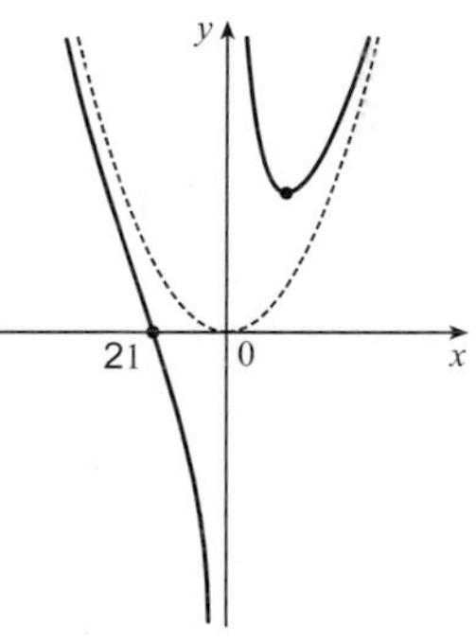

75. $\displaystyle\lim_{x\to\pm\infty}\left[f(x) - x^3\right] = \lim_{x\to\pm\infty}\frac{x^4+1}{x} - \frac{x^4}{x} = \lim_{x\to\pm\infty}\frac{1}{x} = 0$, so the graph of f is asymptotic to that of $y = x^3$.

A. $D = \{x \mid x \neq 0\}$ **B.** No intercept **C.** f is symmetric about the origin. **D.** $\displaystyle\lim_{x\to 0^-}\left(x^3 + \frac{1}{x}\right) = -\infty$ and

$\displaystyle\lim_{x\to 0^+}\left(x^3 + \frac{1}{x}\right) = \infty$, so $x = 0$ is a vertical asymptote, and as shown above, the graph of f is asymptotic to that of $y = x^3$.

E. $f'(x) = 3x^2 - 1/x^2 > 0 \;\Leftrightarrow\; x^4 > \frac{1}{3} \;\Leftrightarrow\; |x| > \frac{1}{\sqrt[4]{3}}$, so f is increasing on $\left(-\infty, -\dfrac{1}{\sqrt[4]{3}}\right)$ and $\left(\dfrac{1}{\sqrt[4]{3}}, \infty\right)$ and decreasing on $\left(-\dfrac{1}{\sqrt[4]{3}}, 0\right)$ and $\left(0, \dfrac{1}{\sqrt[4]{3}}\right)$. **F.** Local maximum value $f\left(-\dfrac{1}{\sqrt[4]{3}}\right) = -4\cdot 3^{-5/4}$, local minimum value $f\left(\dfrac{1}{\sqrt[4]{3}}\right) = 4\cdot 3^{-5/4}$

G. $f''(x) = 6x + 2/x^3 > 0 \;\Leftrightarrow\; x > 0$, so f is CU on $(0, \infty)$ and CD on $(-\infty, 0)$. No IP

H.

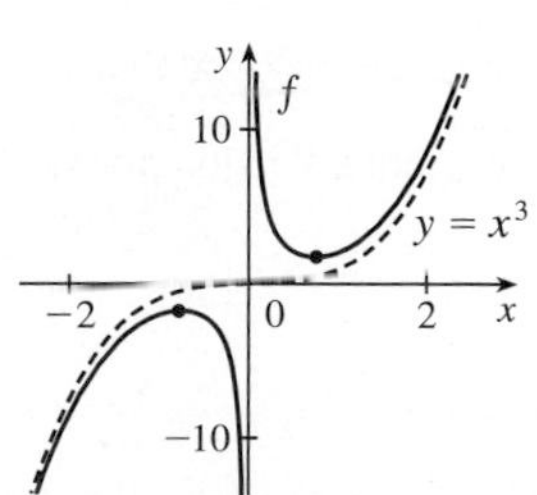

76. $f(x) = \sin x + e^{-x}$. $\lim\limits_{x\to\infty} [f(x) - \sin x] = \lim\limits_{x\to\infty} e^{-x} = 0$, so the graph of f is asymptotic to the graph of $\sin x$ as $x \to \infty$. $\lim\limits_{x\to-\infty} e^{-x} = \infty$, whereas $|\sin x| \le 1$, so for large negative x, the graph of f looks like the graph of e^{-x}.

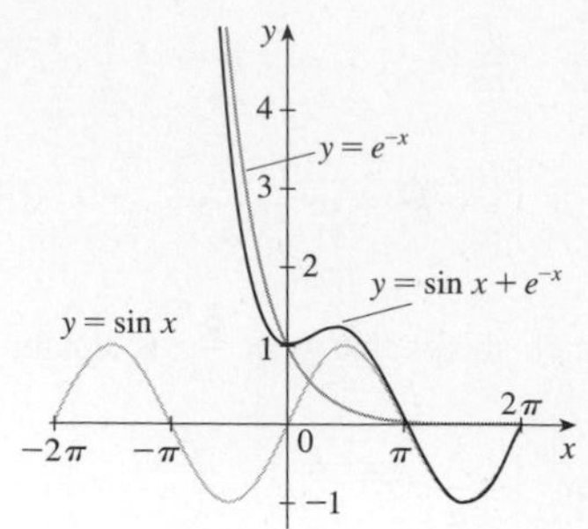

4.6 Graphing with Calculus and Calculators

1. $f(x) = x^5 - 5x^4 - x^3 + 28x^2 - 2x \;\Rightarrow\; f'(x) = 5x^4 - 20x^3 - 3x^2 + 56x - 2 \;\Rightarrow\; f''(x) = 20x^3 - 60x^2 - 6x + 56$. $f(x) = 0 \;\Leftrightarrow\; x = 0$ or $x \approx -2.09, 0.07$; $f'(x) = 0 \;\Leftrightarrow\; x \approx -1.50, 0.04, 2.62, 2.84$; $f''(x) = 0 \;\Leftrightarrow\; x \approx -0.89, 1.15, 2.74$.

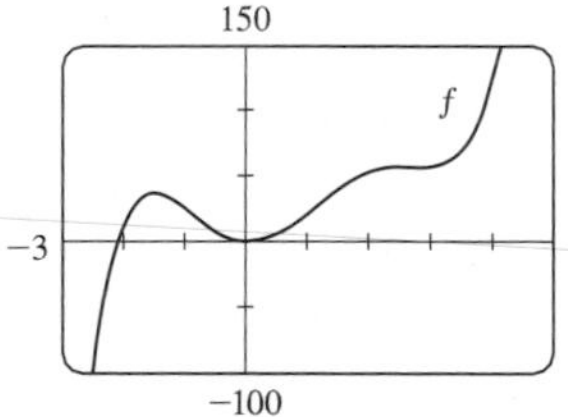

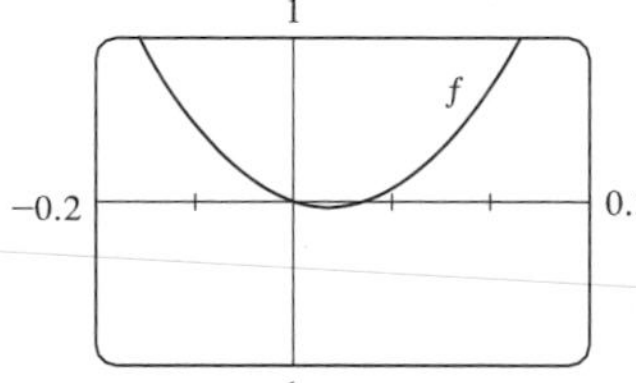

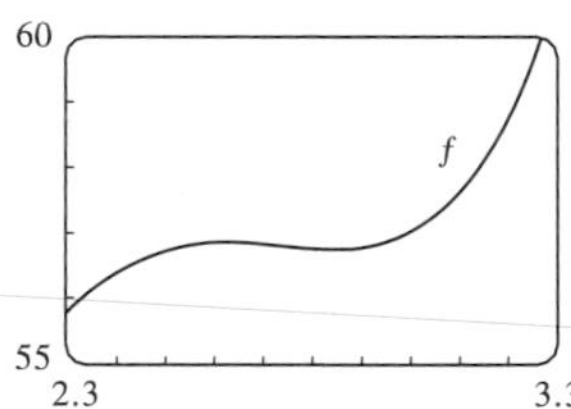

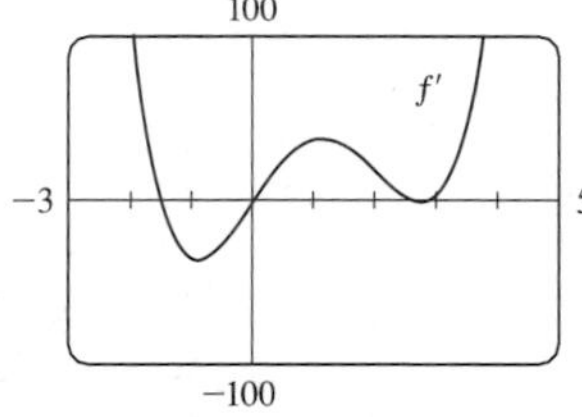

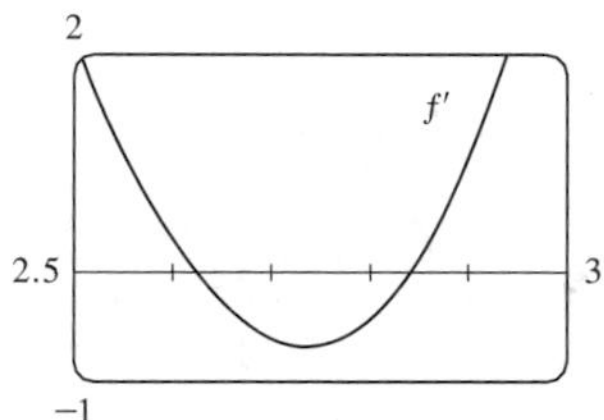

From the graphs of f', we estimate that $f' < 0$ and that f is decreasing on $(-1.50, 0.04)$ and $(2.62, 2.84)$, and that $f' > 0$ and f is increasing on $(-\infty, -1.50)$, $(0.04, 2.62)$, and $(2.84, \infty)$ with local minimum values $f(0.04) \approx -0.04$ and $f(2.84) \approx 56.73$ and local maximum values $f(-1.50) \approx 36.47$ and $f(2.62) \approx 56.83$.

From the graph of f'', we estimate that $f'' > 0$ and that f is CU on $(-0.89, 1.15)$ and $(2.74, \infty)$, and that $f'' < 0$ and f is CD on $(-\infty, -0.89)$ and $(1.15, 2.74)$. There are inflection points at about $(-0.89, 20.90)$, $(1.15, 26.57)$, and $(2.74, 56.78)$.

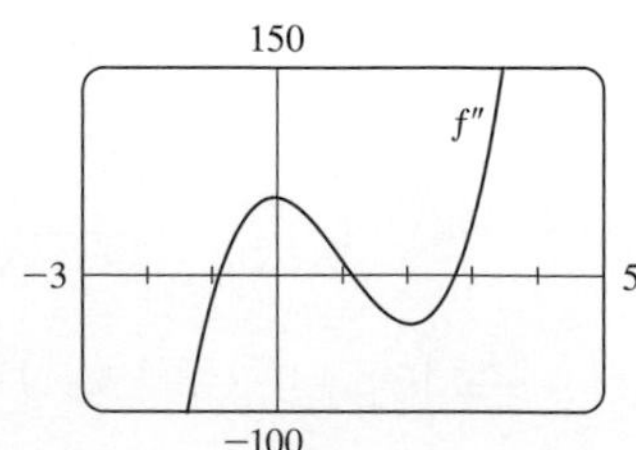

2. $f(x) = -2x^6 + 5x^5 + 140x^3 - 110x^2 \;\Rightarrow\; f'(x) = -12x^5 + 25x^4 + 420x^2 - 220x \;\Rightarrow$ $f''(x) = -60x^4 + 100x^3 + 840x - 220$. $f(x) = 0 \;\Leftrightarrow\; x = 0$ or $x \approx 0.77, 4.93$; $f'(x) = 0 \;\Leftrightarrow\; x = 0$ or

$x \approx 0.52, 3.99$; $f''(x) = 0 \;\Leftrightarrow\; x \approx 0.26, 3.05$.

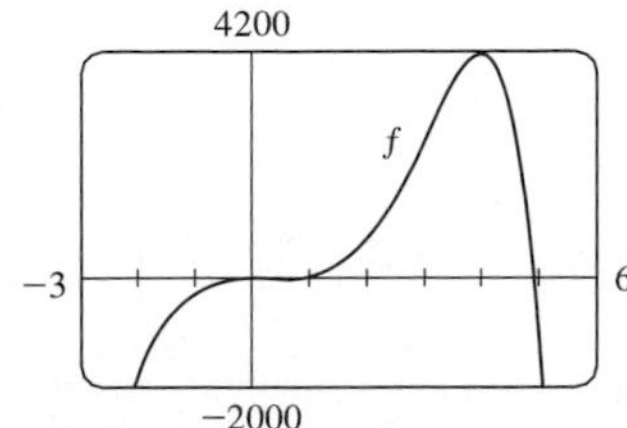

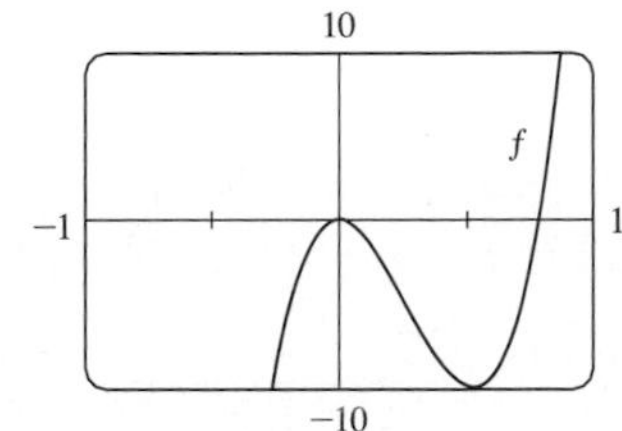

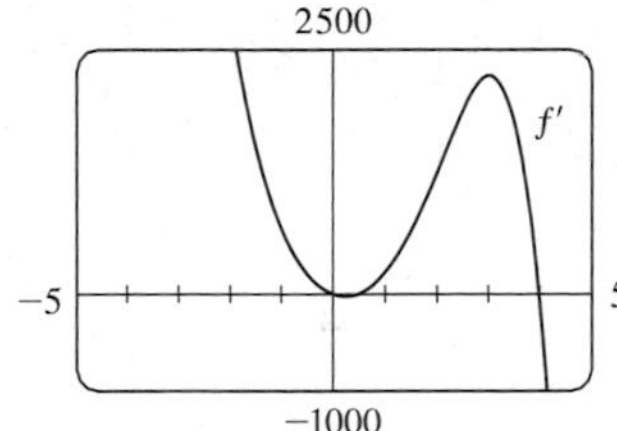

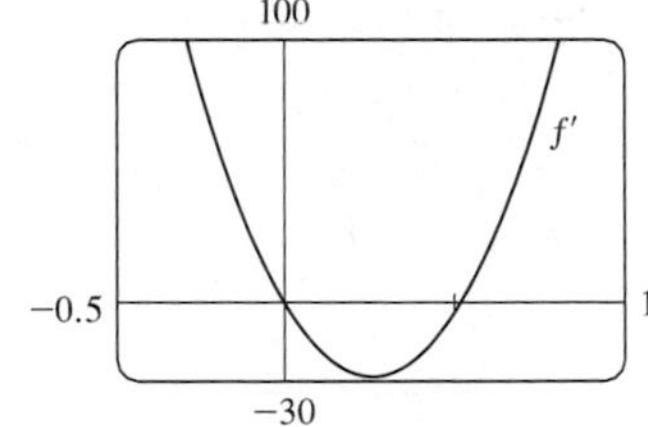

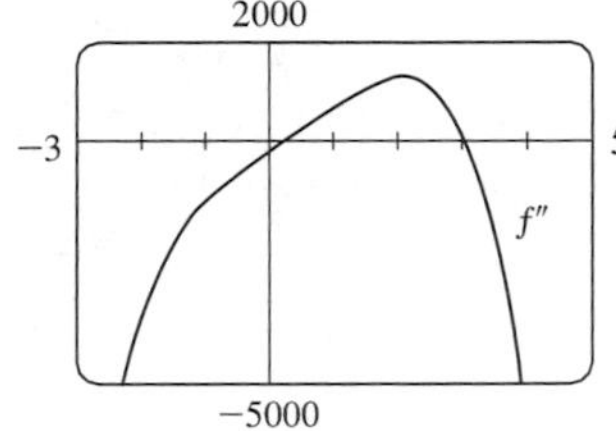

From the graphs of f', we estimate that $f' > 0$ and that f is increasing on $(-\infty, 0)$ and $(0.52, 3.99)$, and that $f' < 0$ and that f is decreasing on $(0, 0.52)$ and $(3.99, \infty)$. f has local maximum values $f(0) = 0$ and $f(3.99) \approx 4128.20$, and f has local minimum value $f(0.52) \approx -9.91$. From the graph of f'', we estimate that $f'' > 0$ and f is CU on $(0.26, 3.05)$, and that $f'' < 0$ and f is CD on $(-\infty, 0.26)$ and $(3.05, \infty)$. There are inflection points at about $(0.26, -4.97)$ and $(3.05, 2649.46)$.

3. $f(x) = x^6 - 5x^5 + 25x^3 - 6x^2 - 48x \;\Rightarrow$

$f'(x) = 6x^5 - 25x^4 + 75x^2 - 12x - 48 \;\Rightarrow$

$f''(x) = 30x^4 - 100x^3 + 150x - 12$. $f(x) = 0 \;\Leftrightarrow\; x = 0$ or $x \approx 3.20$;

$f'(x) = 0 \;\Leftrightarrow\; x \approx -1.31, -0.84, 1.06, 2.50, 2.75$; $f''(x) = 0 \;\Leftrightarrow$

$x \approx -1.10, 0.08, 1.72, 2.64$.

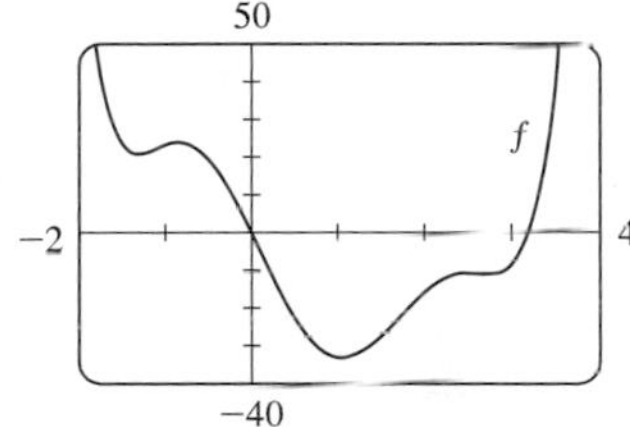

From the graph of f', we estimate that f is decreasing on $(-\infty, -1.31)$, increasing on $(-1.31, -0.84)$, decreasing on $(-0.84, 1.06)$, increasing on $(1.06, 2.50)$, decreasing on $(2.50, 2.75)$, and increasing on $(2.75, \infty)$. f has local minimum values $f(-1.31) \approx 20.72$, $f(1.06) \approx -33.12$, and $f(2.75) \approx -11.33$. f has local maximum values $f(-0.84) \approx 23.71$ and $f(2.50) \approx -11.02$.

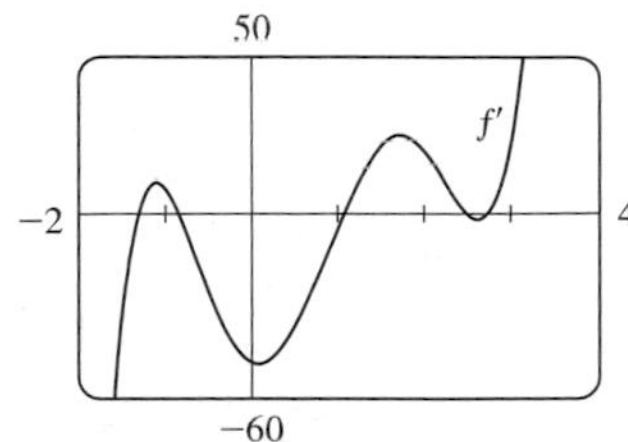

From the graph of f'', we estimate that f is CU on $(-\infty, -1.10)$, CD on $(-1.10, 0.08)$, CU on $(0.08, 1.72)$, CD on $(1.72, 2.64)$, and CU on $(2.64, \infty)$. There are inflection points at about $(-1.10, 22.09)$, $(0.08, -3.88)$, $(1.72, -22.53)$, and $(2.64, -11.18)$.

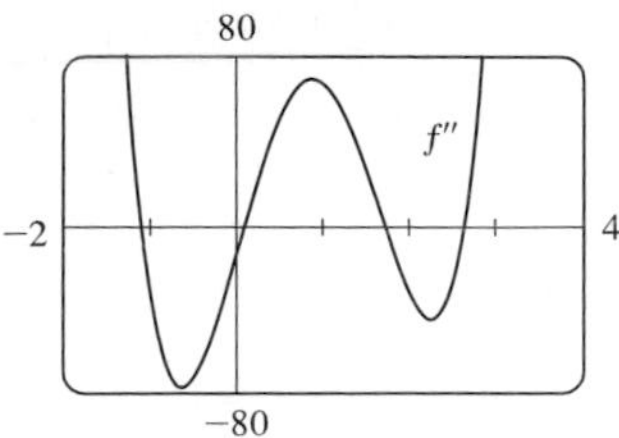

4. $f(x) = \dfrac{x^4 - x^3 - 8}{x^2 - x - 6} \quad \Rightarrow \quad f'(x) = \dfrac{2(x^5 - 2x^4 - 11x^3 + 9x^2 + 8x - 4)}{(x^2 - x - 6)^2} \quad \Rightarrow$

$f''(x) = \dfrac{2(x^6 - 3x^5 - 15x^4 + 41x^3 + 174x^2 - 84x - 56}{(x^2 - x - 6)^3}$. $f(x) = 0 \;\Leftrightarrow\; x \approx -1.48$ or $x = 2$; $f'(x) = 0 \;\Leftrightarrow$

$x \approx -2.74, -0.81, 0.41, 1.08, 4.06$; $f''(x) = 0 \;\Leftrightarrow\; x \approx -0.39, 0.79$. The VAs are $x = -2$ and $x = 3$.

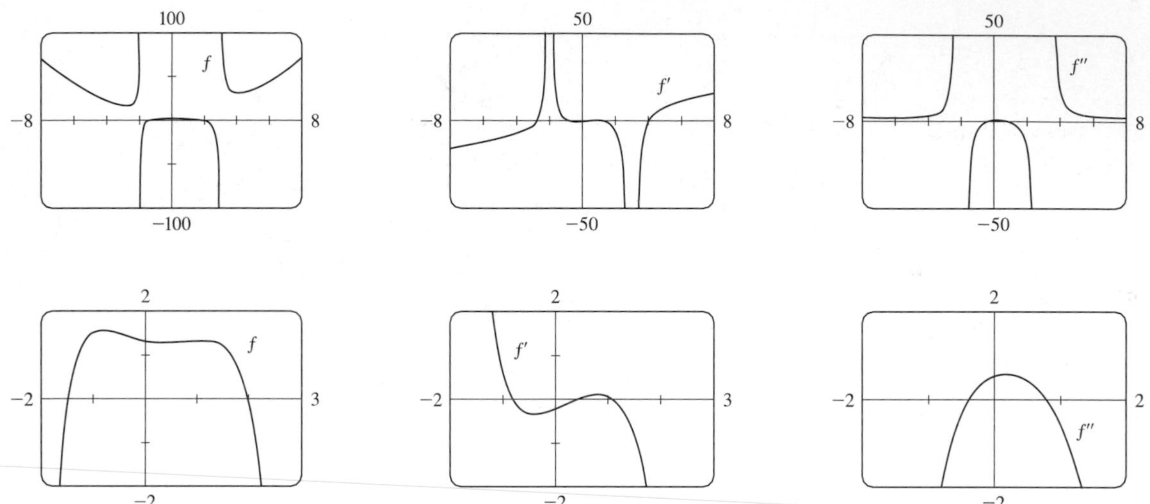

From the graphs of f', we estimate that f is decreasing on $(-\infty, -2.74)$, increasing on $(-2.74, -2)$, increasing on $(-2, -0.81)$, decreasing on $(-0.81, 0.41)$, increasing on $(0.41, 1.08)$, decreasing on $(1.08, 3)$, decreasing on $(3, 4.06)$, and increasing on $(4.06, \infty)$. f has local minimum values $f(-2.74) \approx 16.23$, $f(0.41) \approx 1.29$, and $f(4.06) \approx 30.63$. f has local maximum values $f(-0.81) \approx 1.55$ and $f(1.08) \approx 1.34$.

From the graphs of f'', we estimate that f is CU on $(-\infty, -2)$, CD on $(-2, -0.39)$, CU on $(-0.39, 0.79)$, CD on $(0.79, 3)$, and CU on $(3, \infty)$. There are inflection points at about $(-0.39, 1.45)$ and $(0.79, 1.31)$.

5. $f(x) = \dfrac{x}{x^3 + x^2 + 1} \quad \Rightarrow \quad f'(x) = -\dfrac{2x^3 + x^2 - 1}{(x^3 + x^2 + 1)^2} \quad \Rightarrow \quad f''(x) = \dfrac{2x(3x^4 + 3x^3 + x^2 - 6x - 3)}{(x^3 + x^2 + 1)^3}$

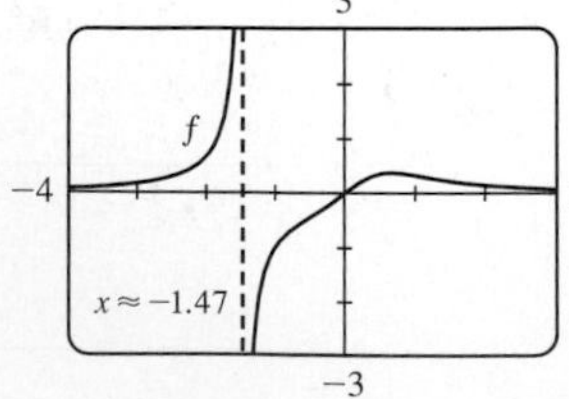

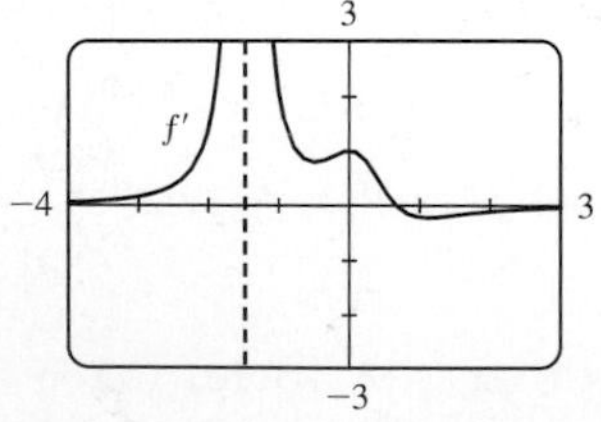

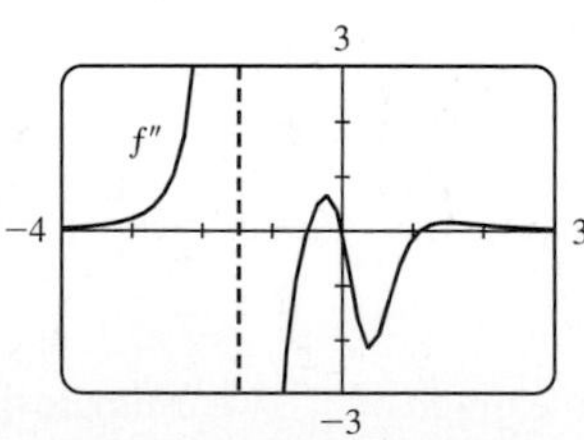

From the graph of f, we see that there is a VA at $x \approx -1.47$. From the graph of f', we estimate that f is increasing on $(-\infty, -1.47)$, increasing on $(-1.47, 0.66)$, and decreasing on $(0.66, \infty)$, with local maximum value $f(0.66) \approx 0.38$.

From the graph of f'', we estimate that f is CU on $(-\infty, -1.47)$, CD on $(-1.47, -0.49)$, CU on $(-0.49, 0)$, CD on $(0, 1.10)$, and CU on $(1.10, \infty)$. There is an inflection point at $(0, 0)$ and at about $(-0.49, -0.44)$ and $(1.10, 0.31)$.

6. $f(x) = 6\sin x - x^2$, $-5 \le x \le 3 \quad \Rightarrow \quad f'(x) = 6\cos x - 2x \quad \Rightarrow \quad f''(x) = -6\sin x - 2$

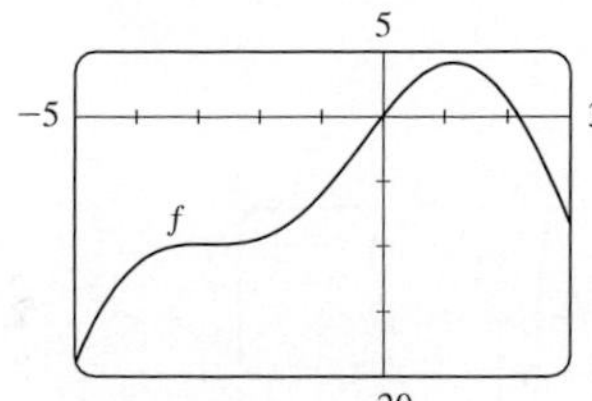

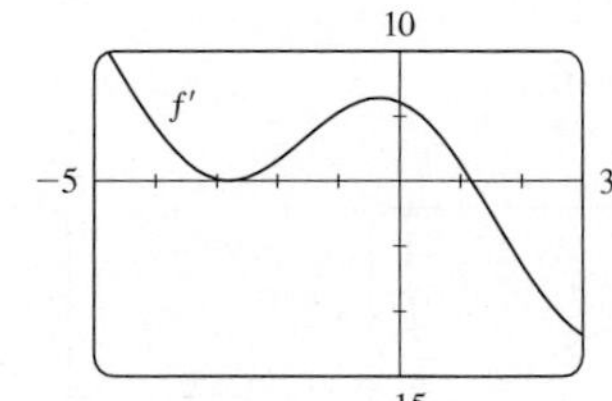

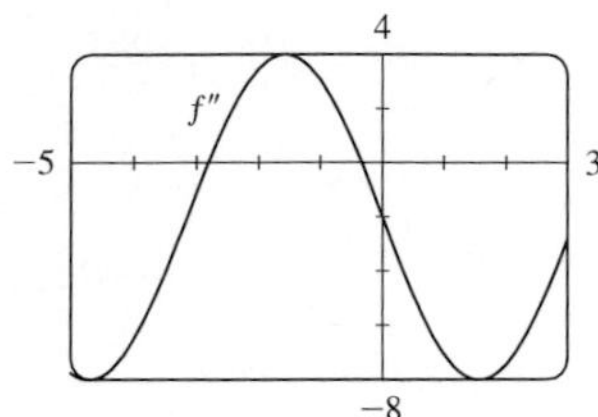

From the graph of f', which has two negative zeros, we estimate that f is increasing on $(-5, -2.94)$, decreasing on $(-2.94, -2.66)$, increasing on $(-2.66, 1.17)$, and decreasing on $(1.17, 3)$, with local maximum values $f(-2.94) \approx -9.84$ and $f(1.17) \approx 4.16$, and local minimum value $f(-2.66) \approx -9.85$.

From the graph of f'', we estimate that f is CD on $(-5, -2.80)$, CU on $(-2.80, -0.34)$, and CD on $(-0.34, 3)$. There are inflection points at about $(-2.80, -9.85)$ and $(-0.34, -2.12)$.

7. $f(x) = 6\sin x + \cot x$, $-\pi \le x \le \pi \quad \Rightarrow \quad f'(x) = 6\cos x - \csc^2 x \quad \Rightarrow \quad f''(x) = -6\sin x + 2\csc^2 x\,\cot x$

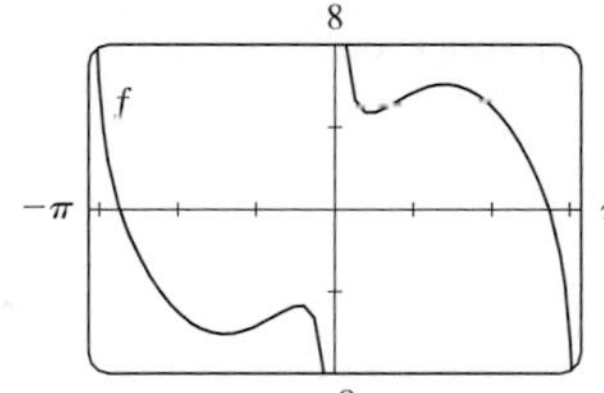

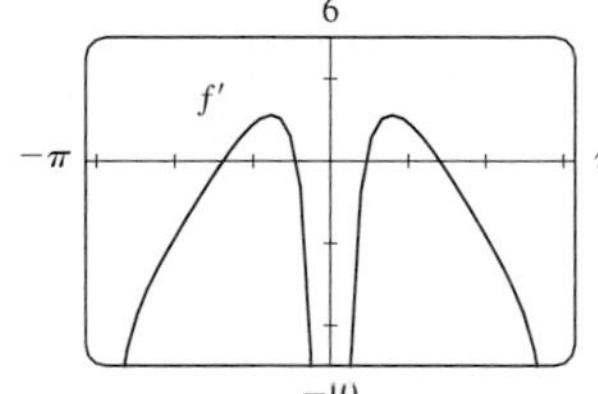

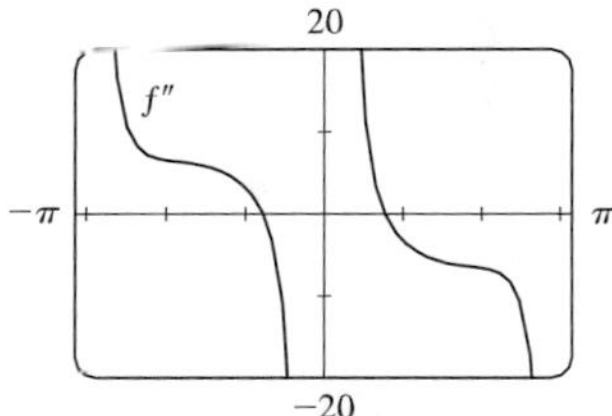

From the graph of f, we see that there are VAs at $x = 0$ and $x = \pm\pi$. f is an odd function, so its graph is symmetric about the origin. From the graph of f', we estimate that f is decreasing on $(-\pi, -1.40)$, increasing on $(-1.40, -0.44)$, decreasing on $(-0.44, 0)$, decreasing on $(0, 0.44)$, increasing on $(0.44, 1.40)$, and decreasing on $(1.40, \pi)$, with local minimum values $f(-1.40) \approx -6.09$ and $f(0.44) \approx 4.68$, and local maximum values $f(-0.44) \approx -4.68$ and $f(1.40) \approx 6.09$.

From the graph of f'', we estimate that f is CU on $(-\pi, -0.77)$, CD on $(-0.77, 0)$, CU on $(0, 0.77)$, and CD on $(0.77, \pi)$. There are IPs at about $(-0.77, -5.22)$ and $(0.77, 5.22)$.

8. $f(x) = e^x - 0.186x^4 \quad \Rightarrow \quad f'(x) = e^x - 0.744x^3 \quad \Rightarrow \quad f''(x) = e^x - 2.232x^2$

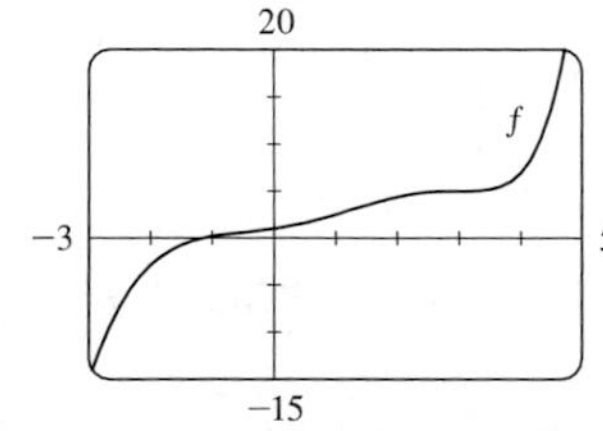

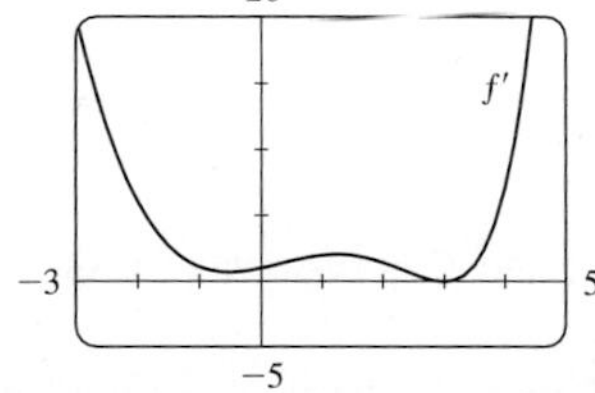

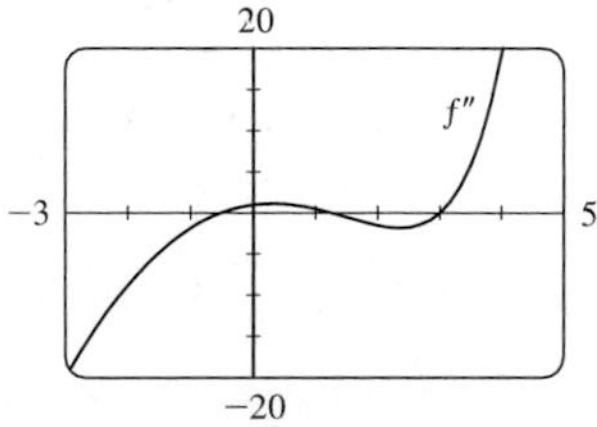

From the graph of f', which has two positive zeros, we estimate that f is increasing on $(-\infty, 2.973)$, decreasing on $(2.973, 3.027)$, and increasing on $(3.027, \infty)$, with local maximum value $f(2.973) \approx 5.01958$ and local minimum value $f(3.027) \approx 5.01949$.

From the graph of f'', we estimate that f is CD on $(-\infty, -0.52)$, CU on $(-0.52, 1.25)$, CD on $(1.25, 3.00)$, and CU on $(3.00, \infty)$. There are inflection points at about $(-0.52, 0.58)$, $(1.25, 3.04)$ and $(3.00, 5.01954)$.

9. $f(x) = 1 + \dfrac{1}{x} + \dfrac{8}{x^2} + \dfrac{1}{x^3} \quad \Rightarrow \quad f'(x) = -\dfrac{1}{x^2} - \dfrac{16}{x^3} - \dfrac{3}{x^4} = -\dfrac{1}{x^4}(x^2 + 16x + 3) \quad \Rightarrow$

$f''(x) = \dfrac{2}{x^3} + \dfrac{48}{x^4} + \dfrac{12}{x^5} = \dfrac{2}{x^5}(x^2 + 24x + 6).$

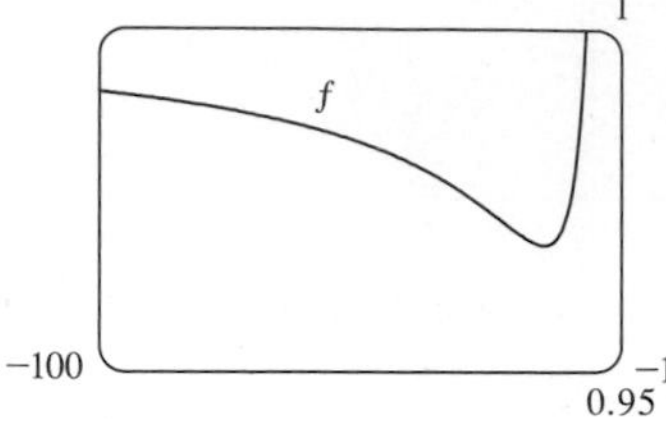

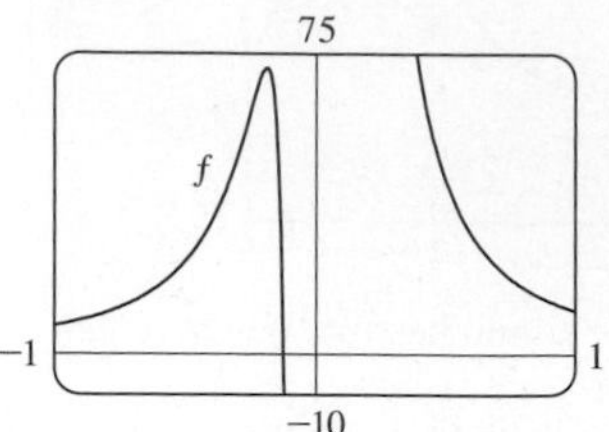

From the graphs, it appears that f increases on $(-15.8, -0.2)$ and decreases on $(-\infty, -15.8)$, $(-0.2, 0)$, and $(0, \infty)$; that f has a local minimum value of $f(-15.8) \approx 0.97$ and a local maximum value of $f(-0.2) \approx 72$; that f is CD on $(-\infty, -24)$ and $(-0.25, 0)$ and is CU on $(-24, -0.25)$ and $(0, \infty)$; and that f has IPs at $(-24, 0.97)$ and $(-0.25, 60)$.

To find the exact values, note that $f' = 0 \quad \Rightarrow \quad x = \dfrac{-16 \pm \sqrt{256 - 12}}{2} = -8 \pm \sqrt{61} \quad [\approx -0.19 \text{ and } -15.81]$.
f' is positive (f is increasing) on $\left(-8 - \sqrt{61}, -8 + \sqrt{61}\right)$ and f' is negative (f is decreasing) on $\left(-\infty, -8 - \sqrt{61}\right)$, $\left(-8 + \sqrt{61}, 0\right)$, and $(0, \infty)$. $\quad f'' = 0 \quad \Rightarrow \quad x = \dfrac{-24 \pm \sqrt{576 - 24}}{2} = -12 \pm \sqrt{138} \quad [\approx -0.25 \text{ and } -23.75]$. $\quad f''$ is positive (f is CU) on $\left(-12 - \sqrt{138}, -12 + \sqrt{138}\right)$ and $(0, \infty)$ and f'' is negative (f is CD) on $\left(-\infty, -12 - \sqrt{138}\right)$ and $\left(-12 + \sqrt{138}, 0\right)$.

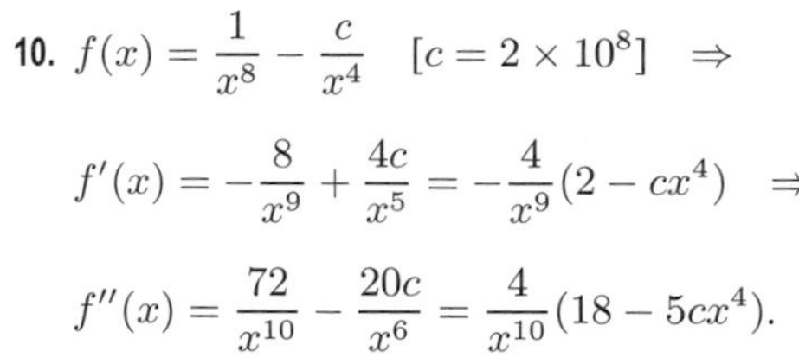

10. $f(x) = \dfrac{1}{x^8} - \dfrac{c}{x^4} \quad [c = 2 \times 10^8] \quad \Rightarrow$

$f'(x) = -\dfrac{8}{x^9} + \dfrac{4c}{x^5} = -\dfrac{4}{x^9}(2 - cx^4) \quad \Rightarrow$

$f''(x) = \dfrac{72}{x^{10}} - \dfrac{20c}{x^6} = \dfrac{4}{x^{10}}(18 - 5cx^4).$

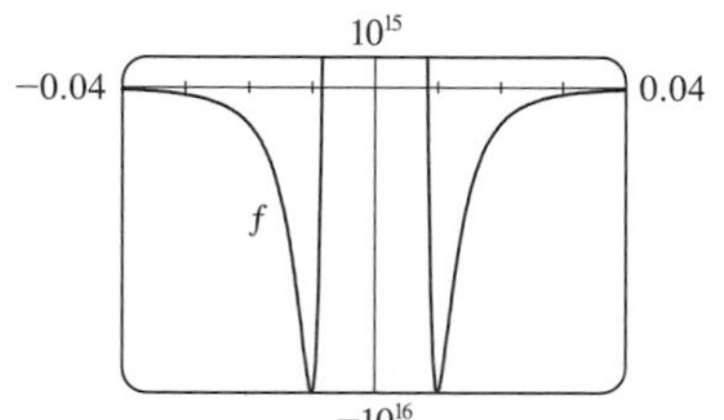

From the graph, it appears that f increases on $(-0.01, 0)$ and $(0.01, \infty)$ and decreases on $(-\infty, -0.01)$ and $(0, 0.01)$; that f has a local minimum value of $f(\pm 0.01) = -10^{16}$; and that f is CU on $(-0.012, 0)$ and $(0, 0.012)$ and f is CD on $(-\infty, -0.012)$ and $(0.012, \infty)$.

To find the exact values, note that $f' = 0 \quad \Rightarrow \quad x^4 = \frac{2}{c} \quad \Rightarrow \quad x \pm \sqrt[4]{\frac{2}{c}} = \pm\frac{1}{100} \quad [c = 2 \times 10^8]$. $\quad f'$ is positive (f is increasing) on $(-0.01, 0)$ and $(0.01, \infty)$ and f' is negative (f is decreasing) on $(-\infty, -0.01)$ and $(0, 0.01)$.
$f'' = 0 \quad \Rightarrow \quad x^4 = \dfrac{18}{5c} \quad \Rightarrow \quad x = \pm\sqrt[4]{\dfrac{18}{5c}} = \pm\dfrac{1}{100}\sqrt[4]{1.8} \quad [\approx \pm 0.0116]$. $\quad f''$ is positive (f is CU) on $\left(-\frac{1}{100}\sqrt[4]{1.8}, 0\right)$ and $\left(0, \frac{1}{100}\sqrt[4]{1.8}\right)$ and f'' is negative (f is CD) on $\left(-\infty, -\frac{1}{100}\sqrt[4]{1.8}\right)$ and $\left(\frac{1}{100}\sqrt[4]{1.8}, \infty\right)$.

11. (a) $f(x) = x^2 \ln x$. The domain of f is $(0, \infty)$.

(b) $\displaystyle\lim_{x \to 0^+} x^2 \ln x = \lim_{x \to 0^+} \frac{\ln x}{1/x^2} \overset{\text{H}}{=} \lim_{x \to 0^+} \frac{1/x}{-2/x^3} = \lim_{x \to 0^+} \left(-\frac{x^2}{2}\right) = 0.$

There is a hole at $(0, 0)$.

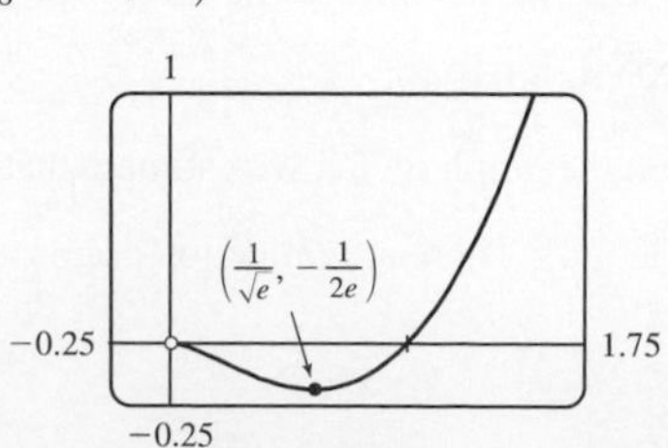

(c) It appears that there is an IP at about $(0.2, -0.06)$ and a local minimum at $(0.6, -0.18)$. $f(x) = x^2 \ln x \Rightarrow$

$f'(x) = x^2(1/x) + (\ln x)(2x) = x(2\ln x + 1) > 0 \Leftrightarrow \ln x > -\frac{1}{2} \Leftrightarrow x > e^{-1/2}$, so f is increasing on

$\left(1/\sqrt{e}, \infty\right)$, decreasing on $\left(0, 1/\sqrt{e}\right)$. By the FDT, $f\left(1/\sqrt{e}\right) = -1/(2e)$ is a local minimum value. This point is approximately $(0.6065, -0.1839)$, which agrees with our estimate.

$f''(x) = x(2/x) + (2\ln x + 1) = 2\ln x + 3 > 0 \Leftrightarrow \ln x > -\frac{3}{2} \Leftrightarrow x > e^{-3/2}$, so f is CU on $(e^{-3/2}, \infty)$ and CD on $(0, e^{-3/2})$. IP is $(e^{-3/2}, -3/(2e^3)) \approx (0.2231, -0.0747)$.

12. (a) $f(x) = xe^{1/x}$. The domain of f is $(-\infty, 0) \cup (0, \infty)$.

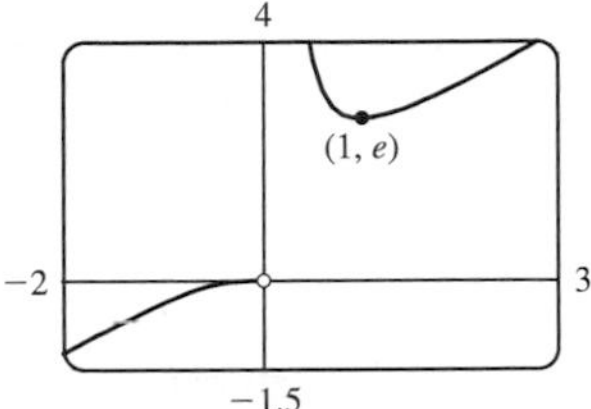

(b) $\lim\limits_{x\to 0^+} xe^{1/x} = \lim\limits_{x\to 0^+} \dfrac{e^{1/x}}{1/x} \overset{\text{H}}{=} \lim\limits_{x\to 0^+} \dfrac{e^{1/x}(-1/x^2)}{-1/x^2} = \lim\limits_{x\to 0^+} e^{1/x} = \infty$,

so $x = 0$ is a VA.

Also $\lim\limits_{x\to 0^-} xe^{1/x} = 0$ since $1/x \to -\infty \Rightarrow e^{1/x} \to 0$.

(c) It appears that there is a local minimum at $(1, 2.7)$. There are no IP and f is CD on $(-\infty, 0)$ and CU on $(0, \infty)$.

$f(x) = xe^{1/x} \Rightarrow f'(x) = xe^{1/x}\left(-\dfrac{1}{x^2}\right) + e^{1/x} = e^{1/x}\left(1 - \dfrac{1}{x}\right) > 0 \Leftrightarrow \dfrac{1}{x} < 1 \Leftrightarrow x < 0$ or $x > 1$,

so f is increasing on $(-\infty, 0)$ and $(1, \infty)$, and decreasing on $(0, 1)$. By the FDT, $f(1) = e$ is a local minimum value, which agrees with our estimate.

$f''(x) = e^{1/x}(1/x^2) + (1 - 1/x)e^{1/x}(-1/x^2) = (e^{1/x}/x^2)(1 - 1 + 1/x) - e^{1/x}/x^3 > 0 \Leftrightarrow x > 0$, so f is CU on $(0, \infty)$ and CD on $(-\infty, 0)$. No IP

13.

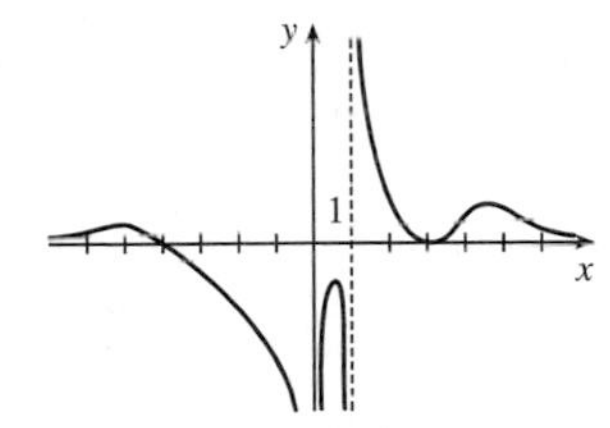

$f(x) = \dfrac{(x+4)(x-3)^2}{x^4(x-1)}$ has VA at $x = 0$ and at $x = 1$ since $\lim\limits_{x\to 0} f(x) = -\infty$,

$\lim\limits_{x\to 1^-} f(x) = -\infty$ and $\lim\limits_{x\to 1^+} f(x) = \infty$.

$$f(x) = \frac{\dfrac{x+4}{x} \cdot \dfrac{(x-3)^2}{x^2}}{\dfrac{x^4}{x^3} \cdot (x-1)} \quad \begin{bmatrix}\text{dividing numerator} \\ \text{and denominator by } x^3\end{bmatrix} = \frac{(1+4/x)(1-3/x)^2}{x(x-1)} \to 0$$

as $x \to \pm\infty$, so f is asymptotic to the x-axis.

Since f is undefined at $x = 0$, it has no y-intercept. $f(x) = 0 \Rightarrow (x+4)(x-3)^2 = 0 \Rightarrow x = -4$ or $x = 3$, so f has x-intercepts -4 and 3. Note, however, that the graph of f is only tangent to the x-axis and does not cross it at $x = 3$, since f is positive as $x \to 3^-$ and as $x \to 3^+$.

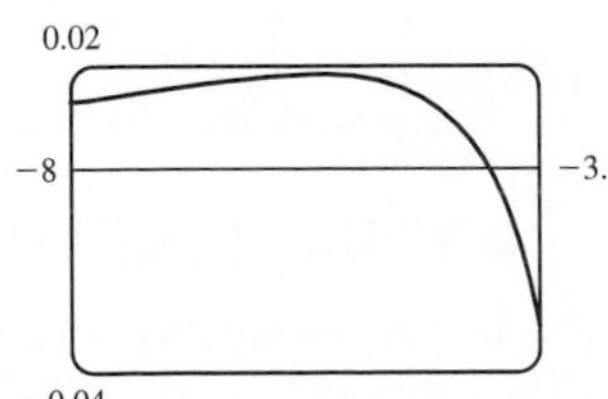

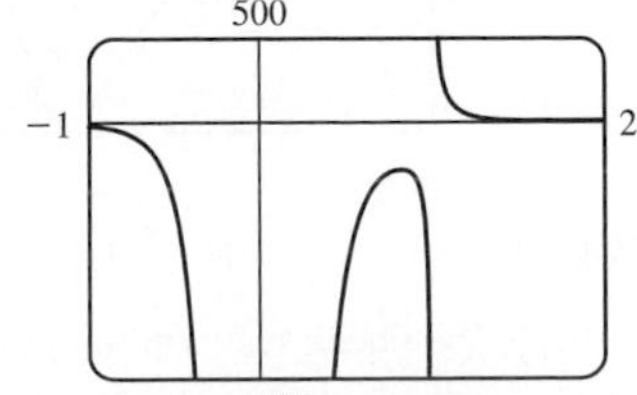

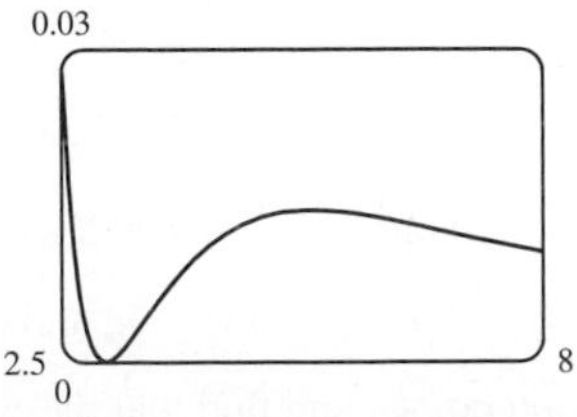

From these graphs, it appears that f has three maximum values and one minimum value. The maximum values are approximately $f(-5.6) = 0.0182$, $f(0.82) = -281.5$ and $f(5.2) = 0.0145$ and we know (since the graph is tangent to the x-axis at $x = 3$) that the minimum value is $f(3) = 0$.

14. 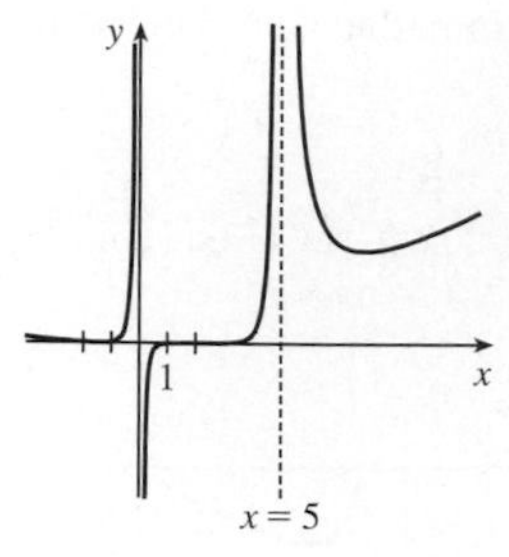

$f(x) = \dfrac{(2x+3)^2(x-2)^5}{x^3(x-5)^2}$ has VAs at $x = 0$ and $x = 5$ since $\lim\limits_{x\to 0^-} f(x) = \infty$, $\lim\limits_{x\to 0^+} f(x) = -\infty$, and $\lim\limits_{x\to 5} f(x) = \infty$. No HA since $\lim\limits_{x\to\pm\infty} f(x) = \infty$. Since f is undefined at $x = 0$, it has no y-intercept. $f(x) = 0 \Leftrightarrow (2x+3)^2(x-2)^5 = 0 \Leftrightarrow x = -\frac{3}{2}$ or $x = 2$, so f has x-intercepts at $-\frac{3}{2}$ and 2. Note, however, that the graph of f is only tangent to the x-axis and does not cross it at $x = -\frac{3}{2}$, since f is positive as $x \to \left(-\frac{3}{2}\right)^-$ and as $x \to \left(-\frac{3}{2}\right)^+$. There is a local minimum value of $f\left(-\frac{3}{2}\right) = 0$.

The only "mystery" feature is the local minimum to the right of the VA $x = 5$. From the graph, we see that $f(7.98) \approx 609$ is a local minimum value.

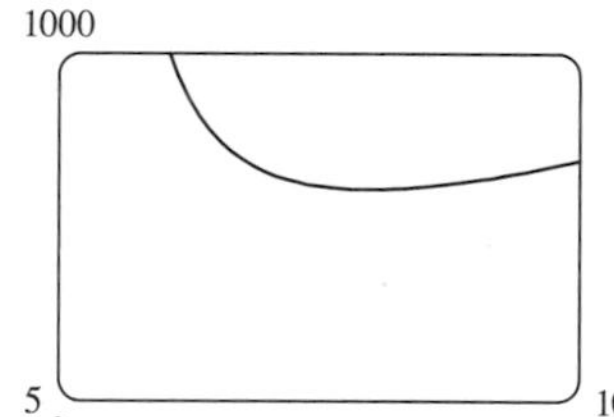

15. $f(x) = \dfrac{x^2(x+1)^3}{(x-2)^2(x-4)^4} \Rightarrow f'(x) = -\dfrac{x(x+1)^2(x^3+18x^2-44x-16)}{(x-2)^3(x-4)^5}$ [from CAS].

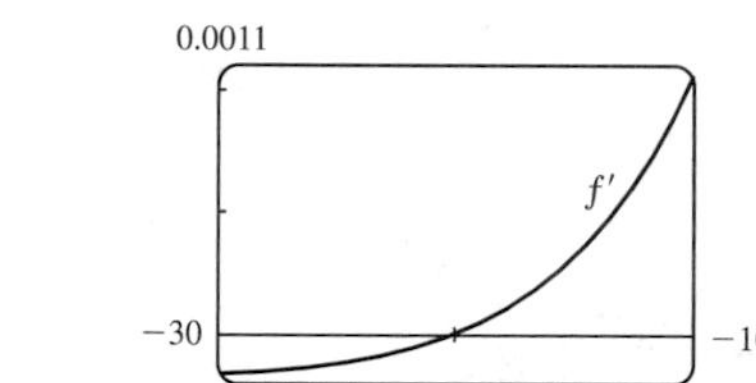

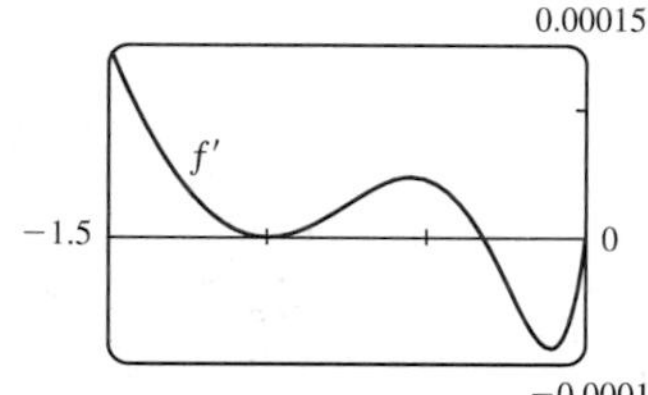

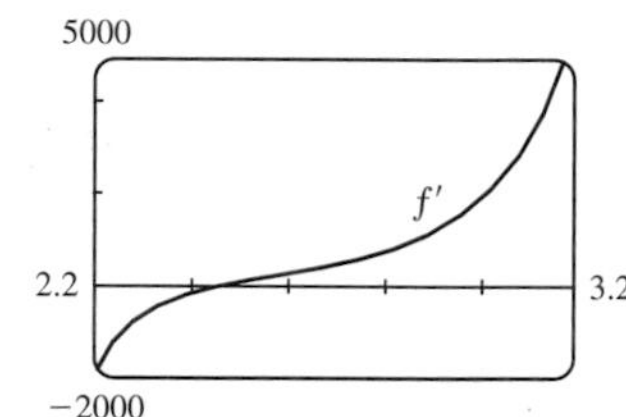

From the graphs of f', it seems that the critical points which indicate extrema occur at $x \approx -20$, -0.3, and 2.5, as estimated in Example 3. (There is another critical point at $x = -1$, but the sign of f' does not change there.) We differentiate again, obtaining $f''(x) = 2\dfrac{(x+1)(x^6+36x^5+6x^4-628x^3+684x^2+672x+64)}{(x-2)^4(x-4)^6}$.

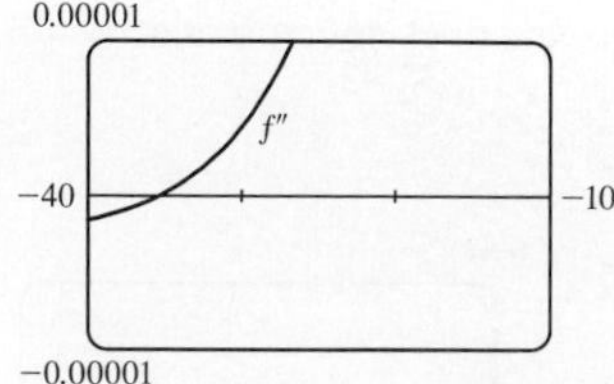

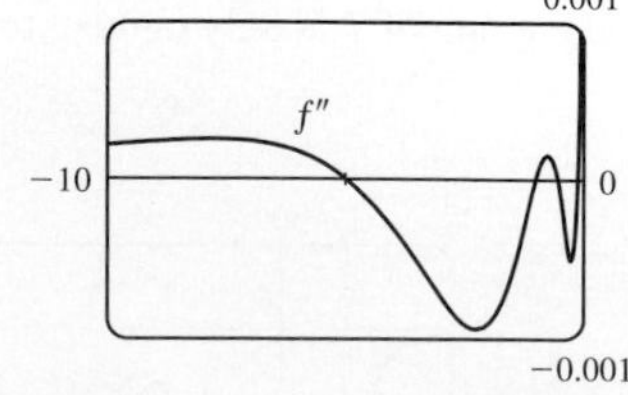

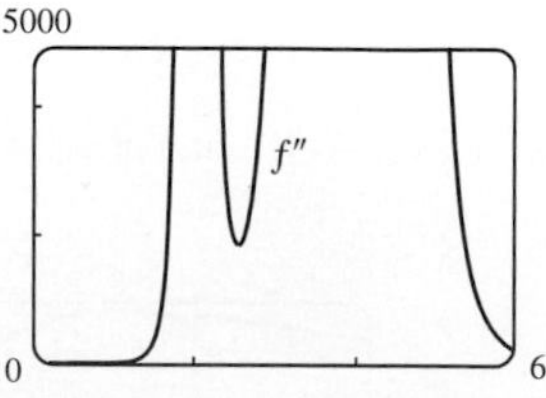

From the graphs of f'', it appears that f is CU on $(-35.3, -5.0)$, $(-1, -0.5)$, $(-0.1, 2)$, $(2, 4)$ and $(4, \infty)$ and CD on $(-\infty, -35.3)$, $(-5.0, -1)$ and $(-0.5, -0.1)$. We check back on the graphs of f to find the y-coordinates of the inflection points, and find that these points are approximately $(-35.3, -0.015)$, $(-5.0, -0.005)$, $(-1, 0)$, $(-0.5, 0.00001)$, and $(-0.1, 0.0000066)$.

16. From a CAS, $$f'(x) = \frac{2(x-2)^4(2x+3)(2x^3-14x^2-10x-45)}{x^4(x-5)^3}$$

and $$f''(x) = \frac{2(x-2)^3(4x^6-56x^5+216x^4+460x^3+805x^2+1710x+5400)}{x^5(x-5)^4}$$

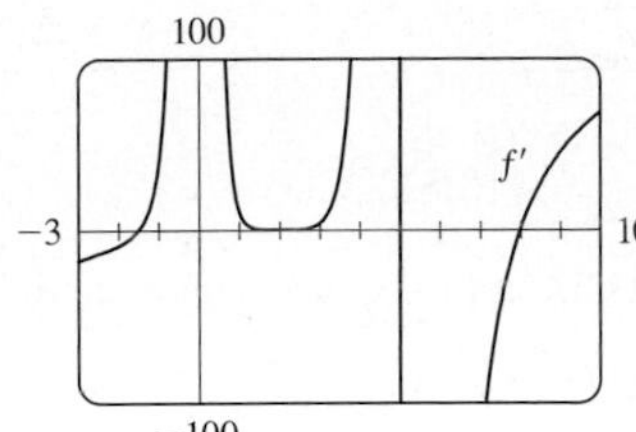

From Exercise 14 and $f'(x)$ above, we know that the zeros of f' are -1.5, 2, and 7.98. From the graph of f', we conclude that f is decreasing on $(-\infty, -1.5)$, increasing on $(-1.5, 0)$ and $(0, 5)$, decreasing on $(5, 7.98)$, and increasing on $(7.98, \infty)$.

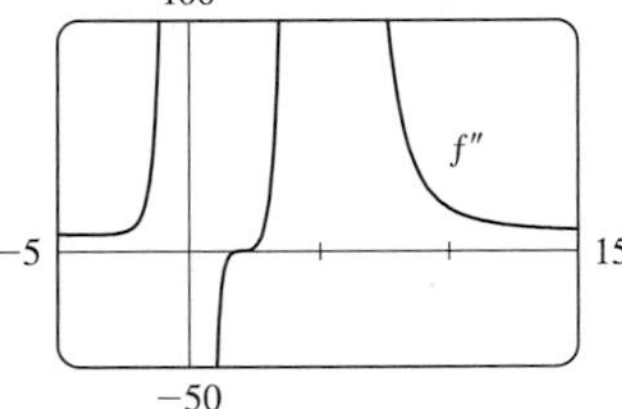

From $f''(x)$, we know that $x = 2$ is a zero, and the graph of f'' shows us that $x = 2$ is the only zero of f''. Thus, f is CU on $(-\infty, 0)$, CD on $(0, 2)$, CU on $(2, 5)$, and CU on $(5, \infty)$.

17. $f(x) = \dfrac{x^3+5x^2+1}{x^4+x^3-x^2+2}$. From a CAS, $f'(x) = \dfrac{-x(x^5+10x^4+6x^3+4x^2-3x-22)}{(x^4+x^3-x^2+2)^2}$ and

$$f''(x) = \frac{2(x^9+15x^8+18x^7+21x^6-9x^5-135x^4-76x^3+21x^2+6x+22)}{(x^4+x^3-x^2+2)^3}$$

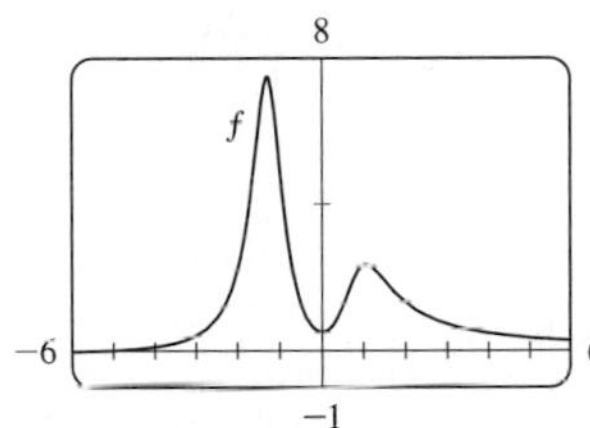

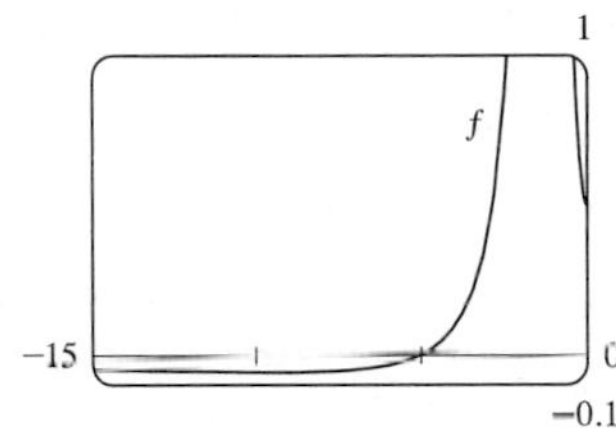

The first graph of f shows that $y = 0$ is a HA. As $x \to \infty$, $f(x) \to 0$ through positive values. As $x \to -\infty$, it is not clear if $f(x) \to 0$ through positive or negative values. The second graph of f shows that f has an x-intercept near -5, and will have a local minimum and inflection point to the left of -5.

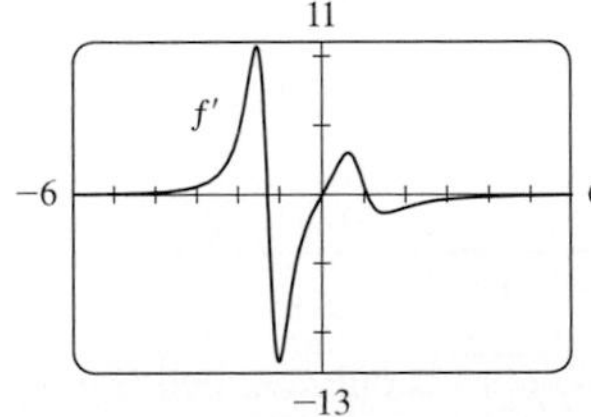

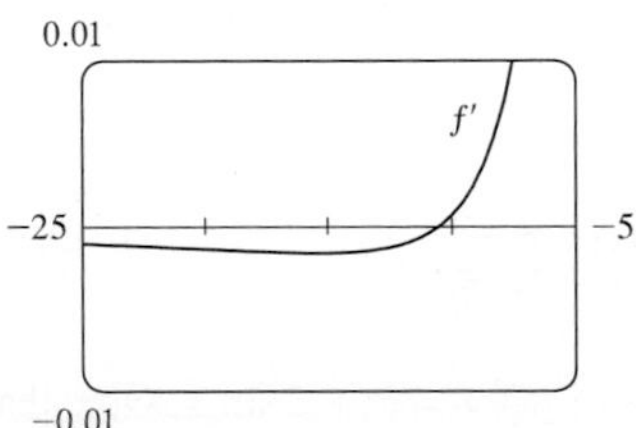

From the two graphs of f', we see that f' has four zeros. We conclude that f is decreasing on $(-\infty, -9.41)$, increasing on $(-9.41, -1.29)$, decreasing on $(-1.29, 0)$, increasing on $(0, 1.05)$, and decreasing on $(1.05, \infty)$. We have local minimum values $f(-9.41) \approx -0.056$ and $f(0) = 0.5$, and local maximum values $f(-1.29) \approx 7.49$ and $f(1.05) \approx 2.35$.

[continued]

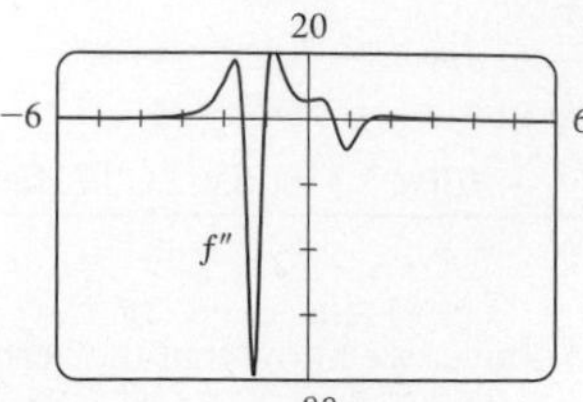

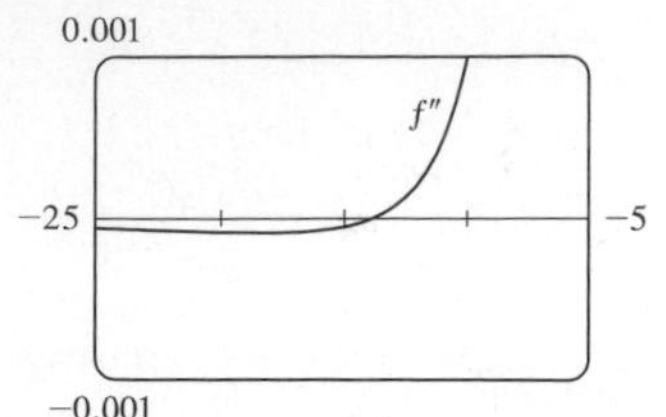

From the two graphs of f'', we see that f'' has five zeros. We conclude that f is CD on $(-\infty, -13.81)$, CU on $(-13.81, -1.55)$, CD on $(-1.55, -1.03)$, CU on $(-1.03, 0.60)$, CD on $(0.60, 1.48)$, and CU on $(1.48, \infty)$. There are five inflection points: $(-13.81, -0.05)$, $(-1.55, 5.64)$, $(-1.03, 5.39)$, $(0.60, 1.52)$, and $(1.48, 1.93)$.

18. $y = f(x) = \dfrac{x^{2/3}}{1 + x + x^4}$. From a CAS, $y' = -\dfrac{10x^4 + x - 2}{3x^{1/3}(x^4 + x + 1)^2}$ and $y'' = \dfrac{2(65x^8 - 14x^5 - 80x^4 + 2x^2 - 8x - 1)}{9x^{4/3}(x^4 + x + 1)^3}$

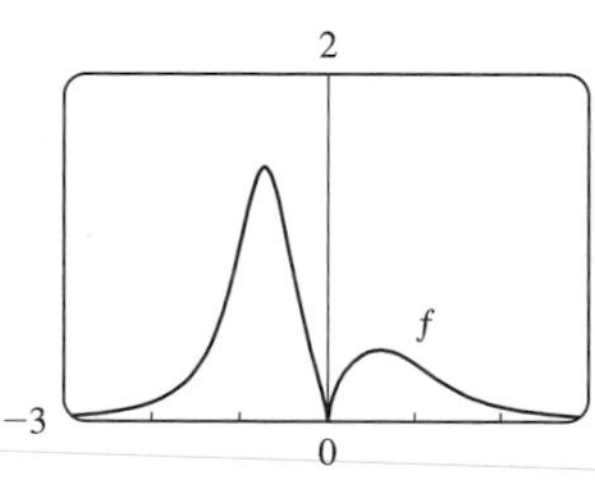

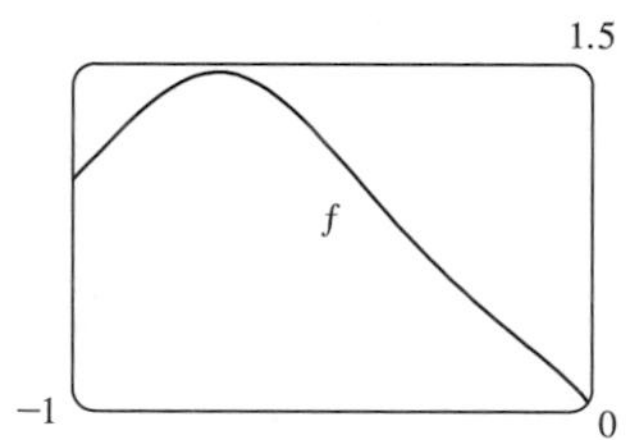

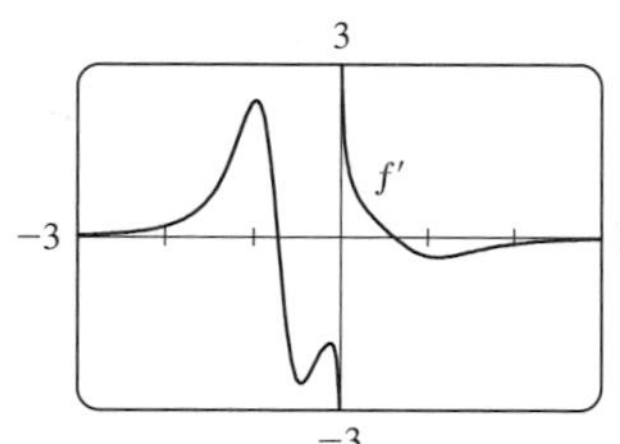

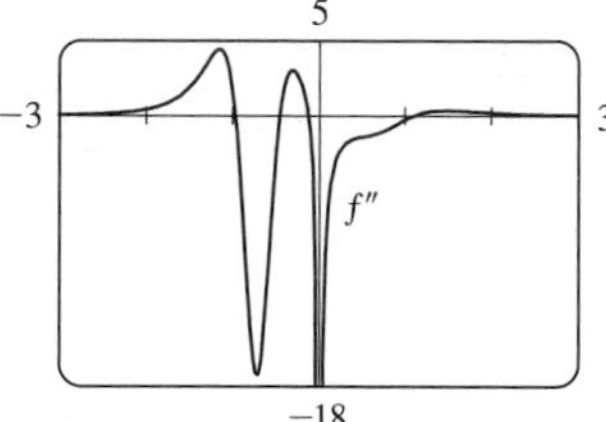

$f'(x)$ does not exist at $x = 0$ and $f'(x) = 0 \quad \Leftrightarrow \quad x \approx -0.72$ and 0.61, so f is increasing on $(-\infty, -0.72)$, decreasing on $(-0.72, 0)$, increasing on $(0, 0.61)$, and decreasing on $(0.61, \infty)$. There is a local maximum value of $f(-0.72) \approx 1.46$ and a local minimum value of $f(0.61) \approx 0.41$. $f''(x)$ does not exist at $x = 0$ and $f''(x) = 0 \quad \Leftrightarrow \quad x \approx -0.97, -0.46, -0.12$, and 1.11, so f is CU on $(-\infty, -0.97)$, CD on $(-0.97, -0.46)$, CU on $(-0.46, -0.12)$, CD on $(-0.12, 0)$, CD on $(0, 1.11)$, and CU on $(1.11, \infty)$. There are inflection points at $(-0.97, 1.08)$, $(-0.46, 1.01)$, $(-0.12, 0.28)$, and $(1.11, 0.29)$.

19. $y = f(x) = \sqrt{x + 5\sin x}$, $x \le 20$.

From a CAS, $y' = \dfrac{5\cos x + 1}{2\sqrt{x + 5\sin x}}$ and $y'' = -\dfrac{10\cos x + 25\sin^2 x + 10x\sin x + 26}{4(x + 5\sin x)^{3/2}}$.

We'll start with a graph of $g(x) = x + 5\sin x$. Note that $f(x) = \sqrt{g(x)}$ is only defined if $g(x) \ge 0$. $g(x) = 0 \quad \Leftrightarrow \quad x = 0$ or $x \approx -4.91, -4.10, 4.10$, and 4.91. Thus, the domain of f is $[-4.91, -4.10] \cup [0, 4.10] \cup [4.91, 20]$.

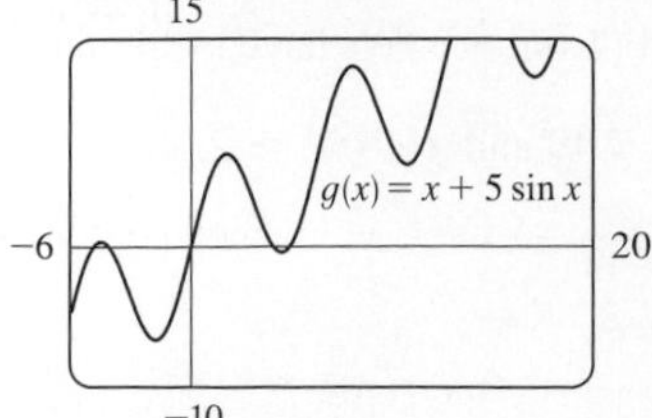

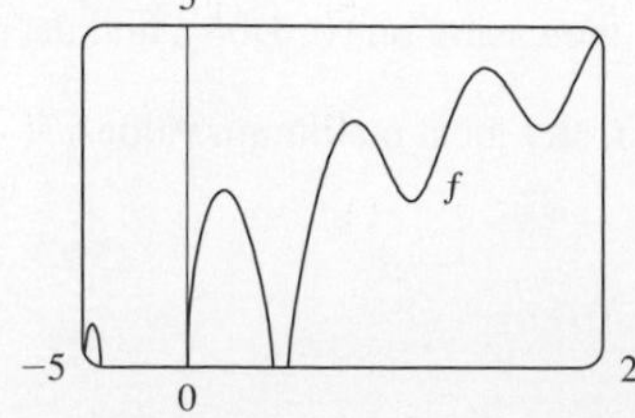

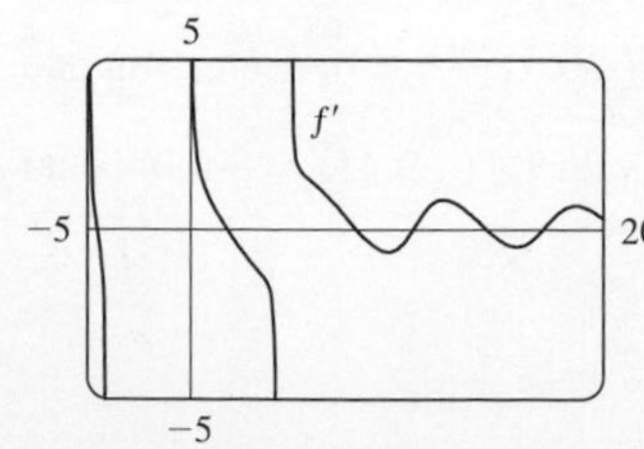

From the expression for y', we see that $y' = 0 \Leftrightarrow 5\cos x + 1 = 0 \Rightarrow x_1 = \cos^{-1}\left(-\frac{1}{5}\right) \approx 1.77$ and $x_2 = 2\pi - x_1 \approx -4.51$ (not in the domain of f). The leftmost zero of f' is $x_1 - 2\pi \approx -4.51$. Moving to the right, the zeros of f' are x_1, $x_1 + 2\pi$, $x_2 + 2\pi$, $x_1 + 4\pi$, and $x_2 + 4\pi$. Thus, f is increasing on $(-4.91, -4.51)$, decreasing on $(-4.51, -4.10)$, increasing on $(0, 1.77)$, decreasing on $(1.77, 4.10)$, increasing on $(4.91, 8.06)$, decreasing on $(8.06, 10.79)$, increasing on $(10.79, 14.34)$, decreasing on $(14.34, 17.08)$, and increasing on $(17.08, 20)$. The local maximum values are $f(-4.51) \approx 0.62$, $f(1.77) \approx 2.58$, $f(8.06) \approx 3.60$, and $f(14.34) \approx 4.39$. The local minimum values are $f(10.79) \approx 2.43$ and $f(17.08) \approx 3.49$.

f is CD on $(-4.91, -4.10)$, $(0, 4.10)$, $(4.91, 9.60)$, CU on $(9.60, 12.25)$, CD on $(12.25, 15.81)$, CU on $(15.81, 18.65)$, and CD on $(18.65, 20)$. There are inflection points at $(9.60, 2.95)$, $(12.25, 3.27)$, $(15.81, 3.91)$, and $(18.65, 4.20)$.

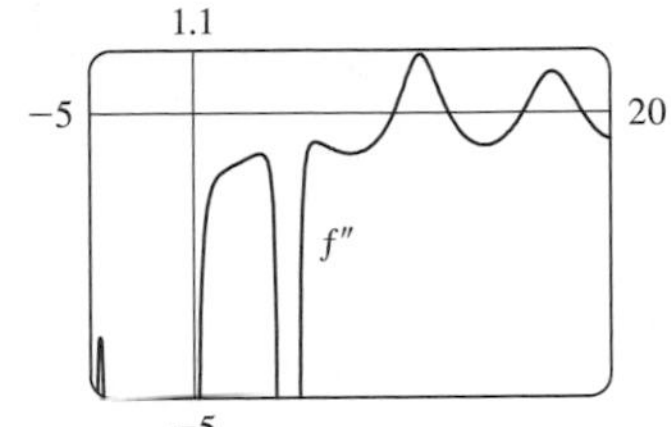

20. $y = f(x) = x - \tan^{-1} x^2$. From a CAS, $y' = \dfrac{x^4 - 2x + 1}{x^4 + 1}$ and $y'' = \dfrac{2(3x^4 - 1)}{(x^4 + 1)^2}$. $y' = 0 \Leftrightarrow x \approx 0.54$ or $x = 1$.

$y'' = 0 \Leftrightarrow x \approx \pm 0.76$.

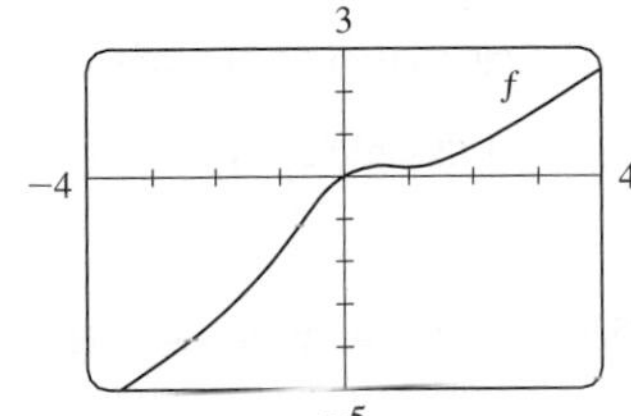

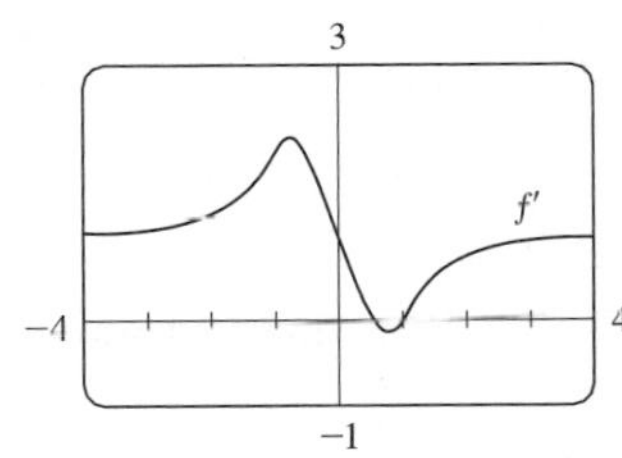

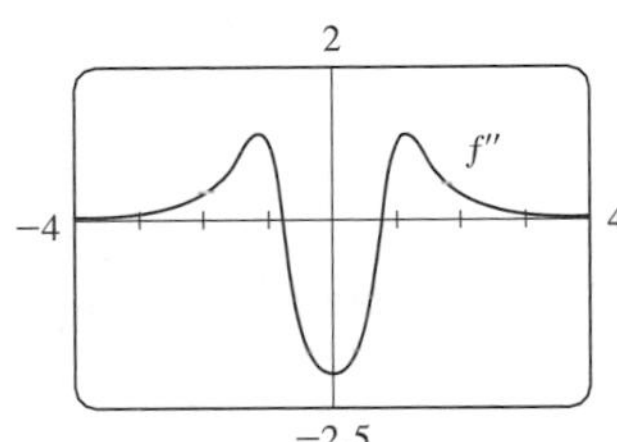

From the graphs of f and f', we estimate that f is increasing on $(-\infty, 0.54)$, decreasing on $(0.54, 1)$, and increasing on $(1, \infty)$. f has local maximum value $f(0.54) \approx 0.26$ and local minimum value $f(1) \approx 0.21$.

From the graph of f'', we estimate that f is CU on $(-\infty, -0.76)$, CD on $(-0.76, 0.76)$, and CU on $(0.76, \infty)$. There are inflection points at about $(-0.76, -1.28)$ and $(0.76, 0.24)$.

21. $y = f(x) = \dfrac{1 - e^{1/x}}{1 + e^{1/x}}$. From a CAS, $y' = \dfrac{2e^{1/x}}{x^2(1 + e^{1/x})^2}$ and $y'' = \dfrac{-2e^{1/x}(1 - e^{1/x} + 2x + 2xe^{1/x})}{x^4(1 + e^{1/x})^3}$.

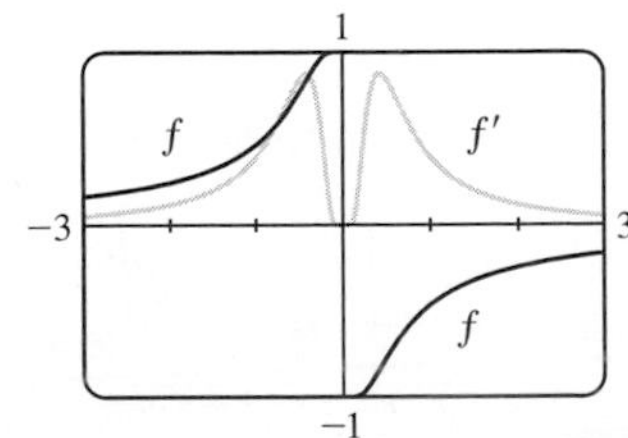

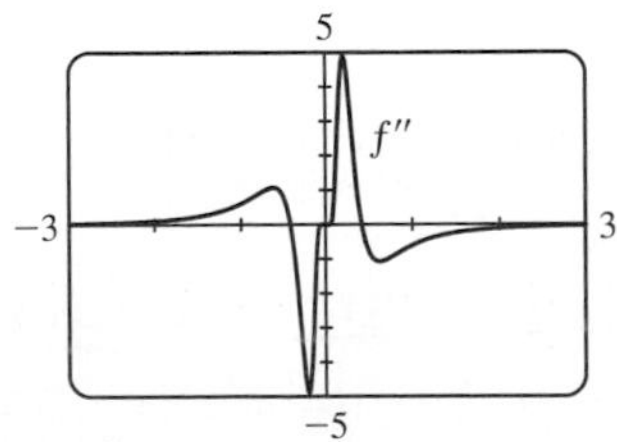

f is an odd function defined on $(-\infty, 0) \cup (0, \infty)$. Its graph has no x- or y-intercepts. Since $\lim\limits_{x \to \pm\infty} f(x) = 0$, the x-axis is a HA. $f'(x) > 0$ for $x \neq 0$, so f is increasing on $(-\infty, 0)$ and $(0, \infty)$. It has no local extreme values.

$f''(x) = 0$ for $x \approx \pm 0.417$, so f is CU on $(-\infty, -0.417)$, CD on $(-0.417, 0)$, CU on $(0, 0.417)$, and CD on $(0.417, \infty)$. f has IPs at $(-0.417, 0.834)$ and $(0.417, -0.834)$.

22. $y = f(x) = \dfrac{3}{3 + 2\sin x}$. From a CAS, $y' = -\dfrac{6\cos x}{(3 + 2\sin x)^2}$ and $y'' = \dfrac{6(2\sin^2 x + 4\cos^2 x + 3\sin x)}{(3 + 2\sin x)^3}$. Since f is periodic with period 2π, we'll restrict our attention to the interval $[0, 2\pi)$. $y' = 0 \Leftrightarrow 6\cos x = 0 \Leftrightarrow x = \frac{\pi}{2}$ or $\frac{3\pi}{2}$. $y'' = 0 \Leftrightarrow x \approx 4.16$ or 5.27.

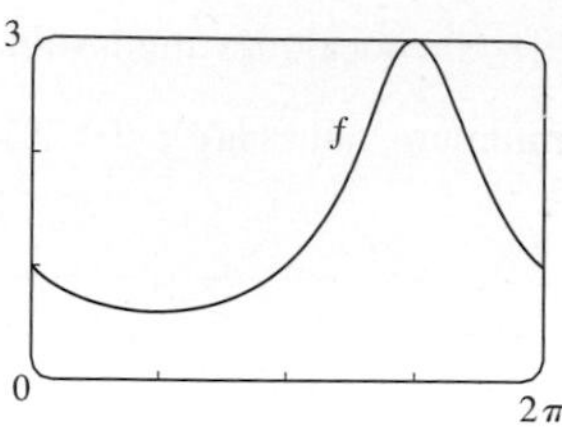

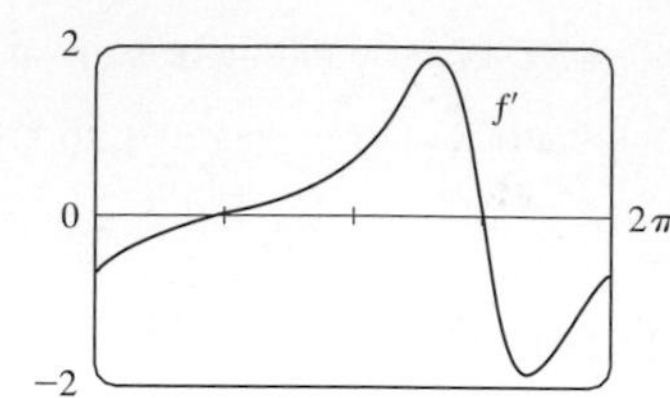

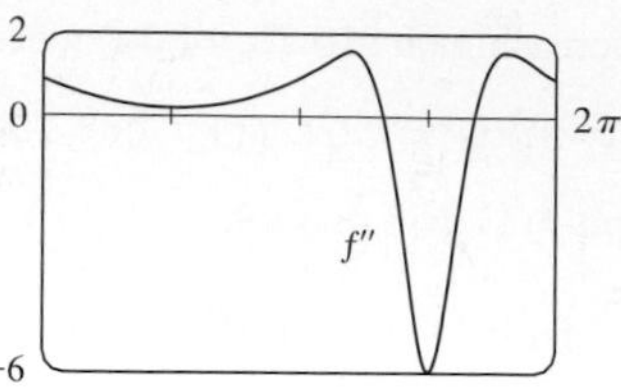

From the graphs of f and f', we conclude that f is decreasing on $\left(0, \frac{\pi}{2}\right)$, increasing on $\left(\frac{\pi}{2}, \frac{3\pi}{2}\right)$, and decreasing on $\left(\frac{3\pi}{2}, 2\pi\right)$. f has local minimum value $f\left(\frac{\pi}{2}\right) = \frac{3}{5}$ and local maximum value $f\left(\frac{3\pi}{2}\right) = 3$.

From the graph of f'', we conclude that f is CU on $(0, 4.16)$, CD on $(4.16, 5.27)$, and CU on $(5.27, 2\pi)$. There are inflection points at about $(4.16, 2.31)$ and $(5.27, 2.31)$.

23. $f(x) = \dfrac{1 - \cos(x^4)}{x^8} \geq 0$. f is an even function, so its graph is symmetric with respect to the y-axis. The first graph shows that f levels off at $y = \frac{1}{2}$ for $|x| < 0.7$. It also shows that f then drops to the x-axis. Your graphing utility may show some severe oscillations near the origin, but there are none. See the discussion in Section 2.2 after Example 2, as well as "Lies My Calculator and Computer Told Me" on the website.

The second graph indicates that as $|x|$ increases, f has progressively smaller humps.

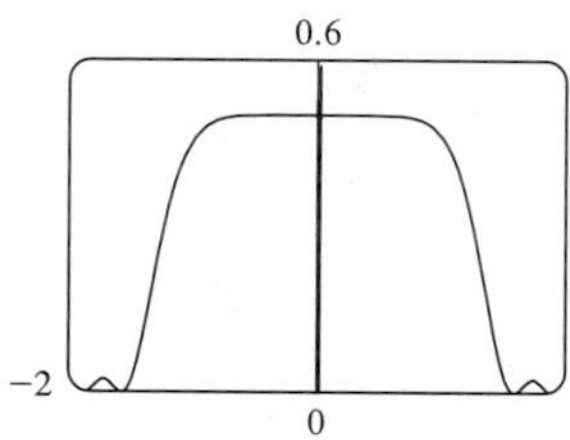

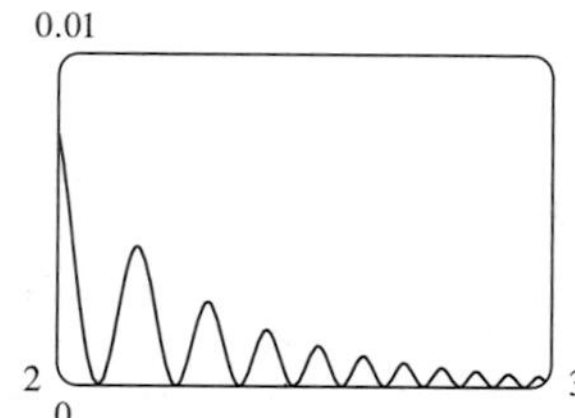

24. $f(x) = e^x + \ln|x - 4|$. The first graph shows the big picture of f but conceals hidden behavior.

The second graph shows that for large negative values of x, f looks like $g(x) = \ln|x|$. It also shows a minimum value and a point of inflection.

The third graph hints at the vertical asymptote that we know exists at $x = 4$ because $\lim\limits_{x \to 4} (e^x + \ln|x - 4|) = -\infty$.

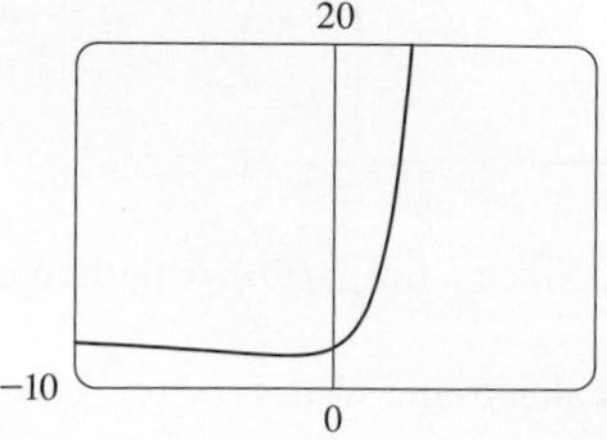

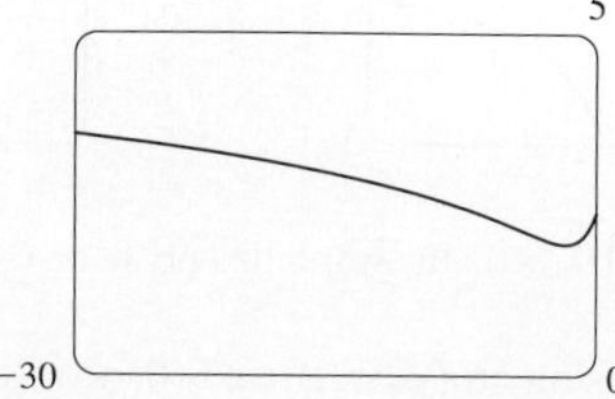

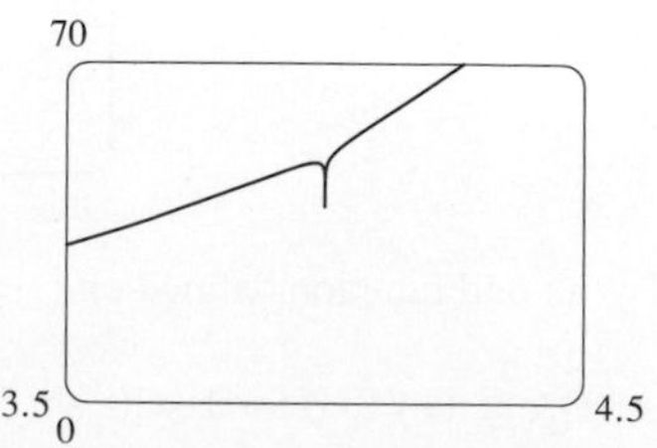

A graphing calculator is unable to show much of the dip of the curve toward the vertical asymptote because of limited resolution. A computer can show more if we restrict ourselves to a narrow interval around $x = 4$. See the solution to Exercise 2.2.48 for a hand-drawn graph of this function.

25. (a) $f(x) = x^{1/x}$

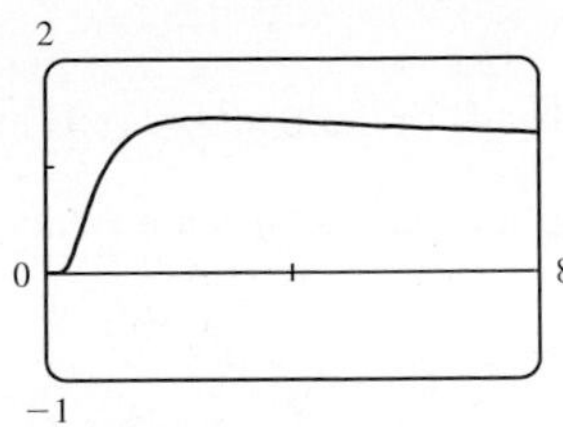

(b) Recall that $a^b = e^{b \ln a}$. $\lim\limits_{x\to 0^+} x^{1/x} = \lim\limits_{x\to 0^+} e^{(1/x)\ln x}$. As $x \to 0^+$, $\dfrac{\ln x}{x} \to -\infty$, so $x^{1/x} = e^{(1/x)\ln x} \to 0$. This indicates that there is a hole at $(0, 0)$. As $x \to \infty$, we have the indeterminate form ∞^0. $\lim\limits_{x\to\infty} x^{1/x} = \lim\limits_{x\to\infty} e^{(1/x)\ln x}$, but $\lim\limits_{x\to\infty} \dfrac{\ln x}{x} \overset{\text{H}}{=} \lim\limits_{x\to\infty} \dfrac{1/x}{1} = 0$, so $\lim\limits_{x\to\infty} x^{1/x} = e^0 = 1$. This indicates that $y = 1$ is a HA.

(c) Estimated maximum: $(2.72, 1.45)$. No estimated minimum. We use logarithmic differentiation to find any critical numbers. $y = x^{1/x} \Rightarrow \ln y = \dfrac{1}{x}\ln x \Rightarrow \dfrac{y'}{y} = \dfrac{1}{x}\cdot\dfrac{1}{x} + (\ln x)\left(-\dfrac{1}{x^2}\right) \Rightarrow y' = x^{1/x}\left(\dfrac{1-\ln x}{x^2}\right) = 0 \Rightarrow$ $\ln x = 1 \Rightarrow x = e$. For $0 < x < e$, $y' > 0$ and for $x > e$, $y' < 0$, so $f(e) = e^{1/e}$ is a local maximum value. This point is approximately $(2.7183, 1.4447)$, which agrees with our estimate.

(d) From the graph, we see that $f''(x) = 0$ at $x \approx 0.58$ and $x \approx 4.37$. Since f'' changes sign at these values, they are x-coordinates of inflection points.

26. (a) $f(x) = (\sin x)^{\sin x}$ is continuous where $\sin x > 0$, that is, on intervals of the form $(2n\pi, (2n+1)\pi)$, so we have graphed f on $(0, \pi)$.

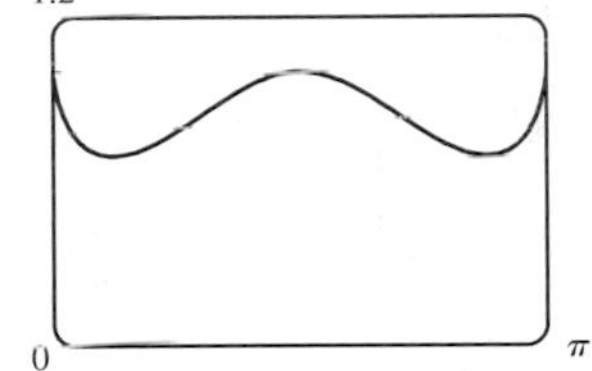

(b) $y = (\sin x)^{\sin x} \Rightarrow \ln y = \sin x \ln \sin x$, so

$$\lim_{x\to 0^+} \ln y = \lim_{x\to 0^+} \sin x \ln \sin x = \lim_{x\to 0^+} \frac{\ln \sin x}{\csc x} \overset{\text{H}}{=} \lim_{x\to 0^+} \frac{\cot x}{-\csc x \cot x}$$
$$= \lim_{x\to 0^+} (-\sin x) = 0 \quad\Rightarrow\quad \lim_{x\to 0^+} y = e^0 = 1.$$

(c) It appears that we have a local maximum at $(1.57, 1)$ and local minima at $(0.38, 0.69)$ and $(2.76, 0.69)$.

$y = (\sin x)^{\sin x} \Rightarrow \ln y = \sin x \ln \sin x \Rightarrow \dfrac{y'}{y} = (\sin x)\left(\dfrac{\cos x}{\sin x}\right) + (\ln \sin x)\cos x = \cos x\,(1 + \ln \sin x) \Rightarrow$ $y' = (\sin x)^{\sin x}(\cos x)(1 + \ln \sin x)$. $y' = 0 \Rightarrow \cos x = 0$ or $\ln \sin x = -1 \Rightarrow x_2 = \frac{\pi}{2}$ or $\sin x = e^{-1}$. On $(0, \pi)$, $\sin x = e^{-1} \Rightarrow x_1 = \sin^{-1}(e^{-1})$ and $x_3 = \pi - \sin^{-1}(e^{-1})$. Approximating these points gives us $(x_1, f(x_1)) \approx (0.3767, 0.6922)$, $(x_2, f(x_2)) \approx (1.5708, 1)$, and $(x_3, f(x_3)) \approx (2.7649, 0.6922)$. The approximations confirm our estimates.

(d) 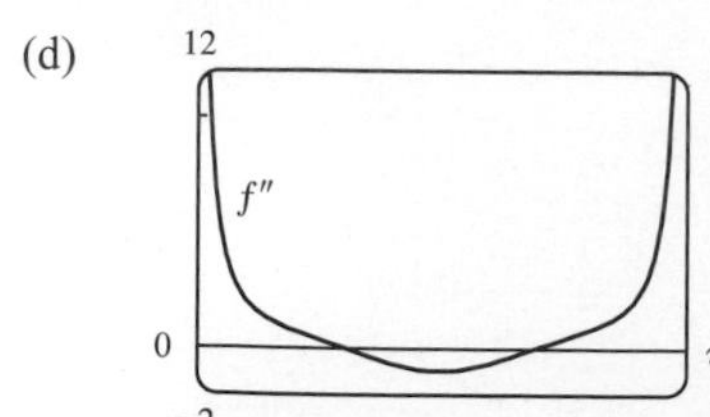

From the graph, we see that $f''(x) = 0$ at $x \approx 0.94$ and $x \approx 2.20$. Since f'' changes sign at these values, they are x-coordinates of inflection points.

27. 

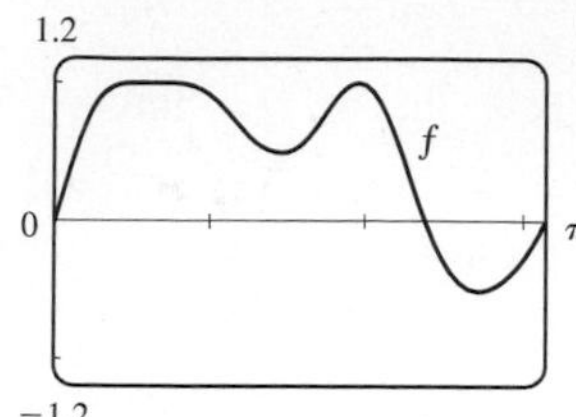

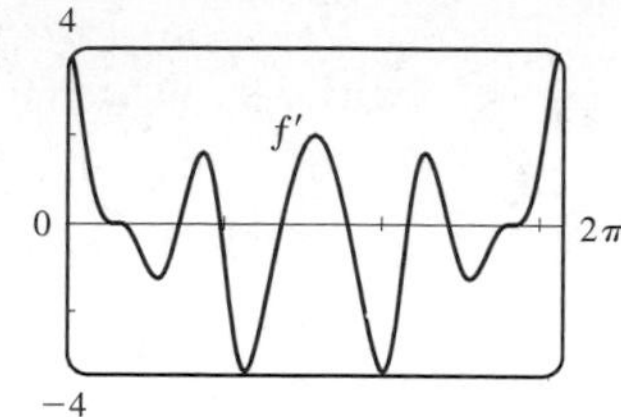

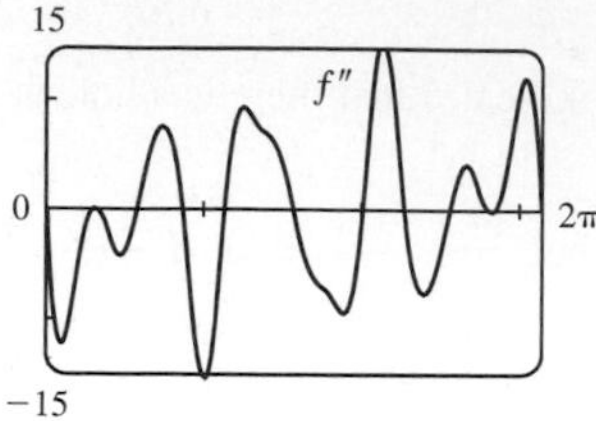

From the graph of $f(x) = \sin(x + \sin 3x)$ in the viewing rectangle $[0, \pi]$ by $[-1.2, 1.2]$, it looks like f has two maxima and two minima. If we calculate and graph $f'(x) = [\cos(x + \sin 3x)]\,(1 + 3\cos 3x)$ on $[0, 2\pi]$, we see that the graph of f' appears to be almost tangent to the x-axis at about $x = 0.7$. The graph of

$$f'' = -\,[\sin(x + \sin 3x)]\,(1 + 3\cos 3x)^2 + \cos(x + \sin 3x)(-9\sin 3x)$$

is even more interesting near this x-value: it seems to just touch the x-axis.

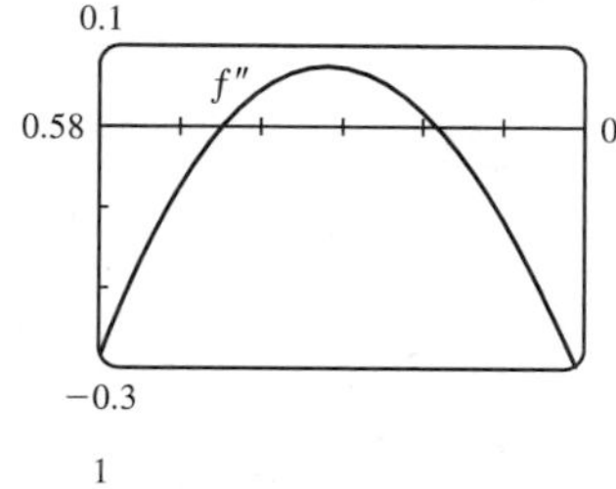

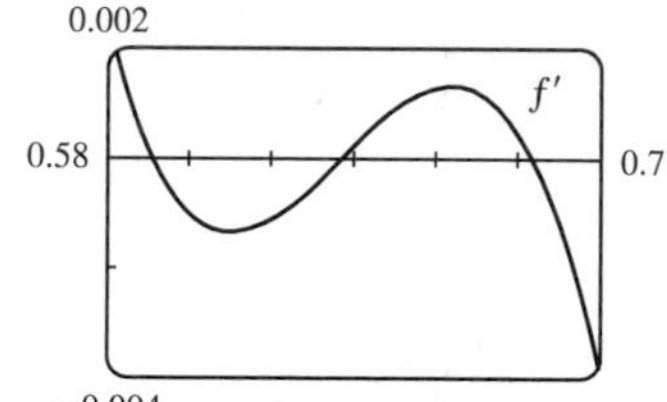

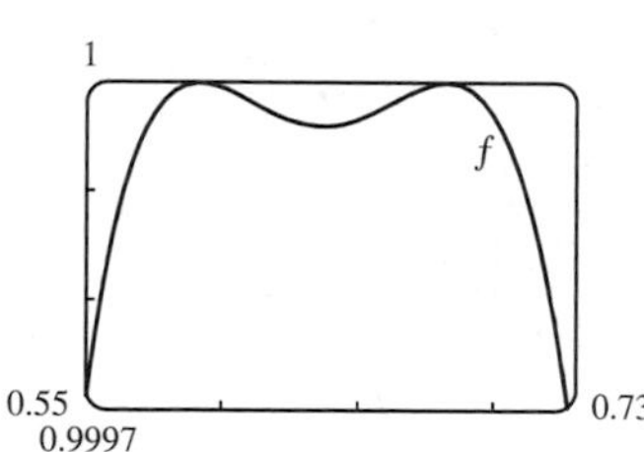

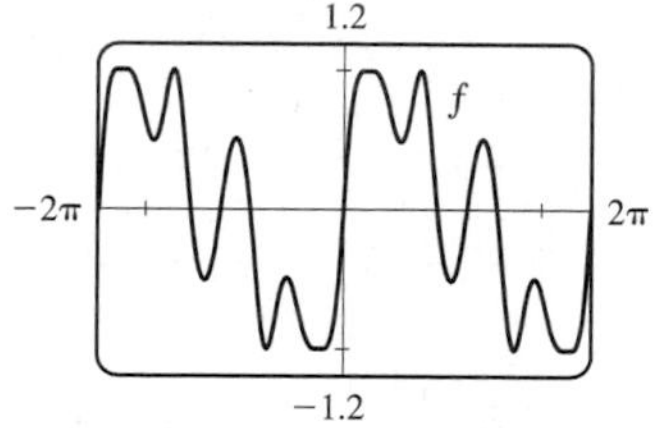

If we zoom in on this place on the graph of f'', we see that f'' actually does cross the axis twice near $x = 0.65$, indicating a change in concavity for a very short interval. If we look at the graph of f' on the same interval, we see that it changes sign three times near $x = 0.65$, indicating that what we had thought was a broad extremum at about $x = 0.7$ actually consists of three extrema (two maxima and a minimum). These maximum values are roughly $f(0.59) = 1$ and $f(0.68) = 1$, and the minimum value is roughly $f(0.64) = 0.99996$. There are also a maximum value of about $f(1.96) = 1$ and minimum values of about $f(1.46) = 0.49$ and $f(2.73) = -0.51$. The points of inflection on $(0, \pi)$ are about $(0.61, 0.99998)$, $(0.66, 0.99998)$, $(1.17, 0.72)$, $(1.75, 0.77)$, and $(2.28, 0.34)$. On $(\pi, 2\pi)$, they are about $(4.01, -0.34)$, $(4.54, -0.77)$, $(5.11, -0.72)$, $(5.62, -0.99998)$, and $(5.67, -0.99998)$. There are also IP at $(0, 0)$ and $(\pi, 0)$. Note that the function is odd and periodic with period 2π, and it is also rotationally symmetric about all points of the form $((2n + 1)\pi, 0)$, n an integer.

28. $f(x) = x^3 + cx = x(x^2 + c) \;\Rightarrow\; f'(x) = 3x^2 + c \;\Rightarrow\; f''(x) = 6x$

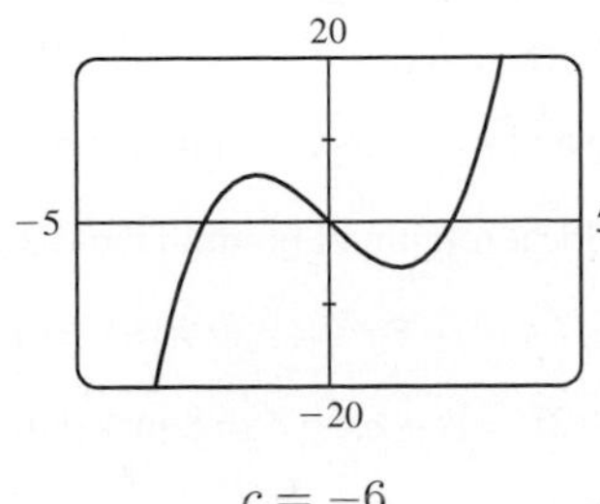

$c = -6$

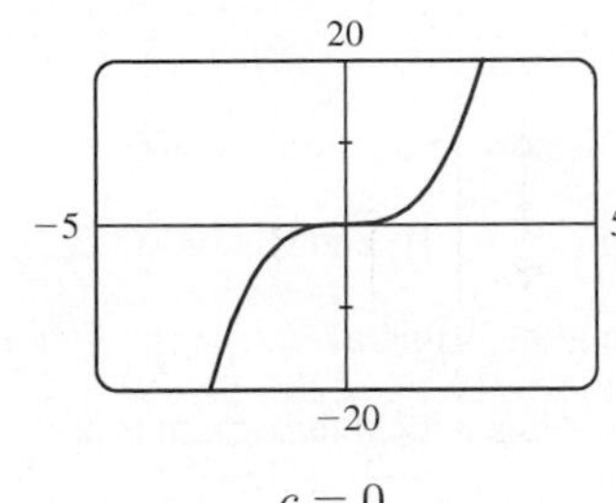

$c = 0$

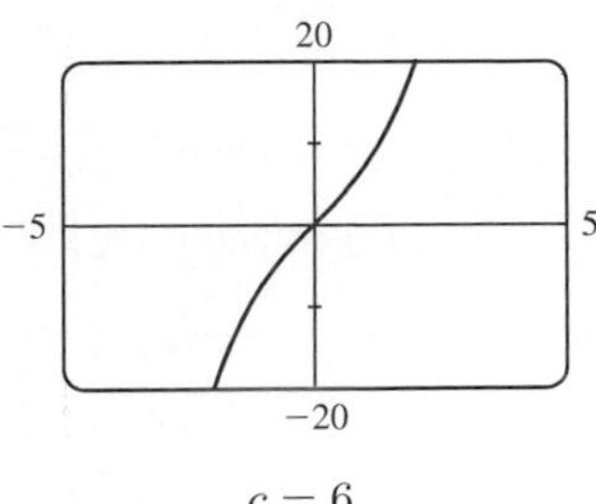

$c = 6$

x-intercepts: When $c \geq 0$, 0 is the only x-intercept. When $c < 0$, the x-intercepts are 0 and $\pm\sqrt{-c}$.

y-intercept $= f(0) = 0$. f is odd, so the graph is symmetric with respect to the origin. $f''(x) < 0$ for $x < 0$ and $f''(x) > 0$ for $x > 0$, so f is CD on $(-\infty, 0)$ and CU on $(0, \infty)$. The origin is the only inflection point.

If $c > 0$, then $f'(x) > 0$ for all x, so f is increasing and has no local maximum or minimum.

If $c = 0$, then $f'(x) \geq 0$ with equality at $x = 0$, so again f is increasing and has no local maximum or minimum.

If $c < 0$, then $f'(x) = 3[x^2 - (-c/3)] = 3\left(x + \sqrt{-c/3}\right)\left(x - \sqrt{-c/3}\right)$, so $f'(x) > 0$ on $\left(-\infty, -\sqrt{-c/3}\right)$ and $\left(\sqrt{-c/3}, \infty\right)$; $f'(x) < 0$ on $\left(-\sqrt{-c/3}, \sqrt{-c/3}\right)$. It follows that $f\left(-\sqrt{-c/3}\right) = -\frac{2}{3}c\sqrt{-c/3}$ is a local maximum value and $f\left(\sqrt{-c/3}\right) = \frac{2}{3}c\sqrt{-c/3}$ is a local minimum value. As c decreases (toward more negative values), the local maximum and minimum move further apart.

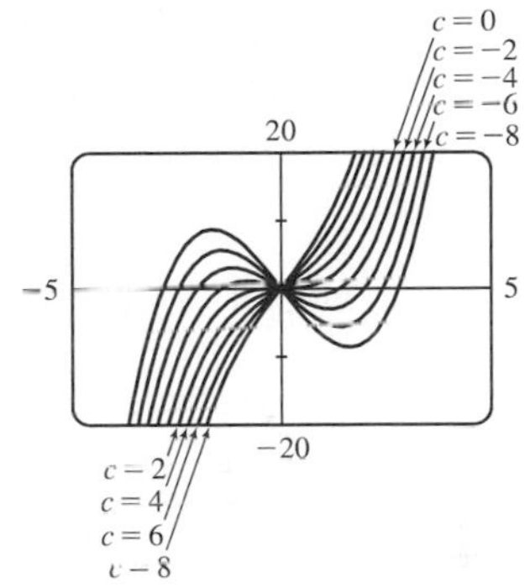

There is no absolute maximum or minimum value. The only transitional value of c corresponding to a change in character of the graph is $c = 0$.

29. $f(x) = x^2 + 6x + c/x \;\Rightarrow\; f'(x) = 2x + 6 - c/x^2 \;\Rightarrow\; f''(x) = 2 + 2c/x^3$

$\mathbf{c = 0}$: The graph is the parabola $y = x^2 + 6x$, which has x-intercepts -6 and 0, vertex $(-3, -9)$, and opens upward.

$\mathbf{c \neq 0}$: The parabola $y = x^2 + 6x$ is an asymptote that the graph of f approaches as $x \to \pm\infty$. The y-axis is a vertical asymptote.

$\mathbf{c < 0}$: The x-intercepts are found by solving $f(x) = 0 \;\Leftrightarrow\; x^3 + 6x^2 + c = g(x) = 0$. Now $g'(x) = 0 \;\Leftrightarrow\; x = -4$ or 0, and g (*not* f) has a local maximum at $x = -4$. $g(-4) = 32 + c$, so if $c < -32$, the maximum is negative and there are no negative x-intercepts; if $c = -32$, the maximum is 0 and there is one negative x-intercept; if $-32 < c < 0$, the maximum is positive and there are two negative x-intercepts. In all cases, there is one positive x-intercept.

As $c \to 0^-$, the local minimum point moves down and right, approaching $(-3, -9)$. [Note that since $f'(x) = \dfrac{2x^3 + 6x^2 - c}{x^2}$, Descartes' Rule of Signs implies that f' has no positive roots and one negative root when $c < 0$. $f''(x) = \dfrac{2(x^3 + c)}{x^3} > 0$ at that negative root, so that critical point yields a local minimum value. This tells us that there are no local maximums when $c < 0$.] $f'(x) > 0$ for $x > 0$, so f is increasing on $(0, \infty)$. From $f''(x) = \dfrac{2(x^3 + c)}{x^3}$, we see that f

has an inflection point at $\left(\sqrt[3]{-c}, 6\sqrt[3]{-c}\right)$. This inflection point moves down and left, approaching the origin as $c \to 0^-$. f is CU on $(-\infty, 0)$, CD on $\left(0, \sqrt[3]{-c}\right)$, and CU on $\left(\sqrt[3]{-c}, \infty\right)$.

$c > 0$: The inflection point $\left(\sqrt[3]{-c}, 6\sqrt[3]{-c}\right)$ is now in the third quadrant and moves up and right, approaching the origin as $c \to 0^+$. f is CU on $\left(-\infty, \sqrt[3]{-c}\right)$, CD on $\left(\sqrt[3]{-c}, 0\right)$, and CU on $(0, \infty)$. f has a local minimum point in the first quadrant. It moves down and left, approaching the origin as $c \to 0^+$. $f'(x) = 0 \Leftrightarrow 2x^3 + 6x^2 - c = h(x) = 0$. Now $h'(x) = 0 \Leftrightarrow x = -2$ or 0, and h (*not* f) has a local maximum at $x = -2$. $h(-2) = 8 - c$, so $c = 8$ makes $h(x) = 0$, and hence, $f'(x) = 0$. When $c > 8$, $f'(x) < 0$ and f is decreasing on $(-\infty, 0)$. For $0 < c < 8$, there is a local minimum that moves toward $(-3, -9)$ and a local maximum that moves toward the origin as c decreases.

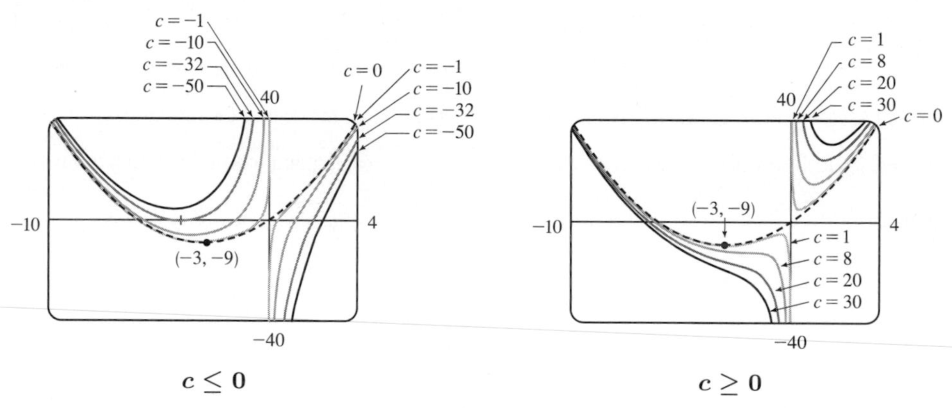

30. With $c = 0$ in $y = f(x) = x\sqrt{c^2 - x^2}$, the graph of f is just the point $(0, 0)$. Since $(-c)^2 = c^2$, we only consider $c > 0$. Since $f(-x) = -f(x)$, the graph is symmetric about the origin. The domain of f is found by solving $c^2 - x^2 \geq 0 \Leftrightarrow x^2 \leq c^2 \Leftrightarrow |x| \leq c$, which gives us $[-c, c]$.

$$f'(x) = x \cdot \tfrac{1}{2}(c^2 - x^2)^{-1/2}(-2x) + (c^2 - x^2)^{1/2}(1) = (c^2 - x^2)^{-1/2}[-x^2 + (c^2 - x^2)] = \frac{c^2 - 2x^2}{\sqrt{c^2 - x^2}}.$$

$f'(x) > 0 \Leftrightarrow c^2 - 2x^2 > 0 \Leftrightarrow x^2 < c^2/2 \Leftrightarrow |x| < c/\sqrt{2}$, so f is increasing on $\left(-c/\sqrt{2}, c/\sqrt{2}\right)$ and decreasing on $\left(-c, -c/\sqrt{2}\right)$ and $\left(c/\sqrt{2}, c\right)$. There is a local minimum value of $f\left(-c/\sqrt{2}\right) = \left(-c/\sqrt{2}\right)\sqrt{c^2 - c^2/2} = \left(-c/\sqrt{2}\right)\left(c/\sqrt{2}\right) = -c^2/2$ and a local maximum value of $f\left(c/\sqrt{2}\right) = c^2/2$.

$$f''(x) = \frac{(c^2 - x^2)^{1/2}(-4x) - (c^2 - 2x^2)\frac{1}{2}\left(c^2 - x^2\right)^{-1/2}(-2x)}{[(c^2 - x^2)^{1/2}]^2}$$

$$= \frac{x(c^2 - x^2)^{-1/2}[(c^2 - x^2)(-4) + (c^2 - 2x^2)]}{(c^2 - x^2)^1} = \frac{2x\left(2x^2 - 3c^2\right)}{(c^2 - x^2)^{3/2}},$$

so $f''(x) = 0 \Leftrightarrow x = 0$ or $x = \pm\sqrt{\frac{3}{2}}\,c$, but only 0 is in the domain of f. $f''(x) < 0$ for $0 < x < c$ and $f''(x) > 0$ for $-c < x < 0$, so f is CD on $(0, c)$ and CU on $(-c, 0)$. There is an IP at $(0, 0)$. So as $|c|$ gets larger, the maximum and minimum values increase in magnitude. The value of c does not affect the concavity of f.

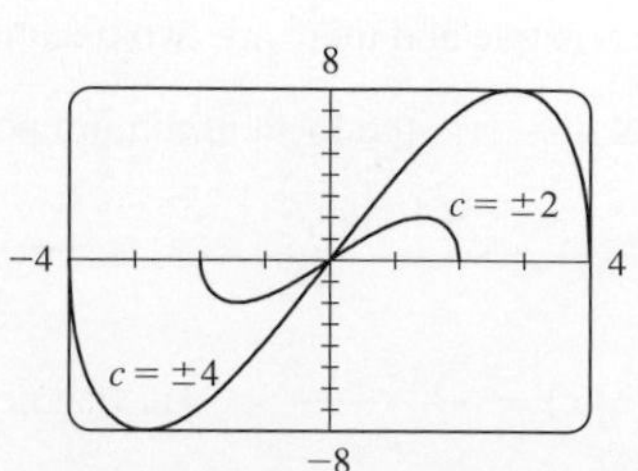

31. $f(x) = e^x + ce^{-x}$. $f = 0 \Rightarrow ce^{-x} = -e^x \Rightarrow c = -e^{2x} \Rightarrow 2x = \ln(-c) \Rightarrow x = \frac{1}{2}\ln(-c)$.

$f'(x) = e^x - ce^{-x}$. $f' = 0 \Rightarrow ce^{-x} = e^x \Rightarrow c = e^{2x} \Rightarrow 2x = \ln c \Rightarrow x = \frac{1}{2}\ln c$.

$f''(x) = e^x + ce^{-x} = f(x)$.

The only transitional value of c is 0. As c increases from $-\infty$ to 0, $\frac{1}{2}\ln(-c)$ is both the the x-intercept and inflection point, and this decreases from ∞ to $-\infty$. Also $f' > 0$, so f is increasing. When $c = 0$, $f(x) = f'(x) = f''(x) = e^x$, f is positive, increasing, and concave upward. As c increases from 0 to ∞, the absolute minimum occurs at $x = \frac{1}{2}\ln c$, which increases from $-\infty$ to ∞. Also, $f = f'' > 0$, so f is positive and concave upward. The value of the y-intercept is $f(0) = 1 + c$, and this increases as c increases from $-\infty$ to ∞.

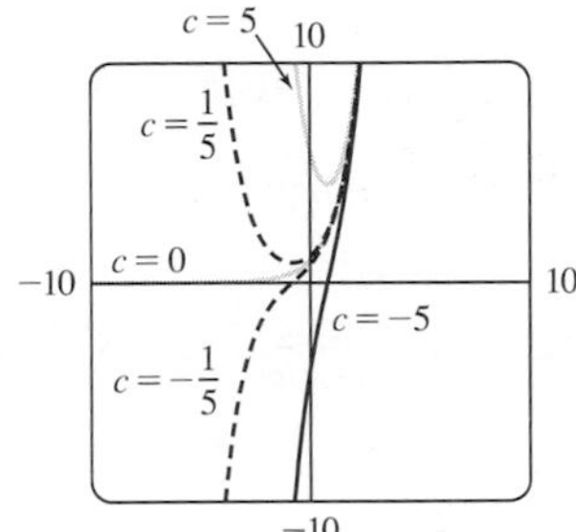

Note: The minimum point $\left(\frac{1}{2}\ln c, 2\sqrt{c}\right)$ can be parameterized by $x = \frac{1}{2}\ln c$, $y = 2\sqrt{c}$, and after eliminating the parameter c, we see that the minimum point lies on the graph of $y = 2e^x$.

32. We see that if $c \le 0$, $f(x) = \ln(x^2 + c)$ is only defined for $x^2 > -c \Rightarrow |x| > \sqrt{-c}$, and $\lim\limits_{x \to \sqrt{-c}^+} f(x) = \lim\limits_{x \to -\sqrt{-c}^-} f(x) = -\infty$, since $\ln y \to -\infty$ as $y \to 0$. Thus, for $c < 0$, there are vertical asymptotes at $x = \pm\sqrt{c}$, and as c decreases (that is, $|c|$ increases), the asymptotes get further apart. For $c = 0$, $\lim\limits_{x \to 0} f(x) = -\infty$, so there is a vertical asymptote at $x = 0$. If $c > 0$, there are no asymptotes. To find the extrema and inflection points, we differentiate: $f(x) = \ln(x^2 + c) \Rightarrow f'(x) = \dfrac{1}{x^2 + c}(2x)$, so by the First Derivative Test there is a local and absolute minimum at $x = 0$. Differentiating again, we get $f''(x) = \dfrac{1}{x^2 + c}(2) + 2x\left[-(x^2 + c)^{-2}(2x)\right] = \dfrac{2(c - x^2)}{(x^2 + c)^2}$.

Now if $c \le 0$, f'' is always negative, so f is concave down on both of the intervals on which it is defined. If $c > 0$, then f'' changes sign when $c = x^2 \Leftrightarrow x = \pm\sqrt{c}$. So for $c > 0$ there are inflection points at $x = \pm\sqrt{c}$, and as c increases, the inflection points get further apart.

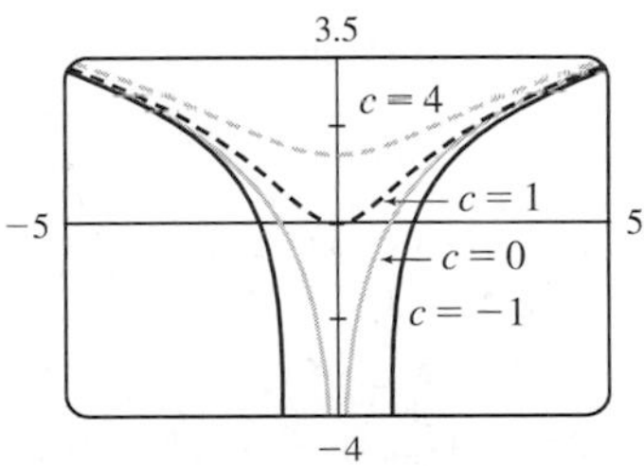

33. Note that $c = 0$ is a transitional value at which the graph consists of the x-axis. Also, we can see that if we substitute $-c$ for c, the function $f(x) = \dfrac{cx}{1 + c^2x^2}$ will be reflected in the x-axis, so we investigate only positive values of c (except $c = -1$, as a demonstration of this reflective property). Also, f is an odd function. $\lim\limits_{x \to \pm\infty} f(x) = 0$, so $y = 0$ is a horizontal asymptote for all c. We calculate $f'(x) = \dfrac{(1 + c^2x^2)c - cx(2c^2x)}{(1 + c^2x^2)^2} = -\dfrac{c(c^2x^2 - 1)}{(1 + c^2x^2)^2}$. $f'(x) = 0 \Leftrightarrow c^2x^2 - 1 = 0 \Leftrightarrow$

[continued]

$x = \pm 1/c$. So there is an absolute maximum value of $f(1/c) = \frac{1}{2}$ and an absolute minimum value of $f(-1/c) = -\frac{1}{2}$. These extrema have the same value regardless of c, but the maximum points move closer to the y-axis as c increases.

$$f''(x) = \frac{(-2c^3x)(1+c^2x^2)^2 - (-c^3x^2+c)[2(1+c^2x^2)(2c^2x)]}{(1+c^2x^2)^4}$$

$$= \frac{(-2c^3x)(1+c^2x^2) + (c^3x^2-c)(4c^2x)}{(1+c^2x^2)^3} = \frac{2c^3x(c^2x^2-3)}{(1+c^2x^2)^3}$$

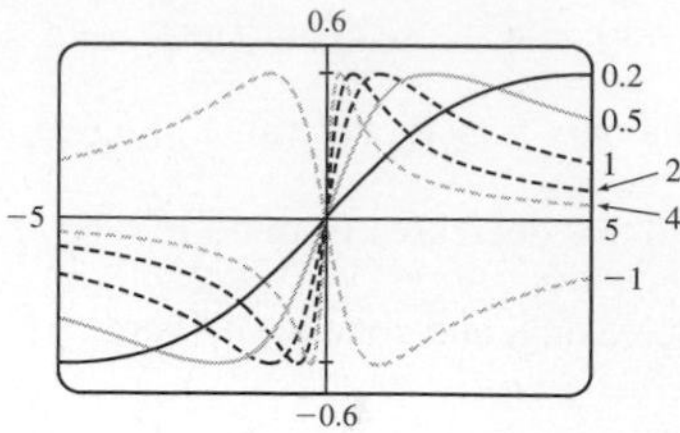

$f''(x) = 0 \quad\Leftrightarrow\quad x = 0$ or $\pm\sqrt{3}/c$, so there are inflection points at $(0, 0)$ and at $\left(\pm\sqrt{3}/c, \pm\sqrt{3}/4\right)$. Again, the y-coordinate of the inflection points does not depend on c, but as c increases, both inflection points approach the y-axis.

34. $f(x) = \dfrac{\sin x}{c + \cos x} \quad\Rightarrow\quad f'(x) = \dfrac{1 + c\cos x}{\cos^2 x + 2c\cos x + c^2} \quad\Rightarrow\quad f''(x) = \dfrac{\sin x(c\cos x - c^2 + 2)}{\cos^3 x + 3c\cos^2 x + 3c^2\cos x + c^3}$. Notice that f is an odd function and has period 2π. We will graph f for $0 \le x \le 2\pi$.

$|\mathbf{c}| \le \mathbf{1}$: See the first figure. f has VAs when the denominator is zero, that is, at $x = \cos^{-1}(-c)$ and $x = 2\pi - \cos^{-1}(-c)$. So for $c = -1$, there are VAs at $x = 0$ and $x = 2\pi$, and as c increases, they move closer to $x = \pi$, which is the single VA when $c = 1$. Note that if $c = 0$, then $f(x) = \tan x$. There are no extreme points (on the entire domain) and inflection points occur at multiples of π.

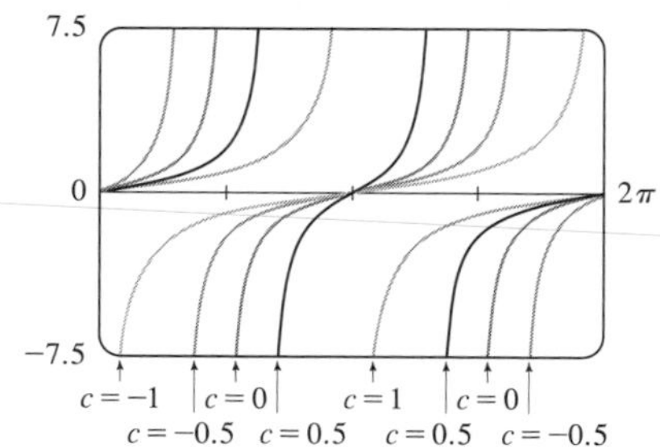

$\mathbf{c > 1}$: See the second figure. $f'(x) = 0 \quad\Leftrightarrow\quad x = \cos^{-1}\left(\dfrac{-1}{c}\right)$ or $x = 2\pi - \cos^{-1}\left(\dfrac{-1}{c}\right)$. The VA disappears and there is now a local maximum and a local minimum. As $c \to 1^+$, the coordinates of the local maximum approach π and ∞, and the coordinates of the local minimum approach π and $-\infty$.

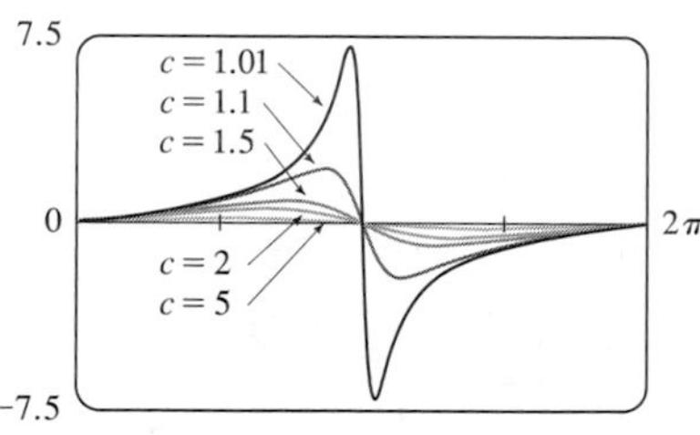

As $c \to \infty$, the graph of f looks like a graph of $y = \sin x$ that is vertically compressed, and the local maximum and local minimum approach $\left(\frac{\pi}{2}, 0\right)$ and $\left(\frac{3\pi}{2}, 0\right)$, respectively.

$f''(x) = 0 \quad\Leftrightarrow\quad \sin x = 0$ (IPs at $x = n\pi$) or $c\cos x - c^2 + 2 = 0$. The second condition is true if $\cos x = \dfrac{c^2 - 2}{c}$ $[c \ne 0]$. The last equation has two solutions if $-1 < \dfrac{c^2-2}{c} < 1 \quad\Rightarrow\quad -c < c^2 - 2 < c \quad\Rightarrow\quad -c < c^2 - 2$ and $c^2 - 2 < c \quad\Rightarrow\quad c^2 + c - 2 > 0$ and $c^2 - c - 2 < 0 \quad\Rightarrow\quad (c+2)(c-1) > 0$ and $(c-2)(c+1) < 0 \quad\Rightarrow\quad c - 1 > 0$

[continued]

[since $c > 1$] and $c - 2 < 0 \;\Rightarrow\; c > 1$ and $c < 2$. Thus, for $1 < c < 2$, we have 2 nontrivial IPs at $x = \cos^{-1}\left(\dfrac{c^2-2}{c}\right)$ and $x = 2\pi - \cos^{-1}\left(\dfrac{c^2-2}{c}\right)$.

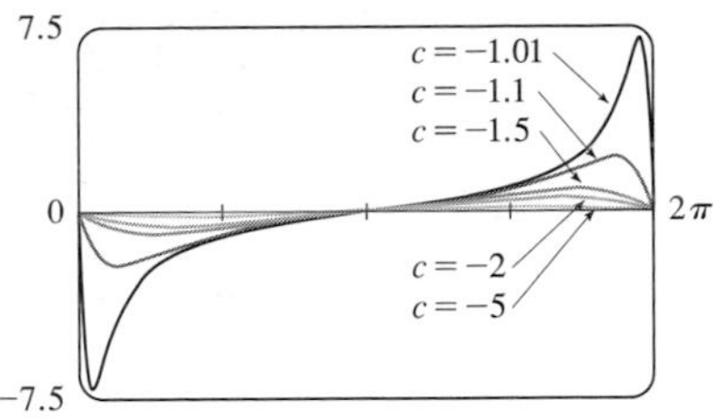

c < −1: See the third figure. The VAs for $c = -1$ at $x = 0$ and $x = 2\pi$ in the first figure disappear and we now have a local minimum and a local maximum. As $c \to -1^+$, the coordinates of the local minimum approach 0 and $-\infty$, and the coordinates of the local maximum approach 2π and ∞. As $c \to -\infty$, the graph of f looks like a graph of $y = \sin x$ that is vertically compressed, and the local minimum and local maximum approach $\left(\frac{\pi}{2}, 0\right)$ and $\left(\frac{3\pi}{2}, 0\right)$, respectively. As above, we have two nontrivial IPs for $-2 < c < -1$.

35. $f(x) = cx + \sin x \;\Rightarrow\; f'(x) = c + \cos x \;\Rightarrow\; f''(x) = -\sin x$

$f(-x) = -f(x)$, so f is an odd function and its graph is symmetric with respect to the origin.

$f(x) = 0 \;\Leftrightarrow\; \sin x = -cx$, so 0 is always an x-intercept.

$f'(x) = 0 \;\Leftrightarrow\; \cos x = -c$, so there is no critical number when $|c| > 1$. If $|c| \le 1$, then there are infinitely many critical numbers. If x_1 is the unique solution of $\cos x = -c$ in the interval $[0, \pi]$, then the critical numbers are $2n\pi \pm x_1$, where n ranges over the integers. (Special cases: When $c = -1$, $x_1 = 0$; when $c = 0$, $x = \frac{\pi}{2}$; and when $c = 1$, $x_1 = \pi$.)

$f''(x) < 0 \;\Leftrightarrow\; \sin x > 0$, so f is CD on intervals of the form $(2n\pi, (2n+1)\pi)$. f is CU on intervals of the form $((2n-1)\pi, 2n\pi)$. The inflection points of f are the points $(n\pi, n\pi c)$, where n is an integer.

If $c \ge 1$, then $f'(x) \ge 0$ for all x, so f is increasing and has no extremum. If $c \le -1$, then $f'(x) \le 0$ for all x, so f is decreasing and has no extremum. If $|c| < 1$, then $f'(x) > 0 \;\Leftrightarrow\; \cos x > -c \;\Leftrightarrow\; x$ is in an interval of the form $(2n\pi - x_1, 2n\pi + x_1)$ for some integer n. These are the intervals on which f is increasing. Similarly, we find that f is decreasing on the intervals of the form $(2n\pi + x_1, 2(n+1)\pi - x_1)$. Thus, f has local maxima at the points $2n\pi + x_1$, where f has the values $c(2n\pi + x_1) + \sin x_1 = c(2n\pi + x_1) + \sqrt{1-c^2}$, and f has local minima at the points $2n\pi - x_1$, where we have $f(2n\pi - x_1) = c(2n\pi - x_1) - \sin x_1 = c(2n\pi - x_1) - \sqrt{1-c^2}$.

The transitional values of c are -1 and 1. The inflection points move vertically, but not horizontally, when c changes. When $|c| \ge 1$, there is no extremum. For $|c| < 1$, the maxima are spaced 2π apart horizontally, as are the minima. The horizontal spacing between maxima and adjacent minima is regular (and equals π) when $c = 0$, but the horizontal space between a local maximum and the nearest local minimum shrinks as $|c|$ approaches 1.

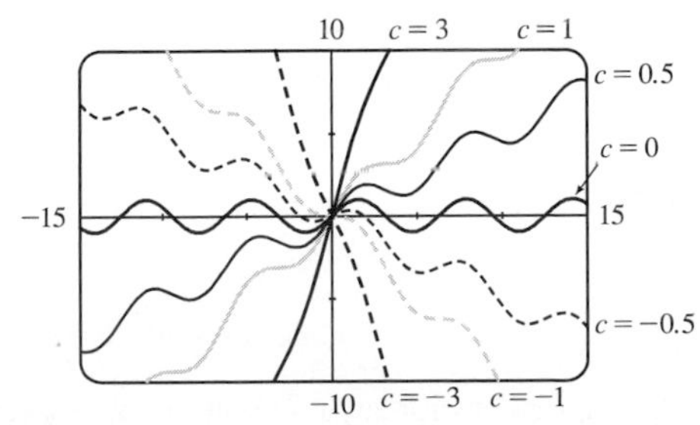

36. For $f(t) = C(e^{-at} - e^{-bt})$, C affects only vertical stretching, so we let $C = 1$. From the first figure, we notice that the graphs all pass through the origin, approach the t-axis as t increases, and approach $-\infty$ as $t \to -\infty$. Next we let $a = 2$ and produce the second figure.

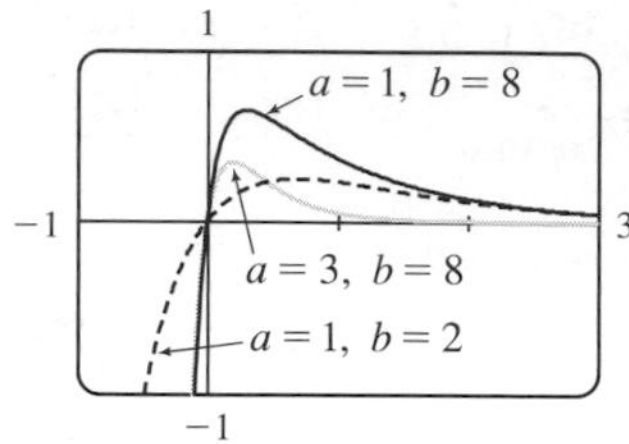

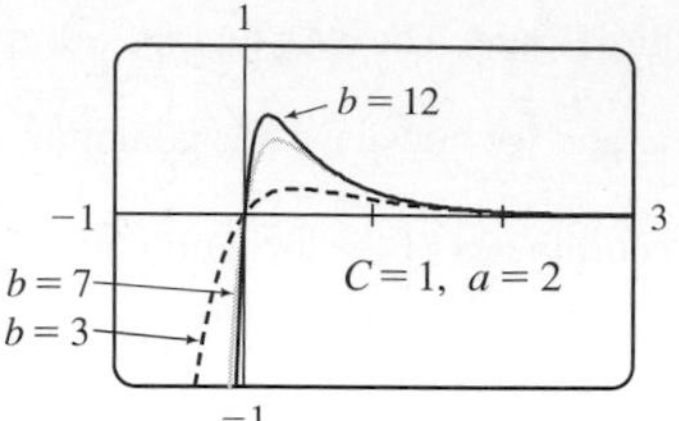

Here, as b increases, the slope of the tangent at the origin increases and the local maximum value increases.

$f(t) = e^{-2t} - e^{-bt} \Rightarrow f'(t) = be^{-bt} - 2e^{-2t}$. $f'(0) = b - 2$, which increases as b increases.

$f'(t) = 0 \Rightarrow be^{-bt} = 2e^{-2t} \Rightarrow \dfrac{b}{2} = e^{(b-2)t} \Rightarrow \ln\dfrac{b}{2} = (b-2)t \Rightarrow t = t_1 = \dfrac{\ln b - \ln 2}{b - 2}$, which decreases as b increases (the maximum is getting closer to the y-axis). $f(t_1) = \dfrac{(b-2)2^{2/(b-2)}}{b^{1+2/(b-2)}}$. We can show that this value increases as b increases by considering it to be a function of b and graphing its derivative with respect to b, which is always positive.

37. If $c < 0$, then $\displaystyle\lim_{x\to-\infty} f(x) = \lim_{x\to-\infty} xe^{-cx} = \lim_{x\to-\infty} \frac{x}{e^{cx}} \overset{\text{H}}{=} \lim_{x\to-\infty} \frac{1}{ce^{cx}} = 0$, and $\displaystyle\lim_{x\to\infty} f(x) = \infty$.

If $c > 0$, then $\displaystyle\lim_{x\to-\infty} f(x) = -\infty$, and $\displaystyle\lim_{x\to\infty} f(x) \overset{\text{H}}{=} \lim_{x\to\infty} \frac{1}{ce^{cx}} = 0$.

If $c = 0$, then $f(x) = x$, so $\displaystyle\lim_{x\to\pm\infty} f(x) = \pm\infty$, respectively.

So we see that $c = 0$ is a transitional value. We now exclude the case $c = 0$, since we know how the function behaves in that case. To find the maxima and minima of f, we differentiate: $f(x) = xe^{-cx} \Rightarrow$

$f'(x) = x(-ce^{-cx}) + e^{-cx} = (1 - cx)e^{-cx}$. This is 0 when $1 - cx = 0 \Leftrightarrow x = 1/c$. If $c < 0$ then this represents a minimum value of $f(1/c) = 1/(ce)$, since $f'(x)$ changes from negative to positive at $x = 1/c$; and if $c > 0$, it represents a maximum value. As $|c|$ increases, the maximum or minimum point gets closer to the origin. To find the inflection points, we differentiate again: $f'(x) = e^{-cx}(1 - cx) \Rightarrow$

$f''(x) = e^{-cx}(-c) + (1 - cx)(-ce^{-cx}) = (cx - 2)ce^{-cx}$. This changes sign when $cx - 2 = 0 \Leftrightarrow x = 2/c$. So as $|c|$ increases, the points of inflection get closer to the origin.

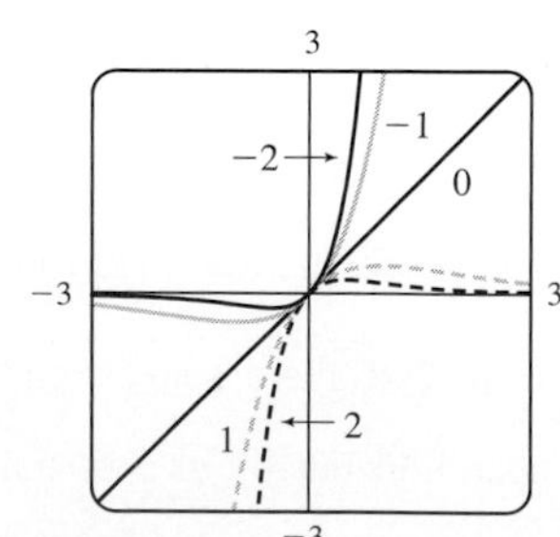

38. For $c = 0$, there is no inflection point; the curve is CU everywhere. If c increases, the curve simply becomes steeper, and there are still no inflection points. If c starts at 0 and decreases, a slight upward bulge appears near $x = 0$, so that there are two inflection points for any $c < 0$. This can be seen algebraically by calculating the second derivative:

$f(x) = x^4 + cx^2 + x \;\Rightarrow\; f'(x) = 4x^3 + 2cx + 1 \;\Rightarrow\; f''(x) = 12x^2 + 2c$. Thus, $f''(x) > 0$ when $c > 0$. For $c < 0$, there are inflection points when $x = \pm\sqrt{-\frac{1}{6}c}$. For $c = 0$, the graph has one critical number, at the absolute minimum somewhere around $x = -0.6$. As c increases, the number of critical points does not change. If c instead decreases from 0, we see that the graph eventually sprouts another local minimum, to the right of the origin, somewhere between $x = 1$ and $x = 2$. Consequently, there is also a maximum near $x = 0$.

After a bit of experimentation, we find that at $c = -1.5$, there appear to be two critical numbers: the absolute minimum at about $x = -1$, and a horizontal tangent with no extremum at about $x = 0.5$. For any c smaller than this there will be 3 critical points, as shown in the graphs with $c = -3$ and with $c = -5$.

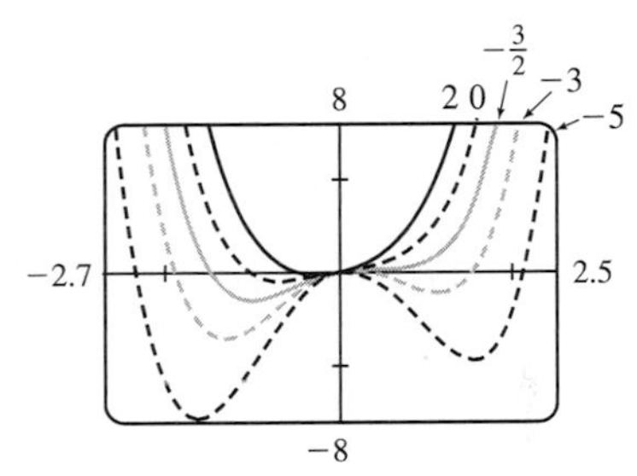

To prove this algebraically, we calculate $f'(x) = 4x^3 + 2cx + 1$. Now if we substitute our value of $c = -1.5$, the formula for $f'(x)$ becomes $4x^3 - 3x + 1 = (x+1)(2x-1)^2$. This has a double root at $x = \frac{1}{2}$, indicating that the function has two critical points: $x = -1$ and $x = \frac{1}{2}$, just as we had guessed from the graph.

39. (a) $f(x) = cx^4 - 2x^2 + 1$. For $c = 0$, $f(x) = -2x^2 + 1$, a parabola whose vertex, $(0, 1)$, is the absolute maximum. For $c > 0$, $f(x) = cx^4 - 2x^2 + 1$ opens upward with two minimum points. As $c \to 0$, the minimum points spread apart and move downward; they are below the x-axis for $0 < c < 1$ and above for $c > 1$. For $c < 0$, the graph opens downward, and has an absolute maximum at $x = 0$ and no local minimum.

(b) $f'(x) = 4cx^3 - 4x = 4cx(x^2 - 1/c)$ $[c \neq 0]$. If $c \leq 0$, 0 is the only critical number. $f''(x) = 12cx^2 - 4$, so $f''(0) = -4$ and there is a local maximum at $(0, f(0)) = (0, 1)$, which lies on $y = 1 - x^2$. If $c > 0$, the critical numbers are 0 and $\pm 1/\sqrt{c}$. As before, there is a local maximum at $(0, f(0)) = (0, 1)$, which lies on $y = 1 - x^2$. $f''\left(\pm 1/\sqrt{c}\right) = 12 - 4 = 8 > 0$, so there is a local minimum at $x = \pm 1/\sqrt{c}$. Here $f\left(\pm 1/\sqrt{c}\right) = c(1/c^2) - 2/c + 1 = -1/c + 1$. But $\left(\pm 1/\sqrt{c}, -1/c + 1\right)$ lies on $y = 1 - x^2$ since $1 - \left(\pm 1/\sqrt{c}\right)^2 = 1 - 1/c$.

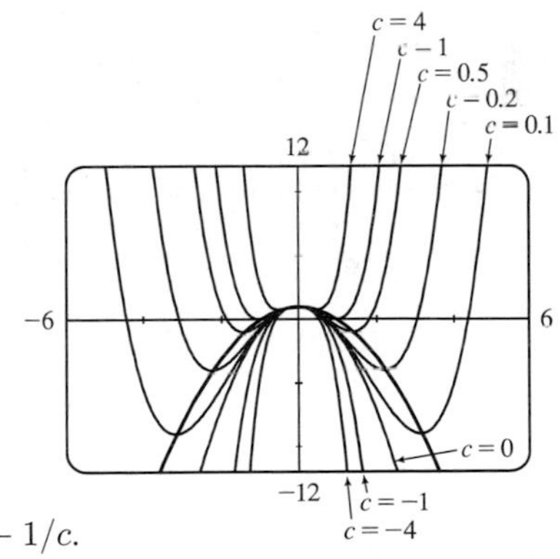

40. (a) $f(x) = 2x^3 + cx^2 + 2x \;\Rightarrow\; f'(x) = 6x^2 + 2cx + 2 = 2(3x^2 + cx + 1)$. $f'(x) = 0 \;\Leftrightarrow\; x = \dfrac{-c \pm \sqrt{c^2 - 12}}{6}$.

So f has critical points $\;\Leftrightarrow\; c^2 - 12 \geq 0 \;\Leftrightarrow\; |c| \geq 2\sqrt{3}$. For $c = \pm 2\sqrt{3}$, $f'(x) \geq 0$ on $(-\infty, \infty)$, so f' does not change signs at $-c/6$, and there is no extremum. If $c^2 - 12 > 0$, then f' changes from positive to negative at $x = \dfrac{-c - \sqrt{c^2 - 12}}{6}$ and from negative to positive at $x = \dfrac{-c + \sqrt{c^2 - 12}}{6}$. So f has a local maximum at $x = \dfrac{-c - \sqrt{c^2 - 12}}{6}$ and a local minimum at $x = \dfrac{-c + \sqrt{c^2 - 12}}{6}$.

(b) Let x_0 be a critical number for $f(x)$. Then $f'(x_0) = 0 \Rightarrow$

$3x_0^2 + cx_0 + 1 = 0 \Leftrightarrow c = \dfrac{-1 - 3x_0^2}{x_0}$. Now

$$f(x_0) = 2x_0^3 + cx_0^2 + 2x_0 = 2x_0^3 + x_0^2\left(\frac{-1 - 3x_0^2}{x_0}\right) + 2x_0$$
$$= 2x_0^3 - x_0 - 3x_0^3 + 2x_0 = x_0 - x_0^3$$

So the point is $(x_0, y_0) = \left(x_0, x_0 - x_0^3\right)$; that is, the point lies on the curve $y = x - x^3$.

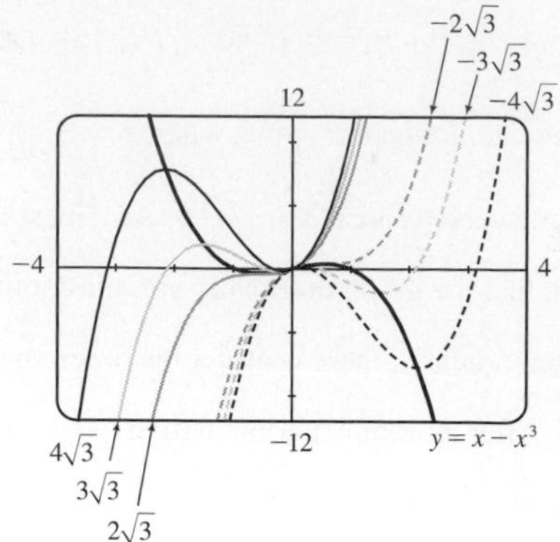

4.7 Optimization Problems

1. (a)

First Number	Second Number	Product
1	22	22
2	21	42
3	20	60
4	19	76
5	18	90
6	17	102
7	16	112
8	15	120
9	14	126
10	13	130
11	12	132

We needn't consider pairs where the first number is larger than the second, since we can just interchange the numbers in such cases. The answer appears to be 11 and 12, but we have considered only integers in the table.

(b) Call the two numbers x and y. Then $x + y = 23$, so $y = 23 - x$. Call the product P. Then $P = xy = x(23 - x) = 23x - x^2$, so we wish to maximize the function $P(x) = 23x - x^2$. Since $P'(x) = 23 - 2x$, we see that $P'(x) = 0 \Leftrightarrow x = \frac{23}{2} = 11.5$. Thus, the maximum value of P is $P(11.5) = (11.5)^2 = 132.25$ and it occurs when $x = y = 11.5$.

Or: Note that $P''(x) = -2 < 0$ for all x, so P is everywhere concave downward and the local maximum at $x = 11.5$ must be an absolute maximum.

2. The two numbers are $x + 100$ and x. Minimize $f(x) = (x + 100)x = x^2 + 100x$. $f'(x) = 2x + 100 = 0 \Rightarrow x = -50$. Since $f''(x) = 2 > 0$, there is an absolute minimum at $x = -50$. The two numbers are 50 and -50.

3. The two numbers are x and $\dfrac{100}{x}$, where $x > 0$. Minimize $f(x) = x + \dfrac{100}{x}$. $f'(x) = 1 - \dfrac{100}{x^2} = \dfrac{x^2 - 100}{x^2}$. The critical number is $x = 10$. Since $f'(x) < 0$ for $0 < x < 10$ and $f'(x) > 0$ for $x > 10$, there is an absolute minimum at $x = 10$. The numbers are 10 and 10.

4. Call the two numbers x and y. Then $x + y = 16$, so $y = 16 - x$. Call the sum of their squares S. Then

$S = x^2 + y^2 = x^2 + (16 - x)^2 \quad\Rightarrow\quad S' = 2x + 2(16 - x)(-1) = 2x - 32 + 2x = 4x - 32.$ $S' = 0 \quad\Rightarrow\quad x = 8.$

Since $S'(x) < 0$ for $0 < x < 8$ and $S'(x) > 0$ for $x > 8$, there is an absolute minimum at $x = 8$. Thus, $y = 16 - 8 = 8$ and $S = 8^2 + 8^2 = 128$.

5. Let the vertical distance be given by $v(x) = (x + 2) - x^2$, $-1 \le x \le 2$.

$v'(x) = 1 - 2x = 0 \quad\Leftrightarrow\quad x = \frac{1}{2}$. $v(-1) = 0$, $v\left(\frac{1}{2}\right) = \frac{9}{4}$, and $v(2) = 0$, so there is an absolute maximum at $x = \frac{1}{2}$. The maximum distance is $v\left(\frac{1}{2}\right) = \frac{1}{2} + 2 - \frac{1}{4} = \frac{9}{4}$.

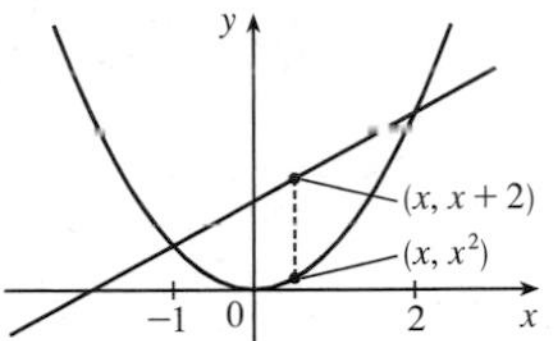

6. Let the vertical distance be given by

$v(x) = (x^2 + 1) - (x - x^2) = 2x^2 - x + 1.$ $v'(x) = 4x - 1 = 0 \quad\Leftrightarrow$

$x = \frac{1}{4}$. $v'(x) < 0$ for $x < \frac{1}{4}$ and $v'(x) > 0$ for $x > \frac{1}{4}$, so there is an absolute minimum at $x = \frac{1}{4}$. The minimum distance is $v\left(\frac{1}{4}\right) = \frac{1}{8} - \frac{1}{4} + 1 = \frac{7}{8}$.

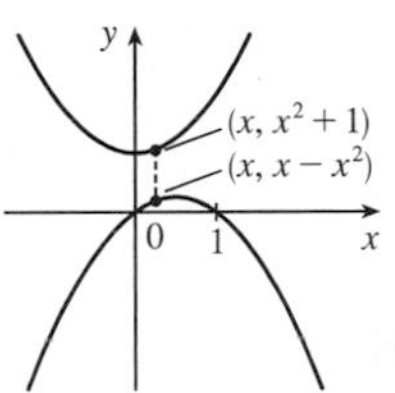

7. If the rectangle has dimensions x and y, then its perimeter is $2x + 2y = 100$ m, so $y = 50 - x$. Thus, the area is $A = xy = x(50 - x)$. We wish to maximize the function $A(x) = x(50 - x) = 50x - x^2$, where $0 < x < 50$. Since $A'(x) = 50 - 2x = -2(x - 25)$, $A'(x) > 0$ for $0 < x < 25$ and $A'(x) < 0$ for $25 < x < 50$. Thus, A has an absolute maximum at $x = 25$, and $A(25) = 25^2 = 625\text{ m}^2$. The dimensions of the rectangle that maximize its area are $x = y = 25$ m. (The rectangle is a square.)

8. If the rectangle has dimensions x and y, then its area is $xy = 1000\text{ m}^2$, so $y = 1000/x$. The perimeter $P = 2x + 2y = 2x + 2000/x$. We wish to minimize the function $P(x) = 2x + 2000/x$ for $x > 0$.

$P'(x) = 2 - 2000/x^2 = (2/x^2)(x^2 - 1000)$, so the only critical number in the domain of P is $x = \sqrt{1000}$.

$P''(x) = 4000/x^3 > 0$, so P is concave upward throughout its domain and $P\left(\sqrt{1000}\,\right) = 4\sqrt{1000}$ is an absolute minimum value. The dimensions of the rectangle with minimal perimeter are $x = y = \sqrt{1000} = 10\sqrt{10}$ m. (The rectangle is a square.)

9. We need to maximize Y for $N \ge 0$. $Y(N) = \dfrac{kN}{1 + N^2} \quad\Rightarrow$

$Y'(N) = \dfrac{(1 + N^2)k - kN(2N)}{(1 + N^2)^2} = \dfrac{k(1 - N^2)}{(1 + N^2)^2} = \dfrac{k(1 + N)(1 - N)}{(1 + N^2)^2}$. $Y'(N) > 0$ for $0 < N < 1$ and $Y'(N) < 0$ for $N > 1$. Thus, Y has an absolute maximum of $Y(1) = \frac{1}{2}k$ at $N = 1$.

10. We need to maximize P for $I \ge 0$. $P(I) = \dfrac{100I}{I^2 + I + 4} \quad\Rightarrow$

$$P'(I) = \frac{(I^2 + I + 4)(100) - 100I(2I + 1)}{(I^2 + I + 4)^2} = \frac{100(I^2 + I + 4 - 2I^2 - I)}{(I^2 + I + 4)^2} = \frac{-100(I^2 - 4)}{(I^2 + I + 4)^2} = \frac{-100(I + 2)(I - 2)}{(I^2 + I + 4)^2}.$$

$P'(I) > 0$ for $0 < I < 2$ and $P'(I) < 0$ for $I > 2$. Thus, P has an absolute maximum of $P(2) = 20$ at $I = 2$.

11. (a)

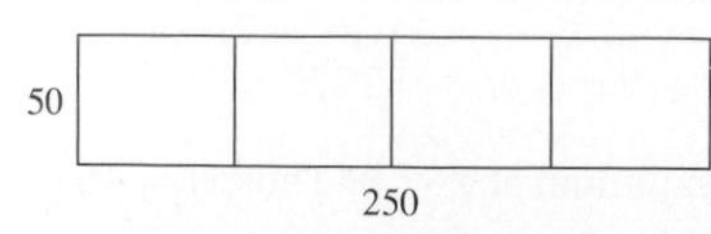

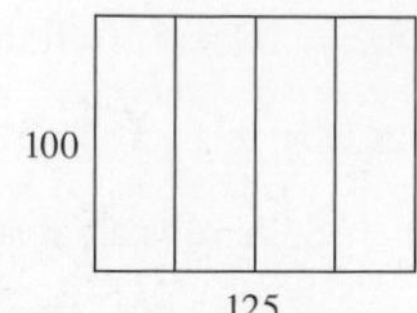

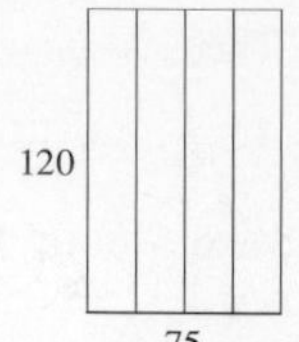

The areas of the three figures are 12,500, 12,500, and 9000 ft^2. There appears to be a maximum area of at least 12,500 ft^2.

(b) Let x denote the length of each of two sides and three dividers.

Let y denote the length of the other two sides.

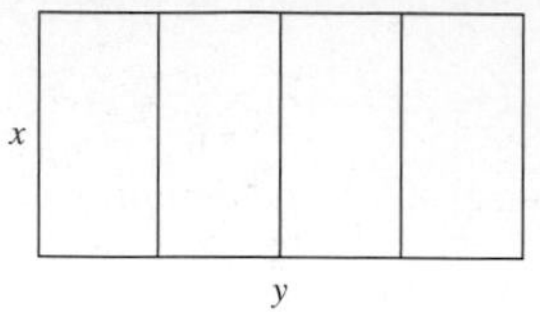

(c) Area $A = \text{length} \times \text{width} = y \cdot x$

(d) Length of fencing $= 750 \quad \Rightarrow \quad 5x + 2y = 750$

(e) $5x + 2y = 750 \quad \Rightarrow \quad y = 375 - \frac{5}{2}x \quad \Rightarrow \quad A(x) = \left(375 - \frac{5}{2}x\right)x = 375x - \frac{5}{2}x^2$

(f) $A'(x) = 375 - 5x = 0 \quad \Rightarrow \quad x = 75$. Since $A''(x) = -5 < 0$ there is an absolute maximum when $x = 75$. Then $y = \frac{375}{2} = 187.5$. The largest area is $75\left(\frac{375}{2}\right) = 14{,}062.5$ ft^2. These values of x and y are between the values in the first and second figures in part (a). Our original estimate was low.

12. (a)

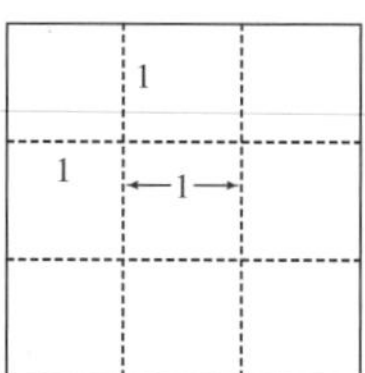

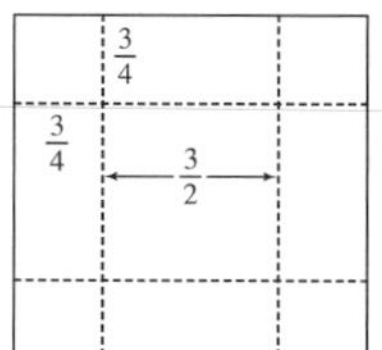

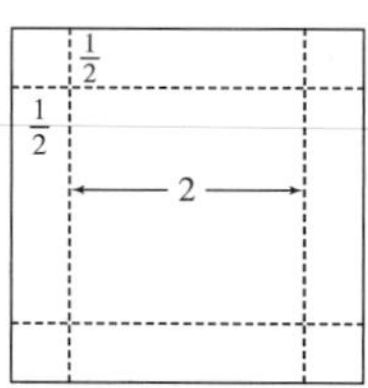

The volumes of the resulting boxes are 1, 1.6875, and 2 ft^3. There appears to be a maximum volume of at least 2 ft^3.

(b) Let x denote the length of the side of the square being cut out. Let y denote the length of the base.

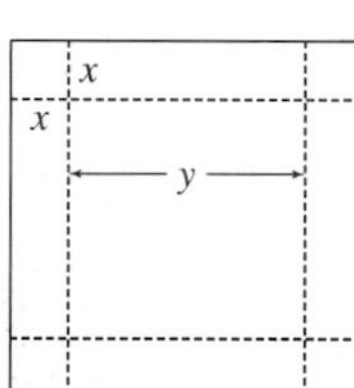

(c) Volume $V = \text{length} \times \text{width} \times \text{height} \quad \Rightarrow \quad V = y \cdot y \cdot x = xy^2$

(d) Length of cardboard $= 3 \quad \Rightarrow \quad x + y + x = 3 \quad \Rightarrow \quad y + 2x = 3$

(e) $y + 2x = 3 \quad \Rightarrow \quad y = 3 - 2x \quad \Rightarrow \quad V(x) = x(3 - 2x)^2$

(f) $V(x) = x(3 - 2x)^2 \quad \Rightarrow$

$V'(x) = x \cdot 2(3 - 2x)(-2) + (3 - 2x)^2 \cdot 1 = (3 - 2x)[-4x + (3 - 2x)] = (3 - 2x)(-6x + 3)$,

so the critical numbers are $x = \frac{3}{2}$ and $x = \frac{1}{2}$. Now $0 \le x \le \frac{3}{2}$ and $V(0) = V\left(\frac{3}{2}\right) = 0$, so the maximum is $V\left(\frac{1}{2}\right) = \left(\frac{1}{2}\right)(2)^2 = 2$ ft^3, which is the value found from our third figure in part (a).

13.

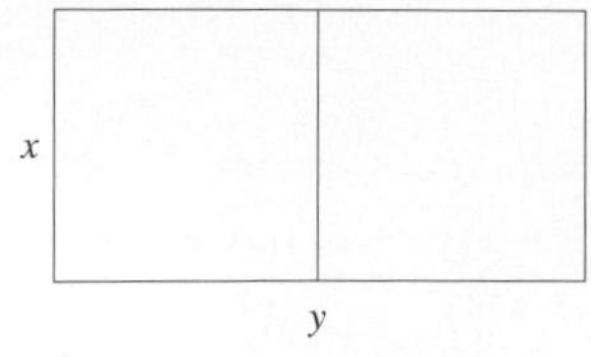

$xy = 1.5 \times 10^6$, so $y = 1.5 \times 10^6/x$. Minimize the amount of fencing, which is $3x + 2y = 3x + 2(1.5 \times 10^6/x) = 3x + 3 \times 10^6/x = F(x)$.

$F'(x) = 3 - 3 \times 10^6/x^2 = 3(x^2 - 10^6)/x^2$. The critical number is $x = 10^3$ and $F'(x) < 0$ for $0 < x < 10^3$ and $F'(x) > 0$ if $x > 10^3$, so the absolute minimum occurs when $x = 10^3$ and $y = 1.5 \times 10^3$.

The field should be 1000 feet by 1500 feet with the middle fence parallel to the short side of the field.

14. Let b be the length of the base of the box and h the height. The volume is $32{,}000 = b^2h \quad\Rightarrow\quad h = 32{,}000/b^2$.

The surface area of the open box is $S = b^2 + 4hb = b^2 + 4(32{,}000/b^2)b = b^2 + 4(32{,}000)/b$.

So $S'(b) = 2b - 4(32{,}000)/b^2 = 2\left(b^3 - 64{,}000\right)/b^2 = 0 \quad\Leftrightarrow\quad b = \sqrt[3]{64{,}000} = 40$. This gives an absolute minimum since $S'(b) < 0$ if $0 < b < 40$ and $S'(b) > 0$ if $b > 40$. The box should be $40 \times 40 \times 20$.

15. Let b be the length of the base of the box and h the height. The surface area is $1200 = b^2 + 4hb \quad\Rightarrow\quad h = (1200 - b^2)/(4b)$.

The volume is $V = b^2h = b^2(1200 - b^2)/4b = 300b - b^3/4 \quad\Rightarrow\quad V'(b) = 300 - \frac{3}{4}b^2$.

$V'(b) = 0 \quad\Rightarrow\quad 300 = \frac{3}{4}b^2 \quad\Rightarrow\quad b^2 = 400 \quad\Rightarrow\quad b = \sqrt{400} = 20$. Since $V'(b) > 0$ for $0 < b < 20$ and $V'(b) < 0$ for $b > 20$, there is an absolute maximum when $b = 20$ by the First Derivative Test for Absolute Extreme Values (see page 328). If $b = 20$, then $h = (1200 - 20^2)/(4 \cdot 20) = 10$, so the largest possible volume is $b^2h = (20)^2(10) = 4000\text{ cm}^3$.

16. 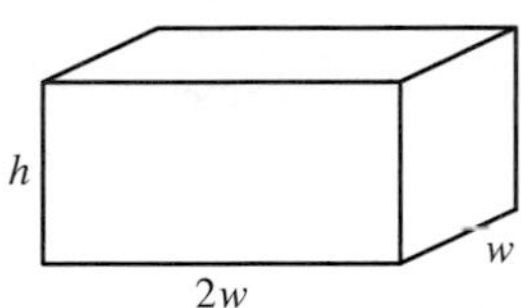

$V = lwh \quad\Rightarrow\quad 10 = (2w)(w)h = 2w^2h$, so $h = 5/w^2$.

The cost is $10(2w^2) + 6[2(2wh) + 2(hw)] = 20w^2 + 36wh$, so

$C(w) = 20w^2 + 36w(5/w^2) = 20w^2 + 180/w$.

$C'(w) = 40w - 180/w^2 = (40w^3 - 180)/w^2 = 40\left(w^3 - \frac{9}{2}\right)/w^2 \quad\Rightarrow\quad w = \sqrt[3]{\frac{9}{2}}$ is the critical number. There is an absolute minimum for C when $w = \sqrt[3]{\frac{9}{2}}$ since $C'(w) < 0$ for $0 < w < \sqrt[3]{\frac{9}{2}}$ and $C'(w) > 0$ for $w > \sqrt[3]{\frac{9}{2}}$. The minimum cost is $C\left(\sqrt[3]{\frac{9}{2}}\right) = 20\left(\sqrt[3]{\frac{9}{2}}\right)^2 + \dfrac{180}{\sqrt[3]{9/2}} \approx \$163.54$.

17. 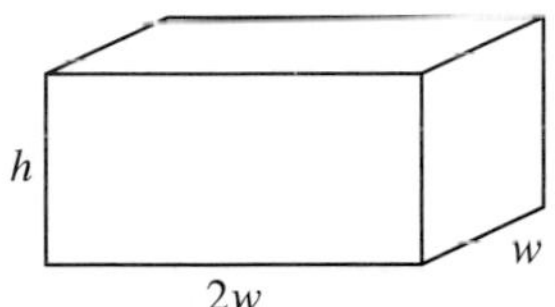

$V = lwh \quad\Rightarrow\quad 10 = (2w)(w)h = 2w^2h$, so $h = 5/w^2$.

The cost is $10(2w^2) + 6[2(2wh) + 2(hw)] + 6(2w^2) = 32w^2 + 36wh$, so

$C(w) = 32w^2 + 36w(5/w^2) = 32w^2 + 180/w$.

$C'(w) = 64w - 180/w^2 = (64w^3 - 180)/w^2 = 4(16w^3 - 45)/w^2 \quad\Rightarrow\quad w = \sqrt[3]{\frac{45}{16}}$ is the critical number. There is an absolute minimum for C when $w = \sqrt[3]{\frac{45}{16}}$ since $C'(w) < 0$ for $0 < w < \sqrt[3]{\frac{45}{16}}$ and $C'(w) > 0$ for $w > \sqrt[3]{\frac{45}{16}}$. The minimum cost is $C\left(\sqrt[3]{\frac{45}{16}}\right) = 32\left(\sqrt[3]{\frac{45}{16}}\right)^2 + \dfrac{180}{\sqrt[3]{45/16}} \approx \$191.28$.

18. 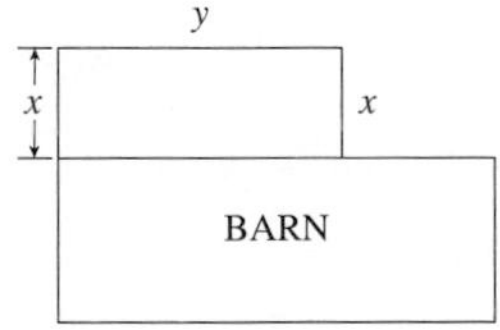

See the figure. The fencing cost \$20 per linear foot to install and the cost of the fencing on the west side will be split with the neighbor, so the farmer's cost C will be $C = \frac{1}{2}(20x) + 20y + 20x = 20y + 30x$. The area A will be maximized when $C = 5000$, so $5000 = 20y + 30x \quad\Leftrightarrow\quad 20y = 5000 - 30x \quad\Leftrightarrow$

$y = 250 - \frac{3}{2}x$. Now $A = xy = x\left(250 - \frac{3}{2}x\right) = 250x - \frac{3}{2}x^2 \quad\Rightarrow\quad A' = 250 - 3x$. $\quad A' = 0 \quad\Leftrightarrow\quad x = \frac{250}{3}$ and since $A'' = -3 < 0$, we have a maximum for A when $x = \frac{250}{3}$ ft and $y = 250 - \frac{3}{2}\left(\frac{250}{3}\right) = 125$ ft. [The maximum area is $125\left(\frac{250}{3}\right) = 10{,}416.\overline{6}\text{ ft}^2$.]

19. 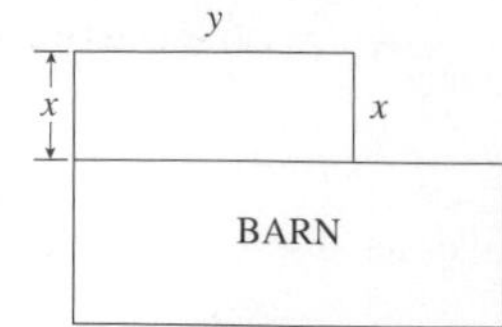

See the figure. The fencing cost \$20 per linear foot to install and the cost of the fencing on the west side will be split with the neighbor, so the farmer's cost C will be $C = \frac{1}{2}(20x) + 20y + 20x = 20y + 30x$. The area A to be enclosed is 8000 ft^2, so $A = xy = 8000 \;\Rightarrow\; y = \dfrac{8000}{x}$.

Now $C = 20y + 30x = 20\left(\dfrac{8000}{x}\right) + 30x = \dfrac{160{,}000}{x} + 30x \;\Rightarrow\; C' = -\dfrac{160{,}000}{x^2} + 30.$ $C' = 0 \;\Leftrightarrow$ $30 = \dfrac{160{,}000}{x^2} \;\Leftrightarrow\; x^2 = \dfrac{16{,}000}{3} \;\Rightarrow\; x = \sqrt{\dfrac{16{,}000}{3}} = 40\sqrt{\dfrac{10}{3}} = \dfrac{40}{3}\sqrt{30}$. Since $C'' = \dfrac{320{,}000}{x^3} > 0$ [for $x > 0$], we have a minimum for C when $x = \dfrac{40}{3}\sqrt{30}$ ft and $y = \dfrac{8000}{x} = \dfrac{8000}{40} \cdot \dfrac{3}{\sqrt{30}} \cdot \dfrac{\sqrt{30}}{\sqrt{30}} = 20\sqrt{30}$ ft. [The minimum cost is $20(20\sqrt{30}\,) + 30\left(\frac{40}{3}\sqrt{30}\right) = 800\sqrt{30} \approx \$4381.78$.]

20. (a) Let the rectangle have sides x and y and area A, so $A = xy$ or $y = A/x$. The problem is to minimize the perimeter $= 2x + 2y = 2x + 2A/x = P(x)$. Now $P'(x) = 2 - 2A/x^2 = 2(x^2 - A)/x^2$. So the critical number is $x = \sqrt{A}$. Since $P'(x) < 0$ for $0 < x < \sqrt{A}$ and $P'(x) > 0$ for $x > \sqrt{A}$, there is an absolute minimum at $x = \sqrt{A}$. The sides of the rectangle are $\sqrt{A}$ and $A/\sqrt{A} = \sqrt{A}$, so the rectangle is a square.

(b) Let p be the perimeter and x and y the lengths of the sides, so $p = 2x + 2y \;\Rightarrow\; 2y = p - 2x \;\Rightarrow\; y = \frac{1}{2}p - x$. The area is $A(x) = x\left(\frac{1}{2}p - x\right) = \frac{1}{2}px - x^2$. Now $A'(x) = 0 \;\Rightarrow\; \frac{1}{2}p - 2x = 0 \;\Rightarrow\; 2x = \frac{1}{2}p \;\Rightarrow\; x = \frac{1}{4}p$. Since $A''(x) = -2 < 0$, there is an absolute maximum for A when $x = \frac{1}{4}p$ by the Second Derivative Test. The sides of the rectangle are $\frac{1}{4}p$ and $\frac{1}{2}p - \frac{1}{4}p = \frac{1}{4}p$, so the rectangle is a square.

21. The distance d from the origin $(0,0)$ to a point $(x, 2x+3)$ on the line is given by $d = \sqrt{(x-0)^2 + (2x+3-0)^2}$ and the square of the distance is $S = d^2 = x^2 + (2x+3)^2$. $S' = 2x + 2(2x+3)2 = 10x + 12$ and $S' = 0 \;\Leftrightarrow\; x = -\frac{6}{5}$. Now $S'' = 10 > 0$, so we know that S has a minimum at $x = -\frac{6}{5}$. Thus, the y-value is $2\left(-\frac{6}{5}\right) + 3 = \frac{3}{5}$ and the point is $\left(-\frac{6}{5}, \frac{3}{5}\right)$.

22. The distance d from the point $(3,0)$ to a point $(x, \sqrt{x}\,)$ on the curve is given by $d = \sqrt{(x-3)^2 + (\sqrt{x} - 0)^2}$ and the square of the distance is $S = d^2 = (x-3)^2 + x$. $S' = 2(x-3) + 1 = 2x - 5$ and $S' = 0 \;\Leftrightarrow\; x = \frac{5}{2}$. Now $S'' = 2 > 0$, so we know that S has a minimum at $x = \frac{5}{2}$. Thus, the y-value is $\sqrt{\frac{5}{2}}$ and the point is $\left(\frac{5}{2}, \sqrt{\frac{5}{2}}\right)$.

23. 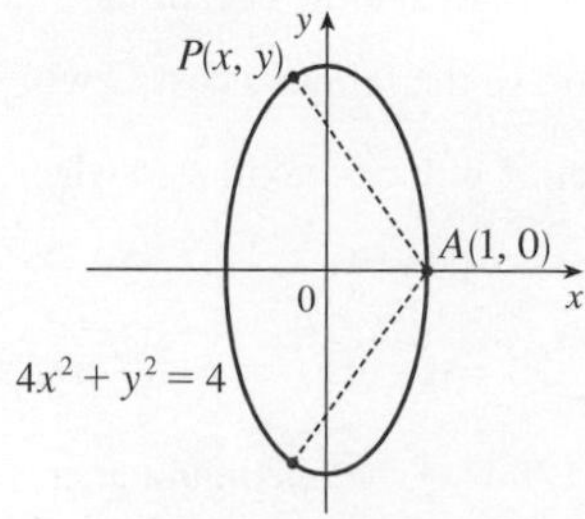

From the figure, we see that there are two points that are farthest away from $A(1,0)$. The distance d from A to an arbitrary point $P(x,y)$ on the ellipse is $d = \sqrt{(x-1)^2 + (y-0)^2}$ and the square of the distance is $S = d^2 = x^2 - 2x + 1 + y^2 = x^2 - 2x + 1 + (4 - 4x^2) = -3x^2 - 2x + 5$. $S' = -6x - 2$ and $S' = 0 \;\Rightarrow\; x = -\frac{1}{3}$. Now $S'' = -6 < 0$, so we know that S has a maximum at $x = -\frac{1}{3}$. Since $-1 \le x \le 1$, $S(-1) = 4$,

$S(-\frac{1}{3}) = \frac{16}{3}$, and $S(1) = 0$, we see that the maximum distance is $\sqrt{\frac{16}{3}}$. The corresponding y-values are

$y = \pm\sqrt{4 - 4(-\frac{1}{3})^2} = \pm\sqrt{\frac{32}{9}} = \pm\frac{4}{3}\sqrt{2} \approx \pm 1.89$. The points are $\left(-\frac{1}{3}, \pm\frac{4}{3}\sqrt{2}\right)$.

24. The distance d from the point $(4, 2)$ to a point $(x, \sin x)$ on the curve is given by $d = \sqrt{(x-4)^2 + (\sin x - 2)^2}$ and the square of the distance is $S = d^2 = (x-4)^2 + (\sin x - 2)^2$. $S' = 2(x-4) + 2(\sin x - 2)\cos x$. Using a calculator, it is clear that S has a minimum between 0 and 5, and from a graph of S', we find that $S' = 0 \Rightarrow x \approx 2.65$, so the point is about $(2.65, 0.47)$.

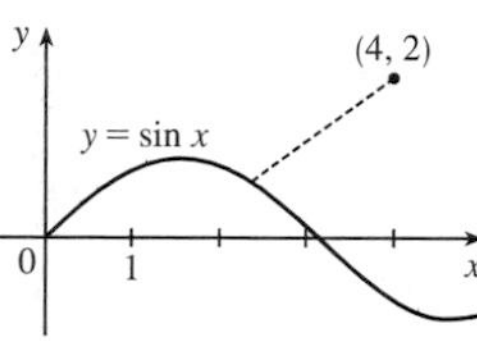

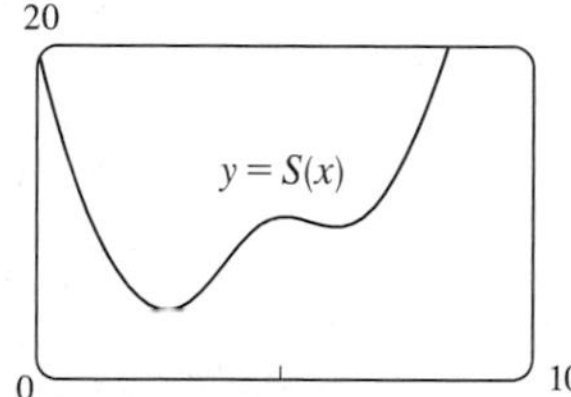

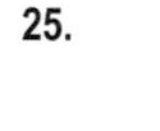

25.

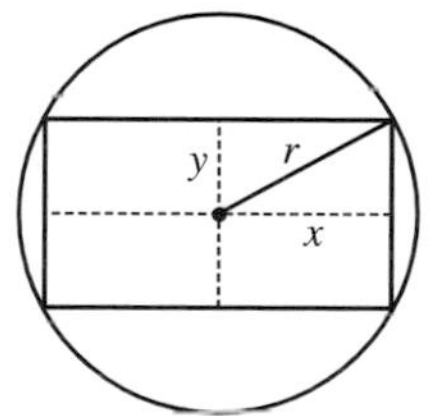

The area of the rectangle is $(2x)(2y) = 4xy$. Also $r^2 = x^2 + y^2$ so $y = \sqrt{r^2 - x^2}$, so the area is $A(x) = 4x\sqrt{r^2 - x^2}$. Now

$A'(x) = 4\left(\sqrt{r^2 - x^2} - \frac{x^2}{\sqrt{r^2 - x^2}}\right) = 4\frac{r^2 - 2x^2}{\sqrt{r^2 - x^2}}$. The critical number is $x = \frac{1}{\sqrt{2}}r$. Clearly this gives a maximum.

$y = \sqrt{r^2 - \left(\frac{1}{\sqrt{2}}r\right)^2} = \sqrt{\frac{1}{2}r^2} = \frac{1}{\sqrt{2}}r = x$, which tells us that the rectangle is a square. The dimensions are $2x = \sqrt{2}\,r$ and $2y = \sqrt{2}\,r$.

26.

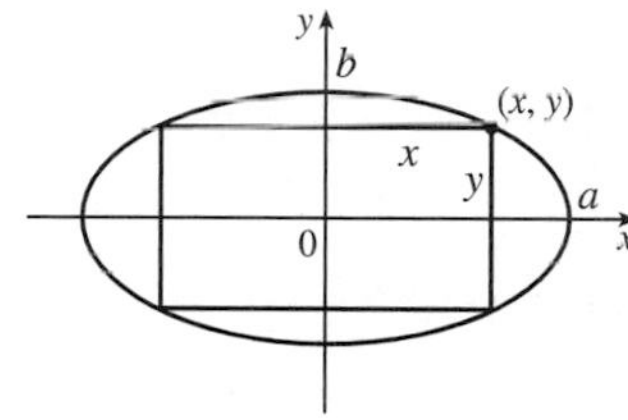

The area of the rectangle is $(2x)(2y) = 4xy$. Now $\frac{x^2}{a^2} + \frac{y^2}{b^2} = 1$ gives $y = \frac{b}{a}\sqrt{a^2 - x^2}$, so we maximize $A(x) = 4\frac{b}{a}x\sqrt{a^2 - x^2}$.

$$A'(x) = \frac{4b}{a}\left[x \cdot \tfrac{1}{2}(a^2 - x^2)^{-1/2}(-2x) + (a^2 - x^2)^{1/2} \cdot 1\right]$$

$$= \frac{4b}{a}(a^2 - x^{2-1/2}[-x^2 + a^2 - x^2] = \frac{4b}{a\sqrt{a^2 - x^2}}[a^2 - 2x^2]$$

So the critical number is $x = \frac{1}{\sqrt{2}}a$, and this clearly gives a maximum. Then $y = \frac{1}{\sqrt{2}}b$, so the maximum area is $4\left(\frac{1}{\sqrt{2}}a\right)\left(\frac{1}{\sqrt{2}}b\right) = 2ab$.

27.

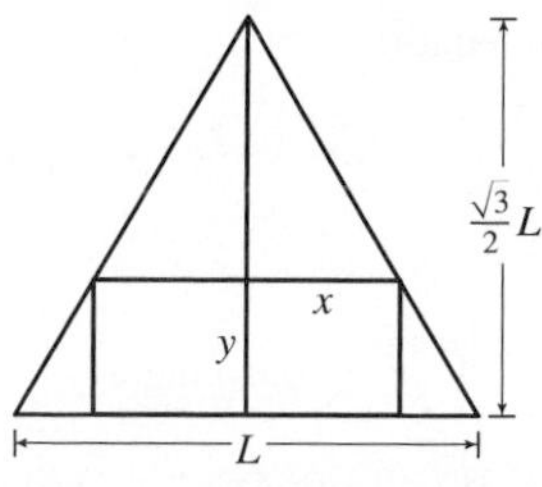

The height h of the equilateral triangle with sides of length L is $\frac{\sqrt{3}}{2}L$, since $h^2 + (L/2)^2 = L^2 \Rightarrow h^2 = L^2 - \frac{1}{4}L^2 = \frac{3}{4}L^2 \Rightarrow$ $h = \frac{\sqrt{3}}{2}L$. Using similar triangles, $\frac{\frac{\sqrt{3}}{2}L - y}{x} = \frac{\frac{\sqrt{3}}{2}L}{L/2} = \sqrt{3} \Rightarrow$ $\sqrt{3}x = \frac{\sqrt{3}}{2}L - y \Rightarrow y = \frac{\sqrt{3}}{2}L - \sqrt{3}x \Rightarrow y = \frac{\sqrt{3}}{2}(L - 2x)$.

[continued]

The area of the inscribed rectangle is $A(x) = (2x)y = \sqrt{3}\,x(L - 2x) = \sqrt{3}\,Lx - 2\sqrt{3}\,x^2$, where $0 \le x \le L/2$. Now $0 = A'(x) = \sqrt{3}\,L - 4\sqrt{3}\,x \;\Rightarrow\; x = \sqrt{3}\,L/(4\sqrt{3}) = L/4$. Since $A(0) = A(L/2) = 0$, the maximum occurs when $x = L/4$, and $y = \frac{\sqrt{3}}{2}L - \frac{\sqrt{3}}{4}L = \frac{\sqrt{3}}{4}L$, so the dimensions are $L/2$ and $\frac{\sqrt{3}}{4}L$.

28.

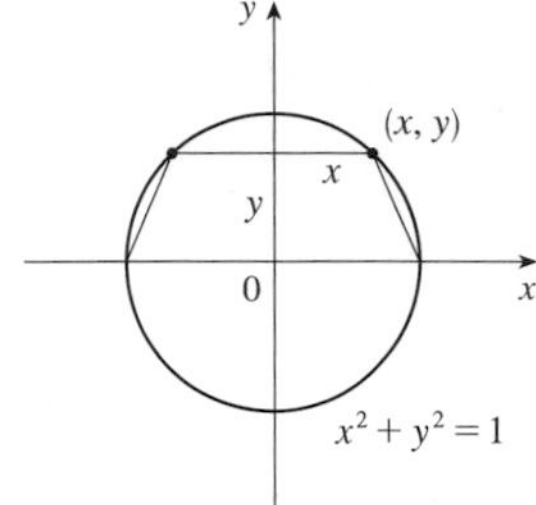

The area A of a trapezoid is given by $A = \frac{1}{2}h(B + b)$. From the diagram, $h = y$, $B = 2$, and $b = 2x$, so $A = \frac{1}{2}y(2 + 2x) = y(1 + x)$. Since it's easier to substitute for y^2, we'll let $T = A^2 = y^2(1 + x)^2 = (1 - x^2)(1 + x)^2$. Now

$$T' = (1 - x^2)2(1 + x) + (1 + x)^2(-2x) = -2(1 + x)[-(1 - x^2) + (1 + x)x]$$
$$= -2(1 + x)(2x^2 + x - 1) = -2(1 + x)(2x - 1)(x + 1)$$

$T' = 0 \;\Leftrightarrow\; x = -1$ or $x = \frac{1}{2}$. $T' > 0$ if $x < \frac{1}{2}$ and $T' < 0$ if $x > \frac{1}{2}$, so we get a maximum at $x = \frac{1}{2}$ [$x = -1$ gives us $A = 0$]. Thus, $y = \sqrt{1 - \left(\frac{1}{2}\right)^2} = \frac{\sqrt{3}}{2}$ and the maximum area is $A = y(1 + x) = \frac{\sqrt{3}}{2}\left(1 + \frac{1}{2}\right) = \frac{3\sqrt{3}}{4}$.

29.

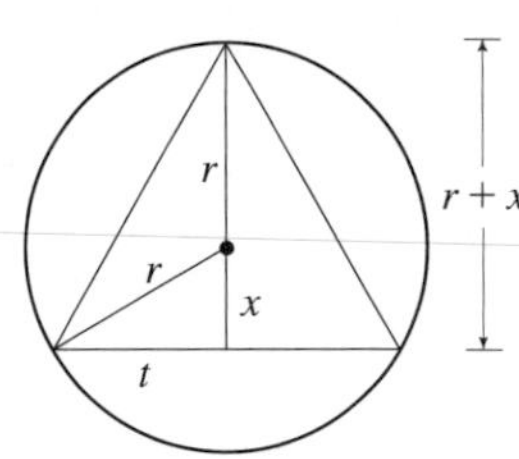

The area of the triangle is

$A(x) = \frac{1}{2}(2t)(r + x) = t(r + x) = \sqrt{r^2 - x^2}\,(r + x)$. Then

$$0 = A'(x) = r\frac{-2x}{2\sqrt{r^2 - x^2}} + \sqrt{r^2 - x^2} + x\frac{-2x}{2\sqrt{r^2 - x^2}}$$
$$= -\frac{x^2 + rx}{\sqrt{r^2 - x^2}} + \sqrt{r^2 - x^2} \quad\Rightarrow$$

$$\frac{x^2 + rx}{\sqrt{r^2 - x^2}} = \sqrt{r^2 - x^2} \;\Rightarrow\; x^2 + rx = r^2 - x^2 \;\Rightarrow\; 0 = 2x^2 + rx - r^2 = (2x - r)(x + r) \;\Rightarrow$$

$x = \frac{1}{2}r$ or $x = -r$. Now $A(r) = 0 = A(-r) \;\Rightarrow\;$ the maximum occurs where $x = \frac{1}{2}r$, so the triangle has height $r + \frac{1}{2}r = \frac{3}{2}r$ and base $2\sqrt{r^2 - \left(\frac{1}{2}r\right)^2} = 2\sqrt{\frac{3}{4}r^2} = \sqrt{3}\,r$.

30.

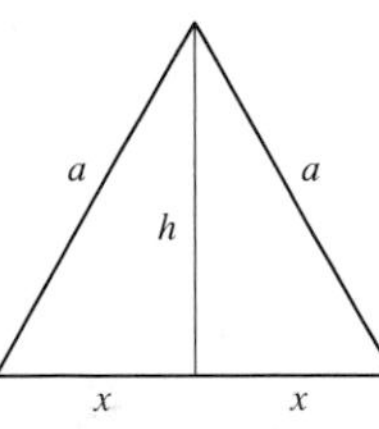

From the figure, we have $x^2 + h^2 = a^2 \;\Rightarrow\; h = \sqrt{a^2 - x^2}$. The area of the isosceles triangle is $A = \frac{1}{2}(2x)h = xh = x\sqrt{a^2 - x^2}$ with $0 \le x \le a$. Now

$$A' = x \cdot \tfrac{1}{2}(a^2 - x^2)^{-1/2}(-2x) + (a^2 - x^2)^{1/2}(1)$$
$$= (a^2 - x^2)^{-1/2}[-x^2 + (a^2 - x^2)] = \frac{a^2 - 2x^2}{\sqrt{a^2 - x^2}}$$

$A' = 0 \;\Leftrightarrow\; x^2 = \frac{1}{2}a^2 \;\Rightarrow\; x = a/\sqrt{2}$. Since $A(0) = 0$, $A(a) = 0$, and $A(a/\sqrt{2}) = (a/\sqrt{2})\sqrt{a^2/2} = \frac{1}{2}a^2$, we see that $x = a/\sqrt{2}$ gives us the maximum area and the length of the base is $2x = 2(a/\sqrt{2}) = \sqrt{2}\,a$. Note that the triangle has sides a, a, and $\sqrt{2}\,a$, which form a *right* triangle, with the right angle between the two sides of equal length.

31.

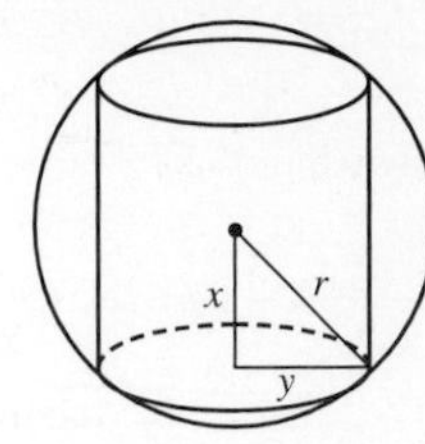

The cylinder has volume $V = \pi y^2(2x)$. Also $x^2 + y^2 = r^2 \;\Rightarrow\; y^2 = r^2 - x^2$, so $V(x) = \pi(r^2 - x^2)(2x) = 2\pi(r^2x - x^3)$, where $0 \le x \le r$.

$V'(x) = 2\pi(r^2 - 3x^2) = 0 \;\Rightarrow\; x = r/\sqrt{3}$. Now $V(0) = V(r) = 0$, so there is a maximum when $x = r/\sqrt{3}$ and $V(r/\sqrt{3}) = \pi(r^2 - r^2/3)(2r/\sqrt{3}) = 4\pi r^3/(3\sqrt{3})$.

32.

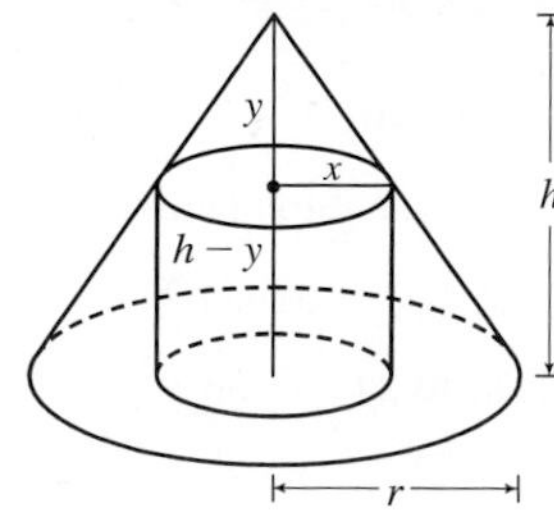

By similar triangles, $y/x = h/r$, so $y = hx/r$. The volume of the cylinder is

$\pi x^2(h - y) = \pi hx^2 - (\pi h/r)x^3 = V(x)$. Now

$V'(x) = 2\pi hx - (3\pi h/r)x^2 = \pi hx(2 - 3x/r)$.

So $V'(x) = 0 \;\Rightarrow\; x = 0$ or $x = \frac{2}{3}r$. The maximum clearly occurs when

$x = \frac{2}{3}r$ and then the volume is

$\pi hx^2 - (\pi h/r)x^3 = \pi hx^2(1 - x/r) = \pi\left(\frac{2}{3}r\right)^2 h\left(1 - \frac{2}{3}\right) = \frac{4}{27}\pi r^2 h$.

33.

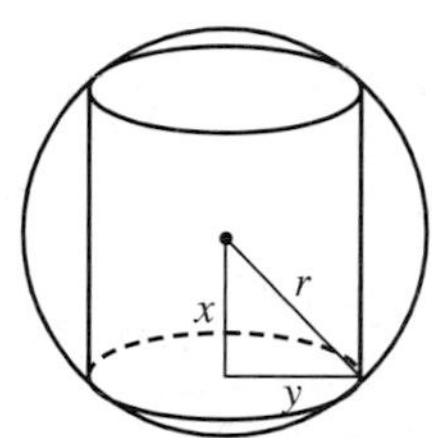

The cylinder has surface area

$$2(\text{area of the base}) + (\text{lateral surface area}) = 2\pi(\text{radius})^2 + 2\pi(\text{radius})(\text{height})$$
$$= 2\pi y^2 + 2\pi y(2x)$$

Now $x^2 + y^2 = r^2 \;\Rightarrow\; y^2 = r^2 - x^2 \;\Rightarrow\; y = \sqrt{r^2 - x^2}$, so the surface area is

$$S(x) = 2\pi(r^2 - x^2) + 4\pi x\sqrt{r^2 - x^2}, \quad 0 \le x \le r$$
$$= 2\pi r^2 - 2\pi x^2 + 4\pi\left(x\sqrt{r^2 - x^2}\right)$$

Thus,
$$S'(x) = 0 - 4\pi x + 4\pi\left[x \cdot \tfrac{1}{2}(r^2 - x^2)^{-1/2}(-2x) + (r^2 - x^2)^{1/2} \cdot 1\right]$$
$$= 4\pi\left[-x - \frac{x^2}{\sqrt{r^2 - x^2}} + \sqrt{r^2 - x^2}\right] = 4\pi \cdot \frac{-x\sqrt{r^2 - x^2} - x^2 + r^2 - x^2}{\sqrt{r^2 - x^2}}$$

$S'(x) = 0 \;\Rightarrow\; x\sqrt{r^2 - x^2} = r^2 - 2x^2 \quad (\star) \;\Rightarrow\; \left(x\sqrt{r^2 - x^2}\right)^2 = (r^2 - 2x^2)^2 \;\Rightarrow$

$x^2(r^2 - x^2) = r^4 - 4r^2x^2 + 4x^4 \;\Rightarrow\; r^2x^2 - x^4 = r^4 - 4r^2x^2 + 4x^4 \;\Rightarrow\; 5x^4 - 5r^2x^2 + r^4 = 0$.

This is a quadratic equation in x^2. By the quadratic formula, $x^2 = \frac{5 \pm \sqrt{5}}{10}r^2$, but we reject the root with the + sign since it doesn't satisfy ($\star$). [The right side is negative and the left side is positive.] So $x = \sqrt{\frac{5 - \sqrt{5}}{10}}\,r$. Since $S(0) = S(r) = 0$, the maximum surface area occurs at the critical number and $x^2 = \frac{5 - \sqrt{5}}{10}r^2 \;\Rightarrow\; y^2 = r^2 - \frac{5 - \sqrt{5}}{10}r^2 = \frac{5 + \sqrt{5}}{10}r^2 \;\Rightarrow$ the surface area is

$$2\pi\left(\tfrac{5+\sqrt{5}}{10}\right)r^2 + 4\pi\sqrt{\tfrac{5-\sqrt{5}}{10}}\sqrt{\tfrac{5+\sqrt{5}}{10}}\,r^2 = \pi r^2\left[2 \cdot \tfrac{5+\sqrt{5}}{10} + 4\tfrac{\sqrt{(5-\sqrt{5})(5+\sqrt{5})}}{10}\right] = \pi r^2\left[\tfrac{5+\sqrt{5}}{5} + \tfrac{2\sqrt{20}}{5}\right]$$
$$= \pi r^2\left[\tfrac{5+\sqrt{5}+2\cdot 2\sqrt{5}}{5}\right] = \pi r^2\left[\tfrac{5+5\sqrt{5}}{5}\right] = \pi r^2\left(1 + \sqrt{5}\right).$$

34.

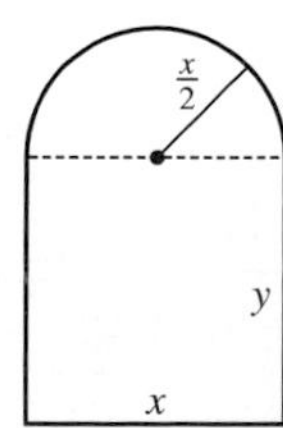

Perimeter $= 30 \;\Rightarrow\; 2y + x + \pi\left(\dfrac{x}{2}\right) = 30 \;\Rightarrow$

$y = \dfrac{1}{2}\left(30 - x - \dfrac{\pi x}{2}\right) = 15 - \dfrac{x}{2} - \dfrac{\pi x}{4}$. The area is the area of the rectangle plus the area of the semicircle, or $xy + \frac{1}{2}\pi\left(\dfrac{x}{2}\right)^2$, so $A(x) = x\left(15 - \dfrac{x}{2} - \dfrac{\pi x}{4}\right) + \frac{1}{8}\pi x^2 = 15x - \frac{1}{2}x^2 - \frac{\pi}{8}x^2$.

$A'(x) = 15 - \left(1 + \frac{\pi}{4}\right)x = 0 \;\Rightarrow\; x = \dfrac{15}{1 + \pi/4} = \dfrac{60}{4 + \pi}$. $A''(x) = -\left(1 + \dfrac{\pi}{4}\right) < 0$, so this gives a maximum.

The dimensions are $x = \dfrac{60}{4 + \pi}$ ft and $y = 15 - \dfrac{30}{4 + \pi} - \dfrac{15\pi}{4 + \pi} = \dfrac{60 + 15\pi - 30 - 15\pi}{4 + \pi} = \dfrac{30}{4 + \pi}$ ft, so the height of the rectangle is half the base.

35.

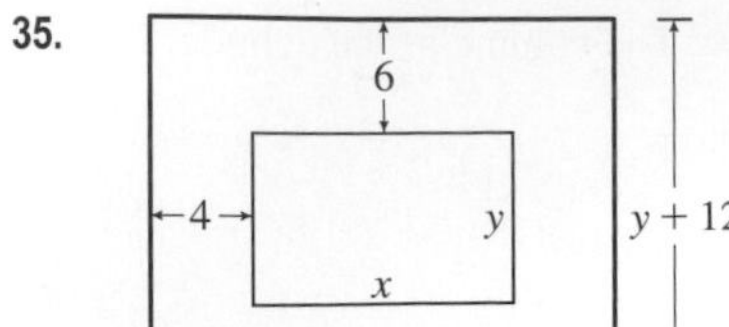

$xy = 384 \;\Rightarrow\; y = 384/x$. Total area is $A(x) = (8 + x)(12 + 384/x) = 12(40 + x + 256/x)$, so $A'(x) = 12(1 - 256/x^2) = 0 \;\Rightarrow\; x = 16$. There is an absolute minimum when $x = 16$ since $A'(x) < 0$ for $0 < x < 16$ and $A'(x) > 0$ for $x > 16$. When $x = 16$, $y = 384/16 = 24$, so the dimensions are 24 cm and 36 cm.

36.

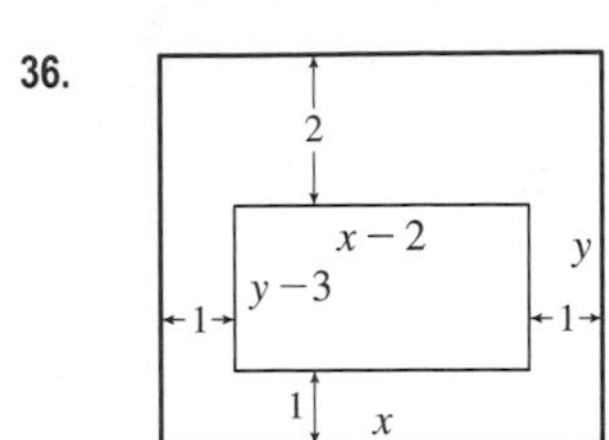

$xy = 180$, so $y = 180/x$. The printed area is $(x - 2)(y - 3) = (x - 2)(180/x - 3) = 186 - 3x - 360/x = A(x)$. $A'(x) = -3 + 360/x^2 = 0$ when $x^2 = 120 \;\Rightarrow\; x = 2\sqrt{30}$. This gives an absolute maximum since $A'(x) > 0$ for $0 < x < 2\sqrt{30}$ and $A'(x) < 0$ for $x > 2\sqrt{30}$. When $x = 2\sqrt{30}$, $y = 180/(2\sqrt{30})$, so the dimensions are $2\sqrt{30}$ in. and $90/\sqrt{30}$ in.

37.

10
x $\quad 10 - x$
$\frac{x}{4}$ $\quad \frac{\sqrt{3}}{2}\left(\frac{10-x}{3}\right)$
$\frac{10-x}{3}$

Let x be the length of the wire used for the square. The total area is

$$A(x) = \left(\frac{x}{4}\right)^2 + \frac{1}{2}\left(\frac{10-x}{3}\right)\frac{\sqrt{3}}{2}\left(\frac{10-x}{3}\right)$$
$$= \tfrac{1}{16}x^2 + \tfrac{\sqrt{3}}{36}(10 - x)^2,\; 0 \le x \le 10$$

$$A'(x) = \tfrac{1}{8}x - \tfrac{\sqrt{3}}{18}(10 - x) = 0 \;\Leftrightarrow\; \tfrac{9}{72}x + \tfrac{4\sqrt{3}}{72}x - \tfrac{40\sqrt{3}}{72} = 0 \;\Leftrightarrow\; x = \frac{40\sqrt{3}}{9 + 4\sqrt{3}}.$$

Now $A(0) = \left(\frac{\sqrt{3}}{36}\right)100 \approx 4.81$, $A(10) = \frac{100}{16} = 6.25$ and $A\left(\frac{40\sqrt{3}}{9+4\sqrt{3}}\right) \approx 2.72$, so

(a) The maximum area occurs when $x = 10$ m, and all the wire is used for the square.

(b) The minimum area occurs when $x = \frac{40\sqrt{3}}{9 + 4\sqrt{3}} \approx 4.35$ m.

38.

10
x $\quad 10 - x$
$\frac{x}{4}$ $\quad r$
$r = \frac{10-x}{2\pi}$

Total area is $A(x) = \left(\frac{x}{4}\right)^2 + \pi\left(\frac{10-x}{2\pi}\right)^2 = \frac{x^2}{16} + \frac{(10-x)^2}{4\pi}$, $0 \le x \le 10$.

$$A'(x) = \frac{x}{8} - \frac{10 - x}{2\pi} = \left(\frac{1}{2\pi} + \frac{1}{8}\right)x - \frac{5}{\pi} = 0 \;\Rightarrow\; x = 40/(4 + \pi).$$

$A(0) = 25/\pi \approx 7.96$, $A(10) = 6.25$, and $A(40/(4 + \pi)) \approx 3.5$, so the maximum occurs when $x = 0$ m and the minimum occurs when $x = 40/(4 + \pi)$ m.

39.

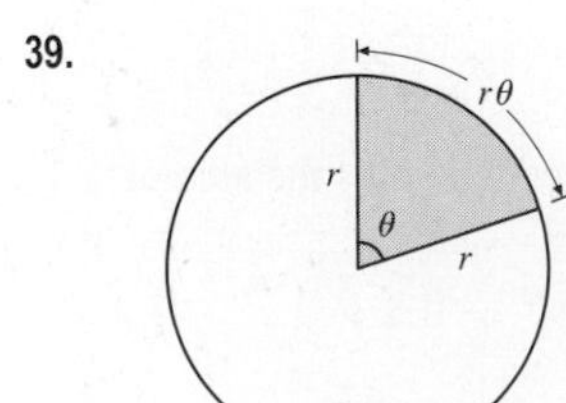

From the figure, the perimeter of the slice is $2r + r\theta = 32$, so $\theta = \dfrac{32 - 2r}{r}$. The area A of the slice is $A = \frac{1}{2}r^2\theta = \frac{1}{2}r^2\left(\frac{32 - 2r}{r}\right) = r(16 - r) = 16r - r^2$ for $0 \le r \le 16$. $A'(r) = 16 - 2r$, so $A' = 0$ when $r = 8$. Since $A(0) = 0$, $A(16) = 0$, and $A(8) = 64$ in.^{2}, the largest piece comes from a pizza with radius 8 in. and diameter 16 in. Note that $\theta = 2$ radians $\approx 114.6^\circ$, which is about 32% of the whole pizza.

40. 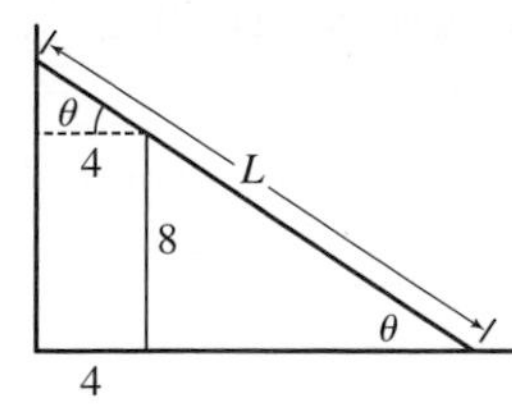

$L = 8 \csc\theta + 4 \sec\theta$, $0 < \theta < \frac{\pi}{2}$, $\dfrac{dL}{d\theta} = -8 \csc\theta \cot\theta + 4 \sec\theta \tan\theta = 0$ when

$\sec\theta \tan\theta = 2 \csc\theta \cot\theta \;\Leftrightarrow\; \tan^3\theta = 2 \;\Leftrightarrow\; \tan\theta = \sqrt[3]{2} \;\Leftrightarrow\; \theta = \tan^{-1}\sqrt[3]{2}$.

$dL/d\theta < 0$ when $0 < \theta < \tan^{-1}\sqrt[3]{2}$, $dL/d\theta > 0$ when $\tan^{-1}\sqrt[3]{2} < \theta < \frac{\pi}{2}$, so L has an absolute minimum when $\theta = \tan^{-1}\sqrt[3]{2}$, and the shortest ladder has length

$$L = 8\frac{\sqrt{1 + 2^{2/3}}}{2^{1/3}} + 4\sqrt{1 + 2^{2/3}} \approx 16.65 \text{ ft.}$$

Another method: Minimize $L^2 = x^2 + (4 + y)^2$, where $\dfrac{x}{4 + y} = \dfrac{8}{y}$.

41. 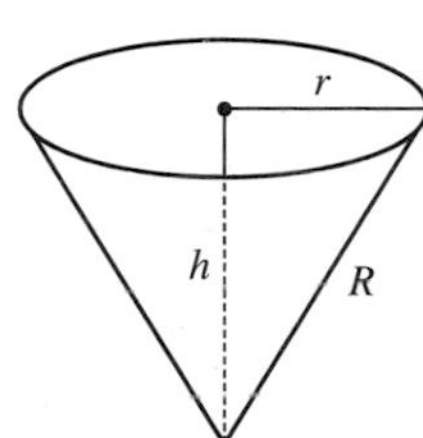

$h^2 + r^2 = R^2 \;\Rightarrow\; V = \frac{\pi}{3} r^2 h = \frac{\pi}{3}(R^2 - h^2)h = \frac{\pi}{3}(R^2 h - h^3)$.

$V'(h) = \frac{\pi}{3}(R^2 - 3h^2) = 0$ when $h = \frac{1}{\sqrt{3}}R$. This gives an absolute maximum, since

$V'(h) > 0$ for $0 < h < \frac{1}{\sqrt{3}}R$ and $V'(h) < 0$ for $h > \frac{1}{\sqrt{3}}R$. The maximum volume is

$$V\left(\tfrac{1}{\sqrt{3}}R\right) = \tfrac{\pi}{3}\left(\tfrac{1}{\sqrt{3}}R^3 - \tfrac{1}{3\sqrt{3}}R^3\right) = \tfrac{2}{9\sqrt{3}}\pi R^3.$$

42. The volume and surface area of a cone with radius r and height h are given by $V = \frac{1}{3}\pi r^2 h$ and $S = \pi r\sqrt{r^2 + h^2}$.

We'll minimize $A = S^2$ subject to $V = 27$. $V = 27 \;\Rightarrow\; \frac{1}{3}\pi r^2 h = 27 \;\Rightarrow\; r^2 = \dfrac{81}{\pi h}$ **(1)**.

$$A = \pi^2 r^2 (r^2 + h^2) = \pi^2\left(\frac{81}{\pi h}\right)\left(\frac{81}{\pi h} + h^2\right) = \frac{81^2}{h^2} + 81\pi h, \text{ so } A' = 0 \;\Rightarrow\; \frac{-2 \cdot 81^2}{h^3} + 81\pi = 0 \;\Rightarrow$$

$$81\pi = \frac{2 \cdot 81^2}{h^3} \;\Rightarrow\; h^3 = \frac{162}{\pi} \;\Rightarrow\; h = \sqrt[3]{\frac{162}{\pi}} = 3\sqrt[3]{\frac{6}{\pi}} \approx 3.722. \text{ From (1), } r^2 = \frac{81}{\pi h} = \frac{81}{\pi \cdot 3\sqrt[3]{6/\pi}} = \frac{27}{\sqrt[3]{6\pi^2}} \;\Rightarrow$$

$r = \dfrac{3\sqrt{3}}{\sqrt[6]{6\pi^2}} \approx 2.632$. $A'' = 6 \cdot 81^2/h^4 > 0$, so A and hence S has an absolute minimum at these values of r and h.

43. 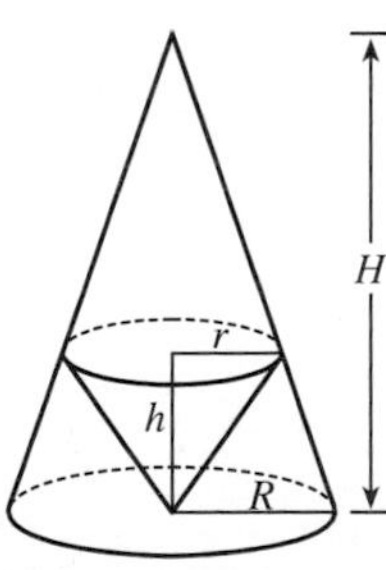

By similar triangles, $\dfrac{H}{R} = \dfrac{H - h}{r}$ **(1)**. The volume of the inner cone is $V = \frac{1}{3}\pi r^2 h$,

so we'll solve **(1)** for h. $\dfrac{Hr}{R} = H - h \;\Rightarrow$

$$h = H - \frac{Hr}{R} = \frac{HR - Hr}{R} = \frac{H}{R}(R - r) \quad \textbf{(2)}.$$

Thus, $V(r) = \dfrac{\pi}{3}r^2 \cdot \dfrac{H}{R}(R - r) = \dfrac{\pi H}{3R}(Rr^2 - r^3) \;\Rightarrow$

$$V'(r) = \frac{\pi H}{3R}(2Rr - 3r^2) = \frac{\pi H}{3R}r(2R - 3r).$$

$V'(r) = 0 \;\Rightarrow\; r = 0$ or $2R = 3r \;\Rightarrow\; r = \frac{2}{3}R$ and from **(2)**, $h = \dfrac{H}{R}\left(R - \frac{2}{3}R\right) = \dfrac{H}{R}\left(\frac{1}{3}R\right) = \frac{1}{3}H$.

$V'(r)$ changes from positive to negative at $r = \frac{2}{3}R$, so the inner cone has a maximum volume of

$V = \frac{1}{3}\pi r^2 h = \frac{1}{3}\pi\left(\frac{2}{3}R\right)^2\left(\frac{1}{3}H\right) = \frac{4}{27} \cdot \frac{1}{3}\pi R^2 H$, which is approximately 15% of the volume of the larger cone.

44. We need to minimize F for $0 \le \theta < \pi/2$. $F(\theta) = \dfrac{\mu W}{\mu \sin\theta + \cos\theta} \Rightarrow F'(\theta) = \dfrac{-\mu W\,(\mu\cos\theta - \sin\theta)}{(\mu\sin\theta + \cos\theta)^2}$ [by the Reciprocal Rule]. $F'(\theta) > 0 \Rightarrow \mu\cos\theta - \sin\theta < 0 \Rightarrow \mu\cos\theta < \sin\theta \Rightarrow \mu < \tan\theta \Rightarrow \theta > \tan^{-1}\mu$.
So F is decreasing on $(0, \tan^{-1}\mu)$ and increasing on $\left(\tan^{-1}\mu, \frac{\pi}{2}\right)$. Thus, F attains its minimum value at $\theta = \tan^{-1}\mu$.
This maximum value is $F(\tan^{-1}\mu) = \dfrac{\mu W}{\sqrt{\mu^2+1}}$.

45. $P(R) = \dfrac{E^2 R}{(R+r)^2} \Rightarrow$

$$P'(R) = \frac{(R+r)^2 \cdot E^2 - E^2 R \cdot 2(R+r)}{[(R+r)^2]^2} = \frac{(R^2+2Rr+r^2)E^2 - 2E^2R^2 - 2E^2Rr}{(R+r)^4}$$
$$= \frac{E^2r^2 - E^2R^2}{(R+r)^4} = \frac{E^2(r^2-R^2)}{(R+r)^4} = \frac{E^2(r+R)(r-R)}{(R+r)^4} = \frac{E^2(r-R)}{(R+r)^3}$$

$P'(R) = 0 \Rightarrow R = r \Rightarrow P(r) = \dfrac{E^2 r}{(r+r)^2} = \dfrac{E^2 r}{4r^2} = \dfrac{E^2}{4r}$.

The expression for $P'(R)$ shows that $P'(R) > 0$ for $R < r$ and $P'(R) < 0$ for $R > r$. Thus, the maximum value of the power is $E^2/(4r)$, and this occurs when $R = r$.

46. (a) $E(v) = \dfrac{aLv^3}{v-u} \Rightarrow E'(v) = aL\dfrac{(v-u)3v^2 - v^3}{(v-u)^2} = 0$ when

$2v^3 = 3uv^2 \Rightarrow 2v = 3u \Rightarrow v = \frac{3}{2}u$.

The First Derivative Test shows that this value of v gives the minimum value of E.

(b)

47. $S = 6sh - \frac{3}{2}s^2\cot\theta + 3s^2\frac{\sqrt{3}}{2}\csc\theta$

(a) $\dfrac{dS}{d\theta} = \frac{3}{2}s^2\csc^2\theta - 3s^2\frac{\sqrt{3}}{2}\csc\theta\,\cot\theta$ or $\frac{3}{2}s^2\csc\theta\left(\csc\theta - \sqrt{3}\cot\theta\right)$.

(b) $\dfrac{dS}{d\theta} = 0$ when $\csc\theta - \sqrt{3}\cot\theta = 0 \Rightarrow \dfrac{1}{\sin\theta} - \sqrt{3}\dfrac{\cos\theta}{\sin\theta} = 0 \Rightarrow \cos\theta = \frac{1}{\sqrt{3}}$. The First Derivative Test shows that the minimum surface area occurs when $\theta = \cos^{-1}\left(\frac{1}{\sqrt{3}}\right) \approx 55^\circ$.

(c)

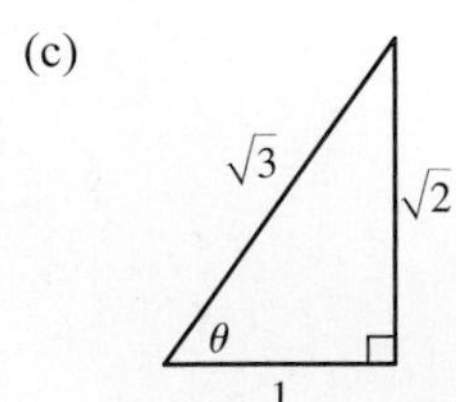

If $\cos\theta = \frac{1}{\sqrt{3}}$, then $\cot\theta = \frac{1}{\sqrt{2}}$ and $\csc\theta = \frac{\sqrt{3}}{\sqrt{2}}$, so the surface area is

$$S = 6sh - \tfrac{3}{2}s^2\tfrac{1}{\sqrt{2}} + 3s^2\tfrac{\sqrt{3}}{2}\tfrac{\sqrt{3}}{\sqrt{2}} = 6sh - \tfrac{3}{2\sqrt{2}}s^2 + \tfrac{9}{2\sqrt{2}}s^2$$
$$= 6sh + \tfrac{6}{2\sqrt{2}}s^2 = 6s\left(h + \tfrac{1}{2\sqrt{2}}s\right)$$

48. Let t be the time, in hours, after 2:00 PM. The position of the boat heading south at time t is $(0, -20t)$. The position of the boat heading east at time t is $(-15+15t, 0)$. If $D(t)$ is the distance between the boats at time t, we minimize $f(t) = [D(t)]^2 = 20^2t^2 + 15^2(t-1)^2$.

$f'(t) = 800t + 450(t-1) = 1250t - 450 = 0$ when $t = \frac{450}{1250} = 0.36$ h.

$0.36 \text{ h} \times \frac{60 \text{ min}}{\text{h}} = 21.6 \text{ min} = 21 \text{ min } 36 \text{ s}$. Since $f''(t) > 0$, this gives a minimum, so the boats are closest together at 2:21:36 PM.

49. Here $T(x) = \dfrac{\sqrt{x^2+25}}{6} + \dfrac{5-x}{8}$, $0 \le x \le 5 \;\Rightarrow\; T'(x) = \dfrac{x}{6\sqrt{x^2+25}} - \dfrac{1}{8} = 0 \;\Leftrightarrow\; 8x = 6\sqrt{x^2+25} \;\Leftrightarrow$

$16x^2 = 9(x^2+25) \;\Leftrightarrow\; x = \frac{15}{\sqrt{7}}$. But $\frac{15}{\sqrt{7}} > 5$, so T has no critical number. Since $T(0) \approx 1.46$ and $T(5) \approx 1.18$, he should row directly to B.

50.

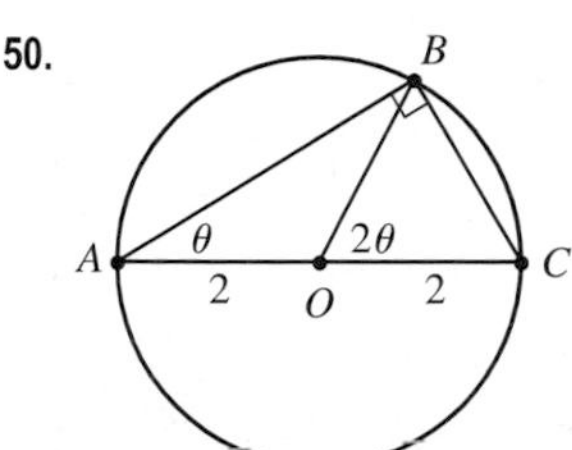

In isosceles triangle AOB, $\angle O = 180° - \theta - \theta$, so $\angle BOC = 2\theta$. The distance rowed is $4\cos\theta$ while the distance walked is the length of arc $BC = 2(2\theta) = 4\theta$. The time taken is given by $T(\theta) = \dfrac{4\cos\theta}{2} + \dfrac{4\theta}{4} = 2\cos\theta + \theta$, $0 \le \theta \le \frac{\pi}{2}$.

$T'(\theta) = -2\sin\theta + 1 = 0 \;\Leftrightarrow\; \sin\theta = \frac{1}{2} \;\Rightarrow\; \theta = \frac{\pi}{6}$.

Check the value of T at $\theta = \frac{\pi}{6}$ and at the endpoints of the domain of T; that is, $\theta = 0$ and $\theta = \frac{\pi}{2}$.

$T(0) = 2$, $T\left(\frac{\pi}{6}\right) = \sqrt{3} + \frac{\pi}{6} \approx 2.26$, and $T\left(\frac{\pi}{2}\right) = \frac{\pi}{2} \approx 1.57$. Therefore, the minimum value of T is $\frac{\pi}{2}$ when $\theta = \frac{\pi}{2}$; that is, the woman should walk all the way. Note that $T''(\theta) = -2\cos\theta < 0$ for $0 \le \theta < \frac{\pi}{2}$, so $\theta = \frac{\pi}{6}$ gives a maximum time.

51. There are $(6-x)$ km over land and $\sqrt{x^2+4}$ km under the river.

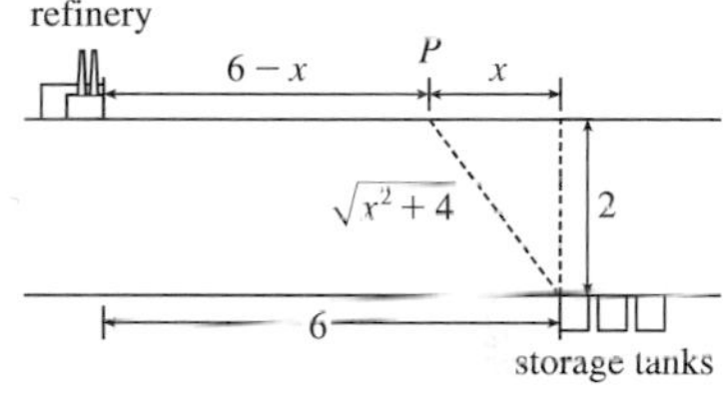

We need to minimize the cost C (measured in \$100,000) of the pipeline.

$C(x) = (6-x)(4) + \left(\sqrt{x^2+4}\right)(8) \;\Rightarrow$

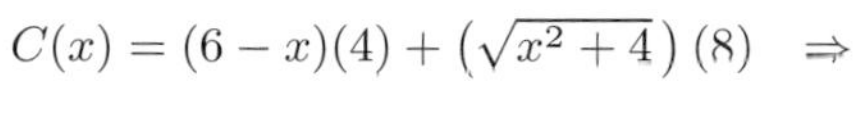

$C'(x) = -4 + 8 \cdot \frac{1}{2}(x^2+4)^{-1/2}(2x) = -4 + \dfrac{8x}{\sqrt{x^2+4}}$.

$C'(x) = 0 \;\Rightarrow\; 4 = \dfrac{8x}{\sqrt{x^2+4}} \;\Rightarrow\; \sqrt{x^2+4} = 2x \;\Rightarrow\; x^2 + 4 = 4x^2 \;\Rightarrow\; 4 = 3x^2 \;\Rightarrow\; x^2 = \frac{4}{3} \;\Rightarrow$

$x = 2/\sqrt{3}$ $[0 \le x \le 6]$. Compare the costs for $x = 0, 2/\sqrt{3}$, and 6. $C(0) = 24 + 16 = 40$,

$C\left(2/\sqrt{3}\right) = 24 - 8/\sqrt{3} + 32/\sqrt{3} = 24 + 24/\sqrt{3} \approx 37.9$, and $C(6) = 0 + 8\sqrt{40} \approx 50.6$. So the minimum cost is about \$3.79 million when P is $6 - 2/\sqrt{3} \approx 4.85$ km east of the refinery.

52. The distance from the refinery to P is now $\sqrt{(6-x)^2 + 1^2} = \sqrt{x^2 - 12x + 37}$.

Thus, $C(x) = 4\sqrt{x^2 - 12x + 37} + 8\sqrt{x^2+4} \;\Rightarrow$

$C'(x) = 4 \cdot \frac{1}{2}(x^2 - 12x + 37)^{-1/2}(2x - 12) + 8 \cdot \frac{1}{2}(x^2+4)^{-1/2}(2x) = \dfrac{4(x-6)}{\sqrt{x^2 - 12x + 37}} + \dfrac{8x}{\sqrt{x^2+4}}$.

$C'(x) = 0 \;\Rightarrow\; x \approx 1.12$ [from a graph of C' or a numerical rootfinder]. $C(0) \approx 40.3$, $C(1.12) \approx 38.3$, and $C(6) \approx 54.6$. So the minimum cost is slightly higher (than in the previous exercise) at about \$3.83 million when P is approximately 4.88 km from the point on the bank 1 km south of the refinery.

53. The total illumination is $I(x) = \dfrac{3k}{x^2} + \dfrac{k}{(10-x)^2}$, $0 < x < 10$. Then

$$I'(x) = \frac{-6k}{x^3} + \frac{2k}{(10-x)^3} = 0 \;\Rightarrow\; 6k(10-x)^3 = 2kx^3 \;\Rightarrow$$

$$3(10-x)^3 = x^3 \;\Rightarrow\; \sqrt[3]{3}\,(10-x) = x \;\Rightarrow\; 10\sqrt[3]{3} - \sqrt[3]{3}\,x = x \;\Rightarrow\; 10\sqrt[3]{3} = x + \sqrt[3]{3}\,x \;\Rightarrow$$

$10\sqrt[3]{3} = \left(1+\sqrt[3]{3}\right)x \;\Rightarrow\; x = \dfrac{10\sqrt[3]{3}}{1+\sqrt[3]{3}} \approx 5.9$ ft. This gives a minimum since $I''(x) > 0$ for $0 < x < 10$.

54.

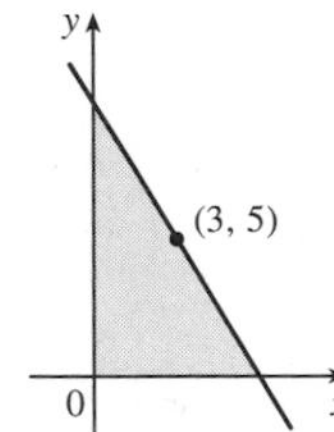

The line with slope m (where $m < 0$) through $(3, 5)$ has equation $y - 5 = m(x-3)$ or $y = mx + (5 - 3m)$. The y-intercept is $5 - 3m$ and the x-intercept is $-5/m + 3$. So the triangle has area $A(m) = \frac{1}{2}(5-3m)(-5/m+3) = 15 - 25/(2m) - \frac{9}{2}m$. Now

$$A'(m) = \frac{25}{2m^2} - \frac{9}{2} = 0 \;\Leftrightarrow\; m^2 = \tfrac{25}{9} \;\Rightarrow\; m = -\tfrac{5}{3} \text{ (since } m < 0\text{)}.$$

$A''(m) = -\dfrac{25}{m^3} > 0$, so there is an absolute minimum when $m = -\frac{5}{3}$. Thus, an equation of the line is $y - 5 = -\frac{5}{3}(x-3)$ or $y = -\frac{5}{3}x + 10$.

55.

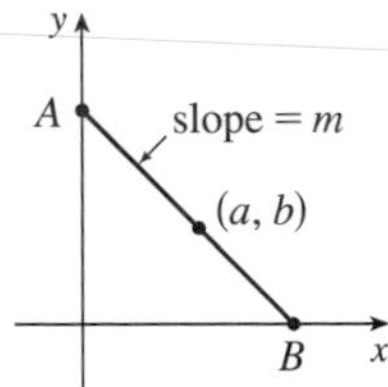

Every line segment in the first quadrant passing through (a, b) with endpoints on the x- and y-axes satisfies an equation of the form $y - b = m(x-a)$, where $m < 0$. By setting $x = 0$ and then $y = 0$, we find its endpoints, $A(0, b - am)$ and $B\left(a - \frac{b}{m}, 0\right)$. The distance d from A to B is given by $d = \sqrt{\left[\left(a - \frac{b}{m}\right) - 0\right]^2 + [0 - (b - am)]^2}$.

It follows that the square of the length of the line segment, as a function of m, is given by

$$S(m) = \left(a - \frac{b}{m}\right)^2 + (am-b)^2 = a^2 - \frac{2ab}{m} + \frac{b^2}{m^2} + a^2m^2 - 2abm + b^2. \text{ Thus,}$$

$$\begin{aligned} S'(m) &= \frac{2ab}{m^2} - \frac{2b^2}{m^3} + 2a^2m - 2ab = \frac{2}{m^3}(abm - b^2 + a^2m^4 - abm^3) \\ &= \frac{2}{m^3}[b(am-b) + am^3(am-b)] = \frac{2}{m^3}(am-b)(b+am^3) \end{aligned}$$

Thus, $S'(m) = 0 \;\Leftrightarrow\; m = b/a$ or $m = -\sqrt[3]{\frac{b}{a}}$. Since $b/a > 0$ and $m < 0$, m must equal $-\sqrt[3]{\frac{b}{a}}$. Since $\dfrac{2}{m^3} < 0$, we see that $S'(m) < 0$ for $m < -\sqrt[3]{\frac{b}{a}}$ and $S'(m) > 0$ for $m > -\sqrt[3]{\frac{b}{a}}$. Thus, S has its absolute minimum value when $m = -\sqrt[3]{\frac{b}{a}}$. That value is

$$\begin{aligned} S\left(-\sqrt[3]{\tfrac{b}{a}}\right) &= \left(a + b\sqrt[3]{\tfrac{a}{b}}\right)^2 + \left(-a\sqrt[3]{\tfrac{b}{a}} - b\right)^2 = \left(a + \sqrt[3]{ab^2}\right)^2 + \left(\sqrt[3]{a^2b} + b\right)^2 \\ &= a^2 + 2a^{4/3}b^{2/3} + a^{2/3}b^{4/3} + a^{4/3}b^{2/3} + 2a^{2/3}b^{4/3} + b^2 = a^2 + 3a^{4/3}b^{2/3} + 3a^{2/3}b^{4/3} + b^2 \end{aligned}$$

The last expression is of the form $x^3 + 3x^2y + 3xy^2 + y^3$ $[= (x+y)^3]$ with $x = a^{2/3}$ and $y = b^{2/3}$, so we can write it as $(a^{2/3} + b^{2/3})^3$ and the shortest such line segment has length $\sqrt{S} = (a^{2/3} + b^{2/3})^{3/2}$.

56. $y = 1 + 40x^3 - 3x^5 \;\Rightarrow\; y' = 120x^2 - 15x^4$, so the tangent line to the curve at $x = a$ has slope $m(a) = 120a^2 - 15a^4$. Now $m'(a) = 240a - 60a^3 = -60a(a^2 - 4) = -60a(a+2)(a-2)$, so $m'(a) > 0$ for $a < -2$, and $0 < a < 2$, and

$m'(a) < 0$ for $-2 < a < 0$ and $a > 2$. Thus, m is increasing on $(-\infty, -2)$, decreasing on $(-2, 0)$, increasing on $(0, 2)$, and decreasing on $(2, \infty)$. Clearly, $m(a) \to -\infty$ as $a \to \pm\infty$, so the maximum value of $m(a)$ must be one of the two local maxima, $m(-2)$ or $m(2)$. But both $m(-2)$ and $m(2)$ equal $120 \cdot 2^2 - 15 \cdot 2^4 = 480 - 240 = 240$. So 240 is the largest slope, and it occurs at the points $(-2, -223)$ and $(2, 225)$. *Note:* $a = 0$ corresponds to a local *minimum* of m.

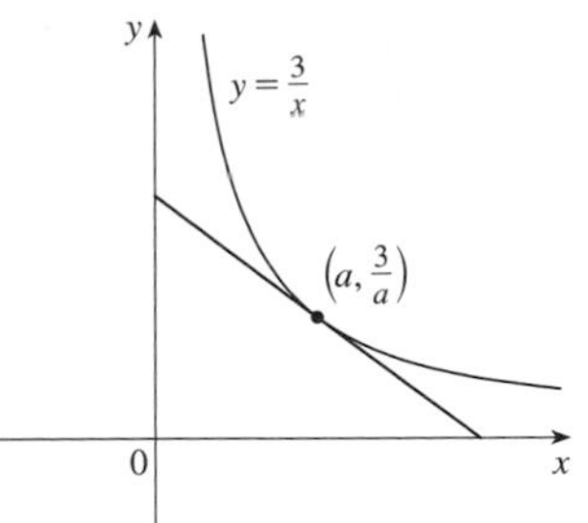

57. $y = \dfrac{3}{x} \Rightarrow y' = -\dfrac{3}{x^2}$, so an equation of the tangent line at the point $(a, \frac{3}{a})$ is $y - \dfrac{3}{a} = -\dfrac{3}{a^2}(x - a)$, or $y = -\dfrac{3}{a^2}x + \dfrac{6}{a}$. The y-intercept $[x = 0]$ is $6/a$. The x-intercept $[y = 0]$ is $2a$. The distance d of the line segment that has endpoints at the intercepts is $d = \sqrt{(2a - 0)^2 + (0 - 6/a)^2}$. Let $S = d^2$, so $S = 4a^2 + \dfrac{36}{a^2} \Rightarrow$

$S' = 8a - \dfrac{72}{a^3}$. $S' = 0 \Leftrightarrow \dfrac{72}{a^3} = 8a \Leftrightarrow a^4 = 9 \Leftrightarrow a^2 = 3 \Rightarrow a = \sqrt{3}$.

$S'' = 8 + \dfrac{216}{a^4} > 0$, so there is an absolute minimum at $a = \sqrt{3}$. Thus, $S = 4(3) + \frac{36}{3} = 12 + 12 = 24$ and hence, $d = \sqrt{24} = 2\sqrt{6}$.

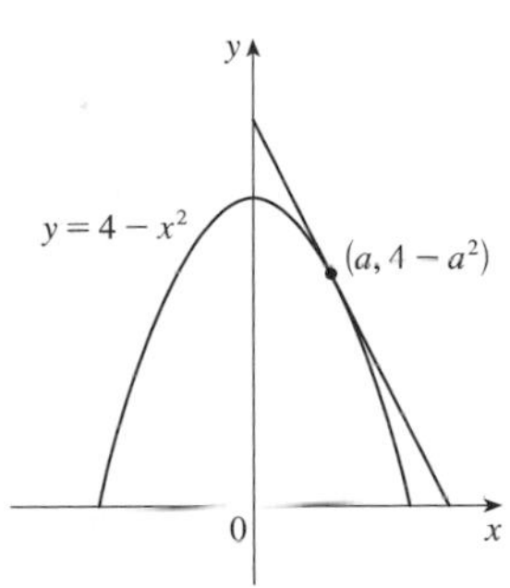

58. $y = 4 - x^2 \Rightarrow y' = -2x$, so an equation of the tangent line at $(a, 4 - a^2)$ is $y - (4 - a^2) = -2a(x - a)$, or $y = -2ax + a^2 + 4$. The y-intercept $[x = 0]$ is $a^2 + 4$. The x-intercept $[y = 0]$ is $\dfrac{a^2 + 4}{2a}$. The area A of the triangle is

$$A = \frac{1}{2}(\text{base})(\text{height}) = \frac{1}{2} \cdot \frac{a^2 + 4}{2a}(a^2 + 4) = \frac{1}{4}\frac{a^4 + 8a^2 + 16}{a} = \frac{1}{4}\left(a^3 + 8a + \frac{16}{a}\right).$$

$A' = 0 \Rightarrow \dfrac{1}{4}\left(3a^2 + 8 - \dfrac{16}{a^2}\right) - 0 \Rightarrow 3a^4 + 8a^2 - 16 = 0 \Rightarrow$

$(3a^2 - 4)(a^2 + 4) = 0 \Rightarrow a^2 = \dfrac{4}{3} \Rightarrow a = \dfrac{2}{\sqrt{3}}$. $A'' = \dfrac{1}{4}\left(6a + \dfrac{32}{a^3}\right) > 0$, so there is an absolute minimum at

$a = \dfrac{2}{\sqrt{3}}$. Thus, $A = \dfrac{1}{2} \cdot \dfrac{4/3 + 4}{2(2/\sqrt{3})}\left(\dfrac{4}{3} + 4\right) = \dfrac{1}{2} \cdot \dfrac{4\sqrt{3}}{3} \cdot \dfrac{16}{3} = \dfrac{32}{9}\sqrt{3}$.

59. (a) If $c(x) = \dfrac{C(x)}{x}$, then, by the Quotient Rule, we have $c'(x) = \dfrac{xC'(x) - C(x)}{x^2}$. Now $c'(x) = 0$ when $xC'(x) - C(x) = 0$ and this gives $C'(x) = \dfrac{C(x)}{x} = c(x)$. Therefore, the marginal cost equals the average cost.

(b) (i) $C(x) = 16{,}000 + 200x + 4x^{3/2}$, $C(1000) = 16{,}000 + 200{,}000 + 40{,}000\sqrt{10} \approx 216{,}000 + 126{,}491$, so $C(1000) \approx \$342{,}491$. $c(x) = C(x)/x = \dfrac{16{,}000}{x} + 200 + 4x^{1/2}$, $c(1000) \approx \$342.49/\text{unit}$. $C'(x) = 200 + 6x^{1/2}$, $C'(1000) = 200 + 60\sqrt{10} \approx \$389.74/\text{unit}$.

(ii) We must have $C'(x) = c(x) \Leftrightarrow 200 + 6x^{1/2} = \dfrac{16{,}000}{x} + 200 + 4x^{1/2} \Leftrightarrow 2x^{3/2} = 16{,}000 \Leftrightarrow$ $x = (8{,}000)^{2/3} = 400$ units. To check that this is a minimum, we calculate

$c'(x) = \dfrac{-16{,}000}{x^2} + \dfrac{2}{\sqrt{x}} = \dfrac{2}{x^2}(x^{3/2} - 8000)$. This is negative for $x < (8000)^{2/3} = 400$, zero at $x = 400$, and positive for $x > 400$, so c is decreasing on $(0, 400)$ and increasing on $(400, \infty)$. Thus, c has an absolute minimum at $x = 400$. [*Note:* $c''(x)$ is *not* positive for all $x > 0$.]

(iii) The minimum average cost is $c(400) = 40 + 200 + 80 = \$320/\text{unit}$.

60. (a) The total profit is $P(x) = R(x) - C(x)$. In order to maximize profit we look for the critical numbers of P, that is, the numbers where the marginal profit is 0. But if $P'(x) = R'(x) - C'(x) = 0$, then $R'(x) = C'(x)$. Therefore, if the profit is a maximum, then the marginal revenue equals the marginal cost.

(b) $C(x) = 16{,}000 + 500x - 1.6x^2 + 0.004x^3$, $p(x) = 1700 - 7x$. Then $R(x) = xp(x) = 1700x - 7x^2$. If the profit is maximum, then $R'(x) = C'(x) \quad\Leftrightarrow\quad 1700 - 14x = 500 - 3.2x + 0.012x^2 \quad\Leftrightarrow\quad 0.012x^2 + 10.8x - 1200 = 0 \quad\Leftrightarrow$ $x^2 + 900x - 100{,}000 = 0 \quad\Leftrightarrow\quad (x + 1000)(x - 100) = 0 \quad\Leftrightarrow\quad x = 100$ (since $x > 0$). The profit is maximized if $P''(x) < 0$, but since $P''(x) = R''(x) - C''(x)$, we can just check the condition $R''(x) < C''(x)$. Now $R''(x) = -14 < -3.2 + 0.024x = C''(x)$ for $x > 0$, so there is a maximum at $x = 100$.

61. (a) We are given that the demand function p is linear and $p(27{,}000) = 10$, $p(33{,}000) = 8$, so the slope is $\frac{10 - 8}{27{,}000 - 33{,}000} = -\frac{1}{3000}$ and an equation of the line is $y - 10 = \left(-\frac{1}{3000}\right)(x - 27{,}000) \quad\Rightarrow$ $y = p(x) = -\frac{1}{3000}x + 19 = 19 - (x/3000)$.

(b) The revenue is $R(x) = xp(x) = 19x - (x^2/3000) \quad\Rightarrow\quad R'(x) = 19 - (x/1500) = 0$ when $x = 28{,}500$. Since $R''(x) = -1/1500 < 0$, the maximum revenue occurs when $x = 28{,}500 \quad\Rightarrow\quad$ the price is $p(28{,}500) = \$9.50$.

62. (a) Let $p(x)$ be the demand function. Then $p(x)$ is linear and $y = p(x)$ passes through $(20, 10)$ and $(18, 11)$, so the slope is $-\frac{1}{2}$ and an equation of the line is $y - 10 = -\frac{1}{2}(x - 20) \quad\Leftrightarrow\quad y = -\frac{1}{2}x + 20$. Thus, the demand is $p(x) = -\frac{1}{2}x + 20$ and the revenue is $R(x) = xp(x) = -\frac{1}{2}x^2 + 20x$.

(b) The cost is $C(x) = 6x$, so the profit is $P(x) = R(x) - C(x) = -\frac{1}{2}x^2 + 14x$. Then $0 = P'(x) = -x + 14 \quad\Rightarrow$ $x = 14$. Since $P''(x) = -1 < 0$, the selling price for maximum profit is $p(14) = -\frac{1}{2}(14) + 20 = \$13$.

63. (a) As in Example 6, we see that the demand function p is linear. We are given that $p(1200) = 350$ and deduce that $p(1280) = 340$, since a \$10 reduction in price increases sales by 80 per week. The slope for p is $\dfrac{340 - 350}{1280 - 1200} = -\dfrac{1}{8}$, so an equation is $p - 350 = -\frac{1}{8}(x - 1200)$ or $p(x) = -\frac{1}{8}x + 500$, where $x \geq 1200$.

(b) $R(x) = x\,p(x) = -\frac{1}{8}x^2 + 500x$. $\quad R'(x) = -\frac{1}{4}x + 500 = 0$ when $x = 4(500) = 2000$. $\quad p(2000) = 250$, so the price should be set at \$250 to maximize revenue.

(c) $C(x) = 35{,}000 + 120x \quad\Rightarrow\quad P(x) = R(x) - C(x) = -\frac{1}{8}x^2 + 500x - 35{,}000 - 120x = -\frac{1}{8}x^2 + 380x - 35{,}000$. $P'(x) = -\frac{1}{4}x + 380 = 0$ when $x = 4(380) = 1520$. $\quad p(1520) = 310$, so the price should be set at \$310 to maximize profit.

64. Let w denote the number of operating wells. Then the amount of daily oil production for each well is $240 - 8(w - 16) = 368 - 8w$, where $w \geq 16$. The total daily oil production P for all wells is given by $P(w) = w(368 - 8w) = 368w - 8w^2$. Now $P'(w) = 368 - 16w$ and $P'(w) = 0 \Leftrightarrow w = \frac{368}{16} = 23$. $P''(w) = -16 < 0$, so the daily production is maximized when the company adds $23 - 16 = 7$ wells.

65. Here $s^2 = h^2 + b^2/4$, so $h^2 = s^2 - b^2/4$. The area is $A = \frac{1}{2}b\sqrt{s^2 - b^2/4}$.

Let the perimeter be p, so $2s + b = p$ or $s = (p - b)/2 \Rightarrow$

$A(b) = \frac{1}{2}b\sqrt{(p-b)^2/4 - b^2/4} = b\sqrt{p^2 - 2pb}/4$. Now

$$A'(b) = \frac{\sqrt{p^2 - 2pb}}{4} - \frac{bp/4}{\sqrt{p^2 - 2pb}} = \frac{-3pb + p^2}{4\sqrt{p^2 - 2pb}}.$$

Therefore, $A'(b) = 0 \Rightarrow -3pb + p^2 = 0 \Rightarrow b = p/3$. Since $A'(b) > 0$ for $b < p/3$ and $A'(b) < 0$ for $b > p/3$, there is an absolute maximum when $b = p/3$. But then $2s + p/3 = p$, so $s = p/3 \Rightarrow s = b \Rightarrow$ the triangle is equilateral.

66. From Exercise 51, with K replacing 8 for the "under river" cost (measured in \$100,000), we see that $C'(x) = 0 \Leftrightarrow$ $4\sqrt{x^2 + 4} = Kx \Leftrightarrow 16x^2 + 64 = K^2x^2 \Leftrightarrow 64 = (K^2 - 16)x^2 \Leftrightarrow x = \dfrac{8}{\sqrt{K^2 - 16}}$. Also from Exercise 51, we have $C(x) = (6 - x)4 + \sqrt{x^2 + 4}\,K$. We now compare costs for using the minimum distance possible under the river $[x = 0]$ and using the critical number above. $C(0) = 24 + 2K$ and

$$\begin{aligned} C\left(\frac{8}{\sqrt{K^2 - 16}}\right) &= 24 - \frac{32}{\sqrt{K^2 - 16}} + \sqrt{\frac{64}{K^2 - 16} + 4}\,K = 24 - \frac{32}{\sqrt{K^2 - 16}} + \sqrt{\frac{4K^2}{K^2 - 16}}\,K \\ &= 24 - \frac{32}{\sqrt{K^2 - 16}} + \frac{2K^2}{\sqrt{K^2 - 16}} = 24 + \frac{2(K^2 - 16)}{\sqrt{K^2 - 16}} = 24 + 2\sqrt{K^2 - 16} \end{aligned}$$

Since $\sqrt{K^2 - 16} < K$, we see that $C\left(\dfrac{8}{\sqrt{K^2 - 16}}\right) < C(0)$ for any cost K, so the minimum distance possible for the "under river" portion of the pipeline should *never* be used.

67. (a)

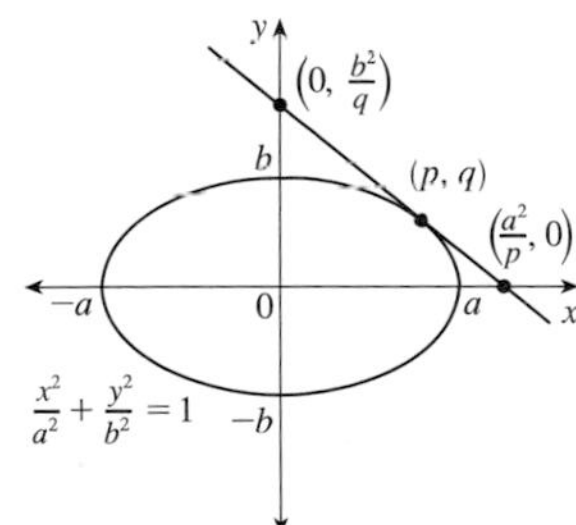

Using implicit differentiation, $\dfrac{x^2}{a^2} + \dfrac{y^2}{b^2} = 1 \Rightarrow \dfrac{2x}{a^2} + \dfrac{2y\,y'}{b^2} = 0 \Rightarrow$ $\dfrac{2y\,y'}{b^2} = -\dfrac{2x}{a^2} \Rightarrow y' = -\dfrac{b^2x}{a^2y}$. At (p, q), $y' = -\dfrac{b^2p}{a^2q}$, and an equation of the tangent line is $y - q = -\dfrac{b^2p}{a^2q}(x - p) \Leftrightarrow y = -\dfrac{b^2p}{a^2q}x + \dfrac{b^2p^2}{a^2q} + q \Leftrightarrow$ $y = -\dfrac{b^2p}{a^2q}x + \dfrac{b^2p^2 + a^2q^2}{a^2q}$. The last term is the y-intercept, but not the term we want, namely b^2/q. Since (p, q) is on the ellipse, we know $\dfrac{p^2}{a^2} + \dfrac{q^2}{b^2} = 1$. To use that relationship we must divide b^2p^2 in the y-intercept by a^2b^2, so divide all terms by a^2b^2. $\dfrac{(b^2p^2 + a^2q^2)/a^2b^2}{(a^2q)/a^2b^2} = \dfrac{p^2/a^2 + q^2/b^2}{q/b^2} = \dfrac{1}{q/b^2} = \dfrac{b^2}{q}$. So the tangent line has equation $y = -\dfrac{b^2p}{a^2q}x + \dfrac{b^2}{q}$. Let $y = 0$ and solve for x to find that x-intercept: $\dfrac{b^2p}{a^2q}x = \dfrac{b^2}{q} \Leftrightarrow$ $x = \dfrac{b^2a^2q}{qb^2p} = \dfrac{a^2}{p}$.

(b) The portion of the tangent line cut off by the coordinate axes is the distance between the intercepts, $(a^2/p, 0)$ and $(0, b^2/q)$: $\sqrt{\left(\frac{a^2}{p}\right)^2 + \left(-\frac{b^2}{q}\right)^2} = \sqrt{\frac{a^4}{p^2} + \frac{b^4}{q^2}}$. To eliminate p or q, we turn to the relationship $\frac{p^2}{a^2} + \frac{q^2}{b^2} = 1 \Leftrightarrow$ $\frac{q^2}{b^2} = 1 - \frac{p^2}{a^2} \Leftrightarrow q^2 = b^2 - \frac{b^2p^2}{a^2} \Leftrightarrow q^2 = \frac{b^2(a^2 - p^2)}{a^2}$. Now substitute for q^2 and use the square S of the distance. $S(p) = \frac{a^4}{p^2} + \frac{b^4a^2}{b^2(a^2 - p^2)} = \frac{a^4}{p^2} + \frac{a^2b^2}{a^2 - p^2}$ for $0 < p < a$. Note that as $p \to 0$ or $p \to a$, $S(p) \to \infty$, so the minimum value of S must occur at a critical number. Now $S'(p) = -\frac{2a^4}{p^3} + \frac{2a^2b^2p}{(a^2 - p^2)^2}$ and $S'(p) = 0 \Leftrightarrow$

$$\frac{2a^4}{p^3} = \frac{2a^2b^2p}{(a^2 - p^2)^2} \Leftrightarrow a^2(a^2 - p^2)^2 = b^2p^4 \Rightarrow a(a^2 - p^2) = bp^2 \Leftrightarrow a^3 = (a + b)p^2 \Leftrightarrow p^2 = \frac{a^3}{a + b}.$$

Substitute for p^2 in $S(p)$:

$$\frac{a^4}{\frac{a^3}{a+b}} + \frac{a^2b^2}{a^2 - \frac{a^3}{a+b}} = \frac{a^4(a + b)}{a^3} + \frac{a^2b^2(a + b)}{a^2(a + b) - a^3} = \frac{a(a + b)}{1} + \frac{a^2b^2(a + b)}{a^2b}$$
$$= a(a + b) + b(a + b) = (a + b)(a + b) = (a + b)^2$$

Taking the square root gives us the desired minimum length of $a + b$.

(c) The triangle formed by the tangent line and the coordinate axes has area $A = \frac{1}{2}\left(\frac{a^2}{p}\right)\left(\frac{b^2}{q}\right)$. As in part (b), we'll use the square of the area and substitute for q^2. $S = \frac{a^4b^4}{4p^2q^2} = \frac{a^4b^4a^2}{4p^2b^2(a^2 - p^2)} = \frac{a^6b^2}{4p^2(a^2 - p^2)}$. Minimizing S (and hence A) is equivalent to maximizing $p^2(a^2 - p^2)$. Let $f(p) = p^2(a^2 - p^2) = a^2p^2 - p^4$ for $0 < p < a$. As in part (b), the minimum value of S must occur at a critical number. Now $f'(p) = 2a^2p - 4p^3 = 2p(a^2 - 2p^2)$. $f'(p) = 0 \Rightarrow$ $p^2 = a^2/2 \Rightarrow p = a/\sqrt{2}$ $[p > 0]$. Substitute for p^2 in $S(p)$: $\frac{a^6b^2}{4\left(\frac{a^2}{2}\right)\left(a^2 - \frac{a^2}{2}\right)} = \frac{a^6b^2}{a^4} = a^2b^2 = (ab)^2$. Taking the square root gives us the desired minimum area of ab.

68. See the figure. The area is given by

$$A(x) = \tfrac{1}{2}\left(2\sqrt{a^2 - x^2}\right)x + \tfrac{1}{2}\left(2\sqrt{a^2 - x^2}\right)\left(\sqrt{x^2 + b^2 - a^2}\right)$$
$$= \sqrt{a^2 - x^2}\left(x + \sqrt{x^2 + b^2 - a^2}\right)$$

for $0 \le x \le a$. Now

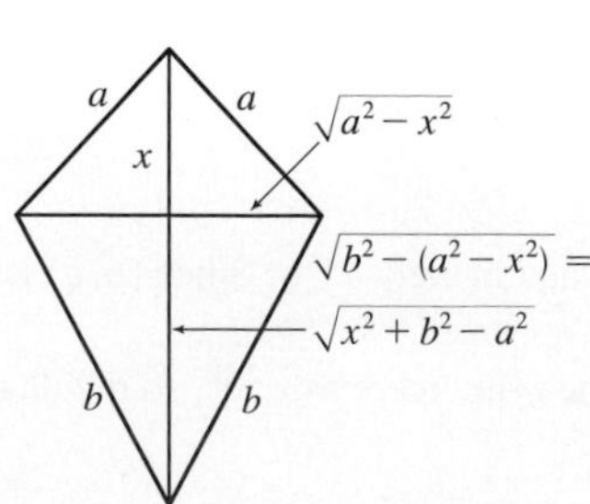

$$A'(x) = \sqrt{a^2 - x^2}\left(1 + \frac{x}{\sqrt{x^2 + b^2 - a^2}}\right) + \left(x + \sqrt{x^2 + b^2 - a^2}\right)\frac{-x}{\sqrt{a^2 - x^2}}$$
$$= 0 \Leftrightarrow$$

$$\frac{x}{\sqrt{a^2-x^2}}\left(x+\sqrt{x^2+b^2-a^2}\right)=\sqrt{a^2-x^2}\left(\frac{x+\sqrt{x^2+b^2-a^2}}{\sqrt{x^2+b^2-a^2}}\right).$$

Except for the trivial case where $x=0$, $a=b$ and $A(x)=0$, we have $x+\sqrt{x^2+b^2-a^2}>0$. Hence, cancelling this factor gives $\dfrac{x}{\sqrt{a^2-x^2}}=\dfrac{\sqrt{a^2-x^2}}{\sqrt{x^2+b^2-a^2}} \Rightarrow x\sqrt{x^2+b^2-a^2}=a^2-x^2 \Rightarrow$

$$x^2(x^2+b^2-a^2)=a^4-2a^2x^2+x^4 \Rightarrow x^2(b^2-a^2)=a^4-2a^2x^2 \Rightarrow x^2(b^2+a^2)=a^4 \Rightarrow x=\frac{a^2}{\sqrt{a^2+b^2}}.$$

Now we must check the value of A at this point as well as at the endpoints of the domain to see which gives the maximum value. $A(0)=a\sqrt{b^2-a^2}$, $A(a)=0$ and

$$A\left(\frac{a^2}{\sqrt{a^2+b^2}}\right)=\sqrt{a^2-\left(\frac{a^2}{\sqrt{a^2+b^2}}\right)^2}\left[\frac{a^2}{\sqrt{a^2+b^2}}+\sqrt{\left(\frac{a^2}{\sqrt{a^2+b^2}}\right)^2+b^2-a^2}\right]$$

$$=\frac{ab}{\sqrt{a^2+b^2}}\left[\frac{a^2}{\sqrt{a^2+b^2}}+\frac{b^2}{\sqrt{a^2+b^2}}\right]=\frac{ab(a^2+b^2)}{a^2+b^2}=ab$$

Since $b\geq\sqrt{b^2-a^2}$, $A\left(a^2/\sqrt{a^2+b^2}\right)\geq A(0)$. So there is an absolute maximum when $x=\dfrac{a^2}{\sqrt{a^2+b^2}}$. In this case the horizontal piece should be $\dfrac{2ab}{\sqrt{a^2+b^2}}$ and the vertical piece should be $\dfrac{a^2+b^2}{\sqrt{a^2+b^2}}=\sqrt{a^2+b^2}$.

69. Note that $|AD|=|AP|+|PD| \Rightarrow 5=x+|PD| \Rightarrow |PD|=5-x$.

Using the Pythagorean Theorem for ΔPDB and ΔPDC gives us

$$L(x)=|AP|+|BP|+|CP|=x+\sqrt{(5-x)^2+2^2}+\sqrt{(5-x)^2+3^2}$$
$$=x+\sqrt{x^2-10x+29}+\sqrt{x^2-10x+34} \Rightarrow$$

$L'(x)=1+\dfrac{x-5}{\sqrt{x^2-10x+29}}+\dfrac{x-5}{\sqrt{x^2-10x+34}}$. From the graphs of L and L', it seems that the minimum value of L is about $L(3.59)=9.35$ m.

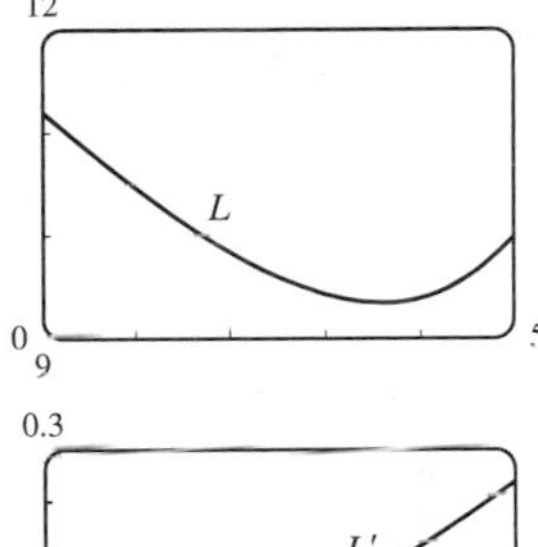

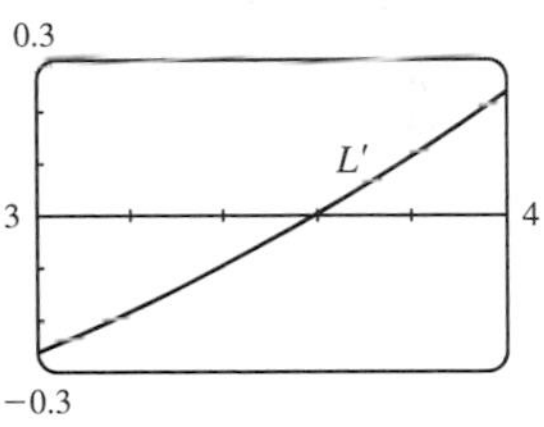

70. We note that since c is the consumption in gallons per hour, and v is the velocity in miles per hour, then $\dfrac{c}{v}=\dfrac{\text{gallons/hour}}{\text{miles/hour}}=\dfrac{\text{gallons}}{\text{mile}}$ gives us the consumption in gallons per mile, that is, the quantity G. To find the minimum, we calculate $\dfrac{dG}{dv}=\dfrac{d}{dv}\left(\dfrac{c}{v}\right)=\dfrac{v\dfrac{dc}{dv}-c\dfrac{dv}{dv}}{v^2}=\dfrac{v\dfrac{dc}{dv}-c}{v^2}$.

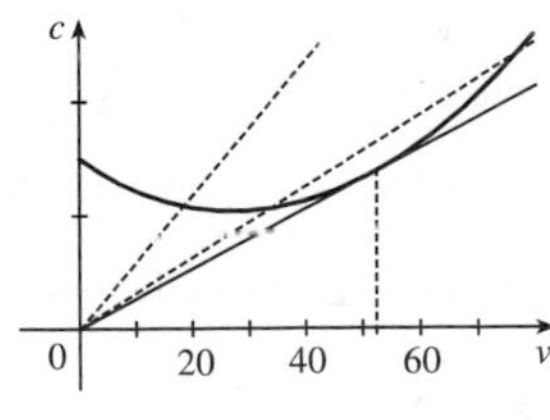

This is 0 when $v\dfrac{dc}{dv}-c=0 \Leftrightarrow \dfrac{dc}{dv}=\dfrac{c}{v}$. This implies that the tangent line of $c(v)$ passes through the origin, and this occurs when $v\approx 53$ mi/h. Note that the slope of the secant line through the origin and a point $(v,c(v))$ on the graph is equal to $G(v)$, and it is intuitively clear that G is minimized in the case where the secant is in fact a tangent.

71. 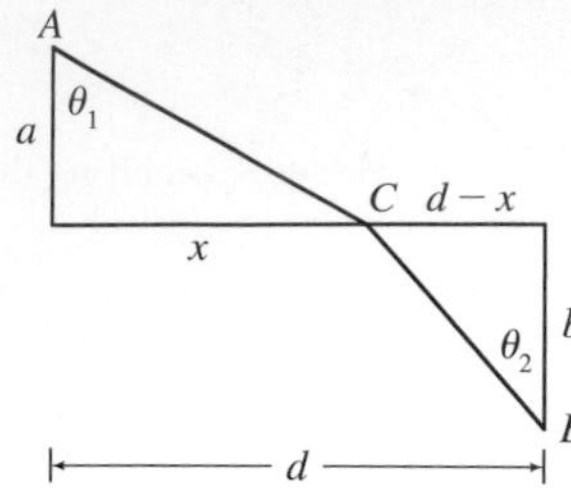

The total time is

$$T(x) = (\text{time from } A \text{ to } C) + (\text{time from } C \text{ to } B) = \frac{\sqrt{a^2+x^2}}{v_1} + \frac{\sqrt{b^2+(d-x)^2}}{v_2}, \; 0 < x < d$$

$$T'(x) = \frac{x}{v_1\sqrt{a^2+x^2}} - \frac{d-x}{v_2\sqrt{b^2+(d-x)^2}} = \frac{\sin\theta_1}{v_1} - \frac{\sin\theta_2}{v_2}$$

The minimum occurs when $T'(x) = 0 \;\Rightarrow\; \dfrac{\sin\theta_1}{v_1} = \dfrac{\sin\theta_2}{v_2}$.

[*Note:* $T''(x) > 0$]

72. 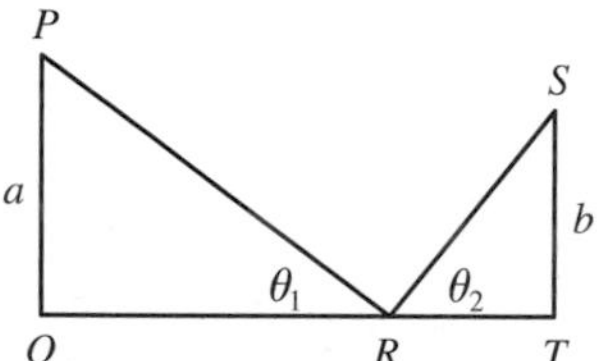

If $d = |QT|$, we minimize $f(\theta_1) = |PR| + |RS| = a\csc\theta_1 + b\csc\theta_2$.

Differentiating with respect to θ_1, and setting $\dfrac{df}{d\theta_1}$ equal to 0, we get

$$\frac{df}{d\theta_1} = 0 = -a\csc\theta_1\,\cot\theta_1 - b\csc\theta_2\,\cot\theta_2\,\frac{d\theta_2}{d\theta_1}.$$

So we need to find an expression for $\dfrac{d\theta_2}{d\theta_1}$. We can do this by observing that $|QT| = \text{constant} = a\cot\theta_1 + b\cot\theta_2$.

Differentiating this equation implicitly with respect to θ_1, we get $-a\csc^2\theta_1 - b\csc^2\theta_2\,\dfrac{d\theta_2}{d\theta_1} = 0 \;\Rightarrow$

$\dfrac{d\theta_2}{d\theta_1} = -\dfrac{a\csc^2\theta_1}{b\csc^2\theta_2}$. We substitute this into the expression for $\dfrac{df}{d\theta_1}$ to get

$$-a\csc\theta_1\,\cot\theta_1 - b\csc\theta_2\,\cot\theta_2\left(-\frac{a\csc^2\theta_1}{b\csc^2\theta_2}\right) = 0 \;\Leftrightarrow\; -a\csc\theta_1\,\cot\theta_1 + a\frac{\csc^2\theta_1\,\cot\theta_2}{\csc\theta_2} = 0 \;\Leftrightarrow$$

$\cot\theta_1\,\csc\theta_2 = \csc\theta_1\,\cot\theta_2 \;\Leftrightarrow\; \dfrac{\cot\theta_1}{\csc\theta_1} = \dfrac{\cot\theta_2}{\csc\theta_2} \;\Leftrightarrow\; \cos\theta_1 = \cos\theta_2$. Since θ_1 and θ_2 are both acute, we have $\theta_1 = \theta_2$.

73. 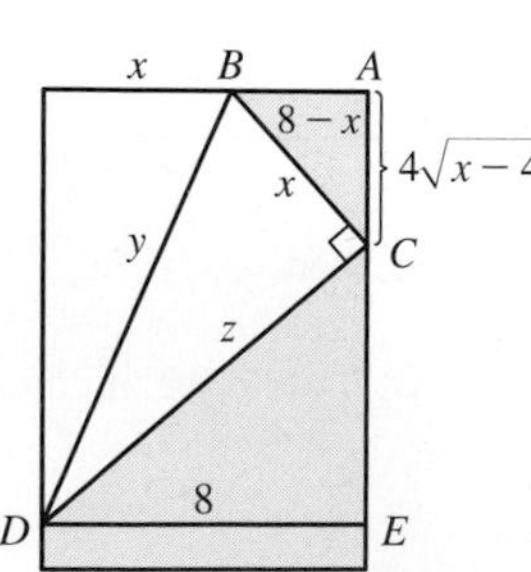

$y^2 = x^2 + z^2$, but triangles CDE and BCA are similar, so $z/8 = x/\left(4\sqrt{x-4}\right) \;\Rightarrow\; z = 2x/\sqrt{x-4}$. Thus, we minimize $f(x) = y^2 = x^2 + 4x^2/(x-4) = x^3/(x-4)$, $4 < x \le 8$.

$$f'(x) = \frac{(x-4)(3x^2) - x^3}{(x-4)^2} = \frac{x^2[3(x-4) - x]}{(x-4)^2} = \frac{2x^2(x-6)}{(x-4)^2} = 0$$

when $x = 6$. $f'(x) < 0$ when $x < 6$, $f'(x) > 0$ when $x > 6$, so the minimum occurs when $x = 6$ in.

74. 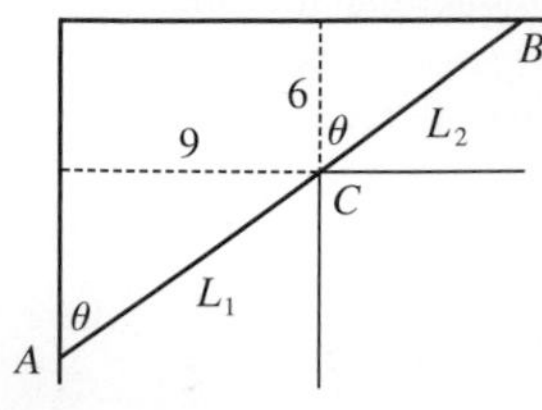

Paradoxically, we solve this maximum problem by solving a minimum problem. Let L be the length of the line ACB going from wall to wall touching the inner corner C. As $\theta \to 0$ or $\theta \to \frac{\pi}{2}$, we have $L \to \infty$ and there will be an angle that makes L a minimum. A pipe of this length will just fit around the corner.

From the diagram, $L = L_1 + L_2 = 9\csc\theta + 6\sec\theta \;\Rightarrow\; dL/d\theta = -9\csc\theta\,\cot\theta + 6\sec\theta\,\tan\theta = 0$ when $6\sec\theta\,\tan\theta = 9\csc\theta\,\cot\theta \;\Leftrightarrow\; \tan^3\theta = \frac{9}{6} = 1.5 \;\Leftrightarrow\; \tan\theta = \sqrt[3]{1.5}$. Then $\sec^2\theta = 1 + \left(\frac{3}{2}\right)^{2/3}$ and $\csc^2\theta = 1 + \left(\frac{3}{2}\right)^{-2/3}$, so the longest pipe has length $L = 9\left[1 + \left(\frac{3}{2}\right)^{-2/3}\right]^{1/2} + 6\left[1 + \left(\frac{3}{2}\right)^{2/3}\right]^{1/2} \approx 21.07$ ft.

Or, use $\theta = \tan^{-1}\left(\sqrt[3]{1.5}\right) \approx 0.853 \;\Rightarrow\; L = 9\csc\theta + 6\sec\theta \approx 21.07$ ft.

75. 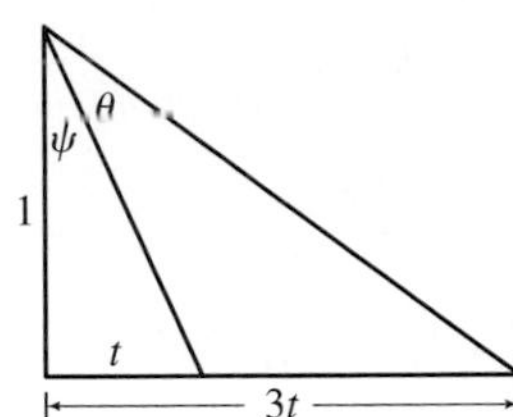

$\theta = (\theta + \psi) - \psi = \arctan\dfrac{3t}{1} - \arctan\dfrac{t}{1} \;\Rightarrow\; \theta' = \dfrac{3}{1 + 9t^2} - \dfrac{1}{1 + t^2}$.

$\theta' = 0 \;\Rightarrow\; \dfrac{3}{1 + 9t^2} = \dfrac{1}{1 + t^2} \;\Rightarrow\; 3 + 3t^2 = 1 + 9t^2 \;\Rightarrow\; 2 = 6t^2 \;\Rightarrow$

$t^2 = \frac{1}{3} \;\Rightarrow\; t = 1/\sqrt{3}$. Thus,

$\theta = \arctan 3/\sqrt{3} - \arctan 1/\sqrt{3} = \frac{\pi}{3} - \frac{\pi}{6} = \frac{\pi}{6}$.

76. We maximize the cross-sectional area

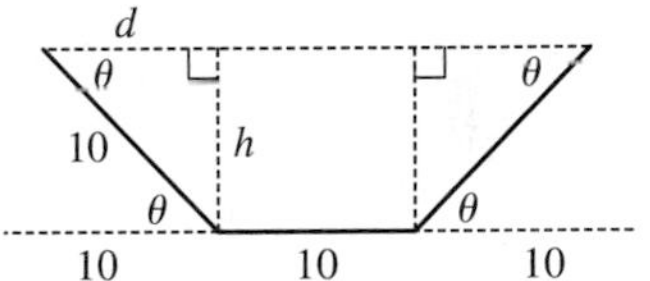

$$A(\theta) = 10h + 2\left(\tfrac{1}{2}dh\right) = 10h + dh = 10(10\sin\theta) + (10\cos\theta)(10\sin\theta)$$
$$= 100(\sin\theta + \sin\theta\,\cos\theta),\quad 0 \le \theta \le \tfrac{\pi}{2}$$

$$A'(\theta) = 100(\cos\theta + \cos^2\theta - \sin^2\theta) = 100(\cos\theta + 2\cos^2\theta - 1)$$
$$= 100(2\cos\theta - 1)(\cos\theta + 1) = 0 \text{ when } \cos\theta = \tfrac{1}{2} \;\Leftrightarrow\; \theta = \tfrac{\pi}{3} \qquad [\cos\theta \ne -1 \text{ since } 0 \le \theta \le \tfrac{\pi}{2}.]$$

Now $A(0) = 0$, $A\left(\frac{\pi}{2}\right) = 100$ and $A\left(\frac{\pi}{3}\right) = 75\sqrt{3} \approx 129.9$, so the maximum occurs when $\theta = \frac{\pi}{3}$.

77. 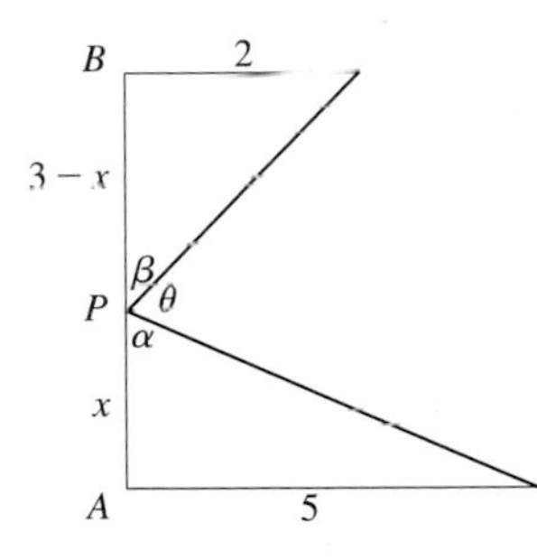

From the figure, $\tan\alpha = \dfrac{5}{x}$ and $\tan\beta = \dfrac{2}{3 - x}$. Since

$\alpha + \beta + \theta = 180^\circ = \pi$, $\theta = \pi - \tan^{-1}\left(\dfrac{5}{x}\right) - \tan^{-1}\left(\dfrac{2}{3 - x}\right) \;\Rightarrow$

$$\frac{d\theta}{dx} = -\frac{1}{1 + \left(\dfrac{5}{x}\right)^2}\left(-\frac{5}{x^2}\right) - \frac{1}{1 + \left(\dfrac{2}{3 - x}\right)^2}\left[\frac{2}{(3 - x)^2}\right]$$
$$= \frac{x^2}{x^2 + 25}\cdot\frac{5}{x^2} - \frac{(3 - x)^2}{(3 - x)^2 + 4}\cdot\frac{2}{(3 - x)^2}.$$

Now $\dfrac{d\theta}{dx} = 0 \;\Rightarrow\; \dfrac{5}{x^2 + 25} = \dfrac{2}{x^2 - 6x + 13} \;\Rightarrow\; 2x^2 + 50 = 5x^2 - 30x + 65 \;\Rightarrow$

$3x^2 - 30x + 15 = 0 \;\Rightarrow\; x^2 - 10x + 5 = 0 \;\Rightarrow\; x = 5 \pm 2\sqrt{5}$. We reject the root with the $+$ sign, since it is larger than 3. $d\theta/dx > 0$ for $x < 5 - 2\sqrt{5}$ and $d\theta/dx < 0$ for $x > 5 - 2\sqrt{5}$, so θ is maximized when $|AP| = x = 5 - 2\sqrt{5} \approx 0.53$.

78. Let x be the distance from the observer to the wall. Then, from the given figure,

$$\theta = \tan^{-1}\left(\frac{h+d}{x}\right) - \tan^{-1}\left(\frac{d}{x}\right), \quad x > 0 \quad \Rightarrow$$

$$\frac{d\theta}{dx} = \frac{1}{1+[(h+d)/x]^2}\left[-\frac{h+d}{x^2}\right] - \frac{1}{1+(d/x)^2}\left[-\frac{d}{x^2}\right] = -\frac{h+d}{x^2+(h+d)^2} + \frac{d}{x^2+d^2}$$

$$= \frac{d[x^2+(h+d)^2]-(h+d)(x^2+d^2)}{[x^2+(h+d)^2](x^2+d^2)} = \frac{h^2d+hd^2-hx^2}{[x^2+(h+d)^2](x^2+d^2)} = 0 \quad \Leftrightarrow$$

$hx^2 = h^2d + hd^2 \quad \Leftrightarrow \quad x^2 = hd + d^2 \quad \Leftrightarrow \quad x = \sqrt{d(h+d)}$. Since $d\theta/dx > 0$ for all $x < \sqrt{d(h+d)}$ and $d\theta/dx < 0$ for all $x > \sqrt{d(h+d)}$, the absolute maximum occurs when $x = \sqrt{d(h+d)}$.

79.

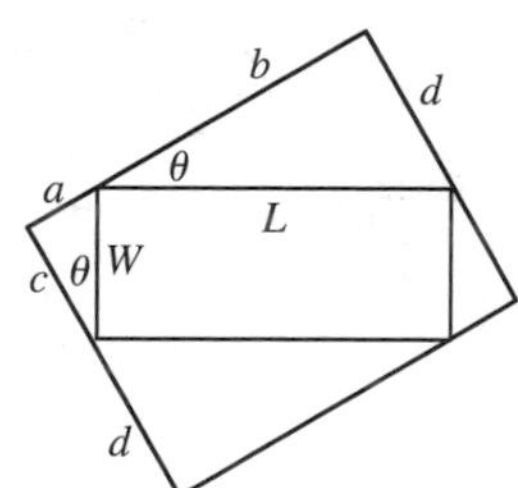

In the small triangle with sides a and c and hypotenuse W, $\sin\theta = \dfrac{a}{W}$ and $\cos\theta = \dfrac{c}{W}$. In the triangle with sides b and d and hypotenuse L, $\sin\theta = \dfrac{d}{L}$ and $\cos\theta = \dfrac{b}{L}$. Thus, $a = W\sin\theta$, $c = W\cos\theta$, $d = L\sin\theta$, and $b = L\cos\theta$, so the area of the circumscribed rectangle is

$$\begin{aligned} A(\theta) &= (a+b)(c+d) = (W\sin\theta + L\cos\theta)(W\cos\theta + L\sin\theta) \\ &= W^2\sin\theta\cos\theta + WL\sin^2\theta + LW\cos^2\theta + L^2\sin\theta\cos\theta \\ &= LW\sin^2\theta + LW\cos^2\theta + (L^2+W^2)\sin\theta\cos\theta \\ &= LW(\sin^2\theta+\cos^2\theta) + (L^2+W^2)\cdot\tfrac{1}{2}\cdot 2\sin\theta\cos\theta = LW + \tfrac{1}{2}(L^2+W^2)\sin 2\theta, \ 0 \le \theta \le \tfrac{\pi}{2} \end{aligned}$$

This expression shows, without calculus, that the maximum value of $A(\theta)$ occurs when $\sin 2\theta = 1 \quad \Leftrightarrow \quad 2\theta = \frac{\pi}{2} \quad \Rightarrow$ $\theta = \frac{\pi}{4}$. So the maximum area is $A\left(\frac{\pi}{4}\right) = LW + \frac{1}{2}(L^2+W^2) = \frac{1}{2}(L^2+2LW+W^2) = \frac{1}{2}(L+W)^2$.

80. (a) Let D be the point such that $a = |AD|$. From the figure, $\sin\theta = \dfrac{b}{|BC|} \quad \Rightarrow \quad |BC| = b\csc\theta$ and

$\cos\theta = \dfrac{|BD|}{|BC|} = \dfrac{a-|AB|}{|BC|} \quad \Rightarrow \quad |BC| = (a-|AB|)\sec\theta$. Eliminating $|BC|$ gives

$(a-|AB|)\sec\theta = b\csc\theta \quad \Rightarrow \quad b\cot\theta = a-|AB| \quad \Rightarrow \quad |AB| = a - b\cot\theta$. The total resistance is

$$R(\theta) = C\frac{|AB|}{r_1^4} + C\frac{|BC|}{r_2^4} = C\left(\frac{a-b\cot\theta}{r_1^4} + \frac{b\csc\theta}{r_2^4}\right).$$

(b) $R'(\theta) = C\left(\dfrac{b\csc^2\theta}{r_1^4} - \dfrac{b\csc\theta\cot\theta}{r_2^4}\right) = bC\csc\theta\left(\dfrac{\csc\theta}{r_1^4} - \dfrac{\cot\theta}{r_2^4}\right).$

$R'(\theta) = 0 \quad \Leftrightarrow \quad \dfrac{\csc\theta}{r_1^4} = \dfrac{\cot\theta}{r_2^4} \quad \Leftrightarrow \quad \dfrac{r_2^4}{r_1^4} = \dfrac{\cot\theta}{\csc\theta} = \cos\theta.$

$R'(\theta) > 0 \quad \Leftrightarrow \quad \dfrac{\csc\theta}{r_1^4} > \dfrac{\cot\theta}{r_2^4} \quad \Rightarrow \quad \cos\theta < \dfrac{r_2^4}{r_1^4}$ and $R'(\theta) < 0$ when $\cos\theta > \dfrac{r_2^4}{r_1^4}$, so there is an absolute minimum when $\cos\theta = r_2^4/r_1^4$.

(c) When $r_2 = \frac{2}{3}r_1$, we have $\cos\theta = \left(\frac{2}{3}\right)^4$, so $\theta = \cos^{-1}\left(\frac{2}{3}\right)^4 \approx 79°$.

81. (a) If $k =$ energy/km over land, then energy/km over water $= 1.4k$.

So the total energy is $E = 1.4k\sqrt{25 + x^2} + k(13 - x),\ 0 \le x \le 13$,

and so $\dfrac{dE}{dx} = \dfrac{1.4kx}{(25 + x^2)^{1/2}} - k$.

Set $\dfrac{dE}{dx} = 0$: $1.4kx = k(25 + x^2)^{1/2} \Rightarrow 1.96x^2 = x^2 + 25 \Rightarrow 0.96x^2 = 25 \Rightarrow x = \frac{5}{\sqrt{0.96}} \approx 5.1$.

Testing against the value of E at the endpoints: $E(0) = 1.4k(5) + 13k = 20k$, $E(5.1) \approx 17.9k$, $E(13) \approx 19.5k$.

Thus, to minimize energy, the bird should fly to a point about 5.1 km from B.

(b) If W/L is large, the bird would fly to a point C that is closer to B than to D to minimize the energy used flying over water. If W/L is small, the bird would fly to a point C that is closer to D than to B to minimize the distance of the flight.

$E = W\sqrt{25 + x^2} + L(13 - x) \Rightarrow \dfrac{dE}{dx} = \dfrac{Wx}{\sqrt{25 + x^2}} - L = 0$ when $\dfrac{W}{L} = \dfrac{\sqrt{25 + x^2}}{x}$. By the same sort of argument as in part (a), this ratio will give the minimal expenditure of energy if the bird heads for the point x km from B.

(c) For flight direct to D, $x = 13$, so from part (b), $W/L = \frac{\sqrt{25 + 13^2}}{13} \approx 1.07$. There is no value of W/L for which the bird should fly directly to B. But note that $\lim\limits_{x \to 0^+} (W/L) = \infty$, so if the point at which E is a minimum is close to B, then W/L is large.

(d) Assuming that the birds instinctively choose the path that minimizes the energy expenditure, we can use the equation for $dE/dx = 0$ from part (a) with $1.4k = c$, $x = 4$, and $k = 1$: $c(4) = 1 \cdot (25 + 4^2)^{1/2} \Rightarrow c = \sqrt{41}/4 \approx 1.6$.

82. (a) $I(x) \propto \dfrac{\text{strength of source}}{(\text{distance from source})^2}$. Adding the intensities from the left and right lightbulbs,

$$I(x) = \frac{k}{x^2 + d^2} + \frac{k}{(10 - x)^2 + d^2} = \frac{k}{x^2 + d^2} + \frac{k}{x^2 - 20x + 100 + d^2}.$$

(b) The magnitude of the constant k won't affect the location of the point of maximum intensity, so for convenience we take

$k = 1$. $I'(x) = -\dfrac{2x}{(x^2 + d^2)^2} - \dfrac{2(x - 10)}{(x^2 - 20x + 100 + d^2)^2}$.

Substituting $d = 5$ into the equations for $I(x)$and $I'(x)$, we get

$$I_5(x) = \frac{1}{x^2 + 25} + \frac{1}{x^2 - 20x + 125} \quad \text{and} \quad I_5'(x) = -\frac{2x}{(x^2 + 25)^2} - \frac{2(x - 10)}{(x^2 - 20x + 125)^2}$$

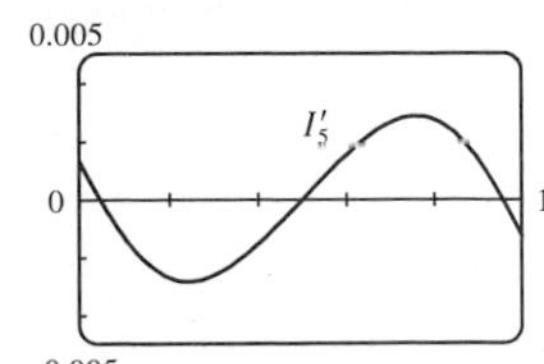

From the graphs, it appears that $I_5(x)$ has a minimum at $x = 5$ m.

(c) Substituting $d = 10$ into the equations for $I(x)$ and $I'(x)$ gives

$$I_{10}(x) = \frac{1}{x^2 + 100} + \frac{1}{x^2 - 20x + 200} \quad \text{and} \quad I'_{10}(x) = -\frac{2x}{(x^2 + 100)^2} - \frac{2\,(x - 10)}{(x^2 - 20x + 200)^2}$$

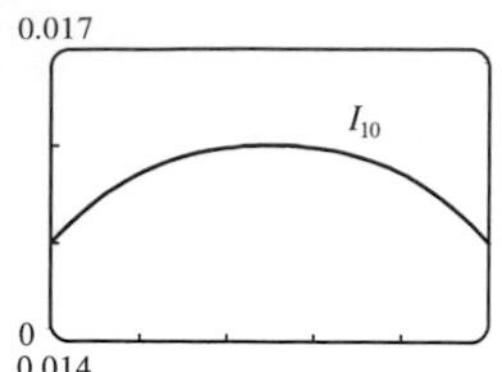

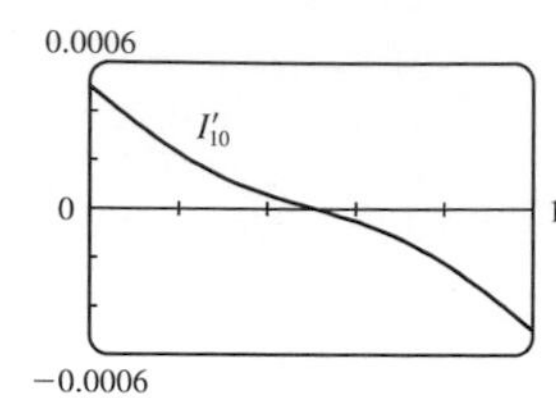

From the graphs, it seems that for $d = 10$, the intensity is minimized at the endpoints, that is, $x = 0$ and $x = 10$. The midpoint is now the most brightly lit point!

(d) From the first figures in parts (b) and (c), we see that the minimal illumination changes from the midpoint ($x = 5$ with $d = 5$) to the endpoints ($x = 0$ and $x = 10$ with $d = 10$).

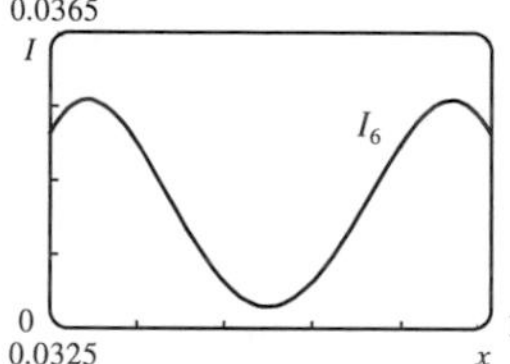

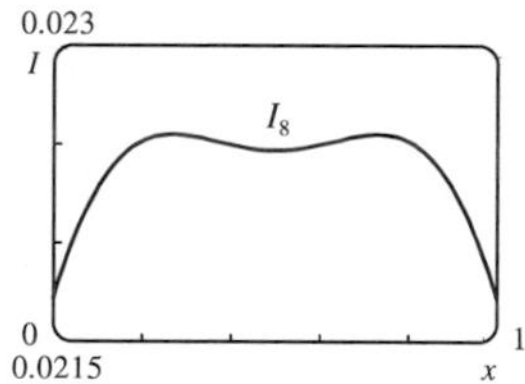

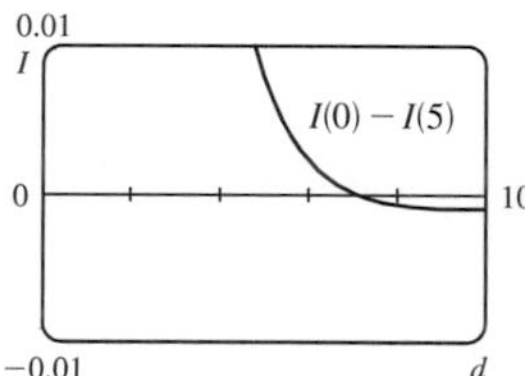

So we try $d = 6$ (see the first figure) and we see that the minimum value still occurs at $x = 5$. Next, we let $d = 8$ (see the second figure) and we see that the minimum value occurs at the endpoints. It appears that for some value of d between 6 and 8, we must have minima at both the midpoint and the endpoints, that is, $I(5)$ must equal $I(0)$. To find this value of d, we solve $I(0) = I(5)$ (with $k = 1$):

$$\frac{1}{d^2} + \frac{1}{100 + d^2} = \frac{1}{25 + d^2} + \frac{1}{25 + d^2} = \frac{2}{25 + d^2} \quad \Rightarrow \quad (25 + d^2)(100 + d^2) + d^2(25 + d^2) = 2d^2(100 + d^2) \quad \Rightarrow$$

$$2500 + 125d^2 + d^4 + 25d^2 + d^4 = 200d^2 + 2d^4 \quad \Rightarrow \quad 2500 = 50d^2 \quad \Rightarrow \quad d^2 = 50 \quad \Rightarrow \quad d = 5\sqrt{2} \approx 7.071$$

[for $0 \le d \le 10$]. The third figure, a graph of $I(0) - I(5)$ with d independent, confirms that $I(0) - I(5) = 0$, that is, $I(0) = I(5)$, when $d = 5\sqrt{2}$. Thus, the point of minimal illumination changes abruptly from the midpoint to the endpoints when $d = 5\sqrt{2}$.

APPLIED PROJECT The Shape of a Can

1. In this case, the amount of metal used in the making of each top or bottom is $(2r)^2 = 4r^2$. So the quantity we want to minimize is $A = 2\pi rh + 2(4r^2)$. But $V = \pi r^2 h \;\Leftrightarrow\; h = V/\pi r^2$. Substituting this expression for h in A gives $A = 2V/r + 8r^2$. Differentiating A with respect to r, we get $dA/dr = -2V/r^2 + 16r = 0 \;\Rightarrow$

$$16r^3 = 2V = 2\pi r^2 h \quad \Leftrightarrow \quad \frac{h}{r} = \frac{8}{\pi} \approx 2.55.$$ This gives a minimum because $\dfrac{d^2A}{dr^2} = 16 + \dfrac{4V}{r^3} > 0$.

2.

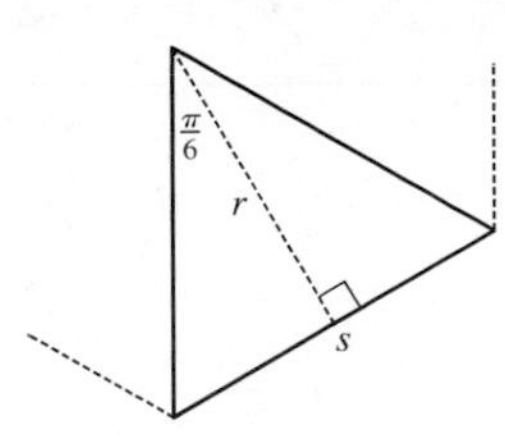

We need to find the area of metal used up by each end, that is, the area of each hexagon. We subdivide the hexagon into six congruent triangles, each sharing one side (s in the diagram) with the hexagon. We calculate the length of $s = 2r\tan\frac{\pi}{6} = \frac{2}{\sqrt{3}}r$, so the area of each triangle is $\frac{1}{2}sr = \frac{1}{\sqrt{3}}r^2$, and the total area of the hexagon is $6 \cdot \frac{1}{\sqrt{3}}r^2 = 2\sqrt{3}\,r^2$. So the quantity we want to minimize is $A = 2\pi rh + 2 \cdot 2\sqrt{3}\,r^2$. Substituting for h as in Problem 1 and differentiating, we get $\dfrac{dA}{dr} = -\dfrac{2V}{r^2} + 8\sqrt{3}\,r$.

Setting this equal to 0, we get $8\sqrt{3}\,r^3 = 2V = 2\pi r^2 h \;\Rightarrow\; \dfrac{h}{r} = \dfrac{4\sqrt{3}}{\pi} \approx 2.21$. Again this minimizes A because $\dfrac{d^2A}{dr^2} = 8\sqrt{3} + \dfrac{4V}{r^3} > 0$.

3. Let $C = 4\sqrt{3}\,r^2 + 2\pi rh + k(4\pi r + h) = 4\sqrt{3}\,r^2 + 2\pi r\left(\dfrac{V}{\pi r^2}\right) + k\left(4\pi r + \dfrac{V}{\pi r^2}\right)$. Then

$\dfrac{dC}{dr} = 8\sqrt{3}\,r - \dfrac{2V}{r^2} + 4k\pi - \dfrac{2kV}{\pi r^3}$. Setting this equal to 0, dividing by 2 and substituting $\dfrac{V}{r^2} = \pi h$ and $\dfrac{V}{\pi r^3} = \dfrac{h}{r}$ in the second and fourth terms respectively, we get $0 = 4\sqrt{3}\,r - \pi h + 2k\pi - \dfrac{kh}{r} \;\Leftrightarrow$

$k\left(2\pi - \dfrac{h}{r}\right) = \pi h - 4\sqrt{3}\,r \;\Rightarrow\; \dfrac{k}{r}\dfrac{2\pi - h/r}{\pi h/r - 4\sqrt{3}} = 1$. We now multiply by $\dfrac{\sqrt[3]{V}}{k}$, noting that $\dfrac{\sqrt[3]{V}}{k}\dfrac{k}{r} = \sqrt[3]{\dfrac{V}{r^3}} = \sqrt[3]{\dfrac{\pi h}{r}}$,

and get $\dfrac{\sqrt[3]{V}}{k} = \sqrt[3]{\dfrac{\pi h}{r}} \cdot \dfrac{2\pi - h/r}{\pi h/r - 4\sqrt{3}}$.

4. Let $\sqrt[3]{V}/k = T$ and $h/r = x$ so that $T(x) = \sqrt[3]{\pi x} \cdot \dfrac{2\pi - x}{\pi x - 4\sqrt{3}}$. We see from the graph of T that when the ratio $\sqrt[3]{V}/k$ is large; that is, either the volume of the can is large or the cost of joining (proportional to k) is small, the optimum value of h/r is about 2.21, but when $\sqrt[3]{V}/k$ is small, indicating small volume or expensive joining, the optimum value of h/r is larger. (The part of the graph for $\sqrt[3]{V}/k < 0$ has no physical meaning, but confirms the location of the asymptote.)

5. Our conclusion is usually true in practice. But there are exceptions, such as cans of tuna, which may have to do with the shape of a reasonable slice of tuna. And for a comfortable grip on a soda or beer can, the geometry of the human hand is a restriction on the radius. Other possible considerations are packaging, transportation and stocking constraints, aesthetic appeal and other marketing concerns. Also, there may be better models than ours which prescribe a differently shaped can in special circumstances.

APPLIED PROJECT Planes and Birds: Minimizing Energy

1. $P(v) = Av^3 + \dfrac{BL^2}{v} \Rightarrow P'(v) = 3Av^2 - \dfrac{BL^2}{v^2}$. $P'(v) = 0 \Leftrightarrow 3Av^2 = \dfrac{BL^2}{v^2} \Leftrightarrow v^4 = \dfrac{BL^2}{3A} \Rightarrow$

$v = \sqrt[4]{\dfrac{BL^2}{3A}}$. $P''(v) = 6Av + \dfrac{2BL^2}{v^3} > 0$, so the speed that minimizes the required power is $v_P = \left(\dfrac{BL^2}{3A}\right)^{1/4}$.

2. $E(v) = \dfrac{P(v)}{v} = Av^2 + \dfrac{BL^2}{v^2} \Rightarrow E'(v) = 2Av - \dfrac{2BL^2}{v^3}$. $E'(v) = 0 \Leftrightarrow 2Av = \dfrac{2BL^2}{v^3} \Leftrightarrow v^4 = \dfrac{BL^2}{A} \Rightarrow$

$v = \sqrt[4]{\dfrac{BL^2}{A}}$. $E''(v) = 2A + \dfrac{6BL^2}{v^4} > 0$, so the speed that minimizes the energy needed to propel the plane is

$v_E = \left(\dfrac{BL^2}{A}\right)^{1/4}$.

3. $\dfrac{v_E}{v_P} = \dfrac{\left(\dfrac{BL^2}{A}\right)^{1/4}}{\left(\dfrac{BL^2}{3A}\right)^{1/4}} = \left(\dfrac{\dfrac{BL^2}{A}}{\dfrac{BL^2}{3A}}\right)^{1/4} = 3^{1/4} \approx 1.316$. Thus, $v_E \approx 1.316\, v_P$, so the speed for minimum energy is about

31.6% greater (faster) than the speed for minimum power.

4. Since x is the fraction of flying time spent in flapping mode, $1 - x$ is the fraction of time spent in folded mode. The average power $\overline{P}$ is the weighted average of P_{flap} and P_{fold}, so

$$\overline{P} = xP_{\text{flap}} + (1-x)P_{\text{fold}} = x\left[(A_b + A_w)\,v^3 + \frac{B(mg/x)^2}{v}\right] + (1-x)A_bv^3$$

$$= xA_bv^3 + xA_wv^3 + x\frac{Bm^2g^2}{x^2v} + A_bv^3 - xA_bv^3 = A_bv^3 + xA_wv^3 + \frac{Bm^2g^2}{xv}$$

5. $\overline{P}(x) = A_bv^3 + xA_wv^3 + \dfrac{Bm^2g^2}{xv} \Rightarrow \overline{P}'(x) = A_wv^3 - \dfrac{Bm^2g^2}{x^2v}$. $\overline{P}'(x) = 0 \Leftrightarrow A_wv^3 = \dfrac{Bm^2g^2}{x^2v} \Leftrightarrow$

$x^2 = \dfrac{Bm^2g^2}{A_wv^4} \Rightarrow x = \dfrac{mg}{v^2}\sqrt{\dfrac{B}{A_w}}$. Since $\overline{P}''(x) = \dfrac{2Bm^2g^2}{x^3v} > 0$, this critical number, call it $x_{\overline{P}}$, gives an absolute minimum for the average power. If the bird flies slowly, then v is smaller and $x_{\overline{P}}$ increases, and the bird spends a larger fraction of its flying time flapping. If the bird flies faster and faster, then v is larger and $x_{\overline{P}}$ decreases, and the bird spends a smaller fraction of its flying time flapping, while still minimizing average power.

6. $\overline{E}(x) = \dfrac{\overline{P}(x)}{v} \Rightarrow \overline{E}'(x) = \dfrac{1}{v}\overline{P}'(x)$, so $\overline{E}'(x) = 0 \Leftrightarrow \overline{P}'(x) = 0$. The value of x that minimizes $\overline{E}$ is the same value of x that minimizes $\overline{P}$, namely $x_{\overline{P}} = \dfrac{mg}{v^2}\sqrt{\dfrac{B}{A_w}}$.

4.8 Newton's Method

1. (a) The tangent line at $x_1 = 6$ intersects the x-axis at $x \approx 7.3$, so $x_2 = 7.3$. The tangent line at $x = 7.3$ intersects the x-axis at $x \approx 6.8$, so $x_3 \approx 6.8$.

(b) $x_1 = 8$ would be a better first approximation because the tangent line at $x = 8$ intersects the x-axis closer to s than does the first approximation $x_1 = 6$.

2. 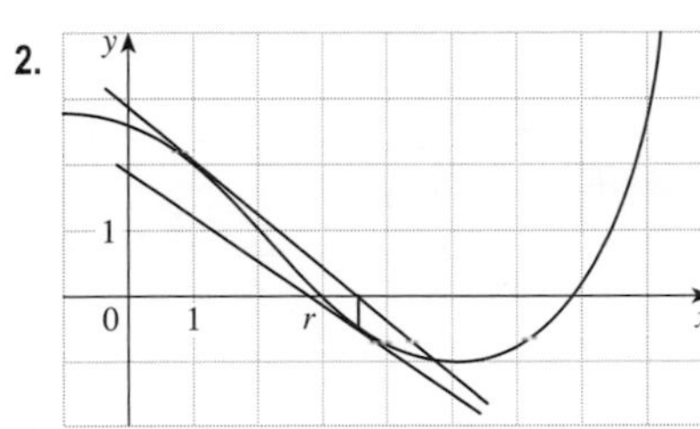

The tangent line at $x_1 = 1$ intersects the x-axis at $x \approx 3.5$, so $x_2 = 3.5$. The tangent line at $x = 3.5$ intersects the x-axis at $x \approx 2.8$, so $x_3 = 2.8$.

3. Since the tangent line $y = 9 - 2x$ is tangent to the curve $y = f(x)$ at the point $(2, 5)$, we have $x_1 = 2$, $f(x_1) = 5$, and $f'(x_1) = -2$ [the slope of the tangent line]. Thus, by Equation 2,

$$x_2 = x_1 - \frac{f(x_1)}{f'(x_1)} = 2 - \frac{5}{-2} = \frac{9}{2}$$

Note that geometrically $\frac{9}{2}$ represents the x-intercept of the tangent line $y = 9 - 2x$.

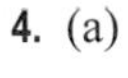

4. (a)

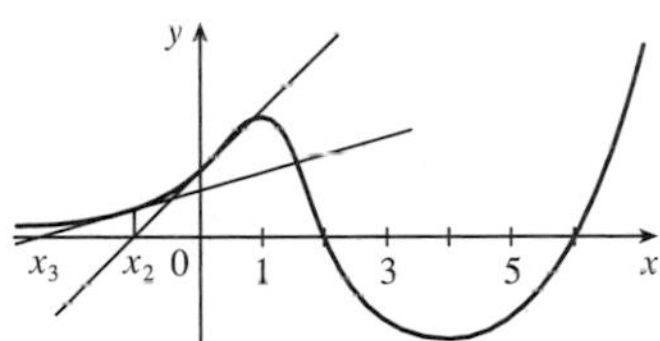

If $x_1 = 0$, then x_2 is negative, and x_3 is even more negative. The sequence of approximations does not converge, that is, Newton's method fails.

(b)

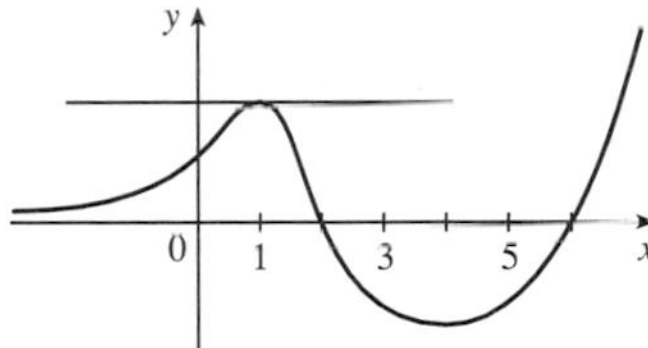

If $x_1 = 1$, the tangent line is horizontal and Newton's method fails.

(c)

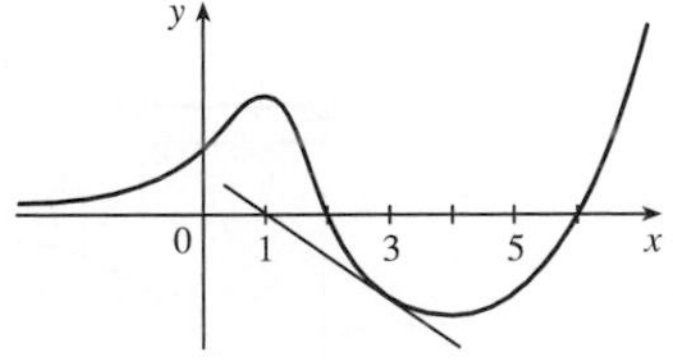

If $x_1 = 3$, then $x_2 = 1$ and we have the same situation as in part (b). Newton's method fails again.

(d)

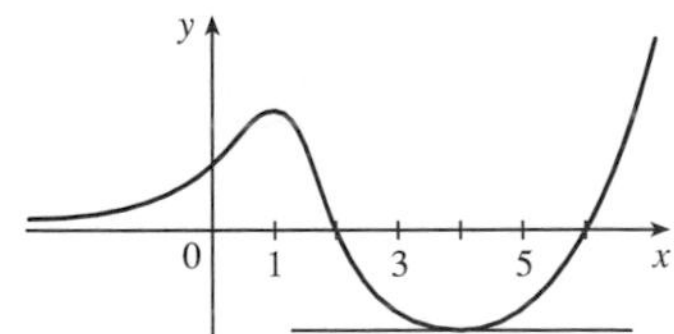

If $x_1 = 4$, the tangent line is horizontal and Newton's method fails.

(e)

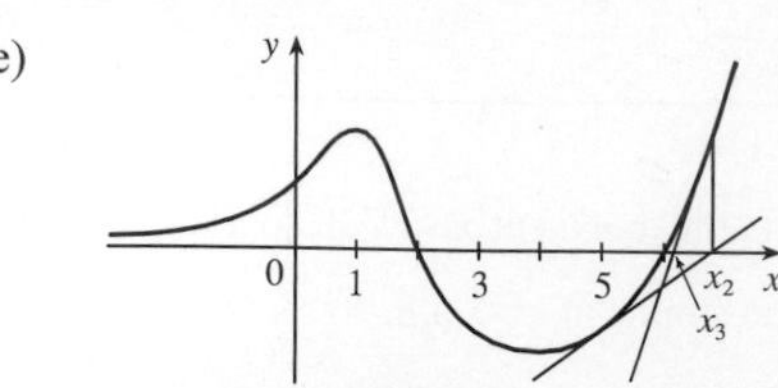

If $x_1 = 5$, then x_2 is greater than 6, x_3 gets closer to 6, and the sequence of approximations converges to 6. Newton's method succeeds!

5. The initial approximations $x_1 = a$, b, and c will work, resulting in a second approximation closer to the origin, and lead to the root of the equation $f(x) = 0$, namely, $x = 0$. The initial approximation $x_1 = d$ will not work because it will result in successive approximations farther and farther from the origin.

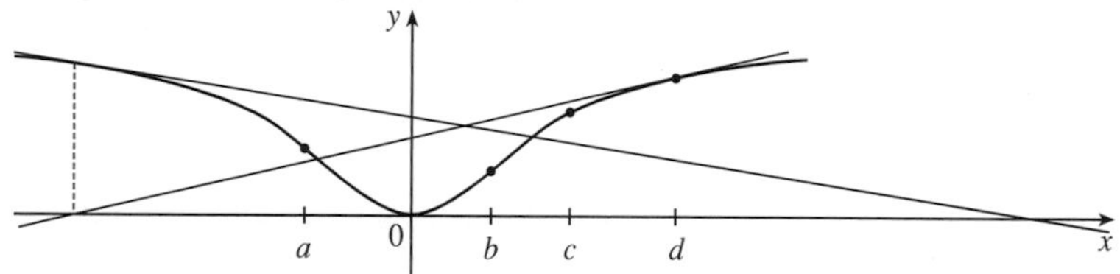

6. $f(x) = 2x^3 - 3x^2 + 2 \Rightarrow f'(x) = 6x^2 - 6x$, so $x_{n+1} = x_n - \dfrac{2x_n^3 - 3x_n^2 + 2}{6x_n^2 - 6x_n}$. Now $x_1 = -1 \Rightarrow$

$$x_2 = -1 - \frac{2(-1)^3 - 3(-1)^2 + 2}{6(-1)^2 - 6(-1)} = -1 - \frac{-3}{12} = -\frac{3}{4} \Rightarrow$$

$$x_3 = -\frac{3}{4} - \frac{2\left(-\frac{3}{4}\right)^3 - 3\left(-\frac{3}{4}\right)^2 + 2}{6\left(-\frac{3}{4}\right)^2 - 6\left(-\frac{3}{4}\right)} = -\frac{3}{4} - \frac{-17/32}{63/8} = -\frac{43}{63} \approx -0.6825.$$

7. $f(x) = \dfrac{2}{x} - x^2 + 1 \Rightarrow f'(x) = -\dfrac{2}{x^2} - 2x$, so $x_{n+1} = x_n - \dfrac{2/x_n - x_n^2 + 1}{-2/x_n^2 - 2x_n}$. Now $x_1 = 2 \Rightarrow$

$$x_2 = 2 - \frac{1 - 4 + 1}{-1/2 - 4} = 2 - \frac{-2}{-9/2} = \frac{14}{9} \Rightarrow x_3 = \frac{14}{9} - \frac{2/(14/9) - (14/9)^2 + 1}{-2(14/9)^2 - 2(14/9)} \approx 1.5215.$$

8. $f(x) = x^7 + 4 \Rightarrow f'(x) = 7x^6$, so $x_{n+1} = x_n - \dfrac{x_n^7 + 4}{7x_n^6}$. Now $x_1 = -1 \Rightarrow$

$$x_2 = -1 - \frac{(-1)^7 + 4}{7(-1)^6} = -1 - \frac{3}{7} = -\frac{10}{7} \Rightarrow x_3 = -\frac{10}{7} - \frac{\left(-\frac{10}{7}\right)^7 + 4}{7\left(-\frac{10}{7}\right)^6} \approx -1.2917.$$

9. $f(x) = x^3 + x + 3 \Rightarrow f'(x) = 3x^2 + 1$, so $x_{n+1} = x_n - \dfrac{x_n^3 + x_n + 3}{3x_n^2 + 1}$.

Now $x_1 = -1 \Rightarrow$

$$x_2 = -1 - \frac{(-1)^3 + (-1) + 3}{3(-1)^2 + 1} = -1 - \frac{-1 - 1 + 3}{3 + 1} = -1 - \frac{1}{4} = -1.25.$$

Newton's method follows the tangent line at $(-1, 1)$ down to its intersection with the x-axis at $(-1.25, 0)$, giving the second approximation $x_2 = -1.25$.

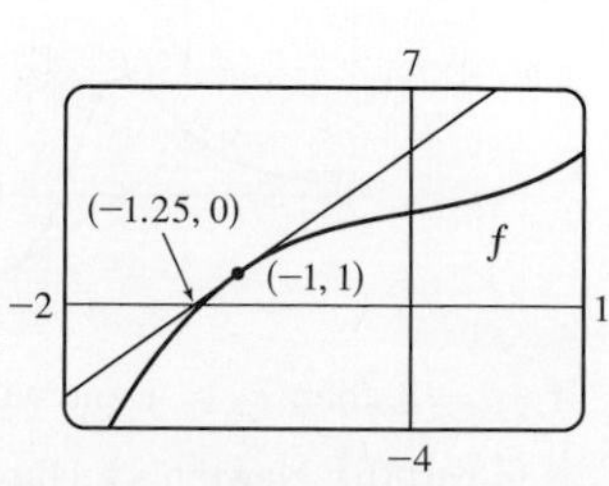

10. $f(x) = x^4 - x - 1 \Rightarrow f'(x) = 4x^3 - 1$, so $x_{n+1} = x_n - \dfrac{x_n^4 - x_n - 1}{4x_n^3 - 1}$.

Now $x_1 = 1 \Rightarrow x_2 = 1 - \dfrac{1^4 - 1 - 1}{4 \cdot 1^3 - 1} = 1 - \dfrac{-1}{3} = \dfrac{4}{3}$. Newton's method follows the tangent line at $(1, -1)$ up to its intersection with the x-axis at $\left(\frac{4}{3}, 0\right)$, giving the second approximation $x_2 = \frac{4}{3}$.

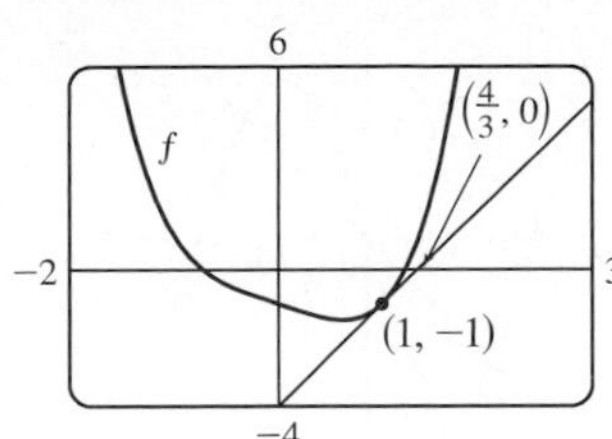

11. To approximate $x = \sqrt[4]{75}$ (so that $x^4 = 75$), we can take $f(x) = x^4 - 75$. So $f'(x) = 4x^3$, and thus, $x_{n+1} = x_n - \dfrac{x_n^4 - 75}{4x_n^3}$. Since $\sqrt[4]{81} = 3$ and 81 is reasonably close to 75, we'll use $x_1 = 3$. We need to find approximations until they agree to eight decimal places. $x_1 = 3 \Rightarrow x_2 = 2.9\overline{4}$, $x_3 \approx 2.94283228$, $x_4 \approx 2.94283096 \approx x_5$. So $\sqrt[4]{75} \approx 2.94283096$, to eight decimal places.

To use Newton's method on a calculator, assign f to Y_1 and f' to Y_2. Then store x_1 in X and enter $X - Y_1/Y_2 \to X$ to get x_2 and further approximations (repeatedly press ENTER).

12. $f(x) = x^8 - 500 \Rightarrow f'(x) = 8x^7$, so $x_{n+1} = x_n - \dfrac{x_n^8 - 500}{8x_n^7}$. Since $\sqrt[8]{256} = 2$ and 256 is reasonably close to 500, we'll use $x_1 = 2$. We need to find approximations until they agree to eight decimal places. $x_1 = 2 \Rightarrow x_2 \approx 2.23828125$, $x_3 \approx 2.18055972$, $x_4 \approx 2.17461675$, $x_5 \approx 2.17455928 \approx x_6$. So $\sqrt[8]{500} \approx 2.17455928$, to eight decimal places.

13. (a) Let $f(x) = 3x^4 - 8x^3 + 2$. The polynomial f is continuous on $[2, 3]$, $f(2) = -14 < 0$, and $f(3) = 29 > 0$, so by the Intermediate Value Theorem, there is a number c in $(2, 3)$ such that $f(c) = 0$. In other words, the equation $3x^4 - 8x^3 + 2 = 0$ has a root in $[2, 3]$.

(b) $f'(x) = 12x^3 - 24x^2 \Rightarrow x_{n+1} = x_n - \dfrac{3x_n^4 - 8x_n^3 + 2}{12x_n^3 - 24x_n^2}$. Taking $x_1 = 2.5$, we get $x_2 = 2.655$, $x_3 \approx 2.630725$, $x_4 \approx 2.630021$, $x_5 \approx 2.630020 \approx x_6$. To six decimal places, the root is 2.630020. Note that taking $x_1 = 2$ is not allowed since $f'(2) = 0$.

14. (a) Let $f(x) = -2x^5 + 9x^4 - 7x^3 - 11x$. The polynomial f is continuous on $[3, 4]$, $f(3) = 21 > 0$, and $f(4) = -236 < 0$, so by the Intermediate Value Theorem, there is a number c in $(3, 4)$ such that $f(c) = 0$. In other words, the equation $-2x^5 + 9x^4 - 7x^3 - 11x = 0$ has a root in $[3, 4]$.

(b) $f'(x) = -10x^4 + 36x^3 - 21x^2 - 11$. $x_{n+1} = x_n - \dfrac{-2x_n^5 + 9x_n^4 - 7x_n^3 - 11x_n}{-10x_n^4 + 36x_n^3 - 21x_n^2 - 11}$. Taking $x_1 = 3.5$, we get $x_2 \approx 3.329174$, $x_3 = 3.278706$, $x_4 \approx 3.274501$, and $x_5 \approx 3.274473 \approx x_6$. To six decimal places, the root is 3.274473.

15.

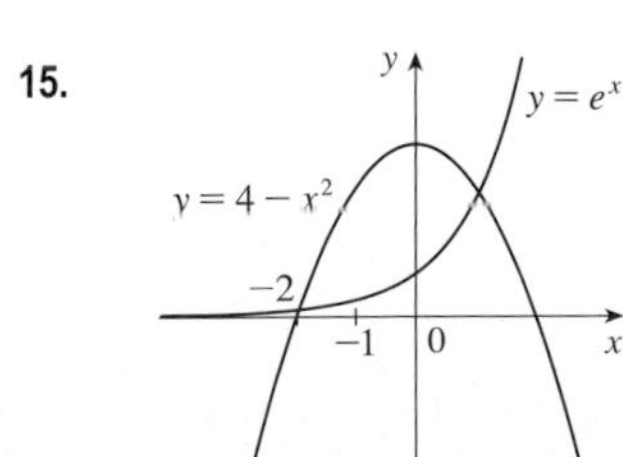

$e^x = 4 - x^2$, so $f(x) = e^x - 4 + x^2 \Rightarrow x_{n+1} = x_n - \dfrac{e^{x_n} - 4 + x_n^2}{e^{x_n} + 2x_n}$.

From the figure, the negative root of $e^x = 4 - x^2$ is near -2.

$x_1 = -2 \Rightarrow x_2 \approx -1.964981$, $x_3 \approx -1.964636 \approx x_4$. So the negative root is -1.964636, to six decimal places.

16.

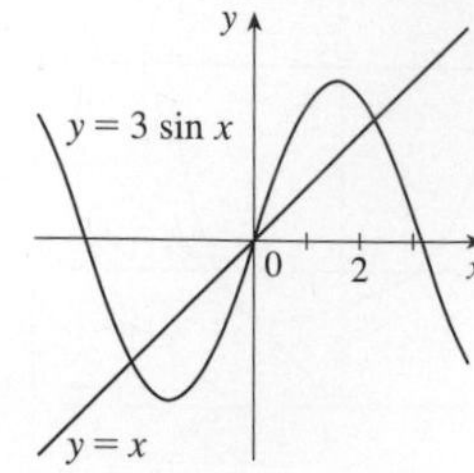

$3\sin x = x$, so $f(x) = 3\sin x - x \;\Rightarrow\; f'(x) = 3\cos x - 1 \;\Rightarrow$

$x_{n+1} = x_n - \dfrac{3\sin x_n - x_n}{3\cos x_n - 1}$. From the figure, the positive root of

$3\sin x = x$ is near 2. $x_1 = 2 \;\Rightarrow\; x_2 \approx 2.323732$, $x_3 \approx 2.279595$,

$x_4 \approx 2.278863 \approx x_5$. So the positive root is 2.278863, to six decimal places.

17.

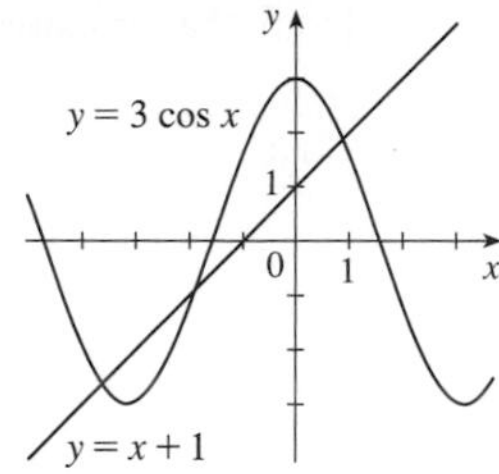

From the graph, we see that there appear to be points of intersection near $x = -4$, $x = -2$, and $x = 1$. Solving $3\cos x = x + 1$ is the same as solving $f(x) = 3\cos x - x - 1 = 0$. $f'(x) = -3\sin x - 1$, so

$$x_{n+1} = x_n - \frac{3\cos x_n - x_n - 1}{-3\sin x_n - 1}.$$

$x_1 = -4$	$x_1 = -2$	$x_1 = 1$
$x_2 \approx -3.682281$	$x_2 \approx -1.856218$	$x_2 \approx 0.892438$
$x_3 \approx -3.638960$	$x_3 \approx -1.862356$	$x_3 \approx 0.889473$
$x_4 \approx -3.637959$	$x_4 \approx -1.862365 \approx x_5$	$x_4 \approx 0.889470 \approx x_5$
$x_5 \approx -3.637958 \approx x_6$		

To six decimal places, the roots of the equation are -3.637958, -1.862365, and 0.889470.

18.

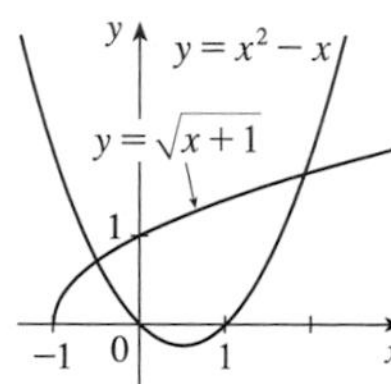

From the graph, we see that there appear to be points of intersection near $x = -0.5$ and $x = 2$. Solving $\sqrt{x+1} = x^2 - x$ is the same as solving $f(x) = \sqrt{x+1} - x^2 + x = 0$. $f'(x) = \dfrac{1}{2\sqrt{x+1}} - 2x + 1$, so

$$x_{n+1} = x_n - \frac{\sqrt{x_n+1} - x_n^2 + x_n}{\dfrac{1}{2\sqrt{x_n+1}} - 2x_n + 1}.$$

$x_1 = -0.5$	$x_1 = 2$
$x_2 \approx -0.484155$	$x_2 \approx 1.901174$
$x_3 \approx -0.484028 \approx x_4$	$x_3 \approx 1.897186$
	$x_4 \approx 1.897179 \approx x_5$

To six decimal places, the roots of the equation are -0.484028 and 0.897179.

19.

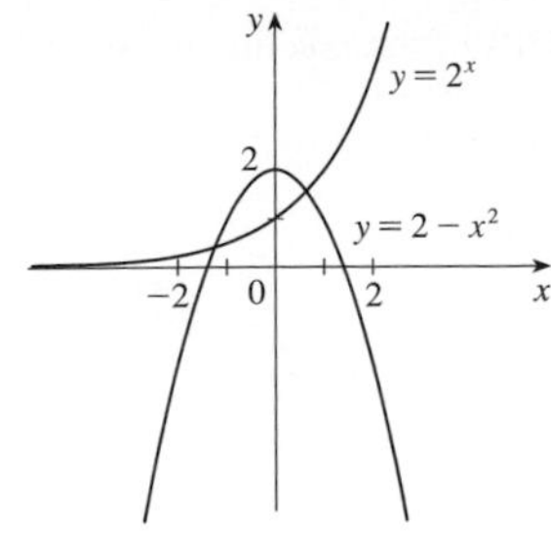

From the figure, we see that the graphs intersect between -2 and -1 and between 0 and 1. Solving $2^x = 2 - x^2$ is the same as solving $f(x) = 2^x - 2 + x^2 = 0$. $f'(x) = 2^x \ln 2 + 2x$, so

$$x_{n+1} = x_n - \frac{2^{x_n} - 2 + x_n^2}{2^{x_n} \ln 2 + 2x_n}.$$

$x_1 = -1$	$x_1 = 1$
$x_2 \approx -1.302402$	$x_2 \approx 0.704692$
$x_3 \approx -1.258636$	$x_3 \approx 0.654915$
$x_4 \approx -1.257692$	$x_4 \approx 0.653484$
$x_5 \approx -1.257691 \approx x_6$	$x_5 \approx 0.653483 \approx x_6$

To six decimal places, the roots of the equation are -1.257691 and 0.653483.

20.

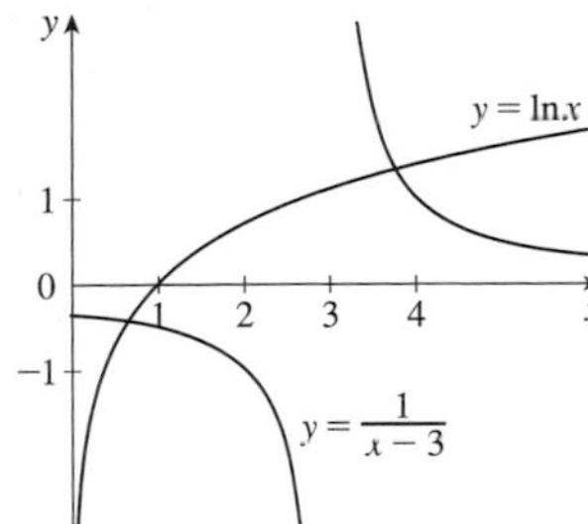

From the figure, we see that the graphs intersect between 0 and 1 and between 3 and 4. Solving $\ln x = \dfrac{1}{x-3}$ is the same as solving $f(x) = \ln x - \dfrac{1}{x-3} = 0$.

$$f'(x) = \frac{1}{x} + \frac{1}{(x-3)^2}, \text{ so } x_{n+1} = x_n - \frac{\ln x_n - 1/(x_n - 3)}{(1/x_n) + 1/(x_n - 3)^2}.$$

$x_1 = 1$	$x_1 = 4$
$x_2 \approx 0.6$	$x_2 \approx 3.690965$
$x_3 \approx 0.651166$	$x_3 \approx 3.750726$
$x_4 \approx 0.653057$	$x_4 \approx 3.755672$
$x_5 \approx 0.653060 \approx x_6$	$x_5 \approx 3.755701 \approx x_6$

To six decimal places, the roots of the equation are 0.653060 and 3.755701.

21.

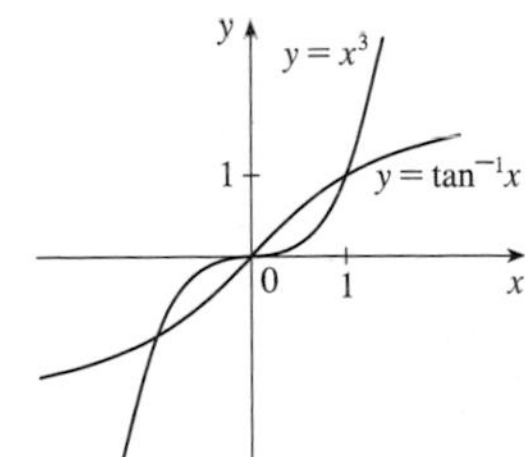

From the figure, we see that the graphs intersect at 0 and at $x = \pm a$, where $a \approx 1$. [Both functions are odd, so the roots are negatives of each other.] Solving $x^3 = \tan^{-1} x$ is the same as solving $f(x) = x^3 - \tan^{-1} x = 0$.

$$f'(x) = 3x^2 - \frac{1}{1+x^2}, \text{ so } x_{n+1} = x_n - \frac{x_n^3 - \tan^{-1} x_n}{3x_n^2 - \dfrac{1}{1+x_n^2}}.$$

Now $x_1 = 1 \;\Rightarrow\; x_2 \approx 0.914159$, $x_3 \approx 0.902251$, $x_4 \approx 0.902026$, $x_5 \approx 0.902025 \approx x_6$. To six decimal places, the nonzero roots of the equation are ± 0.902025.

22. 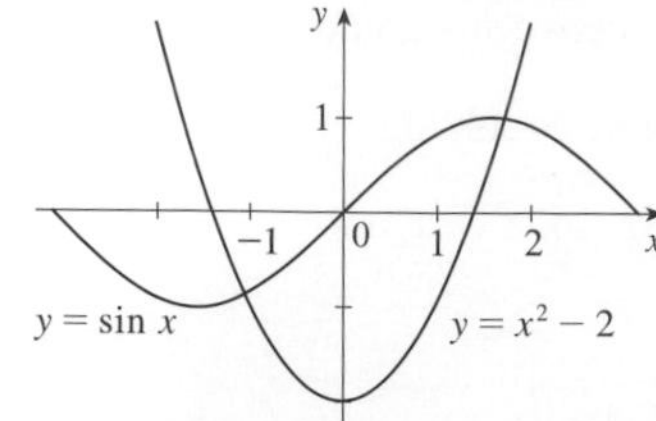

From the graph, we see that there appear to be points of intersection near $x = -1$ and $x = 2$. Solving $\sin x = x^2 - 2$ is the same as solving $f(x) = \sin x - x^2 + 2 = 0$. $f'(x) = \cos x - 2x$, so

$$x_{n+1} = x_n - \frac{\sin x_n - x_n^2 + 2}{\cos x_n - 2x_n}.$$

$x_1 = -1$	$x_1 = 2$
$x_2 \approx -1.062406$	$x_2 \approx 1.753019$
$x_3 \approx -1.061550 \approx x_4$	$x_3 \approx 1.728710$
	$x_4 \approx 1.728466 \approx x_5$

To six decimal places, the roots of the equation are -1.061550 and 1.728466.

23. 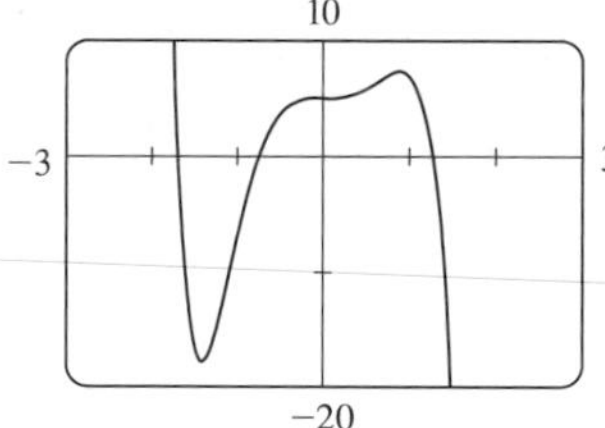

$f(x) = -2x^7 - 5x^4 + 9x^3 + 5 \;\Rightarrow\; f'(x) = -14x^6 - 20x^3 + 27x^2 \;\Rightarrow$

$$x_{n+1} = x_n - \frac{-2x_n^7 - 5x_n^4 + 9x_n^3 + 5}{-14x_n^6 - 20x_n^3 + 27x_n^2}.$$

From the graph of f, there appear to be roots near -1.7, -0.7, and 1.3.

$x_1 = -1.7$	$x_1 = -0.7$	$x_1 = 1.3$
$x_2 = -1.693255$	$x_2 \approx -0.74756345$	$x_2 = 1.268776$
$x_3 \approx -1.69312035$	$x_3 \approx -0.74467752$	$x_3 \approx 1.26589387$
$x_4 \approx -1.69312029 \approx x_5$	$x_4 \approx -0.74466668 \approx x_5$	$x_4 \approx 1.26587094 \approx x_5$

To eight decimal places, the roots of the equation are -1.69312029, -0.74466668, and 1.26587094.

24. 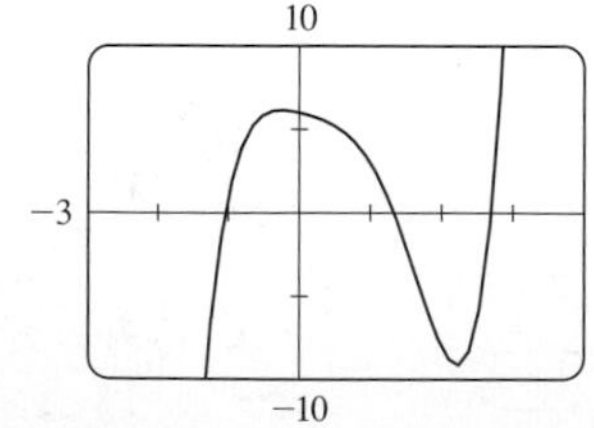

$f(x) = x^5 - 3x^4 + x^3 - x^2 - x + 6 \;\Rightarrow$

$f'(x) = 5x^4 - 12x^3 + 3x^2 - 2x - 1 \;\Rightarrow$

$x_{n+1} = x_n - \dfrac{x_n^5 - 3x_n^4 + x_n^3 - x_n^2 - x_n + 6}{5x_n^4 - 12x_n^3 + 3x_n^2 - 2x_n - 1}$. From the graph of f, there appear to be roots near -1, 1.3, and 2.7.

$x_1 = -1$	$x_1 = 1.3$	$x_1 = 2.7$
$x_2 \approx -1.04761905$	$x_2 \approx 1.33313045$	$x_2 \approx 2.70556135$
$x_3 \approx -1.04451724$	$x_3 \approx 1.33258330$	$x_3 \approx 2.70551210$
$x_4 \approx -1.04450307 \approx x_5$	$x_4 \approx 1.33258316 \approx x_5$	$x_4 \approx 2.70551209 \approx x_5$

To eight decimal places, the roots of the equation are -1.04450307, 1.33258316, and 2.70551209.

25.

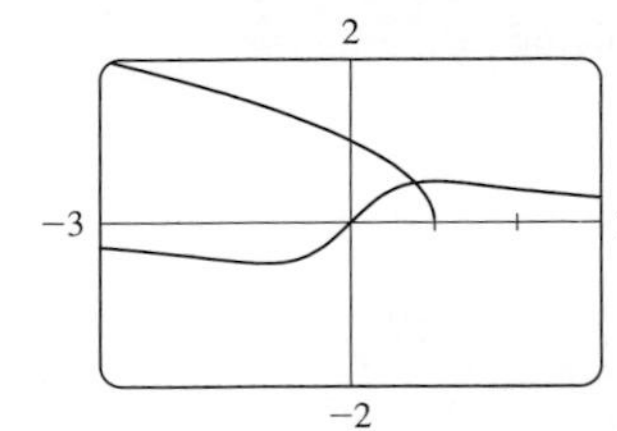

Solving $\dfrac{x}{x^2+1} = \sqrt{1-x}$ is the same as solving

$f(x) = \dfrac{x}{x^2+1} - \sqrt{1-x} = 0.$ $f'(x) = \dfrac{1-x^2}{(x^2+1)^2} + \dfrac{1}{2\sqrt{1-x}}$ $\Rightarrow$

$$x_{n+1} = x_n - \frac{\dfrac{x_n}{x_n^2+1} - \sqrt{1-x_n}}{\dfrac{1-x_n^2}{(x_n^2+1)^2} + \dfrac{1}{2\sqrt{1-x_n}}}.$$

From the graph, we see that the curves intersect at about 0.8. $x_1 = 0.8 \Rightarrow x_2 \approx 0.76757581$, $x_3 \approx 0.76682610$, $x_4 \approx 0.76682579 \approx x_5$. To eight decimal places, the root of the equation is 0.76682579.

26.

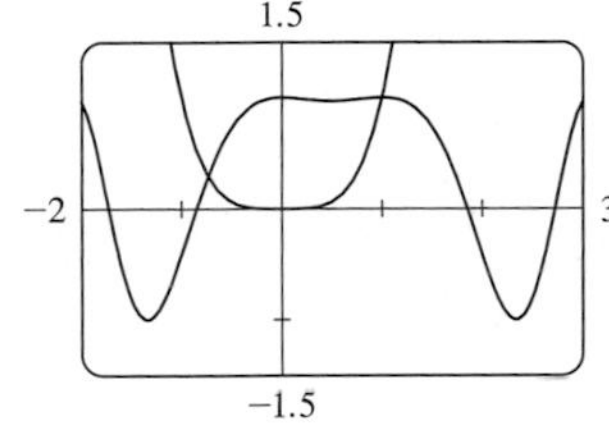

Solving $\cos(x^2 - x) = x^4$ is the same as solving

$f(x) = \cos(x^2 - x) - x^4 = 0.$ $f'(x) = -(2x-1)\sin(x^2-x) - 4x^3 \Rightarrow$

$x_{n+1} = x_n - \dfrac{\cos(x_n^2 - x_n) - x_n^4}{-(2x_n - 1)\sin(x_n^2 - x_n) - 4x_n^3}$. From the equations $y = \cos(x^2 - x)$ and $y = x^4$ and the graph, we deduce that one root of the equation $\cos(x^2 - x) = x^4$ is $x = 1$. We also see that the graphs intersect at approximately $x = -0.7$. $x_1 = -0.7 \Rightarrow x_2 \approx -0.73654354$, $x_3 \approx -0.73486274$, $x_4 \approx -0.73485910 \approx x_5$.

To eight decimal places, one root of the equation is -0.73485910; the other root is 1.

27.

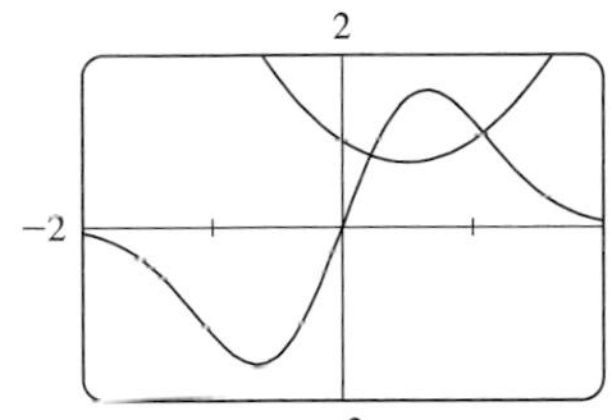

Solving $4e^{-x^2}\sin x = x^2 - x + 1$ is the same as solving

$f(x) = 4e^{-x^2}\sin x - x^2 + x - 1 = 0.$

$f'(x) = 4e^{-x^2}(\cos x - 2x\sin x) - 2x + 1 \Rightarrow$

$$x_{n+1} = x_n - \frac{4e^{-x_n^2}\sin x_n - x_n^2 + x_n - 1}{4e^{-x_n^2}(\cos x_n - 2x_n\sin x_n) - 2x_n + 1}.$$

From the figure, we see that the graphs intersect at approximately $x = 0.2$ and $x = 1.1$.

$x_1 = 0.2$	$x_1 = 1.1$
$x_2 \approx 0.21883273$	$x_2 \approx 1.08432830$
$x_3 \approx 0.21916357$	$x_3 \approx 1.08422462 \approx x_4$
$x_4 \approx 0.21916368 \approx x_5$	

To eight decimal places, the roots of the equation are 0.21916368 and 1.08422462.

28. 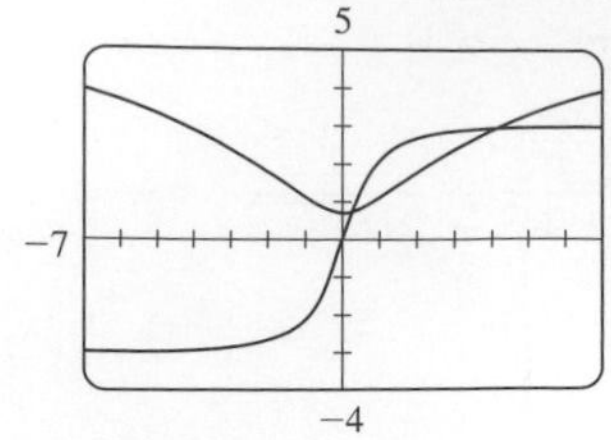

Solving $\ln(x^2+2) = \dfrac{3x}{\sqrt{x^2+1}}$ is the same as solving

$$f(x) = \ln(x^2+2) - \frac{3x}{\sqrt{x^2+1}} = 0.$$

$$f'(x) = \frac{2x}{x^2+2} - \frac{(x^2+1)^{1/2}(3) - (3x)\frac{1}{2}(x^2+1)^{-1/2}(2x)}{[(x^2+1)^{1/2}]^2}$$

$$= \frac{2x}{x^2+2} - \frac{(x^2+1)^{-1/2}[3(x^2+1) - 3x^2]}{(x^2+1)^1}$$

$$= \frac{2x}{x^2+2} - \frac{3}{(x^2+1)^{3/2}} \quad \Rightarrow \quad x_{n+1} = x_n - \frac{\ln(x_n^2+2) - \dfrac{3x_n}{\sqrt{x_n^2+1}}}{\dfrac{2x_n}{x_n^2+2} - \dfrac{3}{(x_n^2+1)^{3/2}}}.$$

From the figure, we see that the graphs intersect at approximately $x = 0.2$ and $x = 4$.

$x_1 = 0.2$	$x_1 = 4$
$x_2 \approx 0.24733161$	$x_2 \approx 4.04993412$
$x_3 \approx 0.24852333$	$x_3 \approx 4.05010983$
$x_4 \approx 0.24852414 \approx x_5$	$x_4 \approx 4.05010984 \approx x_5$

To eight decimal places, the roots of the equation are 0.24852414 and 4.05010984.

29. (a) $f(x) = x^2 - a \quad \Rightarrow \quad f'(x) = 2x$, so Newton's method gives

$$x_{n+1} = x_n - \frac{x_n^2 - a}{2x_n} = x_n - \frac{1}{2}x_n + \frac{a}{2x_n} = \frac{1}{2}x_n + \frac{a}{2x_n} = \frac{1}{2}\left(x_n + \frac{a}{x_n}\right).$$

(b) Using (a) with $a = 1000$ and $x_1 = \sqrt{900} = 30$, we get $x_2 \approx 31.666667$, $x_3 \approx 31.622807$, and $x_4 \approx 31.622777 \approx x_5$. So $\sqrt{1000} \approx 31.622777$.

30. (a) $f(x) = \dfrac{1}{x} - a \quad \Rightarrow \quad f'(x) = -\dfrac{1}{x^2}$, so $x_{n+1} = x_n - \dfrac{1/x_n - a}{-1/x_n^2} = x_n + x_n - ax_n^2 = 2x_n - ax_n^2$.

(b) Using (a) with $a = 1.6894$ and $x_1 = \frac{1}{2} = 0.5$, we get $x_2 = 0.5754$, $x_3 \approx 0.588485$, and $x_4 \approx 0.588789 \approx x_5$. So $1/1.6984 \approx 0.588789$.

31. $f(x) = x^3 - 3x + 6 \quad \Rightarrow \quad f'(x) = 3x^2 - 3$. If $x_1 = 1$, then $f'(x_1) = 0$ and the tangent line used for approximating x_2 is horizontal. Attempting to find x_2 results in trying to divide by zero.

32. $x^3 - x = 1 \quad \Leftrightarrow \quad x^3 - x - 1 = 0$. $f(x) = x^3 - x - 1 \quad \Rightarrow \quad f'(x) = 3x^2 - 1$, so $x_{n+1} = x_n - \dfrac{x_n^3 - x_n - 1}{3x_n^2 - 1}$.

(a) $x_1 = 1$, $x_2 = 1.5$, $x_3 \approx 1.347826$, $x_4 \approx 1.325200$, $x_5 \approx 1.324718 \approx x_6$

(b) $x_1 = 0.6$, $x_2 = 17.9$, $x_3 \approx 11.946802$, $x_4 \approx 7.985520$, $x_5 \approx 5.356909$, $x_6 \approx 3.624996$, $x_7 \approx 2.505589$, $x_8 \approx 1.820129$, $x_9 \approx 1.461044$, $x_{10} \approx 1.339323$, $x_{11} \approx 1.324913$, $x_{12} \approx 1.324718 \approx x_{13}$

(c) $x_1 = 0.57$, $x_2 \approx -54.165455$, $x_3 \approx -36.114293$, $x_4 \approx -24.082094$, $x_5 \approx -16.063387$, $x_6 \approx -10.721483$, $x_7 \approx -7.165534$, $x_8 \approx -4.801704$, $x_9 \approx -3.233425$, $x_{10} \approx -2.193674$, $x_{11} \approx -1.496867$, $x_{12} \approx -0.997546$,

$x_{13} \approx -0.496305$, $x_{14} \approx -2.894162$, $x_{15} \approx -1.967962$, $x_{16} \approx -1.341355$, $x_{17} \approx -0.870187$, $x_{18} \approx -0.249949$, $x_{19} \approx -1.192219$, $x_{20} \approx -0.731952$, $x_{21} \approx 0.355213$, $x_{22} \approx -1.753322$, $x_{23} \approx -1.189420$, $x_{24} \approx -0.729123$, $x_{25} \approx 0.377844$, $x_{26} \approx -1.937872$, $x_{27} \approx -1.320350$, $x_{28} \approx -0.851919$, $x_{29} \approx -0.200959$, $x_{30} \approx -1.119386$, $x_{31} \approx -0.654291$, $x_{32} \approx 1.547010$, $x_{33} \approx 1.360051$, $x_{34} \approx 1.325828$, $x_{35} \approx 1.324719$, $x_{36} \approx 1.324718 \approx x_{37}$.

(d)

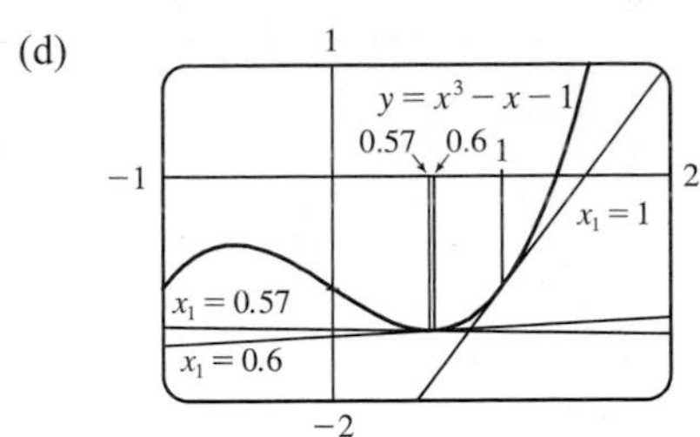

From the figure, we see that the tangent line corresponding to $x_1 = 1$ results in a sequence of approximations that converges quite quickly ($x_5 \approx x_6$). The tangent line corresponding to $x_1 = 0.6$ is close to being horizontal, so x_2 is quite far from the root. But the sequence still converges — just a little more slowly ($x_{12} \approx x_{13}$). Lastly, the tangent line corresponding to $x_1 = 0.57$ is very nearly horizontal, x_2 is farther away from the root, and the sequence takes more iterations to converge ($x_{36} \approx x_{37}$).

33. For $f(x) = x^{1/3}$, $f'(x) = \frac{1}{3}x^{-2/3}$ and

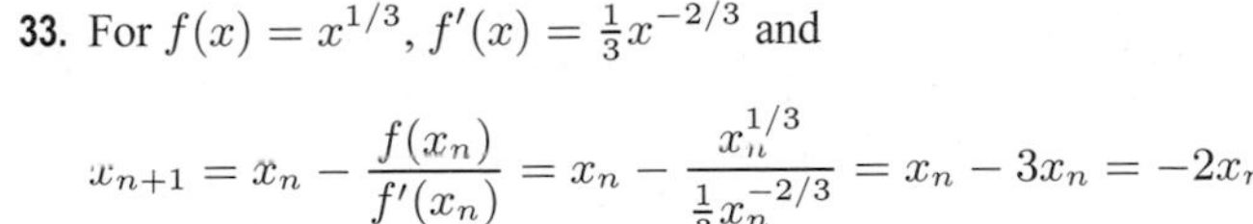

$$x_{n+1} = x_n - \frac{f(x_n)}{f'(x_n)} = x_n - \frac{x_n^{1/3}}{\frac{1}{3}x_n^{-2/3}} = x_n - 3x_n = -2x_n.$$

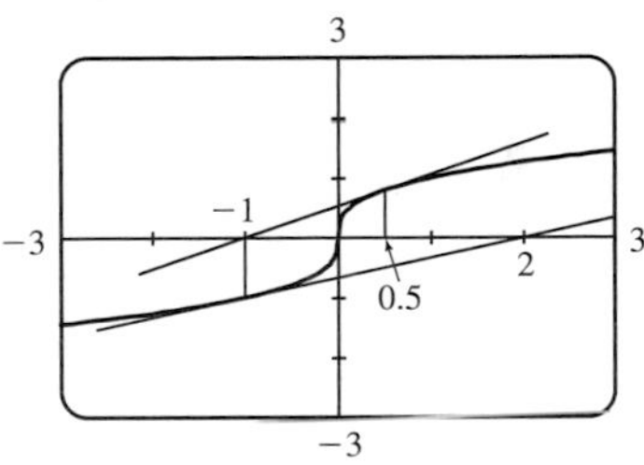

Therefore, each successive approximation becomes twice as large as the previous one in absolute value, so the sequence of approximations fails to converge to the root, which is 0. In the figure, we have $x_1 = 0.5$, $x_2 = -2(0.5) = -1$, and $x_3 = -2(-1) = 2$.

34. According to Newton's Method, for $x_n > 0$,

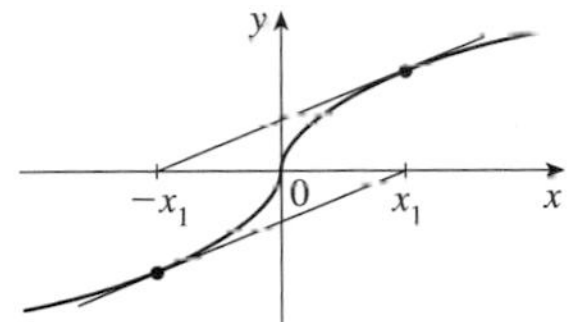

$$x_{n+1} = x_n - \frac{\sqrt{x_n}}{1/\left(2\sqrt{x_n}\right)} = x_n - 2x_n = -x_n \text{ and for } x_n < 0,$$

$$x_{n+1} = x_n - \frac{-\sqrt{-x_n}}{1/\left(2\sqrt{-x_n}\right)} = x_n - [-2(-x_n)] = -x_n. \text{ So we can see that}$$

after choosing any value x_1 the subsequent values will alternate between $-x_1$ and x_1 and never approach the root.

35. (a) $f(x) = x^6 - x^4 + 3x^3 - 2x \;\Rightarrow\; f'(x) = 6x^5 - 4x^3 + 9x^2 - 2 \;\Rightarrow$ $f''(x) = 30x^4 - 12x^2 + 18x$. To find the critical numbers of f, we'll find the zeros of f'. From the graph of f', it appears there are zeros at approximately $x = -1.3$, -0.4, and 0.5. Try $x_1 = -1.3 \;\Rightarrow$

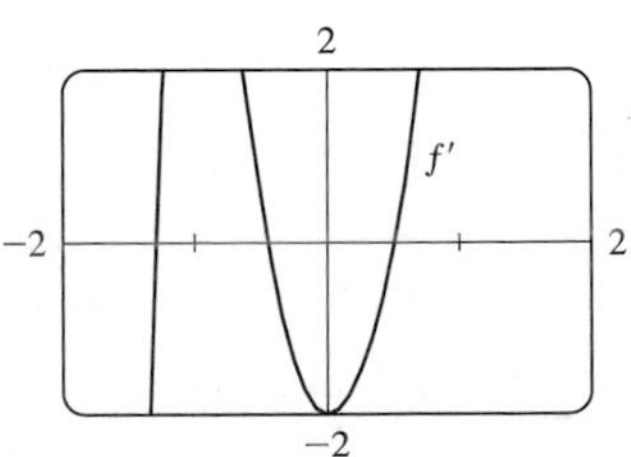

$$x_2 = x_1 - \frac{f'(x_1)}{f''(x_1)} \approx -1.293344 \;\Rightarrow\; x_3 \approx -1.293227 \approx x_4.$$

Now try $x_1 = -0.4 \;\Rightarrow\; x_2 \approx -0.443755 \;\Rightarrow\; x_3 \approx -0.441735 \;\Rightarrow\; x_4 \approx -0.441731 \approx x_5$. Finally try $x_1 = 0.5 \;\Rightarrow\; x_2 \approx 0.507937 \;\Rightarrow\; x_3 \approx 0.507854 \approx x_4$. Therefore, $x = -1.293227$, -0.441731, and 0.507854 are all the critical numbers correct to six decimal places.

(b) There are two critical numbers where f' changes from negative to positive, so f changes from decreasing to increasing. $f(-1.293227) \approx -2.0212$ and $f(0.507854) \approx -0.6721$, so -2.0212 is the absolute minimum value of f correct to four decimal places.

36. $f(x) = x\cos x \;\Rightarrow\; f'(x) = \cos x - x\sin x$. $f'(x)$ exists for all x, so to find the maximum of f, we can examine the zeros of f'. From the graph of f', we see that a good choice for x_1 is $x_1 = 0.9$. Use $g(x) = \cos x - x\sin x$ and $g'(x) = -2\sin x - x\cos x$ to obtain $x_2 \approx 0.860781$, $x_3 \approx 0.860334 \approx x_4$. Now we have $f(0) = 0$, $f(\pi) = -\pi$, and $f(0.860334) \approx 0.561096$, so 0.561096 is the absolute maximum value of f correct to six decimal places.

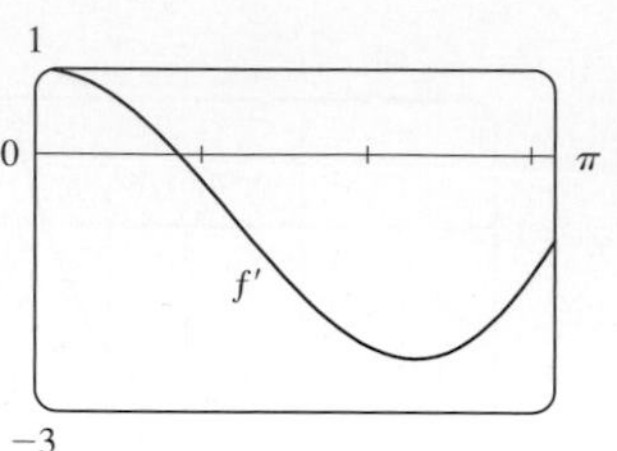

37.

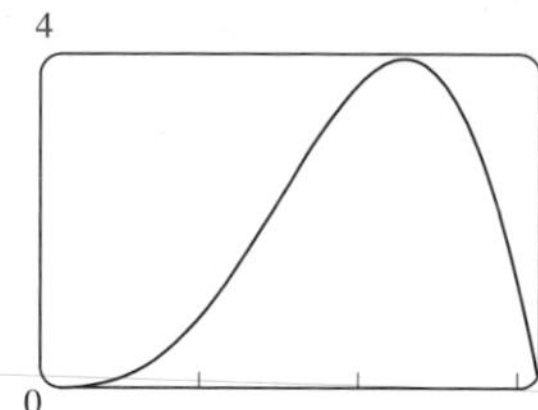

$$y = x^2\sin x \;\Rightarrow\; y' = x^2\cos x + (\sin x)(2x) \;\Rightarrow$$

$$\begin{aligned} y'' &= x^2(-\sin x) + (\cos x)(2x) + (\sin x)(2) + 2x\cos x \\ &= -x^2\sin x + 4x\cos x + 2\sin x \;\Rightarrow \end{aligned}$$

$$\begin{aligned} y''' &= -x^2\cos x + (\sin x)(-2x) + 4x(-\sin x) + (\cos x)(4) + 2\cos x \\ &= -x^2\cos x - 6x\sin x + 6\cos x. \end{aligned}$$

From the graph of $y = x^2\sin x$, we see that $x = 1.5$ is a reasonable guess for the x-coordinate of the inflection point. Using Newton's method with $g(x) = y''$ and $g'(x) = y'''$, we get $x_1 = 1.5 \;\Rightarrow\; x_2 \approx 1.520092$, $x_3 \approx 1.519855 \approx x_4$. The inflection point is about $(1.519855, 2.306964)$.

38.

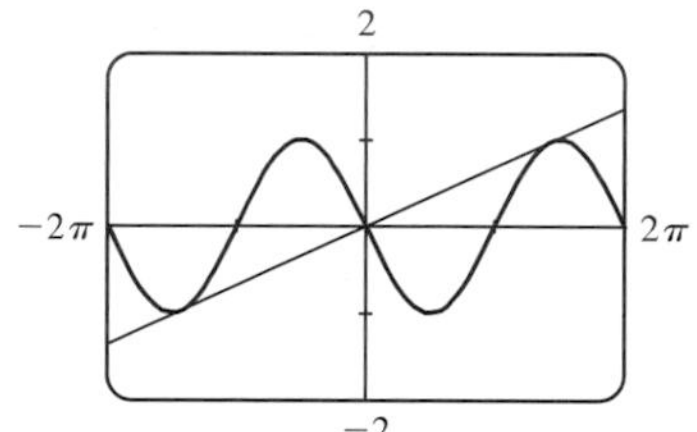

$f(x) = -\sin x \;\Rightarrow\; f'(x) = -\cos x$. At $x = a$, the slope of the tangent line is $f'(a) = -\cos a$. The line through the origin and $(a, f(a))$ is $y = \dfrac{-\sin a - 0}{a - 0}x$. If this line is to be tangent to f at $x = a$, then its slope must equal $f'(a)$. Thus, $\dfrac{-\sin a}{a} = -\cos a \;\Rightarrow\; \tan a = a$.

To solve this equation using Newton's method, let $g(x) = \tan x - x$, $g'(x) = \sec^2 x - 1$, and $x_{n+1} = x_n - \dfrac{\tan x_n - x_n}{\sec^2 x_n - 1}$ with $x_1 = 4.5$ (estimated from the figure). $x_2 \approx 4.493614$, $x_3 \approx 4.493410$, $x_4 \approx 4.493409 \approx x_5$. Thus, the slope of the line that has the largest slope is $f'(x_5) \approx 0.217234$.

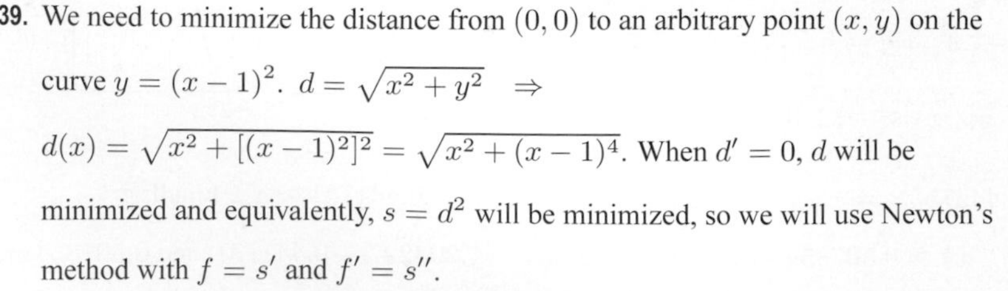

39. We need to minimize the distance from $(0,0)$ to an arbitrary point (x,y) on the curve $y = (x-1)^2$. $d = \sqrt{x^2 + y^2} \;\Rightarrow$ $d(x) = \sqrt{x^2 + [(x-1)^2]^2} = \sqrt{x^2 + (x-1)^4}$. When $d' = 0$, d will be minimized and equivalently, $s = d^2$ will be minimized, so we will use Newton's method with $f = s'$ and $f' = s''$.

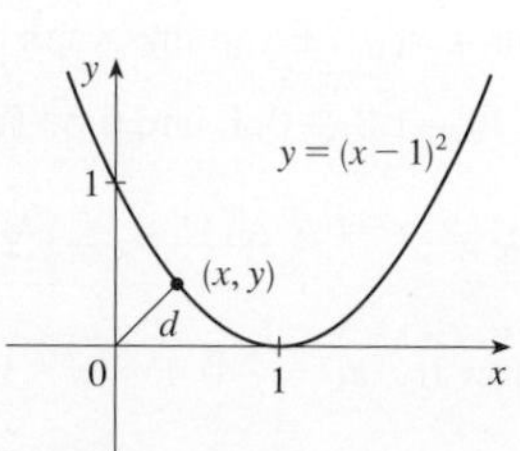

$f(x) = 2x + 4(x-1)^3 \quad \Rightarrow \quad f'(x) = 2 + 12(x-1)^2$, so $x_{n+1} = x_n - \dfrac{2x_n + 4(x_n - 1)^3}{2 + 12(x_n - 1)^2}$. Try $x_1 = 0.5 \quad \Rightarrow$ $x_2 = 0.4$, $x_3 \approx 0.410127$, $x_4 \approx 0.410245 \approx x_5$. Now $d(0.410245) \approx 0.537841$ is the minimum distance and the point on the parabola is $(0.410245, 0.347810)$, correct to six decimal places.

40. Let the radius of the circle be r. Using $s = r\theta$, we have $5 = r\theta$ and so $r = 5/\theta$. From the Law of Cosines we get $4^2 = r^2 + r^2 - 2 \cdot r \cdot r \cdot \cos\theta \quad \Leftrightarrow \quad 16 = 2r^2(1 - \cos\theta) = 2(5/\theta)^2(1 - \cos\theta)$. Multiplying by θ^2 gives $16\theta^2 = 50(1 - \cos\theta)$, so we take $f(\theta) = 16\theta^2 + 50\cos\theta - 50$ and $f'(\theta) = 32\theta - 50\sin\theta$. The formula for Newton's method is $\theta_{n+1} = \theta_n - \dfrac{16\theta_n^2 + 50\cos\theta_n - 50}{32\theta_n - 50\sin\theta_n}$. From the graph of f, we can use $\theta_1 = 2.2$, giving us $\theta_2 \approx 2.2662$, $\theta_3 \approx 2.2622 \approx \theta_4$. So correct to four decimal places, the angle is 2.2622 radians $\approx 130°$.

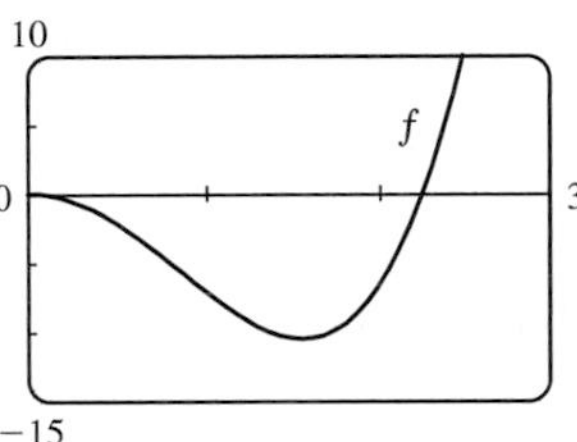

41. In this case, $A = 18{,}000$, $R = 375$, and $n = 5(12) = 60$. So the formula $A = \dfrac{R}{i}[1 - (1+i)^{-n}]$ becomes

$18{,}000 = \dfrac{375}{x}[1 - (1+x)^{-60}] \quad \Leftrightarrow \quad 48x = 1 - (1+x)^{-60}$ [multiply each term by $(1+x)^{60}$] $\Leftrightarrow$

$48x(1+x)^{60} - (1+x)^{60} + 1 = 0$. Let the LHS be called $f(x)$, so that

$$\begin{aligned} f'(x) &= 48x(60)(1+x)^{59} + 48(1+x)^{60} - 60(1+x)^{59} \\ &= 12(1+x)^{59}[4x(60) + 4(1+x) - 5] = 12(1+x)^{59}(244x - 1) \end{aligned}$$

$x_{n+1} = x_n - \dfrac{48x_n(1+x_n)^{60} - (1+x_n)^{60} + 1}{12(1+x_n)^{59}(244x_n - 1)}$. An interest rate of 1% per month seems like a reasonable estimate for $x = i$. So let $x_1 = 1\% = 0.01$, and we get $x_2 \approx 0.0082202$, $x_3 \approx 0.0076802$, $x_4 \approx 0.0076291$, $x_5 \approx 0.0076286 \approx x_6$. Thus, the dealer is charging a monthly interest rate of 0.76286% (or 9.55% per year, compounded monthly).

42. (a) $p(x) = x^5 - (2+r)x^4 + (1+2r)x^3 - (1-r)x^2 + 2(1-r)x + r - 1 \quad \Rightarrow$

$p'(x) = 5x^4 - 4(2+r)x^3 + 3(1+2r)x^2 - 2(1-r)x + 2(1-r)$. So we use

$$x_{n+1} = x_n - \frac{x_n^5 - (2+r)x_n^4 + (1+2r)x_n^3 - (1-r)x_n^2 + 2(1-r)x_n + r - 1}{5x_n^4 - 4(2+r)x_n^3 + 3(1+2r)x_n^2 - 2(1-r)x_n + 2(1-r)}.$$

We substitute in the value $r \approx 3.04042 \times 10^{-6}$ in order to evaluate the approximations numerically. The libration point L_1 is slightly less than 1 AU from the sun, so we take $x_1 = 0.95$ as our first approximation, and get $x_2 \approx 0.96682$, $x_3 \approx 0.97770$, $x_4 \approx 0.98451$, $x_5 \approx 0.98830$, $x_6 \approx 0.98976$, $x_7 \approx 0.98998$, $x_8 \approx 0.98999 \approx x_9$. So, to five decimal places, L_1 is located 0.98999 AU from the sun (or 0.01001 AU from the earth)

(b) In this case we use Newton's method with the function

$p(x) - 2rx^2 = x^5 - (2+r)x^4 + (1+2r)x^3 - (1+r)x^2 + 2(1-r)x + r - 1 \quad \Rightarrow$

$[p(x) - 2rx^2]' = 5x^4 - 4(2+r)x^3 + 3(1+2r)x^2 - 2(1+r)x + 2(1-r)$. So

$x_{n+1} = x_n - \dfrac{x_n^5 - (2+r)x_n^4 + (1+2r)x_n^3 - (1+r)x_n^2 + 2(1-r)x_n + r - 1}{5x_n^4 - 4(2+r)x_n^3 + 3(1+2r)x_n^2 - 2(1+r)x_n + 2(1-r)}$. Again, we substitute $r \approx 3.04042 \times 10^{-6}$. L_2 is slightly more than 1 AU from the sun and, judging from the result of part (a), probably less than 0.02 AU from earth. So we take $x_1 = 1.02$ and get $x_2 \approx 1.01422$, $x_3 \approx 1.01118$, $x_4 \approx 1.01018$, $x_5 \approx 1.01008 \approx x_6$. So, to five decimal places, L_2 is located 1.01008 AU from the sun (or 0.01008 AU from the earth).

4.9 Antiderivatives

1. $f(x) = 4x + 7 = 4x^1 + 7 \quad\Rightarrow\quad F(x) = 4\dfrac{x^{1+1}}{1+1} + 7x + C = 2x^2 + 7x + C$

Check: $F'(x) = 2(2x) + 7 + 0 = 4x + 7 = f(x)$

2. $f(x) = x^2 - 3x + 2 \quad\Rightarrow\quad F(x) = \dfrac{x^3}{3} - 3\dfrac{x^2}{2} + 2x + C = \frac{1}{3}x^3 - \frac{3}{2}x^2 + 2x + C$

Check: $F'(x) = \frac{1}{3}(3x^2) - \frac{3}{2}(2x) + 2 + 0 = x^2 - 3x + 2 = f(x)$

3. $f(x) = 2x^3 - \frac{2}{3}x^2 + 5x \quad\Rightarrow\quad F(x) = 2\dfrac{x^{3+1}}{3+1} - \dfrac{2}{3}\dfrac{x^{2+1}}{2+1} + 5\dfrac{x^{1+1}}{1+1} = \frac{1}{2}x^4 - \frac{2}{9}x^3 + \frac{5}{2}x^2 + C$

Check: $F'(x) = \frac{1}{2}(4x^3) - \frac{2}{9}(3x^2) + \frac{5}{2}(2x) + 0 = 2x^3 - \frac{2}{3}x^2 + 5x = f(x)$

4. $f(x) = 6x^5 - 8x^4 - 9x^2 \quad\Rightarrow\quad F(x) = 6\dfrac{x^6}{6} - 8\dfrac{x^5}{5} - 9\dfrac{x^3}{3} + C = x^6 - \frac{8}{5}x^5 - 3x^3 + C$

5. $f(x) = x(12x + 8) = 12x^2 + 8x \quad\Rightarrow\quad F(x) = 12\dfrac{x^3}{3} + 8\dfrac{x^2}{2} + C = 4x^3 + 4x^2 + C$

6. $f(x) = (x - 5)^2 = x^2 - 10x + 25 \quad\Rightarrow\quad F(x) = \dfrac{x^3}{3} - 10\dfrac{x^2}{2} + 25x + C = \frac{1}{3}x^3 - 5x^2 + 25x + C$

7. $f(x) = 7x^{2/5} + 8x^{-4/5} \quad\Rightarrow\quad F(x) = 7\left(\frac{5}{7}x^{7/5}\right) + 8(5x^{1/5}) + C = 5x^{7/5} + 40x^{1/5} + C$

8. $f(x) = x^{3.4} - 2x^{\sqrt{2}-1} \quad\Rightarrow\quad F(x) = \dfrac{x^{4.4}}{4.4} - 2\left(\dfrac{x^{\sqrt{2}}}{\sqrt{2}}\right) + C = \dfrac{5}{22}x^{4.4} - \sqrt{2}\,x^{\sqrt{2}} + C$

9. $f(x) = \sqrt{2}$ is a constant function, so $F(x) = \sqrt{2}\,x + C$.

10. $f(x) = e^2$ is a constant function, so $F(x) = e^2x + C$.

11. $f(x) = 3\sqrt{x} - 2\sqrt[3]{x} = 3x^{1/2} - 2x^{1/3} \quad\Rightarrow\quad F(x) = 3\left(\frac{2}{3}x^{3/2}\right) - 2\left(\frac{3}{4}x^{4/3}\right) + C = 2x^{3/2} - \frac{3}{2}x^{4/3} + C$

12. $f(x) = \sqrt[3]{x^2} + x\sqrt{x} = x^{2/3} + x^{3/2} \quad\Rightarrow\quad F(x) = \frac{3}{5}x^{5/3} + \frac{2}{5}x^{5/2} + C$

13. $f(x) = \dfrac{1}{5} - \dfrac{2}{x} = \dfrac{1}{5} - 2\left(\dfrac{1}{x}\right)$ has domain $(-\infty, 0) \cup (0, \infty)$, so $F(x) = \begin{cases} \frac{1}{5}x - 2\ln|x| + C_1 & \text{if } x < 0 \\ \frac{1}{5}x - 2\ln|x| + C_2 & \text{if } x > 0 \end{cases}$

See Example 1(b) for a similar problem.

14. $f(t) = \dfrac{3t^4 - t^3 + 6t^2}{t^4} = 3 - \dfrac{1}{t} + \dfrac{6}{t^2}$ has domain $(-\infty, 0) \cup (0, \infty)$, so $F(t) = \begin{cases} 3t - \ln|t| - \dfrac{6}{t} + C_1 & \text{if } t < 0 \\ 3t - \ln|t| - \dfrac{6}{t} + C_2 & \text{if } t > 0 \end{cases}$

See Example 1(b) for a similar problem.

15. $g(t) = \dfrac{1 + t + t^2}{\sqrt{t}} = t^{-1/2} + t^{1/2} + t^{3/2} \quad\Rightarrow\quad G(t) = 2t^{1/2} + \frac{2}{3}t^{3/2} + \frac{2}{5}t^{5/2} + C$

16. $r(\theta) = \sec\theta\,\tan\theta - 2e^{\theta} \quad\Rightarrow\quad R(\theta) = \sec\theta - 2e^{\theta} + C_n$ on the interval $\left(n\pi - \frac{\pi}{2}, n\pi + \frac{\pi}{2}\right)$.

17. $h(\theta) = 2\sin\theta - \sec^2\theta \quad\Rightarrow\quad H(\theta) = -2\cos\theta - \tan\theta + C_n$ on the interval $\left(n\pi - \frac{\pi}{2}, n\pi + \frac{\pi}{2}\right)$.

18. $g(v) = 2\cos v - \dfrac{3}{\sqrt{1 - v^2}} \quad\Rightarrow\quad G(v) = 2\sin v - 3\sin^{-1} v + C$

19. $f(x) = 2^x + 4\sinh x \quad\Rightarrow\quad F(x) = \dfrac{2^x}{\ln 2} + 4\cosh x + C$

20. $f(x) = 1 + 2\sin x + 3/\sqrt{x} = 1 + 2\sin x + 3x^{-1/2} \quad\Rightarrow\quad F(x) = x - 2\cos x + 3\dfrac{x^{1/2}}{1/2} + C = x - 2\cos x + 6\sqrt{x} + C$

21. $f(x) = \dfrac{2x^4 + 4x^3 - x}{x^3}$, $x > 0$; $f(x) = 2x + 4 - x^{-2} \quad\Rightarrow$

$F(x) = 2\dfrac{x^2}{2} + 4x - \dfrac{x^{-2+1}}{-2+1} + C = x^2 + 4x + \dfrac{1}{x} + C$, $x > 0$

22. $f(x) = \dfrac{2x^2 + 5}{x^2 + 1} = \dfrac{2(x^2 + 1) + 3}{x^2 + 1} = 2 + \dfrac{3}{x^2 + 1} \quad\Rightarrow\quad F(x) = 2x + 3\tan^{-1} x + C$

23. $f(x) = 5x^4 - 2x^5 \quad\Rightarrow\quad F(x) = 5 \cdot \dfrac{x^5}{5} - 2 \cdot \dfrac{x^6}{6} + C = x^5 - \frac{1}{3}x^6 + C$.

$F(0) = 4 \quad\Rightarrow\quad 0^5 - \frac{1}{3} \cdot 0^6 + C = 4 \quad\Rightarrow\quad C = 4$, so $F(x) = x^5 - \frac{1}{3}x^6 + 4$.

The graph confirms our answer since $f(x) = 0$ when F has a local maximum, f is positive when F is increasing, and f is negative when F is decreasing.

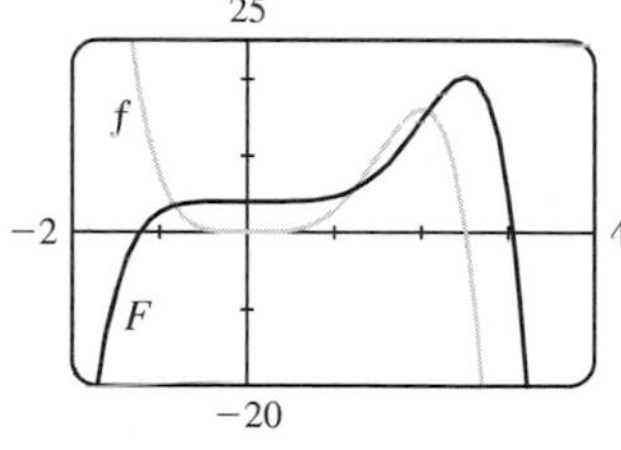

24. $f(x) = 4 - 3(1 + x^2)^{-1} = 4 - \dfrac{3}{1 + x^2} \quad\Rightarrow\quad F(x) = 4x - 3\tan^{-1} x + C$.

$F(1) = 0 \quad\Rightarrow\quad 4 - 3\left(\frac{\pi}{4}\right) + C = 0 \quad\Rightarrow\quad C = \frac{3\pi}{4} - 4$, so

$F(x) = 4x - 3\tan^{-1} x + \frac{3\pi}{4} - 4$. Note that f is positive and F is increasing on $\mathbb{R}$. Also, f has smaller values where the slopes of the tangent lines of F are smaller.

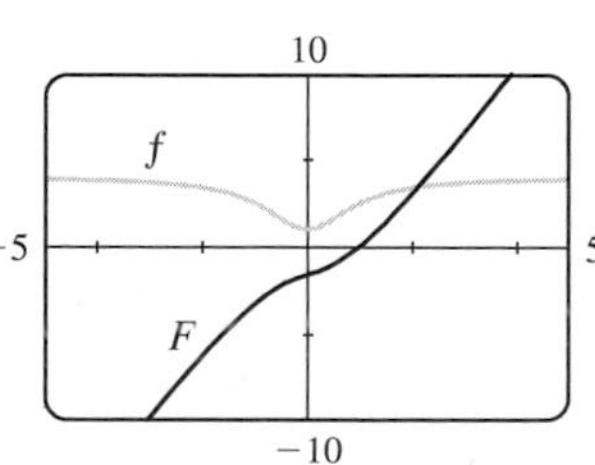

25. $f''(x) = 20x^3 - 12x^2 + 6x \quad\Rightarrow\quad f'(x) = 20\left(\dfrac{x^4}{4}\right) - 12\left(\dfrac{x^3}{3}\right) + 6\left(\dfrac{x^2}{2}\right) + C = 5x^4 - 4x^3 + 3x^2 + C \quad\Rightarrow$

$f(x) = 5\left(\dfrac{x^5}{5}\right) - 4\left(\dfrac{x^4}{4}\right) + 3\left(\dfrac{x^3}{3}\right) + Cx + D = x^5 - x^4 + x^3 + Cx + D$

26. $f''(x) = x^6 - 4x^4 + x + 1 \;\Rightarrow\; f'(x) = \frac{1}{7}x^7 - \frac{4}{5}x^5 + \frac{1}{2}x^2 + x + C \;\Rightarrow$

$f(x) = \frac{1}{56}x^8 - \frac{2}{15}x^6 + \frac{1}{6}x^3 + \frac{1}{2}x^2 + Cx + D$

27. $f''(x) = 2x + 3e^x \;\Rightarrow\; f'(x) = x^2 + 3e^x + C \;\Rightarrow\; f(x) = \frac{1}{3}x^3 + 3e^x + Cx + D$

28. $f''(x) = 1/x^2 = x^{-2} \;\Rightarrow\; f'(x) = \begin{cases} -1/x + C_1 & \text{if } x < 0 \\ -1/x + C_2 & \text{if } x > 0 \end{cases} \;\Rightarrow\; f(x) = \begin{cases} -\ln(-x) + C_1x + D_1 & \text{if } x < 0 \\ -\ln x + C_2x + D_2 & \text{if } x > 0 \end{cases}$

29. $f'''(t) = 12 + \sin t \;\Rightarrow\; f''(t) = 12t - \cos t + C_1 \;\Rightarrow\; f'(t) = 6t^2 - \sin t + C_1t + D \;\Rightarrow$

$f(t) = 2t^3 + \cos t + Ct^2 + Dt + E$, where $C = \frac{1}{2}C_1$.

30. $f'''(t) = \sqrt{t} - 2\cos t = t^{1/2} - 2\cos t \;\Rightarrow\; f''(t) = \frac{2}{3}t^{3/2} - 2\sin t + C_1 \;\Rightarrow\; f'(t) = \frac{4}{15}t^{5/2} + 2\cos t + C_1t + D \;\Rightarrow$

$f(t) = \frac{8}{105}t^{7/2} + 2\sin t + Ct^2 + Dt + E$, where $C = \frac{1}{2}C_1$.

31. $f'(x) = 1 + 3\sqrt{x} \;\Rightarrow\; f(x) = x + 3\left(\frac{2}{3}x^{3/2}\right) + C = x + 2x^{3/2} + C.$ $f(4) = 4 + 2(8) + C$ and $f(4) = 25 \;\Rightarrow$

$20 + C = 25 \;\Rightarrow\; C = 5$, so $f(x) = x + 2x^{3/2} + 5$.

32. $f'(x) = 5x^4 - 3x^2 + 4 \;\Rightarrow\; f(x) = x^5 - x^3 + 4x + C.$ $f(-1) = -1 + 1 - 4 + C$ and $f(-1) = 2 \;\Rightarrow$

$-4 + C = 2 \;\Rightarrow\; C = 6$, so $f(x) = x^5 - x^3 + 4x + 6$.

33. $f'(t) = \dfrac{4}{1+t^2} \;\Rightarrow\; f(t) = 4\arctan t + C.$ $f(1) = 4\left(\dfrac{\pi}{4}\right) + C$ and $f(1) = 0 \;\Rightarrow\; \pi + C = 0 \;\Rightarrow\; C = -\pi$,

so $f(t) = 4\arctan t - \pi$.

34. $f'(t) = t + \dfrac{1}{t^3},\ t > 0 \;\Rightarrow\; f(t) = \dfrac{1}{2}t^2 - \dfrac{1}{2t^2} + C.$ $f(1) = \dfrac{1}{2} - \dfrac{1}{2} + C$ and $f(1) = 6 \;\Rightarrow\; C = 6$, so

$f(t) = \dfrac{1}{2}t^2 - \dfrac{1}{2t^2} + 6.$

35. $f'(x) = 5x^{2/3} \;\Rightarrow\; f(x) = 5\left(\frac{3}{5}x^{5/3}\right) + C = 3x^{5/3} + C.$

$f(8) = 3 \cdot 32 + C$ and $f(8) = 21 \;\Rightarrow\; 96 + C = 21 \;\Rightarrow\; C = -75$, so $f(x) = 3x^{5/3} - 75$.

36. $f'(x) = \dfrac{x+1}{\sqrt{x}} = x^{1/2} + x^{-1/2} \;\Rightarrow\; f(x) = \frac{2}{3}x^{3/2} + 2x^{1/2} + C.$ $f(1) = \frac{2}{3} + 2 + C = \frac{8}{3} + C$ and $f(1) = 5 \;\Rightarrow$

$C = 5 - \frac{8}{3} = \frac{7}{3}$, so $f(x) = \frac{2}{3}x^{3/2} + 2\sqrt{x} + \frac{7}{3}$.

37. $f'(t) = \sec t(\sec t + \tan t) = \sec^2 t + \sec t\,\tan t,\ -\frac{\pi}{2} < t < \frac{\pi}{2} \;\Rightarrow\; f(t) = \tan t + \sec t + C.$ $f\left(\frac{\pi}{4}\right) = 1 + \sqrt{2} + C$

and $f\left(\frac{\pi}{4}\right) = -1 \;\Rightarrow\; 1 + \sqrt{2} + C = -1 \;\Rightarrow\; C = -2 - \sqrt{2}$, so $f(t) = \tan t + \sec t - 2 - \sqrt{2}$.

Note: The fact that f is defined and continuous on $\left(-\frac{\pi}{2}, \frac{\pi}{2}\right)$ means that we have only one constant of integration.

38. $f'(t) = 3^t - \dfrac{3}{t} \Rightarrow f(t) = \begin{cases} 3^t/\ln 3 - 3\ln(-t) + C & \text{if } t < 0 \\ 3^t/\ln 3 - 3\ln t + D & \text{if } t > 0 \end{cases}$

$f(-1) = \dfrac{1}{3\ln 3} - 3\ln 1 + C$ and $f(-1) = 1 \Rightarrow C = 1 - \dfrac{1}{3\ln 3}$.

$f(1) = \dfrac{3}{\ln 3} - 3\ln 1 + D$ and $f(1) = 2 \Rightarrow D = 2 - \dfrac{3}{\ln 3}$.

Thus, $f(t) = \begin{cases} 3^t/\ln 3 - 3\ln(-t) + 1 - 1/(3\ln 3) & \text{if } t < 0 \\ 3^t/\ln 3 - 3\ln t + 2 - 3/\ln 3 & \text{if } t > 0 \end{cases}$

39. $f''(x) = -2 + 12x - 12x^2 \Rightarrow f'(x) = -2x + 6x^2 - 4x^3 + C$. $f'(0) = C$ and $f'(0) = 12 \Rightarrow C = 12$, so $f'(x) = -2x + 6x^2 - 4x^3 + 12$ and hence, $f(x) = -x^2 + 2x^3 - x^4 + 12x + D$. $f(0) = D$ and $f(0) = 4 \Rightarrow D = 4$, so $f(x) = -x^2 + 2x^3 - x^4 + 12x + 4$.

40. $f''(x) = 8x^3 + 5 \Rightarrow f'(x) = 2x^4 + 5x + C$. $f'(1) = 2 + 5 + C$ and $f'(1) = 8 \Rightarrow C = 1$, so $f'(x) = 2x^4 + 5x + 1$. $f(x) = \frac{2}{5}x^5 + \frac{5}{2}x^2 + x + D$. $f(1) = \frac{2}{5} + \frac{5}{2} + 1 + D = D + \frac{39}{10}$ and $f(1) = 0 \Rightarrow D = -\frac{39}{10}$, so $f(x) = \frac{2}{5}x^5 + \frac{5}{2}x^2 + x - \frac{39}{10}$.

41. $f''(\theta) = \sin\theta + \cos\theta \Rightarrow f'(\theta) = -\cos\theta + \sin\theta + C$. $f'(0) = -1 + C$ and $f'(0) = 4 \Rightarrow C = 5$, so $f'(\theta) = -\cos\theta + \sin\theta + 5$ and hence, $f(\theta) = -\sin\theta - \cos\theta + 5\theta + D$. $f(0) = -1 + D$ and $f(0) = 3 \Rightarrow D = 4$, so $f(\theta) = -\sin\theta - \cos\theta + 5\theta + 4$.

42. $f''(t) = t^2 + \dfrac{1}{t^2} = t^2 + t^{-2},\ t > 0 \Rightarrow f'(t) = \frac{1}{3}t^3 - \dfrac{1}{t} + C$. $f'(1) = \frac{1}{3} - 1 + C$ and $f'(1) = 2 \Rightarrow$ $C - \frac{2}{3} = 2 \Rightarrow C = \frac{8}{3}$, so $f'(t) = \frac{1}{3}t^3 - \dfrac{1}{t} + \frac{8}{3}$ and hence, $f(t) = \frac{1}{12}t^4 - \ln t + \frac{8}{3}t + D$. $f(2) = \frac{4}{3} - \ln 2 + \frac{16}{3} + D$ and $f(2) = 3 \Rightarrow \frac{20}{3} - \ln 2 + D = 3 \Rightarrow D = \ln 2 - \frac{11}{3}$, so $f(t) = \frac{1}{12}t^4 - \ln t + \frac{8}{3}t + \ln 2 - \frac{11}{3}$.

43. $f''(x) = 4 + 6x + 24x^2 \Rightarrow f'(x) = 4x + 3x^2 + 8x^3 + C \Rightarrow f(x) = 2x^2 + x^3 + 2x^4 + Cx + D$. $f(0) = D$ and $f(0) = 3 \Rightarrow D = 3$, so $f(x) = 2x^2 + x^3 + 2x^4 + Cx + 3$. $f(1) = 8 + C$ and $f(1) = 10 \Rightarrow C = 2$, so $f(x) = 2x^2 + x^3 + 2x^4 + 2x + 3$.

44. $f''(x) = x^3 + \sinh x \Rightarrow f'(x) = \frac{1}{4}x^4 + \cosh x + C \Rightarrow f(x) = \frac{1}{20}x^5 + \sinh x + Cx + D$. $f(0) = D$ and $f(0) = 1 \Rightarrow D = 1$, so $f(x) = \frac{1}{20}x^5 + \sinh x + Cx + 1$. $f(2) = \frac{32}{20} + \sinh 2 + 2C + 1$ and $f(2) = 2.6 \Rightarrow$ $\sinh 2 + 2C = 0 \Rightarrow C = -\frac{1}{2}\sinh 2$, so $f(x) = \frac{1}{20}x^5 + \sinh x - \frac{1}{2}(\sinh 2)x + 1$.

45. $f''(x) = e^x - 2\sin x \Rightarrow f'(x) = e^x + 2\cos x + C \Rightarrow f(x) = e^x + 2\sin x + Cx + D$. $f(0) = 1 + 0 + D$ and $f(0) = 3 \Rightarrow D = 2$, so $f(x) = e^x + 2\sin x + Cx + 2$. $f(\frac{\pi}{2}) = e^{\pi/2} + 2 + \frac{\pi}{2}C + 2$ and $f(\frac{\pi}{2}) = 0 \Rightarrow e^{\pi/2} + 4 + \frac{\pi}{2}C = 0 \Rightarrow \frac{\pi}{2}C = -e^{\pi/2} - 4 \Rightarrow C = -\frac{2}{\pi}(e^{\pi/2} + 4)$, so $f(x) = e^x + 2\sin x + -\frac{2}{\pi}(e^{\pi/2} + 4)x + 2$.

46. $f''(t) = \sqrt[3]{t} - \cos t = t^{1/3} - \cos t \Rightarrow f'(t) = \frac{3}{4}t^{4/3} - \sin t + C \Rightarrow f(t) = \frac{9}{28}t^{7/3} + \cos t + Ct + D$.

$f(0) = 0 + 1 + 0 + D$ and $f(0) = 2 \Rightarrow D = 1$, so $f(t) = \frac{9}{28}t^{7/3} + \cos t + Ct + 1$. $f(1) = \frac{9}{28} + \cos 1 + C + 1$ and

$f(1) = 2 \Rightarrow C = 2 - \frac{9}{28} - \cos 1 - 1 = \frac{19}{28} - \cos 1$, so $f(t) = \frac{9}{28}t^{7/3} + \cos t + \left(\frac{19}{28} - \cos 1\right)t + 1$.

47. $f''(x) = x^{-2}, x > 0 \Rightarrow f'(x) = -1/x + C \Rightarrow f(x) = -\ln|x| + Cx + D = -\ln x + Cx + D$ [since $x > 0$].

$f(1) = 0 \Rightarrow C + D = 0$ and $f(2) = 0 \Rightarrow -\ln 2 + 2C + D = 0 \Rightarrow -\ln 2 + 2C - C = 0$ [since $D = -C$] $\Rightarrow$

$-\ln 2 + C = 0 \Rightarrow C = \ln 2$ and $D = -\ln 2$. So $f(x) = -\ln x + (\ln 2)x - \ln 2$.

48. $f'''(x) = \cos x \Rightarrow f''(x) = \sin x + C$. $f''(0) = C$ and $f''(0) = 3 \Rightarrow C = 3$. $f''(x) = \sin x + 3 \Rightarrow$

$f'(x) = -\cos x + 3x + D$. $f'(0) = -1 + D$ and $f'(0) = 2 \Rightarrow D = 3$. $f'(x) = -\cos x + 3x + 3 \Rightarrow$

$f(x) = -\sin x + \frac{3}{2}x^2 + 3x + E$. $f(0) = E$ and $f(0) = 1 \Rightarrow E = 1$. Thus, $f(x) = -\sin x + \frac{3}{2}x^2 + 3x + 1$.

49. "The slope of its tangent line at $(x, f(x))$ is $3 - 4x$" means that $f'(x) = 3 - 4x$, so $f(x) = 3x - 2x^2 + C$.

"The graph of f passes through the point $(2, 5)$" means that $f(2) = 5$, but $f(2) = 3(2) - 2(2)^2 + C$, so $5 = 6 - 8 + C \Rightarrow$

$C = 7$. Thus, $f(x) = 3x - 2x^2 + 7$ and $f(1) = 3 - 2 + 7 = 8$.

50. $f'(x) = x^3 \Rightarrow f(x) = \frac{1}{4}x^4 + C$. $x + y = 0 \Rightarrow y = -x \Rightarrow m = -1$. Now $m = f'(x) \Rightarrow -1 = x^3 \Rightarrow$

$x = -1 \Rightarrow y = 1$ (from the equation of the tangent line), so $(-1, 1)$ is a point on the graph of f. From f,

$1 = \frac{1}{4}(-1)^4 + C \Rightarrow C = \frac{3}{4}$. Therefore, the function is $f(x) = \frac{1}{4}x^4 + \frac{3}{4}$.

51. b is the antiderivative of f. For small x, f is negative, so the graph of its antiderivative must be decreasing. But both a and c are increasing for small x, so only b can be f's antiderivative. Also, f is positive where b is increasing, which supports our conclusion.

52. We know right away that c cannot be f's antiderivative, since the slope of c is not zero at the x-value where $f = 0$. Now f is positive when a is increasing and negative when a is decreasing, so a is the antiderivative of f.

53.

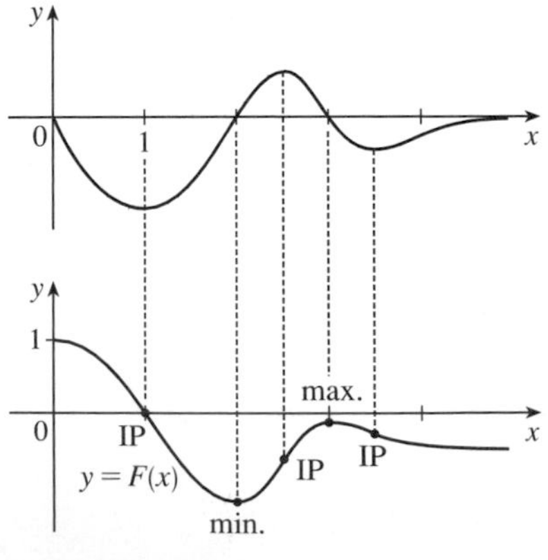

The graph of F must start at $(0, 1)$. Where the given graph, $y = f(x)$, has a local minimum or maximum, the graph of F will have an inflection point.

Where f is negative (positive), F is decreasing (increasing).

Where f changes from negative to positive, F will have a minimum.

Where f changes from positive to negative, F will have a maximum.

Where f is decreasing (increasing), F is concave downward (upward).

54. 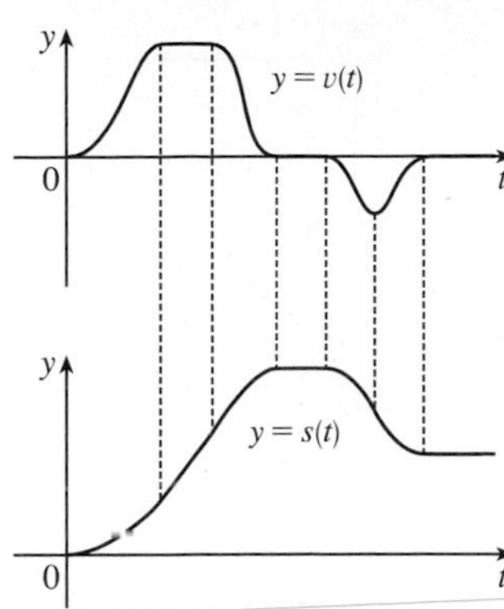

Where v is positive (negative), s is increasing (decreasing).

Where v is increasing (decreasing), s is concave upward (downward).

Where v is horizontal (a steady velocity), s is linear.

55. 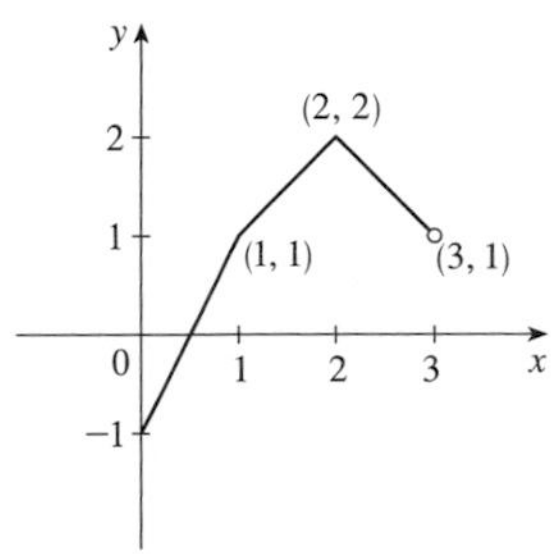

$$f'(x) = \begin{cases} 2 & \text{if } 0 \le x < 1 \\ 1 & \text{if } 1 < x < 2 \\ -1 & \text{if } 2 < x < 3 \end{cases} \quad \Rightarrow \quad f(x) = \begin{cases} 2x + C & \text{if } 0 \le x < 1 \\ x + D & \text{if } 1 < x < 2 \\ -x + E & \text{if } 2 < x < 3 \end{cases}$$

$f(0) = -1 \quad \Rightarrow \quad 2(0) + C = -1 \quad \Rightarrow \quad C = -1$. Starting at the point $(0, -1)$ and moving to the right on a line with slope 2 gets us to the point $(1, 1)$.

The slope for $1 < x < 2$ is 1, so we get to the point $(2, 2)$. Here we have used the fact that f is continuous. We can include the point $x = 1$ on either the first or the second part of f. The line connecting $(1, 1)$ to $(2, 2)$ is $y = x$, so $D = 0$. The slope for $2 < x < 3$ is -1, so we get to $(3, 1)$. $\quad f(2) = 2 \quad \Rightarrow \quad -2 + E = 2 \quad \Rightarrow \quad E = 4$. Thus,

$$f(x) = \begin{cases} 2x - 1 & \text{if } 0 \le x \le 1 \\ x & \text{if } 1 < x < 2 \\ -x + 4 & \text{if } 2 \le x < 3 \end{cases}$$

Note that $f'(x)$ does not exist at $x = 1$, 2, or 3.

56. (a)

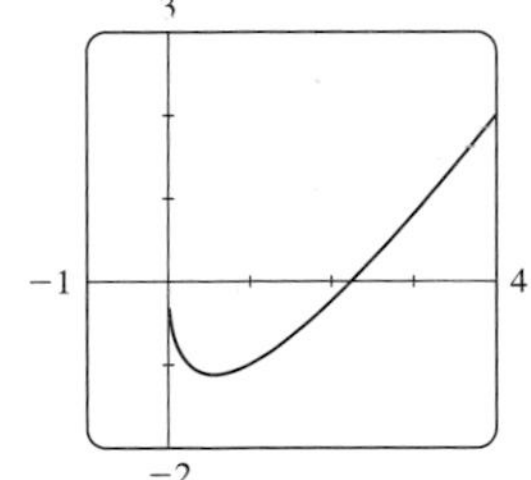

(b) Since $F(0) = 1$, we can start our graph at $(0, 1)$. f has a minimum at about $x = 0.5$, so its derivative is zero there. f is decreasing on $(0, 0.5)$, so its derivative is negative and hence, F is CD on $(0, 0.5)$ and has an IP at $x \approx 0.5$. On $(0.5, 2.2)$, f is negative and increasing (f' is positive), so F is decreasing and CU. On $(2.2, \infty)$, f is positive and increasing, so F is increasing and CU.

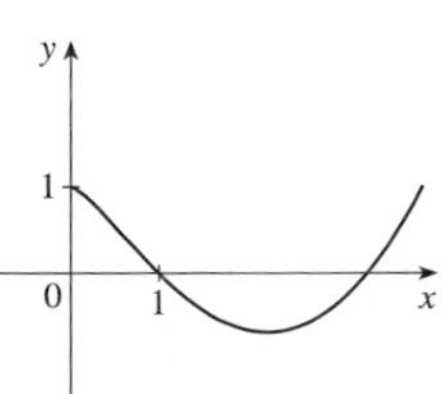

(c) $f(x) = 2x - 3\sqrt{x} \;\Rightarrow\; F(x) = x^2 - 3 \cdot \frac{2}{3}x^{3/2} + C.$

$F(0) = C$ and $F(0) = 1 \;\Rightarrow\; C = 1$, so $F(x) = x^2 - 2x^{3/2} + 1$.

(d)

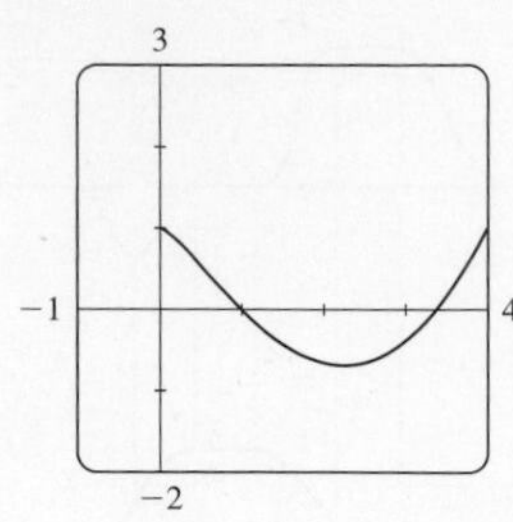

57. $f(x) = \dfrac{\sin x}{1 + x^2}$, $-2\pi \le x \le 2\pi$

Note that the graph of f is one of an odd function, so the graph of F will be one of an even function.

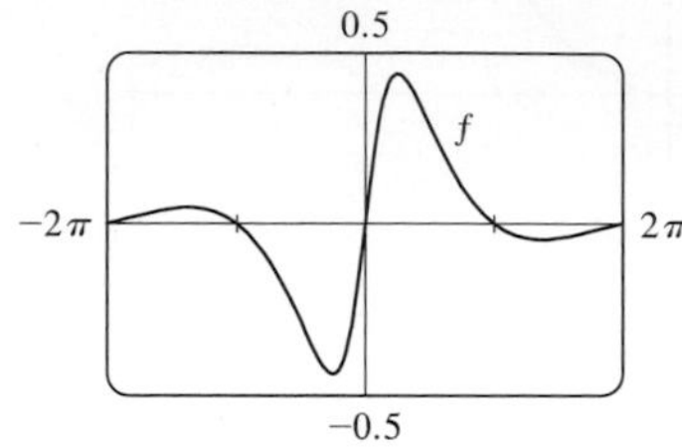

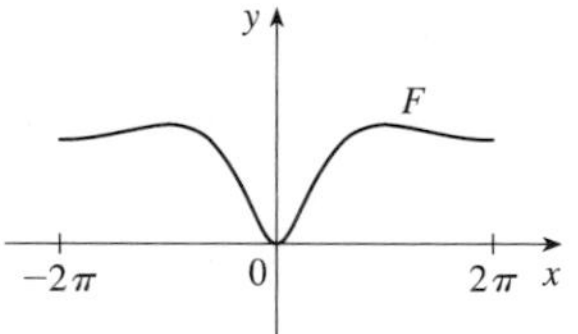

58. $f(x) = \sqrt{x^4 - 2x^2 + 2} - 2$, $-3 \le x \le 3$

Note that the graph of f is one of an even function, so the graph of F will be one of an odd function.

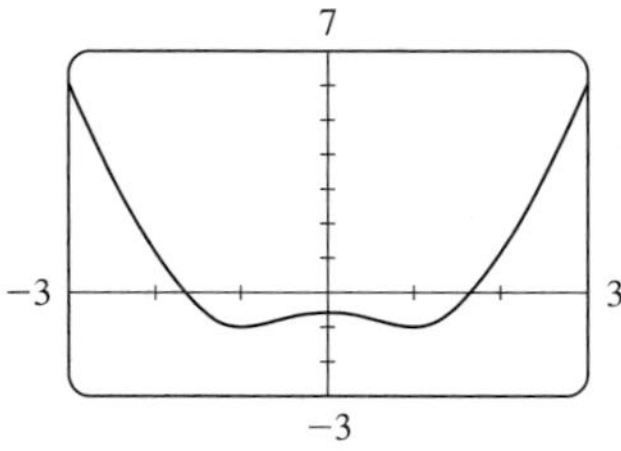

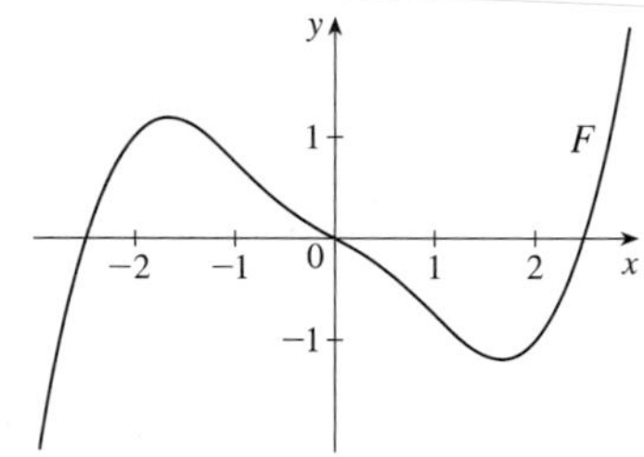

59. $v(t) = s'(t) = \sin t - \cos t \;\Rightarrow\; s(t) = -\cos t - \sin t + C$. $s(0) = -1 + C$ and $s(0) = 0 \;\Rightarrow\; C = 1$, so $s(t) = -\cos t - \sin t + 1$.

60. $v(t) = s'(t) = t^2 - 3\sqrt{t} = t^2 - 3t^{1/2} \;\Rightarrow\; s(t) = \frac{1}{3}t^3 - 2t^{3/2} + C$. $s(4) = \frac{64}{3} - 16 + C$ and $s(4) = 8 \;\Rightarrow\; C = 8 - \frac{64}{3} + 16 = \frac{8}{3}$, so $s(t) = \frac{1}{3}t^3 - 2t^{3/2} + \frac{8}{3}$.

61. $a(t) = v'(t) = 2t + 1 \;\Rightarrow\; v(t) = t^2 + t + C$. $v(0) = C$ and $v(0) = -2 \;\Rightarrow\; C = -2$, so $v(t) = t^2 + t - 2$ and $s(t) = \frac{1}{3}t^3 + \frac{1}{2}t^2 - 2t + D$. $s(0) = D$ and $s(0) = 3 \;\Rightarrow\; D = 3$, so $s(t) = \frac{1}{3}t^3 + \frac{1}{2}t^2 - 2t + 3$.

62. $a(t) = v'(t) = 3\cos t - 2\sin t \;\Rightarrow\; v(t) = 3\sin t + 2\cos t + C$. $v(0) = 2 + C$ and $v(0) = 4 \;\Rightarrow\; C = 2$, so $v(t) = 3\sin t + 2\cos t + 2$ and $s(t) = -3\cos t + 2\sin t + 2t + D$. $s(0) = -3 + D$ and $s(0) = 0 \;\Rightarrow\; D = 3$, so $s(t) = -3\cos t + 2\sin t + 2t + 3$.

63. $a(t) = v'(t) = 10\sin t + 3\cos t \;\Rightarrow\; v(t) = -10\cos t + 3\sin t + C \;\Rightarrow\; s(t) = -10\sin t - 3\cos t + Ct + D$. $s(0) = -3 + D = 0$ and $s(2\pi) = -3 + 2\pi C + D = 12 \;\Rightarrow\; D = 3$ and $C = \frac{6}{\pi}$. Thus, $s(t) = -10\sin t - 3\cos t + \frac{6}{\pi}t + 3$.

64. $a(t) = t^2 - 4t + 6 \;\Rightarrow\; v(t) = \frac{1}{3}t^3 - 2t^2 + 6t + C \;\Rightarrow\; s(t) = \frac{1}{12}t^4 - \frac{2}{3}t^3 + 3t^2 + Ct + D$. $s(0) = D$ and $s(0) = 0 \;\Rightarrow\; D = 0$. $s(1) = \frac{29}{12} + C$ and $s(1) = 20 \;\Rightarrow\; C = \frac{211}{12}$. Thus, $s(t) = \frac{1}{12}t^4 - \frac{2}{3}t^3 + 3t^2 + \frac{211}{12}t$.

65. (a) We first observe that since the stone is dropped 450 m above the ground, $v(0) = 0$ and $s(0) = 450$.

$v'(t) = a(t) = -9.8 \;\Rightarrow\; v(t) = -9.8t + C$. Now $v(0) = 0 \;\Rightarrow\; C = 0$, so $v(t) = -9.8t \;\Rightarrow$ $s(t) = -4.9t^2 + D$. Last, $s(0) = 450 \;\Rightarrow\; D = 450 \;\Rightarrow\; s(t) = 450 - 4.9t^2$.

(b) The stone reaches the ground when $s(t) = 0$. $450 - 4.9t^2 = 0 \;\Rightarrow\; t^2 = 450/4.9 \;\Rightarrow\; t_1 = \sqrt{450/4.9} \approx 9.58$ s.

(c) The velocity with which the stone strikes the ground is $v(t_1) = -9.8\sqrt{450/4.9} \approx -93.9$ m/s.

(d) This is just reworking parts (a) and (b) with $v(0) = -5$. Using $v(t) = -9.8t + C$, $v(0) = -5 \;\Rightarrow\; 0 + C = -5 \;\Rightarrow$ $v(t) = -9.8t - 5$. So $s(t) = -4.9t^2 - 5t + D$ and $s(0) = 450 \;\Rightarrow\; D = 450 \;\Rightarrow\; s(t) = -4.9t^2 - 5t + 450$. Solving $s(t) = 0$ by using the quadratic formula gives us $t = \left(5 \pm \sqrt{8845}\,\right)/(-9.8) \;\Rightarrow\; t_1 \approx 9.09$ s.

66. $v'(t) = a(t) = a \;\Rightarrow\; v(t) = at + C$ and $v_0 = v(0) = C \;\Rightarrow\; v(t) = at + v_0 \;\Rightarrow$ $s(t) = \frac{1}{2}at^2 + v_0t + D \;\Rightarrow\; s_0 = s(0) = D \;\Rightarrow\; s(t) = \frac{1}{2}at^2 + v_0t + s_0$

67. By Exercise 66 with $a = -9.8$, $s(t) = -4.9t^2 + v_0t + s_0$ and $v(t) = s'(t) = -9.8t + v_0$. So $[v(t)]^2 = (-9.8t + v_0)^2 = (9.8)^2t^2 - 19.6v_0t + v_0^2 = v_0^2 + 96.04t^2 - 19.6v_0t = v_0^2 - 19.6\left(-4.9t^2 + v_0t\right)$. But $-4.9t^2 + v_0t$ is just $s(t)$ without the s_0 term; that is, $s(t) - s_0$. Thus, $[v(t)]^2 = v_0^2 - 19.6\,[s(t) - s_0]$.

68. For the first ball, $s_1(t) = -16t^2 + 48t + 432$ from Example 7. For the second ball, $a(t) = -32 \;\Rightarrow\; v(t) = -32t + C$, but $v(1) = -32(1) + C = 24 \;\Rightarrow\; C = 56$, so $v(t) = -32t + 56 \;\Rightarrow\; s(t) = -16t^2 + 56t + D$, but $s(1) = -16(1)^2 + 56(1) + D = 432 \;\Rightarrow\; D = 392$, and $s_2(t) = -16t^2 + 56t + 392$. The balls pass each other when $s_1(t) = s_2(t) \;\Rightarrow\; -16t^2 + 48t + 432 = -16t^2 + 56t + 392 \;\Leftrightarrow\; 8t = 40 \;\Leftrightarrow\; t = 5$ s.

Another solution: From Exercise 66, we have $s_1(t) = -16t^2 + 48t + 432$ and $s_2(t) = -16t^2 + 24t + 432$. We now want to solve $s_1(t) = s_2(t - 1) \;\Rightarrow\; -16t^2 + 48t + 432 = -16(t-1)^2 + 24(t-1) + 432 \;\Rightarrow$ $48t = 32t - 16 + 24t - 24 \;\Rightarrow\; 40 = 8t \;\Rightarrow\; t = 5$ s.

69. Using Exercise 66 with $a = -32$, $v_0 = 0$, and $s_0 = h$ (the height of the cliff), we know that the height at time t is $s(t) = -16t^2 + h$. $v(t) = s'(t) = -32t$ and $v(t) = -120 \;\Rightarrow\; -32t = -120 \;\Rightarrow\; t = 3.75$, so $0 = s(3.75) = -16(3.75)^2 + h \;\Rightarrow\; h = 16(3.75)^2 = 225$ ft.

70. (a) $EIy'' = mg(L - x) + \frac{1}{2}\rho g(L - x)^2 \;\Rightarrow\; EIy' = -\frac{1}{2}mg(L - x)^2 - \frac{1}{6}\rho g(L - x)^3 + C \;\Rightarrow$ $EIy = \frac{1}{6}mg(L - x)^3 + \frac{1}{24}\rho g(L - x)^4 + Cx + D$. Since the left end of the board is fixed, we must have $y = y' = 0$ when $x = 0$. Thus, $0 = -\frac{1}{2}mgL^2 - \frac{1}{6}\rho gL^3 + C$ and $0 = \frac{1}{6}mgL^3 + \frac{1}{24}\rho gL^4 + D$. It follows that

$EIy = \frac{1}{6}mg(L - x)^3 + \frac{1}{24}\rho g(L - x)^4 + \left(\frac{1}{2}mgL^2 + \frac{1}{6}\rho gL^3\right)x - \left(\frac{1}{6}mgL^3 + \frac{1}{24}\rho gL^4\right)$ and

$$f(x) = y = \frac{1}{EI}\left[\tfrac{1}{6}mg(L - x)^3 + \tfrac{1}{24}\rho g(L - x)^4 + \left(\tfrac{1}{2}mgL^2 + \tfrac{1}{6}\rho gL^3\right)x - \left(\tfrac{1}{6}mgL^3 + \tfrac{1}{24}\rho gL^4\right)\right]$$

(b) $f(L) < 0$, so the end of the board is a *distance* approximately $-f(L)$ below the horizontal. From our result in (a), we calculate

$$-f(L) = \frac{-1}{EI}\left[\tfrac{1}{2}mgL^3 + \tfrac{1}{6}\rho gL^4 - \tfrac{1}{6}mgL^3 - \tfrac{1}{24}\rho gL^4\right] = \frac{-1}{EI}\left(\tfrac{1}{3}mgL^3 + \tfrac{1}{8}\rho gL^4\right) = -\frac{gL^3}{EI}\left(\frac{m}{3} + \frac{\rho L}{8}\right)$$

Note: This is positive because g is negative.

71. Marginal cost $= 1.92 - 0.002x = C'(x) \Rightarrow C(x) = 1.92x - 0.001x^2 + K$. But $C(1) = 1.92 - 0.001 + K = 562 \Rightarrow K = 560.081$. Therefore, $C(x) = 1.92x - 0.001x^2 + 560.081 \Rightarrow C(100) = 742.081$, so the cost of producing 100 items is \$742.08.

72. Let the mass, measured from one end, be $m(x)$. Then $m(0) = 0$ and $\rho = \dfrac{dm}{dx} = x^{-1/2} \Rightarrow m(x) = 2x^{1/2} + C$ and $m(0) = C = 0$, so $m(x) = 2\sqrt{x}$. Thus, the mass of the 100-centimeter rod is $m(100) = 2\sqrt{100} = 20$ g.

73. Taking the upward direction to be positive we have that for $0 \le t \le 10$ (using the subscript 1 to refer to $0 \le t \le 10$),

$a_1(t) = -(9 - 0.9t) = v_1'(t) \Rightarrow v_1(t) = -9t + 0.45t^2 + v_0$, but $v_1(0) = v_0 = -10 \Rightarrow$

$v_1(t) = -9t + 0.45t^2 - 10 = s_1'(t) \Rightarrow s_1(t) = -\frac{9}{2}t^2 + 0.15t^3 - 10t + s_0$. But $s_1(0) = 500 = s_0 \Rightarrow$

$s_1(t) = -\frac{9}{2}t^2 + 0.15t^3 - 10t + 500$. $s_1(10) = -450 + 150 - 100 + 500 = 100$, so it takes more than 10 seconds for the raindrop to fall. Now for $t > 10$, $a(t) = 0 = v'(t) \Rightarrow$

$v(t) = \text{constant} = v_1(10) = -9(10) + 0.45(10)^2 - 10 = -55 \Rightarrow v(t) = -55$.

At 55 m/s, it will take $100/55 \approx 1.8$ s to fall the last 100 m. Hence, the total time is $10 + \frac{100}{55} = \frac{130}{11} \approx 11.8$ s.

74. $v'(t) = a(t) = -22$. The initial velocity is 50 mi/h $= \frac{50 \cdot 5280}{3600} = \frac{220}{3}$ ft/s, so $v(t) = -22t + \frac{220}{3}$.

The car stops when $v(t) = 0 \Leftrightarrow t = \frac{220}{3 \cdot 22} = \frac{10}{3}$. Since $s(t) = -11t^2 + \frac{220}{3}t$, the distance covered is

$s\left(\frac{10}{3}\right) = -11\left(\frac{10}{3}\right)^2 + \frac{220}{3} \cdot \frac{10}{3} = \frac{1100}{9} = 122.\overline{2}$ ft.

75. $a(t) = k$, the initial velocity is 30 mi/h $= 30 \cdot \frac{5280}{3600} = 44$ ft/s, and the final velocity (after 5 seconds) is 50 mi/h $= 50 \cdot \frac{5280}{3600} = \frac{220}{3}$ ft/s. So $v(t) = kt + C$ and $v(0) = 44 \Rightarrow C = 44$. Thus, $v(t) = kt + 44 \Rightarrow v(5) = 5k + 44$. But $v(5) = \frac{220}{3}$, so $5k + 44 = \frac{220}{3} \Rightarrow 5k = \frac{88}{3} \Rightarrow k = \frac{88}{15} \approx 5.87$ ft/s^2.

76. $a(t) = -16 \Rightarrow v(t) = -16t + v_0$ where v_0 is the car's speed (in ft/s) when the brakes were applied. The car stops when $-16t + v_0 = 0 \Leftrightarrow t = \frac{1}{16}v_0$. Now $s(t) = \frac{1}{2}(-16)t^2 + v_0t = -8t^2 + v_0t$. The car travels 200 ft in the time that it takes to stop, so $s\left(\frac{1}{16}v_0\right) = 200 \Rightarrow 200 = -8\left(\frac{1}{16}v_0\right)^2 + v_0\left(\frac{1}{16}v_0\right) = \frac{1}{32}v_0^2 \Rightarrow v_0^2 = 32 \cdot 200 = 6400 \Rightarrow v_0 = 80$ ft/s $[54.\overline{54}$ mi/h$]$.

77. Let the acceleration be $a(t) = k$ km/h^2. We have $v(0) = 100$ km/h and we can take the initial position $s(0)$ to be 0. We want the time t_f for which $v(t) = 0$ to satisfy $s(t) < 0.08$ km. In general, $v'(t) = a(t) = k$, so $v(t) = kt + C$, where $C = v(0) = 100$. Now $s'(t) = v(t) = kt + 100$, so $s(t) = \frac{1}{2}kt^2 + 100t + D$, where $D = s(0) = 0$.

Thus, $s(t) = \frac{1}{2}kt^2 + 100t$. Since $v(t_f) = 0$, we have $kt_f + 100 = 0$ or $t_f = -100/k$, so

$s(t_f) = \frac{1}{2}k\left(-\frac{100}{k}\right)^2 + 100\left(-\frac{100}{k}\right) = 10{,}000\left(\frac{1}{2k} - \frac{1}{k}\right) = -\frac{5{,}000}{k}$. The condition $s(t_f)$ must satisfy is

$-\frac{5{,}000}{k} < 0.08 \quad\Rightarrow\quad -\frac{5{,}000}{0.08} > k \quad [k \text{ is negative}] \quad\Rightarrow\quad k < -62{,}500 \text{ km/h}^2$, or equivalently,

$k < -\frac{3125}{648} \approx -4.82 \text{ m/s}^2$.

78 (a) For $0 \le t \le 3$ we have $a(t) = 60t \quad\Rightarrow\quad v(t) = 30t^2 + C \quad\Rightarrow\quad v(0) = 0 = C \quad\Rightarrow\quad v(t) = 30t^2$, so

$s(t) = 10t^3 + C \quad\Rightarrow\quad s(0) = 0 = C \quad\Rightarrow\quad s(t) = 10t^3$. Note that $v(3) = 270$ and $s(3) = 270$.

For $3 < t \le 17$: $a(t) = -g = -32$ ft/s $\quad\Rightarrow\quad v(t) = -32(t-3) + C \quad\Rightarrow\quad v(3) = 270 = C \quad\Rightarrow$

$v(t) = -32(t-3) + 270 \quad\Rightarrow\quad s(t) = -16(t-3)^2 + 270(t-3) + C \quad\Rightarrow\quad s(3) = 270 = C \quad\Rightarrow$

$s(t) = -16(t-3)^2 + 270(t-3) + 270$. Note that $v(17) = -178$ and $s(17) = 914$.

For $17 < t \le 22$: The velocity increases linearly from -178 ft/s to -18 ft/s during this period, so

$\frac{\Delta v}{\Delta t} = \frac{-18-(-178)}{22-17} = \frac{160}{5} = 32$. Thus, $v(t) = 32(t-17) - 178 \quad\Rightarrow$

$s(t) = 16(t-17)^2 - 178(t-17) + 914$ and $s(22) = 424$ ft.

For $t > 22$: $v(t) = -18 \quad\Rightarrow\quad s(t) = -18(t-22) + C$. But $s(22) = 424 = C \quad\Rightarrow\quad s(t) = -18(t-22) + 424$.

Therefore, until the rocket lands, we have

$$v(t) = \begin{cases} 30t^2 & \text{if } 0 \le t \le 3 \\ -32(t-3) + 270 & \text{if } 3 < t \le 17 \\ 32(t-17) - 178 & \text{if } 17 < t \le 22 \\ -18 & \text{if } t > 22 \end{cases}$$

and

$$s(t) = \begin{cases} 10t^3 & \text{if } 0 \le t \le 3 \\ -16(t-3)^2 + 270(t-3) + 270 & \text{if } 3 < t \le 17 \\ 16(t-17)^2 - 178(t-17) + 914 & \text{if } 17 < t \le 22 \\ -18(t-22) + 424 & \text{if } t > 22 \end{cases}$$

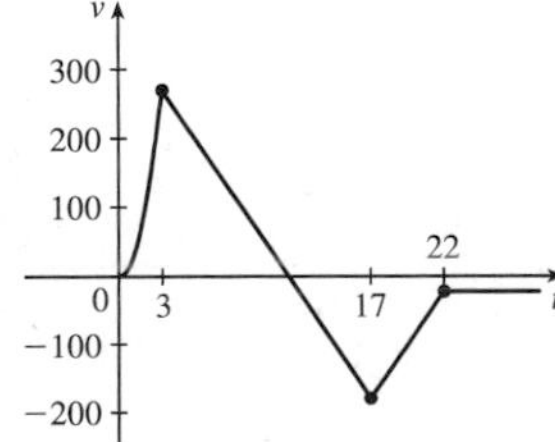

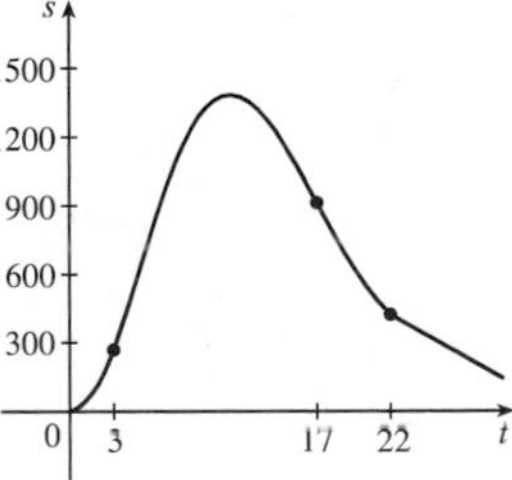

(b) To find the maximum height, set $v(t)$ on $3 < t \le 17$ equal to 0. $-32(t-3) + 270 = 0 \quad\Rightarrow\quad t_1 = 11.4375$ s and the maximum height is $s(t_1) = -16(t_1-3)^2 + 270(t_1-3) + 270 = 1409.0625$ ft.

(c) To find the time to land, set $s(t) = -18(t-22) + 424 = 0$. Then $t - 22 = \frac{424}{18} = 23.\overline{5}$, so $t \approx 45.6$ s.

79. (a) First note that 90 mi/h $= 90 \times \frac{5280}{3600}$ ft/s $= 132$ ft/s. Then $a(t) = 4\text{ ft/s}^2 \Rightarrow v(t) = 4t + C$, but $v(0) = 0 \Rightarrow C = 0$. Now $4t = 132$ when $t = \frac{132}{4} = 33$ s, so it takes 33 s to reach 132 ft/s. Therefore, taking $s(0) = 0$, we have $s(t) = 2t^2$, $0 \le t \le 33$. So $s(33) = 2178$ ft. 15 minutes $= 15(60) = 900$ s, so for $33 < t \le 933$ we have $v(t) = 132\text{ ft/s} \Rightarrow s(933) = 132(900) + 2178 = 120{,}978\text{ ft} = 22.9125$ mi.

(b) As in part (a), the train accelerates for 33 s and travels 2178 ft while doing so. Similarly, it decelerates for 33 s and travels 2178 ft at the end of its trip. During the remaining $900 - 66 = 834$ s it travels at 132 ft/s, so the distance traveled is $132 \cdot 834 = 110{,}088$ ft. Thus, the total distance is $2178 + 110{,}088 + 2178 = 114{,}444\text{ ft} = 21.675$ mi.

(c) 45 mi $= 45(5280) = 237{,}600$ ft. Subtract $2(2178)$ to take care of the speeding up and slowing down, and we have 233,244 ft at 132 ft/s for a trip of $233{,}244/132 = 1767$ s at 90 mi/h. The total time is $1767 + 2(33) = 1833\text{ s} = 30\text{ min }33\text{ s} = 30.55$ min.

(d) $37.5(60) = 2250$ s. $2250 - 2(33) = 2184$ s at maximum speed. $2184(132) + 2(2178) = 292{,}644$ total feet or $292{,}644/5280 = 55.425$ mi.

4 Review

TRUE-FALSE QUIZ

1. False. For example, take $f(x) = x^3$, then $f'(x) = 3x^2$ and $f'(0) = 0$, but $f(0) = 0$ is not a maximum or minimum; $(0, 0)$ is an inflection point.

2. False. For example, $f(x) = |x|$ has an absolute minimum at 0, but $f'(0)$ does not exist.

3. False. For example, $f(x) = x$ is continuous on $(0, 1)$ but attains neither a maximum nor a minimum value on $(0, 1)$. Don't confuse this with f being continuous on the *closed* interval $[a, b]$, which would make the statement true.

4. True. By the Mean Value Theorem, $f'(c) = \dfrac{f(1) - f(-1)}{1 - (-1)} = \dfrac{0}{2} = 0$. Note that $|c| < 1 \Leftrightarrow c \in (-1, 1)$.

5. True. This is an example of part (b) of the I/D Test.

6. False. For example, the curve $y = f(x) = 1$ has no inflection points but $f''(c) = 0$ for all c.

7. False. $f'(x) = g'(x) \Rightarrow f(x) = g(x) + C$. For example, if $f(x) = x + 2$ and $g(x) = x + 1$, then $f'(x) = g'(x) = 1$, but $f(x) \ne g(x)$.

8. False. Assume there is a function f such that $f(1) = -2$ and $f(3) = 0$. Then by the Mean Value Theorem there exists a number $c \in (1, 3)$ such that $f'(c) = \dfrac{f(3) - f(1)}{3 - 1} = \dfrac{0 - (-2)}{2} = 1$. But $f'(x) > 1$ for all x, a contradiction.

9. True. The graph of one such function is sketched.

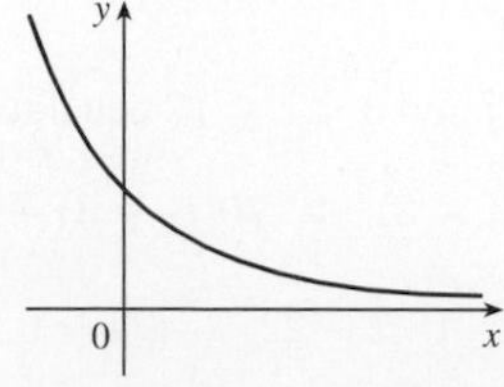

10. False. At any point $(a, f(a))$, we know that $f'(a) < 0$. So since the tangent line at $(a, f(a))$ is not horizontal, it must cross the x-axis—at $x = b$, say. But since $f''(x) > 0$ for all x, the graph of f must lie above all of its tangents; in particular, $f(b) > 0$. But this is a contradiction, since we are given that $f(x) < 0$ for all x.

11. True. Let $x_1 < x_2$ where $x_1, x_2 \in I$. Then $f(x_1) < f(x_2)$ and $g(x_1) < g(x_2)$ [since f and g are increasing on I], so $(f + g)(x_1) = f(x_1) + g(x_1) < f(x_2) + g(x_2) = (f + g)(x_2)$.

12. False. $f(x) = x$ and $g(x) = 2x$ are both increasing on $(0, 1)$, but $f(x) - g(x) = -x$ is not increasing on $(0, 1)$

13. False. Take $f(x) = x$ and $g(x) = x - 1$. Then both f and g are increasing on $(0, 1)$. But $f(x)\,g(x) = x(x - 1)$ is not increasing on $(0, 1)$.

14. True. Let $x_1 < x_2$ where $x_1, x_2 \in I$. Then $0 < f(x_1) < f(x_2)$ and $0 < g(x_1) < g(x_2)$ [since f and g are both positive and increasing]. Hence, $f(x_1)\,g(x_1) < f(x_2)\,g(x_1) < f(x_2)\,g(x_2)$. So fg is increasing on I.

15. True. Let $x_1, x_2 \in I$ and $x_1 < x_2$. Then $f(x_1) < f(x_2)$ [f is increasing] $\Rightarrow$ $\dfrac{1}{f(x_1)} > \dfrac{1}{f(x_2)}$ [f is positive] $\Rightarrow$ $g(x_1) > g(x_2)$ $\Rightarrow$ $g(x) = 1/f(x)$ is decreasing on I.

16. False. If f is even, then $f(x) = f(-x)$. Using the Chain Rule to differentiate this equation, we get $f'(x) = f'(-x)\,\dfrac{d}{dx}(-x) = -f'(-x)$. Thus, $f'(-x) = -f'(x)$, so f' is odd.

17. True. If f is periodic, then there is a number p such that $f(x + p) = f(p)$ for all x. Differentiating gives $f'(x) = f'(x + p) \cdot (x + p)' = f'(x + p) \cdot 1 = f'(x + p)$, so f' is periodic.

18. False. The most general antiderivative of $f(x) = x^{-2}$ is $F(x) = -1/x + C_1$ for $x < 0$ and $F(x) = -1/x + C_2$ for $x > 0$ [see Example 4.9.1(b)].

19. True. By the Mean Value Theorem, there exists a number c in $(0, 1)$ such that $f(1) - f(0) = f'(c)(1 - 0) = f'(c)$. Since $f'(c)$ is nonzero, $f(1) - f(0) \neq 0$, so $f(1) \neq f(0)$.

20. False. Let $f(x) = 1 + \dfrac{1}{x}$ and $g(x) = x$. Then $\lim\limits_{x\to\infty} f(x) = 1$ and $\lim\limits_{x\to\infty} g(x) = \infty$, but

$$\lim_{x\to\infty} [f(x)]^{g(x)} = \lim_{x\to\infty} \left(1 + \frac{1}{x}\right)^x = e, \text{ not } 1.$$

21. False. $$\lim_{x\to 0} \frac{x}{e^x} = \frac{\lim\limits_{x\to 0} x}{\lim\limits_{x\to 0} e^x} = \frac{0}{1} = 0, \text{ not } 1.$$

EXERCISES

1. $f(x) = x^3 - 9x^2 + 24x - 2$, $[0, 5]$. $f'(x) = 3x^2 - 18x + 24 = 3(x^2 - 6x + 8) = 3(x - 2)(x - 4)$. $f'(x) = 0 \Leftrightarrow x = 2$ or $x = 4$. $f'(x) > 0$ for $0 < x < 2$, $f'(x) < 0$ for $2 < x < 4$, and $f'(x) > 0$ for $4 < x < 5$, so $f(2) = 18$ is a local maximum value and $f(4) = 14$ is a local minimum value. Checking the endpoints, we find $f(0) = -2$ and $f(5) = 18$. Thus, $f(0) = -2$ is the absolute minimum value and $f(2) = f(5) = 18$ is the absolute maximum value.

2. $f(x) = x\sqrt{1-x}$, $[-1, 1]$. $f'(x) = x \cdot \frac{1}{2}(1-x)^{-1/2}(-1) + (1-x)^{1/2}(1) = (1-x)^{-1/2}\left[-\frac{1}{2}x + (1-x)\right] = \dfrac{1 - \frac{3}{2}x}{\sqrt{1-x}}$.

$f'(x) = 0 \Rightarrow x = \frac{2}{3}$. $f'(x)$ does not exist $\Leftrightarrow x = 1$. $f'(x) > 0$ for $-1 < x < \frac{2}{3}$ and $f'(x) < 0$ for $\frac{2}{3} < x < 1$, so $f\left(\frac{2}{3}\right) = \frac{2}{3}\sqrt{\frac{1}{3}} = \frac{2}{9}\sqrt{3}$ [≈ 0.38] is a local maximum value. Checking the endpoints, we find $f(-1) = -\sqrt{2}$ and $f(1) = 0$. Thus, $f(-1) = -\sqrt{2}$ is the absolute minimum value and $f\left(\frac{2}{3}\right) = \frac{2}{9}\sqrt{3}$ is the absolute maximum value.

3. $f(x) = \dfrac{3x-4}{x^2+1}$, $[-2, 2]$. $f'(x) = \dfrac{(x^2+1)(3) - (3x-4)(2x)}{(x^2+1)^2} = \dfrac{-(3x^2 - 8x - 3)}{(x^2+1)^2} = \dfrac{-(3x+1)(x-3)}{(x^2+1)^2}$.

$f'(x) = 0 \Rightarrow x = -\frac{1}{3}$ or $x = 3$, but 3 is not in the interval. $f'(x) > 0$ for $-\frac{1}{3} < x < 2$ and $f'(x) < 0$ for $-2 < x < -\frac{1}{3}$, so $f\left(-\frac{1}{3}\right) = \frac{-5}{10/9} = -\frac{9}{2}$ is a local minimum value. Checking the endpoints, we find $f(-2) = -2$ and $f(2) = \frac{2}{5}$. Thus, $f\left(-\frac{1}{3}\right) = -\frac{9}{2}$ is the absolute minimum value and $f(2) = \frac{2}{5}$ is the absolute maximum value.

4. $f(x) = \sqrt{x^2+x+1}$, $[-2, 1]$. $f'(x) = \frac{1}{2}(x^2+x+1)^{-1/2}(2x+1) = \dfrac{2x+1}{2\sqrt{x^2+x+1}}$. $f'(x) = 0 \Rightarrow x = -\dfrac{1}{2}$.

$f'(x) > 0$ for $-\frac{1}{2} < x < 1$ and $f'(x) < 0$ for $-2 < x < -\frac{1}{2}$, so $f\left(-\frac{1}{2}\right) = \sqrt{3}/2$ is a local minimum value. Checking the endpoints, we find $f(-2) = f(1) = \sqrt{3}$. Thus, $f\left(-\frac{1}{2}\right) = \sqrt{3}/2$ is the absolute minimum value and $f(-2) = f(1) = \sqrt{3}$ is the absolute maximum value.

5. $f(x) = x + 2\cos x$, $[-\pi, \pi]$. $f'(x) = 1 - 2\sin x$. $f'(x) = 0 \Rightarrow \sin x = \frac{1}{2} \Rightarrow x = \frac{\pi}{6}, \frac{5\pi}{6}$. $f'(x) > 0$ for $\left(-\pi, \frac{\pi}{6}\right)$ and $\left(\frac{5\pi}{6}, \pi\right)$, and $f'(x) < 0$ for $\left(\frac{\pi}{6}, \frac{5\pi}{6}\right)$, so $f\left(\frac{\pi}{6}\right) = \frac{\pi}{6} + \sqrt{3} \approx 2.26$ is a local maximum value and $f\left(\frac{5\pi}{6}\right) = \frac{5\pi}{6} - \sqrt{3} \approx 0.89$ is a local minimum value. Checking the endpoints, we find $f(-\pi) = -\pi - 2 \approx -5.14$ and $f(\pi) = \pi - 2 \approx 1.14$. Thus, $f(-\pi) = -\pi - 2$ is the absolute minimum value and $f\left(\frac{\pi}{6}\right) = \frac{\pi}{6} + \sqrt{3}$ is the absolute maximum value.

6. $f(x) = x^2 e^{-x}$, $[-1, 3]$. $f'(x) = x^2(-e^{-x}) + e^{-x}(2x) = xe^{-x}(-x+2)$. $f'(x) = 0 \Rightarrow x = 0$ or $x = 2$.

$f'(x) > 0$ for $0 < x < 2$ and $f'(x) < 0$ for $-1 < x < 0$ and $2 < x < 3$, so $f(0) = 0$ is a local minimum value and $f(2) = 4e^{-2} \approx 0.54$ is a local maximum value. Checking the endpoints, we find $f(-1) = e \approx 2.72$ and $f(3) = 9e^{-3} \approx 0.45$. Thus, $f(0) = 0$ is the absolute minimum value and $f(-1) = e$ is the absolute maximum value.

7. This limit has the form $\frac{0}{0}$. $\displaystyle\lim_{x \to 0} \frac{e^x - 1}{\tan x} \overset{\text{H}}{=} \lim_{x \to 0} \frac{e^x}{\sec^2 x} = \frac{1}{1} = 1$

8. This limit has the form $\frac{0}{0}$. $\displaystyle\lim_{x \to 0} \frac{\tan 4x}{x + \sin 2x} \overset{\text{H}}{=} \lim_{x \to 0} \frac{4\sec^2 4x}{1 + 2\cos 2x} = \frac{4(1)}{1 + 2(1)} = \frac{4}{3}$

9. This limit has the form $\frac{0}{0}$. $\displaystyle\lim_{x \to 0} \frac{e^{2x} - e^{-2x}}{\ln(x+1)} \overset{\text{H}}{=} \lim_{x \to 0} \frac{2e^{2x} + 2e^{-2x}}{1/(x+1)} = \frac{2+2}{1} = 4$

10. This limit has the form $\frac{\infty}{\infty}$. $\displaystyle\lim_{x \to \infty} \frac{e^{2x} - e^{-2x}}{\ln(x+1)} \overset{\text{H}}{=} \lim_{x \to \infty} \frac{2e^{2x} + 2e^{-2x}}{1/(x+1)} = \lim_{x \to \infty} 2(x+1)(e^{2x} + e^{-2x}) = \infty$

since $2(x+1) \to \infty$ and $(e^{2x} + e^{-2x}) \to \infty$ as $x \to \infty$.

11. This limit has the form $\infty \cdot 0$.

$$\lim_{x\to-\infty}(x^2-x^3)e^{2x} = \lim_{x\to-\infty}\frac{x^2-x^3}{e^{-2x}}\quad\left[\tfrac{\infty}{\infty}\text{ form}\right] \stackrel{\text{H}}{=} \lim_{x\to-\infty}\frac{2x-3x^2}{-2e^{-2x}}\quad\left[\tfrac{\infty}{\infty}\text{ form}\right]$$
$$\stackrel{\text{H}}{=} \lim_{x\to-\infty}\frac{2-6x}{4e^{-2x}}\quad\left[\tfrac{\infty}{\infty}\text{ form}\right] \stackrel{\text{H}}{=} \lim_{x\to-\infty}\frac{-6}{-8e^{-2x}} = 0$$

12. This limit has the form $0\cdot\infty$. $\lim\limits_{x\to\pi^-}(x-\pi)\csc x = \lim\limits_{x\to\pi^-}\dfrac{x-\pi}{\sin x}\ \left[\tfrac{0}{0}\text{ form}\right] \stackrel{\text{H}}{=} \lim\limits_{x\to\pi^-}\dfrac{1}{\cos x} = \dfrac{1}{-1} = -1$

13. This limit has the form $\infty - \infty$.

$$\lim_{x\to1^+}\left(\frac{x}{x-1}-\frac{1}{\ln x}\right) = \lim_{x\to1^+}\left(\frac{x\ln x - x + 1}{(x-1)\ln x}\right) \stackrel{\text{H}}{=} \lim_{x\to1^+}\frac{x\cdot(1/x)+\ln x - 1}{(x-1)\cdot(1/x)+\ln x} = \lim_{x\to1^+}\frac{\ln x}{1-1/x+\ln x}$$
$$\stackrel{\text{H}}{=} \lim_{x\to1^+}\frac{1/x}{1/x^2+1/x} = \frac{1}{1+1} = \frac{1}{2}$$

14. $y = (\tan x)^{\cos x} \Rightarrow \ln y = \cos x \ln\tan x$, so

$$\lim_{x\to(\pi/2)^-}\ln y = \lim_{x\to(\pi/2)^-}\frac{\ln\tan x}{\sec x} \stackrel{\text{H}}{=} \lim_{x\to(\pi/2)^-}\frac{(1/\tan x)\sec^2 x}{\sec x\,\tan x} = \lim_{x\to(\pi/2)^-}\frac{\sec x}{\tan^2 x} = \lim_{x\to(\pi/2)^-}\frac{\cos x}{\sin^2 x} = \frac{0}{1^2} = 0,$$

so $\lim\limits_{x\to(\pi/2)^-}(\tan x)^{\cos x} = \lim\limits_{x\to(\pi/2)^-}e^{\ln y} = e^0 = 1$.

15. $f(0)=0$, $f'(-2)=f'(1)=f'(9)=0$, $\lim\limits_{x\to\infty}f(x)=0$, $\lim\limits_{x\to6}f(x)=-\infty$,

$f'(x)<0$ on $(-\infty,-2)$, $(1,6)$, and $(9,\infty)$, $f'(x)>0$ on $(-2,1)$ and $(6,9)$,

$f''(x)>0$ on $(-\infty,0)$ and $(12,\infty)$, $f''(x)<0$ on $(0,6)$ and $(6,12)$

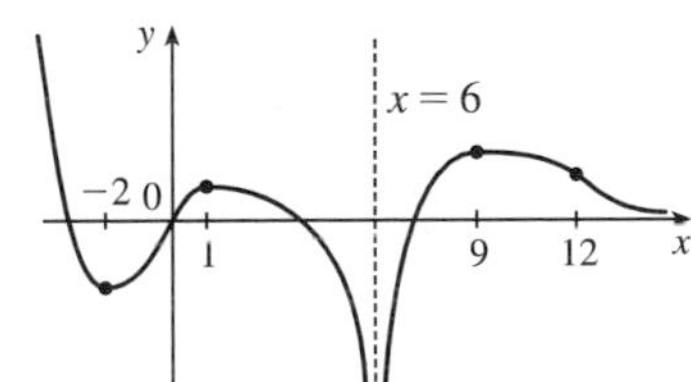

16. For $0<x<1$, $f'(x)=2x$, so $f(x)=x^2+C$. Since $f(0)=0$,

$f(x)=x^2$ on $[0,1]$. For $1<x<3$, $f'(x)=-1$, so $f(x)=-x+D$.

$1=f(1)=-1+D \Rightarrow D=2$, so $f(x)=2-x$. For $x>3$, $f'(x)=1$,

so $f(x)=x+E$. $-1=f(3)=3+E \Rightarrow E=-4$, so $f(x)=x-4$.

Since f is even, its graph is symmetric about the y-axis.

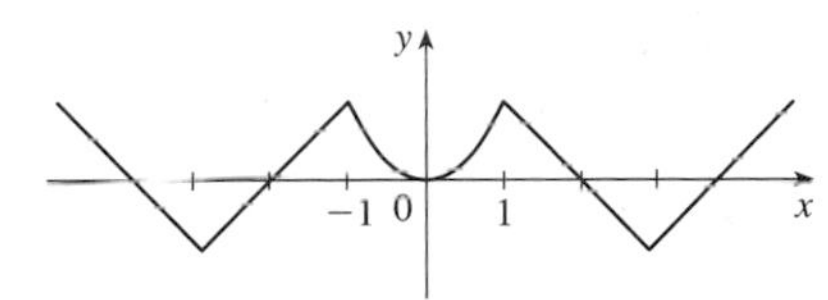

17. f is odd, $f'(x)<0$ for $0<x<2$, $f'(x)>0$ for $x>2$,

$f''(x)>0$ for $0<x<3$, $f''(x)<0$ for $x>3$, $\lim_{x\to\infty}f(x)=-2$

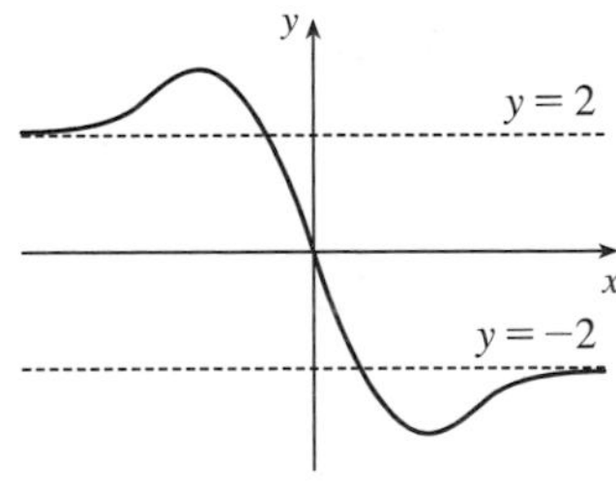

18. (a) Using the Test for Monotonic Functions we know that f is increasing on $(-2,0)$ and $(4,\infty)$ because $f'>0$ on $(-2,0)$ and $(4,\infty)$, and that f is decreasing on $(-\infty,-2)$ and $(0,4)$ because $f'<0$ on $(-\infty,-2)$ and $(0,4)$.

(b) Using the First Derivative Test, we know that f has a local maximum at $x=0$ because f' changes from positive to negative at $x=0$, and that f has a local minimum at $x=4$ because f' changes from negative to positive at $x=4$.

(c)

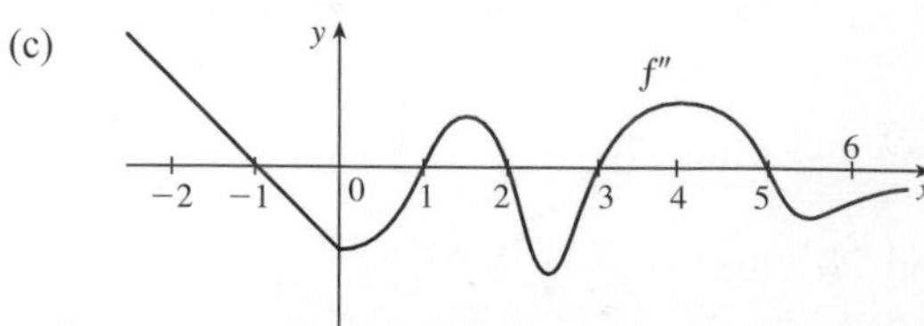

(d)

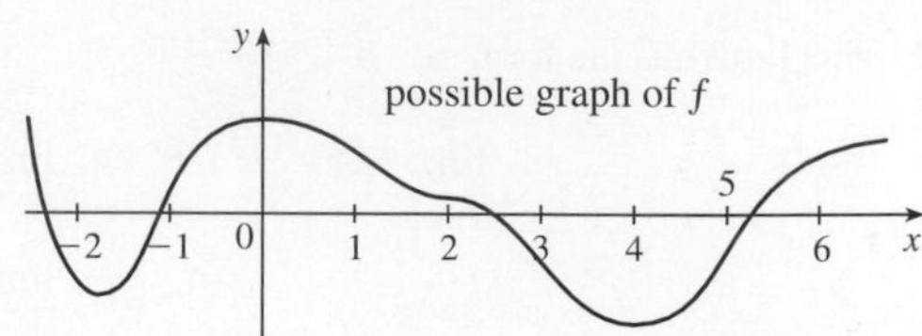

19. $y = f(x) = 2 - 2x - x^3$ **A.** $D = \mathbb{R}$ **B.** y-intercept: $f(0) = 2$. The x-intercept (approximately 0.770917) can be found using Newton's Method. **C.** No symmetry **D.** No asymptote **E.** $f'(x) = -2 - 3x^2 = -(3x^2 + 2) < 0$, so f is decreasing on $\mathbb{R}$. **F.** No extreme value **G.** $f''(x) = -6x < 0$ on $(0, \infty)$ and $f''(x) > 0$ on $(-\infty, 0)$, so f is CD on $(0, \infty)$ and CU on $(-\infty, 0)$. There is an IP at $(0, 2)$.

H.

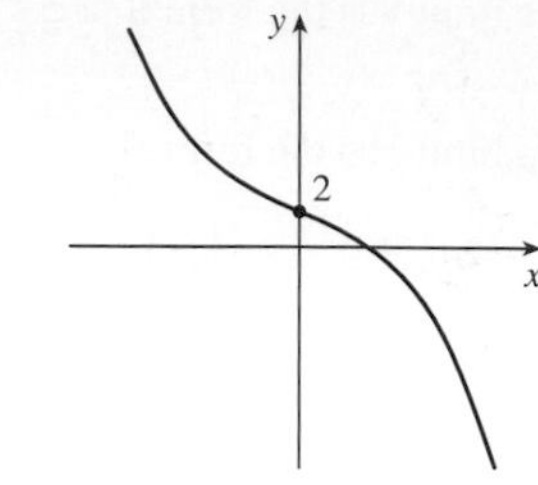

20. $y = f(x) = -2x^3 - 3x^2 + 12x + 5$ **A.** $D = \mathbb{R}$ **B.** y-intercept: $f(0) = 5$; x-intercept: $f(x) = 0 \Leftrightarrow x \approx -3.15, -0.39, 2.04$ **C.** No symmetry **D.** No asymptote **E.** $f'(x) = -6x^2 - 6x + 12 = -6(x^2 + x - 2) = -6(x + 2)(x - 1)$. $f'(x) > 0$ for $-2 < x < 1$, so f is increasing on $(-2, 1)$ and decreasing on $(-\infty, -2)$ and $(1, \infty)$. **F.** Local minimum value $f(-2) = -15$, local maximum value $f(1) = 12$ **G.** $f''(x) = -12x - 6 = -12\left(x + \frac{1}{2}\right)$. $f''(x) > 0$ for $x < -\frac{1}{2}$, so f is CU on $\left(-\infty, -\frac{1}{2}\right)$ and CD on $\left(-\frac{1}{2}, \infty\right)$. There is an IP at $\left(-\frac{1}{2}, -\frac{3}{2}\right)$.

H.

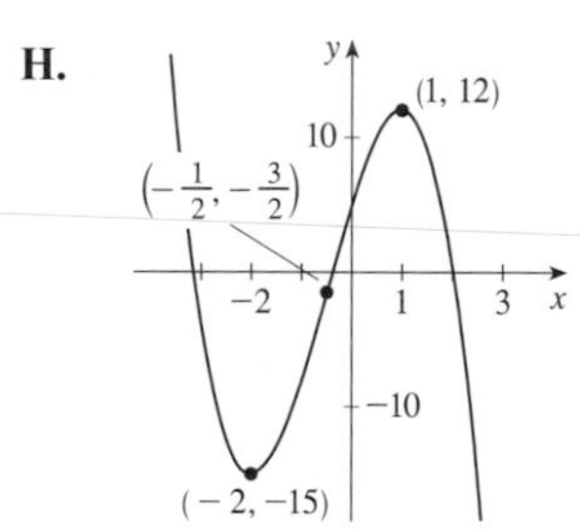

21. $y = f(x) = 3x^4 - 4x^3 + 2$ **A.** $D = \mathbb{R}$ **B.** y-intercept: $f(0) = 2$; no x-intercept **C.** No symmetry **D.** No asymptote **E.** $f'(x) = 12x^3 - 12x^2 = 12x^2(x - 1)$. $f'(x) > 0$ for $x > 1$, so f is increasing on $(1, \infty)$ and decreasing on $(-\infty, 1)$. **F.** $f'(x)$ does not change sign at $x = 0$, so there is no local extremum there. $f(1) = 1$ is a local minimum value. **G.** $f''(x) = 36x^2 - 24x = 12x(3x - 2)$. $f''(x) < 0$ for $0 < x < \frac{2}{3}$, so f is CD on $\left(0, \frac{2}{3}\right)$ and f is CU on $(-\infty, 0)$ and $\left(\frac{2}{3}, \infty\right)$. There are inflection points at $(0, 2)$ and $\left(\frac{2}{3}, \frac{38}{27}\right)$.

H.

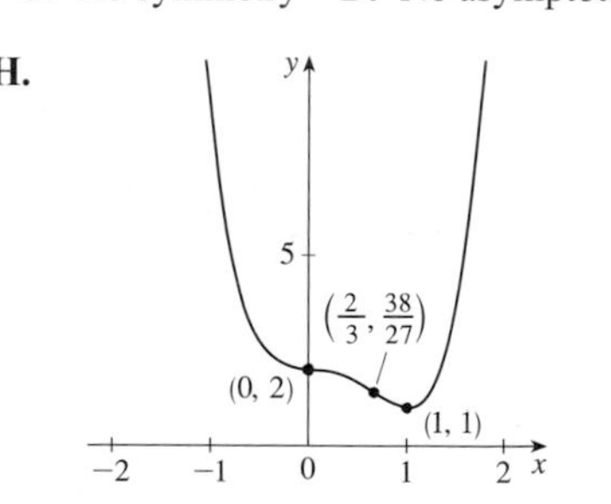

22. $y = f(x) = \dfrac{x}{1 - x^2}$ **A.** $D = (-\infty, -1) \cup (-1, 1) \cup (1, \infty)$ **B.** y-intercept: $f(0) = 0$; x-intercept: 0

C. $f(-x) = -f(x)$, so f is odd and the graph is symmetric about the origin. **D.** $\displaystyle\lim_{x \to \pm\infty} \frac{x}{1 - x^2} = 0$, so $y = 0$ is a HA.

$\displaystyle\lim_{x \to -1^-} \frac{x}{1 - x^2} = \infty$ and $\displaystyle\lim_{x \to -1^+} \frac{x}{1 - x^2} = -\infty$, so $x = -1$ is a VA. Similarly, $\displaystyle\lim_{x \to 1^-} \frac{x}{1 - x^2} = \infty$ and

$\displaystyle\lim_{x \to 1^+} \frac{x}{1 - x^2} = -\infty$, so $x = 1$ is a VA. **E.** $f'(x) = \dfrac{(1 - x^2)(1) - x(-2x)}{(1 - x^2)^2} = \dfrac{1 + x^2}{(1 - x^2)^2} > 0$ for $x \neq \pm 1$, so f is increasing on $(-\infty, -1)$, $(-1, 1)$, and $(1, \infty)$. **F.** No local extrema

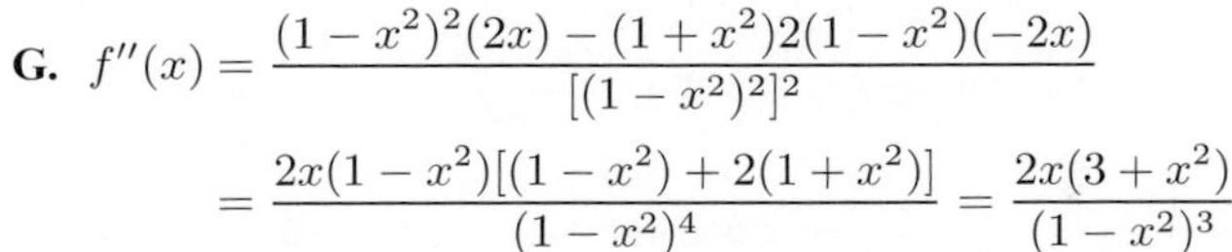

G. $f''(x) = \dfrac{(1-x^2)^2(2x) - (1+x^2)2(1-x^2)(-2x)}{[(1-x^2)^2]^2}$

$= \dfrac{2x(1-x^2)[(1-x^2) + 2(1+x^2)]}{(1-x^2)^4} = \dfrac{2x(3+x^2)}{(1-x^2)^3}$

H.

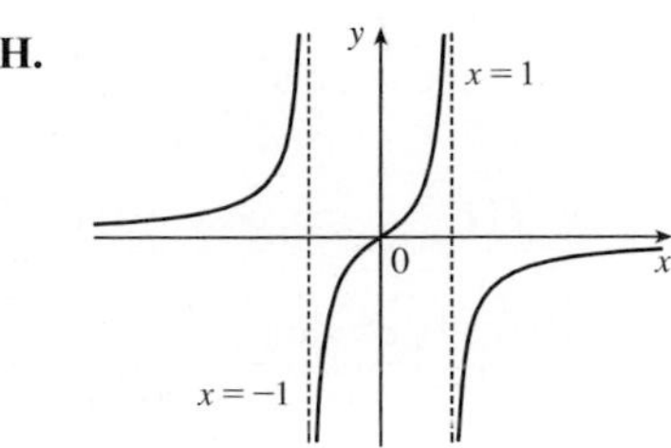

$f''(x) > 0$ for $x < -1$ and $0 < x < 1$, and $f''(x) < 0$ for $-1 < x < 0$ and $x > 1$, so f is CU on $(-\infty, -1)$ and $(0, 1)$, and f is CD on $(-1, 0)$ and $(1, \infty)$. $(0, 0)$ is an IP.

23. $y = f(x) = \dfrac{1}{x(x-3)^2}$ **A.** $D = \{x \mid x \neq 0, 3\} = (-\infty, 0) \cup (0, 3) \cup (3, \infty)$ **B.** No intercepts. **C.** No symmetry.

D. $\displaystyle\lim_{x\to\pm\infty} \frac{1}{x(x-3)^2} = 0$, so $y = 0$ is a HA. $\displaystyle\lim_{x\to 0^+} \frac{1}{x(x-3)^2} = \infty$, $\displaystyle\lim_{x\to 0^-} \frac{1}{x(x-3)^2} = -\infty$, $\displaystyle\lim_{x\to 3} \frac{1}{x(x-3)^2} = \infty$,

so $x = 0$ and $x = 3$ are VA. **E.** $f'(x) = -\dfrac{(x-3)^2 + 2x(x-3)}{x^2(x-3)^4} = \dfrac{3(1-x)}{x^2(x-3)^3} \Rightarrow f'(x) > 0 \Leftrightarrow 1 < x < 3$,

so f is increasing on $(1, 3)$ and decreasing on $(-\infty, 0)$, $(0, 1)$, and $(3, \infty)$.

F. Local minimum value $f(1) = \frac{1}{4}$ **G.** $f''(x) = \dfrac{6(2x^2 - 4x + 3)}{x^3(x-3)^4}$.

Note that $2x^2 - 4x + 3 > 0$ for all x since it has negative discriminant.

So $f''(x) > 0 \Leftrightarrow x > 0 \Rightarrow f$ is CU on $(0, 3)$ and $(3, \infty)$ and CD on $(-\infty, 0)$. No IP

H.

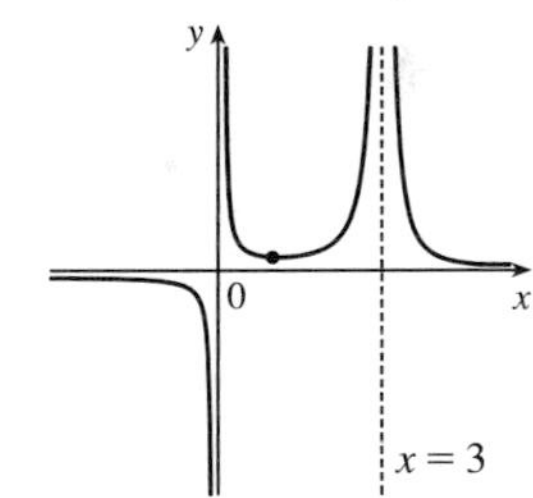

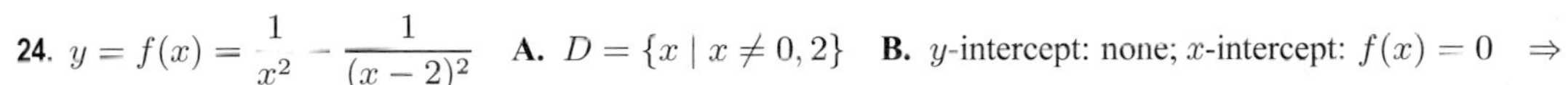

24. $y = f(x) = \dfrac{1}{x^2} - \dfrac{1}{(x-2)^2}$ **A.** $D = \{x \mid x \neq 0, 2\}$ **B.** y-intercept: none; x-intercept: $f(x) = 0 \Rightarrow$

$\dfrac{1}{x^2} = \dfrac{1}{(x-2)^2} \Leftrightarrow (x-2)^2 = x^2 \Leftrightarrow x^2 - 4x + 4 = x^2 \Leftrightarrow 4x = 4 \Leftrightarrow x = 1$ **C.** No symmetry

D. $\displaystyle\lim_{x\to 0} f(x) = \infty$ and $\displaystyle\lim_{x\to 2} f(x) = -\infty$, so $x = 0$ and $x = 2$ are VA; $\displaystyle\lim_{x\to\pm\infty} f(x) = 0$, so $y = 0$ is a HA

E. $f'(x) = -\dfrac{2}{x^3} + \dfrac{2}{(x-2)^3} > 0 \Rightarrow \dfrac{-(x-2)^3 + x^3}{x^3(x-2)^3} > 0 \Leftrightarrow \dfrac{-x^3 + 6x^2 - 12x + 8 + x^3}{x^3(x-2)^3} > 0 \Leftrightarrow$

$\dfrac{2(3x^2 - 6x + 4)}{x^3(x-2)^3} > 0$. The numerator is positive (the discriminant of the quadratic is negative), so $f'(x) > 0$ if $x < 0$ or $x > 2$, and hence, f is increasing on $(-\infty, 0)$ and $(2, \infty)$ and decreasing on $(0, 2)$.

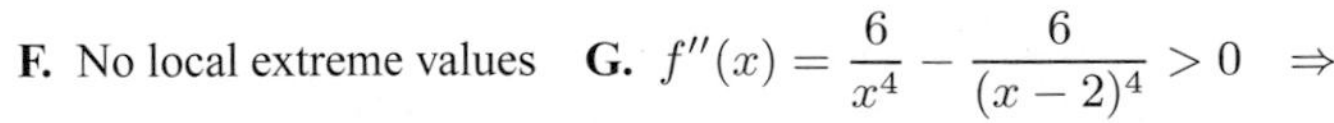

F. No local extreme values **G.** $f''(x) = \dfrac{6}{x^4} - \dfrac{6}{(x-2)^4} > 0 \Rightarrow$

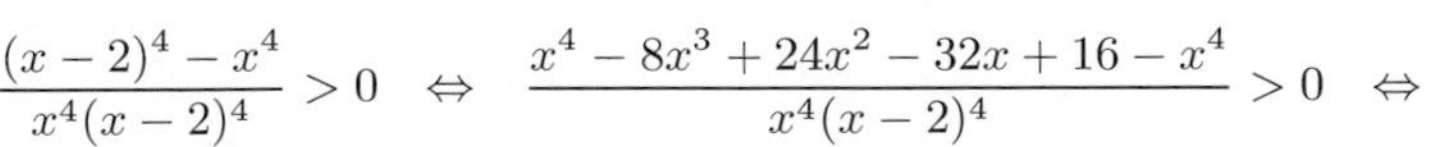

$\dfrac{(x-2)^4 - x^4}{x^4(x-2)^4} > 0 \Leftrightarrow \dfrac{x^4 - 8x^3 + 24x^2 - 32x + 16 - x^4}{x^4(x-2)^4} > 0 \Leftrightarrow$

$\dfrac{-8(x^3 - 3x^2 + 4x - 2)}{x^4(x-2)^4} > 0 \Leftrightarrow \dfrac{-8(x-1)(x^2 - 2x + 2)}{x^4(x-2)^4} > 0$. So f'' is

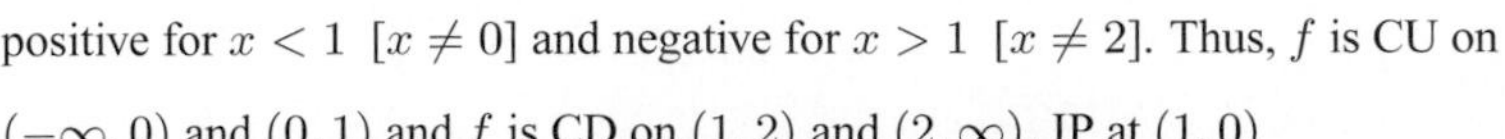

positive for $x < 1$ $[x \neq 0]$ and negative for $x > 1$ $[x \neq 2]$. Thus, f is CU on $(-\infty, 0)$ and $(0, 1)$ and f is CD on $(1, 2)$ and $(2, \infty)$. IP at $(1, 0)$

H.

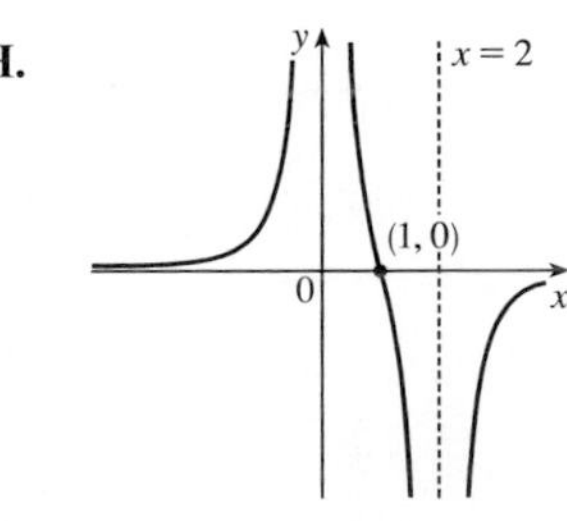

25. $y = f(x) = \dfrac{(x-1)^3}{x^2} = \dfrac{x^3 - 3x^2 + 3x - 1}{x^2} = x - 3 + \dfrac{3x-1}{x^2}$ **A.** $D = \{x \mid x \neq 0\} = (-\infty, 0) \cup (0, \infty)$

B. y-intercept: none; x-intercept: $f(x) = 0 \iff x = 1$ **C.** No symmetry **D.** $\lim\limits_{x \to 0^-} \dfrac{(x-1)^3}{x^2} = -\infty$ and

$\lim\limits_{x \to 0^+} f(x) = -\infty$, so $x = 0$ is a VA. $f(x) - (x-3) = \dfrac{3x-1}{x^2} \to 0$ as $x \to \pm\infty$, so $y = x - 3$ is a SA.

E. $f'(x) = \dfrac{x^2 \cdot 3(x-1)^2 - (x-1)^3(2x)}{(x^2)^2} = \dfrac{x(x-1)^2[3x - 2(x-1)]}{x^4} = \dfrac{(x-1)^2(x+2)}{x^3}$. $f'(x) < 0$ for $-2 < x < 0$,

so f is increasing on $(-\infty, -2)$, decreasing on $(-2, 0)$, and increasing on $(0, \infty)$.

F. Local maximum value $f(-2) = -\frac{27}{4}$ **G.** $f(x) = x - 3 + \dfrac{3}{x} - \dfrac{1}{x^2} \Rightarrow$

$f'(x) = 1 - \dfrac{3}{x^2} + \dfrac{2}{x^3} \Rightarrow f''(x) = \dfrac{6}{x^3} - \dfrac{6}{x^4} = \dfrac{6x-6}{x^4} = \dfrac{6(x-1)}{x^4}$.

$f''(x) > 0$ for $x > 1$, so f is CD on $(-\infty, 0)$ and $(0, 1)$, and f is CU on $(1, \infty)$.

There is an inflection point at $(1, 0)$.

H.

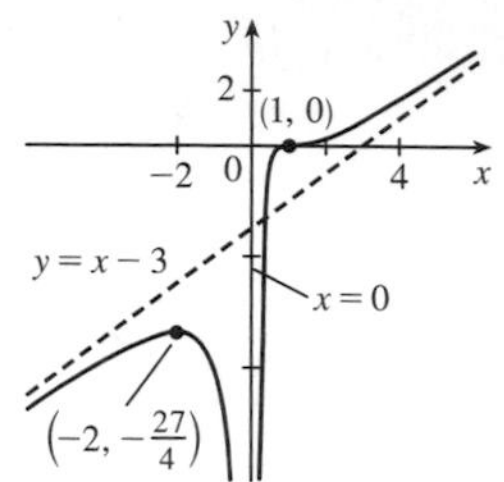

26. $y = f(x) = \sqrt{1-x} + \sqrt{1+x}$ **A.** $1 - x \geq 0$ and $1 + x \geq 0 \Rightarrow x \leq 1$ and $x \geq -1$, so $D = [-1, 1]$.

B. y-intercept: $f(0) = 1 + 1 = 2$; no x-intercept because $f(x) > 0$ for all x.

C. $f(-x) = f(x)$, so the curve is symmetric about the y-axis. **D.** No asymptote

E. $f'(x) = \frac{1}{2}(1-x)^{-1/2}(-1) + \frac{1}{2}(1+x)^{-1/2} = \dfrac{-1}{2\sqrt{1-x}} + \dfrac{1}{2\sqrt{1+x}} = \dfrac{-\sqrt{1+x} + \sqrt{1-x}}{2\sqrt{1-x}\sqrt{1+x}} > 0 \Rightarrow$

$-\sqrt{1+x} + \sqrt{1-x} > 0 \Rightarrow \sqrt{1-x} > \sqrt{1+x} \Rightarrow 1 - x > 1 + x \Rightarrow -2x > 0 \Rightarrow x < 0$, so $f'(x) > 0$ for

$-1 < x < 0$ and $f'(x) < 0$ for $0 < x < 1$. Thus, f is increasing on $(-1, 0)$

and decreasing on $(0, 1)$. **F.** Local maximum value $f(0) = 2$

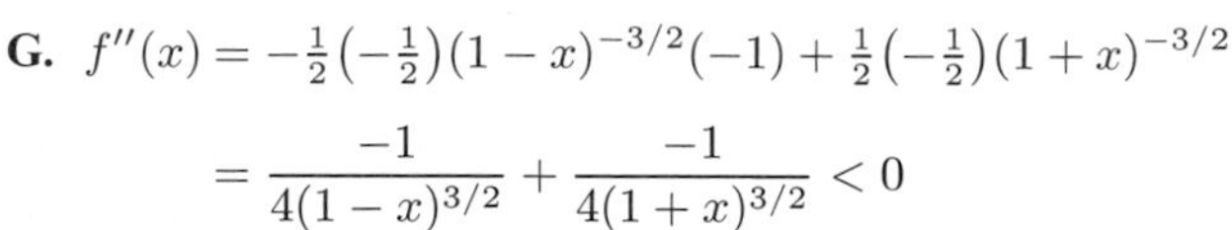

G. $f''(x) = -\frac{1}{2}\left(-\frac{1}{2}\right)(1-x)^{-3/2}(-1) + \frac{1}{2}\left(-\frac{1}{2}\right)(1+x)^{-3/2}$

$= \dfrac{-1}{4(1-x)^{3/2}} + \dfrac{-1}{4(1+x)^{3/2}} < 0$

for all x in the domain, so f is CD on $(-1, 1)$. No IP

H.

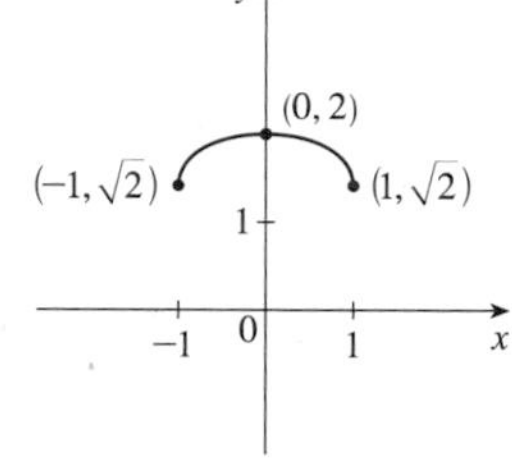

27. $y = f(x) = x\sqrt{2+x}$ **A.** $D = [-2, \infty)$ **B.** y-intercept: $f(0) = 0$; x-intercepts: -2 and 0 **C.** No symmetry

D. No asymptote **E.** $f'(x) = \dfrac{x}{2\sqrt{2+x}} + \sqrt{2+x} = \dfrac{1}{2\sqrt{2+x}}[x + 2(2+x)] = \dfrac{3x+4}{2\sqrt{2+x}} = 0$ when $x = -\frac{4}{3}$, so f is

decreasing on $\left(-2, -\frac{4}{3}\right)$ and increasing on $\left(-\frac{4}{3}, \infty\right)$. **F.** Local minimum value $f\left(-\frac{4}{3}\right) = -\frac{4}{3}\sqrt{\frac{2}{3}} = -\frac{4\sqrt{6}}{9} \approx -1.09$,

no local maximum

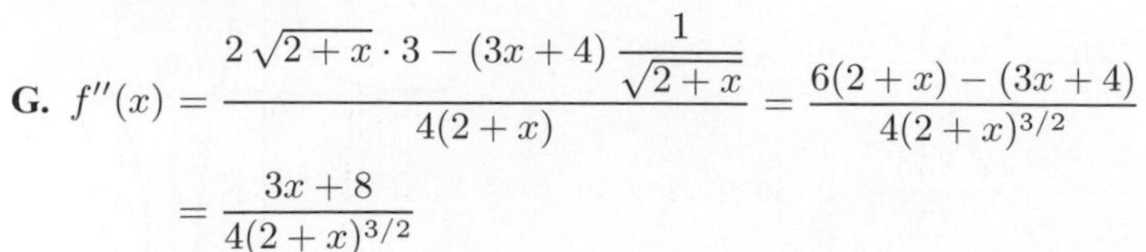

G. $f''(x) = \dfrac{2\sqrt{2+x} \cdot 3 - (3x+4)\dfrac{1}{\sqrt{2+x}}}{4(2+x)} = \dfrac{6(2+x) - (3x+4)}{4(2+x)^{3/2}}$

$= \dfrac{3x+8}{4(2+x)^{3/2}}$

$f''(x) > 0$ for $x > -2$, so f is CU on $(-2, \infty)$. No IP

H.

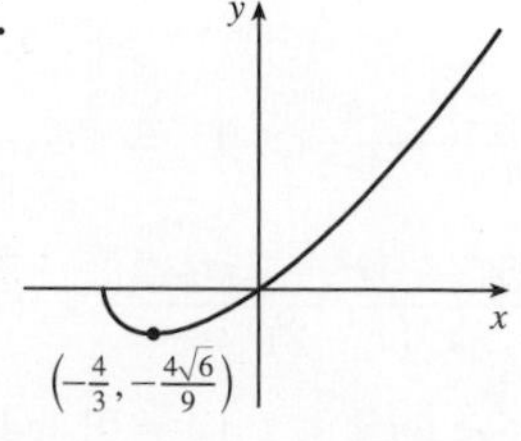

28. $y = f(x) = x^{2/3}(x-3)^2$ **A.** $D = \mathbb{R}$ **B.** y-intercept: $f(0) = 0$; x-intercepts: $f(x) = 0 \Leftrightarrow x = 0, 3$

C. No symmetry **D.** No asymptote

E. $f'(x) = x^{2/3} \cdot 2(x-3) + (x-3)^2 \cdot \frac{2}{3}x^{-1/3} = \frac{2}{3}x^{-1/3}(x-3)[3x+(x-3)] = \frac{2}{3}x^{-1/3}(x-3)(4x-3)$.

$f'(x) > 0 \Leftrightarrow 0 < x < \frac{3}{4}$ or $x > 3$, so f is decreasing on $(-\infty, 0)$, increasing on $\left(0, \frac{3}{4}\right)$, decreasing on $\left(\frac{3}{4}, 3\right)$, and increasing on $(3, \infty)$. **F.** Local minimum value $f(0) = f(3) = 0$; local maximum value

$f\left(\frac{3}{4}\right) = \left(\frac{3}{4}\right)^{2/3}\left(-\frac{9}{4}\right)^2 = \frac{81}{16}\sqrt[3]{\frac{9}{16}} = \frac{81}{32}\sqrt[3]{\frac{9}{2}}$ $[\approx 4.18]$

G. $f'(x) = \left(\frac{2}{3}x^{-1/3}\right)(4x^2 - 15x + 9) \Rightarrow$

$$f''(x) = \left(\tfrac{2}{3}x^{-1/3}\right)(8x-15) + (4x^2-15x+9)\left(-\tfrac{2}{9}x^{-4/3}\right)$$
$$= \tfrac{2}{9}x^{-4/3}[3x(8x-15) - (4x^2-15x+9)]$$
$$= \tfrac{2}{9}x^{-4/3}(20x^2 - 30x - 9)$$

$f''(x) = 0 \Leftrightarrow x \approx -0.26$ or 1.76. $f''(x)$ does not exist at $x = 0$.

f is CU on $(-\infty, -0.26)$, CD on $(-0.26, 0)$, CD on $(0, 1.76)$, and CU on $(1.76, \infty)$. There are inflection points at $(-0.26, 4.28)$ and $(1.76, 2.25)$.

H.

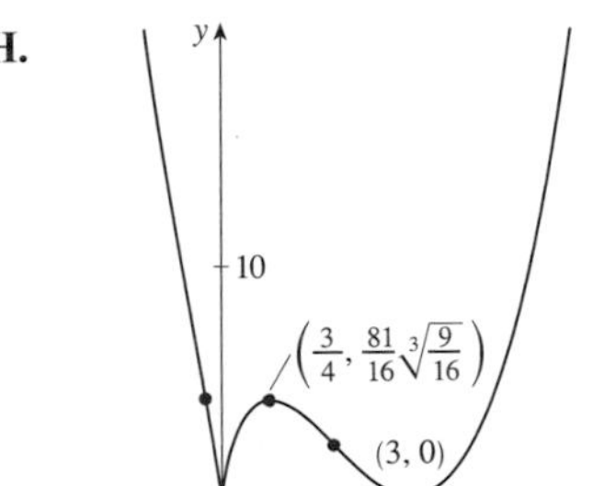

29. $y = f(x) = e^x \sin x$, $-\pi \le x \le \pi$ **A.** $D = [-\pi, \pi]$ **B.** y-intercept: $f(0) = 0$; $f(x) = 0 \Leftrightarrow \sin x = 0 \Rightarrow x = -\pi, 0, \pi$. **C.** No symmetry **D.** No asymptote **E.** $f'(x) = e^x \cos x + \sin x \cdot e^x = e^x(\cos x + \sin x)$.

$f'(x) = 0 \Leftrightarrow -\cos x = \sin x \Leftrightarrow -1 = \tan x \Rightarrow x = -\frac{\pi}{4}, \frac{3\pi}{4}$. $f'(x) > 0$ for $-\frac{\pi}{4} < x < \frac{3\pi}{4}$ and $f'(x) < 0$ for $-\pi < x < -\frac{\pi}{4}$ and $\frac{3\pi}{4} < x < \pi$, so f is increasing on $\left(-\frac{\pi}{4}, \frac{3\pi}{4}\right)$ and f is decreasing on $\left(-\pi, -\frac{\pi}{4}\right)$ and $\left(\frac{3\pi}{4}, \pi\right)$.

F. Local minimum value $f\left(-\frac{\pi}{4}\right) = \left(-\sqrt{2}/2\right)e^{-\pi/4} \approx -0.32$ and local maximum value $f\left(\frac{3\pi}{4}\right) = \left(\sqrt{2}/2\right)e^{3\pi/4} \approx 7.46$

G. $f''(x) = e^x(-\sin x + \cos x) + (\cos x + \sin x)e^x = e^x(2\cos x) > 0 \Rightarrow -\frac{\pi}{2} < x < \frac{\pi}{2}$ and $f''(x) < 0 \Rightarrow -\pi < x < -\frac{\pi}{2}$ and $\frac{\pi}{2} < x < \pi$, so f is CU on $\left(-\frac{\pi}{2}, \frac{\pi}{2}\right)$, and f is CD on $\left(-\pi, -\frac{\pi}{2}\right)$ and $\left(\frac{\pi}{2}, \pi\right)$. There are inflection points at $\left(-\frac{\pi}{2}, -e^{-\pi/2}\right)$ and $\left(\frac{\pi}{2}, e^{\pi/2}\right)$.

H.

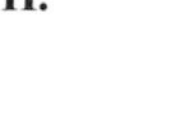

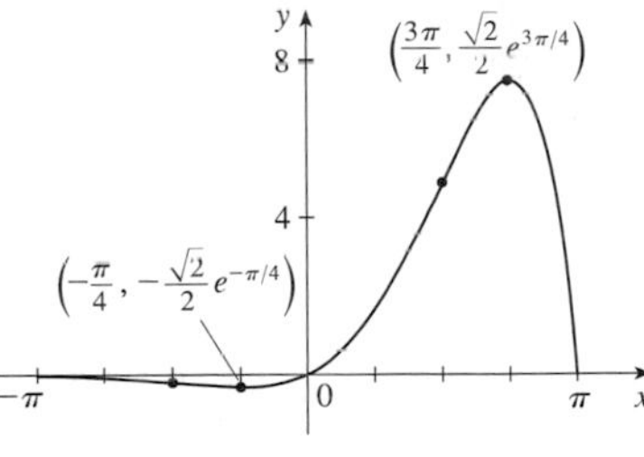

30. $y = f(x) = 4x - \tan x$, $-\frac{\pi}{2} < x < \frac{\pi}{2}$ **A.** $D = \left(-\frac{\pi}{2}, \frac{\pi}{2}\right)$. **B.** y-intercept $= f(0) = 0$ **C.** $f(-x) = -f(x)$, so the curve is symmetric about $(0, 0)$. **D.** $\lim\limits_{x \to \pi/2^-}(4x - \tan x) = -\infty$, $\lim\limits_{x \to -\pi/2^+}(4x - \tan x) = \infty$, so $x = \frac{\pi}{2}$ and $x = -\frac{\pi}{2}$ are VA. **E.** $f'(x) = 4 - \sec^2 x > 0 \Leftrightarrow \sec x < 2 \Leftrightarrow \cos x > \frac{1}{2} \Leftrightarrow -\frac{\pi}{3} < x < \frac{\pi}{3}$, so f is increasing on $\left(-\frac{\pi}{3}, \frac{\pi}{3}\right)$ and decreasing on $\left(-\frac{\pi}{2}, -\frac{\pi}{3}\right)$ and $\left(\frac{\pi}{3}, \frac{\pi}{2}\right)$. **F.** $f\left(\frac{\pi}{3}\right) = \frac{4\pi}{3} - \sqrt{3}$ is a local maximum value, $f\left(-\frac{\pi}{3}\right) = \sqrt{3} - \frac{4\pi}{3}$ is a local minimum value.

G. $f''(x) = -2\sec^2 x \tan x > 0 \Leftrightarrow \tan x < 0 \Leftrightarrow -\frac{\pi}{2} < x < 0$, so f is CU on $\left(-\frac{\pi}{2}, 0\right)$ and CD on $\left(0, \frac{\pi}{2}\right)$. IP at $(0, 0)$

H.

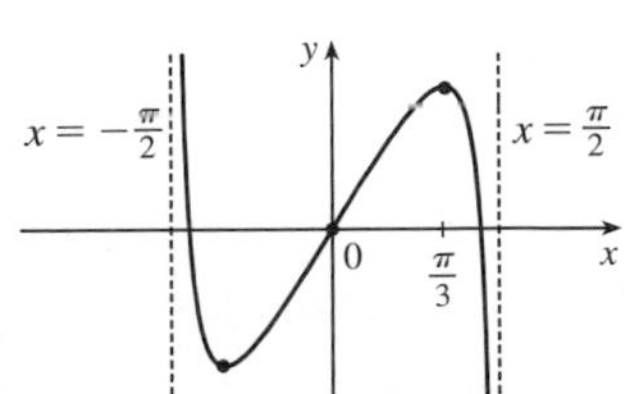

31. $y = f(x) = \sin^{-1}(1/x)$ **A.** $D = \{x \mid -1 \leq 1/x \leq 1\} = (-\infty, -1] \cup [1, \infty)$. **B.** No intercept
C. $f(-x) = -f(x)$, symmetric about the origin **D.** $\lim\limits_{x \to \pm\infty} \sin^{-1}(1/x) = \sin^{-1}(0) = 0$, so $y = 0$ is a HA.

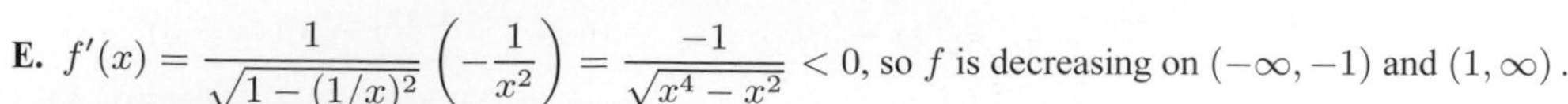

E. $f'(x) = \dfrac{1}{\sqrt{1 - (1/x)^2}} \left(-\dfrac{1}{x^2}\right) = \dfrac{-1}{\sqrt{x^4 - x^2}} < 0$, so f is decreasing on $(-\infty, -1)$ and $(1, \infty)$.

F. No local extreme value, but $f(1) = \frac{\pi}{2}$ is the absolute maximum value and $f(-1) = -\frac{\pi}{2}$ is the absolute minimum value.

H.

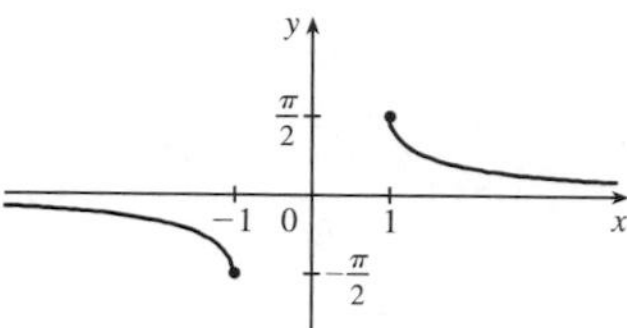

G. $f''(x) = \dfrac{4x^3 - 2x}{2(x^4 - x^2)^{3/2}} = \dfrac{x(2x^2 - 1)}{(x^4 - x^2)^{3/2}} > 0$ for $x > 1$ and $f''(x) < 0$ for $x < -1$, so f is CU on $(1, \infty)$ and CD on $(-\infty, -1)$. No IP

32. $y = f(x) = e^{2x - x^2}$ **A.** $D = \mathbb{R}$ **B.** y-intercept 1; no x-intercept **C.** No symmetry **D.** $\lim\limits_{x \to \pm\infty} e^{2x - x^2} = 0$, so $y = 0$ is a HA. **E.** $y = f(x) = e^{2x - x^2} \Rightarrow f'(x) = 2(1 - x)e^{2x - x^2} > 0 \Leftrightarrow x < 1$, so f is increasing on $(-\infty, 1)$ and decreasing on $(1, \infty)$. **F.** $f(1) = e$ is a local and absolute maximum value.

G. $f''(x) = 2(2x^2 - 4x + 1)e^{2x - x^2} = 0 \Leftrightarrow x = 1 \pm \frac{\sqrt{2}}{2}$.
$f''(x) > 0 \Leftrightarrow x < 1 - \frac{\sqrt{2}}{2}$ or $x > 1 + \frac{\sqrt{2}}{2}$, so f is CU on $\left(-\infty, 1 - \frac{\sqrt{2}}{2}\right)$ and $\left(1 + \frac{\sqrt{2}}{2}, \infty\right)$, and CD on $\left(1 - \frac{\sqrt{2}}{2}, 1 + \frac{\sqrt{2}}{2}\right)$. IP at $\left(1 \pm \frac{\sqrt{2}}{2}, \sqrt{e}\right)$

H.

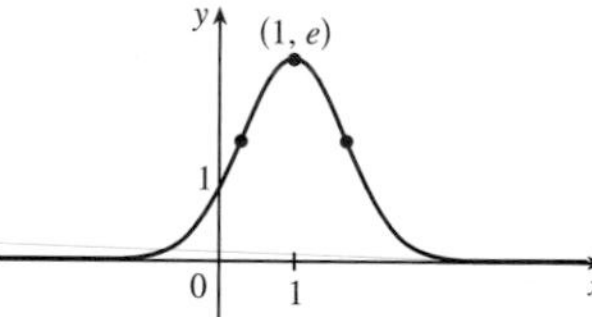

33. $y = f(x) = (x - 2)e^{-x}$ **A.** $D = \mathbb{R}$ **B.** y-intercept: $f(0) = -2$; x-intercept: $f(x) = 0 \Leftrightarrow x = 2$
C. No symmetry **D.** $\lim\limits_{x \to \infty} \dfrac{x - 2}{e^x} \overset{\text{H}}{=} \lim\limits_{x \to \infty} \dfrac{1}{e^x} = 0$, so $y = 0$ is a HA. No VA

E. $f'(x) = (x - 2)(-e^{-x}) + e^{-x}(1) = e^{-x}[-(x - 2) + 1] = (3 - x)e^{-x}$.
$f'(x) > 0$ for $x < 3$, so f is increasing on $(-\infty, 3)$ and decreasing on $(3, \infty)$.
F. Local maximum value $f(3) = e^{-3}$, no local minimum value
G. $f''(x) = (3 - x)(-e^{-x}) + e^{-x}(-1) = e^{-x}[-(3 - x) + (-1)]$
$= (x - 4)e^{-x} > 0$
for $x > 4$, so f is CU on $(4, \infty)$ and CD on $(-\infty, 4)$. IP at $(4, 2e^{-4})$

H.

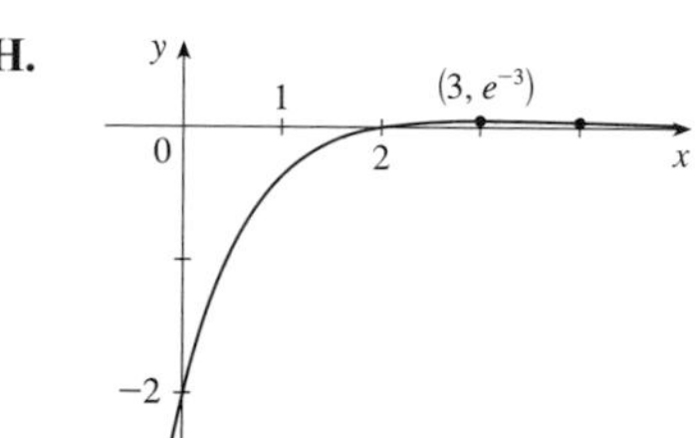

34. $y = f(x) = x + \ln(x^2 + 1)$ **A.** $D = \mathbb{R}$ **B.** y-intercept: $f(0) = 0 + \ln 1 = 0$; x-intercept: $f(x) = 0 \Leftrightarrow \ln(x^2 + 1) = -x \Leftrightarrow x^2 + 1 = e^{-x} \Rightarrow x = 0$ since the graphs of $y = x^2 + 1$ and $y = e^{-x}$ intersect only at $x = 0$.
C. No symmetry **D.** No asymptote **E.** $f'(x) = 1 + \dfrac{2x}{x^2 + 1} = \dfrac{x^2 + 2x + 1}{x^2 + 1} = \dfrac{(x + 1)^2}{x^2 + 1}$. $f'(x) > 0$ if $x \neq -1$ and f is increasing on $\mathbb{R}$. **F.** No local extreme values

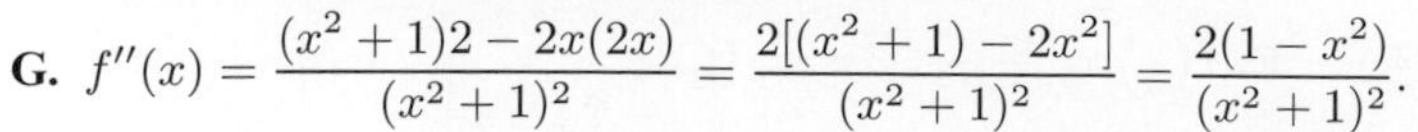

G. $f''(x) = \dfrac{(x^2 + 1)2 - 2x(2x)}{(x^2 + 1)^2} = \dfrac{2[(x^2 + 1) - 2x^2]}{(x^2 + 1)^2} = \dfrac{2(1 - x^2)}{(x^2 + 1)^2}$.

$f''(x) > 0 \Leftrightarrow -1 < x < 1$ and $f''(x) < 0 \Leftrightarrow x < -1$ or $x > 1$, so f is CU on $(-1, 1)$ and f is CD on $(-\infty, -1)$ and $(1, \infty)$. IP at $(-1, -1 + \ln 2)$ and $(1, 1 + \ln 2)$

H.

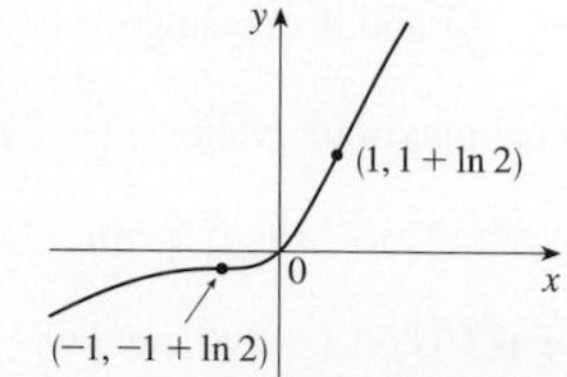

35. $f(x) = \dfrac{x^2-1}{x^3} \quad\Rightarrow\quad f'(x) = \dfrac{x^3(2x)-(x^2-1)3x^2}{x^6} = \dfrac{3-x^2}{x^4} \quad\Rightarrow$

$f''(x) = \dfrac{x^4(-2x)-(3-x^2)4x^3}{x^8} = \dfrac{2x^2-12}{x^5}$

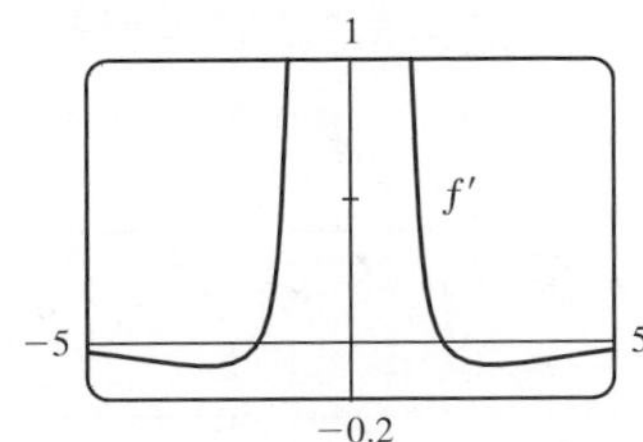

Estimates: From the graphs of f' and f'', it appears that f is increasing on $(-1.73, 0)$ and $(0, 1.73)$ and decreasing on $(-\infty, -1.73)$ and $(1.73, \infty)$; f has a local maximum of about $f(1.73) = 0.38$ and a local minimum of about $f(-1.7) = -0.38$; f is CU on $(-2.45, 0)$ and $(2.45, \infty)$, and CD on $(-\infty, -2.45)$ and $(0, 2.45)$; and f has inflection points at about $(-2.45, -0.34)$ and $(2.45, 0.34)$.

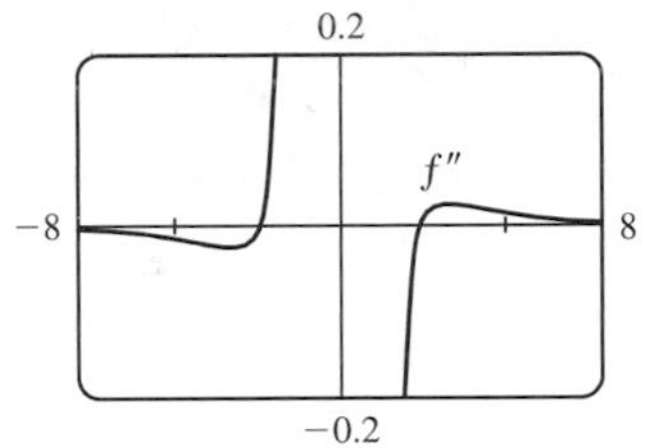

Exact: Now $f'(x) = \dfrac{3-x^2}{x^4}$ is positive for $0 < x^2 < 3$, that is, f is increasing on $(-\sqrt{3}, 0)$ and $(0, \sqrt{3})$; and $f'(x)$ is negative (and so f is decreasing) on $(-\infty, -\sqrt{3})$ and $(\sqrt{3}, \infty)$. $f'(x) = 0$ when $x = \pm\sqrt{3}$.

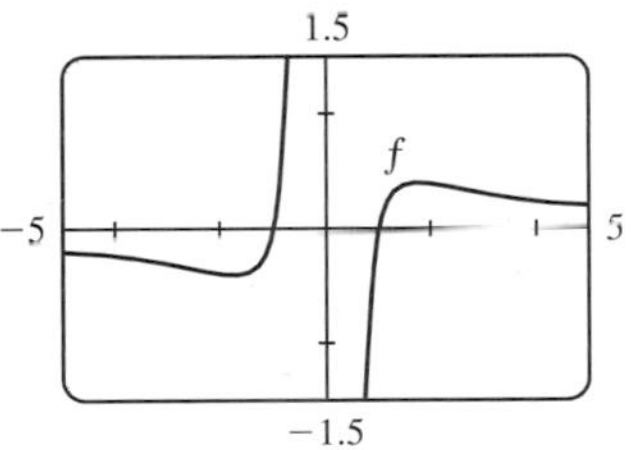

f' goes from positive to negative at $x = \sqrt{3}$, so f has a local maximum of $f(\sqrt{3}) = \frac{(\sqrt{3})^2-1}{(\sqrt{3})^3} = \frac{2\sqrt{3}}{9}$; and since f is odd, we know that maxima on the interval $(0, \infty)$ correspond to minima on $(-\infty, 0)$, so f has a local minimum of $f(-\sqrt{3}) = -\frac{2\sqrt{3}}{9}$. Also, $f''(x) = \dfrac{2x^2-12}{x^5}$ is positive (so f is CU) on $(-\sqrt{6}, 0)$ and $(\sqrt{6}, \infty)$, and negative (so f is CD) on $(-\infty, -\sqrt{6})$ and $(0, \sqrt{6})$. There are IP at $\left(\sqrt{6}, \frac{5\sqrt{6}}{36}\right)$ and $\left(-\sqrt{6}, -\frac{5\sqrt{6}}{36}\right)$.

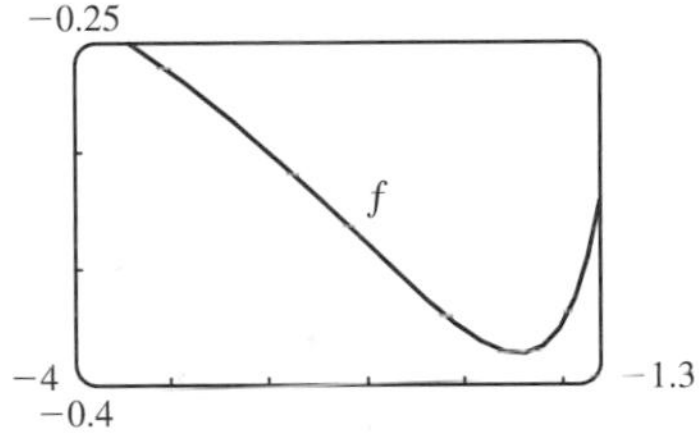

36. $f(x) = \dfrac{x^3+1}{x^6+1} \quad\Rightarrow\quad f'(x) = -\dfrac{3x^2(x^6+2x^3-1)}{(x^6+1)^2} \quad\Rightarrow\quad f''(x) = \dfrac{6x(2x^{12}+7x^9-9x^6-5x^3+1)}{(x^6+1)^3}$.

$f(x) = 0 \;\Leftrightarrow\; x = -1$. $f'(x) = 0 \;\Leftrightarrow\; x = 0$ or $x \approx -1.34, 0.75$. $f''(x) = 0 \;\Leftrightarrow\; x = 0$ or $x \approx -1.64, -0.82, 0.54, 1.09$. From the graphs of f and f', it appears that f is decreasing on $(-\infty, -1.34)$, increasing on $(-1.34, 0.75)$, and decreasing on $(0.75, \infty)$. f has a local minimum value of $f(-1.34) \approx -0.21$ and a local maximum value of $f(0.75) \approx 1.21$. From the graphs of f and f'', it appears that f is CD on $(-\infty, -1.64)$, CU on $(-1.64, -0.82)$, CD on $(-0.82, 0)$, CU on $(0, 0.54)$, CD on $(0.54, 1.09)$ and CU on $(1.09, \infty)$. There are inflection points at about $(-1.64, -0.17)$, $(-0.82, 0.34)$, $(0.54, 1.13)$, $(1.09, 0.86)$, and at $(0, 1)$.

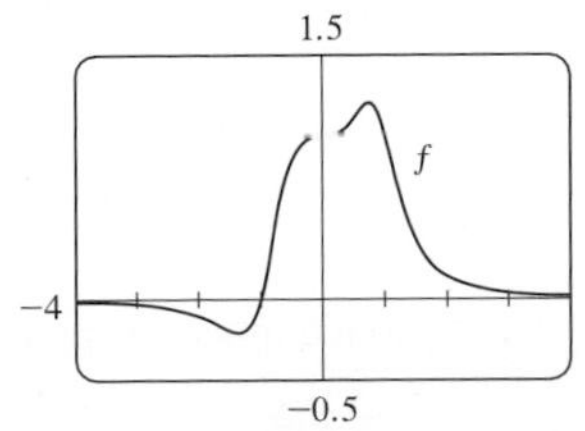

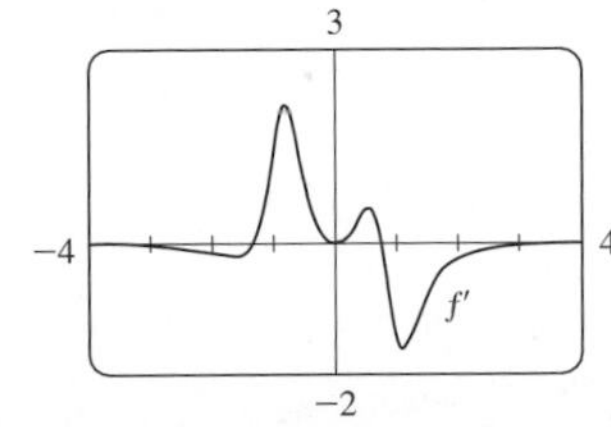

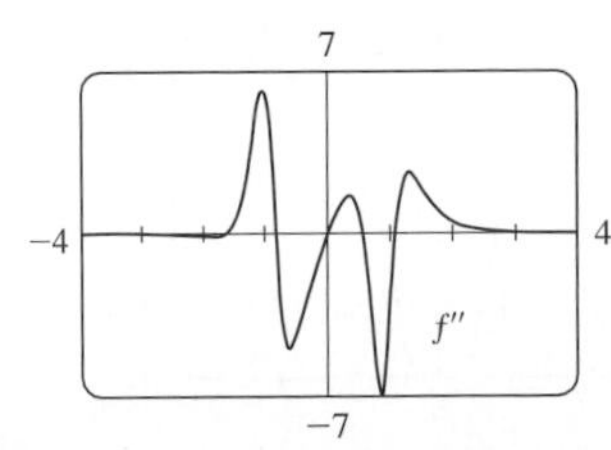

37. $f(x) = 3x^6 - 5x^5 + x^4 - 5x^3 - 2x^2 + 2 \;\Rightarrow\; f'(x) = 18x^5 - 25x^4 + 4x^3 - 15x^2 - 4x \;\Rightarrow$
$f''(x) = 90x^4 - 100x^3 + 12x^2 - 30x - 4$

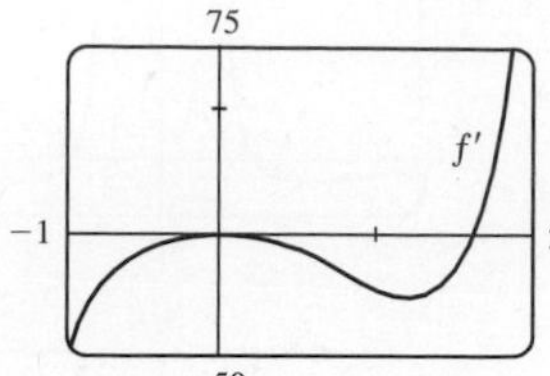

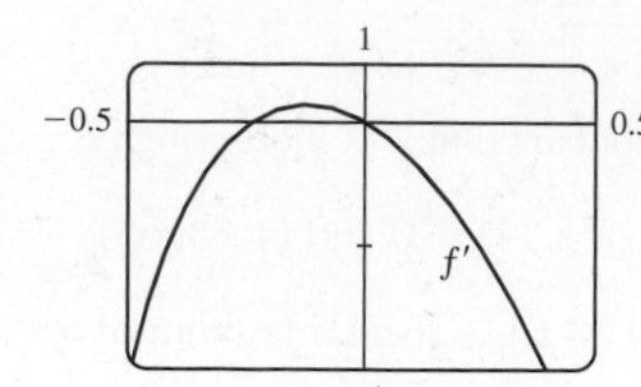

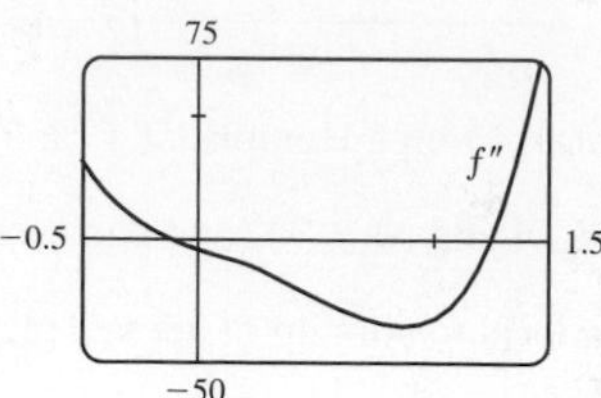

From the graphs of f' and f'', it appears that f is increasing on $(-0.23, 0)$ and $(1.62, \infty)$ and decreasing on $(-\infty, -0.23)$ and $(0, 1.62)$; f has a local maximum of $f(0) = 2$ and local minima of about $f(-0.23) = 1.96$ and $f(1.62) = -19.2$; f is CU on $(-\infty, -0.12)$ and $(1.24, \infty)$ and CD on $(-0.12, 1.24)$; and f has inflection points at about $(-0.12, 1.98)$ and $(1.24, -12.1)$.

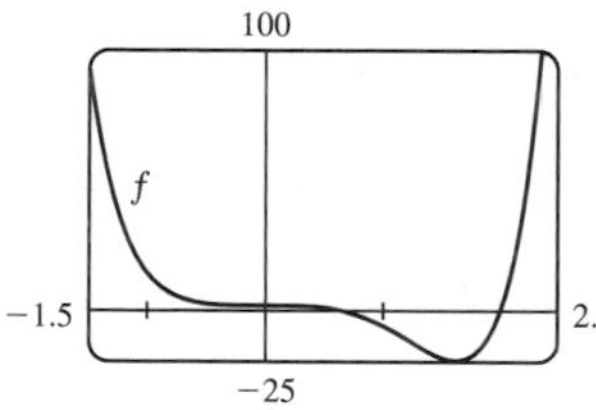

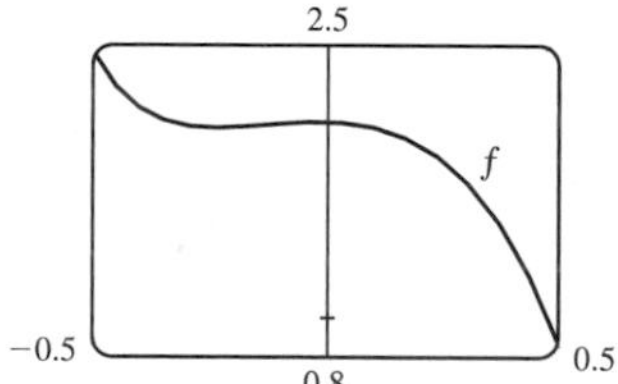

38. $f(x) = x^2 + 6.5 \sin x,\; -5 \le x \le 5 \;\Rightarrow\; f'(x) = 2x + 6.5 \cos x \;\Rightarrow\; f''(x) = 2 - 6.5 \sin x.\;\; f(x) = 0 \;\Leftrightarrow$
$x \approx -2.25$ and $x = 0$; $f'(x) = 0 \;\Leftrightarrow\; x \approx -1.19, 2.40, 3.24$; $f''(x) = 0 \;\Leftrightarrow\; x \approx -3.45, 0.31, 2.83$.

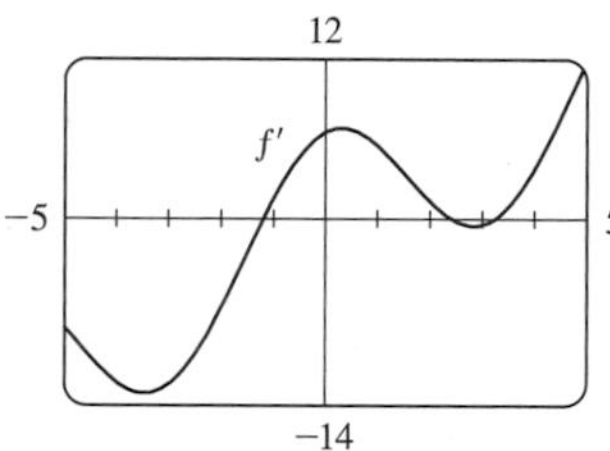

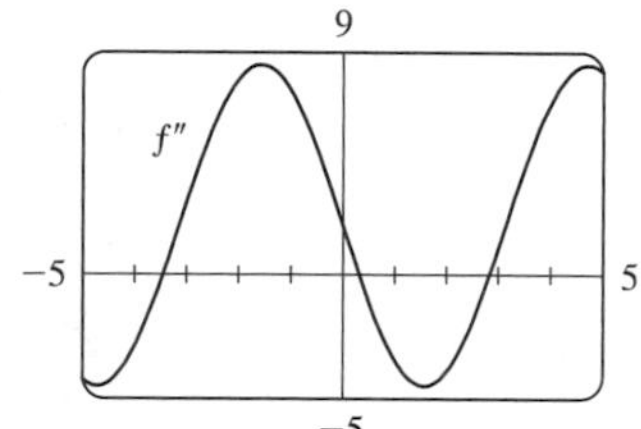

From the graphs of f' and f'', it appears that f is decreasing on $(-5, -1.19)$ and $(2.40, 3.24)$ and increasing on $(-1.19, 2.40)$ and $(3.24, 5)$; f has a local maximum of about $f(2.40) = 10.15$ and local minima of about $f(-1.19) = -4.62$ and $f(3.24) = 9.86$; f is CU on $(-3.45, 0.31)$ and $(2.83, 5)$ and CD on $(-5, -3.45)$ and $(0.31, 2.83)$; and f has inflection points at about $(-3.45, 13.93)$, $(0.31, 2.10)$, and $(2.83, 10.00)$.

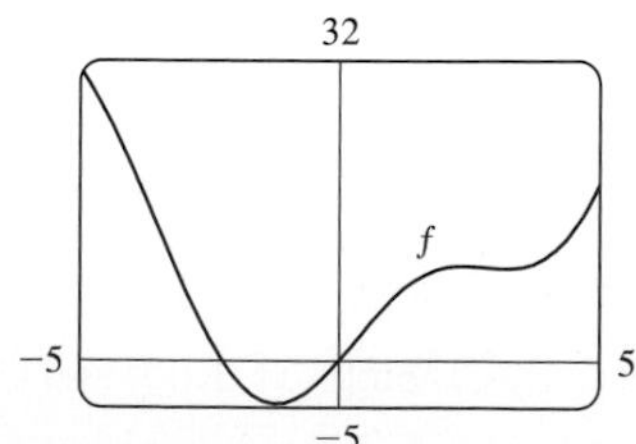

39.

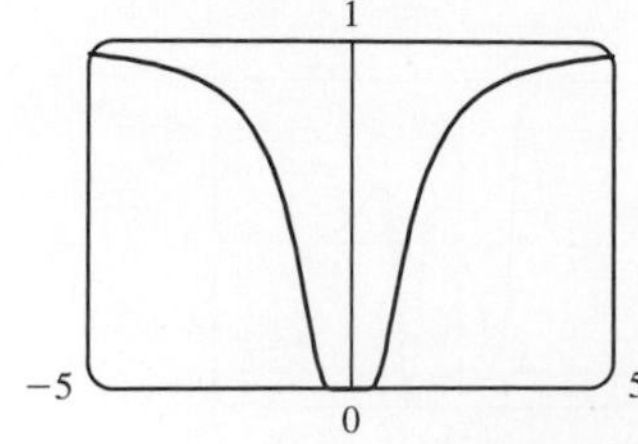

From the graph, we estimate the points of inflection to be about $(\pm 0.82, 0.22)$.

$f(x) = e^{-1/x^2} \;\Rightarrow\; f'(x) = 2x^{-3}e^{-1/x^2} \;\Rightarrow$

$f''(x) = 2[x^{-3}(2x^{-3})e^{-1/x^2} + e^{-1/x^2}(-3x^{-4})] = 2x^{-6}e^{-1/x^2}\,(2 - 3x^2)$.

This is 0 when $2 - 3x^2 = 0 \;\Leftrightarrow\; x = \pm\sqrt{\frac{2}{3}}$, so the inflection points are $\left(\pm\sqrt{\frac{2}{3}}, e^{-3/2}\right)$.

40. (a)

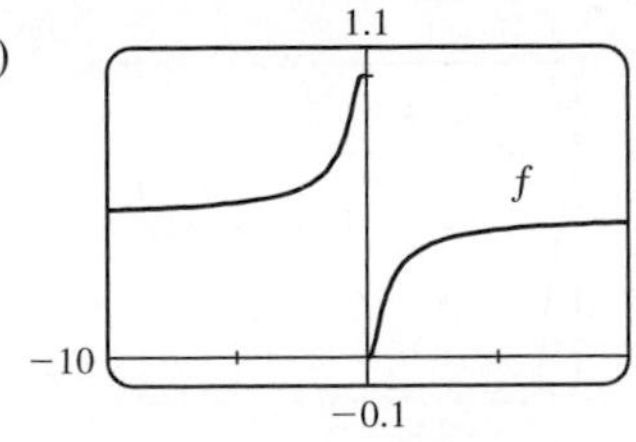

(b) $f(x) = \dfrac{1}{1+e^{1/x}}$.

$\lim\limits_{x\to\infty} f(x) = \dfrac{1}{1+1} = \dfrac{1}{2}$, $\lim\limits_{x\to-\infty} f(x) = \dfrac{1}{1+1} = \dfrac{1}{2}$,

as $x \to 0^+$, $1/x \to \infty$, so $e^{1/x} \to \infty \;\Rightarrow\; \lim\limits_{x\to 0^+} f(x) = 0$,

as $x \to 0^-$, $1/x \to -\infty$, so $e^{1/x} \to 0 \;\Rightarrow\; \lim\limits_{x\to 0^-} f(x) = \dfrac{1}{1+0} = 1$

(c) From the graph of f, estimates for the IP are $(\;0.4, 0.9)$ and $(0.4, 0.08)$.

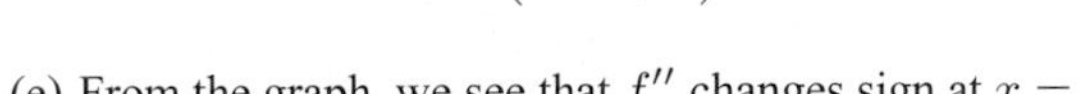

(d) $f''(x) = -\dfrac{e^{1/x}\left[e^{1/x}(2x-1)+2x+1\right]}{x^4(e^{1/x}+1)^3}$

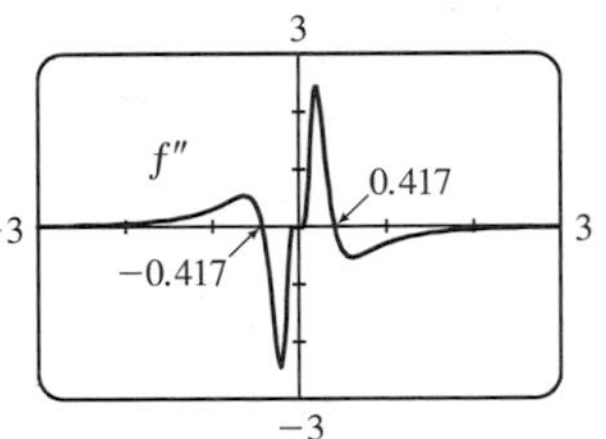

(e) From the graph, we see that f'' changes sign at $x = \pm 0.417$ ($x = 0$ is not in the domain of f). IP are approximately $(0.417, 0.083)$ and $(-0.417, 0.917)$.

41. $f(x) = \dfrac{\cos^2 x}{\sqrt{x^2+x+1}}$, $-\pi \le x \le \pi \;\Rightarrow\; f'(x) = -\dfrac{\cos x\left[(2x+1)\cos x + 4(x^2+x+1)\sin x\right]}{2(x^2+x+1)^{3/2}} \;\Rightarrow$

$$f''(x) = -\frac{(8x^4+16x^3+16x^2+8x+9)\cos^2 x - 8(x^2+x+1)(2x+1)\sin x\,\cos x - 8(x^2+x+1)^2\sin^2 x}{4(x^2+x+1)^{5/2}}$$

$f(x) = 0 \;\Leftrightarrow\; x = \pm\frac{\pi}{2}$; $f'(x) = 0 \;\Leftrightarrow\; x \approx -2.96, -1.57, -0.18, 1.57, 3.01$;

$f''(x) = 0 \;\Leftrightarrow\; x \approx -2.16, -0.75, 0.46$, and 2.21.

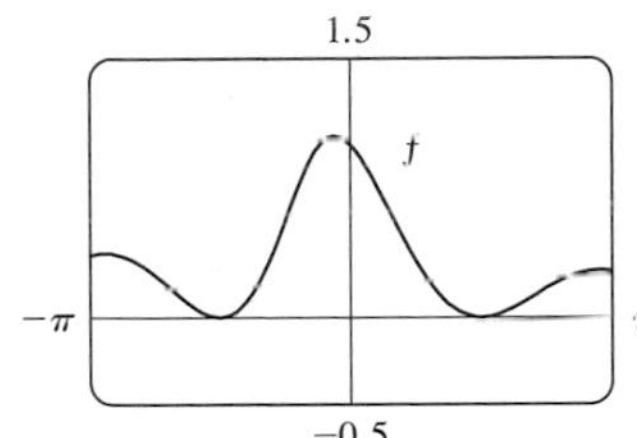

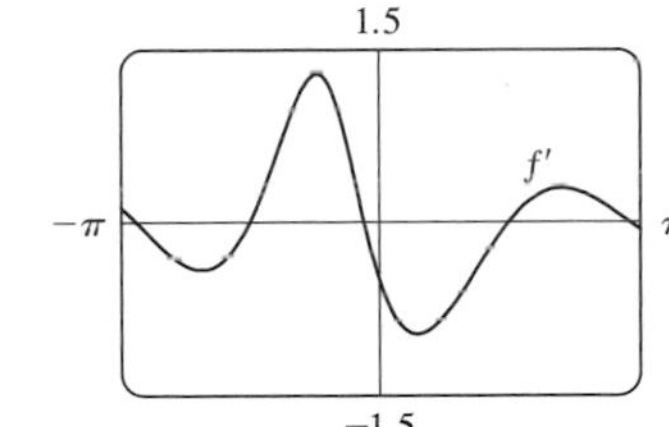

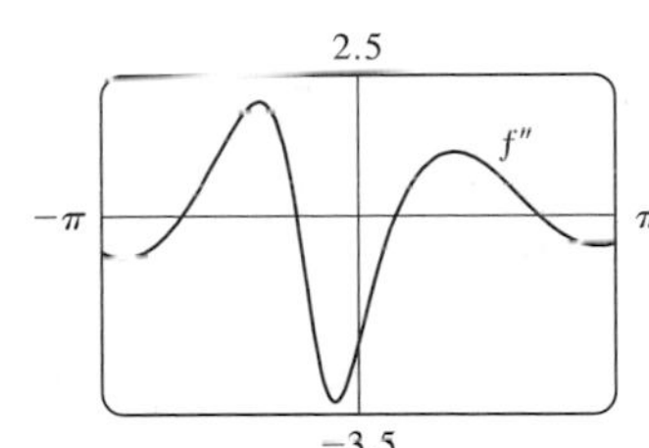

The x-coordinates of the maximum points are the values at which f' changes from positive to negative, that is, -2.96, -0.18, and 3.01. The x-coordinates of the minimum points are the values at which f' changes from negative to positive, that is, -1.57 and 1.57. The x-coordinates of the inflection points are the values at which f'' changes sign, that is, -2.16, -0.75, 0.46, and 2.21.

42. $f(x) = e^{-0.1x}\ln(x^2-1) \;\Rightarrow\; f'(x) = \dfrac{e^{-0.1x}\left[(x^2-1)\ln(x^2-1) - 20x\right]}{10(1-x^2)} \;\Rightarrow$

$$f''(x) = \frac{e^{-0.1x}\left[(x^2-1)^2\ln(x^2-1) - 40(x^3+5x^2-x+5)\right]}{100(x^2-1)^2}.$$

The domain of f is $(-\infty,-1)\cup(1,\infty)$. $f(x) = 0 \;\Leftrightarrow\; x = \pm\sqrt{2}$; $f'(x) = 0 \;\Leftrightarrow\; x \approx 5.87$;

$f''(x) = 0 \;\Leftrightarrow\; x \approx -4.31$ and 11.74.

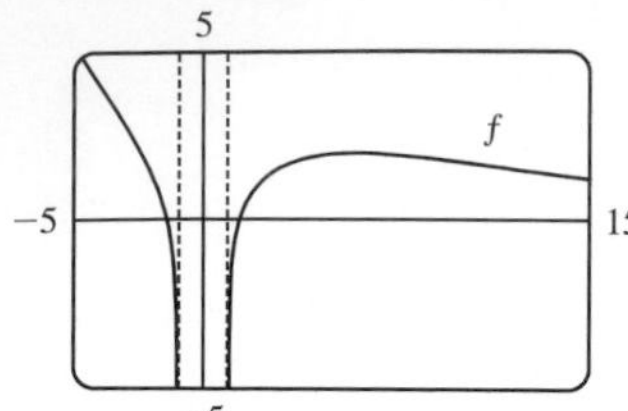

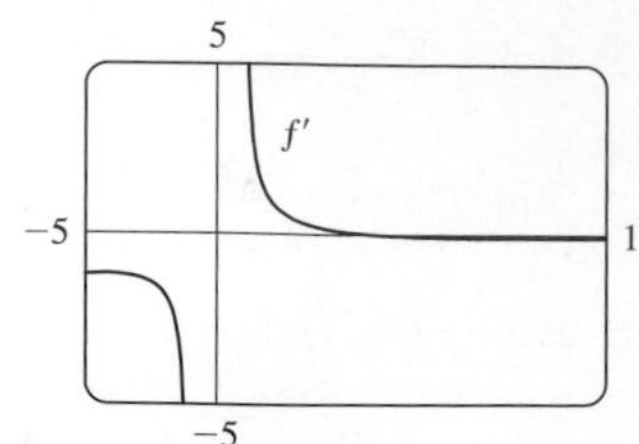

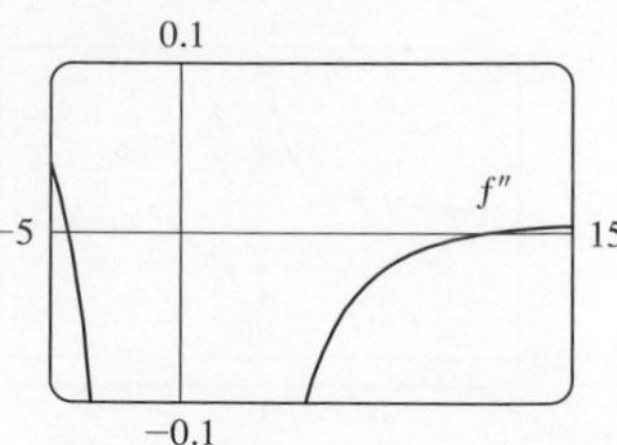

f' changes from positive to negative at $x \approx 5.87$, so 5.87 is the x-coordinate of the maximum point. There is no minimum point. The x-coordinates of the inflection points are the values at which f'' changes sign, that is, -4.31 and 11.74.

43. The family of functions $f(x) = \ln(\sin x + C)$ all have the same period and all have maximum values at $x = \frac{\pi}{2} + 2\pi n$. Since the domain of ln is $(0, \infty)$, f has a graph only if $\sin x + C > 0$ somewhere. Since $-1 \leq \sin x \leq 1$, this happens if $C > -1$, that is, f has no graph if $C \leq -1$. Similarly, if $C > 1$, then $\sin x + C > 0$ and f is continuous on $(-\infty, \infty)$. As C increases, the graph of f is shifted vertically upward and flattens out. If $-1 < C \leq 1$, f is defined where $\sin x + C > 0 \;\Leftrightarrow\; \sin x > -C \;\Leftrightarrow\; \sin^{-1}(-C) < x < \pi - \sin^{-1}(-C)$. Since the period is 2π, the domain of f is $\left(2n\pi + \sin^{-1}(-C), (2n+1)\pi - \sin^{-1}(-C)\right)$, n an integer.

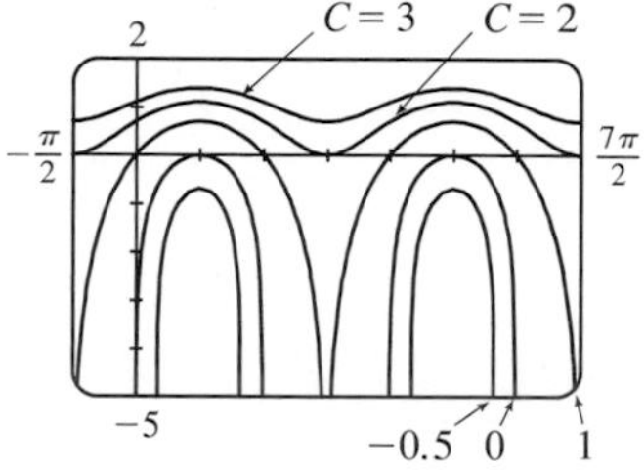

44. We exclude the case $c = 0$, since in that case $f(x) = 0$ for all x. To find the maxima and minima, we differentiate:

$$f(x) = cxe^{-cx^2} \quad\Rightarrow\quad f'(x) = c\left[xe^{-cx^2}(-2cx) + e^{-cx^2}(1)\right] = ce^{-cx^2}(-2cx^2 + 1)$$

This is 0 where $-2cx^2 + 1 = 0 \;\Leftrightarrow\; x = \pm 1/\sqrt{2c}$. So if $c > 0$, there are two maxima or minima, whose x-coordinates approach 0 as c increases. The negative root gives a minimum and the positive root gives a maximum, by the First Derivative Test. By substituting back into the equation, we see that $f\left(\pm 1/\sqrt{2c}\right) = c\left(\pm 1/\sqrt{2c}\right)e^{-c\left(\pm 1/\sqrt{2c}\right)^2} = \pm\sqrt{c/2e}$. So as c increases, the extreme points become more pronounced. Note that if $c > 0$, then $\lim\limits_{x \to \pm\infty} f(x) = 0$. If $c < 0$, then there are no extreme values, and $\lim\limits_{x \to \pm\infty} f(x) = \mp\infty$.

To find the points of inflection, we differentiate again: $f'(x) = ce^{-cx^2}(-2cx^2 + 1) \;\Rightarrow$ $f''(x) = c\left[e^{-cx^2}(-4cx) + (-2cx^2 + 1)(-2cxe^{-cx^2})\right] = -2c^2xe^{-cx^2}(3 - 2cx^2)$. This is 0 at $x = 0$ and where $3 - 2cx^2 = 0 \;\Leftrightarrow\; x = \pm\sqrt{3/(2c)} \;\Rightarrow\;$ IP at $\left(\pm\sqrt{3/(2c)}, \pm\sqrt{3c/2}\,e^{-3/2}\right)$. If $c > 0$ there are three inflection points, and as c increases, the x-coordinates of the nonzero inflection points approach 0. If $c < 0$, there is only one inflection point, the origin.

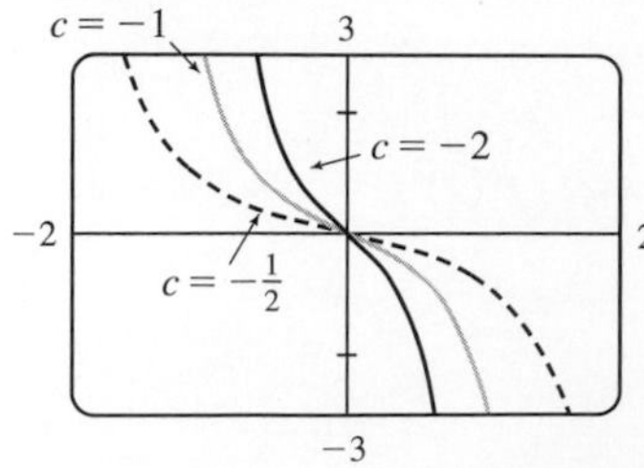

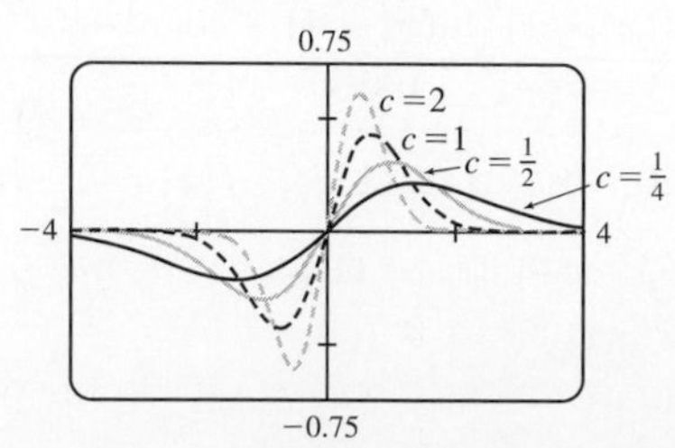

45. Let $f(x) = 3x + 2\cos x + 5$. Then $f(0) = 7 > 0$ and $f(-\pi) = -3\pi - 2 + 5 = -3\pi + 3 = -3(\pi - 1) < 0$, and since f is continuous on $\mathbb{R}$ (hence on $[-\pi, 0]$), the Intermediate Value Theorem assures us that there is at least one zero of f in $[-\pi, 0]$. Now $f'(x) = 3 - 2\sin x > 0$ implies that f is increasing on $\mathbb{R}$, so there is exactly one zero of f, and hence, exactly one real root of the equation $3x + 2\cos x + 5 = 0$.

46. By the Mean Value Theorem, $f'(c) = \dfrac{f(4) - f(0)}{4 - 0} \quad \Leftrightarrow \quad 4f'(c) = f(4) - 1$ for some c with $0 < c < 4$. Since $2 \le f'(c) \le 5$, we have $4(2) \le 4f'(c) \le 4(5) \quad \Leftrightarrow \quad 4(2) \le f(4) - 1 \le 4(5) \quad \Leftrightarrow \quad 8 \le f(4) - 1 \le 20 \quad \Leftrightarrow$ $9 \le f(4) \le 21$.

47. Since f is continuous on $[32, 33]$ and differentiable on $(32, 33)$, then by the Mean Value Theorem there exists a number c in $(32, 33)$ such that $f'(c) = \frac{1}{5}c^{-4/5} = \dfrac{\sqrt[5]{33} - \sqrt[5]{32}}{33 - 32} = \sqrt[5]{33} - 2$, but $\frac{1}{5}c^{-4/5} > 0 \quad \Rightarrow \quad \sqrt[5]{33} - 2 > 0 \quad \Rightarrow \quad \sqrt[5]{33} > 2$. Also f' is decreasing, so that $f'(c) < f'(32) = \frac{1}{5}(32)^{-4/5} = 0.0125 \quad \Rightarrow \quad 0.0125 > f'(c) = \sqrt[5]{33} - 2 \quad \Rightarrow \quad \sqrt[5]{33} < 2.0125$. Therefore, $2 < \sqrt[5]{33} < 2.0125$.

48. Since the point $(1, 3)$ is on the curve $y = ax^3 + bx^2$, we have $3 = a(1)^3 + b(1)^2 \quad \Rightarrow \quad 3 = a + b$ **(1)**.
$y' = 3ax^2 + 2bx \quad \Rightarrow \quad y'' = 6ax + 2b$. $y'' = 0$ [for inflection points] $\Leftrightarrow \quad x = \dfrac{-2b}{6a} = -\dfrac{b}{3a}$. Since we want $x = 1$, $1 = -\dfrac{b}{3a} \quad \Rightarrow \quad b = -3a$. Combining with **(1)** gives us $3 = a - 3a \quad \Leftrightarrow \quad 3 = -2a \quad \Leftrightarrow \quad a = -\frac{3}{2}$. Hence, $b = -3\left(-\frac{3}{2}\right) = \frac{9}{2}$ and the curve is $y = -\frac{3}{2}x^3 + \frac{9}{2}x^2$.

49. (a) $g(x) = f(x^2) \quad \Rightarrow \quad g'(x) = 2xf'(x^2)$ by the Chain Rule. Since $f'(x) > 0$ for all $x \ne 0$, we must have $f'(x^2) > 0$ for $x \ne 0$, so $g'(x) = 0 \quad \Leftrightarrow \quad x = 0$. Now $g'(x)$ changes sign (from negative to positive) at $x = 0$, since one of its factors, $f'(x^2)$, is positive for all x, and its other factor, $2x$, changes from negative to positive at this point, so by the First Derivative Test, f has a local and absolute minimum at $x = 0$.

(b) $g'(x) = 2xf'(x^2) \quad \Rightarrow \quad g''(x) = 2[xf''(x^2)(2x) + f'(x^2)] = 4x^2 f''(x^2) + 2f'(x^2)$ by the Product Rule and the Chain Rule. But $x^2 > 0$ for all $x \ne 0$, $f''(x^2) > 0$ [since f is CU for $x > 0$], and $f'(x^2) > 0$ for all $x \ne 0$, so since all of its factors are positive, $g''(x) > 0$ for $x \ne 0$. Whether $g''(0)$ is positive or 0 doesn't matter [since the sign of g'' does not change there]; g is concave upward on $\mathbb{R}$.

50. Call the two integers x and y. Then $x + 4y = 1000$, so $x = 1000 - 4y$. Their product is $P = xy = (1000 - 4y)y$, so our problem is to maximize the function $P(y) = 1000y - 4y^2$, where $0 < y < 250$ and y is an integer. $P'(y) = 1000 - 8y$, so $P'(y) = 0 \quad \Leftrightarrow \quad y = 125$. $P''(y) = -8 < 0$, so $P(125) = 62{,}500$ is an absolute maximum. Since the optimal y turned out to be an integer, we have found the desired pair of numbers, namely $x = 1000 - 4(125) = 500$ and $y = 125$.

51. If $B = 0$, the line is vertical and the distance from $x = -\dfrac{C}{A}$ to (x_1, y_1) is $\left|x_1 + \dfrac{C}{A}\right| = \dfrac{|Ax_1 + By_1 + C|}{\sqrt{A^2 + B^2}}$, so assume $B \ne 0$. The square of the distance from (x_1, y_1) to the line is $f(x) = (x - x_1)^2 + (y - y_1)^2$ where $Ax + By + C = 0$, so we minimize $f(x) = (x - x_1)^2 + \left(-\frac{A}{B}x - \frac{C}{B} - y_1\right)^2 \Rightarrow f'(x) = 2(x - x_1) + 2\left(-\frac{A}{B}x - \frac{C}{B} - y_1\right)\left(-\frac{A}{B}\right)$.

$f'(x) = 0 \Rightarrow x = \dfrac{B^2x_1 - ABy_1 - AC}{A^2 + B^2}$ and this gives a minimum since $f''(x) = 2\left(1 + \dfrac{A^2}{B^2}\right) > 0$. Substituting this value of x into $f(x)$ and simplifying gives $f(x) = \dfrac{(Ax_1 + By_1 + C)^2}{A^2 + B^2}$, so the minimum distance is $\sqrt{f(x)} = \dfrac{|Ax_1 + By_1 + C|}{\sqrt{A^2 + B^2}}$.

52. On the hyperbola $xy = 8$, if $d(x)$ is the distance from the point $(x, y) = (x, 8/x)$ to the point $(3, 0)$, then $[d(x)]^2 = (x - 3)^2 + 64/x^2 = f(x)$. $f'(x) = 2(x - 3) - 128/x^3 = 0 \Rightarrow x^4 - 3x^3 - 64 = 0 \Rightarrow (x - 4)(x^3 + x^2 + 4x + 16) = 0 \Rightarrow x = 4$ since the solution must have $x > 0$. Then $y = \frac{8}{4} = 2$, so the point is $(4, 2)$.

53.

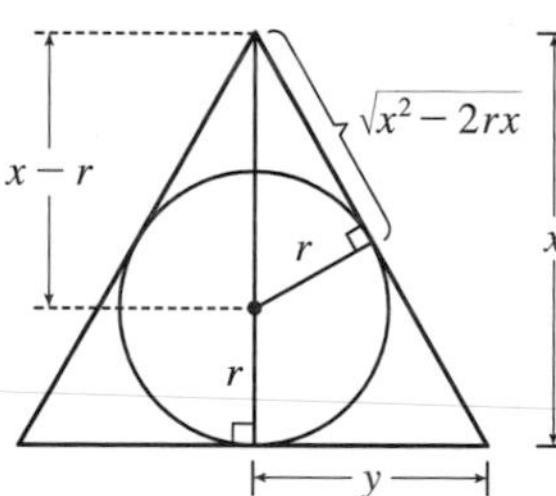

By similar triangles, $\dfrac{y}{x} = \dfrac{r}{\sqrt{x^2 - 2rx}}$, so the area of the triangle is

$$A(x) = \tfrac{1}{2}(2y)x = xy = \frac{rx^2}{\sqrt{x^2 - 2rx}} \Rightarrow$$

$$A'(x) = \frac{2rx\sqrt{x^2 - 2rx} - rx^2(x - r)/\sqrt{x^2 - 2rx}}{x^2 - 2rx} = \frac{rx^2(x - 3r)}{(x^2 - 2rx)^{3/2}} = 0$$

when $x = 3r$.

$A'(x) < 0$ when $2r < x < 3r$, $A'(x) > 0$ when $x > 3r$. So $x = 3r$ gives a minimum and $A(3r) = \dfrac{r(9r^2)}{\sqrt{3}\,r} = 3\sqrt{3}\,r^2$.

54.

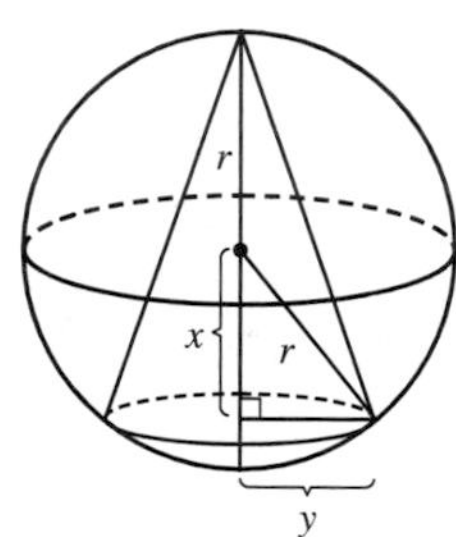

The volume of the cone is $V = \frac{1}{3}\pi y^2(r + x) = \frac{1}{3}\pi(r^2 - x^2)(r + x)$, $-r \le x \le r$.

$$V'(x) = \tfrac{\pi}{3}[(r^2 - x^2)(1) + (r + x)(-2x)] = \tfrac{\pi}{3}[(r + x)(r - x - 2x)]$$
$$= \tfrac{\pi}{3}(r + x)(r - 3x) = 0 \text{ when } x = -r \text{ or } x = r/3.$$

Now $V(r) = 0 = V(-r)$, so the maximum occurs at $x = r/3$ and the volume is

$$V\left(\frac{r}{3}\right) = \frac{\pi}{3}\left(r^2 - \frac{r^2}{9}\right)\left(\frac{4r}{3}\right) = \frac{32\pi r^3}{81}.$$

55.

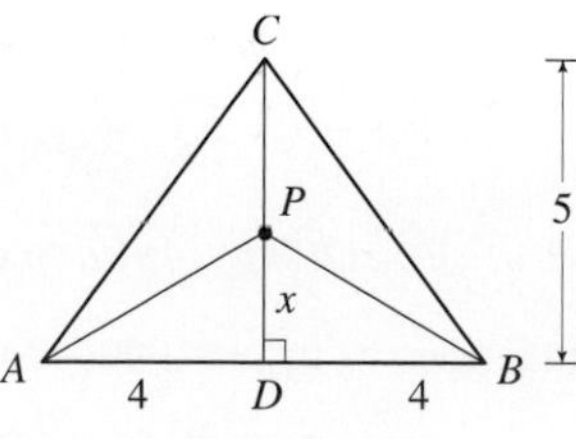

We minimize $L(x) = |PA| + |PB| + |PC| = 2\sqrt{x^2 + 16} + (5 - x)$, $0 \le x \le 5$. $L'(x) = 2x/\sqrt{x^2 + 16} - 1 = 0 \Leftrightarrow 2x = \sqrt{x^2 + 16} \Leftrightarrow 4x^2 = x^2 + 16 \Leftrightarrow x = \frac{4}{\sqrt{3}}$. $L(0) = 13$, $L\left(\frac{4}{\sqrt{3}}\right) \approx 11.9$, $L(5) \approx 12.8$, so the minimum occurs when $x = \frac{4}{\sqrt{3}} \approx 2.3$.

56.

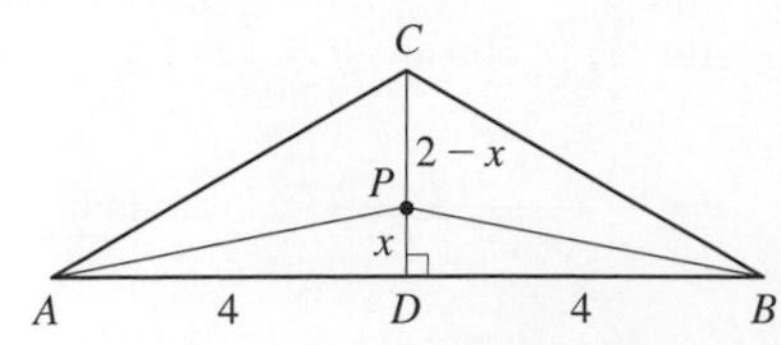

If $|CD| = 2$, the last part of $L(x)$ changes from $(5 - x)$ to $(2 - x)$ with $0 \le x \le 2$. But we still get $L'(x) = 0 \Leftrightarrow x = \frac{4}{\sqrt{3}}$, which isn't in the interval $[0, 2]$. Now $L(0) = 10$ and $L(2) = 2\sqrt{20} = 4\sqrt{5} \approx 8.9$. The minimum occurs when $P = C$.

57. $v = K\sqrt{\dfrac{L}{C} + \dfrac{C}{L}} \;\Rightarrow\; \dfrac{dv}{dL} = \dfrac{K}{2\sqrt{(L/C) + (C/L)}}\left(\dfrac{1}{C} - \dfrac{C}{L^2}\right) = 0 \;\Leftrightarrow\; \dfrac{1}{C} = \dfrac{C}{L^2} \;\Leftrightarrow\; L^2 = C^2 \;\Leftrightarrow\; L = C.$

This gives the minimum velocity since $v' < 0$ for $0 < L < C$ and $v' > 0$ for $L > C$.

58.

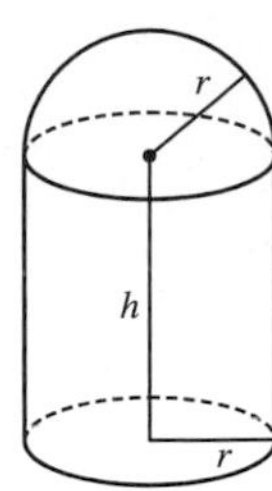

We minimize the surface area $S = \pi r^2 + 2\pi rh + \frac{1}{2}(4\pi r^2) = 3\pi r^2 + 2\pi rh$.

Solving $V = \pi r^2 h + \frac{2}{3}\pi r^3$ for h, we get $h = \dfrac{V - \frac{2}{3}\pi r^3}{\pi r^2} = \dfrac{V}{\pi r^2} - \frac{2}{3}r$, so

$S(r) = 3\pi r^2 + 2\pi r\left[\dfrac{V}{\pi r^2} - \frac{2}{3}r\right] = \frac{5}{3}\pi r^2 + \dfrac{2V}{r}.$

$S'(r) = -\dfrac{2V}{r^2} + \frac{10}{3}\pi r = \dfrac{\frac{10}{3}\pi r^3 - 2V}{r^2} = 0 \;\Leftrightarrow\; \frac{10}{3}\pi r^3 = 2V \;\Leftrightarrow\; r^3 = \dfrac{3V}{5\pi} \;\Leftrightarrow\; r = \sqrt[3]{\dfrac{3V}{5\pi}}.$

This gives an absolute minimum since $S'(r) < 0$ for $0 < r < \sqrt[3]{\dfrac{3V}{5\pi}}$ and $S'(r) > 0$ for $r > \sqrt[3]{\dfrac{3V}{5\pi}}$. Thus,

$$h = \frac{V - \frac{2}{3}\pi \cdot \dfrac{3V}{5\pi}}{\pi\sqrt[3]{\dfrac{(3V)^2}{(5\pi)^2}}} = \frac{\left(V - \frac{2}{5}V\right)\sqrt[3]{(5\pi)^2}}{\pi\sqrt[3]{(3V)^2}} = \frac{3V\sqrt[3]{(5\pi)^2}}{5\pi\sqrt[3]{(3V)^2}} = \sqrt[3]{\frac{3V}{5\pi}} = r$$

59. Let x denote the number of \$1 decreases in ticket price. Then the ticket price is $\$12 - \$1(x)$, and the average attendance is $11{,}000 + 1000(x)$. Now the revenue per game is

$$\begin{aligned} R(x) &= (\text{price per person}) \times (\text{number of people per game}) \\ &= (12 - x)(11{,}000 + 1000x) = -1000x^2 + 1000x + 132{,}000 \end{aligned}$$

for $0 \le x \le 4$ [since the seating capacity is 15,000] $\Rightarrow\; R'(x) = -2000x + 1000 = 0 \;\Leftrightarrow\; x = 0.5$. This is a maximum since $R''(x) = -2000 < 0$ for all x. Now we must check the value of $R(x) = (12 - x)(11{,}000 + 1000x)$ at $x = 0.5$ and at the endpoints of the domain to see which value of x gives the maximum value of R.

$R(0) = (12)(11{,}000) = 132{,}000$, $R(0.5) = (11.5)(11{,}500) = 132{,}250$, and $R(4) = (8)(15{,}000) = 120{,}000$. Thus, the maximum revenue of \$132,250 per game occurs when the average attendance is 11,500 and the ticket price is \$11.50.

60. (a) $C(x) = 1800 + 25x - 0.2x^2 + 0.001x^3$ and

$R(x) = xp(x) = 48.2x - 0.03x^2$.

The profit is maximized when $C'(x) = R'(x)$.

From the figure, we estimate that the tangents are parallel when $x \approx 160$.

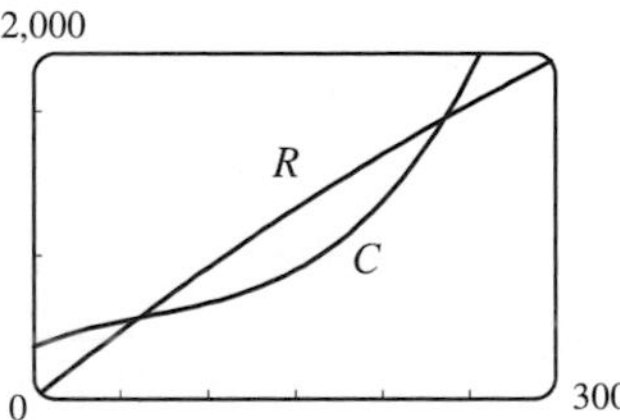

(b) $C'(x) = 25 - 0.4x + 0.003x^2$ and $R'(x) = 48.2 - 0.06x$. $C'(x) = R'(x) \;\Rightarrow\; 0.003x^2 - 0.34x - 23.2 = 0 \;\Rightarrow$ $x_1 \approx 161.3$ $(x > 0)$. $R''(x) = -0.06$ and $C''(x) = -0.4 + 0.006x$, so $R''(x_1) = -0.06 < C''(x_1) \approx 0.57 \;\Rightarrow$ profit is maximized by producing 161 units.

(c) $c(x) = \dfrac{C(x)}{x} = \dfrac{1800}{x} + 25 - 0.2x + 0.001x^2$ is the average cost. Since the average cost is minimized when the marginal cost equals the average cost, we graph $c(x)$ and $C'(x)$ and estimate the point of intersection. From the figure, $C'(x) = c(x) \iff x \approx 144$.

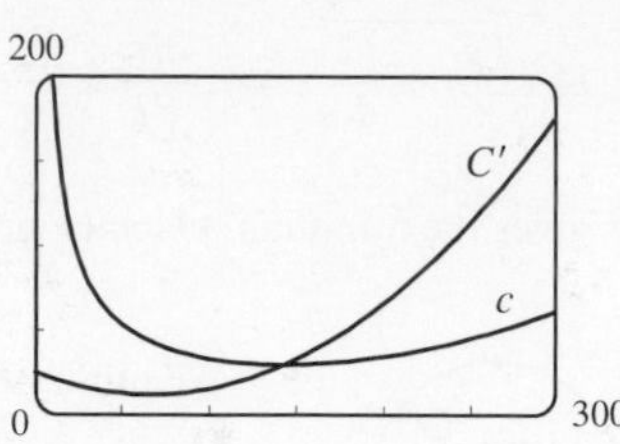

61. $f(x) = x^5 - x^4 + 3x^2 - 3x - 2 \Rightarrow f'(x) = 5x^4 - 4x^3 + 6x - 3$, so $x_{n+1} = x_n - \dfrac{x_n^5 - x_n^4 + 3x_n^2 - 3x_n - 2}{5x_n^4 - 4x_n^3 + 6x_n - 3}$.

Now $x_1 = 1 \Rightarrow x_2 = 1.5 \Rightarrow x_3 \approx 1.343860 \Rightarrow x_4 \approx 1.300320 \Rightarrow x_5 \approx 1.297396 \Rightarrow$ $x_6 \approx 1.297383 \approx x_7$, so the root in $[1, 2]$ is 1.297383, to six decimal places.

62. Graphing $y = \sin x$ and $y = x^2 - 3x + 1$ shows that there are two roots, one about 0.3 and the other about 2.8. $f(x) = \sin x - x^2 + 3x - 1 \Rightarrow$ $f'(x) = \cos x - 2x + 3 \Rightarrow x_{n+1} = x_n - \dfrac{\sin x_n - x_n^2 + 3x_n - 1}{\cos x_n - 2x_n + 3}$.

Now $x_1 = 0.3 \Rightarrow x_2 \approx 0.268552 \Rightarrow x_3 \approx 0.268881 \approx x_4$ and $x_1 = 2.8 \Rightarrow x_2 \approx 2.770354 \Rightarrow x_3 \approx 2.770058 \approx x_4$, so to six decimal places, the roots are 0.268881 and 2.770058.

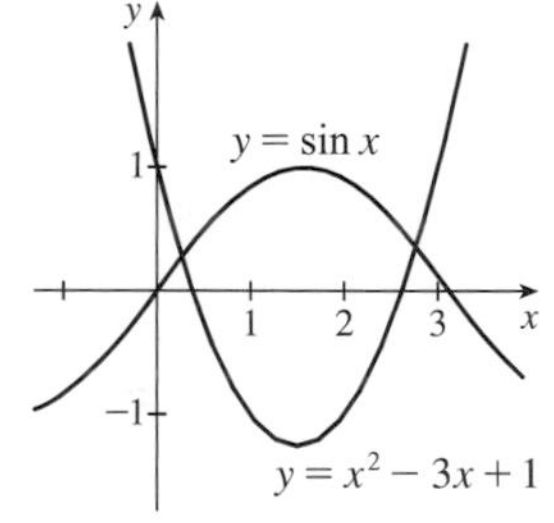

63. $f(t) = \cos t + t - t^2 \Rightarrow f'(t) = -\sin t + 1 - 2t$. $f'(t)$ exists for all t, so to find the maximum of f, we can examine the zeros of f'. From the graph of f', we see that a good choice for t_1 is $t_1 = 0.3$. Use $g(t) = -\sin t + 1 - 2t$ and $g'(t) = -\cos t - 2$ to obtain $t_2 \approx 0.33535293$, $t_3 \approx 0.33541803 \approx t_4$. Since $f''(t) = -\cos t - 2 < 0$ for all t, $f(0.33541803) \approx 1.16718557$ is the absolute maximum.

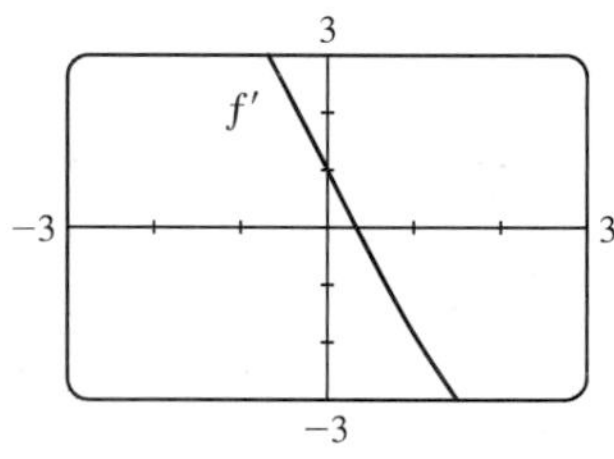

64. $y = f(x) = x \sin x$, $0 \le x \le 2\pi$. **A.** $D = [0, 2\pi]$ **B.** y-intercept: $f(0) = 0$; x-intercepts: $f(x) = 0 \iff x = 0$ or $\sin x = 0 \iff x = 0, \pi$, or 2π. **C.** There is no symmetry on D, but if f is defined for all real numbers x, then f is an even function. **D.** No asymptote **E.** $f'(x) = x \cos x + \sin x$. To find critical numbers in $(0, 2\pi)$, we graph f' and see that there are two critical numbers, about 2 and 4.9. To find them more precisely, we use Newton's method, setting

$g(x) = f'(x) = x \cos x + \sin x$, so that $g'(x) = f''(x) = 2\cos x - x \sin x$ and $x_{n+1} = x_n - \dfrac{x_n \cos x_n + \sin x_n}{2\cos x_n - x_n \sin x_n}$.

$x_1 = 2 \Rightarrow x_2 \approx 2.029048$, $x_3 \approx 2.028758 \approx x_4$ and $x_1 = 4.9 \Rightarrow x_2 \approx 4.913214$, $x_3 \approx 4.913180 \approx x_4$, so the critical numbers, to six decimal places, are $r_1 = 2.028758$ and $r_2 = 4.913180$. By checking sample values of f' in $(0, r_1)$, (r_1, r_2), and $(r_2, 2\pi)$, we see that f is increasing on $(0, r_1)$, decreasing on (r_1, r_2), and increasing on $(r_2, 2\pi)$. **F.** Local maximum value $f(r_1) \approx 1.819706$, local minimum value $f(r_2) \approx -4.814470$. **G.** $f''(x) = 2\cos x - x \sin x$. To find points where $f''(x) = 0$, we graph f'' and find that $f''(x) = 0$ at about 1 and 3.6. To find the values more precisely, we use Newton's method. Set $h(x) = f''(x) = 2\cos x - x\sin x$. Then $h'(x) = -3\sin x - x\cos x$, so

$x_{n+1} = x_n - \dfrac{2\cos x_n - x_n \sin x_n}{-3\sin x_n - x_n \cos x_n}$. $x_1 = 1 \Rightarrow x_2 \approx 1.078028$, $x_3 \approx 1.076874 \approx x_4$ and $x_1 = 3.6 \Rightarrow$

$x_2 \approx 3.643996$, $x_3 \approx 3.643597 \approx x_4$, so the zeros of f'', to six decimal places, are $r_3 = 1.076874$ and $r_4 = 3.643597$.

By checking sample values of f'' in $(0, r_3)$, (r_3, r_4), and $(r_4, 2\pi)$, we see that f is CU on $(0, r_3)$, CD on (r_3, r_4), and CU on $(r_4, 2\pi)$. f has inflection points at $(r_3, f(r_3) \approx 0.948166)$ and $(r_4, f(r_4) \approx -1.753240)$.

H.

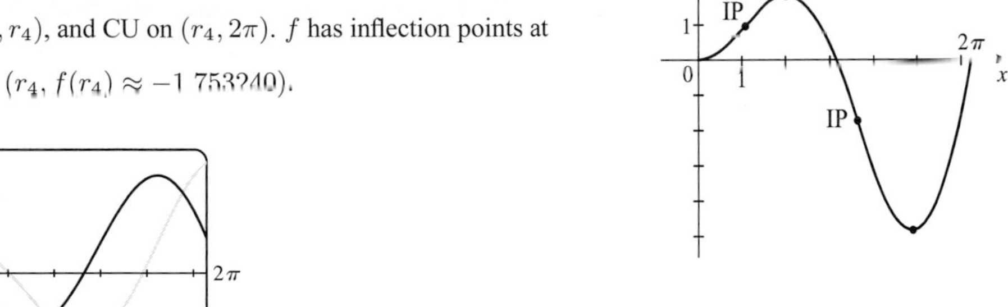

65. $f(x) = 4\sqrt{x} - 6x^2 + 3 = 4x^{1/2} - 6x^2 + 3 \Rightarrow F(x) = 4\left(\frac{2}{3}x^{3/2}\right) \quad 6\left(\frac{1}{3}x^3\right) + 3x + C = \frac{8}{3}x^{3/2} - 2x^3 + 3x + C$

66. $g(x) = \dfrac{1}{x} + \dfrac{1}{x^2+1} \Rightarrow G(x) = \begin{cases} \ln x + \tan^{-1} x + C_1 & \text{if } x > 0 \\ \ln(-x) + \tan^{-1} x + C_2 & \text{if } x < 0 \end{cases}$

67. $f(t) = 2\sin t - 3e^t \Rightarrow F(t) = -2\cos t - 3e^t + C$

68. $f(x) = x^{-3} + \cosh x \Rightarrow F(x) = \begin{cases} -1/(2x^2) + \sinh x + C_1 & \text{if } x > 0 \\ -1/(2x^2) + \sinh x + C_2 & \text{if } x < 0 \end{cases}$

69. $f'(t) = 2t - 3\sin t \Rightarrow f(t) = t^2 + 3\cos t + C$.

$f(0) = 3 + C$ and $f(0) = 5 \Rightarrow C = 2$, so $f(t) = t^2 + 3\cos t + 2$.

70. $f'(u) = \dfrac{u^2 + \sqrt{u}}{u} = u + u^{-1/2} \Rightarrow f(u) = \frac{1}{2}u^2 + 2u^{1/2} + C$.

$f(1) = \frac{1}{2} + 2 + C$ and $f(1) = 3 \Rightarrow C = \frac{1}{2}$, so $f(u) = \frac{1}{2}u^2 + 2\sqrt{u} + \frac{1}{2}$.

71. $f''(x) = 1 - 6x + 48x^2 \Rightarrow f'(x) = x - 3x^2 + 16x^3 + C$. $f'(0) = C$ and $f'(0) = 2 \Rightarrow C = 2$, so

$f'(x) = x - 3x^2 + 16x^3 + 2$ and hence, $f(x) = \frac{1}{2}x^2 - x^3 + 4x^4 + 2x + D$.

$f(0) = D$ and $f(0) = 1 \Rightarrow D = 1$, so $f(x) = \frac{1}{2}x^2 - x^3 + 4x^4 + 2x + 1$.

72. $f''(x) = 5x^3 + 6x^2 + 2 \Rightarrow f'(x) = \frac{5}{4}x^4 + 2x^3 + 2x + C \Rightarrow f(x) = \frac{1}{4}x^5 + \frac{1}{2}x^4 + x^2 + Cx + D$. Now $f(0) = D$ and $f(0) = 3$, so $D = 3$. Also, $f(1) = \frac{1}{4} + \frac{1}{2} + 1 + C + 3 = C + \frac{19}{4}$ and $f(1) = -2$, so $C + \frac{19}{4} = -2 \Rightarrow C = \frac{27}{4}$.

Thus, $f(x) = \frac{1}{4}x^5 + \frac{1}{2}x^4 + x^2 - \frac{27}{4}x + 3$.

73. $v(t) = s'(t) = 2t - \dfrac{1}{1+t^2} \Rightarrow s(t) = t^2 - \tan^{-1} t + C$.

$s(0) = 0 - 0 + C = C$ and $s(0) = 1 \Rightarrow C = 1$, so $s(t) = t^2 - \tan^{-1} t + 1$.

74. $a(t) = v'(t) = \sin t + 3\cos t \quad \Rightarrow \quad v(t) = -\cos t + 3\sin t + C.$

$v(0) = -1 + 0 + C$ and $v(0) = 2 \quad \Rightarrow \quad C = 3$, so $v(t) = -\cos t + 3\sin t + 3$ and $s(t) = -\sin t - 3\cos t + 3t + D$.

$s(0) = -3 + D$ and $s(0) = 0 \quad \Rightarrow \quad D = 3$, and $s(t) = -\sin t - 3\cos t + 3t + 3$.

75. (a) Since f is 0 just to the left of the y-axis, we must have a minimum of F at the same place since we are increasing through $(0,0)$ on F. There must be a local maximum to the left of $x = -3$, since f changes from positive to negative there.

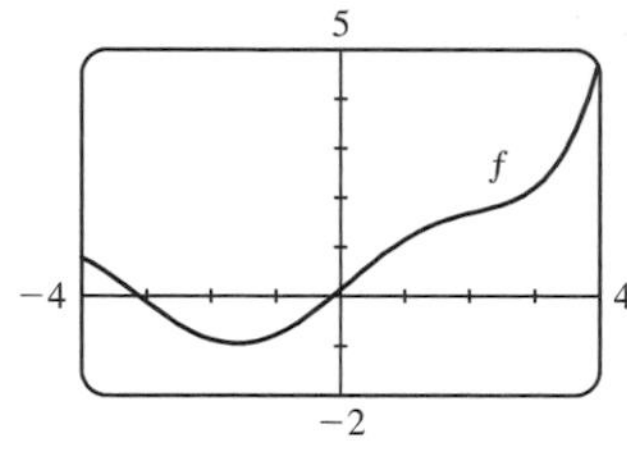

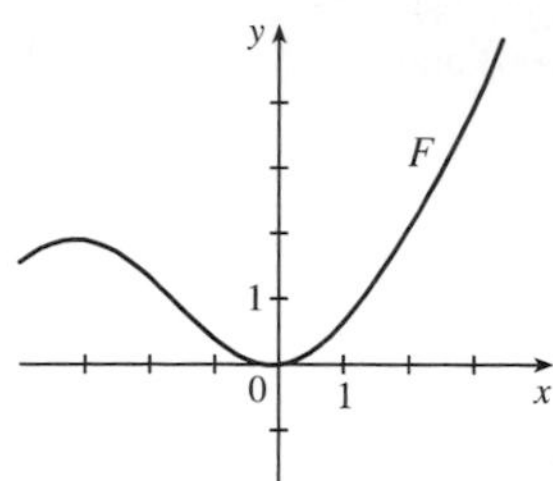

(b) $f(x) = 0.1e^x + \sin x \quad \Rightarrow$

$F(x) = 0.1e^x - \cos x + C$. $F(0) = 0 \quad \Rightarrow$

$0.1 - 1 + C = 0 \quad \Rightarrow \quad C = 0.9$, so

$F(x) = 0.1e^x - \cos x + 0.9$.

(c)

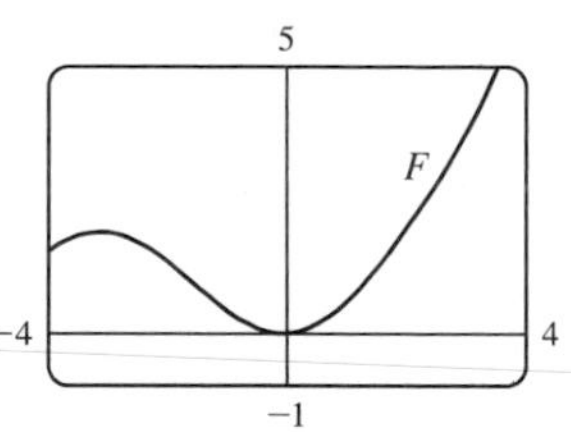

76. $f(x) = x^4 + x^3 + cx^2 \quad \Rightarrow \quad f'(x) = 4x^3 + 3x^2 + 2cx$. This is 0 when $x(4x^2 + 3x + 2c) = 0 \quad \Leftrightarrow \quad x = 0$ or $4x^2 + 3x + 2c = 0$. Using the quadratic formula, we find that the roots of this last equation are $x = \dfrac{-3 \pm \sqrt{9 - 32c}}{8}$.

Now if $9 - 32c < 0 \quad \Leftrightarrow \quad c > \frac{9}{32}$, then $(0,0)$ is the only critical point, a minimum. If $c = \frac{9}{32}$, then there are two critical points (a minimum at $x = 0$, and a horizontal tangent with no maximum or minimum at $x = -\frac{3}{8}$) and if $c < \frac{9}{32}$, then there are three critical points except when $c = 0$, in which case the root with the $+$ sign coincides with the critical point at $x = 0$. For $0 < c < \frac{9}{32}$, there is a minimum at $x = -\dfrac{3}{8} - \dfrac{\sqrt{9 - 32c}}{8}$, a maximum at $x = -\dfrac{3}{8} + \dfrac{\sqrt{9 - 32c}}{8}$, and a minimum at $x = 0$. For $c = 0$, there is a minimum at $x = -\frac{3}{4}$ and a horizontal tangent with no extremum at $x = 0$, and for $c < 0$, there is a maximum at $x = 0$, and there are minima at $x = -\dfrac{3}{8} \pm \dfrac{\sqrt{9 - 32c}}{8}$. Now we calculate $f''(x) = 12x^2 + 6x + 2c$.

The roots of this equation are $x = \dfrac{-6 \pm \sqrt{36 - 4 \cdot 12 \cdot 2c}}{24}$. So if $36 - 96c \le 0 \quad \Leftrightarrow \quad c \ge \frac{3}{8}$, then there is no inflection point. If $c < \frac{3}{8}$, then there are two inflection points at $x = -\dfrac{1}{4} \pm \dfrac{\sqrt{9 - 24c}}{12}$.

[continued]

Value of c	No. of CP	No. of IP
$c < 0$	3	2
$c = 0$	2	2
$0 < c < \frac{9}{32}$	3	2
$c = \frac{9}{32}$	2	2
$\frac{9}{32} < c < \frac{3}{8}$	1	2
$c \geq \frac{3}{8}$	1	0

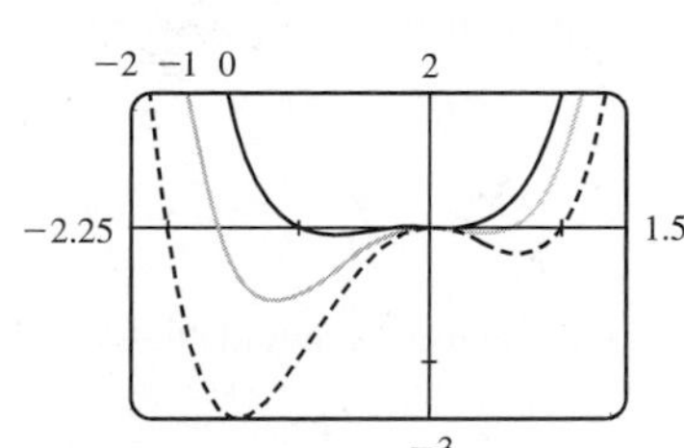

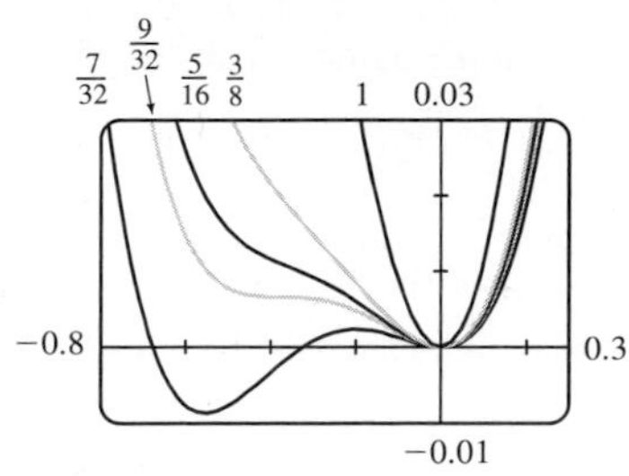

77. Choosing the positive direction to be upward, we have $a(t) = -9.8 \Rightarrow v(t) = -9.8t + v_0$, but $v(0) = 0 = v_0 \Rightarrow$ $v(t) = -9.8t = s'(t) \Rightarrow s(t) = -4.9t^2 + s_0$, but $s(0) = s_0 = 500 \Rightarrow s(t) = -4.9t^2 + 500$. When $s = 0$, $-4.9t^2 + 500 = 0 \Rightarrow t_1 = \sqrt{\frac{500}{4.9}} \approx 10.1 \Rightarrow v(t_1) = -9.8\sqrt{\frac{500}{4.9}} \approx -98.995 \text{ m/s}$. Since the canister has been designed to withstand an impact velocity of 100 m/s, the canister will *not burst*.

78. Let $s_A(t)$ and $s_B(t)$ be the position functions for cars A and B and let $f(t) = s_A(t) - s(t)$. Since A passed B twice, there must be three values of t such that $f(t) = 0$. Then by three applications of Rolle's Theorem (see Exercise 4.2.22), there is a number c such that $f''(c) = 0$. So $s_A''(c) = s_B''(c)$; that is, A and B had equal accelerations at $t = c$. We assume that f is continuous on $[0, T]$ and twice differentiable on $(0, T)$, where T is the total time of the race.

79. (a)

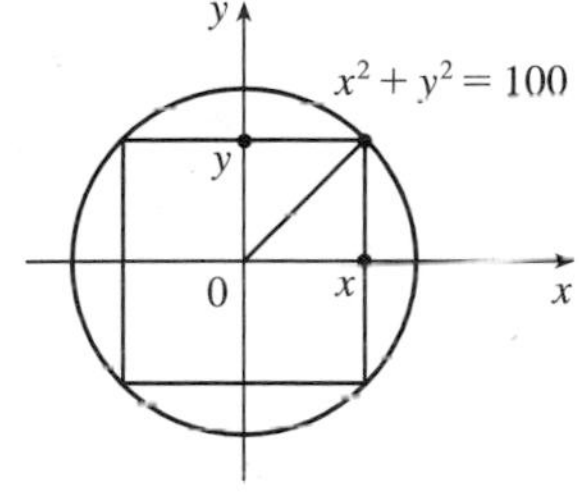

The cross-sectional area of the rectangular beam is

$A = 2x \cdot 2y = 4xy = 4x\sqrt{100 - x^2}$, $0 \leq x \leq 10$, so

$$\frac{dA}{dx} = 4x\left(\tfrac{1}{2}\right)(100 - x^2)^{-1/2}(-2x) + (100 - x^2)^{1/2} \cdot 4$$
$$= \frac{-4x^2}{(100 - x^2)^{1/2}} + 4(100 - x^2)^{1/2} = \frac{4[-x^2 + (100 - x^2)]}{(100 - x^2)^{1/2}}.$$

$\frac{dA}{dx} = 0$ when $-x^2 + (100 - x^2) = 0 \Rightarrow x^2 = 50 \Rightarrow x = \sqrt{50} \approx 7.07 \Rightarrow y = \sqrt{100 - (\sqrt{50})^2} = \sqrt{50}$.

Since $A(0) = A(10) = 0$, the rectangle of maximum area is a square.

(b)

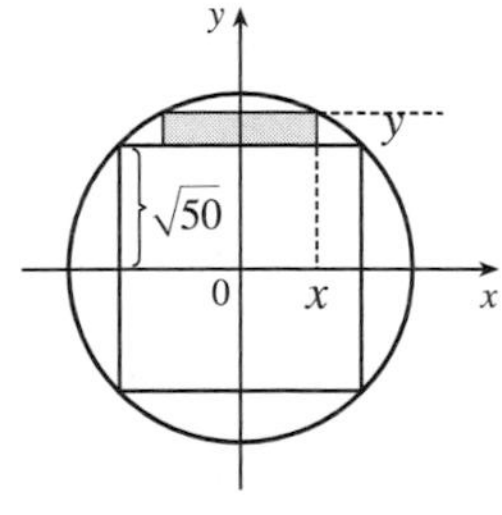

The cross-sectional area of each rectangular plank (shaded in the figure) is

$A = 2x(y - \sqrt{50}) = 2x\left[\sqrt{100 - x^2} - \sqrt{50}\right]$, $0 \leq x \leq \sqrt{50}$, so

$$\frac{dA}{dx} = 2\left(\sqrt{100 - x^2} - \sqrt{50}\right) + 2x\left(\tfrac{1}{2}\right)(100 - x^2)^{-1/2}(-2x)$$
$$= 2(100 - x^2)^{1/2} - 2\sqrt{50} - \frac{2x^2}{(100 - x^2)^{1/2}}$$

Set $\frac{dA}{dx} = 0$: $(100 - x^2) - \sqrt{50}\,(100 - x^2)^{1/2} - x^2 = 0 \Rightarrow 100 - 2x^2 = \sqrt{50}\,(100 - x^2)^{1/2} \Rightarrow$

$10{,}000 - 400x^2 + 4x^4 = 50(100 - x^2) \Rightarrow 4x^4 - 350x^2 + 5000 = 0 \Rightarrow 2x^4 - 175x^2 + 2500 = 0 \Rightarrow$

$x^2 = \dfrac{175 \pm \sqrt{10{,}625}}{4} \approx 69.52$ or $17.98 \Rightarrow x \approx 8.34$ or 4.24. But $8.34 > \sqrt{50}$, so $x_1 \approx 4.24 \Rightarrow$

$y - \sqrt{50} = \sqrt{100 - x_1^2} - \sqrt{50} \approx 1.99$. Each plank should have dimensions about $8\frac{1}{2}$ inches by 2 inches.

(c) From the figure in part (a), the width is $2x$ and the depth is $2y$, so the strength is

$S = k(2x)(2y)^2 = 8kxy^2 = 8kx(100 - x^2) = 800kx - 8kx^3$, $0 \le x \le 10$. $dS/dx = 800k - 24kx^2 = 0$ when

$24kx^2 = 800k \Rightarrow x^2 = \frac{100}{3} \Rightarrow x = \frac{10}{\sqrt{3}} \Rightarrow y = \sqrt{\frac{200}{3}} = \frac{10\sqrt{2}}{\sqrt{3}} = \sqrt{2}\,x$. Since $S(0) = S(10) = 0$, the maximum strength occurs when $x = \frac{10}{\sqrt{3}}$. The dimensions should be $\frac{20}{\sqrt{3}} \approx 11.55$ inches by $\frac{20\sqrt{2}}{\sqrt{3}} \approx 16.33$ inches.

80. (a)

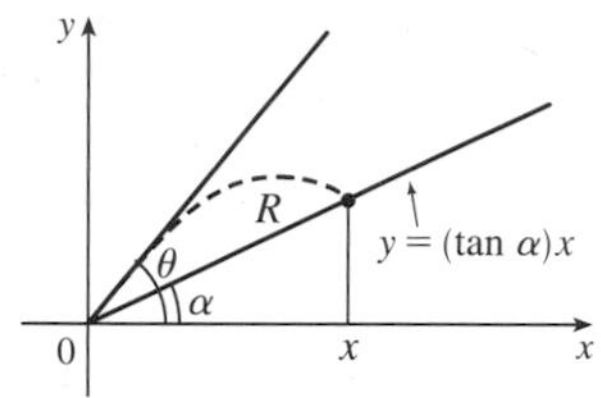

$y = (\tan\theta)x - \dfrac{g}{2v^2\cos^2\theta}x^2$. The parabola intersects the line when

$$(\tan\alpha)x = (\tan\theta)x - \frac{g}{2v^2\cos^2\theta}x^2 \Rightarrow$$

$$x = \frac{(\tan\theta - \tan\alpha)2v^2\cos^2\theta}{g} \Rightarrow$$

$$R(\theta) = \frac{x}{\cos\alpha} = \left(\frac{\sin\theta}{\cos\theta} - \frac{\sin\alpha}{\cos\alpha}\right)\frac{2v^2\cos^2\theta}{g\cos\alpha} = \left(\frac{\sin\theta}{\cos\theta} - \frac{\sin\alpha}{\cos\alpha}\right)(\cos\theta\,\cos\alpha)\frac{2v^2\cos\theta}{g\cos^2\alpha}$$

$$= (\sin\theta\,\cos\alpha - \sin\alpha\,\cos\theta)\frac{2v^2\cos\theta}{g\cos^2\alpha} = \sin(\theta - \alpha)\frac{2v^2\cos\theta}{g\cos^2\alpha}$$

(b) $$R'(\theta) = \frac{2v^2}{g\cos^2\alpha}[\cos\theta\cdot\cos(\theta-\alpha) + \sin(\theta-\alpha)(-\sin\theta)] = \frac{2v^2}{g\cos^2\alpha}\cos[\theta + (\theta - \alpha)]$$

$$= \frac{2v^2}{g\cos^2\alpha}\cos(2\theta - \alpha) = 0$$

when $\cos(2\theta - \alpha) = 0 \Rightarrow 2\theta - \alpha = \frac{\pi}{2} \Rightarrow \theta = \dfrac{\pi/2 + \alpha}{2} = \dfrac{\pi}{4} + \dfrac{\alpha}{2}$. The First Derivative Test shows that this gives a maximum value for $R(\theta)$. [This could be done without calculus by applying the formula for $\sin x\,\cos y$ to $R(\theta)$.]

(c) Replacing α by $-\alpha$ in part (a), we get $R(\theta) = \dfrac{2v^2\cos\theta\,\sin(\theta+\alpha)}{g\cos^2\alpha}$. Proceeding as in part (b), or simply by replacing α by $-\alpha$ in the result of part (b), we see that $R(\theta)$ is maximized when $\theta = \dfrac{\pi}{4} - \dfrac{\alpha}{2}$.

81. $$\lim_{E\to 0^+} P(E) = \lim_{E\to 0^+}\left(\frac{e^E + e^{-E}}{e^E - e^{-E}} - \frac{1}{E}\right)$$

$$= \lim_{E\to 0^+}\frac{E(e^E + e^{-E}) - 1(e^E - e^{-E})}{(e^E - e^{-E})E} = \lim_{E\to 0^+}\frac{Ee^E + Ee^{-E} - e^E + e^{-E}}{Ee^E - Ee^{-E}} \quad \left[\text{form is } \tfrac{0}{0}\right]$$

$$\overset{\text{H}}{=} \lim_{E\to 0^+}\frac{Ee^E + e^E\cdot 1 + E(-e^{-E}) + e^{-E}\cdot 1 - e^E + (-e^{-E})}{Ee^E + e^E\cdot 1 - [E(-e^{-E}) + e^{-E}\cdot 1]}$$

$$= \lim_{E\to 0^+}\frac{Ee^E - Ee^{-E}}{Ee^E + e^E + Ee^{-E} - e^{-E}} = \lim_{E\to 0^+}\frac{e^E - e^{-E}}{e^E + \dfrac{e^E}{E} + e^{-E} - \dfrac{e^{-E}}{E}} \quad [\text{divide by } E]$$

$$= \frac{0}{2 + L}, \text{ where } L = \lim_{E\to 0^+}\frac{e^E - e^{-E}}{E} \quad \left[\text{form is } \tfrac{0}{0}\right] \overset{\text{H}}{=} \lim_{E\to 0^+}\frac{e^E + e^{-E}}{1} = \frac{1+1}{1} = 2$$

Thus, $\lim\limits_{E\to 0^+} P(E) = \dfrac{0}{2+2} = 0$.

82. $\lim\limits_{c\to 0^+} s(t) = \lim\limits_{c\to 0^+}\left(\dfrac{m}{c}\ln\cosh\sqrt{\dfrac{gc}{mt}}\right) = m\lim\limits_{c\to 0^+}\dfrac{\ln\cosh\sqrt{ac}}{c}$ [let $a = g/(mt)$]

$$\overset{\text{H}}{=} m\lim_{c\to 0^+}\frac{\dfrac{1}{\cosh\sqrt{ac}}(\sinh\sqrt{ac})\left(\dfrac{\sqrt{a}}{2\sqrt{c}}\right)}{1} = \frac{m\sqrt{a}}{2}\lim_{c\to 0^+}\frac{\tanh\sqrt{ac}}{\sqrt{c}}$$

$$\overset{\text{H}}{=} \frac{m\sqrt{a}}{2}\lim_{c\to 0^+}\frac{\operatorname{sech}^2\sqrt{ac}\left[\sqrt{a}/\left(2\sqrt{c}\right)\right]}{1/\left(2\sqrt{c}\right)} = \frac{ma}{2}\lim_{c\to 0^+}\operatorname{sech}^2\sqrt{ac} = \frac{ma}{2}(1)^2 = \frac{mg}{2mt} = \frac{g}{2t}$$

83. We first show that $\dfrac{x}{1+x^2} < \tan^{-1}x$ for $x > 0$. Let $f(x) = \tan^{-1}x - \dfrac{x}{1+x^2}$. Then

$f'(x) = \dfrac{1}{1+x^2} - \dfrac{1(1+x^2) - x(2x)}{(1+x^2)^2} = \dfrac{(1+x^2)-(1-x^2)}{(1+x^2)^2} = \dfrac{2x^2}{(1+x^2)^2} > 0$ for $x > 0$. So $f(x)$ is increasing on $(0,\infty)$. Hence, $0 < x \;\Rightarrow\; 0 = f(0) < f(x) = \tan^{-1}x - \dfrac{x}{1+x^2}$. So $\dfrac{x}{1+x^2} < \tan^{-1}x$ for $0 < x$. We next show that $\tan^{-1}x < x$ for $x > 0$. Let $h(x) = x - \tan^{-1}x$. Then $h'(x) = 1 - \dfrac{1}{1+x^2} = \dfrac{x^2}{1+x^2} > 0$. Hence, $h(x)$ is increasing on $(0,\infty)$. So for $0 < x$, $0 = h(0) < h(x) = x - \tan^{-1}x$. Hence, $\tan^{-1}x < x$ for $x > 0$, and we conclude that $\dfrac{x}{1+x^2} < \tan^{-1}x < x$ for $x > 0$.

84. If $f'(x) < 0$ for all x, $f''(x) > 0$ for $|x| > 1$, $f''(x) < 0$ for $|x| < 1$, and $\lim\limits_{x\to\pm\infty}[f(x) + x] = 0$, then f is decreasing everywhere, concave up on $(-\infty,-1)$ and $(1,\infty)$, concave down on $(-1,1)$, and approaches the line $y = -x$ as $x \to \pm\infty$. An example of such a graph is sketched.

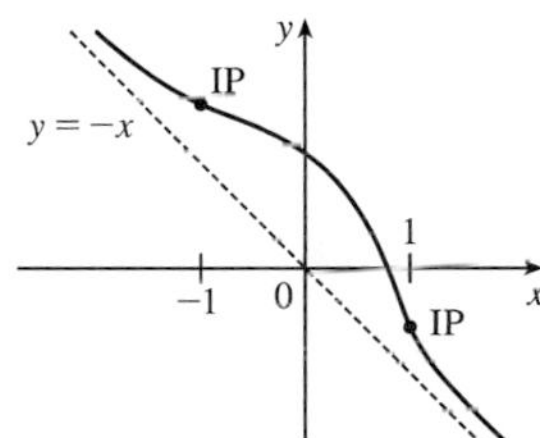

85. (a) $I = \dfrac{k\cos\theta}{d^2} = \dfrac{k(h/d)}{d^2} = k\dfrac{h}{d^3} = k\dfrac{h}{\left(\sqrt{40^2+h^2}\right)^3} = k\dfrac{h}{(1600+h^2)^{3/2}} \;\Rightarrow$

$$\frac{dI}{dh} = k\,\frac{(1600+h^2)^{3/2} - h\frac{3}{2}(1600+h^2)^{1/2}\cdot 2h}{[(1600+h^2)^{3/2}]^2} = \frac{k(1600+h^2)^{1/2}(1600+h^2-3h^2)}{(1600+h^2)^3}$$

$$= \frac{k(1600-2h^2)}{(1600+h^2)^{5/2}} \quad [k \text{ is the constant of proportionality}]$$

Set $dI/dh = 0$: $1600 - 2h^2 = 0 \;\Rightarrow\; h^2 = 800 \;\Rightarrow\; h = \sqrt{800} = 20\sqrt{2}$. By the First Derivative Test, I has a local maximum at $h = 20\sqrt{2} \approx 28$ ft.

(b)

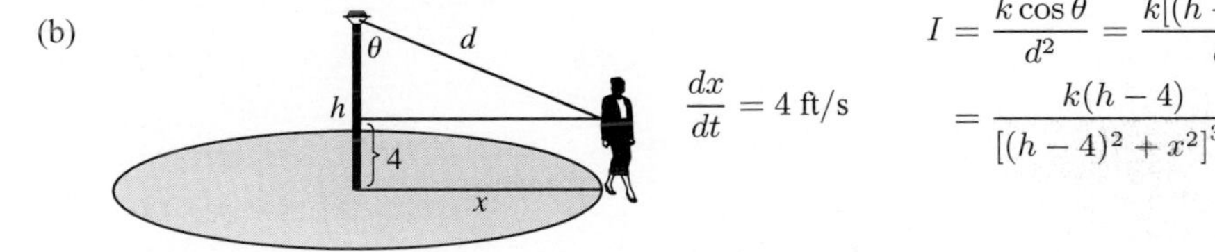

$\dfrac{dx}{dt} = 4$ ft/s

$$I = \frac{k\cos\theta}{d^2} = \frac{k[(h-4)/d]}{d^2} = \frac{k(h-4)}{d^3}$$
$$= \frac{k(h-4)}{[(h-4)^2+x^2]^{3/2}} = k(h-4)\left[(h-4)^2 + x^2\right]^{-3/2}$$

[continued]

$$\frac{dI}{dt} = \frac{dI}{dx} \cdot \frac{dx}{dt} = k(h-4)\left(-\tfrac{3}{2}\right)\left[(h-4)^2 + x^2\right]^{-5/2} \cdot 2x \cdot \frac{dx}{dt}$$

$$= k(h-4)(-3x)\left[(h-4)^2 + x^2\right]^{-5/2} \cdot 4 = \frac{-12xk(h-4)}{[(h-4)^2 + x^2]^{5/2}}$$

$$\left.\frac{dI}{dt}\right|_{x=40} = -\frac{480k(h-4)}{[(h-4)^2 + 1600]^{5/2}}$$

86. (a) $V'(t)$ is the rate of change of the volume of the water with respect to time. $H'(t)$ is the rate of change of the height of the water with respect to time. Since the volume and the height are increasing, $V'(t)$ and $H'(t)$ are positive.

(b) $V'(t)$ is constant, so $V''(t)$ is zero (the slope of a constant function is 0).

(c) At first, the height H of the water increases quickly because the tank is narrow. But as the sphere widens, the rate of increase of the height slows down, reaching a minimum at $t = t_2$. Thus, the height is increasing at a decreasing rate on $(0, t_2)$, so its graph is concave downward and $H''(t_1) < 0$. As the sphere narrows for $t > t_2$, the rate of increase of the height begins to increase, and the graph of H is concave upward. Therefore, $H''(t_2) = 0$ and $H''(t_3) > 0$.

PROBLEMS PLUS

1. Let $y = f(x) = e^{-x^2}$. The area of the rectangle under the curve from $-x$ to x is $A(x) = 2xe^{-x^2}$ where $x \geq 0$. We maximize $A(x)$: $A'(x) = 2e^{-x^2} - 4x^2e^{-x^2} = 2e^{-x^2}(1 - 2x^2) = 0 \Rightarrow x = \frac{1}{\sqrt{2}}$. This gives a maximum since $A'(x) > 0$ for $0 \leq x < \frac{1}{\sqrt{2}}$ and $A'(x) < 0$ for $x > \frac{1}{\sqrt{2}}$. We next determine the points of inflection of $f(x)$. Notice that $f'(x) = -2xe^{-x^2} = -A(x)$. So $f''(x) = -A'(x)$ and hence, $f''(x) < 0$ for $-\frac{1}{\sqrt{2}} < x < \frac{1}{\sqrt{2}}$ and $f''(x) > 0$ for $x < -\frac{1}{\sqrt{2}}$ and $x > \frac{1}{\sqrt{2}}$. So $f(x)$ changes concavity at $x = \pm\frac{1}{\sqrt{2}}$, and the two vertices of the rectangle of largest area are at the inflection points.

2. Let $f(x) = \sin x - \cos x$ on $[0, 2\pi]$ since f has period 2π. $f'(x) = \cos x + \sin x = 0 \quad \Leftrightarrow \quad \cos x = -\sin x \quad \Leftrightarrow$ $\tan x = -1 \quad \Leftrightarrow \quad x = \frac{3\pi}{4}$ or $\frac{7\pi}{4}$. Evaluating f at its critical numbers and endpoints, we get $f(0) = -1$, $f\left(\frac{3\pi}{4}\right) = \sqrt{2}$, $f\left(\frac{7\pi}{4}\right) = -\sqrt{2}$, and $f(2\pi) = -1$. So f has absolute maximum value $\sqrt{2}$ and absolute minimum value $-\sqrt{2}$. Thus, $-\sqrt{2} \leq \sin x - \cos x \leq \sqrt{2} \quad \Rightarrow \quad |\sin x - \cos x| \leq \sqrt{2}$.

3. $f(x)$ has the form $e^{g(x)}$, so it will have an absolute maximum (minimum) where g has an absolute maximum (minimum).

$$g(x) = 10|x-2| - x^2 = \begin{cases} 10(x-2) - x^2 & \text{if } x - 2 > 0 \\ 10[-(x-2)] - x^2 & \text{if } x - 2 < 0 \end{cases} = \begin{cases} -x^2 + 10x - 20 & \text{if } x > 2 \\ -x^2 - 10x + 20 & \text{if } x < 2 \end{cases} \quad \Rightarrow$$

$$g'(x) = \begin{cases} -2x + 10 & \text{if } x > 2 \\ -2x - 10 & \text{if } x < 2 \end{cases}$$

$g'(x) = 0$ if $x = -5$ or $x = 5$, and $g'(2)$ does not exist, so the critical numbers of g are -5, 2, and 5. Since $g''(x) = -2$ for all $x \neq 2$, g is concave downward on $(-\infty, 2)$ and $(2, \infty)$, and g will attain its absolute maximum at one of the critical numbers. Since $g(-5) = 45$, $g(2) = -4$, and $g(5) = 5$, we see that $f(-5) = e^{45}$ is the absolute maximum value of f. Also, $\lim\limits_{x \to \infty} g(x) = -\infty$, so $\lim\limits_{x \to \infty} f(x) = \lim\limits_{x \to \infty} e^{g(x)} = 0$. But $f(x) > 0$ for all x, so there is no absolute minimum value of f.

4. $x^2y^2(4 - x^2)(4 - y^2) = x^2(4 - x^2)y^2(4 - y^2) = f(x)f(y)$, where $f(t) = t^2(4 - t^2)$. We will show that $0 \leq f(t) \leq 4$ for $|t| \leq 2$, which gives $0 \leq f(x)f(y) \leq 16$ for $|x| \leq 2$ and $|y| \leq 2$.

$f(t) = 4t^2 - t^4 \quad \Rightarrow \quad f'(t) = 8t - 4t^3 = 4t(2 - t^2) = 0 \quad \Rightarrow \quad t = 0$ or $\pm\sqrt{2}$.

$f(0) = 0$, $f(\pm\sqrt{2}) = 2(4 - 2) = 4$, and $f(2) = 0$. So 0 is the absolute minimum value of $f(t)$ on $[-2, 2]$ and 4 is the absolute maximum value of $f(t)$ on $[-2, 2]$. We conclude that $0 \leq f(t) \leq 4$ for $|t| \leq 2$ and hence, $0 \leq f(x)f(y) \leq 4^2$ or $0 \leq x^2(4 - x^2)y^2(4 - y^2) \leq 16$.

5. $y = \dfrac{\sin x}{x} \;\Rightarrow\; y' = \dfrac{x\cos x - \sin x}{x^2} \;\Rightarrow\; y'' = \dfrac{-x^2\sin x - 2x\cos x + 2\sin x}{x^3}$. If (x, y) is an inflection point,

then $y'' = 0 \;\Rightarrow\; (2 - x^2)\sin x = 2x\cos x \;\Rightarrow\; (2 - x^2)^2\sin^2 x = 4x^2\cos^2 x \;\Rightarrow$

$(2 - x^2)^2\sin^2 x = 4x^2(1 - \sin^2 x) \;\Rightarrow\; (4 - 4x^2 + x^4)\sin^2 x = 4x^2 - 4x^2\sin^2 x \;\Rightarrow$

$(4 + x^4)\sin^2 x = 4x^2 \;\Rightarrow\; (x^4 + 4)\dfrac{\sin^2 x}{x^2} = 4 \;\Rightarrow\; y^2(x^4 + 4) = 4$ since $y = \dfrac{\sin x}{x}$.

6. Let $P(a, 1 - a^2)$ be the point of contact. The equation of the tangent line at P is $y - (1 - a^2) = (-2a)(x - a) \;\Rightarrow$

$y - 1 + a^2 = -2ax + 2a^2 \;\Rightarrow\; y = -2ax + a^2 + 1$. To find the x-intercept, put $y = 0$: $2ax = a^2 + 1 \;\Rightarrow$

$x = \dfrac{a^2 + 1}{2a}$. To find the y-intercept, put $x = 0$: $y = a^2 + 1$. Therefore, the area of the triangle is

$\dfrac{1}{2}\left(\dfrac{a^2 + 1}{2a}\right)(a^2 + 1) = \dfrac{(a^2 + 1)^2}{4a}$. Therefore, we minimize the function $A(a) = \dfrac{(a^2 + 1)^2}{4a}$, $a > 0$.

$$A'(a) = \frac{(4a)2(a^2 + 1)(2a) - (a^2 + 1)^2(4)}{16a^2} = \frac{(a^2 + 1)[4a^2 - (a^2 + 1)]}{4a^2} = \frac{(a^2 + 1)(3a^2 - 1)}{4a^2}.$$

$A'(a) = 0$ when $3a^2 - 1 = 0 \;\Rightarrow\; a = \frac{1}{\sqrt{3}}$. $A'(a) < 0$ for $a < \frac{1}{\sqrt{3}}$, $A'(a) > 0$ for $a > \frac{1}{\sqrt{3}}$. So by the First Derivative Test, there is an absolute minimum when $a = \frac{1}{\sqrt{3}}$. The required point is $\left(\frac{1}{\sqrt{3}}, \frac{2}{3}\right)$ and the corresponding minimum area is $A\left(\frac{1}{\sqrt{3}}\right) = \frac{4\sqrt{3}}{9}$.

7. Let $L = \lim\limits_{x\to 0} \dfrac{ax^2 + \sin bx + \sin cx + \sin dx}{3x^2 + 5x^4 + 7x^6}$. Now L has the indeterminate form of type $\frac{0}{0}$, so we can apply l'Hospital's Rule. $L = \lim\limits_{x\to 0} \dfrac{2ax + b\cos bx + c\cos cx + d\cos dx}{6x + 20x^3 + 42x^5}$. The denominator approaches 0 as $x \to 0$, so the numerator must also approach 0 (because the limit exists). But the numerator approaches $0 + b + c + d$, so $b + c + d = 0$. Apply l'Hospital's Rule again. $L = \lim\limits_{x\to 0} \dfrac{2a - b^2\sin bx - c^2\sin cx - d^2\sin dx}{6 + 60x^2 + 210x^4} = \dfrac{2a - 0}{6 + 0} = \dfrac{2a}{6}$, which must equal 8.

$\dfrac{2a}{6} = 8 \;\Rightarrow\; a = 24$. Thus, $a + b + c + d = a + (b + c + d) = 24 + 0 = 24$.

8. We first present some preliminary results that we will invoke when calculating the limit.

(1) If $y = (1 + ax)^x$, then $\ln y = x\ln(1 + ax)$, and $\lim\limits_{x\to 0^+} \ln y = \lim\limits_{x\to 0^+} x\ln(1 + ax) = 0$. Thus, $\lim\limits_{x\to 0^+} (1 + ax)^x = e^0 = 1$.

(2) If $y = (1 + ax)^x$, then $\ln y = x\ln(1 + ax)$, and implicitly differentiating gives us $\dfrac{y'}{y} = x \cdot \dfrac{a}{1 + ax} + \ln(1 + ax) \;\Rightarrow$

$y' = y\left[\dfrac{ax}{1 + ax} + \ln(1 + ax)\right]$. Thus, $y = (1 + ax)^x \;\Rightarrow\; y' = (1 + ax)^x\left[\dfrac{ax}{1 + ax} + \ln(1 + ax)\right]$.

(3) If $y = \dfrac{ax}{1 + ax}$, then $y' = \dfrac{(1 + ax)a - ax(a)}{(1 + ax)^2} = \dfrac{a + a^2x - a^2x}{(1 + ax)^2} = \dfrac{a}{(1 + ax)^2}$.

$$\lim_{x\to\infty}\frac{(x+2)^{1/x}-x^{1/x}}{(x+3)^{1/x}-x^{1/x}}=\lim_{x\to\infty}\frac{x^{1/x}[(1+2/x)^{1/x}-1]}{x^{1/x}[(1+3/x)^{1/x}-1]} \qquad \text{[factor out } x^{1/x}\text{]}$$

$$=\lim_{x\to\infty}\frac{(1+2/x)^{1/x}-1}{(1+3/x)^{1/x}-1}$$

$$=\lim_{t\to 0^+}\frac{(1+2t)^t-1}{(1+3t)^t-1} \qquad \text{[let } t=1/x\text{, form } 0/0 \text{ by } \mathbf{(1)}\text{]}$$

$$\overset{\text{H}}{=}\lim_{t\to 0^+}\frac{(1+2t)^t\left[\dfrac{2t}{1+2t}+\ln(1+2t)\right]}{(1+3t)^t\left[\dfrac{3t}{1+3t}+\ln(1+3t)\right]} \qquad \text{[by } \mathbf{(2)}\text{]}$$

$$=\lim_{t\to 0^+}\frac{(1+2t)^t}{(1+3t)^t}\cdot\lim_{t\to 0^+}\frac{\dfrac{2t}{1+2t}+\ln(1+2t)}{\dfrac{3t}{1+3t}+\ln(1+3t)}$$

$$=\frac{1}{1}\cdot\lim_{t\to 0^+}\frac{\dfrac{2t}{1+2t}+\ln(1+2t)}{\dfrac{3t}{1+3t}+\ln(1+3t)} \qquad \text{[by } \mathbf{(1)}\text{, now form } 0/0\text{]}$$

$$\overset{\text{H}}{=}\lim_{t\to 0^+}\frac{\dfrac{2}{(1+2t)^2}+\dfrac{2}{1+2t}}{\dfrac{3}{(1+3t)^2}+\dfrac{3}{1+3t}} \qquad \text{[by } \mathbf{(3)}\text{]}$$

$$=\frac{2+2}{3+3}=\frac{4}{6}=\frac{2}{3}$$

9. Differentiating $x^2+xy+y^2=12$ implicitly with respect to x gives $2x+y+x\dfrac{dy}{dx}+2y\dfrac{dy}{dx}=0$, so $\dfrac{dy}{dx}=-\dfrac{2x+y}{x+2y}$.

At a highest or lowest point, $\dfrac{dy}{dx}=0 \quad\Leftrightarrow\quad y=-2x$. Substituting $-2x$ for y in the original equation gives $x^2+x(-2x)+(-2x)^2=12$, so $3x^2=12$ and $x=\pm 2$. If $x=2$, then $y=-2x=-4$, and if $x=-2$ then $y=4$. Thus, the highest and lowest points are $(-2,4)$ and $(2,-4)$.

10. *Case (i) (first graph):* For $x+y\geq 0$, that is, $y\geq -x$, $|x+y|=x+y\leq e^x \quad\Rightarrow\quad y\leq e^x-x$.

Note that $y=e^x-x$ is always above the line $y=-x$ and that $y=-x$ is a slant asymptote.

Case (ii) (second graph): For $x+y<0$, that is, $y<-x$, $|x+y|=-x-y\leq e^x \quad\Rightarrow\quad y\geq -x-e^x$.

Note that $-x-e^x$ is always below the line $y=-x$ and $y=-x$ is a slant asymptote.

Putting the two pieces together gives the third graph.

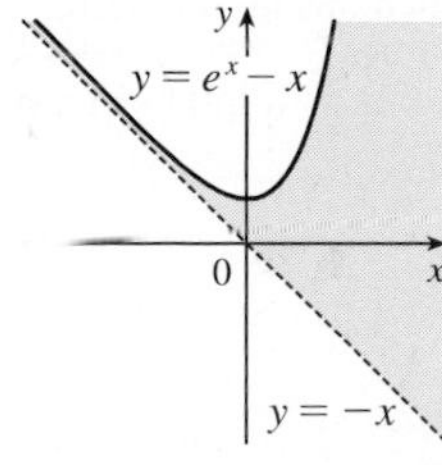

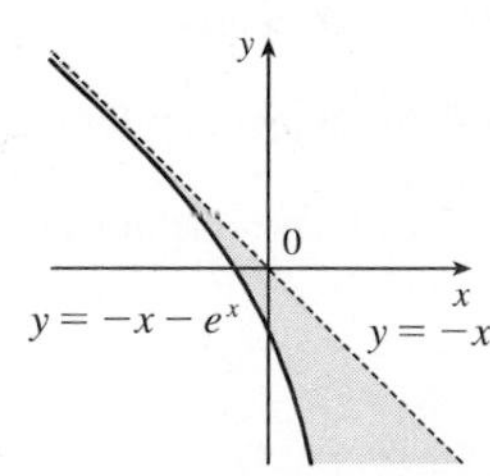

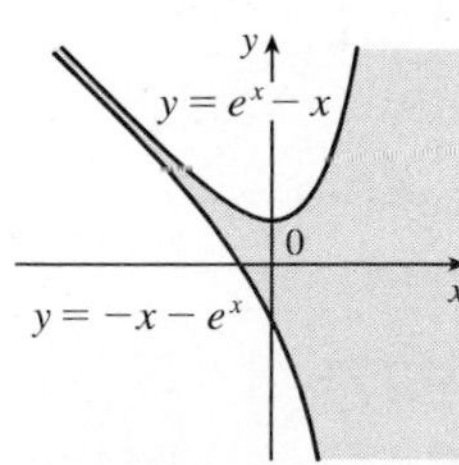

11. (a) $y = x^2 \;\Rightarrow\; y' = 2x$, so the slope of the tangent line at $P(a, a^2)$ is $2a$ and the slope of the normal line is $-\dfrac{1}{2a}$ for $a \neq 0$. An equation of the normal line is $y - a^2 = -\dfrac{1}{2a}(x - a)$. Substitute x^2 for y to find the x-coordinates of the two points of intersection of the parabola and the normal line. $x^2 - a^2 = -\dfrac{x}{2a} + \dfrac{1}{2} \;\Leftrightarrow\; x^2 + \left(\dfrac{1}{2a}\right)x - \dfrac{1}{2} - a^2 = 0$. We know that a is a root of this quadratic equation, so $x - a$ is a factor, and we have $(x - a)\left(x + \dfrac{1}{2a} + a\right) = 0$, and hence, $x = -a - \dfrac{1}{2a}$ is the x-coordinate of the point Q. We want to minimize the y-coordinate of Q, which is

$$\left(-a - \frac{1}{2a}\right)^2 = a^2 + 1 + \frac{1}{4a^2} = y(a). \text{ Now } y'(a) = 2a - \frac{1}{2a^3} = \frac{4a^4 - 1}{2a^3} = \frac{(2a^2+1)(2a^2-1)}{2a^3} = 0 \;\Rightarrow$$

$a = \dfrac{1}{\sqrt{2}}$ for $a > 0$. Since $y''(a) = 2 + \dfrac{3}{2a^4} > 0$, we see that $a = \dfrac{1}{\sqrt{2}}$ gives us the minimum value of the y-coordinate of Q.

(b) The square S of the distance from $P(a, a^2)$ to $Q\left(-a - \dfrac{1}{2a}, \left(-a - \dfrac{1}{2a}\right)^2\right)$ is given by

$$\begin{aligned}
S &= \left(-a - \frac{1}{2a} - a\right)^2 + \left[\left(-a - \frac{1}{2a}\right)^2 - a^2\right]^2 = \left(-2a - \frac{1}{2a}\right)^2 + \left[\left(a^2 + 1 + \frac{1}{4a^2}\right) - a^2\right]^2 \\
&= \left(4a^2 + 2 + \frac{1}{4a^2}\right) + \left(1 + \frac{1}{4a^2}\right)^2 = \left(4a^2 + 2 + \frac{1}{4a^2}\right) + 1 + \frac{2}{4a^2} + \frac{1}{16a^4} \\
&= 4a^2 + 3 + \frac{3}{4a^2} + \frac{1}{16a^4}
\end{aligned}$$

$S' = 8a - \dfrac{6}{4a^3} - \dfrac{4}{16a^5} = 8a - \dfrac{3}{2a^3} - \dfrac{1}{4a^5} = \dfrac{32a^6 - 6a^2 - 1}{4a^5} = \dfrac{(2a^2-1)(4a^2+1)^2}{4a^5}$. The only real positive zero of the equation $S' = 0$ is $a = \dfrac{1}{\sqrt{2}}$. Since $S'' = 8 + \dfrac{9}{2a^4} + \dfrac{5}{4a^6} > 0$, $a = \dfrac{1}{\sqrt{2}}$ corresponds to the shortest possible length of the line segment PQ.

12. $y = cx^3 + e^x \;\Rightarrow\; y' = 3cx^2 + e^x \;\Rightarrow\; y'' = 6cx + e^x$. The curve will have inflection points when y'' changes sign. $y'' = 0 \;\Rightarrow\; -6cx = e^x$, so y'' will change sign when the line $y = -6cx$ intersects the curve $y = e^x$ (but is not tangent to it). Note that if $c = 0$, the curve is just $y = e^x$, which has no inflection point.

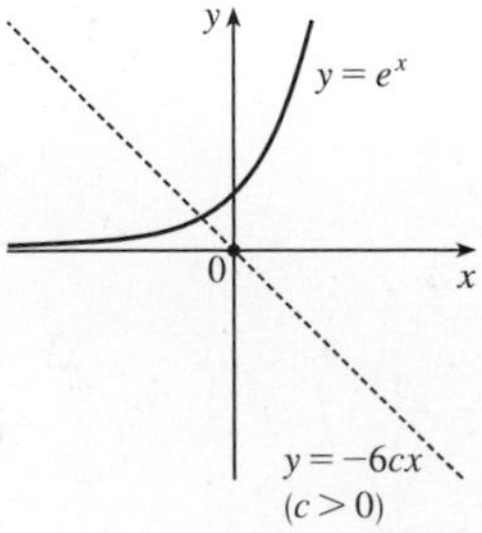

The first figure shows that for $c > 0$, $y = -6cx$ will intersect $y = e^x$ once, so $y = cx^3 + e^x$ will have one inflection point.

[continued]

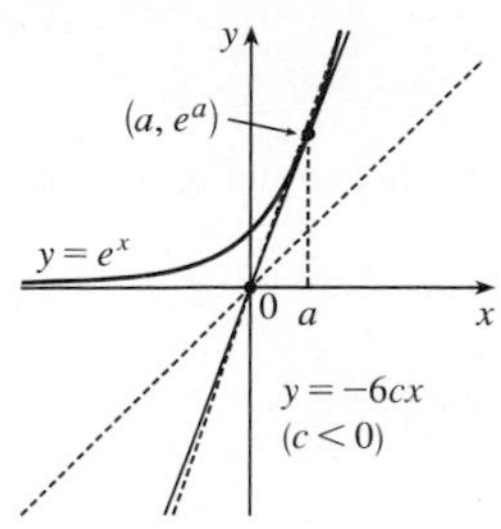

The second figure shows that for $c < 0$, the line $y = -6cx$ can intersect the curve $y = e^x$ in two points (two inflection points), be tangent to it (no inflection point), or not intersect it (no inflection point). The tangent line at (a, e^a) has slope e^a, but from the diagram we see that the slope is $\frac{e^a}{a}$. So $\frac{e^a}{a} = e^a \;\Rightarrow\; a = 1$. Thus, the slope is e. The line $y = -6cx$ must have slope greater than e, so $-6c > e \;\Rightarrow\; c < -e/6$.

Therefore, the curve $y = cx^3 + e^x$ will have one inflection point if $c > 0$ and two inflection points if $c < -e/6$.

13.

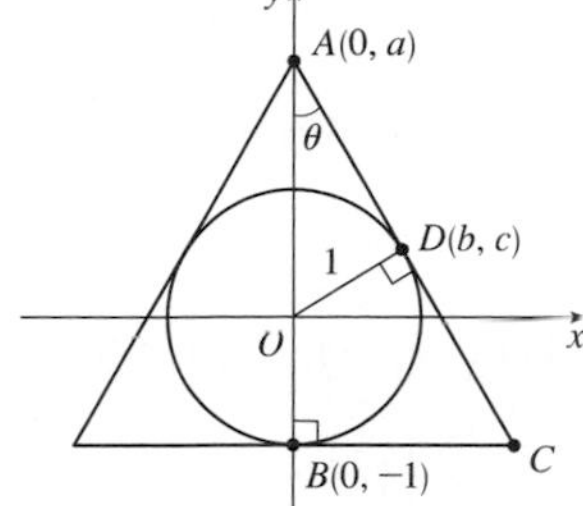

$\overline{AC}$ is tangent to the unit circle at D. To find the slope of $\overline{AC}$ at D, use implicit differentiation. $x^2 + y^2 = 1 \;\Rightarrow\; 2x + 2y\,y' = 0 \;\Rightarrow\; y\,y' = -x \;\Rightarrow\; y' = -\frac{x}{y}$.

Thus, the tangent line at $D(b, c)$ has equation $y = -\frac{b}{c}\,x + a$. At D, $x = b$ and $y = c$, so $c = -\frac{b}{c}(b) + a \;\Rightarrow\; a = c + \frac{b^2}{c} = \frac{c^2 + b^2}{c} = \frac{1}{c}$, and hence $c = \frac{1}{a}$.

Since $b^2 + c^2 = 1$, $b = \sqrt{1 - c^2} = \sqrt{1 - 1/a^2} = \sqrt{\frac{a^2 - 1}{a^2}} = \frac{\sqrt{a^2 - 1}}{a}$, and now we have both b and c in terms of a. At C, $y = -1$, so $-1 = -\frac{b}{c}\,x + a \;\Rightarrow\; \frac{b}{c}\,x = a + 1 \;\Rightarrow$

$x = \frac{c}{b}(a + 1) = \frac{1/a}{\sqrt{a^2 - 1}/a}(a + 1) = \frac{a + 1}{\sqrt{(a + 1)(a - 1)}} = \sqrt{\frac{a + 1}{a - 1}}$, and C has coordinates $\left(\sqrt{\frac{a + 1}{a - 1}}, -1\right)$. Let S be the square of the distance from A to C. Then $S(a) = \left(0 - \sqrt{\frac{a + 1}{a - 1}}\right)^2 + (a + 1)^2 = \frac{a + 1}{a - 1} + (a + 1)^2 \;\Rightarrow$

$$S'(a) = \frac{(a - 1)(1) - (a + 1)(1)}{(a - 1)^2} + 2(a + 1) = \frac{-2 + 2(a + 1)(a - 1)^2}{(a - 1)^2}$$

$$= \frac{-2 + 2(a^3 - a^2 - a + 1)}{(a - 1)^2} = \frac{2a^3 - 2a^2 - 2a}{(a - 1)^2} = \frac{2a(a^2 - a - 1)}{(a - 1)^2}$$

Using the quadratic formula, we find that the solutions of $a^2 - a - 1 = 0$ are $a = \frac{1 \pm \sqrt{5}}{2}$, so $a_1 = \frac{1 + \sqrt{5}}{2}$ (the "golden mean") since $a > 0$. For $1 < a < a_1$, $S'(a) < 0$, and for $a > a_1$, $S'(a) > 0$, so a_1 minimizes S.

Note: The minimum length of the equal sides is $\sqrt{S(a_1)} = \cdots = \sqrt{\frac{11 + 5\sqrt{5}}{2}} \approx 3.33$ and the corresponding length of the third side is $2\sqrt{\frac{a_1 + 1}{a_1 - 1}} = \cdots = 2\sqrt{2 + \sqrt{5}} \approx 4.12$, so the triangle is *not* equilateral.

Another method: In ΔABC, $\cos\theta = \frac{a + 1}{\overline{AC}}$, so $\overline{AC} = \frac{a + 1}{\cos\theta}$. In ΔADO, $\sin\theta = \frac{1}{a}$, so

$\cos\theta = \sqrt{1 - \sin^2\theta} = \sqrt{1 - 1/a^2} = \frac{1}{a}\sqrt{a^2 - 1}$. Thus $\overline{AC} = \frac{a + 1}{(1/a)\sqrt{a^2 - 1}} = \frac{a(a + 1)}{\sqrt{a^2 - 1}} = f(a)$. Now find the minimum of f.

14. To sketch the region $\{(x,y) \mid 2xy \le |x-y| \le x^2+y^2\}$, we consider two cases.

Case 1: $x \ge y$ This is the case in which (x,y) lies on or below the line $y=x$. The double inequality becomes $2xy \le x-y \le x^2+y^2$. The right-hand inequality holds if and only if $x^2-x+y^2+y \ge 0 \iff (x-\frac{1}{2})^2+(y+\frac{1}{2})^2 \ge \frac{1}{2} \iff (x,y)$ lies on or outside the circle with radius $\frac{1}{\sqrt{2}}$ centered at $(\frac{1}{2},-\frac{1}{2})$. The left-hand inequality holds if and only if $2xy-x+y \le 0 \iff xy-\frac{1}{2}x+\frac{1}{2}y \le 0 \iff (x+\frac{1}{2})(y-\frac{1}{2}) \le -\frac{1}{4} \iff (x,y)$ lies on or below the hyperbola $(x+\frac{1}{2})(y-\frac{1}{2})=-\frac{1}{4}$, which passes through the origin and approaches the lines $y=\frac{1}{2}$ and $x=-\frac{1}{2}$ asymptotically.

Case 2: $y \ge x$ This is the case in which (x,y) lies on or above the line $y=x$. The double inequality becomes $2xy \le y-x \le x^2+y^2$. The right-hand inequality holds if and only if $x^2+x+y^2-y \ge 0 \iff (x+\frac{1}{2})^2+(y-\frac{1}{2})^2 \ge \frac{1}{2} \iff (x,y)$ lies on or outside the circle of radius $\frac{1}{\sqrt{2}}$ centered at $(-\frac{1}{2},\frac{1}{2})$. The left-hand inequality holds if and only if $2xy+x-y \le 0 \iff xy+\frac{1}{2}x-\frac{1}{2}y \le 0 \iff (x-\frac{1}{2})(y+\frac{1}{2}) \le -\frac{1}{4} \iff (x,y)$ lies on or above the left-hand branch of the hyperbola $(x-\frac{1}{2})(y+\frac{1}{2})=-\frac{1}{4}$, which passes through the origin and approaches the lines $y=-\frac{1}{2}$ and $x=\frac{1}{2}$ asymptotically. Therefore, the region of interest consists of the points on or above the left branch of the hyperbola $(x-\frac{1}{2})(y+\frac{1}{2})=-\frac{1}{4}$ that are on or outside the circle $(x+\frac{1}{2})^2+(y-\frac{1}{2})^2=\frac{1}{2}$, together with the points on or below the right branch of the hyperbola $(x+\frac{1}{2})(y-\frac{1}{2})=-\frac{1}{4}$ that are on or outside the circle $(x-\frac{1}{2})^2+(y+\frac{1}{2})^2=\frac{1}{2}$. Note that the inequalities are unchanged when x and y are interchanged, so the region is symmetric about the line $y=x$. So we need only have analyzed case 1 and then reflected that region about the line $y=x$, instead of considering case 2.

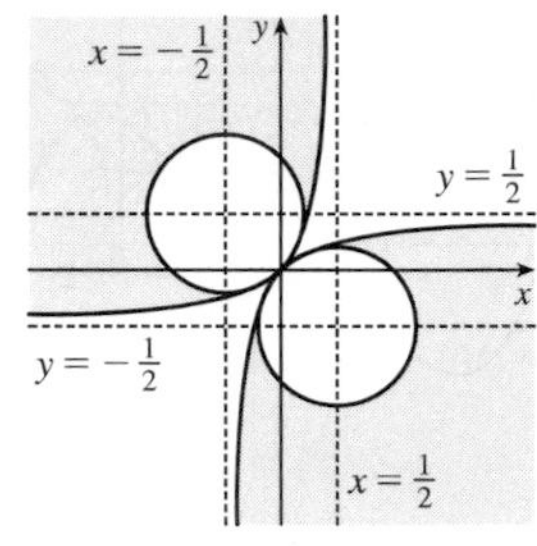

15. $A=(x_1,x_1^2)$ and $B=(x_2,x_2^2)$, where x_1 and x_2 are the solutions of the quadratic equation $x^2=mx+b$. Let $P=(x,x^2)$ and set $A_1=(x_1,0)$, $B_1=(x_2,0)$, and $P_1=(x,0)$. Let $f(x)$ denote the area of triangle PAB. Then $f(x)$ can be expressed in terms of the areas of three trapezoids as follows:

$$\begin{aligned} f(x) &= \text{area}\,(A_1ABB_1) - \text{area}\,(A_1APP_1) - \text{area}\,(B_1BPP_1) \\ &= \tfrac{1}{2}(x_1^2+x_2^2)(x_2-x_1) - \tfrac{1}{2}(x_1^2+x^2)(x-x_1) - \tfrac{1}{2}(x^2+x_2^2)(x_2-x) \end{aligned}$$

After expanding and canceling terms, we get

$f(x)=\frac{1}{2}(x_2x_1^2-x_1x_2^2-xx_1^2+x_1x^2-x_2x^2+xx_2^2)=\frac{1}{2}\left[x_1^2(x_2-x)+x_2^2(x-x_1)+x^2(x_1-x_2)\right]$

$f'(x)=\frac{1}{2}\left[-x_1^2+x_2^2+2x(x_1-x_2)\right]$. $f''(x)=\frac{1}{2}[2(x_1-x_2)]=x_1-x_2<0$ since $x_2>x_1$.

$f'(x)=0 \Rightarrow 2x(x_1-x_2)=x_1^2-x_2^2 \Rightarrow x_P=\frac{1}{2}(x_1+x_2)$.

$$\begin{aligned} f(x_P) &= \tfrac{1}{2}\left(x_1^2\left[\tfrac{1}{2}(x_2-x_1)\right] + x_2^2\left[\tfrac{1}{2}(x_2-x_1)\right] + \tfrac{1}{4}(x_1+x_2)^2(x_1-x_2)\right) \\ &= \tfrac{1}{2}\left[\tfrac{1}{2}(x_2-x_1)\left(x_1^2+x_2^2\right) - \tfrac{1}{4}(x_2-x_1)(x_1+x_2)^2\right] = \tfrac{1}{8}(x_2-x_1)\left[2\left(x_1^2+x_2^2\right) - \left(x_1^2+2x_1x_2+x_2^2\right)\right] \\ &= \tfrac{1}{8}(x_2-x_1)\left(x_1^2-2x_1x_2+x_2^2\right) = \tfrac{1}{8}(x_2-x_1)(x_1-x_2)^2 = \tfrac{1}{8}(x_2-x_1)(x_2-x_1)^2 = \tfrac{1}{8}(x_2-x_1)^3 \end{aligned}$$

To put this in terms of m and b, we solve the system $y = x_1^2$ and $y = mx_1 + b$, giving us $x_1^2 - mx_1 - b = 0 \Rightarrow$ $x_1 = \frac{1}{2}\left(m - \sqrt{m^2+4b}\right)$. Similarly, $x_2 = \frac{1}{2}\left(m + \sqrt{m^2+4b}\right)$. The area is then $\frac{1}{8}(x_2-x_1)^3 = \frac{1}{8}\left(\sqrt{m^2+4b}\right)^3$, and is attained at the point $P\left(x_P, x_P^2\right) = P\left(\frac{1}{2}m, \frac{1}{4}m^2\right)$.

Note: Another way to get an expression for $f(x)$ is to use the formula for an area of a triangle in terms of the coordinates of the vertices: $f(x) = \frac{1}{2}\left[\left(x_2x_1^2 - x_1x_2^2\right) + \left(x_1x^2 - xx_1^2\right) + \left(xx_2^2 - x_2x^2\right)\right]$.

16. Let $x = |AE|$, $y = |AF|$ as shown. The area $\mathcal{A}$ of the ΔAEF is $\mathcal{A} = \frac{1}{2}xy$. We need to find a relationship between x and y, so that we can take the derivative $d\mathcal{A}/dx$ and then find the maximum and minimum areas. Now let A' be the point on which A ends up after the fold has been performed, and let P be the intersection of AA' and EF. Note that AA' is perpendicular to EF since we are reflecting A through the line EF to get to A', and that $|AP| = |PA'|$ for the same reason.

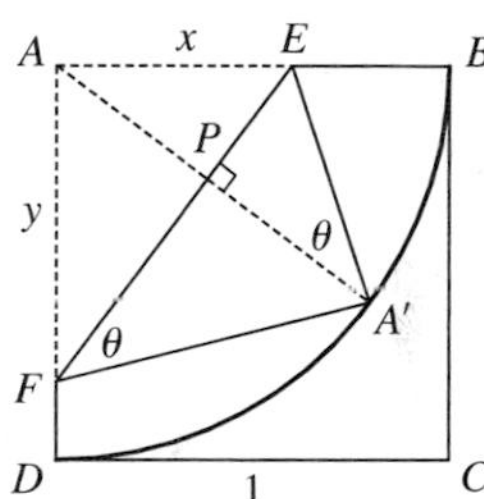

But $|AA'| = 1$, since AA' is a radius of the circle. Since $|AP| + |PA'| = |AA'|$, we have $|AP| = \frac{1}{2}$. Another way to express the area of the triangle is $\mathcal{A} = \frac{1}{2}|EF|\,|AP| = \frac{1}{2}\sqrt{x^2+y^2}\left(\frac{1}{2}\right) = \frac{1}{4}\sqrt{x^2+y^2}$. Equating the two expressions for $\mathcal{A}$, we get $\frac{1}{2}xy = \frac{1}{4}\sqrt{x^2+y^2} \Rightarrow 4x^2y^2 = x^2+y^2 \Rightarrow y^2\left(4x^2-1\right) = x^2 \Rightarrow y = x/\sqrt{4x^2-1}$.

(Note that we could also have derived this result from the similarity of $\triangle A'PE$ and $\triangle A'FE$; that is,

$$\frac{|A'P|}{|PE|} = \frac{|A'F|}{|A'E|} \Rightarrow \frac{\frac{1}{2}}{\sqrt{x^2-\left(\frac{1}{2}\right)^2}} = \frac{y}{x} \Rightarrow y = \frac{\frac{1}{2}x}{\sqrt{4x^2-1}/2} = \frac{x}{\sqrt{4x^2-1}}.$$) Now we can substitute for y and calculate $\dfrac{d\mathcal{A}}{dx}$: $\mathcal{A} = \dfrac{1}{2}\dfrac{x^2}{\sqrt{4x^2-1}} \Rightarrow \dfrac{d\mathcal{A}}{dx} = \dfrac{1}{2}\left[\dfrac{\sqrt{4x^2-1}\,(2x) - x^2\left(\frac{1}{2}\right)\left(4x^2-1\right)^{-1/2}(8x)}{4x^2-1}\right]$. This is 0 when

$2x\sqrt{4x^2-1} - 4x^3\left(4x^2-1\right)^{-1/2} = 0 \Leftrightarrow 2x\left(4x^2-1\right)^{-1/2}\left[\left(4x^2-1\right) - 2x^2\right] = 0 \Rightarrow \left(4x^2-1\right) - 2x^2 = 0$ $(x > 0) \Leftrightarrow 2x^2 = 1 \Rightarrow x = \frac{1}{\sqrt{2}}$. So this is one possible value for an extremum. We must also test the endpoints of the interval over which x ranges. The largest value that x can attain is 1, and the smallest value of x occurs when $y = 1 \Leftrightarrow$ $1 = x/\sqrt{4x^2-1} \Leftrightarrow x^2 = 4x^2-1 \Leftrightarrow 3x^2 = 1 \Leftrightarrow x = \frac{1}{\sqrt{3}}$. This will give the same value of $\mathcal{A}$ as will $x = 1$, since the geometric situation is the same (reflected through the line $y = x$). We calculate

$\mathcal{A}\left(\frac{1}{\sqrt{2}}\right) = \dfrac{1}{2}\dfrac{\left(1/\sqrt{2}\right)^2}{\sqrt{4\left(1/\sqrt{2}\right)^2-1}} = \dfrac{1}{4} = 0.25$, and $\mathcal{A}(1) = \dfrac{1}{2}\dfrac{1^2}{\sqrt{4(1)^2-1}} = \dfrac{1}{2\sqrt{3}} \approx 0.29$. So the maximum area is

$\mathcal{A}(1) = \mathcal{A}\left(\frac{1}{\sqrt{3}}\right) = \frac{1}{2\sqrt{3}}$ and the minimum area is $\mathcal{A}\left(\frac{1}{\sqrt{2}}\right) = \frac{1}{4}$.

[continued]

Another method: Use the angle θ (see diagram above) as a variable:

$\mathcal{A} = \frac{1}{2}xy = \frac{1}{2}\left(\frac{1}{2}\sec\theta\right)\left(\frac{1}{2}\csc\theta\right) = \dfrac{1}{8\sin\theta\cos\theta} = \dfrac{1}{4\sin 2\theta}$. $\mathcal{A}$ is minimized when $\sin 2\theta$ is maximal, that is, when $\sin 2\theta = 1 \Rightarrow 2\theta = \frac{\pi}{2} \Rightarrow \theta = \frac{\pi}{4}$. Also note that $A'E = x = \frac{1}{2}\sec\theta \le 1 \Rightarrow \sec\theta \le 2 \Rightarrow \cos\theta \ge \frac{1}{2} \Rightarrow \theta \le \frac{\pi}{3}$, and similarly, $A'F = y = \frac{1}{2}\csc\theta \le 1 \Rightarrow \csc\theta \le 2 \Rightarrow \sin\theta \le \frac{1}{2} \Rightarrow \theta \ge \frac{\pi}{6}$.

As above, we find that $\mathcal{A}$ is maximized at these endpoints: $\mathcal{A}\left(\frac{\pi}{6}\right) = \dfrac{1}{4\sin\frac{\pi}{3}} = \dfrac{1}{2\sqrt{3}} = \dfrac{1}{4\sin\frac{2\pi}{3}} = \mathcal{A}\left(\frac{\pi}{3}\right)$; and minimized at $\theta = \frac{\pi}{4}$: $\mathcal{A}\left(\frac{\pi}{4}\right) = \dfrac{1}{4\sin\frac{\pi}{2}} = \dfrac{1}{4}$.

17. Suppose that the curve $y = a^x$ intersects the line $y = x$. Then $a^{x_0} = x_0$ for some $x_0 > 0$, and hence $a = x_0^{1/x_0}$. We find the maximum value of $g(x) = x^{1/x}$, $x > 0$, because if a is larger than the maximum value of this function, then the curve $y = a^x$ does not intersect the line $y = x$. $g'(x) = e^{(1/x)\ln x}\left(-\dfrac{1}{x^2}\ln x + \dfrac{1}{x}\cdot\dfrac{1}{x}\right) = x^{1/x}\left(\dfrac{1}{x^2}\right)(1 - \ln x)$. This is 0 only where $x = e$, and for $0 < x < e$, $f'(x) > 0$, while for $x > e$, $f'(x) < 0$, so g has an absolute maximum of $g(e) = e^{1/e}$. So if $y = a^x$ intersects $y = x$, we must have $0 < a \le e^{1/e}$. Conversely, suppose that $0 < a \le e^{1/e}$. Then $a^e \le e$, so the graph of $y = a^x$ lies below or touches the graph of $y = x$ at $x = e$. Also $a^0 = 1 > 0$, so the graph of $y = a^x$ lies above that of $y = x$ at $x = 0$. Therefore, by the Intermediate Value Theorem, the graphs of $y = a^x$ and $y = x$ must intersect somewhere between $x = 0$ and $x = e$.

18. If $L = \lim\limits_{x\to\infty}\left(\dfrac{x+a}{x-a}\right)^x$, then L has the indeterminate form 1^∞, so

$$\ln L = \lim_{x\to\infty}\ln\left(\frac{x+a}{x-a}\right)^x = \lim_{x\to\infty} x\ln\left(\frac{x+a}{x-a}\right) = \lim_{x\to\infty}\frac{\ln(x+a) - \ln(x-a)}{1/x} \overset{\text{H}}{=} \lim_{x\to\infty}\frac{\dfrac{1}{x+a} - \dfrac{1}{x-a}}{-1/x^2}$$

$$= \lim_{x\to\infty}\left[\frac{(x-a)-(x+a)}{(x+a)(x-a)}\cdot\frac{-x^2}{1}\right] = \lim_{x\to\infty}\frac{2ax^2}{x^2 - a^2} = \lim_{x\to\infty}\frac{2a}{1 - a^2/x^2} = 2a$$

Hence, $\ln L = 2a$, so $L = e^{2a}$. From the original equation, we want $L = e^1 \Rightarrow 2a = 1 \Rightarrow a = \frac{1}{2}$.

19. Note that $f(0) = 0$, so for $x \ne 0$, $\left|\dfrac{f(x) - f(0)}{x - 0}\right| = \left|\dfrac{f(x)}{x}\right| = \dfrac{|f(x)|}{|x|} \le \dfrac{|\sin x|}{|x|} = \dfrac{\sin x}{x}$.

Therefore, $|f'(0)| = \left|\lim\limits_{x\to 0}\dfrac{f(x) - f(0)}{x - 0}\right| = \lim\limits_{x\to 0}\left|\dfrac{f(x) - f(0)}{x - 0}\right| \le \lim\limits_{x\to 0}\dfrac{\sin x}{x} = 1$.

But $f(x) = a_1\sin x + a_2\sin 2x + \cdots + a_n\sin nx \Rightarrow f'(x) = a_1\cos x + 2a_2\cos 2x + \cdots + na_n\cos nx$, so $|f'(0)| = |a_1 + 2a_2 + \cdots + na_n| \le 1$.

Another solution: We are given that $\left|\sum_{k=1}^n a_k\sin kx\right| \le |\sin x|$. So for x close to 0, and $x \ne 0$, we have

$\left|\sum\limits_{k=1}^n a_k\dfrac{\sin kx}{\sin x}\right| \le 1 \Rightarrow \lim\limits_{x\to 0}\left|\sum\limits_{k=1}^n a_k\dfrac{\sin kx}{\sin x}\right| \le 1 \Rightarrow \left|\sum\limits_{k=1}^n a_k\lim\limits_{x\to 0}\dfrac{\sin kx}{\sin x}\right| \le 1$. But by l'Hospital's Rule, $\lim\limits_{x\to 0}\dfrac{\sin kx}{\sin x} = \lim\limits_{x\to 0}\dfrac{k\cos kx}{\cos x} = k$, so $\left|\sum\limits_{k=1}^n ka_k\right| \le 1$.

20. Let the circle have radius r, so $|OP| = |OQ| = r$, where O is the center of the circle. Now $\angle POR$ has measure $\frac{1}{2}\theta$, and $\angle OPR$ is a right angle, so $\tan\frac{1}{2}\theta = \dfrac{|PR|}{r}$ and the area of $\triangle OPR$ is $\frac{1}{2}|OP|\,|PR| = \frac{1}{2}r^2\tan\frac{1}{2}\theta$. The area of the sector cut by OP and OR is $\frac{1}{2}r^2\left(\frac{1}{2}\theta\right) = \frac{1}{4}r^2\theta$. Let S be the intersection of PQ and OR. Then $\sin\frac{1}{2}\theta = \dfrac{|PS|}{r}$ and $\cos\frac{1}{2}\theta = \dfrac{|OS|}{r}$, and the area of $\triangle OSP$ is $\frac{1}{2}|OS|\,|PS| = \frac{1}{2}\left(r\cos\frac{1}{2}\theta\right)\left(r\sin\frac{1}{2}\theta\right) = \frac{1}{2}r^2\sin\frac{1}{2}\theta\cos\frac{1}{2}\theta = \frac{1}{4}r^2\sin\theta$.

So $B(\theta) = 2\left(\frac{1}{2}r^2\tan\frac{1}{2}\theta - \frac{1}{4}r^2\theta\right) = r^2\left(\tan\frac{1}{2}\theta - \frac{1}{2}\theta\right)$ and $A(\theta) = 2\left(\frac{1}{4}r^2\theta - \frac{1}{4}r^2\sin\theta\right) = \frac{1}{2}r^2(\theta - \sin\theta)$. Thus,

$$\lim_{\theta\to 0^+}\frac{A(\theta)}{B(\theta)} = \lim_{\theta\to 0^+}\frac{\frac{1}{2}r^2(\theta-\sin\theta)}{r^2\left(\tan\frac{1}{2}\theta - \frac{1}{2}\theta\right)} = \lim_{\theta\to 0^+}\frac{\theta-\sin\theta}{2\left(\tan\frac{1}{2}\theta-\frac{1}{2}\theta\right)} \overset{\text{H}}{=} \lim_{\theta\to 0^+}\frac{1-\cos\theta}{2\left(\frac{1}{2}\sec^2\frac{1}{2}\theta - \frac{1}{2}\right)}$$

$$= \lim_{\theta\to 0^+}\frac{1-\cos\theta}{\sec^2\frac{1}{2}\theta - 1} = \lim_{\theta\to 0^+}\frac{1-\cos\theta}{\tan^2\frac{1}{2}\theta} \overset{\text{H}}{=} \lim_{\theta\to 0^+}\frac{\sin\theta}{2\left(\tan\frac{1}{2}\theta\right)\left(\sec^2\frac{1}{2}\theta\right)\frac{1}{2}}$$

$$= \lim_{\theta\to 0^+}\frac{\sin\theta\cos^3\frac{1}{2}\theta}{\sin\frac{1}{2}\theta} = \lim_{\theta\to 0^+}\frac{\left(2\sin\frac{1}{2}\theta\cos\frac{1}{2}\theta\right)\cos^3\frac{1}{2}\theta}{\sin\frac{1}{2}\theta} = 2\lim_{\theta\to 0^+}\cos^4\left(\tfrac{1}{2}\theta\right) = 2(1)^4 = 2$$

21. (a) Distance $=$ rate $\times$ time, so time $=$ distance/rate. $T_1 = \dfrac{D}{c_1}$, $T_2 = \dfrac{2|PR|}{c_1} + \dfrac{|RS|}{c_2} = \dfrac{2h\sec\theta}{c_1} + \dfrac{D - 2h\tan\theta}{c_2}$,

$$T_3 = \frac{2\sqrt{h^2 + D^2/4}}{c_1} = \frac{\sqrt{4h^2+D^2}}{c_1}.$$

(b) $\dfrac{dT_2}{d\theta} = \dfrac{2h}{c_1}\cdot\sec\theta\tan\theta - \dfrac{2h}{c_2}\sec^2\theta = 0$ when $2h\sec\theta\left(\dfrac{1}{c_1}\tan\theta - \dfrac{1}{c_2}\sec\theta\right) = 0 \;\rightarrow$

$\dfrac{1}{c_1}\dfrac{\sin\theta}{\cos\theta} - \dfrac{1}{c_2}\dfrac{1}{\cos\theta} = 0 \;\rightarrow\; \dfrac{\sin\theta}{c_1\cos\theta} = \dfrac{1}{c_2\cos\theta} \;\Rightarrow\; \sin\theta = \dfrac{c_1}{c_2}$. The First Derivative Test shows that this gives a minimum.

(c) Using part (a) with $D = 1$ and $T_1 = 0.26$, we have $T_1 = \dfrac{D}{c_1} \;\Rightarrow\; c_1 = \frac{1}{0.26} \approx 3.85$ km/s. $T_3 = \dfrac{\sqrt{4h^2+D^2}}{c_1} \;\Rightarrow$ $4h^2 + D^2 - T_3^2c_1^2 \;\Rightarrow\; h = \frac{1}{2}\sqrt{T_3^2c_1^2 - D^2} = \frac{1}{2}\sqrt{(0.34)^2(1/0.26)^2 - 1^2} \approx 0.42$ km. To find c_2, we use $\sin\theta = \dfrac{c_1}{c_2}$ from part (b) and $T_2 = \dfrac{2h\sec\theta}{c_1} + \dfrac{D-2h\tan\theta}{c_2}$ from part (a). From the figure,

$\sin\theta = \dfrac{c_1}{c_2} \;\Rightarrow\; \sec\theta = \dfrac{c_2}{\sqrt{c_2^2 - c_1^2}}$ and $\tan\theta = \dfrac{c_1}{\sqrt{c_2^2-c_1^2}}$, so

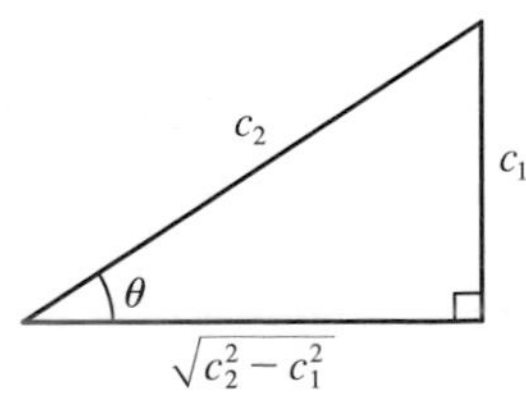

$T_2 = \dfrac{2hc_2}{c_1\sqrt{c_2^2-c_1^2}} + \dfrac{D\sqrt{c_2^2-c_1^2} - 2hc_1}{c_2\sqrt{c_2^2-c_1^2}}$. Using the values for T_2 [given as 0.32], h, c_1, and D, we can graph $Y_1 = T_2$ and $Y_2 = \dfrac{2hc_2}{c_1\sqrt{c_2^2-c_1^2}} + \dfrac{D\sqrt{c_2^2-c_1^2}-2hc_1}{c_2\sqrt{c_2^2-c_1^2}}$ and find their intersection points. Doing so gives us $c_2 \approx 4.10$ and 7.66, but if $c_2 = 4.10$, then $\theta = \arcsin(c_1/c_2) \approx 69.6°$, which implies that point S is to the left of point R in the diagram. So $c_2 = 7.66$ km/s.

22. A straight line intersects the curve $y = f(x) = x^4 + cx^3 + 12x^2 - 5x + 2$ in four distinct points if and only if the graph of f has two inflection points. $f'(x) = 4x^3 + 3cx^2 + 24x - 5$ and $f''(x) = 12x^2 + 6cx + 24$.

$f''(x) = 0 \quad\Leftrightarrow\quad x = \dfrac{-6c \pm \sqrt{(6c)^2 - 4(12)(24)}}{2(12)}$. There are two distinct roots for $f''(x) = 0$ (and hence two inflection points) if and only if the discriminant is positive; that is, $36c^2 - 1152 > 0 \quad\Leftrightarrow\quad c^2 > 32 \quad\Leftrightarrow\quad |c| > \sqrt{32}$. Thus, the desired values of c are $c < -4\sqrt{2}$ or $c > 4\sqrt{2}$.

23.

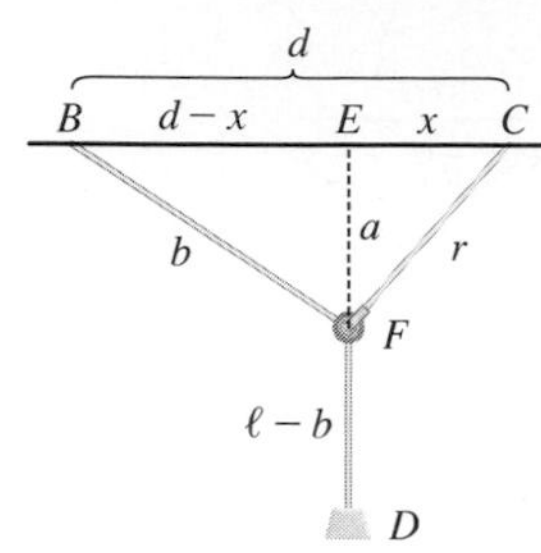

Let $a = |EF|$ and $b = |BF|$ as shown in the figure.

Since $\ell = |BF| + |FD|$, $|FD| = \ell - b$. Now

$$\begin{aligned} |ED| &= |EF| + |FD| = a + \ell - b \\ &\sqrt{r^2 - x^2} + \ell - \sqrt{(d-x)^2 + a^2} \\ &= \sqrt{r^2 - x^2} + \ell - \sqrt{(d-x)^2 + \left(\sqrt{r^2 - x^2}\right)^2} \\ &= \sqrt{r^2 - x^2} + \ell - \sqrt{d^2 - 2dx + x^2 + r^2 - x^2} \end{aligned}$$

Let $f(x) = \sqrt{r^2 - x^2} + \ell - \sqrt{d^2 + r^2 - 2dx}$.

$$f'(x) = \tfrac{1}{2}\left(r^2 - x^2\right)^{-1/2}(-2x) - \tfrac{1}{2}\left(d^2 + r^2 - 2dx\right)^{-1/2}(-2d) = \frac{-x}{\sqrt{r^2 - x^2}} + \frac{d}{\sqrt{d^2 + r^2 - 2dx}}.$$

$$f'(x) = 0 \quad\Rightarrow\quad \frac{x}{\sqrt{r^2 - x^2}} = \frac{d}{\sqrt{d^2 + r^2 - 2dx}} \quad\Rightarrow\quad \frac{x^2}{r^2 - x^2} = \frac{d^2}{d^2 + r^2 - 2dx} \quad\Rightarrow$$

$$d^2x^2 + r^2x^2 - 2dx^3 = d^2r^2 - d^2x^2 \quad\Rightarrow\quad 0 = 2dx^3 - 2d^2x^2 - r^2x^2 + d^2r^2 \quad\Rightarrow$$

$$0 = 2dx^2(x-d) - r^2\left(x^2 - d^2\right) \quad\Rightarrow\quad 0 = 2dx^2(x-d) - r^2(x+d)(x-d) \quad\Rightarrow\quad 0 = (x-d)\left[2dx^2 - r^2(x+d)\right]$$

But $d > r > x$, so $x \neq d$. Thus, we solve $2dx^2 - r^2x - dr^2 = 0$ for x:

$x = \dfrac{-(-r^2) \pm \sqrt{(-r^2)^2 - 4(2d)(-dr^2)}}{2(2d)} = \dfrac{r^2 \pm \sqrt{r^4 + 8d^2r^2}}{4d}$. Because $\sqrt{r^4 + 8d^2r^2} > r^2$, the "negative" can be discarded. Thus, $x = \dfrac{r^2 + \sqrt{r^2}\sqrt{r^2 + 8d^2}}{4d} = \dfrac{r^2 + r\sqrt{r^2 + 8d^2}}{4d} \quad [r > 0] \quad = \dfrac{r}{4d}\left(r + \sqrt{r^2 + 8d^2}\right)$. The maximum value of $|ED|$ occurs at this value of x.

24.

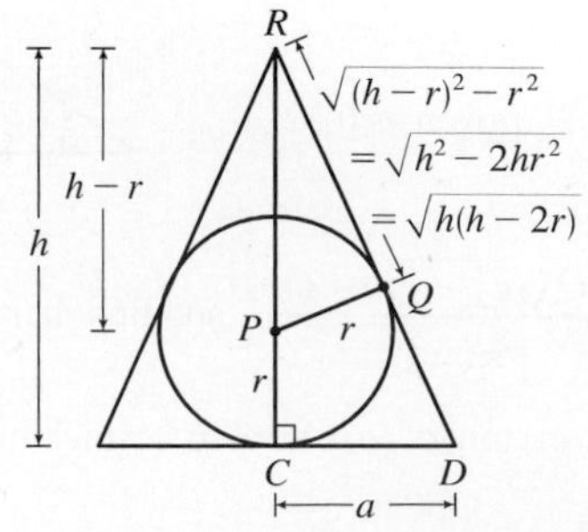

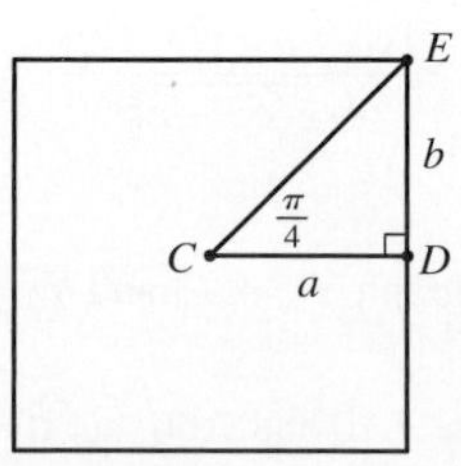

Let $a = \overline{CD}$ denote the distance from the center C of the base to the midpoint D of a side of the base.

Since ΔPQR is similar to ΔDCR, $\dfrac{a}{h} = \dfrac{r}{\sqrt{h(h-2r)}} \;\Rightarrow\; a = \dfrac{rh}{\sqrt{h(h-2r)}} = r\dfrac{\sqrt{h}}{\sqrt{h-2r}}$.

Let b denote one-half the length of a side of the base. The area A of the base is

$A = 8(\text{area of } \Delta CDE) = 8\left(\frac{1}{2}ab\right) = 4a\left(a\tan\frac{\pi}{4}\right) = 4a^2$.

The volume of the pyramid is $V = \frac{1}{3}Ah = \frac{1}{3}\left(4a^2\right)h = \dfrac{4}{3}\left(r\dfrac{\sqrt{h}}{\sqrt{h-2r}}\right)^2 h = \frac{4}{3}r^2\dfrac{h^2}{h-2r}$, with domain $h > 2r$.

Now $\dfrac{dV}{dh} = \dfrac{4}{3}r^2 \cdot \dfrac{(h-2r)(2h)-h^2(1)}{(h-2r)^2} = \dfrac{4}{3}r^2\dfrac{h^2-4hr}{(h-2r)^2} = \dfrac{4}{3}r^2\dfrac{h(h-4r)}{(h-2r)^2}$

and
$$\begin{aligned}\frac{d^2V}{dh^2} &= \frac{4}{3}r^2 \cdot \frac{(h-2r)^2(2h-4r)-\left(h^2-4hr\right)(2)(h-2r)(1)}{[(h-2r)^2]^2} \\ &= \frac{4}{3}r^2 \cdot \frac{2(h-2r)\left[\left(h^2-4hr+4r^2\right)-\left(h^2-4hr\right)\right]}{(h-2r)^2} \\ &= \frac{8}{3}r^2 \cdot \frac{4r^2}{(h-2r)^3} = \frac{32}{3}r^4 \cdot \frac{1}{(h-2r)^3}.\end{aligned}$$

The first derivative is equal to zero for $h = 4r$ and the second derivative is positive for $h > 2r$, so the volume of the pyramid is minimized when $h = 4r$.

To extend our solution to a regular n-gon, we make the following changes:

(1) the number of sides of the base is n

(2) the number of triangles in the base is $2n$

(3) $\angle DCE = \dfrac{\pi}{n}$

(4) $b = a\tan\dfrac{\pi}{n}$

We then obtain the following results: $A = na^2\tan\dfrac{\pi}{n}$, $V = \dfrac{nr^2}{3}\cdot\tan\left(\dfrac{\pi}{n}\right)\cdot\dfrac{h^2}{h-2r}$, $\dfrac{dV}{dh} = \dfrac{nr^2}{3}\cdot\tan\left(\dfrac{\pi}{n}\right)\cdot\dfrac{h(h-4r)}{(h-2r)^2}$,

and $\dfrac{d^2V}{dh^2} = \dfrac{8nr^4}{3}\cdot\tan\left(\dfrac{\pi}{n}\right)\cdot\dfrac{1}{(h-2r)^3}$. Notice that the answer, $h = 4r$, is independent of the number of sides of the base of the polygon!

25. $V = \frac{4}{3}\pi r^3 \;\Rightarrow\; \dfrac{dV}{dt} = 4\pi r^2\dfrac{dr}{dt}$. But $\dfrac{dV}{dt}$ is proportional to the surface area, so $\dfrac{dV}{dt} = k\cdot 4\pi r^2$ for some constant k.

Therefore, $4\pi r^2\dfrac{dr}{dt} = k\cdot 4\pi r^2 \;\Leftrightarrow\; \dfrac{dr}{dt} = k =$ constant. An antiderivative of k with respect to t is kt, so $r = kt + C$.

When $t = 0$, the radius r must equal the original radius r_0, so $C = r_0$, and $r = kt + r_0$. To find k we use the fact that when $t = 3$, $r = 3k + r_0$ and $V = \frac{1}{2}V_0 \;\Rightarrow\; \frac{4}{3}\pi(3k+r_0)^3 = \frac{1}{2}\cdot\frac{4}{3}\pi r_0^3 \;\Rightarrow\; (3k+r_0)^3 = \frac{1}{2}r_0^3 \;\Rightarrow$

$3k + r_0 = \dfrac{1}{\sqrt[3]{2}}r_0 \;\Rightarrow\; k = \frac{1}{3}r_0\left(\dfrac{1}{\sqrt[3]{2}} - 1\right)$. Since $r = kt + r_0$, $r = \frac{1}{3}r_0\left(\dfrac{1}{\sqrt[3]{2}} - 1\right)t + r_0$. When the snowball has melted completely we have $r = 0 \;\Rightarrow\; \frac{1}{3}r_0\left(\dfrac{1}{\sqrt[3]{2}} - 1\right)t + r_0 = 0$ which gives $t = \dfrac{3\sqrt[3]{2}}{\sqrt[3]{2}-1}$. Hence, it takes

$\dfrac{3\sqrt[3]{2}}{\sqrt[3]{2}-1} - 3 = \dfrac{3}{\sqrt[3]{2}-1} \approx 11$ h 33 min longer.

26. By ignoring the bottom hemisphere of the initial spherical bubble, we can rephrase the problem as follows: Prove that the maximum height of a stack of n hemispherical bubbles is $\sqrt{n}$ if the radius of the bottom hemisphere is 1. We proceed by induction. The case $n = 1$ is obvious since $\sqrt{1}$ is the height of the first hemisphere. Suppose the assertion is true for $n = k$ and let's suppose we have $k + 1$ hemispherical bubbles forming a stack of maximum height. Suppose the second hemisphere (counting from the bottom) has radius r. Then by our induction hypothesis (scaled to the setting of a bottom hemisphere of radius r), the height of the stack formed by the top k bubbles is $\sqrt{k}\,r$. (If it were shorter, then the total stack of $k + 1$ bubbles wouldn't have maximum height.)

The height of the whole stack is $H(r) = \sqrt{k}\,r + \sqrt{1 - r^2}$. (See the figure.) We want to choose r so as to maximize $H(r)$. Note that $0 < r < 1$.

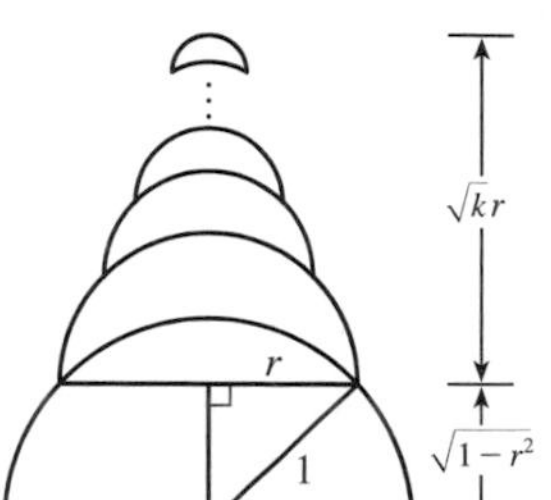

We calculate $H'(r) = \sqrt{k} - \dfrac{r}{\sqrt{1 - r^2}}$ and $H''(r) = \dfrac{-1}{(1 - r^2)^{3/2}}$.

$$H'(r) = 0 \quad \Leftrightarrow \quad r^2 = k(1 - r^2) \quad \Leftrightarrow \quad (k + 1)r^2 = k \quad \Leftrightarrow \quad r = \sqrt{\frac{k}{k + 1}}.$$

This is the only critical number in $(0, 1)$ and it represents a local maximum (hence an absolute maximum) since $H''(r) < 0$ on $(0, 1)$. When $r = \sqrt{\dfrac{k}{k + 1}}$,

$H(r) = \sqrt{k}\dfrac{\sqrt{k}}{\sqrt{k + 1}} + \sqrt{1 - \dfrac{k}{k + 1}} = \dfrac{k}{\sqrt{k + 1}} + \dfrac{1}{\sqrt{k + 1}} = \sqrt{k + 1}$. Thus, the assertion is true for $n = k + 1$ when it is true for $n = k$. By induction, it is true for all positive integers n.

Note: In general, a maximally tall stack of n hemispherical bubbles consists of bubbles with radii

$$1, \sqrt{\frac{n - 1}{n}}, \sqrt{\frac{n - 2}{n}}, \ldots, \sqrt{\frac{2}{n}}, \sqrt{\frac{1}{n}}.$$

5 □ INTEGRALS

5.1 Areas and Distances

1. (a) Since f is *decreasing*, we can obtain a *lower* estimate by using *right* endpoints. We are instructed to use five rectangles, so $n = 5$.

$$\begin{aligned} R_5 &= \sum_{i=1}^{5} f(x_i)\,\Delta x \qquad \left[\Delta x = \frac{b-a}{n} = \frac{10-0}{5} = 2\right] \\ &= f(x_1)\cdot 2 + f(x_2)\cdot 2 + f(x_3)\cdot 2 + f(x_4)\cdot 2 + f(x_5)\cdot 2 \\ &= 2[f(2) + f(4) + f(6) + f(8) + f(10)] \\ &\approx 2(3.2 + 1.8 + 0.8 + 0.2 + 0) \\ &= 2(6) = 12 \end{aligned}$$

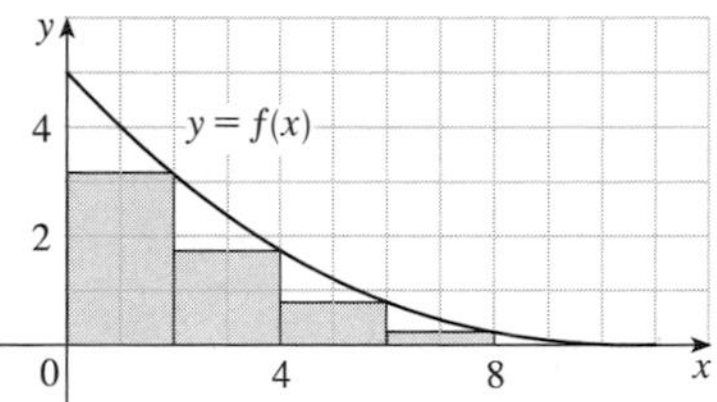

Since f is *decreasing*, we can obtain an *upper* estimate by using *left* endpoints.

$$\begin{aligned} L_5 &= \sum_{i=1}^{5} f(x_{i-1})\,\Delta x \\ &= f(x_0)\cdot 2 + f(x_1)\cdot 2 + f(x_2)\cdot 2 + f(x_3)\cdot 2 + f(x_4)\cdot 2 \\ &= 2[f(0) + f(2) + f(4) + f(6) + f(8)] \\ &\approx 2(5 + 3.2 + 1.8 + 0.8 + 0.2) \\ &= 2(11) = 22 \end{aligned}$$

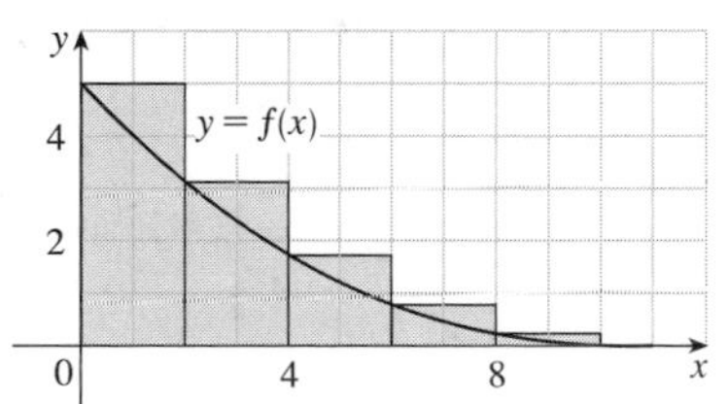

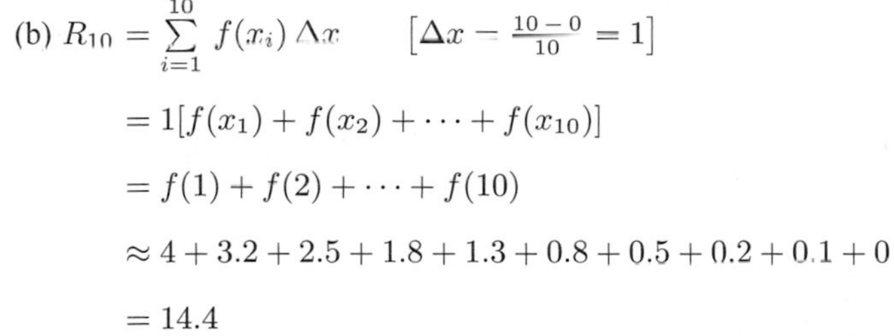

(b) $$\begin{aligned} R_{10} &= \sum_{i=1}^{10} f(x_i)\,\Delta x \qquad \left[\Delta x = \tfrac{10-0}{10} = 1\right] \\ &= 1[f(x_1) + f(x_2) + \cdots + f(x_{10})] \\ &= f(1) + f(2) + \cdots + f(10) \\ &\approx 4 + 3.2 + 2.5 + 1.8 + 1.3 + 0.8 + 0.5 + 0.2 + 0.1 + 0 \\ &= 14.4 \end{aligned}$$

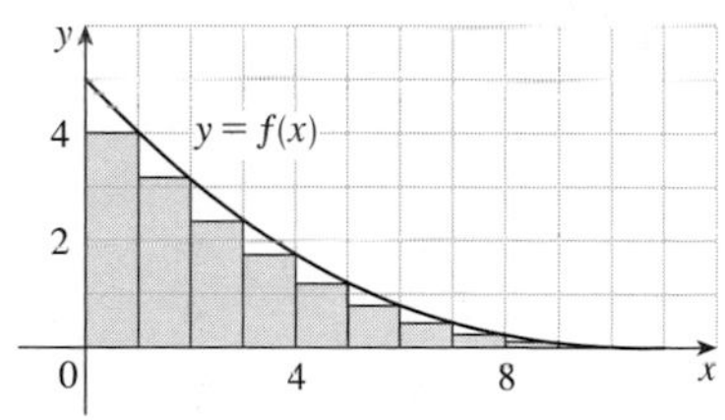

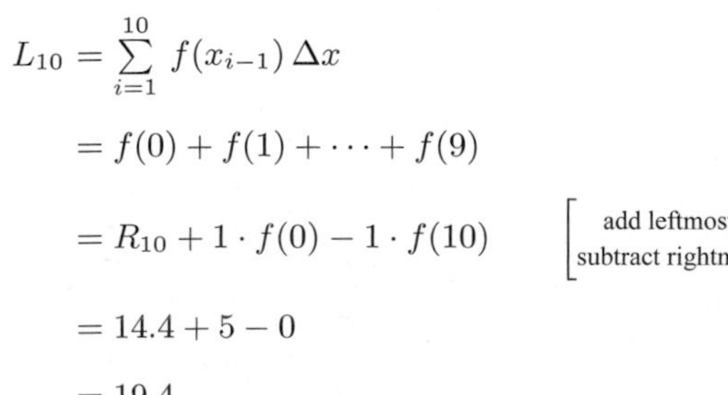

$$\begin{aligned} L_{10} &= \sum_{i=1}^{10} f(x_{i-1})\,\Delta x \\ &= f(0) + f(1) + \cdots + f(9) \\ &= R_{10} + 1\cdot f(0) - 1\cdot f(10) \qquad \left[\begin{array}{c}\text{add leftmost upper rectangle,}\\ \text{subtract rightmost lower rectangle}\end{array}\right] \\ &= 14.4 + 5 - 0 \\ &= 19.4 \end{aligned}$$

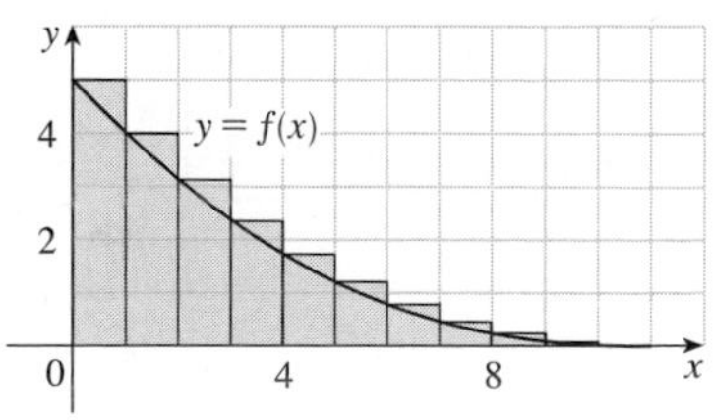

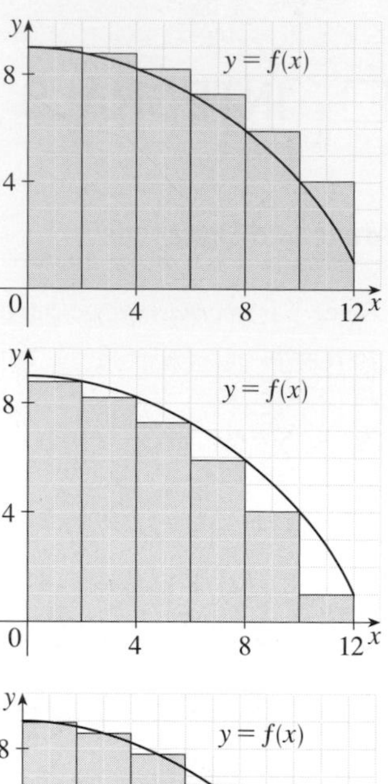

2. (a) (i) $L_6 = \sum_{i=1}^{6} f(x_{i-1})\,\Delta x \qquad [\Delta x = \frac{12-0}{6} = 2]$

$= 2[f(x_0) + f(x_1) + f(x_2) + f(x_3) + f(x_4) + f(x_5)]$

$= 2[f(0) + f(2) + f(4) + f(6) + f(8) + f(10)]$

$\approx 2(9 + 8.8 + 8.2 + 7.3 + 5.9 + 4.1)$

$= 2(43.3) = 86.6$

(ii) $R_6 = L_6 + 2 \cdot f(12) - 2 \cdot f(0)$

$\approx 86.6 + 2(1) - 2(9) = 70.6$

[Add area of rightmost lower rectangle and subtract area of leftmost upper rectangle.]

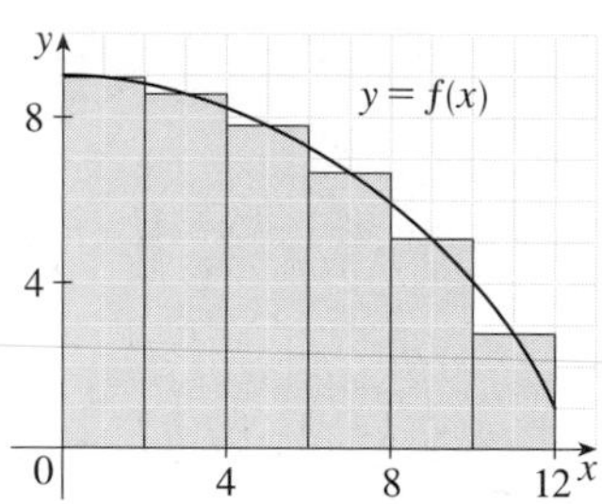

(iii) $M_6 = \sum_{i=1}^{6} f(\overline{x}_i)\,\Delta x$

$= 2[f(1) + f(3) + f(5) + f(7) + f(9) + f(11)]$

$\approx 2(8.9 + 8.5 + 7.8 + 6.6 + 5.1 + 2.8)$

$= 2(39.7) = 79.4$

(b) Since f is *decreasing*, we obtain an *overestimate* by using *left* endpoints; that is, L_6.

(c) Since f is *decreasing*, we obtain an *underestimate* by using *right* endpoints; that is, R_6.

(d) M_6 gives the best estimate, since the area of each rectangle appears to be closer to the true area than the overestimates and underestimates in L_6 and R_6.

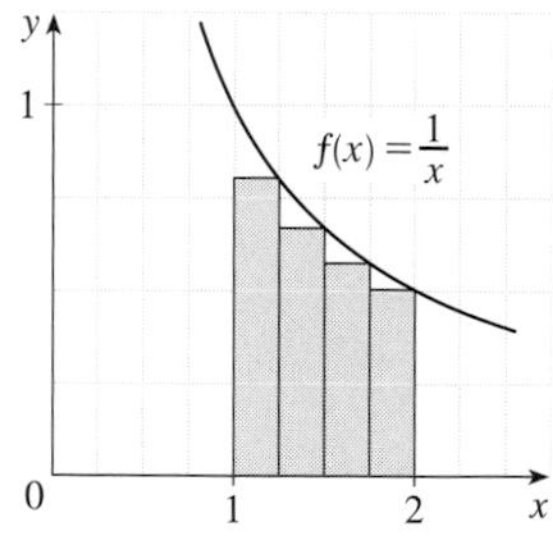

3. (a) $R_4 = \sum_{i=1}^{4} f(x_i)\,\Delta x \qquad \left[\Delta x = \frac{2-1}{4} = \frac{1}{4}\right] = \left[\sum_{i=1}^{4} f(x_i)\right] \Delta x$

$= [f(x_1) + f(x_2) + f(x_3) + f(x_4)]\,\Delta x$

$= \left[\frac{1}{5/4} + \frac{1}{6/4} + \frac{1}{7/4} + \frac{1}{8/4}\right]\frac{1}{4} = \left[\frac{4}{5} + \frac{2}{3} + \frac{4}{7} + \frac{1}{2}\right]\frac{1}{4} \approx 0.6345$

Since f is *decreasing* on $[1, 2]$, an *underestimate* is obtained by using the *right* endpoint approximation, R_4.

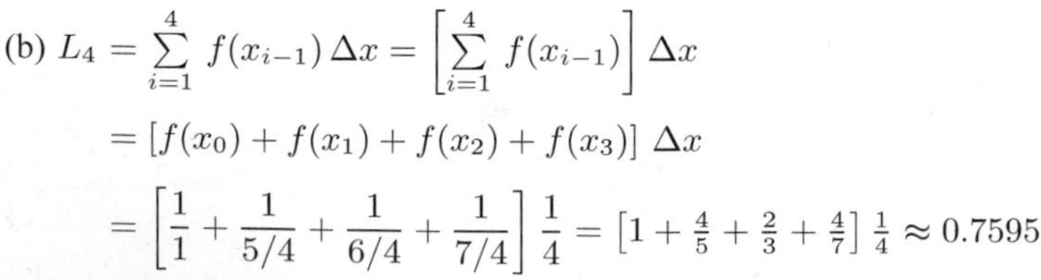

(b) $L_4 = \sum_{i=1}^{4} f(x_{i-1})\,\Delta x = \left[\sum_{i=1}^{4} f(x_{i-1})\right] \Delta x$

$= [f(x_0) + f(x_1) + f(x_2) + f(x_3)]\,\Delta x$

$= \left[\frac{1}{1} + \frac{1}{5/4} + \frac{1}{6/4} + \frac{1}{7/4}\right]\frac{1}{4} = \left[1 + \frac{4}{5} + \frac{2}{3} + \frac{4}{7}\right]\frac{1}{4} \approx 0.7595$

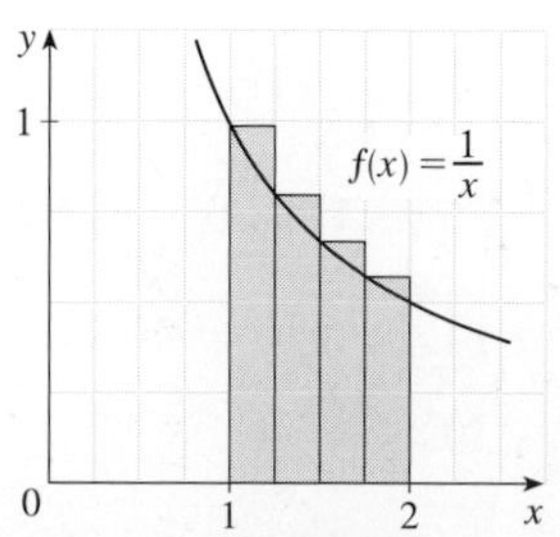

L_4 is an overestimate. Alternatively, we could just add the area of the leftmost upper rectangle and subtract the area of the rightmost lower rectangle; that is, $L_4 = R_4 + f(1) \cdot \frac{1}{4} - f(2) \cdot \frac{1}{4}$.

4. (a) $R_4 = \sum\limits_{i=1}^{4} f(x_i)\,\Delta x \qquad \left[\Delta x = \dfrac{\pi/2 - 0}{4} = \dfrac{\pi}{8}\right] = \left[\sum\limits_{i=1}^{4} f(x_i)\right] \Delta x$

$= [f(x_1) + f(x_2) + f(x_3) + f(x_4)]\,\Delta x$

$= \left[\sin\frac{\pi}{8} + \sin\frac{2\pi}{8} + \sin\frac{3\pi}{8} + \sin\frac{4\pi}{8}\right]\frac{\pi}{8}$

≈ 1.1835

Since f is *increasing* on $\left[0, \frac{\pi}{2}\right]$, R_4 is an *overestimate*.

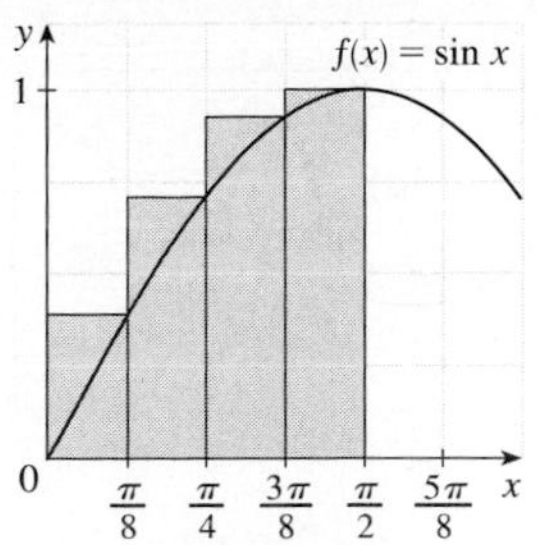

(b) $L_4 = \sum\limits_{i=1}^{4} f(x_{i-1})\,\Delta x = \left[\sum\limits_{i=1}^{4} f(x_{i-1})\right] \Delta x$

$= [f(x_0) + f(x_1) + f(x_2) + f(x_3)]\,\Delta x$

$= \left[\sin 0 + \sin\frac{\pi}{8} + \sin\frac{2\pi}{8} + \sin\frac{3\pi}{8}\right]\frac{\pi}{8}$

≈ 0.7908

Since f is *increasing* on $\left[0, \frac{\pi}{2}\right]$, L_4 is an *underestimate*.

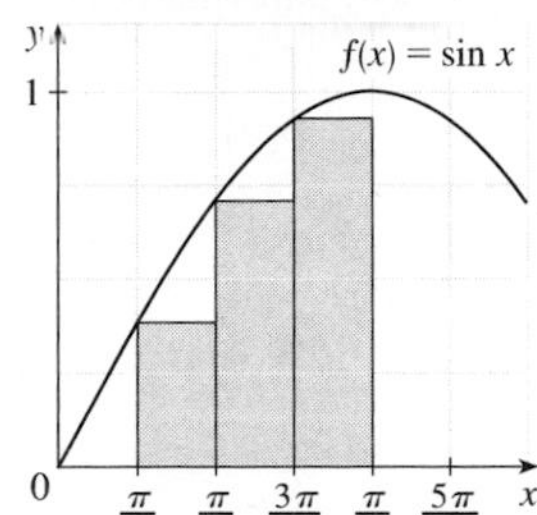

5. (a) $f(x) = 1 + x^2$ and $\Delta x = \dfrac{2-(-1)}{3} = 1 \quad \Rightarrow$

$R_3 = 1 \cdot f(0) + 1 \cdot f(1) + 1 \cdot f(2) = 1 \cdot 1 + 1 \cdot 2 + 1 \cdot 5 = 8.$

$\Delta x = \dfrac{2-(-1)}{6} = 0.5 \quad \Rightarrow$

$R_6 = 0.5[f(-0.5) + f(0) + f(0.5) + f(1) + f(1.5) + f(2)]$

$= 0.5(1.25 + 1 + 1.25 + 2 + 3.25 + 5)$

$= 0.5(13.75) = 6.875$

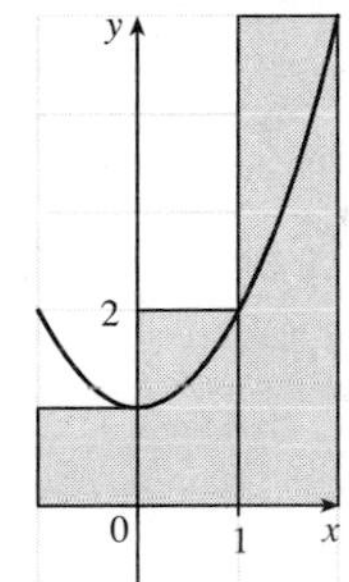

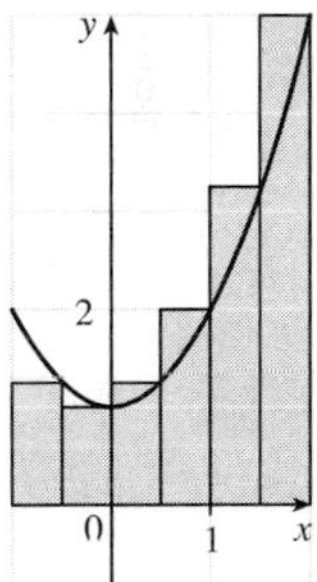

(b) $L_3 = 1 \cdot f(-1) + 1 \cdot f(0) + 1 \cdot f(1) = 1 \cdot 2 + 1 \cdot 1 + 1 \cdot 2 = 5$

$L_6 = 0.5[f(-1) + f(-0.5) + f(0) + f(0.5) + f(1) + f(1.5)]$

$= 0.5(2 + 1.25 + 1 + 1.25 + 2 + 3.25)$

$= 0.5(10.75) = 5.375$

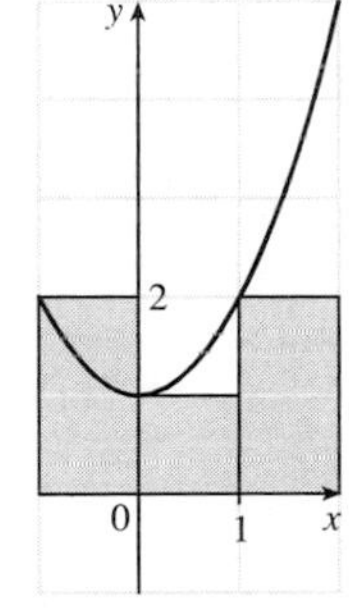

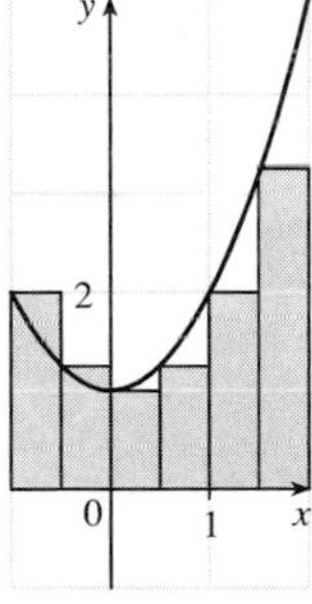

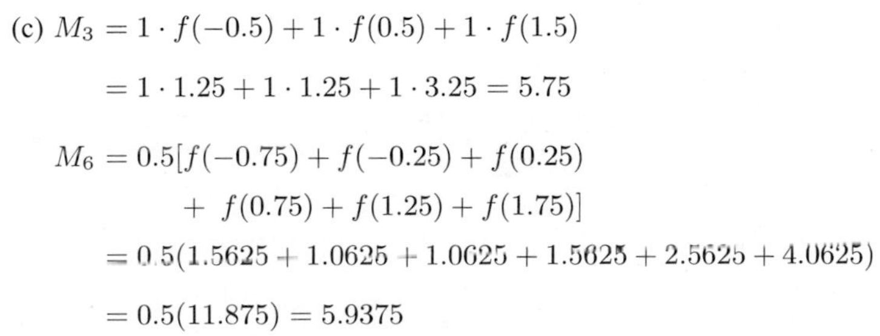

(c) $M_3 = 1 \cdot f(-0.5) + 1 \cdot f(0.5) + 1 \cdot f(1.5)$

$= 1 \cdot 1.25 + 1 \cdot 1.25 + 1 \cdot 3.25 = 5.75$

$M_6 = 0.5[f(-0.75) + f(-0.25) + f(0.25)$
$\quad + f(0.75) + f(1.25) + f(1.75)]$

$= 0.5(1.5625 + 1.0625 + 1.0625 + 1.5625 + 2.5625 + 4.0625)$

$= 0.5(11.875) = 5.9375$

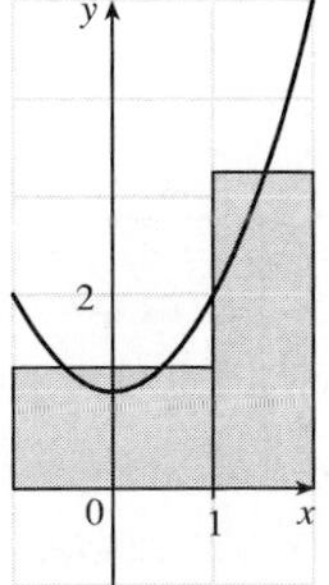

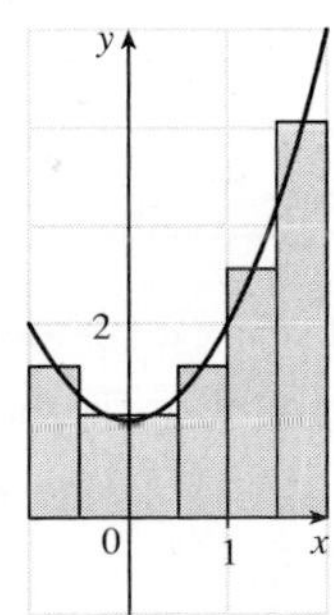

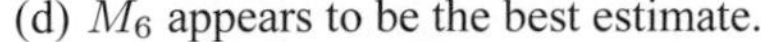

(d) M_6 appears to be the best estimate.

6. (a)

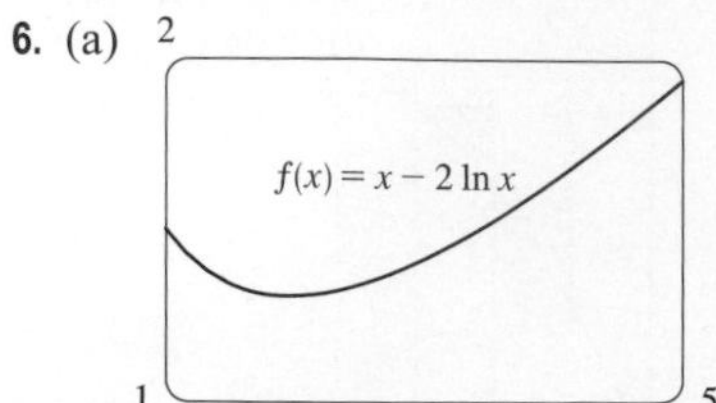

(b) $f(x) = x - 2\ln x$ and $\Delta x = \dfrac{5-1}{4} = 1 \quad \Rightarrow$

(i) $R_4 = 1 \cdot f(2) + 1 \cdot f(3) + 1 \cdot f(4) + 1 \cdot f(5)$

$= (2 - 2\ln 2) + (3 - 2\ln 3) + (4 - 2\ln 4) + (5 - 2\ln 5)$

≈ 4.425

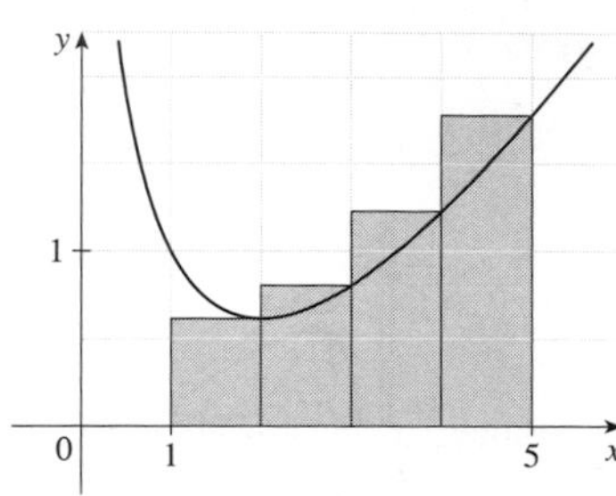

(ii) $M_4 = 1 \cdot f(1.5) + 1 \cdot f(2.5) + 1 \cdot f(3.5) + 1 \cdot f(4.5)$

$= (1.5 - 2\ln 1.5) + (2.5 - 2\ln 2.5)$

$+ (3.5 - 2\ln 3.5) + (4.5 - 2\ln 4.5)$

≈ 3.843

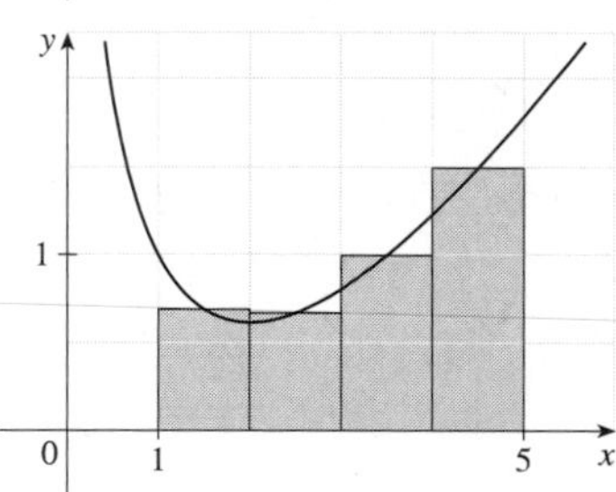

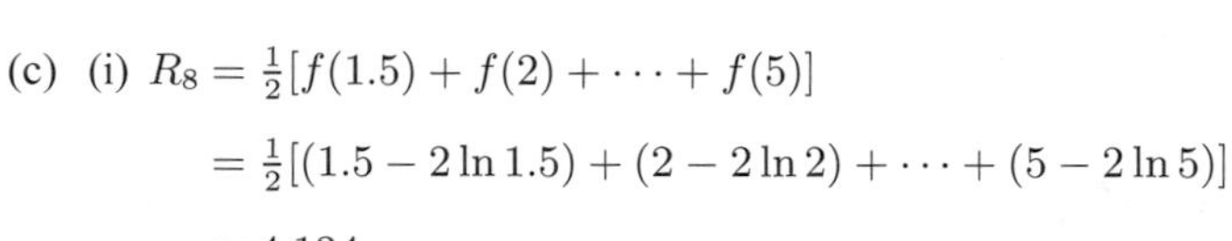

(c) (i) $R_8 = \frac{1}{2}[f(1.5) + f(2) + \cdots + f(5)]$

$= \frac{1}{2}[(1.5 - 2\ln 1.5) + (2 - 2\ln 2) + \cdots + (5 - 2\ln 5)]$

≈ 4.134

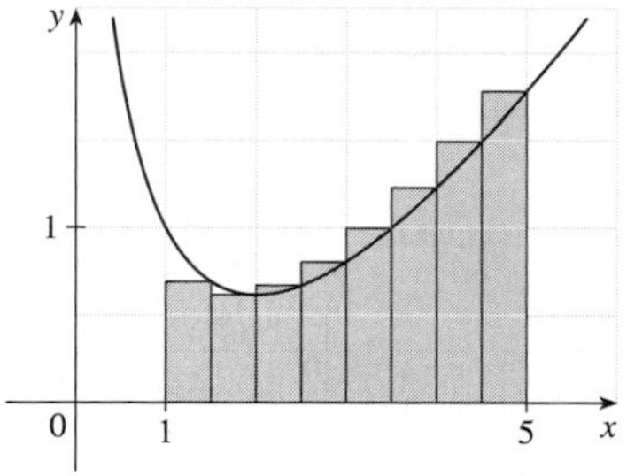

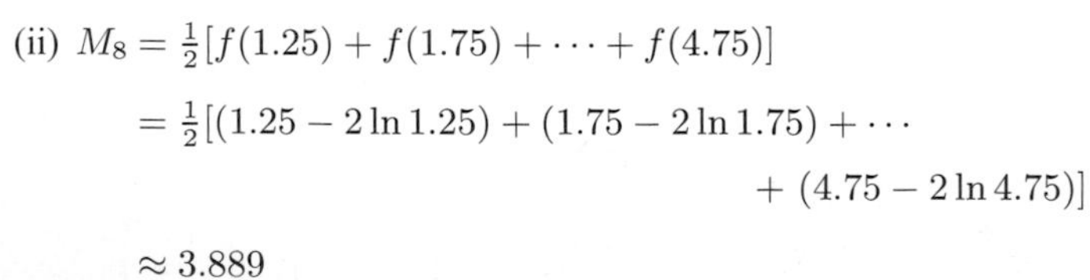

(ii) $M_8 = \frac{1}{2}[f(1.25) + f(1.75) + \cdots + f(4.75)]$

$= \frac{1}{2}[(1.25 - 2\ln 1.25) + (1.75 - 2\ln 1.75) + \cdots$

$+ (4.75 - 2\ln 4.75)]$

≈ 3.889

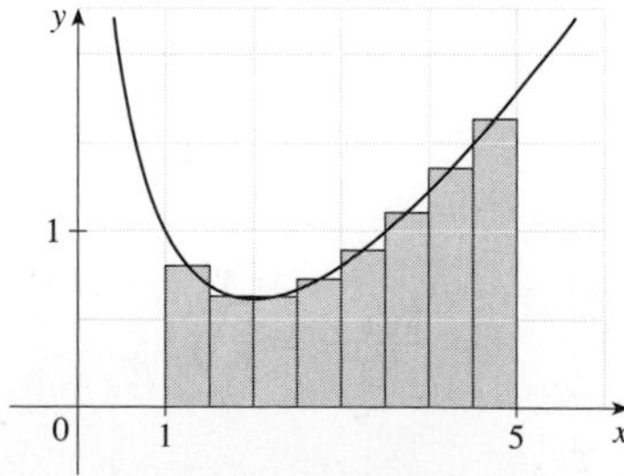

7. $f(x) = 2 + \sin x,\ 0 \le x \le \pi,\ \Delta x = \pi/n.$

$n = 2$: The maximum values of f on both subintervals occur at $x = \frac{\pi}{2}$, so

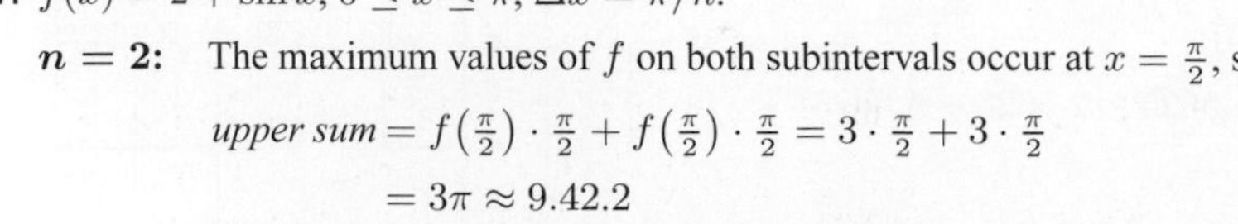

upper sum $= f\left(\frac{\pi}{2}\right) \cdot \frac{\pi}{2} + f\left(\frac{\pi}{2}\right) \cdot \frac{\pi}{2} = 3 \cdot \frac{\pi}{2} + 3 \cdot \frac{\pi}{2}$

$= 3\pi \approx 9.42.2$

The minimum values of f on the subintervals occur at $x = 0$ and $x = \pi$, so

lower sum $= f(0) \cdot \frac{\pi}{2} + f(\pi) \cdot \frac{\pi}{2} = 2 \cdot \frac{\pi}{2} + 2 \cdot \frac{\pi}{2} = 2\pi \approx 6.28.$

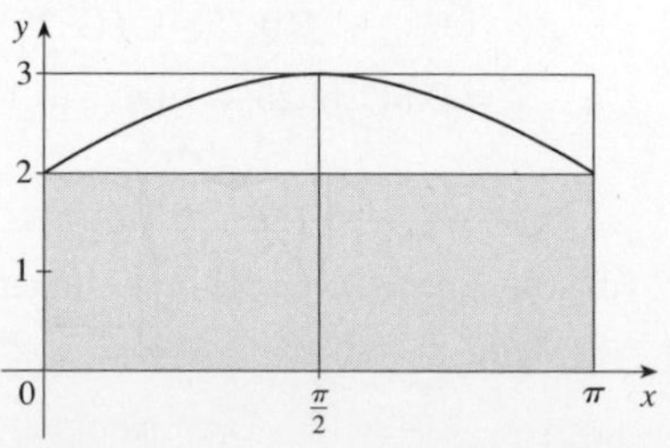

$n = 4$: $\quad$ *upper sum* $= \left[f\left(\frac{\pi}{4}\right) + f\left(\frac{\pi}{2}\right) + f\left(\frac{\pi}{2}\right) + f\left(\frac{3\pi}{4}\right)\right]\left(\frac{\pi}{4}\right)$

$$= \left[\left(2 + \tfrac{1}{2}\sqrt{2}\right) + (2+1) + (2+1) + \left(2 + \tfrac{1}{2}\sqrt{2}\right)\right]\left(\tfrac{\pi}{4}\right)$$

$$= \left(10 + \sqrt{2}\right)\left(\tfrac{\pi}{4}\right) \approx 8.96$$

lower sum $= \left[f(0) + f\left(\frac{\pi}{4}\right) + f\left(\frac{3\pi}{4}\right) + f(\pi)\right]\left(\frac{\pi}{4}\right)$

$$= \left[(2+0) + \left(2 + \tfrac{1}{2}\sqrt{2}\right) + \left(2 + \tfrac{1}{2}\sqrt{2}\right) + (2+0)\right]\left(\tfrac{\pi}{4}\right)$$

$$= \left(8 + \sqrt{2}\right)\left(\tfrac{\pi}{4}\right) \approx 7.39$$

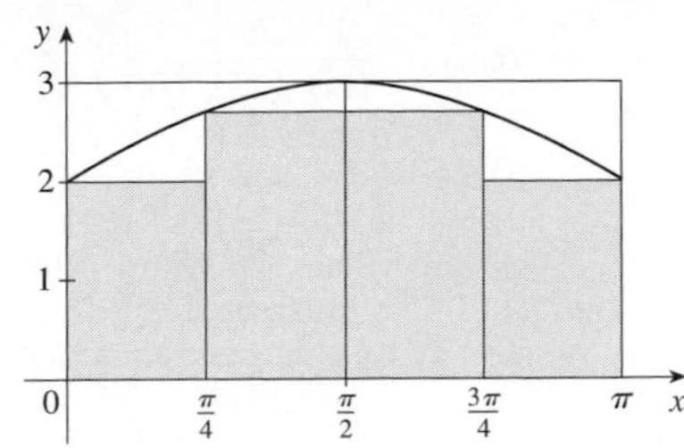

$n = 8$: $\quad$ *upper sum* $= \left[f\left(\frac{\pi}{8}\right) + f\left(\frac{\pi}{4}\right) + f\left(\frac{3\pi}{8}\right) + f\left(\frac{\pi}{2}\right) + f\left(\frac{\pi}{2}\right) + f\left(\frac{5\pi}{8}\right) + f\left(\frac{3\pi}{4}\right) + f\left(\frac{7\pi}{8}\right)\right]\left(\frac{\pi}{8}\right)$

$$\approx 8.65$$

lower sum $= \left[f(0) + f\left(\frac{\pi}{8}\right) + f\left(\frac{\pi}{4}\right) + f\left(\frac{3\pi}{8}\right) + f\left(\frac{5\pi}{8}\right) + f\left(\frac{3\pi}{4}\right) + f\left(\frac{7\pi}{8}\right) + f(\pi)\right]\left(\frac{\pi}{8}\right)$

$$\approx 7.86$$

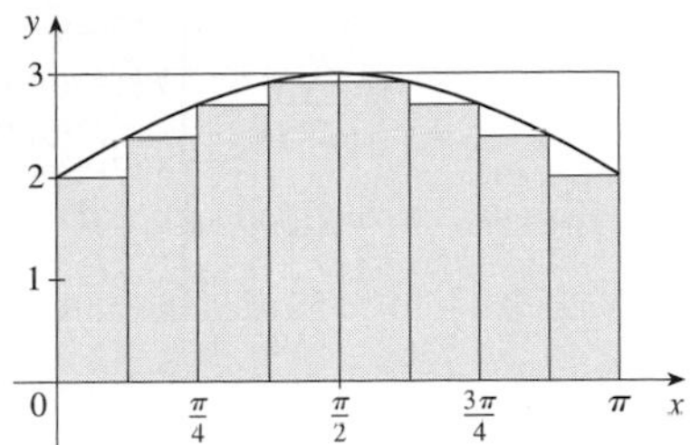

8. $f(x) = 1 + x^2$, $-1 \le x \le 1$, $\Delta x = 2/n$.

$n = 3$: $\quad$ *upper sum* $= \left[f(-1) + f\left(\frac{1}{3}\right) + f(1)\right]\left(\frac{2}{3}\right)$

$$= \left(2 + \tfrac{10}{9} + 2\right)\left(\tfrac{2}{3}\right)$$

$$= \tfrac{92}{27} \approx 3.41$$

lower sum $= \left[f\left(-\frac{1}{3}\right) + f(0) + f\left(\frac{1}{3}\right)\right]\left(\frac{2}{3}\right)$

$$= \left(\tfrac{10}{9} + 1 + \tfrac{10}{9}\right)\left(\tfrac{2}{3}\right)$$

$$= \tfrac{58}{27} \approx 2.15$$

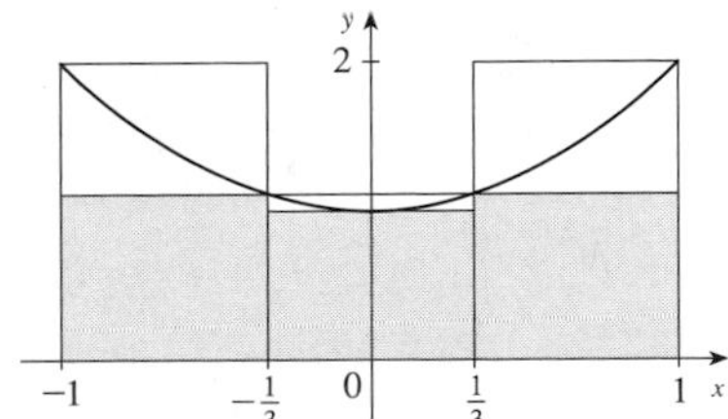

$n = 4$: $\quad$ *upper sum* $= \left[f(-1) + f\left(-\frac{1}{2}\right) + f\left(\frac{1}{2}\right) + f(1)\right]\left(\frac{2}{4}\right)$

$$= \left(2 + \tfrac{5}{4} + \tfrac{5}{4} + 2\right)\left(\tfrac{1}{2}\right)$$

$$= \tfrac{13}{4} = 3.25$$

lower sum $= \left[f\left(-\frac{1}{2}\right) + f(0) + f(0) + f\left(\frac{1}{2}\right)\right]\left(\frac{2}{4}\right)$

$$= \left(\tfrac{5}{4} + 1 + 1 + \tfrac{5}{4}\right)\left(\tfrac{1}{2}\right)$$

$$= \tfrac{9}{4} = 2.25$$

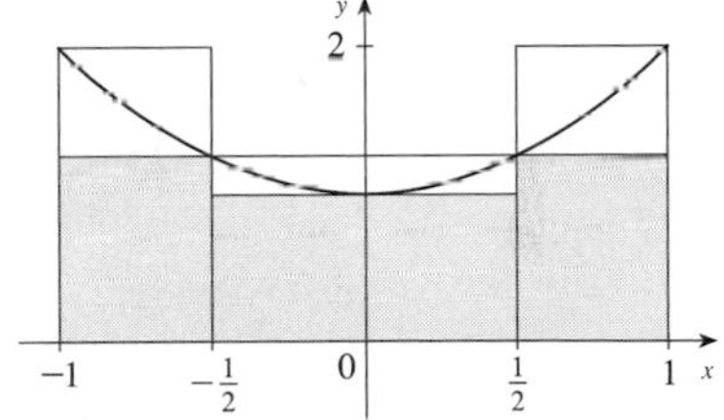

9. Here is one possible algorithm (ordered sequence of operations) for calculating the sums:

1 Let SUM = 0, X_MIN = 0, X_MAX = 1, N = 10 (depending on which sum we are calculating),

DELTA_X = (X_MAX - X_MIN)/N, and RIGHT_ENDPOINT = X_MIN + DELTA_X.

2 Repeat steps 2a, 2b in sequence until RIGHT_ENDPOINT > X_MAX.

2a Add (RIGHT_ENDPOINT)^4 to SUM.

Add DELTA_X to RIGHT_ENDPOINT.

At the end of this procedure, (DELTA_X)·(SUM) is equal to the answer we are looking for. We find that

$R_{10} = \dfrac{1}{10}\displaystyle\sum_{i=1}^{10}\left(\frac{i}{10}\right)^4 \approx 0.2533$, $R_{30} = \dfrac{1}{30}\displaystyle\sum_{i=1}^{30}\left(\frac{i}{30}\right)^4 \approx 0.2170$, $R_{50} = \dfrac{1}{50}\displaystyle\sum_{i=1}^{50}\left(\frac{i}{50}\right)^4 \approx 0.2101$, and

$R_{100} = \dfrac{1}{100}\sum\limits_{i=1}^{100}\left(\dfrac{i}{100}\right)^4 \approx 0.2050$. It appears that the exact area is 0.2. The following display shows the program SUMRIGHT and its output from a TI-83/4 Plus calculator. To generalize the program, we have input (rather than assign) values for Xmin, Xmax, and N. Also, the function, x^4, is assigned to Y_1, enabling us to evaluate any right sum merely by changing Y_1 and running the program.

```
PROGRAM:SUMRIGHT
:0→S
:Prompt Xmin
:Prompt Xmax
:Prompt N
:(Xmax-Xmin)/N→D
:Xmin+D→R
:For(I,1,N)
:S+Y1(R)→S
:R+D→R
:End
:D*S→Z
:Disp Z
```

```
prgmSUMRIGHT
Xmin=?0
Xmax=?1
N=?10
                .25333
                  Done
```

10. We can use the algorithm from Exercise 9 with X_MIN = 0, X_MAX = $\pi/2$, and cos(RIGHT_ENDPOINT) instead of (RIGHT_ENDPOINT)^4 in step 2a. We find that $R_{10} = \dfrac{\pi/2}{10}\sum\limits_{i=1}^{10}\cos\left(\dfrac{i\pi}{20}\right) \approx 0.9194$, $R_{30} = \dfrac{\pi/2}{30}\sum\limits_{i=1}^{30}\cos\left(\dfrac{i\pi}{60}\right) \approx 0.9736$, and $R_{50} = \dfrac{\pi/2}{50}\sum\limits_{i=1}^{50}\cos\left(\dfrac{i\pi}{100}\right) \approx 0.9842$, and $R_{100} = \dfrac{\pi/2}{100}\sum\limits_{i=1}^{100}\cos\left(\dfrac{i\pi}{200}\right) \approx 0.9921$. It appears that the exact area is 1.

11. In Maple, we have to perform a number of steps before getting a numerical answer. After loading the student package [command: `with(student);`] we use the command `left_sum:=leftsum(1/(x^2+1),x=0..1,10` [or `30`, or `50`]`);` which gives us the expression in summation notation. To get a numerical approximation to the sum, we use `evalf(left_sum);`. Mathematica does not have a special command for these sums, so we must type them in manually. For example, the first left sum is given by `(1/10)*Sum[1/(((i-1)/10)^2+1)],{i,1,10}]`, and we use the `N` command on the resulting output to get a numerical approximation.

In Derive, we use the `LEFT_RIEMANN` command to get the left sums, but must define the right sums ourselves. (We can define a new function using `LEFT_RIEMANN` with k ranging from 1 to n instead of from 0 to $n-1$.)

(a) With $f(x) = \dfrac{1}{x^2+1}$, $0 \le x \le 1$, the left sums are of the form $L_n = \dfrac{1}{n}\sum\limits_{i=1}^{n}\dfrac{1}{\left(\frac{i-1}{n}\right)^2+1}$. Specifically, $L_{10} \approx 0.8100$, $L_{30} \approx 0.7937$, and $L_{50} \approx 0.7904$. The right sums are of the form $R_n = \dfrac{1}{n}\sum\limits_{i=1}^{n}\dfrac{1}{\left(\frac{i}{n}\right)^2+1}$. Specifically, $R_{10} \approx 0.7600$, $R_{30} \approx 0.7770$, and $R_{50} \approx 0.7804$.

(b) In Maple, we use the `leftbox` (with the same arguments as `left_sum`) and `rightbox` commands to generate the graphs.

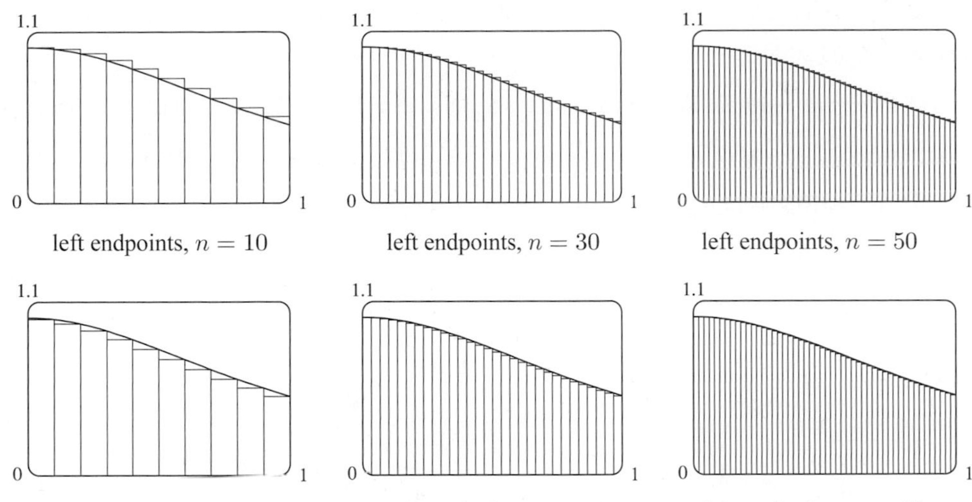

(c) We know that since $y = 1/(x^2 + 1)$ is a decreasing function on $(0, 1)$, all of the left sums are larger than the actual area, and all of the right sums are smaller than the actual area. Since the left sum with $n = 50$ is about $0.7904 < 0.791$ and the right sum with $n = 50$ is about $0.7804 > 0.780$, we conclude that $0.780 < R_{50} <$ exact area $< L_{50} < 0.791$, so the exact area is between 0.780 and 0.791.

12. See the solution to Exercise 11 for the CAS commands for evaluating the sums.

(a) With $f(x) = \ln x$, $1 \leq x \leq 4$, the left sums are of the form $L_n = \dfrac{3}{n} \sum\limits_{i=1}^{n} \ln\left(1 + \dfrac{3(i-1)}{n}\right)$. In particular, $L_{10} \approx 2.3316$, $L_{30} \approx 2.4752$, and $L_{50} \approx 2.5034$. The right sums are of the form $R_n = \dfrac{3}{n} \sum\limits_{i=1}^{n} \ln\left(1 + \dfrac{3i}{n}\right)$. In particular, $R_{10} \approx 2.7475$, $R_{30} \approx 2.6139$, and $R_{50} \approx 2.5865$.

(b) In Maple, we use the `leftbox` (with the same arguments as `left_sum`) and `rightbox` commands to generate the graphs.

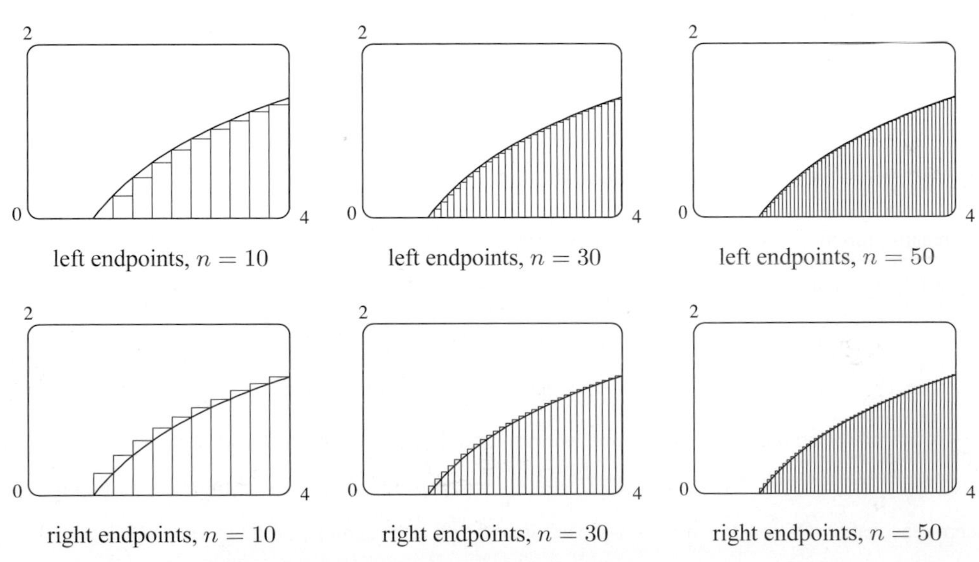

(c) We know that since $y = \ln x$ is an increasing function on $(1, 4)$, all of the left sums are smaller than the actual area, and all of the right sums are larger than the actual area. Since the left sum with $n = 50$ is about $2.503 > 2.50$ and the right sum with $n = 50$ is about $2.587 < 2.59$, we conclude that $2.50 < L_{50} < \text{exact area} < R_{50} < 2.59$, so the exact area is between 2.50 and 2.59.

13. Since v is an increasing function, L_6 will give us a lower estimate and R_6 will give us an upper estimate.

$$L_6 = (0 \text{ ft/s})(0.5 \text{ s}) + (6.2)(0.5) + (10.8)(0.5) + (14.9)(0.5) + (18.1)(0.5) + (19.4)(0.5) = 0.5(69.4) = 34.7 \text{ ft}$$

$$R_6 = 0.5(6.2 + 10.8 + 14.9 + 18.1 + 19.4 + 20.2) = 0.5(89.6) = 44.8 \text{ ft}$$

14. (a) The velocities are given with units mi/h, so we must convert the 10-second intervals to hours:

$$10 \text{ seconds} = \frac{10 \text{ seconds}}{3600 \text{ seconds/h}} = \frac{1}{360} \text{ h}$$

$$\begin{aligned}\text{distance} \approx L_6 &= (182.9 \text{ mi/h})\left(\tfrac{1}{360} \text{ h}\right) + (168.0)\left(\tfrac{1}{360}\right) + (106.6)\left(\tfrac{1}{360}\right) + (99.8)\left(\tfrac{1}{360}\right) \\ &\quad + (124.5)\left(\tfrac{1}{360}\right) + (176.1)\left(\tfrac{1}{360}\right) \\ &= \frac{857.9}{360} \approx 2.383 \text{ miles}\end{aligned}$$

(b) $\text{Distance} \approx R_6 = \left(\frac{1}{360}\right)(168.0 + 106.6 + 99.8 + 124.5 + 176.1 + 175.6) = \dfrac{850.6}{360} \approx 2.363 \text{ miles}$

(c) The velocity is neither increasing nor decreasing on the given interval, so the estimates in parts (a) and (b) are neither upper nor lower estimates.

15. Lower estimate for oil leakage: $R_5 = (7.6 + 6.8 + 6.2 + 5.7 + 5.3)(2) = (31.6)(2) = 63.2$ L.

Upper estimate for oil leakage: $L_5 = (8.7 + 7.6 + 6.8 + 6.2 + 5.7)(2) = (35)(2) = 70$ L.

16. We can find an upper estimate by using the final velocity for each time interval. Thus, the distance d traveled after 62 seconds can be approximated by

$$d = \sum_{i=1}^{6} v(t_i)\, \Delta t_i = (185 \text{ ft/s})(10 \text{ s}) + 319 \cdot 5 + 447 \cdot 5 + 742 \cdot 12 + 1325 \cdot 27 + 1445 \cdot 3 = 54{,}694 \text{ ft}$$

17. For a decreasing function, using left endpoints gives us an overestimate and using right endpoints results in an underestimate. We will use M_6 to get an estimate. $\Delta t = 1$, so

$$M_6 = 1[v(0.5) + v(1.5) + v(2.5) + v(3.5) + v(4.5) + v(5.5)] \approx 55 + 40 + 28 + 18 + 10 + 4 = 155 \text{ ft}$$

For a very rough check on the above calculation, we can draw a line from $(0, 70)$ to $(6, 0)$ and calculate the area of the triangle: $\frac{1}{2}(70)(6) = 210$. This is clearly an overestimate, so our midpoint estimate of 155 is reasonable.

18. For an increasing function, using left endpoints gives us an underestimate and using right endpoints results in an overestimate. We will use M_6 to get an estimate. $\Delta t = \frac{30-0}{6} = 5 \text{ s} = \frac{5}{3600} \text{ h} = \frac{1}{720} \text{ h}$.

$$\begin{aligned}M_6 &= \tfrac{1}{720}[v(2.5) + v(7.5) + v(12.5) + v(17.5) + v(22.5) + v(27.5)] \\ &= \tfrac{1}{720}(31.25 + 66 + 88 + 103.5 + 113.75 + 119.25) = \tfrac{1}{720}(521.75) \approx 0.725 \text{ km}\end{aligned}$$

For a very rough check on the above calculation, we can draw a line from $(0, 0)$ to $(30, 120)$ and calculate the area of the triangle: $\frac{1}{2}(30)(120) = 1800$. Divide by 3600 to get 0.5, which is clearly an underestimate, making our midpoint estimate of 0.725 seem reasonable. Of course, answers will vary due to different readings of the graph.

19. $f(t) = -t(t-21)(t+1)$ and $\Delta t = \frac{12-0}{6} = 2$

$$\begin{aligned} M_6 &= 2 \cdot f(1) + 2 \cdot f(3) + 2 \cdot f(5) + 2 \cdot f(7) + 2 \cdot f(9) + 2 \cdot f(11) \\ &= 2 \cdot 40 + 2 \cdot 216 + 2 \cdot 480 + 2 \cdot 784 + 2 \cdot 1080 + 2 \cdot 1320 \\ &= 7840 \text{ (infected cells/mL)} \cdot \text{days} \end{aligned}$$

Thus, the total amount of infection needed to develop symptoms of measles is about 7840 infected cells per mL of blood plasma.

20. (a) Use $\Delta t = 14$ days. The number of people who died of SARS in Singapore between March 1 and May 24, 2003, using left endpoints is

$$L_6 = 14(0.0079 + 0.0638 + 0.1944 + 0.4435 + 0.5620 + 0.4630) = 14(1.7346) = 24.2844 \approx 24 \text{ people}$$

Using right endpoints,

$$R_6 = 14(0.0638 + 0.1944 + 0.4435 + 0.5620 + 0.4630 + 0.2897) = 14(2.0164) = 28.2296 \approx 28 \text{ people}$$

(b) Let t be the number of days since March 1, 2003, $f(t)$ be the number of deaths per day on day t, and the graph of $y = f(t)$ be a reasonable continuous function on the interval $[0, 84]$. Then the number of SARS deaths from $t = a$ to $t = b$ is approximately equal to the area under the curve $y = f(t)$ from $t = a$ to $t = b$.

21. $f(x) = \dfrac{2x}{x^2+1}$, $1 \le x \le 3$. $\Delta x = (3-1)/n = 2/n$ and $x_i = 1 + i\Delta x = 1 + 2i/n$.

$$A = \lim_{n\to\infty} R_n = \lim_{n\to\infty} \sum_{i=1}^{n} f(x_i)\Delta x = \lim_{n\to\infty} \sum_{i=1}^{n} \frac{2(1+2i/n)}{(1+2i/n)^2+1} \cdot \frac{2}{n}.$$

22. $f(x) = x^2 + \sqrt{1+2x}$, $4 \le x \le 7$. $\Delta x = (7-4)/n = 3/n$ and $x_i = 4 + i\,\Delta x = 4 + 3i/n$.

$$A = \lim_{n\to\infty} R_n = \lim_{n\to\infty} \sum_{i=1}^{n} f(x_i)\,\Delta x = \lim_{n\to\infty} \sum_{i=1}^{n} \left[(4+3i/n)^2 + \sqrt{1+2(4+3i/n)}\right] \cdot \frac{3}{n}.$$

23. $f(x) = \sqrt{\sin x}$, $0 \le x \le \pi$. $\Delta x = (\pi - 0)/n = \pi/n$ and $x_i = 0 + i\,\Delta x = \pi i/n$.

$$A = \lim_{n\to\infty} R_n = \lim_{n\to\infty} \sum_{i=1}^{n} f(x_i)\,\Delta x = \lim_{n\to\infty} \sum_{i=1}^{n} \sqrt{\sin(\pi i/n)} \cdot \frac{\pi}{n}.$$

24. $\displaystyle\lim_{n\to\infty} \sum_{i=1}^{n} \frac{3}{n}\sqrt{1+\frac{3i}{n}}$ can be interpreted as the area of the region lying under the graph of $y = \sqrt{1+x}$ on the interval $[0,3]$, since for $y = \sqrt{1+x}$ on $[0,3]$ with $\Delta x = \dfrac{3-0}{n} = \dfrac{3}{n}$, $x_i = 0 + i\,\Delta x = \dfrac{3i}{n}$, and $x_i^* = x_i$, the expression for the area is $A = \displaystyle\lim_{n\to\infty} \sum_{i=1}^{n} f(x_i^*)\,\Delta x = \lim_{n\to\infty} \sum_{i=1}^{n} \sqrt{1+\frac{3i}{n}}\,\frac{3}{n}$. Note that this answer is not unique. We could use $y = \sqrt{x}$ on $[1,4]$ or, in general, $y = \sqrt{x-n}$ on $[n+1, n+4]$, where n is any real number.

25. $\displaystyle\lim_{n\to\infty} \sum_{i=1}^{n} \frac{\pi}{4n} \tan\frac{i\pi}{4n}$ can be interpreted as the area of the region lying under the graph of $y = \tan x$ on the interval $\left[0, \frac{\pi}{4}\right]$, since for $y = \tan x$ on $\left[0, \frac{\pi}{4}\right]$ with $\Delta x = \dfrac{\pi/4 - 0}{n} = \dfrac{\pi}{4n}$, $x_i = 0 + i\,\Delta x = \dfrac{i\pi}{4n}$, and $x_i^* = x_i$, the expression for the area is $A = \displaystyle\lim_{n\to\infty} \sum_{i=1}^{n} f(x_i^*)\,\Delta x = \lim_{n\to\infty} \sum_{i=1}^{n} \tan\left(\frac{i\pi}{4n}\right)\frac{\pi}{4n}$. Note that this answer is not unique, since the expression for the area is the same for the function $y = \tan(x - k\pi)$ on the interval $\left[k\pi, k\pi + \frac{\pi}{4}\right]$, where k is any integer.

26. (a) $\Delta x = \dfrac{1-0}{n} = \dfrac{1}{n}$ and $x_i = 0 + i\,\Delta x = \dfrac{i}{n}$. $A = \lim\limits_{n\to\infty} R_n = \lim\limits_{n\to\infty} \sum\limits_{i=1}^{n} f(x_i)\,\Delta x = \lim\limits_{n\to\infty} \sum\limits_{i=1}^{n} \left(\dfrac{i}{n}\right)^3 \cdot \dfrac{1}{n}$.

(b) $\lim\limits_{n\to\infty} \sum\limits_{i=1}^{n} \dfrac{i^3}{n^3} \cdot \dfrac{1}{n} = \lim\limits_{n\to\infty} \dfrac{1}{n^4} \sum\limits_{i=1}^{n} i^3 = \lim\limits_{n\to\infty} \dfrac{1}{n^4} \left[\dfrac{n(n+1)}{2}\right]^2 = \lim\limits_{n\to\infty} \dfrac{(n+1)^2}{4n^2} = \dfrac{1}{4} \lim\limits_{n\to\infty} \left(1 + \dfrac{1}{n}\right)^2 = \dfrac{1}{4}$

27. (a) Since f is an increasing function, L_n is an underestimate of A [lower sum] and R_n is an overestimate of A [upper sum]. Thus, A, L_n, and R_n are related by the inequality $L_n < A < R_n$.

(b)

$$\begin{aligned} R_n &= f(x_1)\Delta x + f(x_2)\Delta x + \cdots + f(x_n)\Delta x \\ L_n &= f(x_0)\Delta x + f(x_1)\Delta x + \cdots + f(x_{n-1})\Delta x \\ R_n - L_n &= f(x_n)\Delta x - f(x_0)\Delta x \\ &= \Delta x[f(x_n) - f(x_0)] \\ &= \frac{b-a}{n}[f(b) - f(a)] \end{aligned}$$

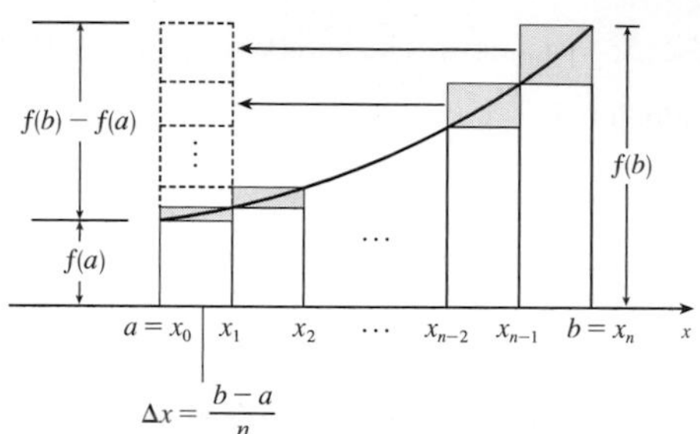

In the diagram, $R_n - L_n$ is the sum of the areas of the shaded rectangles. By sliding the shaded rectangles to the left so that they stack on top of the leftmost shaded rectangle, we form a rectangle of height $f(b) - f(a)$ and width $\dfrac{b-a}{n}$.

(c) $A > L_n$, so $R_n - A < R_n - L_n$; that is, $R_n - A < \dfrac{b-a}{n}[f(b) - f(a)]$.

28. $R_n - A < \dfrac{b-a}{n}[f(b) - f(a)] = \dfrac{3-1}{n}[f(3) - f(1)] = \dfrac{2}{n}(e^3 - e)$

Solving $\dfrac{2}{n}(e^3 - e) < 0.0001$ for n gives us $2(e^3 - e) < 0.0001n \;\Rightarrow\; n > \dfrac{2(e^3 - e)}{0.0001} \;\Rightarrow\; n > 347{,}345.1$. Thus, a value of n that assures us that $R_n - A < 0.0001$ is $n = 347{,}346$. [This is not the *least* value of n.]

29. (a) $y = f(x) = x^5$. $\Delta x = \dfrac{2-0}{n} = \dfrac{2}{n}$ and $x_i = 0 + i\,\Delta x = \dfrac{2i}{n}$.

$$A = \lim_{n\to\infty} R_n = \lim_{n\to\infty} \sum_{i=1}^{n} f(x_i)\,\Delta x = \lim_{n\to\infty} \sum_{i=1}^{n} \left(\frac{2i}{n}\right)^5 \cdot \frac{2}{n} = \lim_{n\to\infty} \sum_{i=1}^{n} \frac{32i^5}{n^5} \cdot \frac{2}{n} = \lim_{n\to\infty} \frac{64}{n^6} \sum_{i=1}^{n} i^5.$$

(b) $\sum\limits_{i=1}^{n} i^5 \overset{\text{CAS}}{=} \dfrac{n^2(n+1)^2(2n^2+2n-1)}{12}$

(c)

$$\begin{aligned} \lim_{n\to\infty} \frac{64}{n^6} \cdot \frac{n^2(n+1)^2(2n^2+2n-1)}{12} &= \frac{64}{12} \lim_{n\to\infty} \frac{(n^2+2n+1)(2n^2+2n-1)}{n^2 \cdot n^2} \\ &= \frac{16}{3} \lim_{n\to\infty} \left(1 + \frac{2}{n} + \frac{1}{n^2}\right)\left(2 + \frac{2}{n} - \frac{1}{n^2}\right) = \tfrac{16}{3} \cdot 1 \cdot 2 = \tfrac{32}{3} \end{aligned}$$

30. From Example 3(a), we have $A = \lim\limits_{n\to\infty} \dfrac{2}{n} \sum\limits_{i=1}^{n} e^{-2i/n}$. Using a CAS, $\sum\limits_{i=1}^{n} e^{-2i/n} = \dfrac{e^{-2}(e^2-1)}{e^{2/n}-1}$ and

$\lim\limits_{n\to\infty} \dfrac{2}{n} \cdot \dfrac{e^{-2}(e^2-1)}{e^{2/n}-1} = e^{-2}(e^2-1) \approx 0.8647$, whereas the estimate from Example 3(b) using M_{10} was 0.8632.

31. $y = f(x) = \cos x$. $\Delta x = \dfrac{b-0}{n} = \dfrac{b}{n}$ and $x_i = 0 + i\,\Delta x = \dfrac{bi}{n}$.

$$A = \lim_{n\to\infty} R_n = \lim_{n\to\infty} \sum_{i=1}^{n} f(x_i)\,\Delta x = \lim_{n\to\infty} \sum_{i=1}^{n} \cos\left(\frac{bi}{n}\right) \cdot \frac{b}{n}$$

$$\overset{\text{CAS}}{=} \lim_{n\to\infty} \left[\frac{b \sin\left(b\left(\dfrac{1}{2n}+1\right)\right)}{2n \sin\left(\dfrac{b}{2n}\right)} - \frac{b}{2n} \right] \overset{\text{CAS}}{=} \sin b$$

If $b = \frac{\pi}{2}$, then $A = \sin\frac{\pi}{2} = 1$.

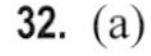

32. (a)

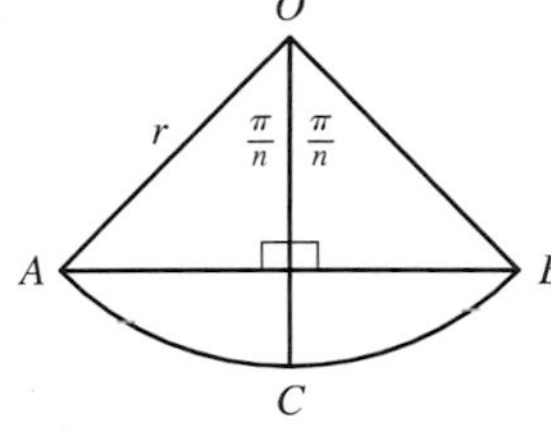

The diagram shows one of the n congruent triangles, $\triangle AOB$, with central angle $2\pi/n$. O is the center of the circle and AB is one of the sides of the polygon. Radius OC is drawn so as to bisect $\angle AOB$. It follows that OC intersects AB at right angles and bisects AB. Thus, $\triangle AOB$ is divided into two right triangles with legs of length $\frac{1}{2}(AB) = r\sin(\pi/n)$ and $r\cos(\pi/n)$. $\triangle AOB$ has area $2 \cdot \frac{1}{2}[r\sin(\pi/n)][r\cos(\pi/n)] = r^2 \sin(\pi/n)\cos(\pi/n) = \frac{1}{2}r^2\sin(2\pi/n)$, so $A_n = n \cdot \text{area}(\triangle AOB) = \frac{1}{2}nr^2\sin(2\pi/n)$.

(b) To use Equation 3.3.2, $\lim\limits_{\theta\to 0} \dfrac{\sin\theta}{\theta} = 1$, we need to have the same expression in the denominator as we have in the argument of the sine function—in this case, $2\pi/n$.

$$\lim_{n\to\infty} A_n = \lim_{n\to\infty} \tfrac{1}{2}nr^2\sin(2\pi/n) = \lim_{n\to\infty} \tfrac{1}{2}nr^2 \frac{\sin(2\pi/n)}{2\pi/n} \cdot \frac{2\pi}{n} = \lim_{n\to\infty} \frac{\sin(2\pi/n)}{2\pi/n}\pi r^2. \text{ Let } \theta = \frac{2\pi}{n}.$$

Then as $n\to\infty$, $\theta\to 0$, so $\lim\limits_{n\to\infty} \dfrac{\sin(2\pi/n)}{2\pi/n}\pi r^2 = \lim\limits_{\theta\to 0} \dfrac{\sin\theta}{\theta}\pi r^2 = (1)\,\pi r^2 = \pi r^2$.

5.2 The Definite Integral

1. $f(x) = x - 1$, $-6 \le x \le 4$. $\Delta x = \dfrac{b-a}{n} = \dfrac{4-(-6)}{5} = 2$.

Since we are using right endpoints, $x_i^* = x_i$.

$$\begin{aligned} R_5 &= \sum_{i=1}^{5} f(x_i)\,\Delta x \\ &= (\Delta x)[f(x_1) + f(x_2) + f(x_3) + f(x_4) + f(x_5) + f(x_6)] \\ &= 2[f(-4) + f(-2) + f(0) + f(2) + f(4)] \\ &= 2[-5 + (-3) + (-1) + 1 + 3] \\ &= 2(-5) = -10 \end{aligned}$$

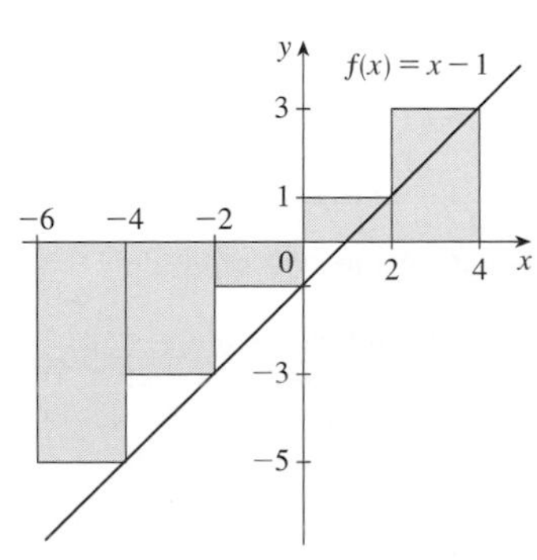

The Riemann sum represents the sum of the areas of the two rectangles above the x-axis minus the sum of the areas of the three rectangles below the x-axis; that is, the *net area* of the rectangles with respect to the x-axis.

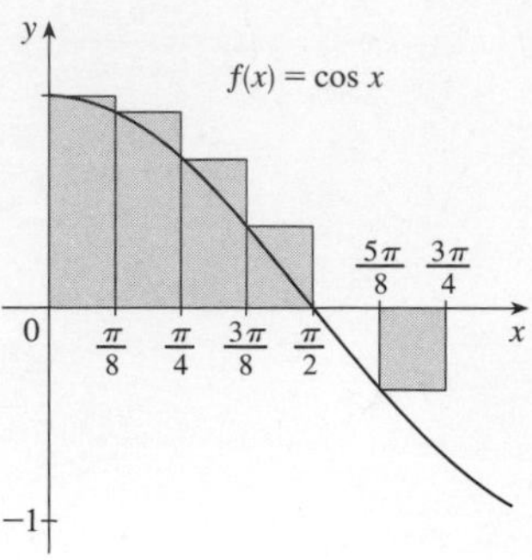

2. $f(x) = \cos x$, $0 \le x \le \frac{3\pi}{4}$. $\Delta x = \dfrac{b-a}{n} = \dfrac{3\pi/4 - 0}{6} = \dfrac{\pi}{8}$.

Since we are using left endpoints, $x_i^* = x_{i-1}$.

$$\begin{aligned} L_6 &= \sum_{i=1}^{6} f(x_{i-1})\,\Delta x \\ &= (\Delta x)[f(x_0) + f(x_1) + f(x_2) + f(x_3) + f(x_4) + f(x_5)] \\ &= \tfrac{\pi}{8}[f(0) + f(\tfrac{\pi}{8}) + f(\tfrac{2\pi}{8}) + f(\tfrac{3\pi}{8}) + f(\tfrac{4\pi}{8}) + f(\tfrac{5\pi}{8})] \\ &\approx 1.033186 \end{aligned}$$

The Riemann sum represents the sum of the areas of the four rectangles above the x-axis minus the area of the rectangle below the x-axis; that is, the *net area* of the rectangles with respect to the x-axis. A sixth rectangle is degenerate, with height 0, and has no area.

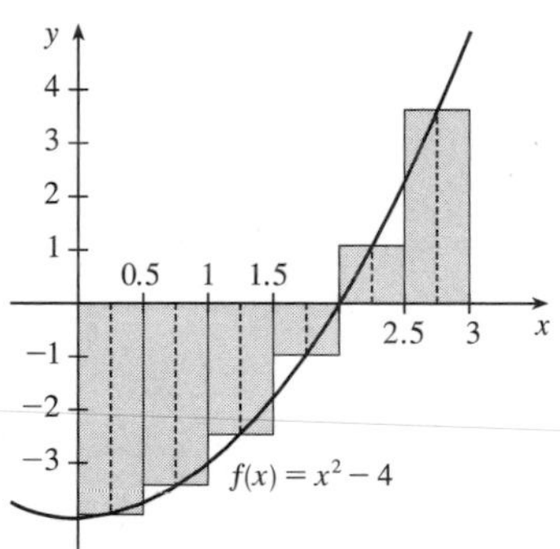

3. $f(x) = x^2 - 4$, $0 \le x \le 3$. $\Delta x = \dfrac{b-a}{n} = \dfrac{3-0}{6} = \dfrac{1}{2}$.

Since we are using midpoints, $x_i^* = \overline{x}_i = \frac{1}{2}(x_{i-1} + x_i)$.

$$\begin{aligned} M_6 &= \sum_{i=1}^{6} f(\overline{x}_i)\,\Delta x \\ &= (\Delta x)[f(\overline{x}_1) + f(\overline{x}_2) + f(\overline{x}_3) + f(\overline{x}_4) + f(\overline{x}_5) + f(\overline{x}_6)] \\ &= \tfrac{1}{2}\left[f(\tfrac{1}{4}) + f(\tfrac{3}{4}) + f(\tfrac{5}{4}) + f(\tfrac{7}{4}) + f(\tfrac{9}{4}) + f(\tfrac{11}{4})\right] \\ &= \tfrac{1}{2}\left(-\tfrac{63}{16} - \tfrac{55}{16} - \tfrac{39}{16} - \tfrac{15}{16} + \tfrac{17}{16} + \tfrac{57}{16}\right) = \tfrac{1}{2}\left(-\tfrac{98}{16}\right) = -\tfrac{49}{16} \end{aligned}$$

The Riemann sum represents the sum of the areas of the two rectangles above the x-axis minus the sum of the areas of the four rectangles below the x-axis; that is, the *net area* of the rectangles with respect to the x-axis.

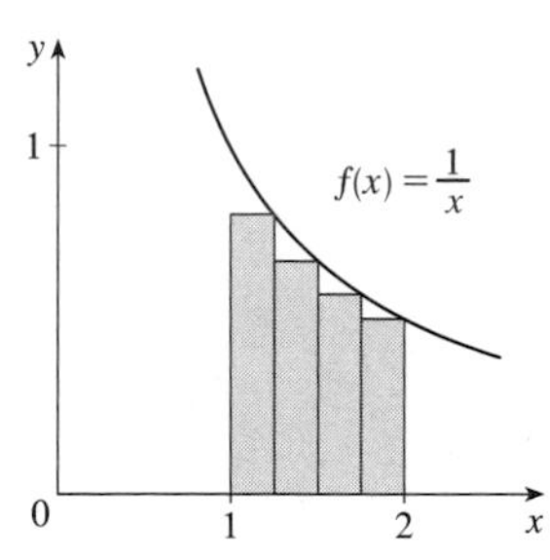

4. (a) $f(x) = \dfrac{1}{x}$, $1 \le x \le 2$. $\Delta x = \dfrac{b-a}{n} = \dfrac{2-1}{4} = \dfrac{1}{4}$.

Since we are using right endpoints, $x_i^* = x_i$.

$$\begin{aligned} R_4 &= \sum_{i=1}^{4} f(x_i)\,\Delta x \\ &= (\Delta x)[f(x_1) + f(x_2) + f(x_3) + f(x_4)] \\ &= \tfrac{1}{4}[f(\tfrac{5}{4}) + f(\tfrac{6}{4}) + f(\tfrac{7}{4}) + f(\tfrac{8}{4})] \\ &= \tfrac{1}{4}\left[\tfrac{4}{5} + \tfrac{2}{3} + \tfrac{4}{7} + \tfrac{1}{2}\right] \\ &\approx 0.634524 \end{aligned}$$

The Riemann sum represents the sum of the areas of the four rectangles.

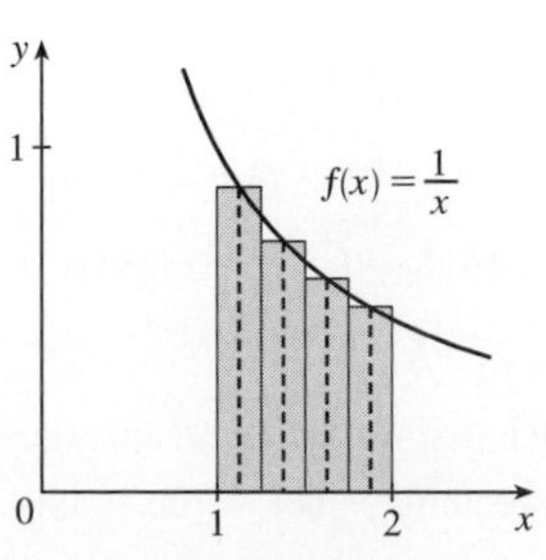

(b) Since we are using midpoints, $x_i^* = \overline{x}_i = \frac{1}{2}(x_{i-1} + x_i)$.

$$\begin{aligned} M_4 &= \sum_{i=1}^{4} f(\overline{x}_i)\,\Delta x \\ &= (\Delta x)[f(\overline{x}_1) + f(\overline{x}_2) + f(\overline{x}_3) + f(\overline{x}_4)] \\ &= \tfrac{1}{4}\left[f(\tfrac{9}{8}) + f(\tfrac{11}{8}) + f(\tfrac{13}{8}) + f(\tfrac{15}{8})\right] \\ &= \tfrac{1}{4}\left(\tfrac{8}{9} + \tfrac{8}{11} + \tfrac{8}{13} + \tfrac{8}{15}\right) \approx 0.691220 \end{aligned}$$

The Riemann sum represents the sum of the areas of the four rectangles.

5. (a) $\displaystyle\int_0^{10} f(x)\,dx \approx R_5 = [f(2) + f(4) + f(6) + f(8) + f(10)]\,\Delta x$

$$= [-1 + 0 + (-2) + 2 + 4](2) = 3(2) = 6$$

(b) $\displaystyle\int_0^{10} f(x)\,dx \approx L_5 = [f(0) + f(2) + f(4) + f(6) + f(8)]\,\Delta x$

$$= [3 + (-1) + 0 + (-2) + 2](2) = 2(2) = 4$$

(c) $\displaystyle\int_0^{10} f(x)\,dx \approx M_5 = [f(1) + f(3) + f(5) + f(7) + f(9)]\,\Delta x$

$$= [0 + (-1) + (-1) + 0 + 3](2) = 1(2) = 2$$

6. (a) $\displaystyle\int_{-2}^{4} g(x)\,dx \approx R_6 = [g(-1) + g(0) + g(1) + g(2) + g(3) + g(4)]\,\Delta x$

$$= \left[-\tfrac{3}{2} + 0 + \tfrac{3}{2} + \tfrac{1}{2} + (-1) + \tfrac{1}{2}\right](1) = 0$$

(b) $\displaystyle\int_{-2}^{4} g(x)\,dx \approx L_6 = [g(-2) + g(-1) + g(0) + g(1) + g(2) + g(3)]\,\Delta x$

$$= \left[0 + \left(-\tfrac{3}{2}\right) + 0 + \tfrac{3}{2} + \tfrac{1}{2} + (-1)\right](1) = -\tfrac{1}{2}$$

(c) $\displaystyle\int_{-2}^{4} g(x)\,dx \approx M_6 = \left[g\left(-\tfrac{3}{2}\right) + g\left(-\tfrac{1}{2}\right) + g\left(\tfrac{1}{2}\right) + g\left(\tfrac{3}{2}\right) + g\left(\tfrac{5}{2}\right) + g\left(\tfrac{7}{2}\right)\right]\Delta x$

$$= \left[-1 + (-1) + 1 + 1 + 0 + \left(-\tfrac{1}{2}\right)\right](1) = -\tfrac{1}{2}$$

7. Since f is increasing, $L_5 \le \int_{10}^{30} f(x)\,dx \le R_5$.

Lower estimate $\displaystyle = L_5 = \sum_{i=1}^{5} f(x_{i-1})\Delta x = 4[f(10) + f(14) + f(18) + f(22) + f(26)]$

$$= 4[-12 + (-6) + (-2) + 1 + 3] = 4(-16) = -64$$

Upper estimate $\displaystyle = R_5 = \sum_{i=1}^{5} f(x)\Delta x = 4[f(14) + f(18) + f(22) + f(26) + f(30)]$

$$= 4[-6 + (-2) + 1 + 3 + 8] = 4(4) = 16$$

8. (a) Using the right endpoints to approximate $\int_3^9 f(x)\,dx$, we have

$\displaystyle\sum_{i=1}^{3} f(x_i)\,\Delta x = 2[f(5) + f(7) + f(9)] = 2\,(-0.6 + 0.9 + 1.8) = 4.2.$

Since f is *increasing*, using *right* endpoints gives an *overestimate*.

(b) Using the left endpoints to approximate $\int_3^9 f(x)\,dx$, we have

$\displaystyle\sum_{i=1}^{3} f(x_{i-1})\,\Delta x = 2[f(3) + f(5) + f(7)] = 2\,(-3.4 - 0.6 + 0.9) = -6.2.$

Since f is *increasing*, using *left* endpoints gives an *underestimate*.

(c) Using the midpoint of each interval to approximate $\int_3^9 f(x)\,dx$, we have

$\displaystyle\sum_{i=1}^{3} f(\overline{x}_i)\,\Delta x = 2[f(4) + f(6) + f(8)] = 2\,(-2.1 + 0.3 + 1.4) = -0.8.$

We cannot say anything about the midpoint estimate compared to the exact value of the integral.

9. $\Delta x = (8-0)/4 = 2$, so the endpoints are 0, 2, 4, 6, and 8, and the midpoints are 1, 3, 5, and 7. The Midpoint Rule gives

$\int_0^8 \sin\sqrt{x}\,dx \approx \sum_{i=1}^{4} f(\bar{x}_i)\,\Delta x = 2\left(\sin\sqrt{1} + \sin\sqrt{3} + \sin\sqrt{5} + \sin\sqrt{7}\right) \approx 2(3.0910) = 6.1820.$

10. $\Delta x = (1-0)/5 = \frac{1}{5}$, so the endpoints are 0, $\frac{1}{5}$, $\frac{2}{5}$, $\frac{3}{5}$, $\frac{4}{5}$, and 1, and the midpoints are $\frac{1}{10}$, $\frac{3}{10}$, $\frac{5}{10}$, $\frac{7}{10}$, and $\frac{9}{10}$. The Midpoint Rule gives

$$\int_0^1 \sqrt{x^3+1}\,dx \approx \sum_{i=1}^{5} f(\bar{x}_i)\,\Delta x = \frac{1}{5}\left(\sqrt{\left(\tfrac{1}{10}\right)^3+1} + \sqrt{\left(\tfrac{3}{10}\right)^3+1} + \sqrt{\left(\tfrac{5}{10}\right)^3+1} + \sqrt{\left(\tfrac{7}{10}\right)^3+1} + \sqrt{\left(\tfrac{9}{10}\right)^3+1}\right)$$
$$\approx 1.1097$$

11. $\Delta x = (2-0)/5 = \frac{2}{5}$, so the endpoints are 0, $\frac{2}{5}$, $\frac{4}{5}$, $\frac{6}{5}$, $\frac{8}{5}$, and 2, and the midpoints are $\frac{1}{5}$, $\frac{3}{5}$, $\frac{5}{5}$, $\frac{7}{5}$ and $\frac{9}{5}$. The Midpoint Rule gives

$$\int_0^2 \frac{x}{x+1}\,dx \approx \sum_{i=1}^{5} f(\bar{x}_i)\,\Delta x = \frac{2}{5}\left(\frac{\frac{1}{5}}{\frac{1}{5}+1} + \frac{\frac{3}{5}}{\frac{3}{5}+1} + \frac{\frac{5}{5}}{\frac{5}{5}+1} + \frac{\frac{7}{5}}{\frac{7}{5}+1} + \frac{\frac{9}{5}}{\frac{9}{5}+1}\right) = \frac{2}{5}\left(\frac{127}{56}\right) = \frac{127}{140} \approx 0.9071.$$

12. $\Delta x = (\pi-0)/4 = \frac{\pi}{4}$, so the endpoints are $\frac{\pi}{4}$, $\frac{2\pi}{4}$, $\frac{3\pi}{4}$, and $\frac{4\pi}{4}$, and the midpoints are $\frac{\pi}{8}$, $\frac{3\pi}{8}$, $\frac{5\pi}{8}$, and $\frac{7\pi}{8}$. The Midpoint Rule gives

$$\int_0^\pi x\sin^2 x\,dx \approx \sum_{i=1}^{5} f(\bar{x}_i)\,\Delta x = \frac{\pi}{4}\left(\frac{\pi}{8}\sin^2\frac{\pi}{8} + \frac{3\pi}{8}\sin^2\frac{3\pi}{8} + \frac{5\pi}{8}\sin^2\frac{5\pi}{8} + \frac{7\pi}{8}\sin^2\frac{7\pi}{8}\right) \approx 2.4674$$

13. In Maple 14, use the commands `with(Student[Calculus1])` and `ReimannSum(x/(x+1),0..2,partition=5,method=midpoint,output=plot)`. In some older versions of Maple, use `with(student)` to load the sum and box commands, then `m:=middlesum(x/(x+1),x=0..2)`, which gives us the sum in summation notation, then `M:=evalf(m)` to get the numerical approximation, and finally `middlebox(x/(x+1),x=0..2)` to generate the graph. The values obtained for $n = 5$, 10, and 20 are 0.9071, 0.9029, and 0.9018, respectively.

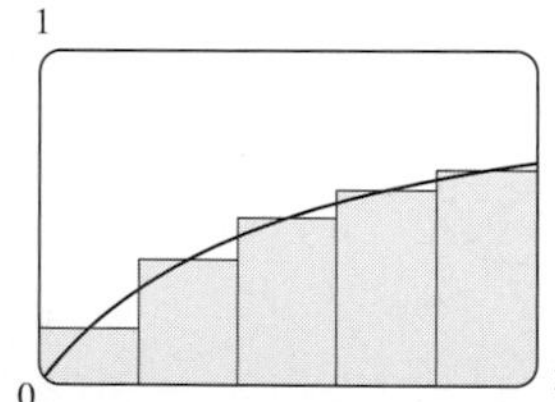

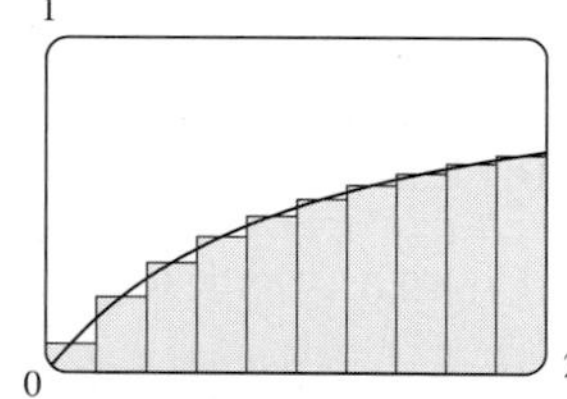

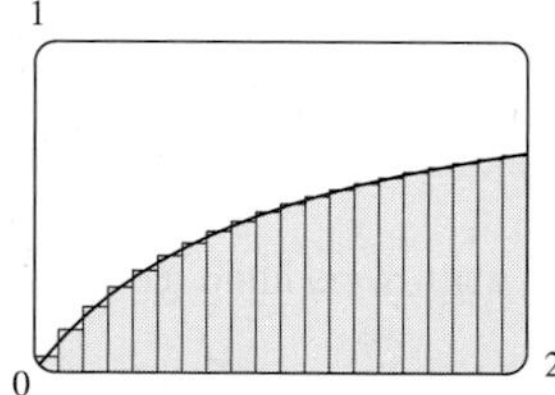

14. For $f(x) = x/(x+1)$ on $[0,2]$, we calculate $L_{100} \approx 0.89469$ and $R_{100} \approx 0.90802$. Since f is increasing on $[0,2]$, L_{100} is an underestimate of $\int_0^2 \frac{x}{x+1}\,dx$ and R_{100} is an overestimate. Thus, $0.8946 < \int_0^2 \frac{x}{x+1}\,dx < 0.9081$.

15. We'll create the table of values to approximate $\int_0^\pi \sin x\,dx$ by using the program in the solution to Exercise 5.1.9 with $\text{Y}_1 = \sin x$, Xmin $= 0$, Xmax $= \pi$, and $n = 5$, 10, 50, and 100.

The values of R_n appear to be approaching 2.

n	R_n
5	1.933766
10	1.983524
50	1.999342
100	1.999836

16. $\int_0^2 e^{-x^2}\,dx$ with $n = 5$, 10, 50, and 100.

n	L_n	R_n
5	1.077467	0.684794
10	0.980007	0.783670
50	0.901705	0.862438
100	0.891896	0.872262

The value of the integral lies between 0.872 and 0.892. Note that $f(x) = e^{-x^2}$ is decreasing on $(0, 2)$. We cannot make a similar statement for $\int_{-1}^2 e^{-x^2}\,dx$ since f is increasing on $(-1, 0)$.

17. On $[0, 1]$, $\displaystyle\lim_{n\to\infty}\sum_{i=1}^n \frac{e^{x_i}}{1+x_i}\,\Delta x = \int_0^1 \frac{e^x}{1+x}\,dx$.

18. On $[2, 5]$, $\displaystyle\lim_{n\to\infty}\sum_{i=1}^n x_i\sqrt{1+x_i^3}\,\Delta x = \int_2^5 x\sqrt{1+x^3}\,dx$.

19. On $[2, 7]$, $\displaystyle\lim_{n\to\infty}\sum_{i=1}^n [5(x_i^*)^3 - 4x_i^*]\,\Delta x = \int_2^7 (5x^3 - 4x)\,dx$.

20. On $[1, 3]$, $\displaystyle\lim_{n\to\infty}\sum_{i=1}^n \frac{x_i^*}{(x_i^*)^2+4}\,\Delta x = \int_1^3 \frac{x}{x^2+4}\,dx$.

21. Note that $\Delta x = \dfrac{5-2}{n} = \dfrac{3}{n}$ and $x_i = 2 + i\,\Delta x = 2 + \dfrac{3i}{n}$.

$$\begin{aligned}
\int_2^5 (4-2x)\,dx &= \lim_{n\to\infty}\sum_{i=1}^n f(x_i)\,\Delta x = \lim_{n\to\infty}\sum_{i=1}^n f\left(2+\frac{3i}{n}\right)\frac{3}{n} = \lim_{n\to\infty}\frac{3}{n}\sum_{i=1}^n\left[4-2\left(2+\frac{3i}{n}\right)\right] \\
&= \lim_{n\to\infty}\frac{3}{n}\sum_{i=1}^n\left[-\frac{6i}{n}\right] = \lim_{n\to\infty}\frac{3}{n}\left(-\frac{6}{n}\right)\sum_{i=1}^n i = \lim_{n\to\infty}\left(-\frac{18}{n^2}\right)\left[\frac{n(n+1)}{2}\right] \\
&= \lim_{n\to\infty}\left(-\frac{18}{2}\right)\left(\frac{n+1}{n}\right) = -9\lim_{n\to\infty}\left(1+\frac{1}{n}\right) = -9(1) = -9
\end{aligned}$$

22. Note that $\Delta x = \dfrac{4-1}{n} = \dfrac{3}{n}$ and $x_i = 1 + i\,\Delta x = 1 + \dfrac{3i}{n}$.

$$\begin{aligned}
\int_1^4 (x^2-4x+2)\,dx &= \lim_{n\to\infty}\sum_{i=1}^n f(x_i)\,\Delta x = \lim_{n\to\infty}\sum_{i=1}^n f\left(1+\frac{3i}{n}\right)\frac{3}{n} = \lim_{n\to\infty}\frac{3}{n}\sum_{i=1}^n\left[\left(1+\frac{3i}{n}\right)^2 - 4\left(1+\frac{3i}{n}\right)+2\right] \\
&= \lim_{n\to\infty}\frac{3}{n}\sum_{i=1}^n\left[1+\frac{6i}{n}+\frac{9i^2}{n^2}-4-\frac{12i}{n}+2\right] = \lim_{n\to\infty}\frac{3}{n}\sum_{i=1}^n\left[\frac{9i^2}{n^2}-\frac{6i}{n}-1\right] \\
&= \lim_{n\to\infty}\frac{3}{n}\left[\frac{9}{n^2}\sum_{i=1}^n i^2 - \frac{6}{n}\sum_{i=1}^n i - \sum_{i=1}^n 1\right] \\
&= \lim_{n\to\infty}\left[\frac{27}{n^3}\frac{n(n+1)(2n+1)}{6} - \frac{18}{n^2}\frac{n(n+1)}{2} - \frac{3}{n}\cdot n(1)\right] \\
&= \lim_{n\to\infty}\left[\frac{9}{2}\frac{(n+1)(2n+1)}{n^2} - 9\frac{n+1}{n} - 3\right] = \lim_{n\to\infty}\left[\frac{9}{2}\frac{n+1}{n}\frac{2n+1}{n} - 9\left(1+\frac{1}{n}\right) - 3\right] \\
&= \lim_{n\to\infty}\left[\frac{9}{2}\left(1+\frac{1}{n}\right)\left(2+\frac{1}{n}\right) - 9\left(1+\frac{1}{n}\right) - 3\right] = \frac{9}{2}(1)(2) - 9(1) - 3 = -3
\end{aligned}$$

23. Note that $\Delta x = \dfrac{0-(-2)}{n} = \dfrac{2}{n}$ and $x_i = -2 + i\,\Delta x = -2 + \dfrac{2i}{n}$.

$$\int_{-2}^{0}(x^2+x)\,dx = \lim_{n\to\infty}\sum_{i=1}^{n} f(x_i)\,\Delta x = \lim_{n\to\infty}\sum_{i=1}^{n} f\left(-2+\frac{2i}{n}\right)\frac{2}{n} = \lim_{n\to\infty}\frac{2}{n}\sum_{i=1}^{n}\left[\left(-2+\frac{2i}{n}\right)^2+\left(-2+\frac{2i}{n}\right)\right]$$

$$= \lim_{n\to\infty}\frac{2}{n}\sum_{i=1}^{n}\left[4-\frac{8i}{n}+\frac{4i^2}{n^2}-2+\frac{2i}{n}\right] = \lim_{n\to\infty}\frac{2}{n}\sum_{i=1}^{n}\left(\frac{4i^2}{n^2}-\frac{6i}{n}+2\right)$$

$$= \lim_{n\to\infty}\frac{2}{n}\left[\frac{4}{n^2}\sum_{i=1}^{n} i^2 - \frac{6}{n}\sum_{i=1}^{n} i + \sum_{i=1}^{n} 2\right] = \lim_{n\to\infty}\left[\frac{8}{n^3}\frac{n(n+1)(2n+1)}{6} - \frac{12}{n^2}\frac{n(n+1)}{2} + \frac{2}{n}\cdot n(2)\right]$$

$$= \lim_{n\to\infty}\left[\frac{4}{3}\frac{(n+1)(2n+1)}{n^2} - 6\,\frac{n+1}{n} + 4\right] = \lim_{n\to\infty}\left[\frac{4}{3}\frac{n+1}{n}\frac{2n+1}{n} - 6\left(1+\frac{1}{n}\right)+4\right]$$

$$= \lim_{n\to\infty}\left[\frac{4}{3}\left(1+\frac{1}{n}\right)\left(2+\frac{1}{n}\right) - 6\left(1+\frac{1}{n}\right)+4\right] = \frac{4}{3}(1)(2) - 6(1) + 4 = \frac{2}{3}$$

24. Note that $\Delta x = \dfrac{2-0}{n} = \dfrac{2}{n}$ and $x_i = 0 + i\,\Delta x = \dfrac{2i}{n}$.

$$\int_{0}^{2}(2x-x^3)\,dx = \lim_{n\to\infty}\sum_{i=1}^{n} f(x_i)\,\Delta x = \lim_{n\to\infty}\sum_{i=1}^{n} f\left(\frac{2i}{n}\right)\frac{2}{n} = \lim_{n\to\infty}\frac{2}{n}\sum_{i=1}^{n}\left[2\left(\frac{2i}{n}\right)-\left(\frac{2i}{n}\right)^3\right]$$

$$= \lim_{n\to\infty}\frac{2}{n}\sum_{i=1}^{n}\left[\frac{4i}{n}-\frac{8i^3}{n^3}\right] = \lim_{n\to\infty}\frac{2}{n}\left[\frac{4}{n}\sum_{i=1}^{n} i - \frac{8}{n^3}\sum_{i=1}^{n} i^3\right]$$

$$= \lim_{n\to\infty}\left\{\frac{8}{n^2}\frac{n(n+1)}{2} - \frac{16}{n^4}\left[\frac{n(n+1)}{2}\right]^2\right\} = \lim_{n\to\infty}\left[4\frac{n+1}{n} - 4\,\frac{(n+1)^2}{n^2}\right]$$

$$= \lim_{n\to\infty}\left[4\left(1+\frac{1}{n}\right) - 4\,\frac{n+1}{n}\,\frac{n+1}{n}\right] = \lim_{n\to\infty}\left[4\left(1+\frac{1}{n}\right) - 4\left(1+\frac{1}{n}\right)\left(1+\frac{1}{n}\right)\right]$$

$$= 4(1) - 4(1)(1) = 0$$

25. Note that $\Delta x = \dfrac{1-0}{n} = \dfrac{1}{n}$ and $x_i = 0 + i\,\Delta x = \dfrac{i}{n}$.

$$\int_{0}^{1}(x^3-3x^2)\,dx = \lim_{n\to\infty}\sum_{i=1}^{n} f(x_i)\,\Delta x = \lim_{n\to\infty}\sum_{i=1}^{n} f\left(\frac{i}{n}\right)\Delta x = \lim_{n\to\infty}\sum_{i=1}^{n}\left[\left(\frac{i}{n}\right)^3 - 3\left(\frac{i}{n}\right)^2\right]\frac{1}{n}$$

$$= \lim_{n\to\infty}\frac{1}{n}\sum_{i=1}^{n}\left[\frac{i^3}{n^3}-\frac{3i^2}{n^2}\right] = \lim_{n\to\infty}\frac{1}{n}\left[\frac{1}{n^3}\sum_{i=1}^{n} i^3 - \frac{3}{n^2}\sum_{i=1}^{n} i^2\right]$$

$$= \lim_{n\to\infty}\left\{\frac{1}{n^4}\left[\frac{n(n+1)}{2}\right]^2 - \frac{3}{n^3}\,\frac{n(n+1)(2n+1)}{6}\right\} = \lim_{n\to\infty}\left[\frac{1}{4}\,\frac{n+1}{n}\,\frac{n+1}{n} - \frac{1}{2}\,\frac{n+1}{n}\,\frac{2n+1}{n}\right]$$

$$= \lim_{n\to\infty}\left[\frac{1}{4}\left(1+\frac{1}{n}\right)\left(1+\frac{1}{n}\right) - \frac{1}{2}\left(1+\frac{1}{n}\right)\left(2+\frac{1}{n}\right)\right] = \frac{1}{4}(1)(1) - \frac{1}{2}(1)(2) = -\frac{3}{4}$$

26. (a) $\Delta x = (4-0)/8 = 0.5$ and $x_i^* = x_i = 0.5i$.

$$\textstyle\int_0^4 (x^2-3x)\,dx \approx \sum_{i=1}^{8} f(x_i^*)\,\Delta x$$

$$= 0.5\{[0.5^2 - 3(0.5)] + [1.0^2 - 3(1.0)] + \cdots + [3.5^2 - 3(3.5)] + [4.0^2 - 3(4.0)]\}$$

$$= \tfrac{1}{2}\left(-\tfrac{5}{4} - 2 - \tfrac{9}{4} - 2 - \tfrac{5}{4} + 0 + \tfrac{7}{4} + 4\right) = -1.5$$

(b)

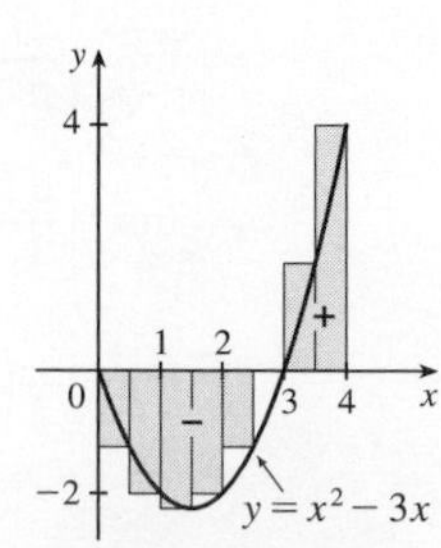

(c) $\displaystyle\int_0^4 (x^2-3x)\,dx = \lim_{n\to\infty}\sum_{i=1}^{n}\left[\left(\frac{4i}{n}\right)^2 - 3\left(\frac{4i}{n}\right)\right]\left(\frac{4}{n}\right)$

$$= \lim_{n\to\infty}\frac{4}{n}\left[\frac{16}{n^2}\sum_{i=1}^{n} i^2 - \frac{12}{n}\sum_{i=1}^{n} i\right]$$

$$= \lim_{n\to\infty}\left[\frac{64}{n^3}\cdot\frac{n(n+1)(2n+1)}{6} - \frac{48}{n^2}\cdot\frac{n(n+1)}{2}\right]$$

$$= \lim_{n\to\infty}\left[\frac{32}{3}\left(1+\frac{1}{n}\right)\left(2+\frac{1}{n}\right) - 24\left(1+\frac{1}{n}\right)\right]$$

$$= \tfrac{32}{3}\cdot 2 - 24 = -\tfrac{8}{3}$$

(d) $\int_0^4 (x^2-3x)\,dx = A_1 - A_2$, where A_1 is the area marked $+$ and A_2 is the area marked $-$.

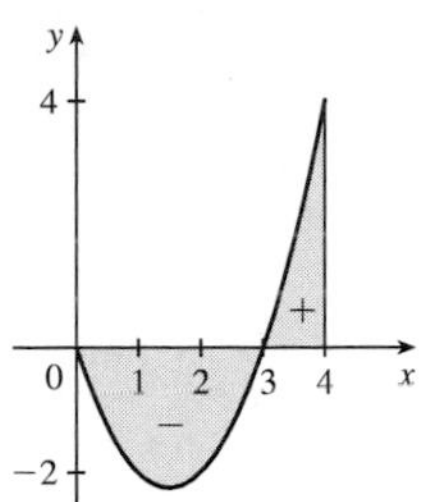

27. $\displaystyle\int_a^b x\,dx = \lim_{n\to\infty}\frac{b-a}{n}\sum_{i=1}^{n}\left[a+\frac{b-a}{n}i\right] = \lim_{n\to\infty}\left[\frac{a(b-a)}{n}\sum_{i=1}^{n}1 + \frac{(b-a)^2}{n^2}\sum_{i=1}^{n}i\right]$

$$= \lim_{n\to\infty}\left[\frac{a(b-a)}{n}n + \frac{(b-a)^2}{n^2}\cdot\frac{n(n+1)}{2}\right] = a(b-a) + \lim_{n\to\infty}\frac{(b-a)^2}{2}\left(1+\frac{1}{n}\right)$$

$$= a(b-a) + \tfrac{1}{2}(b-a)^2 = (b-a)\left(a+\tfrac{1}{2}b-\tfrac{1}{2}a\right) = (b-a)\tfrac{1}{2}(b+a) = \tfrac{1}{2}\left(b^2-a^2\right)$$

28. $\displaystyle\int_a^b x^2\,dx = \lim_{n\to\infty}\frac{b-a}{n}\sum_{i=1}^{n}\left[a+\frac{b-a}{n}i\right]^2 = \lim_{n\to\infty}\frac{b-a}{n}\sum_{i=1}^{n}\left[a^2+2a\frac{b-a}{n}i+\frac{(b-a)^2}{n^2}i^2\right]$

$$= \lim_{n\to\infty}\left[\frac{(b-a)^3}{n^3}\sum_{i=1}^{n}i^2 + \frac{2a(b-a)^2}{n^2}\sum_{i=1}^{n}i + \frac{a^2(b-a)}{n}\sum_{i=1}^{n}1\right]$$

$$= \lim_{n\to\infty}\left[\frac{(b-a)^3}{n^3}\frac{n(n+1)(2n+1)}{6} + \frac{2a(b-a)^2}{n^2}\frac{n(n+1)}{2} + \frac{a^2(b-a)}{n}n\right]$$

$$= \lim_{n\to\infty}\left[\frac{(b-a)^3}{6}\cdot 1\cdot\left(1+\frac{1}{n}\right)\left(2+\frac{1}{n}\right) + a(b-a)^2\cdot 1\cdot\left(1+\frac{1}{n}\right) + a^2(b-a)\right]$$

$$= \frac{(b-a)^3}{3} + a(b-a)^2 + a^2(b-a) = \frac{b^3-3ab^2+3a^2b-a^3}{3} + ab^2 - 2a^2b + a^3 + a^2b - a^3$$

$$= \frac{b^3}{3} - \frac{a^3}{3} - ab^2 + a^2b + ab^2 - a^2b = \frac{b^3-a^3}{3}$$

29. $f(x) = \sqrt{4+x^2}$, $a=1$, $b=3$, and $\Delta x = \dfrac{3-1}{n} = \dfrac{2}{n}$. Using Theorem 4, we get $x_i^* = x_i = 1 + i\,\Delta x = 1 + \dfrac{2i}{n}$, so

$$\int_1^3 \sqrt{4+x^2}\,dx = \lim_{n\to\infty}R_n = \lim_{n\to\infty}\sum_{i=1}^{n}\sqrt{4+\left(1+\frac{2i}{n}\right)^2}\cdot\frac{2}{n}.$$

30. $f(x) = x^2 + \dfrac{1}{x}$, $a=2$, $b=5$, and $\Delta x = \dfrac{5-2}{n} = \dfrac{3}{n}$. Using Theorem 4, we get $x_i^* = x_i = 2 + i\,\Delta x = 2 + \dfrac{3i}{n}$, so

$$\int_2^5 \left(x^2+\frac{1}{x}\right)dx = \lim_{n\to\infty}R_n = \lim_{n\to\infty}\sum_{i=1}^{n}\left[\left(2+\frac{3i}{n}\right)^2 + \frac{1}{2+\frac{3i}{n}}\right]\cdot\frac{3}{n}.$$

31. $\Delta x = (\pi-0)/n = \pi/n$ and $x_i^* = x_i = \pi i/n$.

$$\int_0^\pi \sin 5x\,dx = \lim_{n\to\infty}\sum_{i=1}^{n}(\sin 5x_i)\left(\frac{\pi}{n}\right) = \lim_{n\to\infty}\sum_{i=1}^{n}\left(\sin\frac{5\pi i}{n}\right)\frac{\pi}{n} \overset{\text{CAS}}{=} \pi\lim_{n\to\infty}\frac{1}{n}\cot\left(\frac{5\pi}{2n}\right) \overset{\text{CAS}}{=} \pi\left(\frac{2}{5\pi}\right) = \frac{2}{5}$$

32. $\Delta x = (10 - 2)/n = 8/n$ and $x_i^* = x_i = 2 + 8i/n$.

$$\int_2^{10} x^6\,dx = \lim_{n\to\infty} \sum_{i=1}^{n} \left(2 + \frac{8i}{n}\right)^6 \left(\frac{8}{n}\right) = 8 \lim_{n\to\infty} \frac{1}{n} \sum_{i=1}^{n} \left(2 + \frac{8i}{n}\right)^6$$

$$\stackrel{\text{CAS}}{=} 8 \lim_{n\to\infty} \frac{1}{n} \cdot \frac{64(58{,}593n^6 + 164{,}052n^5 + 131{,}208n^4 - 27{,}776n^2 + 2048)}{21n^5}$$

$$\stackrel{\text{CAS}}{=} 8\left(\frac{1{,}249{,}984}{7}\right) = \frac{9{,}999{,}872}{7} \approx 1{,}428{,}553.1$$

33. (a) Think of $\int_0^2 f(x)\,dx$ as the area of a trapezoid with bases 1 and 3 and height 2. The area of a trapezoid is $A = \frac{1}{2}(b + B)h$, so $\int_0^2 f(x)\,dx = \frac{1}{2}(1 + 3)2 = 4$.

(b) $\int_0^5 f(x)\,dx = \int_0^2 f(x)\,dx + \int_2^3 f(x)\,dx + \int_3^5 f(x)\,dx$

trapezoid   rectangle   triangle

$= \frac{1}{2}(1 + 3)2 + 3 \cdot 1 + \frac{1}{2} \cdot 2 \cdot 3 = 4 + 3 + 3 = 10$

(c) $\int_5^7 f(x)\,dx$ is the negative of the area of the triangle with base 2 and height 3. $\int_5^7 f(x)\,dx = -\frac{1}{2} \cdot 2 \cdot 3 = -3$.

(d) $\int_7^9 f(x)\,dx$ is the negative of the area of a trapezoid with bases 3 and 2 and height 2, so it equals $-\frac{1}{2}(B + b)h = -\frac{1}{2}(3 + 2)2 = -5$. Thus,

$\int_0^9 f(x)\,dx = \int_0^5 f(x)\,dx + \int_5^7 f(x)\,dx + \int_7^9 f(x)\,dx = 10 + (-3) + (-5) = 2$.

34. (a) $\int_0^2 g(x)\,dx = \frac{1}{2} \cdot 4 \cdot 2 = 4$ [area of a triangle]

(b) $\int_2^6 g(x)\,dx = -\frac{1}{2}\pi(2)^2 = -2\pi$ [negative of the area of a semicircle]

(c) $\int_6^7 g(x)\,dx = \frac{1}{2} \cdot 1 \cdot 1 = \frac{1}{2}$ [area of a triangle]

$\int_0^7 g(x)\,dx = \int_0^2 g(x)\,dx + \int_2^6 g(x)\,dx + \int_6^7 g(x)\,dx = 4 - 2\pi + \frac{1}{2} = 4.5 - 2\pi$

35. $\int_{-1}^2 (1 - x)\,dx$ can be interpreted as the difference of the areas of the two shaded triangles; that is, $\frac{1}{2}(2)(2) - \frac{1}{2}(1)(1) = 2 - \frac{1}{2} = \frac{3}{2}$.

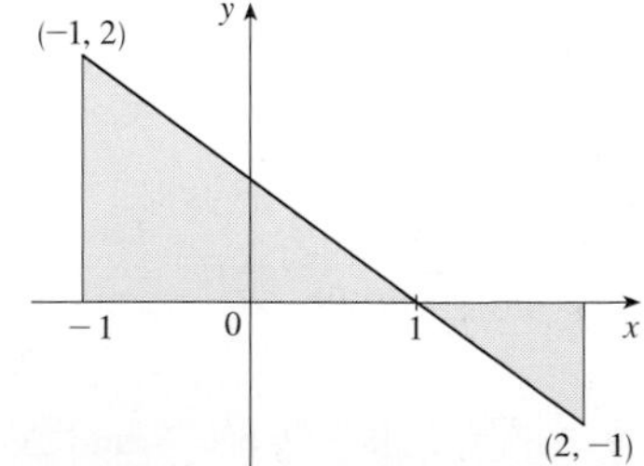

36. $\int_0^9 \left(\frac{1}{3}x - 2\right) dx$ can be interpreted as the difference of the areas of the two shaded triangles; that is, $-\frac{1}{2}(6)(2) + \frac{1}{2}(3)(1) = -6 + \frac{3}{2} = -\frac{9}{2}$.

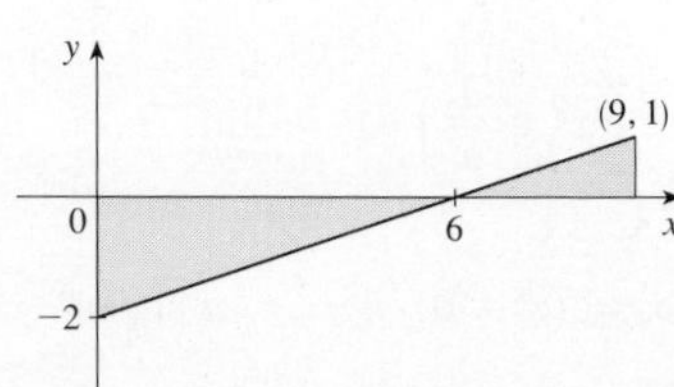

37. $\int_{-3}^{0} \left(1 + \sqrt{9 - x^2}\,\right) dx$ can be interpreted as the area under the graph of $f(x) = 1 + \sqrt{9 - x^2}$ between $x = -3$ and $x = 0$. This is equal to one-quarter the area of the circle with radius 3, plus the area of the rectangle, so $\int_{-3}^{0} \left(1 + \sqrt{9 - x^2}\,\right) dx = \frac{1}{4}\pi \cdot 3^2 + 1 \cdot 3 = 3 + \frac{9}{4}\pi$.

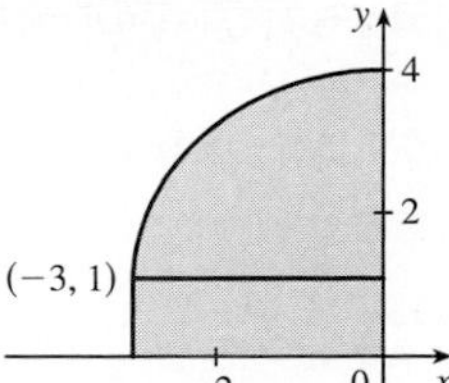

38. $\int_{-5}^{5} \left(x - \sqrt{25 - x^2}\,\right) dx = \int_{-5}^{5} x\,dx - \int_{-5}^{5} \sqrt{25 - x^2}\,dx$. By symmetry, the value of the first integral is 0 since the shaded area above the x-axis equals the shaded area below the x-axis. The second integral can be interpreted as one half the area of a circle with radius 5; that is, $\frac{1}{2}\pi(5)^2 = \frac{25}{2}\pi$. Thus, the value of the original integral is $0 - \frac{25}{2}\pi = -\frac{25}{2}\pi$.

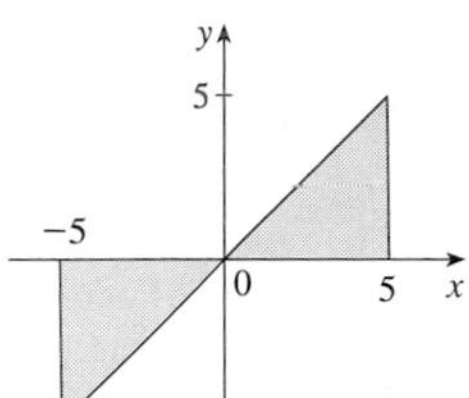

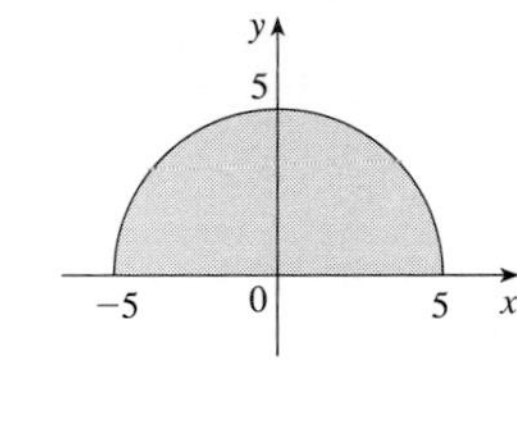

39. $\int_{-4}^{3} \left|\frac{1}{2}x\right| dx$ can be interpreted as the sum of the areas of the two shaded triangles; that is, $\frac{1}{2}(4)(2) + \frac{1}{2}(3)\left(\frac{3}{2}\right) = 4 + \frac{9}{4} = \frac{25}{4}$.

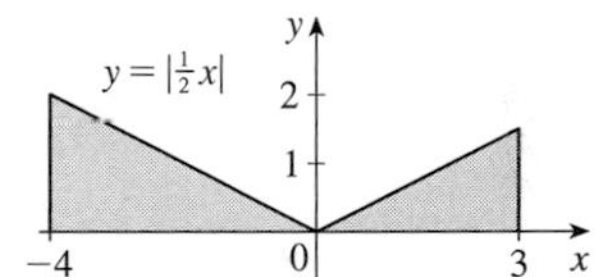

40. $\int_{0}^{1} |2x - 1|\,dx$ can be interpreted as the sum of the areas of the two shaded triangles; that is, $2\left(\frac{1}{2}\right)\left(\frac{1}{2}\right)(1) = \frac{1}{2}$.

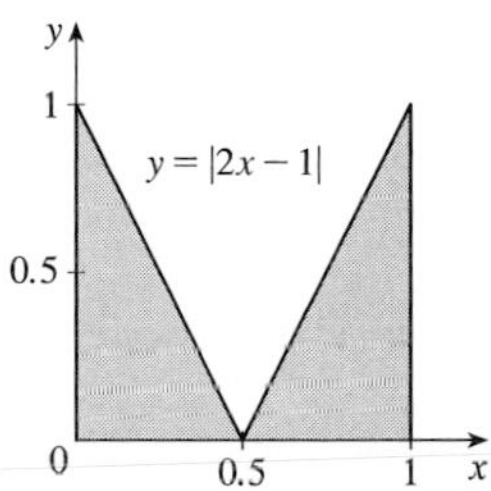

41. $\int_{1}^{1} \sqrt{1 + x^4}\,dx = 0$ since the limits of integration are equal.

42. $\int_{\pi}^{0} \sin^4 \theta\,d\theta = -\int_{0}^{\pi} \sin^4 \theta\,d\theta$ [because we reversed the limits of integration]

$= -\int_{0}^{\pi} \sin^4 x\,dx$ [we can use any letter without changing the value of the integral]

$= -\frac{3}{8}\pi$ [given value]

43. $\int_{0}^{1} (5 - 6x^2)\,dx = \int_{0}^{1} 5\,dx - 6\int_{0}^{1} x^2\,dx = 5(1 - 0) - 6\left(\frac{1}{3}\right) = 5 - 2 = 3$

44. $\int_{1}^{3} (2e^x - 1)\,dx = 2\int_{1}^{3} e^x\,dx - \int_{1}^{3} 1\,dx = 2(e^3 - e) - 1(3 - 1) = 2e^3 - 2e - 2$

45. $\int_{1}^{3} e^{x+2}\,dx = \int_{1}^{3} e^x \cdot e^2\,dx = e^2 \int_{1}^{3} e^x\,dx = e^2(e^3 - e) = e^5 - e^3$

46. $\int_{0}^{\pi/2} (2\cos x - 5x)\,dx = \int_{0}^{\pi/2} 2\cos x\,dx - \int_{0}^{\pi/2} 5x\,dx = 2\int_{0}^{\pi/2} \cos x\,dx - 5\int_{0}^{\pi/2} x\,dx$

$$= 2(1) - 5\frac{(\pi/2)^2 - 0^2}{2} = 2 - \frac{5\pi^2}{8}$$

47. $\int_{-2}^{2} f(x)\,dx + \int_{2}^{5} f(x)\,dx - \int_{-2}^{-1} f(x)\,dx = \int_{-2}^{5} f(x)\,dx + \int_{-1}^{-2} f(x)\,dx$ [by Property 5 and reversing limits]

$= \int_{-1}^{5} f(x)\,dx$ [Property 5]

48. $\int_2^4 f(x)\,dx + \int_4^8 f(x)\,dx = \int_2^8 f(x)\,dx$, so $\int_4^8 f(x)\,dx = \int_2^8 f(x)\,dx - \int_2^4 f(x)\,dx = 7.3 - 5.9 = 1.4$.

49. $\int_0^9 [2f(x) + 3g(x)]\,dx = 2\int_0^9 f(x)\,dx + 3\int_0^9 g(x)\,dx = 2(37) + 3(16) = 122$

50. If $f(x) = \begin{cases} 3 & \text{for } x < 3 \\ x & \text{for } x \geq 3 \end{cases}$, then $\int_0^5 f(x)\,dx$ can be interpreted as the area of the shaded region, which consists of a 5-by-3 rectangle surmounted by an isosceles right triangle whose legs have length 2. Thus, $\int_0^5 f(x)\,dx = 5(3) + \frac{1}{2}(2)(2) = 17$.

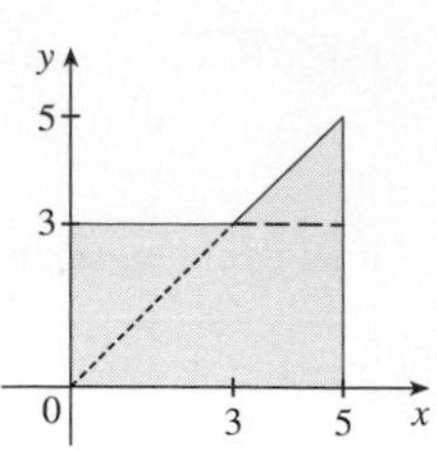

51. $\int_0^3 f(x)\,dx$ is clearly less than -1 and has the smallest value. The slope of the tangent line of f at $x = 1$, $f'(1)$, has a value between -1 and 0, so it has the next smallest value. The largest value is $\int_3^8 f(x)\,dx$, followed by $\int_4^8 f(x)\,dx$, which has a value about 1 unit less than $\int_3^8 f(x)\,dx$. Still positive, but with a smaller value than $\int_4^8 f(x)\,dx$, is $\int_0^8 f(x)\,dx$. Ordering these quantities from smallest to largest gives us

$$\int_0^3 f(x)\,dx < f'(1) < \int_0^8 f(x)\,dx < \int_4^8 f(x)\,dx < \int_3^8 f(x)\,dx \quad \text{or} \quad \text{B} < \text{E} < \text{A} < \text{D} < \text{C}$$

52. $F(0) = \int_2^0 f(t)\,dt = -\int_0^2 f(t)\,dt$, so $F(0)$ is negative, and similarly, so is $F(1)$. $F(3)$ and $F(4)$ are negative since they represent negatives of areas below the x-axis. Since $F(2) = \int_2^2 f(t)\,dt = 0$ is the only non-negative value, choice C is the largest.

53. $I = \int_{-4}^2 [f(x) + 2x + 5]\,dx = \int_{-4}^2 f(x)\,dx + 2\int_{-4}^2 x\,dx + \int_{-4}^2 5\,dx = I_1 + 2I_2 + I_3$

$I_1 = -3$ [area below x-axis] $+3 - 3 = -3$

$I_2 = -\frac{1}{2}(4)(4)$ [area of triangle, see figure] $+\frac{1}{2}(2)(2)$

$= -8 + 2 = -6$

$I_3 = 5[2 - (-4)] = 5(6) = 30$

Thus, $I = -3 + 2(-6) + 30 = 15$.

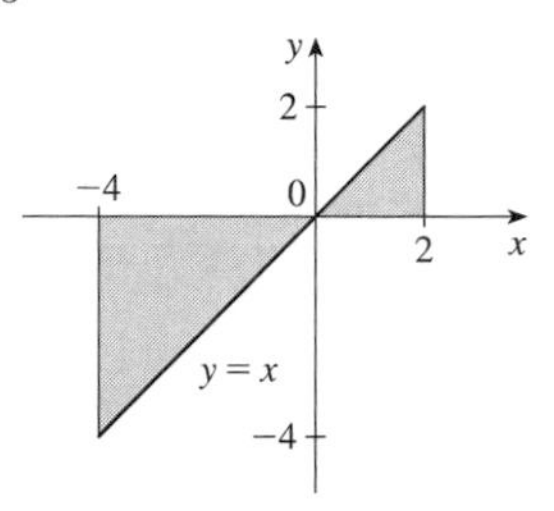

54. Using Integral Comparison Property 8, $m \leq f(x) \leq M \;\Rightarrow\; m(2 - 0) \leq \int_0^2 f(x)\,dx \leq M(2 - 0) \;\Rightarrow$ $2m \leq \int_0^2 f(x)\,dx \leq 2M$.

55. $x^2 - 4x + 4 = (x - 2)^2 \geq 0$ on $[0, 4]$, so $\int_0^4 (x^2 - 4x + 4)\,dx \geq 0$ [Property 6].

56. $x^2 \leq x$ on $[0, 1]$, so $\sqrt{1 + x^2} \leq \sqrt{1 + x}$ on $[0, 1]$. Hence, $\int_0^1 \sqrt{1 + x^2}\,dx \leq \int_0^1 \sqrt{1 + x}\,dx$ [Property 7].

57. If $-1 \leq x \leq 1$, then $0 \leq x^2 \leq 1$ and $1 \leq 1 + x^2 \leq 2$, so $1 \leq \sqrt{1 + x^2} \leq \sqrt{2}$ and $1[1 - (-1)] \leq \int_{-1}^1 \sqrt{1 + x^2}\,dx \leq \sqrt{2}\,[1 - (-1)]$ [Property 8]; that is, $2 \leq \int_{-1}^1 \sqrt{1 + x^2}\,dx \leq 2\sqrt{2}$.

58. If $\dfrac{\pi}{6} \leq x \leq \dfrac{\pi}{3}$, then $\dfrac{1}{2} \leq \sin x \leq \dfrac{\sqrt{3}}{2}$ ($\sin x$ is increasing on $\left[\frac{\pi}{6}, \frac{\pi}{3}\right]$), so

$$\frac{1}{2}\left(\frac{\pi}{3} - \frac{\pi}{6}\right) \leq \int_{\pi/6}^{\pi/3} \sin x\,dx \leq \frac{\sqrt{3}}{2}\left(\frac{\pi}{3} - \frac{\pi}{6}\right) \text{ [Property 8]; that is, } \frac{\pi}{12} \leq \int_{\pi/6}^{\pi/3} \sin x\,dx \leq \frac{\sqrt{3}\,\pi}{12}.$$

59. If $0 \le x \le 1$, then $0 \le x^3 \le 1$, so $0(1-0) \le \int_0^1 x^3\,dx \le 1(1-0)$ [Property 8]; that is, $0 \le \int_0^1 x^3\,dx \le 1$.

60. If $0 \le x \le 3$, then $4 \le x+4 \le 7$ and $\frac{1}{7} \le \frac{1}{x+4} \le \frac{1}{4}$, so $\frac{1}{7}(3-0) \le \int_0^3 \frac{1}{x+4}\,dx \le \frac{1}{4}(3-0)$ [Property 8]; that is,

$\frac{3}{7} \le \int_0^3 \frac{1}{x+4}\,dx \le \frac{3}{4}$.

61. If $\frac{\pi}{4} \le x \le \frac{\pi}{3}$, then $1 \le \tan x \le \sqrt{3}$, so $1\left(\frac{\pi}{3}-\frac{\pi}{4}\right) \le \int_{\pi/4}^{\pi/3} \tan x\,dx \le \sqrt{3}\left(\frac{\pi}{3}-\frac{\pi}{4}\right)$ or $\frac{\pi}{12} \le \int_{\pi/4}^{\pi/3} \tan x\,dx \le \frac{\pi}{12}\sqrt{3}$.

62. Let $f(x) = x^3 - 3x + 3$ for $0 \le x \le 2$. Then $f'(x) = 3x^2 - 3 = 3(x+1)(x-1)$, so f is decreasing on $(0,1)$ and increasing on $(1,2)$. f has the absolute minimum value $f(1) = 1$. Since $f(0) = 3$ and $f(2) = 5$, the absolute maximum value of f is $f(2) = 5$. Thus, $1 \le x^3 - 3x + 3 \le 5$ for x in $[0,2]$. It follows from Property 8 that

$1 \cdot (2-0) \le \int_0^2 (x^3 - 3x + 3)\,dx \le 5 \cdot (2-0)$; that is, $2 \le \int_0^2 (x^3 - 3x + 3)\,dx \le 10$.

63. The only critical number of $f(x) = xe^{-x}$ on $[0,2]$ is $x = 1$. Since $f(0) = 0$, $f(1) = e^{-1} \approx 0.368$, and $f(2) = 2e^{-2} \approx 0.271$, we know that the absolute minimum value of f on $[0,2]$ is 0, and the absolute maximum is e^{-1}. By Property 8, $0 \le xe^{-x} \le e^{-1}$ for $0 \le x \le 2 \;\Rightarrow\; 0(2-0) \le \int_0^2 xe^{-x}\,dx \le e^{-1}(2-0) \;\Rightarrow\; 0 \le \int_0^2 xe^{-x}\,dx \le 2/e$.

64. Let $f(x) = x - 2\sin x$ for $\pi \le x \le 2\pi$. Then $f'(x) = 1 - 2\cos x$ and $f'(x) = 0 \;\Rightarrow\; \cos x = \frac{1}{2} \;\Rightarrow\; x = \frac{5\pi}{3}$.

f has the absolute maximum value $f\left(\frac{5\pi}{3}\right) = \frac{5\pi}{3} - 2\sin\frac{5\pi}{3} = \frac{5\pi}{3} + \sqrt{3} \approx 6.97$ since $f(\pi) = \pi$ and $f(2\pi) = 2\pi$ are both smaller than 6.97. Thus, $\pi \le f(x) \le \frac{5\pi}{3} + \sqrt{3} \;\Rightarrow\; \pi(2\pi - \pi) \le \int_\pi^{2\pi} f(x)\,dx \le \left(\frac{5\pi}{3} + \sqrt{3}\right)(2\pi - \pi)$; that is,

$\pi^2 \le \int_\pi^{2\pi} (x - 2\sin x)\,dx \le \frac{5}{3}\pi^2 + \sqrt{3}\,\pi$.

65. $\sqrt{x^4+1} \ge \sqrt{x^4} = x^2$, so $\int_1^3 \sqrt{x^4+1}\,dx \ge \int_1^3 x^2\,dx = \frac{1}{3}(3^3 - 1^3) = \frac{26}{3}$.

66. $0 \le \sin x \le 1$ for $0 \le x \le \frac{\pi}{2}$, so $x\sin x \le x \;\Rightarrow\; \int_0^{\pi/2} x\sin x\,dx \le \int_0^{\pi/2} x\,dx = \frac{1}{2}\left[\left(\frac{\pi}{2}\right)^2 - 0^2\right] = \frac{\pi^2}{8}$.

67. $\sin x < \sqrt{x} < x$ for $1 \le x \le 2$ and arctan is an increasing function, so $\arctan(\sin x) < \arctan\sqrt{x} < \arctan x$, and hence, $\int_1^2 \arctan(\sin x)\,dx < \int_1^2 \arctan\sqrt{x}\,dx < \int_1^2 \arctan x\,dx$. Thus, $\int_1^2 \arctan x\,dx$ has the largest value.

68. $x^2 < \sqrt{x}$ for $0 < x \le 0.5$ and cosine is a decreasing function on $[0, 0.5]$, so $\cos(x^2) > \cos\sqrt{x}$, and hence, $\int_0^{0.5} \cos(x^2)\,dx > \int_0^{0.5} \cos\sqrt{x}\,dx$. Thus, $\int_0^{0.5} \cos(x^2)\,dx$ is larger.

69. Using right endpoints as in the proof of Property 2, we calculate

$\int_a^b cf(x)\,dx = \lim_{n\to\infty} \sum_{i=1}^n cf(x_i)\,\Delta x = \lim_{n\to\infty} c\sum_{i=1}^n f(x_i)\,\Delta x = c\lim_{n\to\infty} \sum_{i=1}^n f(x_i)\,\Delta x = c\int_a^b f(x)\,dx$.

70. (a) Since $-|f(x)| \le f(x) \le |f(x)|$, it follows from Property 7 that

$$-\int_a^b |f(x)|\,dx \le \int_a^b f(x)\,dx \le \int_a^b |f(x)|\,dx \;\Rightarrow\; \left|\int_a^b f(x)\,dx\right| \le \int_a^b |f(x)|\,dx$$

Note that the definite integral is a real number, and so the following property applies: $-a \le b \le a \;\Rightarrow\; |b| \le a$ for all real numbers b and nonnegative numbers a.

(b) $\left|\int_0^{2\pi} f(x)\sin 2x\,dx\right| \le \int_0^{2\pi} |f(x)\sin 2x|\,dx$ [by part (a)] $= \int_0^{2\pi} |f(x)|\,|\sin 2x|\,dx \le \int_0^{2\pi} |f(x)|\,dx$ by Property 7, since $|\sin 2x| \le 1 \;\Rightarrow\; |f(x)|\,|\sin 2x| \le |f(x)|$.

71. Suppose that f is integrable on $[0,1]$, that is, $\lim\limits_{n\to\infty}\sum\limits_{i=1}^{n} f(x_i^*)\,\Delta x$ exists for any choice of x_i^* in $[x_{i-1}, x_i]$. Let n denote a positive integer and divide the interval $[0,1]$ into n equal subintervals $\left[0, \frac{1}{n}\right], \left[\frac{1}{n}, \frac{2}{n}\right], \ldots, \left[\frac{n-1}{n}, 1\right]$. If we choose x_i^* to be a rational number in the ith subinterval, then we obtain the Riemann sum $\sum\limits_{i=1}^{n} f(x_i^*)\cdot\frac{1}{n} = 0$, so $\lim\limits_{n\to\infty}\sum\limits_{i=1}^{n} f(x_i^*)\cdot\frac{1}{n} = \lim\limits_{n\to\infty} 0 = 0$. Now suppose we choose x_i^* to be an irrational number. Then we get $\sum\limits_{i=1}^{n} f(x_i^*)\cdot\frac{1}{n} = \sum\limits_{i=1}^{n} 1\cdot\frac{1}{n} = n\cdot\frac{1}{n} = 1$ for each n, so $\lim\limits_{n\to\infty}\sum\limits_{i=1}^{n} f(x_i^*)\cdot\frac{1}{n} = \lim\limits_{n\to\infty} 1 = 1$. Since the value of $\lim\limits_{n\to\infty}\sum\limits_{i=1}^{n} f(x_i^*)\,\Delta x$ depends on the choice of the sample points x_i^*, the limit does not exist, and f is not integrable on $[0,1]$.

72. Partition the interval $[0,1]$ into n equal subintervals and choose $x_1^* = \frac{1}{n^2}$. Then with $f(x) = \frac{1}{x}$, $\sum\limits_{i=1}^{n} f(x_i^*)\,\Delta x \ge f(x_1^*)\Delta x = \frac{1}{1/n^2}\cdot\frac{1}{n} = n$. Thus, $\sum\limits_{i=1}^{n} f(x_i^*)\,\Delta x$ can be made arbitrarily large and hence, f is not integrable on $[0,1]$.

73. $\lim\limits_{n\to\infty}\sum\limits_{i=1}^{n}\frac{i^4}{n^5} = \lim\limits_{n\to\infty}\sum\limits_{i=1}^{n}\frac{i^4}{n^4}\cdot\frac{1}{n} = \lim\limits_{n\to\infty}\sum\limits_{i=1}^{n}\left(\frac{i}{n}\right)^4\frac{1}{n}$. At this point, we need to recognize the limit as being of the form $\lim\limits_{n\to\infty}\sum\limits_{i=1}^{n} f(x_i)\,\Delta x$, where $\Delta x = (1-0)/n = 1/n$, $x_i = 0 + i\,\Delta x = i/n$, and $f(x) = x^4$. Thus, the definite integral is $\int_0^1 x^4\,dx$.

74. $\lim\limits_{n\to\infty}\frac{1}{n}\sum\limits_{i=1}^{n}\frac{1}{1+(i/n)^2} = \lim\limits_{n\to\infty}\sum\limits_{i=1}^{n}\frac{1}{1+(i/n)^2}\cdot\frac{1}{n} = \lim\limits_{n\to\infty}\sum\limits_{i=1}^{n} f(x_i)\,\Delta x$, where $\Delta x = (1-0)/n = 1/n$, $x_i = 0 + i\,\Delta x = i/n$, and $f(x) = \frac{1}{1+x^2}$. Thus, the definite integral is $\int_0^1 \frac{dx}{1+x^2}$.

75. Choose $x_i = 1 + \frac{i}{n}$ and $x_i^* = \sqrt{x_{i-1}x_i} = \sqrt{\left(1+\frac{i-1}{n}\right)\left(1+\frac{i}{n}\right)}$. Then

$$\begin{aligned}
\int_1^2 x^{-2}\,dx &= \lim_{n\to\infty}\frac{1}{n}\sum_{i=1}^{n}\frac{1}{\left(1+\frac{i-1}{n}\right)\left(1+\frac{i}{n}\right)} = \lim_{n\to\infty} n\sum_{i=1}^{n}\frac{1}{(n+i-1)(n+i)} \\
&= \lim_{n\to\infty} n\sum_{i=1}^{n}\left(\frac{1}{n+i-1}-\frac{1}{n+i}\right) \quad \text{[by the hint]} = \lim_{n\to\infty} n\left(\sum_{i=0}^{n-1}\frac{1}{n+i}-\sum_{i=1}^{n}\frac{1}{n+i}\right) \\
&= \lim_{n\to\infty} n\left(\left[\frac{1}{n}+\frac{1}{n+1}+\cdots+\frac{1}{2n-1}\right]-\left[\frac{1}{n+1}+\cdots+\frac{1}{2n-1}+\frac{1}{2n}\right]\right) \\
&= \lim_{n\to\infty} n\left(\frac{1}{n}-\frac{1}{2n}\right) = \lim_{n\to\infty}\left(1-\tfrac{1}{2}\right) = \tfrac{1}{2}
\end{aligned}$$

DISCOVERY PROJECT Area Functions

1. (a)

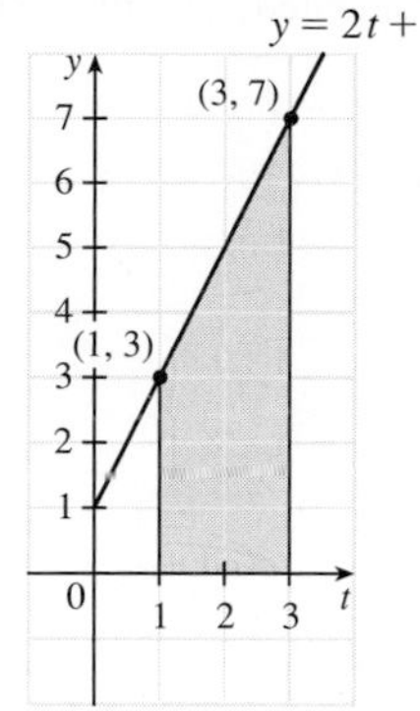

Area of trapezoid $= \frac{1}{2}(b_1 + b_2)h = \frac{1}{2}(3 + 7)2$

$= 10$ square units

Or:

Area of rectangle + area of triangle

$= b_r h_r + \frac{1}{2} b_t h_t = (2)(3) + \frac{1}{2}(2)(4) = 10$ square units

(b)

As in part (a),

$$A(x) = \tfrac{1}{2}[3 + (2x + 1)](x - 1) = \tfrac{1}{2}(2x + 4)(x - 1)$$
$$= (x + 2)(x - 1) = x^2 + x - 2 \text{ square units}$$

(c) $A'(x) = 2x + 1$. This is the y-coordinate of the point $(x, 2x + 1)$ on the given line.

2. (a)

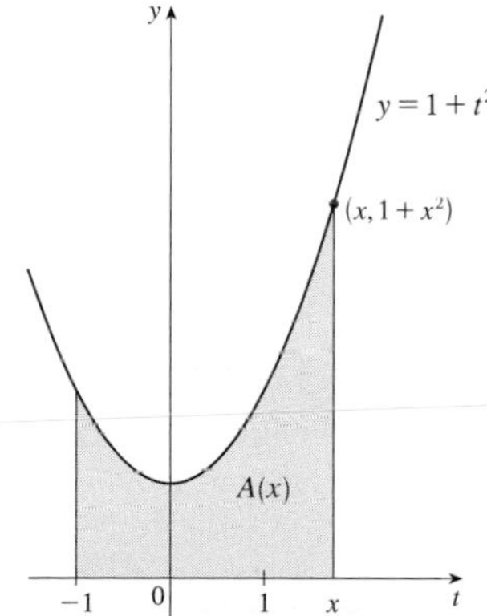

(b) $A(x) = \int_{-1}^{x} (1 + t^2)\, dt = \int_{-1}^{x} 1\, dt + \int_{-1}^{x} t^2\, dt$ [Property 2]

$$= 1[x - (-1)] + \frac{x^3 - (-1)^3}{3} \qquad \left[\begin{array}{c}\text{Property 1 and}\\ \text{Exercise 5.2.28}\end{array}\right]$$
$$= x + 1 + \tfrac{1}{3}x^3 + \tfrac{1}{3}$$
$$= \tfrac{1}{3}x^3 + x + \tfrac{4}{3}$$

(c) $A'(x) = x^2 + 1$. This is the y-coordinate of the point $(x, 1 + x^2)$ on the given curve.

(d)

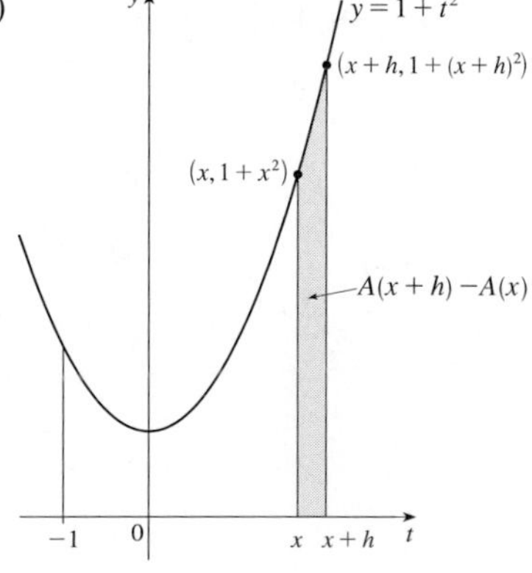

$A(x + h) - A(x)$ is the area under the curve $y = 1 + t^2$ from $t = x$ to $t = x + h$.

(e)

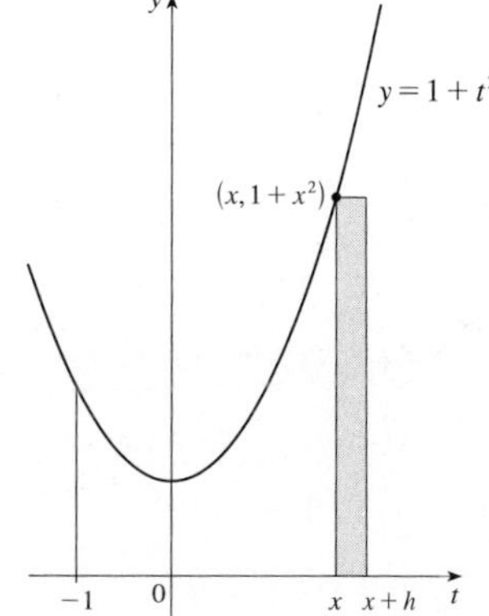

An approximating rectangle is shown in the figure.

It has height $1 + x^2$, width h, and area $h(1 + x^2)$, so

$$A(x + h) - A(x) \approx h(1 + x^2) \quad \Rightarrow \quad \frac{A(x + h) - A(x)}{h} \approx 1 + x^2.$$

(f) Part (e) says that the average rate of change of A is approximately $1 + x^2$. As h approaches 0, the quotient approaches the instantaneous rate of change—namely, $A'(x)$. So the result of part (c), $A'(x) = x^2 + 1$, is geometrically plausible.

3. (a) $f(x) = \cos(x^2)$

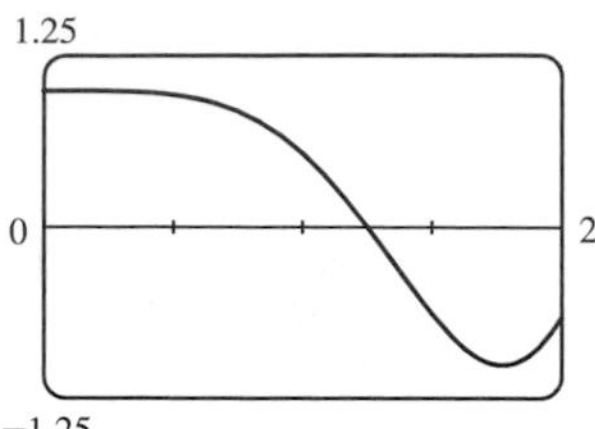

(b) $g(x)$ starts to decrease at that value of x where $\cos(t^2)$ changes from positive to negative; that is, at about $x = 1.25$.

(c) $g(x) = \int_0^x \cos(t^2)\,dt$. Using an integration command, we find that

$g(0) = 0$, $g(0.2) \approx 0.200$, $g(0.4) \approx 0.399$, $g(0.6) \approx 0.592$,

$g(0.8) \approx 0.768$, $g(1.0) \approx 0.905$, $g(1.2) \approx 0.974$, $g(1.4) \approx 0.950$,

$g(1.6) \approx 0.826$, $g(1.8) \approx 0.635$, and $g(2.0) \approx 0.461$.

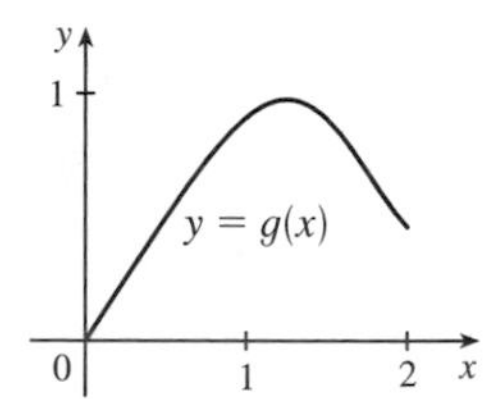

(d) We sketch the graph of g' using the method of Example 1 in Section 2.8. The graphs of $g'(x)$ and $f(x)$ look alike, so we guess that $g'(x) = f(x)$.

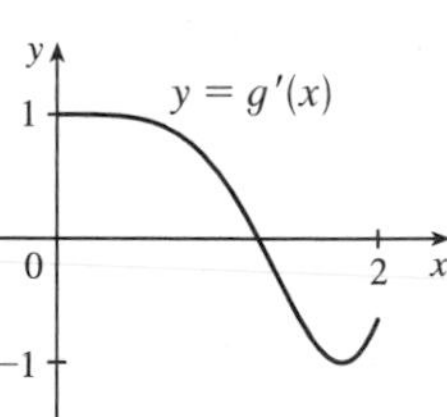

4. In Problems 1 and 2, we showed that if $g(x) = \int_a^x f(t)\,dt$, then $g'(x) = f(x)$, for the functions $f(t) = 2t + 1$ and $f(t) = 1 + t^2$. In Problem 3 we guessed that the same is true for $f(t) = \cos(t^2)$, based on visual evidence. So we conjecture that $g'(x) = f(x)$ for any continuous function f. This turns out to be true and is proved in Section 5.3 (the Fundamental Theorem of Calculus).

5.3 The Fundamental Theorem of Calculus

1. One process undoes what the other one does. The precise version of this statement is given by the Fundamental Theorem of Calculus. See the statement of this theorem and the paragraph that follows it on page 398.

2. (a) $g(x) = \int_0^x f(t)\,dt$, so $g(0) = \int_0^0 f(t)\,dt = 0$.

$g(1) = \int_0^1 f(t)\,dt = \frac{1}{2} \cdot 1 \cdot 1$ [area of triangle] $= \frac{1}{2}$.

$g(2) = \int_0^2 f(t)\,dt = \int_0^1 f(t)\,dt + \int_1^2 f(t)\,dt$ [below the t-axis]
$= \frac{1}{2} - \frac{1}{2} \cdot 1 \cdot 1 = 0$.

$g(3) = g(2) + \int_2^3 f(t)\,dt = 0 - \frac{1}{2} \cdot 1 \cdot 1 = -\frac{1}{2}$.

$g(4) = g(3) + \int_3^4 f(t)\,dt = -\frac{1}{2} + \frac{1}{2} \cdot 1 \cdot 1 = 0$.

$g(5) = g(4) + \int_4^5 f(t)\,dt = 0 + 1.5 = 1.5$.

$g(6) = g(5) + \int_5^6 f(t)\,dt = 1.5 + 2.5 = 4$.

(b) $g(7) = g(6) + \int_6^7 f(t)\,dt \approx 4 + 2.2$ [estimate from the graph] $= 6.2$.

(c) The answers from part (a) and part (b) indicate that g has a minimum at $x = 3$ and a maximum at $x = 7$. This makes sense from the graph of f since we are subtracting area on $1 < x < 3$ and adding area on $3 < x < 7$.

(d)

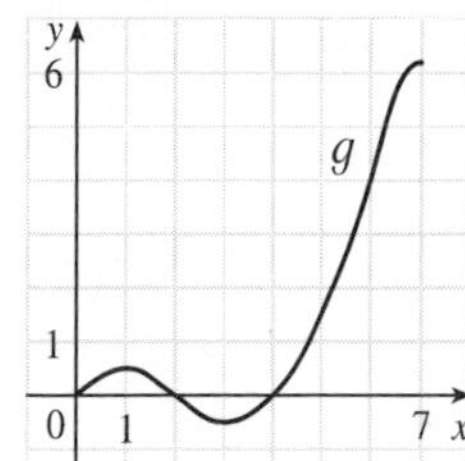

3. (a) $g(x) = \int_0^x f(t)\,dt$.

$g(0) = \int_0^0 f(t)\,dt = 0$

$g(1) = \int_0^1 f(t)\,dt = 1 \cdot 2 = 2$ [rectangle],

$g(2) = \int_0^2 f(t)\,dt = \int_0^1 f(t)\,dt + \int_1^2 f(t)\,dt = g(1) + \int_1^2 f(t)\,dt$
$= 2 + 1 \cdot 2 + \frac{1}{2} \cdot 1 \cdot 2 = 5$ [rectangle plus triangle],

$g(3) = \int_0^3 f(t)\,dt = g(2) + \int_2^3 f(t)\,dt = 5 + \frac{1}{2} \cdot 1 \cdot 4 = 7$,

$g(6) = g(3) + \int_3^6 f(t)\,dt$ [the integral is negative since f lies under the t-axis]
$= 7 + \left[-\left(\frac{1}{2} \cdot 2 \cdot 2 + 1 \cdot 2\right)\right] = 7 - 4 = 3$

(b) g is increasing on $(0, 3)$ because as x increases from 0 to 3, we keep adding more area.

(c) g has a maximum value when we start subtracting area; that is, at $x = 3$.

(d)

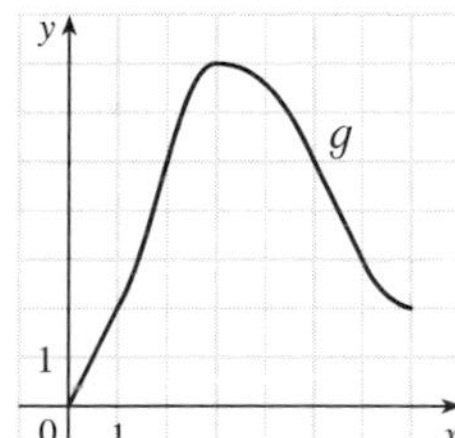

4. (a) $g(x) = \int_0^x f(t)\,dt$, so $g(0) = 0$ since the limits of integration are equal and $g(6) = 0$ since the areas above and below the t-axis are equal.

(b) $g(1)$ is the area under the curve from 0 to 1, which includes two unit squares and about 80% to 90% of a third unit square, so $g(1) \approx 2.8$. Similarly, $g(2) \approx 4.9$ and $g(3) \approx 5.7$. Now $g(3) - g(2) \approx 0.8$, so $g(4) \approx g(3) - 0.8 \approx 4.9$ by the symmetry of f about $x = 3$. Likewise, $g(5) \approx 2.8$.

(c) As we go from $x = 0$ to $x = 3$, we are adding area, so g increases on the interval $(0, 3)$.

(d) g increases on $(0, 3)$ and decreases on $(3, 6)$ [where we are subtracting area], so g has a maximum value at $x = 3$.

(e) A graph of g must have a maximum at $x = 3$, be symmetric about $x = 3$, and have zeros at $x = 0$ and $x = 6$.

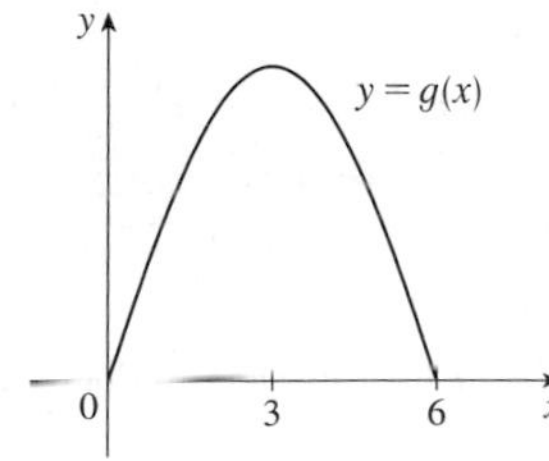

(f) If we sketch the graph of g' by estimating slopes on the graph of g (as in Section 2.8), we get a graph that looks like f (as indicated by FTC1).

5. 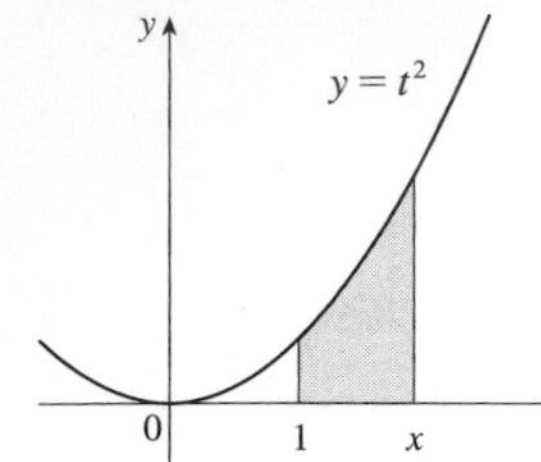

(a) By FTC1 with $f(t) = t^2$ and $a = 1$, $g(x) = \int_1^x t^2\,dt \;\Rightarrow$

$g'(x) = f(x) = x^2$.

(b) Using FTC2, $g(x) = \int_1^x t^2\,dt = \left[\frac{1}{3}t^3\right]_1^x = \frac{1}{3}x^3 - \frac{1}{3} \;\Rightarrow\; g'(x) = x^2$.

6. 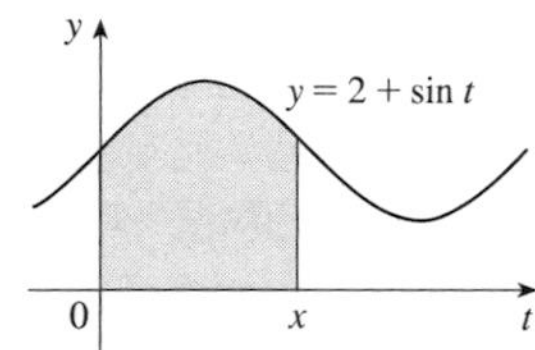

(a) By FTC1 with $f(t) = 2 + \sin t$ and $a = 0$, $g(x) = \int_0^x (2 + \sin t)\,dt \;\Rightarrow$

$g'(x) = f(x) = 2 + \sin x$.

(b) Using FTC2,

$$g(x) = \int_0^x (2 + \sin t)\,dt = [2t - \cos t]_0^x = (2x - \cos x) - (0 - 1)$$
$$= 2x - \cos x + 1 \;\Rightarrow$$

$$g'(x) = 2 - (-\sin x) + 0 = 2 + \sin x$$

7. $f(t) = \sqrt{t + t^3}$ and $g(x) = \int_0^x \sqrt{t + t^3}\,dt$, so by FTC1, $g'(x) = f(x) = \sqrt{x + x^3}$.

8. $f(t) = \ln(1 + t^2)$ and $g(x) = \int_1^x \ln(1 + t^2)\,dt$, so by FTC1, $g'(x) = f(x) = \ln(1 + x^2)$.

9. $f(t) = (t - t^2)^8$ and $g(s) = \int_5^s (t - t^2)^8\,dt$, so by FTC1, $g'(s) = f(s) = (s - s^2)^8$.

10. $f(t) = \dfrac{\sqrt{t}}{t + 1}$ and $h(u) = \displaystyle\int_0^u \frac{\sqrt{t}}{t + 1}\,dt$, so by FTC1, $h'(u) = f(u) = \dfrac{\sqrt{u}}{u + 1}$.

11. $F(x) = \displaystyle\int_x^0 \sqrt{1 + \sec t}\,dt = -\int_0^x \sqrt{1 + \sec t}\,dt \;\Rightarrow\; F'(x) = -\frac{d}{dx}\int_0^x \sqrt{1 + \sec t}\,dt = -\sqrt{1 + \sec x}$

12. $R(y) = \displaystyle\int_y^2 t^3 \sin t\,dt = -\int_2^y t^3 \sin t\,dt \;\Rightarrow\; R'(y) = -\frac{d}{dy}\int_2^y t^3 \sin t\,dt = -y^3 \sin y$

13. Let $u = e^x$. Then $\dfrac{du}{dx} = e^x$. Also, $\dfrac{dh}{dx} = \dfrac{dh}{du}\dfrac{du}{dx}$, so

$$h'(x) = \frac{d}{dx}\int_1^{e^x} \ln t\,dt = \frac{d}{du}\int_1^u \ln t\,dt \cdot \frac{du}{dx} = \ln u \frac{du}{dx} = (\ln e^x) \cdot e^x = xe^x.$$

14. Let $u = \sqrt{x}$. Then $\dfrac{du}{dx} = \dfrac{1}{2\sqrt{x}}$. Also, $\dfrac{dh}{dx} = \dfrac{dh}{du}\dfrac{du}{dx}$, so

$$h'(x) = \frac{d}{dx}\int_1^{\sqrt{x}} \frac{z^2}{z^4 + 1}\,dz = \frac{d}{du}\int_1^u \frac{z^2}{z^4 + 1}\,dz \cdot \frac{du}{dx} = \frac{u^2}{u^4 + 1}\frac{du}{dx} = \frac{x}{x^2 + 1}\frac{1}{2\sqrt{x}} = \frac{\sqrt{x}}{2(x^2 + 1)}.$$

15. Let $u = 3x + 2$. Then $\dfrac{du}{dx} = 3$. Also, $\dfrac{dy}{dx} = \dfrac{dy}{du}\dfrac{du}{dx}$, so

$$y' = \frac{d}{dx}\int_1^{3x+2} \frac{t}{1 + t^3}\,dt = \frac{d}{du}\int_1^u \frac{t}{1 + t^3}\,dt \cdot \frac{du}{dx} = \frac{u}{1 + u^3}\frac{du}{dx} = \frac{3x + 2}{1 + (3x + 2)^3} \cdot 3 = \frac{3(3x + 2)}{1 + (3x + 2)^3}$$

16. Let $u = x^4$. Then $\dfrac{du}{dx} = 4x^3$. Also, $\dfrac{dy}{dx} = \dfrac{dy}{du}\dfrac{du}{dx}$, so

$$y' = \frac{d}{dx}\int_0^{x^4} \cos^2\theta\,d\theta = \frac{d}{du}\int_0^u \cos^2\theta\,d\theta \cdot \frac{du}{dx} = \cos^2 u\,\frac{du}{dx} = \cos^2(x^4) \cdot 4x^3.$$

17. Let $u = \sqrt{x}$. Then $\dfrac{du}{dx} = \dfrac{1}{2\sqrt{x}}$. Also, $\dfrac{dy}{dx} = \dfrac{dy}{du}\dfrac{du}{dx}$, so

$$y' = \frac{d}{dx}\int_{\sqrt{x}}^{\pi/4} \theta \tan\theta \, d\theta = -\frac{d}{du}\int_{\pi/4}^{\sqrt{x}} \theta\tan\theta\,d\theta \cdot \frac{du}{dx} = -u\tan u\,\frac{du}{dx} = -\sqrt{x}\tan\sqrt{x}\cdot\frac{1}{2\sqrt{x}} = -\frac{1}{2}\tan\sqrt{x}$$

18. Let $u = \sin x$. Then $\dfrac{du}{dx} = \cos x$. Also, $\dfrac{dy}{dx} = \dfrac{dy}{du}\dfrac{du}{dx}$, so

$$y' = \frac{d}{dx}\int_{\sin x}^{1} \sqrt{1+t^2}\,dt = \frac{d}{du}\int_{u}^{1}\sqrt{1+t^2}\,dt\cdot\frac{du}{dx} = -\frac{d}{du}\int_{1}^{u}\sqrt{1+t^2}\,dt\cdot\frac{du}{dx}$$
$$= -\sqrt{1+u^2}\,\cos x = -\sqrt{1+\sin^2 x}\,\cos x$$

19. $\int_1^3 (x^2+2x-4)\,dx = \left[\frac{1}{3}x^3 + x^2 - 4x\right]_1^3 = (9+9-12) - \left(\frac{1}{3}+1-4\right) = 6 + \frac{8}{3} = \frac{26}{3}$

20. $\int_{-1}^1 x^{100}\,dx = \left[\frac{1}{101}x^{101}\right]_{-1}^1 = \frac{1}{101} - \left(-\frac{1}{101}\right) = \frac{2}{101}$

21. $\int_0^2 \left(\frac{4}{5}t^3 - \frac{3}{4}t^2 + \frac{2}{5}t\right)dt = \left[\frac{1}{5}t^4 - \frac{1}{4}t^3 + \frac{1}{5}t^2\right]_0^2 = \left(\frac{16}{5} - 2 + \frac{4}{5}\right) - 0 = 2$

22. $\int_0^1 (1 - 8v^3 + 16v^7)\,dv = \left[v - 2v^4 + 2v^8\right]_0^1 = (1-2+2) - 0 = 1$

23. $$\int_1^9 \sqrt{x}\,dx = \int_1^9 x^{1/2}\,dx = \left[\frac{x^{3/2}}{3/2}\right]_1^9 = \frac{2}{3}\left[x^{3/2}\right]_1^9 = \frac{2}{3}(9^{3/2} - 1^{3/2}) = \frac{2}{3}(27-1) = \frac{52}{3}$$

24. $$\int_1^8 x^{-2/3}\,dx = \left[\frac{x^{1/3}}{1/3}\right]_1^8 = 3\left[x^{1/3}\right]_1^8 = 3(8^{1/3} - 1^{1/3}) = 3(2-1) = 3$$

25. $$\int_{\pi/6}^{\pi} \sin\theta\,d\theta = \left[-\cos\theta\right]_{\pi/6}^{\pi} = -\cos\pi - \left(-\cos\frac{\pi}{6}\right) = -(-1) - \left(-\sqrt{3}/2\right) = 1 + \sqrt{3}/2$$

26. $\int_{-5}^5 e\,dx = \left[ex\right]_{-5}^5 = 5e - (-5e) = 10e$

27. $$\int_0^1 (u+2)(u-3)\,du = \int_0^1 (u^2 - u - 6)\,du = \left[\frac{1}{3}u^3 - \frac{1}{2}u^2 - 6u\right]_0^1 = \left(\frac{1}{3} - \frac{1}{2} - 6\right) - 0 = -\frac{37}{6}$$

28. $$\int_0^4 (4-t)\sqrt{t}\,dt = \int_0^4 (4-t)t^{1/2}\,dt = \int_0^4 (4t^{1/2} - t^{3/2})\,dt = \left[\frac{8}{3}t^{3/2} - \frac{2}{5}t^{5/2}\right]_0^4 = \frac{8}{3}(8) - \frac{2}{5}(32) = \frac{320-192}{15} = \frac{128}{15}$$

29. $$\int_1^4 \frac{2+x^2}{\sqrt{x}}\,dx = \int_1^4 \left(\frac{2}{\sqrt{x}} + \frac{x^2}{\sqrt{x}}\right)dx = \int_1^4 (2x^{-1/2} + x^{3/2})\,dx$$
$$= \left[4x^{1/2} + \frac{2}{5}x^{5/2}\right]_1^4 = \left[4(2) + \frac{2}{5}(32)\right] - \left(4 + \frac{2}{5}\right) = 8 + \frac{64}{5} - 4 - \frac{2}{5} = \frac{82}{5}$$

30. $\int_{-1}^2 (3u-2)(u+1)\,du = \int_{-1}^2 (3u^2 + u - 2)\,du = \left[u^3 + \frac{1}{2}u^2 - 2u\right]_{-1}^2 = (8+2-4) - \left(-1 + \frac{1}{2} + 2\right) = 6 - \frac{3}{2} = \frac{9}{2}$

31. $$\int_{\pi/6}^{\pi/2} \csc t\,\cot t\,dt = \left[-\csc t\right]_{\pi/6}^{\pi/2} = \left(-\csc\frac{\pi}{2}\right) - \left(-\csc\frac{\pi}{6}\right) = -1 - (-2) = 1$$

32. $$\int_{\pi/4}^{\pi/3} \csc^2\theta\,d\theta = \left[-\cot\theta\right]_{\pi/4}^{\pi/3} = \left(-\cot\frac{\pi}{3}\right) - \left(-\cot\frac{\pi}{4}\right) = -\frac{1}{\sqrt{3}} - (-1) = 1 - \frac{1}{\sqrt{3}}$$

33. $\int_0^1 (1+r)^3\,dr = \int_0^1 (1+3r+3r^2+r^3)\,dr = \left[r + \frac{3}{2}r^2 + r^3 + \frac{1}{4}r^4\right]_0^1 = \left(1 + \frac{3}{2} + 1 + \frac{1}{4}\right) - 0 = \frac{15}{4}$

34. $\displaystyle\int_0^3 (2\sin x - e^x)\,dx = \left[-2\cos x - e^x\right]_0^3 = (-2\cos 3 - e^3) - (-2-1) = 3 - 2\cos 3 - e^3$

35. $\displaystyle\int_1^2 \frac{v^3+3v^6}{v^4} = \int_1^2 \left(\frac{1}{v} + 3v^2\right) dv = \left[\ln|v| + v^3\right]_1^2 = (\ln 2 + 8) - (\ln 1 + 1) = \ln 2 + 7$

36. $\displaystyle\int_1^{18} \sqrt{\frac{3}{z}}\,dz = \int_1^{18} \sqrt{3}z^{-1/2}\,dz = \sqrt{3}\left[2z^{1/2}\right]_1^{18} = 2\sqrt{3}(18^{1/2} - 1^{1/2}) = 2\sqrt{3}(3\sqrt{2}-1)$

37. $\displaystyle\int_0^1 (x^e + e^x)\,dx = \left[\frac{x^{e+1}}{e+1} + e^x\right]_0^1 = \left(\frac{1}{e+1} + e\right) - (0+1) = \frac{1}{e+1} + e - 1$

38. $\int_0^1 \cosh t\,dt = \left[\sinh t\right]_0^1 = \sinh 1 - \sinh 0 = \sinh 1 \quad \left[\text{or } \frac{1}{2}(e - e^{-1})\right]$

39. $\displaystyle\int_{1/\sqrt{3}}^{\sqrt{3}} \frac{8}{1+x^2}\,dx = \left[8\arctan x\right]_{1/\sqrt{3}}^{\sqrt{3}} = 8\left(\frac{\pi}{3} - \frac{\pi}{6}\right) = 8\left(\frac{\pi}{6}\right) = \frac{4\pi}{3}$

40. $\displaystyle\int_1^3 \frac{y^3 - 2y^2 - y}{y^2}\,dy = \int_1^3 \left(y - 2 - \frac{1}{y}\right) dy = \left[\frac{1}{2}y^2 - 2y - \ln|y|\right]_1^3 = \left(\frac{9}{2} - 6 - \ln 3\right) - \left(\frac{1}{2} - 2 - 0\right) = -\ln 3$

41. $\displaystyle\int_0^4 2^s\,ds = \left[\frac{1}{\ln 2}2^s\right]_0^4 = \frac{16}{\ln 2} - \frac{1}{\ln 2} = \frac{15}{\ln 2}$

42. $\displaystyle\int_{1/2}^{1/\sqrt{2}} \frac{4}{\sqrt{1-x^2}}\,dx = \left[4\arcsin x\right]_{1/2}^{1/\sqrt{2}} = 4\left(\frac{\pi}{4} - \frac{\pi}{6}\right) = 4\left(\frac{\pi}{12}\right) = \frac{\pi}{3}$

43. If $f(x) = \begin{cases} \sin x & \text{if } 0 \le x < \pi/2 \\ \cos x & \text{if } \pi/2 \le x \le \pi \end{cases}$ then

$$\begin{aligned}\int_0^\pi f(x)\,dx &= \int_0^{\pi/2} \sin x\,dx + \int_{\pi/2}^{\pi} \cos x\,dx = \left[-\cos x\right]_0^{\pi/2} + \left[\sin x\right]_{\pi/2}^{\pi} = -\cos\tfrac{\pi}{2} + \cos 0 + \sin \pi - \sin\tfrac{\pi}{2} \\ &= -0 + 1 + 0 - 1 = 0\end{aligned}$$

Note that f is integrable by Theorem 3 in Section 5.2.

44. If $f(x) = \begin{cases} 2 & \text{if } -2 \le x \le 0 \\ 4 - x^2 & \text{if } 0 < x \le 2 \end{cases}$ then

$$\int_{-2}^2 f(x)\,dx = \int_{-2}^0 2\,dx + \int_0^2 (4 - x^2)\,dx = \left[2x\right]_{-2}^0 + \left[4x - \tfrac{1}{3}x^3\right]_0^2 = [0 - (-4)] + \left(\tfrac{16}{3} - 0\right) = \tfrac{28}{3}$$

Note that f is integrable by Theorem 3 in Section 5.2.

45. Area $\displaystyle= \int_0^4 \sqrt{x}\,dx = \int_0^4 x^{1/2}\,dx = \left[\frac{2}{3}x^{3/2}\right]_0^4 = \frac{2}{3}(8) - 0 = \frac{16}{3}$

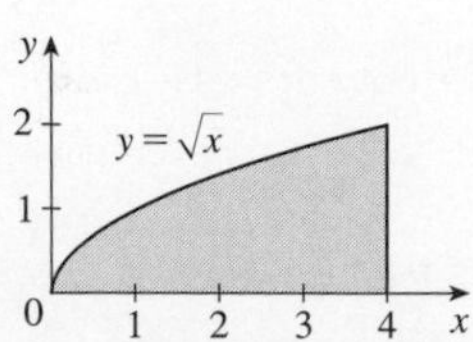

46. Area $= \int_0^1 x^3\,dx = \left[\frac{1}{4}x^4\right]_0^1 = \frac{1}{4} - 0 = \frac{1}{4}$

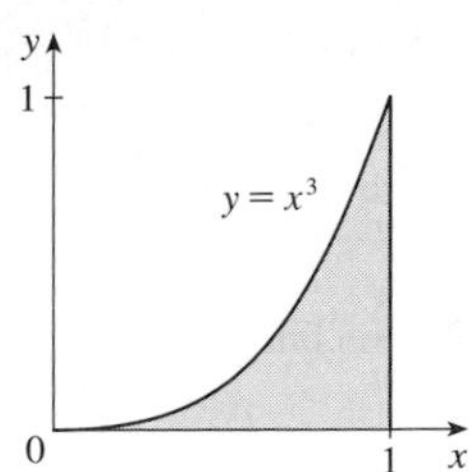

47. Area $= \int_{-2}^{2} (4 - x^2)\,dx = \left[4x - \frac{1}{3}x^3\right]_{-2}^{2} = \left(8 - \frac{8}{3}\right) - \left(-8 + \frac{8}{3}\right) = \frac{32}{3}$

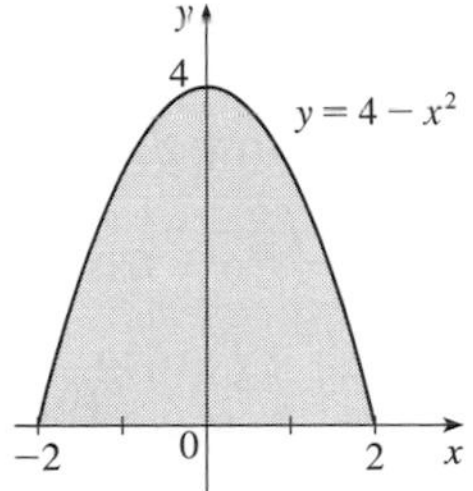

48. Area $= \int_0^2 (2x - x^2)\,dx = \left[x^2 - \frac{1}{3}x^3\right]_0^2 = \left(4 - \frac{8}{3}\right) - 0 = \frac{4}{3}$

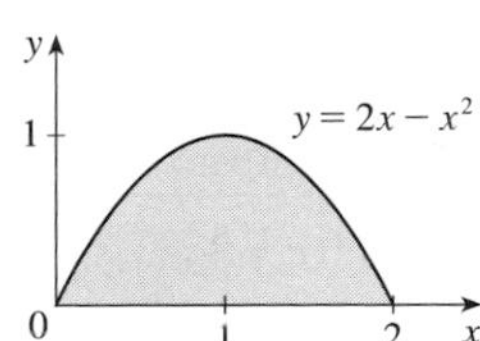

49. From the graph, it appears that the area is about 60. The actual area is

$\int_0^{27} x^{1/3}dx = \left[\frac{3}{4}x^{4/3}\right]_0^{27} = \frac{3}{4} \cdot 81 - 0 = \frac{243}{4} = 60.75$. This is $\frac{3}{4}$ of the area of the viewing rectangle.

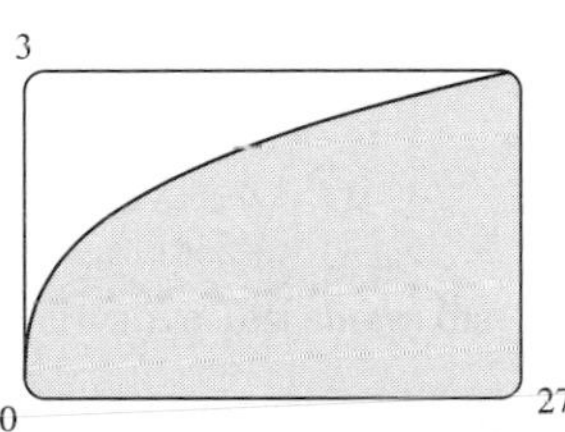

50. From the graph, it appears that the area is about $\frac{1}{3}$. The actual area is

$$\int_1^6 x^{-4}\,dx = \left[\frac{x^{-3}}{-3}\right]_1^6 = \left[\frac{-1}{3x^3}\right]_1^6 = -\frac{1}{3 \cdot 216} + \frac{1}{3} = \frac{215}{648} \approx 0.3318.$$

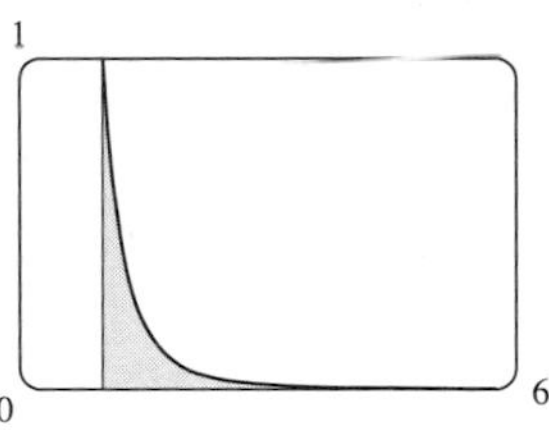

51. It appears that the area under the graph is about $\frac{2}{3}$ of the area of the viewing rectangle, or about $\frac{2}{3}\pi \approx 2.1$. The actual area is

$\int_0^{\pi} \sin x\,dx = [-\cos x]_0^{\pi} = (-\cos \pi) - (-\cos 0) = -(-1) + 1 = 2.$

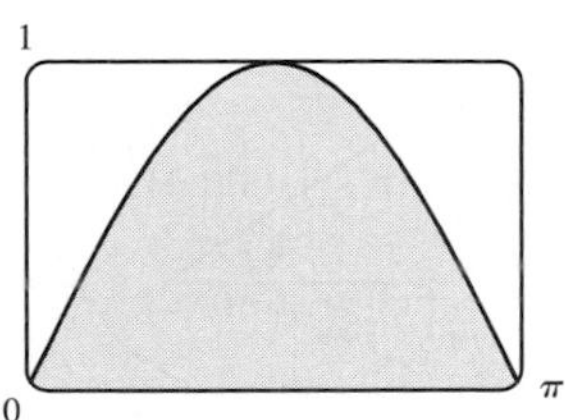

52. Splitting up the region as shown, we estimate that the area under the graph is $\frac{\pi}{3} + \frac{1}{4}\left(3 \cdot \frac{\pi}{3}\right) \approx 1.8$. The actual area is

$\int_0^{\pi/3} \sec^2 x\,dx = [\tan x]_0^{\pi/3} = \sqrt{3} - 0 = \sqrt{3} \approx 1.73.$

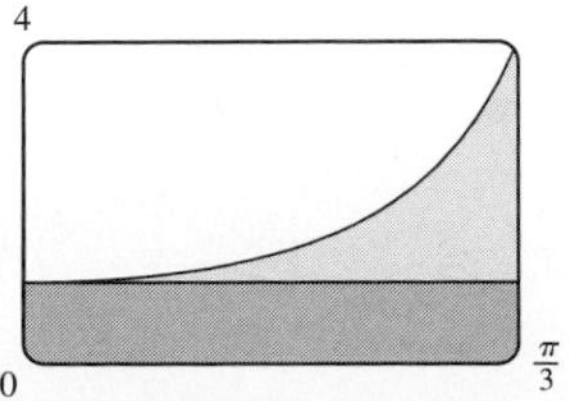

53. $\int_{-1}^{2} x^3\,dx = \left[\frac{1}{4}x^4\right]_{-1}^{2} = 4 - \frac{1}{4} - \frac{15}{4} = 3.75$

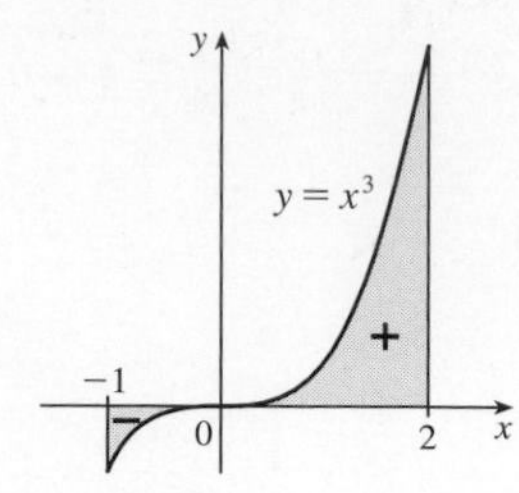

54. $\int_{\pi/6}^{2\pi} \cos x\,dx = \Big[\sin x\Big]_{\pi/6}^{2\pi} = 0 - \frac{1}{2} = -\frac{1}{2}$

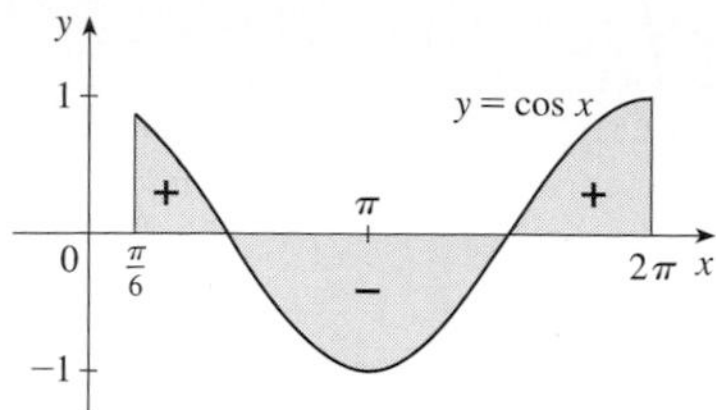

55. $f(x) = x^{-4}$ is not continuous on the interval $[-2, 1]$, so FTC2 cannot be applied. In fact, f has an infinite discontinuity at $x = 0$, so $\int_{-2}^{1} x^{-4}\,dx$ does not exist.

56. $f(x) = \dfrac{4}{x^3}$ is not continuous on the interval $[-1, 2]$, so FTC2 cannot be applied. In fact, f has an infinite discontinuity at $x = 0$, so $\displaystyle\int_{-1}^{2} \frac{4}{x^3}\,dx$ does not exist.

57. $f(\theta) = \sec\theta\,\tan\theta$ is not continuous on the interval $[\pi/3, \pi]$, so FTC2 cannot be applied. In fact, f has an infinite discontinuity at $x = \pi/2$, so $\int_{\pi/3}^{\pi} \sec\theta\,\tan\theta\,d\theta$ does not exist.

58. $f(x) = \sec^2 x$ is not continuous on the interval $[0, \pi]$, so FTC2 cannot be applied. In fact, f has an infinite discontinuity at $x = \pi/2$, so $\int_{0}^{\pi} \sec^2 x\,dx$ does not exist.

59. $$g(x) = \int_{2x}^{3x} \frac{u^2-1}{u^2+1}\,du = \int_{2x}^{0} \frac{u^2-1}{u^2+1}\,du + \int_{0}^{3x} \frac{u^2-1}{u^2+1}\,du = -\int_{0}^{2x} \frac{u^2-1}{u^2+1}\,du + \int_{0}^{3x} \frac{u^2-1}{u^2+1}\,du \;\Rightarrow$$

$$g'(x) = -\frac{(2x)^2-1}{(2x)^2+1}\cdot\frac{d}{dx}(2x) + \frac{(3x)^2-1}{(3x)^2+1}\cdot\frac{d}{dx}(3x) = -2\cdot\frac{4x^2-1}{4x^2+1} + 3\cdot\frac{9x^2-1}{9x^2+1}$$

60. $$g(x) = \int_{1-2x}^{1+2x} t\sin t\,dt = \int_{1-2x}^{0} t\sin t\,dt + \int_{0}^{1+2x} t\sin t\,dt = -\int_{0}^{1-2x} t\sin t\,dt + \int_{0}^{1+2x} t\sin t\,dt \;\Rightarrow$$

$$\begin{aligned} g'(x) &= -(1-2x)\sin(1-2x)\cdot\frac{d}{dx}(1-2x) + (1+2x)\sin(1+2x)\cdot\frac{d}{dx}(1+2x) \\ &= 2(1-2x)\sin(1-2x) + 2(1+2x)\sin(1+2x) \end{aligned}$$

61. $$F(x) = \int_{x}^{x^2} e^{t^2}\,dt = \int_{x}^{0} e^{t^2}\,dt + \int_{0}^{x^2} e^{t^2}\,dt = -\int_{0}^{x} e^{t^2}\,dt + \int_{0}^{x^2} e^{t^2}\,dt \;\Rightarrow$$

$$F'(x) = -e^{x^2} + e^{(x^2)^2}\cdot\frac{d}{dx}(x^2) = -e^{x^2} + 2xe^{x^4}$$

62. $F(x) = \int_{\sqrt{x}}^{2x} \arctan t\, dt = \int_{\sqrt{x}}^{0} \arctan t\, dt + \int_{0}^{2x} \arctan t\, dt = -\int_{0}^{\sqrt{x}} \arctan t\, dt + \int_{0}^{2x} \arctan t\, dt \quad \Rightarrow$

$F'(x) = -\arctan\sqrt{x} \cdot \frac{d}{dx}(\sqrt{x}) + \arctan 2x \cdot \frac{d}{dx}(2x) = -\frac{1}{2\sqrt{x}} \arctan\sqrt{x} + 2\arctan 2x$

63. $y = \int_{\cos x}^{\sin x} \ln(1+2v)\, dv = \int_{\cos x}^{0} \ln(1+2v)\, dv + \int_{0}^{\sin x} \ln(1+2v)\, dv$

$= -\int_{0}^{\cos x} \ln(1+2v)\, dv + \int_{0}^{\sin x} \ln(1+2v)\, dv \quad \Rightarrow$

$y' = -\ln(1+2\cos x) \cdot \frac{d}{dx}\cos x + \ln(1+2\sin x) \cdot \frac{d}{dx}\sin x = \sin x\, \ln(1+2\cos x) + \cos x\, \ln(1+2\sin x)$

64. $f(x) = \int_0^x (1-t^2)e^{t^2}\, dt$ is increasing when $f'(x) = (1-x^2)e^{x^2}$ is positive.

Since $e^{x^2} > 0$, $f'(x) > 0 \quad \Leftrightarrow \quad 1 - x^2 > 0 \quad \Leftrightarrow \quad |x| < 1$, so f is increasing on $(-1, 1)$.

65. $y = \int_0^x \frac{t^2}{t^2+t+2}\, dt \quad \Rightarrow \quad y' = \frac{x^2}{x^2+x+2} \quad \Rightarrow$

$y'' = \frac{(x^2+x+2)(2x) - x^2(2x+1)}{(x^2+x+2)^2} = \frac{2x^3+2x^2+4x-2x^3-x^2}{(x^2+x+2)^2} = \frac{x^2+4x}{(x^2+x+2)^2} = \frac{x(x+4)}{(x^2+x+2)^2}.$

The curve y is concave downward when $y'' < 0$; that is, on the interval $(-4, 0)$.

66. If $F(x) = \int_1^x f(t)\, dt$, then by FTC1, $F'(x) = f(x)$, and also, $F''(x) = f'(x)$. F is concave downward where F'' is negative; that is, where f' is negative. The given graph shows that f is decreasing ($f' < 0$) on the interval $(-1, 1)$.

67. $F(x) = \int_2^x e^{t^2}\, dt \quad \Rightarrow \quad F'(x) = e^{x^2}$, so the slope at $x = 2$ is $e^{2^2} = e^4$. The y-coordinate of the point on F at $x = 2$ is $F(2) = \int_2^2 e^{t^2}\, dt = 0$ since the limits are equal. An equation of the tangent line is $y - 0 = e^4(x-2)$, or $y = e^4 x - 2e^4$.

68. $g(y) = \int_3^y f(x)\, dx \quad \Rightarrow \quad g'(y) = f(y)$. Since $f(x) = \int_0^{\sin x} \sqrt{1+t^2}\, dt$, $g''(y) = f'(y) = \sqrt{1+\sin^2 y} \cdot \cos y$,

so $g''(\frac{\pi}{6}) = \sqrt{1+\sin^2(\frac{\pi}{6})} \cdot \cos\frac{\pi}{6} = \sqrt{1+(\frac{1}{2})^2} \cdot \frac{\sqrt{3}}{2} = \frac{\sqrt{5}}{2} \cdot \frac{\sqrt{3}}{2} = \frac{\sqrt{15}}{4}$.

69. By FTC2, $\int_1^4 f'(x)\, dx = f(4) - f(1)$, so $17 = f(4) - 12 \quad \Rightarrow \quad f(4) = 17 + 12 = 29$.

70. (a) $\operatorname{erf}(x) = \frac{2}{\sqrt{\pi}} \int_0^x e^{-t^2}\, dt \quad \Rightarrow \quad \int_0^x e^{-t^2}\, dt = \frac{\sqrt{\pi}}{2} \operatorname{erf}(x)$. By Property 5 of definite integrals in Section 5.2,

$\int_0^b e^{-t^2}\, dt = \int_0^a e^{-t^2}\, dt + \int_a^b e^{-t^2}\, dt$, so

$\int_a^b e^{-t^2}\, dt = \int_0^b e^{-t^2}\, dt - \int_0^a e^{-t^2}\, dt = \frac{\sqrt{\pi}}{2} \operatorname{erf}(b) - \frac{\sqrt{\pi}}{2} \operatorname{erf}(a) = \frac{1}{2}\sqrt{\pi}\, [\operatorname{erf}(b) - \operatorname{erf}(a)].$

(b) $y = e^{x^2} \operatorname{erf}(x) \quad \Rightarrow \quad y' = 2xe^{x^2} \operatorname{erf}(x) + e^{x^2} \operatorname{erf}'(x) = 2xy + e^{x^2} \cdot \frac{2}{\sqrt{\pi}} e^{-x^2}$ [by FTC1] $= 2xy + \frac{2}{\sqrt{\pi}}$.

71. (a) The Fresnel function $S(x) = \int_0^x \sin\left(\frac{\pi}{2}t^2\right) dt$ has local maximum values where $0 = S'(x) = \sin\left(\frac{\pi}{2}t^2\right)$ and S' changes from positive to negative. For $x > 0$, this happens when $\frac{\pi}{2}x^2 = (2n-1)\pi$ [odd multiples of π] $\Leftrightarrow$

$x^2 = 2(2n-1) \quad \Leftrightarrow \quad x = \sqrt{4n-2}$, n any positive integer. For $x < 0$, S' changes from positive to negative where $\frac{\pi}{2}x^2 = 2n\pi$ [even multiples of π] $\Leftrightarrow \quad x^2 = 4n \quad \Leftrightarrow \quad x = -2\sqrt{n}$. S' does not change sign at $x = 0$.

(b) S is concave upward on those intervals where $S''(x) > 0$. Differentiating our expression for $S'(x)$, we get $S''(x) = \cos\left(\frac{\pi}{2}x^2\right)\left(2\frac{\pi}{2}x\right) = \pi x \cos\left(\frac{\pi}{2}x^2\right)$. For $x > 0$, $S''(x) > 0$ where $\cos\left(\frac{\pi}{2}x^2\right) > 0 \quad \Leftrightarrow \quad 0 < \frac{\pi}{2}x^2 < \frac{\pi}{2}$ or $\left(2n - \frac{1}{2}\right)\pi < \frac{\pi}{2}x^2 < \left(2n + \frac{1}{2}\right)\pi$, n any integer $\Leftrightarrow \quad 0 < x < 1$ or $\sqrt{4n-1} < x < \sqrt{4n+1}$, n any positive integer.

For $x < 0$, $S''(x) > 0$ where $\cos\left(\frac{\pi}{2}x^2\right) < 0 \quad \Leftrightarrow \quad \left(2n - \frac{3}{2}\right)\pi < \frac{\pi}{2}x^2 < \left(2n - \frac{1}{2}\right)\pi$, n any integer $\Leftrightarrow$

$4n - 3 < x^2 < 4n - 1 \quad \Leftrightarrow \quad \sqrt{4n-3} < |x| < \sqrt{4n-1} \quad \Rightarrow \quad \sqrt{4n-3} < -x < \sqrt{4n-1} \quad \Rightarrow$

$-\sqrt{4n-3} > x > -\sqrt{4n-1}$, so the intervals of upward concavity for $x < 0$ are $\left(-\sqrt{4n-1}, -\sqrt{4n-3}\right)$, n any positive integer. To summarize: S is concave upward on the intervals $(0, 1)$, $\left(-\sqrt{3}, -1\right)$, $\left(\sqrt{3}, \sqrt{5}\right)$, $\left(-\sqrt{7}, -\sqrt{5}\right)$, $\left(\sqrt{7}, 3\right)$, $\ldots$.

(c) In Maple, we use `plot({int(sin(Pi*t^2/2),t=0..x),0.2},x=0..2);`. Note that Maple recognizes the Fresnel function, calling it `FresnelS(x)`. In Mathematica, we use `Plot[{Integrate[Sin[Pi*t^2/2],{t,0,x}],0.2},{x,0,2}]`. In Derive, we load the utility file `FRESNEL` and plot `FRESNEL_SIN(x)`. From the graphs, we see that $\int_0^x \sin\left(\frac{\pi}{2}t^2\right) dt = 0.2$ at $x \approx 0.74$.

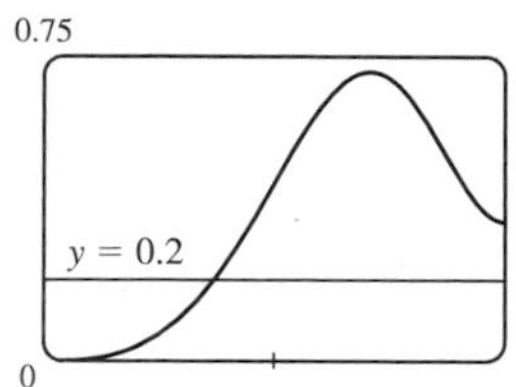

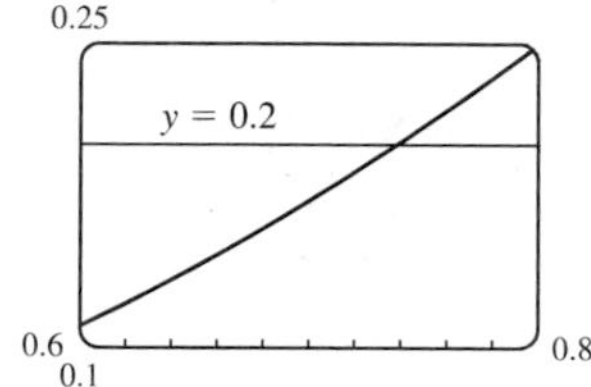

72. (a) In Maple, we should start by setting `si:=int(sin(t)/t,t=0..x);`. In Mathematica, the command is `si=Integrate[Sin[t]/t,{t,0,x}]`. Note that both systems recognize this function; Maple calls it `Si(x)` and Mathematica calls it `SinIntegral[x]`. In Maple, the command to generate the graph is `plot(si,x=-4*Pi..4*Pi);`. In Mathematica, it is `Plot[si,{x,-4*Pi,4*Pi}]`. In Derive, we load the utility file `EXP_INT` and plot `SI(x)`.

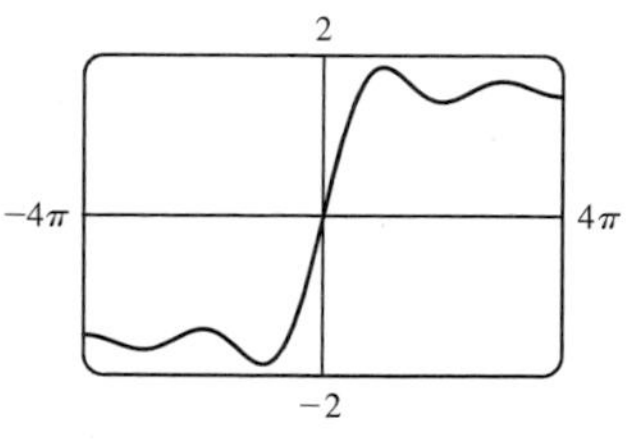

(b) $\text{Si}(x)$ has local maximum values where $\text{Si}'(x)$ changes from positive to negative, passing through 0. From the Fundamental Theorem we know that $\text{Si}'(x) = \dfrac{d}{dx}\displaystyle\int_0^x \frac{\sin t}{t}\,dt = \frac{\sin x}{x}$, so we must have $\sin x = 0$ for a maximum, and for $x > 0$ we must have $x = (2n-1)\pi$, n any positive integer, for Si' to be changing from positive to negative at x. For $x < 0$, we must have $x = 2n\pi$, n any positive integer, for a maximum, since the denominator of $\text{Si}'(x)$ is negative for $x < 0$. Thus, the local maxima occur at $x = \pi, -2\pi, 3\pi, -4\pi, 5\pi, -6\pi, \ldots$.

(c) To find the first inflection point, we solve $\text{Si}''(x) = \dfrac{\cos x}{x} - \dfrac{\sin x}{x^2} = 0$. We can see from the graph that the first inflection point lies somewhere between $x = 3$ and $x = 5$. Using a rootfinder gives the value $x \approx 4.4934$. To find the y-coordinate of the inflection point, we evaluate $\text{Si}(4.4934) \approx 1.6556$. So the coordinates of the first inflection point to the right of the origin are about $(4.4934, 1.6556)$. Alternatively, we could graph $S''(x)$ and estimate the first positive x-value at which it changes sign.

(d) It seems from the graph that the function has horizontal asymptotes at $y \approx 1.5$, with $\lim\limits_{x\to\pm\infty} \text{Si}(x) \approx \pm 1.5$ respectively. Using the limit command, we get $\lim\limits_{x\to\infty} \text{Si}(x) = \frac{\pi}{2}$. Since $\text{Si}(x)$ is an odd function, $\lim\limits_{x\to-\infty} \text{Si}(x) = -\frac{\pi}{2}$. So $\text{Si}(x)$ has the horizontal asymptotes $y = \pm\frac{\pi}{2}$.

(e) We use the `fsolve` command in Maple (or `FindRoot` in Mathematica) to find that the solution is $x \approx 1.1$. Or, as in Exercise 65(c), we graph $y = \text{Si}(x)$ and $y = 1$ on the same screen to see where they intersect.

73. (a) By FTC1, $g'(x) = f(x)$. So $g'(x) = f(x) = 0$ at $x = 1, 3, 5, 7$, and 9. g has local maxima at $x = 1$ and 5 (since $f = g'$ changes from positive to negative there) and local minima at $x = 3$ and 7. There is no local maximum or minimum at $x = 9$, since f is not defined for $x > 9$.

(b) We can see from the graph that $\left|\int_0^1 f\,dt\right| < \left|\int_1^3 f\,dt\right| < \left|\int_3^5 f\,dt\right| < \left|\int_5^7 f\,dt\right| < \left|\int_7^9 f\,dt\right|$. So $g(1) = \left|\int_0^1 f\,dt\right|$, $g(5) = \int_0^5 f\,dt = g(1) - \left|\int_1^3 f\,dt\right| + \left|\int_3^5 f\,dt\right|$, and $g(9) = \int_0^9 f\,dt = g(5) - \left|\int_5^7 f\,dt\right| + \left|\int_7^9 f\,dt\right|$. Thus, $g(1) < g(5) < g(9)$, and so the absolute maximum of $g(x)$ occurs at $x = 9$.

(c) g is concave downward on those intervals where $g'' < 0$. But $g'(x) = f(x)$, so $g''(x) = f'(x)$, which is negative on (approximately) $\left(\frac{1}{2}, 2\right)$, $(4, 6)$ and $(8, 9)$. So g is concave downward on these intervals.

(d)

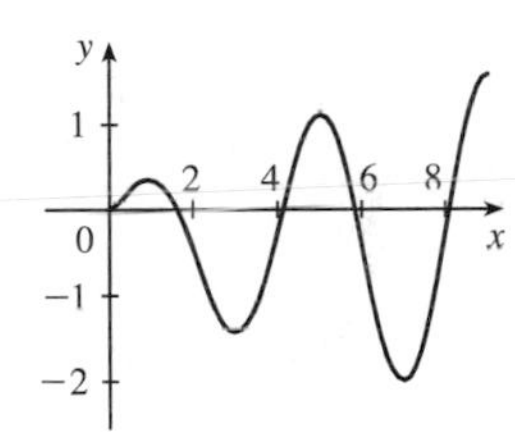

74. (a) By FTC1, $g'(x) = f(x)$. So $g'(x) = f(x) = 0$ at $x = 2, 4, 6, 8$, and 10. g has local maxima at $x = 2$ and 6 (since $f = g'$ changes from positive to negative there) and local minima at $x = 4$ and 8. There is no local maximum or minimum at $x = 10$, since f is not defined for $x > 10$.

(b) We can see from the graph that $\left|\int_0^2 f\,dt\right| > \left|\int_2^4 f\,dt\right| > \left|\int_4^6 f\,dt\right| > \left|\int_6^8 f\,dt\right| > \left|\int_8^{10} f\,dt\right|$. So $g(2) = \left|\int_0^2 f\,dt\right|$, $g(6) = \int_0^6 f\,dt = g(2) - \left|\int_2^4 f\,dt\right| + \left|\int_4^6 f\,dt\right|$, and $g(10) = \int_0^{10} f\,dt = g(6) - \left|\int_6^8 f\,dt\right| + \left|\int_8^{10} f\,dt\right|$. Thus, $g(2) > g(6) > g(10)$, and so the absolute maximum of $g(x)$ occurs at $x = 2$.

(c) g is concave downward on those intervals where $g'' < 0$. But $g'(x) = f(x)$, so $g''(x) = f'(x)$, which is negative on $(1, 3)$, $(5, 7)$ and $(9, 10)$. So g is concave downward on these intervals.

(d)

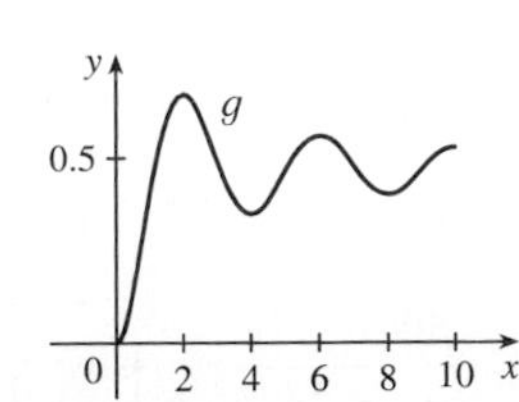

75. $\displaystyle\lim_{n\to\infty}\sum_{i=1}^{n}\left(\frac{i^4}{n^5}+\frac{i}{n^2}\right)=\lim_{n\to\infty}\sum_{i=1}^{n}\left(\frac{i^4}{n^4}+\frac{i}{n}\right)\frac{1}{n}=\lim_{n\to\infty}\frac{1-0}{n}\sum_{i=1}^{n}\left[\left(\frac{i}{n}\right)^4+\frac{i}{n}\right]=\int_0^1(x^4+x)\,dx$

$$=\left[\tfrac{1}{5}x^5+\tfrac{1}{2}x^2\right]_0^1=\left(\tfrac{1}{5}+\tfrac{1}{2}\right)-0=\tfrac{7}{10}$$

76. $\displaystyle\lim_{n\to\infty}\frac{1}{n}\left(\sqrt{\frac{1}{n}}+\sqrt{\frac{2}{n}}+\cdots+\sqrt{\frac{n}{n}}\right)=\lim_{n\to\infty}\frac{1-0}{n}\sum_{i=1}^{n}\sqrt{\frac{i}{n}}=\int_0^1\sqrt{x}\,dx=\left[\frac{2x^{3/2}}{3}\right]_0^1=\frac{2}{3}-0=\frac{2}{3}$

77. Suppose $h<0$. Since f is continuous on $[x+h,x]$, the Extreme Value Theorem says that there are numbers u and v in $[x+h,x]$ such that $f(u)=m$ and $f(v)=M$, where m and M are the absolute minimum and maximum values of f on $[x+h,x]$. By Property 8 of integrals, $m(-h)\le\int_{x+h}^{x}f(t)\,dt\le M(-h)$; that is, $f(u)(-h)\le-\int_x^{x+h}f(t)\,dt\le f(v)(-h)$.

Since $-h>0$, we can divide this inequality by $-h$: $f(u)\le\dfrac{1}{h}\displaystyle\int_x^{x+h}f(t)\,dt\le f(v)$. By Equation 2,

$\dfrac{g(x+h)-g(x)}{h}=\dfrac{1}{h}\displaystyle\int_x^{x+h}f(t)\,dt$ for $h\ne0$, and hence $f(u)\le\dfrac{g(x+h)-g(x)}{h}\le f(v)$, which is Equation 3 in the case where $h<0$.

78. $\displaystyle\frac{d}{dx}\int_{g(x)}^{h(x)}f(t)\,dt=\frac{d}{dx}\left[\int_{g(x)}^{a}f(t)\,dt+\int_a^{h(x)}f(t)\,dt\right]$ [where a is in the domain of f]

$$=\frac{d}{dx}\left[-\int_a^{g(x)}f(t)\,dt\right]+\frac{d}{dx}\left[\int_a^{h(x)}f(t)\,dt\right]=-f(g(x))\,g'(x)+f(h(x))\,h'(x)$$

$$=f(h(x))\,h'(x)-f(g(x))\,g'(x)$$

79. (a) Let $f(x)=\sqrt{x}\;\Rightarrow\;f'(x)=1/(2\sqrt{x})>0$ for $x>0\;\Rightarrow\;f$ is increasing on $(0,\infty)$. If $x\ge0$, then $x^3\ge0$, so $1+x^3\ge1$ and since f is increasing, this means that $f(1+x^3)\ge f(1)\;\Rightarrow\;\sqrt{1+x^3}\ge1$ for $x\ge0$. Next let $g(t)=t^2-t\;\Rightarrow\;g'(t)=2t-1\;\Rightarrow\;g'(t)>0$ when $t\ge1$. Thus, g is increasing on $(1,\infty)$. And since $g(1)=0$, $g(t)\ge0$ when $t\ge1$. Now let $t=\sqrt{1+x^3}$, where $x\ge0$. $\sqrt{1+x^3}\ge1$ (from above) $\Rightarrow\;t\ge1\;\Rightarrow\;g(t)\ge0\;\Rightarrow$ $(1+x^3)-\sqrt{1+x^3}\ge0$ for $x\ge0$. Therefore, $1\le\sqrt{1+x^3}\le1+x^3$ for $x\ge0$.

(b) From part (a) and Property 7: $\int_0^1 1\,dx\le\int_0^1\sqrt{1+x^3}\,dx\le\int_0^1(1+x^3)\,dx\;\Leftrightarrow$

$\left[x\right]_0^1\le\int_0^1\sqrt{1+x^3}\,dx\le\left[x+\tfrac{1}{4}x^4\right]_0^1\;\Leftrightarrow\;1\le\int_0^1\sqrt{1+x^3}\,dx\le1+\tfrac{1}{4}=1.25.$

80. (a) For $0\le x\le1$, we have $x^2\le x$. Since $f(x)=\cos x$ is a decreasing function on $[0,1]$, $\cos(x^2)\ge\cos x$.

(b) $\pi/6<1$, so by part (a), $\cos(x^2)\ge\cos x$ on $[0,\pi/6]$. Thus,

$\int_0^{\pi/6}\cos(x^2)\,dx\ge\int_0^{\pi/6}\cos x\,dx=\left[\sin x\right]_0^{\pi/6}=\sin(\pi/6)-\sin0=\tfrac{1}{2}-0=\tfrac{1}{2}.$

81. $0<\dfrac{x^2}{x^4+x^2+1}<\dfrac{x^2}{x^4}=\dfrac{1}{x^2}$ on $[5,10]$, so

$$0\le\int_5^{10}\frac{x^2}{x^4+x^2+1}\,dx<\int_5^{10}\frac{1}{x^2}\,dx=\left[-\frac{1}{x}\right]_5^{10}=-\frac{1}{10}-\left(-\frac{1}{5}\right)=\frac{1}{10}=0.1.$$

82. (a) If $x < 0$, then $g(x) = \int_0^x f(t)\,dt = \int_0^x 0\,dt = 0$.

If $0 \le x \le 1$, then $g(x) = \int_0^x f(t)\,dt = \int_0^x t\,dt = \left[\frac{1}{2}t^2\right]_0^x = \frac{1}{2}x^2$.

If $1 < x \le 2$, then

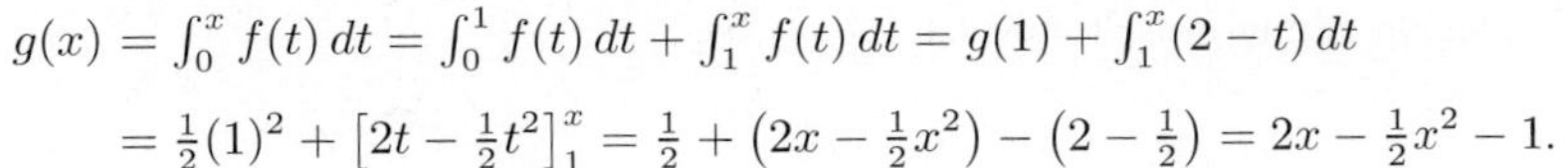

$$g(x) = \int_0^x f(t)\,dt = \int_0^1 f(t)\,dt + \int_1^x f(t)\,dt = g(1) + \int_1^x (2-t)\,dt$$
$$= \tfrac{1}{2}(1)^2 + \left[2t - \tfrac{1}{2}t^2\right]_1^x = \tfrac{1}{2} + \left(2x - \tfrac{1}{2}x^2\right) - \left(2 - \tfrac{1}{2}\right) = 2x - \tfrac{1}{2}x^2 - 1.$$

If $x > 2$, then $g(x) = \int_0^x f(t)\,dt = g(2) + \int_2^x 0\,dt = 1 + 0 = 1$. So

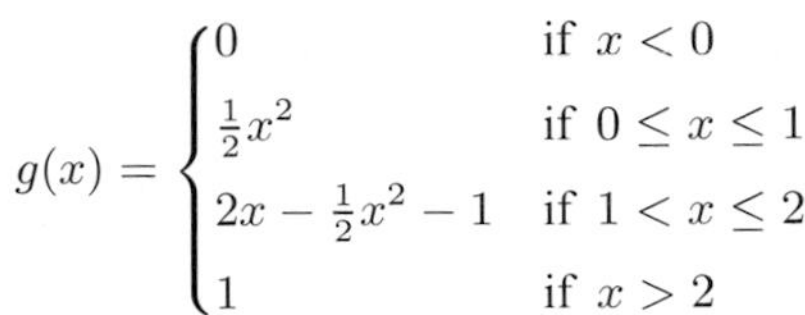

$$g(x) = \begin{cases} 0 & \text{if } x < 0 \\ \frac{1}{2}x^2 & \text{if } 0 \le x \le 1 \\ 2x - \frac{1}{2}x^2 - 1 & \text{if } 1 < x \le 2 \\ 1 & \text{if } x > 2 \end{cases}$$

(b)

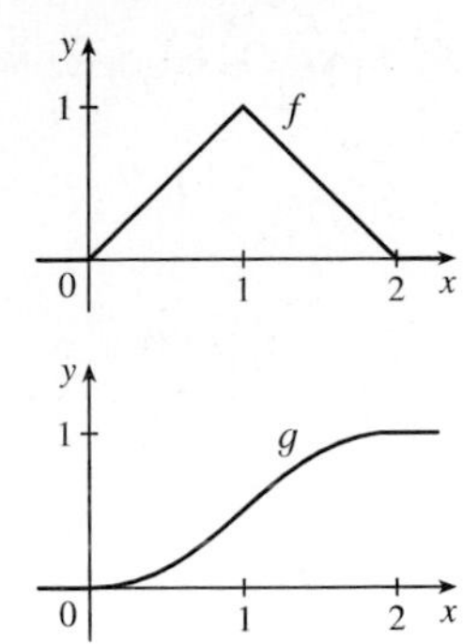

(c) f is not differentiable at its corners at $x = 0$, 1, and 2. f is differentiable on $(-\infty, 0)$, $(0, 1)$, $(1, 2)$ and $(2, \infty)$.

g is differentiable on $(-\infty, \infty)$.

83. Using FTC1, we differentiate both sides of $6 + \int_a^x \frac{f(t)}{t^2}\,dt = 2\sqrt{x}$ to get $\frac{f(x)}{x^2} = 2\frac{1}{2\sqrt{x}} \Rightarrow f(x) = x^{3/2}$.

To find a, we substitute $x = a$ in the original equation to obtain $6 + \int_a^a \frac{f(t)}{t^2}\,dt = 2\sqrt{a} \Rightarrow 6 + 0 = 2\sqrt{a} \Rightarrow$

$3 = \sqrt{a} \Rightarrow a = 9$.

84. $B = 3A \Rightarrow \int_0^b e^x\,dx = 3\int_0^a e^x\,dx \Rightarrow \left[e^x\right]_0^b = 3\left[e^x\right]_0^a \Rightarrow e^b - 1 = 3(e^a - 1) \Rightarrow e^b = 3e^a - 2 \Rightarrow$

$b = \ln(3e^a - 2)$

85. (a) Let $F(t) = \int_0^t f(s)\,ds$. Then, by FTC1, $F'(t) = f(t) =$ rate of depreciation, so $F(t)$ represents the loss in value over the interval $[0, t]$.

(b) $C(t) = \frac{1}{t}\left[A + \int_0^t f(s)\,ds\right] = \frac{A + F(t)}{t}$ represents the average expenditure per unit of t during the interval $[0, t]$, assuming that there has been only one overhaul during that time period. The company wants to minimize average expenditure.

(c) $C(t) = \frac{1}{t}\left[A + \int_0^t f(s)\,ds\right]$. Using FTC1, we have $C'(t) = -\frac{1}{t^2}\left[A + \int_0^t f(s)\,ds\right] + \frac{1}{t}f(t)$.

$C'(t) = 0 \Rightarrow t\,f(t) = A + \int_0^t f(s)\,ds \Rightarrow f(t) = \frac{1}{t}\left[A + \int_0^t f(s)\,ds\right] = C(t)$.

86. (a) $C(t) = \frac{1}{t}\int_0^t [f(s) + g(s)]\,ds$. Using FTC1 and the Product Rule, we have

$C'(t) = \frac{1}{t}[f(t) + g(t)] - \frac{1}{t^2}\int_0^t [f(s) + g(s)]\,ds$. Set $C'(t) = 0$: $\frac{1}{t}[f(t) + g(t)] - \frac{1}{t^2}\int_0^t [f(s) + g(s)]\,ds = 0 \Rightarrow$

$[f(t) + g(t)] - \frac{1}{t}\int_0^t [f(s) + g(s)]\,ds = 0 \Rightarrow [f(t) + g(t)] - C(t) = 0 \Rightarrow C(t) = f(t) + g(t)$.

(b) For $0 \le t \le 30$, we have $D(t) = \int_0^t \left(\frac{V}{15} - \frac{V}{450}s\right) ds = \left[\frac{V}{15}s - \frac{V}{900}s^2\right]_0^t = \frac{V}{15}t - \frac{V}{900}t^2$.

So $D(t) = V \Rightarrow \frac{V}{15}t - \frac{V}{900}t^2 = V \Rightarrow 60t - t^2 = 900 \Rightarrow t^2 - 60t + 900 = 0 \Rightarrow$

$(t-30)^2 = 0 \Rightarrow t = 30$. So the length of time T is 30 months.

(c) $C(t) = \frac{1}{t}\int_0^t \left(\frac{V}{15} - \frac{V}{450}s + \frac{V}{12{,}900}s^2\right) ds = \frac{1}{t}\left[\frac{V}{15}s - \frac{V}{900}s^2 + \frac{V}{38{,}700}s^3\right]_0^t$

$$= \frac{1}{t}\left(\frac{V}{15}t - \frac{V}{900}t^2 + \frac{V}{38{,}700}t^3\right) = \frac{V}{15} - \frac{V}{900}t + \frac{V}{38{,}700}t^2 \Rightarrow$$

$C'(t) = -\frac{V}{900} + \frac{V}{19{,}350}t = 0$ when $\frac{1}{19{,}350}t = \frac{1}{900} \Rightarrow t = 21.5$.

$C(21.5) = \frac{V}{15} - \frac{V}{900}(21.5) + \frac{V}{38{,}700}(21.5)^2 \approx 0.05472V$, $C(0) = \frac{V}{15} \approx 0.06667V$, and

$C(30) = \frac{V}{15} - \frac{V}{900}(30) + \frac{V}{38{,}700}(30)^2 \approx 0.05659V$, so the absolute minimum is $C(21.5) \approx 0.05472V$.

(d) As in part (c), we have $C(t) = \frac{V}{15} - \frac{V}{900}t + \frac{V}{38{,}700}t^2$, so $C(t) = f(t) + g(t) \Leftrightarrow$

$$\frac{V}{15} - \frac{V}{900}t + \frac{V}{38{,}700}t^2 = \frac{V}{15} - \frac{V}{450}t + \frac{V}{12{,}900}t^2 \Leftrightarrow$$

$$t^2\left(\frac{1}{12{,}900} - \frac{1}{38{,}700}\right) = t\left(\frac{1}{450} - \frac{1}{900}\right) \Leftrightarrow t = \frac{1/900}{2/38{,}700} = \frac{43}{2} = 21.5.$$

This is the value of t that we obtained as the critical number of C in part (c), so we have verified the result of (a) in this case.

y
V/15
y = f(t) + g(t)
y = C(t)
0
21.5
30
t

5.4 Indefinite Integrals and the Net Change Theorem

1. $\frac{d}{dx}\left[-\frac{\sqrt{1+x^2}}{x} + C\right] = \frac{d}{dx}\left[-\frac{(1+x^2)^{1/2}}{x} + C\right] = -\frac{x \cdot \frac{1}{2}(1+x^2)^{-1/2}(2x) - (1+x^2)^{1/2} \cdot 1}{(x)^2} + 0$

$$= -\frac{(1+x^2)^{-1/2}\left[x^2 - (1+x^2)\right]}{x^2} = -\frac{-1}{(1+x^2)^{1/2}x^2} = \frac{1}{x^2\sqrt{1+x^2}}$$

2. $\frac{d}{dx}\left(\frac{1}{2}x + \frac{1}{4}\sin 2x + C\right) = \frac{1}{2} + \frac{1}{4}\cos 2x \cdot 2 + 0 = \frac{1}{2} + \frac{1}{2}\cos 2x$

$$= \tfrac{1}{2} + \tfrac{1}{2}(2\cos^2 x - 1) = \tfrac{1}{2} + \cos^2 x - \tfrac{1}{2} = \cos^2 x$$

3. $\frac{d}{dx}(\tan x - x + C) = \sec^2 x - 1 + 0 = \tan^2 x$

4. $\frac{d}{dx}\left[\frac{2}{15b^2}(3bx - 2a)(a+bx)^{3/2} + C\right] = \frac{2}{15b^2}\left[(3bx-2a)\tfrac{3}{2}(a+bx)^{1/2}(b) + (a+bx)^{3/2}(3b) + 0\right]$

$$= \frac{2}{15b^2}(3b)(a+bx)^{1/2}\left[(3bx-2a)\tfrac{1}{2} + (a+bx)\right]$$

$$= \frac{2}{5b}(a+bx)^{1/2}\left(\frac{5}{2}bx\right) = x\sqrt{a+bx}$$

5. $\displaystyle\int (x^{1.3} + 7x^{2.5})\,dx = \frac{1}{2.3}x^{2.3} + \frac{7}{3.5}x^{3.5} + C = \frac{1}{2.3}x^{2.3} + 2x^{3.5} + C$

6. $\int \sqrt[4]{x^5}\,dx = \int x^{5/4}\,dx = \frac{4}{9}x^{9/4} + C$

7. $\int \left(5 + \frac{2}{3}x^2 + \frac{3}{4}x^3\right)\,dx = 5x + \frac{2}{3}\cdot\frac{1}{3}x^3 + \frac{3}{4}\cdot\frac{1}{4}x^4 + C = 5x + \frac{2}{9}x^3 + \frac{3}{16}x^4 + C$

8. $\int (u^6 - 2u^5 - u^3 + \frac{2}{7})\,du = \frac{1}{7}u^7 - 2\cdot\frac{1}{6}u^6 - \frac{1}{4}u^4 + \frac{2}{7}u + C = \frac{1}{7}u^7 - \frac{1}{3}u^6 - \frac{1}{4}u^4 + \frac{2}{7}u + C$

9. $\displaystyle\int (u + 4)(2u + 1)\,du = \int (2u^2 + 9u + 4)\,du = 2\,\frac{u^3}{3} + 9\,\frac{u^2}{2} + 4u + C = \frac{2}{3}u^3 + \frac{9}{2}u^2 + 4u + C$

10. $\int \sqrt{t}(t^2 + 3t + 2)\,dt = \int t^{1/2}(t^2 + 3t + 2)\,dt = \int (t^{5/2} + 3t^{3/2} + 2t^{1/2})\,dt$

$= \frac{2}{7}t^{7/2} + 3\cdot\frac{2}{5}t^{5/2} + 2\cdot\frac{2}{3}t^{3/2} + C = \frac{2}{7}t^{7/2} + \frac{6}{5}t^{5/2} + \frac{4}{3}t^{3/2} + C$

11. $\displaystyle\int \frac{1 + \sqrt{x} + x}{x}\,dx = \int \left(\frac{1}{x} + \frac{\sqrt{x}}{x} + \frac{x}{x}\right)dx = \int \left(\frac{1}{x} + x^{-1/2} + 1\right)dx$

$= \ln|x| + 2x^{1/2} + x + C = \ln|x| + 2\sqrt{x} + x + C$

12. $\displaystyle\int \left(x^2 + 1 + \frac{1}{x^2 + 1}\right)dx = \frac{x^3}{3} + x + \tan^{-1}x + C$

13. $\int (\sin x + \sinh x)\,dx = -\cos x + \cosh x + C$

14. $\displaystyle\int \left(\frac{1 + r}{r}\right)^2 dr = \int \frac{1 + 2r + r^2}{r^2}\,dr = \int (r^{-2} + 2r^{-1} + 1)\,dr = -r^{-1} + 2\ln|r| + r + C = -\frac{1}{r} + 2\ln|r| + r + C$

15. $\int (2 + \tan^2\theta)\,d\theta = \int [2 + (\sec^2\theta - 1)]\,d\theta = \int (1 + \sec^2\theta)\,d\theta = \theta + \tan\theta + C$

16. $\int \sec t\,(\sec t + \tan t)\,dt = \int (\sec^2 t + \sec t\,\tan t)\,dt = \tan t + \sec t + C$

17. $\displaystyle\int 2^t(1 + 5^t)\,dt = \int (2^t + 2^t\cdot 5^t)\,dt = \int (2^t + 10^t)\,dt = \frac{2^t}{\ln 2} + \frac{10^t}{\ln 10} + C$

18. $\displaystyle\int \frac{\sin 2x}{\sin x}\,dx = \int \frac{2\sin x\,\cos x}{\sin x}\,dx = \int 2\cos x\,dx = 2\sin x + C$

19. $\int \left(\cos x + \frac{1}{2}x\right)dx = \sin x + \frac{1}{4}x^2 + C$. The members of the family in the figure correspond to $C = -5$, 0, 5, and 10.

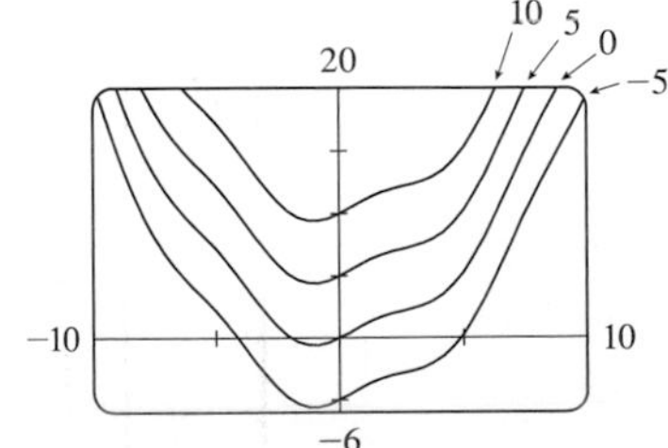

20. $\int (e^x - 2x^2)\,dx = e^x - \frac{2}{3}x^3 + C$. The members of the family in the figure correspond to $C = -5$, 0, 2, and 5.

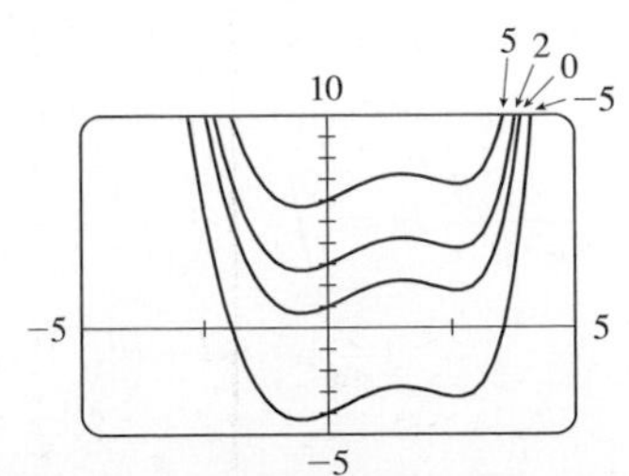

21. $\int_{-2}^{3}(x^2-3)\,dx=\left[\frac{1}{3}x^3-3x\right]_{-2}^{3}=(9-9)-\left(-\frac{8}{3}+6\right)=\frac{8}{3}-\frac{18}{3}=-\frac{10}{3}$

22. $\int_1^2(4x^3-3x^2+2x)\,dx=\left[x^4-x^3+x^2\right]_1^2=(16-8+4)-(1-1+1)=12-1=11$

23. $\int_{-2}^{0}\left(\frac{1}{2}t^4+\frac{1}{4}t^3-t\right)dt=\left[\frac{1}{10}t^5+\frac{1}{16}t^4-\frac{1}{2}t^2\right]_{-2}^{0}=0-\left[\frac{1}{10}(-32)+\frac{1}{16}(16)-\frac{1}{2}(4)\right]=-\left(-\frac{16}{5}+1-2\right)=\frac{21}{5}$

24. $\int_0^3(1+6w^2-10w^4)\,dw=\left[w+2w^3-2w^5\right]_0^3=(3+54-486)-0=-429$

25. $\int_0^2(2x-3)(4x^2+1)\,dx=\int_0^2(8x^3-12x^2+2x-3)\,dx=\left[2x^4-4x^3+x^2-3x\right]_0^2=(32-32+4-6)-0=-2$

26. $\int_{-1}^{1}t(1-t)^2\,dt=\int_{-1}^{1}t(1-2t+t^2)\,dt=\int_{-1}^{1}(t-2t^2+t^3)\,dt=\left[\frac{1}{2}t^2-\frac{2}{3}t^3+\frac{1}{4}t^4\right]_{-1}^{1}$

$$=\left(\tfrac{1}{2}-\tfrac{2}{3}+\tfrac{1}{4}\right)-\left(\tfrac{1}{2}+\tfrac{2}{3}+\tfrac{1}{4}\right)=-\tfrac{4}{3}$$

27. $\int_0^{\pi}(5e^x+3\sin x)\,dx=\left[5e^x-3\cos x\right]_0^{\pi}=\left[5e^{\pi}-3(-1)\right]-\left[5(1)-3(1)\right]=5e^{\pi}+1$

28. $$\int_1^2\left(\frac{1}{x^2}-\frac{4}{x^3}\right)dx=\int_1^2(x^{-2}-4x^{-3})\,dx=\left[\frac{x^{-1}}{-1}-\frac{4x^{-2}}{-2}\right]_1^2=\left[-\frac{1}{x}+\frac{2}{x^2}\right]_1^2=\left(-\frac{1}{2}+\frac{1}{2}\right)-(-1+2)=-1$$

29. $$\int_1^4\left(\frac{4+6u}{\sqrt{u}}\right)du=\int_1^4\left(\frac{4}{\sqrt{u}}+\frac{6u}{\sqrt{u}}\right)du=\int_1^4(4u^{-1/2}+6u^{1/2})\,du=\left[8u^{1/2}+4u^{3/2}\right]_1^4=(16+32)-(8+4)=36$$

30. $$\int_0^1\frac{4}{1+p^2}\,dp=\left[4\arctan p\right]_0^1=4\arctan 1-4\arctan 0=4\left(\frac{\pi}{4}\right)-4(0)=\pi$$

31. $\int_0^1 x\left(\sqrt[3]{x}+\sqrt[4]{x}\right)dx=\int_0^1(x^{4/3}+x^{5/4})\,dx=\left[\frac{3}{7}x^{7/3}+\frac{4}{9}x^{9/4}\right]_0^1=\left(\frac{3}{7}+\frac{4}{9}\right)-0=\frac{55}{63}$

32. $$\int_1^4\frac{\sqrt{y}-y}{y^2}\,dy=\int_1^4\left(\frac{\sqrt{y}}{y^2}-\frac{y}{y^2}\right)dy=\int_1^4(y^{-3/2}-y^{-1})\,dy=\left[-2y^{-1/2}-\ln|y|\right]_1^4=\left[-\frac{2}{\sqrt{y}}-\ln|y|\right]_1^4$$

$$=(-1-\ln 4)-(-2-\ln 1)=1-\ln 4$$

33. $$\int_1^2\left(\frac{x}{2}-\frac{2}{x}\right)dx=\left[\frac{1}{4}x^2-2\ln|x|\right]_1^2=(1-2\ln 2)-\left(\frac{1}{4}-2\ln 1\right)=\frac{3}{4}-2\ln 2$$

34. $$\int_0^1(5x-5^x)\,dx=\left[\frac{5}{2}x^2-\frac{5^x}{\ln 5}\right]_0^1=\left(\frac{5}{2}-\frac{5}{\ln 5}\right)-\left(0-\frac{1}{\ln 5}\right)=\frac{5}{2}-\frac{4}{\ln 5}$$

35. $$\int_0^1(x^{10}+10^x)\,dx=\left[\frac{x^{11}}{11}+\frac{10^x}{\ln 10}\right]_0^1=\left(\frac{1}{11}+\frac{10}{\ln 10}\right)-\left(0+\frac{1}{\ln 10}\right)=\frac{1}{11}+\frac{9}{\ln 10}$$

36. $$\int_0^{\pi/4}\sec\theta\,\tan\theta\,d\theta=\left[\sec\theta\right]_0^{\pi/4}=\sec\frac{\pi}{4}-\sec 0=\sqrt{2}-1$$

37. $$\int_0^{\pi/4}\frac{1+\cos^2\theta}{\cos^2\theta}\,d\theta=\int_0^{\pi/4}\left(\frac{1}{\cos^2\theta}+\frac{\cos^2\theta}{\cos^2\theta}\right)d\theta=\int_0^{\pi/4}(\sec^2\theta+1)\,d\theta$$

$$=\left[\tan\theta+\theta\right]_0^{\pi/4}=\left(\tan\tfrac{\pi}{4}+\tfrac{\pi}{4}\right)-(0+0)=1+\tfrac{\pi}{4}$$

38. $\displaystyle\int_0^{\pi/3} \frac{\sin\theta + \sin\theta\,\tan^2\theta}{\sec^2\theta}\,d\theta = \int_0^{\pi/3} \frac{\sin\theta\,(1+\tan^2\theta)}{\sec^2\theta}\,d\theta = \int_0^{\pi/3} \frac{\sin\theta\,\sec^2\theta}{\sec^2\theta}\,d\theta = \int_0^{\pi/3} \sin\theta\,d\theta$

$\displaystyle = \Big[-\cos\theta\Big]_0^{\pi/3} = -\tfrac{1}{2} - (-1) = \tfrac{1}{2}$

39. $\displaystyle\int_1^8 \frac{2+t}{\sqrt[3]{t^2}}\,dt = \int_1^8 \left(\frac{2}{t^{2/3}} + \frac{t}{t^{2/3}}\right) dt = \int_1^8 (2t^{-2/3} + t^{1/3})\,dt = \Big[2\cdot 3t^{1/3} + \tfrac{3}{4}t^{4/3}\Big]_1^8 = (12+12) - (6 + \tfrac{3}{4}) = \tfrac{69}{4}$

40. $\displaystyle\int_{-10}^{10} \frac{2e^x}{\sinh x + \cosh x}\,dx = \int_{-10}^{10} \frac{2e^x}{\dfrac{e^x - e^{-x}}{2} + \dfrac{e^x + e^{-x}}{2}}\,dx = \int_{-10}^{10} \frac{2e^x}{e^x}\,dx = \int_{-10}^{10} 2\,dx = \Big[2x\Big]_{-10}^{10} = 20 - (-20) = 40$

41. $\displaystyle\int_0^{\sqrt{3}/2} \frac{dr}{\sqrt{1-r^2}} = \Big[\arcsin r\Big]_0^{\sqrt{3}/2} = \arcsin\!\big(\sqrt{3}/2\big) - \arcsin 0 = \frac{\pi}{3} - 0 = \frac{\pi}{3}$

42. $\displaystyle\int_1^2 \frac{(x-1)^3}{x^2}\,dx = \int_1^2 \frac{x^3 - 3x^2 + 3x - 1}{x^2}\,dx = \int_1^2 \left(x - 3 + \frac{3}{x} - \frac{1}{x^2}\right) dx = \left[\tfrac{1}{2}x^2 - 3x + 3\ln|x| + \frac{1}{x}\right]_1^2$

$\displaystyle = \big(2 - 6 + 3\ln 2 + \tfrac{1}{2}\big) - \big(\tfrac{1}{2} - 3 + 0 + 1\big) = 3\ln 2 - 2$

43. $\displaystyle\int_0^{1/\sqrt{3}} \frac{t^2-1}{t^4-1}\,dt = \int_0^{1/\sqrt{3}} \frac{t^2-1}{(t^2+1)(t^2-1)}\,dt = \int_0^{1/\sqrt{3}} \frac{1}{t^2+1}\,dt = \Big[\arctan t\Big]_0^{1/\sqrt{3}} = \arctan\!\big(1/\sqrt{3}\big) - \arctan 0$

$\displaystyle = \tfrac{\pi}{6} - 0 = \tfrac{\pi}{6}$

44. $|2x-1| = \begin{cases} 2x-1 & \text{if } 2x-1 \geq 0 \\ -(2x-1) & \text{if } 2x-1 < 0 \end{cases} = \begin{cases} 2x-1 & \text{if } x \geq \frac{1}{2} \\ 1-2x & \text{if } x < \frac{1}{2} \end{cases}$

Thus, $\int_0^2 |2x-1|\,dx = \int_0^{1/2}(1-2x)\,dx + \int_{1/2}^2 (2x-1)\,dx = \big[x - x^2\big]_0^{1/2} + \big[x^2 - x\big]_{1/2}^2$

$= \big(\tfrac{1}{2} - \tfrac{1}{4}\big) - 0 + (4-2) - \big(\tfrac{1}{4} - \tfrac{1}{2}\big) = \tfrac{1}{4} + 2 - \big(-\tfrac{1}{4}\big) = \tfrac{5}{2}$

45. $\int_{-1}^2 (x - 2|x|)\,dx = \int_{-1}^0 [x - 2(-x)]\,dx + \int_0^2 [x - 2(x)]\,dx = \int_{-1}^0 3x\,dx + \int_0^2 (-x)\,dx = 3\big[\tfrac{1}{2}x^2\big]_{-1}^0 - \big[\tfrac{1}{2}x^2\big]_0^2$

$= 3\big(0 - \tfrac{1}{2}\big) - (2 - 0) = -\tfrac{7}{2} = -3.5$

46. $\int_0^{3\pi/2} |\sin x|\,dx = \int_0^{\pi} \sin x\,dx + \int_{\pi}^{3\pi/2}(-\sin x)\,dx = \big[-\cos x\big]_0^{\pi} + \big[\cos x\big]_{\pi}^{3\pi/2} = [1 - (-1)] + [0 - (-1)] = 2 + 1 = 3$

47. The graph shows that $y = 1 - 2x - 5x^4$ has x-intercepts at $x = a \approx -0.86$ and at $x = b \approx 0.42$. So the area of the region that lies under the curve and above the x-axis is

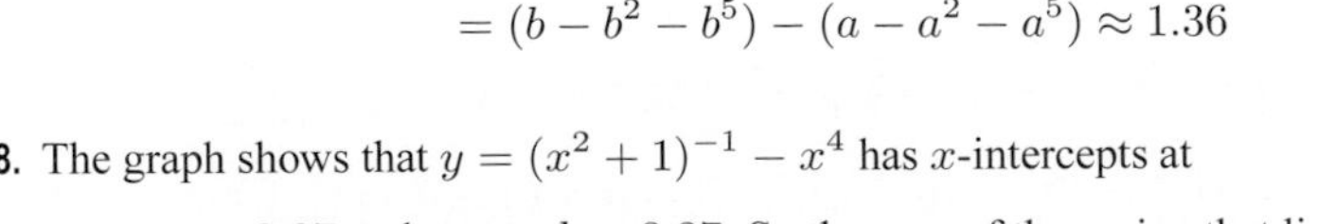

$\int_a^b (1 - 2x - 5x^4)\,dx = \big[x - x^2 - x^5\big]_a^b$

$= (b - b^2 - b^5) - (a - a^2 - a^5) \approx 1.36$

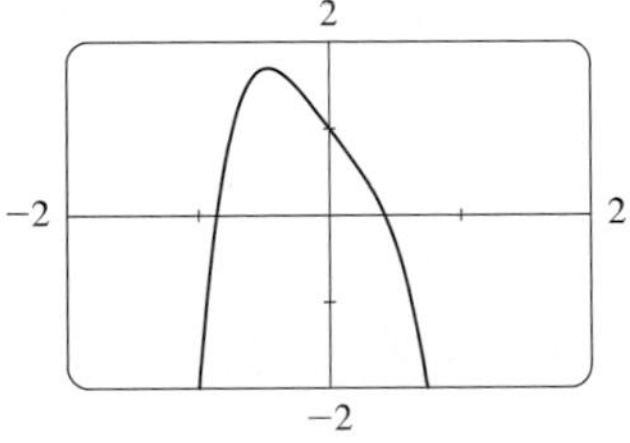

48. The graph shows that $y = (x^2+1)^{-1} - x^4$ has x-intercepts at $x = a \approx -0.87$ and at $x = b \approx 0.87$. So the area of the region that lies under the curve and above the x-axis is

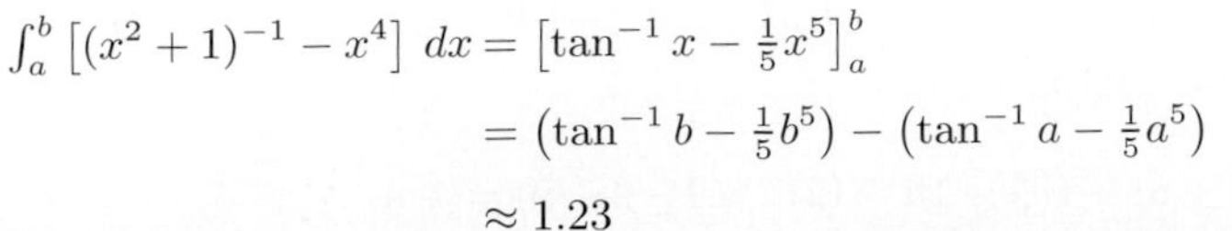

$\int_a^b \big[(x^2+1)^{-1} - x^4\big]\,dx = \big[\tan^{-1}x - \tfrac{1}{5}x^5\big]_a^b$

$= \big(\tan^{-1}b - \tfrac{1}{5}b^5\big) - \big(\tan^{-1}a - \tfrac{1}{5}a^5\big)$

≈ 1.23

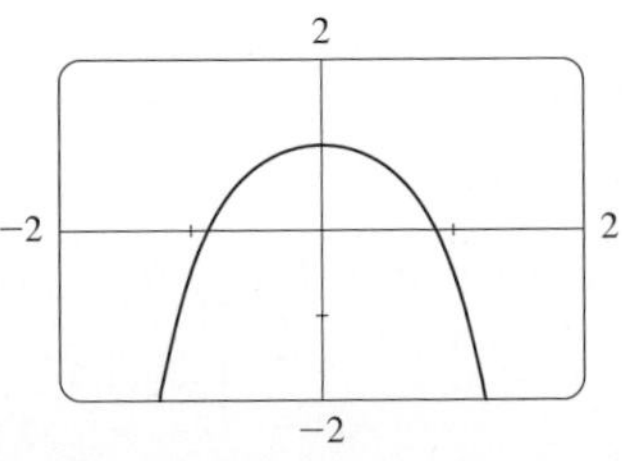

49. $A = \int_0^2 (2y - y^2)\,dy = \left[y^2 - \frac{1}{3}y^3\right]_0^2 = \left(4 - \frac{8}{3}\right) - 0 = \frac{4}{3}$

50. $y = \sqrt[4]{x} \;\Rightarrow\; x = y^4$, so $A = \int_0^1 y^4\,dy = \left[\frac{1}{5}y^5\right]_0^1 = \frac{1}{5}$.

51. If $w'(t)$ is the rate of change of weight in pounds per year, then $w(t)$ represents the weight in pounds of the child at age t. We know from the Net Change Theorem that $\int_5^{10} w'(t)\,dt = w(10) - w(5)$, so the integral represents the increase in the child's weight (in pounds) between the ages of 5 and 10.

52. $\int_a^b I(t)\,dt = \int_a^b Q'(t)\,dt = Q(b) - Q(a)$ by the Net Change Theorem, so it represents the change in the charge Q from time $t = a$ to $t = b$.

53. Since $r(t)$ is the rate at which oil leaks, we can write $r(t) = -V'(t)$, where $V(t)$ is the volume of oil at time t. [Note that the minus sign is needed because V is decreasing, so $V'(t)$ is negative, but $r(t)$ is positive.] Thus, by the Net Change Theorem, $\int_0^{120} r(t)\,dt = -\int_0^{120} V'(t)\,dt = -\left[V(120) - V(0)\right] = V(0) - V(120)$, which is the number of gallons of oil that leaked from the tank in the first two hours (120 minutes).

54. By the Net Change Theorem, $\int_0^{15} n'(t)\,dt = n(15) - n(0) = n(15) - 100$ represents the increase in the bee population in 15 weeks. So $100 + \int_0^{15} n'(t)\,dt = n(15)$ represents the total bee population after 15 weeks.

55. By the Net Change Theorem, $\int_{1000}^{5000} R'(x)\,dx = R(5000) - R(1000)$, so it represents the increase in revenue when production is increased from 1000 units to 5000 units.

56. The slope of the trail is the rate of change of the elevation E, so $f(x) = E'(x)$. By the Net Change Theorem, $\int_3^5 f(x)\,dx = \int_3^5 E'(x)\,dx = E(5) - E(3)$ is the change in the elevation E between $x = 3$ miles and $x = 5$ miles from the start of the trail.

57. In general, the unit of measurement for $\int_a^b f(x)\,dx$ is the product of the unit for $f(x)$ and the unit for x. Since $f(x)$ is measured in newtons and x is measured in meters, the units for $\int_0^{100} f(x)\,dx$ are newton-meters (or joules). (A newton-meter is abbreviated N·m.)

58. The units for $a(x)$ are pounds per foot and the units for x are feet, so the units for da/dx are pounds per foot per foot, denoted (lb/ft)/ft. The unit of measurement for $\int_2^8 a(x)\,dx$ is the product of pounds per foot and feet; that is, pounds.

59. (a) Displacement $= \int_0^3 (3t - 5)\,dt = \left[\frac{3}{2}t^2 - 5t\right]_0^3 = \frac{27}{2} - 15 = -\frac{3}{2}$ m

(b) Distance traveled $= \int_0^3 |3t - 5|\,dt = \int_0^{5/3} (5 - 3t)\,dt + \int_{5/3}^3 (3t - 5)\,dt$

$$= \left[5t - \tfrac{3}{2}t^2\right]_0^{5/3} + \left[\tfrac{3}{2}t^2 - 5t\right]_{5/3}^3 = \tfrac{25}{3} - \tfrac{3}{2}\cdot\tfrac{25}{9} + \tfrac{27}{2} - 15 - \left(\tfrac{3}{2}\cdot\tfrac{25}{9} - \tfrac{25}{3}\right) = \tfrac{41}{6}\text{ m}$$

60. (a) Displacement $= \int_2^4 (t^2 - 2t - 3)\,dt = \left[\frac{1}{3}t^3 - t^2 - 3t\right]_2^4 = \left(\frac{64}{3} - 16 - 12\right) - \left(\frac{8}{3} - 4 - 6\right) = \frac{2}{3}$ m

(b) $v(t) = t^2 - 2t - 3 = (t + 1)(t - 3)$, so $v(t) < 0$ for $-1 < t < 3$, but on the interval $[2, 4]$, $v(t) < 0$ for $2 \le t < 3$.

Distance traveled $= \int_2^4 \left|t^2 - 2t - 3\right|\,dt = \int_2^3 -(t^2 - 2t - 3)\,dt + \int_3^4 (t^2 - 2t - 3)\,dt$

$$= \left[-\tfrac{1}{3}t^3 + t^2 + 3t\right]_2^3 + \left[\tfrac{1}{3}t^3 - t^2 - 3t\right]_3^4$$

$$= (-9 + 9 + 9) - \left(-\tfrac{8}{3} + 4 + 6\right) + \left(\tfrac{64}{3} - 16 - 12\right) - (9 - 9 - 9) = 4\text{ m}$$

61. (a) $v'(t) = a(t) = t + 4 \Rightarrow v(t) = \frac{1}{2}t^2 + 4t + C \Rightarrow v(0) = C = 5 \Rightarrow v(t) = \frac{1}{2}t^2 + 4t + 5$ m/s

(b) Distance traveled $= \int_0^{10} |v(t)|\,dt = \int_0^{10} \left|\frac{1}{2}t^2 + 4t + 5\right| dt = \int_0^{10} \left(\frac{1}{2}t^2 + 4t + 5\right) dt = \left[\frac{1}{6}t^3 + 2t^2 + 5t\right]_0^{10}$

$= \frac{500}{3} + 200 + 50 = 416\frac{2}{3}$ m

62. (a) $v'(t) = a(t) = 2t + 3 \Rightarrow v(t) = t^2 + 3t + C \Rightarrow v(0) = C = -4 \Rightarrow v(t) = t^2 + 3t - 4$

(b) Distance traveled $= \int_0^3 |t^2 + 3t - 4|\,dt = \int_0^3 |(t+4)(t-1)|\,dt = \int_0^1 (-t^2 - 3t + 4)\,dt + \int_1^3 (t^2 + 3t - 4)\,dt$

$= \left[-\frac{1}{3}t^3 - \frac{3}{2}t^2 + 4t\right]_0^1 + \left[\frac{1}{3}t^3 + \frac{3}{2}t^2 - 4t\right]_1^3$

$= \left(-\frac{1}{3} - \frac{3}{2} + 4\right) + \left(9 + \frac{27}{2} - 12\right) - \left(\frac{1}{3} + \frac{3}{2} - 4\right) = \frac{89}{6}$ m

63. Since $m'(x) = \rho(x)$, $m = \int_0^4 \rho(x)\,dx = \int_0^4 \left(9 + 2\sqrt{x}\right) dx = \left[9x + \frac{4}{3}x^{3/2}\right]_0^4 = 36 + \frac{32}{3} - 0 = \frac{140}{3} = 46\frac{2}{3}$ kg.

64. By the Net Change Theorem, the amount of water that flows from the tank during the first 10 minutes is

$\int_0^{10} r(t)\,dt = \int_0^{10} (200 - 4t)\,dt = \left[200t - 2t^2\right]_0^{10} = (2000 - 200) - 0 = 1800$ liters.

65. Let s be the position of the car. We know from Equation 2 that $s(100) - s(0) = \int_0^{100} v(t)\,dt$. We use the Midpoint Rule for $0 \le t \le 100$ with $n = 5$. Note that the length of each of the five time intervals is 20 seconds $= \frac{20}{3600}$ hour $= \frac{1}{180}$ hour. So the distance traveled is

$\int_0^{100} v(t)\,dt \approx \frac{1}{180}[v(10) + v(30) + v(50) + v(70) + v(90)] = \frac{1}{180}(38 + 58 + 51 + 53 + 47) = \frac{247}{180} \approx 1.4$ miles.

66. (a) By the Net Change Theorem, the total amount spewed into the atmosphere is $Q(6) - Q(0) = \int_0^6 r(t)\,dt = Q(6)$ since $Q(0) = 0$. The rate $r(t)$ is positive, so Q is an increasing function. Thus, an upper estimate for $Q(6)$ is R_6 and a lower estimate for $Q(6)$ is L_6. $\Delta t = \dfrac{b-a}{n} = \dfrac{6-0}{6} = 1.$

$$R_6 = \sum_{i=1}^{6} r(t_i)\,\Delta t = 10 + 24 + 36 + 46 + 54 + 60 = 230 \text{ tonnes.}$$

$$L_6 = \sum_{i=1}^{6} r(t_{i-1})\,\Delta t = R_6 + r(0) - r(6) = 230 + 2 - 60 = 172 \text{ tonnes.}$$

(b) $\Delta t = \dfrac{b-a}{n} = \dfrac{6-0}{3} = 2.$ $Q(6) \approx M_3 = 2[r(1) + r(3) + r(5)] = 2(10 + 36 + 54) = 2(100) = 200$ tonnes.

67. From the Net Change Theorem, the increase in cost if the production level is raised from 2000 yards to 4000 yards is $C(4000) - C(2000) = \int_{2000}^{4000} C'(x)\,dx$.

$$\int_{2000}^{4000} C'(x)\,dx = \int_{2000}^{4000} (3 - 0.01x + 0.000006x^2)\,dx = \left[3x - 0.005x^2 + 0.000002x^3\right]_{2000}^{4000}$$

$$= 60{,}000 - 2{,}000 = \$58{,}000$$

68. By the Net Change Theorem, the amount of water after four days is

$$25{,}000 + \int_0^4 r(t)\,dt \approx 25{,}000 + M_4 = 25{,}000 + \tfrac{4-0}{4}\,[r(0.5) + r(1.5) + r(2.5) + r(3.5)]$$

$$\approx 25{,}000 + [1500 + 1770 + 740 + (-690)] = 28{,}320 \text{ liters}$$

69. To use the Midpoint Rule, we'll use the midpoint of each of three 2-second intervals.

$v(6) - v(0) = \int_0^6 a(t)\,dt \approx [a(1) + a(3) + a(5)]\,\dfrac{6-0}{3} \approx (0.6 + 10 + 9.3)(2) = 39.8\text{ ft/s}$

70. Use the midpoint of each of four 2-day intervals. Let $t = 0$ correspond to July 18 and note that the inflow rate, $r(t)$, is in ft^3/s.

Amount of water $= \int_0^8 r(t)\,dt \approx [r(1) + r(3) + r(5) + r(7)]\,\dfrac{8-0}{4} \approx [6401 + 4249 + 3821 + 2628](2) = 34{,}198.$

Now multiply by the number of seconds in a day, $24 \cdot 60^2$, to get 2,954,707,200 ft^3.

71. Let $P(t)$ denote the bacteria population at time t (in hours). By the Net Change Theorem,

$$P(1) - P(0) = \int_0^1 P'(t)\,dt = \int_0^1 (1000 \cdot 2^t)\,dt = \left[1000\,\frac{2^t}{\ln 2}\right]_0^1 = \frac{1000}{\ln 2}\,(2^1 - 2^0) = \frac{1000}{\ln 2} \approx 1443.$$

Thus, the population after one hour is $4000 + 1443 = 5443$.

72. Let $M(t)$ denote the number of megabits transmitted at time t (in hours) [note that $D(t)$ is measured in megabits/second]. By the Net Change Theorem and the Midpoint Rule,

$$\begin{aligned} M(8) - M(0) &= \textstyle\int_0^8 3600D(t)\,dt \approx 3600 \cdot \frac{8-0}{4}[D(1) + D(3) + D(5) + D(7)] \\ &\approx 7200(0.32 + 0.50 + 0.56 + 0.83) = 7200(2.21) = 15{,}912 \text{ megabits} \end{aligned}$$

73. Power is the rate of change of energy with respect to time; that is, $P(t) = E'(t)$. By the Net Change Theorem and the Midpoint Rule,

$$\begin{aligned} E(24) - E(0) &= \int_0^{24} P(t)\,dt \approx \frac{24-0}{12}[P(1) + P(3) + P(5) + \cdots + P(21) + P(23)] \\ &\approx 2(16{,}900 + 16{,}400 + 17{,}000 + 19{,}800 + 20{,}700 + 21{,}200 \\ &\qquad + 20{,}500 + 20{,}500 + 21{,}700 + 22{,}300 + 21{,}700 + 18{,}900) \\ &= 2(237{,}600) = 475{,}200 \end{aligned}$$

Thus, the energy used on that day was approximately 4.75×10^5 megawatt-hours.

74. (a) From Exercise 4.1.74(a), $v(t) = 0.00146t^3 - 0.11553t^2 + 24.98169t - 21.26872$.

(b) $h(125) - h(0) = \int_0^{125} v(t)\,dt = \left[0.000365t^4 - 0.03851t^3 + 12.490845t^2 - 21.26872t\right]_0^{125} \approx 206{,}407$ ft

5.5 The Substitution Rule

1. Let $u = 2x$. Then $du = 2\,dx$ and $dx = \frac{1}{2}\,du$, so $\int \cos 2x\,dx = \int \cos u\,\left(\frac{1}{2}\,du\right) = \frac{1}{2}\sin u + C = \frac{1}{2}\sin 2x + C$.

2. Let $u = -x^2$. Then $du = -2x\,dx$ and $x\,dx = -\frac{1}{2}\,du$, so $\int xe^{-x^2}\,dx = \int e^u\left(-\frac{1}{2}\,du\right) = -\frac{1}{2}e^u + C = -\frac{1}{2}e^{-x^2} + C$.

3. Let $u = x^3 + 1$. Then $du = 3x^2\,dx$ and $x^2\,dx = \frac{1}{3}\,du$, so

$$\int x^2\sqrt{x^3+1}\,dx = \int \sqrt{u}\left(\tfrac{1}{3}\,du\right) = \frac{1}{3}\frac{u^{3/2}}{3/2} + C = \frac{1}{3}\cdot\frac{2}{3}u^{3/2} + C = \tfrac{2}{9}(x^3+1)^{3/2} + C.$$

4. Let $u = \sin\theta$. Then $du = \cos\theta\,d\theta$, so $\int \sin^2\theta\,\cos\theta\,d\theta = \int u^2\,du = \frac{1}{3}u^3 + C = \frac{1}{3}\sin^3\theta + C$.

5. Let $u = x^4 - 5$. Then $du = 4x^3\,dx$ and $x^3\,dx = \frac{1}{4}\,du$, so

$$\int \frac{x^3}{x^4 - 5}\,dx = \int \frac{1}{u}\left(\frac{1}{4}\,du\right) = \frac{1}{4}\ln|u| + C = \frac{1}{4}\ln|x^4 - 5| + C.$$

6. Let $u = 2t + 1$. Then $du = 2\,dt$ and $dt = \frac{1}{2}\,du$, so $\int \sqrt{2t+1}\,dt = \int \sqrt{u}\,\left(\frac{1}{2}\,du\right) = \frac{1}{2} \cdot \frac{2}{3}u^{3/2} + C = \frac{1}{3}(2t+1)^{3/2} + C$.

7. Let $u = 1 - x^2$. Then $du = -2x\,dx$ and $x\,dx = -\frac{1}{2}\,du$, so

$\int x\sqrt{1 - x^2}\,dx = \int \sqrt{u}\,\left(-\frac{1}{2}\,du\right) = -\frac{1}{2} \cdot \frac{2}{3}u^{3/2} + C = -\frac{1}{3}(1 - x^2)^{3/2} + C$.

8. Let $u = x^3$. Then $du = 3x^2\,dx$ and $x^2\,dx = \frac{1}{3}\,du$, so $\int x^2 e^{x^3}\,dx = \int e^u\,\left(\frac{1}{3}\,du\right) = \frac{1}{3}e^u + C = \frac{1}{3}e^{x^3} + C$.

9. Let $u = 1 - 2x$. Then $du = -2\,dx$ and $dx = -\frac{1}{2}\,du$, so

$\int (1 - 2x)^9\,dx = \int u^9\,\left(-\frac{1}{2}\,du\right) = -\frac{1}{2} \cdot \frac{1}{10}u^{10} + C = -\frac{1}{20}(1 - 2x)^{10} + C$.

10. Let $u = 1 + \cos t$. Then $du = -\sin t\,dt$ and $\sin t\,dt = -du$, so

$\int \sin t\,\sqrt{1 + \cos t}\,dt = \int \sqrt{u}\,(-du) = -\frac{2}{3}u^{3/2} + C = -\frac{2}{3}(1 + \cos t)^{3/2} + C$.

11. Let $u = \frac{\pi}{2}t$. Then $du = \frac{\pi}{2}\,dt$ and $dt = \frac{2}{\pi}\,du$, so $\int \cos\left(\frac{\pi}{2}t\right)dt = \int \cos u\,\left(\frac{2}{\pi}\,du\right) = \frac{2}{\pi}\sin u + C = \frac{2}{\pi}\sin\left(\frac{\pi}{2}t\right) + C$.

12. Let $u = 2\theta$. Then $du = 2\,d\theta$ and $d\theta = \frac{1}{2}\,du$, so $\int \sec^2 2\theta\,d\theta = \int \sec^2 u\,\left(\frac{1}{2}\,du\right) = \frac{1}{2}\tan u + C = \frac{1}{2}\tan 2\theta + C$.

13. Let $u = 5 - 3x$. Then $du = -3\,dx$ and $dx = -\frac{1}{3}\,du$, so

$$\int \frac{dx}{5 - 3x} = \int \frac{1}{u}\left(-\frac{1}{3}\,du\right) = -\frac{1}{3}\ln|u| + C = -\frac{1}{3}\ln|5 - 3x| + C.$$

14. Let $u = 4 - y^3$. Then $du = -3y^2\,dy$ and $y^2\,dy = -\frac{1}{3}\,du$, so

$\int y^2(4 - y^3)^{2/3}\,dy = \int u^{2/3}\,\left(-\frac{1}{3}\,du\right) = -\frac{1}{3} \cdot \frac{3}{5}u^{5/3} + C = -\frac{1}{5}(4 - y^3)^{5/3} + C$.

15. Let $u = \cos\theta$. Then $du = -\sin\theta\,d\theta$ and $\sin\theta\,d\theta = -du$, so

$\int \cos^3\theta\,\sin\theta\,d\theta = \int u^3(-du) = -\frac{1}{4}u^4 + C = -\frac{1}{4}\cos^4\theta + C$.

16. Let $u = -5r$. Then $du = -5\,dr$ and $dr = -\frac{1}{5}\,du$, so $\int e^{-5r}\,dr = \int e^u\left(-\frac{1}{5}\,du\right) = -\frac{1}{5}e^u + C = -\frac{1}{5}e^{-5r} + C$.

17. Let $x = 1 - e^u$. Then $dx = -e^u\,du$ and $e^u\,du = -dx$, so

$$\int \frac{e^u}{(1 - e^u)^2}\,du = \int \frac{1}{x^2}(-dx) = -\int x^{-2}\,dx = -(-x^{-1}) + C = \frac{1}{x} + C = \frac{1}{1 - e^u} + C.$$

18. Let $u = \sqrt{x}$. Then $du = \dfrac{1}{2\sqrt{x}}\,dx$ and $2\,du = \dfrac{1}{\sqrt{x}}\,dx$, so

$$\int \frac{\sin\sqrt{x}}{\sqrt{x}}\,dx = \int \sin u\,(2\,du) = -2\cos u + C = -2\cos\sqrt{x} + C.$$

19. Let $u = 3ax + bx^3$. Then $du = (3a + 3bx^2)\,dx = 3(a + bx^2)\,dx$, so

$$\int \frac{a + bx^2}{\sqrt{3ax + bx^3}}\,dx = \int \frac{\frac{1}{3}\,du}{u^{1/2}} = \frac{1}{3}\int u^{-1/2}\,du = \frac{1}{3} \cdot 2u^{1/2} + C = \frac{2}{3}\sqrt{3ax + bx^3} + C.$$

20. Let $u = z^3 + 1$. Then $du = 3z^2\,dz$ and $\frac{1}{3}\,du = z^2\,dz + C$, so

$$\int \frac{z^2}{z^3+1}\,dz = \int \frac{1}{u}\left(\tfrac{1}{3}\,du\right) = \tfrac{1}{3}\ln|u| + C = \tfrac{1}{3}\ln|z^3+1| + C.$$

21. Let $u = \ln x$. Then $du = \dfrac{dx}{x}$, so $\displaystyle\int \frac{(\ln x)^2}{x}\,dx = \int u^2\,du = \tfrac{1}{3}u^3 + C = \tfrac{1}{3}(\ln x)^3 + C.$

22. Let $u = \cos x$. Then $du = -\sin x\,dx$ and $-du = \sin x\,dx$, so

$\int \sin x\,\sin(\cos x)\,dx = \int \sin u\,(-du) = (-\cos u)(-1) + C = \cos(\cos x) + C.$

23. Let $u = \tan\theta$. Then $du = \sec^2\theta\,d\theta$, so $\int \sec^2\theta\,\tan^3\theta\,d\theta = \int u^3\,du = \frac{1}{4}u^4 + C = \frac{1}{4}\tan^4\theta + C.$

24. Let $u = x + 2$. Then $du = dx$ and $x = u - 2$, so

$\int x\sqrt{x+2}\,dx = \int (u-2)\sqrt{u}\,du = \int (u^{3/2} - 2u^{1/2})\,du = \frac{2}{5}u^{5/2} - 2\cdot\frac{2}{3}u^{3/2} + C = \frac{2}{5}(x+2)^{5/2} - \frac{4}{3}(x+2)^{3/2} + C.$

25. Let $u = 1 + e^x$. Then $du = e^x\,dx$, so $\int e^x\sqrt{1+e^x}\,dx = \int \sqrt{u}\,du = \frac{2}{3}u^{3/2} + C = \frac{2}{3}(1+e^x)^{3/2} + C.$

Or: Let $u = \sqrt{1+e^x}$. Then $u^2 = 1 + e^x$ and $2u\,du = e^x\,dx$, so

$\int e^x\sqrt{1+e^x}\,dx = \int u\cdot 2u\,du = \frac{2}{3}u^3 + C = \frac{2}{3}(1+e^x)^{3/2} + C.$

26. Let $u = ax + b$. Then $du = a\,dx$ and $dx = (1/a)\,du$, so

$$\int \frac{dx}{ax+b} = \int \frac{(1/a)\,du}{u} = \frac{1}{a}\int \frac{1}{u}\,du = \frac{1}{a}\ln|u| + C = \frac{1}{a}\ln|ax+b| + C.$$

27. Let $u = x^3 + 3x$. Then $du = (3x^2+3)\,dx$ and $\frac{1}{3}\,du = (x^2+1)\,dx$, so

$\int (x^2+1)(x^3+3x)^4\,dx = \int u^4\left(\frac{1}{3}\,du\right) = \frac{1}{3}\cdot\frac{1}{5}u^5 + C = \frac{1}{15}(x^3+3x)^5 + C.$

28. Let $u = \cos t$. Then $du = -\sin t\,dt$ and $\sin t\,dt = -du$, so $\int e^{\cos t}\sin t\,dt = \int e^u\,(-du) = -e^u + C = -e^{\cos t} + C.$

29. Let $u = 5^t$. Then $du = 5^t\ln 5\,dt$ and $5^t\,dt = \dfrac{1}{\ln 5}\,du$, so

$$\int 5^t\sin(5^t)\,dt = \int \sin u\left(\frac{1}{\ln 5}\,du\right) = -\frac{1}{\ln 5}\cos u + C = -\frac{1}{\ln 5}\cos(5^t) + C.$$

30. Let $u = \tan x$. Then $du = \sec^2 x\,dx$, so

$$\int \frac{\sec^2 x}{\tan^2 x}\,dx = \int \frac{1}{u^2}\,du = \int u^{-2}\,du = -1u^{-1} + C = -\frac{1}{\tan x} + C = -\cot x + C.$$

Or: $\displaystyle\int \frac{\sec^2 x}{\tan^2 x}\,dx = \int\left(\frac{1}{\cos^2 x}\cdot\frac{\cos^2 x}{\sin^2 x}\right)dx = \int \csc^2 x\,dx = -\cot x + C$

31. Let $u = \arctan x$. Then $du = \dfrac{1}{x^2+1}\,dx$, so $\displaystyle\int \frac{(\arctan x)^2}{x^2+1}\,dx = \int u^2\,du = \tfrac{1}{3}u^3 + C = \tfrac{1}{3}(\arctan x)^3 + C.$

32. Let $u = x^2 + 4$. Then $du = 2x\,dx$ and $x\,dx = \frac{1}{2}\,du$, so

$$\int \frac{x}{x^2+4}\,dx = \int \frac{1}{u}\left(\frac{1}{2}\,du\right) = \frac{1}{2}\ln|u| + C = \frac{1}{2}\ln|x^2+4| + C = \frac{1}{2}\ln(x^2+4) + C \quad [\text{since } x^2+4>0].$$

33. Let $u = 1 + 5t$. Then $du = 5\,dt$ and $dt = \frac{1}{5}\,du$, so

$\int \cos(1+5t)\,dt = \int \cos u\left(\frac{1}{5}\,du\right) = \frac{1}{5}\sin u + C = \frac{1}{5}\sin(1+5t) + C.$

34. Let $u = \dfrac{\pi}{x}$. Then $du = -\dfrac{\pi}{x^2}\,dx$ and $\dfrac{1}{x^2}\,dx = -\dfrac{1}{\pi}\,du$, so

$$\int \frac{\cos(\pi/x)}{x^2}\,dx = \int \cos u\left(-\frac{1}{\pi}\,du\right) = -\frac{1}{\pi}\sin u + C = -\frac{1}{\pi}\sin\frac{\pi}{x} + C$$

35. Let $u = \cot x$. Then $du = -\csc^2 x\,dx$ and $\csc^2 x\,dx = -du$, so

$$\int \sqrt{\cot x}\csc^2 x\,dx = \int \sqrt{u}\,(-du) = -\frac{u^{3/2}}{3/2} + C = -\tfrac{2}{3}(\cot x)^{3/2} + C.$$

36. Let $u = 2^t + 3$. Then $du = 2^t \ln 2\,dt$ and $2^t\,dt = \dfrac{1}{\ln 2}\,du$, so

$$\int \frac{2^t}{2^t+3}\,dt = \int \frac{1}{u}\left(\frac{1}{\ln 2}\,du\right) = \frac{1}{\ln 2}\ln|u| + C = \frac{1}{\ln 2}\ln(2^t+3) + C.$$

37. Let $u = \sinh x$. Then $du = \cosh x\,dx$, so $\int \sinh^2 x\,\cosh x\,dx = \int u^2\,du = \frac{1}{3}u^3 + C = \frac{1}{3}\sinh^3 x + C$.

38. Let $u = 1 + \tan t$. Then $du = \sec^2 t\,dt$, so

$$\int \frac{dt}{\cos^2 t\sqrt{1+\tan t}} = \int \frac{\sec^2 t\,dt}{\sqrt{1+\tan t}} = \int \frac{du}{\sqrt{u}} = \int u^{-1/2}\,du = \frac{u^{1/2}}{1/2} + C = 2\sqrt{1+\tan t} + C.$$

39. $\displaystyle\int \frac{\sin 2x}{1+\cos^2 x}\,dx = 2\int \frac{\sin x\cos x}{1+\cos^2 x}\,dx = 2I$. Let $u = \cos x$. Then $du = -\sin x\,dx$, so

$$2I = -2\int \frac{u\,du}{1+u^2} = -2\cdot\tfrac{1}{2}\ln(1+u^2) + C = -\ln(1+u^2) + C = -\ln(1+\cos^2 x) + C.$$

Or: Let $u = 1 + \cos^2 x$.

40. Let $u = \cos x$. Then $du = -\sin x\,dx$ and $\sin x\,dx = -du$, so

$$\int \frac{\sin x}{1+\cos^2 x}\,dx = \int \frac{-du}{1+u^2} = -\tan^{-1} u + C = -\tan^{-1}(\cos x) + C.$$

41. $\displaystyle\int \cot x\,dx = \int \frac{\cos x}{\sin x}\,dx$. Let $u = \sin x$. Then $du = \cos x\,dx$, so $\displaystyle\int \cot x\,dx = \int \frac{1}{u}\,du = \ln|u| + C = \ln|\sin x| + C$.

42. Let $u = \ln t$. Then $du = \dfrac{1}{t}\,dt$, so $\displaystyle\int \frac{\cos(\ln t)}{t}\,dt = \int \cos u\,du = \sin u + C = \sin(\ln t) + C$.

43. Let $u = \sin^{-1} x$. Then $du = \dfrac{1}{\sqrt{1-x^2}}\,dx$, so $\displaystyle\int \frac{dx}{\sqrt{1-x^2}\sin^{-1} x} = \int \frac{1}{u}\,du = \ln|u| + C = \ln\left|\sin^{-1} x\right| + C$.

44. Let $u = x^2$. Then $du = 2x\,dx$, so $\displaystyle\int \frac{x}{1+x^4}\,dx = \int \frac{\frac{1}{2}\,du}{1+u^2} = \tfrac{1}{2}\tan^{-1} u + C = \tfrac{1}{2}\tan^{-1}(x^2) + C$.

45. Let $u = 1 + x^2$. Then $du = 2x\,dx$, so

$$\int \frac{1+x}{1+x^2}\,dx = \int \frac{1}{1+x^2}\,dx + \int \frac{x}{1+x^2}\,dx = \tan^{-1} x + \int \frac{\frac{1}{2}\,du}{u} = \tan^{-1} x + \tfrac{1}{2}\ln|u| + C$$

$$= \tan^{-1} x + \tfrac{1}{2}\ln\left|1+x^2\right| + C = \tan^{-1} x + \tfrac{1}{2}\ln\left(1+x^2\right) + C \quad [\text{since } 1+x^2 > 0].$$

46. Let $u = 2 + x$. Then $du = dx$, $x = u - 2$, and $x^2 = (u - 2)^2$, so

$$\begin{aligned}\int x^2\sqrt{2+x}\,dx &= \int (u-2)^2\sqrt{u}\,du = \int (u^2 - 4u + 4)u^{1/2}\,du = \int (u^{5/2} - 4u^{3/2} + 4u^{1/2})\,du \\ &= \tfrac{2}{7}u^{7/2} - \tfrac{8}{5}u^{5/2} + \tfrac{8}{3}u^{3/2} + C = \tfrac{2}{7}(2+x)^{7/2} - \tfrac{8}{5}(2+x)^{5/2} + \tfrac{8}{3}(2+x)^{3/2} + C\end{aligned}$$

47. Let $u = 2x + 5$. Then $du = 2\,dx$ and $x = \frac{1}{2}(u - 5)$, so

$$\begin{aligned}\int x(2x+5)^8\,dx &= \int \tfrac{1}{2}(u-5)u^8\left(\tfrac{1}{2}\,du\right) = \tfrac{1}{4}\int (u^9 - 5u^8)\,du \\ &= \tfrac{1}{4}\left(\tfrac{1}{10}u^{10} - \tfrac{5}{9}u^9\right) + C = \tfrac{1}{40}(2x+5)^{10} - \tfrac{5}{36}(2x+5)^9 + C\end{aligned}$$

48. Let $u = x^2 + 1$ [so $x^2 = u - 1$]. Then $du = 2x\,dx$ and $x\,dx = \frac{1}{2}\,du$, so

$$\begin{aligned}\int x^3\sqrt{x^2+1}\,dx &= \int x^2\sqrt{x^2+1}\,x\,dx = \int (u-1)\sqrt{u}\left(\tfrac{1}{2}\,du\right) = \tfrac{1}{2}\int (u^{3/2} - u^{1/2})\,du \\ &= \tfrac{1}{2}\left(\tfrac{2}{5}u^{5/2} - \tfrac{2}{3}u^{3/2}\right) + C = \tfrac{1}{5}(x^2+1)^{5/2} - \tfrac{1}{3}(x^2+1)^{3/2} + C\end{aligned}$$

Or: Let $u = \sqrt{x^2+1}$. Then $u^2 = x^2 + 1 \;\Rightarrow\; 2u\,du = 2x\,dx \;\Rightarrow\; u\,du = x\,dx$, so

$$\begin{aligned}\int x^3\sqrt{x^2+1}\,dx &= \int x^2\sqrt{x^2+1}\,x\,dx = \int (u^2-1)\,u\cdot u\,du = \int (u^4 - u^2)\,du \\ &= \tfrac{1}{5}u^5 - \tfrac{1}{3}u^3 + C = \tfrac{1}{5}(x^2+1)^{5/2} - \tfrac{1}{3}(x^2+1)^{3/2} + C\end{aligned}$$

Note: This answer can be written as $\frac{1}{15}\sqrt{x^2+1}(3x^4 + x^2 - 2) + C$.

49. $f(x) = x(x^2-1)^3$. $\;u = x^2 - 1 \;\Rightarrow\; du = 2x\,dx$, so

$$\int x(x^2-1)^3\,dx = \int u^3\left(\tfrac{1}{2}\,du\right) = \tfrac{1}{8}u^4 + C = \tfrac{1}{8}(x^2-1)^4 + C$$

Where f is positive (negative), F is increasing (decreasing). Where f changes from negative to positive (positive to negative), F has a local minimum (maximum).

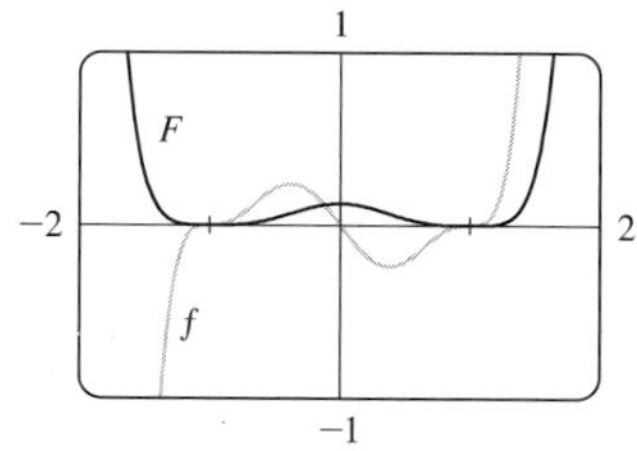

50. $f(\theta) = \tan^2\theta\,\sec^2\theta$. $u = \tan\theta \;\Rightarrow\; du = \sec^2\theta\,d\theta$, so

$$\int \tan^2\theta\,\sec^2\theta\,d\theta = \int u^2\,du = \tfrac{1}{3}u^3 + C = \tfrac{1}{3}\tan^3\theta + C$$

Note that f is positive and F is increasing. At $x = 0$, $f = 0$ and F has a horizontal tangent.

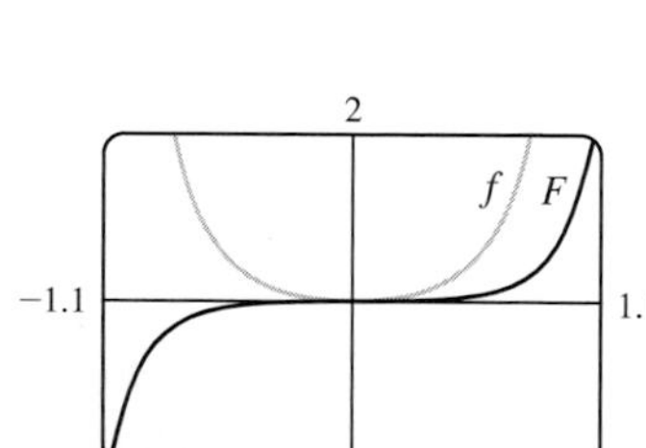

51. $f(x) = e^{\cos x}\sin x$. $u = \cos x \;\Rightarrow\; du = -\sin x\,dx$, so

$$\int e^{\cos x}\sin x\,dx = \int e^u\,(-du) = -e^u + C = -e^{\cos x} + C$$

Note that at $x = \pi$, f changes from positive to negative and F has a local maximum. Also, both f and F are periodic with period 2π, so at $x = 0$ and at $x = 2\pi$, f changes from negative to positive and F has a local minimum.

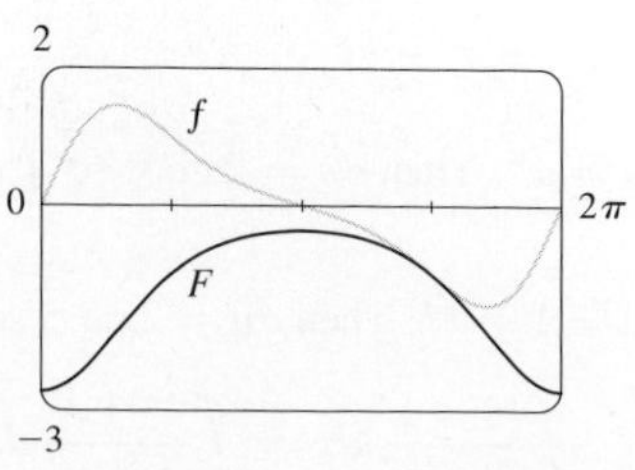

52. $f(x) = \sin x \cos^4 x$. $u = \cos x \quad \Rightarrow \quad du = -\sin x\, dx$, so

$$\int \sin x \cos^4 x\, dx = \int u^4\,(-du) = -\tfrac{1}{5}u^5 + C = -\tfrac{1}{5}\cos^5 x + C$$

Note that at $x = \pi$, f changes from positive to negative and F has a local maximum. Also, both f and F are periodic with period 2π, so at $x = 0$ and at $x = 2\pi$, f changes from negative to positive and F has a local minimum.

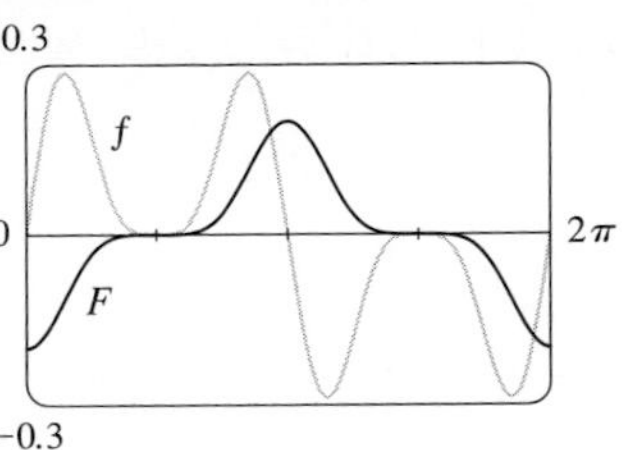

53. Let $u = \frac{\pi}{2}t$, so $du = \frac{\pi}{2}\,dt$. When $t = 0$, $u = 0$; when $t = 1$, $u = \frac{\pi}{2}$. Thus,

$$\int_0^1 \cos(\pi t/2)\,dt = \int_0^{\pi/2} \cos u\,\left(\tfrac{2}{\pi}\,du\right) = \tfrac{2}{\pi}\left[\sin u\right]_0^{\pi/2} = \tfrac{2}{\pi}\left(\sin\tfrac{\pi}{2} - \sin 0\right) = \tfrac{2}{\pi}(1 - 0) = \tfrac{2}{\pi}$$

54. Let $u = 3t - 1$, so $du = 3\,dt$. When $t = 0$, $u = -1$; when $t = 1$, $u = 2$. Thus,

$$\int_0^1 (3t - 1)^{50}\,dt = \int_{-1}^2 u^{50}\,\left(\tfrac{1}{3}\,du\right) = \tfrac{1}{3}\left[\tfrac{1}{51}u^{51}\right]_{-1}^2 = \tfrac{1}{153}\left[2^{51} - (-1)^{51}\right] = \tfrac{1}{153}(2^{51} + 1)$$

55. Let $u = 1 + 7x$, so $du = 7\,dx$. When $x = 0$, $u = 1$; when $x = 1$, $u = 8$. Thus,

$$\int_0^1 \sqrt[3]{1 + 7x}\,dx = \int_1^8 u^{1/3}\left(\tfrac{1}{7}\,du\right) = \tfrac{1}{7}\left[\tfrac{3}{4}u^{4/3}\right]_1^8 = \tfrac{3}{28}(8^{4/3} - 1^{4/3}) = \tfrac{3}{28}(16 - 1) = \tfrac{45}{28}$$

56. Let $u = 5x + 1$, so $du = 5\,dx$. When $x = 0$, $u = 1$; when $x = 3$, $u = 16$. Thus,

$$\int_0^3 \frac{dx}{5x + 1} = \int_1^{16} \frac{1}{u}\left(\frac{1}{5}\,du\right) = \frac{1}{5}\Big[\ln|u|\Big]_1^{16} = \frac{1}{5}(\ln 16 - \ln 1) = \frac{1}{5}\ln 16.$$

57. Let $u = \cos t$, so $du = -\sin t\,dt$. When $t = 0$, $u = 1$; when $t = \frac{\pi}{6}$, $u = \sqrt{3}/2$. Thus,

$$\int_0^{\pi/6} \frac{\sin t}{\cos^2 t}\,dt = \int_1^{\sqrt{3}/2} \frac{1}{u^2}\,(-du) = \left[\frac{1}{u}\right]_1^{\sqrt{3}/2} = \frac{2}{\sqrt{3}} - 1.$$

58. Let $u = \frac{1}{2}t$, so $du = \frac{1}{2}\,dt$. When $t = \frac{\pi}{3}$, $u = \frac{\pi}{6}$; when $t = \frac{2\pi}{3}$, $u = \frac{\pi}{3}$. Thus,

$$\int_{\pi/3}^{2\pi/3} \csc^2\left(\tfrac{1}{2}t\right)dt = \int_{\pi/6}^{\pi/3} \csc^2 u\,(2\,du) = 2\Big[-\cot u\Big]_{\pi/6}^{\pi/3} = -2\left(\cot\frac{\pi}{3} - \cot\frac{\pi}{6}\right)$$

$$= -2\left(\frac{1}{\sqrt{3}} - \sqrt{3}\right) = -2\left(\tfrac{1}{3}\sqrt{3} - \sqrt{3}\right) = \tfrac{4}{3}\sqrt{3}$$

59. Let $u = 1/x$, so $du = -1/x^2\,dx$. When $x = 1$, $u = 1$; when $x = 2$, $u = \frac{1}{2}$. Thus,

$$\int_1^2 \frac{e^{1/x}}{x^2}\,dx = \int_1^{1/2} e^u\,(-du) = -\left[e^u\right]_1^{1/2} = -(e^{1/2} - e) = e - \sqrt{e}.$$

60. Let $u = -x^2$, so $du = -2x\,dx$. When $x = 0$, $u = 0$; when $x = 1$, $u = -1$. Thus,

$$\int_0^1 xe^{-x^2}\,dx = \int_0^{-1} e^u\left(-\tfrac{1}{2}\,du\right) = -\tfrac{1}{2}\left[e^u\right]_0^{-1} = -\tfrac{1}{2}(e^{-1} - e^0) = \tfrac{1}{2}(1 - 1/e).$$

61. $\int_{-\pi/4}^{\pi/4}(x^3 + x^4\tan x)\,dx = 0$ by Theorem 7(b), since $f(x) = x^3 + x^4\tan x$ is an odd function.

62. Let $u = \sin x$, so $du = \cos x\,dx$. When $x = 0$, $u = 0$; when $x = \frac{\pi}{2}$, $u = 1$. Thus,

$$\int_0^{\pi/2} \cos x\,\sin(\sin x)\,dx = \int_0^1 \sin u\,du = \Big[-\cos u\Big]_0^1 = -(\cos 1 - 1) = 1 - \cos 1.$$

63. Let $u = 1 + 2x$, so $du = 2\,dx$. When $x = 0$, $u = 1$; when $x = 13$, $u = 27$. Thus,

$$\int_0^{13} \frac{dx}{\sqrt[3]{(1+2x)^2}} = \int_1^{27} u^{-2/3}\left(\tfrac{1}{2}\,du\right) = \left[\tfrac{1}{2} \cdot 3u^{1/3}\right]_1^{27} = \tfrac{3}{2}(3-1) = 3.$$

64. Assume $a > 0$. Let $u = a^2 - x^2$, so $du = -2x\,dx$. When $x = 0$, $u = a^2$; when $x = a$, $u = 0$. Thus,

$$\int_0^a x\sqrt{a^2 - x^2}\,dx = \int_{a^2}^0 u^{1/2}\left(-\tfrac{1}{2}\,du\right) = \tfrac{1}{2}\int_0^{a^2} u^{1/2}\,du = \tfrac{1}{2} \cdot \left[\tfrac{2}{3}u^{3/2}\right]_0^{a^2} = \tfrac{1}{3}a^3.$$

65. Let $u = x^2 + a^2$, so $du = 2x\,dx$ and $x\,dx = \frac{1}{2}\,du$. When $x = 0$, $u = a^2$; when $x = a$, $u = 2a^2$. Thus,

$$\int_0^a x\sqrt{x^2 + a^2}\,dx = \int_{a^2}^{2a^2} u^{1/2}\left(\tfrac{1}{2}\,du\right) = \tfrac{1}{2}\left[\tfrac{2}{3}u^{3/2}\right]_{a^2}^{2a^2} = \left[\tfrac{1}{3}u^{3/2}\right]_{a^2}^{2a^2} = \tfrac{1}{3}\left[(2a^2)^{3/2} - (a^2)^{3/2}\right] = \tfrac{1}{3}\left(2\sqrt{2} - 1\right)a^3$$

66. $\int_{-\pi/3}^{\pi/3} x^4 \sin x\,dx = 0$ by Theorem 7(b), since $f(x) = x^4 \sin x$ is an odd function.

67. Let $u = x - 1$, so $u + 1 = x$ and $du = dx$. When $x = 1$, $u = 0$; when $x = 2$, $u = 1$. Thus,

$$\int_1^2 x\sqrt{x-1}\,dx = \int_0^1 (u+1)\sqrt{u}\,du = \int_0^1 \left(u^{3/2} + u^{1/2}\right)du = \left[\tfrac{2}{5}u^{5/2} + \tfrac{2}{3}u^{3/2}\right]_0^1 = \tfrac{2}{5} + \tfrac{2}{3} = \tfrac{16}{15}.$$

68. Let $u = 1 + 2x$, so $x = \frac{1}{2}(u-1)$ and $du = 2\,dx$. When $x = 0$, $u = 1$; when $x = 4$, $u = 9$. Thus,

$$\int_0^4 \frac{x\,dx}{\sqrt{1+2x}} = \int_1^9 \frac{\frac{1}{2}(u-1)}{\sqrt{u}}\frac{du}{2} = \tfrac{1}{4}\int_1^9 \left(u^{1/2} - u^{-1/2}\right)du = \tfrac{1}{4}\left[\tfrac{2}{3}u^{3/2} - 2u^{1/2}\right]_1^9 = \tfrac{1}{4} \cdot \tfrac{2}{3}\left[u^{3/2} - 3u^{1/2}\right]_1^9$$
$$= \tfrac{1}{6}[(27-9) - (1-3)] = \tfrac{20}{6} = \tfrac{10}{3}$$

69. Let $u = \ln x$, so $du = \dfrac{dx}{x}$. When $x = e$, $u = 1$; when $x = e^4$; $u = 4$. Thus,

$$\int_e^{e^4} \frac{dx}{x\sqrt{\ln x}} = \int_1^4 u^{-1/2}\,du = 2\left[u^{1/2}\right]_1^4 = 2(2-1) = 2.$$

70. Let $u = (x-1)^2$, so $du = 2(x-1)\,dx$. When $x = 0$, $u = 1$; when $x = 2$, $u = 1$. Thus,

$$\int_0^2 (x-1)e^{(x-1)^2}\,dx = \int_1^1 e^u\left(\tfrac{1}{2}\,du\right) = 0 \text{ since the limits are equal.}$$

71. Let $u = e^z + z$, so $du = (e^z + 1)\,dz$. When $z = 0$, $u = 1$; when $z = 1$, $u = e + 1$. Thus,

$$\int_0^1 \frac{e^z + 1}{e^z + z}\,dz = \int_1^{e+1} \frac{1}{u}\,du = \left[\ln|u|\right]_1^{e+1} = \ln|e+1| - \ln|1| = \ln(e+1).$$

72. Let $u = \dfrac{2\pi t}{T} - \alpha$, so $du = \dfrac{2\pi}{T}\,dt$. When $t = 0$, $u = -\alpha$; when $t = \dfrac{T}{2}$, $u = \pi - \alpha$. Thus,

$$\int_0^{T/2} \sin\left(\frac{2\pi t}{T} - \alpha\right)dt = \int_{-\alpha}^{\pi-\alpha} \sin u\left(\frac{T}{2\pi}\,du\right) = \frac{T}{2\pi}\left[-\cos u\right]_{-\alpha}^{\pi-\alpha} = -\frac{T}{2\pi}[\cos(\pi - \alpha) - \cos(-\alpha)]$$
$$= -\frac{T}{2\pi}(-\cos\alpha - \cos\alpha) = -\frac{T}{2\pi}(-2\cos\alpha) = \frac{T}{\pi}\cos\alpha$$

73. Let $u = 1 + \sqrt{x}$, so $du = \dfrac{1}{2\sqrt{x}}\,dx \;\Rightarrow\; 2\sqrt{x}\,du = dx \;\Rightarrow\; 2(u-1)\,du = dx$. When $x = 0$, $u = 1$; when $x = 1$, $u = 2$. Thus,

$$\int_0^1 \frac{dx}{(1+\sqrt{x})^4} = \int_1^2 \frac{1}{u^4}\cdot[2(u-1)\,du] = 2\int_1^2\left(\frac{1}{u^3}-\frac{1}{u^4}\right)du = 2\left[-\frac{1}{2u^2}+\frac{1}{3u^3}\right]_1^2$$
$$= 2\left[\left(-\tfrac{1}{8}+\tfrac{1}{24}\right)-\left(-\tfrac{1}{2}+\tfrac{1}{3}\right)\right] = 2\left(\tfrac{1}{12}\right) = \tfrac{1}{6}$$

74. If $f(x) = \sin\sqrt[3]{x}$, then $f(-x) = \sin\sqrt[3]{-x} = \sin(-\sqrt[3]{x}) = -\sin\sqrt[3]{x} = -f(x)$, so f is an odd function. Now $I = \int_{-2}^{3}\sin\sqrt[3]{x}\,dx = \int_{-2}^{2}\sin\sqrt[3]{x}\,dx + \int_{2}^{3}\sin\sqrt[3]{x}\,dx = I_1 + I_2$. $I_1 = 0$ by Theorem 7(b). To estimate I_2, note that $2 \le x \le 3 \;\Rightarrow\; \sqrt[3]{2} \le \sqrt[3]{x} \le \sqrt[3]{3}$ [≈ 1.44] $\Rightarrow\; 0 \le \sqrt[3]{x} \le \frac{\pi}{2}$ [≈ 1.57] $\Rightarrow\; \sin 0 \le \sin\sqrt[3]{x} \le \sin\frac{\pi}{2}$ [since sine is increasing on this interval] $\Rightarrow\; 0 \le \sin\sqrt[3]{x} \le 1$. By comparison property 8, $0(3-2) \le I_2 \le 1(3-2) \;\Rightarrow\; 0 \le I_2 \le 1 \;\Rightarrow\; 0 \le I \le 1$.

75. From the graph, it appears that the area under the curve is about $1 + \left(\text{a little more than } \frac{1}{2}\cdot 1\cdot 0.7\right)$, or about 1.4. The exact area is given by $A = \int_0^1 \sqrt{2x+1}\,dx$. Let $u = 2x+1$, so $du = 2\,dx$. The limits change to $2\cdot 0 + 1 = 1$ and $2\cdot 1 + 1 = 3$, and

$A = \int_1^3 \sqrt{u}\left(\frac{1}{2}\,du\right) = \frac{1}{2}\left[\frac{2}{3}u^{3/2}\right]_1^3 = \frac{1}{3}\left(3\sqrt{3}-1\right) = \sqrt{3} - \frac{1}{3} \approx 1.399$.

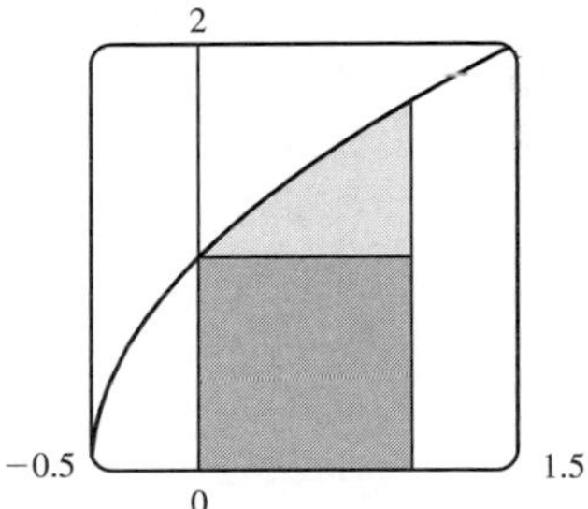

76. From the graph, it appears that the area under the curve is almost $\frac{1}{2}\cdot\pi\cdot 2.6$, or about 4. The exact area is given by

$$A = \int_0^\pi (2\sin x - \sin 2x)\,dx = -2\left[\cos x\right]_0^\pi - \int_0^\pi \sin 2x\,dx$$
$$= -2(-1-1) - 0 = 4$$

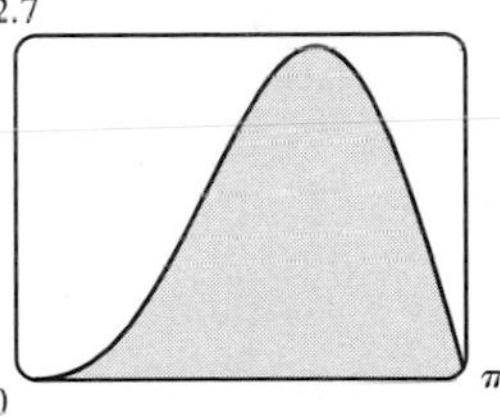

Note: $\int_0^\pi \sin 2x\,dx = 0$ since it is clear from the graph of $y = \sin 2x$ that $\int_{\pi/2}^{\pi}\sin 2x\,dx = -\int_0^{\pi/2}\sin 2x\,dx$.

77. First write the integral as a sum of two integrals:

$I = \int_{-2}^{2}(x+3)\sqrt{4-x^2}\,dx = I_1 + I_2 = \int_{-2}^{2} x\sqrt{4-x^2}\,dx + \int_{-2}^{2} 3\sqrt{4-x^2}\,dx$. $I_1 = 0$ by Theorem 7(b), since $f(x) = x\sqrt{4-x^2}$ is an odd function and we are integrating from $x = -2$ to $x = 2$. We interpret I_2 as three times the area of a semicircle with radius 2, so $I = 0 + 3\cdot\frac{1}{2}\left(\pi\cdot 2^2\right) = 6\pi$.

78. Let $u = x^2$. Then $du = 2x\,dx$ and the limits are unchanged ($0^2 = 0$ and $1^2 = 1$), so $I = \int_0^1 x\sqrt{1-x^4}\,dx = \frac{1}{2}\int_0^1\sqrt{1-u^2}\,du$. But this integral can be interpreted as the area of a quarter-circle with radius 1. So $I = \frac{1}{2}\cdot\frac{1}{4}\left(\pi\cdot 1^2\right) = \frac{1}{8}\pi$.

79. First Figure Let $u = \sqrt{x}$, so $x = u^2$ and $dx = 2u\,du$. When $x = 0$, $u = 0$; when $x = 1$, $u = 1$. Thus,
$A_1 = \int_0^1 e^{\sqrt{x}}\,dx = \int_0^1 e^u (2u\,du) = 2\int_0^1 ue^u\,du$.

Second Figure $A_2 = \int_0^1 2xe^x\,dx = 2\int_0^1 ue^u\,du$.

Third Figure Let $u = \sin x$, so $du = \cos x\,dx$. When $x = 0$, $u = 0$; when $x = \frac{\pi}{2}$, $u = 1$. Thus,
$A_3 = \int_0^{\pi/2} e^{\sin x} \sin 2x\,dx = \int_0^{\pi/2} e^{\sin x}(2 \sin x\,\cos x)\,dx = \int_0^1 e^u(2u\,du) = 2\int_0^1 ue^u\,du$.

Since $A_1 = A_2 = A_3$, all three areas are equal.

80. Let $u = \dfrac{\pi t}{12}$. Then $du = \dfrac{\pi}{12}\,dt$ and

$$\int_0^{24} R(t)\,dt = \int_0^{24} \left[85 - 0.18\cos\left(\frac{\pi t}{12}\right)\right] dt = \int_0^{2\pi} (85 - 0.18\cos u)\left(\frac{12}{\pi}\,du\right) = \frac{12}{\pi}\left[85u - 0.18\sin u\right]_0^{2\pi}$$
$$= \frac{12}{\pi}\left[(85 \cdot 2\pi - 0) - (0 - 0)\right] = 2040 \text{ kcal}$$

81. The rate is measured in liters per minute. Integrating from $t = 0$ minutes to $t = 60$ minutes will give us the total amount of oil that leaks out (in liters) during the first hour.

$$\int_0^{60} r(t)\,dt = \int_0^{60} 100e^{-0.01t}\,dt \qquad [u = -0.01t,\, du = -0.01dt]$$
$$= 100\int_0^{-0.6} e^u(-100\,du) = -10{,}000\left[e^u\right]_0^{-0.6} = -10{,}000(e^{-0.6} - 1) \approx 4511.9 \approx 4512 \text{ liters}$$

82. Let $r(t) = ae^{bt}$ with $a = 450.268$ and $b = 1.12567$, and $n(t) =$ population after t hours. Since $r(t) = n'(t)$, $\int_0^3 r(t)\,dt = n(3) - n(0)$ is the total change in the population after three hours. Since we start with 400 bacteria, the population will be

$$n(3) = 400 + \int_0^3 r(t)\,dt = 400 + \int_0^3 ae^{bt}\,dt = 400 + \frac{a}{b}\left[e^{bt}\right]_0^3 = 400 + \frac{a}{b}(e^{3b} - 1)$$
$$\approx 400 + 11{,}313 = 11{,}713 \text{ bacteria}$$

83. The volume of inhaled air in the lungs at time t is

$$V(t) = \int_0^t f(u)\,du = \int_0^t \frac{1}{2}\sin\left(\frac{2\pi}{5}u\right)du = \int_0^{2\pi t/5} \frac{1}{2}\sin v\left(\frac{5}{2\pi}\,dv\right) \qquad \left[\text{substitute } v = \tfrac{2\pi}{5}u,\ dv = \tfrac{2\pi}{5}\,du\right]$$
$$= \frac{5}{4\pi}\left[-\cos v\right]_0^{2\pi t/5} = \frac{5}{4\pi}\left[-\cos\left(\frac{2\pi}{5}t\right) + 1\right] = \frac{5}{4\pi}\left[1 - \cos\left(\frac{2\pi}{5}t\right)\right] \text{ liters}$$

84. The rate G is measured in kilograms per year. Integrating from $t = 0$ years (2000) to $t = 20$ years (2020) will give us the net change in biomass from 2000 to 2020.

$$\int_0^{20} \frac{60{,}000e^{-0.6t}}{(1 + 5e^{-0.6t})^2}\,dt = \int_6^{1+5e^{-12}} \frac{60{,}000}{u^2}\left(-\tfrac{1}{3}\,du\right) \qquad \left[\begin{array}{l} u = 1 + 5e^{-0.6t}, \\ du = -3e^{-0.6t}\,dt \end{array}\right]$$
$$= \left[\frac{20{,}000}{u}\right]_6^{1+5e^{-12}} = \frac{20{,}000}{1 + 5e^{-12}} - \frac{20{,}000}{6} \approx 16{,}666$$

Thus, the predicted biomass for the year 2020 is approximately $25{,}000 + 16{,}666 = 41{,}666$ kg.

85. $\displaystyle\int_0^{30} u(t)\,dt = \int_0^{30} \frac{r}{V} C_0 e^{-rt/V}\,dt = C_0 \int_1^{e^{-30r/V}} (-dx)$ $\quad\left[\begin{array}{l} x = e^{-rt/V}, \\ dx = -\frac{r}{V}e^{-rt/V}\,dt \end{array}\right]$

$$= C_0\Big[-x\Big]_1^{e^{-30r/V}} = C_0(-e^{-30r/V}+1)$$

The integral $\int_0^{30} u(t)\,dt$ represents the total amount of urea removed from the blood in the first 30 minutes of dialysis.

86. Number of calculators $= x(4) - x(2) = \int_2^4 5000\left[1 - 100(t+10)^{-2}\right]dt$

$$= 5000\left[t + 100(t+10)^{-1}\right]_2^4 = 5000\left[\left(4 + \tfrac{100}{14}\right) - \left(2 + \tfrac{100}{12}\right)\right] \approx 4048$$

87. Let $u = 2x$. Then $du = 2\,dx$, so $\int_0^2 f(2x)\,dx = \int_0^4 f(u)\left(\frac{1}{2}\,du\right) = \frac{1}{2}\int_0^4 f(u)\,du = \frac{1}{2}(10) = 5$.

88. Let $u = x^2$. Then $du = 2x\,dx$, so $\int_0^3 xf(x^2)\,dx = \int_0^9 f(u)\left(\frac{1}{2}\,du\right) = \frac{1}{2}\int_0^9 f(u)\,du = \frac{1}{2}(4) = 2$.

89. Let $u = -x$. Then $du = -dx$, so

$$\int_a^b f(-x)\,dx = \int_{-a}^{-b} f(u)(-du) = \int_{-b}^{-a} f(u)\,du = \int_{-b}^{-a} f(x)\,dx$$

From the diagram, we see that the equality follows from the fact that we are reflecting the graph of f, and the limits of integration, about the y-axis.

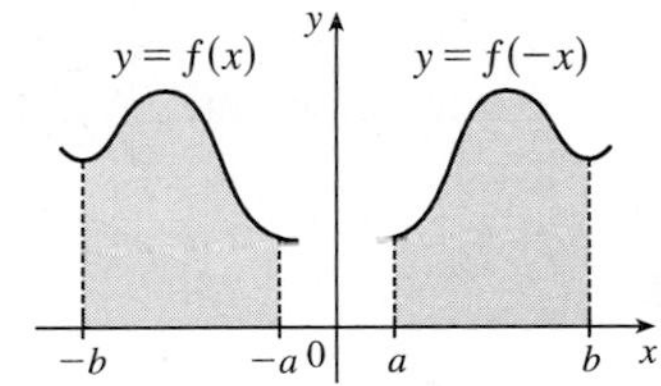

90. Let $u = x + c$. Then $du = dx$, so

$$\int_a^b f(x+c)\,dx = \int_{a+c}^{b+c} f(u)\,du = \int_{a+c}^{b+c} f(x)\,dx$$

From the diagram, we see that the equality follows from the fact that we are translating the graph of f, and the limits of integration, by a distance c.

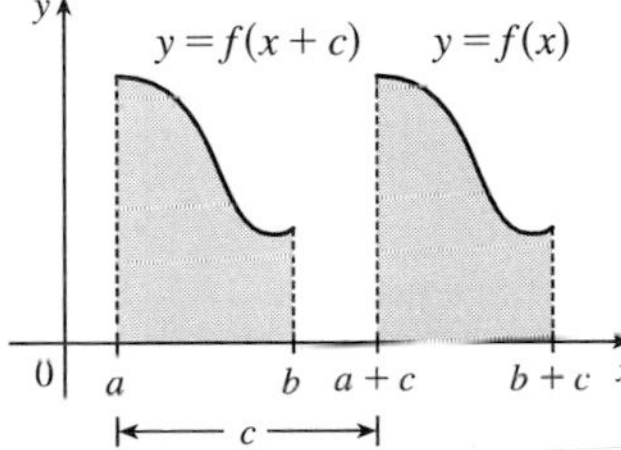

91. Let $u = 1 - x$. Then $x = 1 - u$ and $dx = -du$, so

$\int_0^1 x^a(1-x)^b\,dx = \int_1^0 (1-u)^a u^b(-du) = \int_0^1 u^b(1-u)^a\,du = \int_0^1 x^b(1-x)^a\,dx$.

92. Let $u = \pi - x$. Then $du = -dx$. When $x = \pi$, $u = 0$ and when $x = 0$, $u = \pi$. So

$$\int_0^\pi xf(\sin x)\,dx = -\int_\pi^0 (\pi - u)\,f(\sin(\pi - u))\,du = \int_0^\pi (\pi - u)\,f(\sin u)\,du$$

$$= \pi\int_0^\pi f(\sin u)\,du - \int_0^\pi u\,f(\sin u)\,du = \pi\int_0^\pi f(\sin x)\,dx - \int_0^\pi x\,f(\sin x)\,dx \quad\Rightarrow$$

$2\int_0^\pi x\,f(\sin x)\,dx = \pi\int_0^\pi f(\sin x)\,dx \quad\Rightarrow\quad \int_0^\pi x\,f(\sin x)\,dx = \frac{\pi}{2}\int_0^\pi f(\sin x)\,dx$.

93. $\dfrac{x\sin x}{1+\cos^2 x} = x\cdot\dfrac{\sin x}{2-\sin^2 x} = x\,f(\sin x)$, where $f(t) = \dfrac{t}{2-t^2}$. By Exercise 92,

$$\int_0^\pi \frac{x\sin x}{1+\cos^2 x}\,dx = \int_0^\pi x\,f(\sin x)\,dx = \frac{\pi}{2}\int_0^\pi f(\sin x)\,dx = \frac{\pi}{2}\int_0^\pi \frac{\sin x}{1+\cos^2 x}\,dx$$

Let $u = \cos x$. Then $du = -\sin x\,dx$. When $x = \pi$, $u = -1$ and when $x = 0$, $u = 1$. So

$$\frac{\pi}{2}\int_0^\pi \frac{\sin x}{1+\cos^2 x}\,dx = -\frac{\pi}{2}\int_1^{-1}\frac{du}{1+u^2} = \frac{\pi}{2}\int_{-1}^1 \frac{du}{1+u^2} = \frac{\pi}{2}\left[\tan^{-1}u\right]_{-1}^1$$

$$= \frac{\pi}{2}[\tan^{-1}1 - \tan^{-1}(-1)] = \frac{\pi}{2}\left[\frac{\pi}{4} - \left(-\frac{\pi}{4}\right)\right] = \frac{\pi^2}{4}$$

94. (a) $\int_0^{\pi/2} f(\cos x)\,dx = \int_0^{\pi/2} f\left[\sin\left(\frac{\pi}{2} - x\right)\right] dx \qquad \left[u = \frac{\pi}{2} - x,\ du = -dx\right]$

$$= \int_{\pi/2}^{0} f(\sin u)(-du) = \int_0^{\pi/2} f(\sin u)\,du = \int_0^{\pi/2} f(\sin x)\,dx$$

Continuity of f is needed in order to apply the substitution rule for definite integrals.

(b) In part (a), take $f(x) = x^2$, so $\int_0^{\pi/2} \cos^2 x\,dx = \int_0^{\pi/2} \sin^2 x\,dx$. Now

$\int_0^{\pi/2} \cos^2 x\,dx + \int_0^{\pi/2} \sin^2 x\,dx = \int_0^{\pi/2} (\cos^2 x + \sin^2 x)\,dx = \int_0^{\pi/2} 1\,dx = \left[x\right]_0^{\pi/2} = \frac{\pi}{2}$,

so $2\int_0^{\pi/2} \cos^2 x\,dx = \frac{\pi}{2} \Rightarrow \int_0^{\pi/2} \cos^2 x\,dx = \frac{\pi}{4} \quad \left[= \int_0^{\pi/2} \sin^2 x\,dx\right]$.

5 Review

TRUE-FALSE QUIZ

1. True by Property 2 of the Integral in Section 5.2.

2. False. Try $a = 0$, $b = 2$, $f(x) = g(x) = 1$ as a counterexample.

3. True by Property 3 of the Integral in Section 5.2.

4. False. You can't take a variable outside the integral sign. For example, using $f(x) = 1$ on $[0, 1]$, $\int_0^1 x\,f(x)\,dx = \int_0^1 x\,dx = \left[\frac{1}{2}x^2\right]_0^1 = \frac{1}{2}$ (a constant) while $x\int_0^1 1\,dx = x\,[x]_0^1 = x \cdot 1 = x$ (a variable).

5. False. For example, let $f(x) = x^2$. Then $\int_0^1 \sqrt{x^2}\,dx = \int_0^1 x\,dx = \frac{1}{2}$, but $\sqrt{\int_0^1 x^2\,dx} = \sqrt{\frac{1}{3}} = \frac{1}{\sqrt{3}}$.

6. True by the Net Change Theorem.

7. True by Comparison Property 7 of the Integral in Section 5.2.

8. False. For example, let $a = 0$, $b = 1$, $f(x) = 3$, $g(x) = x$. $f(x) > g(x)$ for each x in $(0, 1)$, but $f'(x) = 0 < 1 = g'(x)$ for $x \in (0, 1)$.

9. True. The integrand is an odd function that is continuous on $[-1, 1]$.

10. True. $\int_{-5}^{5} (ax^2 + bx + c)\,dx = \int_{-5}^{5} (ax^2 + c)\,dx + \int_{-5}^{5} bx\,dx$

$$= 2\int_0^5 (ax^2 + c)\,dx + 0 \quad [\text{because } ax^2 + c \text{ is even and } bx \text{ is odd}]$$

11. False. For example, the function $y = |x|$ is continuous on $\mathbb{R}$, but has no derivative at $x = 0$.

12. True by FTC1.

13. True by Property 5 of Integrals.

14. False. For example, $\int_0^1 \left(x - \frac{1}{2}\right) dx = \left[\frac{1}{2}x^2 - \frac{1}{2}x\right]_0^1 = \left(\frac{1}{2} - \frac{1}{2}\right) - (0 - 0) = 0$, but $f(x) = x - \frac{1}{2} \neq 0$.

15. False. $\int_a^b f(x)\,dx$ is a constant, so $\dfrac{d}{dx}\left(\int_a^b f(x)\,dx\right) = 0$, not $f(x)$ [unless $f(x) = 0$]. Compare the given statement carefully with FTC1, in which the upper limit in the integral is x.

16. False. See the paragraph before Note 4 and Figure 4 in Section 5.2, and notice that $y = x - x^3 < 0$ for $1 < x \le 2$.

17. False. The function $f(x) = 1/x^4$ is not bounded on the interval $[-2, 1]$. It has an infinite discontinuity at $x = 0$, so it is not integrable on the interval. (If the integral were to exist, a positive value would be expected, by Comparison Property 6 of Integrals.)

18. False. For example, if $f(x) = \begin{cases} 1 & \text{if } 0 \le x \le 1 \\ 0 & \text{if } -1 \le x < 0 \end{cases}$ then f has a jump discontinuity at 0, but $\int_{-1}^1 f(x)\,dx$ exists and is equal to 1.

EXERCISES

1. (a)

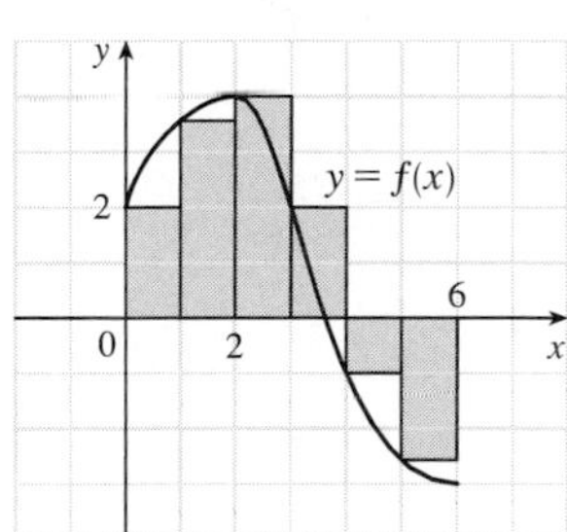

$$L_6 = \sum_{i=1}^{6} f(x_{i-1})\,\Delta x \quad [\Delta x = \tfrac{6-0}{6} = 1]$$

$$= f(x_0)\cdot 1 + f(x_1)\cdot 1 + f(x_2)\cdot 1 + f(x_3)\cdot 1 + f(x_4)\cdot 1 + f(x_5)\cdot 1$$

$$\approx 2 + 3.5 + 4 + 2 + (-1) + (-2.5) = 8$$

The Riemann sum represents the sum of the areas of the four rectangles above the x-axis minus the sum of the areas of the two rectangles below the x-axis.

(b)

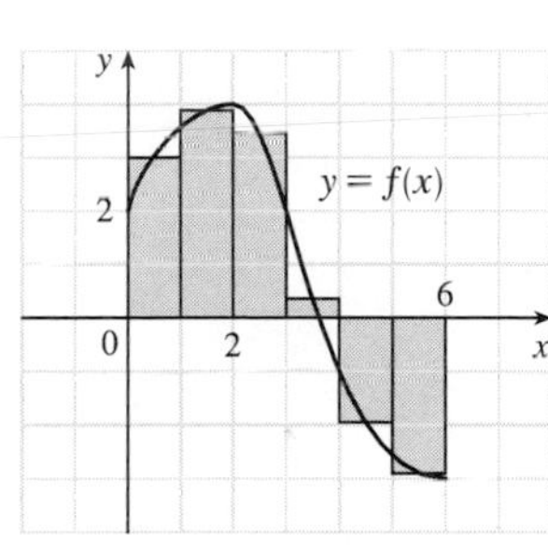

$$M_6 = \sum_{i=1}^{6} f(\overline{x}_i)\,\Delta x \quad [\Delta x = \tfrac{6-0}{6} = 1]$$

$$= f(\overline{x}_1)\cdot 1 + f(\overline{x}_2)\cdot 1 + f(\overline{x}_3)\cdot 1 + f(\overline{x}_4)\cdot 1 + f(\overline{x}_5)\cdot 1 + f(\overline{x}_6)\cdot 1$$

$$= f(0.5) + f(1.5) + f(2.5) + f(3.5) + f(4.5) + f(5.5)$$

$$\approx 3 + 3.9 + 3.4 + 0.3 + (-2) + (-2.9) = 5.7$$

The Riemann sum represents the sum of the areas of the four rectangles above the x-axis minus the sum of the areas of the two rectangles below the x-axis.

2. (a)

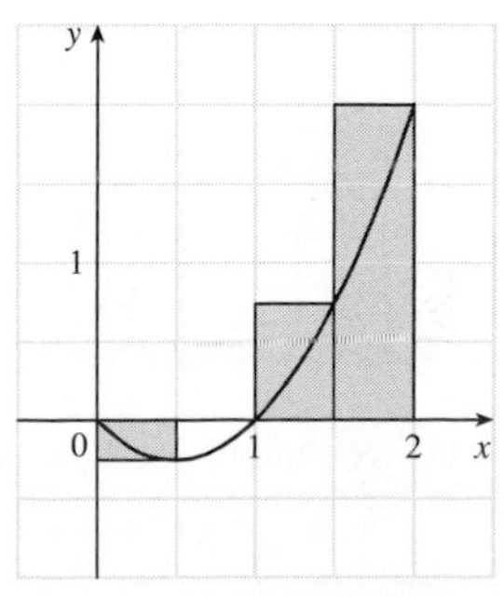

$f(x) = x^2 - x$ and $\Delta x = \frac{2-0}{4} = 0.5 \;\Rightarrow$

$$R_4 = 0.5f(0.5) + 0.5f(1) + 0.5f(1.5) + 0.5f(2)$$

$$= 0.5(-0.25 + 0 + 0.75 + 2) = 1.25$$

The Riemann sum represents the sum of the areas of the two rectangles above the x-axis minus the area of the rectangle below the x-axis. (The second rectangle vanishes.)

(b) $\int_0^2 (x^2 - x)\,dx = \lim\limits_{n\to\infty} \sum\limits_{i=1}^{n} f(x_i)\,\Delta x \qquad [\Delta x = 2/n \text{ and } x_i = 2i/n]$

$$= \lim_{n\to\infty} \sum_{i=1}^{n} \left(\frac{4i^2}{n^2} - \frac{2i}{n}\right)\left(\frac{2}{n}\right) = \lim_{n\to\infty} \frac{2}{n}\left[\frac{4}{n^2}\sum_{i=1}^{n} i^2 - \frac{2}{n}\sum_{i=1}^{n} i\right]$$

$$= \lim_{n\to\infty} \left[\frac{8}{n^3}\cdot\frac{n(n+1)(2n+1)}{6} - \frac{4}{n^2}\cdot\frac{n(n+1)}{2}\right] = \lim_{n\to\infty} \left[\frac{4}{3}\cdot\frac{n+1}{n}\cdot\frac{2n+1}{n} - 2\cdot\frac{n+1}{n}\right]$$

$$= \lim_{n\to\infty} \left[\frac{4}{3}\left(1+\frac{1}{n}\right)\left(2+\frac{1}{n}\right) - 2\left(1+\frac{1}{n}\right)\right] = \tfrac{4}{3}\cdot 1\cdot 2 - 2\cdot 1 = \tfrac{2}{3}$$

(c) $\int_0^2 (x^2 - x)\,dx = \left[\frac{1}{3}x^3 - \frac{1}{2}x^2\right]_0^2 = \left(\frac{8}{3} - 2\right) = \frac{2}{3}$

(d)

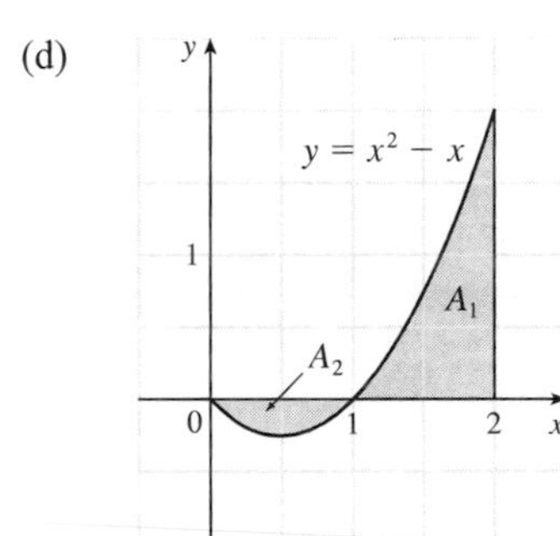

$\int_0^2 (x^2 - x)\,dx = A_1 - A_2$, where A_1 and A_2 are the areas shown in the diagram.

3. $\int_0^1 \left(x + \sqrt{1 - x^2}\right)dx = \int_0^1 x\,dx + \int_0^1 \sqrt{1 - x^2}\,dx = I_1 + I_2.$

I_1 can be interpreted as the area of the triangle shown in the figure and I_2 can be interpreted as the area of the quarter-circle.

Area $= \frac{1}{2}(1)(1) + \frac{1}{4}(\pi)(1)^2 = \frac{1}{2} + \frac{\pi}{4}$.

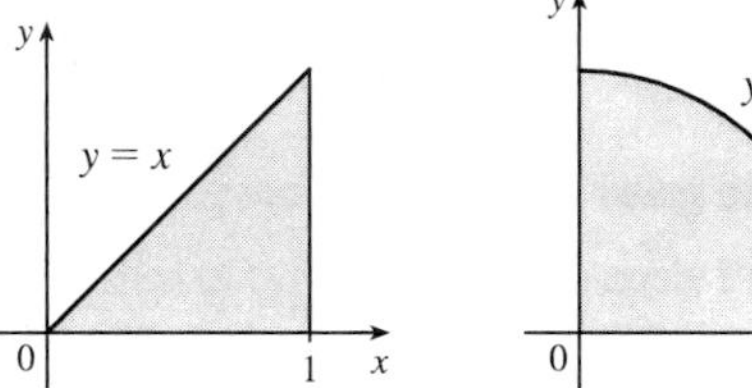

4. On $[0, \pi]$, $\lim\limits_{n\to\infty} \sum\limits_{i=1}^{n} \sin x_i\,\Delta x = \int_0^\pi \sin x\,dx = [-\cos x]_0^\pi = -(-1) - (-1) = 2.$

5. $\int_0^6 f(x)\,dx = \int_0^4 f(x)\,dx + \int_4^6 f(x)\,dx \;\Rightarrow\; 10 = 7 + \int_4^6 f(x)\,dx \;\Rightarrow\; \int_4^6 f(x)\,dx = 10 - 7 = 3$

6. (a) $\int_1^5 (x + 2x^5)\,dx = \lim\limits_{n\to\infty} \sum\limits_{i=1}^{n} f(x_i)\,\Delta x \qquad \left[\Delta x = \frac{5-1}{n} = \frac{4}{n},\ x_i = 1 + \frac{4i}{n}\right]$

$$= \lim_{n\to\infty} \sum_{i=1}^{n} \left[\left(1 + \frac{4i}{n}\right) + 2\left(1 + \frac{4i}{n}\right)^5\right]\cdot\frac{4}{n} \overset{\text{CAS}}{=} \lim_{n\to\infty} \frac{1305n^4 + 3126n^3 + 2080n^2 - 256}{n^3}\cdot\frac{4}{n}$$

$$= 5220$$

(b) $\int_1^5 (x + 2x^5)\,dx = \left[\frac{1}{2}x^2 + \frac{2}{6}x^6\right]_1^5 = \left(\frac{25}{2} + \frac{15{,}625}{3}\right) - \left(\frac{1}{2} + \frac{1}{3}\right) = 12 + 5208 = 5220$

7. First note that either a or b must be the graph of $\int_0^x f(t)\,dt$, since $\int_0^0 f(t)\,dt = 0$, and $c(0) \neq 0$. Now notice that $b > 0$ when c is increasing, and that $c > 0$ when a is increasing. It follows that c is the graph of $f(x)$, b is the graph of $f'(x)$, and a is the graph of $\int_0^x f(t)\,dt$.

8. (a) By the Net Change Theorem (FTC2), $\displaystyle\int_0^1 \frac{d}{dx}\left(e^{\arctan x}\right) dx = \left[e^{\arctan x}\right]_0^1 = e^{\pi/4} - 1$

(b) $\displaystyle\frac{d}{dx}\int_0^1 e^{\arctan x}\, dx = 0$ since this is the derivative of a constant.

(c) By FTC1, $\displaystyle\frac{d}{dx}\int_0^x e^{\arctan t}\, dt = e^{\arctan x}$.

9. $g(4) = \int_0^4 f(t)\, dt = \int_0^1 f(t)\, dt + \int_1^2 f(t)\, dt + \int_2^3 f(t)\, dt + \int_3^4 f(t)\, dt$

$= -\frac{1}{2}\cdot 1\cdot 2 \left[\begin{array}{l}\text{area of triangle,}\\ \textit{below } t\text{-axis}\end{array}\right] + \frac{1}{2}\cdot 1\cdot 2 + 1\cdot 2 + \frac{1}{2}\cdot 1\cdot 2 = 3$

By FTC1, $g'(x) = f(x)$, so $g'(4) = f(4) = 0$.

10. $g(x) = \int_0^x f(t)\, dt \;\Rightarrow\; g'(x) = f(x)$ [by FTC1] $\Rightarrow\; g''(x) = f'(x)$, so $g''(4) = f'(4) = -2$, which is the slope of the line segment at $x = 4$.

11. $\int_1^2 (8x^3 + 3x^2)\, dx = \left[8\cdot\frac{1}{4}x^4 + 3\cdot\frac{1}{3}x^3\right]_1^2 = \left[2x^4 + x^3\right]_1^2 = (2\cdot 2^4 + 2^3) - (2+1) = 40 - 3 = 37$

12. $\displaystyle\int_0^T (x^4 - 8x + 7)\, dx = \left[\tfrac{1}{5}x^5 - 4x^2 + 7x\right]_0^T = \left(\tfrac{1}{5}T^5 - 4T^2 + 7T\right) - 0 = \tfrac{1}{5}T^5 - 4T^2 + 7T$

13. $\int_0^1 (1 - x^9)\, dx = \left[x - \frac{1}{10}x^{10}\right]_0^1 = \left(1 - \frac{1}{10}\right) - 0 = \frac{9}{10}$

14. Let $u = 1 - x$, so $du = -dx$ and $dx = -du$. When $x = 0$, $u = 1$; when $x = 1$, $u = 0$. Thus,

$\int_0^1 (1-x)^9\, dx = \int_1^0 u^9(-du) = \int_0^1 u^9\, du = \frac{1}{10}\left[u^{10}\right]_0^1 = \frac{1}{10}(1-0) = \frac{1}{10}$.

15. $\displaystyle\int_1^9 \frac{\sqrt{u} - 2u^2}{u}\, du = \int_1^9 (u^{-1/2} - 2u)\, du = \left[2u^{1/2} - u^2\right]_1^9 = (6 - 81) - (2 - 1) = -76$

16. $\int_0^1 \left(\sqrt[4]{u} + 1\right)^2 du = \int_0^1 (u^{1/2} + 2u^{1/4} + 1)\, du = \left[\frac{2}{3}u^{3/2} + \frac{8}{5}u^{5/4} + u\right]_0^1 = \left(\frac{2}{3} + \frac{8}{5} + 1\right) - 0 = \frac{49}{15}$

17. Let $u = y^2 + 1$, so $du = 2y\, dy$ and $y\, dy = \frac{1}{2}\, du$. When $y = 0$, $u = 1$; when $y = 1$, $u = 2$. Thus,

$\int_0^1 y(y^2+1)^5\, dy = \int_1^2 u^5\left(\frac{1}{2}\, du\right) = \frac{1}{2}\left[\frac{1}{6}u^6\right]_1^2 = \frac{1}{12}(64 - 1) = \frac{63}{12} = \frac{21}{4}$.

18. Let $u = 1 + y^3$, so $du = 3y^2\, dy$ and $y^2\, dy = \frac{1}{3}\, du$. When $y = 0$, $u = 1$; when $y = 2$, $u = 9$. Thus,

$\int_0^2 y^2\sqrt{1 + y^3}\, dy = \int_1^9 u^{1/2}\left(\frac{1}{3}\, du\right) = \frac{1}{3}\left[\frac{2}{3}u^{3/2}\right]_1^9 = \frac{2}{9}(27 - 1) = \frac{52}{9}$.

19. $\displaystyle\int_1^5 \frac{dt}{(t-4)^2}$ does not exist because the function $f(t) = \dfrac{1}{(t-4)^2}$ has an infinite discontinuity at $t = 4$; that is, f is discontinuous on the interval $[1, 5]$.

20. Let $u = 3\pi t$, so $du = 3\pi\, dt$. When $t = 0$, $u = 1$; when $t = 1$, $u = 3\pi$. Thus,

$$\int_0^1 \sin(3\pi t)\, dt = \int_0^{3\pi} \sin u\left(\frac{1}{3\pi}\, du\right) = \frac{1}{3\pi}\left[-\cos u\right]_0^{3\pi} = -\frac{1}{3\pi}(-1 - 1) = \frac{2}{3\pi}.$$

21. Let $u = v^3$, so $du = 3v^2\,dv$. When $v = 0$, $u = 0$; when $v = 1$, $u = 1$. Thus,

$\int_0^1 v^2 \cos(v^3)\,dv = \int_0^1 \cos u\,(\frac{1}{3}\,du) = \frac{1}{3}\left[\sin u\right]_0^1 = \frac{1}{3}(\sin 1 - 0) = \frac{1}{3}\sin 1.$

22. $\displaystyle\int_{-1}^{1} \frac{\sin x}{1+x^2}\,dx = 0$ by Theorem 5.5.7(b), since $f(x) = \dfrac{\sin x}{1+x^2}$ is an odd function.

23. $\displaystyle\int_{-\pi/4}^{\pi/4} \frac{t^4 \tan t}{2+\cos t}\,dt = 0$ by Theorem 5.5.7(b), since $f(t) = \dfrac{t^4 \tan t}{2+\cos t}$ is an odd function.

24. Let $u = e^x$, so $du = e^x\,dx$. When $x = 0, u = 1$; when $x = 1, u = e$. Thus,

$$\int_0^1 \frac{e^x}{1+e^{2x}}\,dx = \int_1^e \frac{1}{1+u^2}\,du = \left[\arctan u\right]_1^e = \arctan e - \arctan 1 = \arctan e - \tfrac{\pi}{4}.$$

25. $\displaystyle\int\left(\frac{1-x}{x}\right)^2 dx = \int\left(\frac{1}{x}-1\right)^2 dx = \int\left(\frac{1}{x^2}-\frac{2}{x}+1\right) dx = -\frac{1}{x} - 2\ln|x| + x + C$

26. $\displaystyle\int_1^{10} \frac{x}{x^2-4}\,dx$ does not exist because the function $f(x) = \dfrac{x}{x^2-4}$ has an infinite discontinuity at $x = 2$;

that is, f is discontinuous on the interval $[1, 10]$.

27. Let $u = x^2 + 4x$. Then $du = (2x+4)\,dx = 2(x+2)\,dx$, so

$$\int \frac{x+2}{\sqrt{x^2+4x}}\,dx = \int u^{-1/2}\left(\tfrac{1}{2}\,du\right) = \tfrac{1}{2}\cdot 2u^{1/2} + C = \sqrt{u} + C = \sqrt{x^2+4x} + C.$$

28. Let $u = 1 + \cot x$. Then $du = -\csc^2 x\,dx$, so $\displaystyle\int \frac{\csc^2 x}{1+\cot x}\,dx = \int \frac{1}{u}(-du) = -\ln|u| + C = -\ln|1+\cot x| + C.$

29. Let $u = \sin \pi t$. Then $du = \pi\,\cos \pi t\,dt$, so $\int \sin \pi t\,\cos \pi t\,dt = \int u\left(\frac{1}{\pi}\,du\right) = \frac{1}{\pi}\cdot\frac{1}{2}u^2 + C = \frac{1}{2\pi}(\sin \pi t)^2 + C.$

30. Let $u = \cos x$. Then $du = -\sin x\,dx$, so $\int \sin x\,\cos(\cos x)\,dx = -\int \cos u\,du = -\sin u + C = -\sin(\cos x) + C.$

31. Let $u = \sqrt{x}$. Then $du = \dfrac{dx}{2\sqrt{x}}$, so $\displaystyle\int \frac{e^{\sqrt{x}}}{\sqrt{x}}\,dx = 2\int e^u\,du = 2e^u + C = 2e^{\sqrt{x}} + C.$

32. Let $u = \ln x$. Then $du = \dfrac{1}{x}\,dx$, so $\displaystyle\int \frac{\sin(\ln x)}{x}\,dx = \int \sin u\,du = -\cos u + C = -\cos(\ln x) + C.$

33. Let $u = \ln(\cos x)$. Then $du = \dfrac{-\sin x}{\cos x}\,dx = -\tan x\,dx$, so

$\int \tan x\,\ln(\cos x)\,dx = -\int u\,du = -\frac{1}{2}u^2 + C = -\frac{1}{2}[\ln(\cos x)]^2 + C.$

34. Let $u = x^2$. Then $du = 2x\,dx$, so $\displaystyle\int \frac{x}{\sqrt{1-x^4}}\,dx = \frac{1}{2}\int \frac{du}{\sqrt{1-u^2}} = \tfrac{1}{2}\sin^{-1}u + C = \tfrac{1}{2}\sin^{-1}(x^2) + C.$

35. Let $u = 1 + x^4$. Then $du = 4x^3\,dx$, so $\displaystyle\int \frac{x^3}{1+x^4}\,dx = \frac{1}{4}\int \frac{1}{u}\,du = \tfrac{1}{4}\ln|u| + C = \tfrac{1}{4}\ln(1+x^4) + C.$

36. Let $u = 1 + 4x$. Then $du = 4\,dx$, so $\int \sinh(1+4x)\,dx = \frac{1}{4}\int \sinh u\,du = \frac{1}{4}\cosh u + C = \frac{1}{4}\cosh(1+4x) + C.$

37. Let $u = 1 + \sec\theta$. Then $du = \sec\theta\,\tan\theta\,d\theta$, so

$$\int \frac{\sec\theta\,\tan\theta}{1+\sec\theta}\,d\theta = \int \frac{1}{1+\sec\theta}\,(\sec\theta\,\tan\theta\,d\theta) = \int \frac{1}{u}\,du = \ln|u| + C = \ln|1+\sec\theta| + C.$$

38. Let $u = 1 + \tan t$, so $du = \sec^2 t\,dt$. When $t = 0$, $u = 1$; when $t = \frac{\pi}{4}$, $u = 2$. Thus,

$$\int_0^{\pi/4} (1+\tan t)^3 \sec^2 t\,dt = \int_1^2 u^3\,du = \left[\tfrac{1}{4}u^4\right]_1^2 = \tfrac{1}{4}(2^4 - 1^4) = \tfrac{1}{4}(16-1) = \tfrac{15}{4}.$$

39. Since $x^2 - 4 < 0$ for $0 \le x < 2$ and $x^2 - 4 > 0$ for $2 < x \le 3$, we have $|x^2 - 4| = -(x^2 - 4) = 4 - x^2$ for $0 \le x < 2$ and $|x^2 - 4| = x^2 - 4$ for $2 < x \le 3$. Thus,

$$\begin{aligned}\int_0^3 |x^2 - 4|\,dx &= \int_0^2 (4 - x^2)\,dx + \int_2^3 (x^2 - 4)\,dx = \left[4x - \frac{x^3}{3}\right]_0^2 + \left[\frac{x^3}{3} - 4x\right]_2^3 \\ &= \left(8 - \tfrac{8}{3}\right) - 0 + (9 - 12) - \left(\tfrac{8}{3} - 8\right) = \tfrac{16}{3} - 3 + \tfrac{16}{3} = \tfrac{32}{3} - \tfrac{9}{3} = \tfrac{23}{3}\end{aligned}$$

40. Since $\sqrt{x} - 1 < 0$ for $0 \le x < 1$ and $\sqrt{x} - 1 > 0$ for $1 < x \le 4$, we have $\left|\sqrt{x} - 1\right| = -\left(\sqrt{x} - 1\right) = 1 - \sqrt{x}$ for $0 \le x < 1$ and $\left|\sqrt{x} - 1\right| = \sqrt{x} - 1$ for $1 < x \le 4$. Thus,

$$\begin{aligned}\int_0^4 \left|\sqrt{x} - 1\right|\,dx &= \int_0^1 \left(1 - \sqrt{x}\right)dx + \int_1^4 \left(\sqrt{x} - 1\right)dx = \left[x - \tfrac{2}{3}x^{3/2}\right]_0^1 + \left[\tfrac{2}{3}x^{3/2} - x\right]_1^4 \\ &= \left(1 - \tfrac{2}{3}\right) - 0 + \left(\tfrac{16}{3} - 4\right) - \left(\tfrac{2}{3} - 1\right) = \tfrac{1}{3} + \tfrac{16}{3} - 4 + \tfrac{1}{3} = 6 - 4 = 2\end{aligned}$$

41. Let $u = 1 + \sin x$. Then $du = \cos x\,dx$, so

$$\int \frac{\cos x\,dx}{\sqrt{1+\sin x}} = \int u^{-1/2}\,du = 2u^{1/2} + C = 2\sqrt{1+\sin x} + C.$$

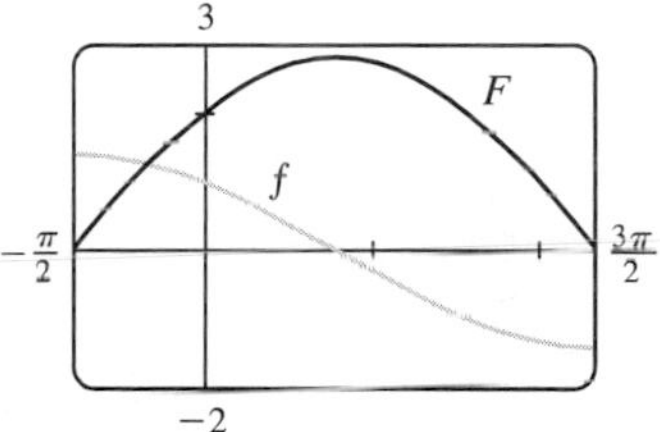

42. Let $u = x^2 + 1$. Then $x^2 = u - 1$ and $x\,dx = \frac{1}{2}\,du$, so

$$\begin{aligned}\int \frac{x^3}{\sqrt{x^2+1}}\,dx &= \int \frac{(u-1)}{\sqrt{u}}\left(\tfrac{1}{2}\,du\right) = \frac{1}{2}\int \left(u^{1/2} - u^{-1/2}\right)du \\ &= \tfrac{1}{2}\left(\tfrac{2}{3}u^{3/2} - 2u^{1/2}\right) + C = \tfrac{1}{3}(x^2+1)^{3/2} - (x^2+1)^{1/2} + C \\ &= \tfrac{1}{3}(x^2+1)^{1/2}\left[(x^2+1) - 3\right] + C = \tfrac{1}{3}\sqrt{x^2+1}\,(x^2 - 2) + C\end{aligned}$$

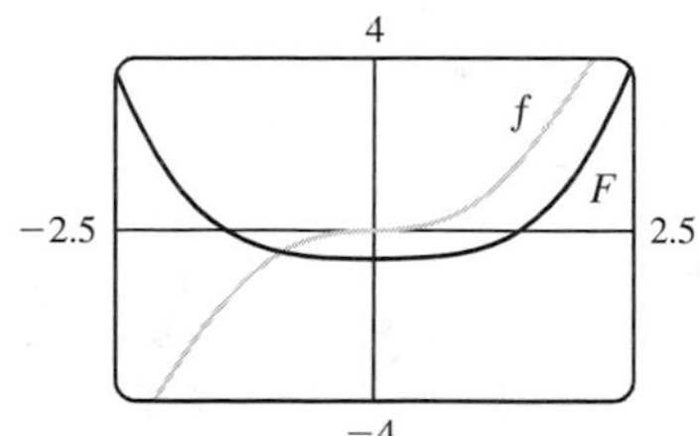

43. From the graph, it appears that the area under the curve $y = x\sqrt{x}$ between $x = 0$ and $x = 4$ is somewhat less than half the area of an 8×4 rectangle, so perhaps about 13 or 14. To find the exact value, we evaluate

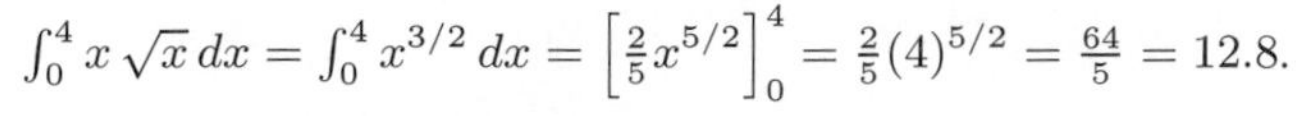

$$\int_0^4 x\sqrt{x}\,dx = \int_0^4 x^{3/2}\,dx = \left[\tfrac{2}{5}x^{5/2}\right]_0^4 = \tfrac{2}{5}(4)^{5/2} = \tfrac{64}{5} = 12.8.$$

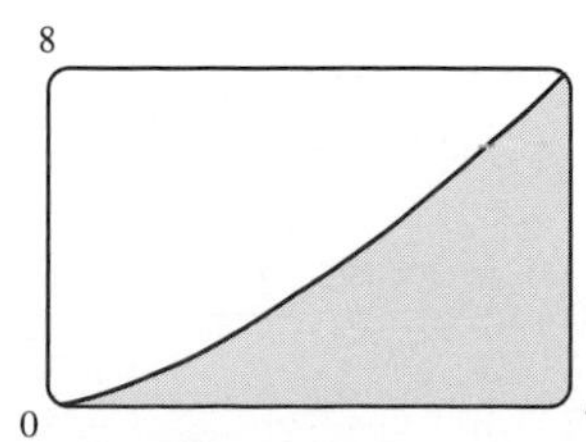

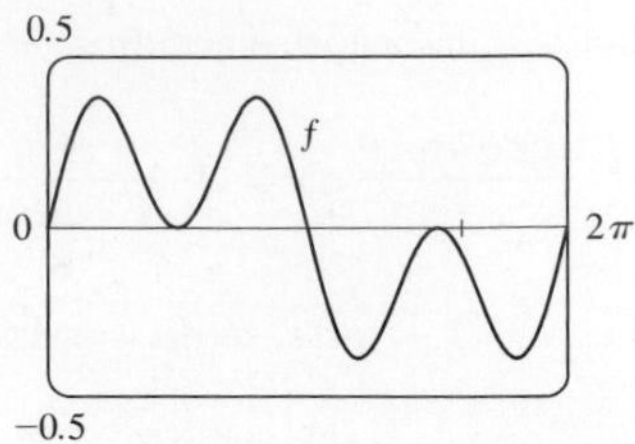

44. From the graph, it seems as though $\int_0^{2\pi} \cos^2 x \, \sin x \, dx$ is equal to 0. To evaluate the integral, let $u = \cos x \;\Rightarrow\; du = -\sin x \, dx$. Thus, $I = \int_1^1 u^2 \, (-du) = 0$.

45. $F(x) = \displaystyle\int_0^x \frac{t^2}{1+t^3}\,dt \;\Rightarrow\; F'(x) = \frac{d}{dx}\int_0^x \frac{t^2}{1+t^3}\,dt = \frac{x^2}{1+x^3}$

46. $F(x) = \displaystyle\int_x^1 \sqrt{t+\sin t}\,dt = -\int_1^x \sqrt{t+\sin t}\,dt \;\Rightarrow\; F'(x) = -\frac{d}{dx}\int_1^x \sqrt{t+\sin t}\,dt = -\sqrt{x+\sin x}$

47. Let $u = x^4$. Then $\dfrac{du}{dx} = 4x^3$. Also, $\dfrac{dg}{dx} = \dfrac{dg}{du}\dfrac{du}{dx}$, so

$$g'(x) = \frac{d}{dx}\int_0^{x^4} \cos(t^2)\,dt = \frac{d}{du}\int_0^u \cos(t^2)\,dt \cdot \frac{du}{dx} = \cos(u^2)\,\frac{du}{dx} = 4x^3\cos(x^8).$$

48. Let $u = \sin x$. Then $\dfrac{du}{dx} = \cos x$. Also, $\dfrac{dg}{dx} = \dfrac{dg}{du}\dfrac{du}{dx}$, so

$$g'(x) = \frac{d}{dx}\int_1^{\sin x} \frac{1-t^2}{1+t^4}\,dt = \frac{d}{du}\int_1^u \frac{1-t^2}{1+t^4}\,dt \cdot \frac{du}{dx} = \frac{1-u^2}{1+u^4}\cdot\frac{du}{dx} = \frac{1-\sin^2 x}{1+\sin^4 x}\cdot\cos x = \frac{\cos^3 x}{1+\sin^4 x}$$

49. $y = \displaystyle\int_{\sqrt{x}}^x \frac{e^t}{t}\,dt = \int_{\sqrt{x}}^1 \frac{e^t}{t}\,dt + \int_1^x \frac{e^t}{t}\,dt = -\int_1^{\sqrt{x}} \frac{e^t}{t}\,dt + \int_1^x \frac{e^t}{t}\,dt \;\Rightarrow$

$\dfrac{dy}{dx} = -\dfrac{d}{dx}\left(\displaystyle\int_1^{\sqrt{x}} \frac{e^t}{t}\,dt\right) + \frac{d}{dx}\left(\int_1^x \frac{e^t}{t}\,dt\right)$. Let $u = \sqrt{x}$. Then

$$\frac{d}{dx}\int_1^{\sqrt{x}} \frac{e^t}{t}\,dt = \frac{d}{dx}\int_1^u \frac{e^t}{t}\,dt = \frac{d}{du}\left(\int_1^u \frac{e^t}{t}\,dt\right)\frac{du}{dx} = \frac{e^u}{u}\cdot\frac{1}{2\sqrt{x}} = \frac{e^{\sqrt{x}}}{\sqrt{x}}\cdot\frac{1}{2\sqrt{x}} = \frac{e^{\sqrt{x}}}{2x},$$

so $\dfrac{dy}{dx} = -\dfrac{e^{\sqrt{x}}}{2x} + \dfrac{e^x}{x} = \dfrac{2e^x - e^{\sqrt{x}}}{2x}$.

50. $y = \int_{2x}^{3x+1} \sin(t^4)dt = \int_{2x}^0 \sin(t^4)\,dt + \int_0^{3x+1} \sin(t^4)\,dt = \int_0^{3x+1} \sin(t^4)\,dt - \int_0^{2x} \sin(t^4)\,dt \;\Rightarrow$

$y' = \sin\left[(3x+1)^4\right]\cdot\dfrac{d}{dx}(3x+1) - \sin\left[(2x)^4\right]\cdot\dfrac{d}{dx}(2x) = 3\sin\left[(3x+1)^4\right] - 2\sin\left[(2x)^4\right]$

51. If $1 \le x \le 3$, then $\sqrt{1^2+3} \le \sqrt{x^2+3} \le \sqrt{3^2+3} \;\Rightarrow\; 2 \le \sqrt{x^2+3} \le 2\sqrt{3}$, so $2(3-1) \le \int_1^3 \sqrt{x^2+3}\,dx \le 2\sqrt{3}(3-1)$; that is, $4 \le \int_1^3 \sqrt{x^2+3}\,dx \le 4\sqrt{3}$.

52. If $3 \le x \le 5$, then $4 \le x+1 \le 6$ and $\dfrac{1}{6} \le \dfrac{1}{x+1} \le \dfrac{1}{4}$, so $\dfrac{1}{6}(5-3) \le \displaystyle\int_3^5 \frac{1}{x+1}\,dx \le \frac{1}{4}(5-3)$;

that is, $\dfrac{1}{3} \le \displaystyle\int_3^5 \frac{1}{x+1}\,dx \le \frac{1}{2}$.

53. $0 \le x \le 1 \;\Rightarrow\; 0 \le \cos x \le 1 \;\Rightarrow\; x^2\cos x \le x^2 \;\Rightarrow\; \int_0^1 x^2\cos x\,dx \le \int_0^1 x^2\,dx = \frac{1}{3}\left[x^3\right]_0^1 = \frac{1}{3}$ [Property 7].

54. On the interval $\left[\frac{\pi}{4}, \frac{\pi}{2}\right]$, x is increasing and $\sin x$ is decreasing, so $\frac{\sin x}{x}$ is decreasing. Therefore, the largest value of $\frac{\sin x}{x}$ on $\left[\frac{\pi}{4}, \frac{\pi}{2}\right]$ is $\frac{\sin(\pi/4)}{\pi/4} = \frac{\sqrt{2}/2}{\pi/4} = \frac{2\sqrt{2}}{\pi}$. By Property 8 with $M = \frac{2\sqrt{2}}{\pi}$ we get $\int_{\pi/4}^{\pi/2} \frac{\sin x}{x}\,dx \le \frac{2\sqrt{2}}{\pi}\left(\frac{\pi}{2} - \frac{\pi}{4}\right) = \frac{\sqrt{2}}{2}$.

55. $\cos x \le 1 \;\Rightarrow\; e^x \cos x \le e^x \;\Rightarrow\; \int_0^1 e^x \cos x\,dx \le \int_0^1 e^x\,dx = [e^x]_0^1 = e - 1$

56. For $0 \le x \le 1$, $0 \le \sin^{-1} x \le \frac{\pi}{2}$, so $\int_0^1 x \sin^{-1} x\,dx \le \int_0^1 x\left(\frac{\pi}{2}\right) dx = \left[\frac{\pi}{4}x^2\right]_0^1 = \frac{\pi}{4}$.

57. $\Delta x = (3 - 0)/6 = \frac{1}{2}$, so the endpoints are 0, $\frac{1}{2}$, 1, $\frac{3}{2}$, 2, $\frac{5}{2}$, and 3, and the midpoints are $\frac{1}{4}$, $\frac{3}{4}$, $\frac{5}{4}$, $\frac{7}{4}$, $\frac{9}{4}$, and $\frac{11}{4}$.

The Midpoint Rule gives

$$\int_0^3 \sin(x^3)\,dx \approx \sum_{i=1}^{6} f(\overline{x}_i)\,\Delta x = \frac{1}{2}\left[\sin\left(\tfrac{1}{4}\right)^3 + \sin\left(\tfrac{3}{4}\right)^3 + \sin\left(\tfrac{5}{4}\right)^3 + \sin\left(\tfrac{7}{4}\right)^3 + \sin\left(\tfrac{9}{4}\right)^3 + \sin\left(\tfrac{11}{4}\right)^3\right] \approx 0.280981.$$

58. (a) Displacement $= \int_0^5 (t^2 - t)\,dt = \left[\frac{1}{3}t^3 - \frac{1}{2}t^2\right]_0^5 = \frac{125}{3} - \frac{25}{2} = \frac{175}{6} = 29.1\overline{6}$ meters

(b) Distance traveled $= \int_0^5 |t^2 - t|\,dt = \int_0^5 |t(t-1)|\,dt = \int_0^1 (t - t^2)\,dt + \int_1^5 (t^2 - t)\,dt$

$$= \left[\tfrac{1}{2}t^2 - \tfrac{1}{3}t^3\right]_0^1 + \left[\tfrac{1}{3}t^3 - \tfrac{1}{2}t^2\right]_1^5 = \tfrac{1}{2} - \tfrac{1}{3} - 0 + \left(\tfrac{125}{3} - \tfrac{25}{2}\right) - \left(\tfrac{1}{3} - \tfrac{1}{2}\right) = \tfrac{177}{6} = 29.5 \text{ meters}$$

59. Note that $r(t) = b'(t)$, where $b(t) =$ the number of barrels of oil consumed up to time t. So, by the Net Change Theorem, $\int_0^8 r(t)\,dt = b(8) - b(0)$ represents the number of barrels of oil consumed from Jan. 1, 2000, through Jan. 1, 2008.

60. Distance covered $= \int_0^{5.0} v(t)\,dt \approx M_5 = \frac{5.0 - 0}{5}[v(0.5) + v(1.5) + v(2.5) + v(3.5) + v(4.5)]$

$$= 1(4.67 + 8.86 + 10.22 + 10.67 + 10.81) = 45.23 \text{ m}$$

61. We use the Midpoint Rule with $n = 6$ and $\Delta t = \frac{24 - 0}{6} = 4$. The increase in the bee population was

$$\int_0^{24} r(t)\,dt \approx M_6 = 4[r(2) + r(6) + r(10) + r(14) + r(18) + r(22)]$$
$$\approx 4[50 + 1000 + 7000 + 8550 + 1350 + 150] = 4(18{,}100) = 72{,}400$$

62. $A_1 = \frac{1}{2}bh = \frac{1}{2}(2)(2) = 2$, $A_2 = \frac{1}{2}bh = \frac{1}{2}(1)(1) = \frac{1}{2}$, and since $y = -\sqrt{1 - x^2}$ for $0 \le x \le 1$ represents a quarter-circle with radius 1, $A_3 = \frac{1}{4}\pi r^2 = \frac{1}{4}\pi(1)^2 = \frac{\pi}{4}$. So

$$\int_{-3}^{1} f(x)\,dx = A_1 - A_2 - A_3 = 2 - \tfrac{1}{2} - \tfrac{\pi}{4} = \tfrac{1}{4}(6 - \pi)$$

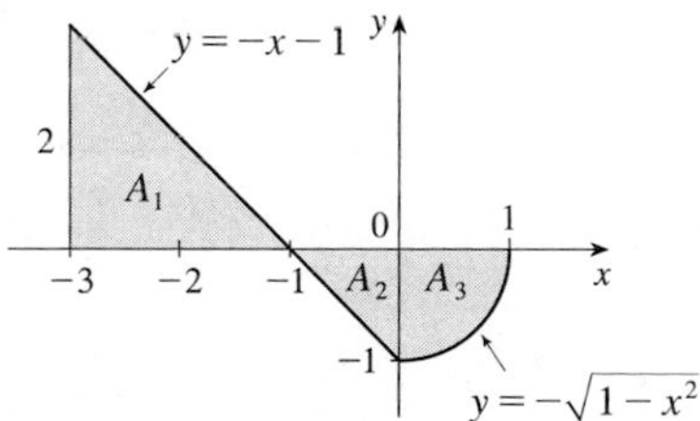

63. Let $u = 2\sin\theta$. Then $du = 2\cos\theta\,d\theta$ and when $\theta = 0$, $u = 0$; when $\theta = \frac{\pi}{2}$, $u = 2$. Thus,

$\int_0^{\pi/2} f(2\sin\theta)\cos\theta\,d\theta = \int_0^2 f(u)\left(\frac{1}{2}\,du\right) = \frac{1}{2}\int_0^2 f(u)\,du = \frac{1}{2}\int_0^2 f(x)\,dx = \frac{1}{2}(6) = 3$.

64. (a) C is increasing on those intervals where C' is positive. By the Fundamental Theorem of Calculus, $C'(x) = \frac{d}{dx}\left[\int_0^x \cos\left(\frac{\pi}{2}t^2\right)dt\right] = \cos\left(\frac{\pi}{2}x^2\right)$. This is positive when $\frac{\pi}{2}x^2$ is in the interval $\left(\left(2n - \frac{1}{2}\right)\pi, \left(2n + \frac{1}{2}\right)\pi\right)$, n any integer. This implies that $\left(2n - \frac{1}{2}\right)\pi < \frac{\pi}{2}x^2 < \left(2n + \frac{1}{2}\right)\pi \;\Leftrightarrow\; 0 \le |x| < 1$ or $\sqrt{4n - 1} < |x| < \sqrt{4n + 1}$, n any positive integer. So C is increasing on the intervals $(-1, 1)$, $\left(\sqrt{3}, \sqrt{5}\right)$, $\left(-\sqrt{5}, -\sqrt{3}\right)$, $\left(\sqrt{7}, 3\right)$, $\left(-3, -\sqrt{7}\right)$, ...

(b) C is concave upward on those intervals where $C'' > 0$. We differentiate C' to find C'': $C'(x) = \cos\left(\frac{\pi}{2}x^2\right) \Rightarrow$ $C''(x) = -\sin\left(\frac{\pi}{2}x^2\right)\left(\frac{\pi}{2} \cdot 2x\right) = -\pi x \sin\left(\frac{\pi}{2}x^2\right)$. For $x > 0$, this is positive where $(2n-1)\pi < \frac{\pi}{2}x^2 < 2n\pi$, n any positive integer $\Leftrightarrow$ $\sqrt{2(2n-1)} < x < 2\sqrt{n}$, n any positive integer. Since there is a factor of $-x$ in C'', the intervals of upward concavity for $x < 0$ are $\left(-\sqrt{2(2n+1)}, -2\sqrt{n}\right)$, n any nonnegative integer. That is, C is concave upward on $(-\sqrt{2}, 0)$, $(\sqrt{2}, 2)$, $(-\sqrt{6}, -2)$, $(\sqrt{6}, 2\sqrt{2})$, ...

(c)

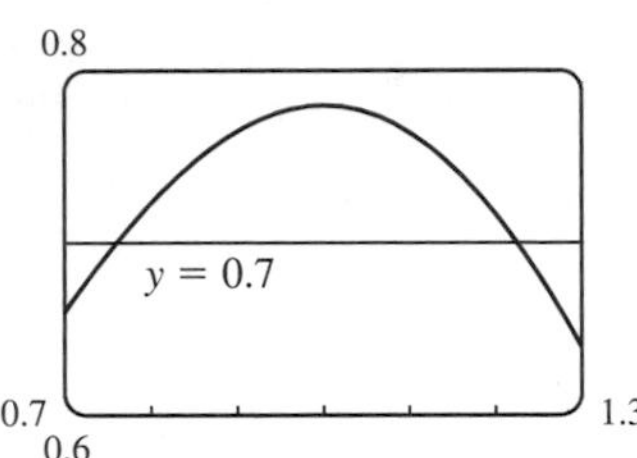

From the graphs, we can determine that $\int_0^x \cos\left(\frac{\pi}{2}t^2\right) dt = 0.7$ at $x \approx 0.76$ and $x \approx 1.22$.

(d)

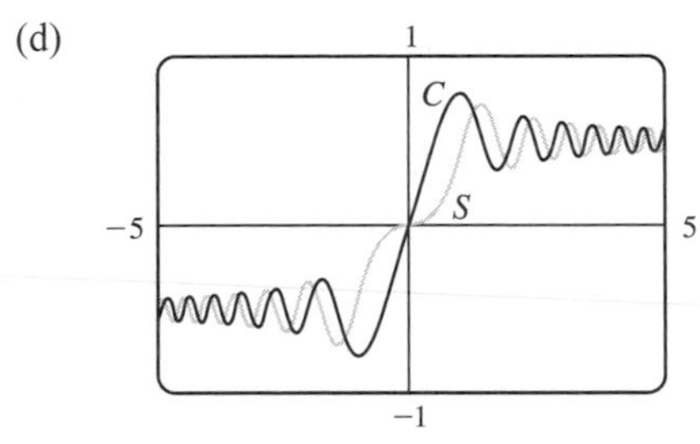

The graphs of $S(x)$ and $C(x)$ have similar shapes, except that S's flattens out near the origin, while C's does not. Note that for $x > 0$, C is increasing where S is concave up, and C is decreasing where S is concave down. Similarly, S is increasing where C is concave down, and S is decreasing where C is concave up. For $x < 0$, these relationships are reversed; that is, C is increasing where S is concave down, and S is increasing where C is concave up. See Example 5.3.3 and Exercise 5.3.65 for a discussion of $S(x)$.

65. Area under the curve $y = \sinh cx$ between $x = 0$ and $x = 1$ is equal to 1 $\Rightarrow$ $\int_0^1 \sinh cx\, dx = 1 \Rightarrow \frac{1}{c}\left[\cosh cx\right]_0^1 = 1 \Rightarrow \frac{1}{c}(\cosh c - 1) = 1 \Rightarrow$ $\cosh c - 1 = c \Rightarrow \cosh c = c + 1$. From the graph, we get $c = 0$ and $c \approx 1.6161$, but $c = 0$ isn't a solution for this problem since the curve $y = \sinh cx$ becomes $y = 0$ and the area under it is 0. Thus, $c \approx 1.6161$.

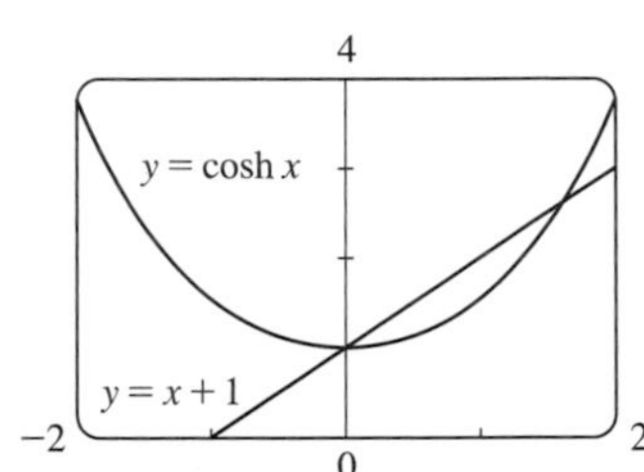

66. Both numerator and denominator approach 0 as $a \to 0$, so we use l'Hospital's Rule. (Note that we are differentiating *with respect to* a, since that is the quantity which is changing.) We also use FTC1:

$$\lim_{a \to 0} T(x,t) = \lim_{a \to 0} \frac{C\int_0^a e^{-(x-u)^2/(4kt)}\,du}{a\sqrt{4\pi kt}} \overset{\text{H}}{=} \lim_{a \to 0} \frac{Ce^{-(x-a)^2/(4kt)}}{\sqrt{4\pi kt}} = \frac{Ce^{-x^2/(4kt)}}{\sqrt{4\pi kt}}$$

67. Using FTC1, we differentiate both sides of the given equation, $\int_1^x f(t)\,dt = (x-1)e^{2x} + \int_1^x e^{-t} f(t)\,dt$, and get

$$f(x) = e^{2x} + 2(x-1)e^{2x} + e^{-x}f(x) \Rightarrow f(x)(1 - e^{-x}) = e^{2x} + 2(x-1)e^{2x} \Rightarrow f(x) = \frac{e^{2x}(2x-1)}{1 - e^{-x}}.$$

68. The second derivative is the derivative of the first derivative, so we'll apply the Net Change Theorem with $F = h'$. $\int_1^2 h''(u)\,du = \int_1^2 (h')'(u)\,du = h'(2) - h'(1) = 5 - 2 = 3$. The other information is unnecessary.

69. Let $u = f(x)$ and $du = f'(x)\,dx$. So $2\int_a^b f(x)f'(x)\,dx = 2\int_{f(a)}^{f(b)} u\,du = \left[u^2\right]_{f(a)}^{f(b)} = [f(b)]^2 - [f(a)]^2$.

70. Let $F(x) = \displaystyle\int_2^x \sqrt{1+t^3}\,dt$. Then $F'(2) = \displaystyle\lim_{h\to 0} \frac{F(2+h) - F(2)}{h} = \lim_{h\to 0} \frac{1}{h}\int_2^{2+h} \sqrt{1+t^3}\,dt$, and $F'(x) = \sqrt{1+x^3}$, so

$$\lim_{h\to 0} \frac{1}{h}\int_2^{2+h} \sqrt{1+t^3}\,dt = F'(2) = \sqrt{1+2^3} = \sqrt{9} = 3.$$

71. Let $u = 1 - x$. Then $du = -dx$, so $\int_0^1 f(1-x)\,dx = \int_1^0 f(u)(-du) = \int_0^1 f(u)\,du = \int_0^1 f(x)\,dx$.

72. $$\lim_{n\to\infty} \frac{1}{n}\left[\left(\frac{1}{n}\right)^9 + \left(\frac{2}{n}\right)^9 + \left(\frac{3}{n}\right)^9 + \cdots + \left(\frac{n}{n}\right)^9\right] = \lim_{n\to\infty} \frac{1-0}{n}\sum_{i=1}^{n}\left(\frac{i}{n}\right)^9 = \int_0^1 x^9\,dx = \left[\frac{x^{10}}{10}\right]_0^1 = \frac{1}{10}$$

The limit is based on Riemann sums using right endpoints and subintervals of equal length.

73. The shaded region has area $\int_0^1 f(x)\,dx = \frac{1}{3}$. The integral $\int_0^1 f^{-1}(y)\,dy$ gives the area of the unshaded region, which we know to be $1 - \frac{1}{3} = \frac{2}{3}$. So $\int_0^1 f^{-1}(y)\,dy = \frac{2}{3}$.

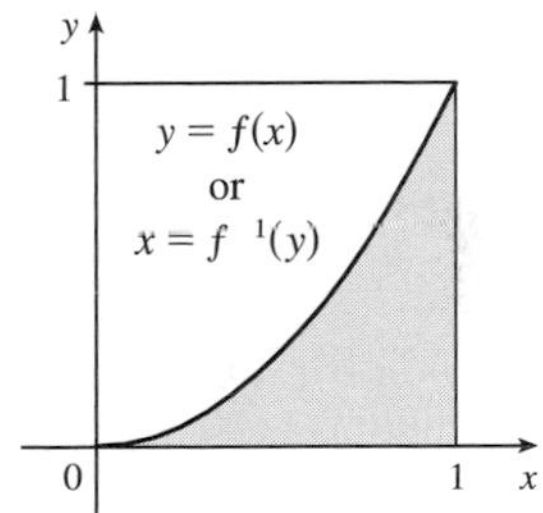

PROBLEMS PLUS

1. Differentiating both sides of the equation $x \sin \pi x = \int_0^{x^2} f(t)\,dt$ (using FTC1 and the Chain Rule for the right side) gives $\sin \pi x + \pi x \cos \pi x = 2x f(x^2)$. Letting $x = 2$ so that $f(x^2) = f(4)$, we obtain $\sin 2\pi + 2\pi \cos 2\pi = 4f(4)$, so $f(4) = \frac{1}{4}(0 + 2\pi \cdot 1) = \frac{\pi}{2}$.

2. The area A under the curve $y = x + 1/x$ from $x = a$ to $x = a + 1.5$ is given by $A(a) = \displaystyle\int_a^{a+1.5} \left(x + \frac{1}{x}\right) dx$.

To find the minimum value of A, we'll differentiate A using FTC1 and set the derivative equal to 0.

$$\begin{aligned} A'(a) &= \frac{d}{da}\int_a^{a+1.5}\left(x+\frac{1}{x}\right)dx \\ &= \frac{d}{da}\int_a^{1}\left(x+\frac{1}{x}\right)dx + \frac{d}{da}\int_1^{a+1.5}\left(x+\frac{1}{x}\right)dx \\ &= -\frac{d}{da}\int_1^{a}\left(x+\frac{1}{x}\right)dx + \frac{d}{da}\int_1^{a+1.5}\left(x+\frac{1}{x}\right)dx \\ &= -\left(a+\frac{1}{a}\right) + \left(a+1.5+\frac{1}{a+1.5}\right) = 1.5 + \frac{1}{a+1.5} - \frac{1}{a} \end{aligned}$$

$A'(a) = 0 \;\Leftrightarrow\; 1.5 + \dfrac{1}{a+1.5} - \dfrac{1}{a} = 0 \;\Leftrightarrow\; 1.5a(a+1.5) + a - (a+1.5) = 0 \;\Leftrightarrow\;$ $1.5a^2 + 2.25a - 1.5 = 0$ $\left[\text{multiply by } \frac{4}{3}\right]$ $\Leftrightarrow\; 2a^2 + 3a - 2 = 0 \;\Leftrightarrow\; (2a-1)(a+2) = 0 \;\Leftrightarrow\; a = \frac{1}{2}$ or $a = -2$. Since $a > 0$, $a = \frac{1}{2}$. $A''(a) = -\dfrac{1}{(a+1.5)^2} + \dfrac{1}{a^2} > 0$, so

$A\left(\frac{1}{2}\right) = \displaystyle\int_{1/2}^{2}\left(x+\frac{1}{x}\right)dx = \left[\tfrac{1}{2}x^2 + \ln|x|\right]_{1/2}^{2} = (2+\ln 2) - \left(\tfrac{1}{8} - \ln 2\right) = \tfrac{15}{8} + 2\ln 2$ is the minimum value of A.

3. For $I = \int_0^4 x e^{(x-2)^4}\,dx$, let $u = x - 2$ so that $x = u + 2$ and $dx = du$. Then $I = \int_{-2}^{2}(u+2)e^{u^4}\,du = \int_{-2}^{2} u e^{u^4}\,du + \int_{-2}^{2} 2e^{u^4}\,du = 0$ [by 5.5.7(b)] $+ 2\int_0^4 e^{(x-2)^4}\,dx = 2k$.

4. (a)

From the graph of $f(x) = \dfrac{2cx - x^2}{c^3}$, it appears that the areas are equal; that is, the area enclosed is independent of c.

(b) We first find the x-intercepts of the curve, to determine the limits of integration: $y = 0 \;\Leftrightarrow\; 2cx - x^2 = 0 \;\Leftrightarrow\; x = 0$ or $x = 2c$. Now we integrate the function between these limits to find the enclosed area:

$$A = \int_0^{2c} \frac{2cx - x^2}{c^3}\,dx = \frac{1}{c^3}\left[cx^2 - \tfrac{1}{3}x^3\right]_0^{2c} = \frac{1}{c^3}\left[c(2c)^2 - \tfrac{1}{3}(2c)^3\right] = \frac{1}{c^3}\left[4c^3 - \tfrac{8}{3}c^3\right] = \tfrac{4}{3}, \text{ a constant.}$$

(c)

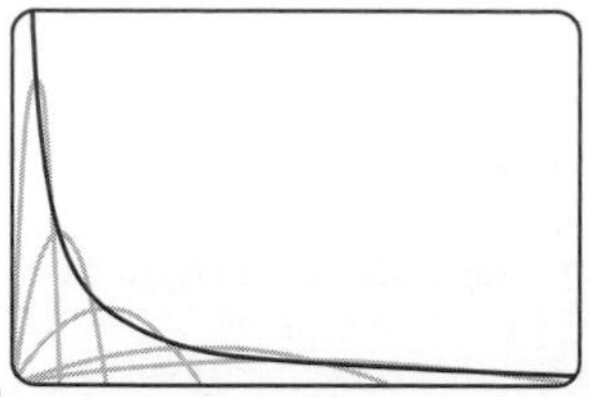

The vertices of the family of parabolas seem to determine a branch of a hyperbola.

(d) For a particular c, the vertex is the point where the maximum occurs. We have seen that the x-intercepts are 0 and $2c$, so by symmetry, the maximum occurs at $x = c$, and its value is $\dfrac{2c(c) - c^2}{c^3} = \dfrac{1}{c}$. So we are interested in the curve consisting of all points of the form $\left(c, \dfrac{1}{c}\right)$, $c > 0$. This is the part of the hyperbola $y = 1/x$ lying in the first quadrant.

5. $f(x) = \displaystyle\int_0^{g(x)} \frac{1}{\sqrt{1+t^3}}\,dt$, where $g(x) = \displaystyle\int_0^{\cos x} [1 + \sin(t^2)]\,dt$. Using FTC1 and the Chain Rule (twice) we have

$$f'(x) = \frac{1}{\sqrt{1 + [g(x)]^3}}\,g'(x) = \frac{1}{\sqrt{1 + [g(x)]^3}}\,[1 + \sin(\cos^2 x)](-\sin x).\text{ Now } g\left(\tfrac{\pi}{2}\right) = \int_0^0 [1 + \sin(t^2)]\,dt = 0, \text{ so}$$

$$f'\left(\tfrac{\pi}{2}\right) = \frac{1}{\sqrt{1+0}}\,(1 + \sin 0)(-1) = 1 \cdot 1 \cdot (-1) = -1.$$

6. If $f(x) = \int_0^x x^2 \sin(t^2)\,dt = x^2 \int_0^x \sin(t^2)\,dt$, then $f'(x) = x^2 \sin(x^2) + 2x \int_0^x \sin(t^2)\,dt$, by the Product Rule and FTC1.

7. By l'Hospital's Rule and the Fundamental Theorem, using the notation $\exp(y) = e^y$,

$$\lim_{x\to 0} \frac{\int_0^x (1 - \tan 2t)^{1/t}\,dt}{x} \overset{\text{H}}{=} \lim_{x\to 0} \frac{(1 - \tan 2x)^{1/x}}{1} = \exp\left(\lim_{x\to 0} \frac{\ln(1 - \tan 2x)}{x}\right)$$

$$\overset{\text{H}}{=} \exp\left(\lim_{x\to 0} \frac{-2\sec^2 2x}{1 - \tan 2x}\right) = \exp\left(\frac{-2 \cdot 1^2}{1 - 0}\right) = e^{-2}$$

8. The area $A(t) = \int_0^t \sin(x^2)\,dx$, and the area $B(t) = \frac{1}{2} t \sin(t^2)$. Since $\lim\limits_{t\to 0^+} A(t) = 0 = \lim\limits_{t\to 0^+} B(t)$, we can use l'Hospital's Rule:

$$\lim_{t\to 0^+} \frac{A(t)}{B(t)} \overset{\text{H}}{=} \lim_{t\to 0^+} \frac{\sin(t^2)}{\frac{1}{2}\sin(t^2) + \frac{1}{2}t[2t\cos(t^2)]} \qquad \text{[by FTC1 and the Product Rule]}$$

$$\overset{\text{H}}{=} \lim_{t\to 0^+} \frac{2t\cos(t^2)}{t\cos(t^2) - 2t^3\sin(t^2) + 2t\cos(t^2)} = \lim_{t\to 0^+} \frac{2\cos(t^2)}{3\cos(t^2) - 2t^2\sin(t^2)} = \frac{2}{3 - 0} = \frac{2}{3}$$

9. $f(x) = 2 + x - x^2 = (-x + 2)(x + 1) = 0 \;\Leftrightarrow\; x = 2$ or $x = -1$. $f(x) \geq 0$ for $x \in [-1, 2]$ and $f(x) < 0$ everywhere else. The integral $\int_a^b (2 + x - x^2)\,dx$ has a maximum on the interval where the integrand is positive, which is $[-1, 2]$. So $a = -1$, $b = 2$. (Any larger interval gives a smaller integral since $f(x) < 0$ outside $[-1, 2]$. Any smaller interval also gives a smaller integral since $f(x) \geq 0$ in $[-1, 2]$.)

10. This sum can be interpreted as a Riemann sum, with the right endpoints of the subintervals as sample points and with $a = 0$, $b = 10{,}000$, and $f(x) = \sqrt{x}$. So we approximate

$$\sum_{i=1}^{10{,}000} \sqrt{i} \approx \lim_{n\to\infty} \frac{10{,}000}{n} \sum_{i=1}^{n} \sqrt{\frac{10{,}000i}{n}} = \int_0^{10{,}000} \sqrt{x}\,dx = \left[\tfrac{2}{3}x^{3/2}\right]_0^{10{,}000} = \tfrac{2}{3}(1{,}000{,}000) \approx 666{,}667.$$

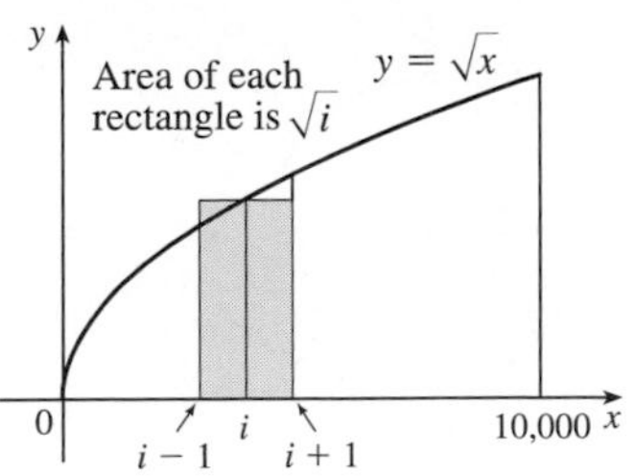

Alternate method: We can use graphical methods as follows:

From the figure we see that $\int_{i-1}^{i} \sqrt{x}\,dx < \sqrt{i} < \int_{i}^{i+1} \sqrt{x}\,dx$, so

$\int_0^{10{,}000} \sqrt{x}\,dx < \sum\limits_{i=1}^{10{,}000} \sqrt{i} < \int_1^{10{,}001} \sqrt{x}\,dx$. Since

$\int \sqrt{x}\,dx = \frac{2}{3}x^{3/2} + C$, we get $\int_0^{10{,}000} \sqrt{x}\,dx = 666{,}666.\overline{6}$ and

$\int_1^{10{,}001} \sqrt{x}\,dx = \frac{2}{3}[(10{,}001)^{3/2} - 1] \approx 666{,}766$.

Hence, $666{,}666.\overline{6} < \sum\limits_{i=1}^{10{,}000} \sqrt{i} < 666{,}766$. We can estimate the sum by averaging these bounds:

$\sum\limits_{i=1}^{10{,}000} \approx \dfrac{666{,}666.\overline{6} + 666{,}766}{2} \approx 666{,}716$. The actual value is about 666,716.46.

11. (a) We can split the integral $\int_0^n [\![x]\!]\,dx$ into the sum $\sum\limits_{i=1}^{n} \left[\int_{i-1}^{i} [\![x]\!]\,dx\right]$. But on each of the intervals $[i-1, i)$ of integration, $[\![x]\!]$ is a constant function, namely $i-1$. So the ith integral in the sum is equal to $(i-1)[i-(i-1)] = (i-1)$. So the original integral is equal to $\sum\limits_{i=1}^{n} (i-1) = \sum\limits_{i=1}^{n-1} i = \dfrac{(n-1)n}{2}$.

(b) We can write $\int_a^b [\![x]\!]\,dx = \int_0^b [\![x]\!]\,dx - \int_0^a [\![x]\!]\,dx$.

Now $\int_0^b [\![x]\!]\,dx = \int_0^{[\![b]\!]} [\![x]\!]\,dx + \int_{[\![b]\!]}^b [\![x]\!]\,dx$. The first of these integrals is equal to $\frac{1}{2}([\![b]\!]-1)\,[\![b]\!]$, by part (a), and since $[\![x]\!] = [\![b]\!]$ on $[[\![b]\!], b]$, the second integral is just $[\![b]\!]\,(b - [\![b]\!])$. So

$\int_0^b [\![x]\!]\,dx = \frac{1}{2}([\![b]\!]-1)\,[\![b]\!] + [\![b]\!]\,(b-[\![b]\!]) = \frac{1}{2}\,[\![b]\!]\,(2b - [\![b]\!] - 1)$ and similarly $\int_0^a [\![x]\!]\,dx = \frac{1}{2}\,[\![a]\!]\,(2a - [\![a]\!] - 1)$.

Therefore, $\int_a^b [\![x]\!]\,dx = \frac{1}{2}\,[\![b]\!]\,(2b - [\![b]\!] - 1) - \frac{1}{2}\,[\![a]\!]\,(2a - [\![a]\!] - 1)$.

12. By FTC1, $\dfrac{d}{dx}\displaystyle\int_0^x \left(\int_1^{\sin t} \sqrt{1+u^4}\,du\right)dt = \int_1^{\sin x} \sqrt{1+u^4}\,du$. Again using FTC1,

$$\frac{d^2}{dx^2}\int_0^x \left(\int_1^{\sin t} \sqrt{1+u^4}\,du\right)dt = \frac{d}{dx}\int_1^{\sin x} \sqrt{1+u^4}\,du = \sqrt{1+\sin^4 x}\,\cos x.$$

13. Let $Q(x) = \displaystyle\int_0^x P(t)\,dt = \left[at + \frac{b}{2}t^2 + \frac{c}{3}t^3 + \frac{d}{4}t^4\right]_0^x = ax + \frac{b}{2}x^2 + \frac{c}{3}x^3 + \frac{d}{4}x^4$. Then $Q(0) = 0$, and $Q(1) = 0$ by the given condition, $a + \dfrac{b}{2} + \dfrac{c}{3} + \dfrac{d}{4} = 0$. Also, $Q'(x) = P(x) = a + bx + cx^2 + dx^3$ by FTC1. By Rolle's Theorem, applied to Q on $[0,1]$, there is a number r in $(0,1)$ such that $Q'(r) = 0$, that is, such that $P(r) = 0$. Thus, the equation $P(x) = 0$ has a root between 0 and 1.

More generally, if $P(x) = a_0 + a_1 x + a_2 x^2 + \cdots + a_n x^n$ and if $a_0 + \dfrac{a_1}{2} + \dfrac{a_2}{3} + \cdots + \dfrac{a_n}{n+1} = 0$, then the equation $P(x) = 0$ has a root between 0 and 1. The proof is the same as before:

Let $Q(x) = \displaystyle\int_0^x P(t)\,dt = a_0 x + \frac{a_1}{2}x^2 + \frac{a_2}{3}x^3 + \cdots + \frac{a_n}{n+1}x^n$. Then $Q(0) = Q(1) = 0$ and $Q'(x) = P(x)$. By Rolle's Theorem applied to Q on $[0,1]$, there is a number r in $(0,1)$ such that $Q'(r) = 0$, that is, such that $P(r) = 0$.

14. 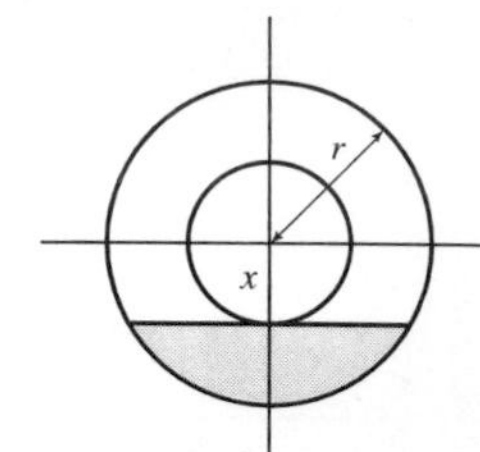

Let x be the distance between the center of the disk and the surface of the liquid. The wetted circular region has area $\pi r^2 - \pi x^2$ while the unexposed wetted region (shaded in the diagram) has area $2\int_x^r \sqrt{r^2 - t^2}\,dt$, so the exposed wetted region has area $A(x) = \pi r^2 - \pi x^2 - 2\int_x^r \sqrt{r^2 - t^2}\,dt$, $0 \le x \le r$. By FTC1, we have $A'(x) = -2\pi x + 2\sqrt{r^2 - x^2}$.

Now $A'(x) > 0 \;\Rightarrow\; -2\pi x + 2\sqrt{r^2 - x^2} > 0 \;\Rightarrow\; \sqrt{r^2 - x^2} > \pi x \;\Rightarrow\; r^2 - x^2 > \pi^2 x^2 \;\Rightarrow$

$r^2 > \pi^2 x^2 + x^2 \;\Rightarrow\; r^2 > x^2(\pi^2 + 1) \;\Rightarrow\; x^2 < \dfrac{r^2}{\pi^2 + 1} \;\Rightarrow\; x < \dfrac{r}{\sqrt{\pi^2 + 1}}$, and we'll call this value x^*.

Since $A'(x) > 0$ for $0 < x < x^*$ and $A'(x) < 0$ for $x^* < x < r$, we have an absolute maximum when $x = x^*$.

15. Note that $\dfrac{d}{dx}\left(\displaystyle\int_0^x \left[\int_0^u f(t)\,dt\right] du\right) = \displaystyle\int_0^x f(t)\,dt$ by FTC1, while

$$\begin{aligned}\frac{d}{dx}\left[\int_0^x f(u)(x-u)\,du\right] &= \frac{d}{dx}\left[x\int_0^x f(u)\,du\right] - \frac{d}{dx}\left[\int_0^x f(u)u\,du\right]\\ &= \int_0^x f(u)\,du + xf(x) - f(x)x = \int_0^x f(u)\,du\end{aligned}$$

Hence, $\int_0^x f(u)(x-u)\,du = \int_0^x \left[\int_0^u f(t)\,dt\right] du + C$. Setting $x = 0$ gives $C = 0$.

16. The parabola $y = 4 - x^2$ and the line $y = x + 2$ intersect when $4 - x^2 = x + 2 \;\Leftrightarrow\; x^2 + x - 2 = 0 \;\Leftrightarrow\; (x+2)(x-1) = 0 \;\Leftrightarrow$ $x = -2$ or 1. So the point A is $(-2, 0)$ and B is $(1, 3)$. The slope of the line $y = x + 2$ is 1 and the slope of the parabola $y = 4 - x^2$ at x-coordinate x is $-2x$. These slopes are equal when $x = -\frac{1}{2}$, so the point C is $\left(-\frac{1}{2}, \frac{15}{4}\right)$.

The area A_1 of the parabolic segment is the area under the parabola from $x = -2$ to $x = 1$, minus the area under the line $y = x + 2$ from -2 to 1. Thus,

$$\begin{aligned}A_1 &= \int_{-2}^{1}(4 - x^2)\,dx - \int_{-2}^{1}(x+2)\,dx = \left[4x - \tfrac{1}{3}x^3\right]_{-2}^{1} - \left[\tfrac{1}{2}x^2 + 2x\right]_{-2}^{1}\\ &= \left[\left(4 - \tfrac{1}{3}\right) - \left(-8 + \tfrac{8}{3}\right)\right] - \left[\left(\tfrac{1}{2} + 2\right) - (2 - 4)\right] = 9 - \tfrac{9}{2} = \tfrac{9}{2}.\end{aligned}$$

The area A_2 of the inscribed triangle is the area under the line segment AC plus the area under the line segment CB minus the area under the line segment AB. The line through A and C has slope $\dfrac{15/4 - 0}{-1/2 + 2} = \dfrac{5}{2}$ and equation $y - 0 = \frac{5}{2}(x + 2)$, or $y = \frac{5}{2}x + 5$. The line through C and B has slope $\dfrac{3 - 15/4}{1 + 1/2} = -\dfrac{1}{2}$ and equation $y - 3 = -\frac{1}{2}(x - 1)$, or $y = -\frac{1}{2}x + \frac{7}{2}$.

Thus,

$$\begin{aligned}A_2 &= \int_{-2}^{-1/2}\left(\tfrac{5}{2}x + 5\right)dx + \int_{-1/2}^{1}\left(-\tfrac{1}{2}x + \tfrac{7}{2}\right)dx - \int_{-2}^{1}(x+2)\,dx = \left[\tfrac{5}{4}x^2 + 5x\right]_{-2}^{-1/2} + \left[-\tfrac{1}{4}x^2 + \tfrac{7}{2}x\right]_{-1/2}^{1} - \tfrac{9}{2}\\ &= \left[\left(\tfrac{5}{16} - \tfrac{5}{2}\right) - (5 - 10)\right] + \left[\left(-\tfrac{1}{4} + \tfrac{7}{2}\right) - \left(-\tfrac{1}{16} - \tfrac{7}{4}\right)\right] - \tfrac{9}{2} = \tfrac{45}{16} + \tfrac{81}{16} - \tfrac{72}{16} = \tfrac{54}{16} = \tfrac{27}{8}\end{aligned}$$

Archimedes' result states that $A_1 = \frac{4}{3}A_2$, which is verified in this case since $\frac{4}{3} \cdot \frac{27}{8} = \frac{9}{2}$.

17. 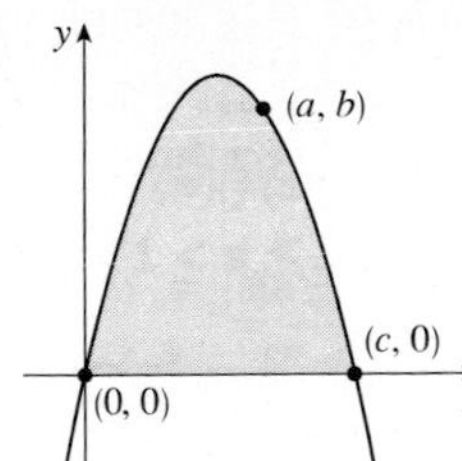

Let c be the nonzero x-intercept so that the parabola has equation $f(x) = kx(x-c)$, or $y = kx^2 - ckx$, where $k < 0$. The area A under the parabola is

$$A = \int_0^c kx(x-c)\,dx = k\int_0^c (x^2 - cx)\,dx = k\left[\tfrac{1}{3}x^3 - \tfrac{1}{2}cx^2\right]_0^c$$
$$= k\left(\tfrac{1}{3}c^3 - \tfrac{1}{2}c^3\right) = -\tfrac{1}{6}kc^3$$

The point (a, b) is on the parabola, so $f(a) = b \;\Rightarrow\; b = ka(a-c) \;\Rightarrow$

$k = \dfrac{b}{a(a-c)}$. Substituting for k in A gives $A(c) = -\dfrac{b}{6a}\cdot\dfrac{c^3}{a-c} \;\Rightarrow$

$$A'(c) = -\frac{b}{6a}\cdot\frac{(a-c)(3c^2) - c^3(-1)}{(a-c)^2} = -\frac{b}{6a}\cdot\frac{c^2[3(a-c)+c]}{(a-c)^2} = -\frac{bc^2(3a-2c)}{6a(a-c)^2}$$

Now $A' = 0 \;\Rightarrow\; c = \frac{3}{2}a$. Since $A'(c) < 0$ for $a < c < \frac{3}{2}a$ and $A'(c) > 0$ for $c > \frac{3}{2}a$, so A has an absolute minimum when $c = \frac{3}{2}a$. Substituting for c in k gives us $k = \dfrac{b}{a\left(a - \frac{3}{2}a\right)} = -\dfrac{2b}{a^2}$, so $f(x) = -\dfrac{2b}{a^2}\,x\left(x - \frac{3}{2}a\right)$, or $f(x) = -\dfrac{2b}{a^2}\,x^2 + \dfrac{3b}{a}\,x$. Note that the vertex of the parabola is $\left(\frac{3}{4}a, \frac{9}{8}b\right)$ and the minimal area under the parabola is $A\left(\frac{3}{2}a\right) = \frac{9}{8}ab$.

18. 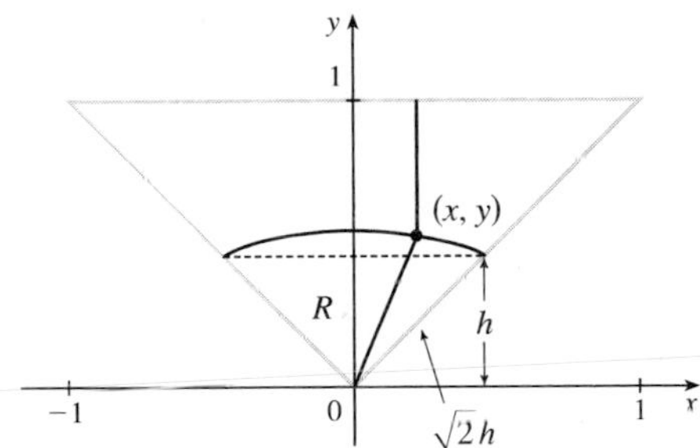

We restrict our attention to the triangle shown. A point in this triangle is closer to the side shown than to any other side, so if we find the area of the region R consisting of all points in the triangle that are closer to the center than to that side, we can multiply this area by 4 to find the total area. We find the equation of the set of points which are equidistant from the center and the side: the distance of the point (x, y) from the side is $1 - y$, and its distance from the center is $\sqrt{x^2 + y^2}$.

So the distances are equal if $\sqrt{x^2 + y^2} = 1 - y \;\Leftrightarrow\; x^2 + y^2 = 1 - 2y + y^2 \;\Leftrightarrow\; y = \frac{1}{2}(1 - x^2)$. Note that the area we are interested in is equal to the area of a triangle plus a crescent-shaped area. To find these areas, we have to find the y-coordinate h of the horizontal line separating them. From the diagram, $1 - h = \sqrt{2}\,h \;\Leftrightarrow\; h = \frac{1}{1+\sqrt{2}} = \sqrt{2} - 1$. We calculate the areas in terms of h, and substitute afterward.

The area of the triangle is $\frac{1}{2}(2h)(h) = h^2$, and the area of the crescent-shaped section is

$$\int_{-h}^{h}\left[\tfrac{1}{2}(1 - x^2) - h\right]dx = 2\int_0^h\left(\tfrac{1}{2} - h - \tfrac{1}{2}x^2\right)dx = 2\left[\left(\tfrac{1}{2} - h\right)x - \tfrac{1}{6}x^3\right]_0^h = h - 2h^2 - \tfrac{1}{3}h^3.$$

So the area of the whole region is

$$4\left[\left(h - 2h^2 - \tfrac{1}{3}h^3\right) + h^2\right] = 4h\left(1 - h - \tfrac{1}{3}h^2\right) = 4\left(\sqrt{2} - 1\right)\left[1 - \left(\sqrt{2} - 1\right) - \tfrac{1}{3}\left(\sqrt{2} - 1\right)^2\right]$$
$$= 4\left(\sqrt{2} - 1\right)\left(1 - \tfrac{1}{3}\sqrt{2}\right) = \tfrac{4}{3}\left(4\sqrt{2} - 5\right)$$

19. $$\lim_{n\to\infty}\left(\frac{1}{\sqrt{n}\sqrt{n+1}}+\frac{1}{\sqrt{n}\sqrt{n+2}}+\cdots+\frac{1}{\sqrt{n}\sqrt{n+n}}\right)$$
$$=\lim_{n\to\infty}\frac{1}{n}\left(\sqrt{\frac{n}{n+1}}+\sqrt{\frac{n}{n+2}}+\cdots+\sqrt{\frac{n}{n+n}}\right)$$
$$=\lim_{n\to\infty}\frac{1}{n}\left(\frac{1}{\sqrt{1+1/n}}+\frac{1}{\sqrt{1+2/n}}+\cdots+\frac{1}{\sqrt{1+1}}\right)$$
$$=\lim_{n\to\infty}\frac{1}{n}\sum_{i=1}^{n}f\left(\frac{i}{n}\right)\qquad\left[\text{where } f(x)=\frac{1}{\sqrt{1+x}}\right]$$
$$=\int_0^1\frac{1}{\sqrt{1+x}}\,dx=\left[2\sqrt{1+x}\right]_0^1=2\left(\sqrt{2}-1\right)$$

20. Note that the graphs of $(x-c)^2$ and $[(x-c)-2]^2$ intersect when $|x-c|=|x-c-2|$ $\Leftrightarrow$ $c-x=x-c-2$ $\Leftrightarrow$ $x=c+1$. The integration will proceed differently depending on the value of c.

Case 1: $-2\le c<-1$

In this case, $f_c(x)=(x-c-2)^2$ for $x\in[0,1]$, so

$$g(c)=\int_0^1(x-c-2)^2\,dx=\tfrac{1}{3}\left[(x-c-2)^3\right]_0^1=\tfrac{1}{3}\left[(-c-1)^3-(-c-2)^3\right]$$
$$=\tfrac{1}{3}(3c^2+9c+7)=c^2+3c+\tfrac{7}{3}=\left(c+\tfrac{3}{2}\right)^2+\tfrac{1}{12}$$

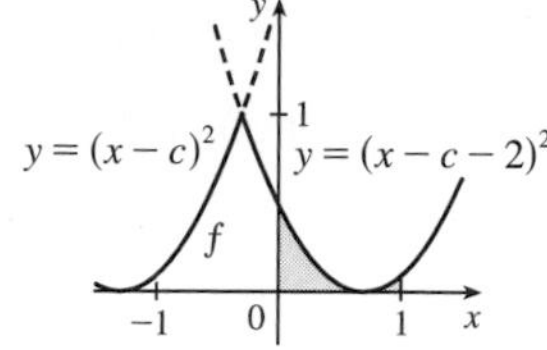

This is a parabola; its maximum value for $-2\le c<-1$ is $g(-2)=\frac{1}{3}$, and its minimum value is $g\left(-\frac{3}{2}\right)=\frac{1}{12}$.

Case 2: $-1\le c<0$

In this case, $f_c(x)=\begin{cases}(x-c)^2 & \text{if } 0\le x\le c+1\\(x-c-2)^2 & \text{if } c+1<x\le 1\end{cases}$

Therefore,

$$g(c)=\int_0^1 f_c(x)\,dx=\int_0^{c+1}(x-c)^2\,dx+\int_{c+1}^1(x-c-2)^2\,dx$$
$$=\tfrac{1}{3}\left[(x-c)^3\right]_0^{c+1}+\tfrac{1}{3}\left[(x-c-2)^3\right]_{c+1}^1=\tfrac{1}{3}\left[1+c^3+(-c-1)^3-(-1)\right]$$
$$=-c^2-c+\tfrac{1}{3}=-\left(c+\tfrac{1}{2}\right)^2+\tfrac{7}{12}$$

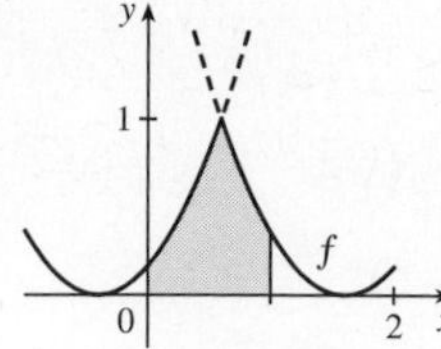

Again, this is a parabola, whose maximum value for $-1\le c<0$ is $g\left(-\frac{1}{2}\right)=\frac{7}{12}$, and whose minimum value on this c-interval is $g(-1)=\frac{1}{3}$.

Case 3: $0 \le c \le 2$

In this case, $f_c(x) = (x - c)^2$ for $x \in [0, 1]$, so

$$\begin{aligned} g(c) &= \int_0^1 (x-c)^2\,dx = \tfrac{1}{3}\left[(x-c)^3\right]_0^1 = \tfrac{1}{3}\left[(1-c)^3 - (-c)^3\right] \\ &= c^2 - c + \tfrac{1}{3} = \left(c - \tfrac{1}{2}\right)^2 + \tfrac{1}{12} \end{aligned}$$

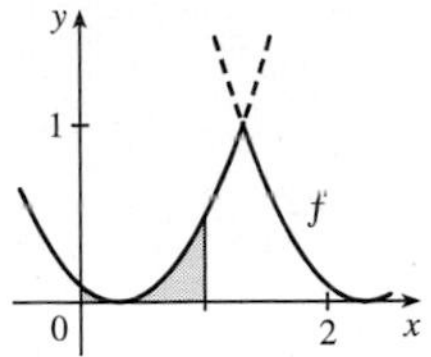

This parabola has a maximum value of $g(2) = \frac{7}{3}$ and a minimum value of $g\left(\frac{1}{2}\right) = \frac{1}{12}$.

We conclude that $g(c)$ has an absolute maximum value of $g(2) = \frac{7}{3}$, and absolute minimum values of $g\left(-\frac{3}{2}\right) = g\left(\frac{1}{2}\right) = \frac{1}{12}$.

6 □ APPLICATIONS OF INTEGRATION

6.1 Areas Between Curves

1. $A = \int_{x=1}^{x=8} (y_T - y_B)\,dx = \int_1^8 \left(\sqrt[3]{x} - \frac{1}{x}\right) dx = \left[\frac{3}{4}x^{4/3} - \ln|x|\right]_1^8 = (12 - \ln 8) - \left(\frac{3}{4} - \ln 1\right) = \frac{45}{4} - \ln 8$

2. $A = \int_0^1 \left(e^x - xe^{x^2}\right) dx = \left[e^x - \frac{1}{2}e^{x^2}\right]_0^1 = (e - \frac{1}{2}e) - (1 - \frac{1}{2}) = \frac{1}{2}e - \frac{1}{2} = \frac{1}{2}(e - 1)$

3. $A = \int_{y=-1}^{y=1} (x_R - x_L)\,dy = \int_{-1}^1 \left[e^y - (y^2 - 2)\right] dy = \int_{-1}^1 (e^y - y^2 + 2)\,dy$

$= \left[e^y - \frac{1}{3}y^3 + 2y\right]_{-1}^1 = (e^1 - \frac{1}{3} + 2) - (e^{-1} + \frac{1}{3} - 2) = e - \frac{1}{e} + \frac{10}{3}$

4. $A = \int_0^3 \left[(2y - y^2) - (y^2 - 4y)\right] dy = \int_0^3 (-2y^2 + 6y)\,dy = \left[-\frac{2}{3}y^3 + 3y^2\right]_0^3 = (-18 + 27) - 0 = 9$

5. $A = \int_{-1}^1 \left[e^x - (x^2 - 1)\right] dx = \left[e^x - \frac{1}{3}x^3 + x\right]_{-1}^1$

$= (e - \frac{1}{3} + 1) - (e^{-1} + \frac{1}{3} - 1) = e - \frac{1}{e} + \frac{4}{3}$

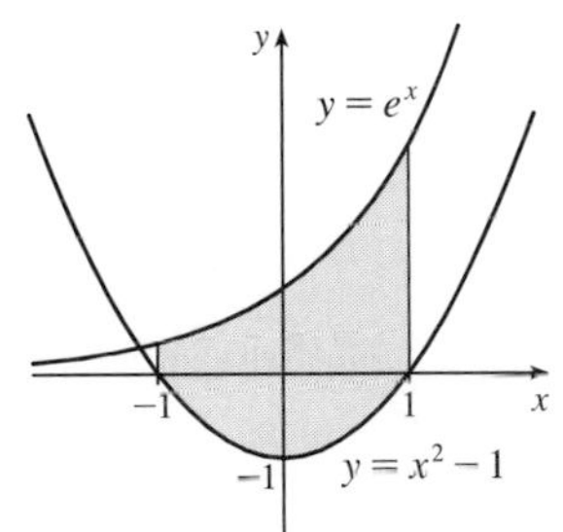

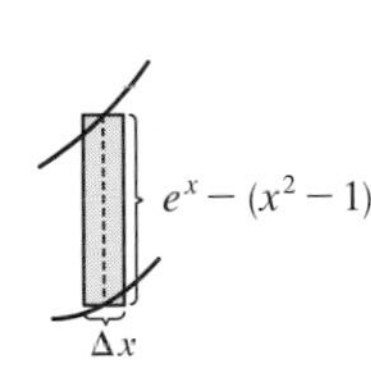

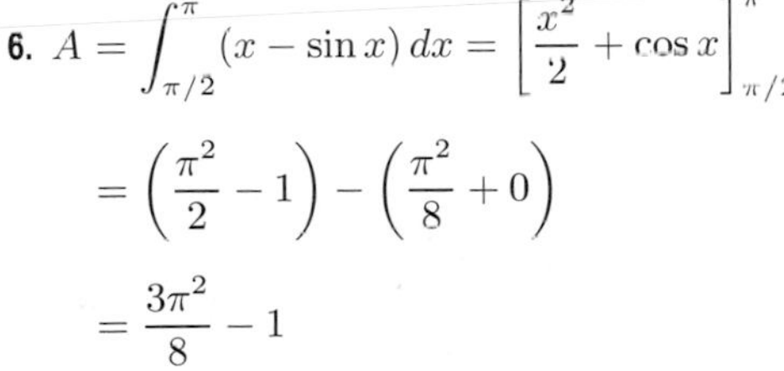

6. $A = \int_{\pi/2}^{\pi} (x - \sin x)\,dx = \left[\frac{x^2}{2} + \cos x\right]_{\pi/2}^{\pi}$

$= \left(\frac{\pi^2}{2} - 1\right) - \left(\frac{\pi^2}{8} + 0\right)$

$= \frac{3\pi^2}{8} - 1$

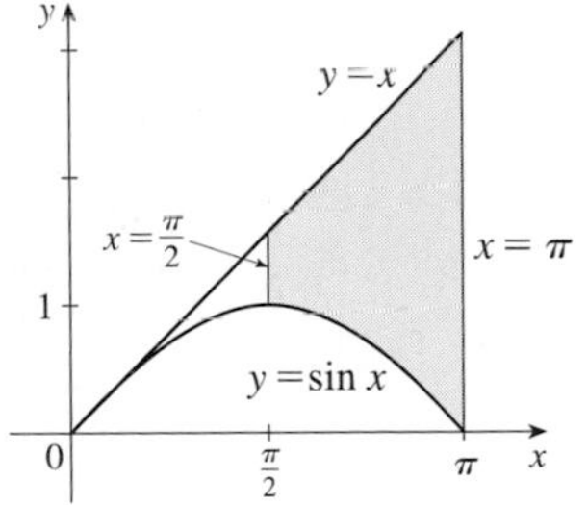

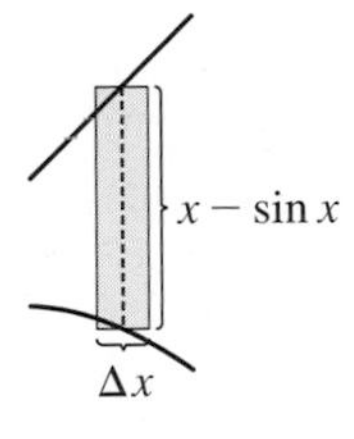

7. The curves intersect when $(x - 2)^2 = x \Leftrightarrow x^2 - 4x + 4 = x \Leftrightarrow x^2 - 5x + 4 = 0 \Leftrightarrow$ $(x - 1)(x - 4) = 0 \Leftrightarrow x = 1$ or 4.

$A = \int_1^4 \left[x - (x - 2)^2\right] dx = \int_1^4 (-x^2 + 5x - 4)\,dx$

$= \left[-\frac{1}{3}x^3 + \frac{5}{2}x^2 - 4x\right]_1^4$

$= (-\frac{64}{3} + 40 - 16) - (-\frac{1}{3} + \frac{5}{2} - 4)$

$= \frac{9}{2}$

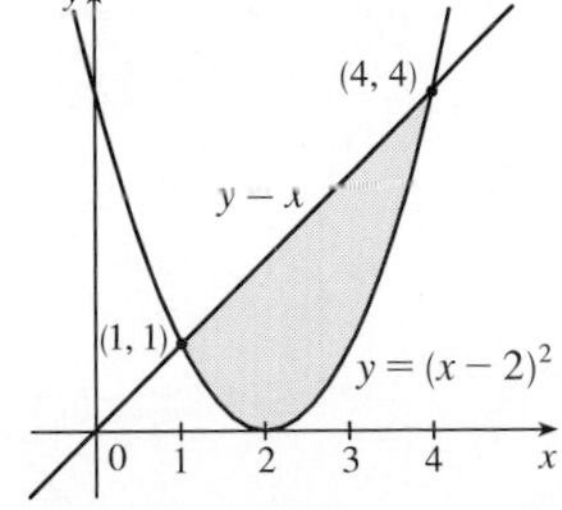

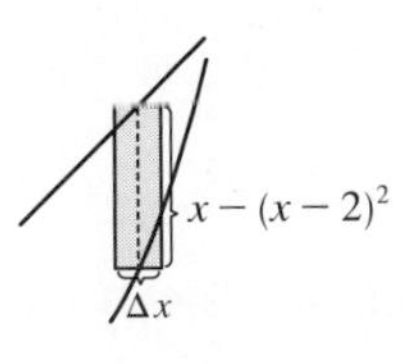

8. The curves intesect when $x^2 - 4x = 2x \Rightarrow x^2 - 6x = 0 \Rightarrow x(x-6) = 0 \Rightarrow x = 0$ or 6.

$$A = \int_0^6 [2x - (x^2 - 4x)]\,dx$$
$$= \int_0^6 (6x - x^2)\,dx = \left[3x^2 - \tfrac{1}{3}x^3\right]_0^6$$
$$= \left[3(6)^2 - \tfrac{1}{3}(6)^3\right] - (0 - 0)$$
$$= 108 - 72 = 36$$

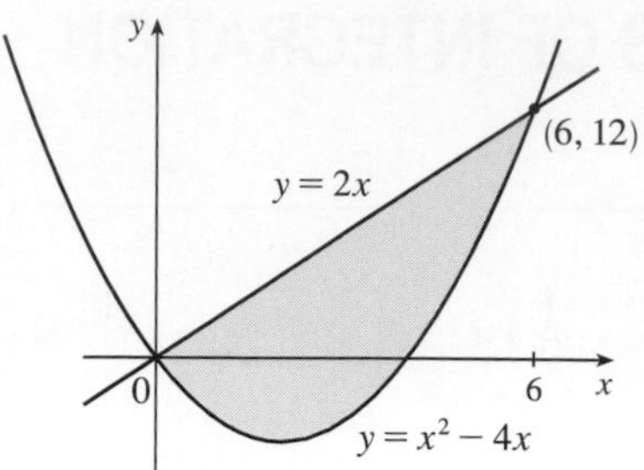

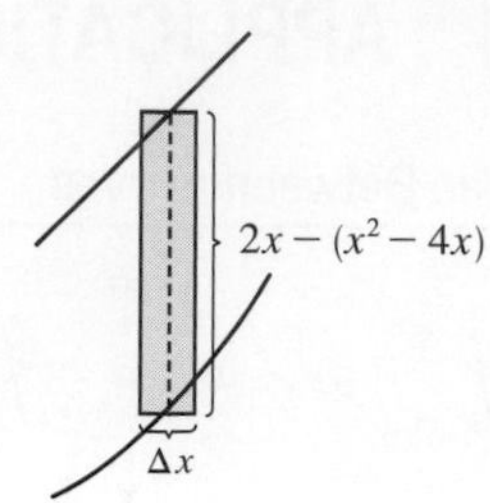

9. $A = \displaystyle\int_1^2 \left(\frac{1}{x} - \frac{1}{x^2}\right) dx = \left[\ln x + \frac{1}{x}\right]_1^2$

$$= \left(\ln 2 + \tfrac{1}{2}\right) - (\ln 1 + 1)$$
$$= \ln 2 - \tfrac{1}{2} \approx 0.19$$

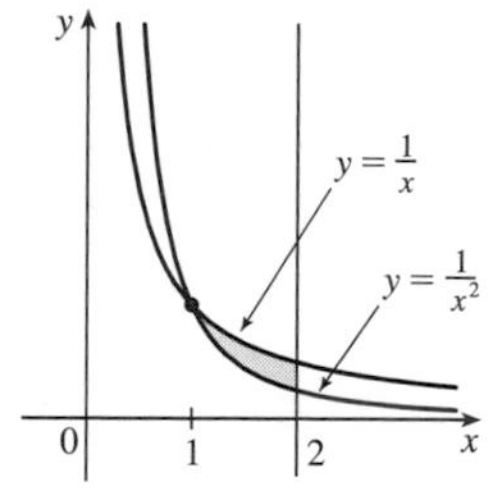

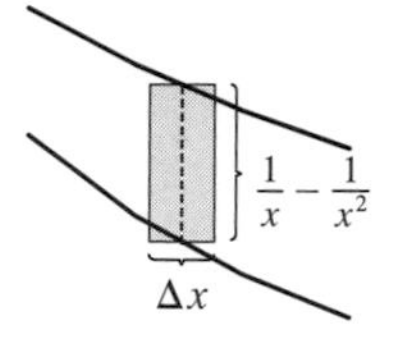

10. By observation, $y = \sin x$ and $y = 2x/\pi$ intersect at $(0, 0)$ and $(\pi/2, 1)$ for $x \geq 0$.

$$A = \int_0^{\pi/2} \left(\sin x - \frac{2x}{\pi}\right) dx = \left[-\cos x - \frac{1}{\pi}x^2\right]_0^{\pi/2} = \left(0 - \frac{\pi}{4}\right) - (-1) = 1 - \frac{\pi}{4}$$

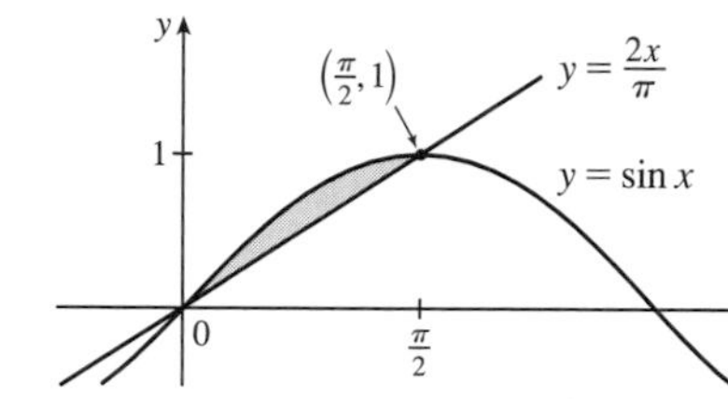

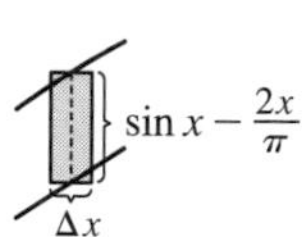

11. The curves intersect when $1 - y^2 = y^2 - 1 \Leftrightarrow 2 = 2y^2 \Leftrightarrow y^2 = 1 \Leftrightarrow y = \pm 1$.

$$A = \int_{-1}^{1} \left[(1 - y^2) - (y^2 - 1)\right] dy$$
$$= \int_{-1}^{1} 2(1 - y^2)\,dy$$
$$= 2 \cdot 2 \int_0^1 (1 - y^2)\,dy$$
$$= 4\left[y - \tfrac{1}{3}y^3\right]_0^1 = 4\left(1 - \tfrac{1}{3}\right) = \tfrac{8}{3}$$

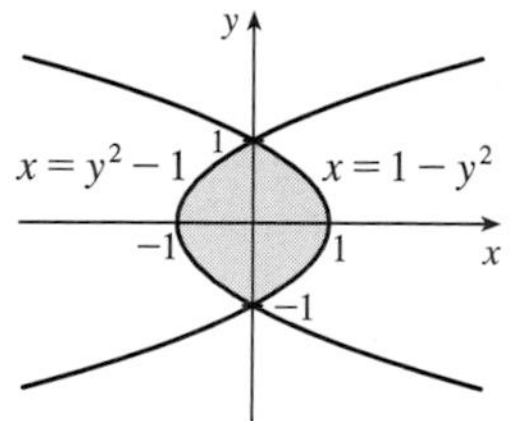

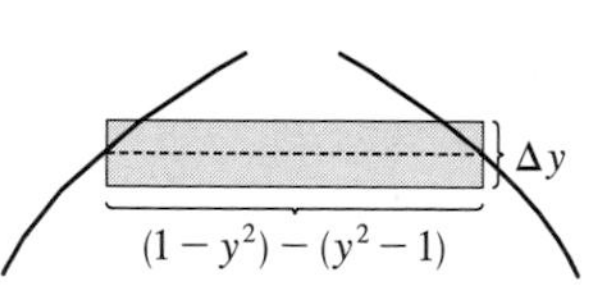

12. $4x + x^2 = 12 \Leftrightarrow (x + 6)(x - 2) = 0 \Leftrightarrow$
$x = -6$ or $x = 2$, so $y = -6$ or $y = 2$ and

$$A = \int_{-6}^{2} \left[\left(-\tfrac{1}{4}y^2 + 3\right) - y\right] dy$$
$$= \left[-\tfrac{1}{12}y^3 - \tfrac{1}{2}y^2 + 3y\right]_{-6}^{2}$$
$$= \left(-\tfrac{2}{3} - 2 + 6\right) - (18 - 18 - 18)$$
$$= 22 - \tfrac{2}{3} = \tfrac{64}{3}$$

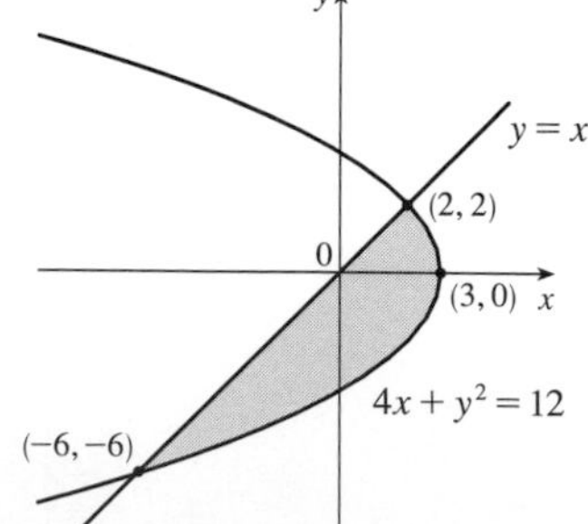

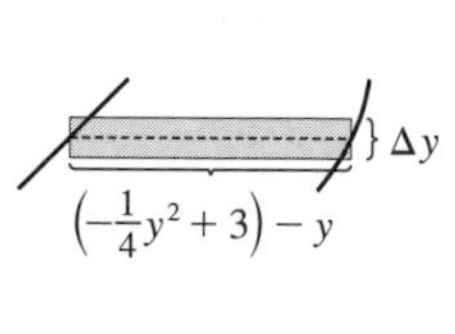

13. $12 - x^2 = x^2 - 6 \;\Leftrightarrow\; 2x^2 = 18 \;\Leftrightarrow$
$x^2 = 9 \;\Leftrightarrow\; x = \pm 3$, so

$$A = \int_{-3}^{3} \left[(12 - x^2) - (x^2 - 6)\right] dx$$

$$= 2\int_0^3 \left(18 - 2x^2\right) dx \qquad \text{[by symmetry]}$$

$$= 2\left[18x - \tfrac{2}{3}x^3\right]_0^3 = 2\left[(54 - 18) - 0\right]$$

$$= 2(36) = 72$$

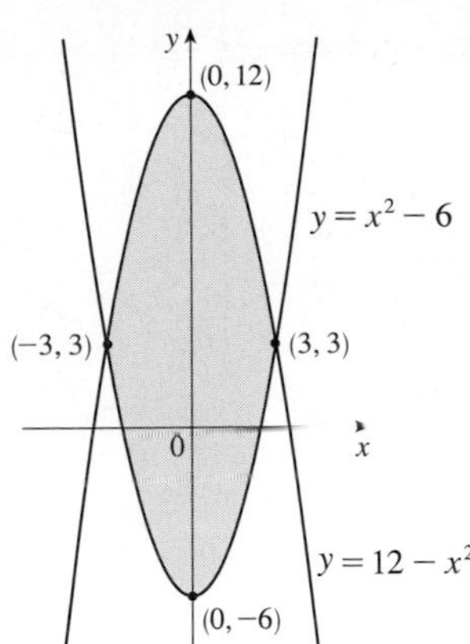

14. $x^2 = 4x - x^2 \;\Leftrightarrow\; 2x^2 - 4x = 0 \;\Leftrightarrow$
$2x(x - 2) = 0 \;\Leftrightarrow\; x = 0$ or 2, so

$$A = \int_0^2 \left[(4x - x^2) - x^2\right] dx = \int_0^2 (4x - 2x^2)\, dx$$

$$= \left[2x^2 - \tfrac{2}{3}x^3\right]_0^2 = 8 - \tfrac{16}{3} = \tfrac{8}{3}$$

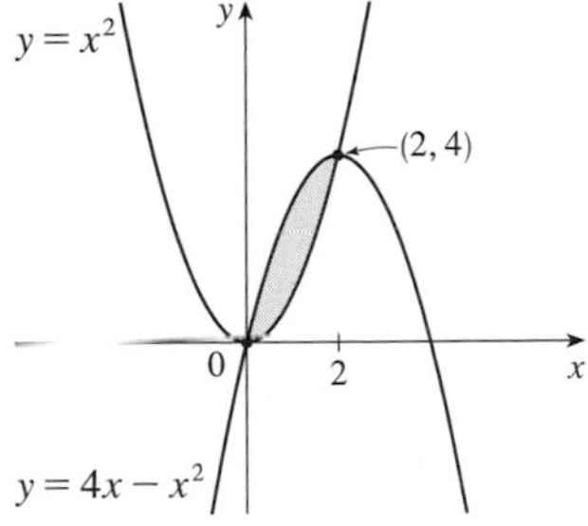

15. The curves intersect when $8\cos x = \sec^2 x \;\Rightarrow\; 8\cos^3 x = 1 \;\Rightarrow\; \cos^3 x = \frac{1}{8} \;\Rightarrow\; \cos x = \frac{1}{2} \;\Rightarrow$
$x = \frac{\pi}{3}$ for $0 < x < \frac{\pi}{2}$. By symmetry,

$$A = 2\int_0^{\pi/3} (8\cos x - \sec^2 x)\, dx$$

$$= 2\left[8\sin x - \tan x\right]_0^{\pi/3}$$

$$= 2\left(8 \cdot \tfrac{\sqrt{3}}{2} - \sqrt{3}\right) = 2\left(3\sqrt{3}\right)$$

$$= 6\sqrt{3}$$

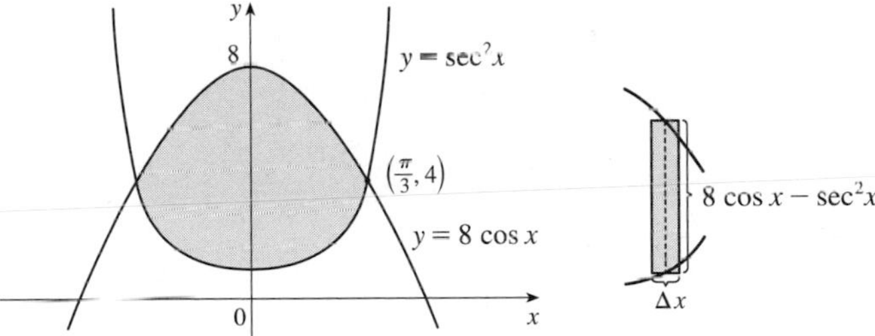

16. $$A = \int_0^{2\pi} \left[(2 - \cos x) - \cos x\right] dx$$

$$= \int_0^{2\pi} (2 - 2\cos x)\, dx$$

$$= \Big[2x - 2\sin x\Big]_0^{2\pi}$$

$$= (4\pi - 0) - 0 = 4\pi$$

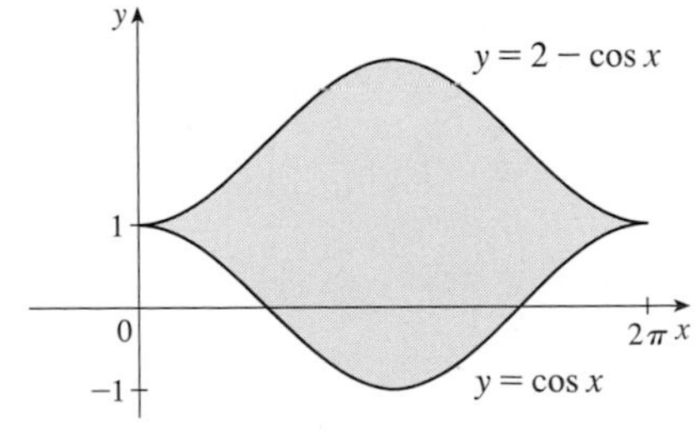

17. $2y^2 = 4 + y^2 \;\Leftrightarrow\; y^2 = 4 \;\Leftrightarrow\; y = \pm 2$, so

$$A = \int_{-2}^{2} \left[(4 + y^2) - 2y^2\right] dy$$

$$= 2\int_0^2 (4 - y^2)\, dy \qquad \text{[by symmetry]}$$

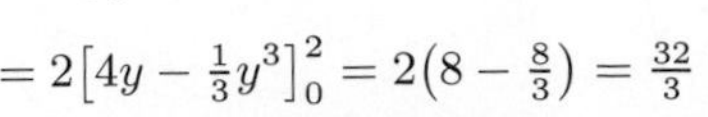

$$= 2\left[4y - \tfrac{1}{3}y^3\right]_0^2 = 2\left(8 - \tfrac{8}{3}\right) = \tfrac{32}{3}$$

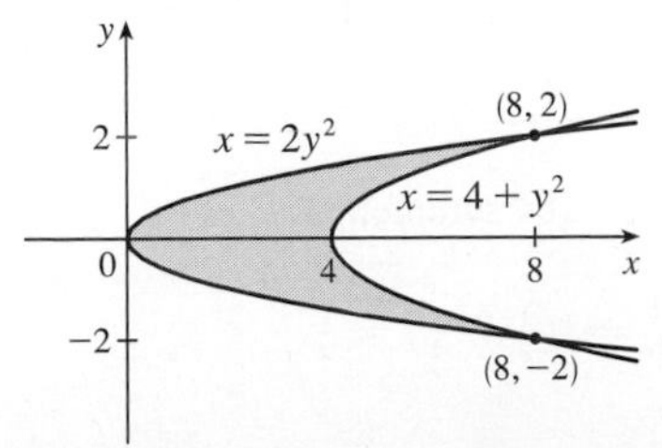

18. The curves intersect when $\sqrt{x-1} = x - 1 \;\Rightarrow$

$x - 1 = x^2 - 2x + 1 \;\Leftrightarrow\; 0 = x^2 - 3x + 2 \;\Leftrightarrow$

$0 = (x-1)(x-2) \;\Leftrightarrow\; x = 1$ or 2.

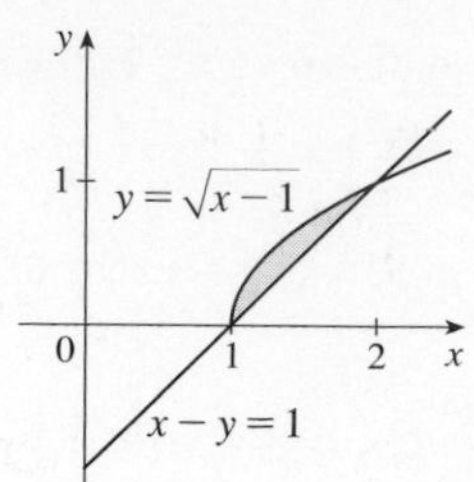

$$A = \int_1^2 \left[\sqrt{x-1} - (x-1)\right] dx$$

$$= \left[\tfrac{2}{3}(x-1)^{3/2} - \tfrac{1}{2}(x-1)^2\right]_1^2 = \left(\tfrac{2}{3} - \tfrac{1}{2}\right) - (0 - 0) = \tfrac{1}{6}$$

19. By inspection, the curves intersect at $x = \pm\frac{1}{2}$.

$$A = \int_{-1/2}^{1/2} [\cos\pi x - (4x^2 - 1)]\, dx$$

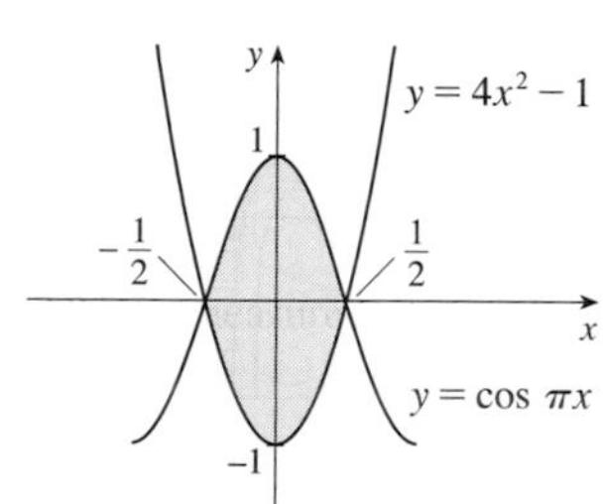

$$= 2\int_0^{1/2} (\cos\pi x - 4x^2 + 1)\, dx \quad \text{[by symmetry]}$$

$$= 2\left[\tfrac{1}{\pi}\sin\pi x - \tfrac{4}{3}x^3 + x\right]_0^{1/2} = 2\left[\left(\tfrac{1}{\pi} - \tfrac{1}{6} + \tfrac{1}{2}\right) - 0\right]$$

$$= 2\left(\tfrac{1}{\pi} + \tfrac{1}{3}\right) = \tfrac{2}{\pi} + \tfrac{2}{3}$$

20. $y = \sqrt{2-x} \;\Rightarrow\; y^2 = 2 - x \;\Leftrightarrow\; x = 2 - y^2$, so the curves intersect when $y^4 = 2 - y^2 \;\Leftrightarrow\; y^4 + y^2 - 2 = 0 \;\Leftrightarrow$ $(y^2+2)(y^2-1) = 0 \;\Leftrightarrow\; y = 1$ [since $y \geq 0$].

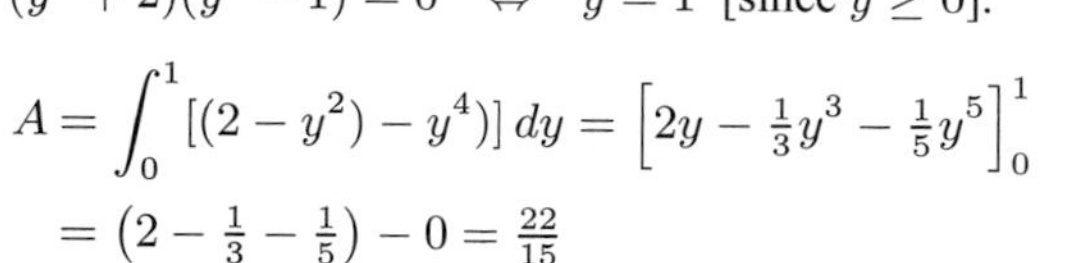

$$A = \int_0^1 [(2-y^2) - y^4]\, dy = \left[2y - \tfrac{1}{3}y^3 - \tfrac{1}{5}y^5\right]_0^1$$

$$= \left(2 - \tfrac{1}{3} - \tfrac{1}{5}\right) - 0 = \tfrac{22}{15}$$

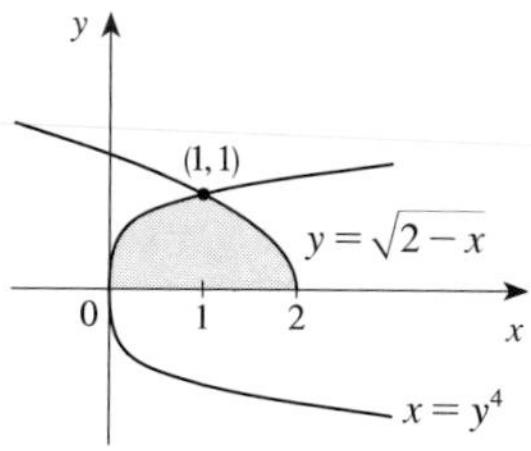

21. The curves intersect when $\tan x = 2\sin x$ (on $[-\pi/3, \pi/3]$) $\Leftrightarrow\; \sin x = 2\sin x\cos x \;\Leftrightarrow$

$2\sin x\cos x - \sin x = 0 \;\Leftrightarrow\; \sin x\,(2\cos x - 1) = 0 \;\Leftrightarrow\; \sin x = 0$ or $\cos x = \frac{1}{2} \;\Leftrightarrow\; x = 0$ or $x = \pm\frac{\pi}{3}$.

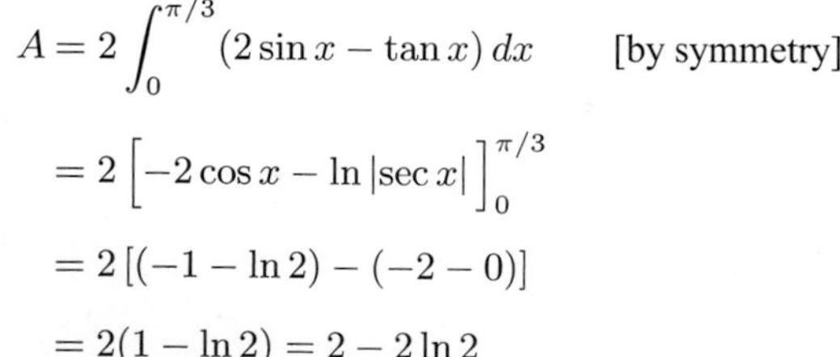

$$A = 2\int_0^{\pi/3} (2\sin x - \tan x)\, dx \qquad \text{[by symmetry]}$$

$$= 2\Big[-2\cos x - \ln|\sec x|\Big]_0^{\pi/3}$$

$$= 2\left[(-1 - \ln 2) - (-2 - 0)\right]$$

$$= 2(1 - \ln 2) = 2 - 2\ln 2$$

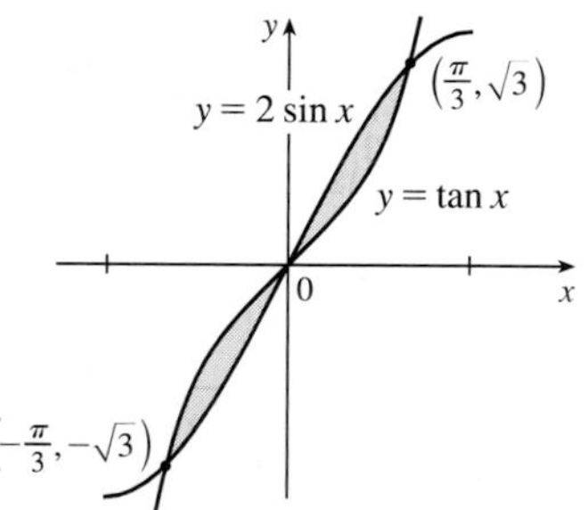

22. The curves intersect when $x^3 = x \;\Leftrightarrow\; x^3 - x = 0 \;\Leftrightarrow$

$x(x^2 - 1) = 0 \;\Leftrightarrow\; x(x+1)(x-1) = 0 \;\Leftrightarrow$

$x = 0$ or $x = \pm 1$.

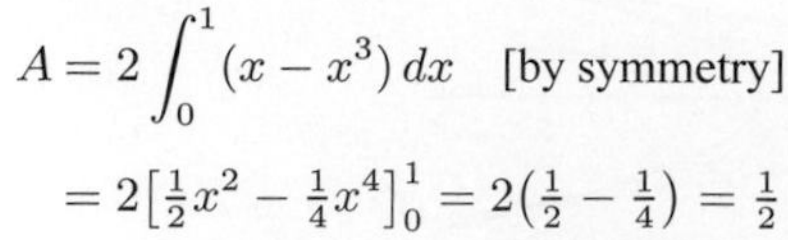

$$A = 2\int_0^1 (x - x^3)\, dx \quad \text{[by symmetry]}$$

$$= 2\left[\tfrac{1}{2}x^2 - \tfrac{1}{4}x^4\right]_0^1 = 2\left(\tfrac{1}{2} - \tfrac{1}{4}\right) = \tfrac{1}{2}$$

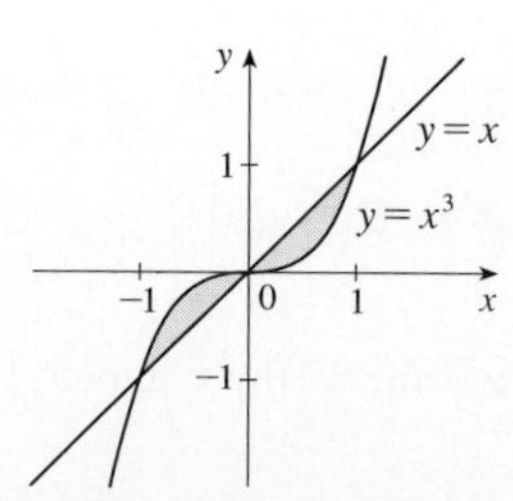

23. The curves intersect when $\sqrt[3]{2x} = \frac{1}{8}x^2 \quad \Leftrightarrow \quad 2x = \dfrac{1}{(2^3)^3}x^6 \quad \Leftrightarrow \quad 2^{10}x = x^6 \quad \Leftrightarrow \quad x^6 - 2^{10}x = 0 \quad \Leftrightarrow$

$x(x^5 - 2^{10}) = 0 \quad \Leftrightarrow \quad x = 0$ or $x^5 = 2^{10} \quad \Leftrightarrow \quad x = 0$ or $x = 2^2 = 4$, so for $0 \le x \le 6$,

$$\begin{aligned} A &= \int_0^4 \left(\sqrt[3]{2x} - \tfrac{1}{8}x^2\right) dx + \int_4^6 \left(\tfrac{1}{8}x^2 - \sqrt[3]{2x}\right) dx = \left[\tfrac{3}{4}\sqrt[3]{2}\,x^{4/3} - \tfrac{1}{24}x^3\right]_0^4 + \left[\tfrac{1}{24}x^3 - \tfrac{3}{4}\sqrt[3]{2}\,x^{4/3}\right]_4^6 \\ &= \left(\tfrac{3}{4}\sqrt[3]{2}\cdot 4\sqrt[3]{4} - \tfrac{64}{24}\right) - (0 - 0) + \left(\tfrac{216}{24} - \tfrac{3}{4}\sqrt[3]{2}\cdot 6\sqrt[3]{6}\right) - \left(\tfrac{64}{24} - \tfrac{3}{4}\sqrt[3]{2}\cdot 4\sqrt[3]{4}\right) \\ &= 6 - \tfrac{8}{3} + 9 - \tfrac{9}{2}\sqrt[3]{12} - \tfrac{8}{3} + 6 = \tfrac{47}{3} - \tfrac{9}{2}\sqrt[3]{12} \end{aligned}$$

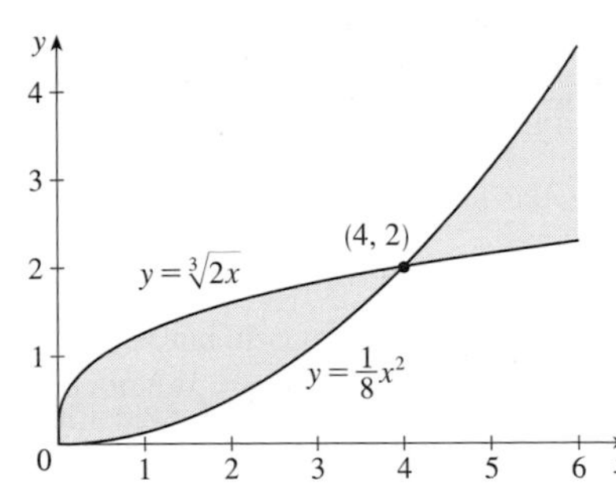

24. The curves intersect when $\cos x = 1 - \cos x$ (on $[0, \pi]$) $\quad \Leftrightarrow \quad 2\cos x = 1 \quad \Leftrightarrow \quad \cos x = \frac{1}{2} \quad \Leftrightarrow \quad x = \frac{\pi}{3}$.

$$\begin{aligned} A &= \int_0^{\pi/3} [\cos x - (1 - \cos x)]\,dx + \int_{\pi/3}^{\pi} [(1 - \cos x) - \cos x]\,dx \\ &= \int_0^{\pi/3} (2\cos x - 1)\,dx + \int_{\pi/3}^{\pi} (1 - 2\cos x)\,dx \\ &= \Big[2\sin x - x\Big]_0^{\pi/3} + \Big[x - 2\sin x\Big]_{\pi/3}^{\pi} \\ &= \left(\sqrt{3} - \frac{\pi}{3}\right) - 0 + (\pi - 0) - \left(\frac{\pi}{3} - \sqrt{3}\right) \\ &= 2\sqrt{3} + \frac{\pi}{3} \end{aligned}$$

25. By inspection, we see that the curves intersect at $x = \pm 1$ and that the area of the region enclosed by the curves is twice the area enclosed in the first quadrant.

$$\begin{aligned} A &= 2\int_0^1 [(2 - x) - x^4]\,dx = 2\left[2x - \tfrac{1}{2}x^2 - \tfrac{1}{5}x^5\right]_0^1 \\ &= 2\left[\left(2 - \tfrac{1}{2} - \tfrac{1}{5}\right) - 0\right] = 2\left(\tfrac{13}{10}\right) = \tfrac{13}{5} \end{aligned}$$

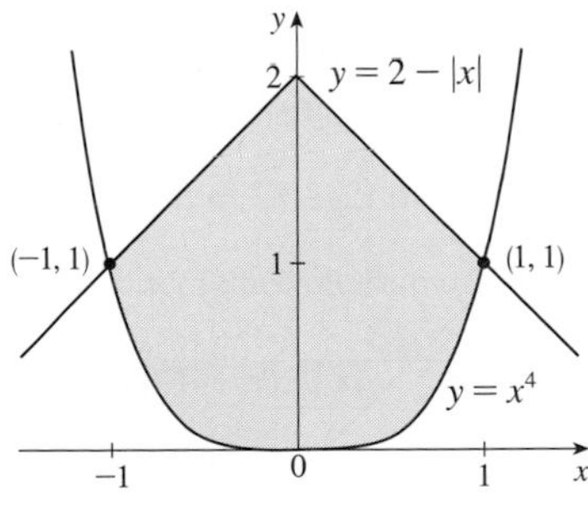

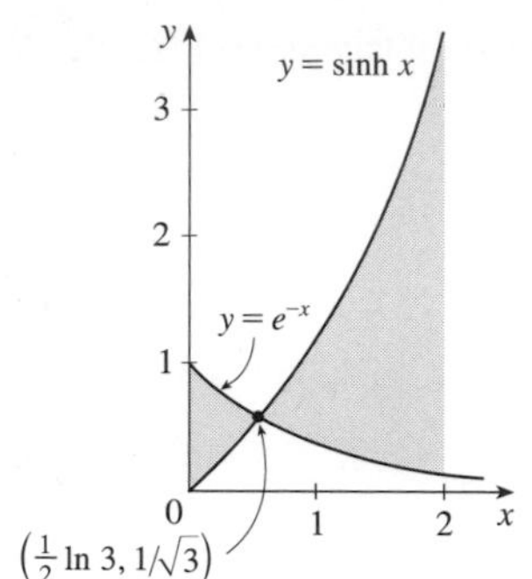

26. $\sinh x = e^{-x} \;\Leftrightarrow\; \frac{1}{2}(e^x - e^{-x}) = e^{-x} \;\Leftrightarrow\; \frac{1}{2}e^x = \frac{3}{2}e^{-x} \;\Leftrightarrow$

$e^{2x} = 3 \;\Leftrightarrow\; 2x = \ln 3 \;\Leftrightarrow\; x = \frac{1}{2}\ln 3$ (or $\ln\sqrt{3}$).

$$A = \int_0^{\ln\sqrt{3}} (e^{-x} - \sinh x)\,dx + \int_{\ln\sqrt{3}}^2 (\sinh x - e^{-x})\,dx$$

$$= \Big[-e^{-x} - \cosh x\Big]_0^{\ln\sqrt{3}} + \Big[\cosh x + e^{-x}\Big]_{\ln\sqrt{3}}^2$$

$$= \left(-\frac{1}{\sqrt{3}} - \frac{2}{\sqrt{3}}\right) - (-1-1) + (\cosh 2 + e^{-2}) - \left(\frac{2}{\sqrt{3}} + \frac{1}{\sqrt{3}}\right)$$

$$= -2\sqrt{3} + 2 + \cosh 2 + e^{-2}, \text{ or } 2 - 2\sqrt{3} + \tfrac{1}{2}e^2 + \tfrac{3}{2}e^{-2}$$

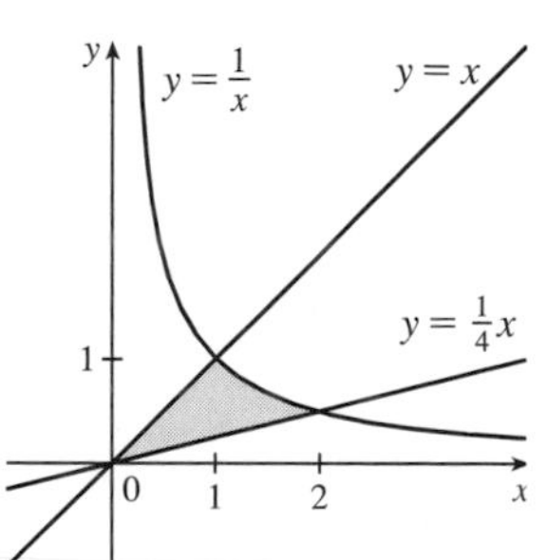

27. $1/x = x \;\Leftrightarrow\; 1 = x^2 \;\Leftrightarrow\; x = \pm 1$ and $1/x = \frac{1}{4}x \;\Leftrightarrow$

$4 = x^2 \;\Leftrightarrow\; x = \pm 2$, so for $x > 0$,

$$A = \int_0^1 \left(x - \frac{1}{4}x\right)dx + \int_1^2 \left(\frac{1}{x} - \frac{1}{4}x\right)dx$$

$$= \int_0^1 \left(\frac{3}{4}x\right)dx + \int_1^2 \left(\frac{1}{x} - \frac{1}{4}x\right)dx$$

$$= \left[\tfrac{3}{8}x^2\right]_0^1 + \left[\ln|x| - \tfrac{1}{8}x^2\right]_1^2$$

$$= \tfrac{3}{8} + \left(\ln 2 - \tfrac{1}{2}\right) - \left(0 - \tfrac{1}{8}\right) = \ln 2$$

28. $\frac{1}{4}x^2 = -x + 3 \;\Leftrightarrow\; x^2 + 4x - 12 = 0 \;\Leftrightarrow\; (x+6)(x-2) = 0 \;\Leftrightarrow\; x = -6$ or 2 and $2x^2 = -x + 3 \;\Leftrightarrow$

$2x^2 + x - 3 = 0 \;\Leftrightarrow\; (2x+3)(x-1) = 0 \;\Leftrightarrow\; x = -\frac{3}{2}$ or 1, so for $x \geq 0$,

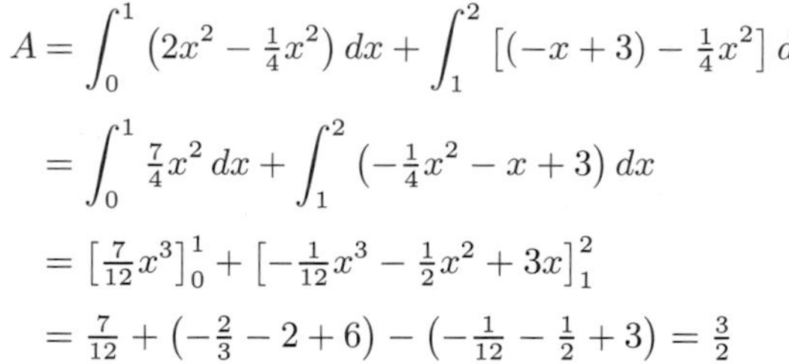

$$A = \int_0^1 \left(2x^2 - \tfrac{1}{4}x^2\right)dx + \int_1^2 \left[(-x+3) - \tfrac{1}{4}x^2\right]dx$$

$$= \int_0^1 \tfrac{7}{4}x^2\,dx + \int_1^2 \left(-\tfrac{1}{4}x^2 - x + 3\right)dx$$

$$= \left[\tfrac{7}{12}x^3\right]_0^1 + \left[-\tfrac{1}{12}x^3 - \tfrac{1}{2}x^2 + 3x\right]_1^2$$

$$= \tfrac{7}{12} + \left(-\tfrac{2}{3} - 2 + 6\right) - \left(-\tfrac{1}{12} - \tfrac{1}{2} + 3\right) = \tfrac{3}{2}$$

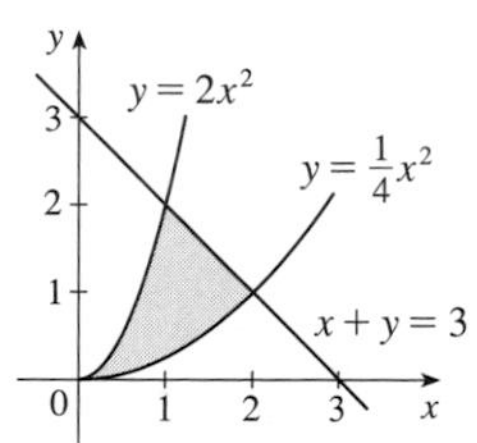

29. (a) Total area $= 12 + 27 = 39$.

(b) $f(x) \leq g(x)$ for $0 \leq x \leq 2$ and $f(x) \geq g(x)$ for $2 \leq x \leq 5$, so

$$\int_0^5 [f(x) - g(x)]\,dx = \int_0^2 [f(x) - g(x)]\,dx + \int_2^5 [f(x) - g(x)]\,dx = -\int_0^2 [g(x) - f(x)]\,dx + \int_2^5 [f(x) - g(x)]\,dx$$
$$= -(12) + 27 = 15$$

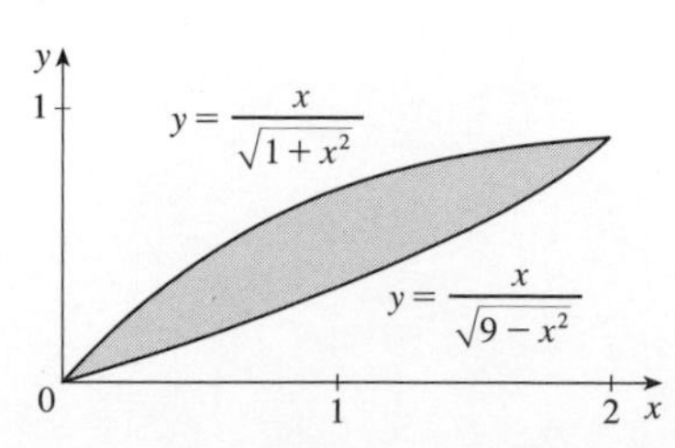

30. $\dfrac{x}{\sqrt{1+x^2}} = \dfrac{x}{\sqrt{9-x^2}} \;\Leftrightarrow\; x = 0$ or $\sqrt{1+x^2} = \sqrt{9-x^2} \;\Rightarrow$

$1 + x^2 = 9 - x^2 \;\Rightarrow\; 2x^2 = 8 \;\Rightarrow\; x^2 = 4 \;\Rightarrow\; x = 2 \;(x \geq 0)$.

$$A = \int_0^2 \left(\frac{x}{\sqrt{1+x^2}} - \frac{x}{\sqrt{9-x^2}}\right)dx = \left[\sqrt{1+x^2} + \sqrt{9-x^2}\right]_0^2$$

$$= (\sqrt{5} + \sqrt{5}) - (1+3) = 2\sqrt{5} - 4$$

31. $\dfrac{x}{1+x^2} = \dfrac{x^2}{1+x^3} \;\Leftrightarrow\; x + x^4 = x^2 + x^4 \;\Leftrightarrow\; x = x^2 \;\Leftrightarrow$

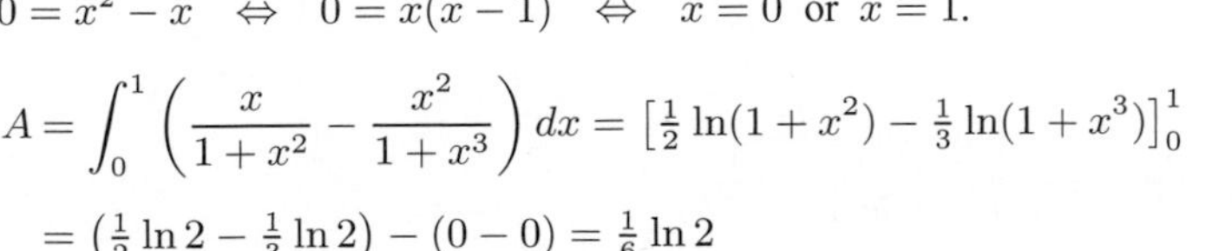

$0 = x^2 - x \;\Leftrightarrow\; 0 = x(x-1) \;\Leftrightarrow\; x = 0$ or $x = 1$.

$$A = \int_0^1 \left(\frac{x}{1+x^2} - \frac{x^2}{1+x^3} \right) dx = \left[\tfrac{1}{2}\ln(1+x^2) - \tfrac{1}{3}\ln(1+x^3) \right]_0^1$$

$$= \left(\tfrac{1}{2}\ln 2 - \tfrac{1}{3}\ln 2 \right) - (0 - 0) = \tfrac{1}{6}\ln 2$$

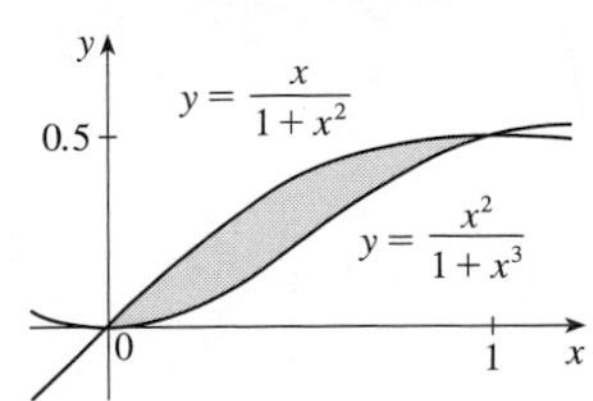

32. $\dfrac{\ln x}{x} = \dfrac{(\ln x)^2}{x} \;\Leftrightarrow\; \ln x = (\ln x)^2 \;\Leftrightarrow\; 0 = (\ln x)^2 - \ln x \;\Leftrightarrow$

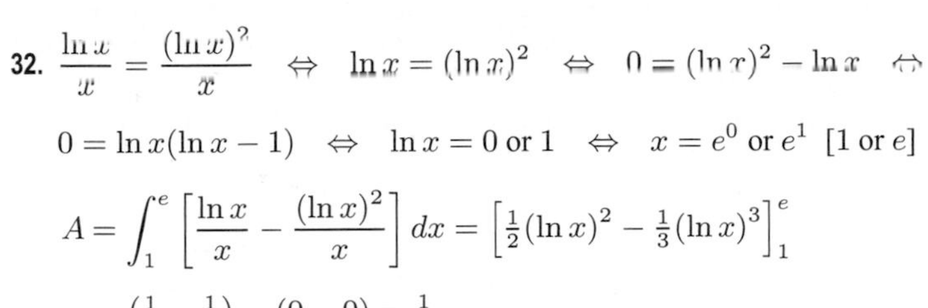

$0 = \ln x(\ln x - 1) \;\Leftrightarrow\; \ln x = 0$ or $1 \;\Leftrightarrow\; x = e^0$ or e^1 $[1$ or $e]$

$$A = \int_1^e \left[\frac{\ln x}{x} - \frac{(\ln x)^2}{x} \right] dx = \left[\tfrac{1}{2}(\ln x)^2 - \tfrac{1}{3}(\ln x)^3 \right]_1^e$$

$$= \left(\tfrac{1}{2} - \tfrac{1}{3} \right) - (0 - 0) = \tfrac{1}{6}$$

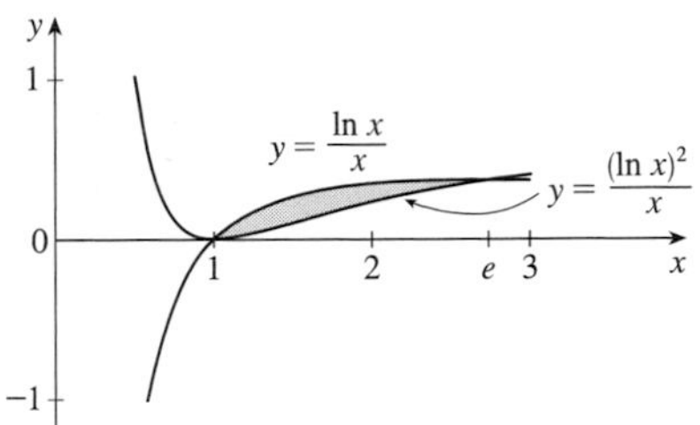

33. An equation of the line through $(0,0)$ and $(3,1)$ is $y = \frac{1}{3}x$; through $(0,0)$ and $(1,2)$ is $y = 2x$; through $(3,1)$ and $(1,2)$ is $y = -\frac{1}{2}x + \frac{5}{2}$.

$$A = \int_0^1 \left(2x - \tfrac{1}{3}x\right) dx + \int_1^3 \left[\left(-\tfrac{1}{2}x + \tfrac{5}{2}\right) - \tfrac{1}{3}x\right] dx$$

$$= \int_0^1 \tfrac{5}{3}x\,dx + \int_1^3 \left(-\tfrac{5}{6}x + \tfrac{5}{2}\right) dx = \left[\tfrac{5}{6}x^2\right]_0^1 + \left[-\tfrac{5}{12}x^2 + \tfrac{5}{2}x\right]_1^3$$

$$= \tfrac{5}{6} + \left(-\tfrac{15}{4} + \tfrac{15}{2}\right) - \left(-\tfrac{5}{12} + \tfrac{5}{2}\right) = \tfrac{5}{2}$$

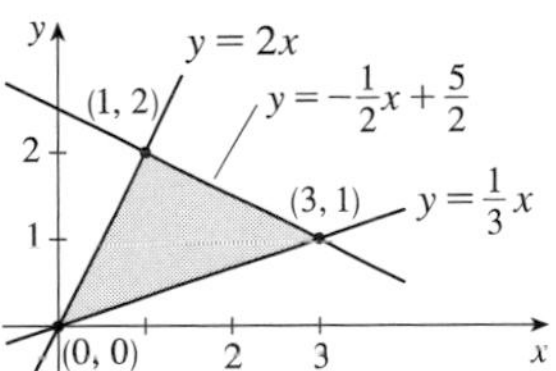

34. An equation of the line through $(2,0)$ and $(0,2)$ is $y = -x + 2$; through $(2,0)$ and $(-1,1)$ is $y = -\frac{1}{3}x + \frac{2}{3}$; through $(0,2)$ and $(-1,1)$ is $y = x + 2$.

$$A = \int_{-1}^0 \left[(x+2) - \left(-\tfrac{1}{3}x + \tfrac{2}{3}\right)\right] dx + \int_0^2 \left[(-x+2) - \left(-\tfrac{1}{3}x + \tfrac{2}{3}\right)\right] dx$$

$$= \int_{-1}^0 \left(\tfrac{4}{3}x + \tfrac{4}{3}\right) dx + \int_0^2 \left(-\tfrac{2}{3}x + \tfrac{4}{3}\right) dx$$

$$= \left[\tfrac{2}{3}x^2 + \tfrac{4}{3}x\right]_{-1}^0 + \left[-\tfrac{1}{3}x^2 + \tfrac{4}{3}x\right]_0^2$$

$$= 0 - \left(\tfrac{2}{3} - \tfrac{4}{3}\right) + \left(-\tfrac{4}{3} + \tfrac{8}{3}\right) - 0 = 2$$

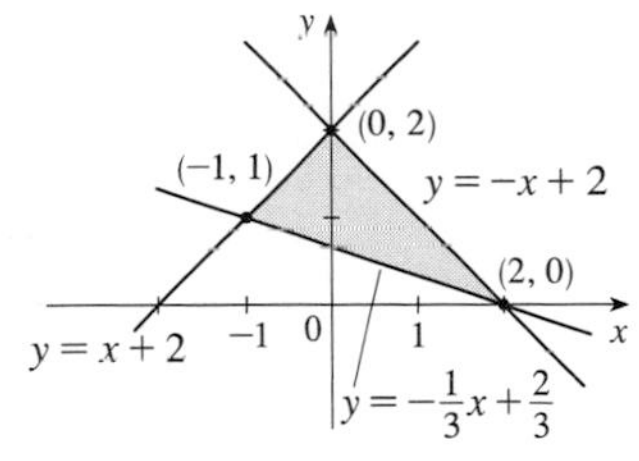

35. The curves intersect when $\sin x = \cos 2x$ (on $[0, \pi/2]$) $\Leftrightarrow$ $\sin x = 1 - 2\sin^2 x$ $\Leftrightarrow$ $2\sin^2 x + \sin x - 1 = 0$ $\Leftrightarrow$ $(2\sin x - 1)(\sin x + 1) = 0 \;\Rightarrow\; \sin x = \frac{1}{2} \;\Rightarrow\; x = \frac{\pi}{6}$.

$$A = \int_0^{\pi/2} |\sin x - \cos 2x|\,dx$$

$$= \int_0^{\pi/6} (\cos 2x - \sin x)\,dx + \int_{\pi/6}^{\pi/2} (\sin x - \cos 2x)\,dx$$

$$= \left[\tfrac{1}{2}\sin 2x + \cos x\right]_0^{\pi/6} + \left[-\cos x - \tfrac{1}{2}\sin 2x\right]_{\pi/6}^{\pi/2}$$

$$= \left(\tfrac{1}{4}\sqrt{3} + \tfrac{1}{2}\sqrt{3}\right) - (0 + 1) + (0 - 0) - \left(-\tfrac{1}{2}\sqrt{3} - \tfrac{1}{4}\sqrt{3}\right)$$

$$= \tfrac{3}{2}\sqrt{3} - 1$$

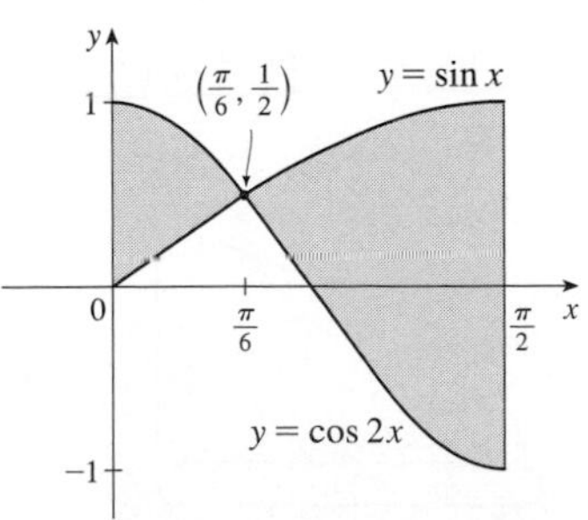

36. $A = \int_{-1}^{1} |3^x - 2^x|\,dx = \int_{-1}^{0} (2^x - 3^x)\,dx + \int_0^1 (3^x - 2^x)\,dx$

$$= \left[\frac{2^x}{\ln 2} - \frac{3^x}{\ln 3}\right]_{-1}^{0} + \left[\frac{3^x}{\ln 3} - \frac{2^x}{\ln 2}\right]_0^1$$

$$= \left(\frac{1}{\ln 2} - \frac{1}{\ln 3}\right) - \left(\frac{1}{2\ln 2} - \frac{1}{3\ln 3}\right) + \left(\frac{3}{\ln 3} - \frac{2}{\ln 2}\right) - \left(\frac{1}{\ln 3} - \frac{1}{\ln 2}\right)$$

$$= \frac{2-1-4+2}{2\ln 2} + \frac{-3+1+9-3}{3\ln 3} = \frac{4}{3\ln 3} - \frac{1}{2\ln 2}$$

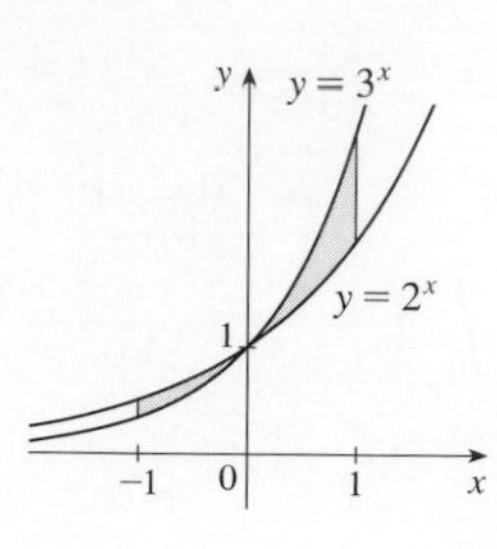

37.

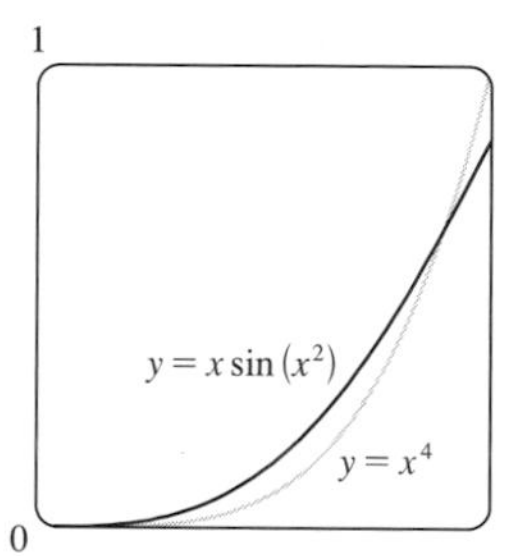

From the graph, we see that the curves intersect at $x = 0$ and $x = a \approx 0.896$, with $x\sin(x^2) > x^4$ on $(0, a)$. So the area A of the region bounded by the curves is

$$A = \int_0^a \left[x\sin(x^2) - x^4\right] dx = \left[-\tfrac{1}{2}\cos(x^2) - \tfrac{1}{5}x^5\right]_0^a$$

$$= -\tfrac{1}{2}\cos(a^2) - \tfrac{1}{5}a^5 + \tfrac{1}{2} \approx 0.037$$

38. From the graph, we see that the curves intersect (with $x \geq 0$) at $x = 0$ and $x = a$, where $a \approx 1.052$, with $x/(x^2+1)^2 > x^5 - x$ on $(0, a)$. The area A of the region bounded by the curves is

$$A = \int_0^a \left[\frac{x}{(x^2+1)^2} - (x^5 - x)\right] dx = \left[-\frac{1}{2}\cdot\frac{1}{x^2+1} - \frac{1}{6}x^6 + \frac{1}{2}x^2\right]_0^a$$

≈ 0.59

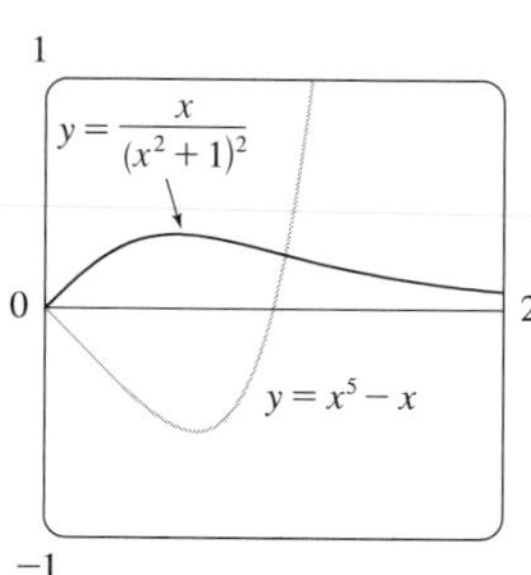

39.

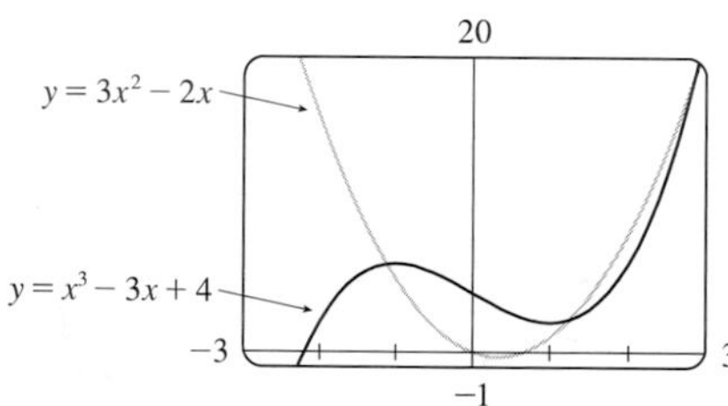

From the graph, we see that the curves intersect at $x = a \approx -1.11$, $x = b \approx 1.25$, and $x = c \approx 2.86$, with $x^3 - 3x + 4 > 3x^2 - 2x$ on (a, b) and $3x^2 - 2x > x^3 - 3x + 4$ on (b, c). So the area of the region bounded by the curves is

$$A = \int_a^b \left[(x^3 - 3x + 4) - (3x^2 - 2x)\right] dx + \int_b^c \left[(3x^2 - 2x) - (x^3 - 3x + 4)\right] dx$$

$$= \int_a^b (x^3 - 3x^2 - x + 4)\,dx + \int_b^c (-x^3 + 3x^2 + x - 4)\,dx$$

$$= \left[\tfrac{1}{4}x^4 - x^3 - \tfrac{1}{2}x^2 + 4x\right]_a^b + \left[-\tfrac{1}{4}x^4 + x^3 + \tfrac{1}{2}x^2 - 4x\right]_b^c \approx 8.38$$

40.

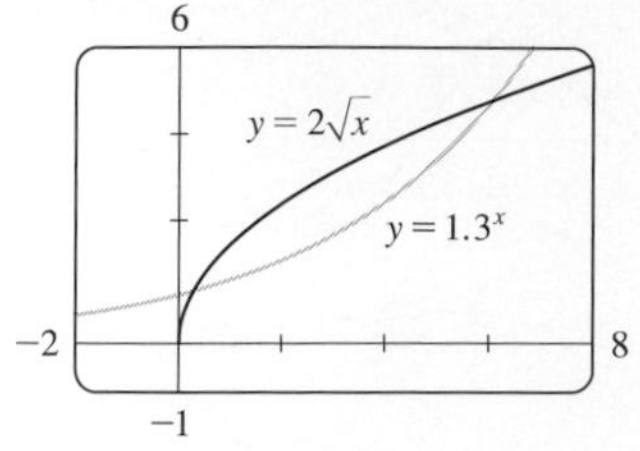

From the graph, we see that the curves intersect at $x = a \approx 0.29$ and $x = b \approx 6.08$. $y = 2\sqrt{x}$ is the upper curve, so the area of the region bounded by the curves is

$$A \approx \int_a^b \left(2\sqrt{x} - 1.3^x\right) dx = \left[\tfrac{4}{3}x^{3/2} - \frac{1}{\ln 1.3}1.3^x\right]_a^b \approx 5.11$$

41. 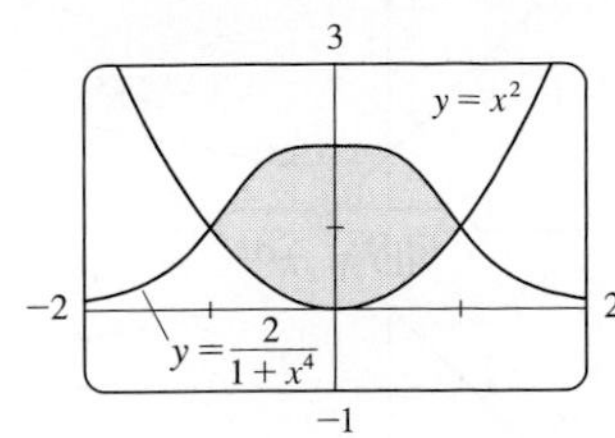

Graph `Y1=2/(1+x^4)` and `Y2=x^2`. We see that $Y_1 > Y_2$ on $(-1, 1)$, so the area is given by $\int_{-1}^{1} \left(\frac{2}{1+x^4} - x^2 \right) dx$. Evaluate the integral with a command such as `fnInt(Y1-Y2,x,-1,1)` to get 2.80123 to five decimal places.

Another method: Graph $f(x) =$ `Y1=2/(1+x^4)-x^2` and from the graph evaluate $\int f(x)\,dx$ from -1 to 1.

42. 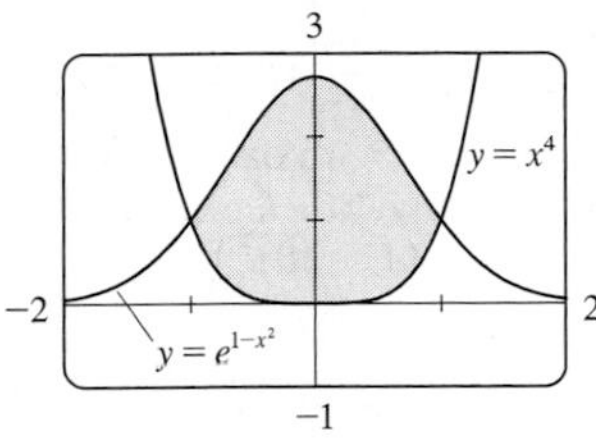

The curves intersect at $x = \pm 1$.

$$A = \int_{-1}^{1} (e^{1-x^2} - x^4)\, dx \approx 3.66016$$

43. 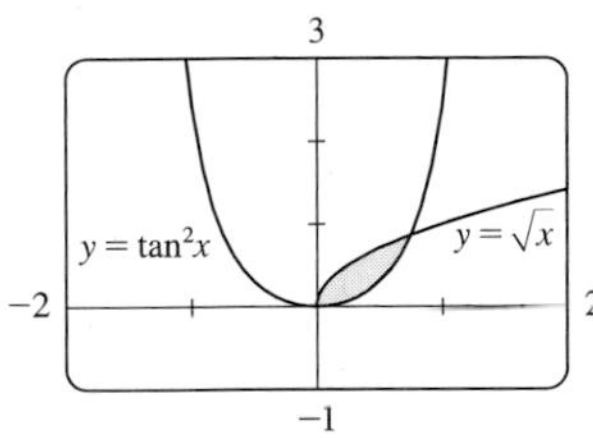

The curves intersect at $x = 0$ and $x = a \approx 0.749363$.

$$A = \int_0^a \left(\sqrt{x} - \tan^2 x \right) dx \approx 0.25142$$

44. 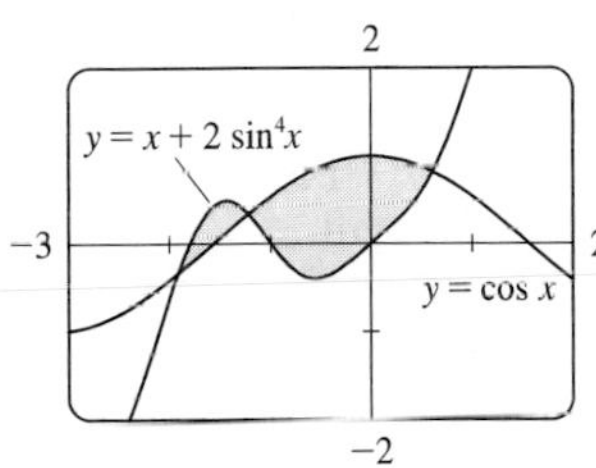

The curves intersect at $x = a \approx -1.911917$, $x = b \approx -1.223676$, and $x = c \approx 0.607946$.

$$A = \int_a^b \left[(x + 2\sin^4 x) - \cos x \right] dx + \int_b^c \left[\cos x - (x + 2\sin^4 x) \right] dx$$
$$\approx 1.70413$$

45. As the figure illustrates, the curves $y = x$ and $y = x^5 - 6x^3 + 4x$ enclose a four-part region symmetric about the origin (since $x^5 - 6x^3 + 4x$ and x are odd functions of x). The curves intersect at values of x where $x^5 - 6x^3 + 4x = x$; that is, where $x(x^4 - 6x^2 + 3) = 0$. That happens at $x = 0$ and where

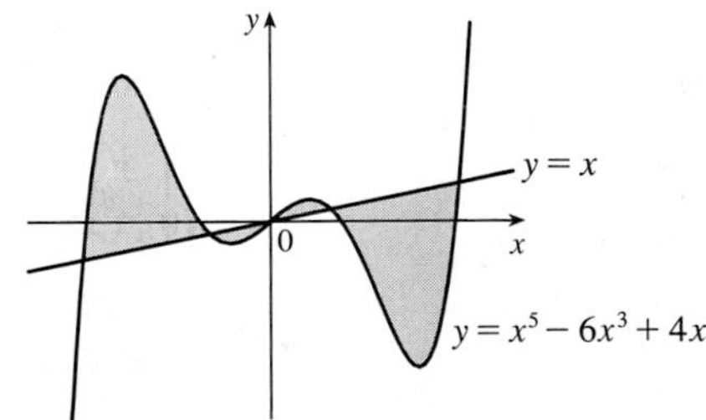

$x^2 = \dfrac{6 \pm \sqrt{36 - 12}}{2} = 3 \pm \sqrt{6}$; that is, at $x = -\sqrt{3 + \sqrt{6}}, -\sqrt{3 - \sqrt{6}}, 0, \sqrt{3 - \sqrt{6}}$, and $\sqrt{3 + \sqrt{6}}$. The exact area is

$$2\int_0^{\sqrt{3+\sqrt{6}}} \left| (x^5 - 6x^3 + 4x) - x \right| dx = 2\int_0^{\sqrt{3+\sqrt{6}}} \left| x^5 - 6x^3 + 3x \right| dx$$
$$= 2\int_0^{\sqrt{3-\sqrt{6}}} (x^5 - 6x^3 + 3x)\, dx + 2\int_{\sqrt{3-\sqrt{6}}}^{\sqrt{3+\sqrt{6}}} (-x^5 + 6x^3 - 3x)\, dx$$
$$\overset{\text{CAS}}{=} 12\sqrt{6} - 9$$

46. The inequality $x \geq 2y^2$ describes the region that lies on, or to the right of, the parabola $x = 2y^2$. The inequality $x \leq 1 - |y|$ describes the region that lies on, or to the left of, the curve $x = 1 - |y| = \begin{cases} 1 - y & \text{if } y \geq 0 \\ 1 + y & \text{if } y < 0 \end{cases}$.

So the given region is the shaded region that lies between the curves.

The graphs of $x = 1 - y$ and $x = 2y^2$ intersect when $1 - y = 2y^2 \Leftrightarrow$ $2y^2 + y - 1 = 0 \Leftrightarrow (2y - 1)(y + 1) = 0 \Rightarrow y = \frac{1}{2}$ [for $y \geq 0$]. By symmetry,

$A = 2\int_0^{1/2} [(1 - y) - 2y^2]\, dy = 2\left[-\frac{2}{3}y^3 - \frac{1}{2}y^2 + y\right]_0^{1/2} = 2\left[\left(-\frac{1}{12} - \frac{1}{8} + \frac{1}{2}\right) - 0\right] = 2\left(\frac{7}{24}\right) = \frac{7}{12}.$

47. 1 second $= \frac{1}{3600}$ hour, so 10 s $= \frac{1}{360}$ h. With the given data, we can take $n = 5$ to use the Midpoint Rule.

$\Delta t = \frac{1/360 - 0}{5} = \frac{1}{1800}$, so

$$\begin{aligned} \text{distance}_{\text{Kelly}} - \text{distance}_{\text{Chris}} &= \int_0^{1/360} v_K\, dt - \int_0^{1/360} v_C\, dt = \int_0^{1/360} (v_K - v_C)\, dt \\ &\approx M_5 = \tfrac{1}{1800}\left[(v_K - v_C)(1) + (v_K - v_C)(3) + (v_K - v_C)(5) \right. \\ &\qquad \left. + (v_K - v_C)(7) + (v_K - v_C)(9)\right] \\ &= \tfrac{1}{1800}[(22 - 20) + (52 - 46) + (71 - 62) + (86 - 75) + (98 - 86)] \\ &= \tfrac{1}{1800}(2 + 6 + 9 + 11 + 12) = \tfrac{1}{1800}(40) = \tfrac{1}{45} \text{ mile, or } 117\tfrac{1}{3} \text{ feet} \end{aligned}$$

48. If x = distance from left end of pool and $w = w(x)$ = width at x, then the Midpoint Rule with $n = 4$ and

$\Delta x = \dfrac{b - a}{n} = \dfrac{8 \cdot 2 - 0}{4} = 4$ gives Area $= \int_0^{16} w\, dx \approx 4(6.2 + 6.8 + 5.0 + 4.8) = 4(22.8) = 91.2\text{ m}^2$.

49. Let $h(x)$ denote the height of the wing at x cm from the left end.

$$\begin{aligned} A \approx M_5 &= \frac{200 - 0}{5}[h(20) + h(60) + h(100) + h(140) + h(180)] \\ &= 40(20.3 + 29.0 + 27.3 + 20.5 + 8.7) = 40(105.8) = 4232\text{ cm}^2 \end{aligned}$$

50. For $0 \leq t \leq 10$, $b(t) > d(t)$, so the area between the curves is given by

$$\begin{aligned} \int_0^{10} [b(t) - d(t)]\, dt &= \int_0^{10} (2200e^{0.024t} - 1460e^{0.018t})\, dt = \left[\frac{2200}{0.024}e^{0.024t} - \frac{1460}{0.018}e^{0.018t}\right]_0^{10} \\ &= \left(\frac{275{,}000}{3}e^{0.24} - \frac{730{,}000}{9}e^{0.18}\right) - \left(\frac{275{,}000}{3} - \frac{730{,}000}{9}\right) \approx 8868 \text{ people} \end{aligned}$$

This area A represents the increase in population over a 10-year period.

51. (a) From Example 5(a), the infectiousness concentration is 1210 cells/mL. $g(t) = 1210 \Leftrightarrow 0.9f(t) = 1210 \Leftrightarrow$ $0.9(-t)(t - 21)(t + 1) = 1210$. Using a calculator to solve the last equation for $t > 0$ gives us two solutions with the lesser being $t = t_3 \approx 11.26$ days, or the 12th day.

(b) From Example 5(b), the slope of the line through P_1 and P_2 is -23. From part (a), $P_3 = (t_3, 1210)$. An equation of the line through P_3 that is parallel to $\overline{P_1P_2}$ is $N - 1210 = -23(t - t_3)$, or $N = -23t + 23t_3 + 1210$. Using a calculator, we find that this line intersects g at $t = t_4 \approx 17.18$, or the 18th day. So in the patient with some immunity, the infection lasts about 2 days less than in the patient without immunity.

(c) The level of infectiousness for this patient is the area between the graph of g and the line in part (b). This area is

$$\int_{t_3}^{t_4} \Big[g(t) - (-23t + 23t_3 + 1210) \Big] \, dt \approx \int_{11.26}^{17.18} (-0.9t^3 + 18t^2 + 41.9t - 1468.94) \, dt$$

$$= \Big[-0.225t^4 + 6t^3 + 20.95t^2 - 1468.94t \Big]_{11.26}^{17.18} \approx 706$$

52. From the figure, $g(t) > f(t)$ for $0 \le t \le 2$. The area between the curves is given by

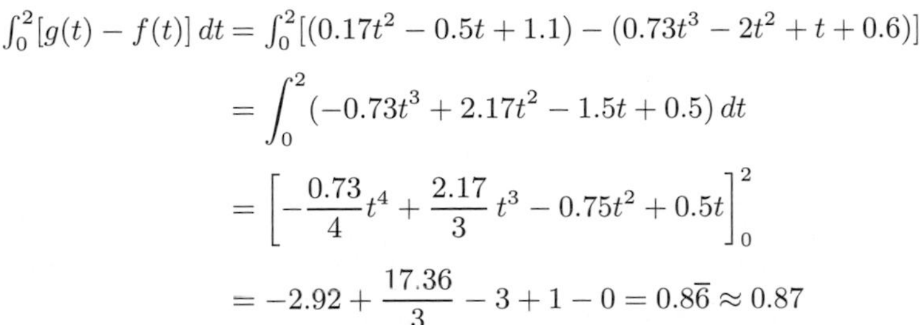

$$\int_0^2 [g(t) - f(t)] \, dt = \int_0^2 [(0.17t^2 - 0.5t + 1.1) - (0.73t^3 - 2t^2 + t + 0.6)] \, dt$$

$$= \int_0^2 (-0.73t^3 + 2.17t^2 - 1.5t + 0.5) \, dt$$

$$= \left[-\frac{0.73}{4}t^4 + \frac{2.17}{3}t^3 - 0.75t^2 + 0.5t \right]_0^2$$

$$= -2.92 + \frac{17.36}{3} - 3 + 1 - 0 = 0.8\overline{6} \approx 0.87$$

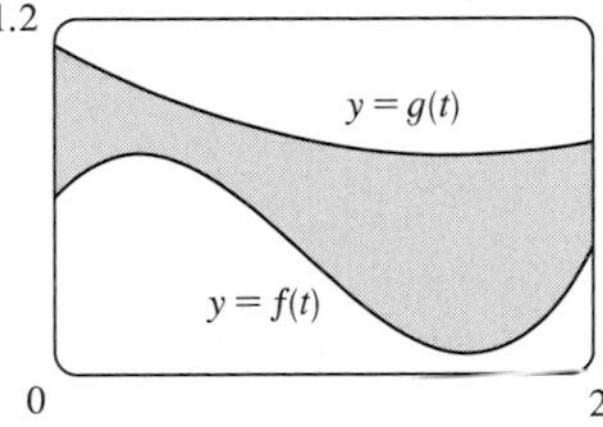

Thus, about 0.87 more inches of rain fell at the second location than at the first during the first two hours of the storm.

53. We know that the area under curve A between $t = 0$ and $t = x$ is $\int_0^x v_A(t) \, dt = s_A(x)$, where $v_A(t)$ is the velocity of car A and s_A is its displacement. Similarly, the area under curve B between $t = 0$ and $t = x$ is $\int_0^x v_B(t) \, dt = s_B(x)$.

(a) After one minute, the area under curve A is greater than the area under curve B. So car A is ahead after one minute.

(b) The area of the shaded region has numerical value $s_A(1) - s_B(1)$, which is the distance by which A is ahead of B after 1 minute.

(c) After two minutes, car B is traveling faster than car A and has gained some ground, but the area under curve A from $t = 0$ to $t = 2$ is still greater than the corresponding area for curve B, so car A is still ahead.

(d) From the graph, it appears that the area between curves A and B for $0 \le t \le 1$ (when car A is going faster), which corresponds to the distance by which car A is ahead, seems to be about 3 squares. Therefore, the cars will be side by side at the time x where the area between the curves for $1 \le t \le x$ (when car B is going faster) is the same as the area for $0 \le t \le 1$. From the graph, it appears that this time is $x \approx 2.2$. So the cars are side by side when $t \approx 2.2$ minutes.

54. The area under $R'(x)$ from $x = 50$ to $x = 100$ represents the change in revenue, and the area under $C'(x)$ from $x = 50$ to $x = 100$ represents the change in cost. The shaded region represents the difference between these two values; that is, the increase in profit as the production level increases from 50 units to 100 units. We use the Midpoint Rule with $n = 5$ and $\Delta x = 10$:

$$M_5 = \Delta x\{[R'(55) - C'(55)] + [R'(65) - C'(65)] + [R'(75) - C'(75)] + [R'(85) - C'(85)] + [R'(95) - C'(95)]\}$$

$$\approx 10(2.40 - 0.85 + 2.20 - 0.90 + 2.00 - 1.00 + 1.80 - 1.10 + 1.70 - 1.20)$$

$$= 10(5.05) = 50.5 \text{ thousand dollars}$$

Using M_1 would give us $50(2 - 1) = 50$ thousand dollars.

55. 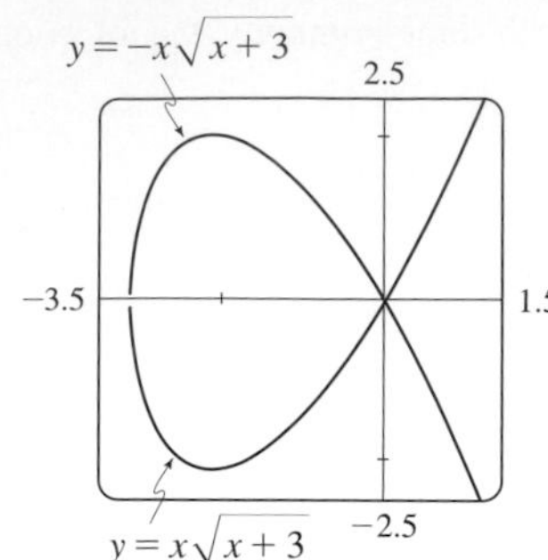

To graph this function, we must first express it as a combination of explicit functions of y; namely, $y = \pm x\sqrt{x+3}$. We can see from the graph that the loop extends from $x = -3$ to $x = 0$, and that by symmetry, the area we seek is just twice the area under the top half of the curve on this interval, the equation of the top half being $y = -x\sqrt{x+3}$. So the area is $A = 2\int_{-3}^{0}\left(-x\sqrt{x+3}\right)dx$. We substitute $u = x + 3$, so $du = dx$ and the limits change to 0 and 3, and we get

$$A = -2\int_0^3 \left[(u-3)\sqrt{u}\right]du = -2\int_0^3 (u^{3/2} - 3u^{1/2})\,du$$

$$= -2\left[\tfrac{2}{5}u^{5/2} - 2u^{3/2}\right]_0^3 = -2\left[\tfrac{2}{5}(3^2\sqrt{3}) - 2(3\sqrt{3})\right] = \tfrac{24}{5}\sqrt{3}$$

56. 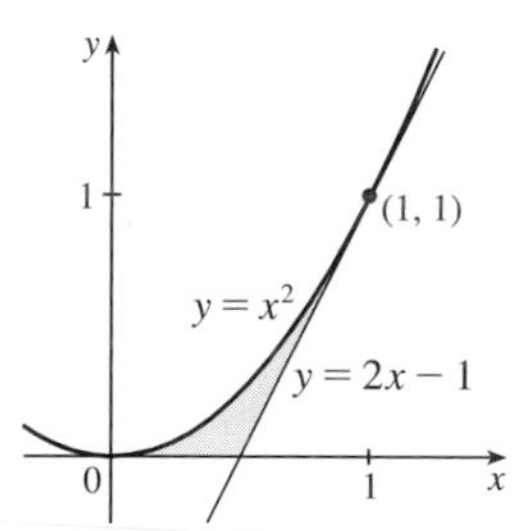

We start by finding the equation of the tangent line to $y = x^2$ at the point $(1, 1)$: $y' = 2x$, so the slope of the tangent is $2(1) = 2$, and its equation is $y - 1 = 2(x - 1)$, or $y = 2x - 1$. We would need two integrals to integrate with respect to x, but only one to integrate with respect to y.

$$A = \int_0^1 \left[\tfrac{1}{2}(y+1) - \sqrt{y}\right]dy = \left[\tfrac{1}{4}y^2 + \tfrac{1}{2}y - \tfrac{2}{3}y^{3/2}\right]_0^1$$

$$= \tfrac{1}{4} + \tfrac{1}{2} - \tfrac{2}{3} = \tfrac{1}{12}$$

57. 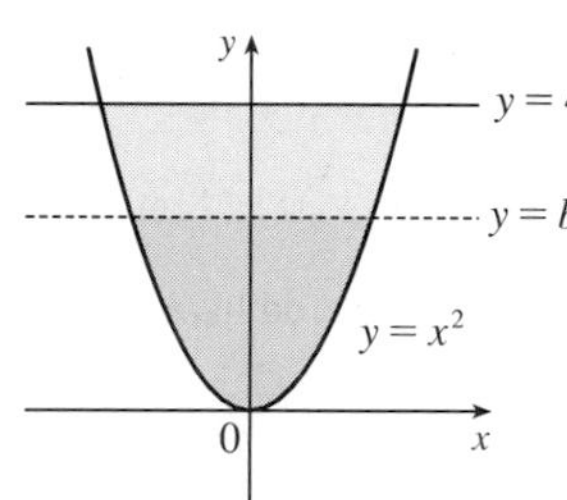

By the symmetry of the problem, we consider only the first quadrant, where $y = x^2 \;\Rightarrow\; x = \sqrt{y}$. We are looking for a number b such that

$$\int_0^b \sqrt{y}\,dy = \int_b^4 \sqrt{y}\,dy \;\Rightarrow\; \tfrac{2}{3}\left[y^{3/2}\right]_0^b = \tfrac{2}{3}\left[y^{3/2}\right]_b^4 \;\Rightarrow$$

$$b^{3/2} = 4^{3/2} - b^{3/2} \;\Rightarrow\; 2b^{3/2} = 8 \;\Rightarrow\; b^{3/2} = 4 \;\Rightarrow\; b = 4^{2/3} \approx 2.52.$$

58. (a) We want to choose a so that

$$\int_1^a \frac{1}{x^2}\,dx = \int_a^4 \frac{1}{x^2}\,dx \;\Rightarrow\; \left[\frac{-1}{x}\right]_1^a = \left[\frac{-1}{x}\right]_a^4 \;\Rightarrow\; -\frac{1}{a} + 1 = -\frac{1}{4} + \frac{1}{a} \;\Rightarrow\; \frac{5}{4} = \frac{2}{a} \;\Rightarrow\; a = \frac{8}{5}.$$

(b) The area under the curve $y = 1/x^2$ from $x = 1$ to $x = 4$ is $\frac{3}{4}$ [take $a = 4$ in the first integral in part (a)]. Now the line $y = b$ must intersect the curve $x = 1/\sqrt{y}$ and not the line $x = 4$, since the area under the line $y = 1/4^2$ from $x = 1$ to $x = 4$ is only $\frac{3}{16}$, which is less than half of $\frac{3}{4}$. We want to choose b so that the upper area in the diagram is half of the total area under the curve $y = 1/x^2$ from $x = 1$ to $x = 4$. This implies that

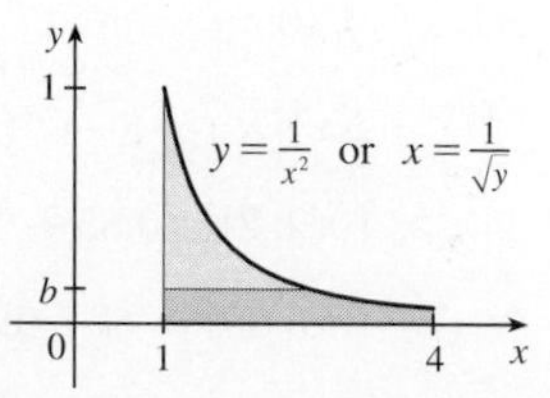

$\int_b^1 (1/\sqrt{y} - 1)\,dy = \frac{1}{2}\cdot\frac{3}{4} \;\Rightarrow\; \left[2\sqrt{y} - y\right]_b^1 = \frac{3}{8} \;\Rightarrow\; 1 - 2\sqrt{b} + b = \frac{3}{8} \;\Rightarrow$
$b - 2\sqrt{b} + \frac{5}{8} = 0$. Letting $c = \sqrt{b}$, we get $c^2 - 2c + \frac{5}{8} = 0 \;\Rightarrow$
$8c^2 - 16c + 5 = 0$. Thus, $c = \frac{16 \pm \sqrt{256 - 160}}{16} = 1 \pm \frac{\sqrt{6}}{4}$. But $c = \sqrt{b} < 1 \;\Rightarrow$
$c = 1 - \frac{\sqrt{6}}{4} \;\Rightarrow\; b = c^2 = 1 + \frac{3}{8} - \frac{\sqrt{6}}{2} = \frac{1}{8}(11 - 4\sqrt{6}) \approx 0.1503.$

59. We first assume that $c > 0$, since c can be replaced by $-c$ in both equations without changing the graphs, and if $c = 0$ the curves do not enclose a region. We see from the graph that the enclosed area A lies between $x = -c$ and $x = c$, and by symmetry, it is equal to four times the area in the first quadrant. The enclosed area is

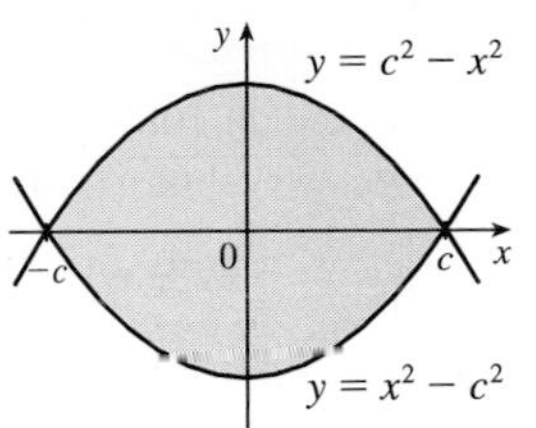

$A = 4\int_0^c (c^2 - x^2)\,dx = 4\left[c^2x - \frac{1}{3}x^3\right]_0^c = 4\left(c^3 - \frac{1}{3}c^3\right) = 4\left(\frac{2}{3}c^3\right) = \frac{8}{3}c^3$

So $A = 576 \;\Leftrightarrow\; \frac{8}{3}c^3 = 576 \;\Leftrightarrow\; c^3 = 216 \;\Leftrightarrow\; c = \sqrt[3]{216} = 6.$

Note that $c = -6$ is another solution, since the graphs are the same.

60.

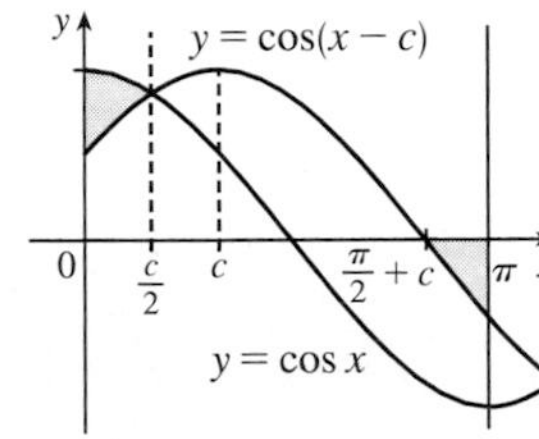

It appears from the diagram that the curves $y = \cos x$ and $y = \cos(x - c)$ intersect halfway between 0 and c, namely, when $x = c/2$. We can verify that this is indeed true by noting that $\cos(c/2 - c) = \cos(-c/2) = \cos(c/2)$. The point where $\cos(x - c)$ crosses the x-axis is $x = \frac{\pi}{2} + c$. So we require that $\int_0^{c/2} [\cos x - \cos(x - c)]\,dx = -\int_{\pi/2+c}^{\pi} \cos(x - c)\,dx$ [the negative sign on the RHS is needed since the second area is beneath the x-axis] $\Leftrightarrow$ $[\sin x - \sin(x - c)]_0^{c/2} = -[\sin(x - c)]_{\pi/2+c}^{\pi}$ $\Rightarrow$ $[\sin(c/2) - \sin(-c/2)] - [-\sin(-c)] = -\sin(\pi - c) + \sin\left[\left(\frac{\pi}{2} + c\right) - c\right]$ $\Leftrightarrow$ $2\sin(c/2) - \sin c = -\sin c + 1.$ [Here we have used the oddness of the sine function, and the fact that $\sin(\pi - c) = \sin c$]. So $2\sin(c/2) = 1$ $\Leftrightarrow$ $\sin(c/2) = \frac{1}{2}$ $\Leftrightarrow$ $c/2 = \frac{\pi}{6}$ $\Leftrightarrow$ $c = \frac{\pi}{3}.$

61. The curve and the line will determine a region when they intersect at two or more points. So we solve the equation $x/(x^2 + 1) = mx$ $\Rightarrow$ $x = x(mx^2 + m)$ $\Rightarrow$ $x(mx^2 + m) - x = 0$ $\Rightarrow$ $x(mx^2 + m - 1) = 0$ $\Rightarrow$ $x = 0$ or $mx^2 + m - 1 = 0$ $\Rightarrow$

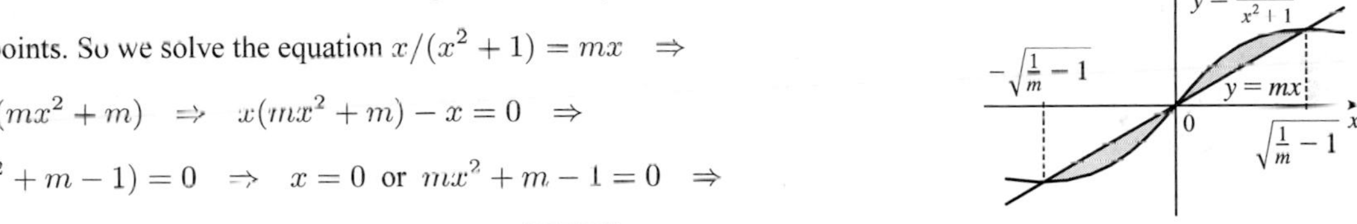

$x = 0$ or $x^2 = \dfrac{1 - m}{m}$ $\Rightarrow$ $x = 0$ or $x = \pm\sqrt{\dfrac{1}{m} - 1}$. Note that if $m = 1$, this has only the solution $x = 0$, and no region is determined. But if $1/m - 1 > 0$ $\Leftrightarrow$ $1/m > 1$ $\Leftrightarrow$ $0 < m < 1$, then there are two solutions. [Another way of seeing this is to observe that the slope of the tangent to $y = x/(x^2 + 1)$ at the origin is $y'(0) = 1$ and therefore we must have $0 < m < 1$.] Note that we cannot just integrate between the positive and negative roots, since the curve and the line cross at the origin. Since mx and $x/(x^2 + 1)$ are both odd functions, the total area is twice the area between the curves on the interval $\left[0, \sqrt{1/m - 1}\right]$. So the total area enclosed is

$$2\int_0^{\sqrt{1/m-1}} \left[\frac{x}{x^2 + 1} - mx\right] dx = 2\left[\tfrac{1}{2}\ln(x^2 + 1) - \tfrac{1}{2}mx^2\right]_0^{\sqrt{1/m-1}} = [\ln(1/m - 1 + 1) - m(1/m - 1)] - (\ln 1 - 0)$$

$$= \ln(1/m) - 1 + m = m - \ln m - 1$$

APPLIED PROJECT The Gini Index

1. (a) $G = \dfrac{\text{area between } L \text{ and } y = x}{\text{area under } y = x} = \dfrac{\int_0^1 [x - L(x)]\,dx}{\frac{1}{2}} = 2\displaystyle\int_0^1 [x - L(x)]\,dx$

(b) For a perfectly egalitarian society, $L(x) = x$, so $G = 2\int_0^1 [x - x]\,dx = 0$. For a perfectly totalitarian society,

$$L(x) = \begin{cases} 1 & \text{if } x = 1 \\ 0 & \text{if } 0 \le x < 1 \end{cases} \quad \text{so } G = 2\int_0^1 (x - 0)\,dx = 2\left[\tfrac{1}{2}x^2\right]_0^1 = 2\left(\tfrac{1}{2}\right) = 1.$$

2. (a) The richest 20% of the population in 2010 received $1 - L(0.8) = 1 - 0.498 = 0.502$, or 50.2%, of the total US income.

(b) A quadratic model has the form $Q(x) = ax^2 + bx + c$. Rounding to six decimal places, we get $a = 1.305\,357$, $b = -0.371\,357$, and $c = 0.026\,714$. The quadratic model appears to be a reasonable fit, but note that $Q(0) \neq 0$ and Q is both decreasing and increasing.

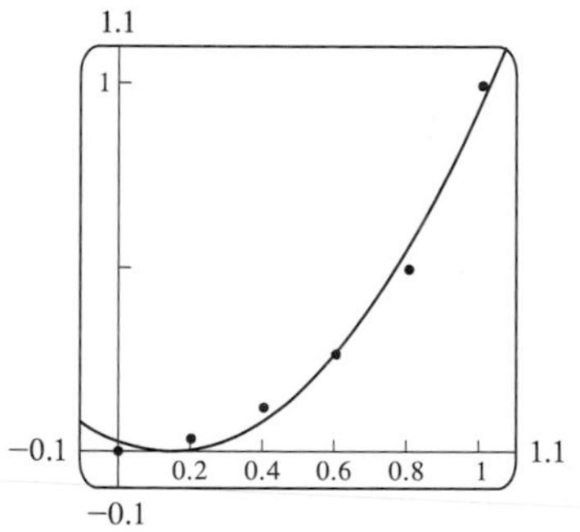

(c) $G = 2\displaystyle\int_0^1 [x - Q(x)]\,dx \approx 0.4477$

3.

Year	$Q(x) = ax^2 + bx + c$: a	b	c	Gini
1970	1.117 411	−0.152 411	0.013 321	0.3808
1980	1.149 554	−0.189 696	0.016 179	0.3910
1990	1.216 071	−0.268 214	0.020 714	0.4161
2000	1.280 804	−0.345 232	0.025 821	0.4397

The Gini index has risen steadily from 1970 to 2010. The trend is toward a less egalitarian society.

4. Using Maple's `PowerFit` or TI's `PwrReg` command and omitting the point $(0, 0)$ gives us $P(x) = 0.845\,446x^{2.050\,379}$ and a Gini index $2\int_0^1 [x - P(x)]\,dx \approx 0.4457$. Note that the power function is nearly quadratic.

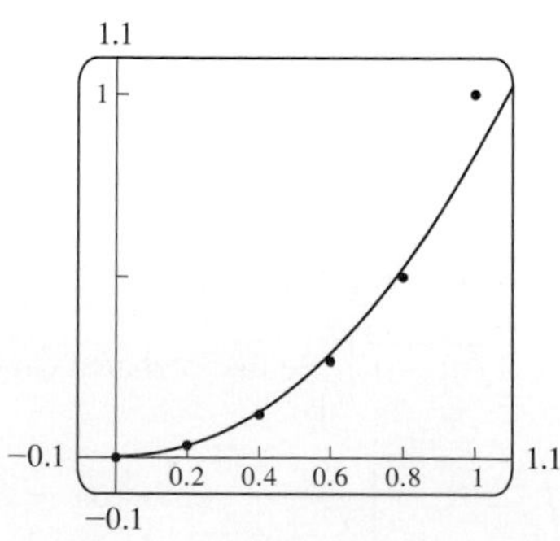

6.2 Volumes

1. A cross-section is a disk with radius $x+1$, so its area is

$A(x) = \pi(x+1)^2 = \pi(x^2+2x+1)$.

$V = \int_0^2 A(x)\,dx = \int_0^2 \pi(x^2+2x+1)\,dx$

$= \pi\left[\frac{1}{3}x^3 + x^2 + x\right]_0^2$

$= \pi\left(\frac{8}{3}+4+2\right) = \frac{26\pi}{3}$

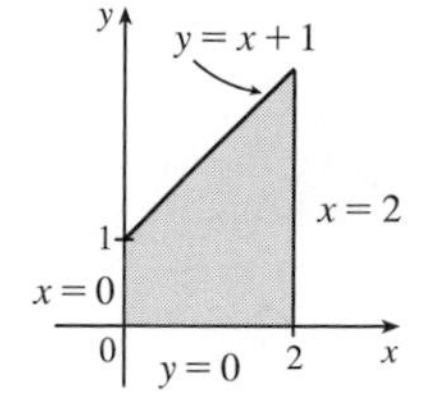

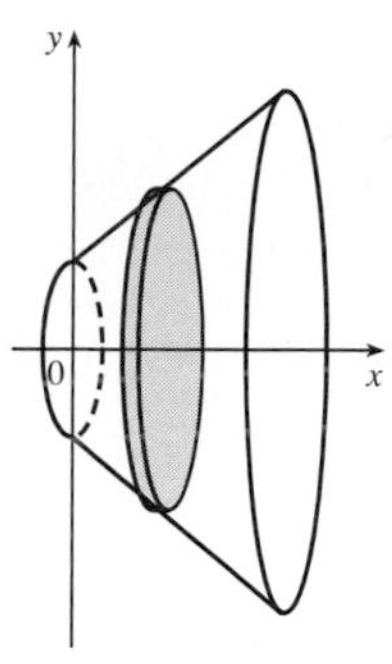

2. A cross-section is a disk with radius $\dfrac{1}{x}$, so

its area is $A(x) = \pi\left(\dfrac{1}{x}\right)^2 = \pi x^{-2}$.

$V = \int_1^4 A(x)\,dx = \int_1^4 \pi x^{-2}\,dx$

$= \pi\Big[-x^{-1}\Big]_1^4 = \pi\left(-\frac{1}{4}+1\right)$

$= \frac{3\pi}{4}$

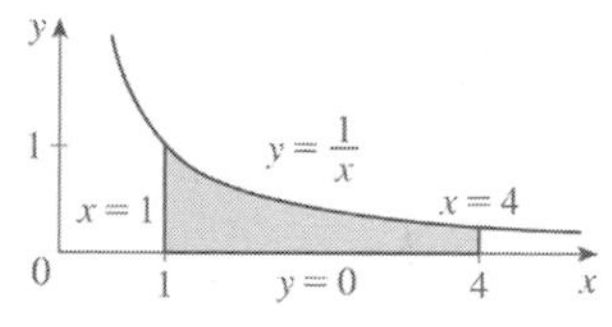

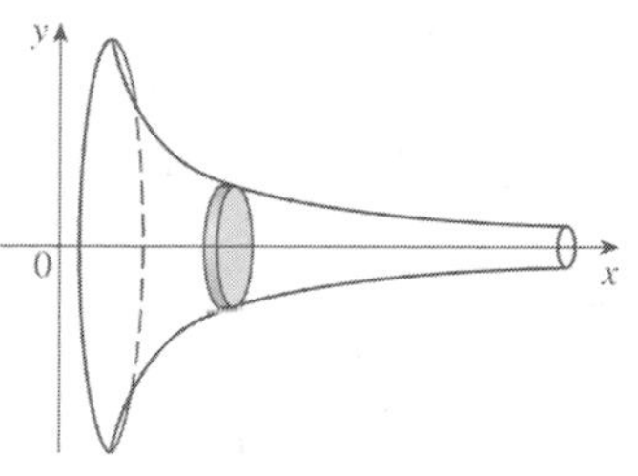

3. A cross-section is a disk with radius $\sqrt{x-1}$, so its area is $A(x) = \pi\left(\sqrt{x-1}\right)^2 = \pi(x-1)$.

$$V = \int_1^5 A(x)\,dx = \int_1^5 \pi(x-1)\,dx = \pi\left[\tfrac{1}{2}x^2 - x\right]_1^5 = \pi\left[\left(\tfrac{25}{2}-5\right) - \left(\tfrac{1}{2}-1\right)\right] = 8\pi$$

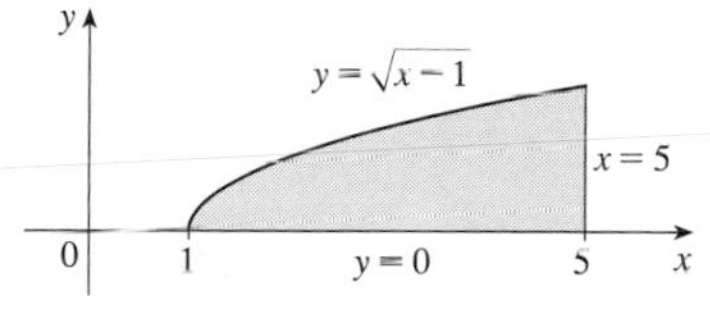

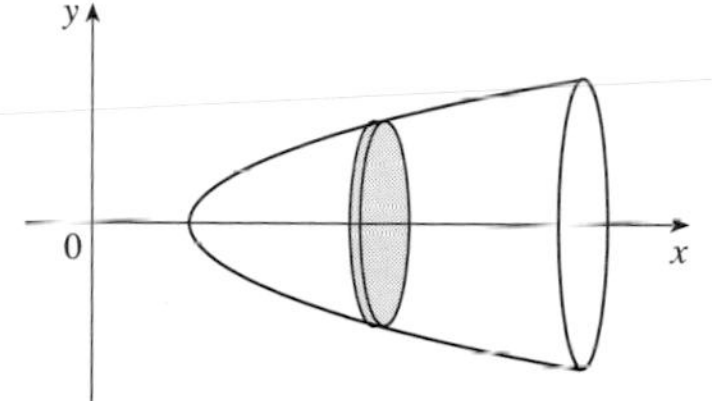

4. A cross-section is a disk with radius e^x, so

its area is $A(x) = \pi(e^x)^2 = \pi e^{2x}$.

$V = \displaystyle\int_{-1}^{1} A(x)\,dx = \int_{-1}^{1} \pi e^{2x}\,dx$

$= \pi\left[\frac{1}{2}e^{2x}\right]_{-1}^{1} = \frac{\pi}{2}(e^2 - e^{-2})$

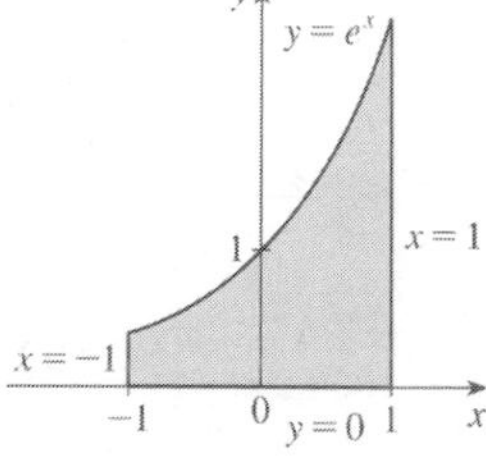

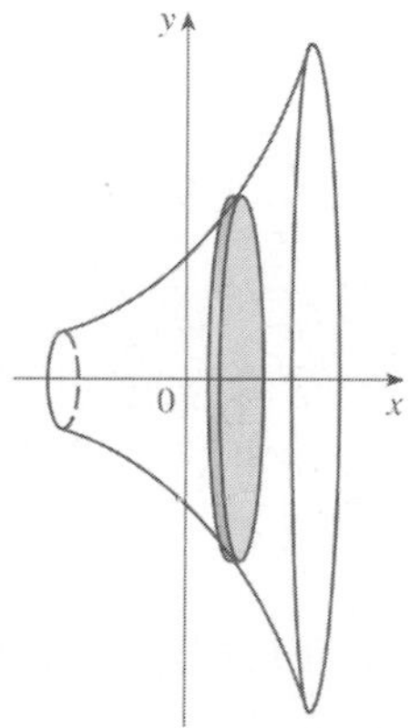

5. A cross-section is a disk with radius $2\sqrt{y}$, so its area is $A(y) = \pi\left(2\sqrt{y}\right)^2$.

$$V = \int_0^9 A(y)\,dy = \int_0^9 \pi\left(2\sqrt{y}\right)^2 dy = 4\pi \int_0^9 y\,dy$$

$$= 4\pi\left[\tfrac{1}{2}y^2\right]_0^9 = 2\pi(81) = 162\pi$$

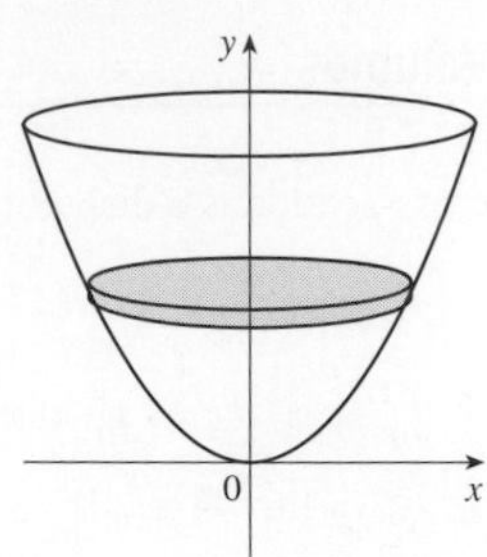

6. A cross-section is a disk with radius $\frac{1}{2}y^2$, so its area is $A(y) = \pi\left(\frac{1}{2}y^2\right)^2 = \frac{1}{4}\pi y^4$.

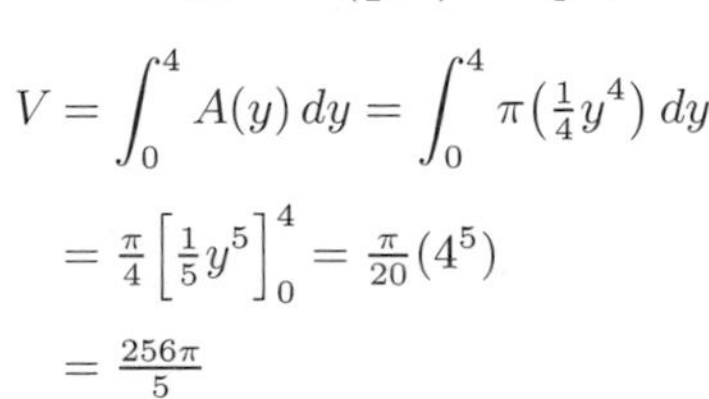

$$V = \int_0^4 A(y)\,dy = \int_0^4 \pi\left(\tfrac{1}{4}y^4\right) dy$$

$$= \tfrac{\pi}{4}\left[\tfrac{1}{5}y^5\right]_0^4 = \tfrac{\pi}{20}(4^5)$$

$$= \tfrac{256\pi}{5}$$

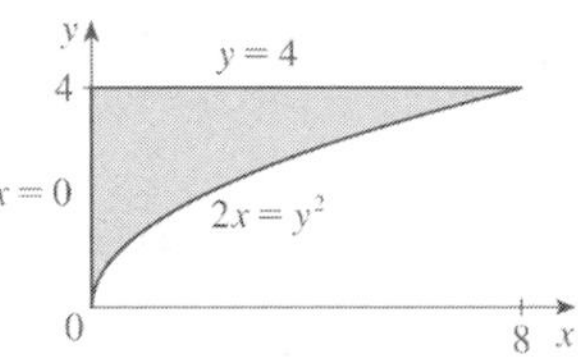

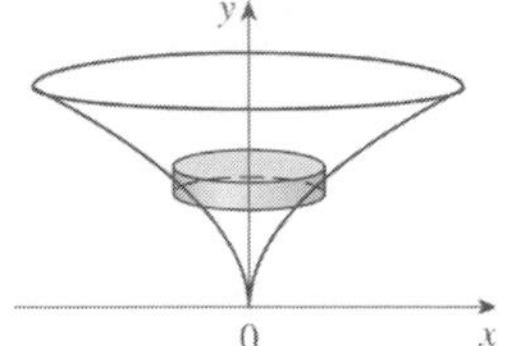

7. A cross-section is a washer (annulus) with inner radius x^3 and outer radius x, so its area is $A(x) = \pi(x)^2 - \pi(x^3)^2 = \pi(x^2 - x^6)$.

$$V = \int_0^1 A(x)\,dx = \int_0^1 \pi(x^2 - x^6)\,dx$$

$$= \pi\left[\tfrac{1}{3}x^3 - \tfrac{1}{7}x^7\right]_0^1 = \pi\left(\tfrac{1}{3} - \tfrac{1}{7}\right) = \tfrac{4}{21}\pi$$

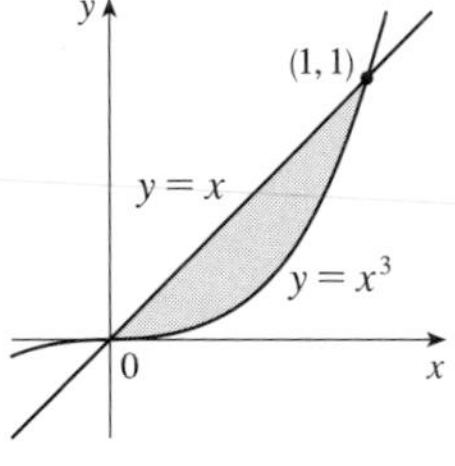

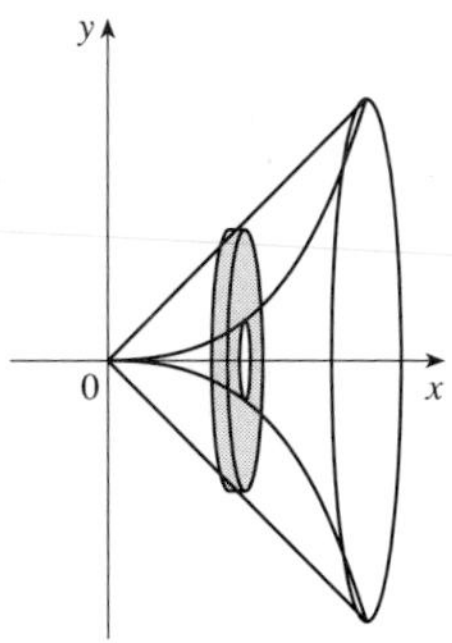

8. A cross-section is a washer (annulus) with inner radius 2 and outer radius $6 - x^2$, so its area is $A(x) = \pi[(6 - x^2)^2 - 2^2] = \pi(x^4 - 12x^2 + 32)$.

$$V = \int_{-2}^2 A(x)\,dx = 2\int_0^2 \pi(x^4 - 12x^2 + 32)\,dx$$

$$= 2\pi\left[\tfrac{1}{5}x^5 - 4x^3 + 32x\right]_0^2$$

$$= 2\pi\left(\tfrac{32}{5} - 32 + 64\right) = 2\pi\left(\tfrac{192}{5}\right) = \tfrac{384\pi}{5}$$

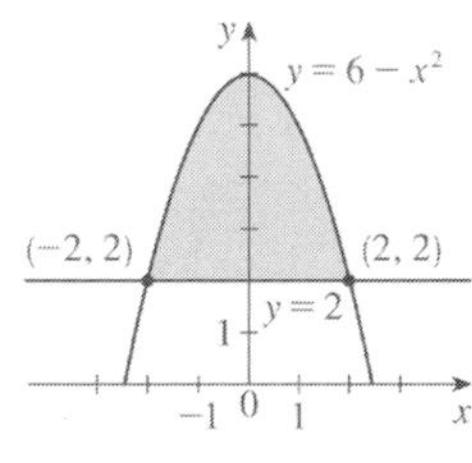

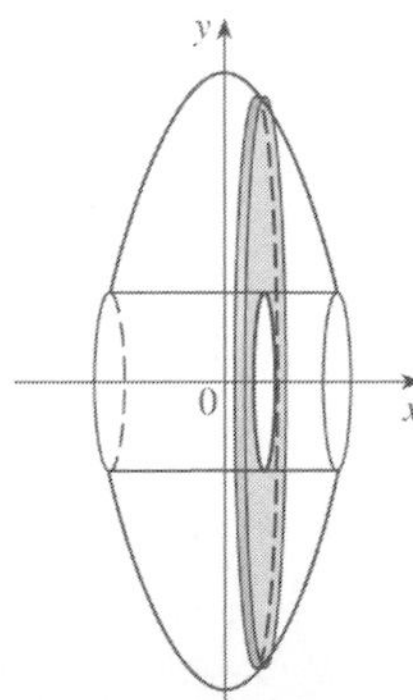

9. A cross-section is a washer with inner radius y^2 and outer radius $2y$, so its area is $A(y) = \pi(2y)^2 - \pi(y^2)^2 = \pi(4y^2 - y^4)$.

$$V = \int_0^2 A(y)\,dy = \pi\int_0^2 (4y^2 - y^4)\,dy$$

$$= \pi\left[\tfrac{4}{3}y^3 - \tfrac{1}{5}y^5\right]_0^2 = \pi\left(\tfrac{32}{3} - \tfrac{32}{5}\right) = \tfrac{64}{15}\pi$$

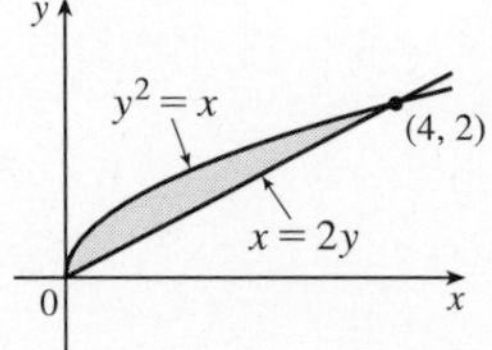

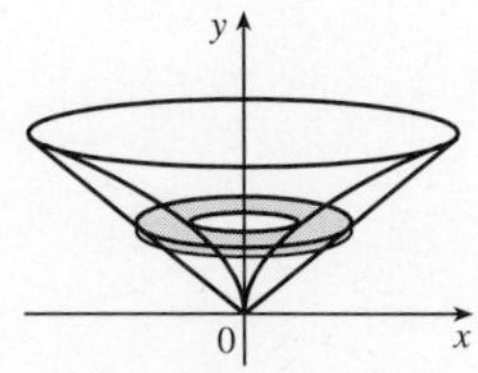

10. A cross-section is a washer with inner radius y^4 and outer radius $2 - y^2$, so its area is

$A(y) = \pi(2 - y^2)^2 - \pi(y^4)^2 = \pi(4 - 4y^2 + y^4 - y^8)$.

$V = \int_{-1}^{1} A(y)\,dy = 2\int_0^1 \pi(4 - 4y^2 + y^4 - y^8)\,dy$

$= 2\pi\left[4y - \frac{4}{3}y^3 + \frac{1}{5}y^5 - \frac{1}{9}y^9\right]_0^1$

$= 2\pi\left(4 - \frac{4}{3} + \frac{1}{5} - \frac{1}{9}\right) = 2\pi\left(\frac{124}{45}\right) = \frac{248\pi}{45}$

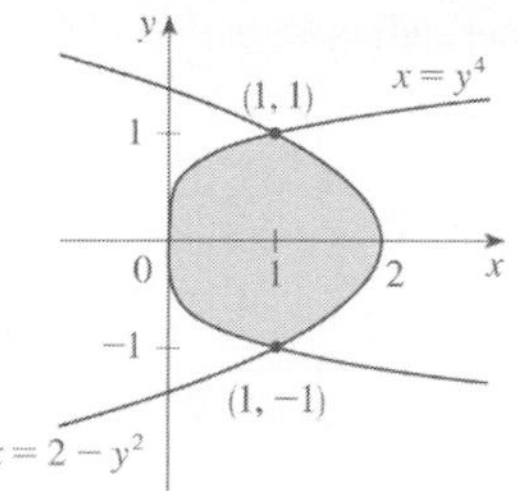

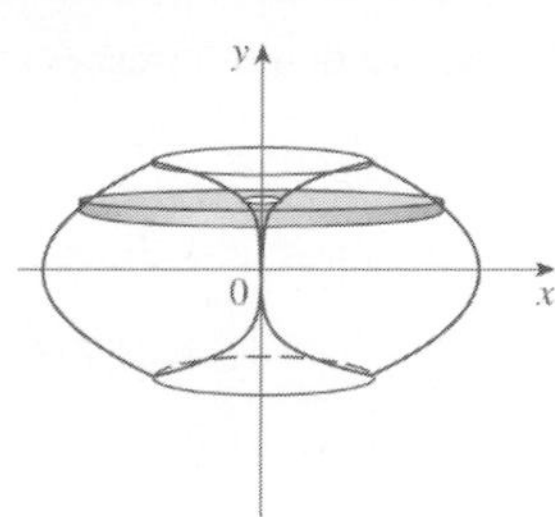

11. A cross-section is a washer with inner radius $1 - \sqrt{x}$ and outer radius $1 - x^2$, so its area is

$A(x) = \pi\left[(1 - x^2)^2 - (1 - \sqrt{x})^2\right]$

$= \pi\left[(1 - 2x^2 + x^4) - (1 - 2\sqrt{x} + x)\right]$

$= \pi\left(x^4 - 2x^2 + 2\sqrt{x} - x\right)$.

$V = \int_0^1 A(x)\,dx = \int_0^1 \pi(x^4 - 2x^2 + 2x^{1/2} - x)\,dx$

$= \pi\left[\frac{1}{5}x^5 - \frac{2}{3}x^3 + \frac{4}{3}x^{3/2} - \frac{1}{2}x^2\right]_0^1$

$= \pi\left(\frac{1}{5} - \frac{2}{3} + \frac{4}{3} - \frac{1}{2}\right) = \frac{11}{30}\pi$

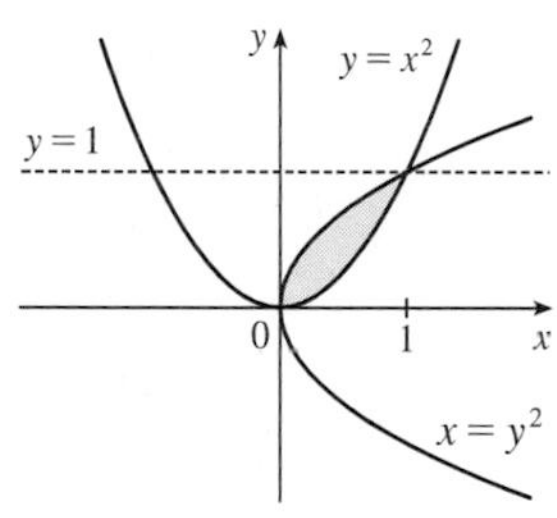

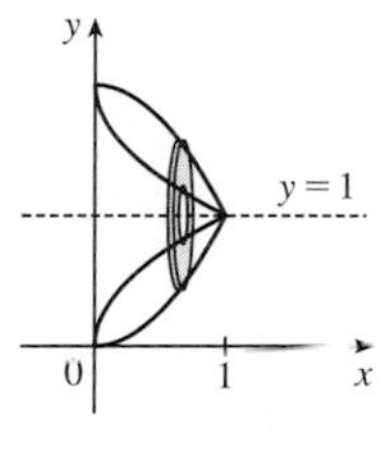

12. A cross-section is a washer with inner radius $1 - (-3) = 4$ and outer radius $x^3 - (-3) = x^3 + 3$, so its area is

$A(x) = \pi(x^3 + 3)^2 - \pi(4)^2 = \pi(x^6 + 6x^3 - 7)$.

$V = \int_1^2 A(x)\,dx = \int_1^2 \pi(x^6 + 6x^3 - 7)\,dx$

$= \pi\left[\frac{1}{7}x^7 + \frac{3}{2}x^4 - 7x\right]_1^2$

$= \pi\left[\left(\frac{128}{7} + 24 - 14\right) - \left(\frac{1}{7} + \frac{3}{2} - 7\right)\right] = \frac{471\pi}{14}$

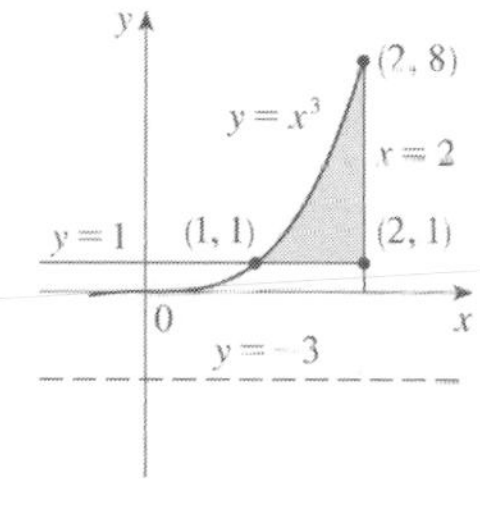

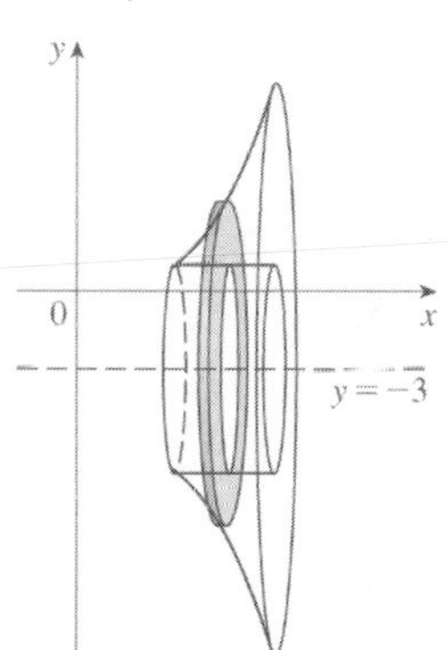

13. A cross-section is a washer with inner radius $(1 + \sec x) - 1 = \sec x$ and outer radius $3 - 1 = 2$, so its area is

$A(x) = \pi\left[2^2 - (\sec x)^2\right] = \pi(4 - \sec^2 x)$.

$V = \int_{-\pi/3}^{\pi/3} A(x)\,dx = \int_{-\pi/3}^{\pi/3} \pi(4 - \sec^2 x)\,dx$

$= 2\pi\int_0^{\pi/3} (4 - \sec^2 x)\,dx$ [by symmetry]

$= 2\pi\left[4x - \tan x\right]_0^{\pi/3} = 2\pi\left[\left(\frac{4\pi}{3} - \sqrt{3}\right) - 0\right]$

$= 2\pi\left(\frac{4\pi}{3} - \sqrt{3}\right)$

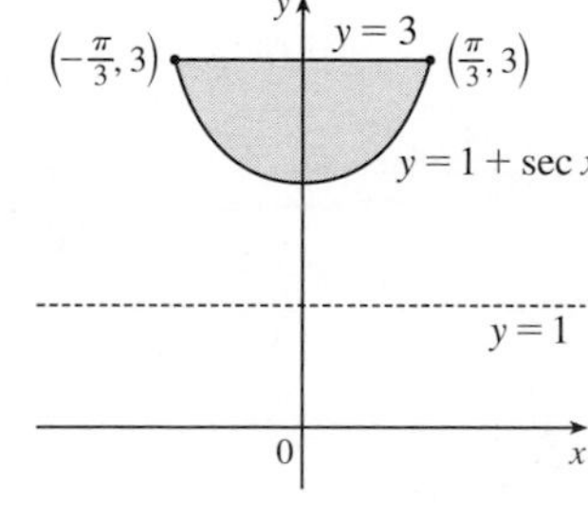

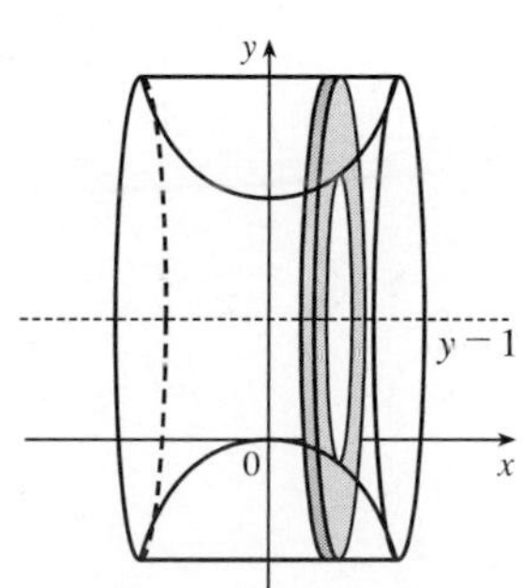

14. A cross-section is a washer with inner radius $\sin x - (-1)$ and outer radius $\cos x - (-1)$, so its area is

$$\begin{aligned} A(x) &= \pi\left[(\cos x + 1)^2 - (\sin x + 1)^2\right] \\ &= \pi(\cos^2 x + 2\cos x - \sin^2 x - 2\sin x) \\ &= \pi(\cos 2x + 2\cos x - 2\sin x). \end{aligned}$$

$$\begin{aligned} V &= \textstyle\int_0^{\pi/4} A(x)\,dx = \int_0^{\pi/4} \pi(\cos 2x + 2\cos x - 2\sin x)\,dx \\ &= \pi\left[\tfrac{1}{2}\sin 2x + 2\sin x + 2\cos x\right]_0^{\pi/4} \\ &= \pi\left[\left(\tfrac{1}{2} + \sqrt{2} + \sqrt{2}\right) - (0 + 0 + 2)\right] = \left(2\sqrt{2} - \tfrac{3}{2}\right)\pi \end{aligned}$$

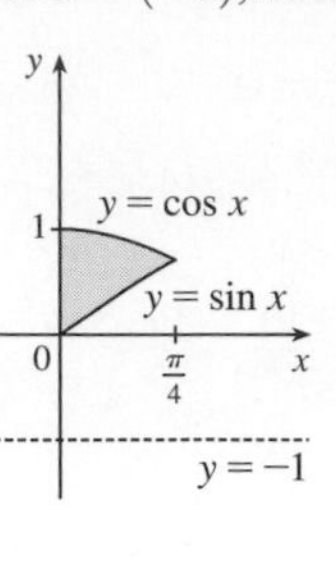

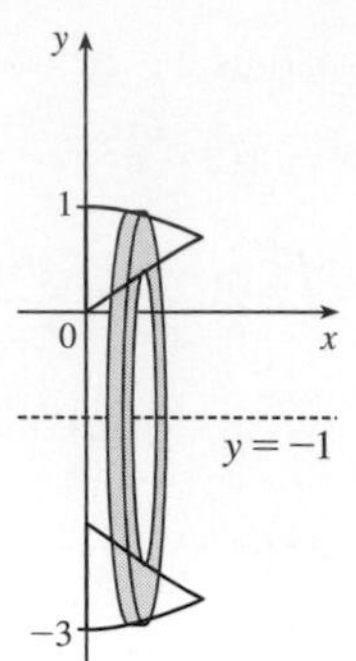

15. A cross-section is a washer with inner radius $2 - 1$ and outer radius $2 - \sqrt[3]{y}$, so its area is

$$A(y) = \pi\left[(2 - \sqrt[3]{y})^2 - (2 - 1)^2\right] = \pi\left[4 - 4\sqrt[3]{y} + \sqrt[3]{y^2} - 1\right].$$

$$V = \int_0^1 A(y)\,dy = \int_0^1 \pi(3 - 4y^{1/3} + y^{2/3})\,dy = \pi\left[3y - 3y^{4/3} + \tfrac{3}{5}y^{2/3}\right]_0^1 = \pi\left(3 - 3 + \tfrac{3}{5}\right) = \tfrac{3}{5}\pi.$$

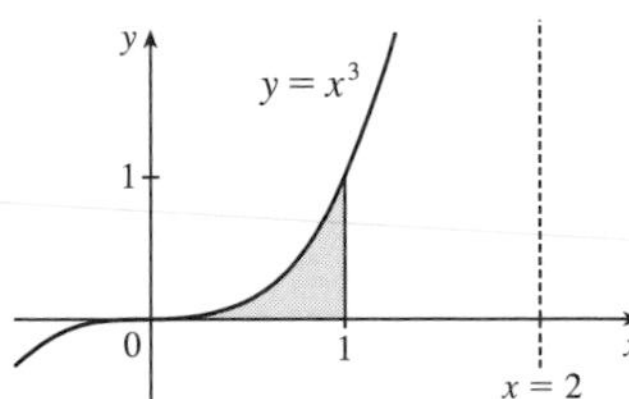

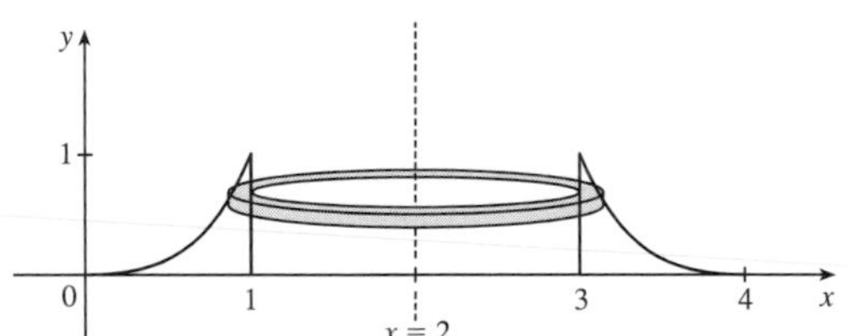

16. For $0 \le y < \frac{1}{2}$, a cross-section is a washer with inner radius $1 - (-1)$ and outer radius $2 - (-1)$, so its area is $A(y) = \pi(3^2 - 2^2) = 5\pi$. For $\frac{1}{2} \le y \le 1$, a cross-section is a washer with inner radius $1 - (-1)$ and outer radius $1/y - (-1)$, so its area is $A(y) = \pi\left[(1/y + 1)^2 - (2)^2\right] = \pi(1/y^2 + 2/y + 1 - 4)$.

$$\begin{aligned} V &= \int_0^{1/2} 5\pi\,dy + \int_{1/2}^1 \pi\left(\frac{1}{y^2} + \frac{2}{y} - 3\right)dy = 5\pi\left[y\right]_0^{1/2} + \pi\left[-\frac{1}{y} + 2\ln y - 3y\right]_{1/2}^1 \\ &= 5\pi\left(\tfrac{1}{2} - 0\right) + \pi\left[(-1 + 0 - 3) - \left(-2 + 2\ln\tfrac{1}{2} - \tfrac{3}{2}\right)\right] = \tfrac{5}{2}\pi + \pi\left(-\tfrac{1}{2} + 2\ln 2\right) \\ &= (2 + 2\ln 2)\pi = 2\pi(1 + \ln 2) \end{aligned}$$

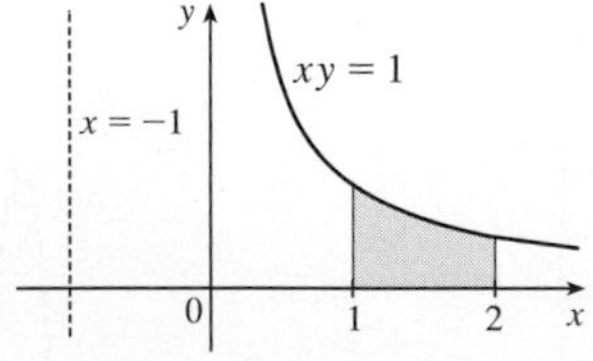

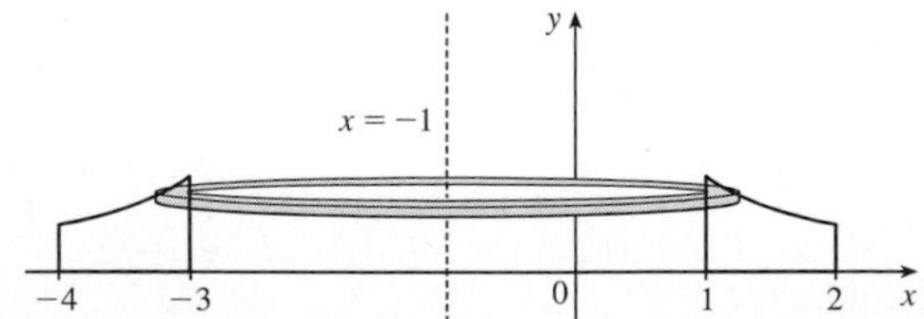

17. From the symmetry of the curves, we see they intersect at $x = \frac{1}{2}$ and so $y^2 = \frac{1}{2} \;\Leftrightarrow\; y = \pm\sqrt{\frac{1}{2}}$. A cross-section is a washer with inner radius $3 - (1 - y^2)$ and outer radius $3 - y^2$, so its area is

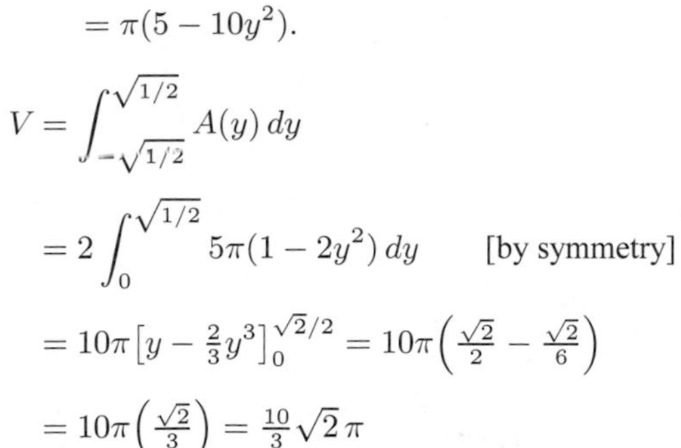

$$\begin{aligned} A(y) &= \pi\left[(3 - y^2)^2 - (2 + y^2)^2\right] \\ &= \pi\left[(9 - 6y^2 + y^4) - (4 + 4y^2 + y^4)\right] \\ &= \pi(5 - 10y^2). \end{aligned}$$

$$\begin{aligned} V &= \int_{-\sqrt{1/2}}^{\sqrt{1/2}} A(y)\,dy \\ &= 2\int_0^{\sqrt{1/2}} 5\pi(1 - 2y^2)\,dy \qquad \text{[by symmetry]} \\ &= 10\pi\left[y - \tfrac{2}{3}y^3\right]_0^{\sqrt{2}/2} = 10\pi\left(\tfrac{\sqrt{2}}{2} - \tfrac{\sqrt{2}}{6}\right) \\ &= 10\pi\left(\tfrac{\sqrt{2}}{3}\right) = \tfrac{10}{3}\sqrt{2}\,\pi \end{aligned}$$

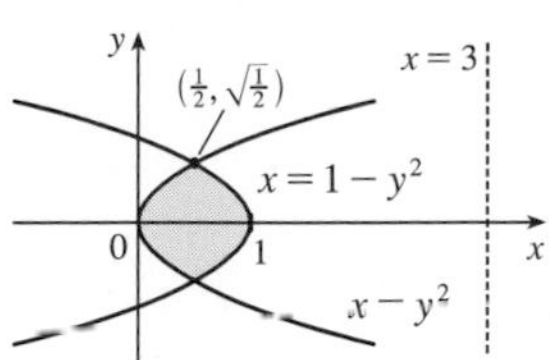

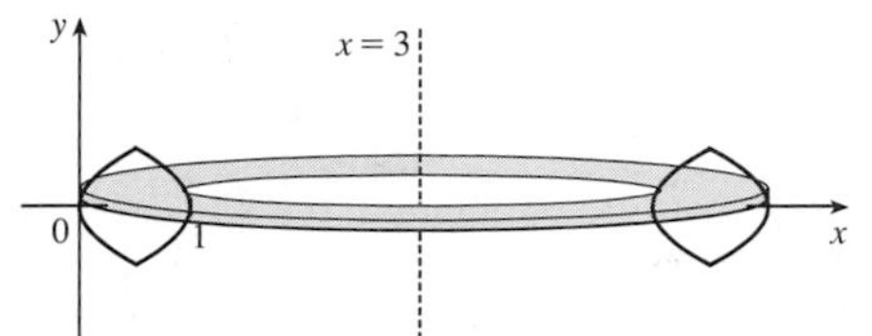

18. For $0 \le y < 2$, a cross-section is an annulus with inner radius $2 - 1$ and outer radius $4 - 1$, the area of which is $A_1(y) = \pi(4 - 1)^2 - \pi(2 - 1)^2$. For $2 \le y \le 4$, a cross-section is an annulus with inner radius $y - 1$ and outer radius $4 - 1$, the area of which is $A_2(y) = \pi(4 - 1)^2 - \pi(y - 1)^2$.

$$\begin{aligned} V &= \textstyle\int_0^4 A(y)\,dy = \pi \int_0^2 \left[(4 - 1)^2 - (2 - 1)^2\right] dy + \pi \int_2^4 \left[(4 - 1)^2 - (y - 1)^2\right] dy \\ &= \pi\left[8y\right]_0^2 + \pi \textstyle\int_2^4 (8 + 2y - y^2)\,dy \\ &= 16\pi + \pi\left[8y + y^2 - \tfrac{1}{3}y^3\right]_2^4 \\ &= 16\pi + \pi\left[\left(32 + 16 - \tfrac{64}{3}\right) - \left(16 + 4 - \tfrac{8}{3}\right)\right] \\ &= \tfrac{76}{3}\pi \end{aligned}$$

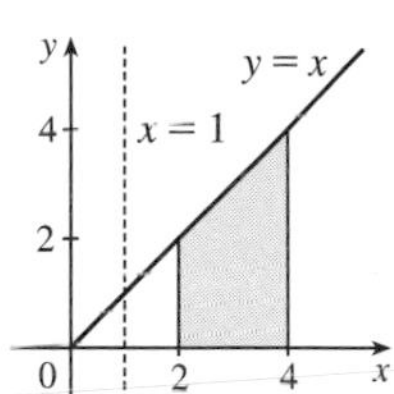

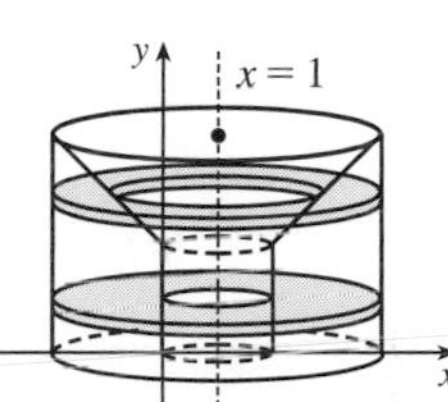

19. $\mathscr{R}_1$ about OA (the line $y = 0$):

$$V = \int_0^1 A(x)\,dx = \int_0^1 \pi(x)^2\,dx = \pi\left[\tfrac{1}{3}x^3\right]_0^1 = \tfrac{1}{3}\pi$$

20. $\mathscr{R}_1$ about OC (the line $x = 0$):

$$V = \int_0^1 A(y)\,dy = \int_0^1 \pi(1^2 - y^2)\,dy = \pi\left[y - \tfrac{1}{3}y^3\right]_0^1 = \pi\left(1 - \tfrac{1}{3}\right) = \tfrac{2}{3}\pi$$

21. $\mathscr{R}_1$ about AB (the line $x = 1$):

$$V = \int_0^1 A(y)\,dy = \int_0^1 \pi(1 - y)^2\,dy = \pi\int_0^1 (1 - 2y + y^2)\,dy = \pi\left[y - y^2 + \tfrac{1}{3}y^3\right]_0^1 = \tfrac{1}{3}\pi$$

22. $\mathscr{R}_1$ about BC (the line $y = 1$):

$$\begin{aligned} V &= \int_0^1 A(x)\,dx = \int_0^1 \pi\left[(1 - 0)^2 - (1 - x)^2\right] dx = \pi\int_0^1 \left[1 - (1 - 2x + x^2)\right] dx \\ &= \pi\int_0^1 (-x^2 + 2x)\,dx = \pi\left[-\tfrac{1}{3}x^3 + x^2\right]_0^1 = \pi\left(-\tfrac{1}{3} + 1\right) = \tfrac{2}{3}\pi \end{aligned}$$

23. $\mathcal{R}_2$ about OA (the line $y = 0$):

$$V = \int_0^1 A(x)\,dx = \int_0^1 \pi\left[1^2 - \left(\sqrt[4]{x}\right)^2\right] dx = \pi \int_0^1 (1 - x^{1/2})\,dx = \pi\left[x - \tfrac{2}{3}x^{3/2}\right]_0^1 = \pi\left(1 - \tfrac{2}{3}\right) = \tfrac{1}{3}\pi$$

24. $\mathcal{R}_2$ about OC (the line $x = 0$):

$$V = \int_0^1 A(y)\,dy = \int_0^1 \pi[(y^4)^2]\,dy = \pi \int_0^1 y^8\,dy = \pi\left[\tfrac{1}{9}y^9\right]_0^1 = \tfrac{1}{9}\pi$$

25. $\mathcal{R}_2$ about AB (the line $x = 1$):

$$V = \int_0^1 A(y)\,dy = \int_0^1 \pi[1^2 - (1 - y^4)^2]\,dy = \pi \int_0^1 [1 - (1 - 2y^4 + y^8)]\,dy$$
$$= \pi \int_0^1 (2y^4 - y^8)\,dy = \pi\left[\tfrac{2}{5}y^5 - \tfrac{1}{9}y^9\right]_0^1 = \pi\left(\tfrac{2}{5} - \tfrac{1}{9}\right) = \tfrac{13}{45}\pi$$

26. $\mathcal{R}_2$ about BC (the line $y = 1$):

$$V = \int_0^1 A(x)\,dx = \int_0^1 \pi(1 - \sqrt[4]{x})^2\,dx = \pi \int_0^1 (1 - 2x^{1/4} + x^{1/2})\,dx$$
$$= \pi\left[x - \tfrac{8}{5}x^{5/4} + \tfrac{2}{3}x^{3/2}\right]_0^1 = \pi\left(1 - \tfrac{8}{5} + \tfrac{2}{3}\right) = \tfrac{1}{15}\pi$$

27. $\mathcal{R}_3$ about OA (the line $y = 0$):

$$V = \int_0^1 A(x)\,dx = \int_0^1 \pi\left[\left(\sqrt[4]{x}\right)^2 - x^2\right] dx = \pi \int_0^1 (x^{1/2} - x^2)\,dx = \pi\left[\tfrac{2}{3}x^{3/2} - \tfrac{1}{3}x^3\right]_0^1 = \pi\left(\tfrac{2}{3} - \tfrac{1}{3}\right) = \tfrac{1}{3}\pi$$

Note: Let $\mathcal{R} = \mathcal{R}_1 \cup \mathcal{R}_2 \cup \mathcal{R}_3$. If we rotate $\mathcal{R}$ about any of the segments OA, OC, AB, or BC, we obtain a right circular cylinder of height 1 and radius 1. Its volume is $\pi r^2 h = \pi(1)^2 \cdot 1 = \pi$. As a check for Exercises 19, 23, and 27, we can add the answers, and that sum must equal π. Thus, $\frac{1}{3}\pi + \frac{1}{3}\pi + \frac{1}{3}\pi = \pi$.

28. $\mathcal{R}_3$ about OC (the line $x = 0$):

$$V = \int_0^1 A(y)\,dy = \int_0^1 \pi[y^2 - (y^4)^2]\,dy = \pi \int_0^1 (y^2 - y^8)\,dy = \pi\left[\tfrac{1}{3}y^3 - \tfrac{1}{9}y^9\right]_0^1 = \pi\left(\tfrac{1}{3} - \tfrac{1}{9}\right) = \tfrac{2}{9}\pi$$

Note: See the note in Exercise 27. For Exercises 20, 24, and 28, we have $\frac{2}{3}\pi + \frac{1}{9}\pi + \frac{2}{9}\pi = \pi$.

29. $\mathcal{R}_3$ about AB (the line $x = 1$):

$$V = \int_0^1 A(y)\,dy = \int_0^1 \pi[(1 - y^4)^2 - (1 - y)^2]\,dy = \pi \int_0^1 [(1 - 2y^4 + y^8) - (1 - 2y + y^2)]\,dy$$
$$= \pi \int_0^1 (y^8 - 2y^4 - y^2 + 2y)\,dy = \pi\left[\tfrac{1}{9}y^9 - \tfrac{2}{5}y^5 - \tfrac{1}{3}y^3 + y^2\right]_0^1 = \pi\left(\tfrac{1}{9} - \tfrac{2}{5} - \tfrac{1}{3} + 1\right) = \tfrac{17}{45}\pi$$

Note: See the note in Exercise 27. For Exercises 21, 25, and 29, we have $\frac{1}{3}\pi + \frac{13}{45}\pi + \frac{17}{45}\pi = \pi$.

30. $\mathcal{R}_3$ about BC (the line $y = 1$):

$$V = \int_0^1 A(x)\,dx = \int_0^1 \pi[(1-x)^2 - (1-\sqrt[4]{x}\,)^2]\,dx = \pi\int_0^1 [(1-2x+x^2) - (1-2x^{1/4}+x^{1/2})]\,dx$$

$$= \pi\int_0^1 (x^2 - 2x - x^{1/2} + 2x^{1/4})\,dx = \pi\left[\tfrac{1}{3}x^3 - x^2 - \tfrac{2}{3}x^{3/2} + \tfrac{8}{5}x^{5/4}\right]_0^1 = \pi\left(\tfrac{1}{3} - 1 - \tfrac{2}{3} + \tfrac{8}{5}\right) = \tfrac{4}{15}\pi$$

Note: See the note in Exercise 27. For Exercises 22, 26, and 30, we have $\frac{2}{3}\pi + \frac{1}{15}\pi + \frac{4}{15}\pi = \pi$.

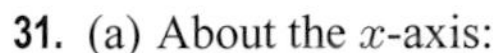

31. (a) About the x-axis:

$$V = \int_{-1}^1 \pi(e^{-x^2})^2\,dx = 2\pi\int_0^1 e^{-2x^2}\,dx \quad \text{[by symmetry]}$$

$$\approx 3.75825$$

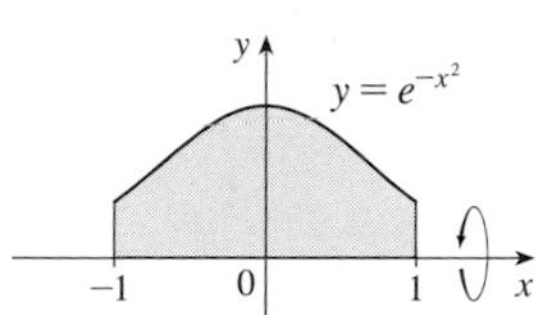

(b) About $y = -1$:

$$V = \int_{-1}^1 \pi\left\{[e^{-x^2} - (-1)]^2 - [0-(-1)]^2\right\}\,dx$$

$$= 2\pi\int_0^1 [(e^{-x^2}+1)^2 - 1]\,dx = 2\pi\int_0^1 (c^{-2x^2} + 2e^{-x^2})\,dx$$

$$\approx 13.14312$$

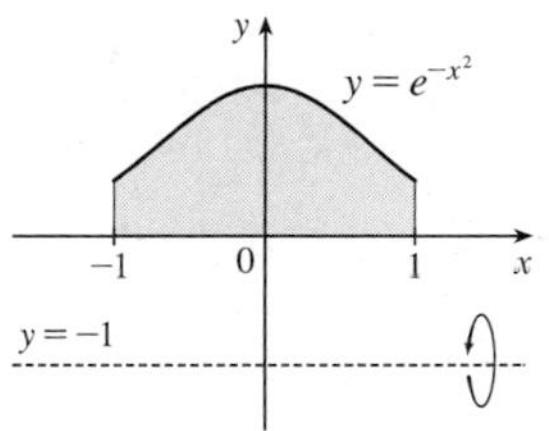

32. (a) About the x-axis:

$$V = \int_{-\pi/2}^{\pi/2} \pi(\cos^2 x)^2\,dx = 2\pi\int_0^{\pi/2} \cos^4 x\,dx \quad \text{[by symmetry]}$$

$$\approx 3.70110$$

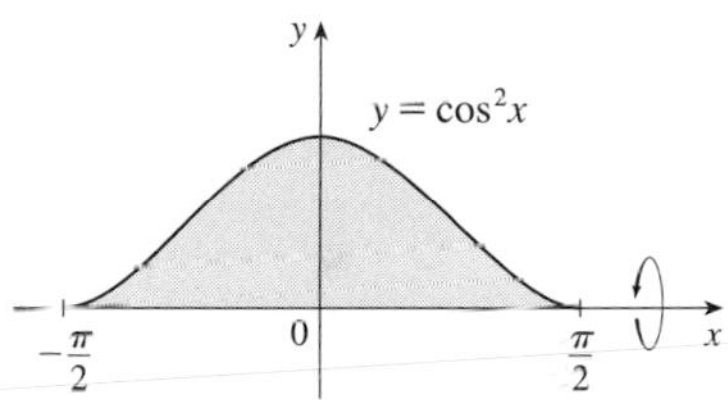

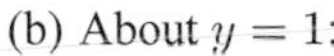

(b) About $y = 1$:

$$V = \int_{-\pi/2}^{\pi/2} \pi[(1-0)^2 - (1-\cos^2 x)^2]\,dx$$

$$= 2\pi\int_0^{\pi/2} [1 - (1 - 2\cos^2 x + \cos^4 x)]\,dx$$

$$= 2\pi\int_0^{\pi/2} (2\cos^2 x - \cos^4 x)\,dx \approx 6.16850$$

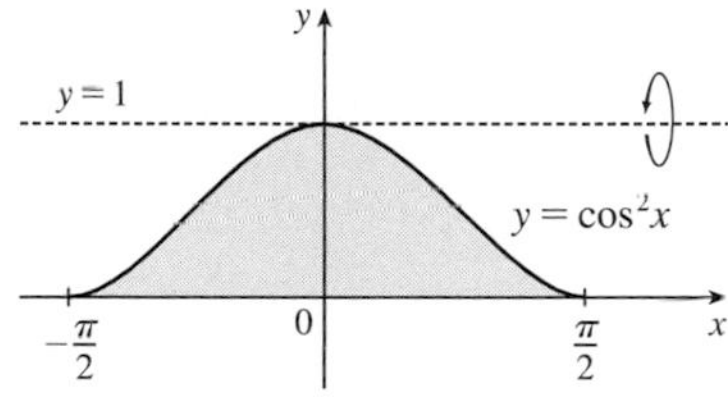

33. (a) About $y = 2$:

$x^2 + 4y^2 = 4 \;\Rightarrow\; 4y^2 = 4 - x^2 \;\Rightarrow\; y^2 = 1 - x^2/4 \;\Rightarrow$

$y = \pm\sqrt{1 - x^2/4}$

$$V = \int_{-2}^2 \pi\left\{\left[2 - \left(-\sqrt{1-x^2/4}\right)\right]^2 - \left(2 - \sqrt{1-x^2/4}\right)^2\right\}\,dx$$

$$= 2\pi\int_0^2 8\sqrt{1-x^2/4}\,dx \approx 78.95684$$

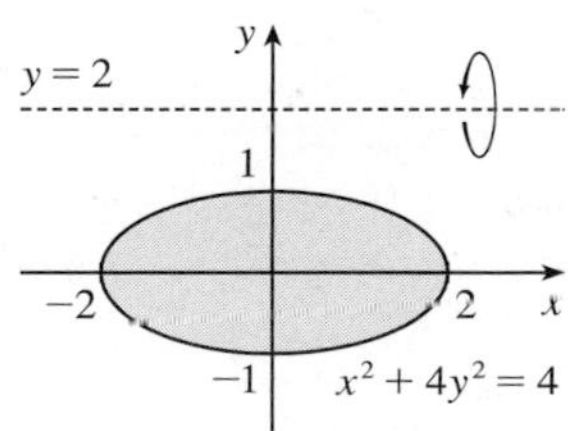

(b) About $x = 2$:

$x^2 + 4y^2 = 4 \quad\Rightarrow\quad x^2 = 4 - 4y^2 \quad\Rightarrow\quad x = \pm\sqrt{4 - 4y^2}$

$$V = \int_{-1}^{1} \pi \left\{ \left[2 - \left(-\sqrt{4 - 4y^2}\right)\right]^2 - \left(2 - \sqrt{4 - 4y^2}\right)^2 \right\} dy$$

$$= 2\pi \int_{0}^{1} 8\sqrt{4 - 4y^2}\, dy \approx 78.95684$$

[Notice that this is the same approximation as in part (a). This can be explained by Pappus's Theorem in Section 8.3.]

34. (a) About the x-axis:

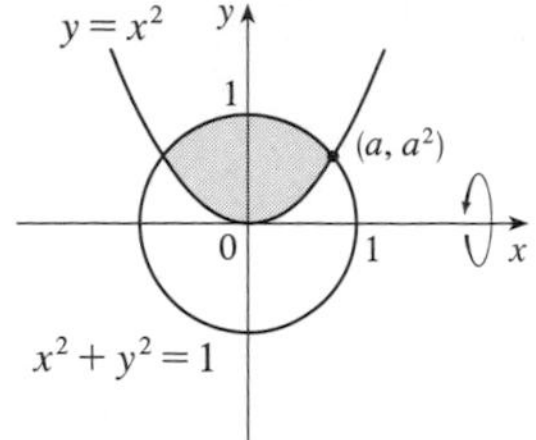

$y = x^2$ and $x^2 + y^2 = 1 \quad\Rightarrow\quad x^2 + x^4 = 1 \quad\Rightarrow\quad x^4 + x^2 - 1 = 0 \quad\Rightarrow$

$$x^2 = \frac{-1 + \sqrt{5}}{2} \approx 0.618 \quad\Rightarrow\quad x = \pm a = \pm\sqrt{\frac{-1 + \sqrt{5}}{2}} \approx \pm 0.786.$$

$$V = \int_{-a}^{a} \pi \left[\left(\sqrt{1 - x^2}\right)^2 - (x^2)^2 \right] dx = 2\pi \int_{0}^{a} (1 - x^2 - x^4)\, dx$$

$$\approx 3.54459$$

(b) About the y-axis:

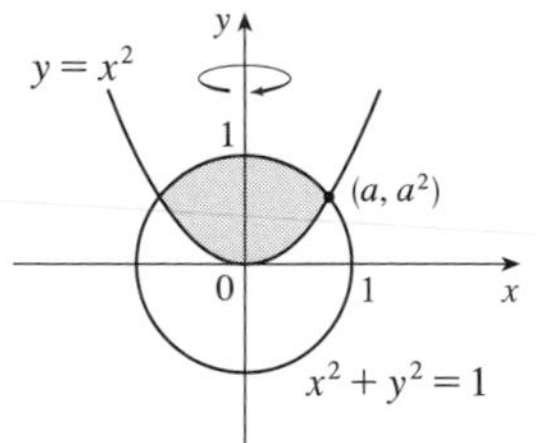

$$V = \int_{0}^{a^2} \pi \left(\sqrt{y}\right)^2 dy + \int_{a^2}^{1} \pi \left(\sqrt{1 - y^2}\right)^2 dy$$

$$= \pi \int_{0}^{a^2} y\, dy + \pi \int_{a^2}^{1} (1 - y^2)\, dy \approx 0.99998$$

35. $y = \ln(x^6 + 2)$ and $y = \sqrt{3 - x^3}$ intersect at $x = a \approx -4.091$, $x = b \approx -1.467$, and $x = c \approx 1.091$.

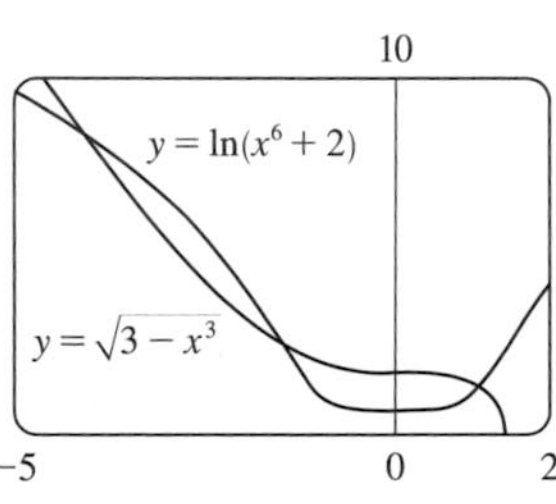

$$V = \pi \int_{a}^{b} \left\{ \left[\ln(x^6 + 2)\right]^2 - \left(\sqrt{3 - x^3}\right)^2 \right\} dx + \pi \int_{b}^{c} \left\{ \left(\sqrt{3 - x^3}\right)^2 - \left[\ln(x^6 + 2)\right]^2 \right\} dx \approx 89.023$$

36. $y = 1 + xe^{-x^3}$ and $y = \arctan x^2$ intersect at $x = a \approx -0.570$ and $x = b \approx 1.391$.

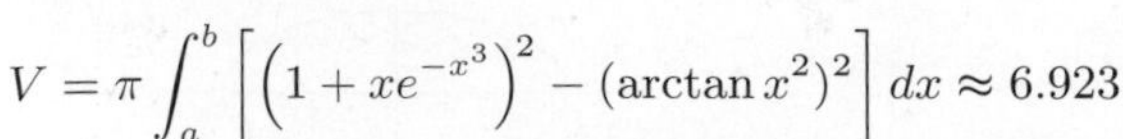

$$V = \pi \int_{a}^{b} \left[\left(1 + xe^{-x^3}\right)^2 - (\arctan x^2)^2 \right] dx \approx 6.923$$

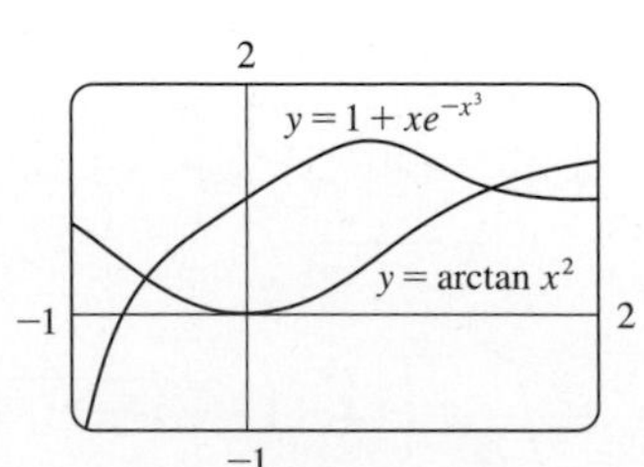

37. $V = \pi \int_0^{\pi} \left\{ \left[\sin^2 x - (-1)\right]^2 - \left[0 - (-1)\right]^2 \right\} dx$

$\overset{\text{CAS}}{=} \frac{11}{8}\pi^2$

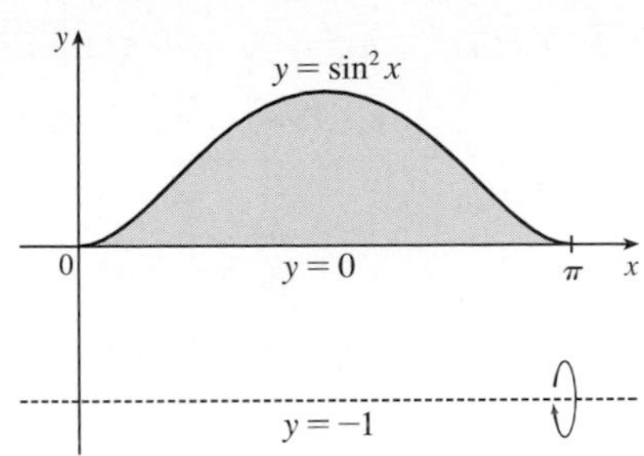

38. $V = \pi \int_0^2 \left[(3-x)^2 - (3 - xe^{1-x/2})^2 \right] dx$

$\overset{\text{CAS}}{=} \pi\left(-2e^2 + 24e - \frac{142}{3}\right)$

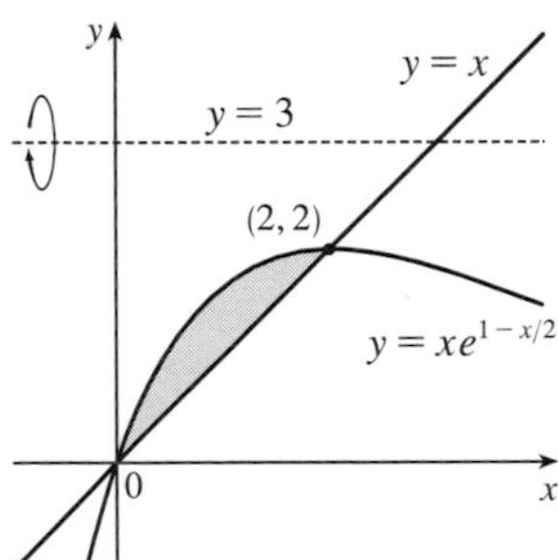

39. $\pi \int_0^{\pi} \sin x\, dx = \pi \int_0^{\pi} \left(\sqrt{\sin x}\right)^2 dx$ describes the volume of solid obtained by rotating the region

$\mathcal{R} = \left\{ (x, y) \mid 0 \le x \le \pi, 0 \le y \le \sqrt{\sin x} \right\}$ of the xy-plane about the x-axis.

40. $\pi \int_{-1}^{1} (1 - y^2)^2\, dy$ describes the volume of the solid obtained by rotating the region

$\mathcal{R} = \{(x, y) \mid -1 \le y \le 1, 0 \le x \le 1 - y^2\}$ of the xy-plane about the y-axis.

41. $\pi \int_0^1 (y^4 - y^8)\, dy = \pi \int_0^1 \left[(y^2)^2 - (y^4)^2 \right] dy$ describes the volume of the solid obtained by rotating the region

$\mathcal{R} = \{(x, y) \mid 0 \le y \le 1, y^4 \le x \le y^2\}$ of the xy-plane about the y-axis.

42. $\pi \int_1^4 [3^2 - (3 - \sqrt{x})^2\, dx$ describes the volume of the solid obtained by rotating the region

$\mathcal{R} = \{(x, y) \mid 1 \le x \le 4, 3 - \sqrt{x} \le y \le 3\}$ of the xy-plane about the x-axis.

43. There are 10 subintervals over the 15-cm length, so we'll use $n = 10/2 = 5$ for the Midpoint Rule.

$$V = \int_0^{15} A(x)\, dx \approx M_5 = \frac{15-0}{5}[A(1.5) + A(4.5) + A(7.5) + A(10.5) + A(13.5)]$$
$$= 3(18 + 79 + 106 + 128 + 39) = 3 \cdot 370 = 1110 \text{ cm}^3$$

44.
$$V = \int_0^{10} A(x)\, dx \approx M_5 = \frac{10-0}{5}[A(1) + A(3) + A(5) + A(7) + A(9)]$$
$$= 2(0.65 + 0.61 + 0.59 + 0.55 + 0.50) = 2(2.90) = 5.80 \text{ m}^3$$

45. (a) $V = \int_2^{10} \pi [f(x)]^2\, dx \approx \pi \frac{10-2}{4} \{[f(3)]^2 + [f(5)]^2 + [f(7)]^2 + [f(9)]^2\}$

$\approx 2\pi \left[(1.5)^2 + (2.2)^2 + (3.8)^2 + (3.1)^2 \right] \approx 196 \text{ units}^3$

(b) $V = \int_0^4 \pi \left[(\text{outer radius})^2 - (\text{inner radius})^2 \right] dy$

$\approx \pi \frac{4-0}{4} \left\{ \left[(9.9)^2 - (2.2)^2\right] + \left[(9.7)^2 - (3.0)^2\right] + \left[(9.3)^2 - (5.6)^2\right] + \left[(8.7)^2 - (6.5)^2\right] \right\}$

$\approx 838 \text{ units}^3$

46. (a) $V = \int_{-1}^{1} \pi\left[\left(ax^3+bx^2+cx+d\right)\sqrt{1-x^2}\,\right]^2 dx \overset{\text{CAS}}{=} \dfrac{4\left\{5a^2+18ac+3\left[3b^2+14bd+7(c^2+5d^2)\right]\right\}\pi}{315}$

(b) $y = (-0.06x^3+0.04x^2+0.1x+0.54)\sqrt{1-x^2}$ is graphed in the figure. Substitute $a=-0.06$, $b=0.04$, $c=0.1$, and $d=0.54$ in the answer for part (a) to get $V \overset{\text{CAS}}{=} \dfrac{3769\pi}{9375} \approx 1.263$.

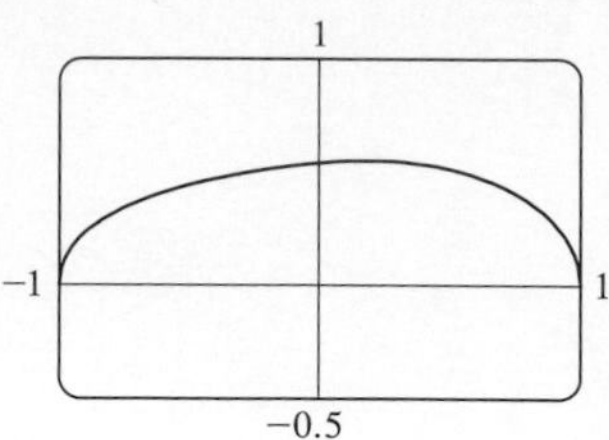

47. We'll form a right circular cone with height h and base radius r by revolving the line $y = \frac{r}{h}x$ about the x-axis.

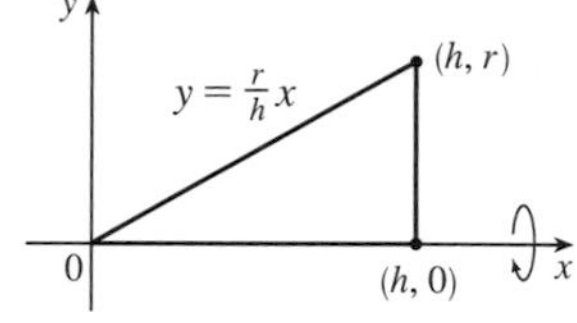

$$V = \pi\int_0^h \left(\frac{r}{h}x\right)^2 dx = \pi\int_0^h \frac{r^2}{h^2}x^2\,dx = \pi\frac{r^2}{h^2}\left[\frac{1}{3}x^3\right]_0^h$$
$$= \pi\frac{r^2}{h^2}\left(\frac{1}{3}h^3\right) = \frac{1}{3}\pi r^2 h$$

Another solution: Revolve $x = -\dfrac{r}{h}\,y + r$ about the y-axis.

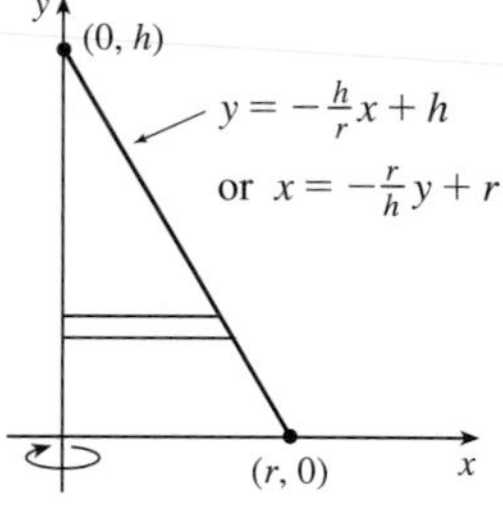

$$V = \pi\int_0^h \left(-\frac{r}{h}y+r\right)^2 dy \overset{*}{=} \pi\int_0^h \left[\frac{r^2}{h^2}y^2 - \frac{2r^2}{h}y + r^2\right] dy$$
$$= \pi\left[\frac{r^2}{3h^2}y^3 - \frac{r^2}{h}y^2 + r^2y\right]_0^h = \pi\left(\tfrac{1}{3}r^2h - r^2h + r^2h\right) = \tfrac{1}{3}\pi r^2 h$$

* Or use substitution with $u = r - \dfrac{r}{h}\,y$ and $du = -\dfrac{r}{h}\,dy$ to get

$$\pi\int_r^0 u^2\left(-\frac{h}{r}\,du\right) = -\pi\frac{h}{r}\left[\frac{1}{3}u^3\right]_r^0 = -\pi\frac{h}{r}\left(-\frac{1}{3}r^3\right) = \frac{1}{3}\pi r^2 h.$$

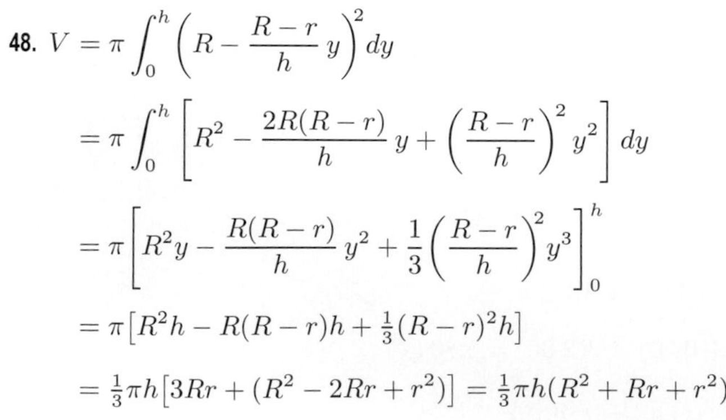

48. $V = \pi\int_0^h \left(R - \dfrac{R-r}{h}\,y\right)^2 dy$

$$= \pi\int_0^h \left[R^2 - \frac{2R(R-r)}{h}\,y + \left(\frac{R-r}{h}\right)^2 y^2\right] dy$$
$$= \pi\left[R^2y - \frac{R(R-r)}{h}\,y^2 + \frac{1}{3}\left(\frac{R-r}{h}\right)^2 y^3\right]_0^h$$
$$= \pi\left[R^2h - R(R-r)h + \tfrac{1}{3}(R-r)^2h\right]$$
$$= \tfrac{1}{3}\pi h\left[3Rr + (R^2 - 2Rr + r^2)\right] = \tfrac{1}{3}\pi h(R^2 + Rr + r^2)$$

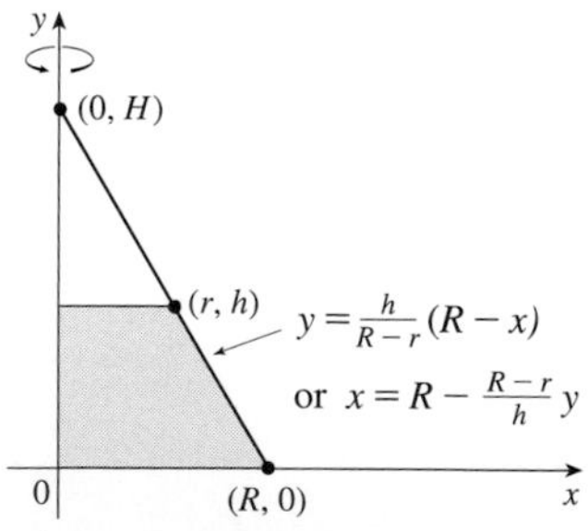

Another solution: $\dfrac{H}{R} = \dfrac{H-h}{r}$ by similar triangles. Therefore, $Hr = HR - hR \;\Rightarrow\; hR = H(R-r) \;\Rightarrow$

$H = \dfrac{hR}{R-r}$. Now

$$V = \tfrac{1}{3}\pi R^2 H - \tfrac{1}{3}\pi r^2 (H-h) \qquad \text{[by Exercise 47]}$$

$$= \tfrac{1}{3}\pi R^2 \frac{hR}{R-r} - \tfrac{1}{3}\pi r^2 \frac{rh}{R-r} \qquad \left[H - h = \frac{rH}{R} = \frac{rhR}{R(R-r)}\right]$$

$$= \frac{1}{3}\pi h \frac{R^3 - r^3}{R-r} = \tfrac{1}{3}\pi h \left(R^2 + Rr + r^2\right)$$

$$= \tfrac{1}{3}\left[\pi R^2 + \pi r^2 + \sqrt{(\pi R^2)(\pi r^2)}\,\right] h = \tfrac{1}{3}\left(A_1 + A_2 + \sqrt{A_1 A_2}\,\right) h$$

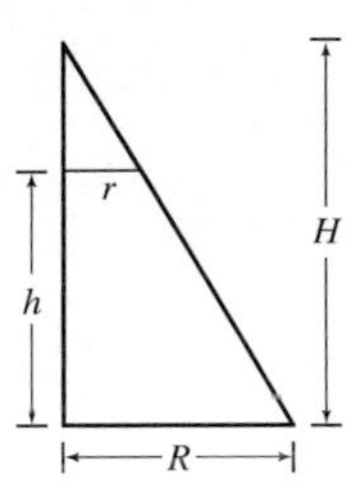

where A_1 and A_2 are the areas of the bases of the frustum. (See Exercise 50 for a related result.)

49. $x^2 + y^2 = r^2 \quad \Leftrightarrow \quad x^2 = r^2 - y^2$

$$V = \pi \int_{r-h}^{r} \left(r^2 - y^2\right) dy = \pi \left[r^2 y - \frac{y^3}{3}\right]_{r-h}^{r} = \pi \left\{\left[r^3 - \frac{r^3}{3}\right] - \left[r^2(r-h) - \frac{(r-h)^3}{3}\right]\right\}$$

$$= \pi \left\{\tfrac{2}{3} r^3 - \tfrac{1}{3}(r-h)\left[3r^2 - (r-h)^2\right]\right\}$$

$$= \tfrac{1}{3}\pi \left\{2r^3 - (r-h)\left[3r^2 - \left(r^2 - 2rh + h^2\right)\right]\right\}$$

$$= \tfrac{1}{3}\pi \left\{2r^3 - (r-h)\left[2r^2 + 2rh - h^2\right]\right\}$$

$$= \tfrac{1}{3}\pi \left(2r^3 - 2r^3 - 2r^2 h + rh^2 + 2r^2 h + 2rh^2 - h^3\right)$$

$$= \tfrac{1}{3}\pi \left(3rh^2 - h^3\right) = \tfrac{1}{3}\pi h^2 (3r - h), \text{ or, equivalently, } \pi h^2 \left(r - \frac{h}{3}\right)$$

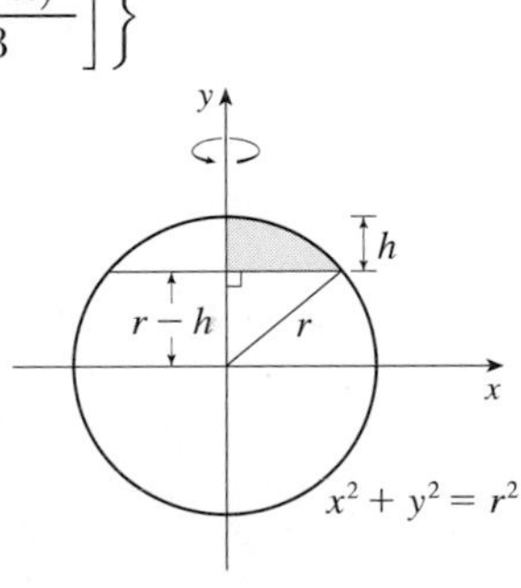

50. An equation of the line is $x = \dfrac{\Delta x}{\Delta y}\, y + (x\text{-intercept}) = \dfrac{a/2 - b/2}{h - 0}\, y + \dfrac{b}{2} = \dfrac{a-b}{2h}\, y + \dfrac{b}{2}$.

$$V = \int_0^h A(y)\, dy = \int_0^h (2x)^2\, dy$$

$$= \int_0^h \left[2\left(\frac{a-b}{2h} y + \frac{b}{2}\right)\right]^2 dy = \int_0^h \left[\frac{a-b}{h} y + b\right]^2 dy$$

$$= \int_0^h \left[\frac{(a-b)^2}{h^2} y^2 + \frac{2b(a-b)}{h} y + b^2\right] dy$$

$$= \left[\frac{(a-b)^2}{3h^2} y^3 + \frac{b(a-b)}{h} y^2 + b^2 y\right]_0^h$$

$$= \tfrac{1}{3}(a-b)^2 h + b(a-b)h + b^2 h = \tfrac{1}{3}\left(a^2 - 2ab + b^2 + 3ab\right) h$$

$$= \tfrac{1}{3}\left(a^2 + ab + b^2\right) h$$

[Note that this can be written as $\tfrac{1}{3}\left(A_1 + A_2 + \sqrt{A_1 A_2}\,\right) h$, as in Exercise 48.]

If $a = b$, we get a rectangular solid with volume $b^2 h$. If $a = 0$, we get a square pyramid with volume $\tfrac{1}{3} b^2 h$.

51. For a cross-section at height y, we see from similar triangles that $\dfrac{\alpha/2}{b/2} = \dfrac{h-y}{h}$, so $\alpha = b\left(1 - \frac{y}{h}\right)$.

Similarly, for cross-sections having $2b$ as their base and β replacing α, $\beta = 2b\left(1 - \frac{y}{h}\right)$. So

$$V = \int_0^h A(y)\,dy = \int_0^h \left[b\left(1 - \tfrac{y}{h}\right)\right]\left[2b\left(1 - \tfrac{y}{h}\right)\right] dy$$

$$= \int_0^h 2b^2\left(1 - \tfrac{y}{h}\right)^2 dy = 2b^2 \int_0^h \left(1 - \frac{2y}{h} + \frac{y^2}{h^2}\right) dy$$

$$= 2b^2\left[y - \frac{y^2}{h} + \frac{y^3}{3h^2}\right]_0^h = 2b^2\left[h - h + \tfrac{1}{3}h\right]$$

$$= \tfrac{2}{3}b^2 h \quad \left[= \tfrac{1}{3}Bh \text{ where } B \text{ is the area of the base, as with any pyramid.}\right]$$

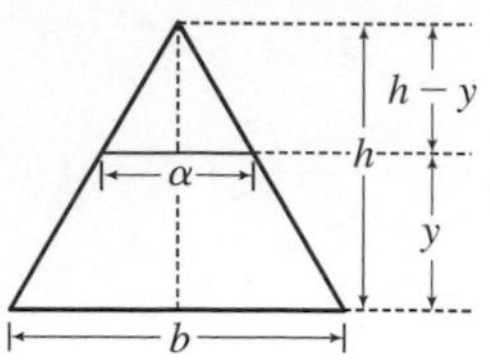

52. Consider the triangle consisting of two vertices of the base and the center of the base. This triangle is similar to the corresponding triangle at a height y, so $a/b = \alpha/\beta \;\Rightarrow\; \alpha = a\beta/b$. Also by similar triangles, $b/h = \beta/(h-y) \Rightarrow \beta = b(h-y)/h$. These two equations imply that $\alpha = a(1 - y/h)$, and since the cross-section is an equilateral triangle, it has area

$$A(y) = \frac{1}{2}\cdot\alpha\cdot\frac{\sqrt{3}}{2}\alpha = \frac{a^2(1-y/h)^2}{4}\sqrt{3}, \text{ so}$$

$$V = \int_0^h A(y)\,dy = \frac{a^2\sqrt{3}}{4}\int_0^h \left(1 - \tfrac{y}{h}\right)^2 dy$$

$$= \frac{a^2\sqrt{3}}{4}\left[-\frac{h}{3}\left(1 - \frac{y}{h}\right)^3\right]_0^h = -\frac{\sqrt{3}}{12}a^2 h(-1) = \frac{\sqrt{3}}{12}a^2 h$$

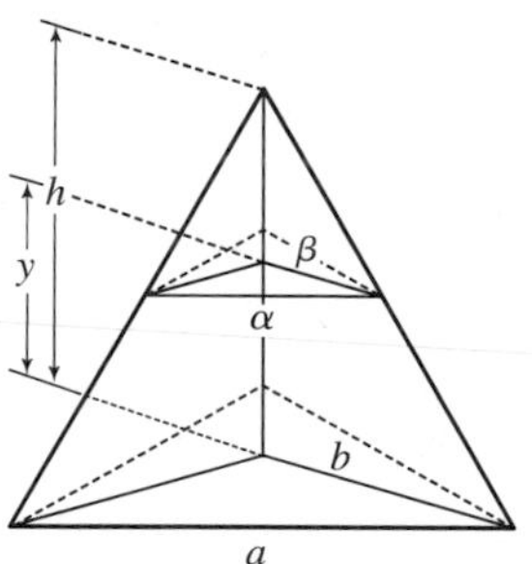

53. A cross-section at height z is a triangle similar to the base, so we'll multiply the legs of the base triangle, 3 and 4, by a proportionality factor of $(5-z)/5$. Thus, the triangle at height z has area

$$A(z) = \frac{1}{2}\cdot 3\left(\frac{5-z}{5}\right)\cdot 4\left(\frac{5-z}{5}\right) = 6\left(1 - \frac{z}{5}\right)^2, \text{ so}$$

$$V = \int_0^5 A(z)\,dz = 6\int_0^5 \left(1 - \frac{z}{5}\right)^2 dz = 6\int_1^0 u^2(-5\,du) \qquad \begin{bmatrix} u = 1 - z/5, \\ du = -\frac{1}{5}\,dz \end{bmatrix}$$

$$= -30\left[\tfrac{1}{3}u^3\right]_1^0 = -30\left(-\tfrac{1}{3}\right) = 10 \text{ cm}^3$$

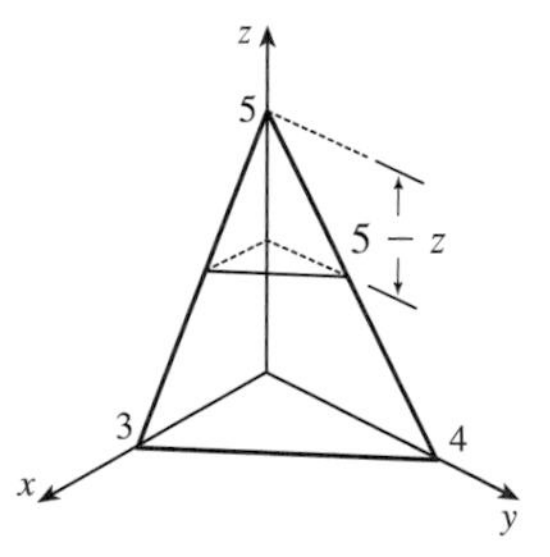

54. A cross-section is shaded in the diagram.

$$A(x) = (2y)^2 = \left(2\sqrt{r^2 - x^2}\right)^2, \text{ so}$$

$$V = \textstyle\int_{-r}^r A(x)\,dx = 2\int_0^r 4(r^2 - x^2)\,dx$$

$$= 8\left[r^2 x - \tfrac{1}{3}x^3\right]_0^r = 8\left(\tfrac{2}{3}r^3\right) = \tfrac{16}{3}r^3$$

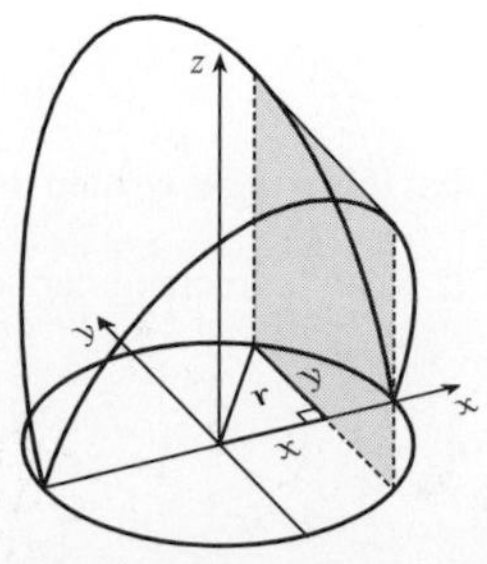

55. If l is a leg of the isosceles right triangle and $2y$ is the hypotenuse, then $l^2 + l^2 = (2y)^2 \Rightarrow 2l^2 = 4y^2 \Rightarrow l^2 = 2y^2$.

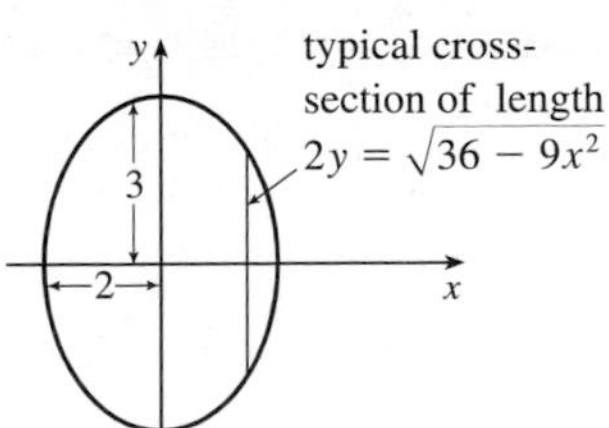

$$V = \int_{-2}^{2} A(x)\,dx = 2\int_0^2 A(x)\,dx = 2\int_0^2 \tfrac{1}{2}(l)(l)\,dx = 2\int_0^2 y^2\,dx$$
$$= 2\int_0^2 \tfrac{1}{4}(36 - 9x^2)\,dx = \tfrac{9}{2}\int_0^2 (4 - x^2)\,dx$$
$$= \tfrac{9}{2}\left[4x - \tfrac{1}{3}x^3\right]_0^2 = \tfrac{9}{2}\left(8 - \tfrac{8}{3}\right) = 24$$

56. The cross-section of the base corresponding to the coordinate y has length $x = 1 - y$. The corresponding equilateral triangle with side s has area $A(y) = s^2\left(\frac{\sqrt{3}}{4}\right) = (1-y)^2\left(\frac{\sqrt{3}}{4}\right)$. Therefore,

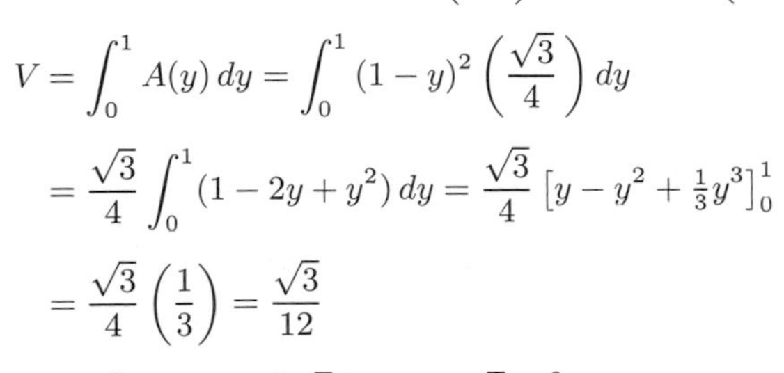

$$V = \int_0^1 A(y)\,dy = \int_0^1 (1-y)^2\left(\frac{\sqrt{3}}{4}\right)dy$$
$$= \frac{\sqrt{3}}{4}\int_0^1 (1 - 2y + y^2)\,dy = \frac{\sqrt{3}}{4}\left[y - y^2 + \tfrac{1}{3}y^3\right]_0^1$$
$$= \frac{\sqrt{3}}{4}\left(\frac{1}{3}\right) = \frac{\sqrt{3}}{12}$$

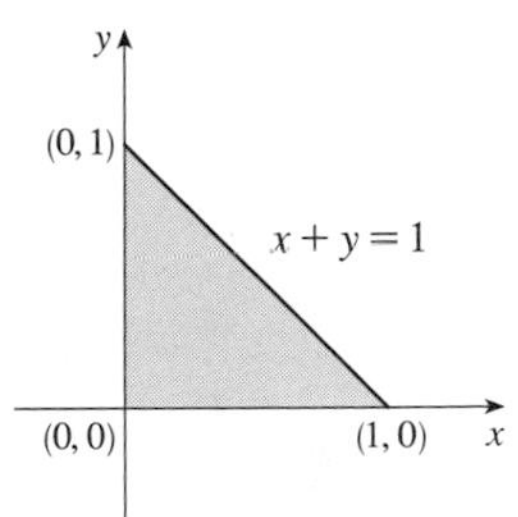

Or: $\displaystyle\int_0^1 (1-y)^2\left(\frac{\sqrt{3}}{4}\right)dy = \frac{\sqrt{3}}{4}\int_1^0 u^2(-du)\quad [u = 1 - y] = \frac{\sqrt{3}}{4}\left[\frac{1}{3}u^3\right]_0^1 = \frac{\sqrt{3}}{12}$

57. The cross-section of the base corresponding to the coordinate x has length $y = 1 - x$. The corresponding square with side s has area $A(x) = s^2 = (1-x)^2 = 1 - 2x + x^2$. Therefore,

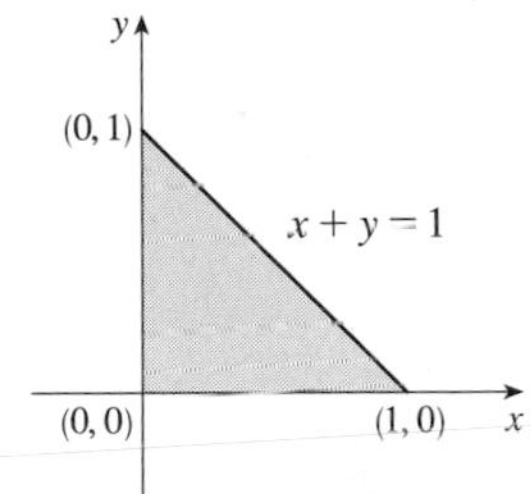

$$V = \int_0^1 A(x)\,dx = \int_0^1 (1 - 2x + x^2)\,dx$$
$$= \left[x - x^2 + \tfrac{1}{3}x^3\right]_0^1 = \left(1 - 1 + \tfrac{1}{3}\right) - 0 = \tfrac{1}{3}$$

Or: $\displaystyle\int_0^1 (1-x)^2\,dx = \int_1^0 u^2(-du)\quad [u = 1 - x] = \left[\tfrac{1}{3}u^3\right]_0^1 = \tfrac{1}{3}$

58. The cross-section of the base corresponding to the coordinate y has length $2x = 2\sqrt{1-y}$. $\left[y = 1 - x^2 \Leftrightarrow x = \pm\sqrt{1-y}\right]$ The corresponding square with side s has area $A(x) = s^2 = \left(2\sqrt{1-y}\right)^2 = 4(1-y)$. Therefore,

$$V = \int_0^1 A(y)\,dy = \int_0^1 4(1-y)\,dy = 4\left[y - \tfrac{1}{2}y^2\right]_0^1 = 4\left[\left(1 - \tfrac{1}{2}\right) - 0\right] = 2.$$

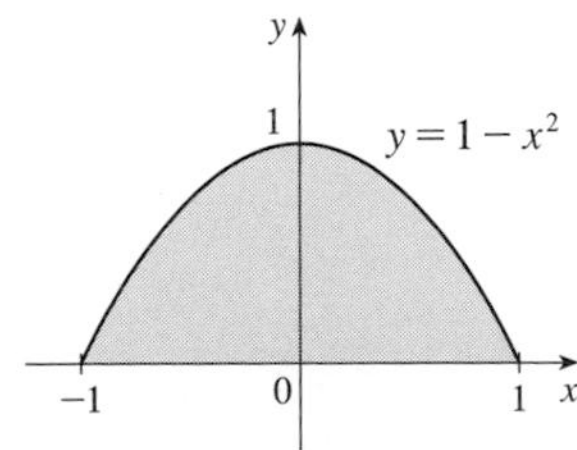

59. The cross-section of the base b corresponding to the coordinate x has length $1 - x^2$. The height h also has length $1 - x^2$, so the corresponding isosceles triangle has area $A(x) = \frac{1}{2}bh = \frac{1}{2}(1 - x^2)^2$. Therefore,

$$V = \int_{-1}^{1} A(x)\,dx = \int_{-1}^{1} \tfrac{1}{2}(1 - x^2)^2\,dx$$
$$= 2 \cdot \tfrac{1}{2}\int_0^1 (1 - 2x^2 + x^4)\,dx \qquad \text{[by symmetry]}$$
$$= \left[x - \tfrac{2}{3}x^3 + \tfrac{1}{5}x^5\right]_0^1 = \left(1 - \tfrac{2}{3} + \tfrac{1}{5}\right) - 0 = \tfrac{8}{15}$$

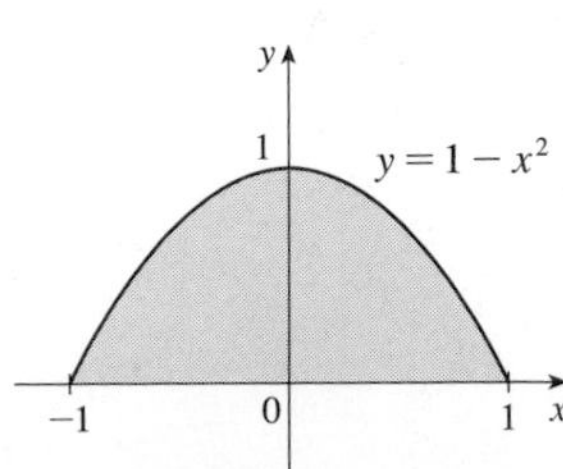

60. The cross-section of the base corresponding to the coordinate y has length $2x = 2\sqrt{2-y}$. $[y = 2 - x^2 \Leftrightarrow x = \pm\sqrt{2-y}\,]$ The corresponding cross-section of the solid S is a quarter-circle with radius $2\sqrt{2-y}$ and area $A(y) = \frac{1}{4}\pi(2\sqrt{2-y}\,)^2 = \pi(2-y)$. Therefore,

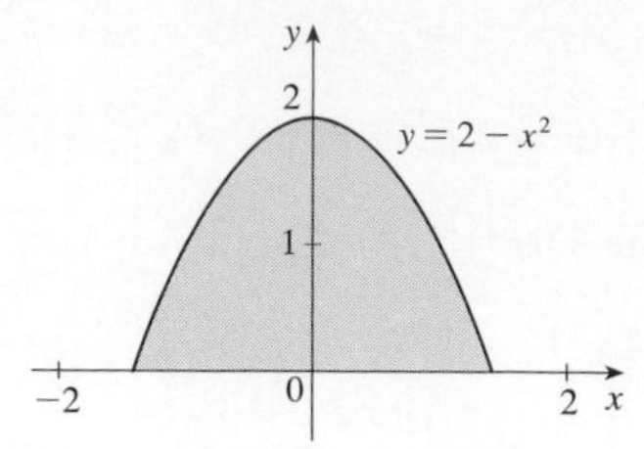

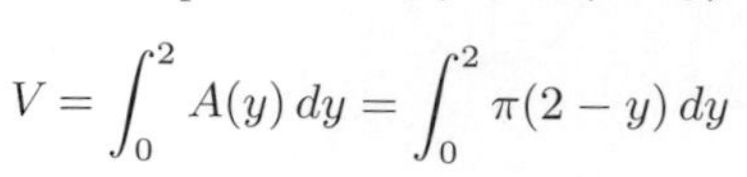

$$V = \int_0^2 A(y)\,dy = \int_0^2 \pi(2-y)\,dy$$

$$= \pi\left[2y - \tfrac{1}{2}y^2\right]_0^2 = \pi(4-2) = 2\pi$$

61. The cross-section of S at coordinate x, $-1 \le x \le 1$, is a circle centered at the point $\left(x, \frac{1}{2}(1-x^2)\right)$ with radius $\frac{1}{2}(1-x^2)$.

The area of the cross-section is

$$A(x) = \pi\left[\tfrac{1}{2}(1-x^2)\right]^2 = \tfrac{\pi}{4}(1 - 2x^2 + x^4)$$

The volume of S is

$$V = \int_{-1}^{1} A(x)\,dx = 2\int_0^1 \tfrac{\pi}{4}(1 - 2x^2 + x^4)\,dx = \tfrac{\pi}{2}\left[x - \tfrac{2}{3}x^3 + \tfrac{1}{5}x^5\right]_0^1 = \tfrac{\pi}{2}\left(1 - \tfrac{2}{3} + \tfrac{1}{5}\right) = \tfrac{\pi}{2}\left(\tfrac{8}{15}\right) = \tfrac{4\pi}{15}$$

62. (a) $V = \int_{-r}^{r} A(x)\,dx = 2\int_0^r A(x)\,dx = 2\int_0^r \frac{1}{2}h\left(2\sqrt{r^2-x^2}\,\right)dx = 2h\int_0^r \sqrt{r^2-x^2}\,dx$

(b) Observe that the integral represents one quarter of the area of a circle of radius r, so $V = 2h \cdot \frac{1}{4}\pi r^2 = \frac{1}{2}\pi h r^2$.

63. (a) The torus is obtained by rotating the circle $(x-R)^2 + y^2 = r^2$ about the y-axis. Solving for x, we see that the right half of the circle is given by $x = R + \sqrt{r^2-y^2} = f(y)$ and the left half by $x = R - \sqrt{r^2-y^2} = g(y)$. So

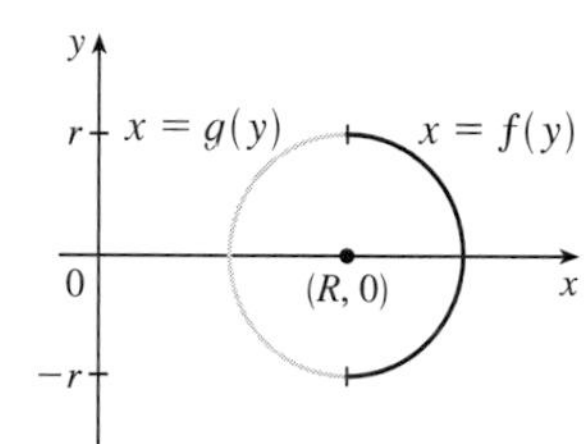

$$V = \pi\int_{-r}^{r} \left\{[f(y)]^2 - [g(y)]^2\right\}dy$$

$$= 2\pi\int_0^r \left[\left(R^2 + 2R\sqrt{r^2-y^2} + r^2 - y^2\right) - \left(R^2 - 2R\sqrt{r^2-y^2} + r^2 - y^2\right)\right]dy$$

$$= 2\pi\int_0^r 4R\sqrt{r^2-y^2}\,dy = 8\pi R\int_0^r \sqrt{r^2-y^2}\,dy$$

(b) Observe that the integral represents a quarter of the area of a circle with radius r, so $8\pi R\int_0^r \sqrt{r^2-y^2}\,dy = 8\pi R \cdot \frac{1}{4}\pi r^2 = 2\pi^2 r^2 R$.

64. The cross-sections perpendicular to the y-axis in Figure 17 are rectangles. The rectangle corresponding to the coordinate y has a base of length $2\sqrt{16-y^2}$ in the xy-plane and a height of $\frac{1}{\sqrt{3}}\,y$, since $\angle BAC = 30°$ and $|BC| = \frac{1}{\sqrt{3}}|AB|$. Thus, $A(y) = \frac{2}{\sqrt{3}}\,y\sqrt{16-y^2}$ and

$$V = \int_0^4 A(y)\,dy = \tfrac{2}{\sqrt{3}}\int_0^4 \sqrt{16-y^2}\,y\,dy = \tfrac{2}{\sqrt{3}}\int_{16}^0 u^{1/2}\left(-\tfrac{1}{2}\,du\right) \qquad [\text{Put } u = 16 - y^2, \text{ so } du = -2y\,dy]$$

$$= \tfrac{1}{\sqrt{3}}\int_0^{16} u^{1/2}\,du = \tfrac{1}{\sqrt{3}}\,\tfrac{2}{3}\left[u^{3/2}\right]_0^{16} = \tfrac{2}{3\sqrt{3}}(64) = \tfrac{128}{3\sqrt{3}}$$

65. (a) Volume$(S_1) = \int_0^h A(z)\,dz =$ Volume(S_2) since the cross-sectional area $A(z)$ at height z is the same for both solids.

(b) By Cavalieri's Principle, the volume of the cylinder in the figure is the same as that of a right circular cylinder with radius r and height h, that is, $\pi r^2 h$.

66. Each cross-section of the solid S in a plane perpendicular to the x-axis is a square (since the edges of the cut lie on the cylinders, which are perpendicular). One-quarter of this square and one-eighth of S are shown. The area of this quarter-square is $|PQ|^2 = r^2 - x^2$. Therefore, $A(x) = 4(r^2 - x^2)$ and the volume of S is

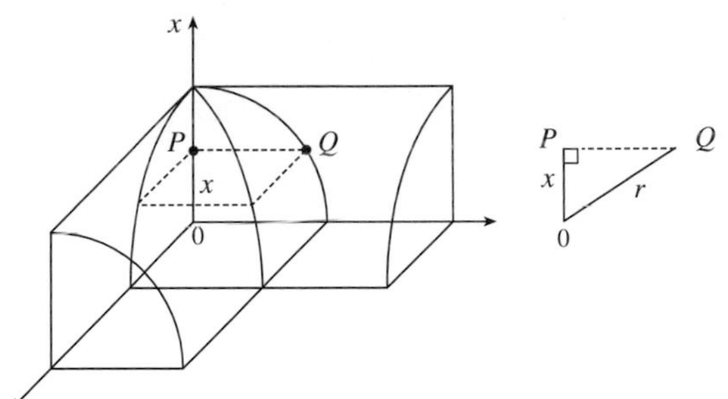

$$V = \int_{-r}^{r} A(x)\,dx = 4\int_{-r}^{r}(r^2 - x^2)\,dx$$
$$= 8(r^2 - x^2)\,dx = 8\left[r^2x - \tfrac{1}{3}x^3\right]_0^r = \tfrac{16}{3}r^3$$

67. The volume is obtained by rotating the area common to two circles of radius r, as shown. The volume of the right half is

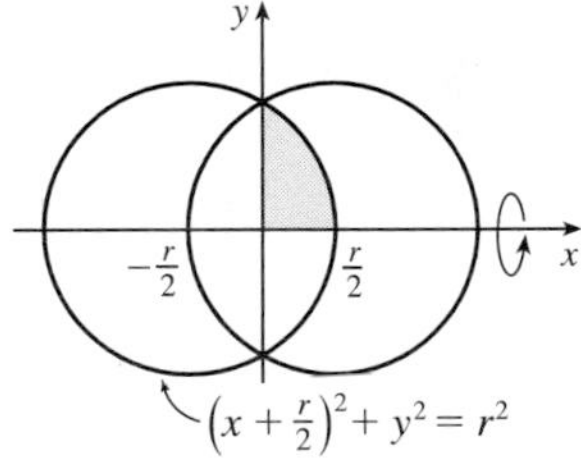

$$V_{\text{right}} = \pi\int_0^{r/2} y^2\,dx = \pi\int_0^{r/2}\left[r^2 - \left(\tfrac{1}{2}r + x\right)^2\right]dx$$
$$= \pi\left[r^2x - \tfrac{1}{3}\left(\tfrac{1}{2}r + x\right)^3\right]_0^{r/2} = \pi\left[\left(\tfrac{1}{2}r^3 - \tfrac{1}{3}r^3\right) - \left(0 - \tfrac{1}{24}r^3\right)\right] = \tfrac{5}{24}\pi r^3$$

So by symmetry, the total volume is twice this, or $\frac{5}{12}\pi r^3$.

Another solution: We observe that the volume is the twice the volume of a cap of a sphere, so we can use the formula from Exercise 49 with $h = \frac{1}{2}r$: $V = 2\cdot\frac{1}{3}\pi h^2(3r - h) - \frac{2}{3}\pi\left(\frac{1}{2}r\right)^2\left(3r - \frac{1}{2}r\right) = \frac{5}{12}\pi r^3$.

68. We consider two cases: one in which the ball is not completely submerged and the other in which it is.

Case 1: $0 \le h \le 10$ The ball will not be completely submerged, and so a cross-section of the water parallel to the surface will be the shaded area shown in the first diagram. We can find the area of the cross-section at height x above the bottom of the bowl by using the Pythagorean Theorem: $R^2 = 15^2 - (15 - x)^2$ and $r^2 = 5^2 - (x - 5)^2$, so $A(x) = \pi(R^2 - r^2) = 20\pi x$. The volume of water when it has depth h is then $V(h) = \int_0^h A(x)\,dx = \int_0^h 20\pi x\,dx = \left[10\pi x^2\right]_0^h = 10\pi h^2\ \text{cm}^3$, $0 \le h \le 10$.

Case 2: $10 < h \le 15$ In this case we can find the volume by simply subtracting the volume displaced by the ball from the total volume inside the bowl underneath the surface of the water. The total volume underneath the surface is just the volume of a cap of the bowl, so we use the formula from Exercise 49: $V_{\text{cap}}(h) = \frac{1}{3}\pi h^2(45 - h)$. The volume of the small sphere is $V_{\text{ball}} = \frac{4}{3}\pi(5)^3 = \frac{500}{3}\pi$, so the total volume is $V_{\text{cap}} - V_{\text{ball}} = \frac{1}{3}\pi(45h^2 - h^3 - 500)\ \text{cm}^3$.

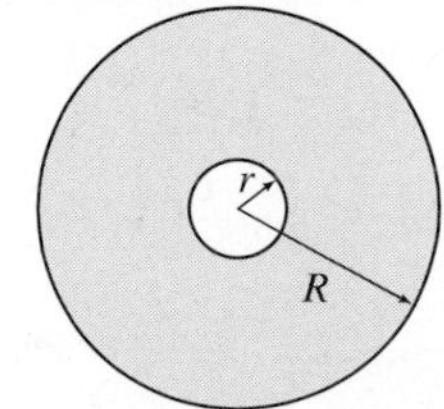

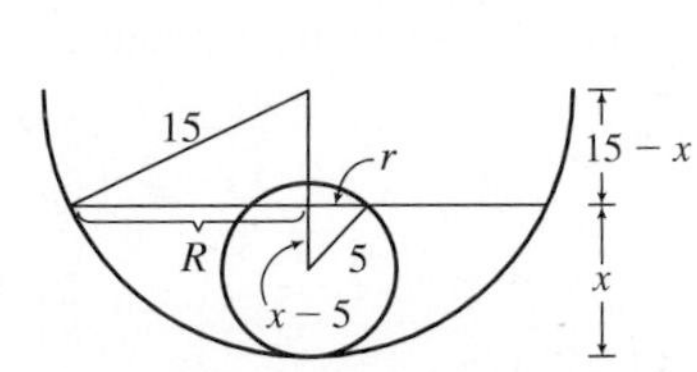

69. Take the x-axis to be the axis of the cylindrical hole of radius r.

A quarter of the cross-section through y, perpendicular to the y-axis, is the rectangle shown. Using the Pythagorean Theorem twice, we see that the dimensions of this rectangle are

$x = \sqrt{R^2 - y^2}$ and $z = \sqrt{r^2 - y^2}$, so

$\frac{1}{4}A(y) = xz = \sqrt{r^2 - y^2}\,\sqrt{R^2 - y^2}$, and

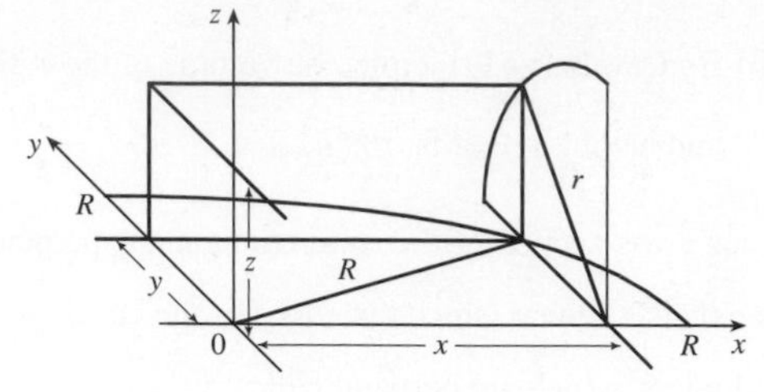

$$V = \int_{-r}^{r} A(y)\,dy = \int_{-r}^{r} 4\sqrt{r^2 - y^2}\,\sqrt{R^2 - y^2}\,dy = 8\int_0^r \sqrt{r^2 - y^2}\,\sqrt{R^2 - y^2}\,dy$$

70. The line $y = r$ intersects the semicircle $y = \sqrt{R^2 - x^2}$ when $r = \sqrt{R^2 - x^2} \Rightarrow r^2 = R^2 - x^2 \Rightarrow$ $x^2 = R^2 - r^2 \Rightarrow x = \pm\sqrt{R^2 - r^2}$. Rotating the shaded region about the x-axis gives us

$$V = \int_{-\sqrt{R^2-r^2}}^{\sqrt{R^2-r^2}} \pi\left[\left(\sqrt{R^2 - x^2}\right)^2 - r^2\right] dx = 2\pi \int_0^{\sqrt{R^2-r^2}} (R^2 - x^2 - r^2)\,dx \qquad \text{[by symmetry]}$$

$$= 2\pi \int_0^{\sqrt{R^2-r^2}} \left[(R^2 - r^2) - x^2\right] dx = 2\pi\left[(R^2 - r^2)x - \tfrac{1}{3}x^3\right]_0^{\sqrt{R^2-r^2}}$$

$$= 2\pi\left[(R^2 - r^2)^{3/2} - \tfrac{1}{3}(R^2 - r^2)^{3/2}\right] = 2\pi \cdot \tfrac{2}{3}(R^2 - r^2)^{3/2} = \tfrac{4\pi}{3}(R^2 - r^2)^{3/2}$$

Our answer makes sense in limiting cases. As $r \to 0$, $V \to \frac{4}{3}\pi R^3$, which is the volume of the full sphere. As $r \to R$, $V \to 0$, which makes sense because the hole's radius is approaching that of the sphere.

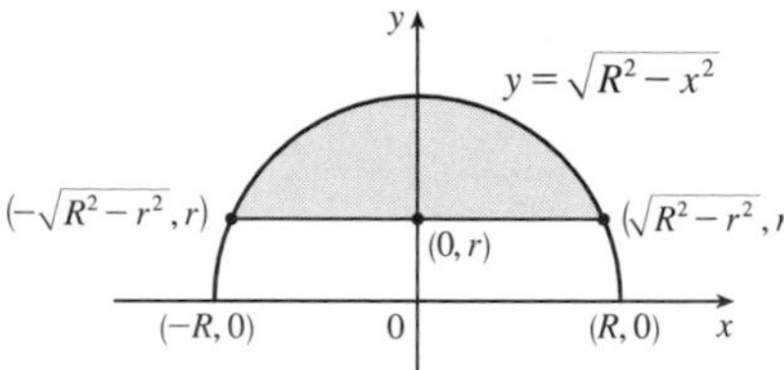

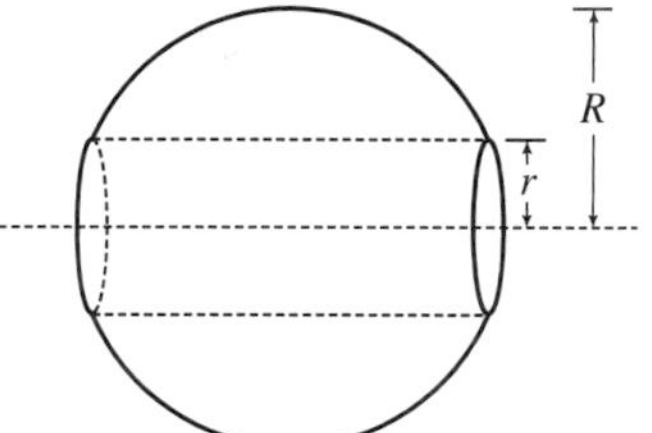

71. (a) The radius of the barrel is the same at each end by symmetry, since the function $y = R - cx^2$ is even. Since the barrel is obtained by rotating the graph of the function y about the x-axis, this radius is equal to the value of y at $x = \frac{1}{2}h$, which is $R - c\left(\frac{1}{2}h\right)^2 = R - d = r$.

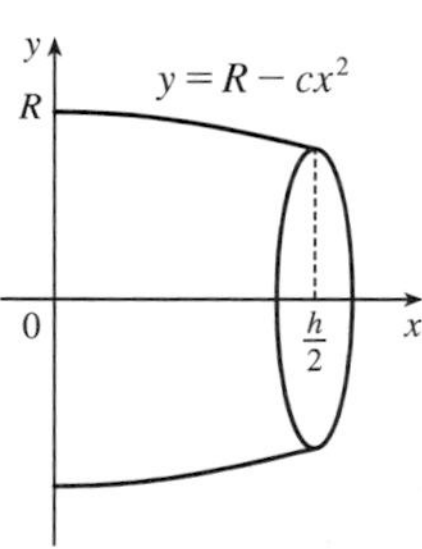

(b) The barrel is symmetric about the y-axis, so its volume is twice the volume of that part of the barrel for $x > 0$. Also, the barrel is a volume of rotation, so

$$V = 2\int_0^{h/2} \pi y^2\,dx = 2\pi \int_0^{h/2} \left(R - cx^2\right)^2 dx = 2\pi\left[R^2x - \tfrac{2}{3}Rcx^3 + \tfrac{1}{5}c^2x^5\right]_0^{h/2}$$

$$= 2\pi\left(\tfrac{1}{2}R^2h - \tfrac{1}{12}Rch^3 + \tfrac{1}{160}c^2h^5\right)$$

[continued]

Trying to make this look more like the expression we want, we rewrite it as $V = \frac{1}{3}\pi h\left[2R^2 + \left(R^2 - \frac{1}{2}Rch^2 + \frac{3}{80}c^2h^4\right)\right]$.

But $R^2 - \frac{1}{2}Rch^2 + \frac{3}{80}c^2h^4 = \left(R - \frac{1}{4}ch^2\right)^2 - \frac{1}{40}c^2h^4 = (R-d)^2 - \frac{2}{5}\left(\frac{1}{4}ch^2\right)^2 = r^2 - \frac{2}{5}d^2$.

Substituting this back into V, we see that $V = \frac{1}{3}\pi h\left(2R^2 + r^2 - \frac{2}{5}d^2\right)$, as required.

72. It suffices to consider the case where $\mathcal{R}$ is bounded by the curves $y = f(x)$ and $y = g(x)$ for $a \le x \le b$, where $g(x) \le f(x)$ for all x in $[a, b]$, since other regions can be decomposed into subregions of this type. We are concerned with the volume obtained when $\mathcal{R}$ is rotated about the line $y = -k$, which is equal to

$$\begin{aligned} V_2 &= \pi \int_a^b \left([f(x)+k]^2 - [g(x)+k]^2\right) dx \\ &= \pi \int_a^b \left([f(x)]^2 - [g(x)]^2\right) dx + 2\pi k \int_a^b [f(x) - g(x)]\, dx = V_1 + 2\pi k A \end{aligned}$$

6.3 Volumes by Cylindrical Shells

1.

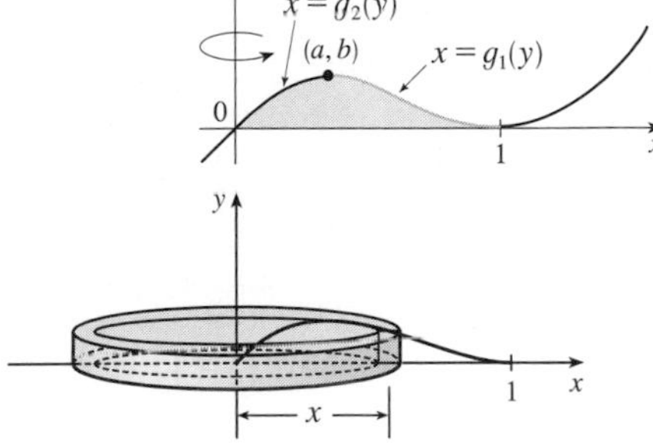

If we were to use the "washer" method, we would first have to locate the local maximum point (a, b) of $y = x(x-1)^2$ using the methods of Chapter 4. Then we would have to solve the equation $y = x(x-1)^2$ for x in terms of y to obtain the functions $x = g_1(y)$ and $x = g_2(y)$ shown in the first figure. This step would be difficult because it involves the cubic formula. Finally we would find the volume using $V = \pi \int_0^b \left\{[g_1(y)]^2 - [g_2(y)]^2\right\} dy$.

Using shells, we find that a typical approximating shell has radius x, so its circumference is $2\pi x$. Its height is y, that is, $x(x-1)^2$. So the total volume is

$$V = \int_0^1 2\pi x\left[x(x-1)^2\right] dx = 2\pi \int_0^1 \left(x^4 - 2x^3 + x^2\right) dx = 2\pi\left[\frac{x^5}{5} - 2\frac{x^4}{4} + \frac{x^3}{3}\right]_0^1 = \frac{\pi}{15}$$

2.

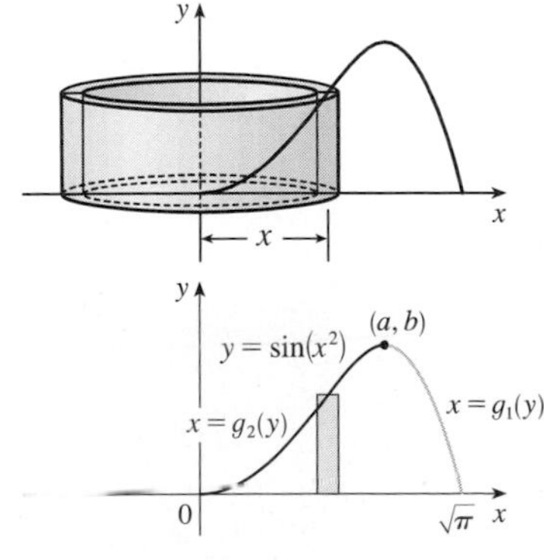

A typical cylindrical shell has circumference $2\pi x$ and height $\sin(x^2)$. $V = \int_0^{\sqrt{\pi}} 2\pi x \sin(x^2)\, dx$. Let $u = x^2$. Then $du = 2x\, dx$, so $V = \pi \int_0^{\pi} \sin u\, du = \pi[-\cos u]_0^{\pi} = \pi[1 - (-1)] = 2\pi$. For slicing, we would first have to locate the local maximum point (a, b) of $y = \sin(x^2)$ using the methods of Chapter 4. Then we would have to solve the equation $y = \sin(x^2)$ for x in terms of y to obtain the functions $x = g_1(y)$ and $x = g_2(y)$ shown in the second figure. Finally we would find the volume using $V = \pi \int_0^b \left\{[g_1(y)]^2 - [g_2(y)]^2\right\} dy$. Using shells is definitely preferable to slicing.

3. $V = \int_0^1 2\pi x \sqrt[3]{x}\, dx = 2\pi \int_0^1 x^{4/3}\, dx$

$= 2\pi \left[\frac{3}{7}x^{7/3}\right]_0^1 = 2\pi\left(\frac{3}{7}\right) = \frac{6}{7}\pi$

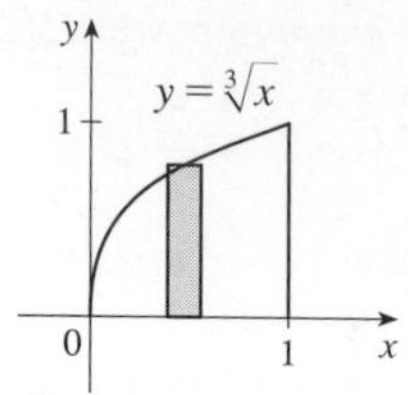

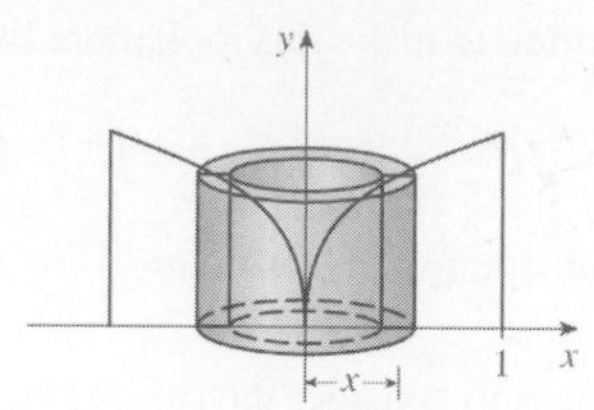

4. $V = \int_1^2 2\pi x \cdot x^3\, dx = 2\pi \int_1^2 x^4\, dx$

$= 2\pi\left[\frac{1}{5}x^5\right]_1^2 = 2\pi\left(\frac{32}{5} - \frac{1}{5}\right) = \frac{62}{5}\pi$

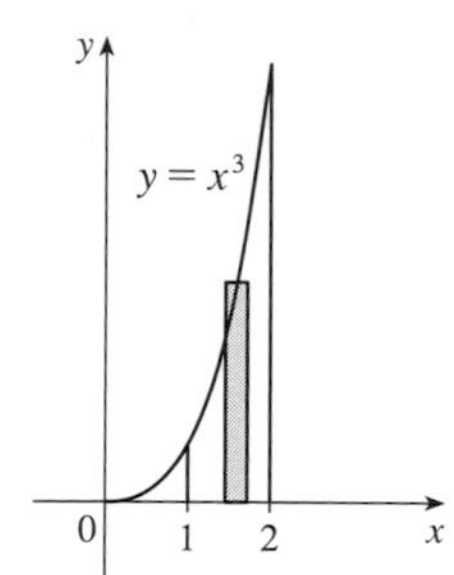

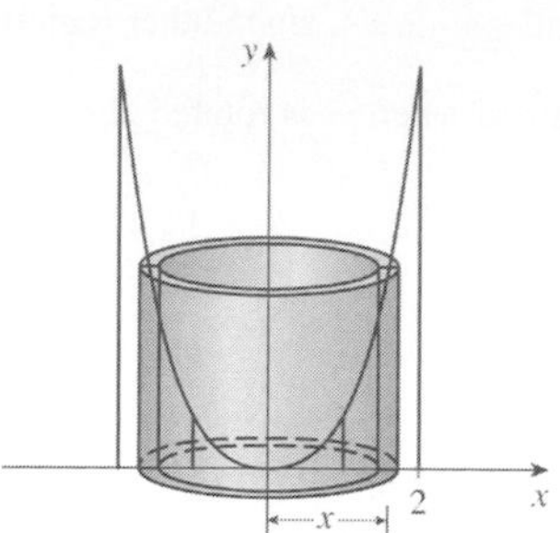

5. $V = \int_0^1 2\pi x e^{-x^2}\, dx$. Let $u = x^2$.

Thus, $du = 2x\, dx$, so

$V = \pi \int_0^1 e^{-u}\, du = \pi\left[-e^{-u}\right]_0^1 = \pi(1 - 1/e)$.

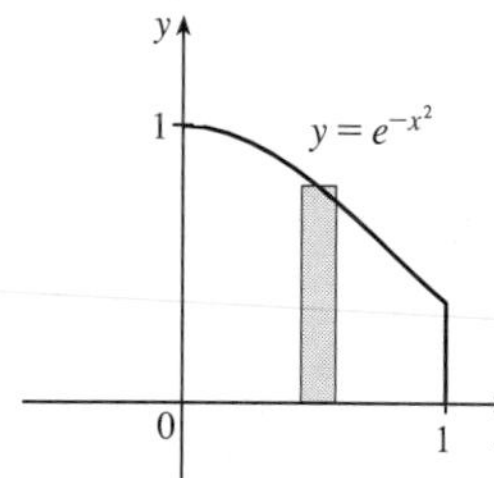

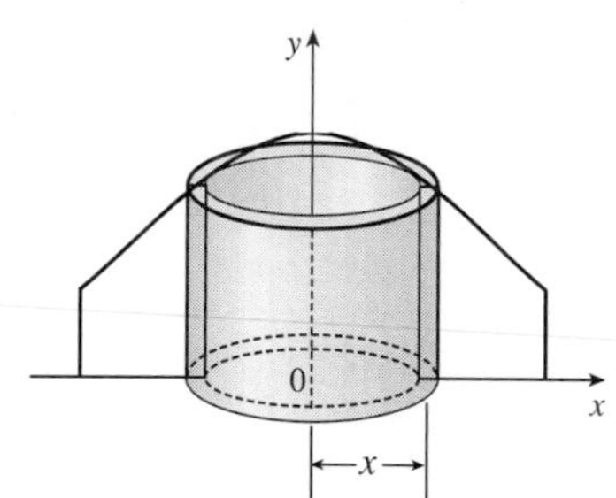

6. $4x - x^2 = x \quad \Leftrightarrow \quad 0 = x^2 - 3x \quad \Leftrightarrow \quad 0 = x(x-3) \quad \Leftrightarrow \quad x = 0$ or 3.

$V = \int_0^3 2\pi x[(4x - x^2) - x]\, dx$

$= 2\pi \int_0^3 (-x^3 + 3x^2)\, dx$

$= 2\pi\left[-\frac{1}{4}x^4 + x^3\right]_0^3$

$= 2\pi\left(-\frac{81}{4} + 27\right) = 2\pi\left(\frac{27}{4}\right) = \frac{27}{2}\pi$

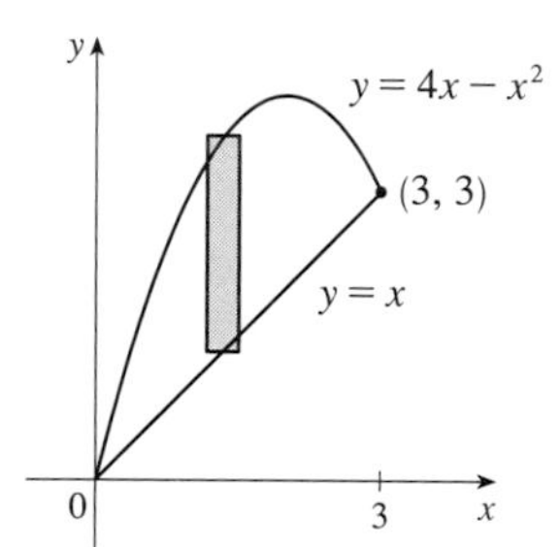

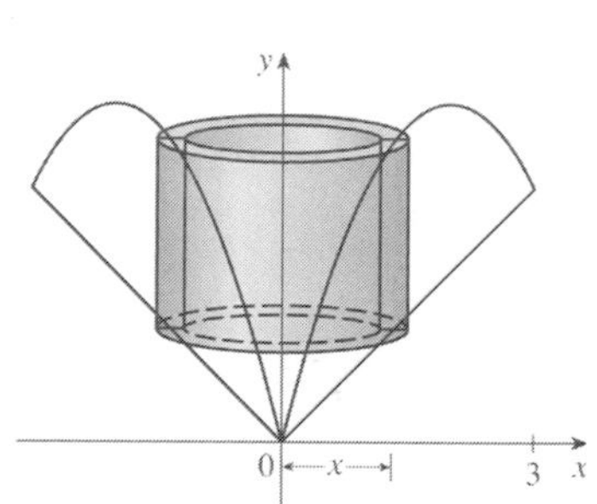

7. $x^2 = 6x - 2x^2 \quad \Leftrightarrow \quad 3x^2 - 6x = 0 \quad \Leftrightarrow \quad 3x(x-2) = 0 \quad \Leftrightarrow \quad x = 0$ or 2.

$V = \int_0^2 2\pi x[(6x - 2x^2) - x^2]\, dx$

$= 2\pi \int_0^2 (-3x^3 + 6x^2)\, dx$

$= 2\pi\left[-\frac{3}{4}x^4 + 2x^3\right]_0^2$

$= 2\pi(-12 + 16) = 8\pi$

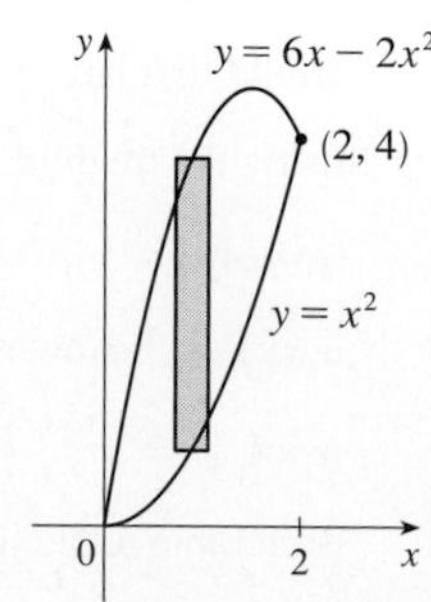

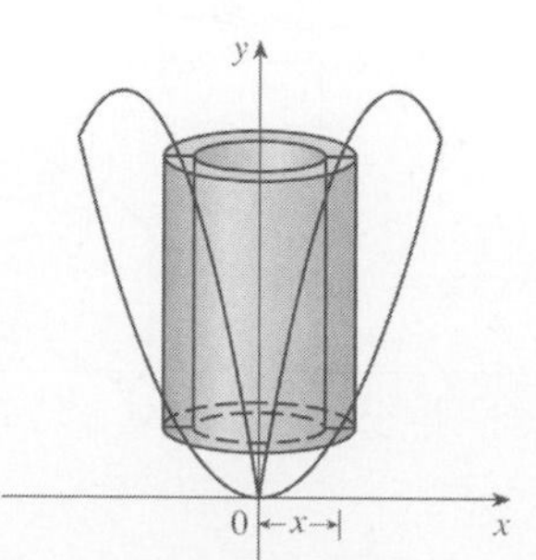

8. By slicing:

$$V = \int_0^1 \pi\left[\left(\sqrt{y}\right)^2 - (y^2)^2\right] dy = \pi \int_0^1 (y - y^4)\, dy$$

$$= \pi\left[\tfrac{1}{2}y^2 - \tfrac{1}{5}y^5\right]_0^1 = \pi\left(\tfrac{1}{2} - \tfrac{1}{5}\right) = \tfrac{3}{10}\pi$$

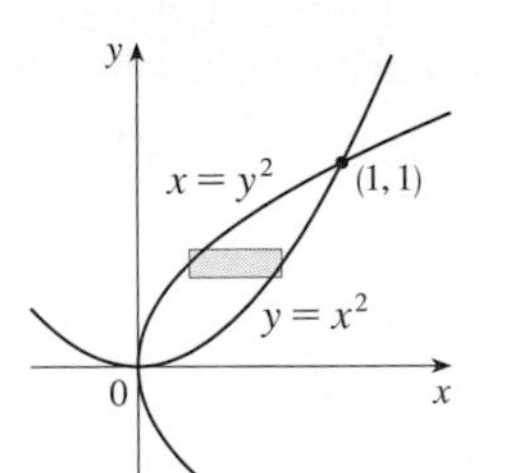

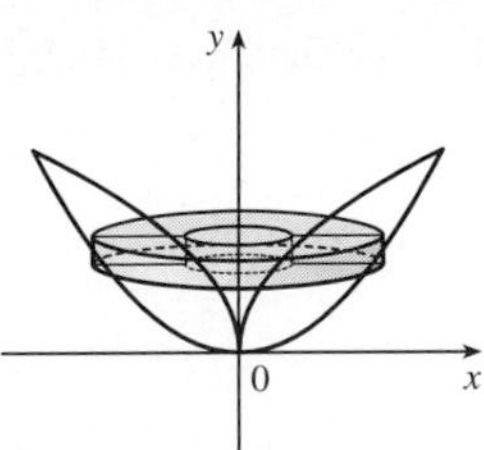

By cylindrical shells:

$$V = \int_0^1 2\pi x\left(\sqrt{x} - x^2\right) dx = 2\pi \int_0^1 (x^{3/2} - x^3)\, dx = 2\pi\left[\tfrac{2}{5}x^{5/2} - \tfrac{1}{4}x^4\right]_0^1$$

$$= 2\pi\left(\tfrac{2}{5} - \tfrac{1}{4}\right) = 2\pi\left(\tfrac{3}{20}\right) = \tfrac{3}{10}\pi$$

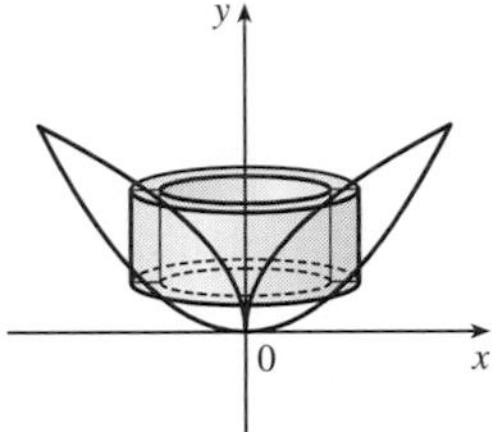

9. $xy = 1 \;\Rightarrow\; x = \frac{1}{y}$. The shell has radius y, circumference $2\pi y$, and height $1/y$, so

$$V = \int_1^3 2\pi y\left(\frac{1}{y}\right) dy$$

$$= 2\pi \int_1^3 dy = 2\pi\left[y\right]_1^3$$

$$= 2\pi(3 - 1) = 4\pi$$

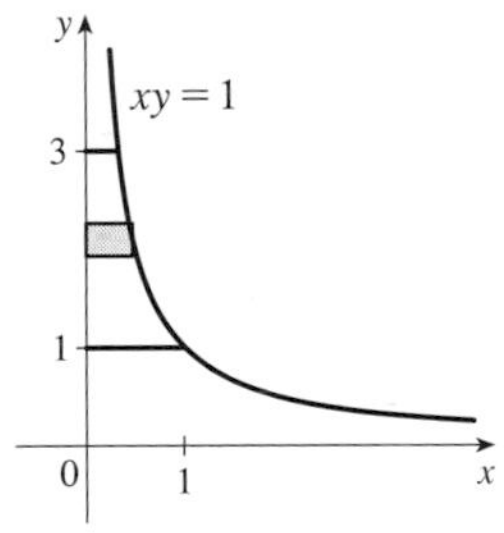

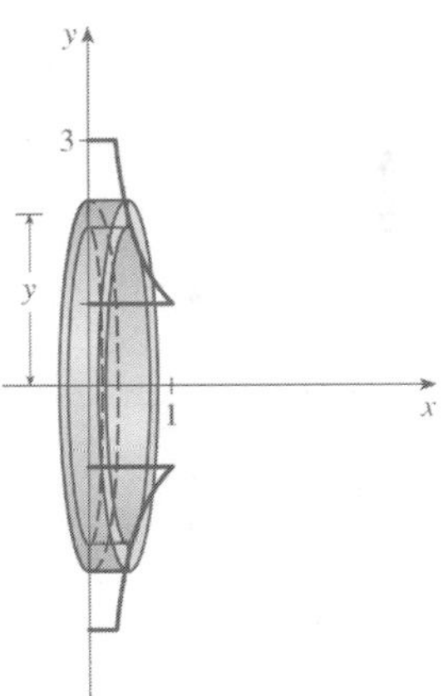

10. $y = \sqrt{x} \;\Rightarrow\; x = y^2$. The shell has radius y, circumference $2\pi y$, and height y^2, so

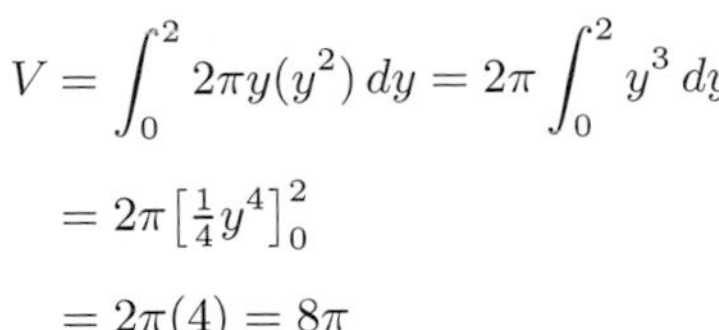

$$V = \int_0^2 2\pi y(y^2)\, dy = 2\pi \int_0^2 y^3\, dy$$

$$= 2\pi\left[\tfrac{1}{4}y^4\right]_0^2$$

$$= 2\pi(4) = 8\pi$$

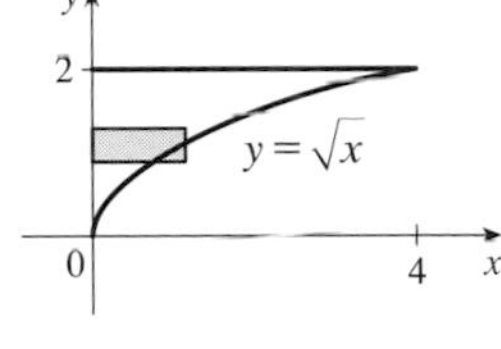

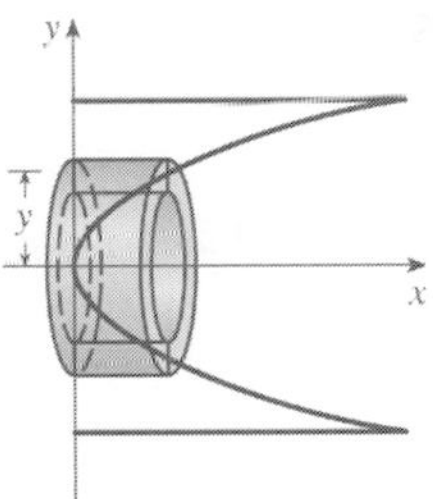

11. $y = x^{3/2} \;\Rightarrow\; x = y^{2/3}$. The shell has radius y, circumference $2\pi y$, and height $y^{2/3}$, so

$$V = \int_0^8 2\pi y(y^{2/3})\, dy = 2\pi \int_0^8 y^{5/3}\, dy$$

$$= 2\pi\left[\tfrac{3}{8}y^{8/3}\right]_0^8$$

$$= 2\pi \cdot \tfrac{3}{8} \cdot 256 = 192\pi$$

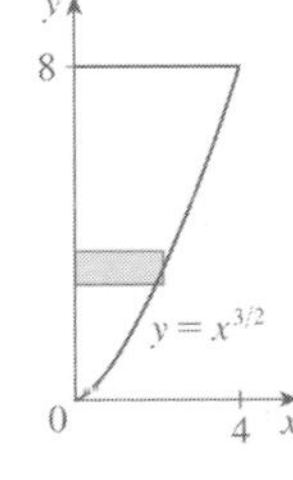

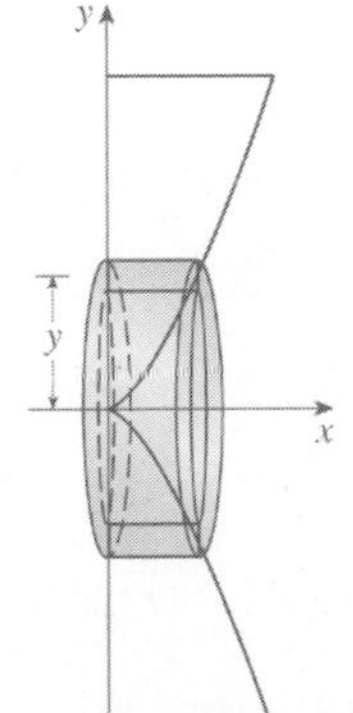

12. The shell has radius y, circumference $2\pi y$, and height $-3y^2 + 12y - 9$, so

$$\begin{aligned} V &= \int_1^3 2\pi y(-3y^2 + 12y - 9)\,dy \\ &= 2\pi \int_1^3 (-3y^3 + 12y^2 - 9y)\,dy \\ &= -6\pi \int_1^3 (y^3 - 4y^2 + 3y)\,dy \\ &= -6\pi \left[\tfrac{1}{4}y^4 - \tfrac{4}{3}y^3 + \tfrac{3}{2}y^2\right]_1^3 \\ &= -6\pi\left[\left(\tfrac{81}{4} - 36 + \tfrac{27}{2}\right) - \left(\tfrac{1}{4} - \tfrac{4}{3} + \tfrac{3}{2}\right)\right] \\ &= -6\pi\left(-\tfrac{8}{3}\right) = 16\pi \end{aligned}$$

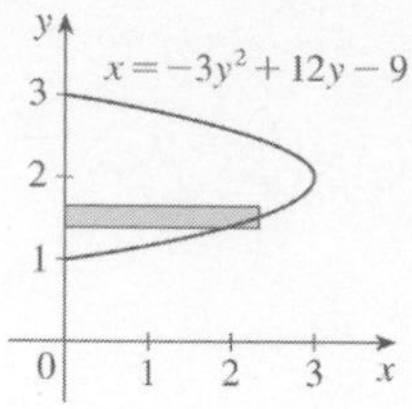

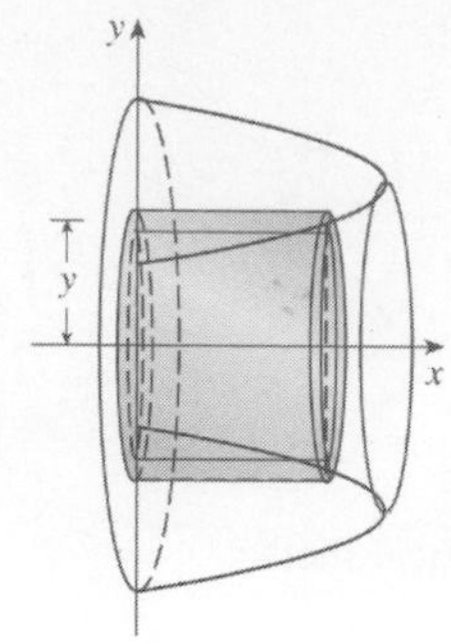

13. The shell has radius y, circumference $2\pi y$, and height

$2 - \left[1 + (y-2)^2\right] = 1 - (y-2)^2 = 1 - (y^2 - 4y + 4) = -y^2 + 4y - 3$, so

$$\begin{aligned} V &= \int_1^3 2\pi y(-y^2 + 4y - 3)\,dy \\ &= 2\pi \int_1^3 (-y^3 + 4y^2 - 3y)\,dy \\ &= 2\pi\left[-\tfrac{1}{4}y^4 + \tfrac{4}{3}y^3 - \tfrac{3}{2}y^2\right]_1^3 \\ &= 2\pi\left[\left(-\tfrac{81}{4} + 36 - \tfrac{27}{2}\right) - \left(-\tfrac{1}{4} + \tfrac{4}{3} - \tfrac{3}{2}\right)\right] \\ &= 2\pi\left(\tfrac{8}{3}\right) = \tfrac{16}{3}\pi \end{aligned}$$

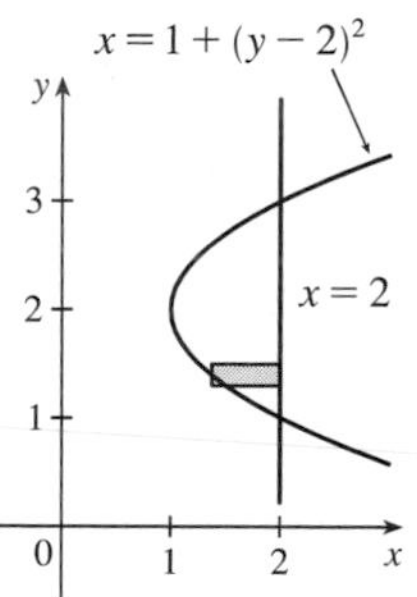

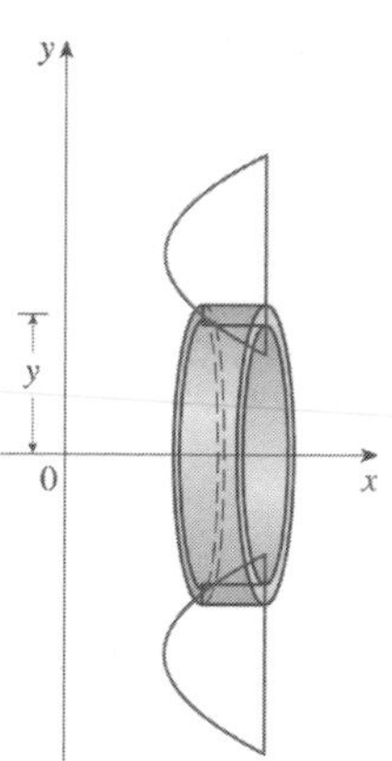

14. The curves intersect when $4 - y = y^2 - 4y + 4 \Leftrightarrow 0 = y^2 - 3y \Leftrightarrow 0 = y(y-3) \Leftrightarrow y = 0$ or 3. The shell has radius y, circumference $2\pi y$, and height $(4 - y) - (y^2 - 4y + 4) = -y^2 + 3y$, so

$$\begin{aligned} V &= \int_0^3 2\pi y(-y^2 + 3y)\,dy = 2\pi \int_0^3 (3y^2 - y^3)\,dy \\ &= 2\pi\left[y^3 - \tfrac{1}{4}y^4\right]_0^3 = 2\pi\left(27 - \tfrac{81}{4}\right) = 2\pi\left(\tfrac{27}{4}\right) = \tfrac{27\pi}{2} \end{aligned}$$

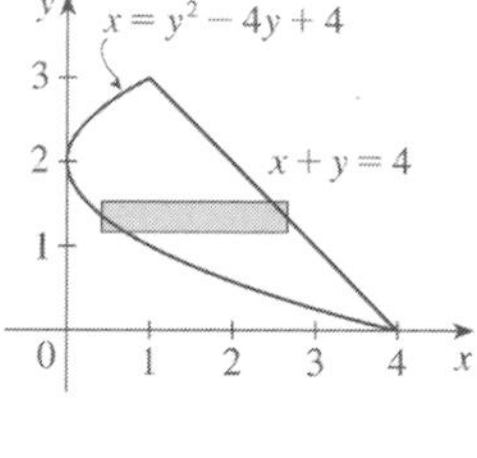

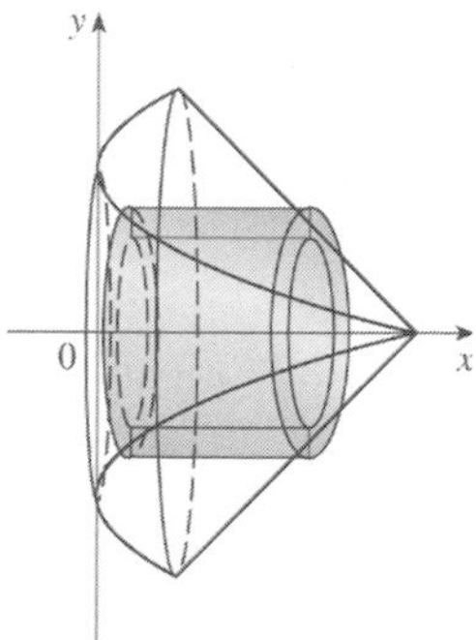

15. The shell has radius $3 - x$, circumference $2\pi(3 - x)$, and height $8 - x^3$.

$$\begin{aligned} V &= \int_0^2 2\pi(3 - x)(8 - x^3)\,dx \\ &= 2\pi \int_0^2 (x^4 - 3x^3 - 8x + 24)\,dx \\ &= 2\pi\left[\tfrac{1}{5}x^5 - \tfrac{3}{4}x^4 - 4x^2 + 24x\right]_0^2 \\ &= 2\pi\left(\tfrac{32}{5} - 12 - 16 + 48\right) = 2\pi\left(\tfrac{132}{5}\right) = \tfrac{264\pi}{5} \end{aligned}$$

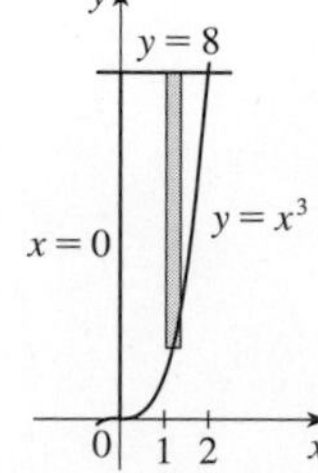

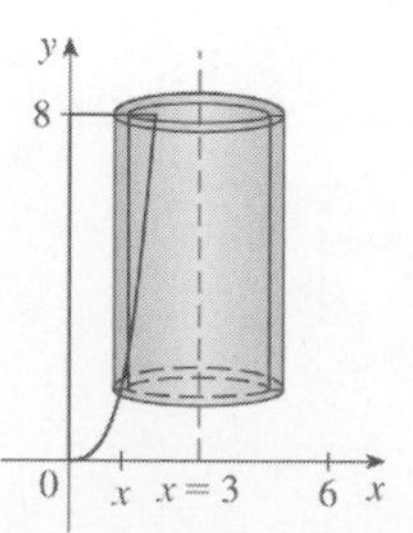

16. The shell has radius $x - (-1) = x + 1$, circumference $2\pi(x+1)$, and height $4 - 2x$.

$$\begin{aligned} V &= \int_0^2 2\pi(x+1)(4-2x)\,dx \\ &= 4\pi \int_0^2 (x+1)(2-x)\,dx \\ &= 4\pi \int_0^2 (-x^2+x+2)\,dx \\ &= 4\pi \left[-\tfrac{1}{3}x^3 + \tfrac{1}{2}x^2 + 2x\right]_0^2 \\ &= 4\pi\left(-\tfrac{8}{3} + 2 + 4\right) = 4\pi\left(\tfrac{10}{3}\right) = \tfrac{40\pi}{3} \end{aligned}$$

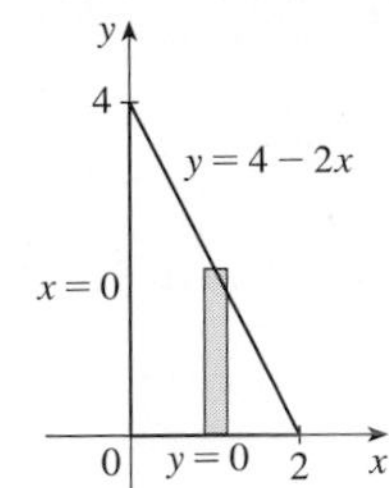

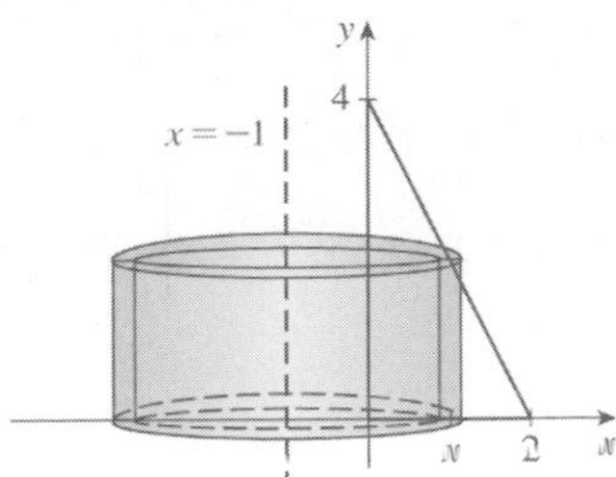

17. The shell has radius $x - 1$, circumference $2\pi(x-1)$, and height $(4x - x^2) - 3 = -x^2 + 4x - 3$.

$$\begin{aligned} V &= \int_1^3 2\pi(x-1)(-x^2+4x-3)\,dx \\ &= 2\pi \int_1^3 (-x^3+5x^2-7x+3)\,dx \\ &= 2\pi\left[-\tfrac{1}{4}x^4 + \tfrac{5}{3}x^3 - \tfrac{7}{2}x^2 + 3x\right]_1^3 \\ &= 2\pi\left[\left(-\tfrac{81}{4} + 45 - \tfrac{63}{2} + 9\right) - \left(-\tfrac{1}{4} + \tfrac{5}{3} - \tfrac{7}{2} + 3\right)\right] \\ &= 2\pi\left(\tfrac{4}{3}\right) = \tfrac{8}{3}\pi \end{aligned}$$

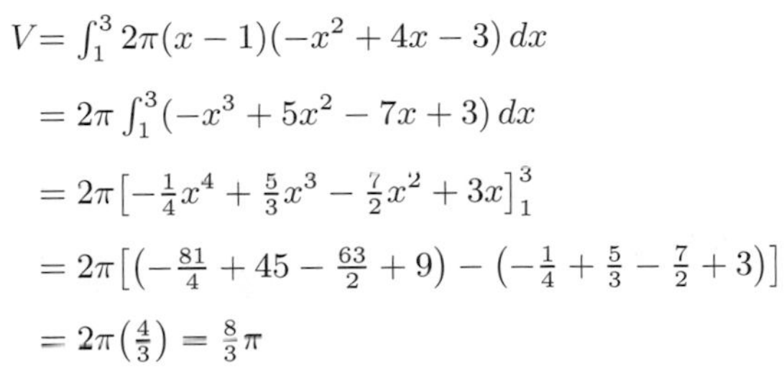

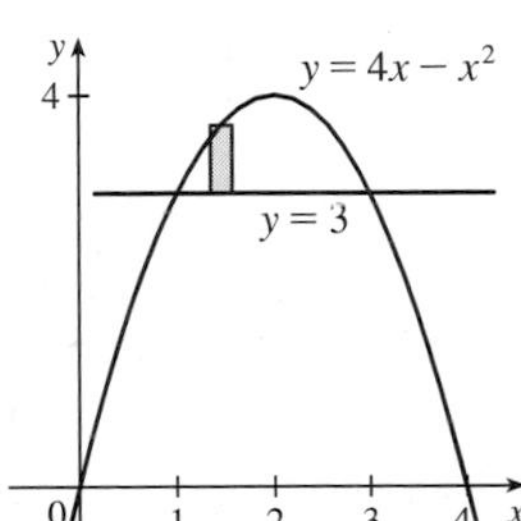

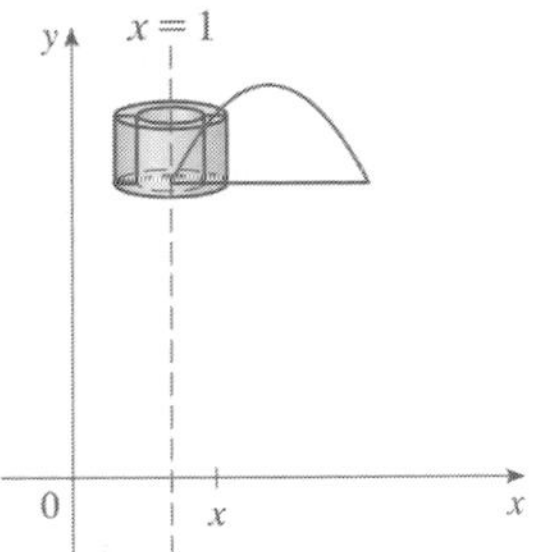

18. The shell has radius $5 - x$, circumference $2\pi(5-x)$, and height $\sqrt{x} - \frac{1}{2}x$.

$$\begin{aligned} V &= \int_0^4 2\pi(5-x)\left(\sqrt{x} - \tfrac{1}{2}x\right)dx = 2\pi\int_0^4 \left(5x^{1/2} - \tfrac{5}{2}x - x^{3/2} + \tfrac{1}{2}x^2\right)dx \\ &= 2\pi\left[\tfrac{10}{3}x^{3/2} - \tfrac{5}{4}x^2 - \tfrac{2}{5}x^{5/2} + \tfrac{1}{6}x^3\right]_0^4 = 2\pi\left(\tfrac{80}{3} - 20 - \tfrac{64}{5} + \tfrac{32}{3}\right) \\ &= 2\pi\left(\tfrac{68}{15}\right) = \tfrac{136\pi}{15} \end{aligned}$$

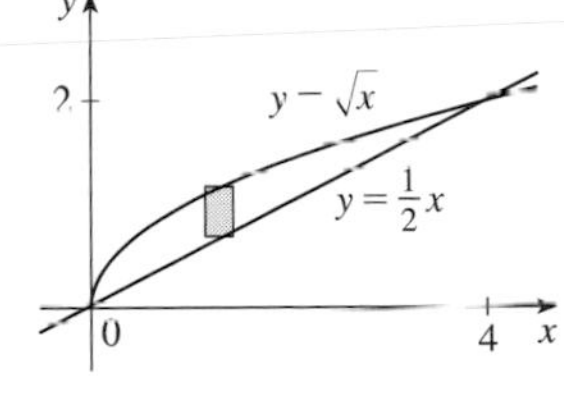

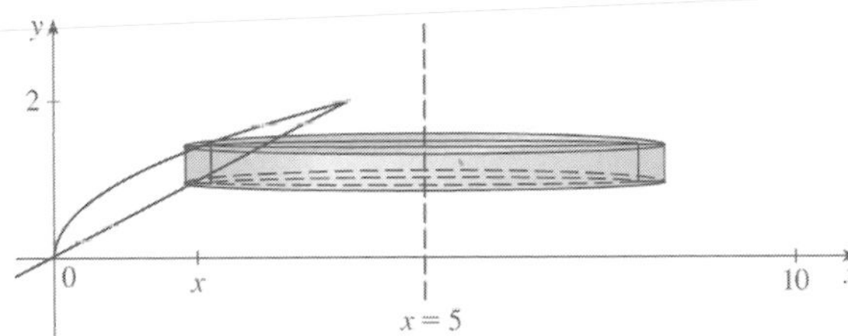

19. The shell has radius $2 - y$, circumference $2\pi(2-y)$, and height $2 - 2y^2$.

$$\begin{aligned} V &= \int_0^1 2\pi(2-y)(2-2y^2)\,dy \\ &= 4\pi\int_0^1 (2-y)(1-y^2)\,dy \\ &= 4\pi\int_0^1 (y^3 - 2y^2 - y + 2)\,dy \\ &= 4\pi\left[\tfrac{1}{4}y^4 - \tfrac{2}{3}y^3 - \tfrac{1}{2}y^2 + 2y\right]_0^1 \\ &= 4\pi\left(\tfrac{1}{4} - \tfrac{2}{3} - \tfrac{1}{2} + 2\right) \\ &= 4\pi\left(\tfrac{13}{12}\right) = \tfrac{13\pi}{3} \end{aligned}$$

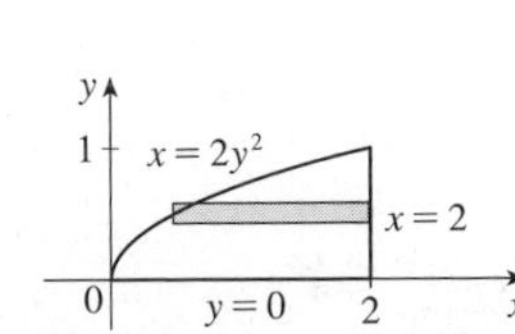

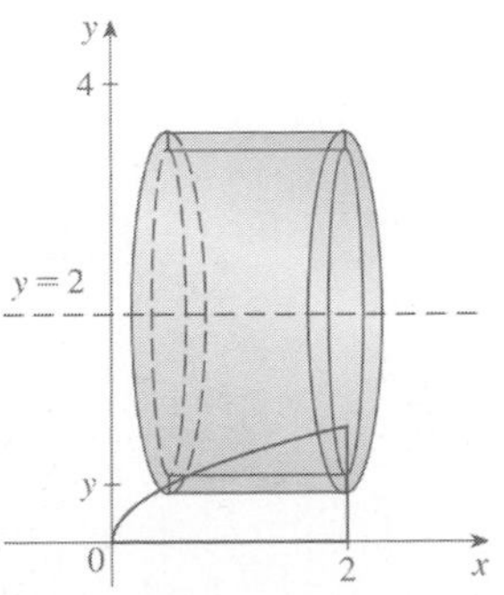

20. The shell has radius $y-(-2)=y+2$, circumference $2\pi(y+2)$, and height $(y^2+1)-2y^2=1-y^2$.

$$\begin{aligned} V &= \int_{-1}^{1} 2\pi(y+2)(1-y^2)\,dy \\ &= 2\pi\int_{-1}^{1}(-y^3-2y^2+y+2)\,dy \\ &= 4\pi\int_0^1(-2y^2+2)\,dy \qquad \text{[by Theorem 5.5.7]} \\ &= 8\pi\int_0^1(1-y^2)\,dy = 8\pi\left[y-\tfrac{1}{3}y^3\right]_0^1 \\ &= 8\pi\left(1-\tfrac{1}{3}\right) = 8\pi\left(\tfrac{2}{3}\right) = \tfrac{16\pi}{3} \end{aligned}$$

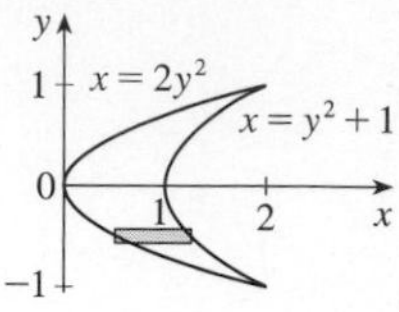

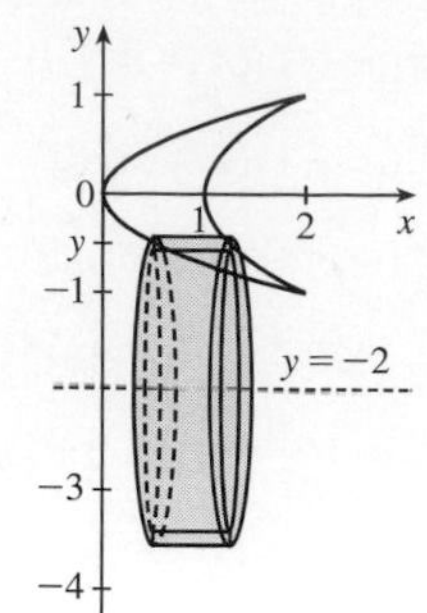

21. (a) $V = 2\pi\displaystyle\int_0^2 x(xe^{-x})\,dx = 2\pi\int_0^2 x^2e^{-x}\,dx$

(b) $V \approx 4.06300$

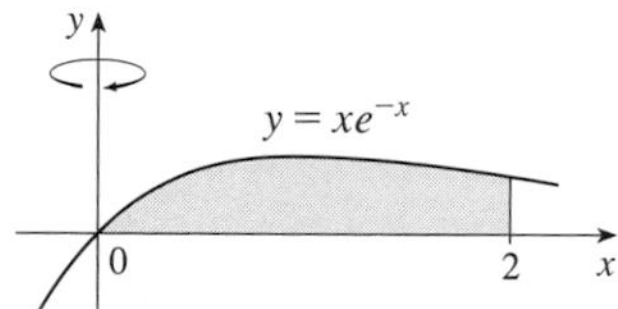

22. (a) $V = 2\pi\displaystyle\int_0^{\pi/4}\left(\frac{\pi}{2}-x\right)\tan x\,dx$

(b) $V \approx 2.25323$

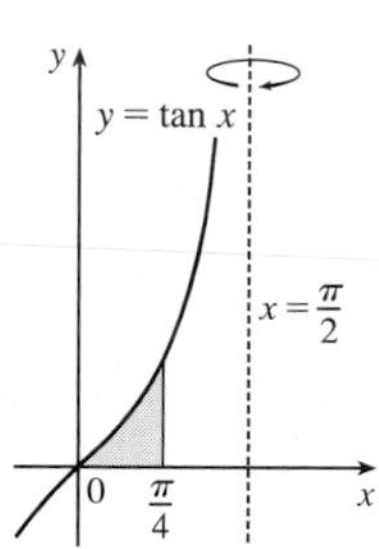

23. (a) $V = 2\pi\displaystyle\int_{-\pi/2}^{\pi/2}(\pi-x)[\cos^4 x-(-\cos^4 x)]\,dx$

$$= 4\pi\int_{-\pi/2}^{\pi/2}(\pi-x)\cos^4 x\,dx$$

[or $8\pi^2\int_0^{\pi/2}\cos^4 x\,dx$ using Theorem 5.5.7]

(b) $V \approx 46.50942$

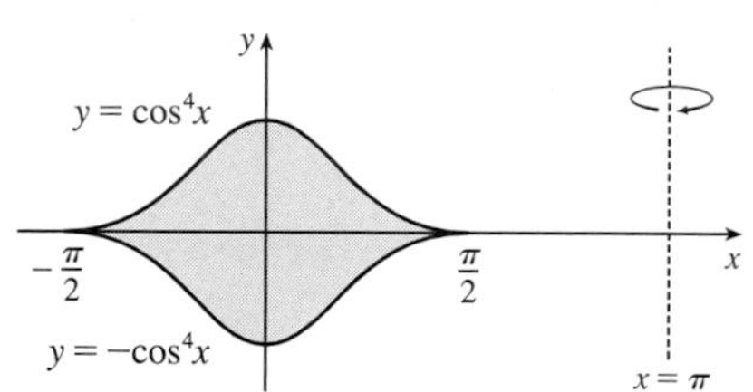

24. (a) $x = \dfrac{2x}{1+x^3} \;\Rightarrow\; x+x^4=2x \;\Rightarrow\; x^4-x=0 \;\Rightarrow$

$x(x^3-1)=0 \;\Rightarrow\; x(x-1)(x^2+x+1)=0 \;\Rightarrow\; x=0$ or 1

$$V = 2\pi\int_0^1[x-(-1)]\left(\frac{2x}{1+x^3}-x\right)dx$$

(b) $V \approx 2.36164$

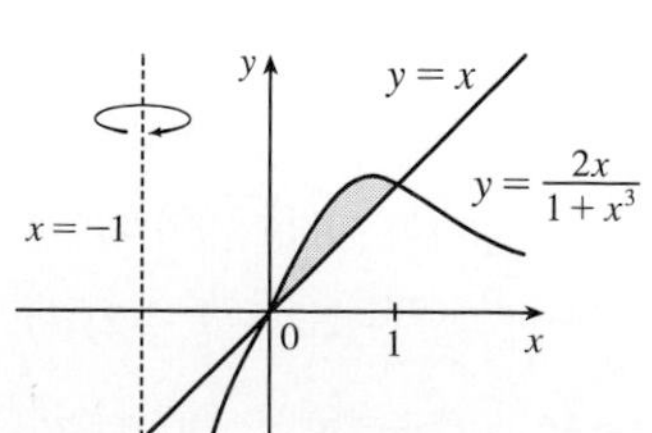

25. (a) $V = \int_0^{\pi} 2\pi(4 - y)\sqrt{\sin y}\,dy$ (b) $V \approx 36.57476$

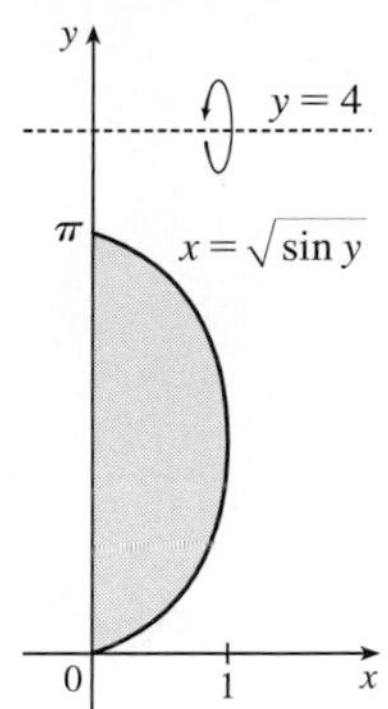

26. (a) $V = \int_{-3}^{3} 2\pi(5 - y)\left(4 - \sqrt{y^2 + 7}\right) dy$ (b) $V \approx 163.02712$

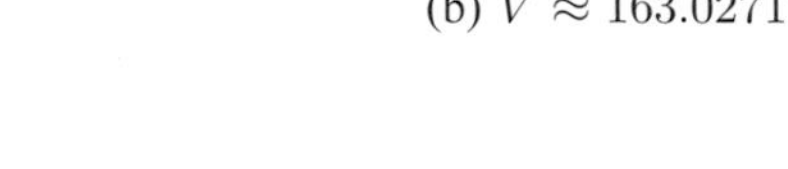

27. $V = \int_0^1 2\pi x\sqrt{1 + x^3}\,dx$. Let $f(x) = x\sqrt{1 + x^3}$.

Then the Midpoint Rule with $n = 5$ gives

$$\int_0^1 f(x)\,dx \approx \tfrac{1-0}{5}\left[f(0.1) + f(0.3) + f(0.5) + f(0.7) + f(0.9)\right]$$
$$\approx 0.2(2.9290)$$

Multiplying by 2π gives $V \approx 3.68$.

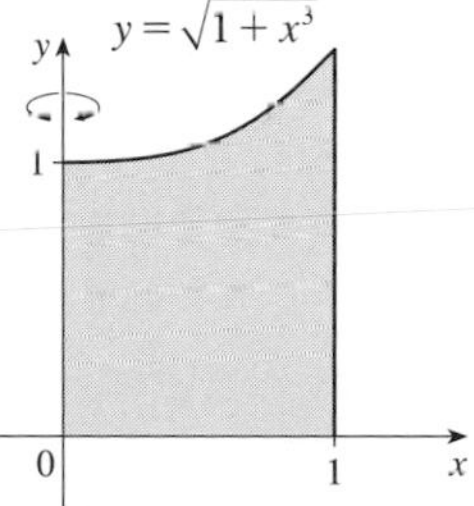

28. $V = \int_0^{10} 2\pi x f(x)\,dx$. Let $g(x) = x f(x)$, where the values of f are obtained from the graph.
Using the Midpoint Rule with $n = 5$ gives

$$\begin{aligned}\int_0^{10} g(x)\,dx &\approx \tfrac{10-0}{5}[g(1) + g(3) + g(5) + g(7) + g(9)] \\ &= 2[1f(1) + 3f(3) + 5f(5) + 7f(7) + 9f(9)] \\ &= 2[1(4-2) + 3(5-1) + 5(4-1) + 7(4-2) + 9(4-2)] \\ &= 2(2 + 12 + 15 + 14 + 18) = 2(61) = 122\end{aligned}$$

Multiplying by 2π gives $V \approx 244\pi \approx 766.5$.

29. $\int_0^3 2\pi x^5\,dx = 2\pi\int_0^3 x(x^4)\,dx$. The solid is obtained by rotating the region $0 \le y \le x^4$, $0 \le x \le 3$ about the y-axis using cylindrical shells.

30. $\int_1^3 2\pi y \ln y\,dy$. The solid is obtained by rotating the region $0 \le x \le \ln y$, $1 \le y \le 3$ about the x-axis using cylindrical shells.

31. $2\pi \int_1^4 \frac{y+2}{y^2}\,dy = 2\pi \int_1^4 (y+2)\left(\frac{1}{y^2}\right) dy$. The solid is obtained by rotating the region $0 \leq x \leq 1/y^2$, $1 \leq y \leq 4$ about the line $y = -2$ using cylindrical shells.

32. $\int_0^1 2\pi(2-x)(3^x - 2^x)\,dx$. The solid is obtained by rotating the region $2^x \leq y \leq 3^x$, $0 \leq x \leq 1$ about the line $x = 2$ using cylindrical shells.

33. From the graph, the curves intersect at $x = 0$ and $x = a \approx 2.175$, with $\frac{x}{x^2+1} > x^2 - 2x$ on the interval $(0, a)$. So the volume of the solid obtained by rotating the region about the y-axis is

$$V = 2\pi \int_0^a x\left[\frac{x}{x^2+1} - (x^2 - 2x)\right] dx \approx 14.450$$

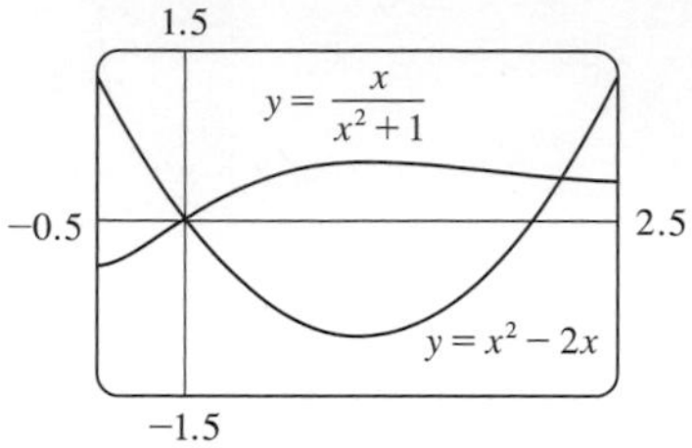

34. From the graph, the curves intersect at $x = a \approx 0.906$ and $x = b \approx 2.715$, with $e^{\sin x} > x^2 - 4x + 5$ on the interval (a, b). So the volume of the solid obtained by rotating the region about the y-axis is

$$V = 2\pi \int_a^b x\left[e^{\sin x} - (x^2 - 4x + 5)\right] dx \approx 21.253$$

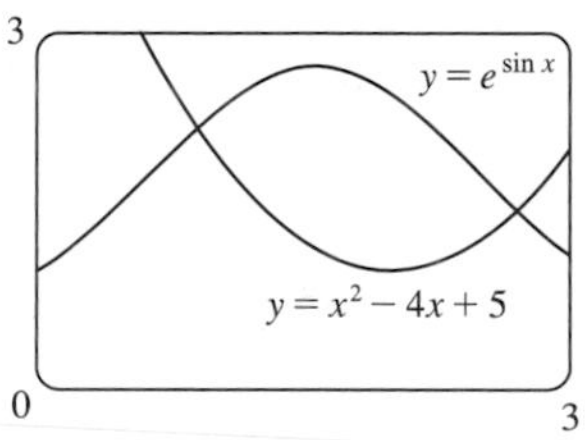

35. $V = 2\pi \int_0^{\pi/2} \left[\left(\frac{\pi}{2} - x\right)\left(\sin^2 x - \sin^4 x\right)\right] dx$

$\overset{\text{CAS}}{=} \frac{1}{32}\pi^3$

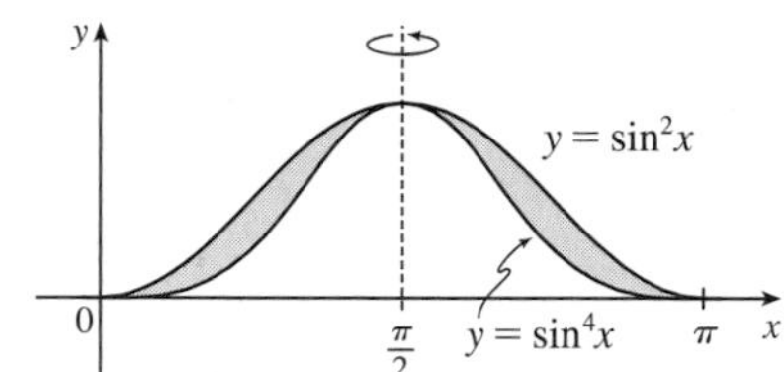

36. $V = 2\pi \int_0^\pi \{[x - (-1)](x^3 \sin x)\}\,dx$

$\overset{\text{CAS}}{=} 2\pi(\pi^4 + \pi^3 - 12\pi^2 - 6\pi + 48)$

$= 2\pi^5 + 2\pi^4 - 24\pi^3 - 12\pi^2 + 96\pi$

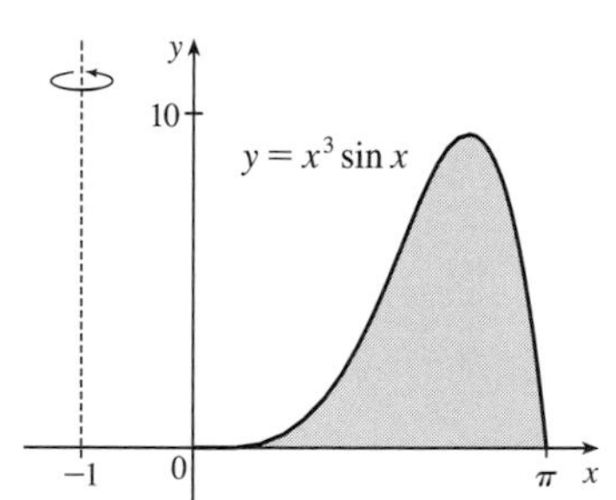

37. Use shells:

$V = \int_2^4 2\pi x(-x^2 + 6x - 8)\,dx = 2\pi \int_2^4 (-x^3 + 6x^2 - 8x)\,dx$

$= 2\pi\left[-\frac{1}{4}x^4 + 2x^3 - 4x^2\right]_2^4$

$= 2\pi[(-64 + 128 - 64) - (-4 + 16 - 16)]$

$= 2\pi(4) = 8\pi$

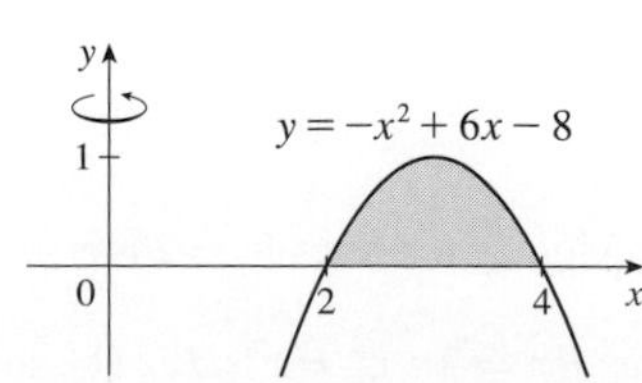

38. Use disks:

$$
\begin{aligned}
V &= \textstyle\int_2^4 \pi(-x^2+6x-8)^2\,dx \\
&= \pi \textstyle\int_2^4 (x^4 - 12x^3 + 52x^2 - 96x + 64)\,dx \\
&= \pi\left[\tfrac{1}{5}x^5 - 3x^4 + \tfrac{52}{3}x^3 - 48x^2 + 64x\right]_2^4 \\
&= \pi\left(\tfrac{512}{15} - \tfrac{496}{15}\right) = \tfrac{16}{15}\pi
\end{aligned}
$$

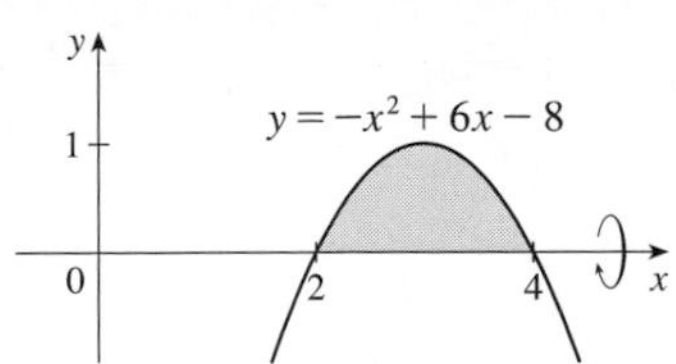

39. Use washers: $y^2 \quad x^2 - 1 \quad \Rightarrow \quad y = \pm\sqrt{x^2 \pm 1}$

$$
\begin{aligned}
V &= \int_{-\sqrt{3}}^{\sqrt{3}} \pi\left[(2-0)^2 - \left(\sqrt{x^2+1} - 0\right)^2\right] dx \\
&= 2\pi \int_0^{\sqrt{3}} [4 - (x^2+1)]\,dx \qquad \text{[by symmetry]} \\
&= 2\pi \int_0^{\sqrt{3}} (3 - x^2)\,dx = 2\pi\left[3x - \tfrac{1}{3}x^3\right]_0^{\sqrt{3}} \\
&= 2\pi\left(3\sqrt{3} - \sqrt{3}\right) = 4\sqrt{3}\,\pi
\end{aligned}
$$

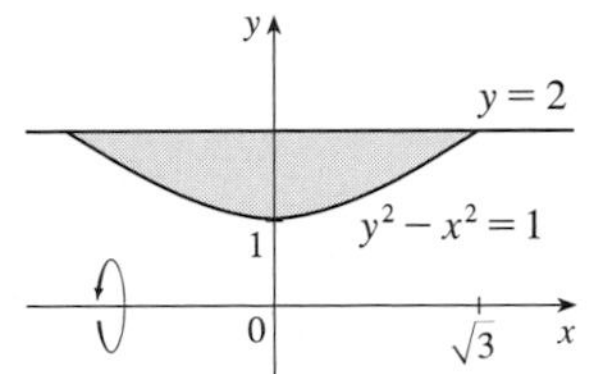

40. Use disks: $y^2 - x^2 = 1 \quad \Rightarrow \quad x = \pm\sqrt{y^2 - 1}$

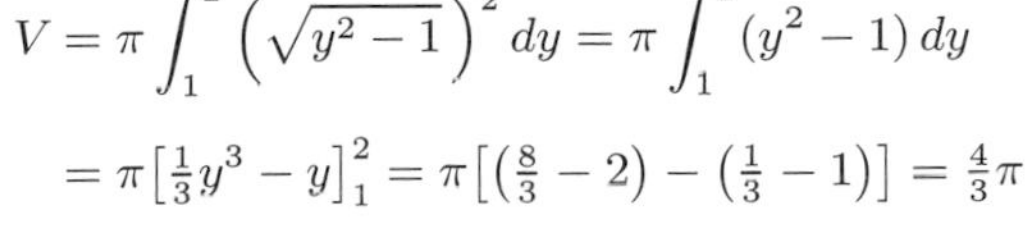

$$
\begin{aligned}
V &= \pi \int_1^2 \left(\sqrt{y^2-1}\right)^2 dy = \pi \int_1^2 (y^2 - 1)\,dy \\
&= \pi\left[\tfrac{1}{3}y^3 - y\right]_1^2 = \pi\left[\left(\tfrac{8}{3} - 2\right) - \left(\tfrac{1}{3} - 1\right)\right] = \tfrac{4}{3}\pi
\end{aligned}
$$

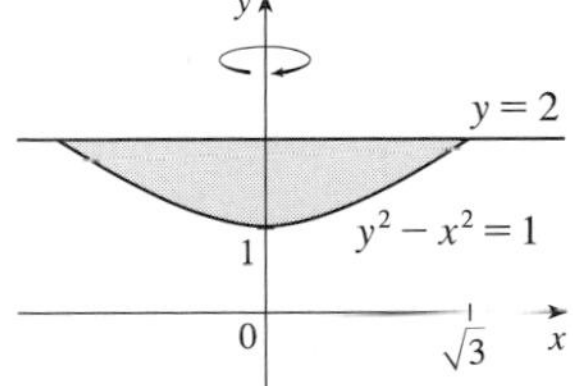

41. Use disks: $x^2 + (y-1)^2 = 1 \quad \Leftrightarrow \quad x = \pm\sqrt{1 - (y-1)^2}$

$$
\begin{aligned}
V &= \pi \int_0^2 \left[\sqrt{1-(y-1)^2}\right]^2 dy = \pi \int_0^2 (2y - y^2)\,dy \\
&= \pi\left[y^2 - \tfrac{1}{3}y^3\right]_0^2 = \pi\left(4 - \tfrac{8}{3}\right) = \tfrac{4}{3}\pi
\end{aligned}
$$

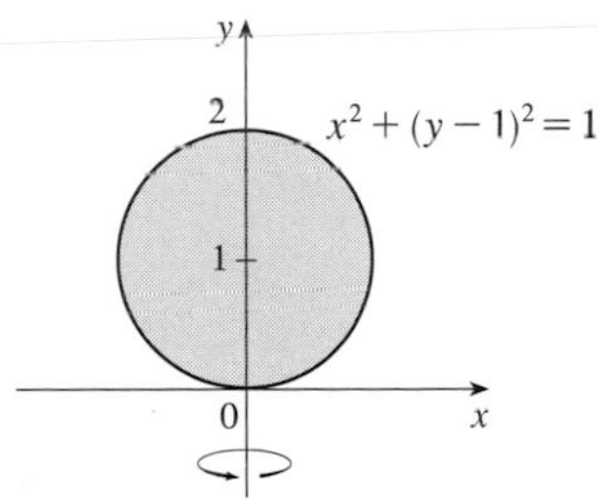

42. Use shells:

$$
\begin{aligned}
V &= \textstyle\int_1^5 2\pi(y-1)[4-(y-3)^2]\,dy \\
&= 2\pi \textstyle\int_1^5 (y-1)(-y^2+6y-5)\,dy \\
&= 2\pi \textstyle\int_1^5 (-y^3 + 7y^2 - 11y + 5)\,dy \\
&= 2\pi\left[-\tfrac{1}{4}y^4 + \tfrac{7}{3}y^3 \quad \tfrac{11}{2}y^2 + 5y\right]_1^5 \\
&= 2\pi\left(\tfrac{275}{12} - \tfrac{19}{12}\right) = \tfrac{128}{3}\pi
\end{aligned}
$$

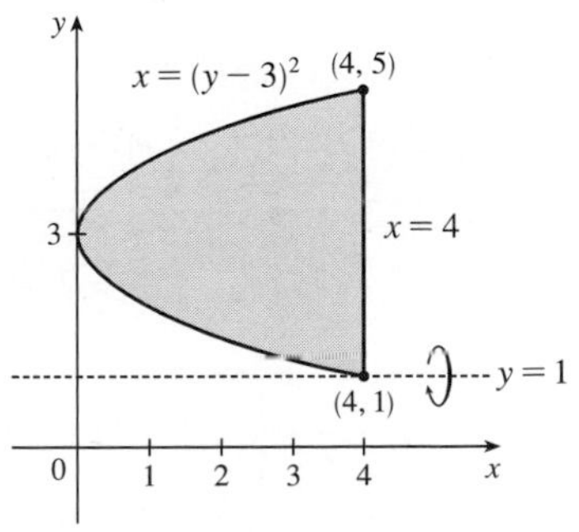

43. $y+1=(y-1)^2 \;\Leftrightarrow\; y+1=y^2-2y+1 \;\Leftrightarrow\; 0=y^2-3y \;\Leftrightarrow$
$0=y(y-3) \;\Leftrightarrow\; y=0$ or 3.

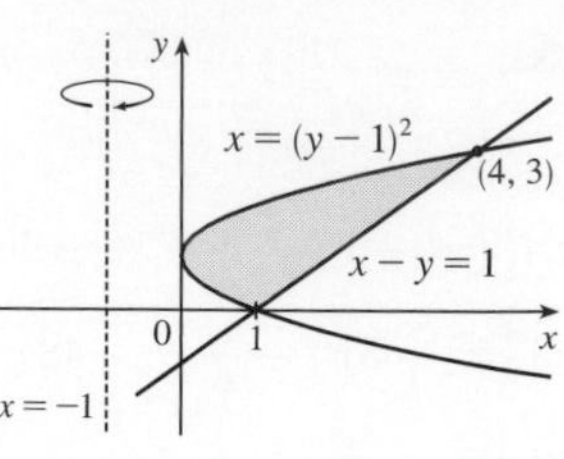

Use disks:

$$V=\pi\int_0^3 \{[(y+1)-(-1)]^2-[(y-1)^2-(-1)]^2\}\,dy$$
$$=\pi\int_0^3 [(y+2)^2-(y^2-2y+2)^2]\,dy$$
$$=\pi\int_0^3 [(y^2+4y+4)-(y^4-4y^3+8y^2-8y+4)]\,dy=\pi\int_0^3 (-y^4+4y^3-7y^2+12y)\,dy$$
$$=\pi\left[-\tfrac{1}{5}y^5+y^4-\tfrac{7}{3}y^3+6y^2\right]_0^3=\pi\left(-\tfrac{243}{5}+81-63+54\right)=\tfrac{117}{5}\pi$$

44. Use cylindrical shells to find the volume V.

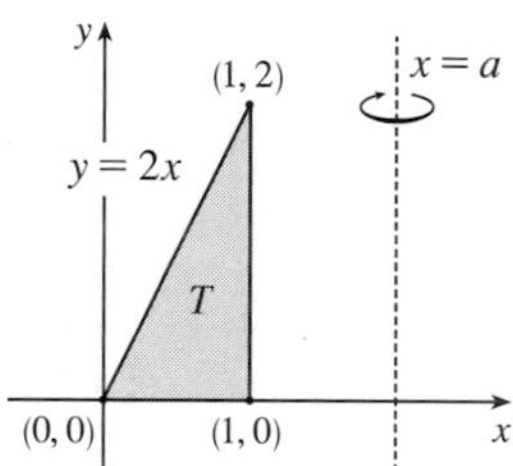

$$V=\int_0^1 2\pi(a-x)(2x)\,dx=4\pi\int_0^1 (ax-x^2)\,dx$$
$$=4\pi\left[\tfrac{1}{2}ax^2-\tfrac{1}{3}x^3\right]_0^1=4\pi\left(\tfrac{1}{2}a-\tfrac{1}{3}\right)$$

Now solve for a in terms of V:

$$V=4\pi\left(\tfrac{1}{2}a-\tfrac{1}{3}\right) \;\Leftrightarrow\; \frac{V}{4\pi}=\frac{1}{2}a-\frac{1}{3} \;\Leftrightarrow\; \frac{1}{2}a=\frac{V}{4\pi}+\frac{1}{3} \;\Leftrightarrow$$
$$a=\frac{V}{2\pi}+\frac{2}{3}$$

45. Use shells:

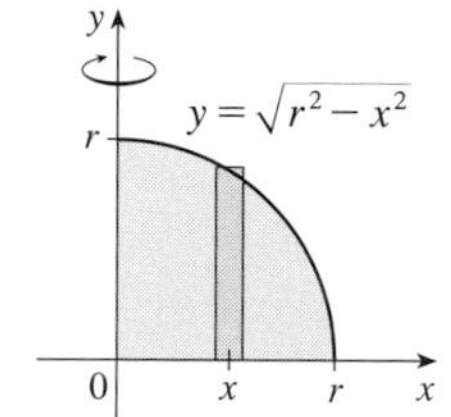

$$V=2\int_0^r 2\pi x\sqrt{r^2-x^2}\,dx=-2\pi\int_0^r (r^2-x^2)^{1/2}(-2x)\,dx$$
$$=\left[-2\pi\cdot\tfrac{2}{3}(r^2-x^2)^{3/2}\right]_0^r=-\tfrac{4}{3}\pi(0-r^3)=\tfrac{4}{3}\pi r^3$$

46. $V=\int_{R-r}^{R+r} 2\pi x\cdot 2\sqrt{r^2-(x-R)^2}\,dx$

$=\int_{-r}^{r} 4\pi(u+R)\sqrt{r^2-u^2}\,du \qquad$ [let $u=x-R$]

$=4\pi R\int_{-r}^{r}\sqrt{r^2-u^2}\,du+4\pi\int_{-r}^{r} u\sqrt{r^2-u^2}\,du$

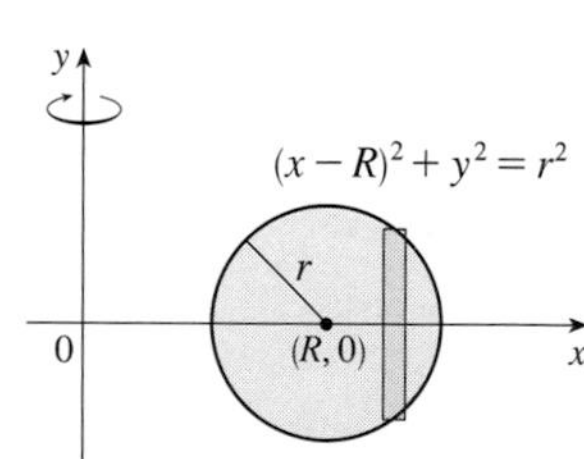

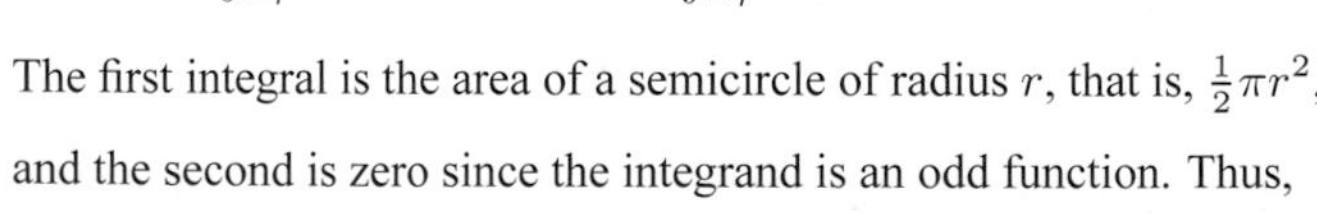

The first integral is the area of a semicircle of radius r, that is, $\tfrac{1}{2}\pi r^2$, and the second is zero since the integrand is an odd function. Thus,

$V=4\pi R\left(\tfrac{1}{2}\pi r^2\right)+4\pi\cdot 0=2\pi^2Rr^2$.

47. $V=2\pi\displaystyle\int_0^r x\left(-\frac{h}{r}x+h\right)dx=2\pi h\int_0^r\left(-\frac{x^2}{r}+x\right)dx$

$$=2\pi h\left[-\frac{x^3}{3r}+\frac{x^2}{2}\right]_0^r=2\pi h\,\frac{r^2}{6}=\frac{\pi r^2h}{3}$$

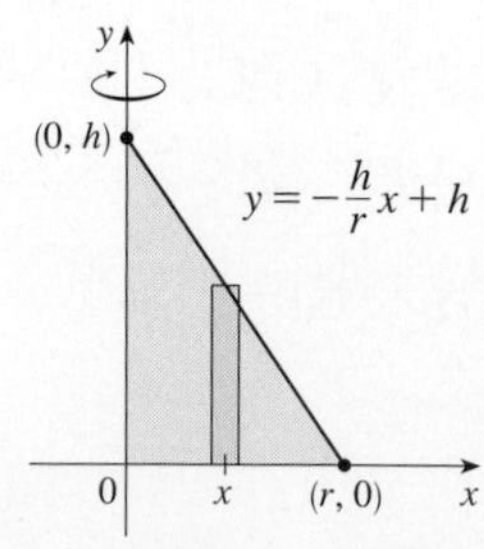

48. By symmetry, the volume of a napkin ring obtained by drilling a hole of radius r through a sphere with radius R is twice the volume obtained by rotating the area above the x-axis and below the curve $y = \sqrt{R^2 - x^2}$ (the equation of the top half of the cross-section of the sphere), between $x = r$ and $x = R$, about the y-axis. This volume is equal to

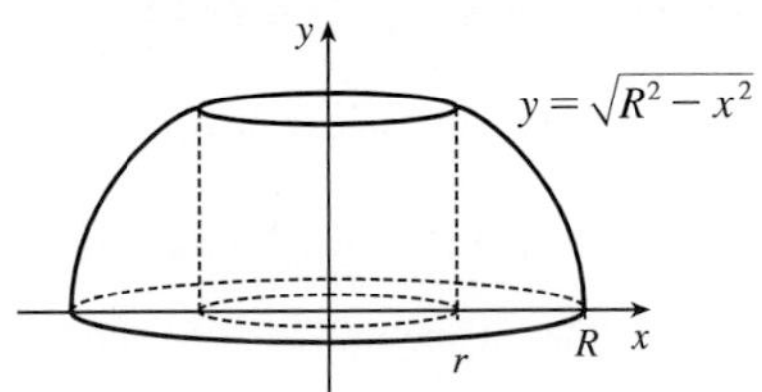

$$2\int_{\text{inner radius}}^{\text{outer radius}} 2\pi rh\,dx = 2\cdot 2\pi\int_r^R x\sqrt{R^2 - x^2}\,dx = 4\pi\left[-\tfrac{1}{3}\left(R^2 - x^2\right)^{3/2}\right]_r^R = \tfrac{4}{3}\pi(R^2 - r^2)^{3/2}$$

But by the Pythagorean Theorem, $R^2 - r^2 = \left(\frac{1}{2}h\right)^2$, so the volume of the napkin ring is $\frac{4}{3}\pi\left(\frac{1}{2}h\right)^3 = \frac{1}{6}\pi h^3$, which is independent of both R and r; that is, the amount of wood in a napkin ring of height h is the same regardless of the size of the sphere used. Note that most of this calculation has been done already, but with more difficulty, in Exercise 6.2.70.

Another solution: The height of the missing cap is the radius of the sphere minus half the height of the cut-out cylinder, that is, $R - \frac{1}{2}h$. Using Exercise 6.2.49,

$$V_{\text{napkin ring}} = V_{\text{sphere}} - V_{\text{cylinder}} - 2V_{\text{cap}} = \tfrac{4}{3}\pi R^3 - \pi r^2 h - 2\cdot\tfrac{\pi}{3}\left(R - \tfrac{1}{2}h\right)^2\left[3R - \left(R - \tfrac{1}{2}h\right)\right] = \tfrac{1}{6}\pi h^3$$

6.4 Work

1. (a) The work done by the gorilla in lifting its weight of 360 pounds to a height of 20 feet is $W = Fd = (360\text{ lb})(20\text{ ft}) = 7200$ ft-lb.

(b) The amount of time it takes the gorilla to climb the tree doesn't change the amount of work done, so the work done is still 7200 ft-lb.

2. $W = Fd = (mg)d = [(200\text{ kg})(9.8\text{ m/s}^2)](3\text{ m}) = (1960\text{ N})(3\text{ m}) = 5880\text{ J}$

3. $W = \int_a^b f(x)\,dx = \int_1^{10} 5x^{-2}\,dx = 5\left[-x^{-1}\right]_1^{10} = 5\left(-\frac{1}{10} + 1\right) = 4.5$ ft-lb

4. $W = \int_1^2 \cos\left(\frac{1}{3}\pi x\right)dx = \frac{3}{\pi}\left[\sin\left(\frac{1}{3}\pi x\right)\right]_1^2 = \frac{3}{\pi}\left(\frac{\sqrt{3}}{2} - \frac{\sqrt{3}}{2}\right) = 0\text{ N·m} = 0\text{ J}.$

Interpretation: From $x = 1$ to $x = \frac{3}{2}$, the force does work equal to $\int_1^{3/2}\cos\left(\frac{1}{3}\pi x\right)dx = \frac{3}{\pi}\left(1 - \frac{\sqrt{3}}{2}\right)$ J in accelerating the particle and increasing its kinetic energy. From $x = \frac{3}{2}$ to $x = 2$, the force opposes the motion of the particle, decreasing its kinetic energy. This is negative work, equal in magnitude but opposite in sign to the work done from $x = 1$ to $x = \frac{3}{2}$.

5. The force function is given by $F(x)$ (in newtons) and the work (in joules) is the area under the curve, given by $\int_0^8 F(x)\,dx = \int_0^4 F(x)\,dx + \int_4^8 F(x)\,dx = \frac{1}{2}(4)(30) + (4)(30) = 180$ J.

6. $W = \int_4^{20} f(x)\,dx \approx M_4 = \Delta x[f(6) + f(10) + f(14) + f(18)] = \frac{20-4}{4}[5.8 + 8.8 + 8.2 + 5.2] = 4(28) = 112\text{ J}$

7. According to Hooke's Law, the force required to maintain a spring stretched x units beyond its natural length (or compressed x units less than its natural length) is proportional to x, that is, $f(x) = kx$. Here, the amount stretched is 4 in. $= \frac{1}{3}$ ft and the force is 10 lb. Thus, $10 = k\left(\frac{1}{3}\right) \Rightarrow k = 30$ lb/ft, and $f(x) = 30x$. The work done in stretching the spring from its natural length to 6 in. $= \frac{1}{2}$ ft beyond its natural length is $W = \int_0^{1/2} 30x\,dx = \left[15x^2\right]_0^{1/2} = \frac{15}{4}$ ft-lb.

8. According to Hooke's Law, the force required to maintain a spring stretched x units beyond its natural length (or compressed x units less than its natural length) is proportional to x, that is, $f(x) = kx$. Here, the amount compressed is $40 - 30 = 10$ cm $= 0.1$ m and the force is 60 N. Thus, $60 = k(0.1) \Rightarrow k = 600$ N/m, and $f(x) = 600x$. The work required to compress the spring 0.1 m is $W = \int_0^{0.1} 600x\,dx = \left[300x^2\right]_0^{0.1} = 300(0.01) = 3$ N-m (or J). The work required to compress the spring $40 - 25 = 15$ cm $= 0.15$ m is $W = \int_0^{0.15} 600x\,dx = \left[300x^2\right]_0^{0.15} = 300(0.0225) = 6.75$ J.

9. (a) If $\int_0^{0.12} kx\,dx = 2$ J, then $2 = \left[\frac{1}{2}kx^2\right]_0^{0.12} = \frac{1}{2}k(0.0144) = 0.0072k$ and $k = \frac{2}{0.0072} = \frac{2500}{9} \approx 277.78$ N/m.

Thus, the work needed to stretch the spring from 35 cm to 40 cm is

$\int_{0.05}^{0.10} \frac{2500}{9}x\,dx = \left[\frac{1250}{9}x^2\right]_{1/20}^{1/10} = \frac{1250}{9}\left(\frac{1}{100} - \frac{1}{400}\right) = \frac{25}{24} \approx 1.04$ J.

(b) $f(x) = kx$, so $30 = \frac{2500}{9}x$ and $x = \frac{270}{2500}$ m $= 10.8$ cm

10. If $12 = \int_0^1 kx\,dx = \left[\frac{1}{2}kx^2\right]_0^1 = \frac{1}{2}k$, then $k = 24$ lb/ft and the work required is

$\int_0^{3/4} 24x\,dx = \left[12x^2\right]_0^{3/4} = 12 \cdot \frac{9}{16} = \frac{27}{4} = 6.75$ ft-lb.

11. The distance from 20 cm to 30 cm is 0.1 m, so with $f(x) = kx$, we get $W_1 = \int_0^{0.1} kx\,dx = k\left[\frac{1}{2}x^2\right]_0^{0.1} = \frac{1}{200}k$.

Now $W_2 = \int_{0.1}^{0.2} kx\,dx = k\left[\frac{1}{2}x^2\right]_{0.1}^{0.2} = k\left(\frac{4}{200} - \frac{1}{200}\right) = \frac{3}{200}k$. Thus, $W_2 = 3W_1$.

12. Let L be the natural length of the spring in meters. Then

$6 = \int_{0.10-L}^{0.12-L} kx\,dx = \left[\frac{1}{2}kx^2\right]_{0.10-L}^{0.12-L} = \frac{1}{2}k\left[(0.12 - L)^2 - (0.10 - L)^2\right]$ and

$10 = \int_{0.12-L}^{0.14-L} kx\,dx = \left[\frac{1}{2}kx^2\right]_{0.12-L}^{0.14-L} = \frac{1}{2}k\left[(0.14 - L)^2 - (0.12 - L)^2\right]$.

Simplifying gives us $12 = k(0.0044 - 0.04L)$ and $20 = k(0.0052 - 0.04L)$. Subtracting the first equation from the second gives $8 = 0.0008k$, so $k = 10{,}000$. Now the second equation becomes $20 = 52 - 400L$, so $L = \frac{32}{400}$ m $= 8$ cm.

In Exercises 13–22, n is the number of subintervals of length Δx, and x_i^* is a sample point in the ith subinterval $[x_{i-1}, x_i]$.

13. (a) The portion of the rope from x ft to $(x + \Delta x)$ ft below the top of the building weighs $\frac{1}{2}\,\Delta x$ lb and must be lifted x_i^* ft, so its contribution to the total work is $\frac{1}{2}x_i^*\,\Delta x$ ft-lb. The total work is

$$W = \lim_{n\to\infty} \sum_{i=1}^{n} \tfrac{1}{2}x_i^*\,\Delta x = \int_0^{50} \tfrac{1}{2}x\,dx = \left[\tfrac{1}{4}x^2\right]_0^{50} = \tfrac{2500}{4} = 625 \text{ ft-lb}$$

Notice that the exact height of the building does not matter (as long as it is more than 50 ft).

(b) When half the rope is pulled to the top of the building, the work to lift the top half of the rope is $W_1 = \int_0^{25} \frac{1}{2}x\,dx = \left[\frac{1}{4}x^2\right]_0^{25} = \frac{625}{4}$ ft-lb. The bottom half of the rope is lifted 25 ft and the work needed to accomplish that is $W_2 = \int_{25}^{50} \frac{1}{2} \cdot 25\,dx = \frac{25}{2}\left[x\right]_{25}^{50} = \frac{625}{2}$ ft-lb. The total work done in pulling half the rope to the top of the building is $W = W_1 + W_2 = \frac{625}{2} + \frac{625}{4} = \frac{3}{4} \cdot 625 = \frac{1875}{4}$ ft-lb.

14. (a) The 60 ft cable weighs 180 lb, or 3 lb/ft. If we divide the cable into n equal parts of length $\Delta x = 60/n$ ft, then for large n, all points in the ith part are lifted by approximately the same amount. Choose a representative distance from the winch in the ith part of the cable, say x_i^*. If $x_i^* < 25$ ft, then the ith part has to be lifted roughly x_i^* ft. If $x_i^* \geq 25$ ft, then the ith part has to be lifted 25 ft. The ith part weighs $(3\text{ lb/ft})(\Delta x\text{ ft}) = 3\,\Delta x$ lb, so the work done in lifting it is $(3\,\Delta x)x_i^*$ if $x_i^* < 25$ ft and $(3\,\Delta x)(25) = 75\,\Delta x$ if $x_i^* \geq 25$ ft. The work of lifting the top 25 ft of the cable is

$W_1 = \lim\limits_{n\to\infty} \sum\limits_{i=1}^{n_1} 3x_i^*\,\Delta x = \int_0^{25} 3x\,dx = \left[\frac{3}{2}x^2\right]_0^{25} = \frac{3}{2}(625) = 937.5$ ft-lb. Here n_1 represents the number of parts of the cable in the top 25 ft. The work of lifting the bottom 35 ft of the cable is

$W_2 = \lim\limits_{n\to\infty} \sum\limits_{i=1}^{n_2} 75\,\Delta x = \int_{25}^{60} 75\,dx = 75(60-25) = 2625$ ft-lb, where n_2 represents the number of small parts in the bottom 35 feet of the cable. The total work done is $W = W_1 + W_2 = 937.5 + 2625 = 3562.5$ ft-lb.

(b) Once x feet of cable have been wound up by the winch, there is $(60 - x)$ ft of cable still hanging from the winch. That portion of the cable weighs $3(60 - x)$ lb. Lifting it Δx feet requires $3(60 - x)\,\Delta x$ ft-lb of work. Thus, the total work needed to lift the cable 25 ft is $W = \int_0^{25} 3(60 - x)\,dx = \left[180x - \frac{3}{2}x^2\right]_0^{25} = 4500 - 937.5 = 3562.5$ ft-lb.

15. The work needed to lift the cable is $\lim\limits_{n\to\infty} \sum_{i=1}^{n} 2x_i^*\,\Delta x = \int_0^{500} 2x\,dx = \left[x^2\right]_0^{500} = 250{,}000$ ft-lb. The work needed to lift the coal is $800\text{ lb} \cdot 500\text{ ft} = 400{,}000$ ft-lb. Thus, the total work required is $250{,}000 + 400{,}000 = 650{,}000$ ft-lb.

16. *Assumptions*:

1. After lifting, the chain is L-shaped, with 4 m of the chain lying along the ground.
2. The chain slides effortlessly and without friction along the ground while its end is lifted.
3. The weight density of the chain is constant throughout its length and therefore equals $(8\text{ kg/m})(9.8\text{ m/s}^2) = 78.4\text{ N/m}$.

The part of the chain x m from the lifted end is raised $6 - x$ m if $0 \leq x \leq 6$ m, and it is lifted 0 m if $x > 6$ m. Thus, the work needed is

$$W = \lim_{n\to\infty} \sum_{i=1}^{n} (6 - x_i^*) \cdot 78.4\,\Delta x = \int_0^6 (6 - x)78.4\,dx = 78.4\left[6x - \tfrac{1}{2}x^2\right]_0^6 = (78.4)(18) = 1411.2\text{ J}$$

17. At a height of x meters $(0 \leq x \leq 12)$, the mass of the rope is $(0.8\text{ kg/m})(12 - x\text{ m}) = (9.6 - 0.8x)$ kg and the mass of the water is $\left(\frac{36}{12}\text{ kg/m}\right)(12 - x\text{ m}) = (36 - 3x)$ kg. The mass of the bucket is 10 kg, so the total mass is $(9.6 - 0.8x) + (36 - 3x) + 10 = (55.6 - 3.8x)$ kg, and hence, the total force is $9.8(55.6 - 3.8x)$ N. The work needed to lift the bucket Δx m through the ith subinterval of $[0, 12]$ is $9.8(55.6 - 3.8x_i^*)\Delta x$, so the total work is

$$W = \lim_{n\to\infty} \sum_{i=1}^{n} 9.8(55.6 - 3.8x_i^*)\,\Delta x = \int_0^{12} (9.8)(55.6 - 3.8x)\,dx = 9.8\left[55.6x - 1.9x^2\right]_0^{12} = 9.8(393.6) \approx 3857\text{ J}$$

18. The work needed to lift the bucket itself is 4 lb · 80 ft = 320 ft-lb. At time t (in seconds) the bucket is $x_i^* = 2t$ ft above its original 80 ft depth, but it now holds only $(40 - 0.2t)$ lb of water. In terms of distance, the bucket holds $\left[40 - 0.2\left(\frac{1}{2}x_i^*\right)\right]$ lb of water when it is x_i^* ft above its original 80 ft depth. Moving this amount of water a distance Δx requires $\left(40 - \frac{1}{10}x_i^*\right)\Delta x$ ft-lb of work. Thus, the work needed to lift the water is

$$W = \lim_{n\to\infty}\sum_{i=1}^{n}\left(40 - \tfrac{1}{10}x_i^*\right)\Delta x = \int_0^{80}\left(40 - \tfrac{1}{10}x\right)dx = \left[40x - \tfrac{1}{20}x^2\right]_0^{80} = (3200 - 320)\text{ ft-lb}$$

Adding the work of lifting the bucket gives a total of 3200 ft-lb of work.

19. The chain's weight density is $\dfrac{25\text{ lb}}{10\text{ ft}} = 2.5\text{ lb/ft}$. The part of the chain x ft below the ceiling (for $5 \le x \le 10$) has to be lifted $2(x-5)$ ft, so the work needed to lift the ith subinterval of the chain is $2(x_i^* - 5)(2.5\,\Delta x)$. The total work needed is

$$\begin{aligned} W &= \lim_{n\to\infty}\sum_{i=1}^{n} 2(x_i^* - 5)(2.5)\,\Delta x = \int_5^{10}[2(x-5)(2.5)]\,dx = 5\int_5^{10}(x-5)\,dx \\ &= 5\left[\tfrac{1}{2}x^2 - 5x\right]_5^{10} = 5\left[(50-50) - \left(\tfrac{25}{2} - 25\right)\right] = 5\left(\tfrac{25}{2}\right) = 62.5\text{ ft-lb} \end{aligned}$$

20. A horizontal cylindrical slice of water Δx ft thick has a volume of $\pi r^2 h = \pi \cdot 12^2 \cdot \Delta x$ ft^3 and weighs about $\left(62.5\text{ lb/ft}^3\right)\left(144\pi\,\Delta x\text{ ft}^3\right) = 9000\pi\,\Delta x$ lb. If the slice lies x_i^* ft below the edge of the pool (where $1 \le x_i^* \le 5$), then the work needed to pump it out is about $9000\pi x_i^*\,\Delta x$. Thus,

$$W = \lim_{n\to\infty}\sum_{i=1}^{n} 9000\pi x_i^*\,\Delta x = \int_1^5 9000\pi x\,dx = \left[4500\pi x^2\right]_1^5 = 4500\pi(25-1) = 108{,}000\pi\text{ ft-lb}$$

21. A "slice" of water Δx m thick and lying at a depth of x_i^* m (where $0 \le x_i^* \le \frac{1}{2}$) has volume $(2 \times 1 \times \Delta x)$ m^3, a mass of $2000\,\Delta x$ kg, weighs about $(9.8)(2000\,\Delta x) = 19{,}600\,\Delta x$ N, and thus requires about $19{,}600x_i^*\,\Delta x$ J of work for its removal.

So $W = \lim\limits_{n\to\infty}\sum\limits_{i=1}^{n} 19{,}600x_i^*\,\Delta x = \int_0^{1/2} 19{,}600x\,dx = \left[9800x^2\right]_0^{1/2} = 2450$ J.

22. We use a vertical coodinate x measured from the center of the water tank. The top and bottom of the tank have coordinates $x = -12$ ft and $x = 12$ ft, respectively.

A thin horizontal slice of water at coordinate x is a disk of radius $\sqrt{12^2 - x^2}$ as shown in the figure. The disk has area $\pi r^2 = \pi(12^2 - x^2)$, so if the slice has thickness Δx, the slice has volume $\pi(12^2 - x^2)\,\Delta x$ and weight $62.5\pi(12^2 - x^2)\,\Delta x$. The work needed to raise this water from ground level (coordinate 72) to coordinate x, a distance of $(72 - x)$ ft, is $62.5\pi(12^2 - x^2)(72 - x)\,\Delta x$ ft-lb. The total work needed to fill the tank is

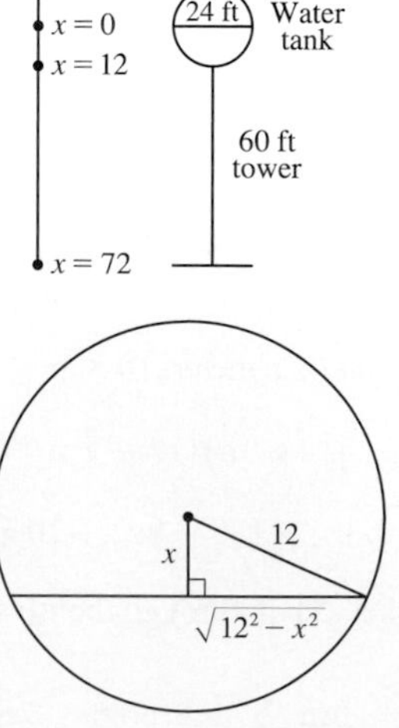

approximated by a Riemann sum $\sum_{i=1}^{n} 62.5\pi[(12^2 - (x_i^*)^2)](72 - x_i^*)\,\Delta x$. Thus, the total work is

$$W = \lim_{n\to\infty} \sum_{i=1}^{n} 62.5\pi[(12^2 - (x_i^*)^2)](72 - x_i^*)\,\Delta x = \int_{-12}^{12} 62.5\pi(12^2 - x^2)(72 - x)\,dx$$

$$= 62.5\pi \int_{-12}^{12}[\underbrace{72(12^2 - x^2)}_{\text{even function}} - \underbrace{x(12^2 - x^2)}_{\text{odd function}}]\,dx = 62.5\pi(2)\int_0^{12} 72(12^2 - x^2)\,dx \qquad \text{[by Theorem 5.5.7]}$$

$$= 125\pi(72)\left[12^2 x - \tfrac{1}{3}x^3\right]_0^{12} = 9000\pi\left(12^3 - \tfrac{1}{3}\cdot 12^3\right) = 9000\pi\left(\tfrac{2}{3}\cdot 12^3\right)$$

$$= 10{,}368{,}000\pi \text{ ft-lb}$$

The 1.5 horsepower pump does $1.5(550) = 825$ ft-lb of work per second. To fill the tank, it will take

$$\frac{10{,}368{,}000\pi \text{ ft-lb}}{825 \text{ ft-lb/s}} \approx 39{,}481 \text{ s} \approx 10.97 \text{ hours.}$$

23. A rectangular "slice" of water Δx m thick and lying x m above the bottom has width x m and volume $8x\,\Delta x$ m^3. It weighs about $(9.8\times 1000)(8x\,\Delta x)$ N, and must be lifted $(5 - x)$ m by the pump, so the work needed is about $(9.8\times 10^3)(5 - x)(8x\,\Delta x)$ J. The total work required is

$$W \approx \int_0^3 (9.8\times 10^3)(5 - x)8x\,dx = (9.8\times 10^3)\int_0^3 (40x - 8x^2)\,dx = (9.8\times 10^3)\left[20x^2 - \tfrac{8}{3}x^3\right]_0^3$$

$$= (9.8\times 10^3)(180 - 72) = (9.8\times 10^3)(108) = 1058.4\times 10^3 \approx 1.06\times 10^6 \text{ J}$$

24. Let y measure depth (in meters) below the center of the spherical tank, so that $y = -3$ at the top of the tank and $y = -4$ at the spigot. A horizontal disk-shaped "slice" of water Δy m thick and lying at coordinate y has radius $\sqrt{9 - y^2}$ m and volume $\pi r^2\Delta y = \pi(9 - y^2)\,\Delta y$ m^3. It weighs about $(9.8\times 1000)\pi(9 - y^2)\,\Delta y$ N and must be lifted $(y + 4)$ m by the pump, so the work needed to pump it out is about $(9.8\times 10^3)(y + 4)\pi(9 - y^2)\,\Delta y$ J. The total work required is

$$W \approx \int_{-3}^{3}(9.8\times 10^3)(y + 4)\pi(9 - y^2)\,dy = (9.8\times 10^3)\pi\int_{-3}^{3}[y(9 - y^2) + 4(9 - y^2)]\,dy$$

$$= (9.8\times 10^3)\pi(2)(4)\int_0^3 (9 - y^2)\,dy \qquad \text{[by Theorem 5.5.7]}$$

$$= (78.4\times 10^3)\pi\left[9y - \tfrac{1}{3}y^3\right]_0^3 = (78.4\times 10^3)\pi(18) = 1{,}411{,}200\pi \approx 4.43\times 10^6 \text{ J}$$

25. Let x measure depth (in feet) below the spout at the top of the tank. A horizontal disk-shaped "slice" of water Δx ft thick and lying at coordinate x has radius $\frac{3}{8}(16 - x)$ ft $(\star)$ and volume $\pi r^2\Delta x = \pi\cdot\frac{9}{64}(16 - x)^2\,\Delta x$ ft^3. It weighs about $(62.5)\frac{9\pi}{64}(16 - x)^2\,\Delta x$ lb and must be lifted x ft by the pump, so the work needed to pump it out is about $(62.5)x\,\frac{9\pi}{64}(16 - x)^2\,\Delta x$ ft-lb. The total work required is

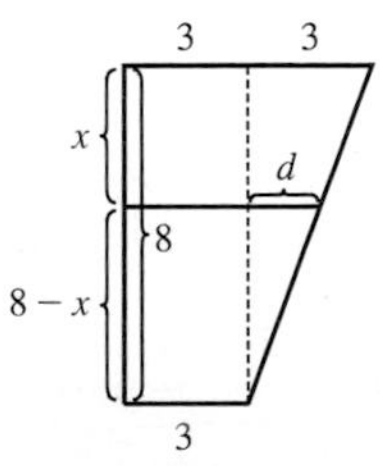

$(\star)$ From similar triangles, $\dfrac{d}{8 - x} = \dfrac{3}{8}$.

So $r = 3 + d = 3 + \frac{3}{8}(8 - x)$

$= \dfrac{3(8)}{8} + \dfrac{3}{8}(8 - x)$

$= \frac{3}{8}(16 - x)$

$$W \approx \int_0^8 (62.5)x\,\tfrac{9\pi}{64}(16 - x)^2\,dx = (62.5)\tfrac{9\pi}{64}\int_0^8 x(256 - 32x + x^2)\,dx$$

$$= (62.5)\tfrac{9\pi}{64}\int_0^8 (256x - 32x^2 + x^3)\,dx = (62.5)\tfrac{9\pi}{64}\left[128x^2 - \tfrac{32}{3}x^3 + \tfrac{1}{4}x^4\right]_0^8$$

$$= (62.5)\frac{9\pi}{64}\left(\frac{11{,}264}{3}\right) = 33{,}000\pi \approx 1.04\times 10^5 \text{ ft-lb}$$

26. Let x measure the distance (in feet) above the bottom of the tank. A horizontal "slice" of water Δx ft thick and lying at coordinate x has volume $10(2x)\,\Delta x$ ft^3. It weighs about $(62.5)20x\,\Delta x$ lb and must be lifted $(6-x)$ ft by the pump, so the work needed to pump it out is about $(62.5)(6-x)20x\,\Delta x$ ft-lb. The total work required is

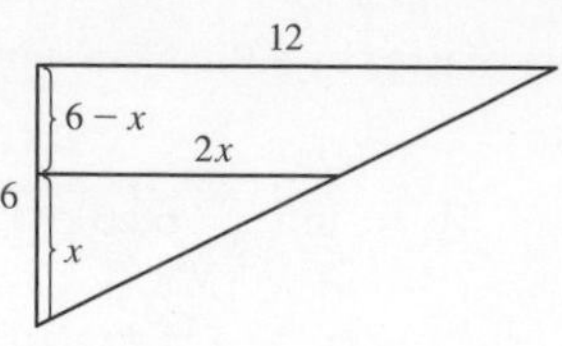

$W \approx \int_0^6 (62.5)(6-x)20x\,dx = 1250\int_0^6 (6x-x^2)\,dx = 1250\left[3x^2 - \frac{1}{3}x^3\right]_0^6 = 1250(36) = 45{,}000$ ft-lb.

27. If only 4.7×10^5 J of work is done, then only the water above a certain level (call it h) will be pumped out. So we use the same formula as in Exercise 23, except that the work is fixed, and we are trying to find the lower limit of integration:

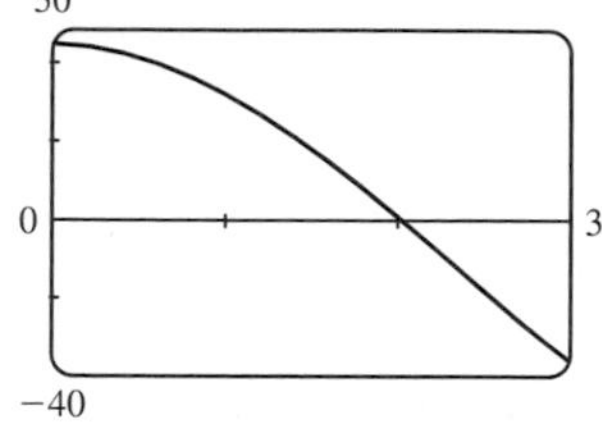

$4.7 \times 10^5 \approx \int_h^3 (9.8 \times 10^3)(5-x)8x\,dx = (9.8 \times 10^3)\left[20x^2 - \frac{8}{3}x^3\right]_h^3 \quad \Leftrightarrow$

$\frac{4.7}{9.8} \times 10^2 \approx 48 = \left(20 \cdot 3^2 - \frac{8}{3} \cdot 3^3\right) - \left(20h^2 - \frac{8}{3}h^3\right) \quad \Leftrightarrow$

$2h^3 - 15h^2 + 45 = 0$. To find the solution of this equation, we plot $2h^3 - 15h^2 + 45$ between $h = 0$ and $h = 3$. We see that the equation is satisfied for $h \approx 2.0$. So the depth of water remaining in the tank is about 2.0 m.

28. The only changes needed in the solution for Exercise 24 are: (1) change the lower limit from -3 to 0 and (2) change 1000 to 900.

$$\begin{aligned} W &\approx \int_0^3 (9.8 \times 900)(y+4)\pi(9-y^2)\,dy = (9.8 \times 900)\,\pi \int_0^3 (9y - y^3 + 36 - 4y^2)\,dy \\ &= (9.8 \times 900)\pi\left[\tfrac{9}{2}y^2 - \tfrac{1}{4}y^4 + 36y - \tfrac{4}{3}y^3\right]_0^3 = (9.8 \times 900)\pi(92.25) = 813{,}645\pi \\ &\approx 2.56 \times 10^6 \text{ J} \quad [\text{about } 58\% \text{ of the work in Exercise 24}] \end{aligned}$$

29. $V = \pi r^2 x$, so V is a function of x and P can also be regarded as a function of x. If $V_1 = \pi r^2 x_1$ and $V_2 = \pi r^2 x_2$, then

$$\begin{aligned} W &= \int_{x_1}^{x_2} F(x)\,dx = \int_{x_1}^{x_2} \pi r^2 P(V(x))\,dx = \int_{x_1}^{x_2} P(V(x))\,dV(x) \qquad [\text{Let } V(x) = \pi r^2 x, \text{ so } dV(x) = \pi r^2\,dx.] \\ &= \int_{V_1}^{V_2} P(V)\,dV \quad \text{by the Substitution Rule.} \end{aligned}$$

30. $160 \text{ lb/in}^2 = 160 \cdot 144 \text{ lb/ft}^2$, $100 \text{ in}^3 = \frac{100}{1728} \text{ ft}^3$, and $800 \text{ in}^3 = \frac{800}{1728} \text{ ft}^3$.

$k = PV^{1.4} = (160 \cdot 144)\left(\frac{100}{1728}\right)^{1.4} = 23{,}040\left(\frac{25}{432}\right)^{1.4} \approx 426.5$. Therefore, $P \approx 426.5V^{-1.4}$ and

$$W = \int_{100/1728}^{800/1728} 426.5V^{-1.4}\,dV = 426.5\left[\frac{1}{-0.4}V^{-0.4}\right]_{25/432}^{25/54} = (426.5)(2.5)\left[\left(\tfrac{432}{25}\right)^{0.4} - \left(\tfrac{54}{25}\right)^{0.4}\right] \approx 1.88 \times 10^3 \text{ ft-lb.}$$

31. (a)
$$\begin{aligned} W &= \int_{x_1}^{x_2} f(x)\,dx = \int_{t_1}^{t_2} f(s(t))\,v(t)\,dt \quad \left[\begin{array}{l} x = s(t), \\ dx = v(t)\,dt \end{array}\right] \\ &= \int_{t_1}^{t_2} m\,a(t)\,v(t)\,dt = \int_{v_1}^{v_2} m\,u\,du \quad \left[\begin{array}{l} u = v(t), \\ du = a(t)\,dt \end{array}\right] \\ &= \left[\tfrac{1}{2}mu^2\right]_{v_1}^{v_2} = \tfrac{1}{2}mv_2^2 - \tfrac{1}{2}mv_1^2 \end{aligned}$$

(b) The mass of the bowling ball is $\dfrac{12 \text{ lb}}{32 \text{ ft/s}^2} = \dfrac{3}{8}$ slug. Converting 20 mi/h to ft/s^2 gives us

$\dfrac{20 \text{ mi}}{\text{h}} \cdot \dfrac{5280 \text{ ft}}{1 \text{ mi}} \cdot \dfrac{1 \text{ h}}{3600 \text{ s}^2} = \dfrac{88}{3}$ ft/s^2. From part (a) with $v_1 = 0$ and $v_2 = \frac{88}{3}$, the work required to hurl the bowling ball is $W = \frac{1}{2} \cdot \frac{3}{8}\left(\frac{88}{3}\right)^2 - \frac{1}{2} \cdot \frac{3}{8}(0)^2 = \frac{484}{3} = 161.\overline{3}$ ft-lb.

32. The work required to move the 800 kg roller coaster car is

$W = \int_0^{60} (5.7x^2 + 1.5x)\,dx = \left[1.9x^3 + 0.75x^2\right]_0^{60} = 410{,}400 + 2700 = 413{,}100$ J.

Using Exercise 31(a) with $v_1 = 0$, we get $W = \frac{1}{2}mv_2^2 \;\Rightarrow\; v_2 = \sqrt{\dfrac{2W}{m}} = \sqrt{\dfrac{2(413{,}100)}{800}} \approx 32.14$ m/s.

33. (a) $W = \displaystyle\int_a^b F(r)\,dr = \int_a^b G\frac{m_1 m_2}{r^2}\,dr = Gm_1m_2\left[\frac{-1}{r}\right]_a^b = Gm_1m_2\left(\frac{1}{a} - \frac{1}{b}\right)$

(b) By part (a), $W = GMm\left(\dfrac{1}{R} - \dfrac{1}{R + 1{,}000{,}000}\right)$ where M = mass of the earth in kg, R = radius of the earth in m, and m = mass of satellite in kg. (Note that 1000 km = 1,000,000 m.) Thus,

$$W = (6.67 \times 10^{-11})(5.98 \times 10^{24})(1000) \times \left(\frac{1}{6.37 \times 10^6} - \frac{1}{7.37 \times 10^6}\right) \approx 8.50 \times 10^9 \text{ J}$$

34. (a) Assume the pyramid has smooth sides. From the figure for $0 \le x \le 378$, an equation for the side is $y = \frac{-481}{378}x + 481 \;\Leftrightarrow\; x = -\frac{378}{481}(y - 481)$. The horizontal length of a cross-section is $2x$ and the area of a cross-section is $A = (2x)^2 = 4x^2 = 4\dfrac{378^2}{481^2}(y - 481)^2$. A slice of thickness Δy at height y has volume $\Delta V = A\,\Delta y$ ft^3 and weight $150\,\Delta V$ lb, so the work needed to build the pyramid was

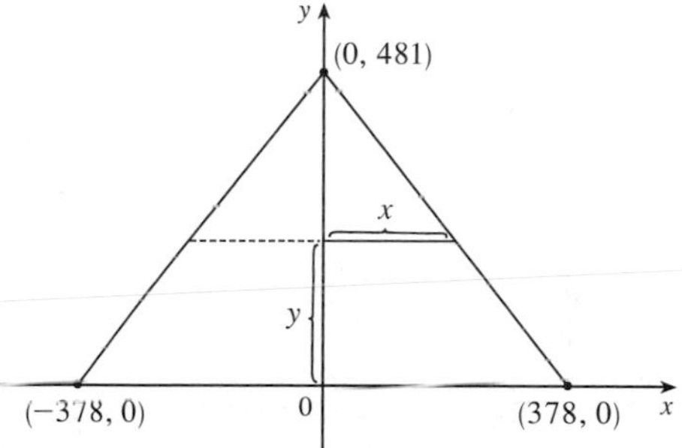

$$W_1 = \int_0^{481} 150y \cdot 4\frac{378^2}{481^2}(y - 481)^2\,dy = 600\,\frac{378^2}{481^2}\int_0^{481}(y^3 - 2 \cdot 481y^2 + 481^2y)\,dy$$

$$= 600\,\frac{378^2}{481^2}\left[\frac{1}{4}y^4 - \frac{2 \cdot 481}{3}y^3 + \frac{481^2}{2}y^2\right]_0^{481} = 600\,\frac{378^2}{481^2}\left(\frac{481^4}{4} - \frac{2 \cdot 481^4}{3} + \frac{481^4}{2}\right)$$

$$= 600\,\frac{378^2}{481^2}\,\frac{481^4}{12} = 50 \cdot 378^2 \cdot 481^2 \approx 1.653 \times 10^{12} \text{ ft-lb}$$

(b) Work done $= W_2 = \dfrac{10 \text{ h}}{\text{day}} \cdot \dfrac{340 \text{ days}}{\text{year}} \cdot \dfrac{20 \text{ yr}}{1 \text{ laborer}} \cdot \dfrac{200 \text{ ft-lb}}{\text{hour}} = 1.36 \times 10^7 \dfrac{\text{ft-lb}}{\text{laborer}}$. Dividing W_1 by W_2 gives us about 121,536 laborers.

6.5 Average Value of a Function

1. $f_{\text{ave}} = \frac{1}{b-a}\int_a^b f(x)\,dx = \frac{1}{2-(-1)}\int_{-1}^2 (3x^2+8x)\,dx = \frac{1}{3}[x^3+4x^2]_{-1}^2 = \frac{1}{3}[(8+16)-(-1+4)] = 7$

2. $f_{\text{ave}} = \frac{1}{b-a}\int_a^b f(x)\,dx = \frac{1}{4-0}\int_0^4 \sqrt{x}\,dx = \frac{1}{4}\left[\frac{2}{3}x^{3/2}\right]_0^4 = \frac{1}{4}\left(\frac{2}{3}\cdot 8\right) = \frac{4}{3}$

3. $g_{\text{ave}} = \dfrac{1}{b-a}\displaystyle\int_a^b g(x)\,dx = \dfrac{1}{\pi/2-(-\pi/2)}\displaystyle\int_{-\pi/2}^{\pi/2} 3\cos x\,dx = \dfrac{3\cdot 2}{\pi}\displaystyle\int_0^{\pi/2}\cos x\,dx$ [by Theorem 5.5.7]

$= \frac{6}{\pi}\Big[\sin x\Big]_0^{\pi/2} = \frac{6}{\pi}(1-0) = \frac{6}{\pi}$

4. $g_{\text{ave}} = \dfrac{1}{b-a}\displaystyle\int_a^b g(t)\,dt = \dfrac{1}{3-1}\displaystyle\int_1^3 \dfrac{t}{\sqrt{3+t^2}}\,dt = \frac{1}{2}\left[(3+t^2)^{1/2}\right]_1^3 = \frac{1}{2}\left(2\sqrt{3}-2\right) = \sqrt{3}-1$

5. $f_{\text{ave}} = \frac{1}{b-a}\int_a^b f(t)\,dt = \frac{1}{\pi/2-0}\int_0^{\pi/2} e^{\sin t}\cos t\,dt = \frac{2}{\pi}[e^{\sin t}]_0^{\pi/2} = \frac{2}{\pi}(e-1)$

6. $f_{\text{ave}} = \dfrac{1}{b-a}\displaystyle\int_a^b f(x)\,dx = \dfrac{1}{1-(-1)}\displaystyle\int_{-1}^1 \dfrac{x^2}{(x^3+3)^2}\,dx = \dfrac{1}{2}\displaystyle\int_2^4 \dfrac{1}{u^2}\left(\dfrac{1}{3}\,du\right)$ $\left[\begin{array}{c} u = x^3+3, \\ du = 3x^2\,dx \end{array}\right]$

$= \dfrac{1}{6}\left[-\dfrac{1}{u}\right]_2^4 = \dfrac{1}{6}\left(-\dfrac{1}{4}+\dfrac{1}{2}\right) = \dfrac{1}{24}$

7. $h_{\text{ave}} = \frac{1}{b-a}\int_a^b h(x)\,dx = \frac{1}{\pi-0}\int_0^\pi \cos^4 x\,\sin x\,dx = \frac{1}{\pi}\int_1^{-1} u^4(-du)$ $[u=\cos x,\ du=-\sin x\,dx]$

$= \frac{1}{\pi}\int_{-1}^1 u^4\,du = \frac{1}{\pi}\cdot 2\int_0^1 u^4\,du$ [by Theorem 5.5.7] $= \frac{2}{\pi}\left[\frac{1}{5}u^5\right]_0^1 = \frac{2}{5\pi}$

8. $h_{\text{ave}} = \dfrac{1}{b-a}\displaystyle\int_a^b h(u)\,du = \dfrac{1}{5-1}\displaystyle\int_1^5 \dfrac{\ln u}{u}\,du = \dfrac{1}{4}\displaystyle\int_0^{\ln 5} y\,dy$ $\left[\begin{array}{c} y = \ln u, \\ dy = 1/u\,du \end{array}\right]$

$= \frac{1}{4}\left[\frac{1}{2}y^2\right]_0^{\ln 5} = \frac{1}{8}(\ln 5)^2$

9. (a) $f_{\text{ave}} = \dfrac{1}{5-2}\displaystyle\int_2^5 (x-3)^2\,dx = \dfrac{1}{3}\left[\dfrac{1}{3}(x-3)^3\right]_2^5$

$= \frac{1}{9}\left[2^3-(-1)^3\right] = \frac{1}{9}(8+1) = 1$

(b) $f(c) = f_{\text{ave}} \iff (c-3)^2 = 1 \iff$

$c-3 = \pm 1 \iff c = 2$ or 4

(c)

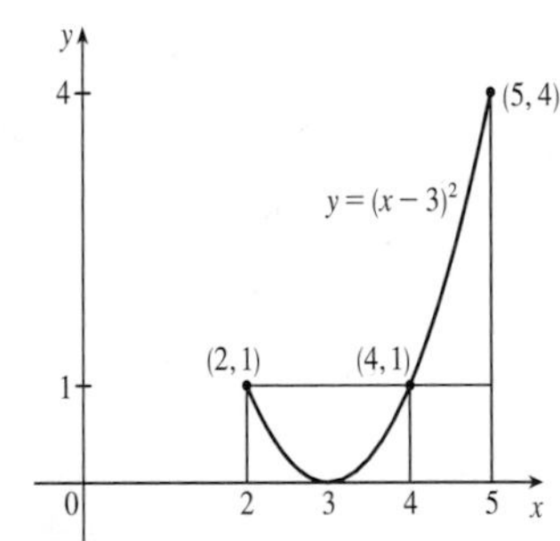

10. (a) $f_{\text{ave}} = \dfrac{1}{3-1}\displaystyle\int_1^3 \dfrac{1}{x}\,dx = \dfrac{1}{2}\Big[\ln|x|\Big]_1^3$

$= \frac{1}{2}(\ln 3 - \ln 1) = \frac{1}{2}\ln 3$

(b) $f(c) = f_{\text{ave}} \iff \frac{1}{c} = \frac{1}{2}\ln 3 \iff c = 2/\ln 3 \approx 1.820$

(c)

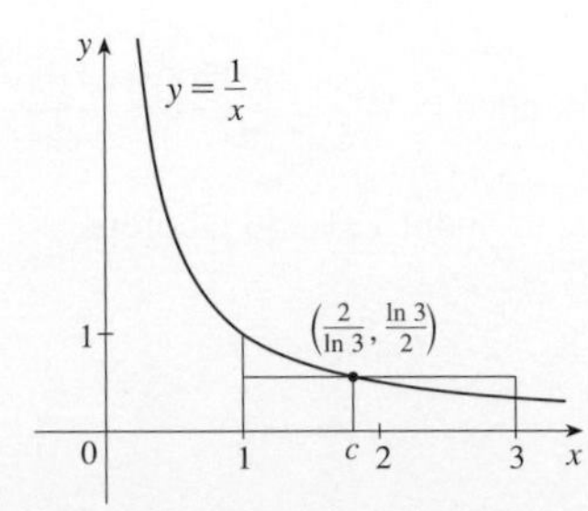

11. (a) $f_{\text{ave}} = \dfrac{1}{\pi - 0}\displaystyle\int_0^\pi (2\sin x - \sin 2x)\,dx$

$= \frac{1}{\pi}\left[-2\cos x + \frac{1}{2}\cos 2x\right]_0^\pi$

$= \frac{1}{\pi}\left[\left(2 + \frac{1}{2}\right) - \left(-2 + \frac{1}{2}\right)\right] = \frac{4}{\pi}$

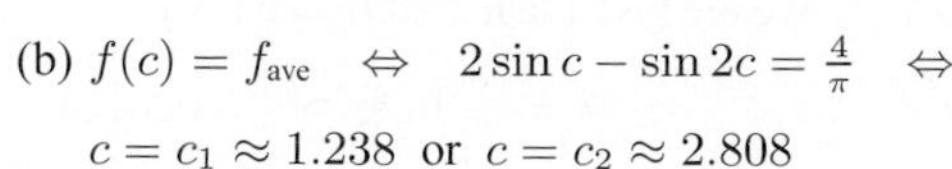

(b) $f(c) = f_{\text{ave}} \;\Leftrightarrow\; 2\sin c - \sin 2c = \frac{4}{\pi} \;\Leftrightarrow$

$c = c_1 \approx 1.238$ or $c = c_2 \approx 2.808$

(c)

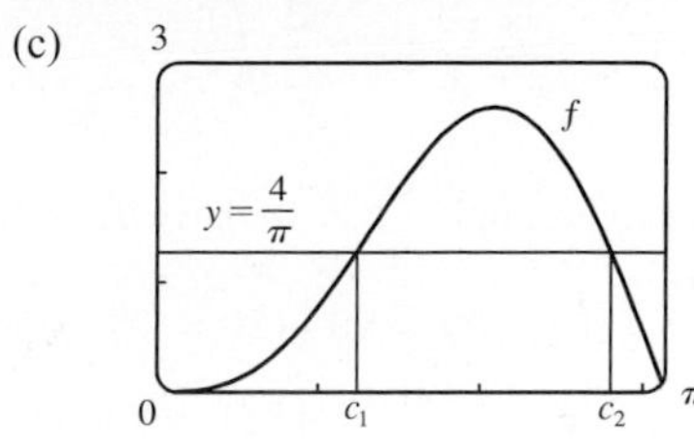

12. (a) $f_{\text{ave}} = \frac{1}{2-0}\int_0^2 2xe^{-x^2}\,dx$

$= \frac{1}{2}\left[-e^{-x^2}\right]_0^2 = \frac{1}{2}(-e^{-4} + 1)$

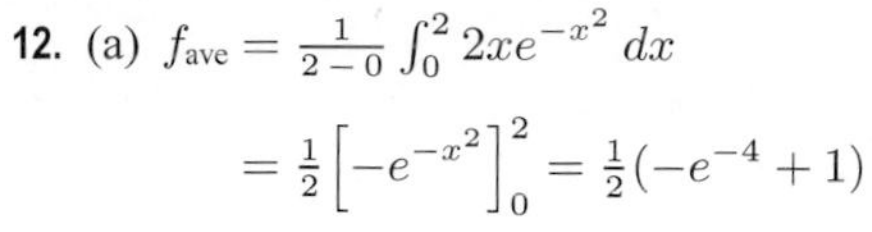

(b) $f(c) = f_{\text{ave}} \;\Leftrightarrow\; 2ce^{-c^2} = \frac{1}{2}(1 - e^{-4}) \;\Leftrightarrow$

$c = c_1 \approx 0.263$ or $c = c_2 \approx 1.287$

(c)

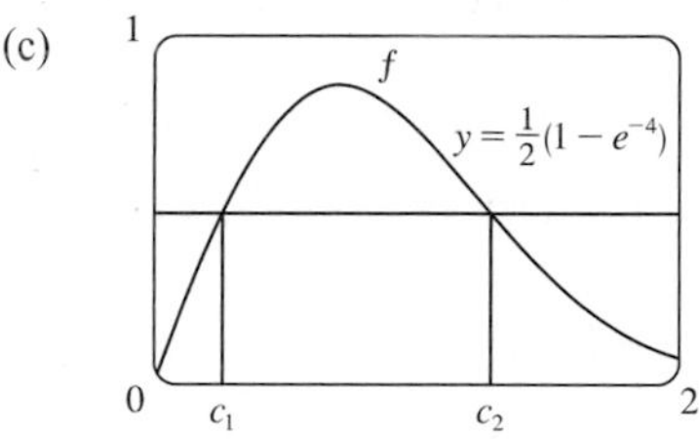

13. f is continuous on $[1, 3]$, so by the Mean Value Theorem for Integrals there exists a number c in $[1, 3]$ such that

$\int_1^3 f(x)\,dx = f(c)(3-1) \;\Rightarrow\; 8 = 2f(c)$; that is, there is a number c such that $f(c) = \frac{8}{2} = 4$.

14. The requirement is that $\dfrac{1}{b-0}\displaystyle\int_0^b f(x)\,dx = 3$. The LHS of this equation is equal to

$\dfrac{1}{b}\displaystyle\int_0^b (2 + 6x - 3x^2)\,dx = \dfrac{1}{b}\left[2x + 3x^2 - x^3\right]_0^b = 2 + 3b - b^2$, so we solve the equation $2 + 3b - b^2 = 3 \;\Leftrightarrow$

$b^2 - 3b + 1 = 0 \;\Leftrightarrow\; b = \dfrac{3 \pm \sqrt{(-3)^2 - 4 \cdot 1 \cdot 1}}{2 \cdot 1} = \dfrac{3 \pm \sqrt{5}}{2}$. Both roots are valid since they are positive.

15. Use geometric interpretations to find the values of the integrals.

$$\begin{aligned}\int_0^8 f(x)\,dx &= \int_0^1 f(x)\,dx + \int_1^2 f(x)\,dx + \int_2^3 f(x)\,dx + \int_3^4 f(x)\,dx + \int_4^6 f(x)\,dx + \int_6^7 f(x)\,dx + \int_7^8 f(x)\,dx \\ &= -\tfrac{1}{2} + \tfrac{1}{2} + \tfrac{1}{2} + 1 + 4 + \tfrac{3}{2} + 2 = 9\end{aligned}$$

Thus, the average value of f on $[0, 8] = f_{\text{ave}} = \frac{1}{8-0}\int_0^8 f(x)\,dx = \frac{1}{8}(9) = \frac{9}{8}$.

16. (a) $v_{\text{ave}} = \frac{1}{12-0}\int_0^{12} v(t)\,dt = \frac{1}{12}I$. Use the Midpoint Rule with $n = 3$ and $\Delta t = \frac{12-0}{3} = 4$ to estimate I.

$I \approx M_3 = 4[v(2) + v(6) + v(10)] = 4[21 + 50 + 66] = 4(137) = 548$. Thus, $v_{\text{ave}} \approx \frac{1}{12}(548) = 45\frac{2}{3}$ km/h.

(b) Estimating from the graph, $v(t) = 45\frac{2}{3}$ when $t \approx 5.2$ s.

17. Let $t = 0$ and $t = 12$ correspond to 9 AM and 9 PM, respectively.

$$\begin{aligned}T_{\text{ave}} &= \tfrac{1}{12-0}\int_0^{12}\left[50 + 14\sin\tfrac{1}{12}\pi t\right]dt = \tfrac{1}{12}\left[50t - 14 \cdot \tfrac{12}{\pi}\cos\tfrac{1}{12}\pi t\right]_0^{12} \\ &= \tfrac{1}{12}\left[50 \cdot 12 + 14 \cdot \tfrac{12}{\pi} + 14 \cdot \tfrac{12}{\pi}\right] = \left(50 + \tfrac{28}{\pi}\right)°\text{F} \approx 59°\text{F}\end{aligned}$$

18. $v_{\text{ave}} = \dfrac{1}{R-0}\displaystyle\int_0^R v(r)\,dr = \dfrac{1}{R}\int_0^R \dfrac{P}{4\eta l}(R^2 - r^2)\,dr = \dfrac{P}{4\eta lR}\left[R^2 r - \tfrac{1}{3}r^3\right]_0^R = \dfrac{P}{4\eta lR}\left(\tfrac{2}{3}\right)R^3 = \dfrac{PR^2}{6\eta l}$.

Since $v(r)$ is decreasing on $(0, R]$, $v_{\max} = v(0) = \dfrac{PR^2}{4\eta l}$. Thus, $v_{\text{ave}} = \frac{2}{3}v_{\max}$.

19. $\rho_{\text{ave}} = \frac{1}{8}\int_0^8 \frac{12}{\sqrt{x+1}}\,dx = \frac{3}{2}\int_0^8 (x+1)^{-1/2}\,dx = \left[3\sqrt{x+1}\right]_0^8 = 9 - 3 = 6 \text{ kg/m}$

20. (a) Similar to Example 3.8.3, we have $T_s = 20°\text{C}$ and hence $\frac{dT}{dt} = c(T - 20)$. Let $y = T - 20$, so that $y(0) = T(0) - 20 = 95 - 20 = 75$. Now y satisfies (3.8.2), so $y = 75e^{ct}$. We are given that $T(30) = 61$, so $y(30) = 61 - 20 = 41$ and $41 = 75e^{c(30)} \Rightarrow \frac{41}{75} = e^{30c} \Rightarrow 30c = \ln\frac{41}{75} \Rightarrow c = \frac{1}{30}\ln\frac{41}{75} \approx -0.020131$. Thus, $T(t) = 20 + 75e^{-kt}$, where $k = -c \approx 0.02$.

(b) $T_{\text{ave}} = \frac{1}{30-0}\int_0^{30} T(t)\,dt = \frac{1}{30}\int_0^{30}(20 + 75e^{-kt})\,dt = \frac{1}{30}\left[20t - \frac{75}{k}e^{-kt}\right]_0^{30} = \frac{1}{30}\left[\left(600 - \frac{75}{k}e^{-30k}\right) - \left(0 - \frac{75}{k}\right)\right]$

$= \frac{1}{30}\left(600 - \frac{75}{k}\cdot\frac{41}{75} + \frac{75}{k}\right) = \frac{1}{30}\left(600 + \frac{34}{k}\right) = 20 + \frac{34}{30k} \approx 76.3°\text{C}$

21. $P_{\text{ave}} = \frac{1}{50-0}\int_0^{50} P(t)\,dt = \frac{1}{50}\int_0^{50} 2560e^{bt}\,dt \quad [\text{with } b = 0.017185]$

$= \frac{2560}{50}\left[\frac{1}{b}e^{bt}\right]_0^{50} = \frac{2560}{50b}(e^{50b} - 1) \approx 4056$ million, or about 4 billion people

22. $s = \frac{1}{2}gt^2 \Rightarrow t = \sqrt{2s/g}$ [since $t \geq 0$]. Now $v = ds/dt = gt = g\sqrt{2s/g} = \sqrt{2gs} \Rightarrow v^2 = 2gs \Rightarrow s = \frac{v^2}{2g}$.

We see that v can be regarded as a function of t or of s: $v = F(t) = gt$ and $v = G(s) = \sqrt{2gs}$. Note that $v_T = F(T) = gT$.

Displacement can be viewed as a function of t: $s = s(t) = \frac{1}{2}gt^2$; also $s(t) = \frac{v^2}{2g} = \frac{[F(t)]^2}{2g}$. When $t = T$, these two formulas for $s(t)$ imply that

$$\sqrt{2gs(T)} = F(T) = v_T = gT = 2\left(\tfrac{1}{2}gT^2\right)/T = 2s(T)/T \qquad (\star)$$

The average of the velocities with respect to time t during the interval $[0, T]$ is

$$v_{t\text{-ave}} = F_{\text{ave}} = \frac{1}{T-0}\int_0^T F(t)\,dt = \frac{1}{T}[s(T) - s(0)] \quad [\text{by FTC}] \quad = \frac{s(T)}{T} \quad [\text{since } s(0) = 0] \quad = \frac{1}{2}v_T \quad [\text{by } (\star)]$$

But the average of the velocities with respect to displacement s during the corresponding displacement interval $[s(0), s(T)] = [0, s(T)]$ is

$$v_{s\text{-ave}} = G_{\text{ave}} = \frac{1}{s(T)-0}\int_0^{s(T)} G(s)\,ds = \frac{1}{s(T)}\int_0^{s(T)}\sqrt{2gs}\,ds = \frac{\sqrt{2g}}{s(T)}\int_0^{s(T)} s^{1/2}\,ds$$

$$= \frac{\sqrt{2g}}{s(T)}\cdot\frac{2}{3}\left[s^{3/2}\right]_0^{s(T)} = \frac{2}{3}\cdot\frac{\sqrt{2g}}{s(T)}\cdot\left[s(T)\right]^{3/2} = \frac{2}{3}\sqrt{2gs(T)} = \frac{2}{3}v_T \quad [\text{by } (\star)]$$

23. $V_{\text{ave}} = \frac{1}{5}\int_0^5 V(t)\,dt = \frac{1}{5}\int_0^5 \frac{5}{4\pi}\left[1 - \cos\left(\frac{2}{5}\pi t\right)\right]dt = \frac{1}{4\pi}\int_0^5\left[1 - \cos\left(\frac{2}{5}\pi t\right)\right]dt$

$= \frac{1}{4\pi}\left[t - \frac{5}{2\pi}\sin\left(\frac{2}{5}\pi t\right)\right]_0^5 = \frac{1}{4\pi}[(5-0) - 0] = \frac{5}{4\pi} \approx 0.4 \text{ L}$

24. $f_{\text{ave}} = \frac{1}{b-a}\int_a^b f(x)\,dx$

$> \frac{1}{b-a}$ (area of trapezoid $ABDF$)

$= \frac{1}{b-a}$ (area of rectangle $ACEF$)

$= \frac{1}{b-a}\left[f\left(\frac{a+b}{2}\right)\cdot(b-a)\right]$

$= f\left(\frac{a+b}{2}\right)$

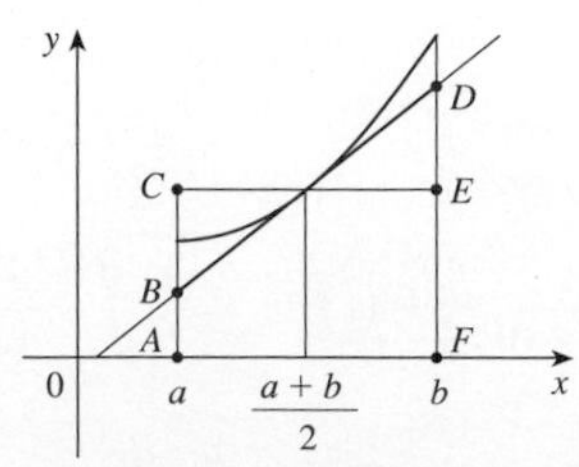

25. Let $F(x) = \int_a^x f(t)\,dt$ for x in $[a, b]$. Then F is continuous on $[a, b]$ and differentiable on (a, b), so by the Mean Value Theorem there is a number c in (a, b) such that $F(b) - F(a) = F'(c)(b - a)$. But $F'(x) = f(x)$ by the Fundamental Theorem of Calculus. Therefore, $\int_a^b f(t)\,dt - 0 = f(c)(b - a)$.

26. $$f_{\text{ave}}\,[a,b] = \frac{1}{b-a}\int_a^b f(x)\,dx = \frac{1}{b-a}\int_a^c f(x)\,dx + \frac{1}{b-a}\int_c^b f(x)\,dx$$

$$= \frac{c-a}{b-a}\left[\frac{1}{c-a}\int_a^c f(x)\,dx\right] + \frac{b-c}{b-a}\left[\frac{1}{b-c}\int_c^b f(x)\,dx\right] = \frac{c-a}{b-a}\,f_{\text{ave}}\,[a,c] + \frac{b-c}{b-a}\,f_{\text{ave}}\,[c,b]$$

APPLIED PROJECT Calculus and Baseball

1. (a) $F = ma = m\,\dfrac{dv}{dt}$, so by the Substitution Rule we have

$$\int_{t_0}^{t_1} F(t)\,dt = \int_{t_0}^{t_1} m\left(\frac{dv}{dt}\right)dt = m\int_{v_0}^{v_1} dv = \left[mv\right]_{v_0}^{v_1} = mv_1 - mv_0 = p(t_1) - p(t_0)$$

(b) (i) We have $v_1 = 110$ mi/h $= \frac{110(5280)}{3600}$ ft/s $= 161.\overline{3}$ ft/s, $v_0 = -90$ mi/h $= -132$ ft/s, and the mass of the baseball is $m = \dfrac{w}{g} = \frac{5/16}{32} = \frac{5}{512}$. So the change in momentum is

$p(t_1) - p(t_0) = mv_1 - mv_0 = \frac{5}{512}\,[161.\overline{3} - (-132)] \approx 2.86$ slug-ft/s.

(ii) From part (a) and part (b)(i), we have $\int_0^{0.001} F(t)\,dt = p(0.001) - p(0) \approx 2.86$, so the average force over the interval $[0, 0.001]$ is $\frac{1}{0.001}\int_0^{0.001} F(t)\,dt \approx \frac{1}{0.001}(2.86) = 2860$ lb.

2. (a) $W = \displaystyle\int_{s_0}^{s_1} F(s)\,ds$, where $F(s) = m\,\dfrac{dv}{dt} = m\,\dfrac{dv}{ds}\dfrac{ds}{dt} = mv\,\dfrac{dv}{ds}$ and so, by the Substitution Rule,

$$W = \int_{s_0}^{s_1} F(s)\,ds = \int_{s_0}^{s_1} mv\,\frac{dv}{ds}\,ds = \int_{v(s_0)}^{v(s_1)} mv\,dv = \left[\tfrac{1}{2}mv^2\right]_{v_0}^{v_1} = \tfrac{1}{2}mv_1^2 - \tfrac{1}{2}mv_0^2$$

(b) From part (b)(i), 90 mi/h $= 132$ ft/s. Assume $v_0 = v(s_0) = 0$ and $v_1 = v(s_1) = 132$ ft/s [note that s_1 is the point of release of the baseball]. $m = \frac{5}{512}$, so the work done is $W = \frac{1}{2}mv_1^2 - \frac{1}{2}mv_0^2 = \frac{1}{2}\cdot\frac{5}{512}\cdot(132)^2 \approx 85$ ft-lb.

3. (a) Here we have a differential equation of the form $dv/dt = kv$, so by Theorem 3.8.2, the solution is $v(t) = v(0)e^{kt}$. In this case $k = -\frac{1}{10}$ and $v(0) = 100$ ft/s, so $v(t) = 100e^{-t/10}$. We are interested in the time t that the ball takes to travel 280 ft, so we find the distance function

$$s(t) = \textstyle\int_0^t v(x)\,dx = \int_0^t 100e^{-x/10}\,dx = 100\left[-10e^{-x/10}\right]_0^t = -1000(e^{-t/10} - 1) = 1000(1 - e^{-t/10})$$

Now we set $s(t) = 280$ and solve for t: $280 = 1000(1 - e^{-t/10}) \Rightarrow 1 - e^{-t/10} = \frac{7}{25} \Rightarrow -\frac{1}{10}t = \ln\left(1 - \frac{7}{25}\right) \Rightarrow t \approx 3.285$ seconds.

(b) Let x be the distance of the shortstop from home plate. We calculate the time for the ball to reach home plate as a function of x, then differentiate with respect to x to find the value of x which corresponds to the minimum time. The total time that it takes the ball to reach home is the sum of the times of the two throws, plus the relay time $\left(\frac{1}{2}\text{ s}\right)$. The distance from the fielder to the shortstop is $280 - x$, so to find the time t_1 taken by the first throw, we solve the equation

$s_1(t_1) = 280 - x \quad\Leftrightarrow\quad 1 - e^{-t_1/10} = \dfrac{280 - x}{1000} \quad\Leftrightarrow\quad t_1 = -10\ln\dfrac{720 + x}{1000}$. We find the time t_2 taken by the second throw if the shortstop throws with velocity w, since we see that this velocity varies in the rest of the problem. We use $v = we^{-t/10}$ and isolate t_2 in the equation $s(t_2) = 10w(1 - e^{-t_2/10}) = x \quad\Leftrightarrow\quad e^{-t_2/10} = 1 - \dfrac{x}{10w} \quad\Leftrightarrow$

$t_2 = -10\ln\dfrac{10w - x}{10w}$, so the total time is $t_w(x) = \dfrac{1}{2} - 10\left[\ln\dfrac{720 + x}{1000} + \ln\dfrac{10w - x}{10w}\right]$.

To find the minimum, we differentiate: $\dfrac{dt_w}{dx} = -10\left[\dfrac{1}{720 + x} - \dfrac{1}{10w - x}\right]$, which changes from negative to positive when $720 + x = 10w - x \quad\Leftrightarrow\quad x = 5w - 360$. By the First Derivative Test, t_w has a minimum at this distance from the shortstop to home plate. So if the shortstop throws at $w = 105$ ft/s from a point $x = 5(105) - 360 = 165$ ft from home plate, the minimum time is $t_{105}(165) = \frac{1}{2} - 10\left(\ln\frac{720 + 165}{1000} + \ln\frac{1050 - 165}{1050}\right) \approx 3.431$ seconds. This is longer than the time taken in part (a), so in this case the manager should encourage a direct throw. If $w = 115$ ft/s, then $x = 215$ ft from home, and the minimum time is $t_{115}(215) = \frac{1}{2} - 10\left(\ln\frac{720 + 215}{1000} + \ln\frac{1150 - 215}{1150}\right) \approx 3.242$ seconds. This is less than the time taken in part (a), so in this case, the manager should encourage a relayed throw.

(c) In general, the minimum time is $t_w(5w - 360) = \dfrac{1}{2} - 10\left[\ln\dfrac{360 + 5w}{1000} + \ln\dfrac{360 + 5w}{10w}\right] = \dfrac{1}{2} - 10\ln\dfrac{(w + 72)^2}{400w}$.

We want to find out when this is about 3.285 seconds, the same time as the direct throw. From the graph, we estimate that this is the case for $w \approx 112.8$ ft/s. So if the shortstop can throw the ball with this velocity, then a relayed throw takes the same time as a direct throw.

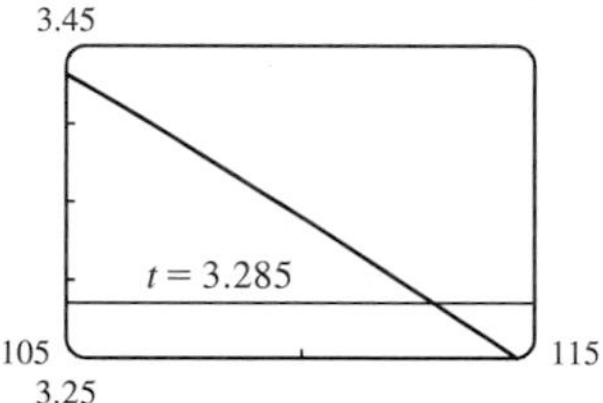

APPLIED PROJECT Where to Sit at the Movies

1. $|VP| = 9 + x\cos\alpha$, $|PT| = 35 - (4 + x\sin\alpha) = 31 - x\sin\alpha$, and $|PB| = (4 + x\sin\alpha) - 10 = x\sin\alpha - 6$. So using the Pythagorean Theorem, we have $|VT| = \sqrt{|VP|^2 + |PT|^2} = \sqrt{(9 + x\cos\alpha)^2 + (31 - x\sin\alpha)^2} = a$,

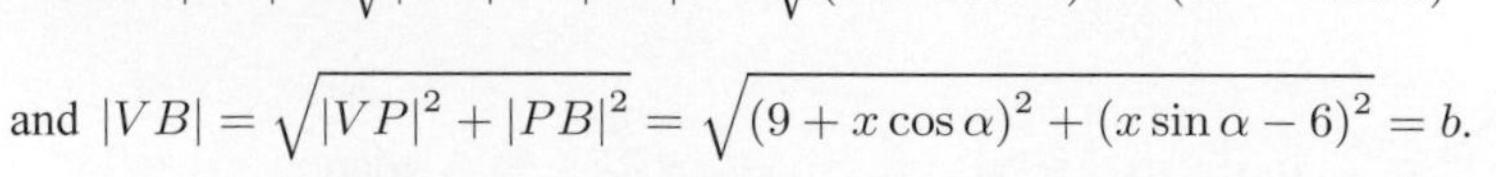

and $|VB| = \sqrt{|VP|^2 + |PB|^2} = \sqrt{(9 + x\cos\alpha)^2 + (x\sin\alpha - 6)^2} = b$.

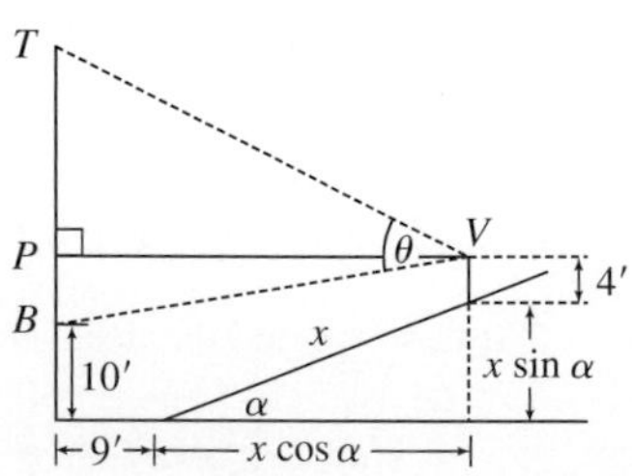

Using the Law of Cosines on $\triangle VBT$, we get $25^2 = a^2 + b^2 - 2ab\cos\theta \quad\Leftrightarrow\quad \cos\theta = \dfrac{a^2 + b^2 - 625}{2ab} \quad\Leftrightarrow$

$\theta = \arccos\left(\dfrac{a^2 + b^2 - 625}{2ab}\right)$, as required.

2. From the graph of θ, it appears that the value of x which maximizes θ is $x \approx 8.25$ ft. Assuming that the first row is at $x = 0$, the row closest to this value of x is the fourth row, at $x = 9$ ft, and from the graph, the viewing angle in this row seems to be about 0.85 radians, or about $49°$.

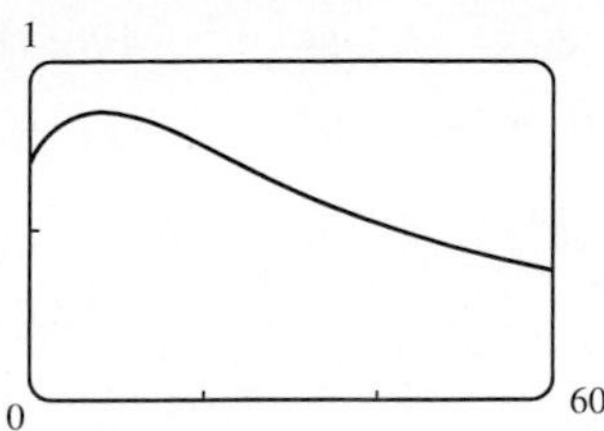

3. With a CAS, we type in the definition of θ, substitute in the proper values of a and b in terms of x and $\alpha = 20° = \frac{\pi}{9}$ radians, and then use the differentiation command to find the derivative. We use a numerical rootfinder and find that the root of the equation $d\theta/dx = 0$ is $x \approx 8.253062$, as approximated in Problem 2.

4. From the graph in Problem 2, it seems that the average value of the function on the interval $[0, 60]$ is about 0.6. We can use a CAS to approximate $\frac{1}{60}\int_0^{60} \theta(x)\,dx \approx 0.625 \approx 36°$. (The calculation is much faster if we reduce the number of digits of accuracy required.) The minimum value is $\theta(60) \approx 0.38$ and, from Problem 2, the maximum value is about 0.85.

6 Review

EXERCISES

1. The curves intersect when $x^2 = 4x - x^2 \;\Leftrightarrow\; 2x^2 - 4x = 0 \;\Leftrightarrow\; 2x(x-2) = 0 \;\Leftrightarrow\; x = 0$ or 2.

$$A = \int_0^2 \left[(4x - x^2) - x^2\right] dx = \int_0^2 (4x - 2x^2)\,dx$$
$$= \left[2x^2 - \tfrac{2}{3}x^3\right]_0^2 = \left[\left(8 - \tfrac{16}{3}\right) - 0\right] = \tfrac{8}{3}$$

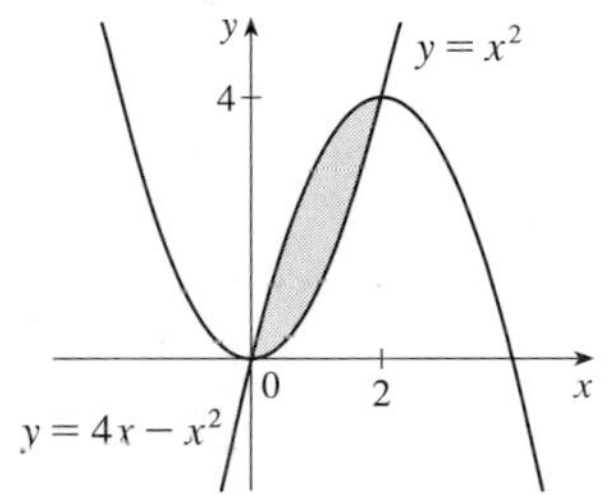

2. The line $y = x - 2$ intersects the curve $y = \sqrt{x}$ at $(4, 2)$ and it intersects the curve $y = -\sqrt[3]{x}$ at $(1, -1)$.

$$A = \int_0^1 \left[\sqrt{x} - \left(-\sqrt[3]{x}\right)\right] dx + \int_1^4 \left[\sqrt{x} - (x-2)\right] dx$$
$$= \left[\tfrac{2}{3}x^{3/2} + \tfrac{3}{4}x^{4/3}\right]_0^1 + \left[\tfrac{2}{3}x^{3/2} - \tfrac{1}{2}x^2 + 2x\right]_1^4$$
$$= \left(\tfrac{2}{3} + \tfrac{3}{4}\right) - 0 + \left(\tfrac{16}{3} - 8 + 8\right) - \left(\tfrac{2}{3} - \tfrac{1}{2} + 2\right)$$
$$= \tfrac{16}{3} + \tfrac{3}{4} - \tfrac{3}{2} = \tfrac{55}{12}$$

Or, integrating with respect to y: $A = \int_{-1}^0 \left[(y+2) - (-y^3)\right] dy + \int_0^2 \left[(y+2) - y^2\right] dy$

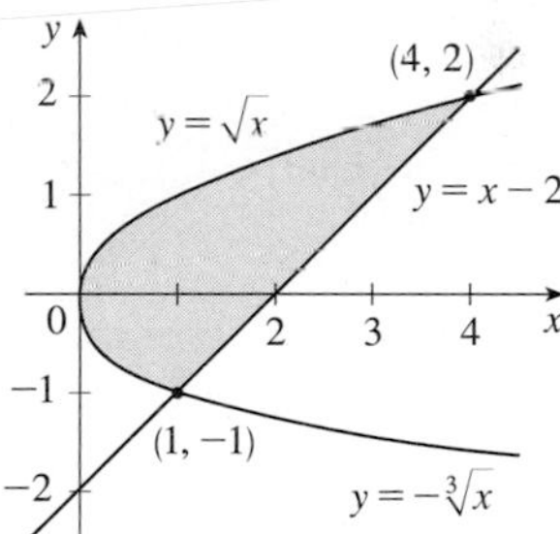

3. If $x \geq 0$, then $|x| = x$, and the graphs intersect when $x = 1 - 2x^2 \;\Leftrightarrow\; 2x^2 + x - 1 = 0 \;\Leftrightarrow\; (2x-1)(x+1) = 0 \;\Leftrightarrow\; x = \frac{1}{2}$ or -1, but $-1 < 0$. By symmetry, we can double the area from $x = 0$ to $x = \frac{1}{2}$.

$$A = 2\int_0^{1/2} \left[(1 - 2x^2) - x\right] dx = 2\int_0^{1/2} (-2x^2 - x + 1)\,dx$$
$$= 2\left[-\tfrac{2}{3}x^3 - \tfrac{1}{2}x^2 + x\right]_0^{1/2} = 2\left[\left(-\tfrac{1}{12} - \tfrac{1}{8} + \tfrac{1}{2}\right) - 0\right]$$
$$= 2\left(\tfrac{7}{24}\right) = \tfrac{7}{12}$$

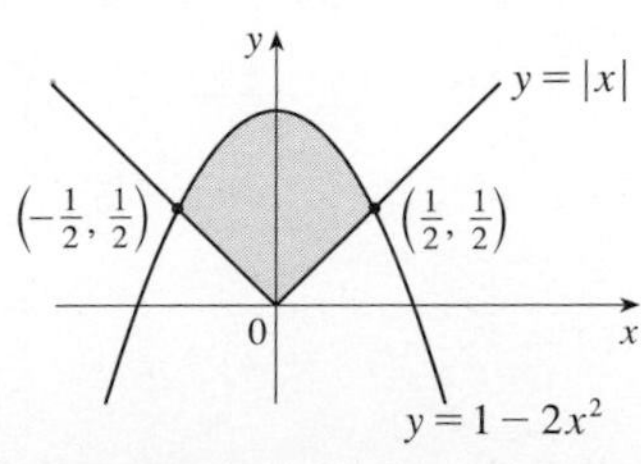

4. $y^2 + 3y = -y \quad \Leftrightarrow \quad y^2 + 4y = 0 \quad \Leftrightarrow \quad y(y+4) = 0 \quad \Leftrightarrow$
$y = 0$ or -4.

$A = \int_{-4}^{0} [-y - (y^2 + 3y)]\,dy = \int_{-4}^{0}(-y^2 - 4y)\,dy$
$= \left[-\frac{1}{3}y^3 - 2y^2\right]_{-4}^{0} = 0 - \left(\frac{64}{3} - 32\right) = \frac{32}{3}$

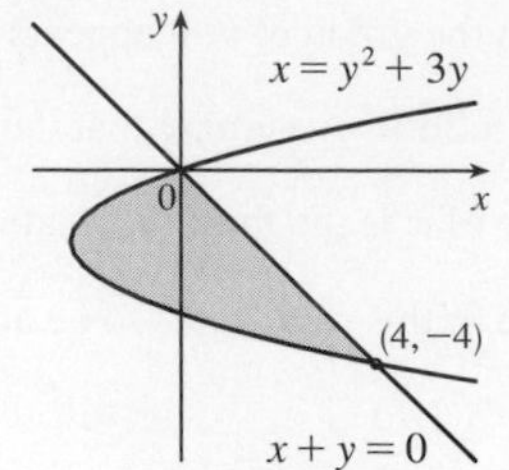

5. $A = \displaystyle\int_0^2 \left[\sin\left(\frac{\pi x}{2}\right) - (x^2 - 2x)\right] dx$
$= \left[-\dfrac{2}{\pi}\cos\left(\dfrac{\pi x}{2}\right) - \dfrac{1}{3}x^3 + x^2\right]_0^2$
$= \left(\frac{2}{\pi} - \frac{8}{3} + 4\right) - \left(-\frac{2}{\pi} - 0 + 0\right) = \frac{4}{3} + \frac{4}{\pi}$

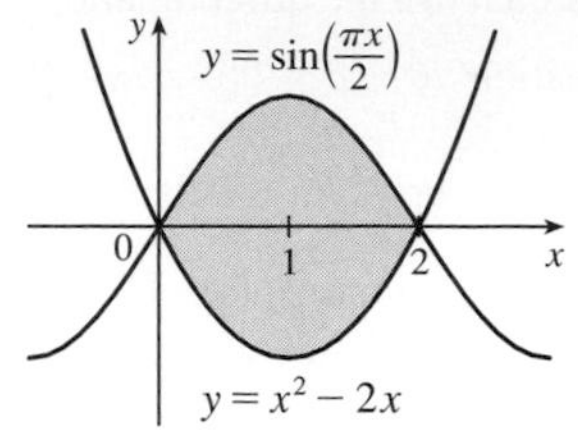

6. $A = \int_0^1 \left(\sqrt{x} - x^2\right) dx + \int_1^2 \left(x^2 - \sqrt{x}\right) dx$
$= \left[\frac{2}{3}x^{3/2} - \frac{1}{3}x^3\right]_0^1 + \left[\frac{1}{3}x^3 - \frac{2}{3}x^{3/2}\right]_1^2$
$= \left[\left(\frac{2}{3} - \frac{1}{3}\right) - 0\right] + \left[\left(\frac{8}{3} - \frac{4}{3}\sqrt{2}\right) - \left(\frac{1}{3} - \frac{2}{3}\right)\right]$
$= \frac{10}{3} - \frac{4}{3}\sqrt{2}$

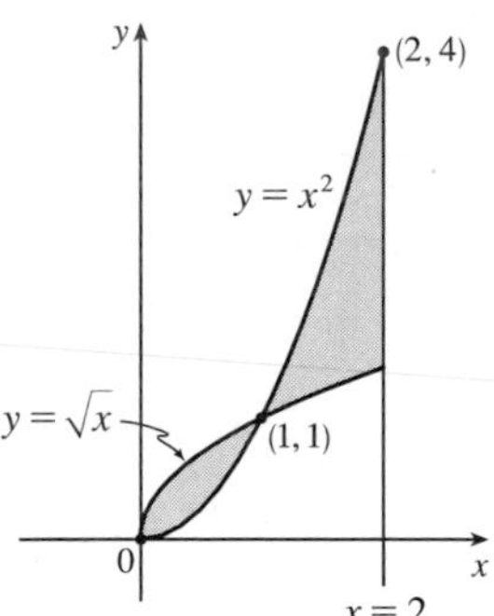

7. Using washers with inner radius x^2 and outer radius $2x$, we have
$V = \pi \int_0^2 \left[(2x)^2 - (x^2)^2\right] dx = \pi \int_0^2 (4x^2 - x^4)\,dx$
$= \pi\left[\frac{4}{3}x^3 - \frac{1}{5}x^5\right]_0^2 = \pi\left(\frac{32}{3} - \frac{32}{5}\right)$
$= 32\pi \cdot \frac{2}{15} = \frac{64}{15}\pi$

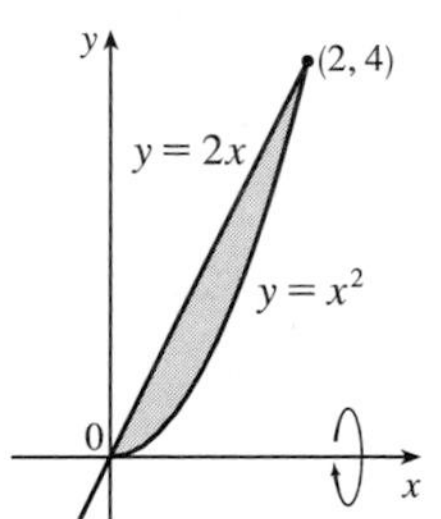

8. $1 + y^2 = y + 3 \quad \Leftrightarrow \quad y^2 - y - 2 = 0 \quad \Leftrightarrow \quad (y-2)(y+1) = 0 \quad \Leftrightarrow$
$y = 2$ or -1.
$V = \pi \int_{-1}^{2} \left[(y+3)^2 - (1+y^2)^2\right] dy = \pi \int_{-1}^{2} (y^2 + 6y + 9 - 1 - 2y^2 - y^4)\,dy$
$= \pi \int_{-1}^{2} (8 + 6y - y^2 - y^4)\,dy = \pi\left[8y + 3y^2 - \frac{1}{3}y^3 - \frac{1}{5}y^5\right]_{-1}^{2}$
$= \pi\left[\left(16 + 12 - \frac{8}{3} - \frac{32}{5}\right) - \left(-8 + 3 + \frac{1}{3} + \frac{1}{5}\right)\right] = \pi\left(33 - \frac{9}{3} - \frac{33}{5}\right) = \frac{117}{5}\pi$

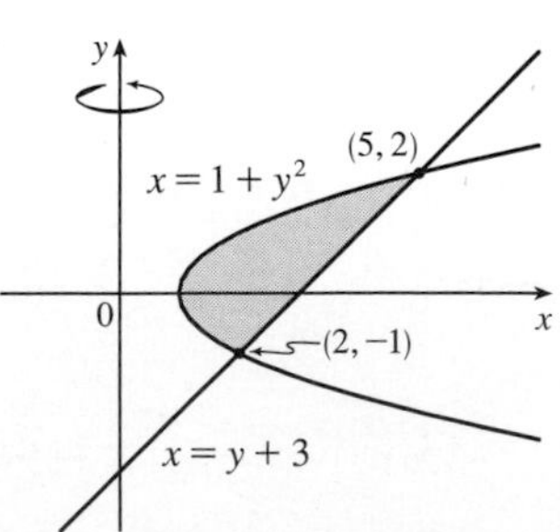

9. $V = \pi \int_{-3}^{3} \left\{ \left[(9 - y^2) - (-1)\right]^2 - \left[0 - (-1)\right]^2 \right\} dy$

$= 2\pi \int_0^3 \left[(10 - y^2)^2 - 1\right] dy = 2\pi \int_0^3 (100 - 20y^2 + y^4 - 1)\, dy$

$= 2\pi \int_0^3 (99 - 20y^2 + y^4)\, dy = 2\pi \left[99y - \frac{20}{3}y^3 + \frac{1}{5}y^5\right]_0^3$

$= 2\pi \left(297 - 180 + \frac{243}{5}\right) = \frac{1656}{5}\pi$

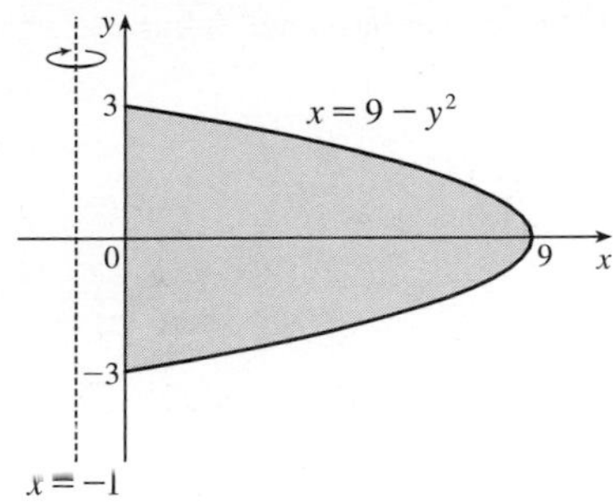

10. $V = \pi \int_{-2}^{2} \left\{ \left[(9 - x^2) - (-1)\right]^2 - \left[(x^2 + 1) - (-1)\right]^2 \right\} dx$

$= \pi \int_{-2}^{2} \left[(10 - x^2)^2 - (x^2 + 2)^2\right] dx$

$= 2\pi \int_0^2 (96 - 24x^2)\, dx = 48\pi \int_0^2 (4 - x^2)\, dx$

$= 48\pi \left[4x - \frac{1}{3}x^3\right]_0^2 = 48\pi \left(8 - \frac{8}{3}\right) = 256\pi$

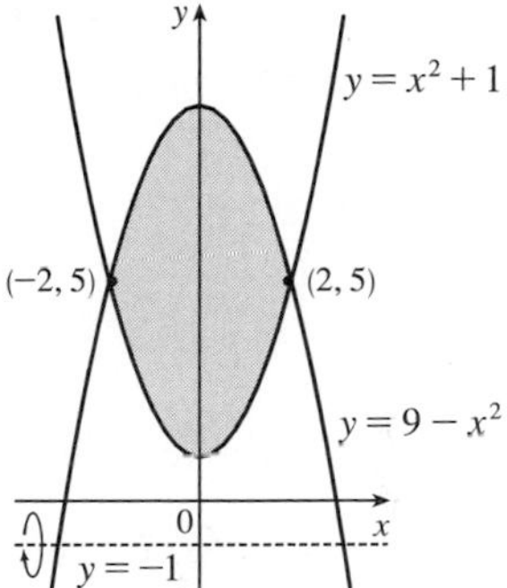

11. The graph of $x^2 - y^2 = a^2$ is a hyperbola with right and left branches.

Solving for y gives us $y^2 = x^2 - a^2 \quad \Rightarrow \quad y = \pm\sqrt{x^2 - a^2}$.

We'll use shells and the height of each shell is

$\sqrt{x^2 - a^2} - \left(-\sqrt{x^2 - a^2}\right) = 2\sqrt{x^2 - a^2}$.

The volume is $V = \int_a^{a+h} 2\pi x \cdot 2\sqrt{x^2 - a^2}\, dx$. To evaluate, let $u = x^2 - a^2$,

so $du = 2x\, dx$ and $x\, dx = \frac{1}{2}\, du$. When $x = a$, $u = 0$, and when $x = a + h$,

$u = (a + h)^2 - a^2 = a^2 + 2ah + h^2 - a^2 = 2ah + h^2$.

Thus, $V = 4\pi \displaystyle\int_0^{2ah+h^2} \sqrt{u} \left(\frac{1}{2}\, du\right) = 2\pi \left[\frac{2}{3}u^{3/2}\right]_0^{2ah+h^2} = \frac{4}{3}\pi \left(2ah + h^2\right)^{3/2}$.

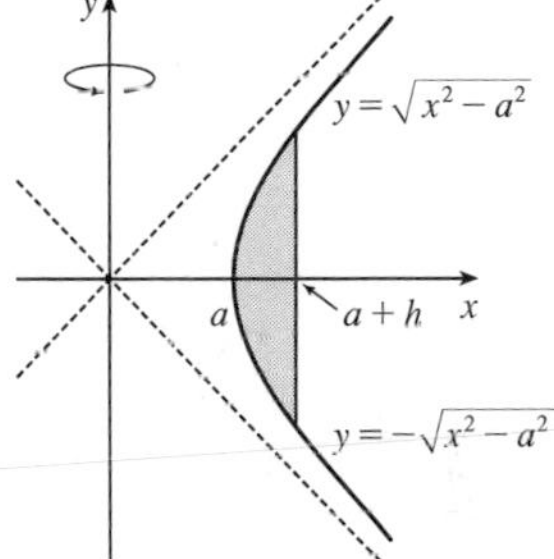

12. A shell has radius x, circumference $2\pi x$, and height $\tan x - x$.

$V = \int_0^{\pi/3} 2\pi x\, (\tan x - x)\, dx$

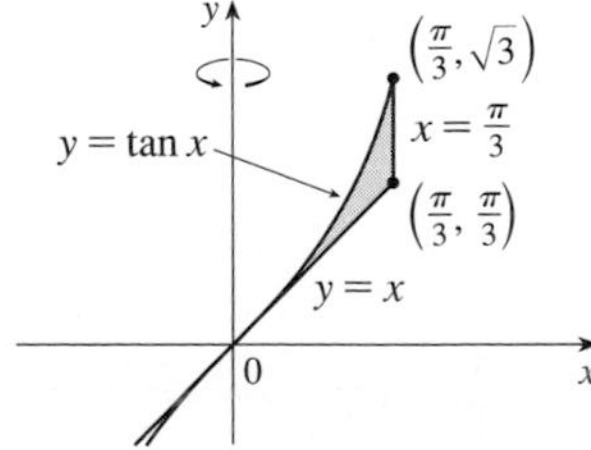

13. A shell has radius $\frac{\pi}{2} - x$, circumference $2\pi\left(\frac{\pi}{2} - x\right)$, and height $\cos^2 x - \frac{1}{4}$.

$y = \cos^2 x$ intersects $y = \frac{1}{4}$ when $\cos^2 x = \frac{1}{4} \quad \Leftrightarrow$

$\cos x = \pm\frac{1}{2} \quad [\,|x| \leq \pi/2] \quad \Leftrightarrow \quad x = \pm\frac{\pi}{3}$.

$$V = \int_{-\pi/3}^{\pi/3} 2\pi\left(\frac{\pi}{2} - x\right)\left(\cos^2 x - \frac{1}{4}\right) dx$$

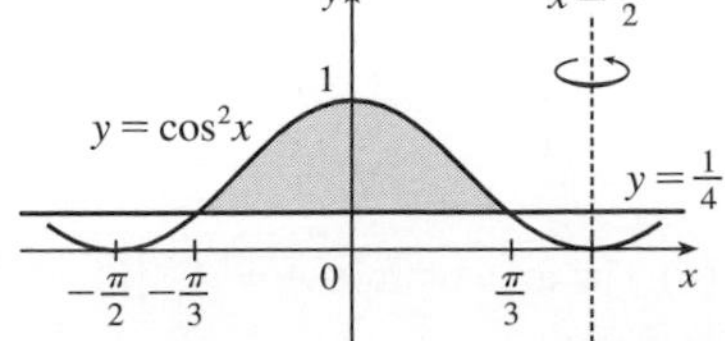

14. A washer has outer radius $2 - x^2$ and inner radius $2 - \sqrt{x}$.

$$V = \int_0^1 \pi\left[(2-x^2)^2 - \left(2-\sqrt{x}\right)^2\right] dx$$

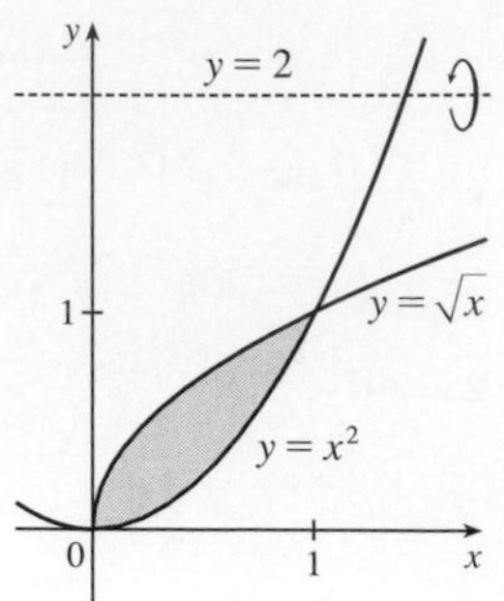

15. (a) A cross-section is a washer with inner radius x^2 and outer radius x.

$$V = \int_0^1 \pi\left[(x)^2 - (x^2)^2\right] dx = \int_0^1 \pi(x^2 - x^4)\, dx = \pi\left[\tfrac{1}{3}x^3 - \tfrac{1}{5}x^5\right]_0^1 = \pi\left[\tfrac{1}{3} - \tfrac{1}{5}\right] = \tfrac{2}{15}\pi$$

(b) A cross-section is a washer with inner radius y and outer radius $\sqrt{y}$.

$$V = \int_0^1 \pi\left[\left(\sqrt{y}\right)^2 - y^2\right] dy = \int_0^1 \pi(y - y^2)\, dy = \pi\left[\tfrac{1}{2}y^2 - \tfrac{1}{3}y^3\right]_0^1 = \pi\left[\tfrac{1}{2} - \tfrac{1}{3}\right] = \tfrac{\pi}{6}$$

(c) A cross-section is a washer with inner radius $2 - x$ and outer radius $2 - x^2$.

$$V = \int_0^1 \pi\left[(2-x^2)^2 - (2-x)^2\right] dx = \int_0^1 \pi(x^4 - 5x^2 + 4x)\, dx = \pi\left[\tfrac{1}{5}x^5 - \tfrac{5}{3}x^3 + 2x^2\right]_0^1 = \pi\left[\tfrac{1}{5} - \tfrac{5}{3} + 2\right] = \tfrac{8}{15}\pi$$

16. (a) $A = \int_0^1 (2x - x^2 - x^3)\, dx = \left[x^2 - \tfrac{1}{3}x^3 - \tfrac{1}{4}x^4\right]_0^1 = 1 - \tfrac{1}{3} - \tfrac{1}{4} = \tfrac{5}{12}$

(b) A cross-section is a washer with inner radius x^3 and outer radius $2x - x^2$, so its area is $\pi(2x - x^2)^2 - \pi(x^3)^2$.

$$\begin{aligned} V &= \int_0^1 A(x)\, dx = \int_0^1 \pi[(2x-x^2)^2 - (x^3)^2]\, dx = \int_0^1 \pi(4x^2 - 4x^3 + x^4 - x^6)\, dx \\ &= \pi\left[\tfrac{4}{3}x^3 - x^4 + \tfrac{1}{5}x^5 - \tfrac{1}{7}x^7\right]_0^1 = \pi\left(\tfrac{4}{3} - 1 + \tfrac{1}{5} - \tfrac{1}{7}\right) = \tfrac{41}{105}\pi \end{aligned}$$

(c) Using the method of cylindrical shells,

$$V = \int_0^1 2\pi x(2x - x^2 - x^3)\, dx = \int_0^1 2\pi(2x^2 - x^3 - x^4)\, dx = 2\pi\left[\tfrac{2}{3}x^3 - \tfrac{1}{4}x^4 - \tfrac{1}{5}x^5\right]_0^1 = 2\pi\left(\tfrac{2}{3} - \tfrac{1}{4} - \tfrac{1}{5}\right) = \tfrac{13}{30}\pi.$$

17. (a) Using the Midpoint Rule on $[0,1]$ with $f(x) = \tan(x^2)$ and $n = 4$, we estimate

$$A = \int_0^1 \tan(x^2)\, dx \approx \tfrac{1}{4}\left[\tan\left(\left(\tfrac{1}{8}\right)^2\right) + \tan\left(\left(\tfrac{3}{8}\right)^2\right) + \tan\left(\left(\tfrac{5}{8}\right)^2\right) + \tan\left(\left(\tfrac{7}{8}\right)^2\right)\right] \approx \tfrac{1}{4}(1.53) \approx 0.38$$

(b) Using the Midpoint Rule on $[0,1]$ with $f(x) = \pi\tan^2(x^2)$ (for disks) and $n = 4$, we estimate

$$V = \int_0^1 f(x)\, dx \approx \tfrac{1}{4}\pi\left[\tan^2\left(\left(\tfrac{1}{8}\right)^2\right) + \tan^2\left(\left(\tfrac{3}{8}\right)^2\right) + \tan^2\left(\left(\tfrac{5}{8}\right)^2\right) + \tan^2\left(\left(\tfrac{7}{8}\right)^2\right)\right] \approx \tfrac{\pi}{4}(1.114) \approx 0.87$$

18. (a)

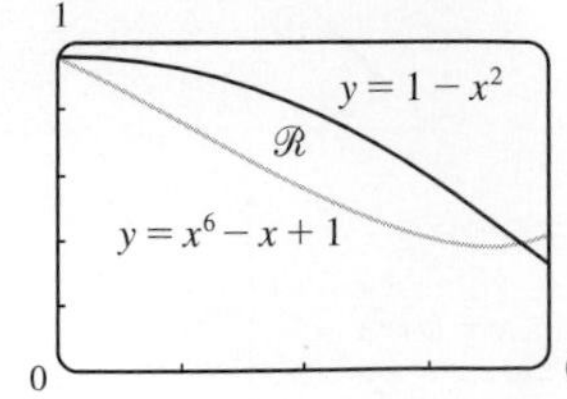

From the graph, we see that the curves intersect at $x = 0$ and at $x = a \approx 0.75$, with $1 - x^2 > x^6 - x + 1$ on $(0, a)$.

(b) The area of $\mathscr{R}$ is $A = \int_0^a \left[(1 - x^2) - (x^6 - x + 1)\right] dx = \left[-\tfrac{1}{3}x^3 - \tfrac{1}{7}x^7 + \tfrac{1}{2}x^2\right]_0^a \approx 0.12.$

(c) Using washers, the volume generated when $\mathcal{R}$ is rotated about the x-axis is

$$V = \pi \int_0^a [(1-x^2)^2 - (x^6 - x + 1)^2]\,dx = \pi \int_0^a (-x^{12} + 2x^7 - 2x^6 + x^4 - 3x^2 + 2x)\,dx$$
$$= \pi\left[-\tfrac{1}{13}x^{13} + \tfrac{1}{4}x^8 - \tfrac{2}{7}x^7 + \tfrac{1}{5}x^5 - x^3 + x^2\right]_0^a \approx 0.54$$

(d) Using shells, the volume generated when $\mathcal{R}$ is rotated about the y-axis is

$$V = \int_0^a 2\pi x[(1-x^2) - (x^6 - x + 1)]\,dx = 2\pi \int_0^a (-x^3 - x^7 + x^2)\,dx = 2\pi\left[-\tfrac{1}{4}x^4 - \tfrac{1}{8}x^8 + \tfrac{1}{3}x^3\right]_0^a \approx 0.31.$$

19. $\int_0^{\pi/2} 2\pi x \cos x\,dx = \int_0^{\pi/2} (2\pi x)\cos x\,dx$

The solid is obtained by rotating the region $\mathcal{R} = \left\{(x,y) \mid 0 \le x \le \frac{\pi}{2}, 0 \le y \le \cos x\right\}$ about the y-axis.

20. $\int_0^{\pi/2} 2\pi \cos^2 x\,dx = \int_0^{\pi/2} \pi\left(\sqrt{2}\cos x\right)^2\,dx$

The solid is obtained by rotating the region $\mathcal{R} = \left\{(x,y) \mid 0 \le x \le \frac{\pi}{2}, 0 \le y \le \sqrt{2}\,\cos x\right\}$ about the x-axis.

21. $\int_0^{\pi} \pi(2 - \sin x)^2\,dx$

The solid is obtained by rotating the region $\mathcal{R} = \{(x,y) \mid 0 \le x \le \pi, 0 \le y \le 2 - \sin x\}$ about the x-axis.

22. $\int_0^4 2\pi(6-y)(4y - y^2)\,dy$

The solid is obtained by rotating the region $\mathcal{R} = \left\{(x,y) \mid 0 \le x \le 4y - y^2, 0 \le y \le 4\right\}$ about the line $y = 6$.

23. Take the base to be the disk $x^2 + y^2 \le 9$. Then $V = \int_{-3}^3 A(x)\,dx$, where $A(x_0)$ is the area of the isosceles right triangle whose hypotenuse lies along the line $x = x_0$ in the xy-plane. The length of the hypotenuse is $2\sqrt{9 - x^2}$ and the length of each leg is $\sqrt{2}\sqrt{9 - x^2}$. $A(x) = \frac{1}{2}\left(\sqrt{2}\sqrt{9-x^2}\right)^2 = 9 - x^2$, so

$$V = 2\int_0^3 A(x)\,dx = 2\int_0^3 (9 - x^2)\,dx = 2\left[9x - \tfrac{1}{3}x^3\right]_0^3 = 2(27 - 9) = 36$$

24. $V = \int_{-1}^1 A(x)\,dx = 2\int_0^1 A(x)\,dx = 2\int_0^1 \left[(2 - x^2) - x^2\right]^2\,dx = 2\int_0^1 \left[2(1 - x^2)\right]^2\,dx$

$$= 8\int_0^1 (1 - 2x^2 + x^4)\,dx = 8\left[x - \tfrac{2}{3}x^3 + \tfrac{1}{5}x^5\right]_0^1 = 8\left(1 - \tfrac{2}{3} + \tfrac{1}{5}\right) = \tfrac{64}{15}$$

25. Equilateral triangles with sides measuring $\frac{1}{4}x$ meters have height $\frac{1}{4}x\sin 60^\circ = \frac{\sqrt{3}}{8}x$. Therefore,

$$A(x) = \tfrac{1}{2}\cdot\tfrac{1}{4}x\cdot\tfrac{\sqrt{3}}{8}x = \tfrac{\sqrt{3}}{64}x^2. \quad V = \int_0^{20} A(x)\,dx = \tfrac{\sqrt{3}}{64}\int_0^{20} x^2\,dx = \tfrac{\sqrt{3}}{64}\left[\tfrac{1}{3}x^3\right]_0^{20} = \tfrac{8000\sqrt{3}}{64\cdot 3} = \tfrac{125\sqrt{3}}{3}\text{ m}^3.$$

26. (a) By the symmetry of the problem, we consider only the solid to the right of the origin. The semicircular cross-sections perpendicular to the x-axis have radius $1 - x$, so $A(x) = \frac{1}{2}\pi(1-x)^2$. Now we can calculate

$$V = 2\int_0^1 A(x)\,dx = 2\int_0^1 \tfrac{1}{2}\pi(1-x)^2\,dx = \int_0^1 \pi(1-x)^2\,dx = -\tfrac{\pi}{3}\left[(1-x)^3\right]_0^1 = \tfrac{\pi}{3}.$$

(b) Cut the solid with a plane perpendicular to the x-axis and passing through the y-axis. Fold the half of the solid in the region $x \le 0$ under the xy-plane so that the point $(-1, 0)$ comes around and touches the point $(1, 0)$. The resulting solid is a right circular cone of radius 1 with vertex at $(x, y, z) = (1, 0, 0)$ and with its base in the yz-plane, centered at the origin. The volume of this cone is $\frac{1}{3}\pi r^2 h = \frac{1}{3}\pi\cdot 1^2\cdot 1 = \frac{\pi}{3}$.

27. $f(x) = kx \Rightarrow 30\text{ N} = k(15 - 12)\text{ cm} \Rightarrow k = 10\text{ N/cm} = 1000\text{ N/m}$. $20\text{ cm} - 12\text{ cm} = 0.08\text{ m} \Rightarrow$

$W = \int_0^{0.08} kx\,dx = 1000\int_0^{0.08} x\,dx = 500\left[x^2\right]_0^{0.08} = 500(0.08)^2 = 3.2\text{ N·m} = 3.2\text{ J}$.

28. The work needed to raise the elevator alone is 1600 lb × 30 ft = 48,000 ft-lb. The work needed to raise the bottom 170 ft of cable is 170 ft × 10 lb/ft × 30 ft = 51,000 ft-lb. The work needed to raise the top 30 ft of cable is $\int_0^{30} 10x\,dx = \left[5x^2\right]_0^{30} = 5 \cdot 900 = 4500$ ft-lb. Adding these, we see that the total work needed is $48{,}000 + 51{,}000 + 4{,}500 = 103{,}500$ ft-lb.

29. (a) The parabola has equation $y = ax^2$ with vertex at the origin and passing through $(4, 4)$. $4 = a \cdot 4^2 \Rightarrow a = \frac{1}{4} \Rightarrow y = \frac{1}{4}x^2 \Rightarrow x^2 = 4y \Rightarrow$ $x = 2\sqrt{y}$. Each circular disk has radius $2\sqrt{y}$ and is moved $4 - y$ ft.

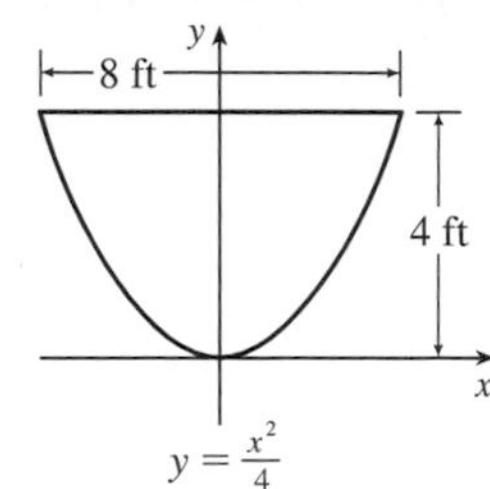

$$W = \int_0^4 \pi\left(2\sqrt{y}\right)^2 62.5(4 - y)\,dy = 250\pi\int_0^4 y(4 - y)\,dy$$

$$= 250\pi\left[2y^2 - \tfrac{1}{3}y^3\right]_0^4 = 250\pi\left(32 - \tfrac{64}{3}\right) = \tfrac{8000\pi}{3} \approx 8378\text{ ft-lb}$$

(b) In part (a) we knew the final water level (0) but not the amount of work done. Here we use the same equation, except with the work fixed, and the lower limit of integration (that is, the final water level—call it h) unknown: $W = 4000 \Leftrightarrow$ $250\pi\left[2y^2 - \frac{1}{3}y^3\right]_h^4 = 4000 \Leftrightarrow \frac{16}{\pi} = \left[\left(32 - \frac{64}{3}\right) - \left(2h^2 - \frac{1}{3}h^3\right)\right] \Leftrightarrow$ $h^3 - 6h^2 + 32 - \frac{48}{\pi} = 0$. We graph the function $f(h) = h^3 - 6h^2 + 32 - \frac{48}{\pi}$ on the interval $[0, 4]$ to see where it is 0. From the graph, $f(h) = 0$ for $h \approx 2.1$. So the depth of water remaining is about 2.1 ft.

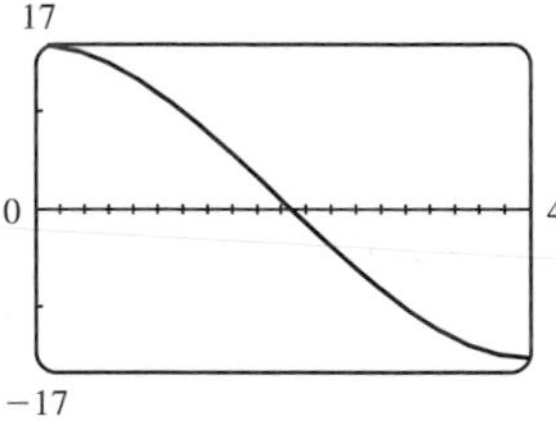

30. A horizontal slice of cooking oil Δx m thick has a volume of $\pi r^2 h = \pi \cdot 2^2 \cdot \Delta x\text{ m}^3$, a mass of $920(4\pi\,\Delta x)$ kg, weighs about $(9.8)(3680\pi\,\Delta x) = 36{,}064\pi\Delta x$ N, and thus requires about $36{,}064\pi x_i^*\Delta x$ J of work for its removal (where $3 \le x_i^* \le 6$). The total work needed to empty the tank is

$$W = \lim_{n\to\infty}\sum_{i=1}^{n} 36{,}064\pi x_i^*\,\Delta x = \int_3^6 36{,}064\pi x\,dx = 36{,}064\pi\left[\tfrac{1}{2}x^2\right]_3^6 = 18{,}032\pi(36 - 9) = 486{,}864\pi \approx 1.53 \times 10^6\text{ J}.$$

31. $f_{\text{ave}} = \dfrac{1}{b - a}\displaystyle\int_a^b f(t)\,dt = \dfrac{1}{\pi/4 - 0}\int_0^{\pi/4}\sec^2 t\,dt = \dfrac{4}{\pi}\Big[\tan t\Big]_0^{\pi/4} = \dfrac{4}{\pi}(1 - 0) = \dfrac{4}{\pi}$

32. (a) $f_{\text{ave}} = \dfrac{1}{b - a}\displaystyle\int_a^b f(x)\,dx = \dfrac{1}{4 - 1}\int_1^4 \dfrac{1}{\sqrt{x}}\,dx$

$= \frac{1}{3}\displaystyle\int_1^4 x^{-1/2}\,dx = \frac{1}{3}\Big[2\sqrt{x}\Big]_1^4$

$= \frac{2}{3}(2 - 1) = \frac{2}{3}$

(b) $f(c) = f_{\text{ave}} \Leftrightarrow \dfrac{1}{\sqrt{c}} = \dfrac{2}{3} \Leftrightarrow \sqrt{c} = \dfrac{3}{2} \Leftrightarrow c = \dfrac{9}{4}$

(c)

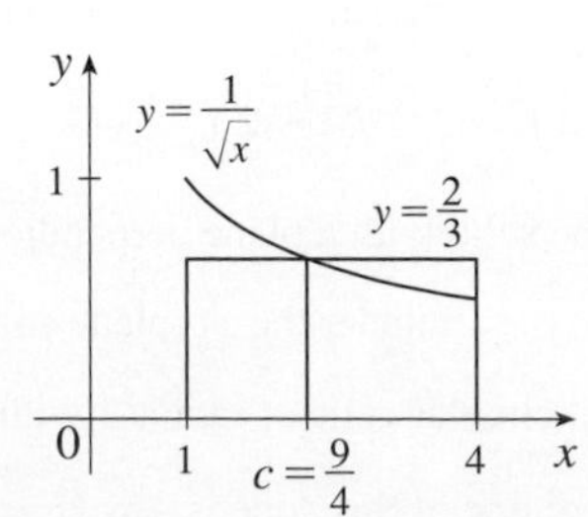

33. $\lim_{h\to 0} f_{\text{ave}} = \lim_{h\to 0} \dfrac{1}{(x+h)-x}\displaystyle\int_x^{x+h} f(t)\,dt = \lim_{h\to 0}\dfrac{F(x+h)-F(x)}{h}$, where $F(x) = \int_a^x f(t)\,dt$. But we recognize this limit as being $F'(x)$ by the definition of a derivative. Therefore, $\lim_{h\to 0} f_{\text{ave}} = F'(x) = f(x)$ by FTC1.

34. (a) $\mathcal{R}_1$ is the region below the graph of $y = x^2$ and above the x-axis between $x = 0$ and $x = b$, and $\mathcal{R}_2$ is the region to the left of the graph of $x = \sqrt{y}$ and to the right of the y-axis between $y = 0$ and $y = b^2$. So the area of $\mathcal{R}_1$ is $A_1 = \int_0^b x^2\,dx = \left[\frac{1}{3}x^3\right]_0^b = \frac{1}{3}b^3$, and the area of $\mathcal{R}_2$ is $A_2 = \int_0^{b^2} \sqrt{y}\,dy = \left[\frac{2}{3}y^{3/2}\right]_0^{b^2} = \frac{2}{3}b^3$. So there is no solution to $A_1 = A_2$ for $b \neq 0$.

(b) Using disks, we calculate the volume of rotation of $\mathcal{R}_1$ about the x-axis to be $V_{1,x} = \pi\int_0^b (x^2)^2\,dx = \frac{1}{5}\pi b^5$. Using cylindrical shells, we calculate the volume of rotation of $\mathcal{R}_1$ about the y-axis to be $V_{1,y} = 2\pi\int_0^b x(x^2)\,dx = 2\pi\left[\frac{1}{4}x^4\right]_0^b = \frac{1}{2}\pi b^4$. So $V_{1,x} = V_{1,y} \Leftrightarrow \frac{1}{5}\pi b^5 = \frac{1}{2}\pi b^4 \Leftrightarrow 2b = 5 \Leftrightarrow b = \frac{5}{2}$. So the volumes of rotation about the x- and y-axes are the same for $b = \frac{5}{2}$.

(c) We use cylindrical shells to calculate the volume of rotation of $\mathcal{R}_2$ about the x-axis:

$$\mathcal{R}_{2,x} = 2\pi\int_0^{b^2} y\left(\sqrt{y}\right)dy = 2\pi\left[\tfrac{2}{5}y^{5/2}\right]_0^{b^2} = \tfrac{4}{5}\pi b^5.$$

We already know the volume of rotation of $\mathcal{R}_1$ about the x-axis from part (b), and $\mathcal{R}_{1,x} = \mathcal{R}_{2,x} \Leftrightarrow \frac{1}{5}\pi b^5 = \frac{4}{5}\pi b^5$, which has no solution for $b \neq 0$.

(d) We use disks to calculate the volume of rotation of $\mathcal{R}_2$ about the y-axis: $\mathcal{R}_{2,y} = \pi\int_0^{b^2}\left(\sqrt{y}\right)^2 dy = \pi\left[\frac{1}{2}y^2\right]_0^{b^2} = \frac{1}{2}\pi b^4$. We know the volume of rotation of $\mathcal{R}_1$ about the y-axis from part (b), and $\mathcal{R}_{1,y} = \mathcal{R}_{2,y} \Leftrightarrow \frac{1}{2}\pi b^4 = \frac{1}{2}\pi b^4$. But this equation is true for all b, so the volumes of rotation about the y-axis are equal for all values of b.

PROBLEMS PLUS

1. (a) The area under the graph of f from 0 to t is equal to $\int_0^t f(x)\,dx$, so the requirement is that $\int_0^t f(x)\,dx = t^3$ for all t. We differentiate both sides of this equation with respect to t (with the help of FTC1) to get $f(t) = 3t^2$. This function is positive and continuous, as required.

(b) The volume generated from $x = 0$ to $x = b$ is $\int_0^b \pi[f(x)]^2\,dx$. Hence, we are given that $b^2 = \int_0^b \pi[f(x)]^2\,dx$ for all $b > 0$. Differentiating both sides of this equation with respect to b using the Fundamental Theorem of Calculus gives $2b = \pi[f(b)]^2 \;\Rightarrow\; f(b) = \sqrt{2b/\pi}$, since f is positive. Therefore, $f(x) = \sqrt{2x/\pi}$.

2. The total area of the region bounded by the parabola $y = x - x^2 = x(1 - x)$ and the x-axis is $\int_0^1 (x - x^2)\,dx = \left[\frac{1}{2}x^2 - \frac{1}{3}x^3\right]_0^1 = \frac{1}{6}$. Let the slope of the line we are looking for be m. Then the area above this line but below the parabola is $\int_0^a \left[(x - x^2) - mx\right]dx$, where a is the x-coordinate of the point of intersection of the line and the parabola. We find the point of intersection by solving the equation $x - x^2 = mx \;\Leftrightarrow\; 1 - x = m \;\Leftrightarrow\; x = 1 - m$. So the value of a is $1 - m$, and

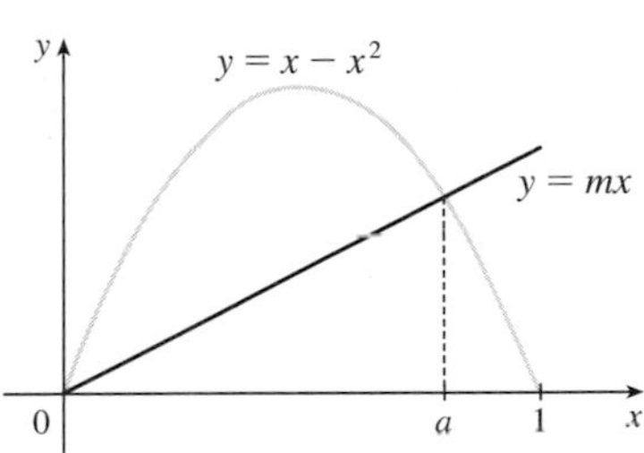

$$\begin{aligned}\int_0^{1-m}\left[(x - x^2) - mx\right]dx &= \int_0^{1-m}\left[(1-m)x - x^2\right]dx = \left[\tfrac{1}{2}(1-m)x^2 - \tfrac{1}{3}x^3\right]_0^{1-m}\\ &= \tfrac{1}{2}(1-m)(1-m)^2 - \tfrac{1}{3}(1-m)^3 = \tfrac{1}{6}(1-m)^3\end{aligned}$$

We want this to be half of $\frac{1}{6}$, so $\frac{1}{6}(1-m)^3 = \frac{1}{12} \;\Leftrightarrow\; (1-m)^3 = \frac{6}{12} \;\Leftrightarrow\; 1 - m = \sqrt[3]{\frac{1}{2}} \;\Leftrightarrow\; m = 1 - \frac{1}{\sqrt[3]{2}}$. So the slope of the required line is $1 - \frac{1}{\sqrt[3]{2}} \approx 0.206$.

3. Let a and b be the x-coordinates of the points where the line intersects the curve. From the figure, $R_1 = R_2 \;\Rightarrow$

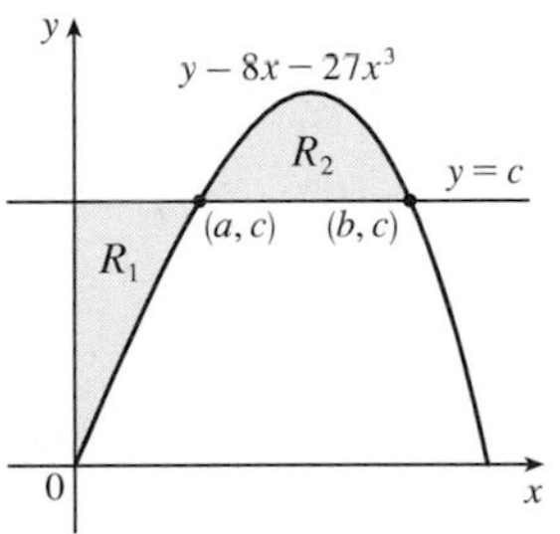

$$\begin{aligned}\int_0^a \left[c - (8x - 27x^3)\right]dx &= \int_a^b \left[(8x - 27x^3) - c\right]dx\\ \left[cx - 4x^2 + \tfrac{27}{4}x^4\right]_0^a &= \left[4x^2 - \tfrac{27}{4}x^4 - cx\right]_a^b\\ ac - 4a^2 + \tfrac{27}{4}a^4 &= \left(4b^2 - \tfrac{27}{4}b^4 - bc\right) - \left(4a^2 - \tfrac{27}{4}a^4 - ac\right)\\ 0 &= 4b^2 - \tfrac{27}{4}b^4 - bc = 4b^2 - \tfrac{27}{4}b^4 - b(8b - 27b^3)\\ &= 4b^2 - \tfrac{27}{4}b^4 - 8b^2 + 27b^4 = \tfrac{81}{4}b^4 - 4b^2\\ &= b^2\left(\tfrac{81}{4}b^2 - 4\right)\end{aligned}$$

So for $b > 0$, $b^2 = \frac{16}{81} \;\Rightarrow\; b = \frac{4}{9}$. Thus, $c = 8b - 27b^3 = 8\left(\frac{4}{9}\right) - 27\left(\frac{64}{729}\right) = \frac{32}{9} - \frac{64}{27} = \frac{32}{27}$.

4. (a) Take slices perpendicular to the line through the center C of the bottom of the glass and the point P where the top surface of the water meets the bottom of the glass.

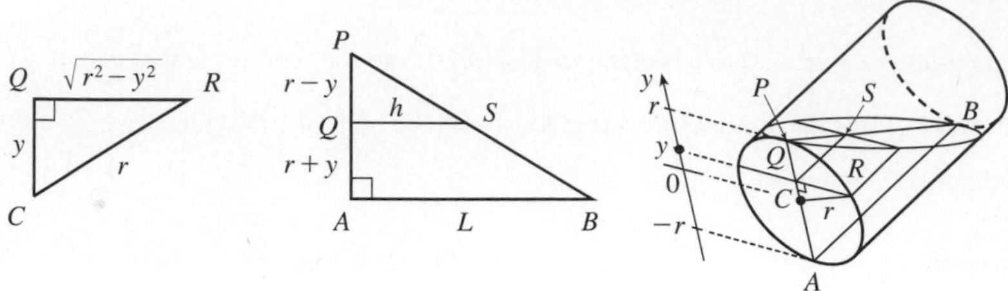

A typical rectangular cross-section y units above the axis of the glass has width $2\,|QR| = 2\sqrt{r^2-y^2}$ and length $h = |QS| = \dfrac{L}{2r}(r-y)$. [Triangles PQS and PAB are similar, so $\dfrac{h}{L} = \dfrac{|PQ|}{|PA|} = \dfrac{r-y}{2r}$.] Thus,

$$\begin{aligned} V &= \int_{-r}^{r} 2\sqrt{r^2-y^2}\cdot\frac{L}{2r}(r-y)\,dy = L\int_{-r}^{r}\left(1-\frac{y}{r}\right)\sqrt{r^2-y^2}\,dy \\ &= L\int_{-r}^{r}\sqrt{r^2-y^2}\,dy - \frac{L}{r}\int_{-r}^{r} y\sqrt{r^2-y^2}\,dy \\ &= L\cdot\frac{\pi r^2}{2} - \frac{L}{r}\cdot 0 \quad \left[\begin{array}{c}\text{the first integral is the area of a semicircle of radius } r,\\ \text{and the second has an odd integrand}\end{array}\right] \quad = \frac{\pi r^2 L}{2} \end{aligned}$$

(b) Slice parallel to the plane through the axis of the glass and the point of contact P. (This is the plane determined by P, B, and C in the figure.) $STUV$ is a typical trapezoidal slice. With respect to an x-axis with origin at C as shown, if S and V have x-coordinate x, then $|SV| = 2\sqrt{r^2-x^2}$. Projecting the trapezoid $STUV$ onto the plane of the triangle PAB (call the projection $S'T'U'V'$), we see that $|AP| = 2r$, $|SV| = 2\sqrt{r^2-x^2}$, and $|S'P| = |V'A| = \frac{1}{2}(|AP|-|SV|) = r - \sqrt{r^2-x^2}$.

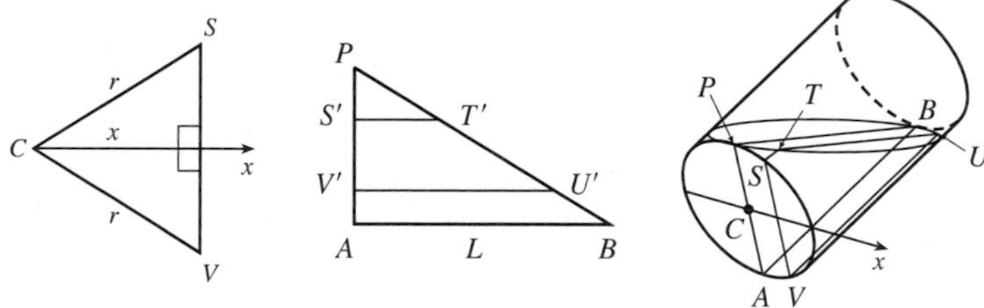

By similar triangles, $\dfrac{|ST|}{|S'P|} = \dfrac{|AB|}{|AP|}$, so $|ST| = \left(r-\sqrt{r^2-x^2}\right)\cdot\dfrac{L}{2r}$. In the same way, we find that $\dfrac{|VU|}{|V'P|} = \dfrac{|AB|}{|AP|}$, so $|VU| = |V'P|\cdot\dfrac{L}{2r} = (|AP|-|V'A|)\cdot\dfrac{L}{2r} = \left(r+\sqrt{r^2-x^2}\right)\cdot\dfrac{L}{2r}$. The area $A(x)$ of the trapezoid $STUV$ is $\frac{1}{2}|SV|\cdot(|ST|+|VU|)$; that is,

$A(x) = \frac{1}{2}\cdot 2\sqrt{r^2-x^2}\cdot\left[\left(r-\sqrt{r^2-x^2}\right)\cdot\dfrac{L}{2r} + \left(r+\sqrt{r^2-x^2}\right)\cdot\dfrac{L}{2r}\right] = L\sqrt{r^2-x^2}$. Thus,

$V = \displaystyle\int_{-r}^{r} A(x)\,dx = L\int_{-r}^{r}\sqrt{r^2-x^2}\,dx = L\cdot\frac{\pi r^2}{2} = \frac{\pi r^2 L}{2}$.

(c) See the computation of V in part (a) or part (b).

(d) The volume of the water is exactly half the volume of the cylindrical glass, so $V = \frac{1}{2}\pi r^2 L$.

(e) Choose x-, y-, and z-axes as shown in the figure. Then slices perpendicular to the x-axis are triangular, slices perpendicular to the y-axis are rectangular, and slices perpendicular to the z-axis are segments of circles. Using triangular slices, we find that the area $A(x)$ of a typical slice DEF, where D has x-coordinate x, is given by

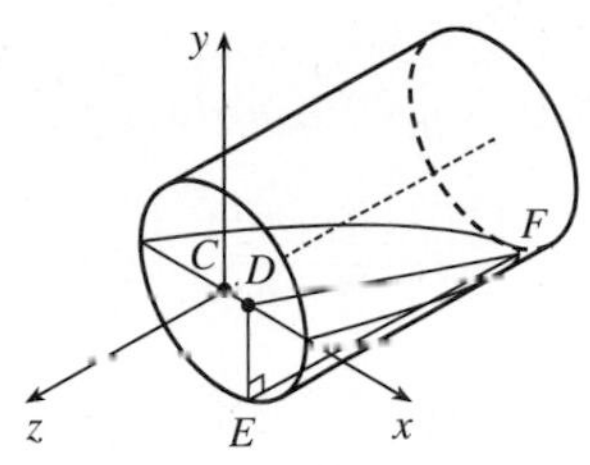

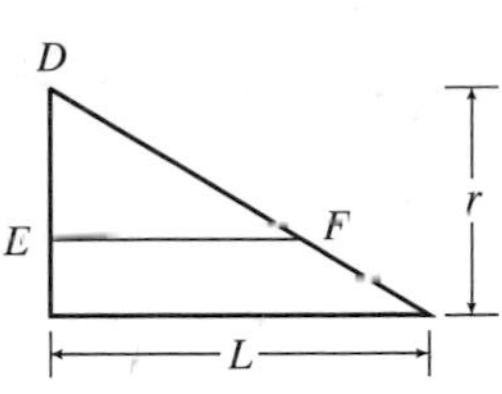

$$A(x) = \tfrac{1}{2}|DE| \cdot |EF| = \tfrac{1}{2}|DE| \cdot \left(\frac{L}{r}|DE|\right) = \frac{L}{2r}|DE|^2 = \frac{L}{2r}\left(r^2 - x^2\right). \text{ Thus,}$$

$$V = \int_{-r}^{r} A(x)\,dx = \frac{L}{2r}\int_{-r}^{r}\left(r^2 - x^2\right)dx = \frac{L}{r}\int_{-r}^{r}\left(r^2 - x^2\right)dx = \frac{L}{r}\left[r^2 x - \frac{x^3}{3}\right]_0^r$$

$$= \frac{L}{r}\left(r^3 - \frac{r^3}{3}\right) = \frac{L}{r}\cdot\tfrac{2}{3}r^3 = \tfrac{2}{3}r^2 L \qquad \text{[This is } 2/(3\pi) \approx 0.21 \text{ of the volume of the glass.]}$$

5. (a) $V = \pi h^2(r - h/3) = \frac{1}{3}\pi h^2(3r - h)$. See the solution to Exercise 6.2.49.

(b) The smaller segment has height $h = 1 - x$ and so by part (a) its volume is

$V = \frac{1}{3}\pi(1-x)^2\,[3(1) - (1-x)] = \frac{1}{3}\pi(x-1)^2(x+2)$. This volume must be $\frac{1}{3}$ of the total volume of the sphere, which is $\frac{4}{3}\pi(1)^3$. So $\frac{1}{3}\pi(x-1)^2(x+2) = \frac{1}{3}\left(\frac{4}{3}\pi\right) \Rightarrow (x^2 - 2x + 1)(x+2) = \frac{4}{3} \Rightarrow x^3 - 3x + 2 = \frac{4}{3} \Rightarrow$ $3x^3 - 9x + 2 = 0$. Using Newton's method with $f(x) = 3x^3 - 9x + 2$, $f'(x) = 9x^2 - 9$, we get

$x_{n+1} = x_n - \dfrac{3x_n^3 - 9x_n + 2}{9x_n^2 - 9}$. Taking $x_1 = 0$, we get $x_2 \approx 0.2222$, and $x_3 \approx 0.2261 \approx x_4$, so, correct to four decimal places, $x \approx 0.2261$.

(c) With $r = 0.5$ and $s = 0.75$, the equation $x^3 - 3rx^2 + 4r^3 s = 0$ becomes $x^3 - 3(0.5)x^2 + 4(0.5)^3(0.75) = 0 \Rightarrow$ $x^3 - \frac{3}{2}x^2 + 4\left(\frac{1}{8}\right)\frac{3}{4} = 0 \Rightarrow 8x^3 - 12x^2 + 3 = 0$. We use Newton's method with $f(x) = 8x^3 - 12x^2 + 3$,

$f'(x) = 24x^2 - 24x$, so $x_{n+1} = x_n - \dfrac{8x_n^3 - 12x_n^2 + 3}{24x_n^2 - 24x_n}$. Take $x_1 = 0.5$. Then $x_2 \approx 0.6667$, and $x_3 \approx 0.6736 \approx x_4$.

So to four decimal places the depth is 0.6736 m.

(d) (i) From part (a) with $r = 5$ in., the volume of water in the bowl is

$V = \frac{1}{3}\pi h^2(3r - h) = \frac{1}{3}\pi h^2(15 - h) = 5\pi h^2 - \frac{1}{3}\pi h^3$. We are given that $\dfrac{dV}{dt} = 0.2 \text{ in}^3\text{/s}$ and we want to find $\dfrac{dh}{dt}$

when $h = 3$. Now $\dfrac{dV}{dt} = 10\pi h\dfrac{dh}{dt} - \pi h^2\dfrac{dh}{dt}$, so $\dfrac{dh}{dt} = \dfrac{0.2}{\pi(10h - h^2)}$. When $h = 3$, we have

$$\frac{dh}{dt} = \frac{0.2}{\pi(10\cdot 3 - 3^2)} = \frac{1}{105\pi} \approx 0.003 \text{ in/s.}$$

(ii) From part (a), the volume of water required to fill the bowl from the instant that the water is 4 in. deep is $V = \frac{1}{2} \cdot \frac{4}{3}\pi(5)^3 - \frac{1}{3}\pi(4)^2(15-4) = \frac{2}{3} \cdot 125\pi - \frac{16}{3} \cdot 11\pi = \frac{74}{3}\pi$. To find the time required to fill the bowl we divide this volume by the rate: Time $= \frac{74\pi/3}{0.2} = \frac{370\pi}{3} \approx 387$ s ≈ 6.5 min.

6. (a) The volume above the surface is $\int_0^{L-h} A(y)\,dy = \int_{-h}^{L-h} A(y)\,dy - \int_{-h}^{0} A(y)\,dy$. So the proportion of volume above the surface is $\dfrac{\int_0^{L-h} A(y)\,dy}{\int_{-h}^{L-h} A(y)\,dy} = \dfrac{\int_{-h}^{L-h} A(y)\,dy - \int_{-h}^{0} A(y)\,dy}{\int_{-h}^{L-h} A(y)\,dy}$. Now by Archimedes' Principle, we have $F = W \Rightarrow$ $\rho_f g \int_{-h}^{0} A(y)\,dy = \rho_0 g \int_{-h}^{L-h} A(y)\,dy$, so $\int_{-h}^{0} A(y)\,dy = (\rho_0/\rho_f)\int_{-h}^{L-h} A(y)\,dy$. Therefore, $\dfrac{\int_0^{L-h} A(y)\,dy}{\int_{-h}^{L-h} A(y)\,dy} = \dfrac{\int_{-h}^{L-h} A(y)\,dy - (\rho_0/\rho_f)\int_{-h}^{L-h} A(y)\,dy}{\int_{-h}^{L-h} A(y)\,dy} = \dfrac{\rho_f - \rho_0}{\rho_f}$, so the percentage of volume above the surface is $100\left(\dfrac{\rho_f - \rho_0}{\rho_f}\right)\%$.

(b) For an iceberg, the percentage of volume above the surface is $100\left(\frac{1030 - 917}{1030}\right)\% \approx 11\%$.

(c) No, the water does not overflow. Let V_i be the volume of the ice cube, and let V_w be the volume of the water which results from the melting. Then by the formula derived in part (a), the volume of ice above the surface of the water is $\left[(\rho_f - \rho_0)/\rho_f\right]V_i$, so the volume below the surface is $V_i - \left[(\rho_f - \rho_0)/\rho_f\right]V_i = (\rho_0/\rho_f)V_i$. Now the mass of the ice cube is the same as the mass of the water which is created when it melts, namely $m = \rho_0 V_i = \rho_f V_w \Rightarrow$ $V_w = (\rho_0/\rho_f)V_i$. So when the ice cube melts, the volume of the resulting water is the same as the underwater volume of the ice cube, and so the water does not overflow.

(d) The figure shows the instant when the height of the exposed part of the ball is y. Using the formula in Problem 5(a) with $r = 0.4$ and $h = 0.8 - y$, we see that the volume of the submerged part of the sphere is $\frac{1}{3}\pi(0.8-y)^2[1.2 - (0.8-y)]$, so its weight is $1000g \cdot \frac{1}{3}\pi s^2(1.2 - s)$, where $s = 0.8 - y$. Then the work done to submerge the sphere is

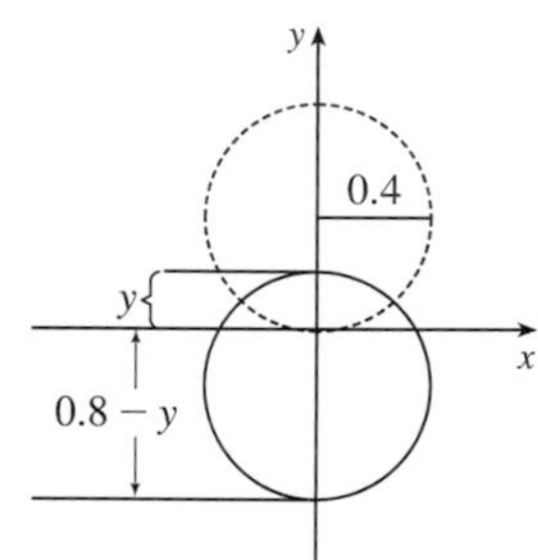

$$\begin{aligned} W &= \int_0^{0.8} g\tfrac{1000}{3}\pi s^2(1.2 - s)\,ds = g\tfrac{1000}{3}\pi \int_0^{0.8}(1.2s^2 - s^3)\,ds \\ &= g\tfrac{1000}{3}\pi\left[0.4s^3 - \tfrac{1}{4}s^4\right]_0^{0.8} = g\tfrac{1000}{3}\pi(0.2048 - 0.1024) = 9.8\,\tfrac{1000}{3}\,\pi(0.1024) \approx 1.05 \times 10^3 \text{ J} \end{aligned}$$

7. We are given that the rate of change of the volume of water is $\dfrac{dV}{dt} = -kA(x)$, where k is some positive constant and $A(x)$ is the area of the surface when the water has depth x. Now we are concerned with the rate of change of the depth of the water with respect to time, that is, $\dfrac{dx}{dt}$. But by the Chain Rule, $\dfrac{dV}{dt} = \dfrac{dV}{dx}\dfrac{dx}{dt}$, so the first equation can be written $\dfrac{dV}{dx}\dfrac{dx}{dt} = -kA(x)$ $(\star)$. Also, we know that the total volume of water up to a depth x is $V(x) = \int_0^x A(s)\,ds$, where $A(s)$ is the area of a cross-section of the water at a depth s. Differentiating this equation with respect to x, we get $dV/dx = A(x)$. Substituting this into equation $\star$, we get $A(x)(dx/dt) = -kA(x) \Rightarrow dx/dt = -k$, a constant.

8. A typical sphere of radius r is shown in the figure. We wish to maximize the shaded volume V, which can be thought of as the volume of a hemisphere of radius r minus the volume of the spherical cap with height $h = 1 - \sqrt{1 - r^2}$ and radius 1.

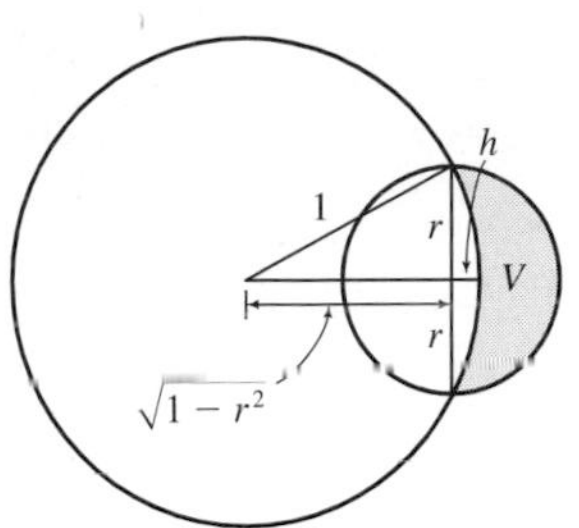

$$V = \tfrac{1}{2} \cdot \tfrac{4}{3}\pi r^3 - \tfrac{1}{3}\pi\left(1 - \sqrt{1 - r^2}\right)^2 \left[3(1) - \left(1 - \sqrt{1 - r^2}\right)\right] \quad \text{[by Problem 5(a)]}$$

$$= \tfrac{1}{3}\pi\left[2r^3 - \left(2 - 2\sqrt{1 - r^2} - r^2\right)\left(2 + \sqrt{1 - r^2}\right)\right]$$

$$= \tfrac{1}{3}\pi\left[2r^3 - 2 + \left(r^2 + 2\right)\sqrt{1 - r^2}\right]$$

$$V' = \tfrac{1}{3}\pi\left[6r^2 + \frac{(r^2 + 2)(-r)}{\sqrt{1 - r^2}} + \sqrt{1 - r^2}(2r)\right] = \tfrac{1}{3}\pi\left[\frac{6r^2\sqrt{1 - r^2} - r(r^2 + 2) + 2r(1 - r^2)}{\sqrt{1 - r^2}}\right]$$

$$= \tfrac{1}{3}\pi\left(\frac{6r^2\sqrt{1 - r^2} - 3r^3}{\sqrt{1 - r^2}}\right) = \frac{\pi r^2\left(2\sqrt{1 - r^2} - r\right)}{\sqrt{1 - r^2}}$$

$V'(r) = 0 \Leftrightarrow 2\sqrt{1 - r^2} = r \Leftrightarrow 4 - 4r^2 = r^2 \Leftrightarrow r^2 = \frac{4}{5} \Leftrightarrow r = \frac{2}{\sqrt{5}} \approx 0.89.$

Since $V'(r) > 0$ for $0 < r < \frac{2}{\sqrt{5}}$ and $V'(r) < 0$ for $\frac{2}{\sqrt{5}} < r < 1$, we know that V attains a maximum at $r = \frac{2}{\sqrt{5}}$.

9. We must find expressions for the areas A and B, and then set them equal and see what this says about the curve C. If $P = (a, 2a^2)$, then area A is just $\int_0^a (2x^2 - x^2)\,dx = \int_0^a x^2\,dx = \frac{1}{3}a^3$. To find area B, we use y as the variable of integration. So we find the equation of the middle curve as a function of y: $y = 2x^2 \Leftrightarrow x = \sqrt{y/2}$, since we are concerned with the first quadrant only. We can express area B as

$$\int_0^{2a^2}\left[\sqrt{y/2} - C(y)\right]dy = \left[\frac{4}{3}(y/2)^{3/2}\right]_0^{2a^2} - \int_0^{2a^2} C(y)\,dy = \frac{4}{3}a^3 - \int_0^{2a^2} C(y)\,dy$$

where $C(y)$ is the function with graph C. Setting $A = B$, we get $\frac{1}{3}a^3 = \frac{4}{3}a^3 - \int_0^{2a^2} C(y)\,dy \Leftrightarrow \int_0^{2a^2} C(y)\,dy = a^3$. Now we differentiate this equation with respect to a using the Chain Rule and the Fundamental Theorem: $C(2a^2)(4a) = 3a^2 \Rightarrow C(y) = \frac{3}{4}\sqrt{y/2}$, where $y = 2a^2$. Now we can solve for y: $x = \frac{3}{4}\sqrt{y/2} \Rightarrow x^2 = \frac{9}{16}(y/2) \Rightarrow y = \frac{32}{9}x^2$.

10. We want to find the volume of that part of the sphere which is below the surface of the water. As we can see from the diagram, this region is a cap of a sphere with radius r and height $r + d$. If we can find an expression for d in terms of h, r and θ, then we can determine the volume of the region [see Problem 5(a)], and then differentiate with respect to r to find the maximum. We see that

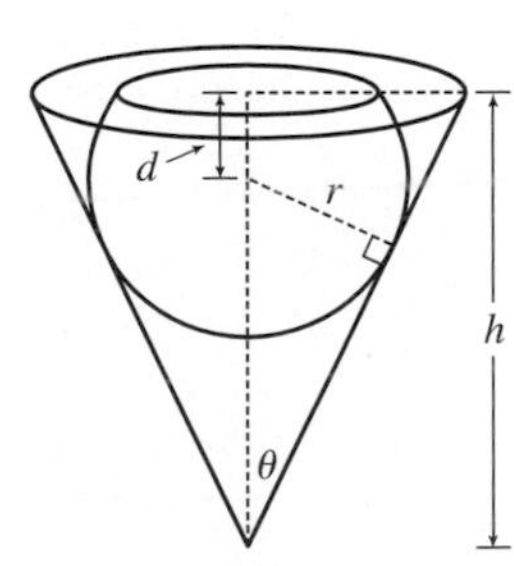

$$\sin\theta = \frac{r}{h - d} \Leftrightarrow h - d = \frac{r}{\sin\theta} \Leftrightarrow d = h - r\csc\theta.$$

[continued]

Now we can use the formula from Problem 5(a) to find the volume of water displaced:

$$V = \tfrac{1}{3}\pi h^2(3r-h) = \tfrac{1}{3}\pi(r+d)^2\,[3r-(r+d)] = \tfrac{1}{3}\pi(r+h-r\csc\theta)^2(2r-h+r\csc\theta)$$
$$= \tfrac{\pi}{3}[r(1-\csc\theta)+h]^2[r(2+\csc\theta)-h]$$

Now we differentiate with respect to r:

$$dV/dr = \tfrac{\pi}{3}\left([r(1-\csc\theta)+h]^2(2+\csc\theta)+2[r(1-\csc\theta)+h](1-\csc\theta)[r(2+\csc\theta)-h]\right)$$
$$= \tfrac{\pi}{3}[r(1-\csc\theta)+h]([r(1-\csc\theta)+h](2+\csc\theta)+2(1-\csc\theta)[r(2+\csc\theta)-h])$$
$$= \tfrac{\pi}{3}[r(1-\csc\theta)+h](3(2+\csc\theta)(1-\csc\theta)r+[(2+\csc\theta)-2(1-\csc\theta)]h)$$
$$= \tfrac{\pi}{3}[r(1-\csc\theta)+h][3(2+\csc\theta)(1-\csc\theta)r+3h\csc\theta]$$

This is 0 when $r = \dfrac{h}{\csc\theta - 1}$ and when $r = \dfrac{h\csc\theta}{(\csc\theta+2)(\csc\theta-1)}$. Now since $V\left(\dfrac{h}{\csc\theta-1}\right) = 0$ (the first factor vanishes; this corresponds to $d=-r$), the maximum volume of water is displaced when $r = \dfrac{h\csc\theta}{(\csc\theta-1)(\csc\theta+2)}$.

(Our intuition tells us that a maximum value does exist, and it must occur at a critical number.) Multiplying numerator and denominator by $\sin^2\theta$, we get an alternative form of the answer: $r = \dfrac{h\sin\theta}{\sin\theta+\cos 2\theta}$.

11. (a) Stacking disks along the y-axis gives us $V = \int_0^h \pi\,[f(y)]^2\,dy$.

(b) Using the Chain Rule, $\dfrac{dV}{dt} = \dfrac{dV}{dh}\cdot\dfrac{dh}{dt} = \pi\,[f(h)]^2\,\dfrac{dh}{dt}$.

(c) $kA\sqrt{h} = \pi[f(h)]^2\,\dfrac{dh}{dt}$. Set $\dfrac{dh}{dt} = C$: $\pi[f(h)]^2\,C = kA\sqrt{h} \;\Rightarrow\; [f(h)]^2 = \dfrac{kA}{\pi C}\sqrt{h} \;\Rightarrow\; f(h) = \sqrt{\dfrac{kA}{\pi C}}\,h^{1/4}$; that is, $f(y) = \sqrt{\dfrac{kA}{\pi C}}\,y^{1/4}$. The advantage of having $\dfrac{dh}{dt} = C$ is that the markings on the container are equally spaced.

12. (a) We first use the cylindrical shell method to express the volume V in terms of h, r, and ω:

$$V = \int_0^r 2\pi xy\,dx = \int_0^r 2\pi x\left[h+\frac{\omega^2x^2}{2g}\right]dx = 2\pi\int_0^r\left(hx+\frac{\omega^2x^3}{2g}\right)dx$$
$$= 2\pi\left[\frac{hx^2}{2}+\frac{\omega^2x^4}{8g}\right]_0^r = 2\pi\left[\frac{hr^2}{2}+\frac{\omega^2r^4}{8g}\right] = \pi hr^2+\frac{\pi\omega^2r^4}{4g} \quad\Rightarrow$$

$$h = \frac{V-(\pi\omega^2r^4)/(4g)}{\pi r^2} = \frac{4gV-\pi\omega^2r^4}{4\pi gr^2}.$$

(b) The surface touches the bottom when $h=0 \;\Rightarrow\; 4gV-\pi\omega^2r^4=0 \;\Rightarrow\; \omega^2 = \dfrac{4gV}{\pi r^4} \;\Rightarrow\; \omega = \dfrac{2\sqrt{gV}}{\sqrt{\pi}r^2}$.

To spill over the top, $y(r) > L \;\Leftrightarrow$

$$L < h+\frac{\omega^2r^2}{2g} = \frac{4gV-\pi\omega^2r^4}{4\pi gr^2}+\frac{\omega^2r^2}{2g} = \frac{4gV}{4\pi gr^2}-\frac{\pi\omega^2r^2}{4\pi gr^2}+\frac{\omega^2r^2}{2g}$$
$$= \frac{V}{\pi r^2}-\frac{\omega^2r^2}{4g}+\frac{\omega^2r^2}{2g} = \frac{V}{\pi r^2}+\frac{\omega^2r^2}{4g} \quad\Leftrightarrow$$

$\dfrac{\omega^2r^2}{4g} > L-\dfrac{V}{\pi r^2} = \dfrac{\pi r^2L-V}{\pi r^2} \;\Leftrightarrow\; \omega^2 > \dfrac{4g(\pi r^2L-V)}{\pi r^4}$. So for spillage, the angular speed should be $\omega > \dfrac{2\sqrt{g(\pi r^2L-V)}}{r^2\sqrt{\pi}}$.

(c) (i) Here we have $r = 2$, $L = 7$, $h = 7 - 5 = 2$. When $x = 1$, $y = 7 - 4 = 3$. Therefore, $3 = 2 + \dfrac{\omega^2 \cdot 1^2}{2 \cdot 32} \Rightarrow$

$$1 = \frac{\omega^2}{2 \cdot 32} \quad \Rightarrow \quad \omega^2 = 64 \quad \Rightarrow \quad \omega = 8 \text{ rad/s. } V = \pi(2)(2)^2 + \frac{\pi \cdot 8^2 \cdot 2^4}{4g} = 8\pi + 8\pi = 16\pi \text{ ft}^2.$$

(ii) At the wall, $x = 2$, so $y = 2 + \dfrac{8^2 \cdot 2^2}{2 \cdot 32} = 6$ and the surface is $7 - 6 = 1$ ft below the top of the tank.

13. The cubic polynomial passes through the origin, so let its equation be $y = px^3 + qx^2 + rx$. The curves intersect when $px^3 + qx^2 + rx = x^2 \Leftrightarrow$ $px^3 + (q-1)x^2 + rx = 0$. Call the left side $f(x)$. Since $f(a) = f(b) = 0$, another form of f is

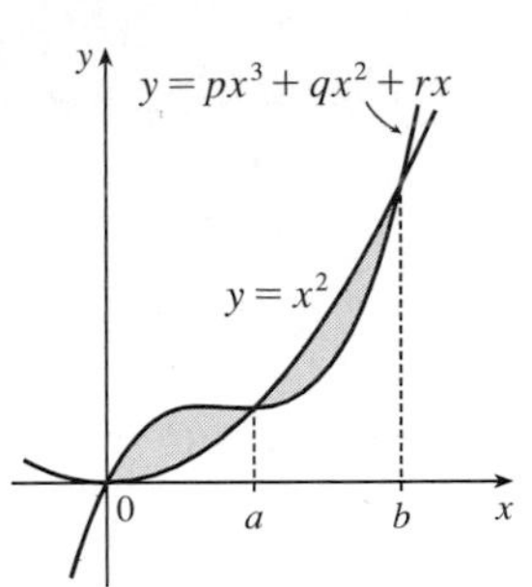

$$\begin{aligned} f(x) &= px(x-a)(x-b) = px[x^2 - (a+b)x + ab] \\ &= p[x^3 - (a+b)x^2 + abx] \end{aligned}$$

Since the two areas are equal, we must have $\int_0^a f(x)\,dx = -\int_a^b f(x)\,dx \quad \Rightarrow$

$[F(x)]_0^a = [F(x)]_b^a \quad \Rightarrow \quad F(a) - F(0) = F(a) - F(b) \quad \Rightarrow \quad F(0) = F(b)$, where F is an antiderivative of f.

Now $F(x) = \int f(x)\,dx = \int p[x^3 - (a+b)x^2 + abx]\,dx = p\left[\frac{1}{4}x^4 - \frac{1}{3}(a+b)x^3 + \frac{1}{2}abx^2\right] + C$, so

$F(0) = F(b) \quad \Rightarrow \quad C = p\left[\frac{1}{4}b^4 - \frac{1}{3}(a+b)b^3 + \frac{1}{2}ab^3\right] + C \quad \Rightarrow \quad 0 = p\left[\frac{1}{4}b^4 - \frac{1}{3}(a+b)b^3 + \frac{1}{2}ab^3\right] \quad \Rightarrow$

$0 = 3b - 4(a+b) + 6a \quad [\text{multiply by } 12/(pb^3),\, b \neq 0] \quad \Rightarrow \quad 0 = 3b - 4a - 4b + 6a \quad \Rightarrow \quad b = 2a.$

Hence, b is twice the value of a.

14. (a) Place the round flat tortilla on an xy-coordinate system as shown in the first figure. An equation of the circle is $x^2 + y^2 = 4^2$ and the height of a cross-section is $2\sqrt{16 - x^2}$.

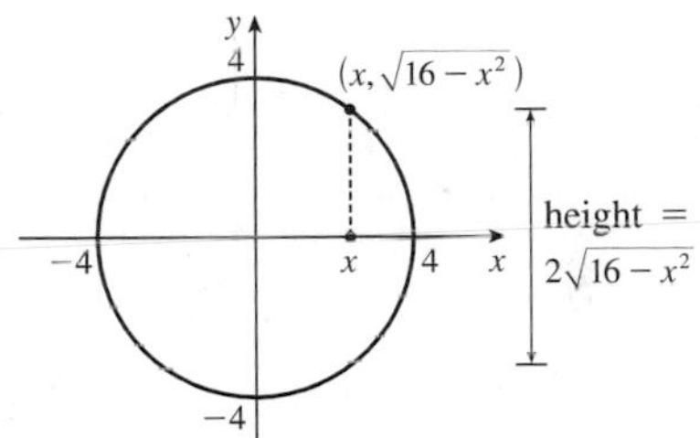

Now look at a cross-section with central angle θ_x as shown in the second figure (r is the radius of the circular cylinder). The filled area $A(x)$ is equal to the area $A_1(x)$ of the sector minus the area $A_2(x)$ of the triangle.

$$\begin{aligned} A(x) &= A_1(x) - A_2(x) = \tfrac{1}{2}r^2\theta_x - \tfrac{1}{2}r^2 \sin\theta_x && [\text{area formulas from trigonometry}] \\ &= \tfrac{1}{2}r(r\theta_x) - \tfrac{1}{2}r^2 \sin\left(\frac{s}{r}\right) && [\text{arc length } s = r\theta_x \quad \Rightarrow \quad \theta_x = s/r] \\ &= \tfrac{1}{2}r \cdot 2\sqrt{16 - x^2} - \tfrac{1}{2}r^2 \sin\left(\frac{2\sqrt{16 - x^2}}{r}\right) && [s = 2\sqrt{16 - x^2}\,] \\ &= r\sqrt{16 - x^2} - \tfrac{1}{2}r^2 \sin\left(\frac{2}{r}\sqrt{16 - x^2}\right) \quad (\star) \end{aligned}$$

Note that the central angle θ_x will be small near the ends of the tortilla; that is, when $|x| \approx 4$. But near the center of the tortilla (when $|x| \approx 0$), the central angle θ_x may exceed $180°$. Thus, the sine of θ_x will be negative and the second term in $(\star)$ will be positive (actually adding area to the area of the sector). The volume of the taco can be found by integrating the cross-sectional areas from $x = -4$ to $x = 4$. Thus,

$$V(x) = \int_{-4}^{4} A(x)\,dx = \int_{-4}^{4} \left[r\sqrt{16 - x^2} - \tfrac{1}{2}r^2 \sin\left(\frac{2}{r}\sqrt{16 - x^2}\right)\right] dx$$

(b) To find the value of r that maximizes the volume of the taco, we can define the function

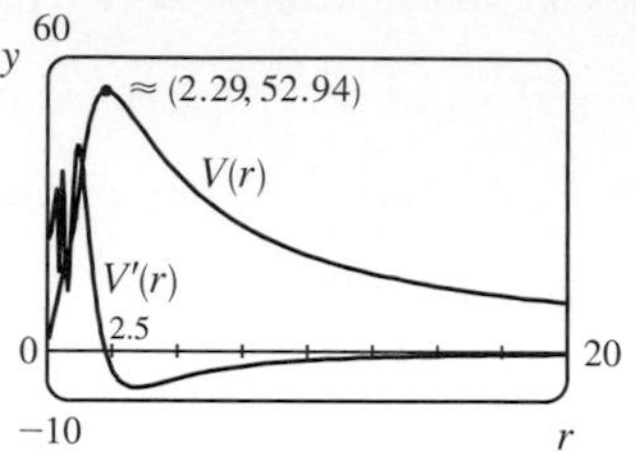

$$V(r) = \int_{-4}^{4} \left[r\sqrt{16 - x^2} - \tfrac{1}{2}r^2 \sin\left(\frac{2}{r}\sqrt{16 - x^2}\right)\right] dx$$

The figure shows a graph of $y = V(r)$ and $y = V'(r)$. The maximum volume of about 52.94 occurs when $r \approx 2.2912$.

15. We assume that P lies in the region of positive x. Since $y = x^3$ is an odd function, this assumption will not affect the result of the calculation. Let $P = (a, a^3)$. The slope of the tangent to the curve $y = x^3$ at P is $3a^2$, and so the equation of the tangent is $y - a^3 = 3a^2(x - a) \quad\Leftrightarrow\quad y = 3a^2x - 2a^3$. We solve this simultaneously with $y = x^3$ to find the other point of intersection: $x^3 = 3a^2x - 2a^3 \quad\Leftrightarrow\quad (x - a)^2(x + 2a) = 0$. So $Q = (-2a, -8a^3)$ is the other point of intersection. The equation of the tangent at Q is $y - (-8a^3) = 12a^2[x - (-2a)] \quad\Leftrightarrow\quad y = 12a^2x + 16a^3$. By symmetry, this tangent will intersect the curve again at $x = -2(-2a) = 4a$. The curve lies above the first tangent, and below the second, so we are looking for a relationship between $A = \int_{-2a}^{a} \left[x^3 - (3a^2x - 2a^3)\right] dx$ and $B = \int_{-2a}^{4a} \left[(12a^2x + 16a^3) - x^3\right] dx$. We calculate $A = \left[\frac{1}{4}x^4 - \frac{3}{2}a^2x^2 + 2a^3x\right]_{-2a}^{a} = \frac{3}{4}a^4 - (-6a^4) = \frac{27}{4}a^4$, and $B = \left[6a^2x^2 + 16a^3x - \frac{1}{4}x^4\right]_{-2a}^{4a} = 96a^4 - (-12a^4) = 108a^4$. We see that $B = 16A = 2^4A$. This is because our calculation of area B was essentially the same as that of area A, with a replaced by $-2a$, so if we replace a with $-2a$ in our expression for A, we get $\frac{27}{4}(-2a)^4 = 108a^4 = B$.

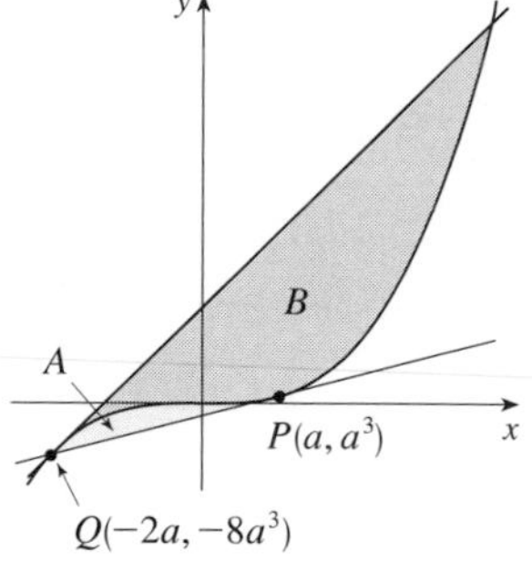

7 ☐ TECHNIQUES OF INTEGRATION

7.1 Integration by Parts

1. Let $u = x$, $dv = e^{2x}\,dx \Rightarrow du = dx$, $v = \frac{1}{2}e^{2x}$. Then by Equation 2,

$\int xe^{2x}\,dx = \frac{1}{2}xe^{2x} - \int \frac{1}{2}e^{2x}\,dx = \frac{1}{2}xe^{2x} - \frac{1}{4}e^{2x} + C.$

2. Let $u = \ln x$, $dv = \sqrt{x}\,dx \Rightarrow du = \dfrac{1}{x}\,dx$, $v = \frac{2}{3}x^{3/2}$. Then by Equation 2,

$$\int \sqrt{x}\ln x\,dx = \tfrac{2}{3}x^{3/2}\ln x - \int \tfrac{2}{3}x^{3/2}\cdot\frac{1}{x}\,dx = \tfrac{2}{3}x^{3/2}\ln x - \int \tfrac{2}{3}x^{1/2}\,dx = \tfrac{2}{3}x^{3/2}\ln x - \tfrac{4}{9}x^{3/2} + C.$$

Note: A mnemonic device which is helpful for selecting u when using integration by parts is the LIATE principle of precedence for u:

Logarithmic
Inverse trigonometric
Algebraic
Trigonometric
Exponential

If the integrand has several factors, then we try to choose among them a u which appears as high as possible on the list. For example, in $\int xe^{2x}\,dx$ the integrand is xe^{2x}, which is the product of an algebraic function (x) and an exponential function (e^{2x}). Since Algebraic appears before Exponential, we choose $u = x$. Sometimes the integration turns out to be similar regardless of the selection of u and dv, but it is advisable to refer to LIATE when in doubt.

3. Let $u = x$, $dv = \cos 5x\,dx \Rightarrow du = dx$, $v = \frac{1}{5}\sin 5x$. Then by Equation 2,

$\int x\cos 5x\,dx = \frac{1}{5}x\sin 5x - \int \frac{1}{5}\sin 5x\,dx = \frac{1}{5}x\sin 5x + \frac{1}{25}\cos 5x + C.$

4. Let $u = y$, $dv = e^{0.2y}dy \Rightarrow du = dy$, $v = \frac{1}{0.2}e^{0.2y}$. Then by Equation 2,

$\int ye^{0.2y}dy = 5ye^{0.2y} - \int 5e^{0.2y}dy = 5ye^{0.2y} - 25e^{0.2y} + C.$

5. Let $u = t$, $dv = e^{-3t}dt \Rightarrow du = dt$, $v = -\frac{1}{3}e^{-3t}$. Then by Equation 2,

$\int te^{-3t}dt = -\frac{1}{3}te^{-3t} - \int -\frac{1}{3}e^{-3t}dt = -\frac{1}{3}te^{-3t} + \frac{1}{3}\int e^{-3t}dt = -\frac{1}{3}te^{-3t} - \frac{1}{9}e^{-3t} + C.$

6. Let $u = x - 1$, $dv = \sin \pi x\,dx \Rightarrow du = dx$, $v = -\frac{1}{\pi}\cos \pi x$. Then by Equation 2,

$$\begin{aligned}\int (x-1)\sin \pi x\,dx &= -\frac{1}{\pi}(x-1)\cos \pi x - \int -\frac{1}{\pi}\cos \pi x\,dx = -\frac{1}{\pi}(x-1)\cos \pi x + \frac{1}{\pi}\int \cos \pi x\,dx \\ &= -\frac{1}{\pi}(x-1)\cos \pi x + \frac{1}{\pi^2}\sin \pi x + C\end{aligned}$$

7. First let $u = x^2 + 2x$, $dv = \cos x\,dx \Rightarrow du = (2x+2)\,dx$, $v = \sin x$. Then by Equation 2,

$I = \int (x^2+2x)\cos x\,dx = (x^2+2x)\sin x - \int (2x+2)\sin x\,dx$. Next let $U = 2x+2$, $dV = \sin x\,dx \Rightarrow dU = 2\,dx$, $V = -\cos x$, so $\int (2x+2)\sin x\,dx = -(2x+2)\cos x - \int -2\cos x\,dx = -(2x+2)\cos x + 2\sin x$. Thus,

$I = (x^2+2x)\sin x + (2x+2)\cos x - 2\sin x + C.$

8. First let $u = t^2$, $dv = \sin\beta t\,dt \Rightarrow du = 2t\,dt$, $v = -\frac{1}{\beta}\cos\beta t$. Then by Equation 2,

$I = \int t^2 \sin\beta t\,dt = -\frac{1}{\beta}t^2\cos\beta t - \int -\frac{2}{\beta}t\cos\beta t\,dt$. Next let $U = t$, $dV = \cos\beta t\,dt \Rightarrow dU = dt$,

$V = \dfrac{1}{\beta}\sin\beta t$, so $\int t\cos\beta t\,dt = \dfrac{1}{\beta}t\sin\beta t - \displaystyle\int \frac{1}{\beta}\sin\beta t\,dt = \frac{1}{\beta}t\sin\beta t + \frac{1}{\beta^2}\cos\beta t$. Thus,

$$I = -\frac{1}{\beta}t^2\cos\beta t + \frac{2}{\beta}\left(\frac{1}{\beta}t\sin\beta t + \frac{1}{\beta^2}\cos\beta t\right) + C = -\frac{1}{\beta}t^2\cos\beta t + \frac{2}{\beta^2}t\sin\beta t + \frac{2}{\beta^3}\cos\beta t + C.$$

9. Let $u = \cos^{-1}x$, $dv = dx \Rightarrow du = \dfrac{-1}{\sqrt{1-x^2}}\,dx$, $v = x$. Then by Equation 2,

$$\begin{aligned}\int \cos^{-1}x\,dx &= x\cos^{-1}x - \int \frac{-x}{\sqrt{1-x^2}}\,dx = x\cos^{-1}x - \int \frac{1}{\sqrt{t}}\left(\frac{1}{2}\,dt\right) \qquad \begin{bmatrix} t = 1-x^2, \\ dt = -2x\,dx\end{bmatrix} \\ &= x\cos^{-1}x - \tfrac{1}{2}\cdot 2t^{1/2} + C = x\cos^{-1}x - \sqrt{1-x^2} + C\end{aligned}$$

10. Let $u = \ln\sqrt{x}$, $dv = dx \Rightarrow du = \dfrac{1}{\sqrt{x}}\cdot\dfrac{1}{2\sqrt{x}}\,dx = \dfrac{1}{2x}\,dx$, $v = x$. Then by Equation 2,

$$\int \ln\sqrt{x}\,dx = x\ln\sqrt{x} - \int x\cdot\frac{1}{2x}\,dx = x\ln\sqrt{x} - \int \frac{1}{2}\,dx = x\ln\sqrt{x} - \frac{1}{2}x + C.$$

Note: We could start by using $\ln\sqrt{x} = \frac{1}{2}\ln x$.

11. Let $u = \ln t$, $dv = t^4\,dt \Rightarrow du = \dfrac{1}{t}\,dt$, $v = \dfrac{1}{5}t^5$. Then by Equation 2,

$$\int t^4\ln t\,dt = \frac{1}{5}t^5\ln t - \int \frac{1}{5}t^5\cdot\frac{1}{t}\,dt = \frac{1}{5}t^5\ln t - \int \frac{1}{5}t^4\,dt = \frac{1}{5}t^5\ln t - \frac{1}{25}t^5 + C.$$

12. Let $u = \tan^{-1}2y$, $dv = dy \Rightarrow du = \dfrac{2}{1+4y^2}\,dy$, $v = y$. Then by Equation 2,

$$\begin{aligned}\int \tan^{-1}2y\,dy &= y\tan^{-1}2y - \int \frac{2y}{1+4y^2}\,dy = y\tan^{-1}2y - \int \frac{1}{t}\left(\frac{1}{4}\,dt\right) \qquad \begin{bmatrix} t = 1+4y^2, \\ dt = 8y\,dy\end{bmatrix} \\ &= y\tan^{-1}2y - \tfrac{1}{4}\ln|t| + C = y\tan^{-1}2y - \tfrac{1}{4}\ln(1+4y^2) + C\end{aligned}$$

13. Let $u = t$, $dv = \csc^2 t\,dt \Rightarrow du = dt$, $v = -\cot t$. Then by Equation 2,

$$\begin{aligned}\int t\csc^2 t\,dt &= -t\cot t - \int -\cot t\,dt = -t\cot t + \int \frac{\cos t}{\sin t}\,dt = -t\cot t + \int \frac{1}{z}\,dz \qquad \begin{bmatrix} z = \sin t, \\ dz = \cos t\,dt\end{bmatrix} \\ &= -t\cot t + \ln|z| + C = -t\cot t + \ln|\sin t| + C\end{aligned}$$

14. Let $u = x$, $dv = \cosh ax\,dx \Rightarrow du = dx$, $v = \dfrac{1}{a}\sinh ax$. Then by Equation 2,

$$\int x\cosh ax\,dx = \frac{1}{a}x\sinh ax - \int \frac{1}{a}\sinh ax\,dx = \frac{1}{a}x\sinh ax - \frac{1}{a^2}\cosh ax + C.$$

15. First let $u = (\ln x)^2$, $dv = dx \Rightarrow du = 2\ln x\cdot\frac{1}{x}\,dx$, $v = x$. Then by Equation 2,

$I = \int(\ln x)^2\,dx = x(\ln x)^2 - 2\int x\ln x\cdot\frac{1}{x}\,dx = x(\ln x)^2 - 2\int \ln x\,dx$. Next let $U = \ln x$, $dV = dx \Rightarrow$ $dU = 1/x\,dx$, $V = x$ to get $\int \ln x\,dx = x\ln x - \int x\cdot(1/x)\,dx = x\ln x - \int dx = x\ln x - x + C_1$. Thus, $I = x(\ln x)^2 - 2(x\ln x - x + C_1) = x(\ln x)^2 - 2x\ln x + 2x + C$, where $C = -2C_1$.

16. $\displaystyle\int \frac{z}{10^z}\,dz = \int z\,10^{-z}\,dz$. Let $u = z$, $dv = 10^{-z}\,dz \Rightarrow du = dz$, $v = \dfrac{-10^{-z}}{\ln 10}$. Then by Equation 2,

$$\int z\,10^{-z}\,dz = \frac{-z\,10^{-z}}{\ln 10} - \int \frac{-10^{-z}}{\ln 10}\,dz = \frac{-z}{10^z \ln 10} - \frac{10^{-z}}{(\ln 10)(\ln 10)} + C = -\frac{z}{10^z \ln 10} - \frac{1}{10^z(\ln 10)^2} + C.$$

17. First let $u = \sin 3\theta$, $dv = e^{2\theta}\,d\theta \Rightarrow du = 3\cos 3\theta\,d\theta$, $v = \frac{1}{2}e^{2\theta}$. Then

$I = \int e^{2\theta}\sin 3\theta\,d\theta = \frac{1}{2}e^{2\theta}\sin 3\theta - \frac{3}{2}\int e^{2\theta}\cos 3\theta\,d\theta$. Next let $U = \cos 3\theta$, $dV = e^{2\theta}\,d\theta \Rightarrow dU = -3\sin 3\theta\,d\theta$,

$V = \frac{1}{2}e^{2\theta}$ to get $\int e^{2\theta}\cos 3\theta\,d\theta = \frac{1}{2}e^{2\theta}\cos 3\theta + \frac{3}{2}\int e^{2\theta}\sin 3\theta\,d\theta$. Substituting in the previous formula gives

$I = \frac{1}{2}e^{2\theta}\sin 3\theta - \frac{3}{4}e^{2\theta}\cos 3\theta - \frac{9}{4}\int e^{2\theta}\sin 3\theta\,d\theta = \frac{1}{2}e^{2\theta}\sin 3\theta - \frac{3}{4}e^{2\theta}\cos 3\theta - \frac{9}{4}I \Rightarrow$

$\frac{13}{4}I = \frac{1}{2}e^{2\theta}\sin 3\theta - \frac{3}{4}e^{2\theta}\cos 3\theta + C_1$. Hence, $I = \frac{1}{13}e^{2\theta}(2\sin 3\theta - 3\cos 3\theta) + C$, where $C = \frac{4}{13}C_1$.

18. First let $u = e^{-\theta}$, $dv = \cos 2\theta\,d\theta \Rightarrow du = -e^{-\theta}\,d\theta$, $v = \frac{1}{2}\sin 2\theta$. Then

$I = \int e^{-\theta}\cos 2\theta\,d\theta = \frac{1}{2}e^{-\theta}\sin 2\theta - \int \frac{1}{2}\sin 2\theta\,(-e^{-\theta}\,d\theta) = \frac{1}{2}e^{-\theta}\sin 2\theta + \frac{1}{2}\int e^{-\theta}\sin 2\theta\,d\theta$.

Next let $U = e^{-\theta}$, $dV = \sin 2\theta\,d\theta \Rightarrow dU = -e^{-\theta}\,d\theta$, $V = -\frac{1}{2}\cos 2\theta$, so

$\int e^{-\theta}\sin 2\theta\,d\theta = -\frac{1}{2}e^{-\theta}\cos 2\theta - \int(-\frac{1}{2})\cos 2\theta(-e^{-\theta}\,d\theta) = -\frac{1}{2}e^{-\theta}\cos 2\theta - \frac{1}{2}\int e^{-\theta}\cos 2\theta\,d\theta$.

So $I = \frac{1}{2}e^{-\theta}\sin 2\theta + \frac{1}{2}\left[(-\frac{1}{2}e^{-\theta}\cos 2\theta) - \frac{1}{2}I\right] = \frac{1}{2}e^{-\theta}\sin 2\theta - \frac{1}{4}e^{-\theta}\cos 2\theta - \frac{1}{4}I \Rightarrow$

$\frac{5}{4}I = \frac{1}{2}e^{-\theta}\sin 2\theta - \frac{1}{4}e^{-\theta}\cos 2\theta + C_1 \Rightarrow I = \frac{4}{5}(\frac{1}{2}e^{-\theta}\sin 2\theta - \frac{1}{4}e^{-\theta}\cos 2\theta + C_1) = \frac{2}{5}e^{-\theta}\sin 2\theta - \frac{1}{5}e^{-\theta}\cos 2\theta + C$.

19. First let $u = z^3$, $dv = e^z dz \Rightarrow du = 3z^2 dz$, $v = e^z$. Then $I_1 = \int z^3 e^z dz = z^3 e^z - 3\int z^2 e^z dz$. Next let $u_1 = z^2$,

$dv_1 = e^z dz \Rightarrow du_1 = 2z\,dz$, $v_1 = e^z$. Then $I_2 = z^2 e^z - 2\int z e^z dz$. Finally, let $u_2 = z$, $dv_2 = e^z dz \Rightarrow du_2 = dz$,

$v_2 = e^z$. Then $\int z e^z dz = z e^z - \int e^z dz = z e^z - e^z + C_1$. Substituting in the expression for I_2, we get

$I_2 = z^2 e^z - 2(z e^z - e^z + C_1) = z^2 e^z - 2z e^z + 2e^z - 2C_1$. Substituting the last expression for I_2 into I_1 gives

$I_1 = z^3 e^z - 3(z^2 e^z - 2z e^z + 2e^z - 2C_1) = z^3 e^z - 3z^2 e^z + 6z e^z - 6e^z + C$, where $C = 6C_1$.

20. $\int x\tan^2 x\,dx = \int x(\sec^2 x - 1)\,dx = \int x\sec^2 x\,dx - \int x\,dx$. Let $u = x$, $dv = \sec^2 x\,dx \Rightarrow du = dx$, $v = \tan x$.

Then by Equation 2, $\int x\sec^2 x\,dx = x\tan x - \int \tan x\,dx = x\tan x - \ln|\sec x|$, and thus,

$\int x\tan^2 x\,dx = x\tan x - \ln|\sec x| - \frac{1}{2}x^2 + C$.

21. Let $u = xe^{2x}$, $dv = \dfrac{1}{(1+2x)^2}\,dx \Rightarrow du = (x\cdot 2e^{2x} + e^{2x}\cdot 1)\,dx = e^{2x}(2x+1)\,dx$, $v = -\dfrac{1}{2(1+2x)}$.

Then by Equation 2,

$$\int \frac{xe^{2x}}{(1+2x)^2}\,dx = -\frac{xe^{2x}}{2(1+2x)} + \frac{1}{2}\int \frac{e^{2x}(2x+1)}{1+2x}\,dx = -\frac{xe^{2x}}{2(1+2x)} + \frac{1}{2}\int e^{2x}\,dx = -\frac{xe^{2x}}{2(1+2x)} + \frac{1}{4}e^{2x} + C.$$

The answer could be written as $\dfrac{e^{2x}}{4(2x+1)} + C$.

22. First let $u = (\arcsin x)^2$, $dv = dx \Rightarrow du = 2\arcsin x \cdot \dfrac{1}{\sqrt{1-x^2}}\,dx$, $v = x$. Then

$I = \displaystyle\int (\arcsin x)^2 dx = x(\arcsin x)^2 - 2\int \frac{x\arcsin x}{\sqrt{1-x^2}}\,dx$. To simplify the last integral, let $t = \arcsin x$ $[x = \sin t]$, so

$dt = \dfrac{1}{\sqrt{1-x^2}}\,dx$, and $\displaystyle\int \frac{x \arcsin x}{\sqrt{1-x^2}}\,dx = \int t \sin t\,dt$. To evaluate just the last integral, now let $U = t$, $dV = \sin t\,dt \;\Rightarrow$ $dU = dt$, $V = -\cos t$. Thus,

$$\int t \sin t\,dt = -t\cos t + \int \cos t\,dt = -t\cos t + \sin t + C$$

$$= -\arcsin x \cdot \frac{\sqrt{1-x^2}}{1} + x + C_1 \quad \text{[refer to the figure]}$$

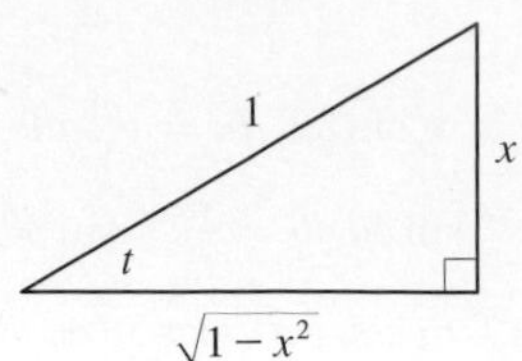

Returning to I, we get $I = x(\arcsin x)^2 + 2\sqrt{1-x^2}\,\arcsin x - 2x + C$, where $C = -2C_1$.

23. Let $u = x$, $dv = \cos \pi x\,dx \;\Rightarrow\; du = dx$, $v = \frac{1}{\pi}\sin \pi x$. By (6),

$$\int_0^{1/2} x\cos \pi x\,dx = \left[\frac{1}{\pi}x \sin \pi x\right]_0^{1/2} - \int_0^{1/2} \frac{1}{\pi}\sin \pi x\,dx = \frac{1}{2\pi} - 0 - \frac{1}{\pi}\left[-\frac{1}{\pi}\cos \pi x\right]_0^{1/2}$$

$$= \frac{1}{2\pi} + \frac{1}{\pi^2}(0-1) = \frac{1}{2\pi} - \frac{1}{\pi^2} \text{ or } \frac{\pi - 2}{2\pi^2}$$

24. First let $u = x^2 + 1$, $dv = e^{-x}\,dx \;\Rightarrow\; du = 2x\,dx$, $v = -e^{-x}$. By (6),

$\int_0^1 (x^2+1)e^{-x}\,dx = \left[-(x^2+1)e^{-x}\right]_0^1 + \int_0^1 2xe^{-x}\,dx = -2e^{-1} + 1 + 2\int_0^1 xe^{-x}\,dx$.

Next let $U = x$, $dV = e^{-x}\,dx \;\Rightarrow\; dU = dx$, $V = -e^{-x}$. By (6) again,

$\int_0^1 xe^{-x}\,dx = \left[-xe^{-x}\right]_0^1 + \int_0^1 e^{-x}\,dx = -e^{-1} + \left[-e^{-x}\right]_0^1 = -e^{-1} - e^{-1} + 1 = -2e^{-1} + 1$. So

$\int_0^1 (x^2+1)e^{-x}\,dx = -2e^{-1} + 1 + 2(-2e^{-1} + 1) = -2e^{-1} + 1 - 4e^{-1} + 2 = -6e^{-1} + 3$.

25. Let $u = y$, $dv = \sinh y\,dy \;\Rightarrow\; du = dy$, $v = \cosh y$. By (6),

$\int_0^2 y \sinh y\,dy = \Big[y\cosh y\Big]_0^2 - \int_0^2 \cosh y\,dy = 2\cosh 2 - 0 - \Big[\sinh y\Big]_0^2 = 2\cosh 2 - \sinh 2$.

26. Let $u = \ln w$, $dv = w^2\,dw \;\Rightarrow\; du = \dfrac{1}{w}\,dw$, $v = \frac{1}{3}w^3$. By (6),

$\int_1^2 w^2 \ln w\,dw = \left[\frac{1}{3}w^3 \ln w\right]_1^2 - \int_1^2 \frac{1}{3}w^2\,dw = \frac{8}{3}\ln 2 - 0 - \left[\frac{1}{9}w^3\right]_1^2 = \frac{8}{3}\ln 2 - \left(\frac{8}{9} - \frac{1}{9}\right) = \frac{8}{3}\ln 2 - \frac{7}{9}$.

27. Let $u = \ln R$, $dv = \dfrac{1}{R^2}\,dR \;\Rightarrow\; du = \dfrac{1}{R}\,dR$, $v = -\dfrac{1}{R}$. By (6),

$$\int_1^5 \frac{\ln R}{R^2}\,dR = \left[-\frac{1}{R}\ln R\right]_1^5 - \int_1^5 -\frac{1}{R^2}\,dR = -\tfrac{1}{5}\ln 5 - 0 - \left[\frac{1}{R}\right]_1^5 = -\tfrac{1}{5}\ln 5 - \left(\tfrac{1}{5} - 1\right) = \tfrac{4}{5} - \tfrac{1}{5}\ln 5.$$

28. First let $u = t^2$, $dv = \sin 2t\,dt \;\Rightarrow\; du = 2t\,dt$, $v = -\frac{1}{2}\cos 2t$. By (6),

$\int_0^{2\pi} t^2 \sin 2t\,dt = \left[-\frac{1}{2}t^2 \cos 2t\right]_0^{2\pi} + \int_0^{2\pi} t\cos 2t\,dt = -2\pi^2 + \int_0^{2\pi} t \cos 2t\,dt$. Next let $U = t$, $dV = \cos 2t\,dt \;\Rightarrow$ $dU = dt$, $V = \frac{1}{2}\sin 2t$. By (6) again,

$\int_0^{2\pi} t\cos 2t\,dt = \left[\frac{1}{2}t\sin 2t\right]_0^{2\pi} - \int_0^{2\pi} \frac{1}{2}\sin 2t\,dt = 0 - \left[-\frac{1}{4}\cos 2t\right]_0^{2\pi} = \frac{1}{4} - \frac{1}{4} = 0$. Thus, $\int_0^{2\pi} t^2 \sin 2t\,dt = -2\pi^2$.

29. $\sin 2x = 2\sin x\,\cos x$, so $\int_0^{\pi} x\,\sin x\,\cos x\,dx = \frac{1}{2}\int_0^{\pi} x\,\sin 2x\,dx$. Let $u = x$, $dv = \sin 2x\,dx \;\Rightarrow\; du = dx$, $v = -\frac{1}{2}\cos 2x$. By (6), $\frac{1}{2}\int_0^{\pi} x\,\sin 2x\,dx = \frac{1}{2}\left[-\frac{1}{2}x\cos 2x\right]_0^{\pi} - \frac{1}{2}\int_0^{\pi} -\frac{1}{2}\cos 2x\,dx = -\frac{1}{4}\pi - 0 + \frac{1}{4}\left[\frac{1}{2}\sin 2x\right]_0^{\pi} = -\dfrac{\pi}{4}$.

30. Let $u = \arctan(1/x)$, $dv = dx \quad\Rightarrow\quad du = \dfrac{1}{1+(1/x)^2}\cdot\dfrac{-1}{x^2}\,dx = \dfrac{-dx}{x^2+1}$, $v = x$. By (6),

$$\int_1^{\sqrt{3}} \arctan\left(\frac{1}{x}\right) dx = \left[x\arctan\left(\frac{1}{x}\right)\right]_1^{\sqrt{3}} + \int_1^{\sqrt{3}} \frac{x\,dx}{x^2+1} = \sqrt{3}\,\frac{\pi}{6} - 1\cdot\frac{\pi}{4} + \frac{1}{2}\Big[\ln(x^2+1)\Big]_1^{\sqrt{3}}$$

$$= \frac{\pi\sqrt{3}}{6} - \frac{\pi}{4} + \frac{1}{2}(\ln 4 - \ln 2) = \frac{\pi\sqrt{3}}{6} - \frac{\pi}{4} + \frac{1}{2}\ln\frac{4}{2} = \frac{\pi\sqrt{3}}{6} - \frac{\pi}{4} + \frac{1}{2}\ln 2$$

31. Let $u = M$, $dv = e^{-M}\,dM \quad\Rightarrow\quad du = dM$, $v = -e^{-M}$. By (6),

$$\int_1^5 \frac{M}{e^M}\,dM = \int_1^5 Me^{-M}\,dM = \Big[-Me^{-M}\Big]_1^5 - \int_1^5 -e^{-M}\,dM = -5e^{-5} + e^{-1} - \Big[e^{-M}\Big]_1^5$$

$$= -5e^{-5} + e^{-1} - (e^{-5} - e^{-1}) = 2e^{-1} - 6e^{-5}$$

32. Let $u = (\ln x)^2$, $dv = x^{-3}\,dx \quad\Rightarrow\quad du = \dfrac{2\ln x}{x}\,dx$, $v = -\frac{1}{2}x^{-2}$. By (6),

$I = \displaystyle\int_1^2 \frac{(\ln x)^2}{x^3}\,dx = \left[-\frac{(\ln x)^2}{2x^2}\right]_1^2 + \int_1^2 \frac{\ln x}{x^3}\,dx$. Now let $U = \ln x$, $dV = x^{-3}\,dx \quad\Rightarrow\quad dU = \dfrac{1}{x}\,dx$, $V = -\frac{1}{2}x^{-2}$.

Then

$$\int_1^2 \frac{\ln x}{x^3}\,dx = \left[-\frac{\ln x}{2x^2}\right]_1^2 + \tfrac{1}{2}\int_1^2 x^{-3}dx = -\tfrac{1}{8}\ln 2 + 0 + \tfrac{1}{2}\left[-\tfrac{1}{2x^2}\right]_1^2 = -\tfrac{1}{8}\ln 2 + \tfrac{1}{2}\left(-\tfrac{1}{8} + \tfrac{1}{2}\right) = \tfrac{3}{16} - \tfrac{1}{8}\ln 2.$$

Thus $I = \left(-\frac{1}{8}(\ln 2)^2 + 0\right) + \left(\frac{3}{16} - \frac{1}{8}\ln 2\right) = -\frac{1}{8}(\ln 2)^2 - \frac{1}{8}\ln 2 + \frac{3}{16}$.

33. Let $u = \ln(\cos x)$, $dv = \sin x\,dx \quad\Rightarrow\quad du = \dfrac{1}{\cos x}(-\sin x)\,dx$, $v = -\cos x$. By (6),

$$\textstyle\int_0^{\pi/3} \sin x\,\ln(\cos x)\,dx = \Big[-\cos x\,\ln(\cos x)\Big]_0^{\pi/3} - \int_0^{\pi/3} \sin x\,dx = -\frac{1}{2}\ln\frac{1}{2} - 0 - \Big[-\cos x\Big]_0^{\pi/3}$$

$$= -\tfrac{1}{2}\ln\tfrac{1}{2} + \left(\tfrac{1}{2} - 1\right) = \tfrac{1}{2}\ln 2 - \tfrac{1}{2}$$

34. Let $u = r^2$, $dv = \dfrac{r}{\sqrt{4+r^2}}\,dr \quad\Rightarrow\quad du = 2r\,dr$, $v = \sqrt{4+r^2}$. By (6),

$$\int_0^1 \frac{r^3}{\sqrt{4+r^2}}\,dr = \left[r^2\sqrt{4+r^2}\right]_0^1 - 2\int_0^1 r\sqrt{4+r^2}\,dr = \sqrt{5} - \tfrac{2}{3}\left[(4+r^2)^{3/2}\right]_0^1$$

$$= \sqrt{5} - \tfrac{2}{3}(5)^{3/2} + \tfrac{2}{3}(8) = \sqrt{5}\left(1 - \tfrac{10}{3}\right) + \tfrac{16}{3} = \tfrac{16}{3} - \tfrac{7}{3}\sqrt{5}$$

35. Let $u = (\ln x)^2$, $dv = x^4\,dx \quad\Rightarrow\quad du = 2\dfrac{\ln x}{x}\,dx$, $v = \dfrac{x^5}{5}$. By (6),

$$\int_1^2 x^4(\ln x)^2\,dx = \left[\frac{x^5}{5}(\ln x)^2\right]_1^2 - 2\int_1^2 \frac{x^4}{5}\ln x\,dx = \tfrac{32}{5}(\ln 2)^2 - 0 - 2\int_1^2 \frac{x^4}{5}\ln x\,dx.$$

Let $U = \ln x$, $dV = \dfrac{x^4}{5}\,dx \quad\Rightarrow\quad dU = \dfrac{1}{x}\,dx$, $V = \dfrac{x^5}{25}$.

Then $\displaystyle\int_1^2 \frac{x^4}{5}\ln x\,dx = \left[\frac{x^5}{25}\ln x\right]_1^2 - \int_1^2 \frac{x^4}{25}\,dx = \tfrac{32}{25}\ln 2 - 0 - \left[\frac{x^5}{125}\right]_1^2 = \tfrac{32}{25}\ln 2 - \left(\tfrac{32}{125} - \tfrac{1}{125}\right).$

So $\int_1^2 x^4(\ln x)^2\,dx = \frac{32}{5}(\ln 2)^2 - 2\left(\frac{32}{25}\ln 2 - \frac{31}{125}\right) = \frac{32}{5}(\ln 2)^2 - \frac{64}{25}\ln 2 + \frac{62}{125}$.

36. Let $u = \sin(t-s)$, $dv = e^s\,ds \;\Rightarrow\; du = -\cos(t-s)\,ds$, $v = e^s$. Then

$I = \int_0^t e^s \sin(t-s)\,ds = \Big[e^s \sin(t-s)\Big]_0^t + \int_0^t e^s \cos(t-s)\,ds = e^t \sin 0 - e^0 \sin t + I_1$. For I_1, let $U = \cos(t-s)$,

$dV = e^s\,ds \;\Rightarrow\; dU = \sin(t-s)\,ds$, $V = e^s$. So $I_1 = \big[e^s \cos(t-s)\big]_0^t - \int_0^t e^s \sin(t-s)\,ds = e^t \cos 0 - e^0 \cos t - I$.

Thus, $I = -\sin t + e^t - \cos t - I \;\Rightarrow\; 2I = e^t - \cos t - \sin t \;\Rightarrow\; I = \frac{1}{2}(e^t - \cos t - \sin t)$.

37. Let $t = \sqrt{x}$, so that $t^2 = x$ and $2t\,dt = dx$. Thus, $\int e^{\sqrt{x}}\,dx = \int e^t (2t)\,dt$. Now use parts with $u = t$, $dv = e^t\,dt$, $du = dt$, and $v = e^t$ to get $2\int te^t\,dt = 2t\,e^t - 2\int e^t\,dt = 2t\,e^t - 2e^t + C = 2\sqrt{x}\,e^{\sqrt{x}} - 2e^{\sqrt{x}} + C$.

38. Let $t = \ln x$, so that $e^t = x$ and $e^t\,dt = dx$. Thus, $\int \cos(\ln x)\,dx = \int \cos t \cdot e^t\,dt = I$. Now use parts with $u = \cos t$, $dv = e^t\,dt$, $du = -\sin t\,dt$, and $v = e^t$ to get $\int e^t \cos t\,dt = e^t \cos t - \int -e^t \sin t\,dt = e^t \cos t + \int e^t \sin t\,dt$. Now use parts with $U = \sin t$, $dV = e^t\,dt$, $dU = \cos t\,dt$, and $V = e^t$ to get

$\int e^t \sin t\,dt = e^t \sin t - \int e^t \cos t\,dt$. Thus, $I = e^t \cos t + e^t \sin t - I \;\Rightarrow\; 2I = e^t \cos t + e^t \sin t \;\Rightarrow$

$I = \frac{1}{2}e^t \cos t + \frac{1}{2}e^t \sin t + C = \frac{1}{2}x\cos(\ln x) + \frac{1}{2}x\sin(\ln x) + C$.

39. Let $x = \theta^2$, so that $dx = 2\theta\,d\theta$. Thus, $\displaystyle\int_{\sqrt{\pi/2}}^{\sqrt{\pi}} \theta^3 \cos\!\left(\theta^2\right)\,d\theta = \int_{\sqrt{\pi/2}}^{\sqrt{\pi}} \theta^2 \cos\!\left(\theta^2\right)\cdot \tfrac{1}{2}(2\theta\,d\theta) = \tfrac{1}{2}\int_{\pi/2}^{\pi} x\cos x\,dx$. Now use parts with $u = x$, $dv = \cos x\,dx$, $du = dx$, $v = \sin x$ to get

$$\begin{aligned}\tfrac{1}{2}\int_{\pi/2}^{\pi} x\cos x\,dx &= \tfrac{1}{2}\left(\big[x\sin x\big]_{\pi/2}^{\pi} - \int_{\pi/2}^{\pi} \sin x\,dx\right) = \tfrac{1}{2}\big[x\sin x + \cos x\big]_{\pi/2}^{\pi} \\ &= \tfrac{1}{2}(\pi \sin \pi + \cos \pi) - \tfrac{1}{2}\left(\tfrac{\pi}{2}\sin\tfrac{\pi}{2} + \cos\tfrac{\pi}{2}\right) = \tfrac{1}{2}(\pi\cdot 0 - 1) - \tfrac{1}{2}\left(\tfrac{\pi}{2}\cdot 1 + 0\right) = -\tfrac{1}{2} - \tfrac{\pi}{4}\end{aligned}$$

40. Let $x = \cos t$, so that $dx = -\sin t\,dt$. Thus,

$\int_0^\pi e^{\cos t} \sin 2t\,dt = \int_0^\pi e^{\cos t}(2\sin t\,\cos t)\,dt = \int_1^{-1} e^x \cdot 2x\,(-dx) = 2\int_{-1}^1 xe^x\,dx$. Now use parts with $u = x$, $dv = e^x\,dx$, $du = dx$, $v = e^x$ to get

$2\int_{-1}^1 xe^x\,dx = 2\left(\big[xe^x\big]_{-1}^1 - \int_{-1}^1 e^x\,dx\right) = 2\left(e^1 + e^{-1} - \big[e^x\big]_{-1}^1\right) = 2(e + e^{-1} - [e^1 - e^{-1}]) = 2(2e^{-1}) = 4/e$.

41. Let $y = 1 + x$, so that $dy = dx$. Thus, $\int x\ln(1+x)\,dx = \int (y-1)\ln y\,dy$. Now use parts with $u = \ln y$, $dv = (y-1)\,dy$, $du = \frac{1}{y}\,dy$, $v = \frac{1}{2}y^2 - y$ to get

$$\begin{aligned}\int (y-1)\ln y\,dy &= \left(\tfrac{1}{2}y^2 - y\right)\ln y - \int\left(\tfrac{1}{2}y - 1\right)dy = \tfrac{1}{2}y(y-2)\ln y - \tfrac{1}{4}y^2 + y + C \\ &= \tfrac{1}{2}(1+x)(x-1)\ln(1+x) - \tfrac{1}{4}(1+x)^2 + 1 + x + C,\end{aligned}$$

which can be written as $\frac{1}{2}(x^2 - 1)\ln(1+x) - \frac{1}{4}x^2 + \frac{1}{2}x + \frac{3}{4} + C$.

42. Let $y = \ln x$, so that $dy = \dfrac{1}{x}\,dx$. Thus, $\displaystyle\int \frac{\arcsin(\ln x)}{x}\,dx = \int \arcsin y\,dy$. Now use parts with $u = \arcsin y$, $dv = dy$, $du = \dfrac{1}{\sqrt{1-y^2}}\,dy$, and $v = y$ to get

$$\int \arcsin y\,dy = y\arcsin y - \int \frac{y}{\sqrt{1-y^2}}\,dy = y\arcsin y + \sqrt{1-y^2} + C = (\ln x)\arcsin(\ln x) + \sqrt{1-(\ln x)^2} + C.$$

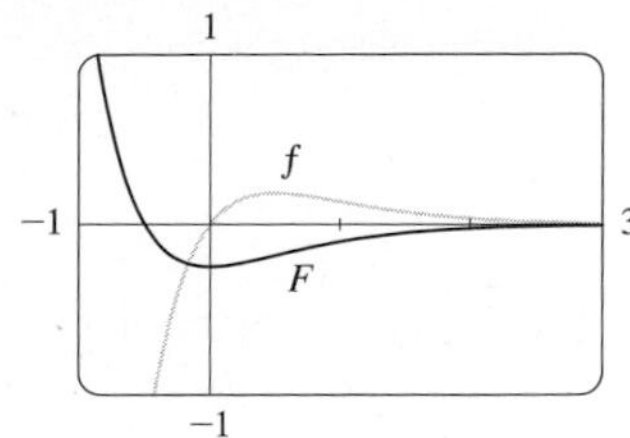

43. Let $u = x$, $dv = e^{-2x}\,dx \;\Rightarrow\; du = dx$, $v = -\frac{1}{2}e^{-2x}$. Then

$\int xe^{-2x}\,dx = -\frac{1}{2}xe^{-2x} + \int \frac{1}{2}e^{-2x}\,dx = -\frac{1}{2}xe^{-2x} - \frac{1}{4}e^{-2x} + C$. We see from the graph that this is reasonable, since F has a minimum where f changes from negative to positive. Also, F increases where f is positive and F decreases where f is negative.

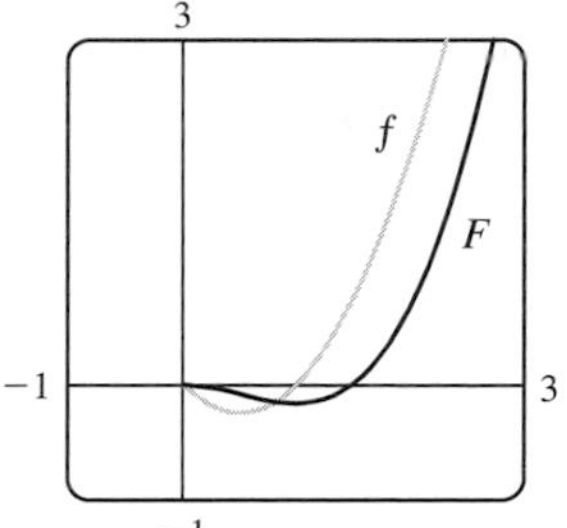

44. Let $u = \ln x$, $dv = x^{3/2}\,dx \;\Rightarrow\; du = \frac{1}{x}\,dx$, $v = \frac{2}{5}x^{5/2}$. Then

$$\begin{aligned}\int x^{3/2}\ln x\,dx &= \tfrac{2}{5}x^{5/2}\ln x - \tfrac{2}{5}\int x^{3/2}\,dx = \tfrac{2}{5}x^{5/2}\ln x - \left(\tfrac{2}{5}\right)^2 x^{5/2} + C \\ &= \tfrac{2}{5}x^{5/2}\ln x - \tfrac{4}{25}x^{5/2} + C\end{aligned}$$

We see from the graph that this is reasonable, since F has a minimum where f changes from negative to positive.

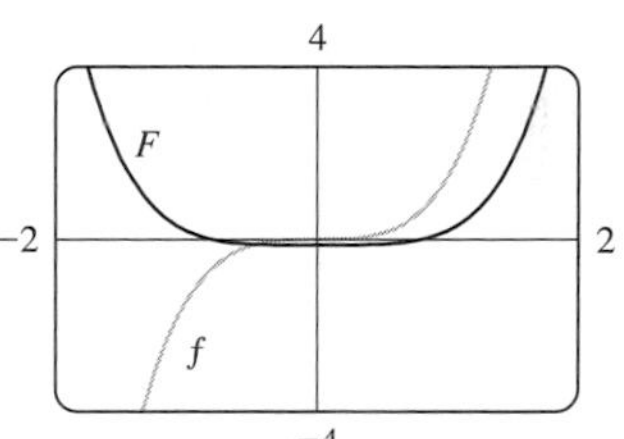

45. Let $u = \frac{1}{2}x^2$, $dv = 2x\sqrt{1+x^2}\,dx \;\Rightarrow\; du = x\,dx$, $v = \frac{2}{3}(1+x^2)^{3/2}$.

Then

$$\begin{aligned}\int x^3\sqrt{1+x^2}\,dx &= \tfrac{1}{2}x^2\left[\tfrac{2}{3}(1+x^2)^{3/2}\right] - \tfrac{2}{3}\int x(1+x^2)^{3/2}dx \\ &= \tfrac{1}{3}x^2(1+x^2)^{3/2} - \tfrac{2}{3}\cdot\tfrac{2}{5}\cdot\tfrac{1}{2}(1+x^2)^{5/2} + C \\ &= \tfrac{1}{3}x^2(1+x^2)^{3/2} - \tfrac{2}{15}(1+x^2)^{5/2} + C\end{aligned}$$

We see from the graph that this is reasonable, since F increases where f is positive and F decreases where f is negative. Note also that f is an odd function and F is an even function.

Another method: Use substitution with $u = 1 + x^2$ to get $\frac{1}{5}(1+x^2)^{5/2} - \frac{1}{3}(1+x^2)^{3/2} + C$.

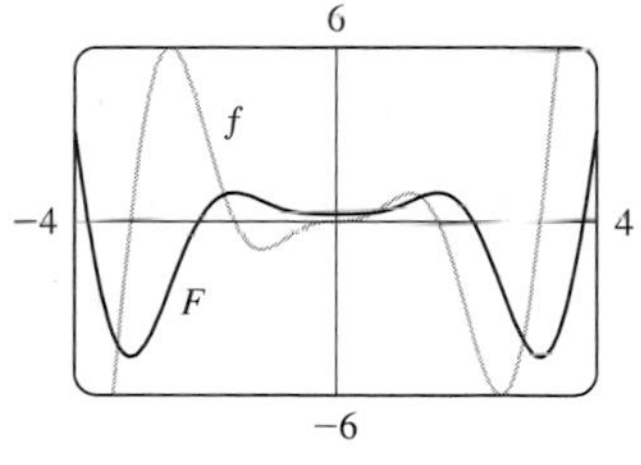

46. First let $u = x^2$, $dv = \sin 2x\,dx \;\Rightarrow\; du = 2x\,dx$, $v = -\frac{1}{2}\cos 2x$.

Then $I = \int x^2\sin 2x\,dx = -\frac{1}{2}x^2\cos 2x + \int x\cos 2x\,dx$.

Next let $U = x$, $dV = \cos 2x\,dx \;\Rightarrow\; dU = dx$, $V = \frac{1}{2}\sin 2x$, so

$\int x\cos 2x\,dx = \frac{1}{2}x\sin 2x - \int \frac{1}{2}\sin 2x\,dx = \frac{1}{2}x\sin 2x + \frac{1}{4}\cos 2x + C$.

Thus, $I = -\frac{1}{2}x^2\cos 2x + \frac{1}{2}x\sin 2x + \frac{1}{4}\cos 2x + C$.

We see from the graph that this is reasonable, since F increases where f is positive and F decreases where f is negative. Note also that f is an odd function and F is an even function.

47. (a) Take $n = 2$ in Example 6 to get $\displaystyle\int \sin^2 x\,dx = -\frac{1}{2}\cos x\sin x + \frac{1}{2}\int 1\,dx = \frac{x}{2} - \frac{\sin 2x}{4} + C$.

(b) $\int \sin^4 x\,dx = -\frac{1}{4}\cos x\sin^3 x + \frac{3}{4}\int \sin^2 x\,dx = -\frac{1}{4}\cos x\sin^3 x + \frac{3}{8}x - \frac{3}{16}\sin 2x + C$.

48. (a) Let $u = \cos^{n-1}x$, $dv = \cos x\,dx \;\Rightarrow\; du = -(n-1)\cos^{n-2}x\,\sin x\,dx$, $v = \sin x$ in (2):

$$\begin{aligned}\int \cos^n x\,dx &= \cos^{n-1}x\,\sin x + (n-1)\int \cos^{n-2}x\,\sin^2 x\,dx \\ &= \cos^{n-1}x\,\sin x + (n-1)\int \cos^{n-2}x\,(1-\cos^2 x)dx \\ &= \cos^{n-1}x\,\sin x + (n-1)\int \cos^{n-2}x\,dx - (n-1)\int \cos^n x\,dx\end{aligned}$$

[continued]

Rearranging terms gives $n\int \cos^n x\,dx = \cos^{n-1} x\,\sin x + (n-1)\int \cos^{n-2} x\,dx$ or

$$\int \cos^n x\,dx = \frac{1}{n}\cos^{n-1} x\,\sin x + \frac{n-1}{n}\int \cos^{n-2} x\,dx$$

(b) Take $n = 2$ in part (a) to get $\int \cos^2 x\,dx = \frac{1}{2}\cos x\,\sin x + \frac{1}{2}\int 1\,dx = \dfrac{x}{2} + \dfrac{\sin 2x}{4} + C$.

(c) $\int \cos^4 x\,dx = \frac{1}{4}\cos^3 x\,\sin x + \frac{3}{4}\int \cos^2 x\,dx = \frac{1}{4}\cos^3 x\,\sin x + \frac{3}{8}x + \frac{3}{16}\sin 2x + C$

49. (a) From Example 6, $\displaystyle\int \sin^n x\,dx = -\frac{1}{n}\cos x\,\sin^{n-1} x + \frac{n-1}{n}\int \sin^{n-2} x\,dx$. Using (6),

$$\begin{aligned}\int_0^{\pi/2} \sin^n x\,dx &= \left[-\frac{\cos x\,\sin^{n-1} x}{n}\right]_0^{\pi/2} + \frac{n-1}{n}\int_0^{\pi/2} \sin^{n-2} x\,dx \\ &= (0-0) + \frac{n-1}{n}\int_0^{\pi/2} \sin^{n-2} x\,dx = \frac{n-1}{n}\int_0^{\pi/2} \sin^{n-2} x\,dx\end{aligned}$$

(b) Using $n = 3$ in part (a), we have $\int_0^{\pi/2} \sin^3 x\,dx = \frac{2}{3}\int_0^{\pi/2} \sin x\,dx = \left[-\frac{2}{3}\cos x\right]_0^{\pi/2} = \frac{2}{3}$.

Using $n = 5$ in part (a), we have $\int_0^{\pi/2} \sin^5 x\,dx = \frac{4}{5}\int_0^{\pi/2} \sin^3 x\,dx = \frac{4}{5}\cdot\frac{2}{3} = \frac{8}{15}$.

(c) The formula holds for $n = 1$ (that is, $2n+1 = 3$) by (b). Assume it holds for some $k \geq 1$. Then

$\displaystyle\int_0^{\pi/2} \sin^{2k+1} x\,dx = \frac{2\cdot 4\cdot 6\cdot\cdots\cdot(2k)}{3\cdot 5\cdot 7\cdot\cdots\cdot(2k+1)}$. By Example 6,

$$\begin{aligned}\int_0^{\pi/2} \sin^{2k+3} x\,dx &= \frac{2k+2}{2k+3}\int_0^{\pi/2} \sin^{2k+1} x\,dx = \frac{2k+2}{2k+3}\cdot\frac{2\cdot 4\cdot 6\cdot\cdots\cdot(2k)}{3\cdot 5\cdot 7\cdot\cdots\cdot(2k+1)} \\ &= \frac{2\cdot 4\cdot 6\cdot\cdots\cdot(2k)[2(k+1)]}{3\cdot 5\cdot 7\cdot\cdots\cdot(2k+1)[2(k+1)+1]},\end{aligned}$$

so the formula holds for $n = k+1$. By induction, the formula holds for all $n \geq 1$.

50. Using Exercise 49(a), we see that the formula holds for $n = 1$, because $\int_0^{\pi/2} \sin^2 x\,dx = \frac{1}{2}\int_0^{\pi/2} 1\,dx = \frac{1}{2}\left[x\right]_0^{\pi/2} = \frac{1}{2}\cdot\frac{\pi}{2}$.

Now assume it holds for some $k \geq 1$. Then $\displaystyle\int_0^{\pi/2} \sin^{2k} x\,dx = \frac{1\cdot 3\cdot 5\cdot\cdots\cdot(2k-1)}{2\cdot 4\cdot 6\cdot\cdots\cdot(2k)}\,\frac{\pi}{2}$. By Exercise 49(a),

$$\begin{aligned}\int_0^{\pi/2} \sin^{2(k+1)} x\,dx &= \frac{2k+1}{2k+2}\int_0^{\pi/2} \sin^{2k} x\,dx = \frac{2k+1}{2k+2}\cdot\frac{1\cdot 3\cdot 5\cdot\cdots\cdot(2k-1)}{2\cdot 4\cdot 6\cdot\cdots\cdot(2k)}\,\frac{\pi}{2} \\ &= \frac{1\cdot 3\cdot 5\cdot\cdots\cdot(2k-1)(2k+1)}{2\cdot 4\cdot 6\cdot\cdots\cdot(2k)(2k+2)}\cdot\frac{\pi}{2},\end{aligned}$$

so the formula holds for $n = k+1$. By induction, the formula holds for all $n \geq 1$.

51. Let $u = (\ln x)^n$, $dv = dx$ $\Rightarrow$ $du = n(\ln x)^{n-1}(dx/x)$, $v = x$. By Equation 2,

$\int (\ln x)^n\,dx = x(\ln x)^n - \int nx(\ln x)^{n-1}(dx/x) = x(\ln x)^n - n\int (\ln x)^{n-1}\,dx$.

52. Let $u = x^n$, $dv = e^x\,dx$ $\Rightarrow$ $du = nx^{n-1}\,dx$, $v = e^x$. By Equation 2, $\int x^n e^x\,dx = x^n e^x - n\int x^{n-1}e^x\,dx$.

53. $\int \tan^n x\,dx = \int \tan^{n-2} x \tan^2 x\,dx = \int \tan^{n-2} x\,(\sec^2 x - 1)\,dx = \int \tan^{n-2} x \sec^2 x\,dx - \int \tan^{n-2} x\,dx$

$= I - \int \tan^{n-2} x\,dx.$

Let $u = \tan^{n-2} x$, $dv = \sec^2 x\,dx \quad \Rightarrow \quad du = (n-2)\tan^{n-3} x \sec^2 x\,dx$, $v = \tan x$. Then, by Equation 2,

$$\begin{aligned} I &= \tan^{n-1} x - (n-2) \textstyle\int \tan^{n-2} x \sec^2 x\,dx \\ 1I &= \tan^{n-1} x - (n-2)I \\ (n-1)I &= \tan^{n-1} x \\ I &= \frac{\tan^{n-1} x}{n-1} \end{aligned}$$

Returning to the original integral, $\int \tan^n x\,dx = \dfrac{\tan^{n-1} x}{n-1} - \int \tan^{n-2} x\,dx.$

54. Let $u = \sec^{n-2} x$, $dv = \sec^2 x\,dx \quad \Rightarrow \quad du = (n-2)\sec^{n-3} x \sec x \tan x\,dx$, $v = \tan x$. Then, by Equation 2,

$$\begin{aligned} \textstyle\int \sec^n x\,dx &= \tan x \sec^{n-2} x - (n-2) \textstyle\int \sec^{n-2} x \tan^2 x\,dx \\ &= \tan x \sec^{n-2} x - (n-2) \textstyle\int \sec^{n-2} x\,(\sec^2 x - 1)\,dx \\ &= \tan x \sec^{n-2} x - (n-2) \textstyle\int \sec^n x\,dx + (n-2) \textstyle\int \sec^{n-2} x\,dx \end{aligned}$$

so $(n-1)\int \sec^n x\,dx = \tan x \sec^{n-2} x + (n-2)\int \sec^{n-2} x\,dx$. If $n - 1 \neq 0$, then

$$\int \sec^n x\,dx = \frac{\tan x \sec^{n-2} x}{n-1} + \frac{n-2}{n-1}\int \sec^{n-2} x\,dx.$$

55. By repeated applications of the reduction formula in Exercise 51,

$$\begin{aligned} \textstyle\int (\ln x)^3\,dx &= x(\ln x)^3 - 3\textstyle\int (\ln x)^2\,dx = x(\ln x)^3 - 3\left[x(\ln x)^2 - 2\textstyle\int (\ln x)^1\,dx\right] \\ &= x(\ln x)^3 - 3x(\ln x)^2 + 6\left[x(\ln x)^1 - 1\textstyle\int (\ln x)^0\,dx\right] \\ &= x(\ln x)^3 - 3x(\ln x)^2 + 6x\ln x - 6\textstyle\int 1\,dx = x(\ln x)^3 - 3x(\ln x)^2 + 6x\ln x - 6x + C \end{aligned}$$

56. By repeated applications of the reduction formula in Exercise 52,

$$\begin{aligned} \textstyle\int x^4 e^x\,dx &= x^4e^x - 4\textstyle\int x^3e^x\,dx = x^4e^x - 4\left(x^3e^x - 3\textstyle\int x^2e^x\,dx\right) \\ &= x^4e^x - 4x^3e^x + 12\left(x^2e^x - 2\textstyle\int x^1e^x\,dx\right) = x^4e^x - 4x^3e^x + 12x^2e^x - 24\left(x^1e^x - \textstyle\int x^0e^x\,dx\right) \\ &= x^4e^x - 4x^3e^x + 12x^2e^x - 24xe^x + 24e^x + C \quad \left[\text{or } e^x(x^4 - 4x^3 + 12x^2 - 24x + 24) + C\right] \end{aligned}$$

57. The curves $y = x^2 \ln x$ and $y = 4\ln x$ intersect when $x^2 \ln x = 4\ln x \quad \Leftrightarrow$

$x^2 \ln x - 4\ln x = 0 \quad \Leftrightarrow \quad (x^2 - 4)\ln x = 0 \quad \Leftrightarrow$

$x = 1$ or 2 [since $x > 0$]. For $1 < x < 2$, $4\ln x > x^2 \ln x$. Thus,

area $= \int_1^2 (4\ln x - x^2\ln x)\,dx = \int_1^2 [(4 - x^2)\ln x]\,dx$. Let $u = \ln x$,

$dv = (4 - x^2)\,dx \quad \Rightarrow \quad du = \frac{1}{x}\,dx$, $v = 4x - \frac{1}{3}x^3$. Then

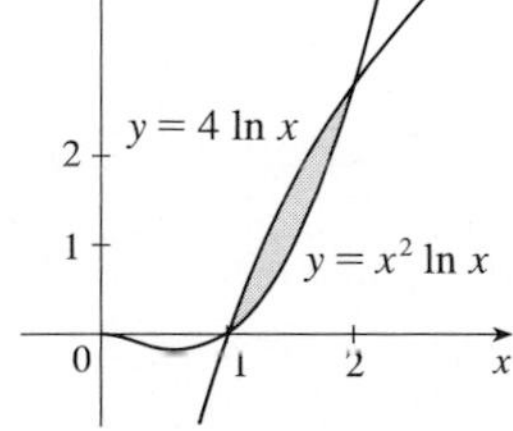

$$\begin{aligned} \text{area} &= \left[(\ln x)\left(4x - \tfrac{1}{3}x^3\right)\right]_1^2 - \int_1^2 \left[\left(4x - \tfrac{1}{3}x^3\right)\frac{1}{x}\right]dx = (\ln 2)\left(\tfrac{16}{3}\right) - 0 - \int_1^2 \left(4 - \tfrac{1}{3}x^2\right)dx \\ &= \tfrac{16}{3}\ln 2 - \left[4x - \tfrac{1}{9}x^3\right]_1^2 = \tfrac{16}{3}\ln 2 - \left(\tfrac{64}{9} - \tfrac{35}{9}\right) = \tfrac{16}{3}\ln 2 - \tfrac{29}{9} \end{aligned}$$

58. The curves $y = x^2e^{-x}$ and $y = xe^{-x}$ intersect when $x^2e^{-x} = xe^{-x} \Leftrightarrow$ $x^2 - x = 0 \Leftrightarrow x(x-1) = 0 \Leftrightarrow x = 0$ or 1.

For $0 < x < 1$, $xe^{-x} > x^2e^{-x}$. Thus,

area $= \int_0^1 (xe^{-x} - x^2e^{-x})\,dx = \int_0^1 (x - x^2)e^{-x}\,dx$. Let $u = x - x^2$, $dv = e^{-x}dx \Rightarrow du = (1-2x)\,dx$, $v = -e^{-x}$. Then

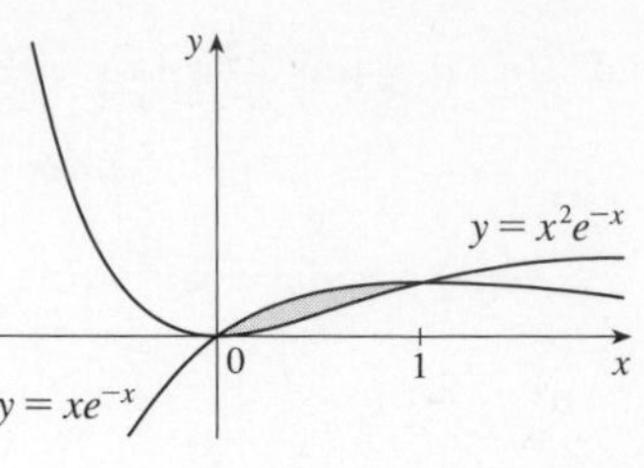

area $= \left[(x - x^2)(-e^{-x})\right]_0^1 - \int_0^1 [-e^{-x}(1-2x)]\,dx = 0 + \int_0^1 (1-2x)e^{-x}\,dx$.

Now let $U = 1 - 2x$, $dV = e^{-x}dx \Rightarrow dU = -2\,dx$, $V = -e^{-x}$. Now

area $= \left[(1-2x)(-e^{-x})\right]_0^1 - \int_0^1 2e^{-x}\,dx = e^{-1} + 1 - \left[-2e^{-x}\right]_0^1 = e^{-1} + 1 + 2(e^{-1} - 1) = 3e^{-1} - 1$.

59. The curves $y = \arcsin\left(\frac{1}{2}x\right)$ and $y = 2 - x^2$ intersect at $x = a \approx -1.75119$ and $x = b \approx 1.17210$. From the figure, the area bounded by the curves is given by

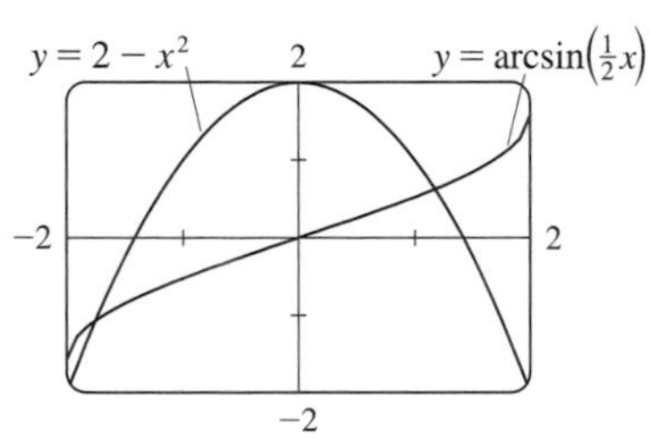

$A = \int_a^b [(2 - x^2) - \arcsin\left(\frac{1}{2}x\right)]\,dx = \left[2x - \frac{1}{3}x^3\right]_a^b - \int_a^b \arcsin\left(\frac{1}{2}x\right)dx$.

Let $u = \arcsin\left(\frac{1}{2}x\right)$, $dv = dx \Rightarrow du = \dfrac{1}{\sqrt{1 - \left(\frac{1}{2}x\right)^2}} \cdot \dfrac{1}{2}\,dx$, $v = x$.

Then

$$A = \left[2x - \frac{1}{3}x^3\right]_a^b - \left\{\left[x \arcsin\left(\frac{1}{2}x\right)\right]_a^b - \int_a^b \frac{x}{2\sqrt{1 - \frac{1}{4}x^2}}\,dx\right\}$$

$$= \left[2x - \tfrac{1}{3}x^3 - x\arcsin\left(\tfrac{1}{2}x\right) - 2\sqrt{1 - \tfrac{1}{4}x^2}\right]_a^b \approx 3.99926$$

60. The curves $y = x\ln(x+1)$ and $y = 3x - x^2$ intersect at $x = 0$ and $x = a \approx 1.92627$. From the figure, the area bounded by the curves is given by

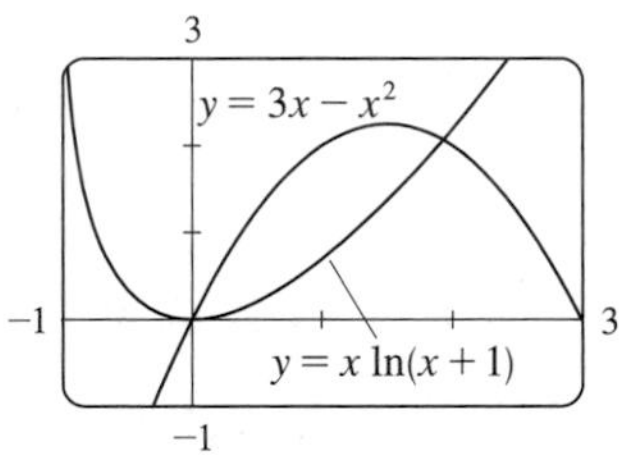

$A = \int_0^a [(3x - x^2) - x\ln(x+1)]\,dx = \left[\frac{3}{2}x^2 - \frac{1}{3}x^3\right]_0^a - \int_0^a x\ln(x+1)\,dx$.

Let $u = \ln(x+1)$, $dv = x\,dx \Rightarrow du = \dfrac{1}{x+1}\,dx$, $v = \frac{1}{2}x^2$. Then

$$A = \left[\frac{3}{2}x^2 - \frac{1}{3}x^3\right]_0^a - \left\{\left[\frac{1}{2}x^2\ln(x+1)\right]_0^a - \frac{1}{2}\int_0^a \frac{x^2}{x+1}\,dx\right\}$$

$$= \left[\frac{3}{2}x^2 - \frac{1}{3}x^3\right]_0^a - \left[\frac{1}{2}x^2\ln(x+1)\right]_0^a + \frac{1}{2}\int_0^a \left(x - 1 + \frac{1}{x+1}\right)dx$$

$$= \left[\tfrac{3}{2}x^2 - \tfrac{1}{3}x^3 - \tfrac{1}{2}x^2\ln(x+1) + \tfrac{1}{4}x^2 - \tfrac{1}{2}x + \tfrac{1}{2}\ln|x+1|\right]_0^a \approx 1.69260$$

61. Volume $= \int_0^1 2\pi x\cos(\pi x/2)\,dx$. Let $u = x$, $dv = \cos(\pi x/2)\,dx \Rightarrow du = dx$, $v = \frac{2}{\pi}\sin(\pi x/2)$.

$$V = 2\pi\left[\frac{2}{\pi}x\sin\left(\frac{\pi x}{2}\right)\right]_0^1 - 2\pi\cdot\frac{2}{\pi}\int_0^1 \sin\left(\frac{\pi x}{2}\right)dx = 2\pi\left(\frac{2}{\pi} - 0\right) - 4\left[-\frac{2}{\pi}\cos\left(\frac{\pi x}{2}\right)\right]_0^1 = 4 + \frac{8}{\pi}(0 - 1) = 4 - \frac{8}{\pi}.$$

62. Volume $= \int_0^1 2\pi x(e^x - e^{-x})\,dx = 2\pi\int_0^1 (xe^x - xe^{-x})\,dx = 2\pi\left[\int_0^1 xe^x\,dx - \int_0^1 xe^{-x}\,dx\right]$ [both integrals by parts]

$= 2\pi\left[(xe^x - e^x) - (-xe^{-x} - e^{-x})\right]_0^1 = 2\pi[2/e - 0] = 4\pi/e$

63. Volume $= \int_{-1}^{0} 2\pi(1-x)e^{-x}\,dx$. Let $u = 1-x$, $dv = e^{-x}\,dx \Rightarrow du = -\,dx$, $v = -e^{-x}$.

$$V = 2\pi\left[(1-x)(-e^{-x})\right]_{-1}^{0} - 2\pi\int_{-1}^{0} e^{-x}\,dx = 2\pi\left[(x-1)(e^{-x}) + e^{-x}\right]_{-1}^{0} = 2\pi\left[xe^{-x}\right]_{-1}^{0} = 2\pi(0+e) = 2\pi e.$$

64. $y = e^x \Leftrightarrow x = \ln y$. Volume $= \int_1^3 2\pi y \ln y\,dy$. Let $u = \ln y$, $dv = y\,dy \Rightarrow du = \dfrac{1}{y}\,dy$, $v = \frac{1}{2}y^2$.

$$\begin{aligned} V &= 2\pi\left[\tfrac{1}{2}y^2\ln y\right]_1^3 - 2\pi\int_1^3 \tfrac{1}{2}y\,dy = 2\pi\left[\tfrac{1}{2}y^2\ln y - \tfrac{1}{4}y^2\right]_1^3 \\ &= 2\pi\left[\left(\tfrac{9}{2}\ln 3 - \tfrac{9}{4}\right) - \left(0 - \tfrac{1}{4}\right)\right] = 2\pi\left(\tfrac{9}{2}\ln 3 - 2\right) = (9\ln 3 - 4)\,\pi \end{aligned}$$

65. (a) Use shells about the y-axis:

$$V = \int_1^2 2\pi x\ln x\,dx \qquad \begin{bmatrix} u = \ln x, & dv = x\,dx \\ du = \frac{1}{x}\,dx, & v = \frac{1}{2}x^2 \end{bmatrix}$$

$$= 2\pi\left\{\left[\tfrac{1}{2}x^2\ln x\right]_1^2 - \int_1^2 \tfrac{1}{2}x\,dx\right\} = 2\pi\left\{(2\ln 2 - 0) - \left[\tfrac{1}{4}x^2\right]_1^2\right\} = 2\pi\left(2\ln 2 - \tfrac{3}{4}\right)$$

(b) Use disks about the x-axis:

$$V = \int_1^2 \pi(\ln x)^2\,dx \qquad \begin{bmatrix} u = (\ln x)^2, & dv = dx \\ du = 2\ln x\cdot\frac{1}{x}\,dx, & v = x \end{bmatrix}$$

$$= \pi\left\{\left[x(\ln x)^2\right]_1^2 - \int_1^2 2\ln x\,dx\right\} \qquad \begin{bmatrix} u = \ln x, & dv = dx \\ du = \frac{1}{x}\,dx, & v = x \end{bmatrix}$$

$$= \pi\left\{2(\ln 2)^2 - 2\left(\left[x\ln x\right]_1^2 - \int_1^2 dx\right)\right\} = \pi\left\{2(\ln 2)^2 - 4\ln 2 + 2\left[x\right]_1^2\right\}$$

$$= \pi\left[2(\ln 2)^2 - 4\ln 2 + 2\right] = 2\pi\left[(\ln 2)^2 - 2\ln 2 + 1\right]$$

66.
$$f_{\text{ave}} = \frac{1}{b-a}\int_a^b f(x)\,dx = \frac{1}{\pi/4 - 0}\int_0^{\pi/4} x\sec^2 x\,dx \qquad \begin{bmatrix} u = x, & dv = \sec^2 x\,dx \\ du = dx, & v = \tan x \end{bmatrix}$$

$$= \frac{4}{\pi}\left\{\left[x\tan x\right]_0^{\pi/4} - \int_0^{\pi/4}\tan x\,dx\right\} = \frac{4}{\pi}\left\{\frac{\pi}{4} - \left[\ln|\sec x|\right]_0^{\pi/4}\right\} = \frac{4}{\pi}\left(\frac{\pi}{4} - \ln\sqrt{2}\right)$$

$$= 1 - \tfrac{4}{\pi}\ln\sqrt{2} \text{ or } 1 - \tfrac{2}{\pi}\ln 2$$

67. $S(x) = \displaystyle\int_0^x \sin\left(\tfrac{1}{2}\pi t^2\right)dt \Rightarrow \int S(x)\,dx = \int\left[\int_0^x \sin\left(\tfrac{1}{2}\pi t^2\right)dt\right]dx.$

Let $u = \displaystyle\int_0^x \sin\left(\tfrac{1}{2}\pi t^2\right)dt = S(x)$, $dv = dx \Rightarrow du = \sin\left(\tfrac{1}{2}\pi x^2\right)dx$, $v = x$. Thus,

$$\int S(x)\,dx = xS(x) - \int x\sin\left(\tfrac{1}{2}\pi x^2\right)dx = xS(x) - \int \sin y\left(\tfrac{1}{\pi}\,dy\right) \qquad \begin{bmatrix} u = \frac{1}{2}\pi x^2, \\ du = \pi x\,dx \end{bmatrix}$$

$$= xS(x) + \tfrac{1}{\pi}\cos y + C = xS(x) + \tfrac{1}{\pi}\cos\left(\tfrac{1}{2}\pi x^2\right) + C$$

68. The rocket will have height $H = \int_0^{60} v(t)\,dt$ after 60 seconds.

$$\begin{aligned} H &= \int_0^{60}\left[-gt - v_e\ln\left(\frac{m-rt}{m}\right)\right]dt = -g\left[\tfrac{1}{2}t^2\right]_0^{60} - v_e\left[\int_0^{60}\ln(m-rt)\,dt - \int_0^{60}\ln m\,dt\right] \\ &= -g(1800) + v_e(\ln m)(60) - v_e\int_0^{60}\ln(m-rt)\,dt \end{aligned}$$

Let $u = \ln(m-rt)$, $dv = dt \Rightarrow du = \dfrac{1}{m-rt}(-r)\,dt$, $v = t$. Then

$$\int_0^{60} \ln(m - rt)\,dt = \Big[t\ln(m - rt)\Big]_0^{60} + \int_0^{60} \frac{rt}{m - rt}\,dt = 60\ln(m - 60r) + \int_0^{60}\left(-1 + \frac{m}{m - rt}\right)dt$$

$$= 60\ln(m - 60r) + \left[-t - \frac{m}{r}\ln(m - rt)\right]_0^{60} = 60\ln(m - 60r) - 60 - \frac{m}{r}\ln(m - 60r) + \frac{m}{r}\ln m$$

So $H = -1800g + 60v_e \ln m - 60v_e \ln(m - 60r) + 60v_e + \dfrac{m}{r}v_e \ln(m - 60r) - \dfrac{m}{r}v_e \ln m$. Substituting $g = 9.8$, $m = 30{,}000$, $r = 160$, and $v_e = 3000$ gives us $H \approx 14{,}844$ m.

69. Since $v(t) > 0$ for all t, the desired distance is $s(t) = \int_0^t v(w)\,dw = \int_0^t w^2 e^{-w}\,dw$.

First let $u = w^2$, $dv = e^{-w}\,dw \;\Rightarrow\; du = 2w\,dw$, $v = -e^{-w}$. Then $s(t) = \left[-w^2 e^{-w}\right]_0^t + 2\int_0^t we^{-w}\,dw$.

Next let $U = w$, $dV = e^{-w}\,dw \;\Rightarrow\; dU = dw$, $V = -e^{-w}$. Then

$$s(t) = -t^2e^{-t} + 2\left(\left[-we^{-w}\right]_0^t + \int_0^t e^{-w}\,dw\right) = -t^2e^{-t} + 2\left(-te^{-t} + 0 + \left[-e^{-w}\right]_0^t\right)$$

$$= -t^2e^{-t} + 2(-te^{-t} - e^{-t} + 1) = -t^2e^{-t} - 2te^{-t} - 2e^{-t} + 2 = 2 - e^{-t}(t^2 + 2t + 2) \text{ meters}$$

70. Suppose $f(0) = g(0) = 0$ and let $u = f(x)$, $dv = g''(x)\,dx \;\Rightarrow\; du = f'(x)\,dx$, $v = g'(x)$.

Then $\int_0^a f(x)\,g''(x)\,dx = \Big[f(x)\,g'(x)\Big]_0^a - \int_0^a f'(x)\,g'(x)\,dx = f(a)\,g'(a) - \int_0^a f'(x)\,g'(x)\,dx$.

Now let $U = f'(x)$, $dV = g'(x)\,dx \;\Rightarrow\; dU = f''(x)\,dx$ and $V = g(x)$, so

$\int_0^a f'(x)\,g'(x)\,dx = \Big[f'(x)\,g(x)\Big]_0^a - \int_0^a f''(x)\,g(x)\,dx = f'(a)\,g(a) - \int_0^a f''(x)\,g(x)\,dx$.

Combining the two results, we get $\int_0^a f(x)\,g''(x)\,dx = f(a)\,g'(a) - f'(a)\,g(a) + \int_0^a f''(x)\,g(x)\,dx$.

71. For $I = \int_1^4 xf''(x)\,dx$, let $u = x$, $dv = f''(x)\,dx \;\Rightarrow\; du = dx$, $v = f'(x)$. Then

$I = \left[xf'(x)\right]_1^4 - \int_1^4 f'(x)\,dx = 4f'(4) - 1\cdot f'(1) - [f(4) - f(1)] = 4\cdot 3 - 1\cdot 5 - (7 - 2) = 12 - 5 - 5 = 2$.

We used the fact that f'' is continuous to guarantee that I exists.

72. (a) Take $g(x) = x$ and $g'(x) = 1$ in Equation 1.

(b) By part (a), $\int_a^b f(x)\,dx = bf(b) - a\,f(a) - \int_a^b x\,f'(x)\,dx$. Now let $y = f(x)$, so that $x = g(y)$ and $dy = f'(x)\,dx$. Then $\int_a^b x\,f'(x)\,dx = \int_{f(a)}^{f(b)} g(y)\,dy$. The result follows.

(c) Part (b) says that the area of region $ABFC$ is

$$= \quad bf(b) \quad - \quad af(a) \quad - \quad \int_{f(a)}^{f(b)} g(y)\,dy$$

$= $ (area of rectangle $OBFE$) $-$ (area of rectangle $OACD$) $-$ (area of region $DCFE$)

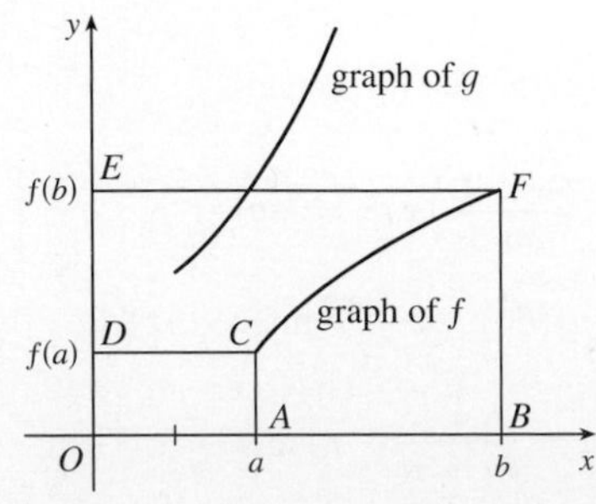

(d) We have $f(x) = \ln x$, so $f^{-1}(x) = e^x$, and since $g = f^{-1}$, we have $g(y) = e^y$. By part (b),

$$\int_1^e \ln x\,dx = e\ln e - 1\ln 1 - \int_{\ln 1}^{\ln e} e^y\,dy = e - \int_0^1 e^y\,dy = e - \left[e^y\right]_0^1 = e - (e-1) = 1.$$

73. Using the formula for volumes of rotation and the figure, we see that

Volume $= \int_0^d \pi b^2\,dy - \int_0^c \pi a^2\,dy - \int_c^d \pi[g(y)]^2\,dy = \pi b^2 d - \pi a^2 c - \int_c^d \pi[g(y)]^2\,dy$. Let $y = f(x)$, which gives $dy = f'(x)\,dx$ and $g(y) = x$, so that $V = \pi b^2 d - \pi a^2 c - \pi\int_a^b x^2 f'(x)\,dx$.

Now integrate by parts with $u = x^2$, and $dv = f'(x)\,dx \;\Rightarrow\; du = 2x\,dx$, $v = f(x)$, and

$\int_a^b x^2 f'(x)\,dx = \left[x^2 f(x)\right]_a^b - \int_a^b 2x\,f(x)\,dx = b^2 f(b) - a^2 f(a) - \int_a^b 2x\,f(x)\,dx$, but $f(a) = c$ and $f(b) = d \;\Rightarrow$

$V = \pi b^2 d - \pi a^2 c - \pi\left[b^2 d - a^2 c - \int_a^b 2x\,f(x)\,dx\right] = \int_a^b 2\pi x\,f(x)\,dx$.

74. (a) We note that for $0 \le x \le \frac{\pi}{2}$, $0 \le \sin x \le 1$, so $\sin^{2n+2} x \le \sin^{2n+1} x \le \sin^{2n} x$. So by the second Comparison Property of the Integral, $I_{2n+2} \le I_{2n+1} \le I_{2n}$.

(b) Substituting directly into the result from Exercise 50, we get

$$\frac{I_{2n+2}}{I_{2n}} = \frac{\dfrac{1\cdot 3\cdot 5\cdot\cdots\cdot[2(n+1)-1]}{2\cdot 4\cdot 6\cdot\cdots\cdot[2(n+1)]}\dfrac{\pi}{2}}{\dfrac{1\cdot 3\cdot 5\cdot\cdots\cdot(2n-1)}{2\cdot 4\cdot 6\cdot\cdots\cdot(2n)}\dfrac{\pi}{2}} = \frac{2(n+1)-1}{2(n+1)} = \frac{2n+1}{2n+2}$$

(c) We divide the result from part (a) by I_{2n}. The inequalities are preserved since I_{2n} is positive: $\dfrac{I_{2n+2}}{I_{2n}} \le \dfrac{I_{2n+1}}{I_{2n}} \le \dfrac{I_{2n}}{I_{2n}}$.

Now from part (b), the left term is equal to $\dfrac{2n+1}{2n+2}$, so the expression becomes $\dfrac{2n+1}{2n+2} \le \dfrac{I_{2n+1}}{I_{2n}} \le 1$. Now $\lim\limits_{n\to\infty} \dfrac{2n+1}{2n+2} = \lim\limits_{n\to\infty} 1 = 1$, so by the Squeeze Theorem, $\lim\limits_{n\to\infty} \dfrac{I_{2n+1}}{I_{2n}} = 1$.

(d) We substitute the results from Exercises 49 and 50 into the result from part (c):

$$1 = \lim_{n\to\infty} \frac{I_{2n+1}}{I_{2n}} = \lim_{n\to\infty} \frac{\dfrac{2\cdot 4\cdot 6\cdot\cdots\cdot(2n)}{3\cdot 5\cdot 7\cdot\cdots\cdot(2n+1)}}{\dfrac{1\cdot 3\cdot 5\cdot\cdots\cdot(2n-1)}{2\cdot 4\cdot 6\cdot\cdots\cdot(2n)}\dfrac{\pi}{2}} = \lim_{n\to\infty} \left[\frac{2\cdot 4\cdot 6\cdot\cdots\cdot(2n)}{3\cdot 5\cdot 7\cdot\cdots\cdot(2n+1)}\right]\left[\frac{2\cdot 4\cdot 6\cdot\cdots\cdot(2n)}{1\cdot 3\cdot 5\cdot\cdots\cdot(2n-1)}\left(\frac{2}{\pi}\right)\right]$$

$$= \lim_{n\to\infty} \frac{2}{1}\cdot\frac{2}{3}\cdot\frac{4}{3}\cdot\frac{4}{5}\cdot\frac{6}{5}\cdot\frac{6}{7}\cdot\cdots\cdot\frac{2n}{2n-1}\cdot\frac{2n}{2n+1}\cdot\frac{2}{\pi} \qquad \text{[rearrange terms]}$$

Multiplying both sides by $\frac{\pi}{2}$ gives us the *Wallis product*:

$$\frac{\pi}{2} = \frac{2}{1}\cdot\frac{2}{3}\cdot\frac{4}{3}\cdot\frac{4}{5}\cdot\frac{6}{5}\cdot\frac{6}{7}\cdot\cdots$$

(e) The area of the kth rectangle is k. At the $2n$th step, the area is increased from $2n-1$ to $2n$ by multiplying the width by $\dfrac{2n}{2n-1}$, and at the $(2n+1)$th step, the area is increased from $2n$ to $2n+1$ by multiplying the height by $\dfrac{2n+1}{2n}$. These two steps multiply the ratio of width to height by $\dfrac{2n}{2n-1}$ and $\dfrac{1}{(2n+1)/(2n)} = \dfrac{2n}{2n+1}$ respectively. So, by part (d), the limiting ratio is $\dfrac{2}{1}\cdot\dfrac{2}{3}\cdot\dfrac{4}{3}\cdot\dfrac{4}{5}\cdot\dfrac{6}{5}\cdot\dfrac{6}{7}\cdot\cdots = \dfrac{\pi}{2}$.

7.2 Trigonometric Integrals

The symbols $\overset{s}{=}$ and $\overset{c}{=}$ indicate the use of the substitutions $\{u = \sin x, du = \cos x\,dx\}$ and $\{u = \cos x, du = -\sin x\,dx\}$, respectively.

1. $\int \sin^2 x \cos^3 x\,dx = \int \sin^2 x \cos^2 x \cos x\,dx = \int \sin^2 x\,(1 - \sin^2 x)\cos x\,dx$

$$\overset{s}{=} \int u^2(1-u^2)\,du = \int (u^2 - u^4)\,du = \tfrac{1}{3}u^3 - \tfrac{1}{5}u^5 + C = \tfrac{1}{3}\sin^3 x - \tfrac{1}{5}\sin^5 x + C$$

2. $\int \sin^3\theta \cos^4\theta\,d\theta = \int \sin^2\theta\cos^4\theta\sin\theta\,d\theta = \int (1-\cos^2\theta)\cos^4\theta\sin\theta\,d\theta$

$$\overset{c}{=} \int (1-u^2)u^4(-du) = \int (u^6 - u^4)\,du = \tfrac{1}{7}u^7 - \tfrac{1}{5}u^5 + C = \tfrac{1}{7}\cos^7\theta - \tfrac{1}{5}\cos^5\theta + C$$

3. $\int_0^{\pi/2} \sin^7\theta\cos^5\theta\,d\theta = \int_0^{\pi/2}\sin^7\theta\cos^4\theta\cos\theta\,d\theta = \int_0^{\pi/2}\sin^7\theta\,(1-\sin^2\theta)^2\cos\theta\,d\theta$

$$\overset{s}{=} \int_0^1 u^7(1-u^2)^2\,du = \int_0^1 u^7(1-2u^2+u^4)\,du = \int_0^1 (u^7 - 2u^9 + u^{11})\,du$$

$$= \left[\frac{1}{8}u^8 - \frac{1}{5}u^{10} + \frac{1}{12}u^{12}\right]_0^1 = \left(\frac{1}{8} - \frac{1}{5} + \frac{1}{12}\right) - 0 = \frac{15 - 24 + 10}{120} = \frac{1}{120}$$

4. $\int_0^{\pi/2}\sin^5 x\,dx = \int_0^{\pi/2}\sin^4 x\sin x\,dx = \int_0^{\pi/2}(1-\cos^2 x)^2\sin x\,dx \overset{c}{=} \int_1^0 (1-u^2)^2(-du)$

$$= \int_0^1 (1 - 2u^2 + u^4)\,du = \left[u - \frac{2}{3}u^3 + \frac{1}{5}u^5\right]_0^1 = \left(1 - \frac{2}{3} + \frac{1}{5}\right) - 0 = \frac{15 - 10 + 3}{15} = \frac{8}{15}$$

5. $\int \sin^5(2t)\cos^2(2t)\,dt = \int \sin^4(2t)\cos^2(2t)\sin(2t)\,dt = \int [1-\cos^2(2t)]^2\cos^2(2t)\sin(2t)\,dt$

$$= \int (1-u^2)^2 u^2\left(-\tfrac{1}{2}\,du\right) \qquad [u = \cos(2t),\ du = -2\sin(2t)\,dt]$$

$$= -\tfrac{1}{2}\int (u^4 - 2u^2 + 1)u^2\,du = -\tfrac{1}{2}\int (u^6 - 2u^4 + u^2)\,du$$

$$= -\tfrac{1}{2}\left(\tfrac{1}{7}u^7 - \tfrac{2}{5}u^5 + \tfrac{1}{3}u^3\right) + C = -\tfrac{1}{14}\cos^7(2t) + \tfrac{1}{5}\cos^5(2t) - \tfrac{1}{6}\cos^3(2t) + C$$

6. $\int t\cos^5(t^2)\,dt = \int t\cos^4(t^2)\cos(t^2)\,dt = \int t[1 - \sin^2(t^2)]^2\cos(t^2)\,dt$

$$= \int \tfrac{1}{2}(1-u^2)^2\,du \qquad [u = \sin(t^2),\ du = 2t\cos(t^2)\,dt]$$

$$= \tfrac{1}{2}\int (u^4 - 2u^2 + 1)\,du = \tfrac{1}{2}\left(\tfrac{1}{5}u^5 - \tfrac{2}{3}u^3 + u\right) + C = \tfrac{1}{10}\sin^5(t^2) - \tfrac{1}{3}\sin^3(t^2) + \tfrac{1}{2}\sin(t^2) + C$$

7. $\int_0^{\pi/2}\cos^2\theta\,d\theta = \int_0^{\pi/2}\tfrac{1}{2}(1 + \cos 2\theta)\,d\theta$ [half-angle identity]

$$= \tfrac{1}{2}\left[\theta + \tfrac{1}{2}\sin 2\theta\right]_0^{\pi/2} = \tfrac{1}{2}\left[\left(\tfrac{\pi}{2} + 0\right) - (0 + 0)\right] = \tfrac{\pi}{4}$$

8. $\int_0^{2\pi}\sin^2\left(\tfrac{1}{3}\theta\right)d\theta = \int_0^{2\pi}\tfrac{1}{2}\left[1 - \cos\left(2\cdot\tfrac{1}{3}\theta\right)\right]d\theta$ [half-angle identity]

$$= \frac{1}{2}\left[\theta - \frac{3}{2}\sin\left(\frac{2}{3}\theta\right)\right]_0^{2\pi} = \frac{1}{2}\left[\left(2\pi - \frac{3}{2}\left(-\frac{\sqrt{3}}{2}\right)\right) - 0\right] = \pi + \frac{3}{8}\sqrt{3}$$

9. $\int_0^{\pi}\cos^4(2t)\,dt = \int_0^{\pi}[\cos^2(2t)]^2\,dt = \int_0^{\pi}\left[\tfrac{1}{2}(1 + \cos(2\cdot 2t))\right]^2 dt$ [half-angle identity]

$$= \tfrac{1}{4}\int_0^{\pi}[1 + 2\cos 4t + \cos^2(4t)]\,dt = \tfrac{1}{4}\int_0^{\pi}\left[1 + 2\cos 4t + \tfrac{1}{2}(1 + \cos 8t)\right]dt$$

$$= \tfrac{1}{4}\int_0^{\pi}\left(\tfrac{3}{2} + 2\cos 4t + \tfrac{1}{2}\cos 8t\right)dt = \tfrac{1}{4}\left[\tfrac{3}{2}t + \tfrac{1}{2}\sin 4t + \tfrac{1}{16}\sin 8t\right]_0^{\pi} = \tfrac{1}{4}\left[\left(\tfrac{3}{2}\pi + 0 + 0\right) - 0\right] = \tfrac{3}{8}\pi$$

10. $\int_0^{\pi}\sin^2 t\cos^4 t\,dt = \tfrac{1}{4}\int_0^{\pi}(4\sin^2 t\cos^2 t)\cos^2 t\,dt = \tfrac{1}{4}\int_0^{\pi}(2\sin t\cos t)^2\,\tfrac{1}{2}(1 + \cos 2t)\,dt$

$$= \tfrac{1}{8}\int_0^{\pi}(\sin 2t)^2(1 + \cos 2t)\,dt = \tfrac{1}{8}\int_0^{\pi}(\sin^2 2t + \sin^2 2t\cos 2t)\,dt$$

$$= \tfrac{1}{8}\int_0^{\pi}\sin^2 2t\,dt + \tfrac{1}{8}\int_0^{\pi}\sin^2 2t\cos 2t\,dt = \tfrac{1}{8}\int_0^{\pi}\tfrac{1}{2}(1 - \cos 4t)\,dt + \tfrac{1}{8}\left[\tfrac{1}{3}\cdot\tfrac{1}{2}\sin^3 2t\right]_0^{\pi}$$

$$= \tfrac{1}{16}\left[t - \tfrac{1}{4}\sin 4t\right]_0^{\pi} + \tfrac{1}{8}(0 - 0) = \tfrac{1}{16}[(\pi - 0) - 0] = \tfrac{\pi}{16}$$

11. $\int_0^{\pi/2} \sin^2 x \cos^2 x\,dx = \int_0^{\pi/2} \frac{1}{4}(4\sin^2 x \cos^2 x)\,dx = \int_0^{\pi/2} \frac{1}{4}(2\sin x \cos x)^2 dx = \frac{1}{4}\int_0^{\pi/2} \sin^2 2x\,dx$

$= \frac{1}{4}\int_0^{\pi/2} \frac{1}{2}(1-\cos 4x)\,dx = \frac{1}{8}\int_0^{\pi/2}(1-\cos 4x)\,dx = \frac{1}{8}\left[x - \frac{1}{4}\sin 4x\right]_0^{\pi/2} = \frac{1}{8}\left(\frac{\pi}{2}\right) = \frac{\pi}{16}$

12. $\int_0^{\pi/2}(2-\sin\theta)^2\,d\theta = \int_0^{\pi/2}(4-4\sin\theta+\sin^2\theta)\,d\theta = \int_0^{\pi/2}\left[4-4\sin\theta+\frac{1}{2}(1-\cos 2\theta)\right]d\theta$

$= \int_0^{\pi/2}\left(\frac{9}{2}-4\sin\theta-\frac{1}{2}\cos 2\theta\right)d\theta = \left[\frac{9}{2}\theta+4\cos\theta-\frac{1}{4}\sin 2\theta\right]_0^{\pi/2}$

$= \left(\frac{9\pi}{4}+0-0\right)-(0+4-0) = \frac{9}{4}\pi - 4$

13. $\int \sqrt{\cos\theta}\sin^3\theta\,d\theta = \int\sqrt{\cos\theta}\sin^2\theta\,\sin\theta\,d\theta = \int(\cos\theta)^{1/2}(1-\cos^2\theta)\sin\theta\,d\theta$

$\overset{c}{=} \int u^{1/2}(1-u^2)\,(-du) = \int(u^{5/2}-u^{1/2})\,du$

$= \frac{2}{7}u^{7/2} - \frac{2}{3}u^{3/2} + C = \frac{2}{7}(\cos\theta)^{7/2} - \frac{2}{3}(\cos\theta)^{3/2} + C$

14. $\displaystyle\int \frac{\sin^2(1/t)}{t^2}\,dt = \int \sin^2 u\,(-du) \qquad \left[u = \frac{1}{t},\ du = -\frac{1}{t^2}\,dt\right]$

$\displaystyle = -\int\frac{1}{2}(1-\cos 2u)\,du = -\frac{1}{2}\left(u - \frac{1}{2}\sin 2u\right) + C = -\frac{1}{2t} + \frac{1}{4}\sin\left(\frac{2}{t}\right) + C$

15. $\displaystyle\int \cot x\cos^2 x\,dx = \int\frac{\cos x}{\sin x}(1-\sin^2 x)\,dx$

$\displaystyle \overset{s}{=} \int\frac{1-u^2}{u}\,du = \int\left(\frac{1}{u}-u\right)du = \ln|u| - \frac{1}{2}u^2 + C = \ln|\sin x| - \frac{1}{2}\sin^2 x + C$

16. $\displaystyle\int\tan^2 x\cos^3 x\,dx = \int\frac{\sin^2 x}{\cos^2 x}\cos^3 x\,dx = \int\sin^2 x\cos x\,dx \overset{s}{=} \int u^2\,du = \frac{1}{3}u^3 + C = \frac{1}{3}\sin^3 x + C$

17. $\int\sin^2 x\sin 2x\,dx = \int\sin^2 x\,(2\sin x\cos x)\,dx \overset{s}{=} \int 2u^3\,du = \frac{1}{2}u^4 + C = \frac{1}{2}\sin^4 x + C$

18. $\int\sin x\cos\left(\frac{1}{2}x\right)dx = \int\sin\left(2\cdot\frac{1}{2}x\right)\cos\left(\frac{1}{2}x\right)dx = \int 2\sin\left(\frac{1}{2}x\right)\cos^2\left(\frac{1}{2}x\right)dx$

$= \int 2u^2\,(-2\,du) \qquad \left[u = \cos\left(\frac{1}{2}x\right),\ du = -\frac{1}{2}\sin\left(\frac{1}{2}x\right)dx\right]$

$= -\frac{4}{3}u^3 + C = -\frac{4}{3}\cos^3\left(\frac{1}{2}x\right) + C$

19. $\int t\sin^2 t\,dt = \int t\left[\frac{1}{2}(1-\cos 2t)\right]dt = \frac{1}{2}\int(t - t\cos 2t)\,dt = \frac{1}{2}\int t\,dt - \frac{1}{2}\int t\cos 2t\,dt$

$= \frac{1}{2}\left(\frac{1}{2}t^2\right) - \frac{1}{2}\left(\frac{1}{2}t\sin 2t - \int\frac{1}{2}\sin 2t\,dt\right) \qquad \left[\begin{matrix}u = t, & dv = \cos 2t\,dt\\ du = dt, & v = \frac{1}{2}\sin 2t\end{matrix}\right]$

$= \frac{1}{4}t^2 - \frac{1}{4}t\sin 2t + \frac{1}{2}\left(-\frac{1}{4}\cos 2t\right) + C = \frac{1}{4}t^2 - \frac{1}{4}t\sin 2t - \frac{1}{8}\cos 2t + C$

20. $I = \int x\sin^3 x\,dx$. First, evaluate

$\int\sin^3 x\,dx = \int(1-\cos^2 x)\sin x\,dx \overset{c}{=} \int(1-u^2)(-du) = \int(u^2-1)\,du = \frac{1}{3}u^3 - u + C_1 = \frac{1}{3}\cos^3 x - \cos x + C_1.$

Now for I, let $u = x$, $dv = \sin^3 x \Rightarrow du = dx$, $v = \frac{1}{3}\cos^3 x - \cos x$, so

$I = \frac{1}{3}x\cos^3 x - x\cos x - \int\left(\frac{1}{3}\cos^3 x - \cos x\right)dx = \frac{1}{3}x\cos^3 x - x\cos x - \frac{1}{3}\int\cos^3 x\,dx + \sin x$

$= \frac{1}{3}x\cos^3 x - x\cos x - \frac{1}{3}\left(\sin x - \frac{1}{3}\sin^3 x\right) + \sin x + C$ [by Example 1]

$= \frac{1}{3}x\cos^3 x - x\cos x + \frac{2}{3}\sin x + \frac{1}{9}\sin^3 x + C$

21. $\int\tan x\sec^3 x\,dx = \int\tan x\sec x\sec^2 x\,dx = \int u^2\,du \qquad [u = \sec x,\ du = \sec x\tan x\,dx]$

$= \frac{1}{3}u^3 + C = \frac{1}{3}\sec^3 x + C$

22. $\int \tan^2\theta \sec^4\theta\, d\theta = \int \tan^2\theta \sec^2\theta \sec^2\theta\, d\theta = \int \tan^2\theta\,(\tan^2\theta + 1)\sec^2\theta\, d\theta$

$= \int u^2(u^2+1)\,du \quad [u = \tan\theta,\ du = \sec^2\theta\, d\theta]$

$= \int (u^4 + u^2)\,du = \frac{1}{5}u^5 + \frac{1}{3}u^3 + C = \frac{1}{5}\tan^5\theta + \frac{1}{3}\tan^3\theta + C$

23. $\int \tan^2 x\, dx = \int (\sec^2 x - 1)\,dx = \tan x - x + C$

24. $\int (\tan^2 x + \tan^4 x)\,dx = \int \tan^2 x\,(1+\tan^2 x)\,dx = \int \tan^2 x \sec^2 x\, dx = \int u^2\,du \quad [u = \tan x,\ du = \sec^2 x\, dx]$

$= \frac{1}{3}u^3 + C = \frac{1}{3}\tan^3 x + C$

25. Let $u = \tan x$. Then $du = \sec^2 x\, dx$, so

$$\begin{aligned}\int \tan^4 x \sec^6 x\, dx &= \int \tan^4 x \sec^4 x\,(\sec^2 x\, dx) = \int \tan^4 x(1+\tan^2 x)^2\,(\sec^2 x\, dx)\\ &= \int u^4(1+u^2)^2\,du = \int (u^8 + 2u^6 + u^4)\,du\\ &= \tfrac{1}{9}u^9 + \tfrac{2}{7}u^7 + \tfrac{1}{5}u^5 + C = \tfrac{1}{9}\tan^9 x + \tfrac{2}{7}\tan^7 x + \tfrac{1}{5}\tan^5 x + C\end{aligned}$$

26. $\int_0^{\pi/4} \sec^6\theta \tan^6\theta\, d\theta = \int_0^{\pi/4} \tan^6\theta \sec^4\theta \sec^2\theta\, d\theta = \int_0^{\pi/4} \tan^6\theta(1+\tan^2\theta)^2\sec^2\theta\, d\theta$

$= \int_0^1 u^6(1+u^2)^2\,du \quad \begin{bmatrix} u = \tan\theta, \\ du = \sec^2\theta\, d\theta \end{bmatrix}$

$= \int_0^1 u^6(u^4 + 2u^2 + 1)\,du = \int_0^1 (u^{10} + 2u^8 + u^6)\,du$

$= \left[\frac{1}{11}u^{11} + \frac{2}{9}u^9 + \frac{1}{7}u^7\right]_0^1 = \frac{1}{11} + \frac{2}{9} + \frac{1}{7} = \frac{63 + 154 + 99}{693} = \frac{316}{693}$

27. $\int \tan^3 x \sec x\, dx = \int \tan^2 x \sec x \tan x\, dx = \int (\sec^2 x - 1)\sec x \tan x\, dx$

$= \int (u^2 - 1)\,du \quad [u = \sec x,\ du = \sec x \tan x\, dx] \quad = \frac{1}{3}u^3 - u + C = \frac{1}{3}\sec^3 x - \sec x + C$

28. Let $u = \sec x$, so $du = \sec x \tan x\, dx$. Thus,

$$\begin{aligned}\int \tan^5 x \sec^3 x\, dx &= \int \tan^4 x \sec^2 x\,(\sec x \tan x)\,dx = \int (\sec^2 x - 1)^2 \sec^2 x\,(\sec x \tan x\, dx)\\ &= \int (u^2-1)^2 u^2\,du = \int (u^6 - 2u^4 + u^2)\,du\\ &= \tfrac{1}{7}u^7 - \tfrac{2}{5}u^5 + \tfrac{1}{3}u^3 + C = \tfrac{1}{7}\sec^7 x - \tfrac{2}{5}\sec^5 x + \tfrac{1}{3}\sec^3 x + C\end{aligned}$$

29. $\int \tan^3 x \sec^6 x\, dx = \int \tan^3 x \sec^4 x \sec^2 x\, dx = \int \tan^3 x\,(1+\tan^2 x)^2 \sec^2 x\, dx$

$= \int u^3(1+u^2)^2\,du \quad \begin{bmatrix} u = \tan x, \\ du = \sec^2 x\, dx \end{bmatrix}$

$= \int u^3(u^4 + 2u^2 + 1)\,du = \int (u^7 + 2u^5 + u^3)\,du$

$= \frac{1}{8}u^8 + \frac{1}{3}u^6 + \frac{1}{4}u^4 + C = \frac{1}{8}\tan^8 x + \frac{1}{3}\tan^6 x + \frac{1}{4}\tan^4 x + C$

30. $\int_0^{\pi/4} \tan^4 t\, dt = \int_0^{\pi/4} \tan^2 t\,(\sec^2 t - 1)\,dt = \int_0^{\pi/4} \tan^2 t \sec^2 t\, dt - \int_0^{\pi/4} \tan^2 t\, dt$

$= \int_0^1 u^2\,du \ [u = \tan t] \ - \int_0^{\pi/4} (\sec^2 t - 1)\,dt = \left[\frac{1}{3}u^3\right]_0^1 - \Big[\tan t - t\Big]_0^{\pi/4}$

$= \frac{1}{3} - \left[\left(1 - \frac{\pi}{4}\right) - 0\right] = \frac{\pi}{4} - \frac{2}{3}$

31. $\int \tan^5 x\, dx = \int (\sec^2 x - 1)^2 \tan x\, dx = \int \sec^4 x \tan x\, dx - 2\int \sec^2 x \tan x\, dx + \int \tan x\, dx$

$= \int \sec^3 x \sec x \tan x\, dx - 2\int \tan x \sec^2 x\, dx + \int \tan x\, dx$

$= \frac{1}{4}\sec^4 x - \tan^2 x + \ln|\sec x| + C \quad [\text{or } \frac{1}{4}\sec^4 x - \sec^2 x + \ln|\sec x| + C]$

32. $\int \tan^2 x \sec x\,dx = \int (\sec^2 x - 1)\sec x\,dx = \int \sec^3 x\,dx - \int \sec x\,dx$

$$= \tfrac{1}{2}(\sec x\tan x + \ln|\sec x + \tan x|) - \ln|\sec x + \tan x| + C \qquad \text{[by Example 8 and (1)]}$$

$$= \tfrac{1}{2}(\sec x\tan x - \ln|\sec x + \tan x|) + C$$

33. Let $u = x$, $dv = \sec x\tan x\,dx \Rightarrow du = dx$, $v = \sec x$. Then

$\int x\sec x\tan x\,dx = x\sec x - \int \sec x\,dx = x\sec x - \ln|\sec x + \tan x| + C.$

34. $$\int \frac{\sin\phi}{\cos^3\phi}\,d\phi = \int \frac{\sin\phi}{\cos\phi}\cdot\frac{1}{\cos^2\phi}\,d\phi = \int \tan\phi\sec^2\phi\,d\phi = \int u\,du \qquad [u = \tan\phi,\ du = \sec^2\phi\,d\phi]$$

$$= \tfrac{1}{2}u^2 + C = \tfrac{1}{2}\tan^2\phi + C$$

Alternate solution: Let $u = \cos\phi$ to get $\frac{1}{2}\sec^2\phi + C$.

35. $\int_{\pi/6}^{\pi/2}\cot^2 x\,dx = \int_{\pi/6}^{\pi/2}(\csc^2 x - 1)\,dx = \left[-\cot x - x\right]_{\pi/6}^{\pi/2} = \left(0 - \frac{\pi}{2}\right) - \left(-\sqrt{3} - \frac{\pi}{6}\right) = \sqrt{3} - \frac{\pi}{3}$

36. $$\int_{\pi/4}^{\pi/2}\cot^3 x\,dx = \int_{\pi/4}^{\pi/2}\cot x(\csc^2 x - 1)\,dx = \int_{\pi/4}^{\pi/2}\cot x\csc^2 x\,dx - \int_{\pi/4}^{\pi/2}\frac{\cos x}{\sin x}\,dx$$

$$= \left[-\tfrac{1}{2}\cot^2 x - \ln|\sin x|\right]_{\pi/4}^{\pi/2} = (0 - \ln 1) - \left[-\tfrac{1}{2} - \ln\tfrac{1}{\sqrt{2}}\right] = \tfrac{1}{2} + \ln\tfrac{1}{\sqrt{2}} = \tfrac{1}{2}(1 - \ln 2)$$

37. $\int_{\pi/4}^{\pi/2}\cot^5\phi\csc^3\phi\,d\phi = \int_{\pi/4}^{\pi/2}\cot^4\phi\csc^2\phi\csc\phi\cot\phi\,d\phi = \int_{\pi/4}^{\pi/2}(\csc^2\phi - 1)^2\csc^2\phi\csc\phi\cot\phi\,d\phi$

$$= \int_{\sqrt{2}}^{1}(u^2 - 1)^2u^2(-du) \qquad [u = \csc\phi,\ du = -\csc\phi\cot\phi\,d\phi]$$

$$= \int_{1}^{\sqrt{2}}(u^6 - 2u^4 + u^2)\,du = \left[\tfrac{1}{7}u^7 - \tfrac{2}{5}u^5 + \tfrac{1}{3}u^3\right]_1^{\sqrt{2}} = \left(\tfrac{8}{7}\sqrt{2} - \tfrac{8}{5}\sqrt{2} + \tfrac{2}{3}\sqrt{2}\right) - \left(\tfrac{1}{7} - \tfrac{2}{5} + \tfrac{1}{3}\right)$$

$$= \frac{120 - 168 + 70}{105}\sqrt{2} - \frac{15 - 42 + 35}{105} = \frac{22}{105}\sqrt{2} - \frac{8}{105}$$

38. $\int_{\pi/4}^{\pi/2}\csc^4\theta\cot^4\theta\,d\theta = \int_{\pi/4}^{\pi/2}\cot^4\theta\csc^2\theta\csc^2\theta\,d\theta = \int_{\pi/4}^{\pi/2}\cot^4\theta(\cot^2\theta + 1)\csc^2\theta\,d\theta$

$$= \int_1^0 u^4(u^2 + 1)(-du) \qquad \begin{bmatrix} u = \cot\theta, \\ du = -\csc^2\theta\,d\theta \end{bmatrix}$$

$$= \int_0^1 (u^6 + u^4)\,du$$

$$= \left[\tfrac{1}{7}u^7 + \tfrac{1}{5}u^5\right]_0^1 = \tfrac{1}{7} + \tfrac{1}{5} = \tfrac{12}{35}$$

39. $I = \int \csc x\,dx = \int \dfrac{\csc x(\csc x - \cot x)}{\csc x - \cot x}\,dx = \int \dfrac{-\csc x\cot x + \csc^2 x}{\csc x - \cot x}\,dx$. Let $u = \csc x - \cot x \Rightarrow$

$du = (-\csc x\cot x + \csc^2 x)\,dx$. Then $I = \int du/u = \ln|u| = \ln|\csc x - \cot x| + C$.

40. Let $u = \csc x$, $dv = \csc^2 x\,dx$. Then $du = -\csc x\cot x\,dx$, $v = -\cot x \Rightarrow$

$$\int \csc^3 x\,dx = -\csc x\cot x - \int \csc x\cot^2 x\,dx = -\csc x\cot x - \int \csc x(\csc^2 x - 1)\,dx$$

$$= -\csc x\cot x + \int \csc x\,dx - \int \csc^3 x\,dx$$

Solving for $\int \csc^3 x\,dx$ and using Exercise 39, we get

$\int \csc^3 x\,dx = -\frac{1}{2}\csc x\,\cot x + \frac{1}{2}\int \csc x\,dx = -\frac{1}{2}\csc x\,\cot x + \frac{1}{2}\ln|\csc x - \cot x| + C$. Thus,

$$\begin{aligned}\int_{\pi/6}^{\pi/3} \csc^3 x\,dx &= \left[-\tfrac{1}{2}\csc x\,\cot x + \tfrac{1}{2}\ln|\csc x - \cot x|\right]_{\pi/6}^{\pi/3} \\ &= -\tfrac{1}{2}\cdot\tfrac{2}{\sqrt{3}}\cdot\tfrac{1}{\sqrt{3}} + \tfrac{1}{2}\ln\left|\tfrac{2}{\sqrt{3}} - \tfrac{1}{\sqrt{3}}\right| + \tfrac{1}{2}\cdot 2\cdot\sqrt{3} - \tfrac{1}{2}\ln\left|2 - \sqrt{3}\right| \\ &= -\tfrac{1}{3} + \sqrt{3} + \tfrac{1}{2}\ln\tfrac{1}{\sqrt{3}} - \tfrac{1}{2}\ln(2 - \sqrt{3}) \approx 1.7825\end{aligned}$$

41. $\int \sin 8x\,\cos 5x\,dx \overset{2a}{=} \int \frac{1}{2}[\sin(8x - 5x) + \sin(8x + 5x)]\,dx = \frac{1}{2}\int (\sin 3x + \sin 13x)\,dx$

$= \frac{1}{2}(-\frac{1}{3}\cos 3x - \frac{1}{13}\cos 13x) + C = -\frac{1}{6}\cos 3x - \frac{1}{26}\cos 13x + C$

42. $\int \sin 2\theta\,\sin 6\theta\,d\theta \overset{2b}{=} \int \frac{1}{2}[\cos(2\theta - 6\theta) - \cos(2\theta + 6\theta)]\,d\theta$

$= \frac{1}{2}\int [\cos(-4\theta) - \cos 8\theta]\,d\theta = \frac{1}{2}\int (\cos 4\theta - \cos 8\theta)\,d\theta$

$= \frac{1}{2}\left(\frac{1}{4}\sin 4\theta - \frac{1}{8}\sin 8\theta\right) + C = \frac{1}{8}\sin 4\theta - \frac{1}{16}\sin 8\theta + C$

43. $\int_0^{\pi/2} \cos 5t\,\cos 10t\,dt \overset{2c}{=} \int_0^{\pi/2} \frac{1}{2}[\cos(5t - 10t) + \cos(5t + 10t)]\,dt$

$= \frac{1}{2}\int_0^{\pi/2} [\cos(-5t) + \cos 15t]\,dt = \frac{1}{2}\int_0^{\pi/2} (\cos 5t + \cos 15t)\,dt$

$= \frac{1}{2}\left[\frac{1}{5}\sin 5t + \frac{1}{15}\sin 15t\right]_0^{\pi/2} = \frac{1}{2}\left(\frac{1}{5} - \frac{1}{15}\right) = \frac{1}{15}$

44. $\displaystyle\int \sin x\,\sec^5 x\,dx = \int \frac{\sin x}{\cos^5 x}\,dx \overset{c}{=} \int \frac{1}{u^5}\,(-du) = \frac{1}{4u^4} + C = \frac{1}{4\cos^4 x} + C = \frac{1}{4}\sec^4 x + C$

45. $\int_0^{\pi/6} \sqrt{1 + \cos 2x}\,dx = \int_0^{\pi/6} \sqrt{1 + (2\cos^2 x - 1)}\,dx = \int_0^{\pi/6} \sqrt{2\cos^2 x}\,dx = \sqrt{2}\int_0^{\pi/6} \sqrt{\cos^2 x}\,dx$

$= \sqrt{2}\int_0^{\pi/6} |\cos x|\,dx = \sqrt{2}\int_0^{\pi/6} \cos x\,dx \qquad$ [since $\cos x > 0$ for $0 \leq x \leq \pi/6$]

$= \sqrt{2}\left[\sin x\right]_0^{\pi/6} = \sqrt{2}\left(\frac{1}{2} - 0\right) = \frac{1}{2}\sqrt{2}$

46. $\int_0^{\pi/4} \sqrt{1 - \cos 4\theta}\,d\theta = \int_0^{\pi/4} \sqrt{1 - (1 - 2\sin^2(2\theta))}\,d\theta = \int_0^{\pi/4} \sqrt{2\sin^2(2\theta)}\,d\theta = \sqrt{2}\int_0^{\pi/4} \sqrt{\sin^2(2\theta)}\,d\theta$

$= \sqrt{2}\int_0^{\pi/4} |\sin 2\theta|\,d\theta = \sqrt{2}\int_0^{\pi/4} \sin 2\theta\,d\theta \qquad$ [since $\sin 2\theta \geq 0$ for $0 \leq \theta \leq \pi/4$]

$= \sqrt{2}\left[-\frac{1}{2}\cos 2\theta\right]_0^{\pi/4} = -\frac{1}{2}\sqrt{2}\,(0 - 1) = \frac{1}{2}\sqrt{2}$

47. $\displaystyle\int \frac{1 - \tan^2 x}{\sec^2 x}\,dx = \int \left(\cos^2 x - \sin^2 x\right)dx = \int \cos 2x\,dx = \frac{1}{2}\sin 2x + C$

48. $\displaystyle\int \frac{dx}{\cos x - 1} = \int \frac{1}{\cos x - 1}\cdot\frac{\cos x + 1}{\cos x + 1}\,dx = \int \frac{\cos x + 1}{\cos^2 x - 1}\,dx = \int \frac{\cos x + 1}{-\sin^2 x}\,dx$

$= \int \left(-\cot x\,\csc x - \csc^2 x\right)dx = \csc x + \cot x + C$

49. $\int x\tan^2 x\,dx = \int x(\sec^2 x - 1)\,dx = \int x\sec^2 x\,dx - \int x\,dx$

$= x\tan x - \int \tan x\,dx - \frac{1}{2}x^2 \qquad \begin{bmatrix} u = x, & dv = \sec^2 x\,dx \\ du = dx, & v = \tan x \end{bmatrix}$

$= x\tan x - \ln|\sec x| - \frac{1}{2}x^2 + C$

50. Let $u = \tan^7 x$, $dv = \sec x \tan x\,dx \quad \Rightarrow \quad du = 7\tan^6 x \sec^2 x\,dx$, $v = \sec x$. Then

$$\begin{aligned}\int \tan^8 x \sec x\,dx &= \int \tan^7 x \cdot \sec x \tan x\,dx = \tan^7 x \sec x - \int 7\tan^6 x \sec^2 x \sec x\,dx \\ &= \tan^7 x \sec x - 7\int \tan^6 x\,(\tan^2 x + 1)\sec x\,dx \\ &= \tan^7 x \sec x - 7\int \tan^8 x \sec x\,dx - 7\int \tan^6 x \sec x\,dx.\end{aligned}$$

Thus, $8\int \tan^8 x \sec x\,dx = \tan^7 x \sec x - 7\int \tan^6 x \sec x\,dx$ and

$$\int_0^{\pi/4} \tan^8 x \sec x\,dx = \frac{1}{8}\big[\tan^7 x \sec x\big]_0^{\pi/4} - \frac{7}{8}\int_0^{\pi/4} \tan^6 x \sec x\,dx = \frac{\sqrt{2}}{8} - \frac{7}{8}I.$$

In Exercises 51–54, let $f(x)$ denote the integrand and $F(x)$ its antiderivative (with $C = 0$).

51. Let $u = x^2$, so that $du = 2x\,dx$. Then

$$\begin{aligned}\int x\sin^2(x^2)\,dx &= \int \sin^2 u\left(\tfrac{1}{2}\,du\right) = \tfrac{1}{2}\int \tfrac{1}{2}(1 - \cos 2u)\,du \\ &= \tfrac{1}{4}\left(u - \tfrac{1}{2}\sin 2u\right) + C = \tfrac{1}{4}u - \tfrac{1}{4}\left(\tfrac{1}{2}\cdot 2\sin u \cos u\right) + C \\ &= \tfrac{1}{4}x^2 - \tfrac{1}{4}\sin(x^2)\cos(x^2) + C\end{aligned}$$

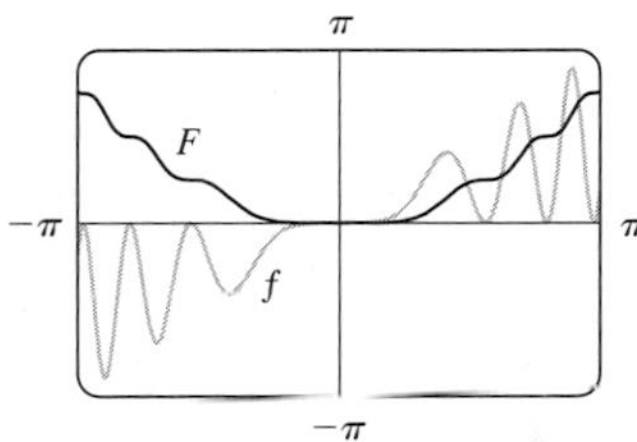

We see from the graph that this is reasonable, since F increases where f is positive and F decreases where f is negative. Note also that f is an odd function and F is an even function.

52.
$$\begin{aligned}\int \sin^5 x \cos^3 x\,dx &= \int \sin^5 x \cos^2 x \cos x\,dx \\ &= \int \sin^5 x\,(1 - \sin^2 x)\cos x\,dx \\ &\overset{s}{=} \int u^5(1 - u^2)\,du = \int (u^5 - u^7)\,du \\ &= \tfrac{1}{6}\sin^6 x - \tfrac{1}{8}\sin^8 x + C\end{aligned}$$

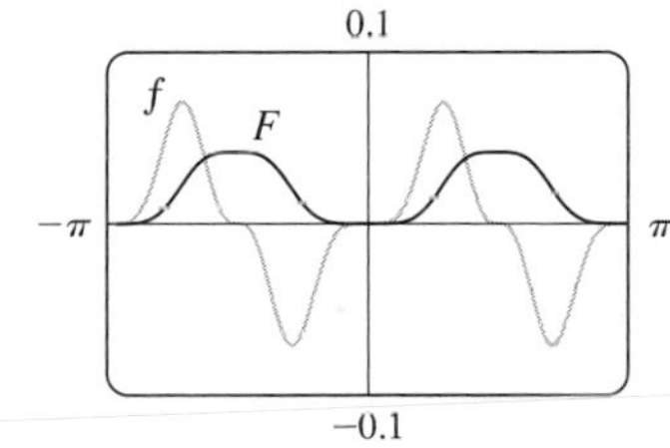

We see from the graph that this is reasonable, since F increases where f is positive and F decreases where f is negative. Note also that f is an odd function and F is an even function.

53.
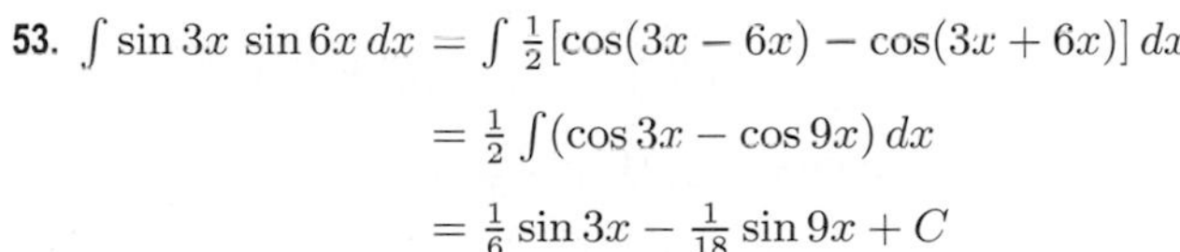
$$\begin{aligned}\int \sin 3x \sin 6x\,dx &= \int \tfrac{1}{2}[\cos(3x - 6x) - \cos(3x + 6x)]\,dx \\ &= \tfrac{1}{2}\int (\cos 3x - \cos 9x)\,dx \\ &= \tfrac{1}{6}\sin 3x - \tfrac{1}{18}\sin 9x + C\end{aligned}$$

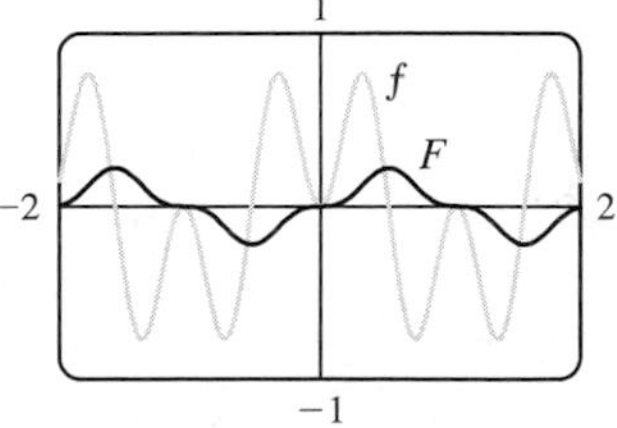

Notice that $f(x) = 0$ whenever F has a horizontal tangent.

54.
$$\begin{aligned}\int \sec^4\left(\tfrac{1}{2}x\right)dx &= \int \left(\tan^2\tfrac{x}{2} + 1\right)\sec^2\tfrac{x}{2}\,dx \\ &= \int (u^2 + 1)\,2\,du \quad \left[u = \tan\tfrac{x}{2},\ du = \tfrac{1}{2}\sec^2\tfrac{x}{2}\,dx\right] \\ &= \tfrac{2}{3}u^3 + 2u + C = \tfrac{2}{3}\tan^3\tfrac{x}{2} + 2\tan\tfrac{x}{2} + C\end{aligned}$$

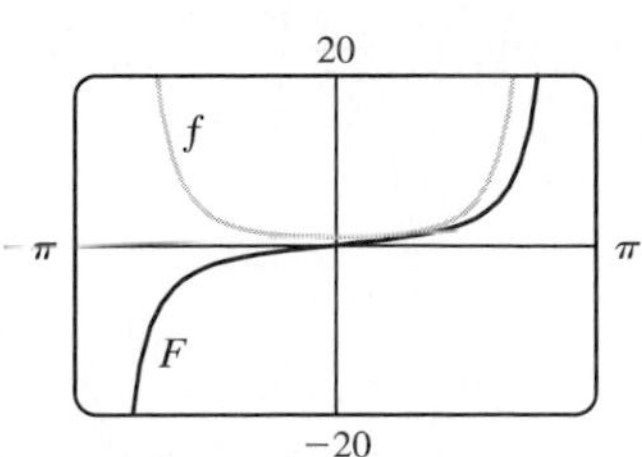

Notice that F is increasing and f is positive on the intervals on which they are defined. Also, F has no horizontal tangent and f is never zero.

55. $f_{\text{ave}} = \frac{1}{2\pi}\int_{-\pi}^{\pi} \sin^2 x \cos^3 x\, dx = \frac{1}{2\pi}\int_{-\pi}^{\pi} \sin^2 x\,(1-\sin^2 x)\,\cos x\, dx$

$= \frac{1}{2\pi}\int_0^0 u^2(1-u^2)\,du$ [where $u = \sin x$] $= 0$

56. (a) Let $u = \cos x$. Then $du = -\sin x\, dx \;\Rightarrow\; \int \sin x \cos x\, dx = \int u(-du) = -\frac{1}{2}u^2 + C = -\frac{1}{2}\cos^2 x + C_1$.

(b) Let $u = \sin x$. Then $du = \cos x\, dx \;\Rightarrow\; \int \sin x \cos x\, dx = \int u\, du = \frac{1}{2}u^2 + C = \frac{1}{2}\sin^2 x + C_2$.

(c) $\int \sin x \cos x\, dx = \int \frac{1}{2}\sin 2x\, dx = -\frac{1}{4}\cos 2x + C_3$

(d) Let $u = \sin x$, $dv = \cos x\, dx$. Then $du = \cos x\, dx$, $v = \sin x$, so $\int \sin x \cos x\, dx = \sin^2 x - \int \sin x \cos x\, dx$, by Equation 7.1.2, so $\int \sin x \cos x\, dx = \frac{1}{2}\sin^2 x + C_4$.

Using $\cos^2 x = 1 - \sin^2 x$ and $\cos 2x = 1 - 2\sin^2 x$, we see that the answers differ only by a constant.

57. $A = \int_0^{\pi}(\sin^2 x - \sin^3 x)\,dx = \int_0^{\pi}\left[\frac{1}{2}(1-\cos 2x) - \sin x\,(1-\cos^2 x)\right]dx$

$= \int_0^{\pi}\left(\frac{1}{2} - \frac{1}{2}\cos 2x\right)dx + \int_1^{-1}(1-u^2)\,du \qquad \left[\begin{array}{l} u = \cos x, \\ du = -\sin x\, dx \end{array}\right]$

$= \left[\frac{1}{2}x - \frac{1}{4}\sin 2x\right]_0^{\pi} + 2\int_0^1 (u^2-1)\,du$

$= \left(\frac{1}{2}\pi - 0\right) - (0-0) + 2\left[\frac{1}{3}u^3 - u\right]_0^1$

$= \frac{1}{2}\pi + 2\left(\frac{1}{3} - 1\right) = \frac{1}{2}\pi - \frac{4}{3}$

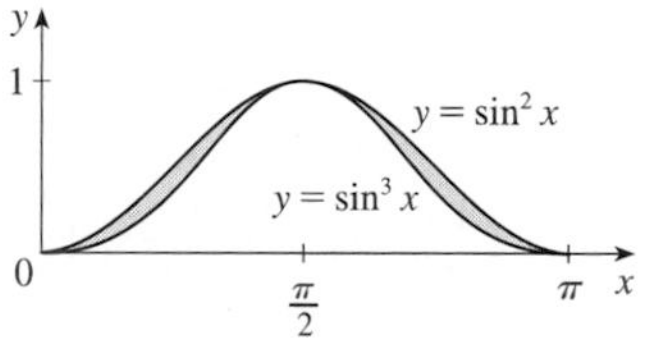

58. 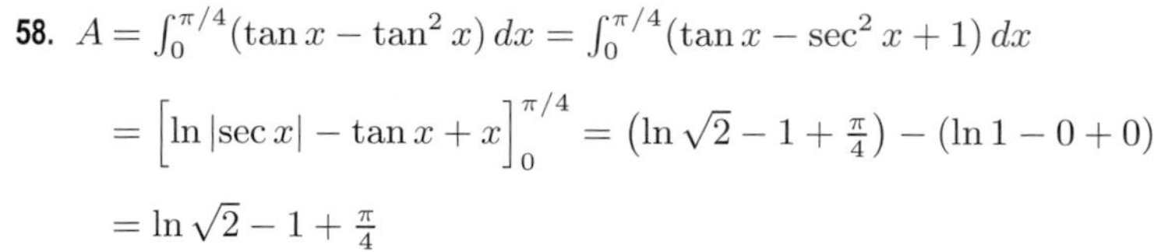

$A = \int_0^{\pi/4}(\tan x - \tan^2 x)\,dx = \int_0^{\pi/4}(\tan x - \sec^2 x + 1)\,dx$

$= \Big[\ln|\sec x| - \tan x + x\Big]_0^{\pi/4} = \left(\ln\sqrt{2} - 1 + \frac{\pi}{4}\right) - (\ln 1 - 0 + 0)$

$= \ln\sqrt{2} - 1 + \frac{\pi}{4}$

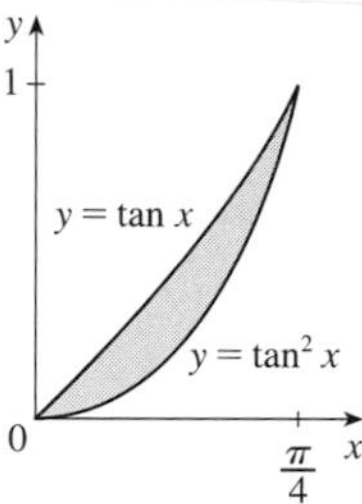

59. 

It seems from the graph that $\int_0^{2\pi}\cos^3 x\, dx = 0$, since the area below the x-axis and above the graph looks about equal to the area above the axis and below the graph. By Example 1, the integral is $\left[\sin x - \frac{1}{3}\sin^3 x\right]_0^{2\pi} = 0$. Note that due to symmetry, the integral of any odd power of $\sin x$ or $\cos x$ between limits which differ by $2n\pi$ (n any integer) is 0.

**60.** 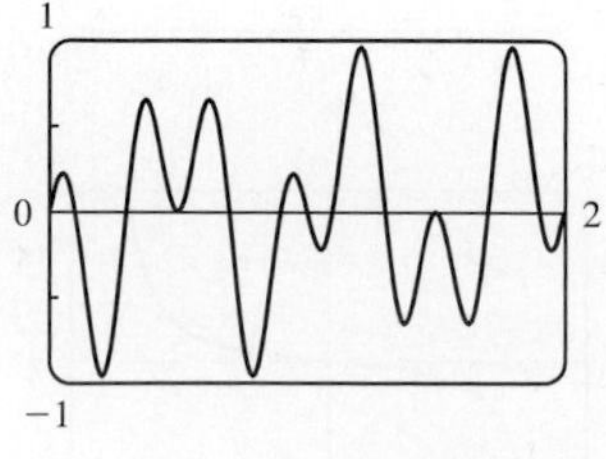

It seems from the graph that $\int_0^2 \sin 2\pi x \cos 5\pi x\, dx = 0$, since each bulge above the x-axis seems to have a corresponding depression below the x-axis. To evaluate the integral, we use a trigonometric identity:

$\int_0^1 \sin 2\pi x \cos 5\pi x\, dx = \frac{1}{2}\int_0^2[\sin(2\pi x - 5\pi x) + \sin(2\pi x + 5\pi x)]\,dx$

$= \frac{1}{2}\int_0^2[\sin(-3\pi x) + \sin 7\pi x]\,dx$

$= \frac{1}{2}\left[\frac{1}{3\pi}\cos(-3\pi x) - \frac{1}{7\pi}\cos 7\pi x\right]_0^2$

$= \frac{1}{2}\left[\frac{1}{3\pi}(1-1) - \frac{1}{7\pi}(1-1)\right] = 0$

61. Using disks, $V = \int_{\pi/2}^{\pi} \pi \sin^2 x\,dx = \pi \int_{\pi/2}^{\pi} \frac{1}{2}(1 - \cos 2x)\,dx = \pi\left[\frac{1}{2}x - \frac{1}{4}\sin 2x\right]_{\pi/2}^{\pi} = \pi\left(\frac{\pi}{2} - 0 - \frac{\pi}{4} + 0\right) = \frac{\pi^2}{4}$

62. Using disks,

$$\begin{aligned} V &= \int_0^{\pi} \pi(\sin^2 x)^2\,dx = 2\pi \int_0^{\pi/2} \left[\tfrac{1}{2}(1 - \cos 2x)\right]^2 dx \\ &= \tfrac{\pi}{2} \int_0^{\pi/2} (1 - 2\cos 2x + \cos^2 2x)\,dx \\ &= \tfrac{\pi}{2} \int_0^{\pi/2} \left[1 - 2\cos 2x + \tfrac{1}{2}(1 - \cos 4x)\right] dx \\ &= \tfrac{\pi}{2} \int_0^{\pi/2} \left(\tfrac{3}{2} - 2\cos 2x - \tfrac{1}{2}\cos 4x\right) dx = \tfrac{\pi}{2}\left[\tfrac{3}{2}x - \sin 2x + \tfrac{1}{8}\sin 4x\right]_0^{\pi/2} \\ &= \tfrac{\pi}{2}\left[\left(\tfrac{3\pi}{4} - 0 + 0\right) - (0 - 0 + 0)\right] = \tfrac{3}{8}\pi^2 \end{aligned}$$

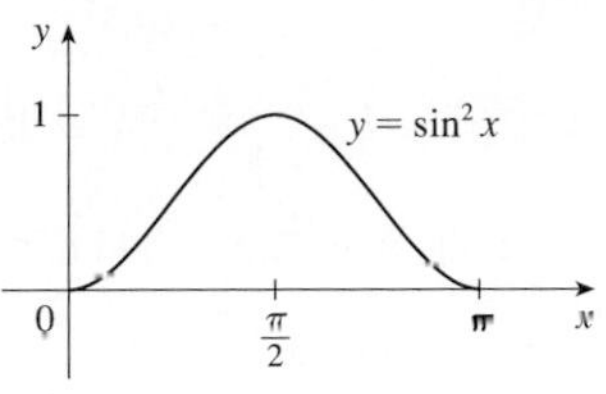

63. Using washers,

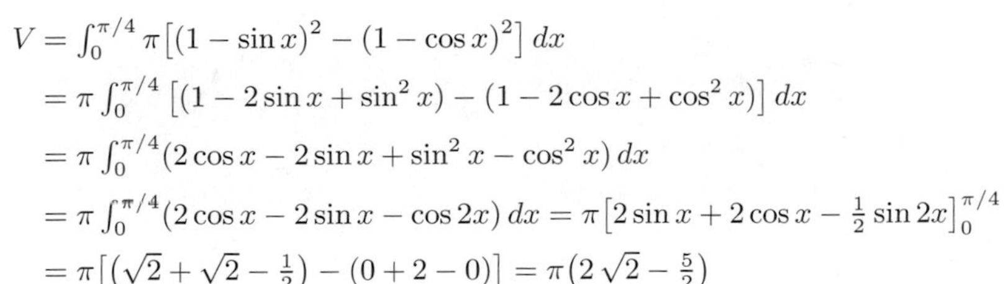

$$\begin{aligned} V &= \int_0^{\pi/4} \pi\left[(1 - \sin x)^2 - (1 - \cos x)^2\right] dx \\ &= \pi \int_0^{\pi/4} \left[(1 - 2\sin x + \sin^2 x) - (1 - 2\cos x + \cos^2 x)\right] dx \\ &= \pi \int_0^{\pi/4} (2\cos x - 2\sin x + \sin^2 x - \cos^2 x)\,dx \\ &= \pi \int_0^{\pi/4} (2\cos x - 2\sin x - \cos 2x)\,dx = \pi\left[2\sin x + 2\cos x - \tfrac{1}{2}\sin 2x\right]_0^{\pi/4} \\ &= \pi\left[\left(\sqrt{2} + \sqrt{2} - \tfrac{1}{2}\right) - (0 + 2 - 0)\right] = \pi\left(2\sqrt{2} - \tfrac{5}{2}\right) \end{aligned}$$

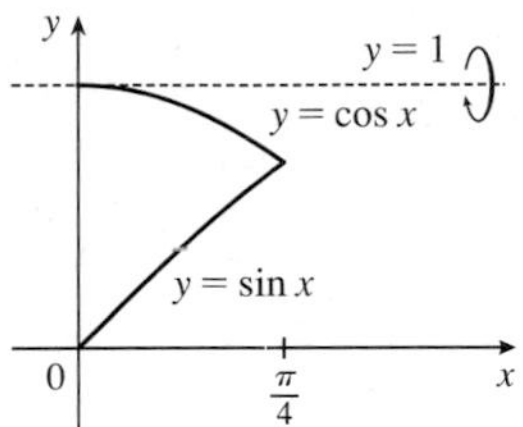

64. Using washers,

$$\begin{aligned} V &= \int_0^{\pi/3} \pi\left\{[\sec x - (-1)]^2 - [\cos x - (-1)]^2\right\} dx \\ &= \pi \int_0^{\pi/3} [(\sec^2 x + 2\sec x + 1) - (\cos^2 x + 2\cos x + 1)]\,dx \\ &= \pi \int_0^{\pi/3} \left[\sec^2 x + 2\sec x - \tfrac{1}{2}(1 + \cos 2x) - 2\cos x\right] dx \\ &= \pi\left[\tan x + 2\ln|\sec x + \tan x| - \tfrac{1}{2}x - \tfrac{1}{4}\sin 2x - 2\sin x\right]_0^{\pi/3} \\ &= \pi\left[\left(\sqrt{3} + 2\ln(2 + \sqrt{3}) - \tfrac{\pi}{6} - \tfrac{1}{8}\sqrt{3} - \sqrt{3}\right) - 0\right] \\ &= 2\pi\ln(2 + \sqrt{3}) - \tfrac{1}{6}\pi^2 - \tfrac{1}{8}\pi\sqrt{3} \end{aligned}$$

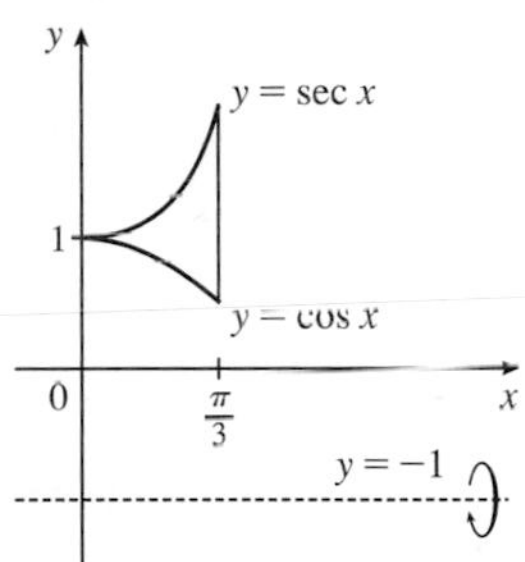

65. $s = f(t) = \int_0^t \sin\omega u \cos^2\omega u\,du$. Let $y = \cos\omega u \Rightarrow dy = -\omega\sin\omega u\,du$. Then

$s = -\frac{1}{\omega}\int_1^{\cos\omega t} y^2 dy = -\frac{1}{\omega}\left[\frac{1}{3}y^3\right]_1^{\cos\omega t} = \frac{1}{3\omega}(1 - \cos^3\omega t)$.

66. (a) We want to calculate the square root of the average value of $[E(t)]^2 = [155\sin(120\pi t)]^2 = 155^2\sin^2(120\pi t)$. First, we calculate the average value itself, by integrating $[E(t)]^2$ over one cycle (between $t = 0$ and $t = \frac{1}{60}$, since there are 60 cycles per second) and dividing by $\left(\frac{1}{60} - 0\right)$:

$$\begin{aligned} [E(t)]^2_{\text{ave}} &= \tfrac{1}{1/60}\int_0^{1/60} [155^2\sin^2(120\pi t)]\,dt = 60 \cdot 155^2 \int_0^{1/60} \tfrac{1}{2}[1 - \cos(240\pi t)]\,dt \\ &= 60 \cdot 155^2\left(\tfrac{1}{2}\right)\left[t - \tfrac{1}{240\pi}\sin(240\pi t)\right]_0^{1/60} = 60 \cdot 155^2\left(\tfrac{1}{2}\right)\left[\left(\tfrac{1}{60} - 0\right) - (0 - 0)\right] = \tfrac{155^2}{2} \end{aligned}$$

The RMS value is just the square root of this quantity, which is $\frac{155}{\sqrt{2}} \approx 110$ V.

(b) $220 = \sqrt{[E(t)]^2_{\text{ave}}} \quad \Rightarrow$

$$220^2 = [E(t)]^2_{\text{ave}} = \tfrac{1}{1/60}\int_0^{1/60} A^2 \sin^2(120\pi t)\,dt = 60A^2 \int_0^{1/60} \tfrac{1}{2}[1 - \cos(240\pi t)]\,dt$$

$$= 30A^2\left[t - \tfrac{1}{240\pi}\sin(240\pi t)\right]_0^{1/60} = 30A^2\left[\left(\tfrac{1}{60} - 0\right) - (0 - 0)\right] = \tfrac{1}{2}A^2$$

Thus, $220^2 = \frac{1}{2}A^2 \quad \Rightarrow \quad A = 220\sqrt{2} \approx 311$ V.

67. Just note that the integrand is odd $[f(-x) = -f(x)]$.

Or: If $m \neq n$, calculate

$$\int_{-\pi}^{\pi} \sin mx\, \cos nx\, dx = \int_{-\pi}^{\pi} \tfrac{1}{2}[\sin(m-n)x + \sin(m+n)x]\,dx = \tfrac{1}{2}\left[-\frac{\cos(m-n)x}{m-n} - \frac{\cos(m+n)x}{m+n}\right]_{-\pi}^{\pi} = 0$$

If $m = n$, then the first term in each set of brackets is zero.

68. $\int_{-\pi}^{\pi} \sin mx\, \sin nx\, dx = \int_{-\pi}^{\pi} \frac{1}{2}[\cos(m-n)x - \cos(m+n)x]\,dx$.

If $m \neq n$, this is equal to $\dfrac{1}{2}\left[\dfrac{\sin(m-n)x}{m-n} - \dfrac{\sin(m+n)x}{m+n}\right]_{-\pi}^{\pi} = 0.$

If $m = n$, we get $\int_{-\pi}^{\pi} \frac{1}{2}[1 - \cos(m+n)x]\,dx = \left[\frac{1}{2}x\right]_{-\pi}^{\pi} - \left[\dfrac{\sin(m+n)x}{2(m+n)}\right]_{-\pi}^{\pi} = \pi - 0 = \pi.$

69. $\int_{-\pi}^{\pi} \cos mx\, \cos nx\, dx = \int_{-\pi}^{\pi} \frac{1}{2}[\cos(m-n)x + \cos(m+n)x]\,dx$.

If $m \neq n$, this is equal to $\dfrac{1}{2}\left[\dfrac{\sin(m-n)x}{m-n} + \dfrac{\sin(m+n)x}{m+n}\right]_{-\pi}^{\pi} = 0.$

If $m = n$, we get $\int_{-\pi}^{\pi} \frac{1}{2}[1 + \cos(m+n)x]\,dx = \left[\frac{1}{2}x\right]_{-\pi}^{\pi} + \left[\dfrac{\sin(m+n)x}{2(m+n)}\right]_{-\pi}^{\pi} = \pi + 0 = \pi.$

70. $\dfrac{1}{\pi}\displaystyle\int_{-\pi}^{\pi} f(x)\, \sin mx\, dx = \dfrac{1}{\pi}\int_{-\pi}^{\pi}\left[\left(\sum_{n=1}^{m} a_n \sin nx\right)\sin mx\right]dx = \sum_{n=1}^{m}\dfrac{a_n}{\pi}\int_{-\pi}^{\pi} \sin mx\, \sin nx\, dx$. By Exercise 68, every term is zero except the mth one, and that term is $\dfrac{a_m}{\pi} \cdot \pi = a_m$.

7.3 Trigonometric Substitution

1. Let $x = 2\sin\theta$, where $-\pi/2 \le \theta \le \pi/2$. Then $dx = 2\cos\theta\, d\theta$ and

$\sqrt{4 - x^2} = \sqrt{4 - 4\sin^2\theta} = \sqrt{4\cos^2\theta} = 2\,|\cos\theta| = 2\cos\theta$.

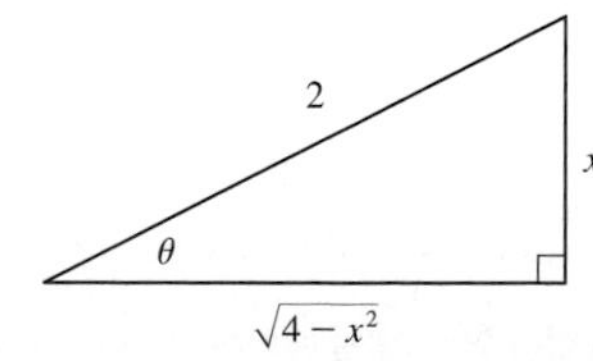

Thus, $\displaystyle\int \frac{dx}{x^2\sqrt{4 - x^2}} = \int \frac{2\cos\theta}{4\sin^2\theta(2\cos\theta)}\,d\theta = \frac{1}{4}\int \csc^2\theta\, d\theta$

$$= -\frac{1}{4}\cot\theta + C = -\frac{\sqrt{4 - x^2}}{4x} + C \quad \text{[see figure]}$$

2. Let $x = 2\tan\theta$, where $-\frac{\pi}{2} < \theta < \frac{\pi}{2}$. Then $dx = 2\sec^2\theta\, d\theta$ and

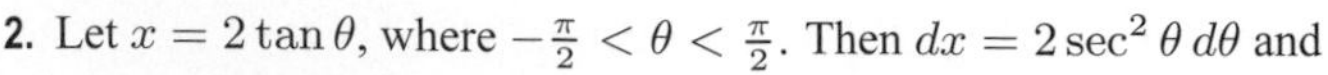

$\sqrt{x^2 + 4} = \sqrt{4\tan^2\theta + 4} = \sqrt{4(\tan^2\theta + 1)} = \sqrt{4\sec^2\theta} = 2\,|\sec\theta|$

$= 2\sec\theta \quad$ for the relevant values of θ.

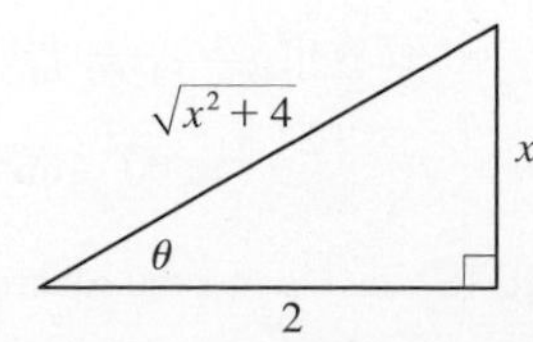

[continued]

$$\int \frac{x^3}{\sqrt{x^2+4}}\,dx = \int \frac{8\tan^3\theta}{2\sec\theta}\,2\sec^2\theta\,d\theta = 8\int \tan^2\theta\,\sec\theta\,\tan\theta\,d\theta$$

$$= 8\int(\sec^2\theta - 1)\,\sec\theta\,\tan\theta\,d\theta = 8\int(u^2-1)\,du \qquad [u=\sec\theta]$$

$$= 8\left(\frac{1}{3}u^3 - u\right) + C = \frac{8}{3}\sec^3\theta - 8\sec\theta + C = \frac{8}{3}\left(\frac{\sqrt{x^2+4}}{2}\right)^3 - 8\left(\frac{\sqrt{x^2+4}}{2}\right) + C$$

$$= \tfrac{1}{3}(x^2+4)^{3/2} - 4\sqrt{x^2+4} + C$$

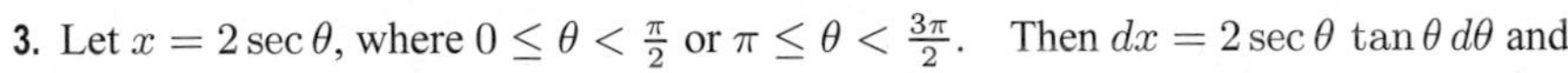
3. Let $x = 2\sec\theta$, where $0 \le \theta < \frac{\pi}{2}$ or $\pi \le \theta < \frac{3\pi}{2}$. Then $dx = 2\sec\theta\,\tan\theta\,d\theta$ and

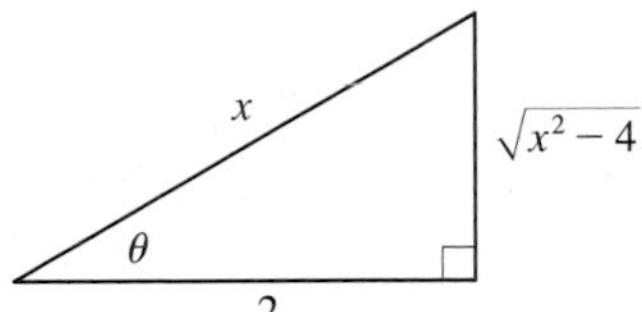

$$\sqrt{x^2-4} = \sqrt{4\sec^2\theta - 4} = \sqrt{4(\sec^2\theta-1)}$$

$$= \sqrt{4\tan^2\theta} = 2\,|\tan\theta| = 2\tan\theta \quad \text{for the relevant values of } \theta$$

$$\int \frac{\sqrt{x^2-4}}{x}\,dx = \int \frac{2\tan\theta}{2\sec\theta}\,2\sec\theta\,\tan\theta\,d\theta = 2\int \tan^2\theta\,d\theta$$

$$= 2\int(\sec^2\theta - 1)\,d\theta = 2\,(\tan\theta - \theta) + C = 2\left[\frac{\sqrt{x^2-4}}{2} - \sec^{-1}\left(\frac{x}{2}\right)\right] + C$$

$$= \sqrt{x^2-4} - 2\sec^{-1}\left(\frac{x}{2}\right) + C$$

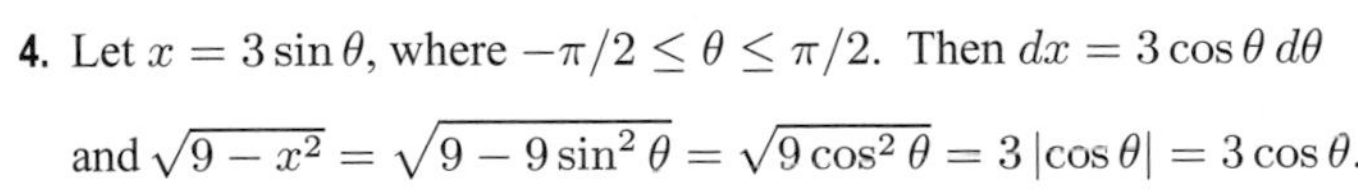
4. Let $x = 3\sin\theta$, where $-\pi/2 \le \theta \le \pi/2$. Then $dx = 3\cos\theta\,d\theta$ and $\sqrt{9-x^2} = \sqrt{9-9\sin^2\theta} = \sqrt{9\cos^2\theta} = 3\,|\cos\theta| = 3\cos\theta$.

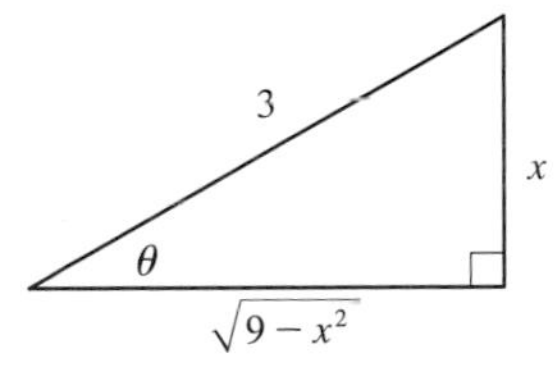

$$\int \frac{x^2}{\sqrt{9-x^2}}\,dx = \int \frac{9\sin^2\theta}{3\cos\theta}\,3\cos\theta\,d\theta = 9\int \sin^2\theta\,d\theta$$

$$= 9\int \tfrac{1}{2}(1-\cos 2\theta)\,d\theta = \tfrac{9}{2}\left(\theta - \tfrac{1}{2}\sin 2\theta\right) + C = \tfrac{9}{2}\theta - \tfrac{9}{4}(2\sin\theta\,\cos\theta) + C$$

$$= \frac{9}{2}\sin^{-1}\left(\frac{x}{3}\right) - \frac{9}{2}\cdot\frac{x}{3}\cdot\frac{\sqrt{9-x^2}}{3} + C = \frac{9}{2}\sin^{-1}\left(\frac{x}{3}\right) - \frac{1}{2}x\sqrt{9-x^2} + C$$

5. Let $x = \sec\theta$, where $0 \le \theta \le \frac{\pi}{2}$ or $\pi \le \theta < \frac{3\pi}{2}$. Then $dx = \sec\theta\,\tan\theta\,d\theta$ and $\sqrt{x^2-1} = \sqrt{\sec^2\theta - 1} = \sqrt{\tan^2\theta} = |\tan\theta| = \tan\theta$ for the relevant values of θ, so

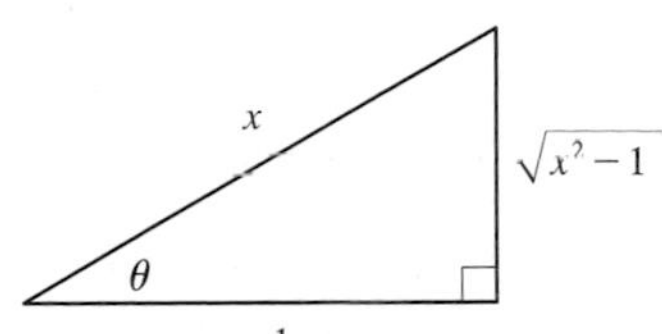

$$\int \frac{\sqrt{x^2-1}}{x^4}\,dx = \int \frac{\tan\theta}{\sec^4\theta}\,\sec\theta\,\tan\theta\,d\theta = \int \tan^2\theta\,\cos^3\theta\,d\theta$$

$$= \int \sin^2\theta\,\cos\theta\,d\theta \overset{s}{=} \int u^2\,du = \tfrac{1}{3}u^3 + C = \tfrac{1}{3}\sin^3\theta + C$$

$$= \frac{1}{3}\left(\frac{\sqrt{x^2-1}}{x}\right)^3 + C = \frac{1}{3}\frac{(x^2-1)^{3/2}}{x^3} + C$$

6. Let $u = 36 - x^2$, so $du = -2x\,dx$. When $x = 0$, $u = 36$; when $x = 3$, $u = 27$. Thus,

$$\int_0^3 \frac{x}{\sqrt{36-x^2}}\,dx = \int_{36}^{27} \frac{1}{\sqrt{u}}\left(-\frac{1}{2}\,du\right) = -\frac{1}{2}\Big[2\sqrt{u}\Big]_{36}^{27} = -\left(\sqrt{27}-\sqrt{36}\right) = 6 - 3\sqrt{3}$$

[continued]

Another method: Let $x = 6\sin\theta$, so $dx = 6\cos\theta\,d\theta$, $x = 0 \;\Rightarrow\; \theta = 0$, and $x = 3 \;\Rightarrow\; \theta = \frac{\pi}{6}$. Then

$$\int_0^3 \frac{x}{\sqrt{36-x^2}}\,dx = \int_0^{\pi/6} \frac{6\sin\theta}{\sqrt{36(1-\sin^2\theta)}}\,6\cos\theta\,d\theta = \int_0^{\pi/6} \frac{6\sin\theta}{6\cos\theta}\,6\cos\theta\,d\theta = 6\int_0^{\pi/6} \sin\theta\,d\theta$$

$$= 6\Big[-\cos\theta\Big]_0^{\pi/6} = 6\left(-\tfrac{\sqrt{3}}{2}+1\right) = 6 - 3\sqrt{3}$$

7. Let $x = a\tan\theta$, where $a > 0$ and $-\frac{\pi}{2} < \theta < \frac{\pi}{2}$. Then $dx = a\sec^2\theta\,d\theta$, $x = 0 \;\Rightarrow\; \theta = 0$, and $x = a \;\Rightarrow\; \theta = \frac{\pi}{4}$. Thus,

$$\int_0^a \frac{dx}{(a^2+x^2)^{3/2}} = \int_0^{\pi/4} \frac{a\sec^2\theta\,d\theta}{[a^2(1+\tan^2\theta)]^{3/2}} = \int_0^{\pi/4} \frac{a\sec^2\theta\,d\theta}{a^3\sec^3\theta} = \frac{1}{a^2}\int_0^{\pi/4}\cos\theta\,d\theta = \frac{1}{a^2}\Big[\sin\theta\Big]_0^{\pi/4}$$

$$= \frac{1}{a^2}\left(\frac{\sqrt{2}}{2} - 0\right) = \frac{1}{\sqrt{2}\,a^2}.$$

8. Let $t = 4\sec\theta$, where $0 \le \theta < \frac{\pi}{2}$ or $\pi \le \theta < \frac{3\pi}{2}$. Then $dt = 4\sec\theta\,\tan\theta\,d\theta$ and $\sqrt{t^2-16} = \sqrt{16\sec^2\theta - 16} = \sqrt{16\tan^2\theta} = 4\tan\theta$ for the relevant values of θ, so

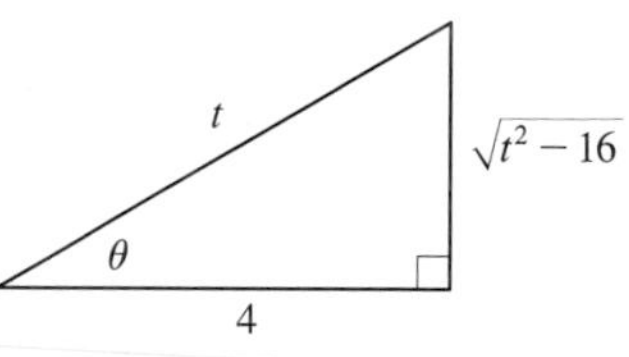

$$\int \frac{dt}{t^2\sqrt{t^2-16}} = \int \frac{4\sec\theta\,\tan\theta\,d\theta}{16\sec^2\theta\cdot 4\tan\theta} = \frac{1}{16}\int \frac{1}{\sec\theta}\,d\theta = \frac{1}{16}\int \cos\theta\,d\theta$$

$$= \frac{1}{16}\sin\theta + C = \frac{1}{16}\,\frac{\sqrt{t^2-16}}{t} + C = \frac{\sqrt{t^2-16}}{16t} + C$$

9. Let $x = \sec\theta$, so $dx = \sec\theta\,\tan\theta\,d\theta$, $x = 2 \;\Rightarrow\; \theta = \frac{\pi}{3}$, and $x = 3 \;\Rightarrow\; \theta = \sec^{-1}3$. Then

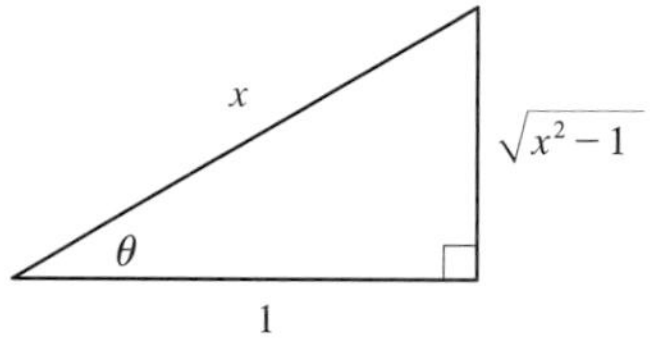

$$\int_2^3 \frac{dx}{(x^2-1)^{3/2}} = \int_{\pi/3}^{\sec^{-1}3} \frac{\sec\theta\,\tan\theta\,d\theta}{\tan^3\theta} = \int_{\pi/3}^{\sec^{-1}3} \frac{\cos\theta}{\sin^2\theta}\,d\theta$$

$$\overset{s}{=} \int_{\sqrt{3}/2}^{\sqrt{8}/3} \frac{1}{u^2}\,du = \left[-\frac{1}{u}\right]_{\sqrt{3}/2}^{\sqrt{8}/3} = \frac{-3}{\sqrt{8}} + \frac{2}{\sqrt{3}} = -\frac{3}{4}\sqrt{2} + \frac{2}{3}\sqrt{3}$$

10. Let $x = \frac{2}{3}\sin\theta$, so $dx = \frac{2}{3}\cos\theta\,d\theta$, $x = 0 \;\Rightarrow\; \theta = 0$, and $x = \frac{2}{3} \;\Rightarrow\; \theta = \frac{\pi}{2}$. Thus,

$$\int_0^{2/3} \sqrt{4-9x^2}\,dx = \int_0^{\pi/2} \sqrt{4 - 9\cdot\frac{4}{9}\sin^2\theta}\;\frac{2}{3}\cos\theta\,d\theta$$

$$= \int_0^{\pi/2} 2\cos\theta\cdot\frac{2}{3}\cos\theta\,d\theta = \frac{4}{3}\int_0^{\pi/2}\cos^2\theta\,d\theta$$

$$= \frac{4}{3}\int_0^{\pi/2}\frac{1}{2}(1+\cos 2\theta)\,d\theta = \frac{2}{3}\left[\theta + \frac{1}{2}\sin 2\theta\right]_0^{\pi/2} = \frac{2}{3}\left[\left(\frac{\pi}{2}+0\right) - (0+0)\right] = \frac{\pi}{3}$$

11. $\int_0^{1/2} x\sqrt{1-4x^2}\,dx = \int_1^0 u^{1/2}\left(-\frac{1}{8}\,du\right)$ $\quad\left[\begin{array}{c} u = 1-4x^2, \\ du = -8x\,dx \end{array}\right]$

$$= \tfrac{1}{8}\left[\tfrac{2}{3}u^{3/2}\right]_0^1 = \tfrac{1}{12}(1-0) = \tfrac{1}{12}$$

12. Let $t = 2\tan\theta$, so $dt = 2\sec^2\theta\,d\theta$, $t = 0 \;\Rightarrow\; \theta = 0$, and $t = 2 \;\Rightarrow\; \theta = \frac{\pi}{4}$. Thus,

$$\int_0^2 \frac{dt}{\sqrt{4+t^2}} = \int_0^{\pi/4} \frac{2\sec^2\theta\,d\theta}{\sqrt{4+4\tan^2\theta}} = \int_0^{\pi/4} \frac{2\sec^2\theta\,d\theta}{2\sec\theta} = \int_0^{\pi/4} \sec\theta\,d\theta = \Big[\ln|\sec\theta+\tan\theta|\Big]_0^{\pi/4}$$

$$= \ln\left|\sqrt{2}+1\right| - \ln|1+0| = \ln(\sqrt{2}+1)$$

13. Let $x = 3\sec\theta$, where $0 \le \theta < \frac{\pi}{2}$ or $\pi \le \theta < \frac{3\pi}{2}$. Then

$dx = 3\sec\theta\,\tan\theta\,d\theta$ and $\sqrt{x^2-9} = 3\tan\theta$, so

$$\int \frac{\sqrt{x^2-9}}{x^3}\,dx = \int \frac{3\tan\theta}{27\sec^3\theta}\,3\sec\theta\,\tan\theta\,d\theta = \frac{1}{3}\int \frac{\tan^2\theta}{\sec^2\theta}\,d\theta$$

$$= \tfrac{1}{3}\int \sin^2\theta\,d\theta = \tfrac{1}{3}\int \tfrac{1}{2}(1-\cos 2\theta)\,d\theta = \tfrac{1}{6}\theta - \tfrac{1}{12}\sin 2\theta + C = \tfrac{1}{6}\theta - \tfrac{1}{6}\sin\theta\,\cos\theta + C$$

$$= \frac{1}{6}\sec^{-1}\left(\frac{x}{3}\right) - \frac{1}{6}\,\frac{\sqrt{x^2-9}}{x}\,\frac{3}{x} + C = \frac{1}{6}\sec^{-1}\left(\frac{x}{3}\right) - \frac{\sqrt{x^2-9}}{2x^2} + C$$

14. Let $x = \tan\theta$, so $dx = \sec^2\theta\,d\theta$, $x = 0 \;\Rightarrow\; \theta = 0$, and $x = 1 \;\Rightarrow\; \theta = \frac{\pi}{4}$. Then

$$\int_0^1 \frac{dx}{(x^2+1)^2} = \int_0^{\pi/4} \frac{\sec^2\theta\,d\theta}{(\tan^2\theta+1)^2} = \int_0^{\pi/4} \frac{\sec^2\theta\,d\theta}{(\sec^2\theta)^2}$$

$$= \int_0^{\pi/4} \cos^2\theta\,d\theta = \int_0^{\pi/4} \frac{1}{2}(1+\cos 2\theta)\,d\theta$$

$$= \tfrac{1}{2}\left[\theta + \tfrac{1}{2}\sin 2\theta\right]_0^{\pi/4} = \tfrac{1}{2}\left[\left(\tfrac{\pi}{4}+\tfrac{1}{2}\right) - 0\right] = \tfrac{\pi}{8} + \tfrac{1}{4}$$

15. Let $x = a\sin\theta$, $dx = a\cos\theta\,d\theta$, $x = 0 \;\Rightarrow\; \theta = 0$ and $x = a \;\Rightarrow\; \theta = \frac{\pi}{2}$. Then

$$\int_0^a x^2\sqrt{a^2-x^2}\,dx = \int_0^{\pi/2} a^2\sin^2\theta\,(a\cos\theta)\,a\cos\theta\,d\theta = a^4\int_0^{\pi/2}\sin^2\theta\,\cos^2\theta\,d\theta$$

$$= a^4\int_0^{\pi/2}\left[\tfrac{1}{2}(2\sin\theta\,\cos\theta)\right]^2 d\theta = \frac{a^4}{4}\int_0^{\pi/2}\sin^2 2\theta\,d\theta = \frac{a^4}{4}\int_0^{\pi/2}\tfrac{1}{2}(1-\cos 4\theta)\,d\theta$$

$$= \frac{a^4}{8}\left[\theta - \tfrac{1}{4}\sin 4\theta\right]_0^{\pi/2} = \frac{a^4}{8}\left[\left(\frac{\pi}{2}-0\right) - 0\right] = \frac{\pi}{16}a^4$$

16. Let $x = \frac{1}{3}\sec\theta$, so $dx = \frac{1}{3}\sec\theta\,\tan\theta\,d\theta$, $x = \sqrt{2}/3 \;\Rightarrow\; \theta = \frac{\pi}{4}$, $x = \frac{2}{3} \;\Rightarrow\; \theta = \frac{\pi}{3}$. Then

$$\int_{\sqrt{2}/3}^{2/3} \frac{dx}{x^5\sqrt{9x^2-1}} = \int_{\pi/4}^{\pi/3} \frac{\frac{1}{3}\sec\theta\,\tan\theta\,d\theta}{\left(\frac{1}{3}\right)^5\sec^5\theta\,\tan\theta} = 3^4\int_{\pi/4}^{\pi/3}\cos^4\theta\,d\theta = 81\int_{\pi/4}^{\pi/3}\left[\tfrac{1}{2}(1+\cos 2\theta)\right]^2 d\theta$$

$$= \tfrac{81}{4}\int_{\pi/4}^{\pi/3}(1+2\cos 2\theta+\cos^2 2\theta)\,d\theta = \tfrac{81}{4}\int_{\pi/4}^{\pi/3}\left[1+2\cos 2\theta+\tfrac{1}{2}(1+\cos 4\theta)\right]d\theta$$

$$= \tfrac{81}{4}\int_{\pi/4}^{\pi/3}\left(\tfrac{3}{2}+2\cos 2\theta+\tfrac{1}{2}\cos 4\theta\right)d\theta = \tfrac{81}{4}\left[\tfrac{3}{2}\theta+\sin 2\theta+\tfrac{1}{8}\sin 4\theta\right]_{\pi/4}^{\pi/3}$$

$$= \tfrac{81}{4}\left[\left(\tfrac{\pi}{2}+\tfrac{\sqrt{3}}{2}-\tfrac{\sqrt{3}}{16}\right) - \left(\tfrac{3\pi}{8}+1+0\right)\right] = \tfrac{81}{4}\left(\tfrac{\pi}{8}+\tfrac{7}{16}\sqrt{3}-1\right)$$

17. Let $u = x^2-7$, so $du = 2x\,dx$. Then $\displaystyle\int \frac{x}{\sqrt{x^2-7}}\,dx = \frac{1}{2}\int\frac{1}{\sqrt{u}}\,du = \tfrac{1}{2}\cdot 2\sqrt{u}+C = \sqrt{x^2-7}+C.$

18. Let $ax = b\sec\theta$, so $(ax)^2 = b^2\sec^2\theta \Rightarrow$

$(ax)^2 - b^2 = b^2\sec^2\theta - b^2 = b^2(\sec^2\theta - 1) = b^2\tan^2\theta.$

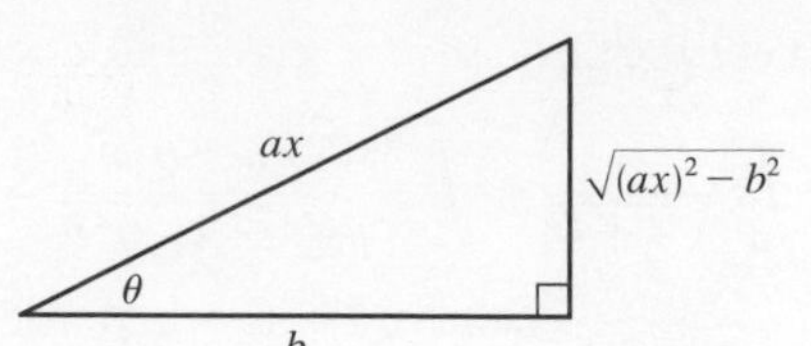

So $\sqrt{(ax)^2 - b^2} = b\tan\theta$, $dx = \dfrac{b}{a}\sec\theta\,\tan\theta\,d\theta$, and

$$\int \frac{dx}{[(ax)^2 - b^2]^{3/2}} = \int \frac{\frac{b}{a}\sec\theta\,\tan\theta}{b^3\tan^3\theta}\,d\theta = \frac{1}{ab^2}\int \frac{\sec\theta}{\tan^2\theta}\,d\theta$$
$$= \frac{1}{ab^2}\int \frac{\cos\theta}{\sin^2\theta}\,d\theta = \frac{1}{ab^2}\int \csc\theta\,\cot\theta\,d\theta$$
$$= -\frac{1}{ab^2}\csc\theta + C = -\frac{1}{ab^2}\frac{ax}{\sqrt{(ax)^2 - b^2}} + C$$
$$= -\frac{x}{b^2\sqrt{(ax)^2 - b^2}} + C$$

19. Let $x = \tan\theta$, where $-\frac{\pi}{2} < \theta < \frac{\pi}{2}$. Then $dx = \sec^2\theta\,d\theta$ and $\sqrt{1 + x^2} = \sec\theta$, so

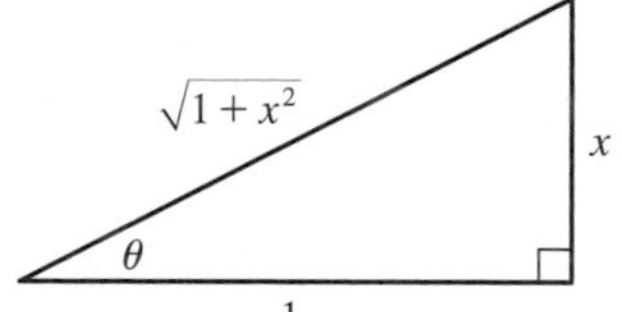

$$\int \frac{\sqrt{1 + x^2}}{x}\,dx = \int \frac{\sec\theta}{\tan\theta}\sec^2\theta\,d\theta = \int \frac{\sec\theta}{\tan\theta}(1 + \tan^2\theta)\,d\theta$$
$$= \int(\csc\theta + \sec\theta\,\tan\theta)\,d\theta$$
$$= \ln|\csc\theta - \cot\theta| + \sec\theta + C \quad \text{[by Exercise 7.2.39]}$$
$$= \ln\left|\frac{\sqrt{1 + x^2}}{x} - \frac{1}{x}\right| + \frac{\sqrt{1 + x^2}}{1} + C = \ln\left|\frac{\sqrt{1 + x^2} - 1}{x}\right| + \sqrt{1 + x^2} + C$$

20. Let $u = 1 + x^2$, so $du = 2x\,dx$. Then

$$\int \frac{x}{\sqrt{1 + x^2}}\,dx = \int \frac{1}{\sqrt{u}}\left(\frac{1}{2}\,du\right) = \frac{1}{2}\int u^{-1/2}\,du = \frac{1}{2}\cdot 2u^{1/2} + C = \sqrt{1 + x^2} + C$$

21. Let $x = \frac{3}{5}\sin\theta$, so $dx = \frac{3}{5}\cos\theta\,d\theta$, $x = 0 \Rightarrow \theta = 0$, and $x = 0.6 \Rightarrow \theta = \frac{\pi}{2}$. Then

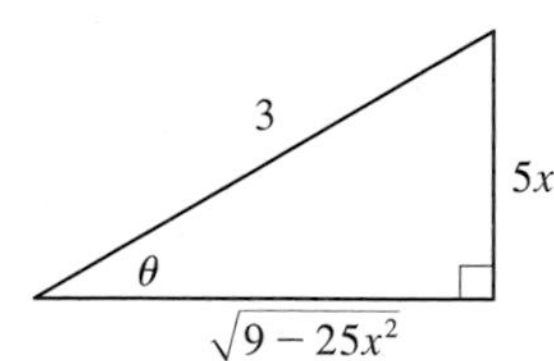

$$\int_0^{0.6} \frac{x^2}{\sqrt{9 - 25x^2}}\,dx = \int_0^{\pi/2} \frac{\left(\frac{3}{5}\right)^2\sin^2\theta}{3\cos\theta}\left(\tfrac{3}{5}\cos\theta\,d\theta\right) = \tfrac{9}{125}\int_0^{\pi/2}\sin^2\theta\,d\theta$$
$$= \tfrac{9}{125}\int_0^{\pi/2}\tfrac{1}{2}(1 - \cos 2\theta)\,d\theta = \tfrac{9}{250}\left[\theta - \tfrac{1}{2}\sin 2\theta\right]_0^{\pi/2}$$
$$= \tfrac{9}{250}\left[\left(\tfrac{\pi}{2} - 0\right) - 0\right] = \tfrac{9}{500}\pi$$

22. Let $x = \tan\theta$, where $-\frac{\pi}{2} < \theta < \frac{\pi}{2}$. Then $dx = \sec^2\theta\,d\theta$, $\sqrt{x^2 + 1} = \sec\theta$ and $x = 0 \Rightarrow \theta = 0$, $x = 1 \Rightarrow \theta = \frac{\pi}{4}$, so

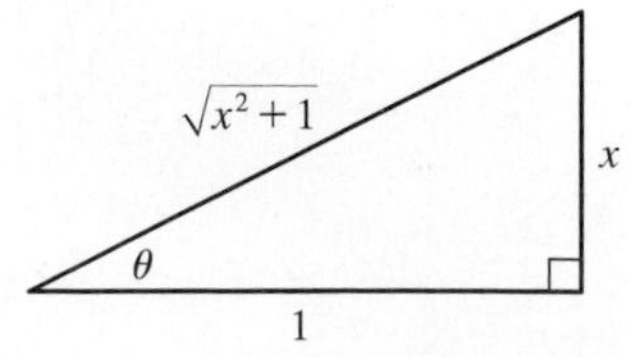

$$\int_0^1 \sqrt{x^2 + 1}\,dx = \int_0^{\pi/4}\sec\theta\,\sec^2\theta\,d\theta = \int_0^{\pi/4}\sec^3\theta\,d\theta$$
$$= \tfrac{1}{2}\Big[\sec\theta\,\tan\theta + \ln|\sec\theta + \tan\theta|\Big]_0^{\pi/4} \quad \text{[by Example 7.2.8]}$$
$$= \tfrac{1}{2}\left[\sqrt{2}\cdot 1 + \ln\left(1 + \sqrt{2}\right) - 0 - \ln(1 + 0)\right] = \tfrac{1}{2}\left[\sqrt{2} + \ln\left(1 + \sqrt{2}\right)\right]$$

23. $\displaystyle\int \frac{dx}{\sqrt{x^2+2x+5}} = \int \frac{dx}{\sqrt{(x+1)^2+4}} = \int \frac{2\sec^2\theta\,d\theta}{\sqrt{4\tan^2\theta+4}} \qquad \begin{bmatrix} x+1 = 2\tan\theta, \\ dx = 2\sec^2\theta\,d\theta \end{bmatrix}$

$\displaystyle = \int \frac{2\sec^2\theta\,d\theta}{2\sec\theta} = \int \sec\theta\,d\theta = \ln|\sec\theta+\tan\theta| + C_1$

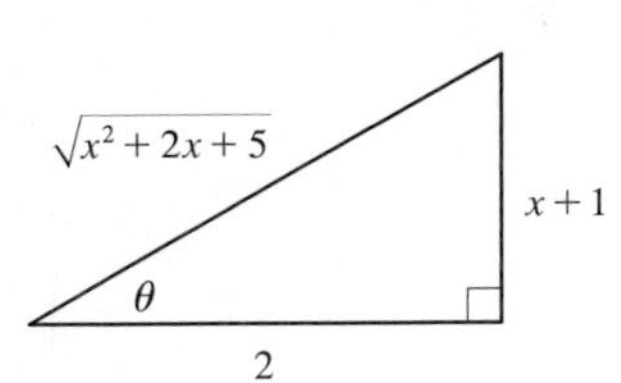

$\displaystyle = \ln\left|\frac{\sqrt{x^2+2x+5}}{2} + \frac{x+1}{2}\right| + C_1,$

or $\ln\left|\sqrt{x^2+2x+5}+x+1\right| + C$, where $C = C_1 - \ln 2$.

24. $\displaystyle\int_0^1 \sqrt{x-x^2}\,dx = \int_0^1 \sqrt{\tfrac14 - \left(x^2 - x + \tfrac14\right)}\,dx = \int_0^1 \sqrt{\tfrac14 - \left(x-\tfrac12\right)^2}\,dx$

$\displaystyle = \int_{-\pi/2}^{\pi/2} \sqrt{\tfrac14 - \tfrac14\sin^2\theta}\;\tfrac12\cos\theta\,d\theta \qquad \begin{bmatrix} x-\frac12 = \frac12\sin\theta, \\ dx = \frac12\cos\theta\,d\theta \end{bmatrix}$

$\displaystyle = 2\int_0^{\pi/2} \tfrac12\cos\theta\,\tfrac12\cos\theta\,d\theta = \tfrac12\int_0^{\pi/2}\cos^2\theta\,d\theta = \tfrac12\int_0^{\pi/2}\tfrac12(1+\cos 2\theta)\,d\theta$

$\displaystyle = \tfrac14\left[\theta + \tfrac12\sin 2\theta\right]_0^{\pi/2} = \tfrac14\left(\tfrac{\pi}{2}\right) = \tfrac{\pi}{8}$

25. $\int x^2\sqrt{3+2x-x^2}\,dx = \int x^2\sqrt{4-(x^2+2x+1)}\,dx = \int x^2\sqrt{2^2-(x-1)^2}\,dx$

$= \int (1+2\sin\theta)^2\sqrt{4\cos^2\theta}\,2\cos\theta\,d\theta \qquad \begin{bmatrix} x-1 = 2\sin\theta, \\ dx = 2\cos\theta\,d\theta \end{bmatrix}$

$= \int (1+4\sin\theta+4\sin^2\theta)\,4\cos^2\theta\,d\theta$

$= 4\int(\cos^2\theta + 4\sin\theta\,\cos^2\theta + 4\sin^2\theta\,\cos^2\theta)\,d\theta$

$= 4\int \tfrac12(1+\cos 2\theta)\,d\theta + 4\int 4\sin\theta\,\cos^2\theta\,d\theta + 4\int(2\sin\theta\,\cos\theta)^2\,d\theta$

$= 2\int(1+\cos 2\theta)\,d\theta + 16\int\sin\theta\,\cos^2\theta\,d\theta + 4\int\sin^2 2\theta\,d\theta$

$= 2\left(\theta+\tfrac12\sin 2\theta\right) + 16\left(-\tfrac13\cos^3\theta\right) + 4\int\tfrac12(1-\cos 4\theta)\,d\theta$

$= 2\theta + \sin 2\theta - \tfrac{16}{3}\cos^3\theta + 2\left(\theta - \tfrac14\sin 4\theta\right) + C$

$= 4\theta - \tfrac12\sin 4\theta + \sin 2\theta - \tfrac{16}{3}\cos^3\theta + C$

$= 4\theta - \tfrac12(2\sin 2\theta\,\cos 2\theta) + \sin 2\theta - \tfrac{16}{3}\cos^3\theta + C$

$= 4\theta + \sin 2\theta(1-\cos 2\theta) - \tfrac{16}{3}\cos^3\theta + C$

$= 4\theta + (2\sin\theta\,\cos\theta)(2\sin^2\theta) - \tfrac{16}{3}\cos^3\theta + C$

$= 4\theta + 4\sin^3\theta\,\cos\theta - \tfrac{16}{3}\cos^3\theta + C$

$\displaystyle = 4\sin^{-1}\left(\frac{x-1}{2}\right) + 4\left(\frac{x-1}{2}\right)^3\frac{\sqrt{3+2x-x^2}}{2} - \frac{16}{3}\,\frac{(3+2x-x^2)^{3/2}}{2^3} + C$

$\displaystyle = 4\sin^{-1}\left(\frac{x-1}{2}\right) + \frac14(x-1)^3\sqrt{3+2x-x^2} - \frac23(3+2x-x^2)^{3/2} + C$

26. $3+4x-4x^2 = -(4x^2-4x+1)+4 = 2^2-(2x-1)^2.$

Let $2x-1 = 2\sin\theta$, so $2\,dx = 2\cos\theta\,d\theta$ and $\sqrt{3+4x-4x^2} = 2\cos\theta$.

Then

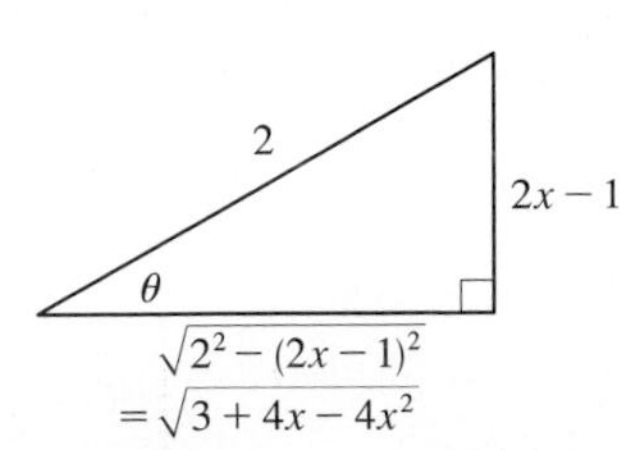

$$\int \frac{x^2}{(3+4x-4x^2)^{3/2}}\,dx = \int \frac{\left[\frac{1}{2}(1+2\sin\theta)\right]^2}{(2\cos\theta)^3}\cos\theta\,d\theta$$

$$= \frac{1}{32}\int \frac{1+4\sin\theta+4\sin^2\theta}{\cos^2\theta}\,d\theta = \frac{1}{32}\int(\sec^2\theta+4\tan\theta\,\sec\theta+4\tan^2\theta)\,d\theta$$

$$= \tfrac{1}{32}\int[\sec^2\theta+4\tan\theta\,\sec\theta+4(\sec^2\theta-1)]\,d\theta$$

$$= \tfrac{1}{32}\int(5\sec^2\theta+4\tan\theta\,\sec\theta-4)\,d\theta = \tfrac{1}{32}(5\tan\theta+4\sec\theta-4\theta)+C$$

$$= \frac{1}{32}\left[5\cdot\frac{2x-1}{\sqrt{3+4x-4x^2}}+4\cdot\frac{2}{\sqrt{3+4x-4x^2}}-4\cdot\sin^{-1}\left(\frac{2x-1}{2}\right)\right]+C$$

$$= \frac{10x+3}{32\sqrt{3+4x-4x^2}}-\frac{1}{8}\sin^{-1}\left(\frac{2x-1}{2}\right)+C$$

27. $x^2+2x=(x^2+2x+1)-1=(x+1)^2-1$. Let $x+1=1\sec\theta$,

so $dx=\sec\theta\,\tan\theta\,d\theta$ and $\sqrt{x^2+2x}=\tan\theta$. Then

$x+1$; $\sqrt{(x+1)^2-1^2}=\sqrt{x^2+2x}$; θ; 1

$$\int\sqrt{x^2+2x}\,dx=\int\tan\theta\,(\sec\theta\,\tan\theta\,d\theta)=\int\tan^2\theta\,\sec\theta\,d\theta$$

$$=\int(\sec^2\theta-1)\,\sec\theta\,d\theta=\int\sec^3\theta\,d\theta-\int\sec\theta\,d\theta$$

$$=\tfrac{1}{2}\sec\theta\,\tan\theta+\tfrac{1}{2}\ln|\sec\theta+\tan\theta|-\ln|\sec\theta+\tan\theta|+C$$

$$=\tfrac{1}{2}\sec\theta\,\tan\theta-\tfrac{1}{2}\ln|\sec\theta+\tan\theta|+C=\tfrac{1}{2}(x+1)\sqrt{x^2+2x}-\tfrac{1}{2}\ln\left|x+1+\sqrt{x^2+2x}\,\right|+C$$

28. $x^2-2x+2=(x^2-2x+1)+1=(x-1)^2+1$. Let $x-1=1\tan\theta$,

so $dx=\sec^2\theta\,d\theta$ and $\sqrt{x^2-2x+2}=\sec\theta$. Then

$\sqrt{1^2+(x-1)^2}=\sqrt{x^2-2x+2}$; $x-1$; θ; 1

$$\int\frac{x^2+1}{(x^2-2x+2)^2}\,dx=\int\frac{(\tan\theta+1)^2+1}{\sec^4\theta}\sec^2\theta\,d\theta$$

$$=\int\frac{\tan^2\theta+2\tan\theta+2}{\sec^2\theta}\,d\theta$$

$$=\int(\sin^2\theta+2\sin\theta\,\cos\theta+2\cos^2\theta)\,d\theta=\int(1+2\sin\theta\,\cos\theta+\cos^2\theta)\,d\theta$$

$$=\int\left[1+2\sin\theta\,\cos\theta+\tfrac{1}{2}(1+\cos2\theta)\right]\,d\theta=\int\left(\tfrac{3}{2}+2\sin\theta\,\cos\theta+\tfrac{1}{2}\cos2\theta\right)\,d\theta$$

$$=\tfrac{3}{2}\theta+\sin^2\theta+\tfrac{1}{4}\sin2\theta+C=\tfrac{3}{2}\theta+\sin^2\theta+\tfrac{1}{2}\sin\theta\,\cos\theta+C$$

$$=\frac{3}{2}\tan^{-1}\left(\frac{x-1}{1}\right)+\frac{(x-1)^2}{x^2-2x+2}+\frac{1}{2}\,\frac{x-1}{\sqrt{x^2-2x+2}}\,\frac{1}{\sqrt{x^2-2x+2}}+C$$

$$=\frac{3}{2}\tan^{-1}(x-1)+\frac{2(x^2-2x+1)+x-1}{2(x^2-2x+2)}+C=\frac{3}{2}\tan^{-1}(x-1)+\frac{2x^2-3x+1}{2(x^2-2x+2)}+C$$

We can write the answer as

$$\frac{3}{2}\tan^{-1}(x-1)+\frac{(2x^2-4x+4)+x-3}{2(x^2-2x+2)}+C=\frac{3}{2}\tan^{-1}(x-1)+1+\frac{x-3}{2(x^2-2x+2)}+C$$

$$=\frac{3}{2}\tan^{-1}(x-1)+\frac{x-3}{2(x^2-2x+2)}+C_1,\ \text{where } C_1=1+C$$

29. Let $u = x^2$, $du = 2x\,dx$. Then

$$\int x\sqrt{1-x^4}\,dx = \int \sqrt{1-u^2}\left(\tfrac{1}{2}\,du\right) = \tfrac{1}{2}\int \cos\theta \cdot \cos\theta\,d\theta \qquad \left[\begin{array}{c}\text{where } u = \sin\theta,\ du = \cos\theta\,d\theta, \\ \text{and } \sqrt{1-u^2} = \cos\theta\end{array}\right]$$

$$= \tfrac{1}{2}\int \tfrac{1}{2}(1+\cos 2\theta)\,d\theta = \tfrac{1}{4}\theta + \tfrac{1}{8}\sin 2\theta + C = \tfrac{1}{4}\theta + \tfrac{1}{4}\sin\theta\,\cos\theta + C$$

$$= \tfrac{1}{4}\sin^{-1} u + \tfrac{1}{4}u\sqrt{1-u^2} + C = \tfrac{1}{4}\sin^{-1}(x^2) + \tfrac{1}{4}x^2\sqrt{1-x^4} + C$$

30. Let $u = \sin t$, $du = \cos t\,dt$. Then

$$\int_0^{\pi/2} \frac{\cos t}{\sqrt{1+\sin^2 t}}\,dt = \int_0^1 \frac{1}{\sqrt{1+u^2}}\,du = \int_0^{\pi/4} \frac{1}{\sec\theta}\sec^2\theta\,d\theta \qquad \left[\begin{array}{c}\text{where } u = \tan\theta,\ du = \sec^2\theta\,d\theta, \\ \text{and } \sqrt{1+u^2} = \sec\theta\end{array}\right]$$

$$= \int_0^{\pi/4} \sec\theta\,d\theta = \Big[\ln|\sec\theta + \tan\theta|\Big]_0^{\pi/4} \qquad \text{[by (1) in Section 7.2]}$$

$$= \ln(\sqrt{2}+1) - \ln(1+0) = \ln(\sqrt{2}+1)$$

31. (a) Let $x = a\tan\theta$, where $-\frac{\pi}{2} < \theta < \frac{\pi}{2}$. Then $\sqrt{x^2+a^2} = a\sec\theta$ and

$$\int \frac{dx}{\sqrt{x^2+a^2}} = \int \frac{a\sec^2\theta\,d\theta}{a\sec\theta} = \int \sec\theta\,d\theta = \ln|\sec\theta + \tan\theta| + C_1 = \ln\left|\frac{\sqrt{x^2+a^2}}{a} + \frac{x}{a}\right| + C_1$$

$$= \ln\left(x + \sqrt{x^2+a^2}\right) + C \quad \text{where } C = C_1 - \ln|a|$$

(b) Let $x = a\sinh t$, so that $dx = a\cosh t\,dt$ and $\sqrt{x^2+a^2} = a\cosh t$. Then

$$\int \frac{dx}{\sqrt{x^2+a^2}} = \int \frac{a\cosh t\,dt}{a\cosh t} = t + C = \sinh^{-1}\frac{x}{a} + C.$$

32. (a) Let $x = a\tan\theta$, $-\frac{\pi}{2} < \theta < \frac{\pi}{2}$. Then

$$I = \int \frac{x^2}{(x^2+a^2)^{3/2}}\,dx = \int \frac{a^2\tan^2\theta}{a^3\sec^3\theta}\,a\sec^2\theta\,d\theta = \int \frac{\tan^2\theta}{\sec\theta}\,d\theta = \int \frac{\sec^2\theta - 1}{\sec\theta}\,d\theta$$

$$= \int (\sec\theta - \cos\theta)\,d\theta = \ln|\sec\theta + \tan\theta| - \sin\theta + C$$

$$= \ln\left|\frac{\sqrt{x^2+a^2}}{a} + \frac{x}{a}\right| - \frac{x}{\sqrt{x^2+a^2}} + C = \ln\left(x + \sqrt{x^2+a^2}\right) - \frac{x}{\sqrt{x^2+a^2}} + C_1$$

(b) Let $x = a\sinh t$. Then

$$I = \int \frac{a^2\sinh^2 t}{a^3\cosh^3 t}\,a\cosh t\,dt = \int \tanh^2 t\,dt = \int (1 - \operatorname{sech}^2 t)\,dt = t - \tanh t + C$$

$$= \sinh^{-1}\frac{x}{a} - \frac{x}{\sqrt{a^2+x^2}} + C$$

33. The average value of $f(x) = \sqrt{x^2-1}/x$ on the interval $[1, 7]$ is

$$\frac{1}{7-1}\int_1^7 \frac{\sqrt{x^2-1}}{x}\,dx = \frac{1}{6}\int_0^\alpha \frac{\tan\theta}{\sec\theta}\cdot\sec\theta\tan\theta\,d\theta \qquad \left[\begin{array}{c}\text{where } x = \sec\theta,\ dx = \sec\theta\tan\theta\,d\theta, \\ \sqrt{x^2-1} = \tan\theta, \text{ and } \alpha = \sec^{-1} 7\end{array}\right]$$

$$= \tfrac{1}{6}\int_0^\alpha \tan^2\theta\,d\theta = \tfrac{1}{6}\int_0^\alpha (\sec^2\theta - 1)\,d\theta = \tfrac{1}{6}\Big[\tan\theta - \theta\Big]_0^\alpha$$

$$= \tfrac{1}{6}(\tan\alpha - \alpha) = \tfrac{1}{6}\left(\sqrt{48} - \sec^{-1} 7\right)$$

34. $9x^2 - 4y^2 = 36 \quad\Rightarrow\quad y = \pm\frac{3}{2}\sqrt{x^2-4} \quad\Rightarrow$

$\text{area} = 2\int_2^3 \frac{3}{2}\sqrt{x^2-4}\,dx = 3\int_2^3 \sqrt{x^2-4}\,dx$

$= 3\int_0^\alpha 2\tan\theta\, 2\sec\theta\,\tan\theta\,d\theta \qquad \left[\begin{array}{c}\text{where } x = 2\sec\theta, \\ dx = 2\sec\theta\,\tan\theta\,d\theta, \\ \alpha = \sec^{-1}\left(\frac{3}{2}\right)\end{array}\right]$

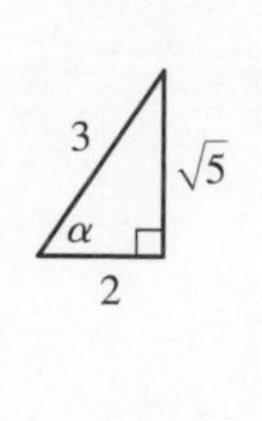

$= 12\int_0^\alpha (\sec^2\theta - 1)\sec\theta\,d\theta = 12\int_0^\alpha (\sec^3\theta - \sec\theta)\,d\theta$

$= 12\left[\frac{1}{2}(\sec\theta\,\tan\theta + \ln|\sec\theta+\tan\theta|) - \ln|\sec\theta+\tan\theta|\right]_0^\alpha$

$= 6\left[\sec\theta\,\tan\theta - \ln|\sec\theta+\tan\theta|\right]_0^\alpha = 6\left[\frac{3\sqrt{5}}{4} - \ln\left(\frac{3}{2}+\frac{\sqrt{5}}{2}\right)\right] = \frac{9\sqrt{5}}{2} - 6\ln\left(\frac{3+\sqrt{5}}{2}\right)$

35. Area of $\triangle POQ = \frac{1}{2}(r\cos\theta)(r\sin\theta) = \frac{1}{2}r^2\sin\theta\,\cos\theta$. Area of region $PQR = \int_{r\cos\theta}^{r}\sqrt{r^2-x^2}\,dx$. Let $x = r\cos u \;\Rightarrow\; dx = -r\sin u\,du$ for $\theta \le u \le \frac{\pi}{2}$. Then we obtain

$$\begin{aligned}\int\sqrt{r^2-x^2}\,dx &= \int r\sin u\,(-r\sin u)\,du = -r^2\int\sin^2 u\,du = -\tfrac{1}{2}r^2(u - \sin u\,\cos u) + C \\ &= -\tfrac{1}{2}r^2\cos^{-1}(x/r) + \tfrac{1}{2}x\sqrt{r^2-x^2} + C\end{aligned}$$

so

$$\begin{aligned}\text{area of region } PQR &= \tfrac{1}{2}\left[-r^2\cos^{-1}(x/r) + x\sqrt{r^2-x^2}\right]_{r\cos\theta}^{r} \\ &= \tfrac{1}{2}\left[0 - (-r^2\theta + r\cos\theta\, r\sin\theta)\right] = \tfrac{1}{2}r^2\theta - \tfrac{1}{2}r^2\sin\theta\,\cos\theta\end{aligned}$$

and thus, (area of sector POR) = (area of $\triangle POQ$) + (area of region PQR) $= \frac{1}{2}r^2\theta$.

36. Let $x = \sqrt{2}\sec\theta$, where $0 \le \theta < \frac{\pi}{2}$ or $\pi \le \theta < \frac{3\pi}{2}$, so $dx = \sqrt{2}\sec\theta\tan\theta\,d\theta$. Then

$$\begin{aligned}\int\frac{dx}{x^4\sqrt{x^2-2}} &= \int\frac{\sqrt{2}\sec\theta\,\tan\theta\,d\theta}{4\sec^4\theta\,\sqrt{2}\,\tan\theta} \\ &= \tfrac{1}{4}\int\cos^3\theta\,d\theta = \tfrac{1}{4}\int(1-\sin^2\theta)\cos\theta\,d\theta \\ &= \tfrac{1}{4}\left[\sin\theta - \tfrac{1}{3}\sin^3\theta\right] + C \qquad [\text{substitute } u = \sin\theta] \\ &= \frac{1}{4}\left[\frac{\sqrt{x^2-2}}{x} - \frac{(x^2-2)^{3/2}}{3x^3}\right] + C\end{aligned}$$

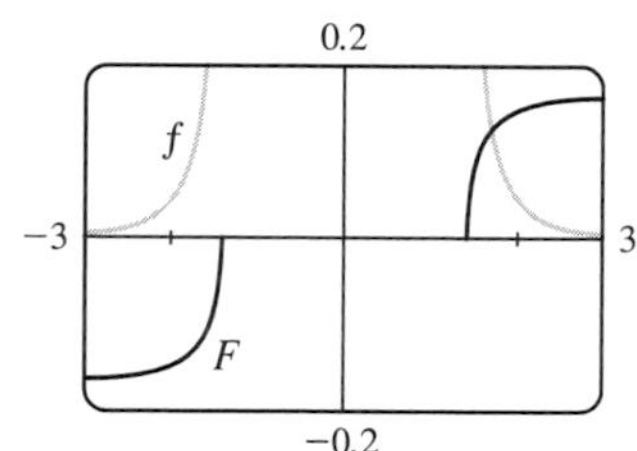

From the graph, it appears that our answer is reasonable. [Notice that $f(x)$ is large when F increases rapidly and small when F levels out.]

37. Use disks about the x-axis:

$$V = \int_0^3 \pi\left(\frac{9}{x^2+9}\right)^2 dx = 81\pi\int_0^3 \frac{1}{(x^2+9)^2}\,dx$$

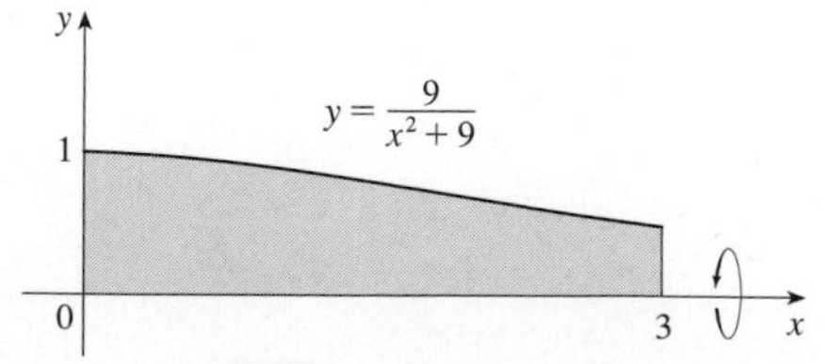

Let $x = 3\tan\theta$, so $dx = 3\sec^2\theta\,d\theta$, $x = 0 \;\Rightarrow\; \theta = 0$ and $x = 3 \;\Rightarrow\; \theta = \frac{\pi}{4}$. Thus,

$$\begin{aligned}V &= 81\pi\int_0^{\pi/4}\frac{1}{(9\sec^2\theta)^2}\,3\sec^2\theta\,d\theta = 3\pi\int_0^{\pi/4}\cos^2\theta\,d\theta = 3\pi\int_0^{\pi/4}\frac{1}{2}(1+\cos 2\theta)\,d\theta \\ &= \tfrac{3\pi}{2}\left[\theta + \tfrac{1}{2}\sin 2\theta\right]_0^{\pi/4} = \tfrac{3\pi}{2}\left[\left(\tfrac{\pi}{4}+\tfrac{1}{2}\right) - 0\right] = \tfrac{3}{8}\pi^2 + \tfrac{3}{4}\pi\end{aligned}$$

38. Use shells about $x = 1$:

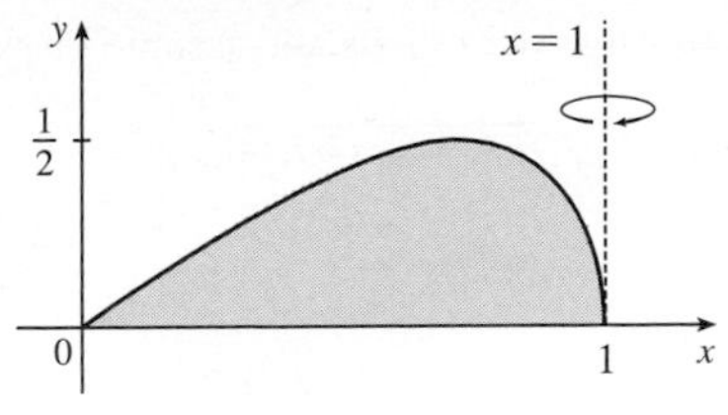

$V = \int_0^1 2\pi(1-x)\,x\sqrt{1-x^2}\,dx$

$= 2\pi\int_0^1 x\sqrt{1-x^2}\,dx - 2\pi\int_0^1 x^2\sqrt{1-x^2}\,dx = 2\pi V_1 - 2\pi V_2$

For V_1, let $u = 1-x^2$, so $du = -2x\,dx$, and

$V_1 = \int_1^0 \sqrt{u}\left(-\frac{1}{2}\,du\right) = \frac{1}{2}\int_0^1 u^{1/2}\,du = \frac{1}{2}\left[\frac{2}{3}u^{3/2}\right]_0^1 = \frac{1}{2}\left(\frac{2}{3}\right) = \frac{1}{3}.$

For V_2, let $x = \sin\theta$, so $dx = \cos\theta\,d\theta$, and

$$V_2 = \int_0^{\pi/2} \sin^2\theta\sqrt{\cos^2\theta}\,\cos\theta\,d\theta = \int_0^{\pi/2} \sin^2\theta\,\cos^2\theta\,d\theta = \int_0^{\pi/2} \frac{1}{4}(2\sin\theta\,\cos\theta)^2\,d\theta$$

$$= \frac{1}{4}\int_0^{\pi/2}\sin^2 2\theta\,d\theta = \frac{1}{4}\int_0^{\pi/2}\frac{1}{2}(1-\cos 2\theta)\,d\theta = \frac{1}{8}\left[\theta - \frac{1}{2}\sin 2\theta\right]_0^{\pi/2} = \frac{1}{8}\left(\frac{\pi}{2}\right) = \frac{\pi}{16}$$

Thus, $V = 2\pi\left(\frac{1}{3}\right) - 2\pi\left(\frac{\pi}{16}\right) = \frac{2}{3}\pi - \frac{1}{8}\pi^2.$

39. (a) Let $t = a\sin\theta$, $dt = a\cos\theta\,d\theta$, $t = 0 \Rightarrow \theta = 0$ and $t = x \Rightarrow$ $\theta = \sin^{-1}(x/a)$. Then

$$\int_0^x \sqrt{a^2-t^2}\,dt = \int_0^{\sin^{-1}(x/a)} a\cos\theta\,(a\cos\theta\,d\theta) = a^2\int_0^{\sin^{-1}(x/a)}\cos^2\theta\,d\theta$$

$$= \frac{a^2}{2}\int_0^{\sin^{-1}(x/a)}(1+\cos 2\theta)\,d\theta = \frac{a^2}{2}\left[\theta + \frac{1}{2}\sin 2\theta\right]_0^{\sin^{-1}(x/a)} = \frac{a^2}{2}\left[\theta + \sin\theta\,\cos\theta\right]_0^{\sin^{-1}(x/a)}$$

$$= \frac{a^2}{2}\left[\left(\sin^{-1}\left(\frac{x}{a}\right) + \frac{x}{a}\cdot\frac{\sqrt{a^2-x^2}}{a}\right) - 0\right] = \frac{1}{2}a^2\sin^{-1}(x/a) + \frac{1}{2}x\sqrt{a^2-x^2}$$

(b) The integral $\int_0^x \sqrt{a^2-t^2}\,dt$ represents the area under the curve $y = \sqrt{a^2-t^2}$ between the vertical lines $t = 0$ and $t = x$. The figure shows that this area consists of a triangular region and a sector of the circle $t^2 + y^2 = a^2$. The triangular region has base x and height $\sqrt{a^2-x^2}$, so its area is $\frac{1}{2}x\sqrt{a^2-x^2}$. The sector has area $\frac{1}{2}a^2\theta = \frac{1}{2}a^2\sin^{-1}(x/a)$.

40. The curves intersect when $x^2 + \left(\frac{1}{2}x^2\right)^2 = 8 \Leftrightarrow x^2 + \frac{1}{4}x^4 = 8 \Leftrightarrow x^4 + 4x^2 - 32 = 0 \Leftrightarrow$ $(x^2+8)(x^2-4) = 0 \Leftrightarrow x = \pm 2$. The area inside the circle and above the parabola is given by

$A_1 = \int_{-2}^2\left(\sqrt{8-x^2} - \frac{1}{2}x^2\right)dx = 2\int_0^2\sqrt{8-x^2}\,dx - 2\int_0^2\frac{1}{2}x^2\,dx$

$= 2\left[\frac{1}{2}(8)\sin^{-1}\left(\frac{2}{\sqrt{8}}\right) + \frac{1}{2}(2)\sqrt{8-2^2} - \frac{1}{2}\left[\frac{1}{3}x^3\right]_0^2\right]$ [by Exercise 39]

$= 8\sin^{-1}\left(\frac{1}{\sqrt{2}}\right) + 2\sqrt{4} - \frac{8}{3} = 8\left(\frac{\pi}{4}\right) + 4 - \frac{8}{3} = 2\pi + \frac{4}{3}$

Since the area of the disk is $\pi\left(\sqrt{8}\right)^2 = 8\pi$, the area inside the circle and below the parabola is $A_2 = 8\pi - \left(2\pi + \frac{4}{3}\right) = 6\pi - \frac{4}{3}$.

41. We use cylindrical shells and assume that $R > r$. $x^2 = r^2 - (y-R)^2 \Rightarrow x = \pm\sqrt{r^2-(y-R)^2}$, so $g(y) = 2\sqrt{r^2-(y-R)^2}$ and

$$V = \int_{R-r}^{R+r} 2\pi y \cdot 2\sqrt{r^2-(y-R)^2}\,dy = \int_{-r}^{r} 4\pi(u+R)\sqrt{r^2-u^2}\,du \qquad [\text{where } u = y-R]$$

$$= 4\pi\int_{-r}^{r} u\sqrt{r^2-u^2}\,du + 4\pi R\int_{-r}^{r}\sqrt{r^2-u^2}\,du \qquad \left[\begin{array}{c}\text{where } u = r\sin\theta,\ du = r\cos\theta\,d\theta\\ \text{in the second integral}\end{array}\right]$$

$$= 4\pi\left[-\tfrac{1}{3}(r^2-u^2)^{3/2}\right]_{-r}^{r} + 4\pi R\int_{-\pi/2}^{\pi/2} r^2\cos^2\theta\,d\theta = -\tfrac{4\pi}{3}(0-0) + 4\pi Rr^2\int_{-\pi/2}^{\pi/2}\cos^2\theta\,d\theta$$

$$= 2\pi Rr^2\int_{-\pi/2}^{\pi/2}(1+\cos 2\theta)\,d\theta = 2\pi Rr^2\left[\theta + \tfrac{1}{2}\sin 2\theta\right]_{-\pi/2}^{\pi/2} = 2\pi^2 Rr^2$$

Another method: Use washers instead of shells, so $V = 8\pi R\int_0^r\sqrt{r^2-y^2}\,dy$ as in Exercise 6.2.63(a), but evaluate the integral using $y = r\sin\theta$.

42. Let $x = b\tan\theta$, so that $dx = b\sec^2\theta\,d\theta$ and $\sqrt{x^2+b^2} = b\sec\theta$.

$$E(P) = \int_{-a}^{L-a}\frac{\lambda b}{4\pi\varepsilon_0(x^2+b^2)^{3/2}}\,dx = \frac{\lambda b}{4\pi\varepsilon_0}\int_{\theta_1}^{\theta_2}\frac{1}{(b\sec\theta)^3}\,b\sec^2\theta\,d\theta$$

$$= \frac{\lambda}{4\pi\varepsilon_0 b}\int_{\theta_1}^{\theta_2}\frac{1}{\sec\theta}\,d\theta = \frac{\lambda}{4\pi\varepsilon_0 b}\int_{\theta_1}^{\theta_2}\cos\theta\,d\theta = \frac{\lambda}{4\pi\varepsilon_0 b}\Big[\sin\theta\Big]_{\theta_1}^{\theta_2}$$

$$= \frac{\lambda}{4\pi\varepsilon_0 b}\left[\frac{x}{\sqrt{x^2+b^2}}\right]_{-a}^{L-a} = \frac{\lambda}{4\pi\varepsilon_0 b}\left(\frac{L-a}{\sqrt{(L-a)^2+b^2}} + \frac{a}{\sqrt{a^2+b^2}}\right)$$

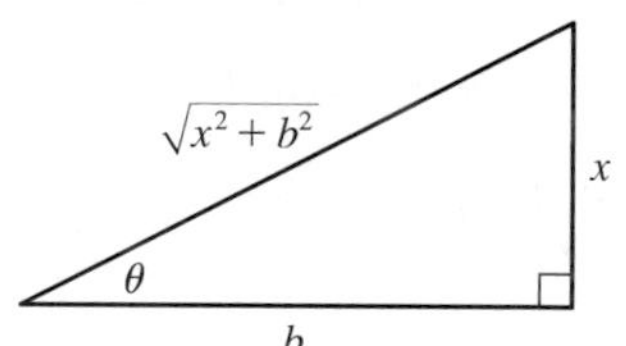

43. Let the equation of the large circle be $x^2+y^2 = R^2$. Then the equation of the small circle is $x^2+(y-b)^2 = r^2$, where $b = \sqrt{R^2-r^2}$ is the distance between the centers of the circles. The desired area is

$$A = \int_{-r}^{r}\left[\left(b+\sqrt{r^2-x^2}\right) - \sqrt{R^2-x^2}\right]dx$$

$$= 2\int_0^r\left(b+\sqrt{r^2-x^2}-\sqrt{R^2-x^2}\right)dx$$

$$= 2\int_0^r b\,dx + 2\int_0^r\sqrt{r^2-x^2}\,dx - 2\int_0^r\sqrt{R^2-x^2}\,dx$$

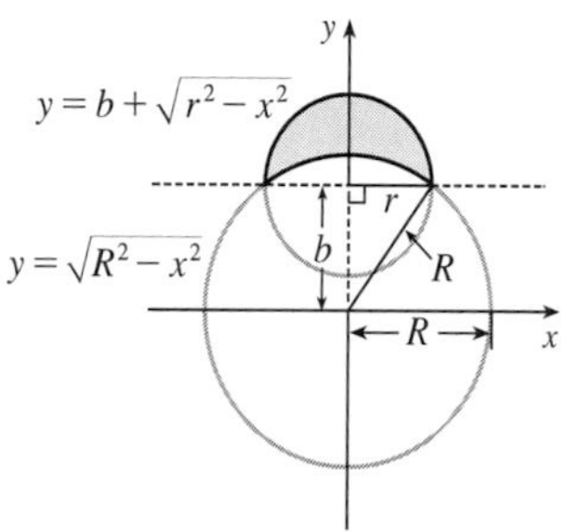

The first integral is just $2br = 2r\sqrt{R^2-r^2}$. The second integral represents the area of a quarter-circle of radius r, so its value is $\frac{1}{4}\pi r^2$. To evaluate the other integral, note that

$$\int\sqrt{a^2-x^2}\,dx = \int a^2\cos^2\theta\,d\theta \quad [x = a\sin\theta,\ dx = a\cos\theta\,d\theta] \;=\left(\tfrac{1}{2}a^2\right)\int(1+\cos 2\theta)\,d\theta$$

$$= \tfrac{1}{2}a^2\left(\theta+\tfrac{1}{2}\sin 2\theta\right) + C = \tfrac{1}{2}a^2(\theta+\sin\theta\,\cos\theta) + C$$

$$= \frac{a^2}{2}\arcsin\left(\frac{x}{a}\right) + \frac{a^2}{2}\left(\frac{x}{a}\right)\frac{\sqrt{a^2-x^2}}{a} + C = \frac{a^2}{2}\arcsin\left(\frac{x}{a}\right) + \frac{x}{2}\sqrt{a^2-x^2} + C$$

Thus, the desired area is

$$A = 2r\sqrt{R^2-r^2} + 2\left(\tfrac{1}{4}\pi r^2\right) - \left[R^2\arcsin(x/R) + x\sqrt{R^2-x^2}\right]_0^r$$

$$= 2r\sqrt{R^2-r^2} + \tfrac{1}{2}\pi r^2 - \left[R^2\arcsin(r/R) + r\sqrt{R^2-r^2}\right] = r\sqrt{R^2-r^2} + \tfrac{\pi}{2}r^2 - R^2\arcsin(r/R)$$

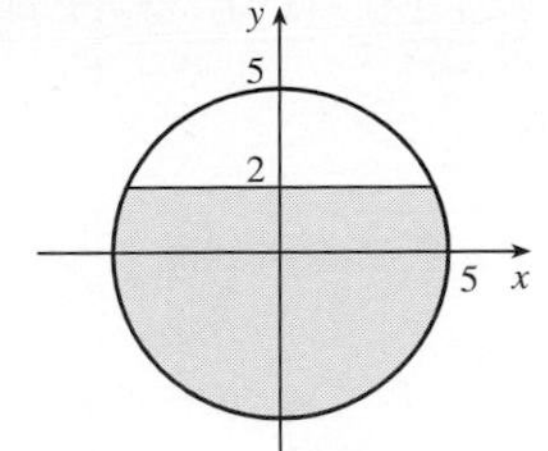

44. Note that the circular cross-sections of the tank are the same everywhere, so the percentage of the total capacity that is being used is equal to the percentage of any cross-section that is under water. The underwater area is

$$\begin{aligned} A &= 2\int_{-5}^{2}\sqrt{25-y^2}\,dy \\ &= \left[25\arcsin(y/5) + y\sqrt{25-y^2}\right]_{-5}^{2} \qquad [\text{substitute } y = 5\sin\theta] \\ &= 25\arcsin\tfrac{2}{5} + 2\sqrt{21} + \tfrac{25}{2}\pi \approx 58.72 \text{ ft}^2 \end{aligned}$$

so the fraction of the total capacity in use is $\dfrac{A}{\pi(5)^2} \approx \dfrac{58.72}{25\pi} \approx 0.748$ or 74.8%.

7.4 Integration of Rational Functions by Partial Fractions

1. (a) $\dfrac{4+x}{(1+2x)(3-x)} = \dfrac{A}{1+2x} + \dfrac{B}{3-x}$

(b) $\dfrac{1-x}{x^3+x^4} = \dfrac{1-x}{x^3(1+x)} = \dfrac{A}{x} + \dfrac{B}{x^2} + \dfrac{C}{x^3} + \dfrac{D}{1+x}$

2. (a) $\dfrac{x-6}{x^2+x-6} = \dfrac{x-6}{(x+3)(x-2)} = \dfrac{A}{x+3} + \dfrac{B}{x-2}$

(b) $\dfrac{x^2}{x^2+x+6} = \dfrac{(x^2+x+6)-(x+6)}{x^2+x+6} = 1 - \dfrac{x+6}{x^2+x+6}$

Notice that x^2+x+6 can't be factored because its discriminant is $b^2-4ac = -23 < 0$.

3. (a) $\dfrac{1}{x^2+x^4} = \dfrac{1}{x^2(1+x^2)} = \dfrac{A}{x} + \dfrac{B}{x^2} + \dfrac{Cx+D}{1+x^2}$

(b) $$\begin{aligned} \frac{x^3+1}{x^3-3x^2+2x} &= \frac{(x^3-3x^2+2x)+3x^2-2x+1}{x^3-3x^2+2x} = 1 + \frac{3x^2-2x+1}{x(x^2-3x+2)} \qquad [\text{or use long division}] \\ &= 1 + \frac{3x^2-2x+1}{x(x-1)(x-2)} = 1 + \frac{A}{x} + \frac{B}{x-1} + \frac{C}{x-2} \end{aligned}$$

4. (a) $$\begin{aligned} \frac{x^4-2x^3+x^2+2x-1}{x^2-2x+1} &= \frac{x^2(x^2-2x+1)+2x-1}{x^2-2x+1} = x^2 + \frac{2x-1}{(x-1)^2} \qquad [\text{or use long division}] \\ &= x^2 + \frac{A}{x-1} + \frac{B}{(x-1)^2} \end{aligned}$$

(b) $\dfrac{x^2-1}{x^3+x^2+x} = \dfrac{x^2-1}{x(x^2+x+1)} = \dfrac{A}{x} + \dfrac{Bx+C}{x^2+x+1}$

5. (a) $$\begin{aligned} \frac{x^6}{x^2-4} &= x^4 + 4x^2 + 16 + \frac{64}{(x+2)(x-2)} \qquad [\text{by long division}] \\ &= x^4 + 4x^2 + 16 + \frac{A}{x+2} + \frac{B}{x-2} \end{aligned}$$

(b) $\dfrac{x^4}{(x^2-x+1)(x^2+2)^2} = \dfrac{Ax+B}{x^2-x+1} + \dfrac{Cx+D}{x^2+2} + \dfrac{Ex+F}{(x^2+2)^2}$

6. (a) $\dfrac{t^6+1}{t^6+t^3} = \dfrac{(t^6+t^3)-t^3+1}{t^6+t^3} = 1+\dfrac{-t^3+1}{t^3(t^3+1)} = 1+\dfrac{-t^3+1}{t^3(t+1)(t^2-t+1)} = 1+\dfrac{A}{t}+\dfrac{B}{t^2}+\dfrac{C}{t^3}+\dfrac{D}{t+1}+\dfrac{Ex+F}{t^2-t+1}$

(b) $\dfrac{x^5+1}{(x^2-x)(x^4+2x^2+1)} = \dfrac{x^5+1}{x(x-1)(x^2+1)^2} = \dfrac{A}{x}+\dfrac{B}{x-1}+\dfrac{Cx+D}{x^2+1}+\dfrac{Ex+F}{(x^2+1)^2}$

7. $\displaystyle\int \frac{x^4}{x-1}\,dx = \int\left(x^3+x^2+x+1+\frac{1}{x-1}\right)dx$ [by division] $= \frac{1}{4}x^4+\frac{1}{3}x^3+\frac{1}{2}x^2+x+\ln|x-1|+C$

8. $\displaystyle\int \frac{3t-2}{t+1}\,dt = \int\left(3-\frac{5}{t+1}\right)dt = 3t-5\ln|t+1|+C$

9. $\dfrac{5x+1}{(2x+1)(x-1)} = \dfrac{A}{2x+1}+\dfrac{B}{x-1}$. Multiply both sides by $(2x+1)(x-1)$ to get $5x+1 = A(x-1)+B(2x+1) \;\Rightarrow$

$5x+1 = Ax-A+2Bx+B \;\Rightarrow\; 5x+1 = (A+2B)x+(-A+B)$.

The coefficients of x must be equal and the constant terms are also equal, so $A+2B=5$ and $-A+B=1$. Adding these equations gives us $3B=6 \;\Leftrightarrow\; B=2$, and hence, $A=1$. Thus,

$$\int \frac{5x+1}{(2x+1)(x-1)}\,dx = \int\left(\frac{1}{2x+1}+\frac{2}{x-1}\right)dx = \tfrac{1}{2}\ln|2x+1|+2\ln|x-1|+C.$$

Another method: Substituting 1 for x in the equation $5x+1=A(x-1)+B(2x+1)$ gives $6=3B \;\Leftrightarrow\; B=2$.

Substituting $-\frac{1}{2}$ for x gives $-\frac{3}{2} = -\frac{3}{2}A \;\Leftrightarrow\; A=1$.

10. $\dfrac{y}{(y+4)(2y-1)} = \dfrac{A}{y+4}+\dfrac{B}{2y-1}$. Multiply both sides by $(y+4)(2y-1)$ to get $y = A(2y-1)+B(y+4) \;\Rightarrow$

$y = 2Ay-A+By+4B \;\Rightarrow\; y=(2A+B)y+(-A+4B)$. The coefficients of y must be equal and the constant terms are also equal, so $2A+B=1$ and $-A+4B=0$. Adding 2 times the second equation and the first equation gives us $9B=1 \;\Leftrightarrow\; B=\frac{1}{9}$ and hence, $A=\frac{4}{9}$. Thus,

$$\begin{aligned}\int \frac{y\,dy}{(y+4)(2y-1)} &= \int\left(\frac{\frac{4}{9}}{y+4}+\frac{\frac{1}{9}}{2y-1}\right)dy = \frac{4}{9}\ln|y+4|+\frac{1}{9}\cdot\frac{1}{2}\ln|2y-1|+C\\ &= \tfrac{4}{9}\ln|y+4|+\tfrac{1}{18}\ln|2y-1|+C\end{aligned}$$

Another method: Substituting $\frac{1}{2}$ for y in the equation $y=A(2y-1)+B(y+4)$ gives $\frac{1}{2}=\frac{9}{2}B \;\Leftrightarrow\; B=\frac{1}{9}$.

Substituting -4 for y gives $-4=-9A \;\Leftrightarrow\; A=\frac{4}{9}$.

11. $\dfrac{2}{2x^2+3x+1} = \dfrac{2}{(2x+1)(x+1)} = \dfrac{A}{2x+1}+\dfrac{B}{x+1}$. Multiply both sides by $(2x+1)(x+1)$ to get

$2=A(x+1)+B(2x+1)$. The coefficients of x must be equal and the constant terms are also equal, so $A+2B=0$ and $A+B=2$. Subtracting the second equation from the first gives $B=-2$, and hence, $A=4$. Thus,

$$\int_0^1 \frac{2}{2x^2+3x+1}\,dx = \int_0^1\left(\frac{4}{2x+1}-\frac{2}{x+1}\right)dx = \left[\frac{4}{2}\ln|2x+1|-2\ln|x+1|\right]_0^1 = (2\ln 3-2\ln 2)-0 = 2\ln\frac{3}{2}.$$

Another method: Substituting -1 for x in the equation $2=A(x+1)+B(2x+1)$ gives $2=-B \;\Leftrightarrow\; B=-2$.

Substituting $-\frac{1}{2}$ for x gives $2=\frac{1}{2}A \;\Leftrightarrow\; A=4$.

12. $\dfrac{x-4}{x^2-5x+6} = \dfrac{A}{x-2} + \dfrac{B}{x-3}$. Multiply both sides by $(x-2)(x-3)$ to get $x-4 = A(x-3) + B(x-2) \;\Rightarrow$

$x-4 = Ax - 3A + Bx - 2B \;\Rightarrow\; x-4 = (A+B)x + (-3A-2B)$.

The coefficients of x must be equal and the constant terms are also equal, so $A+B=1$ and $-3A-2B=-4$.

Adding twice the first equation to the second gives us $-A=-2 \;\Leftrightarrow\; A=2$, and hence, $B=-1$.Thus,

$$\int_0^1 \frac{x-4}{x^2-5x+6}\,dx = \int_0^1 \left(\frac{2}{x-2} - \frac{1}{x-3}\right) dx = \left[2\ln|x-2| - \ln|x-3|\right]_0^1$$

$$= (0 - \ln 2) - (2\ln 2 - \ln 3) = -3\ln 2 + \ln 3 \;\;\left[\text{or } \ln\tfrac{3}{8}\right]$$

Another method: Substituting 3 for x in the equation $x-4 = A(x-3)+B(x-2)$ gives $-1=B$. Substituting 2 for x gives $-2=-A \;\Leftrightarrow\; A=2$.

13. $\displaystyle\int \frac{ax}{x^2-bx}\,dx = \int \frac{ax}{x(x-b)}\,dx = \int \frac{a}{x-b}\,dx = a\ln|x-b| + C$

14. If $a \neq b$, $\dfrac{1}{(x+a)(x+b)} = \dfrac{1}{b-a}\left(\dfrac{1}{x+a} - \dfrac{1}{x+b}\right)$, so if $a \neq b$, then

$$\int \frac{dx}{(x+a)(x+b)} = \frac{1}{b-a}\left(\ln|x+a| - \ln|x+b|\right) + C = \frac{1}{b-a}\ln\left|\frac{x+a}{x+b}\right| + C$$

If $a=b$, then $\displaystyle\int \frac{dx}{(x+a)^2} = -\frac{1}{x+a} + C$.

15. $\dfrac{x^3-4x+1}{x^2-3x+2} = x+3+\dfrac{3x-5}{(x-1)(x-2)}$. Write $\dfrac{3x-5}{(x-1)(x-2)} = \dfrac{A}{x-1} + \dfrac{B}{x-2}$. Multiplying both sides by $(x-1)(x-2)$ gives $3x-5 = A(x-2)+B(x-1)$. Substituting 2 for x gives $1=B$. Substituting 1 for x gives $-2=-A \;\Leftrightarrow\; A=2$. Thus,

$$\int_{-1}^{0} \frac{x^3-4x+1}{x^2-3x+2}\,dx = \int_{-1}^{0}\left(x+3+\frac{2}{x-1}+\frac{1}{x-2}\right)dx = \left[\tfrac{1}{2}x^2 + 3x + 2\ln|x-1| + \ln|x-2|\right]_{-1}^{0}$$

$$= (0+0+0+\ln 2) - \left(\tfrac{1}{2} - 3 + 2\ln 2 + \ln 3\right) = \tfrac{5}{2} - \ln 2 - \ln 3, \text{ or } \tfrac{5}{2} - \ln 6$$

16. $\dfrac{x^3+4x^2+x-1}{x^3+x^2} = 1 + \dfrac{3x^2+x-1}{x^2(x+1)}$. Write $\dfrac{3x^2+x-1}{x^2(x+1)} = \dfrac{A}{x} + \dfrac{B}{x^2} + \dfrac{C}{x+1}$. Multiplying both sides by $x^2(x+1)$ gives $3x^2+x-1 = Ax(x+1)+B(x+1)+Cx^2$. Substituting 0 for x gives $-1=B$. Substituting -1 for x gives $1=C$. Equating coefficients of x^2 gives $3=A+C=A+1$, so $A=2$. Thus,

$$\int_1^2 \frac{x^3+4x^2+x-1}{x^3+x^2}\,dx = \int_1^2\left(1+\frac{2}{x}-\frac{1}{x^2}+\frac{1}{x+1}\right)dx = \left[x + 2\ln|x| + \frac{1}{x} + \ln|x+1|\right]_1^2$$

$$= \left(2 + 2\ln 2 + \tfrac{1}{2} + \ln 3\right) - (1+0+1+\ln 2) = \tfrac{1}{2} + \ln 2 + \ln 3, \text{ or } \tfrac{1}{2} + \ln 6.$$

17. $\dfrac{4y^2-7y-12}{y(y+2)(y-3)} = \dfrac{A}{y} + \dfrac{B}{y+2} + \dfrac{C}{y-3} \;\Rightarrow\; 4y^2-7y-12 = A(y+2)(y-3)+By(y-3)+Cy(y+2)$. Setting $y=0$ gives $-12=-6A$, so $A=2$. Setting $y=-2$ gives $18=10B$, so $B=\frac{9}{5}$. Setting $y=3$ gives $3=15C$, so $C=\frac{1}{5}$.

Now

$$\int_1^2 \frac{4y^2-7y-12}{y(y+2)(y-3)}\,dy = \int_1^2 \left(\frac{2}{y}+\frac{9/5}{y+2}+\frac{1/5}{y-3}\right)dy = \left[2\ln|y|+\tfrac{9}{5}\ln|y+2|+\tfrac{1}{5}\ln|y-3|\right]_1^2$$
$$= 2\ln 2+\tfrac{9}{5}\ln 4+\tfrac{1}{5}\ln 1-2\ln 1-\tfrac{9}{5}\ln 3-\tfrac{1}{5}\ln 2$$
$$= 2\ln 2+\tfrac{18}{5}\ln 2-\tfrac{1}{5}\ln 2-\tfrac{9}{5}\ln 3=\tfrac{27}{5}\ln 2-\tfrac{9}{5}\ln 3=\tfrac{9}{5}(3\ln 2-\ln 3)=\tfrac{9}{5}\ln\tfrac{8}{3}$$

18. $\dfrac{3x^2+6x+2}{x^2+3x+2}=3+\dfrac{-3x-4}{(x+1)(x+2)}$. Write $\dfrac{-3x-4}{(x+1)(x+2)}=\dfrac{A}{x+1}+\dfrac{B}{x+2}$. Multiplying both sides by $(x+1)(x+2)$ gives $-3x-4=A(x+2)+B(x+1)$. Substituting -2 for x gives $2=-B \quad\Leftrightarrow\quad B=-2$. Substituting -1 for x gives $-1=A$. Thus,

$$\int_1^2 \frac{3x^2+6x+2}{x^2+3x+2}\,dx=\int_1^2\left(3-\frac{1}{x+1}-\frac{2}{x+2}\right)dx=\Big[3x-\ln|x+1|-2\ln|x+2|\Big]_1^2$$
$$=(6-\ln 3-2\ln 4)-(3-\ln 2-2\ln 3)=3+\ln 2+\ln 3-2\ln 4,\text{ or } 3+\ln\tfrac{3}{8}$$

19. $\dfrac{x^2+x+1}{(x+1)^2(x+2)}=\dfrac{A}{x+1}+\dfrac{B}{(x+1)^2}+\dfrac{C}{x+2}$. Multiplying both sides by $(x+1)^2(x+2)$ gives $x^2+x+1=A(x+1)(x+2)+B(x+2)+C(x+1)^2$. Substituting -1 for x gives $1=B$. Substituting -2 for x gives $3=C$. Equating coefficients of x^2 gives $1=A+C=A+3$, so $A=-2$. Thus,

$$\int_0^1 \frac{x^2+x+1}{(x+1)^2(x+2)}\,dx=\int_0^1\left(\frac{-2}{x+1}+\frac{1}{(x+1)^2}+\frac{3}{x+2}\right)dx=\left[-2\ln|x+1|-\frac{1}{x+1}+3\ln|x+2|\right]_0^1$$
$$=\left(-2\ln 2-\tfrac{1}{2}+3\ln 3\right)-(0-1+3\ln 2)=\tfrac{1}{2}-5\ln 2+3\ln 3,\text{ or } \tfrac{1}{2}+\ln\tfrac{27}{32}$$

20. $\dfrac{x(3-5x)}{(3x-1)(x-1)^2}=\dfrac{A}{3x-1}+\dfrac{B}{x-1}+\dfrac{C}{(x-1)^2}$. Multiplying both sides by $(3x-1)(x-1)^2$ gives $x(3-5x)=A(x-1)^2+B(x-1)(3x-1)+C(3x-1)$. Substituting 1 for x gives $-2=2C \quad\Leftrightarrow\quad C=-1$. Substituting $\frac{1}{3}$ for x gives $\frac{4}{9}=\frac{4}{9}A \quad\Leftrightarrow\quad A=1$. Substituting 0 for x gives $0=A+B-C=1+B+1$, so $B=-2$. Thus,

$$\int_2^3 \frac{x(3-5x)}{(3x-1)(x-1)^2}\,dx=\int_2^3\left[\frac{1}{3x-1}-\frac{2}{x-1}-\frac{1}{(x-1)^2}\right]dx=\left[\frac{1}{3}\ln|3x-1|-2\ln|x-1|+\frac{1}{x-1}\right]_2^3$$
$$=\left(\tfrac{1}{3}\ln 8-2\ln 2+\tfrac{1}{2}\right)-\left(\tfrac{1}{3}\ln 5-0+1\right)=-\ln 2-\tfrac{1}{3}\ln 5-\tfrac{1}{2}$$

21. $\dfrac{1}{(t^2-1)^2}=\dfrac{1}{(t+1)^2(t-1)^2}=\dfrac{A}{t+1}+\dfrac{B}{(t+1)^2}+\dfrac{C}{t-1}+\dfrac{D}{(t-1)^2}$. Multiplying both sides by $(t+1)^2(t-1)^2$ gives $1=A(t+1)(t-1)^2+B(t-1)^2+C(t-1)(t+1)^2+D(t+1)^2$. Substituting 1 for t gives $1=4D \quad\Leftrightarrow\quad D=\frac{1}{4}$. Substituting -1 for t gives $1=4B \quad\Leftrightarrow\quad B=\frac{1}{4}$. Substituting 0 for t gives $1=A+B-C+D=A+\frac{1}{4}-C+\frac{1}{4}$, so $\frac{1}{2}=A-C$. Equating coefficients of t^3 gives $0=A+C$. Adding the last two equations gives $2A=\frac{1}{2} \quad\Leftrightarrow\quad A=\frac{1}{4}$, and so $C=-\frac{1}{4}$. Thus,

$$\int\frac{dt}{(t^2-1)^2}=\int\left[\frac{1/4}{t+1}+\frac{1/4}{(t+1)^2}-\frac{1/4}{t-1}+\frac{1/4}{(t-1)^2}\right]dt$$
$$=\frac{1}{4}\left[\ln|t+1|-\frac{1}{t+1}-\ln|t-1|-\frac{1}{t-1}\right]+C,\text{ or } \frac{1}{4}\left(\ln\left|\frac{t+1}{t-1}\right|+\frac{2t}{1-t^2}\right)+C$$

22. $\displaystyle\int \frac{x^4+9x^2+x+2}{x^2+9}\,dx = \int\left(x^2+\frac{x+2}{x^2+9}\right)dx = \int\left(x^2+\frac{x}{x^2+9}+\frac{2}{x^2+9}\right)dx$

$$= \frac{1}{3}x^3 + \frac{1}{2}\ln(x^2+9) + \frac{2}{3}\tan^{-1}\frac{x}{3} + C$$

23. $\dfrac{10}{(x-1)(x^2+9)} = \dfrac{A}{x-1} + \dfrac{Bx+C}{x^2+9}$. Multiply both sides by $(x-1)(x^2+9)$ to get

$10 = A(x^2+9) + (Bx+C)(x-1)$ $(\star)$. Substituting 1 for x gives $10 = 10A \;\Leftrightarrow\; A = 1$. Substituting 0 for x gives $10 = 9A - C \;\Rightarrow\; C = 9(1) - 10 = -1$. The coefficients of the x^2-terms in $(\star)$ must be equal, so $0 = A + B \;\Rightarrow\; B = -1$. Thus,

$$\int \frac{10}{(x-1)(x^2+9)}\,dx = \int\left(\frac{1}{x-1}+\frac{-x-1}{x^2+9}\right)dx = \int\left(\frac{1}{x-1}-\frac{x}{x^2+9}-\frac{1}{x^2+9}\right)dx$$
$$= \ln|x-1| - \tfrac{1}{2}\ln(x^2+9) - \tfrac{1}{3}\tan^{-1}\left(\tfrac{x}{3}\right) + C$$

In the second term we used the substitution $u = x^2+9$ and in the last term we used Formula 10.

24. $\dfrac{x^2-x+6}{x^3+3x} = \dfrac{x^2-x+6}{x(x^2+3)} = \dfrac{A}{x} + \dfrac{Bx+C}{x^2+3}$. Multiply by $x(x^2+3)$ to get $x^2 - x + 6 = A(x^2+3) + (Bx+C)x$.

Substituting 0 for x gives $6 = 3A \;\Leftrightarrow\; A = 2$. The coefficients of the x^2-terms must be equal, so $1 = A + B \;\Rightarrow\; B = 1 - 2 = -1$. The coefficients of the x-terms must be equal, so $-1 = C$. Thus,

$$\int \frac{x^2-x+6}{x^3+3x}\,dx = \int\left(\frac{2}{x}+\frac{-x-1}{x^2+3}\right)dx = \int\left(\frac{2}{x}-\frac{x}{x^2+3}-\frac{1}{x^2+3}\right)dx$$
$$= 2\ln|x| - \frac{1}{2}\ln(x^2+3) - \frac{1}{\sqrt{3}}\tan^{-1}\frac{x}{\sqrt{3}} + C$$

25. $\dfrac{4x}{x^3+x^2+x+1} = \dfrac{4x}{x^2(x+1)+1(x+1)} = \dfrac{4x}{(x+1)(x^2+1)} = \dfrac{A}{x+1} + \dfrac{Bx+C}{x^2+1}$. Multiply both sides by $(x+1)(x^2+1)$ to get $4x = A(x^2+1) + (Bx+C)(x+1) \;\Leftrightarrow\; 4x = Ax^2 + A + Bx^2 + Bx + Cx + C \;\Leftrightarrow\; 4x = (A+B)x^2 + (B+C)x + (A+C)$. Comparing coefficients gives us the following system of equations:

$$A + B = 0 \quad \textbf{(1)} \qquad B + C = 4 \quad \textbf{(2)} \qquad A + C = 0 \quad \textbf{(3)}$$

Subtracting equation **(1)** from equation **(2)** gives us $-A + C = 4$, and adding that equation to equation **(3)** gives us $2C = 4 \;\Leftrightarrow\; C = 2$, and hence $A = -2$ and $B = 2$. Thus,

$$\int \frac{4x}{x^3+x^2+x+1}\,dx = \int\left(\frac{-2}{x+1}+\frac{2x+2}{x^2+1}\right)dx = \int\left(\frac{-2}{x+1}+\frac{2x}{x^2+1}+\frac{2}{x^2+1}\right)dx$$
$$= -2\ln|x+1| + \ln(x^2+1) + 2\tan^{-1}x + C$$

26. $\displaystyle\int \frac{x^2+x+1}{(x^2+1)^2}\,dx = \int \frac{x^2+1}{(x^2+1)^2}\,dx + \int \frac{x}{(x^2+1)^2}\,dx = \int \frac{1}{x^2+1}\,dx + \frac{1}{2}\int \frac{1}{u^2}\,du \quad [u = x^2+1,\ du = 2x\,dx]$

$$= \tan^{-1}x + \frac{1}{2}\left(-\frac{1}{u}\right) + C = \tan^{-1}x - \frac{1}{2(x^2+1)} + C$$

27. $\dfrac{x^3+4x+3}{x^4+5x^2+4} = \dfrac{x^3+4x+3}{(x^2+1)(x^2+4)} = \dfrac{Ax+B}{x^2+1} + \dfrac{Cx+D}{x^2+4}$. Multiply both sides by $(x^2+1)(x^2+4)$

to get $x^3+4x+3 = (Ax+B)(x^2+4)+(Cx+D)(x^2+1) \quad \Leftrightarrow$

$x^3+4x+3 = Ax^3+Bx^2+4Ax+4B+Cx^3+Dx^2+Cx+D \quad \Leftrightarrow$

$x^3+4x+3 = (A+C)x^3+(B+D)x^2+(4A+C)x+(4B+D)$. Comparing coefficients gives us the following system of equations:

$A+C=1$ **(1)** $\qquad B+D=0$ **(2)** $\qquad 4A+C=4$ **(3)** $\qquad 4B+D=3$ **(4)**

Subtracting equation **(1)** from equation **(3)** gives us $A=1$ and hence, $C=0$. Subtracting equation **(2)** from equation **(4)** gives us $B=1$ and hence, $D=-1$. Thus,

$$\begin{aligned}\int \frac{x^3+4x+3}{x^4+5x^2+4}\,dx &= \int\left(\frac{x+1}{x^2+1}+\frac{-1}{x^2+4}\right)dx = \int\left(\frac{x}{x^2+1}+\frac{1}{x^2+1}-\frac{1}{x^2+4}\right)dx \\ &= \frac{1}{2}\ln(x^2+1)+\tan^{-1}x-\frac{1}{2}\tan^{-1}\left(\frac{x}{2}\right)+C\end{aligned}$$

28. $\dfrac{x^3+6x-2}{x^4+6x^2} = \dfrac{x^3+6x-2}{x^2(x^2+6)} = \dfrac{A}{x}+\dfrac{B}{x^2}+\dfrac{Cx+D}{x^2+6}$. Multiply both sides by $x^2(x^2+6)$ to get

$x^3+6x-2 = Ax(x^2+6)+B(x^2+6)+(Cx+D)x^2 \quad \Leftrightarrow$

$x^3+6x-2 = Ax^3+6Ax+Bx^2+6B+Cx^3+Dx^2 \quad \Leftrightarrow \quad x^3+6x-2 = (A+C)x^3+(B+D)x^2+6Ax+6B$.

Substituting 0 for x gives $-2=6B \quad \Leftrightarrow \quad B=-\frac{1}{3}$. Equating coefficients of x^2 gives $0=B+D$, so $D=\frac{1}{3}$. Equating coefficients of x gives $6=6A \quad \Leftrightarrow \quad A=1$. Equating coefficients of x^3 gives $1=A+C$, so $C=0$. Thus,

$$\int \frac{x^3+6x-2}{x^4+6x^2}\,dx = \int\left(\frac{1}{x}+\frac{-1/3}{x^2}+\frac{1/3}{x^2+6}\right)dx = \ln|x|+\frac{1}{3x}+\frac{1}{3\sqrt{6}}\tan^{-1}\left(\frac{x}{\sqrt{6}}\right)+C.$$

29. $$\begin{aligned}\int\frac{x+4}{x^2+2x+5}\,dx &= \int\frac{x+1}{x^2+2x+5}\,dx+\int\frac{3}{x^2+2x+5}\,dx = \frac{1}{2}\int\frac{(2x+2)\,dx}{x^2+2x+5}+\int\frac{3\,dx}{(x+1)^2+4} \\ &= \frac{1}{2}\ln\left|x^2+2x+5\right|+3\int\frac{2\,du}{4(u^2+1)} \qquad \left[\begin{array}{l}\text{where } x+1=2u,\\ \text{and } dx=2\,du\end{array}\right] \\ &= \frac{1}{2}\ln(x^2+2x+5)+\frac{3}{2}\tan^{-1}u+C = \frac{1}{2}\ln(x^2+2x+5)+\frac{3}{2}\tan^{-1}\left(\frac{x+1}{2}\right)+C\end{aligned}$$

30. $\dfrac{x^3-2x^2+2x-5}{x^4+4x^2+3} = \dfrac{x^3-2x^2+2x-5}{(x^2+1)(x^2+3)} = \dfrac{Ax+B}{x^2+1}+\dfrac{Cx+D}{x^2+3}$. Multiply both sides by $(x^2+1)(x^2+3)$ to get

$x^3-2x^2+2x-5 = (Ax+B)(x^2+3)+(Cx+D)(x^2+1) \quad \Leftrightarrow$

$x^3-2x^2+2x-5 = Ax^3+Bx^2+3Ax+3B+Cx^3+Dx^2+Cx+D \quad \Leftrightarrow$

$x^3-2x^2+2x-5 = (A+C)x^3+(B+D)x^2+(3A+C)x+(3B+D)$. Comparing coefficients gives us the following system of equations:

$A+C=1$ **(1)** $\qquad B+D=-2$ **(2)** $\qquad 3A+C=2$ **(3)** $\qquad 3B+D=-5$ **(4)**

Subtracting equation **(1)** from equation **(3)** gives us $2A=1 \quad \Leftrightarrow \quad A=\frac{1}{2}$, and hence, $C=\frac{1}{2}$. Subtracting equation **(2)** from equation **(4)** gives us $2B=-3 \quad \Leftrightarrow \quad B=-\frac{3}{2}$, and hence, $D=-\frac{1}{2}$.

Thus,

$$\int \frac{x^3 - 2x^2 + 2x - 5}{x^4 + 4x^2 + 3}\,dx = \int \left(\frac{\frac{1}{2}x - \frac{3}{2}}{x^2 + 1} + \frac{\frac{1}{2}x - \frac{1}{2}}{x^2 + 3} \right) dx = \int \left(\frac{\frac{1}{2}x}{x^2 + 1} - \frac{\frac{3}{2}}{x^2 + 1} + \frac{\frac{1}{2}x}{x^2 + 3} - \frac{\frac{1}{2}}{x^2 + 3} \right) dx$$

$$= \tfrac{1}{4}\ln(x^2 + 1) - \tfrac{3}{2}\tan^{-1} x + \tfrac{1}{4}\ln(x^2 + 3) - \frac{1}{2\sqrt{3}}\tan^{-1}\left(\frac{x}{\sqrt{3}}\right) + C$$

31. $\dfrac{1}{x^3 - 1} = \dfrac{1}{(x - 1)(x^2 + x + 1)} = \dfrac{A}{x - 1} + \dfrac{Bx + C}{x^2 + x + 1} \quad \Rightarrow \quad 1 = A(x^2 + x + 1) + (Bx + C)(x - 1).$

Take $x = 1$ to get $A = \frac{1}{3}$. Equating coefficients of x^2 and then comparing the constant terms, we get $0 = \frac{1}{3} + B$, $1 = \frac{1}{3} - C$, so $B = -\frac{1}{3}$, $C = -\frac{2}{3} \quad \Rightarrow$

$$\int \frac{1}{x^3 - 1}\,dx = \int \frac{\frac{1}{3}}{x - 1}\,dx + \int \frac{-\frac{1}{3}x - \frac{2}{3}}{x^2 + x + 1}\,dx = \tfrac{1}{3}\ln|x - 1| - \frac{1}{3}\int \frac{x + 2}{x^2 + x + 1}\,dx$$

$$= \tfrac{1}{3}\ln|x - 1| - \frac{1}{3}\int \frac{x + 1/2}{x^2 + x + 1}\,dx - \frac{1}{3}\int \frac{(3/2)\,dx}{(x + 1/2)^2 + 3/4}$$

$$= \tfrac{1}{3}\ln|x - 1| - \tfrac{1}{6}\ln(x^2 + x + 1) - \tfrac{1}{2}\left(\frac{2}{\sqrt{3}}\right)\tan^{-1}\left(\frac{x + \frac{1}{2}}{\sqrt{3}/2}\right) + K$$

$$= \tfrac{1}{3}\ln|x - 1| - \tfrac{1}{6}\ln(x^2 + x + 1) - \tfrac{1}{\sqrt{3}}\tan^{-1}\left(\tfrac{1}{\sqrt{3}}(2x + 1)\right) + K$$

32. $\displaystyle\int_0^1 \frac{x}{x^2 + 4x + 13}\,dx = \int_0^1 \frac{\frac{1}{2}(2x + 4)}{x^2 + 4x + 13}\,dx - 2\int_0^1 \frac{dx}{(x + 2)^2 + 9}$

$$= \frac{1}{2}\int_{13}^{18} \frac{dy}{y} - 2\int_{2/3}^{1} \frac{3\,du}{9u^2 + 9} \qquad \left[\begin{matrix}\text{where } y = x^2 + 4x + 13,\ dy = (2x + 4)\,dx, \\ x + 2 = 3u, \text{ and } dx = 3\,du\end{matrix}\right]$$

$$= \tfrac{1}{2}\left[\ln y\right]_{13}^{18} - \tfrac{2}{3}\left[\tan^{-1} u\right]_{2/3}^{1} = \tfrac{1}{2}\ln\tfrac{18}{13} - \tfrac{2}{3}\left(\tfrac{\pi}{4} - \tan^{-1}\left(\tfrac{2}{3}\right)\right)$$

$$= \tfrac{1}{2}\ln\tfrac{18}{13} - \tfrac{\pi}{6} + \tfrac{2}{3}\tan^{-1}\left(\tfrac{2}{3}\right)$$

33. Let $u = x^4 + 4x^2 + 3$, so that $du = (4x^3 + 8x)\,dx = 4(x^3 + 2x)\,dx$, $x = 0 \quad \Rightarrow \quad u = 3$, and $x = 1 \quad \Rightarrow \quad u = 8$.

Then $\displaystyle\int_0^1 \frac{x^3 + 2x}{x^4 + 4x^2 + 3}\,dx = \int_3^8 \frac{1}{u}\left(\frac{1}{4}\,du\right) = \frac{1}{4}\left[\ln|u|\right]_3^8 = \frac{1}{4}(\ln 8 - \ln 3) = \frac{1}{4}\ln\frac{8}{3}.$

34. $\dfrac{x^5 + x - 1}{x^3 + 1} = x^2 + \dfrac{-x^2 + x - 1}{x^3 + 1} = x^2 + \dfrac{-x^2 + x - 1}{(x + 1)(x^2 - x + 1)} = x^2 + \dfrac{-1}{x + 1}$, so

$$\int \frac{x^5 + x - 1}{x^3 + 1}\,dx = \int \left(x^2 - \frac{1}{x + 1} \right) dx = \frac{1}{3}x^3 - \ln|x + 1| + C$$

35. $\dfrac{5x^4 + 7x^2 + x + 2}{x(x^2 + 1)^2} = \dfrac{A}{x} + \dfrac{Bx + C}{x^2 + 1} + \dfrac{Dx + E}{(x^2 + 1)^2}$. Multiply by $x(x^2 + 1)^2$ to get

$5x^4 + 7x^2 + x + 2 = A(x^2 + 1)^2 + (Bx + C)x(x^2 + 1) + (Dx + E)x \quad \Leftrightarrow$

$5x^4 + 7x^2 + x + 2 = A(x^4 + 2x^2 + 1) + (Bx^2 + Cx)(x^2 + 1) + Dx^2 + Ex \quad \Leftrightarrow$

$5x^4 + 7x^2 + x + 2 = Ax^4 + 2Ax^2 + A + Bx^4 + Cx^3 + Bx^2 + Cx + Dx^2 + Ex \quad \Leftrightarrow$

$5x^4 + 7x^2 + x + 2 = (A + B)x^4 + Cx^3 + (2A + B + D)x^2 + (C + E)x + A$. Equating coefficients gives us $C = 0$,

$A = 2, A + B = 5 \Rightarrow B = 3, C + E = 1 \Rightarrow E = 1$, and $2A + B + D = 7 \Rightarrow D = 0$. Thus,

$$\int \frac{5x^4 + 7x^2 + x + 2}{x(x^2+1)^2}\,dx = \int \left[\frac{2}{x} + \frac{3x}{x^2+1} + \frac{1}{(x^2+1)^2}\right] dx = I. \text{ Now}$$

$$\begin{aligned}
\int \frac{dx}{(x^2+1)^2} &= \int \frac{\sec^2\theta\,d\theta}{(\tan^2\theta+1)^2} \qquad \begin{bmatrix} x = \tan\theta, \\ dx = \sec^2\theta\,d\theta \end{bmatrix} \\
&= \int \frac{\sec^2\theta}{\sec^4\theta}\,d\theta = \int \cos^2\theta\,d\theta = \int \tfrac{1}{2}(1+\cos 2\theta)\,d\theta \\
&= \tfrac{1}{2}\theta + \tfrac{1}{4}\sin 2\theta + C = \tfrac{1}{2}\theta + \tfrac{1}{2}\sin\theta\,\cos\theta + C \\
&= \frac{1}{2}\tan^{-1}x + \frac{1}{2}\frac{x}{\sqrt{x^2+1}}\frac{1}{\sqrt{x^2+1}} + C
\end{aligned}$$

Therefore, $I = 2\ln|x| + \tfrac{3}{2}\ln(x^2+1) + \tfrac{1}{2}\tan^{-1}x + \dfrac{x}{2(x^2+1)} + C.$

36. Let $u = x^5 + 5x^3 + 5x$, so that $du = (5x^4 + 15x^2 + 5)dx = 5(x^4 + 3x^2 + 1)dx$. Then

$$\int \frac{x^4 + 3x^2 + 1}{x^5 + 5x^3 + 5x}\,dx = \int \frac{1}{u}\left(\frac{1}{5}\,du\right) = \frac{1}{5}\ln|u| + C = \frac{1}{5}\ln\left|x^5 + 5x^3 + 5x\right| + C$$

37. $\dfrac{x^2 - 3x + 7}{(x^2 - 4x + 6)^2} = \dfrac{Ax + B}{x^2 - 4x + 6} + \dfrac{Cx + D}{(x^2 - 4x + 6)^2} \Rightarrow x^2 - 3x + 7 = (Ax + B)(x^2 - 4x + 6) + Cx + D \Rightarrow$

$x^2 - 3x + 7 = Ax^3 + (-4A + B)x^2 + (6A - 4B + C)x + (6B + D)$. So $A = 0, -4A + B = 1 \Rightarrow B = 1$,

$6A - 4B + C = -3 \Rightarrow C = 1, 6B + D = 7 \Rightarrow D = 1$. Thus,

$$\begin{aligned}
I &= \int \frac{x^2 - 3x + 7}{(x^2 - 4x + 6)^2}\,dx = \int \left(\frac{1}{x^2 - 4x + 6} + \frac{x + 1}{(x^2 - 4x + 6)^2}\right) dx \\
&= \int \frac{1}{(x-2)^2 + 2}\,dx + \int \frac{x - 2}{(x^2 - 4x + 6)^2}\,dx + \int \frac{3}{(x^2 - 4x + 6)^2}\,dx \\
&= I_1 + I_2 + I_3.
\end{aligned}$$

$$I_1 = \int \frac{1}{(x-2)^2 + \left(\sqrt{2}\right)^2}\,dx = \frac{1}{\sqrt{2}}\tan^{-1}\left(\frac{x-2}{\sqrt{2}}\right) + C_1$$

$$I_2 = \frac{1}{2}\int \frac{2x - 4}{(x^2 - 4x + 6)^2}\,dx = \tfrac{1}{2}\int \frac{1}{u^2}\,du = \tfrac{1}{2}\left(-\frac{1}{u}\right) + C_2 = -\frac{1}{2(x^2 - 4x + 6)} + C_2$$

$$\begin{aligned}
I_3 &= 3\int \frac{1}{\left[(x-2)^2 + \left(\sqrt{2}\right)^2\right]^2}\,dx = 3\int \frac{1}{[2(\tan^2\theta + 1)]^2}\sqrt{2}\sec^2\theta\,d\theta \qquad \begin{bmatrix} x - 2 = \sqrt{2}\tan\theta, \\ dx = \sqrt{2}\sec^2\theta\,d\theta \end{bmatrix} \\
&= \frac{3\sqrt{2}}{4}\int \frac{\sec^2\theta}{\sec^4\theta}\,d\theta = \frac{3\sqrt{2}}{4}\int \cos^2\theta\,d\theta = \frac{3\sqrt{2}}{4}\int \tfrac{1}{2}(1 + \cos 2\theta)\,d\theta \\
&= \frac{3\sqrt{2}}{8}\left(\theta + \tfrac{1}{2}\sin 2\theta\right) + C_3 = \frac{3\sqrt{2}}{8}\tan^{-1}\left(\frac{x-2}{\sqrt{2}}\right) + \frac{3\sqrt{2}}{8}\left(\tfrac{1}{2}\cdot 2\sin\theta\,\cos\theta\right) + C_3 \\
&= \frac{3\sqrt{2}}{8}\tan^{-1}\left(\frac{x-2}{\sqrt{2}}\right) + \frac{3\sqrt{2}}{8}\cdot\frac{x-2}{\sqrt{x^2 - 4x + 6}}\cdot\frac{\sqrt{2}}{\sqrt{x^2 - 4x + 6}} + C_3 \\
&= \frac{3\sqrt{2}}{8}\tan^{-1}\left(\frac{x-2}{\sqrt{2}}\right) + \frac{3(x-2)}{4(x^2 - 4x + 6)} + C_3
\end{aligned}$$

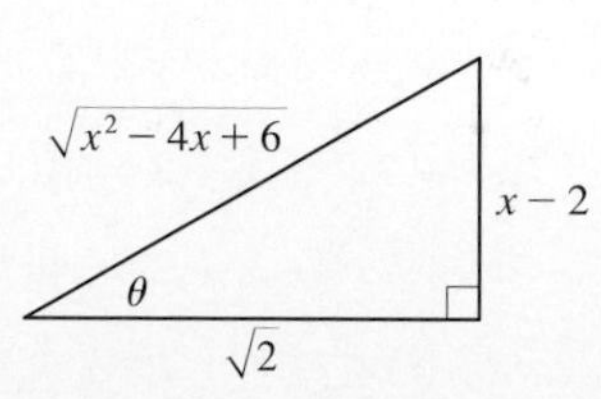

So $I = I_1 + I_2 + I_3 \qquad [C = C_1 + C_2 + C_3]$

$$= \frac{1}{\sqrt{2}}\tan^{-1}\left(\frac{x-2}{\sqrt{2}}\right) + \frac{-1}{2(x^2-4x+6)} + \frac{3\sqrt{2}}{8}\tan^{-1}\left(\frac{x-2}{\sqrt{2}}\right) + \frac{3(x-2)}{4(x^2-4x+6)} + C$$

$$= \left(\frac{4\sqrt{2}}{8} + \frac{3\sqrt{2}}{8}\right)\tan^{-1}\left(\frac{x-2}{\sqrt{2}}\right) + \frac{3(x-2)-2}{4(x^2-4x+6)} + C = \frac{7\sqrt{2}}{8}\tan^{-1}\left(\frac{x-2}{\sqrt{2}}\right) + \frac{3x-8}{4(x^2-4x+6)} + C$$

38. $\dfrac{x^3+2x^2+3x-2}{(x^2+2x+2)^2} = \dfrac{Ax+B}{x^2+2x+2} + \dfrac{Cx+D}{(x^2+2x+2)^2} \quad \Rightarrow$

$x^3+2x^2+3x-2 = (Ax+B)(x^2+2x+2) + Cx + D \quad \Rightarrow$

$x^3+2x^2+3x-2 = Ax^3 + (2A+B)x^2 + (2A+2B+C)x + 2B + D.$

So $A = 1$, $2A + B = 2 \quad \Rightarrow \quad B = 0$, $2A + 2B + C = 3 \quad \Rightarrow \quad C = 1$, and $2B + D = -2 \quad \Rightarrow \quad D = -2$. Thus,

$$I = \int \frac{x^3+2x^2+3x-2}{(x^2+2x+2)^2}\,dx = \int\left(\frac{x}{x^2+2x+2} + \frac{x-2}{(x^2+2x+2)^2}\right)dx$$

$$= \int \frac{x+1}{x^2+2x+2}\,dx + \int \frac{-1}{x^2+2x+2}\,dx + \int \frac{x+1}{(x^2+2x+2)^2}\,dx + \int \frac{-3}{(x^2+2x+2)^2}\,dx$$

$$= I_1 + I_2 + I_3 + I_4.$$

$$I_1 = \int \frac{x+1}{x^2+2x+2}\,dx = \int \frac{1}{u}\left(\frac{1}{2}\,du\right) \quad \begin{bmatrix} u = x^2+2x+2, \\ du = 2(x+1)\,dx \end{bmatrix} \quad = \frac{1}{2}\ln\left|x^2+2x+2\right| + C_1$$

$$I_2 = -\int \frac{1}{(x+1)^2+1}\,dx = -\frac{1}{1}\tan^{-1}\left(\frac{x+1}{1}\right) + C_2 = -\tan^{-1}(x+1) + C_2$$

$$I_3 = \int \frac{x+1}{(x^2+2x+2)^2}\,dx = \int \frac{1}{u^2}\left(\frac{1}{2}\,du\right) = -\frac{1}{2u} + C_3 = -\frac{1}{2(x^2+2x+2)} + C_3$$

$$I_4 = -3\int \frac{1}{[(x+1)^2+1]^2}\,dx = -3\int \frac{1}{(\tan^2\theta+1)^2}\sec^2\theta\,d\theta \quad \begin{bmatrix} x+1 = 1\tan\theta, \\ dx = \sec^2\theta\,d\theta \end{bmatrix}$$

$$= -3\int \frac{1}{\sec^2\theta}\,d\theta = -3\int \cos^2\theta\,d\theta = -\frac{3}{2}\int(1+\cos 2\theta)\,d\theta$$

$$= -\tfrac{3}{2}\left(\theta + \tfrac{1}{2}\sin 2\theta\right) + C_4 = -\tfrac{3}{2}\theta - \tfrac{3}{2}\left(\tfrac{1}{2}\cdot 2\sin\theta\,\cos\theta\right) + C_4$$

$$= -\frac{3}{2}\tan^{-1}\left(\frac{x+1}{1}\right) - \frac{3}{2}\cdot\frac{x+1}{\sqrt{x^2+2x+2}}\cdot\frac{1}{\sqrt{x^2+2x+2}} + C_4$$

$$= -\frac{3}{2}\tan^{-1}(x+1) - \frac{3(x+1)}{2(x^2+2x+2)} + C_4$$

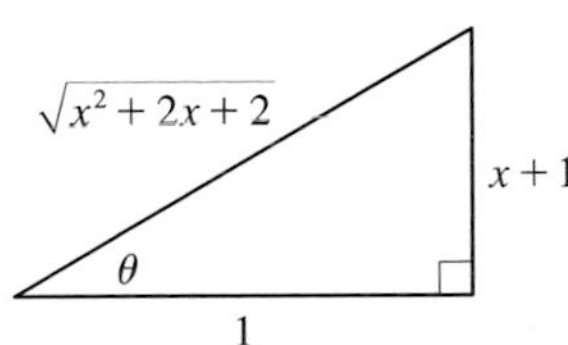

So $I = I_1 + I_2 + I_3 + I_4 \qquad [C = C_1 + C_2 + C_3 + C_4]$

$$= \frac{1}{2}\ln(x^2+2x+2) - \tan^{-1}(x+1) - \frac{1}{2(x^2+2x+2)} - \frac{3}{2}\tan^{-1}(x+1) - \frac{3(x+1)}{2(x^2+2x+2)} + C$$

$$= \frac{1}{2}\ln(x^2+2x+2) - \frac{5}{2}\tan^{-1}(x+1) - \frac{3x+4}{2(x^2+2x+2)} + C$$

39. $\displaystyle\int \frac{dx}{x\sqrt{x-1}} = \int \frac{2u}{u(u^2+1)}\,du \qquad \left[\begin{matrix} u=\sqrt{x-1},\ x=u^2+1 \\ u^2=x-1,\ \ dx=2u\,du \end{matrix}\right]$

$$= 2\int \frac{1}{u^2+1}\,du = 2\tan^{-1}u + C = 2\tan^{-1}\sqrt{x-1}+C$$

40. Let $u=\sqrt{x+3}$, so $u^2=x+3$ and $2u\,du=dx$. Then

$\displaystyle\int \frac{dx}{2\sqrt{x+3}+x} = \int \frac{2u\,du}{2u+(u^2-3)} = \int \frac{2u}{u^2+2u-3}\,du = \int \frac{2u}{(u+3)(u-1)}\,du.$ Now

$\displaystyle\frac{2u}{(u+3)(u-1)} = \frac{A}{u+3}+\frac{B}{u-1} \quad\Rightarrow\quad 2u = A(u-1)+B(u+3).$ Setting $u=1$ gives $2=4B$, so $B=\frac{1}{2}$.

Setting $u=-3$ gives $-6=-4A$, so $A=\frac{3}{2}$. Thus,

$$\int \frac{2u}{(u+3)(u-1)}\,du = \int \left(\frac{\frac{3}{2}}{u+3}+\frac{\frac{1}{2}}{u-1}\,du\right)$$
$$= \tfrac{3}{2}\ln|u+3| + \tfrac{1}{2}\ln|u-1| + C = \tfrac{3}{2}\ln\left(\sqrt{x+3}+3\right) + \tfrac{1}{2}\ln\left|\sqrt{x+3}-1\right| + C$$

41. Let $u=\sqrt{x}$, so $u^2=x$ and $2u\,du=dx$. Then $\displaystyle\int \frac{dx}{x^2+x\sqrt{x}} = \int \frac{2u\,du}{u^4+u^3} = \int \frac{2\,du}{u^3+u^2} = \int \frac{2\,du}{u^2(u+1)}.$

$\displaystyle\frac{2}{u^2(u+1)} = \frac{A}{u}+\frac{B}{u^2}+\frac{C}{u+1} \quad\Rightarrow\quad 2 = Au(u+1)+B(u+1)+Cu^2.$ Setting $u=0$ gives $B=2$. Setting $u=-1$

gives $C=2$. Equating coefficients of u^2, we get $0=A+C$, so $A=-2$. Thus,

$$\int \frac{2\,du}{u^2(u+1)} = \int \left(\frac{-2}{u}+\frac{2}{u^2}+\frac{2}{u+1}\right) du = -2\ln|u| - \frac{2}{u} + 2\ln|u+1| + C = -2\ln\sqrt{x} - \frac{2}{\sqrt{x}} + 2\ln\left(\sqrt{x}+1\right) + C.$$

42. Let $u=\sqrt[3]{x}$. Then $x=u^3$, $dx=3u^2\,du \quad\Rightarrow$

$$\int_0^1 \frac{1}{1+\sqrt[3]{x}}\,dx = \int_0^1 \frac{3u^2\,du}{1+u} = \int_0^1 \left(3u-3+\frac{3}{1+u}\right) du = \left[\tfrac{3}{2}u^2 - 3u + 3\ln(1+u)\right]_0^1 = 3\left(\ln 2 - \tfrac{1}{2}\right).$$

43. Let $u=\sqrt[3]{x^2+1}$. Then $x^2=u^3-1$, $2x\,dx=3u^2\,du \quad\Rightarrow$

$$\int \frac{x^3\,dx}{\sqrt[3]{x^2+1}} = \int \frac{(u^3-1)\frac{3}{2}u^2\,du}{u} = \frac{3}{2}\int (u^4-u)\,du$$
$$= \tfrac{3}{10}u^5 - \tfrac{3}{4}u^2 + C = \tfrac{3}{10}(x^2+1)^{5/3} - \tfrac{3}{4}(x^2+1)^{2/3} + C$$

44. $\displaystyle\int \frac{dx}{(1+\sqrt{x})^2} = \int \frac{2(u-1)}{u^2}\,du \qquad \left[\begin{matrix} u=1+\sqrt{x}, \\ x=(u-1)^2,\ dx=2(u-1)\,du \end{matrix}\right]$

$$= 2\int \left(\frac{1}{u}-\frac{1}{u^2}\right) du = 2\ln|u| + \frac{2}{u} + C = 2\ln(1+\sqrt{x}) + \frac{2}{1+\sqrt{x}} + C$$

45. If we were to substitute $u=\sqrt{x}$, then the square root would disappear but a cube root would remain. On the other hand, the substitution $u=\sqrt[3]{x}$ would eliminate the cube root but leave a square root. We can eliminate both roots by means of the substitution $u=\sqrt[6]{x}$. (Note that 6 is the least common multiple of 2 and 3.)

Let $u=\sqrt[6]{x}$. Then $x=u^6$, so $dx=6u^5\,du$ and $\sqrt{x}=u^3$, $\sqrt[3]{x}=u^2$. Thus,

$$\int \frac{dx}{\sqrt{x} - \sqrt[3]{x}} = \int \frac{6u^5\,du}{u^3 - u^2} = 6 \int \frac{u^5}{u^2(u-1)}\,du = 6 \int \frac{u^3}{u-1}\,du$$

$$= 6 \int \left(u^2 + u + 1 + \frac{1}{u-1} \right) du \qquad \text{[by long division]}$$

$$= 6\left(\tfrac{1}{3}u^3 + \tfrac{1}{2}u^2 + u + \ln|u-1|\right) + C = 2\sqrt{x} + 3\sqrt[3]{x} + 6\sqrt[6]{x} + 6\ln\left|\sqrt[6]{x} - 1\right| + C$$

46. Let $u = \sqrt{1+\sqrt{x}}$, so that $u^2 = 1 + \sqrt{x}$, $x = (u^2-1)^2$, and $dx = 2(u^2-1) \cdot 2u\,du = 4u(u^2-1)\,du$. Then

$\displaystyle\int \frac{\sqrt{1+\sqrt{x}}}{x}\,dx = \int \frac{u}{(u^2-1)^2} \cdot 4u(u^2-1)\,du = \int \frac{4u^2}{u^2-1}\,du = \int \left(4 + \frac{4}{u^2-1}\right) du$. Now

$\displaystyle\frac{4}{u^2-1} = \frac{A}{u+1} + \frac{B}{u-1} \quad\Rightarrow\quad 4 = A(u-1) + B(u+1)$. Setting $u = 1$ gives $4 = 2B$, so $B = 2$. Setting $u = -1$ gives $4 = -2A$, so $A = -2$. Thus,

$$\int \left(4 + \frac{4}{u^2-1}\right) du = \int \left(4 - \frac{2}{u+1} + \frac{2}{u-1}\right) du = 4u - 2\ln|u+1| + 2\ln|u-1| + C$$

$$= 4\sqrt{1+\sqrt{x}} - 2\ln\left(\sqrt{1+\sqrt{x}} + 1\right) + 2\ln\left(\sqrt{1+\sqrt{x}} - 1\right) + C$$

47. Let $u = e^x$. Then $x = \ln u$, $\displaystyle dx = \frac{du}{u} \quad\Rightarrow$

$$\int \frac{e^{2x}\,dx}{e^{2x} + 3e^x + 2} = \int \frac{u^2\,(du/u)}{u^2 + 3u + 2} = \int \frac{u\,du}{(u+1)(u+2)} = \int \left[\frac{-1}{u+1} + \frac{2}{u+2}\right] du$$

$$= 2\ln|u+2| - \ln|u+1| + C = \ln\frac{(e^x+2)^2}{e^x+1} + C$$

48. Let $u = \cos x$, so that $du = -\sin x\,dx$. Then $\displaystyle\int \frac{\sin x}{\cos^2 x - 3\cos x}\,dx = \int \frac{1}{u^2 - 3u}(-du) = \int \frac{-1}{u(u-3)}\,du$.

$\displaystyle\frac{-1}{u(u-3)} = \frac{A}{u} + \frac{B}{u-3} \quad\Rightarrow\quad -1 = A(u-3) + Bu$. Setting $u = 3$ gives $B = -\frac{1}{3}$. Setting $u = 0$ gives $A = \frac{1}{3}$.

Thus, $\displaystyle\int \frac{-1}{u(u-3)}\,du = \int \left(\frac{\frac{1}{3}}{u} - \frac{\frac{1}{3}}{u-3}\right) du = \frac{1}{3}\ln|u| - \frac{1}{3}\ln|u-3| + C = \tfrac{1}{3}\ln|\cos x| - \tfrac{1}{3}\ln|\cos x - 3| + C$.

49. Let $u = \tan t$, so that $du = \sec^2 t\,dt$. Then $\displaystyle\int \frac{\sec^2 t}{\tan^2 t + 3\tan t + 2}\,dt = \int \frac{1}{u^2 + 3u + 2}\,du = \int \frac{1}{(u+1)(u+2)}\,du$.

Now $\displaystyle\frac{1}{(u+1)(u+2)} = \frac{A}{u+1} + \frac{B}{u+2} \quad\Rightarrow\quad 1 = A(u+2) + B(u+1)$.

Setting $u = -2$ gives $1 = -B$, so $B = -1$. Setting $u = -1$ gives $1 = A$.

Thus, $\displaystyle\int \frac{1}{(u+1)(u+2)}\,du = \int \left(\frac{1}{u+1} - \frac{1}{u+2}\right) du = \ln|u+1| - \ln|u+2| + C = \ln|\tan t + 1| - \ln|\tan t + 2| + C$.

50. Let $u = e^x$, so that $du = e^x\,dx$. Then $\displaystyle\int \frac{e^x}{(e^x-2)(e^{2x}+1)}\,dx = \int \frac{1}{(u-2)(u^2+1)}\,du$. Now

$\displaystyle\frac{1}{(u-2)(u^2+1)} = \frac{A}{u-2} + \frac{Bu+C}{u^2+1} \quad\Rightarrow\quad 1 = A(u^2+1) + (Bu+C)(u-2)$. Setting $u = 2$ gives $1 = 5A$, so $A = \frac{1}{5}$.

Setting $u = 0$ gives $1 = \frac{1}{5} - 2C$, so $C = -\frac{2}{5}$. Comparing coefficients of u^2 gives $0 = \frac{1}{5} + B$, so $B = -\frac{1}{5}$. Thus,

$$\begin{aligned}\int \frac{1}{(u-2)(u^2+1)}\,du &= \int \left(\frac{\frac{1}{5}}{u-2} + \frac{-\frac{1}{5}u - \frac{2}{5}}{u^2+1}\right) du = \frac{1}{5}\int \frac{1}{u-2}\,du - \frac{1}{5}\int \frac{u}{u^2+1}\,du - \frac{2}{5}\int \frac{1}{u^2+1}\,du \\ &= \frac{1}{5}\ln|u-2| - \frac{1}{5}\cdot\frac{1}{2}\ln|u^2+1| - \frac{2}{5}\tan^{-1} u + C \\ &= \frac{1}{5}\ln|e^x - 2| - \frac{1}{10}\ln(e^{2x}+1) - \frac{2}{5}\tan^{-1} e^x + C\end{aligned}$$

51. Let $u = e^x$, so that $du = e^x\,dx$ and $dx = \dfrac{du}{u}$. Then $\displaystyle\int \frac{dx}{1+e^x} = \int \frac{du}{(1+u)u}$. $\quad \dfrac{1}{u(u+1)} = \dfrac{A}{u} + \dfrac{B}{u+1} \quad\Rightarrow$

$1 = A(u+1) + Bu$. Setting $u = -1$ gives $B = -1$. Setting $u = 0$ gives $A = 1$. Thus,

$$\int \frac{du}{u(u+1)} = \int \left(\frac{1}{u} - \frac{1}{u+1}\right) du = \ln|u| - \ln|u+1| + C = \ln e^x - \ln(e^x+1) + C = x - \ln(e^x+1) + C.$$

52. Let $u = \sinh t$, so that $du = \cosh t\,dt$. Then $\displaystyle\int \frac{\cosh t}{\sinh^2 t + \sinh^4 t}\,dt = \int \frac{1}{u^2+u^4}\,du = \int \frac{1}{u^2(u^2+1)}\,du.$

$\dfrac{1}{u^2(u^2+1)} = \dfrac{A}{u} + \dfrac{B}{u^2} + \dfrac{Cu+D}{u^2+1} \quad\Rightarrow\quad 1 = Au(u^2+1) + B(u^2+1) + (Cu+D)u^2$. Setting $u = 0$ gives $B = 1$.

Comparing coefficients of u^2, we get $0 = B + D$, so $D = -1$. Comparing coefficients of u, we get $0 = A$. Comparing coefficients of u^3, we get $0 = A + C$, so $C = 0$. Thus,

$$\begin{aligned}\int \frac{1}{u^2(u^2+1)}\,du &= \int \left(\frac{1}{u^2} - \frac{1}{u^2+1}\right) du = -\frac{1}{u} - \tan^{-1} u + C = -\frac{1}{\sinh t} - \tan^{-1}(\sinh t) + C \\ &= -\operatorname{csch} t - \tan^{-1}(\sinh t) + C\end{aligned}$$

53. Let $u = \ln(x^2 - x + 2)$, $dv = dx$. Then $du = \dfrac{2x-1}{x^2-x+2}\,dx$, $v = x$, and (by integration by parts)

$$\begin{aligned}\int \ln(x^2-x+2)\,dx &= x\ln(x^2-x+2) - \int \frac{2x^2-x}{x^2-x+2}\,dx = x\ln(x^2-x+2) - \int \left(2 + \frac{x-4}{x^2-x+2}\right) dx \\ &= x\ln(x^2-x+2) - 2x - \int \frac{\frac{1}{2}(2x-1)}{x^2-x+2}\,dx + \frac{7}{2}\int \frac{dx}{(x-\frac{1}{2})^2 + \frac{7}{4}} \\ &= x\ln(x^2-x+2) - 2x - \frac{1}{2}\ln(x^2-x+2) + \frac{7}{2}\int \frac{\frac{\sqrt{7}}{2}\,du}{\frac{7}{4}(u^2+1)} \qquad \left[\begin{array}{c}\text{where } x - \frac{1}{2} = \frac{\sqrt{7}}{2}u, \\ dx = \frac{\sqrt{7}}{2}\,du, \\ (x-\frac{1}{2})^2 + \frac{7}{4} = \frac{7}{4}(u^2+1)\end{array}\right] \\ &= (x - \tfrac{1}{2})\ln(x^2-x+2) - 2x + \sqrt{7}\tan^{-1} u + C \\ &= (x - \tfrac{1}{2})\ln(x^2-x+2) - 2x + \sqrt{7}\tan^{-1}\frac{2x-1}{\sqrt{7}} + C\end{aligned}$$

54. Let $u = \tan^{-1} x$, $dv = x\,dx \quad\Rightarrow\quad du = dx/(1+x^2)$, $v = \frac{1}{2}x^2$.

Then $\displaystyle\int x\tan^{-1} x\,dx = \frac{1}{2}x^2\tan^{-1} x - \frac{1}{2}\int \frac{x^2}{1+x^2}\,dx$. To evaluate the last integral, use long division or observe that

$\displaystyle\int \frac{x^2}{1+x^2}\,dx = \int \frac{(1+x^2)-1}{1+x^2}\,dx = \int 1\,dx - \int \frac{1}{1+x^2}\,dx = x - \tan^{-1} x + C_1$. So

$\int x\tan^{-1} x\,dx = \frac{1}{2}x^2\tan^{-1} x - \frac{1}{2}(x - \tan^{-1} x + C_1) = \frac{1}{2}(x^2\tan^{-1} x + \tan^{-1} x - x) + C.$

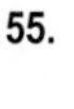

55.

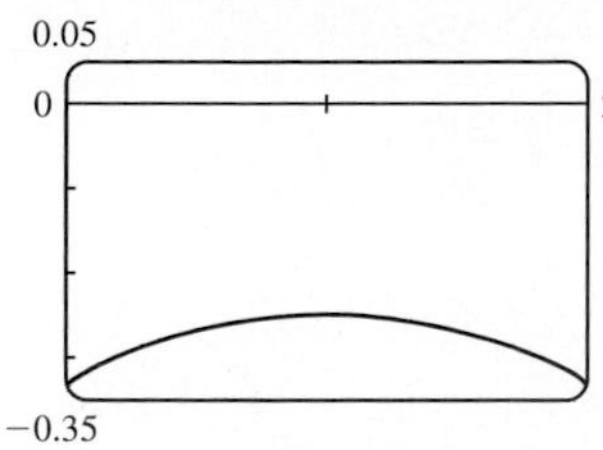

From the graph, we see that the integral will be negative, and we guess that the area is about the same as that of a rectangle with width 2 and height 0.3, so we estimate the integral to be $-(2\cdot 0.3)=-0.6$. Now

$$\frac{1}{x^2-2x-3}=\frac{1}{(x-3)(x+1)}=\frac{A}{x-3}+\frac{B}{x+1}\quad\Leftrightarrow$$

$1=(A+B)x+A-3B$, so $A=-B$ and $A-3B=1 \quad\Leftrightarrow\quad A=\frac{1}{4}$ and $B=-\frac{1}{4}$, so the integral becomes

$$\int_0^2\frac{dx}{x^2-2x-3}=\frac{1}{4}\int_0^2\frac{dx}{x-3}-\frac{1}{4}\int_0^2\frac{dx}{x+1}=\frac{1}{4}\Big[\ln|x-3|-\ln|x+1|\Big]_0^2=\frac{1}{4}\left[\ln\left|\frac{x-3}{x+1}\right|\right]_0^2$$

$$=\tfrac{1}{4}\left(\ln\tfrac{1}{3}-\ln 3\right)=-\tfrac{1}{2}\ln 3\approx -0.55$$

56. $k=0$: $\displaystyle\int\frac{dx}{x^2+k}=\int\frac{dx}{x^2}=-\frac{1}{x}+C$

$k>0$: $\displaystyle\int\frac{dx}{x^2+k}=\int\frac{dx}{x^2+(\sqrt{k}\,)^2}=\frac{1}{\sqrt{k}}\tan^{-1}\left(\frac{x}{\sqrt{k}}\right)+C$

$k<0$: $\displaystyle\int\frac{dx}{x^2+k}=\int\frac{dx}{x^2-(-k)}=\int\frac{dx}{x^2-\left(\sqrt{-k}\right)^2}=\frac{1}{2\sqrt{-k}}\ln\left|\frac{x-\sqrt{-k}}{x+\sqrt{-k}}\right|+C$ [by Example 3]

57. $\displaystyle\int\frac{dx}{x^2-2x}=\int\frac{dx}{(x-1)^2-1}=\int\frac{du}{u^2-1}$ [put $u=x-1$]

$\displaystyle=\frac{1}{2}\ln\left|\frac{u-1}{u+1}\right|+C$ [by Equation 6] $\displaystyle=\frac{1}{2}\ln\left|\frac{x-2}{x}\right|+C$

58. $\displaystyle\int\frac{(2x+1)\,dx}{4x^2+12x-7}=\frac{1}{4}\int\frac{(8x+12)\,dx}{4x^2+12x-7}-\int\frac{2\,dx}{(2x+3)^2-16}$

$\displaystyle=\tfrac{1}{4}\ln\left|4x^2+12x-7\right|-\int\frac{du}{u^2-16}$ [put $u=2x+3$]

$=\tfrac{1}{4}\ln\left|4x^2+12x-7\right|-\tfrac{1}{8}\ln|(u-4)/(u+4)|+C$ [by Equation 6]

$=\tfrac{1}{4}\ln\left|4x^2+12x-7\right|-\tfrac{1}{8}\ln|(2x-1)/(2x+7)|+C$

59. (a) If $t=\tan\left(\dfrac{x}{2}\right)$, then $\dfrac{x}{2}=\tan^{-1}t$. The figure gives

$$\cos\left(\frac{x}{2}\right)=\frac{1}{\sqrt{1+t^2}}\text{ and }\sin\left(\frac{x}{2}\right)=\frac{t}{\sqrt{1+t^2}}.$$

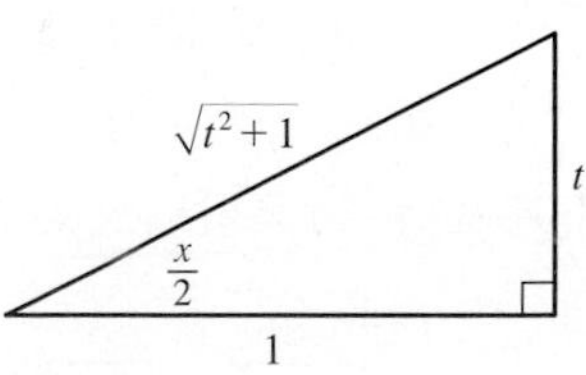

(b) $\displaystyle\cos x=\cos\left(2\cdot\frac{x}{2}\right)=2\cos^2\left(\frac{x}{2}\right)-1$

$$=2\left(\frac{1}{\sqrt{1+t^2}}\right)^2-1=\frac{2}{1+t^2}-1=\frac{1-t^2}{1+t^2}$$

(c) $\displaystyle\frac{x}{2}=\arctan t\quad\Rightarrow\quad x=2\arctan t\quad\Rightarrow\quad dx=\frac{2}{1+t^2}\,dt$

60. Let $t = \tan(x/2)$. Then, by using the expressions in Exercise 59, we have

$$\int \frac{dx}{1-\cos x} = \int \frac{2\,dt/(1+t^2)}{1-(1-t^2)/(1+t^2)} = \int \frac{2\,dt}{(1+t^2)-(1-t^2)} = \int \frac{2\,dt}{2t^2} = \int \frac{1}{t^2}\,dt$$

$$= -\frac{1}{t} + C = -\frac{1}{\tan(x/2)} + C = -\cot(x/2) + C$$

Another method: $\displaystyle\int \frac{dx}{1-\cos x} = \int \left(\frac{1}{1-\cos x}\cdot\frac{1+\cos x}{1+\cos x}\right)dx = \int \frac{1+\cos x}{1-\cos^2 x}\,dx = \int \frac{1+\cos x}{\sin^2 x}\,dx$

$$= \int \left(\frac{1}{\sin^2 x} + \frac{\cos x}{\sin^2 x}\right)dx = \int (\csc^2 x + \csc x\,\cot x)\,dx = -\cot x - \csc x + C$$

61. Let $t = \tan(x/2)$. Then, using the expressions in Exercise 59, we have

$$\int \frac{1}{3\sin x - 4\cos x}\,dx = \int \frac{1}{3\left(\frac{2t}{1+t^2}\right) - 4\left(\frac{1-t^2}{1+t^2}\right)}\,\frac{2\,dt}{1+t^2} = 2\int \frac{dt}{3(2t)-4(1-t^2)} = \int \frac{dt}{2t^2+3t-2}$$

$$= \int \frac{dt}{(2t-1)(t+2)} = \int \left[\frac{2}{5}\frac{1}{2t-1} - \frac{1}{5}\frac{1}{t+2}\right]dt \qquad \text{[using partial fractions]}$$

$$= \tfrac{1}{5}\Big[\ln|2t-1| - \ln|t+2|\Big] + C = \frac{1}{5}\ln\left|\frac{2t-1}{t+2}\right| + C = \frac{1}{5}\ln\left|\frac{2\tan(x/2)-1}{\tan(x/2)+2}\right| + C$$

62. Let $t = \tan(x/2)$. Then, by Exercise 59,

$$\int_{\pi/3}^{\pi/2} \frac{dx}{1+\sin x - \cos x} = \int_{1/\sqrt{3}}^{1} \frac{2\,dt/(1+t^2)}{1+2t/(1+t^2)-(1-t^2)/(1+t^2)} = \int_{1/\sqrt{3}}^{1} \frac{2\,dt}{1+t^2+2t-1+t^2}$$

$$= \int_{1/\sqrt{3}}^{1} \left[\frac{1}{t} - \frac{1}{t+1}\right]dt = \Big[\ln t - \ln(t+1)\Big]_{1/\sqrt{3}}^{1} = \ln\frac{1}{2} - \ln\frac{1}{\sqrt{3}+1} = \ln\frac{\sqrt{3}+1}{2}$$

63. Let $t = \tan(x/2)$. Then, by Exercise 59,

$$\int_0^{\pi/2} \frac{\sin 2x}{2+\cos x}\,dx = \int_0^{\pi/2} \frac{2\sin x\,\cos x}{2+\cos x}\,dx = \int_0^1 \frac{2\cdot\frac{2t}{1+t^2}\cdot\frac{1-t^2}{1+t^2}}{2+\frac{1-t^2}{1+t^2}}\,\frac{2}{1+t^2}\,dt = \int_0^1 \frac{\frac{8t(1-t^2)}{(1+t^2)^2}}{2(1+t^2)+(1-t^2)}\,dt$$

$$= \int_0^1 8t\cdot\frac{1-t^2}{(t^2+3)(t^2+1)^2}\,dt = I$$

If we now let $u = t^2$, then $\displaystyle\frac{1-t^2}{(t^2+3)(t^2+1)^2} = \frac{1-u}{(u+3)(u+1)^2} = \frac{A}{u+3} + \frac{B}{u+1} + \frac{C}{(u+1)^2} \quad\Rightarrow$

$1-u = A(u+1)^2 + B(u+3)(u+1) + C(u+3)$. Set $u=-1$ to get $2 = 2C$, so $C = 1$. Set $u = -3$ to get $4 = 4A$, so $A = 1$. Set $u = 0$ to get $1 = 1+3B+3$, so $B = -1$. So

$$I = \int_0^1 \left[\frac{8t}{t^2+3} - \frac{8t}{t^2+1} + \frac{8t}{(t^2+1)^2}\right]dt = \left[4\ln(t^2+3) - 4\ln(t^2+1) - \frac{4}{t^2+1}\right]_0^1$$

$$= (4\ln 4 - 4\ln 2 - 2) - (4\ln 3 - 0 - 4) = 8\ln 2 - 4\ln 2 - 4\ln 3 + 2 = 4\ln\tfrac{2}{3} + 2$$

64. $\displaystyle\frac{1}{x^3+x} = \frac{1}{x(x^2+1)} = \frac{A}{x} + \frac{Bx+C}{x^2+1} \quad\Rightarrow\quad 1 = A(x^2+1) + (Bx+C)x$. Set $x = 0$ to get $1 = A$. So

$1 = (1+B)x^2 + Cx + 1 \quad\Rightarrow\quad B+1 = 0 \quad [B = -1]$ and $C = 0$. Thus, the area is

$$\int_1^2 \frac{1}{x^3+x}\,dx = \int_1^2 \left(\frac{1}{x} - \frac{x}{x^2+1}\right) dx = \left[\ln|x| - \tfrac{1}{2}\ln|x^2+1|\right]_1^2 = \left(\ln 2 - \tfrac{1}{2}\ln 5\right) - \left(0 - \tfrac{1}{2}\ln 2\right)$$
$$= \tfrac{3}{2}\ln 2 - \tfrac{1}{2}\ln 5 \quad \left[\text{or } \tfrac{1}{2}\ln\tfrac{8}{5}\right]$$

65. By long division, $\dfrac{x^2+1}{3x-x^2} = -1 + \dfrac{3x+1}{3x-x^2}$. Now

$\dfrac{3x+1}{3x-x^2} = \dfrac{3x+1}{x(3-x)} = \dfrac{A}{x} + \dfrac{B}{3-x} \quad\Rightarrow\quad 3x+1 = A(3-x) + Bx$. Set $x=3$ to get $10 = 3B$, so $B = \frac{10}{3}$. Set $x = 0$ to get $1 = 3A$, so $A = \frac{1}{3}$. Thus, the area is

$$\int_1^2 \frac{x^2+1}{3x-x^2}\,dx = \int_1^2 \left(-1 + \frac{\frac{1}{3}}{x} + \frac{\frac{10}{3}}{3-x}\right) dx = \left[-x + \tfrac{1}{3}\ln|x| - \tfrac{10}{3}\ln|3-x|\right]_1^2$$
$$= \left(-2 + \tfrac{1}{3}\ln 2 - 0\right) - \left(-1 + 0 - \tfrac{10}{3}\ln 2\right) = -1 + \tfrac{11}{3}\ln 2$$

66. (a) We use disks, so the volume is $V = \pi\displaystyle\int_0^1 \left[\frac{1}{x^2+3x+2}\right]^2 dx = \pi\int_0^1 \frac{dx}{(x+1)^2(x+2)^2}$. To evaluate the integral, we use partial fractions: $\dfrac{1}{(x+1)^2(x+2)^2} = \dfrac{A}{x+1} + \dfrac{B}{(x+1)^2} + \dfrac{C}{x+2} + \dfrac{D}{(x+2)^2} \quad\Rightarrow$

$1 = A(x+1)(x+2)^2 + B(x+2)^2 + C(x+1)^2(x+2) + D(x+1)^2$. We set $x=-1$, giving $B=1$, then set $x=-2$, giving $D=1$. Now equating coefficients of x^3 gives $A=-C$, and then equating constants gives $1 = 4A + 4 + 2(-A) + 1 \quad\Rightarrow\quad A=-2 \quad\Rightarrow\quad C=2$. So the expression becomes

$$V = \pi\int_0^1 \left[\frac{-2}{x+1} + \frac{1}{(x+1)^2} + \frac{2}{(x+2)} + \frac{1}{(x+2)^2}\right] dx = \pi\left[2\ln\left|\frac{x+2}{x+1}\right| - \frac{1}{x+1} - \frac{1}{x+2}\right]_0^1$$
$$= \pi\left[\left(2\ln\tfrac{3}{2} - \tfrac{1}{2} - \tfrac{1}{3}\right) - \left(2\ln 2 - 1 - \tfrac{1}{2}\right)\right] = \pi\left(2\ln\tfrac{3/2}{2} + \tfrac{2}{3}\right) = \pi\left(\tfrac{2}{3} + \ln\tfrac{9}{16}\right)$$

(b) In this case, we use cylindrical shells, so the volume is $V = 2\pi\displaystyle\int_0^1 \frac{x\,dx}{x^2+3x+2} = 2\pi\int_0^1 \frac{x\,dx}{(x+1)(x+2)}$. We use partial fractions to simplify the integrand: $\dfrac{x}{(x+1)(x+2)} = \dfrac{A}{x+1} + \dfrac{B}{x+2} \quad\Rightarrow\quad x = (A+B)x + 2A + B$. So $A + B = 1$ and $2A + B = 0 \quad\Rightarrow\quad A = -1$ and $B = 2$. So the volume is

$$2\pi\int_0^1 \left[\frac{-1}{x+1} + \frac{2}{x+2}\right] dx = 2\pi\Big[-\ln|x+1| + 2\ln|x+2|\Big]_0^1$$
$$= 2\pi(-\ln 2 + 2\ln 3 + \ln 1 - 2\ln 2) = 2\pi(2\ln 3 - 3\ln 2) = 2\pi\ln\tfrac{9}{8}$$

67. $t = \displaystyle\int \frac{P+S}{P[(r-1)P-S]}\,dP = \int \frac{P+S}{P(0.1P-S)}\,dP \quad [r = 1.1]$. Now $\dfrac{P+S}{P(0.1P-S)} = \dfrac{A}{P} + \dfrac{B}{0.1P-S} \quad\Rightarrow$

$P + S = A(0.1P - S) + BP$. Substituting 0 for P gives $S = -AS \quad\Rightarrow\quad A = -1$. Substituting $10S$ for P gives

$11S = 10BS \quad\Rightarrow\quad B = \frac{11}{10}$. Thus, $t = \displaystyle\int \left(\frac{-1}{P} + \frac{11/10}{0.1P-S}\right) dP \quad\Rightarrow\quad t = -\ln P + 11\ln(0.1P - S) + C$.

When $t = 0$, $P = 10{,}000$ and $S = 900$, so $0 = -\ln 10{,}000 + 11\ln(1000 - 900) + C \quad\Rightarrow$

$C = \ln 10{,}000 - 11\ln 100 \quad [= \ln 10^{-18} \approx -41.45]$.

Therefore, $t = -\ln P + 11\ln\left(\frac{1}{10}P - 900\right) + \ln 10{,}000 - 11\ln 100 \quad\Rightarrow\quad t = \ln\dfrac{10{,}000}{P} + 11\ln\dfrac{P - 9000}{1000}$.

68. If we subtract and add $2x^2$, we get

$$x^4+1=x^4+2x^2+1-2x^2=(x^2+1)^2-2x^2=(x^2+1)^2-(\sqrt{2}\,x)^2$$
$$=\left[(x^2+1)-\sqrt{2}\,x\right]\left[(x^2+1)+\sqrt{2}\,x\right]=(x^2-\sqrt{2}\,x+1)(x^2+\sqrt{2}\,x+1)$$

So we can decompose $\dfrac{1}{x^4+1}=\dfrac{Ax+B}{x^2+\sqrt{2}\,x+1}+\dfrac{Cx+D}{x^2-\sqrt{2}\,x+1}\quad\Rightarrow$

$1=(Ax+B)(x^2-\sqrt{2}\,x+1)+(Cx+D)(x^2+\sqrt{2}\,x+1)$. Setting the constant terms equal gives $B+D=1$, then from the coefficients of x^3 we get $A+C=0$. Now from the coefficients of x we get $A+C+(B-D)\sqrt{2}=0\quad\Leftrightarrow$ $[(1-D)-D]\sqrt{2}=0\quad\Rightarrow\quad D=\frac{1}{2}\quad\Rightarrow\quad B=\frac{1}{2}$, and finally, from the coefficients of x^2 we get $\sqrt{2}\,(C-A)+B+D=0\quad\Rightarrow\quad C-A=-\frac{1}{\sqrt{2}}\quad\Rightarrow\quad C=-\frac{\sqrt{2}}{4}$ and $A=\frac{\sqrt{2}}{4}$. So we rewrite the integrand, splitting the terms into forms which we know how to integrate:

$$\frac{1}{x^4+1}=\frac{\frac{\sqrt{2}}{4}x+\frac{1}{2}}{x^2+\sqrt{2}\,x+1}+\frac{-\frac{\sqrt{2}}{4}x+\frac{1}{2}}{x^2-\sqrt{2}\,x+1}=\frac{1}{4\sqrt{2}}\left[\frac{2x+2\sqrt{2}}{x^2+\sqrt{2}\,x+1}-\frac{2x-2\sqrt{2}}{x^2-\sqrt{2}\,x+1}\right]$$
$$=\frac{\sqrt{2}}{8}\left[\frac{2x+\sqrt{2}}{x^2+\sqrt{2}\,x+1}-\frac{2x-\sqrt{2}}{x^2-\sqrt{2}\,x+1}\right]+\frac{1}{4}\left[\frac{1}{\left(x+\frac{1}{\sqrt{2}}\right)^2+\frac{1}{2}}+\frac{1}{\left(x-\frac{1}{\sqrt{2}}\right)^2+\frac{1}{2}}\right]$$

Now we integrate: $\displaystyle\int\frac{dx}{x^4+1}=\frac{\sqrt{2}}{8}\ln\left(\frac{x^2+\sqrt{2}\,x+1}{x^2-\sqrt{2}\,x+1}\right)+\frac{\sqrt{2}}{4}\left[\tan^{-1}\left(\sqrt{2}\,x+1\right)+\tan^{-1}\left(\sqrt{2}\,x-1\right)\right]+C.$

69. (a) In Maple, we define $f(x)$, and then use `convert(f,parfrac,x);` to obtain

$$f(x)=\frac{24{,}110/4879}{5x+2}-\frac{668/323}{2x+1}-\frac{9438/80{,}155}{3x-7}+\frac{(22{,}098x+48{,}935)/260{,}015}{x^2+x+5}$$

In Mathematica, we use the command `Apart`, and in Derive, we use `Expand`.

(b) $\displaystyle\int f(x)\,dx=\frac{24{,}110}{4879}\cdot\frac{1}{5}\ln|5x+2|-\frac{668}{323}\cdot\frac{1}{2}\ln|2x+1|-\frac{9438}{80{,}155}\cdot\frac{1}{3}\ln|3x-7|$

$$+\frac{1}{260{,}015}\int\frac{22{,}098\left(x+\frac{1}{2}\right)+37{,}886}{\left(x+\frac{1}{2}\right)^2+\frac{19}{4}}\,dx+C$$
$$=\frac{24{,}110}{4879}\cdot\frac{1}{5}\ln|5x+2|-\frac{668}{323}\cdot\frac{1}{2}\ln|2x+1|-\frac{9438}{80{,}155}\cdot\frac{1}{3}\ln|3x-7|$$
$$+\frac{1}{260{,}015}\left[22{,}098\cdot\frac{1}{2}\ln(x^2+x+5)+37{,}886\cdot\sqrt{\frac{4}{19}}\tan^{-1}\left(\frac{1}{\sqrt{19/4}}\left(x+\frac{1}{2}\right)\right)\right]+C$$
$$=\frac{4822}{4879}\ln|5x+2|-\frac{334}{323}\ln|2x+1|-\frac{3146}{80{,}155}\ln|3x-7|+\frac{11{,}049}{260{,}015}\ln(x^2+x+5)$$
$$+\frac{75{,}772}{260{,}015\sqrt{19}}\tan^{-1}\left[\frac{1}{\sqrt{19}}(2x+1)\right]+C$$

Using a CAS, we get

$$\frac{4822\ln(5x+2)}{4879}-\frac{334\ln(2x+1)}{323}-\frac{3146\ln(3x-7)}{80{,}155}$$
$$+\frac{11{,}049\ln(x^2+x+5)}{260{,}015}+\frac{3988\sqrt{19}}{260{,}015}\tan^{-1}\left[\frac{\sqrt{19}}{19}(2x+1)\right]$$

The main difference in this answer is that the absolute value signs and the constant of integration have been omitted. Also, the fractions have been reduced and the denominators rationalized.

70. (a) In Maple, we define $f(x)$, and then use `convert(f,parfrac,x);` to get

$$f(x) = \frac{5828/1815}{(5x-2)^2} - \frac{59{,}096/19{,}965}{5x-2} + \frac{2(2843x+816)/3993}{2x^2+1} + \frac{(313x-251)/363}{(2x^2+1)^2}.$$

In Mathematica, we use the command `Apart`, and in Derive, we use `Expand`.

(b) As we saw in Exercise 69, computer algebra systems omit the absolute value signs in $\int (1/y)\,dy = \ln|y|$. So we use the CAS to integrate the expression in part (a) and add the necessary absolute value signs and constant of integration to get

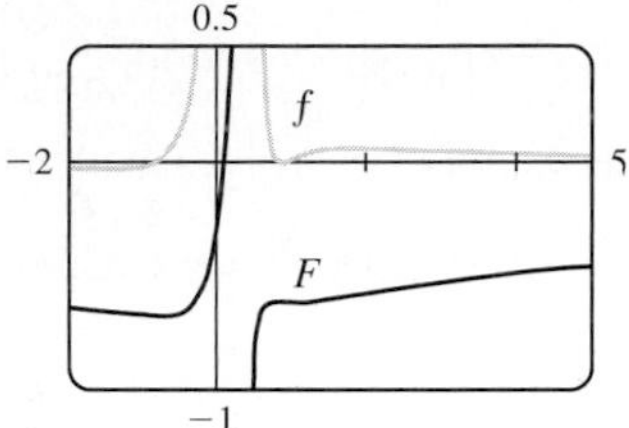

$$\int f(x)\,dx = -\frac{5828}{9075(5x-2)} - \frac{59{,}096\ln|5x-2|}{99{,}825} + \frac{2843\ln(2x^2+1)}{7986}$$
$$+\frac{503}{15{,}972}\sqrt{2}\tan^{-1}(\sqrt{2}\,x) - \frac{1}{2904}\frac{1004x+626}{2x^2+1} + C$$

(c) From the graph, we see that f goes from negative to positive at $x \approx -0.78$, then back to negative at $x \approx 0.8$, and finally back to positive at $x = 1$. Also, $\lim_{x\to 0.4} f(x) = \infty$. So we see (by the First Derivative Test) that $\int f(x)\,dx$ has minima at $x \approx -0.78$ and $x = 1$, and a maximum at $x \approx 0.80$, and that $\int f(x)\,dx$ is unbounded as $x \to 0.4$. Note also that just to the right of $x = 0.4$, f has large values, so $\int f(x)\,dx$ increases rapidly, but slows down as f drops toward 0. $\int f(x)\,dx$ decreases from about 0.8 to 1, then increases slowly since f stays small and positive.

71. $\dfrac{x^4(1-x)^4}{1+x^2} = \dfrac{x^4(1-4x+6x^2-4x^3+x^4)}{1+x^2} = \dfrac{x^8-4x^7+6x^6-4x^5+x^4}{1+x^2} = x^6-4x^5+5x^4-4x^2+4-\dfrac{4}{1+x^2}$, so

$$\int_0^1 \frac{x^4(1-x)^4}{1+x^2}\,dx = \left[\frac{1}{7}x^7 - \frac{2}{3}x^6 + x^5 - \frac{4}{3}x^3 + 4x - 4\tan^{-1}x\right]_0^1 = \left(\frac{1}{7} - \frac{2}{3} + 1 - \frac{4}{3} + 4 - 4\cdot\frac{\pi}{4}\right) - 0 = \frac{22}{7} - \pi.$$

72. (a) Let $u = (x^2+a^2)^{-n}$, $dv = dx \;\Rightarrow\; du = -n(x^2+a^2)^{-n-1}\,2x\,dx$, $v = x$.

$$I_n = \int \frac{dx}{(x^2+a^2)^n} = \frac{x}{(x^2+a^2)^n} - \int \frac{-2nx^2}{(x^2+a^2)^{n+1}}\,dx \qquad \text{[by parts]}$$
$$= \frac{x}{(x^2+a^2)^n} + 2n\int \frac{(x^2+a^2)-a^2}{(x^2+a^2)^{n+1}}\,dx$$
$$= \frac{x}{(x^2+a^2)^n} + 2n\int \frac{dx}{(x^2+a^2)^n} - 2na^2\int \frac{dx}{(x^2+a^2)^{n+1}}$$

Recognizing the last two integrals as I_n and I_{n+1}, we can solve for I_{n+1} in terms of I_n.

$$2na^2 I_{n+1} = \frac{x}{(x^2+a^2)^n} + 2nI_n - I_n \;\Rightarrow\; I_{n+1} = \frac{x}{2a^2n(x^2+a^2)^n} + \frac{2n-1}{2a^2n}I_n \;\Rightarrow$$

$$I_n = \frac{x}{2a^2(n-1)(x^2+a^2)^{n-1}} + \frac{2n-3}{2a^2(n-1)}I_{n-1}$$ [decrease n-values by 1], which is the desired result.

(b) Using part (a) with $a = 1$ and $n = 2$, we get

$$\int \frac{dx}{(x^2+1)^2} = \frac{x}{2(x^2+1)} + \frac{1}{2}\int \frac{dx}{x^2+1} = \frac{x}{2(x^2+1)} + \frac{1}{2}\tan^{-1}x + C$$

Using part (a) with $a = 1$ and $n = 3$, we get

$$\int \frac{dx}{(x^2+1)^3} = \frac{x}{2(2)(x^2+1)^2} + \frac{3}{2(2)} \int \frac{dx}{(x^2+1)^2} = \frac{x}{4(x^2+1)^2} + \frac{3}{4}\left[\frac{x}{2(x^2+1)} + \frac{1}{2}\tan^{-1} x\right] + C$$

$$= \frac{x}{4(x^2+1)^2} + \frac{3x}{8(x^2+1)} + \frac{3}{8}\tan^{-1} x + C$$

73. There are only finitely many values of x where $Q(x) = 0$ (assuming that Q is not the zero polynomial). At all other values of x, $F(x)/Q(x) = G(x)/Q(x)$, so $F(x) = G(x)$. In other words, the values of F and G agree at all except perhaps finitely many values of x. By continuity of F and G, the polynomials F and G must agree at those values of x too.

More explicitly: if a is a value of x such that $Q(a) = 0$, then $Q(x) \neq 0$ for all x sufficiently close to a. Thus,

$$\begin{aligned} F(a) &= \lim_{x\to a} F(x) && \text{[by continuity of } F\text{]} \\ &= \lim_{x\to a} G(x) && \text{[whenever } Q(x) \neq 0\text{]} \\ &= G(a) && \text{[by continuity of } G\text{]} \end{aligned}$$

74. Let $f(x) = ax^2 + bx + c$. We calculate the partial fraction decomposition of $\dfrac{f(x)}{x^2(x+1)^3}$. Since $f(0) = 1$, we must have $c = 1$, so $\dfrac{f(x)}{x^2(x+1)^3} = \dfrac{ax^2+bx+1}{x^2(x+1)^3} = \dfrac{A}{x} + \dfrac{B}{x^2} + \dfrac{C}{x+1} + \dfrac{D}{(x+1)^2} + \dfrac{E}{(x+1)^3}$. Now in order for the integral not to contain any logarithms (that is, in order for it to be a rational function), we must have $A = C = 0$, so $ax^2 + bx + 1 = B(x+1)^3 + Dx^2(x+1) + Ex^2$. Equating constant terms gives $B = 1$, then equating coefficients of x gives $3B = b \;\Rightarrow\; b = 3$. This is the quantity we are looking for, since $f'(0) = b$.

75. If $a \neq 0$ and n is a positive integer, then $f(x) = \dfrac{1}{x^n(x-a)} = \dfrac{A_1}{x} + \dfrac{A_2}{x^2} + \cdots + \dfrac{A_n}{x^n} + \dfrac{B}{x-a}$. Multiply both sides by $x^n(x-a)$ to get $1 = A_1x^{n-1}(x-a) + A_2x^{n-2}(x-a) + \cdots + A_n(x-a) + Bx^n$. Let $x = a$ in the last equation to get $1 = Ba^n \;\Rightarrow\; B = 1/a^n$. So

$$\begin{aligned} f(x) - \frac{B}{x-a} &= \frac{1}{x^n(x-a)} - \frac{1}{a^n(x-a)} = \frac{a^n - x^n}{x^na^n(x-a)} = -\frac{x^n - a^n}{a^nx^n(x-a)} \\ &= -\frac{(x-a)(x^{n-1} + x^{n-2}a + x^{n-3}a^2 + \cdots + xa^{n-2} + a^{n-1})}{a^nx^n(x-a)} \\ &= -\left(\frac{x^{n-1}}{a^nx^n} + \frac{x^{n-2}a}{a^nx^n} + \frac{x^{n-3}a^2}{a^nx^n} + \cdots + \frac{xa^{n-2}}{a^nx^n} + \frac{a^{n-1}}{a^nx^n}\right) \\ &= -\frac{1}{a^nx} - \frac{1}{a^{n-1}x^2} - \frac{1}{a^{n-2}x^3} - \cdots - \frac{1}{a^2x^{n-1}} - \frac{1}{ax^n} \end{aligned}$$

Thus, $f(x) = \dfrac{1}{x^n(x-a)} = -\dfrac{1}{a^nx} - \dfrac{1}{a^{n-1}x^2} - \cdots - \dfrac{1}{ax^n} + \dfrac{1}{a^n(x-a)}$.

7.5 Strategy for Integration

1. Let $u = 1 - \sin x$. Then $du = -\cos x\,dx \;\Rightarrow$

$$\int \frac{\cos x}{1-\sin x}\,dx = \int \frac{1}{u}(-du) = -\ln|u| + C = -\ln|1-\sin x| + C = -\ln(1-\sin x) + C$$

2. Let $u = 3x + 1$. Then $du = 3\,dx \;\Rightarrow$

$$\int_0^1 (3x+1)^{\sqrt{2}}\,dx = \int_1^4 u^{\sqrt{2}}\left(\frac{1}{3}\,du\right) = \frac{1}{3}\left[\frac{1}{\sqrt{2}+1}\,u^{\sqrt{2}+1}\right]_1^4 = \frac{1}{3(\sqrt{2}+1)}\left(4^{\sqrt{2}+1} - 1\right)$$

3. Let $u = \ln y$, $dv = \sqrt{y}\,dy \;\Rightarrow\; du = \dfrac{1}{y}\,dy$, $v = \dfrac{2}{3}y^{3/2}$. Then

$$\int_1^4 \sqrt{y}\ln y\,dy = \left[\frac{2}{3}y^{3/2}\ln y\right]_1^4 - \int_1^4 \frac{2}{3}y^{1/2}\,dy = \frac{2}{3}\cdot 8\ln 4 - 0 - \left[\frac{4}{9}y^{3/2}\right]_1^4 = \frac{16}{3}(2\ln 2) - \left(\frac{4}{9}\cdot 8 - \frac{4}{9}\right) = \frac{32}{3}\ln 2 - \frac{28}{9}$$

4. $$\int \frac{\sin^3 x}{\cos x}\,dx = \int \frac{\sin^2 x\,\sin x}{\cos x}\,dx = \int \frac{(1-\cos^2 x)\sin x}{\cos x}\,dx = \int \frac{1-u^2}{u}(-du) \qquad \left[\begin{array}{l} u = \cos x \\ du = -\sin x\,dx \end{array}\right]$$

$$= \int \left(u - \tfrac{1}{u}\right)du = \tfrac{1}{2}u^2 - \ln|u| + C = \tfrac{1}{2}\cos^2 x - \ln|\cos x| + C$$

5. Let $u = t^2$. Then $du = 2t\,dt \;\Rightarrow$

$$\int \frac{t}{t^4+2}\,dt = \int \frac{1}{u^2+2}\left(\frac{1}{2}\,du\right) = \frac{1}{2}\frac{1}{\sqrt{2}}\tan^{-1}\left(\frac{u}{\sqrt{2}}\right) + C \;\text{[by Formula 17]} = \frac{1}{2\sqrt{2}}\tan^{-1}\left(\frac{t^2}{\sqrt{2}}\right) + C$$

6. Let $u = 2x + 1$. Then $du = 2\,dx \;\Rightarrow$

$$\int_0^1 \frac{x}{(2x+1)^3}\,dx = \int_1^3 \frac{(u-1)/2}{u^3}\left(\frac{1}{2}\,du\right) = \frac{1}{4}\int_1^3\left(\frac{1}{u^2} - \frac{1}{u^3}\right)du = \frac{1}{4}\left[-\frac{1}{u} + \frac{1}{2u^2}\right]_1^3$$

$$= \tfrac{1}{4}\left[\left(-\tfrac{1}{3} + \tfrac{1}{18}\right) - \left(-1 + \tfrac{1}{2}\right)\right] = \tfrac{1}{4}\left(\tfrac{2}{9}\right) = \tfrac{1}{18}$$

7. Let $u = \arctan y$. Then $du = \dfrac{dy}{1+y^2} \;\Rightarrow\; \displaystyle\int_{-1}^1 \frac{e^{\arctan y}}{1+y^2}\,dy = \int_{-\pi/4}^{\pi/4} e^u\,du = \left[e^u\right]_{-\pi/4}^{\pi/4} = e^{\pi/4} - e^{-\pi/4}$.

8. $\int t\sin t\,\cos t\,dt = \int t\cdot\frac{1}{2}(2\sin t\,\cos t)\,dt = \frac{1}{2}\int t\sin 2t\,dt$

$$= \tfrac{1}{2}\left(-\tfrac{1}{2}t\cos 2t - \int -\tfrac{1}{2}\cos 2t\,dt\right) \qquad \left[\begin{array}{ll} u = t, & dv = \sin 2t\,dt \\ du = dt, & v = -\tfrac{1}{2}\cos 2t \end{array}\right]$$

$$= -\tfrac{1}{4}t\cos 2t + \tfrac{1}{4}\int \cos 2t\,dt = -\tfrac{1}{4}t\cos 2t + \tfrac{1}{8}\sin 2t + C$$

9. $\dfrac{x+2}{x^2+3x-4} = \dfrac{x+2}{(x+4)(x-1)} = \dfrac{A}{x+4} + \dfrac{B}{x-1}$. Multiply by $(x+4)(x-1)$ to get $x + 2 = A(x-1) + B(x+4)$.

Substituting 1 for x gives $3 = 5B \;\Leftrightarrow\; B = \frac{3}{5}$. Substituting -4 for x gives $-2 = -5A \;\Leftrightarrow\; A = \frac{2}{5}$. Thus,

$$\int_2^4 \frac{x+2}{x^2+3x-4}\,dx = \int_2^4\left(\frac{2/5}{x+4} + \frac{3/5}{x-1}\right)dx = \left[\frac{2}{5}\ln|x+4| + \frac{3}{5}\ln|x-1|\right]_2^4$$

$$= \left(\tfrac{2}{5}\ln 8 + \tfrac{3}{5}\ln 3\right) - \left(\tfrac{2}{5}\ln 6 + 0\right) = \tfrac{2}{5}(3\ln 2) + \tfrac{3}{5}\ln 3 - \tfrac{2}{5}(\ln 2 + \ln 3)$$

$$= \tfrac{4}{5}\ln 2 + \tfrac{1}{5}\ln 3, \text{ or } \tfrac{1}{5}\ln 48$$

10. Let $u = \dfrac{1}{x}$, $dv = \dfrac{\cos(1/x)}{x^2} \;\Rightarrow\; du = -\dfrac{1}{x^2}\,dx$, $v = -\sin\left(\dfrac{1}{x}\right)$. Then

$$\int \frac{\cos(1/x)}{x^3}\,dx = -\frac{1}{x}\sin\left(\frac{1}{x}\right) - \int \frac{1}{x^2}\sin\left(\frac{1}{x}\right)dx = -\frac{1}{x}\sin\left(\frac{1}{x}\right) - \cos\left(\frac{1}{x}\right) + C.$$

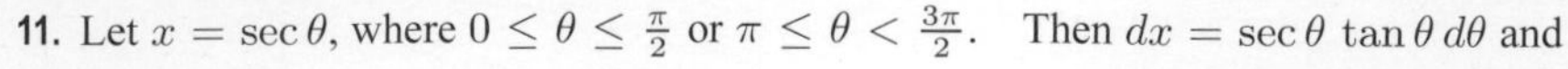

11. Let $x = \sec\theta$, where $0 \le \theta \le \frac{\pi}{2}$ or $\pi \le \theta < \frac{3\pi}{2}$. Then $dx = \sec\theta\,\tan\theta\,d\theta$ and

$\sqrt{x^2-1} = \sqrt{\sec^2\theta - 1} = \sqrt{\tan^2\theta} = |\tan\theta| = \tan\theta$ for the relevant values of θ, so

$$\int \frac{1}{x^3\sqrt{x^2-1}}\,dx = \int \frac{\sec\theta\,\tan\theta}{\sec^3\theta\,\tan\theta}\,d\theta = \int \cos^2\theta\,d\theta = \int \tfrac{1}{2}(1+\cos 2\theta)\,d\theta$$

$$= \tfrac{1}{2}\theta + \tfrac{1}{4}\sin 2\theta + C = \tfrac{1}{2}\theta + \tfrac{1}{2}\sin\theta\,\cos\theta + C$$

$$= \frac{1}{2}\sec^{-1}x + \frac{1}{2}\,\frac{\sqrt{x^2-1}}{x}\,\frac{1}{x} + C = \frac{1}{2}\sec^{-1}x + \frac{\sqrt{x^2-1}}{2x^2} + C$$

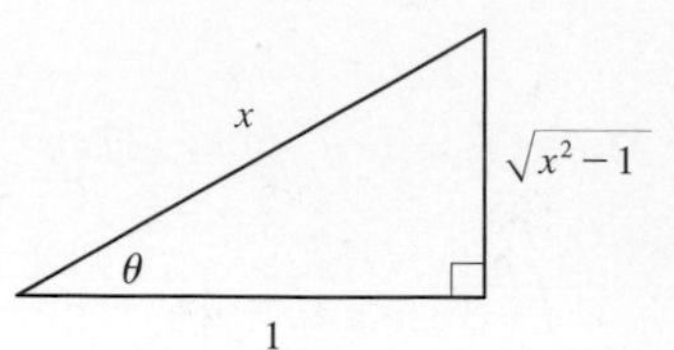

12. $\dfrac{2x-3}{x^3+3x} = \dfrac{2x-3}{x(x^2+3)} = \dfrac{A}{x} + \dfrac{Bx+C}{x^2+3}$. Multiply by $x(x^2+3)$ to get $2x-3 = A(x^2+3) + (Bx+C)x \;\Leftrightarrow$

$2x-3 = (A+B)x^2 + Cx + 3A$. Equating coefficients gives us $C = 2$, $3A = -3 \;\Leftrightarrow\; A = -1$, and $A + B = 0$, so

$B = 1$. Thus,

$$\int \frac{2x-3}{x^3+3x}\,dx = \int\left(\frac{-1}{x} + \frac{x+2}{x^2+3}\right)dx = \int\left(-\frac{1}{x} + \frac{x}{x^2+3} + \frac{2}{x^2+3}\right)dx$$

$$= -\ln|x| + \frac{1}{2}\ln(x^2+3) + \frac{2}{\sqrt{3}}\tan^{-1}\!\left(\frac{x}{\sqrt{3}}\right) + C$$

13. $\int \sin^5 t\,\cos^4 t\,dt = \int \sin^4 t\,\cos^4 t\,\sin t\,dt = \int (\sin^2 t)^2\cos^4 t\,\sin t\,dt$

$$= \int (1-\cos^2 t)^2\cos^4 t\,\sin t\,dt = \int (1-u^2)^2u^4\,(-du) \quad [u = \cos t,\, du = -\sin t\,dt]$$

$$= \int (-u^4 + 2u^6 - u^8)\,du = -\tfrac{1}{5}u^5 + \tfrac{2}{7}u^7 - \tfrac{1}{9}u^9 + C = -\tfrac{1}{5}\cos^5 t + \tfrac{2}{7}\cos^7 t - \tfrac{1}{9}\cos^9 t + C$$

14. Let $u = \ln(1+x^2)$, $dv = dx \;\Rightarrow\; du = \dfrac{2x}{1+x^2}\,dx$, $v = x$. Then

$$\int \ln(1+x^2)\,dx = x\ln(1+x^2) - \int \frac{2x^2}{1+x^2}\,dx = x\ln(1+x^2) - 2\int \frac{(x^2+1)-1}{1+x^2}\,dx$$

$$= x\ln(1+x^2) - 2\int\left(1 - \frac{1}{1+x^2}\right)dx = x\ln(1+x^2) - 2x + 2\tan^{-1}x + C$$

15. Let $u = x$, $dv = \sec x\,\tan x\,dx \;\Rightarrow\; du = dx$, $v = \sec x$. Then

$\int x\sec x\,\tan x\,dx = x\sec x - \int \sec x\,dx = x\sec x - \ln|\sec x + \tan x| + C$.

16. $$\int_0^{\sqrt{2}/2} \frac{x^2}{\sqrt{1-x^2}}\,dx = \int_0^{\pi/4} \frac{\sin^2\theta}{\cos\theta}\cos\theta\,d\theta \quad \begin{bmatrix} u = \sin\theta, \\ du = \cos\theta\,d\theta \end{bmatrix}$$

$$= \int_0^{\pi/4} \tfrac{1}{2}(1-\cos 2\theta)\,d\theta = \tfrac{1}{2}\left[\theta - \tfrac{1}{2}\sin 2\theta\right]_0^{\pi/4} = \tfrac{1}{2}\left[\left(\tfrac{\pi}{4} - \tfrac{1}{2}\right) - (0-0)\right] = \tfrac{\pi}{8} - \tfrac{1}{4}$$

17. $\int_0^\pi t\cos^2 t\,dt = \int_0^\pi t\left[\tfrac{1}{2}(1+\cos 2t)\right]dt = \tfrac{1}{2}\int_0^\pi t\,dt + \tfrac{1}{2}\int_0^\pi t\cos 2t\,dt$

$$= \tfrac{1}{2}\left[\tfrac{1}{2}t^2\right]_0^\pi + \tfrac{1}{2}\left[\tfrac{1}{2}t\sin 2t\right]_0^\pi - \tfrac{1}{2}\int_0^\pi \tfrac{1}{2}\sin 2t\,dt \quad \begin{bmatrix} u = t, & dv = \cos 2t\,dt \\ du = dt, & v = \tfrac{1}{2}\sin 2t \end{bmatrix}$$

$$= \tfrac{1}{4}\pi^2 + 0 - \tfrac{1}{4}\left[-\tfrac{1}{2}\cos 2t\right]_0^\pi = \tfrac{1}{4}\pi^2 + \tfrac{1}{8}(1-1) = \tfrac{1}{4}\pi^2$$

18. Let $u = \sqrt{t}$. Then $du = \dfrac{1}{2\sqrt{t}}\,dt \Rightarrow \displaystyle\int_1^4 \frac{e^{\sqrt{t}}}{\sqrt{t}}\,dt = \int_1^2 e^u\,(2\,du) = 2\Big[e^u\Big]_1^2 = 2(e^2 - e)$.

19. Let $u = e^x$. Then $\int e^{x+e^x}\,dx = \int e^{e^x} e^x\,dx = \int e^u\,du = e^u + C = e^{e^x} + C$.

20. Since e^2 is a constant, $\int e^2\,dx = e^2 x + C$.

21. Let $t = \sqrt{x}$, so that $t^2 = x$ and $2t\,dt = dx$. Then $\int \arctan\sqrt{x}\,dx = \int \arctan t\,(2t\,dt) = I$. Now use parts with

$u = \arctan t$, $dv = 2t\,dt \Rightarrow du = \dfrac{1}{1+t^2}\,dt$, $v = t^2$. Thus,

$$I = t^2 \arctan t - \int \frac{t^2}{1+t^2}\,dt = t^2 \arctan t - \int\left(1 - \frac{1}{1+t^2}\right)dt = t^2 \arctan t - t + \arctan t + C$$

$$= x \arctan\sqrt{x} - \sqrt{x} + \arctan\sqrt{x} + C \quad \left[\text{or } (x+1)\arctan\sqrt{x} - \sqrt{x} + C\right]$$

22. Let $u = 1 + (\ln x)^2$, so that $du = \dfrac{2\ln x}{x}\,dx$. Then

$$\int \frac{\ln x}{x\sqrt{1+(\ln x)^2}}\,dx = \frac{1}{2}\int \frac{1}{\sqrt{u}}\,du = \frac{1}{2}\left(2\sqrt{u}\right) + C = \sqrt{1+(\ln x)^2} + C.$$

23. Let $u = 1 + \sqrt{x}$. Then $x = (u-1)^2$, $dx = 2(u-1)\,du \Rightarrow$

$\int_0^1 \left(1+\sqrt{x}\right)^8 dx = \int_1^2 u^8 \cdot 2(u-1)\,du = 2\int_1^2 (u^9 - u^8)\,du = \left[\frac{1}{5}u^{10} - 2\cdot\frac{1}{9}u^9\right]_1^2 = \frac{1024}{5} - \frac{1024}{9} - \frac{1}{5} + \frac{2}{9} = \frac{4097}{45}$.

24. $\int (1+\tan x)^2 \sec x\,dx = \int (1 + 2\tan x + \tan^2 x)\sec x\,dx$

$= \int [\sec x + 2\sec x\,\tan x + (\sec^2 x - 1)\sec x]\,dx = \int (2\sec x\,\tan x + \sec^3 x)\,dx$

$= 2\sec x + \frac{1}{2}(\sec x\,\tan x + \ln|\sec x + \tan x| + C)$ [by Example 7.2.8]

25. $\displaystyle\int_0^1 \frac{1+12t}{1+3t}\,dt = \int_0^1 \frac{(12t+4)-3}{3t+1}\,dt = \int_0^1\left(4 - \frac{3}{3t+1}\right)dt = \Big[4t - \ln|3t+1|\Big]_0^1 = (4 - \ln 4) - (0 - 0) = 4 - \ln 4$

26. $\dfrac{3x^2+1}{x^3+x^2+x+1} = \dfrac{3x^2+1}{(x^2+1)(x+1)} = \dfrac{A}{x+1} + \dfrac{Bx+C}{x^2+1}$. Multiply by $(x+1)(x^2+1)$ to get

$3x^2 + 1 = A(x^2+1) + (Bx+C)(x+1) \Leftrightarrow 3x^2 + 1 = (A+B)x^2 + (B+C)x + (A+C)$. Substituting -1 for x gives $4 = 2A \Leftrightarrow A = 2$. Equating coefficients of x^2 gives $3 = A + B = 2 + B \Leftrightarrow B = 1$. Equating coefficients of x gives $0 = B + C = 1 + C \Leftrightarrow C = -1$. Thus,

$$\int_0^1 \frac{3x^2+1}{x^3+x^2+x+1}\,dx = \int_0^1\left(\frac{2}{x+1} + \frac{x-1}{x^2+1}\right)dx = \int_0^1\left(\frac{2}{x+1} + \frac{x}{x^2+1} - \frac{1}{x^2+1}\right)dx$$

$$= \left[2\ln|x+1| + \tfrac{1}{2}\ln(x^2+1) - \tan^{-1}x\right]_0^1 = \left(2\ln 2 + \tfrac{1}{2}\ln 2 - \tfrac{\pi}{4}\right) - (0+0-0)$$

$$= \tfrac{5}{2}\ln 2 - \tfrac{\pi}{4}$$

27. Let $u = 1 + e^x$, so that $du = e^x\,dx = (u-1)\,dx$. Then $\displaystyle\int \frac{1}{1+e^x}\,dx = \int \frac{1}{u}\cdot\frac{du}{u-1} = \int \frac{1}{u(u-1)}\,du = I$. Now

$\dfrac{1}{u(u-1)} = \dfrac{A}{u} + \dfrac{B}{u-1} \Rightarrow 1 = A(u-1) + Bu$. Set $u = 1$ to get $1 = B$. Set $u = 0$ to get $1 = -A$, so $A = -1$.

Thus, $I = \int \left(\frac{-1}{u} + \frac{1}{u-1} \right) du = -\ln|u| + \ln|u-1| + C = -\ln(1+e^x) + \ln e^x + C = x - \ln(1+e^x) + C.$

Another method: Multiply numerator and denominator by e^{-x} and let $u = e^{-x} + 1$. This gives the answer in the form $-\ln(e^{-x}+1) + C$.

28. $\int \sin\sqrt{at}\,dt = \int \sin u \cdot \frac{2}{a} u\,du \quad [u = \sqrt{at},\, u^2 = at,\, 2u\,du = a\,dt] \quad = \frac{2}{a}\int u \sin u\,du$

$= \frac{2}{a}[-u\cos u + \sin u] + C \quad$ [integration by parts] $\quad = -\frac{2}{a}\sqrt{at}\,\cos\sqrt{at} + \frac{2}{a}\sin\sqrt{at} + C$

$= -2\sqrt{\frac{t}{a}}\,\cos\sqrt{at} + \frac{2}{a}\sin\sqrt{at} + C$

29. Use integration by parts with $u = \ln\left(x + \sqrt{x^2-1}\right)$, $dv = dx \;\Rightarrow$

$$du = \frac{1}{x+\sqrt{x^2-1}}\left(1 + \frac{x}{\sqrt{x^2-1}}\right)dx = \frac{1}{x+\sqrt{x^2-1}}\left(\frac{\sqrt{x^2-1}+x}{\sqrt{x^2-1}}\right)dx = \frac{1}{\sqrt{x^2-1}}\,dx,\; v = x. \text{ Then}$$

$$\int \ln\left(x+\sqrt{x^2-1}\right)dx = x\ln\left(x+\sqrt{x^2-1}\right) - \int \frac{x}{\sqrt{x^2-1}}\,dx = x\ln\left(x+\sqrt{x^2-1}\right) - \sqrt{x^2-1} + C.$$

30. $|e^x - 1| = \begin{cases} e^x - 1 & \text{if } e^x - 1 \geq 0 \\ -(e^x-1) & \text{if } e^x - 1 < 0 \end{cases} = \begin{cases} e^x - 1 & \text{if } x \geq 0 \\ 1 - e^x & \text{if } x < 0 \end{cases}$

Thus, $\int_{-1}^{2} |e^x - 1|\,dx = \int_{-1}^{0}(1-e^x)\,dx + \int_0^2 (e^x - 1)\,dx = \left[x - e^x\right]_{-1}^{0} + \left[e^x - x\right]_0^2$

$= (0-1) - (-1-e^{-1}) + (e^2 - 2) - (1-0) = e^2 + e^{-1} - 3$

31. As in Example 5,

$$\int \sqrt{\frac{1+x}{1-x}}\,dx = \int \frac{\sqrt{1+x}}{\sqrt{1-x}} \cdot \frac{\sqrt{1+x}}{\sqrt{1+x}}\,dx = \int \frac{1+x}{\sqrt{1-x^2}}\,dx = \int \frac{dx}{\sqrt{1-x^2}} + \int \frac{x\,dx}{\sqrt{1-x^2}} = \sin^{-1}x - \sqrt{1-x^2} + C.$$

Another method: Substitute $u = \sqrt{(1+x)/(1-x)}$.

32. $\displaystyle\int_1^3 \frac{e^{3/x}}{x^2}\,dx = \int_3^1 e^u\left(-\tfrac{1}{3}\,du\right) \quad \left[\begin{array}{c} u = 3/x, \\ du = -3/x^2\,dx \end{array}\right]$

$= -\frac{1}{3}\left[e^u\right]_3^1 = -\frac{1}{3}(e - e^3) = \frac{1}{3}(e^3 - e)$

33. $3 - 2x - x^2 = -(x^2 + 2x + 1) + 4 = 4 - (x+1)^2$. Let $x + 1 = 2\sin\theta$, where $-\frac{\pi}{2} \leq \theta \leq \frac{\pi}{2}$. Then $dx = 2\cos\theta\,d\theta$ and

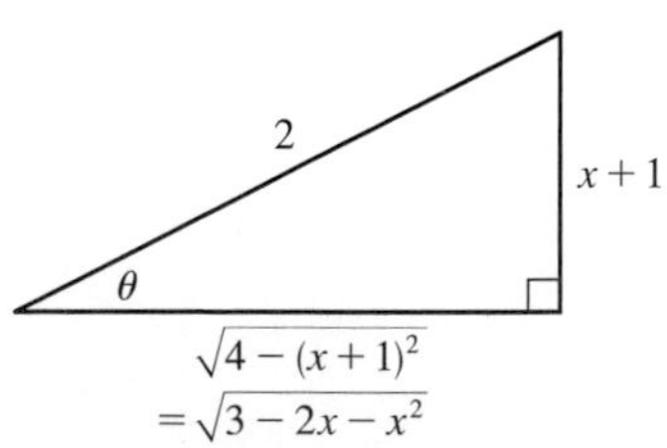

$\int \sqrt{3-2x-x^2}\,dx = \int \sqrt{4-(x+1)^2}\,dx = \int \sqrt{4 - 4\sin^2\theta}\,2\cos\theta\,d\theta$

$= 4\int \cos^2\theta\,d\theta = 2\int(1+\cos 2\theta)\,d\theta$

$= 2\theta + \sin 2\theta + C = 2\theta + 2\sin\theta\cos\theta + C$

$$= 2\sin^{-1}\left(\frac{x+1}{2}\right) + 2 \cdot \frac{x+1}{2} \cdot \frac{\sqrt{3-2x-x^2}}{2} + C$$

$$= 2\sin^{-1}\left(\frac{x+1}{2}\right) + \frac{x+1}{2}\sqrt{3-2x-x^2} + C$$

34. $\displaystyle\int_{\pi/4}^{\pi/2} \frac{1+4\cot x}{4-\cot x}\,dx = \int_{\pi/4}^{\pi/2} \left[\frac{(1+4\cos x/\sin x)}{(4-\cos x/\sin x)}\cdot\frac{\sin x}{\sin x}\right] dx = \int_{\pi/4}^{\pi/2} \frac{\sin x+4\cos x}{4\sin x-\cos x}\,dx$

$$= \int_{3/\sqrt{2}}^{4} \frac{1}{u}\,du \quad \begin{bmatrix} u = 4\sin x - \cos x, \\ du = (4\cos x + \sin x)\,dx \end{bmatrix}$$

$$= \Big[\ln|u|\Big]_{3/\sqrt{2}}^{4} = \ln 4 - \ln\frac{3}{\sqrt{2}} = \ln\frac{4}{3/\sqrt{2}} = \ln\left(\frac{4}{3}\sqrt{2}\right)$$

35. The integrand is an odd function, so $\displaystyle\int_{-\pi/2}^{\pi/2} \frac{x}{1+\cos^2 x}\,dx = 0$ [by 5.5.7(b)].

36. $\displaystyle\int \frac{1+\sin x}{1+\cos x}\,dx = \int \frac{(1+\sin x)(1-\cos x)}{(1+\cos x)(1-\cos x)}\,dx = \int \frac{1-\cos x+\sin x-\sin x\,\cos x}{\sin^2 x}\,dx$

$$= \int \left(\csc^2 x - \frac{\cos x}{\sin^2 x} + \csc x - \frac{\cos x}{\sin x}\right) dx$$

$$\overset{s}{=} -\cot x + \frac{1}{\sin x} + \ln|\csc x - \cot x| - \ln|\sin x| + C \quad \text{[by Exercise 7.2.39]}$$

The answer can be written as $\dfrac{1-\cos x}{\sin x} - \ln(1+\cos x) + C$.

37. Let $u = \tan\theta$. Then $du = \sec^2\theta\,d\theta \;\Rightarrow\; \int_0^{\pi/4} \tan^3\theta\,\sec^2\theta\,d\theta = \int_0^1 u^3\,du = \left[\frac{1}{4}u^4\right]_0^1 = \frac{1}{4}$.

38. $\displaystyle\int_{\pi/6}^{\pi/3} \frac{\sin\theta\,\cot\theta}{\sec\theta}\,d\theta = \int_{\pi/6}^{\pi/3} \cos^2\theta\,d\theta = \frac{1}{2}\int_{\pi/6}^{\pi/3} (1+\cos 2\theta)\,d\theta = \frac{1}{2}\left[\theta + \frac{1}{2}\sin 2\theta\right]_{\pi/6}^{\pi/3}$

$$= \frac{1}{2}\left[\left(\frac{\pi}{3} + \frac{\sqrt{3}}{4}\right) - \left(\frac{\pi}{6} + \frac{\sqrt{3}}{4}\right)\right] = \frac{1}{2}\left(\frac{\pi}{6}\right) = \frac{\pi}{12}$$

39. Let $u = \sec\theta$, so that $du = \sec\theta\,\tan\theta\,d\theta$. Then $\displaystyle\int \frac{\sec\theta\,\tan\theta}{\sec^2\theta - \sec\theta}\,d\theta = \int \frac{1}{u^2 - u}\,du = \int \frac{1}{u(u-1)}\,du = I$. Now

$\dfrac{1}{u(u-1)} = \dfrac{A}{u} + \dfrac{B}{u-1} \;\Rightarrow\; 1 = A(u-1) + Bu$. Set $u = 1$ to get $1 = B$. Set $u = 0$ to get $1 = -A$, so $A = -1$.

Thus, $I = \displaystyle\int \left(\frac{-1}{u} + \frac{1}{u-1}\right) du = -\ln|u| + \ln|u-1| + C = \ln|\sec\theta - 1| - \ln|\sec\theta| + C$ [or $\ln|1-\cos\theta| + C$].

40. Using product formula 2(a) in Section 7.2, $\sin 6x\,\cos 3x = \frac{1}{2}[\sin(6x-3x) + \sin(6x+3x)] = \frac{1}{2}(\sin 3x + \sin 9x)$. Thus,

$$\int_0^\pi \sin 6x\,\cos 3x\,dx = \int_0^\pi \tfrac{1}{2}(\sin 3x + \sin 9x)\,dx = \tfrac{1}{2}\left[-\tfrac{1}{3}\cos 3x - \tfrac{1}{9}\cos 9x\right]_0^\pi$$

$$= \tfrac{1}{2}\left[\left(\tfrac{1}{3} + \tfrac{1}{9}\right) - \left(-\tfrac{1}{3} - \tfrac{1}{9}\right)\right] = \tfrac{1}{2}\left(\tfrac{4}{9} + \tfrac{4}{9}\right) = \tfrac{4}{9}$$

41. Let $u = \theta$, $dv = \tan^2\theta\,d\theta = (\sec^2\theta - 1)\,d\theta \;\Rightarrow\; du = d\theta$ and $v = \tan\theta - \theta$. So

$$\int \theta\tan^2\theta\,d\theta = \theta(\tan\theta - \theta) - \int(\tan\theta - \theta)\,d\theta = \theta\tan\theta - \theta^2 - \ln|\sec\theta| + \tfrac{1}{2}\theta^2 + C$$

$$= \theta\tan\theta - \tfrac{1}{2}\theta^2 - \ln|\sec\theta| + C$$

42. Let $u = \tan^{-1} x$, $dv = \dfrac{1}{x^2}\,dx \;\Rightarrow\; du = \dfrac{1}{1+x^2}\,dx$, $v = -\dfrac{1}{x}$. Then

$$I = \int \frac{\tan^{-1} x}{x^2}\,dx = -\frac{1}{x}\tan^{-1} x - \int\left(-\frac{1}{x(1+x^2)}\right)dx = -\frac{1}{x}\tan^{-1} x + \int\left(\frac{A}{x} + \frac{Bx+C}{1+x^2}\right)dx$$

$\dfrac{1}{x(1+x^2)} = \dfrac{A}{x} + \dfrac{Bx+C}{1+x^2} \;\Rightarrow\; 1 = A(1+x^2) + (Bx+C)x \;\Rightarrow\; 1 = (A+B)x^2 + Cx + A$, so $C = 0$, $A = 1$,

and $A + B = 0 \;\Rightarrow\; B = -1$. Thus,

$$\begin{aligned} I &= -\frac{1}{x}\tan^{-1} x + \int\left(\frac{1}{x} - \frac{x}{1+x^2}\right)dx = -\frac{1}{x}\tan^{-1} x + \ln|x| - \frac{1}{2}\ln\left|1+x^2\right| + C \\ &= -\frac{\tan^{-1} x}{x} + \ln\left|\frac{x}{\sqrt{x^2+1}}\right| + C \end{aligned}$$

Or: Let $x = \tan\theta$, so that $dx = \sec^2\theta\,d\theta$. Then $\displaystyle\int \frac{\tan^{-1} x}{x^2}\,dx = \int \frac{\theta}{\tan^2\theta}\sec^2\theta\,d\theta = \int \theta\csc^2\theta\,d\theta = I$. Now use parts with $u = \theta$, $dv = \csc^2\theta\,d\theta \;\Rightarrow\; du = d\theta$, $v = -\cot\theta$. Thus,

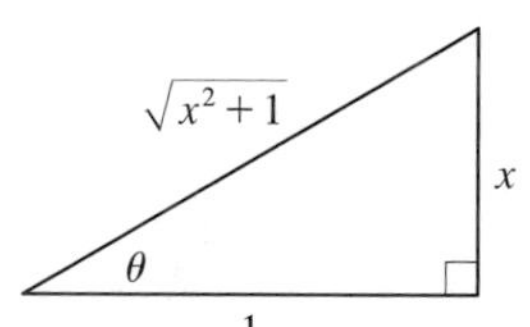

$$\begin{aligned} I &= -\theta\cot\theta - \textstyle\int(-\cot\theta)\,d\theta = -\theta\cot\theta + \ln|\sin\theta| + C \\ &= -\tan^{-1} x \cdot \frac{1}{x} + \ln\left|\frac{x}{\sqrt{x^2+1}}\right| + C = -\frac{\tan^{-1} x}{x} + \ln\left|\frac{x}{\sqrt{x^2+1}}\right| + C \end{aligned}$$

43. Let $u = \sqrt{x}$ so that $du = \dfrac{1}{2\sqrt{x}}\,dx$. Then

$$\begin{aligned} \int \frac{\sqrt{x}}{1+x^3}\,dx &= \int \frac{u}{1+u^6}\,(2u\,du) = 2\int \frac{u^2}{1+(u^3)^2}\,du = 2\int \frac{1}{1+t^2}\left(\frac{1}{3}\,dt\right) \qquad \left[\begin{matrix} t = u^3 \\ dt = 3u^2\,du \end{matrix}\right] \\ &= \tfrac{2}{3}\tan^{-1} t + C = \tfrac{2}{3}\tan^{-1} u^3 + C = \tfrac{2}{3}\tan^{-1}(x^{3/2}) + C \end{aligned}$$

Another method: Let $u = x^{3/2}$ so that $u^2 = x^3$ and $du = \frac{3}{2}x^{1/2}\,dx \;\Rightarrow\; \sqrt{x}\,dx = \frac{2}{3}\,du$. Then

$$\int \frac{\sqrt{x}}{1+x^3}\,dx = \int \frac{\frac{2}{3}}{1+u^2}\,du = \frac{2}{3}\tan^{-1} u + C = \frac{2}{3}\tan^{-1}(x^{3/2}) + C.$$

44. Let $u = \sqrt{1+e^x}$. Then $u^2 = 1 + e^x$, $2u\,du = e^x\,dx = (u^2-1)\,dx$, and $dx = \dfrac{2u}{u^2-1}\,du$, so

$$\begin{aligned} \int \sqrt{1+e^x}\,dx &= \int u \cdot \frac{2u}{u^2-1}\,du = \int \frac{2u^2}{u^2-1}\,du = \int\left(2 + \frac{2}{u^2-1}\right)du = \int\left(2 + \frac{1}{u-1} - \frac{1}{u+1}\right)du \\ &= 2u + \ln|u-1| - \ln|u+1| + C = 2\sqrt{1+e^x} + \ln\left(\sqrt{1+e^x} - 1\right) - \ln\left(\sqrt{1+e^x} + 1\right) + C \end{aligned}$$

45. Let $t = x^3$. Then $dt = 3x^2\,dx \;\Rightarrow\; I = \int x^5 e^{-x^3}\,dx = \frac{1}{3}\int te^{-t}\,dt$. Now integrate by parts with $u = t$, $dv = e^{-t}\,dt$:

$I = -\frac{1}{3}te^{-t} + \frac{1}{3}\int e^{-t}\,dt = -\frac{1}{3}te^{-t} - \frac{1}{3}e^{-t} + C = -\frac{1}{3}e^{-x^3}(x^3+1) + C.$

46. Use integration by parts with $u = (x-1)e^x$, $dv = \dfrac{1}{x^2}\,dx \;\Rightarrow\; du = [(x-1)e^x + e^x]\,dx = xe^x\,dx$, $v = -\dfrac{1}{x}$. Then

$$\int \frac{(x-1)e^x}{x^2}\,dx = (x-1)e^x\left(-\frac{1}{x}\right) - \int -e^x\,dx = -e^x + \frac{e^x}{x} + e^x + C = \frac{e^x}{x} + C.$$

47. Let $u = x - 1$, so that $du = dx$. Then

$$\int x^3(x-1)^{-4}\,dx = \int (u+1)^3 u^{-4}\,du = \int (u^3+3u^2+3u+1)u^{-4}\,du = \int (u^{-1}+3u^{-2}+3u^{-3}+u^{-4})\,du$$
$$= \ln|u| - 3u^{-1} - \tfrac{3}{2}u^{-2} - \tfrac{1}{3}u^{-3} + C = \ln|x-1| - 3(x-1)^{-1} - \tfrac{3}{2}(x-1)^{-2} - \tfrac{1}{3}(x-1)^{-3} + C$$

48. Let $u = \sqrt{1-x^2}$, so $u^2 = 1 - x^2$, and $2u\,du = -2x\,dx$. Then $\int_0^1 x\sqrt{2-\sqrt{1-x^2}}\,dx = \int_1^0 \sqrt{2-u}\,(-u\,du)$.

Now let $v = \sqrt{2-u}$, so $v^2 = 2 - u$, and $2v\,dv = -du$. Thus,

$$\int_1^0 \sqrt{2-u}\,(-u\,du) = \int_1^{\sqrt{2}} v(2-v^2)\,(2v\,dv) = \int_1^{\sqrt{2}} (4v^2 - 2v^4)\,dv = \left[\tfrac{4}{3}v^3 - \tfrac{2}{5}v^5\right]_1^{\sqrt{2}}$$
$$= \left(\tfrac{8}{3}\sqrt{2} - \tfrac{8}{5}\sqrt{2}\right) - \left(\tfrac{4}{3} - \tfrac{2}{5}\right) = \tfrac{16}{15}\sqrt{2} - \tfrac{14}{15}$$

49. Let $u = \sqrt{4x+1} \;\Rightarrow\; u^2 = 4x+1 \;\Rightarrow\; 2u\,du = 4\,dx \;\Rightarrow\; dx = \frac{1}{2}u\,du$. So

$$\int \frac{1}{x\sqrt{4x+1}}\,dx = \int \frac{\frac{1}{2}u\,du}{\frac{1}{4}(u^2-1)\,u} = 2\int \frac{du}{u^2-1} = 2\left(\tfrac{1}{2}\right)\ln\left|\frac{u-1}{u+1}\right| + C \qquad \text{[by Formula 19]}$$
$$= \ln\left|\frac{\sqrt{4x+1}-1}{\sqrt{4x+1}+1}\right| + C$$

50. As in Exercise 49, let $u = \sqrt{4x+1}$. Then $\displaystyle\int \frac{dx}{x^2\sqrt{4x+1}} = \int \frac{\frac{1}{2}u\,du}{\left[\frac{1}{4}(u^2-1)\right]^2 u} = 8\int \frac{du}{(u^2-1)^2}$. Now

$$\frac{1}{(u^2-1)^2} = \frac{1}{(u+1)^2(u-1)^2} = \frac{A}{u+1} + \frac{B}{(u+1)^2} + \frac{C}{u-1} + \frac{D}{(u-1)^2} \quad\Rightarrow$$

$1 = A(u+1)(u-1)^2 + B(u-1)^2 + C(u-1)(u+1)^2 + D(u+1)^2$. $u = 1 \;\Rightarrow\; D = \frac{1}{4}$, $u = -1 \;\Rightarrow\; B = \frac{1}{4}$.

Equating coefficients of u^3 gives $A + C = 0$, and equating coefficients of 1 gives $1 = A + B - C + D \;\Rightarrow$

$1 = A + \frac{1}{4} - C + \frac{1}{4} \;\Rightarrow\; \frac{1}{2} = A - C$. So $A = \frac{1}{4}$ and $C = -\frac{1}{4}$. Therefore,

$$\int \frac{dx}{x^2\sqrt{4x+1}} = 8\int \left[\frac{1/4}{u+1} + \frac{1/4}{(u+1)^2} + \frac{-1/4}{u-1} + \frac{1/4}{(u-1)^2}\right] du$$
$$= \int \left[\frac{2}{u+1} + 2(u+1)^{-2} - \frac{2}{u-1} + 2(u-1)^{-2}\right] du$$
$$= 2\ln|u+1| - \frac{2}{u+1} - 2\ln|u-1| - \frac{2}{u-1} + C$$
$$= 2\ln\left(\sqrt{4x+1}+1\right) - \frac{2}{\sqrt{4x+1}+1} - 2\ln\left|\sqrt{4x+1}-1\right| - \frac{2}{\sqrt{4x+1}-1} + C$$

51. Let $2x = \tan\theta \;\Rightarrow\; x = \frac{1}{2}\tan\theta$, $dx = \frac{1}{2}\sec^2\theta\,d\theta$, $\sqrt{4x^2+1} = \sec\theta$, so

$$\int \frac{dx}{x\sqrt{4x^2+1}} = \int \frac{\frac{1}{2}\sec^2\theta\,d\theta}{\frac{1}{2}\tan\theta\,\sec\theta} = \int \frac{\sec\theta}{\tan\theta}\,d\theta = \int \csc\theta\,d\theta$$
$$= -\ln|\csc\theta + \cot\theta| + C \qquad [\text{or } \ln|\csc\theta - \cot\theta| + C]$$
$$= -\ln\left|\frac{\sqrt{4x^2+1}}{2x} + \frac{1}{2x}\right| + C \qquad \left[\text{or } \ln\left|\frac{\sqrt{4x^2+1}}{2x} - \frac{1}{2x}\right| + C\right]$$

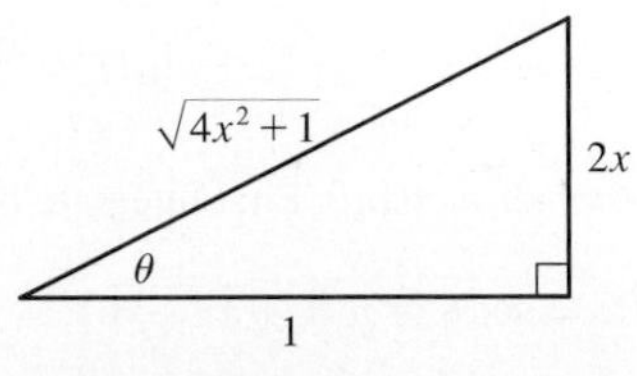

52. Let $u = x^2$. Then $du = 2x\,dx \;\Rightarrow$

$$\int \frac{dx}{x(x^4+1)} = \int \frac{x\,dx}{x^2(x^4+1)} = \frac{1}{2}\int \frac{du}{u(u^2+1)} = \frac{1}{2}\int \left[\frac{1}{u} - \frac{u}{u^2+1}\right] du = \tfrac{1}{2}\ln|u| - \tfrac{1}{4}\ln(u^2+1) + C$$

$$= \tfrac{1}{2}\ln(x^2) - \tfrac{1}{4}\ln(x^4+1) + C = \tfrac{1}{4}\left[\ln(x^4) - \ln(x^4+1)\right] + C = \frac{1}{4}\ln\left(\frac{x^4}{x^4+1}\right) + C$$

Or: Write $I = \displaystyle\int \frac{x^3\,dx}{x^4(x^4+1)}$ and let $u = x^4$.

53. $\displaystyle\int x^2 \sinh(mx)dx = \frac{1}{m}x^2\cosh(mx) - \frac{2}{m}\int x\cosh(mx)\,dx$ $\quad\left[\begin{array}{ll} u = x^2, & dv = \sinh(mx)\,dx, \\ du = 2x\,dx & v = \frac{1}{m}\cosh(mx) \end{array}\right]$

$$= \frac{1}{m}x^2\cosh(mx) - \frac{2}{m}\left(\tfrac{1}{m}x\sinh(mx) - \frac{1}{m}\textstyle\int \sinh(mx)\,dx\right) \quad \left[\begin{array}{ll} U = x, & dV = \cosh(mx)\,dx, \\ dU = dx & V = \frac{1}{m}\sinh(mx) \end{array}\right]$$

$$= \frac{1}{m}x^2\cosh(mx) - \frac{2}{m^2}x\sinh(mx) + \frac{2}{m^3}\cosh(mx) + C$$

54. $\int (x+\sin x)^2\,dx = \int \left(x^2 + 2x\sin x + \sin^2 x\right) dx = \frac{1}{3}x^3 + 2(\sin x - x\cos x) + \frac{1}{2}(x - \sin x\cos x) + C$

$$= \tfrac{1}{3}x^3 + \tfrac{1}{2}x + 2\sin x - \tfrac{1}{2}\sin x\cos x - 2x\cos x + C$$

55. Let $u = \sqrt{x}$, so that $x = u^2$ and $dx = 2u\,du$. Then $\displaystyle\int \frac{dx}{x + x\sqrt{x}} = \int \frac{2u\,du}{u^2 + u^2\cdot u} = \int \frac{2}{u(1+u)}\,du = I.$

Now $\displaystyle\frac{2}{u(1+u)} = \frac{A}{u} + \frac{B}{1+u} \;\Rightarrow\; 2 = A(1+u) + Bu$. Set $u = -1$ to get $2 = -B$, so $B = -2$. Set $u = 0$ to get $2 = A$.

Thus, $I = \displaystyle\int \left(\frac{2}{u} - \frac{2}{1+u}\right) du = 2\ln|u| - 2\ln|1+u| + C = 2\ln\sqrt{x} - 2\ln\left(1+\sqrt{x}\right) + C.$

56. Let $u = \sqrt{x}$, so that $x = u^2$ and $dx = 2u\,du$. Then

$$\int \frac{dx}{\sqrt{x} + x\sqrt{x}} = \int \frac{2u\,du}{u + u^2\cdot u} = \int \frac{2}{1+u^2}\,du = 2\tan^{-1}u + C = 2\tan^{-1}\sqrt{x} + C.$$

57. Let $u = \sqrt[3]{x+c}$. Then $x = u^3 - c \;\Rightarrow$

$$\textstyle\int x\sqrt[3]{x+c}\,dx = \int (u^3 - c)u\cdot 3u^2\,du = 3\int (u^6 - cu^3)\,du = \frac{3}{7}u^7 - \frac{3}{4}cu^4 + C = \frac{3}{7}(x+c)^{7/3} - \frac{3}{4}c(x+c)^{4/3} + C$$

58. Let $t = \sqrt{x^2-1}$. Then $dt = \left(x/\sqrt{x^2-1}\right)dx$, $x^2 - 1 = t^2$, $x = \sqrt{t^2+1}$, so

$I = \displaystyle\int \frac{x\ln x}{\sqrt{x^2-1}}\,dx = \int \ln\sqrt{t^2+1}\,dt = \frac{1}{2}\int \ln(t^2+1)\,dt$. Now use parts with $u = \ln(t^2+1)$, $dv = dt$:

$$I = \tfrac{1}{2}t\ln\left(t^2+1\right) - \int \frac{t^2}{t^2+1}\,dt = \tfrac{1}{2}t\ln\left(t^2+1\right) - \int \left[1 - \frac{1}{t^2+1}\right] dt$$

$$= \tfrac{1}{2}t\ln\left(t^2+1\right) - t + \tan^{-1}t + C = \sqrt{x^2-1}\,\ln x - \sqrt{x^2-1} + \tan^{-1}\sqrt{x^2-1} + C$$

Another method: First integrate by parts with $u = \ln x$, $dv = \left(x/\sqrt{x^2-1}\right)dx$ and then use substitution $\left(x = \sec\theta \text{ or } u = \sqrt{x^2-1}\right)$.

59. $\dfrac{1}{x^4-16} = \dfrac{1}{(x^2-4)(x^2+4)} = \dfrac{1}{(x-2)(x+2)(x^2+4)} = \dfrac{A}{x-2} + \dfrac{B}{x+2} + \dfrac{Cx+D}{x^2+4}$. Multiply by $(x-2)(x+2)(x^2+4)$ to get $1 = A(x+2)(x^2+4) + B(x-2)(x^2+4) + (Cx+D)(x-2)(x+2)$. Substituting 2 for x gives $1 = 32A \Leftrightarrow A = \frac{1}{32}$. Substituting -2 for x gives $1 = -32B \Leftrightarrow B = -\frac{1}{32}$. Equating coefficients of x^3 gives $0 = A + B + C = \frac{1}{32} - \frac{1}{32} + C$, so $C = 0$. Equating constant terms gives $1 = 8A - 8B - 4D = \frac{1}{4} + \frac{1}{4} - 4D$, so $\frac{1}{2} = -4D \Leftrightarrow D = -\frac{1}{8}$. Thus,

$$\int \frac{dx}{x^4-16} = \int \left(\frac{1/32}{x-2} - \frac{1/32}{x+2} - \frac{1/8}{x^2+4} \right) dx = \frac{1}{32}\ln|x-2| - \frac{1}{32}\ln|x+2| - \frac{1}{8}\cdot\frac{1}{2}\tan^{-1}\left(\frac{x}{2}\right) + C$$

$$= \frac{1}{32}\ln\left|\frac{x-2}{x+2}\right| - \frac{1}{16}\tan^{-1}\left(\frac{x}{2}\right) + C$$

60. Let $2x = \sec\theta$, so that $2\,dx = \sec\theta\,\tan\theta\,d\theta$. Then

$$\int \frac{dx}{x^2\sqrt{4x^2-1}} = \int \frac{\frac{1}{2}\sec\theta\,\tan\theta\,d\theta}{\frac{1}{4}\sec^2\theta\sqrt{\sec^2\theta-1}} = \int \frac{2\tan\theta\,d\theta}{\sec\theta\,\tan\theta}$$

$$= 2\int\cos\theta\,d\theta = 2\sin\theta + C$$

$$= 2\cdot\frac{\sqrt{4x^2-1}}{2x} + C = \frac{\sqrt{4x^2-1}}{x} + C$$

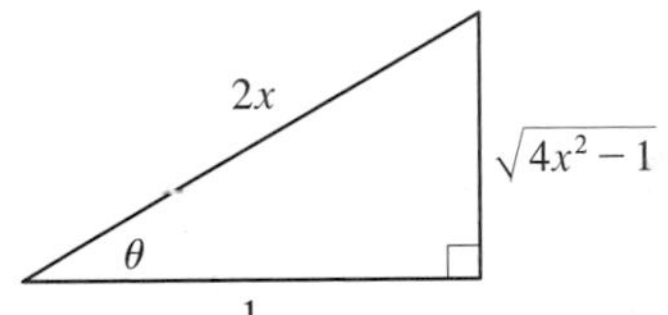

61. $\displaystyle\int \frac{d\theta}{1+\cos\theta} = \int\left(\frac{1}{1+\cos\theta}\cdot\frac{1-\cos\theta}{1-\cos\theta}\right)d\theta = \int\frac{1-\cos\theta}{1-\cos^2\theta}\,d\theta = \int\frac{1-\cos\theta}{\sin^2\theta}\,d\theta = \int\left(\frac{1}{\sin^2\theta} - \frac{\cos\theta}{\sin^2\theta}\right)d\theta$

$$= \int(\csc^2\theta - \cot\theta\,\csc\theta)\,d\theta = -\cot\theta + \csc\theta + C$$

Another method: Use the substitutions in Exercise 7.4.59.

$$\int\frac{d\theta}{1+\cos\theta} = \int\frac{2/(1+t^2)\,dt}{1+(1-t^2)/(1+t^2)} = \int\frac{2\,dt}{(1+t^2)+(1-t^2)} = \int dt = t + C = \tan\left(\frac{\theta}{2}\right) + C$$

62. $\displaystyle\int\frac{d\theta}{1+\cos^2\theta} = \int\frac{(1/\cos^2\theta)\,d\theta}{(1+\cos^2\theta)/\cos^2\theta} = \int\frac{\sec^2\theta}{\sec^2\theta+1}\,d\theta = \int\frac{\sec^2\theta}{\tan^2\theta+2}\,d\theta = \int\frac{1}{u^2+2}\,du \quad \begin{bmatrix} u = \tan\theta, \\ du = \sec^2\theta\,d\theta \end{bmatrix}$

$$= \int\frac{1}{u^2+\left(\sqrt{2}\right)^2}\,du = \frac{1}{\sqrt{2}}\tan^{-1}\left(\frac{u}{\sqrt{2}}\right) + C = \frac{1}{\sqrt{2}}\tan^{-1}\left(\frac{\tan\theta}{\sqrt{2}}\right) + C$$

63. Let $y = \sqrt{x}$ so that $dy = \dfrac{1}{2\sqrt{x}}\,dx \Rightarrow dx = 2\sqrt{x}\,dy = 2y\,dy$. Then

$$\int\sqrt{x}\,e^{\sqrt{x}}\,dx = \int ye^y(2y\,dy) = \int 2y^2e^y\,dy \quad \begin{bmatrix} u = 2y^2, & dv = e^y\,dy, \\ du = 4y\,dy & v = e^y \end{bmatrix}$$

$$= 2y^2e^y - \int 4ye^y\,dy \quad \begin{bmatrix} U = 4y, & dV = e^y\,dy, \\ dU = 4\,dy & V = e^y \end{bmatrix}$$

$$= 2y^2e^y - \left(4ye^y - \textstyle\int 4e^y\,dy\right) = 2y^2e^y - 4ye^y + 4e^y + C$$

$$= 2(y^2 - 2y + 2)e^y + C = 2\left(x - 2\sqrt{x} + 2\right)e^{\sqrt{x}} + C$$

64. Let $u = \sqrt{x} + 1$, so that $x = (u-1)^2$ and $dx = 2(u-1)\,du$. Then

$$\int \frac{1}{\sqrt{\sqrt{x}+1}}\,dx = \int \frac{2(u-1)\,du}{\sqrt{u}} = \int (2u^{1/2} - 2u^{-1/2})\,du = \frac{4}{3}u^{3/2} - 4u^{1/2} + C = \frac{4}{3}(\sqrt{x}+1)^{3/2} - 4\sqrt{\sqrt{x}+1} + C.$$

65. Let $u = \cos^2 x$, so that $du = 2\cos x\,(-\sin x)\,dx$. Then

$$\int \frac{\sin 2x}{1+\cos^4 x}\,dx = \int \frac{2\sin x\,\cos x}{1+(\cos^2 x)^2}\,dx = \int \frac{1}{1+u^2}\,(-du) = -\tan^{-1} u + C = -\tan^{-1}(\cos^2 x) + C.$$

66. Let $u = \tan x$. Then

$$\int_{\pi/4}^{\pi/3} \frac{\ln(\tan x)\,dx}{\sin x\,\cos x} = \int_{\pi/4}^{\pi/3} \frac{\ln(\tan x)}{\tan x}\sec^2 x\,dx = \int_1^{\sqrt{3}} \frac{\ln u}{u}\,du = \left[\tfrac{1}{2}(\ln u)^2\right]_1^{\sqrt{3}} = \tfrac{1}{2}\left(\ln\sqrt{3}\right)^2 = \tfrac{1}{8}(\ln 3)^2.$$

67. $$\int \frac{dx}{\sqrt{x+1}+\sqrt{x}} = \int \left(\frac{1}{\sqrt{x+1}+\sqrt{x}} \cdot \frac{\sqrt{x+1}-\sqrt{x}}{\sqrt{x+1}-\sqrt{x}}\right) dx = \int \left(\sqrt{x+1}-\sqrt{x}\right) dx$$
$$= \tfrac{2}{3}\left[(x+1)^{3/2} - x^{3/2}\right] + C$$

68. $$\int \frac{x^2}{x^6+3x^3+2}\,dx = \int \frac{x^2\,dx}{(x^3+1)(x^3+2)} = \int \frac{\frac{1}{3}\,du}{(u+1)(u+2)} \quad \begin{bmatrix} u = x^3, \\ du = 3x^2\,dx \end{bmatrix}.$$

Now $\dfrac{1}{(u+1)(u+2)} = \dfrac{A}{u+1} + \dfrac{B}{u+2} \Rightarrow 1 = A(u+2) + B(u+1)$. Setting $u = -2$ gives $B = -1$. Setting $u = -1$ gives $A = 1$. Thus,

$$\frac{1}{3}\int \frac{du}{(u+1)(u+2)} = \frac{1}{3}\int \left(\frac{1}{u+1} - \frac{1}{u+2}\right) du = \frac{1}{3}\ln|u+1| - \frac{1}{3}\ln|u+2| + C$$
$$= \tfrac{1}{3}\ln\left|x^3+1\right| - \tfrac{1}{3}\ln\left|x^3+2\right| + C$$

69. Let $x = \tan\theta$, so that $dx = \sec^2\theta\,d\theta$, $x = \sqrt{3} \Rightarrow \theta = \frac{\pi}{3}$, and $x = 1 \Rightarrow \theta = \frac{\pi}{4}$. Then

$$\int_1^{\sqrt{3}} \frac{\sqrt{1+x^2}}{x^2}\,dx = \int_{\pi/4}^{\pi/3} \frac{\sec\theta}{\tan^2\theta}\sec^2\theta\,d\theta = \int_{\pi/4}^{\pi/3} \frac{\sec\theta\,(\tan^2\theta+1)}{\tan^2\theta}\,d\theta = \int_{\pi/4}^{\pi/3} \left(\frac{\sec\theta\,\tan^2\theta}{\tan^2\theta} + \frac{\sec\theta}{\tan^2\theta}\right) d\theta$$
$$= \int_{\pi/4}^{\pi/3} (\sec\theta + \csc\theta\,\cot\theta)\,d\theta = \Big[\ln|\sec\theta + \tan\theta| - \csc\theta\Big]_{\pi/4}^{\pi/3}$$
$$= \left(\ln\left|2+\sqrt{3}\right| - \tfrac{2}{\sqrt{3}}\right) - \left(\ln\left|\sqrt{2}+1\right| - \sqrt{2}\right) = \sqrt{2} - \tfrac{2}{\sqrt{3}} + \ln(2+\sqrt{3}) - \ln(1+\sqrt{2})$$

70. Let $u = e^x$. Then $x = \ln u$, $dx = du/u \Rightarrow$

$$\int \frac{dx}{1+2e^x-e^{-x}} = \int \frac{du/u}{1+2u-1/u} = \int \frac{du}{2u^2+u-1} = \int \left[\frac{2/3}{2u-1} - \frac{1/3}{u+1}\right] du$$
$$= \tfrac{1}{3}\ln|2u-1| - \tfrac{1}{3}\ln|u+1| + C = \tfrac{1}{3}\ln|(2e^x-1)/(e^x+1)| + C$$

71. Let $u = e^x$. Then $x = \ln u$, $dx = du/u$ $\Rightarrow$

$$\int \frac{e^{2x}}{1+e^x}\,dx = \int \frac{u^2}{1+u}\frac{du}{u} = \int \frac{u}{1+u}\,du = \int \left(1 - \frac{1}{1+u}\right) du = u - \ln|1+u| + C = e^x - \ln(1+e^x) + C.$$

72. Use parts with $u = \ln(x+1)$, $dv = dx/x^2$:

$$\int \frac{\ln(x+1)}{x^2}\,dx = -\frac{1}{x}\ln(x+1) + \int \frac{dx}{x(x+1)} = -\frac{1}{x}\ln(x+1) + \int \left[\frac{1}{x} - \frac{1}{x+1}\right] dx$$

$$= -\frac{1}{x}\ln(x+1) + \ln|x| - \ln(x+1) + C = -\left(1 + \frac{1}{x}\right)\ln(x+1) + \ln|x| + C$$

73. Let $\theta = \arcsin x$, so that $d\theta = \dfrac{1}{\sqrt{1-x^2}}\,dx$ and $x = \sin\theta$. Then

$$\int \frac{x + \arcsin x}{\sqrt{1-x^2}}\,dx = \int (\sin\theta + \theta)\,d\theta = -\cos\theta + \tfrac{1}{2}\theta^2 + C$$

$$= -\sqrt{1-x^2} + \tfrac{1}{2}(\arcsin x)^2 + C$$

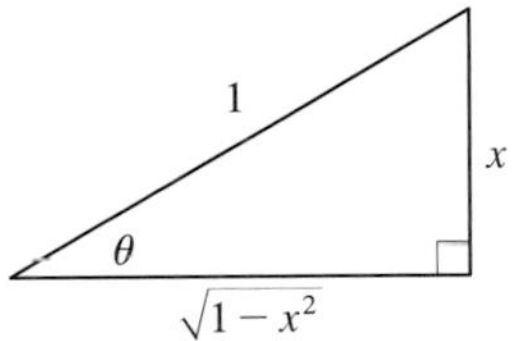

74. $$\int \frac{4^x + 10^x}{2^x}\,dx = \int \left(\frac{4^x}{2^x} + \frac{10^x}{2^x}\right) dx = \int (2^x + 5^x)\,dx = \frac{2^x}{\ln 2} + \frac{5^x}{\ln 5} + C$$

75. $$\int \frac{dx}{x\ln x - x} = \int \frac{dx}{x(\ln x - 1)} = \int \frac{du}{u} \quad \begin{bmatrix} u = \ln x - 1, \\ du = (1/x)\,dx \end{bmatrix}$$

$$= \ln|u| + C = \ln|\ln x - 1| + C$$

76. $$\int \frac{x^2}{\sqrt{x^2+1}}\,dx = \int \frac{\tan^2\theta}{\sec\theta}\sec^2\theta\,d\theta \quad \begin{bmatrix} x = \tan\theta, \\ dx = \sec^2\theta\,d\theta \end{bmatrix}$$

$$= \int \tan^2\theta\,\sec\theta\,d\theta = \int (\sec^2\theta - 1)\sec\theta\,d\theta$$

$$= \int (\sec^3\theta - \sec\theta)\,d\theta$$

$$= \tfrac{1}{2}(\sec\theta\,\tan\theta + \ln|\sec\theta + \tan\theta|) - \ln|\sec\theta + \tan\theta| + C \quad \text{[by (1) and Example 7.2.8]}$$

$$= \tfrac{1}{2}(\sec\theta\,\tan\theta - \ln|\sec\theta + \tan\theta|) + C = \tfrac{1}{2}\left[x\sqrt{x^2+1} - \ln\left(\sqrt{x^2+1} + x\right)\right] + C$$

77. Let $y = \sqrt{1+e^x}$, so that $y^2 = 1 + e^x$, $2y\,dy = e^x\,dx$, $e^x = y^2 - 1$, and $x = \ln(y^2-1)$. Then

$$\int \frac{xe^x}{\sqrt{1+e^x}}\,dx = \int \frac{\ln(y^2-1)}{y}(2y\,dy) = 2\int [\ln(y+1) + \ln(y-1)]\,dy$$

$$= 2[(y+1)\ln(y+1) - (y+1) + (y-1)\ln(y-1) - (y-1)] + C \quad \text{[by Example 7.1.2]}$$

$$= 2[y\ln(y+1) + \ln(y+1) - y - 1 + y\ln(y-1) - \ln(y-1) - y + 1] + C$$

$$= 2[y(\ln(y+1) + \ln(y-1)) + \ln(y+1) - \ln(y-1) - 2y] + C$$

$$= 2\left[y\ln(y^2-1) + \ln\frac{y+1}{y-1} - 2y\right] + C = 2\left[\sqrt{1+e^x}\,\ln(e^x) + \ln\frac{\sqrt{1+e^x}+1}{\sqrt{1+e^x}-1} - 2\sqrt{1+e^x}\right] + C$$

$$= 2x\sqrt{1+e^x} + 2\ln\frac{\sqrt{1+e^x}+1}{\sqrt{1+e^x}-1} - 4\sqrt{1+e^x} + C = 2(x-2)\sqrt{1+e^x} + 2\ln\frac{\sqrt{1+e^x}+1}{\sqrt{1+e^x}-1} + C$$

78. $\dfrac{1+\sin x}{1-\sin x} = \dfrac{1+\sin x}{1-\sin x}\cdot\dfrac{1+\sin x}{1+\sin x} = \dfrac{1+2\sin x+\sin^2 x}{1-\sin^2 x} = \dfrac{1+2\sin x+\sin^2 x}{\cos^2 x} = \dfrac{1}{\cos^2 x}+\dfrac{2\sin x}{\cos^2 x}+\dfrac{\sin^2 x}{\cos^2 x}$

$= \sec^2 x + 2\sec x\,\tan x + \tan^2 x = \sec^2 x + 2\sec x\,\tan x + \sec^2 x - 1 = 2\sec^2 x + 2\sec x\,\tan x - 1$

Thus, $$\int \frac{1+\sin x}{1-\sin x}\,dx = \int (2\sec^2 x + 2\sec x\,\tan x - 1)\,dx = 2\tan x + 2\sec x - x + C$$

79. Let $u = x$, $dv = \sin^2 x\cos x\,dx \;\Rightarrow\; du = dx$, $v = \frac{1}{3}\sin^3 x$. Then

$\int x\sin^2 x\,\cos x\,dx = \frac{1}{3}x\sin^3 x - \int \frac{1}{3}\sin^3 x\,dx = \frac{1}{3}x\sin^3 x - \frac{1}{3}\int(1-\cos^2 x)\sin x\,dx$

$= \dfrac{1}{3}x\sin^3 x + \dfrac{1}{3}\displaystyle\int (1-y^2)\,dy \qquad \left[\begin{matrix} y = \cos x, \\ dy = -\sin x\,dx\end{matrix}\right]$

$= \frac{1}{3}x\sin^3 x + \frac{1}{3}y - \frac{1}{9}y^3 + C = \frac{1}{3}x\sin^3 x + \frac{1}{3}\cos x - \frac{1}{9}\cos^3 x + C$

80. $\displaystyle\int \frac{\sec x\,\cos 2x}{\sin x + \sec x}\,dx = \int \frac{\sec x\,\cos 2x}{\sin x + \sec x}\cdot\frac{2\cos x}{2\cos x}\,dx = \int \frac{2\cos 2x}{2\sin x\cos x + 2}\,dx$

$= \displaystyle\int \frac{2\cos 2x}{\sin 2x + 2}\,dx = \int \frac{1}{u}\,du \qquad \left[\begin{matrix} u = \sin 2x + 2, \\ du = 2\cos 2x\,dx\end{matrix}\right]$

$= \ln|u| + C = \ln|\sin 2x + 2| + C = \ln(\sin 2x + 2) + C$

81. $\displaystyle\int \sqrt{1-\sin x}\,dx = \int \sqrt{\frac{1-\sin x}{1}\cdot\frac{1+\sin x}{1+\sin x}}\,dx = \int \sqrt{\frac{1-\sin^2 x}{1+\sin x}}\,dx$

$= \displaystyle\int \sqrt{\frac{\cos^2 x}{1+\sin x}}\,dx = \int \frac{\cos x\,dx}{\sqrt{1+\sin x}} \qquad [\text{assume } \cos x > 0]$

$= \displaystyle\int \frac{du}{\sqrt{u}} \qquad \left[\begin{matrix} u = 1+\sin x, \\ du = \cos x\,dx\end{matrix}\right]$

$= 2\sqrt{u} + C = 2\sqrt{1+\sin x} + C$

Another method: Let $u = \sin x$ so that $du = \cos x\,dx = \sqrt{1-\sin^2 x}\,dx = \sqrt{1-u^2}\,dx$. Then

$$\int \sqrt{1-\sin x}\,dx = \int \sqrt{1-u}\left(\frac{du}{\sqrt{1-u^2}}\right) = \int \frac{1}{\sqrt{1+u}}\,du = 2\sqrt{1+u} + C = 2\sqrt{1+\sin x} + C.$$

82. $\displaystyle\int \frac{\sin x\,\cos x}{\sin^4 x + \cos^4 x}\,dx = \int \frac{\sin x\,\cos x}{(\sin^2 x)^2 + (\cos^2 x)^2}\,dx = \int \frac{\sin x\,\cos x}{(\sin^2 x)^2 + (1-\sin^2 x)^2}\,dx$

$= \displaystyle\int \frac{1}{u^2 + (1-u)^2}\left(\frac{1}{2}\,du\right) \qquad \left[\begin{matrix} u = \sin^2 x, \\ du = 2\sin x\,\cos x\,dx\end{matrix}\right]$

$= \displaystyle\int \frac{1}{4u^2 - 4u + 2}\,du = \int \frac{1}{(4u^2 - 4u + 1) + 1}\,du$

$= \displaystyle\int \frac{1}{(2u-1)^2 + 1}\,du = \frac{1}{2}\int \frac{1}{y^2+1}\,dy \qquad \left[\begin{matrix} y = 2u - 1, \\ dy = 2\,du\end{matrix}\right]$

$= \frac{1}{2}\tan^{-1} y + C = \frac{1}{2}\tan^{-1}(2u-1) + C = \frac{1}{2}\tan^{-1}(2\sin^2 x - 1) + C$

Another solution:

$$\int \frac{\sin x \cos x}{\sin^4 x + \cos^4 x}\,dx = \int \frac{(\sin x \cos x)/\cos^4 x}{(\sin^4 x + \cos^4 x)/\cos^4 x}\,dx = \int \frac{\tan x \sec^2 x}{\tan^4 x + 1}\,dx$$

$$= \int \frac{1}{u^2+1}\left(\frac{1}{2}\,du\right) \quad \begin{bmatrix} u = \tan^2 x, \\ du = 2\tan x \sec^2 x\,dx \end{bmatrix}$$

$$= \tfrac{1}{2}\tan^{-1} u + C = \tfrac{1}{2}\tan^{-1}(\tan^2 x) + C$$

83. The function $y = 2xe^{x^2}$ *does* have an elementary antiderivative, so we'll use this fact to help evaluate the integral.

$$\int (2x^2+1)e^{x^2}\,dx = \int 2x^2 e^{x^2}\,dx + \int e^{x^2}\,dx = \int x\left(2xe^{x^2}\right)dx + \int e^{x^2}\,dx$$

$$= xe^{x^2} - \int e^{x^2}\,dx + \int e^{x^2}\,dx \quad \begin{bmatrix} u = x, & dv = 2xe^{x^2}\,dx, \\ du = dx & v = e^{x^2} \end{bmatrix} \quad = xe^{x^2} + C$$

84. (a) $$\int_1^2 \frac{e^x}{x}\,dx = \int_0^{\ln 2} \frac{e^{e^t}}{e^t}\,e^t\,dt \quad \begin{bmatrix} x = e^t, \\ dx = e^t dt \end{bmatrix} \quad = \int_0^{\ln 2} e^{e^t}\,dt = F(\ln 2)$$

(b) $$\int_2^3 \frac{1}{\ln x}\,dx = \int_{\ln 2}^{\ln 3} \frac{1}{u}\,(e^u\,du) \quad \begin{bmatrix} u = \ln x, \\ du = \frac{1}{x}\,dx \end{bmatrix} \quad = \int_{\ln\ln 2}^{\ln\ln 3} \frac{e^{e^v}}{e^v}\,e^v\,dv \quad \begin{bmatrix} u = e^v, \\ du = e^v\,dv \end{bmatrix}$$

$$= \int_{\ln\ln 2}^{0} e^{e^v}\,dv + \int_0^{\ln\ln 3} e^{e^v}\,dv \quad [\text{note that } \ln\ln 2 < 0]$$

$$= \int_0^{\ln\ln 3} e^{e^v}\,dv - \int_0^{\ln\ln 2} e^{e^v}\,dv = F(\ln\ln 3) - F(\ln\ln 2)$$

Another method: Substitute $x = e^{e^t}$ in the original integral.

7.6 Integration Using Tables and Computer Algebra Systems

Keep in mind that there are several ways to approach many of these exercises, and different methods can lead to different forms of the answer.

1. $$\int_0^{\pi/2} \cos 5x \cos 2x\,dx \overset{80}{=} \left[\frac{\sin(5-2)x}{2(5-2)} + \frac{\sin(5+2)x}{2(5+2)}\right]_0^{\pi/2} \quad \begin{bmatrix} a = 5, \\ b = 2 \end{bmatrix}$$

$$= \left[\frac{\sin 3x}{6} + \frac{\sin 7x}{14}\right]_0^{\pi/2} = \left(-\frac{1}{6} - \frac{1}{14}\right) - 0 = \frac{-7-3}{42} = -\frac{5}{21}$$

2. $$\int_0^1 \sqrt{x - x^2}\,dx = \int_0^1 \sqrt{2\left(\tfrac{1}{2}\right)x - x^2}\,dx \overset{113}{=} \left[\frac{x - \frac{1}{2}}{2}\sqrt{2\left(\tfrac{1}{2}\right)x - x^2} + \frac{\left(\frac{1}{2}\right)^2}{2}\cos^{-1}\left(\frac{\frac{1}{2} - x}{\frac{1}{2}}\right)\right]_0^1$$

$$= \left[\frac{2x-1}{4}\sqrt{x - x^2} + \frac{1}{8}\cos^{-1}(1-2x)\right]_0^1 = \left(0 + \frac{1}{8}\cdot\pi\right) - \left(0 + \frac{1}{8}\cdot 0\right) = \frac{1}{8}\pi$$

3. $\int_1^2 \sqrt{4x^2 - 3}\,dx = \frac{1}{2}\int_2^4 \sqrt{u^2 - (\sqrt{3})^2}\,du \quad [u = 2x,\ du = 2\,dx]$

$$\overset{39}{=} \frac{1}{2}\left[\frac{u}{2}\sqrt{u^2 - (\sqrt{3})^2} - \frac{(\sqrt{3})^2}{2}\ln\left|u + \sqrt{u^2 - (\sqrt{3})^2}\right|\right]_2^4$$

$$= \tfrac{1}{2}\left[2\sqrt{13} - \tfrac{3}{2}\ln\left(4 + \sqrt{13}\right)\right] - \tfrac{1}{2}\left(1 - \tfrac{3}{2}\ln 3\right) = \sqrt{13} - \tfrac{3}{4}\ln(4 + \sqrt{13}) - \tfrac{1}{2} + \tfrac{3}{4}\ln 3$$

4. $\displaystyle\int_0^1 \tan^3\left(\frac{\pi}{6}x\right) dx = \frac{6}{\pi}\int_0^{\pi/6} \tan^3 u\, du \qquad [u = (\pi/6)x,\ du = (\pi/6)\,dx]$

$$\overset{69}{=} \frac{6}{\pi}\left[\frac{1}{2}\tan^2 u + \ln|\cos u|\right]_0^{\pi/6} = \frac{6}{\pi}\left[\left(\frac{1}{2}\left(\frac{1}{\sqrt{3}}\right)^2 + \ln\frac{\sqrt{3}}{2}\right) - (0 + \ln 1)\right] = \frac{1}{\pi} + \frac{6}{\pi}\ln\frac{\sqrt{3}}{2}$$

5. $\int_0^{\pi/8} \arctan 2x\, dx = \frac{1}{2}\int_0^{\pi/4} \arctan u\, du \qquad [u = 2x,\ du = 2\,dx]$

$$\overset{89}{=} \frac{1}{2}\left[u\arctan u - \frac{1}{2}\ln(1+u^2)\right]_0^{\pi/4} = \frac{1}{2}\left\{\left[\frac{\pi}{4}\arctan\frac{\pi}{4} - \frac{1}{2}\ln\left(1 + \frac{\pi^2}{16}\right)\right] - 0\right\}$$

$$= \frac{\pi}{8}\arctan\frac{\pi}{4} - \frac{1}{4}\ln\left(1 + \frac{\pi^2}{16}\right)$$

6. $\displaystyle\int_0^2 x^2\sqrt{4 - x^2}\, dx \overset{31}{=} \left[\frac{x}{8}(2x^2 - 4)\sqrt{4 - x^2} + \frac{16}{8}\sin^{-1}\left(\frac{x}{2}\right)\right]_0^2 = \left(0 + 2\cdot\frac{\pi}{2}\right) - 0 = \pi$

7. $\displaystyle\int \frac{\cos x}{\sin^2 x - 9}\, dx = \int \frac{1}{u^2 - 9}\, du \quad \begin{bmatrix} u = \sin x, \\ du = \cos x\, dx \end{bmatrix} \quad \overset{20}{=} \frac{1}{2(3)}\ln\left|\frac{u-3}{u+3}\right| + C = \frac{1}{6}\ln\left|\frac{\sin x - 3}{\sin x + 3}\right| + C$

8. $\displaystyle\int \frac{e^x}{4 - e^{2x}}\, dx = \int \frac{1}{4 - u^2}\, du \quad \begin{bmatrix} u = e^x, \\ du = e^x\, dx \end{bmatrix} \quad \overset{19}{=} \frac{1}{2(2)}\ln\left|\frac{u+2}{u-2}\right| + C = \frac{1}{4}\ln\left|\frac{e^x + 2}{e^x - 2}\right| + C$

9. $\displaystyle\int \frac{\sqrt{9x^2 + 4}}{x^2}\, dx = \int \frac{\sqrt{u^2 + 4}}{u^2/9}\left(\frac{1}{3}du\right) \quad \begin{bmatrix} u = 3x, \\ du = 3\,dx \end{bmatrix}$

$$= 3\int \frac{\sqrt{4 + u^2}}{u^2}\, du \overset{24}{=} 3\left[-\frac{\sqrt{4 + u^2}}{u} + \ln\left(u + \sqrt{4 + u^2}\right)\right] + C$$

$$= -\frac{3\sqrt{4 + 9x^2}}{3x} + 3\ln\left(3x + \sqrt{4 + 9x^2}\right) + C = -\frac{\sqrt{9x^2 + 4}}{x} + 3\ln\left(3x + \sqrt{9x^2 + 4}\right) + C$$

10. Let $u = \sqrt{2}\,y$ and $a = \sqrt{3}$. Then $du = \sqrt{2}\,dy$ and

$$\int \frac{\sqrt{2y^2 - 3}}{y^2}\, dy = \int \frac{\sqrt{u^2 - a^2}}{\frac{1}{2}u^2}\,\frac{du}{\sqrt{2}} = \sqrt{2}\int \frac{\sqrt{u^2 - a^2}}{u^2}\, du \overset{42}{=} \sqrt{2}\left(-\frac{\sqrt{u^2 - a^2}}{u} + \ln\left|u + \sqrt{u^2 - a^2}\right|\right) + C$$

$$= \sqrt{2}\left(-\frac{\sqrt{2y^2 - 3}}{\sqrt{2}\,y} + \ln\left|\sqrt{2}\,y + \sqrt{2y^2 - 3}\right|\right) + C$$

$$= -\frac{\sqrt{2y^2 - 3}}{y} + \sqrt{2}\ln\left|\sqrt{2}\,y + \sqrt{2y^2 - 3}\right| + C$$

11. $\int_0^\pi \cos^6\theta\, d\theta \overset{74}{=} \left[\frac{1}{6}\cos^5\theta\,\sin\theta\right]_0^\pi + \frac{5}{6}\int_0^\pi \cos^4\theta\, d\theta \overset{74}{=} 0 + \frac{5}{6}\left\{\left[\frac{1}{4}\cos^3\theta\,\sin\theta\right]_0^\pi + \frac{3}{4}\int_0^\pi \cos^2\theta\, d\theta\right\}$

$$\overset{64}{=} \tfrac{5}{6}\left\{0 + \tfrac{3}{4}\left[\tfrac{1}{2}\theta + \tfrac{1}{4}\sin 2\theta\right]_0^\pi\right\} = \tfrac{5}{6}\cdot\tfrac{3}{4}\cdot\tfrac{\pi}{2} = \tfrac{5\pi}{16}$$

12. $\int x\sqrt{2 + x^4}\, dx = \int \sqrt{2 + u^2}\left(\frac{1}{2}\, du\right) \qquad \begin{bmatrix} u = x^2, \\ du = 2x\, dx \end{bmatrix}$

$$\overset{21}{=} \frac{1}{2}\left[\frac{u}{2}\sqrt{2 + u^2} + \frac{2}{2}\ln\left(u + \sqrt{2 + u^2}\right)\right] + C = \frac{x^2}{4}\sqrt{2 + x^4} + \frac{1}{2}\ln\left(x^2 + \sqrt{2 + x^4}\right) + C$$

13. $\displaystyle\int \frac{\arctan\sqrt{x}}{\sqrt{x}}\,dx = \int \arctan u\,(2\,du) \qquad \begin{bmatrix} u=\sqrt{x}, \\ du = 1/(2\sqrt{x})\,dx \end{bmatrix}$

$\overset{89}{=} 2\left[u\arctan u - \tfrac{1}{2}\ln(1+u^2)\right] + C = 2\sqrt{x}\arctan\sqrt{x} - \ln(1+x) + C$

14. $\int_0^\pi x^3 \sin x\,dx \overset{84}{=} \left[-x^3\cos x\right]_0^\pi + 3\int_0^\pi x^2\cos x\,dx \overset{85}{=} -\pi^3(-1) + 3\left\{\left[x^2\sin x\right]_0^\pi - 2\int_0^\pi x\sin x\,dx\right\}$

$= \pi^3 - 6\int_0^\pi x\sin x\,dx \overset{84}{=} \pi^3 - 6\left\{\left[-x\cos x\right]_0^\pi + \int_0^\pi \cos x\,dx\right\}$

$= \pi^3 - 6[\pi] - 6\left[\sin x\right]_0^\pi = \pi^3 - 6\pi$

15. $\displaystyle\int \frac{\coth(1/y)}{y^2}\,dy = \int \coth u\,(-du) \qquad \begin{bmatrix} u = 1/y, \\ du = -1/y^2\,dy \end{bmatrix}$

$\overset{106}{=} -\ln|\sinh u| + C = -\ln|\sinh(1/y)| + C$

16. $\displaystyle\int \frac{e^{3t}}{\sqrt{e^{2t}-1}}\,dt = \int \frac{e^{2t}}{\sqrt{e^{2t}-1}}\,(e^t\,dt) = \int \frac{u^2}{\sqrt{u^2-1}}\,du \qquad \begin{bmatrix} u = e^t, \\ du = e^t\,dt \end{bmatrix}$

$\displaystyle\overset{44}{=} \frac{u}{2}\sqrt{u^2-1} + \tfrac{1}{2}\ln\left|u+\sqrt{u^2-1}\right| + C = \tfrac{1}{2}e^t\sqrt{e^{2t}-1} + \tfrac{1}{2}\ln\left(e^t + \sqrt{e^{2t}-1}\right) + C$

17. Let $z = 6 + 4y - 4y^2 = 6 - (4y^2 - 4y + 1) + 1 = 7 - (2y-1)^2$, $u = 2y - 1$, and $a = \sqrt{7}$.

Then $z = a^2 - u^2$, $du = 2\,dy$, and

$\int y\sqrt{6+4y-4y^2}\,dy = \int y\sqrt{z}\,dy = \int \tfrac{1}{2}(u+1)\sqrt{a^2-u^2}\,\tfrac{1}{2}\,du = \tfrac{1}{4}\int u\sqrt{a^2-u^2}\,du + \tfrac{1}{4}\int\sqrt{a^2-u^2}\,du$

$= \tfrac{1}{4}\int\sqrt{a^2-u^2}\,du - \tfrac{1}{8}\int(-2u)\sqrt{a^2-u^2}\,du$

$\displaystyle\overset{30}{=} \frac{u}{8}\sqrt{a^2-u^2} + \frac{a^2}{8}\sin^{-1}\!\left(\frac{u}{a}\right) - \frac{1}{8}\int\sqrt{w}\,dw \qquad \begin{bmatrix} w = a^2 - u^2, \\ dw = -2u\,du \end{bmatrix}$

$\displaystyle= \frac{2y-1}{8}\sqrt{6+4y-4y^2} + \frac{7}{8}\sin^{-1}\frac{2y-1}{\sqrt{7}} - \frac{1}{8}\cdot\frac{2}{3}w^{3/2} + C$

$\displaystyle= \frac{2y-1}{8}\sqrt{6+4y-4y^2} + \frac{7}{8}\sin^{-1}\frac{2y-1}{\sqrt{7}} - \frac{1}{12}(6+4y-4y^2)^{3/2} + C$

This can be rewritten as

$\displaystyle\sqrt{6+4y-4y^2}\left[\frac{1}{8}(2y-1) - \frac{1}{12}(6+4y-4y^2)\right] + \frac{7}{8}\sin^{-1}\frac{2y-1}{\sqrt{7}} + C$

$\displaystyle= \left(\frac{1}{3}y^2 - \frac{1}{12}y - \frac{5}{8}\right)\sqrt{6+4y-4y^2} + \frac{7}{8}\sin^{-1}\left(\frac{2y-1}{\sqrt{7}}\right) + C$

$\displaystyle= \frac{1}{24}(8y^2 - 2y - 15)\sqrt{6+4y-4y^2} + \frac{7}{8}\sin^{-1}\left(\frac{2y-1}{\sqrt{7}}\right) + C$

18. $\displaystyle\int \frac{dx}{2x^3 - 3x^2} = \int \frac{dx}{x^2(-3+2x)} \overset{50}{=} -\frac{1}{-3x} + \frac{2}{(-3)^2}\ln\left|\frac{-3+2x}{x}\right| + C = \frac{1}{3x} + \tfrac{2}{9}\ln\left|\frac{2x-3}{x}\right| + C$

19. Let $u = \sin x$. Then $du = \cos x\,dx$, so

$\displaystyle\int \sin^2 x\,\cos x\,\ln(\sin x)\,dx = \int u^2\ln u\,du \overset{101}{=} \frac{u^{2+1}}{(2+1)^2}[(2+1)\ln u - 1] + C = \tfrac{1}{9}u^3(3\ln u - 1) + C$

$= \tfrac{1}{9}\sin^3 x\,[3\ln(\sin x) - 1] + C$

20. Let $u = \sin\theta$, so that $du = \cos\theta\,d\theta$. Then

$$\int \frac{\sin 2\theta}{\sqrt{5-\sin\theta}}\,d\theta = \int \frac{2\sin\theta\,\cos\theta}{\sqrt{5-\sin\theta}}\,d\theta = 2\int \frac{u}{\sqrt{5-u}}\,du \overset{55}{=} 2\cdot\frac{2}{3(-1)^2}\,[-1u-2(5)]\,\sqrt{5-u}+C$$

$$= \tfrac{4}{3}(-u-10)\,\sqrt{5-u}+C = -\tfrac{4}{3}(\sin\theta+10)\,\sqrt{5-\sin\theta}+C$$

21. Let $u = e^x$ and $a = \sqrt{3}$. Then $du = e^x\,dx$ and

$$\int \frac{e^x}{3-e^{2x}}\,dx = \int \frac{du}{a^2-u^2} \overset{19}{=} \frac{1}{2a}\ln\left|\frac{u+a}{u-a}\right|+C = \frac{1}{2\sqrt{3}}\ln\left|\frac{e^x+\sqrt{3}}{e^x-\sqrt{3}}\right|+C.$$

22. Let $u = x^2$ and $a = 2$. Then $du = 2x\,dx$ and

$$\int_0^2 x^3\sqrt{4x^2-x^4}\,dx = \tfrac{1}{2}\int_0^2 x^2\,\sqrt{2\cdot 2\cdot x^2-(x^2)^2}\cdot 2x\,dx = \tfrac{1}{2}\int_0^4 u\,\sqrt{2au-u^2}\,du$$

$$\overset{114}{=} \left[\frac{2u^2-au-3a^2}{12}\sqrt{2au-u^2}+\frac{a^3}{4}\cos^{-1}\!\left(\frac{a-u}{a}\right)\right]_0^4$$

$$= \left[\frac{2u^2-2u-12}{12}\sqrt{4u-u^2}+\frac{8}{4}\cos^{-1}\!\left(\frac{2-u}{2}\right)\right]_0^4$$

$$= \left[\frac{u^2-u-6}{6}\sqrt{4u-u^2}+2\cos^{-1}\!\left(\frac{2-u}{2}\right)\right]_0^4$$

$$= [0+2\cos^{-1}(-1)]-(0+2\cos^{-1}1) = 2\cdot\pi-2\cdot 0 = 2\pi$$

23. $\int \sec^5 x\,dx \overset{77}{=} \frac{1}{4}\tan x\,\sec^3 x+\frac{3}{4}\int\sec^3 x\,dx \overset{77}{=} \frac{1}{4}\tan x\,\sec^3 x+\frac{3}{4}\left(\frac{1}{2}\tan x\,\sec x+\frac{1}{2}\int\sec x\,dx\right)$

$\overset{14}{=} \frac{1}{4}\tan x\,\sec^3 x+\frac{3}{8}\tan x\,\sec x+\frac{3}{8}\ln|\sec x+\tan x|+C$

24. $\int x^3\arcsin(x^2)\,dx = \int u\arcsin u\,\left(\frac{1}{2}du\right)$ $\quad\left[\begin{array}{l} u = x^2, \\ du = 2x\,dx \end{array}\right]$

$$\overset{90}{=} \frac{1}{2}\left[\frac{2u^2-1}{4}\arcsin u+\frac{u\sqrt{1-u^2}}{4}\right]+C = \frac{2x^4-1}{8}\arcsin(x^2)+\frac{x^2\sqrt{1-x^4}}{8}+C$$

25. Let $u = \ln x$ and $a = 2$. Then $du = dx/x$ and

$$\int \frac{\sqrt{4+(\ln x)^2}}{x}\,dx = \int \sqrt{a^2+u^2}\,du \overset{21}{=} \frac{u}{2}\sqrt{a^2+u^2}+\frac{a^2}{2}\ln\left(u+\sqrt{a^2+u^2}\right)+C$$

$$= \tfrac{1}{2}(\ln x)\sqrt{4+(\ln x)^2}+2\ln\left[\ln x+\sqrt{4+(\ln x)^2}\right]+C$$

26. $\int x^4e^{-x}\,dx \overset{97}{=} -x^4e^{-x}+4\int x^3e^{-x}\,dx \overset{97}{=} -x^4e^{-x}+4\left(-x^3e^{-x}+3\int x^2e^{-x}\,dx\right)$

$\overset{97}{=} -(x^4+4x^3)e^{-x}+12\left(-x^2e^{-x}+2\int xe^{-x}\,dx\right)$

$\overset{96}{=} -(x^4+4x^3+12x^2)e^{-x}+24[(-x-1)e^{-x}]+C = -(x^4+4x^3+12x^2+24x+24)e^{-x}+C$

So $\int_0^1 x^4e^{-x}\,dx = \left[-(x^4+4x^3+12x^2+24x+24)e^{-x}\right]_0^1 = -(1+4+12+24+24)e^{-1}+24e^0 = 24-65e^{-1}$.

27. $\displaystyle\int \frac{\cos^{-1}(x^{-2})}{x^3}\,dx = -\frac{1}{2}\int\cos^{-1}u\,du$ $\quad\left[\begin{array}{l} u = x^{-2}, \\ du = -2x^{-3}\,dx \end{array}\right]$

$\overset{88}{=} -\frac{1}{2}\left(u\cos^{-1}u-\sqrt{1-u^2}\right)+C = -\frac{1}{2}x^{-2}\cos^{-1}(x^{-2})+\frac{1}{2}\sqrt{1-x^{-4}}+C$

28. $\displaystyle\int \frac{dx}{\sqrt{1-e^{2x}}} = \int \frac{1}{\sqrt{1-u^2}}\left(\frac{du}{u}\right) \qquad \begin{bmatrix} u = e^x, \\ du = e^x\,dx,\ dx = du/u \end{bmatrix}$

$$\overset{35}{=} -\frac{1}{1}\ln\left|\frac{1+\sqrt{1-u^2}}{u}\right| + C = -\ln\left|\frac{1+\sqrt{1-e^{2x}}}{e^x}\right| + C = -\ln\left(\frac{1+\sqrt{1-e^{2x}}}{e^x}\right) + C$$

29. Let $u = e^x$. Then $x = \ln u$, $dx = du/u$, so

$$\int \sqrt{e^{2x}-1}\,dx = \int \frac{\sqrt{u^2-1}}{u}\,du \overset{41}{=} \sqrt{u^2-1} - \cos^{-1}(1/u) + C = \sqrt{e^{2x}-1} - \cos^{-1}(e^{-x}) + C.$$

30. Let $u = \alpha t - 3$ and assume that $\alpha \neq 0$. Then $du = \alpha\,dt$ and

$$\int e^t \sin(\alpha t - 3)\,dt = \frac{1}{\alpha}\int e^{(u+3)/\alpha}\sin u\,du = \frac{1}{\alpha}e^{3/\alpha}\int e^{(1/\alpha)u}\sin u\,du$$

$$\overset{98}{=} \frac{1}{\alpha}e^{3/\alpha}\frac{e^{(1/\alpha)u}}{(1/\alpha)^2+1^2}\left(\frac{1}{\alpha}\sin u - \cos u\right) + C = \frac{1}{\alpha}e^{3/\alpha}e^{(1/\alpha)u}\frac{\alpha^2}{1+\alpha^2}\left(\frac{1}{\alpha}\sin u - \cos u\right) + C$$

$$= \frac{1}{1+\alpha^2}e^{(u+3)/\alpha}(\sin u - \alpha\cos u) + C = \frac{1}{1+\alpha^2}\,e^t\left[\sin(\alpha t - 3) - \alpha\cos(\alpha t - 3)\right] + C$$

31. $\displaystyle\int \frac{x^4\,dx}{\sqrt{x^{10}-2}} = \int \frac{x^4\,dx}{\sqrt{(x^5)^2-2}} = \frac{1}{5}\int \frac{du}{\sqrt{u^2-2}} \qquad \begin{bmatrix} u = x^5, \\ du = 5x^4\,dx \end{bmatrix}$

$$\overset{43}{=} \tfrac{1}{5}\ln\left|u+\sqrt{u^2-2}\right| + C = \tfrac{1}{5}\ln\left|x^5+\sqrt{x^{10}-2}\right| + C$$

32. Let $u = \tan\theta$ and $a = 3$. Then $du = \sec^2\theta\,d\theta$ and

$$\int \frac{\sec^2\theta\,\tan^2\theta}{\sqrt{9-\tan^2\theta}}\,d\theta = \int \frac{u^2}{\sqrt{a^2-u^2}}\,du \overset{34}{=} -\frac{u}{2}\sqrt{a^2-u^2} + \frac{a^2}{2}\sin^{-1}\left(\frac{u}{a}\right) + C$$

$$= -\frac{1}{2}\tan\theta\,\sqrt{9-\tan^2\theta} + \frac{9}{2}\sin^{-1}\left(\frac{\tan\theta}{3}\right) + C$$

33. Use disks about the x-axis:

$$V = \textstyle\int_0^\pi \pi(\sin^2 x)^2\,dx = \pi\int_0^\pi \sin^4 x\,dx \overset{73}{=} \pi\left\{\left[-\frac{1}{4}\sin^3 x\,\cos x\right]_0^\pi + \frac{3}{4}\int_0^\pi \sin^2 x\,dx\right\}$$

$$\overset{63}{=} \pi\left\{0 + \tfrac{3}{4}\left[\tfrac{1}{2}x - \tfrac{1}{4}\sin 2x\right]_0^\pi\right\} = \pi\left[\tfrac{3}{4}\left(\tfrac{1}{2}\pi - 0\right)\right] = \tfrac{3}{8}\pi^2$$

34. Use shells about the y-axis:

$$V = \int_0^1 2\pi x \arcsin x\,dx \overset{90}{=} 2\pi\left[\frac{2x^2-1}{4}\sin^{-1}x + \frac{x\sqrt{1-x^2}}{4}\right]_0^1 = 2\pi\left[\left(\frac{1}{4}\cdot\frac{\pi}{2}+0\right) - 0\right] = \frac{1}{4}\pi^2$$

35. (a) $\displaystyle\frac{d}{du}\left[\frac{1}{b^3}\left(a+bu-\frac{a^2}{a+bu}-2a\ln|a+bu|\right)+C\right] = \frac{1}{b^3}\left[b + \frac{ba^2}{(a+bu)^2} - \frac{2ab}{(a+bu)}\right]$

$$= \frac{1}{b^3}\left[\frac{b(a+bu)^2 + ba^2 - (a+bu)2ab}{(a+bu)^2}\right]$$

$$= \frac{1}{b^3}\left[\frac{b^3u^2}{(a+bu)^2}\right] = \frac{u^2}{(a+bu)^2}$$

(b) Let $t = a + bu \;\Rightarrow\; dt = b\,du$. Note that $u = \dfrac{t-a}{b}$ and $du = \dfrac{1}{b}\,dt$.

$$\int \frac{u^2\,du}{(a+bu)^2} = \frac{1}{b^3}\int \frac{(t-a)^2}{t^2}\,dt = \frac{1}{b^3}\int \frac{t^2 - 2at + a^2}{t^2}\,dt = \frac{1}{b^3}\int\left(1 - \frac{2a}{t} + \frac{a^2}{t^2}\right)dt$$

$$= \frac{1}{b^3}\left(t - 2a\ln|t| - \frac{a^2}{t}\right) + C = \frac{1}{b^3}\left(a + bu - \frac{a^2}{a+bu} - 2a\ln|a+bu|\right) + C$$

36. (a) $\dfrac{d}{du}\left[\dfrac{u}{8}(2u^2 - a^2)\sqrt{a^2 - u^2} + \dfrac{a^4}{8}\sin^{-1}\dfrac{u}{a} + C\right]$

$$= \frac{u}{8}(2u^2 - a^2)\frac{-u}{\sqrt{a^2 - u^2}} + \sqrt{a^2 - u^2}\left[\frac{u}{8}(4u) + (2u^2 - a^2)\tfrac{1}{8}\right] + \frac{a^4}{8}\frac{1/a}{\sqrt{1 - u^2/a^2}}$$

$$= -\frac{u^2(2u^2 - a^2)}{8\sqrt{a^2 - u^2}} + \sqrt{a^2 - u^2}\left[\frac{u^2}{2} + \frac{2u^2 - a^2}{8}\right] + \frac{a^4}{8\sqrt{a^2 - u^2}}$$

$$= \tfrac{1}{2}(a^2 - u^2)^{-1/2}\left[-\frac{u^2}{4}(2u^2 - a^2) + u^2(a^2 - u^2) + \tfrac{1}{4}(a^2 - u^2)(2u^2 - a^2) + \frac{a^4}{4}\right]$$

$$= \tfrac{1}{2}(a^2 - u^2)^{-1/2}[2u^2a^2 - 2u^4] = \frac{u^2(a^2 - u^2)}{\sqrt{a^2 - u^2}} = u^2\sqrt{a^2 - u^2}$$

(b) Let $u = a\sin\theta \;\Rightarrow\; du = a\cos\theta\,d\theta$. Then

$$\int u^2\sqrt{a^2 - u^2}\,du = \int a^2\sin^2\theta\, a\sqrt{1 - \sin^2\theta}\, a\cos\theta\,d\theta = a^4\int \sin^2\theta\,\cos^2\theta\,d\theta$$

$$= a^4\int \tfrac{1}{2}(1 + \cos 2\theta)\tfrac{1}{2}(1 - \cos 2\theta)\,d\theta = \tfrac{1}{4}a^4\int(1 - \cos^2 2\theta)\,d\theta$$

$$= \tfrac{1}{4}a^4\int\left[1 - \tfrac{1}{2}(1 + \cos 4\theta)\right]d\theta = \tfrac{1}{4}a^4\left(\tfrac{1}{2}\theta - \tfrac{1}{8}\sin 4\theta\right) + C$$

$$= \tfrac{1}{4}a^4\left(\tfrac{1}{2}\theta - \tfrac{1}{8}\cdot 2\sin 2\theta\,\cos 2\theta\right) + C = \tfrac{1}{4}a^4\left[\tfrac{1}{2}\theta - \tfrac{1}{2}\sin\theta\,\cos\theta(1 - 2\sin^2\theta)\right] + C$$

$$= \frac{a^4}{8}\left[\sin^{-1}\frac{u}{a} - \frac{u}{a}\frac{\sqrt{a^2 - u^2}}{a}\left(1 - \frac{2u^2}{a^2}\right)\right] + C = \frac{a^4}{8}\left[\sin^{-1}\frac{u}{a} - \frac{u}{a}\frac{\sqrt{a^2 - u^2}}{a}\frac{a^2 - 2u^2}{a^2}\right] + C$$

$$= \frac{u}{8}(2u^2 - a^2)\sqrt{a^2 - u^2} + \frac{a^4}{8}\sin^{-1}\frac{u}{a} + C$$

37. Maple and Mathematica both give $\int \sec^4 x\,dx = \frac{2}{3}\tan x + \frac{1}{3}\tan x\,\sec^2 x$, while Derive gives the second term as $\dfrac{\sin x}{3\cos^3 x} = \dfrac{1}{3}\dfrac{\sin x}{\cos x}\dfrac{1}{\cos^2 x} = \dfrac{1}{3}\tan x\,\sec^2 x$. Using Formula 77, we get $\int \sec^4 x\,dx = \frac{1}{3}\tan x\,\sec^2 x + \frac{2}{3}\int \sec^2 x\,dx = \frac{1}{3}\tan x\,\sec^2 x + \frac{2}{3}\tan x + C$.

38. Derive gives $\int \csc^5 x\,dx = \dfrac{3}{8}\ln\left(\tan\left(\dfrac{x}{2}\right)\right) - \cos x\left(\dfrac{3}{8\sin^2 x} + \dfrac{1}{4\sin^4 x}\right)$ and Maple gives $-\dfrac{1}{4}\dfrac{\cos x}{\sin^4 x} - \dfrac{3}{8}\dfrac{\cos x}{\sin^2 x} + \dfrac{3}{8}\ln(\csc x - \cot x)$. Using a half-angle identity for tangent, $\tan\dfrac{x}{2} = \dfrac{1 - \cos x}{\sin x}$, we have $\ln\tan\dfrac{x}{2} = \ln\dfrac{1 - \cos x}{\sin x} = \ln\left(\dfrac{1}{\sin x} - \dfrac{\cos x}{\sin x}\right) = \ln(\csc x - \cot x)$, so those two answers are equivalent.

Mathematica gives

$$\begin{aligned}
I &= -\frac{3}{32}\csc^2\frac{x}{2} - \frac{1}{64}\csc^4\frac{x}{2} - \frac{3}{8}\log\cos\frac{x}{2} + \frac{3}{8}\log\sin\frac{x}{2} + \frac{3}{32}\sec^2\frac{x}{2} + \frac{1}{64}\sec^4\frac{x}{2} \\
&= \frac{3}{8}\left(\log\sin\frac{x}{2} - \log\cos\frac{x}{2}\right) + \frac{3}{32}\left(\sec^2\frac{x}{2} - \csc^2\frac{x}{2}\right) + \frac{1}{64}\left(\sec^4\frac{x}{2} - \csc^4\frac{x}{2}\right) \\
&= \frac{3}{8}\log\frac{\sin(x/2)}{\cos(x/2)} + \frac{3}{32}\left[\frac{1}{\cos^2(x/2)} - \frac{1}{\sin^2(x/2)}\right] + \frac{1}{64}\left[\frac{1}{\cos^4(x/2)} - \frac{1}{\sin^4(x/2)}\right] \\
&= \frac{3}{8}\log\tan\frac{x}{2} + \frac{3}{32}\left[\frac{\sin^2(x/2) - \cos^2(x/2)}{\cos^2(x/2)\,\sin^2(x/2)}\right] + \frac{1}{64}\left[\frac{\sin^4(x/2) - \cos^4(x/2)}{\cos^4(x/2)\,\sin^4(x/2)}\right]
\end{aligned}$$

Now $$\frac{\sin^2(x/2) - \cos^2(x/2)}{\cos^2(x/2)\,\sin^2(x/2)} = \frac{\dfrac{1-\cos x}{2} - \dfrac{1+\cos x}{2}}{\dfrac{1+\cos x}{2}\cdot\dfrac{1-\cos x}{2}} = \frac{-\dfrac{2\cos x}{2}}{\dfrac{1-\cos^2 x}{4}} = \frac{-4\cos x}{\sin^2 x}$$

and $$\begin{aligned}
\frac{\sin^4(x/2) - \cos^4(x/2)}{\cos^4(x/2)\,\sin^4(x/2)} &= \frac{\sin^2(x/2) - \cos^2(x/2)}{\cos^2(x/2)\,\sin^2(x/2)}\,\frac{\sin^2(x/2) + \cos^2(x/2)}{\cos^2(x/2)\,\sin^2(x/2)} \\
&= \frac{-4\cos x}{\sin^2 x}\,\frac{1}{\dfrac{1+\cos x}{2}\cdot\dfrac{1-\cos x}{2}} = -\frac{4\cos x}{\sin^2 x}\,\frac{4}{1-\cos^2 x} = -\frac{16\cos x}{\sin^4 x}
\end{aligned}$$

Returning to the expression for I, we get

$$I = \frac{3}{8}\log\tan\frac{x}{2} + \frac{3}{32}\left(\frac{-4\cos x}{\sin^2 x}\right) + \frac{1}{64}\left(\frac{-16\cos x}{\sin^4 x}\right) = \frac{3}{8}\log\tan\frac{x}{2} - \frac{3}{8}\frac{\cos x}{\sin^2 x} - \frac{1}{4}\frac{\cos x}{\sin^4 x},$$

so all are equivalent.

Now use Formula 78 to get

$$\begin{aligned}
\int \csc^5 x\,dx &= \frac{-1}{4}\cot x\csc^3 x + \frac{3}{4}\int \csc^3 x\,dx = -\frac{1}{4}\frac{\cos x}{\sin x}\frac{1}{\sin^3 x} + \frac{3}{4}\left(\frac{-1}{2}\cot x\csc x + \frac{1}{2}\int \csc x\,dx\right) \\
&= -\frac{1}{4}\frac{\cos x}{\sin^4 x} - \frac{3}{8}\frac{\cos x}{\sin x}\frac{1}{\sin x} + \frac{3}{8}\int \csc x\,dx = \frac{1}{4}\frac{\cos x}{\sin^4 x} - \frac{3}{8}\frac{\cos x}{\sin^2 x} + \frac{3}{8}\ln|\csc x - \cot x| + C
\end{aligned}$$

39. Derive gives $\int x^2\sqrt{x^2+4}\,dx = \frac{1}{4}x(x^2+2)\sqrt{x^2+4} - 2\ln\left(\sqrt{x^2+4}+x\right)$. Maple gives $\frac{1}{4}x(x^2+4)^{3/2} - \frac{1}{2}x\sqrt{x^2+4} - 2\,\mathrm{arcsinh}\left(\frac{1}{2}x\right)$. Applying the command `convert(%,ln);` yields

$$\begin{aligned}
\tfrac{1}{4}x(x^2+4)^{3/2} - \tfrac{1}{2}x\sqrt{x^2+4} - 2\ln\left(\tfrac{1}{2}x + \tfrac{1}{2}\sqrt{x^2+4}\right) &= \tfrac{1}{4}x(x^2+4)^{1/2}\left[(x^2+4)-2\right] - 2\ln\left[\left(x+\sqrt{x^2+4}\right)/2\right] \\
&= \tfrac{1}{4}x(x^2+2)\sqrt{x^2+4} - 2\ln\left(\sqrt{x^2+4}+x\right) + 2\ln 2
\end{aligned}$$

Mathematica gives $\frac{1}{4}x(2+x^2)\sqrt{3+x^2} - 2\,\mathrm{arcsinh}(x/2)$. Applying the `TrigToExp` and `Simplify` commands gives $\frac{1}{4}\left[x(2+x^2)\sqrt{4+x^2} - 8\log\left(\frac{1}{2}\left(x+\sqrt{4+x^2}\right)\right)\right] = \frac{1}{4}x(x^2+2)\sqrt{x^2+4} - 2\ln\left(x+\sqrt{4+x^2}\right) + 2\ln 2$, so all are equivalent (without constant).

Now use Formula 22 to get

$$\begin{aligned}
\int x^2\sqrt{2^2+x^2}\,dx &= \frac{x}{8}(2^2+2x^2)\sqrt{2^2+x^2} - \frac{2^4}{8}\ln\left(x+\sqrt{2^2+x^2}\right) + C \\
&= \frac{x}{8}(2)(2+x^2)\sqrt{4+x^2} - 2\ln\left(x+\sqrt{4+x^2}\right) + C \\
&= \tfrac{1}{4}x(x^2+2)\sqrt{x^2+4} - 2\ln\left(\sqrt{x^2+4}+x\right) + C
\end{aligned}$$

40. Derive gives $\int \frac{dx}{e^x(3e^x+2)} = -\frac{e^{-x}}{2} + \frac{3\ln(3e^x+2)}{4} - \frac{3x}{4}$, Maple gives $\frac{3}{4}\ln(3e^x+2) - \frac{1}{2e^x} - \frac{3}{4}\ln(e^x)$, and Mathematica gives

$$-\frac{e^{-x}}{2} + \frac{3}{4}\log(3+2e^{-x}) = -\frac{e^{-x}}{2} + \frac{3}{4}\log\left(\frac{3e^x+2}{e^x}\right) = -\frac{e^{-x}}{2} + \frac{3}{4}\frac{\ln(3e^x+2)}{\ln e^x} = -\frac{e^{-x}}{2} + \frac{3}{4}\ln(3e^x+2) - \frac{3}{4}x,$$

so all are equivalent. Now let $u = e^x$, so $du = e^x\,dx$ and $dx = du/u$. Then

$$\int \frac{1}{e^x(3e^x+2)}\,dx = \int \frac{1}{u(3u+2)}\frac{du}{u} = \int \frac{1}{u^2(2+3u)}\,du \overset{50}{=} -\frac{1}{2u} + \frac{3}{2^2}\ln\left|\frac{2+3u}{u}\right| + C$$

$$= -\frac{1}{2e^x} + \frac{3}{4}\ln(2+3e^x) - \frac{3}{4}\ln e^x + C = -\frac{1}{2e^x} + \frac{3}{4}\ln(3e^x+2) - \frac{3}{4}x + C$$

41. Derive and Maple give $\int \cos^4 x\,dx = \frac{\sin x\cos^3 x}{4} + \frac{3\sin x\cos x}{8} + \frac{3x}{8}$, while Mathematica gives

$$\frac{3x}{8} + \frac{1}{4}\sin(2x) + \frac{1}{32}\sin(4x) = \frac{3x}{8} + \frac{1}{4}(2\sin x\cos x) + \frac{1}{32}(2\sin 2x\cos 2x)$$

$$= \frac{3x}{8} + \frac{1}{2}\sin x\cos x + \frac{1}{16}[2\sin x\cos x\,(2\cos^2 x - 1)]$$

$$= \frac{3x}{8} + \frac{1}{2}\sin x\cos x + \frac{1}{4}\sin x\cos^3 x - \frac{1}{8}\sin x\cos x,$$

so all are equivalent.

Using tables,

$$\int \cos^4 x\,dx \overset{74}{=} \tfrac{1}{4}\cos^3 x\sin x + \tfrac{3}{4}\int \cos^2 x\,dx \overset{64}{=} \tfrac{1}{4}\cos^3 x\sin x + \tfrac{3}{4}\left(\tfrac{1}{2}x + \tfrac{1}{4}\sin 2x\right) + C$$

$$= \tfrac{1}{4}\cos^3 x\sin x + \tfrac{3}{8}x + \tfrac{3}{16}(2\sin x\cos x) + C = \tfrac{1}{4}\cos^3 x\sin x + \tfrac{3}{8}x + \tfrac{3}{8}\sin x\cos x + C$$

42. Derive gives $\int x^2\sqrt{1-x^2}\,dx = \frac{\arcsin x}{8} + \frac{x\sqrt{1-x^2}(2x^2-1)}{8}$, Maple gives

$$-\frac{x}{4}(1-x^2)^{3/2} + \frac{x}{8}\sqrt{1-x^2} + \frac{1}{8}\arcsin x = \frac{x}{8}(1-x^2)^{1/2}[-2(1-x^2)+1] + \frac{1}{8}\arcsin x$$

$$= \frac{x}{8}(1-x^2)^{1/2}(2x^2-1) + \frac{1}{8}\arcsin x,$$

and Mathematica gives $\frac{1}{8}\left(x\sqrt{1-x^2}(-1+2x^2) + \arcsin x\right)$, so all are equivalent.

Now use Formula 31 to get

$$\int x^2\sqrt{1-x^2}\,dx = \frac{x}{8}(2x^2-1)\sqrt{1-x^2} + \frac{1}{8}\sin^{-1}x + C$$

43. Maple gives $\int \tan^5 x\,dx = \frac{1}{4}\tan^4 x - \frac{1}{2}\tan^2 x + \frac{1}{2}\ln(1+\tan^2 x)$, Mathematica gives $\int \tan^5 x\,dx = \frac{1}{4}[-1-2\cos(2x)]\sec^4 x - \ln(\cos x)$, and Derive gives $\int \tan^5 x\,dx = \frac{1}{4}\tan^4 x - \frac{1}{2}\tan^2 x - \ln(\cos x)$. These expressions are equivalent, and none includes absolute value bars or a constant of integration. Note that Mathematica's and Derive's expressions suggest that the integral is undefined where $\cos x < 0$, which is not the case. Using Formula 75, $\int \tan^5 x\,dx = \frac{1}{5-1}\tan^{5-1}x - \int \tan^{5-2}x\,dx = \frac{1}{4}\tan^4 x - \int \tan^3 x\,dx$. Using Formula 69, $\int \tan^3 x\,dx = \frac{1}{2}\tan^2 x + \ln|\cos x| + C$, so $\int \tan^5 x\,dx = \frac{1}{4}\tan^4 x - \frac{1}{2}\tan^2 x - \ln|\cos x| + C$.

44. Derive, Maple, and Mathematica all give $\displaystyle\int \frac{1}{\sqrt{1+\sqrt[3]{x}}}\,dx = \frac{2}{5}\sqrt{\sqrt[3]{x}+1}\left(3\sqrt[3]{x^2}-4\sqrt[3]{x}+8\right)$. [Maple adds a constant of $-\frac{16}{5}$.] We'll change the form of the integral by letting $u=\sqrt[3]{x}$, so that $u^3=x$ and $3u^2\,du=dx$. Then

$$\int \frac{1}{\sqrt{1+\sqrt[3]{x}}}\,dx = \int \frac{3u^2\,du}{\sqrt{1+u}} \overset{56}{=} 3\left[\frac{2}{15(1)^3}\left(8(1)^2+3(1)^2u^2-4(1)(1)u\right)\sqrt{1+u}\right]+C$$

$$= \tfrac{2}{5}(8+3u^2-4u)\sqrt{1+u}+C = \tfrac{2}{5}\left(8+3\sqrt[3]{x^2}-4\sqrt[3]{x}\right)\sqrt{1+\sqrt[3]{x}}+C$$

45. (a) $\displaystyle F(x)=\int f(x)\,dx = \int \frac{1}{x\sqrt{1-x^2}}\,dx \overset{35}{=} -\frac{1}{1}\ln\left|\frac{1+\sqrt{1-x^2}}{x}\right|+C = -\ln\left|\frac{1+\sqrt{1-x^2}}{x}\right|+C.$

f has domain $\{x \mid x\neq 0,\ 1-x^2>0\} = \{x \mid x\neq 0,\ |x|<1\} = (-1,0)\cup(0,1)$. F has the same domain.

(b) Derive gives $F(x)=\ln\left(\sqrt{1-x^2}-1\right)-\ln x$ and Mathematica gives $F(x)=\ln x-\ln\left(1+\sqrt{1-x^2}\right)$. Both are correct if you take absolute values of the logarithm arguments, and both would then have the same domain. Maple gives $F(x)=-\operatorname{arctanh}\left(1/\sqrt{1-x^2}\right)$. This function has domain

$\left\{x \mid |x|<1,\ -1<1/\sqrt{1-x^2}<1\right\} = \left\{x \mid |x|<1,\ 1/\sqrt{1-x^2}<1\right\} = \left\{x \mid |x|<1,\ \sqrt{1-x^2}>1\right\} = \emptyset$,

the empty set! If we apply the command `convert(%,ln);` to Maple's answer, we get

$-\dfrac{1}{2}\ln\left(\dfrac{1}{\sqrt{1-x^2}}+1\right)+\dfrac{1}{2}\ln\left(1-\dfrac{1}{\sqrt{1-x^2}}\right)$, which has the same domain, $\emptyset$.

46. None of Maple, Mathematica and Derive is able to evaluate $\int(1+\ln x)\sqrt{1+(x\ln x)^2}\,dx$. However, if we let $u=x\ln x$, then $du=(1+\ln x)\,dx$ and the integral is simply $\int\sqrt{1+u^2}\,du$, which any CAS can evaluate. The antiderivative is

$\frac{1}{2}\ln\left(x\ln x+\sqrt{1+(x\ln x)^2}\right)+\frac{1}{2}x\ln x\sqrt{1+(x\ln x)^2}+C$.

DISCOVERY PROJECT Patterns in Integrals

1. (a) The CAS results are listed. Note that the absolute value symbols are missing, as is the familiar "$+C$".

(i) $\displaystyle\int\frac{1}{(x+2)(x+3)}\,dx=\ln(x+2)-\ln(x+3)$ (ii) $\displaystyle\int\frac{1}{(x+1)(x+5)}\,dx=\frac{\ln(x+1)}{4}-\frac{\ln(x+5)}{4}$

(iii) $\displaystyle\int\frac{1}{(x+2)(x-5)}\,dx=\frac{\ln(x-5)}{7}-\frac{\ln(x+2)}{7}$ (iv) $\displaystyle\int\frac{1}{(x+2)^2}\,dx=-\frac{1}{x+2}$

(b) If $a\neq b$, it appears that $\ln(x+a)$ is divided by $b-a$ and $\ln(x+b)$ is divided by $a-b$, so we guess that

$\displaystyle\int\frac{1}{(x+a)(x+b)}\,dx=\frac{\ln(x+a)}{b-a}+\frac{\ln(x+b)}{a-b}+C$. If $a=b$, as in part (a)(iv), it appears that

$\displaystyle\int\frac{1}{(x+a)^2}\,dx=-\frac{1}{x+a}+C.$

(c) The CAS verifies our guesses. Now $\dfrac{1}{(x+a)(x+b)}=\dfrac{A}{x+a}+\dfrac{B}{x+b} \Rightarrow 1=A(x+b)+B(x+a)$.

Setting $x=-b$ gives $B=1/(a-b)$ and setting $x=-a$ gives $A=1/(b-a)$. So

$$\int\frac{1}{(x+a)(x+b)}\,dx=\int\left[\frac{1/(b-a)}{x+a}+\frac{1/(a-b)}{x+b}\right]dx=\frac{\ln|x+a|}{b-a}+\frac{\ln|x+b|}{a-b}+C$$

[continued]

and our guess for $a \neq b$ is correct. If $a = b$, then $\dfrac{1}{(x+a)(x+b)} = \dfrac{1}{(x+a)^2} = (x+a)^{-2}$. Letting $u = x + a \Rightarrow$ $du = dx$, we have $\displaystyle\int (x+a)^{-2}\,dx = \int u^{-2}\,du = -\frac{1}{u} + C = -\frac{1}{x+a} + C$, and our guess for $a = b$ is also correct.

2. (a) (i) $\displaystyle\int \sin x \cos 2x\,dx = \frac{\cos x}{2} - \frac{\cos 3x}{6}$ (ii) $\displaystyle\int \sin 3x \cos 7x\,dx = \frac{\cos 4x}{8} - \frac{\cos 10x}{20}$

(iii) $\displaystyle\int \sin 8x \cos 3x\,dx = -\frac{\cos 11x}{22} - \frac{\cos 5x}{10}$

(b) Looking at the sums and differences of a and b in part (a), we guess that

$$\int \sin ax \cos bx\,dx = \frac{\cos((a-b)x)}{2(b-a)} - \frac{\cos((a+b)x)}{2(a+b)} + C$$

Note that $\cos((a-b)x) = \cos((b-a)x)$.

(c) The CAS verifies our guess. Again, we can prove that the guess is correct by differentiating:

$$\begin{aligned}
\frac{d}{dx}\left[\frac{\cos((a-b)x)}{2(b-a)} - \frac{\cos((a+b)x)}{2(a+b)}\right] &= \frac{1}{2(b-a)}[-\sin((a-b)x)](a-b) - \frac{1}{2(a+b)}[-\sin((a+b)x)](a+b) \\
&= \tfrac{1}{2}\sin(ax-bx) + \tfrac{1}{2}\sin(ax+bx) \\
&= \tfrac{1}{2}(\sin ax \cos bx - \cos ax \sin bx) + \tfrac{1}{2}(\sin ax \cos bx + \cos ax \sin bx) \\
&= \sin ax \cos bx
\end{aligned}$$

Our formula is valid for $a \neq b$.

3. (a) (i) $\int \ln x\,dx = x\ln x - x$ (ii) $\int x \ln x\,dx = \frac{1}{2}x^2\ln x - \frac{1}{4}x^2$

(iii) $\int x^2 \ln x\,dx = \frac{1}{3}x^3 \ln x - \frac{1}{9}x^3$ (iv) $\int x^3 \ln x\,dx = \frac{1}{4}x^4 \ln x - \frac{1}{16}x^4$

(v) $\int x^7 \ln x\,dx = \frac{1}{8}x^8 \ln x - \frac{1}{64}x^8$

(b) We guess that $\int x^n \ln x\,dx = \dfrac{1}{n+1}x^{n+1}\ln x - \dfrac{1}{(n+1)^2}x^{n+1}$.

(c) Let $u = \ln x$, $dv = x^n\,dx \Rightarrow du = \dfrac{dx}{x}$, $v = \dfrac{1}{n+1}x^{n+1}$. Then

$$\int x^n \ln x\,dx = \frac{1}{n+1}x^{n+1}\ln x - \frac{1}{n+1}\int x^n\,dx = \frac{1}{n+1}x^{n+1}\ln x - \frac{1}{n+1}\cdot\frac{1}{n+1}x^{n+1},$$

which verifies our guess. We must have $n + 1 \neq 0 \Leftrightarrow n \neq -1$.

4. (a) (i) $\int xe^x\,dx = e^x(x-1)$ (ii) $\int x^2 e^x\,dx = e^x(x^2 - 2x + 2)$

(iii) $\int x^3 e^x\,dx = e^x(x^3 - 3x^2 + 6x - 6)$ (iv) $\int x^4 e^x\,dx = e^x(x^4 - 4x^3 + 12x^2 - 24x + 24)$

(v) $\int x^5 e^x\,dx = e^x(x^5 - 5x^4 + 20x^3 - 60x^2 + 120x - 120)$

(b) Notice from part (a) that we can write

$$\int x^4 e^x\,dx = e^x(x^4 - 4x^3 + 4\cdot 3x^2 - 4\cdot 3\cdot 2x + 4\cdot 3\cdot 2\cdot 1)$$

and $$\int x^5 e^x\,dx = e^x(x^5 - 5x^4 + 5\cdot 4x^3 - 5\cdot 4\cdot 3x^2 + 5\cdot 4\cdot 3\cdot 2x - 5\cdot 4\cdot 3\cdot 2\cdot 1)$$

So we guess that

$$\int x^6 e^x\,dx = e^x(x^6 - 6x^5 + 6\cdot 5x^4 - 6\cdot 5\cdot 4x^3 + 6\cdot 5\cdot 4\cdot 3x^2 - 6\cdot 5\cdot 4\cdot 3\cdot 2x + 6\cdot 5\cdot 4\cdot 3\cdot 2\cdot 1)$$
$$= e^x(x^6 - 6x^5 + 30x^4 - 120x^3 + 360x^2 - 720x + 720)$$

The CAS verifies our guess.

(c) From the results in part (a), as well as our prediction in part (b), we speculate that

$$\int x^n e^x\,dx = e^x\left[x^n - nx^{n-1} + n(n-1)x^{n-2} - n(n-1)(n-2)x^{n-3} + \cdots \pm n!x \mp n!\right] = e^x \sum_{i=0}^{n} (-1)^{n-i}\frac{n!}{i!}\,x^i.$$

(We have reversed the order of the polynomial's terms.)

(d) Let S_n be the statement that $\int x^n e^x\,dx = e^x \sum_{i=0}^{n} (-1)^{n-i}\frac{n!}{i!}\,x^i$.

S_1 is true by part (a)(i). Suppose S_k is true for some k, and consider S_{k+1}. Integrating by parts with $u = x^{k+1}$, $dv = e^x\,dx \Rightarrow du = (k+1)x^k\,dx$, $v = e^x$, we get

$$\int x^{k+1}e^x\,dx = x^{k+1}e^x - (k+1)\int x^k e^x\,dx = x^{k+1}e^x - (k+1)\left[e^x \sum_{i=0}^{k} (-1)^{k-i}\frac{k!}{i!}\,x^i\right]$$
$$= e^x\left[x^{k+1} - (k+1)\sum_{i=0}^{k} (-1)^{k-i}\frac{k!}{i!}\,x^i\right] = e^x\left[x^{k+1} + \sum_{i=0}^{k} (-1)^{k-i+1}\frac{(k+1)k!}{i!}\,x^i\right]$$
$$= e^x \sum_{i=0}^{k+1} (-1)^{(k+1)-i}\frac{(k+1)!}{i!}\,x^i$$

This verifies S_n for $n = k+1$. Thus, by mathematical induction, S_n is true for all n, where n is a positive integer.

7.7 Approximate Integration

1. (a) $\Delta x = (b-a)/n = (4-0)/2 = 2$

$$L_2 = \sum_{i=1}^{2} f(x_{i-1})\,\Delta x = f(x_0)\cdot 2 + f(x_1)\cdot 2 = 2\,[f(0) + f(2)] = 2(0.5 + 2.5) = 6$$

$$R_2 = \sum_{i=1}^{2} f(x_i)\,\Delta x = f(x_1)\cdot 2 + f(x_2)\cdot 2 = 2\,[f(2) + f(4)] = 2(2.5 + 3.5) = 12$$

$$M_2 = \sum_{i=1}^{2} f(\overline{x}_i)\Delta x = f(\overline{x}_1)\cdot 2 + f(\overline{x}_2)\cdot 2 = 2\,[f(1) + f(3)] \approx 2(1.6 + 3.2) = 9.6$$

(b)

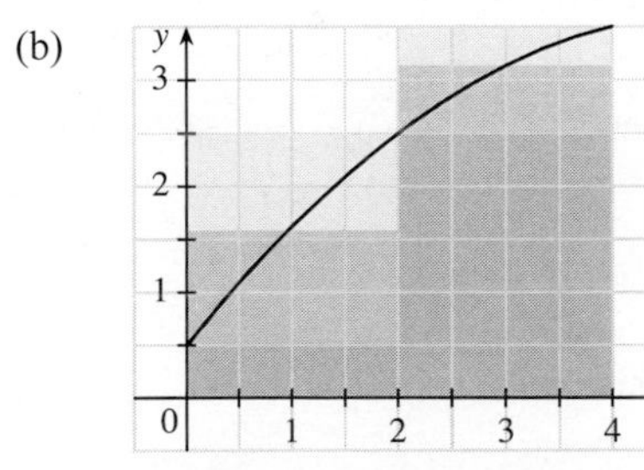

L_2 is an underestimate, since the area under the small rectangles is less than the area under the curve, and R_2 is an overestimate, since the area under the large rectangles is greater than the area under the curve. It appears that M_2 is an overestimate, though it is fairly close to I. See the solution to Exercise 47 for a proof of the fact that if f is concave down on $[a, b]$, then the Midpoint Rule is an overestimate of $\int_a^b f(x)\,dx$.

(c) $T_2 = \left(\frac{1}{2}\,\Delta x\right)[f(x_0) + 2f(x_1) + f(x_2)] = \frac{2}{2}[f(0) + 2f(2) + f(4)] = 0.5 + 2(2.5) + 3.5 = 9.$

This approximation is an underestimate, since the graph is concave down. Thus, $T_2 = 9 < I$. See the solution to Exercise 47 for a general proof of this conclusion.

(d) For any n, we will have $L_n < T_n < I < M_n < R_n$.

2. 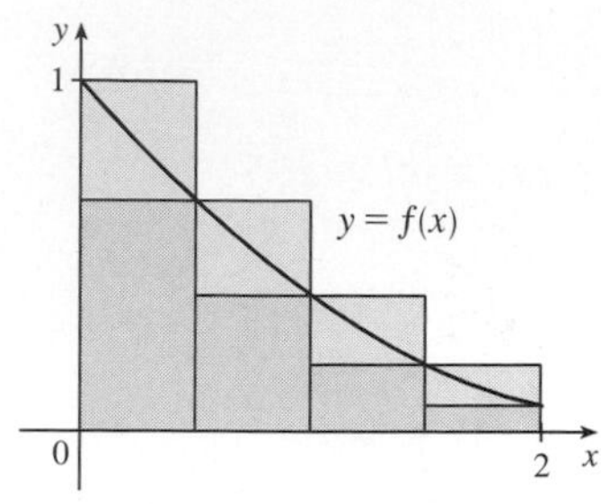

The diagram shows that $L_4 > T_4 > \int_0^2 f(x)\,dx > R_4$, and it appears that M_4 is a bit less than $\int_0^2 f(x)\,dx$. In fact, for any function that is concave upward, it can be shown that $L_n > T_n > \int_0^2 f(x)\,dx > M_n > R_n$.

(a) Since $0.9540 > 0.8675 > 0.8632 > 0.7811$, it follows that $L_n = 0.9540$, $T_n = 0.8675$, $M_n = 0.8632$, and $R_n = 0.7811$.

(b) Since $M_n < \int_0^2 f(x)\,dx < T_n$, we have $0.8632 < \int_0^2 f(x)\,dx < 0.8675$.

3. $f(x) = \cos(x^2)$, $\Delta x = \frac{1-0}{4} = \frac{1}{4}$

(a) $T_4 = \frac{1}{4 \cdot 2}\left[f(0) + 2f\left(\frac{1}{4}\right) + 2f\left(\frac{2}{4}\right) + 2f\left(\frac{3}{4}\right) + f(1)\right] \approx 0.895759$

(b) $M_4 = \frac{1}{4}\left[f\left(\frac{1}{8}\right) + f\left(\frac{3}{8}\right) + f\left(\frac{5}{8}\right) + f\left(\frac{7}{8}\right)\right] \approx 0.908907$

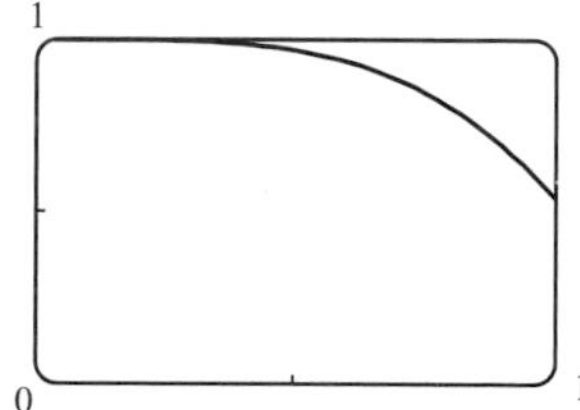

The graph shows that f is concave down on $[0, 1]$. So T_4 is an underestimate and M_4 is an overestimate. We can conclude that $0.895759 < \int_0^1 \cos(x^2)\,dx < 0.908907$.

4. 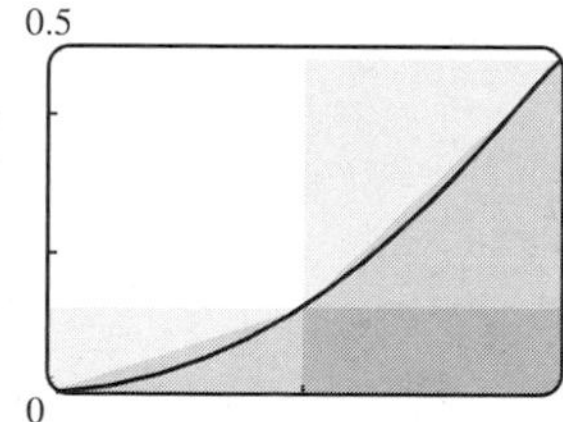

(a) Since f is increasing on $[0, 1]$, L_2 will underestimate I (since the area of the darkest rectangle is less than the area under the curve), and R_2 will overestimate I. Since f is concave upward on $[0, 1]$, M_2 will underestimate I and T_2 will overestimate I (the area under the straight line segments is greater than the area under the curve).

(b) For any n, we will have $L_n < M_n < I < T_n < R_n$.

(c) $L_5 = \sum_{i=1}^{5} f(x_{i-1})\,\Delta x = \frac{1}{5}[f(0.0) + f(0.2) + f(0.4) + f(0.6) + f(0.8)] \approx 0.1187$

$R_5 = \sum_{i=1}^{5} f(x_i)\,\Delta x = \frac{1}{5}[f(0.2) + f(0.4) + f(0.6) + f(0.8) + f(1)] \approx 0.2146$

$M_5 = \sum_{i=1}^{5} f(\overline{x}_i)\,\Delta x = \frac{1}{5}[f(0.1) + f(0.3) + f(0.5) + f(0.7) + f(0.9)] \approx 0.1622$

$T_5 = \left(\frac{1}{2}\,\Delta x\right)[f(0) + 2f(0.2) + 2f(0.4) + 2f(0.6) + 2f(0.8) + f(1)] \approx 0.1666$

From the graph, it appears that the Midpoint Rule gives the best approximation. (This is in fact the case, since $I \approx 0.16371405$.)

5. (a) $f(x) = \dfrac{x}{1+x^2}$, $\Delta x = \dfrac{b-a}{n} = \dfrac{2-0}{10} = \dfrac{1}{5}$

$M_{10} = \frac{1}{5}\left[f\left(\frac{1}{10}\right) + f\left(\frac{3}{10}\right) + f\left(\frac{5}{10}\right) + \cdots + f\left(\frac{19}{10}\right)\right] \approx 0.806598$

(b) $S_{10} = \frac{1}{5 \cdot 3}\left[f(0) + 4f\left(\frac{1}{5}\right) + 2f\left(\frac{2}{5}\right) + 4f\left(\frac{3}{5}\right) + 2f\left(\frac{4}{5}\right) + \cdots + 4f\left(\frac{9}{5}\right) + f(2)\right] \approx 0.804779$

Actual: $I = \displaystyle\int_0^2 \frac{x}{1+x^2}\,dx = \left[\tfrac{1}{2}\ln\left|1+x^2\right|\right]_0^2$ $\quad [u = 1+x^2,\ du = 2x\,dx]$

$= \frac{1}{2}\ln 5 - \frac{1}{2}\ln 1 = \frac{1}{2}\ln 5 \approx 0.804719$

Errors: $E_M = \text{actual} - M_{10} = I - M_{10} \approx -0.001879$

$E_S = \text{actual} - S_{10} = I - S_{10} \approx -0.000060$

6. (a) $f(x) = x\cos x$, $\Delta x = \dfrac{b-a}{n} = \dfrac{\pi - 0}{4} = \dfrac{\pi}{4}$

$M_4 = \frac{\pi}{4}\left[f\left(\frac{\pi}{8}\right) + f\left(\frac{3\pi}{8}\right) + f\left(\frac{5\pi}{8}\right) + f\left(\frac{7\pi}{8}\right)\right] \approx -1.945744$

(b) $S_4 = \frac{\pi}{4 \cdot 3}\left[f(0) + 4f\left(\frac{\pi}{4}\right) + 2f\left(\frac{2\pi}{4}\right) + 4f\left(\frac{3\pi}{4}\right) + f(\pi)\right] \approx -1.985611$

Actual: $I = \int_0^\pi x\cos x\,dx = \Big[x\sin x + \cos x\Big]_0^\pi$ $\quad$ [use parts with $u = x$ and $dv = \cos x\,dx$]

$= (0 + (-1)) - (0 + 1) = -2$

Errors: $E_M = \text{actual} - M_4 = I - M_4 \approx -0.054256$

$E_S = \text{actual} - S_4 = I - S_4 \approx -0.014389$

7. $f(x) = \sqrt{x^3 - 1}$, $\Delta x = \dfrac{b-a}{n} = \dfrac{2-1}{10} = \dfrac{1}{10}$

(a) $T_{10} = \frac{1}{10 \cdot 2}[f(1) + 2f(1.1) + 2f(1.2) + 2f(1.3) + 2f(1.4) + 2f(1.5)$
$\qquad + 2f(1.6) + 2f(1.7) + 2f(1.8) + 2f(1.9) + f(2)]$

≈ 1.506361

(b) $M_{10} = \frac{1}{10}[f(1.05) + f(1.15) + f(1.25) + f(1.35) + f(1.45) + f(1.55) + f(1.65) + f(1.75) + f(1.85) + f(1.95)]$

≈ 1.518362

(c) $S_{10} = \frac{1}{10 \cdot 3}[f(1) + 4f(1.1) + 2f(1.2) + 4f(1.3) + 2f(1.4)$
$\qquad + 4f(1.5) + 2f(1.6) + 4f(1.7) + 2f(1.8) + 4f(1.9) + f(2)]$

≈ 1.511519

8. $f(x) = \dfrac{1}{1+x^6}$, $\Delta x = \dfrac{b-a}{n} = \dfrac{2-0}{8} = \dfrac{1}{4}$

(a) $T_8 = \frac{1}{4 \cdot 2}[f(0) + 2f(0.25) + 2f(0.5) + 2f(0.75) + 2f(1) + 2f(1.25) + 2f(1.5) + 2f(1.75) + f(2)] \approx 1.040756$

(b) $M_8 = \frac{1}{4}[f(0.125) + f(0.375) + f(0.625) + f(0.875) + f(1.125) + f(1.375) + f(1.625) + f(1.875)] \approx 1.041109$

(c) $S_8 = \frac{1}{4 \cdot 3}[f(0) + 4f(0.25) + 2f(0.5) + 4f(0.75) + 2f(1) + 4f(1.25) + 2f(1.5) + 4f(1.75) + f(2)] \approx 1.042172$

9. $f(x) = \dfrac{e^x}{1+x^2}$, $\Delta x = \dfrac{b-a}{n} = \dfrac{2-0}{10} = \dfrac{1}{5}$

(a) $T_{10} = \frac{1}{5 \cdot 2}[f(0) + 2f(0.2) + 2f(0.4) + 2f(0.6) + 2f(0.8) + 2f(1)$
$\qquad + 2f(1.2) + 2f(1.4) + 2f(1.6) + 2f(1.8) + f(2)]$

≈ 2.660833

(b) $M_{10} = \frac{1}{5}[f(0.1) + f(0.3) + f(0.5) + f(0.7) + f(0.9) + f(1.1) + f(1.3) + f(1.5) + f(1.7) + f(1.9)]$
≈ 2.664377

(c) $S_{10} = \frac{1}{5 \cdot 3}[f(0) + 4f(0.2) + 2f(0.4) + 4f(0.6) + 2f(0.8)$
$+ 4f(1) + 2f(1.2) + 4f(1.4) + 2f(1.6) + 4f(1.8) + f(2)] \approx 2.663244$

10. $f(x) = \sqrt[3]{1 + \cos x}$, $\Delta x = \frac{\pi/2 - 0}{4} = \frac{\pi}{8}$

(a) $T_4 = \frac{\pi}{8 \cdot 2}\left[f(0) + 2f\left(\frac{\pi}{8}\right) + 2f\left(\frac{2\pi}{8}\right) + 2f\left(\frac{3\pi}{8}\right) + f\left(\frac{\pi}{2}\right)\right] \approx 1.838967$

(b) $M_4 = \frac{\pi}{8}\left[f\left(\frac{\pi}{16}\right) + f\left(\frac{3\pi}{16}\right) + f\left(\frac{5\pi}{16}\right) + f\left(\frac{7\pi}{16}\right)\right] \approx 1.845390$

(c) $S_4 = \frac{\pi}{8 \cdot 3}\left[f(0) + 4f\left(\frac{\pi}{8}\right) + 2f\left(\frac{2\pi}{8}\right) + 4f\left(\frac{3\pi}{8}\right) + f\left(\frac{\pi}{2}\right)\right] \approx 1.843245$

11. $f(x) = x^3 \sin x$, $\Delta x = \frac{4 - 0}{8} = \frac{1}{2}$

(a) $T_8 = \frac{1}{2 \cdot 2}\left[f(0) + 2f\left(\frac{1}{2}\right) + 2f(1) + 2f\left(\frac{3}{2}\right) + 2f(2) + 2f\left(\frac{5}{2}\right) + 2f(3) + 2f\left(\frac{7}{2}\right) + f(4)\right] \approx -7.276910$

(b) $M_8 = \frac{1}{2}\left[f\left(\frac{1}{4}\right) + f\left(\frac{3}{4}\right) + f\left(\frac{5}{4}\right) + f\left(\frac{7}{4}\right) + f\left(\frac{9}{4}\right) + f\left(\frac{11}{4}\right) + f\left(\frac{13}{4}\right) + f\left(\frac{15}{4}\right)\right] \approx -4.818251$

(c) $S_8 = \frac{1}{2 \cdot 3}\left[f(0) + 4f\left(\frac{1}{2}\right) + 2f(1) + 4f\left(\frac{3}{2}\right) + 2f(2) + 4f\left(\frac{5}{2}\right) + 2f(3) + 4f\left(\frac{7}{2}\right) + f(4)\right] \approx -5.605350$

12. $f(x) = e^{1/x}$, $\Delta x = \frac{3 - 1}{8} = \frac{1}{4}$

(a) $T_8 = \frac{1}{4 \cdot 2}\left[f(1) + 2f\left(\frac{5}{4}\right) + 2f\left(\frac{3}{2}\right) + 2f\left(\frac{7}{4}\right) + 2f(2) + 2f\left(\frac{9}{4}\right) + 2f\left(\frac{5}{2}\right) + 2f\left(\frac{11}{4}\right) + f(3)\right] \approx 3.534934$

(b) $M_8 = \frac{1}{4}\left[f\left(\frac{9}{8}\right) + f\left(\frac{11}{8}\right) + f\left(\frac{13}{8}\right) + f\left(\frac{15}{8}\right) + f\left(\frac{17}{8}\right) + f\left(\frac{19}{8}\right) + f\left(\frac{21}{8}\right) + f\left(\frac{23}{8}\right)\right] \approx 3.515248$

(c) $S_8 = \frac{1}{4 \cdot 3}\left[f(1) + 4f\left(\frac{5}{4}\right) + 2f\left(\frac{3}{2}\right) + 4f\left(\frac{7}{4}\right) + 2f(2) + 4f\left(\frac{9}{4}\right) + 2f\left(\frac{5}{2}\right) + 4f\left(\frac{11}{4}\right) + f(3)\right] \approx 3.522375$

13. $f(y) = \sqrt{y} \cos y$, $\Delta y = \frac{4 - 0}{8} = \frac{1}{2}$

(a) $T_8 = \frac{1}{2 \cdot 2}\left[f(0) + 2f\left(\frac{1}{2}\right) + 2f(1) + 2f\left(\frac{3}{2}\right) + 2f(2) + 2f\left(\frac{5}{2}\right) + 2f(3) + 2f\left(\frac{7}{2}\right) + f(4)\right] \approx -2.364034$

(b) $M_8 = \frac{1}{2}\left[f\left(\frac{1}{4}\right) + f\left(\frac{3}{4}\right) + f\left(\frac{5}{4}\right) + f\left(\frac{7}{4}\right) + f\left(\frac{9}{4}\right) + f\left(\frac{11}{4}\right) + f\left(\frac{13}{4}\right) + f\left(\frac{15}{4}\right)\right] \approx -2.310690$

(c) $S_8 = \frac{1}{2 \cdot 3}\left[f(0) + 4f\left(\frac{1}{2}\right) + 2f(1) + 4f\left(\frac{3}{2}\right) + 2f(2) + 4f\left(\frac{5}{2}\right) + 2f(3) + 4f\left(\frac{7}{2}\right) + f(4)\right] \approx -2.346520$

14. $f(t) = \dfrac{1}{\ln t}$, $\Delta t = \dfrac{3 - 2}{10} = \dfrac{1}{10}$

(a) $T_{10} = \frac{1}{10 \cdot 2}\{f(2) + 2[f(2.1) + f(2.2) + \cdots + f(2.9)] + f(3)\} \approx 1.119061$

(b) $M_{10} = \frac{1}{10}[f(2.05) + f(2.15) + \cdots + f(2.85) + f(2.95)] \approx 1.118107$

(c) $S_{10} = \frac{1}{10 \cdot 3}[f(2) + 4f(2.1) + 2f(2.2) + 4f(2.3) + 2f(2.4) + 4f(2.5) + 2f(2.6)$
$+ 4f(2.7) + 2f(2.8) + 4f(2.9) + f(3)] \approx 1.118428$

15. $f(x) = \dfrac{x^2}{1 + x^4}$, $\Delta x = \dfrac{1 - 0}{10} = \dfrac{1}{10}$

(a) $T_{10} = \frac{1}{10 \cdot 2}\{f(0) + 2[f(0.1 + f(0.2) + \cdots + f(0.9)] + f(1)\} \approx 0.243747$

(b) $M_{10} = \frac{1}{10}[f(0.05) + f(0.15) + \cdots + f(0.85) + f(0.95)] \approx 0.243748$

(c) $S_{10} = \frac{1}{10 \cdot 3}[f(0) + 4f(0.1) + 2f(0.2) + 4f(0.3) + 2f(0.4) + 4f(0.5) + 2f(0.6)$
$+ 4f(0.7) + 2f(0.8) + 4f(0.9) + f(1)] \approx 0.243751$

Note: $\int_0^1 f(x)\,dx \approx 0.24374775$. This is a rare case where the Trapezoidal and Midpoint Rules give better approximations than Simpson's Rule.

16. $f(t) = \dfrac{\sin t}{t}$, $\Delta t = \dfrac{3-1}{4} = \dfrac{1}{2}$

(a) $T_4 = \frac{1}{2 \cdot 2}[f(1) + 2f(1.5) + 2f(2) + 2f(2.5) + f(3)] \approx 0.901645$

(b) $M_4 = \frac{1}{2}[f(1.25) + f(1.75) + f(2.25) + f(2.75)] \approx 0.903031$

(c) $S_4 = \frac{1}{2 \cdot 3}[f(1) + 4f(1.5) + 2f(2) + 4f(2.5) + f(3)] \approx 0.902558$

17. $f(x) = \ln(1 + e^x)$, $\Delta x = \frac{4-0}{8} = \frac{1}{2}$

(a) $T_8 = \frac{1}{2 \cdot 2}\{f(0) + 2[f(0.5) + f(1) + \cdots + f(3) + f(3.5)] + f(4)\} \approx 8.814278$

(b) $M_8 = \frac{1}{2}[f(0.25) + f(0.75) + \cdots + f(3.25) + f(3.75)] \approx 8.799212$

(c) $S_8 = \frac{1}{2 \cdot 3}[f(0) + 4f(0.5) + 2f(1) + 4f(1.5) + 2f(2) + 4f(2.5) + 2f(3) + 4f(3.5) + f(4)] \approx 8.804229$

18. $f(x) = \sqrt{x + x^3}$, $\Delta x = \frac{1-0}{10} = \frac{1}{10}$

(a) $T_{10} = \frac{1}{2 \cdot 2}\{f(0) + 2[f(0.1) + f(0.2) + \cdots + f(0.8) + f(0.9)] + f(1)\} \approx 0.787092$

(b) $M_{10} = \frac{1}{2}[f(0.05) + f(0.15) + \cdots + f(0.85) + f(0.95)] \approx 0.793821$

(c) $S_{10} = \frac{1}{2 \cdot 3}[f(0) + 4f(0.1) + 2f(0.2) + 4f(0.3) + 2f(0.4) + 4f(0.5) + 2f(0.6)$
$+ 4f(0.7) + 2f(0.8) + 4f(0.9) + f(1)]$
≈ 0.789915

19. $f(x) = \cos(x^2)$, $\Delta x = \frac{1-0}{8} = \frac{1}{8}$

(a) $T_8 = \frac{1}{8 \cdot 2}\{f(0) + 2[f(\frac{1}{8}) + f(\frac{2}{8}) + \cdots + f(\frac{7}{8})] + f(1)\} \approx 0.902333$

$M_8 = \frac{1}{8}[f(\frac{1}{16}) + f(\frac{3}{16}) + f(\frac{5}{16}) + \cdots + f(\frac{15}{16})] = 0.905620$

(b) $f(x) = \cos(x^2)$, $f'(x) = -2x\sin(x^2)$, $f''(x) = -2\sin(x^2) - 4x^2\cos(x^2)$. For $0 \le x \le 1$, sin and cos are positive, so $|f''(x)| = 2\sin(x^2) + 4x^2\cos(x^2) \le 2 \cdot 1 + 4 \cdot 1 \cdot 1 = 6$ since $\sin(x^2) \le 1$ and $\cos(x^2) \le 1$ for all x, and $x^2 \le 1$ for $0 \le x \le 1$. So for $n = 8$, we take $K = 6$, $a = 0$, and $b = 1$ in Theorem 3, to get $|E_T| \le 6 \cdot 1^3/(12 \cdot 8^2) = \frac{1}{128} = 0.0078125$ and $|E_M| \le \frac{1}{256} = 0.00390625$. [A better estimate is obtained by noting from a graph of f'' that $|f''(x)| \le 4$ for $0 \le x \le 1$.]

(c) Take $K = 6$ [as in part (b)] in Theorem 3. $|E_T| \le \dfrac{K(b-a)^3}{12n^2} \le 0.0001 \quad \Leftrightarrow \quad \dfrac{6(1-0)^3}{12n^2} \le 10^{-4} \quad \Leftrightarrow$

$\dfrac{1}{2n^2} \le \dfrac{1}{10^4} \quad \Leftrightarrow \quad 2n^2 \ge 10^4 \quad \Leftrightarrow \quad n^2 \ge 5000 \quad \Leftrightarrow \quad n \ge 71$. Take $n = 71$ for T_n. For E_M, again take $K = 6$ in Theorem 3 to get $|E_M| \le 10^{-4} \quad \Leftrightarrow \quad 4n^2 \ge 10^4 \quad \Leftrightarrow \quad n^2 \ge 2500 \quad \Leftrightarrow \quad n \ge 50$. Take $n = 50$ for M_n.

20. $f(x) = e^{1/x}$, $\Delta x = \frac{2-1}{10} = \frac{1}{10}$

(a) $T_{10} = \frac{1}{10 \cdot 2}[f(1) + 2f(1.1) + 2f(1.2) + \cdots + 2f(1.9) + f(2)] \approx 2.021976$

$M_{10} = \frac{1}{10}[f(1.05) + f(1.15) + f(1.25) + \cdots + f(1.95)] \approx 2.019102$

(b) $f(x) = e^{1/x}$, $f'(x) = -\dfrac{1}{x^2}e^{1/x}$, $f''(x) = \dfrac{2x+1}{x^4}\,e^{1/x}$. Now f'' is decreasing on $[1,2]$, so let $x=1$ to take $K = 3e$.

$|E_T| \le \dfrac{3e(2-1)^3}{12(10)^2} = \dfrac{e}{400} \approx 0.006796$. $\quad |E_M| \le \dfrac{|E_T|}{2} = \dfrac{e}{800} \approx 0.003398$.

(c) Take $K = 3e$ [as in part (b)] in Theorem 3. $|E_T| \le \dfrac{K(b-a)^3}{12n^2} \le 0.0001 \;\Leftrightarrow\; \dfrac{3e(2-1)^3}{12n^2} \le 10^{-4} \;\Leftrightarrow$

$\dfrac{e}{4n^2} \le \dfrac{1}{10^4} \;\Leftrightarrow\; n^2 \ge \dfrac{10^4 e}{4} \;\Leftrightarrow\; n \ge 83$. Take $n = 83$ for T_n. For E_M, again take $K = 3e$ in Theorem 3 to get

$|E_M| \le 10^{-4} \;\Leftrightarrow\; n^2 \ge \dfrac{10^4 e}{8} \;\Leftrightarrow\; n \ge 59$. Take $n = 59$ for M_n.

21. $f(x) = \sin x$, $\Delta x = \frac{\pi - 0}{10} = \frac{\pi}{10}$

(a) $T_{10} = \frac{\pi}{10 \cdot 2}\left[f(0) + 2f\left(\frac{\pi}{10}\right) + 2f\left(\frac{2\pi}{10}\right) + \cdots + 2f\left(\frac{9\pi}{10}\right) + f(\pi)\right] \approx 1.983524$

$M_{10} = \frac{\pi}{10}\left[f\left(\frac{\pi}{20}\right) + f\left(\frac{3\pi}{20}\right) + f\left(\frac{5\pi}{20}\right) + \cdots + f\left(\frac{19\pi}{20}\right)\right] \approx 2.008248$

$S_{10} = \frac{\pi}{10 \cdot 3}\left[f(0) + 4f\left(\frac{\pi}{10}\right) + 2f\left(\frac{2\pi}{10}\right) + 4f\left(\frac{3\pi}{10}\right) + \cdots + 4f\left(\frac{9\pi}{10}\right) + f(\pi)\right] \approx 2.000110$

Since $I = \int_0^\pi \sin x\,dx = \big[-\cos x\big]_0^\pi = 1 - (-1) = 2$, $E_T = I - T_{10} \approx 0.016476$, $E_M = I - M_{10} \approx -0.008248$, and $E_S = I - S_{10} \approx -0.000110$.

(b) $f(x) = \sin x \;\Rightarrow\; \left|f^{(n)}(x)\right| \le 1$, so take $K = 1$ for all error estimates.

$|E_T| \le \dfrac{K(b-a)^3}{12n^2} = \dfrac{1(\pi-0)^3}{12(10)^2} = \dfrac{\pi^3}{1200} \approx 0.025839$. $\quad |E_M| \le \dfrac{|E_T|}{2} = \dfrac{\pi^3}{2400} \approx 0.012919$.

$|E_S| \le \dfrac{K(b-a)^5}{180n^4} = \dfrac{1(\pi-0)^5}{180(10)^4} = \dfrac{\pi^5}{1{,}800{,}000} \approx 0.000170$.

The actual error is about 64% of the error estimate in all three cases.

(c) $|E_T| \le 0.00001 \;\Leftrightarrow\; \dfrac{\pi^3}{12n^2} \le \dfrac{1}{10^5} \;\Leftrightarrow\; n^2 \ge \dfrac{10^5\pi^3}{12} \;\Rightarrow\; n \ge 508.3$. Take $n = 509$ for T_n.

$|E_M| \le 0.00001 \;\Leftrightarrow\; \dfrac{\pi^3}{24n^2} \le \dfrac{1}{10^5} \;\Leftrightarrow\; n^2 \ge \dfrac{10^5\pi^3}{24} \;\Rightarrow\; n \ge 359.4$. Take $n = 360$ for M_n.

$|E_S| \le 0.00001 \;\Leftrightarrow\; \dfrac{\pi^5}{180n^4} \le \dfrac{1}{10^5} \;\Leftrightarrow\; n^4 \ge \dfrac{10^5\pi^5}{180} \;\Rightarrow\; n \ge 20.3$.

Take $n = 22$ for S_n (since n must be even).

22. From Example 7(b), we take $K = 76e$ to get $|E_S| \le \dfrac{76e(1)^5}{180n^4} \le 0.00001 \;\Rightarrow\; n^4 \ge \dfrac{76e}{180(0.00001)} \;\Rightarrow\; n \ge 18.4$.

Take $n = 20$ (since n must be even).

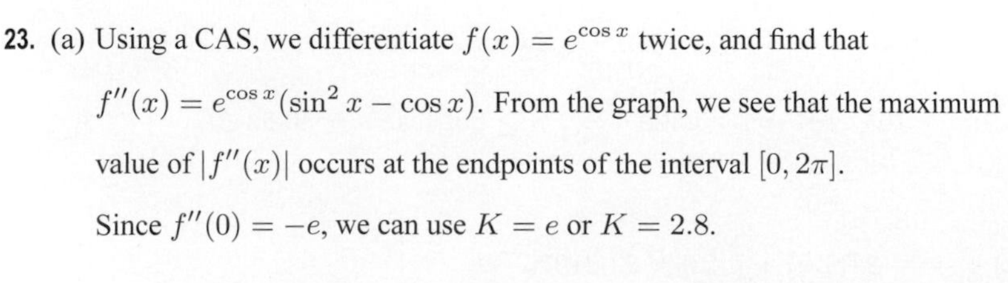

23. (a) Using a CAS, we differentiate $f(x) = e^{\cos x}$ twice, and find that $f''(x) = e^{\cos x}(\sin^2 x - \cos x)$. From the graph, we see that the maximum value of $|f''(x)|$ occurs at the endpoints of the interval $[0, 2\pi]$. Since $f''(0) = -e$, we can use $K = e$ or $K = 2.8$.

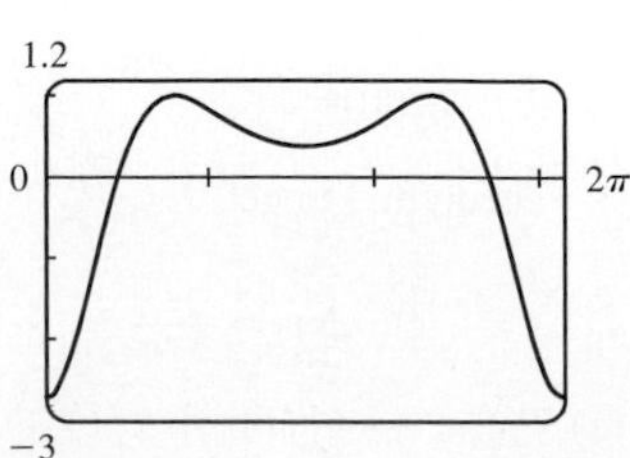

(b) A CAS gives $M_{10} \approx 7.954926518$. (In Maple, use `Student[Calculus1][RiemannSum]` or `Student[Calculus1][ApproximateInt]`.)

(c) Using Theorem 3 for the Midpoint Rule, with $K = e$, we get $|E_M| \le \dfrac{e(2\pi - 0)^3}{24 \cdot 10^2} \approx 0.280945995$.

With $K = 2.8$, we get $|E_M| \le \dfrac{2.8(2\pi - 0)^3}{24 \cdot 10^2} = 0.289391916$.

(d) A CAS gives $I \approx 7.954926521$.

(e) The actual error is only about 3×10^{-9}, much less than the estimate in part (c).

(f) We use the CAS to differentiate twice more, and then graph

$f^{(4)}(x) = e^{\cos x}(\sin^4 x - 6\sin^2 x\,\cos x + 3 - 7\sin^2 x + \cos x)$.

From the graph, we see that the maximum value of $\left|f^{(4)}(x)\right|$ occurs at the endpoints of the interval $[0, 2\pi]$. Since $f^{(4)}(0) = 4e$, we can use $K = 4e$ or $K = 10.9$.

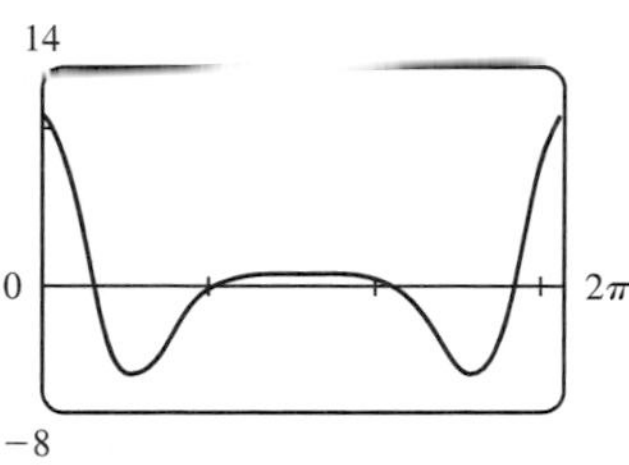

(g) A CAS gives $S_{10} \approx 7.953789422$. (In Maple, use `Student[Calculus1][ApproximateInt]`.)

(h) Using Theorem 4 with $K = 4e$, we get $|E_S| \le \dfrac{4e(2\pi - 0)^5}{180 \cdot 10^4} \approx 0.059153618$.

With $K = 10.9$, we get $|E_S| \le \dfrac{10.9(2\pi - 0)^5}{180 \cdot 10^4} \approx 0.059299814$.

(i) The actual error is about $7.954926521 - 7.953789422 \approx 0.00114$. This is quite a bit smaller than the estimate in part (h), though the difference is not nearly as great as it was in the case of the Midpoint Rule.

(j) To ensure that $|E_S| \le 0.0001$, we use Theorem 4: $|E_S| \le \dfrac{4e(2\pi)^5}{180 \cdot n^4} \le 0.0001 \Rightarrow \dfrac{4e(2\pi)^5}{180 \cdot 0.0001} \le n^4 \Rightarrow$ $n^4 \ge 5{,}915{,}362 \Leftrightarrow n \ge 49.3$. So we must take $n \ge 50$ to ensure that $|I - S_n| \le 0.0001$. ($K = 10.9$ leads to the same value of n.)

24. (a) Using the CAS, we differentiate $f(x) = \sqrt{4 - x^3}$ twice, and find that $f''(x) = -\dfrac{9x^4}{4(4 - x^3)^{3/2}} - \dfrac{3x}{(4 - x^3)^{1/2}}$.

From the graph, we see that $|f''(x)| < 2.2$ on $[-1, 1]$.

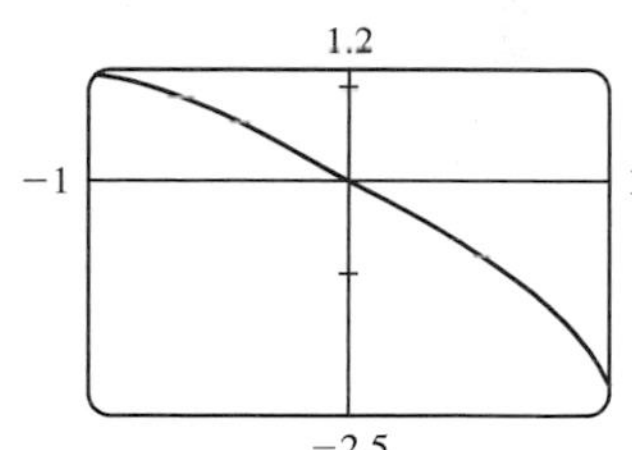

(b) A CAS gives $M_{10} \approx 3.995804152$. (In Maple, use `Student[Calculus1][RiemannSum]` or `Student[Calculus1][ApproximateInt]`.)

(c) Using Theorem 3 for the Midpoint Rule, with $K = 2.2$, we get $|E_M| \le \dfrac{2.2\,[1 - (-1)]^3}{24 \cdot 10^2} \approx 0.00733$.

(d) A CAS gives $I \approx 3.995487677$.

(e) The actual error is about -0.0003165, much less than the estimate in part (c).

(f) We use the CAS to differentiate twice more, and then graph

$f^{(4)}(x) = \dfrac{9}{16}\dfrac{x^2(x^6 - 224x^3 - 1280)}{(4 - x^3)^{7/2}}$.

From the graph, we see that $\left|f^{(4)}(x)\right| < 18.1$ on $[-1, 1]$.

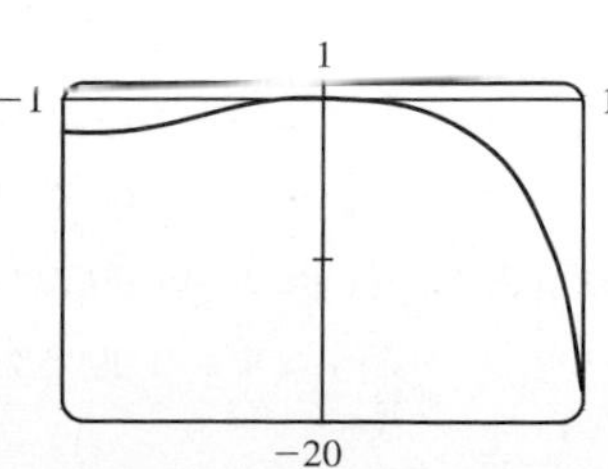

(g) A CAS gives $S_{10} \approx 3.995449790$. (In Maple, use `Student[Calculus1][ApproximateInt]`.)

(h) Using Theorem 4 with $K = 18.1$, we get $|E_S| \le \dfrac{18.1\,[1-(-1)]^5}{180 \cdot 10^4} \approx 0.000322$.

(i) The actual error is about $3.995487677 - 3.995449790 \approx 0.0000379$. This is quite a bit smaller than the estimate in part (h).

(j) To ensure that $|E_S| \le 0.0001$, we use Theorem 4: $|E_S| \le \dfrac{18.1(2)^5}{180 \cdot n^4} \le 0.0001 \;\Rightarrow\; \dfrac{18.1(2)^5}{180 \cdot 0.0001} \le n^4 \;\Rightarrow$ $n^4 \ge 32{,}178 \;\Rightarrow n \ge 13.4$. So we must take $n \ge 14$ to ensure that $|I - S_n| \le 0.0001$.

25. $I = \int_0^1 xe^x\,dx = [(x-1)e^x]_0^1$ [parts or Formula 96] $= 0 - (-1) = 1$, $f(x) = xe^x$, $\Delta x = 1/n$

$n = 5$:

$$L_5 = \tfrac{1}{5}[f(0) + f(0.2) + f(0.4) + f(0.6) + f(0.8)] \approx 0.742943$$
$$R_5 = \tfrac{1}{5}[f(0.2) + f(0.4) + f(0.6) + f(0.8) + f(1)] \approx 1.286599$$
$$T_5 = \tfrac{1}{5 \cdot 2}[f(0) + 2f(0.2) + 2f(0.4) + 2f(0.6) + 2f(0.8) + f(1)] \approx 1.014771$$
$$M_5 = \tfrac{1}{5}[f(0.1) + f(0.3) + f(0.5) + f(0.7) + f(0.9)] \approx 0.992621$$

$$E_L = I - L_5 \approx 1 - 0.742943 = 0.257057$$
$$E_R \approx 1 - 1.286599 = -0.286599$$
$$E_T \approx 1 - 1.014771 = -0.014771$$
$$E_M \approx 1 - 0.992621 = 0.007379$$

$n = 10$:

$$L_{10} = \tfrac{1}{10}[f(0) + f(0.1) + f(0.2) + \cdots + f(0.9)] \approx 0.867782$$
$$R_{10} = \tfrac{1}{10}[f(0.1) + f(0.2) + \cdots + f(0.9) + f(1)] \approx 1.139610$$
$$T_{10} = \tfrac{1}{10 \cdot 2}\{f(0) + 2[f(0.1) + f(0.2) + \cdots + f(0.9)] + f(1)\} \approx 1.003696$$
$$M_{10} = \tfrac{1}{10}[f(0.05) + f(0.15) + \cdots + f(0.85) + f(0.95)] \approx 0.998152$$

$$E_L = I - L_{10} \approx 1 - 0.867782 = 0.132218$$
$$E_R \approx 1 - 1.139610 = -0.139610$$
$$E_T \approx 1 - 1.003696 = -0.003696$$
$$E_M \approx 1 - 0.998152 = 0.001848$$

$n = 20$:

$$L_{20} = \tfrac{1}{20}[f(0) + f(0.05) + f(0.10) + \cdots + f(0.95)] \approx 0.932967$$
$$R_{20} = \tfrac{1}{20}[f(0.05) + f(0.10) + \cdots + f(0.95) + f(1)] \approx 1.068881$$
$$T_{20} = \tfrac{1}{20 \cdot 2}\{f(0) + 2[f(0.05) + f(0.10) + \cdots + f(0.95)] + f(1)\} \approx 1.000924$$
$$M_{20} = \tfrac{1}{20}[f(0.025) + f(0.075) + f(0.125) + \cdots + f(0.975)] \approx 0.999538$$

$$E_L = I - L_{20} \approx 1 - 0.932967 = 0.067033$$
$$E_R \approx 1 - 1.068881 = -0.068881$$
$$E_T \approx 1 - 1.000924 = -0.000924$$
$$E_M \approx 1 - 0.999538 = 0.000462$$

n	L_n	R_n	T_n	M_n
5	0.742943	1.286599	1.014771	0.992621
10	0.867782	1.139610	1.003696	0.998152
20	0.932967	1.068881	1.000924	0.999538

n	E_L	E_R	E_T	E_M
5	0.257057	−0.286599	−0.014771	0.007379
10	0.132218	−0.139610	−0.003696	0.001848
20	0.067033	−0.068881	−0.000924	0.000462

Observations:

1. E_L and E_R are always opposite in sign, as are E_T and E_M.
2. As n is doubled, E_L and E_R are decreased by about a factor of 2, and E_T and E_M are decreased by a factor of about 4.
3. The Midpoint approximation is about twice as accurate as the Trapezoidal approximation.
4. All the approximations become more accurate as the value of n increases.
5. The Midpoint and Trapezoidal approximations are much more accurate than the endpoint approximations.

26. $I = \int_1^2 \frac{1}{x^2}\,dx = \left[-\frac{1}{x}\right]_1^2 = -\frac{1}{2} - (-1) = \frac{1}{2}$, $f(x) = \frac{1}{x^2}$, $\Delta x = \frac{1}{n}$

$n = 5$:

$$L_5 = \tfrac{1}{5}[f(1) + f(1.2) + f(1.4) + f(1.6) + f(1.8)] \approx 0.580783$$

$$R_5 = \tfrac{1}{5}[f(1.2) + f(1.4) + f(1.6) + f(1.8) + f(2)] \approx 0.430783$$

$$T_5 = \tfrac{1}{5\cdot 2}[f(1) + 2f(1.2) + 2f(1.4) + 2f(1.6) + 2f(1.8) + f(2)] \approx 0.505783$$

$$M_5 = \tfrac{1}{5}[f(1.1) + f(1.3) + f(1.5) + f(1.7) + f(1.9)] \approx 0.497127$$

$$E_L = I - L_5 \approx \tfrac{1}{2} - 0.580783 = -0.080783$$

$$E_R \approx \tfrac{1}{2} - 0.430783 = 0.069217$$

$$E_T \approx \tfrac{1}{2} - 0.505783 = -0.005783$$

$$E_M \approx \tfrac{1}{2} - 0.497127 = 0.002873$$

$n = 10$:

$$L_{10} = \tfrac{1}{10}[f(1) + f(1.1) + f(1.2) + \cdots + f(1.9)] \approx 0.538955$$

$$R_{10} = \tfrac{1}{10}[f(1.1) + f(1.2) + \cdots + f(1.9) + f(2)] \approx 0.463955$$

$$T_{10} = \tfrac{1}{10\cdot 2}\{f(1) + 2[f(1.1) + f(1.2) + \cdots + f(1.9)] + f(2)\} \approx 0.501455$$

$$M_{10} = \tfrac{1}{10}[f(1.05) + f(1.15) + \cdots + f(1.85) + f(1.95)] \approx 0.499274$$

$$E_L = I - L_{10} \approx \tfrac{1}{2} - 0.538955 = -0.038955$$

$$E_R \approx \tfrac{1}{2} - 0.463955 = 0.036049$$

$$E_T \approx \tfrac{1}{2} - 0.501455 = -0.001455$$

$$E_M \approx \tfrac{1}{2} - 0.499274 = 0.000726$$

$n = 20$:

$$L_{20} = \tfrac{1}{20}[f(1) + f(1.05) + f(1.10) + \cdots + f(1.95)] \approx 0.519114$$

$$R_{20} = \tfrac{1}{20}[f(1.05) + f(1.10) + \cdots + f(1.95) + f(2)] \approx 0.481614$$

$$T_{20} = \tfrac{1}{20\cdot 2}\{f(1) + 2[f(1.05) + f(1.10) + \cdots + f(1.95)] + f(2)\} \approx 0.500364$$

$$M_{20} = \tfrac{1}{20}[f(1.025) + f(1.075) + f(1.125) + \cdots + f(1.975)] \approx 0.499818$$

$$\begin{aligned} E_L &= I - L_{20} \approx \tfrac{1}{2} - 0.519114 = -0.019114 \\ E_R &\approx \tfrac{1}{2} - 0.481614 = 0.018386 \\ E_T &\approx \tfrac{1}{2} - 0.500364 = -0.000364 \\ E_M &\approx \tfrac{1}{2} - 0.499818 = 0.000182 \end{aligned}$$

n	L_n	R_n	T_n	M_n
5	0.580783	0.430783	0.505783	0.497127
10	0.538955	0.463955	0.501455	0.499274
20	0.519114	0.481614	0.500364	0.499818

n	E_L	E_R	E_T	E_M
5	−0.080783	0.069217	−0.005783	0.002873
10	−0.038955	0.036049	−0.001455	0.000726
20	−0.019114	0.018386	−0.000364	0.000182

Observations:

1. E_L and E_R are always opposite in sign, as are E_T and E_M.
2. As n is doubled, E_L and E_R are decreased by about a factor of 2, and E_T and E_M are decreased by a factor of about 4.
3. The Midpoint approximation is about twice as accurate as the Trapezoidal approximation.
4. All the approximations become more accurate as the value of n increases.
5. The Midpoint and Trapezoidal approximations are much more accurate than the endpoint approximations.

27. $I = \int_0^2 x^4\,dx = \left[\frac{1}{5}x^5\right]_0^2 = \frac{32}{5} - 0 = 6.4$, $f(x) = x^4$, $\Delta x = \frac{2-0}{n} = \frac{2}{n}$

$n = 6$:

$$\begin{aligned} T_6 &= \tfrac{2}{6 \cdot 2}\{f(0) + 2[f(\tfrac{1}{3}) + f(\tfrac{2}{3}) + f(\tfrac{3}{3}) + f(\tfrac{4}{3}) + f(\tfrac{5}{3})] + f(2)\} \approx 6.695473 \\ M_6 &= \tfrac{2}{6}[f(\tfrac{1}{6}) + f(\tfrac{3}{6}) + f(\tfrac{5}{6}) + f(\tfrac{7}{6}) + f(\tfrac{9}{6}) + f(\tfrac{11}{6})] \approx 6.252572 \\ S_6 &= \tfrac{2}{6 \cdot 3}[f(0) + 4f(\tfrac{1}{3}) + 2f(\tfrac{2}{3}) + 4f(\tfrac{3}{3}) + 2f(\tfrac{4}{3}) + 4f(\tfrac{5}{3}) + f(2)] \approx 6.403292 \\ E_T &= I - T_6 \approx 6.4 - 6.695473 = -0.295473 \\ E_M &\approx 6.4 - 6.252572 = 0.147428 \\ E_S &\approx 6.4 - 6.403292 = -0.003292 \end{aligned}$$

$n = 12$:

$$\begin{aligned} T_{12} &= \tfrac{2}{12 \cdot 2}\{f(0) + 2[f(\tfrac{1}{6}) + f(\tfrac{2}{6}) + f(\tfrac{3}{6}) + \cdots + f(\tfrac{11}{6})] + f(2)\} \approx 6.474023 \\ M_6 &= \tfrac{2}{12}[f(\tfrac{1}{12}) + f(\tfrac{3}{12}) + f(\tfrac{5}{12}) + \cdots + f(\tfrac{23}{12})] \approx 6.363008 \\ S_6 &= \tfrac{2}{12 \cdot 3}[f(0) + 4f(\tfrac{1}{6}) + 2f(\tfrac{2}{6}) + 4f(\tfrac{3}{6}) + 2f(\tfrac{4}{6}) + \cdots + 4f(\tfrac{11}{6}) + f(2)] \approx 6.400206 \\ E_T &= I - T_{12} \approx 6.4 - 6.474023 = -0.074023 \\ E_M &\approx 6.4 - 6.363008 = 0.036992 \\ E_S &\approx 6.4 - 6.400206 = -0.000206 \end{aligned}$$

n	T_n	M_n	S_n
6	6.695473	6.252572	6.403292
12	6.474023	6.363008	6.400206

n	E_T	E_M	E_S
6	−0.295473	0.147428	−0.003292
12	−0.074023	0.036992	−0.000206

Observations:

1. E_T and E_M are opposite in sign and decrease by a factor of about 4 as n is doubled.
2. The Simpson's approximation is much more accurate than the Midpoint and Trapezoidal approximations, and E_S seems to decrease by a factor of about 16 as n is doubled.

28. $I = \int_1^4 \frac{1}{\sqrt{x}}\,dx = \left[2\sqrt{x}\right]_1^4 = 4 - 2 = 2$, $f(x) = \frac{1}{\sqrt{x}}$, $\Delta x = \frac{4-1}{n} = \frac{3}{n}$

$n = 6$: $T_6 = \frac{3}{6\cdot 2}\{f(1) + 2[f(\frac{3}{2}) + f(\frac{4}{2}) + f(\frac{5}{2}) + f(\frac{6}{2}) + f(\frac{7}{2})] + f(4)\} \approx 2.008966$

$M_6 = \frac{3}{6}[f(\frac{5}{4}) + f(\frac{7}{4}) + f(\frac{9}{4}) + f(\frac{11}{4}) + f(\frac{13}{4}) + f(\frac{15}{4})] \approx 1.995572$

$S_6 = \frac{3}{6\cdot 3}[f(1) + 4f(\frac{3}{2}) + 2f(\frac{4}{2}) + 4f(\frac{5}{2}) + 2f(\frac{6}{2}) + 4f(\frac{7}{2}) + f(4)] \approx 2.000469$

$E_T = I - T_6 \approx 2 - 2.008966 = -0.008966$,

$E_M \approx 2 - 1.995572 = 0.004428$,

$E_S \approx 2 - 2.000469 = -0.000469$

$n = 12$: $T_{12} = \frac{3}{12\cdot 2}\{f(1) + 2[f(\frac{5}{4}) + f(\frac{6}{4}) + f(\frac{7}{4}) + \cdots + f(\frac{15}{4})] + f(4)\} \approx 2.002269$

$M_{12} = \frac{3}{12}[f(\frac{9}{8}) + f(\frac{11}{8}) + f(\frac{13}{8}) + \cdots + f(\frac{31}{8})] \approx 1.998869$

$S_{12} = \frac{3}{12\cdot 3}[f(1) + 4f(\frac{5}{4}) + 2f(\frac{6}{4}) + 4f(\frac{7}{4}) + 2f(\frac{8}{4}) + \cdots + 4f(\frac{15}{4}) + f(4)] \approx 2.000036$

$E_T = I - T_{12} \approx 2 - 2.002269 = -0.002269$

$E_M \approx 2 - 1.998869 = 0.001131$

$E_S \approx 2 - 2.000036 = -0.000036$

n	T_n	M_n	S_n
6	2.008966	1.995572	2.000469
12	2.002269	1.998869	2.000036

n	E_T	E_M	E_S
6	−0.008966	0.004428	−0.000469
12	−0.002269	0.001131	−0.000036

Observations:

1. E_T and E_M are opposite in sign and decrease by a factor of about 4 as n is doubled.
2. The Simpson's approximation is much more accurate than the Midpoint and Trapezoidal approximations, and E_S seems to decrease by a factor of about 16 as n is doubled.

29. (a) $\Delta x = (b - a)/n = (6 - 0)/6 = 1$

$T_6 = \frac{1}{2}[f(0) + 2f(1) + 2f(2) + 2f(3) + 2f(4) + 2f(5) + f(6)]$

$\approx \frac{1}{2}[2 + 2(1) + 2(3) + 2(5) + 2(4) + 2(3) + 4] = \frac{1}{2}(38) = 19$

(b) $M_6 = 1[f(0.5) + f(1.5) + f(2.5) + f(3.5) + f(4.5) + f(5.5)] \approx 1.3 + 1.5 + 4.6 + 4.7 + 3.3 + 3.2 = 18.6$

(c) $S_6 = \frac{1}{3}[f(0) + 4f(1) + 2f(2) + 4f(3) + 2f(4) + 4f(5) + f(6)]$

$\approx \frac{1}{3}[2 + 4(1) + 2(3) + 4(5) + 2(4) + 4(3) + 4] = \frac{1}{3}(56) = 18.\overline{6}$

30. If x = distance from left end of pool and $w = w(x)$ = width at x, then Simpson's Rule with $n = 8$ and $\Delta x = 2$ gives

Area $= \int_0^{16} w\,dx \approx \frac{2}{3}[0 + 4(6.2) + 2(7.2) + 4(6.8) + 2(5.6) + 4(5.0) + 2(4.8) + 4(4.8) + 0] \approx 84\text{ m}^2$.

31. (a) $\int_1^5 f(x)\,dx \approx M_4 = \frac{5-1}{4}[f(1.5) + f(2.5) + f(3.5) + f(4.5)] = 1(2.9 + 3.6 + 4.0 + 3.9) = 14.4$

(b) $-2 \le f''(x) \le 3 \;\Rightarrow\; |f''(x)| \le 3 \;\Rightarrow\; K = 3$, since $|f''(x)| \le K$. The error estimate for the Midpoint Rule is

$$|E_M| \le \frac{K(b-a)^3}{24n^2} = \frac{3(5-1)^3}{24(4)^2} = \frac{1}{2}.$$

32. (a) $\int_0^{1.6} g(x)\,dx \approx S_8 = \frac{1.6-0}{8\cdot 3}[g(0)+4g(0.2)+2g(0.4)+4g(0.6)+2g(0.8)+4g(1.0)+2g(1.2)+4g(1.4)+g(1.6)]$

$= \frac{1}{15}[12.1+4(11.6)+2(11.3)+4(11.1)+2(11.7)+4(12.2)+2(12.6)+4(13.0)+13.2]$

$= \frac{1}{15}(288.1) = \frac{2881}{150} \approx 19.2$

(b) $-5 \le g^{(4)}(x) \le 2 \;\Rightarrow\; \left|g^{(4)}(x)\right| \le 5 \;\Rightarrow\; K = 5$, since $\left|g^{(4)}(x)\right| \le K$. The error estimate for Simpson's Rule is

$$|E_S| \le \frac{K(b-a)^5}{180n^4} = \frac{5(1.6-0)^5}{180(8)^4} = \frac{2}{28{,}125} = 7.\overline{1} \times 10^{-5}.$$

33. We use Simpson's Rule with $n = 12$ and $\Delta t = \frac{24-0}{12} = 2$.

$$S_{12} = \tfrac{2}{3}[T(0)+4T(2)+2T(4)+4T(6)+2T(8)+4T(10)+2T(12) + 4T(14)+2T(16)+4T(18)+2T(20)+4T(22)+T(24)]$$

$$\approx \tfrac{2}{3}[66.6+4(65.4)+2(64.4)+4(61.7)+2(67.3)+4(72.1)+2(74.9) + 4(77.4)+2(79.1)+4(75.4)+2(75.6)+4(71.4)+67.5] = \tfrac{2}{3}(2550.3) = 1700.2.$$

Thus, $\int_0^{24} T(t)\,dt \approx S_{12}$ and $T_{\text{ave}} = \frac{1}{24-0}\int_0^{24} T(t)\,dt \approx 70.84°\text{F}$.

34. We use Simpson's Rule with $n = 10$ and $\Delta x = \frac{1}{2}$:

$$\text{distance} = \int_0^5 v(t)\,dt \approx S_{10} = \tfrac{1}{2\cdot 3}[f(0)+4f(0.5)+2f(1)+\cdots+4f(4.5)+f(5)]$$

$$= \tfrac{1}{6}[0+4(4.67)+2(7.34)+4(8.86)+2(9.73)+4(10.22) + 2(10.51)+4(10.67)+2(10.76)+4(10.81)+10.81]$$

$$= \tfrac{1}{6}(268.41) = 44.735\text{ m}$$

35. By the Net Change Theorem, the increase in velocity is equal to $\int_0^6 a(t)\,dt$. We use Simpson's Rule with $n = 6$ and $\Delta t = (6-0)/6 = 1$ to estimate this integral:

$$\int_0^6 a(t)\,dt \approx S_6 = \tfrac{1}{3}[a(0)+4a(1)+2a(2)+4a(3)+2a(4)+4a(5)+a(6)]$$

$$\approx \tfrac{1}{3}[0+4(0.5)+2(4.1)+4(9.8)+2(12.9)+4(9.5)+0] = \tfrac{1}{3}(113.2) = 37.7\overline{3}\text{ ft/s}$$

36. By the Net Change Theorem, the total amount of water that leaked out during the first six hours is equal to $\int_0^6 r(t)\,dt$. We use Simpson's Rule with $n = 6$ and $\Delta t = \frac{6-0}{6} = 1$ to estimate this integral:

$$\int_0^6 r(t)\,dt \approx S_6 = \tfrac{1}{3}[r(0)+4r(1)+2r(2)+4r(3)+2r(4)+4r(5)+r(6)]$$

$$\approx \tfrac{1}{3}[4+4(3)+2(2.4)+4(1.9)+2(1.4)+4(1.1)+1] = \tfrac{1}{3}(36.6) = 12.2\text{ liters}$$

37. By the Net Change Theorem, the energy used is equal to $\int_0^6 P(t)\,dt$. We use Simpson's Rule with $n = 12$ and $\Delta t = \frac{6-0}{12} = \frac{1}{2}$ to estimate this integral:

$$\int_0^6 P(t)\,dt \approx S_{12} = \tfrac{1/2}{3}[P(0)+4P(0.5)+2P(1)+4P(1.5)+2P(2)+4P(2.5)+2P(3) + 4P(3.5)+2P(4)+4P(4.5)+2P(5)+4P(5.5)+P(6)]$$

$$= \tfrac{1}{6}[1814+4(1735)+2(1686)+4(1646)+2(1637)+4(1609)+2(1604) + 4(1611)+2(1621)+4(1666)+2(1745)+4(1886)+2052]$$

$$= \tfrac{1}{6}(61{,}064) = 10{,}177.\overline{3}\text{ megawatt-hours}$$

38. By the Net Change Theorem, the total amount of data transmitted is equal to $\int_0^8 D(t)\,dt \times 3600$ [since $D(t)$ is measured in megabits per second and t is in hours]. We use Simpson's Rule with $n = 8$ and $\Delta t = (8-0)/8 = 1$ to estimate this integral:

$$\begin{aligned}\int_0^8 D(t)\,dt \approx S_8 &= \tfrac{1}{3}[D(0) + 4D(1) + 2D(2) + 4D(3) + 2D(4) + 4D(5) + 2D(6) + 4D(7) + D(8)] \\ &\approx \tfrac{1}{3}[0.35 + 4(0.32) + 2(0.41) + 4(0.50) + 2(0.51) + 4(0.56) + 2(0.56) + 4(0.83) + 0.88] \\ &= \tfrac{1}{3}(13.03) = 4.34\overline{3}\end{aligned}$$

Now multiply by 3600 to obtain 15,636 megabits.

39. (a) Let $y = f(x)$ denote the curve. Using disks, $V = \int_2^{10} \pi[f(x)]^2\,dx = \pi \int_2^{10} g(x)\,dx = \pi I_1$.

Now use Simpson's Rule to approximate I_1:

$$\begin{aligned}I_1 \approx S_8 &= \tfrac{10-2}{3(8)}[g(2) + 4g(3) + 2g(4) + 4g(5) + 2g(6) + 4g(7) + g(8)] \\ &\approx \tfrac{1}{3}[0^2 + 4(1.5)^2 + 2(1.9)^2 + 4(2.2)^2 + 2(3.0)^2 + 4(3.8)^2 + 2(4.0)^2 + 4(3.1)^2 + 0^2] \\ &= \tfrac{1}{3}(181.78)\end{aligned}$$

Thus, $V \approx \pi \cdot \frac{1}{3}(181.78) \approx 190.4$ or 190 cubic units.

(b) Using cylindrical shells, $V = \int_2^{10} 2\pi x f(x)\,dx = 2\pi \int_2^{10} x f(x)\,dx = 2\pi I_1$.

Now use Simpson's Rule to approximate I_1:

$$\begin{aligned}I_1 \approx S_8 &= \tfrac{10-2}{3(8)}[2f(2) + 4\cdot 3f(3) + 2\cdot 4f(4) + 4\cdot 5f(5) + 2\cdot 6f(6) \\ &\qquad + 4\cdot 7f(7) + 2\cdot 8f(8) + 4\cdot 9f(9) + 10f(10)] \\ &\approx \tfrac{1}{3}[2(0) + 12(1.5) + 8(1.9) + 20(2.2) + 12(3.0) + 28(3.8) + 16(4.0) + 36(3.1) + 10(0)] \\ &= \tfrac{1}{3}(395.2)\end{aligned}$$

Thus, $V \approx 2\pi \cdot \frac{1}{3}(395.2) \approx 827.7$ or 828 cubic units.

40. $$\begin{aligned}\text{Work} = \int_0^{18} f(x)\,dx \approx S_6 &= \tfrac{18-0}{6\cdot 3}[f(0) + 4f(3) + 2f(6) + 4f(9) + 2f(12) + 4f(15) + f(18)] \\ &= 1\cdot[9.8 + 4(9.1) + 2(8.5) + 4(8.0) + 2(7.7) + 4(7.5) + 7.4] = 148 \text{ joules}\end{aligned}$$

41. The curve is $y = f(x) = 1/(1 + e^{-x})$. Using disks, $V = \int_0^{10} \pi[f(x)]^2\,dx = \pi \int_0^{10} g(x)\,dx = \pi I_1$. Now use Simpson's Rule to approximate I_1:

$$\begin{aligned}I_1 \approx S_{10} &= \tfrac{10-0}{10\cdot 3}[g(0) + 4g(1) + 2g(2) + 4g(3) + 2g(4) + 4g(5) + 2g(6) + 4g(7) + 2g(8) + 4g(9) + g(10)] \\ &\approx 8.80825\end{aligned}$$

Thus, $V \approx \pi I_1 \approx 27.7$ or 28 cubic units.

42. Using Simpson's Rule with $n = 10$, $\Delta x = \frac{\pi/2}{10}$, $L = 1$, $\theta_0 = \frac{42\pi}{180}$ radians, $g = 9.8\text{ m/s}^2$, $k^2 = \sin^2\left(\frac{1}{2}\theta_0\right)$, and $f(x) = 1/\sqrt{1 - k^2\sin^2 x}$, we get

$$\begin{aligned}T &= 4\sqrt{\frac{L}{g}}\int_0^{\pi/2} \frac{dx}{\sqrt{1 - k^2\sin^2 x}} \approx 4\sqrt{\frac{L}{g}}\,S_{10} \\ &= 4\sqrt{\tfrac{1}{9.8}}\left(\tfrac{\pi/2}{10\cdot 3}\right)\left[f(0) + 4f\left(\tfrac{\pi}{20}\right) + 2f\left(\tfrac{2\pi}{20}\right) + \cdots + 4f\left(\tfrac{9\pi}{20}\right) + f\left(\tfrac{\pi}{2}\right)\right] \approx 2.07665\end{aligned}$$

43. $I(\theta) = \dfrac{N^2 \sin^2 k}{k^2}$, where $k = \dfrac{\pi N d \sin\theta}{\lambda}$, $N = 10{,}000$, $d = 10^{-4}$, and $\lambda = 632.8 \times 10^{-9}$. So $I(\theta) = \dfrac{(10^4)^2 \sin^2 k}{k^2}$,

where $k = \dfrac{\pi(10^4)(10^{-4})\sin\theta}{632.8 \times 10^{-9}}$. Now $n = 10$ and $\Delta\theta = \dfrac{10^{-6} - (-10^{-6})}{10} = 2 \times 10^{-7}$, so

$M_{10} = 2 \times 10^{-7}[I(-0.0000009) + I(-0.0000007) + \cdots + I(0.0000009)] \approx 59.4$.

44. $f(x) = \cos(\pi x)$, $\Delta x = \frac{20-0}{10} = 2 \quad \Rightarrow$

$$T_{10} = \tfrac{2}{2}\{f(0) + 2[f(2) + f(4) + \cdots + f(18)] + f(20)\} = 1[\cos 0 + 2(\cos 2\pi + \cos 4\pi + \cdots + \cos 18\pi) + \cos 20\pi]$$
$$= 1 + 2(1+1+1+1+1+1+1+1+1) + 1 = 20$$

The actual value is $\int_0^{20} \cos(\pi x)\,dx = \frac{1}{\pi}\big[\sin \pi x\big]_0^{20} = \frac{1}{\pi}(\sin 20\pi - \sin 0) = 0$. The discrepancy is due to the fact that the function is sampled only at points of the form $2n$, where its value is $f(2n) = \cos(2n\pi) = 1$.

45. Consider the function f whose graph is shown. The area $\int_0^2 f(x)\,dx$ is close to 2. The Trapezoidal Rule gives

$T_2 = \frac{2-0}{2\cdot 2}\,[f(0) + 2f(1) + f(2)] = \frac{1}{2}\,[1 + 2\cdot 1 + 1] = 2$.

The Midpoint Rule gives $M_2 = \frac{2-0}{2}\,[f(0.5) + f(1.5)] = 1[0+0] = 0$, so the Trapezoidal Rule is more accurate.

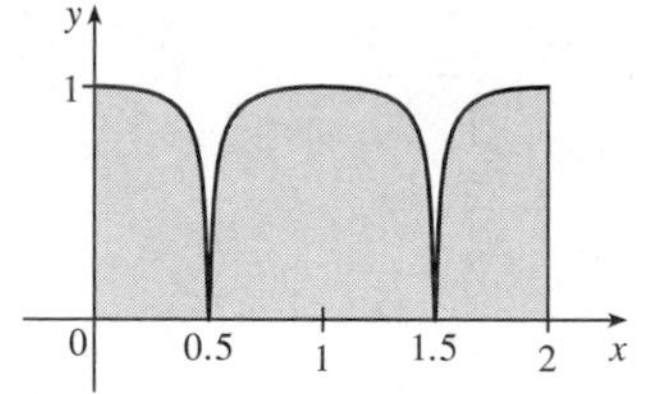

46. Consider the function $f(x) = |x-1|$, $0 \le x \le 2$. The area $\int_0^2 f(x)\,dx$ is exactly 1. So is the right endpoint approximation:

$R_2 = f(1)\,\Delta x + f(2)\,\Delta x = 0\cdot 1 + 1\cdot 1 = 1$. But Simpson's Rule approximates f with the parabola $y = (x-1)^2$, shown dashed, and

$S_2 = \dfrac{\Delta x}{3}\,[f(0) + 4f(1) + f(2)] = \dfrac{1}{3}\,[1 + 4\cdot 0 + 1] = \dfrac{2}{3}$.

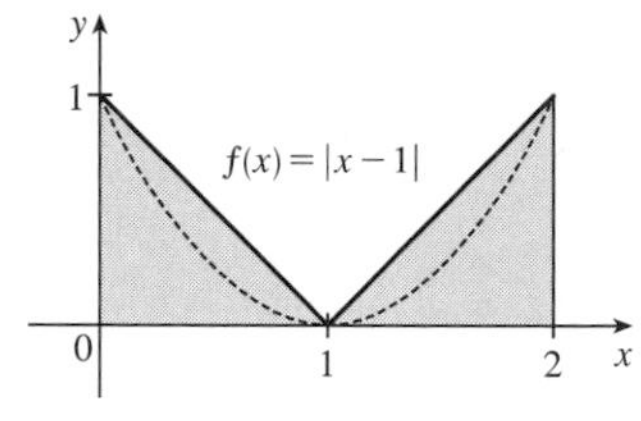

47. Since the Trapezoidal and Midpoint approximations on the interval $[a, b]$ are the sums of the Trapezoidal and Midpoint approximations on the subintervals $[x_{i-1}, x_i]$, $i = 1, 2, \ldots, n$, we can focus our attention on one such interval. The condition $f''(x) < 0$ for $a \le x \le b$ means that the graph of f is concave down as in Figure 5. In that figure, T_n is the area of the trapezoid $AQRD$, $\int_a^b f(x)\,dx$ is the area of the region $AQPRD$, and M_n is the area of the trapezoid $ABCD$, so $T_n < \int_a^b f(x)\,dx < M_n$. In general, the condition $f'' < 0$ implies that the graph of f on $[a, b]$ lies above the chord joining the points $(a, f(a))$ and $(b, f(b))$. Thus, $\int_a^b f(x)\,dx > T_n$. Since M_n is the area under a tangent to the graph, and since $f'' < 0$ implies that the tangent lies above the graph, we also have $M_n > \int_a^b f(x)\,dx$. Thus, $T_n < \int_a^b f(x)\,dx < M_n$.

48. Let f be a polynomial of degree ≤ 3; say $f(x) = Ax^3 + Bx^2 + Cx + D$. It will suffice to show that Simpson's estimate is exact when there are two subintervals ($n = 2$), because for a larger even number of subintervals the sum of exact estimates is exact. As in the derivation of Simpson's Rule, we can assume that $x_0 = -h$, $x_1 = 0$, and $x_2 = h$. Then Simpson's approximation is

$$\int_{-h}^{h} f(x)\,dx \approx \tfrac{1}{3}h[f(-h)+4f(0)+f(h)] = \tfrac{1}{3}h\big[(-Ah^3+Bh^2-Ch+D)+4D+(Ah^3+Bh^2+Ch+D)\big]$$
$$= \tfrac{1}{3}h[2Bh^2+6D] = \tfrac{2}{3}Bh^3+2Dh$$

The exact value of the integral is

$$\int_{-h}^{h}(Ax^3+Bx^2+Cx+D)\,dx = 2\int_0^h (Bx^2+D)\,dx \qquad \text{[by Theorem 5.5.7(a) and (b)]}$$
$$= 2\left[\tfrac{1}{3}Bx^3+Dx\right]_0^h = \tfrac{2}{3}Bh^3+2Dh$$

Thus, Simpson's Rule is exact.

49. $T_n = \frac{1}{2}\,\Delta x\,[f(x_0)+2f(x_1)+\cdots+2f(x_{n-1})+f(x_n)]$ and

$M_n = \Delta x\,[f(\overline{x}_1)+f(\overline{x}_2)+\cdots+f(\overline{x}_{n-1})+f(\overline{x}_n)]$, where $\overline{x}_i = \frac{1}{2}(x_{i-1}+x_i)$. Now

$T_{2n} = \frac{1}{2}\left(\frac{1}{2}\Delta x\right)[f(x_0)+2f(\overline{x}_1)+2f(x_1)+2f(\overline{x}_2)+2f(x_2)+\cdots+2f(\overline{x}_{n-1})+2f(x_{n-1})+2f(\overline{x}_n)+f(x_n)]$ so

$$\tfrac{1}{2}(T_n+M_n) = \tfrac{1}{2}T_n+\tfrac{1}{2}M_n$$
$$= \tfrac{1}{4}\Delta x[f(x_0)+2f(x_1)+\cdots+2f(x_{n-1})+f(x_n)]+\tfrac{1}{4}\Delta x[2f(\overline{x}_1)+2f(\overline{x}_2)+\cdots+2f(\overline{x}_{n-1})+2f(\overline{x}_n)]$$
$$= T_{2n}$$

50. $T_n = \dfrac{\Delta x}{2}\left[f(x_0)+2\sum\limits_{i=1}^{n-1} f(x_i)+f(x_n)\right]$ and $M_n = \Delta x \sum\limits_{i=1}^{n} f\left(x_i-\dfrac{\Delta x}{2}\right)$, so

$$\tfrac{1}{3}T_n+\tfrac{2}{3}M_n = \tfrac{1}{3}(T_n+2M_n) = \frac{\Delta x}{3\cdot 2}\left[f(x_0)+2\sum_{i=1}^{n-1} f(x_i)+f(x_n)+4\sum_{i=1}^{n} f\left(x_i-\frac{\Delta x}{2}\right)\right]$$

where $\Delta x = \dfrac{b-a}{n}$. Let $\delta x = \dfrac{b-a}{2n}$. Then $\Delta x = 2\delta x$, so

$$\tfrac{1}{3}T_n+\tfrac{2}{3}M_n = \frac{\delta x}{3}\left[f(x_0)+2\sum_{i=1}^{n-1} f(x_i)+f(x_n)+4\sum_{i=1}^{n} f(x_i-\delta x)\right]$$
$$= \tfrac{1}{3}\delta x[f(x_0)+4f(x_1-\delta x)+2f(x_1)+4f(x_2-\delta x)$$
$$+2f(x_2)+\cdots+2f(x_{n-1})+4f(x_n-\delta x)+f(x_n)]$$

Since $x_0, x_1-\delta x, x_1, x_2-\delta x, x_2, \ldots, x_{n-1}, x_n-\delta x, x_n$ are the subinterval endpoints for S_{2n}, and since $\delta x = \dfrac{b-a}{2n}$ is the width of the subintervals for S_{2n}, the last expression for $\frac{1}{3}T_n+\frac{2}{3}M_n$ is the usual expression for S_{2n}. Therefore, $\frac{1}{3}T_n+\frac{2}{3}M_n = S_{2n}$.

7.8 Improper Integrals

1. (a) Since $y = \dfrac{x}{x-1}$ has an infinite discontinuity at $x=1$, $\displaystyle\int_1^2 \frac{x}{x-1}\,dx$ is a Type 2 improper integral.

(b) Since $\displaystyle\int_0^\infty \frac{1}{1+x^3}\,dx$ has an infinite interval of integration, it is an improper integral of Type 1.

(c) Since $\displaystyle\int_{-\infty}^\infty x^2e^{-x^2}\,dx$ has an infinite interval of integration, it is an improper integral of Type 1.

(d) Since $y = \cot x$ has an infinite discontinuity at $x=0$, $\int_0^{\pi/4}\cot x\,dx$ is a Type 2 improper integral.

2. (a) Since $y = \tan x$ is defined and continuous on $\left[0, \frac{\pi}{4}\right]$, $\int_0^{\pi/4} \tan x\,dx$ is proper.

(b) Since $y = \tan x$ has an infinite discontinuity at $x = \frac{\pi}{2}$, $\int_0^{\pi} \tan x\,dx$ is a Type 2 improper integral.

(c) Since $y = \dfrac{1}{x^2 - x - 2} = \dfrac{1}{(x-2)(x+1)}$ has an infinite discontinuity at $x = -1$, $\displaystyle\int_{-1}^{1} \frac{dx}{x^2 - x - 2}$ is a Type 2 improper integral.

(d) Since $\int_0^{\infty} e^{-x^3}\,dx$ has an infinite interval of integration, it is an improper integral of Type 1.

3. The area under the graph of $y = 1/x^3 = x^{-3}$ between $x = 1$ and $x = t$ is

$A(t) = \int_1^t x^{-3}\,dx = \left[-\frac{1}{2}x^{-2}\right]_1^t = -\frac{1}{2}t^{-2} - \left(-\frac{1}{2}\right) = \frac{1}{2} - 1/(2t^2)$. So the area for $1 \le x \le 10$ is $A(10) = 0.5 - 0.005 = 0.495$, the area for $1 \le x \le 100$ is $A(100) = 0.5 - 0.00005 = 0.49995$, and the area for $1 \le x \le 1000$ is $A(1000) = 0.5 - 0.0000005 = 0.4999995$. The total area under the curve for $x \ge 1$ is $\lim\limits_{t\to\infty} A(t) = \lim\limits_{t\to\infty} \left[\frac{1}{2} - 1/(2t^2)\right] = \frac{1}{2}$.

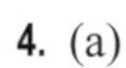
4. (a)

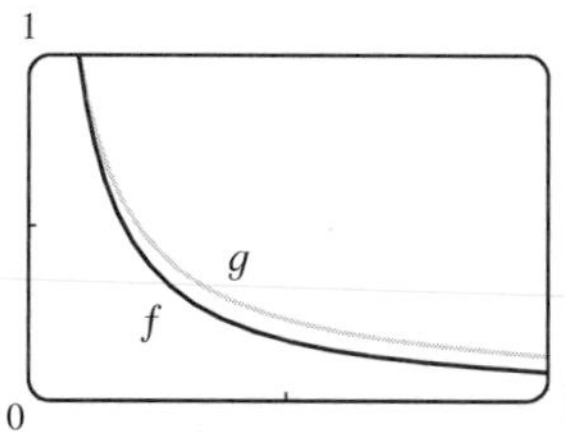

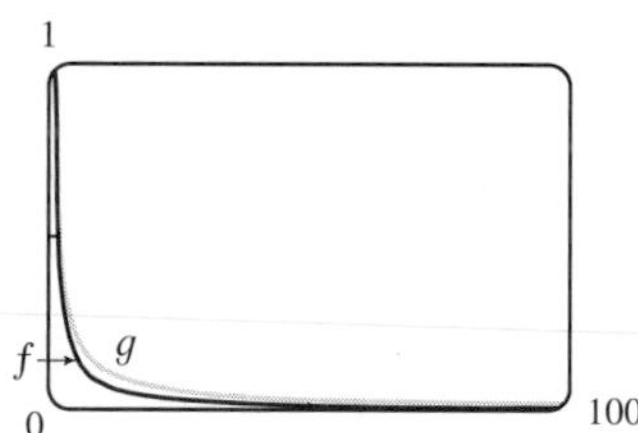

(b) The area under the graph of f from $x = 1$ to $x = t$ is

$$F(t) = \int_1^t f(x)\,dx = \int_1^t x^{-1.1}\,dx = \left[-\tfrac{1}{0.1}x^{-0.1}\right]_1^t$$
$$= -10(t^{-0.1} - 1) = 10(1 - t^{-0.1})$$

and the area under the graph of g is

$$G(t) = \int_1^t g(x)\,dx = \int_1^t x^{-0.9}\,dx = \left[\tfrac{1}{0.1}x^{0.1}\right]_1^t = 10(t^{0.1} - 1).$$

t	$F(t)$	$G(t)$
10	2.06	2.59
100	3.69	5.85
10^4	6.02	15.12
10^6	7.49	29.81
10^{10}	9	90
10^{20}	9.9	990

(c) The total area under the graph of f is $\lim\limits_{t\to\infty} F(t) = \lim\limits_{t\to\infty} 10(1 - t^{-0.1}) = 10$.

The total area under the graph of g does not exist, since $\lim\limits_{t\to\infty} G(t) = \lim\limits_{t\to\infty} 10(t^{0.1} - 1) = \infty$.

5. $$\int_3^{\infty} \frac{1}{(x-2)^{3/2}}\,dx = \lim_{t\to\infty} \int_3^t (x-2)^{-3/2}\,dx = \lim_{t\to\infty} \left[-2(x-2)^{-1/2}\right]_3^t \qquad [u = x - 2,\ du = dx]$$
$$= \lim_{t\to\infty} \left(\frac{-2}{\sqrt{t-2}} + \frac{2}{\sqrt{1}}\right) = 0 + 2 = 2. \qquad \text{Convergent}$$

6. $$\int_0^{\infty} \frac{1}{\sqrt[4]{1+x}}\,dx = \lim_{t\to\infty} \int_0^t (1+x)^{-1/4}\,dx = \lim_{t\to\infty} \left[\tfrac{4}{3}(1+x)^{3/4}\right]_0^t \qquad [u = 1 + x,\ du = dx]$$
$$= \lim_{t\to\infty} \left[\tfrac{4}{3}(1+t)^{3/4} - \tfrac{4}{3}\right] = \infty. \qquad \text{Divergent}$$

7. $\displaystyle\int_{-\infty}^{0} \frac{1}{3-4x}\,dx = \lim_{t\to-\infty} \int_t^0 \frac{1}{3-4x}\,dx = \lim_{t\to-\infty} \left[-\tfrac{1}{4}\ln|3-4x|\right]_t^0 = \lim_{t\to-\infty} \left[-\tfrac{1}{4}\ln 3 + \tfrac{1}{4}\ln|3-4t|\right] = \infty.$

Divergent

8. $\displaystyle\int_1^{\infty} \frac{1}{(2x+1)^3}\,dx = \lim_{t\to\infty} \int_1^t \frac{1}{(2x+1)^3}\,dx = \lim_{t\to\infty} \left[-\frac{1}{4(2x+1)^2}\right]_1^t = \lim_{t\to\infty} \left[-\frac{1}{4(2t+1)^2} + \frac{1}{36}\right] = 0 + \frac{1}{36}.$

Convergent

9. $\int_2^{\infty} e^{-5p}\,dp = \lim\limits_{t\to\infty} \int_2^t e^{-5p}\,dp = \lim\limits_{t\to\infty} \left[-\frac{1}{5}e^{-5p}\right]_2^t = \lim\limits_{t\to\infty} \left(-\frac{1}{5}e^{-5t} + \frac{1}{5}e^{-10}\right) = 0 + \frac{1}{5}e^{-10} = \frac{1}{5}e^{-10}.$ Convergent

10. $\displaystyle\int_{-\infty}^{0} 2^r\,dr = \lim_{t\to-\infty} \int_t^0 2^r\,dr = \lim_{t\to-\infty} \left[\frac{2^r}{\ln 2}\right]_t^0 = \lim_{t\to-\infty} \left(\frac{1}{\ln 2} - \frac{2^t}{\ln 2}\right) = \frac{1}{\ln 2} - 0 = \frac{1}{\ln 2}.$ Convergent

11. $\displaystyle\int_0^{\infty} \frac{x^2}{\sqrt{1+x^3}}\,dx = \lim_{t\to\infty} \int_0^t \frac{x^2}{\sqrt{1+x^3}}\,dx = \lim_{t\to\infty} \left[\frac{2}{3}\sqrt{1+x^3}\right]_0^t = \lim_{t\to\infty} \left(\tfrac{2}{3}\sqrt{1+t^3} - \tfrac{2}{3}\right) = \infty.$ Divergent

12. $I = \int_{-\infty}^{\infty} (y^3 - 3y^2)\,dy = I_1 + I_2 = \int_{-\infty}^{0} (y^3 - 3y^2)\,dy + \int_0^{\infty} (y^3 - 3y^2)\,dy$, but

$I_1 = \lim\limits_{t\to-\infty} \left[\frac{1}{4}y^4 - y^3\right]_t^0 = \lim\limits_{t\to-\infty} \left(t^3 - \frac{1}{4}t^4\right) = -\infty.$ Since I_1 is divergent, I is divergent,

and there is no need to evaluate I_2. Divergent

13. $\int_{-\infty}^{\infty} xe^{-x^2}\,dx = \int_{-\infty}^{0} xe^{-x^2}\,dx + \int_0^{\infty} xe^{-x^2}\,dx.$

$\int_{-\infty}^{0} xe^{-x^2}\,dx = \lim\limits_{t\to-\infty} \left(-\frac{1}{2}\right)\left[e^{-x^2}\right]_t^0 = \lim\limits_{t\to-\infty} \left(-\frac{1}{2}\right)\left(1 - e^{-t^2}\right) = -\frac{1}{2}\cdot 1 = -\frac{1}{2}$, and

$\int_0^{\infty} xe^{-x^2}\,dx = \lim\limits_{t\to\infty} \left(-\frac{1}{2}\right)\left[e^{-x^2}\right]_0^t = \lim\limits_{t\to\infty} \left(-\frac{1}{2}\right)\left(e^{-t^2} - 1\right) = -\frac{1}{2}\cdot(-1) = \frac{1}{2}.$

Therefore, $\int_{-\infty}^{\infty} xe^{-x^2}\,dx = -\frac{1}{2} + \frac{1}{2} = 0.$ Convergent

14. $\displaystyle\int_1^{\infty} \frac{e^{-1/x}}{x^2}\,dx = \lim_{t\to\infty} \int_1^t \frac{e^{-1/x}}{x^2}\,dx = \lim_{t\to\infty} \left[e^{-1/x}\right]_1^t = \lim_{t\to\infty} \left(e^{-1/t} - e^{-1}\right) = 1 - \frac{1}{e}.$ Convergent

15. $\int_0^{\infty} \sin^2\alpha\,d\alpha = \lim\limits_{t\to\infty} \int_0^t \frac{1}{2}(1 - \cos 2\alpha)\,d\alpha = \lim\limits_{t\to\infty} \left[\frac{1}{2}\left(\alpha - \frac{1}{2}\sin 2\alpha\right)\right]_0^t = \lim\limits_{t\to\infty} \left[\frac{1}{2}\left(t - \frac{1}{2}\sin 2t\right) - 0\right] = \infty.$

Divergent

16. $\int_0^{\infty} \sin\theta\, e^{\cos\theta}\,d\theta = \lim\limits_{t\to\infty} \int_0^t \sin\theta\, e^{\cos\theta}\,d\theta = \lim\limits_{t\to\infty} \left[-e^{\cos\theta}\right]_0^t = \lim\limits_{t\to\infty} \left(-e^{\cos t} + e\right)$

This limit does not exist since $\cos t$ oscillates in value between -1 and 1, so $e^{\cos t}$ oscillates in value between e^{-1} and e^1. Divergent

17. $\displaystyle\int_1^{\infty} \frac{1}{x^2+x}\,dx = \lim_{t\to\infty} \int_1^t \frac{1}{x(x+1)}\,dx = \lim_{t\to\infty} \int_1^t \left(\frac{1}{x} - \frac{1}{x+1}\right) dx$ [partial fractions]

$\displaystyle = \lim_{t\to\infty} \Big[\ln|x| - \ln|x+1|\Big]_1^t = \lim_{t\to\infty} \left[\ln\left|\frac{x}{x+1}\right|\right]_1^t = \lim_{t\to\infty} \left(\ln\frac{t}{t+1} - \ln\frac{1}{2}\right) = 0 - \ln\frac{1}{2} = \ln 2.$

Convergent

18. $\displaystyle\int_2^\infty \frac{dv}{v^2+2v-3} = \lim_{t\to\infty}\int_2^t \frac{dv}{(v+3)(v-1)} = \lim_{t\to\infty}\int_2^t \left(\frac{-\frac14}{v+3}+\frac{\frac14}{v-1}\right)dv = \lim_{t\to\infty}\left[-\tfrac14\ln|v+3|+\tfrac14\ln|v-1|\right]_2^t$

$\displaystyle = \frac14\lim_{t\to\infty}\left[\ln\frac{v-1}{v+3}\right]_2^t = \frac14\lim_{t\to\infty}\left(\ln\frac{t-1}{t+3}-\ln\frac15\right) = \frac14(0+\ln 5) = \frac14\ln 5.$ Convergent

19. $\int_{-\infty}^0 ze^{2z}\,dz = \lim_{t\to-\infty}\int_t^0 ze^{2z}\,dz = \lim_{t\to-\infty}\left[\frac12 ze^{2z}-\frac14 e^{2z}\right]_t^0$ $\left[\begin{array}{l}\text{integration by parts with}\\ u=z,\ dv=e^{2z}\,dz\end{array}\right]$

$= \lim_{t\to-\infty}\left[\left(0-\frac14\right)-\left(\frac12 te^{2t}-\frac14 e^{2t}\right)\right] = -\frac14-0+0$ [by l'Hospital's Rule] $= -\frac14$. Convergent

20. $\int_2^\infty ye^{-3y}\,dy = \lim_{t\to\infty}\int_2^t ye^{-3y}\,dy = \lim_{t\to\infty}\left[-\frac13 ye^{-3y}-\frac19 e^{-3y}\right]_2^t$ $\left[\begin{array}{l}\text{integration by parts with}\\ u=y,\ dv=e^{-3y}\,dy\end{array}\right]$

$= \lim_{t\to\infty}\left[\left(-\frac13 te^{-3t}-\frac19 e^{-3t}\right)-\left(-\frac23 e^{-6}-\frac19 e^{-6}\right)\right] = 0-0+\frac79 e^{-6}$ [by l'Hospital's Rule] $= \frac79 e^{-6}$.

Convergent

21. $\displaystyle\int_1^\infty \frac{\ln x}{x}\,dx = \lim_{t\to\infty}\left[\frac{(\ln x)^2}{2}\right]_1^t$ $\left[\begin{array}{l}\text{by substitution with}\\ u=\ln x,\ du=dx/x\end{array}\right]$ $\displaystyle = \lim_{t\to\infty}\frac{(\ln t)^2}{2} = \infty.$ Divergent

22. $\displaystyle\int_1^\infty \frac{\ln x}{x^2}\,dx = \lim_{t\to\infty}\int_1^t \frac{\ln x}{x^2}\,dx = \lim_{t\to\infty}\left[-\frac{\ln x}{x}-\frac1x\right]_1^t$ $\left[\begin{array}{l}\text{integration by parts with}\\ u=\ln x,\ dv=(1/x^2)\,dx\end{array}\right]$

$\displaystyle = \lim_{t\to\infty}\left(-\frac{\ln t}{t}-\frac1t+1\right) \overset{\text{H}}{=} \lim_{t\to\infty}\left(-\frac{1/t}{1}\right)-\lim_{t\to\infty}\frac1t+\lim_{t\to\infty}1 = 0-0+1 = 1.$ Convergent

23. $\displaystyle\int_{-\infty}^0 \frac{z}{z^4+4}\,dz = \lim_{t\to-\infty}\int_t^0 \frac{z}{z^4+4}\,dz = \lim_{t\to-\infty}\frac12\left[\frac12\tan^{-1}\left(\frac{z^2}{2}\right)\right]_t^0$ $\left[\begin{array}{l}u=z^2,\\ du=2z\,dz\end{array}\right]$

$\displaystyle = \lim_{t\to-\infty}\left[0-\frac14\tan^{-1}\left(\frac{t^2}{2}\right)\right] = -\frac14\left(\frac\pi2\right) = -\frac\pi8.$ Convergent

24. $\displaystyle\int_e^\infty \frac{1}{x(\ln x)^2}\,dx = \lim_{t\to\infty}\int_e^t \frac{1}{x(\ln x)^2}\,dx = \lim_{t\to\infty}\left[-\frac{1}{\ln x}\right]_e^t$ $\left[\begin{array}{l}u=\ln x,\\ du=(1/x)\,dx\end{array}\right]$

$\displaystyle = \lim_{t\to\infty}\left(-\frac{1}{\ln t}+1\right) = 0+1 = 1.$ Convergent

25. $\displaystyle\int_0^\infty e^{-\sqrt y}\,dy = \lim_{t\to\infty}\int_0^t e^{-\sqrt y}\,dy = \lim_{t\to\infty}\int_0^{\sqrt t} e^{-x}\,(2x\,dx)$ $\left[\begin{array}{l}x=\sqrt y,\\ dx=1/(2\sqrt y\,)\,dy\end{array}\right]$

$\displaystyle = \lim_{t\to\infty}\left\{\left[-2xe^{-x}\right]_0^{\sqrt t}+\int_0^{\sqrt t} 2e^{-x}\,dx\right\}$ $\left[\begin{array}{ll}u=2x, & dv=e^{-x}\,dx\\ du=2\,dx, & v=-e^{-x}\end{array}\right]$

$\displaystyle = \lim_{t\to\infty}\left(-2\sqrt t\,e^{-\sqrt t}+\left[-2e^{-x}\right]_0^{\sqrt t}\right) = \lim_{t\to\infty}\left(\frac{-2\sqrt t}{e^{\sqrt t}}-\frac{2}{e^{\sqrt t}}+2\right) = 0-0+2 = 2.$

Convergent

Note: $\displaystyle\lim_{t\to\infty}\frac{\sqrt t}{e^{\sqrt t}} \overset{\text{H}}{=} \lim_{t\to\infty}\frac{2\sqrt t}{2\sqrt t\,e^{\sqrt t}} = \lim_{t\to\infty}\frac{1}{e^{\sqrt t}} = 0$

26. $\displaystyle\int_1^\infty \frac{dx}{\sqrt{x}+x\sqrt{x}}\,dx = \lim_{t\to\infty}\int_1^t \frac{dx}{\sqrt{x}\,(1+x)} = \lim_{t\to\infty}\int_1^{\sqrt{t}} \frac{1}{u(1+u^2)}\,(2u\,du)$ $\left[\begin{array}{l} u=\sqrt{x}, \\ du = 1/(2\sqrt{x}\,)\,dx \end{array}\right]$

$\displaystyle= \lim_{t\to\infty}\int_1^{\sqrt{t}} \frac{2}{1+u^2}\,du = \lim_{t\to\infty}\left[2\tan^{-1}u\right]_1^{\sqrt{t}} = \lim_{t\to\infty} 2(\tan^{-1}\sqrt{t}-\tan^{-1}1)$

$= 2(\frac{\pi}{2}-\frac{\pi}{4}) = \frac{\pi}{2}.$ Convergent

27. $\displaystyle\int_0^1 \frac{1}{x}\,dx = \lim_{t\to 0^+}\int_t^1 \frac{1}{x}\,dx = \lim_{t\to 0^+}\Big[\ln|x|\Big]_t^1 = \lim_{t\to 0^+}(-\ln t) = \infty.$ Divergent

28. $\displaystyle\int_0^5 \frac{1}{\sqrt[3]{5-x}}\,dx = \lim_{t\to 5^-}\int_0^t (5-x)^{-1/3}\,dx = \lim_{t\to 5^-}\left[-\tfrac{3}{2}(5-x)^{2/3}\right]_0^t = \lim_{t\to 5^-}\left\{-\tfrac{3}{2}[(5-t)^{2/3}-5^{2/3}]\right\}$

$= \frac{3}{2}5^{2/3}.$ Convergent

29. $\displaystyle\int_{-2}^{14} \frac{dx}{\sqrt[4]{x+2}} = \lim_{t\to -2^+}\int_t^{14}(x+2)^{-1/4}\,dx = \lim_{t\to -2^+}\left[\frac{4}{3}(x+2)^{3/4}\right]_t^{14} = \frac{4}{3}\lim_{t\to -2^+}\left[16^{3/4}-(t+2)^{3/4}\right]$

$= \frac{4}{3}(8-0) = \frac{32}{3}.$ Convergent

30. $\displaystyle\int_{-1}^2 \frac{x}{(x+1)^2}\,dx = \lim_{t\to -1^+}\int_t^2 \frac{x}{(x+1)^2}\,dx = \lim_{t\to -1^+}\int_t^2 \left[\frac{1}{x+1}-\frac{1}{(x+1)^2}\right]dx$ [partial fractions]

$\displaystyle= \lim_{t\to -1^+}\left[\ln|x+1|+\frac{1}{x+1}\right]_t^2 = \lim_{t\to -1^+}\left[\ln 3+\frac{1}{3}-\left(\ln(t+1)+\frac{1}{t+1}\right)\right] = -\infty.$ Divergent

Note: To justify the last step, $\displaystyle\lim_{t\to -1^+}\left[\ln(t+1)+\frac{1}{t+1}\right] = \lim_{x\to 0^+}\left(\ln x+\frac{1}{x}\right)$ $\left[\begin{array}{c}\text{substitute}\\ x \text{ for } t+1\end{array}\right]$ $\displaystyle= \lim_{x\to 0^+}\frac{x\ln x+1}{x} = \infty$

since $\displaystyle\lim_{x\to 0^+}(x\ln x) = \lim_{x\to 0^+}\frac{\ln x}{1/x} \overset{\text{H}}{=} \lim_{x\to 0^+}\frac{1/x}{-1/x^2} = \lim_{x\to 0^+}(-x) = 0.$

31. $\displaystyle\int_{-2}^3 \frac{dx}{x^4} = \int_{-2}^0 \frac{dx}{x^4}+\int_0^3 \frac{dx}{x^4}$, but $\displaystyle\int_{-2}^0 \frac{dx}{x^4} = \lim_{t\to 0^-}\left[-\frac{x^{-3}}{3}\right]_{-2}^t = \lim_{t\to 0^-}\left[-\frac{1}{3t^3}-\frac{1}{24}\right] = \infty.$ Divergent

32. $\displaystyle\int_0^1 \frac{dx}{\sqrt{1-x^2}} = \lim_{t\to 1^-}\int_0^t \frac{dx}{\sqrt{1-x^2}} = \lim_{t\to 1^-}\left[\sin^{-1}x\right]_0^t = \lim_{t\to 1^-}\sin^{-1}t = \frac{\pi}{2}.$ Convergent

33. There is an infinite discontinuity at $x=1$. $\displaystyle\int_0^9 \frac{1}{\sqrt[3]{x-1}}\,dx = \int_0^1 (x-1)^{-1/3}\,dx+\int_1^9 (x-1)^{-1/3}\,dx.$

Here $\int_0^1 (x-1)^{-1/3}\,dx = \lim_{t\to 1^-}\int_0^t (x-1)^{-1/3}\,dx = \lim_{t\to 1^-}\left[\frac{3}{2}(x-1)^{2/3}\right]_0^t = \lim_{t\to 1^-}\left[\frac{3}{2}(t-1)^{2/3}-\frac{3}{2}\right] = -\frac{3}{2}$

and $\int_1^9 (x-1)^{-1/3}\,dx = \lim_{t\to 1^+}\int_t^9 (x-1)^{-1/3}\,dx = \lim_{t\to 1^+}\left[\frac{3}{2}(x-1)^{2/3}\right]_t^9 = \lim_{t\to 1^+}\left[6-\frac{3}{2}(t-1)^{2/3}\right] = 6.$ Thus,

$\displaystyle\int_0^9 \frac{1}{\sqrt[3]{x-1}}\,dx = -\frac{3}{2}+6 = \frac{9}{2}.$ Convergent

34. There is an infinite discontinuity at $w=2$.

$\displaystyle\int_0^2 \frac{w}{w-2}\,dw = \lim_{t\to 2^-}\int_0^t \left(1+\frac{2}{w-2}\right)dw = \lim_{t\to 2^-}\Big[w+2\ln|w-2|\Big]_0^t = \lim_{t\to 2^-}(t+2\ln|t-2|-2\ln 2) = -\infty$, so

$\displaystyle\int_0^2 \frac{w}{w-2}\,dw$ diverges, and hence, $\displaystyle\int_0^5 \frac{w}{w-2}\,dw$ diverges. Divergent

35. $\int_0^{\pi/2} \tan^2\theta\,d\theta = \lim\limits_{t\to(\pi/2)^-} \int_0^t \tan^2\theta\,d\theta = \lim\limits_{t\to(\pi/2)^-} \int_0^t (\sec^2\theta - 1)\,d\theta = \lim\limits_{t\to(\pi/2)^-} \Big[\tan\theta - \theta\Big]_0^t$

$= \lim\limits_{t\to(\pi/2)^-} (\tan t - t) = \infty$ since $\tan t \to \infty$ as $t \to \frac{\pi}{2}^-$. Divergent

36. $\displaystyle\int_0^4 \frac{dx}{x^2 - x - 2} = \int_0^4 \frac{dx}{(x-2)(x+1)} = \int_0^2 \frac{dx}{(x-2)(x+1)} + \int_2^4 \frac{dx}{(x-2)(x+1)}$

Considering only $\displaystyle\int_0^2 \frac{dx}{(x-2)(x+1)}$ and using partial fractions, we have

$$\int_0^2 \frac{dx}{(x-2)(x+1)} = \lim_{t\to 2^-} \int_0^t \left(\frac{\frac{1}{3}}{x-2} - \frac{\frac{1}{3}}{x+1}\right) dx = \lim_{t\to 2^-} \Big[\tfrac{1}{3}\ln|x-2| - \tfrac{1}{3}\ln|x+1|\Big]_0^t$$

$$= \lim_{t\to 2^-} \big[\tfrac{1}{3}\ln|t-2| - \tfrac{1}{3}\ln|t+1| - \tfrac{1}{3}\ln 2 + 0\big] = -\infty \text{ since } \ln|t-2| \to -\infty \text{ as } t \to 2^-.$$

Thus, $\displaystyle\int_0^2 \frac{dx}{x^2 - x - 2}$ is divergent, and hence, $\displaystyle\int_0^4 \frac{dx}{x^2 - x - 2}$ is divergent as well.

37. $\int_0^1 r\ln r\,dr = \lim\limits_{t\to 0^+} \int_t^1 r\ln r\,dr = \lim\limits_{t\to 0^+} \Big[\tfrac{1}{2}r^2\ln r - \tfrac{1}{4}r^2\Big]_t^1$ $\quad\begin{bmatrix} u = \ln r, & dv = r\,dr \\ du = (1/r)\,dr, & v = \frac{1}{2}r^2 \end{bmatrix}$

$= \lim\limits_{t\to 0^+} \big[(0 - \tfrac{1}{4}) - (\tfrac{1}{2}t^2\ln t - \tfrac{1}{4}t^2)\big] = -\tfrac{1}{4} - 0 = -\tfrac{1}{4}$

since $\displaystyle\lim_{t\to 0^+} t^2\ln t = \lim_{t\to 0^+} \frac{\ln t}{1/t^2} \overset{\text{H}}{=} \lim_{t\to 0^+} \frac{1/t}{-2/t^3} = \lim_{t\to 0^+} (-\tfrac{1}{2}t^2) = 0.$ Convergent

38. $\displaystyle\int_0^{\pi/2} \frac{\cos\theta}{\sqrt{\sin\theta}}\,d\theta = \lim_{t\to 0^+} \int_t^{\pi/2} \frac{\cos\theta}{\sqrt{\sin\theta}}\,d\theta = \lim_{t\to 0^+} \Big[2\sqrt{\sin\theta}\Big]_t^{\pi/2}$ $\quad\begin{bmatrix} u = \sin\theta, \\ du = \cos\theta\,d\theta \end{bmatrix}$

$= \lim\limits_{t\to 0^+} (2 - 2\sqrt{\sin t}) = 2 - 0 = 2.$ Convergent

39. $\displaystyle\int_{-1}^0 \frac{e^{1/x}}{x^3}\,dx = \lim_{t\to 0^-} \int_{-1}^t \frac{1}{x}e^{1/x} \cdot \frac{1}{x^2}\,dx = \lim_{t\to 0^-} \int_{-1}^{1/t} ue^u\,(-du)$ $\quad\begin{bmatrix} u = 1/x, \\ du = -dx/x^2 \end{bmatrix}$

$\displaystyle= \lim_{t\to 0^-} \big[(u-1)e^u\big]_{1/t}^{-1}$ $\begin{bmatrix}\text{use parts} \\ \text{or Formula 96}\end{bmatrix}$ $\displaystyle= \lim_{t\to 0^-} \left[-2e^{-1} - \left(\frac{1}{t} - 1\right)e^{1/t}\right]$

$\displaystyle= -\frac{2}{e} - \lim_{s\to -\infty} (s-1)e^s \quad [s = 1/t] \quad = -\frac{2}{e} - \lim_{s\to -\infty} \frac{s-1}{e^{-s}} \overset{\text{H}}{=} -\frac{2}{e} - \lim_{s\to -\infty} \frac{1}{-e^{-s}}$

$\displaystyle= -\frac{2}{e} - 0 = -\frac{2}{e}.$ Convergent

40. $\displaystyle\int_0^1 \frac{e^{1/x}}{x^3}\,dx = \lim_{t\to 0^+} \int_t^1 \frac{1}{x}e^{1/x} \cdot \frac{1}{x^2}\,dx = \lim_{t\to 0^+} \int_{1/t}^1 ue^u\,(-du)$ $\quad\begin{bmatrix} u = 1/x, \\ du = -dx/x^2 \end{bmatrix}$

$\displaystyle= \lim_{t\to 0^+} \big[(u-1)e^u\big]_1^{1/t}$ $\begin{bmatrix}\text{use parts} \\ \text{or Formula 96}\end{bmatrix}$ $\displaystyle= \lim_{t\to 0^+} \left[\left(\frac{1}{t} - 1\right)e^{1/t} - 0\right]$

$\displaystyle= \lim_{s\to\infty} (s-1)e^s \quad [s = 1/t] \quad = \infty.$ Divergent

41.

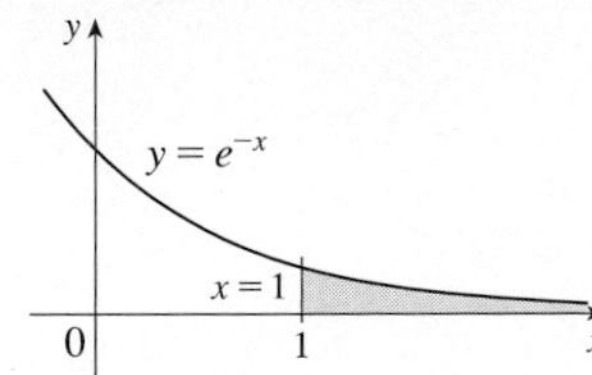

$$\text{Area} = \int_1^\infty e^{-x}\,dx = \lim_{t\to\infty}\int_1^t e^{-x}\,dx = \lim_{t\to\infty}\Big[-e^{-x}\Big]_1^t$$
$$= \lim_{t\to\infty}\left(-e^{-t}+e^{-1}\right) = 0 + e^{-1} = 1/e$$

42.

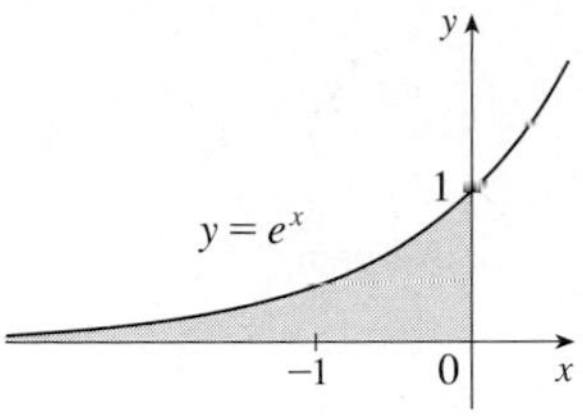

$$\text{Area} = \int_{-\infty}^0 e^x\,dx = \lim_{t\to-\infty}\int_t^0 e^x\,dx = \lim_{t\to-\infty}\Big[e^x\Big]_t^0$$
$$= \lim_{t\to-\infty}\left(e^0 - e^t\right) = 1 - 0 = 1$$

43.

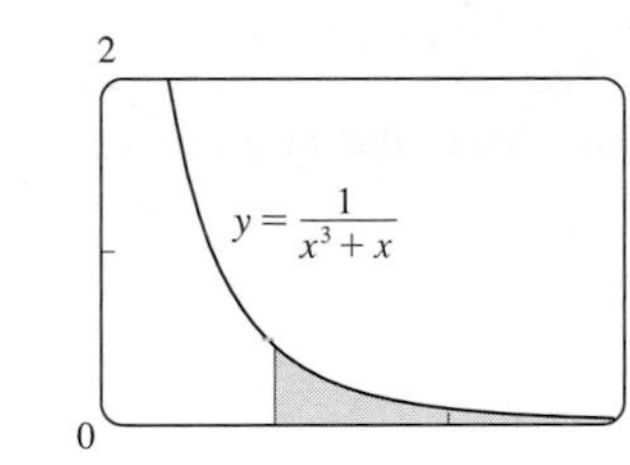

$$\text{Area} = \int_1^\infty \frac{1}{x^3+x}\,dx = \lim_{t\to\infty}\int_1^t \frac{1}{x(x^2+1)}\,dx$$
$$= \lim_{t\to\infty}\int_1^t\left(\frac{1}{x} - \frac{x}{x^2+1}\right)dx \qquad \text{[partial fractions]}$$
$$= \lim_{t\to\infty}\left[\ln|x| - \frac{1}{2}\ln\left|x^2+1\right|\right]_1^t = \lim_{t\to\infty}\left[\ln\frac{x}{\sqrt{x^2+1}}\right]_1^t$$
$$= \lim_{t\to\infty}\left(\ln\frac{t}{\sqrt{t^2+1}} - \ln\frac{1}{\sqrt{2}}\right) = \ln 1 - \ln 2^{-1/2} = \frac{1}{2}\ln 2$$

44.

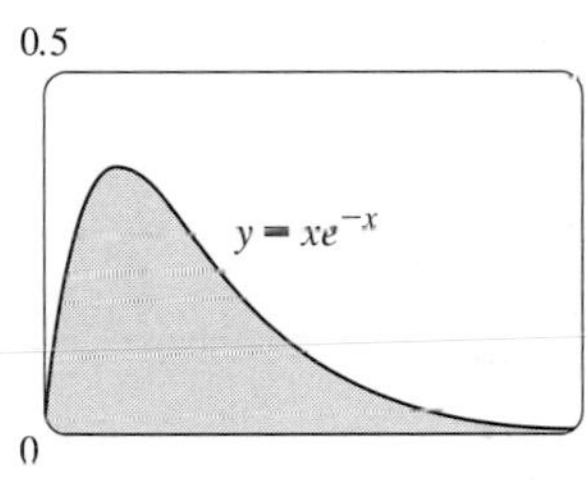

$$\text{Area} = \int_0^\infty xe^{-x}\,dx = \lim_{t\to\infty}\int_0^t xe^{-x}\,dx$$
$$= \lim_{t\to\infty}\Big[-xe^{-x} - e^{-x}\Big]_0^t \qquad \text{[use parts wtih } u = x \text{ and } dv = e^{-x}\,dx\text{]}$$
$$= \lim_{t\to\infty}\left[\left(-te^{-t} - e^{-t}\right) - (-1)\right]$$
$$= 0 \quad \text{[use l'Hospital's Rule]} \quad - 0 + 1 = 1$$

45.

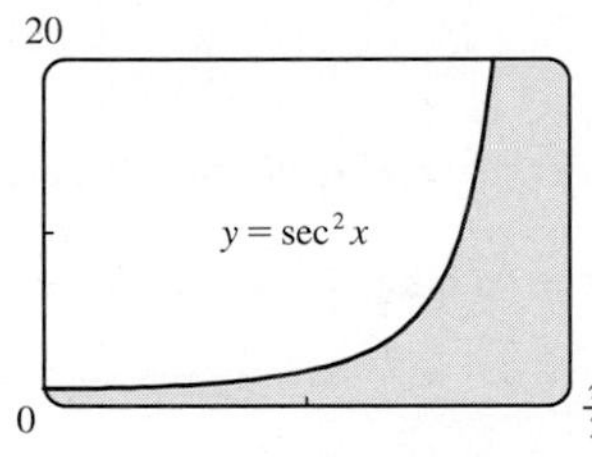

$$\text{Area} = \int_0^{\pi/2}\sec^2 x\,dx = \lim_{t\to(\pi/2)^-}\int_0^t \sec^2 x\,dx = \lim_{t\to(\pi/2)^-}\Big[\tan x\Big]_0^t$$
$$= \lim_{t\to(\pi/2)^-}(\tan t - 0) = \infty$$

Infinite area

46.

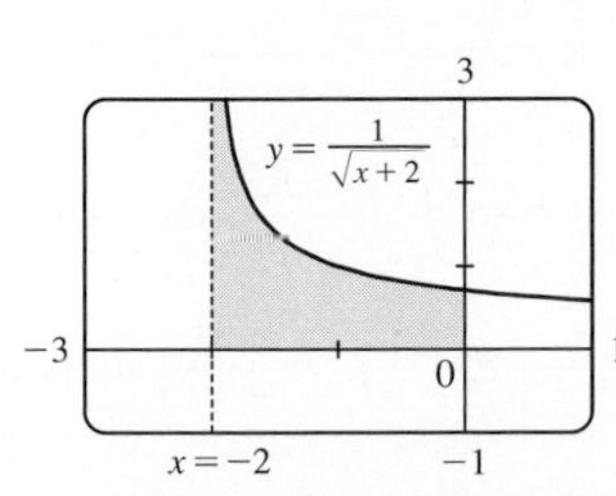

$$\text{Area} = \int_{-2}^0 \frac{1}{\sqrt{x+2}}\,dx = \lim_{t\to-2^+}\int_t^0 \frac{1}{\sqrt{x+2}}\,dx = \lim_{t\to-2^+}\left[2\sqrt{x+2}\right]_t^0$$
$$= \lim_{t\to-2^+}\left(2\sqrt{2} - 2\sqrt{t+2}\right) = 2\sqrt{2} - 0 = 2\sqrt{2}$$

47. (a)

t	$\int_1^t g(x)\,dx$
2	0.447453
5	0.577101
10	0.621306
100	0.668479
1000	0.672957
10,000	0.673407

$g(x) = \dfrac{\sin^2 x}{x^2}$.

It appears that the integral is convergent.

(b) $-1 \le \sin x \le 1 \;\Rightarrow\; 0 \le \sin^2 x \le 1 \;\Rightarrow\; 0 \le \dfrac{\sin^2 x}{x^2} \le \dfrac{1}{x^2}$. Since $\displaystyle\int_1^\infty \frac{1}{x^2}\,dx$ is convergent [Equation 2 with $p = 2 > 1$], $\displaystyle\int_1^\infty \frac{\sin^2 x}{x^2}\,dx$ is convergent by the Comparison Theorem.

(c)

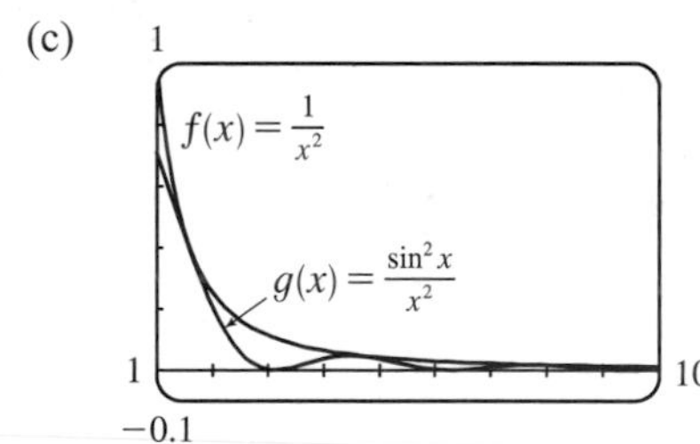

Since $\int_1^\infty f(x)\,dx$ is finite and the area under $g(x)$ is less than the area under $f(x)$ on any interval $[1, t]$, $\int_1^\infty g(x)\,dx$ must be finite; that is, the integral is convergent.

48. (a)

t	$\int_2^t g(x)dx$
5	3.830327
10	6.801200
100	23.328769
1000	69.023361
10,000	208.124560

$g(x) = \dfrac{1}{\sqrt{x} - 1}$.

It appears that the integral is divergent.

(b) For $x \ge 2$, $\sqrt{x} > \sqrt{x} - 1 \;\Rightarrow\; \dfrac{1}{\sqrt{x}} < \dfrac{1}{\sqrt{x} - 1}$. Since $\displaystyle\int_2^\infty \frac{1}{\sqrt{x}}\,dx$ is divergent [Equation 2 with $p = \frac{1}{2} \le 1$], $\displaystyle\int_2^\infty \frac{1}{\sqrt{x} - 1}\,dx$ is divergent by the Comparison Theorem.

(c)

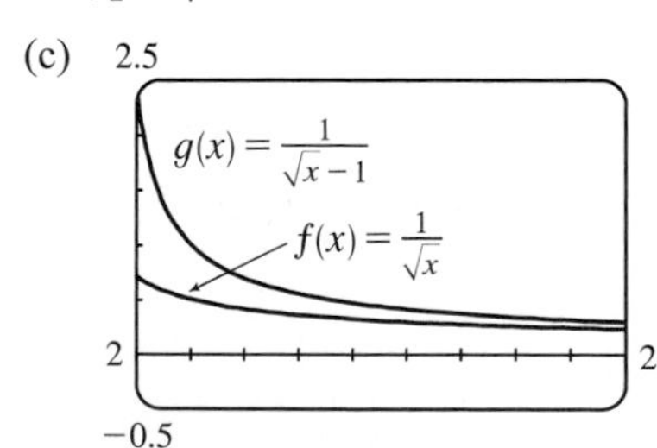

Since $\int_2^\infty f(x)\,dx$ is infinite and the area under $g(x)$ is greater than the area under $f(x)$ on any interval $[2, t]$, $\int_2^\infty g(x)\,dx$ must be infinite; that is, the integral is divergent.

49. For $x > 0$, $\dfrac{x}{x^3 + 1} < \dfrac{x}{x^3} = \dfrac{1}{x^2}$. $\displaystyle\int_1^\infty \frac{1}{x^2}\,dx$ is convergent by Equation 2 with $p = 2 > 1$, so $\displaystyle\int_1^\infty \frac{x}{x^3 + 1}\,dx$ is convergent by the Comparison Theorem. $\displaystyle\int_0^1 \frac{x}{x^3 + 1}\,dx$ is a constant, so $\displaystyle\int_0^\infty \frac{x}{x^3 + 1}\,dx = \int_0^1 \frac{x}{x^3 + 1}\,dx + \int_1^\infty \frac{x}{x^3 + 1}\,dx$ is also convergent.

50. For $x \geq 1$, $\dfrac{1+\sin^2 x}{\sqrt{x}} \geq \dfrac{1}{\sqrt{x}}$. $\displaystyle\int_1^\infty \frac{1}{\sqrt{x}}\,dx$ is divergent by Equation 2 with $p = \frac{1}{2} \leq 1$, so $\displaystyle\int_1^\infty \frac{1+\sin^2 x}{\sqrt{x}}\,dx$ is divergent by the Comparison Theorem.

51. For $x > 1$, $f(x) = \dfrac{x+1}{\sqrt{x^4 - x}} > \dfrac{x+1}{\sqrt{x^4}} > \dfrac{x}{x^2} = \dfrac{1}{x}$, so $\displaystyle\int_2^\infty f(x)\,dx$ diverges by comparison with $\displaystyle\int_2^\infty \frac{1}{x}\,dx$, which diverges by Equation 2 with $p = 1 \leq 1$. Thus, $\int_1^\infty f(x)\,dx = \int_1^2 f(x)\,dx + \int_2^\infty f(x)\,dx$ also diverges.

52. For $x \geq 0$, $\arctan x < \dfrac{\pi}{2} < 2$, so $\dfrac{\arctan x}{2+e^x} < \dfrac{2}{2+e^x} < \dfrac{2}{e^x} = 2e^{-x}$. Now

$I = \displaystyle\int_0^\infty 2e^{-x}\,dx = \lim_{t\to\infty}\int_0^t 2e^{-x}\,dx = \lim_{t\to\infty}\left[-2e^{-x}\right]_0^t = \lim_{t\to\infty}\left(-\frac{2}{e^t}+2\right) = 2$, so I is convergent, and by comparison,

$\displaystyle\int_0^\infty \frac{\arctan x}{2+e^x}\,dx$ is convergent.

53. For $0 < x \leq 1$, $\dfrac{\sec^2 x}{x\sqrt{x}} > \dfrac{1}{x^{3/2}}$. Now

$I = \displaystyle\int_0^1 x^{-3/2}\,dx = \lim_{t\to 0^+}\int_t^1 x^{-3/2}\,dx = \lim_{t\to 0^+}\left[-2x^{-1/2}\right]_t^1 = \lim_{t\to 0^+}\left(-2+\frac{2}{\sqrt{t}}\right) = \infty$, so I is divergent, and by

comparison, $\displaystyle\int_0^1 \frac{\sec^2 x}{x\sqrt{x}}$ is divergent.

54. For $0 < x \leq 1$, $\dfrac{\sin^2 x}{\sqrt{x}} \leq \dfrac{1}{\sqrt{x}}$. Now

$I = \displaystyle\int_0^\pi \frac{1}{\sqrt{x}}\,dx = \lim_{t\to 0^+}\int_t^\pi x^{-1/2}\,dx = \lim_{t\to 0^+}\left[2x^{1/2}\right]_t^\pi = \lim_{t\to 0^+}\left(2\pi - 2\sqrt{t}\right) = 2\pi - 0 = 2\pi$, so I is convergent, and by

comparison, $\displaystyle\int_0^\pi \frac{\sin^2 x}{\sqrt{x}}\,dx$ is convergent.

55. $\displaystyle\int_0^\infty \frac{dx}{\sqrt{x}\,(1+x)} = \int_0^1 \frac{dx}{\sqrt{x}\,(1+x)} + \int_1^\infty \frac{dx}{\sqrt{x}\,(1+x)} = \lim_{t\to 0^+}\int_t^1 \frac{dx}{\sqrt{x}\,(1+x)} + \lim_{t\to\infty}\int_1^t \frac{dx}{\sqrt{x}\,(1+x)}$. Now

$\displaystyle\int \frac{dx}{\sqrt{x}\,(1+x)} = \int \frac{2u\,du}{u(1+u^2)}\quad \begin{bmatrix} u = \sqrt{x},\ x = u^2, \\ dx = 2u\,du \end{bmatrix} \quad = 2\int \frac{du}{1+u^2} = 2\tan^{-1}u + C = 2\tan^{-1}\sqrt{x} + C$, so

$$\begin{aligned}\int_0^\infty \frac{dx}{\sqrt{x}\,(1+x)} &= \lim_{t\to 0^+}\left[2\tan^{-1}\sqrt{x}\right]_t^1 + \lim_{t\to\infty}\left[2\tan^{-1}\sqrt{x}\right]_1^t \\ &= \lim_{t\to 0^+}\left[2\left(\tfrac{\pi}{4}\right) - 2\tan^{-1}\sqrt{t}\right] + \lim_{t\to\infty}\left[2\tan^{-1}\sqrt{t} - 2\left(\tfrac{\pi}{4}\right)\right] = \tfrac{\pi}{2} - 0 + 2\left(\tfrac{\pi}{2}\right) - \tfrac{\pi}{2} = \pi.\end{aligned}$$

56. $\displaystyle\int_2^\infty \frac{dx}{x\sqrt{x^2-4}} = \int_2^3 \frac{dx}{x\sqrt{x^2-4}} + \int_3^\infty \frac{dx}{x\sqrt{x^2-4}} = \lim_{t\to 2^+}\int_t^3 \frac{dx}{x\sqrt{x^2-4}} + \lim_{t\to\infty}\int_3^t \frac{dx}{x\sqrt{x^2-4}}$. Now

$\displaystyle\int \frac{dx}{x\sqrt{x^2-4}} = \int \frac{2\sec\theta\tan\theta\,d\theta}{2\sec\theta\,2\tan\theta}\quad \begin{bmatrix} x = 2\sec\theta, \text{ where} \\ 0 \leq \theta < \pi/2 \text{ or } \pi \leq \theta < 3\pi/2 \end{bmatrix} \quad = \tfrac{1}{2}\theta + C = \tfrac{1}{2}\sec^{-1}\left(\tfrac{1}{2}x\right) + C$, so

$$\int_2^\infty \frac{dx}{x\sqrt{x^2-4}} = \lim_{t\to 2^+}\left[\tfrac{1}{2}\sec^{-1}\left(\tfrac{1}{2}x\right)\right]_t^3 + \lim_{t\to\infty}\left[\tfrac{1}{2}\sec^{-1}\left(\tfrac{1}{2}x\right)\right]_3^t = \tfrac{1}{2}\sec^{-1}\left(\tfrac{3}{2}\right) - 0 + \tfrac{1}{2}\left(\tfrac{\pi}{2}\right) - \tfrac{1}{2}\sec^{-1}\left(\tfrac{3}{2}\right) = \tfrac{\pi}{4}.$$

57. If $p = 1$, then $\displaystyle\int_0^1 \frac{dx}{x^p} = \lim_{t\to 0^+} \int_t^1 \frac{dx}{x} = \lim_{t\to 0^+} [\ln x]_t^1 = \infty.$ Divergent

If $p \neq 1$, then $\displaystyle\int_0^1 \frac{dx}{x^p} = \lim_{t\to 0^+} \int_t^1 \frac{dx}{x^p}$ [note that the integral is not improper if $p < 0$]

$$= \lim_{t\to 0^+} \left[\frac{x^{-p+1}}{-p+1}\right]_t^1 = \lim_{t\to 0^+} \frac{1}{1-p}\left[1 - \frac{1}{t^{p-1}}\right]$$

If $p > 1$, then $p - 1 > 0$, so $\dfrac{1}{t^{p-1}} \to \infty$ as $t \to 0^+$, and the integral diverges.

If $p < 1$, then $p - 1 < 0$, so $\dfrac{1}{t^{p-1}} \to 0$ as $t \to 0^+$ and $\displaystyle\int_0^1 \frac{dx}{x^p} = \frac{1}{1-p}\left[\lim_{t\to 0^+} \left(1 - t^{1-p}\right)\right] = \frac{1}{1-p}.$

Thus, the integral converges if and only if $p < 1$, and in that case its value is $\dfrac{1}{1-p}$.

58. Let $u = \ln x$. Then $du = dx/x \quad\Rightarrow\quad \displaystyle\int_e^\infty \frac{dx}{x\,(\ln x)^p} = \int_1^\infty \frac{du}{u^p}$. By Example 4, this converges to $\dfrac{1}{p-1}$ if $p > 1$ and diverges otherwise.

59. First suppose $p = -1$. Then

$\displaystyle\int_0^1 x^p \ln x\, dx = \int_0^1 \frac{\ln x}{x}\, dx = \lim_{t\to 0^+} \int_t^1 \frac{\ln x}{x}\, dx = \lim_{t\to 0^+} \left[\tfrac{1}{2}(\ln x)^2\right]_t^1 = -\tfrac{1}{2} \lim_{t\to 0^+} (\ln t)^2 = -\infty$, so the integral diverges. Now suppose $p \neq -1$. Then integration by parts gives

$\displaystyle\int x^p \ln x\, dx = \frac{x^{p+1}}{p+1} \ln x - \int \frac{x^p}{p+1}\, dx = \frac{x^{p+1}}{p+1} \ln x - \frac{x^{p+1}}{(p+1)^2} + C$. If $p < -1$, then $p + 1 < 0$, so

$$\int_0^1 x^p \ln x\, dx = \lim_{t\to 0^+} \left[\frac{x^{p+1}}{p+1} \ln x - \frac{x^{p+1}}{(p+1)^2}\right]_t^1 = \frac{-1}{(p+1)^2} - \left(\frac{1}{p+1}\right) \lim_{t\to 0^+} \left[t^{p+1}\left(\ln t - \frac{1}{p+1}\right)\right] = \infty.$$

If $p > -1$, then $p + 1 > 0$ and

$$\textstyle\int_0^1 x^p \ln x\, dx = \displaystyle\frac{-1}{(p+1)^2} - \left(\frac{1}{p+1}\right) \lim_{t\to 0^+} \frac{\ln t - 1/(p+1)}{t^{-(p+1)}} \overset{\text{H}}{=} \frac{-1}{(p+1)^2} - \left(\frac{1}{p+1}\right) \lim_{t\to 0^+} \frac{1/t}{-(p+1)t^{-(p+2)}}$$

$$= \frac{-1}{(p+1)^2} + \frac{1}{(p+1)^2} \lim_{t\to 0^+} t^{p+1} = \frac{-1}{(p+1)^2}$$

Thus, the integral converges to $-\dfrac{1}{(p+1)^2}$ if $p > -1$ and diverges otherwise.

60. (a) $\boldsymbol{n = 0}$: $\displaystyle\int_0^\infty x^n e^{-x}\, dx = \lim_{t\to\infty} \int_0^t e^{-x}\, dx = \lim_{t\to\infty} \left[-e^{-x}\right]_0^t = \lim_{t\to\infty} \left[-e^{-t} + 1\right] = 0 + 1 = 1$

$\boldsymbol{n = 1}$: $\displaystyle\int_0^\infty x^n e^{-x}\, dx = \lim_{t\to\infty} \int_0^t x e^{-x}\, dx$. To evaluate $\displaystyle\int x e^{-x}\, dx$, we'll use integration by parts with $u = x$, $dv = e^{-x}\, dx \quad\Rightarrow\quad du = dx$, $v = -e^{-x}$.

So $\displaystyle\int x e^{-x}\, dx = -x e^{-x} - \int -e^{-x}\, dx = -x e^{-x} - e^{-x} + C = (-x - 1)e^{-x} + C$ and

$$\lim_{t\to\infty}\int_0^t xe^{-x}\,dx = \lim_{t\to\infty}\left[(-x-1)e^{-x}\right]_0^t = \lim_{t\to\infty}\left[(-t-1)e^{-t}+1\right] = \lim_{t\to\infty}\left[-te^{-t}-e^{-t}+1\right]$$

$$= 0-0+1 \quad [\text{use l'Hospital's Rule}] \quad = 1$$

$n = 2$: $\displaystyle\int_0^\infty x^n e^{-x}\,dx = \lim_{t\to\infty}\int_0^t x^2 e^{-x}\,dx$. To evaluate $\int x^2 e^{-x}\,dx$, we could use integration by parts again or Formula 97. Thus,

$$\lim_{t\to\infty}\int_0^t x^2 e^{-x}\,dx = \lim_{t\to\infty}\left[-x^2e^{-x}\right]_0^t + 2\lim_{t\to\infty}\int_0^t xe^{-x}\,dx$$

$$= 0+0+2(1) \quad [\text{use l'Hospital's Rule and the result for } n=1] \quad = 2$$

$n = 3$: $\displaystyle\int_0^\infty x^n e^{-x} dx = \lim_{t\to\infty}\int_0^t x^3 e^{-x}\,dx \overset{97}{=} \lim_{t\to\infty}\left[-x^3e^{-x}\right]_0^t + 3\lim_{t\to\infty}\int_0^t x^2e^{-x}\,dx$

$$= 0+0+3(2) \quad [\text{use l'Hospital's Rule and the result for } n=2] \quad = 6$$

(b) For $n = 1$, 2, and 3, we have $\int_0^\infty x^n e^{-x}\,dx = 1$, 2, and 6. The values for the integral are equal to the factorials for n, so we guess $\int_0^\infty x^n e^{-x}\,dx = n!$.

(c) Suppose that $\int_0^\infty x^k e^{-x}\,dx = k!$ for some positive integer k. Then $\int_0^\infty x^{k+1}e^{-x}\,dx = \lim\limits_{t\to\infty}\int_0^t x^{k+1}e^{-x}\,dx$.

To evaluate $\int x^{k+1}e^{-x}\,dx$, we use parts with $u = x^{k+1}$, $dv = e^{-x}\,dx \;\Rightarrow\; du = (k+1)x^k\,dx$, $v = -e^{-x}$.

So $\int x^{k+1}e^{-x}dx = -x^{k+1}e^{-x} - \int -(k+1)x^k e^{-x}\,dx = -x^{k+1}e^{-x} + (k+1)\int x^k e^{-x}\,dx$ and

$$\lim_{t\to\infty}\int_0^t x^{k+1}e^{-x}\,dx = \lim_{t\to\infty}\left[-x^{k+1}e^{-x}\right]_0^t + (k+1)\lim_{t\to\infty}\int_0^t x^k e^{-x}\,dx$$

$$= \lim_{t\to\infty}\left[-t^{k+1}e^{-t}+0\right] + (k+1)k! = 0+0+(k+1)! = (k+1)!,$$

so the formula holds for $k+1$. By induction, the formula holds for all positive integers. (Since $0! = 1$, the formula holds for $n = 0$, too.)

61. (a) $I = \int_{-\infty}^\infty x\,dx = \int_{-\infty}^0 x\,dx + \int_0^\infty x\,dx$, and $\int_0^\infty x\,dx = \lim\limits_{t\to\infty}\int_0^t x\,dx = \lim\limits_{t\to\infty}\left[\frac12 x^2\right]_0^t = \lim\limits_{t\to\infty}\left[\frac12 t^2 - 0\right] = \infty$, so I is divergent.

(b) $\int_{-t}^t x\,dx = \left[\frac12 x^2\right]_{-t}^t = \frac12 t^2 - \frac12 t^2 = 0$, so $\lim\limits_{t\to\infty}\int_{-t}^t x\,dx = 0$. Therefore, $\int_{-\infty}^\infty x\,dx \neq \lim\limits_{t\to\infty}\int_{-t}^t x\,dx$.

62. Let $k = \dfrac{M}{2RT}$ so that $\overline{v} = \dfrac{4}{\sqrt{\pi}}k^{3/2}\displaystyle\int_0^\infty v^3e^{-kv^2}\,dv$. Let I denote the integral and use parts to integrate I. Let $\alpha = v^2$, $d\beta = ve^{-kv^2}\,dv \quad\Rightarrow\quad d\alpha = 2v\,dv$, $\beta = -\dfrac{1}{2k}e^{-kv^2}$:

$$I = \lim_{t\to\infty}\left[-\frac{1}{2k}v^2e^{-kv^2}\right]_0^t + \frac1k\int_0^\infty ve^{-kv^2}\,dv_0^t = -\frac{1}{2k}\lim_{t\to\infty}\left(t^2e^{-kt^2}\right) + \frac1k\lim_{t\to\infty}\left[-\frac{1}{2k}e^{-kv^2}\right]$$

$$\overset{\text{H}}{=} -\frac{1}{2k}\cdot 0 - \frac{1}{2k^2}(0-1) = \frac{1}{2k^2}$$

Thus, $\overline{v} = \dfrac{4}{\sqrt{\pi}}k^{3/2}\cdot\dfrac{1}{2k^2} = \dfrac{2}{(k\pi)^{1/2}} = \dfrac{2}{[\pi M/(2RT)]^{1/2}} = \dfrac{2\sqrt{2}\sqrt{RT}}{\sqrt{\pi M}} = \sqrt{\dfrac{8RT}{\pi M}}$.

63. Volume $= \int_1^\infty \pi\left(\frac{1}{x}\right)^2 dx = \pi \lim_{t\to\infty}\int_1^t \frac{dx}{x^2} = \pi \lim_{t\to\infty}\left[-\frac{1}{x}\right]_1^t = \pi \lim_{t\to\infty}\left(1-\frac{1}{t}\right) = \pi < \infty.$

64. Work $= \int_R^\infty \frac{GMm}{r^2}\,dr = \lim_{t\to\infty}\int_R^t \frac{GMm}{r^2}\,dr = \lim_{t\to\infty} GMm\left[\frac{-1}{r}\right]_R^t = GMm \lim_{t\to\infty}\left(\frac{-1}{t}+\frac{1}{R}\right) = \frac{GMm}{R}$, where

M = mass of the earth $= 5.98\times 10^{24}$ kg, m = mass of satellite $= 10^3$ kg, R = radius of the earth $= 6.37\times 10^6$ m, and

G = gravitational constant $= 6.67\times 10^{-11}$ N·m²/kg.

Therefore, Work $= \frac{6.67\times 10^{-11}\cdot 5.98\times 10^{24}\cdot 10^3}{6.37\times 10^6} \approx 6.26\times 10^{10}$ J.

65. Work $= \int_R^\infty F\,dr = \lim_{t\to\infty}\int_R^t \frac{GmM}{r^2}\,dr = \lim_{t\to\infty} GmM\left(\frac{1}{R}-\frac{1}{t}\right) = \frac{GmM}{R}$. The initial kinetic energy provides the work,

so $\frac{1}{2}mv_0^2 = \frac{GmM}{R} \Rightarrow v_0 = \sqrt{\frac{2GM}{R}}$.

66. $y(s) = \int_s^R \frac{2r}{\sqrt{r^2-s^2}}\,x(r)\,dr$ and $x(r) = \frac{1}{2}(R-r)^2 \Rightarrow$

$$y(s) = \lim_{t\to s^+}\int_t^R \frac{r(R-r)^2}{\sqrt{r^2-s^2}}\,dr = \lim_{t\to s^+}\int_t^R \frac{r^3-2Rr^2+R^2r}{\sqrt{r^2-s^2}}\,dr$$

$$= \lim_{t\to s^+}\left[\int_t^R \frac{r^3\,dr}{\sqrt{r^2-s^2}} - 2R\int_t^R \frac{r^2\,dr}{\sqrt{r^2-s^2}} + R^2\int_t^R \frac{r\,dr}{\sqrt{r^2-s^2}}\right] = \lim_{t\to s^+}\left(I_1 - 2RI_2 + R^2 I_3\right) = L$$

For I_1: Let $u = \sqrt{r^2-s^2} \Rightarrow u^2 = r^2-s^2$, $r^2 = u^2+s^2$, $2r\,dr = 2u\,du$, so, omitting limits and constant of integration,

$$I_1 = \int \frac{(u^2+s^2)u}{u}\,du = \int (u^2+s^2)\,du = \tfrac{1}{3}u^3 + s^2u = \tfrac{1}{3}u(u^2+3s^2)$$
$$= \tfrac{1}{3}\sqrt{r^2-s^2}\,(r^2-s^2+3s^2) = \tfrac{1}{3}\sqrt{r^2-s^2}\,(r^2+2s^2)$$

For I_2: Using Formula 44, $I_2 = \frac{r}{2}\sqrt{r^2-s^2} + \frac{s^2}{2}\ln\left|r+\sqrt{r^2-s^2}\right|$.

For I_3: Let $u = r^2-s^2 \Rightarrow du = 2r\,dr$. Then $I_3 = \frac{1}{2}\int \frac{du}{\sqrt{u}} = \frac{1}{2}\cdot 2\sqrt{u} = \sqrt{r^2-s^2}$.

Thus,

$$L = \lim_{t\to s^+}\left[\tfrac{1}{3}\sqrt{r^2-s^2}\,(r^2+2s^2) - 2R\left(\frac{r}{2}\sqrt{r^2-s^2} + \frac{s^2}{2}\ln\left|r+\sqrt{r^2-s^2}\right|\right) + R^2\sqrt{r^2-s^2}\right]_t^R$$
$$= \lim_{t\to s^+}\left[\tfrac{1}{3}\sqrt{R^2-s^2}(R^2+2s^2) - 2R\left(\frac{R}{2}\sqrt{R^2-s^2} + \frac{s^2}{2}\ln\left|R+\sqrt{R^2-s^2}\right|\right) + R^2\sqrt{R^2-s^2}\right]$$
$$- \lim_{t\to s^+}\left[\tfrac{1}{3}\sqrt{t^2-s^2}\,(t^2+2s^2) - 2R\left(\frac{t}{2}\sqrt{t^2-s^2} + \frac{s^2}{2}\ln\left|t+\sqrt{t^2-s^2}\right|\right) + R^2\sqrt{t^2-s^2}\right]$$
$$= \left[\tfrac{1}{3}\sqrt{R^2-s^2}\,(R^2+2s^2) - Rs^2\ln\left|R+\sqrt{R^2-s^2}\right|\right] - \left[-Rs^2\ln|s|\right]$$
$$= \tfrac{1}{3}\sqrt{R^2-s^2}\,(R^2+2s^2) - Rs^2\ln\left(\frac{R+\sqrt{R^2-s^2}}{s}\right)$$

67. We would expect a small percentage of bulbs to burn out in the first few hundred hours, most of the bulbs to burn out after close to 700 hours, and a few overachievers to burn on and on.

(a)

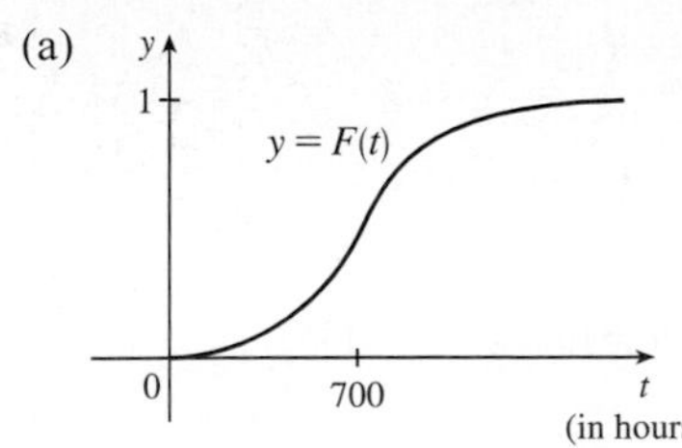

(b) $r(t) = F'(t)$ is the rate at which the fraction $F(t)$ of burnt-out bulbs increases as t increases. This could be interpreted as a fractional burnout rate.

(c) $\int_0^\infty r(t)\,dt = \lim\limits_{x\to\infty} F(x) = 1$, since all of the bulbs will eventually burn out.

68. $I = \displaystyle\int_0^\infty te^{kt}\,dt = \lim_{s\to\infty}\left[\frac{1}{k^2}(kt \quad 1)\,e^{kt}\right]_0^s$ [Formula 96, or parts] $= \displaystyle\lim_{s\to\infty}\left[\left(\frac{1}{k}se^{ks} - \frac{1}{k^2}e^{ks}\right) - \left(-\frac{1}{k^2}\right)\right].$

Since $k < 0$ the first two terms approach 0 (you can verify that the first term does so with l'Hospital's Rule), so the limit is equal to $1/k^2$. Thus, $M = -kI = -k(1/k^2) = -1/k = -1/(-0.000121) \approx 8264.5$ years.

69. $$\gamma = \int_0^\infty \frac{cN(1-e^{-kt})}{k}\,e^{-\lambda t}\,dt = \frac{cN}{k}\lim_{x\to\infty}\int_0^x \left[e^{-\lambda t} - e^{(-k-\lambda)t}\right]dt$$

$$= \frac{cN}{k}\lim_{x\to\infty}\left[\frac{1}{-\lambda}e^{-\lambda t} - \frac{1}{-k-\lambda}e^{(-k-\lambda)t}\right]_0^x = \frac{cN}{k}\lim_{x\to\infty}\left[\frac{1}{-\lambda e^{\lambda x}} + \frac{1}{(k+\lambda)e^{(k+\lambda)x}} - \left(\frac{1}{-\lambda} + \frac{1}{k+\lambda}\right)\right]$$

$$= \frac{cN}{k}\left(\frac{1}{\lambda} - \frac{1}{k+\lambda}\right) = \frac{cN}{k}\left(\frac{k+\lambda-\lambda}{\lambda(k+\lambda)}\right) = \frac{cN}{\lambda(k+\lambda)}$$

70. $$\int_0^\infty u(t)\,dt = \lim_{x\to\infty}\int_0^x \frac{r}{V}C_0\,e^{-rt/V}\,dt = \frac{r}{V}C_0\lim_{x\to\infty}\left[\frac{e^{-rt/V}}{-r/V}\right]_0^x = \frac{r}{V}C_0\left(-\frac{V}{r}\right)\lim_{x\to\infty}\left(e^{-rx/V} - 1\right)$$

$$= -C_0(0-1) = C_0.$$

$\int_0^\infty u(t)\,dt$ represents the total amount of urea removed from the blood if dialysis is continued indefinitely. The fact that $\int_0^\infty u(t)\,dt = C_0$ means that, in the limit, as $t \to \infty$, all the urea in the blood at time $t = 0$ is removed. The calculation says nothing about how rapidly that limit is approached.

71. $I = \displaystyle\int_a^\infty \frac{1}{x^2+1}\,dx = \lim_{t\to\infty}\int_a^t \frac{1}{x^2+1}\,dx = \lim_{t\to\infty}\left[\tan^{-1}x\right]_a^t = \lim_{t\to\infty}\left(\tan^{-1}t - \tan^{-1}a\right) = \frac{\pi}{2} - \tan^{-1}a.$

$I < 0.001 \;\Rightarrow\; \frac{\pi}{2} - \tan^{-1}a < 0.001 \;\Rightarrow\; \tan^{-1}a > \frac{\pi}{2} - 0.001 \;\Rightarrow\; a > \tan\left(\frac{\pi}{2} - 0.001\right) \approx 1000.$

72. $f(x) = e^{-x^2}$ and $\Delta x = \frac{4-0}{8} = \frac{1}{2}.$

$\int_0^4 f(x)\,dx \approx S_8 = \frac{1}{2\cdot 3}[f(0) + 4f(0.5) + 2f(1) + \cdots + 2f(3) + 4f(3.5) + f(4)] \approx \frac{1}{6}(5.31717808) \approx 0.8862$

Now $x > 4 \;\Rightarrow\; -x\cdot x < -x\cdot 4 \;\Rightarrow\; e^{-x^2} < e^{-4x} \;\Rightarrow\; \int_4^\infty e^{-x^2}\,dx < \int_4^\infty e^{-4x}\,dx.$

$\int_4^\infty e^{-4x}\,dx = \lim\limits_{t\to\infty}\left[-\frac{1}{4}e^{-4x}\right]_4^t = -\frac{1}{4}\left(0 - e^{-16}\right) = 1/(4e^{16}) \approx 0.0000000281 < 0.0000001$, as desired.

73. (a) $F(s) = \displaystyle\int_0^\infty f(t)e^{-st}\,dt = \int_0^\infty e^{-st}\,dt = \lim_{n\to\infty}\left[-\frac{e^{-st}}{s}\right]_0^n = \lim_{n\to\infty}\left(\frac{e^{-sn}}{-s} + \frac{1}{s}\right)$. This converges to $\dfrac{1}{s}$ only if $s > 0$.

Therefore $F(s) = \dfrac{1}{s}$ with domain $\{s \mid s > 0\}$.

(b) $F(s) = \int_0^\infty f(t)e^{-st}\,dt = \int_0^\infty e^t e^{-st}\,dt = \lim_{n\to\infty}\int_0^n e^{t(1-s)}\,dt = \lim_{n\to\infty}\left[\frac{1}{1-s}e^{t(1-s)}\right]_0^n$

$= \lim_{n\to\infty}\left(\frac{e^{(1-s)n}}{1-s} - \frac{1}{1-s}\right)$

This converges only if $1 - s < 0 \;\Rightarrow\; s > 1$, in which case $F(s) = \dfrac{1}{s-1}$ with domain $\{s \mid s > 1\}$.

(c) $F(s) = \int_0^\infty f(t)e^{-st}\,dt = \lim_{n\to\infty}\int_0^n te^{-st}\,dt$. Use integration by parts: let $u = t$, $dv = e^{-st}\,dt \;\Rightarrow\; du = dt$, $v = -\dfrac{e^{-st}}{s}$. Then $F(s) = \lim_{n\to\infty}\left[-\frac{t}{s}e^{-st} - \frac{1}{s^2}e^{-st}\right]_0^n = \lim_{n\to\infty}\left(\frac{-n}{se^{sn}} - \frac{1}{s^2e^{sn}} + 0 + \frac{1}{s^2}\right) = \frac{1}{s^2}$ only if $s > 0$.

Therefore, $F(s) = \dfrac{1}{s^2}$ and the domain of F is $\{s \mid s > 0\}$.

74. $0 \le f(t) \le Me^{at} \;\Rightarrow\; 0 \le f(t)e^{-st} \le Me^{at}e^{-st}$ for $t \ge 0$. Now use the Comparison Theorem:

$$\int_0^\infty Me^{at}e^{-st}\,dt = \lim_{n\to\infty} M\int_0^n e^{t(a-s)}\,dt = M\cdot\lim_{n\to\infty}\left[\frac{1}{a-s}e^{t(a-s)}\right]_0^n = M\cdot\lim_{n\to\infty}\frac{1}{a-s}\left[e^{n(a-s)} - 1\right]$$

This is convergent only when $a - s < 0 \;\Rightarrow\; s > a$. Therefore, by the Comparison Theorem, $F(s) = \int_0^\infty f(t)\,e^{-st}\,dt$ is also convergent for $s > a$.

75. $G(s) = \int_0^\infty f'(t)e^{-st}\,dt$. Integrate by parts with $u = e^{-st}$, $dv = f'(t)\,dt \;\Rightarrow\; du = -se^{-st}$, $v = f(t)$:

$$G(s) = \lim_{n\to\infty}\left[f(t)e^{-st}\right]_0^n + s\int_0^\infty f(t)e^{-st}\,dt = \lim_{n\to\infty} f(n)e^{-sn} - f(0) + sF(s)$$

But $0 \le f(t) \le Me^{at} \;\Rightarrow\; 0 \le f(t)e^{-st} \le Me^{at}e^{-st}$ and $\lim_{t\to\infty} Me^{t(a-s)} = 0$ for $s > a$. So by the Squeeze Theorem,

$\lim_{t\to\infty} f(t)e^{-st} = 0$ for $s > a \;\Rightarrow\; G(s) = 0 - f(0) + sF(s) = sF(s) - f(0)$ for $s > a$.

76. Assume without loss of generality that $a < b$. Then

$$\begin{aligned}
\int_{-\infty}^a f(x)\,dx + \int_a^\infty f(x)\,dx &= \lim_{t\to-\infty}\int_t^a f(x)\,dx + \lim_{u\to\infty}\int_a^u f(x)\,dx \\
&= \lim_{t\to-\infty}\int_t^a f(x)\,dx + \lim_{u\to\infty}\left[\int_a^b f(x)\,dx + \int_b^u f(x)\,dx\right] \\
&= \lim_{t\to-\infty}\int_t^a f(x)\,dx + \int_a^b f(x)\,dx + \lim_{u\to\infty}\int_b^u f(x)\,dx \\
&= \lim_{t\to-\infty}\left[\int_t^a f(x)\,dx + \int_a^b f(x)\,dx\right] + \int_b^\infty f(x)\,dx \\
&= \lim_{t\to-\infty}\int_t^b f(x)\,dx + \int_b^\infty f(x)\,dx = \int_{-\infty}^b f(x)\,dx + \int_b^\infty f(x)\,dx
\end{aligned}$$

77. We use integration by parts: let $u = x$, $dv = xe^{-x^2}\,dx \;\Rightarrow\; du = dx$, $v = -\frac{1}{2}e^{-x^2}$. So

$$\int_0^\infty x^2e^{-x^2}\,dx = \lim_{t\to\infty}\left[-\frac{1}{2}xe^{-x^2}\right]_0^t + \frac{1}{2}\int_0^\infty e^{-x^2}\,dx = \lim_{t\to\infty}\left[-\frac{t}{2e^{t^2}}\right] + \frac{1}{2}\int_0^\infty e^{-x^2}\,dx = \frac{1}{2}\int_0^\infty e^{-x^2}\,dx$$

(The limit is 0 by l'Hospital's Rule.)

78. $\int_0^\infty e^{-x^2}\,dx$ is the area under the curve $y = e^{-x^2}$ for $0 \le x < \infty$ and $0 < y \le 1$. Solving $y = e^{-x^2}$ for x, we get $y = e^{-x^2} \Rightarrow \ln y = -x^2 \Rightarrow -\ln y = x^2 \Rightarrow x = \pm\sqrt{-\ln y}$. Since x is positive, choose $x = \sqrt{-\ln y}$, and the area is represented by $\int_0^1 \sqrt{-\ln y}\,dy$. Therefore, each integral represents the same area, so the integrals are equal.

79. For the first part of the integral, let $x = 2\tan\theta \Rightarrow dx = 2\sec^2\theta\,d\theta$.

$$\int \frac{1}{\sqrt{x^2+4}}\,dx = \int \frac{2\sec^2\theta}{2\sec\theta}\,d\theta = \int \sec\theta\,d\theta = \ln|\sec\theta + \tan\theta|.$$

From the figure, $\tan\theta = \dfrac{x}{2}$, and $\sec\theta = \dfrac{\sqrt{x^2+4}}{2}$. So

$$I = \int_0^\infty \left(\frac{1}{\sqrt{x^2+4}} - \frac{C}{x+2}\right)dx = \lim_{t\to\infty}\left[\ln\left|\frac{\sqrt{x^2+4}}{2} + \frac{x}{2}\right| - C\ln|x+2|\right]_0^t$$

$$= \lim_{t\to\infty}\left[\ln\frac{\sqrt{t^2+4}+t}{2} - C\ln(t+2) - (\ln 1 - C\ln 2)\right]$$

$$= \lim_{t\to\infty}\left[\ln\left(\frac{\sqrt{t^2+4}+t}{2\,(t+2)^C}\right) + \ln 2^C\right] = \ln\left(\lim_{t\to\infty}\frac{t+\sqrt{t^2+4}}{(t+2)^C}\right) + \ln 2^{C-1}$$

Now $L = \lim\limits_{t\to\infty}\dfrac{t+\sqrt{t^2+4}}{(t+2)^C} \overset{\text{H}}{=} \lim\limits_{t\to\infty}\dfrac{1+t/\sqrt{t^2+4}}{C\,(t+2)^{C-1}} = \dfrac{2}{C\lim\limits_{t\to\infty}(t+2)^{C-1}}$.

If $C < 1$, $L = \infty$ and I diverges.

If $C = 1$, $L = 2$ and I converges to $\ln 2 + \ln 2^0 = \ln 2$.

If $C > 1$, $L = 0$ and I diverges to $-\infty$.

80. $$I = \int_0^\infty\left(\frac{x}{x^2+1} - \frac{C}{3x+1}\right)dx = \lim_{t\to\infty}\left[\tfrac{1}{2}\ln(x^2+1) - \tfrac{1}{3}C\ln(3x+1)\right]_0^t = \lim_{t\to\infty}\left[\ln(t^2+1)^{1/2} - \ln(3t+1)^{C/3}\right]$$

$$= \lim_{t\to\infty}\left(\ln\frac{(t^2+1)^{1/2}}{(3t+1)^{C/3}}\right) = \ln\left(\lim_{t\to\infty}\frac{\sqrt{t^2+1}}{(3t+1)^{C/3}}\right)$$

For $C \le 0$, the integral diverges. For $C > 0$, we have

$$L = \lim_{t\to\infty}\frac{\sqrt{t^2+1}}{(3t+1)^{C/3}} \overset{\text{H}}{=} \lim_{t\to\infty}\frac{t/\sqrt{t^2+1}}{C(3t+1)^{(C/3)-1}} = \frac{1}{C}\lim_{t\to\infty}\frac{1}{(3t+1)^{(C/3)-1}}$$

For $C/3 < 1 \Leftrightarrow C < 3$, $L = \infty$ and I diverges.

For $C = 3$, $L = \frac{1}{3}$ and $I = \ln\frac{1}{3}$.

For $C > 3$, $L = 0$ and I diverges to $-\infty$.

81. No, $I = \int_0^\infty f(x)\,dx$ must be *divergent*. Since $\lim\limits_{x\to\infty} f(x) = 1$, there must exist an N such that if $x \ge N$, then $f(x) \ge \frac{1}{2}$. Thus, $I = I_1 + I_2 = \int_0^N f(x)\,dx + \int_N^\infty f(x)\,dx$, where I_1 is an ordinary definite integral that has a finite value, and I_2 is improper and diverges by comparison with the divergent integral $\int_N^\infty \frac{1}{2}\,dx$.

82. As in Exercise 55, we let $I = \displaystyle\int_0^\infty \frac{x^a}{1+x^b}\,dx = I_1 + I_2$, where $I_1 = \displaystyle\int_0^1 \frac{x^a}{1+x^b}\,dx$ and $I_2 = \displaystyle\int_1^\infty \frac{x^a}{1+x^b}\,dx$. We will show that I_1 converges for $a > -1$ and I_2 converges for $b > a+1$, so that I converges when $a > -1$ and $b > a+1$.

[continued]

I_1 is improper only when $a < 0$. When $0 \le x \le 1$, we have $\dfrac{1}{1+x^b} \le 1 \;\Rightarrow\; \dfrac{1}{x^{-a}(1+x^b)} \le \dfrac{1}{x^{-a}}$. The integral $\displaystyle\int_0^1 \frac{1}{x^{-a}}\,dx$ converges for $-a < 1$ [or $a > -1$] by Exercise 57, so by the Comparison Theorem, $\displaystyle\int_0^1 \frac{1}{x^{-a}(1+x^b)}\,dx$ converges for $-1 < a < 0$. I_1 is not improper when $a \ge 0$, so it has a finite real value in that case. Therefore, I_1 has a finite real value (converges) when $a > -1$.

I_2 is always improper. When $x \ge 1$, $\dfrac{x^a}{1+x^b} = \dfrac{1}{x^{-a}(1+x^b)} = \dfrac{1}{x^{-a}+x^{b-a}} < \dfrac{1}{x^{b-a}}$. By (2), $\displaystyle\int_1^\infty \frac{1}{x^{b-a}}\,dx$ converges for $b-a > 1$ (or $b > a+1$), so by the Comparison Theorem, $\displaystyle\int_1^\infty \frac{x^a}{1+x^b}\,dx$ converges for $b > a+1$.

Thus, I converges if $a > -1$ and $b > a+1$.

7 Review

TRUE-FALSE QUIZ

1. False. Since the numerator has a higher degree than the denominator, $\dfrac{x(x^2+4)}{x^2-4} = x + \dfrac{8x}{x^2-4} = x + \dfrac{A}{x+2} + \dfrac{B}{x-2}$.

2. True. In fact, $A = -1$, $B = C = 1$.

3. False. It can be put in the form $\dfrac{A}{x} + \dfrac{B}{x^2} + \dfrac{C}{x-4}$.

4. False. The form is $\dfrac{A}{x} + \dfrac{Bx+C}{x^2+4}$.

5. False. This is an improper integral, since the denominator vanishes at $x = 1$.

$$\int_0^4 \frac{x}{x^2-1}\,dx = \int_0^1 \frac{x}{x^2-1}\,dx + \int_1^4 \frac{x}{x^2-1}\,dx \text{ and}$$

$$\int_0^1 \frac{x}{x^2-1}\,dx = \lim_{t\to1^-}\int_0^t \frac{x}{x^2-1}\,dx = \lim_{t\to1^-}\left[\tfrac{1}{2}\ln\left|x^2-1\right|\right]_0^t = \lim_{t\to1^-}\tfrac{1}{2}\ln\left|t^2-1\right| = \infty$$

So the integral diverges.

6. True by Theorem 7.8.2 with $p = \sqrt{2} > 1$.

7. False. See Exercise 61 in Section 7.8.

8. False. For example, with $n = 1$ the Trapezoidal Rule is much more accurate than the Midpoint Rule for the function in the diagram.

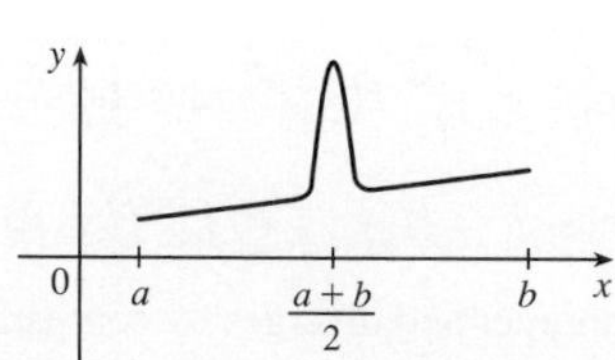

9. (a) True. See the end of Section 7.5.

(b) False. Examples include the functions $f(x) = e^{x^2}$, $g(x) = \sin(x^2)$, and $h(x) = \dfrac{\sin x}{x}$.

10. True. If f is continuous on $[0, \infty)$, then $\int_0^1 f(x)\,dx$ is finite. Since $\int_1^\infty f(x)\,dx$ is finite, so is $\int_0^\infty f(x)\,dx = \int_0^1 f(x)\,dx + \int_1^\infty f(x)\,dx$.

11. False. If $f(x) = 1/x$, then f is continuous and decreasing on $[1, \infty)$ with $\lim\limits_{x\to\infty} f(x) = 0$, but $\int_1^\infty f(x)\,dx$ is divergent.

12. True. $\int_a^\infty [f(x) + g(x)]\,dx = \lim\limits_{t\to\infty} \int_a^t [f(x) + g(x)]\,dx = \lim\limits_{t\to\infty} \left(\int_a^t f(x)\,dx + \int_a^t g(x)\,dx\right)$

$$= \lim_{t\to\infty} \int_a^t f(x)\,dx + \lim_{t\to\infty} \int_a^t g(x)\,dx \quad \left[\begin{array}{l}\text{since both limits}\\ \text{in the sum exist}\end{array}\right]$$

$$= \int_a^\infty f(x)\,dx + \int_a^\infty g(x)\,dx$$

Since the two integrals are finite, so is their sum.

13. False. Take $f(x) = 1$ for all x and $g(x) = -1$ for all x. Then $\int_a^\infty f(x)\,dx = \infty$ [divergent] and $\int_a^\infty g(x)\,dx = -\infty$ [divergent], but $\int_a^\infty [f(x) + g(x)]\,dx = 0$ [convergent].

14. False. $\int_0^\infty f(x)\,dx$ could converge or diverge. For example, if $g(x) = 1$, then $\int_0^\infty f(x)\,dx$ diverges if $f(x) = 1$ and converges if $f(x) = 0$.

EXERCISES

1. $$\int_1^2 \frac{(x+1)^2}{x}\,dx = \int_1^2 \frac{x^2+2x+1}{x}\,dx = \int_1^2 \left(x + 2 + \frac{1}{x}\right)dx = \left[\frac{1}{2}x^2 + 2x + \ln|x|\right]_1^2$$
$$= (2 + 4 + \ln 2) - \left(\tfrac{1}{2} + 2 + 0\right) = \tfrac{7}{2} + \ln 2$$

2. $$\int_1^2 \frac{x}{(x+1)^2}\,dx = \int_2^3 \frac{u-1}{u^2}\,du \quad \left[\begin{array}{l}u = x+1,\\ du = dx\end{array}\right]$$
$$= \int_2^3 \left(\frac{1}{u} - \frac{1}{u^2}\right)du = \left[\ln|u| + \frac{1}{u}\right]_2^3 = \left(\ln 3 + \frac{1}{3}\right) - \left(\ln 2 + \frac{1}{2}\right) = \ln\frac{3}{2} - \frac{1}{6}$$

3. $$\int \frac{e^{\sin x}}{\sec x}\,dx = \int \cos x\, e^{\sin x}\,dx = \int e^u\,du \quad \left[\begin{array}{l}u = \sin x,\\ du = \cos x\,dx\end{array}\right]$$
$$= e^u + C = e^{\sin x} + C$$

4. $$\int_0^{\pi/6} t \sin 2t\,dt = \left[-\tfrac{1}{2}t\cos 2t\right]_0^{\pi/6} - \int_0^{\pi/6} \left(-\tfrac{1}{2}\cos 2t\right)dt \quad \left[\begin{array}{ll}u = t, & dv = \sin 2t\\ du = dt, & v = -\tfrac{1}{2}\cos 2t\end{array}\right]$$
$$= \left(-\tfrac{\pi}{12}\cdot\tfrac{1}{2}\right) - (0) + \left[\tfrac{1}{4}\sin 2t\right]_0^{\pi/6} = -\tfrac{\pi}{24} + \tfrac{1}{8}\sqrt{3}$$

5. $$\int \frac{dt}{2t^2+3t+1} = \int \frac{1}{(2t+1)(t+1)}\,dt = \int \left(\frac{2}{2t+1} - \frac{1}{t+1}\right)dt \quad [\text{partial fractions}] \quad = \ln|2t+1| - \ln|t+1| + C$$

6. $$\int_1^2 x^5 \ln x\,dx = \left[\tfrac{1}{6}x^6 \ln x\right]_1^2 - \int_1^2 \tfrac{1}{6}x^5\,dx \quad \left[\begin{array}{ll}u = \ln x, & dv = x^5\,dx\\ du = \frac{1}{x}\,dx, & v = \frac{1}{6}x^6\end{array}\right]$$
$$= \tfrac{64}{6}\ln 2 - 0 - \left[\tfrac{1}{36}x^6\right]_1^2 = \tfrac{32}{3}\ln 2 - \left(\tfrac{64}{36} - \tfrac{1}{36}\right) = \tfrac{32}{3}\ln 2 - \tfrac{7}{4}$$

7. $$\int_0^{\pi/2} \sin^3\theta\,\cos^2\theta\,d\theta = \int_0^{\pi/2} (1 - \cos^2\theta)\cos^2\theta\,\sin\theta\,d\theta = \int_1^0 (1 - u^2)u^2\,(-du) \quad \left[\begin{array}{l}u = \cos\theta,\\ du = -\sin\theta\,d\theta\end{array}\right]$$
$$= \int_0^1 (u^2 - u^4)\,du = \left[\tfrac{1}{3}u^3 - \tfrac{1}{5}u^5\right]_0^1 = \left(\tfrac{1}{3} - \tfrac{1}{5}\right) - 0 = \tfrac{2}{15}$$

8. Let $u = \sqrt{e^x - 1}$, so that $u^2 = e^x - 1$, $2u\,du = e^x\,dx$, and $e^x = u^2 + 1$. Then

$$\int \frac{1}{\sqrt{e^x - 1}}\,dx = \int \frac{1}{u}\frac{2u\,du}{u^2+1} = 2\int \frac{1}{u^2+1}\,du = 2\tan^{-1} u + C = 2\tan^{-1}\sqrt{e^x - 1} + C.$$

9. Let $u = \ln t$, $du = dt/t$. Then $\displaystyle\int \frac{\sin(\ln t)}{t}\,dt = \int \sin u\,du = -\cos u + C = -\cos(\ln t) + C$.

10. Let $u = \arctan x$, $du = dx/(1 + x^2)$. Then

$$\int_0^1 \frac{\sqrt{\arctan x}}{1+x^2}\,dx = \int_0^{\pi/4} \sqrt{u}\,du = \frac{2}{3}\Big[u^{3/2}\Big]_0^{\pi/4} = \frac{2}{3}\left[\frac{\pi^{3/2}}{4^{3/2}} - 0\right] = \frac{2}{3}\cdot\frac{1}{8}\pi^{3/2} = \frac{1}{12}\pi^{3/2}.$$

11. Let $x = \sec\theta$. Then

$$\int_1^2 \frac{\sqrt{x^2-1}}{x}\,dx = \int_0^{\pi/3} \frac{\tan\theta}{\sec\theta}\sec\theta\,\tan\theta\,d\theta = \int_0^{\pi/3}\tan^2\theta\,d\theta = \int_0^{\pi/3}(\sec^2\theta - 1)\,d\theta = \Big[\tan\theta - \theta\Big]_0^{\pi/3} = \sqrt{3} - \tfrac{\pi}{3}.$$

12. $$\int \frac{e^{2x}}{1+e^{4x}}\,dx = \int \frac{1}{1+u^2}\left(\tfrac{1}{2}\,du\right) \qquad \begin{bmatrix} u = e^{2x}, \\ du = 2e^{2x}\,dx \end{bmatrix}$$
$$= \tfrac{1}{2}\tan^{-1} u + C = \tfrac{1}{2}\tan^{-1} e^{2x} + C$$

13. Let $w = \sqrt[3]{x}$. Then $w^3 = x$ and $3w^2\,dw = dx$, so $\int e^{\sqrt[3]{x}}\,dx = \int e^w \cdot 3w^2\,dw = 3I$. To evaluate I, let $u = w^2$, $dv = e^w\,dw \Rightarrow du = 2w\,dw$, $v = e^w$, so $I = \int w^2 e^w\,dw = w^2 e^w - \int 2we^w\,dw$. Now let $U = w$, $dV = e^w\,dw \Rightarrow dU = dw$, $V = e^w$. Thus, $I = w^2e^w - 2\big[we^w - \int e^w\,dw\big] = w^2e^w - 2we^w + 2e^w + C_1$, and hence

$3I = 3e^w(w^2 - 2w + 2) + C = 3e^{\sqrt[3]{x}}(x^{2/3} - 2x^{1/3} + 2) + C$.

14. $$\int \frac{x^2+2}{x+2}\,dx = \int\left(x - 2 + \frac{6}{x+2}\right)dx = \tfrac{1}{2}x^2 - 2x + 6\ln|x+2| + C$$

15. $\dfrac{x-1}{x^2+2x} = \dfrac{x-1}{x(x+2)} = \dfrac{A}{x} + \dfrac{B}{x+2} \Rightarrow x - 1 = A(x+2) + Bx$. Set $x = -2$ to get $-3 = -2B$, so $B = \frac{3}{2}$. Set $x = 0$ to get $-1 = 2A$, so $A = -\frac{1}{2}$. Thus, $\displaystyle\int \frac{x-1}{x^2+2x}\,dx = \int\left(\frac{-\frac{1}{2}}{x} + \frac{\frac{3}{2}}{x+2}\right)dx = -\tfrac{1}{2}\ln|x| + \tfrac{3}{2}\ln|x+2| + C$.

16. $$\int \frac{\sec^6\theta}{\tan^2\theta}\,d\theta = \int \frac{(\tan^2\theta+1)^2\sec^2\theta}{\tan^2\theta}\,d\theta \quad \begin{bmatrix} u = \tan\theta, \\ du = -\sec^2\theta\,d\theta \end{bmatrix} \quad = \int \frac{(u^2+1)^2}{u^2}\,du = \int \frac{u^4+2u^2+1}{u^2}\,du$$
$$= \int\left(u^2 + 2 + \frac{1}{u^2}\right)du = \frac{u^3}{3} + 2u - \frac{1}{u} + C = \tfrac{1}{3}\tan^3\theta + 2\tan\theta - \cot\theta + C$$

17. $$\int x\cosh x\,dx = x\sinh x - \int \sinh x\,dx \qquad \begin{bmatrix} u = x, & dv = \cosh x\,dx \\ du = dx, & v = \sinh x \end{bmatrix}$$
$$= x\sinh x - \cosh x + C$$

18. $\dfrac{x^2+8x-3}{x^3+3x^2} = \dfrac{x^2+8x-3}{x^2(x+3)} = \dfrac{A}{x} + \dfrac{B}{x^2} + \dfrac{C}{x+3} \Rightarrow x^2 + 8x - 3 = Ax(x+3) + B(x+3) + Cx^2$.

Taking $x = 0$, we get $-3 = 3B$, so $B = -1$. Taking $x = -3$, we get $-18 = 9C$, so $C = -2$.

Taking $x = 1$, we get $6 = 4A + 4B + C = 4A - 4 - 2$, so $4A = 12$ and $A = 3$. Now

$$\int \frac{x^2 + 8x - 3}{x^3 + 3x^2}\,dx = \int \left(\frac{3}{x} - \frac{1}{x^2} - \frac{2}{x+3}\right) dx = 3\ln|x| + \frac{1}{x} - 2\ln|x+3| + C.$$

19. $$\int \frac{x+1}{9x^2 + 6x + 5}\,dx = \int \frac{x+1}{(9x^2 + 6x + 1) + 4}\,dx = \int \frac{x+1}{(3x+1)^2 + 4}\,dx \qquad \begin{bmatrix} u = 3x + 1, \\ du = 3\,dx \end{bmatrix}$$

$$= \int \frac{\left[\frac{1}{3}(u-1)\right] + 1}{u^2 + 4}\left(\frac{1}{3}\,du\right) = \frac{1}{3}\cdot\frac{1}{3}\int \frac{(u-1)+3}{u^2+4}\,du$$

$$= \frac{1}{9}\int \frac{u}{u^2+4}\,du + \frac{1}{9}\int \frac{2}{u^2 + 2^2}\,du = \frac{1}{9}\cdot\frac{1}{2}\ln(u^2+4) + \frac{2}{9}\cdot\frac{1}{2}\tan^{-1}\left(\frac{1}{2}u\right) + C$$

$$= \tfrac{1}{18}\ln(9x^2 + 6x + 5) + \tfrac{1}{9}\tan^{-1}\left[\tfrac{1}{2}(3x+1)\right] + C$$

20. $$\int \tan^5\theta\,\sec^3\theta\,d\theta = \int \tan^4\theta\,\sec^2\theta\,\sec\theta\,\tan\theta\,d\theta = \int (\sec^2\theta - 1)^2\sec^2\theta\,\sec\theta\,\tan\theta\,d\theta \qquad \begin{bmatrix} u = \sec\theta, \\ du = \sec\theta\,\tan\theta\,d\theta \end{bmatrix}$$

$$= \int (u^2 - 1)^2 u^2\,du = \int (u^6 - 2u^4 + u^2)\,du$$

$$= \tfrac{1}{7}u^7 - \tfrac{2}{5}u^5 + \tfrac{1}{3}u^3 + C = \tfrac{1}{7}\sec^7\theta - \tfrac{2}{5}\sec^5\theta + \tfrac{1}{3}\sec^3\theta + C$$

21. $$\int \frac{dx}{\sqrt{x^2 - 4x}} = \int \frac{dx}{\sqrt{(x^2 - 4x + 4) - 4}} = \int \frac{dx}{\sqrt{(x-2)^2 - 2^2}}$$

$$= \int \frac{2\sec\theta\,\tan\theta\,d\theta}{2\tan\theta} \qquad \begin{bmatrix} x - 2 = 2\sec\theta, \\ dx = 2\sec\theta\,\tan\theta\,d\theta \end{bmatrix}$$

$$= \int \sec\theta\,d\theta = \ln|\sec\theta + \tan\theta| + C_1$$

$$= \ln\left|\frac{x-2}{2} + \frac{\sqrt{x^2 - 4x}}{2}\right| + C_1$$

$$= \ln\left|x - 2 + \sqrt{x^2 - 4x}\right| + C, \text{ where } C = C_1 - \ln 2$$

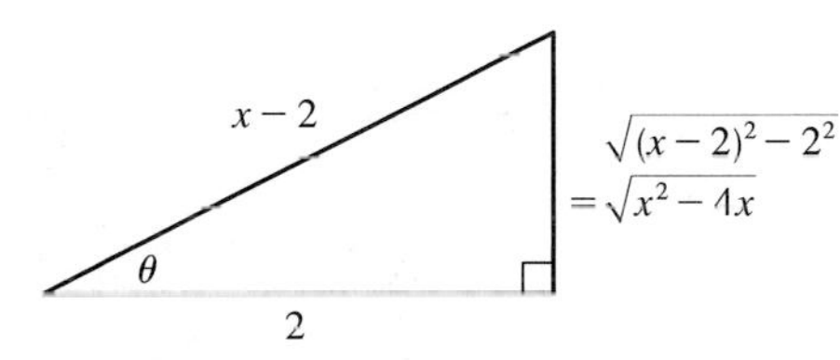

22. $$\int \cos\sqrt{t}\,dt = \int 2x\,\cos x\,dx \qquad \begin{bmatrix} x = \sqrt{t}, \\ x^2 = t, \quad 2x\,dx = dt \end{bmatrix}$$

$$= 2x\,\sin x - \int 2\sin x\,dx \qquad \begin{bmatrix} u = x, & dv = \cos x\,dx \\ du = dx, & v = \sin x \end{bmatrix}$$

$$= 2x\,\sin x + 2\cos x + C = 2\sqrt{t}\,\sin\sqrt{t} + 2\cos\sqrt{t} + C$$

23. Let $x = \tan\theta$, so that $dx = \sec^2\theta\,d\theta$. Then

$$\int \frac{dx}{x\sqrt{x^2+1}} = \int \frac{\sec^2\theta\,d\theta}{\tan\theta\,\sec\theta} = \int \frac{\sec\theta}{\tan\theta}\,d\theta$$

$$= \int \csc\theta\,d\theta = \ln|\csc\theta - \cot\theta| + C$$

$$= \ln\left|\frac{\sqrt{x^2+1}}{x} - \frac{1}{x}\right| + C = \ln\left|\frac{\sqrt{x^2+1} - 1}{x}\right| + C$$

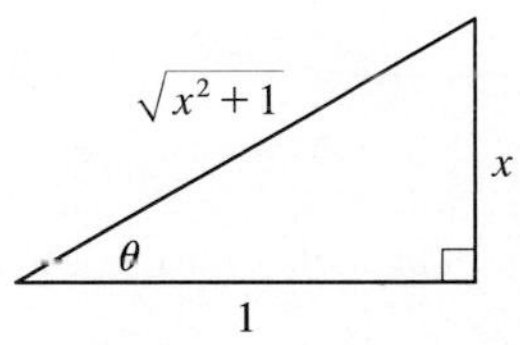

24. Let $u = \cos x$, $dv = e^x\,dx \Rightarrow du = -\sin x\,dx$, $v = e^x$: $(*)\ I = \int e^x \cos x\,dx = e^x \cos x + \int e^x \sin x\,dx$.

To integrate $\int e^x \sin x\,dx$, let $U = \sin x$, $dV = e^x\,dx \Rightarrow dU = \cos x\,dx$, $V = e^x$. Then

$\int e^x \sin x\,dx = e^x \sin x - \int e^x \cos x\,dx = e^x \sin x - I$. By substitution in $(*)$, $I = e^x \cos x + e^x \sin x - I \Rightarrow$

$2I = e^x(\cos x + \sin x) \Rightarrow I = \frac{1}{2}e^x(\cos x + \sin x) + C$.

25. $\dfrac{3x^3 - x^2 + 6x - 4}{(x^2+1)(x^2+2)} = \dfrac{Ax+B}{x^2+1} + \dfrac{Cx+D}{x^2+2} \Rightarrow 3x^3 - x^2 + 6x - 4 = (Ax+B)(x^2+2) + (Cx+D)(x^2+1)$.

Equating the coefficients gives $A + C = 3$, $B + D = -1$, $2A + C = 6$, and $2B + D = -4 \Rightarrow$

$A = 3$, $C = 0$, $B = -3$, and $D = 2$. Now

$$\int \frac{3x^3 - x^2 + 6x - 4}{(x^2+1)(x^2+2)}\,dx = 3\int \frac{x-1}{x^2+1}\,dx + 2\int \frac{dx}{x^2+2} = \frac{3}{2}\ln(x^2+1) - 3\tan^{-1}x + \sqrt{2}\tan^{-1}\left(\frac{x}{\sqrt{2}}\right) + C.$$

26. $\int x \sin x \cos x\,dx = \int \frac{1}{2}x \sin 2x\,dx \quad \begin{bmatrix} u = \frac{1}{2}x, & dv = \sin 2x\,dx, \\ du = \frac{1}{2}\,dx & v = -\frac{1}{2}\cos 2x \end{bmatrix}$

$$= -\tfrac{1}{4}x \cos 2x + \int \tfrac{1}{4}\cos 2x\,dx = -\tfrac{1}{4}x\cos 2x + \tfrac{1}{8}\sin 2x + C$$

27. $\int_0^{\pi/2} \cos^3 x \sin 2x\,dx = \int_0^{\pi/2} \cos^3 x\,(2\sin x \cos x)\,dx = \int_0^{\pi/2} 2\cos^4 x \sin x\,dx = \left[-\frac{2}{5}\cos^5 x\right]_0^{\pi/2} = \frac{2}{5}$

28. Let $u = \sqrt[3]{x}$. Then $x = u^3$, $dx = 3u^2du \Rightarrow$

$$\int \frac{\sqrt[3]{x}+1}{\sqrt[3]{x}-1}\,dx = \int \frac{u+1}{u-1}3u^2du = 3\int \left(u^2 + 2u + 2 + \frac{2}{u-1}\right)du$$

$$= u^3 + 3u^2 + 6u + 6\ln|u-1| + C = x + 3x^{2/3} + 6\sqrt[3]{x} + 6\ln|\sqrt[3]{x} - 1| + C$$

29. The integrand is an odd function, so $\displaystyle\int_{-3}^{3} \frac{x}{1+|x|}\,dx = 0$ [by 5.5.7(b)].

30. Let $u = e^{-x}$, $du = -e^{-x}\,dx$. Then

$$\int \frac{dx}{e^x\sqrt{1-e^{-2x}}} = \int \frac{e^{-x}\,dx}{\sqrt{1-(e^{-x})^2}} = \int \frac{-du}{\sqrt{1-u^2}} = -\sin^{-1}u + C = -\sin^{-1}(e^{-x}) + C.$$

31. Let $u = \sqrt{e^x - 1}$. Then $u^2 = e^x - 1$ and $2u\,du = e^x\,dx$. Also, $e^x + 8 = u^2 + 9$. Thus,

$$\int_0^{\ln 10} \frac{e^x\sqrt{e^x-1}}{e^x+8}\,dx = \int_0^3 \frac{u \cdot 2u\,du}{u^2+9} = 2\int_0^3 \frac{u^2}{u^2+9}\,du = 2\int_0^3 \left(1 - \frac{9}{u^2+9}\right)du$$

$$= 2\left[u - \frac{9}{3}\tan^{-1}\left(\frac{u}{3}\right)\right]_0^3 = 2\left[(3 - 3\tan^{-1}1) - 0\right] = 2\left(3 - 3\cdot\frac{\pi}{4}\right) = 6 - \frac{3\pi}{2}$$

32. $\displaystyle\int_0^{\pi/4} \frac{x\sin x}{\cos^3 x}\,dx = \int_0^{\pi/4} x\tan x \sec^2 x\,dx \quad \begin{bmatrix} u = x, & dv = \tan x \sec^2 x\,dx, \\ du = dx & v = \frac{1}{2}\tan^2 x \end{bmatrix}$

$$= \left[\frac{x}{2}\tan^2 x\right]_0^{\pi/4} - \frac{1}{2}\int_0^{\pi/4}\tan^2 x\,dx = \frac{\pi}{8}\cdot 1^2 - 0 - \frac{1}{2}\int_0^{\pi/4}(\sec^2 x - 1)\,dx$$

$$= \frac{\pi}{8} - \frac{1}{2}\Big[\tan x - x\Big]_0^{\pi/4} = \frac{\pi}{8} - \frac{1}{2}\left(1 - \frac{\pi}{4}\right) = \frac{\pi}{4} - \frac{1}{2}$$

33. Let $x = 2\sin\theta \;\Rightarrow\; (4 - x^2)^{3/2} = (2\cos\theta)^3$, $dx = 2\cos\theta\, d\theta$, so

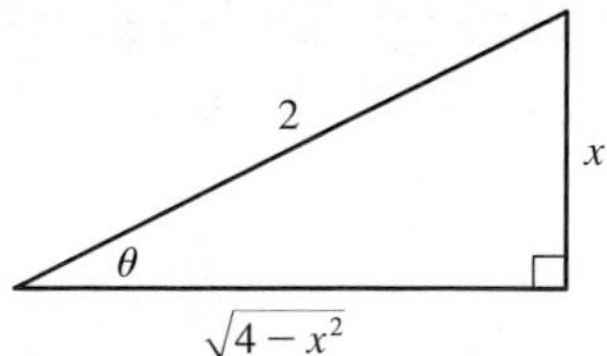

$$\int \frac{x^2}{(4-x^2)^{3/2}}\,dx = \int \frac{4\sin^2\theta}{8\cos^3\theta}\,2\cos\theta\,d\theta = \int \tan^2\theta\,d\theta = \int (\sec^2\theta - 1)\,d\theta$$

$$= \tan\theta - \theta + C = \frac{x}{\sqrt{4-x^2}} - \sin^{-1}\!\left(\frac{x}{2}\right) + C$$

34. Integrate by parts twice, first with $u = (\arcsin x)^2$, $dv = dx$:

$$I = \int (\arcsin x)^2\,dx = x(\arcsin x)^2 - \int 2x\arcsin x\left(\frac{dx}{\sqrt{1-x^2}}\right)$$

Now let $U = \arcsin x$, $dV = \dfrac{x}{\sqrt{1-x^2}}\,dx \;\Rightarrow\; dU = \dfrac{1}{\sqrt{1-x^2}}\,dx$, $V = -\sqrt{1-x^2}$. So

$$I = x(\arcsin x)^2 - 2\left[\arcsin x\left(-\sqrt{1-x^2}\right) + \textstyle\int dx\right] = x(\arcsin x)^2 + 2\sqrt{1-x^2}\arcsin x - 2x + C$$

35. $$\int \frac{1}{\sqrt{x + x^{3/2}}}\,dx = \int \frac{dx}{\sqrt{x\,(1+\sqrt{x}\,)}} = \int \frac{dx}{\sqrt{x}\sqrt{1+\sqrt{x}}} \quad \begin{bmatrix} u = 1+\sqrt{x}, \\ du = \dfrac{dx}{2\sqrt{x}} \end{bmatrix} \quad = \int \frac{2\,du}{\sqrt{u}} = \int 2u^{-1/2}\,du$$

$$= 4\sqrt{u} + C = 4\sqrt{1+\sqrt{x}} + C$$

36. $$\int \frac{1-\tan\theta}{1+\tan\theta}\,d\theta = \int \frac{\dfrac{\cos\theta}{\cos\theta} - \dfrac{\sin\theta}{\cos\theta}}{\dfrac{\cos\theta}{\cos\theta} + \dfrac{\sin\theta}{\cos\theta}}\,d\theta = \int \frac{\cos\theta - \sin\theta}{\cos\theta + \sin\theta}\,d\theta = \ln|\cos\theta + \sin\theta| + C$$

37. $$\textstyle\int (\cos x + \sin x)^2\cos 2x\,dx = \int \left(\cos^2 x + 2\sin x\cos x + \sin^2 x\right)\cos 2x\,dx = \int (1 + \sin 2x)\cos 2x\,dx$$

$$\textstyle = \int \cos 2x\,dx + \frac{1}{2}\int \sin 4x\,dx = \frac{1}{2}\sin 2x - \frac{1}{8}\cos 4x + C$$

Or: $$\textstyle\int (\cos x + \sin x)^2\cos 2x\,dx = \int (\cos x + \sin x)^2(\cos^2 x - \sin^2 x)\,dx$$

$$\textstyle = \int (\cos x + \sin x)^3(\cos x - \sin x)\,dx = \frac{1}{4}(\cos x + \sin x)^4 + C_1$$

38. $$\int \frac{2^{\sqrt{x}}}{\sqrt{x}}\,dx = \textstyle\int 2^u\,(2\,du) \qquad \begin{bmatrix} u = \sqrt{x}, \\ du = 1/(2\sqrt{x}\,)\,dx \end{bmatrix}$$

$$= 2\cdot\frac{2^u}{\ln 2} + C = \frac{2^{\sqrt{x}+1}}{\ln 2} + C$$

39. We'll integrate $I = \displaystyle\int \frac{xe^{2x}}{(1+2x)^2}\,dx$ by parts with $u = xe^{2x}$ and $dv = \dfrac{dx}{(1+2x)^2}$. Then $du = (x\cdot 2e^{2x} + e^{2x}\cdot 1)\,dx$

and $v = -\dfrac{1}{2}\cdot\dfrac{1}{1+2x}$, so

$$I = -\frac{1}{2}\cdot\frac{xe^{2x}}{1+2x} - \int\left[-\frac{1}{2}\cdot\frac{e^{2x}(2x+1)}{1+2x}\right]dx = -\frac{xe^{2x}}{4x+2} + \frac{1}{2}\cdot\frac{1}{2}e^{2x} + C = e^{2x}\left(\frac{1}{4} - \frac{x}{4x+2}\right) + C$$

Thus, $$\int_0^{1/2} \frac{xe^{2x}}{(1+2x)^2}\,dx = \left[e^{2x}\left(\frac{1}{4} - \frac{x}{4x+2}\right)\right]_0^{1/2} = e\left(\frac{1}{4} - \frac{1}{8}\right) - 1\left(\frac{1}{4} - 0\right) = \frac{1}{8}e - \frac{1}{4}.$$

40. $\displaystyle\int_{\pi/4}^{\pi/3} \frac{\sqrt{\tan\theta}}{\sin 2\theta}\,d\theta = \int_{\pi/4}^{\pi/3} \frac{\sqrt{\frac{\sin\theta}{\cos\theta}}}{2\sin\theta\cos\theta}\,d\theta = \int_{\pi/4}^{\pi/3} \frac{1}{2}(\sin\theta)^{-1/2}(\cos\theta)^{-3/2}\,d\theta = \int_{\pi/4}^{\pi/3} \frac{1}{2}\left(\frac{\sin\theta}{\cos\theta}\right)^{-1/2}(\cos\theta)^{-2}\,d\theta$

$$= \int_{\pi/4}^{\pi/3} \frac{1}{2}(\tan\theta)^{-1/2}\sec^2\theta\,d\theta = \Big[\sqrt{\tan\theta}\Big]_{\pi/4}^{\pi/3} = \sqrt{\sqrt{3}} - \sqrt{1} = \sqrt[4]{3} - 1$$

41. $\displaystyle\int_1^\infty \frac{1}{(2x+1)^3}\,dx = \lim_{t\to\infty}\int_1^t \frac{1}{(2x+1)^3}\,dx = \lim_{t\to\infty}\int_1^t \tfrac{1}{2}(2x+1)^{-3}\,2\,dx = \lim_{t\to\infty}\left[-\frac{1}{4(2x+1)^2}\right]_1^t$

$$= -\frac{1}{4}\lim_{t\to\infty}\left[\frac{1}{(2t+1)^2} - \frac{1}{9}\right] = -\frac{1}{4}\left(0 - \frac{1}{9}\right) = \frac{1}{36}$$

42. $\displaystyle\int_1^\infty \frac{\ln x}{x^4}\,dx = \lim_{t\to\infty}\int_1^t \frac{\ln x}{x^4}\,dx \qquad \begin{bmatrix} u = \ln x, & dv = dx/x^4, \\ du = dx/x & v = -1/(3x^3) \end{bmatrix}$

$$= \lim_{t\to\infty}\left[-\frac{\ln x}{3x^3}\right]_1^t + \int_1^t \frac{1}{3x^4}\,dx = \lim_{t\to\infty}\left(-\frac{\ln t}{3t^3} + 0 + \left[\frac{-1}{9x^3}\right]_1^t\right) \overset{\text{H}}{=} \lim_{t\to\infty}\left(-\frac{1}{9t^3} + \left[\frac{-1}{9t^3} + \frac{1}{9}\right]\right)$$

$$= 0 + 0 + \tfrac{1}{9} = \tfrac{1}{9}$$

43. $\displaystyle\int \frac{dx}{x\ln x} \quad \begin{bmatrix} u = \ln x, \\ du = dx/x \end{bmatrix} \quad = \int \frac{du}{u} = \ln|u| + C = \ln|\ln x| + C$, so

$\displaystyle\int_2^\infty \frac{dx}{x\ln x} = \lim_{t\to\infty}\int_2^t \frac{dx}{x\ln x} = \lim_{t\to\infty}\Big[\ln|\ln x|\Big]_2^t = \lim_{t\to\infty}\left[\ln(\ln t) - \ln(\ln 2)\right] = \infty$, so the integral is divergent.

44. Let $u = \sqrt{y-2}$. Then $y = u^2 + 2$ and $dy = 2u\,du$, so

$$\int \frac{y\,dy}{\sqrt{y-2}} = \int \frac{(u^2+2)2u\,du}{u} = 2\int (u^2+2)\,du = 2\left[\tfrac{1}{3}u^3 + 2u\right] + C$$

Thus, $\displaystyle\int_2^6 \frac{y\,dy}{\sqrt{y-2}} = \lim_{t\to 2^+}\int_t^6 \frac{y\,dy}{\sqrt{y-2}} = \lim_{t\to 2^+}\left[\tfrac{2}{3}(y-2)^{3/2} + 4\sqrt{y-2}\right]_t^6$

$$= \lim_{t\to 2^+}\left[\tfrac{16}{3} + 8 - \tfrac{2}{3}(t-2)^{3/2} - 4\sqrt{t-2}\right] = \tfrac{40}{3}.$$

45. $\displaystyle\int_0^4 \frac{\ln x}{\sqrt{x}}\,dx = \lim_{t\to 0^+}\int_t^4 \frac{\ln x}{\sqrt{x}}\,dx \overset{\star}{=} \lim_{t\to 0^+}\left[2\sqrt{x}\ln x - 4\sqrt{x}\right]_t^4$

$$= \lim_{t\to 0^+}\left[(2\cdot 2\ln 4 - 4\cdot 2) - (2\sqrt{t}\ln t - 4\sqrt{t})\right] \overset{\star\star}{=} (4\ln 4 - 8) - (0 - 0) = 4\ln 4 - 8$$

($\star$) Let $u = \ln x$, $dv = \dfrac{1}{\sqrt{x}}\,dx \;\Rightarrow\; du = \dfrac{1}{x}\,dx$, $v = 2\sqrt{x}$. Then

$$\int \frac{\ln x}{\sqrt{x}}\,dx = 2\sqrt{x}\ln x - 2\int \frac{dx}{\sqrt{x}} = 2\sqrt{x}\ln x - 4\sqrt{x} + C$$

($\star\star$) $\displaystyle\lim_{t\to 0^+}(2\sqrt{t}\ln t) = \lim_{t\to 0^+}\frac{2\ln t}{t^{-1/2}} \overset{\text{H}}{=} \lim_{t\to 0^+}\frac{2/t}{-\frac{1}{2}t^{-3/2}} = \lim_{t\to 0^+}(-4\sqrt{t}) = 0$

46. Note that $f(x) = 1/(2 - 3x)$ has an infinite discontinuity at $x = \frac{2}{3}$. Now

$$\int_0^{2/3} \frac{1}{2-3x}\,dx = \lim_{t\to(2/3)^-} \int_0^t \frac{1}{2-3x}\,dx = \lim_{t\to(2/3)^-} \left[-\tfrac{1}{3}\ln|2-3x|\right]_0^t = -\tfrac{1}{3}\lim_{t\to(2/3)^-}\left[\ln|2-3t| - \ln 2\right] = \infty$$

Since $\int_0^{2/3} \frac{1}{2-3x}\,dx$ diverges, so does $\int_0^1 \frac{1}{2-3x}\,dx$.

47. $\int_0^1 \frac{x-1}{\sqrt{x}}\,dx = \lim_{t\to0^+}\int_t^1\left(\frac{x}{\sqrt{x}} - \frac{1}{\sqrt{x}}\right)dx = \lim_{t\to0^+}\int_t^1 \left(x^{1/2} - x^{-1/2}\right)dx = \lim_{t\to0^+}\left[\tfrac{2}{3}x^{3/2} - 2x^{1/2}\right]_t^1$

$$= \lim_{t\to0^+}\left[\left(\tfrac{2}{3} - 2\right) - \left(\tfrac{2}{3}t^{3/2} - 2t^{1/2}\right)\right] = -\tfrac{4}{3} - 0 = -\tfrac{4}{3}$$

48. $I = \int_{-1}^1 \frac{dx}{x^2-2x} = \int_{-1}^1 \frac{dx}{x(x-2)} = \int_{-1}^0 \frac{dx}{x(x-2)} + \int_0^1 \frac{dx}{x(x-2)} = I_1 + I_2$. Now

$\frac{1}{x(x-2)} = \frac{A}{x} + \frac{B}{x-2} \quad\Rightarrow\quad 1 = A(x-2) + Bx$. Set $x = 2$ to get $1 = 2B$, so $B = \frac{1}{2}$. Set $x = 0$ to get $1 = -2A$,

$A = -\frac{1}{2}$. Thus,

$$I_2 = \lim_{t\to0^+}\int_t^1\left(\frac{-\frac{1}{2}}{x} + \frac{\frac{1}{2}}{x-2}\right)dx = \lim_{t\to0^+}\left[-\tfrac{1}{2}\ln|x| + \tfrac{1}{2}\ln|x-2|\right]_t^1 = \lim_{t\to0^+}\left[(0+0) - \left(-\tfrac{1}{2}\ln t + \tfrac{1}{2}\ln|t-2|\right)\right]$$

$$= -\tfrac{1}{2}\ln 2 + \tfrac{1}{2}\lim_{t\to0^+}\ln t = -\infty.$$

Since I_2 diverges, I is divergent.

49. Let $u = 2x + 1$. Then

$$\int_{-\infty}^{\infty}\frac{dx}{4x^2+4x+5} = \int_{-\infty}^{\infty}\frac{\frac{1}{2}\,du}{u^2+4} = \tfrac{1}{2}\int_{-\infty}^0 \frac{du}{u^2+4} + \tfrac{1}{2}\int_0^\infty \frac{du}{u^2+4}$$

$$= \tfrac{1}{2}\lim_{t\to-\infty}\left[\tfrac{1}{2}\tan^{-1}\left(\tfrac{1}{2}u\right)\right]_t^0 + \tfrac{1}{2}\lim_{t\to\infty}\left[\tfrac{1}{2}\tan^{-1}\left(\tfrac{1}{2}u\right)\right]_0^t = \tfrac{1}{4}\left[0 - \left(-\tfrac{\pi}{2}\right)\right] + \tfrac{1}{4}\left[\tfrac{\pi}{2} - 0\right] = \tfrac{\pi}{4}.$$

50. $\int_1^\infty \frac{\tan^{-1}x}{x^2}\,dx = \lim_{t\to\infty}\int_1^t \frac{\tan^{-1}x}{x^2}\,dx$. Integrate by parts:

$$\int\frac{\tan^{-1}x}{x^2}\,dx = \frac{-\tan^{-1}x}{x} + \int\frac{1}{x}\frac{dx}{1+x^2} = \frac{-\tan^{-1}x}{x} + \int\left[\frac{1}{x} - \frac{x}{x^2+1}\right]dx$$

$$= \frac{-\tan^{-1}x}{x} + \ln|x| - \tfrac{1}{2}\ln(x^2+1) + C = \frac{-\tan^{-1}x}{x} + \frac{1}{2}\ln\frac{x^2}{x^2+1} + C$$

Thus,

$$\int_1^\infty \frac{\tan^{-1}x}{x^2}\,dx = \lim_{t\to\infty}\left[-\frac{\tan^{-1}x}{x} + \frac{1}{2}\ln\frac{x^2}{x^2+1}\right]_1^t = \lim_{t\to\infty}\left[-\frac{\tan^{-1}t}{t} + \frac{1}{2}\ln\frac{t^2}{t^2+1} + \frac{\pi}{4} - \frac{1}{2}\ln\frac{1}{2}\right]$$

$$= 0 + \tfrac{1}{2}\ln 1 + \tfrac{\pi}{4} + \tfrac{1}{2}\ln 2 = \tfrac{\pi}{4} + \tfrac{1}{2}\ln 2$$

51. We first make the substitution $t = x + 1$, so $\ln(x^2 + 2x + 2) = \ln\left[(x+1)^2 + 1\right] = \ln(t^2 + 1)$. Then we use parts with $u = \ln(t^2 + 1)$, $dv = dt$:

$$\begin{aligned}\int \ln(t^2+1)\,dt &= t\,\ln(t^2+1) - \int \frac{t(2t)\,dt}{t^2+1} = t\,\ln(t^2+1) - 2\int \frac{t^2\,dt}{t^2+1} = t\,\ln(t^2+1) - 2\int \left(1 - \frac{1}{t^2+1}\right) dt \\ &= t\,\ln(t^2+1) - 2t + 2\arctan t + C \\ &= (x+1)\,\ln(x^2+2x+2) - 2x + 2\arctan(x+1) + K, \text{ where } K = C - 2\end{aligned}$$

[Alternatively, we could have integrated by parts immediately with $u = \ln(x^2 + 2x + 2)$.] Notice from the graph that $f = 0$ where F has a horizontal tangent. Also, F is always increasing, and $f \geq 0$.

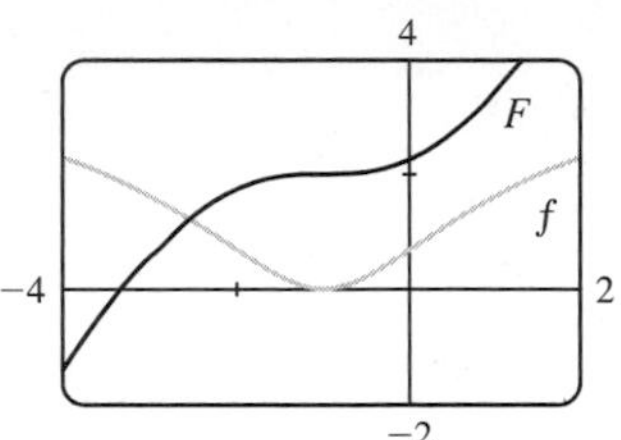

52. Let $u = x^2 + 1$. Then $x^2 = u - 1$ and $x\,dx = \frac{1}{2}\,du$, so

$$\begin{aligned}\int \frac{x^3}{\sqrt{x^2+1}}\,dx &= \int \frac{(u-1)}{\sqrt{u}}\left(\tfrac{1}{2}\,du\right) = \frac{1}{2}\int \left(u^{1/2} - u^{-1/2}\right) du \\ &= \tfrac{1}{2}\left(\tfrac{2}{3}u^{3/2} - 2u^{1/2}\right) + C = \tfrac{1}{3}(x^2+1)^{3/2} - (x^2+1)^{1/2} + C \\ &= \tfrac{1}{3}(x^2+1)^{1/2}\left[(x^2+1) - 3\right] + C = \tfrac{1}{3}\sqrt{x^2+1}\,(x^2-2) + C\end{aligned}$$

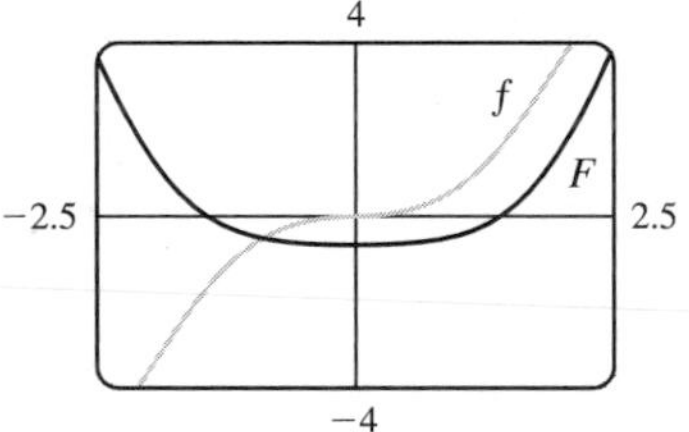

53. From the graph, it seems as though $\int_0^{2\pi} \cos^2 x\,\sin^3 x\,dx$ is equal to 0. To evaluate the integral, we write the integral as $I = \int_0^{2\pi} \cos^2 x\,(1 - \cos^2 x)\,\sin x\,dx$ and let $u = \cos x \Rightarrow du = -\sin x\,dx$. Thus, $I = \int_1^1 u^2(1 - u^2)(-du) = 0$.

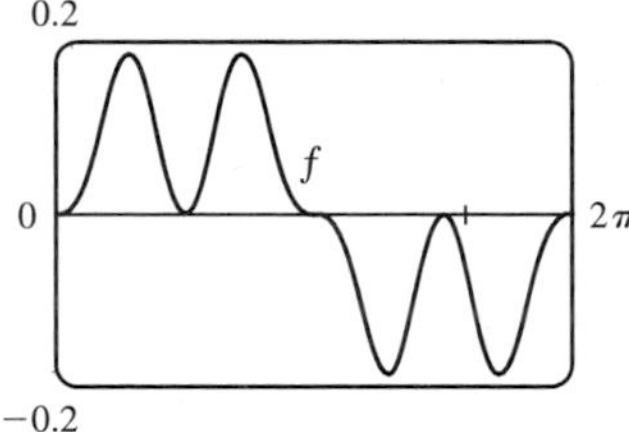

54. (a) To evaluate $\int x^5 e^{-2x}\,dx$ by hand, we would integrate by parts repeatedly, always taking $dv = e^{-2x}$ and starting with $u = x^5$. Each time we would reduce the degree of the x-factor by 1.

(b) To evaluate the integral using tables, we would use Formula 97 (which is proved using integration by parts) until the exponent of x was reduced to 1, and then we would use Formula 96.

(c) $\int x^5 e^{-2x}\,dx = -\frac{1}{8}e^{-2x}\left(4x^5 + 10x^4 + 20x^3 + 30x^2 + 30x + 15\right) + C$

(d)

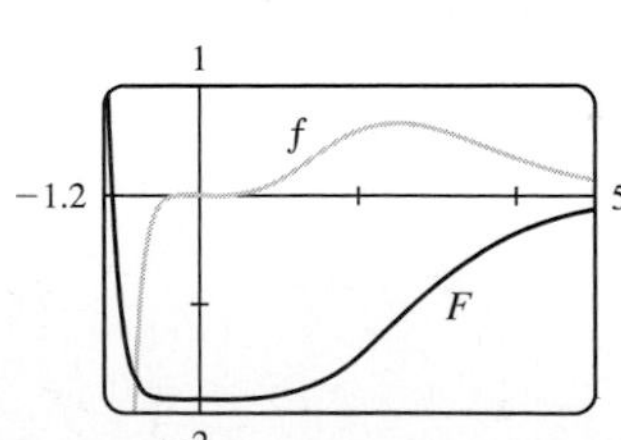

55. $\int \sqrt{4x^2 - 4x - 3}\,dx = \int \sqrt{(2x-1)^2 - 4}\,dx \quad \left[\begin{array}{l} u = 2x - 1, \\ du = 2\,dx \end{array}\right] \quad = \int \sqrt{u^2 - 2^2}\left(\tfrac{1}{2}\,du\right)$

$$\begin{aligned}&\overset{39}{=} \frac{1}{2}\left(\frac{u}{2}\sqrt{u^2-2^2} - \frac{2^2}{2}\ln\left|u + \sqrt{u^2-2^2}\right|\right) + C = \tfrac{1}{4}u\sqrt{u^2-4} - \ln\left|u + \sqrt{u^2-4}\right| + C \\ &= \tfrac{1}{4}(2x-1)\sqrt{4x^2-4x-3} - \ln\left|2x - 1 + \sqrt{4x^2-4x-3}\right| + C\end{aligned}$$

56. $\int \csc^5 t\,dt \overset{78}{=} -\frac{1}{4}\cot t\,\csc^3 t + \frac{3}{4}\int \csc^3 t\,dt \overset{72}{=} -\frac{1}{4}\cot t\,\csc^3 t + \frac{3}{4}\left[-\frac{1}{2}\csc t\,\cot t + \frac{1}{2}\ln|\csc t - \cot t|\right] + C$

$= -\frac{1}{4}\cot t\,\csc^3 t - \frac{3}{8}\csc t\,\cot t + \frac{3}{8}\ln|\csc t - \cot t| + C$

57. Let $u = \sin x$, so that $du = \cos x\,dx$. Then

$$\int \cos x\,\sqrt{4+\sin^2 x}\,dx = \int \sqrt{2^2+u^2}\,du \overset{21}{=} \frac{u}{2}\sqrt{2^2+u^2} + \frac{2^2}{2}\ln\left(u+\sqrt{2^2+u^2}\right) + C$$

$$= \frac{1}{2}\sin x\,\sqrt{4+\sin^2 x} + 2\ln\left(\sin x + \sqrt{4+\sin^2 x}\right) + C$$

58. Let $u = \sin x$. Then $du = \cos x\,dx$, so

$$\int \frac{\cot x\,dx}{\sqrt{1+2\sin x}} = \int \frac{du}{u\sqrt{1+2u}} \overset{\substack{57\text{ with}\\a=1,\,b=2}}{=} \ln\left|\frac{\sqrt{1+2u}-1}{\sqrt{1+2u}+1}\right| + C = \ln\left|\frac{\sqrt{1+2\sin x}-1}{\sqrt{1+2\sin x}+1}\right| + C$$

59. (a) $\frac{d}{du}\left[-\frac{1}{u}\sqrt{a^2-u^2} - \sin^{-1}\left(\frac{u}{a}\right) + C\right] = \frac{1}{u^2}\sqrt{a^2-u^2} + \frac{1}{\sqrt{a^2-u^2}} - \frac{1}{\sqrt{1-u^2/a^2}}\cdot\frac{1}{a}$

$$= \left(a^2-u^2\right)^{-1/2}\left[\frac{1}{u^2}\left(a^2-u^2\right) + 1 - 1\right] = \frac{\sqrt{a^2-u^2}}{u^2}$$

(b) Let $u = a\sin\theta \quad\Rightarrow\quad du = a\cos\theta\,d\theta$, $a^2-u^2 = a^2\left(1-\sin^2\theta\right) = a^2\cos^2\theta$.

$$\int \frac{\sqrt{a^2-u^2}}{u^2}\,du = \int \frac{a^2\cos^2\theta}{a^2\sin^2\theta}\,d\theta = \int \frac{1-\sin^2\theta}{\sin^2\theta}\,d\theta = \int (\csc^2\theta - 1)\,d\theta = -\cot\theta - \theta + C$$

$$= -\frac{\sqrt{a^2-u^2}}{u} - \sin^{-1}\left(\frac{u}{a}\right) + C$$

60. Work backward, and use integration by parts with $U = u^{-(n-1)}$ and $dV = (a+bu)^{-1/2}\,du \quad\Rightarrow$

$dU = \dfrac{-(n-1)\,du}{u^n}$ and $V = \dfrac{2}{b}\sqrt{a+bu}$, to get

$$\int \frac{du}{u^{n-1}\sqrt{a+bu}} = \int U\,dV = UV - \int V\,dU = \frac{2\sqrt{a+bu}}{bu^{n-1}} + \frac{2(n-1)}{b}\int \frac{\sqrt{a+bu}}{u^n}\,du$$

$$= \frac{2\sqrt{a+bu}}{bu^{n-1}} + \frac{2(n-1)}{b}\int \frac{a+bu}{u^n\sqrt{a+bu}}\,du$$

$$= \frac{2\sqrt{a+bu}}{bu^{n-1}} + 2(n-1)\int \frac{du}{u^{n-1}\sqrt{a+bu}} + \frac{2a(n-1)}{b}\int \frac{du}{u^n\sqrt{a+bu}}$$

Rearranging the equation gives $\dfrac{2a(n-1)}{b}\displaystyle\int \frac{du}{u^n\sqrt{a+bu}} = -\frac{2\sqrt{a+bu}}{bu^{n-1}} - (2n-3)\int \frac{du}{u^{n-1}\sqrt{a+bu}} \quad\Rightarrow$

$$\int \frac{du}{u^n\sqrt{a+bu}} = \frac{-\sqrt{a+bu}}{a(n-1)u^{n-1}} - \frac{b(2n-3)}{2a(n-1)}\int \frac{du}{u^{n-1}\sqrt{a+bu}}$$

61. For $n \geq 0$, $\int_0^\infty x^n\,dx = \lim\limits_{t\to\infty}\left[x^{n+1}/(n+1)\right]_0^t = \infty$. For $n < 0$, $\int_0^\infty x^n\,dx = \int_0^1 x^n\,dx + \int_1^\infty x^n\,dx$. Both integrals are improper. By (7.8.2), the second integral diverges if $-1 \leq n < 0$. By Exercise 7.8.57, the first integral diverges if $n \leq -1$. Thus, $\int_0^\infty x^n\,dx$ is divergent for all values of n.

62. $I = \int_0^\infty e^{ax}\cos x\,dx = \lim\limits_{t\to\infty}\int_0^t e^{ax}\cos x\,dx \overset{\substack{99 \text{ with} \\ b=1}}{=} \lim\limits_{t\to\infty}\left[\frac{e^{ax}}{a^2+1}(a\cos x+\sin x)\right]_0^t$

$= \lim\limits_{t\to\infty}\left[\frac{e^{at}}{a^2+1}(a\cos t+\sin t) - \frac{1}{a^2+1}(a)\right] = \frac{1}{a^2+1}\lim\limits_{t\to\infty}\left[e^{at}(a\cos t+\sin t) - a\right].$

For $a \geq 0$, the limit does not exist due to oscillation. For $a < 0$, $\lim\limits_{t\to\infty}\left[e^{at}(a\cos t+\sin t)\right] = 0$ by the Squeeze Theorem, because $\left|e^{at}(a\cos t+\sin t)\right| \leq e^{at}(|a|+1)$, so $I = \frac{1}{a^2+1}(-a) = -\frac{a}{a^2+1}$.

63. $f(x) = \frac{1}{\ln x}$, $\Delta x = \frac{b-a}{n} = \frac{4-2}{10} = \frac{1}{5}$

(a) $T_{10} = \frac{1}{5\cdot 2}\{f(2) + 2[f(2.2) + f(2.4) + \cdots + f(3.8)] + f(4)\} \approx 1.925444$

(b) $M_{10} = \frac{1}{5}[f(2.1) + f(2.3) + f(2.5) + \cdots + f(3.9)] \approx 1.920915$

(c) $S_{10} = \frac{1}{5\cdot 3}[f(2) + 4f(2.2) + 2f(2.4) + \cdots + 2f(3.6) + 4f(3.8) + f(4)] \approx 1.922470$

64. $f(x) = \sqrt{x}\cos x$, $\Delta x = \frac{b-a}{n} = \frac{4-1}{10} = \frac{3}{10}$

(a) $T_{10} = \frac{3}{10\cdot 2}\{f(1) + 2[f(1.3) + f(1.6) + \cdots + f(3.7)] + f(4)\} \approx -2.835151$

(b) $M_{10} = \frac{3}{10}[f(1.15) + f(1.45) + f(1.75) + \cdots + f(3.85)] \approx -2.856809$

(c) $S_{10} = \frac{3}{10\cdot 3}[f(1) + 4f(1.3) + 2f(1.6) + \cdots + 2f(3.4) + 4f(3.7) + f(4)] \approx -2.849672$

65. $f(x) = \frac{1}{\ln x} \;\Rightarrow\; f'(x) = -\frac{1}{x(\ln x)^2} \;\Rightarrow\; f''(x) = \frac{2+\ln x}{x^2(\ln x)^3} = \frac{2}{x^2(\ln x)^3} + \frac{1}{x^2(\ln x)^2}$. Note that each term of $f''(x)$ decreases on $[2,4]$, so we'll take $K = f''(2) \approx 2.022$. $|E_T| \leq \frac{K(b-a)^3}{12n^2} \approx \frac{2.022(4-2)^3}{12(10)^2} = 0.01348$ and $|E_M| \leq \frac{K(b-a)^3}{24n^2} = 0.00674$. $|E_T| \leq 0.00001 \;\Leftrightarrow\; \frac{2.022(8)}{12n^2} \leq \frac{1}{10^5} \;\Leftrightarrow\; n^2 \geq \frac{10^5(2.022)(8)}{12} \;\Rightarrow\; n \geq 367.2$. Take $n = 368$ for T_n. $|E_M| \leq 0.00001 \;\Leftrightarrow\; n^2 \geq \frac{10^5(2.022)(8)}{24} \;\Rightarrow\; n \geq 259.6$. Take $n = 260$ for M_n.

66. $\int_1^4 \frac{e^x}{x}\,dx \approx S_6 = \frac{(4-1)/6}{3}[f(1) + 4f(1.5) + 2f(2) + 4f(2.5) + 2f(3) + 4f(3.5) + f(4)] \approx 17.739438$

67. $\Delta t = \left(\frac{10}{60} - 0\right)/\,10 = \frac{1}{60}$.

Distance traveled $= \int_0^{10} v\,dt \approx S_{10}$

$= \frac{1}{60\cdot 3}[40 + 4(42) + 2(45) + 4(49) + 2(52) + 4(54) + 2(56) + 4(57) + 2(57) + 4(55) + 56]$

$= \frac{1}{180}(1544) = 8.5\overline{7}$ mi

68. We use Simpson's Rule with $n = 6$ and $\Delta t = \frac{24-0}{6} = 4$:

Increase in bee population $= \int_0^{24} r(t)\,dt \approx S_6$

$= \frac{4}{3}[r(0) + 4r(4) + 2r(8) + 4r(12) + 2r(16) + 4r(20) + r(24)]$

$= \frac{4}{3}[0 + 4(300) + 2(3000) + 4(11{,}000) + 2(4000) + 4(400) + 0]$

$= \frac{4}{3}(60{,}800) \approx 81{,}067$ bees

69. (a) $f(x) = \sin(\sin x)$. A CAS gives

$$f^{(4)}(x) = \sin(\sin x)[\cos^4 x + 7\cos^2 x - 3] + \cos(\sin x)\left[6\cos^2 x \sin x + \sin x\right]$$

From the graph, we see that $\left|f^{(4)}(x)\right| < 3.8$ for $x \in [0, \pi]$.

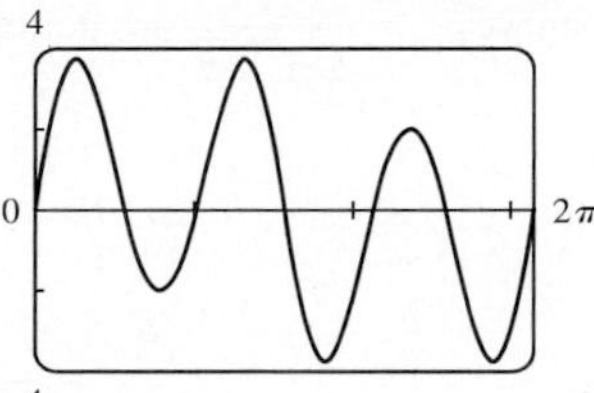

(b) We use Simpson's Rule with $f(x) = \sin(\sin x)$ and $\Delta x = \frac{\pi}{10}$:

$$\int_0^\pi f(x)\,dx \approx \frac{\pi}{10 \cdot 3}\left[f(0) + 4f\left(\frac{\pi}{10}\right) + 2f\left(\frac{2\pi}{10}\right) + \cdots + 4f\left(\frac{9\pi}{10}\right) + f(\pi)\right] \approx 1.786721$$

From part (a), we know that $\left|f^{(4)}(x)\right| < 3.8$ on $[0, \pi]$, so we use Theorem 7.7.4 with $K = 3.8$, and estimate the error as $|E_S| \le \dfrac{3.8(\pi - 0)^5}{180(10)^4} \approx 0.000646$.

(c) If we want the error to be less than 0.00001, we must have $|E_S| \le \dfrac{3.8\pi^5}{180n^4} \le 0.00001$, so $n^4 \ge \dfrac{3.8\pi^5}{180(0.00001)} \approx 646{,}041.6 \quad\Rightarrow\quad n \ge 28.35$. Since n must be even for Simpson's Rule, we must have $n \ge 30$ to ensure the desired accuracy.

70. With an x-axis in the normal position, at $x = 7$ we have $C = 2\pi r = 45 \quad\Rightarrow\quad r(7) = \frac{2\pi}{45}$.

Using Simpson's Rule with $n = 4$ and $\Delta x = 7$, we have

$$V = \int_0^{28} \pi[r(x)]^2\,dx \approx S_4 = \tfrac{7}{3}\left[0 + 4\pi\left(\tfrac{45}{2\pi}\right)^2 + 2\pi\left(\tfrac{53}{2\pi}\right)^2 + 4\pi\left(\tfrac{45}{2\pi}\right)^2 + 0\right] = \tfrac{7}{3}\left(\tfrac{21{,}818}{4\pi}\right) \approx 4051 \text{ cm}^3.$$

71. (a) $\dfrac{2 + \sin x}{\sqrt{x}} \ge \dfrac{1}{\sqrt{x}}$ for x in $[1, \infty)$. $\displaystyle\int_1^\infty \frac{1}{\sqrt{x}}\,dx$ is divergent by (7.8.2) with $p = \dfrac{1}{2} \le 1$. Therefore, $\displaystyle\int_1^\infty \frac{2 + \sin x}{\sqrt{x}}\,dx$ is divergent by the Comparison Theorem.

(b) $\dfrac{1}{\sqrt{1 + x^4}} < \dfrac{1}{\sqrt{x^4}} = \dfrac{1}{x^2}$ for x in $[1, \infty)$. $\displaystyle\int_1^\infty \frac{1}{x^2}\,dx$ is convergent by (7.8.2) with $p = 2 > 1$. Therefore, $\displaystyle\int_1^\infty \frac{1}{\sqrt{1 + x^4}}\,dx$ is convergent by the Comparison Theorem.

72. The line $y = 3$ intersects the hyperbola $y^2 - x^2 = 1$ at two points on its upper branch, namely $\left(-2\sqrt{2}, 3\right)$ and $\left(2\sqrt{2}, 3\right)$. The desired area is

$$A = \int_{-2\sqrt{2}}^{2\sqrt{2}} \left(3 - \sqrt{x^2 + 1}\right) dx = 2\int_0^{2\sqrt{2}} \left(3 - \sqrt{x^2 + 1}\right) dx \overset{21}{=} 2\left[3x - \tfrac{1}{2}x\sqrt{x^2 + 1} - \tfrac{1}{2}\ln\left(x + \sqrt{x^2 + 1}\right)\right]_0^{2\sqrt{2}}$$

$$= \left[6x - x\sqrt{x^2 + 1} - \ln\left(x + \sqrt{x^2 + 1}\right)\right]_0^{2\sqrt{2}} = 12\sqrt{2} - 2\sqrt{2} \cdot 3 - \ln\left(2\sqrt{2} + 3\right) = 6\sqrt{2} - \ln\left(3 + 2\sqrt{2}\right)$$

Another method: $A = 2\int_1^3 \sqrt{y^2 - 1}\,dy$ and use Formula 39.

73. For x in $\left[0, \frac{\pi}{2}\right]$, $0 \le \cos^2 x \le \cos x$. For x in $\left[\frac{\pi}{2}, \pi\right]$, $\cos x \le 0 \le \cos^2 x$. Thus,

$$\text{area} = \int_0^{\pi/2} (\cos x - \cos^2 x)\,dx + \int_{\pi/2}^{\pi} (\cos^2 x - \cos x)\,dx$$

$$= \left[\sin x - \tfrac{1}{2}x - \tfrac{1}{4}\sin 2x\right]_0^{\pi/2} + \left[\tfrac{1}{2}x + \tfrac{1}{4}\sin 2x - \sin x\right]_{\pi/2}^{\pi} = \left[\left(1 - \tfrac{\pi}{4}\right) - 0\right] + \left[\tfrac{\pi}{2} - \left(\tfrac{\pi}{4} - 1\right)\right] = 2$$

74. The curves $y = \dfrac{1}{2 \pm \sqrt{x}}$ are defined for $x \geq 0$. For $x > 0$, $\dfrac{1}{2-\sqrt{x}} > \dfrac{1}{2+\sqrt{x}}$. Thus, the required area is

$$\int_0^1 \left(\frac{1}{2-\sqrt{x}} - \frac{1}{2+\sqrt{x}}\right) dx = \int_0^1 \left(\frac{1}{2-u} - \frac{1}{2+u}\right) 2u\,du \quad \left[u = \sqrt{x}\right] \quad = 2\int_0^1 \left(-\frac{u}{u-2} - \frac{u}{u+2}\right) du$$

$$= 2\int_0^1 \left(-1 - \frac{2}{u-2} - 1 + \frac{2}{u+2}\right) du = 2\left[2\ln\left|\frac{u+2}{u-2}\right| - 2u\right]_0^1 = 4\ln 3 - 4.$$

75. Using the formula for disks, the volume is

$$\begin{aligned} V &= \textstyle\int_0^{\pi/2} \pi\,[f(x)]^2\,dx = \pi \int_0^{\pi/2} (\cos^2 x)^2\,dx = \pi \int_0^{\pi/2} \left[\frac{1}{2}(1+\cos 2x)\right]^2\,dx \\ &= \textstyle\frac{\pi}{4}\int_0^{\pi/2} (1 + \cos^2 2x + 2\cos 2x)\,dx = \frac{\pi}{4}\int_0^{\pi/2} \left[1 + \frac{1}{2}(1+\cos 4x) + 2\cos 2x\right] dx \\ &= \textstyle\frac{\pi}{4}\left[\frac{3}{2}x + \frac{1}{2}\left(\frac{1}{4}\sin 4x\right) + 2\left(\frac{1}{2}\sin 2x\right)\right]_0^{\pi/2} = \frac{\pi}{4}\left[\left(\frac{3\pi}{4} + \frac{1}{8}\cdot 0 + 0\right) - 0\right] = \frac{3}{16}\pi^2 \end{aligned}$$

76. Using the formula for cylindrical shells, the volume is

$$\begin{aligned} V &= \textstyle\int_0^{\pi/2} 2\pi x f(x)\,dx = 2\pi\int_0^{\pi/2} x\cos^2 x\,dx = 2\pi\int_0^{\pi/2} x\left[\frac{1}{2}(1+\cos 2x)\right] dx = 2\left(\frac{1}{2}\right)\pi\int_0^{\pi/2} (x + x\cos 2x)\,dx \\ &= \textstyle\pi\left(\left[\frac{1}{2}x^2\right]_0^{\pi/2} + \left[x\left(\frac{1}{2}\sin 2x\right)\right]_0^{\pi/2} - \int_0^{\pi/2} \frac{1}{2}\sin 2x\,dx\right) \quad \left[\begin{array}{l}\text{parts with } u = x, \\ dv = \cos 2x\,dx\end{array}\right] \\ &= \textstyle\pi\left[\frac{1}{2}\left(\frac{\pi}{2}\right)^2 + 0 - \frac{1}{2}\left[-\frac{1}{2}\cos 2x\right]_0^{\pi/2}\right] = \frac{\pi^3}{8} + \frac{\pi}{4}(-1-1) = \frac{1}{8}(\pi^3 - 4\pi) \end{aligned}$$

77. By the Fundamental Theorem of Calculus,

$$\textstyle\int_0^\infty f'(x)\,dx = \lim\limits_{t\to\infty}\int_0^t f'(x)\,dx = \lim\limits_{t\to\infty}[f(t) - f(0)] = \lim\limits_{t\to\infty} f(t) - f(0) = 0 - f(0) = -f(0).$$

78. (a) $(\tan^{-1} x)_{\text{ave}} = \displaystyle\lim_{t\to\infty}\frac{1}{t-0}\int_0^t \tan^{-1} x\,dx \overset{89}{=} \lim_{t\to\infty}\left\{\frac{1}{t}\left[x\tan^{-1} x - \tfrac{1}{2}\ln(1+x^2)\right]_0^t\right\}$

$$= \lim_{t\to\infty}\left[\frac{1}{t}\left(t\tan^{-1} t - \tfrac{1}{2}\ln(1+t^2)\right)\right] = \lim_{t\to\infty}\left[\tan^{-1} t - \frac{\ln(1+t^2)}{2t}\right]$$

$$\overset{\text{H}}{=} \frac{\pi}{2} - \lim_{t\to\infty}\frac{2t/(1+t^2)}{2} = \frac{\pi}{2} - 0 = \frac{\pi}{2}$$

(b) $f(x) \geq 0$ and $\int_a^\infty f(x)\,dx$ is divergent $\Rightarrow$ $\lim\limits_{t\to\infty}\int_a^t f(x)\,dx = \infty$.

$$f_{\text{ave}} = \lim_{t\to\infty}\frac{\int_a^t f(x)\,dx}{t-a}\,dx \overset{\text{H}}{=} \lim_{t\to\infty}\frac{f(t)}{1} \quad [\text{by FTC1}] \quad = \lim_{x\to\infty} f(x), \text{ if this limit exists.}$$

(c) Suppose $\int_a^\infty f(x)\,dx$ converges; that is, $\lim\limits_{t\to\infty}\int_a^t f(x)\,dx = L < \infty$. Then

$$f_{\text{ave}} = \lim_{t\to\infty}\left[\frac{1}{t-a}\int_a^t f(x)\,dx\right] = \lim_{t\to\infty}\frac{1}{t-a}\cdot\lim_{t\to\infty}\int_a^t f(x)\,dx = 0\cdot L = 0.$$

(d) $(\sin x)_{\text{ave}} = \displaystyle\lim_{t\to\infty}\frac{1}{t}\int_0^t \sin x\,dx = \lim_{t\to\infty}\left(\frac{1}{t}\left[-\cos x\right]_0^t\right) = \lim_{t\to\infty}\left(-\frac{\cos t}{t} + \frac{1}{t}\right) = \lim_{t\to\infty}\frac{1-\cos t}{t} = 0$

79. Let $u = 1/x \Rightarrow x = 1/u \Rightarrow dx = -(1/u^2)\,du$.

$$\int_0^\infty \frac{\ln x}{1+x^2}\,dx = \int_\infty^0 \frac{\ln(1/u)}{1+1/u^2}\left(-\frac{du}{u^2}\right) = \int_\infty^0 \frac{-\ln u}{u^2+1}(-du) = \int_\infty^0 \frac{\ln u}{1+u^2}\,du = -\int_0^\infty \frac{\ln u}{1+u^2}\,du$$

Therefore, $\displaystyle\int_0^\infty \frac{\ln x}{1+x^2}\,dx = -\int_0^\infty \frac{\ln x}{1+x^2}\,dx = 0.$

80. If the distance between P and the point charge is d, then the potential V at P is

$$V = W = \int_\infty^d F\,dr = \int_\infty^d \frac{q}{4\pi\varepsilon_0 r^2}\,dr = \lim_{t\to\infty} \frac{q}{4\pi\varepsilon_0}\left[-\frac{1}{r}\right]_t^d = \frac{q}{4\pi\varepsilon_0}\lim_{t\to\infty}\left(-\frac{1}{d}+\frac{1}{t}\right) = -\frac{q}{4\pi\varepsilon_0 d}.$$

PROBLEMS PLUS

1.

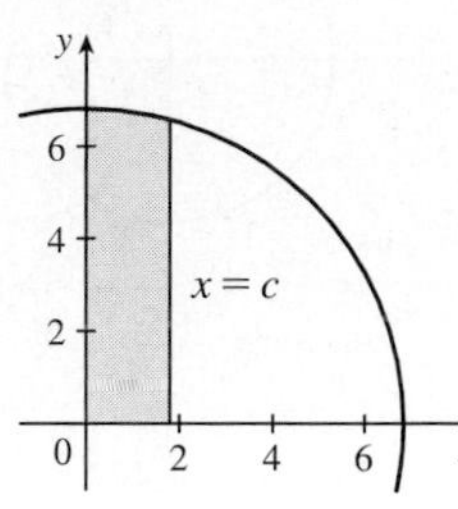

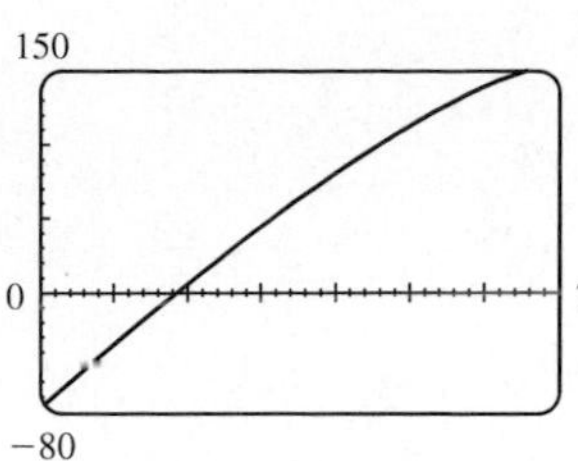

By symmetry, the problem can be reduced to finding the line $x = c$ such that the shaded area is one-third of the area of the quarter-circle. An equation of the semicircle is $y = \sqrt{49 - x^2}$, so we require that $\int_0^c \sqrt{49 - x^2}\,dx = \frac{1}{3} \cdot \frac{1}{4}\pi(7)^2 \quad \Leftrightarrow$
$\left[\frac{1}{2}x\sqrt{49 - x^2} + \frac{49}{2}\sin^{-1}(x/7)\right]_0^c = \frac{49}{12}\pi$ [by Formula 30] $\Leftrightarrow \frac{1}{2}c\sqrt{49 - c^2} + \frac{49}{2}\sin^{-1}(c/7) = \frac{49}{12}\pi$.

This equation would be difficult to solve exactly, so we plot the left-hand side as a function of c, and find that the equation holds for $c \approx 1.85$. So the cuts should be made at distances of about 1.85 inches from the center of the pizza.

2. $\displaystyle\int \frac{1}{x^7 - x}\,dx = \int \frac{dx}{x(x^6 - 1)} = \int \frac{x^5}{x^6(x^6 - 1)}\,dx = \frac{1}{6}\int \frac{1}{u(u - 1)}\,du \quad \begin{bmatrix} u = x^6, \\ du = 6x^5\,dx \end{bmatrix}$

$$= \frac{1}{6}\int\left(\frac{1}{u - 1} - \frac{1}{u}\right)du = \frac{1}{6}(\ln|u - 1| - \ln|u|) + C$$

$$= \frac{1}{6}\ln\left|\frac{u - 1}{u}\right| + C = \frac{1}{6}\ln\left|\frac{x^6 - 1}{x^6}\right| + C$$

Alternate method:

$$\int \frac{1}{x^7 - x}\,dx = \int \frac{x^{-7}}{1 - x^{-6}}\,dx \quad \begin{bmatrix} u = 1 - x^{-6}, \\ du = 6x^{-7}\,dx \end{bmatrix} \quad = \frac{1}{6}\int du/u = \frac{1}{6}\ln|u| + C = \frac{1}{6}\ln\left|1 - x^{-6}\right| + C$$

Other methods: Substitute $u = x^3$ or $x^3 = \sec\theta$.

3. The given integral represents the difference of the shaded areas, which appears to be 0. It can be calculated by integrating with respect to either x or y, so we find x in terms of y for each curve: $y = \sqrt[3]{1 - x^7} \quad \Rightarrow \quad x = \sqrt[7]{1 - y^3}$ and $y = \sqrt[7]{1 - x^3} \quad \Rightarrow \quad x = \sqrt[3]{1 - y^7}$, so
$\int_0^1\left(\sqrt[3]{1 - y^7} - \sqrt[7]{1 - y^3}\right)dy = \int_0^1\left(\sqrt[7]{1 - x^3} - \sqrt[3]{1 - x^7}\right)dx$. But this equation is of the form $z = -z$. So $\int_0^1\left(\sqrt[3]{1 - x^7} - \sqrt[7]{1 - x^3}\right)dx = 0$.

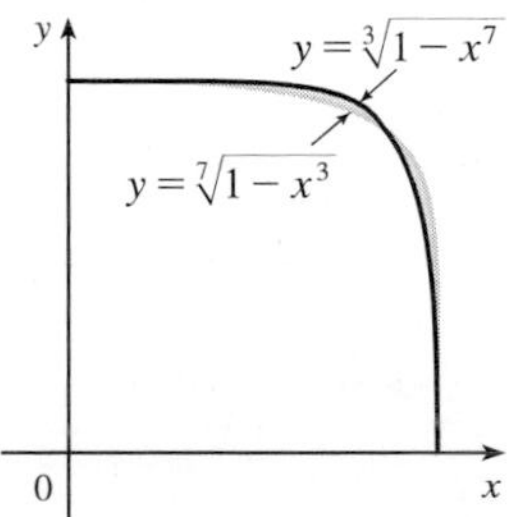

4. The area of each circle is $\pi(1)^2 = \pi$. By symmetry, the area of the union of the two disks is $A = \pi + \pi - 4I$.

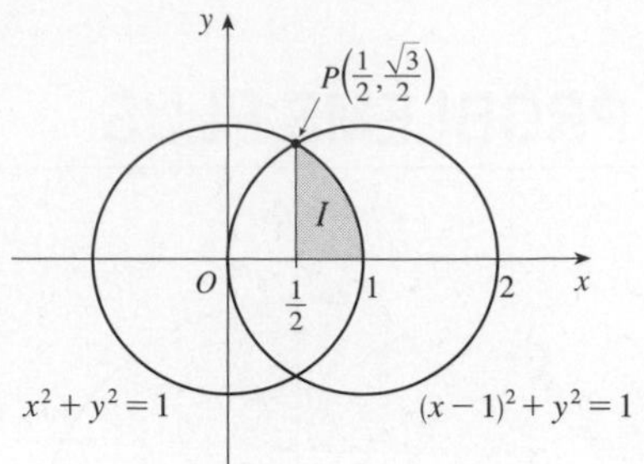

$$I = \int_{1/2}^{1} \sqrt{1-x^2}\, dx$$

$$\overset{30}{=} \left[\tfrac{x}{2}\sqrt{1-x^2} + \tfrac{1}{2}\sin^{-1}\left(\tfrac{x}{1}\right)\right]_{1/2}^{1} \quad \text{[or substitute } x = \sin\theta\text{]}$$

$$= \left(0 + \tfrac{\pi}{4}\right) - \left(\tfrac{1}{4}\tfrac{\sqrt{3}}{2} + \tfrac{1}{2}\tfrac{\pi}{6}\right) = \tfrac{\pi}{4} - \tfrac{\sqrt{3}}{8} - \tfrac{\pi}{12} = \tfrac{\pi}{6} - \tfrac{\sqrt{3}}{8}$$

Thus, $A = 2\pi - 4\left(\frac{\pi}{6} - \frac{\sqrt{3}}{8}\right) = 2\pi - \frac{2\pi}{3} + \frac{\sqrt{3}}{2} = \frac{4\pi}{3} + \frac{\sqrt{3}}{2}$.

Alternate solution (no calculus): The area of the sector, with central angle at the origin, containing I is $\frac{1}{2}r^2\theta = \frac{1}{2}(1)^2\left(\frac{\pi}{3}\right) = \frac{\pi}{6}$. The area of the triangle with hypotenuse OP is $\frac{1}{2}\left(\frac{1}{2}\right)\left(\frac{\sqrt{3}}{2}\right) = \frac{\sqrt{3}}{8}$.

Thus, the area of I is $\frac{\pi}{6} - \frac{\sqrt{3}}{8}$, as calculated above.

5. The area A of the remaining part of the circle is given by

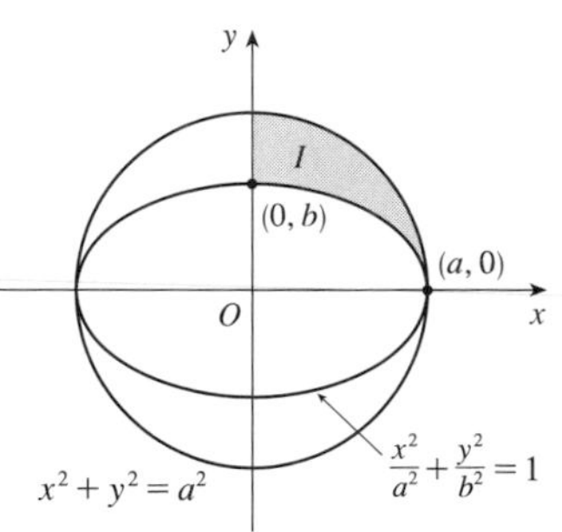

$$A = 4I = 4\int_0^a \left(\sqrt{a^2 - x^2} - \frac{b}{a}\sqrt{a^2-x^2}\right) dx = 4\left(1 - \frac{b}{a}\right)\int_0^a \sqrt{a^2-x^2}\, dx$$

$$\overset{30}{=} \frac{4}{a}(a-b)\left[\frac{x}{2}\sqrt{a^2-x^2} + \frac{a^2}{2}\sin^{-1}\frac{x}{a}\right]_0^a$$

$$= \frac{4}{a}(a-b)\left[\left(0 + \frac{a^2}{2}\frac{\pi}{2}\right) - 0\right] = \frac{4}{a}(a-b)\left(\frac{a^2\pi}{4}\right) = \pi a(a-b),$$

which is the area of an ellipse with semiaxes a and $a - b$.

Alternate solution: Subtracting the area of the ellipse from the area of the circle gives us $\pi a^2 - \pi ab = \pi a\,(a-b)$, as calculated above. (The formula for the area of an ellipse was derived in Example 2 in Section 7.3.)

6. (a) The tangent to the curve $y = f(x)$ at $x = x_0$ has the equation $y - f(x_0) = f'(x_0)(x - x_0)$. The y-intercept of this tangent line is $f(x_0) - f'(x_0)x_0$. Thus, L is the distance from the point $(0, f(x_0) - f'(x_0)x_0)$ to the point $(x_0, f(x_0))$; that is, $L^2 = x_0^2 + [f'(x_0)]^2\, x_0^2$, so $[f'(x_0)]^2 = \dfrac{L^2 - x_0^2}{x_0^2}$ and $f'(x_0) = -\dfrac{\sqrt{L^2 - x_0^2}}{x_0}$ for $0 < x_0 < L$.

(b) $\dfrac{dy}{dx} = -\dfrac{\sqrt{L^2 - x^2}}{x} \quad \Rightarrow \quad y = \displaystyle\int\left(-\frac{\sqrt{L^2-x^2}}{x}\right) dx.$

Let $x = L\sin\theta$. Then $dx = L\cos\theta\, d\theta$ and

$$y = \int \frac{-L\cos\theta\, L\cos\theta\, d\theta}{L\sin\theta} = L\int\frac{\sin^2\theta - 1}{\sin\theta}\, d\theta = L\int(\sin\theta - \csc\theta)\, d\theta$$

$$= -L\cos\theta - L\ln|\csc\theta - \cot\theta| + C = -\sqrt{L^2 - x^2} - L\ln\left(\frac{L}{x} - \frac{\sqrt{L^2-x^2}}{x}\right) + C$$

When $x = L$, $y = 0$, and $0 = -0 - L\ln(1 - 0) + C$, so $C = 0$. Therefore, $y = -\sqrt{L^2 - x^2} - L\ln\left(\dfrac{L - \sqrt{L^2 - x^2}}{x}\right)$.

7. Recall that $\cos A \cos B = \frac{1}{2}[\cos(A + B) + \cos(A - B)]$. So

$$\begin{aligned} f(x) &= \int_0^\pi \cos t \cos(x - t)\,dt = \tfrac{1}{2}\int_0^\pi [\cos(t + x - t) + \cos(t - x + t)]\,dt = \tfrac{1}{2}\int_0^\pi [\cos x + \cos(2t - x)]\,dt \\ &= \tfrac{1}{2}\left[t\cos x + \tfrac{1}{2}\sin(2t - x)\right]_0^\pi = \tfrac{\pi}{2}\cos x + \tfrac{1}{4}\sin(2\pi - x) - \tfrac{1}{4}\sin(-x) \\ &= \tfrac{\pi}{2}\cos x + \tfrac{1}{4}\sin(-x) - \tfrac{1}{4}\sin(-x) = \tfrac{\pi}{2}\cos x \end{aligned}$$

The minimum of $\cos x$ on this domain is -1, so the minimum value of $f(x)$ is $f(\pi) = -\frac{\pi}{2}$.

8. n is a positive integer, so

$$\int (\ln x)^n\,dx = x(\ln x)^n - \int x \cdot n(\ln x)^{n-1}\,(dx/x) \quad \text{[by parts]} = x(\ln x)^n - n\int (\ln x)^{n-1}\,dx$$

Thus,

$$\begin{aligned} \int_0^1 (\ln x)^n\,dx &= \lim_{t\to0^+}\int_t^1 (\ln x)^n\,dx = \lim_{t\to0^+}[x(\ln x)^n]_t^1 - n\lim_{t\to0^+}\int_t^1 (\ln x)^{n-1}\,dx \\ &= -\lim_{t\to0^+}\frac{(\ln t)^n}{1/t} - n\int_0^1 (\ln x)^{n-1}\,dx = -n\int_0^1 (\ln x)^{n-1}\,dx \end{aligned}$$

by repeated application of l'Hospital's Rule. We want to prove that $\int_0^1 (\ln x)^n\,dx = (-1)^n n!$ for every positive integer n. For $n = 1$, we have

$$\int_0^1 (\ln x)^1\,dx = (-1)\int_0^1 (\ln x)^0\,dx = -\int_0^1 dx = -1 \qquad \left[\text{or } \int_0^1 \ln x\,dx = \lim_{t\to0^+}\left[x\ln x - x\right]_t^1 = -1\right]$$

Assuming that the formula holds for n, we find that

$$\int_0^1 (\ln x)^{n+1}\,dx = -(n+1)\int_0^1 (\ln x)^n\,dx = -(n+1)(-1)^n\,n! = (-1)^{n+1}(n+1)!$$

This is the formula for $n + 1$. Thus, the formula holds for all positive integers n by induction.

9. In accordance with the hint, we let $I_k = \int_0^1 (1 - x^2)^k\,dx$, and we find an expression for I_{k+1} in terms of I_k. We integrate I_{k+1} by parts with $u = (1 - x^2)^{k+1} \Rightarrow du = (k+1)(1 - x^2)^k(-2x)$, $dv = dx \Rightarrow v = x$, and then split the remaining integral into identifiable quantities:

$$\begin{aligned} I_{k+1} &= x(1 - x^2)^{k+1}\Big|_0^1 + 2(k+1)\int_0^1 x^2(1 - x^2)^k\,dx = (2k+2)\int_0^1 (1 - x^2)^k[1 - (1 - x^2)]\,dx \\ &= (2k+2)(I_k - I_{k+1}) \end{aligned}$$

So $I_{k+1}[1 + (2k+2)] = (2k+2)I_k \Rightarrow I_{k+1} = \dfrac{2k+2}{2k+3}I_k$. Now to complete the proof, we use induction:

$I_0 = 1 = \dfrac{2^0(0!)^2}{1!}$, so the formula holds for $n = 0$. Now suppose it holds for $n = k$. Then

$$\begin{aligned} I_{k+1} &= \frac{2k+2}{2k+3}I_k = \frac{2k+2}{2k+3}\left[\frac{2^{2k}(k!)^2}{(2k+1)!}\right] = \frac{2(k+1)2^{2k}(k!)^2}{(2k+3)(2k+1)!} = \frac{2(k+1)}{2k+2}\cdot\frac{2(k+1)2^{2k}(k!)^2}{(2k+3)(2k+1)!} \\ &= \frac{[2(k+1)]^2\,2^{2k}(k!)^2}{(2k+3)(2k+2)(2k+1)!} = \frac{2^{2(k+1)}\,[(k+1)!]^2}{[2(k+1)+1]!} \end{aligned}$$

So by induction, the formula holds for all integers $n \geq 0$.

10. (a) Since $-1 \le \sin \le 1$, we have $-f(x) \le f(x)\sin nx \le f(x)$, and the graph of $y = f(x)\,\sin nx$ oscillates between $f(x)$ and $-f(x)$. (The diagram shows the case $f(x) = e^x$ and $n = 10$.) As $n \to \infty$, the graph oscillates more and more frequently; see the graphs in part (b).

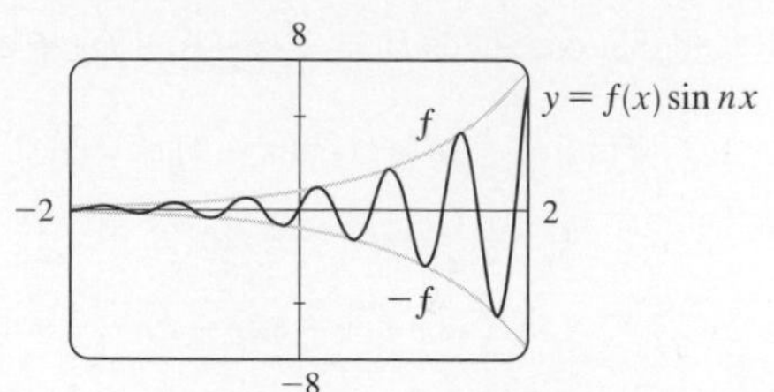

(b) From the graphs of the integrand, it seems that $\lim\limits_{n\to\infty} \int_0^1 f(x)\,\sin nx\,dx = 0$, since as n increases, the integrand oscillates more and more rapidly, and thus (since f' is continuous) it makes sense that the areas above the x-axis and below it during each oscillation approach equality.

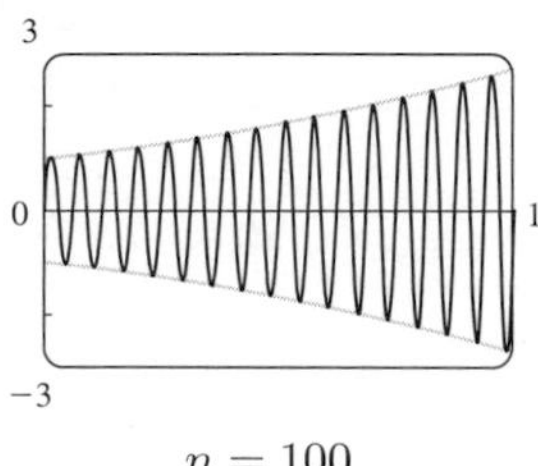

$n = 100$

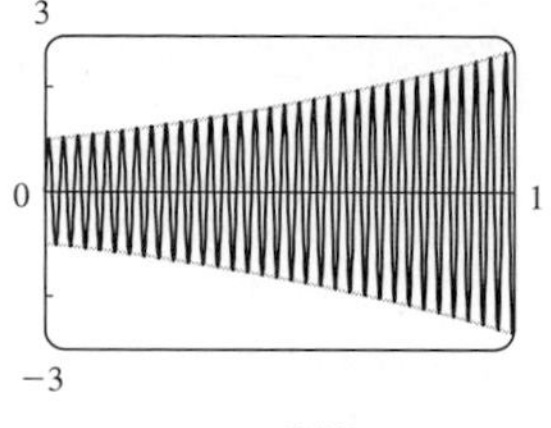

$n = 200$

(c) We integrate by parts with $u = f(x) \;\Rightarrow\; du = f'(x)\,dx$, $dv = \sin nx\,dx \;\Rightarrow\; v = -\dfrac{\cos nx}{n}$:

$$\int_0^1 f(x)\,\sin nx\,dx = \left[-\frac{f(x)\cos nx}{n}\right]_0^1 + \int_0^1 \frac{\cos nx}{n}\,f'(x)\,dx = \frac{1}{n}\left(\int_0^1 \cos nx f'(x)\,dx - \left[f(x)\,\cos nx\right]_0^1\right)$$

$$= \tfrac{1}{n}\left[\int_0^1 \cos nx f'(x)\,dx + f(0) - f(1)\,\cos n\right]$$

Taking absolute values of the first and last terms in this equality, and using the facts that $|\alpha \pm \beta| \le |\alpha| + |\beta|$, $\int_0^1 f(x)\,dx \le \int_0^1 |f(x)|\,dx$, $|f(0)| = f(0)$ [f is positive], $|f'(x)| \le M$ for $0 \le x \le 1$, and $|\cos nx| \le 1$,

$$\left|\int_0^1 f(x)\,\sin nx\,dx\right| \le \tfrac{1}{n}\left[\left|\int_0^1 M\,dx\right| + |f(0)| + |f(1)|\right] = \tfrac{1}{n}\left[M + |f(0)| + |f(1)|\right]$$

which approaches 0 as $n \to \infty$. The result follows by the Squeeze Theorem.

11. $0 < a < b$. Now

$$\int_0^1 [bx + a(1-x)]^t\,dx = \int_a^b \frac{u^t}{(b-a)}\,du \quad [u = bx + a(1-x)] \quad = \left[\frac{u^{t+1}}{(t+1)(b-a)}\right]_a^b = \frac{b^{t+1} - a^{t+1}}{(t+1)(b-a)}.$$

Now let $y = \lim\limits_{t\to 0}\left[\dfrac{b^{t+1} - a^{t+1}}{(t+1)(b-a)}\right]^{1/t}$. Then $\ln y = \lim\limits_{t\to 0}\left[\dfrac{1}{t}\ln \dfrac{b^{t+1} - a^{t+1}}{(t+1)(b-a)}\right]$. This limit is of the form $0/0$, so we can apply l'Hospital's Rule to get

$$\ln y = \lim_{t\to 0}\left[\frac{b^{t+1}\ln b - a^{t+1}\ln a}{b^{t+1} - a^{t+1}} - \frac{1}{t+1}\right] = \frac{b\ln b - a\ln a}{b-a} - 1 = \frac{b\ln b}{b-a} - \frac{a\ln a}{b-a} - \ln e = \ln \frac{b^{b/(b-a)}}{ea^{a/(b-a)}}.$$

Therefore, $y = e^{-1}\left(\dfrac{b^b}{a^a}\right)^{1/(b-a)}$.

12. 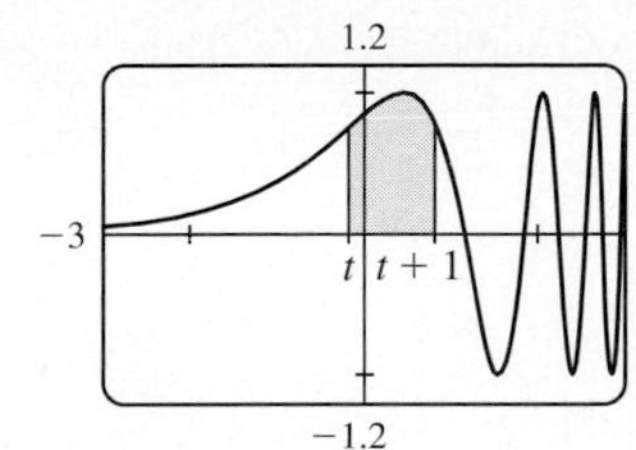

From the graph, it appears that the area under the graph of $f(x) = \sin(e^x)$ on the interval $[t, t+1]$ is greatest when $t \approx -0.2$. To find the exact value, we write the integral as $I = \int_t^{t+1} f(x)\,dx = \int_0^{t+1} f(x)\,dx - \int_0^t f(x)\,dx$, and use FTC1 to find $dI/dt = f(t+1) - f(t) = \sin(e^{t+1}) - \sin(e^t) = 0$ when $\sin(e^{t+1}) = \sin(e^t)$.

Now we have $\sin x = \sin y$ whenever $x - y = 2k\pi$ and also whenever x and y are the same distance from $\left(k + \frac{1}{2}\right)\pi$, k any integer, since $\sin x$ is symmetric about the line $x = \left(k + \frac{1}{2}\right)\pi$ The first possibility is the more obvious one, but if we calculate $e^{t+1} - e^t = 2k\pi$, we get $t = \ln(2k\pi/(e-1))$, which is about 1.3 for $k = 1$ (the least possible value of k). From the graph, this looks unlikely to give the maximum we are looking for. So instead we set $e^{t+1} - \left(k + \frac{1}{2}\right)\pi = \left(k + \frac{1}{2}\right)\pi - e^t \iff$ $e^{t+1} + e^t = (2k+1)\pi \iff e^t(e+1) = (2k+1)\pi \iff t = \ln((2k+1)\pi/(e+1))$. Now $k = 0 \Rightarrow$ $t = \ln(\pi/(e+1)) \approx -0.16853$, which does give the maximum value, as we have seen from the graph of f.

13. Write $I = \displaystyle\int \frac{x^8}{(1+x^6)^2}\,dx = \int x^3 \cdot \frac{x^5}{(1+x^6)^2}\,dx$. Integrate by parts with $u = x^3$, $dv = \dfrac{x^5}{(1+x^6)^2}\,dx$. Then

$du = 3x^2\,dx$, $v = -\dfrac{1}{6(1+x^6)}$ $\Rightarrow$ $I = -\dfrac{x^3}{6(1+x^6)} + \dfrac{1}{2}\displaystyle\int \frac{x^2}{1+x^6}\,dx$. Substitute $t = x^3$ in this latter integral.

$\displaystyle\int \frac{x^2}{1+x^6}\,dx = \frac{1}{3}\int \frac{dt}{1+t^2} = \frac{1}{3}\tan^{-1} t + C = \frac{1}{3}\tan^{-1}(x^3) + C$. Therefore $I = -\dfrac{x^3}{6(1+x^6)} + \dfrac{1}{6}\tan^{-1}(x^3) + C$.

Returning to the improper integral,

$$\begin{aligned}
\int_{-1}^{\infty}\left(\frac{x^4}{1+x^6}\right)^2 dx &= \lim_{t\to\infty}\int_{-1}^{t}\frac{x^8}{(1+x^6)^2}\,dx = \lim_{t\to\infty}\left[-\frac{x^3}{6(1+x^6)} + \frac{1}{6}\tan^{-1}(x^3)\right]_{-1}^{t} \\
&= \lim_{t\to\infty}\left(-\frac{t^3}{6(1+t^6)} + \frac{1}{6}\tan^{-1}(t^3) + \frac{-1}{6(1+1)} - \frac{1}{6}\tan^{-1}(-1)\right) \\
&= 0 + \frac{1}{6}\left(\frac{\pi}{2}\right) - \frac{1}{12} - \frac{1}{6}\left(-\frac{\pi}{4}\right) = \frac{\pi}{12} - \frac{1}{12} + \frac{\pi}{24} = \frac{\pi}{8} - \frac{1}{12}
\end{aligned}$$

14. $\displaystyle\int \sqrt{\tan x}\,dx = \int u\left(\frac{2u}{u^4+1}\,du\right)$ $\left[\begin{array}{ll} u = \sqrt{\tan x}, & u^2 = \tan x \\ 2u\,du = \sec^2 x\,dx, & (\tan^2 x + 1)\,dx = (u^4+1)\,dx \end{array}\right]$

Factoring the denominator, we get

$u^4 + 1 = u^4 + 2u^2 + 1 - 2u^2 = (u^2+1)^2 - \left(\sqrt{2}\,u\right)^2 = \left(u^2 + \sqrt{2}\,u + 1\right)\left(u^2 - \sqrt{2}\,u + 1\right)$. So

$\dfrac{2u^2}{u^4+1} = \dfrac{Au+B}{u^2+\sqrt{2}\,u+1} + \dfrac{Cu+D}{u^2-\sqrt{2}\,u+1}$ $\Rightarrow$ $2u^2 = (Au+B)\left(u^2 - \sqrt{2}\,u + 1\right) + (Cu+D)\left(u^2 + \sqrt{2}\,u + 1\right)$.

Equating coefficients of powers of u, we get $A + C = 0$ (u^3), $B - \sqrt{2}\,A + D + \sqrt{2}\,C = 2$ (u^2), $A - \sqrt{2}\,B + C + \sqrt{2}\,D = 0$ (u), $B + D = 0$ (constants). Substituting $-A$ for C and $-B$ for D in the u-equation leads to

$B = 0$ and $D = 0$, and then substituting those values in the u^2-equation gives us $A = -1/\sqrt{2}$ and $C = 1/\sqrt{2}$. Thus,

$$\begin{aligned}\int \frac{2u^2}{u^4+1}\,du &= -\frac{1}{\sqrt{2}}\int \frac{u}{u^2+\sqrt{2}\,u+1}\,du + \frac{1}{\sqrt{2}}\int \frac{u}{u^2-\sqrt{2}\,u+1}\,du \\ &= \frac{1}{\sqrt{2}}\int \frac{\frac{1}{2}(2u-\sqrt{2})+1/\sqrt{2}}{u^2-\sqrt{2}\,u+1}\,du - \frac{1}{\sqrt{2}}\int \frac{\frac{1}{2}(2u+\sqrt{2})-1/\sqrt{2}}{u^2+\sqrt{2}\,u+1}\,du \\ &= \frac{1}{2\sqrt{2}}\int \frac{2u-\sqrt{2}}{u^2-\sqrt{2}\,u+1}\,du + \frac{1}{2}\int \frac{du}{u^2-\sqrt{2}\,u+1} - \frac{1}{2\sqrt{2}}\int \frac{2u+\sqrt{2}}{u^2+\sqrt{2}\,u+1}\,du + \frac{1}{2}\int \frac{du}{u^2+\sqrt{2}\,u+1} \\ &= \frac{1}{2\sqrt{2}}\ln(u^2-\sqrt{2}\,u+1) + \frac{1}{2}\int \frac{du}{\left(u-1/\sqrt{2}\right)^2+\frac{1}{2}} - \frac{1}{2\sqrt{2}}\ln\left(u^2+\sqrt{2}\,u+1\right) + \frac{1}{2}\int \frac{du}{\left(u+1/\sqrt{2}\right)^2+\frac{1}{2}} \\ &= \frac{1}{2\sqrt{2}}\ln\frac{u^2-\sqrt{2}\,u+1}{u^2+\sqrt{2}\,u+1} + \frac{1}{2}\frac{1}{1/\sqrt{2}}\tan^{-1}\frac{u-1/\sqrt{2}}{1/\sqrt{2}} + \frac{1}{2}\frac{1}{1/\sqrt{2}}\tan^{-1}\frac{u+1/\sqrt{2}}{1/\sqrt{2}} + C \\ &= \frac{\sqrt{2}}{4}\ln\frac{\tan x-\sqrt{2\tan x}+1}{\tan x+\sqrt{2\tan x}+1} + \frac{\sqrt{2}}{2}\tan^{-1}\left(\sqrt{2\tan x}-1\right) + \frac{\sqrt{2}}{2}\tan^{-1}\left(\sqrt{2\tan x}+1\right) + C\end{aligned}$$

15.

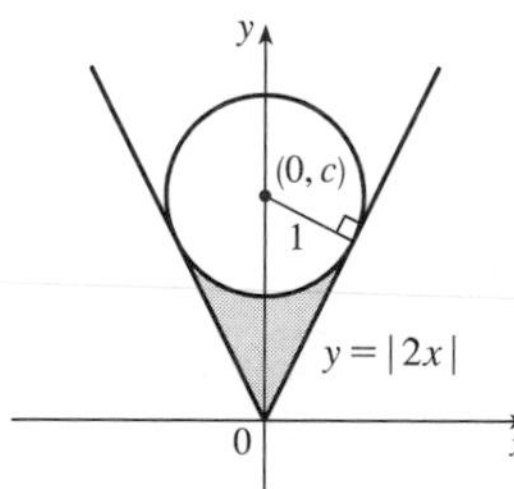

An equation of the circle with center $(0, c)$ and radius 1 is $x^2 + (y-c)^2 = 1^2$, so an equation of the lower semicircle is $y = c - \sqrt{1-x^2}$. At the points of tangency, the slopes of the line and semicircle must be equal. For $x \geq 0$, we must have

$$y' = 2 \;\Rightarrow\; \frac{x}{\sqrt{1-x^2}} = 2 \;\Rightarrow\; x = 2\sqrt{1-x^2} \;\Rightarrow\; x^2 = 4(1-x^2) \;\Rightarrow$$

$5x^2 = 4 \;\Rightarrow\; x^2 = \frac{4}{5} \;\Rightarrow\; x = \frac{2}{5}\sqrt{5}$ and so $y = 2\left(\frac{2}{5}\sqrt{5}\right) = \frac{4}{5}\sqrt{5}$.

The slope of the perpendicular line segment is $-\frac{1}{2}$, so an equation of the line segment is $y - \frac{4}{5}\sqrt{5} = -\frac{1}{2}\left(x - \frac{2}{5}\sqrt{5}\right) \;\Leftrightarrow$ $y = -\frac{1}{2}x + \frac{1}{5}\sqrt{5} + \frac{4}{5}\sqrt{5} \;\Leftrightarrow\; y = -\frac{1}{2}x + \sqrt{5}$, so $c = \sqrt{5}$ and an equation of the lower semicircle is $y = \sqrt{5} - \sqrt{1-x^2}$.

Thus, the shaded area is

$$\begin{aligned}2\int_0^{(2/5)\sqrt{5}}\left[\left(\sqrt{5}-\sqrt{1-x^2}\right)-2x\right]dx &\overset{30}{=} 2\left[\sqrt{5}\,x - \frac{x}{2}\sqrt{1-x^2} - \frac{1}{2}\sin^{-1}x - x^2\right]_0^{(2/5)\sqrt{5}} \\ &= 2\left[2 - \frac{\sqrt{5}}{5}\cdot\frac{1}{\sqrt{5}} - \frac{1}{2}\sin^{-1}\left(\frac{2}{\sqrt{5}}\right) - \frac{4}{5}\right] - 2(0) \\ &= 2\left[1 - \frac{1}{2}\sin^{-1}\left(\frac{2}{\sqrt{5}}\right)\right] = 2 - \sin^{-1}\left(\frac{2}{\sqrt{5}}\right)\end{aligned}$$

16. (a) $M\dfrac{dv}{dt} - ub = -Mg \;\Rightarrow\; (M_0 - bt)\dfrac{dv}{dt} = ub - (M_0 - bt)g \;\Rightarrow\; \dfrac{dv}{dt} = \dfrac{ub}{M_0 - bt} - g \;\Rightarrow$

$v(t) = -u\ln(M_0 - bt) - gt + C$. Now $0 = v(0) = -u\ln M_0 + C$, so $C = u\ln M_0$. Thus

$$v(t) = u\ln M_0 - u\ln(M_0 - bt) - gt = u\ln\frac{M_0}{M_0 - bt} - gt.$$

(b) Burnout velocity $= v\left(\dfrac{M_2}{b}\right) = u\ln\dfrac{M_0}{M_0 - M_2} - g\dfrac{M_2}{b} = u\ln\dfrac{M_0}{M_1} - g\dfrac{M_2}{b}$.

Note: The reason for the term "burnout velocity" is that M_2 kilograms of fuel is used in M_2/b seconds, so $v(M_2/b)$ is the rocket's velocity when the fuel is used up.

(c) Height at burnout time $= y\left(\dfrac{M_2}{b}\right)$. Now $\dfrac{dy}{dt} = v(t) = u \ln M_0 - gt - u\ln(M_0 - bt)$, so

$y(t) = (u \ln M_0)t - \dfrac{gt^2}{2} - \dfrac{u}{b}(M_0 - bt)\ln(M_0 - bt) + ut + C$. Since $0 = y(0) = \dfrac{u}{b}M_0 \ln M_0 + C$, we get

$C = -\dfrac{u}{b}M_0 \ln M_0$ and $y(t) = u(1 + \ln M_0)t - \dfrac{gt^2}{2} + \dfrac{u}{b}(M_0 - bt)\ln(M_0 - bt) - \dfrac{u}{b}M_0 \ln M_0$.

Therefore, the height at burnout is

$$\begin{aligned} y\left(\frac{M_2}{b}\right) &= u(1 + \ln M_0)\frac{M_2}{b} - \frac{g}{2}\left(\frac{M_2}{b}\right)^2 + \frac{u}{b}M_1 \ln M_1 - \frac{u}{b}M_0 \ln M_0 \\ &= \frac{u}{b}M_2 - \frac{u}{b}M_1 \ln M_0 + \frac{u}{b}M_1 \ln M_1 - \frac{g}{2}\left(\frac{M_2}{b}\right)^2 = \frac{u}{b}M_2 + \frac{u}{b}M_1 \ln\frac{M_1}{M_0} - \frac{g}{2}\left(\frac{M_2}{b}\right)^2 \end{aligned}$$

[In the calculation of $y(M_2/b)$, repeated use was made of the relation $M_0 = M_1 + M_2$. In particular, $t = M_2/b \;\Rightarrow\; M_0 - bt = M_1$.]

(d) The formula for $y(t)$ in part (c) holds while there is still fuel. Once the fuel is used up, gravity is the only force acting on the rocket. $-M_1 g = M_1 \dfrac{dv}{dt} \;\Rightarrow\; \dfrac{dv}{dt} = -g \;\Rightarrow\; v(t) = -gt + c_1$, where $c_1 = v\left(\dfrac{M_2}{b}\right) + \dfrac{gM_2}{b} \;\Rightarrow$

$v(t) = v\left(\dfrac{M_2}{b}\right) - g\left(t - \dfrac{M_2}{b}\right) \;\Rightarrow\; y(t) = v\left(\dfrac{M_2}{b}\right)\left(t - \dfrac{M_2}{b}\right) - \dfrac{g}{2}\left(t - \dfrac{M_2}{b}\right)^2 + c_2$, where $c_2 = y\left(\dfrac{M_2}{b}\right)$,

so $y(t) = y\left(\dfrac{M_2}{b}\right) + v\left(\dfrac{M_2}{b}\right)\left(t - \dfrac{M_2}{b}\right) - \dfrac{g}{2}\left(t - \dfrac{M_2}{b}\right)^2$, $t \geq \dfrac{M_2}{b}$.

To summarize: For $0 \leq t \leq \dfrac{M_2}{b}$, $y(t) = u(1 + \ln M_0)t - \dfrac{gt^2}{2} + \dfrac{u}{b}(M_0 - bt)\ln(M_0 - bt) - \dfrac{u}{b}M_0 \ln M_0$

[from part (c)], and for $t \geq \dfrac{M_2}{b}$, $y(t) = y\left(\dfrac{M_2}{b}\right) + v\left(\dfrac{M_2}{b}\right)\left(t - \dfrac{M_2}{b}\right) - \dfrac{g}{2}\left(t - \dfrac{M_2}{b}\right)^2$ [from above].

$y\left(\dfrac{M_2}{b}\right)$ and $v\left(\dfrac{M_2}{b}\right)$ are given in parts (c) and (b), respectively.

8 □ FURTHER APPLICATIONS OF INTEGRATION

8.1 Arc Length

1. $y = 2x - 5 \;\Rightarrow\; L = \int_{-1}^{3} \sqrt{1 + (dy/dx)^2}\,dx = \int_{-1}^{3} \sqrt{1 + (2)^2}\,dx = \sqrt{5}\,[3 - (-1)] = 4\sqrt{5}$.

The arc length can be calculated using the distance formula, since the curve is a line segment, so

$L = [\text{distance from } (-1, -7) \text{ to } (3, 1)] = \sqrt{[3 - (-1)]^2 + [1 - (-7)]^2} = \sqrt{80} = 4\sqrt{5}$

2. Using the arc length formula with $y = \sqrt{2 - x^2} \;\Rightarrow\; \dfrac{dy}{dx} = -\dfrac{x}{\sqrt{2 - x^2}}$, we get

$$L = \int_0^1 \sqrt{1 + \left(\frac{dy}{dx}\right)^2}\,dx = \int_0^1 \sqrt{1 + \frac{x^2}{2 - x^2}}\,dx = \int_0^1 \frac{\sqrt{2}\,dx}{\sqrt{2 - x^2}} = \sqrt{2}\int_0^1 \frac{dx}{\sqrt{\left(\sqrt{2}\right)^2 - x^2}}$$

$$= \sqrt{2}\left[\sin^{-1}\left(\frac{x}{\sqrt{2}}\right)\right]_0^1 = \sqrt{2}\left[\sin^{-1}\left(\frac{1}{\sqrt{2}}\right) - \sin^{-1}0\right] = \sqrt{2}\left[\tfrac{\pi}{4} - 0\right] = \sqrt{2}\,\tfrac{\pi}{4}$$

The curve is a one-eighth of a circle with radius $\sqrt{2}$, so the length of the arc is $\frac{1}{8}\left(2\pi \cdot \sqrt{2}\right) = \sqrt{2}\,\frac{\pi}{4}$, as above.

3. $y = \sin x \;\Rightarrow\; dy/dx = \cos x \;\Rightarrow\; 1 + (dy/dx)^2 = 1 + \cos^2 x$. So $L = \int_0^{\pi} \sqrt{1 + \cos^2 x}\,dx \approx 3.8202$.

4. $y = xe^{-x} \;\Rightarrow\; dy/dx = x(-e^{-x}) + e^{-x}(1) = e^{-x}(1 - x) \;\Rightarrow\; 1 + (dy/dx)^2 = 1 + [e^{-x}(1 - x)]^2$.

So $L = \int_0^2 \sqrt{1 + e^{-2x}(1 - x)^2}\,dx \approx 2.1024$.

5. $y = x - \ln x \;\Rightarrow\; dy/dx = 1 - 1/x \;\Rightarrow\; 1 + (dy/dx)^2 = 1 + (1 - 1/x)^2$. So $L = \int_1^4 \sqrt{1 + (1 - 1/x)^2}\,dx \approx 3.4467$.

6. $x = y^2 - 2y \;\Rightarrow\; dx/dy = 2y - 2 \;\Rightarrow\; 1 + (dx/dy)^2 = 1 + (2y - 2)^2$. So $L = \int_0^2 \sqrt{1 + (2y - 2)^2}\,dy \approx 2.9579$.

7. $x = \sqrt{y} - y \;\Rightarrow\; dx/dy = 1/(2\sqrt{y}) - 1 \;\Rightarrow\; 1 + (dx/dy)^2 = 1 + \left(\dfrac{1}{2\sqrt{y}} - 1\right)^2$.

So $L = \displaystyle\int_1^4 \sqrt{1 + \left(\frac{1}{2\sqrt{y}} - 1\right)^2}\,dy \approx 3.6095$.

8. $y^2 = \ln x \;\Leftrightarrow\; x = e^{y^2} \;\Rightarrow\; dx/dy = 2ye^{y^2} \;\Rightarrow\; 1 + (dx/dy)^2 = 1 + 4y^2e^{2y^2}$.

So $L = \int_{-1}^{1} \sqrt{1 + 4y^2e^{2y^2}}\,dy \approx 4.2552$.

9. $y = 1 + 6x^{3/2} \;\Rightarrow\; dy/dx = 9x^{1/2} \;\Rightarrow\; 1 + (dy/dx)^2 = 1 + 81x$.

So $L = \displaystyle\int_0^1 \sqrt{1 + 81x}\,dx = \int_1^{82} u^{1/2}\left(\tfrac{1}{81}\,du\right) \quad \begin{bmatrix} u = 1 + 81x, \\ du = 81\,dx \end{bmatrix} = \tfrac{1}{81} \cdot \tfrac{2}{3}\left[u^{3/2}\right]_1^{82} = \tfrac{2}{243}\left(82\sqrt{82} - 1\right)$.

10. $36y^2 = (x^2 - 4)^3,\; y \geq 0 \;\Rightarrow\; y = \frac{1}{6}(x^2 - 4)^{3/2} \;\Rightarrow\; dy/dx = \frac{1}{6} \cdot \frac{3}{2}(x^2 - 4)^{1/2}(2x) = \frac{1}{2}x(x^2 - 4)^{1/2} \;\Rightarrow$

$1 + (dy/dx)^2 = 1 + \frac{1}{4}x^2(x^2 - 4) = \frac{1}{4}x^4 - x^2 + 1 = \frac{1}{4}(x^4 - 4x^2 + 4) = \left[\frac{1}{2}(x^2 - 2)\right]^2$. So

$L = \int_2^3 \sqrt{\left[\frac{1}{2}(x^2 - 2)\right]^2}\,dx = \int_2^3 \frac{1}{2}(x^2 - 2)\,dx = \frac{1}{2}\left[\frac{1}{3}x^3 - 2x\right]_2^3 = \frac{1}{2}\left[(9 - 6) - \left(\frac{8}{3} - 4\right)\right] = \frac{1}{2}\left(\frac{13}{3}\right) = \frac{13}{6}$.

11.

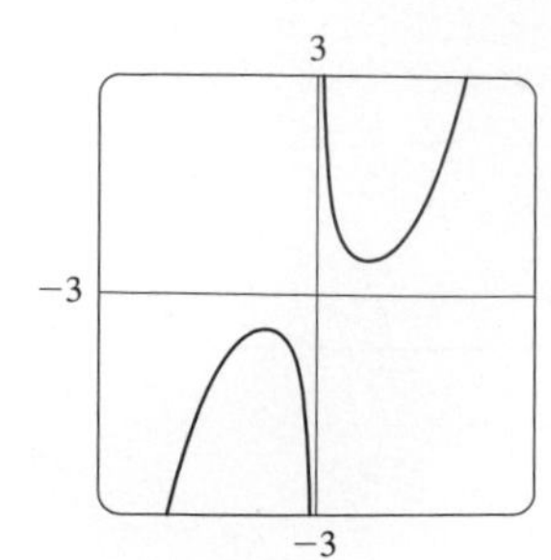

$y = \dfrac{x^3}{3} + \dfrac{1}{4x} \quad\Rightarrow\quad y' = x^2 - \dfrac{1}{4x^2} \quad\Rightarrow$

$1 + (y')^2 = 1 + \left(x^4 - \dfrac{1}{2} + \dfrac{1}{16x^4}\right) = x^4 + \dfrac{1}{2} + \dfrac{1}{16x^4} = \left(x^2 + \dfrac{1}{4x^2}\right)^2$. So

$$L = \int_1^2 \sqrt{1 + (y')^2}\, dx = \int_1^2 \left| x^2 + \frac{1}{4x^2} \right| dx = \int_1^2 \left(x^2 + \frac{1}{4x^2} \right) dx$$

$$= \left[\frac{1}{3}x^3 - \frac{1}{4x} \right]_1^2 = \left(\frac{8}{3} - \frac{1}{8} \right) - \left(\frac{1}{3} - \frac{1}{4} \right) = \frac{7}{3} + \frac{1}{8} = \frac{59}{24}$$

12. $x = \dfrac{y^4}{8} + \dfrac{1}{4y^2} \quad\Rightarrow\quad \dfrac{dx}{dy} = \frac{1}{2}y^3 - \frac{1}{2}y^{-3} \quad\Rightarrow$

$1 + (dx/dy)^2 = 1 + \frac{1}{4}y^6 - \frac{1}{2} + \frac{1}{4}y^{-6} = \frac{1}{4}y^6 + \frac{1}{2} + \frac{1}{4}y^{-6} = \left(\frac{1}{2}y^3 + \frac{1}{2}y^{-3}\right)^2$. So

$L = \int_1^2 \sqrt{\left(\frac{1}{2}y^3 + \frac{1}{2}y^{-3}\right)^2}\, dy = \int_1^2 \left(\frac{1}{2}y^3 + \frac{1}{2}y^{-3}\right) dy = \left[\frac{1}{8}y^4 - \frac{1}{4}y^{-2}\right]_1^2 = \left(2 - \frac{1}{16}\right) - \left(\frac{1}{8} - \frac{1}{4}\right)$

$= 2 + \frac{1}{16} = \frac{33}{16}$.

13. $x = \frac{1}{3}\sqrt{y}\,(y - 3) = \frac{1}{3}y^{3/2} - y^{1/2} \quad\Rightarrow\quad dx/dy = \frac{1}{2}y^{1/2} - \frac{1}{2}y^{-1/2} \quad\Rightarrow$

$1 + (dx/dy)^2 = 1 + \frac{1}{4}y - \frac{1}{2} + \frac{1}{4}y^{-1} = \frac{1}{4}y + \frac{1}{2} + \frac{1}{4}y^{-1} = \left(\frac{1}{2}y^{1/2} + \frac{1}{2}y^{-1/2}\right)^2$. So

$L = \int_1^9 \left(\frac{1}{2}y^{1/2} + \frac{1}{2}y^{-1/2}\right) dy = \frac{1}{2}\left[\frac{2}{3}y^{3/2} + 2y^{1/2}\right]_1^9 = \frac{1}{2}\left[\left(\frac{2}{3}\cdot 27 + 2\cdot 3\right) - \left(\frac{2}{3}\cdot 1 + 2\cdot 1\right)\right]$

$= \frac{1}{2}\left(24 - \frac{8}{3}\right) = \frac{1}{2}\left(\frac{64}{3}\right) = \frac{32}{3}$.

14. $y = \ln(\cos x) \quad\Rightarrow\quad dy/dx = -\tan x \quad\Rightarrow\quad 1 + (dy/dx)^2 = 1 + \tan^2 x = \sec^2 x$. So

$L = \int_0^{\pi/3} \sqrt{\sec^2 x}\, dx = \int_0^{\pi/3} \sec x\, dx = \left[\ln|\sec x + \tan x|\right]_0^{\pi/3} = \ln\left(2 + \sqrt{3}\right) - \ln(1 + 0) = \ln\left(2 + \sqrt{3}\right)$.

15. $y = \ln(\sec x) \quad\Rightarrow\quad \dfrac{dy}{dx} = \dfrac{\sec x\, \tan x}{\sec x} = \tan x \quad\Rightarrow\quad 1 + \left(\dfrac{dy}{dx}\right)^2 = 1 + \tan^2 x = \sec^2 x$, so

$L = \int_0^{\pi/4} \sqrt{\sec^2 x}\, dx = \int_0^{\pi/4} |\sec x|\, dx = \int_0^{\pi/4} \sec x\, dx = \Big[\ln(\sec x + \tan x)\Big]_0^{\pi/4}$

$= \ln\left(\sqrt{2} + 1\right) - \ln(1 + 0) = \ln\left(\sqrt{2} + 1\right)$

16. $y = 3 + \frac{1}{2}\cosh 2x \quad\Rightarrow\quad y' = \sinh 2x \quad\Rightarrow\quad 1 + (dy/dx)^2 = 1 + \sinh^2(2x) = \cosh^2(2x)$. So

$L = \int_0^1 \sqrt{\cosh^2(2x)}\, dx = \int_0^1 \cosh 2x\, dx = \left[\frac{1}{2}\sinh 2x\right]_0^1 = \frac{1}{2}\sinh 2 - 0 = \frac{1}{2}\sinh 2$.

17. $y = \dfrac{1}{4}x^2 - \dfrac{1}{2}\ln x \quad\Rightarrow\quad y' = \dfrac{1}{2}x - \dfrac{1}{2x} \quad\Rightarrow\quad 1 + (y')^2 = 1 + \left(\dfrac{1}{4}x^2 - \dfrac{1}{2} + \dfrac{1}{4x^2}\right) = \dfrac{1}{4}x^2 + \dfrac{1}{2} + \dfrac{1}{4x^2} = \left(\dfrac{1}{2}x + \dfrac{1}{2x}\right)^2$.

So

$$L = \int_1^2 \sqrt{1 + (y')^2}\, dx = \int_1^2 \left| \frac{1}{2}x + \frac{1}{2x} \right| dx = \int_1^2 \left(\frac{1}{2}x + \frac{1}{2x} \right) dx$$

$$= \left[\frac{1}{4}x^2 + \frac{1}{2}\ln|x| \right]_1^2 = \left(1 + \frac{1}{2}\ln 2 \right) - \left(\frac{1}{4} + 0 \right) = \frac{3}{4} + \frac{1}{2}\ln 2$$

18. $y = \sqrt{x - x^2} + \sin^{-1}(\sqrt{x}) \Rightarrow \dfrac{dy}{dx} = \dfrac{1 - 2x}{2\sqrt{x - x^2}} + \dfrac{1}{2\sqrt{x}\sqrt{1 - x}} = \dfrac{2 - 2x}{2\sqrt{x}\sqrt{1 - x}} = \sqrt{\dfrac{1 - x}{x}} \Rightarrow$

$1 + \left(\dfrac{dy}{dx}\right)^2 = 1 + \dfrac{1 - x}{x} = \dfrac{1}{x}$. The curve has endpoints $(0, 0)$ and $(1, \frac{\pi}{2})$,

so $L = \int_0^1 \sqrt{1/x}\,dx = \lim\limits_{t \to 0^+} \int_t^1 \sqrt{1/x}\,dx = \lim\limits_{t \to 0^+} \left[2\sqrt{x}\right]_t^1 = \lim\limits_{t \to 0^+} \left[2\sqrt{1} - 2\sqrt{t}\right] = 2 - 0 = 2$.

19. $y = \ln(1 - x^2) \Rightarrow y' = \dfrac{1}{1 - x^2} \cdot (-2x) \Rightarrow$

$1 + \left(\dfrac{dy}{dx}\right)^2 = 1 + \dfrac{4x^2}{(1 - x^2)^2} = \dfrac{1 - 2x^2 + x^4 + 4x^2}{(1 - x^2)^2} = \dfrac{1 + 2x^2 + x^4}{(1 - x^2)^2} = \dfrac{(1 + x^2)^2}{(1 - x^2)^2} \Rightarrow$

$\sqrt{1 + \left(\dfrac{dy}{dx}\right)^2} = \sqrt{\left(\dfrac{1 + x^2}{1 - x^2}\right)^2} = \dfrac{1 + x^2}{1 - x^2} = -1 + \dfrac{2}{1 - x^2}$ [by division] $= -1 + \dfrac{1}{1 + x} + \dfrac{1}{1 - x}$ [partial fractions].

So $L = \displaystyle\int_0^{1/2} \left(-1 + \frac{1}{1 + x} + \frac{1}{1 - x}\right) dx = \left[-x + \ln|1 + x| - \ln|1 - x|\right]_0^{1/2} = \left(-\tfrac{1}{2} + \ln\tfrac{3}{2} - \ln\tfrac{1}{2}\right) - 0 = \ln 3 - \tfrac{1}{2}$.

20. $y = 1 - e^{-x} \Rightarrow y' = -(-e^{-x}) = e^{-x} \Rightarrow 1 + (dy/dx)^2 = 1 + e^{-2x}$. So

$$L = \int_0^2 \sqrt{1 + e^{-2x}}\,dx = \int_1^{e^{-2}} \sqrt{1 + u^2}\left(-\frac{1}{u}\,du\right) \quad [u = e^{-x}]$$

$$\overset{23}{=} \left[\ln\left|\frac{1 + \sqrt{1 + u^2}}{u}\right| - \sqrt{1 + u^2}\right]_1^{e^{-2}} \quad \text{[or substitute } u = \tan\theta]$$

$$= \ln\left|\frac{1 + \sqrt{1 + e^{-4}}}{e^{-2}}\right| - \sqrt{1 + e^{-4}} - \ln\left|\frac{1 + \sqrt{2}}{1}\right| + \sqrt{2}$$

$$= \ln\left(1 + \sqrt{1 + e^{-4}}\right) - \ln e^{-2} - \sqrt{1 + e^{-4}} - \ln\left(1 + \sqrt{2}\right) + \sqrt{2}$$

$$= \ln\left(1 + \sqrt{1 + e^{-4}}\right) + 2 - \sqrt{1 + e^{-4}} - \ln\left(1 + \sqrt{2}\right) + \sqrt{2}$$

21. $y = \frac{1}{2}x^2 \Rightarrow dy/dx = x \Rightarrow 1 + (dy/dx)^2 = 1 + x^2$. So

$L = \int_{-1}^1 \sqrt{1 + x^2}\,dx = 2\int_0^1 \sqrt{1 + x^2}\,dx$ [by symmetry] $\overset{21}{=} 2\left[\frac{x}{2}\sqrt{1 + x^2} + \frac{1}{2}\ln\left(x + \sqrt{1 + x^2}\right)\right]_0^1$ $\left[\begin{array}{c}\text{or substitute}\\ x = \tan\theta\end{array}\right]$

$= 2\left[\left(\frac{1}{2}\sqrt{2} + \frac{1}{2}\ln\left(1 + \sqrt{2}\right)\right) - \left(0 + \frac{1}{2}\ln 1\right)\right] = \sqrt{2} + \ln\left(1 + \sqrt{2}\right)$

22. $x^2 = (y - 4)^3 \Rightarrow x = (y - 4)^{3/2}$ [for $x > 0$] $\Rightarrow dx/dy = \frac{3}{2}(y - 4)^{1/2} \Rightarrow$

$1 + (dx/dy)^2 = 1 + \frac{9}{4}(y - 4) = \frac{9}{4}y - 8$. So

$$L = \int_5^8 \sqrt{\tfrac{9}{4}y - 8}\,dy = \int_{13/4}^{10} \sqrt{u}\left(\tfrac{4}{9}\,du\right) \quad \left[\begin{array}{l}u = \frac{9}{4}y - 8,\\ du = \frac{9}{4}\,dy\end{array}\right] \quad = \tfrac{4}{9}\left[\tfrac{2}{3}u^{3/2}\right]_{13/4}^{10}$$

$$= \tfrac{8}{27}\left[10^{3/2} - \left(\tfrac{13}{4}\right)^{3/2}\right] \quad \left[\text{or } \tfrac{1}{27}\left(80\sqrt{10} - 13\sqrt{13}\right)\right]$$

23.

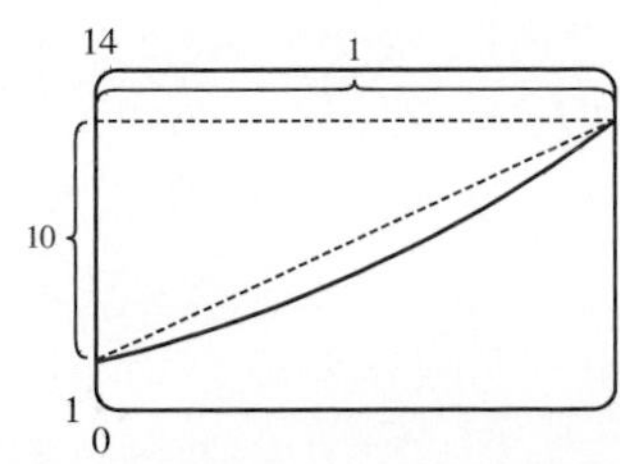

From the figure, the length of the curve is slightly larger than the hypotenuse of the triangle formed by the points $(1, 2)$, $(1, 12)$, and $(2, 12)$. This length is about $\sqrt{10^2 + 1^2} \approx 10$, so we might estimate the length to be 10.

$y = x^2 + x^3 \Rightarrow y' = 2x + 3x^2 \Rightarrow 1 + (y')^2 = 1 + (2x + 3x^2)^2$.

So $L = \int_1^2 \sqrt{1 + (2x + 3x^2)^2}\,dx \approx 10.0556$.

24. 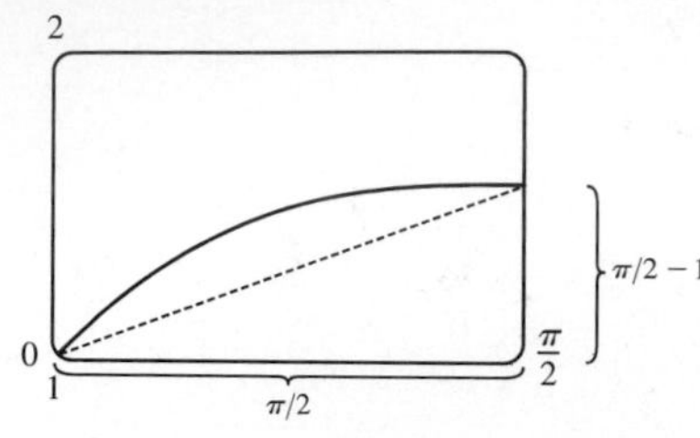

From the figure, the length of the curve is slightly larger than the hypotenuse of the triangle formed by the points $(1, 1)$, $\left(\frac{\pi}{2}, 1\right)$, and $\left(\frac{\pi}{2}, \frac{\pi}{2}\right)$. This length is about $\sqrt{\left(\frac{\pi}{2}\right)^2 + \left(\frac{\pi}{2} - 1\right)^2} \approx 1.7$, so we might estimate the length to be 1.7. $y = x + \cos x \;\Rightarrow\; y' = 1 - \sin x \;\Rightarrow$ $1 + (y')^2 = 1 + (1 - \sin x)^2$. So $L = \int_0^{\pi/2} \sqrt{1 + (1 - \sin x)^2}\,dx \approx 1.7294$.

25. $y = x\sin x \;\Rightarrow\; dy/dx = x\cos x + (\sin x)(1) \;\Rightarrow\; 1 + (dy/dx)^2 = 1 + (x\cos x + \sin x)^2$. Let $f(x) = \sqrt{1 + (dy/dx)^2} = \sqrt{1 + (x\cos x + \sin x)^2}$. Then $L = \int_0^{2\pi} f(x)\,dx$. Since $n = 10$, $\Delta x = \frac{2\pi - 0}{10} = \frac{\pi}{5}$. Now

$$\begin{aligned} L \approx S_{10} &= \tfrac{\pi/5}{3}\left[f(0) + 4f\left(\tfrac{\pi}{5}\right) + 2f\left(\tfrac{2\pi}{5}\right) + 4f\left(\tfrac{3\pi}{5}\right) + 2f\left(\tfrac{4\pi}{5}\right) + 4f\left(\tfrac{5\pi}{5}\right) + 2f\left(\tfrac{6\pi}{5}\right)\right. \\ &\qquad \left. + 4f\left(\tfrac{7\pi}{5}\right) + 2f\left(\tfrac{8\pi}{5}\right) + 4f\left(\tfrac{9\pi}{5}\right) + f(2\pi)\right] \\ &\approx 15.498085 \end{aligned}$$

The value of the integral produced by a calculator is 15.374568 (to six decimal places).

26. $y = \sqrt[3]{x} \;\Rightarrow\; dy/dx = \frac{1}{3}x^{-2/3} \;\Rightarrow\; L = \int_1^6 f(x)\,dx$, where $f(x) = \sqrt{1 + \frac{1}{9}x^{-4/3}}$.

Since $n = 10$, $\Delta x = \frac{6 - 1}{10} = \frac{1}{2}$. Now

$$\begin{aligned} L \approx S_{10} &= \tfrac{1/2}{3}[f(1) + 4f(1.5) + 2f(2) + 4f(2.5) + 2f(3) + 4f(3.5) + 2f(4) \\ &\qquad + 4f(4.5) + 2f(5) + 4f(5.5) + f(6)] \\ &\approx 5.074212 \end{aligned}$$

The value of the integral produced by a calculator is 5.074094 (to six decimal places).

27. $y = \ln(1 + x^3) \;\Rightarrow\; dy/dx = \dfrac{1}{1 + x^3} \cdot 3x^2 \;\Rightarrow\; L = \int_0^5 f(x)\,dx$, where $f(x) = \sqrt{1 + 9x^4/(1 + x^3)^2}$.

Since $n = 10$, $\Delta x = \frac{5 - 0}{10} = \frac{1}{2}$. Now

$$\begin{aligned} L \approx S_{10} &= \tfrac{1/2}{3}[f(0) + 4f(0.5) + 2f(1) + 4f(1.5) + 2f(2) + 4f(2.5) + 2f(3) \\ &\qquad + 4f(3.5) + 2f(4) + 4f(4.5) + f(5)] \\ &\approx 7.094570 \end{aligned}$$

The value of the integral produced by a calculator is 7.118819 (to six decimal places).

28. $y = e^{-x^2} \;\Rightarrow\; dy/dx = e^{-x^2}(-2x) \;\Rightarrow\; L = \int_0^2 f(x)\,dx$, where $f(x) = \sqrt{1 + 4x^2e^{-2x^2}}$.

Since $n = 10$, $\Delta x = \frac{2 - 0}{10} = \frac{1}{5}$. Now

$$\begin{aligned} L \approx S_{10} &= \tfrac{1/5}{3}[f(0) + 4f(0.2) + 2f(0.4) + 4f(0.6) + 2f(0.8) + 4f(1) + 2f(1.2) \\ &\qquad + 4f(1.4) + 2f(1.6) + 4f(1.8) + f(2)] \\ &\approx 2.280559 \end{aligned}$$

The value of the integral produced by a calculator is 2.280526 (to six decimal places).

29. (a) Let $f(x) = y = x\sqrt[3]{4-x}$ with $0 \le x \le 4$.

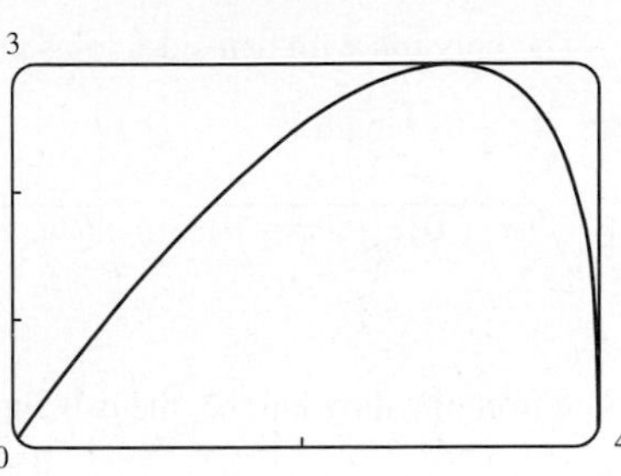

(b) The polygon with one side is just the line segment joining the points $(0, f(0)) = (0, 0)$ and $(4, f(4)) = (4, 0)$, and its length $L_1 = 4$.

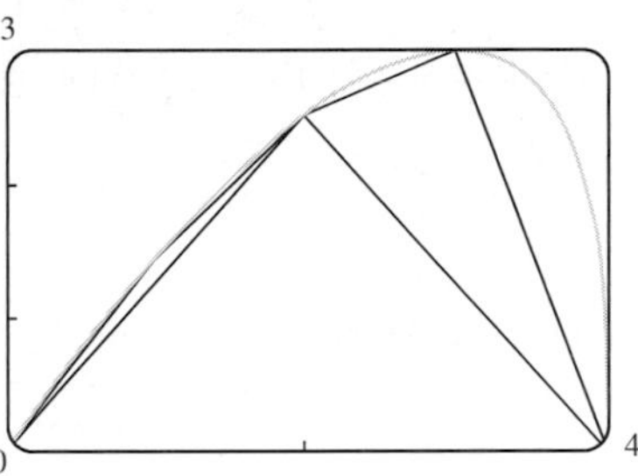

The polygon with two sides joins the points $(0, 0)$, $(2, f(2)) = \left(2, 2\sqrt[3]{2}\right)$ and $(4, 0)$. Its length

$$L_2 = \sqrt{(2-0)^2 + \left(2\sqrt[3]{2} - 0\right)^2} + \sqrt{(4-2)^2 + \left(0 - 2\sqrt[3]{2}\right)^2} = 2\sqrt{4 + 2^{8/3}} \approx 6.43$$

Similarly, the inscribed polygon with four sides joins the points $(0, 0)$, $\left(1, \sqrt[3]{3}\right)$, $\left(2, 2\sqrt[3]{2}\right)$, $(3, 3)$, and $(4, 0)$, so its length

$$L_4 = \sqrt{1 + \left(\sqrt[3]{3}\right)^2} + \sqrt{1 + \left(2\sqrt[3]{2} - \sqrt[3]{3}\right)^2} + \sqrt{1 + \left(3 - 2\sqrt[3]{2}\right)^2} + \sqrt{1+9} \approx 7.50$$

(c) Using the arc length formula with $\dfrac{dy}{dx} = x\left[\frac{1}{3}(4-x)^{-2/3}(-1)\right] + \sqrt[3]{4-x} = \dfrac{12-4x}{3(4-x)^{2/3}}$, the length of the curve is

$$L = \int_0^4 \sqrt{1 + \left(\frac{dy}{dx}\right)^2}\, dx = \int_0^4 \sqrt{1 + \left[\frac{12-4x}{3(4-x)^{2/3}}\right]^2}\, dx.$$

(d) According to a calculator, the length of the curve is $L \approx 7.7988$. The actual value is larger than any of the approximations in part (b). This is always true, since any approximating straight line between two points on the curve is shorter than the length of the curve between the two points.

30. (a) Let $f(x) = y = x + \sin x$ with $0 \le x \le 2\pi$.

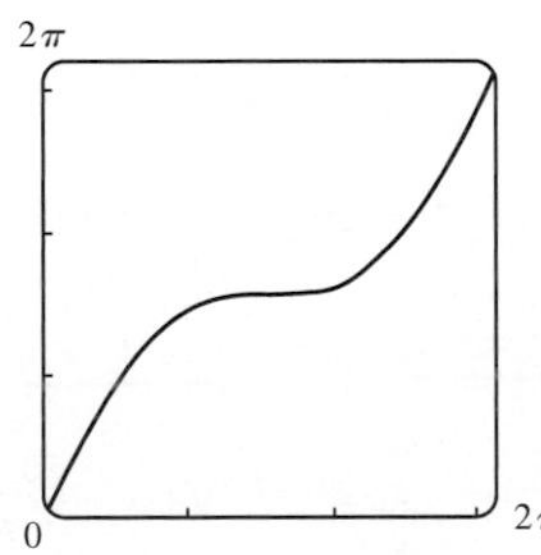

(b) The polygon with one side is just the line segment joining the points $(0, f(0)) = (0, 0)$ and $(2\pi, f(2\pi)) = (2\pi, 2\pi)$, and its length is $\sqrt{(2\pi - 0)^2 + (2\pi - 0)^2} = 2\sqrt{2}\,\pi \approx 8.9$.

[continued]

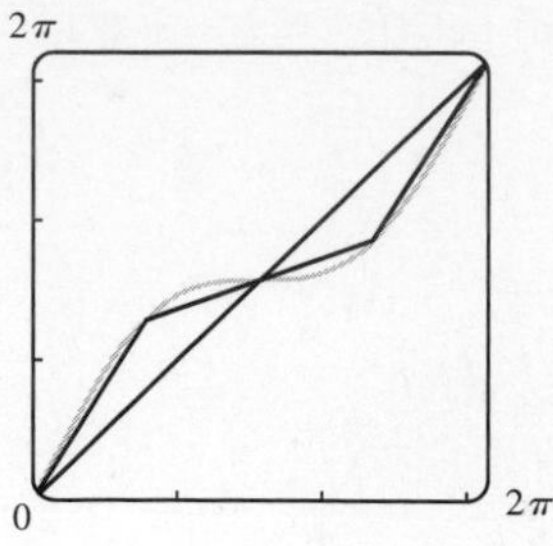

The polygon with two sides joins the points $(0,0)$, $(\pi, f(\pi)) = (\pi, \pi)$, and $(2\pi, 2\pi)$. Its length is

$$\sqrt{(\pi-0)^2+(\pi-0)^2}+\sqrt{(2\pi-\pi)^2+(2\pi-\pi)^2}=\sqrt{2}\,\pi+\sqrt{2}\,\pi$$
$$=2\sqrt{2}\,\pi\approx 8.9$$

Note from the diagram that the two approximations are the same because the sides of the two-sided polygon are in fact on the same line, since $f(\pi)=\pi=\frac{1}{2}f(2\pi)$.

The four-sided polygon joins the points $(0,0)$, $\left(\frac{\pi}{2},\frac{\pi}{2}+1\right)$, (π,π), $\left(\frac{3\pi}{2},\frac{3\pi}{2}-1\right)$, and $(2\pi,2\pi)$, so its length is

$$\sqrt{\left(\tfrac{\pi}{2}\right)^2+\left(\tfrac{\pi}{2}+1\right)^2}+\sqrt{\left(\tfrac{\pi}{2}\right)^2+\left(\tfrac{\pi}{2}-1\right)^2}+\sqrt{\left(\tfrac{\pi}{2}\right)^2+\left(\tfrac{\pi}{2}-1\right)^2}+\sqrt{\left(\tfrac{\pi}{2}\right)^2+\left(\tfrac{\pi}{2}+1\right)^2}\approx 9.4$$

(c) Using the arc length formula with $dy/dx=1+\cos x$, the length of the curve is

$$L=\int_0^{2\pi}\sqrt{1+(1+\cos x)^2}\,dx=\int_0^{2\pi}\sqrt{2+2\cos x+\cos^2 x}\,dx$$

(d) The calculator approximates the integral as 9.5076. The actual length is larger than the approximations in part (b).

31. $y=e^x \;\Rightarrow\; dy/dx=e^x \;\Rightarrow\; 1+(dy/dx)^2 \;\Rightarrow\; 1+e^{2x} \;\Rightarrow$

$$L=\int_0^2\sqrt{1+e^{2x}}\,dx=\int_1^{e^2}\sqrt{1+u^2}\left(\frac{1}{u}\,du\right)\qquad\begin{bmatrix}u=e^x,\\ du=e^x\,dx\end{bmatrix}$$

$$\overset{23}{=}\left[\sqrt{1+u^2}-\ln\left|\frac{1+\sqrt{1+u^2}}{u}\right|\right]_1^{e^2}=\left(\sqrt{1+e^4}-\ln\frac{1+\sqrt{1+e^4}}{e^2}\right)-\left(\sqrt{2}-\ln\frac{1+\sqrt{2}}{1}\right)$$
$$=\sqrt{1+e^4}-\ln(1+\sqrt{1+e^4})+2-\sqrt{2}+\ln(1+\sqrt{2})\approx 6.788651$$

An equivalent answer from a CAS is

$$-\sqrt{2}+\operatorname{arctanh}(\sqrt{2}/2)+\sqrt{e^4+1}-\operatorname{arctanh}(1/\sqrt{e^4+1}).$$

32. $y=x^{4/3} \;\Rightarrow\; dy/dx=\frac{4}{3}x^{1/3} \;\Rightarrow\; 1+(dy/dx)^2=1+\frac{16}{9}x^{2/3} \;\Rightarrow$

$$L=\int_0^1\sqrt{1+\tfrac{16}{9}x^{2/3}}\,dx=\int_0^{4/3}\sqrt{1+u^2}\,\tfrac{81}{64}u^2\,du\qquad\begin{bmatrix}u=\frac{4}{3}x^{1/3},\ du=\frac{4}{9}x^{-2/3}\,dx,\\ dx=\frac{9}{4}x^{2/3}\,du=\frac{9}{4}\cdot\frac{9}{16}u^2\,du=\frac{81}{64}u^2\,du\end{bmatrix}$$

$$\overset{22}{=}\tfrac{81}{64}\left[\tfrac{1}{8}u(1+2u^2)\sqrt{1+u^2}-\tfrac{1}{8}\ln\left(u+\sqrt{1+u^2}\right)\right]_0^{4/3}=\tfrac{81}{64}\left[\tfrac{1}{6}\left(1+\tfrac{32}{9}\right)\sqrt{\tfrac{25}{9}}-\tfrac{1}{8}\ln\left(\tfrac{4}{3}+\sqrt{\tfrac{25}{9}}\right)\right]$$
$$=\tfrac{81}{64}\left(\tfrac{1}{6}\cdot\tfrac{41}{9}\cdot\tfrac{5}{3}-\tfrac{1}{8}\ln 3\right)=\tfrac{205}{128}-\tfrac{81}{512}\ln 3\approx 1.4277586$$

33. $y^{2/3}=1-x^{2/3} \;\Rightarrow\; y=(1-x^{2/3})^{3/2} \;\Rightarrow$

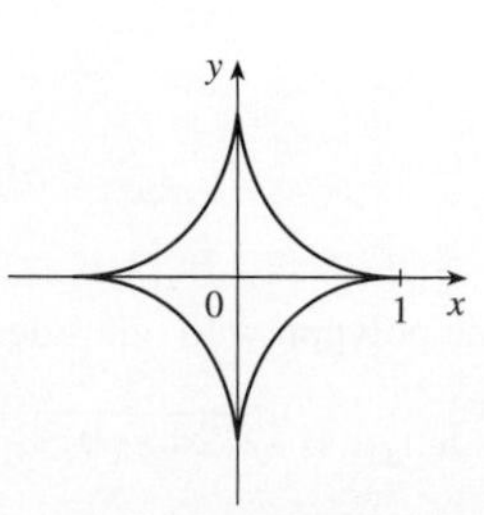

$$\frac{dy}{dx}=\tfrac{3}{2}(1-x^{2/3})^{1/2}\left(-\tfrac{2}{3}x^{-1/3}\right)=-x^{-1/3}(1-x^{2/3})^{1/2}\;\Rightarrow$$

$$\left(\frac{dy}{dx}\right)^2=x^{-2/3}(1-x^{2/3})=x^{-2/3}-1.\text{ Thus}$$

$$L=4\int_0^1\sqrt{1+(x^{-2/3}-1)}\,dx=4\int_0^1 x^{-1/3}\,dx=4\lim_{t\to 0^+}\left[\tfrac{3}{2}x^{2/3}\right]_t^1=6.$$

34. (a)

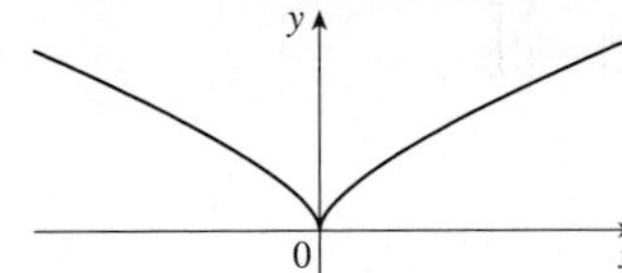

(b) $y = x^{2/3} \;\Rightarrow\; 1 + \left(\dfrac{dy}{dx}\right)^2 = 1 + \left(\frac{2}{3}x^{-1/3}\right)^2 = 1 + \frac{4}{9}x^{-2/3}$. So $L = \int_0^1 \sqrt{1 + \frac{4}{9}x^{-2/3}}\,dx$ [an improper integral].

$x = y^{3/2} \;\Rightarrow\; 1 + \left(\dfrac{dx}{dy}\right)^2 = 1 + \left(\frac{3}{2}y^{1/2}\right)^2 = 1 + \frac{9}{4}y$. So $L = \int_0^1 \sqrt{1 + \frac{9}{4}y}\,dy$.

The second integral equals $\frac{4}{9} \cdot \frac{2}{3}\left[\left(1 + \frac{9}{4}y\right)^{3/2}\right]_0^1 = \frac{8}{27}\left(\frac{13\sqrt{13}}{8} - 1\right) = \frac{13\sqrt{13} - 8}{27}$.

The first integral can be evaluated as follows:

$$\int_0^1 \sqrt{1 + \tfrac{4}{9}x^{-2/3}}\,dx = \lim_{t\to 0^+}\int_t^1 \frac{\sqrt{9x^{2/3} + 4}}{3x^{1/3}}\,dx = \lim_{t\to 0^+}\int_{9t^{2/3}}^9 \frac{\sqrt{u+4}}{18}\,du \qquad \begin{bmatrix} u = 9x^{2/3}, \\ du = 6x^{-1/3}\,dx \end{bmatrix}$$

$$= \int_0^9 \frac{\sqrt{u+4}}{18}\,du = \frac{1}{18}\cdot\left[\frac{2}{3}(u+4)^{3/2}\right]_0^9 = \frac{1}{27}\left(13^{3/2} - 4^{3/2}\right) = \frac{13\sqrt{13} - 8}{27}$$

(c) L = length of the arc of this curve from $(-1, 1)$ to $(8, 4)$

$$= \int_0^1 \sqrt{1 + \frac{9}{4}y}\,dy + \int_0^4 \sqrt{1 + \frac{9}{4}y}\,dy = \frac{13\sqrt{13} - 8}{27} + \frac{8}{27}\left[\left(1 + \frac{9}{4}y\right)^{3/2}\right]_0^4 \qquad \text{[from part (b)]}$$

$$= \frac{13\sqrt{13} - 8}{27} + \frac{8}{27}\left(10\sqrt{10} - 1\right) = \frac{13\sqrt{13} + 80\sqrt{10} - 16}{27}$$

35. $y = 2x^{3/2} \;\Rightarrow\; y' = 3x^{1/2} \;\Rightarrow\; 1 + (y')^2 = 1 + 9x$. The arc length function with starting point $P_0(1, 2)$ is

$s(x) = \int_1^x \sqrt{1 + 9t}\,dt = \left[\frac{2}{27}(1 + 9t)^{3/2}\right]_1^x = \frac{2}{27}\left[(1 + 9x)^{3/2} - 10\sqrt{10}\right]$.

36. (a) $y = f(x) = \ln(\sin x) \;\Rightarrow\; y' = \dfrac{1}{\sin x}\cdot\cos x = \cot x \;\Rightarrow\; 1 + (y')^2 = 1 + \cot^2 x = \csc^2 x \;\Rightarrow$

$\sqrt{1 + (y')^2} = \sqrt{\csc^2 x} = |\csc x|$. Therefore,

$$s(x) = \int_{\pi/2}^x \sqrt{1 + [f'(t)]^2}\,dt = \int_{\pi/2}^x \csc t\,dt = \Big[\ln|\csc t - \cot t|\Big]_{\pi/2}^x$$

$$= \ln|\csc x - \cot x| - \ln|1 - 0| = \ln(\csc x - \cot x)$$

(b) Note that s is increasing on $(0, \pi)$ and that $x = 0$ and $x = \pi$ are vertical asymptotes for both f and s.

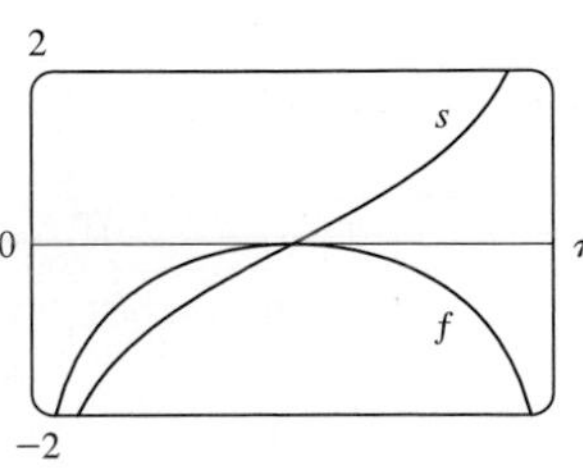

37. $y = \sin^{-1} x + \sqrt{1 - x^2} \;\Rightarrow\; y' = \dfrac{1}{\sqrt{1 - x^2}} - \dfrac{x}{\sqrt{1 - x^2}} = \dfrac{1 - x}{\sqrt{1 - x^2}} \;\Rightarrow$

$$1 + (y')^2 = 1 + \frac{(1 - x)^2}{1 - x^2} = \frac{1 - x^2 + 1 - 2x + x^2}{1 - x^2} = \frac{2 - 2x}{1 - x^2} = \frac{2(1 - x)}{(1 + x)(1 - x)} = \frac{2}{1 + x} \;\Rightarrow$$

$\sqrt{1+(y')^2} = \sqrt{\dfrac{2}{1+x}}$. Thus, the arc length function with starting point $(0,1)$ is given by

$$s(x) = \int_0^x \sqrt{1+[f'(t)]^2}\,dt = \int_0^x \sqrt{\frac{2}{1+t}}\,dt = \sqrt{2}\left[2\sqrt{1+t}\right]_0^x = 2\sqrt{2}\left(\sqrt{1+x}-1\right).$$

38. (a) $s(x) = \int_a^x \sqrt{1+[f'(t)]^2}\,dt$ and $s(x) = \int_0^x \sqrt{3t+5}\,dt \Rightarrow 1+[f'(t)]^2 = 3t+5 \Rightarrow [f'(t)]^2 = 3t+4 \Rightarrow$ $f'(t) = \sqrt{3t+4}$ [since f is increasing]. So $f(t) = \int (3t+4)^{1/2}\,dt = \frac{2}{3}\cdot\frac{1}{3}(3t+4)^{3/2} + C$ and since f has y-intercept 2, $f(0) = \frac{2}{9}\cdot 8 + C$ and $f(0) = 2 \Rightarrow C = 2 - \frac{16}{9} = \frac{2}{9}$. Thus, $f(t) = \frac{2}{9}(3t+4)^{3/2} + \frac{2}{9}$.

(b) $s(x) = \displaystyle\int_0^x \sqrt{3t+5}\,dt = \left[\tfrac{2}{9}(3t+5)^{3/2}\right]_0^x = \tfrac{2}{9}(3x+5)^{3/2} - \tfrac{2}{9}(5)^{3/2}$.

$s(x) = 3 \Leftrightarrow \frac{2}{9}(3x+5)^{3/2} = 3 + \frac{2}{9}(5\sqrt{5}) \Leftrightarrow (3x+5)^{3/2} = \frac{27}{2} + 5\sqrt{5} \Leftrightarrow 3x+5 = \left(\frac{27}{2}+5\sqrt{5}\right)^{2/3} \Rightarrow$ $x_1 = \frac{1}{3}\left[\left(\frac{27}{2}+5\sqrt{5}\right)^{2/3} - 5\right]$. Thus, the point on the graph of f that is 3 units along the curve from the y-intercept is $(x_1, f(x_1)) \approx (1.159, 4.765)$.

39. $f(x) = \frac{1}{4}e^x + e^{-x} \Rightarrow f'(x) = \frac{1}{4}e^x - e^{-x} \Rightarrow$

$1+[f'(x)]^2 = 1 + \left(\frac{1}{4}e^x - e^{-x}\right)^2 = 1 + \frac{1}{16}e^{2x} - \frac{1}{2} + e^{-2x} = \frac{1}{16}e^{2x} + \frac{1}{2} + e^{-2x} = \left(\frac{1}{4}e^x + e^{-x}\right)^2 = [f(x)]^2$. The arc length of the curve $y = f(x)$ on the interval $[a,b]$ is $L = \int_a^b \sqrt{1+[f'(x)]^2}\,dx = \int_a^b \sqrt{[f(x)]^2}\,dx = \int_a^b f(x)\,dx$, which is the area under the curve $y = f(x)$ on the interval $[a,b]$.

40. $y = 150 - \frac{1}{40}(x-50)^2 \Rightarrow y' = -\frac{1}{20}(x-50) \Rightarrow 1+(y')^2 = 1 + \frac{1}{20^2}(x-50)^2$, so the distance traveled by the kite is

$$L = \int_0^{80} \sqrt{1+\frac{1}{20^2}(x-50)^2}\,dx = \int_{-5/2}^{3/2} \sqrt{1+u^2}\,(20\,du) \qquad \begin{bmatrix} u = \frac{1}{20}(x-50), \\ du = \frac{1}{20}\,dx \end{bmatrix}$$

$$\overset{21}{=} 20\left[\tfrac{1}{2}u\sqrt{1+u^2} + \tfrac{1}{2}\ln\left(u+\sqrt{1+u^2}\right)\right]_{-5/2}^{3/2} = 10\left[\tfrac{3}{2}\sqrt{\tfrac{13}{4}} + \ln\left(\tfrac{3}{2}+\sqrt{\tfrac{13}{4}}\right) + \tfrac{5}{2}\sqrt{\tfrac{29}{4}} - \ln\left(-\tfrac{5}{2}+\sqrt{\tfrac{29}{4}}\right)\right]$$

$$= \tfrac{15}{2}\sqrt{13} + \tfrac{25}{2}\sqrt{29} + 10\ln\left(\frac{3+\sqrt{13}}{-5+\sqrt{29}}\right) \approx 122.8 \text{ ft}$$

41. The prey hits the ground when $y = 0 \Leftrightarrow 180 - \frac{1}{45}x^2 = 0 \Leftrightarrow x^2 = 45\cdot 180 \Rightarrow x = \sqrt{8100} = 90$, since x must be positive. $y' = -\frac{2}{45}x \Rightarrow 1+(y')^2 = 1 + \frac{4}{45^2}x^2$, so the distance traveled by the prey is

$$L = \int_0^{90} \sqrt{1+\frac{4}{45^2}x^2}\,dx = \int_0^4 \sqrt{1+u^2}\left(\tfrac{45}{2}\,du\right) \qquad \begin{bmatrix} u = \frac{2}{45}x, \\ du = \frac{2}{45}\,dx \end{bmatrix}$$

$$\overset{21}{=} \tfrac{45}{2}\left[\tfrac{1}{2}u\sqrt{1+u^2} + \tfrac{1}{2}\ln\left(u+\sqrt{1+u^2}\right)\right]_0^4 = \tfrac{45}{2}\left[2\sqrt{17} + \tfrac{1}{2}\ln\left(4+\sqrt{17}\right)\right] = 45\sqrt{17} + \tfrac{45}{4}\ln\left(4+\sqrt{17}\right) \approx 209.1 \text{ m}$$

42. Let $y = a - b\cosh cx$, where $a = 211.49$, $b = 20.96$, and $c = 0.03291765$. Then $y' = -bc\sinh cx \Rightarrow$ $1+(y')^2 = 1 + b^2c^2\sinh^2(cx)$. So $L = \int_{-91.2}^{91.2} \sqrt{1+b^2c^2\sinh^2(cx)}\,dx \approx 451.137 \approx 451$, to the nearest meter.

43. The sine wave has amplitude 1 and period 14, since it goes through two periods in a distance of 28 in., so its equation is $y = 1\sin\left(\frac{2\pi}{14}x\right) = \sin\left(\frac{\pi}{7}x\right)$. The width w of the flat metal sheet needed to make the panel is the arc length of the sine curve from $x = 0$ to $x = 28$. We set up the integral to evaluate w using the arc length formula with $\frac{dy}{dx} = \frac{\pi}{7}\cos\left(\frac{\pi}{7}x\right)$:

$L = \int_0^{28} \sqrt{1 + \left[\frac{\pi}{7}\cos\left(\frac{\pi}{7}x\right)\right]^2}\,dx = 2\int_0^{14} \sqrt{1 + \left[\frac{\pi}{7}\cos\left(\frac{\pi}{7}x\right)\right]^2}\,dx$. This integral would be very difficult to evaluate exactly, so we use a CAS, and find that $L \approx 29.36$ inches.

44. (a) $y = c + a\cosh\left(\frac{x}{a}\right) \;\Rightarrow\; y' = \sinh\left(\frac{x}{a}\right) \;\Rightarrow\; 1 + (y')^2 = 1 + \sinh^2\left(\frac{x}{a}\right) = \cosh^2\left(\frac{x}{a}\right)$. So

$$L = \int_{-b}^{b} \sqrt{\cosh^2\left(\frac{x}{a}\right)}\,dx = 2\int_0^b \cosh\left(\frac{x}{a}\right)dx = 2\left[a\sinh\left(\frac{x}{a}\right)\right]_0^b = 2a\sinh\left(\frac{b}{a}\right).$$

(b) At $x = 0$, $y = c + a$, so $c + a = 20$. The poles are 50 ft apart, so $b = 25$, and $L = 51 \;\Rightarrow\; 51 = 2a\sinh(b/a)$ [from part (a)]. From the figure, we see that $y = 51$ intersects $y = 2x\sinh(25/x)$ at $x \approx 72.3843$ for $x > 0$. So $a \approx 72.3843$ and the wire should be attached at a distance of $y = c + a\cosh(25/a) = 20 - a + a\cosh(25/a) \approx 24.36$ ft above the ground.

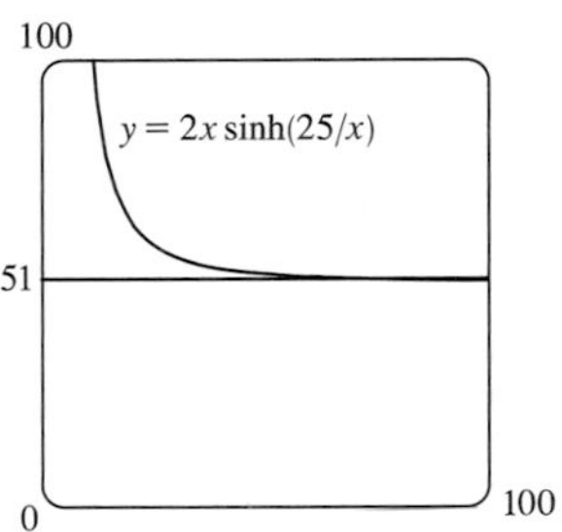

45. $y = \int_1^x \sqrt{t^3 - 1}\,dt \;\Rightarrow\; dy/dx = \sqrt{x^3 - 1}$ [by FTC1] $\;\Rightarrow\; 1 + (dy/dx)^2 = 1 + \left(\sqrt{x^3 - 1}\right)^2 = x^3 \;\Rightarrow$

$$L = \int_1^4 \sqrt{x^3}\,dx = \int_1^4 x^{3/2}\,dx = \tfrac{2}{5}\left[x^{5/2}\right]_1^4 = \tfrac{2}{5}(32 - 1) = \tfrac{62}{5} = 12.4$$

46. By symmetry, the length of the curve in each quadrant is the same, so we'll find the length in the first quadrant and multiply by 4.

$x^{2k} + y^{2k} = 1 \;\Rightarrow\; y^{2k} = 1 - x^{2k} \;\Rightarrow\; y = (1 - x^{2k})^{1/(2k)}$

(in the first quadrant), so we use the arc length formula with

$$\frac{dy}{dx} = \frac{1}{2k}(1 - x^{2k})^{1/(2k)-1}(-2kx^{2k-1}) = -x^{2k-1}(1 - x^{2k})^{1/(2k)-1}$$

The total length is therefore

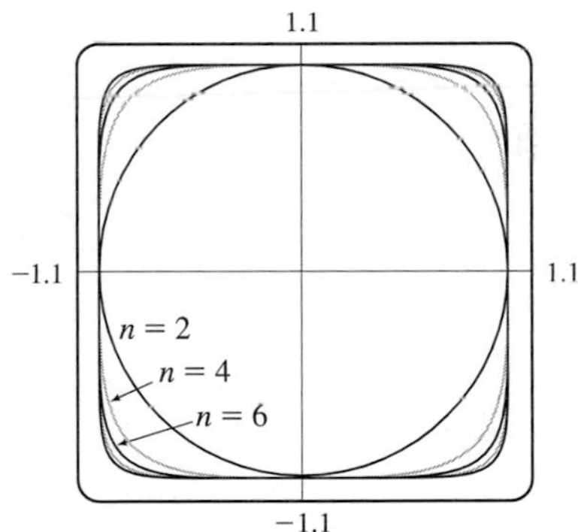

$$L_{2k} = 4\int_0^1 \sqrt{1 + [-x^{2k-1}(1 - x^{2k})^{1/(2k)-1}]^2}\,dx = 4\int_0^1 \sqrt{1 + x^{2(2k-1)}(1 - x^{2k})^{1/k-2}}\,dx$$

Now from the graph, we see that as k increases, the "corners" of these fat circles get closer to the points $(\pm1, \pm1)$ and $(\pm1, \mp1)$, and the "edges" of the fat circles approach the lines joining these four points. It seems plausible that as $k \to \infty$, the total length of the fat circle with $n = 2k$ will approach the length of the perimeter of the square with sides of length 2. This is supported by taking the limit as $k \to \infty$ of the equation of the fat circle in the first quadrant: $\lim\limits_{k\to\infty} (1 - x^{2k})^{1/(2k)} = 1$ for $0 \le x < 1$. So we guess that $\lim\limits_{k\to\infty} L_{2k} = 4 \cdot 2 = 8$.

DISCOVERY PROJECT Arc Length Contest

For advice on how to run the contest and a list of student entries, see the article "Arc Length Contest" by Larry Riddle in *The College Mathematics Journal*, Volume 29, No. 4, September 1998, pages 314–320.

8.2 Area of a Surface of Revolution

1. (a) (i) $y = \tan x \;\Rightarrow\; dy/dx = \sec^2 x \;\Rightarrow\; ds = \sqrt{1 + (dy/dx)^2}\,dx = \sqrt{1 + \sec^4 x}\,dx$. By (7), an integral for the area of the surface obtained by rotating the curve about the x-axis is $S = \int 2\pi y\,ds = \int_0^{\pi/3} 2\pi \tan x\sqrt{1 + \sec^4 x}\,dx$.

(ii) By (8), an integral for the area of the surface obtained by rotating the curve about the y-axis is $S = \int 2\pi x\,ds = \int_0^{\pi/3} 2\pi x\sqrt{1 + \sec^4 x}\,dx$.

(b) (i) 10.5017 (ii) 7.9353

2. (a) (i) $y = x^{-2} \;\Rightarrow\; dy/dx = -2x^{-3} \;\Rightarrow\; ds = \sqrt{1 + (dy/dx)^2}\,dx = \sqrt{1 + 4x^{-6}}\,dx$.

By (7), $S = \int 2\pi y\,ds = \int_1^2 2\pi x^{-2}\sqrt{1 + 4x^{-6}}\,dx$.

(ii) By (8), $S = \int 2\pi x\,ds = \int_1^2 2\pi x\sqrt{1 + 4x^{-6}}\,dx$.

(b) (i) 4.4566 (ii) 11.7299

3. (a) (i) $y = e^{-x^2} \;\Rightarrow\; dy/dx = e^{-x^2}\cdot(-2x) \;\Rightarrow\; ds = \sqrt{1 + (dy/dx)^2}\,dx = \sqrt{1 + 4x^2e^{-2x^2}}\,dx$.

By (7), $S = \int 2\pi y\,ds = \int_{-1}^1 2\pi e^{-x^2}\sqrt{1 + 4x^2e^{-2x^2}}\,dx$.

(ii) By (8), $S = \int 2\pi x\,ds = \int_0^1 2\pi x\sqrt{1 + 4x^2e^{-2x^2}}\,dx$ [symmetric about the y-axis]

(b) (i) 11.0753 (ii) 3.9603

4. (a) (i) $x = \ln(2y + 1) \;\Rightarrow\; dx/dy = \dfrac{2}{2y + 1} \;\Rightarrow\; ds = \sqrt{1 + (dx/dy)^2}\,dy = \sqrt{1 + 4/(2y + 1)^2}\,dy$.

By (7), $S = \int 2\pi y\,ds = \int_0^1 2\pi y\sqrt{1 + 4/(2y + 1)^2}\,dy$.

(ii) By (8), $S = \int 2\pi x\,ds = \int_0^1 2\pi \ln(2y + 1)\sqrt{1 + 4/(2y + 1)^2}\,dy$.

(b) (i) 4.2583 (ii) 5.6053

5. (a) (i) $x = y + y^3 \;\Rightarrow\; dx/dy = 1 + 3y^2 \;\Rightarrow\; ds = \sqrt{1 + (dx/dy)^2}\,dy = \sqrt{1 + (1 + 3y^2)^2}\,dy$.

By (7), $S = \int 2\pi y\,ds = \int_0^1 2\pi y\sqrt{1 + (1 + 3y^2)^2}\,dy$.

(ii) By (8), $S = \int 2\pi x\,ds = \int_0^1 2\pi(y + y^3)\sqrt{1 + (1 + 3y^2)^2}\,dy$.

(b) (i) 8.5302 (ii) 13.5134

6. (a) (i) $y = \tan^{-1} x \;\Rightarrow\; dy/dx = 1/(1 + x^2) \;\Rightarrow\; ds = \sqrt{1 + (dy/dx)^2}\,dx = \sqrt{1 + 1/(1 + x^2)^2}\,dx$.

By (7), $S = \int 2\pi y\,ds = \int_0^2 2\pi \tan^{-1} x\sqrt{1 + 1/(1 + x^2)^2}\,dx$.

(ii) By (8), $S = \int 2\pi x\,ds = \int_0^2 2\pi x\sqrt{1 + 1/(1 + x^2)^2}\,dx$.

(b) (i) 9.7956 (ii) 13.7209

7. $y = x^3 \;\Rightarrow\; y' = 3x^2$. So

$$S = \int_0^2 2\pi y\,\sqrt{1+(y')^2}\,dx = 2\pi \int_0^2 x^3\,\sqrt{1+9x^4}\,dx = \tfrac{2\pi}{36}\int_1^{145}\sqrt{u}\,du \qquad [u = 1+9x^4,\ du = 36x^3\,dx]$$

$$= \tfrac{\pi}{18}\left[\tfrac{2}{3}u^{3/2}\right]_1^{145} = \tfrac{\pi}{27}\left(145\sqrt{145}-1\right)$$

8. $y = \sqrt{5-x} \;\Rightarrow\; y' = \frac{1}{2}(5-x)^{-1/2}(-1) = -1/(2\sqrt{5-x})$. So

$$S = \int_3^5 2\pi y\sqrt{1+(y')^2}\,dx = \int_3^5 2\pi\sqrt{5-x}\sqrt{1+\frac{1}{4(5-x)}}\,dx = 2\pi\int_3^5 \sqrt{5-x+\tfrac{1}{4}}\,dx$$

$$= 2\pi\int_3^5 \sqrt{\tfrac{21}{4}-x}\,dx = 2\pi\int_{9/4}^{1/4}\sqrt{u}\,(-du) \qquad \begin{bmatrix} u = \frac{21}{4}-x, \\ du = -dx \end{bmatrix}$$

$$= 2\pi\int_{1/4}^{9/4} u^{1/2}\,du = 2\pi\left[\tfrac{2}{3}u^{3/2}\right]_{1/4}^{9/4} = \tfrac{4\pi}{3}\left(\tfrac{27}{8}-\tfrac{1}{8}\right) = \tfrac{13\pi}{3}$$

9. $y^2 = x+1 \;\Rightarrow\; y = \sqrt{x+1}$ (for $0 \le x \le 3$ and $1 \le y \le 2$) $\Rightarrow\; y' = 1/(2\sqrt{x+1})$. So

$$S = \int_0^3 2\pi y\sqrt{1+(y')^2}\,dx = 2\pi\int_0^3 \sqrt{x+1}\sqrt{1+\frac{1}{4(x+1)}}\,dx = 2\pi\int_0^3 \sqrt{x+1+\tfrac{1}{4}}\,dx$$

$$= 2\pi\int_0^3 \sqrt{x+\tfrac{5}{4}}\,dx = 2\pi\int_{5/4}^{17/4}\sqrt{u}\,du \qquad \begin{bmatrix} u = x+\frac{5}{4}, \\ du = dx \end{bmatrix}$$

$$= 2\pi\left[\tfrac{2}{3}u^{3/2}\right]_{5/4}^{17/4} = 2\pi\cdot\frac{2}{3}\left(\frac{17^{3/2}}{8}-\frac{5^{3/2}}{8}\right) = \frac{\pi}{6}(17\sqrt{17}-5\sqrt{5}).$$

10. $y = \sqrt{1+e^x} \;\Rightarrow\; y' = \frac{1}{2}(1+e^x)^{-1/2}(e^x) = \dfrac{e^x}{2\sqrt{1+e^x}} \;\Rightarrow$

$$\sqrt{1+(y')^2} = \sqrt{1+\frac{e^{2x}}{4(1+e^x)}} = \sqrt{\frac{4+4e^x+e^{2x}}{4(1+e^x)}} = \sqrt{\frac{(e^x+2)^2}{4(1+e^x)}} = \frac{e^x+2}{2\sqrt{1+e^x}}.\ \text{So}$$

$$S = \int_0^1 2\pi y\,\sqrt{1+(y')^2}\,dx = 2\pi\int_0^1 \sqrt{1+e^x}\,\frac{e^x+2}{2\sqrt{1+e^x}}\,dx = \pi\int_0^1 (e^x+2)\,dx$$

$$= \pi\left[e^x+2x\right]_0^1 = \pi[(e+2)-(1+0)] = \pi(e+1)$$

11. $y = \cos\left(\frac{1}{2}x\right) \;\Rightarrow\; y' = -\frac{1}{2}\sin\left(\frac{1}{2}x\right)$. So

$$S = \int_0^\pi 2\pi y\sqrt{1+(y')^2}\,dx = 2\pi\int_0^\pi \cos\left(\tfrac{1}{2}x\right)\sqrt{1+\tfrac{1}{4}\sin^2\left(\tfrac{1}{2}x\right)}\,dx$$

$$= 2\pi\int_0^1 \sqrt{1+\tfrac{1}{4}u^2}\,(2\,du) \qquad \begin{bmatrix} u = \sin\left(\frac{1}{2}x\right), \\ du = \frac{1}{2}\cos\left(\frac{1}{2}x\right)\,dx \end{bmatrix}$$

$$= 2\pi\int_0^1 \sqrt{4+u^2}\,du \overset{21}{=} 2\pi\left[\frac{u}{2}\sqrt{4+u^2}+2\ln\left(u+\sqrt{4+u^2}\right)\right]_0^1$$

$$= 2\pi\left[\left(\tfrac{1}{2}\sqrt{5}+2\ln(1+\sqrt{5})\right)-(0+2\ln 2)\right] = \pi\sqrt{5}+4\pi\ln\left(\frac{1+\sqrt{5}}{2}\right)$$

12. $y = \dfrac{x^3}{6} + \dfrac{1}{2x} \Rightarrow \dfrac{dy}{dx} = \dfrac{x^2}{2} - \dfrac{1}{2x^2} \Rightarrow \sqrt{1 + \left(\dfrac{dy}{dx}\right)^2} = \sqrt{\dfrac{x^4}{4} + \dfrac{1}{2} + \dfrac{1}{4x^4}} = \sqrt{\left(\dfrac{x^2}{2} + \dfrac{1}{2x^2}\right)^2} = \dfrac{x^2}{2} + \dfrac{1}{2x^2} \Rightarrow$

$$S = \int_{1/2}^{1} 2\pi\left(\frac{x^3}{6} + \frac{1}{2x}\right)\left(\frac{x^2}{2} + \frac{1}{2x^2}\right) dx = 2\pi \int_{1/2}^{1} \left(\frac{x^5}{12} + \frac{x}{12} + \frac{x}{4} + \frac{1}{4x^3}\right) dx$$

$$= 2\pi \int_{1/2}^{1} \left(\frac{x^5}{12} + \frac{x}{3} + \frac{x^{-3}}{4}\right) dx = 2\pi \left[\frac{x^6}{72} + \frac{x^2}{6} - \frac{x^{-2}}{8}\right]_{1/2}^{1}$$

$$= 2\pi\left[\left(\tfrac{1}{72} + \tfrac{1}{6} - \tfrac{1}{8}\right) - \left(\tfrac{1}{64 \cdot 72} + \tfrac{1}{24} - \tfrac{1}{2}\right)\right] = 2\pi\left(\tfrac{263}{512}\right) = \tfrac{263}{256}\pi$$

13. $x = \frac{1}{3}(y^2 + 2)^{3/2} \Rightarrow dx/dy = \frac{1}{2}(y^2 + 2)^{1/2}(2y) = y\sqrt{y^2 + 2} \Rightarrow 1 + (dx/dy)^2 = 1 + y^2(y^2 + 2) = (y^2 + 1)^2$.

So $S = 2\pi \int_1^2 y(y^2 + 1)\, dy = 2\pi\left[\frac{1}{4}y^4 + \frac{1}{2}y^2\right]_1^2 = 2\pi\left(4 + 2 - \frac{1}{4} - \frac{1}{2}\right) = \frac{21\pi}{2}$.

14. $x = 1 + 2y^2 \Rightarrow 1 + (dx/dy)^2 = 1 + (4y)^2 = 1 + 16y^2$.

So $S = 2\pi \int_1^2 y\sqrt{1 + 16y^2}\, dy = \frac{\pi}{16}\int_1^2 (16y^2 + 1)^{1/2} 32y\, dy = \frac{\pi}{16}\left[\frac{2}{3}(16y^2 + 1)^{3/2}\right]_1^2 = \frac{\pi}{24}\left(65\sqrt{65} - 17\sqrt{17}\right)$.

15. $y = \frac{1}{3}x^{3/2} \Rightarrow y' = \frac{1}{2}x^{1/2} \Rightarrow 1 + (y')^2 = 1 + \frac{1}{4}x$. So

$$S = \int_0^{12} 2\pi x\sqrt{1 + (y')^2}\, dx = 2\pi \int_0^{12} x\sqrt{1 + \tfrac{1}{4}x}\, dx = 2\pi \int_0^{12} x\, \tfrac{1}{2}\sqrt{4 + x}\, dx$$

$$= \pi \int_4^{16} (u - 4)\sqrt{u}\, du \qquad \left[\begin{array}{l} u = x + 4, \\ du = dx \end{array}\right]$$

$$= \pi \int_4^{16} \left(u^{3/2} - 4u^{1/2}\right) du = \pi\left[\tfrac{2}{5}u^{5/2} - \tfrac{8}{3}u^{3/2}\right]_4^{16} = \pi\left[\left(\tfrac{2}{5} \cdot 1024 - \tfrac{8}{3} \cdot 64\right) - \left(\tfrac{2}{5} \cdot 32 - \tfrac{8}{3} \cdot 8\right)\right]$$

$$= \pi\left(\tfrac{2}{5} \cdot 992 - \tfrac{8}{3} \cdot 56\right) = \pi\left(\tfrac{5952 - 2240}{15}\right) = \tfrac{3712\pi}{15}$$

16. $x^{2/3} + y^{2/3} = 1$, $0 \le y \le 1$. The curve is symmetric about the y-axis from $x = -1$ to $x = 1$, so we'll use the portion of the curve from $x = 0$ to $x = 1$. $y^{2/3} = 1 - x^{2/3} \Rightarrow y = (1 - x^{2/3})^{3/2} \Rightarrow$

$y' = \frac{3}{2}(1 - x^{2/3})^{1/2}\left(-\frac{2}{3}x^{-1/3}\right) = -\dfrac{\sqrt{1 - x^{2/3}}}{x^{1/3}} \Rightarrow 1 + (y')^2 = 1 + \dfrac{1 - x^{2/3}}{x^{2/3}} = \dfrac{x^{2/3} + 1 - x^{2/3}}{x^{2/3}} = x^{-2/3}$. So

$S = \int_0^1 2\pi x\sqrt{1 + (y')^2}\, dx = 2\pi \int_0^1 x(x^{-1/3})\, dx = 2\pi \int_0^1 x^{2/3}\, dx = 2\pi\left[\frac{3}{5}x^{5/3}\right]_0^1 = 2\pi\left(\frac{3}{5}\right) = \frac{6\pi}{5}$.

17. $x = \sqrt{a^2 - y^2} \Rightarrow dx/dy = \frac{1}{2}(a^2 - y^2)^{-1/2}(-2y) = -y/\sqrt{a^2 - y^2} \Rightarrow$

$$1 + \left(\frac{dx}{dy}\right)^2 = 1 + \frac{y^2}{a^2 - y^2} = \frac{a^2 - y^2}{a^2 - y^2} + \frac{y^2}{a^2 - y^2} = \frac{a^2}{a^2 - y^2} \Rightarrow$$

$$S = \int_0^{a/2} 2\pi\sqrt{a^2 - y^2}\frac{a}{\sqrt{a^2 - y^2}}\, dy = 2\pi \int_0^{a/2} a\, dy = 2\pi a\left[y\right]_0^{a/2} = 2\pi a\left(\frac{a}{2} - 0\right) = \pi a^2.$$

Note that this is $\frac{1}{4}$ the surface area of a sphere of radius a, and the length of the interval $y = 0$ to $y = a/2$ is $\frac{1}{4}$ the length of the interval $y = -a$ to $y = a$.

18. $y = \frac{1}{4}x^2 - \frac{1}{2}\ln x \quad\Rightarrow\quad \dfrac{dy}{dx} = \dfrac{x}{2} - \dfrac{1}{2x} \quad\Rightarrow\quad 1 + \left(\dfrac{dy}{dx}\right)^2 = 1 + \dfrac{x^2}{4} - \dfrac{1}{2} + \dfrac{1}{4x^2} = \dfrac{x^2}{4} + \dfrac{1}{2} + \dfrac{1}{4x^2} = \left(\dfrac{x}{2} + \dfrac{1}{2x}\right)^2$. So

$$\begin{aligned} S &= \int_1^2 2\pi x \sqrt{\left(\frac{x}{2} + \frac{1}{2x}\right)^2}\,dx = 2\pi \int_1^2 x\left(\frac{x}{2} + \frac{1}{2x}\right)dx = \pi \int_1^2 (x^2 + 1)\,dx = \pi\left[\tfrac{1}{3}x^3 + x\right]_1^2 \\ &= \pi\left[\left(\tfrac{8}{3} + 2\right) - \left(\tfrac{1}{3} + 1\right)\right] = \tfrac{10}{3}\pi \end{aligned}$$

19. $y = \frac{1}{5}x^5 \quad\Rightarrow\quad dy/dx = x^4 \quad\Rightarrow\quad 1 + (dy/dx)^2 = 1 + x^8 \quad\Rightarrow\quad S = \int_0^5 2\pi\left(\frac{1}{5}x^5\right)\sqrt{1 + x^8}\,dx.$

Let $f(x) = \frac{2}{5}\pi x^5\sqrt{1 + x^8}$. Since $n = 10$, $\Delta x = \frac{5-0}{10} = \frac{1}{2}$. Then

$$\begin{aligned} S \approx S_{10} &= \tfrac{1/2}{3}[f(0) + 4f(0.5) + 2f(1) + 4f(1.5) + 2f(2) + 4f(2.5) + 2f(3) \\ &\qquad + 4f(3.5) + 2f(4) + 4f(4.5) + f(5)] \\ &\approx 1{,}230{,}507 \end{aligned}$$

The value of the integral produced by a calculator is approximately 1,227,192.

20. $y = x + x^2 \quad\Rightarrow\quad dy/dx = 1 + 2x \quad\Rightarrow\quad 1 + (dy/dx)^2 = 1 + (1 + 2x)^2 \quad\Rightarrow\quad S = \int_0^1 2\pi(x + x^2)\sqrt{1 + (1 + 2x)^2}\,dx.$

Let $f(x) = 2\pi(x + x^2)\sqrt{1 + (1 + 2x)^2}$. Since $n = 10$, $\Delta x = \frac{1-0}{10} = \frac{1}{10}$. Then

$$\begin{aligned} S \approx S_{10} &= \tfrac{1/10}{3}[f(0) + 4f(0.1) + 2f(0.2) + 4f(0.3) + 2f(0.4) + 4f(0.5) + 2f(0.6) \\ &\qquad + 4f(0.7) + 2f(0.8) + 4f(0.9) + f(1)] \\ &\approx 13.649368 \end{aligned}$$

The value of the integral produced by a calculator is 13.649370 (to six decimal places).

21. $y = xe^x \quad\Rightarrow\quad dy/dx = xe^x + e^x \quad\Rightarrow\quad 1 + (dy/dx)^2 = 1 + (xe^x + e^x)^2 \quad\Rightarrow\quad S = \int_0^1 2\pi xe^x\sqrt{1 + (xe^x + e^x)^2}\,dx.$

Let $f(x) = 2\pi xe^x\sqrt{1 + (xe^x + e^x)^2}$. Since $n = 10$, $\Delta x = \frac{1-0}{10} = \frac{1}{10}$. Then

$$\begin{aligned} S \approx S_{10} &= \tfrac{1/10}{3}[f(0) + 4f(0.1) + 2f(0.2) + 4f(0.3) + 2f(0.4) + 4f(0.5) + 2f(0.6) \\ &\qquad + 4f(0.7) + 2f(0.8) + 4f(0.9) + f(1)] \\ &\approx 24.145807 \end{aligned}$$

The value of the integral produced by a calculator is 24.144251 (to six decimal places).

22. $y = x\ln x \quad\Rightarrow\quad dy/dx = x \cdot \dfrac{1}{x} + \ln x = 1 + \ln x \quad\Rightarrow\quad 1 + (dy/dx)^2 = 1 + (1 + \ln x)^2 \quad\Rightarrow$

$S = \int_1^2 2\pi x\ln x\sqrt{1 + (1 + \ln x)^2}\,dx$. Let $f(x) = 2\pi x\ln x\sqrt{1 + (1 + \ln x)^2}$. Since $n = 10$, $\Delta x = \frac{2-1}{10} = \frac{1}{10}$. Then

$$\begin{aligned} S \approx S_{10} &= \tfrac{1/10}{3}[f(1) + 4f(1.1) + 2f(1.2) + 4f(1.3) + 2f(1.4) + 4f(1.5) + 2f(1.6) \\ &\qquad + 4f(1.7) + 2f(1.8) + 4f(1.9) + f(2)] \\ &\approx 7.248933 \end{aligned}$$

The value of the integral produced by a calculator is 7.248934 (to six decimal places).

23. $y = 1/x \quad \Rightarrow \quad ds = \sqrt{1 + (dy/dx)^2}\, dx = \sqrt{1 + (-1/x^2)^2}\, dx = \sqrt{1 + 1/x^4}\, dx \quad \Rightarrow$

$$S = \int_1^2 2\pi \cdot \frac{1}{x}\sqrt{1 + \frac{1}{x^4}}\, dx = 2\pi \int_1^2 \frac{\sqrt{x^4+1}}{x^3}\, dx = 2\pi \int_1^4 \frac{\sqrt{u^2+1}}{u^2}\left(\tfrac{1}{2}\, du\right) \qquad [u = x^2,\, du = 2x\, dx]$$

$$= \pi \int_1^4 \frac{\sqrt{1+u^2}}{u^2}\, du \overset{24}{=} \pi\left[-\frac{\sqrt{1+u^2}}{u} + \ln\left(u + \sqrt{1+u^2}\right)\right]_1^4$$

$$= \pi\left[-\tfrac{\sqrt{17}}{4} + \ln(4 + \sqrt{17}) + \tfrac{\sqrt{2}}{1} - \ln(1 + \sqrt{2})\right] = \frac{\pi}{4}\left[4\ln(\sqrt{17}+4) - 4\ln(\sqrt{2}+1) - \sqrt{17} + 4\sqrt{2}\right]$$

24. $y = \sqrt{x^2+1} \quad \Rightarrow \quad \dfrac{dy}{dx} = \dfrac{x}{\sqrt{x^2+1}} \quad \Rightarrow \quad ds = \sqrt{1 + \left(\dfrac{dy}{dx}\right)^2}\, dx = \sqrt{1 + \dfrac{x^2}{x^2+1}}\, dx \quad \Rightarrow$

$$S = \int_0^3 2\pi\sqrt{x^2+1}\sqrt{1 + \frac{x^2}{x^2+1}}\, dx = 2\pi \int_0^3 \sqrt{2x^2+1}\, dx = 2\sqrt{2}\,\pi \int_0^3 \sqrt{x^2 + \left(\tfrac{1}{\sqrt{2}}\right)^2}\, dx$$

$$\overset{21}{=} 2\sqrt{2}\pi\left[\tfrac{1}{2}x\sqrt{x^2+\tfrac{1}{2}} + \tfrac{1}{4}\ln\left(x + \sqrt{x^2+\tfrac{1}{2}}\right)\right]_0^3 = 2\sqrt{2}\pi\left[\tfrac{3}{2}\sqrt{9+\tfrac{1}{2}} + \tfrac{1}{4}\ln\left(3 + \sqrt{9+\tfrac{1}{2}}\right) - \tfrac{1}{4}\ln\tfrac{1}{\sqrt{2}}\right]$$

$$= 2\sqrt{2}\pi\left[\tfrac{3}{2}\sqrt{\tfrac{19}{2}} + \tfrac{1}{4}\ln\left(3 + \sqrt{\tfrac{19}{2}}\right) + \tfrac{1}{4}\ln\sqrt{2}\right] = 2\sqrt{2}\pi\left[\tfrac{3}{2}\tfrac{\sqrt{19}}{\sqrt{2}} + \tfrac{1}{4}\ln(3\sqrt{2} + \sqrt{19})\right]$$

$$= 3\sqrt{19}\pi + \tfrac{\pi}{\sqrt{2}}\ln(3\sqrt{2} + \sqrt{19})$$

25. $y = x^3$ and $0 \le y \le 1 \quad \Rightarrow \quad y' = 3x^2$ and $0 \le x \le 1$.

$$S = \textstyle\int_0^1 2\pi x\sqrt{1 + (3x^2)^2}\, dx = 2\pi\int_0^3 \sqrt{1+u^2}\,\tfrac{1}{6}\, du \quad \begin{bmatrix} u = 3x^2, \\ du = 6x\, dx \end{bmatrix} \quad = \tfrac{\pi}{3}\int_0^3 \sqrt{1+u^2}\, du$$

$$\overset{21}{=} \text{[or use CAS]} \quad \tfrac{\pi}{3}\left[\tfrac{1}{2}u\sqrt{1+u^2} + \tfrac{1}{2}\ln\left(u + \sqrt{1+u^2}\right)\right]_0^3 = \tfrac{\pi}{3}\left[\tfrac{3}{2}\sqrt{10} + \tfrac{1}{2}\ln(3+\sqrt{10})\right] = \tfrac{\pi}{6}\left[3\sqrt{10} + \ln(3+\sqrt{10})\right]$$

26. $y = \ln(x+1),\, 0 \le x \le 1.\ ds = \sqrt{1 + \left(\dfrac{dy}{dx}\right)^2}\, dx = \sqrt{1 + \left(\dfrac{1}{x+1}\right)^2}\, dx$, so

$$S = \int_0^1 2\pi x\sqrt{1 + \frac{1}{(x+1)^2}}\, dx = \int_1^2 2\pi(u-1)\sqrt{1 + \frac{1}{u^2}}\, du \qquad [u = x+1,\, du = dx]$$

$$= 2\pi\int_1^2 u\frac{\sqrt{1+u^2}}{u}\, du - 2\pi\int_1^2 \frac{\sqrt{1+u^2}}{u}\, du = 2\pi\int_1^2 \sqrt{1+u^2}\, du - 2\pi\int_1^2 \frac{\sqrt{1+u^2}}{u}\, du$$

$$\overset{21,\,23}{=} \text{[or use CAS]} \quad 2\pi\left[\tfrac{1}{2}u\sqrt{1+u^2} + \tfrac{1}{2}\ln\left(u + \sqrt{1+u^2}\right)\right]_1^2 - 2\pi\left[\sqrt{1+u^2} - \ln\left(\frac{1+\sqrt{1+u^2}}{u}\right)\right]_1^2$$

$$= 2\pi\left[\sqrt{5} + \tfrac{1}{2}\ln(2+\sqrt{5}) - \tfrac{1}{2}\sqrt{2} - \tfrac{1}{2}\ln(1+\sqrt{2})\right] - 2\pi\left[\sqrt{5} - \ln\left(\tfrac{1+\sqrt{5}}{2}\right) - \sqrt{2} + \ln(1+\sqrt{2})\right]$$

$$= 2\pi\left[\tfrac{1}{2}\ln(2+\sqrt{5}) + \ln\left(\tfrac{1+\sqrt{5}}{2}\right) + \tfrac{\sqrt{2}}{2} - \tfrac{3}{2}\ln(1+\sqrt{2})\right]$$

27. $S = 2\pi\displaystyle\int_1^\infty y\sqrt{1 + \left(\frac{dy}{dx}\right)^2}\, dx = 2\pi\int_1^\infty \frac{1}{x}\sqrt{1 + \frac{1}{x^4}}\, dx = 2\pi\int_1^\infty \frac{\sqrt{x^4+1}}{x^3}\, dx$. Rather than trying to evaluate this integral, note that $\sqrt{x^4+1} > \sqrt{x^4} = x^2$ for $x > 0$. Thus, if the area is finite,

$S = 2\pi \int_1^\infty \frac{\sqrt{x^4+1}}{x^3}\,dx > 2\pi \int_1^\infty \frac{x^2}{x^3}\,dx = 2\pi \int_1^\infty \frac{1}{x}\,dx$. But we know that this integral diverges, so the area S is infinite.

28. $S = \int_0^\infty 2\pi y\,\sqrt{1+(dy/dx)^2}\,dx = 2\pi \int_0^\infty e^{-x}\sqrt{1+(-e^{-x})^2}\,dx \qquad [y = e^{-x},\, y' = -e^{-x}]$.

Evaluate $I = \int e^{-x}\sqrt{1+(-e^{-x})^2}\,dx$ by using the substitution $u = -e^{-x}$, $du = e^{-x}\,dx$:

$I = \int \sqrt{1+u^2}\,du \overset{21}{=} \frac{1}{2}u\sqrt{1+u^2} + \frac{1}{2}\ln\left(u+\sqrt{1+u^2}\right) + C = \frac{1}{2}(-e^{-x})\sqrt{1+e^{-2x}} + \frac{1}{2}\ln\left(-e^{-x}+\sqrt{1+e^{-2x}}\right) + C.$

Returning to the surface area integral, we have

$$\begin{aligned} S &= 2\pi \lim_{t\to\infty} \int_0^t e^{-x}\sqrt{1+(-e^{-x})^2}\,dx = 2\pi \lim_{t\to\infty} \left[\tfrac{1}{2}(-e^{-x})\sqrt{1+e^{-2x}} + \tfrac{1}{2}\ln\left(-e^{-x}+\sqrt{1+e^{-2x}}\right)\right]_0^t \\ &= 2\pi \lim_{t\to\infty} \left\{\left[\tfrac{1}{2}(-e^{-t})\sqrt{1+e^{-2t}} + \tfrac{1}{2}\ln\left(-e^{-t}+\sqrt{1+e^{-2t}}\right)\right] - \left[\tfrac{1}{2}(-1)\sqrt{1+1} + \tfrac{1}{2}\ln\left(-1+\sqrt{1+1}\right)\right]\right\} \\ &= 2\pi \left\{\left[\tfrac{1}{2}(0)\sqrt{1} + \tfrac{1}{2}\ln\left(0+\sqrt{1}\right)\right] - \left[-\tfrac{1}{2}\sqrt{2} + \tfrac{1}{2}\ln\left(-1+\sqrt{2}\right)\right]\right\} \\ &= 2\pi\left\{[0] + \tfrac{1}{2}\left[\sqrt{2} - \ln\left(\sqrt{2}-1\right)\right]\right\} = \pi\left[\sqrt{2} - \ln\left(\sqrt{2}-1\right)\right] \end{aligned}$$

29. Since $a > 0$, the curve $3ay^2 = x(a-x)^2$ only has points with $x \geq 0$.

$[3ay^2 \geq 0 \;\Rightarrow\; x(a-x)^2 \geq 0 \;\Rightarrow\; x \geq 0.]$

The curve is symmetric about the x-axis (since the equation is unchanged when y is replaced by $-y$). $y = 0$ when $x = 0$ or a, so the curve's loop extends from $x = 0$ to $x = a$.

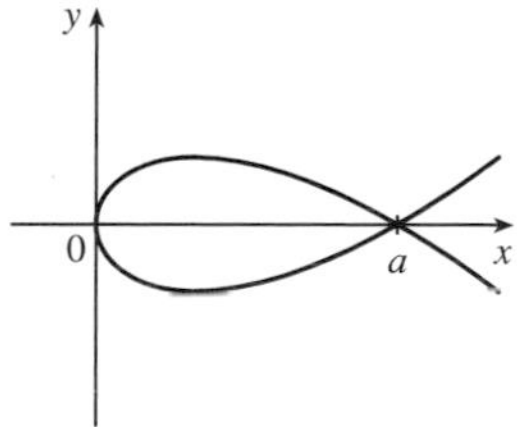

$$\frac{d}{dx}(3ay^2) = \frac{d}{dx}[x(a-x)^2] \;\Rightarrow\; 6ay\frac{dy}{dx} = x\cdot 2(a-x)(-1) + (a-x)^2 \;\Rightarrow\; \frac{dy}{dx} = \frac{(a-x)[-2x+a-x]}{6ay} \;\Rightarrow$$

$$\left(\frac{dy}{dx}\right)^2 = \frac{(a-x)^2(a-3x)^2}{36a^2y^2} = \frac{(a-x)^2(a-3x)^2}{36a^2}\cdot\frac{3a}{x(a-x)^2} \;\begin{bmatrix}\text{the last fraction}\\ \text{is } 1/y^2\end{bmatrix} = \frac{(a-3x)^2}{12ax} \;\Rightarrow$$

$$1 + \left(\frac{dy}{dx}\right)^2 = 1 + \frac{a^2-6ax+9x^2}{12ax} = \frac{12ax}{12ax} + \frac{a^2-6ax+9x^2}{12ax} = \frac{a^2+6ax+9x^2}{12ax} = \frac{(a+3x)^2}{12ax} \quad \text{for } x \neq 0.$$

(a) $$\begin{aligned} S &= \int_{x=0}^a 2\pi y\,ds = 2\pi\int_0^a \frac{\sqrt{x}\,(a-x)}{\sqrt{3a}}\cdot\frac{a+3x}{\sqrt{12ax}}\,dx = 2\pi\int_0^a \frac{(a-x)(a+3x)}{6a}\,dx \\ &= \frac{\pi}{3a}\int_0^a (a^2+2ax-3x^2)\,dx = \frac{\pi}{3a}\left[a^2x+ax^2-x^3\right]_0^a = \frac{\pi}{3a}(a^3+a^3-a^3) = \frac{\pi}{3a}\cdot a^3 = \frac{\pi a^2}{3}. \end{aligned}$$

Note that we have rotated the top half of the loop about the x-axis. This generates the full surface.

(b) We must rotate the full loop about the y-axis, so we get double the area obtained by rotating the top half of the loop:

$$\begin{aligned} S &= 2\cdot 2\pi\int_{x=0}^a x\,ds = 4\pi\int_0^a x\,\frac{a+3x}{\sqrt{12ax}}\,dx = \frac{4\pi}{2\sqrt{3a}}\int_0^a x^{1/2}(a+3x)\,dx = \frac{2\pi}{\sqrt{3a}}\int_0^a (ax^{1/2}+3x^{3/2})\,dx \\ &= \frac{2\pi}{\sqrt{3a}}\left[\frac{2}{3}ax^{3/2}+\frac{6}{5}x^{5/2}\right]_0^a = \frac{2\pi\sqrt{3}}{3\sqrt{a}}\left(\frac{2}{3}a^{5/2}+\frac{6}{5}a^{5/2}\right) = \frac{2\pi\sqrt{3}}{3}\left(\frac{2}{3}+\frac{6}{5}\right)a^2 = \frac{2\pi\sqrt{3}}{3}\left(\frac{28}{15}\right)a^2 \\ &= \frac{56\pi\sqrt{3}\,a^2}{45} \end{aligned}$$

30. In general, if the parabola $y = ax^2$, $-c \le x \le c$, is rotated about the y-axis, the surface area it generates is

$$2\pi \int_0^c x\sqrt{1+(2ax)^2}\,dx = 2\pi \int_0^{2ac} \frac{u}{2a}\sqrt{1+u^2}\,\frac{1}{2a}\,du \quad \begin{bmatrix} u = 2ax, \\ du = 2a\,dx \end{bmatrix} \quad = \frac{\pi}{4a^2}\int_0^{2ac} \left(1+u^2\right)^{1/2} 2u\,du$$

$$= \frac{\pi}{4a^2}\left[\tfrac{2}{3}\left(1+u^2\right)^{3/2}\right]_0^{2ac} = \frac{\pi}{6a^2}\left[\left(1+4a^2c^2\right)^{3/2} - 1\right]$$

Here $2c = 10$ ft and $ac^2 = 2$ ft, so $c = 5$ and $a = \frac{2}{25}$. Thus, the surface area is

$$S = \tfrac{\pi}{6}\tfrac{625}{4}\left[\left(1+4\cdot\tfrac{4}{625}\cdot 25\right)^{3/2} - 1\right] = \tfrac{625\pi}{24}\left[\left(1+\tfrac{16}{25}\right)^{3/2} - 1\right] = \tfrac{625\pi}{24}\left(\tfrac{41\sqrt{41}}{125} - 1\right) = \tfrac{5\pi}{24}\left(41\sqrt{41} - 125\right) \approx 90.01 \text{ ft}^2.$$

31. (a) $\dfrac{x^2}{a^2} + \dfrac{y^2}{b^2} = 1 \;\Rightarrow\; \dfrac{y\,(dy/dx)}{b^2} = -\dfrac{x}{a^2} \;\Rightarrow\; \dfrac{dy}{dx} = -\dfrac{b^2x}{a^2y} \;\Rightarrow$

$$1+\left(\frac{dy}{dx}\right)^2 = 1+\frac{b^4x^2}{a^4y^2} = \frac{b^4x^2+a^4y^2}{a^4y^2} = \frac{b^4x^2+a^4b^2\left(1-x^2/a^2\right)}{a^4b^2\left(1-x^2/a^2\right)} = \frac{a^4b^2+b^4x^2-a^2b^2x^2}{a^4b^2-a^2b^2x^2}$$

$$= \frac{a^4+b^2x^2-a^2x^2}{a^4-a^2x^2} = \frac{a^4-\left(a^2-b^2\right)x^2}{a^2\left(a^2-x^2\right)}$$

The ellipsoid's surface area is twice the area generated by rotating the first-quadrant portion of the ellipse about the x-axis. Thus,

$$S = 2\int_0^a 2\pi y\sqrt{1+\left(\frac{dy}{dx}\right)^2}\,dx = 4\pi\int_0^a \frac{b}{a}\sqrt{a^2-x^2}\,\frac{\sqrt{a^4-\left(a^2-b^2\right)x^2}}{a\sqrt{a^2-x^2}}\,dx = \frac{4\pi b}{a^2}\int_0^a \sqrt{a^4-\left(a^2-b^2\right)x^2}\,dx$$

$$= \frac{4\pi b}{a^2}\int_0^{a\sqrt{a^2-b^2}} \sqrt{a^4-u^2}\,\frac{du}{\sqrt{a^2-b^2}} \quad \left[u = \sqrt{a^2-b^2}\,x\right] \quad \overset{30}{=} \frac{4\pi b}{a^2\sqrt{a^2-b^2}}\left[\frac{u}{2}\sqrt{a^4-u^2}+\frac{a^4}{2}\sin^{-1}\!\left(\frac{u}{a^2}\right)\right]_0^{a\sqrt{a^2-b^2}}$$

$$= \frac{4\pi b}{a^2\sqrt{a^2-b^2}}\left[\frac{a\sqrt{a^2-b^2}}{2}\sqrt{a^4-a^2\left(a^2-b^2\right)}+\frac{a^4}{2}\sin^{-1}\frac{\sqrt{a^2-b^2}}{a}\right] = 2\pi\left[b^2+\frac{a^2b\sin^{-1}\dfrac{\sqrt{a^2-b^2}}{a}}{\sqrt{a^2-b^2}}\right]$$

(b) $\dfrac{x^2}{a^2} + \dfrac{y^2}{b^2} = 1 \;\Rightarrow\; \dfrac{x\,(dx/dy)}{a^2} = -\dfrac{y}{b^2} \;\Rightarrow\; \dfrac{dx}{dy} = -\dfrac{a^2y}{b^2x} \;\Rightarrow$

$$1+\left(\frac{dx}{dy}\right)^2 = 1+\frac{a^4y^2}{b^4x^2} = \frac{b^4x^2+a^4y^2}{b^4x^2} = \frac{b^4a^2(1-y^2/b^2)+a^4y^2}{b^4a^2(1-y^2/b^2)} = \frac{a^2b^4-a^2b^2y^2+a^4y^2}{a^2b^4-a^2b^2y^2}$$

$$= \frac{b^4-b^2y^2+a^2y^2}{b^4-b^2y^2} = \frac{b^4-\left(b^2-a^2\right)y^2}{b^2\left(b^2-y^2\right)}$$

The oblate spheroid's surface area is twice the area generated by rotating the first-quadrant portion of the ellipse about the y-axis. Thus,

$$S = 2\int_0^b 2\pi\,x\sqrt{1+\left(\frac{dx}{dy}\right)^2}\,dy = 4\pi\int_0^b \frac{a}{b}\sqrt{b^2-y^2}\,\frac{\sqrt{b^4-\left(b^2-a^2\right)y^2}}{b\sqrt{b^2-y^2}}\,dy$$

$$= \frac{4\pi a}{b^2}\int_0^b \sqrt{b^4-\left(b^2-a^2\right)y^2}\,dy = \frac{4\pi a}{b^2}\int_0^b \sqrt{b^4+\left(a^2-b^2\right)y^2}\,dy \qquad [\text{since } a > b]$$

$$= \frac{4\pi a}{b^2}\int_0^{b\sqrt{a^2-b^2}} \sqrt{b^4+u^2}\,\frac{du}{\sqrt{a^2-b^2}} \qquad \left[u = \sqrt{a^2-b^2}\,y\right]$$

$$\overset{21}{=} \frac{4\pi a}{b^2\sqrt{a^2-b^2}}\left[\frac{u}{2}\sqrt{b^4+u^2}+\frac{b^4}{2}\ln\!\left(u+\sqrt{b^4+u^2}\right)\right]_0^{b\sqrt{a^2-b^2}}$$

[continued]

$$= \frac{4\pi a}{b^2\sqrt{a^2-b^2}}\left\{\left[\frac{b\sqrt{a^2-b^2}}{2}(ab)+\frac{b^4}{2}\ln\left(b\sqrt{a^2-b^2}+ab\right)\right]-\left[0+\frac{b^4}{2}\ln(b^2)\right]\right\}$$

$$= \frac{4\pi a}{b^2\sqrt{a^2-b^2}}\left[\frac{ab^2\sqrt{a^2-b^2}}{2}+\frac{b^4}{2}\ln\frac{b\sqrt{a^2-b^2}+ab}{b^2}\right] = 2\pi a^2 + \frac{2\pi ab^2}{\sqrt{a^2-b^2}}\ln\frac{\sqrt{a^2-b^2}+a}{b}$$

32. The upper half of the torus is generated by rotating the curve $(x-R)^2+y^2=r^2$, $y>0$, about the y-axis.

$$y\frac{dy}{dx} = -(x-R) \;\Rightarrow\; 1+\left(\frac{dy}{dx}\right)^2 = 1+\frac{(x-R)^2}{y^2} = \frac{y^2+(x-R)^2}{y^2} = \frac{r^2}{r^2-(x-R)^2}.$$ Thus,

$$S = 2\int_{R-r}^{R+r} 2\pi x\sqrt{1+\left(\frac{dy}{dx}\right)^2}\,dx = 4\pi\int_{R-r}^{R+r}\frac{rx}{\sqrt{r^2-(x-R)^2}}\,dx = 4\pi r\int_{-r}^{r}\frac{u+R}{\sqrt{r^2-u^2}}\,du \qquad [u=x-R]$$

$$= 4\pi r\int_{-r}^{r}\frac{u\,du}{\sqrt{r^2-u^2}} + 4\pi Rr\int_{-r}^{r}\frac{du}{\sqrt{r^2-u^2}} = 4\pi r\cdot 0 + 8\pi Rr\int_0^r\frac{du}{\sqrt{r^2-u^2}} \qquad \left[\begin{array}{c}\text{since the first integrand is odd}\\ \text{and the second is even}\end{array}\right]$$

$$= 8\pi Rr\left[\sin^{-1}(u/r)\right]_0^r = 8\pi Rr\left(\tfrac{\pi}{2}\right) = 4\pi^2 Rr$$

33. The analogue of $f(x_i^*)$ in the derivation of (4) is now $c-f(x_i^*)$, so

$$S = \lim_{n\to\infty}\sum_{i=1}^n 2\pi[c-f(x_i^*)]\sqrt{1+[f'(x_i^*)]^2}\,\Delta x = \int_a^b 2\pi[c-f(x)]\sqrt{1+[f'(x)]^2}\,dx.$$

34. $y=x^{1/2} \;\Rightarrow\; y'=\frac{1}{2}x^{-1/2} \;\Rightarrow\; 1+(y')^2=1+1/4x$, so by Exercise 31, $S=\int_0^4 2\pi\left(4-\sqrt{x}\right)\sqrt{1+1/(4x)}\,dx$.

Using a CAS, we get $S=2\pi\ln\left(\sqrt{17}+4\right)+\frac{\pi}{6}\left(31\sqrt{17}+1\right)\approx 80.6095$.

35. For the upper semicircle, $f(x)=\sqrt{r^2-x^2}$, $f'(x)=-x/\sqrt{r^2-x^2}$. The surface area generated is

$$S_1 = \int_{-r}^{r} 2\pi\left(r-\sqrt{r^2-x^2}\right)\sqrt{1+\frac{x^2}{r^2-x^2}}\,dx = 4\pi\int_0^r\left(r-\sqrt{r^2-x^2}\right)\frac{r}{\sqrt{r^2-x^2}}\,dx$$

$$= 4\pi\int_0^r\left(\frac{r^2}{\sqrt{r^2-x^2}}-r\right)dx$$

For the lower semicircle, $f(x)=-\sqrt{r^2-x^2}$ and $f'(x)=\dfrac{x}{\sqrt{r^2-x^2}}$, so $S_2=4\pi\displaystyle\int_0^r\left(\frac{r^2}{\sqrt{r^2-x^2}}+r\right)dx$.

Thus, the total area is $S=S_1+S_2=8\pi\displaystyle\int_0^r\left(\frac{r^2}{\sqrt{r^2-x^2}}\right)dx = 8\pi\left[r^2\sin^{-1}\left(\frac{x}{r}\right)\right]_0^r = 8\pi r^2\left(\frac{\pi}{2}\right) = 4\pi^2 r^2$.

36. (a) Rotate $y=\sqrt{R^2-x^2}$ with $a\le x\le a+h$ about the x-axis to generate a zone of a sphere. $y=\sqrt{R^2-x^2} \;\Rightarrow$

$y'=\frac{1}{2}(R^2-x^2)^{-1/2}(-2x) \;\Rightarrow\; ds=\sqrt{1+\left(\dfrac{-x}{\sqrt{R^2-x^2}}\right)^2}\,dx$. The surface area is

$$S=\int_a^{a+h}2\pi y\,ds = 2\pi\int_a^{a+h}\sqrt{R^2-x^2}\sqrt{1+\frac{x^2}{R^2-x^2}}\,dx$$

$$= 2\pi\int_a^{a+h}\sqrt{R^2-x^2+x^2}\,dx = 2\pi R\left[x\right]_a^{a+h}$$

$$= 2\pi R(a+h-a) = 2\pi Rh$$

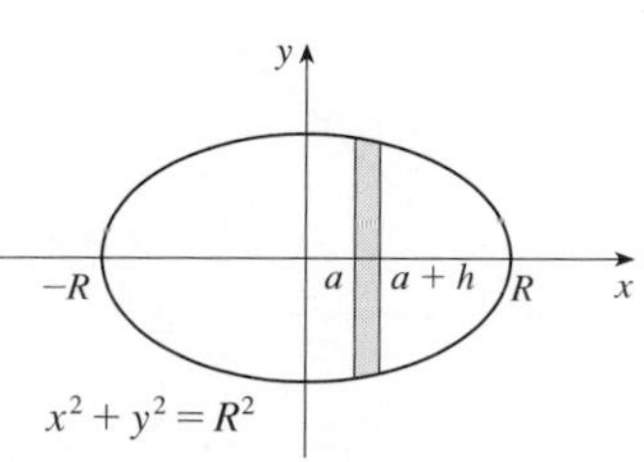

(b) Rotate $y = R$ with $0 \le x \le h$ about the x-axis to generate a zone of a cylinder. $y = R \;\Rightarrow\; y' = 0 \;\Rightarrow$

$ds = \sqrt{1 + 0^2}\,dx = dx$. The surface area is $S = \int_0^h 2\pi y\,ds = 2\pi \int_0^h R\,dx = 2\pi R\left[x\right]_0^h = 2\pi Rh$.

37. $y = e^{x/2} + e^{-x/2} \;\Rightarrow\; y' = \frac{1}{2}e^{x/2} - \frac{1}{2}e^{-x/2} \;\Rightarrow$

$1 + (y')^2 = 1 + \left(\frac{1}{2}e^{x/2} - \frac{1}{2}e^{-x/2}\right)^2 = 1 + \frac{1}{4}e^x - \frac{1}{2} + \frac{1}{4}e^{-x} = \frac{1}{4}e^x + \frac{1}{2} + \frac{1}{4}e^{-x} = \left(\frac{1}{2}e^{x/2} + \frac{1}{2}e^{-x/2}\right)^2$. If we rotate the curve about the x-axis on the interval $a \le x \le b$, the resulting surface area is

$S = \int_a^b 2\pi y\sqrt{1 + (y')^2}\,dx = 2\pi \int_a^b (e^{x/2} + e^{-x/2})\left(\frac{1}{2}e^{x/2} + \frac{1}{2}e^{-x/2}\right)\,dx = \pi \int_a^b (e^{x/2} + e^{-x/2})^2\,dx$, which is the same as the volume obtained by rotating the curve y about the x-axis on the interval $a \le x \le b$, namely, $V = \pi \int_a^b y^2\,dx$.

38. Since $g(x) = f(x) + c$, we have $g'(x) = f'(x)$. Thus,

$$\begin{aligned} S_g &= \int_a^b 2\pi g(x)\,\sqrt{1 + [g'(x)]^2}\,dx = \int_a^b 2\pi[f(x) + c]\,\sqrt{1 + [f'(x)]^2}\,dx \\ &= \int_a^b 2\pi f(x)\,\sqrt{1 + [f'(x)]^2}\,dx + 2\pi c\int_a^b \sqrt{1 + [f'(x)]^2}\,dx = S_f + 2\pi cL \end{aligned}$$

39. In the derivation of (4), we computed a typical contribution to the surface area to be $2\pi\,\dfrac{y_{i-1} + y_i}{2}\,|P_{i-1}P_i|$, the area of a frustum of a cone. When $f(x)$ is not necessarily positive, the approximations $y_i = f(x_i) \approx f(x_i^*)$ and $y_{i-1} = f(x_{i-1}) \approx f(x_i^*)$ must be replaced by $y_i = |f(x_i)| \approx |f(x_i^*)|$ and $y_{i-1} = |f(x_{i-1})| \approx |f(x_i^*)|$. Thus, $2\pi\,\dfrac{y_{i-1} + y_i}{2}\,|P_{i-1}P_i| \approx 2\pi\,|f(x_i^*)|\,\sqrt{1 + [f'(x_i^*)]^2}\,\Delta x$. Continuing with the rest of the derivation as before, we obtain $S = \int_a^b 2\pi\,|f(x)|\,\sqrt{1 + [f'(x)]^2}\,dx$.

DISCOVERY PROJECT Rotating on a Slant

1.

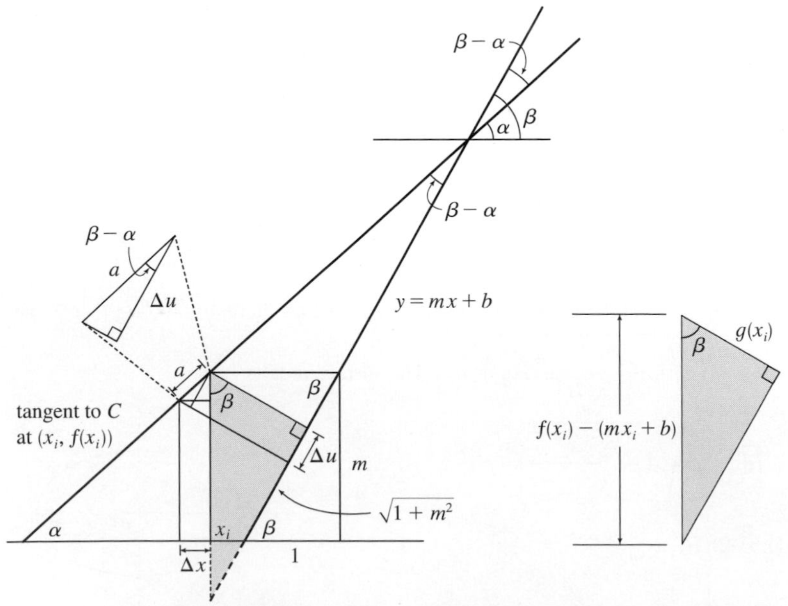

In the figure, the segment a lying above the interval $[x_i - \Delta x, x_i]$ along the tangent to C has length

$\Delta x \sec \alpha = \Delta x \sqrt{1+\tan^2 \alpha} = \sqrt{1+[f'(x_i)]^2}\, \Delta x$. The segment from $(x_i, f(x_i))$ drawn perpendicular to the line $y = mx + b$ has length

$$g(x_i) = [f(x_i) - mx_i - b] \cos \beta = \frac{f(x_i) - mx_i - b}{\sec \beta} = \frac{f(x_i) - mx_i - b}{\sqrt{1+\tan^2 \beta}} = \frac{f(x_i) - mx_i - b}{\sqrt{1+m^2}}$$

Also, $\cos(\beta - \alpha) = \dfrac{\Delta u}{\Delta x \sec \alpha} \quad \Rightarrow$

$$\Delta u = \Delta x \sec \alpha \cos(\beta - \alpha) = \Delta x \frac{\cos \beta \cos \alpha + \sin \beta \sin \alpha}{\cos \alpha} = \Delta x(\cos \beta + \sin \beta \tan \alpha)$$

$$= \Delta x \left[\frac{1}{\sqrt{1+m^2}} + \frac{m}{\sqrt{1+m^2}} f'(x_i) \right] = \frac{1 + mf'(x_i)}{\sqrt{1+m^2}} \Delta x$$

Thus,
$$\text{Area}(\mathscr{R}) = \lim_{n\to\infty} \sum_{i=1}^{n} g(x_i)\, \Delta u = \lim_{n\to\infty} \sum_{i=1}^{n} \frac{f(x_i) - mx_i - b}{\sqrt{1+m^2}} \cdot \frac{1 + mf'(x_i)}{\sqrt{1+m^2}} \Delta x$$

$$= \frac{1}{1+m^2} \int_p^q [f(x) - mx - b][1 + mf'(x)]\, dx$$

2. From Problem 1 with $m = 1$, $f(x) = x + \sin x$, $mx + b = x - 2$, $p = 0$, and $q = 2\pi$,

$$\text{Area} = \frac{1}{1+1^2} \int_0^{2\pi} [x + \sin x - (x-2)]\,[1 + 1(1 + \cos x)]\, dx = \tfrac{1}{2} \int_0^{2\pi} (\sin x + 2)(2 + \cos x)\, dx$$

$$= \tfrac{1}{2} \int_0^{2\pi} (2 \sin x + \sin x \cos x + 4 + 2 \cos x)\, dx = \tfrac{1}{2} \left[-2 \cos x + \tfrac{1}{2} \sin^2 x + 4x + 2 \sin x \right]_0^{2\pi}$$

$$= \tfrac{1}{2} [(-2 + 0 + 8\pi + 0) - (-2 + 0 + 0 + 0)] = \tfrac{1}{2}(8\pi) = 4\pi$$

3. $V = \displaystyle\lim_{n\to\infty} \sum_{i=1}^{n} \pi [g(x_i)]^2\, \Delta u = \lim_{n\to\infty} \sum_{i=1}^{n} \pi \left[\frac{f(x_i) - mx_i - b}{\sqrt{1+m^2}} \right]^2 \frac{1 + mf'(x_i)}{\sqrt{1+m^2}} \Delta x$

$$= \frac{\pi}{(1+m^2)^{3/2}} \int_p^q [f(x) - mx - b]^2 [1 + mf'(x)]\, dx$$

4. $V = \dfrac{\pi}{(1+1^2)^{3/2}} \displaystyle\int_0^{2\pi} (x + \sin x - x + 2)^2 (1 + 1 + \cos x)\, dx$

$$= \frac{\pi}{2\sqrt{2}} \int_0^{2\pi} (\sin x + 2)^2 (\cos x + 2)\, dx = \frac{\pi}{2\sqrt{2}} \int_0^{2\pi} (\sin^2 x + 4 \sin x + 4)(\cos x + 2)\, dx$$

$$= \frac{\pi}{2\sqrt{2}} \int_0^{2\pi} \left(\sin^2 x \cos x + 4 \sin x \cos x + 4 \cos x + 2 \sin^2 x + 8 \sin x + 8\right) dx$$

$$= \frac{\pi}{2\sqrt{2}} \left[\tfrac{1}{3} \sin^3 x + 2 \sin^2 x + 4 \sin x + x - \tfrac{1}{2} \sin 2x - 8 \cos x + 8x \right]_0^{2\pi} \qquad [\text{since } 2 \sin^2 x = 1 - \cos 2x]$$

$$= \frac{\pi}{2\sqrt{2}} [(2\pi - 8 + 16\pi) - (-8)] = \frac{9\sqrt{2}}{2} \pi^2$$

5. $S = \displaystyle\int_p^q 2\pi g(x) \sqrt{1 + [f'(x)]^2}\, dx = \frac{2\pi}{\sqrt{1+m^2}} \int_p^q [f(x) - mx - b] \sqrt{1 + [f'(x)]^2}\, dx$

6. From Problem 5 with $f(x) = \sqrt{x}$, $p = 0$, $q = 4$, $m = \frac{1}{2}$, and $b = 0$,

$$S = \frac{2\pi}{\sqrt{1 + (\frac{1}{2})^2}} \int_0^4 \left(\sqrt{x} - \tfrac{1}{2}x\right) \sqrt{1 + \left(\frac{1}{2\sqrt{x}}\right)^2}\, dx \overset{\text{CAS}}{=} \frac{\pi}{\sqrt{5}} \left[\frac{\ln(\sqrt{17} + 4)}{32} + \frac{37\sqrt{17}}{24} - \frac{1}{3} \right] \approx 8.554$$

8.3 Applications to Physics and Engineering

1. The weight density of water is $\delta = 62.5\ \text{lb/ft}^3$.

(a) $P = \delta d \approx (62.5\ \text{lb/ft}^3)(3\ \text{ft}) = 187.5\ \text{lb/ft}^2$

(b) $F = PA \approx (187.5\ \text{lb/ft}^2)(5\ \text{ft})(2\ \text{ft}) = 1875\ \text{lb}$. ($A$ is the area of the bottom of the tank.)

(c) As in Example 1, the area of the ith strip is $2\,(\Delta x)$ and the pressure is $\delta d = \delta x_i$. Thus,

$$F = \int_0^3 \delta x \cdot 2\,dx \approx (62.5)(2)\int_0^3 x\,dx = 125\left[\tfrac{1}{2}x^2\right]_0^3 = 125\left(\tfrac{9}{2}\right) = 562.5\ \text{lb}.$$

2. (a) $P = \rho g d = (820\ \text{kg/m}^3)(9.8\ \text{m/s}^2)(1.5\ \text{m}) = 12{,}054\ \text{Pa} \approx 12\ \text{kPa}$

(b) $F = PA = (12{,}054\ \text{Pa})(8\ \text{m})(4\ \text{m}) \approx 3.86 \times 10^5\ \text{N}$ (A is the area at the bottom of the tank.)

(c) The area of the ith strip is $4(\Delta x)$ and the pressure is $\rho g d = \rho g\, x_i$. Thus,

$$F = \int_0^{1.5} \rho g x \cdot 4\,dx = (820)(9.8)\cdot 4\int_0^{1.5} x\,dx = 32{,}144\left[\tfrac{1}{2}x^2\right]_0^{3/2} = 16{,}072\left(\tfrac{9}{4}\right) \approx 3.62 \times 10^4\ \text{N}.$$

In Exercises 3–9, n is the number of subintervals of length Δx and x_i^* is a sample point in the ith subinterval $[x_{i-1}, x_i]$.

3. Set up a vertical x-axis as shown, with $x = 0$ at the water's surface and x increasing in the downward direction. Then the area of the ith rectangular strip is $2\,\Delta x$ and the pressure on the strip is δx_i^* (where $\delta \approx 62.5\ \text{lb/ft}^3$). Thus, the hydrostatic force on the strip is $\delta x_i^* \cdot 2\,\Delta x$ and the total hydrostatic force $\approx \sum_{i=1}^{n} \delta x_i^* \cdot 2\,\Delta x$. The total force

$$F = \lim_{n\to\infty}\sum_{i=1}^{n} \delta x_i^* \cdot 2\,\Delta x = \int_3^{11} \delta x \cdot 2\,dx = 2\delta\int_3^{11} x\,dx = 2\delta\left[\tfrac{1}{2}x^2\right]_3^{11} = \delta(121 - 9) = 112\delta \approx 7000\ \text{lb}$$

4. Set up a vertical axis as shown. Then the area of the ith rectangular strip is $2(x_i^* - 2)\,\Delta x$. $\left[\text{By similar triangles, } \dfrac{w_i}{x_i^* - 2} = \dfrac{10}{5}, \text{ so } w_i = 2(x_i^* - 2).\right]$ The pressure on the strip is δx_i^*, so the hydrostatic force on the strip is $\delta x_i^* \cdot 2(x_i^* - 2)\,\Delta x$ and the total hydrostatic force on the plate $\approx \sum_{i=1}^{n} \delta x_i^* \cdot 2(x_i^* - 2)\,\Delta x$. The total force

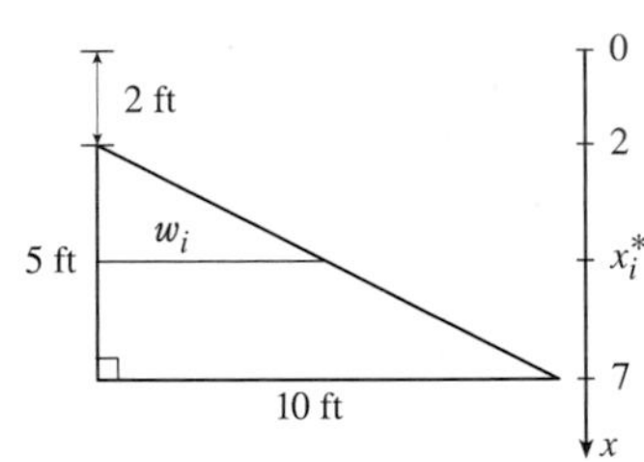

$$F = \lim_{n\to\infty}\sum_{i=1}^{n} \delta x_i^* \cdot 2(x_i^* - 2)\,\Delta x = \int_2^7 \delta x \cdot 2(x - 2)\,dx = 2\delta\int_2^7 (x^2 - 2x)\,dx$$

$$= 2\delta\left[\tfrac{1}{3}x^3 - x^2\right]_2^7 = 2\delta\left[\left(\tfrac{343}{3} - 49\right) - \left(\tfrac{8}{3} - 4\right)\right] = 2\delta\left(\tfrac{200}{3}\right) = \tfrac{400}{3}\delta \approx \tfrac{400}{3}(62.5) = 8333.\overline{3}\ \text{lb}.$$

5. Set up a coordinate system as shown. Then the area of the ith rectangular strip is $2\sqrt{8^2 - (y_i^*)^2}\,\Delta y$. The pressure on the strip is $\delta d_i = \rho g(12 - y_i^*)$, so the hydrostatic force on the strip is $\rho g(12 - y_i^*)\,2\sqrt{64 - (y_i^*)^2}\,\Delta y$ and the total hydrostatic force on the plate $\approx \sum_{i=1}^{n} \rho g(12 - y_i^*)\,2\sqrt{64 - (y_i^*)^2}\,\Delta y$.

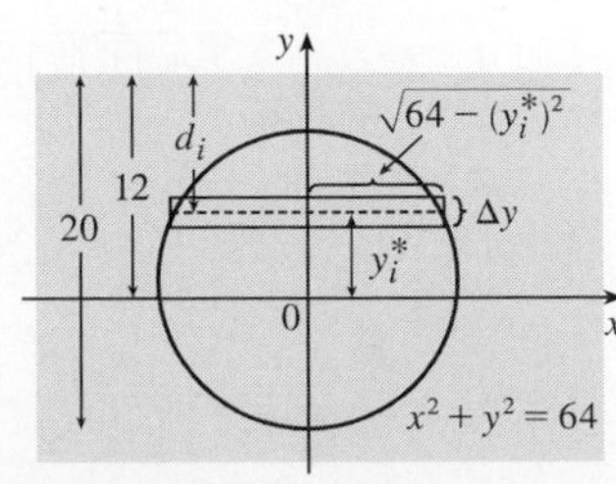

The total force $F = \lim_{n\to\infty} \sum_{i=1}^{n} \rho g(12 - y_i^*)\, 2\sqrt{64 - (y_i^*)^2}\, \Delta y = \int_{-8}^{8} \rho g(12 - y)\, 2\sqrt{64 - y^2}\, dy$

$$= 2\rho g \cdot 12 \int_{-8}^{8} \sqrt{64 - y^2}\, dy - 2\rho g \int_{-8}^{8} y\sqrt{64 - y^2}\, dy.$$

The second integral is 0 because the integrand is an odd function. The first integral is the area of a semicircular disk with radius 8. Thus, $F = 24\rho g\left(\frac{1}{2}\pi(8)^2\right) = 768\pi\rho g \approx 768\pi(1000)(9.8) \approx 2.36 \times 10^7$ N.

6. Set up a coordinate system as shown. Then the area of the ith rectangular strip is $2\sqrt{6^2 - (y_i^*)^2}\, \Delta y$. The pressure on the strip is $\delta d_i = \rho g(4 - y_i^*)$, so the hydrostatic force on the strip is $\rho g(4 - y_i^*)\, 2\sqrt{36 - (y_i^*)^2}\, \Delta y$ and the hydrostatic force on the plate $\approx \sum_{i=1}^{n} \rho g(4 - y_i^*)\, 2\sqrt{36 - (y_i^*)^2}\, \Delta y$. The total force $F = \lim_{n\to\infty} \sum_{i=1}^{n} \rho g(4 - y_i^*)\, 2\sqrt{36 - (y_i^*)^2}\, \Delta y = \int_0^4 \rho g(4 - y)\, 2\sqrt{36 - y^2}\, dy = 8\rho g I_1 - 2\rho g I_2$.

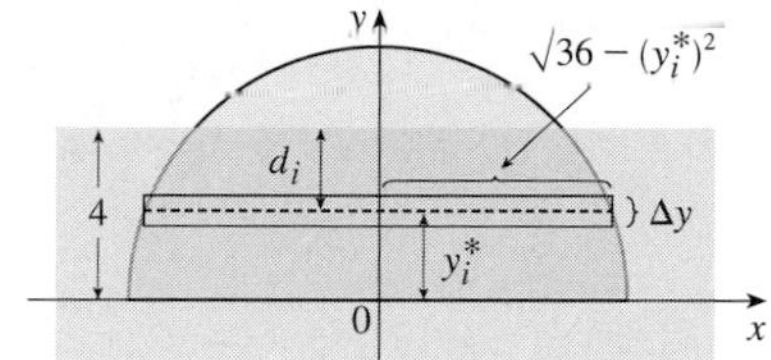

$$I_1 = \int_0^4 \sqrt{36 - y^2}\, dy = \int_0^\alpha \sqrt{36 - 36\sin^2\theta}\,(6\cos\theta\, d\theta) \qquad \begin{bmatrix} y = 6\sin\theta, \\ dy = 6\cos\theta\, d\theta \\ \alpha = \sin^{-1}(2/3) \end{bmatrix}$$

$$= \int_0^\alpha 36\cos^2\theta\, d\theta = \int_0^\alpha 36 \cdot \tfrac{1}{2}(1 + \cos 2\theta)\, d\theta = 18\left[\theta + \tfrac{1}{2}\sin 2\theta\right]_0^\alpha$$

$$= 18\left(\alpha + \tfrac{1}{2}\sin 2\alpha\right) = 18(\alpha + \sin\alpha\, \cos\alpha).$$

$$I_2 = \int_0^4 y\sqrt{36 - y^2}\, dy = \int_{36}^{20} \sqrt{u}\left(-\tfrac{1}{2}du\right) \qquad \begin{bmatrix} u = 36 - y^2, \\ du = -2y\, dy \end{bmatrix}$$

$$= -\tfrac{1}{2}\left[\tfrac{2}{3}u^{3/2}\right]_{36}^{20} = -\tfrac{1}{3}(20^{3/2} - 216) = 72 - \tfrac{40}{3}\sqrt{5}.$$

Thus,

$$F = 8\rho g \cdot 18(\alpha + \sin\alpha\, \cos\alpha) - 2\rho g\left(72 - \tfrac{40}{3}\sqrt{5}\right) = 144\rho g\left(\sin^{-1}\tfrac{2}{3} + \tfrac{2}{3}\tfrac{\sqrt{5}}{3}\right) - 2\rho g\left(72 - \tfrac{40}{3}\sqrt{5}\right)$$

$$= \rho g\left(144\sin^{-1}\tfrac{2}{3} + \tfrac{176}{3}\sqrt{5} - 144\right) \approx 9.04 \times 10^5 \text{ N} \quad [\rho = 1000,\ g \approx 9.8].$$

7. Set up a vertical x-axis as shown. Then the area of the ith rectangular strip is $\left(2 - \frac{2}{\sqrt{3}}x_i^*\right)\Delta x$. $\left[\text{By similar triangles, } \frac{w_i}{2} = \frac{\sqrt{3} - x_i^*}{\sqrt{3}}, \text{ so } w_i = 2 - \frac{2}{\sqrt{3}}x_i^*.\right]$ The pressure on the strip is $\rho g x_i^*$, so the hydrostatic force on the strip is $\rho g x_i^*\left(2 - \frac{2}{\sqrt{3}}x_i^*\right)\Delta x$ and the hydrostatic force on the plate $\approx \sum_{i=1}^{n} \rho g x_i^*\left(2 - \frac{2}{\sqrt{3}}x_i^*\right)\Delta x$. The total force

$$F = \lim_{n\to\infty} \sum_{i=1}^{n} \rho g x_i^*\left(2 - \frac{2}{\sqrt{3}}x_i^*\right)\Delta x = \int_0^{\sqrt{3}} \rho g x\left(2 - \frac{2}{\sqrt{3}}x\right)dx = \rho g \int_0^{\sqrt{3}} \left(2x - \frac{2}{\sqrt{3}}x^2\right)dx$$

$$= \rho g\left[x^2 - \frac{2}{3\sqrt{3}}x^3\right]_0^{\sqrt{3}} = \rho g\,[(3 - 2) - 0] = \rho g \approx 1000 \cdot 9.8 = 9.8 \times 10^3 \text{ N}$$

8. Set up a vertical x-axis as shown. Then the area of the ith rectangular strip is $3x_i^*\,\Delta x$. $\left[\text{By similar triangles, } \frac{w_i}{x_i^*} = \frac{6}{2}, \text{ so } w_i = 3x_i^*.\right]$ The pressure on the strip is $\rho g(x_i^* + 4)$, so the hydrostatic force on the strip is $\rho g(x_i^* + 4)3x_i^*\,\Delta x$ and the hydrostatic force on the plate $\approx \sum_{i=1}^{n} \rho g(x_i^* + 4)3x_i^*\,\Delta x$. The total force

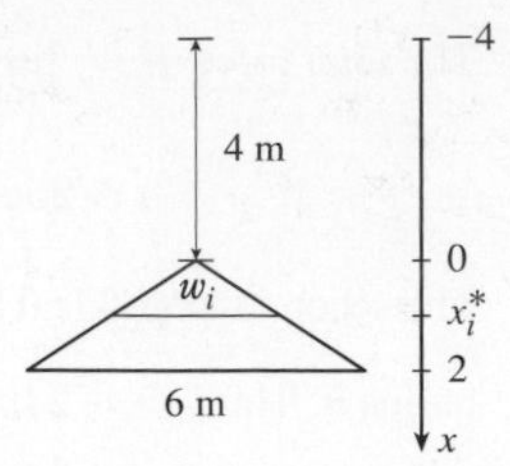

$$F = \lim_{n\to\infty} \sum_{i=1}^{n} \rho g(x_i^* + 4)\,3x_i^*\,\Delta x = \int_0^2 \rho g(x+4)\,3x\,dx = 3\rho g \int_0^2 (x^2 + 4x)\,dx$$

$$= 3\rho g\left[\tfrac{1}{3}x^3 + 2x^2\right]_0^2 = 3\rho g\left(\tfrac{8}{3} + 8\right) = 32\rho g = 313{,}600 \text{ N} \quad [\rho = 1000,\ g \approx 9.8]$$

9. Set up a vertical x-axis as shown. Then the area of the ith rectangular strip is $w_i\,\Delta x = \left(4 + 2\cdot\tfrac{2}{3}x_i^*\right)\Delta x$. The pressure on the strip is $\delta(x_i^* - 1)$, so the hydrostatic force on the strip is $\delta(x_i^* - 1)\left(4 + \tfrac{4}{3}x_i^*\right)\Delta x$ and the hydrostatic force on the plate $\approx \sum_{i=1}^{n} \delta(x_i^* - 1)\left(4 + \tfrac{4}{3}x_i^*\right)\Delta x$. The total force

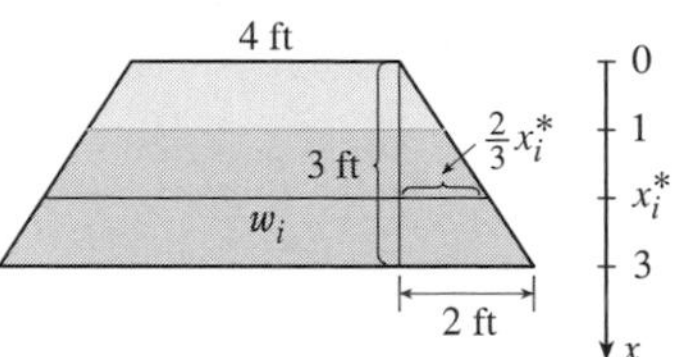

$$F = \lim_{n\to\infty} \sum_{i=1}^{n} \delta(x_i^* - 1)\left(4 + \tfrac{4}{3}x_i^*\right)\Delta x = \int_1^3 \delta(x-1)\left(4 + \tfrac{4}{3}x\right)dx = \delta\int_1^3 \left(\tfrac{4}{3}x^2 + \tfrac{8}{3}x - 4\right)dx$$

$$= \delta\left[\tfrac{4}{9}x^3 + \tfrac{4}{3}x^2 - 4x\right]_1^3 = \delta\left[(12 + 12 - 12) - \left(\tfrac{4}{9} + \tfrac{4}{3} - 4\right)\right] = \delta\left(\tfrac{128}{9}\right) \approx 889 \text{ lb} \quad [\delta \approx 62.5]$$

10. Set up coordinate axes as shown in the figure. For the *top half*, the length of the ith strip is $2\left(a/\sqrt{2} - y_i^*\right)$ and its area is $2\left(a/\sqrt{2} - y_i^*\right)\Delta y$. The pressure on this strip is approximately $\delta d_i = \delta\left(a/\sqrt{2} - y_i^*\right)$ and so the force on the strip is approximately $2\delta\left(a/\sqrt{2} - y_i^*\right)^2\Delta y$. The total force

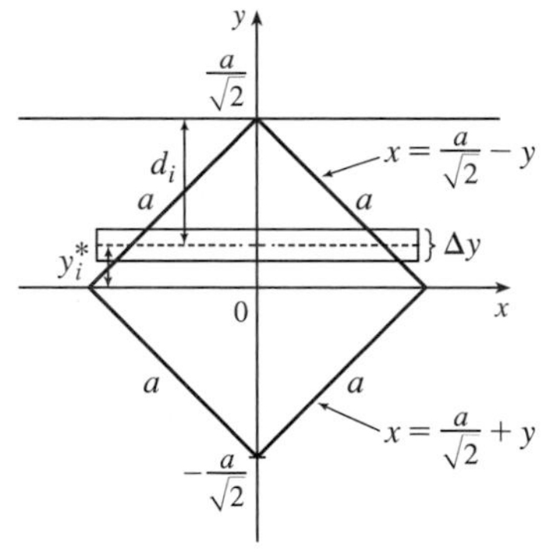

$$F_1 = \lim_{n\to\infty} \sum_{i=1}^{n} 2\delta\left(\frac{a}{\sqrt{2}} - y_i^*\right)^2 \Delta y = 2\delta\int_0^{a/\sqrt{2}} \left(\frac{a}{\sqrt{2}} - y\right)^2 dy$$

$$= 2\delta\left[-\frac{1}{3}\left(\frac{a}{\sqrt{2}} - y\right)^3\right]_0^{a/\sqrt{2}} = -\frac{2}{3}\delta\left[0 - \left(\frac{a}{\sqrt{2}}\right)^3\right] = \frac{2\delta}{3}\frac{a^3}{2\sqrt{2}} = \frac{\sqrt{2}\,a^3\delta}{6}$$

For the *bottom half*, the length is $2\left(a/\sqrt{2} + y_i^*\right)$ and the total force is

$$F_2 = \lim_{n\to\infty} \sum_{i=1}^{n} 2\delta\left(\frac{a}{\sqrt{2}} + y_i^*\right)\left(\frac{a}{\sqrt{2}} - y_i^*\right)\Delta y = 2\delta\int_{-a/\sqrt{2}}^{0} \left(\frac{a^2}{2} - y^2\right)dy = 2\delta\left[\tfrac{1}{2}a^2y - \tfrac{1}{3}y^3\right]_{-a/\sqrt{2}}^{0}$$

$$= 2\delta\left[0 - \left(-\frac{\sqrt{2}\,a^3}{4} + \frac{\sqrt{2}\,a^3}{12}\right)\right] = 2\delta\left(\frac{\sqrt{2}\,a^3}{6}\right) = \frac{2\sqrt{2}\,a^3\delta}{6} \qquad [F_2 = 2F_1]$$

Thus, the total force $F = F_1 + F_2 = \dfrac{3\sqrt{2}\,a^3\delta}{6} = \dfrac{\sqrt{2}\,a^3\delta}{2}$.

11. Set up a vertical x-axis as shown. Then the area of the ith rectangular strip is

$\frac{a}{h}(2h - x_i^*)\,\Delta x$. $\left[\text{By similar triangles, } \frac{w_i}{2h - x_i^*} = \frac{2a}{2h}, \text{ so } w_i = \frac{a}{h}(2h - x_i^*).\right]$

The pressure on the strip is δx_i^*, so the hydrostatic force on the plate

$\approx \sum_{i=1}^{n} \delta x_i^* \frac{a}{h}(2h - x_i^*)\,\Delta x$. The total force

$$F = \lim_{n\to\infty} \sum_{i=1}^{n} \delta x_i^* \frac{a}{h}(2h - x_i^*)\,\Delta x = \delta \frac{a}{h} \int_0^h x(2h - x)\,dx = \frac{a\delta}{h} \int_0^h (2hx - x^2)\,dx$$

$$= \frac{a\delta}{h}\left[hx^2 - \tfrac{1}{3}x^3\right]_0^h = \frac{a\delta}{h}\left(h^3 - \tfrac{1}{3}h^3\right) = \frac{a\delta}{h}\left(\frac{2h^3}{3}\right) = \tfrac{2}{3}\delta a h^2$$

12. (a) The solution is similar to the solution for Example 2. The pressure on a strip is approximately $\delta d_i = 64.6(3 - y_i^*)$ and the total force is

$$F = \lim_{n\to\infty} \sum_{i=1}^{n} 64.6(3 - y_i^*)2\sqrt{9 - (y_i^*)^2}\,\Delta y = 129.2 \int_{-3}^{3} (3 - y)\sqrt{9 - y^2}\,dy$$

$$= 129.2 \cdot 3 \int_{-3}^{3} \sqrt{9 - y^2}\,dy - 129.2 \int_{-3}^{3} y\sqrt{9 - y^2}\,dy$$

$$= 387.6 \cdot \tfrac{1}{2}\pi(3)^2 - 0 \qquad \left[\begin{array}{l}\text{the first integral is the area of a semicircular disk with radius 3 and} \\ \text{the second integral is 0 because the integrand is an odd function}\end{array}\right]$$

$$= (1744.2)\pi \approx 5480 \text{ lb}$$

(b) If the tank is half full, the surface of the milk is $y = 0$, so the pressure on a strip is approximately $\delta d_i = 64.6(0 - y_i^*)$. The upper limit of integration changes from 3 to 0 and the total force is

$$F = 129.2 \int_{-3}^{0} (0 - y)\sqrt{9 - y^2}\,dy = 129.2\left[\tfrac{1}{3}(9 - y^2)^{3/2}\right]_{-3}^{0} = 129.2(9 - 0) = 1162.8 \text{ lb}$$

Note that this is about 21% of the force for a full tank.

13. By similar triangles, $\frac{8}{4\sqrt{3}} = \frac{w_i}{x_i^*} \Rightarrow w_i = \frac{2x_i^*}{\sqrt{3}}$. The area of the ith rectangular strip is $\frac{2x_i^*}{\sqrt{3}}\,\Delta x$ and the pressure on it is $\rho g\left(4\sqrt{3} - x_i^*\right)$.

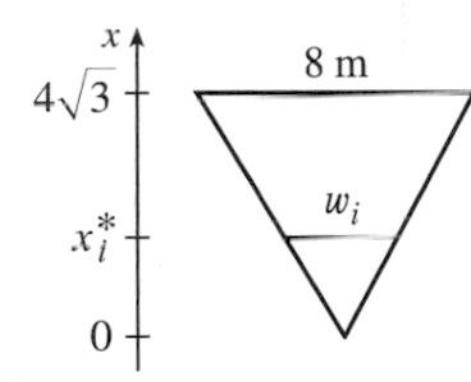

$$F = \int_0^{4\sqrt{3}} \rho g\left(4\sqrt{3} - x\right)\frac{2x}{\sqrt{3}}\,dx = 8\rho g \int_0^{4\sqrt{3}} x\,dx - \frac{2\rho g}{\sqrt{3}} \int_0^{4\sqrt{3}} x^2\,dx$$

$$= 4\rho g\left[x^2\right]_0^{4\sqrt{3}} - \frac{2\rho g}{3\sqrt{3}}\left[x^3\right]_0^{4\sqrt{3}} = 192\rho g - \frac{2\rho g}{3\sqrt{3}}\,64 \cdot 3\sqrt{3} = 192\rho g - 128\rho g = 64\rho g$$

$$\approx 64(840)(9.8) \approx 5.27 \times 10^5 \text{ N}$$

14. $F = \int_0^2 \rho g(10 - x)2\sqrt{4 - x^2}\,dx$

$= 20\rho g \int_0^2 \sqrt{4 - x^2}\,dx - \rho g \int_0^2 \sqrt{4 - x^2}\,2x\,dx$

$= 20\rho g \tfrac{1}{4}\pi(2^2) - \rho g \int_0^4 u^{1/2}\,du \qquad [u = 4 - x^2,\ du = -2x\,dx]$

$= 20\pi\rho g - \tfrac{2}{3}\rho g\left[u^{3/2}\right]_0^4 = 20\pi\rho g - \tfrac{16}{3}\rho g = \rho g\left(20\pi - \tfrac{16}{3}\right)$

$= (1000)(9.8)\left(20\pi - \tfrac{16}{3}\right) \approx 5.63 \times 10^5 \text{ N}$

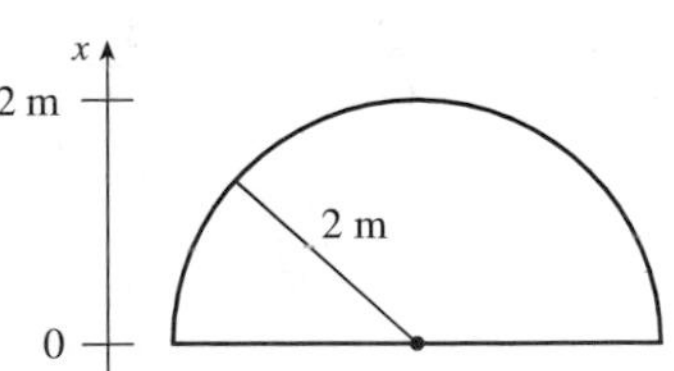

15. (a) The top of the cube has depth $d = 1\text{ m} - 20\text{ cm} = 80\text{ cm} = 0.8\text{ m}$.

$$F = \rho g d A \approx (1000)(9.8)(0.8)(0.2)^2 = 313.6 \approx 314\text{ N}$$

(b) The area of a strip is $0.2\,\Delta x$ and the pressure on it is $\rho g x_i^*$.

$$F = \int_{0.8}^{1} \rho g x(0.2)\,dx = 0.2\rho g\left[\tfrac{1}{2}x^2\right]_{0.8}^{1} = (0.2\rho g)(0.18) = 0.036\rho g = 0.036(1000)(9.8) = 352.8 \approx 353\text{ N}$$

16. The height of the dam is $h = \sqrt{70^2 - 25^2}\cos 30^\circ = 15\sqrt{19}\left(\frac{\sqrt{3}}{2}\right)$.

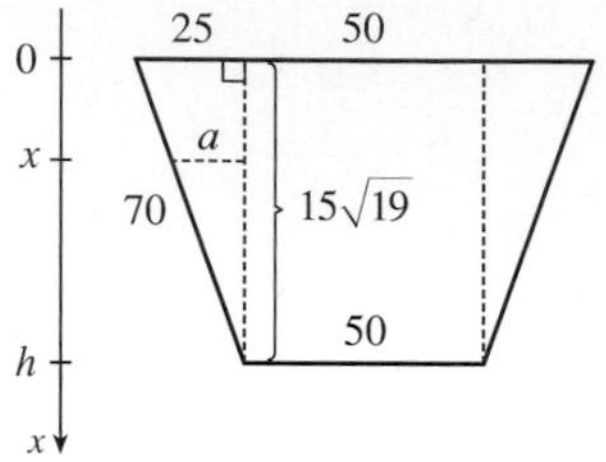

The width of the trapezoid is $w = 50 + 2a$.

By similar triangles, $\dfrac{25}{h} = \dfrac{a}{h-x} \quad\Rightarrow\quad a = \dfrac{25}{h}(h-x)$. Thus,

$$w = 50 + 2\cdot\frac{25}{h}(h-x) = 50 + \frac{50}{h}\cdot h - \frac{50}{h}\cdot x = 50 + 50 - \frac{50x}{h} = 100 - \frac{50x}{h}.$$

From the small triangle in the second figure, $\cos 30^\circ = \dfrac{\Delta x}{z} \quad\Rightarrow$

$z = \Delta x\sec 30^\circ = 2\,\Delta x/\sqrt{3}$.

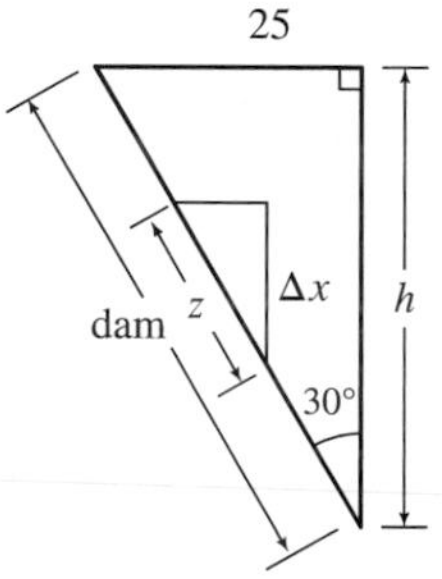

$$F = \int_0^h \delta x\left(100 - \frac{50x}{h}\right)\frac{2}{\sqrt{3}}\,dx = \frac{200\delta}{\sqrt{3}}\int_0^h x\,dx - \frac{100\delta}{h\sqrt{3}}\int_0^h x^2\,dx$$

$$= \frac{200\delta}{\sqrt{3}}\frac{h^2}{2} - \frac{100\delta}{h\sqrt{3}}\frac{h^3}{3} = \frac{200\delta h^2}{3\sqrt{3}} = \frac{200(62.5)}{3\sqrt{3}}\cdot\frac{12{,}825}{4} \approx 7.71\times 10^6\text{ lb}$$

17. (a) The area of a strip is $20\,\Delta x$ and the pressure on it is δx_i.

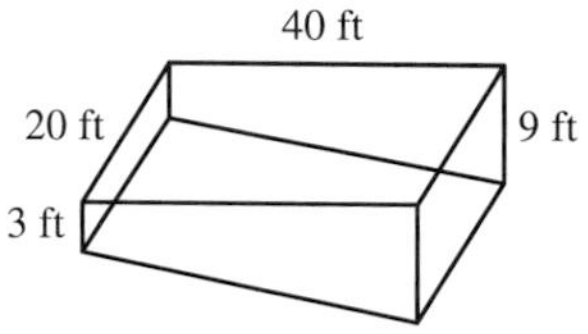

$$F = \int_0^3 \delta x 20\,dx = 20\delta\left[\tfrac{1}{2}x^2\right]_0^3 = 20\delta\cdot\tfrac{9}{2} = 90\delta$$

$$= 90(62.5) = 5625\text{ lb} \approx 5.63\times 10^3\text{ lb}$$

(b) $F = \int_0^9 \delta x 20\,dx = 20\delta\left[\tfrac{1}{2}x^2\right]_0^9 = 20\delta\cdot\tfrac{81}{2} = 810\delta = 810(62.5) = 50{,}625\text{ lb} \approx 5.06\times 10^4\text{ lb}$.

(c) For the first 3 ft, the length of the side is constant at 40 ft. For $3 < x \le 9$, we can use similar triangles to find the length a:

$\dfrac{a}{40} = \dfrac{9-x}{6} \quad\Rightarrow\quad a = 40\cdot\dfrac{9-x}{6}$.

$$F = \int_0^3 \delta x 40\,dx + \int_3^9 \delta x(40)\frac{9-x}{6}\,dx = 40\delta\left[\tfrac{1}{2}x^2\right]_0^3 + \tfrac{20}{3}\delta\int_3^9(9x - x^2)\,dx = 180\delta + \tfrac{20}{3}\delta\left[\tfrac{9}{2}x^2 - \tfrac{1}{3}x^3\right]_3^9$$

$$= 180\delta + \tfrac{20}{3}\delta\left[\left(\tfrac{729}{2} - 243\right) - \left(\tfrac{81}{2} - 9\right)\right] = 180\delta + 600\delta = 780\delta = 780(62.5) = 48{,}750\text{ lb} \approx 4.88\times 10^4\text{ lb}$$

(d) For any right triangle with hypotenuse on the bottom,

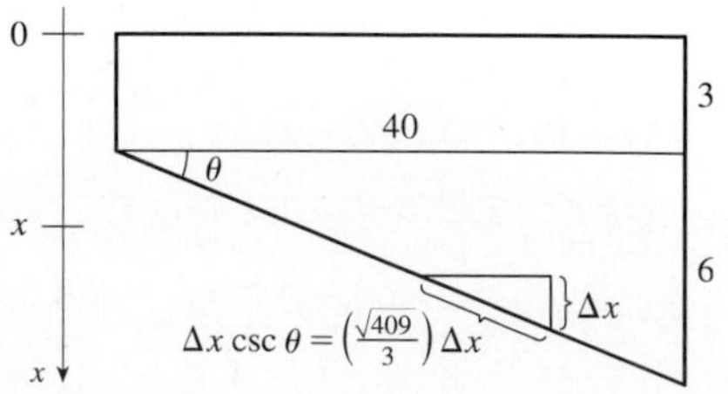

$\sin\theta = \dfrac{\Delta x}{\text{hypotenuse}} \quad\Rightarrow$

$$\text{hypotenuse} = \Delta x\csc\theta = \Delta x\,\frac{\sqrt{40^2+6^2}}{6} = \frac{\sqrt{409}}{3}\,\Delta x.$$

$$F = \int_3^9 \delta x 20\,\frac{\sqrt{409}}{3}\,dx = \tfrac{1}{3}\left(20\sqrt{409}\right)\delta\left[\tfrac{1}{2}x^2\right]_3^9$$

$$= \tfrac{1}{3}\cdot 10\sqrt{409}\,\delta(81-9) \approx 303{,}356\text{ lb} \approx 3.03\times 10^5\text{ lb}$$

18. Partition the interval $[a, b]$ by points x_i as usual and choose $x_i^* \in [x_{i-1}, x_i]$ for each i. The ith horizontal strip of the immersed plate is approximated by a rectangle of height Δx_i and width $w(x_i^*)$, so its area is $A_i \approx w(x_i^*)\,\Delta x_i$. For small Δx_i, the pressure P_i on the ith strip is almost constant and $P_i \approx \rho g x_i^*$ by Equation 1. The hydrostatic force F_i acting on the ith strip is $F_i = P_i A_i \approx \rho g x_i^* w(x_i^*)\,\Delta x_i$. Adding these forces and taking the limit as $n \to \infty$, we obtain the hydrostatic force on the immersed plate:

$$F = \lim_{n\to\infty} \sum_{i=1}^{n} F_i = \lim_{n\to\infty} \sum_{i=1}^{n} \rho g x_i^* w(x_i^*)\,\Delta x_i = \int_a^b \rho g x w(x)\,dx$$

19. From Exercise 18, we have $F = \int_a^b \rho g x w(x)\,dx = \int_{7.0}^{9.4} 64 x w(x)\,dx$. From the table, we see that $\Delta x = 0.4$, so using Simpson's Rule to estimate F, we get

$$\begin{aligned} F &\approx 64\,\tfrac{0.4}{3}[7.0w(7.0) + 4(7.4)w(7.4) + 2(7.8)w(7.8) + 4(8.2)w(8.2) + 2(8.6)w(8.6) + 4(9.0)w(9.0) + 9.4w(9.4)] \\ &= \tfrac{25.6}{3}[7(1.2) + 29.6(1.8) + 15.6(2.9) + 32.8(3.8) + 17.2(3.6) + 36(4.2) + 9.4(4.4)] \\ &= \tfrac{25.6}{3}(486.04) \approx 4148 \text{ lb} \end{aligned}$$

20. (a) From Equation 8, $\overline{x} = \frac{1}{A}\int_a^b x w(x)\,dx \;\Rightarrow\; A\overline{x} = \int_a^b x w(x)\,dx \;\Rightarrow\; \rho g A\overline{x} = \rho g \int_a^b x w(x)\,dx \;\Rightarrow$

$(\rho g \overline{x})A = \int_a^b \rho g\, x w(x)\,dx = F$ by Exercise 18.

(b) For the figure in Exercise 10, let the coordinates of the centroid $(\overline{x}, \overline{y}) = (a/\sqrt{2}, 0)$.

$$F = (\rho g \overline{x})A = \rho g\,\frac{a}{\sqrt{2}}\,a^2 = \delta\,\frac{\sqrt{2}\,a}{2}\,a^2 = \frac{\sqrt{2}\,a^3\delta}{2}.$$

21. The moment M of the system about the origin is $M = \sum_{i=1}^{2} m_i x_i = m_1x_1 + m_2x_2 = 6 \cdot 10 + 9 \cdot 30 = 330$.

The mass m of the system is $m = \sum_{i=1}^{2} m_i = m_1 + m_2 = 6 + 9 = 15$.

The center of mass of the system is $\overline{x} = M/m = \frac{330}{15} = 22$.

22. The moment M is $m_1x_1 + m_2x_2 + m_3x_3 = 12(-3) + 15(2) + 20(8) = 154$. The mass m is $m_1 + m_2 + m_3 = 12 + 15 + 20 = 47$. The center of mass is $\overline{x} = M/m = \frac{154}{47}$.

23. The mass is $m = \sum_{i=1}^{3} m_i = 4 + 2 + 4 = 10$. The moment about the x-axis is $M_x = \sum_{i=1}^{3} m_i y_i = 4(-3) + 2(1) + 4(5) = 10$.

The moment about the y-axis is $M_y = \sum_{i=1}^{3} m_i x_i = 4(2) + 2(-3) + 4(3) = 14$. The center of mass is

$$(\overline{x}, \overline{y}) = \left(\frac{M_y}{m}, \frac{M_x}{m}\right) = \left(\frac{14}{10}, \frac{10}{10}\right) = (1.4, 1).$$

24. The mass is $m = \sum_{i=1}^{4} m_i = 5 + 4 + 3 + 6 = 18$.

The moment about the x-axis is $M_x = \sum_{i=1}^{4} m_i y_i = 5(2) + 4(5) + 3(2) + 6(-2) = 24$.

[continued]

The moment about the y-axis is $M_y = \sum_{i=1}^{4} m_i x_i = 5(-4) + 4(0) + 3(3) + 6(1) = -5$.

The center of mass is $(\overline{x}, \overline{y}) = \left(\frac{M_y}{m}, \frac{M_x}{m}\right) = \left(\frac{-5}{18}, \frac{24}{18}\right) = \left(-\frac{5}{18}, \frac{4}{3}\right)$.

25. The region in the figure is "right-heavy" and "bottom-heavy," so we know that $\overline{x} > 0.5$ and $\overline{y} < 1$, and we might guess that $\overline{x} = 0.7$ and $\overline{y} = 0.7$.

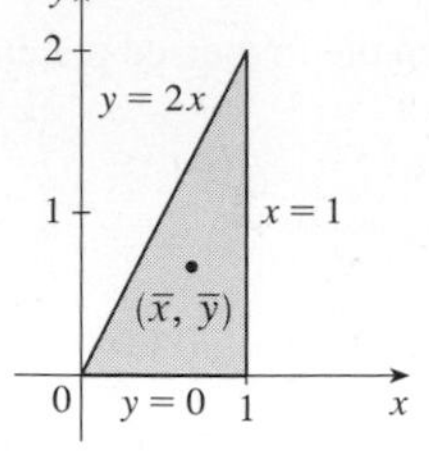

$A = \int_0^1 2x\,dx = \left[x^2\right]_0^1 = 1 - 0 = 1$.

$\overline{x} = \frac{1}{A}\int_0^1 x(2x)\,dx = \frac{1}{1}\left[\frac{2}{3}x^3\right]_0^1 = \frac{2}{3}$.

$\overline{y} = \frac{1}{A}\int_0^1 \frac{1}{2}(2x)^2\,dx = \frac{1}{1}\int_0^1 2x^2\,dx = \left[\frac{2}{3}x^3\right]_0^1 = \frac{2}{3}$.

Thus, the centroid is $(\overline{x}, \overline{y}) = \left(\frac{2}{3}, \frac{2}{3}\right)$.

26. The region in the figure is "right-heavy" and "bottom-heavy," so we know that $\overline{x} > 2$ and $\overline{y} < 1$, and we might guess that $\overline{x} = 2.3$ and $\overline{y} = 0.8$.

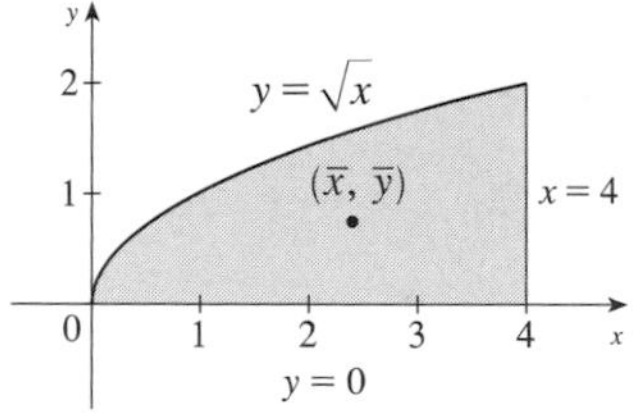

$A = \int_0^4 \sqrt{x}\,dx = \left[\frac{2}{3}x^{3/2}\right]_0^4 = \frac{16}{3}$.

$\overline{x} = \frac{1}{A}\int_0^4 x\left(\sqrt{x}\right)dx = \frac{3}{16}\int_0^4 x^{3/2}\,dx = \frac{3}{16}\left[\frac{2}{5}x^{5/2}\right]_0^4 = \frac{3}{40}(32 - 0) = \frac{12}{5}$.

$\overline{y} = \frac{1}{A}\int_0^4 \frac{1}{2}\left(\sqrt{x}\right)^2\,dx = \frac{3}{16}\int_0^4 \frac{1}{2}x\,dx = \frac{3}{32}\left[\frac{1}{2}x^2\right]_0^4 = \frac{3}{64}(16 - 0) = \frac{3}{4}$.

Thus, the centroid is $(\overline{x}, \overline{y}) = (2.4, 0.75)$.

27. The region in the figure is "right-heavy" and "bottom-heavy," so we know $\overline{x} > 0.5$ and $\overline{y} < 1$, and we might guess that $\overline{x} = 0.6$ and $\overline{y} = 0.9$.

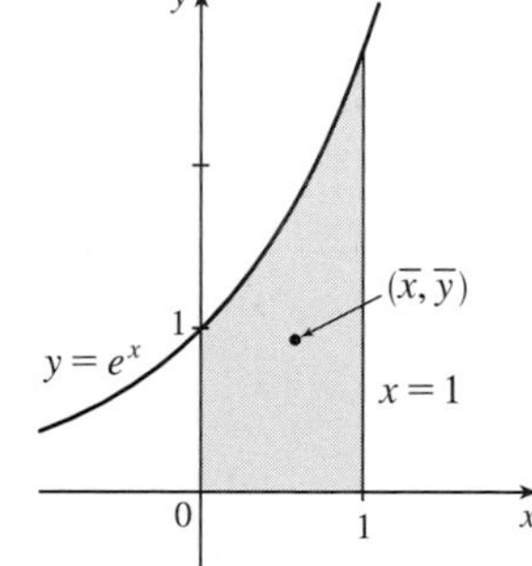

$A = \int_0^1 e^x\,dx = \left[e^x\right]_0^1 = e - 1$.

$\overline{x} = \frac{1}{A}\int_0^1 xe^x\,dx = \frac{1}{e-1}\left[xe^x - e^x\right]_0^1$ [by parts]

$= \frac{1}{e-1}[0 - (-1)] = \frac{1}{e-1}$.

$\overline{y} = \frac{1}{A}\int_0^1 \frac{1}{2}(e^x)^2\,dx = \frac{1}{e-1}\cdot\frac{1}{4}\left[e^{2x}\right]_0^1 = \frac{1}{4(e-1)}\left(e^2 - 1\right) = \frac{e+1}{4}$.

Thus, the centroid is $(\overline{x}, \overline{y}) = \left(\frac{1}{e-1}, \frac{e+1}{4}\right) \approx (0.58, 0.93)$.

28. Since the region in the figure is symmetric about the line $x = \frac{\pi}{2}$, we know that $\overline{x} = \frac{\pi}{2}$. The region is "bottom-heavy," so we know that $\overline{y} < 0.5$, and we might guess that $\overline{y} = 0.4$.

$A = \int_0^\pi \sin x\,dx = \left[-\cos x\right]_0^\pi = 1 - (-1) = 2$.

$\overline{x} = \frac{1}{A}\int_0^\pi x\sin x\,dx \overset{82}{=} \frac{1}{2}\left[\sin x - x\cos x\right]_0^\pi = \frac{1}{2}[(0 + \pi) - (0 - 0)] = \frac{\pi}{2}$.

$\overline{y} = \frac{1}{A}\int_0^\pi \frac{1}{2}(\sin x)^2\,dx = \frac{1}{2}\cdot\frac{1}{2}\int_0^\pi \frac{1}{2}(1 - \cos 2x)\,dx = \frac{1}{8}\left[x - \frac{1}{2}\sin 2x\right]_0^\pi = \frac{1}{8}[(\pi - 0) - (0 - 0)] = \frac{\pi}{8} \approx 0.39$.

Thus, the centroid is $(\overline{x}, \overline{y}) = \left(\frac{\pi}{2}, \frac{\pi}{8}\right)$.

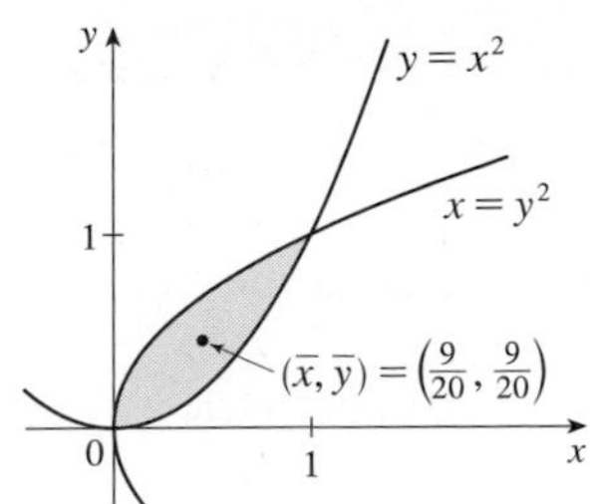

29. $A = \int_0^1 (x^{1/2} - x^2)\,dx = \left[\frac{2}{3}x^{3/2} - \frac{1}{3}x^3\right]_0^1 = \left(\frac{2}{3} - \frac{1}{3}\right) - 0 = \frac{1}{3}$.

$\overline{x} = \frac{1}{A}\int_0^1 x(x^{1/2} - x^2)\,dx = 3\int_0^1 (x^{3/2} - x^3)\,dx$

$= 3\left[\frac{2}{5}x^{5/2} - \frac{1}{4}x^4\right]_0^1 = 3\left(\frac{2}{5} - \frac{1}{4}\right) = 3\left(\frac{3}{20}\right) = \frac{9}{20}$.

$\overline{y} = \frac{1}{A}\int_0^1 \frac{1}{2}\left[(x^{1/2})^2 - (x^2)^2\right]dx = 3\left(\frac{1}{2}\right)\int_0^1 (x - x^4)\,dx$

$= \frac{3}{2}\left[\frac{1}{2}x^2 - \frac{1}{5}x^5\right]_0^1 = \frac{3}{2}\left(\frac{1}{2} - \frac{1}{5}\right) = \frac{3}{2}\left(\frac{3}{10}\right) = \frac{9}{20}$.

Thus, the centroid is $(\overline{x}, \overline{y}) = \left(\frac{9}{20}, \frac{9}{20}\right)$.

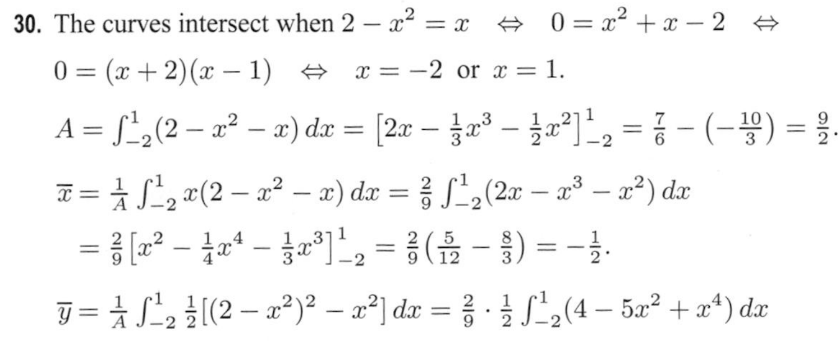

30. The curves intersect when $2 - x^2 = x \;\Leftrightarrow\; 0 = x^2 + x - 2 \;\Leftrightarrow$ $0 = (x + 2)(x - 1) \;\Leftrightarrow\; x = -2$ or $x = 1$.

$A = \int_{-2}^1 (2 - x^2 - x)\,dx = \left[2x - \frac{1}{3}x^3 - \frac{1}{2}x^2\right]_{-2}^1 = \frac{7}{6} - \left(-\frac{10}{3}\right) = \frac{9}{2}$.

$\overline{x} = \frac{1}{A}\int_{-2}^1 x(2 - x^2 - x)\,dx = \frac{2}{9}\int_{-2}^1 (2x - x^3 - x^2)\,dx$

$= \frac{2}{9}\left[x^2 - \frac{1}{4}x^4 - \frac{1}{3}x^3\right]_{-2}^1 = \frac{2}{9}\left(\frac{5}{12} - \frac{8}{3}\right) = -\frac{1}{2}$.

$\overline{y} = \frac{1}{A}\int_{-2}^1 \frac{1}{2}[(2 - x^2)^2 - x^2]\,dx = \frac{2}{9} \cdot \frac{1}{2}\int_{-2}^1 (4 - 5x^2 + x^4)\,dx$

$= \frac{1}{9}\left[4x - \frac{5}{3}x^3 + \frac{1}{5}x^5\right]_{-2}^1 = \frac{1}{9}\left[\frac{38}{15} - \left(-\frac{16}{15}\right)\right] = \frac{2}{5}$.

Thus, the centroid is $(\overline{x}, \overline{y}) = \left(-\frac{1}{2}, \frac{2}{5}\right)$.

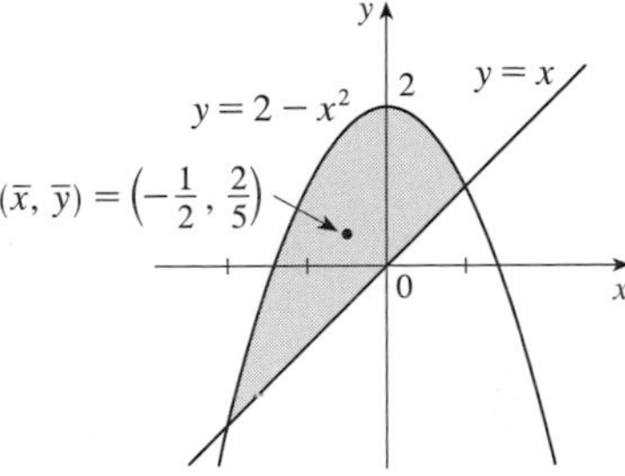

31. $A = \int_0^{\pi/4} (\cos x - \sin x)\,dx = \left[\sin x + \cos x\right]_0^{\pi/4} = \sqrt{2} - 1$.

$\overline{x} = A^{-1}\int_0^{\pi/4} x(\cos x - \sin x)\,dx$

$= A^{-1}\left[x(\sin x + \cos x) + \cos x - \sin x\right]_0^{\pi/4}$ [integration by parts]

$= A^{-1}\left(\frac{\pi}{4}\sqrt{2} - 1\right) = \dfrac{\frac{1}{4}\pi\sqrt{2} - 1}{\sqrt{2} - 1}$.

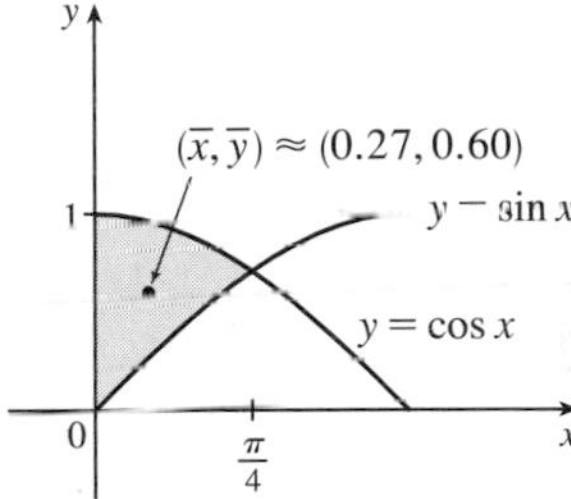

$\overline{y} = A^{-1}\int_0^{\pi/4} \frac{1}{2}(\cos^2 x - \sin^2 x)\,dx = \frac{1}{2A}\int_0^{\pi/4} \cos 2x\,dx = \frac{1}{4A}\left[\sin 2x\right]_0^{\pi/4} = \dfrac{1}{4A} = \dfrac{1}{4\left(\sqrt{2} - 1\right)}$.

Thus, the centroid is $(\overline{x}, \overline{y}) = \left(\dfrac{\pi\sqrt{2} - 4}{4(\sqrt{2} - 1)}, \dfrac{1}{4(\sqrt{2} - 1)}\right) \approx (0.27, 0.60)$.

32. $A = \int_0^1 x^3\,dx + \int_1^2 (2 - x)\,dx = \left[\frac{1}{4}x^4\right]_0^1 + \left[2x - \frac{1}{2}x^2\right]_1^2$

$= \frac{1}{4} + (4 - 2) - \left(2 - \frac{1}{2}\right) = \frac{3}{4}$.

$\overline{x} = \frac{1}{A}\left[\int_0^1 x(x^3)\,dx + \int_1^2 x(2 - x)\,dx\right] = \frac{4}{3}\left[\int_0^1 x^4\,dx + \int_1^2 (2x - x^2)\,dx\right]$

$= \frac{4}{3}\left\{\left[\frac{1}{5}x^5\right]_0^1 + \left[x^2 - \frac{1}{3}x^3\right]_1^2\right\} = \frac{4}{3}\left[\frac{1}{5} + \left(4 - \frac{8}{3}\right) - \left(1 - \frac{1}{3}\right)\right]$

$= \frac{4}{3}\left(\frac{13}{15}\right) = \frac{52}{45}$.

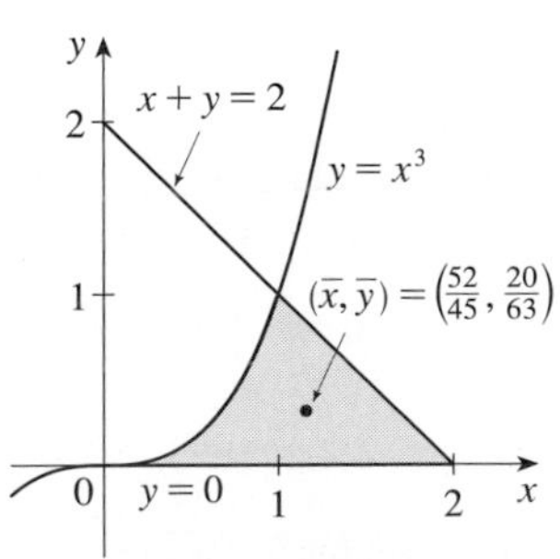

[continued]

$\overline{y} = \frac{1}{A}\left[\int_0^1 \frac{1}{2}(x^3)^2\,dx + \int_1^2 \frac{1}{2}(2-x)^2\,dx\right] = \frac{2}{3}\left[\int_0^1 x^6 dx + \int_1^2 (x-2)^2\,dx\right] = \frac{2}{3}\left\{\left[\frac{1}{7}x^7\right]_0^1 + \left[\frac{1}{3}(x-2)^3\right]_1^2\right\}$

$= \frac{2}{3}\left(\frac{1}{7} - 0 + 0 + \frac{1}{3}\right) = \frac{2}{3}\left(\frac{10}{21}\right) = \frac{20}{63}.$

Thus, the centroid is $(\overline{x}, \overline{y}) = \left(\frac{52}{45}, \frac{20}{63}\right)$.

33. The curves intersect when $2 - y = y^2 \;\Leftrightarrow\; 0 = y^2 + y - 2 \;\Leftrightarrow$
$0 = (y+2)(y-1) \;\Leftrightarrow\; y = -2$ or $y = 1$.

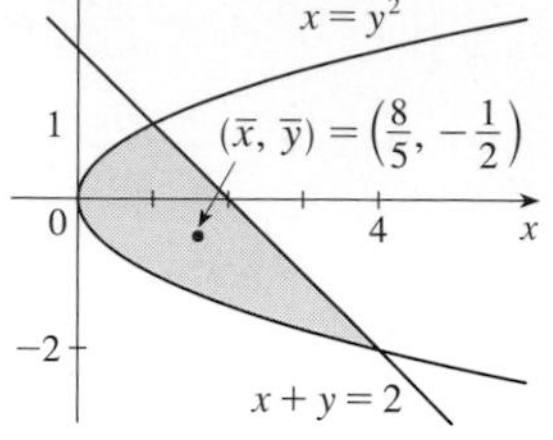

$A = \int_{-2}^1 (2 - y - y^2)\,dy = \left[2y - \frac{1}{2}y^2 - \frac{1}{3}y^3\right]_{-2}^1 = \frac{7}{6} - \left(-\frac{10}{3}\right) = \frac{9}{2}.$

$\overline{x} = \frac{1}{A}\int_{-2}^1 \frac{1}{2}[(2-y)^2 - (y^2)^2]\,dy = \frac{2}{9}\cdot\frac{1}{2}\int_{-2}^1 (4 - 4y + y^2 - y^4)\,dy$

$= \frac{1}{9}\left[4y - 2y^2 + \frac{1}{3}y^3 - \frac{1}{5}y^5\right]_{-2}^1 = \frac{1}{9}\left[\frac{32}{15} - \left(-\frac{184}{15}\right)\right] = \frac{8}{5}.$

$\overline{y} = \frac{1}{A}\int_{-2}^1 y(2 - y - y^2)\,dy = \frac{2}{9}\int_{-2}^1 (2y - y^2 - y^3)\,dy$

$= \frac{2}{9}\left[y^2 - \frac{1}{3}y^3 - \frac{1}{4}y^4\right]_{-2}^1 = \frac{2}{9}\left(\frac{5}{12} - \frac{8}{3}\right) = -\frac{1}{2}.$

Thus, the centroid is $(\overline{x}, \overline{y}) = \left(\frac{8}{5}, -\frac{1}{2}\right)$.

34. An equation of the line is $y = -\frac{3}{2}x + 3$. $A = \frac{1}{2}(2)(3) = 3$, so $m = \rho A = 4(3) = 12$.

$M_x = \rho\int_0^2 \frac{1}{2}\left(-\frac{3}{2}x + 3\right)^2\,dx = \frac{1}{2}\rho\int_0^2 \left(\frac{9}{4}x^2 - 9x + 9\right)dx = \frac{1}{2}(4)\left[\frac{3}{4}x^3 - \frac{9}{2}x^2 + 9x\right]_0^2 = 2(6 - 18 + 18) = 12.$

$M_y = \rho\int_0^2 x\left(-\frac{3}{2}x + 3\right)dx = \rho\int_0^2 \left(-\frac{3}{2}x^2 + 3x\right)dx = 4\left[-\frac{1}{2}x^3 + \frac{3}{2}x^2\right]_0^2 = 4(-4 + 6) = 8.$

$\overline{x} = \dfrac{M_y}{m} = \dfrac{8}{12} = \dfrac{2}{3}$ and $\overline{y} = \dfrac{M_x}{m} = \dfrac{12}{12} = 1$. Thus, the center of mass is $(\overline{x}, \overline{y}) = \left(\frac{2}{3}, 1\right)$. Since ρ is constant, the center of mass is also the centroid.

35. The quarter-circle has equation $y = \sqrt{4^2 - x^2}$ for $0 \le x \le 4$ and the line has equation $y = -2$.

$A = \frac{1}{4}\pi(4)^2 + 2(4) = 4\pi + 8 = 4(\pi + 2)$, so $m = \rho A = 6\cdot 4(\pi + 2) = 24(\pi + 2)$.

$M_x = \rho\int_0^4 \frac{1}{2}\left[\left(\sqrt{16 - x^2}\right)^2 - (-2)^2\right]dx = \frac{1}{2}\rho\int_0^4 (16 - x^2 - 4)\,dx = \frac{1}{2}(6)\left[12x - \frac{1}{3}x^3\right]_0^4 = 3\left(48 - \frac{64}{3}\right) = 80.$

$M_y = \rho\int_0^4 x\left[\sqrt{16 - x^2} - (-2)\right]dx = \rho\int_0^4 x\sqrt{16 - x^2}\,dx + \rho\int_0^4 2x\,dx = 6\left[-\frac{1}{3}(16 - x^2)^{3/2}\right]_0^4 + 6\left[x^2\right]_0^4$

$= 6\left(0 + \frac{64}{3}\right) + 6(16) = 224.$

$$\overline{x} = \frac{M_y}{m} = \frac{224}{24(\pi + 2)} = \frac{28}{3(\pi + 2)} \text{ and } \overline{y} = \frac{M_x}{m} = \frac{80}{24(\pi + 2)} = \frac{10}{3(\pi + 2)}.$$

Thus, the center of mass is $\left(\dfrac{28}{3(\pi + 2)}, \dfrac{10}{3(\pi + 2)}\right) \approx (1.82, 0.65)$.

36. We'll use $n = 8$, so $\Delta x = \frac{b - a}{n} = \frac{8 - 0}{8} = 1$.

$A = \int_0^8 f(x)\,dx \approx S_{10} = \frac{1}{3}[f(0) + 4f(1) + 2f(2) + 4f(3) + 2f(4) + 4f(5) + 2f(6) + 4f(7) + f(8)]$

$\approx \frac{1}{3}[0 + 4(2.0) + 2(2.6) + 4(2.3) + 2(2.2) + 4(3.3) + 2(4.0) + 4(3.2) + 0]$

$= \frac{1}{3}(60.8) = 20.2\overline{6}\;\;\left[\text{or } \frac{304}{15}\right]$

Now $\int_0^8 x\, f(x)\, dx \approx \frac{1}{3}[0 \cdot f(0) + 4 \cdot 1 \cdot f(1) + 2 \cdot 2 \cdot f(2) + 4 \cdot 3 \cdot f(3)$

$$+ 2 \cdot 4 \cdot f(4) + 4 \cdot 5 \cdot f(5) + 2 \cdot 6 \cdot f(6) + 4 \cdot 7 \cdot f(7) + 8 \cdot f(8)]$$

$$\approx \tfrac{1}{3}[0 + 8 + 10.4 + 27.6 + 17.6 + 66 + 48 + 89.6 + 0]$$

$$= \tfrac{1}{3}(267.2) = 89.0\overline{6} \quad \left[\text{or } \tfrac{1336}{15}\right], \text{ so } \overline{x} = \tfrac{1}{A}\int_0^8 x f(x)\, dx \approx 4.39.$$

Also, $\int_0^8 [f(x)]^2\, dx \approx \frac{1}{3}[0^2 + 4(2.0)^2 + 2(2.6)^2 + 4(2.3)^2 + 2(2.2)^2 + 4(3.3)^2 + 2(4.0)^2 + 4(3.2)^2 + 0^2]$

$$= \tfrac{1}{3}(176.88) = 58.96, \text{ so } \overline{y} = \tfrac{1}{A}\int_0^8 \tfrac{1}{2}[f(x)]^2\, dx \approx 1.45.$$

Thus, the centroid is $(\overline{x}, \overline{y}) \approx (4.4, 1.5)$.

37. $A = \int_{-1}^{1}[(x^3 - x) - (x^2 - 1)]\, dx = \int_{-1}^{1}(1 - x^2)\, dx$ $\left[\begin{array}{c}\text{odd-degree terms}\\ \text{drop out}\end{array}\right]$

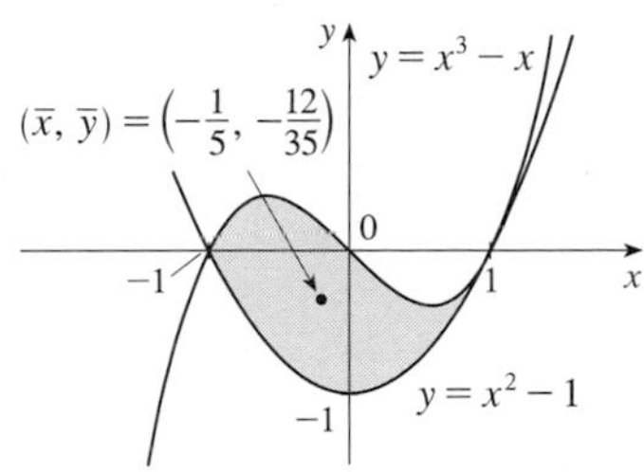

$= 2\int_0^1(1 - x^2)\, dx = 2\left[x - \frac{1}{3}x^3\right]_0^1 = 2\left(\frac{2}{3}\right) = \frac{4}{3}.$

$\overline{x} = \frac{1}{A}\int_{-1}^{1} x(x^3 - x - x^2 + 1)\, dx = \frac{3}{4}\int_{-1}^{1}(x^4 - x^2 - x^3 + x)\, dx$

$= \frac{3}{4}\int_{-1}^{1}(x^4 - x^2)\, dx = \frac{3}{4} \cdot 2\int_0^1(x^4 - x^2)\, dx$

$= \frac{3}{2}\left[\frac{1}{5}x^5 - \frac{1}{3}x^3\right]_0^1 = \frac{3}{2}\left(-\frac{2}{15}\right) = -\frac{1}{5}.$

$\overline{y} = \frac{1}{A}\int_{-1}^{1}\frac{1}{2}[(x^3 - x)^2 - (x^2 - 1)^2]\, dx = \frac{3}{4} \cdot \frac{1}{2}\int_{-1}^{1}(x^6 - 2x^4 + x^2 - x^4 + 2x^2 - 1)\, dx$

$= \frac{3}{8} \cdot 2\int_0^1(x^6 - 3x^4 + 3x^2 - 1)\, dx = \frac{3}{4}\left[\frac{1}{7}x^7 - \frac{3}{5}x^5 + x^3 - x\right]_0^1 = \frac{3}{4}\left(-\frac{16}{35}\right) = -\frac{12}{35}.$

Thus, the centroid is $(\overline{x}, \overline{y}) = \left(-\frac{1}{5}, -\frac{12}{35}\right)$.

38. The curves intersect at $x = a \approx -1.315974$ and $x = b \approx 0.53727445$.

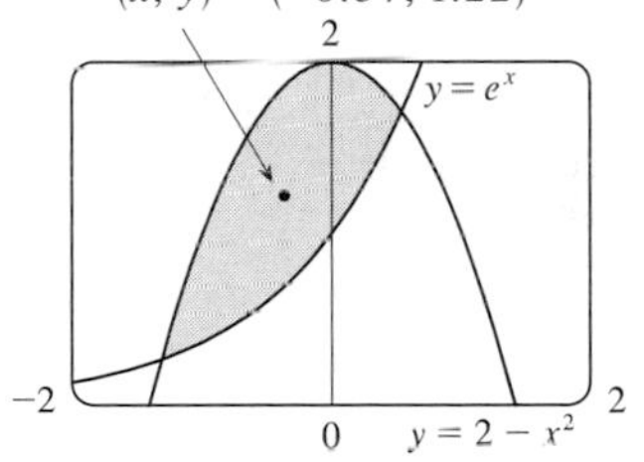

$A = \int_a^b[(2 - x^2) - e^x]\, dx = \left[2x - \frac{1}{3}x^3 - e^x\right]_a^b \approx 1.452014.$

$\overline{x} = \frac{1}{A}\int_a^b x(2 - x^2 - e^x)\, dx = \frac{1}{A}\left[x^2 - \frac{1}{4}x^4 - xe^x + e^x\right]_a^b$

≈ -0.374293

$\overline{y} = \frac{1}{A}\int_a^b \frac{1}{2}[(2 - x^2)^2 - (e^x)^2]\, dx = \frac{1}{2A}\int_a^b(4 - 4x^2 + x^4 - e^{2x})\, dx$

$= \frac{1}{2A}\left[4x - \frac{4}{3}x^3 + \frac{1}{5}x^5 - \frac{1}{2}e^{2x}\right]_a^b \approx 1.218131$

Thus, the centroid is $(\overline{x}, \overline{y}) \approx (-0.37, 1.22)$.

39. Choose x- and y-axes so that the base (one side of the triangle) lies along the x-axis with the other vertex along the positive y-axis as shown. From geometry, we know the medians intersect at a point $\frac{2}{3}$ of the way from each vertex (along the median) to the opposite side. The median from B goes to the midpoint $\left(\frac{1}{2}(a + c), 0\right)$ of side AC, so the point of intersection of the medians is $\left(\frac{2}{3} \cdot \frac{1}{2}(a + c), \frac{1}{3}b\right) = \left(\frac{1}{3}(a + c), \frac{1}{3}b\right)$.

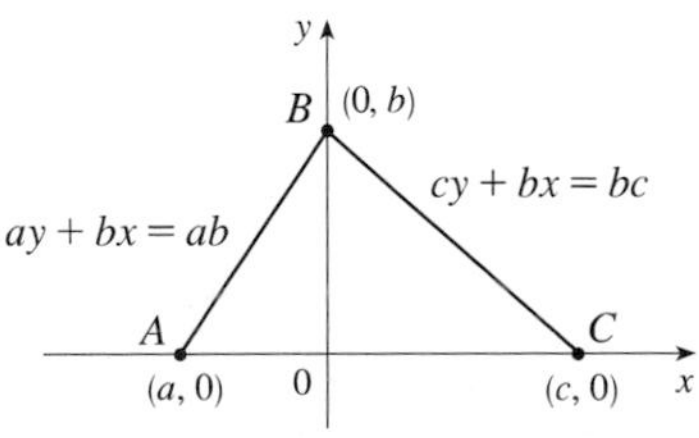

This can also be verified by finding the equations of two medians, and solving them simultaneously to find their point of intersection. Now let us compute the location of the centroid of the triangle. The area is $A = \frac{1}{2}(c - a)b$.

[continued]

$$\overline{x} = \frac{1}{A}\left[\int_a^0 x \cdot \frac{b}{a}(a-x)\,dx + \int_0^c x \cdot \frac{b}{c}(c-x)\,dx\right] = \frac{1}{A}\left[\frac{b}{a}\int_a^0 (ax-x^2)\,dx + \frac{b}{c}\int_0^c (cx-x^2)\,dx\right]$$

$$= \frac{b}{Aa}\left[\frac{1}{2}ax^2 - \frac{1}{3}x^3\right]_a^0 + \frac{b}{Ac}\left[\frac{1}{2}cx^2 - \frac{1}{3}x^3\right]_0^c = \frac{b}{Aa}\left[-\frac{1}{2}a^3 + \frac{1}{3}a^3\right] + \frac{b}{Ac}\left[\frac{1}{2}c^3 - \frac{1}{3}c^3\right]$$

$$= \frac{2}{a\,(c-a)} \cdot \frac{-a^3}{6} + \frac{2}{c\,(c-a)} \cdot \frac{c^3}{6} = \frac{1}{3\,(c-a)}(c^2-a^2) = \frac{a+c}{3}$$

and $$\overline{y} = \frac{1}{A}\left[\int_a^0 \frac{1}{2}\left(\frac{b}{a}(a-x)\right)^2 dx + \int_0^c \frac{1}{2}\left(\frac{b}{c}(c-x)\right)^2 dx\right]$$

$$= \frac{1}{A}\left[\frac{b^2}{2a^2}\int_a^0 (a^2-2ax+x^2)\,dx + \frac{b^2}{2c^2}\int_0^c (c^2-2cx+x^2)\,dx\right]$$

$$= \frac{1}{A}\left[\frac{b^2}{2a^2}\left[a^2x - ax^2 + \tfrac{1}{3}x^3\right]_a^0 + \frac{b^2}{2c^2}\left[c^2x - cx^2 + \tfrac{1}{3}x^3\right]_0^c\right]$$

$$= \frac{1}{A}\left[\frac{b^2}{2a^2}\left(-a^3+a^3-\tfrac{1}{3}a^3\right) + \frac{b^2}{2c^2}\left(c^3-c^3+\tfrac{1}{3}c^3\right)\right] = \frac{1}{A}\left[\frac{b^2}{6}\,(-a+c)\right] = \frac{2}{(c-a)\,b} \cdot \frac{(c-a)b^2}{6} = \frac{b}{3}$$

Thus, the centroid is $(\overline{x}, \overline{y}) = \left(\frac{a+c}{3}, \frac{b}{3}\right)$, as claimed.

Remarks: Actually the computation of $\overline{y}$ is all that is needed. By considering each side of the triangle in turn to be the base, we see that the centroid is $\frac{1}{3}$ of the way from each side to the opposite vertex and must therefore be the intersection of the medians.

The computation of $\overline{y}$ in this problem (and many others) can be simplified by using horizontal rather than vertical approximating rectangles. If the length of a thin rectangle at coordinate y is $\ell(y)$, then its area is $\ell(y)\,\Delta y$, its mass is $\rho\ell(y)\,\Delta y$, and its moment about the x-axis is $\Delta M_x = \rho y\ell(y)\,\Delta y$. Thus,

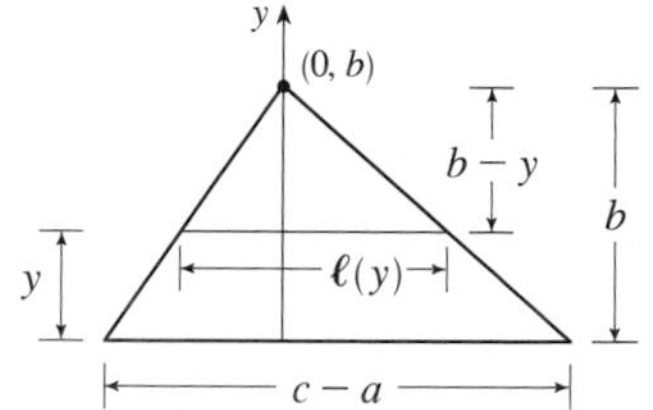

$$M_x = \int \rho y\ell(y)\,dy \quad \text{and} \quad \overline{y} = \frac{\int \rho y\ell(y)\,dy}{\rho A} = \frac{1}{A}\int y\ell(y)\,dy$$

In this problem, $\ell(y) = \dfrac{c-a}{b}(b-y)$ by similar triangles, so

$$\overline{y} = \frac{1}{A}\int_0^b \frac{c-a}{b}\,y(b-y)\,dy = \frac{2}{b^2}\int_0^b (by-y^2)\,dy = \frac{2}{b^2}\left[\tfrac{1}{2}by^2 - \tfrac{1}{3}y^3\right]_0^b = \frac{2}{b^2}\cdot\frac{b^3}{6} = \frac{b}{3}$$

Notice that only one integral is needed when this method is used.

40. The rectangle to the left of the y-axis has centroid $\left(-\frac{1}{2}, 1\right)$ and area 2. The triangle to the right of the y-axis has area 2 and centroid $\left(\frac{2}{3}, \frac{2}{3}\right)$ [by Exercise 39, the centroid is two-thirds of the way from the vertex $(0,0)$ to the point $(1,1)$].

$$\overline{x} = \frac{M_y}{m} = \frac{1}{m}\sum_{i=1}^{2} m_i x_i = \frac{1}{2+2}\left[2\left(-\tfrac{1}{2}\right) + 2\left(\tfrac{2}{3}\right)\right] = \tfrac{1}{4}\left(\tfrac{1}{3}\right) = \tfrac{1}{12}.$$

$$\overline{y} = \frac{M_x}{m} = \frac{1}{m}\sum_{i=1}^{2} m_i y_i = \frac{1}{2+2}\left[2(1) + 2\left(\tfrac{2}{3}\right)\right] = \tfrac{1}{4}\left(\tfrac{10}{3}\right) = \tfrac{5}{6}.$$ Thus, the centroid is $(\overline{x}, \overline{y}) = \left(\frac{1}{12}, \frac{5}{6}\right)$.

41. Divide the lamina into two triangles and one rectangle with respective masses of 2, 2 and 4, so that the total mass is 8. Using the result of Exercise 39, the triangles have centroids $\left(-1, \frac{2}{3}\right)$ and $\left(1, \frac{2}{3}\right)$. The centroid of the rectangle (its center) is $\left(0, -\frac{1}{2}\right)$.

So, using Formulas 5 and 7, we have $\overline{y} = \dfrac{M_x}{m} = \dfrac{1}{m}\sum\limits_{i=1}^{3} m_i\, y_i = \frac{1}{8}\left[2\left(\frac{2}{3}\right) + 2\left(\frac{2}{3}\right) + 4\left(-\frac{1}{2}\right)\right] = \frac{1}{8}\left(\frac{2}{3}\right) = \frac{1}{12}$, and $\overline{x} = 0$,

since the lamina is symmetric about the line $x = 0$. Thus, the centroid is $(\overline{x}, \overline{y}) = \left(0, \frac{1}{12}\right)$.

42. The parabola has equation $y = kx^2$ and passes through (a, b),

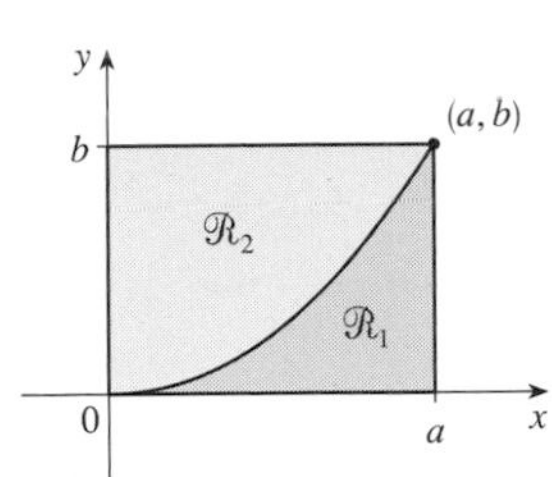

so $b = ka^2 \quad \Rightarrow \quad k = \dfrac{b}{a^2}$ and hence, $y = \dfrac{b}{a^2}x^2$.

$\mathcal{R}_1$ has area $A_1 = \displaystyle\int_0^a \frac{b}{a^2}x^2\,dx = \frac{b}{a^2}\left[\frac{1}{3}x^3\right]_0^a = \frac{b}{a^2}\left(\frac{a^3}{3}\right) = \frac{1}{3}ab.$

Since $\mathcal{R}$ has area ab, $\mathcal{R}_2$ has area $A_2 = ab - \frac{1}{3}ab = \frac{2}{3}ab$.

For $\mathcal{R}_1$:

$$\overline{x}_1 = \frac{1}{A_1}\int_0^a x\left(\frac{b}{a^2}x^2\right)dx = \frac{3}{ab}\frac{b}{a^2}\int_0^a x^3\,dx = \frac{3}{a^3}\left[\frac{1}{4}x^4\right]_0^a = \frac{3}{a^3}\left(\frac{1}{4}a^4\right) = \frac{3}{4}a$$

$$\overline{y}_1 = \frac{1}{A_1}\int_0^a \frac{1}{2}\left(\frac{b}{a^2}x^2\right)^2 dx = \frac{3}{ab}\frac{b^2}{2a^4}\int_0^a x^4\,dx = \frac{3b}{2a^5}\left[\frac{1}{5}x^5\right]_0^a = \frac{3b}{2a^5}\left(\frac{1}{5}a^5\right) = \frac{3}{10}b$$

Thus, the centroid for $\mathcal{R}_1$ is $(\overline{x}_1, \overline{y}_1) = \left(\frac{3}{4}a, \frac{3}{10}b\right)$.

For $\mathcal{R}_2$:

$$\overline{x}_2 = \frac{1}{A_2}\int_0^a x\left(b - \frac{b}{a^2}x^2\right)dx = \frac{3}{2ab}\int_0^a b\left(x - \frac{1}{a^2}x^3\right)dx = \frac{3}{2a}\left[\frac{1}{2}x^2 - \frac{1}{4a^2}x^4\right]_0^a$$

$$= \frac{3}{2a}\left(\frac{a^2}{2} - \frac{a^2}{4}\right) = \frac{3}{2a}\left(\frac{a^2}{4}\right) = \frac{3}{8}a$$

$$\overline{y}_2 = \frac{1}{A_2}\int_0^a \frac{1}{2}\left[(b)^2 - \left(\frac{b}{a^2}x^2\right)^2\right]dx = \frac{3}{2ab}\frac{1}{2}\int_0^a b^2\left(1 - \frac{1}{a^4}x^4\right)dx = \frac{3b}{4a}\left[x - \frac{1}{5a^4}x^5\right]_0^a$$

$$= \frac{3b}{4a}\left(a - \frac{1}{5}a\right) = \frac{3b}{4a}\left(\frac{4a}{5}\right) = \frac{3}{5}b$$

Thus, the centroid for $\mathcal{R}_2$ is $(\overline{x}_2, \overline{y}_2) = \left(\frac{3}{8}a, \frac{3}{5}b\right)$. Note the relationships: $A_2 = 2A_1$, $\overline{x}_1 = 2\,\overline{x}_2$, $\overline{y}_2 = 2\,\overline{y}_1$.

43. $\int_a^b (cx + d)\, f(x)\,dx = \int_a^b cx\, f(x)\,dx + \int_a^b d f(x)\,dx = c\int_a^b x\, f(x)\,dx + d\int_a^b f(x)\,dx = c\overline{x}A + d\int_a^b f(x)\,dx$ [by (8)]

$$= c\overline{x}\int_a^b f(x)\,dx + d\int_a^b f(x)\,dx = (c\overline{x} + d)\int_a^b f(x)\,dx$$

44. A sphere can be generated by rotating a semicircle about its diameter. The center of mass travels a distance

$2\pi\overline{y} = 2\pi\left(\dfrac{4r}{3\pi}\right)$ [from Example 4] $= \dfrac{8r}{3}$, so by the Theorem of Pappus, the volume of the sphere is

$$V = Ad = \frac{\pi r^2}{2}\cdot\frac{8r}{3} = \frac{4}{3}\pi r^3.$$

45. A cone of height h and radius r can be generated by rotating a right triangle about one of its legs as shown. By Exercise 39, $\overline{x} = \frac{1}{3}r$, so by the Theorem of Pappus, the volume of the cone is

$$V = Ad = \left(\tfrac{1}{2} \cdot \text{base} \cdot \text{height}\right) \cdot (2\pi\overline{x}) = \tfrac{1}{2}rh \cdot 2\pi\left(\tfrac{1}{3}r\right) = \tfrac{1}{3}\pi r^2 h.$$

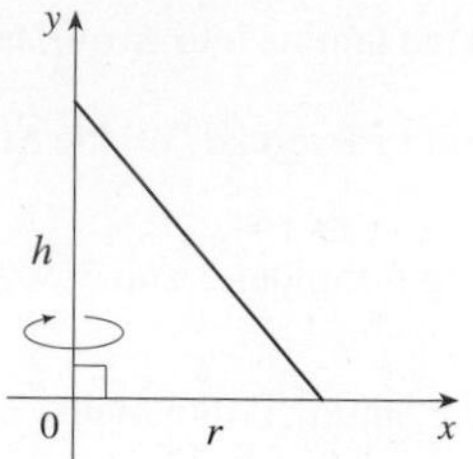

46. From the symmetry in the figure, $\overline{y} = 4$. So the distance traveled by the centroid when rotating the triangle about the x-axis is $d = 2\pi \cdot 4 = 8\pi$. The area of the triangle is $A = \frac{1}{2}bh = \frac{1}{2}(2)(3) = 3$. By the Theorem of Pappus, the volume of the resulting solid is $Ad = 3(8\pi) = 24\pi$.

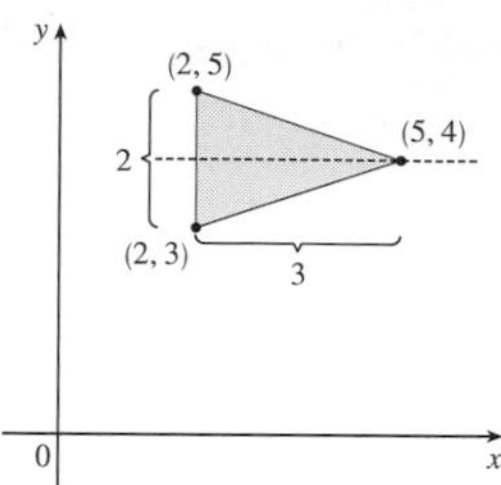

47. The curve C is the quarter-circle $y = \sqrt{16 - x^2}$, $0 \le x \le 4$. Its length L is $\frac{1}{4}(2\pi \cdot 4) = 2\pi$.

Now $y' = \frac{1}{2}(16 - x^2)^{-1/2}(-2x) = \dfrac{-x}{\sqrt{16 - x^2}} \quad\Rightarrow\quad 1 + (y')^2 = 1 + \dfrac{x^2}{16 - x^2} = \dfrac{16}{16 - x^2} \quad\Rightarrow$

$ds = \sqrt{1 + (y')^2}\,dx = \dfrac{4}{\sqrt{16 - x^2}}\,dx$, so

$\overline{x} = \dfrac{1}{L}\displaystyle\int x\,ds = \dfrac{1}{2\pi}\int_0^4 4x(16 - x^2)^{-1/2}\,dx = \dfrac{4}{2\pi}\Big[-(16 - x^2)^{1/2}\Big]_0^4 = \dfrac{2}{\pi}(0 + 4) = \dfrac{8}{\pi}$ and

$\overline{y} = \dfrac{1}{L}\displaystyle\int y\,ds = \dfrac{1}{2\pi}\int_0^4 \sqrt{16 - x^2} \cdot \dfrac{4}{\sqrt{16 - x^2}}\,dx = \dfrac{4}{2\pi}\int_0^4 dx = \dfrac{2}{\pi}\Big[x\Big]_0^4 = \dfrac{2}{\pi}(4 - 0) = \dfrac{8}{\pi}$. Thus, the centroid is $\left(\dfrac{8}{\pi}, \dfrac{8}{\pi}\right)$. Note that the centroid does not lie on the curve, but does lie on the line $y = x$, as expected, due to the symmetry of the curve.

48. (a) From Exercise 47, we have $\overline{y} = (1/L)\int y\,ds \iff \overline{y}L = \int y\,ds$. The surface area is $S = \int 2\pi y\,ds = 2\pi\int y\,ds = 2\pi(\overline{y}L) = L(2\pi\overline{y})$, which is the product of the arc length of C and the distance traveled by the centroid of C.

(b) From Exercise 47, $L = 2\pi$ and $\overline{y} = \frac{8}{\pi}$. By the Second Theorem of Pappus, the surface area is $S = L(2\pi\overline{y}) = 2\pi(2\pi \cdot \frac{8}{\pi}) = 32\pi$.

A geometric formula for the surface area of a half-sphere is $S = 2\pi r^2$. With $r = 4$, we get $S = 32\pi$, which agrees with our first answer.

49. The circle has arc length (circumference) $L = 2\pi r$. As in Example 7, the distance traveled by the centroid during a rotation is $d = 2\pi R$. Therefore, by the Second Theorem of Pappus, the surface area is

$$S = Ld = (2\pi r)(2\pi R) = 4\pi^2 rR$$

50. (a) Let $0 \le x \le 1$. If $n < m$, then $x^n > x^m$; that is, raising x to a larger power produces a smaller number.

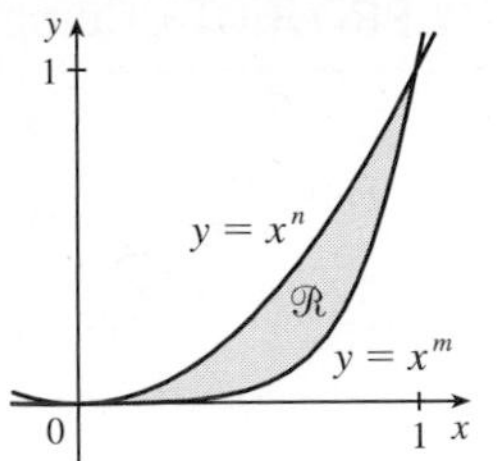

(b) Using Formulas 9 and the fact that the area of $\mathcal{R}$ is

$A = \int_0^1 (x^n - x^m)\,dx = \dfrac{1}{n+1} - \dfrac{1}{m+1} = \dfrac{m-n}{(n+1)(m+1)}$, we get

$$\begin{aligned}\overline{x} &= \frac{(n+1)(m+1)}{m-n}\int_0^1 x[x^n - x^m]\,dx = \frac{(n+1)(m+1)}{m-n}\int_0^1 \left(x^{n+1} - x^{m+1}\right)dx \\ &= \frac{(n+1)(m+1)}{m-n}\left[\frac{1}{n+2} - \frac{1}{m+2}\right] = \frac{(n+1)(m+1)}{(n+2)(m+2)}\end{aligned}$$

and

$$\begin{aligned}\overline{y} &= \frac{(n+1)(m+1)}{m-n}\int_0^1 \tfrac{1}{2}\left[(x^n)^2 - (x^m)^2\right]dx = \frac{(n+1)(m+1)}{2(m-n)}\int_0^1 \left(x^{2n} - x^{2m}\right)dx \\ &= \frac{(n+1)(m+1)}{2(m-n)}\left[\frac{1}{2n+1} - \frac{1}{2m+1}\right] = \frac{(n+1)(m+1)}{(2n+1)(2m+1)}\end{aligned}$$

(c) If we take $n = 3$ and $m = 4$, then

$$(\overline{x}, \overline{y}) = \left(\frac{4\cdot 5}{5\cdot 6}, \frac{4\cdot 5}{7\cdot 9}\right) = \left(\frac{2}{3}, \frac{20}{63}\right)$$

which lies outside $\mathcal{R}$ since $\left(\frac{2}{3}\right)^3 = \frac{8}{27} < \frac{20}{63}$. This is the simplest of many possibilities.

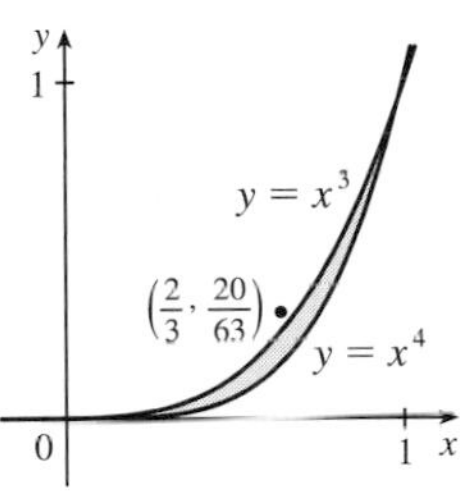

51. Suppose the region lies between two curves $y = f(x)$ and $y = g(x)$ where $f(x) > g(x)$, as illustrated in Figure 13. Choose points x_i with $a = x_0 < x_1 < \cdots < x_n = b$ and choose x_i^* to be the midpoint of the ith subinterval; that is, $x_i^* = \overline{x}_i = \frac{1}{2}(x_{i-1} + x_i)$. Then the centroid of the ith approximating rectangle R_i is its center $C_i = \left(\overline{x}_i, \frac{1}{2}[f(\overline{x}_i) + g(\overline{x}_i)]\right)$. Its area is $[f(\overline{x}_i) - g(\overline{x}_i)]\,\Delta x$, so its mass is

$\rho[f(\overline{x}_i) - g(\overline{x}_i)]\,\Delta x$. Thus, $M_y(R_i) = \rho[f(\overline{x}_i) - g(\overline{x}_i)]\,\Delta x \cdot \overline{x}_i = \rho\overline{x}_i\,[f(\overline{x}_i) - g(\overline{x}_i)]\,\Delta x$ and

$M_x(R_i) = \rho[f(\overline{x}_i) - g(\overline{x}_i)]\,\Delta x \cdot \frac{1}{2}[f(\overline{x}_i) + g(\overline{x}_i)] = \rho\cdot\frac{1}{2}\left[f(\overline{x}_i)^2 - g(\overline{x}_i)^2\right]\Delta x$. Summing over i and taking the limit as $n \to \infty$, we get $M_y = \lim\limits_{n\to\infty}\sum_i \rho\overline{x}_i\,[f(\overline{x}_i) - g(\overline{x}_i)]\,\Delta x = \rho\int_a^b x[f(x) - g(x)]\,dx$ and

$M_x = \lim\limits_{n\to\infty}\sum_i \rho\cdot\frac{1}{2}\left[f(\overline{x}_i)^2 - g(\overline{x}_i)^2\right]\Delta x = \rho\int_a^b \frac{1}{2}\left[f(x)^2 - g(x)^2\right]dx.$

Thus, $\overline{x} = \dfrac{M_y}{m} = \dfrac{M_y}{\rho A} = \dfrac{1}{A}\displaystyle\int_a^b x[f(x) - g(x)]\,dx$ and $\overline{y} = \dfrac{M_x}{m} = \dfrac{M_x}{\rho A} = \dfrac{1}{A}\displaystyle\int_a^b \tfrac{1}{2}\left[f(x)^2 - g(x)^2\right]dx.$

DISCOVERY PROJECT Complementary Coffee Cups

1. Cup A has volume $V_A = \int_0^h \pi[f(y)]^2\,dy$ and cup B has volume

$$V_B = \int_0^h \pi[k - f(y)]^2\,dy = \int_0^h \pi\{k^2 - 2kf(y) + [f(y)]^2\}\,dy$$
$$= \left[\pi k^2 y\right]_0^h - 2\pi k\int_0^h f(y)\,dy + \int_0^h \pi[f(y)]^2\,dy = \pi k^2 h - 2\pi k A_1 + V_A$$

Thus, $V_A = V_B \;\Leftrightarrow\; \pi k(kh - 2A_1) = 0 \;\Leftrightarrow\; k = 2(A_1/h)$; that is, k is twice the average value of f on the interval $[0, h]$.

2. From Problem 1, $V_A = V_B \;\Leftrightarrow\; kh = 2A_1 \;\Leftrightarrow\; A_1 + A_2 = 2A_1 \;\Leftrightarrow\; A_2 = A_1$.

3. Let $\overline{x}_1$ and $\overline{x}_2$ denote the x-coordinates of the centroids of A_1 and A_2, respectively. By Pappus's Theorem,

$V_A = 2\pi\overline{x}_1 A_1$ and $V_B = 2\pi(k - \overline{x}_2)A_2$, so $V_A = V_B \;\Leftrightarrow\; \overline{x}_1 A_1 = kA_2 - \overline{x}_2 A_2 \;\Leftrightarrow\; kA_2 = \overline{x}_1 A_1 + \overline{x}_2 A_2 \;\overset{(\star)}{\Leftrightarrow}$ $kA_2 = \frac{1}{2}k\,(A_1 + A_2) \;\Leftrightarrow\; \frac{1}{2}kA_2 = \frac{1}{2}kA_1 \;\Leftrightarrow\; A_2 = A_1$, as shown in Problem 2. [$(\star)$ The sum of the moments of the regions of areas A_1 and A_2 about the y-axis equals the moment of the entire k-by-h rectangle about the y-axis.]

So, since $A_1 + A_2 = kh$, we have $V_A = V_B \;\Leftrightarrow\; A_1 = A_2 \;\Leftrightarrow\; A_1 = \frac{1}{2}(A_1 + A_2) \;\Leftrightarrow\; A_1 = \frac{1}{2}(kh) \;\Leftrightarrow$ $k = 2\,(A_1/h)$, as shown in Problem 1.

4. We'll use a cup that is $h = 8$ cm high with a diameter of 6 cm on the top and the bottom and symmetrically bulging to a diameter of 8 cm in the middle (all inside dimensions).

For an equation, we'll use a parabola with a vertex at $(4, 4)$; that is, $x = a(y - 4)^2 + 4$. To find a, use the point $(3, 0)$: $3 = a(0 - 4)^2 + 4 \;\Rightarrow\; -1 = 16a \;\Rightarrow\; a = -\frac{1}{16}$. To find k, we'll use the relationship in Problem 1, so we need A_1.

$$A_1 = \int_0^8 \left[-\tfrac{1}{16}(y - 4)^2 + 4\right]dy = \int_{-4}^4 \left(-\tfrac{1}{16}u^2 + 4\right)du \quad [u = y - 4]$$
$$= 2\int_0^4 \left(-\tfrac{1}{16}u^2 + 4\right)du = 2\left[-\tfrac{1}{48}u^3 + 4u\right]_0^4 = 2\left(-\tfrac{4}{3} + 16\right) = \tfrac{88}{3}.$$

Thus, $k = 2(A_1/h) = 2\left(\frac{88/3}{8}\right) = \frac{22}{3}$.

So with $h = 8$ and curve $x = -\frac{1}{16}(y - 4)^2 + 4$, we have

$$V_A = \int_0^8 \pi\left[-\tfrac{1}{16}(y - 4)^2 + 4\right]^2 dy = \pi\int_{-4}^4 \left(-\tfrac{1}{16}u^2 + 4\right)^2 du \quad [u = y - 4] \quad = 2\pi\int_0^4 \left(\tfrac{1}{256}u^4 - \tfrac{1}{2}u^2 + 16\right)du$$
$$= 2\pi\left[\tfrac{1}{1280}u^5 - \tfrac{1}{6}u^3 + 16u\right]_0^4 = 2\pi\left(\tfrac{4}{5} - \tfrac{32}{3} + 64\right) = 2\pi\left(\tfrac{812}{15}\right) = \tfrac{1624}{15}\pi$$

This is approximately 340 cm^3 or 11.5 fl. oz. And with $k = \frac{22}{3}$, we know from Problem 1 that cup B holds the same amount.

8.4 Applications to Economics and Biology

1. By the Net Change Theorem, $C(4000) - C(0) = \int_0^{4000} C'(x)\,dx \Rightarrow$

$$C(4000) = 18{,}000 + \int_0^{4000} (0.82 - 0.000\,03x + 0.000\,000\,003x^2)\,dx$$
$$= 18{,}000 + \left[0.82x - 0.000\,015x^2 + 0.000\,000\,001x^3\right]_0^{4000} = 18{,}000 + 3104 = \$21{,}104$$

2. By the Net Change Theorem,

$$R(10{,}000) - R(5000) = \int_{5000}^{10{,}000} R'(x)\,dx = \int_{5000}^{10{,}000} (48 - 0.0012x)\,dx = \left[48x - 0.0006x^2\right]_{5000}^{10{,}000}$$
$$= 420{,}000 - 225{,}000 = \$195{,}000$$

3. By the Net Change Theorem, $C(50) - C(0) = \int_0^{50} (0.6 + 0.008x)\,dx \Rightarrow$

$C(50) = 100 + \left[0.6x + 0.004x^2\right]_0^{50} = 100 + (40 - 0) = 140$, or \$140,000. Similarly,

$C(100) - C(50) = \left[0.6x + 0.004x^2\right]_{50}^{100} = 100 - 40 = 60$, or \$60,000.

4. Consumer surplus $= \int_0^{400} [p(x) - p(400)]\,dx = \int_0^{400} [(2000 - 46\sqrt{x}) - 1080]\,dx$

$$= \int_0^{400} (920 - 46\sqrt{x})\,dx = 46 \int_0^{400} (20 - x^{1/2})\,dx$$
$$= 46\left[20x - \tfrac{2}{3}x^{3/2}\right]_0^{400} = 46\left(8000 - \tfrac{2}{3} \cdot 8000\right)$$
$$= 46 \cdot \tfrac{1}{3} \cdot 8000 \approx \$122{,}666.67$$

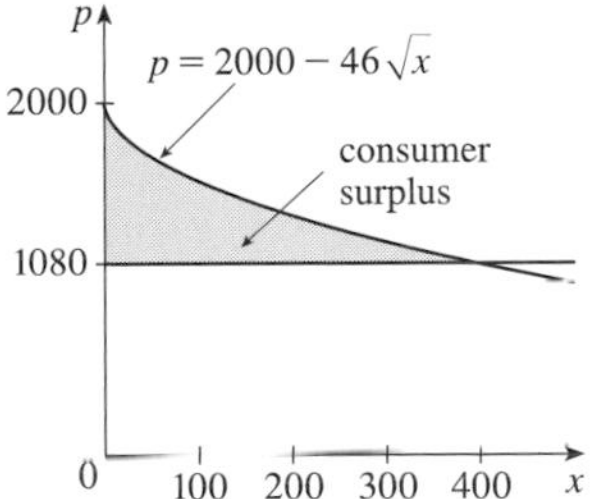

5. $p(x) = 10 \Rightarrow \dfrac{450}{x+8} = 10 \Rightarrow x + 8 = 45 \Rightarrow x = 37.$

$$\text{Consumer surplus} = \int_0^{37} [p(x) - 10]\,dx = \int_0^{37} \left(\frac{450}{x+8} - 10\right) dx$$
$$= \left[450 \ln(x+8) - 10x\right]_0^{37} = (450 \ln 45 - 370) - 450 \ln 8$$
$$= 450 \ln\left(\tfrac{45}{8}\right) - 370 \approx \$407.25$$

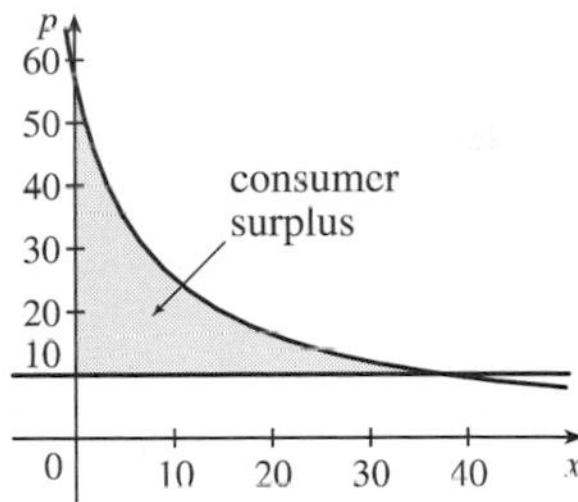

6. $p_S(x) = 3 + 0.01x^2$. $P = p_S(10) = 3 + 1 = 4$.

$$\text{Producer surplus} = \int_0^{10} [P - p_S(x)]\,dx = \int_0^{10} \left[4 - 3 - 0.01x^2\right] dx$$
$$= \left[x - \tfrac{0.01}{3}x^3\right]_0^{10} \approx 10 - 3.33 = \$6.67$$

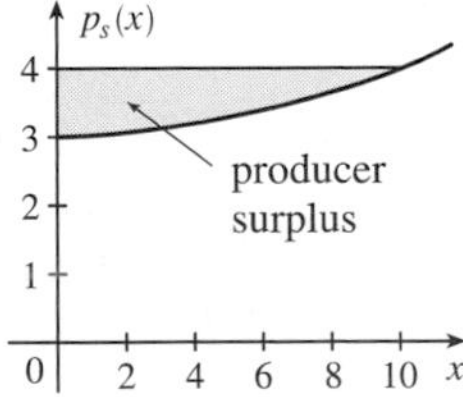

7. $P = p_S(x) \Rightarrow 625 = 125 + 0.002x^2 \Rightarrow 500 = \frac{1}{500}x^2 \Rightarrow x^2 = 500^2 \Rightarrow x = 500.$

$$\text{Producer surplus} = \int_0^{500} [P - p_S(x)]\,dx = \int_0^{500} [625 - (125 + 0.002x^2)]\,dx = \int_0^{500} \left(500 - \tfrac{1}{500}x^2\right) dx$$
$$= \left[500x - \tfrac{1}{1500}x^3\right]_0^{500} = 500^2 - \tfrac{1}{1500}(500^3) \approx \$166{,}666.67$$

8. (a) Demand curve $p_D(x) =$ supply curve $p_S(x) \quad \Leftrightarrow \quad 50 - \frac{1}{20}x = 20 + \frac{1}{10}x \quad \Leftrightarrow \quad 30 = \frac{3}{20}x \quad \Leftrightarrow \quad x = 200.$

$p_D(200) = 50 - \frac{1}{20}(200) = 40$, so the market for this good is in equilibrium when the quantity is 200 and the price is \$40.

(b) At equilibrium, the

$$\text{Consumer surplus} = \int_0^{200} [p_D(x) - 40]\, dx = \int_0^{200} \left(50 - \tfrac{1}{20}x - 40\right) dx = \left[10x - \tfrac{1}{40}x^2\right]_0^{200} = \$1000$$

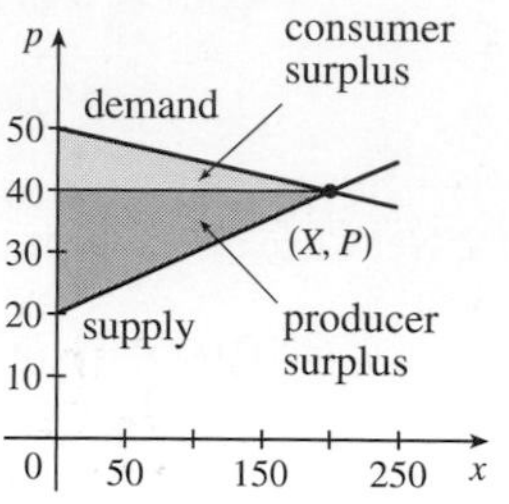

and the

$$\text{Producer surplus} = \int_0^{200} [40 - p_S(x)]\, dx = \int_0^{200} \left(40 - 20 - \tfrac{1}{10}x\right) dx = \left[20x - \tfrac{1}{20}x^2\right]_0^{200} = \$2000$$

9. (a) Demand function $p(x) =$ supply function $p_S(x) \quad \Leftrightarrow \quad 228.4 - 18x = 27x + 57.4 \quad \Leftrightarrow \quad 171 = 45x \quad \Leftrightarrow$

$x = \frac{19}{5}$ [3.8 thousand]. $p(3.8) = 228.4 - 18(3.8) = 160$. The market for the stereos is in equilibrium when the quantity is 3800 and the price is \$160.

(b) $$\text{Consumer surplus} = \int_0^{3.8} [p(x) - 160]\, dx = \int_0^{3.8} (228.4 - 18x - 160)\, dx = \int_0^{3.8} (68.4 - 18x)\, dx = \left[68.4x - 9x^2\right]_0^{3.8} = 68.4(3.8) - 9(3.8)^2 = 129.96$$

$$\text{Producer surplus} = \int_0^{3.8} [160 - p_S(x)]\, dx = \int_0^{3.8} [160 - (27x + 57.4)]\, dx = \int_0^{3.8} (102.6 - 27x)\, dx = \left[102.6x - 13.5x^2\right]_0^{3.8} = 102.6(3.8) - 13.5(3.8)^2 = 194.94$$

Thus, the maximum total surplus for the stereos is $129.96 + 194.94 = 324.9$, or \$324,900.

10. $p(x) = p_S(x) \quad \Leftrightarrow \quad 312e^{-0.14x} = 26e^{0.2x} \quad \Leftrightarrow \quad \dfrac{312}{26} = \dfrac{e^{0.2x}}{e^{-0.14x}} \quad \Leftrightarrow \quad 12 = e^{0.34x} \quad \Leftrightarrow \quad \ln 12 = 0.34x \quad \Leftrightarrow$

$x = X = \dfrac{\ln 12}{0.34}$. $X \approx 7.3085$ (in thousands) and $p(X) \approx 112.1465$.

Consumer surplus $= \int_0^X [p(x) - p(X)]\, dx \approx \int_0^{7.3085} (312e^{-0.14x} - 112.1465)\, dx \approx 607.896$

Producer surplus $= \int_0^X [p_S(X) - p_S(x)]\, dx \approx \int_0^{7.3085} (112.1465 - 26e^{0.2x})\, dx \approx 388.896$

Maximum total surplus $\approx 607.896 + 388.896 = 996.792$, or \$996,792.

Note: Since $p(X) = p_S(X)$, the maximum total surplus could be found by calculating $\int_0^X [p(x) - p_S(x)]\, dx$.

11. $p(x) = \dfrac{800{,}000e^{-x/5000}}{x + 20{,}000} = 16 \quad \Rightarrow \quad x = x_1 \approx 3727.04.$

Consumer surplus $= \int_0^{x_1} [p(x) - 16]\, dx \approx \$37{,}753$

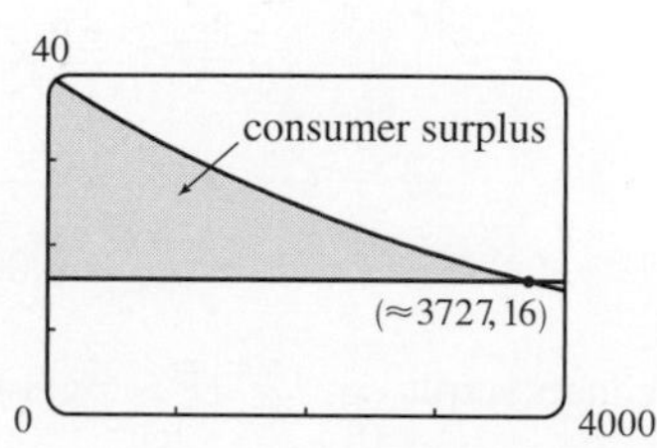

12. The demand function is linear with slope $\frac{-0.5}{50} = -\frac{1}{100}$ and $p(500) = 10$, so an equation is $p - 10 = -\frac{1}{100}(x - 500)$ or $p = -\frac{1}{100}x + 15$. A selling price of \$8 implies that $8 = -\frac{1}{100}x + 15 \Rightarrow \frac{1}{100}x = 7 \Rightarrow x = 700$.

Consumer surplus $= \int_0^{700} \left(-\frac{1}{100}x + 15 - 8\right) dx = \left[-\frac{1}{200}x^2 + 7x\right]_0^{700} = \$2450$.

13. $f(8) - f(4) = \int_4^8 f'(t)\,dt = \int_4^8 \sqrt{t}\,dt = \left[\frac{2}{3}t^{3/2}\right]_4^8 = \frac{2}{3}\left(16\sqrt{2} - 8\right) \approx \$9.75$ million

14. The total revenue R obtained in the first four years is

$$R = \int_0^4 f(t)\,dt = \int_0^4 9000\sqrt{1+2t}\,dt = \int_1^9 9000u^{1/2}\left(\tfrac{1}{2}\,du\right) \qquad [u = 1 + 2t,\ du = 2\,dt]$$

$$= 4500\left[\tfrac{2}{3}u^{3/2}\right]_1^9 = 3000(27 - 1) = \$78{,}000$$

15. Future value $= \int_0^T f(t)\,e^{r(T-t)}\,dt = \int_0^6 8000e^{0.04t}\,e^{0.062(6-t)}\,dt = 8000\int_0^6 e^{0.04t}\,e^{0.372-0.062t}\,dt$

$$= 8000\int_0^6 e^{0.372-0.022t}\,dt = 8000e^{0.372}\int_0^6 e^{-0.022t}\,dt = 8000e^{0.372}\left[\frac{e^{-0.022t}}{-0.022}\right]_0^6$$

$$= \frac{8000e^{0.372}}{-0.022}\left(e^{-0.132} - 1\right) \approx \$65{,}230.48$$

16. Present value $= \int_0^T f(t)\,e^{-rt}\,dt = \int_0^6 8000e^{0.04t}\,e^{-0.062t}\,dt = 8000\int_0^6 e^{-0.022t}\,dt = 8000\left[\frac{e^{-0.022t}}{-0.022}\right]_0^6$

$$= \frac{8000}{-0.022}\left(e^{-0.132} - 1\right) \approx \$44{,}966.91$$

17. $N = \displaystyle\int_a^b Ax^{-k}\,dx = A\left[\frac{x^{-k+1}}{-k+1}\right]_a^b = \frac{A}{1-k}\left(b^{1-k} - a^{1-k}\right)$.

Similarly, $\displaystyle\int_a^b Ax^{1-k}\,dx = A\left[\frac{x^{2-k}}{2-k}\right]_a^b = \frac{A}{2-k}\left(b^{2-k} - a^{2-k}\right)$.

Thus, $\overline{x} = \displaystyle\frac{1}{N}\int_a^b Ax^{1-k}\,dx = \frac{[A/(2-k)](b^{2-k} - a^{2-k})}{[A/(1-k)](b^{1-k} - a^{1-k})} = \frac{(1-k)(b^{2-k} - a^{2-k})}{(2-k)(b^{1-k} - a^{1-k})}$.

18. $n(9) - n(5) = \displaystyle\int_5^9 \left(2200 + 10e^{0.8t}\right)dt = \left[2200t + \frac{10e^{0.8t}}{0.8}\right]_5^9 = \left[2200t\right]_5^9 + \tfrac{25}{2}\left[e^{0.8t}\right]_5^9$

$$= 2200(9 - 5) + 12.5\left(e^{7.2} - e^4\right) \approx 24{,}860$$

19. $F = \dfrac{\pi PR^4}{8\eta l} = \dfrac{\pi(4000)(0.008)^4}{8(0.027)(2)} \approx 1.19 \times 10^{-4}\ \text{cm}^3/\text{s}$

20. If the flux remains constant, then $\dfrac{\pi P_0 R_0^4}{8\eta l} = \dfrac{\pi PR^4}{8\eta l} \Rightarrow P_0 R_0^4 = PR^4 \Rightarrow \dfrac{P}{P_0} = \left(\dfrac{R_0}{R}\right)^4$.

$R = \frac{3}{4}R_0 \Rightarrow \dfrac{P}{P_0} = \left(\dfrac{R_0}{\frac{3}{4}R_0}\right)^4 \Rightarrow P = P_0\left(\frac{4}{3}\right)^4 \approx 3.1605P_0 > 3P_0$; that is, the blood pressure is more than tripled.

21. From (3), $F = \dfrac{A}{\int_0^T c(t)\,dt} = \dfrac{6}{20I}$, where

$$I = \int_0^{10} te^{-0.6t}\,dt = \left[\frac{1}{(-0.6)^2}(-0.6t - 1)\,e^{-0.6t}\right]_0^{10} \quad \begin{bmatrix}\text{integrating} \\ \text{by parts}\end{bmatrix} \quad = \tfrac{1}{0.36}(-7e^{-6} + 1)$$

Thus, $F = \dfrac{6(0.36)}{20(1 - 7e^{-6})} = \dfrac{0.108}{1 - 7e^{-6}} \approx 0.1099$ L/s or 6.594 L/min.

22. As in Example 2, we will estimate the cardiac output using Simpson's Rule with $\Delta t = (16 - 0)/8 = 2$.

$$\begin{aligned}\textstyle\int_0^{16} c(t)\,dt &\approx \tfrac{2}{3}[c(0) + 4c(2) + 2c(4) + 4c(6) + 2c(8) + 4c(10) + 2c(12) + 4c(14) + c(16)] \\ &= \tfrac{2}{3}[0 + 4(4.1) + 2(8.9) + 4(8.5) + 2(6.7) + 4(4.3) + 2(2.5) + 4(1.2) + 0.2] \\ &= \tfrac{2}{3}(108.8) = 72.5\overline{3}\ \text{mg}\cdot\text{s/L}\end{aligned}$$

Therefore, $F \approx \dfrac{A}{72.5\overline{3}} = \dfrac{5.5}{72.5\overline{3}} \approx 0.0758$ L/s or 4.55 L/min.

23. As in Example 2, we will estimate the cardiac output using Simpson's Rule with $\Delta t = (16 - 0)/8 = 2$.

$$\begin{aligned}\textstyle\int_0^{16} c(t)\,dt &\approx \tfrac{2}{3}[c(0) + 4c(2) + 2c(4) + 4c(6) + 2c(8) + 4c(10) + 2c(12) + 4c(14) + c(16)] \\ &\approx \tfrac{2}{3}[0 + 4(6.1) + 2(7.4) + 4(6.7) + 2(5.4) + 4(4.1) + 2(3.0) + 4(2.1) + 1.5] \\ &= \tfrac{2}{3}(109.1) = 72.7\overline{3}\ \text{mg}\cdot\text{s/L}\end{aligned}$$

Therefore, $F \approx \dfrac{A}{72.7\overline{3}} = \dfrac{7}{72.7\overline{3}} \approx 0.0962$ L/s or 5.77 L/min.

8.5 Probability

1. (a) $\int_{30{,}000}^{40{,}000} f(x)\,dx$ is the probability that a randomly chosen tire will have a lifetime between 30,000 and 40,000 miles.

(b) $\int_{25{,}000}^{\infty} f(x)\,dx$ is the probability that a randomly chosen tire will have a lifetime of at least 25,000 miles.

2. (a) The probability that you drive to school in less than 15 minutes is $\int_0^{15} f(t)\,dt$.

(b) The probability that it takes you more than half an hour to get to school is $\int_{30}^{\infty} f(t)\,dt$.

3. (a) In general, we must satisfy the two conditions that are mentioned before Example 1—namely, **(1)** $f(x) \geq 0$ for all x, and **(2)** $\int_{-\infty}^{\infty} f(x)\,dx = 1$. For $0 \leq x \leq 1$, $f(x) = 30x^2(1 - x)^2 \geq 0$ and $f(x) = 0$ for all other values of x, so $f(x) \geq 0$ for all x. Also,

$$\begin{aligned}\textstyle\int_{-\infty}^{\infty} f(x)\,dx &= \textstyle\int_0^1 30x^2(1 - x)^2\,dx = \int_0^1 30x^2(1 - 2x + x^2)\,dx = \int_0^1 (30x^2 - 60x^3 + 30x^4)\,dx \\ &= \left[10x^3 - 15x^4 + 6x^5\right]_0^1 = 10 - 15 + 6 = 1\end{aligned}$$

Therefore, f is a probability density function.

(b) $P(X \leq \frac{1}{3}) = \int_{-\infty}^{1/3} f(x)\,dx = \int_0^{1/3} 30x^2(1 - x)^2\,dx = \left[10x^3 - 15x^4 + 6x^5\right]_0^{1/3} = \frac{10}{27} - \frac{15}{81} + \frac{6}{243} = \frac{17}{81}$

4. (a) In general, we must satisfy the two conditions that are mentioned before Example 1—namely, **(1)** $f(x) \ge 0$ for all x, and **(2)** $\int_{-\infty}^{\infty} f(x)\,dx = 1$. For $f(x) = \dfrac{e^{3-x}}{(1+e^{3-x})^2}$, the numerator and denominator are both positive, so $f(x) \ge 0$ for all x.

Also,

$$\int_{-\infty}^{\infty} f(x)\,dx = \int_{-\infty}^{0} f(x)\,dx + \int_{0}^{\infty} f(x)\,dx = \lim_{t\to-\infty}\int_t^0 \frac{e^{3-x}}{(1+e^{3-x})^2}\,dx + \lim_{s\to\infty}\int_0^s \frac{e^{3-x}}{(1+e^{3-x})^2}\,dx$$

$$= \lim_{t\to-\infty}\int_{x=t}^{0}\frac{-du}{u^2} + \lim_{s\to\infty}\int_{x=0}^{s}\frac{-du}{u^2} \qquad \begin{bmatrix} u = 1+e^{3-x}, \\ du = -e^{3-x}\,dx \end{bmatrix}$$

$$= \lim_{t\to-\infty}\left[\frac{1}{u}\right]_{x=t}^{0} + \lim_{s\to\infty}\left[\frac{1}{u}\right]_{x=0}^{s} = \lim_{t\to-\infty}\left[\frac{1}{1+e^{3-x}}\right]_t^0 + \lim_{s\to\infty}\left[\frac{1}{1+e^{3-x}}\right]_0^s$$

$$= \lim_{t\to-\infty}\left(\frac{1}{1+e^3} - \frac{1}{1+e^{3-t}}\right) + \lim_{s\to\infty}\left(\frac{1}{1+e^{3-s}} - \frac{1}{1+e^3}\right) = \frac{1}{1+e^3} - 0 + 1 - \frac{1}{1+e^3} = 1.$$

Therefore, f is a probability density function.

(b) $P(3 \le X \le 4) = \displaystyle\int_3^4 f(x)\,dx = \left[\frac{1}{1+e^{3-x}}\right]_3^4$ [from part (a)] $= \dfrac{1}{1+e^{-1}} - \dfrac{1}{1+1} \approx 0.231$

(c)

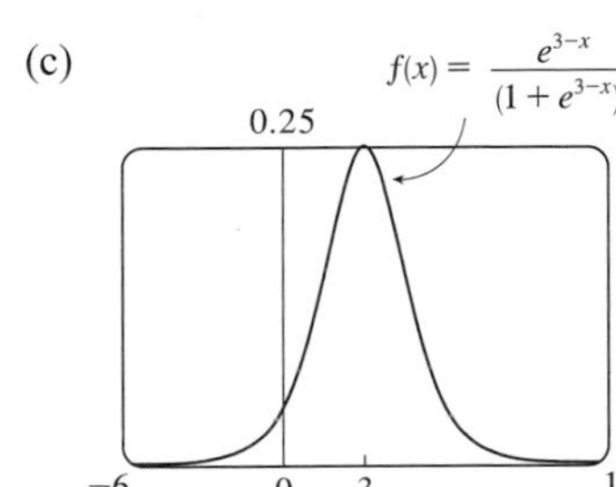

The graph of f appears to be symmetric about the line $x = 3$, so the mean appears to be 3. Similarly, half the area under the graph of f appears to lie to the right of $x = 3$, so the median also appears to be 3.

5. (a) In general, we must satisfy the two conditions that are mentioned before Example 1—namely, (1) $f(x) \ge 0$ for all x, and (2) $\int_{-\infty}^{\infty} f(x)\,dx = 1$. If $c \ge 0$, then $f(x) \ge 0$, so condition (1) is satisfied. For condition (2), we see that

$\displaystyle\int_{-\infty}^{\infty} f(x)\,dx = \int_{-\infty}^{\infty}\frac{c}{1+x^2}\,dx$ and

$$\int_0^{\infty}\frac{c}{1+x^2}\,dx = \lim_{t\to\infty}\int_0^t\frac{c}{1+x^2}\,dx = c\lim_{t\to\infty}\left[\tan^{-1}x\right]_0^t = c\lim_{t\to\infty}\tan^{-1}t = c\left(\frac{\pi}{2}\right)$$

Similarly, $\displaystyle\int_{-\infty}^{0}\frac{c}{1+x^2}\,dx = c\left(\frac{\pi}{2}\right)$, so $\displaystyle\int_{-\infty}^{\infty}\frac{c}{1+x^2}\,dx = 2c\left(\frac{\pi}{2}\right) = c\pi$.

Since $c\pi$ must equal 1, we must have $c = 1/\pi$ so that f is a probability density function.

(b) $P(-1 < X < 1) = \displaystyle\int_{-1}^{1}\frac{1/\pi}{1+x^2}\,dx = \frac{2}{\pi}\int_0^1\frac{1}{1+x^2}\,dx = \frac{2}{\pi}\left[\tan^{-1}x\right]_0^1 = \frac{2}{\pi}\left(\frac{\pi}{4} - 0\right) = \frac{1}{2}$

6. (a) For $0 \le x \le 3$, we have $f(x) = k(3x - x^2)$, which is nonnegative if and only if $k \ge 0$. Also, $\int_{-\infty}^{\infty} f(x)\,dx = \int_0^3 k(3x - x^2)\,dx = k\left[\frac{3}{2}x^2 - \frac{1}{3}x^3\right]_0^3 = k\left(\frac{27}{2} - 9\right) = \frac{9}{2}k$. Now $\frac{9}{2}k = 1 \;\Rightarrow\; k = \frac{2}{9}$. Therefore, f is a probability density function if and only if $k = \frac{2}{9}$.

(b) Let $k = \frac{2}{9}$.

$P(X > 1) = \int_1^\infty f(x)\,dx = \int_1^3 \frac{2}{9}(3x - x^2)\,dx = \frac{2}{9}\left[\frac{3}{2}x^2 - \frac{1}{3}x^3\right]_1^3 = \frac{2}{9}\left[\left(\frac{27}{2} - 9\right) - \left(\frac{3}{2} - \frac{1}{3}\right)\right] = \frac{2}{9}\left(\frac{10}{3}\right) = \frac{20}{27}$.

(c) The mean $\mu = \int_{-\infty}^\infty x f(x)\,dx = \int_0^3 x\left[\frac{2}{9}(3x - x^2)\right]dx = \frac{2}{9}\int_0^3 (3x^2 - x^3)\,dx$
$= \frac{2}{9}\left[x^3 - \frac{1}{4}x^4\right]_0^3 = \frac{2}{9}\left(27 - \frac{81}{4}\right) = \frac{2}{9}\left(\frac{27}{4}\right) = \frac{3}{2}$.

7. (a) In general, we must satisfy the two conditions that are mentioned before Example 1—namely, (1) $f(x) \geq 0$ for all x, and (2) $\int_{-\infty}^\infty f(x)\,dx = 1$. Since $f(x) = 0$ or $f(x) = 0.1$, condition (1) is satisfied. For condition (2), we see that $\int_{-\infty}^\infty f(x)\,dx = \int_0^{10} 0.1\,dx = \left[\frac{1}{10}x\right]_0^{10} = 1$. Thus, $f(x)$ is a probability density function for the spinner's values.

(b) Since all the numbers between 0 and 10 are equally likely to be selected, we expect the mean to be halfway between the endpoints of the interval; that is, $x = 5$.

$$\mu = \int_{-\infty}^\infty x f(x)\,dx = \int_0^{10} x(0.1)\,dx = \left[\tfrac{1}{20}x^2\right]_0^{10} = \tfrac{100}{20} = 5, \quad \text{as expected.}$$

8. (a) As in the preceding exercise, (1) $f(x) \geq 0$ and (2) $\int_{-\infty}^\infty f(x)\,dx = \int_0^{10} f(x)\,dx = \frac{1}{2}(10)(0.2)$ [area of a triangle] $= 1$. So $f(x)$ is a probability density function.

(b) (i) $P(X < 3) = \int_0^3 f(x)\,dx = \frac{1}{2}(3)(0.1) = \frac{3}{20} = 0.15$

(ii) We first compute $P(X > 8)$ and then subtract that value and our answer in (i) from 1 (the total probability).

$P(X > 8) = \int_8^{10} f(x)\,dx = \frac{1}{2}(2)(0.1) = \frac{2}{20} = 0.10$. So $P(3 \leq X \leq 8) = 1 - 0.15 - 0.10 = 0.75$.

(c) We find equations of the lines from $(0, 0)$ to $(6, 0.2)$ and from $(6, 0.2)$ to $(10, 0)$, and find that

$$f(x) = \begin{cases} \frac{1}{30}x & \text{if } 0 \leq x < 6 \\ -\frac{1}{20}x + \frac{1}{2} & \text{if } 6 \leq x < 10 \\ 0 & \text{otherwise} \end{cases}$$

$$\mu = \int_{-\infty}^\infty x f(x)\,dx = \int_0^6 x\left(\tfrac{1}{30}x\right)dx + \int_6^{10} x\left(-\tfrac{1}{20}x + \tfrac{1}{2}\right)dx = \left[\tfrac{1}{90}x^3\right]_0^6 + \left[-\tfrac{1}{60}x^3 + \tfrac{1}{4}x^2\right]_6^{10}$$
$$= \tfrac{216}{90} + \left(-\tfrac{1000}{60} + \tfrac{100}{4}\right) - \left(-\tfrac{216}{60} + \tfrac{36}{4}\right) = \tfrac{16}{3} = 5.\overline{3}$$

9. We need to find m so that $\int_m^\infty f(t)\,dt = \frac{1}{2} \Rightarrow \lim\limits_{x\to\infty} \int_m^x \frac{1}{5}e^{-t/5}\,dt = \frac{1}{2} \Rightarrow \lim\limits_{x\to\infty}\left[\frac{1}{5}(-5)e^{-t/5}\right]_m^x = \frac{1}{2} \Rightarrow$ $(-1)(0 - e^{-m/5}) = \frac{1}{2} \Rightarrow e^{-m/5} = \frac{1}{2} \Rightarrow -m/5 = \ln\frac{1}{2} \Rightarrow m = -5\ln\frac{1}{2} = 5\ln 2 \approx 3.47$ min.

10. (a) $\mu = 1000 \Rightarrow f(t) = \begin{cases} 0 & \text{if } t < 0 \\ \frac{1}{1000}e^{-t/1000} & \text{if } t \geq 0 \end{cases}$

(i) $P(0 \leq X \leq 200) = \int_0^{200} \frac{1}{1000}e^{-t/1000}\,dt = \left[-e^{-t/1000}\right]_0^{200} = -e^{-1/5} + 1 \approx 0.181$

(ii) $P(X > 800) = \int_{800}^\infty \frac{1}{1000}e^{-t/1000}\,dt = \lim\limits_{x\to\infty}\left[-e^{-t/1000}\right]_{800}^x = 0 + e^{-4/5} \approx 0.449$

(b) We need to find m so that $\int_m^\infty f(t)\,dt = \frac{1}{2} \Rightarrow \lim\limits_{x\to\infty}\int_m^x \frac{1}{1000}e^{-t/1000}\,dt = \frac{1}{2} \Rightarrow \lim\limits_{x\to\infty}\left[-e^{-t/1000}\right]_m^x = \frac{1}{2} \Rightarrow$ $0 + e^{-m/1000} = \frac{1}{2} \Rightarrow -m/1000 = \ln\frac{1}{2} \Rightarrow m = -1000\ln\frac{1}{2} = 1000\ln 2 \approx 693.1$ h.

11. (a) An exponential density function with $\mu = 1.6$ is $f(t) = \begin{cases} 0 & \text{if } t < 0 \\ \frac{1}{1.6}e^{-t/1.6} & \text{if } t \geq 0 \end{cases}$.

The probability that a customer waits less than a second is

$$P(X < 1) = \int_0^1 f(t)\,dt = \int_0^1 \tfrac{1}{1.6}e^{-t/1.6}\,dt = \left[-e^{-t/1.6}\right]_0^1 = -e^{-1/1.6} + 1 \approx 0.465.$$

(b) The probability that a customer waits more than 3 seconds is

$$P(X > 3) = \int_3^\infty f(t)\,dt = \lim_{s\to\infty} \int_3^s f(t)\,dt = \lim_{s\to\infty} \left[-e^{-t/1.6}\right]_3^s = \lim_{s\to\infty} (-e^{-s/1.6} + e^{-3/1.6}) = e^{-3/1.6} \approx 0.153.$$

Or: Calculate $1 - \int_0^3 f(t)\,dt$.

(c) We want to find b such that $P(X > b) = 0.05$. From part (b), $P(X > b) = e^{-b/1.6}$. Solving $e^{-b/1.6} = 0.05$ gives us

$-\frac{b}{1.6} = \ln 0.05 \quad \Rightarrow \quad b = -1.6 \ln 0.05 \approx 4.79$ seconds.

Or: Solve $\int_0^b f(t)\,dt = 0.95$ for b.

12. (a) We first find an antiderivative of $g(t) = t^2 e^{at}$.

$$\begin{aligned}
\int t^2 e^{at}\,dt &= \frac{1}{a}t^2 e^{at} - \int \frac{2}{a} t\,e^{at}\,dt && \left[\begin{matrix} u = t^2, & dv = e^{at}\,dt \\ du = 2t\,dt, & v = \frac{1}{a}e^{at} \end{matrix}\right] \\
&= \frac{1}{a}t^2 e^{at} - \frac{2}{a}\left[\frac{1}{a}t\,e^{at} - \int \frac{1}{a}e^{at}\,dt\right] && \left[\begin{matrix} u = t, & dv = e^{at}\,dt \\ du = dt, & v = \frac{1}{a}e^{at} \end{matrix}\right] \\
&= \frac{1}{a}t^2 e^{at} - \frac{2}{a^2}t\,e^{at} + \frac{2}{a^3}e^{at} + C = \frac{1}{a}e^{at}\left(t^2 - \frac{2}{a}t + \frac{2}{a^2}\right) + C \\
&= -20e^{-0.05t}(t^2 + 40t + 800) + C && [\text{with } a = -0.05]
\end{aligned}$$

$$P(0 \leq X \leq 48) = \int_0^{48} f(t)\,dt = \frac{1}{15{,}676}\int_0^{48} g(t)\,dt = \frac{1}{15{,}676}\left[-20e^{-0.05t}(t^2 + 40t + 800)\right]_0^{48}$$

$$= \frac{-20}{15{,}676}(5024e^{-2.4} - 800) \approx 0.439.$$

(b) $$P(X > 36) = P(36 < X \leq 150) = \frac{1}{15{,}676}\int_{36}^{150} g(t)\,dt = \frac{1}{15{,}676}\left[-20e^{-0.05t}(t^2 + 40t + 800)\right]_{36}^{150}$$

$$= \frac{-20}{15{,}676}(29{,}300e^{-7.5} - 3536e^{-1.8}) \approx 0.725$$

13. (a) $f(t) = \begin{cases} \frac{1}{1600}t & \text{if } 0 \leq t \leq 40 \\ \frac{1}{20} - \frac{1}{1600}t & \text{if } 40 < t \leq 80 \\ 0 & \text{otherwise} \end{cases}$

$$P(30 \leq T \leq 60) = \int_{30}^{60} f(t)\,dt = \int_{30}^{40} \frac{t}{1600}\,dt + \int_{40}^{60}\left(\frac{1}{20} - \frac{t}{1600}\right)dt = \left[\frac{t^2}{3200}\right]_{30}^{40} + \left[\frac{t}{20} - \frac{t^2}{3200}\right]_{40}^{60}$$

$$= \left(\frac{1600}{3200} - \frac{900}{3200}\right) + \left(\frac{60}{20} - \frac{3600}{3200}\right) - \left(\frac{40}{20} - \frac{1600}{3200}\right) = -\frac{1300}{3200} + 1 = \frac{19}{32}$$

The probability that the amount of REM sleep is between 30 and 60 minutes is $\frac{19}{32} \approx 59.4\%$.

(b) $\mu = \int_{-\infty}^{\infty} t\,f(t)\,dt = \int_0^{40} t\left(\frac{t}{1600}\right) dt + \int_{40}^{80} t\left(\frac{1}{20} - \frac{t}{1600}\right) dt = \left[\frac{t^3}{4800}\right]_0^{40} + \left[\frac{t^2}{40} - \frac{t^3}{4800}\right]_{40}^{80}$

$$= \frac{64{,}000}{4800} + \left(\frac{6400}{40} - \frac{512{,}000}{4800}\right) - \left(\frac{1600}{40} - \frac{64{,}000}{4800}\right) = -\frac{384{,}000}{4800} + 120 = 40$$

The mean amount of REM sleep is 40 minutes.

14. (a) With $\mu = 69$ and $\sigma = 2.8$, we have $P(65 \le X \le 73) = \int_{65}^{73} \frac{1}{2.8\sqrt{2\pi}} \exp\left(-\frac{(x-69)^2}{2 \cdot 2.8^2}\right) dx \approx 0.847$

(using a calculator or computer to estimate the integral).

(b) $P(X > 6 \text{ feet}) = P(X > 72 \text{ inches}) = 1 - P(0 \le X \le 72) \approx 1 - 0.858 = 0.142$, so 14.2% of the adult male population is more than 6 feet tall.

15. $P(X \ge 10) = \int_{10}^{\infty} \frac{1}{4.2\sqrt{2\pi}} \exp\left(-\frac{(x-9.4)^2}{2 \cdot 4.2^2}\right) dx$. To avoid the improper integral we approximate it by the integral from 10 to 100. Thus, $P(X \ge 10) \approx \int_{10}^{100} \frac{1}{4.2\sqrt{2\pi}} \exp\left(-\frac{(x-9.4)^2}{2 \cdot 4.2^2}\right) dx \approx 0.443$ (using a calculator or computer to estimate the integral), so about 44 percent of the households throw out at least 10 lb of paper a week.

Note: We can't evaluate $1 - P(0 \le X \le 10)$ for this problem since a significant amount of area lies to the left of $X = 0$.

16. (a) $P(0 \le X \le 480) = \int_0^{480} \frac{1}{12\sqrt{2\pi}} \exp\left(-\frac{(x-500)^2}{2 \cdot 12^2}\right) dx \approx 0.0478$ (using a calculator or computer to estimate the integral), so there is about a 4.78% chance that a particular box contains less than 480 g of cereal.

(b) We need to find μ so that $P(0 \le X < 500) = 0.05$. Using our calculator or computer to find $P(0 \le X \le 500)$ for various values of μ, we find that if $\mu = 519.73$, $P = 0.05007$; and if $\mu = 519.74$, $P = 0.04998$. So a good target weight is at least 519.74 g.

17. (a) $P(0 \le X \le 100) = \int_0^{100} \frac{1}{8\sqrt{2\pi}} \exp\left(-\frac{(x-112)^2}{2 \cdot 8^2}\right) dx \approx 0.0668$ (using a calculator or computer to estimate the integral), so there is about a 6.68% chance that a randomly chosen vehicle is traveling at a legal speed.

(b) $P(X \ge 125) = \int_{125}^{\infty} \frac{1}{8\sqrt{2\pi}} \exp\left(-\frac{(x-112)^2}{2 \cdot 8^2}\right) dx = \int_{125}^{\infty} f(x)\,dx$. In this case, we could use a calculator or computer to estimate either $\int_{125}^{300} f(x)\,dx$ or $1 - \int_0^{125} f(x)\,dx$. Both are approximately 0.0521, so about 5.21% of the motorists are targeted.

18. $f(x) = \frac{1}{\sigma\sqrt{2\pi}} e^{-(x-\mu)^2/(2\sigma^2)} \quad\Rightarrow\quad f'(x) = \frac{1}{\sigma\sqrt{2\pi}} e^{-(x-\mu)^2/(2\sigma^2)} \frac{-2(x-\mu)}{2\sigma^2} = \frac{-1}{\sigma^3\sqrt{2\pi}} e^{-(x-\mu)^2/(2\sigma^2)} (x-\mu) \quad\Rightarrow$

$$f''(x) = \frac{-1}{\sigma^3\sqrt{2\pi}} \left[e^{-(x-\mu)^2/(2\sigma^2)} \cdot 1 + (x-\mu)e^{-(x-\mu)^2/(2\sigma^2)} \frac{-2(x-\mu)}{2\sigma^2}\right]$$

$$= \frac{-1}{\sigma^3\sqrt{2\pi}} e^{-(x-\mu)^2/(2\sigma^2)} \left[1 - \frac{(x-\mu)^2}{\sigma^2}\right] = \frac{1}{\sigma^5\sqrt{2\pi}} e^{-(x-\mu)^2/(2\sigma^2)} \left[(x-\mu)^2 - \sigma^2\right]$$

$f''(x) < 0 \quad\Rightarrow\quad (x-\mu)^2 - \sigma^2 < 0 \quad\Rightarrow\quad |x-\mu| < \sigma \quad\Rightarrow\quad -\sigma < x - \mu < \sigma \quad\Rightarrow\quad \mu - \sigma < x < \mu + \sigma$ and similarly, $f''(x) > 0 \quad\Rightarrow\quad x < \mu - \sigma$ or $x > \mu + \sigma$. Thus, f changes concavity and has inflection points at $x = \mu \pm \sigma$.

19. $P(\mu - 2\sigma \le X \le \mu + 2\sigma) = \int_{\mu-2\sigma}^{\mu+2\sigma} \frac{1}{\sigma\sqrt{2\pi}} \exp\left(-\frac{(x-\mu)^2}{2\sigma^2}\right) dx$. Substituting $t = \frac{x-\mu}{\sigma}$ and $dt = \frac{1}{\sigma}\,dx$ gives us

$$\int_{-2}^{2} \frac{1}{\sigma\sqrt{2\pi}} e^{-t^2/2}(\sigma\,dt) = \frac{1}{\sqrt{2\pi}} \int_{-2}^{2} e^{-t^2/2}\,dt \approx 0.9545.$$

20. Let $f(x) = \begin{cases} 0 & \text{if } x < 0 \\ ce^{-cx} & \text{if } x \ge 0 \end{cases}$ where $c = 1/\mu$. By using parts, tables, or a CAS, we find that

(1): $\int xe^{bx}\,dx = (e^{bx}/b^2)(bx - 1)$

(2): $\int x^2e^{bx}\,dx = (e^{bx}/b^3)(b^2x^2 - 2bx + 2)$

Now
$$\begin{aligned}\sigma^2 &= \textstyle\int_{-\infty}^{\infty} (x-\mu)^2 f(x)\,dx = \int_{-\infty}^{0} (x-\mu)^2 f(x)\,dx + \int_{0}^{\infty} (x-\mu)^2 f(x)\,dx \\ &= 0 + \lim_{t\to\infty} c\textstyle\int_0^t (x-\mu)^2 e^{-cx}\,dx = c\cdot \lim_{t\to\infty} \int_0^t \left(x^2e^{-cx} - 2x\mu e^{-cx} + \mu^2 e^{-cx}\right) dx\end{aligned}$$

Next we use (2) and (1) with $b = -c$ to get

$$\sigma^2 = c \lim_{t\to\infty}\left[-\frac{e^{-cx}}{c^3}(c^2x^2 + 2cx + 2) - 2\mu\frac{e^{-cx}}{c^2}(-cx - 1) + \mu^2\frac{e^{-cx}}{-c}\right]_0^t$$

Using l'Hospital's Rule several times, along with the fact that $\mu = 1/c$, we get

$$\sigma^2 = c\left[0 - \left(-\frac{2}{c^3} + \frac{2}{c}\cdot\frac{1}{c^2} + \frac{1}{c^2}\cdot\frac{1}{-c}\right)\right] - c\left(\frac{1}{c^3}\right) = \frac{1}{c^2} \quad\Rightarrow\quad \sigma = \frac{1}{c} = \mu$$

21. (a) First $p(r) = \frac{4}{a_0^3} r^2 e^{-2r/a_0} \ge 0$ for $r \ge 0$. Next,

$$\int_{-\infty}^{\infty} p(r)\,dr = \int_0^{\infty} \frac{4}{a_0^3} r^2 e^{-2r/a_0}\,dr = \frac{4}{a_0^3}\lim_{t\to\infty}\int_0^t r^2 e^{-2r/a_0}\,dr$$

By using parts, tables, or a CAS [or as in Exercise 20], we find that $\int x^2 e^{bx}\,dx = (e^{bx}/b^3)(b^2x^2 - 2bx + 2)$. $(\star)$

Next, we use $(\star)$ (with $b = -2/a_0$) and l'Hospital's Rule to get $\frac{4}{a_0^3}\left[\frac{a_0^3}{-8}(-2)\right] = 1$. This satisfies the second condition for a function to be a probability density function.

(b) Using l'Hospital's Rule, $\frac{4}{a_0^3}\lim_{r\to\infty}\frac{r^2}{e^{2r/a_0}} = \frac{4}{a_0^3}\lim_{r\to\infty}\frac{2r}{(2/a_0)e^{2r/a_0}} = \frac{2}{a_0^2}\lim_{r\to\infty}\frac{2}{(2/a_0)e^{2r/a_0}} = 0.$

To find the maximum of p, we differentiate:

$$p'(r) = \frac{4}{a_0^3}\left[r^2e^{-2r/a_0}\left(-\frac{2}{a_0}\right) + e^{-2r/a_0}(2r)\right] = \frac{4}{a_0^3}e^{-2r/a_0}(2r)\left(-\frac{r}{a_0} + 1\right)$$

$p'(r) = 0 \quad\Leftrightarrow\quad r = 0 \text{ or } 1 = \frac{r}{a_0} \quad\Leftrightarrow\quad r = a_0 \quad [a_0 \approx 5.59 \times 10^{-11}\text{ m}].$

$p'(r)$ changes from positive to negative at $r = a_0$, so $p(r)$ has its maximum value at $r = a_0$.

(c) It is fairly difficult to find a viewing rectangle, but knowing the maximum value from part (b) helps.

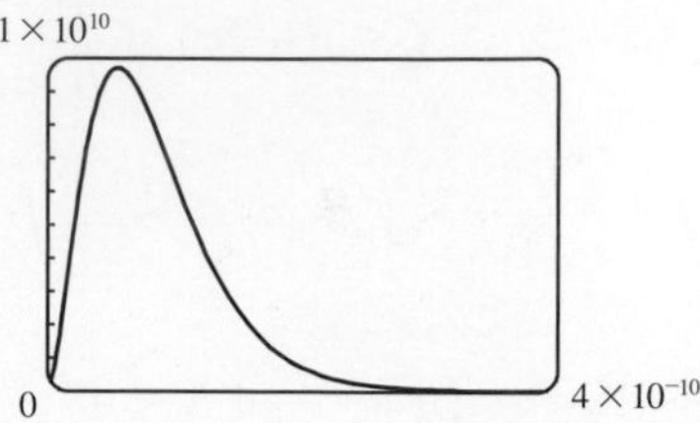

$$p(a_0) = \frac{4}{a_0^3}\,a_0^2 e^{-2a_0/a_0} = \frac{4}{a_0}\,e^{-2} \approx 9{,}684{,}098{,}979$$

With a maximum of nearly 10 billion and a total area under the curve of 1, we know that the "hump" in the graph must be extremely narrow.

(d) $P(r) = \int_0^r \frac{4}{a_0^3} s^2 e^{-2s/a_0}\,ds \;\Rightarrow\; P(4a_0) = \int_0^{4a_0} \frac{4}{a_0^3} s^2 e^{-2s/a_0}\,ds$. Using ($\star$) from part (a) [with $b = -2/a_0$],

$$P(4a_0) = \frac{4}{a_0^3}\left[\frac{e^{-2s/a_0}}{-8/a_0^3}\left(\frac{4}{a_0^2}s^2 + \frac{4}{a_0}s + 2\right)\right]_0^{4a_0} = \frac{4}{a_0^3}\left(\frac{a_0^3}{-8}\right)[e^{-8}(64+16+2) - 1(2)] = -\tfrac{1}{2}(82e^{-8} - 2)$$

$$= 1 - 41e^{-8} \approx 0.986$$

(e) $\mu = \int_{-\infty}^{\infty} r p(r)\,dr = \frac{4}{a_0^3}\lim_{t\to\infty}\int_0^t r^3 e^{-2r/a_0}\,dr$. Integrating by parts three times or using a CAS, we find that

$\int x^3 e^{bx}\,dx = \frac{e^{bx}}{b^4}\left(b^3x^3 - 3b^2x^2 + 6bx - 6\right)$. So with $b = -\frac{2}{a_0}$, we use l'Hospital's Rule, and get

$$\mu = \frac{4}{a_0^3}\left[-\frac{a_0^4}{16}(-6)\right] = \tfrac{3}{2}a_0.$$

8 Review

EXERCISES

1. $y = 4(x-1)^{3/2} \;\Rightarrow\; \frac{dy}{dx} = 6(x-1)^{1/2} \;\Rightarrow\; 1 + \left(\frac{dy}{dx}\right)^2 = 1 + 36(x-1) = 36x - 35$. Thus,

$$L = \int_1^4 \sqrt{36x - 35}\,dx = \int_1^{109} \sqrt{u}\left(\tfrac{1}{36}\,du\right) \qquad \begin{bmatrix} u = 36x - 35, \\ du = 36\,dx \end{bmatrix}$$

$$= \tfrac{1}{36}\left[\tfrac{2}{3}u^{3/2}\right]_1^{109} = \tfrac{1}{54}\left(109\sqrt{109} - 1\right)$$

2. $y = 2\ln\left(\sin\tfrac{1}{2}x\right) \;\Rightarrow\; \frac{dy}{dx} = 2\cdot\frac{1}{\sin\left(\frac{1}{2}x\right)}\cdot\cos\left(\tfrac{1}{2}x\right)\cdot\tfrac{1}{2} = \cot\left(\tfrac{1}{2}x\right) \;\Rightarrow\; 1 + \left(\frac{dy}{dx}\right)^2 = 1 + \cot^2\left(\tfrac{1}{2}x\right) = \csc^2\left(\tfrac{1}{2}x\right)$.

Thus,

$$L = \int_{\pi/3}^{\pi} \sqrt{\csc^2\left(\tfrac{1}{2}x\right)}\,dx = \int_{\pi/3}^{\pi} \left|\csc\left(\tfrac{1}{2}x\right)\right|dx = \int_{\pi/3}^{\pi} \csc\left(\tfrac{1}{2}x\right)dx = \int_{\pi/6}^{\pi/2} \csc u\,(2\,du) \qquad \begin{bmatrix} u = \frac{1}{2}x, \\ du = \frac{1}{2}\,dx \end{bmatrix}$$

$$= 2\Big[\ln|\csc u - \cot u|\Big]_{\pi/6}^{\pi/2} = 2\left[\ln\left|\csc\tfrac{\pi}{2} - \cot\tfrac{\pi}{2}\right| - \ln\left|\csc\tfrac{\pi}{6} - \cot\tfrac{\pi}{6}\right|\right]$$

$$= 2\left[\ln|1 - 0| - \ln\left|2 - \sqrt{3}\right|\right] = -2\ln\left(2 - \sqrt{3}\right) \approx 2.63$$

3. $12x = 4y^3 + 3y^{-1} \;\Rightarrow\; x = \frac{1}{3}y^3 + \frac{1}{4}y^{-1} \;\Rightarrow\; \dfrac{dx}{dy} = y^2 - \frac{1}{4}y^{-2} \;\Rightarrow$

$1 + \left(\dfrac{dx}{dy}\right)^2 = 1 + y^4 - \frac{1}{2} + \frac{1}{16}y^{-4} = y^4 + \frac{1}{2} + \frac{1}{16}y^{-4} = (y^2 + \frac{1}{4}y^{-2})^2$. Thus,

$$L = \int_1^3 \sqrt{\left(y^2 + \tfrac{1}{4}y^{-2}\right)^2}\,dy = \int_1^3 \left|y^2 + \tfrac{1}{4}y^{-2}\right|\,dy = \int_1^3 \left(y^2 + \tfrac{1}{4}y^{-2}\right)\,dy = \left[\tfrac{1}{3}y^3 - \tfrac{1}{4}y^{-1}\right]_1^3$$
$$= \left(9 - \tfrac{1}{12}\right) - \left(\tfrac{1}{3} - \tfrac{1}{4}\right) = \tfrac{106}{12} = \tfrac{53}{6}$$

4. (a) $y = \dfrac{x^4}{16} + \dfrac{1}{2x^2} = \dfrac{1}{16}x^4 + \dfrac{1}{2}x^{-2} \;\Rightarrow\; \dfrac{dy}{dx} = \dfrac{1}{4}x^3 - x^{-3} \;\Rightarrow$

$1 + (dy/dx)^2 = 1 + \left(\frac{1}{4}x^3 - x^{-3}\right)^2 = 1 + \frac{1}{16}x^6 - \frac{1}{2} + x^{-6} = \frac{1}{16}x^6 + \frac{1}{2} + x^{-6} = \left(\frac{1}{4}x^3 + x^{-3}\right)^2$.

Thus, $L = \int_1^2 \left(\frac{1}{4}x^3 + x^{-3}\right)dx = \left[\frac{1}{16}x^4 - \frac{1}{2}x^{-2}\right]_1^2 = \left(1 - \frac{1}{8}\right) - \left(\frac{1}{16} - \frac{1}{2}\right) = \frac{21}{16}$.

(b) $S = \int_1^2 2\pi x\left(\frac{1}{4}x^3 + x^{-3}\right)dx = 2\pi\int_1^2 \left(\frac{1}{4}x^4 + x^{-2}\right)dx = 2\pi\left[\frac{1}{20}x^5 - \frac{1}{x}\right]_1^2$

$= 2\pi\left[\left(\frac{32}{20} - \frac{1}{2}\right) - \left(\frac{1}{20} - 1\right)\right] = 2\pi\left(\frac{8}{5} - \frac{1}{2} - \frac{1}{20} + 1\right) = 2\pi\left(\frac{41}{20}\right) = \frac{41}{10}\pi$

5. (a) $y = \dfrac{2}{x+1} \;\Rightarrow\; y' = \dfrac{-2}{(x+1)^2} \;\Rightarrow\; 1 + (y')^2 = 1 + \dfrac{4}{(x+1)^4}$.

For $0 \le x \le 3$, $L = \int_0^3 \sqrt{1 + (y')^2}\,dx = \int_0^3 \sqrt{1 + 4/(x+1)^4}\,dx \approx 3.5121$.

(b) The area of the surface obtained by rotating C about the x-axis is

$$S = \int_0^3 2\pi y\,ds = 2\pi\int_0^3 \frac{2}{x+1}\sqrt{1 + 4/(x+1)^4}\,dx \approx 22.1391.$$

(c) The area of the surface obtained by rotating C about the y-axis is

$S = \int_0^3 2\pi x\,ds = 2\pi\int_0^3 x\sqrt{1 + 4/(x+1)^4}\,dx \approx 29.8522$.

6. (a) $y = x^2 \;\Rightarrow\; 1 + (y')^2 = 1 + 4x^2$. Rotate about the y-axis for $0 \le x \le 1$:

$$S = \int_0^1 2\pi x\sqrt{1 + 4x^2}\,dx = \int_1^5 \tfrac{\pi}{4}\sqrt{u}\,du \quad [u = 1 + 4x^2] \;=\; \tfrac{\pi}{6}\left[u^{3/2}\right]_1^5 = \tfrac{\pi}{6}(5^{3/2} - 1)$$

(b) $y = x^2 \;\Rightarrow\; 1 + (y')^2 = 1 + 4x^2$. Rotate about the x-axis for $0 \le x \le 1$:

$$S = 2\pi\int_0^1 x^2\sqrt{1 + 4x^2}\,dx = 2\pi\int_0^2 \tfrac{1}{4}u^2\sqrt{1 + u^2}\,\tfrac{1}{2}\,du \quad [u = 2x] \quad = \tfrac{\pi}{4}\int_0^2 u^2\sqrt{1 + u^2}\,du$$
$$= \tfrac{\pi}{4}\left[\tfrac{1}{8}u(1 + 2u^2)\sqrt{1 + u^2} - \tfrac{1}{8}\ln\left|u + \sqrt{1 + u^2}\right|\right]_0^2 \qquad [u = \tan\theta \text{ or use Formula 22}]$$
$$= \tfrac{\pi}{4}\left[\tfrac{1}{4}(9)\sqrt{5} - \tfrac{1}{8}\ln(2 + \sqrt{5}) - 0\right] = \tfrac{\pi}{32}\left[18\sqrt{5} - \ln(2 + \sqrt{5})\right]$$

7. $y = \sin x \;\Rightarrow\; y' = \cos x \;\Rightarrow\; 1 + (y')^2 = 1 + \cos^2 x$. Let $f(x) = \sqrt{1 + \cos^2 x}$. Then

$L = \int_0^\pi f(x)\,dx \approx S_{10}$

$$= \tfrac{(\pi - 0)/10}{3}\left[f(0) + 4f\left(\tfrac{\pi}{10}\right) + 2f\left(\tfrac{2\pi}{10}\right) + 4f\left(\tfrac{3\pi}{10}\right) + 2f\left(\tfrac{4\pi}{10}\right)\right.$$
$$\left. + 4f\left(\tfrac{5\pi}{10}\right) + 2f\left(\tfrac{6\pi}{10}\right) + 4f\left(\tfrac{7\pi}{10}\right) + 2f\left(\tfrac{8\pi}{10}\right) + 4f\left(\tfrac{9\pi}{10}\right) + f(\pi)\right]$$

≈ 3.820188

8. $S = \int_0^\pi 2\pi y\, ds = \int_0^\pi 2\pi \sin x \sqrt{1+\cos^2 x}\, dx$. Let $g(x) = 2\pi \sin x \sqrt{1+\cos^2 x}$. Then

$$\begin{aligned} S = \int_0^\pi g(x)\,dx &\approx S_{10} \\ &= \tfrac{(\pi-0)/10}{3}\left[g(0) + 4g\left(\tfrac{\pi}{10}\right) + 2g\left(\tfrac{2\pi}{10}\right) + 4g\left(\tfrac{3\pi}{10}\right) + 2g\left(\tfrac{4\pi}{10}\right)\right. \\ &\qquad \left. + 4g\left(\tfrac{5\pi}{10}\right) + 2g\left(\tfrac{6\pi}{10}\right) + 4g\left(\tfrac{7\pi}{10}\right) + 2g\left(\tfrac{8\pi}{10}\right) + 4g\left(\tfrac{9\pi}{10}\right) + g(\pi)\right] \\ &\approx 14.426045 \end{aligned}$$

9. $y = \int_1^x \sqrt{\sqrt{t}-1}\, dt \quad\Rightarrow\quad dy/dx = \sqrt{\sqrt{x}-1} \quad\Rightarrow\quad 1 + (dy/dx)^2 = 1 + \left(\sqrt{x}-1\right) = \sqrt{x}$.

Thus, $L = \int_1^{16} \sqrt{\sqrt{x}}\, dx = \int_1^{16} x^{1/4}\, dx = \frac{4}{5}\left[x^{5/4}\right]_1^{16} = \frac{4}{5}(32-1) = \frac{124}{5}$.

10. $S = \int_1^{16} 2\pi x\, ds = 2\pi \int_1^{16} x \cdot x^{1/4}\, dx = 2\pi \int_1^{16} x^{5/4}\, dx = 2\pi \cdot \frac{4}{9}\left[x^{9/4}\right]_1^{16} = \frac{8\pi}{9}(512-1) = \frac{4088}{9}\pi$

11. As in Example 8.3.1, $\dfrac{a}{2-x} = \dfrac{1}{2} \quad\Rightarrow\quad 2a = 2 - x$ and $w = 2(1.5+a) = 3 + 2a = 3 + 2 - x = 5 - x$.

Thus, $F = \int_0^2 \delta x(5-x)\, dx = \delta\left[\frac{5}{2}x^2 - \frac{1}{3}x^3\right]_0^2 = \delta\left(10 - \frac{8}{3}\right) = \frac{22}{3}\delta \approx 458 \text{ lb}$ $[\delta \approx 62.5 \text{ lb/ft}^3]$.

12. $$\begin{aligned} F &= \int_0^4 \delta(4-y)2\left(2\sqrt{y}\right)dy = 4\delta \int_0^4 (4y^{1/2} - y^{3/2})\, dy \\ &= 4\delta\left[\tfrac{8}{3}y^{3/2} - \tfrac{2}{5}y^{5/2}\right]_0^4 = 4\delta\left(\tfrac{64}{3} - \tfrac{64}{5}\right) = 256\delta\left(\tfrac{1}{3} - \tfrac{1}{5}\right) \\ &= \tfrac{512}{15}\delta \approx 2133.3 \text{ lb} \qquad [\delta \approx 62.5 \text{ lb/ft}^3] \end{aligned}$$

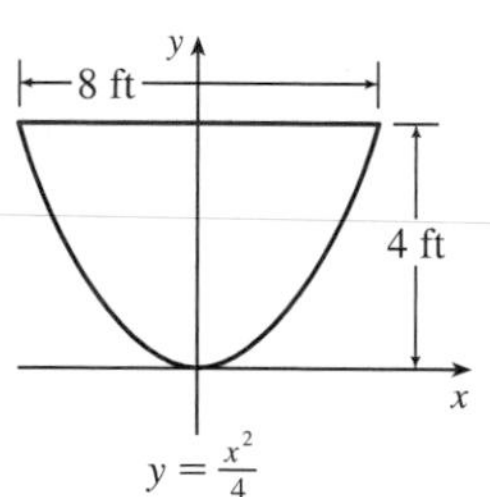

13. $A = \int_0^4 \left(\sqrt{x} - \frac{1}{2}x\right) dx = \left[\frac{2}{3}x^{3/2} - \frac{1}{4}x^2\right]_0^4 = \frac{16}{3} - 4 = \frac{4}{3}$

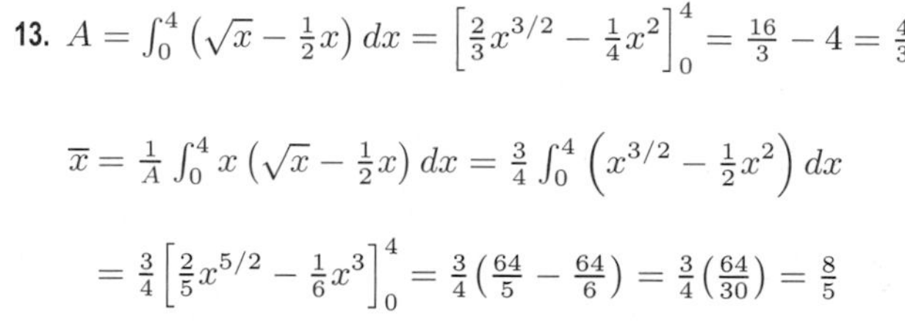

$$\begin{aligned} \overline{x} &= \tfrac{1}{A}\int_0^4 x\left(\sqrt{x} - \tfrac{1}{2}x\right) dx = \tfrac{3}{4}\int_0^4 \left(x^{3/2} - \tfrac{1}{2}x^2\right) dx \\ &= \tfrac{3}{4}\left[\tfrac{2}{5}x^{5/2} - \tfrac{1}{6}x^3\right]_0^4 = \tfrac{3}{4}\left(\tfrac{64}{5} - \tfrac{64}{6}\right) = \tfrac{3}{4}\left(\tfrac{64}{30}\right) = \tfrac{8}{5} \end{aligned}$$

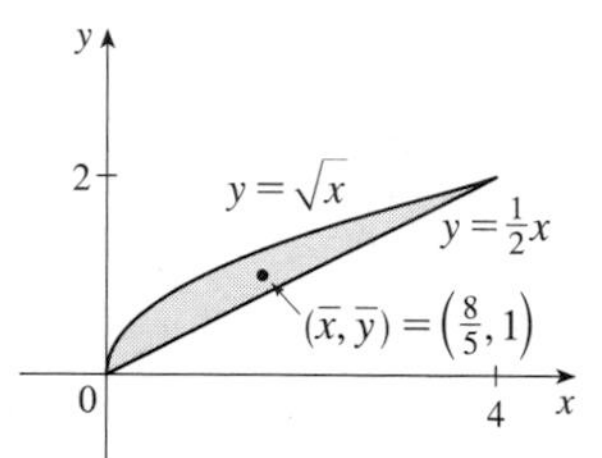

$$\overline{y} = \tfrac{1}{A}\int_0^4 \tfrac{1}{2}\left[\left(\sqrt{x}\right)^2 - \left(\tfrac{1}{2}x\right)^2\right] dx = \tfrac{3}{4}\int_0^4 \tfrac{1}{2}\left(x - \tfrac{1}{4}x^2\right) dx = \tfrac{3}{8}\left[\tfrac{1}{2}x^2 - \tfrac{1}{12}x^3\right]_0^4 = \tfrac{3}{8}\left(8 - \tfrac{16}{3}\right) = \tfrac{3}{8}\left(\tfrac{8}{3}\right) = 1$$

Thus, the centroid is $(\overline{x}, \overline{y}) = \left(\frac{8}{5}, 1\right)$.

14. From the symmetry of the region, $\overline{x} = \frac{\pi}{2}$. $A = \int_{\pi/4}^{3\pi/4} \sin x\, dx = \left[-\cos x\right]_{\pi/4}^{3\pi/4} = \frac{1}{\sqrt{2}} - \left(-\frac{1}{\sqrt{2}}\right) = \sqrt{2}$

$$\begin{aligned} \overline{y} &= \frac{1}{A}\int_{\pi/4}^{3\pi/4} \tfrac{1}{2}\sin^2 x\, dx = \frac{1}{A}\int_{\pi/4}^{3\pi/4} \tfrac{1}{4}(1 - \cos 2x)\, dx \\ &= \frac{1}{4\sqrt{2}}\left[x - \tfrac{1}{2}\sin 2x\right]_{\pi/4}^{3\pi/4} \\ &= \frac{1}{4\sqrt{2}}\left[\tfrac{3\pi}{4} - \tfrac{1}{2}(-1) - \tfrac{\pi}{4} + \tfrac{1}{2}\cdot 1\right] = \tfrac{1}{4\sqrt{2}}\left(\tfrac{\pi}{2} + 1\right) \end{aligned}$$

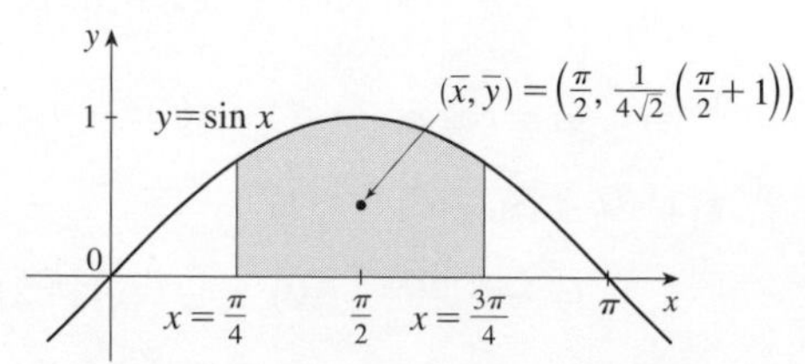

Thus, the centroid is $(\overline{x}, \overline{y}) = \left(\frac{\pi}{2}, \frac{1}{4\sqrt{2}}\left(\frac{\pi}{2} + 1\right)\right) \approx (1.57, 0.45)$.

15. The area of the triangular region is $A = \frac{1}{2}(2)(4) = 4$. An equation of the line is $y = \frac{1}{2}x$ or $x = 2y$.

$$\overline{x} = \frac{1}{A}\int_0^2 \frac{1}{2}[f(y)]^2\,dy = \frac{1}{4}\int_0^2 \frac{1}{2}(2y)^2\,dy = \frac{1}{8}\int_0^2 4y^2\,dy = \frac{1}{8}\left[\frac{4}{3}y^3\right]_0^2 = \frac{1}{6}(8) = \frac{4}{3}$$

$$\overline{y} = \frac{1}{A}\int_0^2 y\,f(y)\,dy = \frac{1}{4}\int_0^2 y(2y)\,dy = \frac{1}{2}\int_0^2 y^2\,dy = \frac{1}{2}\left[\frac{1}{3}y^3\right]_0^2 = \frac{1}{6}(8) = \frac{4}{3}$$

The centroid of the region is $\left(\frac{4}{3}, \frac{4}{3}\right)$.

16. An equation of the line is $y = 8 - x$. An equation of the quarter-circle is $y = -\sqrt{8^2 - x^2}$ with $0 \le x \le 8$. The area of the region is $A = \frac{1}{2}(8)(8) + \frac{1}{4}\pi(8)^2 = 32 + 16\pi = 16(2+\pi)$.

$$\begin{aligned}\overline{x} &= \frac{1}{A}\int_0^8 x[f(x) - g(x)]\,dx = \frac{1}{A}\int_0^8 x\left[(8-x) + \sqrt{64 - x^2}\right]dx \\ &= \frac{1}{A}\int_0^8 \left[8x - x^2 + x(64-x^2)^{1/2}\right]dx = \frac{1}{A}\left[4x^2 - \frac{1}{3}x^3 - \frac{1}{3}(64-x^2)^{3/2}\right]_0^8 \\ &= \frac{1}{A}\left[\left(256 - \frac{512}{3} - 0\right) - \left(0 - 0 - \frac{512}{3}\right)\right] = \frac{256}{16(2+\pi)} = \frac{16}{2+\pi}\end{aligned}$$

$$\begin{aligned}\overline{y} &= \frac{1}{A}\int_0^8 \frac{1}{2}\{[f(x)]^2 - [g(x)]^2\}\,dx = \frac{1}{2A}\int_0^8 \left[(8-x)^2 - \left(-\sqrt{64-x^2}\,\right)^2\right]dx \\ &= \frac{1}{2A}\int_0^8 \left[64 - 16x + x^2 - (64 - x^2)\right]dx = \frac{1}{2A}\int_0^8 (2x^2 - 16x)\,dx \\ &= \frac{1}{A}\int_0^8 (x^2 - 8x)\,dx = \frac{1}{A}\left[\frac{1}{3}x^3 - 4x^2\right]_0^8 = \frac{1}{A}\left(\frac{512}{3} - 256\right) \\ &= \frac{1}{16(2+\pi)}\left(-\frac{256}{3}\right) = -\frac{16}{3(2+\pi)}\end{aligned}$$

The centroid of the region is $\left(\frac{16}{2+\pi}, -\frac{16}{3(2+\pi)}\right) \approx (3.11, -1.04)$.

17. The centroid of this circle, $(1, 0)$, travels a distance $2\pi(1)$ when the lamina is rotated about the y-axis. The area of the circle is $\pi(1)^2$. So by the Theorem of Pappus, $V = A(2\pi\overline{x}) = \pi(1)^2 2\pi(1) = 2\pi^2$.

18. The semicircular region has an area of $\frac{1}{2}\pi r^2$, and sweeps out a sphere of radius r when rotated about the x-axis. $\overline{x} = 0$ because of symmetry about the line $x = 0$. And by the Theorem of Pappus, $V = A(2\pi\overline{y}) \Rightarrow$ $\frac{4}{3}\pi r^3 = \frac{1}{2}\pi r^2(2\pi\overline{y}) \Rightarrow \overline{y} = \frac{4}{3\pi}r$. Thus, the centroid is $(\overline{x}, \overline{y}) = \left(0, \frac{4}{3\pi}r\right)$.

19. $x = 100 \Rightarrow P = 2000 - 0.1(100) - 0.01(100)^2 = 1890$

$$\begin{aligned}\text{Consumer surplus} &= \textstyle\int_0^{100}[p(x) - P]\,dx = \int_0^{100}\left(2000 - 0.1x - 0.01x^2 - 1890\right)dx \\ &= \left[110x - 0.05x^2 - \frac{0.01}{3}x^3\right]_0^{100} = 11{,}000 - 500 - \frac{10{,}000}{3} \approx \$7166.67\end{aligned}$$

20. $\int_0^{24} c(t)\,dt \approx S_{12} = \frac{24-0}{12\cdot 3}[1(0) + 4(1.9) + 2(3.3) + 4(5.1) + 2(7.6) + 4(7.1) + 2(5.8)$

$+ 4(4.7) + 2(3.3) + 4(2.1) + 2(1.1) + 4(0.5) + 1(0)]$

$= \frac{2}{3}(127.8) = 85.2\text{ mg}\cdot\text{s/L}$

Therefore, $F \approx A/85.2 = 6/85.2 \approx 0.0704$ L/s or 4.225 L/min.

21. $f(x) = \begin{cases} \frac{\pi}{20}\sin\left(\frac{\pi}{10}x\right) & \text{if } 0 \le x \le 10 \\ 0 & \text{if } x < 0 \text{ or } x > 10 \end{cases}$

(a) $f(x) \ge 0$ for all real numbers x and

$$\int_{-\infty}^{\infty} f(x)\,dx = \int_0^{10} \frac{\pi}{20}\sin\left(\frac{\pi}{10}x\right)dx = \frac{\pi}{20}\cdot\frac{10}{\pi}\left[-\cos\left(\frac{\pi}{10}x\right)\right]_0^{10} = \frac{1}{2}(-\cos\pi + \cos 0) = \frac{1}{2}(1+1) = 1$$

Therefore, f is a probability density function.

(b) $P(X < 4) = \int_{-\infty}^{4} f(x)\,dx = \int_0^4 \frac{\pi}{20}\sin\left(\frac{\pi}{10}x\right)dx = \frac{1}{2}\left[-\cos\left(\frac{\pi}{10}x\right)\right]_0^4 = \frac{1}{2}\left(-\cos\frac{2\pi}{5} + \cos 0\right)$

$\approx \frac{1}{2}(-0.309017 + 1) \approx 0.3455$

(c) $\mu = \int_{-\infty}^{\infty} xf(x)\,dx = \int_0^{10} \frac{\pi}{20}x\sin\left(\frac{\pi}{10}x\right)dx$

$= \int_0^{\pi} \frac{\pi}{20}\cdot\frac{10}{\pi}u(\sin u)\left(\frac{10}{\pi}\right)du \qquad \left[u = \frac{\pi}{10}x,\ du = \frac{\pi}{10}\,dx\right]$

$= \frac{5}{\pi}\int_0^{\pi} u\sin u\,du \overset{82}{=} \frac{5}{\pi}\left[\sin u - u\cos u\right]_0^{\pi} = \frac{5}{\pi}[0 - \pi(-1)] = 5$

This answer is expected because the graph of f is symmetric about the line $x = 5$.

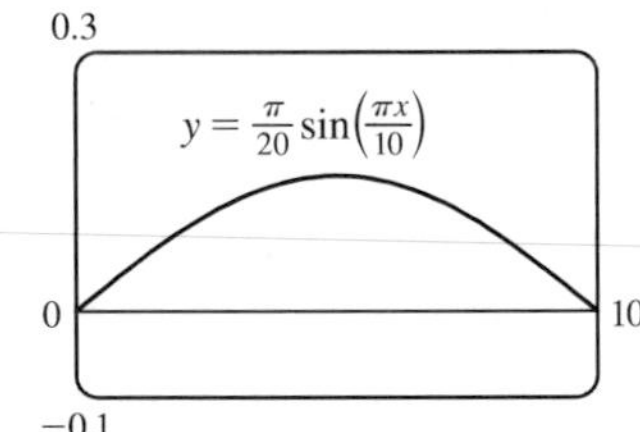

22. $P(250 \le X \le 280) = \displaystyle\int_{250}^{280} \frac{1}{\sigma\sqrt{2\pi}}e^{-(x-\mu)^2/(2\sigma^2)}\,dx = \int_{250}^{280} \frac{1}{15\sqrt{2\pi}}\exp\left(\frac{-(x-268)^2}{2\cdot 15^2}\right)dx \approx 0.673.$

Thus, the percentage of pregnancies that last between 250 and 280 days is about 67.3%.

23. (a) The probability density function is $f(t) = \begin{cases} 0 & \text{if } t < 0 \\ \frac{1}{8}e^{-t/8} & \text{if } t \ge 0 \end{cases}$

$P(0 \le X \le 3) = \int_0^3 \frac{1}{8}e^{-t/8}\,dt = \left[-e^{-t/8}\right]_0^3 = -e^{-3/8} + 1 \approx 0.3127$

(b) $P(X > 10) = \int_{10}^{\infty} \frac{1}{8}e^{-t/8}\,dt = \lim\limits_{x\to\infty}\left[-e^{-t/8}\right]_{10}^{x} = \lim\limits_{x\to\infty}(-e^{-x/8} + e^{-10/8}) = 0 + e^{-5/4} \approx 0.2865$

(c) We need to find m such that $P(X \ge m) = \frac{1}{2} \Rightarrow \int_m^{\infty} \frac{1}{8}e^{-t/8}\,dt = \frac{1}{2} \Rightarrow \lim\limits_{x\to\infty}\left[-e^{-t/8}\right]_m^x = \frac{1}{2} \Rightarrow$

$\lim\limits_{x\to\infty}(-e^{-x/8} + e^{-m/8}) = \frac{1}{2} \Rightarrow e^{-m/8} = \frac{1}{2} \Rightarrow -m/8 = \ln\frac{1}{2} \Rightarrow m = -8\ln\frac{1}{2} = 8\ln 2 \approx 5.55$ minutes.

☐ PROBLEMS PLUS

1. $x^2 + y^2 \le 4y \quad \Leftrightarrow \quad x^2 + (y-2)^2 \le 4$, so S is part of a circle, as shown in the diagram. The area of S is

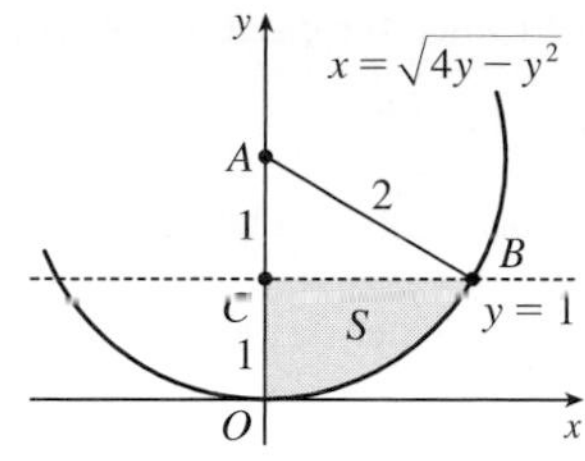

$$\int_0^1 \sqrt{4y - y^2}\,dy \overset{113}{=} \left[\tfrac{y-2}{2}\sqrt{4y-y^2} + 2\cos^{-1}\left(\tfrac{2-y}{2}\right)\right]_0^1 \quad [a = 2]$$

$$= -\tfrac{1}{2}\sqrt{3} + 2\cos^{-1}\left(\tfrac{1}{2}\right) - 2\cos^{-1}1$$

$$= -\tfrac{\sqrt{3}}{2} + 2\left(\tfrac{\pi}{3}\right) - 2(0) = \tfrac{2\pi}{3} - \tfrac{\sqrt{3}}{2}$$

Another method (without calculus): Note that $\theta = \angle CAB = \frac{\pi}{3}$, so the area is

$$(\text{area of sector } OAB) - (\text{area of } \triangle ABC) = \tfrac{1}{2}\left(2^2\right)\tfrac{\pi}{3} - \tfrac{1}{2}(1)\sqrt{3} = \tfrac{2\pi}{3} - \tfrac{\sqrt{3}}{2}$$

2. $y = \pm\sqrt{x^3 - x^4} \quad \Rightarrow$ The loop of the curve is symmetric about $y = 0$, and therefore $\overline{y} = 0$. At each point x where $0 \le x \le 1$, the lamina has a vertical length of $\sqrt{x^3 - x^4} - \left(-\sqrt{x^3 - x^4}\right) = 2\sqrt{x^3 - x^4}$. Therefore,

$$\overline{x} = \frac{\int_0^1 x \cdot 2\sqrt{x^3 - x^4}\,dx}{\int_0^1 2\sqrt{x^3 - x^4}\,dx} = \frac{\int_0^1 x\sqrt{x^3 - x^4}\,dx}{\int_0^1 \sqrt{x^3 - x^4}\,dx}.$$ We evaluate the integrals separately:

$$\int_0^1 x\sqrt{x^3 - x^4}\,dx = \int_0^1 x^{5/2}\sqrt{1-x}\,dx$$

$$= \int_0^{\pi/2} 2\sin^6\theta\cos\theta\sqrt{1 - \sin^2\theta}\,d\theta \qquad \left[\begin{array}{c}\sin\theta = \sqrt{x},\ \cos\theta\,d\theta = dx/(2\sqrt{x}\,), \\ 2\sin\theta\,\cos\theta\,d\theta = dx\end{array}\right]$$

$$= \int_0^{\pi/2} 2\sin^6\theta\cos^2\theta\,d\theta = \int_0^{\pi/2} 2\left[\tfrac{1}{2}(1 - \cos 2\theta)\right]^3 \tfrac{1}{2}(1 + \cos 2\theta)\,d\theta$$

$$= \int_0^{\pi/2} \tfrac{1}{8}(1 - 2\cos 2\theta + 2\cos^3 2\theta - \cos^4 2\theta)\,d\theta$$

$$= \int_0^{\pi/2} \tfrac{1}{8}\left[1 - 2\cos 2\theta + 2\cos 2\theta(1 - \sin^2 2\theta) - \tfrac{1}{4}(1 + \cos 4\theta)^2\right]d\theta$$

$$= \tfrac{1}{8}\left[\theta - \tfrac{1}{3}\sin^3 2\theta\right]_0^{\pi/2} - \tfrac{1}{32}\int_0^{\pi/2}\left(1 + 2\cos 4\theta + \cos^2 4\theta\right)d\theta$$

$$= \tfrac{\pi}{16} - \tfrac{1}{32}\left[\theta + \tfrac{1}{2}\sin 4\theta\right]_0^{\pi/2} - \tfrac{1}{64}\int_0^{\pi/2}(1 + \cos 8\theta)\,d\theta$$

$$= \tfrac{\pi}{16} - \tfrac{1}{32}\left[\theta + \tfrac{1}{2}\sin 4\theta\right]_0^{\pi/2} - \tfrac{1}{64}\int_0^{\pi/2}(1 + \cos 8\theta)\,d\theta$$

$$= \tfrac{3\pi}{64} - \tfrac{1}{64}\left[\theta + \tfrac{1}{8}\sin 8\theta\right]_0^{\pi/2} = \tfrac{5\pi}{128}$$

$$\int_0^1 \sqrt{x^3 - x^4}\,dx = \int_0^1 x^{3/2}\sqrt{1-x}\,dx = \int_0^{\pi/2} 2\sin^4\theta\cos\theta\sqrt{1 - \sin^2\theta}\,d\theta \qquad [\sin\theta = \sqrt{x}\,]$$

$$= \int_0^{\pi/2} 2\sin^4\theta\cos^2\theta\,d\theta = \int_0^{\pi/2} 2 \cdot \tfrac{1}{4}(1 - \cos 2\theta)^2 \cdot \tfrac{1}{2}(1 + \cos 2\theta)\,d\theta$$

$$= \int_0^{\pi/2} \tfrac{1}{4}(1 - \cos 2\theta - \cos^2 2\theta + \cos^3 2\theta)\,d\theta$$

$$= \int_0^{\pi/2} \tfrac{1}{4}\left[1 - \cos 2\theta - \tfrac{1}{2}(1 + \cos 4\theta) + \cos 2\theta(1 - \sin^2 2\theta)\right]d\theta$$

$$= \tfrac{1}{4}\left[\tfrac{\theta}{2} - \tfrac{1}{8}\sin 4\theta - \tfrac{1}{6}\sin^3 2\theta\right]_0^{\pi/2} = \tfrac{\pi}{16}$$

Therefore, $\overline{x} = \dfrac{5\pi/128}{\pi/16} = \dfrac{5}{8}$, and $(\overline{x}, \overline{y}) = \left(\frac{5}{8}, 0\right)$.

3. (a) The two spherical zones, whose surface areas we will call S_1 and S_2, are generated by rotation about the y-axis of circular arcs, as indicated in the figure. The arcs are the upper and lower portions of the circle $x^2 + y^2 = r^2$ that are obtained when the circle is cut with the line $y = d$. The portion of the upper arc in the first quadrant is sufficient to generate the upper spherical zone. That portion of the arc can be described by the relation $x = \sqrt{r^2 - y^2}$ for $d \le y \le r$. Thus, $dx/dy = -y/\sqrt{r^2 - y^2}$ and

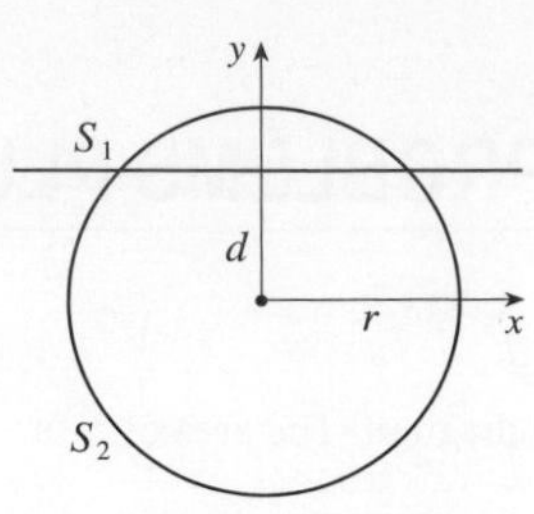

$$ds = \sqrt{1 + \left(\frac{dx}{dy}\right)^2}\, dy = \sqrt{1 + \frac{y^2}{r^2 - y^2}}\, dy = \sqrt{\frac{r^2}{r^2 - y^2}}\, dy = \frac{r\, dy}{\sqrt{r^2 - y^2}}$$

From Formula 8.2.8 we have

$$S_1 = \int_d^r 2\pi x \sqrt{1 + \left(\frac{dx}{dy}\right)^2}\, dy = \int_d^r 2\pi \sqrt{r^2 - y^2} \frac{r\, dy}{\sqrt{r^2 - y^2}} = \int_d^r 2\pi r\, dy = 2\pi r(r - d)$$

Similarly, we can compute $S_2 = \int_{-r}^{d} 2\pi x \sqrt{1 + (dx/dy)^2}\, dy = \int_{-r}^{d} 2\pi r\, dy = 2\pi r(r + d)$. Note that $S_1 + S_2 = 4\pi r^2$, the surface area of the entire sphere.

(b) $r = 3960$ mi and $d = r\,(\sin 75°) \approx 3825$ mi, so the surface area of the Arctic Ocean is about $2\pi r(r - d) \approx 2\pi(3960)(135) \approx 3.36 \times 10^6\ \text{mi}^2$.

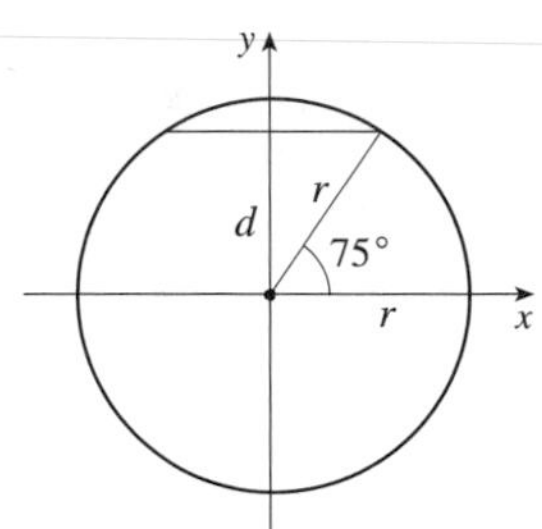

(c) The area on the sphere lies between planes $y = y_1$ and $y = y_2$, where $y_2 - y_1 = h$. Thus, we compute the surface area on the sphere to be $S = \displaystyle\int_{y_1}^{y_2} 2\pi x \sqrt{1 + \left(\frac{dx}{dy}\right)^2}\, dy = \int_{y_1}^{y_2} 2\pi r\, dy = 2\pi r(y_2 - y_1) = 2\pi rh$.

This equals the lateral area of a cylinder of radius r and height h, since such a cylinder is obtained by rotating the line $x = r$ about the y-axis, so the surface area of the cylinder between the planes $y = y_1$ and $y = y_2$ is

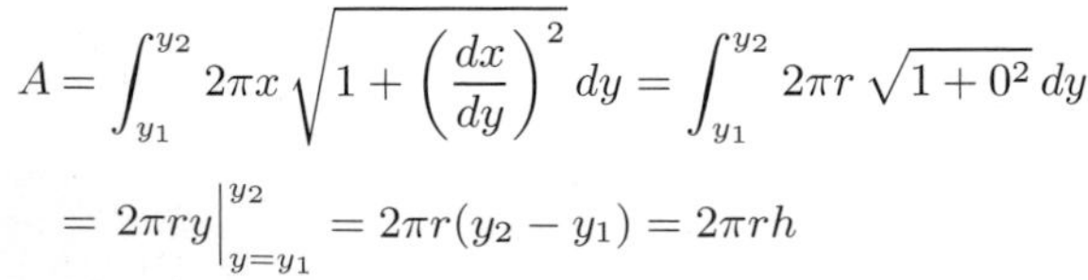

$$A = \int_{y_1}^{y_2} 2\pi x \sqrt{1 + \left(\frac{dx}{dy}\right)^2}\, dy = \int_{y_1}^{y_2} 2\pi r \sqrt{1 + 0^2}\, dy$$

$$= 2\pi r y \Big|_{y=y_1}^{y_2} = 2\pi r(y_2 - y_1) = 2\pi rh$$

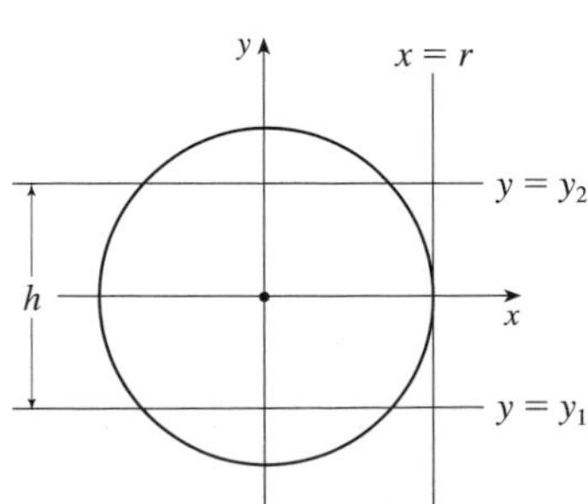

(d) $h = 2r\sin 23.45° \approx 3152$ mi, so the surface area of the Torrid Zone is $2\pi rh \approx 2\pi(3960)(3152) \approx 7.84 \times 10^7$ mi^2.

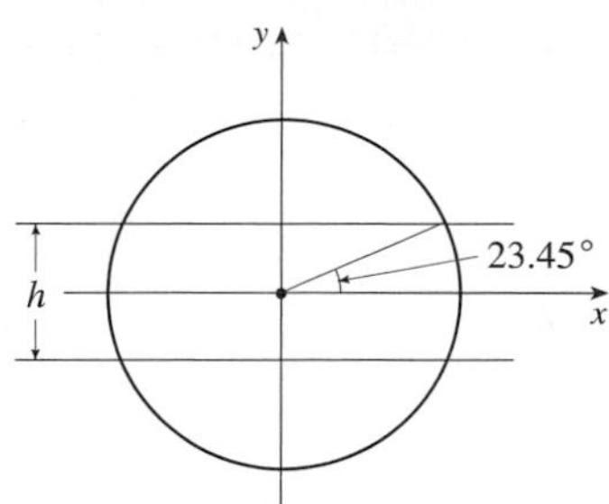

4. (a) Since the right triangles OAT and OTB are similar, we have $\dfrac{r+H}{r} = \dfrac{r}{a} \Rightarrow$

$a = \dfrac{r^2}{r+H}$. The surface area visible from B is $S = \int_a^r 2\pi x\sqrt{1+(dx/dy)^2}\,dy$.

From $x^2+y^2=r^2$, we get $\dfrac{d}{dy}(x^2+y^2) = \dfrac{d}{dy}(r^2) \Rightarrow 2x\dfrac{dx}{dy} + 2y = 0 \Rightarrow$

$\dfrac{dx}{dy} = -\dfrac{y}{x}$ and $1 + \left(\dfrac{dx}{dy}\right)^2 = \dfrac{x^2+y^2}{x^2} = \dfrac{r^2}{x^2}$. Thus,

$$S = \int_a^r 2\pi x \cdot \frac{r}{x}\,dy = 2\pi r(r-a) = 2\pi r\left(r - \frac{r^2}{r+H}\right) = 2\pi r^2\left(1 - \frac{r}{r+H}\right) = 2\pi r^2 \cdot \frac{H}{r+H} = \frac{2\pi r^2 H}{r+H}.$$

(b) Assume $R \geq r$. If a light is placed at point L, at a distance x from the center of the sphere of radius r, then from part (a) we find that the total illuminated area A on the two spheres is [with $r + H = x$ and $r + H = d - x$].

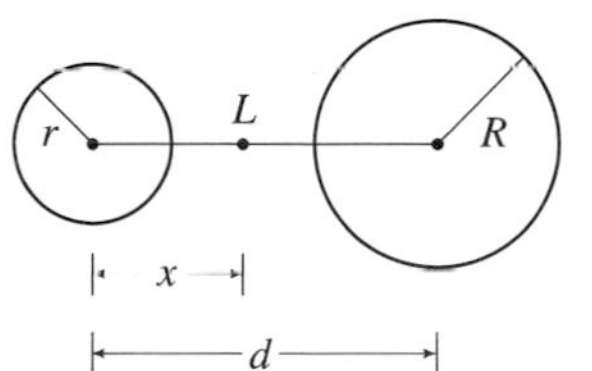

$$A(x) = \frac{2\pi r^2(x-r)}{x} + \frac{2\pi R^2(d-x-R)}{d-x} \quad [r \leq x \leq d-R]. \quad \frac{A(x)}{2\pi} = r^2\left(1 - \frac{r}{x}\right) + R^2\left(1 - \frac{R}{d-x}\right),$$

so $A'(x) = 0 \Leftrightarrow 0 = r^2 \cdot \dfrac{r}{x^2} + R^2 \cdot \dfrac{-R}{(d-x)^2} \Leftrightarrow \dfrac{r^3}{x^2} = \dfrac{R^3}{(d-x)^2} \Leftrightarrow \dfrac{(d-x)^2}{x^2} = \dfrac{R^3}{r^3} \Leftrightarrow$

$$\left(\frac{d}{x} - 1\right)^2 = \left(\frac{R}{r}\right)^3 \Rightarrow \frac{d}{x} - 1 = \left(\frac{R}{r}\right)^{3/2} \Leftrightarrow \frac{d}{x} = 1 + \left(\frac{R}{r}\right)^{3/2} \Leftrightarrow x = x^* = \frac{d}{1+(R/r)^{3/2}}.$$

Now $A'(x) = 2\pi\left(\dfrac{r^3}{x^2} - \dfrac{R^3}{(d-x)^2}\right) \Rightarrow A''(x) = 2\pi\left(-\dfrac{2r^3}{x^3} - \dfrac{2R^3}{(d-x)^3}\right)$ and $A''(x^*) < 0$, so we have a local maximum at $x = x^*$.

However, x^* may not be an allowable value of x—we must show that x^* is between r and $d - R$.

(1) $x^* \geq r \Leftrightarrow \dfrac{d}{1+(R/r)^{3/2}} \geq r \Leftrightarrow d \geq r + R\sqrt{R/r}$

(2) $x^* \leq d - R \Leftrightarrow \dfrac{d}{1+(R/r)^{3/2}} \leq d - R \Leftrightarrow d \leq d - R + d\left(\dfrac{R}{r}\right)^{3/2} - R\left(\dfrac{R}{r}\right)^{3/2} \Leftrightarrow$

$R + R\left(\dfrac{R}{r}\right)^{3/2} \leq d\left(\dfrac{R}{r}\right)^{3/2} \Leftrightarrow d \geq \dfrac{R}{(R/r)^{3/2}} + R = R + r\sqrt{r/R}$, but

$R + r\sqrt{r/R} \leq R + r$, and since $d > r + R$ [given], we conclude that $x^* \leq d - R$.

[continued]

Thus, from (1) and (2), x^* is not an allowable value of x if $d < r + R\sqrt{R/r}$.

So A may have a maximum at $x = r$, x^*, or $d - R$.

$$A(r) = \frac{2\pi R^2(d - r - R)}{d - r} \quad \text{and} \quad A(d - R) = \frac{2\pi r^2(d - r - R)}{d - R}$$

$A(r) > A(d - R) \;\Leftrightarrow\; \dfrac{R^2}{d - r} > \dfrac{r^2}{d - R} \;\Leftrightarrow\; R^2(d - R) > r^2(d - r) \;\Leftrightarrow\; R^2 d - R^3 > r^2 d - r^3 \;\Leftrightarrow$ $R^2 d - r^2 d > R^3 - r^3 \;\Leftrightarrow\; d(R - r)(R + r) > (R - r)(R^2 + Rr + r^2) \;\Leftrightarrow\; d > (R^2 + Rr + r^2)/(R + r) \;\Leftrightarrow$ $d > [(R + r)^2 - Rr]/(R + r) \;\Leftrightarrow\; d > R + r - Rr/(R + r)$. Now $R + r - Rr/(R + r) < R + r$, and we know that $d > R + r$, so we conclude that $A(r) > A(d - R)$.

In conclusion, A has an absolute maximum at $x = x^*$ provided $d \geq r + R\sqrt{R/r}$; otherwise, A has its maximum at $x = r$.

5. (a) Choose a vertical x-axis pointing downward with its origin at the surface. In order to calculate the pressure at depth z, consider n subintervals of the interval $[0, z]$ by points x_i and choose a point $x_i^* \in [x_{i-1}, x_i]$ for each i. The thin layer of water lying between depth x_{i-1} and depth x_i has a density of approximately $\rho(x_i^*)$, so the weight of a piece of that layer with unit cross-sectional area is $\rho(x_i^*)g\,\Delta x$. The total weight of a column of water extending from the surface to depth z (with unit cross-sectional area) would be approximately $\sum_{i=1}^{n} \rho(x_i^*)g\,\Delta x$. The estimate becomes exact if we take the limit as $n \to \infty$; weight (or force) per unit area at depth z is $W = \lim_{n\to\infty} \sum_{i=1}^{n} \rho(x_i^*)g\,\Delta x$. In other words, $P(z) = \int_0^z \rho(x)g\,dx$. More generally, if we make no assumptions about the location of the origin, then $P(z) = P_0 + \int_0^z \rho(x)g\,dx$, where P_0 is the pressure at $x = 0$. Differentiating, we get $dP/dz = \rho(z)g$.

(b)

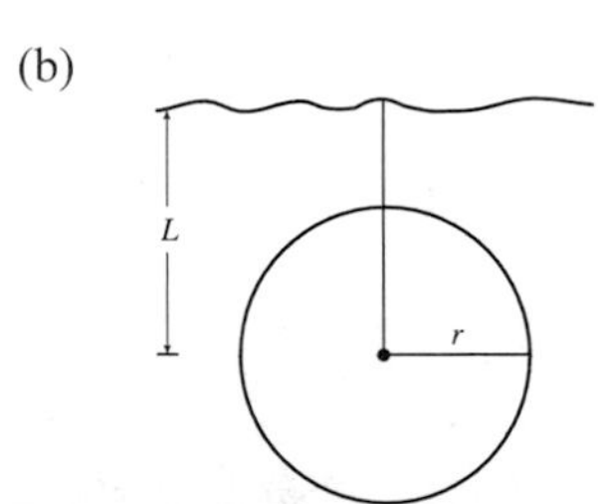

$$\begin{aligned} F &= \textstyle\int_{-r}^{r} P(L + x) \cdot 2\sqrt{r^2 - x^2}\,dx \\ &= \textstyle\int_{-r}^{r} \left(P_0 + \int_0^{L+x} \rho_0 e^{z/H} g\,dz\right) \cdot 2\sqrt{r^2 - x^2}\,dx \\ &= P_0 \textstyle\int_{-r}^{r} 2\sqrt{r^2 - x^2}\,dx + \rho_0 gH \int_{-r}^{r} \left(e^{(L+x)/H} - 1\right) \cdot 2\sqrt{r^2 - x^2}\,dx \\ &= (P_0 - \rho_0 gH) \textstyle\int_{-r}^{r} 2\sqrt{r^2 - x^2}\,dx + \rho_0 gH \int_{-r}^{r} e^{(L+x)/H} \cdot 2\sqrt{r^2 - x^2}\,dx \\ &= (P_0 - \rho_0 gH)(\pi r^2) + \rho_0 gHe^{L/H} \textstyle\int_{-r}^{r} e^{x/H} \cdot 2\sqrt{r^2 - x^2}\,dx \end{aligned}$$

6. The problem can be reduced to finding the line which minimizes the shaded area in the diagram. An equation of the circle in the first quadrant is $x = \sqrt{1 - y^2}$. So the shaded area is

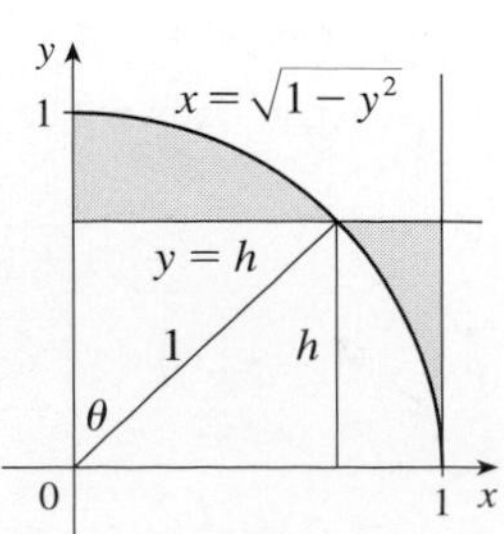

$$\begin{aligned} A(h) &= \int_0^h \left(1 - \sqrt{1 - y^2}\right) dy + \int_h^1 \sqrt{1 - y^2}\,dy \\ &= \int_0^h \left(1 - \sqrt{1 - y^2}\right) dy - \int_1^h \sqrt{1 - y^2}\,dy \end{aligned}$$

$$A'(h) = 1 - \sqrt{1 - h^2} - \sqrt{1 - h^2} \quad [\text{by FTC}] = 1 - 2\sqrt{1 - h^2}$$

$$A' = 0 \;\Leftrightarrow\; \sqrt{1 - h^2} = \tfrac{1}{2} \;\Rightarrow\; 1 - h^2 = \tfrac{1}{4} \;\Rightarrow\; h^2 = \tfrac{3}{4} \;\Rightarrow\; h = \tfrac{\sqrt{3}}{2}.$$

$A''(h) = -2 \cdot \frac{1}{2}(1-h^2)^{-1/2}(-2h) = \dfrac{2h}{\sqrt{1-h^2}} > 0$, so $h = \dfrac{\sqrt{3}}{2}$ gives a minimum value of A.

Note: Another strategy is to use the angle θ as the variable (see the diagram above) and show that $A = \theta + \cos\theta - \frac{\pi}{4} - \frac{1}{2}\sin 2\theta$, which is minimized when $\theta = \frac{\pi}{6}$.

7. To find the height of the pyramid, we use similar triangles. The first figure shows a cross-section of the pyramid passing through the top and through two opposite corners of the square base. Now $|BD| = b$, since it is a radius of the sphere, which has diameter $2b$ since it is tangent to the opposite sides of the square base. Also, $|AD| = b$ since $\triangle ADB$ is isosceles. So the height is $|AB| = \sqrt{b^2 + b^2} = \sqrt{2}\,b$.

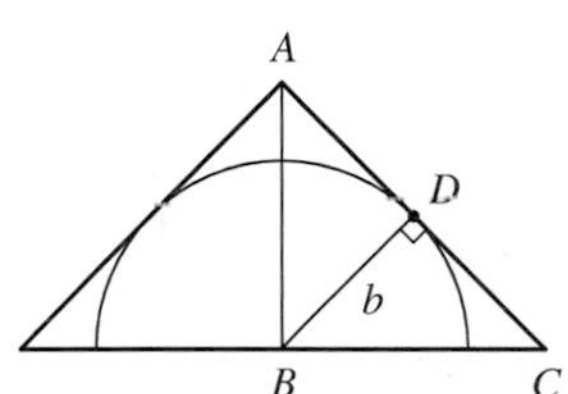

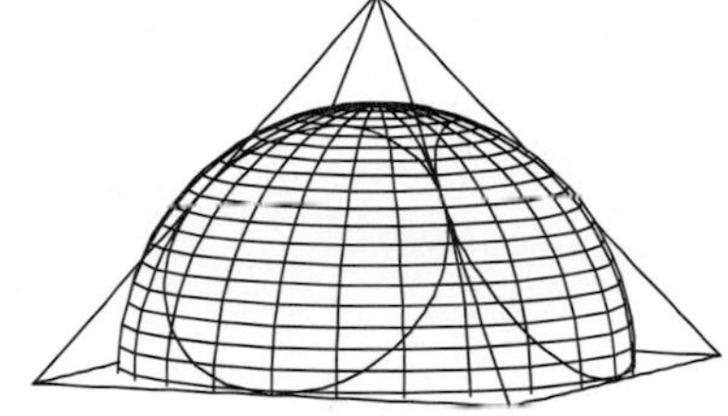
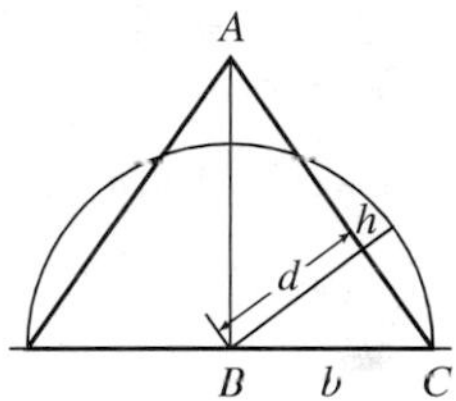

We first observe that the shared volume is equal to half the volume of the sphere, minus the sum of the four equal volumes (caps of the sphere) cut off by the triangular faces of the pyramid. See Exercise 6.2.49 for a derivation of the formula for the volume of a cap of a sphere. To use the formula, we need to find the perpendicular distance h of each triangular face from the surface of the sphere. We first find the distance d from the center of the sphere to one of the triangular faces. The third figure shows a cross-section of the pyramid through the top and through the midpoints of opposite sides of the square base. From similar triangles we find that

$$\frac{d}{b} = \frac{|AB|}{|AC|} = \frac{\sqrt{2}\,b}{\sqrt{b^2 + \left(\sqrt{2}\,b\right)^2}} \quad\Rightarrow\quad d = \frac{\sqrt{2}\,b^2}{\sqrt{3b^2}} = \frac{\sqrt{6}}{3}\,b$$

So $h = b - d = b - \frac{\sqrt{6}}{3}b = \frac{3-\sqrt{6}}{3}b$. So, using the formula $V = \pi h^2(r - h/3)$ from Exercise 6.2.49 with $r = b$, we find that the volume of each of the caps is $\pi\left(\frac{3-\sqrt{6}}{3}b\right)^2\left(b - \frac{3-\sqrt{6}}{3\cdot 3}b\right) = \frac{15-6\sqrt{6}}{9}\cdot\frac{6+\sqrt{6}}{9}\pi b^3 = \left(\frac{2}{3} - \frac{7}{27}\sqrt{6}\right)\pi b^3$. So, using our first observation, the shared volume is $V = \frac{1}{2}\left(\frac{4}{3}\pi b^3\right) - 4\left(\frac{2}{3} - \frac{7}{27}\sqrt{6}\right)\pi b^3 = \left(\frac{28}{27}\sqrt{6} - 2\right)\pi b^3$.

8. Orient the positive x-axis as in the figure. Suppose that the plate has height h and is symmetric about the x-axis. At depth x below the water ($2 \le x \le 2+h$), let the width of the plate be $2f(x)$. Now each of the n horizontal strips has height h/n and the ith strip ($1 \le i \le n$) goes from

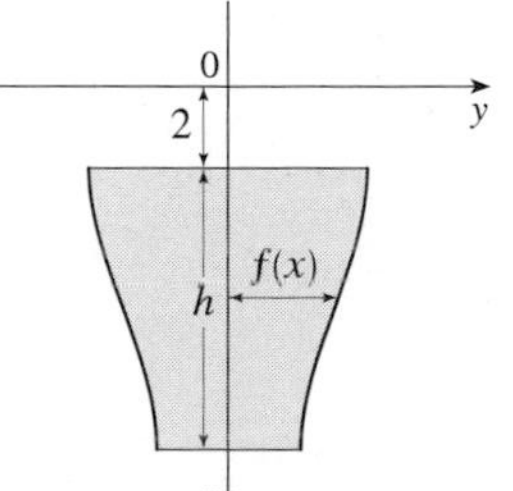

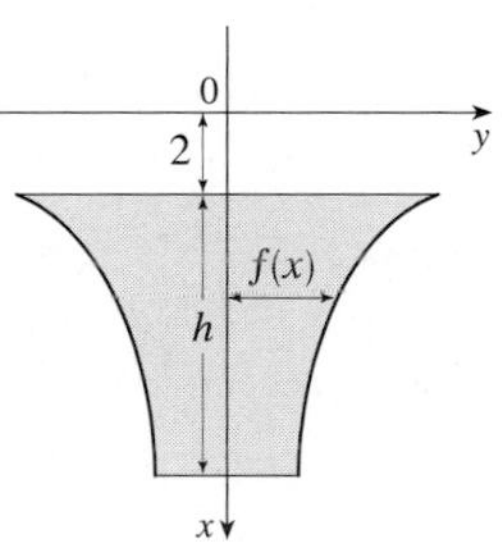

$x = 2 + \left(\dfrac{i-1}{n}\right)h$ to $x = 2 + \left(\dfrac{i}{n}\right)h$. The hydrostatic force on the ith strip is $F(i) = \displaystyle\int_{2+[(i-1)/n]h}^{2+(i/n)h} 62.5x[2f(x)]\,dx$.

[continued]

If we now let $x[2f(x)] = k$ (a constant) so that $f(x) = k/(2x)$, then

$$F(i) = \int_{2+[(i-1)/n]h}^{2+(i/n)h} 62.5k\,dx = 62.5k\left[x\right]_{2+[(i-1)/n]h}^{2+(i/n)h} = 62.5k\left[\left(2+\frac{i}{n}h\right)-\left(2+\frac{i-1}{n}h\right)\right] = 62.5k\left(\frac{h}{n}\right)$$

So the hydrostatic force on the ith strip is independent of i, that is, the force on each strip is the same. So the plate can be shaped as shown in the figure. (In fact, the required condition is satisfied whenever the plate has width C/x at depth x, for some constant C. Many shapes are possible.)

9. We can assume that the cut is made along a vertical line $x = b > 0$, that the disk's boundary is the circle $x^2 + y^2 = 1$, and that the center of mass of the smaller piece (to the right of $x = b$) is $\left(\frac{1}{2}, 0\right)$. We wish to find b to two decimal places. We have $\dfrac{1}{2} = \overline{x} = \dfrac{\int_b^1 x \cdot 2\sqrt{1-x^2}\,dx}{\int_b^1 2\sqrt{1-x^2}\,dx}$. Evaluating the

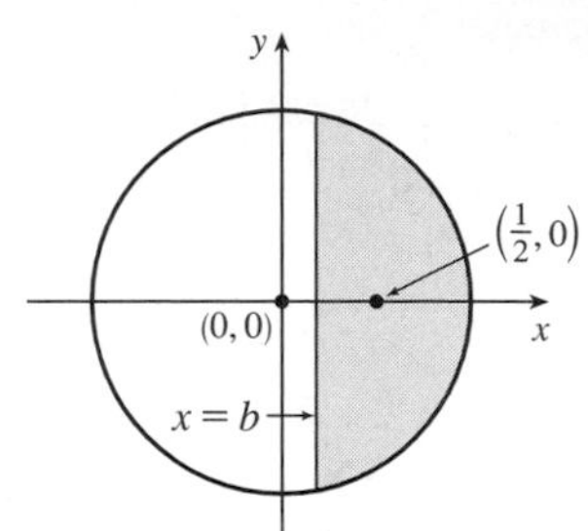

numerator gives us $-\int_b^1 (1-x^2)^{1/2}(-2x)\,dx = -\frac{2}{3}\left[(1-x^2)^{3/2}\right]_b^1 = -\frac{2}{3}\left[0-(1-b^2)^{3/2}\right] = \frac{2}{3}(1-b^2)^{3/2}$.

Using Formula 30 in the table of integrals, we find that the denominator is

$\left[x\sqrt{1-x^2} + \sin^{-1}x\right]_b^1 = \left(0 + \frac{\pi}{2}\right) - \left(b\sqrt{1-b^2} + \sin^{-1}b\right)$. Thus, we have $\dfrac{1}{2} = \overline{x} = \dfrac{\frac{2}{3}(1-b^2)^{3/2}}{\frac{\pi}{2} - b\sqrt{1-b^2} - \sin^{-1}b}$, or,

equivalently, $\frac{2}{3}(1-b^2)^{3/2} = \frac{\pi}{4} - \frac{1}{2}b\sqrt{1-b^2} - \frac{1}{2}\sin^{-1}b$. Solving this equation numerically with a calculator or CAS, we obtain $b \approx 0.138173$, or $b = 0.14$ m to two decimal places.

10. $A_1 = 30 \;\Rightarrow\; \frac{1}{2}bh = 30 \;\Rightarrow\; bh = 60$.

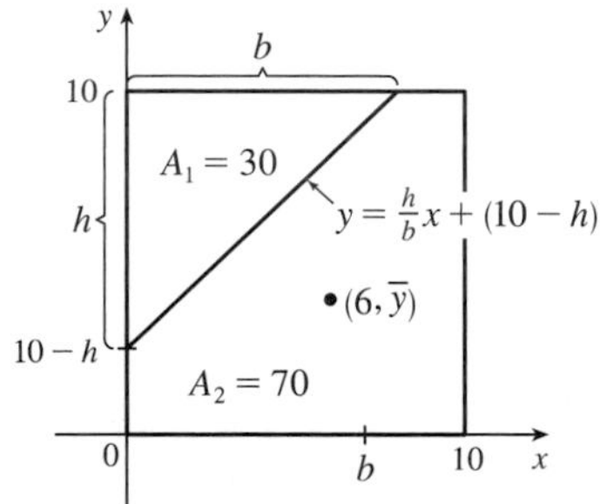

$$\overline{x} = 6 \;\Rightarrow\; \frac{1}{A_2}\int_0^{10} x f(x)\,dx = 6 \;\Rightarrow$$

$$\int_0^b x\left(\frac{h}{b}x + 10 - h\right)dx + \int_b^{10} x(10)\,dx = 6(70) \;\Rightarrow$$

$$\int_0^b \left(\frac{h}{b}x^2 + 10x - hx\right)dx + 10 \cdot \tfrac{1}{2}\left[x^2\right]_b^{10} = 420 \;\Rightarrow$$

$$\left[\frac{h}{3b}x^3 + 5x^2 - \frac{h}{2}x^2\right]_0^b + 5(100 - b^2) = 420 \;\Rightarrow\; \tfrac{1}{3}hb^2 + 5b^2 - \tfrac{1}{2}hb^2 + 500 - 5b^2 = 420 \;\Rightarrow\; 80 = \tfrac{1}{6}hb^2 \;\Rightarrow$$

$480 = (hb)b \;\Rightarrow\; 480 = 60b \;\Rightarrow\; b = 8$. So $h = \frac{60}{8} = \frac{15}{2}$ and an equation of the line is

$y = \dfrac{15/2}{8}x + \left(10 - \dfrac{15}{2}\right) = \dfrac{15}{16}x + \dfrac{5}{2}$. Now

$$\begin{aligned}\overline{y} &= \frac{1}{A_2}\int_0^{10} \tfrac{1}{2}[f(x)]^2\,dx = \frac{1}{70 \cdot 2}\left[\int_0^8 \left(\frac{15}{16}x + \frac{5}{2}\right)^2 dx + \int_8^{10} (10)^2\,dx\right] \\ &= \tfrac{1}{140}\left[\int_0^8 \left(\tfrac{225}{256}x^2 + \tfrac{75}{16}x + \tfrac{25}{4}\right)dx + 100(10-8)\right] = \tfrac{1}{140}\left(\left[\tfrac{225}{768}x^3 + \tfrac{75}{32}x^2 + \tfrac{25}{4}x\right]_0^8 + 200\right) \\ &= \tfrac{1}{140}(150 + 150 + 50 + 200) = \tfrac{550}{140} = \tfrac{55}{14}\end{aligned}$$

[continued]

Another solution: Assume that the right triangle cut from the square has legs a cm and b cm long as shown. The triangle has area 30 cm^2, so $\frac{1}{2}ab = 30$ and $ab = 60$. We place the square in the first quadrant of the xy-plane as shown, and we let T, R, and S denote the triangle, the remaining portion of the square, and the full square, respectively. By symmetry, the centroid of S is $(5, 5)$. By Exercise 8.3.39, the centroid of T is $\left(\frac{b}{3}, 10 - \frac{a}{3}\right)$.

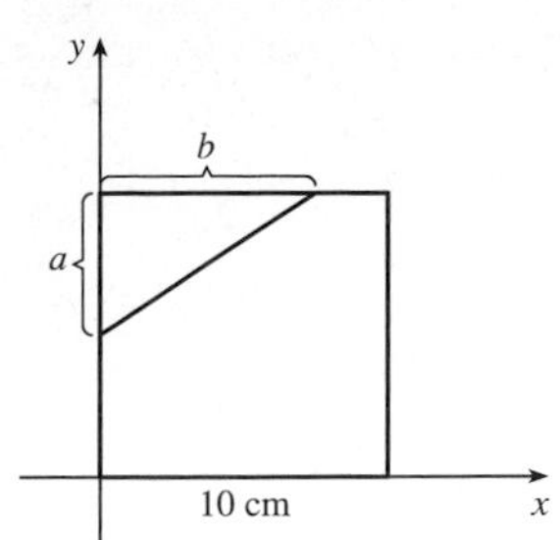

We are given that the centroid of R is $(6, c)$, where c is to be determined. We take the density of the square to be 1, so that areas can be used as masses. Then T has mass $m_T = 30$, S has mass $m_S = 100$, and R has mass $m_R = m_S - m_T = 70$. As in Exercises 40 and 41 of Section 8.3, we view S as consisting of a mass m_T at the centroid $(\overline{x}_T, \overline{y}_T)$ of T and a mass R at the centroid $(\overline{x}_R, \overline{y}_R)$ of R. Then $\overline{x}_S = \dfrac{m_T\overline{x}_T + m_R\overline{x}_R}{m_T + m_R}$ and $\overline{y}_S = \dfrac{m_T\overline{y}_T + m_R\overline{y}_R}{m_T + m_R}$; that is, $5 = \dfrac{30(b/3) + 70(6)}{100}$ and $5 = \dfrac{30(10 - a/3) + 70c}{100}$.

Solving the first equation for b, we get $b = 8$ cm. Since $ab = 60$ cm^2, it follows that $a = \frac{60}{8} = 7.5$ cm. Now the second equation says that $70c = 200 + 10a$, so $7c = 20 + a = \frac{55}{2}$ and $c = \frac{55}{14} = 3.9\overline{285714}$ cm.

The solution is depicted in the figure.

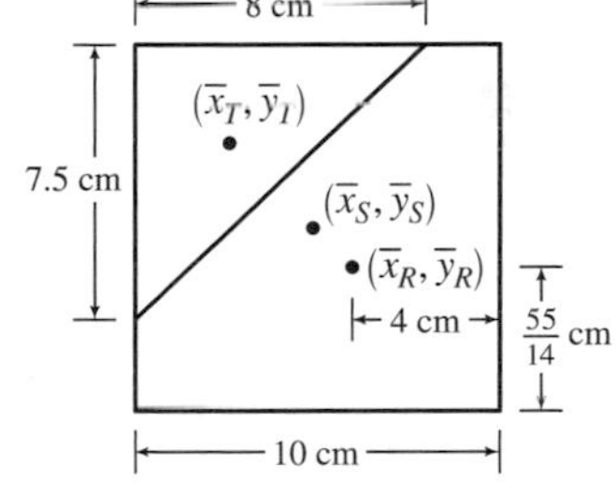

11. If $h = L$, then $P = \dfrac{\text{area under } y = L\sin\theta}{\text{area of rectangle}} = \dfrac{\int_0^\pi L\sin\theta\,d\theta}{\pi L} = \dfrac{[-\cos\theta]_0^\pi}{\pi} = \dfrac{-(-1)+1}{\pi} = \dfrac{2}{\pi}$.

If $h = L/2$, then $P = \dfrac{\text{area under } y = \frac{1}{2}L\sin\theta}{\text{area of rectangle}} = \dfrac{\int_0^\pi \frac{1}{2}L\sin\theta\,d\theta}{\pi L} = \dfrac{[-\cos\theta]_0^\pi}{2\pi} = \dfrac{2}{2\pi} = \dfrac{1}{\pi}$.

12. (a) The total set of possibilities can be identified with the rectangular region $\mathcal{R} = \{(\theta, y) \mid 0 \le y < L, 0 \le \theta < \pi\}$. Even when $h > L$, the needle intersects at least one line if and only if $y \le h\sin\theta$. Let $\mathcal{R}_1 = \{(\theta, y) \mid 0 \le y \le h\sin\theta, 0 \le \theta < \pi\}$. When $h \le L$, $\mathcal{R}_1$ is contained in $\mathcal{R}$, but that is no longer true when $h > L$. Thus, the probability that the needle intersects a line becomes

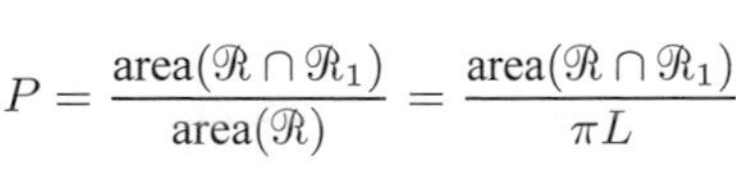

$$P = \frac{\text{area}(\mathcal{R} \cap \mathcal{R}_1)}{\text{area}(\mathcal{R})} = \frac{\text{area}(\mathcal{R} \cap \mathcal{R}_1)}{\pi L}$$

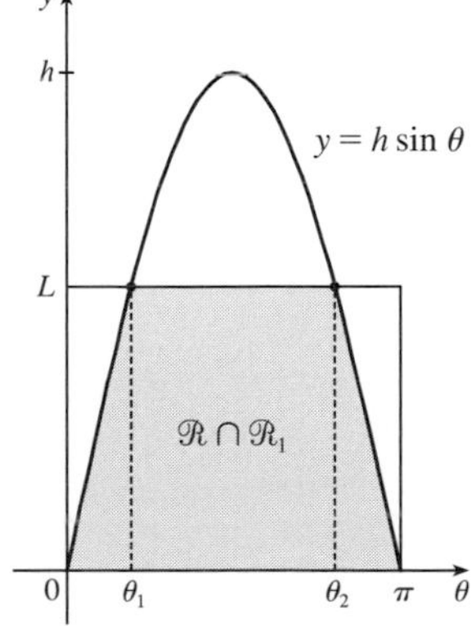

When $h > L$, the curve $y = h\sin\theta$ intersects the line $y = L$ twice—at $\left(\sin^{-1}(L/h), L\right)$ and at $\left(\pi - \sin^{-1}(L/h), L\right)$. Set $\theta_1 = \sin^{-1}(L/h)$ and $\theta_2 = \pi - \theta_1$. Then

$$\begin{aligned}
\text{area}(\mathcal{R} \cap \mathcal{R}_1) &= \int_0^{\theta_1} h \sin\theta \, d\theta + \int_{\theta_1}^{\theta_2} L \, d\theta + \int_{\theta_2}^{\pi} h \sin\theta \, d\theta \\
&= 2\int_0^{\theta_1} h \sin\theta \, d\theta + L(\theta_2 - \theta_1) = 2h\left[-\cos\theta\right]_0^{\theta_1} + L(\pi - 2\theta_1) \\
&= 2h(1 - \cos\theta_1) + L(\pi - 2\theta_1) \\
&= 2h\left(1 - \frac{\sqrt{h^2 - L^2}}{h}\right) + L\left[\pi - 2\sin^{-1}\left(\frac{L}{h}\right)\right] \\
&= 2h - 2\sqrt{h^2 - L^2} + \pi L - 2L\sin^{-1}\left(\frac{L}{h}\right)
\end{aligned}$$

We are told that $L = 4$ and $h = 7$, so $\text{area}(\mathcal{R} \cap \mathcal{R}_1) = 14 - 2\sqrt{33} + 4\pi - 8\sin^{-1}\left(\frac{4}{7}\right) \approx 10.21128$ and $P = \frac{1}{4\pi}\text{area}(\mathcal{R} \cap \mathcal{R}_1) \approx 0.812588$. (By comparison, $P = \frac{2}{\pi} \approx 0.636620$ when $h = L$, as shown in the solution to Problem 11.)

(b) The needle intersects at least two lines when $y + L \leq h \sin\theta$; that is, when $y \leq h\sin\theta - L$. Set $\mathcal{R}_2 = \{(\theta, y) \mid 0 \leq y \leq h\sin\theta - L, 0 \leq \theta < \pi\}$. Then the probability that the needle intersects at least two lines is

$$P_2 = \frac{\text{area}(\mathcal{R} \cap \mathcal{R}_2)}{\text{area}(\mathcal{R})} = \frac{\text{area}(\mathcal{R} \cap \mathcal{R}_2)}{\pi L}$$

When $L = 4$ and $h = 7$, $\mathcal{R}_2$ is contained in $\mathcal{R}$ (see the figure). Thus,

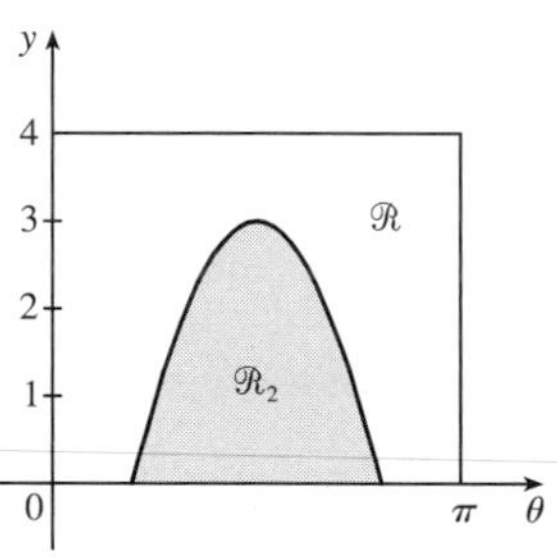

$$\begin{aligned}
P_2 &= \frac{1}{4\pi}\text{area}(\mathcal{R}_2) = \frac{1}{4\pi}\int_{\sin^{-1}(4/7)}^{\pi - \sin^{-1}(4/7)} (7\sin\theta - 4)\, d\theta = \tfrac{1}{4\pi} \cdot 2\int_{\sin^{-1}(4/7)}^{\pi/2} (7\sin\theta - 4)\, d\theta \\
&= \frac{1}{2\pi}\left[-7\cos\theta - 4\theta\right]_{\sin^{-1}(4/7)}^{\pi/2} = \frac{1}{2\pi}\left[0 - 2\pi + 7\frac{\sqrt{33}}{7} + 4\sin^{-1}\left(\tfrac{4}{7}\right)\right] = \frac{\sqrt{33} + 4\sin^{-1}\left(\frac{4}{7}\right) - 2\pi}{2\pi} \\
&\approx 0.301497
\end{aligned}$$

(c) The needle intersects at least three lines when $y + 2L \leq h\sin\theta$: that is, when $y \leq h\sin\theta - 2L$. Set $\mathcal{R}_3 = \{(\theta, y) \mid 0 \leq y \leq h\sin\theta - 2L, 0 \leq \theta < \pi\}$. Then the probability that the needle intersects at least three lines is $P_3 = \dfrac{\text{area}(\mathcal{R} \cap \mathcal{R}_3)}{\text{area}(\mathcal{R})} = \dfrac{\text{area}(\mathcal{R} \cap \mathcal{R}_3)}{\pi L}$. (At this point, the generalization to P_n, n any positive integer, should be clear.)

Under the given assumption,

$$\begin{aligned}
P_3 &= \frac{1}{\pi L}\text{area}(\mathcal{R}_3) = \frac{1}{\pi L}\int_{\sin^{-1}(2L/h)}^{\pi - \sin^{-1}(2L/h)} (h\sin\theta - 2L)\, d\theta = \frac{2}{\pi L}\int_{\sin^{-1}(2L/h)}^{\pi/2} (h\sin\theta - 2L)\, d\theta \\
&= \frac{2}{\pi L}\left[-h\cos\theta - 2L\theta\right]_{\sin^{-1}(2L/h)}^{\pi/2} = \frac{2}{\pi L}\left[-\pi L + \sqrt{h^2 - 4L^2} + 2L\sin^{-1}(2L/h)\right]
\end{aligned}$$

Note that the probability that a needle touches exactly one line is $P_1 - P_2$, the probability that it touches exactly two lines is $P_2 - P_3$, and so on.

13. Solve for y: $x^2 + (x + y + 1)^2 = 1 \;\Rightarrow\; (x + y + 1)^2 = 1 - x^2 \;\Rightarrow\; x + y + 1 = \pm\sqrt{1 - x^2} \;\Rightarrow$

$y = -x - 1 \pm \sqrt{1 - x^2}$.

$$A = \int_{-1}^{1} \left[\left(-x - 1 + \sqrt{1 - x^2}\right) - \left(-x - 1 - \sqrt{1 - x^2}\right)\right] dx$$

$$= \int_{-1}^{1} 2\sqrt{1 - x^2}\, dx = 2\left(\tfrac{\pi}{2}\right) \quad \left[\begin{array}{c}\text{area of}\\ \text{semicircle}\end{array}\right] \quad = \pi$$

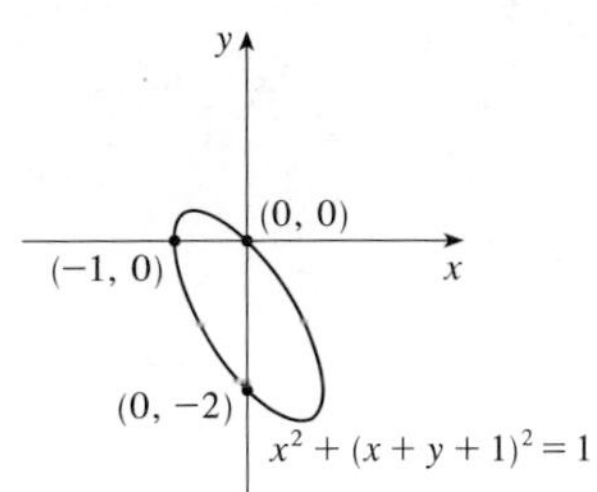

$$\overline{x} = \frac{1}{A}\int_{-1}^{1} x \cdot 2\sqrt{1 - x^2}\, dx = 0 \quad [\text{odd integrand}]$$

$$\overline{y} = \frac{1}{A}\int_{-1}^{1} \frac{1}{2}\left[\left(-x - 1 + \sqrt{1 - x^2}\right)^2 - \left(-x - 1 - \sqrt{1 - x^2}\right)^2\right] dx = \frac{1}{\pi}\int_{-1}^{1} \frac{1}{2}\left(-4x\sqrt{1 - x^2} - 4\sqrt{1 - x^2}\right) dx$$

$$= -\frac{2}{\pi}\int_{-1}^{1}\left(x\sqrt{1 - x^2} + \sqrt{1 - x^2}\right) dx = -\frac{2}{\pi}\int_{-1}^{1} x\sqrt{1 - x^2}\, dx - \frac{2}{\pi}\int_{-1}^{1}\sqrt{1 - x^2}\, dx$$

$$= -\frac{2}{\pi}(0) \quad [\text{odd integrand}] \quad -\frac{2}{\pi}\left(\frac{\pi}{2}\right) \quad \left[\begin{array}{c}\text{area of}\\ \text{semicircle}\end{array}\right] \quad = -1$$

Thus, as expected, the centroid is $(\overline{x}, \overline{y}) = (0, -1)$. We might expect this result since the centroid of an ellipse is located at its center.

9 □ DIFFERENTIAL EQUATIONS

9.1 Modeling with Differential Equations

1. $y = \frac{2}{3}e^x + e^{-2x} \quad \Rightarrow \quad y' = \frac{2}{3}e^x - 2e^{-2x}$. To show that y is a solution of the differential equation, we will substitute the expressions for y and y' in the left-hand side of the equation and show that the left-hand side is equal to the right-hand side.

$$\begin{aligned} \text{LHS} = y' + 2y &= \tfrac{2}{3}e^x - 2e^{-2x} + 2\left(\tfrac{2}{3}e^x + e^{-2x}\right) = \tfrac{2}{3}e^x - 2e^{-2x} + \tfrac{4}{3}e^x + 2e^{-2x} \\ &= \tfrac{6}{3}e^x = 2e^x = \text{RHS} \end{aligned}$$

2. $y = -t\cos t - t \quad \Rightarrow \quad dy/dt = -t(-\sin t) + \cos t(-1) - 1 = t\sin t - \cos t - 1$.

$$\begin{aligned} \text{LHS} = t\,\frac{dy}{dt} &= t(t\sin t - \cos t - 1) = t^2\sin t - t\cos t - t \\ &= t^2\sin t + y = \text{RHS}, \end{aligned}$$

so y is a solution of the differential equation. Also $y(\pi) = -\pi\cos\pi - \pi = -\pi(-1) - \pi = \pi - \pi = 0$, so the initial condition is satisfied.

3. (a) $y = e^{rx} \quad \Rightarrow \quad y' = re^{rx} \quad \Rightarrow \quad y'' = r^2e^{rx}$. Substituting these expressions into the differential equation $2y'' + y' - y = 0$, we get $2r^2e^{rx} + re^{rx} - e^{rx} = 0 \quad \Rightarrow \quad (2r^2 + r - 1)e^{rx} = 0 \quad \Rightarrow$ $(2r - 1)(r + 1) = 0$ [since e^{rx} is never zero] $\Rightarrow \quad r = \frac{1}{2}$ or -1.

(b) Let $r_1 = \frac{1}{2}$ and $r_2 = -1$, so we need to show that every member of the family of functions $y = ae^{x/2} + be^{-x}$ is a solution of the differential equation $2y'' + y' - y = 0$.

$y = ae^{x/2} + be^{-x} \quad \Rightarrow \quad y' = \frac{1}{2}ae^{x/2} - be^{-x} \quad \Rightarrow \quad y'' = \frac{1}{4}ae^{x/2} + be^{-x}$.

$$\begin{aligned} \text{LHS} = 2y'' + y' - y &= 2\left(\tfrac{1}{4}ae^{x/2} + be^{-x}\right) + \left(\tfrac{1}{2}ae^{x/2} - be^{-x}\right) - \left(ae^{x/2} + be^{-x}\right) \\ &= \tfrac{1}{2}ae^{x/2} + 2be^{-x} + \tfrac{1}{2}ae^{x/2} - be^{-x} - ae^{x/2} - be^{-x} \\ &= \left(\tfrac{1}{2}a + \tfrac{1}{2}a - a\right)e^{x/2} + (2b - b - b)e^{-x} \\ &= 0 = \text{RHS} \end{aligned}$$

4. (a) $y = \cos kt \quad \Rightarrow \quad y' = -k\sin kt \quad \Rightarrow \quad y'' = -k^2\cos kt$. Substituting these expressions into the differential equation $4y'' = -25y$, we get $4(-k^2\cos kt) = -25(\cos kt) \quad \Rightarrow \quad (25 - 4k^2)\cos kt = 0 \quad$ [for all t] $\Rightarrow \quad 25 - 4k^2 = 0 \quad \Rightarrow$ $k^2 = \frac{25}{4} \quad \Rightarrow \quad k = \pm\frac{5}{2}$.

(b) $y = A\sin kt + B\cos kt \quad \Rightarrow \quad y' = Ak\cos kt - Bk\sin kt \quad \Rightarrow \quad y'' = -Ak^2\sin kt - Bk^2\cos kt$.

The given differential equation $4y'' = -25y$ is equivalent to $4y'' + 25y = 0$. Thus,

$$\begin{aligned} \text{LHS} = 4y'' + 25y &= 4(-Ak^2\sin kt - Bk^2\cos kt) + 25(A\sin kt + B\cos kt) \\ &= -4Ak^2\sin kt - 4Bk^2\cos kt + 25A\sin kt + 25B\cos kt \\ &= (25 - 4k^2)A\sin kt + (25 - 4k^2)B\cos kt \\ &= 0 \quad \text{since } k^2 = \tfrac{25}{4}. \end{aligned}$$

5. (a) $y = \sin x \;\Rightarrow\; y' = \cos x \;\Rightarrow\; y'' = -\sin x$.

LHS $= y'' + y = -\sin x + \sin x = 0 \neq \sin x$, so $y = \sin x$ **is not** a solution of the differential equation.

(b) $y = \cos x \;\Rightarrow\; y' = -\sin x \;\Rightarrow\; y'' = -\cos x$.

LHS $= y'' + y = -\cos x + \cos x = 0 \neq \sin x$, so $y = \cos x$ **is not** a solution of the differential equation.

(c) $y = \frac{1}{2}x\sin x \;\Rightarrow\; y' = \frac{1}{2}(x\cos x + \sin x) \;\Rightarrow\; y'' = \frac{1}{2}(-x\sin x + \cos x + \cos x)$.

LHS $= y'' + y = \frac{1}{2}(-x\sin x + 2\cos x) + \frac{1}{2}x\sin x = \cos x \neq \sin x$, so $y = \frac{1}{2}x\sin x$ **is not** a solution of the differential equation.

(d) $y = -\frac{1}{2}x\cos x \;\Rightarrow\; y' = -\frac{1}{2}(-x\sin x + \cos x) \;\Rightarrow\; y'' = -\frac{1}{2}(-x\cos x - \sin x - \sin x)$.

LHS $= y'' + y = -\frac{1}{2}(-x\cos x - 2\sin x) + \left(-\frac{1}{2}x\cos x\right) = \sin x =$ RHS, so $y = -\frac{1}{2}x\cos x$ **is** a solution of the differential equation.

6. (a) $y = \dfrac{\ln x + C}{x} \;\Rightarrow\; y' = \dfrac{x\cdot(1/x) - (\ln x + C)}{x^2} = \dfrac{1 - \ln x - C}{x^2}$.

$$\text{LHS} = x^2y' + xy = x^2\cdot\frac{1-\ln x - C}{x^2} + x\cdot\frac{\ln x + C}{x}$$
$$= 1 - \ln x - C + \ln x + C = 1 = \text{RHS}, \text{ so } y \text{ is a solution of the differential equation.}$$

(b)

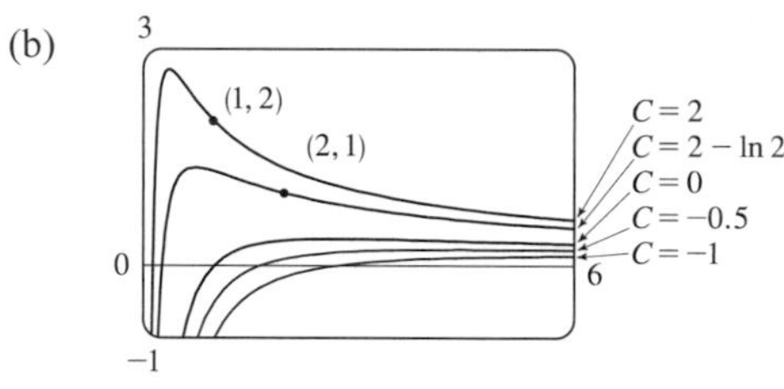

A few notes about the graph of $y = (\ln x + C)/x$:

(1) There is a vertical asymptote of $x = 0$.

(2) There is a horizontal asymptote of $y = 0$.

(3) $y = 0 \;\Rightarrow\; \ln x + C = 0 \;\Rightarrow\; x = e^{-C}$, so there is an x-intercept at e^{-C}.

(4) $y' = 0 \;\Rightarrow\; \ln x = 1 - C \;\Rightarrow\; x = e^{1-C}$, so there is a local maximum at $x = e^{1-C}$.

(c) $y(1) = 2 \;\Rightarrow\; 2 = \dfrac{\ln 1 + C}{1} \;\Rightarrow\; 2 = C$, so the solution is $y = \dfrac{\ln x + 2}{x}$ [shown in part (b)].

(d) $y(2) = 1 \;\Rightarrow\; 1 = \dfrac{\ln 2 + C}{2} \;\Rightarrow\; 2 + \ln 2 + C \;\Rightarrow\; C = 2 - \ln 2$, so the solution is $y = \dfrac{\ln x + 2 - \ln 2}{x}$ [shown in part (b)].

7. (a) Since the derivative $y' = -y^2$ is always negative (or 0 if $y = 0$), the function y must be decreasing (or equal to 0) on any interval on which it is defined.

(b) $y = \dfrac{1}{x + C} \;\Rightarrow\; y' = -\dfrac{1}{(x+C)^2}$. LHS $= y' = -\dfrac{1}{(x+C)^2} = -\left(\dfrac{1}{x+C}\right)^2 = -y^2 =$ RHS

(c) $y = 0$ is a solution of $y' = -y^2$ that is not a member of the family in part (b).

(d) If $y(x) = \dfrac{1}{x+C}$, then $y(0) = \dfrac{1}{0+C} = \dfrac{1}{C}$. Since $y(0) = 0.5$, $\dfrac{1}{C} = \dfrac{1}{2} \;\Rightarrow\; C = 2$, so $y = \dfrac{1}{x+2}$.

8. (a) If x is close to 0, then xy^3 is close to 0, and hence, y' is close to 0. Thus, the graph of y must have a tangent line that is nearly horizontal. If x is large, then xy^3 is large, and the graph of y must have a tangent line that is nearly vertical. (In both cases, we assume reasonable values for y.)

(b) $y = (c - x^2)^{-1/2} \Rightarrow y' = x(c - x^2)^{-3/2}$. RHS $= xy^3 = x[(c - x^2)^{-1/2}]^3 = x(c - x^2)^{-3/2} = y' =$ LHS

(c) When x is close to 0, y' is also close to 0. As x gets larger, so does $|y'|$.

(d) $y(0) = (c - 0)^{-1/2} = 1/\sqrt{c}$ and $y(0) = 2 \Rightarrow \sqrt{c} = \frac{1}{2} \Rightarrow c = \frac{1}{4}$, so $y = \left(\frac{1}{4} - x^2\right)^{-1/2}$.

9. (a) $\dfrac{dP}{dt} = 1.2P\left(1 - \dfrac{P}{4200}\right)$. Now $\dfrac{dP}{dt} > 0 \Rightarrow 1 - \dfrac{P}{4200} > 0$ [assuming that $P > 0$] $\Rightarrow \dfrac{P}{4200} < 1 \Rightarrow$ $P < 4200 \Rightarrow$ the population is increasing for $0 < P < 4200$.

(b) $\dfrac{dP}{dt} < 0 \Rightarrow P > 4200$

(c) $\dfrac{dP}{dt} = 0 \Rightarrow P = 4200$ or $P = 0$

10. (a) $\dfrac{dv}{dt} = -v[v^2 - (1 + a)v + a] = -v(v - a)(v - 1)$, so $\dfrac{dv}{dt} = 0 \Leftrightarrow v = 0, a,$ or 1.

(b) With $0 < a < 1$, $dv/dt = -v(v - a)(v - 1) > 0 \Leftrightarrow v < 0$ or $a < v < 1$, so v is increasing on $(-\infty, 0)$ and $(a, 1)$.

(c) With $0 < a < 1$, $dv/dt = -v(v - a)(v - 1) < 0 \Leftrightarrow 0 < v < a$ or $v > 1$, so v is decreasing on $(0, a)$ and $(1, \infty)$.

11. (a) This function is increasing *and* also decreasing. But $dy/dt = e^t(y - 1)^2 \geq 0$ for all t, implying that the graph of the solution of the differential equation cannot be decreasing on any interval.

(b) When $y = 1$, $dy/dt = 0$, but the graph does not have a horizontal tangent line.

12. The graph for this exercise is shown in the figure at the right.

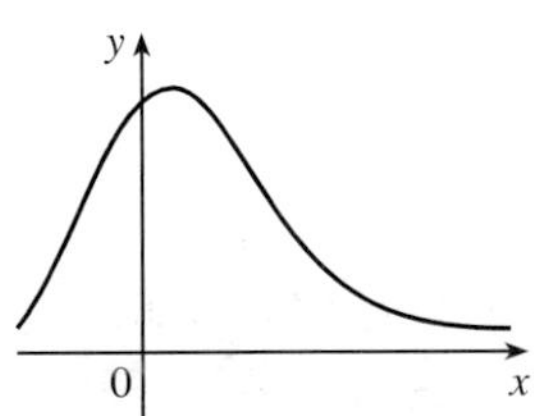

A. $y' = 1 + xy > 1$ for points in the first quadrant, but we can see that $y' < 0$ for some points in the first quadrant.

B. $y' = -2xy = 0$ when $x = 0$, but we can see that $y' > 0$ for $x = 0$. Thus, equations A and B are incorrect, so the correct equation is C.

C. $y' = 1 - 2xy$ seems reasonable since:

(1) When $x = 0$, y' could be 1.

(2) When $x < 0$, y' could be greater than 1.

(3) Solving $y' = 1 - 2xy$ for y gives us $y = \dfrac{1 - y'}{2x}$. If y' takes on small negative values, then as $x \to \infty$, $y \to 0^+$, as shown in the figure.

13. (a) $y' = 1 + x^2 + y^2 \geq 1$ and $y' \to \infty$ as $x \to \infty$. The only curve satisfying these conditions is labeled III.

(b) $y' = xe^{-x^2-y^2} > 0$ if $x > 0$ and $y' < 0$ if $x < 0$. The only curve with negative tangent slopes when $x < 0$ and positive tangent slopes when $x > 0$ is labeled I.

(c) $y' = \dfrac{1}{1 + e^{x^2+y^2}} > 0$ and $y' \to 0$ as $x \to \infty$. The only curve satisfying these conditions is labeled IV.

(d) $y' = \sin(xy)\cos(xy) = 0$ if $y = 0$, which is the solution graph labeled II.

14. (a) The coffee cools most quickly as soon as it is removed from the heat source. The rate of cooling decreases toward 0 since the coffee approaches room temperature.

(b) $\dfrac{dy}{dt} = k(y - R)$, where k is a proportionality constant, y is the temperature of the coffee, and R is the room temperature. The initial condition is $y(0) = 95°\text{C}$. The answer and the model support each other because as y approaches R, dy/dt approaches 0, so the model seems appropriate.

(c)

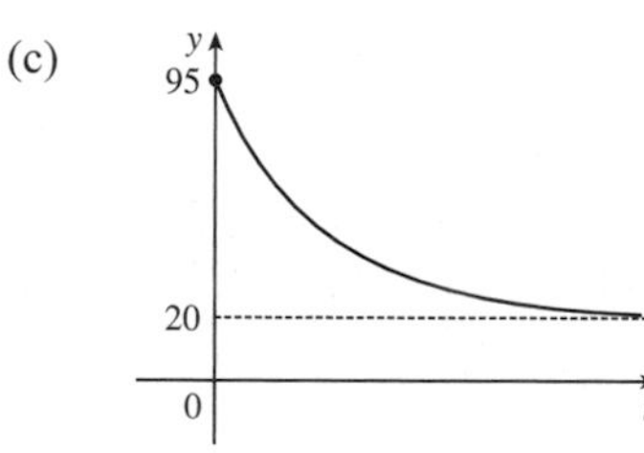

15. (a) P increases most rapidly at the beginning, since there are usually many simple, easily-learned sub-skills associated with learning a skill. As t increases, we would expect dP/dt to remain positive, but decrease. This is because as time progresses, the only points left to learn are the more difficult ones.

(b) $\dfrac{dP}{dt} = k(M - P)$ is always positive, so the level of performance P is increasing. As P gets close to M, dP/dt gets close to 0; that is, the performance levels off, as explained in part (a).

(c)

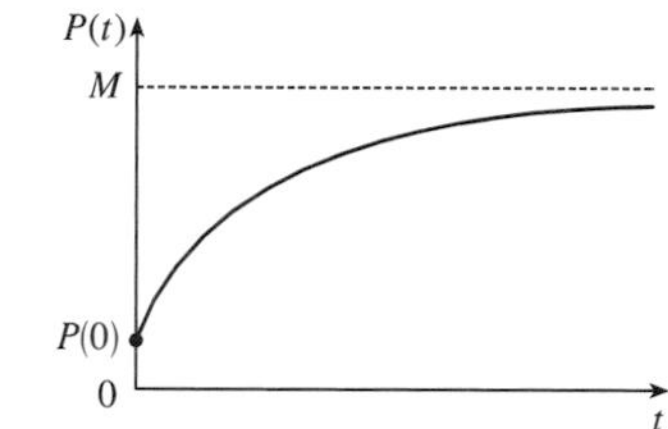

16. (a) $\dfrac{dL}{dt} = k(L_\infty - L)$. Assuming $L_\infty > L$, we have $k > 0$ and $dL/dt > 0$ for all t.

(b)

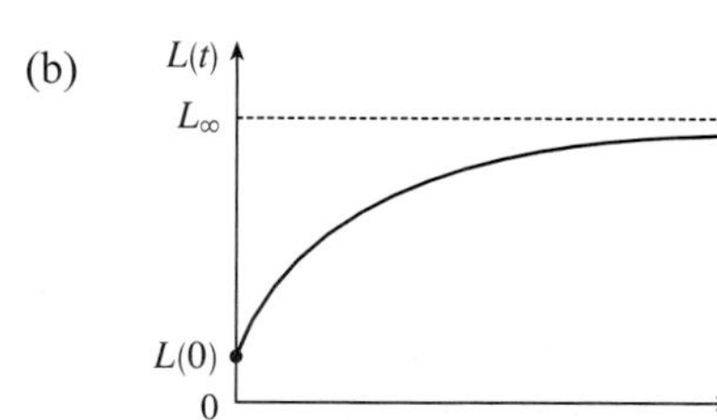

17. If $c(t) = c_s\left(1 - e^{-\alpha t^{1-b}}\right) = c_s - c_s e^{-\alpha t^{1-b}}$ for $t > 0$, where $k > 0$, $c_s > 0$, $0 < b < 1$, and $\alpha = k/(1 - b)$, then

$$\frac{dc}{dt} = c_s\left[0 - e^{-\alpha t^{1-b}} \cdot \frac{d}{dt}\left(-\alpha t^{1-b}\right)\right] = -c_s e^{-\alpha t^{1-b}} \cdot (-\alpha)(1-b)t^{-b} = \frac{\alpha(1-b)}{t^b}\, c_s\, e^{-\alpha t^{1-b}} = \frac{k}{t^b}(c_s - c).$$

The equation for c indicates that as t increases, c approaches c_s. The differential equation indicates that as t increases, the rate of increase of c decreases steadily and approaches 0 as c approaches c_s.

9.2 Direction Fields and Euler's Method

1. (a)

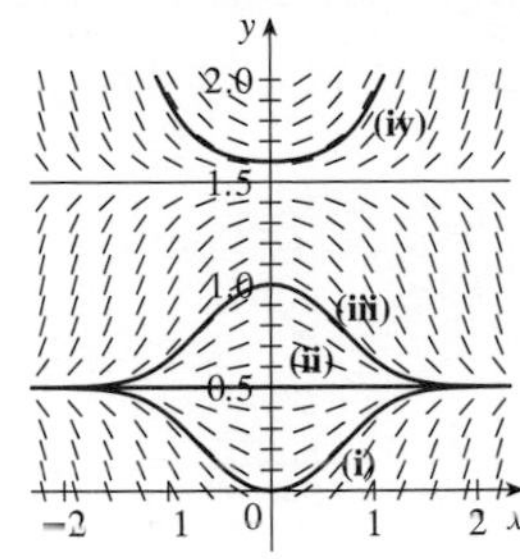

(b) It appears that the constant functions $y = 0.5$ and $y = 1.5$ are equilibrium solutions. Note that these two values of y satisfy the given differential equation $y' = x \cos \pi y$.

2. (a)

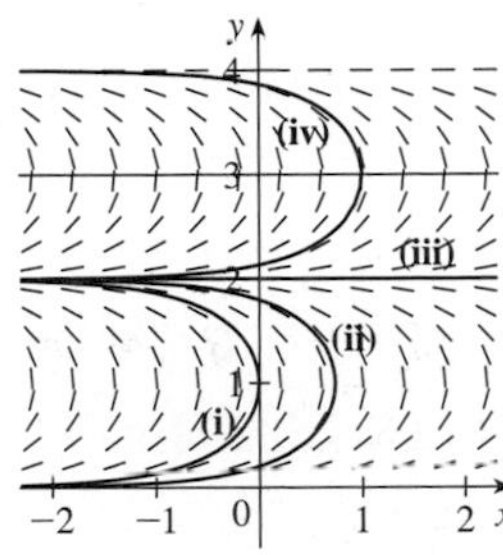

(b) It appears that the constant functions $y = 0$, $y = 2$, and $y = 4$ are equilibrium solutions. Note that these three values of y satisfy the given differential equation $y' = \tan\left(\frac{1}{2}\pi y\right)$.

3. $y' = 2 - y$. The slopes at each point are independent of x, so the slopes are the same along each line parallel to the x-axis. Thus, III is the direction field for this equation. Note that for $y = 2$, $y' = 0$.

4. $y' = x(2 - y) = 0$ on the lines $x = 0$ and $y = 2$. Direction field I satisfies these conditions.

5. $y' = x + y - 1 = 0$ on the line $y = -x + 1$. Direction field IV satisfies this condition. Notice also that on the line $y = -x$ we have $y' - -1$, which is truc in IV.

6. $y' = \sin x \sin y = 0$ on the lines $x = 0$ and $y = 0$, and $y' > 0$ for $0 < x < \pi$, $0 < y < \pi$. Direction field II satisfics these conditions.

7.

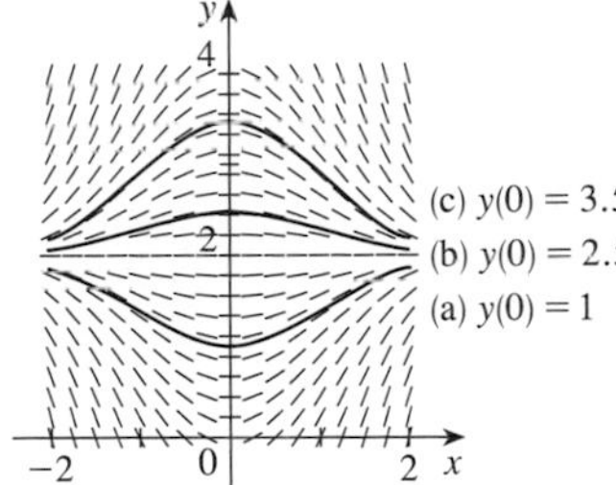

8.

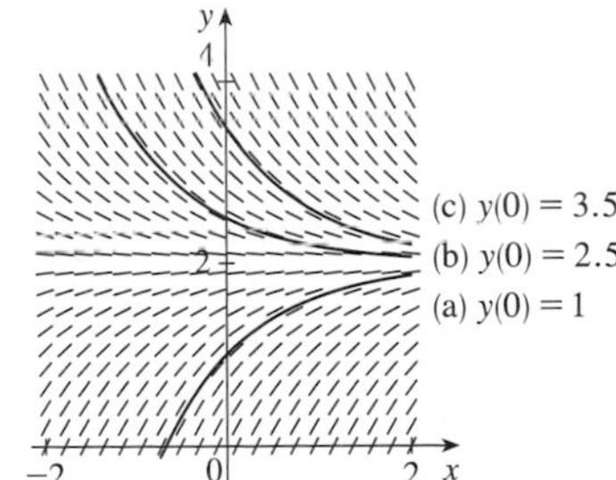

9.

x	y	$y' = \frac{1}{2}y$
0	0	0
0	1	0.5
0	2	1
0	-3	-1.5
0	-2	-1

Note that for $y = 0$, $y' = 0$. The three solution curves sketched go through $(0, 0)$, $(0, 1)$, and $(0, -1)$.

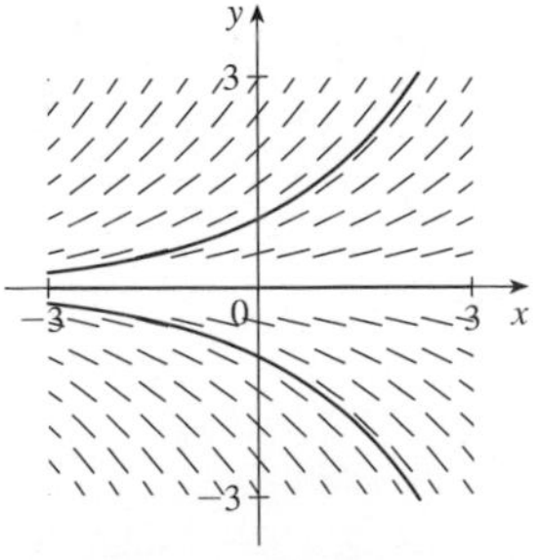

10.

x	y	$y' = x - y + 1$
-1	0	0
-1	-1	1
0	0	1
0	1	0
0	2	-1
0	-1	2
0	-2	3
1	0	2
1	1	1

Note that $y' = 0$ for $y = x + 1$ and that $y' = 1$ for $y = x$. For any constant value of x, y' decreases as y increases and y' increases as y decreases. The three solution curves sketched go through $(0, 0)$, $(0, 1)$, and $(0, -1)$.

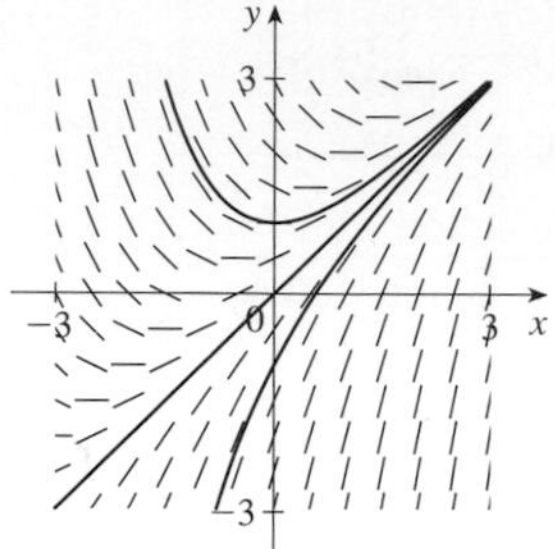

11.

x	y	$y' = y - 2x$
-2	-2	2
-2	2	6
2	2	-2
2	-2	-6

Note that $y' = 0$ for any point on the line $y = 2x$. The slopes are positive to the left of the line and negative to the right of the line. The solution curve in the graph passes through $(1, 0)$.

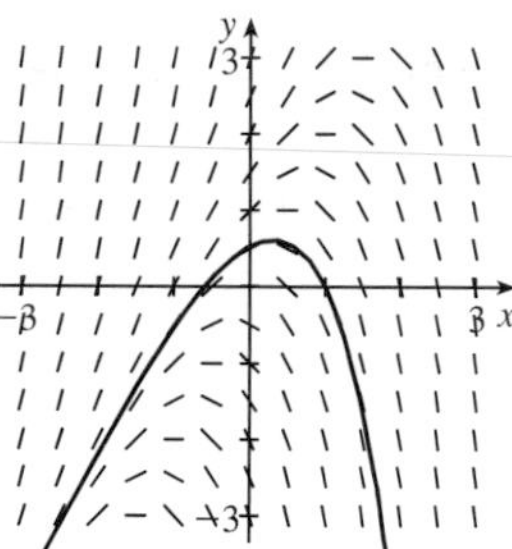

12.

x	y	$y' = xy - x^2$
2	3	2
-2	-3	2
± 2	0	-4
0	0	0
2	2	0

$y' = xy - x^2 = x(y - x)$, so $y' = 0$ for $x = 0$ and $y = x$. The slopes are positive only in the regions in quadrants I and III that are bounded by $x = 0$ and $y = x$. The solution curve in the graph passes through $(0, 1)$.

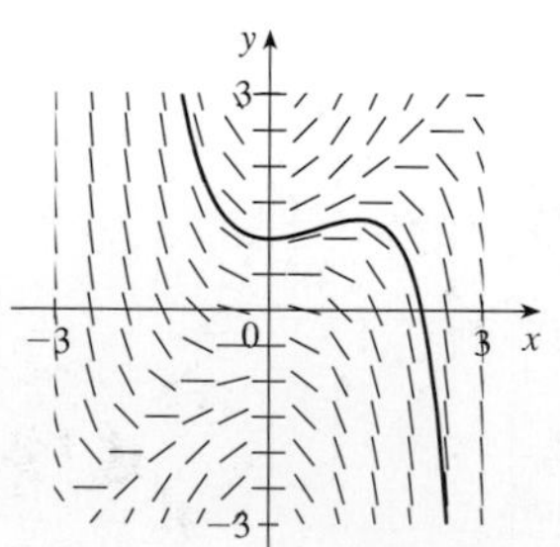

13.

x	y	$y' = y + xy$
0	± 2	± 2
1	± 2	± 4
-3	± 2	∓ 4

Note that $y' = y(x+1) = 0$ for any point on $y = 0$ or on $x = -1$. The slopes are positive when the factors y and $x + 1$ have the same sign and negative when they have opposite signs. The solution curve in the graph passes through $(0, 1)$.

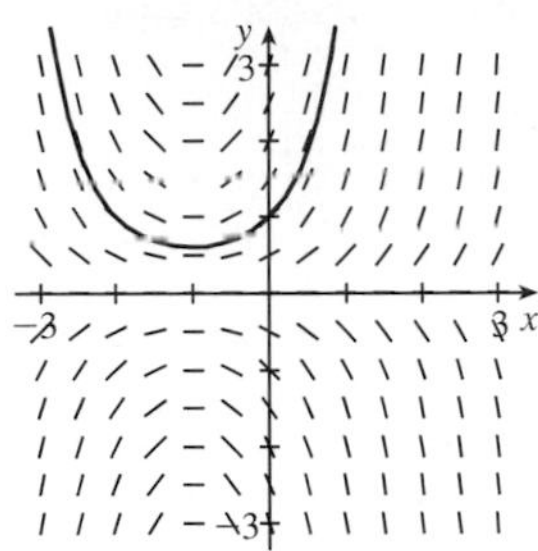

14.

x	y	$y' = x + y^2$
-2	± 1	-1
-2	± 2	2
2	± 1	3
0	± 2	4
0	0	0

Note that $y' = x + y^2 = 0$ only on the parabola $x = -y^2$. The slopes are positive "outside" $x = -y^2$ and negative "inside" $x = -y^2$. The solution curve in the graph passes through $(0, 0)$.

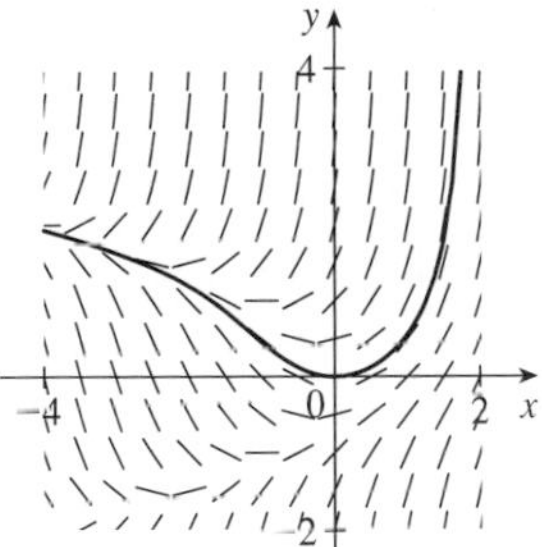

15. $y' = x^2 y - \frac{1}{2} y^2$ and $y(0) = 1$.

In Maple, use the following commands to obtain a similar figure.

```
with(DEtools):
ODE:=diff(y(x),x)-x^2*y(x)-(1/2)*y(x)^2;
ivs:=[y(0)=1];
DEplot({ODE},y(x),x=-3..2,y=0..4,ivs,linecolor=black);
```

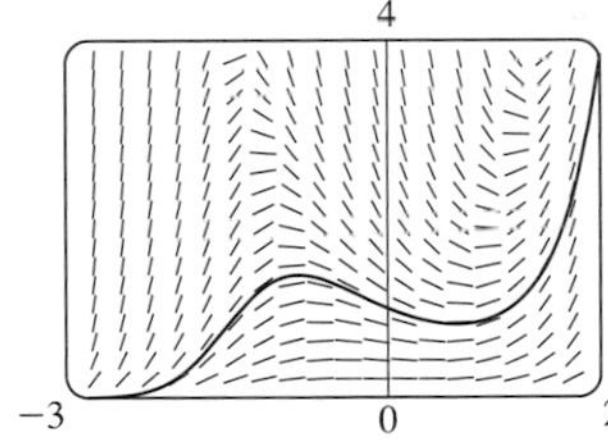

16. $y' = \cos(x + y)$ and $y(0) = 1$.

In Maple, use the following commands to obtain a similar figure.

```
with(DEtools):
ODE:=diff(y(x),x)=cos(x+y(x));
ivs:=[y(0)=1];
DEplot({ODE},y(x),x=-1.5*Pi..1.5*Pi,y=-1.5*Pi..1.5*Pi,
ivs,linecolor=black);
```

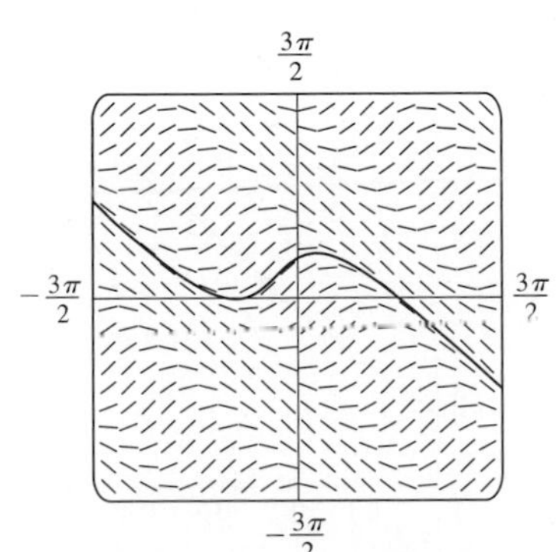

17.

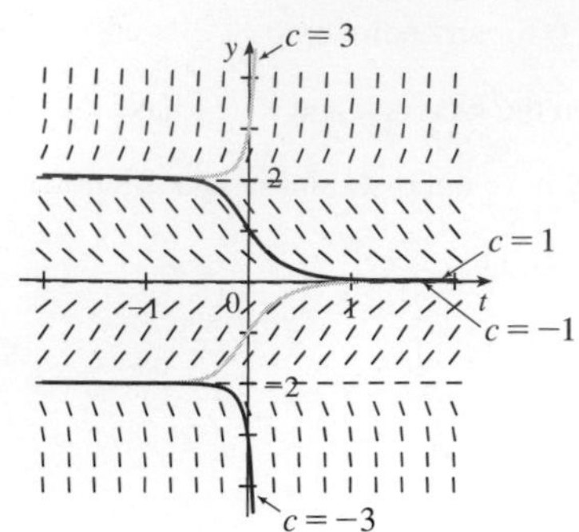

The direction field is for the differential equation $y' = y^3 - 4y$.

$L = \lim_{t \to \infty} y(t)$ exists for $-2 \le c \le 2$;

$L = \pm 2$ for $c = \pm 2$ and $L = 0$ for $-2 < c < 2$.

For other values of c, L does not exist.

18.

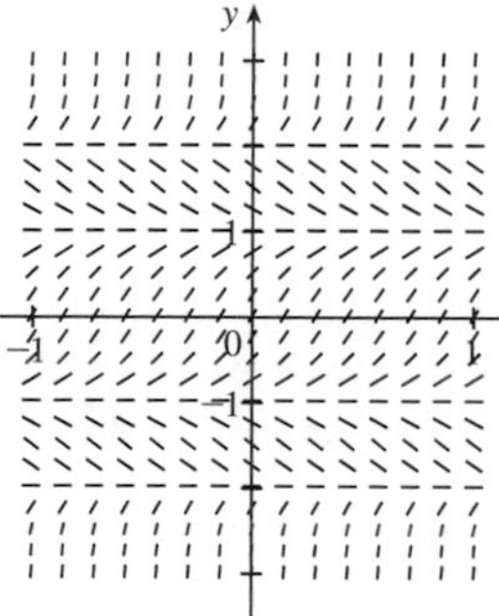

Note that when $f(y) = 0$ on the graph in the text, we have $y' = f(y) = 0$; so we get horizontal segments at $y = \pm 1, \pm 2$. We get segments with negative slopes only for $1 < |y| < 2$. All other segments have positive slope. For the limiting behavior of solutions:

- If $y(0) > 2$, then $\lim_{t \to \infty} y = \infty$ and $\lim_{t \to -\infty} y = 2$.
- If $1 < y(0) < 2$, then $\lim_{t \to \infty} y = 1$ and $\lim_{t \to -\infty} y = 2$.
- If $-1 < y(0) < 1$, then $\lim_{t \to \infty} y = 1$ and $\lim_{t \to -\infty} y = -1$.
- If $-2 < y(0) < -1$, then $\lim_{t \to \infty} y = -2$ and $\lim_{t \to -\infty} y = -1$.
- If $y < -2$, then $\lim_{t \to \infty} y = -2$ and $\lim_{t \to -\infty} y = -\infty$.

19. (a) $y' = F(x, y) = y$ and $y(0) = 1 \quad \Rightarrow \quad x_0 = 0,\ y_0 = 1$.

(i) $h = 0.4$ and $y_1 = y_0 + hF(x_0, y_0) \quad \Rightarrow \quad y_1 = 1 + 0.4 \cdot 1 = 1.4$. $x_1 = x_0 + h = 0 + 0.4 = 0.4$, so $y_1 = y\,(0.4) = 1.4$.

(ii) $h = 0.2 \quad \Rightarrow \quad x_1 = 0.2$ and $x_2 = 0.4$, so we need to find y_2.

$y_1 = y_0 + hF(x_0, y_0) = 1 + 0.2y_0 = 1 + 0.2 \cdot 1 = 1.2$,

$y_2 = y_1 + hF(x_1, y_1) = 1.2 + 0.2y_1 = 1.2 + 0.2 \cdot 1.2 = 1.44$.

(iii) $h = 0.1 \quad \Rightarrow \quad x_4 = 0.4$, so we need to find y_4. $y_1 = y_0 + hF(x_0, y_0) = 1 + 0.1y_0 = 1 + 0.1 \cdot 1 = 1.1$,

$y_2 = y_1 + hF(x_1, y_1) = 1.1 + 0.1y_1 = 1.1 + 0.1 \cdot 1.1 = 1.21$,

$y_3 = y_2 + hF(x_2, y_2) = 1.21 + 0.1y_2 = 1.21 + 0.1 \cdot 1.21 = 1.331$,

$y_4 = y_3 + hF(x_3, y_3) = 1.331 + 0.1y_3 = 1.331 + 0.1 \cdot 1.331 = 1.4641$.

(b)

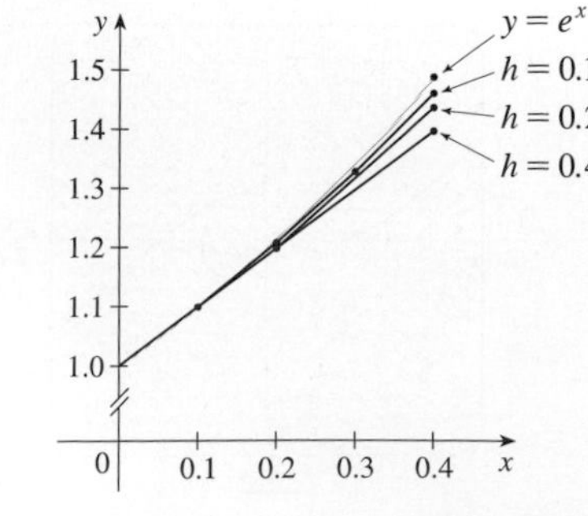

We see that the estimates are underestimates since they are all below the graph of $y = e^x$.

(c) (i) For $h = 0.4$: (exact value) − (approximate value) $= e^{0.4} - 1.4 \approx 0.0918$

(ii) For $h = 0.2$: (exact value) − (approximate value) $= e^{0.4} - 1.44 \approx 0.0518$

(iii) For $h = 0.1$: (exact value) − (approximate value) $= e^{0.4} - 1.4641 \approx 0.0277$

Each time the step size is halved, the error estimate also appears to be halved (approximately).

20. 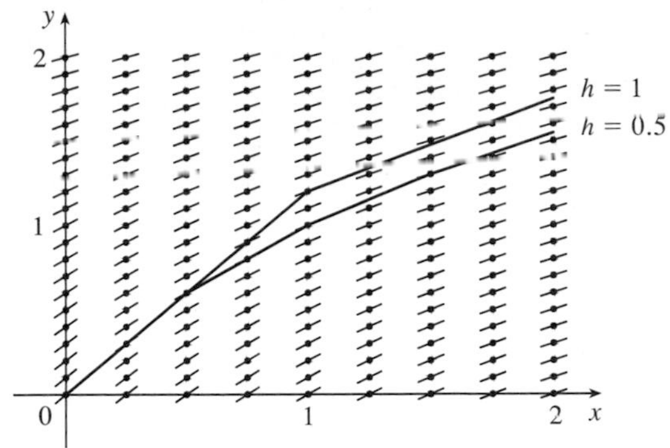

As x increases, the slopes decrease and all of the estimates are above the true values. Thus, all of the estimates are overestimates.

21. $h = 0.5$, $x_0 = 1$, $y_0 = 0$, and $F(x, y) = y - 2x$.

Note that $x_1 = x_0 + h = 1 + 0.5 = 1.5$, $x_2 = 2$, and $x_3 = 2.5$.

$y_1 = y_0 + hF(x_0, y_0) = 0 + 0.5F(1, 0) = 0.5[0 - 2(1)] = -1.$

$y_2 = y_1 + hF(x_1, y_1) = -1 + 0.5F(1.5, -1) = -1 + 0.5[-1 - 2(1.5)] = -3.$

$y_3 = y_2 + hF(x_2, y_2) = -3 + 0.5F(2, -3) = -3 + 0.5[-3 - 2(2)] = -6.5.$

$y_4 = y_3 + hF(x_3, y_3) = -6.5 + 0.5F(2.5, -6.5) = -6.5 + 0.5[-6.5 - 2(2.5)] = -12.25.$

22. $h = 0.2$, $x_0 = 0$, $y_0 = 1$, and $F(x, y) = x^2 y - \frac{1}{2}y^2$. Note that $x_1 = x_0 + h = 0 + 0.2 = 0.2$, $x_2 = 0.4$, $x_3 = 0.6$, $x_4 = 0.8$, and $x_5 = 1$.

$y_1 = y_0 + hF(x_0, y_0) = 1 + 0.2F(0, 1) = 1 + 0.2\left[0^2(1) - \frac{1}{2}(1)^2\right] = 1 + 0.2(-\frac{1}{2}) = 0.9.$

$y_2 = y_1 + hF(x_1, y_1) = 0.9 + 0.2F(0.2, 0.9) = 0.9 + 0.2\left[(0.2)^2(0.9) - \frac{1}{2}(0.9)^2\right] = 0.8262.$

$y_3 = y_2 + hF(x_2, y_2) = 0.8262 + 0.2F(0.4, 0.8262) = 0.8262 + 0.2\left[(0.4)^2(0.8262) - \frac{1}{2}(0.8262)^2\right] = 0.784377756.$

$y_4 = y_3 + hF(x_3, y_3) = 0.784377756 + 0.2F(0.6, 0.784377756) \approx 0.779328108.$

$y_5 = y_4 + hF(x_4, y_4) \approx 0.779328108 + 0.2F(0.8, 0.779328108) \approx 0.818346876.$

Thus, $y(1) \approx 0.8183$.

23. $h = 0.1$, $x_0 = 0$, $y_0 = 1$, and $F(x, y) = y + xy$.

Note that $x_1 = x_0 + h = 0 + 0.1 = 0.1$, $x_2 = 0.2$, $x_3 = 0.3$, and $x_4 = 0.4$.

$y_1 = y_0 + hF(x_0, y_0) = 1 + 0.1F(0, 1) = 1 + 0.1[1 + (0)(1)] = 1.1.$

$y_2 = y_1 + hF(x_1, y_1) = 1.1 + 0.1F(0.1, 1.1) = 1.1 + 0.1[1.1 + (0.1)(1.1)] = 1.221.$

$y_3 = y_2 + hF(x_2, y_2) = 1.221 + 0.1F(0.2, 1.221) = 1.221 + 0.1[1.221 + (0.2)(1.221)] = 1.36752.$

$$y_4 = y_3 + hF(x_3, y_3) = 1.36752 + 0.1F(0.3, 1.36752) = 1.36752 + 0.1[1.36752 + (0.3)(1.36752)] = 1.5452976.$$

$$y_5 = y_4 + hF(x_4, y_4) = 1.5452976 + 0.1F(0.4, 1.5452976) = 1.5452976 + 0.1[1.5452976 + (0.4)(1.5452976)] = 1.761639264.$$

Thus, $y(0.5) \approx 1.7616$.

24. (a) $h = 0.2$, $x_0 = 0$, $y_0 = 0$, and $F(x, y) = \cos(x + y)$. Note that $x_1 = x_0 + h = 0 + 0.2 = 0.2$, $x_2 = 0.4$, and $x_3 = 0.6$.

$y_1 = y_0 + hF(x_0, y_0) = 0 + 0.2F(0, 0) = 0.2\cos(0 + 0) = 0.2(1) = 0.2.$

$y_2 = y_1 + hF(x_1, y_1) = 0.2 + 0.2F(0.2, 0.2) = 0.2 + 0.2\cos(0.4) \approx 0.3842121988.$

$y_3 = y_2 + hF(x_2, y_2) \approx 0.3842 + 0.2F(0.4, 0.3842) \approx 0.5258011763.$

Thus, $y(0.6) \approx 0.5258$.

(b) Now use $h = 0.1$. For $1 \leq n \leq 6$, $x_n = 0.n$.

$y_1 = y_0 + hF(x_0, y_0) = 0 + 0.1\cos(0 + 0) = 0.1(1) = 0.1.$

$y_2 = y_1 + hF(x_1, y_1) = 0.1 + 0.1\cos(0.2) \approx 0.1980.$

$y_3 = y_2 + hF(x_2, y_2) \approx 0.1980 + 0.1\cos(0.3980) \approx 0.2902.$

$y_4 = y_3 + hF(x_3, y_3) \approx 0.2902 + 0.1\cos(0.5902) \approx 0.3733.$

$y_5 = y_4 + hF(x_4, y_4) \approx 0.3733 + 0.1\cos(0.7733) \approx 0.4448.$

$y_6 = y_5 + hF(x_5, y_5) \approx 0.4448 + 0.1\cos(0.9448) \approx 0.5034.$

Thus, $y(0.6) \approx 0.5034$.

25. (a) $dy/dx + 3x^2y = 6x^2 \quad \Rightarrow \quad y' = 6x^2 - 3x^2y$. Store this expression in Y_1 and use the following simple program to evaluate $y(1)$ for each part, using $\text{H} = h = 1$ and $\text{N} = 1$ for part (i), $\text{H} = 0.1$ and $\text{N} = 10$ for part (ii), and so forth.

$h \to$ H: $0 \to$ X: $3 \to$ Y:

For(I, 1, N): Y + H × Y_1 → Y: X + H → X:

End(loop):

Display Y. [To see all iterations, include this statement in the loop.]

(i) $\text{H} = 1, \text{N} = 1 \quad \Rightarrow \quad y(1) = 3$

(ii) $\text{H} = 0.1, \text{N} = 10 \quad \Rightarrow \quad y(1) \approx 2.3928$

(iii) $\text{H} = 0.01, \text{N} = 100 \quad \Rightarrow \quad y(1) \approx 2.3701$

(iv) $\text{H} = 0.001, \text{N} = 1000 \quad \Rightarrow \quad y(1) \approx 2.3681$

(b) $y = 2 + e^{-x^3} \quad \Rightarrow \quad y' = -3x^2e^{-x^3}$

$$\text{LHS} = y' + 3x^2y = -3x^2e^{-x^3} + 3x^2\left(2 + e^{-x^3}\right) = -3x^2e^{-x^3} + 6x^2 + 3x^2e^{-x^3} = 6x^2 = \text{RHS}$$

$y(0) = 2 + e^{-0} = 2 + 1 = 3$

(c) The exact value of $y(1)$ is $2 + e^{-1^3} = 2 + e^{-1}$.

(i) For $h = 1$: (exact value) − (approximate value) $= 2 + e^{-1} - 3 \approx -0.6321$

(ii) For $h = 0.1$: (exact value) − (approximate value) $= 2 + e^{-1} - 2.3928 \approx -0.0249$

(iii) For $h = 0.01$: (exact value) − (approximate value) $= 2 + e^{-1} - 2.3701 \approx -0.0022$

(iv) For $h = 0.001$: (exact value) − (approximate value) $= 2 + e^{-1} - 2.3681 \approx -0.0002$

In (ii)–(iv), it seems that when the step size is divided by 10, the error estimate is also divided by 10 (approximately).

26. (a) We use the program from the solution to Exercise 25 with $\text{Y}_1 = x^3 - y^3$, $\text{H} = 0.01$, and $\text{N} = \frac{2-0}{0.01} = 200$. With $(x_0, y_0) = (0, 1)$, we get $y(2) \approx 1.9000$.

(b)

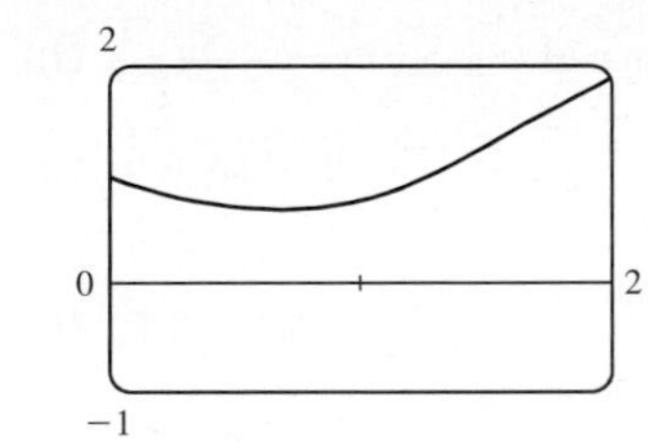

Notice from the graph that $y(2) \approx 1.9$, which serves as a check on our calculation in part (a).

27. (a) $R\dfrac{dQ}{dt} + \dfrac{1}{C}Q = E(t)$ becomes $5Q' + \dfrac{1}{0.05}Q = 60$ or $Q' + 4Q = 12$.

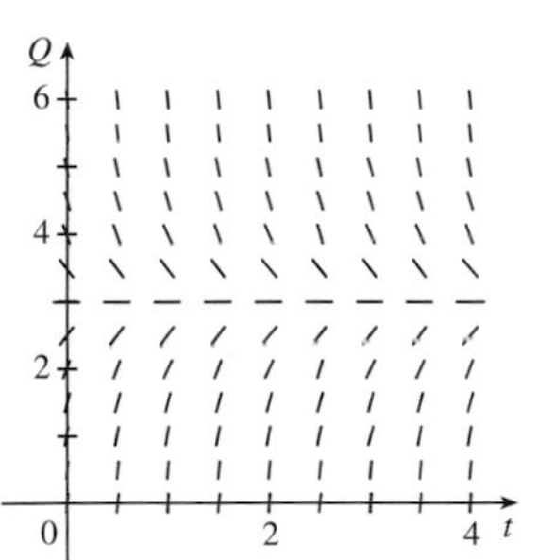

(b) From the graph, it appears that the limiting value of the charge Q is about 3.

(c) If $Q' = 0$, then $4Q = 12 \Rightarrow Q = 3$ is an equilibrium solution.

(d)

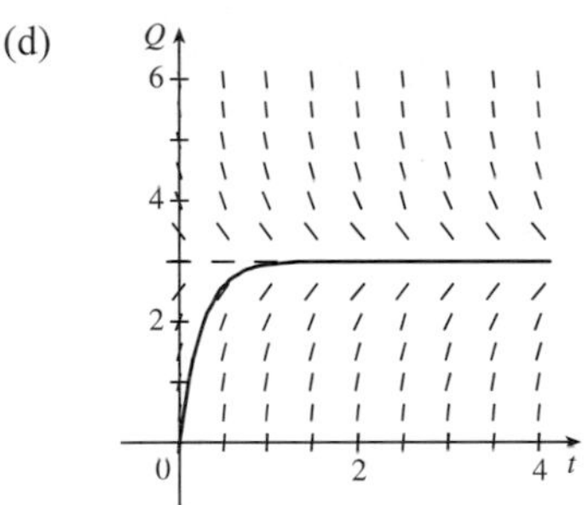

(e) $Q' + 4Q - 12 \Rightarrow Q' = 12 - 4Q$. Now $Q(0) = 0$, so $t_0 = 0$ and $Q_0 = 0$.

$$Q_1 = Q_0 + hF(t_0, Q_0) = 0 + 0.1(12 - 4 \cdot 0) = 1.2$$
$$Q_2 = Q_1 + hF(t_1, Q_1) = 1.2 + 0.1(12 - 4 \cdot 1.2) = 1.92$$
$$Q_3 = Q_2 + hF(t_2, Q_2) = 1.92 + 0.1(12 - 4 \cdot 1.92) = 2.352$$
$$Q_4 = Q_3 + hF(t_3, Q_3) = 2.352 + 0.1(12 - 4 \cdot 2.352) = 2.6112$$
$$Q_5 = Q_4 + hF(t_4, Q_4) = 2.6112 + 0.1(12 - 4 \cdot 2.6112) = 2.76672$$

Thus, $Q_5 = Q(0.5) \approx 2.77$ C.

28. (a) From Exercise 9.1.14, we have $dy/dt = k(y - R)$. We are given that $R = 20°\text{C}$ and $dy/dt = -1°\text{C/min}$ when $y = 70°\text{C}$. Thus, $-1 = k(70 - 20) \Rightarrow k = -\frac{1}{50}$ and the differential equation becomes $dy/dt = -\frac{1}{50}(y - 20)$.

(b)

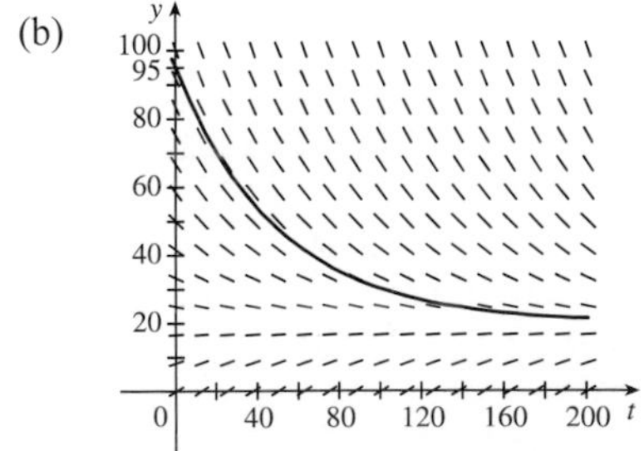

The limiting value of the temperature is 20°C; that is, the temperature of the room.

(c) From part (a), $dy/dt = -\frac{1}{50}(y-20)$. With $t_0 = 0$, $y_0 = 95$, and $h = 2$ min, we get

$$\begin{aligned}
y_1 &= y_0 + hF(t_0, y_0) = 95 + 2\left[-\tfrac{1}{50}(95-20)\right] = 92 \\
y_2 &= y_1 + hF(t_1, y_1) = 92 + 2\left[-\tfrac{1}{50}(92-20)\right] = 89.12 \\
y_3 &= y_2 + hF(t_2, y_2) = 89.12 + 2\left[-\tfrac{1}{50}(89.12-20)\right] = 86.3552 \\
y_4 &= y_3 + hF(t_3, y_3) = 86.3552 + 2\left[-\tfrac{1}{50}(86.3552-20)\right] = 83.700992 \\
y_5 &= y_4 + hF(t_4, y_4) = 83.700992 + 2\left[-\tfrac{1}{50}(83.700992-20)\right] = 81.15295232
\end{aligned}$$

Thus, $y(10) \approx 81.15°\text{C}$.

9.3 Separable Equations

1. $\dfrac{dy}{dx} = 3x^2y^2 \;\Rightarrow\; \dfrac{dy}{y^2} = 3x^2\,dx \;\; [y \neq 0] \;\Rightarrow\; \int y^{-2}\,dy = \int 3x^2\,dx \;\Rightarrow\; -y^{-1} = x^3 + C \;\Rightarrow$

$\dfrac{-1}{y} = x^3 + C \;\Rightarrow\; y = \dfrac{-1}{x^3 + C}$. $y = 0$ is also a solution.

2. $\dfrac{dy}{dx} = x\sqrt{y} \;\Rightarrow\; \dfrac{dy}{\sqrt{y}} = x\,dx \;\; [y \neq 0] \;\Rightarrow\; \int y^{-1/2}\,dy = \int x\,dx \;\Rightarrow\; 2y^{1/2} = \frac{1}{2}x^2 + K \;\Rightarrow$

$\sqrt{y} = \frac{1}{4}x^2 + \frac{1}{2}K \;\Rightarrow\; y = \left(\frac{1}{4}x^2 + C\right)^2$, where $C = \frac{1}{2}K$. $y = 0$ is also a solution.

3. $xyy' = x^2 + 1 \;\Rightarrow\; xy\,\dfrac{dy}{dx} = x^2 + 1 \;\Rightarrow\; y\,dy = \dfrac{x^2+1}{x}\,dx \;\; [x \neq 0] \;\Rightarrow\; \displaystyle\int y\,dy = \int\left(x + \frac{1}{x}\right)dx \;\Rightarrow$

$\frac{1}{2}y^2 = \frac{1}{2}x^2 + \ln|x| + K \;\Rightarrow\; y^2 = x^2 + 2\ln|x| + 2K \;\Rightarrow\; y = \pm\sqrt{x^2 + 2\ln|x| + C}$, where $C = 2K$.

4. $y' + xe^y = 0 \;\Rightarrow\; \dfrac{dy}{dx} = -xe^y \;\Rightarrow\; e^{-y}\,dy = -x\,dx \;\Rightarrow\; \int e^{-y}\,dy = \int -x\,dx \;\Rightarrow\; -e^{-y} = -\frac{1}{2}x^2 + C \;\Rightarrow$

$e^{-y} = \frac{1}{2}x^2 - C \;\Rightarrow\; -y = \ln\left(\frac{1}{2}x^2 - C\right) \;\Rightarrow\; y = -\ln\left(\frac{1}{2}x^2 - C\right)$

5. $(e^y - 1)y' = 2 + \cos x \;\Rightarrow\; (e^y - 1)\dfrac{dy}{dx} = 2 + \cos x \;\Rightarrow\; (e^y - 1)\,dy = (2 + \cos x)\,dx \;\Rightarrow$

$\int(e^y - 1)\,dy = \int(2 + \cos x)\,dx \;\Rightarrow\; e^y - y = 2x + \sin x + C$. We cannot solve explicitly for y.

6. $\dfrac{du}{dt} = \dfrac{1 + t^4}{ut^2 + u^4t^2} \;\Rightarrow\; \dfrac{du}{dt} = \dfrac{1 + t^4}{t^2(u + u^4)} \;\Rightarrow\; (u + u^4)\,du = \dfrac{1 + t^4}{t^2}\,dt \;\Rightarrow\; \int(u + u^4)\,du = \int(t^{-2} + t^2)\,dt \;\Rightarrow$

$\dfrac{1}{2}u^2 + \dfrac{1}{5}u^5 = -\dfrac{1}{t} + \dfrac{1}{3}t^3 + C$. We cannot solve explicitly for u.

7. $\dfrac{d\theta}{dt} = \dfrac{t\sec\theta}{\theta e^{t^2}} \;\Rightarrow\; \theta\cos\theta\,d\theta = te^{-t^2}\,dt \;\Rightarrow\; \int\theta\cos\theta\,d\theta = \int te^{-t^2}\,dt \;\Rightarrow$

$\theta\sin\theta + \cos\theta = -\frac{1}{2}e^{-t^2} + C$ [by parts]. We cannot solve explicitly for θ.

8. $\dfrac{dH}{dR} = \dfrac{RH^2\sqrt{1 + R^2}}{\ln H} \;\Rightarrow\; \dfrac{\ln H}{H^2}\,dH = R\sqrt{1 + R^2}\,dR \;\Rightarrow\; \displaystyle\int\frac{\ln H}{H^2}\,dH = \int R(1 + R^2)^{1/2}\,dR \;\Rightarrow$

$-\dfrac{\ln H}{H} - \dfrac{1}{H} = \dfrac{1}{3}(1 + R^2)^{3/2} + C$ [by parts]. We cannot solve explicitly for H.

9. $\frac{dp}{dt} = t^2p - p + t^2 - 1 = p(t^2-1) + 1(t^2-1) = (p+1)(t^2-1) \Rightarrow \frac{1}{p+1}\,dp = (t^2-1)\,dt \Rightarrow$

$\int \frac{1}{p+1}\,dp = \int (t^2-1)\,dt \Rightarrow \ln|p+1| = \frac{1}{3}t^3 - t + C \Rightarrow |p+1| = e^{t^3/3-t+C} \Rightarrow p+1 = \pm e^C e^{t^3/3-t} \Rightarrow$

$p = Ke^{t^3/3-t} - 1$, where $K = \pm e^C$. Since $p = -1$ is also a solution, K can equal 0, and hence, K can be any real number.

10. $\frac{dz}{dt} + e^{t+z} = 0 \Rightarrow \frac{dz}{dt} = -e^t e^z \Rightarrow \int e^{-z}\,dz = -\int e^t\,dt \Rightarrow -e^{-z} = -e^t + C \Rightarrow e^{-z} = e^t - C \Rightarrow$

$\frac{1}{e^z} = e^t - C \Rightarrow e^z = \frac{1}{e^t - C} \Rightarrow z = \ln\left(\frac{1}{e^t - C}\right) \Rightarrow z = -\ln(e^t - C)$

11. $\frac{dy}{dx} = xe^y \Rightarrow e^{-y}\,dy = x\,dx \Rightarrow \int e^{-y}\,dy = \int x\,dx \Rightarrow -e^{-y} = \frac{1}{2}x^2 + C.$

$y(0) = 0 \Rightarrow -e^{-0} = \frac{1}{2}(0)^2 + C \Rightarrow C = -1$, so $-e^{-y} = \frac{1}{2}x^2 - 1 \Rightarrow e^{-y} = -\frac{1}{2}x^2 + 1 \Rightarrow$

$-y = \ln(1 - \frac{1}{2}x^2) \Rightarrow y = -\ln(1 - \frac{1}{2}x^2).$

12. $\frac{dy}{dx} = \frac{x\sin x}{y} \Rightarrow y\,dy = x\sin x\,dx \Rightarrow \int y\,dy = \int x\sin x\,dx \Rightarrow \frac{1}{2}y^2 = -x\cos x + \sin x + C$ [by parts].

$y(0) = -1 \Rightarrow \frac{1}{2}(-1)^2 = -0\cos 0 + \sin 0 + C \Rightarrow C = \frac{1}{2}$, so $\frac{1}{2}y^2 = -x\cos x + \sin x + \frac{1}{2} \Rightarrow$

$y^2 = -2x\cos x + 2\sin x + 1 \Rightarrow y = -\sqrt{-2x\cos x + 2\sin x + 1}$ since $y(0) = -1 < 0$.

13. $\frac{du}{dt} = \frac{2t + \sec^2 t}{2u}$, $u(0) = -5$. $\int 2u\,du = \int (2t + \sec^2 t)\,dt \Rightarrow u^2 = t^2 + \tan t + C$,

where $[u(0)]^2 = 0^2 + \tan 0 + C \Rightarrow C = (-5)^2 = 25$. Therefore, $u^2 = t^2 + \tan t + 25$, so $u = \pm\sqrt{t^2 + \tan t + 25}$.

Since $u(0) = -5 < 0$, we must have $u = -\sqrt{t^2 + \tan t + 25}$.

14. $x + 3y^2\sqrt{x^2+1}\,\frac{dy}{dx} = 0 \Rightarrow 3y^2\sqrt{x^2+1}\,\frac{dy}{dx} = -x \rightarrow 3y^2\,dy = \frac{-x}{\sqrt{x^2+1}}\,dx \Rightarrow$

$\int 3y^2\,dy = \int -x(x^2+1)^{-1/2}\,dx \Rightarrow y^3 = -(x^2+1)^{1/2} + C$. $y(0) = 1 \Rightarrow 1^3 = -(0^2+1)^{1/2} + C \Rightarrow$

$C = 2$, so $y^3 = -(x^2+1)^{1/2} + 2 \Rightarrow y = (2 - \sqrt{x^2+1})^{1/3}$.

15. $x\ln x = y\left(1 + \sqrt{3+y^2}\right)y'$, $y(1) = 1$. $\int x\ln x\,dx = \int \left(y + y\sqrt{3+y^2}\right)\,dy \Rightarrow \frac{1}{2}x^2\ln x - \int \frac{1}{2}x\,dx$

[use parts with $u = \ln x$, $dv = xdx$] $= \frac{1}{2}y^2 + \frac{1}{3}(3+y^2)^{3/2} \Rightarrow \frac{1}{2}x^2\ln x - \frac{1}{4}x^2 + C = \frac{1}{2}y^2 + \frac{1}{3}(3+y^2)^{3/2}$.

Now $y(1) = 1 \Rightarrow 0 - \frac{1}{4} + C = \frac{1}{2} + \frac{1}{3}(4)^{3/2} \Rightarrow C = \frac{1}{2} + \frac{8}{3} + \frac{1}{4} = \frac{41}{12}$, so

$\frac{1}{2}x^2\ln x - \frac{1}{4}x^2 + \frac{41}{12} = \frac{1}{2}y^2 + \frac{1}{3}(3+y^2)^{3/2}$. We do not solve explicitly for y.

16. $\frac{dP}{dt} = \sqrt{Pt} \Rightarrow dP/\sqrt{P} = \sqrt{t}\,dt \Rightarrow \int P^{-1/2}\,dP = \int t^{1/2}\,dt \Rightarrow 2P^{1/2} = \frac{2}{3}t^{3/2} + C.$

$P(1) = 2 \Rightarrow 2\sqrt{2} = \frac{2}{3} + C \Rightarrow C = 2\sqrt{2} - \frac{2}{3}$, so $2P^{1/2} = \frac{2}{3}t^{3/2} + 2\sqrt{2} - \frac{2}{3} \Rightarrow \sqrt{P} = \frac{1}{3}t^{3/2} + \sqrt{2} - \frac{1}{3} \Rightarrow$

$P = \left(\frac{1}{3}t^{3/2} + \sqrt{2} - \frac{1}{3}\right)^2.$

17. $y' \tan x = a + y,\ 0 < x < \pi/2 \quad \Rightarrow \quad \dfrac{dy}{dx} = \dfrac{a+y}{\tan x} \quad \Rightarrow \quad \dfrac{dy}{a+y} = \cot x\, dx \quad [a + y \neq 0] \quad \Rightarrow$

$\displaystyle\int \frac{dy}{a+y} = \int \frac{\cos x}{\sin x}\, dx \quad \Rightarrow \quad \ln|a+y| = \ln|\sin x| + C \quad \Rightarrow \quad |a+y| = e^{\ln|\sin x| + C} = e^{\ln|\sin x|} \cdot e^C = e^C\,|\sin x| \quad \Rightarrow$

$a + y = K \sin x$, where $K = \pm e^C$. (In our derivation, K was nonzero, but we can restore the excluded case

$y = -a$ by allowing K to be zero.) $\quad y(\pi/3) = a \quad \Rightarrow \quad a + a = K \sin\left(\dfrac{\pi}{3}\right) \quad \Rightarrow \quad 2a = K\dfrac{\sqrt{3}}{2} \quad \Rightarrow \quad K = \dfrac{4a}{\sqrt{3}}.$

Thus, $a + y = \dfrac{4a}{\sqrt{3}} \sin x$ and so $y = \dfrac{4a}{\sqrt{3}} \sin x - a$.

18. $\dfrac{dL}{dt} = kL^2 \ln t \quad \Rightarrow \quad \dfrac{dL}{L^2} = k \ln t\, dt \quad \Rightarrow \quad \displaystyle\int \frac{dL}{L^2} = \int k \ln t\, dt \quad \Rightarrow \quad -\frac{1}{L} = kt \ln t - \int k\, dt$

[by parts with $u = \ln t$, $dv = k\, dt$] $\Rightarrow \quad -\dfrac{1}{L} = kt \ln t - kt + C \quad \Rightarrow \quad L = \dfrac{1}{kt - kt \ln t - C}.$

$L(1) = -1 \quad \Rightarrow \quad -1 = \dfrac{1}{k - k \ln 1 - C} \quad \Rightarrow \quad C - k = 1 \quad \Rightarrow \quad C = k + 1$. Thus, $L = \dfrac{1}{kt - kt \ln t - k - 1}.$

19. $\dfrac{dy}{dx} = \dfrac{x}{y} \quad \Rightarrow \quad y\, dy = x\, dx \quad \Rightarrow \quad \int y\, dy = \int x\, dx \quad \Rightarrow \quad \frac{1}{2}y^2 = \frac{1}{2}x^2 + C. \quad y(0) = 2 \quad \Rightarrow \quad \frac{1}{2}(2)^2 = \frac{1}{2}(0)^2 + C \quad \Rightarrow$

$C = 2$, so $\frac{1}{2}y^2 = \frac{1}{2}x^2 + 2 \quad \Rightarrow \quad y^2 = x^2 + 4 \quad \Rightarrow \quad y = \sqrt{x^2 + 4}$ since $y(0) = 2 > 0$.

20. $f'(x) = x\, f(x) - x \quad \Rightarrow \quad \dfrac{dy}{dx} = xy - x \quad \Rightarrow \quad \dfrac{dy}{dx} = x(y-1) \quad \Rightarrow \quad \dfrac{dy}{y-1} = x\, dx \quad [y \neq 1] \quad \Rightarrow$

$\displaystyle\int \frac{dy}{y-1} = \int x\, dx \quad \Rightarrow \quad \ln|y-1| = \tfrac{1}{2}x^2 + C. \quad f(0) = 2 \quad \Rightarrow \quad \ln|2-1| = \tfrac{1}{2}(0)^2 + C \quad \Rightarrow \quad C = 0$, so

$\ln|y-1| = \frac{1}{2}x^2 \quad \Rightarrow \quad |y-1| = e^{x^2/2} \quad \Rightarrow \quad y - 1 = e^{x^2/2}$ [since $f(0) = 2$] $\Rightarrow \quad y = e^{x^2/2} + 1$.

21. $u = x + y \quad \Rightarrow \quad \dfrac{d}{dx}(u) = \dfrac{d}{dx}(x+y) \quad \Rightarrow \quad \dfrac{du}{dx} = 1 + \dfrac{dy}{dx}$, but $\dfrac{dy}{dx} = x + y = u$, so $\dfrac{du}{dx} = 1 + u \quad \Rightarrow$

$\dfrac{du}{1+u} = dx \quad [u \neq -1] \quad \Rightarrow \quad \displaystyle\int \frac{du}{1+u} = \int dx \quad \Rightarrow \quad \ln|1+u| = x + C \quad \Rightarrow \quad |1+u| = e^{x+C} \quad \Rightarrow$

$1 + u = \pm e^C e^x \quad \Rightarrow \quad u = \pm e^C e^x - 1 \quad \Rightarrow \quad x + y = \pm e^C e^x - 1 \quad \Rightarrow \quad y = Ke^x - x - 1$, where $K = \pm e^C \neq 0$.

If $u = -1$, then $-1 = x + y \quad \Rightarrow \quad y = -x - 1$, which is just $y = Ke^x - x - 1$ with $K = 0$. Thus, the general solution is $y = Ke^x - x - 1$, where $K \in \mathbb{R}$.

22. $xy' = y + xe^{y/x} \quad \Rightarrow \quad y' = y/x + e^{y/x} \quad \Rightarrow \quad \dfrac{dy}{dx} = v + e^v$. Also, $v = y/x \quad \Rightarrow \quad xv = y \quad \Rightarrow \quad \dfrac{dy}{dx} = x\dfrac{dv}{dx} + v$,

so $v + e^v = x\dfrac{dv}{dx} + v \quad \Rightarrow \quad \dfrac{dv}{e^v} = \dfrac{dx}{x} \quad [x \neq 0] \quad \Rightarrow \quad \displaystyle\int \frac{dv}{e^v} = \int \frac{dx}{x} \quad \Rightarrow \quad -e^{-v} = \ln|x| + C \quad \Rightarrow$

$e^{-v} = -\ln|x| - C \quad \Rightarrow \quad -v = \ln(-\ln|x| - C) \quad \Rightarrow \quad y/x = -\ln(-\ln|x| - C) \quad \Rightarrow \quad y = -x \ln(-\ln|x| - C).$

23. (a) $y' = 2x\sqrt{1-y^2} \Rightarrow \dfrac{dy}{dx} = 2x\sqrt{1-y^2} \Rightarrow \dfrac{dy}{\sqrt{1-y^2}} = 2x\,dx \Rightarrow \displaystyle\int \frac{dy}{\sqrt{1-y^2}} = \int 2x\,dx \Rightarrow$

$\sin^{-1} y = x^2 + C$ for $-\frac{\pi}{2} \le x^2 + C \le \frac{\pi}{2}$.

(b) $y(0) = 0 \Rightarrow \sin^{-1} 0 = 0^2 + C \Rightarrow C = 0$,

so $\sin^{-1} y = x^2$ and $y = \sin(x^2)$ for $-\sqrt{\pi/2} \le x \le \sqrt{\pi/2}$.

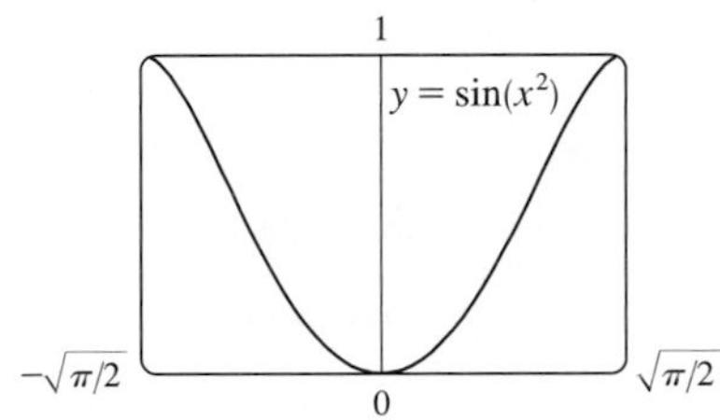

(c) For $\sqrt{1-y^2}$ to be a real number, we must have $-1 \le y \le 1$; that is, $-1 \le y(0) \le 1$. Thus, the initial-value problem $y' = 2x\sqrt{1-y^2}$, $y(0) = 2$ does *not* have a solution.

24. $e^{-y}y' + \cos x = 0 \Leftrightarrow \int e^{-y}\,dy = -\int \cos x\,dx \Leftrightarrow -e^{-y} = -\sin x + C_1 \Leftrightarrow y = -\ln(\sin x + C)$. The solution is periodic, with period 2π. Note that for $C > 1$, the domain of the solution is $\mathbb{R}$, but for $-1 < C \le 1$ it is only defined on the intervals where $\sin x + C > 0$, and it is meaningless for $C \le -1$, since then $\sin x + C \le 0$, and the logarithm is undefined.

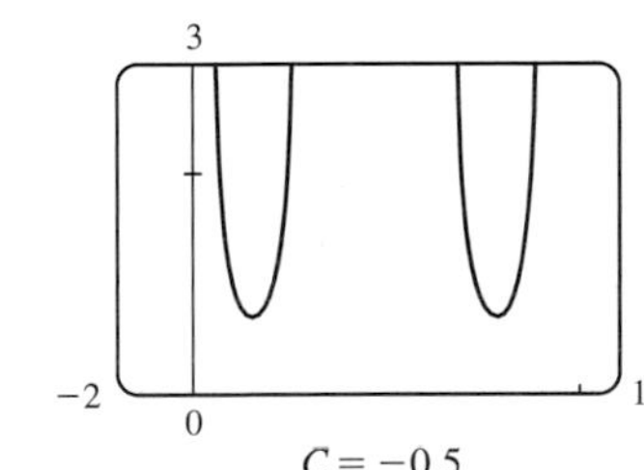

$C = -0.5$

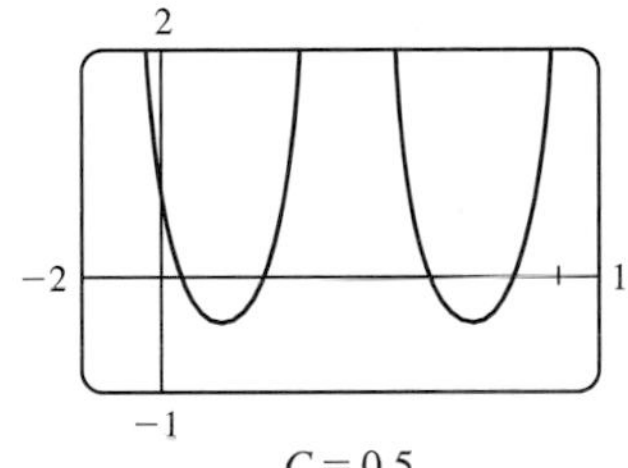

$C = 0.5$

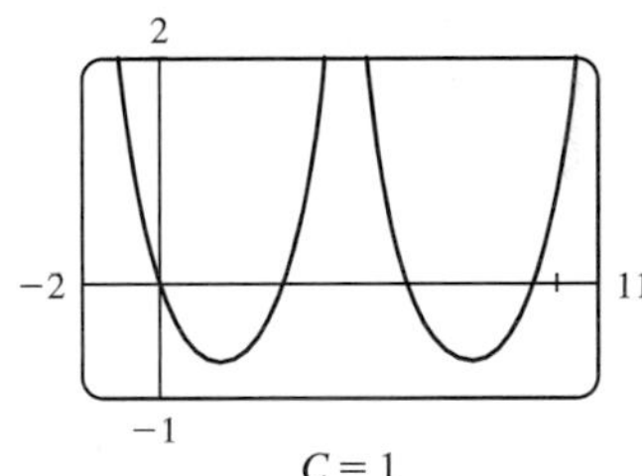

$C = 1$

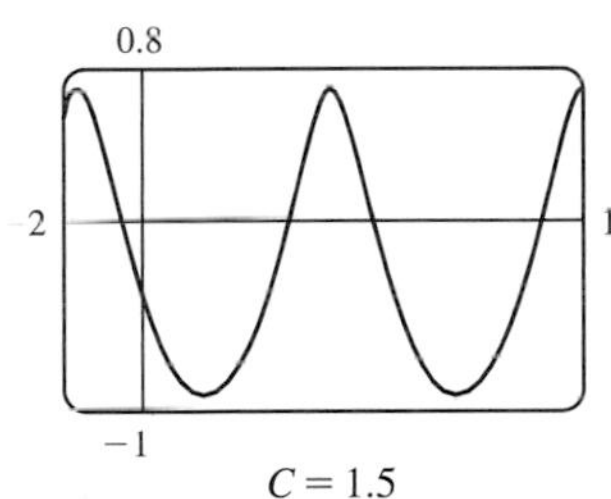

$C = 1.5$

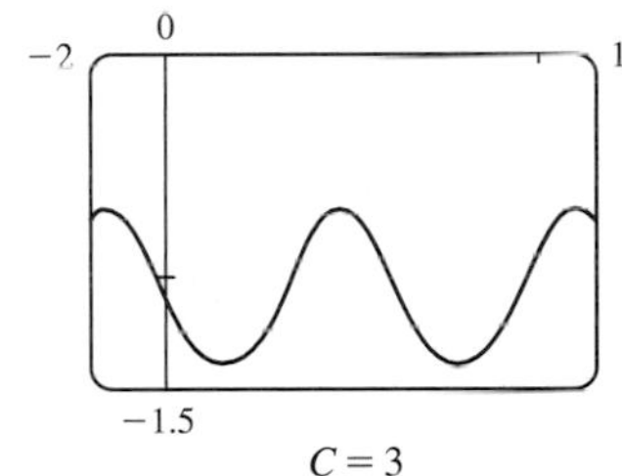

$C = 3$

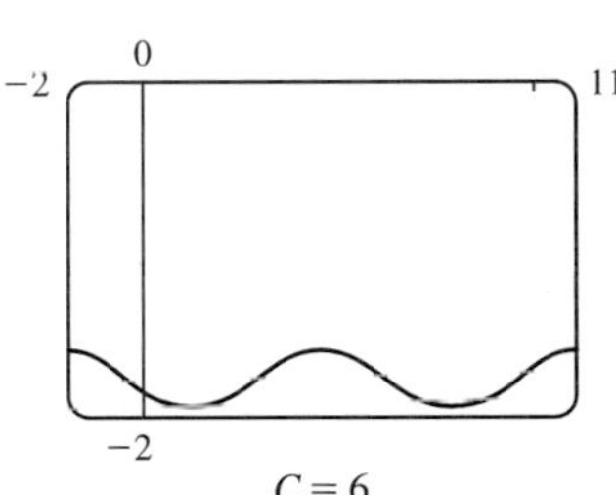

$C = 6$

For $-1 < C < 1$, the solution curve consists of concave-up pieces separated by intervals on which the solution is not defined (where $\sin x + C \le 0$). For $C = 1$, the solution curve consists of concave-up pieces separated by vertical asymptotes at the points where $\sin x + C = 0 \Leftrightarrow \sin x = -1$. For $C > 1$, the curve is continuous, and as C increases, the graph moves downward, and the amplitude of the oscillations decreases.

25. $\dfrac{dy}{dx} = \dfrac{\sin x}{\sin y}$, $y(0) = \dfrac{\pi}{2}$. So $\int \sin y\,dy = \int \sin x\,dx \Leftrightarrow -\cos y = -\cos x + C \Leftrightarrow \cos y = \cos x - C$. From the initial condition, we need $\cos\frac{\pi}{2} = \cos 0 - C \Rightarrow 0 = 1 - C \Rightarrow C = 1$, so the solution is $\cos y = \cos x - 1$. Note that we cannot take $\cos^{-1}$ of both sides, since that would unnecessarily restrict the solution to the case where $-1 \le \cos x - 1 \Leftrightarrow 0 \le \cos x$, as $\cos^{-1}$ is defined only on $[-1, 1]$. Instead we plot the graph using Maple's `plots[implicitplot]` or Mathematica's `Plot[Evaluate[···]]`.

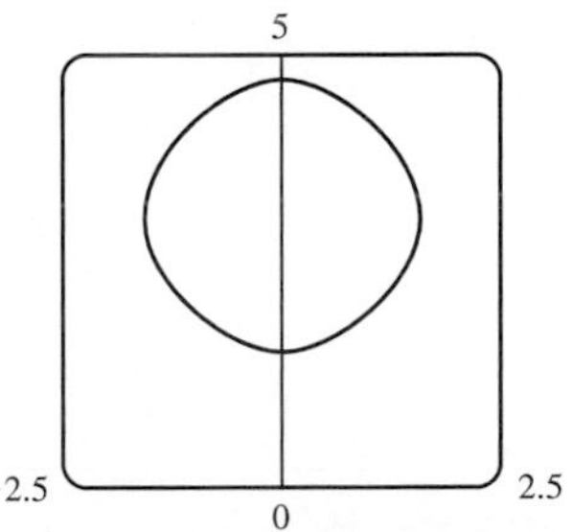

26. $\dfrac{dy}{dx} = \dfrac{x\sqrt{x^2+1}}{ye^y} \quad \Leftrightarrow \quad \int ye^y\,dy = \int x\sqrt{x^2+1}\,dx$. We use parts on the LHS with $u = y$, $dv = e^y\,dy$, and on the RHS we use the substitution $z = x^2 + 1$, so $dz = 2x\,dx$. The equation becomes $ye^y - \int e^y\,dy = \frac{1}{2}\int \sqrt{z}\,dz \quad \Leftrightarrow$ $e^y(y-1) = \frac{1}{3}(x^2+1)^{3/2} + C$, so we see that the curves are symmetric about the y-axis. Every point (x, y) in the plane lies on one of the curves, namely the one for which $C = (y-1)e^y - \frac{1}{3}(x^2+1)^{3/2}$. For example, along the y-axis, $C = (y-1)e^y - \frac{1}{3}$, so the origin lies on the curve with $C = -\frac{4}{3}$. We use Maple's `plots[implicitplot]` command or `Plot[Evaluate[···]]` in Mathematica to plot the solution curves for various values of C.

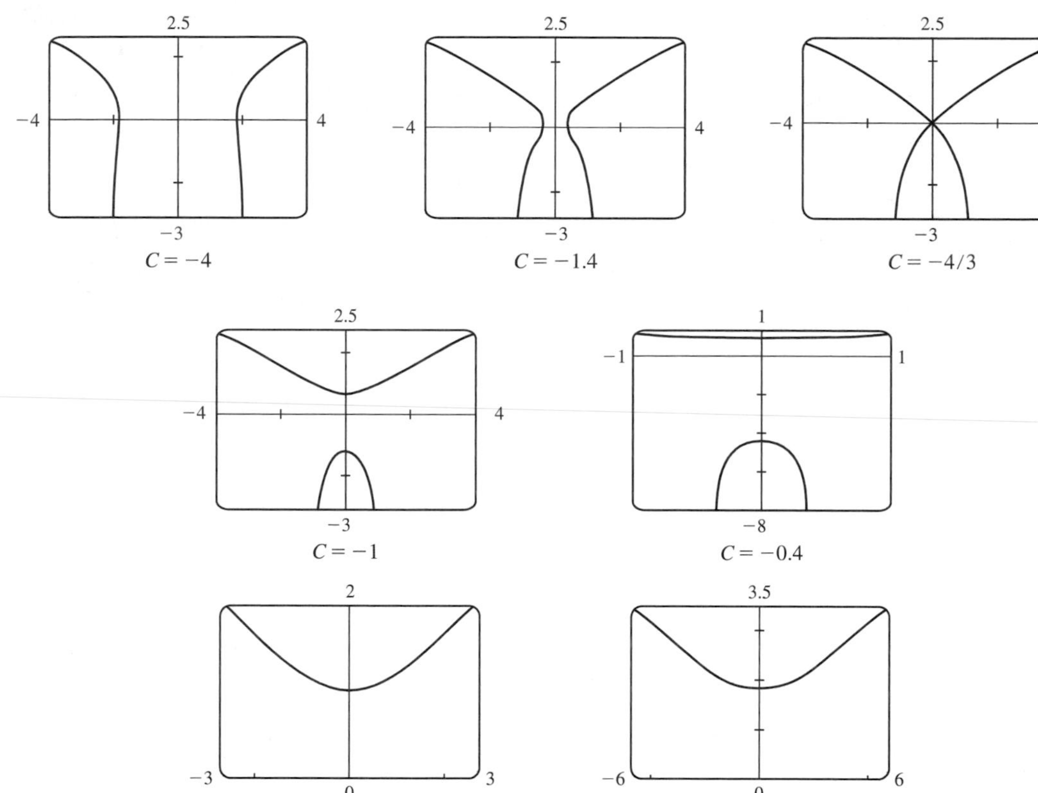

It seems that the transitional values of C are $-\frac{4}{3}$ and $-\frac{1}{3}$. For $C < -\frac{4}{3}$, the graph consists of left and right branches. At $C = -\frac{4}{3}$, the two branches become connected at the origin, and as C increases, the graph splits into top and bottom branches. At $C = -\frac{1}{3}$, the bottom half disappears. As C increases further, the graph moves upward, but doesn't change shape much.

27. (a), (c)

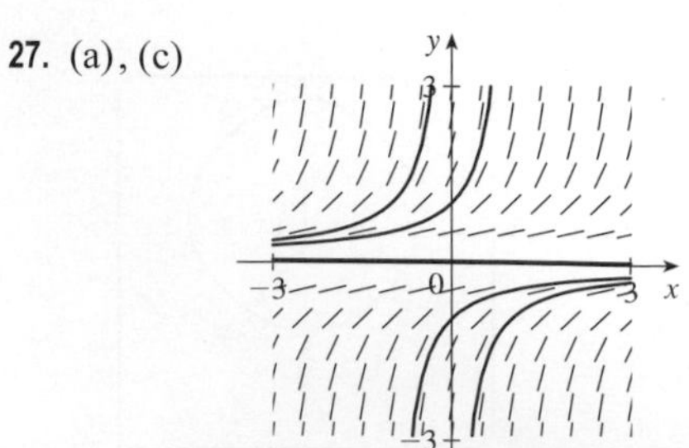

(b) $y' = y^2 \quad \Rightarrow \quad \dfrac{dy}{dx} = y^2 \quad \Rightarrow \quad \displaystyle\int y^{-2}\,dy = \int dx \quad \Rightarrow$

$-y^{-1} = x + C \quad \Rightarrow \quad \dfrac{1}{y} = -x - C \quad \Rightarrow$

$y = \dfrac{1}{K - x}$, where $K = -C$. $y = 0$ is also a solution.

28. (a), (c)

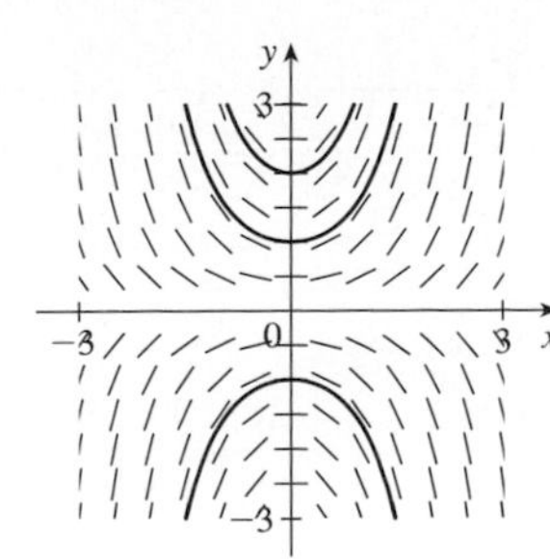

(b) $y' = xy \Rightarrow \dfrac{dy}{dx} = xy \Rightarrow \displaystyle\int \frac{dy}{y} = \int x\,dx \Rightarrow$

$\ln|y| = \frac{1}{2}x^2 + C \Rightarrow |y| = e^{x^2/2 + C} = e^{x^2/2}e^C \Rightarrow$

$y = Ke^{x^2/2}$, where $K = \pm e^C$. Taking $K = 0$ gives us the solution $y = 0$.

29. The curves $x^2 + 2y^2 = k^2$ form a family of ellipses with major axis on the x-axis. Differentiating gives

$\dfrac{d}{dx}(x^2 + 2y^2) = \dfrac{d}{dx}(k^2) \Rightarrow 2x + 4yy' = 0 \Rightarrow 4yy' = -2x \Rightarrow y' = \dfrac{-x}{2y}$. Thus, the slope of the tangent line at any point (x, y) on one of the ellipses is $y' = \dfrac{-x}{2y}$, so the orthogonal trajectories must satisfy $y' = \dfrac{2y}{x} \Leftrightarrow \dfrac{dy}{dx} = \dfrac{2y}{x} \Leftrightarrow \dfrac{dy}{y} = 2 = \dfrac{dx}{x} \Leftrightarrow$

$\displaystyle\int \frac{dy}{y} = 2\int \frac{dx}{x} \Leftrightarrow \ln|y| = 2\ln|x| + C_1 \Leftrightarrow \ln|y| = \ln|x|^2 + C_1 \Leftrightarrow$

$|y| = e^{\ln x^2 + C_1} \Leftrightarrow y = \pm x^2 \cdot e^{C_1} = Cx^2$. This is a family of parabolas.

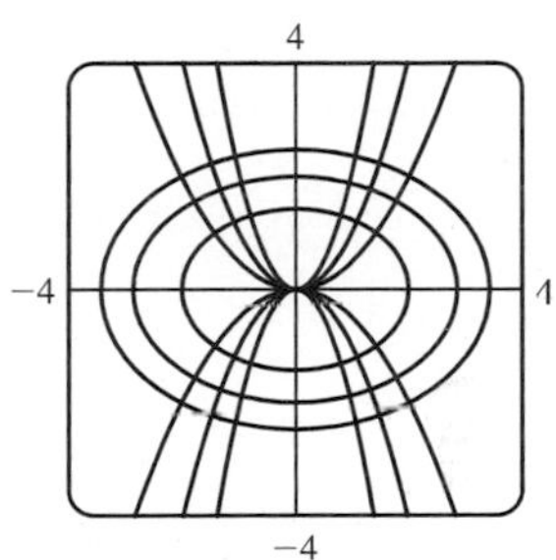

30. The curves $y^2 = kx^3$ form a family of power functions. Differentiating gives $\dfrac{d}{dx}(y^2) = \dfrac{d}{dx}(kx^3) \Rightarrow 2yy' = 3kx^2 \Rightarrow$

$y' = \dfrac{3kx^2}{2y} = \dfrac{3(y^2/x^3)x^2}{2y} = \dfrac{3y}{2x}$, the slope of the tangent line at (x, y) on one of the curves. Thus, the orthogonal trajectories must satisfy $y' = -\dfrac{2x}{3y} \Leftrightarrow \dfrac{dy}{dx} = -\dfrac{2x}{3y} \Leftrightarrow$

$3y\,dy = -2x\,dx \Leftrightarrow \int 3y\,dy = \int -2x\,dx \Leftrightarrow \frac{3}{2}y^2 = -x^2 + C_1 \Leftrightarrow$

$3y^2 = -2x^2 + C_2 \Leftrightarrow 2x^2 + 3y^2 = C$. This is a family of ellipses.

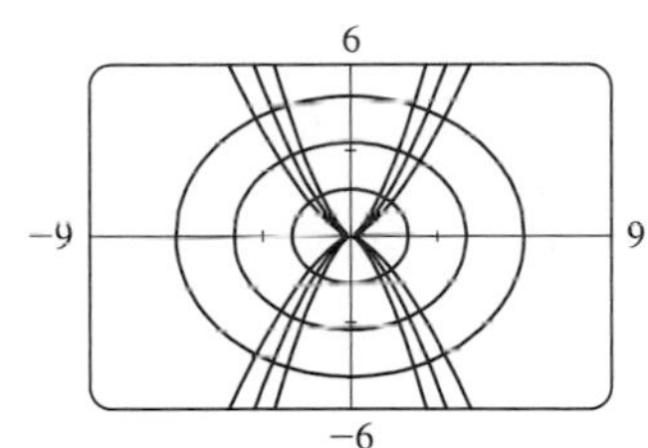

31. The curves $y = k/x$ form a family of hyperbolas with asymptotes $x = 0$ and $y = 0$. Differentiating gives

$\dfrac{d}{dx}(y) = \dfrac{d}{dx}\left(\dfrac{k}{x}\right) \Rightarrow y' = -\dfrac{k}{x^2} \Rightarrow y' = -\dfrac{xy}{x^2}$ [since $y = k/x \Rightarrow xy = k$] $\Rightarrow y' = -\dfrac{y}{x}$. Thus, the slope of the tangent line at any point (x, y) on one of the hyperbolas is $y' = -y/x$, so the orthogonal trajectories must satisfy $y' = x/y \Leftrightarrow \dfrac{dy}{dx} = \dfrac{x}{y} \Leftrightarrow$

$y\,dy = x\,dx \Leftrightarrow \int y\,dy = \int x\,dx \Leftrightarrow \frac{1}{2}y^2 = \frac{1}{2}x^2 + C_1 \Leftrightarrow$

$y^2 = x^2 + C_2 \Leftrightarrow x^2 - y^2 = C$. This is a family of hyperbolas with asymptotes $y = \pm x$.

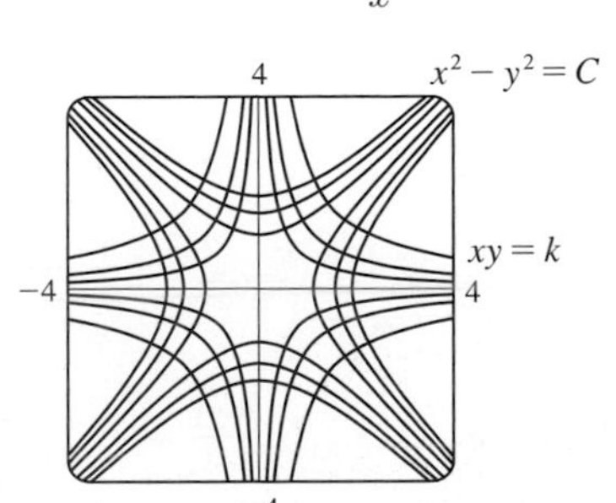

32. The curves $y = 1/(x + k)$ form a family of hyperbolas with asymptotes $x = -k$ and $y = 0$. Differentiating gives

$\dfrac{d}{dx}(y) = \dfrac{d}{dx}\left(\dfrac{1}{x + k}\right) \Rightarrow y' = -\dfrac{1}{(x + k)^2} \Rightarrow y' = -y^2$ [since $y = 1/(x + k)$]. Thus, the slope of the tangent line at any point (x, y) on one of the hyperbolas is $y' = -y^2$, so the orthogonal trajectories must satisfy $y' = 1/y^2 \Leftrightarrow \dfrac{dy}{dx} = \dfrac{1}{y^2} \Leftrightarrow$ $y^2\,dy = dx \Leftrightarrow \int y^2\,dy = \int dx \Leftrightarrow \frac{1}{3}y^3 = x + C_1 \Leftrightarrow$ $y^3 = 3x + C \Leftrightarrow y = (3x + C)^{1/3}$. This is a family of cube root functions with vertical tangents on the x-axis $[y = 0]$.

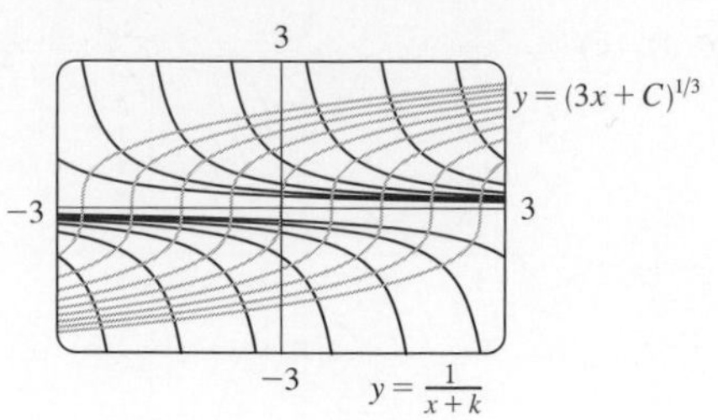

33. $y(x) = 2 + \int_2^x [t - ty(t)]\,dt \Rightarrow y'(x) = x - xy(x)$ [by FTC 1] $\Rightarrow \dfrac{dy}{dx} = x(1 - y) \Rightarrow$ $\int \dfrac{dy}{1 - y} = \int x\,dx \Rightarrow -\ln|1 - y| = \frac{1}{2}x^2 + C$. Letting $x = 2$ in the original integral equation gives us $y(2) = 2 + 0 = 2$. Thus, $-\ln|1 - 2| = \frac{1}{2}(2)^2 + C \Rightarrow 0 = 2 + C \Rightarrow C = -2$. Thus, $-\ln|1 - y| = \frac{1}{2}x^2 - 2 \Rightarrow \ln|1 - y| = 2 - \frac{1}{2}x^2 \Rightarrow |1 - y| = e^{2 - x^2/2} \Rightarrow$ $1 - y = \pm e^{2 - x^2/2} \Rightarrow y = 1 + e^{2 - x^2/2}$ $[y(2) = 2]$.

34. $y(x) = 2 + \int_1^x \dfrac{dt}{ty(t)}, x > 0 \Rightarrow y'(x) = \dfrac{1}{xy(x)} \Rightarrow \dfrac{dy}{dx} = \dfrac{1}{xy} \Rightarrow \int y\,dy = \int \dfrac{1}{x}\,dx \Rightarrow$ $\frac{1}{2}y^2 = \ln x + C$ $[x > 0]$. Letting $x = 1$ in the original integral equation gives us $y(1) = 2 + 0 = 2$. Thus, $\frac{1}{2}(2)^2 = \ln 1 + C \Rightarrow C = 2$. $\frac{1}{2}y^2 = \ln x + 2 \Rightarrow y^2 = 2\ln x + 4$ $[> 0]$ $\Rightarrow y = \sqrt{2\ln x + 4}$.

35. $y(x) = 4 + \int_0^x 2t\sqrt{y(t)}\,dt \Rightarrow y'(x) = 2x\sqrt{y(x)} \Rightarrow \dfrac{dy}{dx} = 2x\sqrt{y} \Rightarrow \int \dfrac{dy}{\sqrt{y}} = \int 2x\,dx \Rightarrow$ $2\sqrt{y} = x^2 + C$. Letting $x = 0$ in the original integral equation gives us $y(0) = 4 + 0 = 4$. Thus, $2\sqrt{4} = 0^2 + C \Rightarrow C = 4$. $2\sqrt{y} = x^2 + 4 \Rightarrow \sqrt{y} = \frac{1}{2}x^2 + 2 \Rightarrow y = \left(\frac{1}{2}x^2 + 2\right)^2$.

36. $(t^2 + 1)f'(t) + [f(t)]^2 + 1 = 0 \Rightarrow (t^2 + 1)\dfrac{dy}{dt} + y^2 + 1 = 0 \Rightarrow \dfrac{dy}{dt} = \dfrac{-y^2 - 1}{t^2 + 1} \Rightarrow$ $\int \dfrac{dy}{y^2 + 1} = -\int \dfrac{dt}{t^2 + 1} \Rightarrow \arctan y = -\arctan t + C \Rightarrow \arctan t + \arctan y = C \Rightarrow$ $\tan(\arctan t + \arctan y) = \tan C \Rightarrow \dfrac{\tan(\arctan t) + \tan(\arctan y)}{1 - \tan(\arctan t)\,\tan(\arctan y)} = \tan C \Rightarrow \dfrac{t + y}{1 - ty} = \tan C = k \Rightarrow$ $t + y = k - kty \Rightarrow y + kty = k - t \Rightarrow y(1 + kt) = k - t \Rightarrow f(t) = y = \dfrac{k - t}{1 + kt}$. Since $f(3) = 2 = \dfrac{k - 3}{1 + 3k} \Rightarrow 2 + 6k = k - 3 \Rightarrow 5k = -5 \Rightarrow k = -1$, we have $y = \dfrac{-1 - t}{1 + (-1)t} = \dfrac{t + 1}{t - 1}$.

37. From Exercise 9.2.27, $\dfrac{dQ}{dt} = 12 - 4Q \Leftrightarrow \int \dfrac{dQ}{12 - 4Q} = \int dt \Leftrightarrow -\frac{1}{4}\ln|12 - 4Q| = t + C \Leftrightarrow$ $\ln|12 - 4Q| = -4t - 4C \Leftrightarrow |12 - 4Q| = e^{-4t - 4C} \Leftrightarrow 12 - 4Q = Ke^{-4t}$ $[K = \pm e^{-4C}] \Leftrightarrow$ $4Q = 12 - Ke^{-4t} \Leftrightarrow Q = 3 - Ae^{-4t}$ $[A = K/4]$. $Q(0) = 0 \Leftrightarrow 0 = 3 - A \Leftrightarrow A = 3 \Leftrightarrow$ $Q(t) = 3 - 3e^{-4t}$. As $t \to \infty$, $Q(t) \to 3 - 0 = 3$ (the limiting value).

38. From Exercise 9.2.28, $\frac{dy}{dt} = -\frac{1}{50}(y - 20) \quad \Leftrightarrow \quad \int \frac{dy}{y - 20} = \int \left(-\frac{1}{50}\right) dt \quad \Leftrightarrow \quad \ln|y - 20| = -\frac{1}{50}t + C \quad \Leftrightarrow$

$y - 20 = Ke^{-t/50} \quad \Leftrightarrow \quad y(t) = Ke^{-t/50} + 20.\;\; y(0) = 95 \quad \Leftrightarrow \quad 95 = K + 20 \quad \Leftrightarrow \quad K = 75 \quad \Leftrightarrow$

$y(t) = 75e^{-t/50} + 20.$

39. $\frac{dP}{dt} = k(M - P) \quad \Leftrightarrow \quad \int \frac{dP}{P - M} = \int (-k)\, dt \quad \Leftrightarrow \quad \ln|P - M| = -kt + C \quad \Leftrightarrow \quad |P - M| = e^{-kt+C} \quad \Leftrightarrow$

$P - M = Ae^{-kt} \;\; [A = \pm e^C] \quad \Leftrightarrow \quad P = M + Ae^{-kt}$. If we assume that performance is at level 0 when $t = 0$, then

$P(0) = 0 \quad \Leftrightarrow \quad 0 = M + A \quad \Leftrightarrow \quad A = -M \quad \Leftrightarrow \quad P(t) = M - Me^{-kt}.\quad \lim\limits_{t \to \infty} P(t) = M - M \cdot 0 = M.$

40. (a) $\frac{dx}{dt} = k(a - x)(b - x),\; a \neq b$. Using partial fractions, $\frac{1}{(a - x)(b - x)} = \frac{1/(b - a)}{a - x} - \frac{1/(b - a)}{b - x}$, so

$$\int \frac{dx}{(a - x)(b - x)} = \int k\, dt \quad \Rightarrow \quad \frac{1}{b - a}\left(-\ln|a - x| + \ln|b - x|\right) = kt + C \quad \Rightarrow \quad \ln\left|\frac{b - x}{a - x}\right| = (b - a)(kt + C).$$

The concentrations [A] $= a - x$ and [B] $= b - x$ cannot be negative, so $\frac{b - x}{a - x} \geq 0$ and $\left|\frac{b - x}{a - x}\right| = \frac{b - x}{a - x}$.

We now have $\ln\left(\frac{b - x}{a - x}\right) = (b - a)(kt + C)$. Since $x(0) = 0$, we get $\ln\left(\frac{b}{a}\right) = (b - a)C$. Hence,

$$\ln\left(\frac{b - x}{a - x}\right) = (b - a)kt + \ln\left(\frac{b}{a}\right) \quad \Rightarrow \quad \frac{b - x}{a - x} = \frac{b}{a}e^{(b-a)kt} \quad \Rightarrow x = \frac{b[e^{(b-a)kt} - 1]}{be^{(b-a)kt}/a - 1} = \frac{ab[e^{(b-a)kt} - 1]}{be^{(b-a)kt} - a}\,\frac{\text{moles}}{\text{L}}.$$

(b) If $b = a$, then $\frac{dx}{dt} = k(a - x)^2$, so $\int \frac{dx}{(a - x)^2} = \int k\, dt$ and $\frac{1}{a - x} = kt + C$. Since $x(0) = 0$, we get $C = \frac{1}{a}$.

Thus, $a - x = \frac{1}{kt + 1/a}$ and $x = a - \frac{a}{akt + 1} = \frac{a^2kt}{akt + 1}\,\frac{\text{moles}}{\text{L}}$. Suppose $x = [\text{C}] = a/2$ when $t = 20$. Then

$$x(20) = a/2 \quad \Rightarrow \quad \frac{a}{2} = \frac{20a^2k}{20ak + 1} \quad \Rightarrow \quad 40a^2k = 20a^2k + a \quad \Rightarrow \quad 20a^2k = a \quad \Rightarrow \quad k = \frac{1}{20a}, \text{ so}$$

$$x = \frac{a^2t/(20a)}{1 + at/(20a)} = \frac{at/20}{1 + t/20} = \frac{at}{t + 20}\,\frac{\text{moles}}{\text{L}}.$$

41. (a) If $a = b$, then $\frac{dx}{dt} = k(a - x)(b - x)^{1/2}$ becomes $\frac{dx}{dt} = k(a - x)^{3/2} \quad \Rightarrow \quad (a - x)^{-3/2}\, dx = k\, dt \quad \Rightarrow$

$\int (a - x)^{-3/2}\, dx = \int k\, dt \quad \Rightarrow \quad 2(a - x)^{-1/2} = kt + C$ [by substitution] $\Rightarrow \quad \frac{2}{kt + C} = \sqrt{a - x} \quad \Rightarrow$

$\left(\frac{2}{kt + C}\right)^2 = a - x \quad \Rightarrow \quad x(t) = a - \frac{4}{(kt + C)^2}$. The initial concentration of HBr is 0, so $x(0) = 0 \quad \Rightarrow$

$0 = a - \frac{4}{C^2} \quad \Rightarrow \quad \frac{4}{C^2} = a \quad \Rightarrow \quad C^2 = \frac{4}{a} \quad \Rightarrow \quad C = 2/\sqrt{a}$ [C is positive since $kt + C = 2(a - x)^{-1/2} > 0$].

Thus, $x(t) = a - \frac{4}{(kt + 2/\sqrt{a})^2}$.

(b) $\dfrac{dx}{dt} = k(a-x)(b-x)^{1/2} \Rightarrow \dfrac{dx}{(a-x)\sqrt{b-x}} = k\,dt \Rightarrow \displaystyle\int \dfrac{dx}{(a-x)\sqrt{b-x}} = \int k\,dt \quad (\star)$.

From the hint, $u = \sqrt{b-x} \Rightarrow u^2 = b - x \Rightarrow 2u\,du = -dx$, so

$$\int \frac{dx}{(a-x)\sqrt{b-x}} = \int \frac{-2u\,du}{[a-(b-u^2)]u} = -2\int \frac{du}{a-b+u^2} = -2\int \frac{du}{\left(\sqrt{a-b}\right)^2 + u^2}$$

$$\overset{17}{=} -2\left(\frac{1}{\sqrt{a-b}}\tan^{-1}\frac{u}{\sqrt{a-b}}\right)$$

So $(\star)$ becomes $\dfrac{-2}{\sqrt{a-b}}\tan^{-1}\dfrac{\sqrt{b-x}}{\sqrt{a-b}} = kt + C$. Now $x(0) = 0 \Rightarrow C = \dfrac{-2}{\sqrt{a-b}}\tan^{-1}\dfrac{\sqrt{b}}{\sqrt{a-b}}$ and we have

$$\frac{-2}{\sqrt{a-b}}\tan^{-1}\frac{\sqrt{b-x}}{\sqrt{a-b}} = kt - \frac{2}{\sqrt{a-b}}\tan^{-1}\frac{\sqrt{b}}{\sqrt{a-b}} \Rightarrow \frac{2}{\sqrt{a-b}}\left(\tan^{-1}\sqrt{\frac{b}{a-b}} - \tan^{-1}\sqrt{\frac{b-x}{a-b}}\right) = kt \Rightarrow$$

$$t(x) = \frac{2}{k\sqrt{a-b}}\left(\tan^{-1}\sqrt{\frac{b}{a-b}} - \tan^{-1}\sqrt{\frac{b-x}{a-b}}\right).$$

42. If $S = \dfrac{dT}{dr}$, then $\dfrac{dS}{dr} = \dfrac{d^2T}{dr^2}$. The differential equation $\dfrac{d^2T}{dr^2} + \dfrac{2}{r}\dfrac{dT}{dr} = 0$ can be written as $\dfrac{dS}{dr} + \dfrac{2}{r}S = 0$. Thus,

$\dfrac{dS}{dr} = \dfrac{-2S}{r} \Rightarrow \dfrac{dS}{S} = -\dfrac{2}{r}\,dr \Rightarrow \displaystyle\int \frac{1}{S}\,dS = \int -\frac{2}{r}\,dr \Rightarrow \ln|S| = -2\ln|r| + C$. Assuming $S = dT/dr > 0$

and $r > 0$, we have $S = e^{-2\ln r + C} = e^{\ln r^{-2}}e^C = r^{-2}k \;\; [k = e^C] \Rightarrow S = \dfrac{1}{r^2}k \Rightarrow \dfrac{dT}{dr} = \dfrac{1}{r^2}k \Rightarrow$

$dT = \dfrac{1}{r^2}k\,dr \Rightarrow \displaystyle\int dT = \int \frac{1}{r^2}k\,dr \Rightarrow T(r) = -\frac{k}{r} + A.$

$T(1) = 15 \Rightarrow 15 = -k + A$ **(1)** and $T(2) = 25 \Rightarrow 25 = -\frac{1}{2}k + A$ **(2)**.

Now solve for k and A: -2**(2)** $+$ **(1)** $\Rightarrow -35 = -A$, so $A = 35$ and $k = 20$, and $T(r) = -20/r + 35$.

43. (a) $\dfrac{dC}{dt} = r - kC \Rightarrow \dfrac{dC}{dt} = -(kC - r) \Rightarrow \displaystyle\int \frac{dC}{kC - r} = \int -dt \Rightarrow (1/k)\ln|kC - r| = -t + M_1 \Rightarrow$

$\ln|kC - r| = -kt + M_2 \Rightarrow |kC - r| = e^{-kt + M_2} \Rightarrow kC - r = M_3e^{-kt} \Rightarrow kC = M_3e^{-kt} + r \Rightarrow$

$C(t) = M_4e^{-kt} + r/k$. $C(0) = C_0 \Rightarrow C_0 = M_4 + r/k \Rightarrow M_4 = C_0 - r/k \Rightarrow$

$C(t) = (C_0 - r/k)e^{-kt} + r/k$.

(b) If $C_0 < r/k$, then $C_0 - r/k < 0$ and the formula for $C(t)$ shows that $C(t)$ increases and $\lim\limits_{t\to\infty} C(t) = r/k$.

As t increases, the formula for $C(t)$ shows how the role of C_0 steadily diminishes as that of r/k increases.

44. (a) Use 1 billion dollars as the x-unit and 1 day as the t-unit. Initially, there is \$10 billion of old currency in circulation, so all of the \$50 million returned to the banks is old. At time t, the amount of new currency is $x(t)$ billion dollars, so $10 - x(t)$ billion dollars of currency is old. The fraction of circulating money that is old is $[10 - x(t)]/10$, and the amount of old currency being returned to the banks each day is $\dfrac{10 - x(t)}{10}\,0.05$ billion dollars. This amount of new currency per day is introduced into circulation, so $\dfrac{dx}{dt} = \dfrac{10 - x}{10} \cdot 0.05 = 0.005(10 - x)$ billion dollars per day.

(b) $\dfrac{dx}{10-x} = 0.005\,dt \;\Rightarrow\; \dfrac{-dx}{10-x} = -0.005\,dt \;\Rightarrow\; \ln(10-x) = -0.005t + c \;\Rightarrow\; 10 - x = Ce^{-0.005t}$,

where $C = e^c \;\Rightarrow\; x(t) = 10 - Ce^{-0.005t}$. From $x(0) = 0$, we get $C = 10$, so $x(t) = 10(1 - e^{-0.005t})$.

(c) The new bills make up 90% of the circulating currency when $x(t) = 0.9 \cdot 10 = 9$ billion dollars.

$9 = 10(1 - e^{-0.005t}) \;\Rightarrow\; 0.9 = 1 - e^{-0.005t} \;\Rightarrow\; e^{-0.005t} = 0.1 \;\Rightarrow\; -0.005t = -\ln 10 \;\Rightarrow$

$t = 200 \ln 10 \approx 460.517$ days ≈ 1.26 years.

45. (a) Let $y(t)$ be the amount of salt (in kg) after t minutes. Then $y(0) = 15$. The amount of liquid in the tank is 1000 L at all times, so the concentration at time t (in minutes) is $y(t)/1000$ kg/L and $\dfrac{dy}{dt} = -\left[\dfrac{y(t)}{1000}\,\dfrac{\text{kg}}{\text{L}}\right]\left(10\,\dfrac{\text{L}}{\text{min}}\right) = -\dfrac{y(t)}{100}\,\dfrac{\text{kg}}{\text{min}}$.

$\displaystyle\int \frac{dy}{y} = -\frac{1}{100}\int dt \;\Rightarrow\; \ln y = -\frac{t}{100} + C$, and $y(0) = 15 \;\Rightarrow\; \ln 15 = C$, so $\ln y = \ln 15 - \dfrac{t}{100}$.

It follows that $\ln\left(\dfrac{y}{15}\right) = -\dfrac{t}{100}$ and $\dfrac{y}{15} = e^{-t/100}$, so $y = 15e^{-t/100}$ kg.

(b) After 20 minutes, $y = 15e^{-20/100} = 15e^{-0.2} \approx 12.3$ kg.

46. Let $y(t)$ be the amount of carbon dioxide in the room after t minutes. Then $y(0) = 0.0015(180) = 0.27\text{ m}^3$. The amount of air in the room is 180 m^3 at all times, so the percentage at time t (in mimutes) is $y(t)/180 \times 100$, and the change in the amount of carbon dioxide with respect to time is

$$\frac{dy}{dt} = (0.0005)\left(2\,\frac{\text{m}^3}{\text{min}}\right) - \frac{y(t)}{180}\left(2\,\frac{\text{m}^3}{\text{min}}\right) = 0.001 - \frac{y}{90} = \frac{9-100y}{9000}\,\frac{\text{m}^3}{\text{min}}$$

Hence, $\displaystyle\int \frac{dy}{9-100y} = \int \frac{dt}{9000}$ and $-\dfrac{1}{100}\ln|9-100y| = \dfrac{1}{9000}t + C$. Because $y(0) = 0.27$, we have

$-\frac{1}{100}\ln 18 = C$, so $-\frac{1}{100}\ln|9-100y| = \frac{1}{9000}t - \frac{1}{100}\ln 18 \;\Rightarrow\; \ln|9-100y| = -\frac{1}{90}t + \ln 18 \;\Rightarrow$

$\ln|9-100y| = \ln e^{-t/90} + \ln 18 \;\Rightarrow\; \ln|9-100y| = \ln(18e^{-t/90})$, and $|9-100y| = 18e^{-t/90}$. Since y is continuous, $y(0) = 0.27$, and the right-hand side is never zero, we deduce that $9 - 100y$ is always negative. Thus, $|9-100y| = 100y - 9$ and we have $100y - 9 = 18e^{-t/90} \;\Rightarrow\; 100y = 9 + 18e^{-t/90} \;\Rightarrow\; y = 0.09 + 0.18e^{-t/90}$. The percentage of carbon dioxide in the room is

$$p(t) = \frac{y}{180} \times 100 = \frac{0.09 + 0.18e^{-t/90}}{180} \times 100 = (0.0005 + 0.001e^{-t/90}) \times 100 = 0.05 + 0.1e^{-t/90}$$

In the long run, we have $\lim\limits_{t\to\infty} p(t) = 0.05 + 0.1(0) = 0.05$; that is, the amount of carbon dioxide approaches 0.05% as time goes on.

47. Let $y(t)$ be the amount of alcohol in the vat after t minutes. Then $y(0) = 0.04(500) = 20$ gal. The amount of beer in the vat is 500 gallons at all times, so the percentage at time t (in minutes) is $y(t)/500 \times 100$, and the change in the amount of alcohol with respect to time t is $\dfrac{dy}{dt} = \text{rate in} - \text{rate out} = 0.06\left(5\,\dfrac{\text{gal}}{\text{min}}\right) - \dfrac{y(t)}{500}\left(5\,\dfrac{\text{gal}}{\text{min}}\right) = 0.3 - \dfrac{y}{100} = \dfrac{30-y}{100}\,\dfrac{\text{gal}}{\text{min}}$.

Hence, $\displaystyle\int \frac{dy}{30-y} = \int \frac{dt}{100}$ and $-\ln|30-y| = \frac{1}{100}t + C$. Because $y(0) = 20$, we have $-\ln 10 = C$, so

$-\ln|30 - y| = \frac{1}{100}t - \ln 10 \;\Rightarrow\; \ln|30 - y| = -t/100 + \ln 10 \;\Rightarrow\; \ln|30 - y| = \ln e^{-t/100} + \ln 10 \;\Rightarrow$ $\ln|30 - y| = \ln(10e^{-t/100}) \;\Rightarrow\; |30 - y| = 10e^{-t/100}$. Since y is continuous, $y(0) = 20$, and the right-hand side is never zero, we deduce that $30 - y$ is always positive. Thus, $30 - y = 10e^{-t/100} \;\Rightarrow\; y = 30 - 10e^{-t/100}$. The percentage of alcohol is $p(t) = y(t)/500 \times 100 = y(t)/5 = 6 - 2e^{-t/100}$. The percentage of alcohol after one hour is $p(60) = 6 - 2e^{-60/100} \approx 4.9$.

48. (a) If $y(t)$ is the amount of salt (in kg) after t minutes, then $y(0) = 0$ and the total amount of liquid in the tank remains constant at 1000 L.

$$\frac{dy}{dt} = \left(0.05\,\frac{\text{kg}}{\text{L}}\right)\left(5\,\frac{\text{L}}{\text{min}}\right) + \left(0.04\,\frac{\text{kg}}{\text{L}}\right)\left(10\,\frac{\text{L}}{\text{min}}\right) - \left(\frac{y(t)}{1000}\,\frac{\text{kg}}{\text{L}}\right)\left(15\,\frac{\text{L}}{\text{min}}\right)$$

$$= 0.25 + 0.40 - 0.015y = 0.65 - 0.015y = \frac{130 - 3y}{200}\,\frac{\text{kg}}{\text{min}}$$

Hence, $\displaystyle\int \frac{dy}{130 - 3y} = \int \frac{dt}{200}$ and $-\frac{1}{3}\ln|130 - 3y| = \frac{1}{200}t + C$. Because $y(0) = 0$, we have $-\frac{1}{3}\ln 130 = C$, so $-\frac{1}{3}\ln|130 - 3y| = \frac{1}{200}t - \frac{1}{3}\ln 130 \;\Rightarrow\; \ln|130 - 3y| = -\frac{3}{200}t + \ln 130 = \ln(130e^{-3t/200})$, and $|130 - 3y| = 130e^{-3t/200}$. Since y is continuous, $y(0) = 0$, and the right-hand side is never zero, we deduce that $130 - 3y$ is always positive. Thus, $130 - 3y = 130e^{-3t/200}$ and $y = \frac{130}{3}(1 - e^{-3t/200})$ kg.

(b) After one hour, $y = \frac{130}{3}(1 - e^{-3\cdot 60/200}) = \frac{130}{3}(1 - e^{-0.9}) \approx 25.7$ kg.

Note: As $t \to \infty$, $y(t) \to \frac{130}{3} = 43\frac{1}{3}$ kg.

49. Assume that the raindrop begins at rest, so that $v(0) = 0$. $dm/dt = km$ and $(mv)' = gm \;\Rightarrow\; mv' + vm' = gm \;\Rightarrow$ $mv' + v(km) = gm \;\Rightarrow\; v' + vk = g \;\Rightarrow\; \dfrac{dv}{dt} = g - kv \;\Rightarrow\; \displaystyle\int \frac{dv}{g - kv} = \int dt \;\Rightarrow$ $-(1/k)\ln|g - kv| = t + C \;\Rightarrow\; \ln|g - kv| = -kt - kC \;\Rightarrow\; g - kv = Ae^{-kt}$. $v(0) = 0 \;\Rightarrow\; A = g$. So $kv = g - ge^{-kt} \;\Rightarrow\; v = (g/k)(1 - e^{-kt})$. Since $k > 0$, as $t \to \infty$, $e^{-kt} \to 0$ and therefore, $\lim\limits_{t\to\infty} v(t) = g/k$.

50. (a) $m\dfrac{dv}{dt} = -kv \;\Rightarrow\; \dfrac{dv}{v} = -\dfrac{k}{m}\,dt \;\Rightarrow\; \ln|v| = -\dfrac{k}{m}t + C$. Since $v(0) = v_0$, $\ln|v_0| = C$. Therefore,

$\ln\left|\dfrac{v}{v_0}\right| = -\dfrac{k}{m}t \;\Rightarrow\; \left|\dfrac{v}{v_0}\right| = e^{-kt/m} \;\Rightarrow\; v(t) = \pm v_0 e^{-kt/m}$. The sign is $+$ when $t = 0$, and we assume v is continuous, so that the sign is $+$ for all t. Thus, $v(t) = v_0 e^{-kt/m}$. $ds/dt = v_0 e^{-kt/m} \;\Rightarrow$ $s(t) = -\dfrac{mv_0}{k}e^{-kt/m} + C'$.

From $s(0) = s_0$, we get $s_0 = -\dfrac{mv_0}{k} + C'$, so $C' = s_0 + \dfrac{mv_0}{k}$ and $s(t) = s_0 + \dfrac{mv_0}{k}(1 - e^{-kt/m})$.

The distance traveled from time 0 to time t is $s(t) - s_0$, so the total distance traveled is $\lim\limits_{t\to\infty}[s(t) - s_0] = \dfrac{mv_0}{k}$.

Note: In finding the limit, we use the fact that $k > 0$ to conclude that $\lim\limits_{t\to\infty} e^{-kt/m} = 0$.

(b) $m\,\dfrac{dv}{dt} = -kv^2 \;\Rightarrow\; \dfrac{dv}{v^2} = -\dfrac{k}{m}\,dt \;\Rightarrow\; \dfrac{-1}{v} = -\dfrac{kt}{m} + C \;\Rightarrow\; \dfrac{1}{v} = \dfrac{kt}{m} - C$. Since $v(0) = v_0$,

$C = -\dfrac{1}{v_0}$ and $\dfrac{1}{v} = \dfrac{kt}{m} + \dfrac{1}{v_0}$. Therefore, $v(t) = \dfrac{1}{kt/m + 1/v_0} = \dfrac{mv_0}{kv_0t + m}$. $\dfrac{ds}{dt} = \dfrac{mv_0}{kv_0t + m} \;\Rightarrow$

$s(t) = \dfrac{m}{k}\displaystyle\int \dfrac{kv_0\,dt}{kv_0t + m} = \dfrac{m}{k}\ln|kv_0t + m| + C'$. Since $s(0) = s_0$, we get $s_0 = \dfrac{m}{k}\ln m + C' \;\Rightarrow$

$C' = s_0 - \dfrac{m}{k}\ln m \;\Rightarrow\; s(t) = s_0 + \dfrac{m}{k}(\ln|kv_0t + m| - \ln m) = s_0 + \dfrac{m}{k}\ln\left|\dfrac{kv_0t + m}{m}\right|$.

We can rewrite the formulas for $v(t)$ and $s(t)$ as $v(t) = \dfrac{v_0}{1 + (kv_0/m)t}$ and $s(t) = s_0 + \dfrac{m}{k}\ln\left|1 + \dfrac{kv_0}{m}t\right|$.

Remarks: This model of horizontal motion through a resistive medium was designed to handle the case in which $v_0 > 0$. Then the term $-kv^2$ representing the resisting force causes the object to decelerate. The absolute value in the expression for $s(t)$ is unnecessary (since k, v_0, and m are all positive), and $\lim\limits_{t\to\infty} s(t) = \infty$. In other words, the object travels infinitely far. However, $\lim\limits_{t\to\infty} v(t) = 0$. When $v_0 < 0$, the term $-kv^2$ increases the magnitude of the object's negative velocity. According to the formula for $s(t)$, the position of the object approaches $-\infty$ as t approaches $m/k(-v_0)$: $\lim\limits_{t\to -m/(kv_0)} s(t) = -\infty$. Again the object travels infinitely far, but this time the feat is accomplished in a finite amount of time. Notice also that $\lim\limits_{t\to -m/(kv_0)} v(t) = -\infty$ when $v_0 < 0$, showing that the speed of the object increases without limit.

51. (a) $\dfrac{1}{L_1}\dfrac{dL_1}{dt} = k\dfrac{1}{L_2}\dfrac{dL_2}{dt} \;\Rightarrow\; \dfrac{d}{dt}(\ln L_1) = \dfrac{d}{dt}(k\ln L_2) \;\Rightarrow\; \displaystyle\int \dfrac{d}{dt}(\ln L_1)\,dt = \int \dfrac{d}{dt}(\ln L_2^k)\,dt \;\Rightarrow$

$\ln L_1 = \ln L_2^k + C \;\Rightarrow\; L_1 = e^{\ln L_2^k + C} = e^{\ln L_2^k}e^C \;\Rightarrow\; L_1 = KL_2^k$, where $K = e^C$.

(b) From part (a) with $L_1 = B$, $L_2 = V$, and $k = 0.0794$, we have $B = KV^{0.0794}$.

52. (a) $\dfrac{dV}{dt} = a(\ln b - \ln V)V \;\Rightarrow\; \dfrac{dV}{dt} = -aV(\ln V - \ln b) \;\Rightarrow\; \dfrac{dV}{V\ln(V/b)} = -a\,dt \;\Rightarrow$

$\displaystyle\int \dfrac{dV}{V\ln(V/b)} = \int -a\,dt \;\Rightarrow\; \int \dfrac{1}{u}\,du = \int -a\,dt \quad \begin{bmatrix} u = \ln(V/b), \\ du = (1/V)\,dV \end{bmatrix} \;\Rightarrow\; \ln|u| = -at + k \;\Rightarrow$

$|u| = e^{-at}e^k \;\Rightarrow\; u = Ce^{-at}$ [where $C = \pm e^k$] $\Rightarrow\; \ln(V/b) = Ce^{-at} \;\Rightarrow\; \dfrac{V}{b} = e^{Ce^{-at}} \;\Rightarrow$

$V = be^{Ce^{-at}}$ with $C \neq 0$.

(b) $V(0) = 1 \;\Rightarrow\; 1 = be^{Ce^{-a(0)}} \;\Rightarrow\; 1 = be^C \;\Rightarrow\; b = e^{-C}$, so $V = e^{-C}e^{Ce^{-at}} = e^{Ce^{-at} - C} = e^{C(e^{-at} - 1)}$.

53. (a) The rate of growth of the area is jointly proportional to $\sqrt{A(t)}$ and $M - A(t)$; that is, the rate is proportional to the product of those two quantities. So for some constant k, $dA/dt = k\sqrt{A}\,(M - A)$. We are interested in the maximum of the function dA/dt (when the tissue grows the fastest), so we differentiate, using the Chain Rule and then substituting for

dA/dt from the differential equation:

$$\frac{d}{dt}\left(\frac{dA}{dt}\right) = k\left[\sqrt{A}\,(-1)\frac{dA}{dt} + (M-A)\cdot\tfrac{1}{2}A^{-1/2}\frac{dA}{dt}\right] = \tfrac{1}{2}kA^{-1/2}\frac{dA}{dt}\,[-2A+(M-A)]$$

$$= \tfrac{1}{2}kA^{-1/2}\left[k\sqrt{A}(M-A)\right][M-3A] = \tfrac{1}{2}k^2(M-A)(M-3A)$$

This is 0 when $M - A = 0$ [this situation never actually occurs, since the graph of $A(t)$ is asymptotic to the line $y = M$, as in the logistic model] and when $M - 3A = 0 \quad\Leftrightarrow\quad A(t) = M/3$. This represents a maximum by the First Derivative Test, since $\dfrac{d}{dt}\left(\dfrac{dA}{dt}\right)$ goes from positive to negative when $A(t) = M/3$.

(b) From the CAS, we get $A(t) = M\left(\dfrac{Ce^{\sqrt{M}kt}-1}{Ce^{\sqrt{M}kt}+1}\right)^2$. To get C in terms of the initial area A_0 and the maximum area M, we substitute $t = 0$ and $A = A_0 = A(0)$: $A_0 = M\left(\dfrac{C-1}{C+1}\right)^2 \quad\Leftrightarrow\quad (C+1)\sqrt{A_0} = (C-1)\sqrt{M} \quad\Leftrightarrow$

$C\sqrt{A_0} + \sqrt{A_0} = C\sqrt{M} - \sqrt{M} \quad\Leftrightarrow\quad \sqrt{M} + \sqrt{A_0} = C\sqrt{M} - C\sqrt{A_0} \quad\Leftrightarrow$

$\sqrt{M} + \sqrt{A_0} = C\left(\sqrt{M} - \sqrt{A_0}\right) \quad\Leftrightarrow\quad C = \dfrac{\sqrt{M}+\sqrt{A_0}}{\sqrt{M}-\sqrt{A_0}}$. [Notice that if $A_0 = 0$, then $C = 1$.]

54. (a) According to the hint we use the Chain Rule: $m\dfrac{dv}{dt} = m\dfrac{dv}{dx}\cdot\dfrac{dx}{dt} = mv\dfrac{dv}{dx} = -\dfrac{mgR^2}{(x+R)^2} \quad\Rightarrow$

$\displaystyle\int v\,dv = \int\frac{-gR^2\,dx}{(x+R)^2} \quad\Rightarrow\quad \frac{v^2}{2} = \frac{gR^2}{x+R} + C$. When $x = 0$, $v = v_0$, so $\dfrac{v_0^2}{2} = \dfrac{gR^2}{0+R} + C \quad\Rightarrow$

$C = \tfrac{1}{2}v_0^2 - gR \quad\Rightarrow\quad \tfrac{1}{2}v^2 - \tfrac{1}{2}v_0^2 = \dfrac{gR^2}{x+R} - gR$. Now at the top of its flight, the rocket's velocity will be 0, and its height will be $x = h$. Solving for v_0: $-\tfrac{1}{2}v_0^2 = \dfrac{gR^2}{h+R} - gR \quad\Rightarrow\quad \dfrac{v_0^2}{2} = g\left[-\dfrac{R^2}{R+h} + \dfrac{R(R+h)}{R+h}\right] = \dfrac{gRh}{R+h} \quad\Rightarrow$

$v_0 = \sqrt{\dfrac{2gRh}{R+h}}$.

(b) $v_e = \displaystyle\lim_{h\to\infty} v_0 = \lim_{h\to\infty}\sqrt{\frac{2gRh}{R+h}} = \lim_{h\to\infty}\sqrt{\frac{2gR}{(R/h)+1}} = \sqrt{2gR}$

(c) $v_e = \sqrt{2\cdot 32\text{ ft/s}^2\cdot 3960\text{ mi}\cdot 5280\text{ ft/mi}} \approx 36{,}581\text{ ft/s} \approx 6.93\text{ mi/s}$

APPLIED PROJECT How Fast Does a Tank Drain?

1. (a) $V = \pi r^2 h \quad\Rightarrow\quad \dfrac{dV}{dt} = \pi r^2\dfrac{dh}{dt}$ [implicit differentiation] $\Rightarrow$

$$\frac{dh}{dt} = \frac{1}{\pi r^2}\frac{dV}{dt} = \frac{1}{\pi r^2}\left(-a\sqrt{2gh}\right) = \frac{1}{\pi 2^2}\left[-\pi\left(\tfrac{1}{12}\right)^2\sqrt{2\cdot 32}\sqrt{h}\right] = -\tfrac{1}{72}\sqrt{h}$$

(b) $\dfrac{dh}{dt} = -\tfrac{1}{72}\sqrt{h} \quad\Rightarrow\quad h^{-1/2}\,dh = -\tfrac{1}{72}\,dt \quad\Rightarrow\quad 2\sqrt{h} = -\tfrac{1}{72}t + C.$

$h(0) = 6 \quad\Rightarrow\quad 2\sqrt{6} = 0 + C \quad\Rightarrow\quad C = 2\sqrt{6} \quad\Rightarrow\quad h(t) = \left(-\tfrac{1}{144}t + \sqrt{6}\right)^2.$

(c) We want to find t when $h = 0$, so we set $h = 0 = \left(-\frac{1}{144}t + \sqrt{6}\right)^2 \Rightarrow t = 144\sqrt{6} \approx 5$ min 53 s.

2. (a) $\dfrac{dh}{dt} = k\sqrt{h} \Rightarrow h^{-1/2}\,dh = k\,dt \quad [h \neq 0] \Rightarrow 2\sqrt{h} = kt + C \Rightarrow$

$h(t) = \frac{1}{4}(kt + C)^2$. Since $h(0) = 10$ cm, the relation $2\sqrt{h(t)} = kt + C$ gives us $2\sqrt{10} = C$. Also, $h(68) = 3$ cm, so $2\sqrt{3} = 68k + 2\sqrt{10}$ and

$k = -\dfrac{\sqrt{10} - \sqrt{3}}{34}$. Thus,

$$h(t) = \frac{1}{4}\left(2\sqrt{10} - \frac{\sqrt{10} - \sqrt{3}}{34}\,t\right)^2 \approx 10 - 0.133t + 0.00044t^2.$$

Here is a table of values of $h(t)$ correct to one decimal place.

t (in s)	$h(t)$ (in cm)
10	8.7
20	7.5
30	6.4
40	5.4
50	4.5
60	3.6

(b) The answers to this part are to be obtained experimentally. See the article by Tom Farmer and Fred Gass, *Physical Demonstrations in the Calculus Classroom,* College Mathematics Journal 1992, pp. 146–148.

3. $V(t) = \pi r^2 h(t) = 100\pi h(t) \Rightarrow \dfrac{dV}{dh} = 100\pi$ and $\dfrac{dV}{dt} = \dfrac{dV}{dh}\dfrac{dh}{dt} = 100\pi\dfrac{dh}{dt}$.

Diameter $= 2.5$ inches $\Rightarrow$ radius $= 1.25$ inches $= \frac{5}{4}\cdot\frac{1}{12}$ foot $= \frac{5}{48}$ foot. Thus, $\dfrac{dV}{dt} = -a\sqrt{2gh} \Rightarrow$

$100\pi\dfrac{dh}{dt} = -\pi\left(\frac{5}{48}\right)^2\sqrt{2\cdot 32h} = -\dfrac{25\pi}{288}\sqrt{h} \Rightarrow \dfrac{dh}{dt} = -\dfrac{\sqrt{h}}{1152} \Rightarrow \int h^{-1/2}\,dh = \int -\frac{1}{1152}\,dt \Rightarrow$

$2\sqrt{h} = -\frac{1}{1152}t + C \Rightarrow \sqrt{h} = -\frac{1}{2304}t + k \Rightarrow h(t) = \left(-\frac{1}{2304}t + k\right)^2$. The water pressure after t seconds is $62.5h(t)$ lb/ft^2, so the condition that the pressure be at least 2160 lb/ft^2 for 10 minutes (600 seconds) is the condition $62.5\cdot h(600) \geq 2160$; that is, $\left(k - \frac{600}{2304}\right)^2 \geq \frac{2160}{62.5} \Rightarrow \left|k - \frac{25}{96}\right| \geq \sqrt{34.56} \Rightarrow k \geq \frac{25}{96} + \sqrt{34.56}$. Now $h(0) = k^2$, so the height of the tank should be at least $\left(\frac{25}{96} + \sqrt{34.56}\right)^2 \approx 37.69$ ft.

4. (a) If the radius of the circular cross-section at height h is r, then the Pythagorean Theorem gives $r^2 = 2^2 - (2 - h)^2$ since the radius of the tank is 2 m. So $A(h) = \pi r^2 = \pi[4 - (2 - h)^2] = \pi(4h - h^2)$. Thus, $A(h)\dfrac{dh}{dt} = -a\sqrt{2gh} \Rightarrow$

$\pi(4h - h^2)\dfrac{dh}{dt} = -\pi(0.01)^2\sqrt{2\cdot 10h} \Rightarrow (4h - h^2)\dfrac{dh}{dt} = -0.0001\sqrt{20h}$.

(b) From part (a) we have $(4h^{1/2} - h^{3/2})\,dh = \left(-0.0001\sqrt{20}\right)dt \Rightarrow \frac{8}{3}h^{3/2} - \frac{2}{5}h^{5/2} = \left(-0.0001\sqrt{20}\right)t + C$.

$h(0) = 2 \Rightarrow \frac{8}{3}(2)^{3/2} - \frac{2}{5}(2)^{5/2} = C \Rightarrow C = \left(\frac{16}{3} - \frac{8}{5}\right)\sqrt{2} = \frac{56}{15}\sqrt{2}$. To find out how long it will take to drain all the water we evaluate t when $h = 0$: $0 = \left(-0.0001\sqrt{20}\right)t + C \Rightarrow$

$$t = \frac{C}{0.0001\sqrt{20}} = \frac{56\sqrt{2}/15}{0.0001\sqrt{20}} = \frac{11{,}200\sqrt{10}}{3} \approx 11{,}806 \text{ s} \approx 3 \text{ h } 17 \text{ min}$$

APPLIED PROJECT Which Is Faster, Going Up or Coming Down?

1. $mv' = -pv - mg \;\Rightarrow\; m\dfrac{dv}{dt} = -(pv + mg) \;\Rightarrow\; \displaystyle\int \frac{dv}{pv + mg} = \int -\frac{1}{m}\,dt \;\Rightarrow$

$\dfrac{1}{p}\ln(pv + mg) = -\dfrac{1}{m}t + C \;\; [pv + mg > 0]$. At $t = 0$, $v = v_0$, so $C = \dfrac{1}{p}\ln(pv_0 + mg)$.

Thus, $\dfrac{1}{p}\ln(pv + mg) = -\dfrac{1}{m}t + \dfrac{1}{p}\ln(pv_0 + mg) \;\Rightarrow\; \ln(pv + mg) = -\dfrac{p}{m}t + \ln(pv_0 + mg) \;\Rightarrow$

$pv + mg = e^{-pt/m}(pv_0 + mg) \;\Rightarrow\; pv = (pv_0 + mg)e^{-pt/m} - mg \;\Rightarrow\; v(t) = \left(v_0 + \dfrac{mg}{p}\right)e^{-pt/m} - \dfrac{mg}{p}$.

2. $y(t) = \int v(t)\,dt = \displaystyle\int \left[\left(v_0 + \frac{mg}{p}\right)e^{-pt/m} - \frac{mg}{p}\right]dt = \left(v_0 + \frac{mg}{p}\right)e^{-pt/m}\left(-\frac{m}{p}\right) - \frac{mg}{p}t + C.$

At $t = 0$, $y = 0$, so $C = \left(v_0 + \dfrac{mg}{p}\right)\dfrac{m}{p}$. Thus,

$$y(t) = \left(v_0 + \frac{mg}{p}\right)\frac{m}{p} - \left(v_0 + \frac{mg}{p}\right)\frac{m}{p}e^{-pt/m} - \frac{mgt}{p} = \left(v_0 + \frac{mg}{p}\right)\frac{m}{p}\left(1 - e^{-pt/m}\right) - \frac{mgt}{p}$$

3. $v(t) = 0 \;\Rightarrow\; \dfrac{mg}{p} = \left(v_0 + \dfrac{mg}{p}\right)e^{-pt/m} \;\Rightarrow\; e^{pt/m} = \dfrac{pv_0}{mg} + 1 \;\Rightarrow\; \dfrac{pt}{m} = \ln\left(\dfrac{pv_0}{mg} + 1\right) \;\Rightarrow$

$t_1 = \dfrac{m}{p}\ln\left(\dfrac{mg + pv_0}{mg}\right)$. With $m = 1$, $v_0 = 20$, $p = \frac{1}{10}$, and $g = 9.8$, we have $t_1 = 10\ln\left(\frac{11.8}{9.8}\right) \approx 1.86$ s.

4.

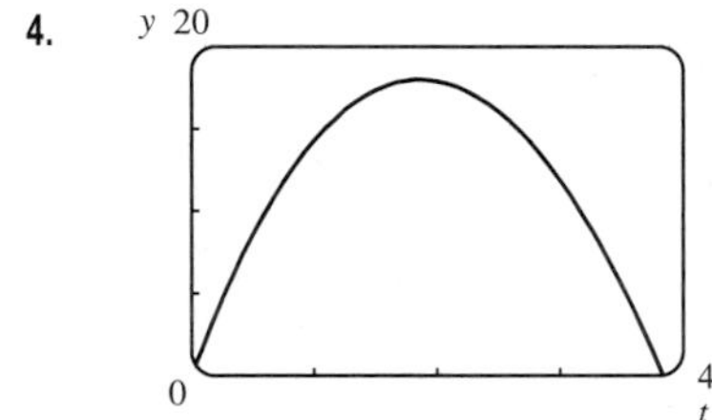

The figure shows the graph of $y = 1180(1 - e^{-0.1t}) - 98t$. The zeros are at $t = 0$ and $t_2 \approx 3.84$. Thus, $t_1 - 0 \approx 1.86$ and $t_2 - t_1 \approx 1.98$. So the time it takes to come down is about 0.12 s longer than the time it takes to go up; hence, going up is faster.

5. $y(2t_1) = \left(v_0 + \dfrac{mg}{p}\right)\dfrac{m}{p}(1 - e^{-2pt_1/m}) - \dfrac{mg}{p}\cdot 2t_1$

$\quad = \left(\dfrac{pv_0 + mg}{p}\right)\dfrac{m}{p}\left[1 - (e^{pt_1/m})^{-2}\right] - \dfrac{mg}{p}\cdot 2\dfrac{m}{p}\ln\left(\dfrac{pv_0 + mg}{mg}\right)$

Substituting $x = e^{pt_1/m} = \dfrac{pv_0}{mg} + 1 = \dfrac{pv_0 + mg}{mg}$ (from Problem 3), we get

$y\,(2t_1) = \left(x\cdot\dfrac{mg}{p}\right)\dfrac{m}{p}(1 - x^{-2}) - \dfrac{m^2g}{p^2}\cdot 2\ln x = \dfrac{m^2g}{p^2}\left(x - \dfrac{1}{x} - 2\ln x\right)$. Now $p > 0$, $m > 0$, $t_1 > 0 \;\Rightarrow$

$x = e^{pt_1/m} > e^0 = 1$. $f(x) = x - \dfrac{1}{x} - 2\ln x \;\Rightarrow\; f'(x) = 1 + \dfrac{1}{x^2} - \dfrac{2}{x} = \dfrac{x^2 - 2x + 1}{x^2} = \dfrac{(x-1)^2}{x^2} > 0$

for $x > 1 \;\Rightarrow\; f(x)$ is increasing for $x > 1$. Since $f(1) = 0$, it follows that $f(x) > 0$ for every $x > 1$. Therefore,

$y(2t_1) = \dfrac{m^2g}{p^2}f(x)$ is positive, which means that the ball has not yet reached the ground at time $2t_1$. This tells us that the time spent going up is always less than the time spent coming down, so *ascent is faster*.

9.4 Models for Population Growth

1. (a) Comparing the given equation, $\dfrac{dP}{dt} = 0.04P\left(1 - \dfrac{P}{M}\right)$, to Equation 4, $\dfrac{dP}{dt} = kP\left(1 - \dfrac{P}{M}\right)$, we see that the carrying capacity is $M = 1200$ and the value of k is 0.04.

(b) By Equation 7, the solution of the equation is $P(t) = \dfrac{M}{1 + Ae^{-kt}}$, where $A = \dfrac{M - P_0}{P_0}$. Since $P(0) = P_0 = 60$, we have $A = \dfrac{1200 - 60}{60} = 19$, and hence, $P(t) = \dfrac{1200}{1 + 19e^{-0.04t}}$

(c) The population after 10 weeks is $P(10) = \dfrac{1200}{1 + 19e^{-0.04(10)}} \approx 87$.

2. (a) $dP/dt = 0.02P - 0.00004P^2 = 0.02P(1 - 0.002P) = 0.02P(1 - P/500)$. Comparing to Equation 4, $dP/dt = kP(1 - P/M)$, we see that the carrying capacity is $M = 500$ and the value of k is 0.02.

(b) By Equation 7, the solution of the equation is $P(t) = \dfrac{M}{1 + Ae^{-kt}}$, where $A = \dfrac{M - P_0}{P_0}$. Since $P(0) = P_0 = 40$, we have $A = \dfrac{500 - 40}{40} = 11.5$, and hence, $P(t) = \dfrac{500}{1 + 11.5e^{-0.02t}}$.

(c) The population after 10 weeks is $P(10) = \dfrac{500}{1 + 11.5e^{-0.02(10)}} \approx 48$.

3. (a) $dP/dt = 0.05P - 0.0005P^2 = 0.05P(1 - 0.01P) = 0.05P(1 - P/100)$. Comparing to Equation 4, $dP/dt = kP(1 - P/M)$, we see that the carrying capacity is $M = 100$ and the value of k is 0.05.

(b) The slopes close to 0 occur where P is near 0 or 100. The largest slopes appear to be on the line $P = 50$. The solutions are increasing for $0 < P_0 < 100$ and decreasing for $P_0 > 100$.

(c)

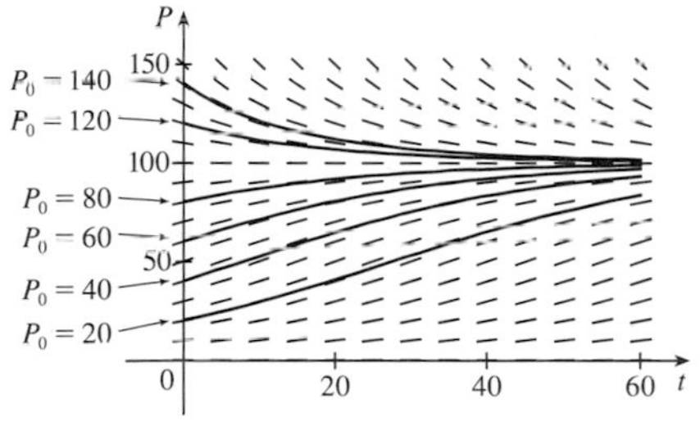

All of the solutions approach $P = 100$ as t increases. As in part (b), the solutions differ since for $0 < P_0 < 100$ they are increasing, and for $P_0 > 100$ they are decreasing. Also, some have an IP and some don't. It appears that the solutions which have $P_0 = 20$ and $P_0 = 40$ have inflection points at $P = 50$.

(d) The equilibrium solutions are $P = 0$ (trivial solution) and $P = 100$. The increasing solutions move away from $P = 0$ and all nonzero solutions approach $P = 100$ as $t \to \infty$.

4. (a) $M = 6000$ and $k = 0.0015 \quad \Rightarrow \quad dP/dt = 0.0015P(1 - P/6000)$.

(b)

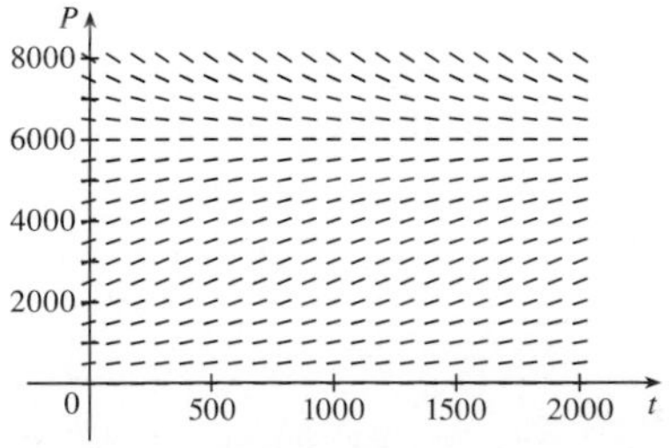

All of the solution curves approach 6000 as $t \to \infty$.

(c) 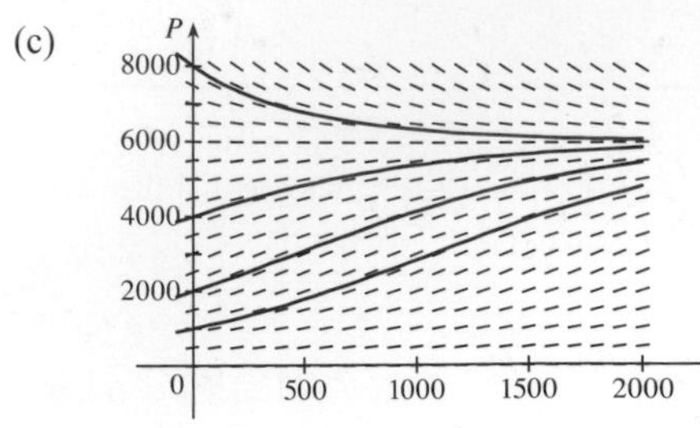

The curves with $P_0 = 1000$ and $P_0 = 2000$ appear to be concave upward at first and then concave downward. The curve with $P_0 = 4000$ appears to be concave downward everywhere. The curve with $P_0 = 8000$ appears to be concave upward everywhere. The inflection points are where the population grows the fastest.

(d) See the solution to Exercise 9.2.25 for a possible program to calculate $P(50)$. [In this case, we use $\mathrm{X} = 0$, $\mathrm{H} = 1$, $\mathrm{N} = 50$, $\mathrm{Y}_1 = 0.0015y(1 - y/6000)$, and $\mathrm{Y} = 1000$.] We find that $P(50) \approx 1064$.

(e) Using Equation 7 with $M = 6000$, $k = 0.0015$, and $P_0 = 1000$, we have $P(t) = \dfrac{M}{1 + Ae^{-kt}} = \dfrac{6000}{1 + Ae^{-0.0015t}}$, where $A = \dfrac{M - P_0}{P_0} = \dfrac{6000 - 1000}{1000} = 5$. Thus, $P(50) = \dfrac{6000}{1 + 5e^{-0.0015(50)}} \approx 1064.1$, which is extremely close to the estimate obtained in part (d).

(f) 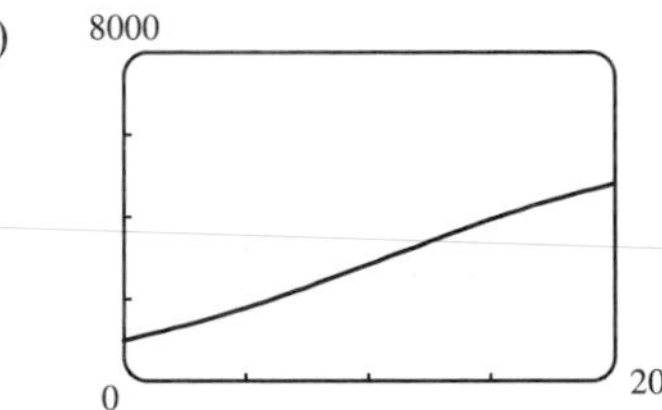

The curves are very similar.

5. (a) $\dfrac{dy}{dt} = ky\left(1 - \dfrac{y}{M}\right) \;\Rightarrow\; y(t) = \dfrac{M}{1 + Ae^{-kt}}$ with $A = \dfrac{M - y(0)}{y(0)}$. With $M = 8 \times 10^7$, $k = 0.71$, and $y(0) = 2 \times 10^7$, we get the model $y(t) = \dfrac{8 \times 10^7}{1 + 3e^{-0.71t}}$, so $y(1) = \dfrac{8 \times 10^7}{1 + 3e^{-0.71}} \approx 3.23 \times 10^7$ kg.

(b) $y(t) = 4 \times 10^7 \;\Rightarrow\; \dfrac{8 \times 10^7}{1 + 3e^{-0.71t}} = 4 \times 10^7 \;\Rightarrow\; 2 = 1 + 3e^{-0.71t} \;\Rightarrow\; e^{-0.71t} = \frac{1}{3} \;\Rightarrow$

$-0.71t = \ln \frac{1}{3} \;\Rightarrow\; t = \dfrac{\ln 3}{0.71} \approx 1.55$ years

6. (a) $\dfrac{dP}{dt} = 0.4P - 0.001P^2 = 0.4P(1 - 0.0025P)$ $\left[\frac{0.001}{0.4} = 0.0025\right]$ $= 0.4P\left(1 - \dfrac{P}{400}\right)$ $\left[0.0025^{-1} = 400\right]$

Thus, by Equation 4, $k = 0.4$ and the carrying capacity is 400.

(b) Using the fact that $P(0) = 50$ and the formula for dP/dt, we get

$$P'(0) = \left.\frac{dP}{dt}\right|_{t=0} = 0.4(50) - 0.001(50)^2 = 20 - 2.5 = 17.5.$$

(c) From Equation 7, $A = \dfrac{M - P_0}{P_0} = \dfrac{400 - 50}{50} = 7$, so $P = \dfrac{400}{1 + 7e^{-0.4t}}$. The population reaches 50% of the carrying capacity, 200, when $200 = \dfrac{400}{1 + 7e^{-0.4t}} \;\Rightarrow\; 1 + 7e^{-0.4t} = 2 \;\Rightarrow\; e^{-0.4t} = \frac{1}{7} \;\Rightarrow\; -0.4t = \ln \frac{1}{7} \;\Rightarrow$

$t = (\ln \frac{1}{7})/(-0.4) \approx 4.86$ years.

7. Using (7), $A = \dfrac{M - P_0}{P_0} = \dfrac{10{,}000 - 1000}{1000} = 9$, so $P(t) = \dfrac{10{,}000}{1 + 9e^{-kt}}$. $P(1) = 2500 \Rightarrow 2500 = \dfrac{10{,}000}{1 + 9e^{-k(1)}} \Rightarrow$

$1 + 9e^{-k} = 4 \Rightarrow 9e^{-k} = 3 \Rightarrow e^{-k} = \frac{1}{3} \Rightarrow -k = \ln\frac{1}{3} \Rightarrow k = \ln 3$. After another three years, $t = 4$,

and $P(4) = \dfrac{10{,}000}{1 + 9e^{-(\ln 3)4}} = \dfrac{10{,}000}{1 + 9\left(e^{\ln 3}\right)^{-4}} = \dfrac{10{,}000}{1 + 9(3)^{-4}} = \dfrac{10{,}000}{1 + \frac{1}{9}} = \dfrac{10{,}000}{\frac{10}{9}} = 9000$.

8. (a) From the graph, we estimate the carrying capacity M for the yeast population to be 680.

(b) An estimate of the initial relative growth rate is $\dfrac{1}{P_0}\dfrac{dP}{dt} = \dfrac{1}{18} \cdot \dfrac{39 - 18}{2 - 0} = \dfrac{7}{12} = 0.58\overline{3}$.

(c) An exponential model is $P(t) = 18e^{7t/12}$. A logistic model is $P(t) = \dfrac{680}{1 + Ae^{-7t/12}}$, where $A = \frac{680 - 18}{18} = \frac{331}{9}$.

(d)

Time in Hours	Observed Values	Exponential Model	Logistic Model
0	18	18	18
2	39	58	55
4	80	186	149
6	171	596	322
8	336	1914	505
10	509	6147	614
12	597	19,739	658
14	640	63,389	673
16	664	203,558	678
18	672	653,679	679

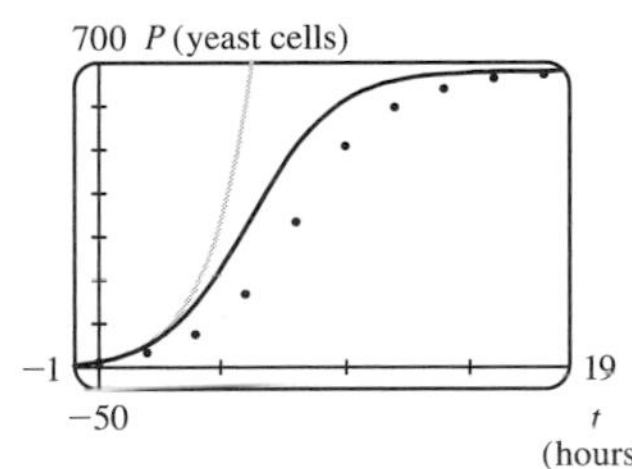

The exponential model is a poor fit for anything beyond the first two observed values. The logistic model varies more for the middle values than it does for the values at either end, but provides a good general fit, as shown in the figure.

(e) $P(7) = \dfrac{680}{1 + \frac{331}{9}e^{-7(7/12)}} \approx 420$ yeast cells

9. (a) We will assume that the difference in birth and death rates is 20 million/year. Let $t = 0$ correspond to the year 2000. Thus,

$$k \approx \frac{1}{P}\frac{dP}{dt} = \frac{1}{6.1 \text{ billion}}\left(\frac{20 \text{ million}}{\text{year}}\right) = \frac{1}{305}, \text{ and } \frac{dP}{dt} = kP\left(1 - \frac{P}{M}\right) = \frac{1}{305}P\left(1 - \frac{P}{20}\right) \text{ with } P \text{ in billions.}$$

(b) $A = \dfrac{M - P_0}{P_0} = \dfrac{20 - 6.1}{6.1} = \dfrac{139}{61} \approx 2.2787$. $P(t) = \dfrac{M}{1 + Ae^{-kt}} = \dfrac{20}{1 + \frac{139}{61}e^{-t/305}}$, so

$P(10) = \dfrac{20}{1 + \frac{139}{61}e^{-10/305}} \approx 6.24$ billion, which underestimates the actual 2010 population of 6.9 billion.

(c) The years 2100 and 2500 correspond to $t = 100$ and $t = 500$, respectively. $P(100) = \dfrac{20}{1 + \frac{139}{61}e^{-100/305}} \approx 7.57$ billion

and $P(500) = \dfrac{20}{1 + \frac{139}{61}e^{-500/305}} \approx 13.87$ billion.

10. (a) Let $t = 0$ correspond to the year 2000. $A = \dfrac{M - P_0}{P_0} = \dfrac{800 - 282}{282} = \dfrac{259}{141} \approx 1.8369$.

$$P(t) = \frac{M}{1 + Ae^{-kt}} = \frac{800}{1 + \frac{259}{141}e^{-kt}} \text{ with } P \text{ in millions.}$$

(b) $P(10) = 309 \;\Leftrightarrow\; \dfrac{800}{1 + \frac{259}{141}e^{-10k}} = 309 \;\Leftrightarrow\; \dfrac{800}{309} = 1 + \dfrac{259}{141}e^{-10k} \;\Leftrightarrow\; \dfrac{491}{309} = \dfrac{259}{141}e^{-10k} \;\Leftrightarrow$

$\dfrac{491 \cdot 141}{309 \cdot 259} = e^{-10k} \;\Leftrightarrow\; -10k = \ln \dfrac{491 \cdot 47}{103 \cdot 259} \;\Leftrightarrow\; k = -\dfrac{1}{10} \ln \dfrac{23{,}077}{26{,}677} \approx 0.0145$.

(c) The years 2100 and 2200 correspond to $t = 100$ and $t = 200$, respectively. $P(100) = \dfrac{800}{1 + \frac{259}{141}e^{-100k}} \approx 559$ million and

$P(200) = \dfrac{800}{1 + \frac{259}{141}e^{-200k}} \approx 727$ million.

(d) $P(t) = 500 \;\Leftrightarrow\; \dfrac{800}{1 + \frac{259}{141}e^{-kt}} = 500 \;\Leftrightarrow\; \dfrac{800}{500} = 1 + \dfrac{259}{141}e^{-kt} \;\Leftrightarrow\; \dfrac{3}{5} = \dfrac{259}{141}e^{-kt} \;\Leftrightarrow\; \dfrac{3 \cdot 141}{5 \cdot 259} = e^{-kt} \;\Leftrightarrow$

$-kt = \ln \dfrac{423}{1295} \;\Leftrightarrow\; t = 10\dfrac{\ln(423/1295)}{\ln(23{,}077/26{,}677)} \approx 77.18$ years. Our logistic model predicts that the US population will exceed 500 million in 77.18 years; that is, in the year 2077.

11. (a) Our assumption is that $\dfrac{dy}{dt} = ky(1 - y)$, where y is the fraction of the population that has heard the rumor.

(b) Using the logistic equation (4), $\dfrac{dP}{dt} = kP\left(1 - \dfrac{P}{M}\right)$, we substitute $y = \dfrac{P}{M}$, $P = My$, and $\dfrac{dP}{dt} = M\dfrac{dy}{dt}$,

to obtain $M\dfrac{dy}{dt} = k(My)(1 - y) \;\Leftrightarrow\; \dfrac{dy}{dt} = ky(1 - y)$, our equation in part (a).

Now the solution to (4) is $P(t) = \dfrac{M}{1 + Ae^{-kt}}$, where $A = \dfrac{M - P_0}{P_0}$.

We use the same substitution to obtain $My = \dfrac{M}{1 + \dfrac{M - My_0}{My_0}e^{-kt}} \;\Rightarrow\; y = \dfrac{y_0}{y_0 + (1 - y_0)e^{-kt}}$.

Alternatively, we could use the same steps as outlined in the solution of Equation 4.

(c) Let t be the number of hours since 8 AM. Then $y_0 = y(0) = \frac{80}{1000} = 0.08$ and $y(4) = \frac{1}{2}$, so

$\dfrac{1}{2} = y(4) = \dfrac{0.08}{0.08 + 0.92e^{-4k}}$. Thus, $0.08 + 0.92e^{-4k} = 0.16$, $e^{-4k} = \frac{0.08}{0.92} = \frac{2}{23}$, and $e^{-k} = \left(\frac{2}{23}\right)^{1/4}$,

so $y = \dfrac{0.08}{0.08 + 0.92(2/23)^{t/4}} = \dfrac{2}{2 + 23(2/23)^{t/4}}$. Solving this equation for t, we get

$$2y + 23y\left(\frac{2}{23}\right)^{t/4} = 2 \;\Rightarrow\; \left(\frac{2}{23}\right)^{t/4} = \frac{2 - 2y}{23y} \;\Rightarrow\; \left(\frac{2}{23}\right)^{t/4} = \frac{2}{23} \cdot \frac{1 - y}{y} \;\Rightarrow\; \left(\frac{2}{23}\right)^{t/4 - 1} = \frac{1 - y}{y}.$$

It follows that $\dfrac{t}{4} - 1 = \dfrac{\ln[(1 - y)/y]}{\ln \frac{2}{23}}$, so $t = 4\left[1 + \dfrac{\ln((1 - y)/y)}{\ln \frac{2}{23}}\right]$.

When $y = 0.9$, $\dfrac{1 - y}{y} = \frac{1}{9}$, so $t = 4\left(1 - \dfrac{\ln 9}{\ln \frac{2}{23}}\right) \approx 7.6$ h or 7 h 36 min. Thus, 90% of the population will have heard the rumor by 3:36 PM.

12. (a) $P(0) = P_0 = 400$, $P(1) = 1200$ and $M = 10{,}000$. From the solution to the logistic differential equation

$$P(t) = \frac{P_0 M}{P_0 + (M - P_0)e^{-kt}}, \text{ we get } P = \frac{400\,(10{,}000)}{400 + (9600)e^{-kt}} = \frac{10{,}000}{1 + 24e^{-kt}}. \quad P(1) = 1200 \;\Rightarrow$$

$$1 + 24e^{-k} = \tfrac{100}{12} \;\Rightarrow\; e^k = \tfrac{288}{88} \;\Rightarrow\; k = \ln\tfrac{36}{11}. \text{ So } P = \frac{10{,}000}{1 + 24e^{-t\ln(36/11)}} = \frac{10{,}000}{1 + 24\cdot(11/36)^t}.$$

(b) $$5000 = \frac{10{,}000}{1 + 24(11/36)^t} \;\Rightarrow\; 24\left(\tfrac{11}{36}\right)^t = 1 \;\Rightarrow\; t\ln\tfrac{11}{36} = \ln\tfrac{1}{24} \;\Rightarrow\; t \approx 2.68 \text{ years.}$$

13. (a) $$\frac{dP}{dt} = kP\left(1 - \frac{P}{M}\right) \;\Rightarrow\; \frac{d^2P}{dt^2} = k\left[P\left(-\frac{1}{M}\frac{dP}{dt}\right) + \left(1 - \frac{P}{M}\right)\frac{dP}{dt}\right] = k\,\frac{dP}{dt}\left(-\frac{P}{M} + 1 - \frac{P}{M}\right)$$

$$= k\left[kP\left(1 - \frac{P}{M}\right)\right]\left(1 - \frac{2P}{M}\right) = k^2P\left(1 - \frac{P}{M}\right)\left(1 - \frac{2P}{M}\right)$$

(b) P grows fastest when P' has a maximum, that is, when $P'' = 0$. From part (a), $P'' = 0 \;\Leftrightarrow\; P = 0, P = M$, or $P = M/2$. Since $0 < P < M$, we see that $P'' = 0 \;\Leftrightarrow\; P = M/2$.

14. First we keep k constant (at 0.1, say) and change P_0 in the function

$$P = \frac{10P_0}{P_0 + (10 - P_0)e^{-0.1t}}.$$ (Notice that P_0 is the P-intercept.) If $P_0 = 0$, the function is 0 everywhere. For $0 < P_0 < 5$, the curve has an inflection point, which moves to the right as P_0 decreases. If $5 < P_0 < 10$, the graph is concave down everywhere. (We are considering only $t \geq 0$.) If $P_0 = 10$, the function is the constant function $P = 10$, and if $P_0 > 10$, the function decreases. For all $P_0 \neq 0$, $\lim_{t\to\infty} P = 10$.

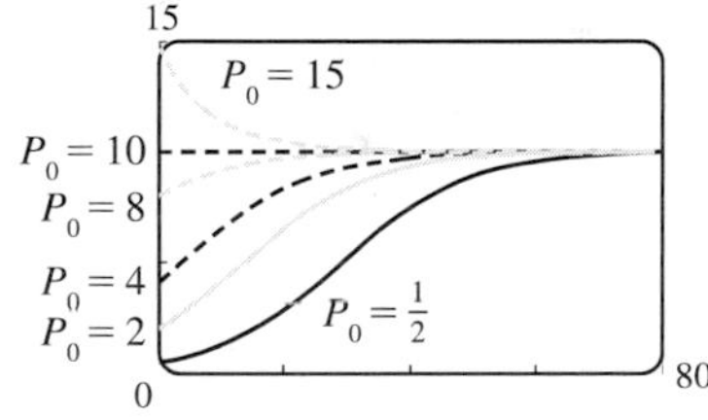

Now we instead keep P_0 constant (at $P_0 = 1$) and change k in the function

$$P = \frac{10}{1 + 9e^{-kt}}.$$ It seems that as k increases, the graph approaches the line $P = 10$ more and more quickly. (Note that the only difference in the shape of the curves is in the horizontal scaling; if we choose suitable x-scales, the graphs all look the same.)

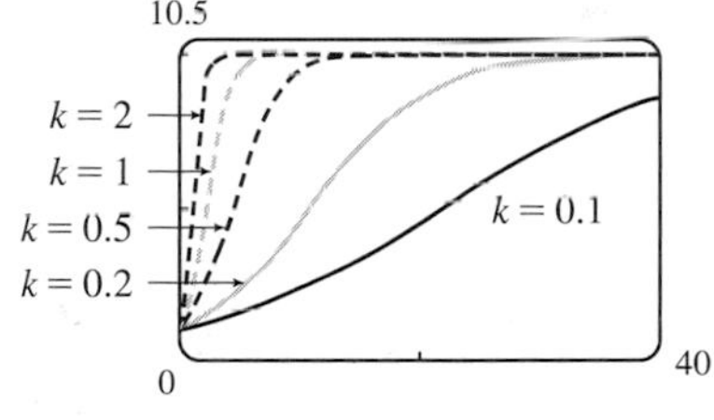

15. Following the hint, we choose $t = 0$ to correspond to 1960 and subtract 94,000 from each of the population figures. We then use a calculator to obtain the models and add 94,000 to get the exponential function $P_E(t) = 1909.7761(1.0796)^t + 94{,}000$ and the logistic function

$$P_L(t) = \frac{33{,}086.4394}{1 + 12.3428e^{-0.1657t}} + 94{,}000.$$ P_L is a reasonably accurate model, while P_E is not, since an exponential model would only be used for the first few data points.

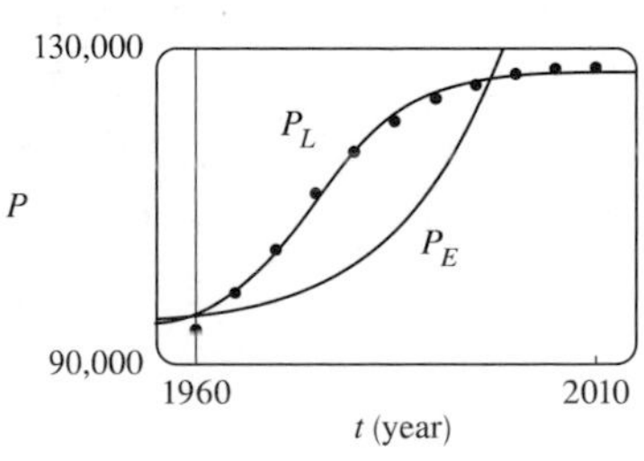

16. Following the hint, we choose $t = 0$ to correspond to 1960 and subtract 3500 from each of the population figures. We then use a calculator to obtain the models and add 3500 to get the exponential function $P_E(t) = 180.9934(1.0445)^t + 3500$ and the logistic function $P_L(t) = \dfrac{1348.9650}{1 + 6.2784e^{-0.0721t}} + 3500$. P_L is a reasonably accurate accurate model, while P_E is not, since an exponential model would only be used for the first few data points.

17. (a) $\dfrac{dP}{dt} = kP - m = k\left(P - \dfrac{m}{k}\right)$. Let $y = P - \dfrac{m}{k}$, so $\dfrac{dy}{dt} = \dfrac{dP}{dt}$ and the differential equation becomes $\dfrac{dy}{dt} = ky$.

The solution is $y = y_0 e^{kt} \Rightarrow P - \dfrac{m}{k} = \left(P_0 - \dfrac{m}{k}\right)e^{kt} \Rightarrow P(t) = \dfrac{m}{k} + \left(P_0 - \dfrac{m}{k}\right)e^{kt}$.

(b) Since $k > 0$, there will be an exponential expansion $\Leftrightarrow P_0 - \dfrac{m}{k} > 0 \Leftrightarrow m < kP_0$.

(c) The population will be constant if $P_0 - \dfrac{m}{k} = 0 \Leftrightarrow m = kP_0$. It will decline if $P_0 - \dfrac{m}{k} < 0 \Leftrightarrow m > kP_0$.

(d) $P_0 = 8{,}000{,}000$, $k = \alpha - \beta = 0.016$, $m = 210{,}000 \Rightarrow m > kP_0\ (= 128{,}000)$, so by part (c), the population was declining.

18. (a) $\dfrac{dy}{dt} = ky^{1+c} \Rightarrow y^{-1-c}\,dy = k\,dt \Rightarrow \dfrac{y^{-c}}{-c} = kt + C$. Since $y(0) = y_0$, we have $C = \dfrac{y_0^{-c}}{-c}$. Thus,

$\dfrac{y^{-c}}{-c} = kt + \dfrac{y_0^{-c}}{-c}$, or $y^{-c} = y_0^{-c} - ckt$. So $y^c = \dfrac{1}{y_0^{-c} - ckt} = \dfrac{y_0^c}{1 - cy_0^c kt}$ and $y(t) = \dfrac{y_0}{(1 - cy_0^c kt)^{1/c}}$.

(b) $y(t) \to \infty$ as $1 - cy_0^c kt \to 0$, that is, as $t \to \dfrac{1}{cy_0^c k}$. Define $T = \dfrac{1}{cy_0^c k}$. Then $\lim\limits_{t \to T^-} y(t) = \infty$.

(c) According to the data given, we have $c = 0.01$, $y(0) = 2$, and $y(3) = 16$, where the time t is given in months. Thus,

$y_0 = 2$ and $16 = y(3) = \dfrac{y_0}{(1 - cy_0^c k \cdot 3)^{1/c}}$. Since $T = \dfrac{1}{cy_0^c k}$, we will solve for $cy_0^c k$. $16 = \dfrac{2}{(1 - 3cy_0^c k)^{100}} \Rightarrow$

$1 - 3cy_0^c k = \left(\frac{1}{8}\right)^{0.01} = 8^{-0.01} \Rightarrow cy_0^c k = \frac{1}{3}\left(1 - 8^{-0.01}\right)$. Thus, doomsday occurs when

$t = T = \dfrac{1}{cy_0^c k} = \dfrac{3}{1 - 8^{-0.01}} \approx 145.77$ months or 12.15 years.

19. (a) The term -15 represents a harvesting of fish at a constant rate—in this case, 15 fish/week. This is the rate at which fish are caught.

(b)

(c) From the graph in part (b), it appears that $P(t) = 250$ and $P(t) = 750$ are the equilibrium solutions. We confirm this analytically by solving the equation $dP/dt = 0$ as follows: $0.08P(1 - P/1000) - 15 = 0 \Rightarrow$

$0.08P - 0.00008P^2 - 15 = 0 \Rightarrow$

$-0.00008(P^2 - 1000P + 187{,}500) = 0 \Rightarrow$

$(P - 250)(P - 750) = 0 \Rightarrow P = 250$ or 750.

(d)

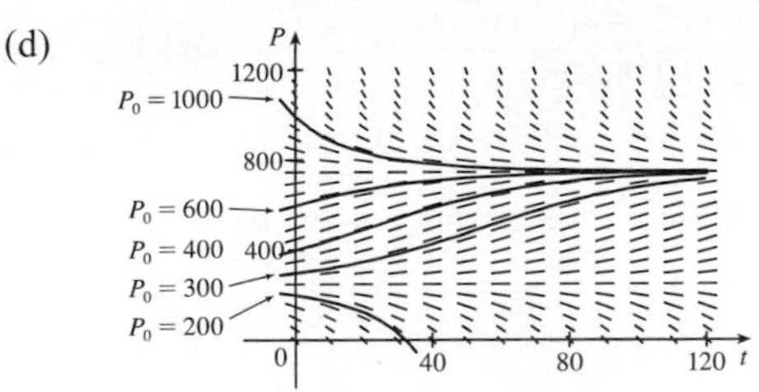

For $0 < P_0 < 250$, $P(t)$ decreases to 0. For $P_0 = 250$, $P(t)$ remains constant. For $250 < P_0 < 750$, $P(t)$ increases and approaches 750. For $P_0 = 750$, $P(t)$ remains constant. For $P_0 > 750$, $P(t)$ decreases and approaches 750.

(e) $\dfrac{dP}{dt} = 0.08P\left(1 - \dfrac{P}{1000}\right) - 15 \;\Leftrightarrow\; -\dfrac{100{,}000}{8}\dfrac{dP}{dt} = (0.08P - 0.00008P^2 - 15)\cdot\left(-\dfrac{100{,}000}{8}\right) \;\Leftrightarrow$

$-12{,}500\,\dfrac{dP}{dt} = P^2 - 1000P + 187{,}500 \;\Leftrightarrow\; \dfrac{dP}{(P-250)(P-750)} = -\dfrac{1}{12{,}500}\,dt \;\Leftrightarrow$

$\displaystyle\int\left(\frac{-1/500}{P-250} + \frac{1/500}{P-750}\right)dP = -\frac{1}{12{,}500}\,dt \;\Leftrightarrow\; \int\left(\frac{1}{P-250} - \frac{1}{P-750}\right)dP = \tfrac{1}{25}\,dt \;\Leftrightarrow$

$\ln|P-250| - \ln|P-750| = \frac{1}{25}t + C \;\Leftrightarrow\; \ln\left|\dfrac{P-250}{P-750}\right| = \frac{1}{25}t + C \;\Leftrightarrow\; \left|\dfrac{P-250}{P-750}\right| = e^{t/25+C} = ke^{t/25} \;\Leftrightarrow$

$\dfrac{P-250}{P-750} = ke^{t/25} \;\Leftrightarrow\; P - 250 = Pke^{t/25} - 750ke^{t/25} \;\Leftrightarrow\; P - Pke^{t/25} = 250 - 750ke^{t/25} \;\Leftrightarrow$

$P(t) = \dfrac{250 - 750ke^{t/25}}{1 - ke^{t/25}}$. If $t = 0$ and $P = 200$, then $200 = \dfrac{250 - 750k}{1-k} \;\Leftrightarrow\; 200 - 200k = 250 - 750k \;\Leftrightarrow$

$550k = 50 \;\Leftrightarrow\; k = \frac{1}{11}$. Similarly, if $t = 0$ and $P = 300$, then $k = -\frac{1}{9}$. Simplifying P with these two values of k gives us

$P(t) = \dfrac{250(3e^{t/25} - 11)}{e^{t/25} - 11}$ and $P(t) = \dfrac{750(e^{t/25} + 3)}{e^{t/25} + 9}$.

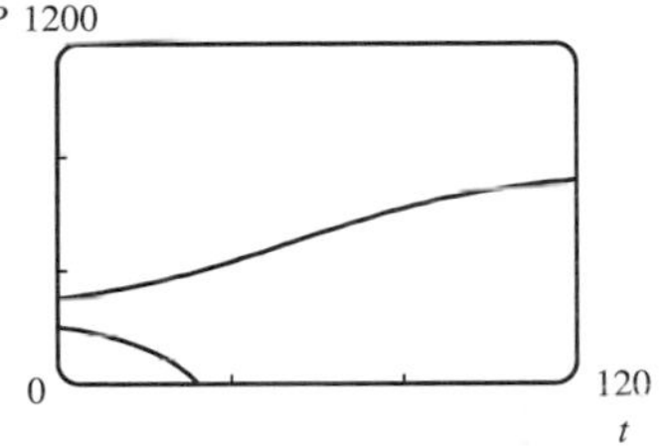

20. (a)

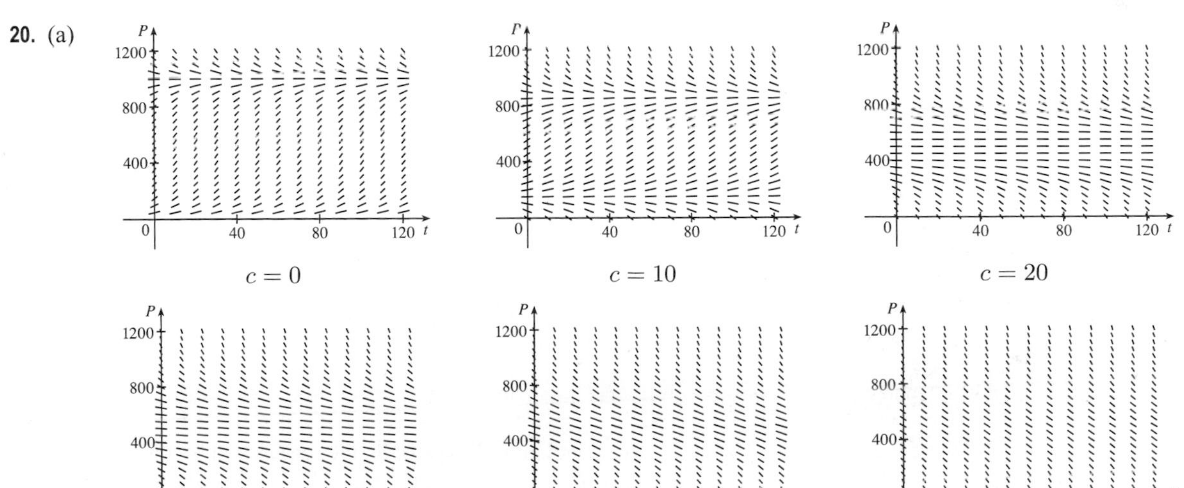

(b) For $0 \le c \le 20$, there is at least one equilibrium solution. For $c > 20$, the population always dies out.

(c) $\dfrac{dP}{dt} = 0.08P - 0.00008P^2 - c$. $\dfrac{dP}{dt} = 0 \Leftrightarrow P = \dfrac{-0.08 \pm \sqrt{(0.08)^2 - 4(-0.00008)(-c)}}{2(-0.00008)}$, which has at least one solution when the discriminant is nonnegative $\Rightarrow$ $0.0064 - 0.00032c \geq 0 \Leftrightarrow c \leq 20$. For $0 \leq c \leq 20$, there is at least one value of P such that $dP/dt = 0$ and hence, at least one equilibrium solution. For $c > 20$, $dP/dt < 0$ and the population always dies out.

(d) The weekly catch should be less than 20 fish per week.

21. (a) $\dfrac{dP}{dt} = (kP)\left(1 - \dfrac{P}{M}\right)\left(1 - \dfrac{m}{P}\right)$. If $m < P < M$, then $dP/dt = (+)(+)(+) = + \Rightarrow P$ is increasing. If $0 < P < m$, then $dP/dt = (+)(+)(-) = - \Rightarrow P$ is decreasing.

(b) 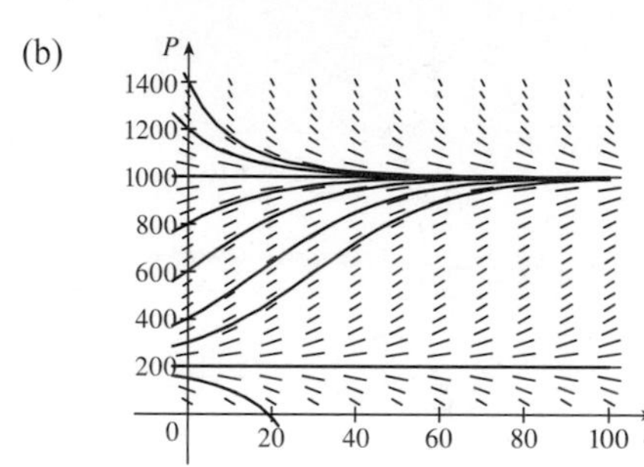

$k = 0.08$, $M = 1000$, and $m = 200 \Rightarrow$

$$\frac{dP}{dt} = 0.08P\left(1 - \frac{P}{1000}\right)\left(1 - \frac{200}{P}\right)$$

For $0 < P_0 < 200$, the population dies out. For $P_0 = 200$, the population is steady. For $200 < P_0 < 1000$, the population increases and approaches 1000. For $P_0 > 1000$, the population decreases and approaches 1000.

The equilibrium solutions are $P(t) = 200$ and $P(t) = 1000$.

(c) $\dfrac{dP}{dt} = kP\left(1 - \dfrac{P}{M}\right)\left(1 - \dfrac{m}{P}\right) = kP\left(\dfrac{M-P}{M}\right)\left(\dfrac{P-m}{P}\right) = \dfrac{k}{M}(M-P)(P-m) \Leftrightarrow$

$\displaystyle\int \frac{dP}{(M-P)(P-m)} = \int \frac{k}{M}\,dt$. By partial fractions, $\dfrac{1}{(M-P)(P-m)} = \dfrac{A}{M-P} + \dfrac{B}{P-m}$, so $A(P-m) + B(M-P) = 1$.

If $P = m$, $B = \dfrac{1}{M-m}$; if $P = M$, $A = \dfrac{1}{M-m}$, so $\displaystyle\frac{1}{M-m}\int\left(\frac{1}{M-P} + \frac{1}{P-m}\right)dP = \int \frac{k}{M}\,dt \Rightarrow$

$\dfrac{1}{M-m}(-\ln|M-P| + \ln|P-m|) = \dfrac{k}{M}t + C \Rightarrow \dfrac{1}{M-m}\ln\left|\dfrac{P-m}{M-P}\right| = \dfrac{k}{M}t + C \Rightarrow$

$\ln\left|\dfrac{P-m}{M-P}\right| = (M-m)\dfrac{k}{M}t + C_1 \Leftrightarrow \dfrac{P-m}{M-P} = De^{(M-m)(k/M)t}$ $\quad [D = \pm e^{C_1}]$.

Let $t = 0$: $\dfrac{P_0 - m}{M - P_0} = D$. So $\dfrac{P-m}{M-P} = \dfrac{P_0 - m}{M - P_0}e^{(M-m)(k/M)t}$.

Solving for P, we get $P(t) = \dfrac{m(M-P_0) + M(P_0 - m)e^{(M-m)(k/M)t}}{M - P_0 + (P_0 - m)e^{(M-m)(k/M)t}}$.

(d) If $P_0 < m$, then $P_0 - m < 0$. Let $N(t)$ be the numerator of the expression for $P(t)$ in part (c). Then $N(0) = P_0(M-m) > 0$, and $P_0 - m < 0 \Leftrightarrow \lim\limits_{t\to\infty} M(P_0 - m)e^{(M-m)(k/M)t} = -\infty \Rightarrow \lim\limits_{t\to\infty} N(t) = -\infty$. Since N is continuous, there is a number t such that $N(t) = 0$ and thus $P(t) = 0$. So the species will become extinct.

22. (a) $\dfrac{dP}{dt} = c\ln\left(\dfrac{M}{P}\right)P \Rightarrow \displaystyle\int \frac{dP}{P\ln(M/P)} = \int c\,dt$. Let $u = \ln\left(\dfrac{M}{P}\right) = \ln M - \ln P \Rightarrow du = -\dfrac{dP}{P} \Rightarrow$

$\displaystyle\int -\frac{du}{u} = ct + D \Rightarrow \ln|u| = -ct - D \Rightarrow |u| = e^{-(ct+D)} \Rightarrow |\ln(M/P)| = e^{-(ct+D)} \Rightarrow$

$\ln(M/P) = \pm e^{-(ct+D)}$. Letting $t = 0$, we get $\ln(M/P_0) = \pm e^{-D}$, so

$\ln(M/P) = \pm e^{-ct-D} = \pm e^{-ct}e^{-D} = \ln(M/P_0)e^{-ct} \quad \Rightarrow \quad M/P = e^{\ln(M/P_0)e^{-ct}} \quad \Rightarrow$

$P(t) = Me^{-\ln(M/P_0)e^{-ct}}$, $c \neq 0$.

(b) $\lim\limits_{t\to\infty} P(t) = \lim\limits_{t\to\infty} Me^{-\ln(M/P_0)e^{-ct}} = Me^{-\ln(M/P_0)\cdot 0} = Me^0 = M$

(c)

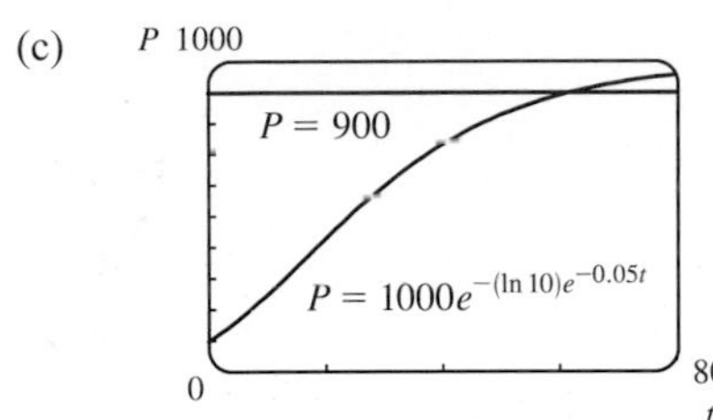

The graphs look very similar. For the Gompertz function, $P(40) \approx 732$, nearly the same as the logistic function. The Gompertz function reaches $P = 900$ at $t \approx 61.7$ and its value at $t = 80$ is about 959, so it doesn't increase quite as fast as the logistic curve.

(d) $\dfrac{dP}{dt} = c\ln\left(\dfrac{M}{P}\right)P = cP(\ln M - \ln P) \quad \Rightarrow$

$$\frac{d^2P}{dt^2} = c\left[P\left(-\frac{1}{P}\frac{dP}{dt}\right) + (\ln M - \ln P)\frac{dP}{dt}\right] = c\frac{dP}{dt}\left[-1 + \ln\left(\frac{M}{P}\right)\right]$$

$$= c[c\ln(M/P)P][\ln(M/P) - 1] = c^2P\ln(M/P)\,[\ln(M/P) - 1]$$

Since $0 < P < M$, $P'' = 0 \;\Leftrightarrow\; \ln(M/P) = 1 \;\Leftrightarrow\; M/P = e \;\Leftrightarrow\; P = M/e$. $P'' > 0$ for $0 < P < M/e$ and $P'' < 0$ for $M/e < P < M$, so P' is a maximum (and P grows fastest) when $P = M/e$.

Note: If $P > M$, then $\ln(M/P) < 0$, so $P''(t) > 0$.

23. (a) $dP/dt = kP\cos(rt - \phi) \;\Rightarrow\; (dP)/P = k\cos(rt-\phi)\,dt \;\Rightarrow\; \int(dP)/P = k\int\cos(rt-\phi)\,dt \;\Rightarrow$

$\ln P = (k/r)\sin(rt-\phi) + C$. (Since this is a growth model, $P > 0$ and we can write $\ln P$ instead of $\ln|P|$.) Since $P(0) = P_0$, we obtain $\ln P_0 = (k/r)\sin(-\phi) + C = -(k/r)\sin\phi + C \;\Rightarrow\; C = \ln P_0 + (k/r)\sin\phi$. Thus, $\ln P = (k/r)\sin(rt-\phi) + \ln P_0 + (k/r)\sin\phi$, which we can rewrite as $\ln(P/P_0) = (k/r)[\sin(rt-\phi) + \sin\phi]$ or, after exponentiation, $P(t) = P_0e^{(k/r)[\sin(rt-\phi)+\sin\phi]}$.

(b) As k increases, the amplitude increases, but the minimum value stays the same.

As r increases, the amplitude and the period decrease.

A change in ϕ produces slight adjustments in the phase shift and amplitude.

Comparing values of k with $P_0 = 1$, $r = 2$, and $\phi = \pi/2$

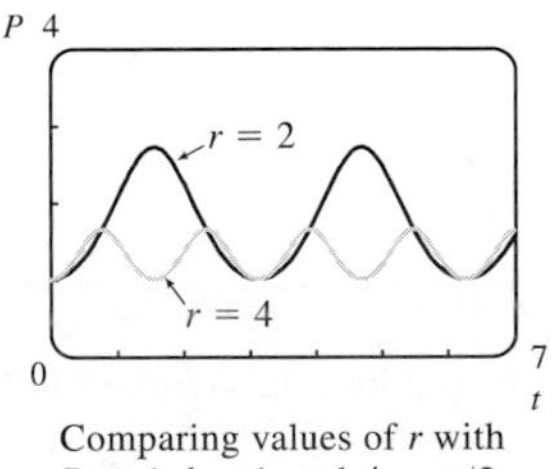

Comparing values of r with $P_0 = 1$, $k = 1$, and $\phi = \pi/2$

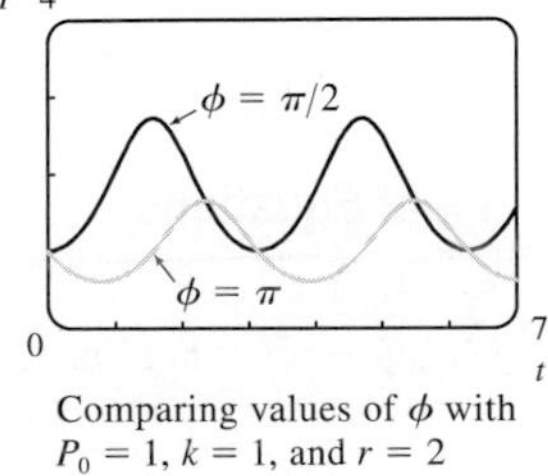

Comparing values of ϕ with $P_0 = 1$, $k = 1$, and $r = 2$

$P(t)$ oscillates between $P_0e^{(k/r)(1+\sin\phi)}$ and $P_0e^{(k/r)(-1+\sin\phi)}$ (the extreme values are attained when $rt - \phi$ is an odd multiple of $\frac{\pi}{2}$), so $\lim\limits_{t\to\infty} P(t)$ does not exist.

24. (a) $dP/dt = kP\cos^2(rt-\phi) \;\Rightarrow\; (dP)/P = k\cos^2(rt-\phi)\,dt \;\Rightarrow\; \int(dP)/P = k\int\cos^2(rt-\phi)\,dt \;\Rightarrow$

$\ln P = k\displaystyle\int\frac{1+\cos(2(rt-\phi))}{2}\,dt = \frac{k}{2}t + \frac{k}{4r}\sin(2(rt-\phi)) + C$. From $P(0)=P_0$, we get

$\ln P_0 = \dfrac{k}{4r}\sin(-2\phi) + C = C - \dfrac{k}{4r}\sin 2\phi$, so $C = \ln P_0 + \dfrac{k}{4r}\sin 2\phi$ and

$\ln P = \dfrac{k}{2}t + \dfrac{k}{4r}\sin(2(rt-\phi)) + \ln P_0 + \dfrac{k}{4r}\sin 2\phi$. Simplifying, we get

$\ln\dfrac{P}{P_0} = \dfrac{k}{2}t + \dfrac{k}{4r}[\sin(2(rt-\phi)) + \sin 2\phi] = f(t)$, or $P(t) = P_0e^{f(t)}$.

(b) An increase in k stretches the graph of P vertically while maintaining $P(0)=P_0$.

An increase in r compresses the graph of P horizontally—similar to changing the period in Exercise 19.

As in Exercise 23, a change in ϕ only makes slight adjustments in the growth of P, as shown in the figure.

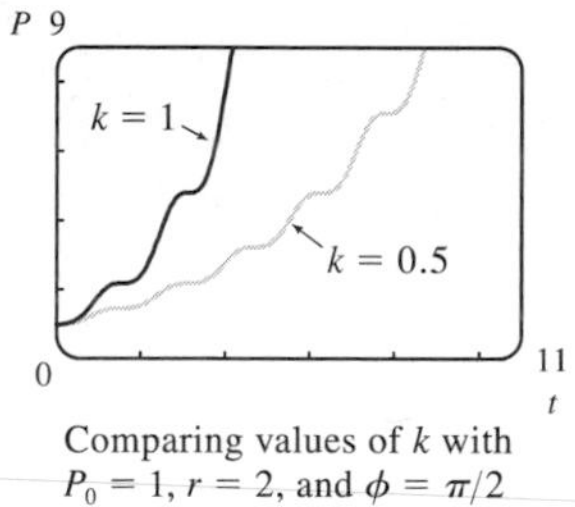

Comparing values of k with $P_0 = 1$, $r = 2$, and $\phi = \pi/2$

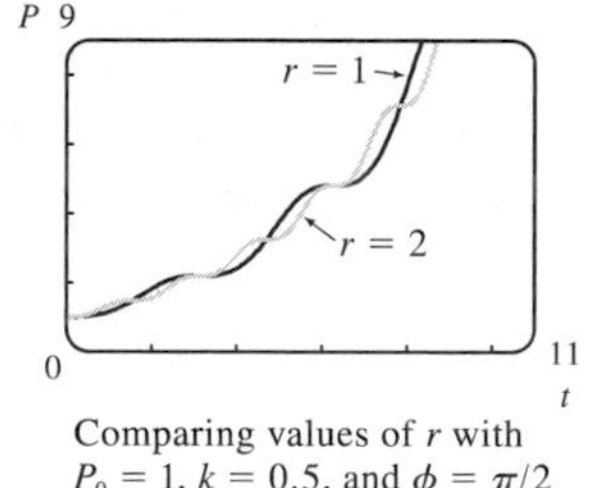

Comparing values of r with $P_0 = 1$, $k = 0.5$, and $\phi = \pi/2$

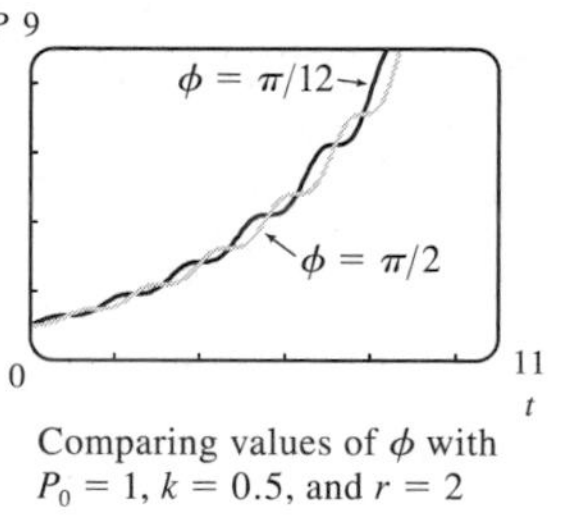

Comparing values of ϕ with $P_0 = 1$, $k = 0.5$, and $r = 2$

$f'(t) = k/2 + [k/(4r)][2r\cos(2(rt-\phi))] = (k/2)[1+\cos(2(rt-\phi))] \ge 0$. Since $P(t) = P_0e^{f(t)}$, we have $P'(t) = P_0f'(t)e^{f(t)} \ge 0$, with equality only when $\cos(2(rt-\phi)) = -1$; that is, when $rt-\phi$ is an odd multiple of $\frac{\pi}{2}$. Therefore, $P(t)$ is an increasing function on $(0,\infty)$. P can also be written as $P(t) = P_0e^{kt/2}e^{(k/4r)[\sin(2(rt-\phi))+\sin 2\phi]}$. The second exponential oscillates between $e^{(k/4r)(1+\sin 2\phi)}$ and $e^{(k/4r)(-1+\sin 2\phi)}$, while the first one, $e^{kt/2}$, grows without bound. So $\lim\limits_{t\to\infty} P(t) = \infty$.

25. By Equation 7, $P(t) = \dfrac{K}{1+Ae^{-kt}}$. By comparison, if $c = (\ln A)/k$ and $u = \frac{1}{2}k(t-c)$, then

$$1+\tanh u = 1 + \frac{e^u - e^{-u}}{e^u + e^{-u}} = \frac{e^u + e^{-u}}{e^u + e^{-u}} + \frac{e^u - e^{-u}}{e^u + e^{-u}} = \frac{2e^u}{e^u + e^{-u}}\cdot\frac{e^{-u}}{e^{-u}} = \frac{2}{1+e^{-2u}}$$

and $e^{-2u} = e^{-k(t-c)} = e^{kc}e^{-kt} = e^{\ln A}e^{-kt} = Ae^{-kt}$, so

$$\tfrac{1}{2}K\left[1+\tanh\left(\tfrac{1}{2}k(t-c)\right)\right] = \frac{K}{2}[1+\tanh u] = \frac{K}{2}\cdot\frac{2}{1+e^{-2u}} = \frac{K}{1+e^{-2u}} = \frac{K}{1+Ae^{-kt}} = P(t).$$

9.5 Linear Equations

1. $y' + x\sqrt{y} = x^2$ is not linear since it cannot be put into the standard form (1), $y' + P(x)\,y = Q(x)$.

2. $y' - x = y\tan x \;\Leftrightarrow\; y' + (-\tan x)y = x$ is linear since it can be put into the standard form (1), $y' + P(x)\,y = Q(x)$.

3. $ue^t = t + \sqrt{t}\,\dfrac{du}{dt} \;\Leftrightarrow\; \sqrt{t}\,u' - e^tu = -t \;\Leftrightarrow\; u' - \dfrac{e^t}{\sqrt{t}}u = -\sqrt{t}$ is linear since it can be put into the standard form, $u' + P(t)\,u = Q(t)$.

4. $\dfrac{dR}{dt} + t\cos R = e^{-t} \quad \Leftrightarrow \quad R' + t\cos R = e^{-t}$ is not linear since it cannot be put into the standard form $R' + P(t)\,R = Q(t)$.

5. Comparing the given equation, $y' + y = 1$, with the general form, $y' + P(x)\,y = Q(x)$, we see that $P(x) = 1$ and the integrating factor is $I(x) = e^{\int P(x)\,dx} = e^{\int 1\,dx} = e^x$. Multiplying the differential equation by $I(x)$ gives

$$e^x y' + e^x y = e^x \quad \Rightarrow \quad (e^x y)' = e^x \quad \Rightarrow \quad e^x y = \int e^x\,dx \quad \Rightarrow \quad e^x y = e^x + C \quad \Rightarrow \quad \frac{e^x y}{e^x} = \frac{e^x}{e^x} + \frac{C}{e^x} \quad \Rightarrow$$

$y = 1 + Ce^{-x}$.

6. $y' - y = e^x \quad \Leftrightarrow \quad y' + (-1)y = e^x \quad \Rightarrow \quad P(x) = -1.\quad I(x) = e^{\int P(x)\,dx} = e^{\int -1\,dx} = e^{-x}$. Multiplying the original differential equation by $I(x)$ gives $e^{-x}y' - e^{-x}y = e^0 \quad \Rightarrow \quad (e^{-x}y)' = 1 \quad \Rightarrow \quad e^{-x}y = \int 1\,dx \quad \Rightarrow$

$$e^{-x}y = x + C \quad \Rightarrow \quad y = \frac{x + C}{e^{-x}} \quad \Rightarrow \quad y = xe^x + Ce^x.$$

7. $y' = x - y \quad \Rightarrow \quad y' + y = x \ (\star)$. $I(x) = e^{\int P(x)\,dx} = e^{\int 1\,dx} = e^x$. Multiplying the differential equation $(\star)$ by $I(x)$ gives $e^x y' + e^x y = xe^x \quad \Rightarrow \quad (e^x y)' = xe^x \quad \Rightarrow \quad e^x y = \int xe^x\,dx \quad \Rightarrow \quad e^x y = xe^x - e^x + C$ [by parts] $\Rightarrow$ $y = x - 1 + Ce^{-x}$ [divide by e^x].

8. $4x^3 y + x^4 y' = \sin^3 x \quad \Rightarrow \quad (x^4 y)' = \sin^3 x \quad \Rightarrow \quad x^4 y = \int \sin^3 x\,dx \quad \Rightarrow$

$$x^4 y = \int \sin x\,(1 - \cos^2 x)\,dx = \int (1 - u^2)(-du) \qquad \begin{bmatrix} u = \cos x, \\ du = -\sin x\,dx \end{bmatrix}$$

$$= \int (u^2 - 1)\,du = \tfrac{1}{3}u^3 - u + C = \tfrac{1}{3}u(u^2 - 3) + C = \tfrac{1}{3}\cos x\,(\cos^2 x - 3) + C \quad \Rightarrow$$

$$y = \frac{1}{3x^4}\cos x\,(\cos^2 x - 3) + \frac{C}{x^4}$$

9. Since $P(x)$ is the derivative of the coefficient of y' [$P(x) = 1$ and the coefficient is x], we can write the differential equation $xy' + y = \sqrt{x}$ in the easily integrable form $(xy)' = \sqrt{x} \quad \Rightarrow \quad xy = \frac{2}{3}x^{3/2} + C \quad \Rightarrow \quad y = \frac{2}{3}\sqrt{x} + C/x$.

10. $2xy' + y = 2\sqrt{x} \quad \Rightarrow \quad y' + \dfrac{1}{2x}\,y = \dfrac{1}{\sqrt{x}} \quad [x > 0] \quad \Rightarrow \quad P(x) = \dfrac{1}{2x}$.

$I(x) = e^{\int P(x)\,dx} = e^{\int 1/(2x)\,dx} = e^{(1/2)\ln|x|} = (e^{\ln x})^{1/2} = \sqrt{x}$. Multiplying the differential equation by $I(x)$ gives

$$\sqrt{x}\,y' + \frac{1}{2\sqrt{x}}\,y = 1 \quad \Rightarrow \quad (\sqrt{x}\,y)' = 1 \quad \Rightarrow \quad \sqrt{x}\,y = \int 1\,dx \quad \Rightarrow \quad \sqrt{x}\,y = x + C \quad \Rightarrow \quad y = \frac{x + C}{\sqrt{x}}.$$

11. $xy' - 2y = x^2 \quad \Rightarrow \quad y' - \dfrac{2}{x}\,y = x \quad \Rightarrow \quad P(x) = -\dfrac{2}{x}$.

$I(x) = e^{\int P(x)\,dx} = e^{\int -2/x\,dx} = e^{-2\ln x} \quad [x > 0] \quad = x^{-2} = \dfrac{1}{x^2}$. Multiplying the differential equation by $I(x)$ gives

$$\frac{1}{x^2}\,y' - \frac{2}{x^3}\,y = \frac{1}{x} \quad \Rightarrow \quad \left(\frac{1}{x^2}\,y\right)' = \frac{1}{x} \quad \Rightarrow \quad \frac{1}{x^2}\,y = \int \frac{1}{x}\,dx \quad \Rightarrow \quad \frac{1}{x^2}\,y = \ln x + C \quad \Rightarrow \quad y = x^2(\ln x + C).$$

12. $y' + 2xy = 1 \quad \Rightarrow \quad P(x) = 2x.\quad I(x) = e^{\int P(x)\,dx} = e^{\int 2x\,dx} = e^{x^2}$. Multiplying the differential equation by $I(x)$ gives $e^{x^2}y' + 2xe^{x^2}y = e^{x^2} \quad \Rightarrow \quad \left(e^{x^2}y\right)' = e^{x^2} \quad \Rightarrow \quad e^{x^2}y = \int_0^x e^{t^2}\,dt + C$ [see page 507] $\Rightarrow$

$y = e^{-x^2}\int_0^x e^{t^2}\,dt + Ce^{-x^2}$.

13. $t^2\dfrac{dy}{dt}+3ty=\sqrt{1+t^2} \;\Rightarrow\; y'+\dfrac{3}{t}y=\dfrac{\sqrt{1+t^2}}{t^2} \;\Rightarrow\; P(t)=\dfrac{3}{t}$.

$I(t)=e^{\int P(t)\,dt}=e^{\int 3/t\,dt}=e^{3\ln t}\;[t>0]\;=t^3$. Multiplying by t^3 gives $t^3y'+3t^2y=t\sqrt{1+t^2} \;\Rightarrow$

$(t^3y)'=t\sqrt{1+t^2} \;\Rightarrow\; t^3y=\int t\sqrt{1+t^2}\,dt \;\Rightarrow\; t^3y=\frac{1}{3}(1+t^2)^{3/2}+C \;\Rightarrow\; y=\frac{1}{3}t^{-3}(1+t^2)^{3/2}+Ct^{-3}$.

14. $t\ln t\,\dfrac{dr}{dt}+r=te^t \;\Rightarrow\; \dfrac{dr}{dt}+\dfrac{1}{t\ln t}r=\dfrac{e^t}{\ln t}$. $I(t)=e^{\int dt/(t\ln t)}=e^{\ln(\ln t)}=\ln t$. Multiplying by $\ln t$ gives

$\ln t\,\dfrac{dr}{dt}+\dfrac{1}{t}r=e^t \;\Rightarrow\; [(\ln t)r]'=e^t \;\Rightarrow\; (\ln t)r=e^t+C \;\Rightarrow\; r=\dfrac{e^t+C}{\ln t}$.

15. $x^2y'+2xy=\ln x \;\Rightarrow\; (x^2y)'=\ln x \;\Rightarrow\; x^2y=\int \ln x\,dx \;\Rightarrow\; x^2y=x\ln x-x+C$ [by parts]. Since $y(1)=2$,

$1^2(2)=1\ln 1-1+C \;\Rightarrow\; 2=-1+C \;\Rightarrow\; C=3$, so $x^2y=x\ln x-x+3$, or $y=\dfrac{1}{x}\ln x-\dfrac{1}{x}+\dfrac{3}{x^2}$.

16. $t^3\dfrac{dy}{dt}+3t^2y=\cos t \;\Rightarrow\; (t^3y)'=\cos t \;\Rightarrow\; t^3y=\int \cos t\,dt \;\Rightarrow\; t^3y=\sin t+C$. Since $y(\pi)=0$,

$\pi^3(0)=\sin\pi+C \;\Rightarrow\; C=0$, so $t^3y=\sin t$, or $y=\dfrac{\sin t}{t^3}$.

17. $t\dfrac{du}{dt}=t^2+3u \;\Rightarrow\; u'-\dfrac{3}{t}u=t\;(\star)$. $I(t)=e^{\int -3/t\,dt}=e^{-3\ln|t|}=(e^{\ln|t|})^{-3}=t^{-3}\;[t>0]\;=\dfrac{1}{t^3}$. Multiplying $(\star)$

by $I(t)$ gives $\dfrac{1}{t^3}u'-\dfrac{3}{t^4}u=\dfrac{1}{t^2} \;\Rightarrow\; \left(\dfrac{1}{t^3}u\right)'=\dfrac{1}{t^2} \;\Rightarrow\; \dfrac{1}{t^3}u=\displaystyle\int\frac{1}{t^2}\,dt \;\Rightarrow\; \dfrac{1}{t^3}u=-\dfrac{1}{t}+C$. Since $u(2)=4$,

$\dfrac{1}{2^3}(4)=-\dfrac{1}{2}+C \;\Rightarrow\; C=1$, so $\dfrac{1}{t^3}u=-\dfrac{1}{t}+1$, or $u=-t^2+t^3$.

18. $xy'+y=x\ln x \;\Rightarrow\; (xy)'=x\ln x \;\Rightarrow\; xy=\int x\ln x\,dx \;\Rightarrow\; xy=\frac{1}{2}x^2\ln x-\frac{1}{4}x^2+C$ $\left[\begin{array}{c}\text{by parts}\\ \text{with } u=\ln x\end{array}\right] \;\Rightarrow$

$y=\dfrac{1}{2}x\ln x-\dfrac{1}{4}x+\dfrac{C}{x}$. $y(1)=0 \;\Rightarrow\; 0=0-\dfrac{1}{4}+C \;\Rightarrow\; C=\dfrac{1}{4}$, so $y=\dfrac{1}{2}x\ln x-\dfrac{1}{4}x+\dfrac{1}{4x}$.

19. $xy'=y+x^2\sin x \;\Rightarrow\; y'-\dfrac{1}{x}y=x\sin x$. $I(x)=e^{\int(-1/x)\,dx}=e^{-\ln x}=e^{\ln x^{-1}}=\dfrac{1}{x}$.

Multiplying by $\dfrac{1}{x}$ gives $\dfrac{1}{x}y'-\dfrac{1}{x^2}y=\sin x \;\Rightarrow\; \left(\dfrac{1}{x}y\right)'=\sin x \;\Rightarrow\; \dfrac{1}{x}y=-\cos x+C \;\Rightarrow\; y=-x\cos x+Cx$.

$y(\pi)=0 \;\Rightarrow\; -\pi\cdot(-1)+C\pi=0 \;\Rightarrow\; C=-1$, so $y=-x\cos x-x$.

20. $(x^2+1)\dfrac{dy}{dx}+3x(y-1)=0 \;\Rightarrow\; (x^2+1)\,y'+3xy=3x \;\Rightarrow\; y'+\dfrac{3x}{x^2+1}y=\dfrac{3x}{x^2+1}$.

$I(x)=e^{\int 3x/(x^2+1)\,dx}=e^{(3/2)\ln|x^2+1|}=\left(e^{\ln(x^2+1)}\right)^{3/2}=(x^2+1)^{3/2}$. Multiplying by $(x^2+1)^{3/2}$ gives

$(x^2+1)^{3/2}y'+3x(x^2+1)^{1/2}y=3x(x^2+1)^{1/2} \;\Rightarrow\; \left[(x^2+1)^{3/2}y\right]'=3x(x^2+1)^{1/2} \;\Rightarrow$

$(x^2+1)^{3/2}y=\int 3x(x^2+1)^{1/2}\,dx=(x^2+1)^{3/2}+C \;\Rightarrow\; y=1+C(x^2+1)^{-3/2}$. Since $y(0)=2$, we have

$2=1+C(1) \;\Rightarrow\; C=1$ and hence, $y=1+(x^2+1)^{-3/2}$.

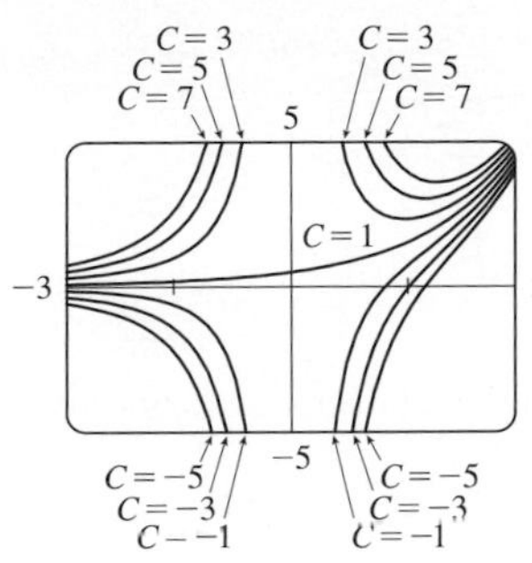

21. $xy' + 2y = e^x \Rightarrow y' + \dfrac{2}{x}y = \dfrac{e^x}{x}$.

$I(x) = e^{\int (2/x)\,dx} = e^{2\ln|x|} = \left(e^{\ln|x|}\right)^2 = |x|^2 = x^2$.

Multiplying by $I(x)$ gives $x^2 y' + 2xy = xe^x \Rightarrow (x^2y)' = xe^x \Rightarrow$

$x^2y = \int xe^x\,dx = (x-1)e^x + C$ [by parts] $\Rightarrow$

$y = [(x-1)e^x + C]/x^2$. The graphs for $C = -5, -3, -1, 1, 3, 5$, and 7 are shown. $C = 1$ is a transitional value. For $C < 1$, there is an inflection point and for $C > 1$, there is a local minimum. As $|C|$ gets larger, the "branches" get further from the origin.

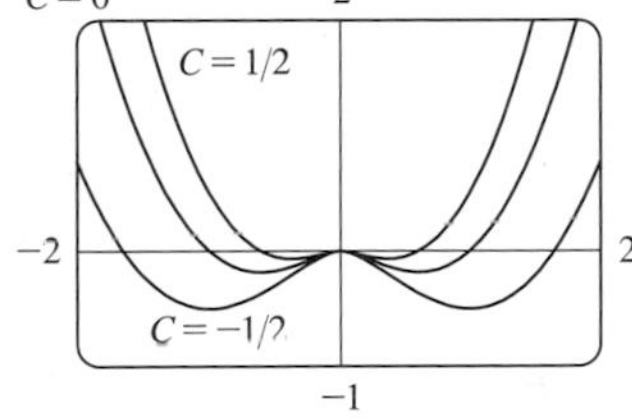

22. $xy' = x^2 + 2y \Leftrightarrow xy' - 2y = x^2 \Leftrightarrow y' - \dfrac{2}{x}y = x$.

$I(x) = e^{\int -2/x\,dx} = e^{-2\ln|x|} = (e^{\ln|x|})^{-2} = |x|^{-2} = \dfrac{1}{x^2}$. Multiplying by

$I(x)$ gives $\dfrac{1}{x^2}y' - \dfrac{2}{x^3}y = \dfrac{1}{x} \Rightarrow \left(\dfrac{1}{x^2}y\right)' = \dfrac{1}{x} \Rightarrow \dfrac{1}{x^2}y = \displaystyle\int \dfrac{1}{x}\,dx \Rightarrow$

$\dfrac{1}{x^2}y = \ln|x| + C \Rightarrow y = (\ln|x| + C)x^2$. For all values of C, as $|x| \to 0$, $y \to 0$, and as $|x| \to \infty$, $y \to \infty$. As $|x|$ increases from 0, the function decreases and attains an absolute minimum. The inflection points, absolute minimums, and x-intercepts all move farther from the origin as C decreases.

23. Setting $u = y^{1-n}$, $\dfrac{du}{dx} = (1-n)\,y^{-n}\dfrac{dy}{dx}$ or $\dfrac{dy}{dx} = \dfrac{y^n}{1-n}\dfrac{du}{dx} = \dfrac{u^{n/(1-n)}}{1-n}\dfrac{du}{dx}$. Then the Bernoulli differential equation becomes $\dfrac{u^{n/(1-n)}}{1-n}\dfrac{du}{dx} + P(x)u^{1/(1-n)} = Q(x)u^{n/(1-n)}$ or $\dfrac{du}{dx} + (1-n)P(x)u = Q(x)(1-n)$.

24. Here $xy' + y = -xy^2 \Rightarrow y' + \dfrac{y}{x} = -y^2$, so $n = 2$, $P(x) = \dfrac{1}{x}$ and $Q(x) = -1$. Setting $u = y^{-1}$, u satisfies $u' - \dfrac{1}{x}u = 1$. Then $I(x) = e^{\int(-1/x)dx} = \dfrac{1}{x}$ (for $x > 0$) and $u = x\left(\displaystyle\int \dfrac{1}{x}\,dx + C\right) = x(\ln|x| + C)$. Thus,

$y = \dfrac{1}{x(C + \ln|x|)}$.

25. Here $y' + \dfrac{2}{x}y = \dfrac{y^3}{x^2}$, so $n = 3$, $P(x) = \dfrac{2}{x}$ and $Q(x) = \dfrac{1}{x^2}$. Setting $u = y^{-2}$, u satisfies $u' - \dfrac{4u}{x} = -\dfrac{2}{x^2}$.

Then $I(x) = e^{\int(-4/x)\,dx} = x^{-4}$ and $u = x^4\left(\displaystyle\int -\dfrac{2}{x^6}\,dx + C\right) = x^4\left(\dfrac{2}{5x^5} + C\right) = Cx^4 + \dfrac{2}{5x}$.

Thus, $y = \pm\left(Cx^4 + \dfrac{2}{5x}\right)^{-1/2}$.

26. $xy'' + 2y' = 12x^2$ and $u = y' \Rightarrow xu' + 2u = 12x^2 \Rightarrow u' + \dfrac{2}{x}u = 12x$.

$I(x) = e^{\int(2/x)\,dx} = e^{2\ln|x|} = \left(e^{\ln|x|}\right)^2 = |x|^2 = x^2$. Multiplying the last differential equation by x^2 gives

$x^2u' + 2xu = 12x^3 \;\Rightarrow\; (x^2u)' = 12x^3 \;\Rightarrow\; x^2u = \int 12x^3\,dx = 3x^4 + C \;\Rightarrow\; u = 3x^2 + C/x^2 \;\Rightarrow$

$y' = 3x^2 + C/x^2 \;\Rightarrow\; y = x^3 - C/x + D.$

27. (a) $2\,\dfrac{dI}{dt} + 10I = 40$ or $\dfrac{dI}{dt} + 5I = 20$. Then the integrating factor is $e^{\int 5\,dt} = e^{5t}$. Multiplying the differential equation by the integrating factor gives $e^{5t}\,\dfrac{dI}{dt} + 5Ie^{5t} = 20e^{5t} \;\Rightarrow\; (e^{5t}I)' = 20e^{5t} \;\Rightarrow$

$I(t) = e^{-5t}\left[\int 20e^{5t}\,dt + C\right] = 4 + Ce^{-5t}$. But $0 = I(0) = 4 + C$, so $I(t) = 4 - 4e^{-5t}$.

(b) $I(0.1) = 4 - 4e^{-0.5} \approx 1.57$ A

28. (a) $\dfrac{dI}{dt} + 20I = 40\sin 60t$, so the integrating factor is e^{20t}. Multiplying the differential equation by the integrating factor gives $e^{20t}\,\dfrac{dI}{dt} + 20Ie^{20t} = 40e^{20t}\sin 60t \;\Rightarrow\; (e^{20t}I)' = 40e^{20t}\sin 60t \;\Rightarrow$

$$I(t) = e^{-20t}\left[\int 40e^{20t}\sin 60t\,dt + C\right] = e^{-20t}\left[40e^{20t}\left(\tfrac{1}{4000}\right)(20\sin 60t - 60\cos 60t)\right] + Ce^{-20t}$$
$$= \frac{\sin 60t - 3\cos 60t}{5} + Ce^{-20t}$$

But $1 = I(0) = -\frac{3}{5} + C$, so $I(t) = \dfrac{\sin 60t - 3\cos 60t + 8e^{-20t}}{5}$.

(b) $I(0.1) = \dfrac{\sin 6 - 3\cos 6 + 8e^{-2}}{5} \approx -0.42$ A

(c)

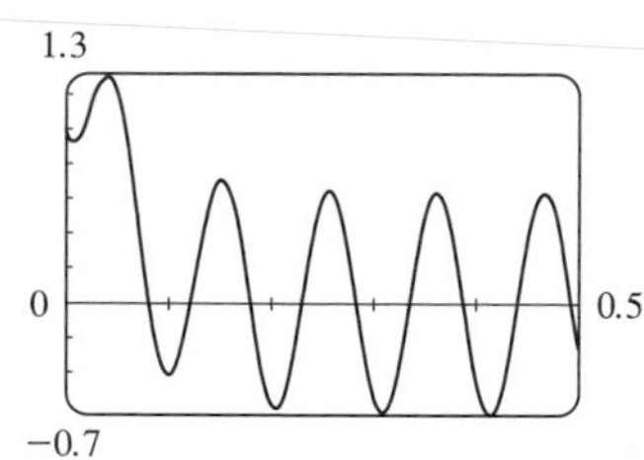

29. $5\,\dfrac{dQ}{dt} + 20Q = 60$ with $Q(0) = 0$ C. Then the integrating factor is $e^{\int 4\,dt} = e^{4t}$, and multiplying the differential equation by the integrating factor gives $e^{4t}\,\dfrac{dQ}{dt} + 4e^{4t}Q = 12e^{4t} \;\Rightarrow\; (e^{4t}Q)' = 12e^{4t} \;\Rightarrow$

$Q(t) = e^{-4t}\left[\int 12e^{4t}\,dt + C\right] = 3 + Ce^{-4t}$. But $0 = Q(0) = 3 + C$ so $Q(t) = 3(1 - e^{-4t})$ is the charge at time t and $I = dQ/dt = 12e^{-4t}$ is the current at time t.

30. $2\,\dfrac{dQ}{dt} + 100Q = 10\sin 60t$ or $\dfrac{dQ}{dt} + 50Q = 5\sin 60t$. Then the integrating factor is $e^{\int 50\,dt} = e^{50t}$, and multiplying the differential equation by the integrating factor gives $e^{50t}\,\dfrac{dQ}{dt} + 50e^{50t}Q = 5e^{50t}\sin 60t \;\Rightarrow\; (e^{50t}Q)' = 5e^{50t}\sin 60t \;\Rightarrow$

$$Q(t) = e^{-50t}\left[\int 5e^{50t}\sin 60t\,dt + C\right] = e^{-50t}\left[5e^{50t}\left(\tfrac{1}{6100}\right)(50\sin 60t - 60\cos 60t)\right] + Ce^{-50t}$$
$$= \tfrac{1}{122}(5\sin 60t - 6\cos 60t) + Ce^{-50t}$$

But $0 = Q(0) = -\frac{6}{122} + C$ so $C = \frac{3}{61}$ and $Q(t) = \dfrac{5\sin 60t - 6\cos 60t}{122} + \dfrac{3e^{-50t}}{61}$ is the charge at time t, while the current is $I(t) = \dfrac{dQ}{dt} = \dfrac{150\cos 60t + 180\sin 60t - 150e^{-50t}}{61}$.

31. $\dfrac{dP}{dt} + kP = kM$, so $I(t) = e^{\int k\,dt} = e^{kt}$. Multiplying the differential equation by $I(t)$ gives $e^{kt}\dfrac{dP}{dt} + kPe^{kt} = kMe^{kt} \;\Rightarrow\; (e^{kt}P)' = kMe^{kt} \;\Rightarrow$ $P(t) = e^{-kt}\left(\int kMe^{kt}dt + C\right) = M + Ce^{-kt}$, $k > 0$. Furthermore, it is reasonable to assume that $0 \le P(0) \le M$, so $-M \le C \le 0$.

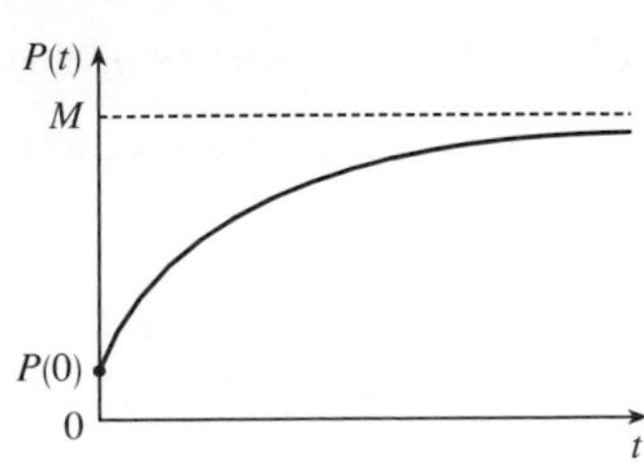

32. Since $P(0) = 0$, we have $P(t) = M(1 - e^{-kt})$. If $P_1(t)$ is Jim's learning curve, then $P_1(1) = 25$ and $P_1(2) = 45$. Hence, $25 = M_1(1 - e^{-k})$ and $45 = M_1(1 - e^{-2k})$, so $1 - 25/M_1 = e^{-k}$ or $k = -\ln\left(1 - \dfrac{25}{M_1}\right) = \ln\left(\dfrac{M_1}{M_1 - 25}\right)$. But $45 = M_1(1 - e^{-2k})$ so $45 = M_1\left[1 - \left(\dfrac{M_1 - 25}{M_1}\right)^2\right]$ or $45 = \dfrac{50M_1 - 625}{M_1}$. Thus, $M_1 = 125$ is the maximum number of units per hour Jim is capable of processing. Similarly, if $P_2(t)$ is Mark's learning curve, then $P_2(1) = 35$ and $P_2(2) = 50$. So $k = \ln\left(\dfrac{M_2}{M_2 - 35}\right)$ and $50 = M_2\left[1 - \left(\dfrac{M_2 - 35}{M_2}\right)^2\right]$ or $M_2 = 61.25$. Hence the maximum number of units per hour for Mark is approximately 61. Another approach would be to use the midpoints of the intervals so that $P_1(0.5) = 25$ and $P_1(1.5) = 45$. Doing so gives us $M_1 \approx 52.6$ and $M_2 \approx 51.8$.

33. $y(0) = 0$ kg. Salt is added at a rate of $\left(0.4\,\dfrac{\text{kg}}{\text{L}}\right)\left(5\,\dfrac{\text{L}}{\text{min}}\right) = 2\,\dfrac{\text{kg}}{\text{min}}$. Since solution is drained from the tank at a rate of 3 L/min, but salt solution is added at a rate of 5 L/min, the tank, which starts out with 100 L of water, contains $(100 + 2t)$ L of liquid after t min. Thus, the salt concentration at time t is $\dfrac{y(t)}{100 + 2t}\,\dfrac{\text{kg}}{\text{L}}$. Salt therefore leaves the tank at a rate of $\left(\dfrac{y(t)}{100 + 2t}\,\dfrac{\text{kg}}{\text{L}}\right)\left(3\,\dfrac{\text{L}}{\text{min}}\right) = \dfrac{3y}{100 + 2t}\,\dfrac{\text{kg}}{\text{min}}$. Combining the rates at which salt enters and leaves the tank, we get $\dfrac{dy}{dt} = 2 - \dfrac{3y}{100 + 2t}$. Rewriting this equation as $\dfrac{dy}{dt} + \left(\dfrac{3}{100 + 2t}\right)y = 2$, we see that it is linear.

$$I(t) = \exp\left(\int \frac{3\,dt}{100 + 2t}\right) = \exp\left(\tfrac{3}{2}\ln(100 + 2t)\right) = (100 + 2t)^{3/2}$$

Multiplying the differential equation by $I(t)$ gives $(100 + 2t)^{3/2}\dfrac{dy}{dt} + 3(100 + 2t)^{1/2}y = 2(100 + 2t)^{3/2} \;\Rightarrow$ $[(100 + 2t)^{3/2}y]' = 2(100 + 2t)^{3/2} \;\Rightarrow\; (100 + 2t)^{3/2}y = \frac{2}{5}(100 + 2t)^{5/2} + C \;\Rightarrow$ $y = \frac{2}{5}(100 + 2t) + C(100 + 2t)^{-3/2}$. Now $0 = y(0) = \frac{2}{5}(100) + C \cdot 100^{-3/2} = 40 + \frac{1}{1000}C \;\Rightarrow\; C = -40{,}000$, so $y = \left[\frac{2}{5}(100 + 2t) - 40{,}000(100 + 2t)^{-3/2}\right]$ kg. From this solution (no pun intended), we calculate the salt concentration at time t to be $C(t) = \dfrac{y(t)}{100 + 2t} = \left[\dfrac{-40{,}000}{(100 + 2t)^{5/2}} + \dfrac{2}{5}\right]\dfrac{\text{kg}}{\text{L}}$. In particular, $C(20) = \dfrac{-40{,}000}{140^{5/2}} + \dfrac{2}{5} \approx 0.2275\,\dfrac{\text{kg}}{\text{L}}$ and $y(20) = \frac{2}{5}(140) - 40{,}000(140)^{-3/2} \approx 31.85$ kg.

34. Let $y(t)$ denote the amount of chlorine in the tank at time t (in seconds). $y(0) = (0.05\text{ g/L})\,(400\text{ L}) = 20\text{ g}$. The amount of liquid in the tank at time t is $(400 - 6t)$ L since 4 L of water enters the tank each second and 10 L of liquid leaves the tank each second. Thus, the concentration of chlorine at time t is $\dfrac{y(t)}{400 - 6t}\,\dfrac{\text{g}}{\text{L}}$. Chlorine doesn't enter the tank, but it leaves at a rate of $\left[\dfrac{y(t)}{400-6t}\,\dfrac{\text{g}}{\text{L}}\right]\left[10\,\dfrac{\text{L}}{\text{s}}\right] = \dfrac{10\,y(t)}{400-6t}\,\dfrac{\text{g}}{\text{s}} = \dfrac{5\,y(t)}{200-3t}\,\dfrac{\text{g}}{\text{s}}$. Therefore, $\dfrac{dy}{dt} = -\dfrac{5y}{200-3t} \;\Rightarrow\; \displaystyle\int \frac{dy}{y} = \int \frac{-5\,dt}{200-3t} \;\Rightarrow$ $\ln y = \frac{5}{3}\ln(200-3t) + C \;\Rightarrow\; y = \exp\!\left(\frac{5}{3}\ln(200-3t) + C\right) = e^C(200-3t)^{5/3}$. Now $20 = y(0) = e^C \cdot 200^{5/3} \;\Rightarrow$ $e^C = \dfrac{20}{200^{5/3}}$, so $y(t) = 20\dfrac{(200-3t)^{5/3}}{200^{5/3}} = 20(1 - 0.015t)^{5/3}$ g for $0 \le t \le 66\frac{2}{3}$ s, at which time the tank is empty.

35. (a) $\dfrac{dv}{dt} + \dfrac{c}{m}v = g$ and $I(t) = e^{\int (c/m)\,dt} = e^{(c/m)t}$, and multiplying the differential equation by $I(t)$ gives $e^{(c/m)t}\dfrac{dv}{dt} + \dfrac{vce^{(c/m)t}}{m} = ge^{(c/m)t} \;\Rightarrow\; \left[e^{(c/m)t}v\right]' = ge^{(c/m)t}$. Hence, $v(t) = e^{-(c/m)t}\left[\int ge^{(c/m)t}\,dt + K\right] = mg/c + Ke^{-(c/m)t}$. But the object is dropped from rest, so $v(0) = 0$ and $K = -mg/c$. Thus, the velocity at time t is $v(t) = (mg/c)[1 - e^{-(c/m)t}]$.

(b) $\displaystyle\lim_{t\to\infty} v(t) = mg/c$

(c) $s(t) = \int v(t)\,dt = (mg/c)[t + (m/c)e^{-(c/m)t}] + c_1$ where $c_1 = s(0) - m^2g/c^2$.

$s(0)$ is the initial position, so $s(0) = 0$ and $s(t) = (mg/c)[t + (m/c)e^{-(c/m)t}] - m^2g/c^2$.

36. $v = (mg/c)(1 - e^{-ct/m}) \;\Rightarrow$

$$\frac{dv}{dm} = \frac{mg}{c}\left(0 - e^{-ct/m}\cdot\frac{ct}{m^2}\right) + \frac{g}{c}(1 - e^{-ct/m})\cdot 1 = -\frac{gt}{m}e^{-ct/m} + \frac{g}{c} - \frac{g}{c}e^{-ct/m}$$
$$= \frac{g}{c}\left(1 - e^{-ct/m} - \frac{ct}{m}e^{-ct/m}\right) \;\Rightarrow$$

$\dfrac{c}{g}\dfrac{dv}{dm} = 1 - \left(1 + \dfrac{ct}{m}\right)e^{-ct/m} = 1 - \dfrac{1 + ct/m}{e^{ct/m}} = 1 - \dfrac{1+Q}{e^Q}$, where $Q = \dfrac{ct}{m} \ge 0$. Since $e^Q > 1 + Q$ for all $Q > 0$, it follows that $dv/dm > 0$ for $t > 0$. In other words, for all $t > 0$, v increases as m increases.

37. (a) $z = \dfrac{1}{P} \;\Rightarrow\; P = \dfrac{1}{z} \;\Rightarrow\; P' = -\dfrac{z'}{z^2}$. Substituting into $P' = kP(1 - P/M)$ gives us $-\dfrac{z'}{z^2} = k\dfrac{1}{z}\left(1 - \dfrac{1}{zM}\right) \;\Rightarrow$ $z' = -kz\left(1 - \dfrac{1}{zM}\right) \;\Rightarrow\; z' = -kz + \dfrac{k}{M} \;\Rightarrow\; z' + kz = \dfrac{k}{M}\;(\star)$.

(b) The integrating factor is $e^{\int k\,dt} = e^{kt}$. Multiplying $(\star)$ by e^{kt} gives $e^{kt}z' + ke^{kt}z = \dfrac{ke^{kt}}{M} \;\Rightarrow\; (e^{kt}z)' = \dfrac{k}{M}e^{kt} \;\Rightarrow$ $e^{kt}z = \displaystyle\int \frac{k}{M}e^{kt}\,dt \;\Rightarrow\; e^{kt}z = \frac{1}{M}e^{kt} + C \;\Rightarrow\; z = \frac{1}{M} + Ce^{-kt}$. Since $P = \dfrac{1}{z}$, we have $P = \dfrac{1}{\frac{1}{M} + Ce^{-kt}} \;\Rightarrow\; P = \dfrac{M}{1 + MCe^{-kt}}$, which agrees with Equation 9.4.7, $P = \dfrac{M}{1 + Ae^{-kt}}$, when $MC = A$.

38. (a) $z = \dfrac{1}{P} \Rightarrow P = \dfrac{1}{z} \Rightarrow P' = \dfrac{-z'}{z^2}$. Substituting into $\dfrac{dP}{dt} = k(t)P\left(1 - \dfrac{P}{M(t)}\right)$ gives us

$$-\frac{z'}{z^2} = \frac{k(t)}{z}\left(1 - \frac{1}{M(t)z}\right) \Rightarrow z' = -k(t)z\left(1 - \frac{1}{M(t)z}\right) \Rightarrow z' = -k(t)z + \frac{k(t)}{M(t)} \Rightarrow$$

$$\frac{dz}{dt} + k(t)z = \frac{k(t)}{M(t)} \quad (\star).$$

(b) The integrating factor is $e^{K(t)}$, where $K(t) = \int_0^t k(s)\,ds$, so that $K'(t) = k(t)$. Multiplying $(\star)$ by

$e^{K(t)}$ gives $e^{K(t)}\dfrac{dz}{dt} + e^{K(t)}k(t)z = \dfrac{e^{K(t)}k(t)}{M(t)} \Rightarrow (e^{K(t)}z)' = \dfrac{K'(t)e^{K(t)}}{M(t)} \Rightarrow$

$e^{K(t)}z = \displaystyle\int_0^t \frac{K'(s)e^{K(s)}}{M(s)}\,ds + C$, so $P = \dfrac{1}{z} = \dfrac{e^{K(t)}}{\displaystyle\int_0^t \frac{K'(s)e^{K(s)}}{M(s)}\,ds + C}$. Now suppose that M is a constant. Then

$P(t) = \dfrac{Me^{K(t)}}{\int_0^t K'(s)e^{K(s)}\,ds + CM} = \dfrac{Me^{K(t)}}{e^{K(t)} + CM} = \dfrac{M}{1 + CMe^{-K(t)}}$. If $\int_0^\infty k(t)\,dt = \infty$, then $\lim\limits_{t\to\infty} K(t) = \infty$, so

$$\lim_{t\to\infty} P(t) = \frac{M}{1 + CM\lim\limits_{t\to\infty} e^{-K(t)}} = \frac{M}{1 + CM \cdot 0} = M.$$

(c) If k is constant, but M varies, then $K(t) = kt$ and we get $e^{kt}z = \displaystyle\int_0^t \frac{ke^{ks}}{M(s)}\,ds + C \Rightarrow$

$z(t) = \dfrac{\displaystyle\int_0^t \frac{ke^{ks}}{M(s)}\,ds + C}{e^{kt}} \Rightarrow z(t) = e^{-kt}\displaystyle\int_0^t \frac{ke^{ks}}{M(s)}\,ds + Ce^{-kt}$. Suppose $M(t)$ has a limit as $t \to \infty$,

say $\lim\limits_{t\to\infty} M(t) = L$. Then

$$\lim_{t\to\infty} P(t) = \lim_{t\to\infty}\frac{1}{z(t)} = \lim_{t\to\infty}\frac{e^{kt}}{\displaystyle\int_0^t \frac{ke^{ks}}{M(s)}\,ds + C} = \lim_{t\to\infty}\frac{ke^{kt}}{\dfrac{ke^{kt}}{M(t)} + 0} \quad \left[\begin{array}{c}\text{l'Hospital's}\\ \text{and FTC 1}\end{array}\right] = \lim_{t\to\infty} M(t) = L.$$

9.6 Predator-Prey Systems

1. (a) $dx/dt = -0.05x + 0.0001xy$. If $y = 0$, we have $dx/dt = -0.05x$, which indicates that in the absence of y, x declines at a rate proportional to itself. So x represents the predator population and y represents the prey population. The growth of the prey population, $0.1y$ (from $dy/dt = 0.1y - 0.005xy$), is restricted only by encounters with predators (the term $-0.005xy$). The predator population increases only through the term $0.0001xy$; that is, by encounters with the prey and not through additional food sources.

(b) $dy/dt = -0.015y + 0.00008xy$. If $x = 0$, we have $dy/dt = -0.015y$, which indicates that in the absence of x, y would decline at a rate proportional to itself. So y represents the predator population and x represents the prey population. The growth of the prey population, $0.2x$ (from $dx/dt = 0.2x - 0.0002x^2 - 0.006xy = 0.2x(1 - 0.001x) - 0.006xy$), is restricted by a carrying capacity of 1000 [from the term $1 - 0.001x = 1 - x/1000$] and by encounters with predators (the term $-0.006xy$). The predator population increases only through the term $0.00008xy$; that is, by encounters with the prey and not through additional food sources.

2. (a) $dx/dt = 0.12x - 0.0006x^2 + 0.00001xy$. $dy/dt = 0.08y + 0.00004xy$.

The xy terms represent encounters between the two species x and y. An increase in y makes dx/dt (the growth rate of x) larger due to the positive term $0.00001xy$. An increase in x makes dy/dt (the growth rate of y) larger due to the positive term $0.00004xy$. Hence, the system describes a cooperation model.

(b) $dx/dt = 0.15x - 0.0002x^2 - 0.0006xy = 0.15x(1 - x/750) - 0.0006xy$.

$dy/dt = 0.2y - 0.00008y^2 - 0.0002xy = 0.2y(1 - y/2500) - 0.0002xy$.

The system shows that x and y have carrying capacities of 750 and 2500. An increase in x reduces the growth rate of y due to the negative term $-0.0002xy$. An increase in y reduces the growth rate of x due to the negative term $-0.0006xy$. Hence, the system describes a competition model.

3. (a) $dx/dt = 0.5x - 0.004x^2 - 0.001xy = 0.5x(1 - x/125) - 0.001xy$.
$dy/dt = 0.4y - 0.001y^2 - 0.002xy = 0.4y(1 - y/400) - 0.002xy$.

The system shows that x and y have carrying capacities of 125 and 400. An increase in x reduces the growth rate of y due to the negative term $-0.002xy$. An increase in y reduces the growth rate of x due to the negative term $-0.001xy$. Hence the system describes a competition model.

(b) $dx/dt = 0 \;\Rightarrow\; x(0.5 - 0.004x - 0.001y) = 0 \;\Rightarrow\; x(500 - 4x - y) = 0$ **(1)** and $dy/dt = 0 \;\Rightarrow$ $y(0.4 - 0.001y - 0.002x) = 0 \;\Rightarrow\; y(400 - y - 2x) = 0$ **(2)**.

From **(1)** and **(2)**, we get four equilibrium solutions.

(i) $x = 0$ and $y = 0$: If the populations are zero, there is no change.

(ii) $x = 0$ and $400 - y - 2x = 0 \;\Rightarrow\; x = 0$ and $y = 400$: In the absence of an x-population, the y-population stabilizes at 400.

(iii) $500 - 4x - y = 0$ and $y = 0 \;\Rightarrow\; x = 125$ and $y = 0$: In the absence of y-population, the x-population stabilizes at 125.

(iv) $500 - 4x - y = 0$ and $400 - y - 2x = 0 \;\Rightarrow\; y = 500 - 4x$ and $y = 400 - 2x \;\Rightarrow\; 500 - 4x = 400 - 2x \;\Rightarrow$ $100 = 2x \;\Rightarrow\; x = 50$ and $y = 300$: A y-population of 300 is just enough to support a constant x-population of 50.

4. Let $L(t)$, $H(t)$, and $W(t)$ represent the populations of lynx, hares, and willows at time t. Let the k_i's and the c_i's denote positive constants, so that a plus sign means an increase and a minus sign means a decrease in the corresponding growth rate. "In the absence of hares, the willow population will grow exponentially and the lynx population will decay exponentially" gives us $dW/dt = +k_1W$ and $dL/dt = -k_2L$. "In the absence of lynx and willow, the hare population will decay exponentially" gives us $dH/dt = -k_3H$. "Lynx eat snowshoe hares and snowshoe hares eat woody plants like willows" gives us encounters that lynx win, hares lose and win, and willows lose. In terms of growth rates, this means that $dL/dt = +c_1LH$, $dH/dt = -c_2LH + c_3HW$, and $dW/dt = -c_4HW$. Putting this information together gives us the following system of differential equations.

$$\begin{aligned} dL/dt &= -k_2L + c_1LH \\ dH/dt &= -k_3H - c_2LH + c_3HW \\ dW/dt &= +k_1W - c_4HW \end{aligned}$$

5. (a) At $t = 0$, there are about 300 rabbits and 100 foxes. At $t = t_1$, the number of foxes reaches a minimum of about 20 while the number of rabbits is about 1000. At $t = t_2$, the number of rabbits reaches a maximum of about 2400, while the number of foxes rebounds to 100. At $t = t_3$, the number of rabbits decreases to about 1000 and the number of foxes reaches a maximum of about 315. As t increases, the number of foxes decreases greatly to 100, and the number of rabbits decreases to 300 (the initial populations), and the cycle starts again.

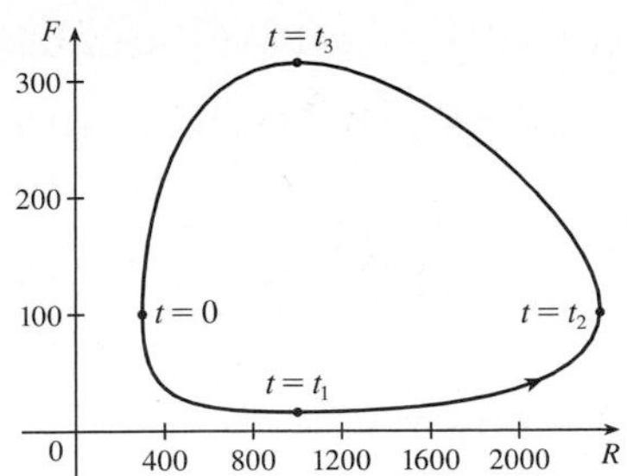

(b)

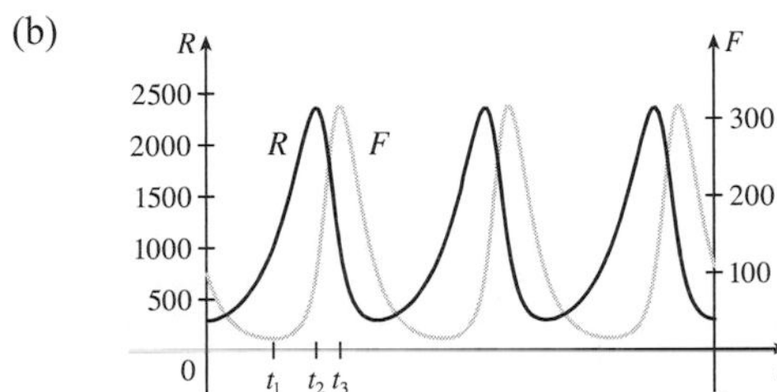

6. (a) At $t = 0$, there are about 600 rabbits and 160 foxes. At $t = t_1$, the number of rabbits reaches a minimum of about 80 and the number of foxes is also 80. At $t = t_2$, the number of foxes reaches a minimum of about 25 while the number of rabbits rebounds to 1000. At $t = t_3$, the number of foxes has increased to 40 and the rabbit population has reached a maximum of about 1750. The curve ends at $t = t_4$, where the number of foxes has increased to 65 and the number of rabbits has decreased to about 950.

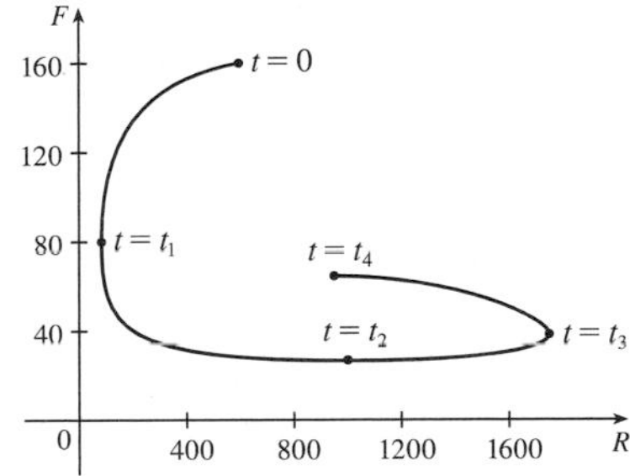

(b)

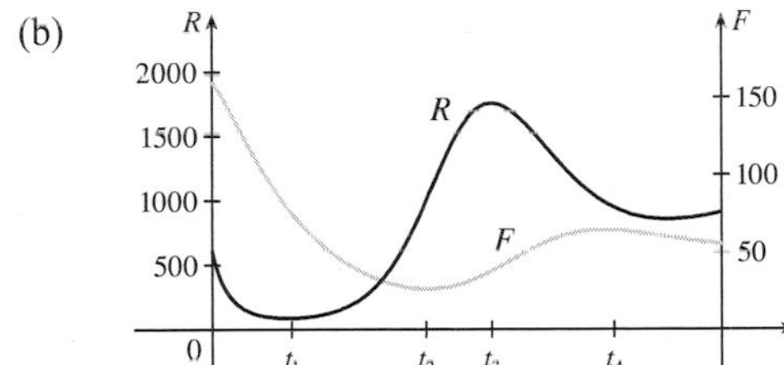

7.

Species 2; Species 1; $t = 0, 5$; $t = 1$; $t = 2$; $t = 3$; $t = 4$

8.

Species 2; Species 1; $t = 0$; $t = 5$; $t = 10$; $t = 15$

9. $\dfrac{dW}{dR} = \dfrac{-0.02W + 0.00002RW}{0.08R - 0.001RW} \Leftrightarrow (0.08 - 0.001W)R\,dW = (-0.02 + 0.00002R)W\,dR \Leftrightarrow$

$$\frac{0.08 - 0.001W}{W}\,dW = \frac{-0.02 + 0.00002R}{R}\,dR \Leftrightarrow \int\left(\frac{0.08}{W} - 0.001\right)dW = \int\left(-\frac{0.02}{R} + 0.00002\right)dR \Leftrightarrow$$

$0.08\ \ln|W| - 0.001W = -0.02\ \ln|R| + 0.00002R + K \quad\Leftrightarrow\quad 0.08\ \ln W + 0.02\ \ln R = 0.001W + 0.00002R + K \quad\Leftrightarrow$

$\ln\left(W^{0.08}R^{0.02}\right) = 0.00002R + 0.001W + K \quad\Leftrightarrow\quad W^{0.08}R^{0.02} = e^{0.00002R+0.001W+K} \quad\Leftrightarrow$

$R^{0.02}W^{0.08} = Ce^{0.00002R}e^{0.001W} \quad\Leftrightarrow\quad \dfrac{R^{0.02}W^{0.08}}{e^{0.00002R}e^{0.001W}} = C$. In general, if $\dfrac{dy}{dx} = \dfrac{-ry+bxy}{kx-axy}$, then $C = \dfrac{x^r y^k}{e^{bx}e^{ay}}$.

10. (a) A and L are constant $\Rightarrow\ A' = 0$ and $L' = 0 \ \Rightarrow \left.\begin{cases} 0 = 2A - 0.01AL \\ 0 = -0.5L + 0.0001AL \end{cases}\right\} \Rightarrow \begin{cases} 0 = A(2 - 0.01L) \\ 0 = L(-0.5 + 0.0001A) \end{cases}$

So either $A = L = 0$ or $L = \frac{2}{0.01} = 200$ and $A = \frac{0.5}{0.0001} = 5000$. The trivial solution $A = L = 0$ just says that if there aren't any aphids or ladybugs, then the populations will not change. The non-trivial solution, $L = 200$ and $A = 5000$, indicates the population sizes needed so that there are no changes in either the number of aphids or the number of ladybugs.

(b) $\dfrac{dL}{dA} = \dfrac{dL/dt}{dA/dt} = \dfrac{-0.5L + 0.0001AL}{2A - 0.01AL}$

(c) The solution curves (phase trajectories) are all closed curves that have the equilibrium point $(5000, 200)$ inside them.

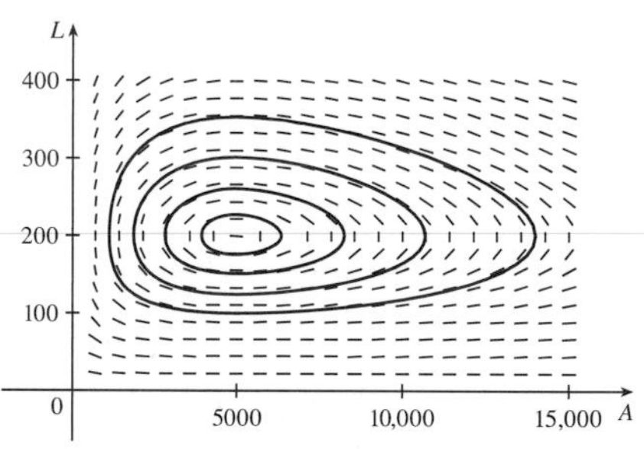

(d)

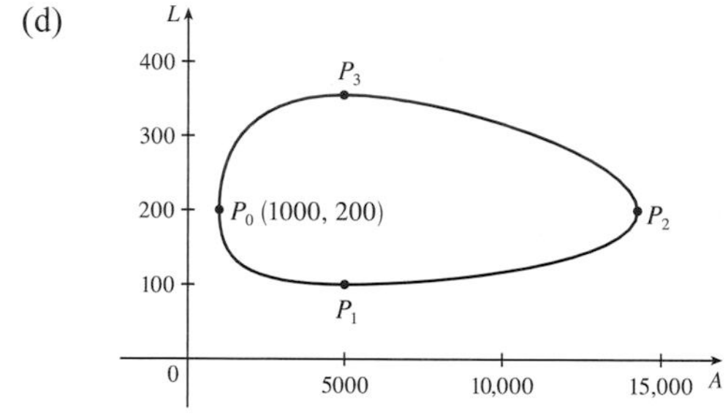

At $P_0(1000, 200)$, $dA/dt = 0$ and $dL/dt = -80 < 0$, so the number of ladybugs is decreasing and hence, we are proceeding in a counterclockwise direction. At P_0, there aren't enough aphids to support the ladybug population, so the number of ladybugs decreases and the number of aphids begins to increase. The ladybug population reaches a minimum at $P_1(5000, 100)$ while the aphid population increases in a dramatic way, reaching its maximum at $P_2(14,250, 200)$.

Meanwhile, the ladybug population is increasing from P_1 to $P_3(5000, 355)$, and as we pass through P_2, the increasing number of ladybugs starts to deplete the aphid population. At P_3 the ladybugs reach a maximum population, and start to decrease due to the reduced aphid population. Both populations then decrease until P_0, where the cycle starts over again.

(e) Both graphs have the same period and the graph of L peaks about a quarter of a cycle after the graph of A.

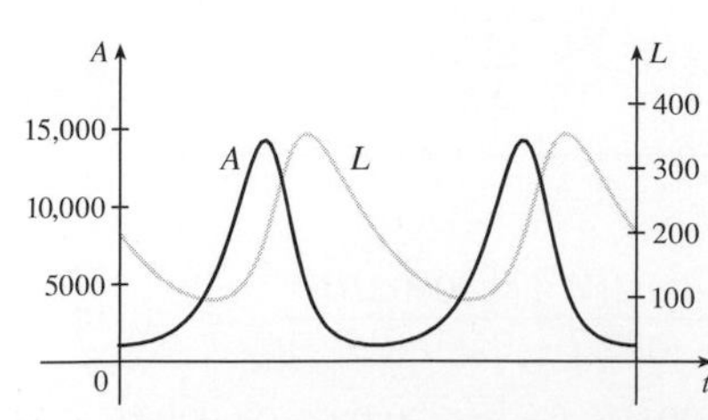

11. (a) Letting $W = 0$ gives us $dR/dt = 0.08R(1 - 0.0002R)$. $dR/dt = 0 \Leftrightarrow R = 0$ or 5000. Since $dR/dt > 0$ for $0 < R < 5000$, we would expect the rabbit population to *increase* to 5000 for these values of R. Since $dR/dt < 0$ for $R > 5000$, we would expect the rabbit population to *decrease* to 5000 for these values of R. Hence, in the absence of wolves, we would expect the rabbit population to stabilize at 5000.

(b) R and W are constant $\Rightarrow$ $R' = 0$ and $W' = 0$ $\Rightarrow$

$$\left.\begin{cases} 0 = 0.08R(1 - 0.0002R) - 0.001\,RW \\ 0 = -0.02W + 0.00002RW \end{cases}\right\} \Rightarrow \begin{cases} 0 = R[0.08(1 - 0.0002R) - 0.001W] \\ 0 = W(-0.02 + 0.00002R) \end{cases}$$

The second equation is true if $W = 0$ or $R = \frac{0.02}{0.00002} = 1000$. If $W = 0$ in the first equation, then either $R = 0$ or $R = \frac{1}{0.0002} = 5000$ [as in part (a)]. If $R = 1000$, then $0 = 1000[0.08(1 - 0.0002 \cdot 1000) - 0.001W] \Leftrightarrow$ $0 = 80(1 - 0.2) - W \Leftrightarrow W = 64$.

Case (i): $W = 0$, $R = 0$: both populations are zero

Case (ii): $W = 0$, $R = 5000$: see part (a)

Case (iii): $R = 1000$, $W = 64$: the predator/prey interaction balances and the populations are stable.

(c) The populations of wolves and rabbits fluctuate around 64 and 1000, respectively, and eventually stabilize at those values.

(d)

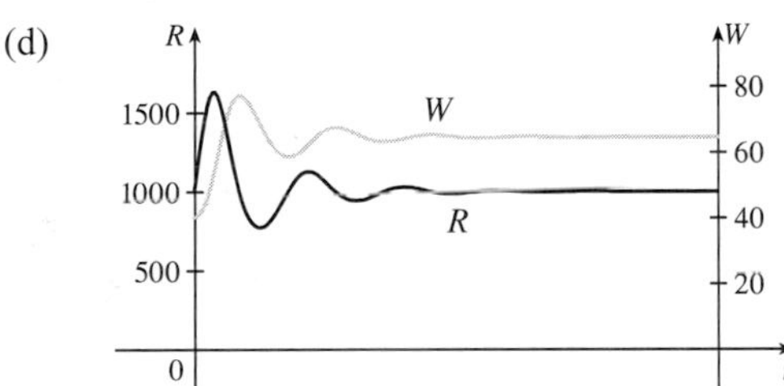

12. (a) If $L = 0$, $dA/dt = 2A\,(1 - 0.0001A)$, so $dA/dt = 0 \Leftrightarrow A = 0$ or $A = \frac{1}{0.0001} = 10{,}000$. Since $dA/dt > 0$ for $0 < A < 10{,}000$, we expect the aphid population to *increase* to 10,000 for these values of A. Since $dA/dt < 0$ for $A > 10{,}000$, we expect the aphid population to *decrease* to 10,000 for these values of A. Hence, in the absence of ladybugs we expect the aphid population to stabilize at 10,000.

(b) A and L are constant $\Rightarrow$ $A' = 0$ and $L' = 0$ $\Rightarrow$

$$\left.\begin{cases} 0 = 2A(1 - 0.0001A) - 0.01AL \\ 0 = -0.5L + 0.0001AL \end{cases}\right\} \Rightarrow \begin{cases} 0 = A[2(1 - 0.0001A) - 0.01L] \\ 0 = L(-0.5 + 0.0001A) \end{cases}$$

The second equation is true if $L = 0$ or $A = \frac{0.5}{0.0001} = 5000$. If $L = 0$ in the first equation, then either $A = 0$ or $A = \frac{1}{0.0001} = 10{,}000$. If $A = 5000$, then $0 = 5000[2(1 - 0.0001 \cdot 5000) - 0.01L] \Leftrightarrow$ $0 = 10{,}000(1 - 0.5) - 50L \Leftrightarrow 50L = 5000 \Leftrightarrow L = 100$.

The equilibrium solutions are: (i) $L = 0$, $A = 0$ (ii) $L = 0$, $A = 10{,}000$ (iii) $A = 5000$, $L = 100$

(c) $$\frac{dL}{dA} = \frac{dL/dt}{dA/dt} = \frac{-0.5L + 0.0001AL}{2A(1 - 0.0001A) - 0.01AL}$$

(d)

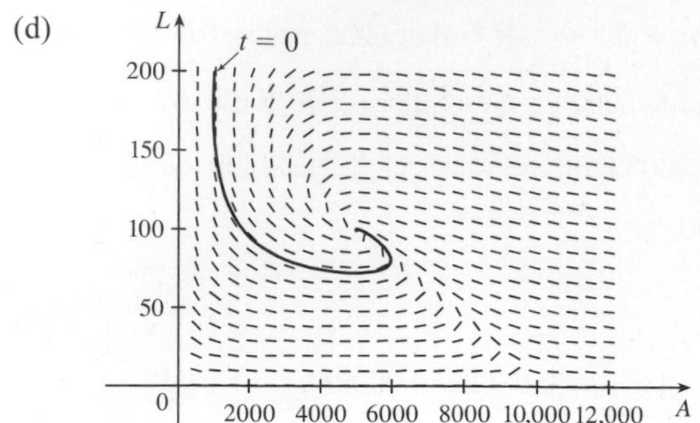

All of the phase trajectories spiral tightly around the equilibrium solution $(5000, 100)$.

(e)

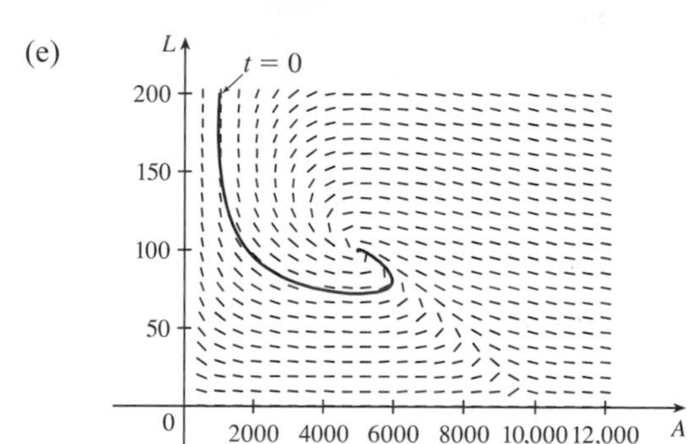

At $t = 0$, the ladybug population decreases rapidly and the aphid population decreases slightly before beginning to increase. As the aphid population continues to increase, the ladybug population reaches a minimum at about $(5000, 75)$. The ladybug population starts to increase and quickly stabilizes at 100, while the aphid population stabilizes at 5000.

(f)

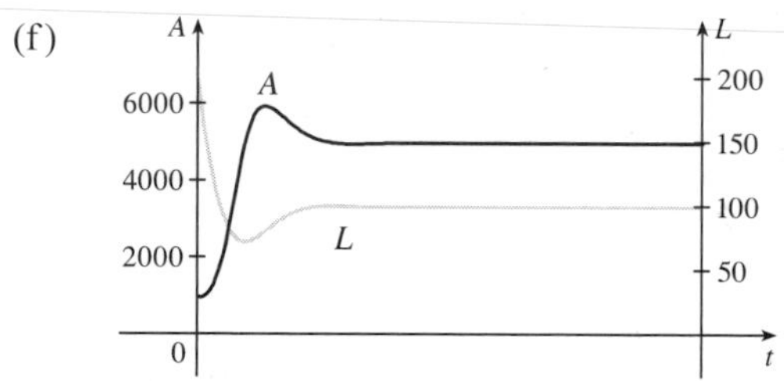

The graph of A peaks just after the graph of L has a minimum.

9 Review

TRUE-FALSE QUIZ

1. True. Since $y^4 \geq 0$, $y' = -1 - y^4 < 0$ and the solutions are decreasing functions.

2. True. $f(x) = y = \dfrac{\ln x}{x} \;\Rightarrow\; y' = \dfrac{1 - \ln x}{x^2}$.

LHS $= x^2 y' + xy = x^2 \cdot \dfrac{1 - \ln x}{x^2} + x \cdot \dfrac{\ln x}{x} = (1 - \ln x) + \ln x = 1 =$ RHS, so $y = \dfrac{\ln x}{x}$ is a solution of $x^2 y' + xy = 1$.

3. False. $x + y$ cannot be written in the form $g(x)f(y)$.

4. True. $y' = 3y - 2x + 6xy - 1 = 6xy - 2x + 3y - 1 = 2x(3y - 1) + 1(3y - 1) = (2x + 1)(3y - 1)$, so y' can be written in the form $g(x)f(y)$, and hence, is separable.

5. True. $e^x y' = y \;\Rightarrow\; y' = e^{-x} y \;\Rightarrow\; y' + (-e^{-x})y = 0$, which is of the form $y' + P(x)\,y = Q(x)$, so the equation is linear.

6. False. $y' + xy = e^y$ cannot be put in the form $y' + P(x)\,y = Q(x)$, so it is not linear.

7. True. By comparing $\dfrac{dy}{dt} = 2y\left(1 - \dfrac{y}{5}\right)$ with the logistic differential equation (9.4.4), we see that the carrying capacity is 5; that is, $\lim\limits_{t\to\infty} y = 5$.

EXERCISES

1. (a)

(b) $\lim\limits_{t\to\infty} y(t)$ appears to be finite for $0 \le c \le 4$. In fact $\lim\limits_{t\to\infty} y(t) = 4$ for $c = 4$, $\lim\limits_{t\to\infty} y(t) = 2$ for $0 < c < 4$, and $\lim\limits_{t\to\infty} y(t) = 0$ for $c = 0$. The equilibrium solutions are $y(t) = 0$, $y(t) = 2$, and $y(t) = 4$.

2. (a) We sketch the direction field and four solution curves, as shown. Note that the slope $y' = x/y$ is not defined on the line $y = 0$.

(b) $y' = x/y \;\Leftrightarrow\; y\,dy = x\,dx \;\Leftrightarrow\; y^2 = x^2 + C$. For $C = 0$, this is the pair of lines $y = \pm x$. For $C \neq 0$, it is the hyperbola $x^2 - y^2 = -C$.

3. (a)

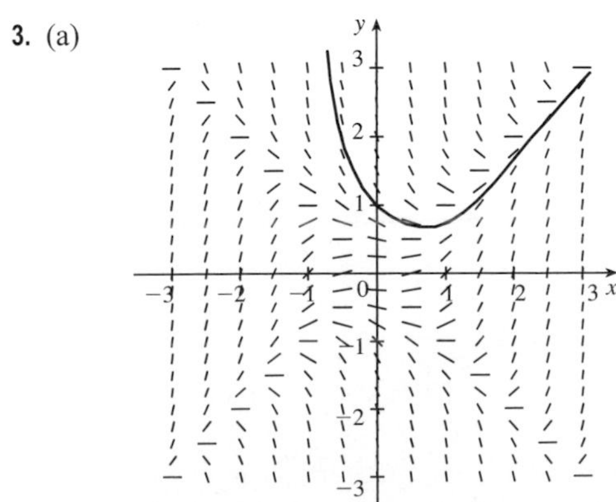

We estimate that when $x = 0.3$, $y = 0.8$, so $y(0.3) \approx 0.8$.

(b) $h = 0.1$, $x_0 = 0$, $y_0 = 1$ and $F(x, y) = x^2 - y^2$. So $y_n = y_{n-1} + 0.1(x_{n-1}^2 - y_{n-1}^2)$. Thus,

$y_1 = 1 + 0.1(0^2 - 1^2) = 0.9$, $y_2 = 0.9 + 0.1(0.1^2 - 0.9^2) = 0.82$, $y_3 = 0.82 + 0.1(0.2^2 - 0.82^2) = 0.75676$.

This is close to our graphical estimate of $y(0.3) \approx 0.8$.

(c) The centers of the horizontal line segments of the direction field are located on the lines $y = x$ and $y = -x$. When a solution curve crosses one of these lines, it has a local maximum or minimum.

4. (a) $h = 0.2$, $x_0 = 0$, $y_0 = 1$ and $F(x, y) = 2xy^2$. We need y_2.

$y_1 = 1 + 0.2(2 \cdot 0 \cdot 1^2) = 1$, $y_2 = 1 + 0.2(2 \cdot 0.2 \cdot 1^2) = 1.08 \approx y(0.4)$.

(b) $h = 0.1$ now, so $y_1 = 1 + 0.1(2 \cdot 0 \cdot 1^2) = 1$, $y_2 = 1 + 0.1(2 \cdot 0.1 \cdot 1^2) = 1.02$,

$y_3 = 1.02 + 0.1(2 \cdot 0.2 \cdot 1.02^2) \approx 1.06162$, $y_4 = 1.06162 + 0.1(2 \cdot 0.3 \cdot 1.06162^2) \approx 1.1292 \approx y(0.4)$.

(c) The equation is separable, so we write $\dfrac{dy}{y^2} = 2x\,dx \;\Rightarrow\; \displaystyle\int \frac{dy}{y^2} = \int 2x\,dx \;\Leftrightarrow\; -\frac{1}{y} = x^2 + C$, but $y(0) = 1$, so

$C = -1$ and $y(x) = \dfrac{1}{1 - x^2} \;\Leftrightarrow\; y(0.4) = \dfrac{1}{1 - 0.16} \approx 1.1905$. From this we see that the approximation was greatly improved by increasing the number of steps, but the approximations were still far off.

5. $y' = xe^{-\sin x} - y\cos x \;\Rightarrow\; y' + (\cos x)\,y = xe^{-\sin x}$ $(\star)$. This is a linear equation and the integrating factor is

$I(x) = e^{\int \cos x\,dx} = e^{\sin x}$. Multiplying $(\star)$ by $e^{\sin x}$ gives $e^{\sin x}\,y' + e^{\sin x}(\cos x)\,y = x \;\Rightarrow\; (e^{\sin x}\,y)' = x \;\Rightarrow$

$e^{\sin x}\,y = \frac{1}{2}x^2 + C \;\Rightarrow\; y = \left(\frac{1}{2}x^2 + C\right)e^{-\sin x}$.

6. $\dfrac{dx}{dt} = 1 - t + x - tx = 1(1 - t) + x(1 - t) = (1 + x)(1 - t) \;\Rightarrow\; \dfrac{dx}{1 + x} = (1 - t)\,dt \;\Rightarrow$

$\displaystyle\int \frac{dx}{1 + x} = \int (1 - t)\,dt \;\Rightarrow\; \ln|1 + x| = t - \tfrac{1}{2}t^2 + C \;\Rightarrow\; |1 + x| = e^{t - t^2/2 + C} \;\Rightarrow$

$1 + x = \pm e^{t - t^2/2} \cdot e^C \;\Rightarrow\; x = -1 + Ke^{t - t^2/2}$, where K is any nonzero constant.

7. $2ye^{y^2}y' = 2x + 3\sqrt{x} \;\Rightarrow\; 2ye^{y^2}\dfrac{dy}{dx} = 2x + 3\sqrt{x} \;\Rightarrow\; 2ye^{y^2}\,dy = \left(2x + 3\sqrt{x}\right)dx \;\Rightarrow$

$\int 2ye^{y^2}\,dy = \int \left(2x + 3\sqrt{x}\right)dx \;\Rightarrow\; e^{y^2} = x^2 + 2x^{3/2} + C \;\Rightarrow\; y^2 = \ln(x^2 + 2x^{3/2} + C) \;\Rightarrow$

$y = \pm\sqrt{\ln(x^2 + 2x^{3/2} + C)}$

8. $x^2y' - y = 2x^3e^{-1/x} \;\Rightarrow\; y' - \dfrac{1}{x^2}\,y = 2xe^{-1/x}$ $(\star)$. This is a linear equation and the integrating factor is

$I(x) = e^{\int(-1/x^2)\,dx} = e^{1/x}$. Multiplying $(\star)$ by $e^{1/x}$ gives $e^{1/x}\,y' - e^{1/x} \cdot \dfrac{1}{x^2}\,y = 2x \;\Rightarrow\; (e^{1/x}\,y)' = 2x \;\Rightarrow$

$e^{1/x}\,y = x^2 + C \;\Rightarrow\; y = e^{-1/x}(x^2 + C)$.

9. $\dfrac{dr}{dt} + 2tr = r \;\Rightarrow\; \dfrac{dr}{dt} = r - 2tr = r(1 - 2t) \;\Rightarrow\; \displaystyle\int \frac{dr}{r} = \int (1 - 2t)\,dt \;\Rightarrow\; \ln|r| = t - t^2 + C \;\Rightarrow$

$|r| = e^{t - t^2 + C} = ke^{t - t^2}$. Since $r(0) = 5$, $5 = ke^0 = k$. Thus, $r(t) = 5e^{t - t^2}$.

10. $(1+\cos x)y' = (1+e^{-y})\sin x \;\Rightarrow\; \dfrac{dy}{1+e^{-y}} = \dfrac{\sin x\,dx}{1+\cos x} \;\Rightarrow\; \displaystyle\int \frac{dy}{1+1/e^y} = \int \frac{\sin x\,dx}{1+\cos x} \;\Rightarrow$

$\displaystyle\int \frac{e^y\,dy}{1+e^y} = \int \frac{\sin x\,dx}{1+\cos x} \;\Rightarrow\; \ln|1+e^y| = -\ln|1+\cos x| + C \;\Rightarrow\; \ln(1+e^y) = -\ln(1+\cos x) + C \;\Rightarrow$

$1+e^y = e^{-\ln(1+\cos x)}\cdot e^C \;\Rightarrow\; e^y = ke^{-\ln(1+\cos x)} - 1 \;\Rightarrow\; y = \ln[ke^{-\ln(1+\cos x)} - 1]$. Since $y(0)=0$,

$0 = \ln[ke^{-\ln 2} - 1] \;\Rightarrow\; e^0 = k\left(\frac{1}{2}\right) - 1 \;\Rightarrow\; k = 4$. Thus, $y(x) = \ln[4e^{-\ln(1+\cos x)} - 1]$. An equivalent form

is $y(x) = \ln\dfrac{3-\cos x}{1+\cos x}$.

11. $xy' - y = x\ln x \;\Rightarrow\; y' - \dfrac{1}{x}y = \ln x$. $I(x) = e^{\int(-1/x)\,dx} = e^{-\ln|x|} = \left(e^{\ln|x|}\right)^{-1} = |x|^{-1} = 1/x$ since the condition $y(1)=2$ implies that we want a solution with $x>0$. Multiplying the last differential equation by $I(x)$ gives

$\dfrac{1}{x}y' - \dfrac{1}{x^2}y = \dfrac{1}{x}\ln x \;\Rightarrow\; \left(\dfrac{1}{x}y\right)' = \dfrac{1}{x}\ln x \;\Rightarrow\; \dfrac{1}{x}y = \displaystyle\int \frac{\ln x}{x}\,dx \;\Rightarrow\; \dfrac{1}{x}y = \frac{1}{2}(\ln x)^2 + C \;\Rightarrow$

$y = \frac{1}{2}x(\ln x)^2 + Cx$. Now $y(1)=2 \;\Rightarrow\; 2 = 0 + C \;\Rightarrow\; C=2$, so $y = \frac{1}{2}x(\ln x)^2 + 2x$.

12. $y' = 3x^2e^y \;\Rightarrow\; \dfrac{dy}{dx} = 3x^2e^y \;\Rightarrow\; e^{-y}\,dy = 3x^2\,dx \;\Rightarrow$

$\int e^{-y}\,dy = \int 3x^2\,dx \;\Rightarrow\; -e^{-y} = x^3 + C$. Now $y(0)=1 \;\Rightarrow$

$-e^{-1} = C$, so $-e^{-y} = x^3 - e^{-1} \;\Rightarrow\; e^{-y} = -x^3 + e^{-1} \;\Rightarrow$

$-y = \ln(-x^3+e^{-1}) \;\Rightarrow\; y = -\ln(-x^3+e^{-1})$. To find the domain,

solve $-x^3 + e^{-1} > 0 \;\Rightarrow\; x^3 < e^{-1} \;\Rightarrow\; x < e^{-1/3}$, so the domain is

$(-\infty, e^{-1/3})$ and $x = e^{-1/3}$ [≈ 0.72] is a vertical asymptote.

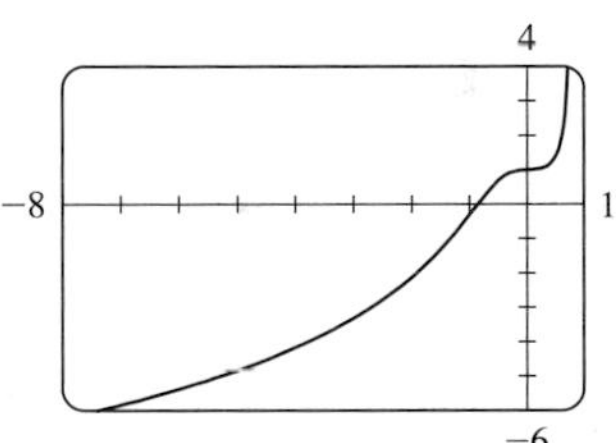

13. $\dfrac{d}{dx}(y) = \dfrac{d}{dx}(ke^x) \;\Rightarrow\; y' = ke^x = y$, so the orthogonal trajectories must have $y' = -\dfrac{1}{y} \;\Rightarrow\; \dfrac{dy}{dx} = -\dfrac{1}{y} \;\Rightarrow$

$y\,dy = -dx \;\Rightarrow\; \int y\,dy = -\int dx \;\Rightarrow\; \frac{1}{2}y^2 = -x + C \;\Rightarrow\; x = C - \frac{1}{2}y^2$, which are parabolas with a horizontal axis.

14. $\dfrac{d}{dx}(y) = \dfrac{d}{dx}(e^{kx}) \;\Rightarrow\; y' = ke^{kx} = ky = \dfrac{\ln y}{x}\cdot y$, so the orthogonal trajectories must have $y' = -\dfrac{x}{y\ln y} \;\Rightarrow$

$\dfrac{dy}{dx} = -\dfrac{x}{y\ln y} \;\Rightarrow\; y\ln y\,dy = -x\,dx \;\Rightarrow\; \int y\ln y\,dy = -\int x\,dx \;\Rightarrow\; \frac{1}{2}y^2\ln y - \frac{1}{4}y^2$ [parts with $u = \ln y$, $dv = y\,dy$] $= -\frac{1}{2}x^2 + C_1 \;\Rightarrow\; 2y^2\ln y - y^2 = C - 2x^2$.

15. (a) Using (4) and (7) in Section 9.4, we see that for $\dfrac{dP}{dt} = 0.1P\left(1 - \dfrac{P}{2000}\right)$ with $P(0) = 100$, we have $k = 0.1$,

$M = 2000$, $P_0 = 100$, and $A = \dfrac{2000-100}{100} = 19$. Thus, the solution of the initial-value problem is

$P(t) = \dfrac{2000}{1+19e^{-0.1t}}$ and $P(20) = \dfrac{2000}{1+19e^{-2}} \approx 560$.

(b) $P = 1200 \;\Leftrightarrow\; 1200 = \dfrac{2000}{1+19e^{-0.1t}} \;\Leftrightarrow\; 1 + 19e^{-0.1t} = \dfrac{2000}{1200} \;\Leftrightarrow\; 19e^{-0.1t} = \dfrac{5}{3} - 1 \;\Leftrightarrow$

$e^{-0.1t} = \left(\frac{2}{3}\right)/19 \;\Leftrightarrow\; -0.1t = \ln\frac{2}{57} \;\Leftrightarrow\; t = -10\ln\frac{2}{57} \approx 33.5$.

16. (a) Let $t = 0$ correspond to the year 2000. An exponential model is $P(t) = ae^{kt}$. $P(0) = 6.1$, so $P(t) = 6.1e^{kt}$.

$P(10) = 6.1e^{10k}$ and $P(10) = 6.9$, so $6.1e^{10k} = 6.9 \quad \Leftrightarrow \quad \dfrac{6.9}{6.1} = e^{10k} \quad \Rightarrow \quad 10k = \ln\dfrac{69}{61} \quad \Rightarrow$

$k = \dfrac{1}{10}\ln\dfrac{69}{61} \approx 0.0123$. Thus, $P(t) = 6.1e^{kt}$ and $P(20) = 6.1e^{20k} \approx 7.8$. Our model predicts that the world population in the year 2020 will be 7.8 billion.

(b) $P(t) = 10 \quad \Leftrightarrow \quad 6.1e^{kt} = 10 \quad \Leftrightarrow \quad e^{kt} = \dfrac{10}{6.1} \quad \Leftrightarrow \quad kt = \ln\dfrac{10}{6.1} \quad \Leftrightarrow \quad t = 10\dfrac{\ln(10/6.1)}{\ln(69/61)} \approx 40.11$ years. Our exponential model predicts that the world population will exceed 10 billion in 40.11 years; that is, in the year 2040.

(c) $A = \dfrac{M - P_0}{P_0} = \dfrac{20 - 6.1}{6.1} = \dfrac{139}{61}$ and from part (a), $k = \dfrac{1}{10}\ln\dfrac{69}{61}$, so $P(t) = \dfrac{M}{1 + Ae^{-kt}} = \dfrac{20}{1 + \frac{139}{61}e^{-kt}}$. Thus,

$P(20) = \dfrac{20}{1 + \frac{139}{61}e^{-20k}} \approx 7.2$ billion, which is less than our prediction of 7.8 billion from the exponential model in part (a).

(d) $P(t) = 10 \quad \Leftrightarrow \quad \dfrac{20}{1 + \frac{139}{61}e^{-kt}} = 10 \quad \Leftrightarrow \quad \dfrac{20}{10} = 1 + \dfrac{139}{61}e^{-kt} \quad \Leftrightarrow \quad 1 = \dfrac{139}{61}e^{-kt} \quad \Leftrightarrow \quad \dfrac{61}{139} = e^{-kt} \quad \Leftrightarrow$

$\ln\dfrac{61}{139} = -kt \quad \Leftrightarrow \quad t = -10\dfrac{\ln(61/139)}{\ln(69/61)} \approx 66.83$ years. Our logistic model predicts that the world population will exceed 10 billion in 66.83 years; that is, in the year 2066, which is considerably later than our prediction of 2040 from the exponential model in part (b).

17. (a) $\dfrac{dL}{dt} \propto L_\infty - L \quad \Rightarrow \quad \dfrac{dL}{dt} = k(L_\infty - L) \quad \Rightarrow \quad \displaystyle\int \frac{dL}{L_\infty - L} = \int k\,dt \quad \Rightarrow \quad -\ln|L_\infty - L| = kt + C \quad \Rightarrow$

$\ln|L_\infty - L| = -kt - C \quad \Rightarrow \quad |L_\infty - L| = e^{-kt-C} \quad \Rightarrow \quad L_\infty - L = Ae^{-kt} \quad \Rightarrow \quad L = L_\infty - Ae^{-kt}$.

At $t = 0$, $L = L(0) = L_\infty - A \quad \Rightarrow \quad A = L_\infty - L(0) \quad \Rightarrow \quad L(t) = L_\infty - [L_\infty - L(0)]e^{-kt}$.

(b) $L_\infty = 53$ cm, $L(0) = 10$ cm, and $k = 0.2 \quad \Rightarrow \quad L(t) = 53 - (53 - 10)e^{-0.2t} = 53 - 43e^{-0.2t}$.

18. Denote the amount of salt in the tank (in kg) by y. $y(0) = 0$ since initially there is only water in the tank.

The rate at which y increases is equal to the rate at which salt flows into the tank minus the rate at which it flows out.

That rate is $\dfrac{dy}{dt} = 0.1\dfrac{\text{kg}}{\text{L}} \times 10\dfrac{\text{L}}{\text{min}} - \dfrac{y}{100}\dfrac{\text{kg}}{\text{L}} \times 10\dfrac{\text{L}}{\text{min}} = 1 - \dfrac{y}{10}\dfrac{\text{kg}}{\text{min}} \quad \Rightarrow \quad \displaystyle\int \frac{dy}{10 - y} = \int \frac{1}{10}\,dt \quad \Rightarrow$

$-\ln|10 - y| = \frac{1}{10}t + C \quad \Rightarrow \quad 10 - y = Ae^{-t/10}$. $y(0) = 0 \quad \Rightarrow \quad 10 = A \quad \Rightarrow \quad y = 10(1 - e^{-t/10})$.

At $t = 6$ minutes, $y = 10(1 - e^{-6/10}) \approx 4.512$ kg.

19. Let P represent the population and I the number of infected people. The rate of spread dI/dt is jointly proportional to I and to $P - I$, so for some constant k, $\dfrac{dI}{dt} = kI(P - I) \quad \Rightarrow \quad I(t) = \dfrac{I_0 P}{I_0 + (P - I_0)e^{-kPt}}$ [from the discussion of logistic growth in Section 9.4].

Now, measuring t in days, we substitute $t = 7$, $P = 5000$, $I_0 = 160$ and $I(7) = 1200$ to find k:

$$1200 = \frac{160 \cdot 5000}{160 + (5000 - 160)e^{-5000 \cdot 7 \cdot k}} \quad \Leftrightarrow \quad 3 = \frac{2000}{160 + 4840e^{-35{,}000k}} \quad \Leftrightarrow \quad 480 + 14{,}520e^{-35{,}000k} = 2000 \quad \Leftrightarrow$$

$e^{-35{,}000k} = \dfrac{2000 - 480}{14{,}520} \quad \Leftrightarrow \quad -35{,}000k = \ln \dfrac{38}{363} \quad \Leftrightarrow \quad k = \dfrac{-1}{35{,}000} \ln \dfrac{38}{363} \approx 0.00006448$. Next, let

$I = 5000 \times 80\% = 4000$, and solve for t: $4000 = \dfrac{160 \cdot 5000}{160 + (5000 - 160)e^{-k \cdot 5000 \cdot t}} \quad \Leftrightarrow \quad 1 = \dfrac{200}{160 + 4840e^{-5000kt}} \quad \Leftrightarrow$

$$160 + 4840e^{-5000kt} = 200 \quad \Leftrightarrow \quad e^{-5000kt} = \frac{200 - 160}{4840} \quad \Leftrightarrow \quad -5000kt = \ln \frac{1}{121} \quad \Leftrightarrow$$

$t = \dfrac{-1}{5000k} \ln \dfrac{1}{121} = \dfrac{1}{\frac{1}{7} \ln \frac{38}{363}} \cdot \ln \dfrac{1}{121} = 7 \cdot \dfrac{\ln 121}{\ln \frac{363}{38}} \approx 14.875$. So it takes about 15 days for 80% of the population to be infected.

20. $\dfrac{1}{R}\dfrac{dR}{dt} = \dfrac{k}{S}\dfrac{dS}{dt} \quad \Rightarrow \quad \dfrac{d}{dt}(\ln R) = \dfrac{d}{dt}(k \ln S) \quad \Rightarrow \quad \ln R = k \ln S + C \quad \Rightarrow$

$R = e^{k \ln S + C} = e^C \left(e^{\ln S}\right)^k \quad \Rightarrow \quad R = AS^k$, where $A = e^C$ is a positive constant.

21. $\dfrac{dh}{dt} = -\dfrac{R}{V}\left(\dfrac{h}{k+h}\right) \quad \Rightarrow \quad \displaystyle\int \frac{k+h}{h}\,dh = \int \left(-\frac{R}{V}\right) dt \quad \Rightarrow \quad \int \left(1 + \frac{k}{h}\right) dh = -\frac{R}{V} \int 1\,dt \quad \Rightarrow$

$h + k \ln h = -\dfrac{R}{V} t + C$. This equation gives a relationship between h and t, but it is not possible to isolate h and express it in terms of t.

22. $dx/dt = 0.4x - 0.002xy$, $dy/dt = -0.2y + 0.000008xy$

(a) The xy terms represent encounters between the birds and the insects. Since the y-population increases from these terms and the x-population decreases, we expect y to represent the birds and x the insects.

(b) x and y are constant $\Rightarrow$ $x' = 0$ and $y' = 0$ $\Rightarrow$

$$\left\{\begin{array}{l} 0 = 0.4x - 0.002xy \\ 0 = -0.2y + 0.000008xy \end{array}\right\} \quad \Rightarrow \quad \left\{\begin{array}{l} 0 = 0.4x(1 - 0.005y) \\ 0 = -0.2y(1 - 0.00004x) \end{array}\right. \quad \Rightarrow \quad y = 0 \text{ and } x = 0 \text{ (zero populations)}$$

or $y = \frac{1}{0.005} = 200$ and $x = \frac{1}{0.00004} = 25{,}000$. The non-trivial solution represents the population sizes needed so that there are no changes in either the number of birds or the number of insects.

(c) $\dfrac{dy}{dx} = \dfrac{dy/dt}{dx/dt} = \dfrac{-0.2y + 0.000008xy}{0.4x - 0.002xy}$

(d)

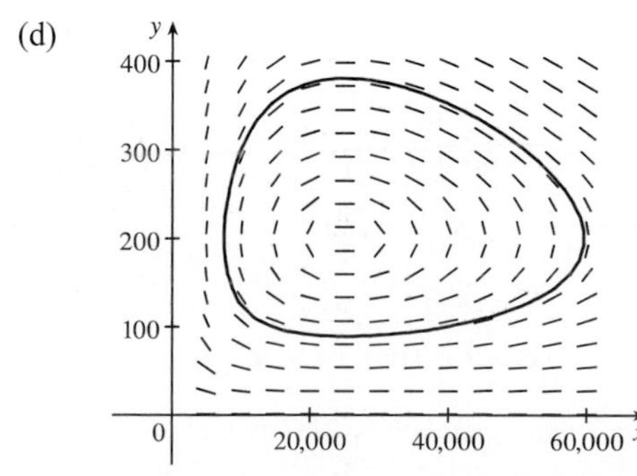

At $(x, y) = (40{,}000, 100)$, $dx/dt = 8000 > 0$, so as t increases we are proceeding in a counterclockwise direction. The populations increase to approximately $(59{,}646, 200)$, at which point the insect population starts to decrease. The birds attain a maximum population of about 380 when the insect population is 25,000. The populations decrease to about $(7370, 200)$, at which point the insect population starts to increase. The birds attain a minimum population of about 88 when the insect population is 25,000, and then the cycle repeats.

(e)

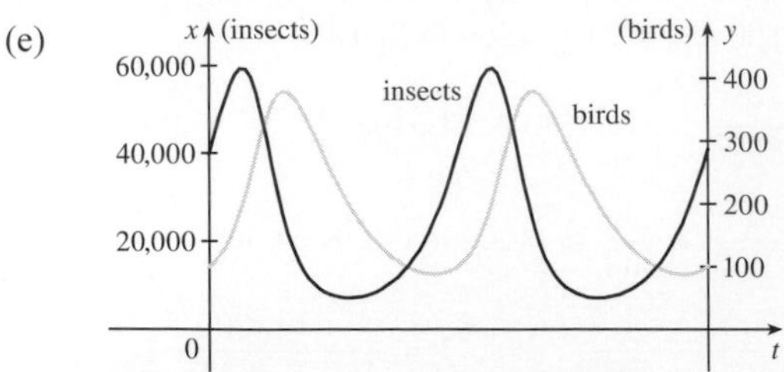

Both graphs have the same period and the bird population peaks about a quarter-cycle after the insect population.

23. (a) $dx/dt = 0.4x(1 - 0.000005x) - 0.002xy$, $dy/dt = -0.2y + 0.000008xy$. If $y = 0$, then $dx/dt = 0.4x(1 - 0.000005x)$, so $dx/dt = 0 \Leftrightarrow x = 0$ or $x = 200{,}000$, which shows that the insect population increases logistically with a carrying capacity of 200,000. Since $dx/dt > 0$ for $0 < x < 200{,}000$ and $dx/dt < 0$ for $x > 200{,}000$, we expect the insect population to stabilize at 200,000.

(b) x and y are constant $\Rightarrow$ $x' = 0$ and $y' = 0$ $\Rightarrow$

$$\left\{\begin{array}{l} 0 = 0.4x(1 - 0.000005x) - 0.002xy \\ 0 = -0.2y + 0.000008xy \end{array}\right\} \Rightarrow \left\{\begin{array}{l} 0 = 0.4x[(1 - 0.000005x) - 0.005y] \\ 0 = y(-0.2 + 0.000008x) \end{array}\right.$$

The second equation is true if $y = 0$ or $x = \frac{0.2}{0.000008} = 25{,}000$. If $y = 0$ in the first equation, then either $x = 0$ or $x = \frac{1}{0.000005} = 200{,}000$. If $x = 25{,}000$, then $0 = 0.4(25{,}000)[(1 - 0.000005 \cdot 25{,}000) - 0.005y]$ $\Rightarrow$ $0 = 10{,}000[(1 - 0.125) - 0.005y]$ $\Rightarrow$ $0 = 8750 - 50y$ $\Rightarrow$ $y = 175$.

Case (i): $y = 0$, $x = 0$: Zero populations

Case (ii): $y = 0$, $x = 200{,}000$: In the absence of birds, the insect population is always 200,000.

Case (iii): $x = 25{,}000$, $y = 175$: The predator/prey interaction balances and the populations are stable.

(c) The populations of the birds and insects fluctuate around 175 and 25,000, respectively, and eventually stabilize at those values.

(d)

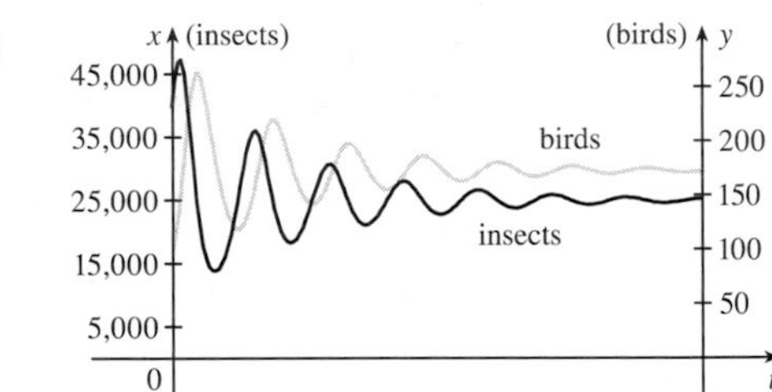

24. First note that, in this question, "weighs" is used in the informal sense, so what we really require is Barbara's mass m in kg as a function of t. Barbara's net intake of calories per day at time t (measured in days) is $c(t) = 1600 - 850 - 15m(t) = 750 - 15m(t)$, where $m(t)$ is her mass at time t. We are given that $m(0) = 60$ kg and $\dfrac{dm}{dt} = \dfrac{c(t)}{10{,}000}$, so $\dfrac{dm}{dt} = \dfrac{750 - 15m}{10{,}000} = \dfrac{150 - 3m}{2000} = \dfrac{-3(m - 50)}{2000}$ with $m(0) = 60$. From $\displaystyle\int \frac{dm}{m - 50} = \int \frac{-3\,dt}{2000}$, we get $\ln|m - 50| = -\frac{3}{2000}t + C$. Since $m(0) = 60$, $C = \ln 10$. Now $\ln\dfrac{|m - 50|}{10} = -\dfrac{3t}{2000}$, so $|m - 50| = 10e^{-3t/2000}$. The quantity $m - 50$ is continuous, initially positive, and the right-hand side is never zero. Thus, $m - 50$ is positive for all t, and $m(t) = 50 + 10e^{-3t/2000}$ kg. As $t \to \infty$, $m(t) \to 50$ kg. Thus, Barbara's mass gradually settles down to 50 kg.

☐ PROBLEMS PLUS

1. We use the Fundamental Theorem of Calculus to differentiate the given equation:

$[f(x)]^2 = 100 + \int_0^x \left\{[f(t)]^2 + [f'(t)]^2\right\} dt \;\Rightarrow\; 2f(x)f'(x) = [f(x)]^2 + [f'(x)]^2 \;\Rightarrow$

$[f(x)]^2 + [f'(x)]^2 - 2f(x)f'(x) = 0 \;\Rightarrow\; [f(x) - f'(x)]^2 = 0 \;\Leftrightarrow\; f(x) = f'(x)$. We can solve this as a separable equation, or else use Theorem 9.4.2 with $k = 1$, which says that the solutions are $f(x) = Ce^x$. Now $[f(0)]^2 = 100$, so $f(0) = C = \pm 10$, and hence $f(x) = \pm 10e^x$ are the only functions satisfying the given equation.

2. $(fg)' = f'g'$, where $f(x) = e^{x^2} \;\Rightarrow\; \left(e^{x^2}g\right)' = 2xe^{x^2}g'$. Since the student's mistake did not affect the answer,

$\left(e^{x^2}g\right)' = e^{x^2}g' + 2xe^{x^2}g = 2xe^{x^2}g'$. So $(2x-1)g' = 2xg$, or $\dfrac{g'}{g} = \dfrac{2x}{2x-1} = 1 + \dfrac{1}{2x-1} \;\Rightarrow$

$\ln|g(x)| = x + \frac{1}{2}\ln(2x-1) + C \;\Rightarrow\; g(x) = Ae^x\sqrt{2x-1}$.

3. $f'(x) = \lim\limits_{h\to 0} \dfrac{f(x+h) - f(x)}{h} = \lim\limits_{h\to 0} \dfrac{f(x)\,[f(h) - 1]}{h}$ [since $f(x+h) = f(x)f(h)$]

$= f(x) \lim\limits_{h\to 0} \dfrac{f(h) - 1}{h} = f(x) \lim\limits_{h\to 0} \dfrac{f(h) - f(0)}{h - 0} = f(x)f'(0) = f(x)$

Therefore, $f'(x) = f(x)$ for all x and from Theorem 9.4.2 we get $f(x) = Ae^x$.

Now $f(0) = 1 \;\Rightarrow\; A = 1 \;\Rightarrow\; f(x) = e^x$.

4. $\left(\displaystyle\int f(x)\,dx\right)\left(\displaystyle\int \frac{dx}{f(x)}\right) = -1 \;\Rightarrow\; \displaystyle\int \frac{dx}{f(x)} = \frac{-1}{\int f(x)\,dx} \;\Rightarrow\; \frac{1}{f(x)} = \frac{f(x)}{\left[\int f(x)\,dx\right]^2}$ [after differentiating] $\Rightarrow$

$\int f(x)\,dx = \pm f(x)$ [after taking square roots] $\Rightarrow\; f(x) = \pm f'(x)$ [after differentiating again] $\Rightarrow\; y = Ae^x$ or $y = Ae^{-x}$ by Theorem 9.4.2. Therefore, $f(x) = Ae^x$ or $f(x) = Ae^{-x}$, for all nonzero constants A, are the functions satisfying the original equation.

5. "The area under the graph of f from 0 to x is proportional to the $(n+1)$st power of $f(x)$" translates to

$\int_0^x f(t)\,dt = k[f(x)]^{n+1}$ for some constant k. By FTC1, $\dfrac{d}{dx}\displaystyle\int_0^x f(t)\,dt = \dfrac{d}{dx}\left\{k[f(x)]^{n+1}\right\} \;\Rightarrow$

$f(x) = k(n+1)[f(x)]^n f'(x) \;\Rightarrow\; 1 = k(n+1)[f(x)]^{n-1}f'(x) \;\Rightarrow\; 1 = k(n+1)y^{n-1}\dfrac{dy}{dx} \;\Rightarrow$

$k(n+1)y^{n-1}\,dy = dx \;\Rightarrow\; \int k(n+1)y^{n-1}\,dy = \int dx \;\Rightarrow\; k(n+1)\dfrac{1}{n}y^n = x + C$.

Now $f(0) = 0 \;\Rightarrow\; 0 = 0 + C \;\Rightarrow\; C = 0$ and then $f(1) = 1 \;\Rightarrow\; k(n+1)\dfrac{1}{n} = 1 \;\Rightarrow\; k = \dfrac{n}{n+1}$,

so $y^n = x$ and $y = f(x) = x^{1/n}$.

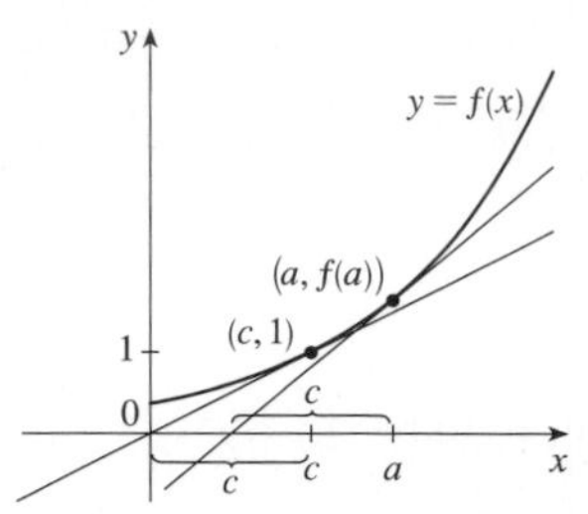

6. Let $y = f(x)$ be a curve that passes through the point $(c, 1)$ and whose subtangents all have length c. The tangent line at $x = a$ has equation $y - f(a) = f'(a)(x - a)$. Assuming $f(a) \neq 0$ and $f'(a) \neq 0$, it has x-intercept $a - \dfrac{f(a)}{f'(a)}$ [let $y = 0$ and solve for x]. Thus, the length of the subtangent is c, so $\left|a - \left(a - \dfrac{f(a)}{f'(a)}\right)\right| = \left|\dfrac{f(a)}{f'(a)}\right| = c \;\Rightarrow\; \dfrac{f'(a)}{f(a)} = \pm\dfrac{1}{c}$.

Now $\dfrac{f'(x)}{f(x)} = \pm\dfrac{1}{c} \;\Rightarrow\; f'(x) = \pm\dfrac{1}{c}f(x) \;\Rightarrow\; \dfrac{dy}{dx} = \pm\dfrac{1}{c}y \;\Rightarrow\; \dfrac{dy}{y} = \pm\dfrac{1}{c}\,dx \;\Rightarrow\; \displaystyle\int \frac{1}{y}\,dy = \pm\frac{1}{c}\int dx \;\Rightarrow$

$\ln|y| = \pm\dfrac{1}{c}x + K$. Since $f(c) = 1$, $\ln 1 = \pm 1 + K \;\Rightarrow\; K = \mp 1$. Thus, $y = e^{\pm x/c \mp 1}$, or $y = e^{\pm(x/c - 1)}$. One curve is an increasing exponential (as shown in the figure) and the other curve is its reflection about the line $x = c$.

7. Let $y(t)$ denote the temperature of the peach pie t minutes after 5:00 PM and R the temperature of the room. Newton's Law of Cooling gives us $dy/dt = k(y - R)$. Solving for y we get $\dfrac{dy}{y - R} = k\,dt \;\Rightarrow\; \ln|y - R| = kt + C \;\Rightarrow$

$|y - R| = e^{kt + C} \;\Rightarrow\; y - R = \pm e^{kt} \cdot e^{C} \;\Rightarrow\; y = Me^{kt} + R$, where M is a nonzero constant. We are given temperatures at three times.

$$y(0) = 100 \;\Rightarrow\; 100 = M + R \;\Rightarrow\; R = 100 - M$$

$$y(10) = 80 \;\Rightarrow\; 80 = Me^{10k} + R \qquad \textbf{(1)}$$

$$y(20) = 65 \;\Rightarrow\; 65 = Me^{20k} + R \qquad \textbf{(2)}$$

Substituting $100 - M$ for R in **(1)** and **(2)** gives us

$$-20 = Me^{10k} - M \;\textbf{(3)} \quad \text{and} \quad -35 = Me^{20k} - M \;\textbf{(4)}$$

Dividing **(3)** by **(4)** gives us $\dfrac{-20}{-35} = \dfrac{M(e^{10k} - 1)}{M(e^{20k} - 1)} \;\Rightarrow\; \dfrac{4}{7} = \dfrac{e^{10k} - 1}{e^{20k} - 1} \;\Rightarrow\; 4e^{20k} - 4 = 7e^{10k} - 7 \;\Rightarrow$

$4e^{20k} - 7e^{10k} + 3 = 0$. This is a quadratic equation in e^{10k}. $(4e^{10k} - 3)(e^{10k} - 1) = 0 \;\Rightarrow\; e^{10k} = \frac{3}{4}$ or $1 \;\Rightarrow$

$10k = \ln\frac{3}{4}$ or $\ln 1 \;\Rightarrow\; k = \frac{1}{10}\ln\frac{3}{4}$ since k is a nonzero constant of proportionality. Substituting $\frac{3}{4}$ for e^{10k} in **(3)** gives us

$-20 = M \cdot \frac{3}{4} - M \;\Rightarrow\; -20 = -\frac{1}{4}M \;\Rightarrow\; M = 80$. Now $R = 100 - M$ so $R = 20°\text{C}$.

8. Let b be the number of hours before noon that it began to snow, t the time measured in hours after noon, and $x = x(t) =$ distance traveled by the plow at time t. Then $dx/dt =$ speed of plow. Since the snow falls steadily, the height at time t is $h(t) = k(t + b)$, where k is a constant. We are given that the rate of removal is constant, say R (in m^3/h).

If the width of the path is w, then $R = \text{height} \times \text{width} \times \text{speed} = h(t) \times w \times \dfrac{dx}{dt} = k(t + b)w\dfrac{dx}{dt}$. Thus, $\dfrac{dx}{dt} = \dfrac{C}{t + b}$, where $C = \dfrac{R}{kw}$ is a constant. This is a separable equation. $\displaystyle\int dx = C\int \frac{dt}{t + b} \;\Rightarrow\; x(t) = C\ln(t + b) + K$.

Put $t = 0$: $0 = C\ln b + K \;\Rightarrow\; K = -C\ln b$, so $x(t) = C\ln(t + b) - C\ln b = C\ln(1 + t/b)$.

Put $t = 1$: $6000 = C\ln(1 + 1/b)$ [$x = 6$ km].

Put $t = 2$: $9000 = C\ln(1 + 2/b)$ [$x = (6 + 3)$ km].

Solve for b: $\dfrac{\ln(1+1/b)}{6000} = \dfrac{\ln(1+2/b)}{9000} \Rightarrow 3\ln\left(1+\dfrac{1}{b}\right) = 2\ln\left(1+\dfrac{2}{b}\right) \Rightarrow \left(1+\dfrac{1}{b}\right)^3 = \left(1+\dfrac{2}{b}\right)^2 \Rightarrow$

$1+\dfrac{3}{b}+\dfrac{3}{b^2}+\dfrac{1}{b^3} = 1+\dfrac{4}{b}+\dfrac{4}{b^2} \Rightarrow \dfrac{1}{b}+\dfrac{1}{b^2}-\dfrac{1}{b^3}=0 \Rightarrow b^2+b-1=0 \Rightarrow b=\frac{-1\pm\sqrt{5}}{2}$.

But $b>0$, so $b=\frac{-1+\sqrt{5}}{2} \approx 0.618$ h ≈ 37 min. The snow began to fall $\frac{\sqrt{5}-1}{2}$ hours before noon; that is, at about 11:23 AM.

9. (a) While running from $(L,0)$ to (x,y), the dog travels a distance $s = \int_x^L \sqrt{1+(dy/dx)^2}\,dx = -\int_L^x \sqrt{1+(dy/dx)^2}\,dx$, so $\dfrac{ds}{dx} = -\sqrt{1+(dy/dx)^2}$. The dog and rabbit run at the same speed, so the rabbit's position when the dog has traveled a distance s is $(0,s)$. Since the dog runs straight for the rabbit, $\dfrac{dy}{dx} = \dfrac{s-y}{0-x}$ (see the figure).

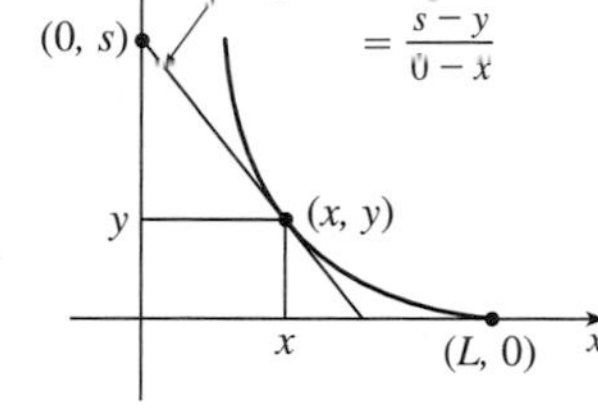

Thus, $s = y - x\dfrac{dy}{dx} \Rightarrow \dfrac{ds}{dx} = \dfrac{dy}{dx} - \left(x\dfrac{d^2y}{dx^2} + 1\dfrac{dy}{dx}\right) = -x\dfrac{d^2y}{dx^2}$. Equating the two expressions for $\dfrac{ds}{dx}$ gives us $x\dfrac{d^2y}{dx^2} = \sqrt{1+\left(\dfrac{dy}{dx}\right)^2}$, as claimed.

(b) Letting $z = \dfrac{dy}{dx}$, we obtain the differential equation $x\dfrac{dz}{dx} = \sqrt{1+z^2}$, or $\dfrac{dz}{\sqrt{1+z^2}} = \dfrac{dx}{x}$. Integrating:

$\ln x = \displaystyle\int \frac{dz}{\sqrt{1+z^2}} \overset{25}{=} \ln\left(z+\sqrt{1+z^2}\right)+C$. When $x=L$, $z=dy/dx=0$, so $\ln L = \ln 1 + C$. Therefore, $C=\ln L$, so $\ln x = \ln\left(\sqrt{1+z^2}+z\right)+\ln L = \ln\left[L\left(\sqrt{1+z^2}+z\right)\right] \Rightarrow x = L\left(\sqrt{1+z^2}+z\right) \Rightarrow$

$\sqrt{1+z^2} = \dfrac{x}{L} - z \Rightarrow 1+z^2 = \left(\dfrac{x}{L}\right)^2 - \dfrac{2xz}{L} + z^2 \Rightarrow \left(\dfrac{x}{L}\right)^2 - 2z\left(\dfrac{x}{L}\right) - 1 = 0 \Rightarrow$

$z = \dfrac{(x/L)^2-1}{2(x/L)} = \dfrac{x^2-L^2}{2Lx} = \dfrac{x}{2L} - \dfrac{L}{2}\dfrac{1}{x}$ [for $x>0$]. Since $z=\dfrac{dy}{dx}$, $y = \dfrac{x^2}{4L} - \dfrac{L}{2}\ln x + C_1$.

Since $y=0$ when $x=L$, $0 = \dfrac{L}{4} - \dfrac{L}{2}\ln L + C_1 \Rightarrow C_1 = \dfrac{L}{2}\ln L - \dfrac{L}{4}$. Thus,

$y = \dfrac{x^2}{4L} - \dfrac{L}{2}\ln x + \dfrac{L}{2}\ln L - \dfrac{L}{4} = \dfrac{x^2-L^2}{4L} - \dfrac{L}{2}\ln\left(\dfrac{x}{L}\right)$.

(c) As $x\to 0^+$, $y\to\infty$, so the dog never catches the rabbit.

10. (a) If the dog runs twice as fast as the rabbit, then the rabbit's position when the dog has traveled a distance s is $(0, s/2)$.

Since the dog runs straight toward the rabbit, the tangent line to the dog's path has slope $\dfrac{dy}{dx} = \dfrac{s/2-y}{0-x}$.

Thus, $s = 2y - 2x\dfrac{dy}{dx} \Rightarrow \dfrac{ds}{dx} = 2\dfrac{dy}{dx} - \left(2x\dfrac{d^2y}{dx^2} + 2\dfrac{dy}{dx}\right) = -2x\dfrac{d^2y}{dx^2}$.

From Problem 9(a), $\dfrac{ds}{dx} = -\sqrt{1+\left(\dfrac{dy}{dx}\right)^2}$, so $2x\dfrac{d^2y}{dx^2} = \sqrt{1+\left(\dfrac{dy}{dx}\right)^2}$.

Letting $z=\dfrac{dy}{dx}$, we obtain the differential equation $2x\dfrac{dz}{dx} = \sqrt{1+z^2}$, or $\dfrac{2\,dz}{\sqrt{1+z^2}} = \dfrac{dx}{x}$. Integrating, we get

$\ln x = \int \frac{2\,dz}{\sqrt{1+z^2}} = 2\ln\left(\sqrt{1+z^2}+z\right) + C$. [See Problem 9(b).] When $x = L$, $z = dy/dx = 0$, so

$\ln L = 2\ln 1 + C = C$. Thus,

$$\ln x = 2\ln\left(\sqrt{1+z^2}+z\right) + \ln L = \ln\left(L\left(\sqrt{1+z^2}+z\right)^2\right) \Rightarrow x = L\left(\sqrt{1+z^2}+z\right)^2 \Rightarrow$$

$$\sqrt{1+z^2} = \sqrt{\frac{x}{L}} - z \Rightarrow 1 + z^2 = \frac{x}{L} - 2\sqrt{\frac{x}{L}}\,z + z^2 \Rightarrow 2\sqrt{\frac{x}{L}}\,z = \frac{x}{L} - 1 \Rightarrow$$

$$\frac{dy}{dx} = z = \frac{1}{2}\sqrt{\frac{x}{L}} - \frac{1}{2\sqrt{x/L}} = \frac{1}{2\sqrt{L}}x^{1/2} - \frac{\sqrt{L}}{2}x^{-1/2} \Rightarrow y = \frac{1}{3\sqrt{L}}x^{3/2} - \sqrt{L}\,x^{1/2} + C_1.$$

When $x = L$, $y = 0$, so $0 = \frac{1}{3\sqrt{L}}L^{3/2} - \sqrt{L}\,L^{1/2} + C_1 = \frac{L}{3} - L + C_1 = C_1 - \frac{2}{3}L$. Therefore, $C_1 = \frac{2}{3}L$ and

$y = \frac{x^{3/2}}{3\sqrt{L}} - \sqrt{L}\,x^{1/2} + \frac{2}{3}L$. As $x \to 0$, $y \to \frac{2}{3}L$, so the dog catches the rabbit when the rabbit is at $\left(0, \frac{2}{3}L\right)$.

(At that point, the dog has traveled a distance of $\frac{4}{3}L$, twice as far as the rabbit has run.)

(b) As in the solutions to part (a) and Problem 9, we get $z = \frac{dy}{dx} = \frac{x^2}{2L^2} - \frac{L^2}{2x^2}$ and hence $y = \frac{x^3}{6L^2} + \frac{L^2}{2x} - \frac{2}{3}L$.

We want to minimize the distance D from the dog at (x, y) to the rabbit at $(0, 2s)$. Now $s = \frac{1}{2}y - \frac{1}{2}x\frac{dy}{dx} \Rightarrow$

$$2s = y - xz \Rightarrow y - 2s = xz = x\left(\frac{x^2}{2L^2} - \frac{L^2}{2x^2}\right) = \frac{x^3}{2L^2} - \frac{L^2}{2x}, \text{ so}$$

$$D = \sqrt{(x-0)^2 + (y-2s)^2} = \sqrt{x^2 + \left(\frac{x^3}{2L^2} - \frac{L^2}{2x}\right)^2} = \sqrt{\frac{x^6}{4L^4} + \frac{x^2}{2} + \frac{L^4}{4x^2}} = \sqrt{\left(\frac{x^3}{2L^2} + \frac{L^2}{2x}\right)^2}$$
$$= \frac{x^3}{2L^2} + \frac{L^2}{2x}$$

$$D' = 0 \Leftrightarrow \frac{3x^2}{2L^2} - \frac{L^2}{2x^2} = 0 \Leftrightarrow \frac{3x^2}{2L^2} = \frac{L^2}{2x^2} \Leftrightarrow x^4 = \frac{L^4}{3} \Leftrightarrow x = \frac{L}{\sqrt[4]{3}},\ x > 0,\ L > 0.$$

Since $D''(x) = \frac{3x}{L^2} + \frac{L^2}{x^3} > 0$ for all $x > 0$, we know that $D\left(\frac{L}{\sqrt[4]{3}}\right) = \frac{(L \cdot 3^{-1/4})^3}{2L^2} + \frac{L^2}{2L \cdot 3^{-1/4}} = \frac{2L}{3^{3/4}}$ is the minimum value of D, that is, the closest the dog gets to the rabbit. The positions at this distance are

$$\text{Dog: } (x, y) = \left(\frac{L}{\sqrt[4]{3}}, \left(\frac{5}{3^{7/4}} - \frac{2}{3}\right)L\right) = \left(\frac{L}{\sqrt[4]{3}}, \frac{5\sqrt[4]{3} - 6}{9}L\right)$$

$$\text{Rabbit: } (0, 2s) = \left(0, \frac{8\sqrt[4]{3}L}{9} - \frac{2L}{3}\right) = \left(0, \frac{8\sqrt[4]{3} - 6}{9}L\right)$$

11. (a) We are given that $V = \frac{1}{3}\pi r^2 h$, $dV/dt = 60{,}000\pi$ ft³/h, and $r = 1.5h = \frac{3}{2}h$. So $V = \frac{1}{3}\pi\left(\frac{3}{2}h\right)^2 h = \frac{3}{4}\pi h^3 \Rightarrow$

$$\frac{dV}{dt} = \frac{3}{4}\pi \cdot 3h^2\frac{dh}{dt} = \frac{9}{4}\pi h^2\frac{dh}{dt}. \text{ Therefore, } \frac{dh}{dt} = \frac{4(dV/dt)}{9\pi h^2} = \frac{240{,}000\pi}{9\pi h^2} = \frac{80{,}000}{3h^2}\ (\star) \Rightarrow$$

$\int 3h^2\,dh = \int 80{,}000\,dt \Rightarrow h^3 = 80{,}000t + C$. When $t = 0$, $h = 60$. Thus, $C = 60^3 = 216{,}000$, so

$h^3 = 80{,}000t + 216{,}000$. Let $h = 100$. Then $100^3 = 1{,}000{,}000 = 80{,}000t + 216{,}000 \Rightarrow$ $80{,}000t = 784{,}000 \Rightarrow t = 9.8$, so the time required is 9.8 hours.

(b) The floor area of the silo is $F = \pi \cdot 200^2 = 40{,}000\pi$ ft^2, and the area of the base of the pile is $A = \pi r^2 = \pi\left(\frac{3}{2}h\right)^2 = \frac{9\pi}{4}h^2$. So the area of the floor which is not covered when $h = 60$ is $F - A = 40{,}000\pi - 8100\pi = 31{,}900\pi \approx 100{,}217$ ft^2. Now $A = \frac{9\pi}{4}h^2 \Rightarrow dA/dt = \frac{9\pi}{4} \cdot 2h\,(dh/dt)$, and from $(\star)$ in part (a) we know that when $h = 60$, $dh/dt = \frac{80{,}000}{3(60)^2} = \frac{200}{27}$ ft/h. Therefore, $dA/dt = \frac{9\pi}{4}(2)(60)\left(\frac{200}{27}\right) = 2000\pi \approx 6283$ ft^2/h.

(c) At $h = 90$ ft, $dV/dt = 60{,}000\pi - 20{,}000\pi = 40{,}000\pi$ ft^3/h. From $(\star)$ in part (a),

$\dfrac{dh}{dt} = \dfrac{4(dV/dt)}{9\pi h^2} = \dfrac{4(40{,}000\pi)}{9\pi h^2} = \dfrac{160{,}000}{9h^2} \Rightarrow \int 9h^2\,dh = \int 160{,}000\,dt \Rightarrow 3h^3 = 160{,}000t + C$. When $t = 0$, $h = 90$; therefore, $C = 3 \cdot 729{,}000 = 2{,}187{,}000$. So $3h^3 = 160{,}000t + 2{,}187{,}000$. At the top, $h = 100 \Rightarrow$ $3(100)^3 = 160{,}000t + 2{,}187{,}000 \Rightarrow t = \frac{813{,}000}{160{,}000} \approx 5.1$. The pile reaches the top after about 5.1 h.

12. Let $P(a, b)$ be any first-quadrant point on the curve $y = f(x)$. The tangent line at P has equation $y - b = f'(a)(x - a)$, or equivalently, $y = mx + b - ma$, where $m = f'(a)$. If $Q(0, c)$ is the y-intercept, then $c = b - am$. If $R(k, 0)$ is the x-intercept, then $k = \dfrac{am - b}{m} = a - \dfrac{b}{m}$. Since the tangent line is bisected at P, we know that $|PQ| = |PR|$; that is, $\sqrt{(a-0)^2 + [b - (b - am)]^2} = \sqrt{[a - (a - b/m)]^2 + (b-0)^2}$. Squaring and simplifying gives us $a^2 + a^2m^2 = b^2/m^2 + b^2 \Rightarrow a^2m^2 + a^2m^4 = b^2 + b^2m^2 \Rightarrow a^2m^4 + (a^2 - b^2)m^2 - b^2 = 0 \Rightarrow$ $(a^2m^2 - b^2)(m^2 + 1) = 0 \Rightarrow m^2 = b^2/a^2$. Since m is the slope of the line from a positive y-intercept to a positive x-intercept, m must be negative. Since a and b are positive, we have $m = -b/a$, so we will solve the equivalent differential equation $\dfrac{dy}{dx} = -\dfrac{y}{x} \Rightarrow \dfrac{dy}{y} = -\dfrac{dx}{x} \Rightarrow \displaystyle\int \frac{dy}{y} = -\int \frac{dx}{x} \Rightarrow \ln y = -\ln x + C\ [x, y > 0] \Rightarrow$ $y = e^{-\ln x + C} = e^{\ln x^{-1}} \cdot e^C = x^{-1} \cdot A \Rightarrow y = A/x$. Since the point $(3, 2)$ is on the curve, $3 = A/2 \Rightarrow A = 6$ and the curve is $y = 6/x$ with $x > 0$.

13. Let $P(a, b)$ be any point on the curve. If m is the slope of the tangent line at P, then $m = y'(a)$, and an equation of the normal line at P is $y - b = -\dfrac{1}{m}(x - a)$, or equivalently, $y = -\dfrac{1}{m}x + b + \dfrac{a}{m}$. The y-intercept is always 6, so $b + \dfrac{a}{m} = 6 \Rightarrow \dfrac{a}{m} = 6 - b \Rightarrow m = \dfrac{a}{6 - b}$. We will solve the equivalent differential equation $\dfrac{dy}{dx} = \dfrac{x}{6 - y} \Rightarrow$ $(6 - y)\,dy = x\,dx \Rightarrow \displaystyle\int (6 - y)\,dy = \int x\,dx \Rightarrow 6y - \tfrac{1}{2}y^2 = \tfrac{1}{2}x^2 + C \Rightarrow 12y - y^2 = x^2 + K$.

Since $(3, 2)$ is on the curve, $12(2) - 2^2 = 3^2 + K \Rightarrow K = 11$. So the curve is given by $12y - y^2 = x^2 + 11 \Rightarrow$ $x^2 + y^2 - 12y + 36 = -11 + 36 \Rightarrow x^2 + (y - 6)^2 = 25$, a circle with center $(0, 6)$ and radius 5.

14. Let $P(x_0, y_0)$ be a point on the curve. Since the midpoint of the line segment determined by the normal line from (x_0, y_0) to its intersection with the x-axis has x-coordinate 0, the x-coordinate of the point of intersection with the x-axis must be $-x_0$. Hence, the normal line has slope $\dfrac{y_0 - 0}{x_0 - (-x_0)} = \dfrac{y_0}{2x_0}$. So the tangent line has slope $-\dfrac{2x_0}{y_0}$. This gives the differential equation $y' = -\dfrac{2x}{y} \Rightarrow y\,dy = -2x\,dx \Rightarrow \int y\,dy = \int(-2x)\,dx \Rightarrow \frac{1}{2}y^2 = -x^2 + C \Rightarrow x^2 + \frac{1}{2}y^2 = C$ $[C > 0]$. This is a family of ellipses.

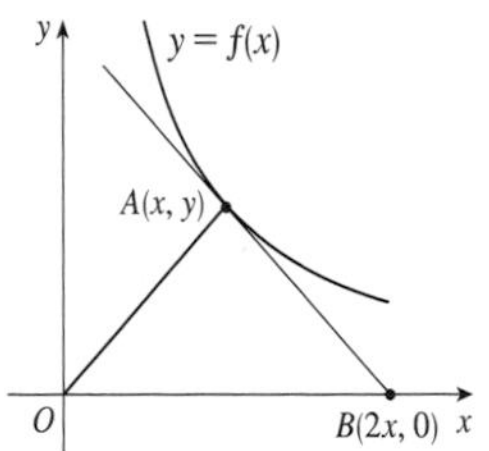

15. From the figure, slope $OA = \dfrac{y}{x}$. If triangle OAB is isosceles, then slope AB must be $-\dfrac{y}{x}$, the negative of slope OA. This slope is also equal to $f'(x)$, so we have $\dfrac{dy}{dx} = -\dfrac{y}{x} \Rightarrow \displaystyle\int \frac{dy}{y} = -\int \frac{dx}{x} \Rightarrow$

$\ln|y| = -\ln|x| + C \Rightarrow |y| = e^{-\ln|x|+C} \Rightarrow$

$|y| = (e^{\ln|x|})^{-1}e^C \Rightarrow |y| = \dfrac{1}{|x|}e^C \Rightarrow y = \dfrac{K}{x}$, $K \neq 0$.

10 □ PARAMETRIC EQUATIONS AND POLAR COORDINATES

10.1 Curves Defined by Parametric Equations

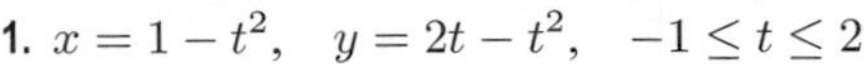

1. $x = 1 - t^2, \quad y = 2t - t^2, \quad -1 \le t \le 2$

t	-1	0	1	2
x	0	1	0	-3
y	-3	0	1	0

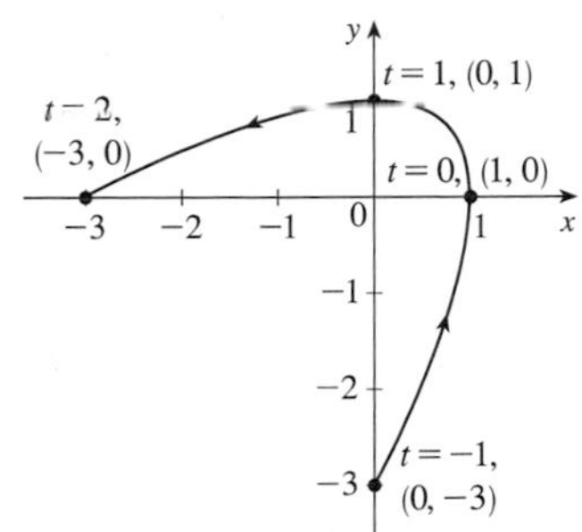

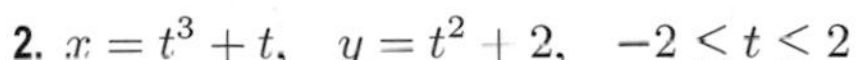

2. $x = t^3 + t, \quad y = t^2 + 2, \quad -2 \le t \le 2$

t	-2	-1	0	1	2
x	-10	-2	0	2	10
y	6	3	2	3	6

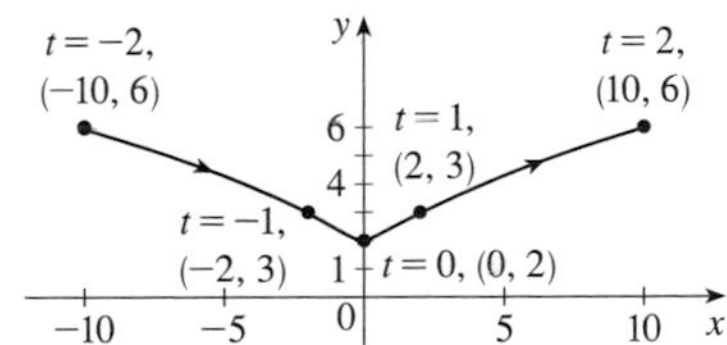

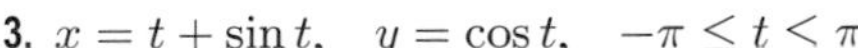

3. $x = t + \sin t, \quad y = \cos t, \quad -\pi \le t \le \pi$

t	$-\pi$	$-\pi/2$	0	$\pi/2$	π
x	$-\pi$	$-\pi/2 + 1$	0	$\pi/2 + 1$	π
y	-1	0	1	0	-1

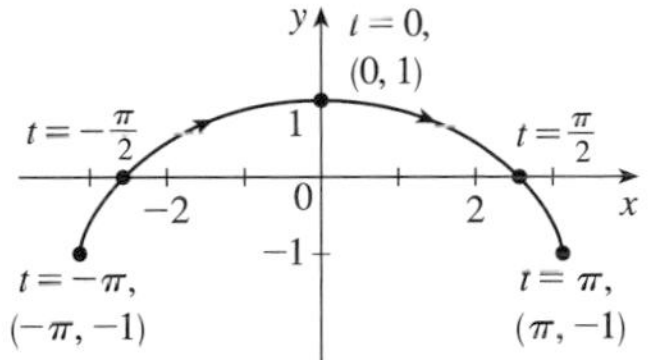

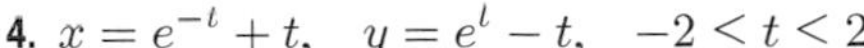

4. $x = e^{-t} + t, \quad y = e^t - t, \quad -2 \le t \le 2$

t	-2	-1	0	1	2
x	$e^2 - 2$	$e - 1$	1	$e^{-1} + 1$	$e^{-2} + 2$
	5.39	1.72		1.37	2.14
y	$e^{-2} + 2$	$e^{-1} + 1$	1	$e - 1$	$e^2 - 2$
	2.14	1.37		1.72	5.39

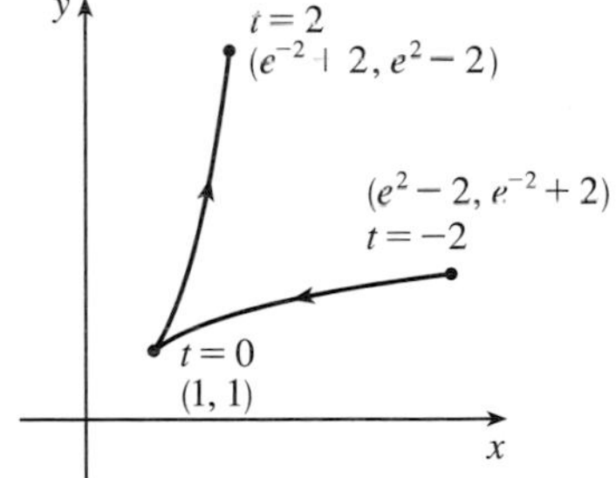

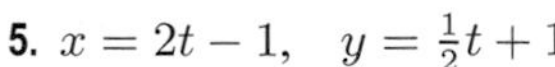

5. $x = 2t - 1, \quad y = \frac{1}{2}t + 1$

(a)

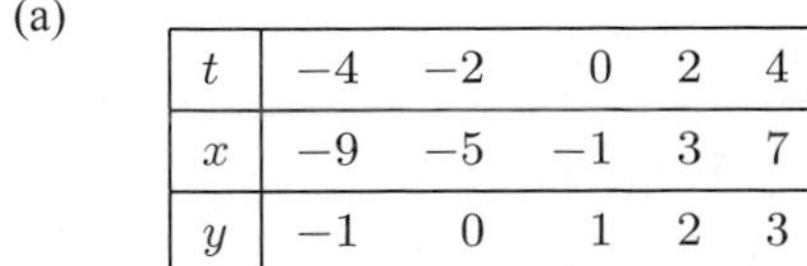

t	-4	-2	0	2	4
x	-9	-5	-1	3	7
y	-1	0	1	2	3

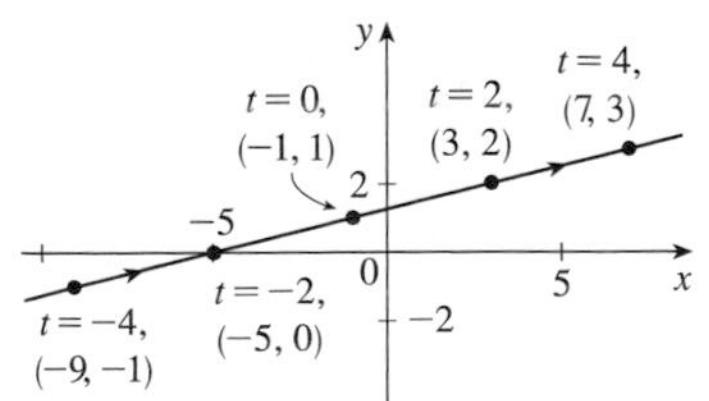

(b) $x = 2t - 1 \;\Rightarrow\; 2t = x + 1 \;\Rightarrow\; t = \frac{1}{2}x + \frac{1}{2}$, so

$$y = \tfrac{1}{2}t + 1 = \tfrac{1}{2}\left(\tfrac{1}{2}x + \tfrac{1}{2}\right) + 1 = \tfrac{1}{4}x + \tfrac{1}{4} + 1 \;\Rightarrow\; y = \tfrac{1}{4}x + \tfrac{5}{4}$$

6. $x = 3t + 2, \quad y = 2t + 3$

(a)

t	-4	-2	0	2	4
x	-10	-4	2	8	14
y	-5	-1	3	7	11

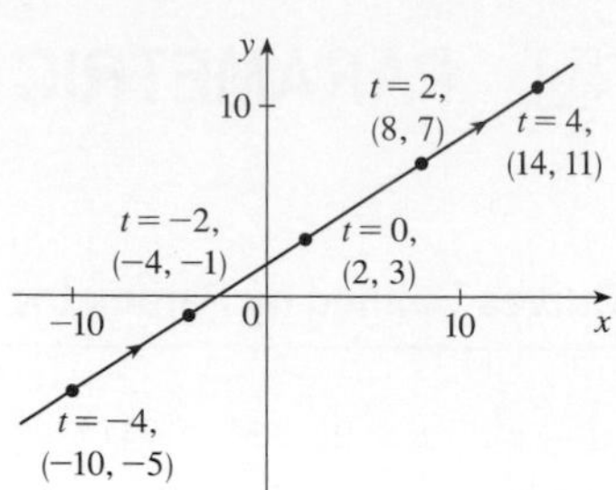

(b) $x = 3t + 2 \;\Rightarrow\; 3t = x - 2 \;\Rightarrow\; t = \frac{1}{3}x - \frac{2}{3}$, so

$y = 2t + 3 = 2\left(\frac{1}{3}x - \frac{2}{3}\right) + 3 = \frac{2}{3}x - \frac{4}{3} + 3 \;\Rightarrow\; y = \frac{2}{3}x + \frac{5}{3}$

7. $x = t^2 - 3, \quad y = t + 2, \quad -3 \le t \le 3$

(a)

t	-3	-1	1	3
x	6	-2	-2	6
y	-1	1	3	5

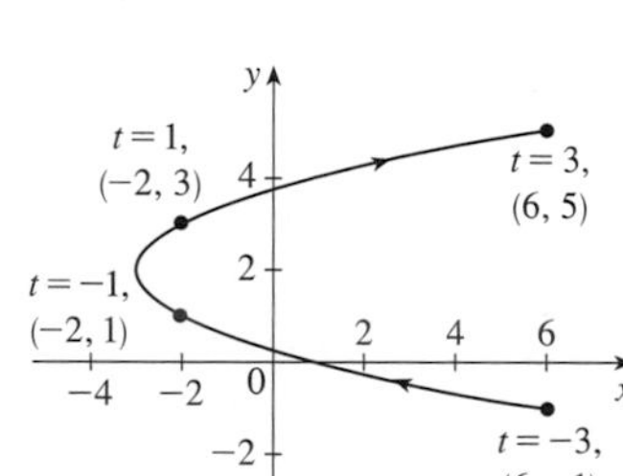

(b) $y = t + 2 \;\Rightarrow\; t = y - 2$, so

$x = t^2 - 3 = (y - 2)^2 - 3 = y^2 - 4y + 4 - 3 \;\Rightarrow$

$x = y^2 - 4y + 1, -1 \le y \le 5$

8. $x = \sin t, \quad y = 1 - \cos t, \quad 0 \le t \le 2\pi$

(a)

t	0	$\pi/2$	π	$3\pi/2$	2π
x	0	1	0	-1	0
y	0	1	2	1	0

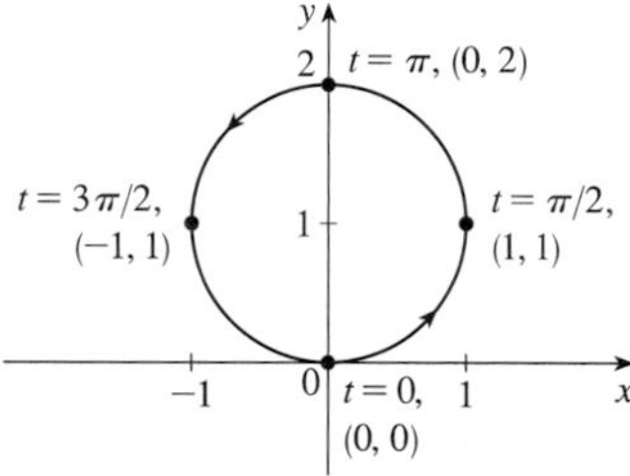

(b) $x = \sin t, y = 1 - \cos t \quad [\text{or } y - 1 = -\cos t] \;\Rightarrow$

$x^2 + (y - 1)^2 = (\sin t)^2 + (-\cos t)^2 \;\Rightarrow\; x^2 + (y - 1)^2 = 1.$

As t varies from 0 to 2π, the circle with center $(0, 1)$ and radius 1 is traced out.

9. $x = \sqrt{t}, \; y = 1 - t$

(a)

t	0	1	2	3	4
x	0	1	1.414	1.732	2
y	1	0	-1	-2	-3

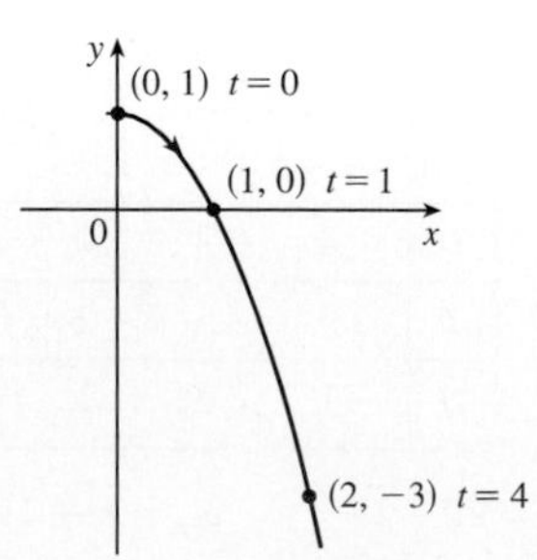

(b) $x = \sqrt{t} \;\Rightarrow\; t = x^2 \;\Rightarrow\; y = 1 - t = 1 - x^2$. Since $t \ge 0$, $x \ge 0$.

So the curve is the right half of the parabola $y = 1 - x^2$.

10. $x = t^2, y = t^3$

(a)

t	-2	-1	0	1	2
x	4	1	0	1	4
y	-8	-1	0	1	8

(b) $y = t^3 \;\Rightarrow\; t = \sqrt[3]{y} \;\Rightarrow\; x = t^2 = \left(\sqrt[3]{y}\right)^2 = y^{2/3}$. $t \in \mathbb{R}, y \in \mathbb{R}, x \geq 0$.

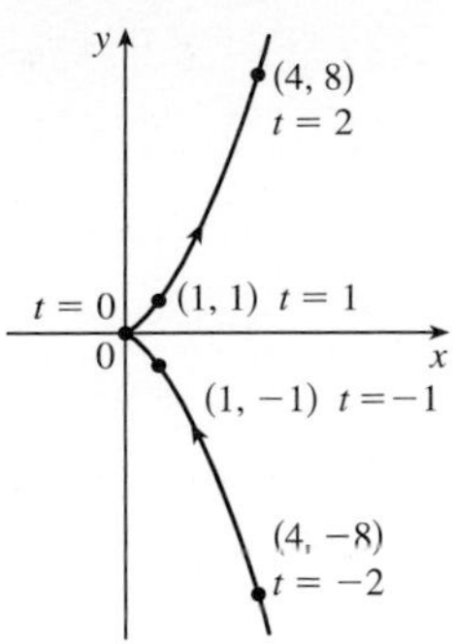

11. (a) $x = \sin \frac{1}{2}\theta, y = \cos \frac{1}{2}\theta, -\pi \leq \theta \leq \pi$.

$x^2 + y^2 = \sin^2 \frac{1}{2}\theta + \cos^2 \frac{1}{2}\theta = 1$. For $-\pi \leq \theta \leq 0$, we have $-1 \leq x \leq 0$ and $0 \leq y \leq 1$. For $0 < \theta \leq \pi$, we have $0 < x \leq 1$ and $1 > y \geq 0$. The graph is a semicircle.

(b)

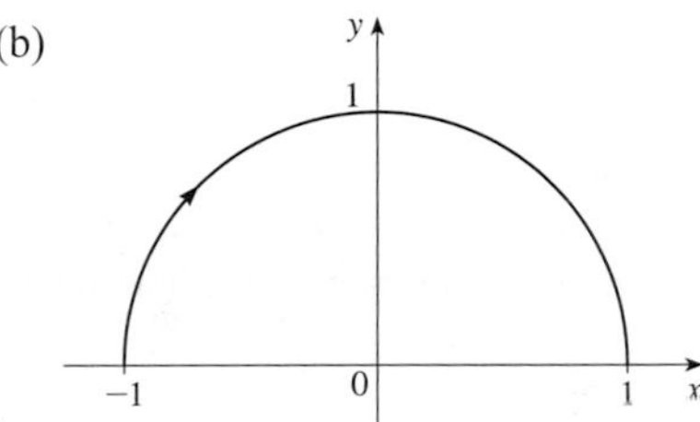

12. (a) $x = \frac{1}{2}\cos\theta, y = 2\sin\theta, 0 \leq \theta \leq \pi$.

$(2x)^2 + \left(\frac{1}{2}y\right)^2 = \cos^2\theta + \sin^2\theta = 1 \;\Rightarrow\; 4x^2 + \frac{1}{4}y^2 = 1 \;\Rightarrow$

$\dfrac{x^2}{(1/2)^2} + \dfrac{y^2}{2^2} = 1$, which is an equation of an ellipse with x-intercepts $\pm\frac{1}{2}$ and y-intercepts ± 2. For $0 \leq \theta \leq \pi/2$, we have $\frac{1}{2} \geq x \geq 0$ and $0 \leq y \leq 2$. For $\pi/2 < \theta \leq \pi$, we have $0 > x \geq -\frac{1}{2}$ and $2 > y \geq 0$. So the graph is the top half of the ellipse.

(b)

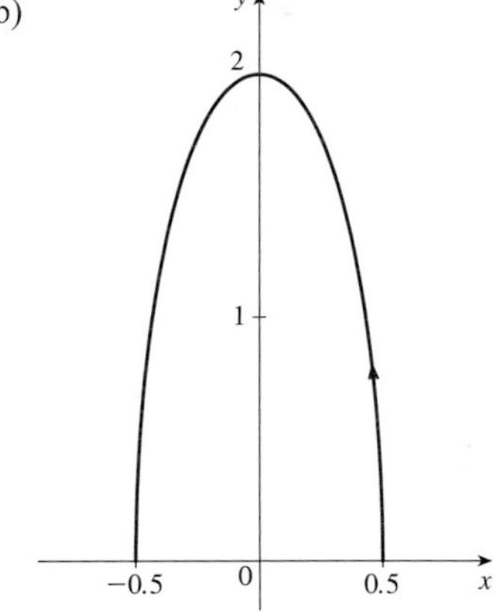

13. (a) $x = \sin t, y = \csc t, 0 < t < \frac{\pi}{2}$. $y = \csc t = \dfrac{1}{\sin t} = \dfrac{1}{x}$.

For $0 < t < \frac{\pi}{2}$, we have $0 < x < 1$ and $y > 1$. Thus, the curve is the portion of the hyperbola $y = 1/x$ with $y > 1$.

(b)

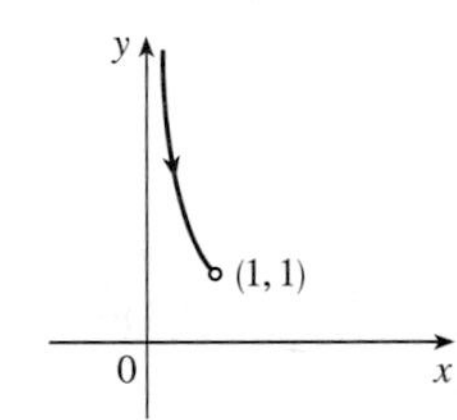

14. (a) $y = e^{-2t} = (e^t)^{-2} = x^{-2} = 1/x^2$ for $x > 0$ since $x = e^t$.

(b)

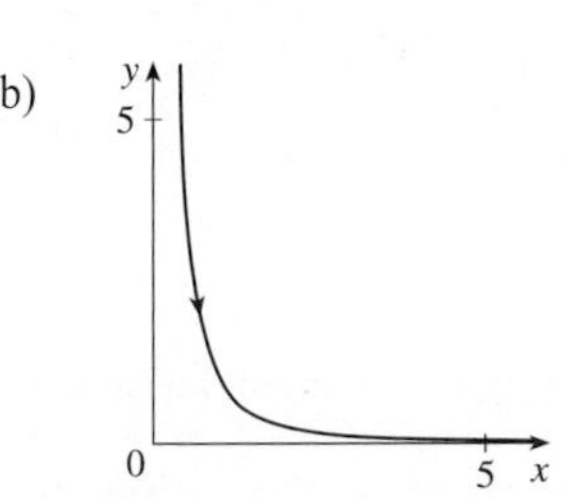

15. (a) $y = \ln t \;\Rightarrow\; t = e^y$, so $x = t^2 = (e^y)^2 = e^{2y}$.

(b)

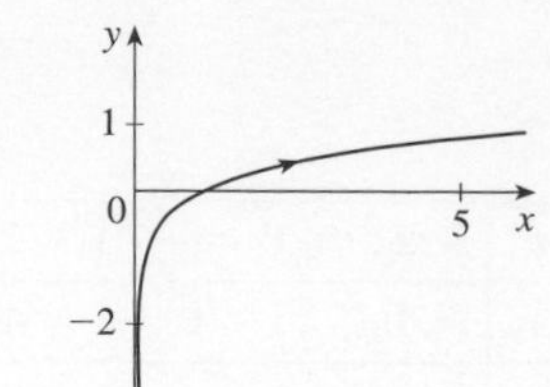

16. (a) $x = \sqrt{t+1} \;\Rightarrow\; x^2 = t+1 \;\Rightarrow\; t = x^2 - 1$.

$y = \sqrt{t-1} = \sqrt{(x^2-1)-1} = \sqrt{x^2-2}$. The curve is the part of the hyperbola $x^2 - y^2 = 2$ with $x \geq \sqrt{2}$ and $y \geq 0$.

(b)

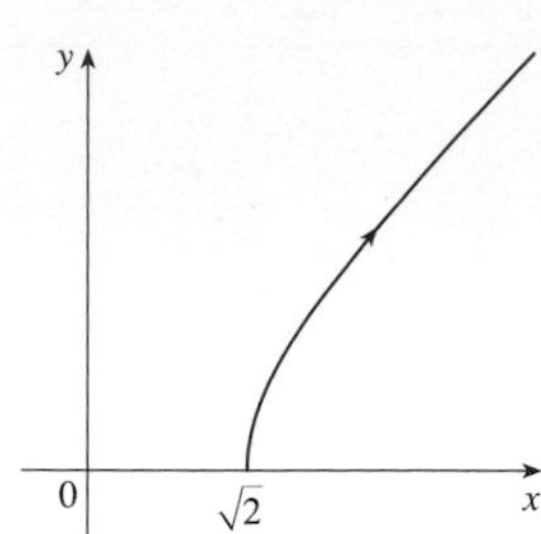

17. (a) $x = \sinh t$, $y = \cosh t \;\Rightarrow\; y^2 - x^2 = \cosh^2 t - \sinh^2 t = 1$.

Since $y = \cosh t \geq 1$, we have the upper branch of the hyperbola $y^2 - x^2 = 1$.

(b)

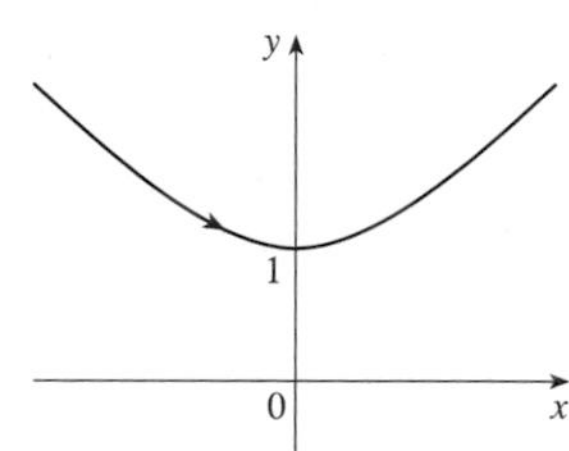

18. (a) $x = \tan^2\theta$, $y = \sec\theta$, $-\pi/2 < \theta < \pi/2$.

$1 + \tan^2\theta = \sec^2\theta \;\Rightarrow\; 1 + x = y^2 \;\Rightarrow\; x = y^2 - 1$. For $-\pi/2 < \theta \leq 0$, we have $x \geq 0$ and $y \geq 1$. For $0 < \theta < \pi/2$, we have $0 < x$ and $1 < y$. Thus, the curve is the portion of the parabola $x = y^2 - 1$ in the first quadrant. As θ increases from $-\pi/2$ to 0, the point (x, y) approaches $(0, 1)$ along the parabola. As θ increases from 0 to $\pi/2$, the point (x, y) retreats from $(0, 1)$ along the parabola.

(b)

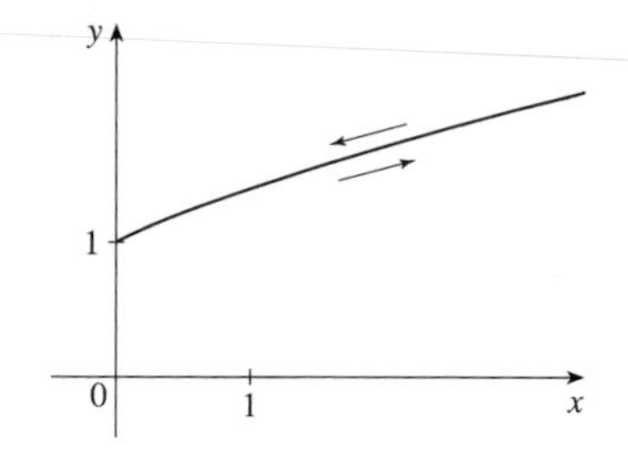

19. $x = 5 + 2\cos\pi t$, $y = 3 + 2\sin\pi t \;\Rightarrow\; \cos\pi t = \dfrac{x-5}{2}$, $\sin\pi t = \dfrac{y-3}{2}$. $\cos^2(\pi t) + \sin^2(\pi t) = 1 \;\Rightarrow$

$\left(\dfrac{x-5}{2}\right)^2 + \left(\dfrac{y-3}{2}\right)^2 = 1$. The motion of the particle takes place on a circle centered at $(5, 3)$ with a radius 2. As t goes from 1 to 2, the particle starts at the point $(3, 3)$ and moves counterclockwise along the circle $\left(\dfrac{x-5}{2}\right)^2 + \left(\dfrac{y-3}{2}\right)^2 = 1$ to $(7, 3)$ [one-half of a circle].

20. $x = 2 + \sin t$, $y = 1 + 3\cos t \;\Rightarrow\; \sin t = x - 2$, $\cos t = \dfrac{y-1}{3}$. $\sin^2 t + \cos^2 t = 1 \;\Rightarrow\; (x-2)^2 + \left(\dfrac{y-1}{3}\right)^2 = 1$.

The motion of the particle takes place on an ellipse centered at $(2, 1)$. As t goes from $\pi/2$ to 2π, the particle starts at the point $(3, 1)$ and moves counterclockwise three-fourths of the way around the ellipse to $(2, 4)$.

21. $x = 5\sin t$, $y = 2\cos t \Rightarrow \sin t = \dfrac{x}{5}$, $\cos t = \dfrac{y}{2}$. $\sin^2 t + \cos^2 t = 1 \Rightarrow \left(\dfrac{x}{5}\right)^2 + \left(\dfrac{y}{2}\right)^2 = 1$. The motion of the particle takes place on an ellipse centered at $(0, 0)$. As t goes from $-\pi$ to 5π, the particle starts at the point $(0, -2)$ and moves clockwise around the ellipse 3 times.

22. $y = \cos^2 t = 1 - \sin^2 t = 1 - x^2$. The motion of the particle takes place on the parabola $y = 1 - x^2$. As t goes from -2π to $-\pi$, the particle starts at the point $(0, 1)$, moves to $(1, 0)$, and goes back to $(0, 1)$. As t goes from $-\pi$ to 0, the particle moves to $(-1, 0)$ and goes back to $(0, 1)$. The particle repeats this motion as t goes from 0 to 2π.

23. We must have $1 \le x \le 4$ and $2 \le y \le 3$. So the graph of the curve must be contained in the rectangle $[1, 4]$ by $[2, 3]$.

24. (a) From the first graph, we have $1 \le x \le 2$. From the second graph, we have $-1 \le y \le 1$. The only choice that satisfies either of those conditions is III.

(b) From the first graph, the values of x cycle through the values from -2 to 2 four times. From the second graph, the values of y cycle through the values from -2 to 2 six times. Choice I satisfies these conditions.

(c) From the first graph, the values of x cycle through the values from -2 to 2 three times. From the second graph, we have $0 \le y \le 2$. Choice IV satisfies these conditions.

(d) From the first graph, the values of x cycle through the values from -2 to 2 two times. From the second graph, the values of y do the same thing. Choice II satisfies these conditions.

25. When $t = -1$, $(x, y) = (1, 1)$. As t increases to 0, x and y both decrease to 0. As t increases from 0 to 1, x increases from 0 to 1 and y decreases from 0 to -1. As t increases beyond 1, x continues to increase and y continues to decrease. For $t < -1$, x and y are both positive and decreasing. We could achieve greater accuracy by estimating x- and y-values for selected values of t from the given graphs and plotting the corresponding points.

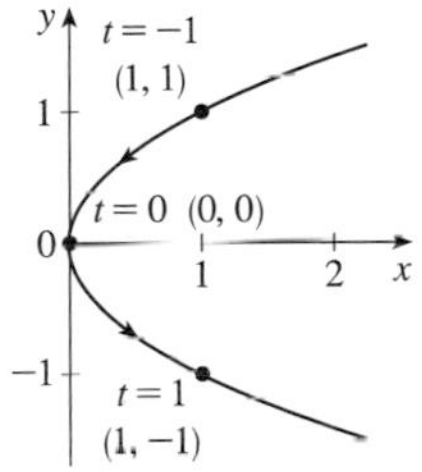

26. When $t = -1$, $(x, y) = (0, 0)$. As t increases to 0, x increases from 0 to 1, while y first decreases to -1 and then increases to 0. As t increases from 0 to 1, x decreases from 1 to 0, while y first increases to 1 and then decreases to 0. We could achieve greater accuracy by estimating x- and y-values for selected values of t from the given graphs and plotting the corresponding points.

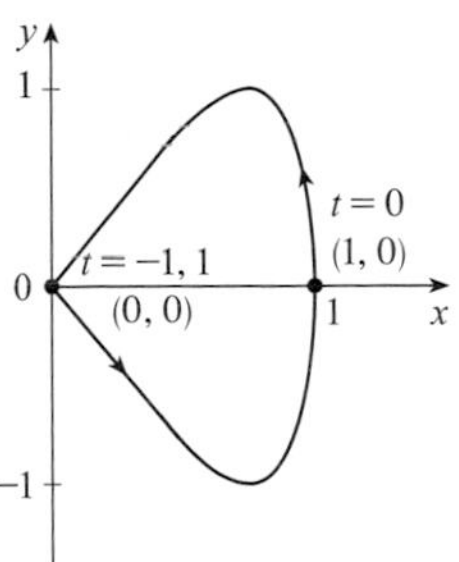

27. When $t = -1$, $(x, y) = (0, 1)$. As t increases to 0, x increases from 0 to 1 and y decreases from 1 to 0. As t increases from 0 to 1, the curve is retraced in the opposite direction with x decreasing from 1 to 0 and y increasing from 0 to 1. We could achieve greater accuracy by estimating x- and y-values for selected values of t from the given graphs and plotting the corresponding points.

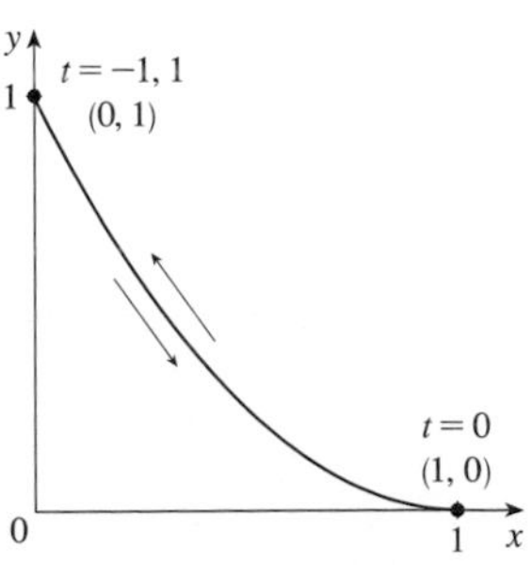

28. (a) $x = t^4 - t + 1 = (t^4 + 1) - t > 0$ [think of the graphs of $y = t^4 + 1$ and $y = t$] and $y = t^2 \geq 0$, so these equations are matched with graph V.

(b) $y = \sqrt{t} \geq 0$. $x = t^2 - 2t = t(t - 2)$ is negative for $0 < t < 2$, so these equations are matched with graph I.

(c) $x = \sin 2t$ has period $2\pi/2 = \pi$. Note that

$y(t + 2\pi) = \sin[t + 2\pi + \sin 2(t + 2\pi)] = \sin(t + 2\pi + \sin 2t) = \sin(t + \sin 2t) = y(t)$, so y has period 2π. These equations match graph II since x cycles through the values -1 to 1 twice as y cycles through those values once.

(d) $x = \cos 5t$ has period $2\pi/5$ and $y = \sin 2t$ has period π, so x will take on the values -1 to 1, and then 1 to -1, before y takes on the values -1 to 1. Note that when $t = 0$, $(x, y) = (1, 0)$. These equations are matched with graph VI.

(e) $x = t + \sin 4t$, $y = t^2 + \cos 3t$. As t becomes large, t and t^2 become the dominant terms in the expressions for x and y, so the graph will look like the graph of $y = x^2$, but with oscillations. These equations are matched with graph IV.

(f) $x = \dfrac{\sin 2t}{4 + t^2}$, $y = \dfrac{\cos 2t}{4 + t^2}$. As $t \to \infty$, x and y both approach 0. These equations are matched with graph III.

29. Use $y = t$ and $x = t - 2\sin \pi t$ with a t-interval of $[-\pi, \pi]$.

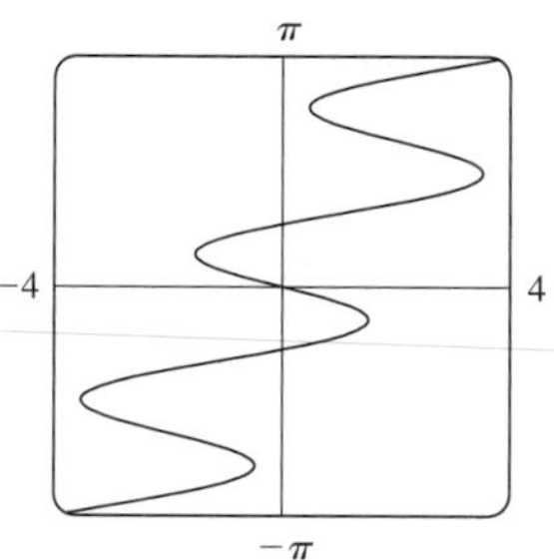

30. Use $x_1 = t$, $y_1 = t^3 - 4t$ and $x_2 = t^3 - 4t$, $y_2 = t$ with a t-interval of $[-3, 3]$. There are 9 points of intersection; $(0, 0)$ is fairly obvious. The point in quadrant I is approximately $(2.2, 2.2)$, and by symmetry, the point in quadrant III is approximately $(-2.2, -2.2)$. The other six points are approximately $(\mp 1.9, \pm 0.5)$, $(\mp 1.7, \pm 1.7)$, and $(\mp 0.5, \pm 1.9)$.

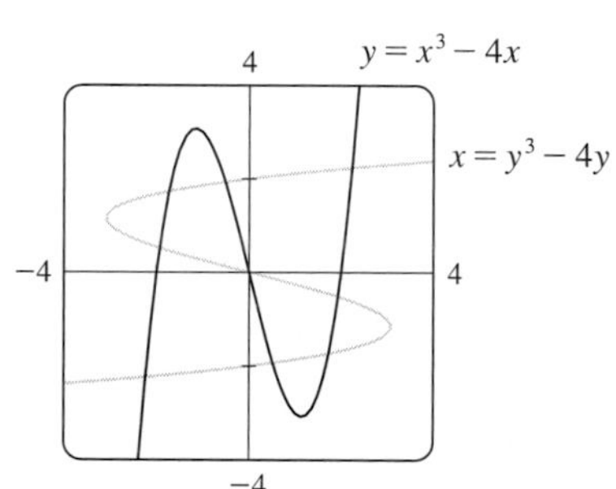

31. (a) $x = x_1 + (x_2 - x_1)t$, $y = y_1 + (y_2 - y_1)t$, $0 \leq t \leq 1$. Clearly the curve passes through $P_1(x_1, y_1)$ when $t = 0$ and through $P_2(x_2, y_2)$ when $t = 1$. For $0 < t < 1$, x is strictly between x_1 and x_2 and y is strictly between y_1 and y_2. For every value of t, x and y satisfy the relation $y - y_1 = \dfrac{y_2 - y_1}{x_2 - x_1}(x - x_1)$, which is the equation of the line through $P_1(x_1, y_1)$ and $P_2(x_2, y_2)$.

Finally, any point (x, y) on that line satisfies $\dfrac{y - y_1}{y_2 - y_1} = \dfrac{x - x_1}{x_2 - x_1}$; if we call that common value t, then the given parametric equations yield the point (x, y); and any (x, y) on the line between $P_1(x_1, y_1)$ and $P_2(x_2, y_2)$ yields a value of t in $[0, 1]$. So the given parametric equations exactly specify the line segment from $P_1(x_1, y_1)$ to $P_2(x_2, y_2)$.

(b) $x = -2 + [3 - (-2)]t = -2 + 5t$ and $y = 7 + (-1 - 7)t = 7 - 8t$ for $0 \leq t \leq 1$.

32. For the side of the triangle from A to B, use $(x_1, y_1) = (1, 1)$ and $(x_2, y_2) = (4, 2)$. Hence, the equations are

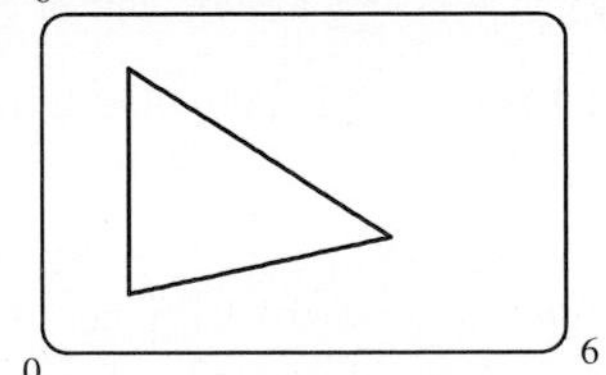

$$x = x_1 + (x_2 - x_1)\,t = 1 + (4 - 1)\,t = 1 + 3t,$$
$$y = y_1 + (y_2 - y_1)\,t = 1 + (2 - 1)\,t = 1 + t.$$

Graphing $x = 1 + 3t$ and $y = 1 + t$ with $0 \le t \le 1$ gives us the side of the triangle from A to B. Similarly, for the side BC we use $x = 4 - 3t$ and $y = 2 + 3t$, and for the side AC we use $x = 1$ and $y = 1 + 4t$.

33. The circle $x^2 + (y - 1)^2 = 4$ has center $(0, 1)$ and radius 2, so by Example 4 it can be represented by $x = 2\cos t$, $y = 1 + 2\sin t$, $0 \le t \le 2\pi$. This representation gives us the circle with a counterclockwise orientation starting at $(2, 1)$.

(a) To get a clockwise orientation, we could change the equations to $x = 2\cos t$, $y = 1 - 2\sin t$, $0 \le t \le 2\pi$.

(b) To get three times around in the counterclockwise direction, we use the original equations $x = 2\cos t$, $y = 1 + 2\sin t$ with the domain expanded to $0 \le t \le 6\pi$.

(c) To start at $(0, 3)$ using the original equations, we must have $x_1 = 0$; that is, $2\cos t = 0$. Hence, $t = \frac{\pi}{2}$. So we use $x = 2\cos t$, $y = 1 + 2\sin t$, $\frac{\pi}{2} \le t \le \frac{3\pi}{2}$.

Alternatively, if we want t to start at 0, we could change the equations of the curve. For example, we could use $x = -2\sin t$, $y = 1 + 2\cos t$, $0 \le t \le \pi$.

34. (a) Let $x^2/a^2 = \sin^2 t$ and $y^2/b^2 = \cos^2 t$ to obtain $x = a\sin t$ and $y = b\cos t$ with $0 \le t \le 2\pi$ as possible parametric equations for the ellipse $x^2/a^2 + y^2/b^2 = 1$.

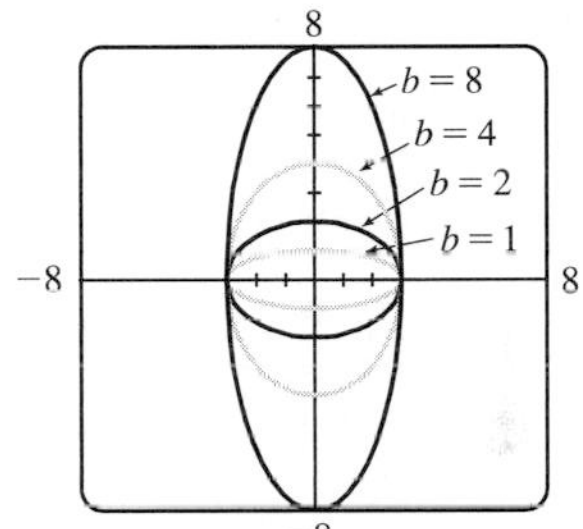

(b) The equations are $x = 3\sin t$ and $y = b\cos t$ for $b \in \{1, 2, 4, 8\}$.

(c) As b increases, the ellipse stretches vertically.

35. *Big circle:* It's centered at $(2, 2)$ with a radius of 2, so by Example 4, parametric equations are

$$x = 2 + 2\cos t, \qquad y = 2 + 2\sin t, \qquad 0 \le t \le 2\pi$$

Small circles: They are centered at $(1, 3)$ and $(3, 3)$ with a radius of 0.1. By Example 4, parametric equations are

	(left)	$x = 1 + 0.1\cos t,$	$y = 3 + 0.1\sin t,$	$0 \le t \le 2\pi$
and	*(right)*	$x = 3 + 0.1\cos t,$	$y = 3 + 0.1\sin t,$	$0 \le t \le 2\pi$

Semicircle: It's the lower half of a circle centered at $(2, 2)$ with radius 1. By Example 4, parametric equations are

$$x = 2 + 1\cos t, \qquad y = 2 + 1\sin t, \qquad \pi \le t \le 2\pi$$

To get all four graphs on the same screen with a typical graphing calculator, we need to change the last t-interval to$[0, 2\pi]$ in order to match the others. We can do this by changing t to $0.5t$. This change gives us the upper half. There are several ways to get the lower half—one is to change the "+" to a "−" in the y-assignment, giving us

$$x = 2 + 1\cos(0.5t), \qquad y = 2 - 1\sin(0.5t), \qquad 0 \le t \le 2\pi$$

36. If you are using a calculator or computer that can overlay graphs (using multiple t-intervals), the following is appropriate.

Left side: $x = 1$ and y goes from 1.5 to 4, so use

$$x = 1, \qquad y = t, \qquad 1.5 \le t \le 4$$

Right side: $x = 10$ and y goes from 1.5 to 4, so use

$$x = 10, \qquad y = t, \qquad 1.5 \le t \le 4$$

Bottom: x goes from 1 to 10 and $y = 1.5$, so use

$$x = t, \qquad y = 1.5, \qquad 1 \le t \le 10$$

Handle: It starts at $(10, 4)$ and ends at $(13, 7)$, so use

$$x = 10 + t, \qquad y = 4 + t, \qquad 0 \le t \le 3$$

Left wheel: It's centered at $(3, 1)$, has a radius of 1, and appears to go about $30°$ above the horizontal, so use

$$x = 3 + 1\cos t, \qquad y = 1 + 1\sin t, \qquad \tfrac{5\pi}{6} \le t \le \tfrac{13\pi}{6}$$

Right wheel: Similar to the left wheel with center $(8, 1)$, so use

$$x = 8 + 1\cos t, \qquad y = 1 + 1\sin t, \qquad \tfrac{5\pi}{6} \le t \le \tfrac{13\pi}{6}$$

If you are using a calculator or computer that cannot overlay graphs (using one t-interval), the following is appropriate. We'll start by picking the t-interval $[0, 2.5]$ since it easily matches the t-values for the two sides. We now need to find parametric equations for all graphs with $0 \le t \le 2.5$.

Left side: $x = 1$ and y goes from 1.5 to 4, so use

$$x = 1, \qquad y = 1.5 + t, \qquad 0 \le t \le 2.5$$

Right side: $x = 10$ and y goes from 1.5 to 4, so use

$$x = 10, \qquad y = 1.5 + t, \qquad 0 \le t \le 2.5$$

Bottom: x goes from 1 to 10 and $y = 1.5$, so use

$$x = 1 + 3.6t, \qquad y = 1.5, \qquad 0 \le t \le 2.5$$

To get the x-assignment, think of creating a linear function such that when $t = 0$, $x = 1$ and when $t = 2.5$, $x = 10$. We can use the point-slope form of a line with $(t_1, x_1) = (0, 1)$ and $(t_2, x_2) = (2.5, 10)$.

$$x - 1 = \frac{10 - 1}{2.5 - 0}(t - 0) \quad \Rightarrow \quad x = 1 + 3.6t.$$

Handle: It starts at $(10, 4)$ and ends at $(13, 7)$, so use

$$x = 10 + 1.2t, \qquad y = 4 + 1.2t, \qquad 0 \le t \le 2.5$$

$(t_1, x_1) = (0, 10)$ and $(t_2, x_2) = (2.5, 13)$ gives us $x - 10 = \dfrac{13 - 10}{2.5 - 0}(t - 0) \quad \Rightarrow \quad x = 10 + 1.2t.$

$(t_1, y_1) = (0, 4)$ and $(t_2, y_2) = (2.5, 7)$ gives us $y - 4 = \dfrac{7 - 4}{2.5 - 0}(t - 0) \quad \Rightarrow \quad y = 4 + 1.2t.$

Left wheel: It's centered at $(3, 1)$, has a radius of 1, and appears to go about $30°$ above the horizontal, so use

$$x = 3 + 1\cos\left(\tfrac{8\pi}{15}t + \tfrac{5\pi}{6}\right), \qquad y = 1 + 1\sin\left(\tfrac{8\pi}{15}t + \tfrac{5\pi}{6}\right), \qquad 0 \le t \le 2.5$$

$(t_1, \theta_1) = \left(0, \frac{5\pi}{6}\right)$ and $(t_2, \theta_2) = \left(\frac{5}{2}, \frac{13\pi}{6}\right)$ gives us $\theta - \frac{5\pi}{6} = \dfrac{\frac{13\pi}{6} - \frac{5\pi}{6}}{\frac{5}{2} - 0}(t - 0) \;\Rightarrow\; \theta = \frac{5\pi}{6} + \frac{8\pi}{15}t.$

Right wheel: Similar to the left wheel with center $(8, 1)$, so use

$$x = 8 + 1\cos\left(\tfrac{8\pi}{15}t + \tfrac{5\pi}{6}\right), \qquad y = 1 + 1\sin\left(\tfrac{8\pi}{15}t + \tfrac{5\pi}{6}\right), \qquad 0 \le t \le 2.5$$

37. (a) $x = t^3 \;\Rightarrow\; t = x^{1/3}$, so $y = t^2 = x^{2/3}$.

We get the entire curve $y = x^{2/3}$ traversed in a left to right direction.

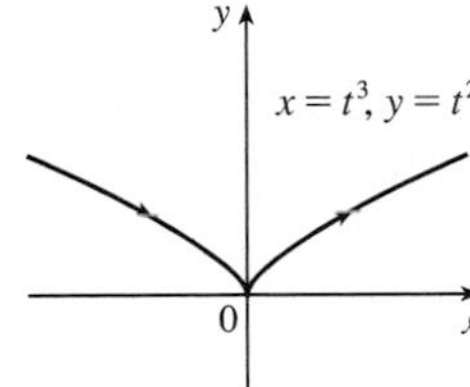

(b) $x = t^6 \;\Rightarrow\; t = x^{1/6}$, so $y = t^4 = x^{4/6} = x^{2/3}$.

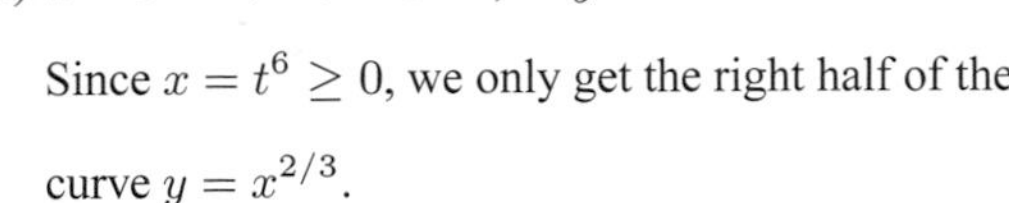

Since $x = t^6 \ge 0$, we only get the right half of the curve $y = x^{2/3}$.

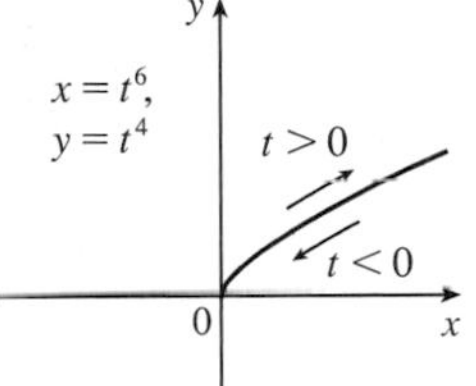

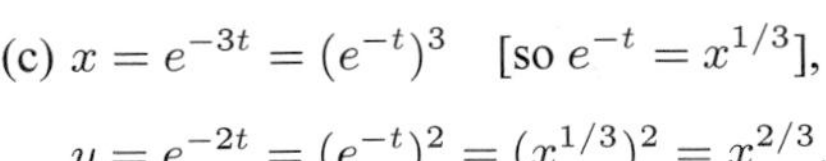

(c) $x = e^{-3t} = (e^{-t})^3$ [so $e^{-t} = x^{1/3}$],

$y = e^{-2t} = (e^{-t})^2 = (x^{1/3})^2 = x^{2/3}$.

If $t < 0$, then x and y are both larger than 1. If $t > 0$, then x and y are between 0 and 1. Since $x > 0$ and $y > 0$, the curve never quite reaches the origin.

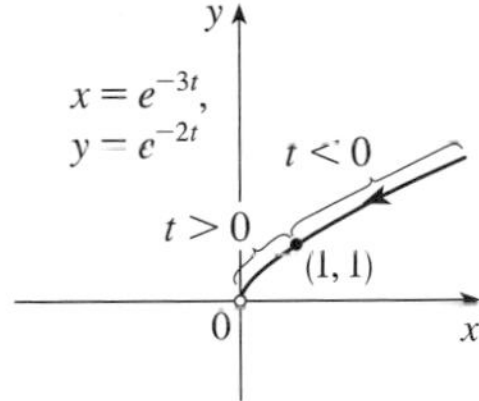

38. (a) $x = t$, so $y = t^{-2} = x^{-2}$. We get the entire curve $y = 1/x^2$ traversed in a left-to-right direction.

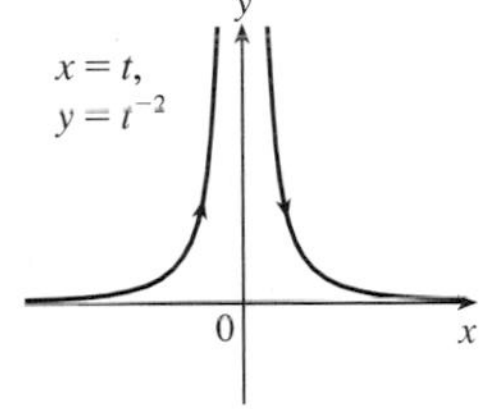

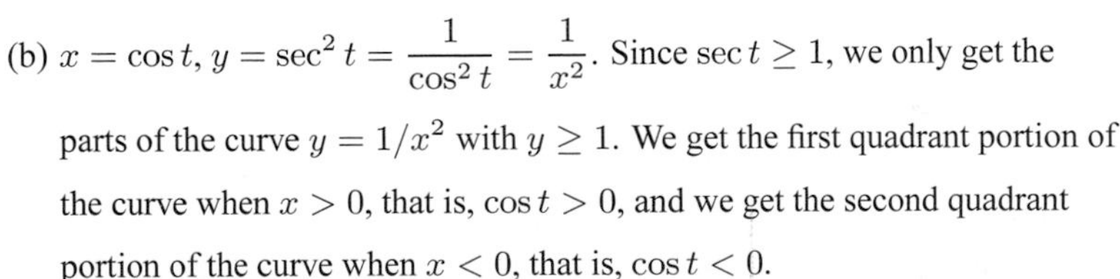

(b) $x = \cos t$, $y = \sec^2 t = \dfrac{1}{\cos^2 t} = \dfrac{1}{x^2}$. Since $\sec t \ge 1$, we only get the parts of the curve $y = 1/x^2$ with $y \ge 1$. We get the first quadrant portion of the curve when $x > 0$, that is, $\cos t > 0$, and we get the second quadrant portion of the curve when $x < 0$, that is, $\cos t < 0$.

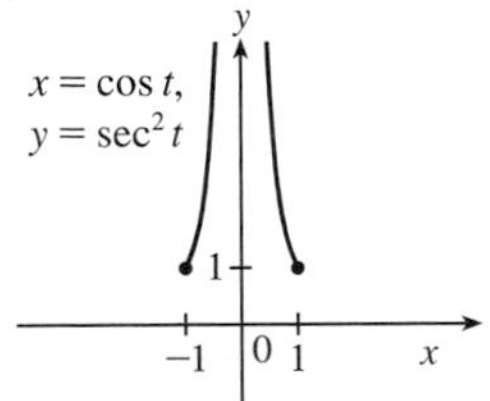

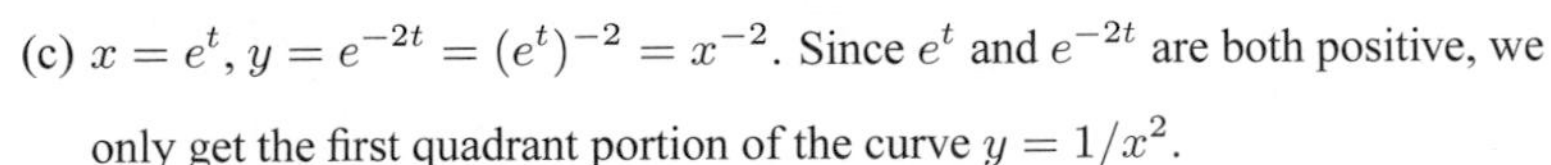

(c) $x = e^t$, $y = e^{-2t} = (e^t)^{-2} = x^{-2}$. Since e^t and e^{-2t} are both positive, we only get the first quadrant portion of the curve $y = 1/x^2$.

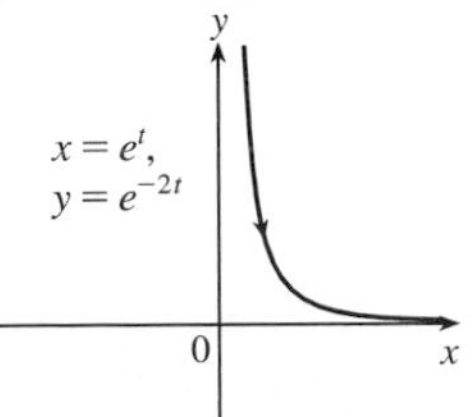

39. The case $\frac{\pi}{2} < \theta < \pi$ is illustrated. C has coordinates $(r\theta, r)$ as in Example 7, and Q has coordinates $(r\theta, r + r\cos(\pi - \theta)) = (r\theta, r(1 - \cos\theta))$ [since $\cos(\pi - \alpha) = \cos\pi\cos\alpha + \sin\pi\sin\alpha = -\cos\alpha$], so P has coordinates $(r\theta - r\sin(\pi - \theta), r(1 - \cos\theta)) = (r(\theta - \sin\theta), r(1 - \cos\theta))$ [since $\sin(\pi - \alpha) = \sin\pi\cos\alpha - \cos\pi\sin\alpha = \sin\alpha$]. Again we have the parametric equations $x = r(\theta - \sin\theta)$, $y = r(1 - \cos\theta)$.

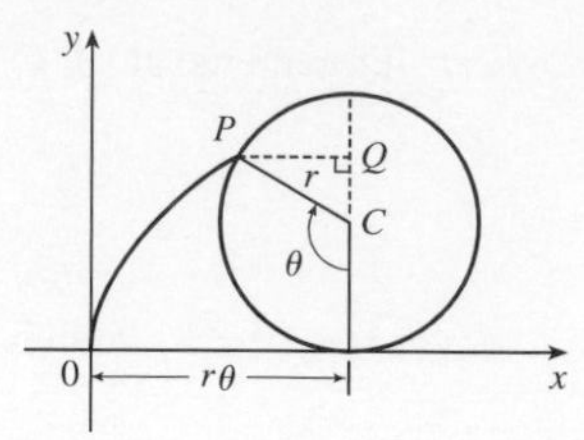

40. The first two diagrams depict the case $\pi < \theta < \frac{3\pi}{2}$, $d < r$. As in Example 7, C has coordinates $(r\theta, r)$. Now Q (in the second diagram) has coordinates $(r\theta, r + d\cos(\theta - \pi)) = (r\theta, r - d\cos\theta)$, so a typical point P of the trochoid has coordinates $(r\theta + d\sin(\theta - \pi), r - d\cos\theta)$. That is, P has coordinates (x, y), where $x = r\theta - d\sin\theta$ and $y = r - d\cos\theta$. When $d = r$, these equations agree with those of the cycloid.

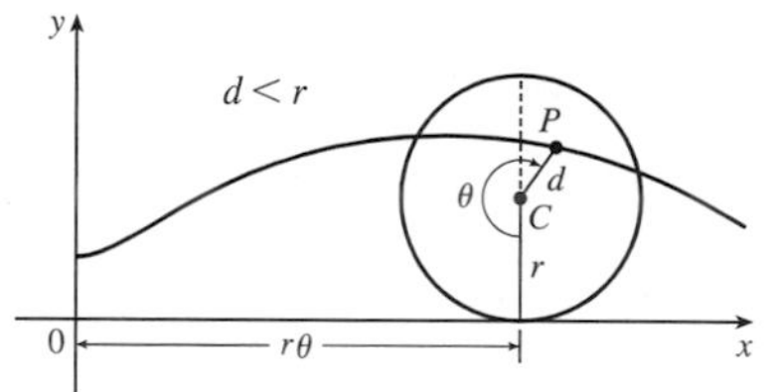

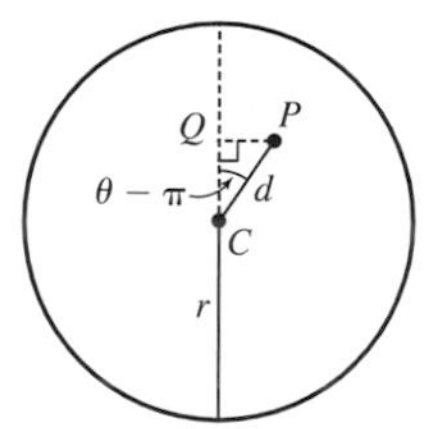

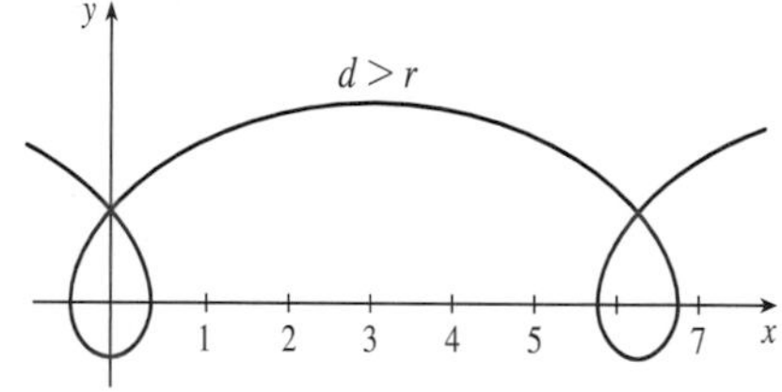

41. It is apparent that $x = |OQ|$ and $y = |QP| = |ST|$. From the diagram, $x = |OQ| = a\cos\theta$ and $y = |ST| = b\sin\theta$. Thus, the parametric equations are $x = a\cos\theta$ and $y = b\sin\theta$. To eliminate θ we rearrange: $\sin\theta = y/b \;\Rightarrow\; \sin^2\theta = (y/b)^2$ and $\cos\theta = x/a \;\Rightarrow\; \cos^2\theta = (x/a)^2$. Adding the two equations: $\sin^2\theta + \cos^2\theta = 1 = x^2/a^2 + y^2/b^2$. Thus, we have an ellipse.

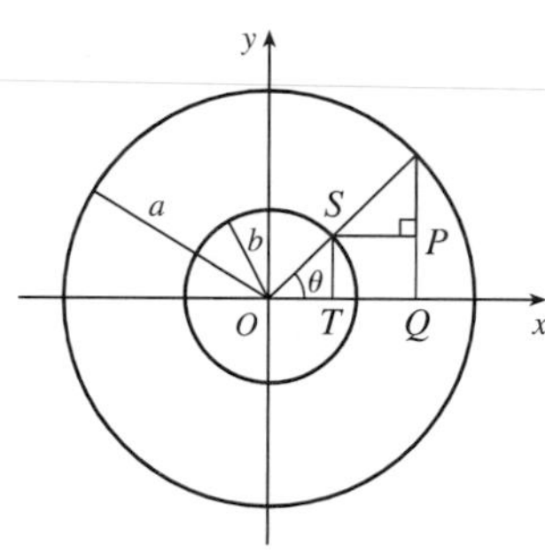

42. A has coordinates $(a\cos\theta, a\sin\theta)$. Since OA is perpendicular to AB, ΔOAB is a right triangle and B has coordinates $(a\sec\theta, 0)$. It follows that P has coordinates $(a\sec\theta, b\sin\theta)$. Thus, the parametric equations are $x = a\sec\theta$, $y = b\sin\theta$.

43. $C = (2a\cot\theta, 2a)$, so the x-coordinate of P is $x = 2a\cot\theta$. Let $B = (0, 2a)$. Then $\angle OAB$ is a right angle and $\angle OBA = \theta$, so $|OA| = 2a\sin\theta$ and $A = ((2a\sin\theta)\cos\theta, (2a\sin\theta)\sin\theta)$. Thus, the y-coordinate of P is $y = 2a\sin^2\theta$.

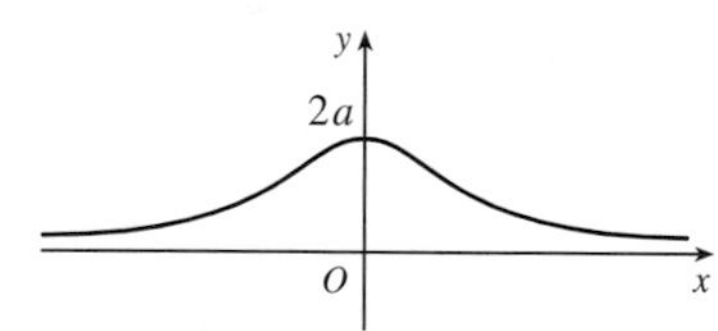

44. (a) Let θ be the angle of inclination of segment OP. Then $|OB| = \dfrac{2a}{\cos\theta}$.

Let $C = (2a, 0)$. Then by use of right triangle OAC we see that $|OA| = 2a\cos\theta$.

Now

$$|OP| = |AB| = |OB| - |OA|$$

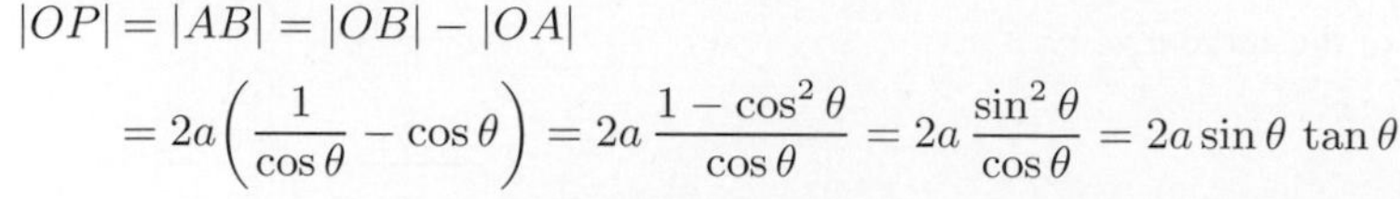

$$= 2a\left(\frac{1}{\cos\theta} - \cos\theta\right) = 2a\,\frac{1 - \cos^2\theta}{\cos\theta} = 2a\,\frac{\sin^2\theta}{\cos\theta} = 2a\sin\theta\,\tan\theta$$

So P has coordinates $x = 2a\sin\theta\,\tan\theta\cdot\cos\theta = 2a\sin^2\theta$ and $y = 2a\sin\theta\,\tan\theta\cdot\sin\theta = 2a\sin^2\theta\,\tan\theta$.

(b)

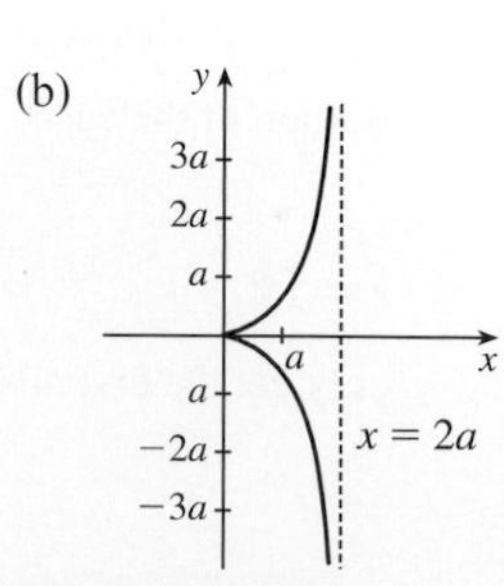

45. (a)

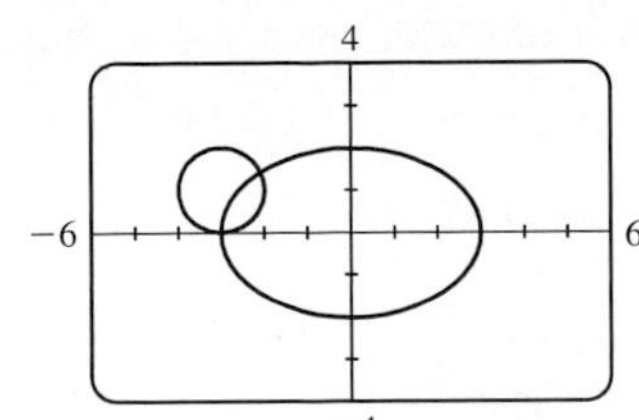

There are 2 points of intersection:

$(-3, 0)$ and approximately $(-2.1, 1.4)$.

(b) A collision point occurs when $x_1 = x_2$ and $y_1 = y_2$ for the same t. So solve the equations:

$$3\sin t = -3 + \cos t \quad \textbf{(1)}$$
$$2\cos t = 1 + \sin t \quad \textbf{(2)}$$

From **(2)**, $\sin t = 2\cos t - 1$. Substituting into **(1)**, we get $3(2\cos t - 1) = -3 + \cos t \;\Rightarrow\; 5\cos t = 0 \;\; (\star) \;\Rightarrow$ $\cos t = 0 \;\Rightarrow\; t = \frac{\pi}{2}$ or $\frac{3\pi}{2}$. We check that $t = \frac{3\pi}{2}$ satisfies **(1)** and **(2)** but $t = \frac{\pi}{2}$ does not. So the only collision point occurs when $t = \frac{3\pi}{2}$, and this gives the point $(-3, 0)$. [We could check our work by graphing x_1 and x_2 together as functions of t and, on another plot, y_1 and y_2 as functions of t. If we do so, we see that the only value of t for which *both* pairs of graphs intersect is $t = \frac{3\pi}{2}$.]

(c) The circle is centered at $(3, 1)$ instead of $(-3, 1)$. There are still 2 intersection points: $(3, 0)$ and $(2.1, 1.4)$, but there are no collision points, since $(\star)$ in part (b) becomes $5\cos t = 6 \;\Rightarrow\; \cos t = \frac{6}{5} > 1$.

46. (a) If $\alpha = 30^\circ$ and $v_0 = 500$ m/s, then the equations become $x = (500\cos 30^\circ)t = 250\sqrt{3}t$ and $y = (500\sin 30^\circ)t - \frac{1}{2}(9.8)t^2 = 250t - 4.9t^2$. $y = 0$ when $t = 0$ (when the gun is fired) and again when $t = \frac{250}{4.9} \approx 51$ s. Then $x = \left(250\sqrt{3}\right)\left(\frac{250}{4.9}\right) \approx 22{,}092$ m, so the bullet hits the ground about 22 km from the gun.

The formula for y is quadratic in t. To find the maximum y-value, we will complete the square:

$$y = -4.9\left(t^2 - \tfrac{250}{4.9}t\right) = -4.9\left[t^2 - \tfrac{250}{4.9}t + \left(\tfrac{125}{4.9}\right)^2\right] + \tfrac{125^2}{4.9} = -4.9\left(t - \tfrac{125}{4.9}\right)^2 + \tfrac{125^2}{4.9} \le \tfrac{125^2}{4.9}$$

with equality when $t = \frac{125}{4.9}$ s, so the maximum height attained is $\frac{125^2}{4.9} \approx 3189$ m.

(b)

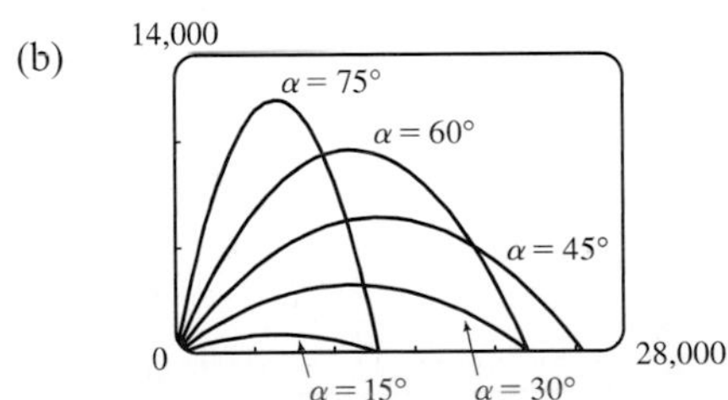

As α $(0^\circ < \alpha < 90^\circ)$ increases up to 45°, the projectile attains a greater height and a greater range. As α increases past 45°, the projectile attains a greater height, but its range decreases.

(c) $x = (v_0\cos\alpha)t \;\Rightarrow\; t = \dfrac{x}{v_0\cos\alpha}$.

$$y = (v_0\sin\alpha)t - \tfrac{1}{2}gt^2 \;\Rightarrow\; y = (v_0\sin\alpha)\frac{x}{v_0\cos\alpha} - \frac{g}{2}\left(\frac{x}{v_0\cos\alpha}\right)^2 = (\tan\alpha)x - \left(\frac{g}{2v_0^2\cos^2\alpha}\right)x^2,$$

which is the equation of a parabola (quadratic in x).

47. $x = t^2, y = t^3 - ct$. We use a graphing device to produce the graphs for various values of c with $-\pi \le t \le \pi$. Note that all the members of the family are symmetric about the x-axis. For $c < 0$, the graph does not cross itself, but for $c = 0$ it has a cusp at $(0, 0)$ and for $c > 0$ the graph crosses itself at $x = c$, so the loop grows larger as c increases.

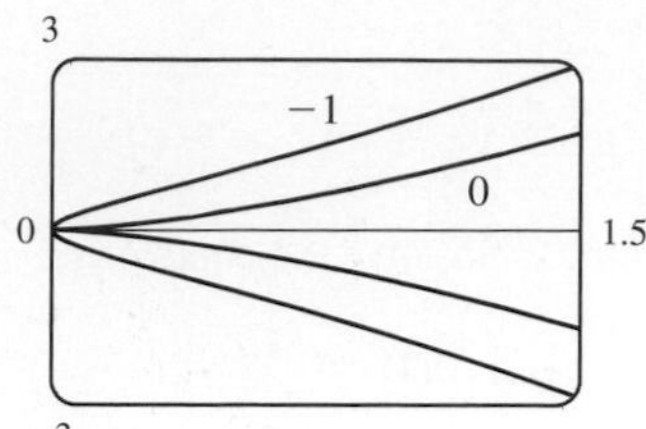

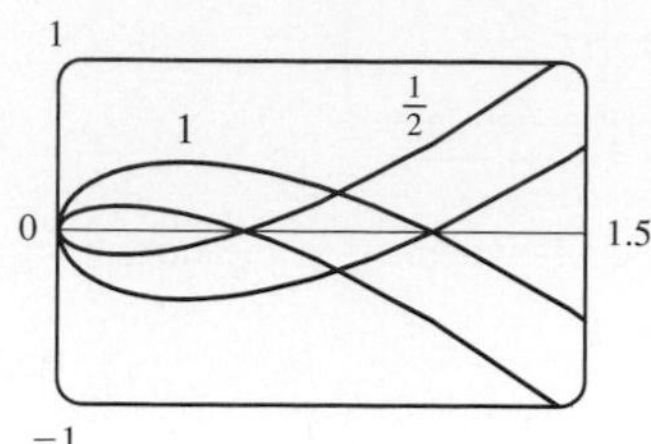

48. $x = 2ct - 4t^3, y = -ct^2 + 3t^4$. We use a graphing device to produce the graphs for various values of c with $-\pi \le t \le \pi$. Note that all the members of the family are symmetric about the y-axis. When $c < 0$, the graph resembles that of a polynomial of even degree, but when $c = 0$ there is a corner at the origin, and when $c > 0$, the graph crosses itself at the origin, and has two cusps below the x-axis. The size of the "swallowtail" increases as c increases.

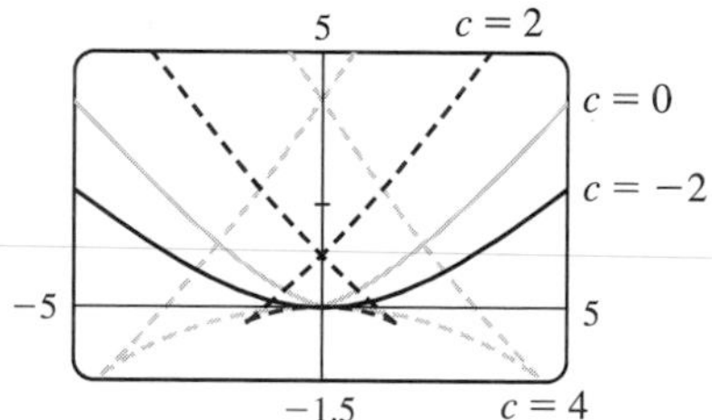

49. $x = t + a\cos t, y = t + a\sin t, a > 0$. From the first figure, we see that curves roughly follow the line $y = x$, and they start having loops when a is between 1.4 and 1.6. The loops increase in size as a increases.

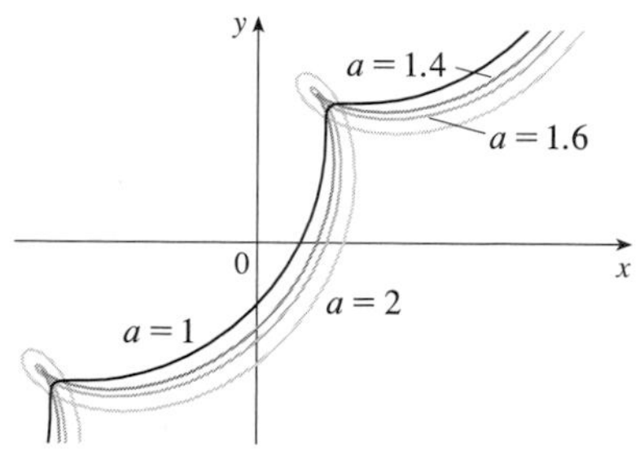

While not required, the following is a solution to determine the *exact* values for which the curve has a loop, that is, we seek the values of a for which there exist parameter values t and u such that $t < u$ and $(t + a\cos t, t + a\sin t) = (u + a\cos u, u + a\sin u)$.

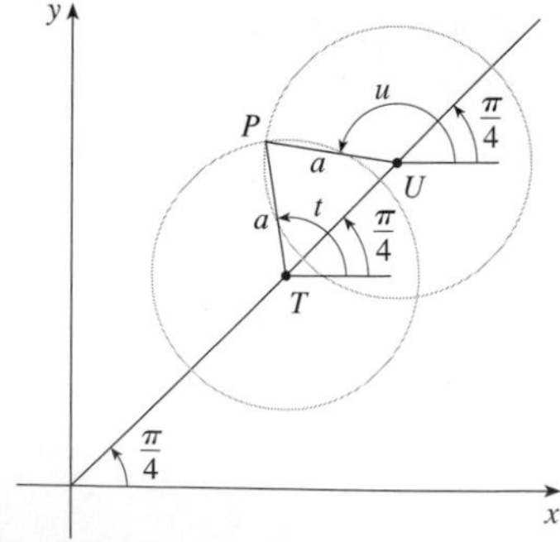

In the diagram at the left, T denotes the point (t, t), U the point (u, u), and P the point $(t + a\cos t, t + a\sin t) = (u + a\cos u, u + a\sin u)$. Since $\overline{PT} = \overline{PU} = a$, the triangle PTU is isosceles. Therefore its base angles, $\alpha = \angle PTU$ and $\beta = \angle PUT$ are equal. Since $\alpha = t - \frac{\pi}{4}$ and $\beta = 2\pi - \frac{3\pi}{4} - u = \frac{5\pi}{4} - u$, the relation $\alpha = \beta$ implies that $u + t = \frac{3\pi}{2}$ **(1)**.

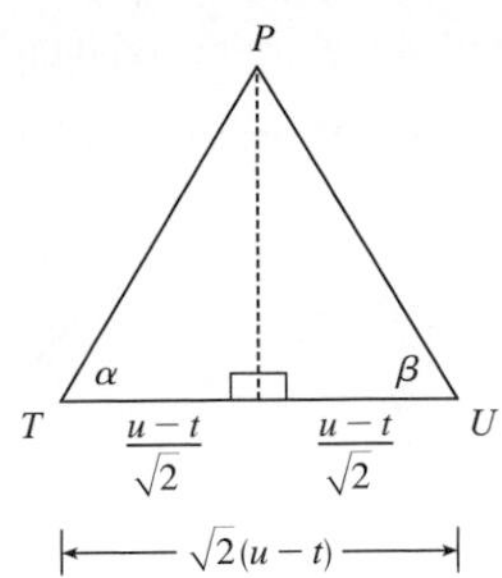

Since $\overline{TU} = \text{distance}((t,t),(u,u)) = \sqrt{2(u-t)^2} = \sqrt{2}\,(u-t)$, we see that

$\cos\alpha = \dfrac{\frac{1}{2}\overline{TU}}{\overline{PT}} = \dfrac{(u-t)/\sqrt{2}}{a}$, so $u - t = \sqrt{2}\,a\cos\alpha$, that is,

$u - t = \sqrt{2}\,a\cos\left(t - \frac{\pi}{4}\right)$ **(2)**. Now $\cos\left(t - \frac{\pi}{4}\right) = \sin\left[\frac{\pi}{2} - \left(t - \frac{\pi}{4}\right)\right] = \sin\left(\frac{3\pi}{4} - t\right)$,

so we can rewrite **(2)** as $u - t = \sqrt{2}\,a\sin\left(\frac{3\pi}{4} - t\right)$ **(2′)**. Subtracting **(2′)** from **(1)** and

dividing by 2, we obtain $t = \frac{3\pi}{4} - \frac{\sqrt{2}}{2}a\sin\left(\frac{3\pi}{4} - t\right)$, or $\frac{3\pi}{4} - t = \frac{a}{\sqrt{2}}\sin\left(\frac{3\pi}{4} - t\right)$ **(3)**.

Since $a > 0$ and $t < u$, it follows from **(2′)** that $\sin\left(\frac{3\pi}{4} - t\right) > 0$. Thus from **(3)** we see that $t < \frac{3\pi}{4}$. [We have implicitly assumed that $0 < t < \pi$ by the way we drew our diagram, but we lost no generality by doing so since replacing t by $t + 2\pi$ merely increases x and y by 2π. The curve's basic shape repeats every time we change t by 2π.] Solving for a in **(3)**, we get $a = \dfrac{\sqrt{2}\left(\frac{3\pi}{4} - t\right)}{\sin\left(\frac{3\pi}{4} - t\right)}$. Write $z = \frac{3\pi}{4} - t$. Then $a = \dfrac{\sqrt{2}\,z}{\sin z}$, where $z > 0$. Now $\sin z < z$ for $z > 0$, so $a > \sqrt{2}$.

$\left[\text{As } z \to 0^+, \text{ that is, as } t \to \left(\frac{3\pi}{4}\right)^-, a \to \sqrt{2}\right]$.

50. Consider the curves $x = \sin t + \sin nt$, $y = \cos t + \cos nt$, where n is a positive integer. For $n = 1$, we get a circle of radius 2 centered at the origin. For $n > 1$, we get a curve lying on or inside that circle that traces out $n - 1$ loops as t ranges from 0 to 2π.

Note:

$$\begin{aligned} x^2 + y^2 &= (\sin t + \sin nt)^2 + (\cos t + \cos nt)^2 \\ &= \sin^2 t + 2\sin t\,\sin nt + \sin^2 nt + \cos^2 t + 2\cos t\,\cos nt + \cos^2 nt \\ &= (\sin^2 t + \cos^2 t) + (\sin^2 nt + \cos^2 nt) + 2(\cos t\,\cos nt + \sin t\,\sin nt) \\ &= 1 + 1 + 2\cos(t - nt) = 2 + 2\cos((1-n)t) \le 4 = 2^2, \end{aligned}$$

with equality for $n = 1$. This shows that each curve lies on or inside the curve for $n = 1$, which is a circle of radius 2 centered at the origin.

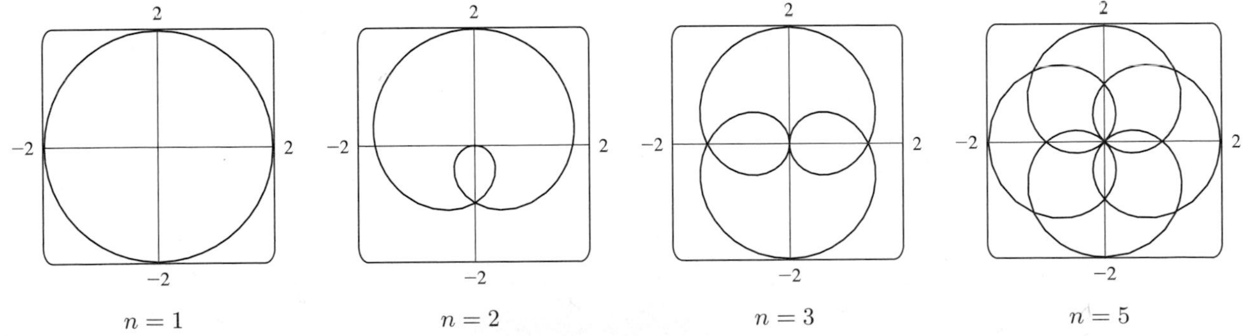

51. Note that all the Lissajous figures are symmetric about the x-axis. The parameters a and b simply stretch the graph in the x- and y-directions respectively. For $a = b = n = 1$ the graph is simply a circle with radius 1. For $n = 2$ the graph crosses itself at the origin and there are loops above and below the x-axis. In general, the figures have $n-1$ points of intersection, all of which are on the y-axis, and a total of n closed loops.

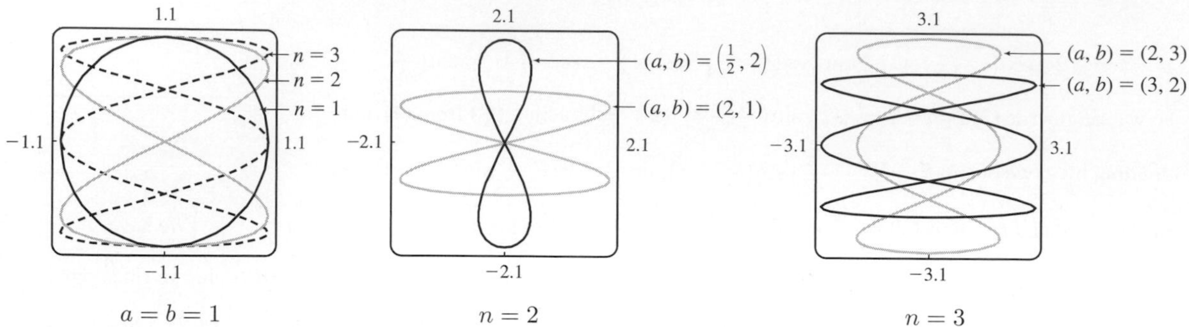

52. $x=\cos t$, $y=\sin t-\sin ct$. If $c=1$, then $y=0$, and the curve is simply the line segment from $(-1,0)$ to $(1,0)$. The graphs are shown for $c=2,3,4$ and 5.

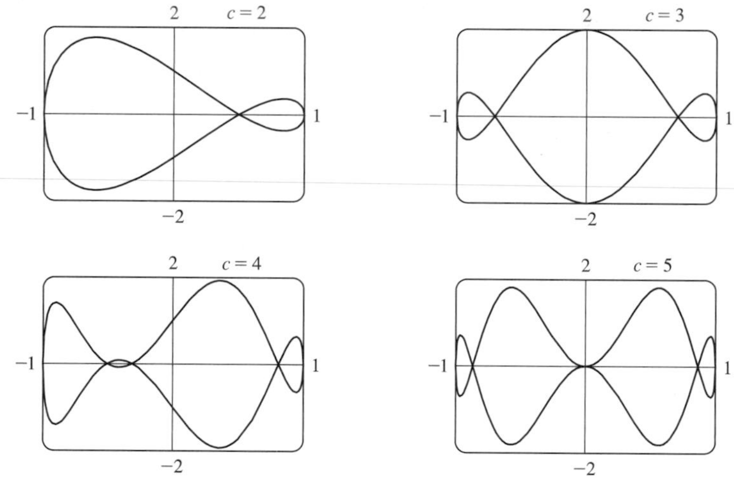

It is easy to see that all the curves lie in the rectangle $[-1,1]$ by $[-2,2]$. When c is an integer, $x(t+2\pi)=x(t)$ and $y(t+2\pi)=y(t)$, so the curve is closed. When c is a positive integer greater than 1, the curve intersects the x-axis $c+1$ times and has c loops (one of which degenerates to a tangency at the origin when c is an odd integer of the form $4k+1$). As c increases, the curve's loops become thinner, but stay in the region bounded by the semicircles $y=\pm\left(1+\sqrt{1-x^2}\,\right)$ and the line segments from $(-1,-1)$ to $(-1,1)$ and from $(1,-1)$ to $(1,1)$. This is true because $|y|=|\sin t-\sin ct|\leq|\sin t|+|\sin ct|\leq\sqrt{1-x^2}+1$. This curve appears to fill the entire region when c is very large, as shown in the figure for $c=1000$.

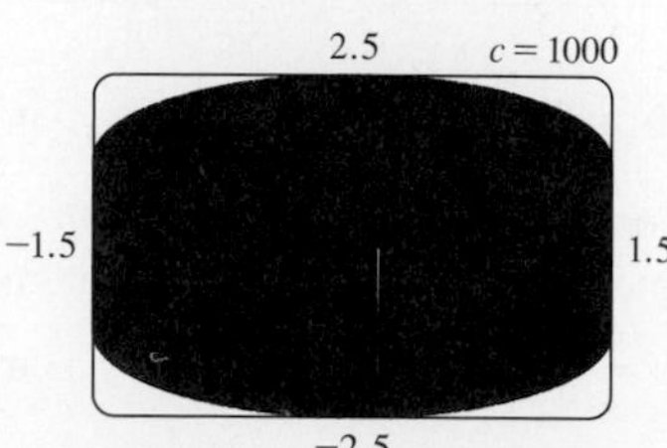

When c is a fraction, we get a variety of shapes with multiple loops, but always within the same region. For some fractional values, such as $c = 2.359$, the curve again appears to fill the region.

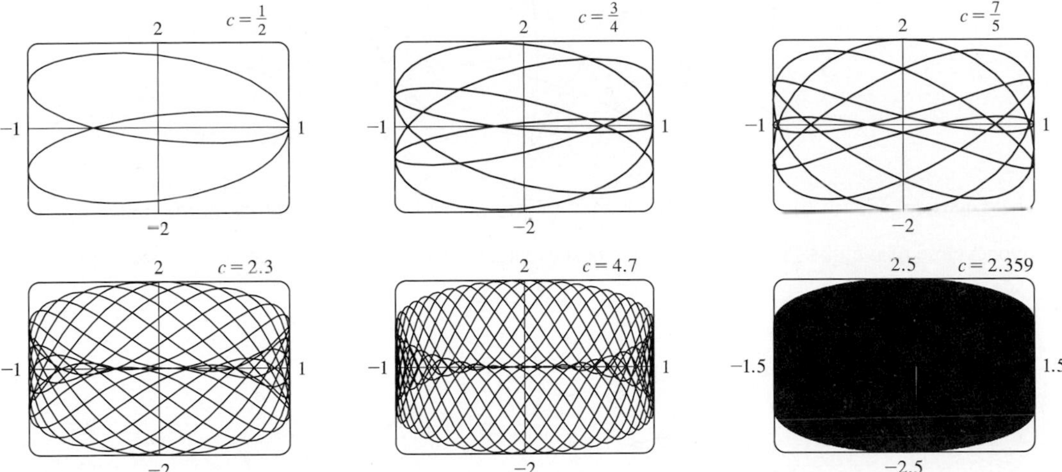

LABORATORY PROJECT Running Circles Around Circles

1. The center Q of the smaller circle has coordinates $((a - b)\cos\theta, (a - b)\sin\theta)$. Arc PS on circle C has length $a\theta$ since it is equal in length to arc AS (the smaller circle rolls without slipping against the larger.)

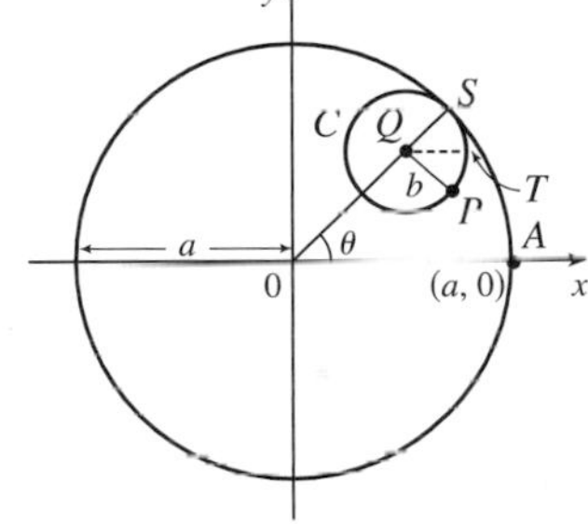

Thus, $\angle PQS = \frac{a}{b}\theta$ and $\angle PQT = \frac{a}{b}\theta - \theta$, so P has coordinates

$$x = (a - b)\cos\theta + b\cos(\angle PQT) = (a - b)\cos\theta + b\cos\left(\frac{a - b}{b}\theta\right)$$

and $$y = (a - b)\sin\theta - b\sin(\angle PQT) = (a - b)\sin\theta - b\sin\left(\frac{a - b}{b}\theta\right).$$

2. With $b = 1$ and a a positive integer greater than 2, we obtain a hypocycloid of a cusps. Shown in the figure is the graph for $a = 4$. Let $a = 4$ and $b = 1$. Using the sum identities to expand $\cos 3\theta$ and $\sin 3\theta$, we obtain

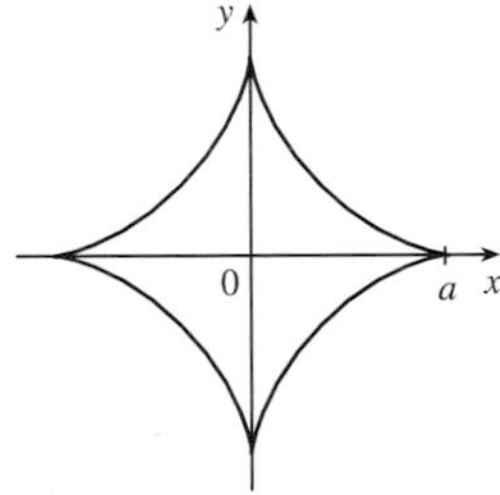

$$x = 3\cos\theta + \cos 3\theta = 3\cos\theta + \left(4\cos^3\theta - 3\cos\theta\right) = 4\cos^3\theta$$

and $$y = 3\sin\theta - \sin 3\theta = 3\sin\theta - \left(3\sin\theta - 4\sin^3\theta\right) = 4\sin^3\theta.$$

3. The graphs at the right are obtained with $b = 1$ and $a = \frac{1}{2}, \frac{1}{3}, \frac{1}{4}$, and $\frac{1}{10}$ with $-2\pi \le \theta \le 2\pi$. We conclude that as the denominator d increases, the graph gets smaller, but maintains the basic shape shown.

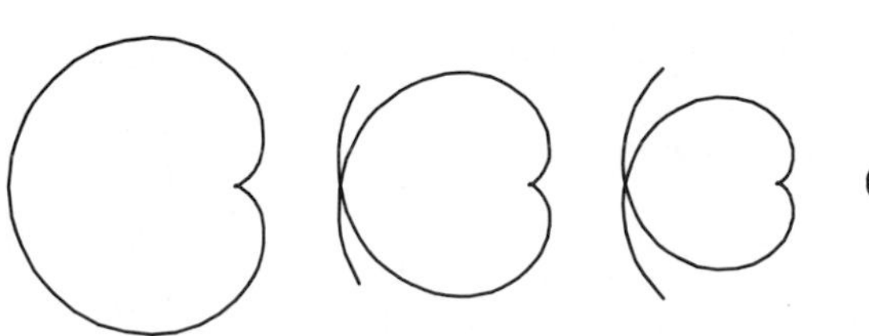

[continued]

Letting $d = 2$ and $n = 3$, 5, and 7 with $-2\pi \le \theta \le 2\pi$ gives us the following:

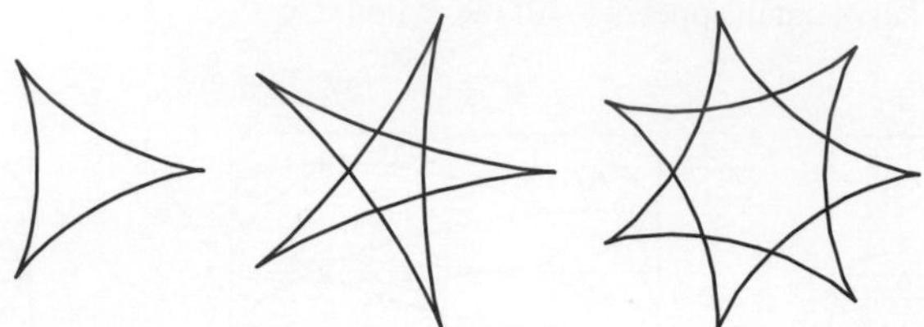

So if d is held constant and n varies, we get a graph with n cusps (assuming n/d is in lowest form). When $n = d + 1$, we obtain a hypocycloid of n cusps. As n increases, we must expand the range of θ in order to get a closed curve. The following graphs have $a = \frac{3}{2}$, $\frac{5}{4}$, and $\frac{11}{10}$.

4. If $b = 1$, the equations for the hypocycloid are

$$x = (a-1)\cos\theta + \cos((a-1)\theta) \qquad y = (a-1)\sin\theta - \sin((a-1)\theta)$$

which is a hypocycloid of a cusps (from Problem 2). In general, if $a > 1$, we get a figure with cusps on the "outside ring" and if $a < 1$, the cusps are on the "inside ring". In any case, as the values of θ get larger, we get a figure that looks more and more like a washer. If we were to graph the hypocycloid for all values of θ, every point on the washer would eventually be arbitrarily close to a point on the curve.

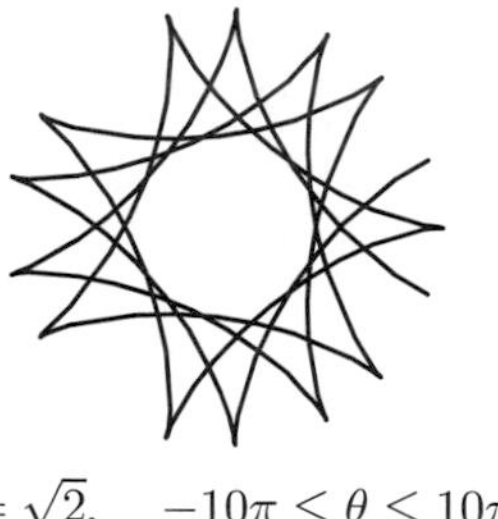

$a = \sqrt{2}, \quad -10\pi \le \theta \le 10\pi$

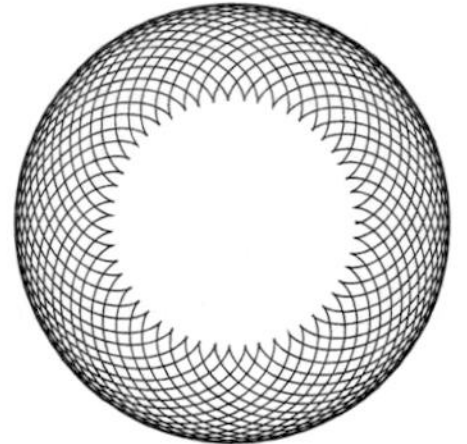

$a = e - 2, \quad 0 \le \theta \le 446$

5. The center Q of the smaller circle has coordinates $((a+b)\cos\theta, (a+b)\sin\theta)$.

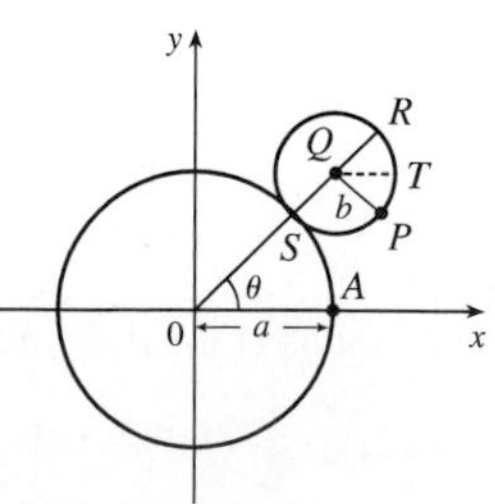

Arc PS has length $a\theta$ (as in Problem 1), so that $\angle PQS = \dfrac{a\theta}{b}$, $\angle PQR = \pi - \dfrac{a\theta}{b}$,

and $\angle PQT = \pi - \dfrac{a\theta}{b} - \theta = \pi - \left(\dfrac{a+b}{b}\right)\theta$ since $\angle RQT = \theta$.

Thus, the coordinates of P are

$$x = (a+b)\cos\theta + b\cos\left(\pi - \frac{a+b}{b}\theta\right) = (a+b)\cos\theta - b\cos\left(\frac{a+b}{b}\theta\right)$$

and $$y = (a+b)\sin\theta - b\sin\left(\pi - \frac{a+b}{b}\theta\right) = (a+b)\sin\theta - b\sin\left(\frac{a+b}{b}\theta\right).$$

6. Let $b = 1$ and the equations become

$$x = (a+1)\cos\theta - \cos((a+1)\theta) \qquad y = (a+1)\sin\theta - \sin((a+1)\theta)$$

If $a = 1$, we have a cardioid. If a is a positive integer greater than 1, we get the graph of an "a-leafed clover", with cusps that are a units from the origin. (Some of the pairs of figures are not to scale.)

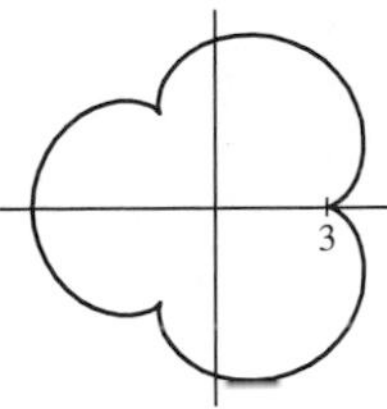

$a = 3$, $-2\pi \le \theta \le 2\pi$

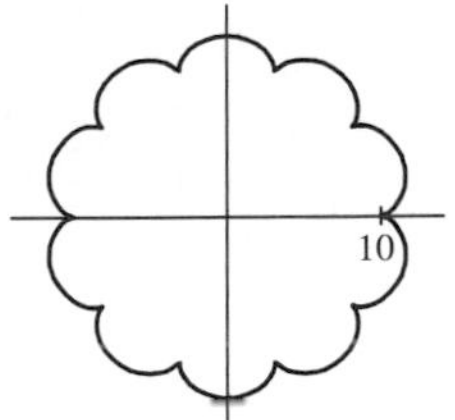

$a = 10$, $-2\pi \le \theta \le 2\pi$

If $a = n/d$ with $n = 1$, we obtain a figure that does not increase in size and requires $-d\pi \le \theta \le d\pi$ to be a closed curve traced exactly once.

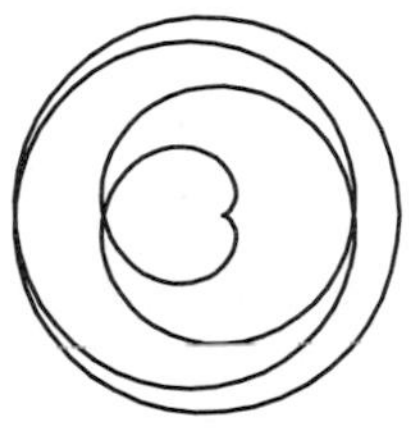

$a = \frac{1}{4}$, $-4\pi \le \theta \le 4\pi$

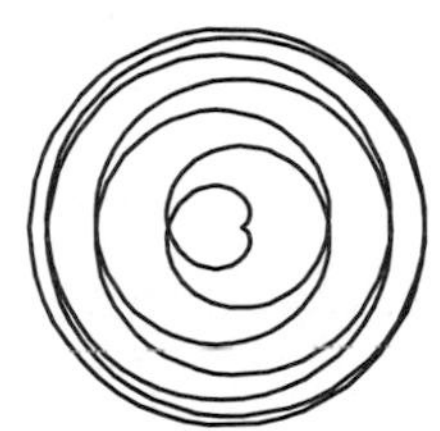

$a = \frac{1}{7}$, $-7\pi \le \theta \le 7\pi$

Next, we keep d constant and let n vary. As n increases, so does the size of the figure. There is an n-pointed star in the middle.

$a = \frac{2}{5}$, $-5\pi \le \theta \le 5\pi$

$a = \frac{7}{5}$, $-5\pi \le \theta \le 5\pi$

Now if $n = d + 1$ we obtain figures similar to the previous ones, but the size of the figure does not increase.

$a = \frac{4}{3}$, $-3\pi \le \theta \le 3\pi$

$a = \frac{7}{6}$, $-6\pi \le \theta \le 6\pi$

If a is irrational, we get washers that increase in size as a increases.

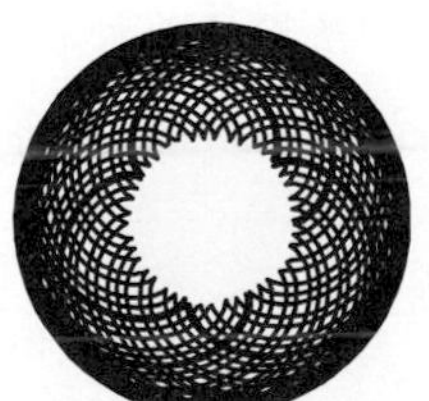

$a = \sqrt{2}$, $0 \le \theta \le 200$

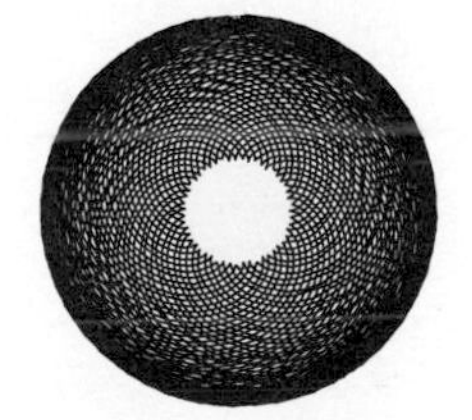

$a = e - 2$, $0 \le \theta \le 446$

10.2 Calculus with Parametric Curves

1. $x = \dfrac{t}{1+t}$, $y = \sqrt{1+t}$ $\Rightarrow$ $\dfrac{dy}{dt} = \dfrac{1}{2}(1+t)^{-1/2} = \dfrac{1}{2\sqrt{1+t}}$, $\dfrac{dx}{dt} = \dfrac{(1+t)(1) - t(1)}{(1+t)^2} = \dfrac{1}{(1+t)^2}$, and

$\dfrac{dy}{dx} = \dfrac{dy/dt}{dx/dt} = \dfrac{1/(2\sqrt{1+t})}{1/(1+t)^2} = \dfrac{(1+t)^2}{2\sqrt{1+t}} = \dfrac{1}{2}(1+t)^{3/2}$.

2. $x = te^t$, $y = t + \sin t$ $\Rightarrow$ $\dfrac{dy}{dt} = 1 + \cos t$, $\dfrac{dx}{dt} = te^t + e^t = e^t(t+1)$, and $\dfrac{dy}{dx} = \dfrac{dy/dt}{dx/dt} = \dfrac{1+\cos t}{e^t(t+1)}$.

3. $x = t^3 + 1$, $y = t^4 + t$; $t = -1$. $\dfrac{dy}{dt} = 4t^3 + 1$, $\dfrac{dx}{dt} = 3t^2$, and $\dfrac{dy}{dx} = \dfrac{dy/dt}{dx/dt} = \dfrac{4t^3+1}{3t^2}$. When $t = -1$, $(x, y) = (0, 0)$ and $dy/dx = -3/3 = -1$, so an equation of the tangent to the curve at the point corresponding to $t = -1$ is $y - 0 = -1(x - 0)$, or $y = -x$.

4. $x = \sqrt{t}$, $y = t^2 - 2t$; $t = 4$. $\dfrac{dy}{dt} = 2t - 2$, $\dfrac{dx}{dt} = \dfrac{1}{2\sqrt{t}}$, and $\dfrac{dy}{dx} = \dfrac{dy/dt}{dx/dt} = (2t-2)2\sqrt{t} = 4(t-1)\sqrt{t}$. When $t = 4$, $(x, y) = (2, 8)$ and $dy/dx = 4(3)(2) = 24$, so an equation of the tangent to the curve at the point corresponding to $t = 4$ is $y - 8 = 24(x - 2)$, or $y = 24x - 40$.

5. $x = t\cos t$, $y = t\sin t$; $t = \pi$. $\dfrac{dy}{dt} = t\cos t + \sin t$, $\dfrac{dx}{dt} = t(-\sin t) + \cos t$, and $\dfrac{dy}{dx} = \dfrac{dy/dt}{dx/dt} = \dfrac{t\cos t + \sin t}{-t\sin t + \cos t}$. When $t = \pi$, $(x, y) = (-\pi, 0)$ and $dy/dx = -\pi/(-1) = \pi$, so an equation of the tangent to the curve at the point corresponding to $t = \pi$ is $y - 0 = \pi[x - (-\pi)]$, or $y = \pi x + \pi^2$.

6. $x = e^t \sin \pi t$, $y = e^{2t}$; $t = 0$. $\dfrac{dy}{dt} = 2e^{2t}$, $\dfrac{dx}{dt} = e^t(\pi\cos\pi t) + (\sin\pi t)e^t = e^t(\pi\cos\pi t + \sin\pi t)$, and

$\dfrac{dy}{dx} = \dfrac{dy/dt}{dx/dt} = \dfrac{2e^{2t}}{e^t(\pi\cos\pi t + \sin\pi t)} = \dfrac{2e^t}{\pi\cos\pi t + \sin\pi t}$. When $t = 0$, $(x, y) = (0, 1)$ and $dy/dx = 2/\pi$, so an equation of the tangent to the curve at the point corresponding to $t = 0$ is $y - 1 = \frac{2}{\pi}(x - 0)$, or $y = \frac{2}{\pi}x + 1$.

7. (a) $x = 1 + \ln t$, $y = t^2 + 2$; $(1, 3)$. $\dfrac{dy}{dt} = 2t$, $\dfrac{dx}{dt} = \dfrac{1}{t}$, and $\dfrac{dy}{dx} = \dfrac{dy/dt}{dx/dt} = \dfrac{2t}{1/t} = 2t^2$. At $(1, 3)$,

$x = 1 + \ln t = 1$ $\Rightarrow$ $\ln t = 0$ $\Rightarrow$ $t = 1$ and $\dfrac{dy}{dx} = 2$, so an equation of the tangent is $y - 3 = 2(x - 1)$, or $y = 2x + 1$.

(b) $x = 1 + \ln t$ $\Rightarrow$ $\ln t = x - 1$ $\Rightarrow$ $t = e^{x-1}$, so $y = t^2 + 2 = (e^{x-1})^2 + 2 = e^{2x-2} + 2$, and $y' = e^{2x-2} \cdot 2$. At $(1, 3)$, $y' = e^{2(1)-2} \cdot 2 = 2$, so an equation of the tangent is $y - 3 = 2(x - 1)$, or $y = 2x + 1$.

8. (a) $x = 1 + \sqrt{t}$, $y = e^{t^2}$; $(2, e)$. $\dfrac{dy}{dt} = e^{t^2} \cdot 2t$, $\dfrac{dx}{dt} = \dfrac{1}{2\sqrt{t}}$, and $\dfrac{dy}{dx} = \dfrac{dy/dt}{dx/dt} = \dfrac{2te^{t^2}}{1/(2\sqrt{t})} = 4t^{3/2}e^{t^2}$. At $(2, e)$,

$x = 1 + \sqrt{t} = 2$ $\Rightarrow$ $\sqrt{t} = 1$ $\Rightarrow$ $t = 1$ and $\dfrac{dy}{dx} = 4e$, so an equation of the tangent is $y - e = 4e(x - 2)$, or $y = 4ex - 7e$.

(b) $x = 1 + \sqrt{t} \;\Rightarrow\; \sqrt{t} = x - 1 \;\Rightarrow\; t = (x-1)^2$, so $y = e^{t^2} = e^{(x-1)^4}$, and $y' = e^{(x-1)^4} \cdot 4(x-1)^3$.

At $(2, e)$, $y' = e \cdot 4 = 4e$, so an equation of the tangent is $y - e = 4e(x - 2)$, or $y = 4ex - 7e$.

9. $x = t^2 - t,\; y = t^2 + t + 1;\; (0, 3)$. $\dfrac{dy}{dx} = \dfrac{dy/dt}{dx/dt} = \dfrac{2t+1}{2t-1}$. To find the value of t corresponding to the point $(0, 3)$, solve $x = 0 \;\Rightarrow$ $t^2 - t = 0 \;\Rightarrow\; t(t-1) = 0 \;\Rightarrow\; t = 0$ or $t = 1$. Only $t = 1$ gives $y = 3$. With $t = 1$, $dy/dx = 3$, and an equation of the tangent is $y - 3 = 3(x - 0)$, or $y = 3x + 3$.

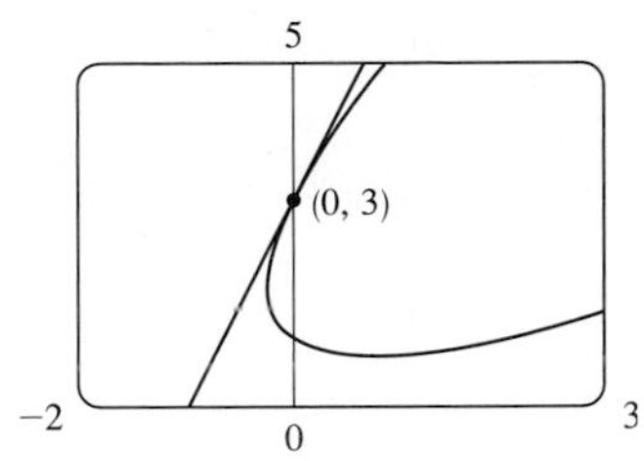

10. $x = \sin \pi t,\; y = t^2 + t;\; (0, 2)$. $\dfrac{dy}{dx} = \dfrac{dy/dt}{dx/dt} = \dfrac{2t+1}{\pi \cos \pi t}$. To find the value of t corresponding to the point $(0, 2)$, solve $y = 2 \;\Rightarrow$ $t^2 + t - 2 = 0 \;\Rightarrow\; (t+2)(t-1) = 0 \;\Rightarrow\; t = -2$ or $t = 1$. Either value gives $dy/dx = -3/\pi$, so an equation of the tangent is $y - 2 = -\frac{3}{\pi}(x - 0)$, or $y = -\frac{3}{\pi}x + 2$.

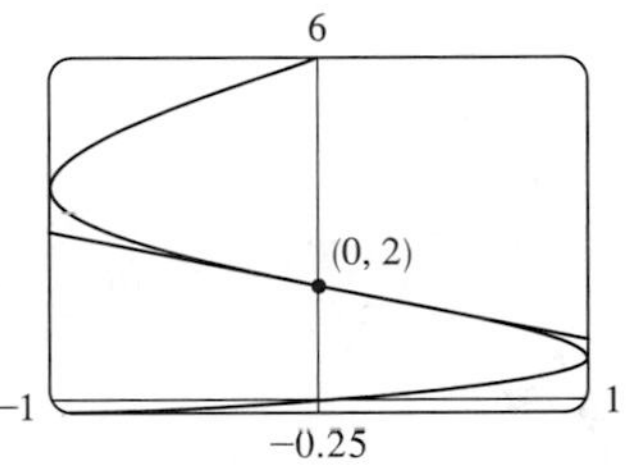

11. $x = t^2 + 1,\; y = t^2 + t \;\Rightarrow\; \dfrac{dy}{dx} = \dfrac{dy/dt}{dx/dt} = \dfrac{2t+1}{2t} = 1 + \dfrac{1}{2t} \;\Rightarrow\; \dfrac{d^2y}{dx^2} = \dfrac{\dfrac{d}{dt}\left(\dfrac{dy}{dx}\right)}{dx/dt} = \dfrac{-1/(2t^2)}{2t} = -\dfrac{1}{4t^3}$.

The curve is CU when $\dfrac{d^2y}{dx^2} > 0$, that is, when $t < 0$.

12. $x = t^3 + 1,\; y = t^2 - t \;\Rightarrow\; \dfrac{dy}{dx} = \dfrac{dy/dt}{dx/dt} = \dfrac{2t-1}{3t^2} = \dfrac{2}{3t} - \dfrac{1}{3t^2} \;\Rightarrow$

$\dfrac{d^2y}{dx^2} = \dfrac{\dfrac{d}{dt}\left(\dfrac{dy}{dx}\right)}{dx/dt} = \dfrac{-\dfrac{2}{3t^2} + \dfrac{2}{3t^3}}{3t^2} = \dfrac{\dfrac{2-2t}{3t^3}}{3t^2} = \dfrac{2(1-t)}{9t^5}$. The curve is CU when $\dfrac{d^2y}{dx^2} > 0$, that is, when $0 < t < 1$.

13. $x = e^t,\; y = te^{-t} \;\Rightarrow\; \dfrac{dy}{dx} = \dfrac{dy/dt}{dx/dt} = \dfrac{-te^{-t} + e^{-t}}{e^t} = \dfrac{e^{-t}(1-t)}{e^t} = e^{-2t}(1-t) \;\Rightarrow$

$\dfrac{d^2y}{dx^2} = \dfrac{\dfrac{d}{dt}\left(\dfrac{dy}{dx}\right)}{dx/dt} = \dfrac{e^{-2t}(-1) + (1-t)(-2e^{-2t})}{e^t} = \dfrac{e^{-2t}(-1-2+2t)}{e^t} = e^{-3t}(2t-3)$. The curve is CU when $\dfrac{d^2y}{dx^2} > 0$, that is, when $t > \frac{3}{2}$.

14. $x = t^2 + 1,\; y = e^t - 1 \;\Rightarrow\; \dfrac{dy}{dx} = \dfrac{dy/dt}{dx/dt} = \dfrac{e^t}{2t} \;\Rightarrow\; \dfrac{d^2y}{dx^2} = \dfrac{\dfrac{d}{dt}\left(\dfrac{dy}{dx}\right)}{dx/dt} = \dfrac{\dfrac{2te^t - e^t \cdot 2}{(2t)^2}}{2t} = \dfrac{2e^t(t-1)}{(2t)^3} = \dfrac{e^t(t-1)}{4t^3}$.

The curve is CU when $\dfrac{d^2y}{dx^2} > 0$, that is, when $t < 0$ or $t > 1$.

15. $x = t - \ln t,\ y = t + \ln t$ [note that $t > 0$] $\Rightarrow \dfrac{dy}{dx} = \dfrac{dy/dt}{dx/dt} = \dfrac{1 + 1/t}{1 - 1/t} = \dfrac{t+1}{t-1} \Rightarrow$

$$\frac{d^2y}{dx^2} = \frac{\dfrac{d}{dt}\left(\dfrac{dy}{dx}\right)}{dx/dt} = \frac{\dfrac{(t-1)(1) - (t+1)(1)}{(t-1)^2}}{(t-1)/t} = \frac{-2t}{(t-1)^3}.$$ The curve is CU when $\dfrac{d^2y}{dx^2} > 0$, that is, when $0 < t < 1$.

16. $x = \cos t,\ y = \sin 2t,\ 0 < t < \pi \Rightarrow \dfrac{dy}{dx} = \dfrac{dy/dt}{dx/dt} = \dfrac{2\cos 2t}{-\sin t} \Rightarrow$

$$\frac{d^2y}{dx^2} = \frac{\dfrac{d}{dt}\left(\dfrac{dy}{dx}\right)}{dx/dt} = \frac{\dfrac{(-\sin t)(-4\sin 2t) - (2\cos 2t)(-\cos t)}{(-\sin t)^2}}{-\sin t} = \frac{(\sin t)(8\sin t\,\cos t) + [2(1 - 2\sin^2 t)](\cos t)}{(-\sin t)\sin^2 t}$$

$$= \frac{(\cos t)(8\sin^2 t + 2 - 4\sin^2 t)}{(-\sin t)\sin^2 t} = -\frac{\cos t}{\sin t}\cdot\frac{4\sin^2 t + 2}{\sin^2 t} \quad [(-\cot t)\cdot\text{positive expression}]$$

The curve is CU when $\dfrac{d^2y}{dx^2} > 0$, that is, when $-\cot t > 0 \Leftrightarrow \cot t < 0 \Leftrightarrow \frac{\pi}{2} < t < \pi$.

17. $x = t^3 - 3t,\ y = t^2 - 3.\ \dfrac{dy}{dt} = 2t$, so $\dfrac{dy}{dt} = 0 \Leftrightarrow t = 0 \Leftrightarrow$ $(x, y) = (0, -3)$. $\dfrac{dx}{dt} = 3t^2 - 3 = 3(t+1)(t-1)$, so $\dfrac{dx}{dt} = 0 \Leftrightarrow$ $t = -1$ or $1 \Leftrightarrow (x, y) = (2, -2)$ or $(-2, -2)$. The curve has a horizontal tangent at $(0, -3)$ and vertical tangents at $(2, -2)$ and $(-2, -2)$.

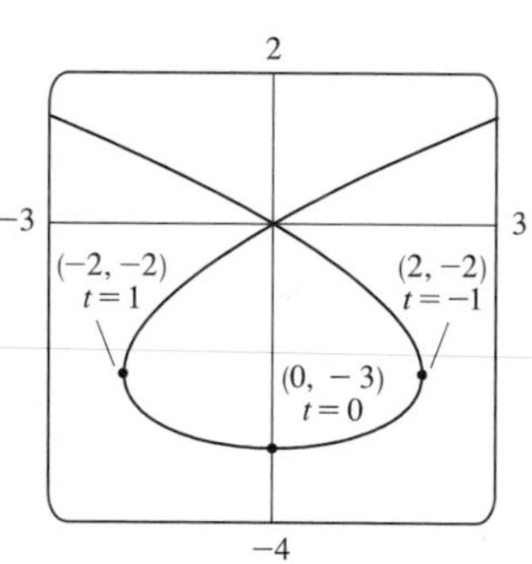

18. $x = t^3 - 3t,\ y = t^3 - 3t^2.\ \dfrac{dy}{dt} = 3t^2 - 6t = 3t(t-2)$, so $\dfrac{dy}{dt} = 0 \Leftrightarrow$ $t = 0$ or $2 \Leftrightarrow (x, y) = (0, 0)$ or $(2, -4)$. $\dfrac{dx}{dt} = 3t^2 - 3 = 3(t+1)(t-1)$, so $\dfrac{dx}{dt} = 0 \Leftrightarrow t = -1$ or $1 \Leftrightarrow (x, y) = (2, -4)$ or $(-2, -2)$. The curve has horizontal tangents at $(0, 0)$ and $(2, -4)$, and vertical tangents at $(2, -4)$ and $(-2, -2)$.

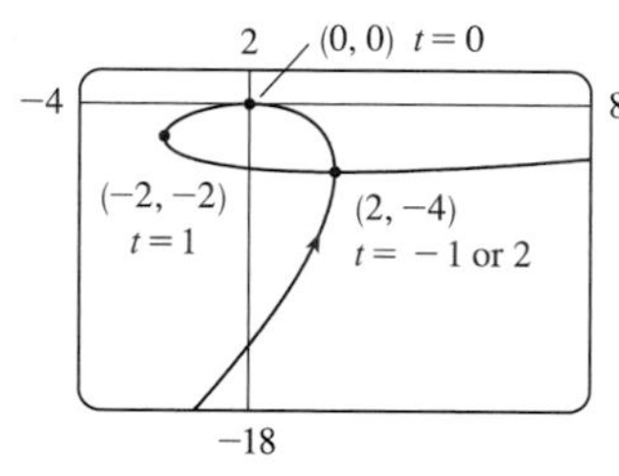

19. $x = \cos\theta,\ y = \cos 3\theta$. The whole curve is traced out for $0 \le \theta \le \pi$.

$\dfrac{dy}{d\theta} = -3\sin 3\theta$, so $\dfrac{dy}{d\theta} = 0 \Leftrightarrow \sin 3\theta = 0 \Leftrightarrow 3\theta = 0, \pi, 2\pi$, or $3\pi \Leftrightarrow$

$\theta = 0, \frac{\pi}{3}, \frac{2\pi}{3}$, or $\pi \Leftrightarrow (x, y) = (1, 1), \left(\frac{1}{2}, -1\right), \left(-\frac{1}{2}, 1\right)$, or $(-1, -1)$.

$\dfrac{dx}{d\theta} = -\sin\theta$, so $\dfrac{dx}{d\theta} = 0 \Leftrightarrow \sin\theta = 0 \Leftrightarrow \theta = 0$ or $\pi \Leftrightarrow$

$(x, y) = (1, 1)$ or $(-1, -1)$. Both $\dfrac{dy}{d\theta}$ and $\dfrac{dx}{d\theta}$ equal 0 when $\theta = 0$ and π.

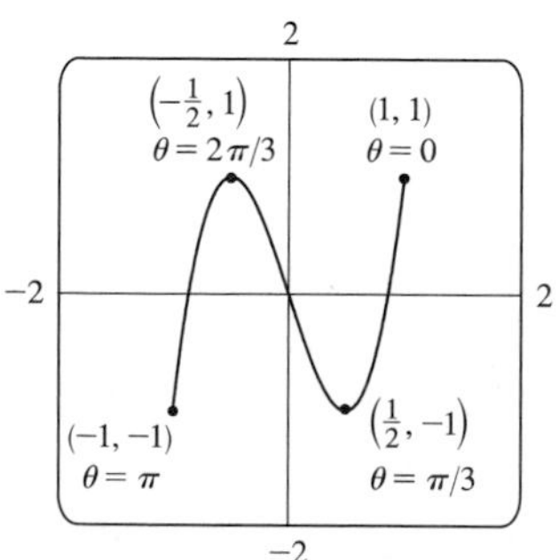

To find the slope when $\theta = 0$, we find $\displaystyle\lim_{\theta\to 0}\frac{dy}{dx} = \lim_{\theta\to 0}\frac{-3\sin 3\theta}{-\sin\theta} \overset{\text{H}}{=} \lim_{\theta\to 0}\frac{-9\cos 3\theta}{-\cos\theta} = 9$, which is the same slope when $\theta = \pi$.

Thus, the curve has horizontal tangents at $\left(\frac{1}{2}, -1\right)$ and $\left(-\frac{1}{2}, 1\right)$, and there are no vertical tangents.

20. $x = e^{\sin\theta}$, $y = e^{\cos\theta}$. The whole curve is traced out for $0 \leq \theta < 2\pi$.

$\dfrac{dy}{d\theta} = -\sin\theta\, e^{\cos\theta}$, so $\dfrac{dy}{d\theta} = \theta \;\Leftrightarrow\; \sin\theta = 0 \;\Leftrightarrow\; \theta = 0$ or $\pi \;\Leftrightarrow$

$(x, y) = (1, e)$ or $(1, 1/e)$. $\dfrac{dx}{d\theta} = \cos\theta\, e^{\sin\theta}$, so $\dfrac{dx}{d\theta} = 0 \;\Leftrightarrow\; \cos\theta = 0 \;\Leftrightarrow$

$\theta = \frac{\pi}{2}$ or $\frac{3\pi}{2} \;\Leftrightarrow\; (x, y) = (e, 1)$ or $(1/e, 1)$. The curve has horizontal tangents at $(1, e)$ and $(1, 1/e)$, and vertical tangents at $(e, 1)$ and $(1/e, 1)$.

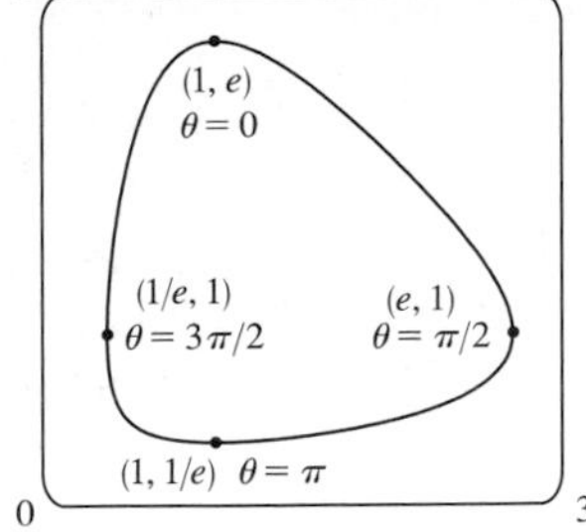

21. From the graph, it appears that the rightmost point on the curve $x = t - t^6$, $y = e^t$ is about $(0.6, 2)$. To find the exact coordinates, we find the value of t for which the graph has a vertical tangent, that is, $0 = dx/dt = 1 - 6t^5 \;\Leftrightarrow\; t = 1/\sqrt[5]{6}$. Hence, the rightmost point is

$$\left(1/\sqrt[5]{6} - 1/\left(6\sqrt[5]{6}\right), e^{1/\sqrt[5]{6}}\right) = \left(5 \cdot 6^{-6/5}, e^{6^{-1/5}}\right) \approx (0.58, 2.01).$$

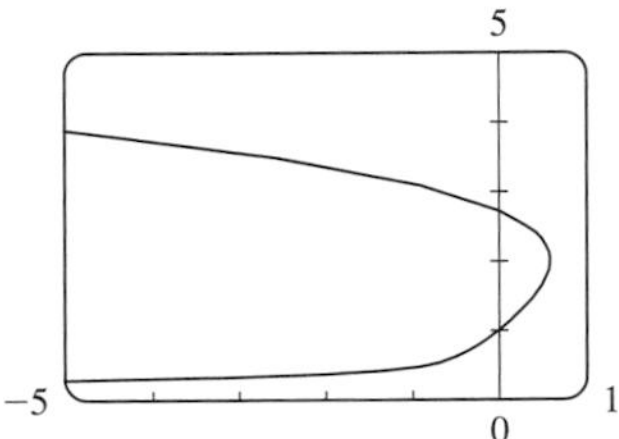

22. From the graph, it appears that the lowest point and the leftmost point on the curve $x = t^4 - 2t$, $y = t + t^4$ are $(1.5, -0.5)$ and $(-1.2, 1.2)$, respectively. To find the exact coordinates, we solve $dy/dt = 0$ (horizontal tangents) and $dx/dt = 0$ (vertical tangents).

$\dfrac{dy}{dt} = 0 \;\Leftrightarrow\; 1 + 4t^3 = 0 \;\Leftrightarrow\; t = -\dfrac{1}{\sqrt[3]{4}}$, so the lowest point is

$$\left(\frac{1}{\sqrt[3]{256}} + \frac{2}{\sqrt[3]{4}}, -\frac{1}{\sqrt[3]{4}} + \frac{1}{\sqrt[3]{256}}\right) = \left(\frac{9}{\sqrt[3]{256}}, -\frac{3}{\sqrt[3]{256}}\right) \approx (1.42, -0.47).$$

$\dfrac{dx}{dt} = 0 \;\Leftrightarrow\; 4t^3 - 2 = 0 \;\Leftrightarrow\; t = \dfrac{1}{\sqrt[3]{2}}$, so the leftmost point is

$$\left(\frac{1}{\sqrt[3]{16}} - \frac{2}{\sqrt[3]{2}}, \frac{1}{\sqrt[3]{2}} + \frac{1}{\sqrt[3]{16}}\right) = \left(-\frac{3}{\sqrt[3]{16}}, \frac{3}{\sqrt[3]{16}}\right) \approx (-1.19, 1.19).$$

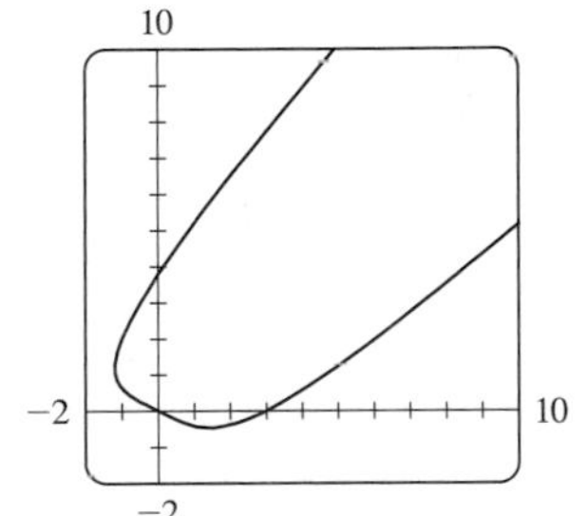

23. We graph the curve $x = t^4 - 2t^3 - 2t^2$, $y = t^3 - t$ in the viewing rectangle $[-2, 1.1]$ by $[-0.5, 0.5]$. This rectangle corresponds approximately to $t \in [-1, 0.8]$.

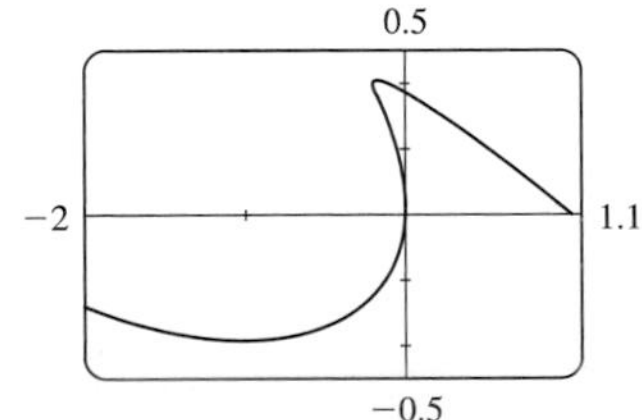

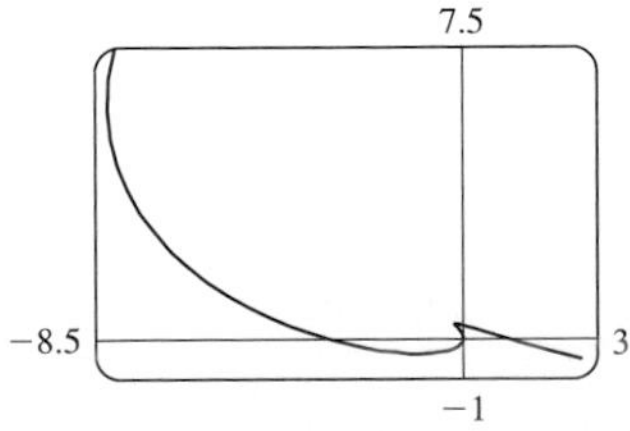

We estimate that the curve has horizontal tangents at about $(-1, -0.4)$ and $(-0.17, 0.39)$ and vertical tangents at about $(0, 0)$ and $(-0.19, 0.37)$. We calculate $\dfrac{dy}{dx} = \dfrac{dy/dt}{dx/dt} = \dfrac{3t^2 - 1}{4t^3 - 6t^2 - 4t}$. The horizontal tangents occur when $dy/dt = 3t^2 - 1 = 0 \;\Leftrightarrow\; t = \pm\frac{1}{\sqrt{3}}$, so both horizontal tangents are shown in our graph. The vertical tangents occur when

$dx/dt = 2t(2t^2 - 3t - 2) = 0 \Leftrightarrow 2t(2t+1)(t-2) = 0 \Leftrightarrow t = 0, -\frac{1}{2}$ or 2. It seems that we have missed one vertical tangent, and indeed if we plot the curve on the t-interval $[-1.2, 2.2]$ we see that there is another vertical tangent at $(-8, 6)$.

24. We graph the curve $x = t^4 + 4t^3 - 8t^2$, $y = 2t^2 - t$ in the viewing rectangle $[-3.7, 0.2]$ by $[-0.2, 1.4]$. It appears that there is a horizontal tangent at about $(-0.4, -0.1)$, and vertical tangents at about $(-3, 1)$ and $(0, 0)$.

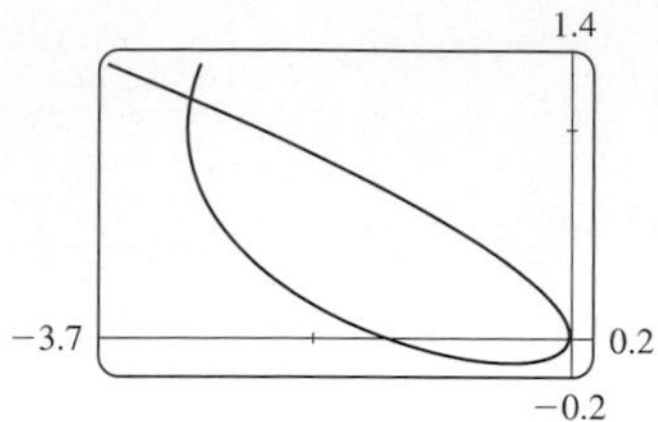

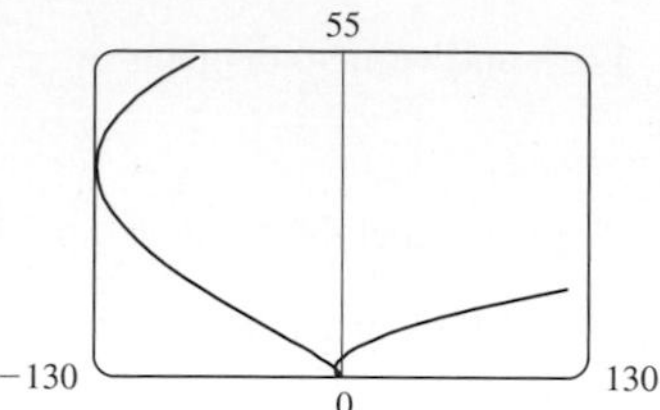

We calculate $\frac{dy}{dx} = \frac{dy/dt}{dx/dt} = \frac{4t - 1}{4t^3 + 12t^2 - 16t}$, so there is a horizontal tangent where $dy/dt = 4t - 1 = 0 \Leftrightarrow t = \frac{1}{4}$.

This point (the lowest point) is shown in the first graph. There are vertical tangents where $dx/dt = 4t^3 + 12t^2 - 16t = 0 \Leftrightarrow$ $4t(t^2 + 3t - 4) = 0 \Leftrightarrow 4t(t+4)(t-1) = 0$. We have missed one vertical tangent corresponding to $t = -4$, and if we plot the graph for $t \in [-5, 3]$, we see that the curve has another vertical tangent line at approximately $(-128, 36)$.

25. $x = \cos t$, $y = \sin t \cos t$. $dx/dt = -\sin t$, $dy/dt = -\sin^2 t + \cos^2 t = \cos 2t$. $(x, y) = (0, 0) \Leftrightarrow \cos t = 0 \Leftrightarrow t$ is an odd multiple of $\frac{\pi}{2}$. When $t = \frac{\pi}{2}$, $dx/dt = -1$ and $dy/dt = -1$, so $dy/dx = 1$. When $t = \frac{3\pi}{2}$, $dx/dt = 1$ and $dy/dt = -1$. So $dy/dx = -1$. Thus, $y = x$ and $y = -x$ are both tangent to the curve at $(0, 0)$.

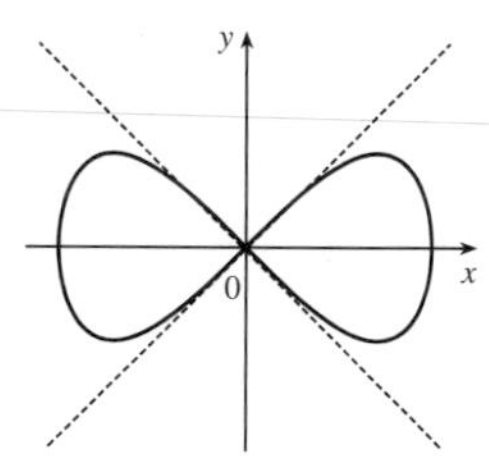

26. $x = -2\cos t$, $y = \sin t + \sin 2t$. From the graph, it appears that the curve crosses itself at the point $(1, 0)$. If this is true, then $x = 1 \Leftrightarrow$ $-2\cos t = 1 \Leftrightarrow \cos t = -\frac{1}{2} \Leftrightarrow t = \frac{2\pi}{3}$ or $\frac{4\pi}{3}$ for $0 \le t \le 2\pi$. Substituting either value of t into y gives $y = 0$, confirming that $(1, 0)$ is the point where the curve crosses itself. $\frac{dy}{dx} = \frac{dy/dt}{dx/dt} = \frac{\cos t + 2\cos 2t}{2\sin t}$.

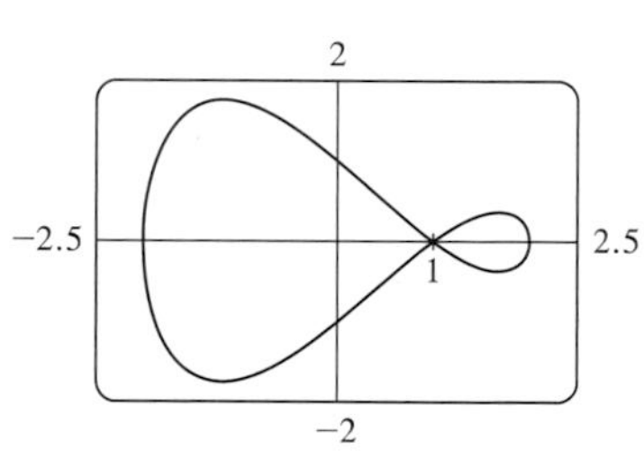

When $t = \frac{2\pi}{3}$, $\frac{dy}{dx} = \frac{-1/2 + 2(-1/2)}{2(\sqrt{3}/2)} = \frac{-3/2}{\sqrt{3}} = -\frac{\sqrt{3}}{2}$, so an equation of the tangent line is $y - 0 = -\frac{\sqrt{3}}{2}(x - 1)$, or $y = -\frac{\sqrt{3}}{2}x + \frac{\sqrt{3}}{2}$. Similarly, when $t = \frac{4\pi}{3}$, an equation of the tangent line is $y = \frac{\sqrt{3}}{2}x - \frac{\sqrt{3}}{2}$.

27. $x = r\theta - d\sin\theta$, $y = r - d\cos\theta$.

(a) $\frac{dx}{d\theta} = r - d\cos\theta$, $\frac{dy}{d\theta} = d\sin\theta$, so $\frac{dy}{dx} = \frac{d\sin\theta}{r - d\cos\theta}$.

(b) If $0 < d < r$, then $|d\cos\theta| \le d < r$, so $r - d\cos\theta \ge r - d > 0$. This shows that $dx/d\theta$ never vanishes, so the trochoid can have no vertical tangent if $d < r$.

28. $x = a\cos^3\theta,\ y = a\sin^3\theta.$

(a) $\dfrac{dx}{d\theta} = -3a\cos^2\theta\sin\theta,\ \dfrac{dy}{d\theta} = 3a\sin^2\theta\cos\theta$, so $\dfrac{dy}{dx} = -\dfrac{\sin\theta}{\cos\theta} = -\tan\theta.$

(b) The tangent is horizontal $\Leftrightarrow$ $dy/dx = 0$ $\Leftrightarrow$ $\tan\theta = 0$ $\Leftrightarrow$ $\theta = n\pi$ $\Leftrightarrow$ $(x, y) = (\pm a, 0)$.

The tangent is vertical $\Leftrightarrow$ $\cos\theta = 0$ $\Leftrightarrow$ θ is an odd multiple of $\frac{\pi}{2}$ $\Leftrightarrow$ $(x, y) = (0, \pm a)$.

(c) $dy/dx = \pm 1$ $\Leftrightarrow$ $\tan\theta = \pm 1$ $\Leftrightarrow$ θ is an odd multiple of $\frac{\pi}{4}$ $\Leftrightarrow$ $(x, y) = \left(\pm\frac{\sqrt{2}}{4}a, \pm\frac{\sqrt{2}}{4}a\right)$

[All sign choices are valid.]

29. $x = 3t^2 + 1,\ y = t^3 - 1$ $\Rightarrow$ $\dfrac{dy}{dx} = \dfrac{dy/dt}{dx/dt} = \dfrac{3t^2}{6t} = \dfrac{t}{2}$. The tangent line has slope $\dfrac{1}{2}$ when $\dfrac{t}{2} = \dfrac{1}{2}$ $\Leftrightarrow$ $t = 1$, so the point is $(4, 0)$.

30. $x = 3t^2 + 1, y = 2t^3 + 1, \dfrac{dx}{dt} = 6t, \dfrac{dy}{dt} = 6t^2$, so $\dfrac{dy}{dx} = \dfrac{6t^2}{6t} = t$ [even where $t = 0$].

So at the point corresponding to parameter value t, an equation of the tangent line is $y - (2t^3 + 1) = t[x - (3t^2 + 1)]$.

If this line is to pass through $(4, 3)$, we must have $3 - (2t^3 + 1) = t[4 - (3t^2 + 1)]$ $\Leftrightarrow$ $2t^3 - 2 = 3t^3 - 3t$ $\Leftrightarrow$ $t^3 - 3t + 2 = 0$ $\Leftrightarrow$ $(t - 1)^2(t + 2) = 0$ $\Leftrightarrow$ $t = 1$ or -2. Hence, the desired equations are $y - 3 = x - 4$, or $y = x - 1$, tangent to the curve at $(4, 3)$, and $y - (-15) = -2(x - 13)$, or $y = -2x + 11$, tangent to the curve at $(13, -15)$.

31. By symmetry of the ellipse about the x- and y-axes,

$$A = 4\int_0^a y\,dx = 4\int_{\pi/2}^0 b\sin\theta\,(-a\sin\theta)\,d\theta = 4ab\int_0^{\pi/2}\sin^2\theta\,d\theta = 4ab\int_0^{\pi/2}\tfrac{1}{2}(1 - \cos 2\theta)\,d\theta$$

$$= 2ab\left[\theta - \tfrac{1}{2}\sin 2\theta\right]_0^{\pi/2} = 2ab\left(\tfrac{\pi}{2}\right) = \pi ab$$

32. The curve $x = t^2 - 2t = t(t - 2)$, $y = \sqrt{t}$ intersects the y-axis when $x = 0$, that is, when $t = 0$ and $t = 2$. The corresponding values of y are 0 and $\sqrt{2}$. The shaded area is given by

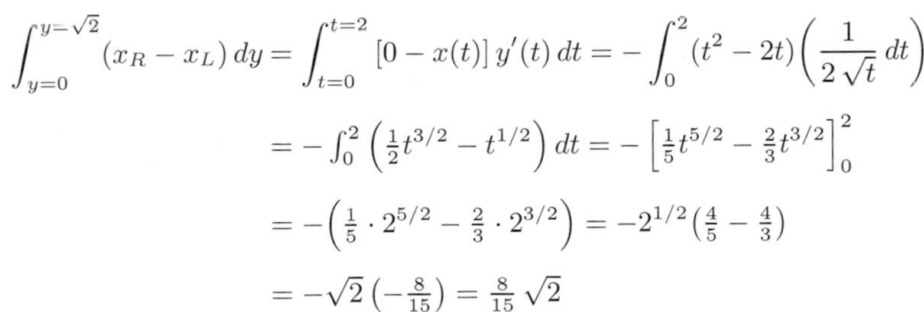

$$\int_{y=0}^{y=\sqrt{2}}(x_R - x_L)\,dy = \int_{t=0}^{t=2}[0 - x(t)]\,y'(t)\,dt = -\int_0^2(t^2 - 2t)\left(\frac{1}{2\sqrt{t}}\,dt\right)$$

$$= -\int_0^2\left(\tfrac{1}{2}t^{3/2} - t^{1/2}\right)dt = -\left[\tfrac{1}{5}t^{5/2} - \tfrac{2}{3}t^{3/2}\right]_0^2$$

$$= -\left(\tfrac{1}{5}\cdot 2^{5/2} - \tfrac{2}{3}\cdot 2^{3/2}\right) = -2^{1/2}\left(\tfrac{4}{5} - \tfrac{4}{3}\right)$$

$$= -\sqrt{2}\left(-\tfrac{8}{15}\right) = \tfrac{8}{15}\sqrt{2}$$

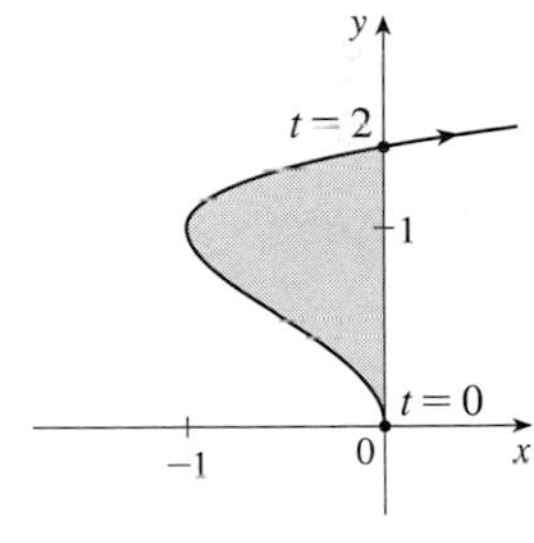

33. The curve $x = t^3 + 1, y = 2t - t^2 = t(2 - t)$ intersects the x-axis when $y = 0$, that is, when $t = 0$ and $t = 2$. The corresponding values of x are 1 and 9. The shaded area is given by

$$\int_{x=1}^{x=9}(y_T - y_B)\,dx = \int_{t=0}^{t=2}[y(t) - 0]\,x'(t)\,dt = \int_0^2(2t - t^2)(3t^2)\,dt$$

$$= 3\int_0^2(2t^3 - t^4)\,dt = 3\left[\tfrac{1}{2}t^4 - \tfrac{1}{5}t^5\right]_0^2 = 3\left(8 - \tfrac{32}{5}\right) = \tfrac{24}{5}$$

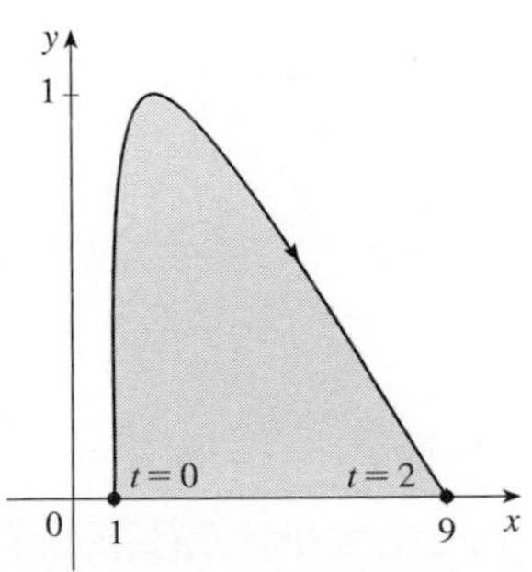

34. By symmetry, $A = 4\int_0^a y\,dx = 4\int_{\pi/2}^0 a\sin^3\theta(-3a\cos^2\theta\,\sin\theta)\,d\theta = 12a^2\int_0^{\pi/2}\sin^4\theta\,\cos^2\theta\,d\theta$. Now

$$\begin{aligned}\int \sin^4\theta\,\cos^2\theta\,d\theta &= \int \sin^2\theta\left(\tfrac{1}{4}\sin^2 2\theta\right)d\theta = \tfrac{1}{8}\int(1-\cos 2\theta)\sin^2 2\theta\,d\theta \\ &= \tfrac{1}{8}\int\left[\tfrac{1}{2}(1-\cos 4\theta) - \sin^2 2\theta\,\cos 2\theta\right]d\theta = \tfrac{1}{16}\theta - \tfrac{1}{64}\sin 4\theta - \tfrac{1}{48}\sin^3 2\theta + C\end{aligned}$$

so $\int_0^{\pi/2}\sin^4\theta\,\cos^2\theta\,d\theta = \left[\frac{1}{16}\theta - \frac{1}{64}\sin 4\theta - \frac{1}{48}\sin^3 2\theta\right]_0^{\pi/2} = \frac{\pi}{32}$. Thus, $A = 12a^2\left(\frac{\pi}{32}\right) = \frac{3}{8}\pi a^2$.

35. $x = r\theta - d\sin\theta,\ y = r - d\cos\theta$.

$$\begin{aligned}A &= \int_0^{2\pi r} y\,dx = \int_0^{2\pi}(r - d\cos\theta)(r - d\cos\theta)\,d\theta = \int_0^{2\pi}(r^2 - 2dr\cos\theta + d^2\cos^2\theta)\,d\theta \\ &= \left[r^2\theta - 2dr\sin\theta + \tfrac{1}{2}d^2\left(\theta + \tfrac{1}{2}\sin 2\theta\right)\right]_0^{2\pi} = 2\pi r^2 + \pi d^2\end{aligned}$$

36. (a) By symmetry, the area of $\mathcal{R}$ is twice the area inside $\mathcal{R}$ above the x-axis. The top half of the loop is described by $x = t^2$, $y = t^3 - 3t$, $-\sqrt{3} \le t \le 0$, so, using the Substitution Rule with $y = t^3 - 3t$ and $dx = 2t\,dt$, we find that

$$\begin{aligned}\text{area} &= 2\int_0^3 y\,dx = 2\int_0^{-\sqrt{3}}(t^3 - 3t)2t\,dt = 2\int_0^{-\sqrt{3}}(2t^4 - 6t^2)\,dt = 2\left[\tfrac{2}{5}t^5 - 2t^3\right]_0^{-\sqrt{3}} \\ &= 2\left[\tfrac{2}{5}(-3^{1/2})^5 - 2(-3^{1/2})^3\right] = 2\left[\tfrac{2}{5}(-9\sqrt{3}) - 2(-3\sqrt{3})\right] = \tfrac{24}{5}\sqrt{3}\end{aligned}$$

(b) Here we use the formula for disks and use the Substitution Rule as in part (a):

$$\begin{aligned}\text{volume} &= \pi\int_0^3 y^2\,dx = \pi\int_0^{-\sqrt{3}}(t^3 - 3t)^2 2t\,dt = 2\pi\int_0^{-\sqrt{3}}(t^6 - 6t^4 + 9t^2)t\,dt = 2\pi\left[\tfrac{1}{8}t^8 - t^6 + \tfrac{9}{4}t^4\right]_0^{-\sqrt{3}} \\ &= 2\pi\left[\tfrac{1}{8}(-3^{1/2})^8 - (-3^{1/2})^6 + \tfrac{9}{4}(-3^{1/2})^4\right] = 2\pi\left[\tfrac{81}{8} - 27 + \tfrac{81}{4}\right] = \tfrac{27}{4}\pi\end{aligned}$$

(c) By symmetry, the y-coordinate of the centroid is 0. To find the x-coordinate, we note that it is the same as the x-coordinate of the centroid of the top half of $\mathcal{R}$, the area of which is $\frac{1}{2}\cdot\frac{24}{5}\sqrt{3} = \frac{12}{5}\sqrt{3}$. So, using Formula 8.3.8 with $A = \frac{12}{5}\sqrt{3}$, we get

$$\begin{aligned}\overline{x} &= \tfrac{5}{12\sqrt{3}}\int_0^3 xy\,dx = \tfrac{5}{12\sqrt{3}}\int_0^{-\sqrt{3}} t^2(t^3 - 3t)2t\,dt = \tfrac{5}{6\sqrt{3}}\left[\tfrac{1}{7}t^7 - \tfrac{3}{5}t^5\right]_0^{-\sqrt{3}} \\ &= \tfrac{5}{6\sqrt{3}}\left[\tfrac{1}{7}(-3^{1/2})^7 - \tfrac{3}{5}(-3^{1/2})^5\right] = \tfrac{5}{6\sqrt{3}}\left[-\tfrac{27}{7}\sqrt{3} + \tfrac{27}{5}\sqrt{3}\right] = \tfrac{9}{7}\end{aligned}$$

So the coordinates of the centroid of $\mathcal{R}$ are $(x, y) = \left(\frac{9}{7}, 0\right)$.

37. $x = t + e^{-t},\ y = t - e^{-t},\ 0 \le t \le 2.\ \ dx/dt = 1 - e^{-t}$ and $dy/dt = 1 + e^{-t}$, so

$(dx/dt)^2 + (dy/dt)^2 = (1 - e^{-t})^2 + (1 + e^{-t})^2 = 1 - 2e^{-t} + e^{-2t} + 1 + 2e^{-t} + e^{-2t} = 2 + 2e^{-2t}$.

Thus, $L = \int_a^b \sqrt{(dx/dt)^2 + (dy/dt)^2}\,dt = \int_0^2 \sqrt{2 + 2e^{-2t}}\,dt \approx 3.1416$.

38. $x = t^2 - t,\ y = t^4,\ 1 \le t \le 4.\ \ dx/dt = 2t - 1$ and $dy/dt = 4t^3$, so

$(dx/dt)^2 + (dy/dt)^2 = (2t - 1)^2 + (4t^3)^2 = 4t^2 - 4t + 1 + 16t^6$.

Thus, $L = \int_a^b \sqrt{(dx/dt)^2 + (dy/dt)^2}\,dt = \int_1^4 \sqrt{16t^6 + 4t^2 - 4t + 1}\,dt \approx 255.3756$.

39. $x = t - 2\sin t,\ y = 1 - 2\cos t,\ 0 \le t \le 4\pi.\ \ dx/dt = 1 - 2\cos t$ and $dy/dt = 2\sin t$, so

$(dx/dt)^2 + (dy/dt)^2 = (1 - 2\cos t)^2 + (2\sin t)^2 = 1 - 4\cos t + 4\cos^2 t + 4\sin^2 t = 5 - 4\cos t$.

Thus, $L = \int_a^b \sqrt{(dx/dt)^2 + (dy/dt)^2}\,dt = \int_0^{4\pi}\sqrt{5 - 4\cos t}\,dt \approx 26.7298$.

40. $x = t + \sqrt{t},\ y = t - \sqrt{t},\ 0 \le t \le 1.\ \ \dfrac{dx}{dt} = 1 + \dfrac{1}{2\sqrt{t}}$ and $\dfrac{dy}{dt} = 1 - \dfrac{1}{2\sqrt{t}}$, so

$$\left(\frac{dx}{dt}\right)^2 + \left(\frac{dy}{dt}\right)^2 = \left(1 + \frac{1}{2\sqrt{t}}\right)^2 + \left(1 - \frac{1}{2\sqrt{t}}\right)^2 = 1 + \frac{1}{\sqrt{t}} + \frac{1}{4t} + 1 - \frac{1}{\sqrt{t}} + \frac{1}{4t} = 2 + \frac{1}{2t}.$$

Thus, $L = \displaystyle\int_a^b \sqrt{(dx/dt)^2 + (dy/dt)^2}\, dt = \int_0^1 \sqrt{2 + \frac{1}{2t}}\, dt = \lim_{t \to 0^+} \int_t^1 \sqrt{2 + \frac{1}{2t}}\, dt \approx 2.0915.$

41. $x = 1 + 3t^2,\ \ y = 4 + 2t^3,\ \ 0 \le t \le 1.\ \ dx/dt = 6t$ and $dy/dt = 6t^2$, so $(dx/dt)^2 + (dy/dt)^2 = 36t^2 + 36t^4$.

Thus, $\ L = \displaystyle\int_0^1 \sqrt{36t^2 + 36t^4}\, dt = \int_0^1 6t\sqrt{1 + t^2}\, dt = 6\int_1^2 \sqrt{u}\left(\tfrac{1}{2}\,du\right)\ \ [u = 1 + t^2,\ du = 2t\,dt]$

$= 3\left[\tfrac{2}{3}u^{3/2}\right]_1^2 = 2(2^{3/2} - 1) = 2(2\sqrt{2} - 1)$

42. $x = e^t - t,\ y = 4e^{t/2},\ 0 \le t \le 2.\ \ dx/dt = e^t - 1$ and $dy/dt = 2e^{t/2}$, so

$(dx/dt)^2 + (dy/dt)^2 = (e^t - 1)^2 + (2e^{t/2})^2 = e^{2t} - 2e^t + 1 + 4e^t = e^{2t} + 2e^t + 1 = (e^t + 1)^2$. Thus,

$$L = \int_0^2 \sqrt{(e^t + 1)^2}\, dt = \int_0^2 |e^t + 1|\, dt = \int_0^2 (e^t + 1)\, dt = \Big[e^t + t\Big]_0^2 = (e^2 + 2) - (1 + 0) = e^2 + 1.$$

43. $x = t\sin t,\ y = t\cos t,\ 0 \le t \le 1.\ \ \dfrac{dx}{dt} = t\cos t + \sin t$ and $\dfrac{dy}{dt} = -t\sin t + \cos t$, so

$$\left(\frac{dx}{dt}\right)^2 + \left(\frac{dy}{dt}\right)^2 = t^2\cos^2 t + 2t\sin t\,\cos t + \sin^2 t + t^2\sin^2 t - 2t\sin t\,\cos t + \cos^2 t$$
$$= t^2(\cos^2 t + \sin^2 t) + \sin^2 t + \cos^2 t = t^2 + 1.$$

Thus, $L = \int_0^1 \sqrt{t^2 + 1}\, dt \overset{21}{=} \left[\tfrac{1}{2}t\sqrt{t^2 + 1} + \tfrac{1}{2}\ln\!\left(t + \sqrt{t^2 + 1}\right)\right]_0^1 = \tfrac{1}{2}\sqrt{2} + \tfrac{1}{2}\ln\!\left(1 + \sqrt{2}\right).$

44. $x = 3\cos t - \cos 3t,\ y = 3\sin t - \sin 3t,\ 0 \le t \le \pi.\ \ \dfrac{dx}{dt} = -3\sin t + 3\sin 3t$ and $\dfrac{dy}{dt} = 3\cos t - 3\cos 3t$, so

$$\left(\frac{dx}{dt}\right)^2 + \left(\frac{dy}{dt}\right)^2 = 9\sin^2 t - 18\sin t\,\sin 3t + 9\sin^2(3t) + 9\cos^2 t - 18\cos t\,\cos 3t + 9\cos^2(3t)$$
$$= 9(\cos^2 t + \sin^2 t) - 18(\cos t\,\cos 3t + \sin t\,\sin 3t) + 9[\cos^2(3t) + \sin^2(3t)]$$
$$= 9(1) - 18\cos(t - 3t) + 9(1) = 18 - 18\cos(-2t) = 18(1 - \cos 2t)$$
$$= 18[1 - (1 - 2\sin^2 t)] = 36\sin^2 t.$$

Thus, $L = \int_0^\pi \sqrt{36\sin^2 t}\, dt = 6\int_0^\pi |\sin t|\, dt = 6\int_0^\pi \sin t\, dt = -6\big[\cos t\big]_0^\pi = -6\,(-1 - 1) = 12.$

45.

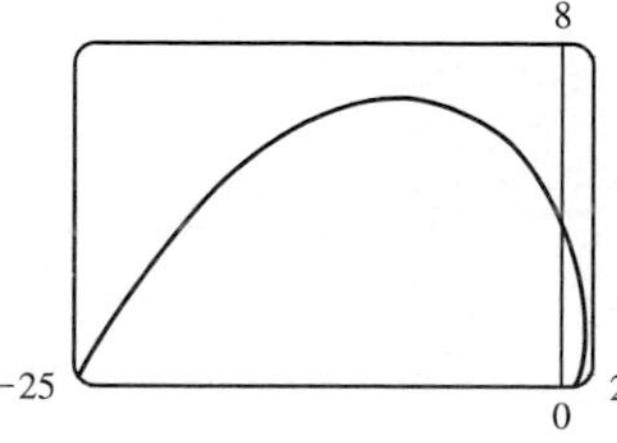

$x = e^t\cos t,\ y = e^t\sin t,\ 0 \le t \le \pi.$

$$\left(\tfrac{dx}{dt}\right)^2 + \left(\tfrac{dy}{dt}\right)^2 = [e^t(\cos t - \sin t)]^2 + [e^t(\sin t + \cos t)]^2$$
$$= (e^t)^2(\cos^2 t - 2\cos t\,\sin t + \sin^2 t)$$
$$+\ (e^t)^2(\sin^2 t + 2\sin t\,\cos t + \cos^2 t$$
$$= e^{2t}(2\cos^2 t + 2\sin^2 t) = 2e^{2t}$$

Thus, $L = \int_0^\pi \sqrt{2e^{2t}}\, dt = \int_0^\pi \sqrt{2}\,e^t\, dt = \sqrt{2}\,\big[e^t\big]_0^\pi = \sqrt{2}\,(e^\pi - 1).$

46. $x = \cos t + \ln(\tan \frac{1}{2}t)$, $y = \sin t$, $\pi/4 \le t \le 3\pi/4$.

$$\frac{dx}{dt} = -\sin t + \frac{\frac{1}{2}\sec^2(t/2)}{\tan(t/2)} = -\sin t + \frac{1}{2\sin(t/2)\cos(t/2)} = -\sin t + \frac{1}{\sin t} \text{ and } \frac{dy}{dt} = \cos t \text{, so}$$

$$\left(\frac{dx}{dt}\right)^2 + \left(\frac{dy}{dt}\right)^2 = \sin^2 t - 2 + \frac{1}{\sin^2 t} + \cos^2 t = 1 - 2 + \csc^2 t = \cot^2 t. \text{ Thus,}$$

$$L = \int_{\pi/4}^{3\pi/4} |\cot t|\, dt = 2\int_{\pi/4}^{\pi/2} \cot t\, dt$$

$$= 2\Big[\ln|\sin t|\Big]_{\pi/4}^{\pi/2} = 2\left(\ln 1 - \ln \frac{1}{\sqrt{2}}\right)$$

$$= 2(0 + \ln\sqrt{2}) = 2(\tfrac{1}{2}\ln 2) = \ln 2.$$

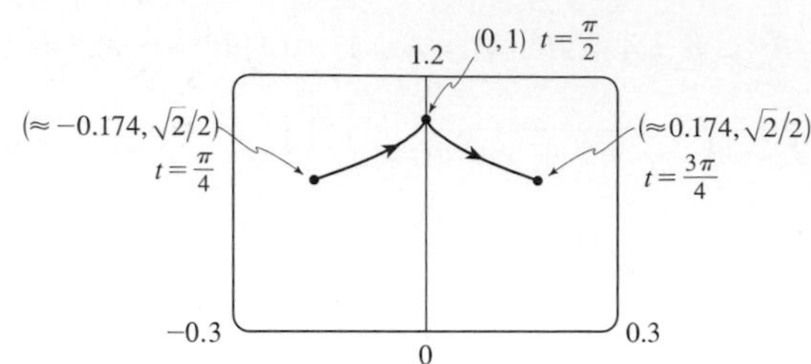

47. The figure shows the curve $x = \sin t + \sin 1.5t$, $y = \cos t$ for $0 \le t \le 4\pi$.

$dx/dt = \cos t + 1.5\cos 1.5t$ and $dy/dt = -\sin t$, so

$(dx/dt)^2 + (dy/dt)^2 = \cos^2 t + 3\cos t\,\cos 1.5t + 2.25\cos^2 1.5t + \sin^2 t$.

Thus, $L = \int_0^{4\pi} \sqrt{1 + 3\cos t\,\cos 1.5t + 2.25\cos^2 1.5t}\, dt \approx 16.7102$.

48. $x = 3t - t^3$, $y = 3t^2$. $dx/dt = 3 - 3t^2$ and $dy/dt = 6t$, so

$$\left(\frac{dx}{dt}\right)^2 + \left(\frac{dy}{dt}\right)^2 = (3 - 3t^2)^2 + (6t)^2 = (3 + 3t^2)^2$$

and the length of the loop is given by

$$L = \int_{-\sqrt{3}}^{\sqrt{3}} (3 + 3t^2)\, dt = 2\int_0^{\sqrt{3}} (3 + 3t^2)\, dt = 2\left[3t + t^3\right]_0^{\sqrt{3}}$$

$$= 2(3\sqrt{3} + 3\sqrt{3}) = 12\sqrt{3}$$

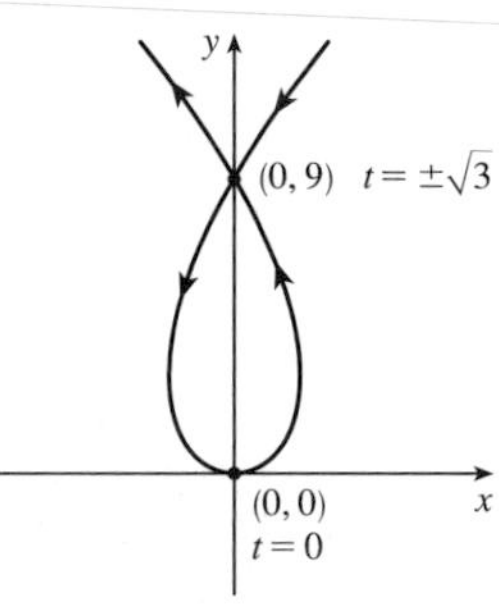

49. $x = t - e^t$, $y = t + e^t$, $-6 \le t \le 6$.

$\left(\frac{dx}{dt}\right)^2 + \left(\frac{dy}{dt}\right)^2 = (1 - e^t)^2 + (1 + e^t)^2 = (1 - 2e^t + e^{2t}) + (1 + 2e^t + e^{2t}) = 2 + 2e^{2t}$, so $L = \int_{-6}^{6} \sqrt{2 + 2e^{2t}}\, dt$.

Set $f(t) = \sqrt{2 + 2e^{2t}}$. Then by Simpson's Rule with $n = 6$ and $\Delta t = \frac{6-(-6)}{6} = 2$, we get

$L \approx \frac{2}{3}[f(-6) + 4f(-4) + 2f(-2) + 4f(0) + 2f(2) + 4f(4) + f(6)] \approx 612.3053$.

50. $x = 2a\cot\theta \;\Rightarrow\; dx/dt = -2a\csc^2\theta$ and $y = 2a\sin^2\theta \;\Rightarrow\; dy/dt = 4a\sin\theta\,\cos\theta = 2a\sin 2\theta$.

So $L = \int_{\pi/4}^{\pi/2} \sqrt{4a^2\csc^4\theta + 4a^2\sin^2 2\theta}\, d\theta = 2a\int_{\pi/4}^{\pi/2} \sqrt{\csc^4\theta + \sin^2 2\theta}\, d\theta$. Using Simpson's Rule with

$n = 4$, $\Delta\theta = \frac{\pi/2 - \pi/4}{4} = \frac{\pi}{16}$, and $f(\theta) = \sqrt{\csc^4\theta + \sin^2 2\theta}$, we get

$L \approx 2a \cdot S_4 = (2a)\frac{\pi}{16 \cdot 3}\left[f(\frac{\pi}{4}) + 4f(\frac{5\pi}{16}) + 2f(\frac{3\pi}{8}) + 4f(\frac{7\pi}{16}) + f(\frac{\pi}{2})\right] \approx 2.2605a$.

51. $x = \sin^2 t$, $y = \cos^2 t$, $0 \le t \le 3\pi$.

$(dx/dt)^2 + (dy/dt)^2 = (2\sin t\,\cos t)^2 + (-2\cos t\sin t)^2 = 8\sin^2 t\cos^2 t = 2\sin^2 2t \;\Rightarrow$

Distance $= \int_0^{3\pi} \sqrt{2}\,|\sin 2t|\,dt = 6\sqrt{2}\int_0^{\pi/2} \sin 2t\,dt$ [by symmetry] $= -3\sqrt{2}\Big[\cos 2t\Big]_0^{\pi/2} = -3\sqrt{2}\,(-1-1) = 6\sqrt{2}.$

The full curve is traversed as t goes from 0 to $\frac{\pi}{2}$, because the curve is the segment of $x + y = 1$ that lies in the first quadrant (since $x, y \geq 0$), and this segment is completely traversed as t goes from 0 to $\frac{\pi}{2}$. Thus, $L = \int_0^{\pi/2} \sin 2t\,dt = \sqrt{2}$, as above.

52. $x = \cos^2 t,\ y = \cos t,\ 0 \leq t \leq 4\pi.\quad \left(\frac{dx}{dt}\right)^2 + \left(\frac{dy}{dt}\right)^2 = (-2\cos t\sin t)^2 + (-\sin t)^2 = \sin^2 t\,(4\cos^2 t + 1)$

$$\begin{aligned}\text{Distance} &= \int_0^{4\pi} |\sin t|\,\sqrt{4\cos^2 t + 1}\,dt = 4\int_0^{\pi} \sin t\,\sqrt{4\cos^2 t + 1}\,dt \\ &= -4\int_1^{-1} \sqrt{4u^2 + 1}\,du \quad [u = \cos t,\ du = -\sin t\,dt] \quad = 4\int_{-1}^{1} \sqrt{4u^2 + 1}\,du \\ &= 8\int_0^1 \sqrt{4u^2 + 1}\,du = 8\int_0^{\tan^{-1} 2} \sec\theta \cdot \tfrac{1}{2}\sec^2\theta\,d\theta \qquad [2u = \tan\theta,\ 2\,du = \sec^2\theta\,d\theta] \\ &= 4\int_0^{\tan^{-1} 2} \sec^3\theta\,d\theta \overset{71}{=} \Big[2\sec\theta\tan\theta + 2\ln|\sec\theta + \tan\theta|\Big]_0^{\tan^{-1} 2} = 4\sqrt{5} + 2\ln\!\left(\sqrt{5} + 2\right)\end{aligned}$$

Thus, $L = \int_0^{\pi} |\sin t|\,\sqrt{4\cos^2 t + 1}\,dt = \sqrt{5} + \frac{1}{2}\ln\!\left(\sqrt{5} + 2\right)$.

53. $x = a\sin\theta,\ y = b\cos\theta,\ 0 \leq \theta \leq 2\pi.$

$$\begin{aligned}\left(\frac{dx}{dt}\right)^2 + \left(\frac{dy}{dt}\right)^2 &= (a\cos\theta)^2 + (-b\sin\theta)^2 = a^2\cos^2\theta + b^2\sin^2\theta = a^2(1 - \sin^2\theta) + b^2\sin^2\theta \\ &= a^2 - (a^2 - b^2)\sin^2\theta = a^2 - c^2\sin^2\theta = a^2\left(1 - \frac{c^2}{a^2}\sin^2\theta\right) = a^2(1 - e^2\sin^2\theta)\end{aligned}$$

So $L = 4\int_0^{\pi/2} \sqrt{a^2\left(1 - e^2\sin^2\theta\right)}\,d\theta$ [by symmetry] $= 4a\int_0^{\pi/2} \sqrt{1 - e^2\sin^2\theta}\,d\theta$.

54. $x = a\cos^3\theta,\ y = a\sin^3\theta.$

$$\begin{aligned}\left(\frac{dx}{dt}\right)^2 + \left(\frac{dy}{dt}\right)^2 &= (-3a\cos^2\theta\,\sin\theta)^2 + (3a\sin^2\theta\,\cos\theta)^2 \\ &= 9a^2\cos^4\theta\,\sin^2\theta + 9a^2\sin^4\theta\,\cos^2\theta \\ &= 9a^2\sin^2\theta\,\cos^2\theta(\cos^2\theta + \sin^2\theta) = 9a^2\sin^2\theta\,\cos^2\theta.\end{aligned}$$

The graph has four-fold symmetry and the curve in the first quadrant corresponds to $0 \leq \theta \leq \pi/2$. Thus,

$$\begin{aligned}L &= 4\int_0^{\pi/2} 3a\sin\theta\,\cos\theta\,d\theta \qquad [\text{since } a > 0 \text{ and } \sin\theta \text{ and } \cos\theta \text{ are positive for } 0 \leq \theta \leq \pi/2] \\ &= 12a\left[\tfrac{1}{2}\sin^2\theta\right]_0^{\pi/2} = 12a\left(\tfrac{1}{2} - 0\right) = 6a\end{aligned}$$

55. (a) $x = 11\cos t - 4\cos(11t/2),\ y = 11\sin t - 4\sin(11t/2).$

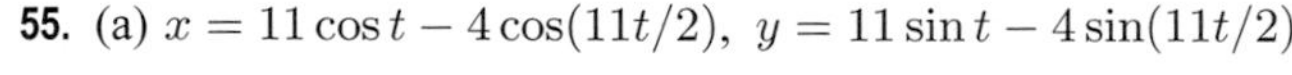

Notice that $0 \leq t \leq 2\pi$ does not give the complete curve because $x(0) \neq x(2\pi)$. In fact, we must take $t \in [0, 4\pi]$ in order to obtain the complete curve, since the first term in each of the parametric equations has period 2π and the second has period $\frac{2\pi}{11/2} = \frac{4\pi}{11}$, and the least common integer multiple of these two numbers is 4π.

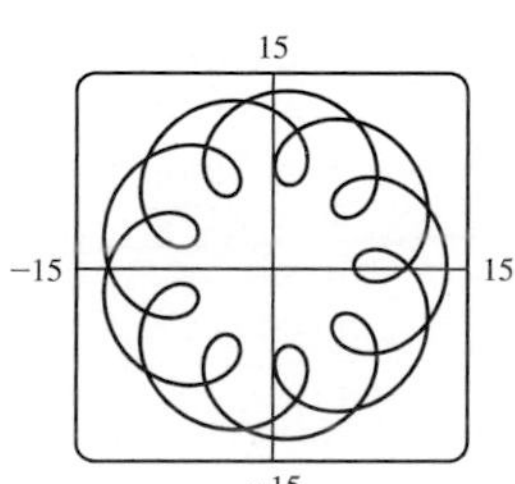

(b) We use the CAS to find the derivatives dx/dt and dy/dt, and then use Theorem 5 to find the arc length. Recent versions of Maple express the integral $\int_0^{4\pi} \sqrt{(dx/dt)^2 + (dy/dt)^2}\, dt$ as $88E\left(2\sqrt{2}\,i\right)$, where $E(x)$ is the elliptic integral $\displaystyle\int_0^1 \frac{\sqrt{1-x^2t^2}}{\sqrt{1-t^2}}\, dt$ and i is the imaginary number $\sqrt{-1}$.

Some earlier versions of Maple (as well as Mathematica) cannot do the integral exactly, so we use the command `evalf(Int(sqrt(diff(x,t)^2+diff(y,t)^2),t=0..4*Pi));` to estimate the length, and find that the arc length is approximately 294.03. Derive's `Para_arc_length` function in the utility file `Int_apps` simplifies the integral to $11\int_0^{4\pi} \sqrt{-4\cos t\, \cos\left(\frac{11t}{2}\right) - 4\sin t\, \sin\left(\frac{11t}{2}\right) + 5}\, dt$.

56. (a) It appears that as $t \to \infty$, $(x, y) \to \left(\frac{1}{2}, \frac{1}{2}\right)$, and as $t \to -\infty$, $(x, y) \to \left(-\frac{1}{2}, -\frac{1}{2}\right)$.

(b) By the Fundamental Theorem of Calculus, $dx/dt = \cos\left(\frac{\pi}{2}t^2\right)$ and $dy/dt = \sin\left(\frac{\pi}{2}t^2\right)$, so by Theorem 5, the length of the curve from the origin to the point with parameter value t is

$$L = \int_0^t \sqrt{\left(\frac{dx}{du}\right)^2 + \left(\frac{dy}{du}\right)^2}\, du = \int_0^t \sqrt{\cos^2\left(\frac{\pi}{2}u^2\right) + \sin^2\left(\frac{\pi}{2}u^2\right)}\, du$$

$$= \int_0^t 1\, du = t \qquad [\text{or } -t \text{ if } t < 0]$$

We have used u as the dummy variable so as not to confuse it with the upper limit of integration.

57. $x = t\sin t,\ y = t\cos t,\ 0 \le t \le \pi/2.\quad dx/dt = t\cos t + \sin t$ and $dy/dt = -t\sin t + \cos t$, so

$$\begin{aligned}(dx/dt)^2 + (dy/dt)^2 &= t^2\cos^2 t + 2t\sin t\, \cos t + \sin^2 t + t^2\sin^2 t - 2t\sin t\, \cos t + \cos^2 t \\ &= t^2(\cos^2 t + \sin^2 t) + \sin^2 t + \cos^2 t = t^2 + 1\end{aligned}$$

$S = \int 2\pi y\, ds = \int_0^{\pi/2} 2\pi t\cos t\sqrt{t^2+1}\, dt \approx 4.7394.$

58. $x = \sin t,\ y = \sin 2t,\ 0 \le t \le \pi/2.\quad dx/dt = \cos t$ and $dy/dt = 2\cos 2t$, so $(dx/dt)^2 + (dy/dt)^2 = \cos^2 t + 4\cos^2 2t$.

$S = \int 2\pi y\, ds = \int_0^{\pi/2} 2\pi \sin 2t\sqrt{\cos^2 t + 4\cos^2 2t}\, dt \approx 8.0285.$

59. $x = t + e^t,\ y = e^{-t},\ 0 \le t \le 1.$

$dx/dt = 1 + e^t$ and $dy/dt = -e^{-t}$, so $(dx/dt)^2 + (dy/dt)^2 = (1+e^t)^2 + (-e^{-t})^2 = 1 + 2e^t + e^{2t} + e^{-2t}$.

$S = \int 2\pi y\, ds = \int_0^1 2\pi e^{-t}\sqrt{1 + 2e^t + e^{2t} + e^{-2t}}\, dt \approx 10.6705.$

60. $x = t^2 - t^3,\ y = t + t^4,\ 0 \le t \le 1.$

$(dx/dt)^2 + (dy/dt)^2 = (2t - 3t^2)^2 + (1 + 4t^3)^2 = 4t^2 - 12t^3 + 9t^4 + 1 + 8t^3 + 16t^6$, so

$S = \int 2\pi y\, ds = \int_0^1 2\pi(t + t^4)\sqrt{16t^6 + 9t^4 - 4t^3 + 4t^2 + 1}\, dt \approx 12.7176.$

61. $x = t^3,\ y = t^2,\ 0 \le t \le 1.\quad \left(\frac{dx}{dt}\right)^2 + \left(\frac{dy}{dt}\right)^2 = \left(3t^2\right)^2 + (2t)^2 = 9t^4 + 4t^2.$

$$S = \int_0^1 2\pi y\sqrt{\left(\tfrac{dx}{dt}\right)^2 + \left(\tfrac{dy}{dt}\right)^2}\,dt = \int_0^1 2\pi t^2\sqrt{9t^4 + 4t^2}\,dt = 2\pi\int_0^1 t^2\sqrt{t^2(9t^2+4)}\,dt$$

$$= 2\pi\int_4^{13}\left(\frac{u-4}{9}\right)\sqrt{u}\left(\tfrac{1}{18}\,du\right)\quad\begin{bmatrix} u = 9t^2+4,\ t^2 = (u-4)/9, \\ du = 18t\,dt,\ \text{so}\ t\,dt = \tfrac{1}{18}\,du\end{bmatrix}\quad = \frac{2\pi}{9\cdot 18}\int_4^{13}\left(u^{3/2} - 4u^{1/2}\right)du$$

$$= \tfrac{\pi}{81}\left[\tfrac{2}{5}u^{5/2} - \tfrac{8}{3}u^{3/2}\right]_4^{13} = \tfrac{\pi}{81}\cdot\tfrac{2}{15}\left[3u^{5/2} - 20u^{3/2}\right]_4^{13}$$

$$= \tfrac{2\pi}{1215}\left[\left(3\cdot 13^2\sqrt{13} - 20\cdot 13\sqrt{13}\right) - (3\cdot 32 - 20\cdot 8)\right] = \tfrac{2\pi}{1215}\left(247\sqrt{13} + 64\right)$$

62. $x = 2t^2 + 1/t,\ y = 8\sqrt{t},\ 1 \le t \le 3.$

$$\left(\frac{dx}{dt}\right)^2 + \left(\frac{dy}{dt}\right)^2 = \left(4t - \frac{1}{t^2}\right)^2 + \left(\frac{4}{\sqrt{t}}\right)^2 = 16t^2 - \frac{8}{t} + \frac{1}{t^4} + \frac{16}{t} = 16t^2 + \frac{8}{t} + \frac{1}{t^4} = \left(4t + \frac{1}{t^2}\right)^2.$$

$$S = \int_1^3 2\pi y\sqrt{\left(\tfrac{dx}{dt}\right)^2 + \left(\tfrac{dy}{dt}\right)^2}\,dt = \int_1^3 2\pi\left(8\sqrt{t}\right)\sqrt{\left(4t + \frac{1}{t^2}\right)^2}\,dt = 16\pi\int_1^3 t^{1/2}\left(4t + t^{-2}\right)dt$$

$$= 16\pi\int_1^3\left(4t^{3/2} + t^{-3/2}\right)dt = 16\pi\left[\tfrac{8}{5}t^{5/2} - 2t^{-1/2}\right]_1^3 = 16\pi\left[\left(\tfrac{72}{5}\sqrt{3} - \tfrac{2}{3}\sqrt{3}\right) - \left(\tfrac{8}{5} - 2\right)\right]$$

$$= 16\pi\left(\tfrac{206}{15}\sqrt{3} + \tfrac{6}{15}\right) = \tfrac{32\pi}{15}\left(103\sqrt{3} + 3\right)$$

63. $x = a\cos^3\theta,\ y = a\sin^3\theta,\ 0 \le \theta \le \frac{\pi}{2}.\quad \left(\frac{dx}{d\theta}\right)^2 + \left(\frac{dy}{d\theta}\right)^2 = \left(-3a\cos^2\theta\,\sin\theta\right)^2 + \left(3a\sin^2\theta\,\cos\theta\right)^2 = 9a^2\sin^2\theta\,\cos^2\theta.$

$$S = \int_0^{\pi/2} 2\pi\cdot a\sin^3\theta\cdot 3a\sin\theta\,\cos\theta\,d\theta = 6\pi a^2\int_0^{\pi/2}\sin^4\theta\,\cos\theta\,d\theta = \tfrac{6}{5}\pi a^2\left[\sin^5\theta\right]_0^{\pi/2} = \tfrac{6}{5}\pi a^2$$

64. $x = 2\cos\theta - \cos 2\theta,\ y = 2\sin\theta - \sin 2\theta\ \Rightarrow$

$$\left(\tfrac{dx}{d\theta}\right)^2 + \left(\tfrac{dy}{d\theta}\right)^2 = (-2\sin\theta + 2\sin 2\theta)^2 + (2\cos\theta - 2\cos 2\theta)^2$$

$$= 4\left[\left(\sin^2\theta - 2\sin\theta\,\sin 2\theta + \sin^2 2\theta\right) + \left(\cos^2\theta - 2\cos\theta\,\cos 2\theta + \cos^2 2\theta\right)\right]$$

$$= 4[1 + 1 - 2(\cos 2\theta\cos\theta + \sin 2\theta\sin\theta)] = 8[1 - \cos(2\theta - \theta)] = 8(1 - \cos\theta)$$

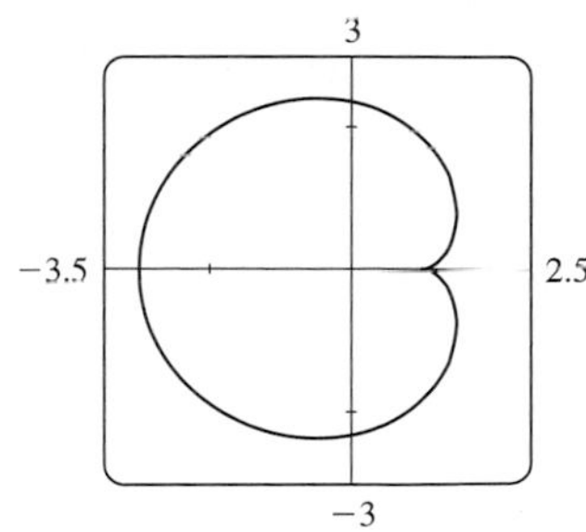

We plot the graph with parameter interval $[0, 2\pi]$, and see that we should only integrate between 0 and π. (If the interval $[0, 2\pi]$ were taken, the surface of revolution would be generated twice.) Also note that $y = 2\sin\theta - \sin 2\theta = 2\sin\theta(1 - \cos\theta)$. So

$$S = \int_0^\pi 2\pi\cdot 2\sin\theta(1 - \cos\theta)\,2\sqrt{2}\sqrt{1 - \cos\theta}\,d\theta$$

$$= 8\sqrt{2}\pi\int_0^\pi (1 - \cos\theta)^{3/2}\sin\theta\,d\theta = 8\sqrt{2}\pi\int_0^2\sqrt{u^3}\,du\quad\begin{bmatrix} u = 1 - \cos\theta, \\ du = \sin\theta\,d\theta\end{bmatrix}$$

$$= 8\sqrt{2}\pi\left[\left(\tfrac{2}{5}\right)u^{5/2}\right]_0^2 = \tfrac{16}{5}\sqrt{2}\pi\left(2^{5/2}\right) = \tfrac{128}{5}\pi$$

65. $x = 3t^2,\ y = 2t^3,\ 0 \le t \le 5\ \Rightarrow\ \left(\frac{dx}{dt}\right)^2 + \left(\frac{dy}{dt}\right)^2 = (6t)^2 + (6t^2)^2 = 36t^2(1 + t^2)\ \Rightarrow$

$$S = \int_0^5 2\pi x\sqrt{(dx/dt)^2 + (dy/dt)^2}\,dt = \int_0^5 2\pi(3t^2)6t\sqrt{1 + t^2}\,dt = 18\pi\int_0^5 t^2\sqrt{1 + t^2}\,2t\,dt$$

$$= 18\pi\int_1^{26}(u - 1)\sqrt{u}\,du\quad\begin{bmatrix} u = 1 + t^2, \\ du = 2t\,dt\end{bmatrix}\quad = 18\pi\int_1^{26}\left(u^{3/2} - u^{1/2}\right)du = 18\pi\left[\tfrac{2}{5}u^{5/2} - \tfrac{2}{3}u^{3/2}\right]_1^{26}$$

$$= 18\pi\left[\left(\tfrac{2}{5}\cdot 676\sqrt{26} - \tfrac{2}{3}\cdot 26\sqrt{26}\right) - \left(\tfrac{2}{5} - \tfrac{2}{3}\right)\right] = \tfrac{24}{5}\pi\left(949\sqrt{26} + 1\right)$$

66. $x = e^t - t,\ y = 4e^{t/2},\ 0 \le t \le 1$. $\left(\frac{dx}{dt}\right)^2 + \left(\frac{dy}{dt}\right)^2 = (e^t - 1)^2 + (2e^{t/2})^2 = e^{2t} + 2e^t + 1 = (e^t + 1)^2$.

$$S = \int_0^1 2\pi(e^t - t)\sqrt{(e^t - 1)^2 + (2e^{t/2})^2}\,dt = \int_0^1 2\pi(e^t - t)(e^t + 1)d$$
$$= 2\pi\left[\tfrac{1}{2}e^{2t} + e^t - (t-1)e^t - \tfrac{1}{2}t^2\right]_0^1 = \pi(e^2 + 2e - 6)$$

67. If f' is continuous and $f'(t) \ne 0$ for $a \le t \le b$, then either $f'(t) > 0$ for all t in $[a, b]$ or $f'(t) < 0$ for all t in $[a, b]$. Thus, f is monotonic (in fact, strictly increasing or strictly decreasing) on $[a, b]$. It follows that f has an inverse. Set $F = g \circ f^{-1}$, that is, define F by $F(x) = g(f^{-1}(x))$. Then $x = f(t) \Rightarrow f^{-1}(x) = t$, so $y = g(t) = g(f^{-1}(x)) = F(x)$.

68. By Formula 8.2.5 with $y = F(x)$, $S = \int_a^b 2\pi F(x)\sqrt{1 + [F'(x)]^2}\,dx$. But by Formula 10.2.1,

$1 + [F'(x)]^2 = 1 + \left(\dfrac{dy}{dx}\right)^2 = 1 + \left(\dfrac{dy/dt}{dx/dt}\right)^2 = \dfrac{(dx/dt)^2 + (dy/dt)^2}{(dx/dt)^2}$. Using the Substitution Rule with $x = x(t)$, where $a = x(\alpha)$ and $b = x(\beta)$, we have $\left[\text{since } dx = \dfrac{dx}{dt}\,dt\right]$

$$S = \int_\alpha^\beta 2\pi\,F(x(t))\sqrt{\frac{(dx/dt)^2 + (dy/dt)^2}{(dx/dt)^2}}\,\frac{dx}{dt}\,dt = \int_\alpha^\beta 2\pi y\sqrt{\left(\frac{dx}{dt}\right)^2 + \left(\frac{dy}{dt}\right)^2}\,dt, \text{ which is Formula 10.2.6.}$$

69. (a) $\phi = \tan^{-1}\left(\dfrac{dy}{dx}\right) \Rightarrow \dfrac{d\phi}{dt} = \dfrac{d}{dt}\tan^{-1}\left(\dfrac{dy}{dx}\right) = \dfrac{1}{1 + (dy/dx)^2}\left[\dfrac{d}{dt}\left(\dfrac{dy}{dx}\right)\right]$. But $\dfrac{dy}{dx} = \dfrac{dy/dt}{dx/dt} = \dfrac{\dot{y}}{\dot{x}} \Rightarrow$

$\dfrac{d}{dt}\left(\dfrac{dy}{dx}\right) = \dfrac{d}{dt}\left(\dfrac{\dot{y}}{\dot{x}}\right) = \dfrac{\ddot{y}\dot{x} - \ddot{x}\dot{y}}{\dot{x}^2} \Rightarrow \dfrac{d\phi}{dt} = \dfrac{1}{1 + (\dot{y}/\dot{x})^2}\left(\dfrac{\ddot{y}\dot{x} - \ddot{x}\dot{y}}{\dot{x}^2}\right) = \dfrac{\dot{x}\ddot{y} - \ddot{x}\dot{y}}{\dot{x}^2 + \dot{y}^2}$. Using the Chain Rule, and the fact that $s = \displaystyle\int_0^t \sqrt{\left(\tfrac{dx}{dt}\right)^2 + \left(\tfrac{dy}{dt}\right)^2}\,dt \Rightarrow \tfrac{ds}{dt} = \sqrt{\left(\tfrac{dx}{dt}\right)^2 + \left(\tfrac{dy}{dt}\right)^2} = \left(\dot{x}^2 + \dot{y}^2\right)^{1/2}$, we have that

$\dfrac{d\phi}{ds} = \dfrac{d\phi/dt}{ds/dt} = \left(\dfrac{\dot{x}\ddot{y} - \ddot{x}\dot{y}}{\dot{x}^2 + \dot{y}^2}\right)\dfrac{1}{(\dot{x}^2 + \dot{y}^2)^{1/2}} = \dfrac{\dot{x}\ddot{y} - \ddot{x}\dot{y}}{(\dot{x}^2 + \dot{y}^2)^{3/2}}$. So $\kappa = \left|\dfrac{d\phi}{ds}\right| = \left|\dfrac{\dot{x}\ddot{y} - \ddot{x}\dot{y}}{(\dot{x}^2 + \dot{y}^2)^{3/2}}\right| = \dfrac{|\dot{x}\ddot{y} - \ddot{x}\dot{y}|}{(\dot{x}^2 + \dot{y}^2)^{3/2}}$.

(b) $x = x$ and $y = f(x) \Rightarrow \dot{x} = 1,\ \ddot{x} = 0$ and $\dot{y} = \dfrac{dy}{dx},\ \ddot{y} = \dfrac{d^2y}{dx^2}$.

So $\kappa = \dfrac{\left|1 \cdot (d^2y/dx^2) - 0 \cdot (dy/dx)\right|}{[1 + (dy/dx)^2]^{3/2}} = \dfrac{\left|d^2y/dx^2\right|}{[1 + (dy/dx)^2]^{3/2}}$.

70. (a) $y = x^2 \Rightarrow \dfrac{dy}{dx} = 2x \Rightarrow \dfrac{d^2y}{dx^2} = 2$. So $\kappa = \dfrac{\left|d^2y/dx^2\right|}{[1 + (dy/dx)^2]^{3/2}} = \dfrac{2}{(1 + 4x^2)^{3/2}}$, and at $(1, 1)$,

$\kappa = \dfrac{2}{5^{3/2}} = \dfrac{2}{5\sqrt{5}}$.

(b) $\kappa' = \dfrac{d\kappa}{dx} = -3(1 + 4x^2)^{-5/2}(8x) = 0 \Leftrightarrow x = 0 \Rightarrow y = 0$. This is a maximum since $\kappa' > 0$ for $x < 0$ and $\kappa' < 0$ for $x > 0$. So the parabola $y = x^2$ has maximum curvature at the origin.

71. $x = \theta - \sin\theta \Rightarrow \dot{x} = 1 - \cos\theta \Rightarrow \ddot{x} = \sin\theta$, and $y = 1 - \cos\theta \Rightarrow \dot{y} = \sin\theta \Rightarrow \ddot{y} = \cos\theta$. Therefore,

$\kappa = \dfrac{\left|\cos\theta - \cos^2\theta - \sin^2\theta\right|}{[(1 - \cos\theta)^2 + \sin^2\theta]^{3/2}} = \dfrac{\left|\cos\theta - (\cos^2\theta + \sin^2\theta)\right|}{(1 - 2\cos\theta + \cos^2\theta + \sin^2\theta)^{3/2}} = \dfrac{|\cos\theta - 1|}{(2 - 2\cos\theta)^{3/2}}$. The top of the arch is characterized by a horizontal tangent, and from Example 2(b) in Section 10.2, the tangent is horizontal when $\theta = (2n-1)\pi$, so take $n = 1$ and substitute $\theta = \pi$ into the expression for κ: $\kappa = \dfrac{|\cos\pi - 1|}{(2 - 2\cos\pi)^{3/2}} = \dfrac{|-1-1|}{[2-2(-1)]^{3/2}} = \dfrac{1}{4}$.

72. (a) Every straight line has parametrizations of the form $x = a + vt$, $y = b + wt$, where a, b are arbitrary and $v, w \neq 0$. For example, a straight line passing through distinct points (a, b) and (c, d) can be described as the parametrized curve $x = a + (c-a)t$, $y = b + (d-b)t$. Starting with $x = a + vt$, $y = b + wt$, we compute $\dot{x} = v$, $\dot{y} = w$, $\ddot{x} = \ddot{y} = 0$, and $\kappa = \dfrac{|v \cdot 0 - w \cdot 0|}{(v^2 + w^2)^{3/2}} = 0$.

(b) Parametric equations for a circle of radius r are $x = r\cos\theta$ and $y = r\sin\theta$. We can take the center to be the origin. So $\dot{x} = -r\sin\theta \;\Rightarrow\; \ddot{x} = -r\cos\theta$ and $\dot{y} = r\cos\theta \;\Rightarrow\; \ddot{y} = -r\sin\theta$. Therefore,

$\kappa = \dfrac{|r^2\sin^2\theta + r^2\cos^2\theta|}{(r^2\sin^2\theta + r^2\cos^2\theta)^{3/2}} = \dfrac{r^2}{r^3} = \dfrac{1}{r}$. And so for any θ (and thus any point), $\kappa = \dfrac{1}{r}$.

73. The coordinates of T are $(r\cos\theta, r\sin\theta)$. Since TP was unwound from arc TA, TP has length $r\theta$. Also $\angle PTQ = \angle PTR - \angle QTR = \frac{1}{2}\pi - \theta$, so P has coordinates $x = r\cos\theta + r\theta\cos\left(\frac{1}{2}\pi - \theta\right) = r(\cos\theta + \theta\sin\theta)$, $y = r\sin\theta - r\theta\sin\left(\frac{1}{2}\pi - \theta\right) = r(\sin\theta - \theta\cos\theta)$.

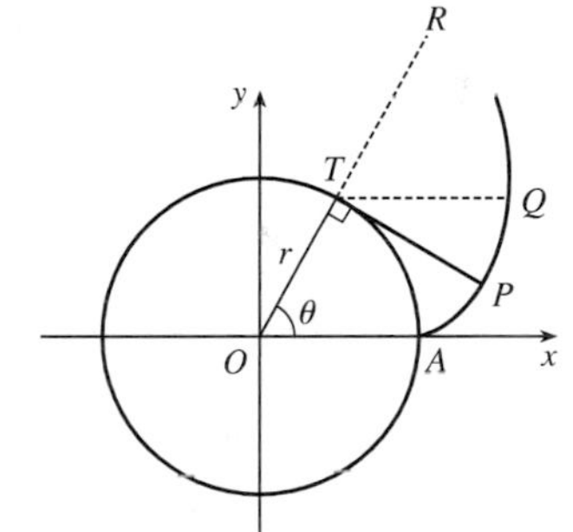

74. If the cow walks with the rope taut, it traces out the portion of the involute in Exercise 73 corresponding to the range $0 \leq \theta \leq \pi$, arriving at the point $(-r, \pi r)$ when $\theta = \pi$. With the rope now fully extended, the cow walks in a semicircle of radius πr, arriving at $(-r, -\pi r)$. Finally, the cow traces out another portion of the involute, namely the reflection about the x-axis of the initial involute path. (This corresponds to the range $-\pi \leq \theta \leq 0$.) Referring to the figure, we see that the total grazing area is $2(A_1 + A_3)$. A_3 is one-quarter of the area of a circle of radius πr, so $A_3 = \frac{1}{4}\pi(\pi r)^2 = \frac{1}{4}\pi^3 r^2$. We will compute $A_1 + A_2$ and then subtract $A_2 = \frac{1}{2}\pi r^2$ to obtain A_1.

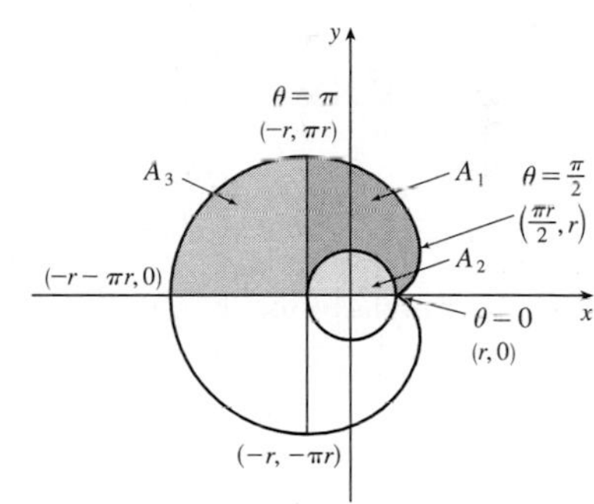

To find $A_1 + A_2$, first note that the rightmost point of the involute is $\left(\frac{\pi}{2}r, r\right)$. [To see this, note that $dx/d\theta = 0$ when $\theta = 0$ or $\frac{\pi}{2}$. $\theta = 0$ corresponds to the cusp at $(r, 0)$ and $\theta = \frac{\pi}{2}$ corresponds to $\left(\frac{\pi}{2}r, r\right)$.] The leftmost point of the involute is $(-r, \pi r)$. Thus, $A_1 + A_2 = \int_{\theta=\pi}^{\pi/2} y\,dx - \int_{\theta=0}^{\pi/2} y\,dx = \int_{\theta=\pi}^{0} y\,dx$.

Now $y\,dx = r(\sin\theta - \theta\cos\theta)\,r\theta\cos\theta\,d\theta = r^2(\theta\sin\theta\,\cos\theta - \theta^2\cos^2\theta)d\theta$. Integrate:

$(1/r^2)\int y\,dx = -\theta\cos^2\theta - \frac{1}{2}(\theta^2 - 1)\sin\theta\,\cos\theta - \frac{1}{6}\theta^3 + \frac{1}{2}\theta + C$. This enables us to compute

$$A_1 + A_2 = r^2\left[-\theta\cos^2\theta - \tfrac{1}{2}(\theta^2 - 1)\sin\theta\cos\theta - \tfrac{1}{6}\theta^3 + \tfrac{1}{2}\theta\right]_\pi^0 = r^2\left[0 - \left(-\pi - \frac{\pi^3}{6} + \frac{\pi}{2}\right)\right] = r^2\left(\frac{\pi}{2} + \frac{\pi^3}{6}\right)$$

Therefore, $A_1 = (A_1 + A_2) - A_2 = \frac{1}{6}\pi^3 r^2$, so the grazing area is $2(A_1 + A_3) = 2\left(\frac{1}{6}\pi^3 r^2 + \frac{1}{4}\pi^3 r^2\right) = \frac{5}{6}\pi^3 r^2$.

LABORATORY PROJECT Bézier Curves

1. The parametric equations for a cubic Bézier curve are

$$x = x_0(1-t)^3 + 3x_1 t(1-t)^2 + 3x_2 t^2(1-t) + x_3 t^3$$
$$y = y_0(1-t)^3 + 3y_1 t(1-t)^2 + 3y_2 t^2(1-t) + y_3 t^3$$

where $0 \le t \le 1$. We are given the points $P_0(x_0, y_0) = (4, 1)$, $P_1(x_1, y_1) = (28, 48)$, $P_2(x_2, y_2) = (50, 42)$, and $P_3(x_3, y_3) = (40, 5)$. The curve is then given by

$$x(t) = 4(1-t)^3 + 3\cdot 28t(1-t)^2 + 3\cdot 50t^2(1-t) + 40t^3$$
$$y(t) = 1(1-t)^3 + 3\cdot 48t(1-t)^2 + 3\cdot 42t^2(1-t) + 5t^3$$

where $0 \le t \le 1$. The line segments are of the form $x = x_0 + (x_1 - x_0)t$, $y = y_0 + (y_1 - y_0)t$:

P_0P_1	$x = 4 + 24t,$	$y = 1 + 47t$
P_1P_2	$x = 28 + 22t,$	$y = 48 - 6t$
P_2P_3	$x = 50 - 10t,$	$y = 42 - 37t$

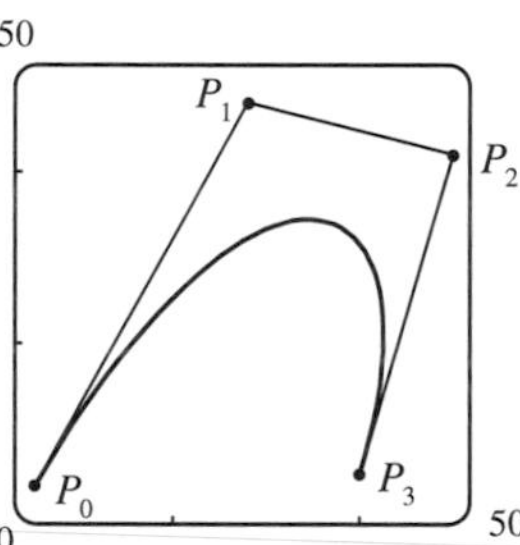

2. It suffices to show that the slope of the tangent at P_0 is the same as that of line segment P_0P_1, namely $\dfrac{y_1 - y_0}{x_1 - x_0}$.

We calculate the slope of the tangent to the Bézier curve:

$$\frac{dy/dt}{dx/dt} = \frac{-3y_0(1-t)^2 + 3y_1\left[-2t(1-t) + (1-t)^2\right] + 3y_2\left[-t^2 + (2t)(1-t)\right] + 3y_3t^2}{-3x_0^2(1-t) + 3x_1\left[-2t(1-t) + (1-t)^2\right] + 3x_2\left[-t^2 + (2t)(1-t)\right] + 3x_3t^2}$$

At point P_0, $t = 0$, so the slope of the tangent is $\dfrac{-3y_0 + 3y_1}{-3x_0 + 3x_1} = \dfrac{y_1 - y_0}{x_1 - x_0}$. So the tangent to the curve at P_0 passes through P_1. Similarly, the slope of the tangent at point P_3 [where $t = 1$] is $\dfrac{-3y_2 + 3y_3}{-3x_2 + 3x_3} = \dfrac{y_3 - y_2}{x_3 - x_2}$, which is also the slope of line P_2P_3.

3. It seems that if P_1 were to the right of P_2, a loop would appear.

We try setting $P_1 = (110, 30)$, and the resulting curve does indeed have a loop.

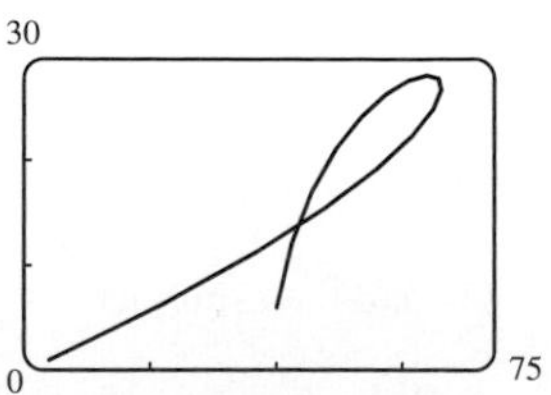

4. Based on the behavior of the Bézier curve in Problems 1–3, we suspect that the four control points should be in an exaggerated C shape. We try $P_0(10, 12)$, $P_1(4, 15)$, $P_2(4, 5)$, and $P_3(10, 8)$, and these produce a decent C. If you are using a CAS, it may be necessary to instruct it to make the x- and y-scales the same so as not to distort the figure (this is called a "constrained projection" in Maple.)

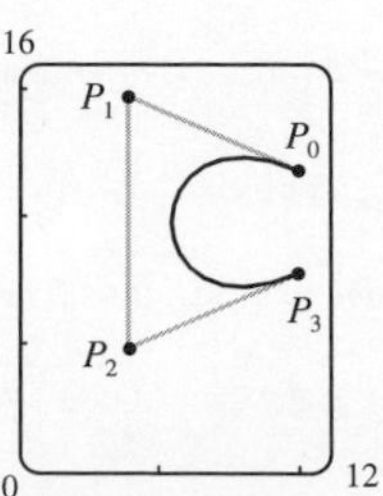

5. We use the same P_0 and P_1 as in Problem 4, and use part of our C as the top of an S. To prevent the center line from slanting up too much, we move P_2 up to $(4, 6)$ and P_3 down and to the left, to $(8, 7)$. In order to have a smooth joint between the top and bottom halves of the S (and a symmetric S), we determine points P_4, P_5, and P_6 by rotating points P_2, P_1, and P_0 about the center of the letter (point P_3). The points are therefore $P_4(12, 8)$, $P_5(12, -1)$, and $P_6(6, 2)$.

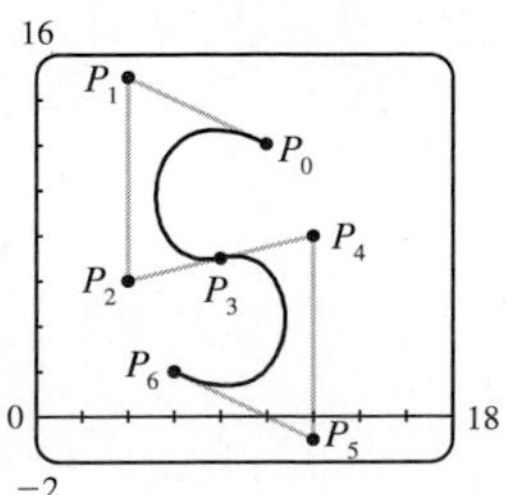

10.3 Polar Coordinates

1. (a) $\left(1, \frac{\pi}{4}\right)$

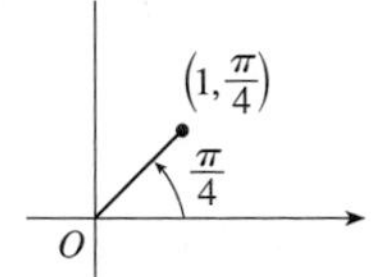

By adding 2π to $\frac{\pi}{4}$, we obtain the point $\left(1, \frac{9\pi}{4}\right)$, which satisfies the $r > 0$ requirement. The direction opposite $\frac{\pi}{4}$ is $\frac{5\pi}{4}$, so $\left(-1, \frac{5\pi}{4}\right)$ is a point that satisfies the $r < 0$ requirement.

(b) $\left(-2, \frac{3\pi}{2}\right)$

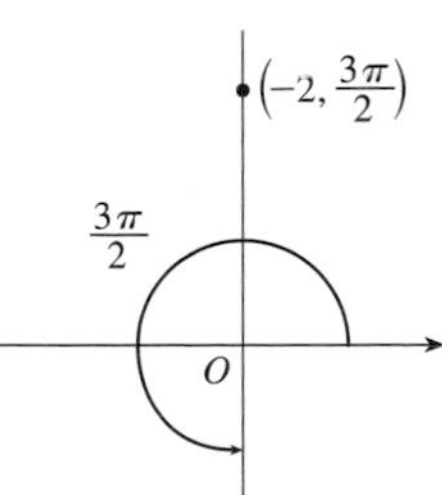

$r > 0$: $\left(-(-2), \frac{3\pi}{2} - \pi\right) = \left(2, \frac{\pi}{2}\right)$

$r < 0$: $\left(-2, \frac{3\pi}{2} + 2\pi\right) = \left(-2, \frac{7\pi}{2}\right)$

(c) $\left(3, -\frac{\pi}{3}\right)$

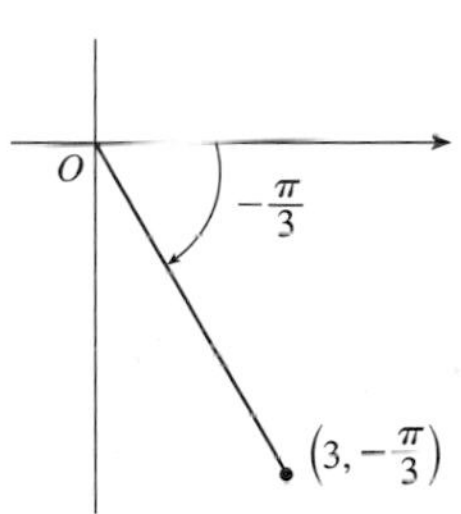

$r > 0$: $\left(3, -\frac{\pi}{3} + 2\pi\right) = \left(3, \frac{5\pi}{3}\right)$

$r < 0$: $\left(-3, -\frac{\pi}{3} + \pi\right) = \left(-3, \frac{2\pi}{3}\right)$

2. (a) $\left(2, \frac{5\pi}{6}\right)$

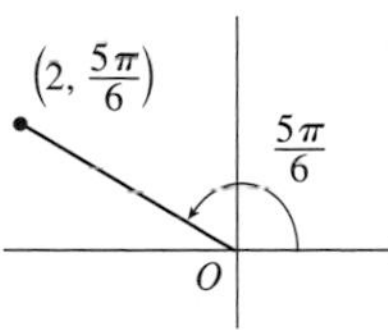

$r > 0$: $\left(2, \frac{5\pi}{6} + 2\pi\right) = \left(2, \frac{17\pi}{6}\right)$

$r < 0$: $\left(-2, \frac{5\pi}{6} - \pi\right) = \left(-2, -\frac{\pi}{6}\right)$

(b) $\left(1, -\frac{2\pi}{3}\right)$

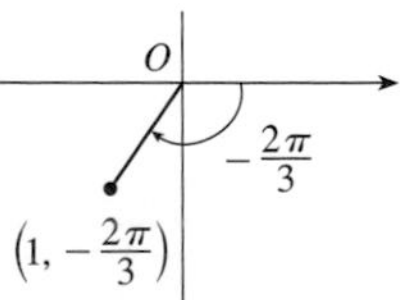

$r > 0$: $\left(1, -\frac{2\pi}{3} + 2\pi\right) = \left(1, \frac{4\pi}{3}\right)$

$r < 0$: $\left(-1, -\frac{2\pi}{3} + \pi\right) = \left(-1, \frac{\pi}{3}\right)$

(c) $\left(-1, \frac{5\pi}{4}\right)$

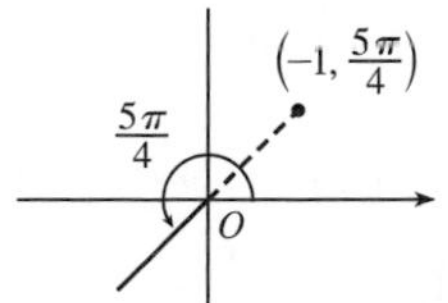

$r > 0$: $\left(-(-1), \frac{5\pi}{4} - \pi\right) = \left(1, \frac{\pi}{4}\right)$

$r < 0$: $\left(-1, \frac{5\pi}{4} - 2\pi\right) = \left(-1, -\frac{3\pi}{4}\right)$

3. (a)

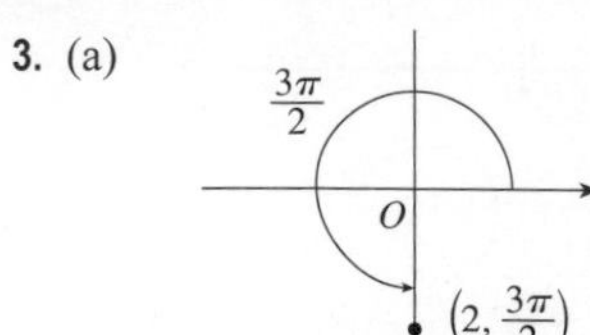

$x = 2\cos\frac{3\pi}{2} = 2(0) = 0$ and $y = 2\sin\frac{3\pi}{2} = 2(-1) = -2$ give us the Cartesian coordinates $(0, -2)$.

(b) $x = \sqrt{2}\cos\dfrac{\pi}{4} = \sqrt{2}\left(\dfrac{1}{\sqrt{2}}\right) = 1$ and $y = \sqrt{2}\sin\dfrac{\pi}{4} = \sqrt{2}\left(\dfrac{1}{\sqrt{2}}\right) = 1$ give us the Cartesian coordinates $(1, 1)$.

(c) $x = -1\cos\left(-\dfrac{\pi}{6}\right) = -1\left(\dfrac{\sqrt{3}}{2}\right) = -\dfrac{\sqrt{3}}{2}$ and

$y = -1\sin\left(-\dfrac{\pi}{6}\right) = -1\left(-\dfrac{1}{2}\right) = \dfrac{1}{2}$ give us the Cartesian

coordinates $\left(-\dfrac{\sqrt{3}}{2}, \dfrac{1}{2}\right)$.

4. (a)

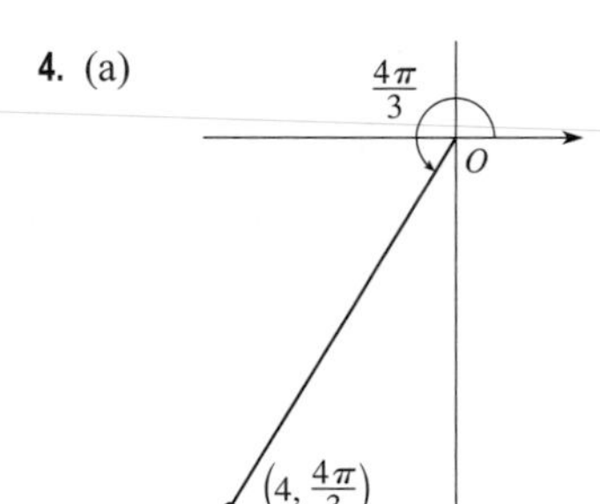

$x = 4\cos\dfrac{4\pi}{3} = 4\left(-\dfrac{1}{2}\right) = -2$ and

$y = 4\sin\dfrac{4\pi}{3} = 4\left(-\dfrac{\sqrt{3}}{2}\right) = -2\sqrt{3}$ give us the Cartesian

coordinates $\left(-2, -2\sqrt{3}\right)$.

(b)

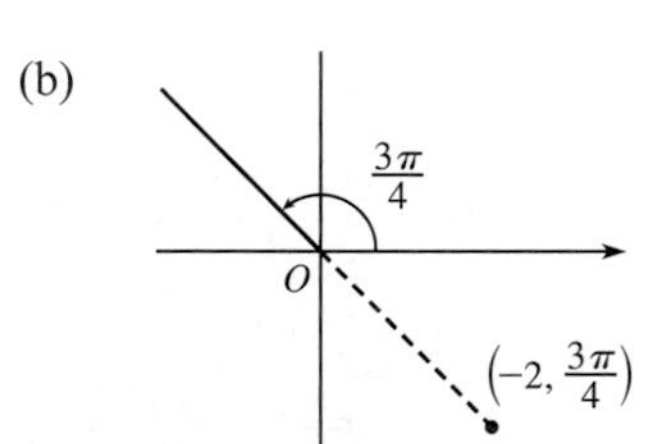

$x = -2\cos\dfrac{3\pi}{4} = -2\left(-\dfrac{\sqrt{2}}{2}\right) = \sqrt{2}$ and

$y = -2\sin\dfrac{3\pi}{4} = -2\left(\dfrac{\sqrt{2}}{2}\right) = -\sqrt{2}$ give us the Cartesian

coordinates $\left(\sqrt{2}, -\sqrt{2}\right)$.

(c)

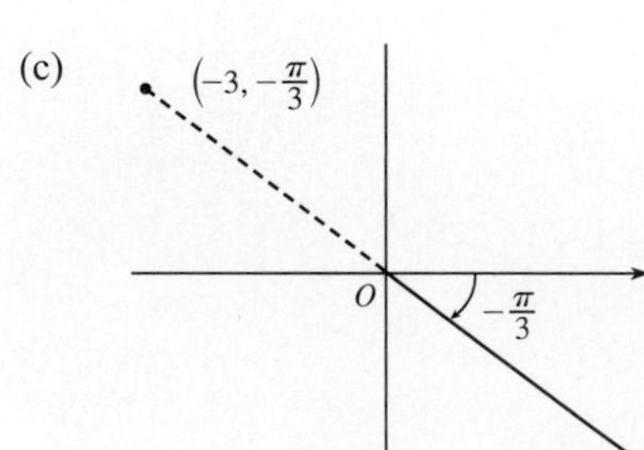

$x = -3\cos\left(-\dfrac{\pi}{3}\right) = -3\left(\dfrac{1}{2}\right) = -\dfrac{3}{2}$ and

$y = -3\sin\left(-\dfrac{\pi}{3}\right) = -3\left(-\dfrac{\sqrt{3}}{2}\right) = \dfrac{3\sqrt{3}}{2}$ give us the Cartesian

coordinates $\left(-\dfrac{3}{2}, \dfrac{3\sqrt{3}}{2}\right)$.

5. (a) $x = -4$ and $y = 4$ $\Rightarrow$ $r = \sqrt{(-4)^2 + 4^2} = 4\sqrt{2}$ and $\tan\theta = \frac{4}{-4} = -1$ $\left[\theta = -\frac{\pi}{4} + n\pi\right]$. Since $(-4, 4)$ is in the second quadrant, the polar coordinates are (i) $\left(4\sqrt{2}, \frac{3\pi}{4}\right)$ and (ii) $\left(-4\sqrt{2}, \frac{7\pi}{4}\right)$.

(b) $x = 3$ and $y = 3\sqrt{3}$ $\Rightarrow$ $r = \sqrt{3^2 + (3\sqrt{3})^2} = \sqrt{9 + 27} = 6$ and $\tan\theta = \dfrac{3\sqrt{3}}{3} = \sqrt{3}$ $[\theta = \frac{\pi}{3} + n\pi]$.

Since $(3, 3\sqrt{3})$ is in the first quadrant, the polar coordinates are (i) $(6, \frac{\pi}{3})$ and (ii) $(-6, \frac{4\pi}{3})$.

6. (a) $x = \sqrt{3}$ and $y = -1$ $\Rightarrow$ $r = \sqrt{(\sqrt{3})^2 + (-1)^2} = 2$ and $\tan\theta = \dfrac{-1}{\sqrt{3}}$ $[\theta = -\frac{\pi}{6} + n\pi]$. Since $(\sqrt{3}, -1)$ is in the fourth quadrant, the polar coordinates are (i) $(2, \frac{11\pi}{6})$ and (ii) $(-2, \frac{5\pi}{6})$.

(b) $x = -6$ and $y = 0$ $\Rightarrow$ $r = \sqrt{(-6)^2 + 0^2} = 6$ and $\tan\theta = \frac{0}{-6} = 0$ $[\theta = n\pi]$. Since $(-6, 0)$ is on the negative x-axis, the polar coordinates are (i) $(6, \pi)$ and (ii) $(-6, 0)$.

7. $r \geq 1$. The curve $r = 1$ represents a circle with center O and radius 1. So $r \geq 1$ represents the region on or outside the circle. Note that θ can take on any value.

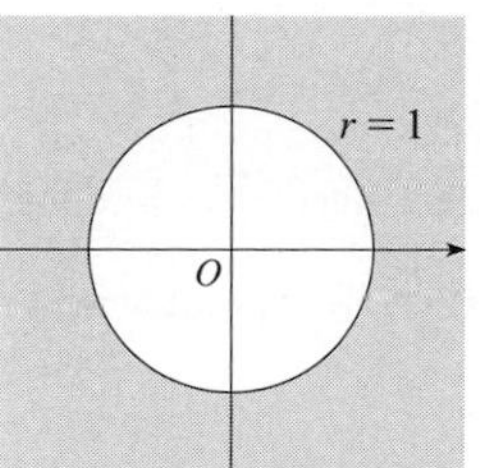

8. $0 \leq r < 2$, $\pi \leq \theta \leq 3\pi/2$. This is the region inside the circle $r = 2$ in the third quadrant.

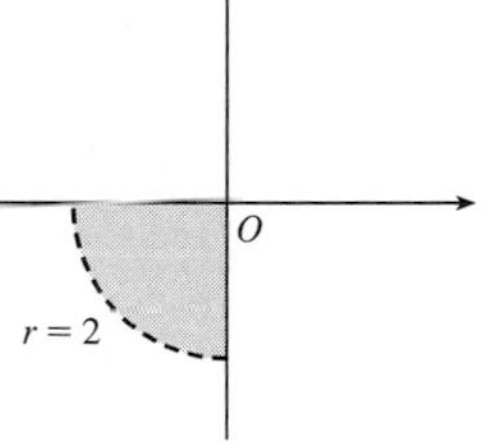

9. $r \geq 0$, $\pi/4 \leq \theta \leq 3\pi/4$.

$\theta = k$ represents a line through O.

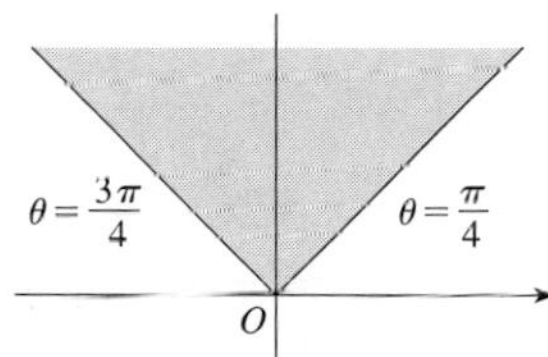

10. $1 \leq r \leq 3$, $\pi/6 < \theta < 5\pi/6$

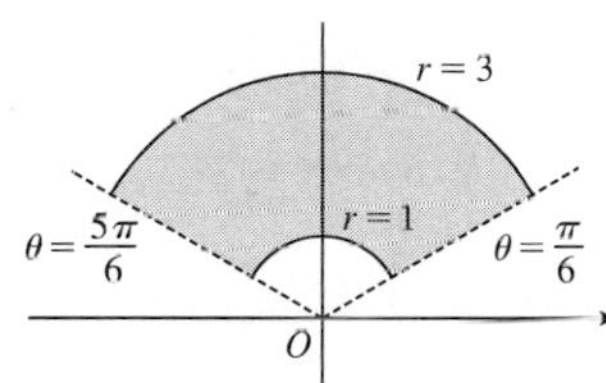

11. $2 < r < 3$, $\frac{5\pi}{3} \leq \theta \leq \frac{7\pi}{3}$

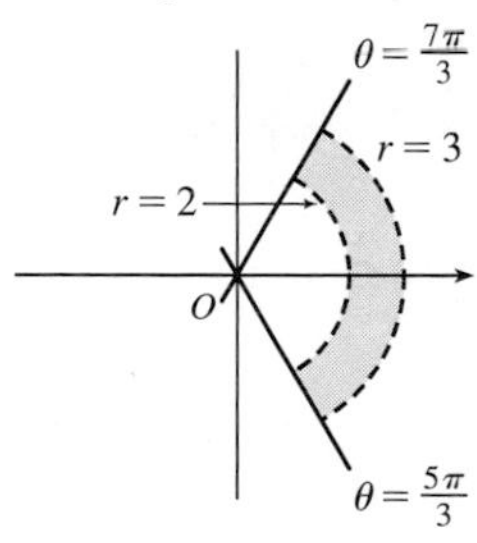

12. $r \geq 1$, $\pi \leq \theta \leq 2\pi$

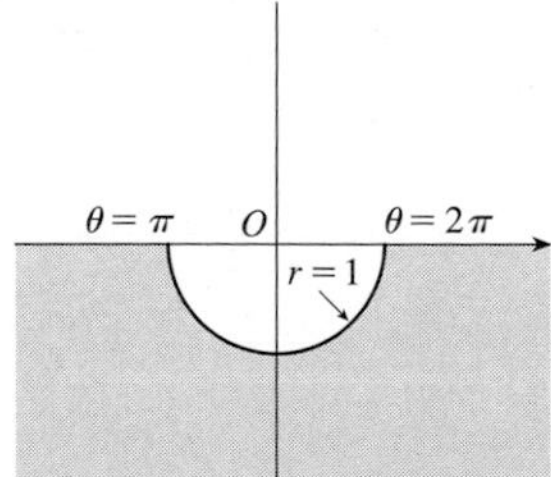

13. Converting the polar coordinates $(4, \frac{4\pi}{3})$ and $(6, \frac{5\pi}{3})$ to Cartesian coordinates gives us $(4\cos\frac{4\pi}{3}, 4\sin\frac{4\pi}{3}) = (-2, -2\sqrt{3})$ and $(6\cos\frac{5\pi}{3}, 6\sin\frac{5\pi}{3}) = (3, -3\sqrt{3})$. Now use the distance formula

$$\begin{aligned} d &= \sqrt{(x_2 - x_1)^2 + (y_2 - y_1)^2} = \sqrt{[3 - (-2)]^2 + \left[-3\sqrt{3} - (-2\sqrt{3})\right]^2} \\ &= \sqrt{5^2 + (-\sqrt{3})^2} = \sqrt{25 + 3} = \sqrt{28} = 2\sqrt{7} \end{aligned}$$

14. The points (r_1, θ_1) and (r_2, θ_2) in Cartesian coordinates are $(r_1 \cos\theta_1, r_1 \sin\theta_1)$ and $(r_2 \cos\theta_2, r_2 \sin\theta_2)$, respectively. The *square* of the distance between them is

$$\begin{aligned}(r_2\cos\theta_2 - r_1\cos\theta_1)^2 &+ (r_2\sin\theta_2 - r_1\sin\theta_1)^2 \\ &= (r_2^2\cos^2\theta_2 - 2r_1r_2\cos\theta_1\cos\theta_2 + r_1^2\cos^2\theta_1) + (r_2^2\sin^2\theta_2 - 2r_1r_2\sin\theta_1\sin\theta_2 + r_1^2\sin^2\theta_1) \\ &= r_1^2(\sin^2\theta_1 + \cos^2\theta_1) + r_2^2(\sin^2\theta_2 + \cos^2\theta_2) - 2r_1r_2(\cos\theta_1\cos\theta_2 + \sin\theta_1\sin\theta_2) \\ &= r_1^2 - 2r_1r_2\cos(\theta_1 - \theta_2) + r_2^2,\end{aligned}$$

so the distance between them is $\sqrt{r_1^2 - 2r_1r_2\cos(\theta_1 - \theta_2) + r_2^2}$.

15. $r^2 = 5 \;\Leftrightarrow\; x^2 + y^2 = 5$, a circle of radius $\sqrt{5}$ centered at the origin.

16. $r = 4\sec\theta \;\Leftrightarrow\; \dfrac{r}{\sec\theta} = 4 \;\Leftrightarrow\; r\cos\theta = 4 \;\Leftrightarrow\; x = 4$, a vertical line.

17. $r = 5\cos\theta \;\Rightarrow\; r^2 = 5r\cos\theta \;\Leftrightarrow\; x^2 + y^2 = 5x \;\Leftrightarrow\; x^2 - 5x + \frac{25}{4} + y^2 = \frac{25}{4} \;\Leftrightarrow\; \left(x - \frac{5}{2}\right)^2 + y^2 = \frac{25}{4}$, a circle of radius $\frac{5}{2}$ centered at $\left(\frac{5}{2}, 0\right)$. The first two equations are actually equivalent since $r^2 = 5r\cos\theta \;\Rightarrow\; r(r - 5\cos\theta) = 0 \;\Rightarrow\; r = 0$ or $r = 5\cos\theta$. But $r = 5\cos\theta$ gives the point $r = 0$ (the pole) when $\theta = 0$. Thus, the equation $r = 5\cos\theta$ is equivalent to the compound condition ($r = 0$ or $r = 5\cos\theta$).

18. $\theta = \dfrac{\pi}{3} \;\Rightarrow\; \tan\theta = \tan\dfrac{\pi}{3} \;\Rightarrow\; \dfrac{y}{x} = \sqrt{3} \;\Leftrightarrow\; y = \sqrt{3}\,x$, a line through the origin.

19. $r^2\cos 2\theta = 1 \;\Leftrightarrow\; r^2(\cos^2\theta - \sin^2\theta) = 1 \;\Leftrightarrow\; (r\cos\theta)^2 - (r\sin\theta)^2 = 1 \;\Leftrightarrow\; x^2 - y^2 = 1$, a hyperbola centered at the origin with foci on the x-axis.

20. $r^2\sin 2\theta = 1 \;\Leftrightarrow\; r^2(2\sin\theta\,\cos\theta) = 1 \;\Leftrightarrow\; 2(r\cos\theta)(r\sin\theta) = 1 \;\Leftrightarrow\; 2xy = 1 \;\Leftrightarrow\; xy = \frac{1}{2}$, a hyperbola centered at the origin with foci on the line $y = x$.

21. $y = 2 \;\Leftrightarrow\; r\sin\theta = 2 \;\Leftrightarrow\; r = \dfrac{2}{\sin\theta} \;\Leftrightarrow\; r = 2\csc\theta$

22. $y = x \;\Rightarrow\; \dfrac{y}{x} = 1\; [x \neq 0] \;\Rightarrow\; \tan\theta = 1 \;\Rightarrow\; \theta = \tan^{-1} 1 \;\Rightarrow\; \theta = \dfrac{\pi}{4}$ or $\theta = \dfrac{5\pi}{4}$ [either includes the pole]

23. $y = 1 + 3x \;\Leftrightarrow\; r\sin\theta = 1 + 3r\cos\theta \;\Leftrightarrow\; r\sin\theta - 3r\cos\theta = 1 \;\Leftrightarrow\; r(\sin\theta - 3\cos\theta) = 1 \;\Leftrightarrow\; r = \dfrac{1}{\sin\theta - 3\cos\theta}$

24. $4y^2 = x \;\Leftrightarrow\; 4(r\sin\theta)^2 = r\cos\theta \;\Leftrightarrow\; 4r^2\sin^2\theta - r\cos\theta = 0 \;\Leftrightarrow\; r(4r\sin^2\theta - \cos\theta) = 0 \;\Leftrightarrow\; r = 0$ or $r = \dfrac{\cos\theta}{4\sin^2\theta} \;\Leftrightarrow\; r = 0$ or $r = \frac{1}{4}\cot\theta\,\csc\theta$. $r = 0$ is included in $r = \frac{1}{4}\cot\theta\,\csc\theta$ when $\theta = \frac{\pi}{2}$, so the curve is represented by the single equation $r = \frac{1}{4}\cot\theta\,\csc\theta$.

25. $x^2 + y^2 = 2cx \;\Leftrightarrow\; r^2 = 2cr\cos\theta \;\Leftrightarrow\; r^2 - 2cr\cos\theta = 0 \;\Leftrightarrow\; r(r - 2c\cos\theta) = 0 \;\Leftrightarrow\; r = 0$ or $r = 2c\cos\theta$. $r = 0$ is included in $r = 2c\cos\theta$ when $\theta = \frac{\pi}{2} + n\pi$, so the curve is represented by the single equation $r = 2c\cos\theta$.

26. $x^2 - y^2 = 4 \quad \Leftrightarrow \quad (r\cos\theta)^2 - (r\sin\theta)^2 = 4 \quad \Leftrightarrow \quad r^2\cos^2\theta - r^2\sin^2\theta = 4 \quad \Leftrightarrow \quad r^2(\cos^2\theta - \sin^2\theta) = 4 \quad \Leftrightarrow$
$r^2 \cos 2\theta = 4$

27. (a) The description leads immediately to the polar equation $\theta = \frac{\pi}{6}$, and the Cartesian equation $y = \tan\left(\frac{\pi}{6}\right)x = \frac{1}{\sqrt{3}}\,x$ is slightly more difficult to derive.

(b) The easier description here is the Cartesian equation $x = 3$.

28. (a) Because its center is not at the origin, it is more easily described by its Cartesian equation, $(x-2)^2 + (y-3)^2 = 5^2$.

(b) This circle is more easily given in polar coordinates: $r = 4$. The Cartesian equation is also simple: $x^2 + y^2 = 16$.

29. $r = -2\sin\theta$

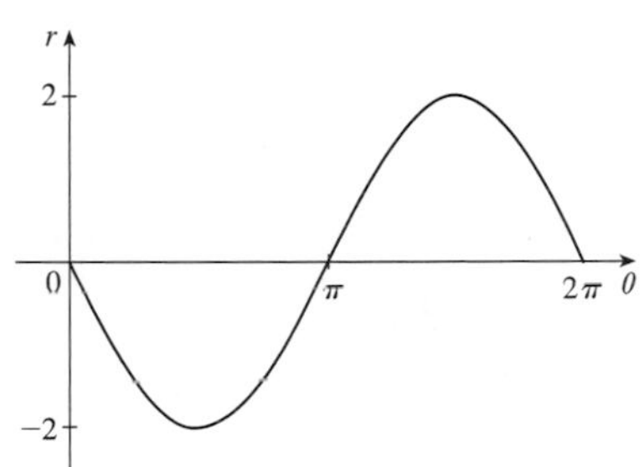

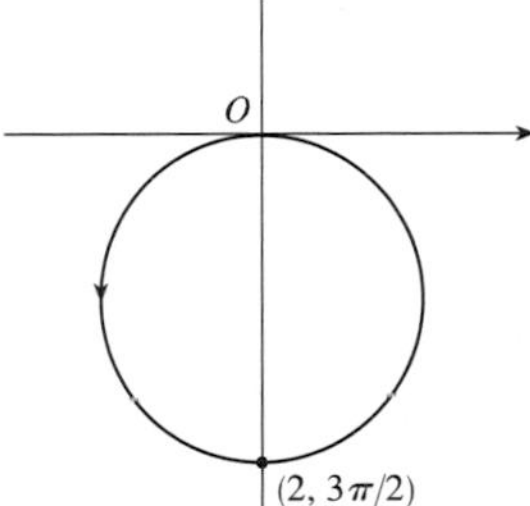

30. $r = 1 - \cos\theta$

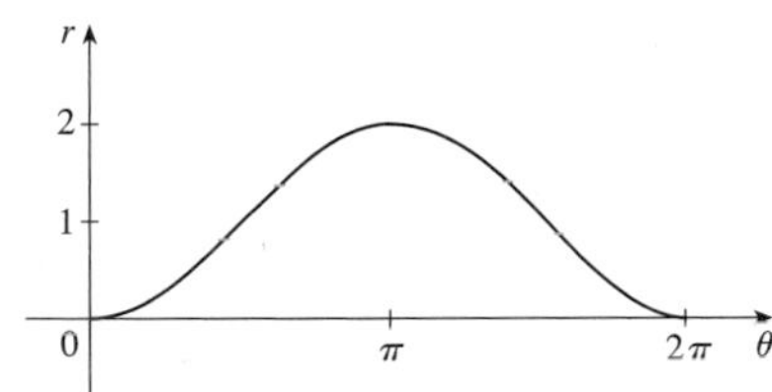

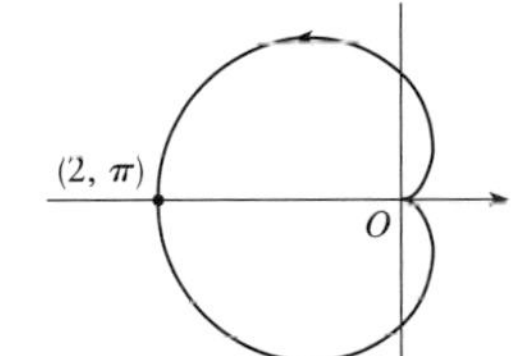

31. $r = 2(1 + \cos\theta)$

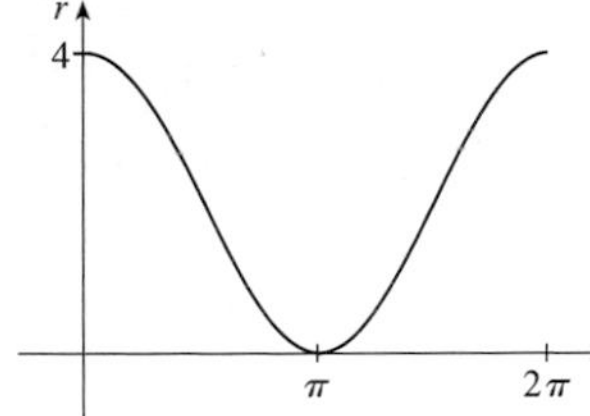

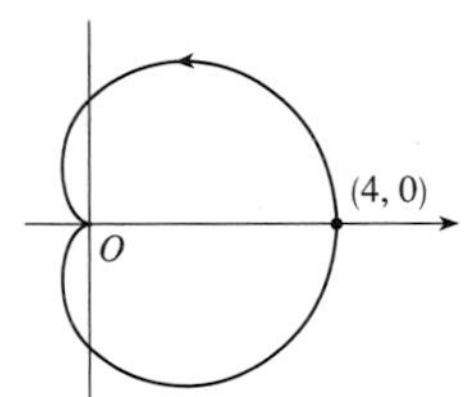

32. $r = 1 + 2\cos\theta$

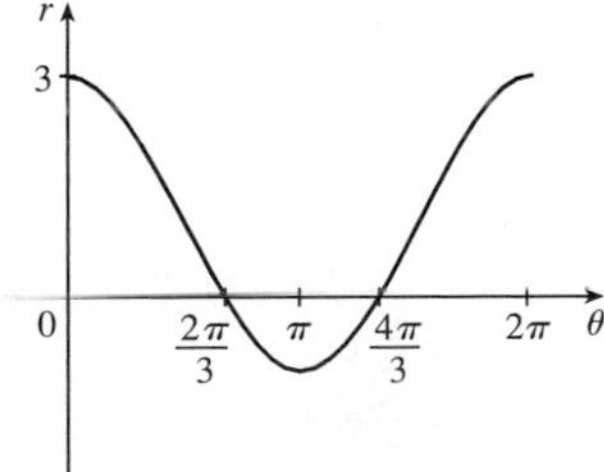

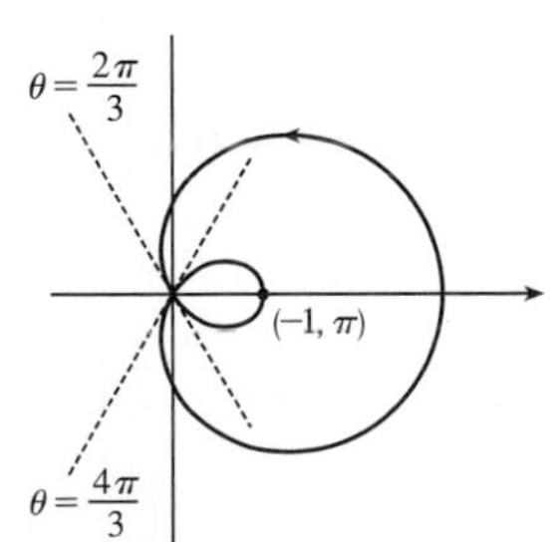

33. $r = \theta, \quad \theta \geq 0$

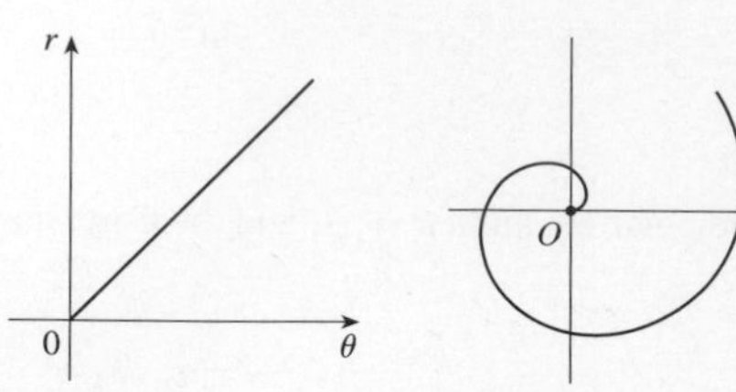

34. $r = \theta^2, \ -2\pi \leq \theta \leq 2\pi$

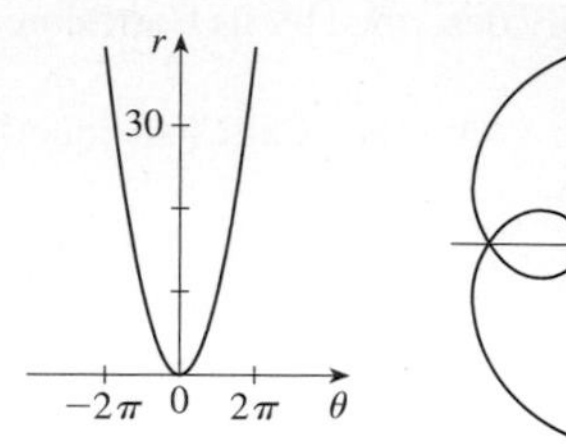

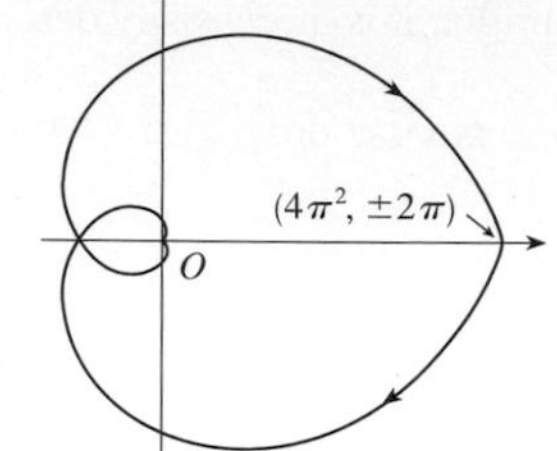

35. $r = 3\cos 3\theta$

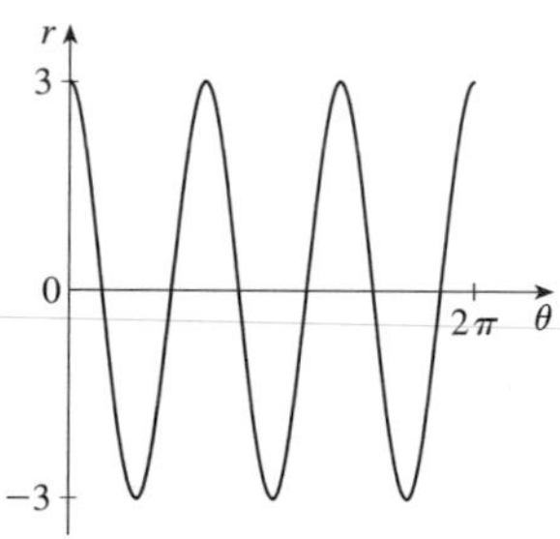

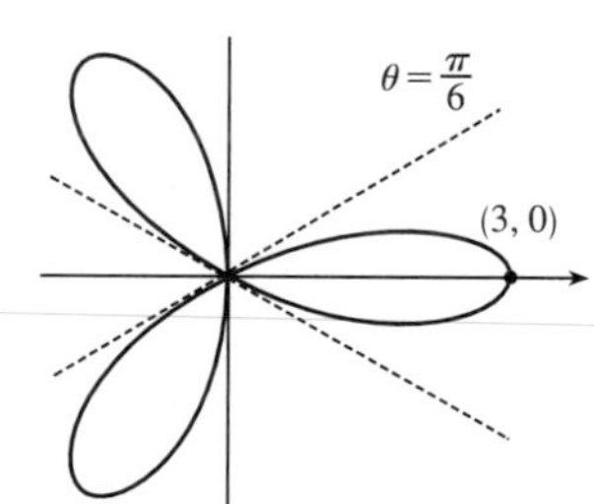

36. $r = -\sin 5\theta$

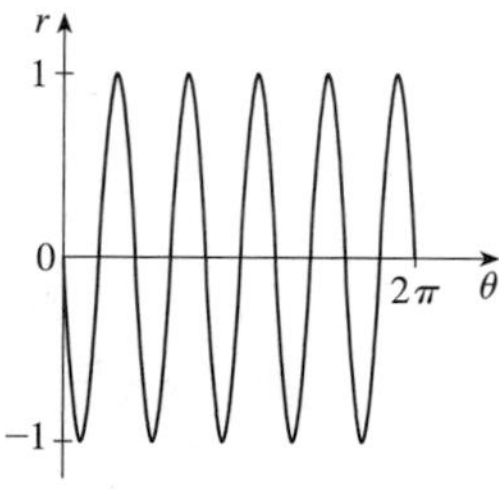

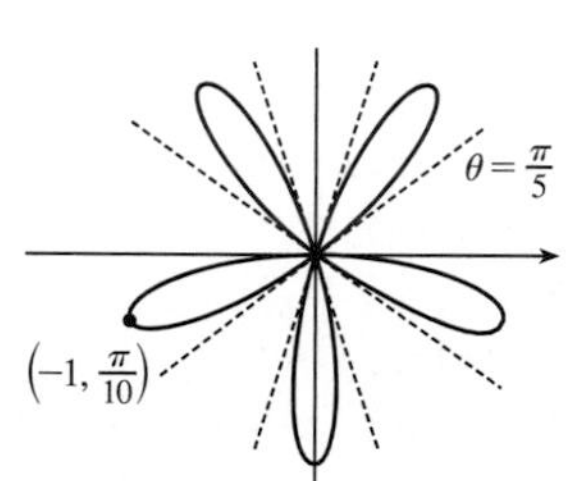

37. $r = 2\cos 4\theta$

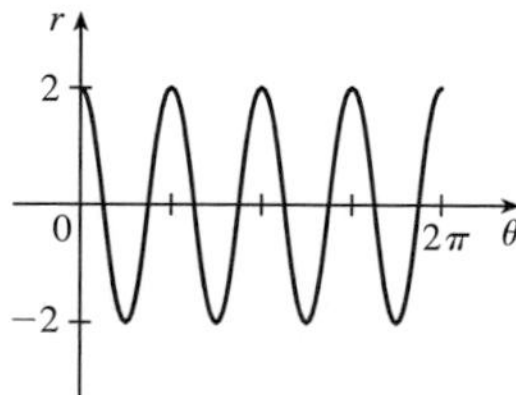

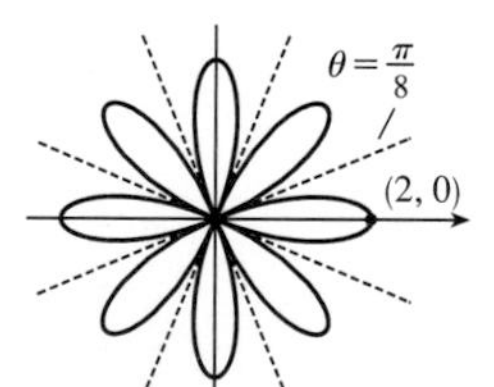

38. $r = 2\sin 6\theta$

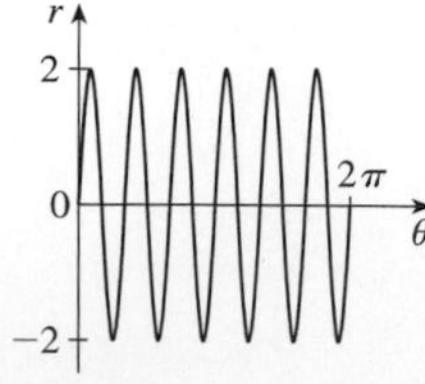

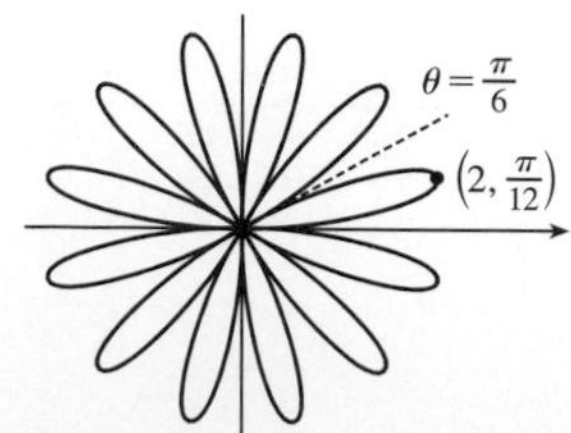

39. $r = 1 + 3\cos\theta$

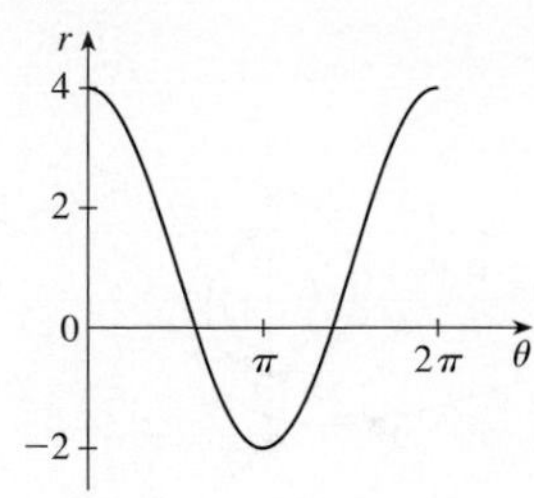

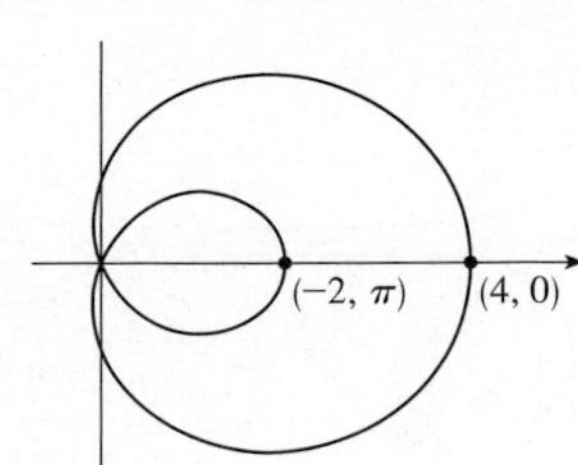

40. $r = 1 + 5\sin\theta$

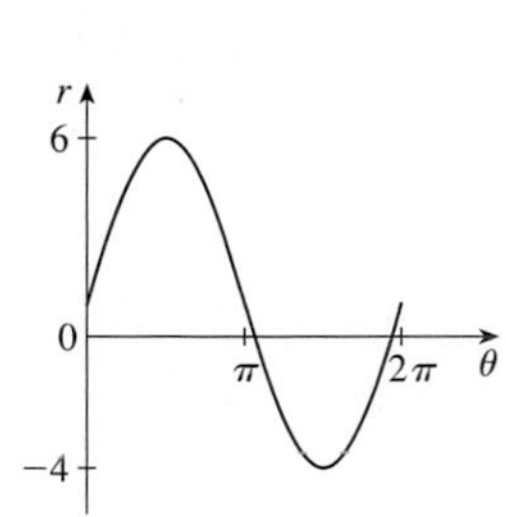

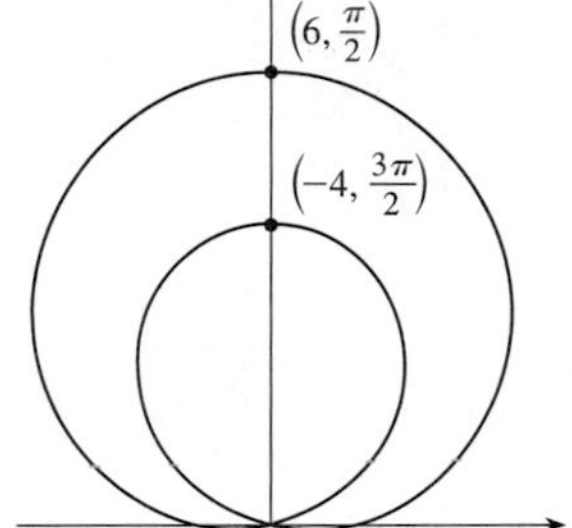

41. $r^2 = 9\sin 2\theta$

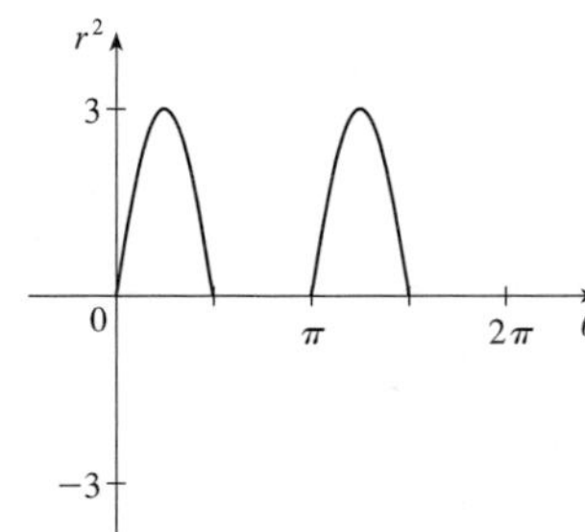

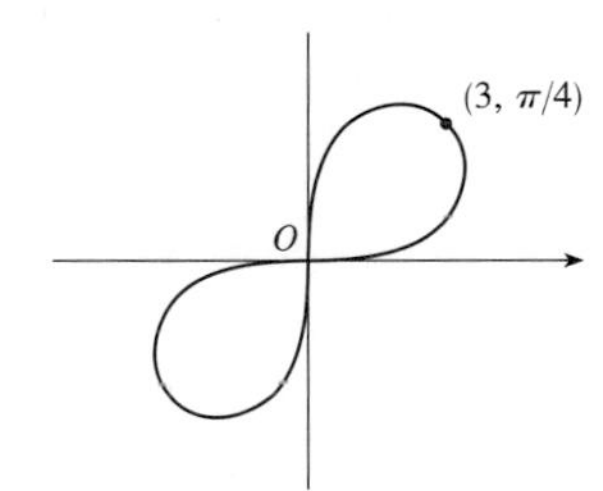

42. $r^2 = \cos 4\theta$

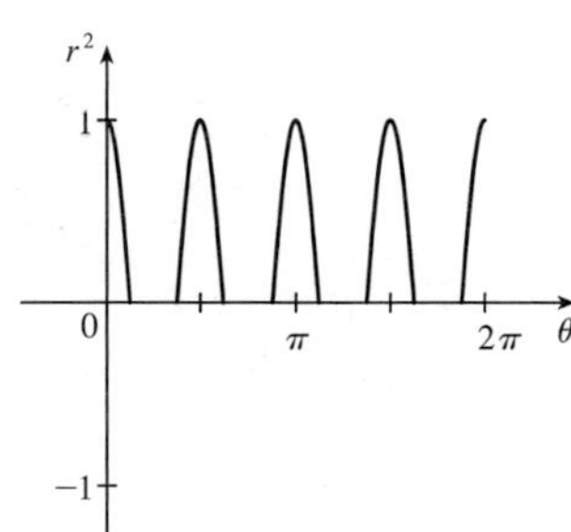

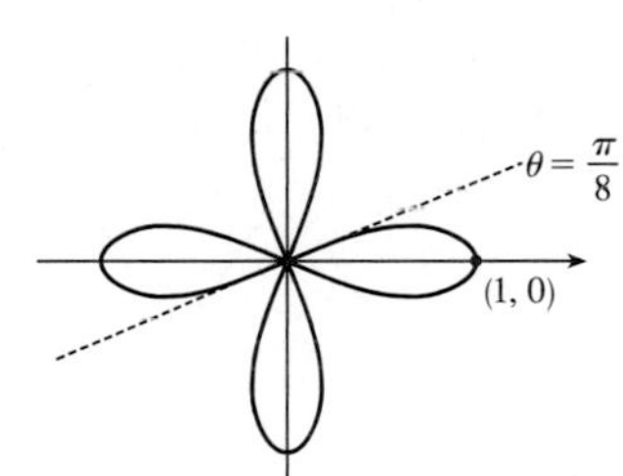

43. $r = 2 + \sin 3\theta$

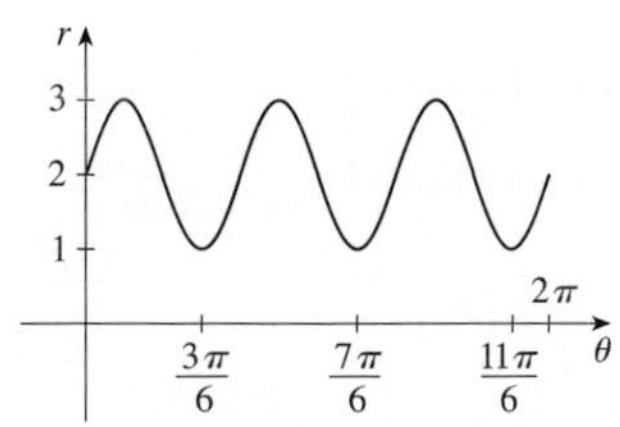

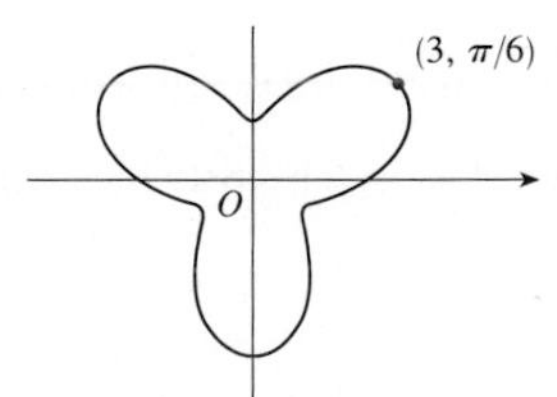

44. $r^2\theta = 1 \quad \Leftrightarrow \quad r = \pm 1/\sqrt{\theta}$ for $\theta > 0$

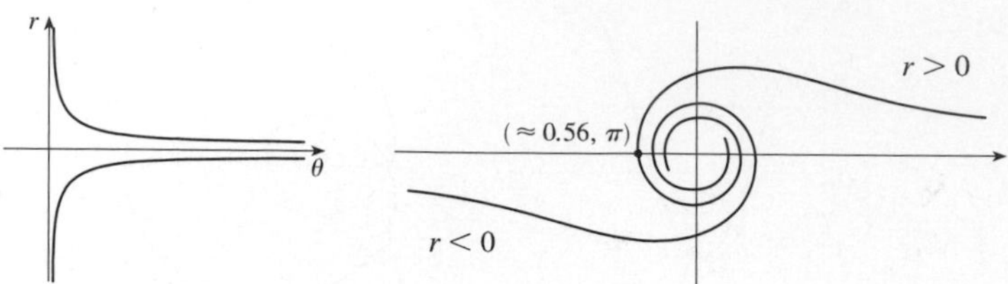

45. $r = \sin(\theta/2)$

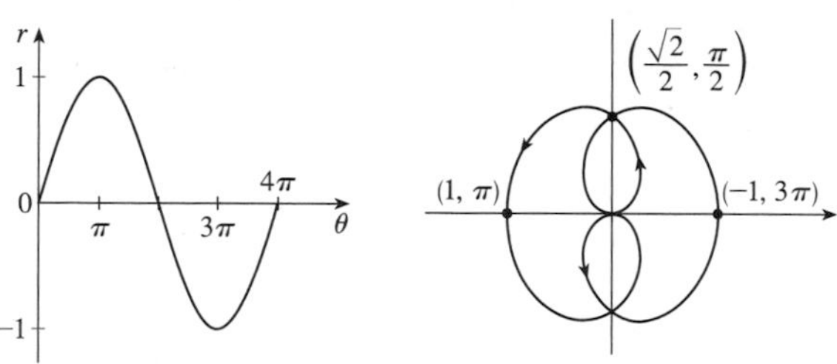

46. $r = \cos(\theta/3)$

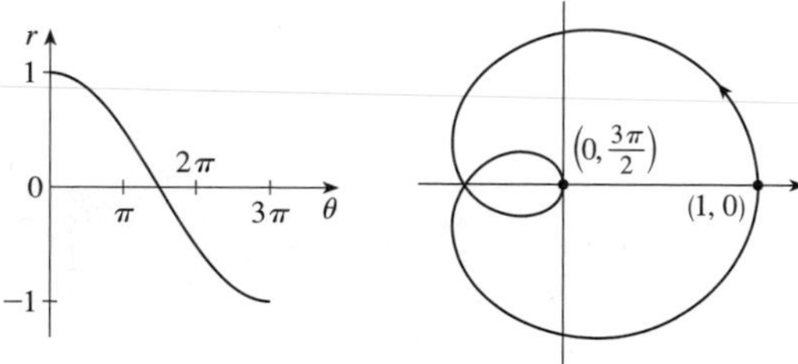

47. For $\theta = 0$, π, and 2π, r has its minimum value of about 0.5. For $\theta = \frac{\pi}{2}$ and $\frac{3\pi}{2}$, r attains its maximum value of 2. We see that the graph has a similar shape for $0 \leq \theta \leq \pi$ and $\pi \leq \theta \leq 2\pi$.

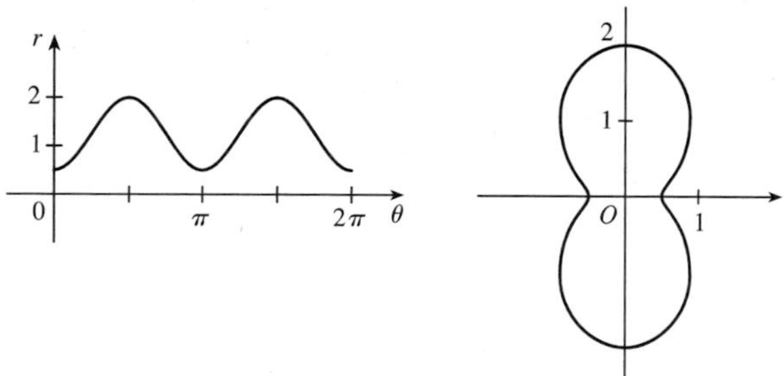

48. The given graph has a maximum of 2 for $\theta = 0$, a minimum of 1 for $\theta = \frac{\pi}{4}$, and then a maximum of 2 for $\theta = \frac{\pi}{2}$. This pattern is repeated 4 times for $0 \leq \theta \leq 2\pi$.

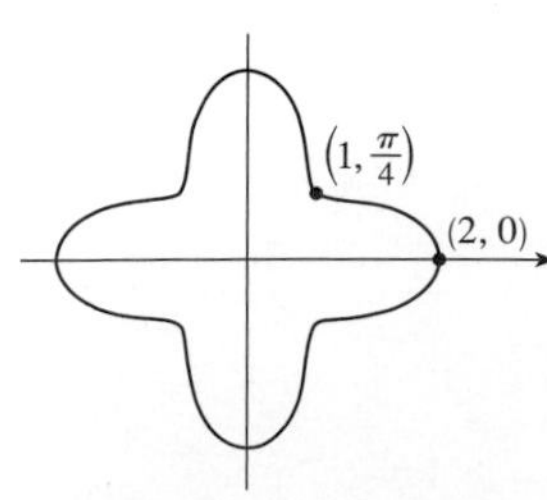

49. $x = r\cos\theta = (4 + 2\sec\theta)\cos\theta = 4\cos\theta + 2$. Now, $r \to \infty \;\Rightarrow$ $(4 + 2\sec\theta) \to \infty \;\Rightarrow\; \theta \to \left(\frac{\pi}{2}\right)^-$ or $\theta \to \left(\frac{3\pi}{2}\right)^+$ [since we need only consider $0 \le \theta < 2\pi$], so $\lim\limits_{r\to\infty} x = \lim\limits_{\theta\to\pi/2^-}(4\cos\theta + 2) = 2$. Also, $r \to -\infty \;\Rightarrow\; (4 + 2\sec\theta) \to -\infty \;\Rightarrow\; \theta \to \left(\frac{\pi}{2}\right)^+$ or $\theta \to \left(\frac{3\pi}{2}\right)^-$, so $\lim\limits_{r\to-\infty} x = \lim\limits_{\theta\to\pi/2^+}(4\cos\theta + 2) = 2$. Therefore, $\lim\limits_{r\to\pm\infty} x = 2 \;\Rightarrow\; x = 2$ is a vertical asymptote.

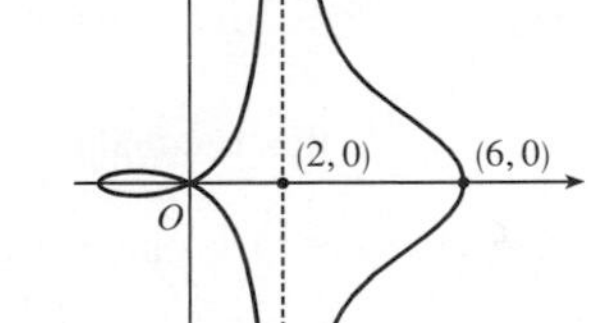

50. $y = r\sin\theta = 2\sin\theta - \csc\theta\sin\theta = 2\sin\theta - 1$.

$r \to \infty \;\Rightarrow\; (2 - \csc\theta) \to \infty \;\Rightarrow$ $\csc\theta \to -\infty \;\Rightarrow\; \theta \to \pi^+$ [since we need only consider $0 \le \theta < 2\pi$] and so $\lim\limits_{r\to\infty} y = \lim\limits_{\theta\to\pi^+} 2\sin\theta - 1 = -1$.

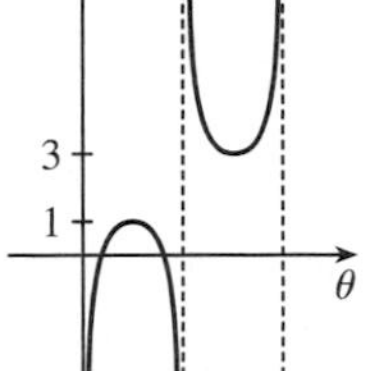

Also $r \to -\infty \;\Rightarrow\; (2 - \csc\theta) \to -\infty \;\Rightarrow\; \csc\theta \to \infty \;\Rightarrow\; \theta \to \pi^-$ and so $\lim\limits_{r\to-\infty} x = \lim\limits_{\theta\to\pi^-} 2\sin\theta - 1 = -1$.

Therefore $\lim\limits_{r\to\pm\infty} y = -1 \;\Rightarrow\; y = -1$ is a horizontal asymptote.

51. To show that $x = 1$ is an asymptote we must prove $\lim\limits_{r\to\pm\infty} x = 1$.

$x = (r)\cos\theta = (\sin\theta\tan\theta)\cos\theta = \sin^2\theta$. Now, $r \to \infty \;\Rightarrow\; \sin\theta\tan\theta \to \infty \;\Rightarrow$ $\theta \to \left(\frac{\pi}{2}\right)^-$, so $\lim\limits_{r\to\infty} x = \lim\limits_{\theta\to\pi/2^-}\sin^2\theta = 1$. Also, $r \to -\infty \;\Rightarrow\; \sin\theta\tan\theta \to -\infty \;\Rightarrow$ $\theta \to \left(\frac{\pi}{2}\right)^+$, so $\lim\limits_{r\to-\infty} x = \lim\limits_{\theta\to\pi/2^+}\sin^2\theta = 1$. Therefore, $\lim\limits_{r\to\pm\infty} x = 1 \;\Rightarrow\; x = 1$ is a vertical asymptote. Also notice that $x = \sin^2\theta \ge 0$ for all θ, and $x = \sin^2\theta \le 1$ for all θ. And $x \ne 1$, since the curve is not defined at odd multiples of $\frac{\pi}{2}$. Therefore, the curve lies entirely within the vertical strip $0 \le x < 1$.

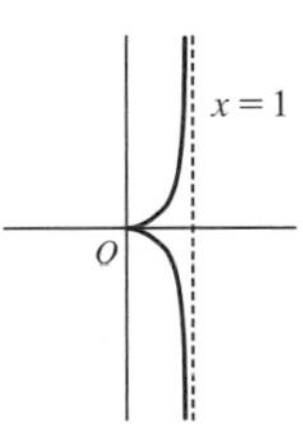

52. The equation is $(x^2 + y^2)^3 = 4x^2y^2$, but using polar coordinates we know that $x^2 + y^2 = r^2$ and $x = r\cos\theta$ and $y = r\sin\theta$. Substituting into the given equation: $r^6 = 4r^2\cos^2\theta\, r^2\sin^2\theta \;\Rightarrow\; r^2 = 4\cos^2\theta\sin^2\theta \;\Rightarrow$ $r = \pm 2\cos\theta\sin\theta = \pm\sin 2\theta$. $r = \pm\sin 2\theta$ is sketched at right.

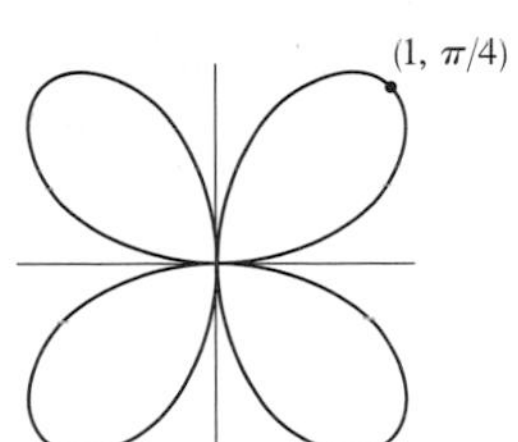

53. (a) We see that the curve $r = 1 + c\sin\theta$ crosses itself at the origin, where $r = 0$ (in fact the inner loop corresponds to negative r-values,) so we solve the equation of the limaçon for $r = 0 \;\Leftrightarrow\; c\sin\theta = -1 \;\Leftrightarrow\; \sin\theta = -1/c$. Now if $|c| < 1$, then this equation has no solution and hence there is no inner loop. But if $c < -1$, then on the interval $(0, 2\pi)$ the equation has the two solutions $\theta = \sin^{-1}(-1/c)$ and $\theta = \pi - \sin^{-1}(-1/c)$, and if $c > 1$, the solutions are $\theta = \pi + \sin^{-1}(1/c)$ and $\theta = 2\pi - \sin^{-1}(1/c)$. In each case, $r < 0$ for θ between the two solutions, indicating a loop.

(b) For $0 < c < 1$, the dimple (if it exists) is characterized by the fact that y has a local maximum at $\theta = \frac{3\pi}{2}$. So we determine for what c-values $\dfrac{d^2y}{d\theta^2}$ is negative at $\theta = \frac{3\pi}{2}$, since by the Second Derivative Test this indicates a maximum:

$y = r\sin\theta = \sin\theta + c\sin^2\theta \quad \Rightarrow \quad \dfrac{dy}{d\theta} = \cos\theta + 2c\sin\theta\,\cos\theta = \cos\theta + c\sin 2\theta \quad \Rightarrow \quad \dfrac{d^2y}{d\theta^2} = -\sin\theta + 2c\cos 2\theta.$

At $\theta = \frac{3\pi}{2}$, this is equal to $-(-1) + 2c(-1) = 1 - 2c$, which is negative only for $c > \frac{1}{2}$. A similar argument shows that for $-1 < c < 0$, y only has a local minimum at $\theta = \frac{\pi}{2}$ (indicating a dimple) for $c < -\frac{1}{2}$.

54. (a) $r = \ln\theta,\ 1 \le \theta \le 6\pi$. r increases as θ increases and there are almost three full revolutions. The graph must be either III or VI. As θ increases, r grows slowly in VI and quickly in III. Since $r = \ln\theta$ grows slowly, its graph must be VI.

(b) $r = \theta^2,\ 0 \le \theta \le 8\pi$. See part (a). This is graph III.

(c) The graph of $r = \cos 3\theta$ is a three-leaved rose, which is graph II.

(d) Since $-1 \le \cos 3\theta \le 1$, $1 \le 2 + \cos 3\theta \le 3$, so $r = 2 + \cos 3\theta$ is never 0; that is, the curve never intersects the pole. The graph must be I or IV. For $0 \le \theta \le 2\pi$, the graph assumes its minimum r-value of 1 three times, at $\theta = \frac{\pi}{3}$, π, and $\frac{5\pi}{3}$, so it must be graph IV.

(e) $r = \cos(\theta/2)$. For $\theta = 0$, $r = 1$, and as θ increases to π, r decreases to 0. Only graph V satisfies those values.

(f) $r = 2 + \cos(3\theta/2)$. As in part (d), this graph never intersects the pole, so it must be graph I.

55. $r = 2\cos\theta \quad \Rightarrow \quad x = r\cos\theta = 2\cos^2\theta,\ y = r\sin\theta = 2\sin\theta\,\cos\theta = \sin 2\theta \quad \Rightarrow$

$$\frac{dy}{dx} = \frac{dy/d\theta}{dx/d\theta} = \frac{2\cos 2\theta}{2\cdot 2\cos\theta(-\sin\theta)} = \frac{\cos 2\theta}{-\sin 2\theta} = -\cot 2\theta$$

When $\theta = \dfrac{\pi}{3}$, $\dfrac{dy}{dx} = -\cot\left(2\cdot\dfrac{\pi}{3}\right) = \cot\dfrac{\pi}{3} = \dfrac{1}{\sqrt{3}}$. [*Another method:* Use Equation 3.]

56. $r = 2 + \sin 3\theta \quad \Rightarrow \quad x = r\cos\theta = (2 + \sin 3\theta)\cos\theta,\ y = r\sin\theta = (2 + \sin 3\theta)\sin\theta \quad \Rightarrow$

$$\frac{dy}{dx} = \frac{dy/d\theta}{dx/d\theta} = \frac{(2 + \sin 3\theta)\cos\theta + \sin\theta(3\cos 3\theta)}{(2 + \sin 3\theta)(-\sin\theta) + \cos\theta(3\cos 3\theta)}$$

When $\theta = \dfrac{\pi}{4}$, $\dfrac{dy}{dx} = \dfrac{\left(2 + \sin\frac{3\pi}{4}\right)\cos\frac{\pi}{4} + \sin\frac{\pi}{4}\left(3\cos\frac{3\pi}{4}\right)}{\left(2 + \sin\frac{3\pi}{4}\right)\left(-\sin\frac{\pi}{4}\right) + \cos\frac{\pi}{4}\left(3\cos\frac{3\pi}{4}\right)} = \dfrac{\left(2 + \frac{\sqrt{2}}{2}\right)\frac{\sqrt{2}}{2} + \frac{\sqrt{2}}{2}\cdot 3\left(-\frac{\sqrt{2}}{2}\right)}{\left(2 + \frac{\sqrt{2}}{2}\right)\left(-\frac{\sqrt{2}}{2}\right) + \frac{\sqrt{2}}{2}\cdot 3\left(-\frac{\sqrt{2}}{2}\right)}$

$$= \frac{\sqrt{2} + \frac{1}{2} - \frac{3}{2}}{-\sqrt{2} - \frac{1}{2} - \frac{3}{2}} = \frac{\sqrt{2} - 1}{-\sqrt{2} - 2}, \text{ or, equivalently, } 2 - \frac{3}{2}\sqrt{2}.$$

57. $r = 1/\theta \quad \Rightarrow \quad x = r\cos\theta = (\cos\theta)/\theta,\ y = r\sin\theta = (\sin\theta)/\theta \quad \Rightarrow$

$$\frac{dy}{dx} = \frac{dy/d\theta}{dx/d\theta} = \frac{\sin\theta(-1/\theta^2) + (1/\theta)\cos\theta}{\cos\theta(-1/\theta^2) - (1/\theta)\sin\theta}\cdot\frac{\theta^2}{\theta^2} = \frac{-\sin\theta + \theta\cos\theta}{-\cos\theta - \theta\sin\theta}$$

When $\theta = \pi$, $\dfrac{dy}{dx} = \dfrac{-0 + \pi(-1)}{-(-1) - \pi(0)} = \dfrac{-\pi}{1} = -\pi$.

58. $r = \cos(\theta/3) \quad \Rightarrow \quad x = r\cos\theta = \cos(\theta/3)\cos\theta,\ y = r\sin\theta = \cos(\theta/3)\sin\theta \quad \Rightarrow$

$$\frac{dy}{dx} = \frac{dy/d\theta}{dx/d\theta} = \frac{\cos(\theta/3)\cos\theta + \sin\theta\left(-\frac{1}{3}\sin(\theta/3)\right)}{\cos(\theta/3)\left(-\sin\theta\right) + \cos\theta\left(-\frac{1}{3}\sin(\theta/3)\right)}$$

When $\theta = \pi$, $\dfrac{dy}{dx} = \dfrac{\frac{1}{2}(-1) + (0)\left(-\sqrt{3}/6\right)}{\frac{1}{2}(0) + (-1)\left(-\sqrt{3}/6\right)} = \dfrac{-1/2}{\sqrt{3}/6} = -\dfrac{3}{\sqrt{3}} = -\sqrt{3}$.

59. $r = \cos 2\theta \quad \Rightarrow \quad x = r\cos\theta = \cos 2\theta\, \cos\theta,\, y = r\sin\theta = \cos 2\theta\, \sin\theta \quad \Rightarrow$

$$\frac{dy}{dx} = \frac{dy/d\theta}{dx/d\theta} = \frac{\cos 2\theta\, \cos\theta + \sin\theta\,(-2\sin 2\theta)}{\cos 2\theta\,(-\sin\theta) + \cos\theta\,(-2\sin 2\theta)}$$

When $\theta = \frac{\pi}{4}$, $\dfrac{dy}{dx} = \dfrac{0(\sqrt{2}/2) + (\sqrt{2}/2)(-2)}{0(-\sqrt{2}/2) + (\sqrt{2}/2)(-2)} = \dfrac{-\sqrt{2}}{-\sqrt{2}} = 1.$

60. $r = 1 + 2\cos\theta \quad \Rightarrow \quad x = r\cos\theta = (1 + 2\cos\theta)\cos\theta,\, y = r\sin\theta = (1 + 2\cos\theta)\sin\theta \quad \Rightarrow$

$$\frac{dy}{dx} = \frac{dy/d\theta}{dx/d\theta} = \frac{(1 + 2\cos\theta)\cos\theta + \sin\theta\,(-2\sin\theta)}{(1 + 2\cos\theta)(-\sin\theta) + \cos\theta\,(-2\sin\theta)}$$

When $\theta = \frac{\pi}{3}$, $\dfrac{dy}{dx} = \dfrac{2(\frac{1}{2}) + (\sqrt{3}/2)(-\sqrt{3})}{2(-\sqrt{3}/2) + \frac{1}{2}(-\sqrt{3})} \cdot \dfrac{2}{2} = \dfrac{2 - 3}{-2\sqrt{3} - \sqrt{3}} = \dfrac{-1}{-3\sqrt{3}} = \dfrac{\sqrt{3}}{9}.$

61. $r = 3\cos\theta \quad \Rightarrow \quad x = r\cos\theta = 3\cos\theta\, \cos\theta,\;\; y = r\sin\theta = 3\cos\theta\, \sin\theta \quad \Rightarrow$

$\frac{dy}{d\theta} = -3\sin^2\theta + 3\cos^2\theta = 3\cos 2\theta = 0 \quad \Rightarrow \quad 2\theta = \frac{\pi}{2}$ or $\frac{3\pi}{2} \quad \Leftrightarrow \quad \theta = \frac{\pi}{4}$ or $\frac{3\pi}{4}$.

So the tangent is horizontal at $\left(\frac{3}{\sqrt{2}}, \frac{\pi}{4}\right)$ and $\left(-\frac{3}{\sqrt{2}}, \frac{3\pi}{4}\right)$ $\left[\text{same as } \left(\frac{3}{\sqrt{2}}, -\frac{\pi}{4}\right)\right]$.

$\frac{dx}{d\theta} = -6\sin\theta\cos\theta = -3\sin 2\theta = 0 \quad \Rightarrow \quad 2\theta = 0$ or $\pi \quad \Leftrightarrow \quad \theta = 0$ or $\frac{\pi}{2}$. So the tangent is vertical at $(3, 0)$ and $\left(0, \frac{\pi}{2}\right)$.

62. $r = 1 - \sin\theta \quad \Rightarrow \quad x = r\cos\theta = \cos\theta\,(1 - \sin\theta),\;\; y = r\sin\theta = \sin\theta\,(1 - \sin\theta) \quad \Rightarrow$

$\frac{dy}{d\theta} = \sin\theta\,(-\cos\theta) + (1 - \sin\theta)\,\cos\theta = \cos\theta\,(1 - 2\sin\theta) = 0 \quad \Rightarrow \quad \cos\theta = 0$ or $\sin\theta = \frac{1}{2} \quad \Rightarrow$

$\theta = \frac{\pi}{6}, \frac{\pi}{2}, \frac{5\pi}{6}$, or $\frac{3\pi}{2} \quad \Rightarrow \quad$ horizontal tangent at $\left(\frac{1}{2}, \frac{\pi}{6}\right)$, $\left(\frac{1}{2}, \frac{5\pi}{6}\right)$, and $\left(2, \frac{3\pi}{2}\right)$.

$$\begin{aligned}\frac{dx}{d\theta} &= \cos\theta\,(-\cos\theta) + (1 - \sin\theta)(-\sin\theta) = -\cos^2\theta - \sin\theta + \sin^2\theta = 2\sin^2\theta - \sin\theta - 1 \\ &= (2\sin\theta + 1)(\sin\theta - 1) = 0 \quad \Rightarrow\end{aligned}$$

$\sin\theta = -\frac{1}{2}$ or $1 \quad \Rightarrow \quad \theta = \frac{7\pi}{6}, \frac{11\pi}{6}$, or $\frac{\pi}{2} \quad \Rightarrow \quad$ vertical tangent at $\left(\frac{3}{2}, \frac{7\pi}{6}\right)$, $\left(\frac{3}{2}, \frac{11\pi}{6}\right)$, and $\left(0, \frac{\pi}{2}\right)$.

Note that the tangent is vertical, not horizontal, when $\theta = \frac{\pi}{2}$, since

$$\lim_{\theta \to (\pi/2)^-} \frac{dy/d\theta}{dx/d\theta} = \lim_{\theta \to (\pi/2)^-} \frac{\cos\theta\,(1 - 2\sin\theta)}{(2\sin\theta + 1)(\sin\theta - 1)} = \infty \text{ and } \lim_{\theta \to (\pi/2)^+} \frac{dy/d\theta}{dx/d\theta} = -\infty.$$

63. $r = 1 + \cos\theta \quad \Rightarrow \quad x = r\cos\theta = \cos\theta\,(1 + \cos\theta),\;\; y = r\sin\theta = \sin\theta\,(1 + \cos\theta) \quad \Rightarrow$

$\frac{dy}{d\theta} = (1 + \cos\theta)\,\cos\theta - \sin^2\theta = 2\cos^2\theta + \cos\theta - 1 = (2\cos\theta - 1)(\cos\theta + 1) = 0 \quad \Rightarrow \quad \cos\theta = \frac{1}{2}$ or $-1 \quad \Rightarrow$

$\theta = \frac{\pi}{3}, \pi$, or $\frac{5\pi}{3} \quad \Rightarrow \quad$ horizontal tangent at $\left(\frac{3}{2}, \frac{\pi}{3}\right)$, $(0, \pi)$, and $\left(\frac{3}{2}, \frac{5\pi}{3}\right)$.

$\frac{dx}{d\theta} = -(1 + \cos\theta)\sin\theta - \cos\theta\sin\theta = -\sin\theta\,(1 + 2\cos\theta) = 0 \quad \Rightarrow \quad \sin\theta = 0$ or $\cos\theta = -\frac{1}{2} \quad \Rightarrow$

$\theta = 0, \pi, \frac{2\pi}{3}$, or $\frac{4\pi}{3} \quad \Rightarrow \quad$ vertical tangent at $(2, 0)$, $\left(\frac{1}{2}, \frac{2\pi}{3}\right)$, and $\left(\frac{1}{2}, \frac{4\pi}{3}\right)$.

Note that the tangent is horizontal, not vertical when $\theta = \pi$, since $\displaystyle\lim_{\theta \to \pi} \frac{dy/d\theta}{dx/d\theta} = 0$.

64. $r = e^\theta \quad \Rightarrow \quad x = r\cos\theta = e^\theta\cos\theta,\;\; y = r\sin\theta = e^\theta\sin\theta \quad \Rightarrow$

$\frac{dy}{d\theta} = e^\theta\sin\theta + e^\theta\cos\theta = e^\theta(\sin\theta + \cos\theta) = 0 \quad \Rightarrow \quad \sin\theta = -\cos\theta \quad \Rightarrow \quad \tan\theta = -1 \quad \Rightarrow$

$\theta = -\frac{1}{4}\pi + n\pi$ [n any integer] $\Rightarrow$ horizontal tangents at $\left(e^{\pi(n-1/4)}, \pi\left(n - \frac{1}{4}\right)\right)$.

$\frac{dx}{d\theta} = e^\theta \cos\theta - e^\theta \sin\theta = e^\theta(\cos\theta - \sin\theta) = 0 \Rightarrow \sin\theta = \cos\theta \Rightarrow \tan\theta = 1 \Rightarrow$

$\theta = \frac{1}{4}\pi + n\pi$ [n any integer] $\Rightarrow$ vertical tangents at $\left(e^{\pi(n+1/4)}, \pi\left(n + \frac{1}{4}\right)\right)$.

65. $r = a\sin\theta + b\cos\theta \Rightarrow r^2 = ar\sin\theta + br\cos\theta \Rightarrow x^2 + y^2 = ay + bx \Rightarrow$

$x^2 - bx + \left(\frac{1}{2}b\right)^2 + y^2 - ay + \left(\frac{1}{2}a\right)^2 = \left(\frac{1}{2}b\right)^2 + \left(\frac{1}{2}a\right)^2 \Rightarrow \left(x - \frac{1}{2}b\right)^2 + \left(y - \frac{1}{2}a\right)^2 = \frac{1}{4}(a^2 + b^2)$, and this is a circle with center $\left(\frac{1}{2}b, \frac{1}{2}a\right)$ and radius $\frac{1}{2}\sqrt{a^2 + b^2}$.

66. These curves are circles which intersect at the origin and at $\left(\frac{1}{\sqrt{2}}a, \frac{\pi}{4}\right)$. At the origin, the first circle has a horizontal tangent and the second a vertical one, so the tangents are perpendicular here. For the first circle [$r = a\sin\theta$], $dy/d\theta = a\cos\theta\,\sin\theta + a\sin\theta\,\cos\theta = a\sin 2\theta = a$ at $\theta = \frac{\pi}{4}$ and $dx/d\theta = a\cos^2\theta - a\sin^2\theta = a\cos 2\theta = 0$ at $\theta = \frac{\pi}{4}$, so the tangent here is vertical. Similarly, for the second circle [$r = a\cos\theta$], $dy/d\theta = a\cos 2\theta = 0$ and $dx/d\theta = -a\sin 2\theta = -a$ at $\theta = \frac{\pi}{4}$, so the tangent is horizontal, and again the tangents are perpendicular.

67. $r = 1 + 2\sin(\theta/2)$. The parameter interval is $[0, 4\pi]$.

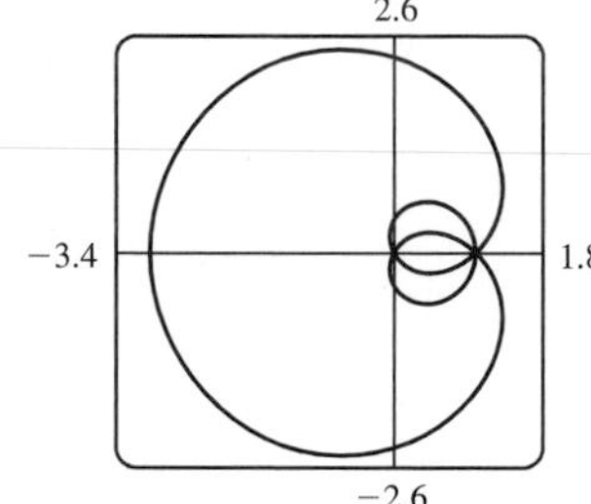

68. $r = \sqrt{1 - 0.8\sin^2\theta}$. The parameter interval is $[0, 2\pi]$.

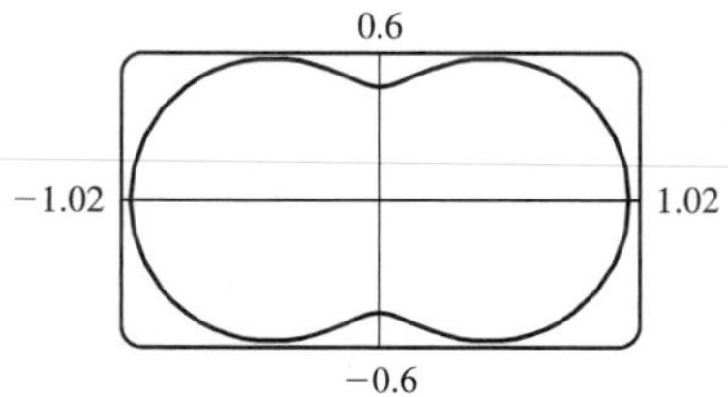

69. $r = e^{\sin\theta} - 2\cos(4\theta)$.

The parameter interval is $[0, 2\pi]$.

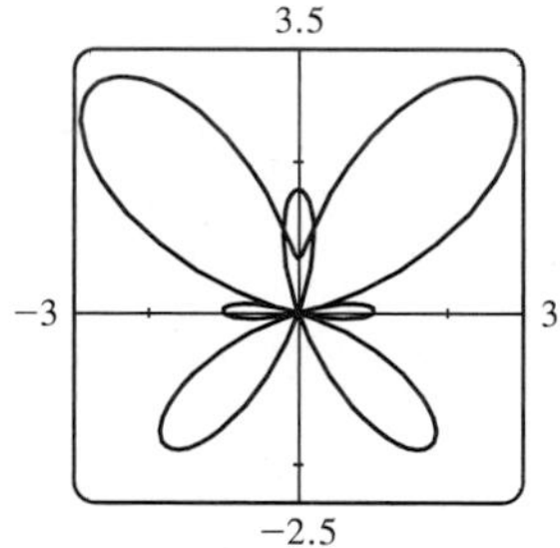

70. $r = |\tan\theta|^{|\cot\theta|}$.

The parameter interval $[0, \pi]$ produces the heart-shaped valentine curve shown in the first window.

The complete curve, including the reflected heart, is produced by the parameter interval $[0, 2\pi]$, but perhaps you'll agree that the first curve is more appropriate.

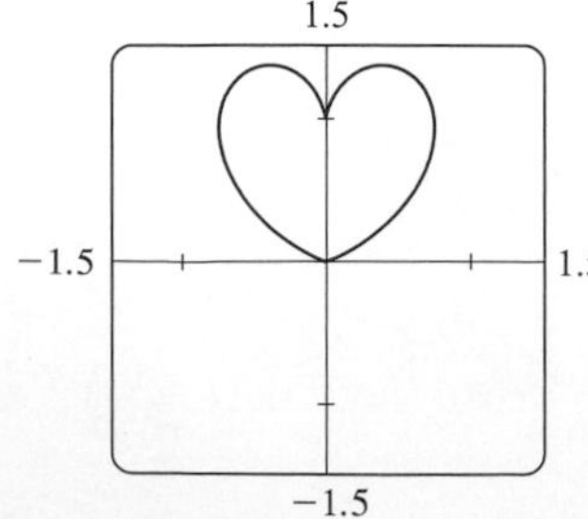

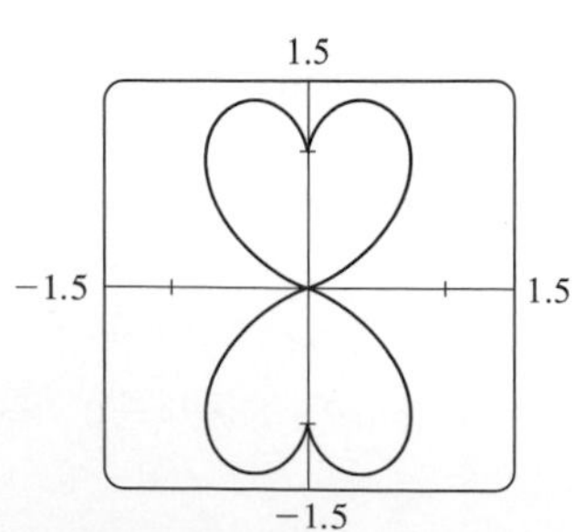

71. $r = 1 + \cos^{999}\theta$. The parameter interval is $[0, 2\pi]$.

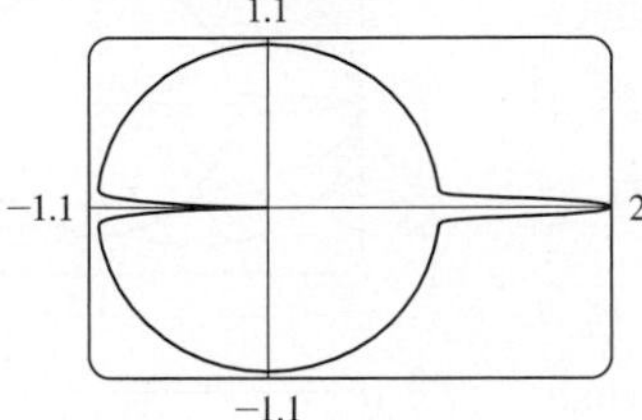

72. $r = 2 + \cos(9\theta/4)$. The parameter interval is $[0, 8\pi]$.

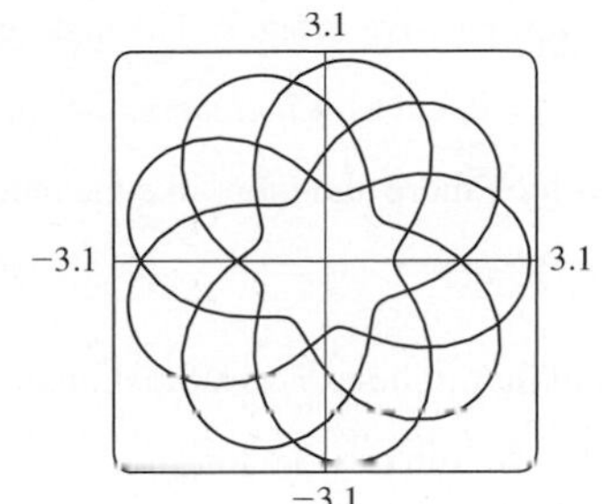

73. It appears that the graph of $r = 1 + \sin\left(\theta - \frac{\pi}{6}\right)$ is the same shape as the graph of $r = 1 + \sin\theta$, but rotated counterclockwise about the origin by $\frac{\pi}{6}$. Similarly, the graph of $r = 1 + \sin\left(\theta - \frac{\pi}{3}\right)$ is rotated by $\frac{\pi}{3}$. In general, the graph of $r = f(\theta - \alpha)$ is the same shape as that of $r = f(\theta)$, but rotated counterclockwise through α about the origin. That is, for any point (r_0, θ_0) on the curve $r = f(\theta)$, the point $(r_0, \theta_0 + \alpha)$ is on the curve $r = f(\theta - \alpha)$, since $r_0 = f(\theta_0) = f((\theta_0 + \alpha) - \alpha)$.

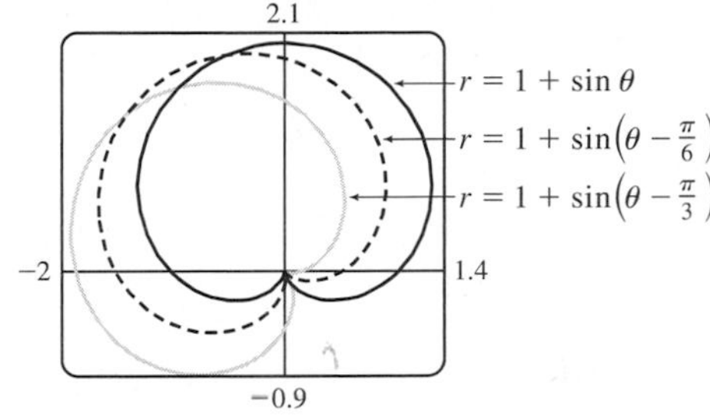

74.

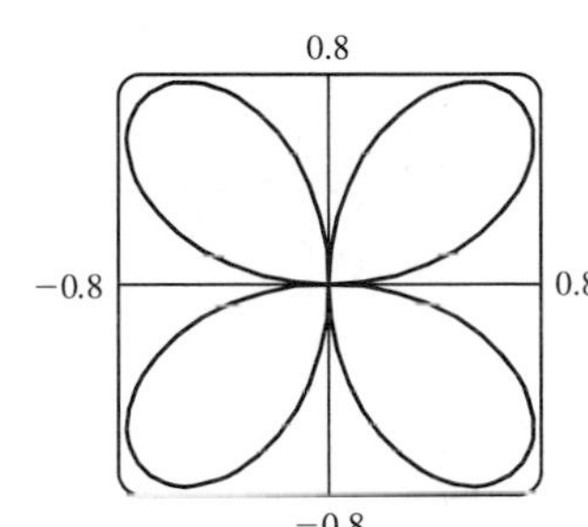

From the graph, the highest points seem to have $y \approx 0.77$. To find the exact value, we solve $dy/d\theta = 0$. $y = r\sin\theta = \sin\theta\,\sin 2\theta \quad \rightarrow$

$$\begin{aligned} dy/d\theta &= 2\sin\theta\,\cos 2\theta + \cos\theta\,\sin 2\theta \\ &= 2\sin\theta\,(2\cos^2\theta - 1) + \cos\theta\,(2\sin\theta\,\cos\theta) \\ &= 2\sin\theta\,(3\cos^2\theta - 1) \end{aligned}$$

In the first quadrant, this is 0 when $\cos\theta = \frac{1}{\sqrt{3}} \quad \Leftrightarrow \quad \sin\theta = \sqrt{\frac{2}{3}} \quad \Leftrightarrow$

$y = 2\sin^2\theta\cos\theta = 2 \cdot \frac{2}{3} \cdot \frac{1}{\sqrt{3}} = \frac{4}{9}\sqrt{3} \approx 0.77.$

75. Consider curves with polar equation $r = 1 + c\cos\theta$, where c is a real number. If $c = 0$, we get a circle of radius 1 centered at the pole. For $0 < c \leq 0.5$, the curve gets slightly larger, moves right, and flattens out a bit on the left side. For $0.5 < c < 1$, the left side has a dimple shape. For $c = 1$, the dimple becomes a cusp. For $c > 1$, there is an internal loop. For $c \geq 0$, the rightmost point on the curve is $(1 + c, 0)$. For $c < 0$, the curves are reflections through the vertical axis of the curves with $c > 0$.

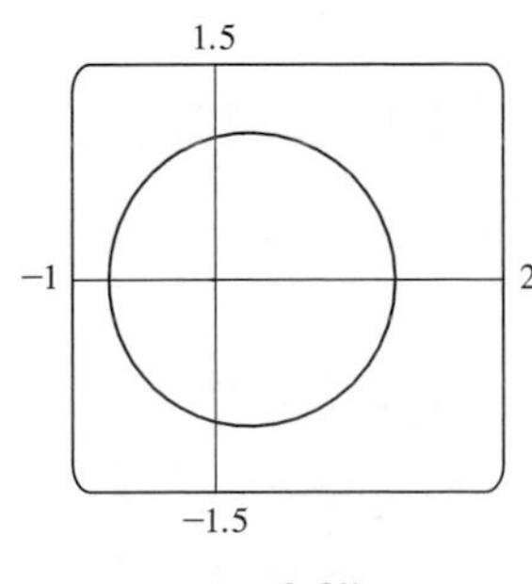

$c = 0.25$

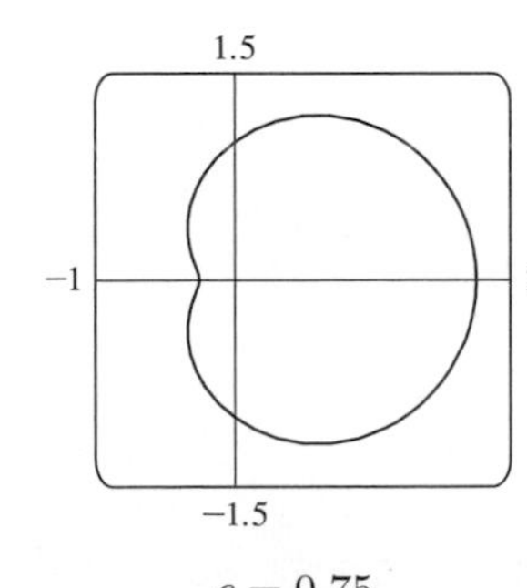

$c = 0.75$

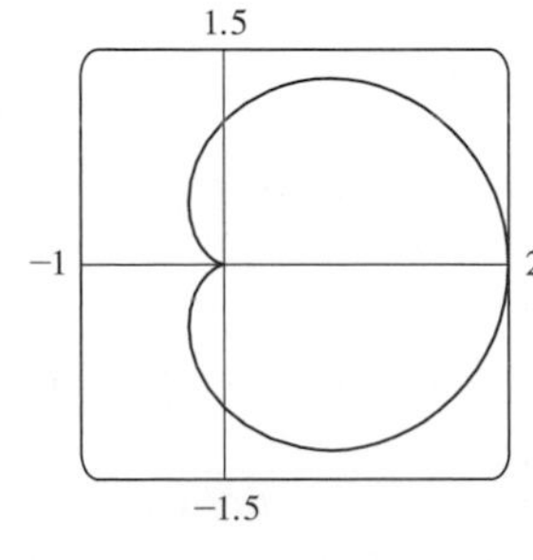

$c = 1$

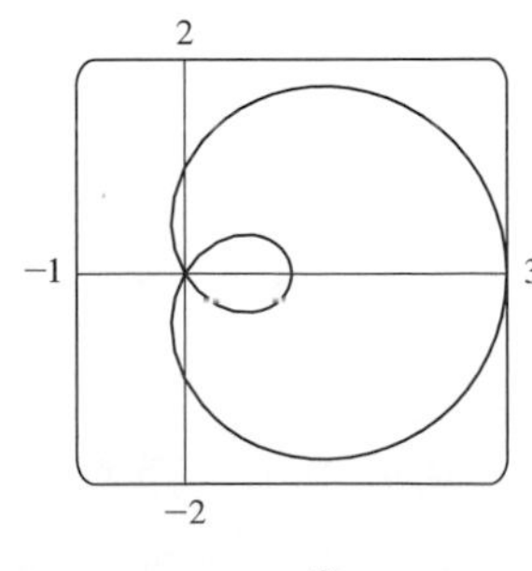

$c = 2$

76. Consider the polar curves $r = 1 + \cos^n \theta$, where n is a positive integer. First, let **n be an even positive integer**. The first figure shows that the curve has a peanut shape for $n = 2$, but as n increases, the ends are squeezed. As n becomes large, the curves look more and more like the unit circle, but with spikes to the points $(2, 0)$ and $(2, \pi)$.

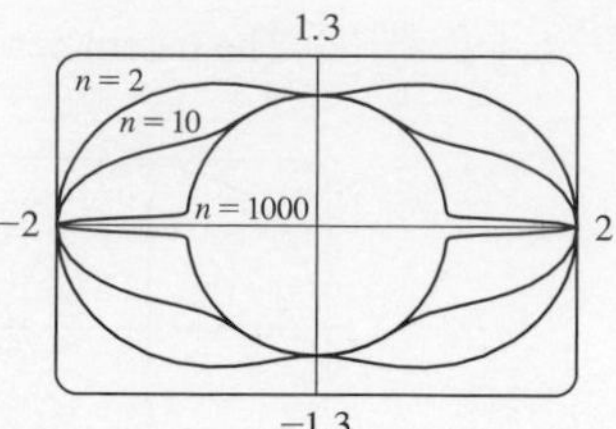

The second figure shows r as a function of θ in Cartesian coordinates for the same values of n. We can see that for large n, the graph is similar to the graph of $y = 1$, but with spikes to $y = 2$ for $x = 0$, π, and 2π. (Note that when $0 < \cos\theta < 1$, $\cos^{1000}\theta$ is very small.)

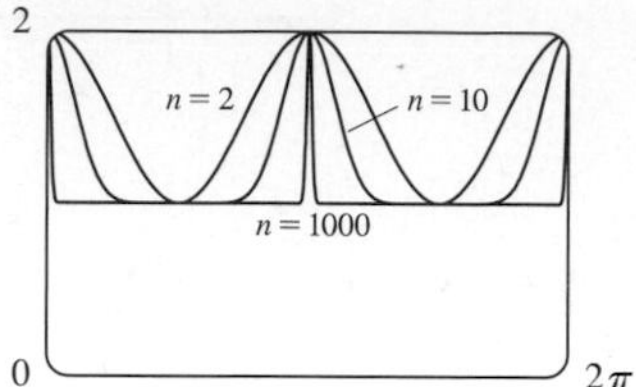

Next, let **n be an odd positive integer**. The third figure shows that the curve is a cardioid for $n = 1$, but as n increases, the heart shape becomes more pronounced. As n becomes large, the curves again look more like the unit circle, but with an outward spike to $(2, 0)$ and an inward spike to $(0, \pi)$.

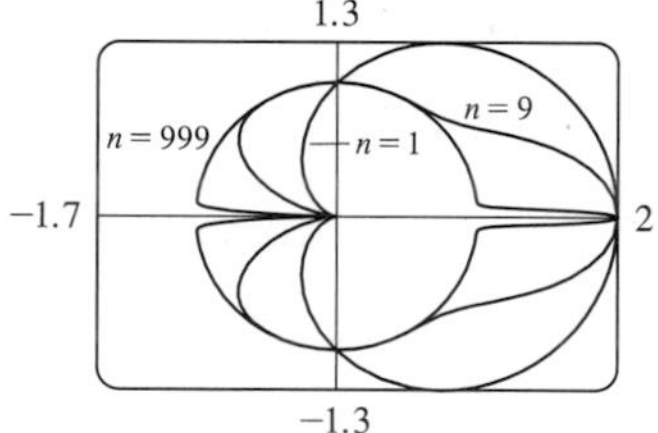

The fourth figure shows r as a function of θ in Cartesian coordinates for the same values of n. We can see that for large n, the graph is similar to the graph of $y = 1$, but spikes to $y = 2$ for $x = 0$ and π, and to $y = 0$ for $x = \pi$.

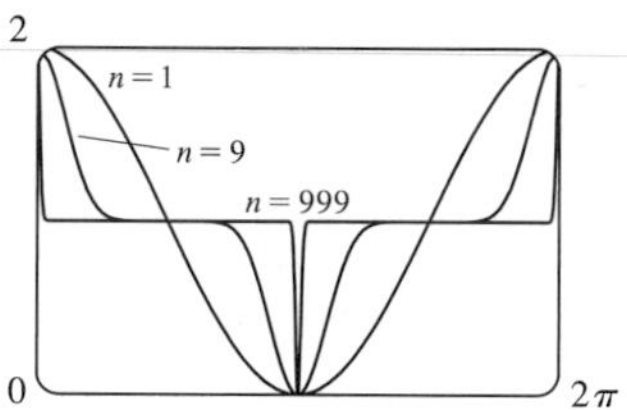

77. $$\tan\psi = \tan(\phi - \theta) = \frac{\tan\phi - \tan\theta}{1 + \tan\phi\,\tan\theta} = \frac{\dfrac{dy}{dx} - \tan\theta}{1 + \dfrac{dy}{dx}\tan\theta} = \frac{\dfrac{dy/d\theta}{dx/d\theta} - \tan\theta}{1 + \dfrac{dy/d\theta}{dx/d\theta}\tan\theta}$$

$$= \frac{\dfrac{dy}{d\theta} - \dfrac{dx}{d\theta}\tan\theta}{\dfrac{dx}{d\theta} + \dfrac{dy}{d\theta}\tan\theta} = \frac{\left(\dfrac{dr}{d\theta}\sin\theta + r\cos\theta\right) - \tan\theta\left(\dfrac{dr}{d\theta}\cos\theta - r\sin\theta\right)}{\left(\dfrac{dr}{d\theta}\cos\theta - r\sin\theta\right) + \tan\theta\left(\dfrac{dr}{d\theta}\sin\theta + r\cos\theta\right)} = \frac{r\cos\theta + r\cdot\dfrac{\sin^2\theta}{\cos\theta}}{\dfrac{dr}{d\theta}\cos\theta + \dfrac{dr}{d\theta}\cdot\dfrac{\sin^2\theta}{\cos\theta}}$$

$$= \frac{r\cos^2\theta + r\sin^2\theta}{\dfrac{dr}{d\theta}\cos^2\theta + \dfrac{dr}{d\theta}\sin^2\theta} = \frac{r}{dr/d\theta}$$

78. (a) $r = e^\theta \;\Rightarrow\; dr/d\theta = e^\theta$, so by Exercise 77, $\tan\psi = r/e^\theta = 1 \;\Rightarrow$ $\psi = \arctan 1 = \frac{\pi}{4}$.

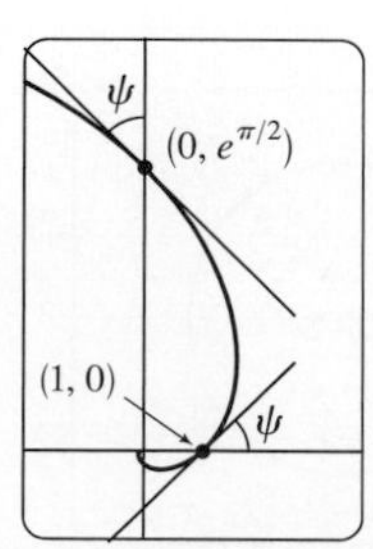

(b) The Cartesian equation of the tangent line at $(1, 0)$ is $y = x - 1$, and that of the tangent line at $(0, e^{\pi/2})$ is $y = e^{\pi/2} - x$.

(c) Let a be the tangent of the angle between the tangent and radial lines, that is, $a = \tan\psi$. Then, by Exercise 77, $a = \dfrac{r}{dr/d\theta} \;\Rightarrow\; \dfrac{dr}{d\theta} = \dfrac{1}{a}r \;\Rightarrow$ $r = Ce^{\theta/a}$ (by Theorem 9.4.2).

LABORATORY PROJECT Families of Polar Curves

1. (a) $r = \sin n\theta$.

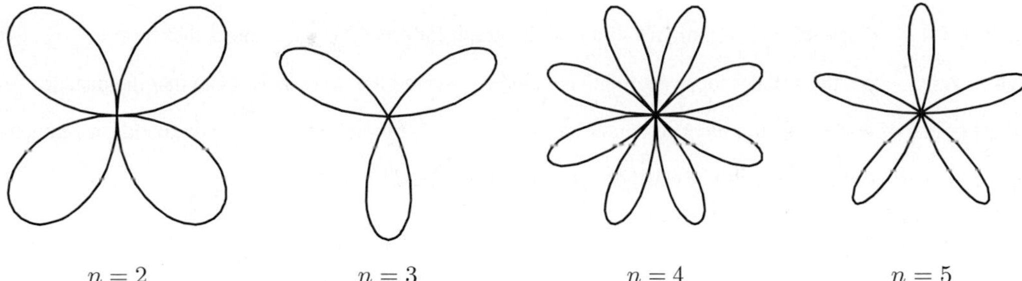

$n = 2$ $n = 3$ $n = 4$ $n = 5$

From the graphs, it seems that when n is even, the number of loops in the curve (called a rose) is $2n$, and when n is odd, the number of loops is simply n. This is because in the case of n odd, every point on the graph is traversed twice, due to the fact that

$$r(\theta + \pi) = \sin[n(\theta + \pi)] = \sin n\theta \, \cos n\pi + \cos n\theta \, \sin n\pi = \begin{cases} \sin n\theta & \text{if } n \text{ is even} \\ -\sin n\theta & \text{if } n \text{ is odd} \end{cases}$$

(b) The graph of $r = |\sin n\theta|$ has $2n$ loops whether n is odd or even, since $r(\theta + \pi) = r(\theta)$.

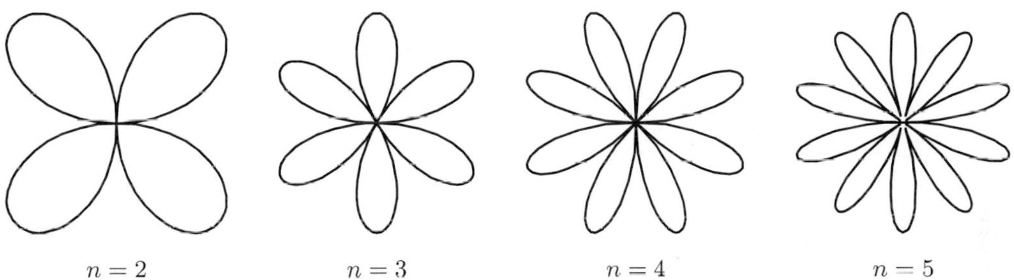

$n = 2$ $n = 3$ $n = 4$ $n = 5$

2. $r = 1 + c \sin n\theta$. We vary n while keeping c constant at 2. As n changes, the curves change in the same way as those in Exercise 1: the number of loops increases. Note that if n is even, the smaller loops are outside the larger ones; if n is odd, they are inside.

$c = 2$

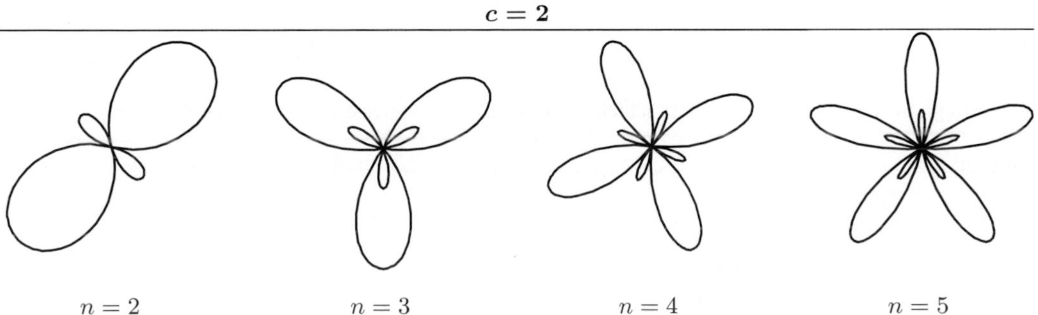

$n = 2$ $n = 3$ $n = 4$ $n = 5$

Now we vary c while keeping $n = 3$. As c increases toward 0, the entire graph gets smaller (the graphs below are not to scale) and the smaller loops shrink in relation to the large ones. At $c = -1$, the small loops disappear entirely, and for $-1 < c < 1$, the graph is a simple, closed curve (at $c = 0$ it is a circle). As c continues to increase, the same changes are seen, but in reverse order, since $1 + (-c)\sin n\theta = 1 + c\sin n(\theta + \pi)$, so the graph for $c = c_0$ is the same as that for $c = -c_0$, with a rotation through π. As $c \to \infty$, the smaller loops get relatively closer in size to the large ones. Note that the distance between the outermost points of corresponding inner and outer loops is always 2. Maple's `animate` command (or Mathematica's `Animate`) is very useful for seeing the changes that occur as c varies.

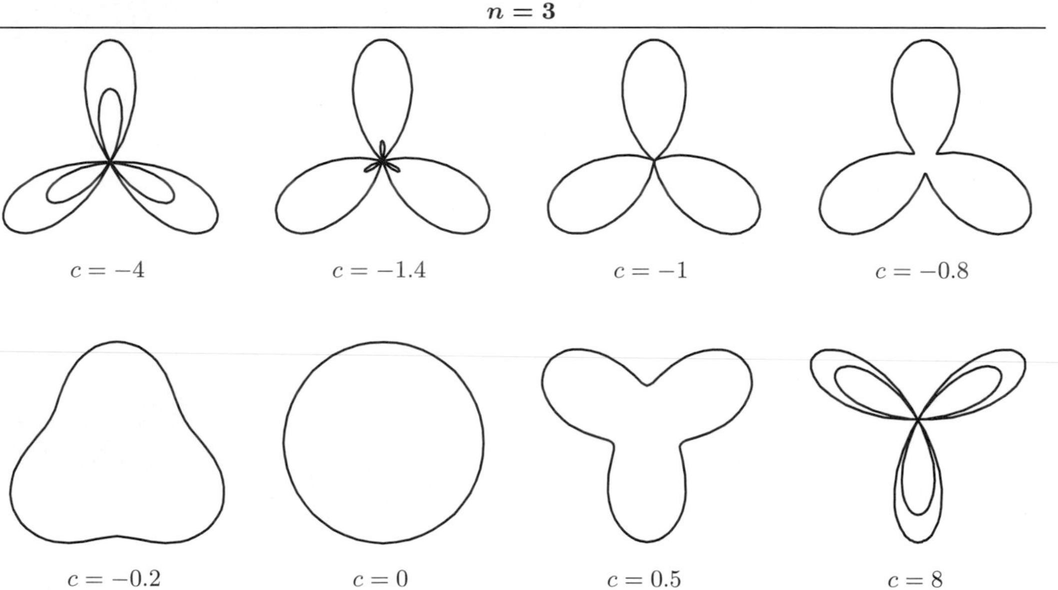

3. $r = \dfrac{1 - a\cos\theta}{1 + a\cos\theta}$. We start with $a = 0$, since in this case the curve is simply the circle $r = 1$.

As a increases, the graph moves to the left, and its right side becomes flattened. As a increases through about 0.4, the right side seems to grow a dimple, which upon closer investigation (with narrower θ-ranges) seems to appear at $a \approx 0.42$ [the actual value is $\sqrt{2} - 1$]. As $a \to 1$, this dimple becomes more pronounced, and the curve begins to stretch out horizontally, until at $a = 1$ the denominator vanishes at $\theta = \pi$, and the dimple becomes an actual cusp. For $a > 1$ we must choose our parameter interval carefully, since $r \to \infty$ as $1 + a\cos\theta \to 0 \quad \Leftrightarrow \quad \theta \to \pm\cos^{-1}(-1/a)$. As a increases from 1, the curve splits into two parts. The left part has a loop, which grows larger as a increases, and the right part grows broader vertically, and its left tip develops a dimple when $a \approx 2.42$ [actually, $\sqrt{2} + 1$]. As a increases, the dimple grows more and more pronounced. If $a < 0$, we get the same graph as we do for the corresponding positive a-value, but with a rotation through π about the pole, as happened when c was replaced with $-c$ in Exercise 2.

[continued]

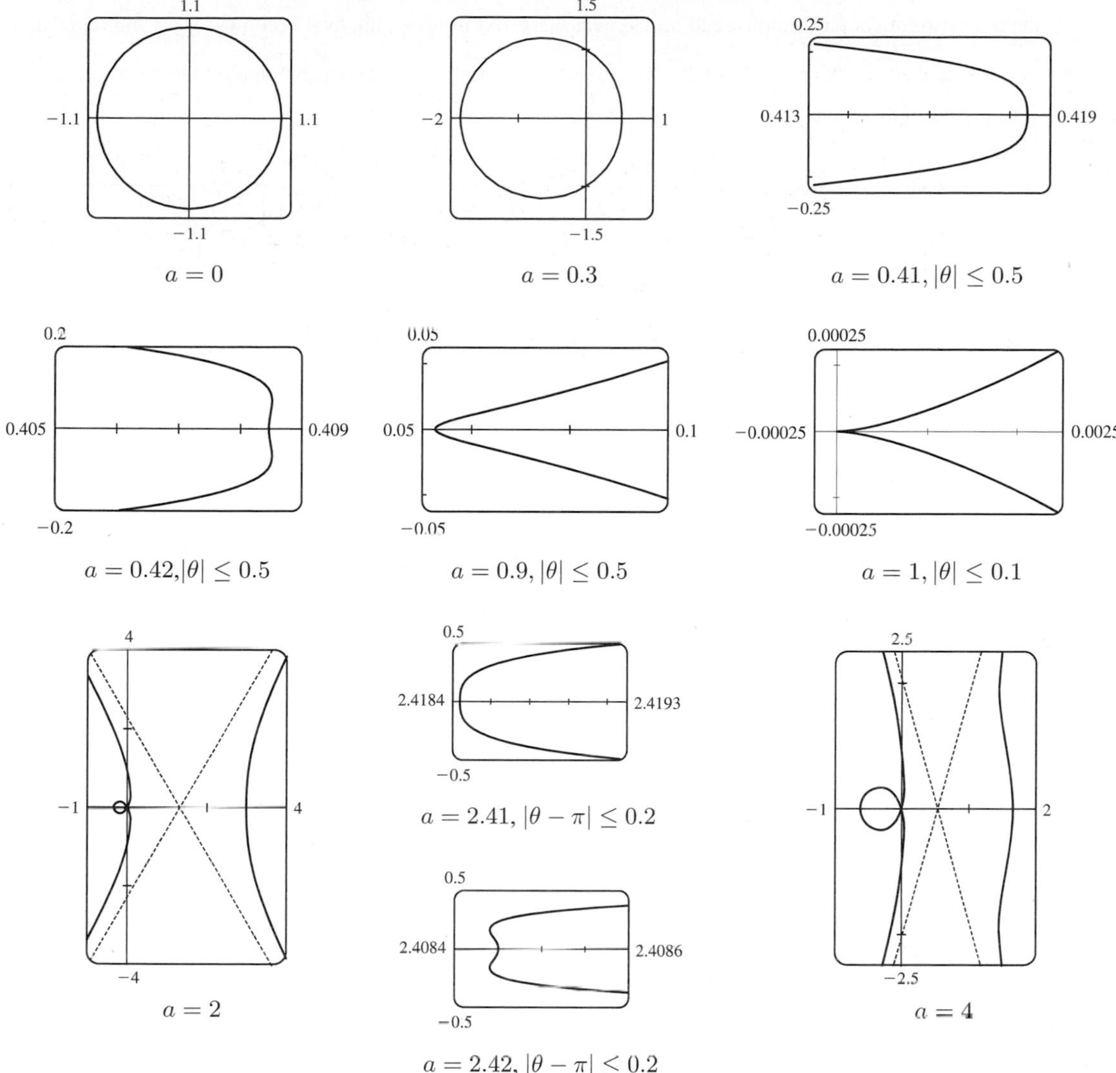

$a = 0$ $\quad a = 0.3$ $\quad a = 0.41, |\theta| \le 0.5$

$a = 0.42, |\theta| \le 0.5$ $\quad a = 0.9, |\theta| \le 0.5$ $\quad a = 1, |\theta| \le 0.1$

$a = 2$ $\quad a = 2.41, |\theta - \pi| \le 0.2$ $\quad a = 2.42, |\theta - \pi| \le 0.2$ $\quad a = 4$

4. Most graphing devices cannot plot implicit polar equations, so we must first find an explicit expression (or expressions) for r in terms of θ, a, and c. We note that the given equation, $r^4 - 2c^2r^2\cos 2\theta + c^4 - a^4 = 0$, is a quadratic in r^2, so we use the quadratic formula and find that

$$r^2 = \frac{2c^2\cos 2\theta \pm \sqrt{4c^4\cos^2 2\theta - 4(c^4 - a^4)}}{2} = c^2\cos 2\theta \pm \sqrt{a^4 - c^4\sin^2 2\theta}$$

so $r = \pm\sqrt{c^2\cos 2\theta \pm \sqrt{a^4 - c^4\sin^2 2\theta}}$. So for each graph, we must plot four curves to be sure of plotting all the points which satisfy the given equation. Note that all four functions have period π.

We start with the case $a = c = 1$, and the resulting curve resembles the symbol for infinity. If we let a decrease, the curve splits into two symmetric parts, and as a decreases further, the parts become smaller, further apart, and rounder. If instead we let a increase from 1, the two lobes of the curve join together, and as a increases further they continue to merge, until at

$a \approx 1.4$, the graph no longer has dimples, and has an oval shape. As $a \to \infty$, the oval becomes larger and rounder, since the c^2 and c^4 terms lose their significance. Note that the shape of the graph seems to depend only on the ratio c/a, while the size of the graph varies as c and a jointly increase.

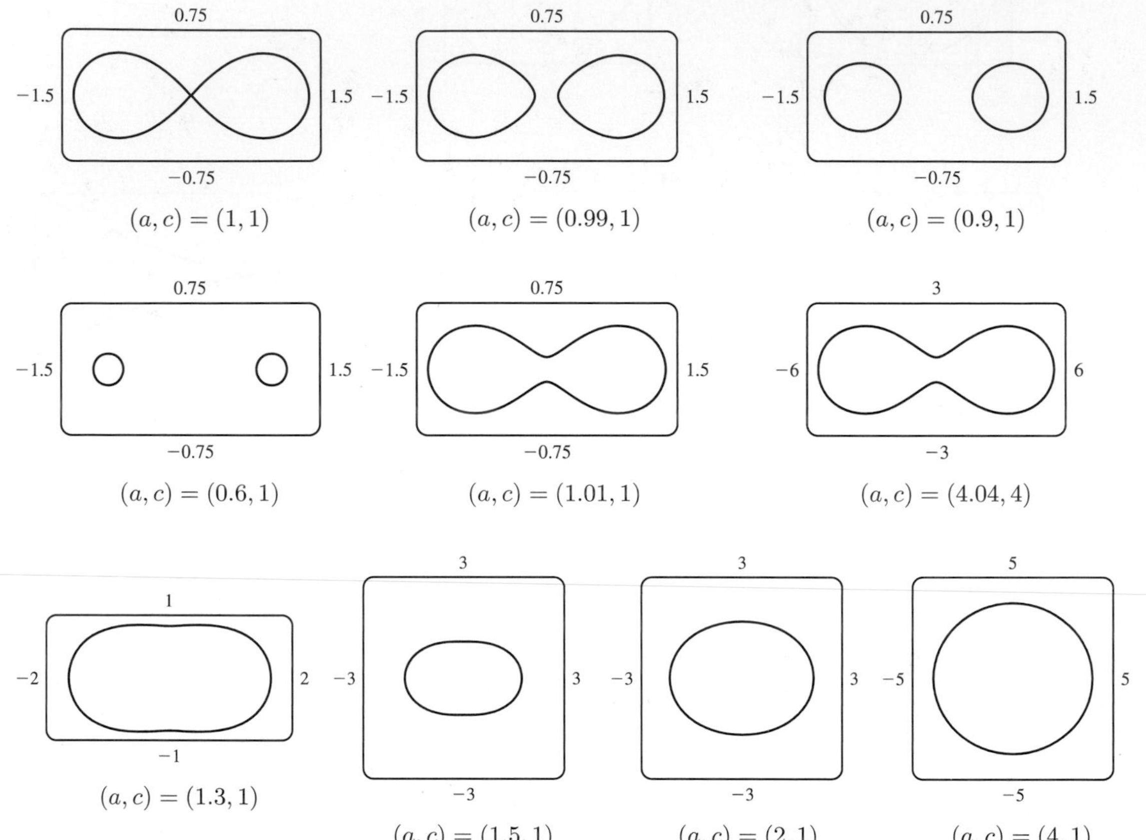

10.4 Areas and Lengths in Polar Coordinates

1. $r = e^{-\theta/4},\ \pi/2 \le \theta \le \pi.$

$$A = \int_{\pi/2}^{\pi} \tfrac{1}{2} r^2\, d\theta = \int_{\pi/2}^{\pi} \tfrac{1}{2}(e^{-\theta/4})^2\, d\theta = \int_{\pi/2}^{\pi} \tfrac{1}{2} e^{-\theta/2}\, d\theta = \tfrac{1}{2}\Big[-2e^{-\theta/2}\Big]_{\pi/2}^{\pi} = -1(e^{-\pi/2} - e^{-\pi/4}) = e^{-\pi/4} - e^{-\pi/2}$$

2. $r = \cos\theta,\ 0 \le \theta \le \pi/6.$

$$A = \int_0^{\pi/6} \tfrac{1}{2} r^2\, d\theta = \int_0^{\pi/6} \tfrac{1}{2}\cos^2\theta\, d\theta = \tfrac{1}{2}\int_0^{\pi/6} \tfrac{1}{2}(1 + \cos 2\theta)\, d\theta = \tfrac{1}{4}\big[\theta + \tfrac{1}{2}\sin 2\theta\big]_0^{\pi/6}$$
$$= \tfrac{1}{4}\big(\tfrac{\pi}{6} + \tfrac{1}{2}\cdot\tfrac{1}{2}\sqrt{3}\big) = \tfrac{\pi}{24} + \tfrac{1}{16}\sqrt{3}$$

3. $r = \sin\theta + \cos\theta,\ 0 \le \theta \le \pi.$

$$A = \int_0^{\pi} \frac{1}{2} r^2\, d\theta = \int_0^{\pi} \frac{1}{2}(\sin\theta + \cos\theta)^2\, d\theta = \int_0^{\pi} \frac{1}{2}(\sin^2\theta + 2\sin\theta\,\cos\theta + \cos^2\theta)\, d\theta = \int_0^{\pi} \frac{1}{2}(1 + \sin 2\theta)\, d\theta$$
$$= \tfrac{1}{2}\Big[\theta - \tfrac{1}{2}\cos 2\theta\Big]_0^{\pi} = \tfrac{1}{2}\big[(\pi - \tfrac{1}{2}) - (0 - \tfrac{1}{2})\big] = \tfrac{\pi}{2}$$

4. $r = 1/\theta,\ \pi/2 \le \theta \le 2\pi.$

$$A = \int_{\pi/2}^{2\pi} \frac{1}{2} r^2\, d\theta = \int_{\pi/2}^{2\pi} \frac{1}{2}\left(\frac{1}{\theta}\right)^2 d\theta = \int_{\pi/2}^{2\pi} \frac{1}{2}\theta^{-2}\, d\theta = \frac{1}{2}\left[-\frac{1}{\theta}\right]_{\pi/2}^{2\pi}$$

$$= \tfrac{1}{2}\left(-\tfrac{1}{2\pi} + \tfrac{2}{\pi}\right) = \tfrac{1}{2}\left(-\tfrac{1}{2\pi} + \tfrac{4}{2\pi}\right) = \tfrac{3}{4\pi}$$

5. $r^2 = \sin 2\theta,\ 0 \le \theta \le \pi/2.$

$$A = \int_0^{\pi/2} \tfrac{1}{2} r^2\, d\theta = \int_0^{\pi/2} \tfrac{1}{2}\sin 2\theta\, d\theta = \left[-\tfrac{1}{4}\cos 2\theta\right]_0^{\pi/2} = -\tfrac{1}{4}(\cos\pi - \cos 0) = -\tfrac{1}{4}(-1-1) = \tfrac{1}{2}$$

6. $r = 2 + \cos\theta,\ \pi/2 \le \theta \le \pi.$

$$A = \int_{\pi/2}^{\pi} \tfrac{1}{2} r^2\, d\theta = \int_{\pi/2}^{\pi} \tfrac{1}{2}(2+\cos\theta)^2\, d\theta = \int_{\pi/2}^{\pi} \tfrac{1}{2}(4 + 4\cos\theta + \cos^2\theta)\, d\theta = \int_{\pi/2}^{\pi} \tfrac{1}{2}[4 + 4\cos\theta + \tfrac{1}{2}(1+\cos 2\theta)]\, d\theta$$

$$= \int_{\pi/2}^{\pi} \left(\tfrac{9}{4} + 2\cos\theta + \tfrac{1}{4}\cos 2\theta\right) d\theta = \left[\tfrac{9}{4}\theta + 2\sin\theta + \tfrac{1}{8}\sin 2\theta\right]_{\pi/2}^{\pi} = \left(\tfrac{9\pi}{4} + 0 + 0\right) - \left(\tfrac{9\pi}{8} + 2 + 0\right) = \tfrac{9\pi}{8} - 2$$

7. $r = 4 + 3\sin\theta,\ -\frac{\pi}{2} \le \theta \le \frac{\pi}{2}.$

$$A = \int_{-\pi/2}^{\pi/2} \tfrac{1}{2}((4+3\sin\theta)^2\, d\theta = \tfrac{1}{2}\int_{-\pi/2}^{\pi/2} (16 + 24\sin\theta + 9\sin^2\theta)\, d\theta$$

$$= \tfrac{1}{2}\int_{-\pi/2}^{\pi/2} (16 + 9\sin^2\theta)\, d\theta \qquad \text{[by Theorem 5.5.7(b)]}$$

$$= \tfrac{1}{2}\cdot 2\int_0^{\pi/2} [16 + 9\cdot\tfrac{1}{2}(1-\cos 2\theta)]\, d\theta \qquad \text{[by Theorem 5.5.7(a)]}$$

$$= \int_0^{\pi/2} \left(\tfrac{41}{2} - \tfrac{9}{2}\cos 2\theta\right) d\theta = \left[\tfrac{41}{2}\theta - \tfrac{9}{4}\sin 2\theta\right]_0^{\pi/2} = \left(\tfrac{41\pi}{4} - 0\right) - (0 - 0) = \tfrac{41\pi}{4}$$

8. $r = \sqrt{\ln\theta},\ 1 \le \theta \le 2\pi.$

$$A = \int_1^{2\pi} \tfrac{1}{2}\left(\sqrt{\ln\theta}\right)^2 d\theta = \int_1^{2\pi} \tfrac{1}{2}\ln\theta\, d\theta = \left[\tfrac{1}{2}\theta\ln\theta\right]_1^{2\pi} - \int_1^{2\pi} \tfrac{1}{2}\, d\theta \qquad \begin{bmatrix} u = \ln\theta, & dv = \tfrac{1}{2}\, d\theta \\ du = (1/\theta)\, d\theta, & v = \tfrac{1}{2}\theta \end{bmatrix}$$

$$= [\pi\ln(2\pi) - 0] - \left[\tfrac{1}{2}\theta\right]_1^{2\pi} = \pi\ln(2\pi) - \pi + \tfrac{1}{2}$$

9. The area is bounded by $r = 2\sin\theta$ for $\theta = 0$ to $\theta = \pi$.

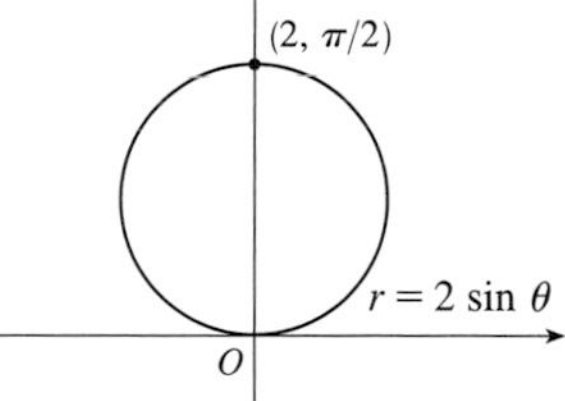

$$A = \int_0^{\pi} \tfrac{1}{2} r^2\, d\theta = \tfrac{1}{2}\int_0^{\pi} (2\sin\theta)^2\, d\theta = \tfrac{1}{2}\int_0^{\pi} 4\sin^2\theta\, d\theta$$

$$= 2\int_0^{\pi} \tfrac{1}{2}(1 - \cos 2\theta) d\theta = \left[\theta - \tfrac{1}{2}\sin 2\theta\right]_0^{\pi} = \pi$$

Also, note that this is a circle with radius 1, so its area is $\pi(1)^2 = \pi$.

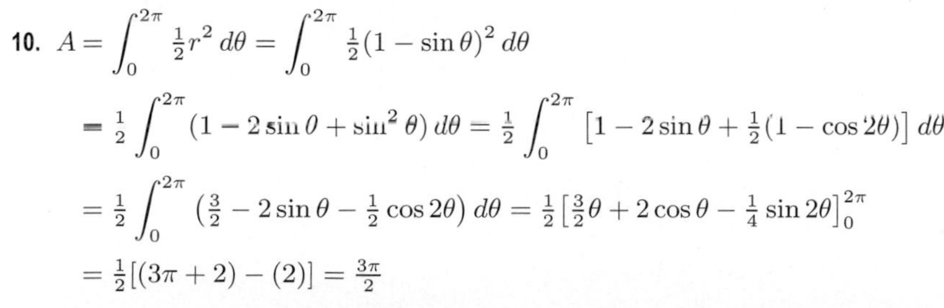

10. $$A = \int_0^{2\pi} \tfrac{1}{2} r^2\, d\theta = \int_0^{2\pi} \tfrac{1}{2}(1 - \sin\theta)^2\, d\theta$$

$$= \tfrac{1}{2}\int_0^{2\pi} (1 - 2\sin\theta + \sin^2\theta)\, d\theta = \tfrac{1}{2}\int_0^{2\pi} [1 - 2\sin\theta + \tfrac{1}{2}(1 - \cos 2\theta)]\, d\theta$$

$$= \tfrac{1}{2}\int_0^{2\pi} \left(\tfrac{3}{2} - 2\sin\theta - \tfrac{1}{2}\cos 2\theta\right) d\theta = \tfrac{1}{2}\left[\tfrac{3}{2}\theta + 2\cos\theta - \tfrac{1}{4}\sin 2\theta\right]_0^{2\pi}$$

$$= \tfrac{1}{2}[(3\pi + 2) - (2)] = \tfrac{3\pi}{2}$$

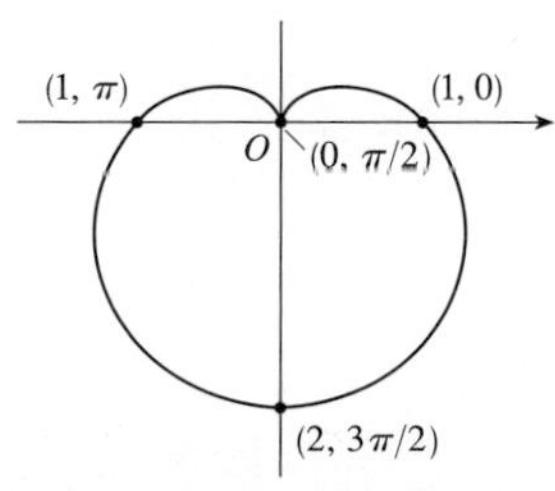

11. $A = \int_0^{2\pi} \frac{1}{2}r^2\,d\theta = \int_0^{2\pi} \frac{1}{2}(3 + 2\cos\theta)^2\,d\theta = \frac{1}{2}\int_0^{2\pi} (9 + 12\cos\theta + 4\cos^2\theta)\,d\theta$

$= \frac{1}{2}\int_0^{2\pi} \left[9 + 12\cos\theta + 4\cdot\frac{1}{2}(1 + \cos 2\theta)\right] d\theta$

$= \frac{1}{2}\int_0^{2\pi} (11 + 12\cos\theta + 2\cos 2\theta)\,d\theta = \frac{1}{2}\left[11\theta + 12\sin\theta + \sin 2\theta\right]_0^{2\pi}$

$= \frac{1}{2}(22\pi) = 11\pi$

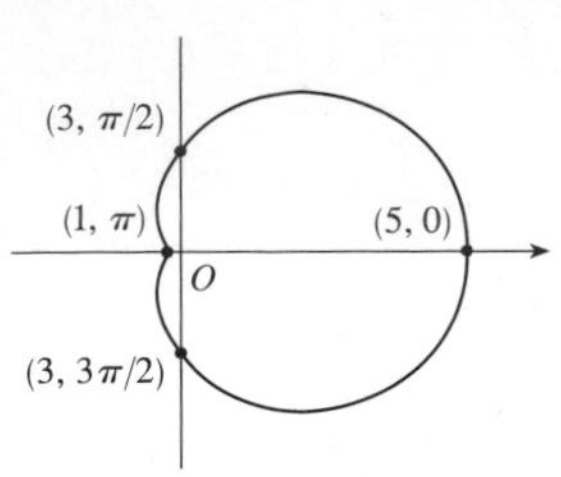

12. $A = \int_0^{2\pi} \frac{1}{2}r^2\,d\theta = \int_0^{2\pi} \frac{1}{2}(2 - \cos\theta)^2\,d\theta = \int_0^{2\pi} \frac{1}{2}(4 - 4\cos\theta + \cos^2\theta)\,d\theta$

$= \int_0^{2\pi} \frac{1}{2}\left[4 - 4\cos\theta + \frac{1}{2}(1 + \cos 2\theta)\right] d\theta = \int_0^{2\pi} \left(\frac{9}{4} - 2\cos\theta + \frac{1}{4}\cos 2\theta\right) d\theta$

$= \left[\frac{9}{4}\theta - 2\sin\theta + \frac{1}{8}\sin 2\theta\right]_0^{2\pi} = \left(\frac{9\pi}{2} - 0 + 0\right) - (0 - 0 + 0) = \frac{9\pi}{2}$

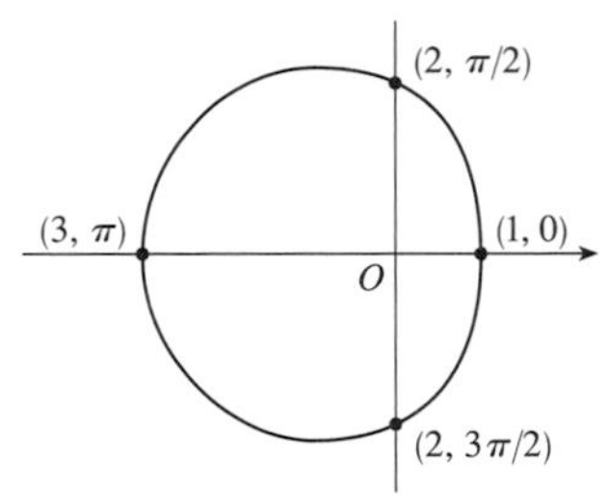

13. $A = \int_0^{2\pi} \frac{1}{2}r^2\,d\theta = \int_0^{2\pi} \frac{1}{2}(2 + \sin 4\theta)^2\,d\theta = \frac{1}{2}\int_0^{2\pi} (4 + 4\sin 4\theta + \sin^2 4\theta)\,d\theta$

$= \frac{1}{2}\int_0^{2\pi} \left[4 + 4\sin 4\theta + \frac{1}{2}(1 - \cos 8\theta)\right] d\theta$

$= \frac{1}{2}\int_0^{2\pi} \left(\frac{9}{2} + 4\sin 4\theta - \frac{1}{2}\cos 8\theta\right) d\theta = \frac{1}{2}\left[\frac{9}{2}\theta - \cos 4\theta - \frac{1}{16}\sin 8\theta\right]_0^{2\pi}$

$= \frac{1}{2}[(9\pi - 1) - (-1)] = \frac{9}{2}\pi$

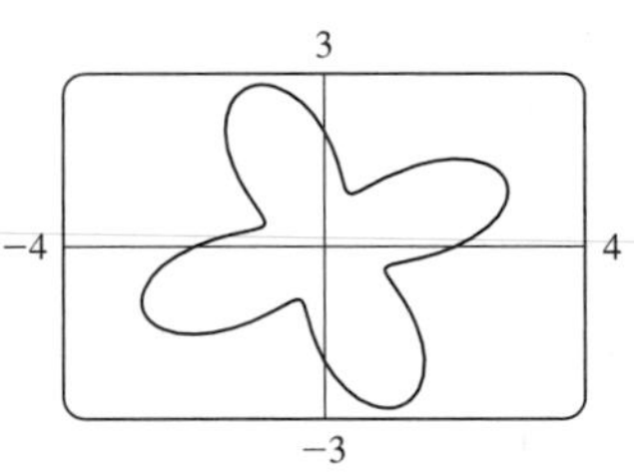

14. $A = \int_0^{2\pi} \frac{1}{2}r^2\,d\theta = \int_0^{2\pi} \frac{1}{2}(3 - 2\cos 4\theta)^2\,d\theta = \frac{1}{2}\int_0^{2\pi} (9 - 12\cos 4\theta + 4\cos^2 4\theta)\,d\theta$

$= \frac{1}{2}\int_0^{2\pi} \left[9 - 12\cos 4\theta + 4\cdot\frac{1}{2}(1 + \cos 8\theta)\right] d\theta$

$= \frac{1}{2}\int_0^{2\pi} (11 - 12\cos 4\theta + 2\cos 8\theta)\,d\theta = \frac{1}{2}\left[11\theta - 3\sin 4\theta + \frac{1}{4}\sin 8\theta\right]_0^{2\pi}$

$= \frac{1}{2}(22\pi) = 11\pi$

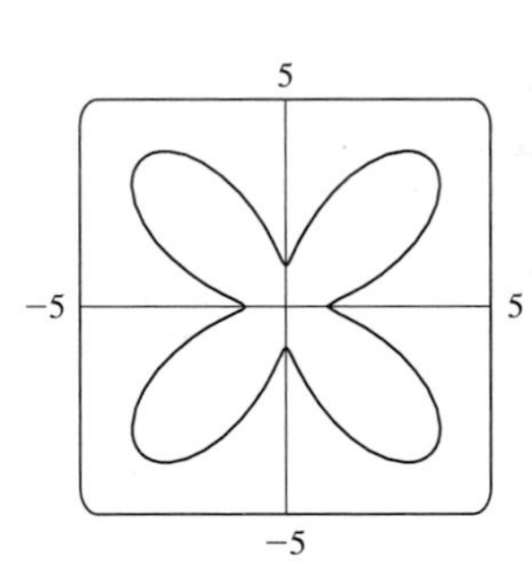

15. $A = \int_0^{2\pi} \frac{1}{2}r^2\,d\theta = \int_0^{2\pi} \frac{1}{2}\left(\sqrt{1 + \cos^2 5\theta}\right)^2 d\theta$

$= \frac{1}{2}\int_0^{2\pi} (1 + \cos^2 5\theta)\,d\theta = \frac{1}{2}\int_0^{2\pi} \left[1 + \frac{1}{2}(1 + \cos 10\theta)\right] d\theta$

$= \frac{1}{2}\left[\frac{3}{2}\theta + \frac{1}{20}\sin 10\theta\right]_0^{2\pi} = \frac{1}{2}(3\pi) = \frac{3}{2}\pi$

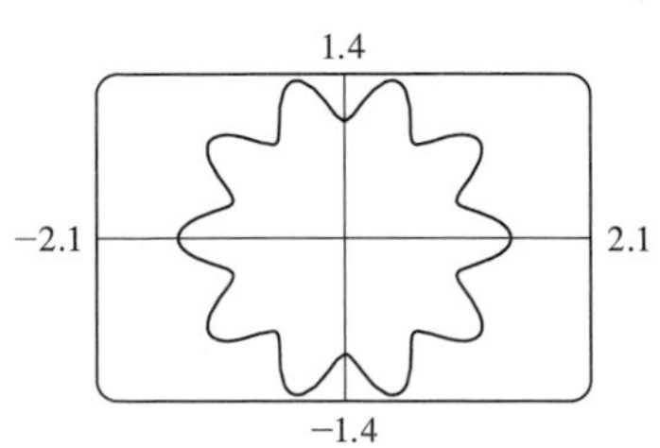

16. $A = \int_0^{2\pi} \frac{1}{2}r^2\,d\theta = \int_0^{2\pi} \frac{1}{2}(1 + 5\sin 6\theta)^2\,d\theta$

$= \frac{1}{2}\int_0^{2\pi} (1 + 10\sin 6\theta + 25\sin^2 6\theta)\,d\theta$

$= \frac{1}{2}\int_0^{2\pi} \left[1 + 10\sin 6\theta + 25 \cdot \frac{1}{2}(1 - \cos 12\theta)\right] d\theta$

$= \frac{1}{2}\int_0^{2\pi} \left[\frac{27}{2} + 10\sin 6\theta - \frac{25}{2}\cos 12\theta\right] d\theta = \frac{1}{2}\left[\frac{27}{2}\theta - \frac{5}{3}\cos 6\theta - \frac{25}{24}\sin 12\theta\right]_0^{2\pi}$

$= \frac{1}{2}\left[\left(27\pi - \frac{5}{3}\right) - \left(-\frac{5}{3}\right)\right] = \frac{27}{2}\pi$

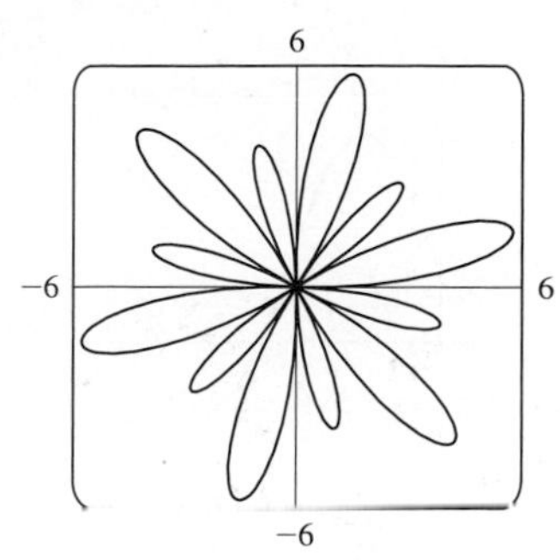

17. The curve passes through the pole when $r = 0 \;\Rightarrow\; 4\cos 3\theta = 0 \;\Rightarrow\; \cos 3\theta = 0 \;\Rightarrow\; 3\theta = \frac{\pi}{2} + \pi n \;\Rightarrow$ $\theta = \frac{\pi}{6} + \frac{\pi}{3}n$. The part of the shaded loop above the polar axis is traced out for $\theta = 0$ to $\theta = \pi/6$, so we'll use $-\pi/6$ and $\pi/6$ as our limits of integration.

$A = \int_{-\pi/6}^{\pi/6} \frac{1}{2}(4\cos 3\theta)^2\,d\theta = 2\int_0^{\pi/6} \frac{1}{2}(16\cos^2 3\theta)\,d\theta$

$= 16\int_0^{\pi/6} \frac{1}{2}(1 + \cos 6\theta)\,d\theta = 8\left[\theta + \frac{1}{6}\sin 6\theta\right]_0^{\pi/6} = 8\left(\frac{\pi}{6}\right) = \frac{4}{3}\pi$

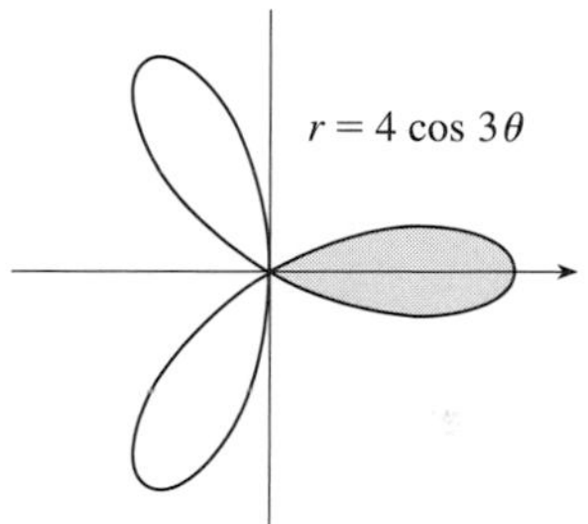

18. The curve given by $r^2 = 4\cos 2\theta$ passes through the pole when $r = 0 \;\Rightarrow\; 4\cos 2\theta = 0 \;\Rightarrow\; \cos 2\theta = 0 \;\Rightarrow$ $2\theta = \frac{\pi}{2} + \pi n \;\Rightarrow\; \theta = \frac{\pi}{4} + \frac{\pi}{2}n$. The part of the shaded loop above the polar axis is traced out for $\theta = 0$ to $\theta = \pi/4$, so we'll use $-\pi/4$ to $\pi/4$ as our limits of integration.

$A = \int_{-\pi/4}^{\pi/4} \frac{1}{2}(4\cos 2\theta)\,d\theta = 2\int_0^{\pi/4} 2\cos 2\theta\,d\theta = 2\Big[\sin 2\theta\Big]_0^{\pi/4}$

$= 2\sin\frac{\pi}{2} = 2(1) = 2$

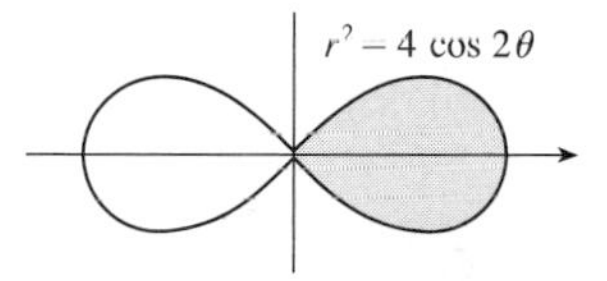

19. $r = 0 \;\Rightarrow\; \sin 4\theta = 0 \;\Rightarrow\; 4\theta = \pi n \;\Rightarrow\; \theta = \frac{\pi}{4}n.$

$A = \int_0^{\pi/4} \frac{1}{2}(\sin 4\theta)^2\,d\theta = \frac{1}{2}\int_0^{\pi/4} \sin^2 4\theta\,d\theta = \frac{1}{2}\int_0^{\pi/4} \frac{1}{2}(1 - \cos 8\theta)\,d\theta$

$= \frac{1}{4}\left[\theta - \frac{1}{8}\sin 8\theta\right]_0^{\pi/4} = \frac{1}{4}\left(\frac{\pi}{4}\right) = \frac{1}{16}\pi$

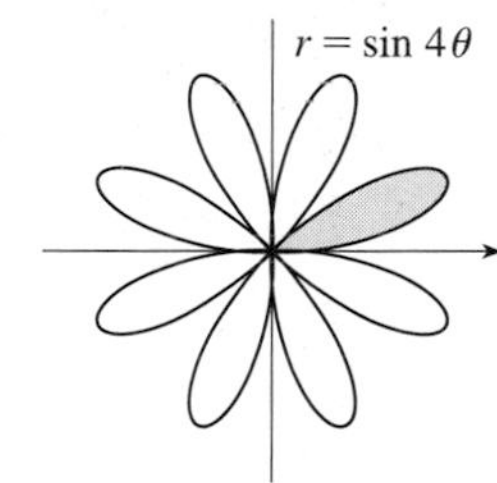

20. $r = 0 \;\Rightarrow\; 2\sin 5\theta = 0 \;\Rightarrow\; \sin 5\theta = 0 \;\Rightarrow\; 5\theta = \pi n \;\Rightarrow\; \theta = \frac{\pi}{5}n.$

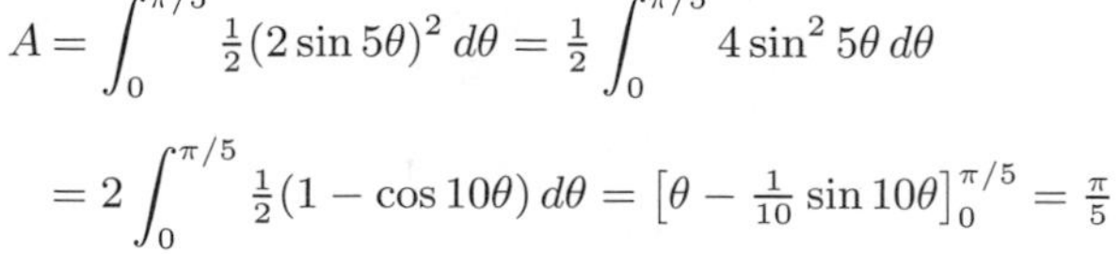

$A = \int_0^{\pi/5} \frac{1}{2}(2\sin 5\theta)^2\,d\theta = \frac{1}{2}\int_0^{\pi/5} 4\sin^2 5\theta\,d\theta$

$= 2\int_0^{\pi/5} \frac{1}{2}(1 - \cos 10\theta)\,d\theta = \left[\theta - \frac{1}{10}\sin 10\theta\right]_0^{\pi/5} = \frac{\pi}{5}$

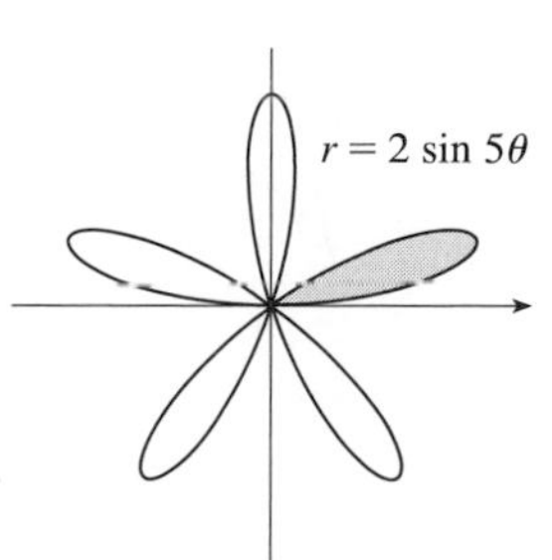

21. 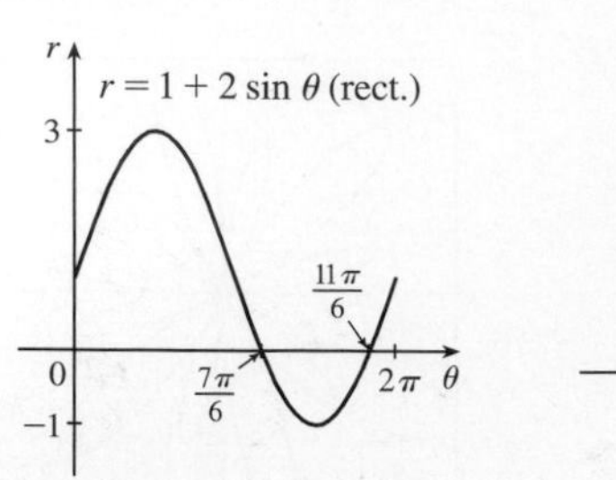

This is a limaçon, with inner loop traced out between $\theta = \frac{7\pi}{6}$ and $\frac{11\pi}{6}$ [found by solving $r = 0$].

$$A = 2\int_{7\pi/6}^{3\pi/2} \tfrac{1}{2}(1 + 2\sin\theta)^2\,d\theta = \int_{7\pi/6}^{3\pi/2} (1 + 4\sin\theta + 4\sin^2\theta)\,d\theta = \int_{7\pi/6}^{3\pi/2} \left[1 + 4\sin\theta + 4 \cdot \tfrac{1}{2}(1 - \cos 2\theta)\right] d\theta$$

$$= \Big[\theta - 4\cos\theta + 2\theta - \sin 2\theta\Big]_{7\pi/6}^{3\pi/2} = \left(\tfrac{9\pi}{2}\right) - \left(\tfrac{7\pi}{2} + 2\sqrt{3} - \tfrac{\sqrt{3}}{2}\right) = \pi - \tfrac{3\sqrt{3}}{2}$$

22. To determine when the strophoid $r = 2\cos\theta - \sec\theta$ passes through the pole, we solve

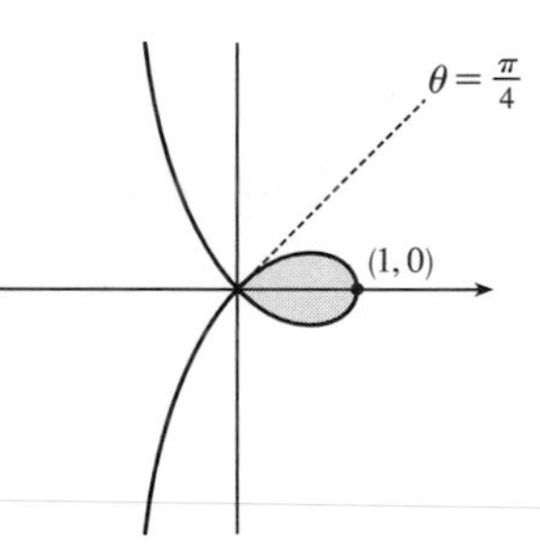

$$r = 0 \;\Rightarrow\; 2\cos\theta - \frac{1}{\cos\theta} = 0 \;\Rightarrow\; 2\cos^2\theta - 1 = 0 \;\Rightarrow\; \cos^2\theta = \frac{1}{2} \;\Rightarrow$$

$$\cos\theta = \pm\frac{1}{\sqrt{2}} \;\Rightarrow\; \theta = \tfrac{\pi}{4} \text{ or } \theta = \tfrac{3\pi}{4} \text{ for } 0 \le \theta \le \pi \text{ with } \theta \ne \tfrac{\pi}{2}.$$

$$A = 2\int_0^{\pi/4} \tfrac{1}{2}(2\cos\theta - \sec\theta)^2\,d\theta = \int_0^{\pi/4} (4\cos^2\theta - 4 + \sec^2\theta)\,d\theta$$

$$= \int_0^{\pi/4} \left[4 \cdot \tfrac{1}{2}(1 + \cos 2\theta) - 4 + \sec^2\theta\right] d\theta = \int_0^{\pi/4} (-2 + 2\cos 2\theta + \sec^2\theta)\,d\theta$$

$$= \big[-2\theta + \sin 2\theta + \tan\theta\big]_0^{\pi/4} = \left(-\tfrac{\pi}{2} + 1 + 1\right) - 0 = 2 - \tfrac{\pi}{2}$$

23. $4\sin\theta = 2 \;\Rightarrow\; \sin\theta = \frac{1}{2} \;\Rightarrow\; \theta = \frac{\pi}{6}$ or $\frac{5\pi}{6} \;\Rightarrow$

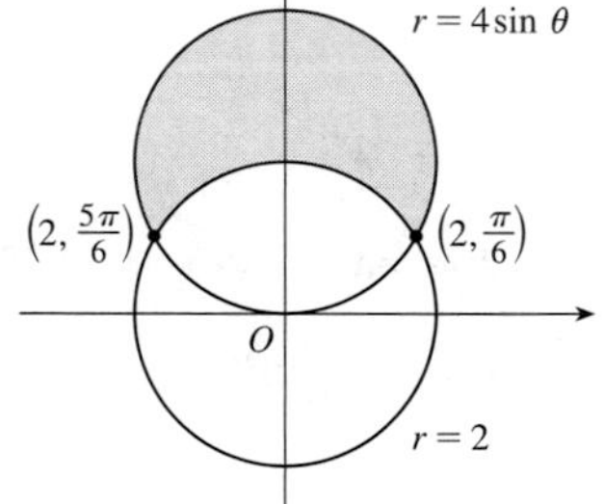

$$A = \int_{\pi/6}^{5\pi/6} \tfrac{1}{2}[(4\sin\theta)^2 - 2^2]\,d\theta = 2\int_{\pi/6}^{\pi/2} \tfrac{1}{2}(16\sin^2\theta - 4)\,d\theta$$

$$= \int_{\pi/6}^{\pi/2} \left[16 \cdot \tfrac{1}{2}(1 - \cos 2\theta) - 4\right] d\theta = \int_{\pi/6}^{\pi/2} (4 - 8\cos 2\theta)\,d\theta$$

$$= \Big[4\theta - 4\sin 2\theta\Big]_{\pi/6}^{\pi/2} = (2\pi - 0) - \left(\tfrac{2\pi}{3} - 2\sqrt{3}\right) = \tfrac{4\pi}{3} + 2\sqrt{3}$$

24. $1 - \sin\theta = 1 \;\Rightarrow\; \sin\theta = 0 \;\Rightarrow\; \theta = 0 \text{ or } \pi \;\Rightarrow$

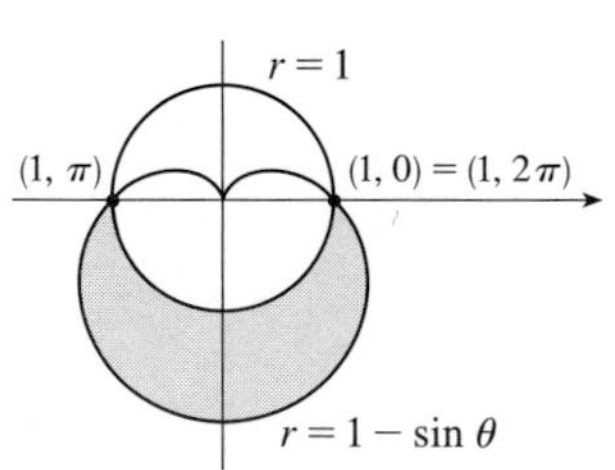

$$A = \int_\pi^{2\pi} \tfrac{1}{2}\left[(1 - \sin\theta)^2 - 1\right] d\theta = \tfrac{1}{2}\int_\pi^{2\pi} (\sin^2\theta - 2\sin\theta)\,d\theta$$

$$= \tfrac{1}{4}\int_\pi^{2\pi} (1 - \cos 2\theta - 4\sin\theta)\,d\theta = \tfrac{1}{4}\Big[\theta - \tfrac{1}{2}\sin 2\theta + 4\cos\theta\Big]_\pi^{2\pi}$$

$$= \tfrac{1}{4}\pi + 2$$

25. To find the area inside the leminiscate $r^2 = 8\cos 2\theta$ and outside the circle $r = 2$, we first note that the two curves intersect when $r^2 = 8\cos 2\theta$ and $r = 2$, that is, when $\cos 2\theta = \frac{1}{2}$. For $-\pi < \theta \le \pi$, $\cos 2\theta = \frac{1}{2} \Leftrightarrow 2\theta = \pm\pi/3$ or $\pm 5\pi/3 \Leftrightarrow \theta = \pm\pi/6$ or $\pm 5\pi/6$. The figure shows that the desired area is 4 times the area between the curves from 0 to $\pi/6$. Thus,

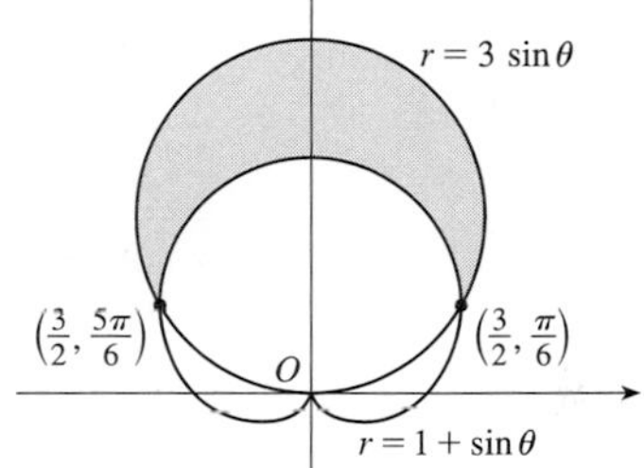

$$A = 4\int_0^{\pi/6}\left[\tfrac{1}{2}(8\cos 2\theta) - \tfrac{1}{2}(2)^2\right]d\theta = 8\int_0^{\pi/6}(2\cos 2\theta - 1)\,d\theta$$

$$= 8\left[\sin 2\theta - \theta\right]_0^{\pi/6} = 8\left(\sqrt{3}/2 - \pi/6\right) = 4\sqrt{3} - 4\pi/3$$

26. $3\sin\theta = 1 + \sin\theta \Rightarrow \sin\theta = \frac{1}{2} \Rightarrow \theta = \frac{\pi}{6}$ or $\frac{5\pi}{6} \Rightarrow$

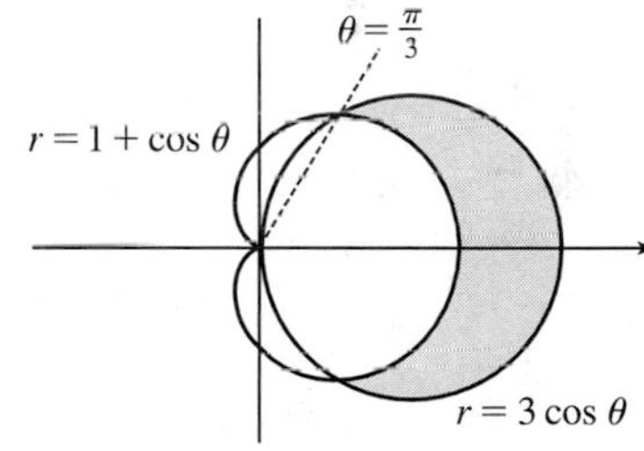

$$A = \int_{\pi/6}^{5\pi/6}\tfrac{1}{2}\left[(3\sin\theta)^2 - (1+\sin\theta)^2\right]d\theta$$

$$= 2\int_{\pi/6}^{\pi/2}\tfrac{1}{2}\left(9\sin^2\theta - 1 - 2\sin\theta - \sin^2\theta\right)d\theta$$

$$= \int_{\pi/6}^{\pi/2}\left(8\sin^2\theta - 1 - 2\sin\theta\right)d\theta$$

$$= \int_{\pi/6}^{\pi/2}\left[8\cdot\tfrac{1}{2}(1-\cos 2\theta) - 1 - 2\sin\theta\right]d\theta = \int_{\pi/6}^{\pi/2}(3 - 4\cos 2\theta - 2\sin\theta)\,d\theta$$

$$= \left[3\theta - 2\sin 2\theta + 2\cos\theta\right]_{\pi/6}^{\pi/2} = \left(\tfrac{3\pi}{2} - 0 + 0\right) - \left(\tfrac{\pi}{2} - \sqrt{3} + \sqrt{3}\right) = \pi$$

27. $3\cos\theta = 1 + \cos\theta \Leftrightarrow \cos\theta = \frac{1}{2} \Rightarrow \theta = \frac{\pi}{3}$ or $-\frac{\pi}{3}$.

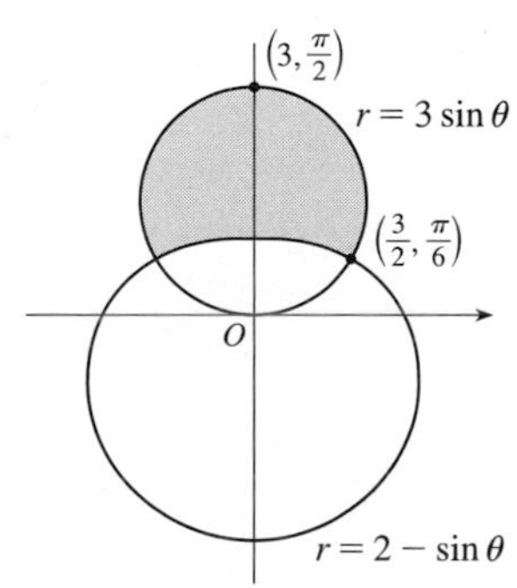

$$A = 2\int_0^{\pi/3}\tfrac{1}{2}\left[(3\cos\theta)^2 - (1+\cos\theta)^2\right]d\theta$$

$$= \int_0^{\pi/3}(8\cos^2\theta - 2\cos\theta - 1)\,d\theta = \int_0^{\pi/3}\left[4(1+\cos 2\theta) - 2\cos\theta - 1\right]d\theta$$

$$= \int_0^{\pi/3}(3 + 4\cos 2\theta - 2\cos\theta)\,d\theta = \left[3\theta + 2\sin 2\theta - 2\sin\theta\right]_0^{\pi/3}$$

$$= \pi + \sqrt{3} - \sqrt{3} = \pi$$

28. $3\sin\theta = 2 - \sin\theta \Rightarrow 4\sin\theta = 2 \Rightarrow \sin\theta = \frac{1}{2} \Rightarrow \theta = \frac{\pi}{6}$ or $\frac{5\pi}{6}$.

(3, π/2)
r = 3 sin θ
(3/2, π/6)
O
r = 2 − sin θ

$$A = 2\int_{\pi/6}^{\pi/2}\tfrac{1}{2}\left[(3\sin\theta)^2 - (2-\sin\theta)^2\right]d\theta$$

$$= \int_{\pi/6}^{\pi/2}\left(9\sin^2\theta - 4 + 4\sin\theta - \sin^2\theta\right]d\theta$$

$$= \int_{\pi/6}^{\pi/2}(8\sin^2\theta + 4\sin\theta - 4)\,d\theta$$

$$= 4\int_{\pi/6}^{\pi/2}\left[2\cdot\tfrac{1}{2}(1-\cos 2\theta) + \sin\theta - 1\right]d\theta$$

$$= 4\int_{\pi/6}^{\pi/2}(\sin\theta - \cos 2\theta)\,d\theta = 4\left[-\cos\theta - \tfrac{1}{2}\sin 2\theta\right]_{\pi/6}^{\pi/2}$$

$$= 4\left[(0-0) - \left(-\tfrac{\sqrt{3}}{2} - \tfrac{\sqrt{3}}{4}\right)\right] = 4\left(\tfrac{3\sqrt{3}}{4}\right) = 3\sqrt{3}$$

29. $3\sin\theta = 3\cos\theta \quad\Rightarrow\quad \dfrac{3\sin\theta}{3\cos\theta} = 1 \quad\Rightarrow\quad \tan\theta = 1 \quad\Rightarrow\quad \theta = \frac{\pi}{4} \quad\Rightarrow$

$$A = 2\int_0^{\pi/4} \tfrac{1}{2}(3\sin\theta)^2\,d\theta = \int_0^{\pi/4} 9\sin^2\theta\,d\theta = \int_0^{\pi/4} 9\cdot\tfrac{1}{2}(1-\cos 2\theta)\,d\theta$$

$$= \int_0^{\pi/4}\left(\tfrac{9}{2} - \tfrac{9}{2}\cos 2\theta\right)d\theta = \left[\tfrac{9}{2}\theta - \tfrac{9}{4}\sin 2\theta\right]_0^{\pi/4} = \left(\tfrac{9\pi}{8} - \tfrac{9}{4}\right) - (0-0)$$

$$= \tfrac{9\pi}{8} - \tfrac{9}{4}$$

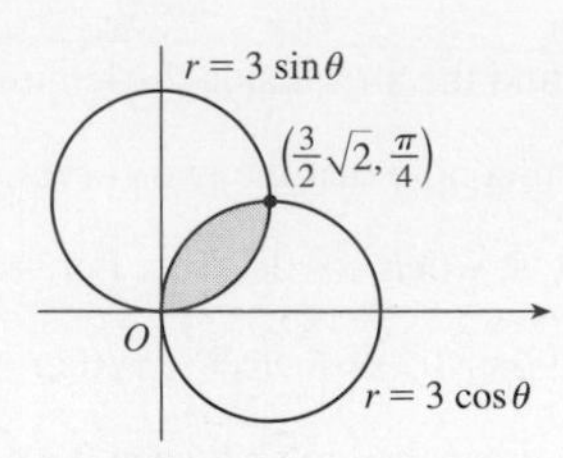

30. $A = 4\int_0^{\pi/2}\frac{1}{2}(1-\cos\theta)^2\,d\theta = 2\int_0^{\pi/2}(1 - 2\cos\theta + \cos^2\theta)\,d\theta$

$$= 2\int_0^{\pi/2}\left[1 - 2\cos\theta + \tfrac{1}{2}(1+\cos 2\theta)\right]d\theta$$

$$= 2\int_0^{\pi/2}\left(\tfrac{3}{2} - 2\cos\theta + \tfrac{1}{2}\cos 2\theta\right)d\theta = \int_0^{\pi/2}(3 - 4\cos\theta + \cos 2\theta)\,d\theta$$

$$= \left[3\theta - 4\sin\theta + \tfrac{1}{2}\sin 2\theta\right]_0^{\pi/2} = \tfrac{3\pi}{2} - 4$$

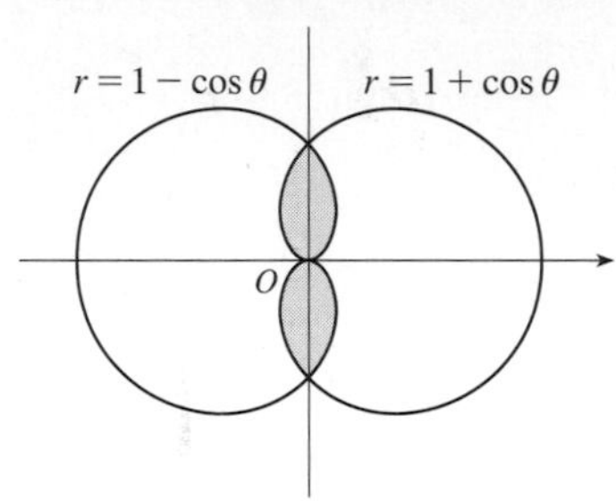

31. $\sin 2\theta = \cos 2\theta \quad\Rightarrow\quad \dfrac{\sin 2\theta}{\cos 2\theta} = 1 \quad\Rightarrow\quad \tan 2\theta = 1 \quad\Rightarrow\quad 2\theta = \frac{\pi}{4} \quad\Rightarrow$

$\theta = \frac{\pi}{8} \quad\Rightarrow$

$$A = 8\cdot 2\int_0^{\pi/8}\tfrac{1}{2}\sin^2 2\theta\,d\theta = 8\int_0^{\pi/8}\tfrac{1}{2}(1-\cos 4\theta)\,d\theta$$

$$= 4\left[\theta - \tfrac{1}{4}\sin 4\theta\right]_0^{\pi/8} = 4\left(\tfrac{\pi}{8} - \tfrac{1}{4}\cdot 1\right) = \tfrac{\pi}{2} - 1$$

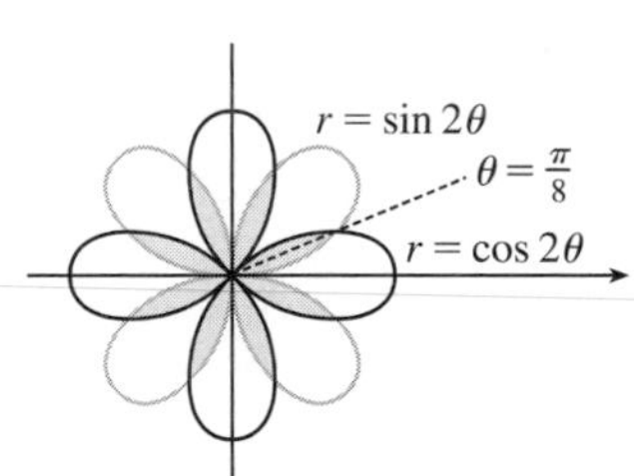

32. $3 + 2\cos\theta = 3 + 2\sin\theta \quad\Rightarrow\quad \cos\theta = \sin\theta \quad\Rightarrow\quad \theta = \frac{\pi}{4}$ or $\frac{5\pi}{4}$.

$$A = 2\int_{\pi/4}^{5\pi/4}\tfrac{1}{2}(3+2\cos\theta)^2\,d\theta = \int_{\pi/4}^{5\pi/4}(9 + 12\cos\theta + 4\cos^2\theta)\,d\theta$$

$$= \int_{\pi/4}^{5\pi/4}\left[9 + 12\cos\theta + 4\cdot\tfrac{1}{2}(1+\cos 2\theta)\right]d\theta$$

$$= \int_{\pi/4}^{5\pi/4}(11 + 12\cos\theta + 2\cos 2\theta)\,d\theta = \left[11\theta + 12\sin\theta + \sin 2\theta\right]_{\pi/4}^{5\pi/4}$$

$$= \left(\tfrac{55\pi}{4} - 6\sqrt{2} + 1\right) - \left(\tfrac{11\pi}{4} + 6\sqrt{2} + 1\right) = 11\pi - 12\sqrt{2}$$

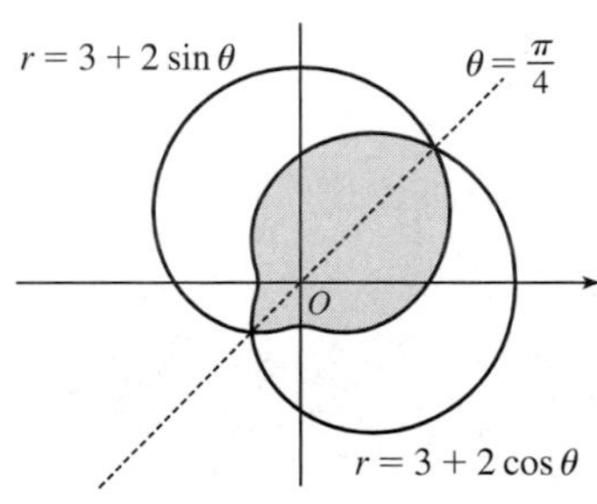

33. From the figure, we see that the shaded region is 4 times the shaded region from $\theta = 0$ to $\theta = \pi/4$. $r^2 = 2\sin 2\theta$ and $r = 1 \quad\Rightarrow$ $2\sin 2\theta = 1^2 \quad\Rightarrow\quad \sin 2\theta = \frac{1}{2} \quad\Rightarrow\quad 2\theta = \frac{\pi}{6} \quad\Rightarrow\quad \theta = \frac{\pi}{12}$.

$$A = 4\int_0^{\pi/12}\tfrac{1}{2}(2\sin 2\theta)\,d\theta + 4\int_{\pi/12}^{\pi/4}\tfrac{1}{2}(1)^2\,d\theta$$

$$= \int_0^{\pi/12} 4\sin 2\theta\,d\theta + \int_{\pi/12}^{\pi/4} 2\,d\theta = \left[-2\cos 2\theta\right]_0^{\pi/12} + \left[2\theta\right]_{\pi/12}^{\pi/4}$$

$$= (-\sqrt{3} + 2) + \left(\tfrac{\pi}{2} - \tfrac{\pi}{6}\right) = -\sqrt{3} + 2 + \tfrac{\pi}{3}$$

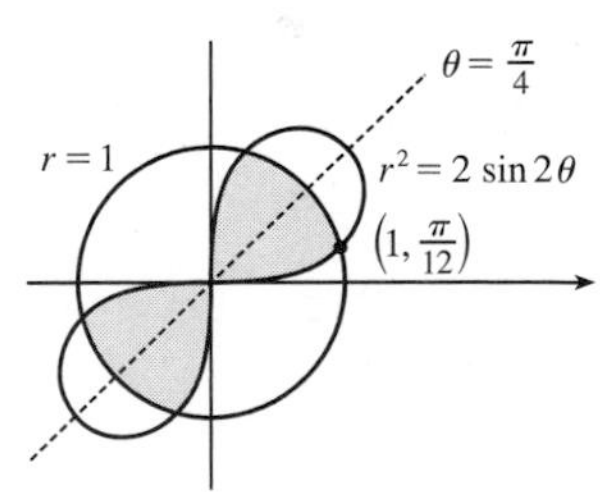

34. Let $\alpha = \tan^{-1}(b/a)$. Then

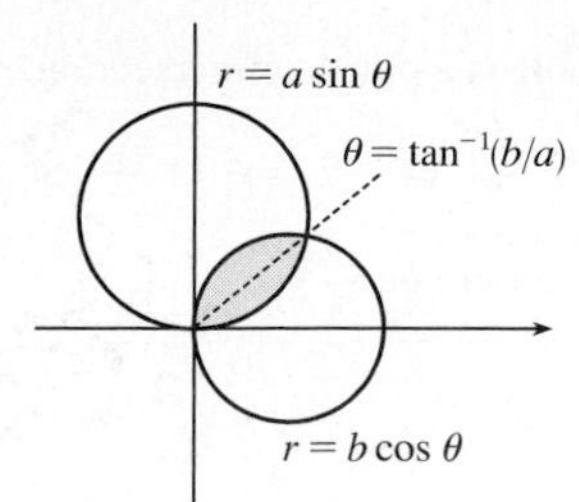

$$\begin{aligned} A &= \int_0^\alpha \tfrac{1}{2}(a\sin\theta)^2\,d\theta + \int_\alpha^{\pi/2} \tfrac{1}{2}(b\cos\theta)^2\,d\theta \\ &= \tfrac{1}{4}a^2\left[\theta - \tfrac{1}{2}\sin 2\theta\right]_0^\alpha + \tfrac{1}{4}b^2\left[\theta + \tfrac{1}{2}\sin 2\theta\right]_\alpha^{\pi/2} \\ &= \tfrac{1}{4}\alpha(a^2 - b^2) + \tfrac{1}{8}\pi b^2 - \tfrac{1}{4}(a^2 + b^2)(\sin\alpha\,\cos\alpha) \\ &= \tfrac{1}{4}(a^2 - b^2)\tan^{-1}(b/a) + \tfrac{1}{8}\pi b^2 - \tfrac{1}{4}ab \end{aligned}$$

35. The darker shaded region (from $\theta = 0$ to $\theta = 2\pi/3$) represents $\frac{1}{2}$ of the desired area plus $\frac{1}{2}$ of the area of the inner loop. From this area, we'll subtract $\frac{1}{2}$ of the area of the inner loop (the lighter shaded region from $\theta = 2\pi/3$ to $\theta = \pi$), and then double that difference to obtain the desired area.

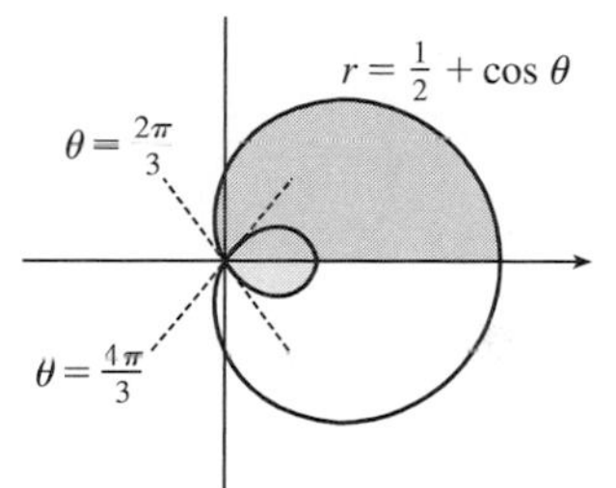

$$\begin{aligned} A &= 2\left[\int_0^{2\pi/3} \tfrac{1}{2}\left(\tfrac{1}{2} + \cos\theta\right)^2 d\theta - \int_{2\pi/3}^{\pi} \tfrac{1}{2}\left(\tfrac{1}{2} + \cos\theta\right)^2 d\theta\right] \\ &= \int_0^{2\pi/3}\left(\tfrac{1}{4} + \cos\theta + \cos^2\theta\right) d\theta - \int_{2\pi/3}^{\pi}\left(\tfrac{1}{4} + \cos\theta + \cos^2\theta\right) d\theta \\ &= \int_0^{2\pi/3}\left[\tfrac{1}{4} + \cos\theta + \tfrac{1}{2}(1 + \cos 2\theta)\right] d\theta \\ &\qquad - \int_{2\pi/3}^{\pi}\left[\tfrac{1}{4} + \cos\theta + \tfrac{1}{2}(1 + \cos 2\theta)\right] d\theta \\ &= \left[\frac{\theta}{4} + \sin\theta + \frac{\theta}{2} + \frac{\sin 2\theta}{4}\right]_0^{2\pi/3} - \left[\frac{\theta}{4} + \sin\theta + \frac{\theta}{2} + \frac{\sin 2\theta}{4}\right]_{2\pi/3}^{\pi} \\ &= \left(\tfrac{\pi}{6} + \tfrac{\sqrt{3}}{2} + \tfrac{\pi}{3} - \tfrac{\sqrt{3}}{8}\right) - \left(\tfrac{\pi}{4} + \tfrac{\pi}{2}\right) + \left(\tfrac{\pi}{6} + \tfrac{\sqrt{3}}{2} + \tfrac{\pi}{3} - \tfrac{\sqrt{3}}{8}\right) \\ &= \tfrac{\pi}{4} + \tfrac{3}{4}\sqrt{3} = \tfrac{1}{4}(\pi + 3\sqrt{3}) \end{aligned}$$

36. $r = 0 \;\Rightarrow\; 1 + 2\cos 3\theta = 0 \;\Rightarrow\; \cos 3\theta = -\frac{1}{2} \;\Rightarrow\; 3\theta = \frac{2\pi}{3}, \frac{4\pi}{3}$ [for $0 \le 3\theta \le 2\pi$] $\Rightarrow\; \theta = \frac{2\pi}{9}, \frac{4\pi}{9}$. The darker shaded region (from $\theta = 0$ to $\theta = 2\pi/9$) represents $\frac{1}{2}$ of the desired area plus $\frac{1}{2}$ of the area of the inner loop. From this area, we'll subtract $\frac{1}{2}$ of the area of the inner loop (the lighter shaded region from $\theta = 2\pi/9$ to $\theta = \pi/3$), and then double that difference to obtain the desired area.

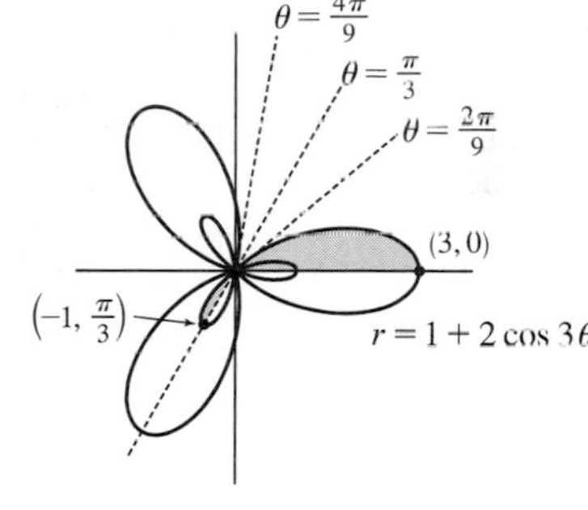

$$A = 2\left[\int_0^{2\pi/9} \tfrac{1}{2}(1 + 2\cos 3\theta)^2\,d\theta - \int_{2\pi/9}^{\pi/3} \tfrac{1}{2}(1 + 2\cos 3\theta)^2\,d\theta\right]$$

Now
$$\begin{aligned} r^2 &= (1 + 2\cos 3\theta)^2 = 1 + 4\cos 3\theta + 4\cos^2 3\theta = 1 + 4\cos 3\theta + 4\cdot\tfrac{1}{2}(1 + \cos 6\theta) \\ &= 1 + 4\cos 3\theta + 2 + 2\cos 6\theta = 3 + 4\cos 3\theta + 2\cos 6\theta \end{aligned}$$

and $\int r^2 d\theta = 3\theta + \frac{4}{3}\sin 3\theta + \frac{1}{3}\sin 6\theta + C$, so

$$\begin{aligned} A &= \left[3\theta + \tfrac{4}{3}\sin 3\theta + \tfrac{1}{3}\sin 6\theta\right]_0^{2\pi/9} - \left[3\theta + \tfrac{4}{3}\sin 3\theta + \tfrac{1}{3}\sin 6\theta\right]_{2\pi/9}^{\pi/3} \\ &= \left[\left(\tfrac{2\pi}{3} + \tfrac{4}{3}\cdot\tfrac{\sqrt{3}}{2} + \tfrac{1}{3}\cdot\tfrac{-\sqrt{3}}{2}\right) - 0\right] - \left[(\pi + 0 + 0) - \left(\tfrac{2\pi}{3} + \tfrac{4}{3}\cdot\tfrac{\sqrt{3}}{2} + \tfrac{1}{3}\cdot\tfrac{-\sqrt{3}}{2}\right)\right] \\ &= \tfrac{4\pi}{3} + \tfrac{4}{3}\sqrt{3} - \tfrac{1}{3}\sqrt{3} - \pi = \tfrac{\pi}{3} + \sqrt{3} \end{aligned}$$

37. The pole is a point of intersection. $\sin\theta = 1 - \sin\theta \Rightarrow 2\sin\theta = 1 \Rightarrow$ $\sin\theta = \frac{1}{2} \Rightarrow \theta = \frac{\pi}{6}$ or $\frac{5\pi}{6}$. So the other points of intersection are $\left(\frac{1}{2}, \frac{\pi}{6}\right)$ and $\left(\frac{1}{2}, \frac{5\pi}{6}\right)$.

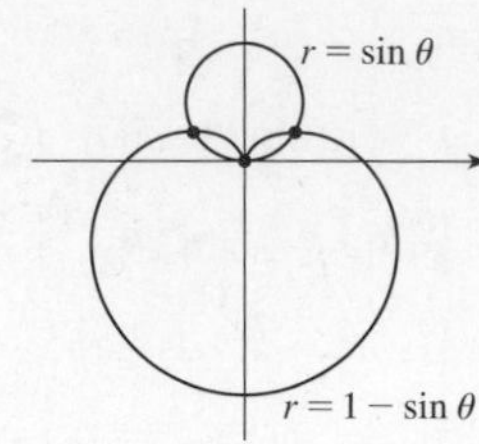

38. The pole is a point of intersection. $1 + \cos\theta = 1 - \sin\theta \Rightarrow$ $\cos\theta = -\sin\theta \Rightarrow \dfrac{\cos\theta}{\sin\theta} = -1 \Rightarrow \cot\theta = -1 \Rightarrow \theta = \frac{3\pi}{4}$ or $\frac{7\pi}{4}$. So the other points of intersection are $\left(1 - \frac{1}{2}\sqrt{2}, \frac{3\pi}{4}\right)$ and $\left(1 + \frac{1}{2}\sqrt{2}, \frac{7\pi}{4}\right)$.

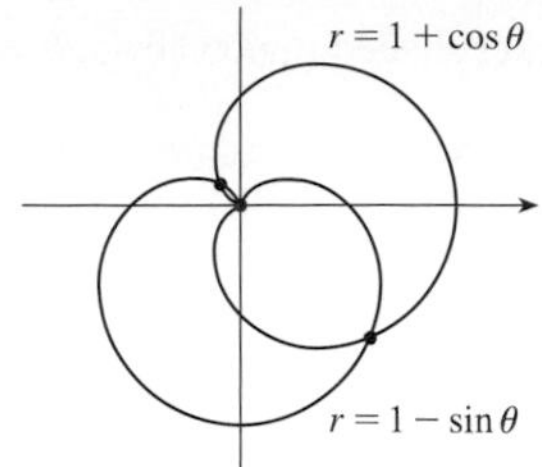

39. $2\sin 2\theta = 1 \Rightarrow \sin 2\theta = \frac{1}{2} \Rightarrow 2\theta = \frac{\pi}{6}, \frac{5\pi}{6}, \frac{13\pi}{6}$, or $\frac{17\pi}{6}$.

By symmetry, the eight points of intersection are given by

$(1, \theta)$, where $\theta = \frac{\pi}{12}, \frac{5\pi}{12}, \frac{13\pi}{12}$, and $\frac{17\pi}{12}$, and

$(-1, \theta)$, where $\theta = \frac{7\pi}{12}, \frac{11\pi}{12}, \frac{19\pi}{12}$, and $\frac{23\pi}{12}$.

[There are many ways to describe these points.]

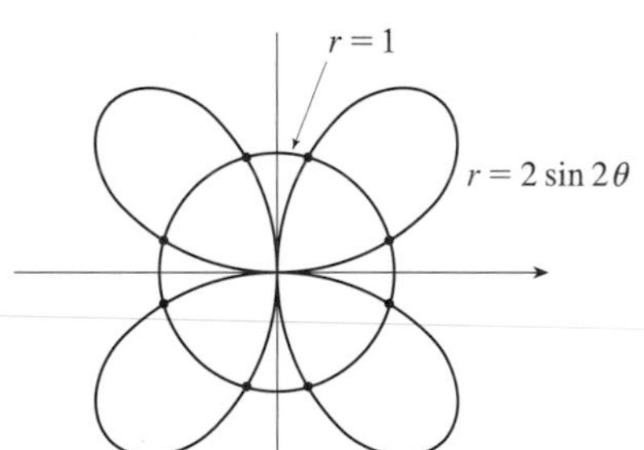

40. Clearly the pole lies on both curves. $\sin 3\theta = \cos 3\theta \Rightarrow \tan 3\theta = 1 \Rightarrow$ $3\theta = \frac{\pi}{4} + n\pi$ [n any integer] $\Rightarrow \theta = \frac{\pi}{12} + \frac{\pi}{3}n \Rightarrow$ $\theta = \frac{\pi}{12}, \frac{5\pi}{12}$, or $\frac{3\pi}{4}$, so the three remaining intersection points are $\left(\frac{1}{\sqrt{2}}, \frac{\pi}{12}\right), \left(-\frac{1}{\sqrt{2}}, \frac{5\pi}{12}\right)$, and $\left(\frac{1}{\sqrt{2}}, \frac{3\pi}{4}\right)$.

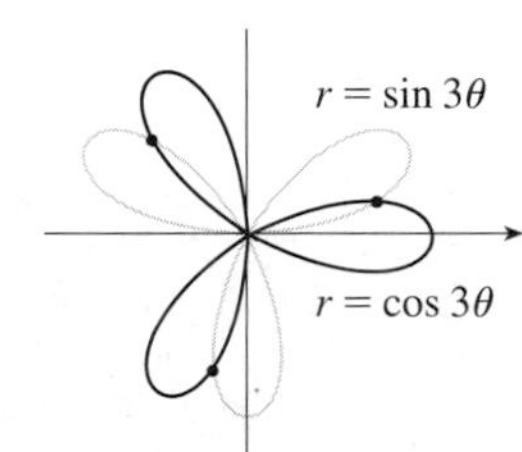

41. The pole is a point of intersection. $\sin\theta = \sin 2\theta = 2\sin\theta\cos\theta \Leftrightarrow$ $\sin\theta\,(1 - 2\cos\theta) = 0 \Leftrightarrow \sin\theta = 0$ or $\cos\theta = \frac{1}{2} \Rightarrow$ $\theta = 0, \pi, \frac{\pi}{3}$, or $-\frac{\pi}{3} \Rightarrow$ the other intersection points are $\left(\frac{\sqrt{3}}{2}, \frac{\pi}{3}\right)$ and $\left(\frac{\sqrt{3}}{2}, \frac{2\pi}{3}\right)$ [by symmetry].

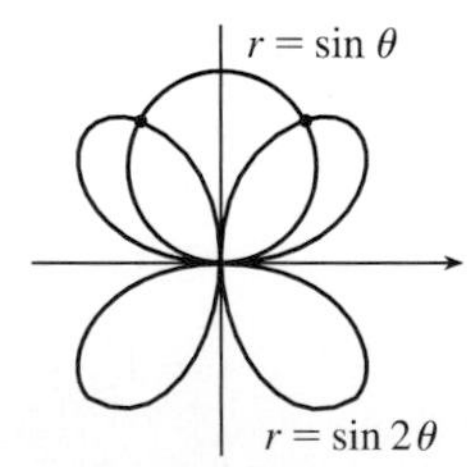

42. Clearly the pole is a point of intersection. $\sin 2\theta = \cos 2\theta \Rightarrow$ $\tan 2\theta = 1 \Rightarrow 2\theta = \frac{\pi}{4} + 2n\pi$ [since $\sin 2\theta$ and $\cos 2\theta$ must be positive in the equations] $\Rightarrow \theta = \frac{\pi}{8} + n\pi \Rightarrow \theta = \frac{\pi}{8}$ or $\frac{9\pi}{8}$. So the curves also intersect at $\left(\frac{1}{\sqrt[4]{2}}, \frac{\pi}{8}\right)$ and $\left(\frac{1}{\sqrt[4]{2}}, \frac{9\pi}{8}\right)$.

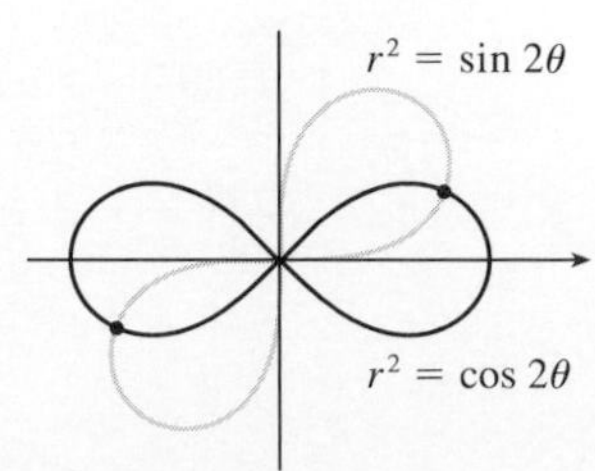

43.

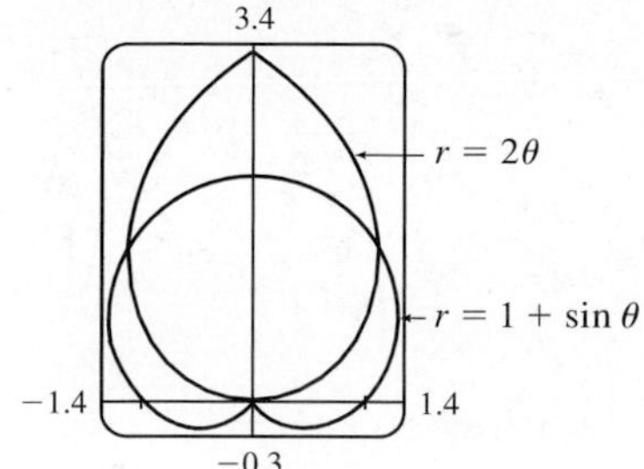

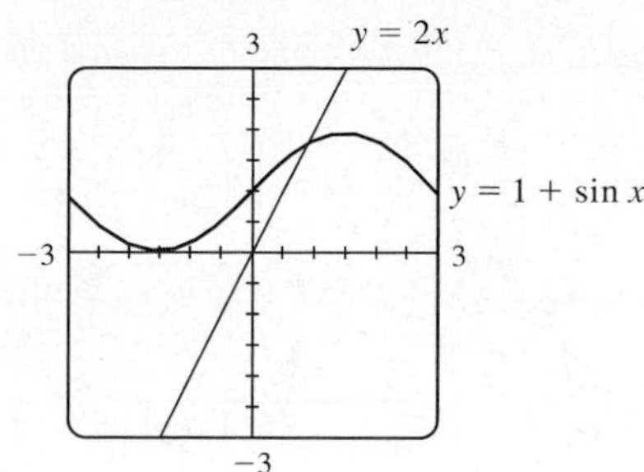

From the first graph, we see that the pole is one point of intersection. By zooming in or using the cursor, we find the θ-values of the intersection points to be $\alpha \approx 0.88786 \approx 0.89$ and $\pi - \alpha \approx 2.25$. (The first of these values may be more easily estimated by plotting $y = 1 + \sin x$ and $y = 2x$ in rectangular coordinates; see the second graph.) By symmetry, the total area contained is twice the area contained in the first quadrant, that is,

$$A = 2\int_0^\alpha \tfrac{1}{2}(2\theta)^2\, d\theta + 2\int_\alpha^{\pi/2} \tfrac{1}{2}(1+\sin\theta)^2\, d\theta = \int_0^\alpha 4\theta^2\, d\theta + \int_\alpha^{\pi/2} \left[1 + 2\sin\theta + \tfrac{1}{2}(1-\cos 2\theta)\right] d\theta$$

$$= \left[\tfrac{4}{3}\theta^3\right]_0^\alpha + \left[\theta - 2\cos\theta + \left(\tfrac{1}{2}\theta - \tfrac{1}{4}\sin 2\theta\right)\right]_\alpha^{\pi/2} = \tfrac{4}{3}\alpha^3 + \left[\left(\tfrac{\pi}{2} + \tfrac{\pi}{4}\right) - \left(\alpha - 2\cos\alpha + \tfrac{1}{2}\alpha - \tfrac{1}{4}\sin 2\alpha\right)\right] \approx 3.4645$$

44.

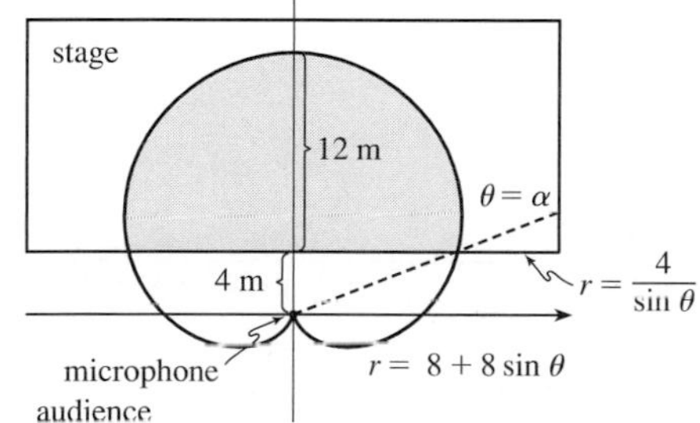

We need to find the shaded area A in the figure. The horizontal line representing the front of the stage has equation $y = 4 \Leftrightarrow r\sin\theta = 4 \Rightarrow r = 4/\sin\theta$. This line intersects the curve $r = 8 + 8\sin\theta$ when $8 + 8\sin\theta = \dfrac{4}{\sin\theta} \Rightarrow 8\sin\theta + 8\sin^2\theta = 4 \Rightarrow 2\sin^2\theta + 2\sin\theta - 1 = 0 \Rightarrow$

$$\sin\theta = \frac{-2 \pm \sqrt{4+8}}{4} = \frac{-2 \pm 2\sqrt{3}}{4} = \frac{-1+\sqrt{3}}{2} \quad \text{[the other value is less than } -1\text{]} \Rightarrow \theta = \sin^{-1}\left(\frac{\sqrt{3}-1}{2}\right).$$

This angle is about 21.5° and is denoted by α in the figure.

$$A = 2\int_\alpha^{\pi/2} \tfrac{1}{2}(8+8\sin\theta)^2\, d\theta - 2\int_\alpha^{\pi/2} \tfrac{1}{2}(4\csc\theta)^2\, d\theta = 64\int_\alpha^{\pi/2}(1 + 2\sin\theta + \sin^2\theta)\, d\theta - 16\int_\alpha^{\pi/2}\csc^2\theta\, d\theta$$

$$= 64\int_\alpha^{\pi/2}\left(1 + 2\sin\theta + \tfrac{1}{2} - \tfrac{1}{2}\cos 2\theta\right) d\theta + 16\int_\alpha^{\pi/2}(-\csc^2\theta)\, d\theta = 64\left[\tfrac{3}{2}\theta - 2\cos\theta - \tfrac{1}{4}\sin 2\theta\right]_\alpha^{\pi/2} + 16\left[\cot\theta\right]_\alpha^{\pi/2}$$

$$= 16\left[6\theta - 8\cos\theta - \sin 2\theta + \cot\theta\right]_\alpha^{\pi/a} = 16[(3\pi - 0 - 0 + 0) - (6\alpha - 8\cos\alpha - \sin 2\alpha + \cot\alpha)]$$

$$= 48\pi - 96\alpha + 128\cos\alpha + 16\sin 2\alpha - 16\cot\alpha$$

From the figure, $x^2 + \left(\sqrt{3}-1\right)^2 = 2^2 \Rightarrow x^2 = 4 - \left(3 - 2\sqrt{3} + 1\right) \Rightarrow$ $x^2 = 2\sqrt{3} = \sqrt{12}$, so $x = \sqrt{2\sqrt{3}} = \sqrt[4]{12}$. Using the trigonometric relationships for a right triangle and the identity $\sin 2\alpha = 2\sin\alpha\cos\alpha$, we continue:

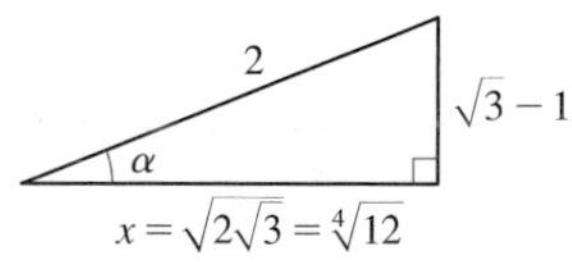

$$A = 48\pi - 96\alpha + 128\cdot\frac{\sqrt[4]{12}}{2} + 16\cdot 2\cdot\frac{\sqrt{3}-1}{2}\cdot\frac{\sqrt[4]{12}}{2} - 16\cdot\frac{\sqrt[4]{12}}{\sqrt{3}-1}\cdot\frac{\sqrt{3}+1}{\sqrt{3}+1}$$

$$= 48\pi - 96\alpha + 64\sqrt[4]{12} + 8\sqrt[4]{12}\left(\sqrt{3}-1\right) - 8\sqrt[4]{12}\left(\sqrt{3}+1\right) = 48\pi + 48\sqrt[4]{12} - 96\sin^{-1}\left(\frac{\sqrt{3}-1}{2}\right)$$

$$\approx 204.16\ \text{m}^2$$

45. $L = \int_a^b \sqrt{r^2 + (dr/d\theta)^2}\, d\theta = \int_0^\pi \sqrt{(2\cos\theta)^2 + (-2\sin\theta)^2}\, d\theta$

$= \int_0^\pi \sqrt{4(\cos^2\theta + \sin^2\theta)}\, d\theta = \int_0^\pi \sqrt{4}\, d\theta = \big[2\theta\big]_0^\pi = 2\pi$

As a check, note that the curve is a circle of radius 1, so its circumference is $2\pi(1) = 2\pi$.

46. $L = \int_a^b \sqrt{r^2 + (dr/d\theta)^2}\, d\theta = \int_0^{2\pi} \sqrt{(5^\theta)^2 + (5^\theta \ln 5)^2}\, d\theta = \int_0^{2\pi} \sqrt{5^{2\theta}[1 + (\ln 5)^2]}\, d\theta$

$= \sqrt{1 + (\ln 5)^2} \int_0^{2\pi} \sqrt{5^{2\theta}}\, d\theta = \sqrt{1 + (\ln 5)^2} \int_0^{2\pi} 5^\theta\, d\theta = \sqrt{1 + (\ln 5)^2} \left[\dfrac{5^\theta}{\ln 5}\right]_0^{2\pi}$

$= \sqrt{1 + (\ln 5)^2} \left(\dfrac{5^{2\pi}}{\ln 5} - \dfrac{1}{\ln 5}\right) = \dfrac{\sqrt{1 + (\ln 5)^2}}{\ln 5}(5^{2\pi} - 1)$

47. $L = \int_a^b \sqrt{r^2 + (dr/d\theta)^2}\, d\theta = \int_0^{2\pi} \sqrt{(\theta^2)^2 + (2\theta)^2}\, d\theta = \int_0^{2\pi} \sqrt{\theta^4 + 4\theta^2}\, d\theta$

$= \int_0^{2\pi} \sqrt{\theta^2(\theta^2 + 4)}\, d\theta = \int_0^{2\pi} \theta\sqrt{\theta^2 + 4}\, d\theta$

Now let $u = \theta^2 + 4$, so that $du = 2\theta\, d\theta$ $\left[\theta\, d\theta = \frac{1}{2}\, du\right]$ and

$\int_0^{2\pi} \theta\sqrt{\theta^2 + 4}\, d\theta = \int_4^{4\pi^2 + 4} \frac{1}{2}\sqrt{u}\, du = \frac{1}{2} \cdot \frac{2}{3}\left[u^{3/2}\right]_4^{4(\pi^2+1)} = \frac{1}{3}[4^{3/2}(\pi^2 + 1)^{3/2} - 4^{3/2}] = \frac{8}{3}[(\pi^2 + 1)^{3/2} - 1]$

48. $L = \int_a^b \sqrt{r^2 + (dr/d\theta)^2}\, d\theta = \int_0^{2\pi} \sqrt{[2(1 + \cos\theta)]^2 + (-2\sin\theta)^2}\, d\theta = \int_0^{2\pi} \sqrt{4 + 8\cos\theta + 4\cos^2\theta + 4\sin^2\theta}\, d\theta$

$= \int_0^{2\pi} \sqrt{8 + 8\cos\theta}\, d\theta = \sqrt{8} \int_0^{2\pi} \sqrt{1 + \cos\theta}\, d\theta = \sqrt{8} \int_0^{2\pi} \sqrt{2 \cdot \frac{1}{2}(1 + \cos\theta)}\, d\theta$

$= \sqrt{8} \int_0^{2\pi} \sqrt{2\cos^2\dfrac{\theta}{2}}\, d\theta = \sqrt{8}\sqrt{2} \int_0^{2\pi} \left|\cos\dfrac{\theta}{2}\right| d\theta = 4 \cdot 2 \int_0^\pi \cos\dfrac{\theta}{2}\, d\theta$ [by symmetry]

$= 8\left[2\sin\dfrac{\theta}{2}\right]_0^\pi = 8(2) = 16$

49. The curve $r = \cos^4(\theta/4)$ is completely traced with $0 \le \theta \le 4\pi$.

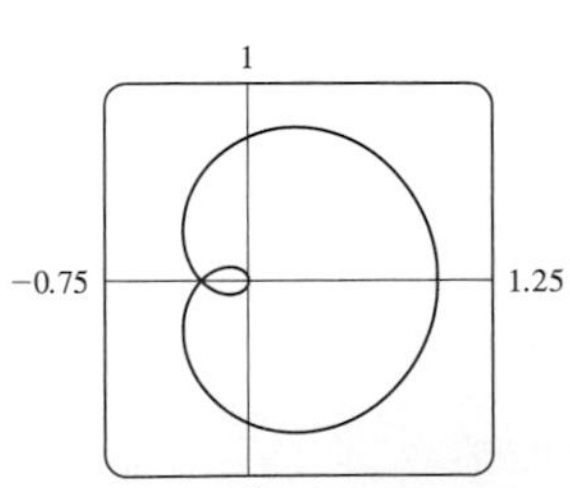

$r^2 + (dr/d\theta)^2 = [\cos^4(\theta/4)]^2 + \left[4\cos^3(\theta/4) \cdot (-\sin(\theta/4)) \cdot \frac{1}{4}\right]^2$

$= \cos^8(\theta/4) + \cos^6(\theta/4)\sin^2(\theta/4)$

$= \cos^6(\theta/4)[\cos^2(\theta/4) + \sin^2(\theta/4)] = \cos^6(\theta/4)$

$L = \int_0^{4\pi} \sqrt{\cos^6(\theta/4)}\, d\theta = \int_0^{4\pi} \left|\cos^3(\theta/4)\right| d\theta$

$= 2\int_0^{2\pi} \cos^3(\theta/4)\, d\theta$ [since $\cos^3(\theta/4) \ge 0$ for $0 \le \theta \le 2\pi$] $= 8\int_0^{\pi/2} \cos^3 u\, du$ $\left[u = \frac{1}{4}\theta\right]$

$= 8\int_0^{\pi/2} (1 - \sin^2 u)\cos u\, du = 8\int_0^1 (1 - x^2)\, dx$ $\left[\begin{array}{c} x = \sin u, \\ dx = \cos u\, du \end{array}\right]$

$= 8\left[x - \frac{1}{3}x^3\right]_0^1 = 8\left(1 - \frac{1}{3}\right) = \frac{16}{3}$

50. The curve $r = \cos^2(\theta/2)$ is completely traced with $0 \le \theta \le 2\pi$.

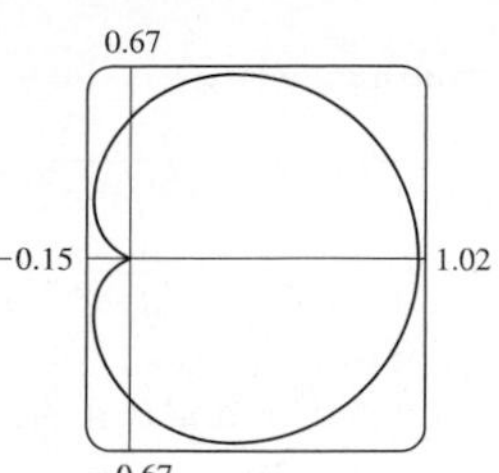

$$\begin{aligned} r^2 + (dr/d\theta)^2 &= [\cos^2(\theta/2)]^2 + \left[2\cos(\theta/2)\cdot(-\sin(\theta/2))\cdot\tfrac{1}{2}\right]^2 \\ &= \cos^4(\theta/2) + \cos^2(\theta/2)\sin^2(\theta/2) \\ &= \cos^2(\theta/2)[\cos^2(\theta/2) + \sin^2(\theta/2)] \\ &= \cos^2(\theta/2) \end{aligned}$$

$L = \int_0^{2\pi} \sqrt{\cos^2(\theta/2)}\,d\theta = \int_0^{2\pi} |\cos(\theta/2)|\,d\theta = 2\int_0^{\pi} \cos(\theta/2)\,d\theta$ [since $\cos(\theta/2) \ge 0$ for $0 \le \theta \le \pi$]

$= 4\int_0^{\pi/2} \cos u\,du$ $\left[u = \tfrac{1}{2}\theta\right]$ $= 4\left[\sin u\right]_0^{\pi/2} = 4(1-0) = 4$

51. One loop of the curve $r = \cos 2\theta$ is traced with $-\pi/4 \le \theta \le \pi/4$.

$$r^2 + \left(\frac{dr}{d\theta}\right)^2 = \cos^2 2\theta + (-2\sin 2\theta)^2 = \cos^2 2\theta + 4\sin^2 2\theta = 1 + 3\sin^2 2\theta \;\Rightarrow$$

$$L = \int_{-\pi/4}^{\pi/4} \sqrt{1 + 3\sin^2 2\theta}\,d\theta \approx 2.4221.$$

52. $r^2 + \left(\dfrac{dr}{d\theta}\right)^2 = \tan^2\theta + (\sec^2\theta)^2 \;\Rightarrow\; L = \displaystyle\int_{\pi/6}^{\pi/3} \sqrt{\tan^2\theta + \sec^4\theta}\,d\theta \approx 1.2789$

53. The curve $r = \sin(6\sin\theta)$ is completely traced with $0 \le \theta \le \pi$. $r = \sin(6\sin\theta) \;\Rightarrow$

$$\frac{dr}{d\theta} = \cos(6\sin\theta)\cdot 6\cos\theta, \text{ so } r^2 + \left(\frac{dr}{d\theta}\right)^2 = \sin^2(6\sin\theta) + 36\cos^2\theta\cos^2(6\sin\theta) \;\Rightarrow$$

$$L = \int_0^{\pi} \sqrt{\sin^2(6\sin\theta) + 36\cos^2\theta\,\cos^2(6\sin\theta)}\,d\theta \approx 8.0091.$$

54. The curve $r = \sin(\theta/4)$ is completely traced with $0 \le \theta \le 8\pi$. $r = \sin(\theta/4) \;\Rightarrow\; \dfrac{dr}{d\theta} = \tfrac{1}{4}\cos(\theta/4)$, so

$$r^2 + \left(\frac{dr}{d\theta}\right)^2 = \sin^2(\theta/4) + \tfrac{1}{16}\cos^2(\theta/4) \;\Rightarrow\; L = \int_0^{8\pi} \sqrt{\sin^2(\theta/4) + \tfrac{1}{16}\cos^2(\theta/4)}\,d\theta \approx 17.1568.$$

55. (a) From (10.2.6),

$$\begin{aligned} S &= \int_a^b 2\pi y\sqrt{(dx/d\theta)^2 + (dy/d\theta)^2}\,d\theta \\ &= \int_a^b 2\pi y\sqrt{r^2 + (dr/d\theta)^2}\,d\theta \qquad \text{[from the derivation of Equation 10.4.5]} \\ &= \int_a^b 2\pi r\sin\theta\sqrt{r^2 + (dr/d\theta)^2}\,d\theta \end{aligned}$$

(b) The curve $r^2 = \cos 2\theta$ goes through the pole when $\cos 2\theta = 0 \;\Rightarrow$ $2\theta = \frac{\pi}{2} \;\Rightarrow\; \theta = \frac{\pi}{4}$. We'll rotate the curve from $\theta = 0$ to $\theta = \frac{\pi}{4}$ and double this value to obtain the total surface area generated.

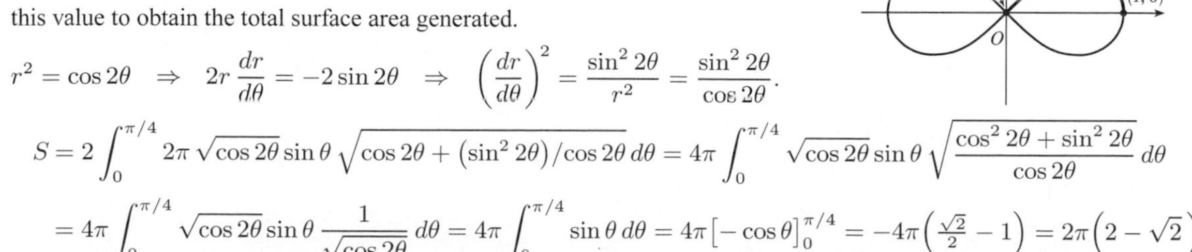

$$r^2 = \cos 2\theta \;\Rightarrow\; 2r\frac{dr}{d\theta} = -2\sin 2\theta \;\Rightarrow\; \left(\frac{dr}{d\theta}\right)^2 = \frac{\sin^2 2\theta}{r^2} = \frac{\sin^2 2\theta}{\cos 2\theta}.$$

$$S = 2\int_0^{\pi/4} 2\pi\sqrt{\cos 2\theta}\sin\theta\sqrt{\cos 2\theta + (\sin^2 2\theta)/\cos 2\theta}\,d\theta = 4\pi\int_0^{\pi/4}\sqrt{\cos 2\theta}\sin\theta\sqrt{\frac{\cos^2 2\theta + \sin^2 2\theta}{\cos 2\theta}}\,d\theta$$

$$= 4\pi\int_0^{\pi/4}\sqrt{\cos 2\theta}\sin\theta\,\frac{1}{\sqrt{\cos 2\theta}}\,d\theta = 4\pi\int_0^{\pi/4}\sin\theta\,d\theta = 4\pi\left[-\cos\theta\right]_0^{\pi/4} = -4\pi\left(\tfrac{\sqrt{2}}{2} - 1\right) = 2\pi\left(2 - \sqrt{2}\right)$$

56. (a) Rotation around $\theta = \frac{\pi}{2}$ is the same as rotation around the y-axis, that is, $S = \int_a^b 2\pi x\, ds$ where

$ds = \sqrt{(dx/dt)^2 + (dy/dt)^2}\, dt$ for a parametric equation, and for the special case of a polar equation, $x = r\cos\theta$ and

$ds = \sqrt{(dx/d\theta)^2 + (dy/d\theta)^2}\, d\theta = \sqrt{r^2 + (dr/d\theta)^2}\, d\theta$ [see the derivation of Equation 10.4.5]. Therefore, for a polar

equation rotated around $\theta = \frac{\pi}{2}$, $S = \int_a^b 2\pi r\cos\theta\sqrt{r^2 + (dr/d\theta)^2}\, d\theta$.

(b) As in the solution for Exercise 55(b), we can double the surface area generated by rotating the curve from $\theta = 0$ to $\theta = \frac{\pi}{4}$ to obtain the total surface area.

$$S = 2\int_0^{\pi/4} 2\pi\sqrt{\cos 2\theta}\cos\theta\sqrt{\cos 2\theta + (\sin^2 2\theta)/\cos 2\theta}\, d\theta = 4\pi\int_0^{\pi/4}\sqrt{\cos 2\theta}\cos\theta\sqrt{\frac{\cos^2 2\theta + \sin^2 2\theta}{\cos 2\theta}}\, d\theta$$

$$= 4\pi\int_0^{\pi/4}\sqrt{\cos 2\theta}\cos\theta\frac{1}{\sqrt{\cos 2\theta}}\, d\theta = 4\pi\int_0^{\pi/4}\cos\theta\, d\theta = 4\pi\big[\sin\theta\big]_0^{\pi/4} = 4\pi\left(\frac{\sqrt{2}}{2} - 0\right) = 2\sqrt{2}\,\pi$$

10.5 Conic Sections

1. $x^2 = 6y$ and $x^2 = 4py \Rightarrow 4p = 6 \Rightarrow p = \frac{3}{2}$.

The vertex is $(0, 0)$, the focus is $\left(0, \frac{3}{2}\right)$, and the directrix is $y = -\frac{3}{2}$.

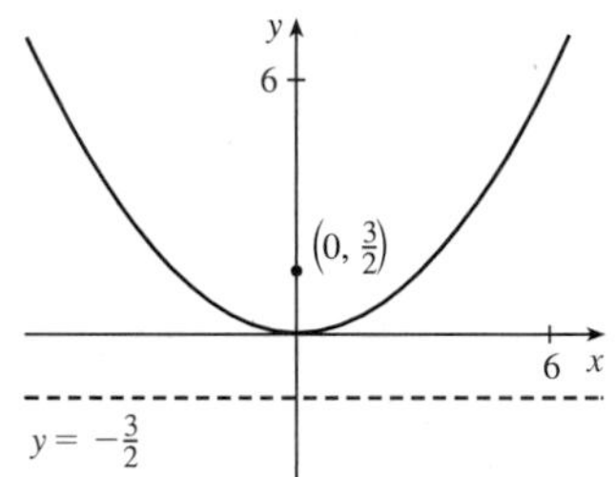
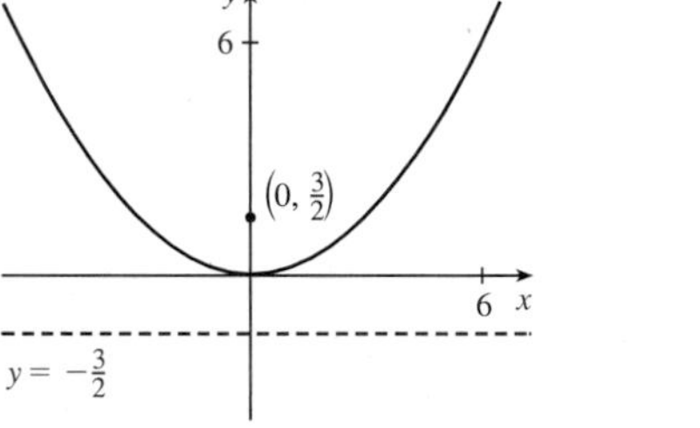

2. $2y^2 = 5x \Rightarrow y^2 = \frac{5}{2}x$. $4p = \frac{5}{2} \Rightarrow p = \frac{5}{8}$.

The vertex is $(0, 0)$, the focus is $\left(\frac{5}{8}, 0\right)$, and the directrix is $x = -\frac{5}{8}$.

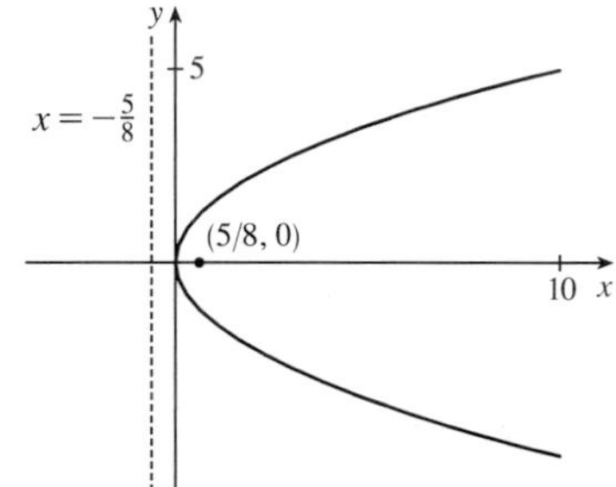

3. $2x = -y^2 \Rightarrow y^2 = -2x$. $4p = -2 \Rightarrow p = -\frac{1}{2}$.

The vertex is $(0, 0)$, the focus is $\left(-\frac{1}{2}, 0\right)$, and the directrix is $x = \frac{1}{2}$.

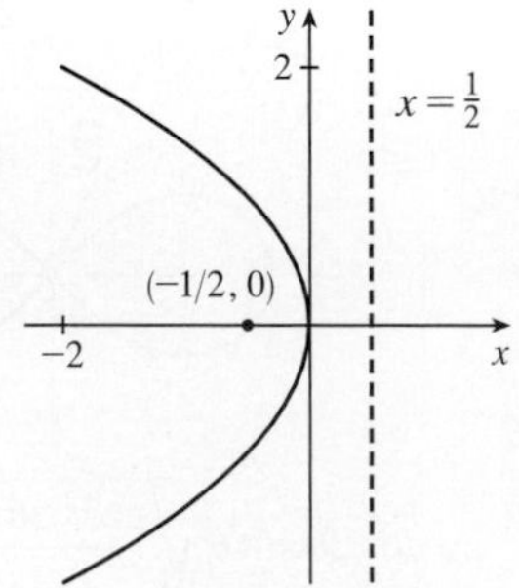

4. $3x^2 + 8y = 0 \Rightarrow 3x^2 = -8y \Rightarrow x^2 = -\frac{8}{3}y$.

$4p = -\frac{8}{3} \Rightarrow p = -\frac{2}{3}$. The vertex is $(0, 0)$, the focus is $\left(0, -\frac{2}{3}\right)$, and the directrix is $y = \frac{2}{3}$.

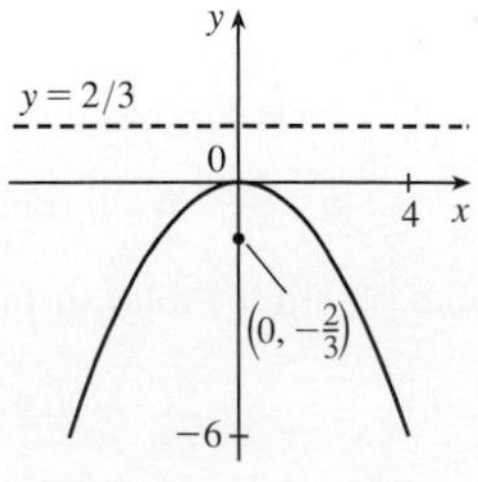

5. $(x+2)^2 = 8(y-3)$. $4p = 8$, so $p = 2$. The vertex is $(-2,3)$, the focus is $(-2,5)$, and the directrix is $y = 1$.

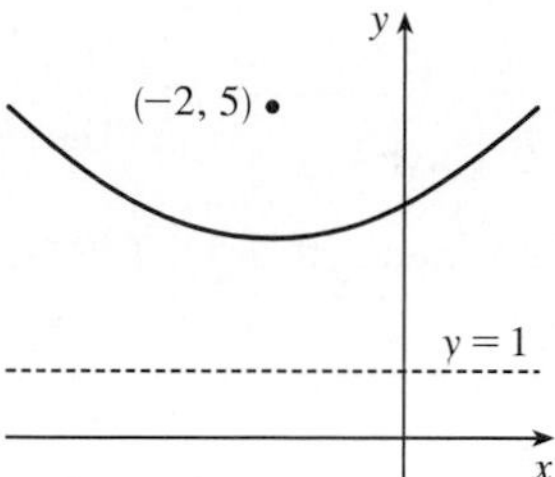

6. $(y-2)^2 = 2x + 1 = 2\left(x + \frac{1}{2}\right)$. $4p = 2$, so $p = \frac{1}{2}$. The vertex is $\left(-\frac{1}{2}, 2\right)$, the focus is $(0,2)$, and the directrix is $x = -1$.

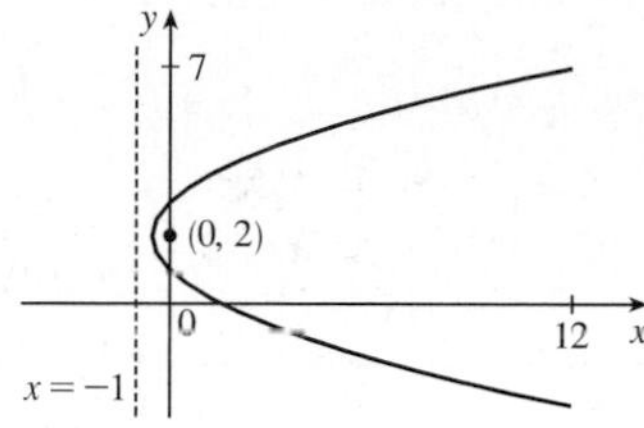

7. $y^2 + 6y + 2x + 1 = 0 \quad \Leftrightarrow \quad y^2 + 6y = -2x - 1$
$\Leftrightarrow \quad y^2 + 6y + 9 = -2x + 8 \quad \Leftrightarrow$
$(y+3)^2 = -2(x-4)$. $4p = -2$, so $p = -\frac{1}{2}$.
The vertex is $(4,-3)$, the focus is $\left(\frac{7}{2}, -3\right)$, and the directrix is $x = \frac{9}{2}$.

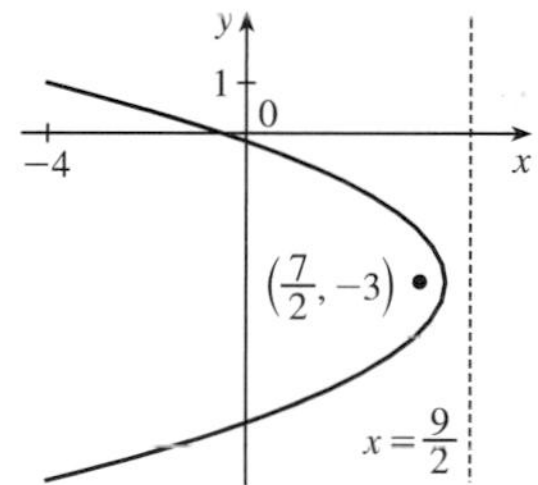

8. $2x^2 - 16x - 3y + 38 = 0 \quad \Leftrightarrow \quad 2x^2 - 16x = 3y - 38$
$\Leftrightarrow \quad 2(x^2 - 8x + 16) = 3y - 38 + 32 \quad \Leftrightarrow$
$2(x-4)^2 = 3y - 6 \quad \Leftrightarrow \quad (x-4)^2 = \frac{3}{2}(y-2)$.
$4p = \frac{3}{2}$, so $p = \frac{3}{8}$. The vertex is $(4,2)$, the focus is $\left(4, \frac{19}{8}\right)$, and the directrix is $y = \frac{13}{8}$.

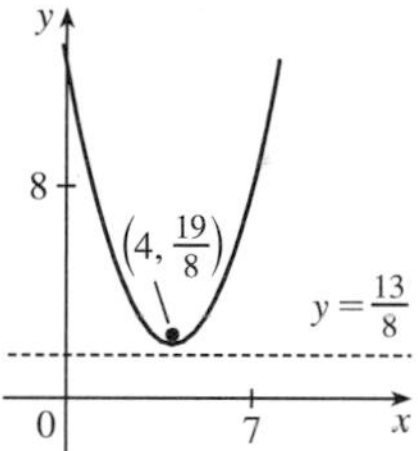

9. The equation has the form $y^2 = 4px$, where $p < 0$. Since the parabola passes through $(-1,1)$, we have $1^2 = 4p(-1)$, so $4p = -1$ and an equation is $y^2 = -x$ or $x = -y^2$. $4p = -1$, so $p = -\frac{1}{4}$ and the focus is $\left(-\frac{1}{4}, 0\right)$ while the directrix is $x = \frac{1}{4}$.

10. The vertex is $(2,-2)$, so the equation is of the form $(x-2)^2 = 4p(y+2)$, where $p > 0$. The point $(0,0)$ is on the parabola, so $4 = 4p(2)$ and $4p = 2$. Thus, an equation is $(x-2)^2 = 2(y+2)$. $4p = 2$, so $p = \frac{1}{2}$ and the focus is $\left(2, -\frac{3}{2}\right)$ while the directrix is $y = -\frac{5}{2}$.

11. $\dfrac{x^2}{2} + \dfrac{y^2}{4} = 1 \quad \Rightarrow \quad a = \sqrt{4} = 2, b = \sqrt{2}$,
$c = \sqrt{a^2 - b^2} = \sqrt{4-2} = \sqrt{2}$. The ellipse is centered at $(0,0)$, with vertices at $(0, \pm 2)$. The foci are $\left(0, \pm\sqrt{2}\right)$.

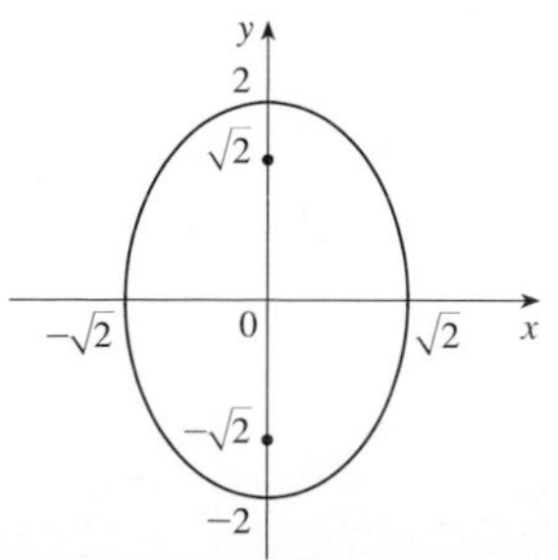

12. $\dfrac{x^2}{36} + \dfrac{y^2}{8} = 1 \quad \Rightarrow \quad a = \sqrt{36} = 6, b = \sqrt{8}$,
$c = \sqrt{a^2 - b^2} = \sqrt{36-8} = \sqrt{28} = 2\sqrt{7}$. The ellipse is centered at $(0,0)$, with vertices at $(\pm 6, 0)$. The foci are $(\pm 2\sqrt{7}, 0)$.

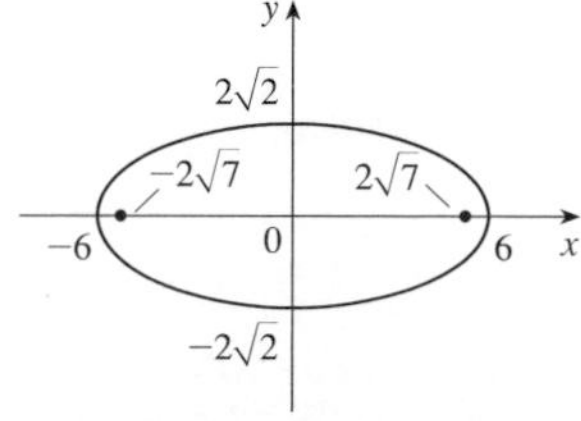

13. $x^2 + 9y^2 = 9 \quad \Leftrightarrow \quad \dfrac{x^2}{9} + \dfrac{y^2}{1} = 1 \quad \Rightarrow \quad a = \sqrt{9} = 3,$

$b = \sqrt{1} = 1,\ c = \sqrt{a^2 - b^2} = \sqrt{9 - 1} = \sqrt{8} = 2\sqrt{2}.$

The ellipse is centered at $(0, 0)$, with vertices $(\pm 3, 0)$.

The foci are $(\pm 2\sqrt{2}, 0)$.

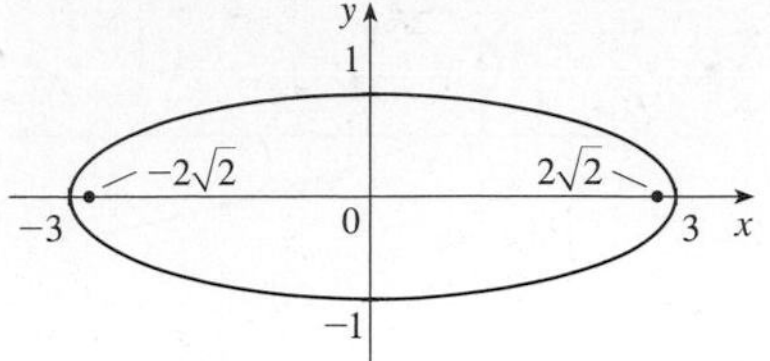

14. $100x^2 + 36y^2 = 225 \quad \Leftrightarrow \quad \dfrac{x^2}{\frac{225}{100}} + \dfrac{y^2}{\frac{225}{36}} = 1 \quad \Leftrightarrow$

$\dfrac{x^2}{\frac{9}{4}} + \dfrac{y^2}{\frac{25}{4}} = 1 \quad \Rightarrow \quad a = \sqrt{\frac{25}{4}} = \frac{5}{2},\ b = \sqrt{\frac{9}{4}} = \frac{3}{2},$

$c = \sqrt{a^2 - b^2} = \sqrt{\frac{25}{4} - \frac{9}{4}} = 2.$ The ellipse is centered

at $(0, 0)$, with vertices $\left(0, \pm\frac{5}{2}\right)$. The foci are $(0, \pm 2)$.

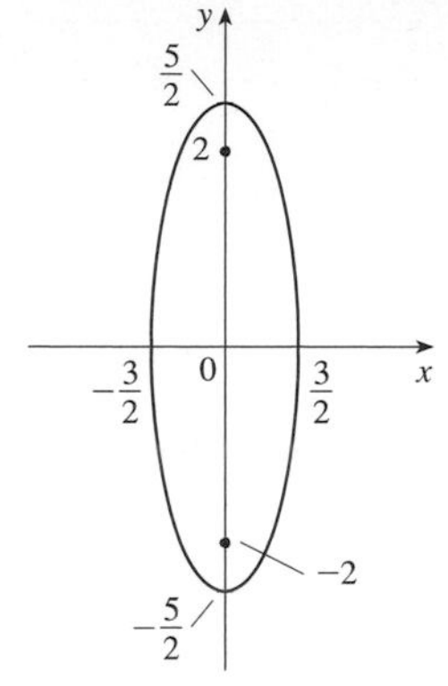

15. $9x^2 - 18x + 4y^2 = 27 \quad \Leftrightarrow$

$9(x^2 - 2x + 1) + 4y^2 = 27 + 9 \quad \Leftrightarrow$

$9(x - 1)^2 + 4y^2 = 36 \quad \Leftrightarrow \quad \dfrac{(x - 1)^2}{4} + \dfrac{y^2}{9} = 1 \quad \Rightarrow$

$a = 3,\ b = 2,\ c = \sqrt{5} \quad \Rightarrow \quad$ center $(1, 0)$,

vertices $(1, \pm 3)$, foci $\left(1, \pm\sqrt{5}\right)$

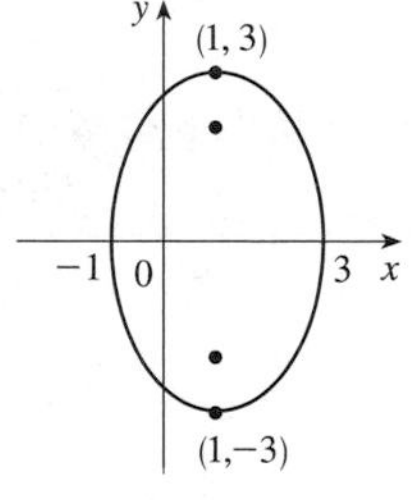

16. $x^2 + 3y^2 + 2x - 12y + 10 = 0 \quad \Leftrightarrow$

$x^2 + 2x + 1 + 3(y^2 - 4y + 4) = -10 + 1 + 12 \quad \Leftrightarrow$

$(x + 1)^2 + 3(y - 2)^2 = 3 \quad \Leftrightarrow$

$\dfrac{(x + 1)^2}{3} + \dfrac{(y - 2)^2}{1} = 1 \quad \Rightarrow \quad a = \sqrt{3},\ b = 1,$

$c = \sqrt{2} \quad \Rightarrow \quad$ center $(-1, 2)$, vertices $\left(-1 \pm \sqrt{3}, 2\right)$,

foci $\left(-1 \pm \sqrt{2}, 2\right)$

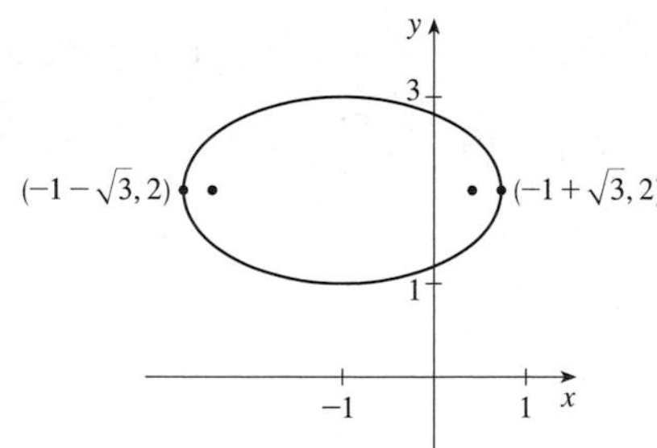

17. The center is $(0, 0)$, $a = 3$, and $b = 2$, so an equation is $\dfrac{x^2}{4} + \dfrac{y^2}{9} = 1$. $c = \sqrt{a^2 - b^2} = \sqrt{5}$, so the foci are $\left(0, \pm\sqrt{5}\right)$.

18. The ellipse is centered at $(2, 1)$, with $a = 3$ and $b = 2$. An equation is $\dfrac{(x - 2)^2}{9} + \dfrac{(y - 1)^2}{4} = 1$. $c = \sqrt{a^2 - b^2} = \sqrt{5}$, so the foci are $\left(2 \pm \sqrt{5}, 1\right)$.

19. $\dfrac{y^2}{25} - \dfrac{x^2}{9} = 1 \;\Rightarrow\; a = 5, b = 3, c = \sqrt{25 + 9} = \sqrt{34} \;\Rightarrow$

center $(0, 0)$, vertices $(0, \pm 5)$, foci $\left(0, \pm\sqrt{34}\right)$, asymptotes $y = \pm\frac{5}{3}x$.

Note: It is helpful to draw a $2a$-by-$2b$ rectangle whose center is the center of the hyperbola. The asymptotes are the extended diagonals of the rectangle.

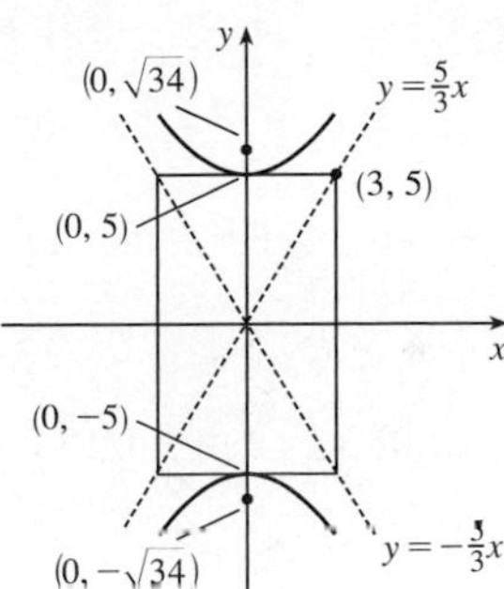

20. $\dfrac{x^2}{36} - \dfrac{y^2}{64} = 1 \;\Rightarrow\; a = 6, b = 8, c = \sqrt{36 + 64} = 10 \;\Rightarrow$

center $(0, 0)$, vertices $(\pm 6, 0)$, foci $(\pm 10, 0)$, asymptotes $y = \pm\frac{8}{6}x = \pm\frac{4}{3}x$

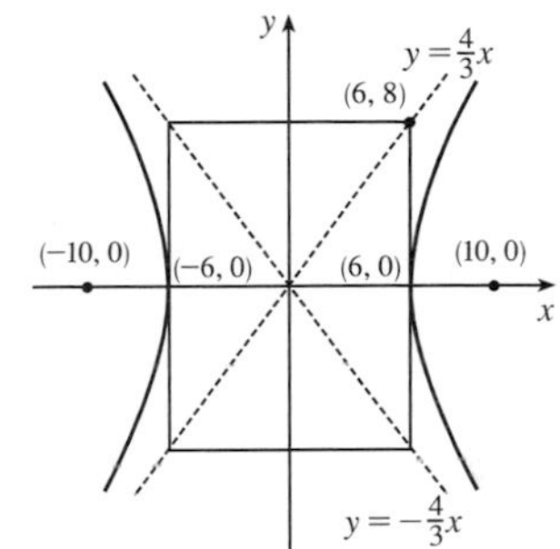

21. $x^2 - y^2 = 100 \;\Leftrightarrow\; \dfrac{x^2}{100} - \dfrac{y^2}{100} = 1 \;\Rightarrow\; a = b = 10,$

$c = \sqrt{100 + 100} = 10\sqrt{2} \;\Rightarrow$ center $(0, 0)$, vertices $(\pm 10, 0)$,

foci $\left(\pm 10\sqrt{2}, 0\right)$, asymptotes $y = \pm\frac{10}{10}x = \pm x$

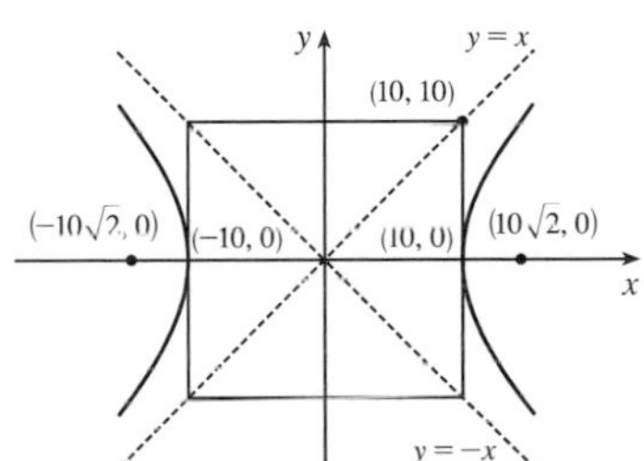

22. $y^2 - 16x^2 = 16 \;\Leftrightarrow\; \dfrac{y^2}{16} - \dfrac{x^2}{1} = 1 \;\Rightarrow\; a = 4, b = 1,$

$c = \sqrt{16 + 1} = \sqrt{17} \;\Rightarrow$ center $(0, 0)$, vertices $(0, \pm 4)$,

foci $\left(0, \pm\sqrt{17}\right)$, asymptotes $y = \pm\frac{4}{1}x = \pm 4x$

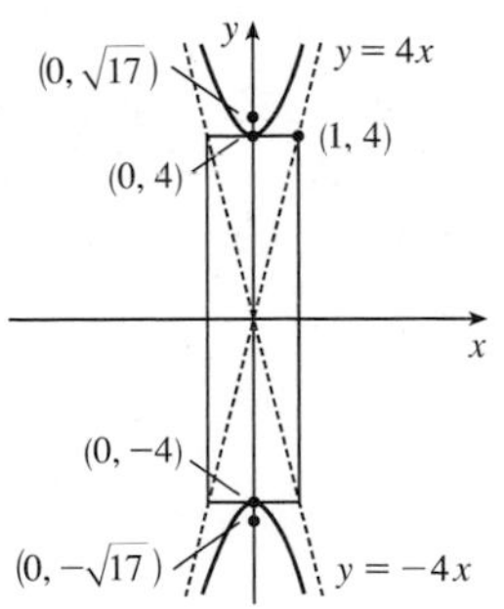

23. $x^2 - y^2 + 2y = 2 \;\Leftrightarrow\; x^2 - (y^2 - 2y + 1) = 2 - 1 \;\Leftrightarrow$

$\dfrac{x^2}{1} - \dfrac{(y - 1)^2}{1} = 1 \;\Rightarrow\; a = b = 1, c = \sqrt{1 + 1} = \sqrt{2} \;\Rightarrow$

center $(0, 1)$, vertices $(\pm 1, 1)$, foci $\left(\pm\sqrt{2}, 1\right)$,

asymptotes $y - 1 = \pm\frac{1}{1}x = \pm x$.

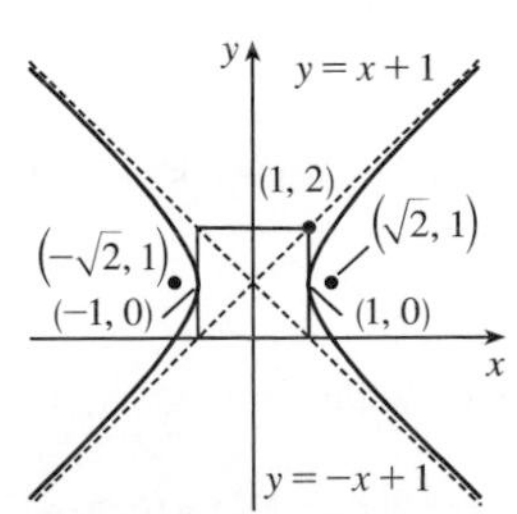

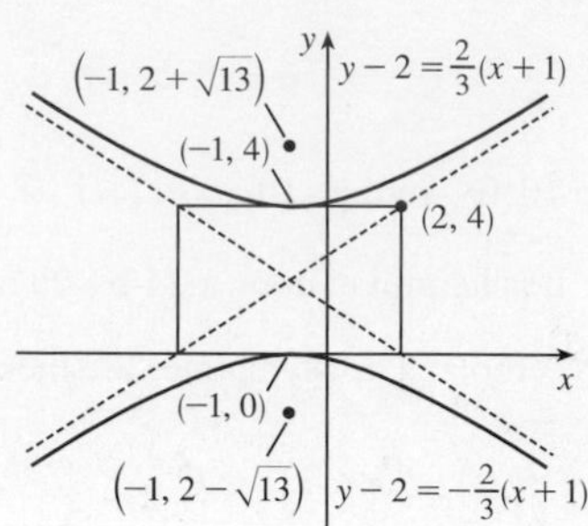

24. $9y^2 - 4x^2 - 36y - 8x = 4 \Leftrightarrow$

$9(y^2 - 4y + 4) - 4(x^2 + 2x + 1) = 4 + 36 - 4 \Leftrightarrow$

$9(y-2)^2 - 4(x+1)^2 = 36 \Leftrightarrow \dfrac{(y-2)^2}{4} - \dfrac{(x+1)^2}{9} = 1 \Rightarrow$

$a = 2, b = 3, c = \sqrt{4+9} = \sqrt{13} \Rightarrow$ center $(-1, 2)$, vertices $(-1, 2 \pm 2)$, foci $(-1, 2 \pm \sqrt{13})$, asymptotes $y - 2 = \pm\frac{2}{3}(x+1)$.

25. $4x^2 = y^2 + 4 \Leftrightarrow 4x^2 - y^2 = 4 \Leftrightarrow \dfrac{x^2}{1} - \dfrac{y^2}{4} = 1$. This is an equation of a *hyperbola* with vertices $(\pm 1, 0)$. The foci are at $(\pm\sqrt{1+4}, 0) = (\pm\sqrt{5}, 0)$.

26. $4x^2 = y + 4 \Leftrightarrow x^2 = \frac{1}{4}(y+4)$. This is an equation of a *parabola* with $4p = \frac{1}{4}$, so $p = \frac{1}{16}$. The vertex is $(0, -4)$ and the focus is $\left(0, -4 + \frac{1}{16}\right) = \left(0, -\frac{63}{16}\right)$.

27. $x^2 = 4y - 2y^2 \Leftrightarrow x^2 + 2y^2 - 4y = 0 \Leftrightarrow x^2 + 2(y^2 - 2y + 1) = 2 \Leftrightarrow x^2 + 2(y-1)^2 = 2 \Leftrightarrow$ $\dfrac{x^2}{2} + \dfrac{(y-1)^2}{1} = 1$. This is an equation of an *ellipse* with vertices at $(\pm\sqrt{2}, 1)$. The foci are at $(\pm\sqrt{2-1}, 1) = (\pm 1, 1)$.

28. $y^2 - 2 = x^2 - 2x \Leftrightarrow y^2 - x^2 + 2x = 2 \Leftrightarrow y^2 - (x^2 - 2x + 1) = 2 - 1 \Leftrightarrow \dfrac{y^2}{1} - \dfrac{(x-1)^2}{1} = 1$. This is an equation of a *hyperbola* with vertices $(1, \pm 1)$. The foci are at $(1, \pm\sqrt{1+1}) = (1, \pm\sqrt{2})$.

29. $3x^2 - 6x - 2y = 1 \Leftrightarrow 3x^2 - 6x = 2y + 1 \Leftrightarrow 3(x^2 - 2x + 1) = 2y + 1 + 3 \Leftrightarrow 3(x-1)^2 = 2y + 4 \Leftrightarrow$ $(x-1)^2 = \frac{2}{3}(y+2)$. This is an equation of a *parabola* with $4p = \frac{2}{3}$, so $p = \frac{1}{6}$. The vertex is $(1, -2)$ and the focus is $\left(1, -2 + \frac{1}{6}\right) = \left(1, -\frac{11}{6}\right)$.

30. $x^2 - 2x + 2y^2 - 8y + 7 = 0 \Leftrightarrow (x^2 - 2x + 1) + 2(y^2 - 4y + 4) = -7 + 1 + 8 \Leftrightarrow (x-1)^2 + 2(y-2)^2 = 2 \Leftrightarrow$ $\dfrac{(x-1)^2}{2} + \dfrac{(y-2)^2}{1} = 1$. This is an equation of an *ellipse* with vertices at $(1 \pm \sqrt{2}, 2)$. The foci are at $(1 \pm \sqrt{2-1}, 2) = (1 \pm 1, 2)$.

31. The parabola with vertex $(0, 0)$ and focus $(1, 0)$ opens to the right and has $p = 1$, so its equation is $y^2 = 4px$, or $y^2 = 4x$.

32. The parabola with focus $(0, 0)$ and directrix $y = 6$ has vertex $(0, 3)$ and opens downward, so $p = -3$ and its equation is $(x-0)^2 = 4p(y-3)$, or $x^2 = -12(y-3)$.

33. The distance from the focus $(-4, 0)$ to the directrix $x = 2$ is $2 - (-4) = 6$, so the distance from the focus to the vertex is $\frac{1}{2}(6) = 3$ and the vertex is $(-1, 0)$. Since the focus is to the left of the vertex, $p = -3$. An equation is $y^2 = 4p(x+1) \Rightarrow$ $y^2 = -12(x+1)$.

34. The parabola with vertex $(2, 3)$ and focus $(2, -1)$ opens downward and has $p = -1 - 3 = -4$, so its equation is $(x-2)^2 = 4p(y-3)$, or $(x-2)^2 = -16(y-3)$.

35. The parabola with vertex $(3, -1)$ having a horizontal axis has equation $[y - (-1)]^2 = 4p(x - 3)$. Since it passes through $(-15, 2)$, $(2 + 1)^2 = 4p(-15 - 3) \Rightarrow 9 = 4p(-18) \Rightarrow 4p = -\frac{1}{2}$. An equation is $(y + 1)^2 = -\frac{1}{2}(x - 3)$.

36. The parabola with vertical axis and passing through $(0, 4)$ has equation $y = ax^2 + bx + 4$. It also passes through $(1, 3)$ and $(-2, -6)$, so

$$\begin{cases} 3 = a + b + 4 \\ -6 = 4a - 2b + 4 \end{cases} \Rightarrow \begin{cases} -1 = a + b \\ -10 = 4a - 2b \end{cases} \Rightarrow \begin{cases} -1 = a + b \\ -5 = 2a - b \end{cases}$$

Adding the last two equations gives us $3a = -6$, or $a = -2$. Since $a + b = -1$, we have $b = 1$, and an equation is $y = -2x^2 + x + 4$.

37. The ellipse with foci $(\pm 2, 0)$ and vertices $(\pm 5, 0)$ has center $(0, 0)$ and a horizontal major axis, with $a = 5$ and $c = 2$, so $b^2 = a^2 - c^2 = 25 - 4 = 21$. An equation is $\dfrac{x^2}{25} + \dfrac{y^2}{21} = 1$.

38. The ellipse with foci $(0, \pm\sqrt{2})$ and vertices $(0, \pm 2)$ has center $(0, 0)$ and a vertical major axis, with $a = 2$ and $c = \sqrt{2}$, so $b^2 = a^2 - c^2 = 4 - 2 = 2$. An equation is $\dfrac{x^2}{2} + \dfrac{y^2}{4} = 1$.

39. Since the vertices are $(0, 0)$ and $(0, 8)$, the ellipse has center $(0, 4)$ with a vertical axis and $a = 4$. The foci at $(0, 2)$ and $(0, 6)$ are 2 units from the center, so $c = 2$ and $b = \sqrt{a^2 - c^2} = \sqrt{4^2 - 2^2} = \sqrt{12}$. An equation is $\dfrac{(x - 0)^2}{b^2} + \dfrac{(y - 4)^2}{a^2} = 1 \Rightarrow$ $\dfrac{x^2}{12} + \dfrac{(y - 4)^2}{16} = 1$.

40. Since the foci are $(0, -1)$ and $(8, -1)$, the ellipse has center $(4, -1)$ with a horizontal axis and $c = 4$.
The vertex $(9, -1)$ is 5 units from the center, so $a = 5$ and $b = \sqrt{a^2 - c^2} = \sqrt{5^2 - 4^2} = \sqrt{9}$. An equation is $\dfrac{(x - 4)^2}{a^2} + \dfrac{(y + 1)^2}{b^2} = 1 \Rightarrow \dfrac{(x - 4)^2}{25} + \dfrac{(y + 1)^2}{9} = 1$.

41. An equation of an ellipse with center $(-1, 4)$ and vertex $(-1, 0)$ is $\dfrac{(x + 1)^2}{b^2} + \dfrac{(y - 4)^2}{4^2} = 1$. The focus $(-1, 6)$ is 2 units from the center, so $c = 2$. Thus, $b^2 + 2^2 = 4^2 \Rightarrow b^2 = 12$, and the equation is $\dfrac{(x + 1)^2}{12} + \dfrac{(y - 4)^2}{16} = 1$.

42. Foci $F_1(-4, 0)$ and $F_2(4, 0) \Rightarrow c = 4$ and an equation is $\dfrac{x^2}{a^2} + \dfrac{y^2}{b^2} = 1$. The ellipse passes through $P(-4, 1.8)$, so $2a = |PF_1| + |PF_2| \Rightarrow 2a = 1.8 + \sqrt{8^2 + (1.8)^2} \Rightarrow 2a = 1.8 + 8.2 \Rightarrow a = 5$.
$b^2 = a^2 - c^2 = 25 - 16 = 9$ and the equation is $\dfrac{x^2}{25} + \dfrac{y^2}{9} = 1$.

43. An equation of a hyperbola with vertices $(\pm 3, 0)$ is $\dfrac{x^2}{3^2} - \dfrac{y^2}{b^2} = 1$. Foci $(\pm 5, 0) \Rightarrow c = 5$ and $3^2 + b^2 = 5^2 \Rightarrow$ $b^2 = 25 - 9 = 16$, so the equation is $\dfrac{x^2}{9} - \dfrac{y^2}{16} = 1$.

44. An equation of a hyperbola with vertices $(0, \pm 2)$ is $\dfrac{y^2}{2^2} - \dfrac{x^2}{b^2} = 1$. Foci $(0, \pm 5)$ $\Rightarrow$ $c = 5$ and $2^2 + b^2 = 5^2$ $\Rightarrow$

$b^2 = 25 - 4 = 21$, so the equation is $\dfrac{y^2}{4} - \dfrac{x^2}{21} = 1$.

45. The center of a hyperbola with vertices $(-3, -4)$ and $(-3, 6)$ is $(-3, 1)$, so $a = 5$ and an equation is

$\dfrac{(y-1)^2}{5^2} - \dfrac{(x+3)^2}{b^2} = 1$. Foci $(-3, -7)$ and $(-3, 9)$ $\Rightarrow$ $c = 8$, so $5^2 + b^2 = 8^2$ $\Rightarrow$ $b^2 = 64 - 25 = 39$ and the

equation is $\dfrac{(y-1)^2}{25} - \dfrac{(x+3)^2}{39} = 1$.

46. The center of a hyperbola with vertices $(-1, 2)$ and $(7, 2)$ is $(3, 2)$, so $a = 4$ and an equation is $\dfrac{(x-3)^2}{4^2} - \dfrac{(y-2)^2}{b^2} = 1$.

Foci $(-2, 2)$ and $(8, 2)$ $\Rightarrow$ $c = 5$, so $4^2 + b^2 = 5^2$ $\Rightarrow$ $b^2 = 25 - 16 = 9$ and the equation is

$\dfrac{(x-3)^2}{16} - \dfrac{(y-2)^2}{9} = 1$.

47. The center of a hyperbola with vertices $(\pm 3, 0)$ is $(0, 0)$, so $a = 3$ and an equation is $\dfrac{x^2}{3^2} - \dfrac{y^2}{b^2} = 1$.

Asymptotes $y = \pm 2x$ $\Rightarrow$ $\dfrac{b}{a} = 2$ $\Rightarrow$ $b = 2(3) = 6$ and the equation is $\dfrac{x^2}{9} - \dfrac{y^2}{36} = 1$.

48. The center of a hyperbola with foci $(2, 0)$ and $(2, 8)$ is $(2, 4)$, so $c = 4$ and an equation is $\dfrac{(y-4)^2}{a^2} - \dfrac{(x-2)^2}{b^2} = 1$.

The asymptote $y = 3 + \frac{1}{2}x$ has slope $\frac{1}{2}$, so $\dfrac{a}{b} = \dfrac{1}{2}$ $\Rightarrow$ $b = 2a$ and $a^2 + b^2 = c^2$ $\Rightarrow$ $a^2 + (2a)^2 = 4^2$ $\Rightarrow$

$5a^2 = 16$ $\Rightarrow$ $a^2 = \frac{16}{5}$ and so $b^2 = 16 - \frac{16}{5} = \frac{64}{5}$. Thus, an equation is $\dfrac{(y-4)^2}{16/5} - \dfrac{(x-2)^2}{64/5} = 1$.

49. In Figure 8, we see that the point on the ellipse closest to a focus is the closer vertex (which is a distance $a - c$ from it) while the farthest point is the other vertex (at a distance of $a + c$). So for this lunar orbit, $(a - c) + (a + c) = 2a = (1728 + 110) + (1728 + 314)$, or $a = 1940$; and $(a + c) - (a - c) = 2c = 314 - 110$, or $c = 102$. Thus, $b^2 = a^2 - c^2 = 3{,}753{,}196$, and the equation is $\dfrac{x^2}{3{,}763{,}600} + \dfrac{y^2}{3{,}753{,}196} = 1$.

50. (a) Choose V to be the origin, with x-axis through V and F. Then F is $(p, 0)$, A is $(p, 5)$, so substituting A into the equation $y^2 = 4px$ gives $25 = 4p^2$ so $p = \frac{5}{2}$ and $y^2 = 10x$.

(b) $x = 11$ $\Rightarrow$ $y = \sqrt{110}$ $\Rightarrow$ $|CD| = 2\sqrt{110}$

51. (a) Set up the coordinate system so that A is $(-200, 0)$ and B is $(200, 0)$.

$|PA| - |PB| = (1200)(980) = 1{,}176{,}000 \text{ ft} = \frac{2450}{11} \text{ mi} = 2a$ $\Rightarrow$ $a = \frac{1225}{11}$, and $c = 200$ so

$b^2 = c^2 - a^2 = \dfrac{3{,}339{,}375}{121}$ $\Rightarrow$ $\dfrac{121x^2}{1{,}500{,}625} - \dfrac{121y^2}{3{,}339{,}375} = 1$.

(b) Due north of B $\Rightarrow$ $x = 200$ $\Rightarrow$ $\dfrac{(121)(200)^2}{1{,}500{,}625} - \dfrac{121y^2}{3{,}339{,}375} = 1$ $\Rightarrow$ $y = \dfrac{133{,}575}{539} \approx 248$ mi

52. $|PF_1| - |PF_2| = \pm 2a \;\Leftrightarrow\; \sqrt{(x+c)^2 + y^2} - \sqrt{(x-c)^2 + y^2} = \pm 2a \;\Leftrightarrow$

$\sqrt{(x+c)^2 + y^2} = \sqrt{(x-c)^2 + y^2} \pm 2a \;\Leftrightarrow\; (x+c)^2 + y^2 = (x-c)^2 + y^2 + 4a^2 \pm 4a\sqrt{(x-c)^2 + y^2} \;\Leftrightarrow$

$4cx - 4a^2 = \pm 4a\sqrt{(x-c)^2 + y^2} \;\Leftrightarrow\; c^2x^2 - 2a^2cx + a^4 = a^2(x^2 - 2cx + c^2 + y^2) \;\Leftrightarrow$

$(c^2 - a^2)x^2 - a^2y^2 = a^2(c^2 - a^2) \;\Leftrightarrow\; b^2x^2 - a^2y^2 = a^2b^2$ [where $b^2 = c^2 - a^2$] $\Leftrightarrow\; \dfrac{x^2}{a^2} - \dfrac{y^2}{b^2} = 1$

53. The function whose graph is the upper branch of this hyperbola is concave upward. The function is

$y = f(x) = a\sqrt{1 + \dfrac{x^2}{b^2}} = \dfrac{a}{b}\sqrt{b^2 + x^2}$, so $y' = \dfrac{a}{b}x(b^2 + x^2)^{-1/2}$ and

$y'' = \dfrac{a}{b}\left[(b^2 + x^2)^{-1/2} - x^2(b^2 + x^2)^{-3/2}\right] = ab(b^2 + x^2)^{-3/2} > 0$ for all x, and so f is concave upward.

54. We can follow exactly the same sequence of steps as in the derivation of Formula 4, except we use the points $(1, 1)$ and $(-1, -1)$ in the distance formula (first equation of that derivation) so $\sqrt{(x-1)^2 + (y-1)^2} + \sqrt{(x+1)^2 + (y+1)^2} = 4$ will lead (after moving the second term to the right, squaring, and simplifying) to $2\sqrt{(x+1)^2 + (y+1)^2} = x + y + 4$, which, after squaring and simplifying again, leads to $3x^2 - 2xy + 3y^2 = 8$.

55. (a) If $k > 16$, then $k - 16 > 0$, and $\dfrac{x^2}{k} + \dfrac{y^2}{k - 16} = 1$ is an *ellipse* since it is the sum of two squares on the left side.

(b) If $0 < k < 16$, then $k - 16 < 0$, and $\dfrac{x^2}{k} + \dfrac{y^2}{k - 16} = 1$ is a *hyperbola* since it is the difference of two squares on the left side.

(c) If $k < 0$, then $k - 16 < 0$, and there is *no curve* since the left side is the sum of two negative terms, which cannot equal 1.

(d) In case (a), $a^2 = k$, $b^2 = k - 16$, and $c^2 = a^2 - b^2 = 16$, so the foci are at $(\pm 4, 0)$. In case (b), $k - 16 < 0$, so $a^2 = k$, $b^2 = 16 - k$, and $c^2 = a^2 + b^2 = 16$, and so again the foci are at $(\pm 4, 0)$.

56. (a) $y^2 = 4px \;\Rightarrow\; 2yy' = 4p \;\Rightarrow\; y' = \dfrac{2p}{y}$, so the tangent line is

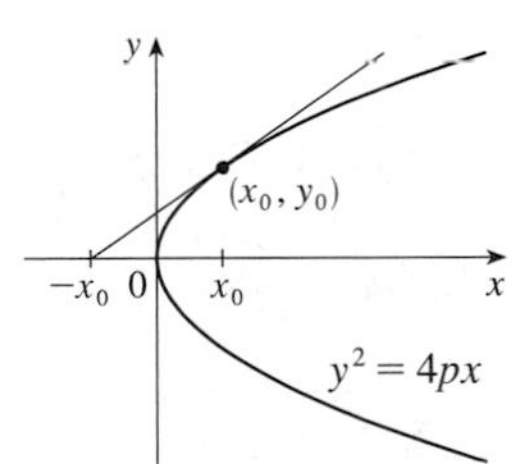

$y - y_0 = \dfrac{2p}{y_0}(x - x_0) \;\Rightarrow\; yy_0 - y_0^2 = 2p(x - x_0) \;\Leftrightarrow$

$yy_0 - 4px_0 = 2px - 2px_0 \;\Rightarrow\; yy_0 = 2p(x + x_0)$.

(b) The x-intercept is $-x_0$.

57. $x^2 = 4py \;\Rightarrow\; 2x = 4py' \;\Rightarrow\; y' = \dfrac{x}{2p}$, so the tangent line at (x_0, y_0) is $y - \dfrac{x_0^2}{4p} = \dfrac{x_0}{2p}(x - x_0)$. This line passes through the point $(a, -p)$ on the directrix, so $-p - \dfrac{x_0^2}{4p} = \dfrac{x_0}{2p}(a - x_0) \;\Rightarrow\; -4p^2 - x_0^2 = 2ax_0 - 2x_0^2 \;\Leftrightarrow$

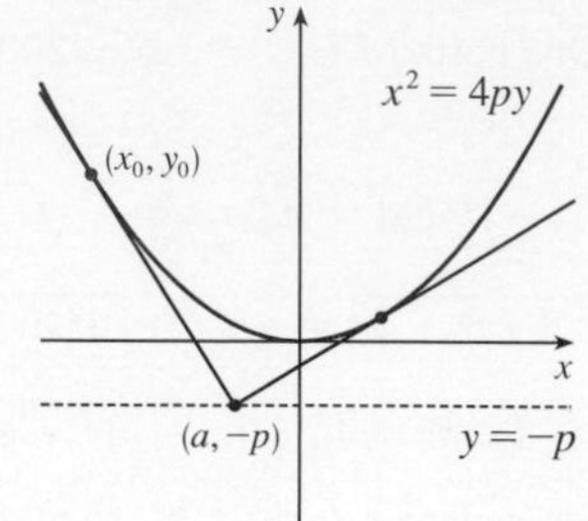

$x_0^2 - 2ax_0 - 4p^2 = 0 \quad \Leftrightarrow \quad x_0^2 - 2ax_0 + a^2 = a^2 + 4p^2 \quad \Leftrightarrow$

$(x_0 - a)^2 = a^2 + 4p^2 \quad \Leftrightarrow \quad x_0 = a \pm \sqrt{a^2 + 4p^2}$. The slopes of the tangent lines at $x = a \pm \sqrt{a^2 + 4p^2}$ are $\dfrac{a \pm \sqrt{a^2 + 4p^2}}{2p}$, so the product of the two slopes is

$$\frac{a + \sqrt{a^2 + 4p^2}}{2p} \cdot \frac{a - \sqrt{a^2 + 4p^2}}{2p} = \frac{a^2 - (a^2 + 4p^2)}{4p^2} = \frac{-4p^2}{4p^2} = -1,$$

showing that the tangent lines are perpendicular.

58. Without a loss of generality, let the ellipse, hyperbola, and foci be as shown in the figure.

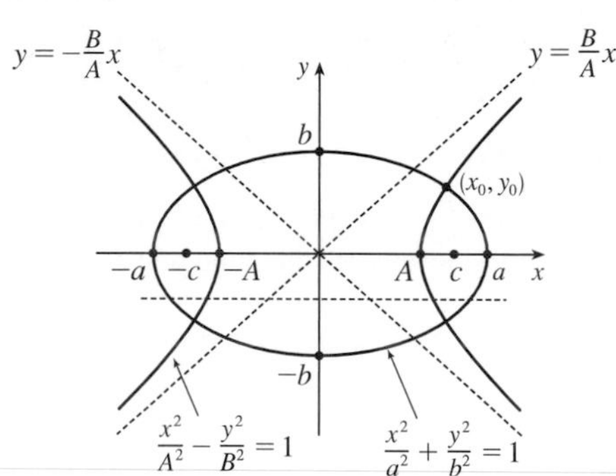

The curves intersect (eliminate y^2) $\Rightarrow$

$$B^2\left(\frac{x^2}{A^2} - \frac{y^2}{B^2}\right) + b^2\left(\frac{x^2}{a^2} + \frac{y^2}{b^2}\right) = B^2 + b^2 \quad \Rightarrow$$

$$\frac{B^2x^2}{A^2} + \frac{b^2x^2}{a^2} = B^2 + b^2 \quad \Rightarrow \quad x^2\left(\frac{B^2}{A^2} + \frac{b^2}{a^2}\right) = B^2 + b^2 \quad \Rightarrow$$

$$x^2 = \frac{B^2 + b^2}{\dfrac{a^2B^2 + b^2A^2}{A^2a^2}} = \frac{A^2a^2(B^2 + b^2)}{a^2B^2 + b^2A^2}.$$

Similarly, $y^2 = \dfrac{B^2b^2(a^2 - A^2)}{b^2A^2 + a^2B^2}$.

Next we find the slopes of the tangent lines of the curves: $\dfrac{x^2}{a^2} + \dfrac{y^2}{b^2} = 1 \;\Rightarrow\; \dfrac{2x}{a^2} + \dfrac{2yy'}{b^2} = 0 \;\Rightarrow\; \dfrac{yy'}{b^2} = -\dfrac{x}{a^2} \;\Rightarrow$

$y'_E = -\dfrac{b^2}{a^2}\dfrac{x}{y}$ and $\dfrac{x^2}{A^2} - \dfrac{y^2}{B^2} = 1 \;\Rightarrow\; \dfrac{2x}{A^2} - \dfrac{2yy'}{B^2} = 0 \;\Rightarrow\; \dfrac{yy'}{B^2} = \dfrac{x}{A^2} \;\Rightarrow\; y'_H = \dfrac{B^2}{A^2}\dfrac{x}{y}$. The product of the slopes

at (x_0, y_0) is $y'_E\,y'_H = -\dfrac{b^2B^2x_0^2}{a^2A^2y_0^2} = -\dfrac{b^2B^2\left[\dfrac{A^2a^2(B^2+b^2)}{a^2B^2 + b^2A^2}\right]}{a^2A^2\left[\dfrac{B^2b^2(a^2 - A^2)}{b^2A^2 + a^2B^2}\right]} = -\dfrac{B^2 + b^2}{a^2 - A^2}$. Since $a^2 - b^2 = c^2$ and $A^2 + B^2 = c^2$,

we have $a^2 - b^2 = A^2 + B^2 \;\Rightarrow\; a^2 - A^2 = b^2 + B^2$, so the product of the slopes is -1, and hence, the tangent lines at each point of intersection are perpendicular.

59. $9x^2 + 4y^2 = 36 \;\Leftrightarrow\; \dfrac{x^2}{4} + \dfrac{y^2}{9} = 1$. We use the parametrization $x = 2\cos t$, $y = 3\sin t$, $0 \le t \le 2\pi$. The circumference is given by

$$\begin{aligned} L &= \textstyle\int_0^{2\pi} \sqrt{(dx/dt)^2 + (dy/dt)^2}\,dt = \int_0^{2\pi} \sqrt{(-2\sin t)^2 + (3\cos t)^2}\,dt \\ &= \textstyle\int_0^{2\pi} \sqrt{4\sin^2 t + 9\cos^2 t}\,dt = \int_0^{2\pi} \sqrt{4 + 5\cos^2 t}\,dt \end{aligned}$$

Now use Simpson's Rule with $n = 8$, $\Delta t = \dfrac{2\pi - 0}{8} = \dfrac{\pi}{4}$, and $f(t) = \sqrt{4 + 5\cos^2 t}$ to get

$$L \approx S_8 = \tfrac{\pi/4}{3}\left[f(0) + 4f\left(\tfrac{\pi}{4}\right) + 2f\left(\tfrac{\pi}{2}\right) + 4f\left(\tfrac{3\pi}{4}\right) + 2f(\pi) + 4f\left(\tfrac{5\pi}{4}\right) + 2f\left(\tfrac{3\pi}{2}\right) + 4f\left(\tfrac{7\pi}{4}\right) + f(2\pi)\right] \approx 15.9.$$

60. The length of the major axis is $2a$, so $a = \frac{1}{2}(1.18 \times 10^{10}) = 5.9 \times 10^9$. The length of the minor axis is $2b$, so $b = \frac{1}{2}(1.14 \times 10^{10}) = 5.7 \times 10^9$. An equation of the ellipse is $\dfrac{x^2}{a^2} + \dfrac{y^2}{b^2} = 1$, or converting into parametric equations,

$x = a\cos\theta$ and $y = b\sin\theta$. So

$$L = 4\int_0^{\pi/2} \sqrt{(dx/d\theta)^2 + (dy/d\theta)^2}\, d\theta = 4\int_0^{\pi/2} \sqrt{a^2\sin^2\theta + b^2\cos^2\theta}\, d\theta$$

Using Simpson's Rule with $n = 10$, $\Delta\theta = \frac{\pi/2 - 0}{10} = \frac{\pi}{20}$, and $f(\theta) = \sqrt{a^2\sin^2\theta + b^2\cos^2\theta}$, we get

$$L \approx 4 \cdot S_{10} = 4 \cdot \tfrac{\pi}{20 \cdot 3}\left[f(0) + 4f\left(\tfrac{\pi}{20}\right) + 2f\left(\tfrac{2\pi}{20}\right) + \cdots + 2f\left(\tfrac{8\pi}{20}\right) + 4f\left(\tfrac{9\pi}{20}\right) + f\left(\tfrac{\pi}{2}\right)\right] \approx 3.64 \times 10^{10} \text{ km}$$

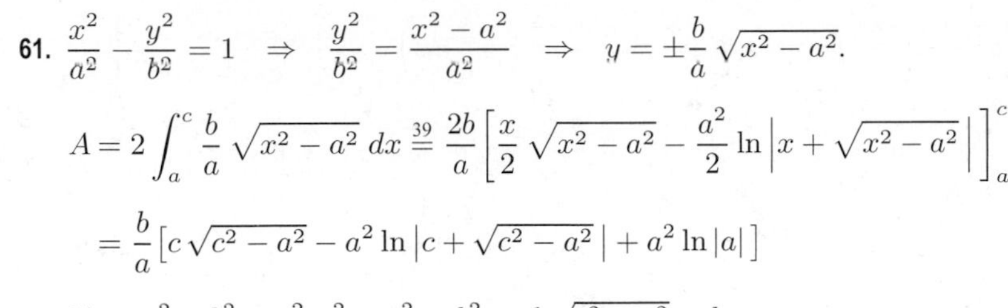

61. $\dfrac{x^2}{a^2} - \dfrac{y^2}{b^2} = 1 \;\Rightarrow\; \dfrac{y^2}{b^2} = \dfrac{x^2 - a^2}{a^2} \;\Rightarrow\; y = \pm\dfrac{b}{a}\sqrt{x^2 - a^2}.$

$$A = 2\int_a^c \frac{b}{a}\sqrt{x^2 - a^2}\, dx \overset{39}{=} \frac{2b}{a}\left[\frac{x}{2}\sqrt{x^2 - a^2} - \frac{a^2}{2}\ln\left|x + \sqrt{x^2 - a^2}\right|\right]_a^c$$

$$= \frac{b}{a}\left[c\sqrt{c^2 - a^2} - a^2\ln\left|c + \sqrt{c^2 - a^2}\right| + a^2\ln|a|\right]$$

Since $a^2 + b^2 = c^2$, $c^2 - a^2 = b^2$, and $\sqrt{c^2 - a^2} = b$.

$$= \frac{b}{a}\left[cb - a^2\ln(c + b) + a^2\ln a\right] = \frac{b}{a}\left[cb + a^2(\ln a - \ln(b + c))\right]$$

$= b^2c/a + ab\ln[a/(b + c)]$, where $c^2 = a^2 + b^2$.

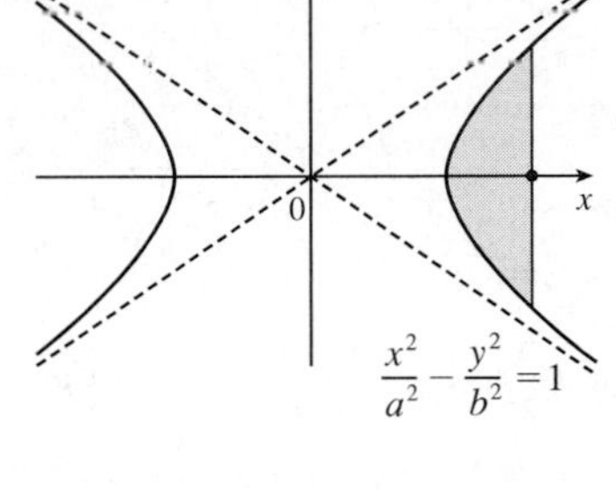

62. (a) $\dfrac{x^2}{a^2} + \dfrac{y^2}{b^2} = 1 \;\Rightarrow\; \dfrac{y^2}{b^2} = \dfrac{a^2 - x^2}{a^2} \;\Rightarrow\; y = \pm\dfrac{b}{a}\sqrt{a^2 - x^2}.$

$$V = \int_{-a}^{a} \pi\left(\frac{b}{a}\sqrt{a^2 - x^2}\right)^2 dx = 2\pi\,\frac{b^2}{a^2}\int_0^a (a^2 - x^2)\, dx$$

$$= \frac{2\pi b^2}{a^2}\left[a^2x - \tfrac{1}{3}x^3\right]_0^a = \frac{2\pi b^2}{a^2}\left(\frac{2a^3}{3}\right) = \frac{4}{3}\pi b^2 a$$

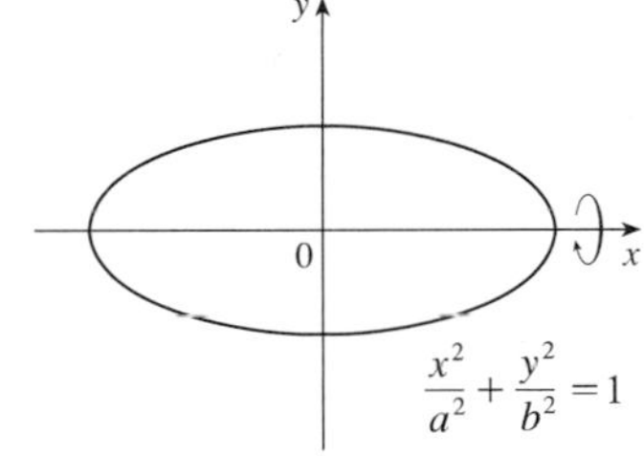

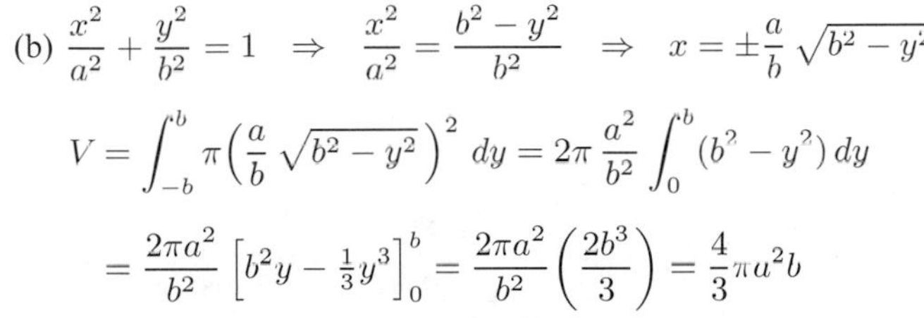

(b) $\dfrac{x^2}{a^2} + \dfrac{y^2}{b^2} = 1 \;\Rightarrow\; \dfrac{x^2}{a^2} = \dfrac{b^2 - y^2}{b^2} \;\Rightarrow\; x = \pm\dfrac{a}{b}\sqrt{b^2 - y^2}.$

$$V = \int_{-b}^{b} \pi\left(\frac{a}{b}\sqrt{b^2 - y^2}\right)^2 dy = 2\pi\,\frac{a^2}{b^2}\int_0^b (b^2 - y^2)\, dy$$

$$= \frac{2\pi a^2}{b^2}\left[b^2y - \tfrac{1}{3}y^3\right]_0^b = \frac{2\pi a^2}{b^2}\left(\frac{2b^3}{3}\right) = \frac{4}{3}\pi a^2 b$$

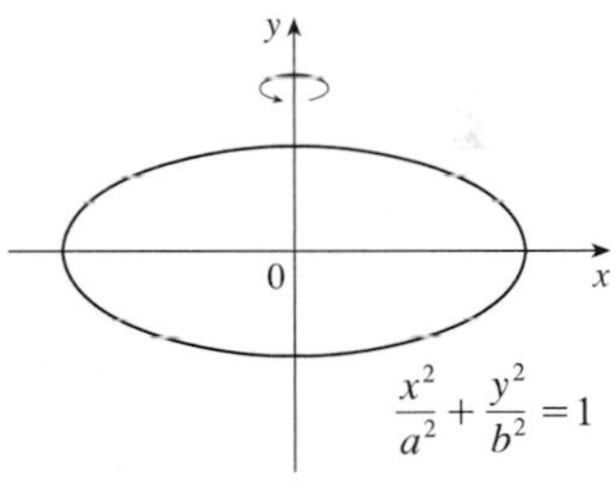

63. $9x^2 + 4y^2 = 36 \;\Leftrightarrow\; \dfrac{x^2}{4} + \dfrac{y^2}{9} = 1 \;\Rightarrow\; a = 3, b = 2$. By symmetry, $\overline{x} = 0$. By Example 2 in Section 7.3, the area of the top half of the ellipse is $\frac{1}{2}(\pi ab) = 3\pi$. Solve $9x^2 + 4y^2 = 36$ for y to get an equation for the top half of the ellipse:

$9x^2 + 4y^2 = 36 \;\Leftrightarrow\; 4y^2 = 36 - 9x^2 \;\Leftrightarrow\; y^2 = \frac{9}{4}(4 - x^2) \;\Rightarrow\; y = \frac{3}{2}\sqrt{4 - x^2}$. Now

$$\overline{y} = \frac{1}{A}\int_a^b \frac{1}{2}[f(x)]^2\, dx = \frac{1}{3\pi}\int_{-2}^{2} \frac{1}{2}\left(\frac{3}{2}\sqrt{4 - x^2}\right)^2 dx = \frac{3}{8\pi}\int_{-2}^{2}(4 - x^2)\, dx$$

$$= \frac{3}{8\pi} \cdot 2\int_0^2 (4 - x^2)\, dx = \frac{3}{4\pi}\left[4x - \frac{1}{3}x^3\right]_0^2 = \frac{3}{4\pi}\left(\frac{16}{3}\right) = \frac{4}{\pi}$$

so the centroid is $(0, 4/\pi)$.

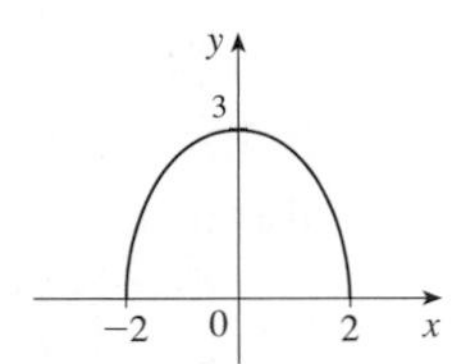

64. (a) Consider the ellipse $\dfrac{x^2}{a^2}+\dfrac{y^2}{b^2}=1$ with $a>b$, so that the major axis is the x-axis. Let the ellipse be parametrized by $x=a\cos t$, $y=b\sin t$, $0\le t\le 2\pi$. Then

$$\left(\frac{dx}{dt}\right)^2+\left(\frac{dy}{dt}\right)^2=a^2\sin^2 t+b^2\cos^2 t=a^2(1-\cos^2 t)+b^2\cos^2 t=a^2+(b^2-a^2)\cos^2 t=a^2-c^2\cos^2 t,$$

where $c^2=a^2-b^2$. Using symmetry and rotating the ellipse about the major axis gives us surface area

$$S=\int 2\pi y\,ds=2\int_0^{\pi/2}2\pi(b\sin t)\sqrt{a^2-c^2\cos^2 t}\,dt=4\pi b\int_c^0\sqrt{a^2-u^2}\left(-\frac{1}{c}\,du\right)\qquad\left[\begin{array}{c}u=c\cos t\\ du=-c\sin t\,dt\end{array}\right]$$

$$=\frac{4\pi b}{c}\int_0^c\sqrt{a^2-u^2}\,du\overset{30}{=}\frac{4\pi b}{c}\left[\frac{u}{2}\sqrt{a^2-u^2}+\frac{a^2}{2}\sin^{-1}\left(\frac{u}{a}\right)\right]_0^c=\frac{2\pi b}{c}\left[c\sqrt{a^2-c^2}+a^2\sin^{-1}\left(\frac{c}{a}\right)\right]$$

$$=\frac{2\pi b}{c}\left[bc+a^2\sin^{-1}\left(\frac{c}{a}\right)\right]$$

(b) As in part (a),

$$\left(\frac{dx}{dt}\right)^2+\left(\frac{dy}{dt}\right)^2=a^2\sin^2 t+b^2\cos^2 t=a^2\sin^2 t+b^2(1-\sin^2 t)=b^2+(a^2-b^2)\sin^2 t=b^2+c^2\sin^2 t.$$

Rotating about the minor axis gives us

$$S=\int 2\pi x\,ds=2\int_0^{\pi/2}2\pi(a\cos t)\sqrt{b^2+c^2\sin^2 t}\,dt=4\pi a\int_0^c\sqrt{b^2+u^2}\left(\frac{1}{c}\,du\right)\qquad\left[\begin{array}{c}u=c\sin t\\ du=c\cos t\,dt\end{array}\right]$$

$$\overset{21}{=}\frac{4\pi a}{c}\left[\frac{u}{2}\sqrt{b^2+u^2}+\frac{b^2}{2}\ln\left(u+\sqrt{b^2+u^2}\right)\right]_0^c=\frac{2\pi a}{c}\left[c\sqrt{b^2+c^2}+b^2\ln\left(c+\sqrt{b^2+c^2}\right)-b^2\ln b\right]$$

$$=\frac{2\pi a}{c}\left[ac+b^2\ln\left(\frac{a+c}{b}\right)\right]$$

65. Differentiating implicitly, $\dfrac{x^2}{a^2}+\dfrac{y^2}{b^2}=1\;\Rightarrow\;\dfrac{2x}{a^2}+\dfrac{2yy'}{b^2}=0\;\Rightarrow\;y'=-\dfrac{b^2x}{a^2y}$ $[y\neq 0]$. Thus, the slope of the tangent line at P is $-\dfrac{b^2x_1}{a^2y_1}$. The slope of F_1P is $\dfrac{y_1}{x_1+c}$ and of F_2P is $\dfrac{y_1}{x_1-c}$. By the formula in Problem 21 on text page 273, we have

$$\tan\alpha=\frac{\dfrac{y_1}{x_1+c}+\dfrac{b^2x_1}{a^2y_1}}{1-\dfrac{b^2x_1y_1}{a^2y_1(x_1+c)}}=\frac{a^2y_1^2+b^2x_1(x_1+c)}{a^2y_1(x_1+c)-b^2x_1y_1}=\frac{a^2b^2+b^2cx_1}{c^2x_1y_1+a^2cy_1}\qquad\left[\begin{array}{c}\text{using } b^2x_1^2+a^2y_1^2=a^2b^2,\\ \text{and } a^2-b^2=c^2\end{array}\right]$$

$$=\frac{b^2(cx_1+a^2)}{cy_1(cx_1+a^2)}=\frac{b^2}{cy_1}$$

and

$$\tan\beta=\frac{-\dfrac{b^2x_1}{a^2y_1}-\dfrac{y_1}{x_1-c}}{1-\dfrac{b^2x_1y_1}{a^2y_1(x_1-c)}}=\frac{-a^2y_1^2-b^2x_1(x_1-c)}{a^2y_1(x_1-c)-b^2x_1y_1}=\frac{-a^2b^2+b^2cx_1}{c^2x_1y_1-a^2cy_1}=\frac{b^2(cx_1-a^2)}{cy_1(cx_1-a^2)}=\frac{b^2}{cy_1}$$

Thus, $\alpha=\beta$.

66. The slopes of the line segments F_1P and F_2P are $\dfrac{y_1}{x_1+c}$ and $\dfrac{y_1}{x_1-c}$, where P is (x_1,y_1). Differentiating implicitly, $\dfrac{2x}{a^2}-\dfrac{2yy'}{b^2}=0\;\Rightarrow\;y'=\dfrac{b^2x}{a^2y}\;\Rightarrow$ the slope of the tangent at P is $\dfrac{b^2x_1}{a^2y_1}$, so by the formula in Problem 21 on text page 273,

$$\tan\alpha = \frac{\dfrac{b^2x_1}{a^2y_1} - \dfrac{y_1}{x_1+c}}{1 + \dfrac{b^2x_1y_1}{a^2y_1(x_1+c)}} = \frac{b^2x_1(x_1+c) - a^2y_1^2}{a^2y_1(x_1+c) + b^2x_1y_1} = \frac{b^2(cx_1+a^2)}{cy_1(cx_1+a^2)} \quad \left[\begin{array}{c}\text{using } x_1^2/a^2 - y_1^2/b^2 = 1, \\ \text{and } a^2+b^2=c^2\end{array}\right] = \frac{b^2}{cy_1}$$

and

$$\tan\beta = \frac{-\dfrac{b^2x_1}{a^2y_1} + \dfrac{y_1}{x_1-c}}{1 + \dfrac{b^2x_1y_1}{a^2y_1(x_1-c)}} = \frac{-b^2x_1(x_1-c) + a^2y_1^2}{a^2y_1(x_1-c) + b^2x_1y_1} = \frac{b^2(cx_1-a^2)}{cy_1(cx_1-a^2)} = \frac{b^2}{cy_1}$$

So $\alpha = \beta$.

10.6 Conic Sections in Polar Coordinates

1. The directrix $x = 4$ is to the right of the focus at the origin, so we use the form with "$+ \,e\cos\theta$" in the denominator. (See Theorem 6 and Figure 2.) An equation of the ellipse is $r = \dfrac{ed}{1+e\cos\theta} = \dfrac{\frac{1}{2}\cdot 4}{1+\frac{1}{2}\cos\theta} = \dfrac{4}{2+\cos\theta}$.

2. The directrix $x = -3$ is to the left of the focus at the origin, so we use the form with "$- \,e\cos\theta$" in the denominator. $e = 1$ for a parabola, so an equation is $r = \dfrac{ed}{1-e\cos\theta} = \dfrac{1\cdot 3}{1-1\cos\theta} = \dfrac{3}{1-\cos\theta}$.

3. The directrix $y = 2$ is above the focus at the origin, so we use the form with "$+ \,e\sin\theta$" in the denominator. An equation of the hyperbola is $r = \dfrac{ed}{1+e\sin\theta} = \dfrac{1.5(2)}{1+1.5\sin\theta} = \dfrac{6}{2+3\sin\theta}$.

4. The directrix $x = 3$ is to the right of the focus at the origin, so we use the form with "$+ \,e\cos\theta$" in the denominator. An equation of the hyperbola is $r = \dfrac{ed}{1+e\cos\theta} = \dfrac{3\cdot 3}{1+3\cos\theta} = \dfrac{9}{1+3\cos\theta}$.

5. The vertex $(2, \pi)$ is to the left of the focus at the origin, so we use the form with "$-e\cos\theta$" in the denominator. An equation of the ellipse is $r = \dfrac{ed}{1-e\cos\theta}$. Using eccentricity $e = \dfrac{2}{3}$ with $\theta = \pi$ and $r = 2$, we get $2 = \dfrac{\frac{2}{3}d}{1-\frac{2}{3}(-1)} \;\Rightarrow$ $2 = \dfrac{2d}{5} \;\Rightarrow\; d = 5$, so we have $r = \dfrac{\frac{2}{3}(5)}{1-\frac{2}{3}\cos\theta} = \dfrac{10}{3-2\cos\theta}$.

6. The directrix $r = 4\csc\theta$ (equivalent to $r\sin\theta = 4$ or $y = 4$) is above the focus at the origin, so we will use the form with "$+e\sin\theta$" in the denominator. The distance from the focus to the directrix is $d = 4$, so an equation of the ellipse is $r = \dfrac{ed}{1+e\sin\theta} = \dfrac{(0.6)(4)}{1+0.6\sin\theta}\cdot\dfrac{5}{5} = \dfrac{12}{5+3\sin\theta}$.

7. The vertex $\left(3, \frac{\pi}{2}\right)$ is 3 units above the focus at the origin, so the directrix is 6 units above the focus ($d = 6$), and we use the form "$+e\sin\theta$" in the denominator. $e = 1$ for a parabola, so an equation is $r = \dfrac{ed}{1+e\sin\theta} = \dfrac{1(6)}{1+1\sin\theta} = \dfrac{6}{1+\sin\theta}$.

8. The directrix $r = -2\sec\theta$ (equivalent to $r\cos\theta = -2$ or $x = -2$) is left of the focus at the origin, so we will use the form with "$-e\cos\theta$" in the denominator. The distance from the focus to the directrix is $d = 2$, so an equation of the hyperbola is $r = \dfrac{ed}{1 - e\cos\theta} = \dfrac{2(2)}{1 - 2\cos\theta} = \dfrac{4}{1 - 2\cos\theta}$.

9. $r = \dfrac{4}{5 - 4\sin\theta} \cdot \dfrac{1/5}{1/5} = \dfrac{4/5}{1 - \frac{4}{5}\sin\theta}$, where $e = \frac{4}{5}$ and $ed = \frac{4}{5} \;\Rightarrow\; d = 1$.

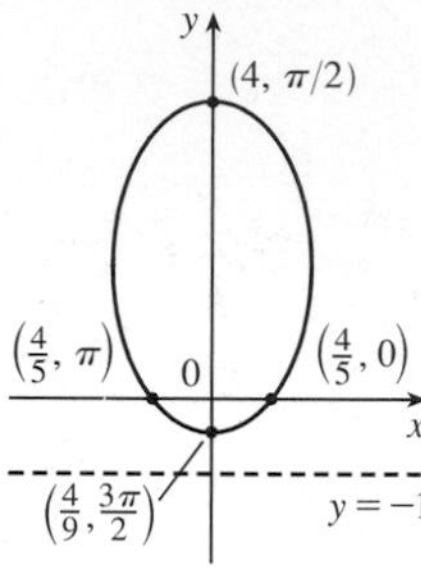

(a) Eccentricity $= e = \frac{4}{5}$

(b) Since $e = \frac{4}{5} < 1$, the conic is an ellipse.

(c) Since "$-\,e\sin\theta$" appears in the denominator, the directrix is below the focus at the origin, $d = |Fl| = 1$, so an equation of the directrix is $y = -1$.

(d) The vertices are $\left(4, \frac{\pi}{2}\right)$ and $\left(\frac{4}{9}, \frac{3\pi}{2}\right)$.

10. $r = \dfrac{1}{2 + \sin\theta} \cdot \dfrac{1/2}{1/2} = \dfrac{1/2}{1 + \frac{1}{2}\sin\theta}$, where $e = \dfrac{1}{2}$ and $ed = \dfrac{1}{2} \;\Rightarrow\; d = 1$.

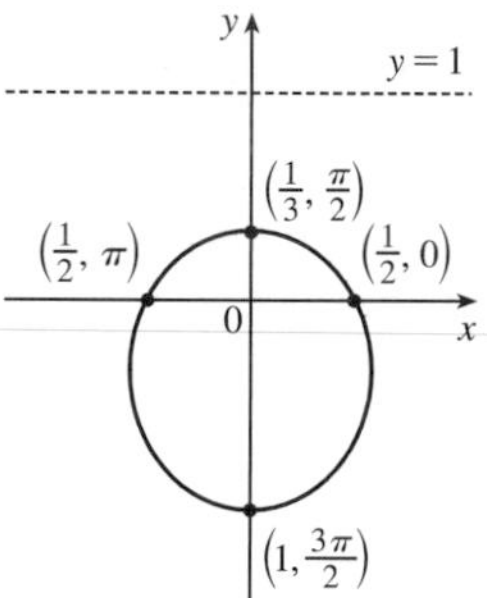

(a) Eccentricity $= e = \dfrac{1}{2}$

(b) Since $e = \dfrac{1}{2} < 1$, the conic is an ellipse.

(c) Since "$+e\sin\theta$" appears in the denominator, the directrix is above the focus at the origin, $d = |Fl| = 1$, so an equation of the directrix is $y = 1$.

(d) The vertices are $\left(\frac{1}{3}, \frac{\pi}{2}\right)$ and $\left(1, \frac{3\pi}{2}\right)$.

11. $r = \dfrac{2}{3 + 3\sin\theta} \cdot \dfrac{1/3}{1/3} = \dfrac{2/3}{1 + 1\sin\theta}$, where $e = 1$ and $ed = \frac{2}{3} \;\Rightarrow\; d = \frac{2}{3}$.

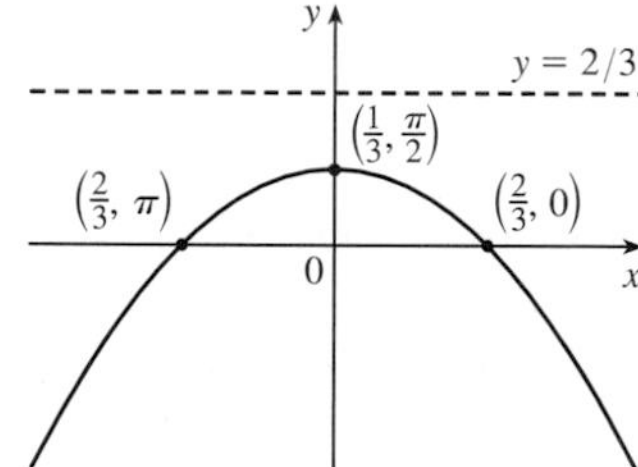

(a) Eccentricity $= e = 1$

(b) Since $e = 1$, the conic is a parabola.

(c) Since "$+\,e\sin\theta$" appears in the denominator, the directrix is above the focus at the origin. $d = |Fl| = \frac{2}{3}$, so an equation of the directrix is $y = \frac{2}{3}$.

(d) The vertex is at $\left(\frac{1}{3}, \frac{\pi}{2}\right)$, midway between the focus and directrix.

12. $r = \dfrac{5}{2 - 4\cos\theta} \cdot \dfrac{1/2}{1/2} = \dfrac{5/2}{1 - 2\cos\theta}$, where $e = 2$ and $ed = \dfrac{5}{2} \;\Rightarrow\; d = \dfrac{5}{4}$.

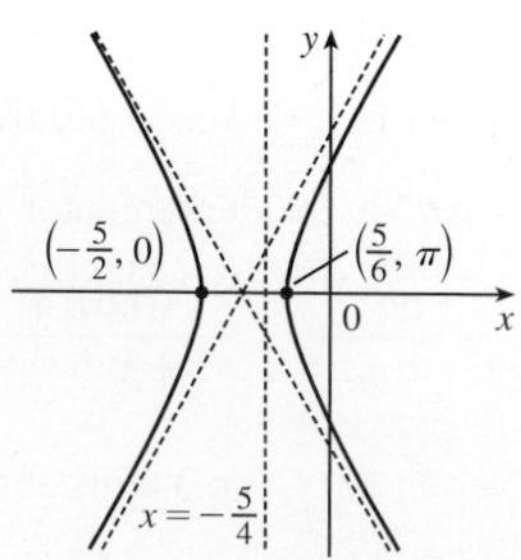

(a) Eccentricity $= e = 2$

(b) Since $e = 2 > 1$, the conic is a hyperbola.

(c) Since "$-e\cos\theta$" appears in the denominator, the directrix is to the left the focus at the origin. $d = |Fl| = \frac{5}{4}$, so an equation of the directrix is $x = -\frac{5}{4}$.

(d) The vertices are $\left(-\frac{5}{2}, 0\right)$ and $\left(\frac{5}{6}, \pi\right)$, so the center is midway between them, that is, $\left(\frac{5}{3}, \pi\right)$.

13. $r = \dfrac{9}{6 + 2\cos\theta} \cdot \dfrac{1/6}{1/6} = \dfrac{3/2}{1 + \frac{1}{3}\cos\theta}$, where $e = \frac{1}{3}$ and $ed = \frac{3}{2} \;\Rightarrow\; d = \frac{9}{2}$.

(a) Eccentricity $= e = \frac{1}{3}$

(b) Since $e = \frac{1}{3} < 1$, the conic is an ellipse.

(c) Since "$+e\cos\theta$" appears in the denominator, the directrix is to the right of the focus at the origin. $d = |Fl| = \frac{9}{2}$, so an equation of the directrix is $x = \frac{9}{2}$.

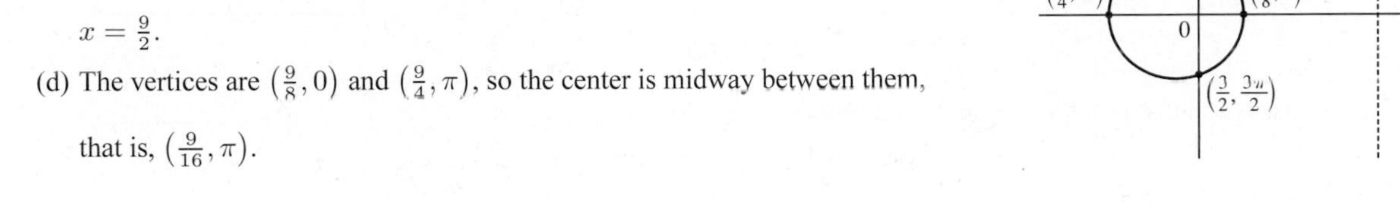

(d) The vertices are $\left(\frac{9}{8}, 0\right)$ and $\left(\frac{9}{4}, \pi\right)$, so the center is midway between them, that is, $\left(\frac{9}{16}, \pi\right)$.

14. $r = \dfrac{1}{3 - 3\sin\theta} \cdot \dfrac{1/3}{1/3} = \dfrac{1/3}{1 - 1\sin\theta}$, where $e = 1$ and $ed = \dfrac{1}{3} \;\Rightarrow\; d = \dfrac{1}{3}$

(a) Eccentricity $= e = 1$

(b) Since $e = 1$, the conic is a parabola.

(c) Since "$-e\sin\theta$" appears in the denominator, the directrix is below the focus at the origin, $d = |Fl| = \frac{1}{3}$, so an equation of the directrix is $y = -\frac{1}{3}$.

(d) The vertex is at $\left(\frac{1}{6}, \frac{3\pi}{2}\right)$, midway between the focus and the directrix.

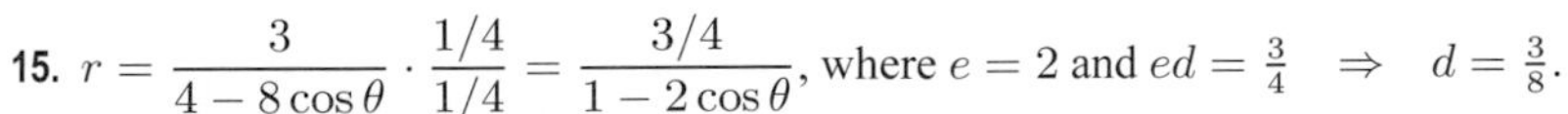

15. $r = \dfrac{3}{4 - 8\cos\theta} \cdot \dfrac{1/4}{1/4} = \dfrac{3/4}{1 - 2\cos\theta}$, where $e = 2$ and $ed = \frac{3}{4} \;\Rightarrow\; d = \frac{3}{8}$.

(a) Eccentricity $= e = 2$

(b) Since $e = 2 > 1$, the conic is a hyperbola.

(c) Since "$-e\cos\theta$" appears in the denominator, the directrix is to the left of the focus at the origin. $d = |Fl| = \frac{3}{8}$, so an equation of the directrix is $x = -\frac{3}{8}$.

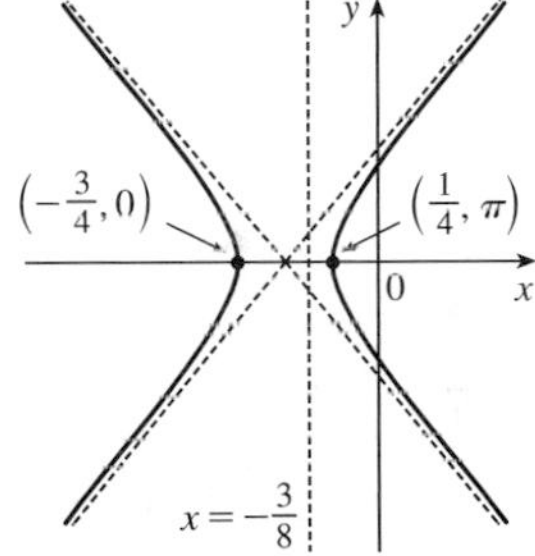

(d) The vertices are $\left(-\frac{3}{4}, 0\right)$ and $\left(\frac{1}{4}, \pi\right)$, so the center is midway between them, that is, $\left(\frac{1}{2}, \pi\right)$.

16. $r = \dfrac{4}{2 + 3\cos\theta} \cdot \dfrac{1/2}{1/2} = \dfrac{2}{1 + \frac{3}{2}\cos\theta}$, where $e = \dfrac{3}{2}$ and $ed = 2 \;\Rightarrow\; d = \dfrac{4}{3}$.

(a) Eccentricity $= e = \dfrac{3}{2}$

(b) Since $e = \dfrac{3}{2} > 1$, the conic is a hyperbola.

(c) Since "$+e\cos\theta$" appears in the denominator, the directrix is to the right of the focus at the origin. $d = |Fl| = \frac{4}{3}$, so an equation of the directrix is $x = \frac{4}{3}$.

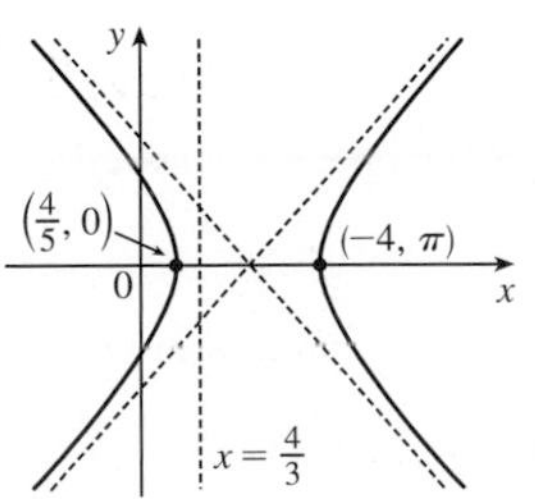

(d) The vertices are $\left(\frac{4}{5}, 0\right)$ and $(-4, \pi)$, so the center is midway between them, that is, $\left(\frac{8}{5}, 0\right)$.

17. (a) $r = \dfrac{1}{1 - 2\sin\theta}$, where $e = 2$ and $ed = 1 \Rightarrow d = \frac{1}{2}$. The eccentricity $e = 2 > 1$, so the conic is a hyperbola. Since "$-e\sin\theta$" appears in the denominator, the directrix is below the focus at the origin. $d = |Fl| = \frac{1}{2}$, so an equation of the directrix is $y = -\frac{1}{2}$. The vertices are $\left(-1, \frac{\pi}{2}\right)$ and $\left(\frac{1}{3}, \frac{3\pi}{2}\right)$, so the center is midway between them, that is, $\left(\frac{2}{3}, \frac{3\pi}{2}\right)$.

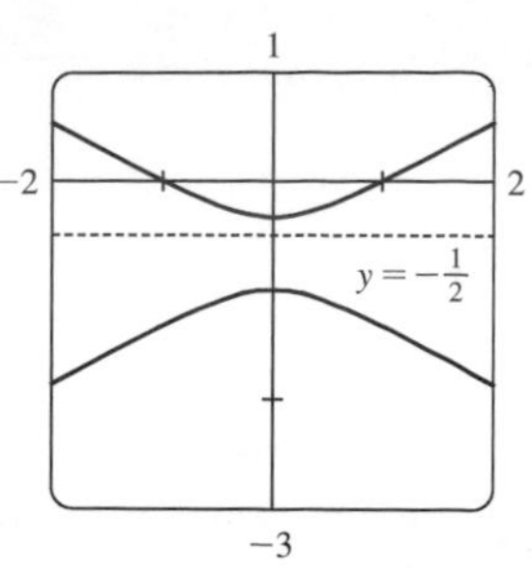

(b) By the discussion that precedes Example 4, the equation is $r = \dfrac{1}{1 - 2\sin\left(\theta - \frac{3\pi}{4}\right)}$.

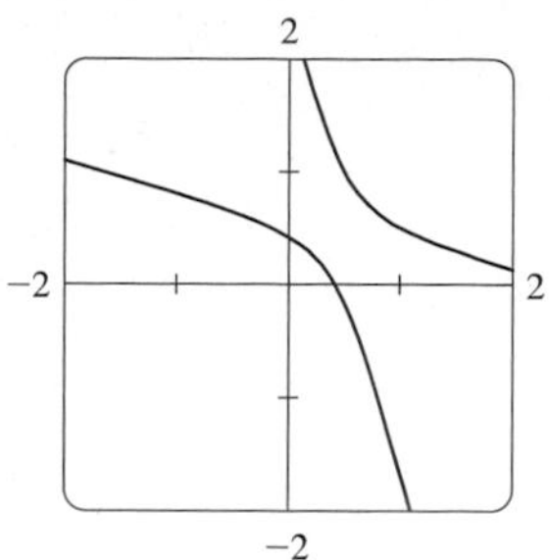

18. $r = \dfrac{4}{5 + 6\cos\theta} = \dfrac{4/5}{1 + \frac{6}{5}\cos\theta}$, so $e = \frac{6}{5}$ and $ed = \frac{4}{5} \Rightarrow d = \frac{2}{3}$.

An equation of the directrix is $x = \frac{2}{3} \Rightarrow r\cos\theta = \frac{2}{3} \Rightarrow r = \dfrac{2}{3\cos\theta}$.

If the hyperbola is rotated about its focus (the origin) through an angle $\pi/3$, its equation is the same as that of the original, with θ replaced by $\theta - \frac{\pi}{3}$ (see Example 4), so $r = \dfrac{4}{5 + 6\cos\left(\theta - \frac{\pi}{3}\right)}$.

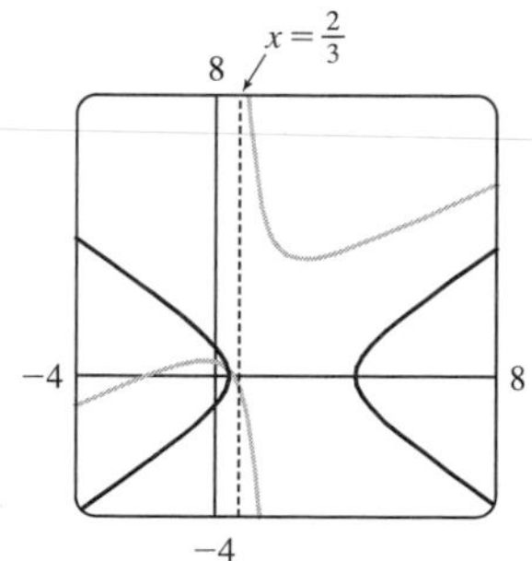

19. For $e < 1$ the curve is an ellipse. It is nearly circular when e is close to 0. As e increases, the graph is stretched out to the right, and grows larger (that is, its right-hand focus moves to the right while its left-hand focus remains at the origin.) At $e = 1$, the curve becomes a parabola with focus at the origin.

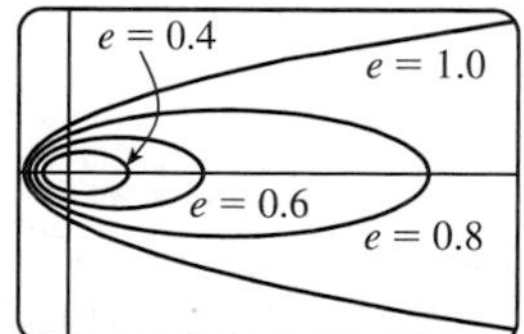

20. (a) The value of d does not seem to affect the shape of the conic (a parabola) at all, just its size, position, and orientation (for $d < 0$ it opens upward, for $d > 0$ it opens downward).

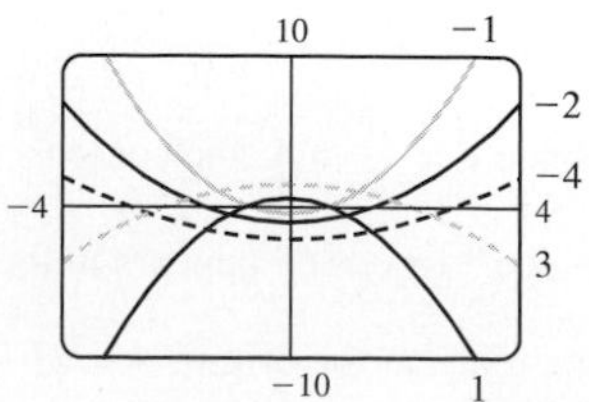

(b) We consider only positive values of e. When $0 < e < 1$, the conic is an ellipse. As $e \to 0^+$, the graph approaches perfect roundness and zero size. As e increases, the ellipse becomes more elongated, until at $e = 1$ it turns into a parabola. For $e > 1$, the conic is a hyperbola, which moves downward and gets broader as e continues to increase.

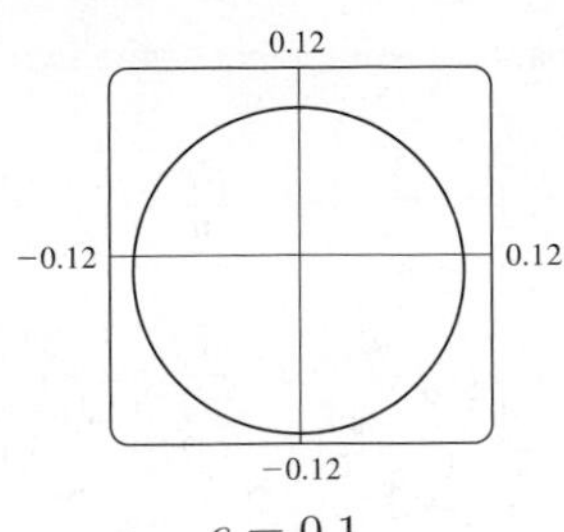

$e = 0.1$

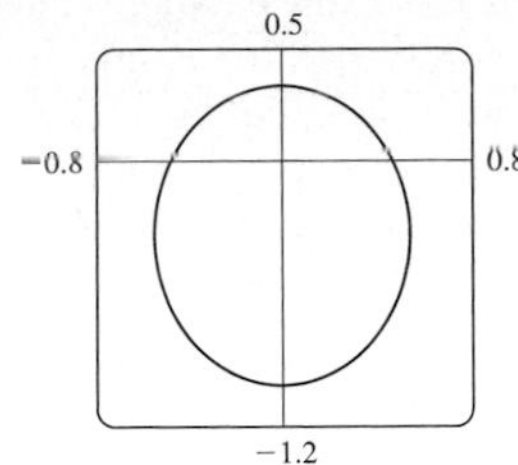

$e = 0.5$

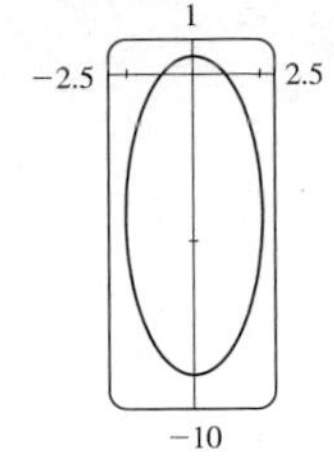

$e = 0.9$

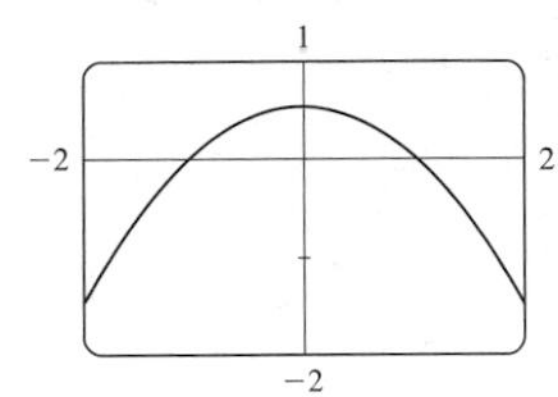

$e = 1$

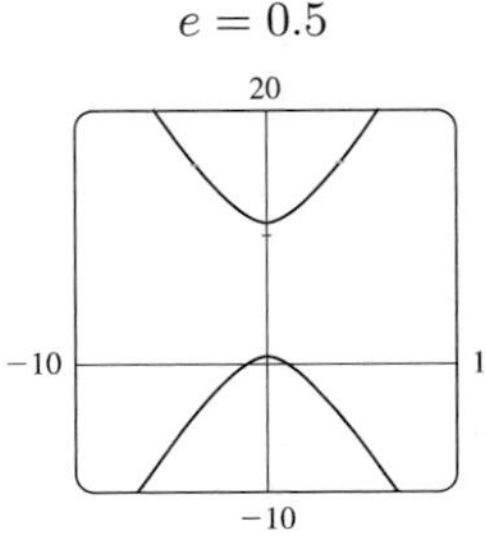

$e = 1.1$

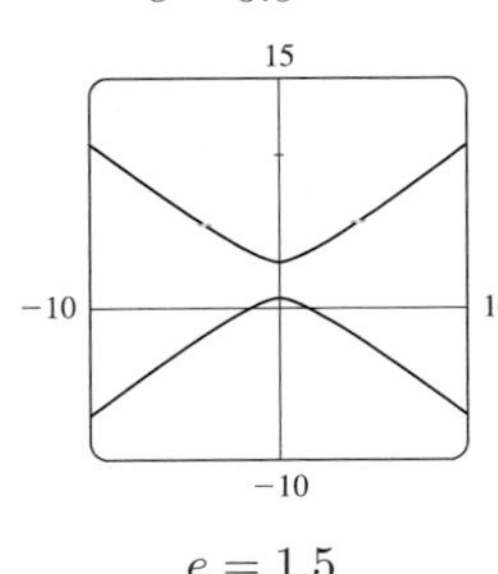

$e = 1.5$

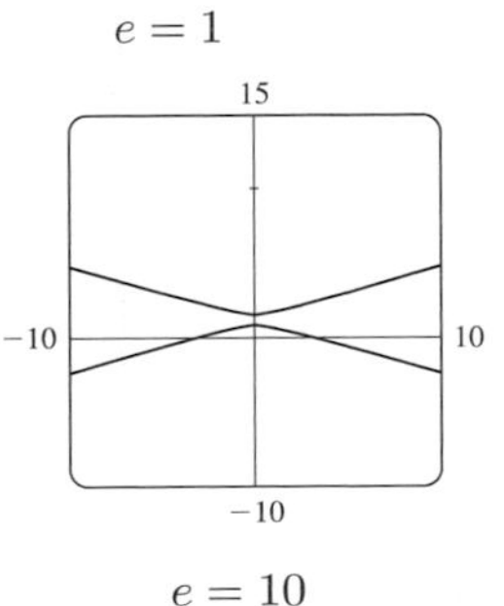

$e = 10$

21. $|PF| = e\,|Pl| \quad\Rightarrow\quad r = e[d - r\cos(\pi - \theta)] = e(d + r\cos\theta) \quad\Rightarrow$

$r(1 - e\cos\theta) = ed \quad\Rightarrow\quad r = \dfrac{ed}{1 - e\cos\theta}$

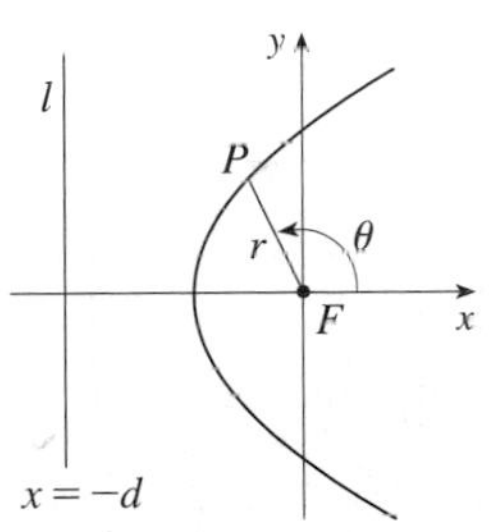

22. $|PF| = e\,|Pl| \quad\Rightarrow\quad r = e[d - r\sin\theta] \quad\Rightarrow\quad r(1 + e\sin\theta) = ed \quad\Rightarrow$

$r = \dfrac{ed}{1 + e\sin\theta}$

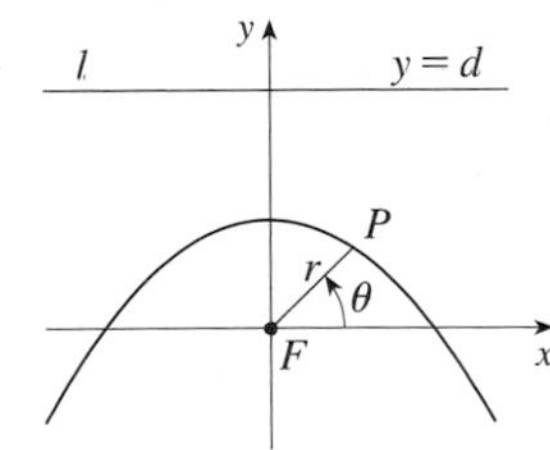

23. $|PF| = e\,|Pl| \quad\Rightarrow\quad r = e[d - r\sin(\theta - \pi)] = e(d + r\sin\theta) \quad\Rightarrow$

$r(1 - e\sin\theta) = ed \quad\Rightarrow\quad r = \dfrac{ed}{1 - e\sin\theta}$

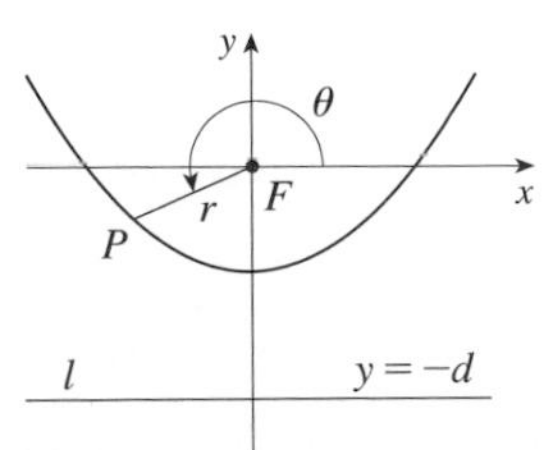

24. The parabolas intersect at the two points where $\dfrac{c}{1+\cos\theta} = \dfrac{d}{1-\cos\theta} \quad\Rightarrow\quad \cos\theta = \dfrac{c-d}{c+d} \quad\Rightarrow\quad r = \dfrac{c+d}{2}$.

For the first parabola, $\dfrac{dr}{d\theta} = \dfrac{c\sin\theta}{(1+\cos\theta)^2}$, so

$$\frac{dy}{dx} = \frac{(dr/d\theta)\sin\theta + r\cos\theta}{(dr/d\theta)\cos\theta - r\sin\theta} = \frac{c\sin^2\theta + c\cos\theta(1+\cos\theta)}{c\sin\theta\cos\theta - c\sin\theta(1+\cos\theta)} = \frac{1+\cos\theta}{-\sin\theta}$$

and similarly for the second, $\dfrac{dy}{dx} = \dfrac{1-\cos\theta}{\sin\theta} = \dfrac{\sin\theta}{1+\cos\theta}$. Since the product of these slopes is -1, the parabolas intersect at right angles.

25. We are given $e = 0.093$ and $a = 2.28 \times 10^8$. By (7), we have

$$r = \frac{a(1-e^2)}{1+e\cos\theta} = \frac{2.28 \times 10^8[1-(0.093)^2]}{1+0.093\cos\theta} \approx \frac{2.26 \times 10^8}{1+0.093\cos\theta}$$

26. We are given $e = 0.048$ and $2a = 1.56 \times 10^9 \quad\Rightarrow\quad a = 7.8 \times 10^8$. By (7), we have

$$r = \frac{a(1-e^2)}{1+e\cos\theta} = \frac{7.8 \times 10^8[1-(0.048)^2]}{1+0.048\cos\theta} \approx \frac{7.78 \times 10^8}{1+0.048\cos\theta}$$

27. Here $2a =$ length of major axis $= 36.18$ AU $\quad\Rightarrow\quad a = 18.09$ AU and $e = 0.97$. By (7), the equation of the orbit is $r = \dfrac{18.09[1-(0.97)^2]}{1+0.97\cos\theta} \approx \dfrac{1.07}{1+0.97\cos\theta}$. By (8), the maximum distance from the comet to the sun is $18.09(1+0.97) \approx 35.64$ AU or about 3.314 billion miles.

28. Here $2a =$ length of major axis $= 356.5$ AU $\quad\Rightarrow\quad a = 178.25$ AU and $e = 0.9951$. By (7), the equation of the orbit is $r = \dfrac{178.25[1-(0.9951)^2]}{1+0.9951\cos\theta} \approx \dfrac{1.7426}{1+0.9951\cos\theta}$. By (8), the minimum distance from the comet to the sun is $178.25(1-0.9951) \approx 0.8734$ AU or about 81 million miles.

29. The minimum distance is at perihelion, where $4.6 \times 10^7 = r = a(1-e) = a(1-0.206) = a(0.794) \quad\Rightarrow$ $a = 4.6 \times 10^7/0.794$. So the maximum distance, which is at aphelion, is $r = a(1+e) = \left(4.6 \times 10^7/0.794\right)(1.206) \approx 7.0 \times 10^7$ km.

30. At perihelion, $r = a(1-e) = 4.43 \times 10^9$, and at aphelion, $r = a(1+e) = 7.37 \times 10^9$. Adding, we get $2a = 11.80 \times 10^9$, so $a = 5.90 \times 10^9$ km. Therefore $1 + e = a(1+e)/a = \frac{7.37}{5.90} \approx 1.249$ and $e \approx 0.249$.

31. From Exercise 29, we have $e = 0.206$ and $a(1-e) = 4.6 \times 10^7$ km. Thus, $a = 4.6 \times 10^7/0.794$. From (7), we can write the equation of Mercury's orbit as $r = a\dfrac{1-e^2}{1+e\cos\theta}$. So since

$$\frac{dr}{d\theta} = \frac{a(1-e^2)e\sin\theta}{(1+e\cos\theta)^2} \quad\Rightarrow$$

$$r^2 + \left(\frac{dr}{d\theta}\right)^2 = \frac{a^2(1-e^2)^2}{(1+e\cos\theta)^2} + \frac{a^2(1-e^2)^2\,e^2\sin^2\theta}{(1+e\cos\theta)^4} = \frac{a^2(1-e^2)^2}{(1+e\cos\theta)^4}\left(1+2e\cos\theta+e^2\right)$$

the length of the orbit is

$$L = \int_0^{2\pi} \sqrt{r^2 + (dr/d\theta)^2}\, d\theta = a(1-e^2) \int_0^{2\pi} \frac{\sqrt{1+e^2+2e\cos\theta}}{(1+e\cos\theta)^2}\, d\theta \approx 3.6 \times 10^8 \text{ km}$$

This seems reasonable, since Mercury's orbit is nearly circular, and the circumference of a circle of radius a is $2\pi a \approx 3.6 \times 10^8$ km.

TRUE-FALSE QUIZ

1. False. Consider the curve defined by $x = f(t) = (t-1)^3$ and $y = g(t) = (t-1)^2$. Then $g'(t) = 2(t-1)$, so $g'(1) = 0$, but its graph has a *vertical* tangent when $t = 1$. *Note:* The statement is true if $f'(1) \neq 0$ when $g'(1) = 0$.

2. False. If $x = f(t)$ and $y = g(t)$ are twice differentiable, then $\dfrac{d^2y}{dx^2} = \dfrac{d}{dx}\left(\dfrac{dy}{dx}\right) = \dfrac{\dfrac{d}{dt}\left(\dfrac{dy}{dx}\right)}{\dfrac{dx}{dt}}$.

3. False. For example, if $f(t) = \cos t$ and $g(t) = \sin t$ for $0 \le t \le 4\pi$, then the curve is a circle of radius 1, hence its length is 2π, but $\int_0^{4\pi} \sqrt{[f'(t)]^2 + [g'(t)]^2}\, dt = \int_0^{4\pi} \sqrt{(-\sin t)^2 + (\cos t)^2}\, dt = \int_0^{4\pi} 1\, dt = 4\pi$, since as t increases from 0 to 4π, the circle is traversed twice.

4. False. If $(r, \theta) = (1, \pi)$, then $(x, y) = (-1, 0)$, so $\tan^{-1}(y/x) = \tan^{-1} 0 = 0 \neq \theta$. The statement is true for points in quadrants I and IV.

5. True. The curve $r = 1 - \sin 2\theta$ is unchanged if we rotate it through $180°$ about O because $1 - \sin 2(\theta + \pi) = 1 - \sin(2\theta + 2\pi) = 1 - \sin 2\theta$. So it's unchanged if we replace r by $-r$. (See the discussion after Example 8 in Section 10.3.) In other words, it's the same curve as $r = -(1 - \sin 2\theta) = \sin 2\theta - 1$.

6. True. The polar equation $r = 2$, the Cartesian equation $x^2 + y^2 = 4$, and the parametric equations $x = 2\sin 3t$, $y = 2\cos 3t$ $[0 \le t \le 2\pi]$ all describe the circle of radius 2 centered at the origin.

7. False. The first pair of equations gives the portion of the parabola $y = x^2$ with $x \ge 0$, whereas the second pair of equations traces out the whole parabola $y = x^2$.

8. True. $y^2 = 2y + 3x \;\Leftrightarrow\; (y-1)^2 = 3x + 1 = 3\left(x + \frac{1}{3}\right) = 4\left(\frac{3}{4}\right)\left(x + \frac{1}{3}\right)$, which is the equation of a parabola with vertex $\left(-\frac{1}{3}, 1\right)$ and focus $\left(-\frac{1}{3} + \frac{3}{4}, 1\right)$, opening to the right.

9. True. By rotating and translating the parabola, we can assume it has an equation of the form $y = cx^2$, where $c > 0$. The tangent at the point (a, ca^2) is the line $y - ca^2 = 2ca(x - a)$; i.e., $y = 2cax - ca^2$. This tangent meets the parabola at the points (x, cx^2) where $cx^2 = 2cax - ca^2$. This equation is equivalent to $x^2 = 2ax - a^2$ [since $c > 0$]. But $x^2 = 2ax - a^2 \;\Leftrightarrow\; x^2 - 2ax + a^2 = 0 \;\Leftrightarrow\; (x-a)^2 = 0 \;\Leftrightarrow\; x = a \;\Leftrightarrow\; (x, cx^2) = (a, ca^2)$. This shows that each tangent meets the parabola at exactly one point.

10. True. Consider a hyperbola with focus at the origin, oriented so that its polar equation is $r = \dfrac{ed}{1 + e\cos\theta}$, where $e > 1$.

The directrix is $x = d$, but along the hyperbola we have $x = r\cos\theta = \dfrac{ed\cos\theta}{1 + e\cos\theta} = d\left(\dfrac{e\cos\theta}{1 + e\cos\theta}\right) \neq d$.

10 Review

EXERCISES

1. $x = t^2 + 4t$, $y = 2 - t$, $-4 \le t \le 1$. $t = 2 - y$, so

$x = (2 - y)^2 + 4(2 - y) = 4 - 4y + y^2 + 8 - 4y = y^2 - 8y + 12 \Leftrightarrow$

$x + 4 = y^2 - 8y + 16 = (y - 4)^2$. This is part of a parabola with vertex $(-4, 4)$, opening to the right.

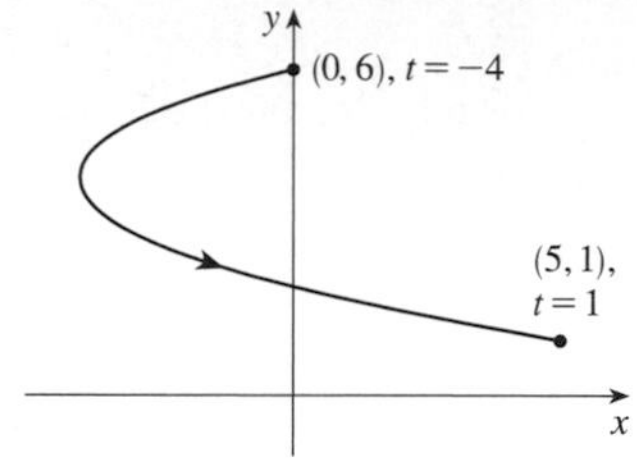

2. $x = 1 + e^{2t}$, $y = e^t$.

$x = 1 + e^{2t} = 1 + (e^t)^2 = 1 + y^2$, $y > 0$.

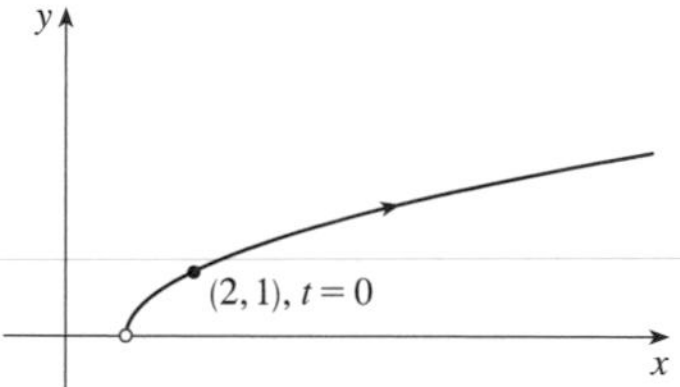

3. $y = \sec\theta = \dfrac{1}{\cos\theta} = \dfrac{1}{x}$. Since $0 \le \theta \le \pi/2$, $0 < x \le 1$ and $y \ge 1$.

This is part of the hyperbola $y = 1/x$.

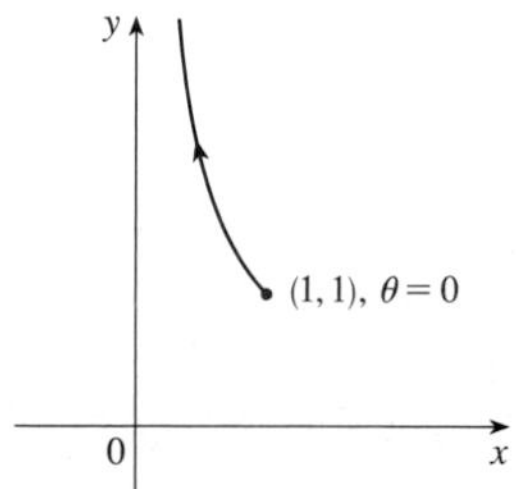

4. $x = 2\cos\theta$, $y = 1 + \sin\theta$, $\cos^2\theta + \sin^2\theta = 1 \Rightarrow$

$\left(\dfrac{x}{2}\right)^2 + (y - 1)^2 = 1 \Rightarrow \dfrac{x^2}{4} + (y - 1)^2 = 1$. This is an ellipse, centered at $(0, 1)$, with semimajor axis of length 2 and semiminor axis of length 1.

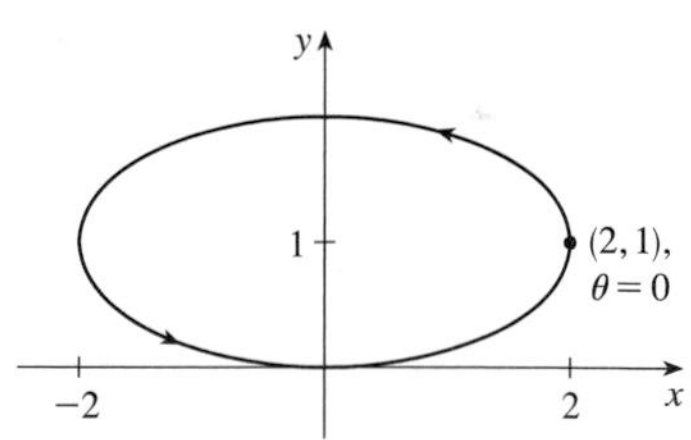

5. Three different sets of parametric equations for the curve $y = \sqrt{x}$ are

(i) $x = t, \quad y = \sqrt{t}$

(ii) $x = t^4, \quad y = t^2$

(iii) $x = \tan^2 t, \quad y = \tan t, \quad 0 \le t < \pi/2$

There are many other sets of equations that also give this curve.

6. For $t < -1$, $x > 0$ and $y < 0$ with x decreasing and y increasing. When $t = -1$, $(x, y) = (0, 0)$. When $-1 < t < 0$, we have $-1 < x < 0$ and $0 < y < 1/2$. When $t = 0$, $(x, y) = (-1, 0)$. When $0 < t < 1$, $-1 < x < 0$ and $-\frac{1}{2} < y < 0$. When $t = 1$, $(x, y) = (0, 0)$ again. When $t > 1$, both x and y are positive and increasing.

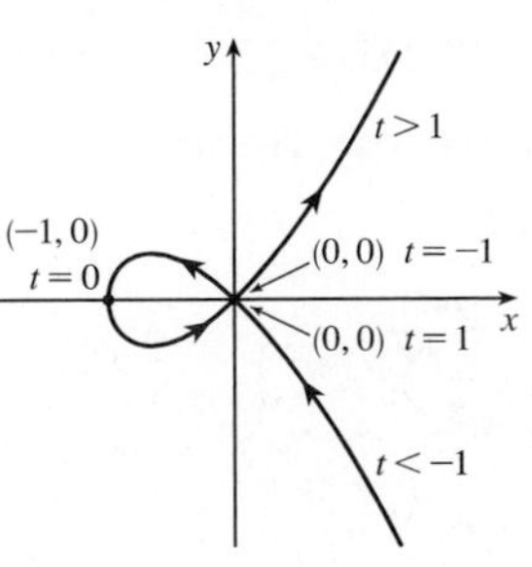

7. (a)

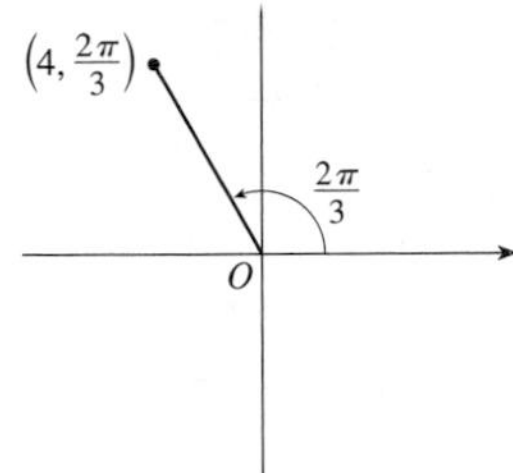

The Cartesian coordinates are $x = 4\cos\frac{2\pi}{3} = 4\left(-\frac{1}{2}\right) = -2$ and $y = 4\sin\frac{2\pi}{3} = 4\left(\frac{\sqrt{3}}{2}\right) = 2\sqrt{3}$, that is, the point $\left(-2, 2\sqrt{3}\right)$.

(b) Given $x = -3$ and $y = 3$, we have $r = \sqrt{(-3)^2 + 3^2} = \sqrt{18} = 3\sqrt{2}$. Also, $\tan\theta = \dfrac{y}{x} \Rightarrow \tan\theta = \dfrac{3}{-3}$, and since $(-3, 3)$ is in the second quadrant, $\theta = \frac{3\pi}{4}$. Thus, one set of polar coordinates for $(-3, 3)$ is $\left(3\sqrt{2}, \frac{3\pi}{4}\right)$, and two others are $\left(3\sqrt{2}, \frac{11\pi}{4}\right)$ and $\left(-3\sqrt{2}, \frac{7\pi}{4}\right)$.

8. $1 \leq r < 2$, $\frac{\pi}{6} \leq \theta \leq \frac{5\pi}{6}$

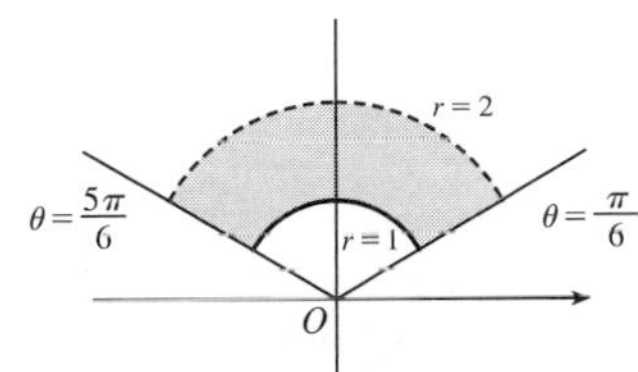

9. $r = 1 + \sin\theta$. This cardioid is symmetric about the $\theta = \pi/2$ axis.

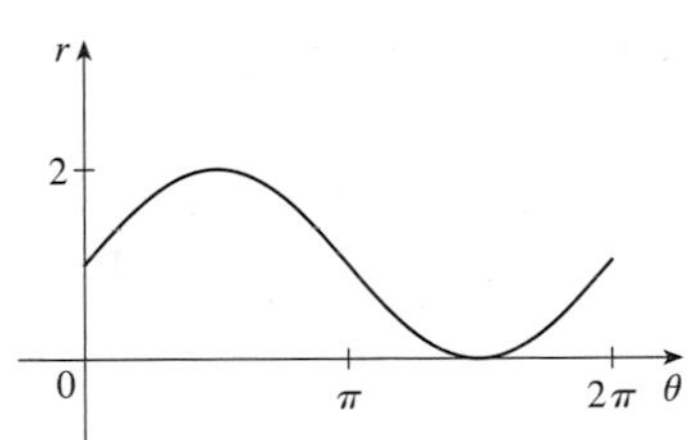

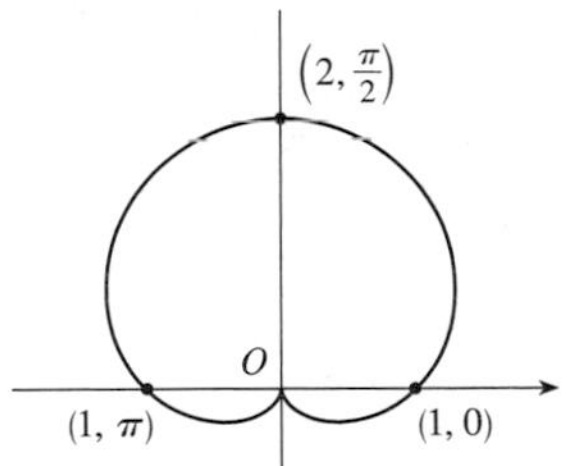

10. $r = \sin 4\theta$. This is an eight-leaved rose.

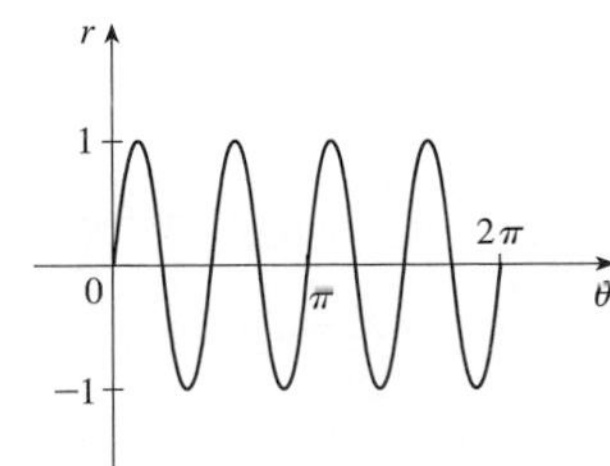

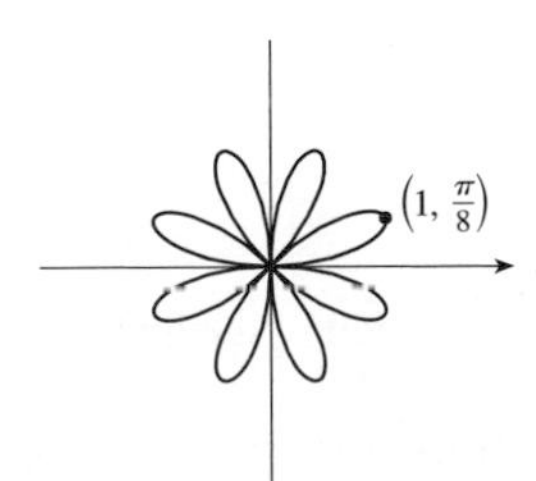

11. $r = \cos 3\theta$. This is a three-leaved rose. The curve is traced twice.

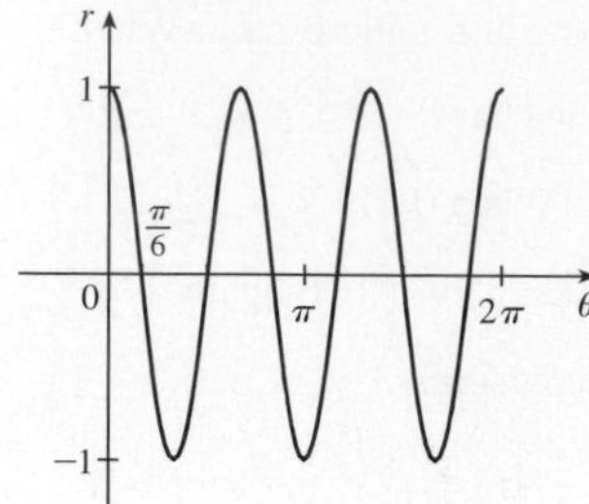

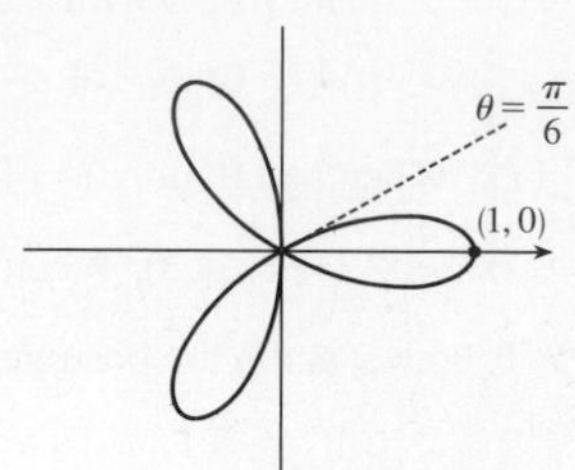

12. $r = 3 + \cos 3\theta$. The curve is symmetric about the horizontal axis.

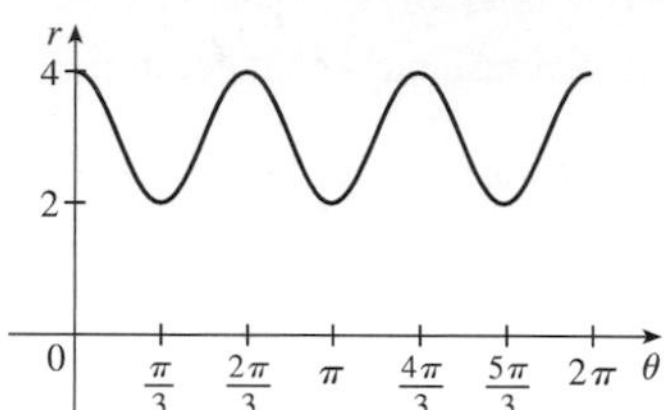

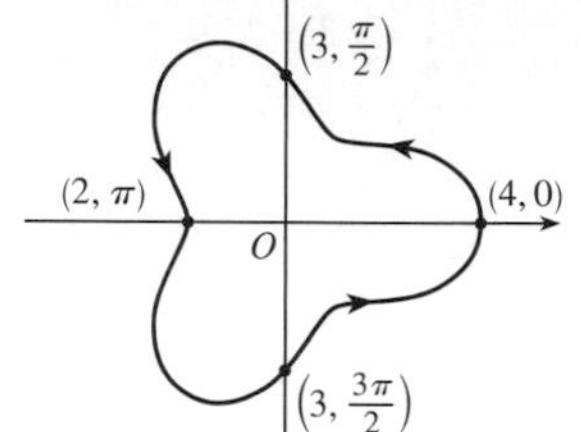

13. $r = 1 + \cos 2\theta$. The curve is symmetric about the pole and both the horizontal and vertical axes.

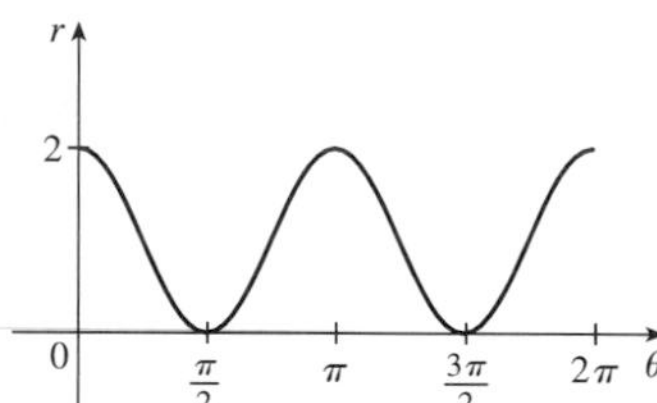

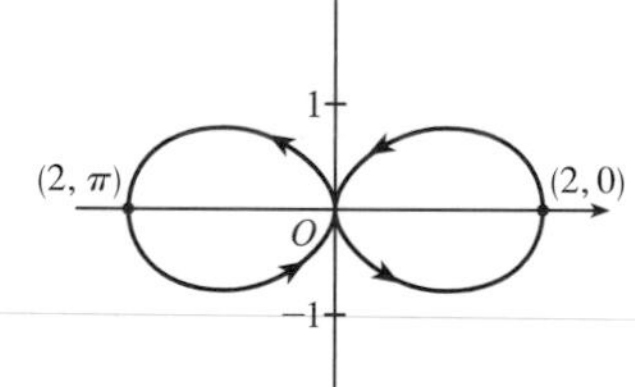

14. $r = 2\cos(\theta/2)$. The curve is symmetric about the pole and both the horizontal and vertical axes.

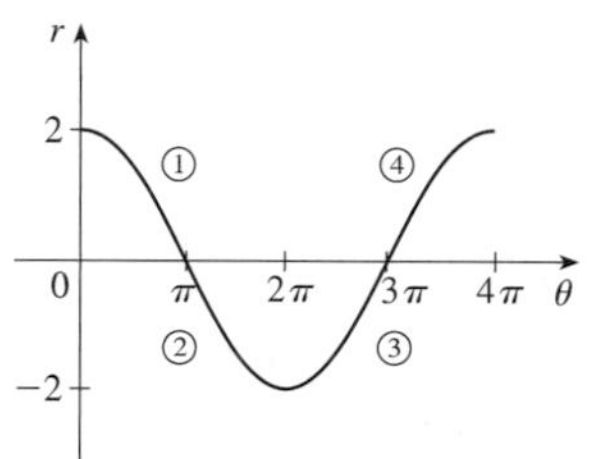

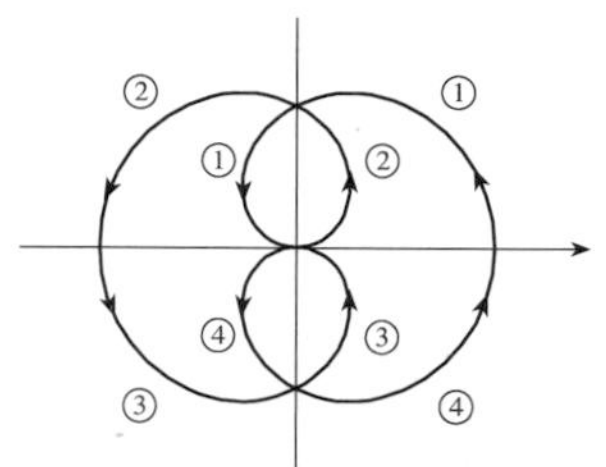

15. $r = \dfrac{3}{1 + 2\sin\theta} \quad \Rightarrow \quad e = 2 > 1$, so the conic is a hyperbola. $de = 3 \quad \Rightarrow$ $d = \frac{3}{2}$ and the form "$+2\sin\theta$" imply that the directrix is above the focus at the origin and has equation $y = \frac{3}{2}$. The vertices are $\left(1, \frac{\pi}{2}\right)$ and $\left(-3, \frac{3\pi}{2}\right)$.

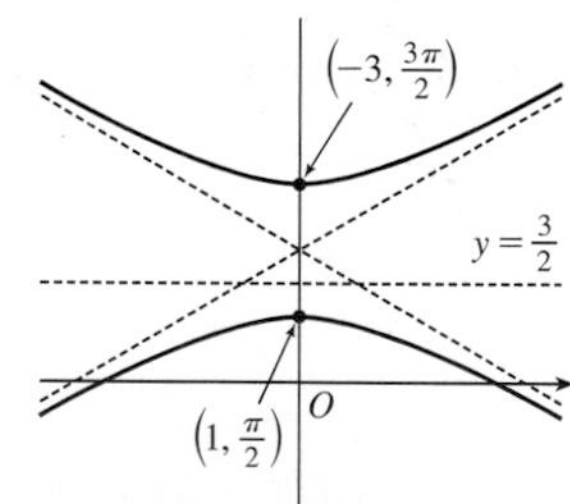

16. $r = \dfrac{3}{2 - 2\cos\theta} \cdot \dfrac{1/2}{1/2} = \dfrac{3/2}{1 - 1\cos\theta} \quad \Rightarrow \quad e = 1$, so the conic is a parabola. $de = \frac{3}{2} \quad \Rightarrow \quad d = \frac{3}{2}$ and the form "$-2\cos\theta$" imply that the directrix is to the left of the focus at the origin and has equation $x = -\frac{3}{2}$. The vertex is $\left(\frac{3}{4}, \pi\right)$.

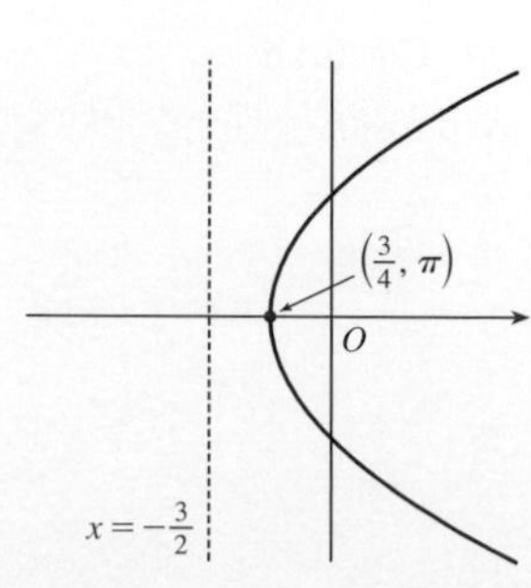

17. $x + y = 2 \;\Leftrightarrow\; r\cos\theta + r\sin\theta = 2 \;\Leftrightarrow\; r(\cos\theta + \sin\theta) = 2 \;\Leftrightarrow\; r = \dfrac{2}{\cos\theta + \sin\theta}$

18. $x^2 + y^2 = 2 \;\Rightarrow\; r^2 = 2 \;\Rightarrow\; r = \sqrt{2}$. [$r = -\sqrt{2}$ gives the same curve.]

19. $r = (\sin\theta)/\theta$. As $\theta \to \pm\infty$, $r \to 0$. As $\theta \to 0$, $r \to 1$. In the first figure, there are an infinite number of x-intercepts at $x = \pi n$, n a nonzero integer. These correspond to pole points in the second figure.

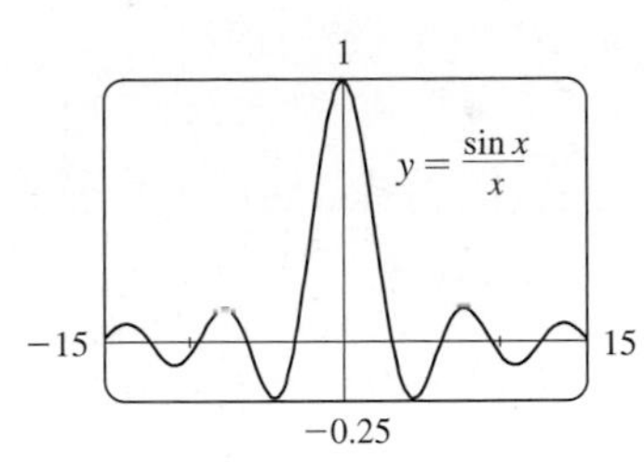

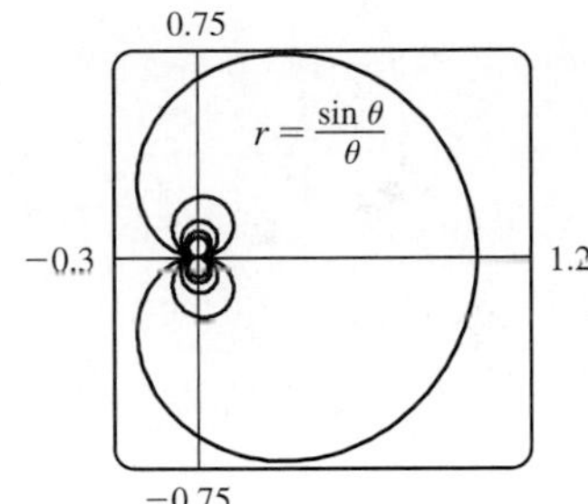

20. $r = \dfrac{2}{4 - 3\cos\theta} = \dfrac{1/2}{1 - \frac{3}{4}\cos\theta} \;\Rightarrow\; e = \frac{3}{4}$ and $d = \frac{2}{3}$. The equation of the directrix is $x = -\frac{2}{3} \;\Rightarrow\; r = -2/(3\cos\theta)$. To obtain the equation of the rotated ellipse, we replace θ in the original equation with $\theta - \frac{2\pi}{3}$, and get $r = \dfrac{2}{4 - 3\cos\left(\theta - \frac{2\pi}{3}\right)}$.

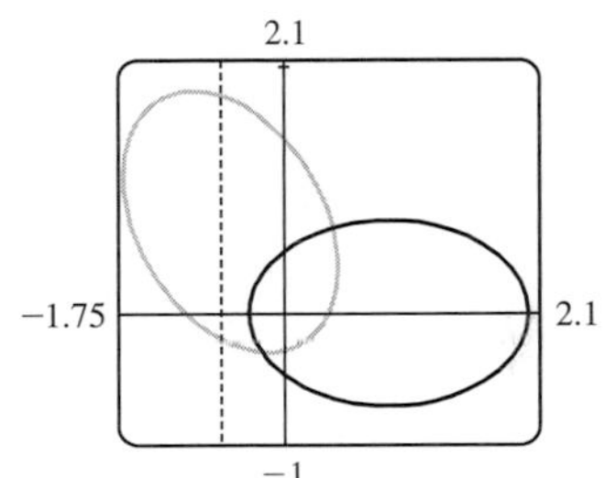

21. $x = \ln t$, $y = 1 + t^2$; $t = 1$. $\dfrac{dy}{dt} = 2t$ and $\dfrac{dx}{dt} = \dfrac{1}{t}$, so $\dfrac{dy}{dx} = \dfrac{dy/dt}{dx/dt} = \dfrac{2t}{1/t} = 2t^2$.

When $t = 1$, $(x, y) = (0, 2)$ and $dy/dx = 2$.

22. $x = t^3 + 6t + 1$, $y = 2t - t^2$; $t = -1$. $\dfrac{dy}{dx} = \dfrac{dy/dt}{dx/dt} = \dfrac{2 - 2t}{3t^2 + 6}$. When $t = -1$, $(x, y) = (-6, -3)$ and $\dfrac{dy}{dx} = \dfrac{4}{9}$.

23. $r = e^{-\theta} \;\Rightarrow\; y = r\sin\theta = e^{-\theta}\sin\theta$ and $x = r\cos\theta = e^{-\theta}\cos\theta \;\Rightarrow$

$$\frac{dy}{dx} = \frac{dy/d\theta}{dx/d\theta} = \frac{\frac{dr}{d\theta}\sin\theta + r\cos\theta}{\frac{dr}{d\theta}\cos\theta - r\sin\theta} = \frac{-e^{-\theta}\sin\theta + e^{-\theta}\cos\theta}{-e^{-\theta}\cos\theta - e^{-\theta}\sin\theta} \cdot \frac{-e^{\theta}}{-e^{\theta}} = \frac{\sin\theta - \cos\theta}{\cos\theta + \sin\theta}.$$

When $\theta = \pi$, $\dfrac{dy}{dx} = \dfrac{0 - (-1)}{-1 + 0} = \dfrac{1}{-1} = -1$.

24. $r = 3 + \cos 3\theta \;\Rightarrow\; \dfrac{dy}{dx} = \dfrac{dy/d\theta}{dx/d\theta} = \dfrac{\frac{dr}{d\theta}\sin\theta + r\cos\theta}{\frac{dr}{d\theta}\cos\theta - r\sin\theta} = \dfrac{-3\sin 3\theta\sin\theta + (3 + \cos 3\theta)\cos\theta}{-3\sin 3\theta\cos\theta - (3 + \cos 3\theta)\sin\theta}$.

When $\theta = \pi/2$, $\dfrac{dy}{dx} = \dfrac{(-3)(-1)(1) + (3 + 0)\cdot 0}{(-3)(-1)(0) - (3 + 0)\cdot 1} = \dfrac{3}{-3} = -1$.

25. $x = t + \sin t$, $y = t - \cos t \;\Rightarrow\; \dfrac{dy}{dx} = \dfrac{dy/dt}{dx/dt} = \dfrac{1 + \sin t}{1 + \cos t} \;\Rightarrow$

$$\frac{d^2y}{dx^2} = \frac{\frac{d}{dt}\left(\frac{dy}{dx}\right)}{dx/dt} = \frac{\frac{(1 + \cos t)\,\cos t - (1 + \sin t)(-\sin t)}{(1 + \cos t)^2}}{1 + \cos t} = \frac{\cos t + \cos^2 t + \sin t + \sin^2 t}{(1 + \cos t)^3} = \frac{1 + \cos t + \sin t}{(1 + \cos t)^3}$$

26. $x = 1 + t^2$, $y = t - t^3$. $\dfrac{dy}{dt} = 1 - 3t^2$ and $\dfrac{dx}{dt} = 2t$, so $\dfrac{dy}{dx} = \dfrac{dy/dt}{dx/dt} = \dfrac{1 - 3t^2}{2t} = \frac{1}{2}t^{-1} - \frac{3}{2}t$.

$$\frac{d^2y}{dx^2} = \frac{d(dy/dx)/dt}{dx/dt} = \frac{-\frac{1}{2}t^{-2} - \frac{3}{2}}{2t} = -\tfrac{1}{4}t^{-3} - \tfrac{3}{4}t^{-1} = -\frac{1}{4t^3}(1 + 3t^2) = -\frac{3t^2 + 1}{4t^3}.$$

27. We graph the curve $x = t^3 - 3t$, $y = t^2 + t + 1$ for $-2.2 \le t \le 1.2$. By zooming in or using a cursor, we find that the lowest point is about $(1.4, 0.75)$. To find the exact values, we find the t-value at which $dy/dt = 2t + 1 = 0 \Leftrightarrow t = -\frac{1}{2} \Leftrightarrow (x, y) = \left(\frac{11}{8}, \frac{3}{4}\right)$.

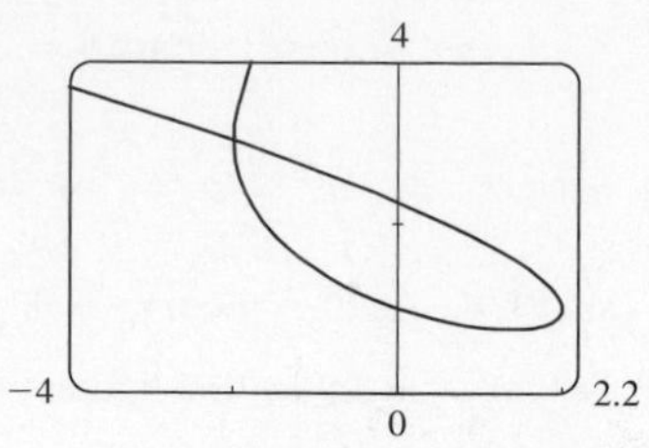

28. We estimate the coordinates of the point of intersection to be $(-2, 3)$. In fact this is exact, since both $t = -2$ and $t = 1$ give the point $(-2, 3)$. So the area enclosed by the loop is

$$\int_{t=-2}^{t=1} y\,dx = \int_{-2}^{1} (t^2 + t + 1)(3t^2 - 3)\,dt = \int_{-2}^{1} (3t^4 + 3t^3 - 3t - 3)\,dt$$

$$= \left[\tfrac{3}{5}t^5 + \tfrac{3}{4}t^4 - \tfrac{3}{2}t^2 - 3t\right]_{-2}^{1} = \left(\tfrac{3}{5} + \tfrac{3}{4} - \tfrac{3}{2} - 3\right) - \left[-\tfrac{96}{5} + 12 - 6 - (-6)\right] = \tfrac{81}{20}$$

29. $x = 2a\cos t - a\cos 2t \Rightarrow \dfrac{dx}{dt} = -2a\sin t + 2a\sin 2t = 2a\sin t(2\cos t - 1) = 0 \Leftrightarrow$

$\sin t = 0$ or $\cos t = \frac{1}{2} \Rightarrow t = 0, \frac{\pi}{3}, \pi,$ or $\frac{5\pi}{3}$.

$y = 2a\sin t - a\sin 2t \Rightarrow \dfrac{dy}{dt} = 2a\cos t - 2a\cos 2t = 2a\left(1 + \cos t - 2\cos^2 t\right) = 2a(1 - \cos t)(1 + 2\cos t) = 0 \Rightarrow$

$t = 0, \frac{2\pi}{3},$ or $\frac{4\pi}{3}$.

Thus the graph has vertical tangents where $t = \frac{\pi}{3}$, π and $\frac{5\pi}{3}$, and horizontal tangents where $t = \frac{2\pi}{3}$ and $\frac{4\pi}{3}$. To determine what the slope is where $t = 0$, we use l'Hospital's Rule to evaluate $\displaystyle\lim_{t\to 0} \frac{dy/dt}{dx/dt} = 0$, so there is a horizontal tangent there.

t	x	y
0	a	0
$\frac{\pi}{3}$	$\frac{3}{2}a$	$\frac{\sqrt{3}}{2}a$
$\frac{2\pi}{3}$	$-\frac{1}{2}a$	$\frac{3\sqrt{3}}{2}a$
π	$-3a$	0
$\frac{4\pi}{3}$	$-\frac{1}{2}a$	$-\frac{3\sqrt{3}}{2}a$
$\frac{5\pi}{3}$	$\frac{3}{2}a$	$-\frac{\sqrt{3}}{2}a$

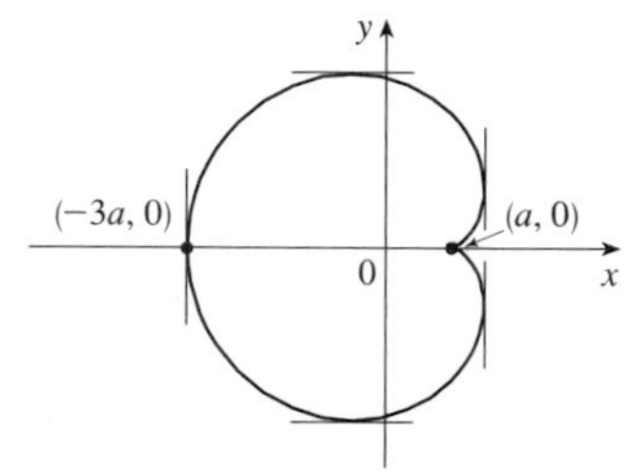

30. From Exercise 29, $x = 2a\cos t - a\cos 2t$, $y = 2a\sin t - a\sin 2t \Rightarrow$

$$A = 2\int_{\pi}^{0} (2a\sin t - a\sin 2t)(-2a\sin t + 2a\sin 2t)\,dt = 4a^2\int_0^{\pi} (2\sin^2 t + \sin^2 2t - 3\sin t\sin 2t)\,dt$$

$$= 4a^2\int_0^{\pi} \left[(1 - \cos 2t) + \tfrac{1}{2}(1 - \cos 4t) - 6\sin^2 t\,\cos t\right] dt = 4a^2\left[t - \tfrac{1}{2}\sin 2t + \tfrac{1}{2}t - \tfrac{1}{8}\sin 4t - 2\sin^3 t\right]_0^{\pi}$$

$$= 4a^2\left(\tfrac{3}{2}\right)\pi = 6\pi a^2$$

31. The curve $r^2 = 9\cos 5\theta$ has 10 "petals." For instance, for $-\frac{\pi}{10} \le \theta \le \frac{\pi}{10}$, there are two petals, one with $r > 0$ and one with $r < 0$.

$$A = 10\int_{-\pi/10}^{\pi/10} \tfrac{1}{2}r^2\,d\theta = 5\int_{-\pi/10}^{\pi/10} 9\cos 5\theta\,d\theta = 5\cdot 9\cdot 2\int_0^{\pi/10}\cos 5\theta\,d\theta = 18\left[\sin 5\theta\right]_0^{\pi/10} = 18$$

32. $r = 1 - 3\sin\theta$. The inner loop is traced out as θ goes from $\alpha = \sin^{-1}\left(\frac{1}{3}\right)$ to $\pi - \alpha$, so

$$A = \int_{\alpha}^{\pi-\alpha} \tfrac{1}{2}r^2\,d\theta = \int_{\alpha}^{\pi/2} (1 - 3\sin\theta)^2\,d\theta = \int_{\alpha}^{\pi/2} \left[1 - 6\sin\theta + \tfrac{9}{2}(1 - \cos 2\theta)\right] d\theta$$

$$= \left[\tfrac{11}{2}\theta + 6\cos\theta - \tfrac{9}{4}\sin 2\theta\right]_{\alpha}^{\pi/2} = \tfrac{11}{4}\pi - \tfrac{11}{2}\sin^{-1}\left(\tfrac{1}{3}\right) - 3\sqrt{2}$$

33. The curves intersect when $4\cos\theta = 2 \Rightarrow \cos\theta = \frac{1}{2} \Rightarrow \theta = \pm\frac{\pi}{3}$ for $-\pi \le \theta \le \pi$. The points of intersection are $\left(2, \frac{\pi}{3}\right)$ and $\left(2, -\frac{\pi}{3}\right)$.

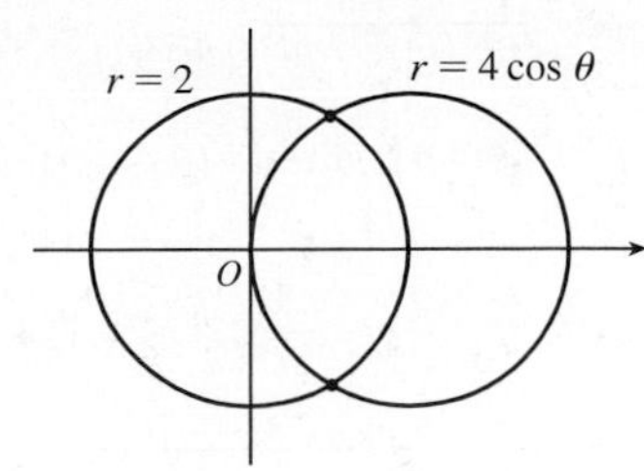

34. The two curves clearly both contain the pole. For other points of intersection, $\cot\theta = 2\cos(\theta + 2n\pi)$ or $-2\cos(\theta + \pi + 2n\pi)$, both of which reduce to $\cot\theta = 2\cos\theta \Leftrightarrow \cos\theta = 2\sin\theta\cos\theta \Leftrightarrow \cos\theta(1 - 2\sin\theta) = 0 \Rightarrow$ $\cos\theta = 0$ or $\sin\theta = \frac{1}{2} \Rightarrow \theta = \frac{\pi}{6}, \frac{\pi}{2}, \frac{5\pi}{6}$ or $\frac{3\pi}{2} \Rightarrow$ intersection points are $\left(0, \frac{\pi}{2}\right)$, $\left(\sqrt{3}, \frac{\pi}{6}\right)$, and $\left(\sqrt{3}, \frac{11\pi}{6}\right)$.

35. The curves intersect where $2\sin\theta = \sin\theta + \cos\theta \Rightarrow$ $\sin\theta = \cos\theta \Rightarrow \theta = \frac{\pi}{4}$, and also at the origin (at which $\theta = \frac{3\pi}{4}$ on the second curve).

$$\begin{aligned} A &= \int_0^{\pi/4} \tfrac{1}{2}(2\sin\theta)^2\,d\theta + \int_{\pi/4}^{3\pi/4} \tfrac{1}{2}(\sin\theta + \cos\theta)^2\,d\theta \\ &= \int_0^{\pi/4} (1 - \cos 2\theta)\,d\theta + \tfrac{1}{2}\int_{\pi/4}^{3\pi/4} (1 + \sin 2\theta)\,d\theta \\ &= \left[\theta - \tfrac{1}{2}\sin 2\theta\right]_0^{\pi/4} + \left[\tfrac{1}{2}\theta - \tfrac{1}{4}\cos 2\theta\right]_{\pi/4}^{3\pi/4} = \tfrac{1}{2}(\pi - 1) \end{aligned}$$

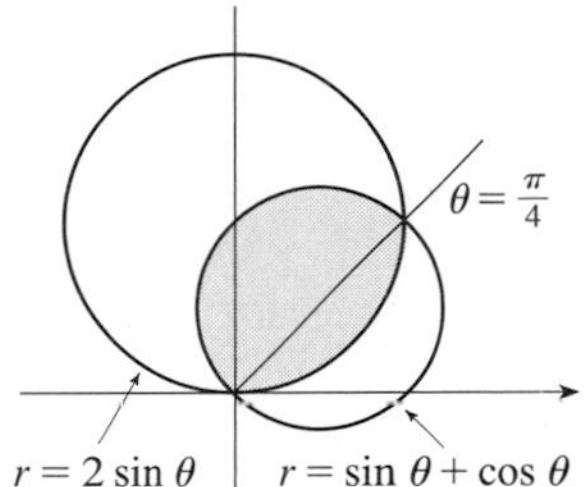

36.
$$\begin{aligned} A &= 2\int_{-\pi/2}^{\pi/6} \tfrac{1}{2}\left[(2 + \cos 2\theta)^2 - (2 + \sin\theta)^2\right] d\theta \\ &= \int_{-\pi/2}^{\pi/6} \left[4\cos 2\theta + \cos^2 2\theta - 4\sin\theta - \sin^2\theta\right] d\theta \\ &= \left[2\sin 2\theta + \tfrac{1}{2}\theta + \tfrac{1}{8}\sin 4\theta + 4\cos\theta - \tfrac{1}{2}\theta + \tfrac{1}{4}\sin 2\theta\right]_{-\pi/2}^{\pi/6} \\ &= \tfrac{51}{16}\sqrt{3} \end{aligned}$$

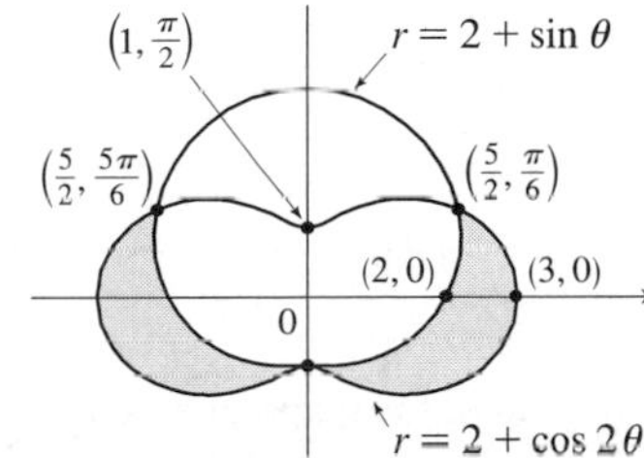

37. $x = 3t^2,\ y = 2t^3.$

$$\begin{aligned} L &= \int_0^2 \sqrt{(dx/dt)^2 + (dy/dt)^2}\,dt = \int_0^2 \sqrt{(6t)^2 + (6t^2)^2}\,dt = \int_0^2 \sqrt{36t^2 + 36t^4}\,dt = \int_0^2 \sqrt{36t^2}\,\sqrt{1 + t^2}\,dt \\ &= \int_0^2 6\,|t|\,\sqrt{1 + t^2}\,dt = 6\int_0^2 t\sqrt{1 + t^2}\,dt = 6\int_1^5 u^{1/2}\left(\tfrac{1}{2}du\right) \qquad \left[u = 1 + t^2,\ du = 2t\,dt\right] \\ &= 6 \cdot \tfrac{1}{2} \cdot \tfrac{2}{3}\left[u^{3/2}\right]_1^5 = 2(5^{3/2} - 1) = 2\left(5\sqrt{5} - 1\right) \end{aligned}$$

38. $x = 2 + 3t,\quad y = \cosh 3t \Rightarrow (dx/dt)^2 + (dy/dt)^2 = 3^2 + (3\sinh 3t)^2 = 9(1 + \sinh^2 3t) = 9\cosh^2 3t$, so $L = \int_0^1 \sqrt{9\cosh^2 3t}\,dt = \int_0^1 |3\cosh 3t|\,dt = \int_0^1 3\cosh 3t\,dt = \left[\sinh 3t\right]_0^1 = \sinh 3 - \sinh 0 = \sinh 3.$

39.
$$\begin{aligned} L &= \int_\pi^{2\pi} \sqrt{r^2 + (dr/d\theta)^2}\,d\theta = \int_\pi^{2\pi} \sqrt{(1/\theta)^2 + (-1/\theta^2)^2}\,d\theta = \int_\pi^{2\pi} \frac{\sqrt{\theta^2 + 1}}{\theta^2}\,d\theta \\ &\overset{24}{=} \left[-\frac{\sqrt{\theta^2 + 1}}{\theta} + \ln\left(\theta + \sqrt{\theta^2 + 1}\right)\right]_\pi^{2\pi} = \frac{\sqrt{\pi^2 + 1}}{\pi} - \frac{\sqrt{4\pi^2 + 1}}{2\pi} + \ln\left(\frac{2\pi + \sqrt{4\pi^2 + 1}}{\pi + \sqrt{\pi^2 + 1}}\right) \\ &= \frac{2\sqrt{\pi^2 + 1} - \sqrt{4\pi^2 + 1}}{2\pi} + \ln\left(\frac{2\pi + \sqrt{4\pi^2 + 1}}{\pi + \sqrt{\pi^2 + 1}}\right) \end{aligned}$$

40. $L = \int_0^\pi \sqrt{r^2 + (dr/d\theta)^2}\,d\theta = \int_0^\pi \sqrt{\sin^6\left(\frac{1}{3}\theta\right) + \sin^4\left(\frac{1}{3}\theta\right)\cos^2\left(\frac{1}{3}\theta\right)}\,d\theta$

$= \int_0^\pi \sin^2\left(\frac{1}{3}\theta\right)d\theta = \left[\frac{1}{2}\left(\theta - \frac{3}{2}\sin\left(\frac{2}{3}\theta\right)\right)\right]_0^\pi = \frac{1}{2}\pi - \frac{3}{8}\sqrt{3}$

41. $x = 4\sqrt{t},\ y = \dfrac{t^3}{3} + \dfrac{1}{2t^2},\ 1 \le t \le 4 \ \Rightarrow$

$S = \int_1^4 2\pi y\sqrt{(dx/dt)^2 + (dy/dt)^2}\,dt = \int_1^4 2\pi\left(\frac{1}{3}t^3 + \frac{1}{2}t^{-2}\right)\sqrt{\left(2/\sqrt{t}\right)^2 + (t^2 - t^{-3})^2}\,dt$

$= 2\pi\int_1^4 \left(\frac{1}{3}t^3 + \frac{1}{2}t^{-2}\right)\sqrt{(t^2 + t^{-3})^2}\,dt = 2\pi\int_1^4 \left(\frac{1}{3}t^5 + \frac{5}{6} + \frac{1}{2}t^{-5}\right)dt = 2\pi\left[\frac{1}{18}t^6 + \frac{5}{6}t - \frac{1}{8}t^{-4}\right]_1^4 = \frac{471{,}295}{1024}\pi$

42. $x = 2 + 3t, \quad y = \cosh 3t \quad \Rightarrow \quad (dx/dt)^2 + (dy/dt)^2 = 3^2 + (3\sinh 3t)^2 = 9(1 + \sinh^2 3t) = 9\cosh^2 3t$, so

$S = \int_0^1 2\pi y\,ds = \int_0^1 2\pi\cosh 3t\sqrt{9\cosh^2 3t}\,dt = \int_0^1 2\pi\cosh 3t\,|3\cosh 3t|\,dt = \int_0^1 2\pi\cosh 3t \cdot 3\cosh 3t\,dt$

$= 6\pi\int_0^1 \cosh^2 3t\,dt = 6\pi\int_0^1 \frac{1}{2}(1 + \cosh 6t)\,dt = 3\pi\left[t + \frac{1}{6}\sinh 6t\right]_0^1 = 3\pi\left(1 + \frac{1}{6}\sinh 6\right) = 3\pi + \frac{\pi}{2}\sinh 6$

43. For all c except -1, the curve is asymptotic to the line $x = 1$. For $c < -1$, the curve bulges to the right near $y = 0$. As c increases, the bulge becomes smaller, until at $c = -1$ the curve is the straight line $x = 1$. As c continues to increase, the curve bulges to the left, until at $c = 0$ there is a cusp at the origin. For $c > 0$, there is a loop to the left of the origin, whose size and roundness increase as c increases. Note that the x-intercept of the curve is always $-c$.

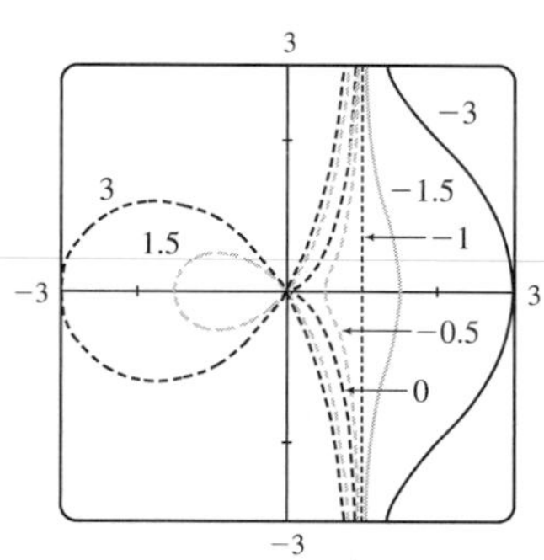

44. For a close to 0, the graph of $r^a = |\sin 2\theta|$ consists of four thin petals. As a increases, the petals get wider, until as $a \to \infty$, each petal occupies almost its entire quarter-circle.

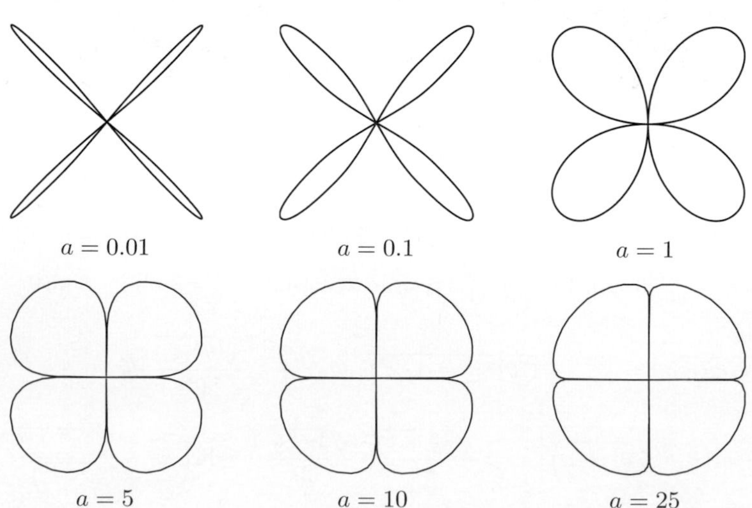

45. $\dfrac{x^2}{9} + \dfrac{y^2}{8} = 1$ is an ellipse with center $(0, 0)$.

$a = 3$, $b = 2\sqrt{2}$, $c = 1$ $\Rightarrow$

foci $(\pm 1, 0)$, vertices $(\pm 3, 0)$.

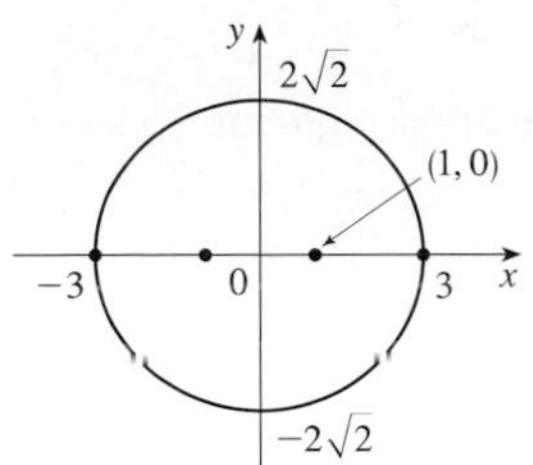

46. $4x^2 - y^2 = 16 \quad \Leftrightarrow \quad \dfrac{x^2}{4} - \dfrac{y^2}{16} = 1$ is a hyperbola

with center $(0, 0)$, vertices $(\pm 2, 0)$, $a = 2$, $b = 4$,

$c = \sqrt{16 + 4} = 2\sqrt{5}$, foci $\left(\pm 2\sqrt{5}, 0\right)$ and

asymptotes $y = \pm 2x$.

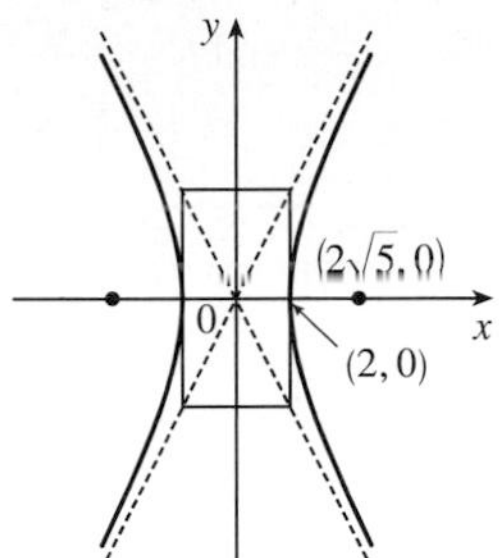

47. $6y^2 + x - 36y + 55 = 0 \quad \Leftrightarrow$

$6(y^2 - 6y + 9) = -(x + 1) \quad \Leftrightarrow$

$(y - 3)^2 = -\frac{1}{6}(x + 1)$, a parabola with vertex $(-1, 3)$,

opening to the left, $p = -\frac{1}{24} \quad \Rightarrow \quad$ focus $\left(-\frac{25}{24}, 3\right)$ and

directrix $x = -\frac{23}{24}$.

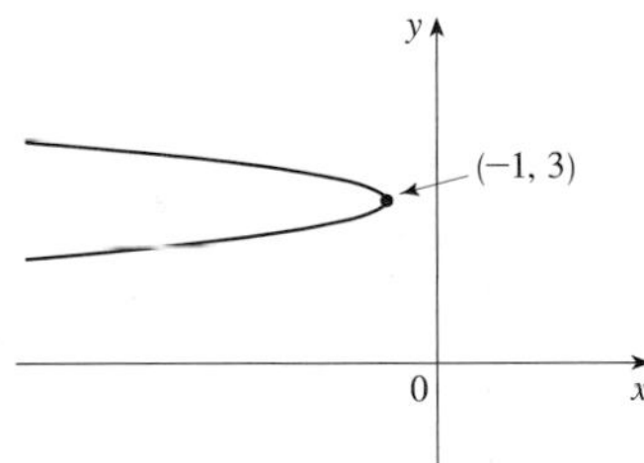

48. $25x^2 + 4y^2 + 50x - 16y = 59 \quad \Leftrightarrow$

$25(x + 1)^2 + 4(y - 2)^2 = 100 \quad \Leftrightarrow$

$\frac{1}{4}(x + 1)^2 + \frac{1}{25}(y - 2)^2 = 1$ is an ellipse centered at

$(-1, 2)$ with foci on the line $x = -1$, vertices $(-1, 7)$

and $(-1, -3)$; $a = 5$, $b = 2 \quad \Rightarrow \quad c = \sqrt{21} \quad \Rightarrow$

foci $\left(-1, 2 \pm \sqrt{21}\,\right)$.

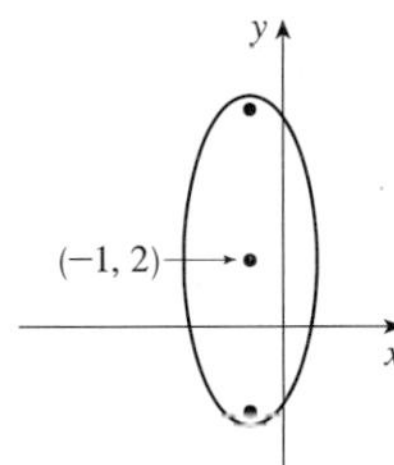

49. The ellipse with foci $(\pm 4, 0)$ and vertices $(\pm 5, 0)$ has center $(0, 0)$ and a horizontal major axis, with $a = 5$ and $c = 4$, so $b^2 = a^2 - c^2 = 5^2 - 4^2 = 9$. An equation is $\dfrac{x^2}{25} + \dfrac{y^2}{9} = 1$.

50. The distance from the focus $(2, 1)$ to the directrix $x = -4$ is $2 - (-4) = 6$, so the distance from the focus to the vertex is $\frac{1}{2}(6) = 3$ and the vertex is $(-1, 1)$. Since the focus is to the right of the vertex, $p = 3$. An equation is $(y - 1)^2 = 4 \cdot 3[x - (-1)]$, or $(y - 1)^2 = 12(x + 1)$.

51. The center of a hyperbola with foci $(0, \pm 4)$ is $(0, 0)$, so $c = 4$ and an equation is $\dfrac{y^2}{a^2} - \dfrac{x^2}{b^2} = 1$.

The asymptote $y = 3x$ has slope 3, so $\dfrac{a}{b} = \dfrac{3}{1} \quad \Rightarrow \quad a = 3b$ and $a^2 + b^2 = c^2 \quad \Rightarrow \quad (3b)^2 + b^2 = 4^2 \quad \Rightarrow$

$10b^2 = 16 \quad \Rightarrow \quad b^2 = \frac{8}{5}$ and so $a^2 = 16 - \frac{8}{5} = \frac{72}{5}$. Thus, an equation is $\dfrac{y^2}{72/5} - \dfrac{x^2}{8/5} = 1$, or $\dfrac{5y^2}{72} - \dfrac{5x^2}{8} = 1$.

52. Center is $(3, 0)$, and $a = \frac{8}{2} = 4$, $c = 2 \quad \Leftrightarrow \quad b = \sqrt{4^2 - 2^2} = \sqrt{12} \quad \Rightarrow$

an equation of the ellipse is $\dfrac{(x - 3)^2}{12} + \dfrac{y^2}{16} = 1$.

53. $x^2 + y = 100 \;\Leftrightarrow\; x^2 = -(y - 100)$ has its vertex at $(0, 100)$, so one of the vertices of the ellipse is $(0, 100)$. Another form of the equation of a parabola is $x^2 = 4p(y - 100)$ so $4p(y - 100) = -(y - 100) \;\Rightarrow\; 4p = -1 \;\Rightarrow\; p = -\frac{1}{4}$. Therefore the shared focus is found at $\left(0, \frac{399}{4}\right)$ so $2c = \frac{399}{4} - 0 \;\Rightarrow\; c = \frac{399}{8}$ and the center of the ellipse is $\left(0, \frac{399}{8}\right)$. So $a = 100 - \frac{399}{8} = \frac{401}{8}$ and $b^2 = a^2 - c^2 = \dfrac{401^2 - 399^2}{8^2} = 25$. So the equation of the ellipse is $\dfrac{x^2}{b^2} + \dfrac{\left(y - \frac{399}{8}\right)^2}{a^2} = 1 \;\Rightarrow$ $\dfrac{x^2}{25} + \dfrac{\left(y - \frac{399}{8}\right)^2}{\left(\frac{401}{8}\right)^2} = 1$, or $\dfrac{x^2}{25} + \dfrac{(8y - 399)^2}{160{,}801} = 1$.

54. $\dfrac{x^2}{a^2} + \dfrac{y^2}{b^2} = 1 \;\Rightarrow\; \dfrac{2x}{a^2} + \dfrac{2y}{b^2}\dfrac{dy}{dx} = 0 \;\Rightarrow\; \dfrac{dy}{dx} = -\dfrac{b^2}{a^2}\dfrac{x}{y}$. Therefore $\dfrac{dy}{dx} = m \;\Leftrightarrow\; y = -\dfrac{b^2}{a^2}\dfrac{x}{m}$. Combining this condition with $\dfrac{x^2}{a^2} + \dfrac{y^2}{b^2} = 1$, we find that $x = \pm\dfrac{a^2 m}{\sqrt{a^2m^2 + b^2}}$. In other words, the two points on the ellipse where the tangent has slope m are $\left(\pm\dfrac{a^2 m}{\sqrt{a^2m^2 + b^2}}, \mp\dfrac{b^2}{\sqrt{a^2m^2 + b^2}}\right)$. The tangent lines at these points have the equations $y \pm \dfrac{b^2}{\sqrt{a^2m^2 + b^2}} = m\left(x \mp \dfrac{a^2 m}{\sqrt{a^2m^2 + b^2}}\right)$ or $y = mx \mp \dfrac{a^2m^2}{\sqrt{a^2m^2 + b^2}} \mp \dfrac{b^2}{\sqrt{a^2m^2 + b^2}} = mx \mp \sqrt{a^2m^2 + b^2}$.

55. Directrix $x = 4 \;\Rightarrow\; d = 4$, so $e = \frac{1}{3} \;\Rightarrow\; r = \dfrac{ed}{1 + e\cos\theta} = \dfrac{4}{3 + \cos\theta}$.

56. See the end of the proof of Theorem 10.6.1. If $e > 1$, then $1 - e^2 < 0$ and Equations 10.6.4 become $a^2 = \dfrac{e^2d^2}{(e^2 - 1)^2}$ and $b^2 = \dfrac{e^2d^2}{e^2 - 1}$, so $\dfrac{b^2}{a^2} = e^2 - 1$. The asymptotes $y = \pm\dfrac{b}{a}x$ have slopes $\pm\dfrac{b}{a} = \pm\sqrt{e^2 - 1}$, so the angles they make with the polar axis are $\pm\tan^{-1}\left[\sqrt{e^2 - 1}\right] = \cos^{-1}(\pm 1/e)$.

57. In polar coordinates, an equation for the circle is $r = 2a\sin\theta$. Thus, the coordinates of Q are $x = r\cos\theta = 2a\sin\theta\,\cos\theta$ and $y = r\sin\theta = 2a\sin^2\theta$. The coordinates of R are $x = 2a\cot\theta$ and $y = 2a$. Since P is the midpoint of QR, we use the midpoint formula to get $x = a(\sin\theta\,\cos\theta + \cot\theta)$ and $y = a(1 + \sin^2\theta)$.

58. (a) If (a, b) lies on the curve, then there is some parameter value t_1 such that $\dfrac{3t_1}{1 + t_1^3} = a$ and $\dfrac{3t_1^2}{1 + t_1^3} = b$. If $t_1 = 0$, the point is $(0, 0)$, which lies on the line $y = x$. If $t_1 \neq 0$, then the point corresponding to $t = \dfrac{1}{t_1}$ is given by $x = \dfrac{3(1/t_1)}{1 + (1/t_1)^3} = \dfrac{3t_1^2}{t_1^3 + 1} = b$, $y = \dfrac{3(1/t_1)^2}{1 + (1/t_1)^3} = \dfrac{3t_1}{t_1^3 + 1} = a$. So (b, a) also lies on the curve. [Another way to see this is to do part (e) first; the result is immediate.] The curve intersects the line $y = x$ when $\dfrac{3t}{1 + t^3} = \dfrac{3t^2}{1 + t^3} \;\Rightarrow$ $t = t^2 \;\Rightarrow\; t = 0$ or 1, so the points are $(0, 0)$ and $\left(\frac{3}{2}, \frac{3}{2}\right)$.

(b) $\dfrac{dy}{dt} = \dfrac{(1+t^3)(6t) - 3t^2(3t^2)}{(1+t^3)^2} = \dfrac{6t - 3t^4}{(1+t^3)^2} = 0$ when $6t - 3t^4 = 3t(2 - t^3) = 0 \Rightarrow t = 0$ or $t = \sqrt[3]{2}$, so there are horizontal tangents at $(0, 0)$ and $\left(\sqrt[3]{2}, \sqrt[3]{4}\right)$. Using the symmetry from part (a), we see that there are vertical tangents at $(0, 0)$ and $\left(\sqrt[3]{4}, \sqrt[3]{2}\right)$.

(c) Notice that as $t \to -1^+$, we have $x \to -\infty$ and $y \to \infty$. As $t \to -1^-$, we have $x \to \infty$ and $y \to -\infty$. Also $y - (-x - 1) = y + x + 1 = \dfrac{3t + 3t^2 + (1 + t^3)}{1 + t^3} = \dfrac{(t+1)^3}{1+t^3} = \dfrac{(t+1)^2}{t^2 - t + 1} \to 0$ as $t \to -1$. So $y = -x - 1$ is a slant asymptote.

(d) $\dfrac{dx}{dt} = \dfrac{(1+t^3)(3) - 3t(3t^2)}{(1+t^3)^2} = \dfrac{3 - 6t^3}{(1+t^3)^2}$ and from part (b) we have $\dfrac{dy}{dt} = \dfrac{6t - 3t^4}{(1+t^3)^2}$. So $\dfrac{dy}{dx} = \dfrac{dy/dt}{dx/dt} = \dfrac{t(2 - t^3)}{1 - 2t^3}$.

Also $\dfrac{d^2y}{dx^2} = \dfrac{\frac{d}{dt}\left(\frac{dy}{dx}\right)}{dx/dt} = \dfrac{2(1+t^3)^4}{3(1 - 2t^3)^3} > 0 \Leftrightarrow t < \dfrac{1}{\sqrt[3]{2}}$.

So the curve is concave upward there and has a minimum point at $(0, 0)$ and a maximum point at $\left(\sqrt[3]{2}, \sqrt[3]{4}\right)$. Using this together with the information from parts (a), (b), and (c), we sketch the curve.

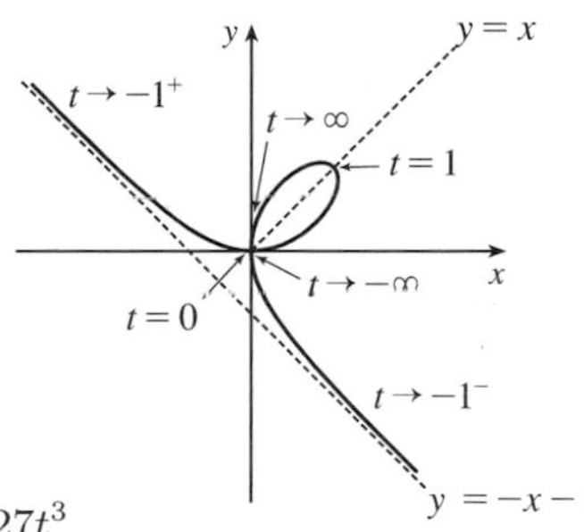

(e) $x^3 + y^3 = \left(\dfrac{3t}{1+t^3}\right)^3 + \left(\dfrac{3t^2}{1+t^3}\right)^3 = \dfrac{27t^3 + 27t^6}{(1+t^3)^3} = \dfrac{27t^3(1+t^3)}{(1+t^3)^3} = \dfrac{27t^3}{(1+t^3)^2}$

and $3xy = 3\left(\dfrac{3t}{1+t^3}\right)\left(\dfrac{3t^2}{1+t^3}\right) = \dfrac{27t^3}{(1+t^3)^2}$, so $x^3 + y^3 = 3xy$.

(f) We start with the equation from part (e) and substitute $x = r\cos\theta$, $y = r\sin\theta$. Then $x^3 + y^3 = 3xy \Rightarrow$ $r^3\cos^3\theta + r^3\sin^3\theta = 3r^2\cos\theta\sin\theta$. For $r \neq 0$, this gives $r = \dfrac{3\cos\theta\sin\theta}{\cos^3\theta + \sin^3\theta}$. Dividing numerator and denominator by $\cos^3\theta$, we obtain $r = \dfrac{3\left(\dfrac{1}{\cos\theta}\right)\dfrac{\sin\theta}{\cos\theta}}{1 + \dfrac{\sin^3\theta}{\cos^3\theta}} = \dfrac{3\sec\theta\tan\theta}{1 + \tan^3\theta}$.

(g) The loop corresponds to $\theta \in \left(0, \frac{\pi}{2}\right)$, so its area is

$$A = \int_0^{\pi/2} \frac{r^2}{2}\,d\theta = \frac{1}{2}\int_0^{\pi/2}\left(\frac{3\sec\theta\tan\theta}{1 + \tan^3\theta}\right)^2 d\theta = \frac{9}{2}\int_0^{\pi/2}\frac{\sec^2\theta\tan^2\theta}{(1 + \tan^3\theta)^2}\,d\theta = \frac{9}{2}\int_0^{\infty}\frac{u^2\,du}{(1+u^3)^2} \quad [\text{let } u = \tan\theta]$$

$$= \lim_{b\to\infty} \tfrac{9}{2}\left[-\tfrac{1}{3}(1+u^3)^{-1}\right]_0^b = \tfrac{3}{2}$$

(h) By symmetry, the area between the folium and the line $y = -x - 1$ is equal to the enclosed area in the third quadrant, plus twice the enclosed area in the fourth quadrant. The area in the third quadrant is $\frac{1}{2}$, and since $y = -x - 1 \Rightarrow$ $r\sin\theta = -r\cos\theta - 1 \Rightarrow r = -\dfrac{1}{\sin\theta + \cos\theta}$, the area in the fourth quadrant is

$$\frac{1}{2}\int_{-\pi/2}^{-\pi/4}\left[\left(-\frac{1}{\sin\theta + \cos\theta}\right)^2 - \left(\frac{3\sec\theta\tan\theta}{1 + \tan^3\theta}\right)^2\right]d\theta \overset{\text{CAS}}{=} \frac{1}{2}.$$ Therefore, the total area is $\frac{1}{2} + 2\left(\frac{1}{2}\right) = \frac{3}{2}$.

PROBLEMS PLUS

1. See the figure. The circle with center $(-1, 0)$ and radius $\sqrt{2}$ has equation $(x+1)^2 + y^2 = 2$ and describes the circular arc from $(0, -1)$ to $(0, 1)$. Converting the equation to polar coordinates gives us

$(r\cos\theta + 1)^2 + (r\sin\theta)^2 = 2 \Rightarrow$

$r^2\cos^2\theta + 2r\cos\theta + 1 + r^2\sin^2\theta = 2 \Rightarrow$

$r^2(\cos^2\theta + \sin^2\theta) + 2r\cos\theta = 1 \Rightarrow r^2 + 2r\cos\theta = 1$. Using the quadratic formula to solve for r gives us

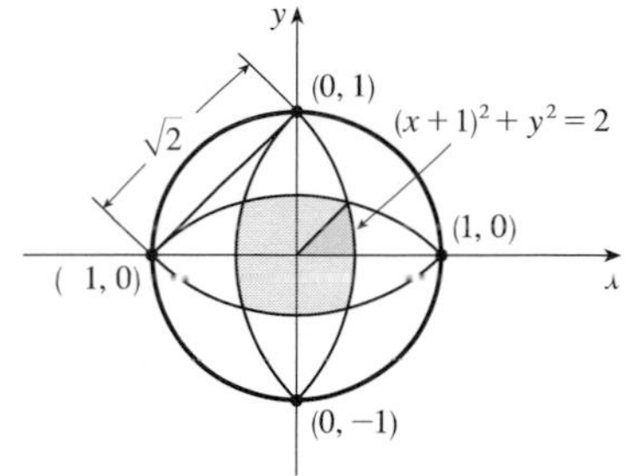

$$r = \frac{-2\cos\theta \pm \sqrt{4\cos^2\theta + 4}}{2} = -\cos\theta + \sqrt{\cos^2\theta + 1} \text{ for } r > 0.$$

The darkest shaded region is $\frac{1}{8}$ of the entire shaded region A, so $\frac{1}{8}A = \int_0^{\pi/4} \frac{1}{2}r^2\,d\theta = \frac{1}{2}\int_0^{\pi/4} (1 - 2r\cos\theta)\,d\theta \Rightarrow$

$$\begin{aligned}
\tfrac{1}{4}A &= \int_0^{\pi/4} \left[1 - 2\cos\theta\left(-\cos\theta + \sqrt{\cos^2\theta+1}\right)\right]d\theta = \int_0^{\pi/4}\left(1 + 2\cos^2\theta - 2\cos\theta\sqrt{\cos^2\theta+1}\right)d\theta \\
&= \int_0^{\pi/4}\left[1 + 2\cdot\tfrac{1}{2}(1+\cos 2\theta) - 2\cos\theta\sqrt{(1-\sin^2\theta)+1}\right]d\theta \\
&= \int_0^{\pi/4}(2+\cos 2\theta)\,d\theta - 2\int_0^{\pi/4}\cos\theta\sqrt{2-\sin^2\theta}\,d\theta \\
&= \left[2\theta + \tfrac{1}{2}\sin 2\theta\right]_0^{\pi/4} - 2\int_0^{1/\sqrt{2}}\sqrt{2-u^2}\,du \qquad \begin{bmatrix} u = \sin\theta, \\ du = \cos\theta\,d\theta \end{bmatrix} \\
&= \left(\frac{\pi}{2} + \frac{1}{2}\right) - (0+0) - 2\left[\frac{u}{2}\sqrt{2-u^2} + \sin^{-1}\frac{u}{\sqrt{2}}\right]_0^{1/\sqrt{2}} \qquad \begin{bmatrix}\text{Formula 30,} \\ a = \sqrt{2}\end{bmatrix} \\
&= \frac{\pi}{2} + \frac{1}{2} - 2\left(\frac{1}{2\sqrt{2}}\cdot\frac{\sqrt{3}}{\sqrt{2}} + \frac{\pi}{6}\right) = \frac{\pi}{2} + \frac{1}{2} - \frac{1}{2}\sqrt{3} - \frac{\pi}{3} = \frac{\pi}{6} + \frac{1}{2} - \frac{1}{2}\sqrt{3}.
\end{aligned}$$

Thus, $A = 4\left(\frac{\pi}{6} + \frac{1}{2} - \frac{1}{2}\sqrt{3}\right) = \frac{2\pi}{3} + 2 - 2\sqrt{3}$.

2. (a) The curve $x^4 + y^4 = x^2 + y^2$ is symmetric about both axes and about the line $y = x$ (since interchanging x and y does not change the equation) so we need only consider $y \geq x \geq 0$ to begin with. Implicit differentiation gives

$4x^3 + 4y^3y' = 2x + 2yy' \Rightarrow y' = \dfrac{x(1-2x^2)}{y(2y^2-1)} \Rightarrow y' = 0$ when $x = 0$ and when $x = \pm\frac{1}{\sqrt{2}}$. If $x = 0$, then

$y^4 = y^2 \Rightarrow y^2(y^2-1) = 0 \Rightarrow y = 0$ or ± 1. The point $(0, 0)$ can't be a highest or lowest point because it is isolated. [If $-1 < x < 1$ and $-1 < y < 1$, then $x^4 < x^2$ and $y^4 < y^2 \Rightarrow x^4 + y^4 < x^2 + y^2$, except for $(0, 0)$.]

If $x = \frac{1}{\sqrt{2}}$, then $x^2 = \frac{1}{2}$, $x^4 = \frac{1}{4}$, so $\frac{1}{4} + y^4 = \frac{1}{2} + y^2 \Rightarrow 4y^4 - 4y^2 - 1 = 0 \Rightarrow y^2 = \frac{4+\sqrt{16+16}}{8} = \frac{1\pm\sqrt{2}}{2}$.

But $y^2 > 0$, so $y^2 = \frac{1+\sqrt{2}}{2} \Rightarrow y = \pm\sqrt{\frac{1}{2}(1+\sqrt{2})}$. Near the point $(0, 1)$, the denominator of y' is positive and the numerator changes from negative to positive as x increases through 0, so $(0, 1)$ is a local minimum point. At

$\left(\frac{1}{\sqrt{2}}, \sqrt{\frac{1+\sqrt{2}}{2}}\right)$, y' changes from positive to negative, so that point gives a maximum. By symmetry, the highest points on the curve are $\left(\pm\frac{1}{\sqrt{2}}, \sqrt{\frac{1+\sqrt{2}}{2}}\right)$ and the lowest points are $\left(\pm\frac{1}{\sqrt{2}}, -\sqrt{\frac{1+\sqrt{2}}{2}}\right)$.

(b) We use the information from part (a), together with symmetry with respect to the axes and the lines $y = \pm x$, to sketch the curve.

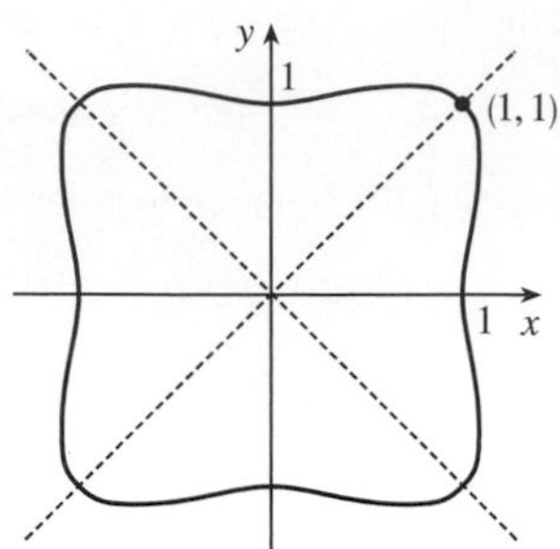

(c) In polar coordinates, $x^4 + y^4 = x^2 + y^2$ becomes $r^4 \cos^4\theta + r^4 \sin^4\theta = r^2$ or $r^2 = \dfrac{1}{\cos^4\theta + \sin^4\theta}$. By the symmetry shown in part (b), the area enclosed by the curve is $A = 8\int_0^{\pi/4} \frac{1}{2} r^2\, d\theta = 4\int_0^{\pi/4} \dfrac{d\theta}{\cos^4\theta + \sin^4\theta} \overset{\text{CAS}}{=} \sqrt{2}\,\pi$.

3. In terms of x and y, we have $x = r\cos\theta = (1 + c\sin\theta)\cos\theta = \cos\theta + c\sin\theta\cos\theta = \cos\theta + \frac{1}{2}c\sin 2\theta$ and $y = r\sin\theta = (1 + c\sin\theta)\sin\theta = \sin\theta + c\sin^2\theta$. Now $-1 \le \sin\theta \le 1 \;\Rightarrow\; -1 \le \sin\theta + c\sin^2\theta \le 1 + c \le 2$, so $-1 \le y \le 2$. Furthermore, $y = 2$ when $c = 1$ and $\theta = \frac{\pi}{2}$, while $y = -1$ for $c = 0$ and $\theta = \frac{3\pi}{2}$. Therefore, we need a viewing rectangle with $-1 \le y \le 2$.

To find the x-values, look at the equation $x = \cos\theta + \frac{1}{2}c\sin 2\theta$ and use the fact that $\sin 2\theta \ge 0$ for $0 \le \theta \le \frac{\pi}{2}$ and $\sin 2\theta \le 0$ for $-\frac{\pi}{2} \le \theta \le 0$. [Because $r = 1 + c\sin\theta$ is symmetric about the y-axis, we only need to consider $-\frac{\pi}{2} \le \theta \le \frac{\pi}{2}$.] So for $-\frac{\pi}{2} \le \theta \le 0$, x has a maximum value when $c = 0$ and then $x = \cos\theta$ has a maximum value of 1 at $\theta = 0$. Thus, the maximum value of x must occur on $\left[0, \frac{\pi}{2}\right]$ with $c = 1$. Then $x = \cos\theta + \frac{1}{2}\sin 2\theta \;\Rightarrow$ $\frac{dx}{d\theta} = -\sin\theta + \cos 2\theta = -\sin\theta + 1 - 2\sin^2\theta \;\Rightarrow\; \frac{dx}{d\theta} = -(2\sin\theta - 1)(\sin\theta + 1) = 0$ when $\sin\theta = -1$ or $\frac{1}{2}$ [but $\sin\theta \ne -1$ for $0 \le \theta \le \frac{\pi}{2}$]. If $\sin\theta = \frac{1}{2}$, then $\theta = \frac{\pi}{6}$ and $x = \cos\frac{\pi}{6} + \frac{1}{2}\sin\frac{\pi}{3} = \frac{3}{4}\sqrt{3}$. Thus, the maximum value of x is $\frac{3}{4}\sqrt{3}$, and, by symmetry, the minimum value is $-\frac{3}{4}\sqrt{3}$. Therefore, the smallest viewing rectangle that contains every member of the family of polar curves $r = 1 + c\sin\theta$, where $0 \le c \le 1$, is $\left[-\frac{3}{4}\sqrt{3}, \frac{3}{4}\sqrt{3}\right] \times [-1, 2]$.

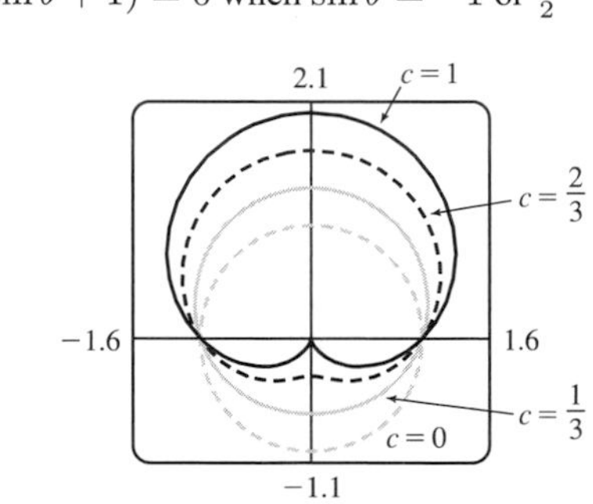

4. (a) Let us find the polar equation of the path of the bug that starts in the upper right corner of the square. If the polar coordinates of this bug, at a particular moment, are (r, θ), then the polar coordinates of the bug that it is crawling toward must be $\left(r, \theta + \frac{\pi}{2}\right)$. (The next bug must be the same distance from the origin and the angle between the lines joining the bugs to the pole must be $\frac{\pi}{2}$.) The Cartesian coordinates of the first bug are $(r\cos\theta, r\sin\theta)$ and for the second bug we have $x = r\cos\left(\theta + \frac{\pi}{2}\right) = -r\sin\theta$, $y = r\sin\left(\theta + \frac{\pi}{2}\right) = r\cos\theta$. So the slope of the line joining the bugs is

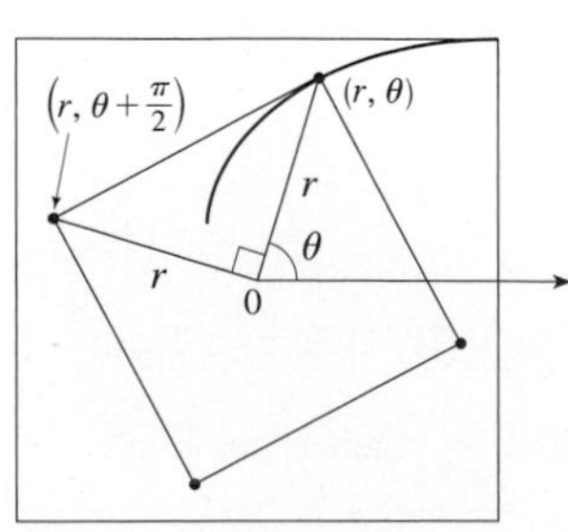

$\dfrac{r\cos\theta - r\sin\theta}{-r\sin\theta - r\cos\theta} = \dfrac{\sin\theta - \cos\theta}{\sin\theta + \cos\theta}$. This must be equal to the slope of the tangent line at (r, θ), so by

Equation 10.3.3 we have $\dfrac{(dr/d\theta)\sin\theta + r\cos\theta}{(dr/d\theta)\cos\theta - r\sin\theta} = \dfrac{\sin\theta - \cos\theta}{\sin\theta + \cos\theta}$. Solving for $\dfrac{dr}{d\theta}$, we get

$\dfrac{dr}{d\theta}\sin^2\theta + \dfrac{dr}{d\theta}\sin\theta\cos\theta + r\sin\theta\cos\theta + r\cos^2\theta = \dfrac{dr}{d\theta}\sin\theta\cos\theta - \dfrac{dr}{d\theta}\cos^2\theta - r\sin^2\theta + r\sin\theta\cos\theta \Rightarrow$

$\dfrac{dr}{d\theta}\left(\sin^2\theta + \cos^2\theta\right) + r\left(\cos^2\theta + \sin^2\theta\right) = 0 \quad\Rightarrow\quad \dfrac{dr}{d\theta} = -r$. Solving this differential equation as a separable equation (as in Section 9.3), or using Theorem 9.4.2 with $k = -1$, we get $r = Ce^{-\theta}$. To determine C we use the fact that, at its starting position, $\theta = \frac{\pi}{4}$ and $r = \frac{1}{\sqrt{2}}a$, so $\frac{1}{\sqrt{2}}a = Ce^{-\pi/4} \quad\Rightarrow\quad C = \frac{1}{\sqrt{2}}ae^{\pi/4}$. Therefore, a polar equation of the bug's path is $r = \frac{1}{\sqrt{2}}ae^{\pi/4}e^{-\theta}$ or $r = \frac{1}{\sqrt{2}}ae^{(\pi/4)-\theta}$.

(b) The distance traveled by this bug is $L = \int_{\pi/4}^{\infty}\sqrt{r^2 + (dr/d\theta)^2}\,d\theta$, where $\dfrac{dr}{d\theta} = \dfrac{a}{\sqrt{2}}\,e^{\pi/4}(-e^{-\theta})$ and so

$r^2 + (dr/d\theta)^2 = \frac{1}{2}a^2e^{\pi/2}e^{-2\theta} + \frac{1}{2}a^2e^{\pi/2}e^{-2\theta} = a^2e^{\pi/2}e^{-2\theta}$. Thus

$$\begin{aligned} L &= \int_{\pi/4}^{\infty} ae^{\pi/4}e^{-\theta}\,d\theta = ae^{\pi/4}\lim_{t\to\infty}\int_{\pi/4}^{t} e^{-\theta}\,d\theta = ae^{\pi/4}\lim_{t\to\infty}\left[-e^{-\theta}\right]_{\pi/4}^{t} \\ &= ae^{\pi/4}\lim_{t\to\infty}\left[e^{-\pi/4} - e^{-t}\right] = ae^{\pi/4}e^{-\pi/4} = a \end{aligned}$$

5. Without loss of generality, assume the hyperbola has equation $\dfrac{x^2}{a^2} - \dfrac{y^2}{b^2} = 1$. Use implicit differentiation to get

$\dfrac{2x}{a^2} - \dfrac{2y\,y'}{b^2} = 0$, so $y' = \dfrac{b^2x}{a^2y}$. The tangent line at the point (c, d) on the hyperbola has equation $y - d = \dfrac{b^2c}{a^2d}(x - c)$.

The tangent line intersects the asymptote $y = \dfrac{b}{a}x$ when $\dfrac{b}{a}x - d = \dfrac{b^2c}{a^2d}(x - c) \quad\Rightarrow\quad abdx - a^2d^2 = b^2cx - b^2c^2 \quad\Rightarrow$

$abdx - b^2cx = a^2d^2 - b^2c^2 \quad\Rightarrow\quad x = \dfrac{a^2d^2 - b^2c^2}{b(ad - bc)} = \dfrac{ad + bc}{b}$ and the y-value is $\dfrac{b}{a}\dfrac{ad + bc}{b} = \dfrac{ad + bc}{a}$.

Similarly, the tangent line intersects $y = -\dfrac{b}{a}x$ at $\left(\dfrac{bc - ad}{b}, \dfrac{ad - bc}{a}\right)$. The midpoint of these intersection points is

$\left(\dfrac{1}{2}\left(\dfrac{ad + bc}{b} + \dfrac{bc - ad}{b}\right), \dfrac{1}{2}\left(\dfrac{ad + bc}{a} + \dfrac{ad - bc}{a}\right)\right) = \left(\dfrac{1}{2}\dfrac{2bc}{b}, \dfrac{1}{2}\dfrac{2ad}{a}\right) = (c, d)$, the point of tangency.

Note: If $y = 0$, then at $(\pm a, 0)$, the tangent line is $x = \pm a$, and the points of intersection are clearly equidistant from the point of tangency.

6. (a) Since the smaller circle rolls without slipping around C, the amount of arc traversed on C ($2r\theta$ in the figure) must equal the amount of arc of the smaller circle that has been in contact with C. Since the smaller circle has radius r, it must have turned through an angle of $2r\theta/r = 2\theta$. In addition to turning through an angle 2θ, the little circle has rolled through an angle θ against C. Thus, P has turned through an angle of 3θ as shown in the figure. (If the little circle had turned through an angle of 2θ with its center pinned to the x-axis,

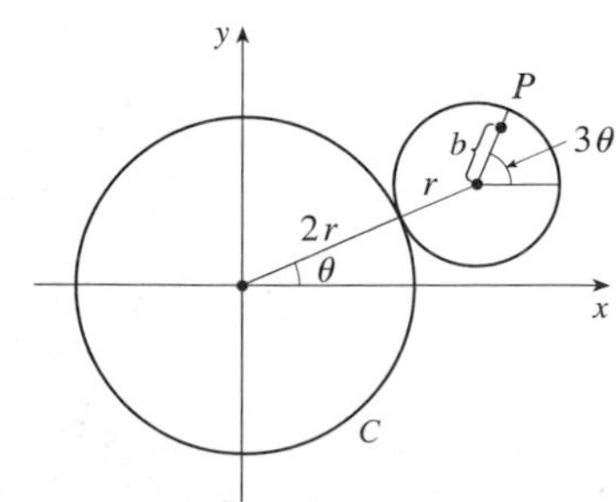

then P would have turned only 2θ instead of 3θ. The movement of the little circle around C adds θ to the angle.) From the figure, we see that the center of the small circle has coordinates $(3r\cos\theta, 3r\sin\theta)$. Thus, P has coordinates (x, y), where $x = b\cos 3\theta + 3r\cos\theta$ and $y = b\sin 3\theta + 3r\sin\theta$.

(b)

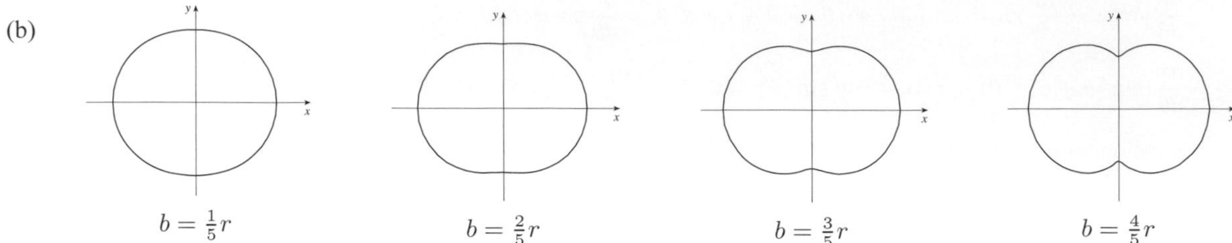

(c) The diagram gives an alternate description of point P on the epitrochoid. Q moves around a circle of radius b, and P rotates one-third as fast with respect to Q at a distance of $3r$. Place an equilateral triangle with sides of length $3\sqrt{3}r$ so that its centroid is at Q and one vertex is at P. (The distance from the centroid to a vertex is $\frac{1}{\sqrt{3}}$ times the length of a side of the equilateral triangle.)

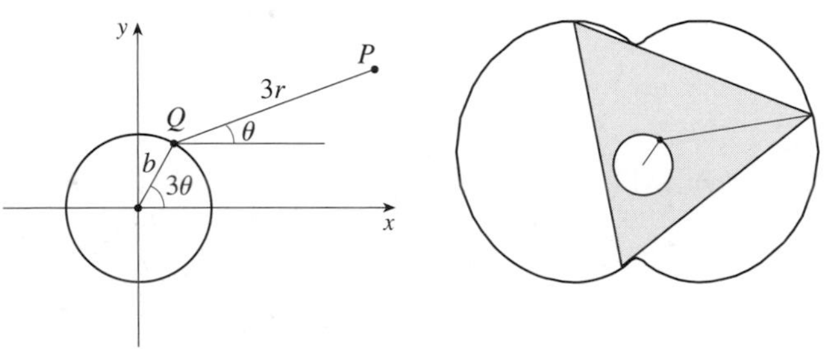

As θ increases by $\frac{2\pi}{3}$, the point Q travels once around the circle of radius b, returning to its original position. At the same time, P (and the rest of the triangle) rotate through an angle of $\frac{2\pi}{3}$ about Q, so P's position is occupied by another vertex. In this way, we see that the epitrochoid traced out by P is simultaneously traced out by the other two vertices as well. The whole equilateral triangle sits inside the epitrochoid (touching it only with its vertices) and each vertex traces out the curve once while the centroid moves around the circle three times.

(d) We view the epitrochoid as being traced out in the same way as in part (c), by a rotor for which the distance from its center to each vertex is $3r$, so it has radius $6r$. To show that the rotor fits inside the epitrochoid, it suffices to show that for any position of the tracing point P, there are no points on the opposite side of the rotor which are outside the epitrochoid. But the most likely case of intersection is when P is on the y-axis, so as long as the diameter of the rotor $\left(\text{which is } 3\sqrt{3}\,r\right)$ is less than the distance between the y-intercepts, the rotor will fit. The y-intercepts occur when $\theta = \frac{\pi}{2}$ or $\theta = \frac{3\pi}{2}$ $\Rightarrow$ $y = -b + 3r$ or $y = b - 3r$, so the distance between the intercepts is $(-b + 3r) - (b - 3r) = 6r - 2b$, and the rotor will fit if $3\sqrt{3}\,r \le 6r - 2b \;\Leftrightarrow\; 2b \le 6r - 3\sqrt{3}\,r \;\Leftrightarrow\; b \le \frac{3}{2}\left(2 - \sqrt{3}\right)r$.

11 □ INFINITE SEQUENCES AND SERIES

11.1 Sequences

1. (a) A sequence is an ordered list of numbers. It can also be defined as a function whose domain is the set of positive integers.

(b) The terms a_n approach 8 as n becomes large. In fact, we can make a_n as close to 8 as we like by taking n sufficiently large.

(c) The terms a_n become large as n becomes large. In fact, we can make a_n as large as we like by taking n sufficiently large.

2. (a) From Definition 1, a convergent sequence is a sequence for which $\lim_{n\to\infty} a_n$ exists. Examples: $\{1/n\}$, $\{1/2^n\}$

(b) A divergent sequence is a sequence for which $\lim_{n\to\infty} a_n$ *does not* exist. Examples: $\{n\}$, $\{\sin n\}$

3. $a_n = \dfrac{2^n}{2n+1}$, so the sequence is $\left\{\dfrac{2^1}{2(1)+1}, \dfrac{2^2}{2(2)+1}, \dfrac{2^3}{2(3)+1}, \dfrac{2^4}{2(4)+1}, \dfrac{2^5}{2(5)+1}, \ldots\right\} = \left\{\dfrac{2}{3}, \dfrac{4}{5}, \dfrac{8}{7}, \dfrac{16}{9}, \dfrac{32}{11}, \ldots\right\}$.

4. $a_n = \dfrac{n^2-1}{n^2+1}$, so the sequence is $\left\{\dfrac{1-1}{1+1}, \dfrac{4-1}{4+1}, \dfrac{9-1}{9+1}, \dfrac{16-1}{16+1}, \dfrac{25-1}{25+1}, \ldots\right\} = \left\{0, \dfrac{3}{5}, \dfrac{8}{10}, \dfrac{15}{17}, \dfrac{24}{26}, \ldots\right\}$.

5. $a_n = \dfrac{(-1)^{n-1}}{5^n}$, so the sequence is $\left\{\dfrac{1}{5^1}, \dfrac{-1}{5^2}, \dfrac{1}{5^3}, \dfrac{-1}{5^4}, \dfrac{1}{5^5}, \ldots\right\} = \left\{\dfrac{1}{5}, -\dfrac{1}{25}, \dfrac{1}{125}, -\dfrac{1}{625}, \dfrac{1}{3125}, \ldots\right\}$.

6. $a_n = \cos\dfrac{n\pi}{2}$, so the sequence is $\left\{\cos\dfrac{\pi}{2}, \cos\pi, \cos\dfrac{3\pi}{2}, \cos 2\pi, \cos\dfrac{5\pi}{2}, \ldots\right\} = \{0, -1, 0, 1, 0, \ldots\}$.

7. $a_n = \dfrac{1}{(n+1)!}$, so the sequence is $\left\{\dfrac{1}{2!}, \dfrac{1}{3!}, \dfrac{1}{4!}, \dfrac{1}{5!}, \dfrac{1}{6!}, \ldots\right\} = \left\{\dfrac{1}{2}, \dfrac{1}{6}, \dfrac{1}{24}, \dfrac{1}{120}, \dfrac{1}{720}, \ldots\right\}$.

8. $a_n = \dfrac{(-1)^n n}{n!+1}$, so $a_1 = \dfrac{(-1)^1 1}{1!+1} = \dfrac{-1}{2}$, and the sequence is

$$\left\{\frac{-1}{2}, \frac{2}{2+1}, \frac{-3}{6+1}, \frac{4}{24+1}, \frac{-5}{120+1}, \ldots\right\} = \left\{-\frac{1}{2}, \frac{2}{3}, -\frac{3}{7}, \frac{4}{25}, -\frac{5}{121}, \ldots\right\}.$$

9. $a_1 = 1$, $a_{n+1} = 5a_n - 3$. Each term is defined in terms of the preceding term. $a_2 = 5a_1 - 3 = 5(1) - 3 = 2$. $a_3 = 5a_2 - 3 = 5(2) - 3 = 7$. $a_4 = 5a_3 - 3 = 5(7) - 3 = 32$. $a_5 = 5a_4 - 3 = 5(32) - 3 = 157$. The sequence is $\{1, 2, 7, 32, 157, \ldots\}$.

10. $a_1 = 6$, $a_{n+1} = \dfrac{a_n}{n}$. $a_2 = \dfrac{a_1}{1} = \dfrac{6}{1} = 6$. $a_3 = \dfrac{a_2}{2} = \dfrac{6}{2} = 3$. $a_4 = \dfrac{a_3}{3} = \dfrac{3}{3} = 1$. $a_5 = \dfrac{a_4}{4} = \dfrac{1}{4}$. The sequence is $\{6, 6, 3, 1, \frac{1}{4}, \ldots\}$.

11. $a_1 = 2$, $a_{n+1} = \dfrac{a_n}{1+a_n}$. $a_2 = \dfrac{a_1}{1+a_1} = \dfrac{2}{1+2} = \dfrac{2}{3}$. $a_3 = \dfrac{a_2}{1+a_2} = \dfrac{2/3}{1+2/3} = \dfrac{2}{5}$. $a_4 = \dfrac{a_3}{1+a_3} = \dfrac{2/5}{1+2/5} = \dfrac{2}{7}$. $a_5 = \dfrac{a_4}{1+a_4} = \dfrac{2/7}{1+2/7} = \dfrac{2}{9}$. The sequence is $\{2, \frac{2}{3}, \frac{2}{5}, \frac{2}{7}, \frac{2}{9}, \ldots\}$.

12. $a_1 = 2, a_2 = 1, a_{n+1} = a_n - a_{n-1}$. Each term is defined in term of the two preceding terms.

$a_3 = a_2 - a_1 = 1 - 2 = -1$. $a_4 = a_3 - a_2 = -1 - 1 = -2$. $a_5 = a_4 - a_3 = -2 - (-1) = -1$.

$a_6 = a_5 - a_4 = -1 - (-2) = 1$. The sequence is $\{2, 1, -1, -2, -1, 1, \ldots\}$.

13. $\{\frac{1}{2}, \frac{1}{4}, \frac{1}{6}, \frac{1}{8}, \frac{1}{10}, \ldots\}$. The denominator is two times the number of the term, n, so $a_n = \dfrac{1}{2n}$.

14. $\{4, -1, \frac{1}{4}, -\frac{1}{16}, \frac{1}{64}, \ldots\}$. The first term is 4 and each term is $-\frac{1}{4}$ times the preceding one, so $a_n = 4\left(-\frac{1}{4}\right)^{n-1}$.

15. $\{-3, 2, -\frac{4}{3}, \frac{8}{9}, -\frac{16}{27}, \ldots\}$. The first term is -3 and each term is $-\frac{2}{3}$ times the preceding one, so $a_n = -3\left(-\frac{2}{3}\right)^{n-1}$.

16. $\{5, 8, 11, 14, 17, \ldots\}$. Each term is larger than the preceding term by 3, so $a_n = a_1 + d(n-1) = 5 + 3(n-1) = 3n + 2$.

17. $\{\frac{1}{2}, -\frac{4}{3}, \frac{9}{4}, -\frac{16}{5}, \frac{25}{6}, \ldots\}$. The numerator of the nth term is n^2 and its denominator is $n + 1$. Including the alternating signs, we get $a_n = (-1)^{n+1}\dfrac{n^2}{n+1}$.

18. $\{1, 0, -1, 0, 1, 0, -1, 0, \ldots\}$. Two possibilities are $a_n = \sin\dfrac{n\pi}{2}$ and $a_n = \cos\dfrac{(n-1)\pi}{2}$.

19.

n	$a_n = \dfrac{3n}{1+6n}$
1	0.4286
2	0.4615
3	0.4737
4	0.4800
5	0.4839
6	0.4865
7	0.4884
8	0.4898
9	0.4909
10	0.4918

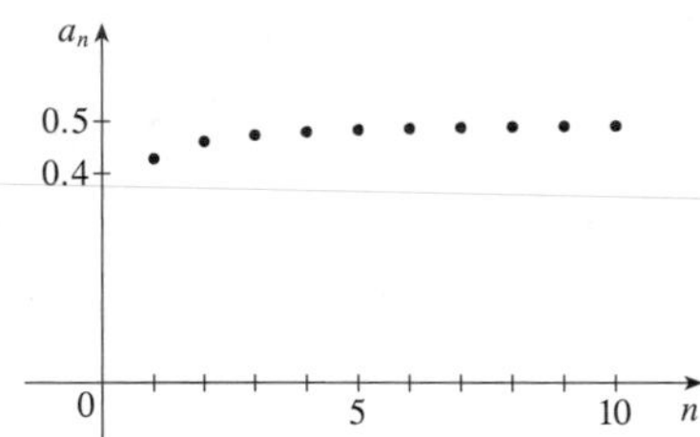

It appears that $\lim\limits_{n\to\infty} a_n = 0.5$.

$$\lim_{n\to\infty} \frac{3n}{1+6n} = \lim_{n\to\infty} \frac{(3n)/n}{(1+6n)/n} = \lim_{n\to\infty} \frac{3}{1/n + 6} = \frac{3}{6} = \frac{1}{2}$$

20.

n	$a_n = 2 + \dfrac{(-1)^n}{n}$
1	1.0000
2	2.5000
3	1.6667
4	2.2500
5	1.8000
6	2.1667
7	1.8571
8	2.1250
9	1.8889
10	2.1000

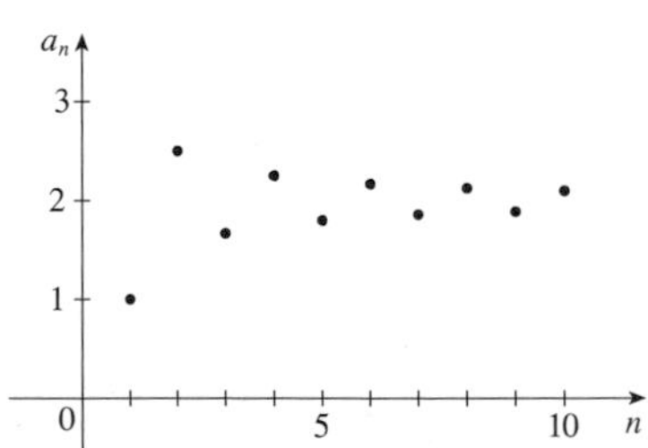

It appears that $\lim\limits_{n\to\infty} a_n = 2$.

$$\lim_{n\to\infty} \left(2 + \frac{(-1)^n}{n}\right) = \lim_{n\to\infty} 2 + \lim_{n\to\infty} \frac{(-1)^n}{n} = 2 + 0 = 2 \text{ since } \lim_{n\to\infty} \frac{1}{n} = 0$$

and by Theorem 6, $\lim\limits_{n\to\infty} \dfrac{(-1)^n}{n} = 0$.

21.

n	$a_n = 1 + \left(-\frac{1}{2}\right)^n$
1	0.5000
2	1.2500
3	0.8750
4	1.0625
5	0.9688
6	1.0156
7	0.9922
8	1.0039
9	0.9980
10	1.0010

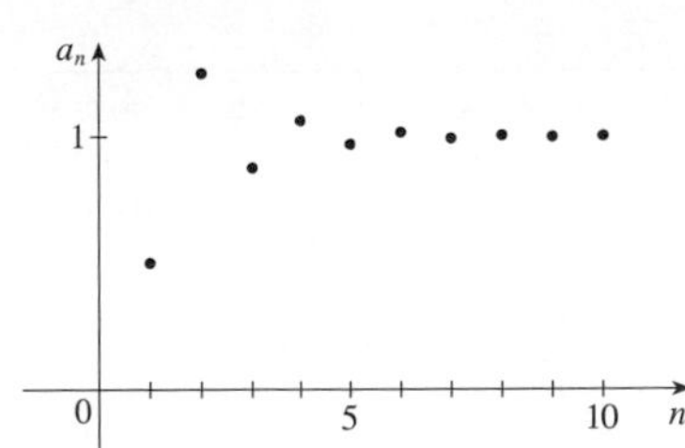

It appears that $\lim\limits_{n\to\infty} a_n = 1$.

$\lim\limits_{n\to\infty} \left(1 + \left(-\frac{1}{2}\right)^n\right) = \lim\limits_{n\to\infty} 1 + \lim\limits_{n\to\infty} \left(-\frac{1}{2}\right)^n = 1 + 0 = 1$ since $\lim\limits_{n\to\infty} \left(-\frac{1}{2}\right)^n = 0$ by (9).

22.

n	$a_n = 1 + \dfrac{10^n}{9^n}$
1	2.1111
2	2.2346
3	2.3717
4	2.5242
5	2.6935
6	2.8817
7	3.0908
8	3.3231
9	3.5812
10	3.8680

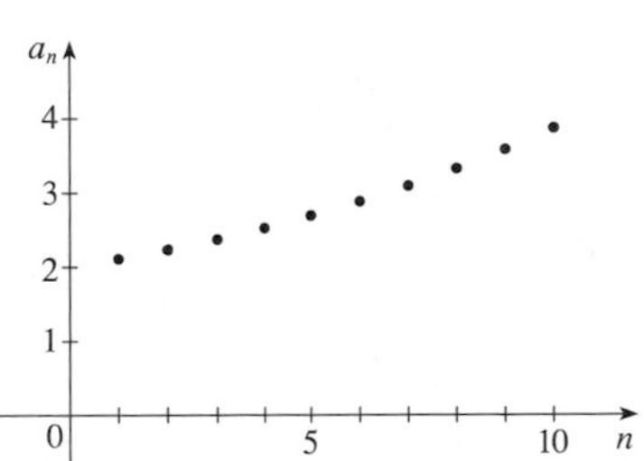

It appears that the sequence does not have a limit.

$\lim\limits_{n\to\infty} \dfrac{10^n}{9^n} = \lim\limits_{n\to\infty} \left(\dfrac{10}{9}\right)^n$, which diverges by (9) since $\frac{10}{9} > 1$.

23. $a_n = \dfrac{3+5n^2}{n+n^2} = \dfrac{(3+5n^2)/n^2}{(n+n^2)/n^2} = \dfrac{5+3/n^2}{1+1/n}$, so $a_n \to \dfrac{5+0}{1+0} = 5$ as $n \to \infty$. Converges

24. $a_n = \dfrac{3+5n^2}{1+n} = \dfrac{(3+5n^2)/n}{(1+n)/n} = \dfrac{3/n+5n}{1/n+1}$, so $a_n \to \infty$ as $n \to \infty$ since $\lim\limits_{n\to\infty} \left(\dfrac{3}{n} + 5n\right) = \infty$ and $\lim\limits_{n\to\infty} \left(\dfrac{1}{n} + 1\right) = 0 + 1 = 1$. Diverges

25. $a_n = \dfrac{n^4}{n^3-2n} = \dfrac{n^4/n^3}{(n^3-2n)/n^3} = \dfrac{n}{1-2/n^2}$, so $a_n \to \infty$ as $n \to \infty$ since $\lim\limits_{n\to\infty} n = \infty$ and $\lim\limits_{n\to\infty} \left(1 - \dfrac{2}{n^2}\right) = 1 - 0 = 1$. Diverges

26. $a_n = 2 + (0.86)^n \to 2 + 0 = 2$ as $n \to \infty$ since $\lim\limits_{n\to\infty} (0.86)^n = 0$ by (9) with $r = 0.86$. Converges

27. $a_n = 3^n 7^{-n} = \dfrac{3^n}{7^n} = \left(\dfrac{3}{7}\right)^n$, so $\lim\limits_{n\to\infty} a_n = 0$ by (9) with $r = \dfrac{3}{7}$. Converges

28. $a_n = \dfrac{3\sqrt{n}}{\sqrt{n}+2} = \dfrac{3\sqrt{n}/\sqrt{n}}{(\sqrt{n}+2)/\sqrt{n}} = \dfrac{3}{1+2/\sqrt{n}} \to \dfrac{3}{1+0} = 3$ as $n \to \infty$. Converges

29. Because the natural exponential function is continuous at 0, Theorem 7 enables us to write

$$\lim_{n\to\infty} a_n = \lim_{n\to\infty} e^{-1/\sqrt{n}} = e^{\lim_{n\to\infty}(-1/\sqrt{n})} = e^0 = 1.$$ Converges

30. $a_n = \dfrac{4^n}{1+9^n} = \dfrac{4^n/9^n}{(1+9^n)/9^n} = \dfrac{(4/9)^n}{(1/9)^n+1} \to \dfrac{0}{0+1} = 0$ as $n \to \infty$ since $\lim\limits_{n\to\infty}\left(\dfrac{4}{9}\right)^n = 0$ and $\lim\limits_{n\to\infty}\left(\dfrac{1}{9}\right)^n = 0$ by (9). Converges

31. $a_n = \sqrt{\dfrac{1+4n^2}{1+n^2}} = \sqrt{\dfrac{(1+4n^2)/n^2}{(1+n^2)/n^2}} = \sqrt{\dfrac{(1/n^2)+4}{(1/n^2)+1}} \to \sqrt{4} = 2$ as $n \to \infty$ since $\lim\limits_{n\to\infty}(1/n^2) = 0$. Converges

32. $a_n = \cos\left(\dfrac{n\pi}{n+1}\right) = \cos\left(\dfrac{n\pi/n}{(n+1)/n}\right) = \cos\left(\dfrac{\pi}{1+1/n}\right)$, so $a_n \to \cos\pi = -1$ as $n \to \infty$ since $\lim\limits_{n\to\infty} 1/n = 0$. Converges

33. $a_n = \dfrac{n^2}{\sqrt{n^3+4n}} = \dfrac{n^2/\sqrt{n^3}}{\sqrt{n^3+4n}/\sqrt{n^3}} = \dfrac{\sqrt{n}}{\sqrt{1+4/n^2}}$, so $a_n \to \infty$ as $n \to \infty$ since $\lim\limits_{n\to\infty}\sqrt{n} = \infty$ and $\lim\limits_{n\to\infty}\sqrt{1+4/n^2} = 1$. Diverges

34. If $b_n = \dfrac{2n}{n+2}$, then $\lim\limits_{n\to\infty} b_n = \lim\limits_{n\to\infty}\dfrac{(2n)/n}{(n+2)/n} = \lim\limits_{n\to\infty}\dfrac{2}{1+2/n} = \dfrac{2}{1} = 2$. Since the natural exponential function is continuous at 2, by Theorem 7, $\lim\limits_{n\to\infty} e^{2n/(n+2)} = e^{\lim_{n\to\infty} b_n} = e^2$. Converges

35. $\lim\limits_{n\to\infty}|a_n| = \lim\limits_{n\to\infty}\left|\dfrac{(-1)^n}{2\sqrt{n}}\right| = \dfrac{1}{2}\lim\limits_{n\to\infty}\dfrac{1}{n^{1/2}} = \dfrac{1}{2}(0) = 0$, so $\lim\limits_{n\to\infty} a_n = 0$ by (6). Converges

36. $\lim\limits_{n\to\infty}\dfrac{n}{n+\sqrt{n}} = \lim\limits_{n\to\infty}\dfrac{n/n}{(n+\sqrt{n})/n} = \lim\limits_{n\to\infty}\dfrac{1}{1+1/\sqrt{n}} = \dfrac{1}{1+0} = 1$. Thus, $a_n = \dfrac{(-1)^{n+1}n}{n+\sqrt{n}}$ has odd-numbered terms that approach 1 and even-numbered terms that approach -1 as $n \to \infty$, and hence, the sequence $\{a_n\}$ is divergent.

37. $a_n = \dfrac{(2n-1)!}{(2n+1)!} = \dfrac{(2n-1)!}{(2n+1)(2n)(2n-1)!} = \dfrac{1}{(2n+1)(2n)} \to 0$ as $n \to \infty$. Converges

38. $a_n = \dfrac{\ln n}{\ln 2n} = \dfrac{\ln n}{\ln 2 + \ln n} = \dfrac{1}{\frac{\ln 2}{\ln n}+1} \to \dfrac{1}{0+1} = 1$ as $n \to \infty$. Converges

39. $a_n = \sin n$. This sequence diverges since the terms don't approach any particular real number as $n \to \infty$. The terms take on values between -1 and 1. Diverges

40. $a_n = \dfrac{\tan^{-1} n}{n}$. $\lim\limits_{n\to\infty}\tan^{-1} n = \lim\limits_{x\to\infty}\tan^{-1} x = \dfrac{\pi}{2}$ by (3), so $\lim\limits_{n\to\infty} a_n = 0$. Converges

41. $a_n = n^2 e^{-n} = \dfrac{n^2}{e^n}$. Since $\lim\limits_{x\to\infty}\dfrac{x^2}{e^x} \overset{\text{H}}{=} \lim\limits_{x\to\infty}\dfrac{2x}{e^x} \overset{\text{H}}{=} \lim\limits_{x\to\infty}\dfrac{2}{e^x} = 0$, it follows from Theorem 3 that $\lim\limits_{n\to\infty} a_n = 0$. Converges

42. $a_n = \ln(n+1) - \ln n = \ln\left(\dfrac{n+1}{n}\right) = \ln\left(1+\dfrac{1}{n}\right) \to \ln(1) = 0$ as $n \to \infty$ because ln is continuous. Converges

43. $0 \le \dfrac{\cos^2 n}{2^n} \le \dfrac{1}{2^n}$ [since $0 \le \cos^2 n \le 1$], so since $\lim\limits_{n\to\infty} \dfrac{1}{2^n} = 0$, $\left\{\dfrac{\cos^2 n}{2^n}\right\}$ converges to 0 by the Squeeze Theorem.

44. $a_n = \sqrt[n]{2^{1+3n}} = (2^{1+3n})^{1/n} = (2^1 2^{3n})^{1/n} = 2^{1/n}2^3 = 8 \cdot 2^{1/n}$, so

$\lim\limits_{n\to\infty} a_n = 8 \lim\limits_{n\to\infty} 2^{1/n} = 8 \cdot 2^{\lim_{n\to\infty}(1/n)} = 8 \cdot 2^0 = 8$ by Theorem 7, since the function $f(x) = 2^x$ is continuous at 0.

Converges

45. $a_n = n\sin(1/n) = \dfrac{\sin(1/n)}{1/n}$. Since $\lim\limits_{x\to\infty} \dfrac{\sin(1/x)}{1/x} = \lim\limits_{t\to 0^+} \dfrac{\sin t}{t}$ [where $t = 1/x$] $= 1$, it follows from Theorem 3

that $\{a_n\}$ converges to 1.

46. $a_n = 2^{-n}\cos n\pi$. $0 \le \left|\dfrac{\cos n\pi}{2^n}\right| \le \dfrac{1}{2^n} = \left(\dfrac{1}{2}\right)^n$, so $\lim\limits_{n\to\infty} |a_n| = 0$ by (9), and $\lim\limits_{n\to\infty} a_n = 0$ by (6). Converges

47. $y = \left(1+\dfrac{2}{x}\right)^x \Rightarrow \ln y = x\ln\left(1+\dfrac{2}{x}\right)$, so

$$\lim_{x\to\infty} \ln y = \lim_{x\to\infty} \frac{\ln(1+2/x)}{1/x} \overset{\text{H}}{=} \lim_{x\to\infty} \frac{\left(\dfrac{1}{1+2/x}\right)\left(-\dfrac{2}{x^2}\right)}{-1/x^2} = \lim_{x\to\infty} \frac{2}{1+2/x} = 2 \Rightarrow$$

$\lim\limits_{x\to\infty}\left(1+\dfrac{2}{x}\right)^x = \lim\limits_{x\to\infty} e^{\ln y} = e^2$, so by Theorem 3, $\lim\limits_{n\to\infty}\left(1+\dfrac{2}{n}\right)^n = e^2$. Converges

48. $y = x^{1/x} \Rightarrow \ln y = \dfrac{1}{x}\ln x$, so $\lim\limits_{x\to\infty} \ln y = \lim\limits_{x\to\infty} \dfrac{\ln x}{x} \overset{\text{H}}{=} \lim\limits_{x\to\infty} \dfrac{1/x}{1} = \lim\limits_{x\to\infty} \dfrac{1}{x} = 0 \Rightarrow$

$\lim\limits_{x\to\infty} x^{1/x} = \lim\limits_{x\to\infty} e^{\ln y} = e^0 = 1$, so by Theorem 3, $\lim\limits_{n\to\infty} \sqrt[n]{n} = 1$. Converges

49. $a_n = \ln(2n^2+1) - \ln(n^2+1) = \ln\left(\dfrac{2n^2+1}{n^2+1}\right) = \ln\left(\dfrac{2+1/n^2}{1+1/n^2}\right) \to \ln 2$ as $n \to \infty$. Converges

50. $\lim\limits_{x\to\infty} \dfrac{(\ln x)^2}{x} \overset{\text{H}}{=} \lim\limits_{x\to\infty} \dfrac{2(\ln x)(1/x)}{1} = 2\lim\limits_{x\to\infty} \dfrac{\ln x}{x} \overset{\text{H}}{=} 2\lim\limits_{x\to\infty} \dfrac{1/x}{1} = 0$, so by Theorem 3, $\lim\limits_{n\to\infty} \dfrac{(\ln n)^2}{n} = 0$. Converges

51. $a_n = \arctan(\ln n)$. Let $f(x) = \arctan(\ln x)$. Then $\lim\limits_{x\to\infty} f(x) = \frac{\pi}{2}$ since $\ln x \to \infty$ as $x \to \infty$ and arctan is continuous.

Thus, $\lim\limits_{n\to\infty} a_n = \lim\limits_{n\to\infty} f(n) = \frac{\pi}{2}$. Converges

52. $a_n = n - \sqrt{n+1}\sqrt{n+3} = n - \sqrt{n^2+4n+3} = \dfrac{n - \sqrt{n^2+4n+3}}{1} \cdot \dfrac{n+\sqrt{n^2+4n+3}}{n+\sqrt{n^2+4n+3}}$

$$= \frac{n^2 - (n^2+4n+3)}{n+\sqrt{n^2+4n+3}} = \frac{-4n-3}{n+\sqrt{n^2+4n+3}} = \frac{(-4n-3)/n}{(n+\sqrt{n^2+4n+3})/n} = \frac{-4-3/n}{1+\sqrt{1+4/n+3/n^2}},$$

so $\lim\limits_{n\to\infty} a_n = \dfrac{-4-0}{1+\sqrt{1+0+0}} = \dfrac{-4}{2} = -2$. Converges

53. $\{0, 1, 0, 0, 1, 0, 0, 0, 1, \ldots\}$ diverges since the sequence takes on only two values, 0 and 1, and never stays arbitrarily close to either one (or any other value) for n sufficiently large.

54. $\left\{\frac{1}{1}, \frac{1}{3}, \frac{1}{2}, \frac{1}{4}, \frac{1}{3}, \frac{1}{5}, \frac{1}{4}, \frac{1}{6}, \ldots\right\}$. $a_{2n-1} = \dfrac{1}{n}$ and $a_{2n} = \dfrac{1}{n+2}$ for all positive integers n. $\lim\limits_{n\to\infty} a_n = 0$ since $\lim\limits_{n\to\infty} a_{2n-1} = \lim\limits_{n\to\infty} \dfrac{1}{n} = 0$ and $\lim\limits_{n\to\infty} a_{2n} = \lim\limits_{n\to\infty} \dfrac{1}{n+2} = 0$. For n sufficiently large, a_n can be made as close to 0 as we like. Converges

55. $a_n = \dfrac{n!}{2^n} = \dfrac{1}{2} \cdot \dfrac{2}{2} \cdot \dfrac{3}{2} \cdot \cdots \cdot \dfrac{(n-1)}{2} \cdot \dfrac{n}{2} \geq \dfrac{1}{2} \cdot \dfrac{n}{2}$ [for $n > 1$] $= \dfrac{n}{4} \to \infty$ as $n \to \infty$, so $\{a_n\}$ diverges.

56. $0 < |a_n| = \dfrac{3^n}{n!} = \dfrac{3}{1} \cdot \dfrac{3}{2} \cdot \dfrac{3}{3} \cdot \cdots \cdot \dfrac{3}{(n-1)} \cdot \dfrac{3}{n} \leq \dfrac{3}{1} \cdot \dfrac{3}{2} \cdot \dfrac{3}{n}$ [for $n > 2$] $= \dfrac{27}{2n} \to 0$ as $n \to \infty$, so by the Squeeze Theorem and Theorem 6, $\{(-3)^n/n!\}$ converges to 0.

57. From the graph, it appears that the sequence $\{a_n\} = \left\{(-1)^n \dfrac{n}{n+1}\right\}$ is divergent, since it oscillates between 1 and -1 (approximately). To prove this, suppose that $\{a_n\}$ converges to L. If $b_n = \dfrac{n}{n+1}$, then $\{b_n\}$ converges to 1, and $\lim\limits_{n\to\infty} \dfrac{a_n}{b_n} = \dfrac{L}{1} = L$. But $\dfrac{a_n}{b_n} = (-1)^n$, so $\lim\limits_{n\to\infty} \dfrac{a_n}{b_n}$ does not exist. This contradiction shows that $\{a_n\}$ diverges.

58. From the graph, it appears that the sequence converges to 0.

$|a_n| = \left|\dfrac{\sin n}{n}\right| = \dfrac{|\sin n|}{|n|} \leq \dfrac{1}{n}$, so $\lim\limits_{n\to\infty} |a_n| = 0$. By (6), it follows that $\lim\limits_{n\to\infty} a_n = 0$.

59. From the graph, it appears that the sequence converges to a number between 0.7 and 0.8.

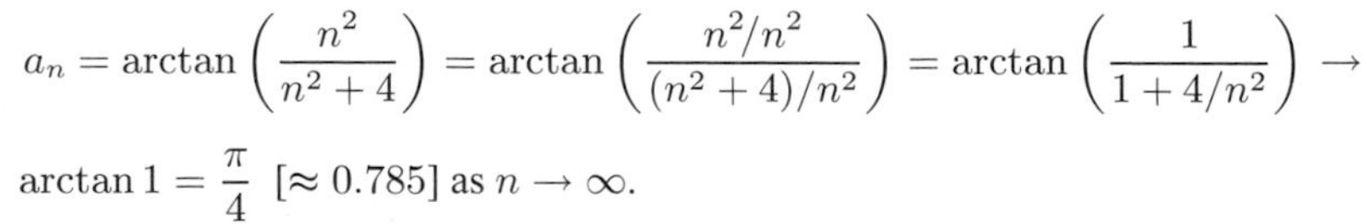

$a_n = \arctan\left(\dfrac{n^2}{n^2+4}\right) = \arctan\left(\dfrac{n^2/n^2}{(n^2+4)/n^2}\right) = \arctan\left(\dfrac{1}{1+4/n^2}\right) \to$ $\arctan 1 = \dfrac{\pi}{4}$ [≈ 0.785] as $n \to \infty$.

60.

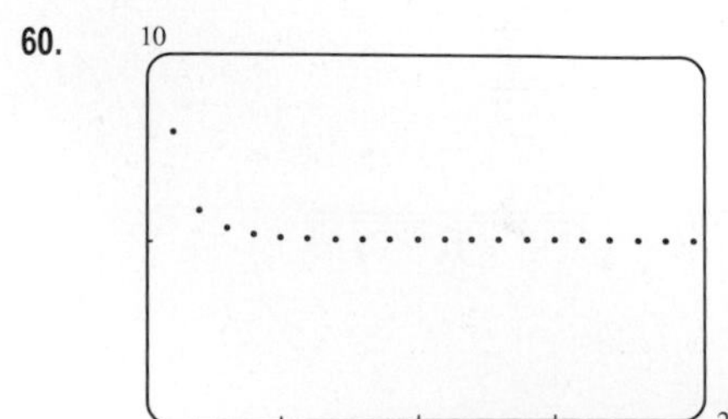

From the graph, it appears that the sequence converges to 5.

$$5 = \sqrt[n]{5^n} \leq \sqrt[n]{3^n + 5^n} \leq \sqrt[n]{5^n + 5^n} = \sqrt[n]{2}\,\sqrt[n]{5^n}$$
$$= \sqrt[n]{2} \cdot 5 \to 5 \text{ as } n \to \infty \quad \left[\lim_{n\to\infty} 2^{1/n} = 2^0 = 1\right]$$

Hence, $a_n \to 5$ by the Squeeze Theorem.

[continued]

Alternate solution: Let $y = (3^x + 5^x)^{1/x}$. Then

$$\lim_{x\to\infty} \ln y = \lim_{x\to\infty} \frac{\ln(3^x + 5^x)}{x} \overset{\text{H}}{=} \lim_{x\to\infty} \frac{3^x \ln 3 + 5^x \ln 5}{3^x + 5^x} = \lim_{x\to\infty} \frac{\left(\frac{3}{5}\right)^x \ln 3 + \ln 5}{\left(\frac{3}{5}\right)^x + 1} = \ln 5,$$

so $\lim\limits_{x\to\infty} y = e^{\ln 5} = 5$, and so $\left\{ \sqrt[n]{3^n + 5^n} \right\}$ converges to 5.

61. 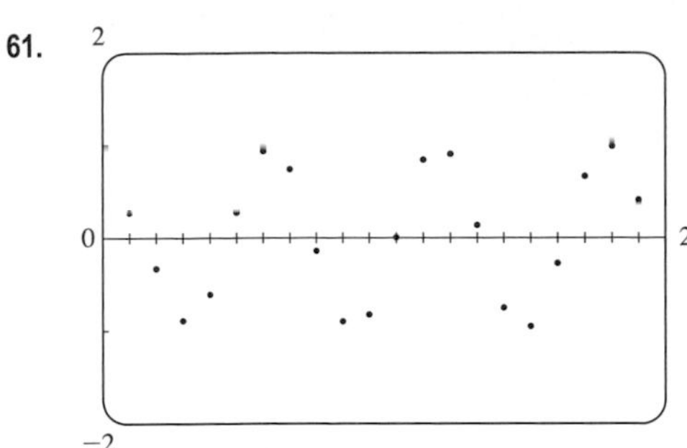

From the graph, it appears that the sequence $\{a_n\} = \left\{ \dfrac{n^2 \cos n}{1 + n^2} \right\}$ is divergent, since it oscillates between 1 and -1 (approximately). To prove this, suppose that $\{a_n\}$ converges to L. If $b_n = \dfrac{n^2}{1 + n^2}$, then $\{b_n\}$ converges to 1, and $\lim\limits_{n\to\infty} \dfrac{a_n}{b_n} = \dfrac{L}{1} = L$. But $\dfrac{a_n}{b_n} = \cos n$, so $\lim\limits_{n\to\infty} \dfrac{a_n}{b_n}$ does not exist. This contradiction shows that $\{a_n\}$ diverges.

62. 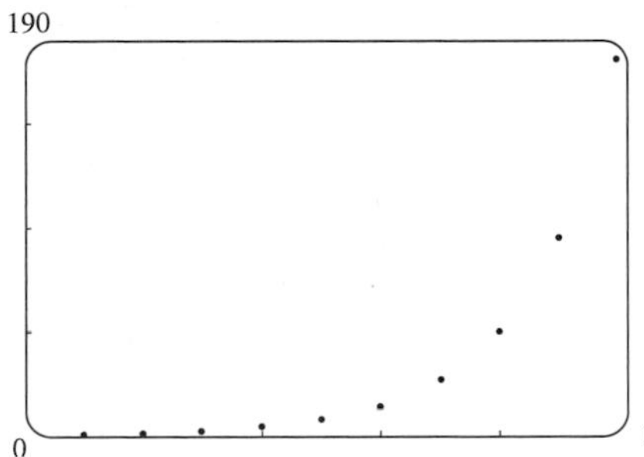

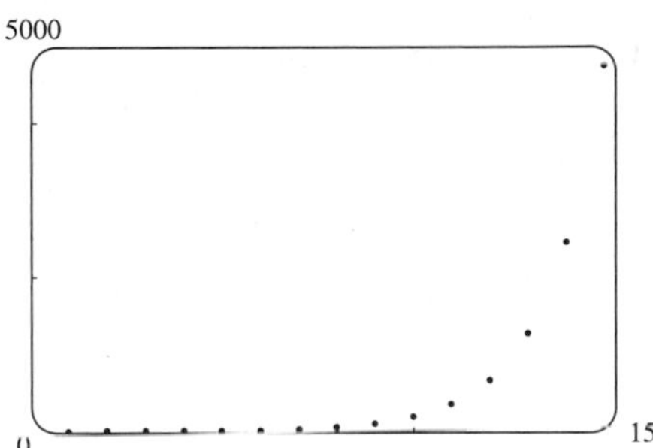

From the graphs, it seems that the sequence diverges. $a_n = \dfrac{1 \cdot 3 \cdot 5 \cdot \cdots \cdot (2n-1)}{n!}$. We first prove by induction that $a_n \geq \left(\dfrac{3}{2}\right)^{n-1}$ for all n. This is clearly true for $n = 1$, so let $P(n)$ be the statement that the above is true for n. We must show it is then true for $n + 1$. $a_{n+1} = a_n \cdot \dfrac{2n+1}{n+1} \geq \left(\dfrac{3}{2}\right)^{n-1} \cdot \dfrac{2n+1}{n+1}$ (induction hypothesis). But $\dfrac{2n+1}{n+1} \geq \dfrac{3}{2}$ [since $2(2n+1) \geq 3(n+1) \;\Leftrightarrow\; 4n + 2 \geq 3n + 3 \;\Leftrightarrow\; n \geq 1$], and so we get that $a_{n+1} \geq \left(\frac{3}{2}\right)^{n-1} \cdot \frac{3}{2} = \left(\frac{3}{2}\right)^n$ which is $P(n+1)$. Thus, we have proved our first assertion, so since $\left\{ \left(\frac{3}{2}\right)^{n-1} \right\}$ diverges [by (9)], so does the given sequence $\{a_n\}$.

63. 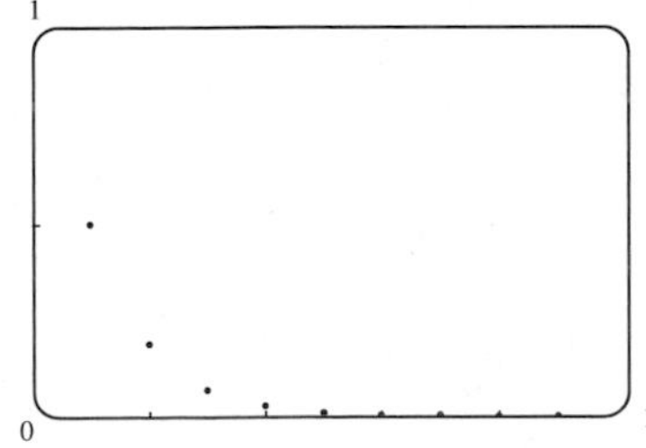

From the graph, it appears that the sequence approaches 0.

$$0 < a_n = \frac{1 \cdot 3 \cdot 5 \cdot \cdots \cdot (2n-1)}{(2n)^n} = \frac{1}{2n} \cdot \frac{3}{2n} \cdot \frac{5}{2n} \cdot \cdots \cdot \frac{2n-1}{2n}$$

$$\leq \frac{1}{2n} \cdot (1) \cdot (1) \cdot \cdots \cdot (1) = \frac{1}{2n} \to 0 \text{ as } n \to \infty$$

So by the Squeeze Theorem, $\left\{ \dfrac{1 \cdot 3 \cdot 5 \cdot \cdots \cdot (2n-1)}{(2n)^n} \right\}$ converges to 0.

64. (a) $a_1 = 1, a_{n+1} = 4 - a_n$ for $n \geq 1$. $a_1 = 1, a_2 = 4 - a_1 = 4 - 1 = 3, a_3 = 4 - a_2 = 4 - 3 = 1$, $a_4 = 4 - a_3 = 4 - 1 = 3, a_5 = 4 - a_4 = 4 - 3 = 1$. Since the terms of the sequence alternate between 1 and 3, the sequence is divergent.

(b) $a_1 = 2, a_2 = 4 - a_1 = 4 - 2 = 2, a_3 = 4 - a_2 = 4 - 2 = 2$. Since all of the terms are 2, $\lim\limits_{n\to\infty} a_n = 2$ and hence, the sequence is convergent.

65. (a) $a_n = 1000(1.06)^n \Rightarrow a_1 = 1060, a_2 = 1123.60, a_3 = 1191.02, a_4 = 1262.48$, and $a_5 = 1338.23$.

(b) $\lim\limits_{n\to\infty} a_n = 1000 \lim\limits_{n\to\infty} (1.06)^n$, so the sequence diverges by (9) with $r = 1.06 > 1$.

66. (a) Substitute 1 to 6 for n in $I_n = 100\left(\dfrac{1.0025^n - 1}{0.0025} - n\right)$ to get $I_1 = \$0, I_2 = \$0.25, I_3 = \$0.75, I_4 = \$1.50$, $I_5 = \$2.51$, and $I_6 = \$3.76$.

(b) For two years, use $2 \cdot 12 = 24$ for n to get $\$70.28$.

67. (a) We are given that the initial population is 5000, so $P_0 = 5000$. The number of catfish increases by 8% per month and is decreased by 300 per month, so $P_1 = P_0 + 8\%P_0 - 300 = 1.08P_0 - 300$, $P_2 = 1.08P_1 - 300$, and so on. Thus, $P_n = 1.08P_{n-1} - 300$.

(b) Using the recursive formula with $P_0 = 5000$, we get $P_1 = 5100$, $P_2 = 5208$, $P_3 = 5325$ (rounding any portion of a catfish), $P_4 = 5451$, $P_5 = 5587$, and $P_6 = 5734$, which is the number of catfish in the pond after six months.

68. $a_{n+1} = \begin{cases} \frac{1}{2}a_n & \text{if } a_n \text{ is an even number} \\ 3a_n + 1 & \text{if } a_n \text{ is an odd number} \end{cases}$ When $a_1 = 11$, the first 40 terms are 11, 34, 17, 52, 26, 13, 40, 20, 10, 5, 16, 8, 4, 2, 1, 4, 2, 1, 4, 2, 1, 4, 2, 1, 4, 2, 1, 4, 2, 1, 4, 2, 1, 4, 2, 1, 4, 2, 1, 4. When $a_1 = 25$, the first 40 terms are 25, 76, 38, 19, 58, 29, 88, 44, 22, 11, 34, 17, 52, 26, 13, 40, 20, 10, 5, 16, 8, 4, 2, 1, 4, 2, 1, 4, 2, 1, 4, 2, 1, 4, 2, 1, 4, 2, 1, 4. The famous Collatz conjecture is that this sequence always reaches 1, regardless of the starting point a_1.

69. If $|r| \geq 1$, then $\{r^n\}$ diverges by (9), so $\{nr^n\}$ diverges also, since $|nr^n| = n\,|r^n| \geq |r^n|$. If $|r| < 1$ then

$\lim\limits_{x\to\infty} xr^x = \lim\limits_{x\to\infty} \dfrac{x}{r^{-x}} \overset{\text{H}}{=} \lim\limits_{x\to\infty} \dfrac{1}{(-\ln r)\,r^{-x}} = \lim\limits_{x\to\infty} \dfrac{r^x}{-\ln r} = 0$, so $\lim\limits_{n\to\infty} nr^n = 0$, and hence $\{nr^n\}$ converges whenever $|r| < 1$.

70. (a) Let $\lim\limits_{n\to\infty} a_n = L$. By Definition 2, this means that for every $\varepsilon > 0$ there is an integer N such that $|a_n - L| < \varepsilon$ whenever $n > N$. Thus, $|a_{n+1} - L| < \varepsilon$ whenever $n + 1 > N \Leftrightarrow n > N - 1$. It follows that $\lim\limits_{n\to\infty} a_{n+1} = L$ and so $\lim\limits_{n\to\infty} a_n = \lim\limits_{n\to\infty} a_{n+1}$.

(b) If $L = \lim\limits_{n\to\infty} a_n$ then $\lim\limits_{n\to\infty} a_{n+1} = L$ also, so L must satisfy $L = 1/(1 + L) \Rightarrow L^2 + L - 1 = 0 \Rightarrow L = \frac{-1+\sqrt{5}}{2}$ (since L has to be nonnegative if it exists).

71. Since $\{a_n\}$ is a decreasing sequence, $a_n > a_{n+1}$ for all $n \geq 1$. Because all of its terms lie between 5 and 8, $\{a_n\}$ is a bounded sequence. By the Monotonic Sequence Theorem, $\{a_n\}$ is convergent; that is, $\{a_n\}$ has a limit L. L must be less than 8 since $\{a_n\}$ is decreasing, so $5 \leq L < 8$.

72. Since $\{a_n\} = \{\cos n\} \approx \{0.54, -0.42, -0.99, -0.65, 0.28, \ldots\}$, the sequence is not monotonic. The sequence is bounded since $-1 \leq \cos n \leq 1$ for all n.

73. $a_n = \dfrac{1}{2n+3}$ is decreasing since $a_{n+1} = \dfrac{1}{2(n+1)+3} = \dfrac{1}{2n+5} < \dfrac{1}{2n+3} = a_n$ for each $n \geq 1$. The sequence is bounded since $0 < a_n \leq \frac{1}{5}$ for all $n \geq 1$. Note that $a_1 = \frac{1}{5}$.

74. $a_n > a_{n+1} \quad \Leftrightarrow \quad \dfrac{1-n}{2+n} > \dfrac{1-(n+1)}{2+(n+1)} \quad \Leftrightarrow \quad \dfrac{1-n}{2+n} > \dfrac{-n}{n+3} \quad \Leftrightarrow \quad -n^2 - 2n + 3 > -n^2 - 2n \quad \Leftrightarrow \quad 3 > 0$, which is true for all $n \geq 1$, so $\{a_n\}$ is decreasing. Since $a_1 = 0$ and $\displaystyle\lim_{n\to\infty} \frac{1-n}{2+n} = \lim_{n\to\infty} \frac{1/n - 1}{2/n + 1} = -1$, the sequence is bounded $(-1 < a_n \leq 0)$.

75. The terms of $a_n = n(-1)^n$ alternate in sign, so the sequence is not monotonic. The first five terms are -1, 2, -3, 4, and -5. Since $\displaystyle\lim_{n\to\infty} |a_n| = \lim_{n\to\infty} n = \infty$, the sequence is not bounded.

76. Since $\{a_n\} = \left\{2 + \dfrac{(-1)^n}{n}\right\} = \{1, 2\frac{1}{2}, 1\frac{2}{3}, \ldots\}$, the sequence is not monotonic. The sequence is bounded since $1 \leq a_n \leq \frac{5}{2}$ for all n.

77. $a_n = 3 - 2ne^{-n}$. Let $f(x) = 3 - 2xe^{-x}$. Then $f'(x) = 0 - 2[x(-e^{-x}) + e^{-x}] = 2e^{-x}(x-1)$, which is positive for $x > 1$, so f is increasing on $(1, \infty)$. It follows that the sequence $\{a_n\} = \{f(n)\}$ is increasing. The sequence is bounded below by $a_1 = 3 - 2e^{-1} \approx 2.26$ and above by 3, so the sequence is bounded.

78. $a_n = n^3 - 3n + 3$. Let $f(x) = x^3 - 3x + 3$. Then $f'(x) = 3x^2 - 3 = 3(x^2 - 1)$, which is positive for $x > 1$, so f is increasing on $(1, \infty)$. It follows that the sequence $\{a_n\} = \{f(n)\}$ is increasing. The sequence is bounded below by $a_1 = 1$, but is not bounded above, so it is not bounded.

79. For $\left\{\sqrt{2}, \sqrt{2\sqrt{2}}, \sqrt{2\sqrt{2\sqrt{2}}}, \ldots\right\}$, $a_1 = 2^{1/2}$, $a_2 = 2^{3/4}$, $a_3 = 2^{7/8}$, …, so $a_n = 2^{(2^n-1)/2^n} = 2^{1-(1/2^n)}$.

$\displaystyle\lim_{n\to\infty} a_n = \lim_{n\to\infty} 2^{1-(1/2^n)} = 2^1 = 2.$

Alternate solution: Let $L = \displaystyle\lim_{n\to\infty} a_n$. (We could show the limit exists by showing that $\{a_n\}$ is bounded and increasing.) Then L must satisfy $L = \sqrt{2 \cdot L} \quad \Rightarrow \quad L^2 = 2L \quad \Rightarrow \quad L(L-2) = 0$. $L \neq 0$ since the sequence increases, so $L = 2$.

80. (a) Let P_n be the statement that $a_{n+1} \geq a_n$ and $a_n \leq 3$. P_1 is obviously true. We will assume that P_n is true and then show that as a consequence P_{n+1} must also be true. $a_{n+2} \geq a_{n+1} \quad \Leftrightarrow \quad \sqrt{2 + a_{n+1}} \geq \sqrt{2 + a_n} \quad \Leftrightarrow$ $2 + a_{n+1} \geq 2 + a_n \quad \Leftrightarrow \quad a_{n+1} \geq a_n$, which is the induction hypothesis. $a_{n+1} \leq 3 \quad \Leftrightarrow \quad \sqrt{2 + a_n} \leq 3 \quad \Leftrightarrow$

$2 + a_n \le 9 \;\Leftrightarrow\; a_n \le 7$, which is certainly true because we are assuming that $a_n \le 3$. So P_n is true for all n, and so $a_1 \le a_n \le 3$ (showing that the sequence is bounded), and hence by the Monotonic Sequence Theorem, $\lim\limits_{n\to\infty} a_n$ exists.

(b) If $L = \lim\limits_{n\to\infty} a_n$, then $\lim\limits_{n\to\infty} a_{n+1} = L$ also, so $L = \sqrt{2+L} \;\Rightarrow\; L^2 = 2 + L \;\Leftrightarrow\; L^2 - L - 2 = 0 \;\Leftrightarrow$ $(L+1)(L-2) = 0 \;\Leftrightarrow\; L = 2$ [since L can't be negative].

81. $a_1 = 1,\, a_{n+1} = 3 - \dfrac{1}{a_n}$. We show by induction that $\{a_n\}$ is increasing and bounded above by 3. Let P_n be the proposition that $a_{n+1} > a_n$ and $0 < a_n < 3$. Clearly P_1 is true. Assume that P_n is true. Then $a_{n+1} > a_n \;\Rightarrow\; \dfrac{1}{a_{n+1}} < \dfrac{1}{a_n} \;\Rightarrow$ $-\dfrac{1}{a_{n+1}} > -\dfrac{1}{a_n}$. Now $a_{n+2} = 3 - \dfrac{1}{a_{n+1}} > 3 - \dfrac{1}{a_n} = a_{n+1} \;\Leftrightarrow\; P_{n+1}$. This proves that $\{a_n\}$ is increasing and bounded above by 3, so $1 = a_1 < a_n < 3$, that is, $\{a_n\}$ is bounded, and hence convergent by the Monotonic Sequence Theorem. If $L = \lim\limits_{n\to\infty} a_n$, then $\lim\limits_{n\to\infty} a_{n+1} = L$ also, so L must satisfy $L = 3 - 1/L \;\Rightarrow\; L^2 - 3L + 1 = 0 \;\Rightarrow\; L = \frac{3 \pm \sqrt{5}}{2}$. But $L > 1$, so $L = \frac{3+\sqrt{5}}{2}$.

82. $a_1 = 2,\, a_{n+1} = \dfrac{1}{3 - a_n}$. We use induction. Let P_n be the statement that $0 < a_{n+1} \le a_n \le 2$. Clearly P_1 is true, since $a_2 = 1/(3-2) = 1$. Now assume that P_n is true. Then $a_{n+1} \le a_n \;\Rightarrow\; -a_{n+1} \ge -a_n \;\Rightarrow\; 3 - a_{n+1} \ge 3 - a_n \;\Rightarrow$ $a_{n+2} = \dfrac{1}{3 - a_{n+1}} \le \dfrac{1}{3 - a_n} = a_{n+1}$. Also $a_{n+2} > 0$ [since $3 - a_{n+1}$ is positive] and $a_{n+1} \le 2$ by the induction hypothesis, so P_{n+1} is true. To find the limit, we use the fact that $\lim\limits_{n\to\infty} a_n = \lim\limits_{n\to\infty} a_{n+1} \;\Rightarrow\; L = \frac{1}{3-L} \;\Rightarrow$ $L^2 - 3L + 1 = 0 \;\Rightarrow\; L = \frac{3 \pm \sqrt{5}}{2}$. But $L \le 2$, so we must have $L = \frac{3 - \sqrt{5}}{2}$.

83. (a) Let a_n be the number of rabbit pairs in the nth month. Clearly $a_1 = 1 = a_2$. In the nth month, each pair that is 2 or more months old (that is, a_{n-2} pairs) will produce a new pair to add to the a_{n-1} pairs already present. Thus, $a_n = a_{n-1} + a_{n-2}$, so that $\{a_n\} = \{f_n\}$, the Fibonacci sequence.

(b) $a_n = \dfrac{f_{n+1}}{f_n} \;\Rightarrow\; a_{n-1} = \dfrac{f_n}{f_{n-1}} = \dfrac{f_{n-1} + f_{n-2}}{f_{n-1}} = 1 + \dfrac{f_{n-2}}{f_{n-1}} = 1 + \dfrac{1}{f_{n-1}/f_{n-2}} = 1 + \dfrac{1}{a_{n-2}}$. If $L = \lim\limits_{n\to\infty} a_n$, then $L = \lim\limits_{n\to\infty} a_{n-1}$ and $L = \lim\limits_{n\to\infty} a_{n-2}$, so L must satisfy $L = 1 + \dfrac{1}{L} \;\Rightarrow\; L^2 - L - 1 = 0 \;\Rightarrow\; L = \frac{1+\sqrt{5}}{2}$ [since L must be positive].

84. (a) If f is continuous, then $f(L) = f\left(\lim\limits_{n\to\infty} a_n\right) = \lim\limits_{n\to\infty} f(a_n) = \lim\limits_{n\to\infty} a_{n+1} = \lim\limits_{n\to\infty} a_n = L$ by Exercise 70(a).

(b) By repeatedly pressing the cosine key on the calculator (that is, taking cosine of the previous answer) until the displayed value stabilizes, we see that $L \approx 0.73909$.

85. (a)

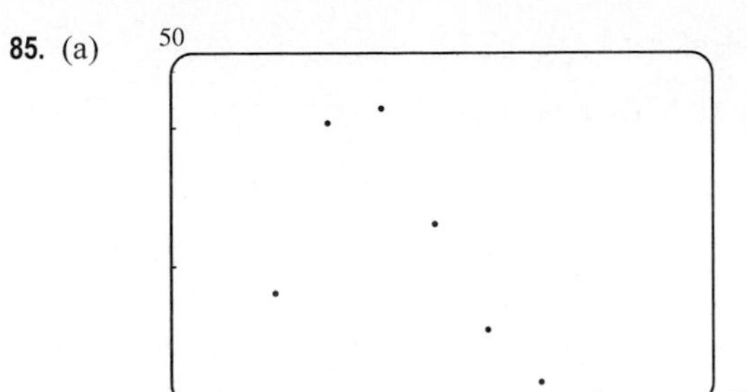

From the graph, it appears that the sequence $\left\{\frac{n^5}{n!}\right\}$ converges to 0, that is, $\lim_{n\to\infty}\frac{n^5}{n!}=0$.

(b)

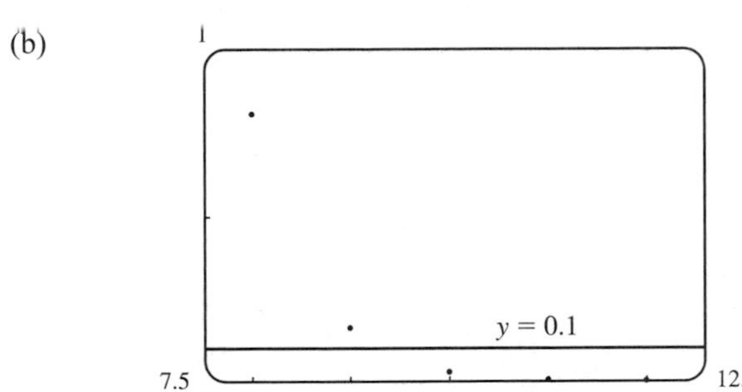

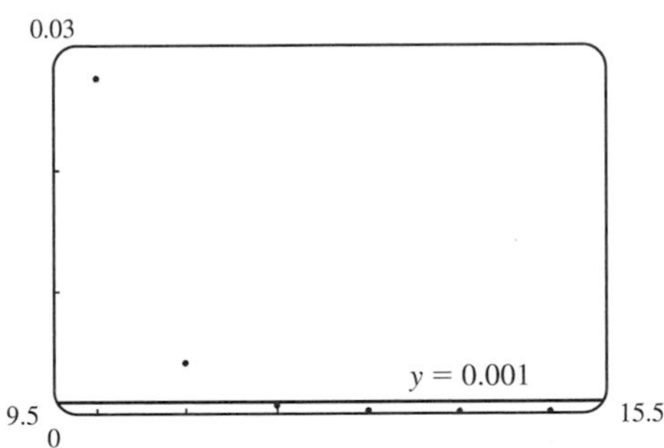

From the first graph, it seems that the smallest possible value of N corresponding to $\varepsilon = 0.1$ is 9, since $n^5/n! < 0.1$ whenever $n \geq 10$, but $9^5/9! > 0.1$. From the second graph, it seems that for $\varepsilon = 0.001$, the smallest possible value for N is 11 since $n^5/n! < 0.001$ whenever $n \geq 12$.

86. Let $\varepsilon > 0$ and let N be any positive integer larger than $\ln(\varepsilon)/\ln|r|$. If $n > N$, then $n > \ln(\varepsilon)/\ln|r| \Rightarrow n\ln|r| < \ln\varepsilon$ [since $|r| < 1 \Rightarrow \ln|r| < 0$] $\Rightarrow \ln(|r|^n) < \ln\varepsilon \Rightarrow |r|^n < \varepsilon \Rightarrow |r^n - 0| < \varepsilon$, and so by Definition 2, $\lim_{n\to\infty} r^n = 0$.

87. Theorem 6: If $\lim_{n\to\infty}|a_n| = 0$ then $\lim_{n\to\infty} -|a_n| = 0$, and since $-|a_n| \leq a_n \leq |a_n|$, we have that $\lim_{n\to\infty} a_n = 0$ by the Squeeze Theorem.

88. Theorem 7: If $\lim_{n\to\infty} a_n = L$ and the function f is continuous at L, then $\lim_{n\to\infty} f(a_n) = f(L)$.

Proof: We must show that, given a number $\varepsilon > 0$, there is an integer N such that $|f(a_n) - f(L)| < \varepsilon$ whenever $n > N$. Suppose $\varepsilon > 0$. Since f is continuous at L, there is a number $\delta > 0$ such that $|f(x) - f(L)| < \varepsilon$ if $|x - L| < \delta$. Since $\lim_{n\to\infty} a_n = L$, there is an integer N such that $|a_n - L| < \delta$ if $n > N$. Suppose $n > N$. Then $0 < |a_n - L| < \delta$, so $|f(a_n) - f(L)| < \varepsilon$.

89. To Prove: If $\lim_{n\to\infty} a_n = 0$ and $\{b_n\}$ is bounded, then $\lim_{n\to\infty}(a_n b_n) = 0$.

Proof: Since $\{b_n\}$ is bounded, there is a positive number M such that $|b_n| \leq M$ and hence, $|a_n|\,|b_n| \leq |a_n|\,M$ for all $n \geq 1$. Let $\varepsilon > 0$ be given. Since $\lim_{n\to\infty} a_n = 0$, there is an integer N such that $|a_n - 0| < \frac{\varepsilon}{M}$ if $n > N$. Then $|a_n b_n - 0| = |a_n b_n| = |a_n|\,|b_n| \leq |a_n|\,M = |a_n - 0|\,M < \frac{\varepsilon}{M}\cdot M = \varepsilon$ for all $n > N$. Since ε was arbitrary, $\lim_{n\to\infty}(a_n b_n) = 0$.

90. (a) $\dfrac{b^{n+1}-a^{n+1}}{b-a} = b^n + b^{n-1}a + b^{n-2}a^2 + b^{n-3}a^3 + \cdots + ba^{n-1} + a^n$

$< b^n + b^{n-1}b + b^{n-2}b^2 + b^{n-3}b^3 + \cdots + bb^{n-1} + b^n = (n+1)b^n$

(b) Since $b-a>0$, we have $b^{n+1}-a^{n+1} < (n+1)b^n(b-a) \;\Rightarrow\; b^{n+1}-(n+1)b^n(b-a) < a^{n+1} \;\Rightarrow$ $b^n[(n+1)a - nb] < a^{n+1}$.

(c) With this substitution, $(n+1)a - nb = 1$, and so $b^n = \left(1+\dfrac{1}{n}\right)^n < a^{n+1} = \left(1+\dfrac{1}{n+1}\right)^{n+1}$.

(d) With this substitution, we get $\left(1+\dfrac{1}{2n}\right)^n\left(\dfrac{1}{2}\right) < 1 \;\Rightarrow\; \left(1+\dfrac{1}{2n}\right)^n < 2 \;\Rightarrow\; \left(1+\dfrac{1}{2n}\right)^{2n} < 4$.

(e) $a_n < a_{2n}$ since $\{a_n\}$ is increasing, so $a_n < a_{2n} < 4$.

(f) Since $\{a_n\}$ is increasing and bounded above by 4, $a_1 \le a_n \le 4$, and so $\{a_n\}$ is bounded and monotonic, and hence has a limit by the Monotonic Sequence Theorem.

91. (a) First we show that $a > a_1 > b_1 > b$.

$a_1 - b_1 = \frac{a+b}{2} - \sqrt{ab} = \frac{1}{2}\left(a - 2\sqrt{ab} + b\right) = \frac{1}{2}\left(\sqrt{a}-\sqrt{b}\right)^2 > 0$ [since $a > b$] $\Rightarrow$ $a_1 > b_1$. Also $a - a_1 = a - \frac{1}{2}(a+b) = \frac{1}{2}(a-b) > 0$ and $b - b_1 = b - \sqrt{ab} = \sqrt{b}\left(\sqrt{b}-\sqrt{a}\right) < 0$, so $a > a_1 > b_1 > b$. In the same way we can show that $a_1 > a_2 > b_2 > b_1$ and so the given assertion is true for $n=1$. Suppose it is true for $n=k$, that is, $a_k > a_{k+1} > b_{k+1} > b_k$. Then

$$a_{k+2} - b_{k+2} = \tfrac{1}{2}(a_{k+1}+b_{k+1}) - \sqrt{a_{k+1}b_{k+1}} = \tfrac{1}{2}\left(a_{k+1} - 2\sqrt{a_{k+1}b_{k+1}} + b_{k+1}\right) = \tfrac{1}{2}\left(\sqrt{a_{k+1}} - \sqrt{b_{k+1}}\right)^2 > 0,$$

$$a_{k+1} - a_{k+2} = a_{k+1} - \tfrac{1}{2}(a_{k+1}+b_{k+1}) = \tfrac{1}{2}(a_{k+1}-b_{k+1}) > 0, \text{ and}$$

$$b_{k+1} - b_{k+2} = b_{k+1} - \sqrt{a_{k+1}b_{k+1}} = \sqrt{b_{k+1}}\left(\sqrt{b_{k+1}} - \sqrt{a_{k+1}}\right) < 0 \quad\Rightarrow\quad a_{k+1} > a_{k+2} > b_{k+2} > b_{k+1},$$

so the assertion is true for $n = k+1$. Thus, it is true for all n by mathematical induction.

(b) From part (a) we have $a > a_n > a_{n+1} > b_{n+1} > b_n > b$, which shows that both sequences, $\{a_n\}$ and $\{b_n\}$, are monotonic and bounded. So they are both convergent by the Monotonic Sequence Theorem.

(c) Let $\lim\limits_{n\to\infty} a_n = \alpha$ and $\lim\limits_{n\to\infty} b_n = \beta$. Then $\lim\limits_{n\to\infty} a_{n+1} = \lim\limits_{n\to\infty} \dfrac{a_n+b_n}{2} \;\Rightarrow\; \alpha = \dfrac{\alpha+\beta}{2} \;\Rightarrow$ $2\alpha = \alpha + \beta \;\Rightarrow\; \alpha = \beta$.

92. (a) Let $\varepsilon > 0$. Since $\lim\limits_{n\to\infty} a_{2n} = L$, there exists N_1 such that $|a_{2n} - L| < \varepsilon$ for $n > N_1$. Since $\lim\limits_{n\to\infty} a_{2n+1} = L$, there exists N_2 such that $|a_{2n+1} - L| < \varepsilon$ for $n > N_2$. Let $N = \max\{2N_1, 2N_2+1\}$ and let $n > N$. If n is even, then $n = 2m$ where $m > N_1$, so $|a_n - L| = |a_{2m} - L| < \varepsilon$. If n is odd, then $n = 2m+1$, where $m > N_2$, so $|a_n - L| = |a_{2m+1} - L| < \varepsilon$. Therefore $\lim\limits_{n\to\infty} a_n = L$.

(b) $a_1 = 1$, $a_2 = 1 + \frac{1}{1+1} = \frac{3}{2} = 1.5$, $a_3 = 1 + \frac{1}{5/2} = \frac{7}{5} = 1.4$, $a_4 = 1 + \frac{1}{12/5} = \frac{17}{12} = 1.41\overline{6}$,

$a_5 = 1 + \frac{1}{29/12} = \frac{41}{29} \approx 1.413793$, $a_6 = 1 + \frac{1}{70/29} = \frac{99}{70} \approx 1.414286$, $a_7 = 1 + \frac{1}{169/70} = \frac{239}{169} \approx 1.414201$,

$a_8 = 1 + \frac{1}{408/169} = \frac{577}{408} \approx 1.414216$. Notice that $a_1 < a_3 < a_5 < a_7$ and $a_2 > a_4 > a_6 > a_8$. It appears that the odd terms are increasing and the even terms are decreasing. Let's prove that $a_{2n-2} > a_{2n}$ and $a_{2n-1} < a_{2n+1}$ by mathematical induction. Suppose that $a_{2k-2} > a_{2k}$. Then $1 + a_{2k-2} > 1 + a_{2k} \Rightarrow \dfrac{1}{1+a_{2k-2}} < \dfrac{1}{1+a_{2k}} \Rightarrow$

$1 + \dfrac{1}{1+a_{2k-2}} < 1 + \dfrac{1}{1+a_{2k}} \Rightarrow a_{2k-1} < a_{2k+1} \Rightarrow 1 + a_{2k-1} < 1 + a_{2k+1} \Rightarrow$

$\dfrac{1}{1+a_{2k-1}} > \dfrac{1}{1+a_{2k+1}} \Rightarrow 1 + \dfrac{1}{1+a_{2k-1}} > 1 + \dfrac{1}{1+a_{2k+1}} \Rightarrow a_{2k} > a_{2k+2}$. We have thus shown, by induction, that the odd terms are increasing and the even terms are decreasing. Also all terms lie between 1 and 2, so both $\{a_n\}$ and $\{b_n\}$ are bounded monotonic sequences and are therefore convergent by the Monotonic Sequence Theorem. Let $\lim\limits_{n\to\infty} a_{2n} = L$. Then $\lim\limits_{n\to\infty} a_{2n+2} = L$ also. We have

$$a_{n+2} = 1 + \frac{1}{1+1+1/(1+a_n)} = 1 + \frac{1}{(3+2a_n)/(1+a_n)} = \frac{4+3a_n}{3+2a_n}$$

so $a_{2n+2} = \dfrac{4+3a_{2n}}{3+2a_{2n}}$. Taking limits of both sides, we get $L = \dfrac{4+3L}{3+2L} \Rightarrow 3L + 2L^2 = 4 + 3L \Rightarrow L^2 = 2 \Rightarrow$ $L = \sqrt{2}$ [since $L > 0$]. Thus, $\lim\limits_{n\to\infty} a_{2n} = \sqrt{2}$. Similarly we find that $\lim\limits_{n\to\infty} a_{2n+1} = \sqrt{2}$. So, by part (a), $\lim\limits_{n\to\infty} a_n = \sqrt{2}$.

93. (a) Suppose $\{p_n\}$ converges to p. Then $p_{n+1} = \dfrac{bp_n}{a+p_n} \Rightarrow \lim\limits_{n\to\infty} p_{n+1} = \dfrac{b \lim\limits_{n\to\infty} p_n}{a + \lim\limits_{n\to\infty} p_n} \Rightarrow p = \dfrac{bp}{a+p} \Rightarrow$

$p^2 + ap = bp \Rightarrow p(p + a - b) = 0 \Rightarrow p = 0$ or $p = b - a$.

(b) $p_{n+1} = \dfrac{bp_n}{a+p_n} = \dfrac{\left(\frac{b}{a}\right)p_n}{1 + \frac{p_n}{a}} < \left(\dfrac{b}{a}\right)p_n$ since $1 + \dfrac{p_n}{a} > 1$.

(c) By part (b), $p_1 < \left(\frac{b}{a}\right)p_0$, $p_2 < \left(\frac{b}{a}\right)p_1 < \left(\frac{b}{a}\right)^2 p_0$, $p_3 < \left(\frac{b}{a}\right)p_2 < \left(\frac{b}{a}\right)^3 p_0$, etc. In general, $p_n < \left(\frac{b}{a}\right)^n p_0$,

so $\lim\limits_{n\to\infty} p_n \le \lim\limits_{n\to\infty} \left(\frac{b}{a}\right)^n \cdot p_0 = 0$ since $b < a$. $\left[\text{By (7), } \lim\limits_{n\to\infty} r^n = 0 \text{ if } -1 < r < 1. \text{ Here } r = \frac{b}{a} \in (0,1).\right]$

(d) Let $a < b$. We first show, by induction, that if $p_0 < b - a$, then $p_n < b - a$ and $p_{n+1} > p_n$.

For $n = 0$, we have $p_1 - p_0 = \dfrac{bp_0}{a+p_0} - p_0 = \dfrac{p_0(b-a-p_0)}{a+p_0} > 0$ since $p_0 < b - a$. So $p_1 > p_0$.

Now we suppose the assertion is true for $n = k$, that is, $p_k < b - a$ and $p_{k+1} > p_k$. Then

$b - a - p_{k+1} = b - a - \dfrac{bp_k}{a+p_k} = \dfrac{a(b-a) + bp_k - ap_k - bp_k}{a+p_k} = \dfrac{a(b-a-p_k)}{a+p_k} > 0$ because $p_k < b - a$. So

$p_{k+1} < b - a$. And $p_{k+2} - p_{k+1} = \dfrac{bp_{k+1}}{a+p_{k+1}} - p_{k+1} = \dfrac{p_{k+1}(b-a-p_{k+1})}{a+p_{k+1}} > 0$ since $p_{k+1} < b - a$. Therefore,

$p_{k+2} > p_{k+1}$. Thus, the assertion is true for $n = k + 1$. It is therefore true for all n by mathematical induction.

A similar proof by induction shows that if $p_0 > b - a$, then $p_n > b - a$ and $\{p_n\}$ is decreasing.

In either case the sequence $\{p_n\}$ is bounded and monotonic, so it is convergent by the Monotonic Sequence Theorem. It then follows from part (a) that $\lim\limits_{n\to\infty} p_n = b - a$.

LABORATORY PROJECT Logistic Sequences

1. To write such a program in Maple it is best to calculate all the points first and then graph them. One possible sequence of commands [taking $p_0 = \frac{1}{2}$ and $k = 1.5$ for the difference equation] is

```
t:='t';p(0):=1/2;k:=1.5;
for j from 1 to 20  do p(j):=k*p(j-1)*(1-p(j-1)) od;
plot([seq([t,p(t)] t=0..20)],t=0..20,p=0..0.5,style=point);
```

In Mathematica, we can use the following program:

```
p[0]=1/2
k=1.5
p[j_]:=k*p[j-1]*(1-p[j-1])
P=Table[p[t],{t,20}]
ListPlot[P]
```

With $p_0 = \frac{1}{2}$ and $k = 1.5$:

n	p_n	n	p_n	n	p_n
0	0.5	7	0.3338465076	14	0.3333373303
1	0.375	8	0.3335895255	15	0.3333353318
2	0.3515625	9	0.3334613309	16	0.3333343326
3	0.3419494629	10	0.3333973076	17	0.3333338329
4	0.3375300416	11	0.3333653143	18	0.3333335831
5	0.3354052689	12	0.3333493223	19	0.3333334582
6	0.3343628617	13	0.3333413274	20	0.3333333958

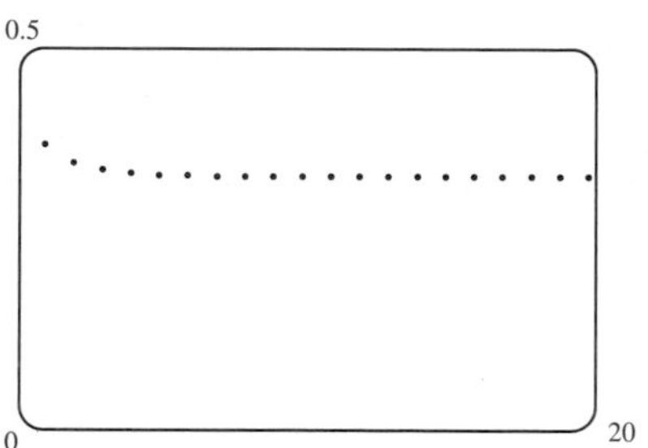

With $p_0 = \frac{1}{2}$ and $k = 2.5$:

n	p_n	n	p_n	n	p_n
0	0.5	7	0.6004164790	14	0.5999967417
1	0.625	8	0.5997913269	15	0.6000016291
2	0.5859375	9	0.6001042277	16	0.5999991854
3	0.6065368651	10	0.5999478590	17	0.6000004073
4	0.5966247409	11	0.6000260637	18	0.5999997964
5	0.6016591486	12	0.5999869664	19	0.6000001018
6	0.5991635437	13	0.6000065164	20	0.5999999491

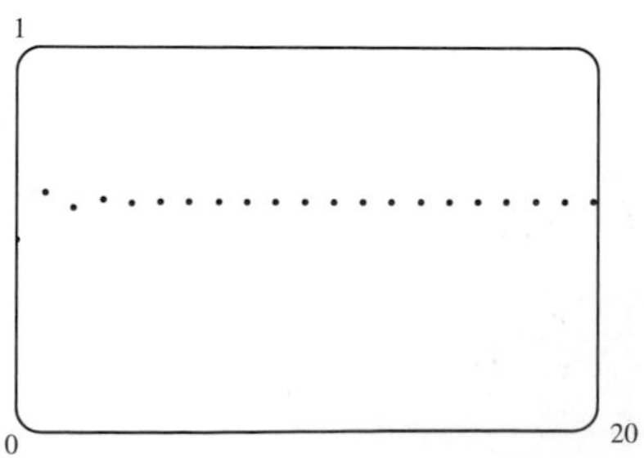

Both of these sequences seem to converge (the first to about $\frac{1}{3}$, the second to about 0.60).

With $p_0 = \frac{7}{8}$ and $k = 1.5$:

n	p_n	n	p_n	n	p_n
0	0.875	7	0.3239166554	14	0.3332554829
1	0.1640625	8	0.3284919837	15	0.3332943990
2	0.2057189941	9	0.3308775005	16	0.3333138639
3	0.2450980344	10	0.3320963702	17	0.3333235980
4	0.2775374819	11	0.3327125567	18	0.3333284655
5	0.3007656421	12	0.3330223670	19	0.3333308994
6	0.3154585059	13	0.3331777051	20	0.3333321164

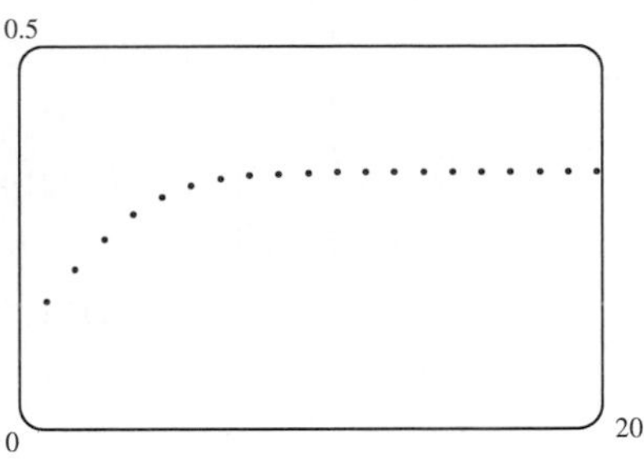

With $p_0 = \frac{7}{8}$ and $k = 2.5$:

n	p_n	n	p_n	n	p_n
0	0.875	7	0.6016572368	14	0.5999869815
1	0.2734375	8	0.5991645155	15	0.6000065088
2	0.4966735840	9	0.6004159972	16	0.5999967455
3	0.6249723374	10	0.5997915688	17	0.6000016272
4	0.5859547872	11	0.6001041070	18	0.5999991864
5	0.6065294364	12	0.5999479194	19	0.6000004068
6	0.5966286980	13	0.6000260335	20	0.5999997966

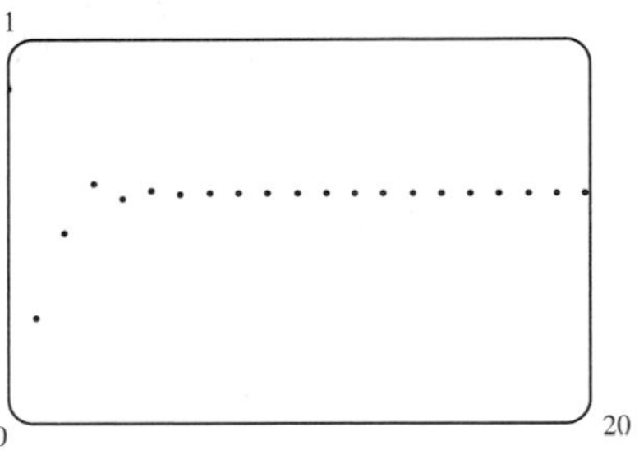

The limit of the sequence seems to depend on k, but not on p_0.

2. With $p_0 - \frac{7}{8}$ and $k - 3.2$:

n	p_n	n	p_n	n	p_n
0	0.875	7	0.5830728495	14	0.7990633827
1	0.35	8	0.7779164854	15	0.5137954979
2	0.728	9	0.5528397669	16	0.7993909896
3	0.6336512	10	0.7910654689	17	0.5131681132
4	0.7428395416	11	0.5288988570	18	0.7994451225
5	0.6112926626	12	0.7973275394	19	0.5130643795
6	0.7603646184	13	0.5171082698	20	0.7994538304

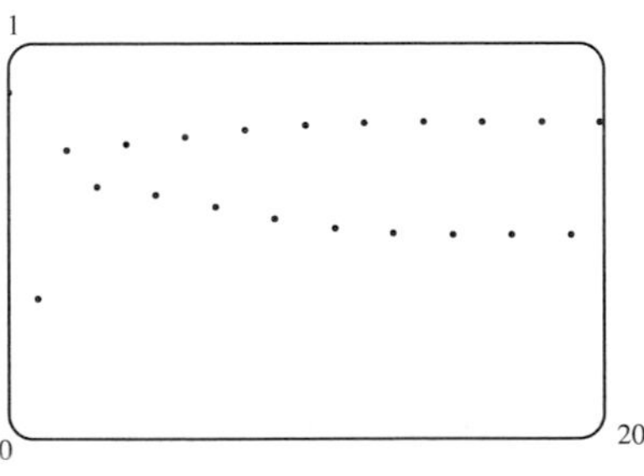

It seems that eventually the terms fluctuate between two values (about 0.5 and 0.8 in this case).

3. With $p_0 = \frac{7}{8}$ and $k = 3.42$:

n	p_n	n	p_n	n	p_n
0	0.875	7	0.4523028596	14	0.8442074951
1	0.3740625	8	0.8472194412	15	0.4498025048
2	0.8007579316	9	0.4426802161	16	0.8463823232
3	0.5456427596	10	0.8437633929	17	0.4446659586
4	0.8478752457	11	0.4508474156	18	0.8445284520
5	0.4411212220	12	0.8467373602	19	0.4490464985
6	0.8431438501	13	0.4438243545	20	0.8461207931

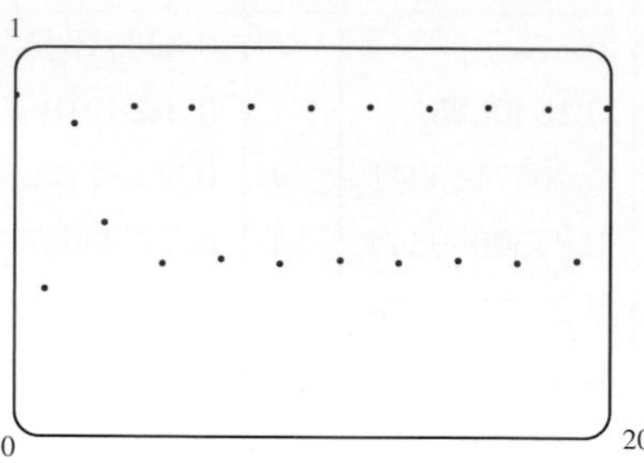

With $p_0 = \frac{7}{8}$ and $k = 3.45$:

n	p_n	n	p_n	n	p_n
0	0.875	7	0.4670259170	14	0.8403376122
1	0.37734375	8	0.8587488490	15	0.4628875685
2	0.8105962830	9	0.4184824586	16	0.8577482026
3	0.5296783241	10	0.8395743720	17	0.4209559716
4	0.8594612299	11	0.4646778983	18	0.8409445432
5	0.4167173034	12	0.8581956045	19	0.4614610237
6	0.8385707740	13	0.4198508858	20	0.8573758782

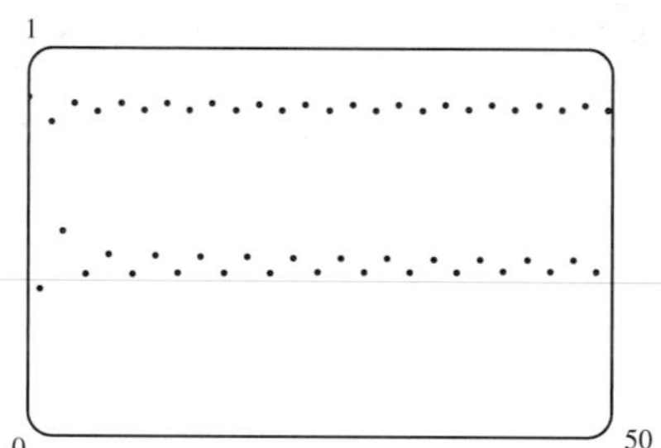

From the graphs above, it seems that for k between 3.4 and 3.5, the terms eventually fluctuate between four values. In the graph below, the pattern followed by the terms is $0.395, 0.832, 0.487, 0.869, 0.395, \ldots$. Note that even for $k = 3.42$ (as in the first graph), there are four distinct "branches"; even after 1000 terms, the first and third terms in the pattern differ by about 2×10^{-9}, while the first and fifth terms differ by only 2×10^{-10}. With $p_0 = \frac{7}{8}$ and $k = 3.48$:

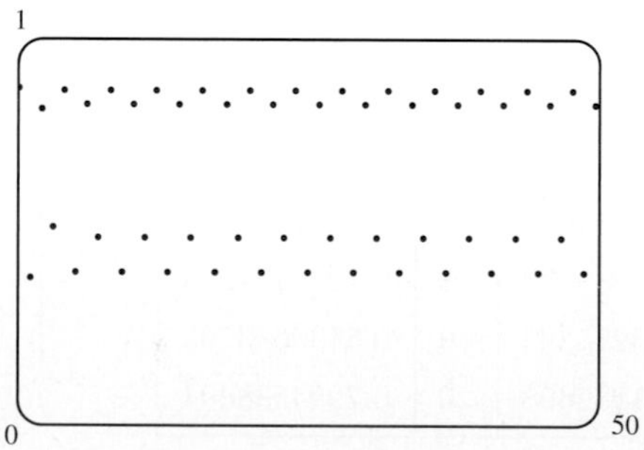

4.

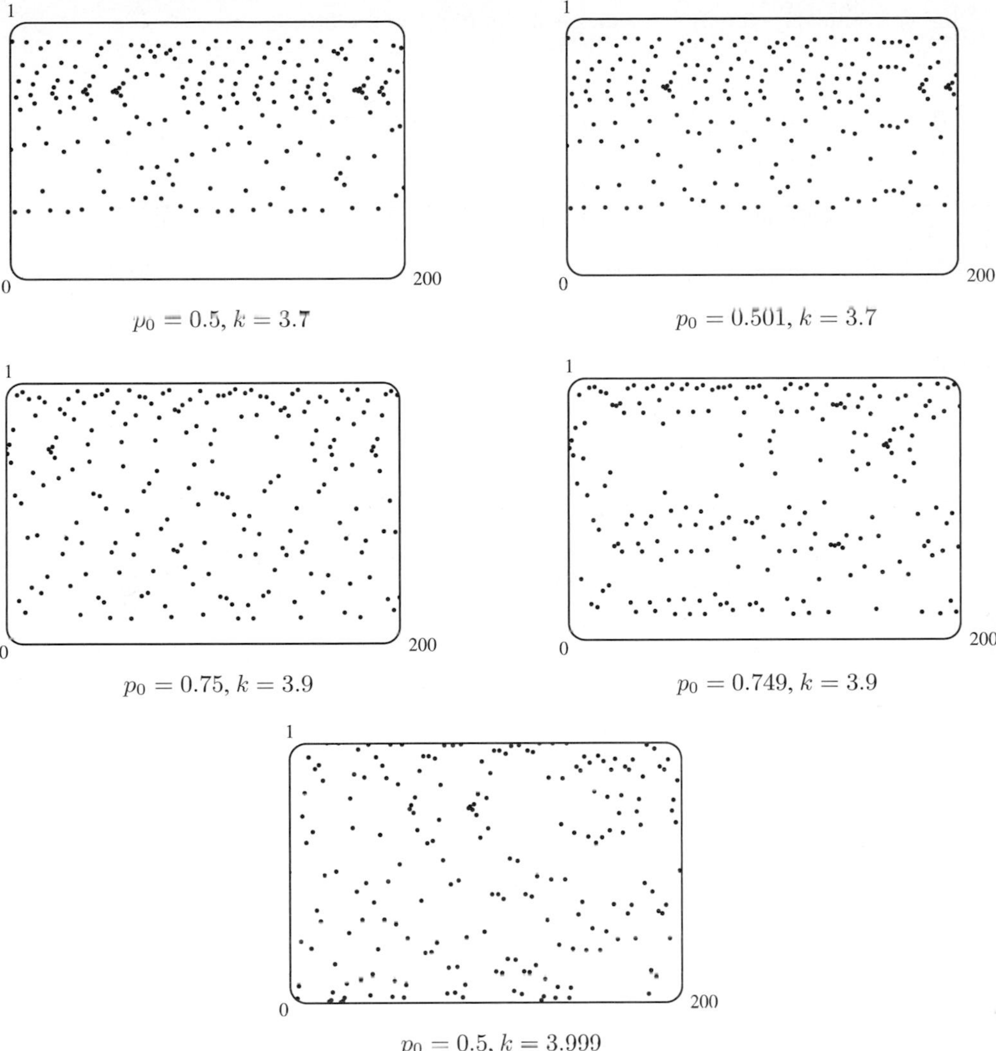

From the graphs, it seems that if p_0 is changed by 0.001, the whole graph changes completely. (Note, however, that this might be partially due to accumulated round-off error in the CAS. These graphs were generated by Maple with 100-digit accuracy, and different degrees of accuracy give different graphs.) There seem to be some some fleeting patterns in these graphs, but on the whole they are certainly very chaotic. As k increases, the graph spreads out vertically, with more extreme values close to 0 or 1.

11.2 Series

1. (a) A sequence is an ordered list of numbers whereas a series is the *sum* of a list of numbers.

(b) A series is convergent if the sequence of partial sums is a convergent sequence. A series is divergent if it is not convergent.

2. $\sum_{n=1}^{\infty} a_n = 5$ means that by adding sufficiently many terms of the series we can get as close as we like to the number 5.

In other words, it means that $\lim_{n\to\infty} s_n = 5$, where s_n is the nth partial sum, that is, $\sum_{i=1}^{n} a_i$.

3. $\sum_{n=1}^{\infty} a_n = \lim_{n\to\infty} s_n = \lim_{n\to\infty} [2 - 3(0.8)^n] = \lim_{n\to\infty} 2 - 3 \lim_{n\to\infty} (0.8)^n = 2 - 3(0) = 2$

4. $\sum_{n=1}^{\infty} a_n = \lim_{n\to\infty} s_n = \lim_{n\to\infty} \frac{n^2-1}{4n^2+1} = \lim_{n\to\infty} \frac{(n^2-1)/n^2}{(4n^2+1)/n^2} = \lim_{n\to\infty} \frac{1-1/n^2}{4+1/n^2} = \frac{1-0}{4+0} = \frac{1}{4}$

5. For $\sum_{n=1}^{\infty} \frac{1}{n^4+n^2}$, $a_n = \frac{1}{n^4+n^2}$. $s_1 = a_1 = \frac{1}{1^4+1^2} = \frac{1}{2} = 0.5$, $s_2 = s_1 + a_2 = \frac{1}{2} + \frac{1}{16+4} = 0.55$,

$s_3 = s_2 + a_3 \approx 0.5611$, $s_4 = s_3 + a_4 \approx 0.5648$, $s_5 = s_4 + a_5 \approx 0.5663$, $s_6 = s_5 + a_6 \approx 0.5671$,

$s_7 = s_6 + a_7 \approx 0.5675$, and $s_8 = s_7 + a_8 \approx 0.5677$. It appears that the series is convergent.

6. For $\sum_{n=1}^{\infty} \frac{1}{\sqrt[3]{n}}$, $a_n = \frac{1}{\sqrt[3]{n}}$. $s_1 = a_1 = \frac{1}{\sqrt[3]{1}} = 1$, $s_2 = s_1 + a_2 = 1 + \frac{1}{\sqrt[3]{2}} \approx 1.7937$,

$s_3 = s_2 + a_3 \approx 2.4871$, $s_4 = s_3 + a_4 \approx 3.1170$, $s_5 = s_4 + a_5 \approx 3.7018$, $s_6 = s_5 + a_6 \approx 4.2521$,

$s_7 = s_6 + a_7 \approx 4.7749$, and $s_8 = s_7 + a_8 \approx 5.2749$. It appears that the series is divergent.

7. For $\sum_{n=1}^{\infty} \sin n$, $a_n = \sin n$. $s_1 = a_1 = \sin 1 \approx 0.8415$, $s_2 = s_1 + a_2 \approx 1.7508$,

$s_3 = s_2 + a_3 \approx 1.8919$, $s_4 = s_3 + a_4 \approx 1.1351$, $s_5 = s_4 + a_5 \approx 0.1762$, $s_6 = s_5 + a_6 \approx -0.1033$,

$s_7 = s_6 + a_7 \approx 0.5537$, and $s_8 = s_7 + a_8 \approx 1.5431$. It appears that the series is divergent.

8. For $\sum_{n=1}^{\infty} \frac{(-1)^{n-1}}{n!}$, $a_n = (-1)^{n-1}\frac{1}{n!}$. $s_1 = a_1 = \frac{1}{1!} = 1$, $s_2 = s_1 + a_2 = 1 - \frac{1}{2!} = 0.5$,

$s_3 = s_2 + a_3 = 0.5 + \frac{1}{3!} \approx 0.6667$, $s_4 = s_3 + a_4 = 0.625$, $s_5 = s_4 + a_5 \approx 0.6333$, $s_6 = s_5 + a_6 \approx 0.6319$,

$s_7 = s_6 + a_7 \approx 0.6321$, and $s_8 = s_7 + a_8 \approx 0.6321$. It appears that the series is convergent.

9.

n	s_n
1	−2.40000
2	−1.92000
3	−2.01600
4	−1.99680
5	−2.00064
6	−1.99987
7	−2.00003
8	−1.99999
9	−2.00000
10	−2.00000

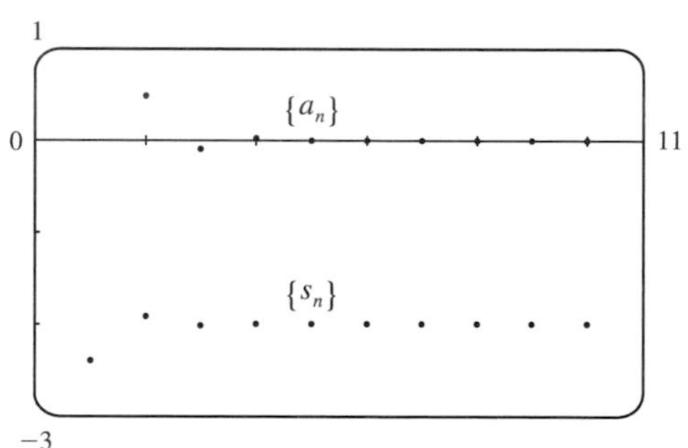

From the graph and the table, it seems that the series converges to -2. In fact, it is a geometric series with $a = -2.4$ and $r = -\frac{1}{5}$, so its sum is $\sum_{n=1}^{\infty} \frac{12}{(-5)^n} = \frac{-2.4}{1-\left(-\frac{1}{5}\right)} = \frac{-2.4}{1.2} = -2$.

Note that the dot corresponding to $n = 1$ is part of both $\{a_n\}$ and $\{s_n\}$.

TI-86 Note: To graph $\{a_n\}$ and $\{s_n\}$, set your calculator to Param mode and DrawDot mode. (DrawDot is under GRAPH, MORE, FORMT (F3).) Now under `E(t)=` make the assignments: `xt1=t`, `yt1=12/(-5)^t`, `xt2=t`, `yt2=sum seq(yt1,t,1,t,1)`. (sum and seq are under LIST, OPS (F5), MORE.) Under WIND use `1,10,1,0,10,1,-3,1,1` to obtain a graph similar to the one above. Then use TRACE (F4) to see the values.

10.

n	s_n
1	0.54030
2	0.12416
3	−0.86584
4	−1.51948
5	−1.23582
6	−0.27565
7	0.47825
8	0.33275
9	−0.57838
10	−1.41745

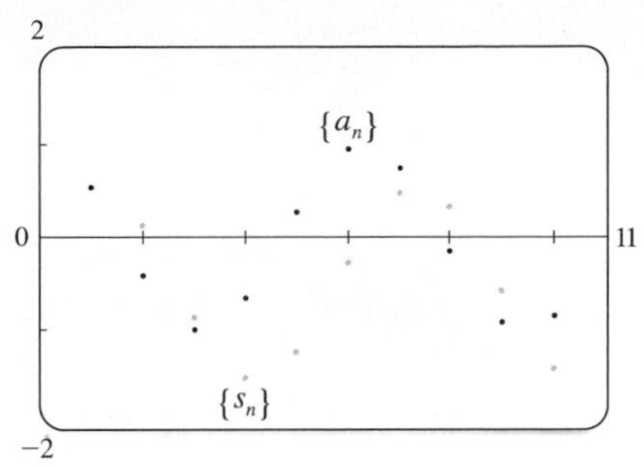

The series $\sum_{n=1}^{\infty} \cos n$ diverges, since its terms do not approach 0.

11.

n	s_n
1	0.44721
2	1.15432
3	1.98637
4	2.88080
5	3.80927
6	4.75796
7	5.71948
8	6.68962
9	7.66581
10	8.64639

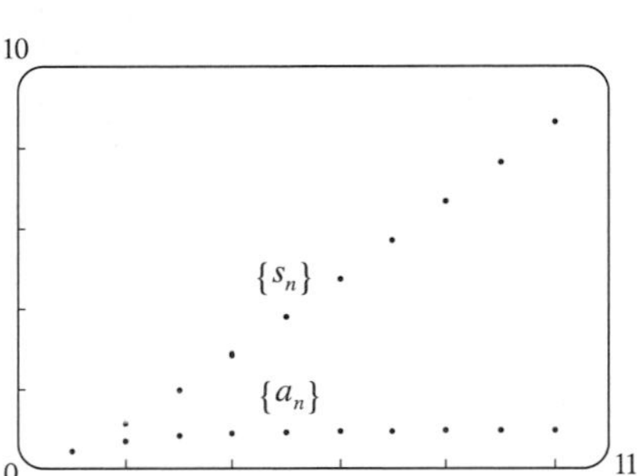

The series $\sum_{n=1}^{\infty} \frac{n}{\sqrt{n^2+4}}$ diverges, since its terms do not approach 0.

12.

n	s_n
1	4.90000
2	8.33000
3	10.73100
4	12.41170
5	13.58819
6	14.41173
7	14.98821
8	15.39175
9	15.67422
10	15.87196

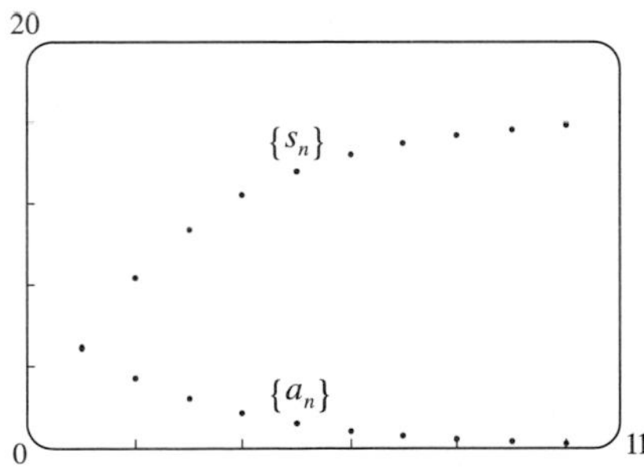

From the graph and the table, we see that the terms are getting smaller and may approach 0, and that the series approaches a value near 16. The series is geometric with $a_1 = 4.9$ and $r = 0.7$, so its sum is $\sum_{n=1}^{\infty} \frac{7^{n+1}}{10^n} = \frac{4.9}{1-0.7} = \frac{4.9}{0.3} = 16.\overline{3}$.

13.

n	s_n
2	1.00000
3	1.33333
4	1.50000
5	1.60000
6	1.66667
7	1.71429
8	1.75000
9	1.77778
10	1.80000
11	1.81818

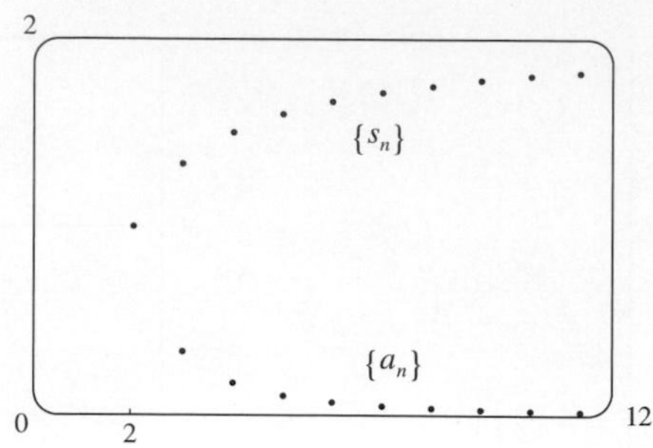

From the graph and the table, we see that the terms are getting smaller and may approach 0, and that the series may approach a number near 2. Using partial fractions, we have

$$\begin{aligned}\sum_{n=2}^{k} \frac{2}{n^2 - n} &= \sum_{n=2}^{k} \left(\frac{2}{n-1} - \frac{2}{n}\right)\\ &= \left(\frac{2}{1} - \frac{2}{2}\right) + \left(\frac{2}{2} - \frac{2}{3}\right) + \left(\frac{2}{3} - \frac{2}{4}\right)\\ &\qquad + \cdots + \left(\frac{2}{k-2} - \frac{2}{k-1}\right) + \left(\frac{2}{k-1} - \frac{2}{k}\right)\\ &= 2 - \frac{2}{k}\end{aligned}$$

As $k \to \infty$, $2 - \dfrac{2}{k} \to 2$, so $\displaystyle\sum_{n=2}^{\infty} \frac{2}{n^2 - n} = 2$.

14.

n	s_n
1	0.36205
2	0.51428
3	0.59407
4	0.64280
5	0.67557
6	0.69910
7	0.71680
8	0.73059
9	0.74164
10	0.75069

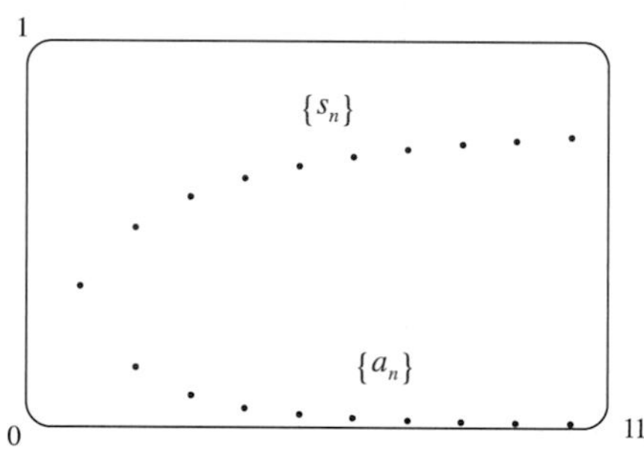

From the graph and the table, we see that the terms are getting smaller and may approach 0, and that the series may approach a number near 1.

$$\begin{aligned}\sum_{n=1}^{k} \left(\sin\frac{1}{n} - \sin\frac{1}{n+1}\right) &= \left(\sin 1 - \sin\frac{1}{2}\right) + \left(\sin\frac{1}{2} - \sin\frac{1}{3}\right)\\ &\qquad + \cdots + \left(\sin\frac{1}{k-1} + \sin\frac{1}{k}\right)\\ &\qquad + \left(\sin\frac{1}{k} - \sin\frac{1}{k+1}\right)\\ &= \sin 1 - \sin\frac{1}{k+1}\end{aligned}$$

As $k \to \infty$, $\sin 1 - \sin\dfrac{1}{k+1} \to \sin 1 - \sin 0 = \sin 1$, so

$$\sum_{n=1}^{\infty} \left(\sin\frac{1}{n} - \sin\frac{1}{n+1}\right) = \sin 1 \approx 0.84147.$$

15. (a) $\lim\limits_{n\to\infty} a_n = \lim\limits_{n\to\infty} \dfrac{2n}{3n+1} = \dfrac{2}{3}$, so the *sequence* $\{a_n\}$ is convergent by (11.1.1).

(b) Since $\lim\limits_{n\to\infty} a_n = \frac{2}{3} \neq 0$, the *series* $\sum\limits_{n=1}^{\infty} a_n$ is divergent by the Test for Divergence.

16. (a) Both $\sum\limits_{i=1}^{n} a_i$ and $\sum\limits_{j=1}^{n} a_j$ represent the sum of the first n terms of the sequence $\{a_n\}$, that is, the nth partial sum.

(b) $\sum\limits_{i=1}^{n} a_j = \underbrace{a_j + a_j + \cdots + a_j}_{n \text{ terms}} = na_j$, which, in general, is not the same as $\sum\limits_{i=1}^{n} a_i = a_1 + a_2 + \cdots + a_n$.

17. $3 - 4 + \frac{16}{3} - \frac{64}{9} + \cdots$ is a geometric series with ratio $r = -\frac{4}{3}$. Since $|r| = \frac{4}{3} > 1$, the series diverges.

18. $4 + 3 + \frac{9}{4} + \frac{27}{16} + \cdots$ is a geometric series with ratio $\frac{3}{4}$. Since $|r| = \frac{3}{4} < 1$, the series converges to $\dfrac{a}{1-r} = \dfrac{4}{1-3/4} = 16$.

19. $10 - 2 + 0.4 - 0.08 + \cdots$ is a geometric series with ratio $-\frac{2}{10} = -\frac{1}{5}$. Since $|r| = \frac{1}{5} < 1$, the series converges to

$\dfrac{a}{1-r} = \dfrac{10}{1-(-1/5)} = \dfrac{10}{6/5} = \dfrac{50}{6} = \dfrac{25}{3}$.

20. $2 + 0.5 + 0.125 + 0.03125 + \cdots$ is a geometric series with ratio $r = \frac{0.5}{2} = \frac{1}{4}$. Since $|r| = \frac{1}{4} < 1$, the series converges

to $\dfrac{a}{1-r} = \dfrac{2}{1-1/4} = \dfrac{2}{3/4} = \dfrac{8}{3}$.

21. $\sum\limits_{n=1}^{\infty} 12\,(0.73)^{n-1}$ is a geometric series with first term $a = 12$ and ratio $r = 0.73$. Since $|r| = 0.73 < 1$, the series converges

to $\dfrac{a}{1-r} = \dfrac{12}{1-0.73} = \dfrac{12}{0.27} = \dfrac{12(100)}{27} = \dfrac{400}{9}$.

22. $\sum\limits_{n=1}^{\infty} \dfrac{5}{\pi^n} = 5\sum\limits_{n=1}^{\infty} \left(\dfrac{1}{\pi}\right)^n$. The latter series is geometric with $a = \dfrac{1}{\pi}$ and ratio $r = \dfrac{1}{\pi}$. Since $|r| = \dfrac{1}{\pi} < 1$, it converges to

$\dfrac{1/\pi}{1-1/\pi} = \dfrac{1}{\pi - 1}$. Thus, the given series converges to $5\left(\dfrac{1}{\pi-1}\right) = \dfrac{5}{\pi-1}$.

23. $\sum\limits_{n=1}^{\infty} \dfrac{(-3)^{n-1}}{4^n} = \dfrac{1}{4}\sum\limits_{n=1}^{\infty}\left(-\dfrac{3}{4}\right)^{n-1}$. The latter series is geometric with $a = 1$ and ratio $r = -\frac{3}{4}$. Since $|r| = \frac{3}{4} < 1$, it

converges to $\dfrac{1}{1-(-3/4)} = \frac{4}{7}$. Thus, the given series converges to $\left(\frac{1}{4}\right)\left(\frac{4}{7}\right) = \frac{1}{7}$.

24. $\sum\limits_{n=0}^{\infty} \dfrac{3^{n+1}}{(-2)^n} = 3\sum\limits_{n=0}^{\infty}\left(-\frac{3}{2}\right)^n$ is a geometric series with ratio $r = -\frac{3}{2}$. Since $|r| = \frac{3}{2} > 1$, the series diverges.

25. $\sum\limits_{n=1}^{\infty} \dfrac{e^{2n}}{6^{n-1}} = \sum\limits_{n=1}^{\infty} \dfrac{(e^2)^n}{6^n 6^{-1}} = 6\sum\limits_{n=1}^{\infty}\left(\dfrac{e^2}{6}\right)^n$ is a geometric series with ratio $r = \dfrac{e^2}{6}$. Since $|r| = \dfrac{e^2}{6}[\approx 1.23] > 1$, the series

diverges.

26. $\displaystyle\sum_{n=1}^{\infty}\frac{6\cdot 2^{2n-1}}{3^n}=\sum_{n=1}^{\infty}\frac{6(2^2)^n\cdot 2^{-1}}{3^n}=3\sum_{n=1}^{\infty}\left(\frac{4}{3}\right)^n$ is a geometric series with ratio $r=\frac{4}{3}$. Since $|r|=\frac{4}{3}>1$, the series diverges.

27. $\frac{1}{3}+\frac{1}{6}+\frac{1}{9}+\frac{1}{12}+\frac{1}{15}+\cdots=\sum_{n=1}^{\infty}\frac{1}{3n}=\frac{1}{3}\sum_{n=1}^{\infty}\frac{1}{n}$. This is a constant multiple of the divergent harmonic series, so it diverges.

28. $\frac{1}{3}+\frac{2}{9}+\frac{1}{27}+\frac{2}{81}+\frac{1}{243}+\frac{2}{729}+\cdots=\left(\frac{1}{3}+\frac{1}{27}+\frac{1}{243}+\cdots\right)+\left(\frac{2}{9}+\frac{2}{81}+\frac{2}{729}+\cdots\right)$, which are both convergent geometric series with sums $\dfrac{1/3}{1-1/9}=\dfrac{3}{8}$ and $\dfrac{2/9}{1-1/9}=\dfrac{1}{4}$, so the original series converges and its sum is $\frac{3}{8}+\frac{1}{4}=\frac{5}{8}$.

29. $\displaystyle\sum_{n=1}^{\infty}\frac{2+n}{1-2n}$ diverges by the Test for Divergence since $\displaystyle\lim_{n\to\infty}a_n=\lim_{n\to\infty}\frac{2+n}{1-2n}=\lim_{n\to\infty}\frac{2/n+1}{1/n-2}=-\frac{1}{2}\neq 0$.

30. $\displaystyle\sum_{k=1}^{\infty}\frac{k^2}{k^2-2k+5}$ diverges by the Test for Divergence since $\displaystyle\lim_{k\to\infty}\frac{k^2}{k^2-2k+5}=\lim_{k\to\infty}\frac{1}{1-2/k+5/k^2}=1\neq 0$.

31. $\displaystyle\sum_{n=1}^{\infty}3^{n+1}4^{-n}=\sum_{n=1}^{\infty}\frac{3^n\cdot 3^1}{4^n}=3\sum_{n=1}^{\infty}\left(\frac{3}{4}\right)^n$. The latter series is geometric with $a=\frac{3}{4}$ and ratio $r=\frac{3}{4}$. Since $|r|=\frac{3}{4}<1$, it converges to $\dfrac{3/4}{1-3/4}=3$. Thus, the given series converges to $3(3)=9$.

32. $\displaystyle\sum_{n=1}^{\infty}\left[(-0.2)^n+(0.6)^{n-1}\right]=\sum_{n=1}^{\infty}(-0.2)^n+\sum_{n=1}^{\infty}(0.6)^{n-1}$ [sum of two geometric series]

$$=\frac{-0.2}{1-(-0.2)}+\frac{1}{1-0.6}=-\frac{1}{6}+\frac{5}{2}=\frac{7}{3}$$

33. $\displaystyle\sum_{n=1}^{\infty}\frac{1}{4+e^{-n}}$ diverges by the Test for Divergence since $\displaystyle\lim_{n\to\infty}\frac{1}{4+e^{-n}}=\frac{1}{4+0}=\frac{1}{4}\neq 0$.

34. $\displaystyle\sum_{n=1}^{\infty}\frac{2^n+4^n}{e^n}$ diverges by the Test for Divergence since $\displaystyle\lim_{n\to\infty}\frac{2^n+4^n}{e^n}=\lim_{n\to\infty}\left(\frac{2^n}{e^n}+\frac{4^n}{e^n}\right)\geq\lim_{n\to\infty}\left(\frac{4}{e}\right)^n=\infty$ since $\frac{4}{e}>1$.

35. $\displaystyle\sum_{k=1}^{\infty}(\sin 100)^k$ is a geometric series with first term $a=\sin 100$ [≈ -0.506] and ratio $r=\sin 100$. Since $|r|<1$, the series converges to $\dfrac{\sin 100}{1-\sin 100}\approx -0.336$.

36. $\displaystyle\sum_{n=1}^{\infty}\frac{1}{1+\left(\frac{2}{3}\right)^n}$ diverges by the Test for Divergence since $\displaystyle\lim_{n\to\infty}\frac{1}{1+\left(\frac{2}{3}\right)^n}=\frac{1}{1+0}=1\neq 0$.

37. $\sum_{n=1}^{\infty} \ln\left(\frac{n^2+1}{2n^2+1}\right)$ diverges by the Test for Divergence since

$$\lim_{n\to\infty} a_n = \lim_{n\to\infty} \ln\left(\frac{n^2+1}{2n^2+1}\right) = \ln\left(\lim_{n\to\infty} \frac{n^2+1}{2n^2+1}\right) = \ln\tfrac{1}{2} \neq 0.$$

38. $\sum_{k=0}^{\infty} (\sqrt{2})^{-k} = \sum_{k=0}^{\infty} \left(\frac{1}{\sqrt{2}}\right)^k$ is a geometric series with first term $a = \left(\frac{1}{\sqrt{2}}\right)^0 = 1$ and ratio $r = \frac{1}{\sqrt{2}}$. Since $|r| < 1$, the series converges to $\frac{1}{1 - 1/\sqrt{2}} = \frac{\sqrt{2}}{\sqrt{2}-1} \approx 3.414$.

39. $\sum_{n=1}^{\infty} \arctan n$ diverges by the Test for Divergence since $\lim_{n\to\infty} a_n = \lim_{n\to\infty} \arctan n = \frac{\pi}{2} \neq 0$.

40. $\sum_{n=1}^{\infty} \left(\frac{3}{5^n} + \frac{2}{n}\right)$ diverges because $\sum_{n=1}^{\infty} \frac{2}{n} = 2\sum_{n=1}^{\infty} \frac{1}{n}$ diverges. (If it converged, then $\frac{1}{2} \cdot 2\sum_{n=1}^{\infty} \frac{1}{n}$ would also converge by Theorem 8(i), but we know from Example 9 that the harmonic series $\sum_{n=1}^{\infty} \frac{1}{n}$ diverges.) If the given series converges, then the difference $\sum_{n=1}^{\infty} \left(\frac{3}{5^n} + \frac{2}{n}\right) - \sum_{n=1}^{\infty} \frac{3}{5^n}$ must converge (since $\sum_{n=1}^{\infty} \frac{3}{5^n}$ is a convergent geometric series) and equal $\sum_{n=1}^{\infty} \frac{2}{n}$, but we have just seen that $\sum_{n=1}^{\infty} \frac{2}{n}$ diverges, so the given series must also diverge.

41. $\sum_{n=1}^{\infty} \frac{1}{e^n} = \sum_{n=1}^{\infty} \left(\frac{1}{e}\right)^n$ is a geometric series with first term $a = \frac{1}{e}$ and ratio $r = \frac{1}{e}$. Since $|r| = \frac{1}{e} < 1$, the series converges to $\frac{1/e}{1-1/e} = \frac{1/e}{1-1/e} \cdot \frac{e}{e} = \frac{1}{e-1}$. By Example 8, $\sum_{n=1}^{\infty} \frac{1}{n(n+1)} = 1$. Thus, by Theorem 8(ii),

$$\sum_{n=1}^{\infty} \left(\frac{1}{e^n} + \frac{1}{n(n+1)}\right) = \sum_{n=1}^{\infty} \frac{1}{e^n} + \sum_{n=1}^{\infty} \frac{1}{n(n+1)} = \frac{1}{e-1} + 1 = \frac{1}{e-1} + \frac{e-1}{e-1} = \frac{e}{e-1}.$$

42. $\sum_{n=1}^{\infty} \frac{e^n}{n^2}$ diverges by the Test for Divergence since $\lim_{n\to\infty} a_n = \lim_{n\to\infty} \frac{e^n}{n^2} = \lim_{x\to\infty} \frac{e^x}{x^2} \overset{\text{H}}{=} \lim_{x\to\infty} \frac{e^x}{2x} \overset{\text{H}}{=} \lim_{x\to\infty} \frac{e^x}{2} = \infty \neq 0$.

43. Using partial fractions, the partial sums of the series $\sum_{n=2}^{\infty} \frac{2}{n^2-1}$ are

$$\begin{aligned} s_n &= \sum_{i=2}^{n} \frac{2}{(i-1)(i+1)} = \sum_{i=2}^{n} \left(\frac{1}{i-1} - \frac{1}{i+1}\right) \\ &= \left(1 - \frac{1}{3}\right) + \left(\frac{1}{2} - \frac{1}{4}\right) + \left(\frac{1}{3} - \frac{1}{5}\right) + \cdots + \left(\frac{1}{n-3} - \frac{1}{n-1}\right) + \left(\frac{1}{n-2} - \frac{1}{n}\right) \end{aligned}$$

This sum is a telescoping series and $s_n = 1 + \frac{1}{2} - \frac{1}{n-1} - \frac{1}{n}$.

Thus, $\sum_{n=2}^{\infty} \frac{2}{n^2-1} = \lim_{n\to\infty} s_n = \lim_{n\to\infty} \left(1 + \frac{1}{2} - \frac{1}{n-1} - \frac{1}{n}\right) = \frac{3}{2}$.

44. For the series $\sum_{n=1}^{\infty} \ln \dfrac{n}{n+1}$,

$$s_n = (\ln 1 - \ln 2) + (\ln 2 - \ln 3) + (\ln 3 - \ln 4) + \cdots + [\ln n - \ln(n+1)] = \ln 1 - \ln(n+1) = -\ln(n+1)$$

[telescoping series]

Thus, $\lim_{n\to\infty} s_n = -\infty$, so the series is divergent.

45. For the series $\sum_{n=1}^{\infty} \dfrac{3}{n(n+3)}$, $s_n = \sum_{i=1}^{n} \dfrac{3}{i(i+3)} = \sum_{i=1}^{n} \left(\dfrac{1}{i} - \dfrac{1}{i+3}\right)$ [using partial fractions]. The latter sum is

$$\left(1 - \tfrac{1}{4}\right) + \left(\tfrac{1}{2} - \tfrac{1}{5}\right) + \left(\tfrac{1}{3} - \tfrac{1}{6}\right) + \left(\tfrac{1}{4} - \tfrac{1}{7}\right) + \cdots + \left(\tfrac{1}{n-3} - \tfrac{1}{n}\right) + \left(\tfrac{1}{n-2} - \tfrac{1}{n+1}\right) + \left(\tfrac{1}{n-1} - \tfrac{1}{n+2}\right) + \left(\tfrac{1}{n} - \tfrac{1}{n+3}\right)$$

$$= 1 + \tfrac{1}{2} + \tfrac{1}{3} - \tfrac{1}{n+1} - \tfrac{1}{n+2} - \tfrac{1}{n+3} \quad \text{[telescoping series]}$$

Thus, $\sum_{n=1}^{\infty} \dfrac{3}{n(n+3)} = \lim_{n\to\infty} s_n = \lim_{n\to\infty} \left(1 + \tfrac{1}{2} + \tfrac{1}{3} - \tfrac{1}{n+1} - \tfrac{1}{n+2} - \tfrac{1}{n+3}\right) = 1 + \tfrac{1}{2} + \tfrac{1}{3} = \tfrac{11}{6}$. Converges

46. For the series $\sum_{n=4}^{\infty} \left(\dfrac{1}{\sqrt{n}} - \dfrac{1}{\sqrt{n+1}}\right)$

$$s_n = \sum_{i=4}^{n} \left(\frac{1}{\sqrt{i}} - \frac{1}{\sqrt{i+1}}\right) = \left(\frac{1}{\sqrt{4}} - \frac{1}{\sqrt{5}}\right) + \left(\frac{1}{\sqrt{5}} - \frac{1}{\sqrt{6}}\right) + \cdots + \left(\frac{1}{\sqrt{n}} - \frac{1}{\sqrt{n+1}}\right) = \frac{1}{\sqrt{4}} - \frac{1}{\sqrt{n+1}}$$

[telescoping series]

Thus, $\sum_{n=4}^{\infty} \left(\dfrac{1}{\sqrt{n}} - \dfrac{1}{\sqrt{n+1}}\right) = \lim_{n\to\infty} s_n = \lim_{n\to\infty} \left(\dfrac{1}{\sqrt{4}} - \dfrac{1}{\sqrt{n+1}}\right) = \dfrac{1}{\sqrt{4}} - 0 = \dfrac{1}{2}$. Converges

47. For the series $\sum_{n=1}^{\infty} \left(e^{1/n} - e^{1/(n+1)}\right)$,

$$s_n = \sum_{i=1}^{n} \left(e^{1/i} - e^{1/(i+1)}\right) = (e^1 - e^{1/2}) + (e^{1/2} - e^{1/3}) + \cdots + \left(e^{1/n} - e^{1/(n+1)}\right) = e - e^{1/(n+1)}$$

[telescoping series]

Thus, $\sum_{n=1}^{\infty} \left(e^{1/n} - e^{1/(n+1)}\right) = \lim_{n\to\infty} s_n = \lim_{n\to\infty} \left(e - e^{1/(n+1)}\right) = e - e^0 = e - 1$. Converges

48. Using partial fractions, the partial sums of the series $\sum_{n=2}^{\infty} \dfrac{1}{n^3 - n}$ are

$$s_n = \sum_{i=2}^{n} \frac{1}{i(i-1)(i+1)} = \sum_{i=2}^{n} \left(-\frac{1}{i} + \frac{1/2}{i-1} + \frac{1/2}{i+1}\right) = \frac{1}{2} \sum_{i=2}^{n} \left(\frac{1}{i-1} - \frac{2}{i} + \frac{1}{i+1}\right)$$

$$= \frac{1}{2}\left[\left(\frac{1}{1} - \frac{2}{2} + \frac{1}{3}\right) + \left(\frac{1}{2} - \frac{2}{3} + \frac{1}{4}\right) + \left(\frac{1}{3} - \frac{2}{4} + \frac{1}{5}\right) + \left(\frac{1}{4} - \frac{2}{5} + \frac{1}{6}\right) + \cdots\right.$$

$$\left. + \left(\frac{1}{n-3} - \frac{2}{n-2} + \frac{1}{n-1}\right) + \left(\frac{1}{n-2} - \frac{2}{n-1} + \frac{1}{n}\right) + \left(\frac{1}{n-1} - \frac{2}{n} + \frac{1}{n+1}\right)\right]$$

Note: In three consecutive expressions in parentheses, the 3rd term in the first expression plus the 2nd term in the second expression plus the 1st term in the third expression sum to 0.

$$= \frac{1}{2}\left(\frac{1}{1} - \frac{2}{2} + \frac{1}{2} + \frac{1}{n} - \frac{2}{n} + \frac{1}{n+1}\right) = \frac{1}{4} - \frac{1}{2n} + \frac{1}{2n+2}$$

Thus, $\sum_{n=2}^{\infty} \dfrac{1}{n^3 - n} = \lim_{n\to\infty} s_n = \lim_{n\to\infty} \left(\dfrac{1}{4} - \dfrac{1}{2n} + \dfrac{1}{2n+2}\right) = \dfrac{1}{4}$.

49. (a) Many people would guess that $x < 1$, but note that x consists of an infinite number of 9s.

(b) $x = 0.99999\ldots = \frac{9}{10} + \frac{9}{100} + \frac{9}{1000} + \frac{9}{10{,}000} + \cdots = \sum_{n=1}^{\infty} \frac{9}{10^n}$, which is a geometric series with $a_1 = 0.9$ and $r = 0.1$. Its sum is $\frac{0.9}{1 - 0.1} = \frac{0.9}{0.9} = 1$, that is, $x = 1$.

(c) The number 1 has two decimal representations, $1.00000\ldots$ and $0.99999\ldots$.

(d) Except for 0, all rational numbers that have a terminating decimal representation can be written in more than one way. For example, 0.5 can be written as $0.49999\ldots$ as well as $0.50000\ldots$.

50. $a_1 = 1$, $a_n = (5-n)a_{n-1} \Rightarrow a_2 = (5-2)a_1 = 3(1) = 3$, $a_3 = (5-3)a_2 = 2(3) = 6$, $a_4 = (5-4)a_3 = 1(6) = 6$, $a_5 = (5-5)a_4 = 0$, and all succeeding terms equal 0. Thus, $\sum_{n=1}^{\infty} a_n = \sum_{n=1}^{4} a_n = 1 + 3 + 6 + 6 = 16$.

51. $0.\overline{8} = \frac{8}{10} + \frac{8}{10^2} + \cdots$ is a geometric series with $a = \frac{8}{10}$ and $r = \frac{1}{10}$. It converges to $\frac{a}{1-r} = \frac{8/10}{1 - 1/10} = \frac{8}{9}$.

52. $0.\overline{46} = \frac{46}{100} + \frac{46}{100^2} + \cdots$ is a geometric series with $a = \frac{46}{100}$ and $r = \frac{1}{100}$. It converges to $\frac{a}{1-r} = \frac{46/100}{1 - 1/100} = \frac{46}{99}$.

53. $2.\overline{516} = 2 + \frac{516}{10^3} + \frac{516}{10^6} + \cdots$. Now $\frac{516}{10^3} + \frac{516}{10^6} + \cdots$ is a geometric series with $a = \frac{516}{10^3}$ and $r = \frac{1}{10^3}$. It converges to $\frac{a}{1-r} = \frac{516/10^3}{1 - 1/10^3} = \frac{516/10^3}{999/10^3} = \frac{516}{999}$. Thus, $2.\overline{516} = 2 + \frac{516}{999} = \frac{2514}{999} = \frac{838}{333}$.

54. $10.1\overline{35} = 10.1 + \frac{35}{10^3} + \frac{35}{10^5} + \cdots$. Now $\frac{35}{10^3} + \frac{35}{10^5} + \cdots$ is a geometric series with $a = \frac{35}{10^3}$ and $r = \frac{1}{10^2}$. It converges to $\frac{a}{1-r} = \frac{35/10^3}{1 - 1/10^2} = \frac{35/10^3}{99/10^2} = \frac{35}{990}$. Thus, $10.1\overline{35} = 10.1 + \frac{35}{990} = \frac{9999 + 35}{990} = \frac{10{,}034}{990} = \frac{5017}{495}$.

55. $1.234\overline{567} = 1.234 + \frac{567}{10^6} + \frac{567}{10^9} + \cdots$. Now $\frac{567}{10^6} + \frac{567}{10^9} + \cdots$ is a geometric series with $a = \frac{567}{10^6}$ and $r = \frac{1}{10^3}$. It converges to $\frac{a}{1-r} = \frac{567/10^6}{1 - 1/10^3} = \frac{567/10^6}{999/10^3} = \frac{567}{999{,}000} = \frac{21}{37{,}000}$. Thus,

$$1.234\overline{567} = 1.234 + \frac{21}{37{,}000} = \frac{1234}{1000} + \frac{21}{37{,}000} = \frac{45{,}658}{37{,}000} + \frac{21}{37{,}000} = \frac{45{,}679}{37{,}000}.$$

56. $5.\overline{71358} = 5 + \frac{71{,}358}{10^5} + \frac{71{,}358}{10^{10}} + \cdots$. Now $\frac{71{,}358}{10^5} + \frac{71{,}358}{10^{10}} + \cdots$ is a geometric series with $a = \frac{71{,}358}{10^5}$ and $r = \frac{1}{10^5}$. It converges to $\frac{a}{1-r} = \frac{71{,}358/10^5}{1 - 1/10^5} = \frac{71{,}358/10^5}{99{,}999/10^5} = \frac{71{,}358}{99{,}999} = \frac{23{,}786}{33{,}333}$. Thus,

$$5.\overline{71358} = 5 + \frac{23{,}786}{33{,}333} = \frac{166{,}665}{33{,}333} + \frac{23{,}786}{33{,}333} = \frac{190{,}451}{33{,}333}.$$

57. $\sum_{n=1}^{\infty}(-5)^n x^n = \sum_{n=1}^{\infty}(-5x)^n$ is a geometric series with $r=-5x$, so the series converges $\Leftrightarrow$ $|r|<1$ $\Leftrightarrow$

$|-5x|<1 \Leftrightarrow |x|<\frac{1}{5}$, that is, $-\frac{1}{5}<x<\frac{1}{5}$. In that case, the sum of the series is $\dfrac{a}{1-r}=\dfrac{-5x}{1-(-5x)}=\dfrac{-5x}{1+5x}$.

58. $\sum_{n=1}^{\infty}(x+2)^n$ is a geometric series with $r=x+2$, so the series converges $\Leftrightarrow$ $|r|<1$ $\Leftrightarrow$ $|x+2|<1$ $\Leftrightarrow$

$-1<x+2<1 \Leftrightarrow -3<x<-1$. In that case, the sum of the series is $\dfrac{a}{1-r}=\dfrac{x+2}{1-(x+2)}=\dfrac{x+2}{-x-1}$.

59. $\sum_{n=0}^{\infty}\dfrac{(x-2)^n}{3^n}=\sum_{n=0}^{\infty}\left(\dfrac{x-2}{3}\right)^n$ is a geometric series with $r=\dfrac{x-2}{3}$, so the series converges $\Leftrightarrow$ $|r|<1$ $\Leftrightarrow$

$\left|\dfrac{x-2}{3}\right|<1 \Leftrightarrow -1<\dfrac{x-2}{3}<1 \Leftrightarrow -3<x-2<3 \Leftrightarrow -1<x<5$. In that case, the sum of the series is

$$\frac{a}{1-r}=\frac{1}{1-\dfrac{x-2}{3}}=\frac{1}{\dfrac{3-(x-2)}{3}}=\frac{3}{5-x}.$$

60. $\sum_{n=0}^{\infty}(-4)^n(x-5)^n=\sum_{n=0}^{\infty}[-4(x-5)]^n$ is a geometric series with $r=-4(x-5)$, so the series converges $\Leftrightarrow$

$|r|<1 \Leftrightarrow |-4(x-5)|<1 \Leftrightarrow |x-5|<\frac{1}{4} \Leftrightarrow -\frac{1}{4}<x-5<\frac{1}{4} \Leftrightarrow \frac{19}{4}<x<\frac{21}{4}$. In that case, the sum of

the series is $\dfrac{a}{1-r}=\dfrac{1}{1-[-4(x-5)]}=\dfrac{1}{4x-19}$.

61. $\sum_{n=0}^{\infty}\dfrac{2^n}{x^n}=\sum_{n=0}^{\infty}\left(\dfrac{2}{x}\right)^n$ is a geometric series with $r=\dfrac{2}{x}$, so the series converges $\Leftrightarrow$ $|r|<1$ $\Leftrightarrow$ $\left|\dfrac{2}{x}\right|<1$ $\Leftrightarrow$

$2<|x| \Leftrightarrow x>2$ or $x<-2$. In that case, the sum of the series is $\dfrac{a}{1-r}=\dfrac{1}{1-2/x}=\dfrac{x}{x-2}$.

62. $\sum_{n=0}^{\infty}\dfrac{\sin^n x}{3^n}=\sum_{n=0}^{\infty}\left(\dfrac{\sin x}{3}\right)^n$ is a geometric series with $r=\dfrac{\sin x}{3}$, so the series converges $\Leftrightarrow$ $|r|<1$ $\Leftrightarrow$

$\left|\dfrac{\sin x}{3}\right|<1 \Leftrightarrow |\sin x|<3$, which is true for all x. Thus, the sum of the series is $\dfrac{a}{1-r}=\dfrac{1}{1-(\sin x)/3}=\dfrac{3}{3-\sin x}$.

63. $\sum_{n=0}^{\infty}e^{nx}=\sum_{n=0}^{\infty}(e^x)^n$ is a geometric series with $r=e^x$, so the series converges $\Leftrightarrow$ $|r|<1$ $\Leftrightarrow$ $|e^x|<1$ $\Leftrightarrow$

$-1<e^x<1 \Leftrightarrow 0<e^x<1 \Leftrightarrow x<0$. In that case, the sum of the series is $\dfrac{a}{1-r}=\dfrac{1}{1-e^x}$.

64. Because $\dfrac{1}{n}\to 0$ and ln is continuous, we have $\lim\limits_{n\to\infty}\ln\left(1+\dfrac{1}{n}\right)=\ln 1=0$.

We now show that the series $\sum_{n=1}^{\infty}\ln\left(1+\dfrac{1}{n}\right)=\sum_{n=1}^{\infty}\ln\left(\dfrac{n+1}{n}\right)=\sum_{n=1}^{\infty}[\ln(n+1)-\ln n]$ diverges.

$s_n=(\ln 2-\ln 1)+(\ln 3-\ln 2)+\cdots+(\ln(n+1)-\ln n)=\ln(n+1)-\ln 1=\ln(n+1)$.

As $n\to\infty$, $s_n=\ln(n+1)\to\infty$, so the series diverges.

65. After defining f, We use `convert(f,parfrac);` in Maple, `Apart` in Mathematica, or `Expand Rational` and `Simplify` in Derive to find that the general term is $\dfrac{3n^2+3n+1}{(n^2+n)^3} = \dfrac{1}{n^3} - \dfrac{1}{(n+1)^3}$. So the nth partial sum is

$$s_n = \sum_{k=1}^{n}\left(\frac{1}{k^3} - \frac{1}{(k+1)^3}\right) = \left(1 - \frac{1}{2^3}\right) + \left(\frac{1}{2^3} - \frac{1}{3^3}\right) + \cdots + \left(\frac{1}{n^3} - \frac{1}{(n+1)^3}\right) = 1 - \frac{1}{(n+1)^3}$$

The series converges to $\lim\limits_{n\to\infty} s_n = 1$. This can be confirmed by directly computing the sum using `sum(f,n=1..infinity);` (in Maple), `Sum[f,{n,1,Infinity}]` (in Mathematica), or `Calculus Sum` (from 1 to ∞) and `Simplify` (in Derive).

66. See Exercise 65 for specific CAS commands.

$\dfrac{1}{n^5 - 5n^3 + 4n} = \dfrac{1}{24(n-2)} + \dfrac{1}{24(n+2)} - \dfrac{1}{6(n-1)} - \dfrac{1}{6(n+1)} + \dfrac{1}{4n}$. So the nth partial sum is

$$\begin{aligned} s_n &= \frac{1}{24}\sum_{k=3}^{n}\left(\frac{1}{k-2} - \frac{4}{k-1} + \frac{6}{k} - \frac{4}{k+1} + \frac{1}{k+2}\right) \\ &= \frac{1}{24}\left[\left(\frac{1}{1} - \frac{4}{2} + \frac{6}{3} - \frac{4}{4} + \frac{1}{5}\right) + \cdots + \left(\frac{1}{n-2} - \frac{4}{n-1} + \frac{6}{n} - \frac{4}{n+1} + \frac{1}{n+2}\right)\right] \end{aligned}$$

The terms with denominator 5 or greater cancel, except for a few terms with n in the denominator. So as $n \to \infty$,

$$s_n \to \frac{1}{24}\left(\frac{1}{1} - \frac{3}{2} + \frac{3}{3} - \frac{1}{4}\right) = \frac{1}{24}\left(\frac{1}{4}\right) = \frac{1}{96}.$$

67. For $n = 1$, $a_1 = 0$ since $s_1 = 0$. For $n > 1$,

$$a_n = s_n - s_{n-1} = \frac{n-1}{n+1} - \frac{(n-1)-1}{(n-1)+1} = \frac{(n-1)n - (n+1)(n-2)}{(n+1)n} = \frac{2}{n(n+1)}$$

Also, $\sum\limits_{n=1}^{\infty} a_n = \lim\limits_{n\to\infty} s_n = \lim\limits_{n\to\infty} \dfrac{1-1/n}{1+1/n} = 1$.

68. $a_1 = s_1 = 3 - \frac{1}{2} = \frac{5}{2}$. For $n \neq 1$,

$$a_n = s_n - s_{n-1} = \left(3 - n2^{-n}\right) - \left[3 - (n-1)2^{-(n-1)}\right] = -\frac{n}{2^n} + \frac{n-1}{2^{n-1}}\cdot\frac{2}{2} = \frac{2(n-1)}{2^n} - \frac{n}{2^n} = \frac{n-2}{2^n}$$

Also, $\sum\limits_{n=1}^{\infty} a_n = \lim\limits_{n\to\infty} s_n = \lim\limits_{n\to\infty}\left(3 - \dfrac{n}{2^n}\right) = 3$ because $\lim\limits_{x\to\infty}\dfrac{x}{2^x} \overset{\text{H}}{=} \lim\limits_{x\to\infty}\dfrac{1}{2^x \ln 2} = 0$.

69. (a) The quantity of the drug in the body after the first tablet is 100 mg. After the second tablet, there is 100 mg plus 20% of the first 100-mg tablet; that is, $100 + 0.20(100) = 120$ mg. After the third tablet, the quantity is $100 + 0.20(120)$ or, equivalently, $100 + 100(0.20) + 100(0.20)^2$. Either expression gives us 124 mg.

(b) From part (a), we see that $Q_{n+1} = 100 + 0.20\,Q_n$.

(c) $Q_n = 100 + 100(0.20)^1 + 100(0.20)^2 + \cdots + 100(0.20)^{n-1}$

$= \sum\limits_{i=1}^{n} 100(0.20)^{i-1}$ [geometric with $a = 100$ and $r = 0.20$].

The quantity of the antibiotic that remains in the body in the long run is $\lim\limits_{n\to\infty} Q_n = \dfrac{100}{1-0.20} = \dfrac{100}{4/5} = 125$ mg.

70. (a) The concentration of the drug after the first injection is 1.5 mg/L. "Reduced by 90%" is the same as 10% remains, so the concentration after the second injection is $1.5 + 0.10(1.5) = 1.65$ mg/L. The concentration after the third injection is $1.5 + 0.10(1.65)$, or, equivalently, $1.5 + 1.5(0.10) + 1.5(0.10)^2$. Either expression gives us 1.665 mg/L.

(b) $C_n = 1.5 + 1.5(0.10)^1 + 1.5(0.10)^2 + \cdots + 1.5(0.10)^{n-1}$

$= \sum_{i=1}^{n} 1.5(0.10)^{i-1}$ [geometric with $a = 1.5$ and $r = 0.10$].

By (3), $C_n = \dfrac{1.5[1-(0.10)^n]}{1-0.10} = \dfrac{1.5}{0.9}[1-(0.10)^n] = \dfrac{5}{3}[1-(0.10)^n]$ mg/L.

(c) The limiting value of the concentration is $\lim\limits_{n\to\infty} C_n = \lim\limits_{n\to\infty} \frac{5}{3}[1-(0.10)^n] = \frac{5}{3}(1-0) = \frac{5}{3}$ mg/L.

71. (a) The quantity of the drug in the body after the first tablet is 150 mg. After the second tablet, there is 150 mg plus 5% of the first 150-mg tablet, that is, $[150 + 150(0.05)]$ mg. After the third tablet, the quantity is $[150 + 150(0.05) + 150(0.05)^2] = 157.875$ mg. After n tablets, the quantity (in mg) is $150 + 150(0.05) + \cdots + 150(0.05)^{n-1}$. We can use Formula 3 to write this as $\dfrac{150(1-0.05^n)}{1-0.05} = \dfrac{3000}{19}(1-0.05^n)$.

(b) The number of milligrams remaining in the body in the long run is $\lim\limits_{n\to\infty}\left[\frac{3000}{19}(1-0.05^n)\right] = \frac{3000}{19}(1-0) \approx 157.895$, only 0.02 mg more than the amount after 3 tablets.

72. (a) The residual concentration just before the second injection is De^{-aT}; before the third, $De^{-aT} + De^{-a2T}$; before the $(n+1)$st, $De^{-aT} + De^{-a2T} + \cdots + De^{-anT}$. This sum is equal to $\dfrac{De^{-aT}(1-e^{-anT})}{1-e^{-aT}}$ [Formula 3].

(b) The limiting pre-injection concentration is $\lim\limits_{n\to\infty} \dfrac{De^{-aT}(1-e^{-anT})}{1-e^{-aT}} = \dfrac{De^{-aT}(1-0)}{1-e^{-aT}} \cdot \dfrac{e^{aT}}{e^{aT}} = \dfrac{D}{e^{aT}-1}$.

(c) $\dfrac{D}{e^{aT}-1} \geq C \;\Rightarrow\; D \geq C(e^{aT}-1)$, so the minimal dosage is $D = C(e^{aT}-1)$.

73. (a) The first step in the chain occurs when the local government spends D dollars. The people who receive it spend a fraction c of those D dollars, that is, Dc dollars. Those who receive the Dc dollars spend a fraction c of it, that is, Dc^2 dollars. Continuing in this way, we see that the total spending after n transactions is

$$S_n = D + Dc + Dc^2 + \cdots + Dc^{n-1} = \frac{D(1-c^n)}{1-c} \text{ by (3).}$$

(b) $\lim\limits_{n\to\infty} S_n = \lim\limits_{n\to\infty} \dfrac{D(1-c^n)}{1-c} = \dfrac{D}{1-c}\lim\limits_{n\to\infty}(1-c^n) = \dfrac{D}{1-c}$ $\left[\text{since } 0 < c < 1 \;\Rightarrow\; \lim\limits_{n\to\infty} c^n = 0\right]$

$= \dfrac{D}{s}$ [since $c + s = 1$] $= kD$ [since $k = 1/s$]

If $c = 0.8$, then $s = 1 - c = 0.2$ and the multiplier is $k = 1/s = 5$.

74. (a) Initially, the ball falls a distance H, then rebounds a distance rH, falls rH, rebounds r^2H, falls r^2H, etc. The total distance it travels is

$$H + 2rH + 2r^2H + 2r^3H + \cdots = H(1 + 2r + 2r^2 + 2r^3 + \cdots) = H[1 + 2r(1 + r + r^2 + \cdots)]$$
$$= H\left[1 + 2r\left(\frac{1}{1-r}\right)\right] = H\left(\frac{1+r}{1-r}\right) \text{ meters}$$

(b) From Example 2.1.3, we know that a ball falls $\frac{1}{2}gt^2$ meters in t seconds, where g is the gravitational acceleration. Thus, a ball falls h meters in $t = \sqrt{2h/g}$ seconds. The total travel time in seconds is

$$\sqrt{\frac{2H}{g}} + 2\sqrt{\frac{2H}{g}r} + 2\sqrt{\frac{2H}{g}r^2} + 2\sqrt{\frac{2H}{g}r^3} + \cdots = \sqrt{\frac{2H}{g}}\left[1 + 2\sqrt{r} + 2\sqrt{r}^2 + 2\sqrt{r}^3 + \cdots\right]$$
$$= \sqrt{\frac{2H}{g}}\left(1 + 2\sqrt{r}\left[1 + \sqrt{r} + \sqrt{r}^2 + \cdots\right]\right)$$
$$= \sqrt{\frac{2H}{g}}\left[1 + 2\sqrt{r}\left(\frac{1}{1-\sqrt{r}}\right)\right] = \sqrt{\frac{2H}{g}}\frac{1+\sqrt{r}}{1-\sqrt{r}}$$

(c) It will help to make a chart of the time for each descent and each rebound of the ball, together with the velocity just before and just after each bounce. Recall that the time in seconds needed to fall h meters is $\sqrt{2h/g}$. The ball hits the ground with velocity $-g\sqrt{2h/g} = -\sqrt{2hg}$ (taking the upward direction to be positive) and rebounds with velocity $kg\sqrt{2h/g} = k\sqrt{2hg}$, taking time $k\sqrt{2h/g}$ to reach the top of its bounce, where its velocity is 0. At that point, its height is k^2h. All these results follow from the formulas for vertical motion with gravitational acceleration $-g$: $\dfrac{d^2y}{dt^2} = -g \;\Rightarrow\; v = \dfrac{dy}{dt} = v_0 - gt \;\Rightarrow\; y = y_0 + v_0t - \frac{1}{2}gt^2$.

number of descent	time of descent	speed before bounce	speed after bounce	time of ascent	peak height
1	$\sqrt{2H/g}$	$\sqrt{2Hg}$	$k\sqrt{2Hg}$	$k\sqrt{2H/g}$	k^2H
2	$\sqrt{2k^2H/g}$	$\sqrt{2k^2Hg}$	$k\sqrt{2k^2Hg}$	$k\sqrt{2k^2H/g}$	k^4H
3	$\sqrt{2k^4H/g}$	$\sqrt{2k^4Hg}$	$k\sqrt{2k^4Hg}$	$k\sqrt{2k^4H/g}$	k^6H
$\cdots$	$\cdots$	$\cdots$	$\cdots$	$\cdots$	$\cdots$

The total travel time in seconds is

$$\sqrt{\frac{2H}{g}} + k\sqrt{\frac{2H}{g}} + k\sqrt{\frac{2H}{g}} + k^2\sqrt{\frac{2H}{g}} + k^2\sqrt{\frac{2H}{g}} + \cdots = \sqrt{\frac{2H}{g}}(1 + 2k + 2k^2 + 2k^3 + \cdots)$$
$$= \sqrt{\frac{2H}{g}}[1 + 2k(1 + k + k^2 + \cdots)]$$
$$= \sqrt{\frac{2H}{g}}\left[1 + 2k\left(\frac{1}{1-k}\right)\right] = \sqrt{\frac{2H}{g}}\frac{1+k}{1-k}$$

Another method: We could use part (b). At the top of the bounce, the height is $k^2h = rh$, so $\sqrt{r} = k$ and the result follows from part (b).

75. $\sum_{n=2}^{\infty}(1+c)^{-n}$ is a geometric series with $a=(1+c)^{-2}$ and $r=(1+c)^{-1}$, so the series converges when

$\left|(1+c)^{-1}\right|<1 \;\Leftrightarrow\; |1+c|>1 \;\Leftrightarrow\; 1+c>1$ or $1+c<-1 \;\Leftrightarrow\; c>0$ or $c<-2$. We calculate the sum of the

series and set it equal to 2: $\dfrac{(1+c)^{-2}}{1-(1+c)^{-1}}=2 \;\Leftrightarrow\; \left(\dfrac{1}{1+c}\right)^2=2-2\left(\dfrac{1}{1+c}\right) \;\Leftrightarrow\; 1=2(1+c)^2-2(1+c) \;\Leftrightarrow$

$2c^2+2c-1=0 \;\Leftrightarrow\; c=\frac{-2\pm\sqrt{12}}{4}=\frac{\pm\sqrt{3}-1}{2}$. However, the negative root is inadmissible because $-2<\frac{-\sqrt{3}-1}{2}<0$.

So $c=\frac{\sqrt{3}-1}{2}$.

76. $\sum_{n=0}^{\infty}e^{nc}=\sum_{n=0}^{\infty}(e^c)^n$ is a geometric series with $a=(e^c)^0=1$ and $r=e^c$. If $e^c<1$, it has sum $\dfrac{1}{1-e^c}$, so $\dfrac{1}{1-e^c}=10 \;\Rightarrow$

$\frac{1}{10}=1-e^c \;\Rightarrow\; e^c=\frac{9}{10} \;\Rightarrow\; c=\ln\frac{9}{10}$.

77. $e^{s_n}=e^{1+\frac12+\frac13+\cdots+\frac1n}=e^1e^{1/2}e^{1/3}\cdots e^{1/n}>(1+1)\left(1+\frac12\right)\left(1+\frac13\right)\cdots\left(1+\frac1n\right)\qquad [e^x>1+x]$

$$=\frac{2}{1}\frac{3}{2}\frac{4}{3}\cdots\frac{n+1}{n}=n+1$$

Thus, $e^{s_n}>n+1$ and $\lim\limits_{n\to\infty}e^{s_n}=\infty$. Since $\{s_n\}$ is increasing, $\lim\limits_{n\to\infty}s_n=\infty$, implying that the harmonic series is divergent.

78. The area between $y=x^{n-1}$ and $y=x^n$ for $0\le x\le 1$ is

$$\int_0^1\left(x^{n-1}-x^n\right)dx=\left[\frac{x^n}{n}-\frac{x^{n+1}}{n+1}\right]_0^1=\frac{1}{n}-\frac{1}{n+1}$$

$$=\frac{(n+1)-n}{n(n+1)}=\frac{1}{n(n+1)}$$

We can see from the diagram that as $n\to\infty$, the sum of the areas between the successive curves approaches the area of the unit square, that is, 1. So $\sum_{n=1}^{\infty}\dfrac{1}{n\,(n+1)}=1$.

79. Let d_n be the diameter of C_n. We draw lines from the centers of the C_i to the center of D (or C), and using the Pythagorean Theorem, we can write

$1^2+\left(1-\frac12 d_1\right)^2=\left(1+\frac12 d_1\right)^2 \;\Leftrightarrow$

$1=\left(1+\frac12 d_1\right)^2-\left(1-\frac12 d_1\right)^2=2d_1$ [difference of squares] $\Rightarrow\; d_1=\frac12$.

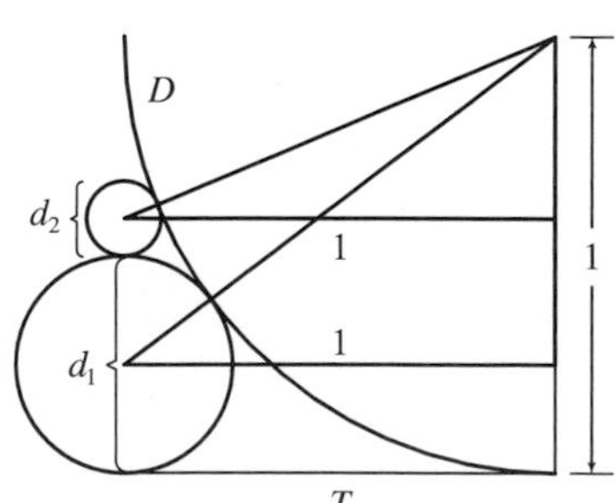

Similarly,

$$1=\left(1+\tfrac12 d_2\right)^2-\left(1-d_1-\tfrac12 d_2\right)^2=2d_2+2d_1-d_1^2-d_1d_2$$
$$=(2-d_1)(d_1+d_2)\;\Leftrightarrow$$

$d_2 = \dfrac{1}{2 - d_1} - d_1 = \dfrac{(1 - d_1)^2}{2 - d_1}$, $1 = \left(1 + \frac{1}{2}d_3\right)^2 - \left(1 - d_1 - d_2 - \frac{1}{2}d_3\right)^2 \Leftrightarrow d_3 = \dfrac{[1 - (d_1 + d_2)]^2}{2 - (d_1 + d_2)}$, and in general,

$d_{n+1} = \dfrac{\left(1 - \sum_{i=1}^{n} d_i\right)^2}{2 - \sum_{i=1}^{n} d_i}$. If we actually calculate d_2 and d_3 from the formulas above, we find that they are $\dfrac{1}{6} = \dfrac{1}{2 \cdot 3}$ and $\dfrac{1}{12} = \dfrac{1}{3 \cdot 4}$ respectively, so we suspect that in general, $d_n = \dfrac{1}{n(n+1)}$. To prove this, we use induction: Assume that for all $k \le n$, $d_k = \dfrac{1}{k(k+1)} = \dfrac{1}{k} - \dfrac{1}{k+1}$. Then $\sum\limits_{i=1}^{n} d_i = 1 - \dfrac{1}{n+1} = \dfrac{n}{n+1}$ [telescoping sum]. Substituting this into our formula for d_{n+1}, we get $d_{n+1} = \dfrac{\left[1 - \dfrac{n}{n+1}\right]^2}{2 - \left(\dfrac{n}{n+1}\right)} = \dfrac{\dfrac{1}{(n+1)^2}}{\dfrac{n+2}{n+1}} = \dfrac{1}{(n+1)(n+2)}$, and the induction is complete.

Now, we observe that the partial sums $\sum_{i=1}^{n} d_i$ of the diameters of the circles approach 1 as $n \to \infty$; that is, $\sum\limits_{n=1}^{\infty} a_n = \sum\limits_{n=1}^{\infty} \dfrac{1}{n(n+1)} = 1$, which is what we wanted to prove.

80. $|CD| = b \sin\theta$, $|DE| = |CD| \sin\theta = b \sin^2\theta$, $|EF| = |DE| \sin\theta = b \sin^3\theta$, ... Therefore,

$|CD| + |DE| + |EF| + |FG| + \cdots = b \sum\limits_{n=1}^{\infty} \sin^n\theta = b\left(\dfrac{\sin\theta}{1 - \sin\theta}\right)$ since this is a geometric series with $r = \sin\theta$ and $|\sin\theta| < 1$ $\left[\text{because } 0 < \theta < \frac{\pi}{2}\right]$.

81. The series $1 - 1 + 1 - 1 + 1 - 1 + \cdots$ diverges (geometric series with $r = -1$) so we cannot say that $0 = 1 - 1 + 1 - 1 + 1 - 1 + \cdots$.

82. If $\sum\limits_{n=1}^{\infty} a_n$ is convergent, then $\lim\limits_{n\to\infty} a_n = 0$ by Theorem 6, so $\lim\limits_{n\to\infty} \dfrac{1}{a_n} \ne 0$, and so $\sum\limits_{n=1}^{\infty} \dfrac{1}{a_n}$ is divergent by the Test for Divergence.

83. $\sum_{n=1}^{\infty} ca_n = \lim\limits_{n\to\infty} \sum_{i=1}^{n} ca_i = \lim\limits_{n\to\infty} c \sum_{i=1}^{n} a_i = c \lim\limits_{n\to\infty} \sum_{i=1}^{n} a_i = c \sum_{n=1}^{\infty} a_n$, which exists by hypothesis.

84. If $\sum ca_n$ were convergent, then $\sum (1/c)(ca_n) = \sum a_n$ would be also, by Theorem 8(i). But this is not the case, so $\sum ca_n$ must diverge.

85. Suppose on the contrary that $\sum (a_n + b_n)$ converges. Then $\sum (a_n + b_n)$ and $\sum a_n$ are convergent series. So by Theorem 8(iii), $\sum [(a_n + b_n) - a_n]$ would also be convergent. But $\sum [(a_n + b_n) - a_n] = \sum b_n$, a contradiction, since $\sum b_n$ is given to be divergent.

86. No. For example, take $\sum a_n = \sum n$ and $\sum b_n = \sum (-n)$, which both diverge, yet $\sum (a_n + b_n) = \sum 0$, which converges with sum 0.

87. The partial sums $\{s_n\}$ form an increasing sequence, since $s_n - s_{n-1} = a_n > 0$ for all n. Also, the sequence $\{s_n\}$ is bounded since $s_n \leq 1000$ for all n. So by the Monotonic Sequence Theorem, the sequence of partial sums converges, that is, the series $\sum a_n$ is convergent.

88. (a) $\text{RHS} = \dfrac{1}{f_{n-1}f_n} - \dfrac{1}{f_nf_{n+1}} = \dfrac{f_nf_{n+1} - f_nf_{n-1}}{f_n^2f_{n-1}f_{n+1}} = \dfrac{f_{n+1} - f_{n-1}}{f_nf_{n-1}f_{n+1}} = \dfrac{(f_{n-1} + f_n) - f_{n-1}}{f_nf_{n-1}f_{n+1}} = \dfrac{1}{f_{n-1}f_{n+1}} = \text{LHS}$

(b) $$\sum_{n=2}^{\infty} \frac{1}{f_{n-1}f_{n+1}} = \sum_{n=2}^{\infty}\left(\frac{1}{f_{n-1}f_n} - \frac{1}{f_nf_{n+1}}\right) \quad \text{[from part (a)]}$$

$$= \lim_{n\to\infty}\left[\left(\frac{1}{f_1f_2} - \frac{1}{f_2f_3}\right) + \left(\frac{1}{f_2f_3} - \frac{1}{f_3f_4}\right) + \left(\frac{1}{f_3f_4} - \frac{1}{f_4f_5}\right) + \cdots + \left(\frac{1}{f_{n-1}f_n} - \frac{1}{f_nf_{n+1}}\right)\right]$$

$$= \lim_{n\to\infty}\left(\frac{1}{f_1f_2} - \frac{1}{f_nf_{n+1}}\right) = \frac{1}{f_1f_2} - 0 = \frac{1}{1\cdot 1} = 1 \quad \text{because } f_n \to \infty \text{ as } n \to \infty.$$

(c) $$\sum_{n=2}^{\infty} \frac{f_n}{f_{n-1}f_{n+1}} = \sum_{n=2}^{\infty}\left(\frac{f_n}{f_{n-1}f_n} - \frac{f_n}{f_nf_{n+1}}\right) \quad \text{[as above]}$$

$$= \sum_{n=2}^{\infty}\left(\frac{1}{f_{n-1}} - \frac{1}{f_{n+1}}\right)$$

$$= \lim_{n\to\infty}\left[\left(\frac{1}{f_1} - \frac{1}{f_3}\right) + \left(\frac{1}{f_2} - \frac{1}{f_4}\right) + \left(\frac{1}{f_3} - \frac{1}{f_5}\right) + \left(\frac{1}{f_4} - \frac{1}{f_6}\right) + \cdots + \left(\frac{1}{f_{n-1}} - \frac{1}{f_{n+1}}\right)\right]$$

$$= \lim_{n\to\infty}\left(\frac{1}{f_1} + \frac{1}{f_2} - \frac{1}{f_n} - \frac{1}{f_{n+1}}\right) = 1 + 1 - 0 - 0 = 2 \quad \text{because } f_n \to \infty \text{ as } n \to \infty.$$

89. (a) At the first step, only the interval $\left(\frac{1}{3}, \frac{2}{3}\right)$ (length $\frac{1}{3}$) is removed. At the second step, we remove the intervals $\left(\frac{1}{9}, \frac{2}{9}\right)$ and $\left(\frac{7}{9}, \frac{8}{9}\right)$, which have a total length of $2 \cdot \left(\frac{1}{3}\right)^2$. At the third step, we remove 2^2 intervals, each of length $\left(\frac{1}{3}\right)^3$. In general, at the nth step we remove 2^{n-1} intervals, each of length $\left(\frac{1}{3}\right)^n$, for a length of $2^{n-1} \cdot \left(\frac{1}{3}\right)^n = \frac{1}{3}\left(\frac{2}{3}\right)^{n-1}$. Thus, the total length of all removed intervals is $\sum\limits_{n=1}^{\infty} \frac{1}{3}\left(\frac{2}{3}\right)^{n-1} = \frac{1/3}{1 - 2/3} = 1$ $\left[\text{geometric series with } a = \frac{1}{3} \text{ and } r = \frac{2}{3}\right]$. Notice that at the nth step, the leftmost interval that is removed is $\left(\left(\frac{1}{3}\right)^n, \left(\frac{2}{3}\right)^n\right)$, so we never remove 0, and 0 is in the Cantor set. Also, the rightmost interval removed is $\left(1 - \left(\frac{2}{3}\right)^n, 1 - \left(\frac{1}{3}\right)^n\right)$, so 1 is never removed. Some other numbers in the Cantor set are $\frac{1}{3}$, $\frac{2}{3}$, $\frac{1}{9}$, $\frac{2}{9}$, $\frac{7}{9}$, and $\frac{8}{9}$.

(b) The area removed at the first step is $\frac{1}{9}$; at the second step, $8 \cdot \left(\frac{1}{9}\right)^2$; at the third step, $(8)^2 \cdot \left(\frac{1}{9}\right)^3$. In general, the area removed at the nth step is $(8)^{n-1}\left(\frac{1}{9}\right)^n = \frac{1}{9}\left(\frac{8}{9}\right)^{n-1}$, so the total area of all removed squares is

$$\sum_{n=1}^{\infty} \frac{1}{9}\left(\frac{8}{9}\right)^{n-1} = \frac{1/9}{1 - 8/9} = 1.$$

90. (a)

a_1	1	2	4	1	1	1000
a_2	2	3	1	4	1000	1
a_3	1.5	2.5	2.5	2.5	500.5	500.5
a_4	1.75	2.75	1.75	3.25	750.25	250.75
a_5	1.625	2.625	2.125	2.875	625.375	375.625
a_6	1.6875	2.6875	1.9375	3.0625	687.813	313.188
a_7	1.65625	2.65625	2.03125	2.96875	656.594	344.406
a_8	1.67188	2.67188	1.98438	3.01563	672.203	328.797
a_9	1.66406	2.66406	2.00781	2.99219	664.398	336.602
a_{10}	1.66797	2.66797	1.99609	3.00391	668.301	332.699
a_{11}	1.66602	2.66602	2.00195	2.99805	666.350	334.650
a_{12}	1.66699	2.66699	1.99902	3.00098	667.325	333.675

The limits seem to be $\frac{5}{3}, \frac{8}{3}$, 2, 3, 667, and 334. Note that the limits appear to be "weighted" more toward a_2. In general, we guess that the limit is $\dfrac{a_1 + 2a_2}{3}$.

(b) $a_{n+1} - a_n = \frac{1}{2}(a_n + a_{n-1}) - a_n = -\frac{1}{2}(a_n - a_{n-1}) = -\frac{1}{2}\left[\frac{1}{2}(a_{n-1} + a_{n-2}) - a_{n-1}\right]$

$$= -\tfrac{1}{2}\left[-\tfrac{1}{2}(a_{n-1} - a_{n-2})\right] = \cdots = \left(-\tfrac{1}{2}\right)^{n-1}(a_2 - a_1)$$

Note that we have used the formula $a_k = \frac{1}{2}(a_{k-1} + a_{k-2})$ a total of $n - 1$ times in this calculation, once for each k between 3 and $n + 1$. Now we can write

$$\begin{aligned} a_n &= a_1 + (a_2 - a_1) + (a_3 - a_2) + \cdots + (a_{n-1} - a_{n-2}) + (a_n - a_{n-1}) \\ &= a_1 + \sum_{k=1}^{n-1}(a_{k+1} - a_k) = a_1 + \sum_{k=1}^{n-1}\left(-\tfrac{1}{2}\right)^{k-1}(a_2 - a_1) \end{aligned}$$

and so

$$\lim_{n\to\infty} a_n = a_1 + (a_2 - a_1)\sum_{k=1}^{\infty}\left(-\tfrac{1}{2}\right)^{k-1} = a_1 + (a_2 - a_1)\left[\frac{1}{1 - (-1/2)}\right] = a_1 + \tfrac{2}{3}(a_2 - a_1) = \frac{a_1 + 2a_2}{3}.$$

91. (a) For $\displaystyle\sum_{n=1}^{\infty}\frac{n}{(n+1)!}$, $s_1 = \dfrac{1}{1\cdot 2} = \dfrac{1}{2}$, $s_2 = \dfrac{1}{2} + \dfrac{2}{1\cdot 2\cdot 3} = \dfrac{5}{6}$, $s_3 = \dfrac{5}{6} + \dfrac{3}{1\cdot 2\cdot 3\cdot 4} = \dfrac{23}{24}$,

$s_4 = \dfrac{23}{24} + \dfrac{4}{1\cdot 2\cdot 3\cdot 4\cdot 5} = \dfrac{119}{120}$. The denominators are $(n+1)!$, so a guess would be $s_n = \dfrac{(n+1)! - 1}{(n+1)!}$.

(b) For $n = 1$, $s_1 = \dfrac{1}{2} = \dfrac{2! - 1}{2!}$, so the formula holds for $n = 1$. Assume $s_k = \dfrac{(k+1)! - 1}{(k+1)!}$. Then

$$\begin{aligned} s_{k+1} &= \frac{(k+1)! - 1}{(k+1)!} + \frac{k+1}{(k+2)!} = \frac{(k+1)! - 1}{(k+1)!} + \frac{k+1}{(k+1)!(k+2)} = \frac{(k+2)! - (k+2) + k + 1}{(k+2)!} \\ &= \frac{(k+2)! - 1}{(k+2)!} \end{aligned}$$

Thus, the formula is true for $n = k + 1$. So by induction, the guess is correct.

(c) $\displaystyle\lim_{n\to\infty} s_n = \lim_{n\to\infty}\frac{(n+1)! - 1}{(n+1)!} = \lim_{n\to\infty}\left[1 - \frac{1}{(n+1)!}\right] = 1$ and so $\displaystyle\sum_{n=1}^{\infty}\frac{n}{(n+1)!} = 1$.

92.

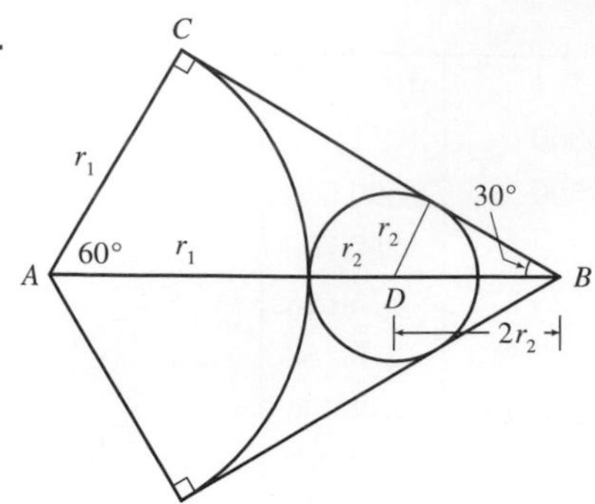

Let r_1 = radius of the large circle, r_2 = radius of next circle, and so on. From the figure we have $\angle BAC = 60°$ and $\cos 60° = r_1/|AB|$, so $|AB| = 2r_1$ and $|DB| = 2r_2$. Therefore, $2r_1 = r_1 + r_2 + 2r_2 \Rightarrow$ $r_1 = 3r_2$. In general, we have $r_{n+1} = \frac{1}{3}r_n$, so the total area is

$$A = \pi r_1^2 + 3\pi r_2^2 + 3\pi r_3^2 + \cdots = \pi r_1^2 + 3\pi r_2^2\left(1 + \frac{1}{3^2} + \frac{1}{3^4} + \frac{1}{3^6} + \cdots\right)$$

$$= \pi r_1^2 + 3\pi r_2^2 \cdot \frac{1}{1 - 1/9} = \pi r_1^2 + \tfrac{27}{8}\pi r_2^2$$

Since the sides of the triangle have length 1, $|BC| = \frac{1}{2}$ and $\tan 30° = \dfrac{r_1}{1/2}$. Thus, $r_1 = \dfrac{\tan 30°}{2} = \dfrac{1}{2\sqrt{3}} \Rightarrow r_2 = \dfrac{1}{6\sqrt{3}}$,

so $A = \pi\left(\dfrac{1}{2\sqrt{3}}\right)^2 + \dfrac{27\pi}{8}\left(\dfrac{1}{6\sqrt{3}}\right)^2 = \dfrac{\pi}{12} + \dfrac{\pi}{32} = \dfrac{11\pi}{96}$. The area of the triangle is $\dfrac{\sqrt{3}}{4}$, so the circles occupy about 83.1% of the area of the triangle.

11.3 The Integral Test and Estimates of Sums

1. The picture shows that $a_2 = \dfrac{1}{2^{1.3}} < \displaystyle\int_1^2 \frac{1}{x^{1.3}}\,dx$, $a_3 = \dfrac{1}{3^{1.3}} < \displaystyle\int_2^3 \frac{1}{x^{1.3}}\,dx$, and so on, so $\displaystyle\sum_{n=2}^{\infty} \frac{1}{n^{1.3}} < \int_1^{\infty} \frac{1}{x^{1.3}}\,dx$. The integral converges by (7.8.2) with $p = 1.3 > 1$, so the series converges.

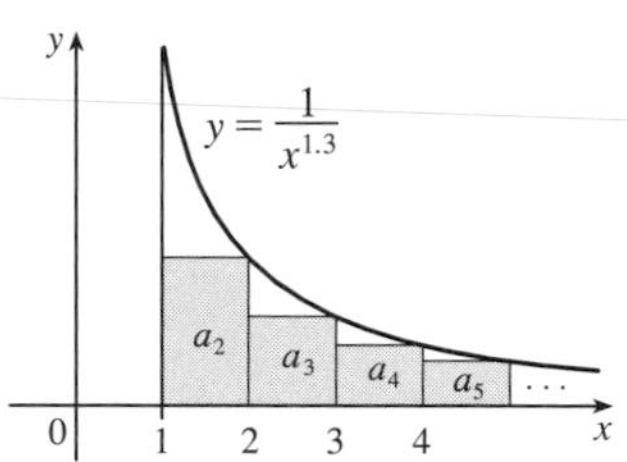

2. From the first figure, we see that $\int_1^6 f(x)\,dx < \sum_{i=1}^{5} a_i$. From the second figure, we see that $\sum_{i=2}^{6} a_i < \int_1^6 f(x)\,dx$. Thus, we have $\sum_{i=2}^{6} a_i < \int_1^6 f(x)\,dx < \sum_{i=1}^{5} a_i$.

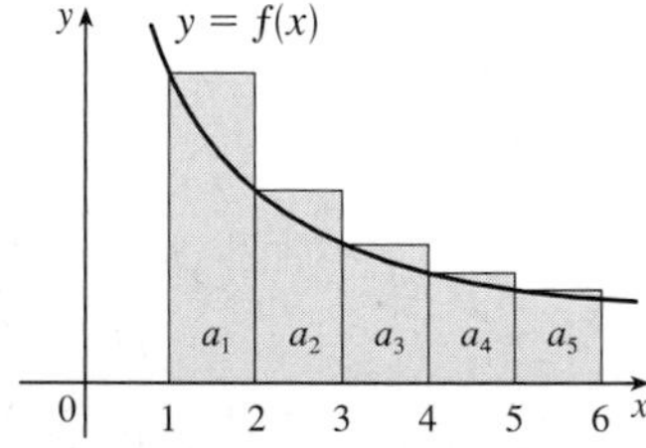

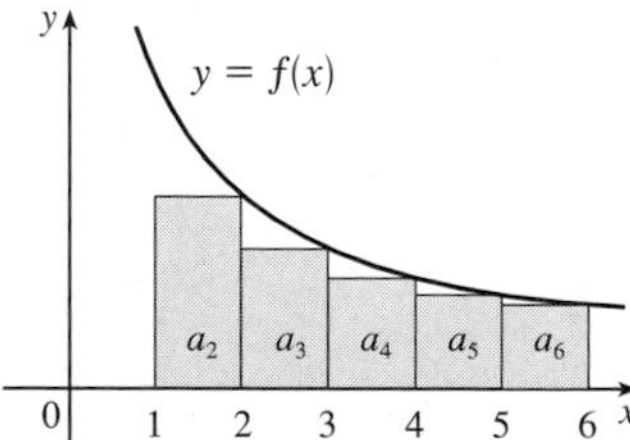

3. The function $f(x) = x^{-3}$ is continuous, positive, and decreasing on $[1, \infty)$, so the Integral Test applies.

$$\int_1^{\infty} x^{-3}\,dx = \lim_{t\to\infty}\int_1^t x^{-3}\,dx = \lim_{t\to\infty}\left[\frac{x^{-2}}{-2}\right]_1^t = \lim_{t\to\infty}\left(-\frac{1}{2t^2} + \frac{1}{2}\right) = \frac{1}{2}.$$

Since this improper integral is convergent, the series $\displaystyle\sum_{n=1}^{\infty} n^{-3}$ is also convergent by the Integral Test.

4. The function $f(x) = x^{-0.3}$ is continuous, positive, and decreasing on $[1, \infty)$, so the Integral Test applies.

$$\int_1^\infty x^{-0.3}\,dx = \lim_{t\to\infty}\int_1^t x^{-0.3}\,dx = \lim_{t\to\infty}\left[\frac{x^{0.7}}{0.7}\right]_1^t = \lim_{t\to\infty}\left(\frac{t^{0.7}}{0.7} - \frac{1}{0.7}\right) = \infty.$$

Since this improper integral is divergent, the series $\sum_{n=1}^{\infty} n^{-0.3}$ is also divergent by the Integral Test.

5. The function $f(x) = \dfrac{2}{5x-1}$ is continuous, positive, and decreasing on $[1, \infty)$, so the Integral Test applies.

$$\int_1^\infty \frac{2}{5x-1}\,dx = \lim_{t\to\infty}\int_1^t \frac{2}{5x-1}\,dx = \lim_{t\to\infty}\left[\frac{2}{5}\ln(5x-1)\right]_1^t = \lim_{t\to\infty}\left[\frac{2}{5}\ln(5t-1) - \frac{2}{5}\ln 4\right] = \infty.$$

Since this improper integral is divergent, the series $\sum_{n=1}^{\infty} \dfrac{2}{5n-1}$ is also divergent by the Integral Test.

6. The function $f(x) = \dfrac{1}{(3x-1)^4}$ is continuous, positive, and decreasing on $[1, \infty)$, so the Integral Test applies.

$$\int_1^\infty \frac{1}{(3x-1)^4}\,dx = \lim_{t\to\infty}\int_1^t (3x-1)^{-4}\,dx = \lim_{t\to\infty}\left[\frac{1}{(-3)3}(3x-1)^{-3}\right]_1^t = \lim_{t\to\infty}\left[-\frac{1}{9(3t-1)^3} + \frac{1}{9\cdot 2^3}\right] = \frac{1}{72}.$$

Since this improper integral is convergent, the series $\sum_{n=1}^{\infty} \dfrac{1}{(3n-1)^4}$ is also convergent by the Integral Test.

7. The function $f(x) = \dfrac{x}{x^2+1}$ is continuous, positive, and decreasing on $[1, \infty)$, so the Integral Test applies.

$\displaystyle\int_1^\infty \frac{x}{x^2+1}\,dx = \lim_{t\to\infty}\int_1^t \frac{x}{x^2+1}\,dx = \lim_{t\to\infty}\left[\frac{1}{2}\ln(x^2+1)\right]_1^t = \frac{1}{2}\lim_{t\to\infty}\left[\ln(t^2+1) - \ln 2\right] = \infty$. Since this improper integral is divergent, the series $\sum_{n=1}^{\infty} \dfrac{n}{n^2+1}$ is also divergent by the Integral Test.

8. The function $f(x) = x^2 e^{-x^3}$ is continuous, positive, and decreasing ($\star$) on $[1, \infty)$, so the Integral Test applies.

$$\int_1^\infty x^2 e^{-x^3}\,dx = \lim_{t\to\infty}\int_1^t x^2 e^{-x^3}\,dx = \lim_{t\to\infty}\left[-\frac{1}{3}e^{-x^3}\right]_1^t = -\frac{1}{3}\lim_{t\to\infty}\left(e^{-t^3} - e^{-1}\right) = -\frac{1}{3}\left(0 - \frac{1}{e}\right) = \frac{1}{3e}.$$

Since this improper integral is convergent, the series $\sum_{n=1}^{\infty} n^2 e^{-n^3}$ is also convergent by the Integral Test.

($\star$): $f'(x) = x^2 e^{-x^3}(-3x^2) + e^{-x^3}(2x) = xe^{-x^3}(-3x^3+2) = \dfrac{x(2-3x^3)}{e^{x^3}} < 0$ for $x > 1$

9. $\sum_{n=1}^{\infty} \dfrac{1}{n^{\sqrt{2}}}$ is a p-series with $p = \sqrt{2} > 1$, so it converges by (1).

10. $\sum_{n=3}^{\infty} n^{-0.9999} = \sum_{n=3}^{\infty} \dfrac{1}{n^{0.9999}}$ is a p-series with $p = 0.9999 \le 1$, so it diverges by (1). The fact that the series begins with $n = 3$ is irrelevant when determining convergence.

11. $1 + \dfrac{1}{8} + \dfrac{1}{27} + \dfrac{1}{64} + \dfrac{1}{125} + \cdots = \sum_{n=1}^{\infty} \dfrac{1}{n^3}$. This is a p-series with $p = 3 > 1$, so it converges by (1).

12. $\frac{1}{5}+\frac{1}{7}+\frac{1}{9}+\frac{1}{11}+\frac{1}{13}+\cdots=\sum_{n=1}^{\infty}\frac{1}{2n+3}$. The function $f(x)=\frac{1}{2x+3}$ is continuous, positive, and decreasing on $[1,\infty)$, so the Integral Test applies.

$$\int_1^{\infty}\frac{1}{2x+3}\,dx=\lim_{t\to\infty}\int_1^t\frac{1}{2x+3}\,dx=\lim_{t\to\infty}\left[\tfrac{1}{2}\ln(2x+3)\right]_1^t=\lim_{t\to\infty}\left[\tfrac{1}{2}\ln(2t+3)-\tfrac{1}{2}\ln 5\right]=\infty,$$ so the series $\sum_{n=1}^{\infty}\frac{1}{2n+3}$ diverges.

13. $\frac{1}{3}+\frac{1}{7}+\frac{1}{11}+\frac{1}{15}+\frac{1}{19}+\cdots=\sum_{n=1}^{\infty}\frac{1}{4n-1}$. The function $f(x)=\frac{1}{4x-1}$ is continuous, positive, and decreasing on $[1,\infty)$, so the Integral Test applies.

$$\int_1^{\infty}\frac{1}{4x-1}\,dx=\lim_{t\to\infty}\int_1^t\frac{1}{4x-1}\,dx=\lim_{t\to\infty}\left[\tfrac{1}{4}\ln(4x-1)\right]_1^t=\lim_{t\to\infty}\left[\tfrac{1}{4}\ln(4t-1)-\tfrac{1}{4}\ln 3\right]=\infty,$$ so the series $\sum_{n=1}^{\infty}\frac{1}{4n-1}$ diverges.

14. $1+\frac{1}{2\sqrt{2}}+\frac{1}{3\sqrt{3}}+\frac{1}{4\sqrt{4}}+\frac{1}{5\sqrt{5}}+\cdots=\sum_{n=1}^{\infty}\frac{1}{n\sqrt{n}}=\sum_{n=1}^{\infty}\frac{1}{n^{3/2}}$. This is a p-series with $p=\frac{3}{2}>1$, so it converges by (1).

15. $\sum_{n=1}^{\infty}\frac{\sqrt{n}+4}{n^2}=\sum_{n=1}^{\infty}\left(\frac{\sqrt{n}}{n^2}+\frac{4}{n^2}\right)=\sum_{n=1}^{\infty}\frac{1}{n^{3/2}}+\sum_{n=1}^{\infty}\frac{4}{n^2}$. $\sum_{n=1}^{\infty}\frac{1}{n^{3/2}}$ is a convergent p-series with $p=\frac{3}{2}>1$.

$\sum_{n=1}^{\infty}\frac{4}{n^2}=4\sum_{n=1}^{\infty}\frac{1}{n^2}$ is a constant multiple of a convergent p-series with $p=2>1$, so it converges. The sum of two convergent series is convergent, so the original series is convergent.

16. The function $f(x)=\frac{\sqrt{x}}{1+x^{3/2}}$ is continuous and positive on $[1,\infty)$.

$$f'(x)=\frac{(1+x^{3/2})\left(\frac{1}{2}x^{-1/2}\right)-x^{1/2}\left(\frac{3}{2}x^{1/2}\right)}{(1+x^{3/2})^2}=\frac{\frac{1}{2}x^{-1/2}+\frac{1}{2}x-\frac{3}{2}x}{(1+x^{3/2})^2}=\frac{1-2x^{3/2}}{2\sqrt{x}(1+x^{3/2})^2}<0$$ for $x\geq 1$, so f is decreasing on $[1,\infty)$, and the Integral Test applies.

$$\int_1^{\infty}\frac{\sqrt{x}}{1+x^{3/2}}\,dx=\lim_{t\to\infty}\int_1^t\frac{\sqrt{x}}{1+x^{3/2}}\,dx=\lim_{t\to\infty}\left[\tfrac{2}{3}\ln(1+x^{3/2})\right]_1^t\qquad\left[\begin{array}{c}\text{substitution}\\ \text{with } u=1+x^{3/2}\end{array}\right]$$

$$=\lim_{t\to\infty}\left[\tfrac{2}{3}\ln(1+t^{3/2})-\tfrac{2}{3}\ln 2\right]=\infty,$$

so the series $\sum_{n=1}^{\infty}\frac{\sqrt{n}}{1+n^{3/2}}$ diverges.

17. The function $f(x)=\frac{1}{x^2+4}$ is continuous, positive, and decreasing on $[1,\infty)$, so we can apply the Integral Test.

$$\int_1^{\infty}\frac{1}{x^2+4}\,dx=\lim_{t\to\infty}\int_1^t\frac{1}{x^2+4}\,dx=\lim_{t\to\infty}\left[\frac{1}{2}\tan^{-1}\frac{x}{2}\right]_1^t=\frac{1}{2}\lim_{t\to\infty}\left[\tan^{-1}\left(\frac{t}{2}\right)-\tan^{-1}\left(\frac{1}{2}\right)\right]$$

$$=\frac{1}{2}\left[\frac{\pi}{2}-\tan^{-1}\left(\frac{1}{2}\right)\right]$$

Therefore, the series $\sum_{n=1}^{\infty}\frac{1}{n^2+4}$ converges.

18. The function $f(x) = \dfrac{1}{x^2+2x+2}$ is continuous, positive, and decreasing on $[1,\infty)$, so the Integral Test applies.

$$\int_1^\infty \frac{1}{x^2+2x+2}\,dx = \lim_{t\to\infty}\int_1^t \frac{1}{(x+1)^2+1}\,dx = \lim_{t\to\infty}\Big[\arctan(x+1)\Big]_1^t$$
$$= \lim_{t\to\infty}[\arctan(t+1) - \arctan 2] = \frac{\pi}{2} - \arctan 2,$$

so the series $\displaystyle\sum_{n=1}^\infty \frac{1}{n^2+2n+2}$ converges.

19. The function $f(x) = \dfrac{x^3}{x^4+4}$ is continuous and positive on $[2,\infty)$, and is also decreasing since

$f'(x) = \dfrac{(x^4+4)(3x^2) - x^3(4x^3)}{(x^4+4)^2} = \dfrac{12x^2 - x^6}{(x^4+4)^2} = \dfrac{x^2(12-x^4)}{(x^4+4)^2} < 0$ for $x > \sqrt[4]{12} \approx 1.86$, so we can use the Integral Test on $[2,\infty)$.

$\displaystyle\int_2^\infty \frac{x^3}{x^4+4}\,dx = \lim_{t\to\infty}\int_2^t \frac{x^3}{x^4+4}\,dx = \lim_{t\to\infty}\Big[\tfrac{1}{4}\ln(x^4+4)\Big]_2^t = \lim_{t\to\infty}\Big[\tfrac{1}{4}\ln(t^4+4) - \tfrac{1}{4}\ln 20\Big] = \infty$, so the series

$\displaystyle\sum_{n=2}^\infty \frac{n^3}{n^4+4}$ diverges, and it follows that $\displaystyle\sum_{n=1}^\infty \frac{n^3}{n^4+4}$ diverges as well.

20. The function $f(x) = \dfrac{3x-4}{x^2-2x} = \dfrac{2}{x} + \dfrac{1}{x-2}$ [by partial fractions] is continuous, positive, and decreasing on $[3,\infty)$ since it is the sum of two such functions, so we can apply the Integral Test.

$\displaystyle\int_3^\infty \frac{3x-4}{x^2-x}\,dx = \lim_{t\to\infty}\int_3^t \left[\frac{2}{x} + \frac{1}{x-2}\right]dx = \lim_{t\to\infty}\Big[2\ln x + \ln(x-2)\Big]_3^t = \lim_{t\to\infty}[2\ln t + \ln(t-2) - 2\ln 3] = \infty.$

The integral is divergent, so the series $\displaystyle\sum_{n=3}^\infty \frac{3n-4}{n^2-n}$ is divergent.

21. $f(x) = \dfrac{1}{x\ln x}$ is continuous and positive on $[2,\infty)$, and also decreasing since $f'(x) = -\dfrac{1+\ln x}{x^2(\ln x)^2} < 0$ for $x > 2$, so we can use the Integral Test. $\displaystyle\int_2^\infty \frac{1}{x\ln x}\,dx = \lim_{t\to\infty}[\ln(\ln x)]_2^t = \lim_{t\to\infty}[\ln(\ln t) - \ln(\ln 2)] = \infty$, so the series $\displaystyle\sum_{n=2}^\infty \frac{1}{n\ln n}$ diverges.

22. The function $f(x) = \dfrac{\ln x}{x^2}$ is continuous and positive on $[2,\infty)$, and also decreasing since

$f'(x) = \dfrac{x^2(1/x) - (\ln x)(2x)}{(x^2)^2} = \dfrac{x - 2x\ln x}{x^4} = \dfrac{1-2\ln x}{x^3} < 0$ for $x > e^{1/2} \approx 1.65$, so we can use the Integral Test on $[2,\infty)$.

$$\int_2^\infty \frac{\ln x}{x^2}\,dx = \lim_{t\to\infty}\int_2^t \frac{\ln x}{x^2}\,dx = \lim_{t\to\infty}\left(\left[-\frac{\ln x}{x}\right]_2^t + \int_2^t \frac{1}{x^2}\,dx\right) \quad \left[\begin{array}{c}\text{by parts with}\\ u = \ln x,\ dv = (1/x^2)\,dx\end{array}\right]$$
$$= \lim_{t\to\infty}\left(-\frac{\ln t}{t} + \frac{\ln 2}{2} + \left[-\frac{1}{x}\right]_2^t\right) \overset{\text{H}}{=} \lim_{t\to\infty}\left(\frac{1/t}{1} + \frac{\ln 2}{2} - \frac{1}{t} + \frac{1}{2}\right) = \frac{\ln 2 + 1}{2},$$

so the series $\displaystyle\sum_{n=2}^\infty \frac{\ln n}{n^2}$ converges.

23. The function $f(x) = xe^{-x} = \dfrac{x}{e^x}$ is continuous and positive on $[1, \infty)$, and also decreasing since

$f'(x) = \dfrac{e^x \cdot 1 - xe^x}{(e^x)^2} = \dfrac{e^x(1-x)}{(e^x)^2} = \dfrac{1-x}{e^x} < 0$ for $x > 1$ [and $f(1) > f(2)$], so we can use the Integral Test on $[1, \infty)$.

$$\begin{aligned}\int_1^\infty xe^{-x}\,dx &= \lim_{t\to\infty}\int_1^t xe^{-x}\,dx = \lim_{t\to\infty}\left(\left[-xe^{-x}\right]_1^t + \int_1^t e^{-x}\,dx\right) \qquad \left[\begin{array}{c}\text{by parts with}\\ u = x,\, dv = e^{-x}\,dx\end{array}\right]\\ &= \lim_{t\to\infty}\left(-te^{-t} + e^{-1} + \left[-e^{-x}\right]_1^t\right) = \lim_{t\to\infty}\left(-\frac{t}{e^t} + \frac{1}{e} - \frac{1}{e^t} + \frac{1}{e}\right)\\ &\overset{\text{H}}{=} \lim_{t\to\infty}\left(-\frac{1}{e^t} + \frac{1}{e} - 0 + \frac{1}{e}\right) = \frac{2}{e},\end{aligned}$$

so the series $\sum\limits_{k=1}^{\infty} ke^{-k}$ converges.

24. The function $f(x) = xe^{-x^2} = \dfrac{x}{e^{x^2}}$ is continuous and positive on $[1, \infty)$, and also decreasing since

$f'(x) = \dfrac{e^{x^2} \cdot 1 - xe^{x^2} \cdot 2x}{(e^{x^2})^2} = \dfrac{1 - 2x^2}{e^{x^2}} < 0$ for $x > \sqrt{\frac{1}{2}} \approx 0.7$, so we can use the Integral Test on $[1, \infty)$.

$\displaystyle\int_1^\infty xe^{-x^2}\,dx = \lim_{t\to\infty}\int_1^t xe^{-x^2}\,dx = \lim_{t\to\infty}\left[-\tfrac{1}{2}e^{-x^2}\right]_1^t = \lim_{t\to\infty}\left(-\tfrac{1}{2}e^{-t^2} + \tfrac{1}{2}e^{-1}\right) = \frac{1}{2e}$, so the series $\sum\limits_{k=1}^{\infty} ke^{-k^2}$ converges.

25. The function $f(x) = \dfrac{1}{x^2 + x^3} = \dfrac{1}{x^2} - \dfrac{1}{x} + \dfrac{1}{x+1}$ [by partial fractions] is continuous, positive and decreasing on $[1, \infty)$, so the Integral Test applies.

$$\begin{aligned}\int_1^\infty f(x)\,dx &= \lim_{t\to\infty}\int_1^t\left(\frac{1}{x^2} - \frac{1}{x} + \frac{1}{x+1}\right)dx = \lim_{t\to\infty}\left[-\frac{1}{x} - \ln x + \ln(x+1)\right]_1^t\\ &= \lim_{t\to\infty}\left[-\frac{1}{t} + \ln\frac{t+1}{t} + 1 - \ln 2\right] = 0 + 0 + 1 - \ln 2\end{aligned}$$

The integral converges, so the series $\sum\limits_{n=1}^{\infty}\dfrac{1}{n^2 + n^3}$ converges.

26. The function $f(x) = \dfrac{x}{x^4 + 1}$ is positive, continuous, and decreasing on $[1, \infty)$. [Note that

$f'(x) = \dfrac{x^4 + 1 - 4x^4}{(x^4+1)^2} = \dfrac{1 - 3x^4}{(x^4+1)^2} < 0$ on $[1, \infty)$.] Thus, we can apply the Integral Test.

$\displaystyle\int_1^\infty \frac{x}{x^4+1}\,dx = \lim_{t\to\infty}\int_1^t \frac{\frac{1}{2}(2x)}{1 + (x^2)^2}\,dx = \lim_{t\to\infty}\left[\frac{1}{2}\tan^{-1}(x^2)\right]_1^t = \frac{1}{2}\lim_{t\to\infty}\left[\tan^{-1}(t^2) - \tan^{-1}1\right] = \frac{1}{2}\left(\frac{\pi}{2} - \frac{\pi}{4}\right) = \frac{\pi}{8}$

so the series $\sum\limits_{n=1}^{\infty}\dfrac{n}{n^4+1}$ converges.

27. The function $f(x) = \dfrac{\cos \pi x}{\sqrt{x}}$ is neither positive nor decreasing on $[1, \infty)$, so the hypotheses of the Integral Test are not satisfied for the series $\sum\limits_{n=1}^{\infty}\dfrac{\cos \pi n}{\sqrt{n}}$.

28. The function $f(x) = \dfrac{\cos^2 x}{1+x^2}$ is not decreasing on $[1, \infty)$, so the hypotheses of the Integral Test are not satisfied for the series $\sum\limits_{n=1}^{\infty} \dfrac{\cos^2 n}{1+n^2}$.

29. We have already shown (in Exercise 21) that when $p = 1$ the series $\sum\limits_{n=2}^{\infty} \dfrac{1}{n(\ln n)^p}$ diverges, so assume that $p \neq 1$.

$f(x) = \dfrac{1}{x(\ln x)^p}$ is continuous and positive on $[2, \infty)$, and $f'(x) = -\dfrac{p + \ln x}{x^2(\ln x)^{p+1}} < 0$ if $x > e^{-p}$, so that f is eventually decreasing and we can use the Integral Test.

$$\int_2^{\infty} \frac{1}{x(\ln x)^p}\,dx = \lim_{t\to\infty}\left[\frac{(\ln x)^{1-p}}{1-p}\right]_2^t \quad [\text{for } p \neq 1] = \lim_{t\to\infty}\left[\frac{(\ln t)^{1-p}}{1-p} - \frac{(\ln 2)^{1-p}}{1-p}\right]$$

This limit exists whenever $1 - p < 0 \quad\Leftrightarrow\quad p > 1$, so the series converges for $p > 1$.

30. $f(x) = \dfrac{1}{x \ln x\,[\ln(\ln x)]^p}$ is positive and continuous on $[3, \infty)$. For $p \geq 0$, f clearly decreases on $[3, \infty)$; and for $p < 0$, it can be verified that f is ultimately decreasing. Thus, we can apply the Integral Test.

$$I = \int_3^{\infty} \frac{dx}{x \ln x\,[\ln(\ln x)]^p} = \lim_{t\to\infty}\int_3^t \frac{[\ln(\ln x)]^{-p}}{x \ln x}\,dx = \lim_{t\to\infty}\left[\frac{[\ln(\ln x)]^{-p+1}}{-p+1}\right]_3^t \qquad [\text{for } p \neq 1]$$

$$= \lim_{t\to\infty}\left[\frac{[\ln(\ln t)]^{-p+1}}{-p+1} - \frac{[\ln(\ln 3)]^{-p+1}}{-p+1}\right],$$

which exists whenever $-p + 1 < 0 \quad\Leftrightarrow\quad p > 1$. If $p = 1$, then $I = \lim\limits_{t\to\infty}\left[\ln(\ln(\ln x))\right]_3^t = \infty$. Therefore, $\sum\limits_{n=3}^{\infty} \dfrac{1}{n \ln n\,[\ln(\ln n)]^p}$ converges for $p > 1$.

31. Clearly the series cannot converge if $p \geq -\frac{1}{2}$, because then $\lim\limits_{n\to\infty} n(1+n^2)^p \neq 0$. So assume $p < -\frac{1}{2}$. Then $f(x) = x(1+x^2)^p$ is continuous, positive, and eventually decreasing on $[1, \infty)$, and we can use the Integral Test.

$$\int_1^{\infty} x(1+x^2)^p\,dx = \lim_{t\to\infty}\left[\frac{1}{2}\cdot\frac{(1+x^2)^{p+1}}{p+1}\right]_1^t = \frac{1}{2(p+1)}\lim_{t\to\infty}\left[(1+t^2)^{p+1} - 2^{p+1}\right].$$

This limit exists and is finite $\Leftrightarrow\quad p + 1 < 0 \quad\Leftrightarrow\quad p < -1$, so the series $\sum\limits_{n=1}^{\infty} n(1+n^2)^p$ converges whenever $p < -1$.

32. If $p \leq 0$, $\lim\limits_{n\to\infty} \dfrac{\ln n}{n^p} = \infty$ and the series diverges, so assume $p > 0$. $f(x) = \dfrac{\ln x}{x^p}$ is positive and continuous and $f'(x) < 0$ for $x > e^{1/p}$, so f is eventually decreasing and we can use the Integral Test. Integration by parts gives

$$\int_1^{\infty} \frac{\ln x}{x^p}\,dx = \lim_{t\to\infty}\left[\frac{x^{1-p}\,[(1-p)\ln x - 1]}{(1-p)^2}\right]_1^t \quad (\text{for } p \neq 1) = \frac{1}{(1-p)^2}\left[\lim_{t\to\infty} t^{1-p}\,[(1-p)\ln t - 1] + 1\right],$$ which exists whenever $1 - p < 0 \quad\Leftrightarrow\quad p > 1$. Thus, $\sum\limits_{n=1}^{\infty} \dfrac{\ln n}{n^p}$ converges $\quad\Leftrightarrow\quad p > 1$.

33. Since this is a p-series with $p = x$, $\zeta(x)$ is defined when $x > 1$. Unless specified otherwise, the domain of a function f is the set of real numbers x such that the expression for $f(x)$ makes sense and defines a real number. So, in the case of a series, it's the set of real numbers x such that the series is convergent.

34. (a) $\sum\limits_{n=2}^{\infty} \frac{1}{n^2} = \sum\limits_{n=1}^{\infty} \frac{1}{n^2} - \frac{1}{1^2}$ [subtract a_1] $= \frac{\pi^2}{6} - 1$

(b) $\sum\limits_{n=3}^{\infty} \frac{1}{(n+1)^2} = \sum\limits_{n=4}^{\infty} \frac{1}{n^2} = \sum\limits_{n=1}^{\infty} \frac{1}{n^2} - \left(\frac{1}{1^2} + \frac{1}{2^2} + \frac{1}{3^2}\right) = \frac{\pi^2}{6} - \frac{49}{36}$

(c) $\sum\limits_{n=1}^{\infty} \frac{1}{(2n)^2} = \sum\limits_{n=1}^{\infty} \frac{1}{4n^2} = \frac{1}{4} \sum\limits_{n=1}^{\infty} \frac{1}{n^2} = \frac{1}{4}\left(\frac{\pi^2}{6}\right) = \frac{\pi^2}{24}$

35. (a) $\sum\limits_{n=1}^{\infty} \left(\frac{3}{n}\right)^4 = \sum\limits_{n=1}^{\infty} \frac{81}{n^4} = 81 \sum\limits_{n=1}^{\infty} \frac{1}{n^4} = 81\left(\frac{\pi^4}{90}\right) = \frac{9\pi^4}{10}$

(b) $\sum\limits_{k=5}^{\infty} \frac{1}{(k-2)^4} = \frac{1}{3^4} + \frac{1}{4^4} + \frac{1}{5^4} + \cdots = \sum\limits_{k=3}^{\infty} \frac{1}{k^4} = \frac{\pi^4}{90} - \left(\frac{1}{1^4} + \frac{1}{2^4}\right)$ [subtract a_1 and a_2] $= \frac{\pi^4}{90} - \frac{17}{16}$

36. (a) $f(x) = 1/x^4$ is positive and continuous and $f'(x) = -4/x^5$ is negative for $x > 0$, and so the Integral Test applies.

$$\sum_{n=1}^{\infty} \frac{1}{n^4} \approx s_{10} = \frac{1}{1^4} + \frac{1}{2^4} + \frac{1}{3^4} + \cdots + \frac{1}{10^4} \approx 1.082037.$$

$R_{10} \le \int_{10}^{\infty} \frac{1}{x^4}\,dx = \lim\limits_{t\to\infty} \left[\frac{1}{-3x^3}\right]_{10}^{t} = \lim\limits_{t\to\infty} \left(-\frac{1}{3t^3} + \frac{1}{3\,(10)^3}\right) = \frac{1}{3000}$, so the error is at most $0.000\overline{3}$.

(b) $s_{10} + \int_{11}^{\infty} \frac{1}{x^4}\,dx \le s \le s_{10} + \int_{10}^{\infty} \frac{1}{x^4}\,dx \quad\Rightarrow\quad s_{10} + \frac{1}{3(11)^3} \le s \le s_{10} + \frac{1}{3(10)^3} \quad\Rightarrow$

$1.082037 + 0.000250 = 1.082287 \le s \le 1.082037 + 0.000333 = 1.082370$, so we get $s \approx 1.08233$ with error ≤ 0.00005.

(c) The estimate in part (b) is $s \approx 1.08233$ with error ≤ 0.00005. The exact value given in Exercise 35 is $\pi^4/90 \approx 1.082323$. The difference is less than 0.00001.

(d) $R_n \le \int_{n}^{\infty} \frac{1}{x^4}\,dx = \frac{1}{3n^3}$. So $R_n < 0.00001 \;\Rightarrow\; \frac{1}{3n^3} < \frac{1}{10^5} \;\Rightarrow\; 3n^3 > 10^5 \;\Rightarrow\; n > \sqrt[3]{(10)^5/3} \approx 32.2$,

that is, for $n > 32$.

37. (a) $f(x) = \frac{1}{x^2}$ is positive and continuous and $f'(x) = -\frac{2}{x^3}$ is negative for $x > 0$, and so the Integral Test applies.

$$\sum_{n=1}^{\infty} \frac{1}{n^2} \approx s_{10} = \frac{1}{1^2} + \frac{1}{2^2} + \frac{1}{3^2} + \cdots + \frac{1}{10^2} \approx 1.549768.$$

$R_{10} \le \int_{10}^{\infty} \frac{1}{x^2}\,dx = \lim\limits_{t\to\infty} \left[\frac{-1}{x}\right]_{10}^{t} = \lim\limits_{t\to\infty} \left(-\frac{1}{t} + \frac{1}{10}\right) = \frac{1}{10}$, so the error is at most 0.1.

(b) $s_{10} + \int_{11}^{\infty} \frac{1}{x^2}\,dx \le s \le s_{10} + \int_{10}^{\infty} \frac{1}{x^2}\,dx \Rightarrow s_{10} + \frac{1}{11} \le s \le s_{10} + \frac{1}{10} \Rightarrow$

$1.549768 + 0.090909 = 1.640677 \le s \le 1.549768 + 0.1 = 1.649768$, so we get $s \approx 1.64522$ (the average of 1.640677 and 1.649768) with error ≤ 0.005 (the maximum of $1.649768 - 1.64522$ and $1.64522 - 1.640677$, rounded up).

(c) The estimate in part (b) is $s \approx 1.64522$ with error ≤ 0.005. The exact value given in Exercise 34 is $\pi^2/6 \approx 1.644934$. The difference is less than 0.0003.

(d) $R_n \le \int_n^{\infty} \frac{1}{x^2}\,dx = \frac{1}{n}$. So $R_n < 0.001$ if $\frac{1}{n} < \frac{1}{1000} \Leftrightarrow n > 1000$.

38. $f(x) = xe^{-2x}$ is continuous, positive, and decreasing on $[1, \infty)$, so the Integral Test applies. Using (2),

$$R_n \le \int_n^{\infty} xe^{-2x}\,dx = \lim_{t\to\infty}\left(\left[-\tfrac{1}{2}xe^{-2x}\right]_n^t + \int_n^t \tfrac{1}{2}e^{-2x}\,dx\right) \quad \left[\begin{array}{c}\text{using parts with}\\ u = x,\ dv = e^{-2x}\,dx\end{array}\right]$$

$$= \lim_{t\to\infty}\left(\frac{-t}{2e^{2t}} + \frac{n}{2e^{2n}} - \frac{1}{4e^{2t}} + \frac{1}{4e^{2n}}\right) \overset{\text{H}}{=} 0 + \frac{n}{2e^{2n}} - 0 + \frac{1}{4e^{2n}} = \frac{2n+1}{4e^{2n}}$$

To be correct to four decimal places, we want $\frac{2n+1}{4e^{2n}} \le \frac{5}{10^5}$. This inequality is true for $n = 6$.

$$s_6 = \sum_{n=1}^{6} \frac{n}{e^{2n}} = \frac{1}{e^2} + \frac{2}{e^4} + \frac{3}{e^6} + \frac{4}{e^8} + \frac{5}{e^{10}} + \frac{6}{e^{12}} \approx 0.1810.$$

39. $f(x) = 1/(2x+1)^6$ is continuous, positive, and decreasing on $[1, \infty)$, so the Integral Test applies. Using (2),

$R_n \le \int_n^{\infty} (2x+1)^{-6}\,dx = \lim_{t\to\infty}\left[\frac{-1}{10(2x+1)^5}\right]_n^t = \frac{1}{10(2n+1)^5}$. To be correct to five decimal places, we want

$$\frac{1}{10(2n+1)^5} \le \frac{5}{10^6} \Leftrightarrow (2n+1)^5 \ge 20{,}000 \Leftrightarrow n \ge \tfrac{1}{2}\left(\sqrt[5]{20{,}000} - 1\right) \approx 3.12, \text{ so use } n = 4.$$

$$s_4 = \sum_{n=1}^{4} \frac{1}{(2n+1)^6} = \frac{1}{3^6} + \frac{1}{5^6} + \frac{1}{7^6} + \frac{1}{9^6} \approx 0.001\,446 \approx 0.00145.$$

40. $f(x) = \frac{1}{x(\ln x)^2}$ is positive and continuous and $f'(x) = -\frac{\ln x + 2}{x^2(\ln x)^3}$ is negative for $x > 1$, so the Integral Test applies.

Using (2), we need $0.01 > \int_n^{\infty} \frac{dx}{x(\ln x)^2} = \lim_{t\to\infty}\left[\frac{-1}{\ln x}\right]_n^t = \frac{1}{\ln n}$. This is true for $n > e^{100}$, so we would have to add this many terms to find the sum of the series $\sum_{n=2}^{\infty} \frac{1}{n(\ln n)^2}$ to within 0.01, which would be problematic because $e^{100} \approx 2.7 \times 10^{43}$.

41. $\sum_{n=1}^{\infty} n^{-1.001} = \sum_{n=1}^{\infty} \frac{1}{n^{1.001}}$ is a convergent p-series with $p = 1.001 > 1$. Using (2), we get

$R_n \le \int_n^{\infty} x^{-1.001}\,dx = \lim_{t\to\infty}\left[\frac{x^{-0.001}}{-0.001}\right]_n^t = -1000 \lim_{t\to\infty}\left[\frac{1}{x^{0.001}}\right]_n^t = -1000\left(-\frac{1}{n^{0.001}}\right) = \frac{1000}{n^{0.001}}$. We want

$R_n < 0.000\,000\,005 \quad \Leftrightarrow \quad \dfrac{1000}{n^{0.001}} < 5 \times 10^{-9} \quad \Leftrightarrow \quad n^{0.001} > \dfrac{1000}{5 \times 10^{-9}} \quad \Leftrightarrow$

$n > \left(2 \times 10^{11}\right)^{1000} = 2^{1000} \times 10^{11{,}000} \approx 1.07 \times 10^{301} \times 10^{11{,}000} = 1.07 \times 10^{11{,}301}$.

42. (a) $f(x) = \left(\dfrac{\ln x}{x}\right)^2$ is continuous and positive for $x > 1$, and since $f'(x) = \dfrac{2\ln x\,(1 - \ln x)}{x^3} < 0$ for $x > e$, we can apply the Integral Test. Using a CAS, we get $\displaystyle\int_1^\infty \left(\frac{\ln x}{x}\right)^2 dx = 2$, so the series $\displaystyle\sum_{n=1}^{\infty} \left(\frac{\ln n}{n}\right)^2$ also converges.

(b) Since the Integral Test applies, the error in $s \approx s_n$ is $R_n \le \displaystyle\int_n^\infty \left(\frac{\ln x}{x}\right)^2 dx = \frac{(\ln n)^2 + 2\ln n + 2}{n}$.

(c) By graphing the functions $y_1 = \dfrac{(\ln x)^2 + 2\ln x + 2}{x}$ and $y_2 = 0.05$, we see that $y_1 < y_2$ for $n \ge 1373$.

(d) Using the CAS to sum the first 1373 terms, we get $s_{1373} \approx 1.94$.

43. (a) From the figure, $a_2 + a_3 + \cdots + a_n \le \int_1^n f(x)\,dx$, so with

$f(x) = \dfrac{1}{x}$, $\dfrac{1}{2} + \dfrac{1}{3} + \dfrac{1}{4} + \cdots + \dfrac{1}{n} \le \displaystyle\int_1^n \frac{1}{x}\,dx = \ln n$.

Thus, $s_n = 1 + \dfrac{1}{2} + \dfrac{1}{3} + \dfrac{1}{4} + \cdots + \dfrac{1}{n} \le 1 + \ln n$.

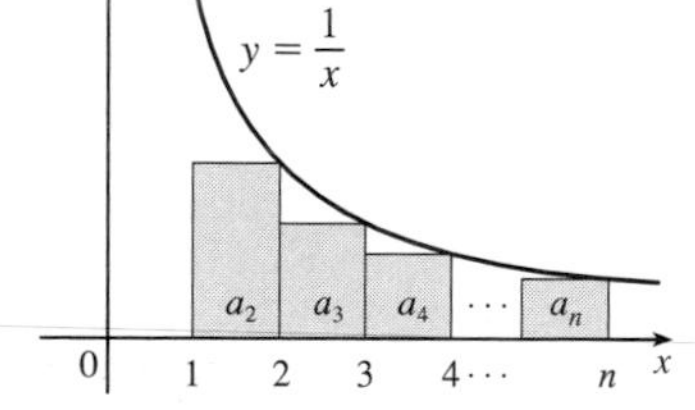

(b) By part (a), $s_{10^6} \le 1 + \ln 10^6 \approx 14.82 < 15$ and

$s_{10^9} \le 1 + \ln 10^9 \approx 21.72 < 22$.

44. (a) The sum of the areas of the n rectangles in the graph to the right is

$1 + \dfrac{1}{2} + \dfrac{1}{3} + \cdots + \dfrac{1}{n}$. Now $\displaystyle\int_1^{n+1} \frac{dx}{x}$ is less than this sum because

the rectangles extend above the curve $y = 1/x$, so

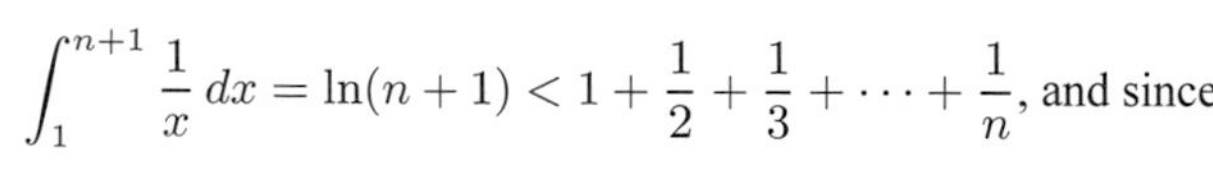

$\displaystyle\int_1^{n+1} \frac{1}{x}\,dx = \ln(n+1) < 1 + \frac{1}{2} + \frac{1}{3} + \cdots + \frac{1}{n}$, and since

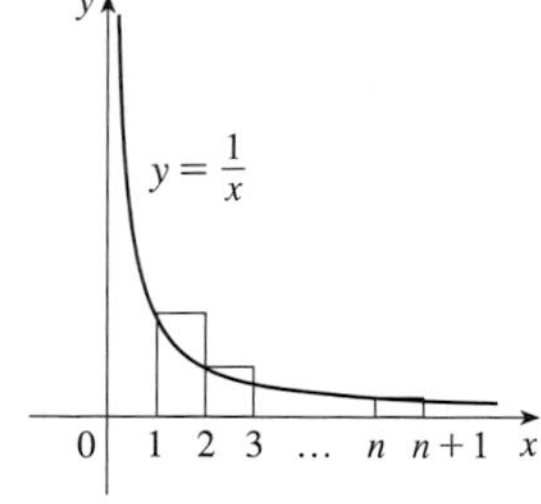

$\ln n < \ln(n+1)$, $0 < 1 + \dfrac{1}{2} + \dfrac{1}{3} + \cdots + \dfrac{1}{n} - \ln n = t_n$.

(b) The area under $f(x) = 1/x$ between $x = n$ and $x = n + 1$ is

$\displaystyle\int_n^{n+1} \frac{dx}{x} = \ln(n+1) - \ln n$, and this is clearly greater than the area of

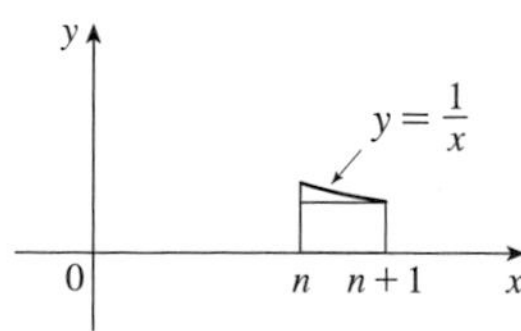

the inscribed rectangle in the figure to the right $\left[\text{which is } \dfrac{1}{n+1}\right]$, so

$t_n - t_{n+1} = [\ln(n+1) - \ln n] - \dfrac{1}{n+1} > 0$, and so $t_n > t_{n+1}$, so $\{t_n\}$ is a decreasing sequence.

(c) We have shown that $\{t_n\}$ is decreasing and that $t_n > 0$ for all n. Thus, $0 < t_n \le t_1 = 1$, so $\{t_n\}$ is a bounded monotonic sequence, and hence converges by the Monotonic Sequence Theorem.

45. $b^{\ln n} = \left(e^{\ln b}\right)^{\ln n} = \left(e^{\ln n}\right)^{\ln b} = n^{\ln b} = \dfrac{1}{n^{-\ln b}}$. This is a p-series, which converges for all b such that $-\ln b > 1 \Leftrightarrow$

$\ln b < -1 \Leftrightarrow b < e^{-1} \Leftrightarrow b < 1/e$ [with $b > 0$].

46. For the series $\sum\limits_{n=1}^{\infty}\left(\dfrac{c}{n} - \dfrac{1}{n+1}\right)$,

$$s_n = \sum_{i=1}^{n}\left(\frac{c}{i} - \frac{1}{i+1}\right) = \left(\frac{c}{1} - \frac{1}{2}\right) + \left(\frac{c}{2} - \frac{1}{3}\right) + \left(\frac{c}{3} - \frac{1}{4}\right) + \cdots + \left(\frac{c}{n} - \frac{1}{n+1}\right)$$

$$= \frac{c}{1} + \frac{c-1}{2} + \frac{c-1}{3} + \frac{c-1}{4} + \cdots + \frac{c-1}{n} - \frac{1}{n+1} = c + (c-1)\left(\frac{1}{2} + \frac{1}{3} + \frac{1}{4} + \cdots + \frac{1}{n}\right) - \frac{1}{n+1}$$

Thus, $\sum\limits_{n=1}^{\infty}\left(\dfrac{c}{n} - \dfrac{1}{n+1}\right) = \lim\limits_{n\to\infty} s_n = \lim\limits_{n\to\infty}\left[c + (c-1)\sum\limits_{i=2}^{n}\dfrac{1}{i} - \dfrac{1}{n+1}\right]$. Since a constant multiple of a divergent series is divergent, the last limit exists only if $c - 1 = 0$, so the original series converges only if $c = 1$.

11.4 The Comparison Tests

1. (a) We cannot say anything about $\sum a_n$. If $a_n > b_n$ for all n and $\sum b_n$ is convergent, then $\sum a_n$ could be convergent or divergent. (See the note after Example 2.)

(b) If $a_n < b_n$ for all n, then $\sum a_n$ is convergent. [This is part (i) of the Comparison Test.]

2. (a) If $a_n > b_n$ for all n, then $\sum a_n$ is divergent. [This is part (ii) of the Comparison Test.]

(b) We cannot say anything about $\sum a_n$. If $a_n < b_n$ for all n and $\sum b_n$ is divergent, then $\sum a_n$ could be convergent or divergent.

3. $\dfrac{1}{n^3+8} < \dfrac{1}{n^3}$ for all $n \geq 1$, so $\sum\limits_{n=1}^{\infty}\dfrac{1}{n^3+8}$ converges by comparison with $\sum\limits_{n=1}^{\infty}\dfrac{1}{n^3}$, which converges because it is a p-series with $p = 3 > 1$.

4. $\dfrac{1}{\sqrt{n}-1} > \dfrac{1}{\sqrt{n}}$ for all $n \geq 2$, so $\sum\limits_{n=2}^{\infty}\dfrac{1}{\sqrt{n}-1}$ diverges by comparison with $\sum\limits_{n=2}^{\infty}\dfrac{1}{\sqrt{n}}$, which diverges because it is a p-series with $p = \frac{1}{2} \leq 1$.

5. $\dfrac{n+1}{n\sqrt{n}} > \dfrac{n}{n\sqrt{n}} = \dfrac{1}{\sqrt{n}}$ for all $n \geq 1$, so $\sum\limits_{n=1}^{\infty}\dfrac{n+1}{n\sqrt{n}}$ diverges by comparison with $\sum\limits_{n=1}^{\infty}\dfrac{1}{\sqrt{n}}$, which diverges because it is a p-series with $p = \frac{1}{2} \leq 1$.

6. $\dfrac{n-1}{n^3+1} < \dfrac{n}{n^3+1} < \dfrac{n}{n^3} = \dfrac{1}{n^2}$ for all $n \geq 1$, so $\sum\limits_{n=1}^{\infty}\dfrac{n-1}{n^3+1}$ converges by comparison with $\sum\limits_{n=1}^{\infty}\dfrac{1}{n^2}$, which converges because it is a p-series with $p = 2 > 1$.

7. $\dfrac{9^n}{3+10^n} < \dfrac{9^n}{10^n} = \left(\dfrac{9}{10}\right)^n$ for all $n \geq 1$. $\sum\limits_{n=1}^{\infty}\left(\frac{9}{10}\right)^n$ is a convergent geometric series $\left(|r| = \frac{9}{10} < 1\right)$, so $\sum\limits_{n=1}^{\infty}\dfrac{9^n}{3+10^n}$ converges by the Comparison Test.

8. $\dfrac{6^n}{5^n - 1} > \dfrac{6^n}{5^n} = \left(\dfrac{6}{5}\right)^n$ for all $n \geq 1$. $\sum\limits_{n=1}^{\infty} \left(\frac{6}{5}\right)^n$ is a divergent geometric series $\left(|r| = \frac{6}{5} > 1\right)$, so $\sum\limits_{n=1}^{\infty} \dfrac{6^n}{5^n - 1}$ diverges by the Comparison Test.

9. $\dfrac{\ln k}{k} > \dfrac{1}{k}$ for all k ≥ 3 [since $\ln k > 1$ for $k \geq 3$], so $\sum\limits_{k=3}^{\infty} \dfrac{\ln k}{k}$ diverges by comparison with $\sum\limits_{k=3}^{\infty} \dfrac{1}{k}$, which diverges because it is a p-series with $p = 1 \leq 1$ (the harmonic series). Thus, $\sum\limits_{k=1}^{\infty} \dfrac{\ln k}{k}$ diverges since a finite number of terms doesn't affect the convergence or divergence of a series.

10. $\dfrac{k \sin^2 k}{1 + k^3} \leq \dfrac{k}{1 + k^3} < \dfrac{k}{k^3} = \dfrac{1}{k^2}$ for all $k \geq 1$, so $\sum\limits_{k=1}^{\infty} \dfrac{k \sin^2 k}{1 + k^3}$ converges by comparison with $\sum\limits_{k=1}^{\infty} \dfrac{1}{k^2}$, which converges because it is a p-series with $p = 2 > 1$.

11. $\dfrac{\sqrt[3]{k}}{\sqrt{k^3 + 4k + 3}} < \dfrac{\sqrt[3]{k}}{\sqrt{k^3}} = \dfrac{k^{1/3}}{k^{3/2}} = \dfrac{1}{k^{7/6}}$ for all $k \geq 1$, so $\sum\limits_{k=1}^{\infty} \dfrac{\sqrt[3]{k}}{\sqrt{k^3 + 4k + 3}}$ converges by comparison with $\sum\limits_{k=1}^{\infty} \dfrac{1}{k^{7/6}}$, which converges because it is a p-series with $p = \frac{7}{6} > 1$.

12. $\dfrac{(2k-1)(k^2-1)}{(k+1)(k^2+4)^2} < \dfrac{2k(k^2)}{k(k^2)^2} = \dfrac{2k^3}{k^5} = \dfrac{2}{k^2}$ for all $k \geq 1$, so $\sum\limits_{k=1}^{\infty} \dfrac{(2k-1)(k^2-1)}{(k+1)(k^2+4)^2}$ converges by comparison with $2 \sum\limits_{k=1}^{\infty} \dfrac{1}{k^2}$, which converges because it is a constant multiple of a p-series with $p = 2 > 1$.

13. $\dfrac{1 + \cos n}{e^n} < \dfrac{2}{e^n}$ for all $n \geq 1$. $\sum\limits_{n=1}^{\infty} \dfrac{2}{e^n}$ is a convergent geometric series ($|r| = \frac{1}{e} < 1$), so $\sum\limits_{n=1}^{\infty} \dfrac{1 + \cos n}{e^n}$ converges by the Comparison Test.

14. $\dfrac{1}{\sqrt[3]{3n^4 + 1}} < \dfrac{1}{\sqrt[3]{3n^4}} < \dfrac{1}{\sqrt[3]{n^4}} = \dfrac{1}{n^{4/3}}$ for all $n \geq 1$, so $\sum\limits_{n=1}^{\infty} \dfrac{1}{\sqrt[3]{3n^4 + 1}}$ converges by comparison with $\sum\limits_{n=1}^{\infty} \dfrac{1}{n^{4/3}}$, which converges because it is a p-series with $p = \frac{4}{3} > 1$.

15. $\dfrac{4^{n+1}}{3^n - 2} > \dfrac{4 \cdot 4^n}{3^n} = 4\left(\dfrac{4}{3}\right)^n$ for all $n \geq 1$. $\sum\limits_{n=1}^{\infty} 4\left(\dfrac{4}{3}\right)^n = 4 \sum\limits_{n=1}^{\infty} \left(\dfrac{4}{3}\right)^n$ is a divergent geometric series $\left(|r| = \frac{4}{3} > 1\right)$, so $\sum\limits_{n=1}^{\infty} \dfrac{4^{n+1}}{3^n - 2}$ diverges by the Comparison Test.

16. $\dfrac{1}{n^n} \leq \dfrac{1}{n^2}$ for all $n \geq 1$, so $\sum\limits_{n=1}^{\infty} \dfrac{1}{n^n}$ converges by comparison with $\sum\limits_{n=1}^{\infty} \dfrac{1}{n^2}$, which converges because it is a p-series with $p = 2 > 1$.

17. Use the Limit Comparison Test with $a_n = \dfrac{1}{\sqrt{n^2 + 1}}$ and $b_n = \dfrac{1}{n}$:

$\lim\limits_{n \to \infty} \dfrac{a_n}{b_n} = \lim\limits_{n \to \infty} \dfrac{n}{\sqrt{n^2 + 1}} = \lim\limits_{n \to \infty} \dfrac{1}{\sqrt{1 + (1/n^2)}} = 1 > 0$. Since the harmonic series $\sum\limits_{n=1}^{\infty} \dfrac{1}{n}$ diverges, so does $\sum\limits_{n=1}^{\infty} \dfrac{1}{\sqrt{n^2 + 1}}$.

18. Use the Limit Comparison Test with $a_n = \dfrac{2}{\sqrt{n}+2}$ and $b_n = \dfrac{1}{\sqrt{n}}$:

$\lim\limits_{n\to\infty} \dfrac{a_n}{b_n} = \lim\limits_{n\to\infty} \dfrac{2\sqrt{n}}{\sqrt{n}+2} = \lim\limits_{n\to\infty} \dfrac{2}{1+2/\sqrt{n}} = 2 > 0$. Since $\sum\limits_{n=1}^{\infty} \dfrac{1}{\sqrt{n}}$ is a divergent p-series $[\,p = \frac{1}{2} \leq 1]$, the series $\sum\limits_{n=1}^{\infty} \dfrac{2}{\sqrt{n}+2}$ is also divergent.

19. Use the Limit Comparison Test with $a_n = \dfrac{n+1}{n^3+n}$ and $b_n = \dfrac{1}{n^2}$:

$\lim\limits_{n\to\infty} \dfrac{a_n}{b_n} = \lim\limits_{n\to\infty} \dfrac{(n+1)n^2}{n(n^2+1)} = \lim\limits_{n\to\infty} \dfrac{n^2+n}{n^2+1} = \lim\limits_{n\to\infty} \dfrac{1+1/n}{1+1/n^2} = 1 > 0$. Since $\sum\limits_{n=1}^{\infty} \dfrac{1}{n^2}$ is a convergent p-series $[p = 2 > 1]$, the series $\sum\limits_{n=1}^{\infty} \dfrac{n+1}{n^3+n}$ also converges.

20. Use the Limit Comparison Test with $a_n = \dfrac{n^2+n+1}{n^4+n^2}$ and $b_n = \dfrac{1}{n^2}$:

$\lim\limits_{n\to\infty} \dfrac{a_n}{b_n} = \lim\limits_{n\to\infty} \dfrac{(n^2+n+1)n^2}{n^2(n^2+1)} = \lim\limits_{n\to\infty} \dfrac{n^2+n+1}{n^2+1} = \lim\limits_{n\to\infty} \dfrac{1+1/n+1/n^2}{1+1/n^2} = 1 > 0$. Since $\sum\limits_{n=1}^{\infty} \dfrac{1}{n^2}$ is a convergent p-series $[p = 2 > 1]$, the series $\sum\limits_{n=1}^{\infty} \dfrac{n^2+n+1}{n^4+n^2}$ also converges.

21. Use the Limit Comparison Test with $a_n = \dfrac{\sqrt{1+n}}{2+n}$ and $b_n = \dfrac{1}{\sqrt{n}}$:

$\lim\limits_{n\to\infty} \dfrac{a_n}{b_n} = \lim\limits_{n\to\infty} \dfrac{\sqrt{1+n}\sqrt{n}}{2+n} = \lim\limits_{n\to\infty} \dfrac{\sqrt{n^2+n}/\sqrt{n^2}}{(2+n)/n} = \lim\limits_{n\to\infty} \dfrac{\sqrt{1+1/n}}{2/n+1} = 1 > 0$. Since $\sum\limits_{n=1}^{\infty} \dfrac{1}{\sqrt{n}}$ is a divergent p-series $[\,p = \frac{1}{2} \leq 1]$, the series $\sum\limits_{n=1}^{\infty} \dfrac{\sqrt{1+n}}{2+n}$ also diverges.

22. Use the Limit Comparison Test with $a_n = \dfrac{n+2}{(n+1)^3}$ and $b_n = \dfrac{1}{n^2}$:

$\lim\limits_{n\to\infty} \dfrac{a_n}{b_n} = \lim\limits_{n\to\infty} \dfrac{n^2(n+2)}{(n+1)^3} = \lim\limits_{n\to\infty} \dfrac{1+\frac{2}{n}}{\left(1+\frac{1}{n}\right)^3} = 1 > 0$. Since $\sum\limits_{n=3}^{\infty} \dfrac{1}{n^2}$ is a convergent (partial) p-series $[p = 2 > 1]$, the series $\sum\limits_{n=3}^{\infty} \dfrac{n+2}{(n+1)^3}$ also converges.

23. Use the Limit Comparison Test with $a_n = \dfrac{5+2n}{(1+n^2)^2}$ and $b_n = \dfrac{1}{n^3}$:

$\lim\limits_{n\to\infty} \dfrac{a_n}{b_n} = \lim\limits_{n\to\infty} \dfrac{n^3(5+2n)}{(1+n^2)^2} = \lim\limits_{n\to\infty} \dfrac{5n^3+2n^4}{(1+n^2)^2} \cdot \dfrac{1/n^4}{1/(n^2)^2} = \lim\limits_{n\to\infty} \dfrac{\frac{5}{n}+2}{\left(\frac{1}{n^2}+1\right)^2} = 2 > 0$. Since $\sum\limits_{n=1}^{\infty} \dfrac{1}{n^3}$ is a convergent p-series $[p = 3 > 1]$, the series $\sum\limits_{n=1}^{\infty} \dfrac{5+2n}{(1+n^2)^2}$ also converges.

24. $\dfrac{n+3^n}{n+2^n} > \dfrac{3^n}{n+2^n} > \dfrac{3^n}{2^n+2^n} = \dfrac{3^n}{2\cdot 2^n} = \dfrac{1}{2}\left(\dfrac{3}{2}\right)^n$, so the series $\displaystyle\sum_{n=1}^{\infty}\frac{n+3^n}{n+2^n}$ diverges by comparison with $\displaystyle\frac{1}{2}\sum_{n=1}^{\infty}\left(\frac{3}{2}\right)^n$, which is a constant multiple of a divergent geometric series $\left[|r| = \frac{3}{2} > 1\right]$. *Or:* Use the Limit Comparison Test with $a_n = \dfrac{n+3^n}{n+2^n}$ and $b_n = \left(\dfrac{3}{2}\right)^n$.

25. $\dfrac{e^n+1}{ne^n+1} \geq \dfrac{e^n+1}{ne^n+n} = \dfrac{e^n+1}{n(e^n+1)} = \dfrac{1}{n}$ for $n \geq 1$, so the series $\displaystyle\sum_{n=1}^{\infty}\frac{e^n+1}{ne^n+1}$ diverges by comparison with the divergent harmonic series $\displaystyle\sum_{n=1}^{\infty}\frac{1}{n}$. *Or:* Use the Limit Comparison Test with $a_n = \dfrac{e^n+1}{ne^n+1}$ and $b_n = \dfrac{1}{n}$.

26. If $a_n = \dfrac{1}{n\sqrt{n^2-1}}$ and $b_n = \dfrac{1}{n^2}$, then

$\displaystyle\lim_{n\to\infty}\frac{a_n}{b_n} = \lim_{n\to\infty}\frac{n^2}{n\sqrt{n^2-1}} = \lim_{n\to\infty}\frac{n/n}{\sqrt{n^2-1}/n} = \lim_{n\to\infty}\frac{1}{\sqrt{1-1/n^2}} = \frac{1}{1} = 1 > 0$, so $\displaystyle\sum_{n=2}^{\infty}\frac{1}{n\sqrt{n^2-1}}$ converges by the Limit Comparison Test with the convergent series $\displaystyle\sum_{n=2}^{\infty}\frac{1}{n^2}$.

27. Use the Limit Comparison Test with $a_n = \left(1+\dfrac{1}{n}\right)^2 e^{-n}$ and $b_n = e^{-n}$: $\displaystyle\lim_{n\to\infty}\frac{a_n}{b_n} = \lim_{n\to\infty}\left(1+\frac{1}{n}\right)^2 = 1 > 0$. Since $\displaystyle\sum_{n=1}^{\infty} e^{-n} = \sum_{n=1}^{\infty}\frac{1}{e^n}$ is a convergent geometric series $\left[|r| = \frac{1}{e} < 1\right]$, the series $\displaystyle\sum_{n=1}^{\infty}\left(1+\frac{1}{n}\right)^2 e^{-n}$ also converges.

28. $\dfrac{e^{1/n}}{n} > \dfrac{1}{n}$ for all $n \geq 1$, so $\displaystyle\sum_{n=1}^{\infty}\frac{e^{1/n}}{n}$ diverges by comparison with the harmonic series $\displaystyle\sum_{n=1}^{\infty}\frac{1}{n}$.

29. Clearly $n! = n(n-1)(n-2)\cdots(3)(2) \geq 2\cdot 2\cdot 2\cdot\cdots\cdot 2\cdot 2 = 2^{n-1}$, so $\dfrac{1}{n!} \leq \dfrac{1}{2^{n-1}}$. $\displaystyle\sum_{n=1}^{\infty}\frac{1}{2^{n-1}}$ is a convergent geometric series $\left[|r| = \frac{1}{2} < 1\right]$, so $\displaystyle\sum_{n=1}^{\infty}\frac{1}{n!}$ converges by the Comparison Test.

30. $\dfrac{n!}{n^n} = \dfrac{1\cdot 2\cdot 3\cdot\cdots\cdot(n-1)n}{n\cdot n\cdot n\cdot\cdots\cdot n\cdot n} \leq \dfrac{1}{n}\cdot\dfrac{2}{n}\cdot 1\cdot 1\cdot\cdots\cdot 1$ for $n \geq 2$, so since $\displaystyle\sum_{n=1}^{\infty}\frac{2}{n^2}$ converges $[p = 2 > 1]$, $\displaystyle\sum_{n=1}^{\infty}\frac{n!}{n^n}$ converges also by the Comparison Test.

31. Use the Limit Comparison Test with $a_n = \sin\left(\dfrac{1}{n}\right)$ and $b_n = \dfrac{1}{n}$. Then $\sum a_n$ and $\sum b_n$ are series with positive terms and

$\displaystyle\lim_{n\to\infty}\frac{a_n}{b_n} = \lim_{n\to\infty}\frac{\sin(1/n)}{1/n} = \lim_{\theta\to 0}\frac{\sin\theta}{\theta} = 1 > 0$. Since $\displaystyle\sum_{n=1}^{\infty} b_n$ is the divergent harmonic series,

$\displaystyle\sum_{n=1}^{\infty}\sin(1/n)$ also diverges. [Note that we could also use l'Hospital's Rule to evaluate the limit:

$$\lim_{x\to\infty}\frac{\sin(1/x)}{1/x} \overset{\text{H}}{=} \lim_{x\to\infty}\frac{\cos(1/x)\cdot\left(-1/x^2\right)}{-1/x^2} = \lim_{x\to\infty}\cos\frac{1}{x} = \cos 0 = 1.]$$

32. Use the Limit Comparison Test with $a_n = \dfrac{1}{n^{1+1/n}}$ and $b_n = \dfrac{1}{n}$. $\displaystyle\lim_{n\to\infty}\frac{a_n}{b_n} = \lim_{n\to\infty}\frac{n}{n^{1+1/n}} = \lim_{n\to\infty}\frac{1}{n^{1/n}} = 1$

$\left[\text{since } \displaystyle\lim_{x\to\infty} x^{1/x} = 1 \text{ by l'Hospital's Rule}\right]$, so $\displaystyle\sum_{n=1}^{\infty}\frac{1}{n}$ diverges [harmonic series] $\Rightarrow$ $\displaystyle\sum_{n=1}^{\infty}\frac{1}{n^{1+1/n}}$ diverges.

33. $\displaystyle\sum_{n=1}^{10}\frac{1}{5+n^5} = \frac{1}{5+1^5} + \frac{1}{5+2^5} + \frac{1}{5+3^5} + \cdots + \frac{1}{5+10^5} \approx 0.19926$. Now $\dfrac{1}{5+n^5} < \dfrac{1}{n^5}$, so the error is

$$R_{10} \le T_{10} \le \int_{10}^{\infty}\frac{1}{x^5}\,dx = \lim_{t\to\infty}\int_{10}^{t} x^{-5}\,dx = \lim_{t\to\infty}\left[\frac{-1}{4x^4}\right]_{10}^{t} = \lim_{t\to\infty}\left(\frac{-1}{4t^4} + \frac{1}{40{,}000}\right) = \frac{1}{40{,}000} = 0.000\,025.$$

34. $\displaystyle\sum_{n=1}^{10}\frac{e^{1/n}}{n^4} = \frac{e^{1/1}}{1^4} + \frac{e^{1/2}}{2^4} + \frac{e^{1/3}}{3^4} + \cdots + \frac{e^{1/10}}{10^4} \approx 2.84748$. Now $\dfrac{e^{1/n}}{n^4} \le \dfrac{e}{n^4}$ for $n \ge 1$, so the error is

$$R_{10} \le T_{10} \le \int_{10}^{\infty}\frac{e}{x^4}\,dx = \lim_{t\to\infty}\int_{10}^{t} ex^{-4}\,dx = \lim_{t\to\infty}\left[\frac{-e}{3x^3}\right]_{10}^{t} = \lim_{t\to\infty}\left(\frac{-e}{3t^3} + \frac{e}{3000}\right) = \frac{e}{3000} \approx 0.000\,906.$$

35. $\displaystyle\sum_{n=1}^{10} 5^{-n}\cos^2 n = \frac{\cos^2 1}{5} + \frac{\cos^2 2}{5^2} + \frac{\cos^2 3}{5^3} + \cdots + \frac{\cos^2 10}{5^{10}} \approx 0.07393$. Now $\dfrac{\cos^2 n}{5^n} \le \dfrac{1}{5^n}$, so the error is

$$R_{10} \le T_{10} \le \int_{10}^{\infty}\frac{1}{5^x}\,dx = \lim_{t\to\infty}\int_{10}^{t} 5^{-x}\,dx = \lim_{t\to\infty}\left[-\frac{5^{-x}}{\ln 5}\right]_{10}^{t} = \lim_{t\to\infty}\left(-\frac{5^{-t}}{\ln 5} + \frac{5^{-10}}{\ln 5}\right) = \frac{1}{5^{10}\ln 5} < 6.4\times 10^{-8}.$$

36. $\displaystyle\sum_{n=1}^{10}\frac{1}{3^n+4^n} = \frac{1}{3^1+4^1} + \frac{1}{3^2+4^2} + \frac{1}{3^3+4^3} + \cdots + \frac{1}{3^{10}+4^{10}} \approx 0.19788$. Now $\dfrac{1}{3^n+4^n} < \dfrac{1}{3^n+3^n} = \dfrac{1}{2\cdot 3^n}$, so the error is

$$R_{10} \le T_{10} \le \int_{10}^{\infty}\frac{1}{2\cdot 3^x}\,dx = \lim_{t\to\infty}\int_{10}^{t}\frac{1}{2}\cdot 3^{-x}\,dx = \lim_{t\to\infty}\left[-\frac{1}{2}\frac{3^{-x}}{\ln 3}\right]_{10}^{t} = \lim_{t\to\infty}\left(-\frac{1}{2}\frac{3^{-t}}{\ln 3} + \frac{1}{2}\frac{3^{-10}}{\ln 3}\right)$$

$$= \frac{1}{2\cdot 3^{10}\ln 3} < 7.7\times 10^{-6}.$$

37. Since $\dfrac{d_n}{10^n} \le \dfrac{9}{10^n}$ for each n, and since $\displaystyle\sum_{n=1}^{\infty}\frac{9}{10^n}$ is a convergent geometric series $\left(|r| = \frac{1}{10} < 1\right)$, $0.d_1d_2d_3\ldots = \displaystyle\sum_{n=1}^{\infty}\frac{d_n}{10^n}$ will always converge by the Comparison Test.

38. Clearly, if $p < 0$ then the series diverges, since $\displaystyle\lim_{n\to\infty}\frac{1}{n^p\ln n} = \infty$. If $0 \le p \le 1$, then $n^p\ln n \le n\ln n$ $\Rightarrow$ $\dfrac{1}{n^p\ln n} \ge \dfrac{1}{n\ln n}$ and $\displaystyle\sum_{n=2}^{\infty}\frac{1}{n\ln n}$ diverges (Exercise 11.3.21), so $\displaystyle\sum_{n=2}^{\infty}\frac{1}{n^p\ln n}$ diverges. If $p > 1$, use the Limit Comparison Test with $a_n = \dfrac{1}{n^p\ln n}$ and $b_n = \dfrac{1}{n^p}$. $\displaystyle\sum_{n=2}^{\infty} b_n$ converges, and $\displaystyle\lim_{n\to\infty}\frac{a_n}{b_n} = \lim_{n\to\infty}\frac{1}{\ln n} = 0$, so $\displaystyle\sum_{n=2}^{\infty}\frac{1}{n^p\ln n}$ also converges.

(Or use the Comparison Test, since $n^p\ln n > n^p$ for $n > e$.) In summary, the series converges if and only if $p > 1$.

39. Since $\sum a_n$ converges, $\displaystyle\lim_{n\to\infty} a_n = 0$, so there exists N such that $|a_n - 0| < 1$ for all $n > N$ $\Rightarrow$ $0 \le a_n < 1$ for all $n > N$ $\Rightarrow$ $0 \le a_n^2 \le a_n$. Since $\sum a_n$ converges, so does $\sum a_n^2$ by the Comparison Test.

40. (a) Since $\lim\limits_{n\to\infty}(a_n/b_n) = 0$, there is a number $N > 0$ such that $|a_n/b_n - 0| < 1$ for all $n > N$, and so $a_n < b_n$ since a_n and b_n are positive. Thus, since $\sum b_n$ converges, so does $\sum a_n$ by the Comparison Test.

(b) (i) If $a_n = \dfrac{\ln n}{n^3}$ and $b_n = \dfrac{1}{n^2}$, then $\lim\limits_{n\to\infty}\dfrac{a_n}{b_n} = \lim\limits_{n\to\infty}\dfrac{\ln n}{n} = \lim\limits_{x\to\infty}\dfrac{\ln x}{x} \overset{H}{=} \lim\limits_{x\to\infty}\dfrac{1/x}{1} = 0$, so $\sum\limits_{n=1}^{\infty}\dfrac{\ln n}{n^3}$ converges by part (a).

(ii) If $a_n = \dfrac{\ln n}{\sqrt{n}e^n}$ and $b_n = \dfrac{1}{e^n}$, then $\lim\limits_{n\to\infty}\dfrac{a_n}{b_n} = \lim\limits_{n\to\infty}\dfrac{\ln n}{\sqrt{n}} = \lim\limits_{x\to\infty}\dfrac{\ln x}{\sqrt{x}} \overset{H}{=} \lim\limits_{x\to\infty}\dfrac{1/x}{1/(2\sqrt{x})} = \lim\limits_{x\to\infty}\dfrac{2}{\sqrt{x}} = 0$. Now $\sum b_n$ is a convergent geometric series with ratio $r = 1/e$ $[|r| < 1]$, so $\sum a_n$ converges by part (a).

41. (a) Since $\lim\limits_{n\to\infty}\dfrac{a_n}{b_n} = \infty$, there is an integer N such that $\dfrac{a_n}{b_n} > 1$ whenever $n > N$. (Take $M = 1$ in Definition 11.1.5.) Then $a_n > b_n$ whenever $n > N$ and since $\sum b_n$ is divergent, $\sum a_n$ is also divergent by the Comparison Test.

(b) (i) If $a_n = \dfrac{1}{\ln n}$ and $b_n = \dfrac{1}{n}$ for $n \geq 2$, then $\lim\limits_{n\to\infty}\dfrac{a_n}{b_n} = \lim\limits_{n\to\infty}\dfrac{n}{\ln n} = \lim\limits_{x\to\infty}\dfrac{x}{\ln x} \overset{H}{=} \lim\limits_{x\to\infty}\dfrac{1}{1/x} = \lim\limits_{x\to\infty} x = \infty$, so by part (a), $\sum\limits_{n=2}^{\infty}\dfrac{1}{\ln n}$ is divergent.

(ii) If $a_n = \dfrac{\ln n}{n}$ and $b_n = \dfrac{1}{n}$, then $\sum\limits_{n=1}^{\infty} b_n$ is the divergent harmonic series and $\lim\limits_{n\to\infty}\dfrac{a_n}{b_n} = \lim\limits_{n\to\infty}\ln n = \lim\limits_{x\to\infty}\ln x = \infty$, so $\sum\limits_{n=1}^{\infty} a_n$ diverges by part (a).

42. Let $a_n = \dfrac{1}{n^2}$ and $b_n = \dfrac{1}{n}$. Then $\lim\limits_{n\to\infty}\dfrac{a_n}{b_n} = \lim\limits_{n\to\infty}\dfrac{1}{n} = 0$, but $\sum b_n$ diverges while $\sum a_n$ converges.

43. $\lim\limits_{n\to\infty} na_n = \lim\limits_{n\to\infty}\dfrac{a_n}{1/n}$, so we apply the Limit Comparison Test with $b_n = \dfrac{1}{n}$. Since $\lim\limits_{n\to\infty} na_n > 0$ we know that either both series converge or both series diverge, and we also know that $\sum\limits_{n=1}^{\infty}\dfrac{1}{n}$ diverges $[p\text{-series with } p = 1]$. Therefore, $\sum a_n$ must be divergent.

44. First we observe that, by l'Hospital's Rule, $\lim\limits_{x\to 0}\dfrac{\ln(1+x)}{x} = \lim\limits_{x\to 0}\dfrac{1}{1+x} = 1$. Also, if $\sum a_n$ converges, then $\lim\limits_{n\to\infty} a_n = 0$ by Theorem 11.2.6. Therefore, $\lim\limits_{n\to\infty}\dfrac{\ln(1+a_n)}{a_n} = \lim\limits_{x\to 0}\dfrac{\ln(1+x)}{x} = 1 > 0$. We are given that $\sum a_n$ is convergent and $a_n > 0$. Thus, $\sum \ln(1+a_n)$ is convergent by the Limit Comparison Test.

45. Yes. Since $\sum a_n$ is a convergent series with positive terms, $\lim\limits_{n\to\infty} a_n = 0$ by Theorem 11.2.6, and $\sum b_n = \sum \sin(a_n)$ is a series with positive terms (for large enough n). We have $\lim\limits_{n\to\infty}\dfrac{b_n}{a_n} = \lim\limits_{n\to\infty}\dfrac{\sin(a_n)}{a_n} = 1 > 0$ by Theorem 3.3.2. Thus, $\sum b_n$ is also convergent by the Limit Comparison Test.

46. Yes. Since $\sum a_n$ converges, its terms approach 0 as $n \to \infty$, so for some integer N, $a_n \leq 1$ for all $n \geq N$. But then $\sum_{n=1}^{\infty} a_n b_n = \sum_{n=1}^{N-1} a_n b_n + \sum_{n=N}^{\infty} a_n b_n \leq \sum_{n=1}^{N-1} a_n b_n + \sum_{n=N}^{\infty} b_n$. The first term is a finite sum, and the second term converges since $\sum_{n=1}^{\infty} b_n$ converges. So $\sum a_n b_n$ converges by the Comparison Test.

11.5 Alternating Series

1. (a) An alternating series is a series whose terms are alternately positive and negative.

(b) An alternating series $\sum_{n=1}^{\infty} a_n = \sum_{n=1}^{\infty} (-1)^{n-1} b_n$, where $b_n = |a_n|$, converges if $0 < b_{n+1} \le b_n$ for all n and $\lim_{n\to\infty} b_n = 0$.

(This is the Alternating Series Test.)

(c) The error involved in using the partial sum s_n as an approximation to the total sum s is the remainder $R_n = s - s_n$ and the size of the error is smaller than b_{n+1}; that is, $|R_n| \le b_{n+1}$. (This is the Alternating Series Estimation Theorem.)

2. $\frac{2}{3} - \frac{2}{5} + \frac{2}{7} - \frac{2}{9} + \frac{2}{11} - \cdots = \sum_{n=1}^{\infty} (-1)^{n+1} \frac{2}{2n+1}$. Now $b_n = \frac{2}{2n+1} > 0$, $\{b_n\}$ is decreasing, and $\lim_{n\to\infty} b_n = 0$, so the series converges by the Alternating Series Test.

3. $-\frac{2}{5} + \frac{4}{6} - \frac{6}{7} + \frac{8}{8} - \frac{10}{9} + \cdots = \sum_{n=1}^{\infty} (-1)^n \frac{2n}{n+4}$. Now $\lim_{n\to\infty} b_n = \lim_{n\to\infty} \frac{2n}{n+4} = \lim_{n\to\infty} \frac{2}{1+4/n} = \frac{2}{1} \ne 0$. Since $\lim_{n\to\infty} a_n \ne 0$ (in fact the limit does not exist), the series diverges by the Test for Divergence.

4. $\frac{1}{\ln 3} - \frac{1}{\ln 4} + \frac{1}{\ln 5} - \frac{1}{\ln 6} + \frac{1}{\ln 7} - \cdots = \sum_{n=1}^{\infty} \frac{(-1)^{n+1}}{\ln(n+2)}$. Now $b_n = \frac{1}{\ln(n+2)} > 0$, $\{b_n\}$ is decreasing, and $\lim_{n\to\infty} b_n = 0$, so the series converges by the Alternating Series Test.

5. $\sum_{n=1}^{\infty} a_n = \sum_{n=1}^{\infty} \frac{(-1)^{n-1}}{3+5n} = \sum_{n=1}^{\infty} (-1)^{n-1} b_n$. Now $b_n = \frac{1}{3+5n} > 0$, $\{b_n\}$ is decreasing, and $\lim_{n\to\infty} b_n = 0$, so the series converges by the Alternating Series Test.

6. $\sum_{n=0}^{\infty} a_n = \sum_{n=0}^{\infty} \frac{(-1)^{n+1}}{\sqrt{n+1}} = \sum_{n=0}^{\infty} (-1)^{n+1} b_n$. Now $b_n = \frac{1}{\sqrt{n+1}} > 0$, $\{b_n\}$ is decreasing, and $\lim_{n\to\infty} b_n = 0$, so the series converges by the Alternating Series Test.

7. $\sum_{n=1}^{\infty} a_n = \sum_{n=1}^{\infty} (-1)^n \frac{3n-1}{2n+1} = \sum_{n=1}^{\infty} (-1)^n b_n$. Now $\lim_{n\to\infty} b_n = \lim_{n\to\infty} \frac{3-1/n}{2+1/n} = \frac{3}{2} \ne 0$. Since $\lim_{n\to\infty} a_n \ne 0$ (in fact the limit does not exist), the series diverges by the Test for Divergence.

8. $\sum_{n=1}^{\infty} a_n = \sum_{n=1}^{\infty} (-1)^n \frac{n^2}{n^2+n+1} = \sum_{n=1}^{\infty} (-1)^n b_n$. Now $\lim_{n\to\infty} b_n = \lim_{n\to\infty} \frac{n^2}{n^2+n+1} = \lim_{n\to\infty} \frac{1}{1+1/n+1/n^2} = 1 \ne 0$. Since $\lim_{n\to\infty} a_n \ne 0$, the series diverges by the Test for Divergence.

9. $\sum_{n=1}^{\infty} a_n = \sum_{n=1}^{\infty} (-1)^n e^{-n} = \sum_{n=1}^{\infty} (-1)^n b_n$. Now $b_n = \frac{1}{e^n} > 0$, $\{b_n\}$ is decreasing, and $\lim_{n\to\infty} b_n = 0$, so the series converges by the Alternating Series Test.

10. $b_n = \dfrac{\sqrt{n}}{2n+3} > 0$ for $n \geq 1$. $\{b_n\}$ is decreasing for $n \geq 2$ since

$$\left(\frac{\sqrt{x}}{2x+3}\right)' = \frac{(2x+3)\left(\frac{1}{2}x^{-1/2}\right) - x^{1/2}(2)}{(2x+3)^2} = \frac{\frac{1}{2}x^{-1/2}[(2x+3)-4x]}{(2x+3)^2} = \frac{3-2x}{2\sqrt{x}\,(2x+3)^2} < 0 \text{ for } x > \tfrac{3}{2}.$$

Also, $\lim\limits_{n\to\infty} b_n = \lim\limits_{n\to\infty} \dfrac{\sqrt{n}/\sqrt{n}}{(2n+3)/\sqrt{n}} = \lim\limits_{n\to\infty} \dfrac{1}{2\sqrt{n}+3/\sqrt{n}} = 0$. Thus, the series $\sum\limits_{n=1}^{\infty} (-1)^n \dfrac{\sqrt{n}}{2n+3}$ converges by the Alternating Series Test.

11. $b_n = \dfrac{n^2}{n^3+4} > 0$ for $n \geq 1$. $\{b_n\}$ is decreasing for $n \geq 2$ since

$$\left(\frac{x^2}{x^3+4}\right)' = \frac{(x^3+4)(2x) - x^2(3x^2)}{(x^3+4)^2} = \frac{x(2x^3+8-3x^3)}{(x^3+4)^2} = \frac{x(8-x^3)}{(x^3+4)^2} < 0 \text{ for } x > 2. \text{ Also,}$$

$\lim\limits_{n\to\infty} b_n = \lim\limits_{n\to\infty} \dfrac{1/n}{1+4/n^3} = 0$. Thus, the series $\sum\limits_{n=1}^{\infty} (-1)^{n+1} \dfrac{n^2}{n^3+4}$ converges by the Alternating Series Test.

12. $b_n = ne^{-n} = \dfrac{n}{e^n} > 0$ for $n \geq 1$. $\{b_n\}$ is decreasing for $n \geq 1$ since $(xe^{-x})' = x(-e^{-x}) + e^{-x} = e^{-x}(1-x) < 0$ for $x > 1$. Also, $\lim\limits_{n\to\infty} b_n = 0$ since $\lim\limits_{x\to\infty} \dfrac{x}{e^x} \overset{H}{=} \lim\limits_{x\to\infty} \dfrac{1}{e^x} = 0$. Thus, the series $\sum\limits_{n=1}^{\infty} (-1)^{n+1} ne^{-n}$ converges by the Alternating Series Test.

13. $\lim\limits_{n\to\infty} b_n = \lim\limits_{n\to\infty} e^{2/n} = e^0 = 1$, so $\lim\limits_{n\to\infty} (-1)^{n-1} e^{2/n}$ does not exist. Thus, the series $\sum\limits_{n=1}^{\infty} (-1)^{n-1} e^{2/n}$ diverges by the Test for Divergence.

14. $\lim\limits_{n\to\infty} b_n = \lim\limits_{n\to\infty} \arctan n = \frac{\pi}{2}$, so $\lim\limits_{n\to\infty} (-1)^{n-1} \arctan n$ does not exist. Thus, the series $\sum\limits_{n=1}^{\infty} (-1)^{n-1} \arctan n$ diverges by the Test for Divergence.

15. $a_n = \dfrac{\sin\left(n+\frac{1}{2}\right)\pi}{1+\sqrt{n}} = \dfrac{(-1)^n}{1+\sqrt{n}}$. Now $b_n = \dfrac{1}{1+\sqrt{n}} > 0$ for $n \geq 0$, $\{b_n\}$ is decreasing, and $\lim\limits_{n\to\infty} b_n = 0$, so the series $\sum\limits_{n=0}^{\infty} \dfrac{\sin\left(n+\frac{1}{2}\right)\pi}{1+\sqrt{n}}$ converges by the Alternating Series Test.

16. $a_n = \dfrac{n\cos n\pi}{2^n} = (-1)^n \dfrac{n}{2^n} = (-1)^n b_n$. $\{b_n\}$ is decreasing for $n \geq 2$ since

$(x2^{-x})' = x(-2^{-x}\ln 2) + 2^{-x} = 2^{-x}(1 - x\ln 2) < 0$ for $x > \dfrac{1}{\ln 2}$ $[\approx 1.4]$. Also, $\lim\limits_{n\to\infty} b_n = 0$ since

$\lim\limits_{x\to\infty} \dfrac{x}{2^x} \overset{H}{=} \lim\limits_{x\to\infty} \dfrac{1}{2^x \ln 2} = 0$. Thus, the series $\sum\limits_{n=1}^{\infty} \dfrac{n\cos n\pi}{2^n}$ converges by the Alternating Series Test.

17. $\sum\limits_{n=1}^{\infty} (-1)^n \sin\left(\dfrac{\pi}{n}\right)$. $b_n = \sin\left(\dfrac{\pi}{n}\right) > 0$ for $n \geq 2$ and $\sin\left(\dfrac{\pi}{n}\right) \geq \sin\left(\dfrac{\pi}{n+1}\right)$, and $\lim\limits_{n\to\infty} \sin\left(\dfrac{\pi}{n}\right) = \sin 0 = 0$, so the series converges by the Alternating Series Test.

18. $\sum_{n=1}^{\infty}(-1)^n\cos\left(\frac{\pi}{n}\right)$. $\lim_{n\to\infty}\cos\left(\frac{\pi}{n}\right)=\cos(0)=1$, so $\lim_{n\to\infty}(-1)^n\cos\left(\frac{\pi}{n}\right)$ does not exist and the series diverges by the Test for Divergence.

19. $\frac{n^n}{n!}=\frac{n\cdot n\cdot\cdots\cdot n}{1\cdot 2\cdot\cdots\cdot n}\geq n \quad\Rightarrow\quad \lim_{n\to\infty}\frac{n^n}{n!}=\infty \quad\Rightarrow\quad \lim_{n\to\infty}\frac{(-1)^n\,n^n}{n!}$ does not exist. So the series $\sum_{n=1}^{\infty}(-1)^n\frac{n^n}{n!}$ diverges by the Test for Divergence.

20. $b_n=\frac{\sqrt{n+1}-\sqrt{n}}{1}\cdot\frac{\sqrt{n+1}+\sqrt{n}}{\sqrt{n+1}+\sqrt{n}}=\frac{(n+1)-n}{\sqrt{n+1}+\sqrt{n}}=\frac{1}{\sqrt{n+1}+\sqrt{n}}>0$ for $n\geq 1$. $\{b_n\}$ is decreasing and $\lim_{n\to\infty}b_n=0$, so the series $\sum_{n=1}^{\infty}(-1)^n\left(\sqrt{n+1}-\sqrt{n}\right)$ converges by the Alternating Series Test.

21. The graph gives us an estimate for the sum of the series $\sum_{n=1}^{\infty}\frac{(-0.8)^n}{n!}$ of -0.55.

$b_8=\frac{(0.8)^n}{8!}\approx 0.000\,004$, so

$$\sum_{n=1}^{\infty}\frac{(-0.8)^n}{n!}\approx s_7=\sum_{n=1}^{7}\frac{(-0.8)^n}{n!}$$

$$\approx -0.8+0.32-0.085\overline{3}+0.0170\overline{6}-0.002\,731+0.000\,364-0.000\,042\approx -0.5507$$

Adding b_8 to s_7 does not change the fourth decimal place of s_7, so the sum of the series, correct to four decimal places, is -0.5507.

22.

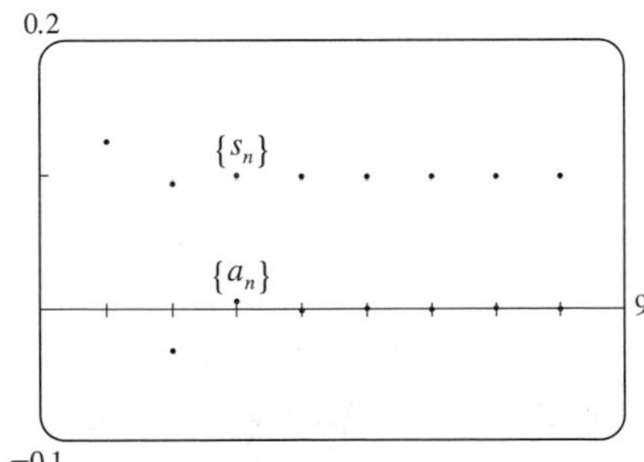

The graph gives us an estimate for the sum of the series $\sum_{n=1}^{\infty}(-1)^{n-1}\frac{n}{8^n}$ of 0.1.

$b_6=\frac{6}{8^6}\approx 0.000\,023$, so

$$\sum_{n=1}^{\infty}(-1)^{n-1}\frac{n}{8^n}\approx s_5=\sum_{n=1}^{5}(-1)^{n-1}\frac{n}{8^n}$$

$$\approx 0.125-0.03125+0.005\,859-0.000\,977+0.000\,153\approx 0.0988$$

Adding b_6 to s_5 does not change the fourth decimal place of s_5, so the sum of the series, correct to four decimal places, is 0.0988.

23. The series $\sum_{n=1}^{\infty} \frac{(-1)^{n+1}}{n^6}$ satisfies (i) of the Alternating Series Test because $\frac{1}{(n+1)^6} < \frac{1}{n^6}$ and (ii) $\lim_{n\to\infty} \frac{1}{n^6} = 0$, so the series is convergent. Now $b_5 = \frac{1}{5^6} = 0.000064 > 0.00005$ and $b_6 = \frac{1}{6^6} \approx 0.00002 < 0.00005$, so by the Alternating Series Estimation Theorem, $n = 5$. (That is, since the 6th term is less than the desired error, we need to add the first 5 terms to get the sum to the desired accuracy.)

24. The series $\sum_{n=1}^{\infty} \frac{\left(-\frac{1}{3}\right)^n}{n} = \sum_{n=1}^{\infty} (-1)^n \frac{1}{n3^n}$ satisfies (i) of the Alternating Series Test because $\frac{1}{(n+1)3^{n+1}} < \frac{1}{n3^n}$ and (ii) $\lim_{n\to\infty} \frac{1}{n3^n} = 0$, so the series is convergent. Now $b_5 = \frac{1}{5 \cdot 3^5} \approx 0.0008 > 0.0005$ and $b_6 = \frac{1}{6 \cdot 3^6} \approx 0.0002 < 0.0005$, so by the Alternating Series Estimation Theorem, $n = 5$. (That is, since the 6th term is less than the desired error, we need to add the first 5 terms to get the sum to the desired accuracy.)

25. The series $\sum_{n=1}^{\infty} \frac{(-1)^{n-1}}{n^2 2^n}$ satisfies (i) of the Alternating Series Test because $\frac{1}{(n+1)^2 2^{n+1}} < \frac{1}{n^2 2^n}$ and (ii) $\lim_{n\to\infty} \frac{1}{n^2 2^n} = 0$, so the series is convergent. Now $b_5 = \frac{1}{5^2 2^5} = 0.00125 > 0.0005$ and $b_6 = \frac{1}{6^2 2^6} \approx 0.0004 < 0.0005$, so by the Alternating Series Estimation Theorem, $n = 5$. (That is, since the 6th term is less than the desired error, we need to add the first 5 terms to get the sum to the desired accuracy.)

26. The series $\sum_{n=1}^{\infty} \left(-\frac{1}{n}\right)^n = \sum_{n=1}^{\infty} (-1)^n \frac{1}{n^n}$ satisfies (i) of the Alternating Series Test because $\frac{1}{(n+1)^{n+1}} < \frac{1}{n^n}$ and (ii) $\lim_{n\to\infty} \frac{1}{n^n} = 0$, so the series is convergent. Now $b_5 = \frac{1}{5^5} = 0.00032 > 0.00005$ and $b_6 = \frac{1}{6^6} \approx 0.00002 < 0.00005$, so by the Alternating Series Estimation Theorem, $n = 5$. (That is, since the 6th term is less than the desired error, we need to add the first 5 terms to get the sum to the desired accuracy.)

27. $b_4 = \frac{1}{8!} = \frac{1}{40{,}320} \approx 0.000\,025$, so

$$\sum_{n=1}^{\infty} \frac{(-1)^n}{(2n)!} \approx s_3 = \sum_{n=1}^{3} \frac{(-1)^n}{(2n)!} = -\frac{1}{2} + \frac{1}{24} - \frac{1}{720} \approx -0.459\,722$$

Adding b_4 to s_3 does not change the fourth decimal place of s_3, so by the Alternating Series Estimation Theorem, the sum of the series, correct to four decimal places, is -0.4597.

28. $\sum_{n=1}^{\infty} \frac{(-1)^{n+1}}{n^6} \approx s_9 = \frac{1}{1^6} - \frac{1}{2^6} + \frac{1}{3^6} - \frac{1}{4^6} + \frac{1}{5^6} - \frac{1}{6^6} + \frac{1}{7^6} - \frac{1}{8^6} + \frac{1}{9^6} \approx 0.985\,552$. Subtracting $b_{10} = 1/10^6$ from s_9 does not change the fourth decimal place of s_9, so by the Alternating Series Estimation Theorem, the sum of the series, correct to four decimal places, is 0.9856.

29. $\sum_{n=1}^{\infty} (-1)^n n e^{-2n} \approx s_5 = -\frac{1}{e^2} + \frac{2}{e^4} - \frac{3}{e^6} + \frac{4}{e^8} - \frac{5}{e^{10}} \approx -0.105\,025$. Adding $b_6 = 6/e^{12} \approx 0.000\,037$ to s_5 does not change the fourth decimal place of s_5, so by the Alternating Series Estimation Theorem, the sum of the series, correct to four decimal places, is -0.1050.

30. $\sum_{n=1}^{\infty} \frac{(-1)^{n-1}}{n4^n} \approx s_6 = \frac{1}{4} - \frac{1}{2 \cdot 4^2} + \frac{1}{3 \cdot 4^3} - \frac{1}{4 \cdot 4^4} + \frac{1}{5 \cdot 4^5} - \frac{1}{6 \cdot 4^6} \approx 0.223136$. Adding $b_7 = \frac{1}{7 \cdot 4^7} \approx 0.000\,0087$ to s_6 does not change the fourth decimal place of s_6, so by the Alternating Series Estimation Theorem, the sum of the series, correct to four decimal places, is 0.2231.

31. $\sum_{n=1}^{\infty} \frac{(-1)^{n-1}}{n} = 1 - \frac{1}{2} + \frac{1}{3} - \frac{1}{4} + \cdots + \frac{1}{49} - \frac{1}{50} + \frac{1}{51} - \frac{1}{52} + \cdots$. The 50th partial sum of this series is an underestimate, since $\sum_{n=1}^{\infty} \frac{(-1)^{n-1}}{n} = s_{50} + \left(\frac{1}{51} - \frac{1}{52}\right) + \left(\frac{1}{53} - \frac{1}{54}\right) + \cdots$, and the terms in parentheses are all positive. The result can be seen geometrically in Figure 1.

32. If $p > 0$, $\frac{1}{(n+1)^p} \leq \frac{1}{n^p}$ ($\{1/n^p\}$ is decreasing) and $\lim_{n \to \infty} \frac{1}{n^p} = 0$, so the series converges by the Alternating Series Test. If $p \leq 0$, $\lim_{n \to \infty} \frac{(-1)^{n-1}}{n^p}$ does not exist, so the series diverges by the Test for Divergence. Thus, $\sum_{n=1}^{\infty} \frac{(-1)^{n-1}}{n^p}$ converges $\Leftrightarrow$ $p > 0$.

33. Clearly $b_n = \frac{1}{n+p}$ is decreasing and eventually positive and $\lim_{n \to \infty} b_n = 0$ for any p. So the series $\sum_{n=1}^{\infty} \frac{(-1)^n}{n+p}$ converges (by the Alternating Series Test) for any p for which every b_n is defined, that is, $n + p \neq 0$ for $n \geq 1$, or p is not a negative integer.

34. Let $f(x) = \frac{(\ln x)^p}{x}$. Then $f'(x) = \frac{(\ln x)^{p-1}(p - \ln x)}{x^2} < 0$ if $x > e^p$ so f is eventually decreasing for every p. Clearly $\lim_{n \to \infty} \frac{(\ln n)^p}{n} = 0$ if $p \leq 0$, and if $p > 0$ we can apply l'Hospital's Rule $[\![p+1]\!]$ times to get a limit of 0 as well. So the series $\sum_{n=2}^{\infty} (-1)^{n-1} \frac{(\ln n)^p}{n}$ converges for all p (by the Alternating Series Test).

35. $\sum b_{2n} = \sum 1/(2n)^2$ clearly converges (by comparison with the p-series for $p = 2$). So suppose that $\sum (-1)^{n-1} b_n$ converges. Then by Theorem 11.2.8(ii), so does $\sum \left[(-1)^{n-1} b_n + b_n\right] = 2\left(1 + \frac{1}{3} + \frac{1}{5} + \cdots\right) = 2 \sum \frac{1}{2n-1}$. But this diverges by comparison with the harmonic series, a contradiction. Therefore, $\sum (-1)^{n-1} b_n$ must diverge. The Alternating Series Test does not apply since $\{b_n\}$ is not decreasing.

36. (a) We will prove this by induction. Let $P(n)$ be the proposition that $s_{2n} = h_{2n} - h_n$. $P(1)$ is the statement $s_2 = h_2 - h_1$, which is true since $1 - \frac{1}{2} = \left(1 + \frac{1}{2}\right) - 1$. So suppose that $P(n)$ is true. We will show that $P(n+1)$ must be true as a consequence.

$$h_{2n+2} - h_{n+1} = \left(h_{2n} + \frac{1}{2n+1} + \frac{1}{2n+2}\right) - \left(h_n + \frac{1}{n+1}\right) = (h_{2n} - h_n) + \frac{1}{2n+1} - \frac{1}{2n+2}$$
$$= s_{2n} + \frac{1}{2n+1} - \frac{1}{2n+2} = s_{2n+2}$$

which is $P(n+1)$, and proves that $s_{2n} = h_{2n} - h_n$ for all n.

(b) We know that $h_{2n} - \ln(2n) \to \gamma$ and $h_n - \ln n \to \gamma$ as $n \to \infty$. So

$s_{2n} = h_{2n} - h_n = [h_{2n} - \ln(2n)] - (h_n - \ln n) + [\ln(2n) - \ln n]$, and

$\lim_{n\to\infty} s_{2n} = \gamma - \gamma + \lim_{n\to\infty} [\ln(2n) - \ln n] = \lim_{n\to\infty} (\ln 2 + \ln n - \ln n) = \ln 2.$

11.6 Absolute Convergence and the Ratio and Root Tests

1. (a) Since $\lim_{n\to\infty} \left|\dfrac{a_{n+1}}{a_n}\right| = 8 > 1$, part (b) of the Ratio Test tells us that the series $\sum a_n$ is divergent.

(b) Since $\lim_{n\to\infty} \left|\dfrac{a_{n+1}}{a_n}\right| = 0.8 < 1$, part (a) of the Ratio Test tells us that the series $\sum a_n$ is absolutely convergent (and therefore convergent).

(c) Since $\lim_{n\to\infty} \left|\dfrac{a_{n+1}}{a_n}\right| = 1$, the Ratio Test fails and the series $\sum a_n$ might converge or it might diverge.

2. $b_n = \dfrac{1}{\sqrt{n}} > 0$ for $n \geq 1$, $\{b_n\}$ is decreasing for $n \geq 1$, and $\lim_{n\to\infty} b_n = 0$, so $\sum_{n=1}^{\infty} \dfrac{(-1)^{n-1}}{\sqrt{n}}$ converges by the Alternating Series Test. To determine absolute convergence, note that $\sum_{n=1}^{\infty} \dfrac{1}{\sqrt{n}}$ diverges because it is a p-series with $p = \frac{1}{2} \leq 1$. Thus, the series $\sum_{n=1}^{\infty} \dfrac{(-1)^{n-1}}{\sqrt{n}}$ is conditionally convergent.

3. $b_n = \dfrac{1}{5n+1} > 0$ for $n \geq 0$, $\{b_n\}$ is decreasing for $n \geq 0$, and $\lim_{n\to\infty} b_n = 0$, so $\sum_{n=0}^{\infty} \dfrac{(-1)^n}{5n+1}$ converges by the Alternating Series Test. To determine absolute convergence, choose $a_n = \dfrac{1}{n}$ to get

$\lim_{n\to\infty} \dfrac{a_n}{b_n} = \lim_{n\to\infty} \dfrac{1/n}{1/(5n+1)} = \lim_{n\to\infty} \dfrac{5n+1}{n} = 5 > 0$, so $\sum_{n=1}^{\infty} \dfrac{1}{5n+1}$ diverges by the Limit Comparison Test with the harmonic series. Thus, the series $\sum_{n=0}^{\infty} \dfrac{(-1)^n}{5n+1}$ is conditionally convergent.

4. $0 < \dfrac{1}{n^3+1} < \dfrac{1}{n^3}$ for $n \geq 1$ and $\sum_{n=1}^{\infty} \dfrac{1}{n^3}$ is a convergent p-series ($p = 3 > 1$), so $\sum_{n=1}^{\infty} \dfrac{1}{n^3+1}$ converges by comparison and the series $\sum_{n=1}^{\infty} \dfrac{(-1)^n}{n^3+1}$ is absolutely convergent.

5. $0 < \left|\dfrac{\sin n}{2^n}\right| < \dfrac{1}{2^n}$ for $n \geq 1$ and $\sum_{n=1}^{\infty} \dfrac{1}{2^n}$ is a convergent geometric series ($r = \frac{1}{2} < 1$), so $\sum_{n=1}^{\infty} \left|\dfrac{\sin n}{2^n}\right|$ converges by comparison and the series $\sum_{n=1}^{\infty} \dfrac{\sin n}{2^n}$ is absolutely convergent.

6. $b_n = \dfrac{n}{n^2+4} > 0$ for $n \geq 1$, $\{b_n\}$ is decreasing for $n \geq 2$, and $\lim_{n\to\infty} b_n = 0$, so $\sum_{n=1}^{\infty} (-1)^{n-1} \dfrac{n}{n^2+4}$ converges by the Alternating Series Test. To determine absolute convergence, choose $a_n = \dfrac{1}{n}$ to get

$\lim\limits_{n\to\infty}\dfrac{a_n}{b_n}=\lim\limits_{n\to\infty}\dfrac{1/n}{n/(n^2+4)}=\lim\limits_{n\to\infty}\dfrac{n^2+4}{n^2}=\lim\limits_{n\to\infty}\dfrac{1+4/n^2}{1}=1>0$, so $\sum\limits_{n=1}^{\infty}\dfrac{n}{n^2+4}$ diverges by the Limit Comparison Test with the harmonic series. Thus, the series $\sum\limits_{n=1}^{\infty}(-1)^{n-1}\dfrac{n}{n^2+4}$ is conditionally convergent.

7. $\lim\limits_{n\to\infty}\left|\dfrac{a_{n+1}}{a_n}\right|=\lim\limits_{n\to\infty}\left|\dfrac{n+1}{5^{n+1}}\cdot\dfrac{5^n}{n}\right|=\lim\limits_{n\to\infty}\left|\dfrac{1}{5}\cdot\dfrac{n+1}{n}\right|=\dfrac{1}{5}\lim\limits_{n\to\infty}\dfrac{1+1/n}{1}=\dfrac{1}{5}(1)=\dfrac{1}{5}<1$, so the series $\sum\limits_{n=1}^{\infty}\dfrac{n}{5^n}$ is absolutely convergent by the Ratio Test.

8. $\lim\limits_{n\to\infty}\left|\dfrac{a_{n+1}}{a_n}\right|=\lim\limits_{n\to\infty}\left|\dfrac{(-2)^{n+1}}{(n+1)^2}\cdot\dfrac{n^2}{(-2)^n}\right|=\lim\limits_{n\to\infty}\left|(-2)\dfrac{n^2}{(n+1)^2}\right|=2\lim\limits_{n\to\infty}\dfrac{1}{(1+1/n)^2}=2(1)=2>1$, so the series $\sum\limits_{n=1}^{\infty}\dfrac{(-2)^n}{n^2}$ is divergent by the Ratio Test.

9. $\lim\limits_{n\to\infty}\left|\dfrac{a_{n+1}}{a_n}\right|=\lim\limits_{n\to\infty}\left|\dfrac{(-1)^n3^{n+1}}{2^{n+1}(n+1)^3}\cdot\dfrac{2^nn^3}{(-1)^{n-1}3^n}\right|=\lim\limits_{n\to\infty}\left|\left(-\dfrac{3}{2}\right)\dfrac{n^3}{(n+1)^3}\right|=\dfrac{3}{2}\lim\limits_{n\to\infty}\dfrac{1}{(1+1/n)^3}=\dfrac{3}{2}(1)=\dfrac{3}{2}>1$,

so the series $\sum\limits_{n=1}^{\infty}(-1)^{n-1}\dfrac{3^n}{2^nn^3}$ is divergent by the Ratio Test.

10. $\lim\limits_{n\to\infty}\left|\dfrac{a_{n+1}}{a_n}\right|=\lim\limits_{n\to\infty}\left|\dfrac{(-3)^{n+1}}{[2(n+1)+1]!}\cdot\dfrac{(2n+1)!}{(-3)^n}\right|=\lim\limits_{n\to\infty}\left|(-3)\dfrac{1}{(2n+3)(2n+2)}\right|=3\lim\limits_{n\to\infty}\dfrac{1}{(2n+3)(2n+2)}$

$=3(0)=0<1$

so the series $\sum\limits_{n=0}^{\infty}\dfrac{(-3)^n}{(2n+1)!}$ is absolutely convergent by the Ratio Test.

11. $\lim\limits_{k\to\infty}\left|\dfrac{a_{k+1}}{a_k}\right|=\lim\limits_{k\to\infty}\left|\dfrac{1}{(k+1)!}\cdot\dfrac{k!}{1}\right|=\lim\limits_{k\to\infty}\dfrac{1}{k+1}=0<1$, so the series $\sum\limits_{k=1}^{\infty}\dfrac{1}{k!}$ is absolutely convergent by the Ratio Test. Since the terms of this series are positive, absolute convergence is the same as convergence.

12. $\lim\limits_{k\to\infty}\left|\dfrac{a_{k+1}}{a_k}\right|=\lim\limits_{k\to\infty}\left|\dfrac{(k+1)e^{-(k+1)}}{ke^{-k}}\right|=\lim\limits_{k\to\infty}\left(\dfrac{k+1}{k}\cdot e^{-1}\right)=\dfrac{1}{e}\lim\limits_{k\to\infty}\dfrac{1+1/k}{1}=\dfrac{1}{e}(1)=\dfrac{1}{e}<1$, so the series $\sum\limits_{k=1}^{\infty}ke^{-k}$ is absolutely convergent by the Ratio Test. Since the terms of this series are positive, absolute convergence is the same as convergence.

13. $\lim\limits_{n\to\infty}\left|\dfrac{a_{n+1}}{a_n}\right|=\lim\limits_{n\to\infty}\left[\dfrac{10^{n+1}}{(n+2)\,4^{2n+3}}\cdot\dfrac{(n+1)\,4^{2n+1}}{10^n}\right]=\lim\limits_{n\to\infty}\left(\dfrac{10}{4^2}\cdot\dfrac{n+1}{n+2}\right)=\dfrac{5}{8}<1$, so the series $\sum\limits_{n=1}^{\infty}\dfrac{10^n}{(n+1)4^{2n+1}}$ is absolutely convergent by the Ratio Test. Since the terms of this series are positive, absolute convergence is the same as convergence.

14. $\lim\limits_{n\to\infty}\left|\dfrac{a_{n+1}}{a_n}\right|=\lim\limits_{n\to\infty}\left[\dfrac{(n+1)!}{100^{n+1}}\cdot\dfrac{100^n}{n!}\right]=\lim\limits_{n\to\infty}\dfrac{n+1}{100}=\infty$, so the series $\sum\limits_{n=1}^{\infty}\dfrac{n!}{100^n}$ diverges by the Ratio Test.

15. $\lim\limits_{n\to\infty}\left|\dfrac{a_{n+1}}{a_n}\right|=\lim\limits_{n\to\infty}\left|\dfrac{(n+1)\pi^{n+1}}{(-3)^n}\cdot\dfrac{(-3)^{n-1}}{n\pi^n}\right|=\lim\limits_{n\to\infty}\left|\dfrac{\pi}{-3}\cdot\dfrac{n+1}{n}\right|=\dfrac{\pi}{3}\lim\limits_{n\to\infty}\dfrac{1+1/n}{1}=\dfrac{\pi}{3}(1)=\dfrac{\pi}{3}>1$, so the series $\sum\limits_{n=1}^{\infty}\dfrac{n\pi^n}{(-3)^{n-1}}$ diverges by the Ratio Test. *Or:* Since $\lim\limits_{n\to\infty}|a_n|=\infty$, the series diverges by the Test for Divergence.

16. $\lim\limits_{n\to\infty}\left|\dfrac{a_{n+1}}{a_n}\right| = \lim\limits_{n\to\infty}\left|\dfrac{(n+1)^{10}}{(-10)^{n+2}}\cdot\dfrac{(-10)^{n+1}}{n^{10}}\right| = \lim\limits_{n\to\infty}\left|\dfrac{1}{-10}\left(\dfrac{n+1}{n}\right)^{10}\right| = \dfrac{1}{10}\lim\limits_{n\to\infty}\left(1+\dfrac{1}{n}\right)^{10} = \dfrac{1}{10}(1) = \dfrac{1}{10} < 1,$

so the series $\sum\limits_{n=1}^{\infty}\dfrac{n^{10}}{(-10)^{n+1}}$ is absolutely convergent by the Ratio Test.

17. $\lim\limits_{n\to\infty}\left|\dfrac{a_{n+1}}{a_n}\right| = \lim\limits_{n\to\infty}\left|\dfrac{\cos[(n+1)\pi/3]}{(n+1)!}\cdot\dfrac{n!}{\cos(n\pi/3)}\right| = \lim\limits_{n\to\infty}\left|\dfrac{\cos[(n+1)\pi/3]}{(n+1)\cos(n\pi/3)}\right| = \lim\limits_{n\to\infty}\dfrac{c}{n+1} = 0 < 1$ (where

$0 < c \le 2$ for all positive integers n), so the series $\sum\limits_{n=1}^{\infty}\dfrac{\cos(n\pi/3)}{n!}$ is absolutely convergent by the Ratio Test.

18. $\lim\limits_{n\to\infty}\left|\dfrac{a_{n+1}}{a_n}\right| = \lim\limits_{n\to\infty}\left|\dfrac{(n+1)!}{(n+1)^{n+1}}\cdot\dfrac{n^n}{n!}\right| = \lim\limits_{n\to\infty}\dfrac{(n+1)n^n}{(n+1)^{n+1}} = \lim\limits_{n\to\infty}\dfrac{n^n}{(n+1)^n} = \lim\limits_{n\to\infty}\dfrac{1}{(1+1/n)^n} = \dfrac{1}{e} < 1$, so the

series $\sum\limits_{n=1}^{\infty}\dfrac{n!}{n^n}$ is absolutely convergent by the Ratio Test.

19. $\lim\limits_{n\to\infty}\left|\dfrac{a_{n+1}}{a_n}\right| = \lim\limits_{n\to\infty}\left|\dfrac{(n+1)^{100}100^{n+1}}{(n+1)!}\cdot\dfrac{n!}{n^{100}100^n}\right| = \lim\limits_{n\to\infty}\dfrac{100}{n+1}\left(\dfrac{n+1}{n}\right)^{100} = \lim\limits_{n\to\infty}\dfrac{100}{n+1}\left(1+\dfrac{1}{n}\right)^{100}$

$= 0\cdot 1 = 0 < 1$

so the series $\sum\limits_{n=1}^{\infty}\dfrac{n^{100}100^n}{n!}$ is absolutely convergent by the Ratio Test.

20. $\lim\limits_{n\to\infty}\left|\dfrac{a_{n+1}}{a_n}\right| = \lim\limits_{n\to\infty}\left|\dfrac{[2(n+1)]!}{[(n+1)!]^2}\cdot\dfrac{(n!)^2}{(2n)!}\right| = \lim\limits_{n\to\infty}\dfrac{(2n+2)(2n+1)}{(n+1)(n+1)} = \lim\limits_{n\to\infty}\dfrac{(2+2/n)(2+1/n)}{(1+1/n)(1+1/n)} = \dfrac{2\cdot 2}{1\cdot 1} = 4 > 1,$

so the series $\sum\limits_{n=1}^{\infty}\dfrac{(2n)!}{(n!)^2}$ diverges by the Ratio Test.

21. $\lim\limits_{n\to\infty}\left|\dfrac{a_{n+1}}{a_n}\right| = \lim\limits_{n\to\infty}\left|\dfrac{(-1)^n(n+1)!}{1\cdot 3\cdot 5\cdot\cdots\cdot(2n-1)(2n+1)}\cdot\dfrac{1\cdot 3\cdot 5\cdot\cdots\cdot(2n-1)}{(-1)^{n-1}n!}\right| = \lim\limits_{n\to\infty}\dfrac{n+1}{2n+1}$

$= \lim\limits_{n\to\infty}\dfrac{1+1/n}{2+1/n} = \dfrac{1}{2} < 1,$

so the series $1 - \dfrac{2!}{1\cdot 3} + \dfrac{3!}{1\cdot 3\cdot 5} - \dfrac{4!}{1\cdot 3\cdot 5\cdot 7} + \cdots + (-1)^{n-1}\dfrac{n!}{1\cdot 3\cdot 5\cdot\cdots\cdot(2n-1)} + \cdots$ is absolutely convergent by the Ratio Test.

22. $\dfrac{2}{3} + \dfrac{2\cdot 5}{3\cdot 5} + \dfrac{2\cdot 5\cdot 8}{3\cdot 5\cdot 7} + \dfrac{2\cdot 5\cdot 8\cdot 11}{3\cdot 5\cdot 7\cdot 9} + \cdots = \sum\limits_{n=1}^{\infty}\dfrac{2\cdot 5\cdot 8\cdot 11\cdot\cdots\cdot(3n-1)}{3\cdot 5\cdot 7\cdot 9\cdot\cdots\cdot(2n+1)}.$

$$\lim_{n\to\infty}\left|\frac{a_{n+1}}{a_n}\right| = \lim_{n\to\infty}\left|\frac{2\cdot 5\cdot 8\cdot\cdots\cdot(3n-1)(3n+2)}{3\cdot 5\cdot 7\cdot\cdots\cdot(2n+1)(2n+3)}\cdot\frac{3\cdot 5\cdot 7\cdot\cdots\cdot(2n+1)}{2\cdot 5\cdot 8\cdot\cdots\cdot(3n-1)}\right|$$
$$= \lim_{n\to\infty}\frac{3n+2}{2n+3} = \lim_{n\to\infty}\frac{3+2/n}{2+3/n} = \frac{3}{2} > 1,$$

so the given series diverges by the Ratio Test.

23. $\lim\limits_{n\to\infty}\left|\dfrac{a_{n+1}}{a_n}\right| = \lim\limits_{n\to\infty}\left|\dfrac{2\cdot 4\cdot 6\cdot\cdots\cdot(2n)(2n+2)}{(n+1)!}\cdot\dfrac{n!}{2\cdot 4\cdot 6\cdot\cdots\cdot(2n)}\right| = \lim\limits_{n\to\infty}\dfrac{2n+2}{n+1} = \lim\limits_{n\to\infty}\dfrac{2(n+1)}{n+1} = 2 > 1$, so

the series $\sum\limits_{n=1}^{\infty}\dfrac{2\cdot 4\cdot 6\cdot\cdots\cdot(2n)}{n!}$ diverges by the Ratio Test.

24. $\lim\limits_{n\to\infty}\left|\dfrac{a_{n+1}}{a_n}\right| = \lim\limits_{n\to\infty}\left|\dfrac{2^{n+1}\,(n+1)!}{5\cdot 8\cdot 11\cdot\cdots\cdot(3n+2)\,(3n+5)}\cdot\dfrac{5\cdot 8\cdot 11\cdot\cdots\cdot(3n+2)}{2^n n!}\right| = \lim\limits_{n\to\infty}\dfrac{2(n+1)}{3n+5} = \dfrac{2}{3} < 1$, so the series $\sum\limits_{n=1}^{\infty}(-1)^n\dfrac{2^n n!}{5\cdot 8\cdot 11\cdot\cdots\cdot(3n+2)}$ is absolutely convergent by the Ratio Test.

25. $\lim\limits_{n\to\infty}\sqrt[n]{|a_n|} = \lim\limits_{n\to\infty}\dfrac{n^2+1}{2n^2+1} = \lim\limits_{n\to\infty}\dfrac{1+1/n^2}{2+1/n^2} = \dfrac{1}{2} < 1$, so the series $\sum\limits_{n=1}^{\infty}\left(\dfrac{n^2+1}{2n^2+1}\right)^n$ is absolutely convergent by the Root Test.

26. $\lim\limits_{n\to\infty}\sqrt[n]{|a_n|} = \lim\limits_{n\to\infty}\sqrt[n]{\left|\dfrac{(-2)^n}{n^n}\right|} = \lim\limits_{n\to\infty}\dfrac{2}{n} = 0 < 1$, so the series $\sum\limits_{n=1}^{\infty}\dfrac{(-2)^n}{n^n}$ is absolutely convergent by the Root Test.

27. $\lim\limits_{n\to\infty}\sqrt[n]{|a_n|} = \lim\limits_{n\to\infty}\sqrt[n]{\left|\dfrac{(-1)^{n-1}}{(\ln n)^n}\right|} = \lim\limits_{n\to\infty}\dfrac{1}{\ln n} = 0 < 1$, so the series $\sum\limits_{n=2}^{\infty}\dfrac{(-1)^{n-1}}{(\ln n)^n}$ is absolutely convergent by the Root Test.

28. $\lim\limits_{n\to\infty}\sqrt[n]{|a_n|} = \lim\limits_{n\to\infty}\sqrt[n]{\left|\left(\dfrac{-2n}{n+1}\right)^{5n}\right|} = \lim\limits_{n\to\infty}\dfrac{2^5\,n^5}{(n+1)^5} = 32\lim\limits_{n\to\infty}\dfrac{1}{\left(\dfrac{n+1}{n}\right)^5} = 32\lim\limits_{n\to\infty}\dfrac{1}{(1+1/n)^5}$

$= 32(1) = 32 > 1,$

so the series $\sum\limits_{n=1}^{\infty}\left(\dfrac{-2n}{n+1}\right)^{5n}$ diverges by the Root Test.

29. $\lim\limits_{n\to\infty}\sqrt[n]{|a_n|} = \lim\limits_{n\to\infty}\sqrt[n]{\left(1+\dfrac{1}{n}\right)^{n^2}} = \lim\limits_{n\to\infty}\left(1+\dfrac{1}{n}\right)^n = e > 1$ [by Equation 3.6.6], so the series $\sum\limits_{n=1}^{\infty}\left(1+\dfrac{1}{n}\right)^{n^2}$ diverges by the Root Test.

30. $\lim\limits_{n\to\infty}\sqrt[n]{|a_n|} = \lim\limits_{n\to\infty}\sqrt[n]{|(\arctan n)^n|} = \lim\limits_{n\to\infty}\arctan n = \frac{\pi}{2} > 1$, so the series $\sum\limits_{n=0}^{\infty}(\arctan n)^n$ diverges by the Root Test.

31. $\sum\limits_{n=2}^{\infty}\dfrac{(-1)^n}{\ln n}$ converges by the Alternating Series Test since $\lim\limits_{n\to\infty}\dfrac{1}{\ln n} = 0$ and $\left\{\dfrac{1}{\ln n}\right\}$ is decreasing. Now $\ln n < n$, so $\dfrac{1}{\ln n} > \dfrac{1}{n}$, and since $\sum\limits_{n=2}^{\infty}\dfrac{1}{n}$ is the divergent (partial) harmonic series, $\sum\limits_{n=2}^{\infty}\dfrac{1}{\ln n}$ diverges by the Comparison Test. Thus, $\sum\limits_{n=2}^{\infty}\dfrac{(-1)^n}{\ln n}$ is conditionally convergent.

32. $\lim\limits_{n\to\infty}\sqrt[n]{|a_n|} = \lim\limits_{n\to\infty}\sqrt[n]{\left|\left(\dfrac{1-n}{2+3n}\right)^n\right|} = \lim\limits_{n\to\infty}\dfrac{n-1}{3n+2} = \lim\limits_{n\to\infty}\dfrac{1-1/n}{3+2/n} = \dfrac{1}{3} < 1$, so the series $\sum\limits_{n=1}^{\infty}\left(\dfrac{1-n}{2+3n}\right)^n$ is absolutely convergent by the Root Test.

33. $\lim\limits_{n\to\infty}\left|\dfrac{a_{n+1}}{a_n}\right| = \lim\limits_{n\to\infty}\left|\dfrac{(-9)^{n+1}}{(n+1)10^{n+2}}\cdot\dfrac{n10^{n+1}}{(-9)^n}\right| = \lim\limits_{n\to\infty}\left|\dfrac{(-9)n}{10(n+1)}\right| = \dfrac{9}{10}\lim\limits_{n\to\infty}\dfrac{1}{1+1/n} = \dfrac{9}{10}(1) = \dfrac{9}{10} < 1$, so the series $\sum\limits_{n=1}^{\infty}\dfrac{(-9)^n}{n10^{n+1}}$ is absolutely convergent by the Ratio Test.

34. $\lim\limits_{n\to\infty}\left|\dfrac{a_{n+1}}{a_n}\right| = \lim\limits_{n\to\infty}\left|\dfrac{(n+1)5^{2n+2}}{10^{n+2}}\cdot\dfrac{10^{n+1}}{n5^{2n}}\right| = \lim\limits_{n\to\infty}\dfrac{5^2(n+1)}{10n} = \dfrac{5}{2}\lim\limits_{n\to\infty}\left(1+\dfrac{1}{n}\right) = \dfrac{5}{2}(1) = \dfrac{5}{2} > 1$, so the series $\sum\limits_{n=1}^{\infty}\dfrac{n5^{2n}}{10^{n+1}}$ diverges by the Ratio Test. *Or:* Since $\lim\limits_{n\to\infty} a_n = \infty$, the series diverges by the Test for Divergence.

35. $\lim\limits_{n\to\infty}\sqrt[n]{|a_n|} = \lim\limits_{n\to\infty}\sqrt[n]{\left|\left(\dfrac{n}{\ln n}\right)^n\right|} = \lim\limits_{n\to\infty}\dfrac{n}{\ln n} = \lim\limits_{x\to\infty}\dfrac{x}{\ln x} \overset{\text{H}}{=} \lim\limits_{x\to\infty}\dfrac{1}{1/x} = \lim\limits_{x\to\infty} x = \infty$, so the series $\sum\limits_{n=2}^{\infty}\left(\dfrac{n}{\ln n}\right)^n$ diverges by the Root Test.

36. $\left|\dfrac{\sin(n\pi/6)}{1+n\sqrt{n}}\right| \le \dfrac{1}{1+n\sqrt{n}} < \dfrac{1}{n^{3/2}}$, so the series $\sum\limits_{n=1}^{\infty}\dfrac{\sin(n\pi/6)}{1+n\sqrt{n}}$ converges by comparison with the convergent p-series $\sum\limits_{n=1}^{\infty}\dfrac{1}{n^{3/2}}$ $\left(p=\frac{3}{2}>1\right)$. It follows that the given series is absolutely convergent.

37. $\left|\dfrac{(-1)^n\arctan n}{n^2}\right| < \dfrac{\pi/2}{n^2}$, so since $\sum\limits_{n=1}^{\infty}\dfrac{\pi/2}{n^2} = \dfrac{\pi}{2}\sum\limits_{n=1}^{\infty}\dfrac{1}{n^2}$ converges ($p=2>1$), the given series $\sum\limits_{n=1}^{\infty}\dfrac{(-1)^n\arctan n}{n^2}$ converges absolutely by the Comparison Test.

38. The function $f(x) = \dfrac{1}{x\ln x}$ is continuous, positive, and decreasing on $[2,\infty)$.

$\displaystyle\int_2^{\infty}\frac{1}{x\ln x}\,dx = \lim_{t\to\infty}\int_2^t\frac{1}{x\ln x}\,dx = \lim_{t\to\infty}\left[\ln(\ln x)\right]_2^t = \lim_{t\to\infty}\left[(\ln(\ln t) - \ln(\ln 2)\right] = \infty$, so the series $\sum\limits_{n=2}^{\infty}\dfrac{(-1)^n}{n\ln n}$ diverges by the Integral Test. Now $\{b_n\} = \left\{\dfrac{1}{n\ln n}\right\}$ with $n\ge 2$ is a decreasing sequence of positive terms and $\lim\limits_{n\to\infty} b_n = 0$. Thus, $\sum\limits_{n=2}^{\infty}\dfrac{(-1)^n}{n\ln n}$ converges by the Alternating Series Test. It follows that $\sum\limits_{n=2}^{\infty}\dfrac{(-1)^n}{n\ln n}$ is conditionally convergent.

39. By the recursive definition, $\lim\limits_{n\to\infty}\left|\dfrac{a_{n+1}}{a_n}\right| = \lim\limits_{n\to\infty}\left|\dfrac{5n+1}{4n+3}\right| = \dfrac{5}{4} > 1$, so the series diverges by the Ratio Test.

40. By the recursive definition, $\lim\limits_{n\to\infty}\left|\dfrac{a_{n+1}}{a_n}\right| = \lim\limits_{n\to\infty}\left|\dfrac{2+\cos n}{\sqrt{n}}\right| = 0 < 1$, so the series converges absolutely by the Ratio Test.

41. The series $\sum\limits_{n=1}^{\infty}\dfrac{b_n^n\cos n\pi}{n} = \sum\limits_{n=1}^{\infty}(-1)^n\dfrac{b_n^n}{n}$, where $b_n > 0$ for $n\ge 1$ and $\lim\limits_{n\to\infty} b_n = \dfrac{1}{2}$.

$\lim\limits_{n\to\infty}\left|\dfrac{a_{n+1}}{a_n}\right| = \lim\limits_{n\to\infty}\left|\dfrac{(-1)^{n+1}b_n^{n+1}}{n+1}\cdot\dfrac{n}{(-1)^n b_n^n}\right| = \lim\limits_{n\to\infty} b_n\dfrac{n}{n+1} = \dfrac{1}{2}(1) = \dfrac{1}{2} < 1$, so the series $\sum\limits_{n=1}^{\infty}\dfrac{b_n^n\cos n\pi}{n}$ is absolutely convergent by the Ratio Test.

42. $\lim\limits_{n\to\infty}\left|\dfrac{a_{n+1}}{a_n}\right| = \lim\limits_{n\to\infty}\left|\dfrac{(-1)^{n+1}(n+1)!}{(n+1)^{n+1}b_1b_2\cdots b_nb_{n+1}}\cdot\dfrac{n^nb_1b_2\cdots b_n}{(-1)^n\,n!}\right| = \lim\limits_{n\to\infty}\left|\dfrac{(-1)(n+1)n^n}{b_{n+1}(n+1)^{n+1}}\right| = \lim\limits_{n\to\infty}\dfrac{n^n}{b_{n+1}(n+1)^n}$

$$= \lim_{n\to\infty}\frac{1}{b_{n+1}}\left(\frac{n}{n+1}\right)^n = \lim_{n\to\infty}\frac{1}{b_{n+1}}\left(\frac{1}{1+1/n}\right)^n = \lim_{n\to\infty}\frac{1}{b_{n+1}(1+1/n)^n} = \frac{1}{\frac{1}{2}e} = \frac{2}{e} < 1$$

so the series $\sum\limits_{n=1}^{\infty}\dfrac{(-1)^n\,n!}{n^nb_1b_2b_3\cdots b_n}$ is absolutely convergent by the Ratio Test.

43. (a) $\lim\limits_{n\to\infty}\left|\dfrac{1/(n+1)^3}{1/n^3}\right| = \lim\limits_{n\to\infty}\dfrac{n^3}{(n+1)^3} = \lim\limits_{n\to\infty}\dfrac{1}{(1+1/n)^3} = 1.$ Inconclusive

(b) $\lim\limits_{n\to\infty}\left|\dfrac{(n+1)}{2^{n+1}}\cdot\dfrac{2^n}{n}\right| = \lim\limits_{n\to\infty}\dfrac{n+1}{2n} = \lim\limits_{n\to\infty}\left(\dfrac{1}{2}+\dfrac{1}{2n}\right) = \dfrac{1}{2}.$ Conclusive (convergent)

(c) $\lim\limits_{n\to\infty}\left|\dfrac{(-3)^n}{\sqrt{n+1}}\cdot\dfrac{\sqrt{n}}{(-3)^{n-1}}\right| = 3\lim\limits_{n\to\infty}\sqrt{\dfrac{n}{n+1}} = 3\lim\limits_{n\to\infty}\sqrt{\dfrac{1}{1+1/n}} = 3.$ Conclusive (divergent)

(d) $\lim\limits_{n\to\infty}\left|\dfrac{\sqrt{n+1}}{1+(n+1)^2}\cdot\dfrac{1+n^2}{\sqrt{n}}\right| = \lim\limits_{n\to\infty}\left[\sqrt{1+\dfrac{1}{n}}\cdot\dfrac{1/n^2+1}{1/n^2+(1+1/n)^2}\right] = 1.$ Inconclusive

44. We use the Ratio Test:

$$\lim_{n\to\infty}\left|\frac{a_{n+1}}{a_n}\right| = \lim_{n\to\infty}\left|\frac{[(n+1)!]^2/[k(n+1)]!}{(n!)^2/(kn)!}\right| = \lim_{n\to\infty}\left|\frac{(n+1)^2}{[k(n+1)]\,[k(n+1)-1]\cdots[kn+1]}\right|$$

Now if $k=1$, then this is equal to $\lim\limits_{n\to\infty}\left|\dfrac{(n+1)^2}{(n+1)}\right| = \infty$, so the series diverges; if $k=2$, the limit is

$\lim\limits_{n\to\infty}\left|\dfrac{(n+1)^2}{(2n+2)(2n+1)}\right| = \dfrac{1}{4} < 1$, so the series converges, and if $k>2$, then the highest power of n in the denominator is larger than 2, and so the limit is 0, indicating convergence. So the series converges for $k\geq 2$.

45. (a) $\lim\limits_{n\to\infty}\left|\dfrac{a_{n+1}}{a_n}\right| = \lim\limits_{n\to\infty}\left|\dfrac{x^{n+1}}{(n+1)!}\cdot\dfrac{n!}{x^n}\right| = \lim\limits_{n\to\infty}\left|\dfrac{x}{n+1}\right| = |x|\lim\limits_{n\to\infty}\dfrac{1}{n+1} = |x|\cdot 0 = 0 < 1$, so by the Ratio Test the series $\sum\limits_{n=0}^{\infty}\dfrac{x^n}{n!}$ converges for all x.

(b) Since the series of part (a) always converges, we must have $\lim\limits_{n\to\infty}\dfrac{x^n}{n!} = 0$ by Theorem 11.2.6.

46. (a) $R_n = a_{n+1}+a_{n+2}+a_{n+3}+a_{n+4}+\cdots = a_{n+1}\left(1+\dfrac{a_{n+2}}{a_{n+1}}+\dfrac{a_{n+3}}{a_{n+1}}+\dfrac{a_{n+4}}{a_{n+1}}+\cdots\right)$

$= a_{n+1}\left(1+\dfrac{a_{n+2}}{a_{n+1}}+\dfrac{a_{n+3}}{a_{n+2}}\dfrac{a_{n+2}}{a_{n+1}}+\dfrac{a_{n+4}}{a_{n+3}}\dfrac{a_{n+3}}{a_{n+2}}\dfrac{a_{n+2}}{a_{n+1}}+\cdots\right)$

$= a_{n+1}(1+r_{n+1}+r_{n+2}r_{n+1}+r_{n+3}r_{n+2}r_{n+1}+\cdots)\quad(\star)$

$\leq a_{n+1}(1+r_{n+1}+r_{n+1}^2+r_{n+1}^3+\cdots)$ [since $\{r_n\}$ is decreasing] $= \dfrac{a_{n+1}}{1-r_{n+1}}$

(b) Note that since $\{r_n\}$ is increasing and $r_n\to L$ as $n\to\infty$, we have $r_n<L$ for all n. So, starting with equation $(\star)$,

$R_n = a_{n+1}(1+r_{n+1}+r_{n+2}r_{n+1}+r_{n+3}r_{n+2}r_{n+1}+\cdots) \leq a_{n+1}(1+L+L^2+L^3+\cdots) = \dfrac{a_{n+1}}{1-L}.$

47. (a) $s_5 = \sum\limits_{n=1}^{5}\dfrac{1}{n2^n} = \dfrac{1}{2}+\dfrac{1}{8}+\dfrac{1}{24}+\dfrac{1}{64}+\dfrac{1}{160} = \dfrac{661}{960} \approx 0.68854.$ Now the ratios

$r_n = \dfrac{a_{n+1}}{a_n} = \dfrac{n2^n}{(n+1)2^{n+1}} = \dfrac{n}{2(n+1)}$ form an increasing sequence, since

$r_{n+1}-r_n = \dfrac{n+1}{2(n+2)} - \dfrac{n}{2(n+1)} = \dfrac{(n+1)^2-n(n+2)}{2(n+1)(n+2)} = \dfrac{1}{2(n+1)(n+2)} > 0.$ So by Exercise 46(b), the error in using s_5 is $R_5 \leq \dfrac{a_6}{1-\lim\limits_{n\to\infty} r_n} = \dfrac{1/(6\cdot 2^6)}{1-1/2} = \dfrac{1}{192} \approx 0.00521.$

(b) The error in using s_n as an approximation to the sum is $R_n = \dfrac{a_{n+1}}{1-\frac{1}{2}} = \dfrac{2}{(n+1)2^{n+1}}$. We want $R_n < 0.00005 \Leftrightarrow$ $\dfrac{1}{(n+1)2^n} < 0.00005 \Leftrightarrow (n+1)2^n > 20{,}000$. To find such an n we can use trial and error or a graph. We calculate $(11+1)2^{11} = 24{,}576$, so $s_{11} = \sum\limits_{n=1}^{11} \dfrac{1}{n2^n} \approx 0.693109$ is within 0.00005 of the actual sum.

48. $s_{10} = \sum\limits_{n=1}^{10} \dfrac{n}{2^n} = \dfrac{1}{2} + \dfrac{2}{4} + \dfrac{3}{8} + \cdots + \dfrac{10}{1024} \approx 1.988$. The ratios $r_n = \dfrac{a_{n+1}}{a_n} = \dfrac{n+1}{2^{n+1}} \cdot \dfrac{2^n}{n} = \dfrac{n+1}{2n} = \dfrac{1}{2}\left(1+\dfrac{1}{n}\right)$ form a decreasing sequence, and $r_{11} = \dfrac{11+1}{2(11)} = \dfrac{12}{22} = \dfrac{6}{11} < 1$, so by Exercise 46(a), the error in using s_{10} to approximate the sum of the series $\sum\limits_{n=1}^{\infty} \dfrac{n}{2^n}$ is $R_{10} \le \dfrac{a_{11}}{1-r_{11}} = \dfrac{\frac{11}{2048}}{1-\frac{6}{11}} = \dfrac{121}{10{,}240} \approx 0.0118$.

49. (i) Following the hint, we get that $|a_n| < r^n$ for $n \ge N$, and so since the geometric series $\sum_{n=1}^{\infty} r^n$ converges $[0 < r < 1]$, the series $\sum_{n=N}^{\infty} |a_n|$ converges as well by the Comparison Test, and hence so does $\sum_{n=1}^{\infty} |a_n|$, so $\sum_{n=1}^{\infty} a_n$ is absolutely convergent.

(ii) If $\lim\limits_{n\to\infty} \sqrt[n]{|a_n|} = L > 1$, then there is an integer N such that $\sqrt[n]{|a_n|} > 1$ for all $n \ge N$, so $|a_n| > 1$ for $n \ge N$. Thus, $\lim\limits_{n\to\infty} a_n \ne 0$, so $\sum_{n=1}^{\infty} a_n$ diverges by the Test for Divergence.

(iii) Consider $\sum\limits_{n=1}^{\infty} \dfrac{1}{n}$ [diverges] and $\sum\limits_{n=1}^{\infty} \dfrac{1}{n^2}$ [converges]. For each sum, $\lim\limits_{n\to\infty} \sqrt[n]{|a_n|} = 1$, so the Root Test is inconclusive.

50. (a)
$$\lim_{n\to\infty} \left|\frac{a_{n+1}}{a_n}\right| = \lim_{n\to\infty} \left| \frac{[4(n+1)]!\,[1103 + 26{,}390(n+1)]}{[(n+1)!]^4\, 396^{4(n+1)}} \cdot \frac{(n!)^4\, 396^{4n}}{(4n)!\,(1103 + 26{,}390n)} \right|$$
$$= \lim_{n\to\infty} \frac{(4n+4)(4n+3)(4n+2)(4n+1)(26{,}390n + 27{,}493)}{(n+1)^4\, 396^4\, (26{,}390n + 1103)} = \frac{4^4}{396^4} = \frac{1}{99^4} < 1,$$
so by the Ratio Test, the series $\sum\limits_{n=0}^{\infty} \dfrac{(4n)!\,(1103 + 26{,}390n)}{(n!)^4\, 396^{4n}}$ converges.

(b) $$\frac{1}{\pi} = \frac{2\sqrt{2}}{9801} \sum_{n=0}^{\infty} \frac{(4n)!\,(1103 + 26{,}390n)}{(n!)^4\, 396^{4n}}$$

With the first term $(n = 0)$, $\dfrac{1}{\pi} \approx \dfrac{2\sqrt{2}}{9801} \cdot \dfrac{1103}{1} \Rightarrow \pi \approx 3.141\,592\,73$, so we get 6 correct decimal places of π, which is $3.141\,592\,653\,589\,793\,238$ to 18 decimal places.

With the second term $(n = 1)$, $\dfrac{1}{\pi} \approx \dfrac{2\sqrt{2}}{9801}\left(\dfrac{1103}{1} + \dfrac{4!\,(1103 + 26{,}390)}{396^4}\right) \Rightarrow \pi \approx 3.141\,592\,653\,589\,793\,878$, so we get 15 correct decimal places of π.

51. (a) Since $\sum a_n$ is absolutely convergent, and since $|a_n^+| \leq |a_n|$ and $|a_n^-| \leq |a_n|$ (because a_n^+ and a_n^- each equal either a_n or 0), we conclude by the Comparison Test that both $\sum a_n^+$ and $\sum a_n^-$ must be absolutely convergent.

Or: Use Theorem 11.2.8.

(b) We will show by contradiction that both $\sum a_n^+$ and $\sum a_n^-$ must diverge. For suppose that $\sum a_n^+$ converged. Then so would $\sum \left(a_n^+ - \frac{1}{2}a_n\right)$ by Theorem 11.2.8. But $\sum \left(a_n^+ - \frac{1}{2}a_n\right) = \sum \left[\frac{1}{2}\left(a_n + |a_n|\right) - \frac{1}{2}a_n\right] = \frac{1}{2}\sum |a_n|$, which diverges because $\sum a_n$ is only conditionally convergent. Hence, $\sum a_n^+$ can't converge. Similarly, neither can $\sum a_n^-$.

52. Let $\sum b_n$ be the rearranged series constructed in the hint. [This series can be constructed by virtue of the result of Exercise 51(b).] This series will have partial sums s_n that oscillate in value back and forth across r. Since $\lim\limits_{n\to\infty} a_n = 0$ (by Theorem 11.2.6), and since the size of the oscillations $|s_n - r|$ is always less than $|a_n|$ because of the way $\sum b_n$ was constructed, we have that $\sum b_n = \lim\limits_{n\to\infty} s_n = r$.

53. Suppose that $\sum a_n$ is conditionally convergent.

(a) $\sum n^2 a_n$ is divergent: Suppose $\sum n^2 a_n$ converges. Then $\lim\limits_{n\to\infty} n^2 a_n = 0$ by Theorem 6 in Section 11.2, so there is an integer $N > 0$ such that $n > N \;\Rightarrow\; n^2 |a_n| < 1$. For $n > N$, we have $|a_n| < \dfrac{1}{n^2}$, so $\sum\limits_{n>N} |a_n|$ converges by comparison with the convergent p-series $\sum\limits_{n>N} \dfrac{1}{n^2}$. In other words, $\sum a_n$ converges absolutely, contradicting the assumption that $\sum a_n$ is conditionally convergent. This contradiction shows that $\sum n^2 a_n$ diverges.

Remark: The same argument shows that $\sum n^p a_n$ diverges for any $p > 1$.

(b) $\sum\limits_{n=2}^{\infty} \dfrac{(-1)^n}{n \ln n}$ is conditionally convergent. It converges by the Alternating Series Test, but does not converge absolutely $\left[\text{by the Integral Test, since the function } f(x) = \dfrac{1}{x \ln x} \text{ is continuous, positive, and decreasing on } [2, \infty) \text{ and } \int_2^\infty \dfrac{dx}{x \ln x} = \lim\limits_{t\to\infty} \int_2^t \dfrac{dx}{x \ln x} = \lim\limits_{t\to\infty} \Big[\ln(\ln x)\Big]_2^t = \infty\right]$. Setting $a_n = \dfrac{(-1)^n}{n \ln n}$ for $n \geq 2$, we find that $\sum\limits_{n=2}^{\infty} n a_n = \sum\limits_{n=2}^{\infty} \dfrac{(-1)^n}{\ln n}$ converges by the Alternating Series Test.

It is easy to find conditionally convergent series $\sum a_n$ such that $\sum n a_n$ diverges. Two examples are $\sum\limits_{n=1}^{\infty} \dfrac{(-1)^{n-1}}{n}$ and $\sum\limits_{n=1}^{\infty} \dfrac{(-1)^{n-1}}{\sqrt{n}}$, both of which converge by the Alternating Series Test and fail to converge absolutely because $\sum |a_n|$ is a p-series with $p \leq 1$. In both cases, $\sum n a_n$ diverges by the Test for Divergence.

11.7 Strategy for Testing Series

1. Use the Limit Comparison Test with $a_n = \dfrac{n^2-1}{n^3+1}$ and $b_n = \dfrac{1}{n}$:

$\lim\limits_{n\to\infty} \dfrac{a_n}{b_n} = \lim\limits_{n\to\infty} \dfrac{(n^2-1)n}{n^3+1} = \lim\limits_{n\to\infty} \dfrac{n^3-n}{n^3+1} = \lim\limits_{n\to\infty} \dfrac{1-1/n^2}{1+1/n^3} = 1 > 0$. Since $\sum\limits_{n=1}^{\infty} \dfrac{1}{n}$ is the divergent harmonic series, the series $\sum\limits_{n=1}^{\infty} \dfrac{n^2-1}{n^3+1}$ also diverges.

2. $\dfrac{n-1}{n^3+1} < \dfrac{n}{n^3+1} < \dfrac{n}{n^3} = \dfrac{1}{n^2}$ for $n \geq 1$, so $\sum\limits_{n=1}^{\infty} \dfrac{n-1}{n^3+1}$ converges by comparison with $\sum\limits_{n=1}^{\infty} \dfrac{1}{n^2}$, which converges because it is a p-series with $p = 2 > 1$.

3. $\sum\limits_{n=1}^{\infty} (-1)^n \dfrac{n^2-1}{n^3+1} = \sum\limits_{n=1}^{\infty} (-1)^n b_n$. Now $b_n = \dfrac{n^2-1}{n^3+1} > 0$ for $n \geq 2$, $\{b_n\}$ is decreasing for $n \geq 2$, and $\lim\limits_{n\to\infty} b_n = 0$, so the series $\sum\limits_{n=1}^{\infty} (-1)^n \dfrac{n^2-1}{n^3+1}$ converges by the Alternating Series Test. By Exercise 1, $\sum\limits_{n=1}^{\infty} \dfrac{n^2-1}{n^3+1}$ diverges, so the series $\sum\limits_{n=1}^{\infty} (-1)^n \dfrac{n^2-1}{n^3+1}$ is conditionally convergent.

4. $\lim\limits_{n\to\infty} |a_n| = \lim\limits_{n\to\infty} \left|(-1)^n \dfrac{n^2-1}{n^2+1}\right| = \lim\limits_{n\to\infty} \dfrac{1-1/n^2}{1+1/n^2} = 1 \neq 0$, so the series $\sum\limits_{n=1}^{\infty} (-1)^n \dfrac{n^2-1}{n^2+1}$ diverges by the Test for Divergence. $\left[\text{Note that } \lim\limits_{n\to\infty} (-1)^n \dfrac{n^2-1}{n^2+1} \text{ does not exist.}\right]$

5. $\lim\limits_{x\to\infty} \dfrac{e^x}{x^2} \overset{\text{H}}{=} \lim\limits_{x\to\infty} \dfrac{e^x}{2x} \overset{\text{H}}{=} \lim\limits_{x\to\infty} \dfrac{e^x}{2} = \infty$, so $\lim\limits_{n\to\infty} \dfrac{e^n}{n^2} = \infty$. Thus, the series $\sum\limits_{n=1}^{\infty} \dfrac{e^n}{n^2}$ diverges by the Test for Divergence.

6. $\lim\limits_{n\to\infty} \sqrt[n]{|a_n|} = \lim\limits_{n\to\infty} \sqrt[n]{\dfrac{n^{2n}}{(1+n)^{3n}}} = \lim\limits_{n\to\infty} \dfrac{n^2}{(1+n)^3} = \lim\limits_{n\to\infty} \dfrac{1/n}{(1/n+1)^3} = \dfrac{0}{1} = 0 < 1$, so the series $\sum\limits_{n=1}^{\infty} \dfrac{n^{2n}}{(1+n)^{3n}}$ converges by the Root Test.

7. Let $f(x) = \dfrac{1}{x\sqrt{\ln x}}$. Then f is positive, continuous, and decreasing on $[2,\infty)$, so we can apply the Integral Test.

Since $\displaystyle\int \frac{1}{x\sqrt{\ln x}}\,dx \left[\begin{matrix} u = \ln x, \\ du = dx/x \end{matrix}\right] = \int u^{-1/2}\,du = 2u^{1/2} + C = 2\sqrt{\ln x} + C$, we find

$\displaystyle\int_2^{\infty} \frac{dx}{x\sqrt{\ln x}} = \lim_{t\to\infty} \int_2^t \frac{dx}{x\sqrt{\ln x}} = \lim_{t\to\infty} \left[2\sqrt{\ln x}\right]_2^t = \lim_{t\to\infty} \left(2\sqrt{\ln t} - 2\sqrt{\ln 2}\right) = \infty$. Since the integral diverges, the given series $\sum\limits_{n=2}^{\infty} \dfrac{1}{n\sqrt{\ln n}}$ diverges.

8. $\lim\limits_{n\to\infty} \left|\dfrac{a_{n+1}}{a_n}\right| = \lim\limits_{n\to\infty} \left|\dfrac{(n+1)^4}{4^{n+1}} \cdot \dfrac{4^n}{n^4}\right| = \lim\limits_{n\to\infty} \dfrac{(n+1)^4}{4n^4} = \dfrac{1}{4} \lim\limits_{n\to\infty} \left(1+\dfrac{1}{n}\right)^4 = \dfrac{1}{4}(1) = \dfrac{1}{4} < 1$, so the series $\sum\limits_{n=1}^{\infty} (-1)^{n-1} \dfrac{n^4}{4^n}$ is absolutely convergent (and therefore convergent) by the Ratio Test.

9. $\lim\limits_{n\to\infty}\left|\dfrac{a_{n+1}}{a_n}\right| = \lim\limits_{n\to\infty}\left|\dfrac{\pi^{2n+2}}{(2n+2)!}\cdot\dfrac{(2n)!}{\pi^{2n}}\right| = \lim\limits_{n\to\infty}\dfrac{\pi^2}{(2n+2)(2n+1)} = 0 < 1$, so the series $\sum\limits_{n=0}^{\infty}(-1)^n\dfrac{\pi^{2n}}{(2n)!}$ is absolutely convergent (and therefore convergent) by the Ratio Test.

10. Let $f(x) = x^2e^{-x^3}$. Then f is continuous and positive on $[1,\infty)$, and $f'(x) = \dfrac{x(2-3x^3)}{e^{x^3}} < 0$ for $x \geq 1$, so f is decreasing on $[1,\infty)$ as well, and we can apply the Integral Test. $\int_1^\infty x^2e^{-x^3}\,dx = \lim\limits_{t\to\infty}\left[-\frac{1}{3}e^{-x^3}\right]_1^t = \frac{1}{3e}$, so the integral converges, and hence, the series converges.

11. $\sum\limits_{n=1}^{\infty}\left(\dfrac{1}{n^3}+\dfrac{1}{3^n}\right) = \sum\limits_{n=1}^{\infty}\dfrac{1}{n^3} + \sum\limits_{n=1}^{\infty}\left(\dfrac{1}{3}\right)^n$. The first series converges since it is a p-series with $p = 3 > 1$ and the second series converges since it is geometric with $|r| = \frac{1}{3} < 1$. The sum of two convergent series is convergent.

12. $\dfrac{1}{k\sqrt{k^2+1}} < \dfrac{1}{k\sqrt{k^2}} = \dfrac{1}{k^2}$, so $\sum\limits_{k=1}^{\infty}\dfrac{1}{k\sqrt{k^2+1}}$ converges by comparison with the convergent p-series $\sum\limits_{k=1}^{\infty}\dfrac{1}{k^2}$ $(p = 2 > 1)$.

13. $\lim\limits_{n\to\infty}\left|\dfrac{a_{n+1}}{a_n}\right| = \lim\limits_{n\to\infty}\left|\dfrac{3^{n+1}(n+1)^2}{(n+1)!}\cdot\dfrac{n!}{3^nn^2}\right| = \lim\limits_{n\to\infty}\dfrac{3(n+1)^2}{(n+1)n^2} = 3\lim\limits_{n\to\infty}\dfrac{n+1}{n^2} = 0 < 1$, so the series $\sum\limits_{n=1}^{\infty}\dfrac{3^nn^2}{n!}$ converges by the Ratio Test.

14. $\left|\dfrac{\sin 2n}{1+2^n}\right| \leq \dfrac{1}{1+2^n} < \dfrac{1}{2^n} = \left(\dfrac{1}{2}\right)^n$, so the series $\sum\limits_{n=1}^{\infty}\left|\dfrac{\sin 2n}{1+2^n}\right|$ converges by comparison with the geometric series $\sum\limits_{n=1}^{\infty}\left(\dfrac{1}{2}\right)^n$ with $|r| = \frac{1}{2} < 1$. Thus, the series $\sum\limits_{n=1}^{\infty}\dfrac{\sin 2n}{1+2^n}$ converges absolutely, implying convergence.

15. $a_k = \dfrac{2^{k-1}3^{k+1}}{k^k} = \dfrac{2^k2^{-1}3^k3^1}{k^k} = \dfrac{3}{2}\left(\dfrac{2\cdot 3}{k}\right)^k$. By the Root Test, $\lim\limits_{k\to\infty}\sqrt[k]{\left(\dfrac{6}{k}\right)^k} = \lim\limits_{k\to\infty}\dfrac{6}{k} = 0 < 1$, so the series $\sum\limits_{k=1}^{\infty}\left(\dfrac{6}{k}\right)^k$ converges. It follows from Theorem 8(i) in Section 11.2 that the given series, $\sum\limits_{k=1}^{\infty}\dfrac{2^{k-1}3^{k+1}}{k^k} = \sum\limits_{k=1}^{\infty}\dfrac{3}{2}\left(\dfrac{6}{k}\right)^k$, also converges.

16. Use the Limit Comparison Test with $a_n = \dfrac{\sqrt{n^4+1}}{n^3+n}$ and $b_n = \dfrac{1}{n}$:

$\lim\limits_{n\to\infty}\dfrac{a_n}{b_n} = \lim\limits_{n\to\infty}\dfrac{n\sqrt{n^4+1}}{n(n^2+1)} = \lim\limits_{n\to\infty}\dfrac{\sqrt{n^4+1}/n^2}{(n^2+1)/n^2} = \lim\limits_{n\to\infty}\dfrac{\sqrt{1+1/n^4}}{1+1/n^2} = 1 > 0$. Since $\sum\limits_{n=1}^{\infty}\dfrac{1}{n}$ is the divergent harmonic series, the series $\sum\limits_{n=1}^{\infty}\dfrac{\sqrt{n^4+1}}{n^3+n}$ also diverges.

17. $\lim\limits_{n\to\infty}\left|\dfrac{a_{n+1}}{a_n}\right| = \lim\limits_{n\to\infty}\left|\dfrac{1\cdot 3\cdot 5\cdot\cdots\cdot(2n-1)(2n+1)}{2\cdot 5\cdot 8\cdot\cdots\cdot(3n-1)(3n+2)}\cdot\dfrac{2\cdot 5\cdot 8\cdot\cdots\cdot(3n-1)}{1\cdot 3\cdot 5\cdot\cdots\cdot(2n-1)}\right| = \lim\limits_{n\to\infty}\dfrac{2n+1}{3n+2}$

$= \lim\limits_{n\to\infty}\dfrac{2+1/n}{3+2/n} = \dfrac{2}{3} < 1,$

so the series $\sum\limits_{n=1}^{\infty}\dfrac{1\cdot 3\cdot 5\cdot\cdots\cdot(2n-1)}{2\cdot 5\cdot 8\cdot\cdots\cdot(3n-1)}$ converges by the Ratio Test.

18. $b_n = \dfrac{1}{\sqrt{n}-1}$ for $n \geq 2$. $\{b_n\}$ is a decreasing sequence of positive numbers and $\lim\limits_{n\to\infty} b_n = 0$, so $\sum\limits_{n=2}^{\infty} \dfrac{(-1)^{n-1}}{\sqrt{n}-1}$ converges by the Alternating Series Test.

19. Let $f(x) = \dfrac{\ln x}{\sqrt{x}}$. Then $f'(x) = \dfrac{2-\ln x}{2x^{3/2}} < 0$ when $\ln x > 2$ or $x > e^2$, so $\dfrac{\ln n}{\sqrt{n}}$ is decreasing for $n > e^2$.

By l'Hospital's Rule, $\lim\limits_{n\to\infty} \dfrac{\ln n}{\sqrt{n}} = \lim\limits_{n\to\infty} \dfrac{1/n}{1/\left(2\sqrt{n}\right)} = \lim\limits_{n\to\infty} \dfrac{2}{\sqrt{n}} = 0$, so the series $\sum\limits_{n=1}^{\infty} (-1)^n \dfrac{\ln n}{\sqrt{n}}$ converges by the Alternating Series Test.

20. $a_k = \dfrac{\sqrt[3]{k}-1}{k(\sqrt{k}+1)} < \dfrac{\sqrt[3]{k}}{k(\sqrt{k}+1)} < \dfrac{\sqrt[3]{k}}{k\sqrt{k}} = \dfrac{k^{1/3}}{k^{3/2}} = \dfrac{1}{k^{7/6}}$, so the series $\sum\limits_{k=1}^{\infty} \dfrac{\sqrt[3]{k}-1}{k(\sqrt{k}+1)}$ converges by comparison with the convergent p-series $\sum\limits_{k=1}^{\infty} \dfrac{1}{k^{7/6}}$ $\left(p = \frac{7}{6} > 1\right)$.

21. $\lim\limits_{n\to\infty} |a_n| = \lim\limits_{n\to\infty} \left|(-1)^n \cos(1/n^2)\right| = \lim\limits_{n\to\infty} \left|\cos(1/n^2)\right| = \cos 0 = 1$, so the series $\sum\limits_{n=1}^{\infty} (-1)^n \cos(1/n^2)$ diverges by the Test for Divergence.

22. $\lim\limits_{k\to\infty} |a_k| = \lim\limits_{k\to\infty} \left|\dfrac{1}{2+\sin k}\right| = \lim\limits_{k\to\infty} \dfrac{1}{2+\sin k}$, which does not exist (the terms vary between $\frac{1}{3}$ and 1). Thus, the series $\sum\limits_{k=1}^{\infty} \dfrac{1}{2+\sin k}$ diverges by the Test for Divergence.

23. Using the Limit Comparison Test with $a_n = \tan\left(\dfrac{1}{n}\right)$ and $b_n = \dfrac{1}{n}$, we have

$\lim\limits_{n\to\infty} \dfrac{a_n}{b_n} = \lim\limits_{n\to\infty} \dfrac{\tan(1/n)}{1/n} = \lim\limits_{x\to\infty} \dfrac{\tan(1/x)}{1/x} \overset{\text{H}}{=} \lim\limits_{x\to\infty} \dfrac{\sec^2(1/x)\cdot(-1/x^2)}{-1/x^2} = \lim\limits_{x\to\infty} \sec^2(1/x) = 1^2 = 1 > 0$. Since $\sum\limits_{n=1}^{\infty} b_n$ is the divergent harmonic series, $\sum\limits_{n=1}^{\infty} a_n$ is also divergent.

24. $\lim\limits_{n\to\infty} a_n = \lim\limits_{n\to\infty} \left(n\sin\dfrac{1}{n}\right) = \lim\limits_{n\to\infty} \dfrac{\sin(1/n)}{1/n} = \lim\limits_{x\to 0^+} \dfrac{\sin x}{x} = 1 \neq 0$, so the series $\sum\limits_{n=1}^{\infty} n\sin(1/n)$ diverges by the Test for Divergence.

25. Use the Ratio Test. $\lim\limits_{n\to\infty} \left|\dfrac{a_{n+1}}{a_n}\right| = \lim\limits_{n\to\infty} \left|\dfrac{(n+1)!}{e^{(n+1)^2}} \cdot \dfrac{e^{n^2}}{n!}\right| = \lim\limits_{n\to\infty} \dfrac{(n+1)n!\cdot e^{n^2}}{e^{n^2+2n+1}n!} = \lim\limits_{n\to\infty} \dfrac{n+1}{e^{2n+1}} = 0 < 1$, so $\sum\limits_{n=1}^{\infty} \dfrac{n!}{e^{n^2}}$ converges.

26. $\lim\limits_{n\to\infty} \left|\dfrac{a_{n+1}}{a_n}\right| = \lim\limits_{n\to\infty} \dfrac{a_{n+1}}{a_n} = \lim\limits_{n\to\infty} \left(\dfrac{n^2+2n+2}{5^{n+1}} \cdot \dfrac{5^n}{n^2+1}\right) = \lim\limits_{n\to\infty} \left(\dfrac{1+2/n+2/n^2}{1+1/n^2} \cdot \dfrac{1}{5}\right) = \dfrac{1}{5} < 1$, so $\sum\limits_{n=1}^{\infty} \dfrac{n^2+1}{5^n}$ converges by the Ratio Test.

27. $\displaystyle\int_2^{\infty} \frac{\ln x}{x^2}\,dx = \lim_{t\to\infty} \left[-\frac{\ln x}{x} - \frac{1}{x}\right]_1^t$ [using integration by parts] $\overset{\text{H}}{=} 1$. So $\sum\limits_{n=1}^{\infty} \dfrac{\ln n}{n^2}$ converges by the Integral Test, and since $\dfrac{k\ln k}{(k+1)^3} < \dfrac{k\ln k}{k^3} = \dfrac{\ln k}{k^2}$, the given series $\sum\limits_{k=1}^{\infty} \dfrac{k\ln k}{(k+1)^3}$ converges by the Comparison Test.

28. Since $\left\{\frac{1}{n}\right\}$ is a decreasing sequence, $e^{1/n} \leq e^{1/1} = e$ for all $n \geq 1$, and $\sum_{n=1}^{\infty} \frac{e}{n^2}$ converges ($p = 2 > 1$), so $\sum_{n=1}^{\infty} \frac{e^{1/n}}{n^2}$ converges by the Comparison Test. (Or use the Integral Test.)

29. $\sum_{n=1}^{\infty} a_n = \sum_{n=1}^{\infty} (-1)^n \frac{1}{\cosh n} = \sum_{n=1}^{\infty} (-1)^n b_n$. Now $b_n = \frac{1}{\cosh n} > 0$, $\{b_n\}$ is decreasing, and $\lim_{n\to\infty} b_n = 0$, so the series converges by the Alternating Series Test.

Or: Write $\frac{1}{\cosh n} = \frac{2}{e^n + e^{-n}} < \frac{2}{e^n}$ and $\sum_{n=1}^{\infty} \frac{1}{e^n}$ is a convergent geometric series, so $\sum_{n=1}^{\infty} \frac{1}{\cosh n}$ is convergent by the Comparison Test. So $\sum_{n=1}^{\infty} (-1)^n \frac{1}{\cosh n}$ is absolutely convergent and therefore convergent.

30. Let $f(x) = \frac{\sqrt{x}}{x+5}$. Then $f(x)$ is continuous and positive on $[1, \infty)$, and since $f'(x) = \frac{5-x}{2\sqrt{x}\,(x+5)^2} < 0$ for $x > 5$, $f(x)$ is eventually decreasing, so we can use the Alternating Series Test. $\lim_{n\to\infty} \frac{\sqrt{n}}{n+5} = \lim_{n\to\infty} \frac{1}{n^{1/2} + 5n^{-1/2}} = 0$, so the series $\sum_{j=1}^{\infty} (-1)^j \frac{\sqrt{j}}{j+5}$ converges.

31. $\lim_{k\to\infty} a_k = \lim_{k\to\infty} \frac{5^k}{3^k + 4^k} =$ [divide by 4^k] $\lim_{k\to\infty} \frac{(5/4)^k}{(3/4)^k + 1} = \infty$ since $\lim_{k\to\infty} \left(\frac{3}{4}\right)^k = 0$ and $\lim_{k\to\infty} \left(\frac{5}{4}\right)^k = \infty$.

Thus, $\sum_{k=1}^{\infty} \frac{5^k}{3^k + 4^k}$ diverges by the Test for Divergence.

32. $$\lim_{n\to\infty} \sqrt[n]{|a_n|} = \lim_{n\to\infty} \sqrt[n]{\left|\frac{(n!)^n}{n^{4n}}\right|} = \lim_{n\to\infty} \frac{n!}{n^4} = \lim_{n\to\infty} \left[\frac{n}{n} \cdot \frac{n-1}{n} \cdot \frac{n-2}{n} \cdot \frac{n-3}{n} \cdot (n-4)!\right]$$
$$= \lim_{n\to\infty} \left[\left(1 - \frac{1}{n}\right)\left(1 - \frac{2}{n}\right)\left(1 - \frac{3}{n}\right)(n-4)!\right] = \infty,$$

so the series $\sum_{n=1}^{\infty} \frac{(n!)^n}{n^{4n}}$ diverges by the Root Test.

33. $\lim_{n\to\infty} \sqrt[n]{|a_n|} = \lim_{n\to\infty} \left(\frac{n}{n+1}\right)^{n^2/n} = \lim_{n\to\infty} \frac{1}{[(n+1)/n]^n} = \frac{1}{\lim_{n\to\infty} (1 + 1/n)^n} = \frac{1}{e} < 1$, so the series $\sum_{n=1}^{\infty} \left(\frac{n}{n+1}\right)^{n^2}$ converges by the Root Test.

34. $0 \leq n\cos^2 n \leq n$, so $\frac{1}{n + n\cos^2 n} \geq \frac{1}{n+n} = \frac{1}{2n}$. Thus, $\sum_{n=1}^{\infty} \frac{1}{n + n\cos^2 n}$ diverges by comparison with $\sum_{n=1}^{\infty} \frac{1}{2n}$, which is a constant multiple of the (divergent) harmonic series.

35. $a_n = \frac{1}{n^{1+1/n}} = \frac{1}{n \cdot n^{1/n}}$, so let $b_n = \frac{1}{n}$ and use the Limit Comparison Test. $\lim_{n\to\infty} \frac{a_n}{b_n} = \lim_{n\to\infty} \frac{1}{n^{1/n}} = 1 > 0$ [see Exercise 4.4.63], so the series $\sum_{n=1}^{\infty} \frac{1}{n^{1+1/n}}$ diverges by comparison with the divergent harmonic series.

36. Note that $(\ln n)^{\ln n} = \left(e^{\ln \ln n}\right)^{\ln n} = \left(e^{\ln n}\right)^{\ln \ln n} = n^{\ln \ln n}$ and $\ln \ln n \to \infty$ as $n \to \infty$, so $\ln \ln n > 2$ for sufficiently large n. For these n we have $(\ln n)^{\ln n} > n^2$, so $\dfrac{1}{(\ln n)^{\ln n}} < \dfrac{1}{n^2}$. Since $\sum\limits_{n=2}^{\infty} \dfrac{1}{n^2}$ converges $[p = 2 > 1]$, so does $\sum\limits_{n=2}^{\infty} \dfrac{1}{(\ln n)^{\ln n}}$ by the Comparison Test.

37. $\lim\limits_{n\to\infty} \sqrt[n]{|a_n|} = \lim\limits_{n\to\infty} (2^{1/n} - 1) = 1 - 1 = 0 < 1$, so the series $\sum\limits_{n=1}^{\infty} \left(\sqrt[n]{2} - 1\right)^n$ converges by the Root Test.

38. Use the Limit Comparison Test with $a_n = \sqrt[n]{2} - 1$ and $b_n = 1/n$. Then

$$\lim_{n\to\infty} \frac{a_n}{b_n} = \lim_{n\to\infty} \frac{2^{1/n} - 1}{1/n} = \lim_{x\to\infty} \frac{2^{1/x} - 1}{1/x} \overset{\text{H}}{=} \lim_{x\to\infty} \frac{2^{1/x} \cdot \ln 2 \cdot (-1/x^2)}{-1/x^2} = \lim_{x\to\infty} (2^{1/x} \cdot \ln 2) = 1 \cdot \ln 2 = \ln 2 > 0.$$

So since $\sum\limits_{n=1}^{\infty} b_n$ diverges (harmonic series), so does $\sum\limits_{n=1}^{\infty} \left(\sqrt[n]{2} - 1\right)$.

Alternate solution: $\sqrt[n]{2} - 1 = \dfrac{1}{2^{(n-1)/n} + 2^{(n-2)/n} + 2^{(n-3)/n} + \cdots + 2^{1/n} + 1}$ [rationalize the numerator] $\geq \dfrac{1}{2n}$,

and since $\sum\limits_{n=1}^{\infty} \dfrac{1}{2n} = \dfrac{1}{2} \sum\limits_{n=1}^{\infty} \dfrac{1}{n}$ diverges (harmonic series), so does $\sum\limits_{n=1}^{\infty} \left(\sqrt[n]{2} - 1\right)$ by the Comparison Test.

11.8 Power Series

1. A power series is a series of the form $\sum_{n=0}^{\infty} c_n x^n = c_0 + c_1 x + c_2 x^2 + c_3 x^3 + \cdots$, where x is a variable and the c_n's are constants called the coefficients of the series.

More generally, a series of the form $\sum_{n=0}^{\infty} c_n (x-a)^n = c_0 + c_1(x-a) + c_2(x-a)^2 + \cdots$ is called a power series in $(x - a)$ or a power series centered at a or a power series about a, where a is a constant.

2. (a) Given the power series $\sum_{n=0}^{\infty} c_n (x-a)^n$, the radius of convergence is:

(i) 0 if the series converges only when $x = a$

(ii) ∞ if the series converges for all x, or

(iii) a positive number R such that the series converges if $|x - a| < R$ and diverges if $|x - a| > R$.

In most cases, R can be found by using the Ratio Test.

(b) The interval of convergence of a power series is the interval that consists of all values of x for which the series converges. Corresponding to the cases in part (a), the interval of convergence is: (i) the single point $\{a\}$, (ii) all real numbers; that is, the real number line $(-\infty, \infty)$, or (iii) an interval with endpoints $a - R$ and $a + R$ which can contain neither, either, or both of the endpoints. In this case, we must test the series for convergence at each endpoint to determine the interval of convergence.

3. If $a_n = (-1)^n n x^n$, then

$$\lim_{n\to\infty} \left|\frac{a_{n+1}}{a_n}\right| = \lim_{n\to\infty} \left|\frac{(-1)^{n+1}(n+1)x^{n+1}}{(-1)^n\, n x^n}\right| = \lim_{n\to\infty} \left|(-1)\frac{n+1}{n}x\right| = \lim_{n\to\infty} \left[\left(1 + \frac{1}{n}\right)|x|\right] = |x|.$$ By the Ratio Test, the series $\sum\limits_{n=1}^{\infty} (-1)^n n x^n$ converges when $|x| < 1$, so the radius of convergence $R = 1$. Now we'll check the endpoints, that is,

$x = \pm 1$. Both series $\sum_{n=1}^{\infty} (-1)^n n(\pm 1)^n = \sum_{n=1}^{\infty} (\mp 1)^n n$ diverge by the Test for Divergence since $\lim_{n\to\infty} |(\mp 1)^n n| = \infty$. Thus, the interval of convergence is $I = (-1, 1)$.

4. If $a_n = \dfrac{(-1)^n x^n}{\sqrt[3]{n}}$, then

$$\lim_{n\to\infty}\left|\frac{a_{n+1}}{a_n}\right| = \lim_{n\to\infty}\left|\frac{(-1)^{n+1}x^{n+1}}{\sqrt[3]{n+1}}\cdot\frac{\sqrt[3]{n}}{(-1)^n x^n}\right| = \lim_{n\to\infty}\left|\frac{(-1)x\sqrt[3]{n}}{\sqrt[3]{n+1}}\right| = \lim_{n\to\infty}\sqrt[3]{\frac{1}{1+1/n}}\,|x| = |x|.$$

By the Ratio Test, the series $\sum_{n=1}^{\infty} \dfrac{(-1)^n x^n}{\sqrt[3]{n}}$ converges when $|x| < 1$, so $R = 1$. When $x = 1$, the series $\sum_{n=1}^{\infty} \dfrac{(-1)^n}{\sqrt[3]{n}}$ converges by the Alternating Series Test. When $x = -1$, the series $\sum_{n=1}^{\infty} \dfrac{1}{\sqrt[3]{n}}$ diverges since it is a p-series $\left(p = \frac{1}{3} \le 1\right)$. Thus, the interval of convergence is $(-1, 1]$.

5. If $a_n = \dfrac{x^n}{2n-1}$, then $\lim_{n\to\infty}\left|\dfrac{a_{n+1}}{a_n}\right| = \lim_{n\to\infty}\left|\dfrac{x^{n+1}}{2n+1}\cdot\dfrac{2n-1}{x^n}\right| = \lim_{n\to\infty}\left(\dfrac{2n-1}{2n+1}|x|\right) = \lim_{n\to\infty}\left(\dfrac{2-1/n}{2+1/n}|x|\right) = |x|$. By the Ratio Test, the series $\sum_{n=1}^{\infty}\dfrac{x^n}{2n-1}$ converges when $|x| < 1$, so $R = 1$. When $x = 1$, the series $\sum_{n=1}^{\infty}\dfrac{1}{2n-1}$ diverges by comparison with $\sum_{n=1}^{\infty}\dfrac{1}{2n}$ since $\dfrac{1}{2n-1} > \dfrac{1}{2n}$ and $\dfrac{1}{2}\sum_{n=1}^{\infty}\dfrac{1}{n}$ diverges since it is a constant multiple of the harmonic series. When $x = -1$, the series $\sum_{n=1}^{\infty}\dfrac{(-1)^n}{2n-1}$ converges by the Alternating Series Test. Thus, the interval of convergence is $[-1, 1)$.

6. If $a_n = \dfrac{(-1)^n x^n}{n^2}$, then

$$\lim_{n\to\infty}\left|\frac{a_{n+1}}{a_n}\right| = \lim_{n\to\infty}\left|\frac{(-1)^{n+1}x^{n+1}}{(n+1)^2}\cdot\frac{n^2}{(-1)^n x^n}\right| = \lim_{n\to\infty}\left|\frac{(-1)xn^2}{(n+1)^2}\right| = \lim_{n\to\infty}\left[\left(\frac{n}{n+1}\right)^2|x|\right] = 1^2\cdot|x| = |x|.$$

By the Ratio Test, the series $\sum_{n=1}^{\infty}\dfrac{(-1)^n x^n}{n^2}$ converges when $|x| < 1$, so $R = 1$. When $x = 1$, the series $\sum_{n=1}^{\infty}\dfrac{(-1)^n}{n^2}$ converges by the Alternating Series Test. When $x = -1$, the series $\sum_{n=1}^{\infty}\dfrac{1}{n^2}$ converges since it is a p-series with $p = 2 > 1$. Thus, the interval of convergence is $[-1, 1]$.

7. If $a_n = \dfrac{x^n}{n!}$, then $\lim_{n\to\infty}\left|\dfrac{a_{n+1}}{a_n}\right| = \lim_{n\to\infty}\left|\dfrac{x^{n+1}}{(n+1)!}\cdot\dfrac{n!}{x^n}\right| = \lim_{n\to\infty}\left|\dfrac{x}{n+1}\right| = |x|\lim_{n\to\infty}\dfrac{1}{n+1} = |x|\cdot 0 = 0 < 1$ for *all* real x. So, by the Ratio Test, $R = \infty$ and $I = (-\infty, \infty)$.

8. Here the Root Test is easier. If $a_n = n^n x^n$, then $\lim_{n\to\infty}\sqrt[n]{|a_n|} = \lim_{n\to\infty} n|x| = \infty$ if $x \ne 0$, so $R = 0$ and $I = \{0\}$.

9. If $a_n = \dfrac{x^n}{n^4 4^n}$, then

$$\lim_{n\to\infty}\left|\frac{a_{n+1}}{a_n}\right| = \lim_{n\to\infty}\left|\frac{x^{n+1}}{(n+1)^4 4^{n+1}}\cdot\frac{n^4 4^n}{x^n}\right| = \lim_{n\to\infty}\left|\frac{n^4}{(n+1)^4}\cdot\frac{x}{4}\right| = \lim_{n\to\infty}\left(\frac{n}{n+1}\right)^4\frac{|x|}{4} = 1^4\cdot\frac{|x|}{4} = \frac{|x|}{4}.$$

By the Ratio Test, the series $\sum_{n=1}^{\infty}\dfrac{x^n}{n^4 4^n}$ converges when $\dfrac{|x|}{4} < 1 \Leftrightarrow |x| < 4$, so $R = 4$. When $x = 4$, the series $\sum_{n=1}^{\infty}\dfrac{1}{n^4}$

converges since it is a p-series ($p = 4 > 1$). When $x = -4$, the series $\sum_{n=1}^{\infty} \frac{(-1)^n}{n^4}$ converges by the Alternating Series Test. Thus, the interval of convergence is $[-4, 4]$.

10. If $a_n = 2^n n^2 x^n$, then $\lim_{n\to\infty} \left|\frac{a_{n+1}}{a_n}\right| = \lim_{n\to\infty} \left|\frac{2^{n+1}(n+1)^2 x^{n+1}}{2^n n^2 x^n}\right| = \lim_{n\to\infty} 2\left(\frac{n+1}{n}\right)^2 |x| = 2\,|x|$. By the Ratio Test, the series $\sum_{n=1}^{\infty} 2^n n^2 x^n$ converges when $2\,|x| < 1 \Leftrightarrow |x| < \frac{1}{2}$, so $R = \frac{1}{2}$. When $x = \pm\frac{1}{2}$, both series $\sum_{n=1}^{\infty} 2^n n^2 \left(\pm\frac{1}{2}\right)^n = \sum_{n=1}^{\infty} (\pm 1)^n n^2$ diverge by the Test for Divergence since $\lim_{n\to\infty} \left|(\pm 1)^n n^2\right| = \infty$. Thus, the interval of convergence is $\left(-\frac{1}{2}, \frac{1}{2}\right)$.

11. If $a_n = \frac{(-1)^n\, 4^n}{\sqrt{n}}\, x^n$, then $\lim_{n\to\infty} \left|\frac{a_{n+1}}{a_n}\right| = \lim_{n\to\infty} \left|\frac{(-1)^{n+1}\, 4^{n+1}\, x^{n+1}}{\sqrt{n+1}} \cdot \frac{\sqrt{n}}{(-1)^n\, 4^n\, x^n}\right| = \lim_{n\to\infty} \sqrt{\frac{n}{n+1}} \cdot 4\,|x| = 4\,|x|$.

By the Ratio Test, the series $\sum_{n=1}^{\infty} \frac{(-1)^n\, 4^n}{\sqrt{n}}\, x^n$ converges when $4\,|x| < 1 \Leftrightarrow |x| < \frac{1}{4}$, so $R = \frac{1}{4}$. When $x = \frac{1}{4}$, the series $\sum_{n=1}^{\infty} \frac{(-1)^n}{\sqrt{n}}$ converges by the Alternating Series Test. When $x = -\frac{1}{4}$, the series $\sum_{n=1}^{\infty} \frac{1}{\sqrt{n}}$ diverges since it is a p-series $\left(p = \frac{1}{2} \le 1\right)$. Thus, the interval of convergence is $\left(-\frac{1}{4}, \frac{1}{4}\right]$.

12. If $a_n = \sum_{n=1}^{\infty} \frac{(-1)^{n-1}}{n5^n}\, x^n$, then $\lim_{n\to\infty} \left|\frac{a_{n+1}}{a_n}\right| = \lim_{n\to\infty} \left|\frac{(-1)^n x^{n+1}}{(n+1)\, 5^{n+1}} \cdot \frac{n5^n}{(-1)^{n-1} x^n}\right| = \lim_{n\to\infty} \left(\frac{n}{n+1}\right) \frac{|x|}{5} = 1 \cdot \frac{|x|}{5} = \frac{|x|}{5}$.

By the Ratio Test, the series $\sum_{n=1}^{\infty} \frac{(-1)^{n-1}}{n5^n}\, x^n$ converges when $\frac{|x|}{5} < 1 \Leftrightarrow |x| < 5$, so $R = 5$. When $x = 5$, the series $\sum_{n=1}^{\infty} \frac{(-1)^{n-1}}{n}$ converges by the Alternating Series Test. When $x = -5$, the series $\sum_{n=1}^{\infty} \frac{-1}{n}$ diverges since it is a constant multiple of the harmonic series. Thus, the interval of convergence is $(-5, 5]$.

13. If $a_n = \frac{n}{2^n(n^2+1)}\, x^n$, then

$$\lim_{n\to\infty} \left|\frac{a_{n+1}}{a_n}\right| = \lim_{n\to\infty} \left|\frac{(n+1)x^{n+1}}{2^{n+1}(n^2+2n+2)} \cdot \frac{2^n(n^2+1)}{n\,x^n}\right| = \lim_{n\to\infty} \frac{n^3+n^2+n+1}{n^3+2n^2+2n} \cdot \frac{|x|}{2}$$
$$= \lim_{n\to\infty} \frac{1+1/n+1/n^2+1/n^3}{1+2/n+2/n^2} \cdot \frac{|x|}{2} = \frac{|x|}{2}$$

By the Ratio Test, the series $\sum_{n=1}^{\infty} \frac{n}{2^n(n^2+1)}\, x^n$ converges when $\frac{|x|}{2} < 1 \Leftrightarrow |x| < 2$, so $R = 2$. When $x = 2$, the series $\sum_{n=1}^{\infty} \frac{n}{n^2+1}$ diverges by the Limit Comparison Test with $b_n = \frac{1}{n}$. When $x = -2$, the series $\sum_{n=1}^{\infty} \frac{(-1)^n n}{n^2+1}$ converges by the Alternating Series Test. Thus, the interval of convergence is $[-2, 2)$.

14. If $a_n = \frac{x^{2n}}{n!}$, then $\lim_{n\to\infty} \left|\frac{a_{n+1}}{a_n}\right| = \lim_{n\to\infty} \left|\frac{x^{2n+2}}{(n+1)!} \cdot \frac{n!}{x^{2n}}\right| = \lim_{n\to\infty} \frac{|x^2|}{n+1} = 0 < 1$ for *all* real x. So, by the Ratio Test, $R = \infty$ and $I = (-\infty, \infty)$.

15. If $a_n = \dfrac{(x-2)^n}{n^2+1}$, then $\displaystyle\lim_{n\to\infty}\left|\frac{a_{n+1}}{a_n}\right| = \lim_{n\to\infty}\left|\frac{(x-2)^{n+1}}{(n+1)^2+1}\cdot\frac{n^2+1}{(x-2)^n}\right| = |x-2|\lim_{n\to\infty}\frac{n^2+1}{(n+1)^2+1} = |x-2|$. By the Ratio Test, the series $\displaystyle\sum_{n=0}^{\infty}\frac{(x-2)^n}{n^2+1}$ converges when $|x-2|<1$ $[R=1]$ $\Leftrightarrow$ $-1<x-2<1$ $\Leftrightarrow$ $1<x<3$. When $x=1$, the series $\displaystyle\sum_{n=0}^{\infty}(-1)^n\frac{1}{n^2+1}$ converges by the Alternating Series Test; when $x=3$, the series $\displaystyle\sum_{n=0}^{\infty}\frac{1}{n^2+1}$ converges by comparison with the p-series $\displaystyle\sum_{n=1}^{\infty}\frac{1}{n^2}$ $[p=2>1]$. Thus, the interval of convergence is $I=[1,3]$.

16. If $a_n = \dfrac{(-1)^n}{(2n-1)2^n}(x-1)^n$, then

$\displaystyle\lim_{n\to\infty}\left|\frac{a_{n+1}}{a_n}\right| = \lim_{n\to\infty}\left|\frac{(-1)^{n+1}(x-1)^{n+1}}{(2n+1)\,2^{n+1}}\cdot\frac{(2n-1)\,2^n}{(-1)^n(x-1)^n}\right| = \lim_{n\to\infty}\frac{2n-1}{2n+1}\cdot\frac{|x-1|}{2} = \frac{|x-1|}{2}$. By the Ratio Test, the series $\displaystyle\sum_{n=1}^{\infty}\frac{(-1)^n}{(2n-1)\,2^n}(x-1)^n$ converges when $\dfrac{|x-1|}{2}<1$ $\Leftrightarrow$ $|x-1|<2$ $[R=2]$ $\Leftrightarrow$ $-2<x-1<2$ $\Leftrightarrow$ $-1<x<3$. When $x=3$, the series $\displaystyle\sum_{n=1}^{\infty}\frac{(-1)^n}{2n-1}$ converges by the Alternating Series Test. When $x=-1$, the series $\displaystyle\sum_{n=1}^{\infty}\frac{1}{2n-1}$ diverges by the Limit Comparison Test with $b_n=\dfrac{1}{n}$. Thus, the interval of convergence is $(-1,3]$.

17. If $a_n = \dfrac{(x+2)^n}{2^n\ln n}$, then $\displaystyle\lim_{n\to\infty}\left|\frac{(x+2)^{n+1}}{2^{n+1}\ln(n+1)}\cdot\frac{2^n\ln n}{(x+2)^n}\right| = \lim_{n\to\infty}\frac{\ln n}{\ln(n+1)}\cdot\frac{|x+2|}{2} = \frac{|x+2|}{2}$ since

$\displaystyle\lim_{n\to\infty}\frac{\ln n}{\ln(n+1)} = \lim_{x\to\infty}\frac{\ln x}{\ln(x+1)} \overset{\text{H}}{=} \lim_{x\to\infty}\frac{1/x}{1/(x+1)} = \lim_{x\to\infty}\frac{x+1}{x} = \lim_{x\to\infty}\left(1+\frac{1}{x}\right) = 1$. By the Ratio Test, the series $\displaystyle\sum_{n=2}^{\infty}\frac{(x+2)^n}{2^n\ln n}$ converges when $\dfrac{|x+2|}{2}<1$ $\Leftrightarrow$ $|x+2|<2$ $[R=2]$ $\Leftrightarrow$ $-2<x+2<2$ $\Leftrightarrow$ $-4<x<0$.

When $x=-4$, the series $\displaystyle\sum_{n=2}^{\infty}\frac{(-1)^n}{\ln n}$ converges by the Alternating Series Test. When $x=0$, the series $\displaystyle\sum_{n=2}^{\infty}\frac{1}{\ln n}$ diverges by the Limit Comparison Test with $b_n=\dfrac{1}{n}$ (or by comparison with the harmonic series). Thus, the interval of convergence is $[-4,0)$.

18. If $a_n = \dfrac{\sqrt{n}}{8^n}(x+6)^n$, then

$$\begin{aligned}\lim_{n\to\infty}\left|\frac{a_{n+1}}{a_n}\right| &= \lim_{n\to\infty}\left|\frac{\sqrt{n+1}\,(x+6)^{n+1}}{8^{n+1}}\cdot\frac{8^n}{\sqrt{n}(x+6)^n}\right| = \lim_{n\to\infty}\sqrt{\frac{n+1}{n}}\cdot\frac{|x+6|}{8}\\ &= \lim_{n\to\infty}\sqrt{1+\frac{1}{n}}\cdot\frac{|x+6|}{8} = \frac{|x+6|}{8}\end{aligned}$$

By the Ratio Test, the series $\displaystyle\sum_{n=1}^{\infty}\frac{\sqrt{n}}{8^n}(x+6)^n$ converges when $\dfrac{|x+6|}{8}<1$ $\Leftrightarrow$ $|x+6|<8$ $[R=8]$ $\Leftrightarrow$ $-8<x+6<8$ $\Leftrightarrow$ $-14<x<2$. When $x=2$, the series $\displaystyle\sum_{n=1}^{\infty}\sqrt{n}$ diverges by the Test for Divergence since $\displaystyle\lim_{n\to\infty}|a_n| = \lim_{n\to\infty}\sqrt{n} = \infty > 0$. Similarly, when $x=-14$, the series $\displaystyle\sum_{n=1}^{\infty}(-1)^n\sqrt{n}$ diverges. Thus, the interval of convergence is $(-14,2)$.

19. If $a_n = \dfrac{(x-2)^n}{n^n}$, then $\lim\limits_{n\to\infty} \sqrt[n]{|a_n|} = \lim\limits_{n\to\infty} \dfrac{|x-2|}{n} = 0$, so the series converges for all x (by the Root Test).

$R = \infty$ and $I = (-\infty, \infty)$.

20. If $a_n = \dfrac{(2x-1)^n}{5^n\sqrt{n}}$, then

$$\lim_{n\to\infty}\left|\frac{a_{n+1}}{a_n}\right| = \lim_{n\to\infty}\left|\frac{(2x-1)^{n+1}}{5^{n+1}\sqrt{n+1}} \cdot \frac{5^n\sqrt{n}}{(2x-1)^n}\right| = \lim_{n\to\infty} \frac{|2x-1|}{5}\sqrt{\frac{n}{n+1}} = \lim_{n\to\infty} \frac{|2x-1|}{5}\sqrt{\frac{1}{1+1/n}} = \frac{|2x-1|}{5}.$$

By the Ratio Test, the series $\sum\limits_{n=1}^{\infty} \dfrac{(2x-1)^n}{5^n\sqrt{n}}$ converges when $\dfrac{|2x-1|}{5} < 1 \iff |2x-1| < 5 \iff \left|x - \frac{1}{2}\right| < \frac{5}{2} \iff$

$-\frac{5}{2} < x - \frac{1}{2} < \frac{5}{2} \iff -2 < x < 3$, so $R = \frac{5}{2}$. When $x = 3$, the series $\sum\limits_{n=1}^{\infty} \dfrac{1}{\sqrt{n}}$ is a divergent p-series $\left(p = \frac{1}{2} \le 1\right)$.

When $x = -2$, the series $\sum\limits_{n=1}^{\infty} \dfrac{(-1)^n}{\sqrt{n}}$ converges by the Alternating Series Test. Thus, the interval of convergence

is $I = [-2, 3)$.

21. $a_n = \dfrac{n}{b^n}(x-a)^n$, where $b > 0$.

$$\lim_{n\to\infty}\left|\frac{a_{n+1}}{a_n}\right| = \lim_{n\to\infty} \frac{(n+1)\,|x-a|^{n+1}}{b^{n+1}} \cdot \frac{b^n}{n\,|x-a|^n} = \lim_{n\to\infty}\left(1+\frac{1}{n}\right)\frac{|x-a|}{b} = \frac{|x-a|}{b}.$$

By the Ratio Test, the series converges when $\dfrac{|x-a|}{b} < 1 \iff |x-a| < b$ [so $R = b$] $\iff -b < x - a < b \iff$

$a - b < x < a + b$. When $|x-a| = b$, $\lim\limits_{n\to\infty} |a_n| = \lim\limits_{n\to\infty} n = \infty$, so the series diverges. Thus, $I = (a-b, a+b)$.

22. $a_n = \dfrac{b^n}{\ln n}(x-a)^n$, where $b > 0$.

$$\lim_{n\to\infty}\left|\frac{a_{n+1}}{a_n}\right| = \lim_{n\to\infty}\left|\frac{b^{n+1}(x-a)^{n+1}}{\ln(n+1)} \cdot \frac{\ln n}{b^n(x-a)^n}\right| = \lim_{n\to\infty} \frac{\ln n}{\ln(n+1)} \cdot b\,|x-a| = b\,|x-a| \text{ since}$$

$\lim\limits_{n\to\infty} \dfrac{\ln n}{\ln(n+1)} = \lim\limits_{x\to\infty} \dfrac{\ln x}{\ln(x+1)} \overset{\text{H}}{=} \lim\limits_{x\to\infty} \dfrac{1/x}{1/(x+1)} = \lim\limits_{x\to\infty} \dfrac{x+1}{x} \overset{\text{H}}{=} \lim\limits_{x\to\infty} \dfrac{1}{1} = 1$. By the Ratio Test, the series

$\sum\limits_{n=2}^{\infty} \dfrac{b^n}{\ln n}(x-a)^n$ converges when $b\,|x-a| < 1 \iff |x-a| < \dfrac{1}{b} \iff -\dfrac{1}{b} < x - a < \dfrac{1}{b} \iff a - \dfrac{1}{b} < x < a + \dfrac{1}{b}$,

so $R = \dfrac{1}{b}$. When $x = a + \dfrac{1}{b}$, the series $\sum\limits_{n=2}^{\infty} \dfrac{1}{\ln n}$ diverges by comparison with the divergent p-series $\sum\limits_{n=2}^{\infty} \dfrac{1}{n}$ since $\dfrac{1}{\ln n} > \dfrac{1}{n}$

for $n \ge 2$. When $x = a - \dfrac{1}{b}$, the series $\sum\limits_{n=2}^{\infty} \dfrac{(-1)^n}{\ln n}$ converges by the Alternating Series Test. Thus, the interval of

convergence is $I = \left[a - \dfrac{1}{b}, a + \dfrac{1}{b}\right)$.

23. If $a_n = n!\,(2x-1)^n$, then $\lim\limits_{n\to\infty}\left|\dfrac{a_{n+1}}{a_n}\right| = \lim\limits_{n\to\infty}\left|\dfrac{(n+1)!\,(2x-1)^{n+1}}{n!(2x-1)^n}\right| = \lim\limits_{n\to\infty} (n+1)\,|2x-1| \to \infty$ as $n \to \infty$

for all $x \ne \frac{1}{2}$. Since the series diverges for all $x \ne \frac{1}{2}$, $R = 0$ and $I = \left\{\frac{1}{2}\right\}$.

24. $a_n = \dfrac{n^2 x^n}{2 \cdot 4 \cdot 6 \cdot \cdots \cdot (2n)} = \dfrac{n^2 x^n}{2^n n!} = \dfrac{n x^n}{2^n (n-1)!}$, so

$\lim\limits_{n\to\infty} \left| \dfrac{a_{n+1}}{a_n} \right| = \lim\limits_{n\to\infty} \dfrac{(n+1)\,|x|^{n+1}}{2^{n+1} n!} \cdot \dfrac{2^n (n-1)!}{n\,|x|^n} = \lim\limits_{n\to\infty} \dfrac{n+1}{n^2} \dfrac{|x|}{2} = 0$. Thus, by the Ratio Test, the series converges for *all* real x and we have $R = \infty$ and $I = (-\infty, \infty)$.

25. If $a_n = \dfrac{(5x-4)^n}{n^3}$, then

$$\lim_{n\to\infty} \left| \frac{a_{n+1}}{a_n} \right| = \lim_{n\to\infty} \left| \frac{(5x-4)^{n+1}}{(n+1)^3} \cdot \frac{n^3}{(5x-4)^n} \right| = \lim_{n\to\infty} |5x-4| \left(\frac{n}{n+1} \right)^3 = \lim_{n\to\infty} |5x-4| \left(\frac{1}{1+1/n} \right)^3$$
$$= |5x-4| \cdot 1 = |5x-4|$$

By the Ratio Test, $\sum\limits_{n=1}^{\infty} \dfrac{(5x-4)^n}{n^3}$ converges when $|5x-4| < 1 \Leftrightarrow \left| x - \frac{4}{5} \right| < \frac{1}{5} \Leftrightarrow -\frac{1}{5} < x - \frac{4}{5} < \frac{1}{5} \Leftrightarrow$ $\frac{3}{5} < x < 1$, so $R = \frac{1}{5}$. When $x = 1$, the series $\sum\limits_{n=1}^{\infty} \dfrac{1}{n^3}$ is a convergent p-series ($p = 3 > 1$). When $x = \frac{3}{5}$, the series $\sum\limits_{n=1}^{\infty} \dfrac{(-1)^n}{n^3}$ converges by the Alternating Series Test. Thus, the interval of convergence is $I = \left[\frac{3}{5}, 1\right]$.

26. If $a_n = \dfrac{x^{2n}}{n\,(\ln n)^2}$, then $\lim\limits_{n\to\infty} \left| \dfrac{a_{n+1}}{a_n} \right| = \lim\limits_{n\to\infty} \left| \dfrac{x^{2n+2}}{(n+1)[\ln(n+1)]^2} \cdot \dfrac{n\,(\ln n)^2}{x^{2n}} \right| = \left| x^2 \right| \lim\limits_{n\to\infty} \dfrac{n\,(\ln n)^2}{(n+1)[\ln(n+1)]^2} = x^2$.

By the Ratio Test, the series $\sum\limits_{n=2}^{\infty} \dfrac{x^{2n}}{n\,(\ln n)^2}$ converges when $x^2 < 1 \Leftrightarrow |x| < 1$, so $R = 1$. When $x = \pm 1$, $x^{2n} = 1$, the series $\sum\limits_{n=2}^{\infty} \dfrac{1}{n\,(\ln n)^2}$ converges by the Integral Test (see Exercise 11.3.22). Thus, the interval of convergence is $I = [-1, 1]$.

27. If $a_n = \dfrac{x^n}{1 \cdot 3 \cdot 5 \cdot \cdots \cdot (2n-1)}$, then

$\lim\limits_{n\to\infty} \left| \dfrac{a_{n+1}}{a_n} \right| = \lim\limits_{n\to\infty} \left| \dfrac{x^{n+1}}{1 \cdot 3 \cdot 5 \cdot \cdots \cdot (2n-1)(2n+1)} \cdot \dfrac{1 \cdot 3 \cdot 5 \cdot \cdots \cdot (2n-1)}{x^n} \right| = \lim\limits_{n\to\infty} \dfrac{|x|}{2n+1} = 0 < 1$. Thus, by the Ratio Test, the series $\sum\limits_{n=1}^{\infty} \dfrac{x^n}{1 \cdot 3 \cdot 5 \cdot \cdots \cdot (2n-1)}$ converges for *all* real x and we have $R = \infty$ and $I = (-\infty, \infty)$.

28. If $a_n = \dfrac{n!\,x^n}{1 \cdot 3 \cdot 5 \cdot \cdots \cdot (2n-1)}$, then

$\lim\limits_{n\to\infty} \left| \dfrac{a_{n+1}}{a_n} \right| = \lim\limits_{n\to\infty} \left| \dfrac{(n+1)!\,x^{n+1}}{1 \cdot 3 \cdot 5 \cdot \cdots \cdot (2n-1)(2n+1)} \cdot \dfrac{1 \cdot 3 \cdot 5 \cdot \cdots \cdot (2n-1)}{n!\,x^n} \right| = \lim\limits_{n\to\infty} \dfrac{(n+1)\,|x|}{2n+1} = \frac{1}{2}\,|x|$.

By the Ratio Test, the series $\sum\limits_{n=1}^{\infty} a_n$ converges when $\frac{1}{2}\,|x| < 1 \Rightarrow |x| < 2$, so $R = 2$. When $x = \pm 2$,

$|a_n| = \dfrac{n!\,2^n}{1 \cdot 3 \cdot 5 \cdot \cdots \cdot (2n-1)} = \dfrac{[1 \cdot 2 \cdot 3 \cdot \cdots \cdot n]\,2^n}{[1 \cdot 3 \cdot 5 \cdot \cdots \cdot (2n-1)]} = \dfrac{2 \cdot 4 \cdot 6 \cdot \cdots \cdot 2n}{1 \cdot 3 \cdot 5 \cdot \cdots \cdot (2n-1)} > 1$, so both endpoint series diverge by the Test for Divergence. Thus, the interval of convergence is $I = (-2, 2)$.

29. (a) We are given that the power series $\sum_{n=0}^{\infty} c_n x^n$ is convergent for $x = 4$. So by Theorem 4, it must converge for at least $-4 < x \le 4$. In particular, it converges when $x = -2$; that is, $\sum_{n=0}^{\infty} c_n (-2)^n$ is convergent.

(b) It does not follow that $\sum_{n=0}^{\infty} c_n(-4)^n$ is necessarily convergent. [See the comments after Theorem 4 about convergence at the endpoint of an interval. An example is $c_n = (-1)^n/(n4^n)$.]

30. We are given that the power series $\sum_{n=0}^{\infty} c_n x^n$ is convergent for $x = -4$ and divergent when $x = 6$. So by Theorem 4 it converges for at least $-4 \leq x < 4$ and diverges for at least $x \geq 6$ and $x < -6$. Therefore:

(a) It converges when $x = 1$; that is, $\sum c_n$ is convergent.

(b) It diverges when $x = 8$; that is, $\sum c_n 8^n$ is divergent.

(c) It converges when $x = -3$; that is, $\sum c_n(-3^n)$ is convergent.

(d) It diverges when $x = -9$; that is, $\sum c_n(-9)^n = \sum(-1)^n c_n 9^n$ is divergent.

31. If $a_n = \dfrac{(n!)^k}{(kn)!}x^n$, then

$$\lim_{n\to\infty}\left|\frac{a_{n+1}}{a_n}\right| = \lim_{n\to\infty}\frac{[(n+1)!]^k\,(kn)!}{(n!)^k\,[k(n+1)]!}|x| = \lim_{n\to\infty}\frac{(n+1)^k}{(kn+k)(kn+k-1)\cdots(kn+2)(kn+1)}|x|$$

$$= \lim_{n\to\infty}\left[\frac{(n+1)}{(kn+1)}\frac{(n+1)}{(kn+2)}\cdots\frac{(n+1)}{(kn+k)}\right]|x|$$

$$= \lim_{n\to\infty}\left[\frac{n+1}{kn+1}\right]\lim_{n\to\infty}\left[\frac{n+1}{kn+2}\right]\cdots\lim_{n\to\infty}\left[\frac{n+1}{kn+k}\right]|x|$$

$$= \left(\frac{1}{k}\right)^k|x| < 1 \quad\Leftrightarrow\quad |x| < k^k \text{ for convergence, and the radius of convergence is } R = k^k.$$

32. (a) Note that the four intervals in parts (a)–(d) have midpoint $m = \frac{1}{2}(p+q)$ and radius of convergence $r = \frac{1}{2}(q-p)$. We also know that the power series $\sum\limits_{n=0}^{\infty} x^n$ has interval of convergence $(-1, 1)$. To change the radius of convergence to r, we can change x^n to $\left(\dfrac{x}{r}\right)^n$. To shift the midpoint of the interval of convergence, we can replace x with $x - m$. Thus, a power series whose interval of convergence is (p, q) is $\sum\limits_{n=0}^{\infty}\left(\dfrac{x-m}{r}\right)^n$, where $m = \frac{1}{2}(p+q)$ and $r = \frac{1}{2}(q-p)$.

(b) Similar to Example 2, we know that $\sum\limits_{n=1}^{\infty}\dfrac{x^n}{n}$ has interval of convergence $[-1, 1)$. By introducing the factor $(-1)^n$ in a_n, the interval of convergence changes to $(-1, 1]$. Now change the midpoint and radius as in part (a) to get $\sum\limits_{n=1}^{\infty}(-1)^n\dfrac{1}{n}\left(\dfrac{x-m}{r}\right)^n$ as a power series whose interval of convergence is $(p, q]$.

(c) As in part (b), $\sum\limits_{n=1}^{\infty}\dfrac{1}{n}\left(\dfrac{x-m}{r}\right)^n$ is a power series whose interval of convergence is $[p, q)$.

(d) If we increase the exponent on n (to say, $n = 2$), in the power series in part (c), then when $x = q$, the power series $\sum\limits_{n=1}^{\infty}\dfrac{1}{n^2}\left(\dfrac{x-m}{r}\right)^n$ will converge by comparison to the p-series with $p = 2 > 1$, and the interval of convergence will be $[p, q]$.

33. No. If a power series is centered at a, its interval of convergence is symmetric about a. If a power series has an infinite radius of convergence, then its interval of convergence must be $(-\infty, \infty)$, not $[0, \infty)$.

34. The partial sums of the series $\sum_{n=0}^{\infty} x^n$ definitely do not converge to $f(x) = 1/(1-x)$ for $x \geq 1$, since f is undefined at $x = 1$ and negative on $(1, \infty)$, while all the partial sums are positive on this interval. The partial sums also fail to converge to f for $x \leq -1$, since $0 < f(x) < 1$ on this interval, while the partial sums are either larger than 1 or less than 0. The partial sums seem to converge to f on $(-1, 1)$. This graphical evidence is consistent with what we know about geometric series: convergence for $|x| < 1$, divergence for $|x| \geq 1$ (see Examples 2 and 7 in Section 11.2).

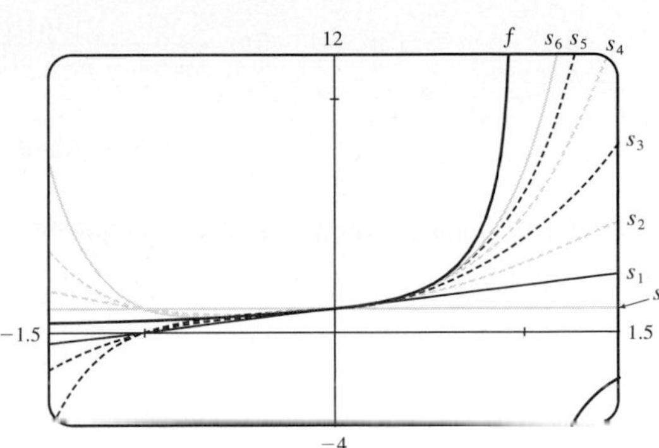

35. (a) If $a_n = \dfrac{(-1)^n x^{2n+1}}{n!(n+1)!\,2^{2n+1}}$, then

$$\lim_{n\to\infty}\left|\frac{a_{n+1}}{a_n}\right| = \lim_{n\to\infty}\left|\frac{x^{2n+3}}{(n+1)!(n+2)!\,2^{2n+3}}\cdot\frac{n!(n+1)!\,2^{2n+1}}{x^{2n+1}}\right| = \left(\frac{x}{2}\right)^2 \lim_{n\to\infty}\frac{1}{(n+1)(n+2)} = 0 \text{ for all } x.$$

So $J_1(x)$ converges for all x and its domain is $(-\infty, \infty)$.

(b), (c) The initial terms of $J_1(x)$ up to $n = 5$ are $a_0 = \dfrac{x}{2}$,

$a_1 = -\dfrac{x^3}{16}$, $a_2 = \dfrac{x^5}{384}$, $a_3 = -\dfrac{x^7}{18{,}432}$, $a_4 = \dfrac{x^9}{1{,}474{,}560}$,

and $a_5 = -\dfrac{x^{11}}{176{,}947{,}200}$. The partial sums seem to approximate $J_1(x)$ well near the origin, but as $|x|$ increases, we need to take a large number of terms to get a good approximation.

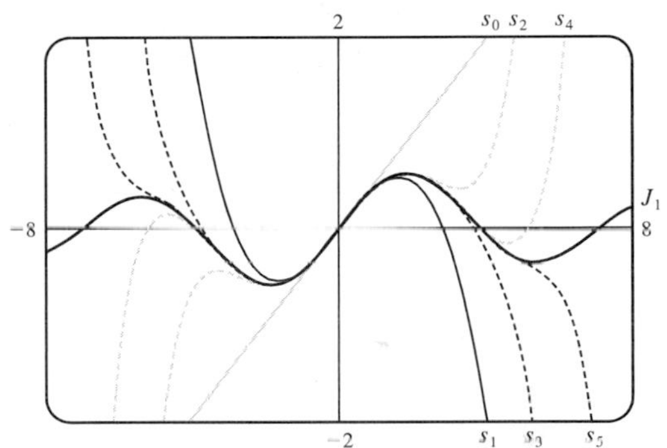

36. (a) $A(x) = 1 + \sum\limits_{n=1}^{\infty} a_n$, where $a_n = \dfrac{x^{3n}}{2\cdot 3\cdot 5\cdot 6\cdot\cdots\cdot(3n-1)(3n)}$, so $\lim\limits_{n\to\infty}\left|\dfrac{a_{n+1}}{a_n}\right| = |x|^3 \lim\limits_{n\to\infty}\dfrac{1}{(3n+2)(3n+3)} = 0$ for all x, so the domain is $\mathbb{R}$.

(b), (c)

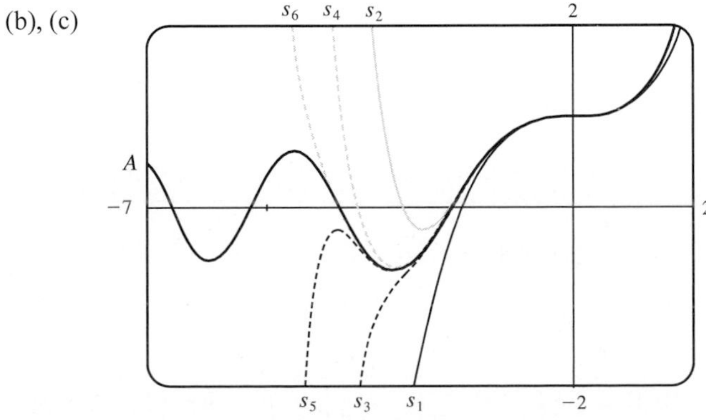

$s_0 = 1$ has been omitted from the graph. The partial sums seem to approximate $A(x)$ well near the origin, but as $|x|$ increases, we need to take a large number of terms to get a good approximation.

To plot A, we must first define $A(x)$ for the CAS. Note that for $n \geq 1$, the denominator of a_n is

$2 \cdot 3 \cdot 5 \cdot 6 \cdot \cdots \cdot (3n-1) \cdot 3n = \dfrac{(3n)!}{1 \cdot 4 \cdot 7 \cdot \cdots \cdot (3n-2)} = \dfrac{(3n)!}{\prod_{k=1}^{n}(3k-2)}$, so $a_n = \dfrac{\prod_{k=1}^{n}(3k-2)}{(3n)!}x^{3n}$ and thus

$A(x) = 1 + \sum\limits_{n=1}^{\infty} \dfrac{\prod_{k=1}^{n}(3k-2)}{(3n)!}x^{3n}$. Both Maple and Mathematica are able to plot A if we define it this way, and Derive is able to produce a similar graph using a suitable partial sum of $A(x)$.

Derive, Maple and Mathematica all have two initially known Airy functions, called `AI·SERIES(z,m)` and `BI·SERIES(z,m)` from `BESSEL.MTH` in Derive and `AiryAi` and `AiryBi` in Maple and Mathematica (just `Ai` and `Bi` in older versions of Maple). However, it is very difficult to solve for A in terms of the CAS's Airy functions, although in fact $A(x) = \dfrac{\sqrt{3}\,\texttt{AiryAi}(x) + \texttt{AiryBi}(x)}{\sqrt{3}\,\texttt{AiryAi}(0) + \texttt{AiryBi}(0)}$.

37. $s_{2n-1} = 1 + 2x + x^2 + 2x^3 + x^4 + 2x^5 + \cdots + x^{2n-2} + 2x^{2n-1}$

$= 1(1+2x) + x^2(1+2x) + x^4(1+2x) + \cdots + x^{2n-2}(1+2x) = (1+2x)(1 + x^2 + x^4 + \cdots + x^{2n-2})$

$= (1+2x)\dfrac{1-x^{2n}}{1-x^2}$ [by (11.2.3) with $r = x^2$] $\to \dfrac{1+2x}{1-x^2}$ as $n \to \infty$ by (11.2.4), when $|x| < 1$.

Also $s_{2n} = s_{2n-1} + x^{2n} \to \dfrac{1+2x}{1-x^2}$ since $x^{2n} \to 0$ for $|x| < 1$. Therefore, $s_n \to \dfrac{1+2x}{1-x^2}$ since s_{2n} and s_{2n-1} both approach $\dfrac{1+2x}{1-x^2}$ as $n \to \infty$. Thus, the interval of convergence is $(-1, 1)$ and $f(x) = \dfrac{1+2x}{1-x^2}$.

38. $s_{4n-1} = c_0 + c_1x + c_2x^2 + c_3x^3 + c_0x^4 + c_1x^5 + c_2x^6 + c_3x^7 + \cdots + c_3x^{4n-1}$

$= \left(c_0 + c_1x + c_2x^2 + c_3x^3\right)\left(1 + x^4 + x^8 + \cdots + x^{4n-4}\right) \to \dfrac{c_0 + c_1x + c_2x^2 + c_3x^3}{1-x^4}$ as $n \to \infty$

[by (11.2.4) with $r = x^4$] for $\left|x^4\right| < 1 \;\Leftrightarrow\; |x| < 1$. Also $s_{4n}, s_{4n+1}, s_{4n+2}$ have the same limits (for example, $s_{4n} = s_{4n-1} + c_0x^{4n}$ and $x^{4n} \to 0$ for $|x| < 1$). So if at least one of c_0, c_1, c_2, and c_3 is nonzero, then the interval of convergence is $(-1, 1)$ and $f(x) = \dfrac{c_0 + c_1x + c_2x^2 + c_3x^3}{1-x^4}$.

39. We use the Root Test on the series $\sum c_nx^n$. We need $\lim\limits_{n\to\infty} \sqrt[n]{|c_nx^n|} = |x| \lim\limits_{n\to\infty} \sqrt[n]{|c_n|} = c\,|x| < 1$ for convergence, or $|x| < 1/c$, so $R = 1/c$.

40. Suppose $c_n \neq 0$. Applying the Ratio Test to the series $\sum c_n(x-a)^n$, we find that

$L = \lim\limits_{n\to\infty}\left|\dfrac{a_{n+1}}{a_n}\right| = \lim\limits_{n\to\infty}\left|\dfrac{c_{n+1}(x-a)^{n+1}}{c_n(x-a)^n}\right| = \lim\limits_{n\to\infty}\dfrac{|x-a|}{|c_n/c_{n+1}|}\,(*) = \dfrac{|x-a|}{\lim\limits_{n\to\infty}|c_n/c_{n+1}|}$ (if $\lim\limits_{n\to\infty}|c_n/c_{n+1}| \neq 0$), so the series converges when $\dfrac{|x-a|}{\lim\limits_{n\to\infty}|c_n/c_{n+1}|} < 1 \;\Leftrightarrow\; |x-a| < \lim\limits_{n\to\infty}\left|\dfrac{c_n}{c_{n+1}}\right|$. Thus, $R = \lim\limits_{n\to\infty}\left|\dfrac{c_n}{c_{n+1}}\right|$. If $\lim\limits_{n\to\infty}\left|\dfrac{c_n}{c_{n+1}}\right| = 0$ and $|x-a| \neq 0$, then $(*)$ shows that $L = \infty$ and so the series diverges, and hence, $R = 0$. Thus, in all cases, $R = \lim\limits_{n\to\infty}\left|\dfrac{c_n}{c_{n+1}}\right|$.

41. For $2 < x < 3$, $\sum c_nx^n$ diverges and $\sum d_nx^n$ converges. By Exercise 11.2.85, $\sum(c_n + d_n)\,x^n$ diverges. Since both series converge for $|x| < 2$, the radius of convergence of $\sum(c_n + d_n)\,x^n$ is 2.

42. Since $\sum c_n x^n$ converges whenever $|x| < R$, $\sum c_n x^{2n} = \sum c_n (x^2)^n$ converges whenever $|x^2| < R \Leftrightarrow |x| < \sqrt{R}$, so the second series has radius of convergence $\sqrt{R}$.

11.9 Representations of Functions as Power Series

1. If $f(x) = \sum\limits_{n=0}^{\infty} c_n x^n$ has radius of convergence 10, then $f'(x) = \sum\limits_{n=1}^{\infty} n c_n x^{n-1}$ also has radius of convergence 10 by Theorem 2.

2. If $f(x) = \sum\limits_{n=0}^{\infty} b_n x^n$ converges on $(-2, 2)$, then $\int f(x)\,dx = C + \sum\limits_{n=0}^{\infty} \dfrac{b_n}{n+1} x^{n+1}$ has the same radius of convergence (by Theorem 2), but may not have the same interval of convergence—it may happen that the integrated series converges at an endpoint (or both endpoints).

3. Our goal is to write the function in the form $\dfrac{1}{1-r}$, and then use Equation (1) to represent the function as a sum of a power series. $f(x) = \dfrac{1}{1+x} = \dfrac{1}{1-(-x)} = \sum\limits_{n=0}^{\infty} (-x)^n = \sum\limits_{n=0}^{\infty} (-1)^n x^n$ with $|-x| < 1 \Leftrightarrow |x| < 1$, so $R = 1$ and $I = (-1, 1)$.

4. $f(x) = \dfrac{5}{1-4x^2} = 5\left(\dfrac{1}{1-4x^2}\right) = 5 \sum\limits_{n=0}^{\infty} (4x^2)^n = 5 \sum\limits_{n=0}^{\infty} 4^n x^{2n}$. The series converges when $|4x^2| < 1 \Leftrightarrow$ $|x|^2 < \frac{1}{4} \Leftrightarrow |x| < \frac{1}{2}$, so $R = \frac{1}{2}$ and $I = \left(-\frac{1}{2}, \frac{1}{2}\right)$.

5. $f(x) = \dfrac{2}{3-x} = \dfrac{2}{3}\left(\dfrac{1}{1-x/3}\right) = \dfrac{2}{3} \sum\limits_{n=0}^{\infty} \left(\dfrac{x}{3}\right)^n$ or, equivalently, $2 \sum\limits_{n=0}^{\infty} \dfrac{1}{3^{n+1}} x^n$. The series converges when $\left|\dfrac{x}{3}\right| < 1$, that is, when $|x| < 3$, so $R = 3$ and $I = (-3, 3)$.

6. $f(x) = \dfrac{4}{2x+3} = \dfrac{4}{3}\left(\dfrac{1}{1+2x/3}\right) = \dfrac{4}{3}\left(\dfrac{1}{1-(-2x/3)}\right) = \dfrac{4}{3} \sum\limits_{n=0}^{\infty} \left(-\dfrac{2x}{3}\right)^n$ or, equivalently, $\sum\limits_{n=0}^{\infty} (-1)^n \dfrac{2^{n+2}}{3^{n+1}} x^n$. The series converges when $\left|-\dfrac{2x}{3}\right| < 1$, that is, when $|x| < \frac{3}{2}$, so $R = \frac{3}{2}$ and $I = \left(-\frac{3}{2}, \frac{3}{2}\right)$.

7. $f(x) = \dfrac{x^2}{x^4+16} = \dfrac{x^2}{16}\left(\dfrac{1}{1+x^4/16}\right) = \dfrac{x^2}{16}\left(\dfrac{1}{1-[-(x/2)]^4}\right) = \dfrac{x^2}{16} \sum\limits_{n=0}^{\infty} \left[-\left(\dfrac{x}{2}\right)^4\right]^n$ or, equivalently, $\sum\limits_{n=0}^{\infty} \dfrac{(-1)^n x^{4n+2}}{2^{4n+4}}$. The series converges when $\left|-\left(\dfrac{x}{2}\right)^4\right| < 1 \Rightarrow \left|\dfrac{x}{2}\right| < 1 \Rightarrow |x| < 2$, so $R = 2$ and $I = (-2, 2)$.

8. $f(x) = \dfrac{x}{2x^2+1} = x\left(\dfrac{1}{1-(-2x^2)}\right) = x \sum\limits_{n=0}^{\infty} (-2x^2)^n$ or, equivalently, $\sum\limits_{n=0}^{\infty} (-1)^n 2^n x^{2n+1}$. The series converges when $|-2x^2| < 1 \Rightarrow |x^2| < \frac{1}{2} \Rightarrow |x| < \dfrac{1}{\sqrt{2}}$, so $R = \dfrac{1}{\sqrt{2}}$ and $I = \left(-\dfrac{1}{\sqrt{2}}, \dfrac{1}{\sqrt{2}}\right)$.

9. $f(x) = \dfrac{x-1}{x+2} = \dfrac{x+2-3}{x+2} = 1 - \dfrac{3}{x+2} = 1 - \dfrac{3/2}{x/2+1} = 1 - \dfrac{3}{2} \cdot \dfrac{1}{1-(-x/2)}$

$$= 1 - \frac{3}{2}\sum_{n=0}^{\infty}\left(-\frac{x}{2}\right)^n = 1 - \frac{3}{2} - \frac{3}{2}\sum_{n=1}^{\infty}\left(-\frac{x}{2}\right)^n = -\frac{1}{2} - \sum_{n=1}^{\infty}\frac{(-1)^n\, 3x^n}{2^{n+1}}.$$

The geometric series $\sum\limits_{n=0}^{\infty}\left(-\frac{x}{2}\right)^n$ converges when $\left|-\dfrac{x}{2}\right| < 1 \iff |x| < 2$, so $R = 2$ and $I = (-2, 2)$.

Alternatively, you could write $f(x) = 1 - 3\left(\dfrac{1}{x+2}\right)$ and use the series for $\dfrac{1}{x+2}$ found in Example 2.

10. $f(x) = \dfrac{a}{x^2+a^2}\quad [a > 0] \quad = \dfrac{a}{a^2}\left[\dfrac{1}{1-(-x^2/a^2)}\right] = \dfrac{1}{a}\sum\limits_{n=0}^{\infty}\left(-\dfrac{x^2}{a^2}\right)^n = \sum\limits_{n=0}^{\infty}\dfrac{(-1)^n x^{2n}}{a^{2n+1}}$. The geometric series $\sum\limits_{n=0}^{\infty}\left(-\dfrac{x^2}{a^2}\right)^n$ converges when $\left|-\dfrac{x^2}{a^2}\right| < 1 \iff |x| < a$, so $R = a$ and $I = (-a, a)$.

11. $f(x) = \dfrac{2x-4}{x^2-4x+3} = \dfrac{2x-4}{(x-1)(x-3)} = \dfrac{A}{x-1} + \dfrac{B}{x-3} \Rightarrow 2x - 4 = A(x-3) + B(x-1)$. Let $x = 1$ to get $-2 = -2A \iff A = 1$ and $x = 3$ to get $2 = 2B \iff B = 1$. Thus,

$$\frac{2x-4}{x^2-4x+3} = \frac{1}{x-1} + \frac{1}{x-3} = \frac{-1}{1-x} + \frac{1}{-3}\left[\frac{1}{1-(x/3)}\right] = -\sum_{n=0}^{\infty}x^n - \frac{1}{3}\sum_{n=0}^{\infty}\left(\frac{x}{3}\right)^n = \sum_{n=0}^{\infty}\left(-1 - \frac{1}{3^{n+1}}\right)x^n.$$

We represented f as the sum of two geometric series; the first converges for $x \in (-1, 1)$ and the second converges for $x \in (-3, 3)$. Thus, the sum converges for $x \in (-1, 1) = I$.

12. $f(x) = \dfrac{2x+3}{x^2+3x+2} = \dfrac{2x+3}{(x+1)(x+2)} = \dfrac{A}{x+1} + \dfrac{B}{x+2} \Rightarrow 2x + 3 = A(x+2) + B(x+1)$. Let $x = -1$ to get $1 = A$ and $x = -2$ to get $-1 = -B \iff B = 1$. Thus,

$$\frac{2x+3}{x^2+3x+2} = \frac{1}{x+1} + \frac{1}{x+2} = \frac{1}{1-(-x)} + \frac{1}{2}\left[\frac{1}{1-(-x/2)}\right]$$

$$= \sum_{n=0}^{\infty}(-x)^n + \frac{1}{2}\sum_{n=0}^{\infty}\left(-\frac{x}{2}\right)^n = \sum_{n=0}^{\infty}\left[(-1)^n\left(1 + \frac{1}{2^{n+1}}\right)\right]x^n$$

We represented f as the sum of two geometric series; the first converges for $x \in (-1, 1)$ and the second converges for $x \in (-2, 2)$. Thus, the sum converges for $x \in (-1, 1) = I$.

13. (a) $f(x) = \dfrac{1}{(1+x)^2} = \dfrac{d}{dx}\left(\dfrac{-1}{1+x}\right) = -\dfrac{d}{dx}\left[\sum\limits_{n=0}^{\infty}(-1)^n x^n\right]$ [from Exercise 3]

$$= \sum_{n=1}^{\infty}(-1)^{n+1}nx^{n-1} \text{ [from Theorem 2(i)] } = \sum_{n=0}^{\infty}(-1)^n(n+1)x^n \text{ with } R = 1.$$

In the last step, note that we *decreased* the initial value of the summation variable n by 1, and then *increased* each occurrence of n in the term by 1 [also note that $(-1)^{n+2} = (-1)^n$].

(b) $f(x) = \dfrac{1}{(1+x)^3} = -\dfrac{1}{2}\dfrac{d}{dx}\left[\dfrac{1}{(1+x)^2}\right] = -\dfrac{1}{2}\dfrac{d}{dx}\left[\sum\limits_{n=0}^{\infty}(-1)^n(n+1)x^n\right]$ [from part (a)]

$$= -\tfrac{1}{2}\sum_{n=1}^{\infty}(-1)^n(n+1)nx^{n-1} = \tfrac{1}{2}\sum_{n=0}^{\infty}(-1)^n(n+2)(n+1)x^n \text{ with } R = 1.$$

(c) $f(x) = \dfrac{x^2}{(1+x)^3} = x^2 \cdot \dfrac{1}{(1+x)^3} = x^2 \cdot \dfrac{1}{2}\sum\limits_{n=0}^{\infty}(-1)^n(n+2)(n+1)x^n$ [from part (b)]

$$= \frac{1}{2}\sum_{n=0}^{\infty}(-1)^n(n+2)(n+1)x^{n+2}$$

[continued]

To write the power series with x^n rather than x^{n+2}, we will *decrease* each occurrence of n in the term by 2 and *increase* the initial value of the summation variable by 2. This gives us $\frac{1}{2}\sum\limits_{n=2}^{\infty}(-1)^n(n)(n-1)x^n$ with $R=1$.

14. (a) $\displaystyle\int \frac{1}{1-x}\,dx = -\ln(1-x)+C$ and

$$\int \frac{1}{1-x}\,dx = \int (1+x+x^2+\cdots)\,dx = \left(x+\frac{x^2}{2}+\frac{x^3}{3}+\cdots\right)+C = \sum_{n=1}^{\infty}\frac{x^n}{n}+C \text{ for } |x|<1.$$

So $-\ln(1-x) = \sum\limits_{n=1}^{\infty}\dfrac{x^n}{n}+C$ and letting $x=0$ gives $0=C$. Thus, $f(x)=\ln(1-x) = -\sum\limits_{n=1}^{\infty}\dfrac{x^n}{n}$ with $R=1$.

(b) $f(x) = x\ln(1-x) = -x\sum\limits_{n=1}^{\infty}\dfrac{x^n}{n} = -\sum\limits_{n=1}^{\infty}\dfrac{x^{n+1}}{n}$.

(c) Letting $x=\frac{1}{2}$ gives $\ln\frac{1}{2} = -\sum\limits_{n=1}^{\infty}\dfrac{(1/2)^n}{n} \;\Rightarrow\; \ln 1-\ln 2 = -\sum\limits_{n=1}^{\infty}\dfrac{1^n}{n2^n} \;\Rightarrow\; \ln 2 = \sum\limits_{n=1}^{\infty}\dfrac{1}{n2^n}$.

15. $f(x) = \ln(5-x) = -\displaystyle\int\frac{dx}{5-x} = -\frac{1}{5}\int\frac{dx}{1-x/5} = -\frac{1}{5}\int\left[\sum_{n=0}^{\infty}\left(\frac{x}{5}\right)^n\right]dx = C-\frac{1}{5}\sum_{n=0}^{\infty}\frac{x^{n+1}}{5^n(n+1)} = C-\sum_{n=1}^{\infty}\frac{x^n}{n5^n}$

Putting $x=0$, we get $C=\ln 5$. The series converges for $|x/5|<1 \;\Leftrightarrow\; |x|<5$, so $R=5$.

16. $f(x) = x^2\tan^{-1}(x^3) = x^2\sum\limits_{n=0}^{\infty}(-1)^n\dfrac{(x^3)^{2n+1}}{2n+1}$ [by Example 7] $= \sum\limits_{n=0}^{\infty}(-1)^n\dfrac{x^{6n+3+2}}{2n+1} = \sum\limits_{n=0}^{\infty}(-1)^n\dfrac{x^{6n+5}}{2n+1}$ for

$|x^3|<1 \;\Leftrightarrow\; |x|<1$, so $R=1$.

17. We know that $\dfrac{1}{1+4x} = \dfrac{1}{1-(-4x)} = \sum\limits_{n=0}^{\infty}(-4x)^n$. Differentiating, we get

$$\frac{-4}{(1+4x)^2} = \sum_{n=1}^{\infty}(-4)^n n x^{n-1} = \sum_{n=0}^{\infty}(-4)^{n+1}(n+1)x^n, \text{ so}$$

$$f(x) = \frac{x}{(1+4x)^2} = \frac{-x}{4}\cdot\frac{-4}{(1+4x)^2} = \frac{-x}{4}\sum_{n=0}^{\infty}(-4)^{n+1}(n+1)x^n = \sum_{n=0}^{\infty}(-1)^n 4^n(n+1)x^{n+1}$$

for $|-4x|<1 \;\Leftrightarrow\; |x|<\frac{1}{4}$, so $R=\frac{1}{4}$.

18. $\dfrac{1}{2-x} = \dfrac{1}{2(1-x/2)} = \dfrac{1}{2}\sum\limits_{n=0}^{\infty}\left(\dfrac{x}{2}\right)^n = \sum\limits_{n=0}^{\infty}\dfrac{1}{2^{n+1}}x^n$. Now $\dfrac{d}{dx}\left(\dfrac{1}{2-x}\right) = \dfrac{d}{dx}\left(\sum\limits_{n=0}^{\infty}\dfrac{1}{2^{n+1}}x^n\right) \;\Rightarrow$

$$\frac{1}{(2-x)^2} = \sum_{n=1}^{\infty}\frac{1}{2^{n+1}}nx^{n-1} \text{ and } \frac{d}{dx}\left(\frac{1}{(2-x)^2}\right) = \frac{d}{dx}\left(\sum_{n=1}^{\infty}\frac{1}{2^{n+1}}nx^{n-1}\right) \;\Rightarrow$$

$$\frac{2}{(2-x)^3} = \sum_{n=2}^{\infty}\frac{1}{2^{n+1}}n(n-1)x^{n-2} = \sum_{n=0}^{\infty}\frac{(n+2)(n+1)}{2^{n+3}}x^n.$$

Thus, $f(x) = \left(\dfrac{x}{2-x}\right)^3 = \dfrac{x^3}{(2-x)^3} = \dfrac{x^3}{2}\cdot\dfrac{2}{(2-x)^3} = \dfrac{x^3}{2}\sum\limits_{n=0}^{\infty}\dfrac{(n+2)(n+1)}{2^{n+3}}x^n = \sum\limits_{n=0}^{\infty}\dfrac{(n+2)(n+1)}{2^{n+4}}x^{n+3}$

for $\left|\dfrac{x}{2}\right|<1 \;\Leftrightarrow\; |x|<2$, so $R=2$.

19. By Example 5, $\dfrac{1}{(1-x)^2} = \sum\limits_{n=0}^{\infty} (n+1)x^n$. Thus,

$$\begin{aligned} f(x) = \frac{1+x}{(1-x)^2} &= \frac{1}{(1-x)^2} + \frac{x}{(1-x)^2} = \sum_{n=0}^{\infty}(n+1)x^n + \sum_{n=0}^{\infty}(n+1)x^{n+1} \\ &= \sum_{n=0}^{\infty}(n+1)x^n + \sum_{n=1}^{\infty} nx^n \qquad \text{[make the starting values equal]} \\ &= 1 + \sum_{n=1}^{\infty}[(n+1)+n]x^n = 1 + \sum_{n=1}^{\infty}(2n+1)x^n = \sum_{n=0}^{\infty}(2n+1)x^n \text{ with } R = 1. \end{aligned}$$

20. By Example 5, $\dfrac{1}{(1-x)^2} = \sum\limits_{n=0}^{\infty} (n+1)x^n$, so

$$\frac{d}{dx}\left(\frac{1}{(1-x)^2}\right) = \frac{d}{dx}\left(\sum_{n=0}^{\infty}(n+1)x^n\right) \quad \Rightarrow \quad \frac{2}{(1-x)^3} = \sum_{n=1}^{\infty}(n+1)nx^{n-1}. \text{ Thus,}$$

$$\begin{aligned} f(x) = \frac{x^2+x}{(1-x)^3} &= \frac{x^2}{(1-x)^3} + \frac{x}{(1-x)^3} = \frac{x^2}{2}\cdot\frac{2}{(1-x)^3} + \frac{x}{2}\cdot\frac{2}{(1-x)^3} \\ &= \frac{x^2}{2}\sum_{n=1}^{\infty}(n+1)nx^{n-1} + \frac{x}{2}\sum_{n=1}^{\infty}(n+1)nx^{n-1} = \sum_{n=1}^{\infty}\frac{(n+1)n}{2}x^{n+1} + \sum_{n=1}^{\infty}\frac{(n+1)n}{2}x^n \\ &= \sum_{n=2}^{\infty}\frac{n(n-1)}{2}x^n + \sum_{n=1}^{\infty}\frac{(n+1)n}{2}x^n \qquad \text{[make the exponents on } x \text{ equal by changing an index]} \\ &= \sum_{n=2}^{\infty}\frac{n^2-n}{2}x^n + x + \sum_{n=2}^{\infty}\frac{n^2+n}{2}x^n \qquad \text{[make the starting values equal]} \\ &= x + \sum_{n=2}^{\infty} n^2x^n = \sum_{n=1}^{\infty} n^2x^n \text{ with } R = 1. \end{aligned}$$

21. $f(x) = \dfrac{x^2}{x^2+1} = x^2\left(\dfrac{1}{1-(-x^2)}\right) = x^2\sum\limits_{n=0}^{\infty}(-x^2)^n = \sum\limits_{n=0}^{\infty}(-1)^n x^{2n+2}$. This series converges when $|-x^2| < 1 \Leftrightarrow$ $x^2 < 1 \Leftrightarrow |x| < 1$, so $R = 1$. The partial sums are $s_1 = x^2$, $s_2 = s_1 - x^4$, $s_3 = s_2 + x^6$, $s_4 = s_3 - x^8$, $s_5 = s_4 + x^{10}$, ... Note that s_1 corresponds to the first term of the infinite sum, regardless of the value of the summation variable and the value of the exponent. As n increases, $s_n(x)$ approximates f better on the interval of convergence, which is $(-1, 1)$.

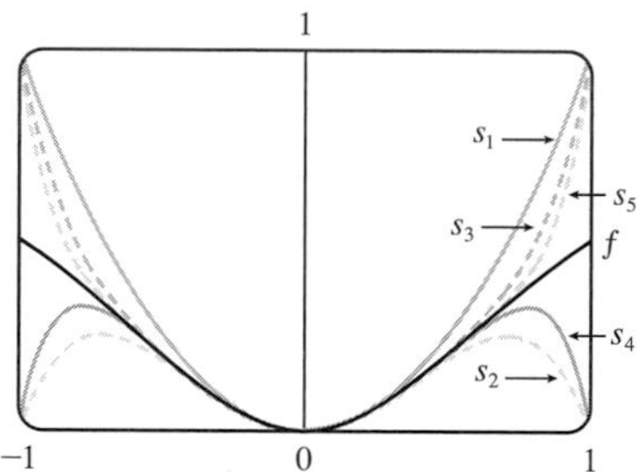

22. From Example 6, we have $\ln(1+x) = \sum\limits_{n=1}^{\infty}(-1)^{n-1}\dfrac{x^n}{n}$ with $|x| < 1$, so $f(x) = \ln(1+x^4) = \sum\limits_{n=1}^{\infty}(-1)^{n-1}\dfrac{x^{4n}}{n}$ with $|x^4| < 1 \Leftrightarrow |x| < 1$ $[R = 1]$. The partial sums are $s_1 = x^4$, $s_2 = s_1 - \frac{1}{2}x^8$, $s_3 = s_2 + \frac{1}{3}x^{12}$, $s_4 = s_3 - \frac{1}{4}x^{16}$, $s_5 = s_4 + \frac{1}{5}x^{20}$, ... Note that s_1 corresponds to the first term of the infinite sum, regardless of the value of the summation variable and the value of the exponent. As n increases, $s_n(x)$ approximates f better on the interval of convergence, which is $[-1, 1]$. (When $x = \pm 1$, the series is the convergent alternating harmonic series.)

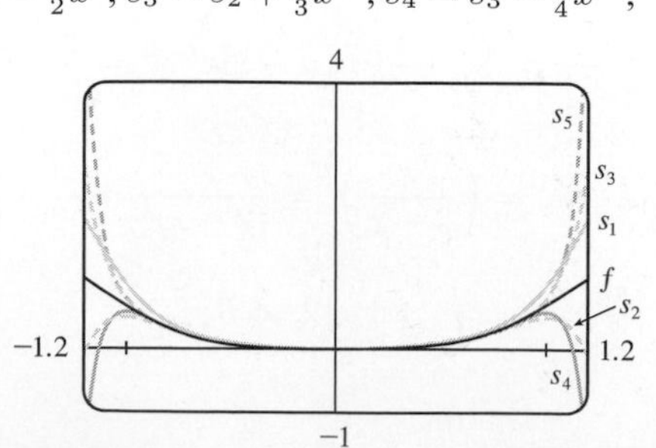

23. $f(x) = \ln\left(\dfrac{1+x}{1-x}\right) = \ln(1+x) - \ln(1-x) = \displaystyle\int \frac{dx}{1+x} + \int \frac{dx}{1-x} = \int \frac{dx}{1-(-x)} + \int \frac{dx}{1-x}$

$$= \int \left[\sum_{n=0}^{\infty} (-1)^n x^n + \sum_{n=0}^{\infty} x^n\right] dx = \int [(1 - x + x^2 - x^3 + x^4 - \cdots) + (1 + x + x^2 + x^3 + x^4 + \cdots)]\, dx$$

$$= \int (2 + 2x^2 + 2x^4 + \cdots)\, dx = \int \sum_{n=0}^{\infty} 2x^{2n}\, dx = C + \sum_{n=0}^{\infty} \frac{2x^{2n+1}}{2n+1}$$

But $f(0) = \ln \frac{1}{1} = 0$, so $C = 0$ and we have $f(x) = \displaystyle\sum_{n=0}^{\infty} \frac{2x^{2n+1}}{2n+1}$ with $R = 1$. If $x = \pm 1$, then $f(x) = \pm 2 \displaystyle\sum_{n=0}^{\infty} \frac{1}{2n+1}$,

which both diverge by the Limit Comparison Test with $b_n = \dfrac{1}{n}$.

The partial sums are $s_1 = \dfrac{2x}{1}$, $s_2 = s_1 + \dfrac{2x^3}{3}$, $s_3 = s_2 + \dfrac{2x^5}{5}$, ...

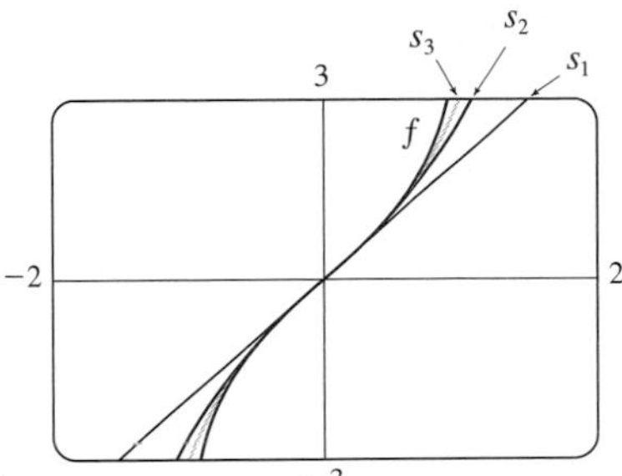

As n increases, $s_n(x)$ approximates f better on the interval of convergence, which is $(-1, 1)$.

24. $f(x) = \tan^{-1}(2x) = 2\displaystyle\int \frac{dx}{1+4x^2} = 2\int \sum_{n=0}^{\infty} (-1)^n \left(4x^2\right)^n dx = 2\int \sum_{n=0}^{\infty} (-1)^n 4^n x^{2n}\, dx$

$$= C + 2\sum_{n=0}^{\infty} \frac{(-1)^n 4^n x^{2n+1}}{2n+1} = \sum_{n=0}^{\infty} \frac{(-1)^n 2^{2n+1} x^{2n+1}}{2n+1} \qquad [f(0) = \tan^{-1} 0 = 0, \text{ so } C = 0]$$

The series converges when $\left|4x^2\right| < 1 \Leftrightarrow |x| < \frac{1}{2}$, so $R = \frac{1}{2}$. If $x = \pm\frac{1}{2}$, then $f(x) = \displaystyle\sum_{n=0}^{\infty} (-1)^n \frac{1}{2n+1}$ and

$f(x) = \displaystyle\sum_{n=0}^{\infty} (-1)^{n+1} \frac{1}{2n+1}$, respectively. Both series converge by the Alternating Series Test. The partial sums are

$s_1 = \dfrac{2x}{1}$, $s_2 = s_1 - \dfrac{2^3 x^3}{3}$, $s_3 = s_2 + \dfrac{2^5 x^5}{5}$, ...

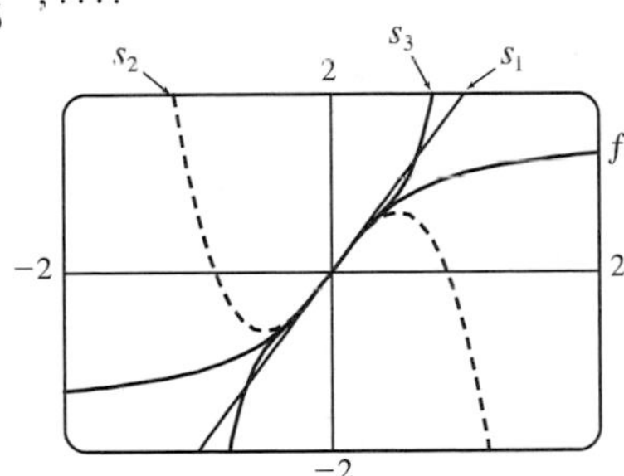

As n increases, $s_n(x)$ approximates f better on the interval of convergence, which is $\left[-\frac{1}{2}, \frac{1}{2}\right]$.

25. $\dfrac{t}{1-t^8} = t \cdot \dfrac{1}{1-t^8} = t\displaystyle\sum_{n=0}^{\infty} (t^8)^n = \sum_{n=0}^{\infty} t^{8n+1} \Rightarrow \int \frac{t}{1-t^8}\, dt = C + \sum_{n=0}^{\infty} \frac{t^{8n+2}}{8n+2}$. The series for $\dfrac{1}{1-t^8}$ converges

when $\left|t^8\right| < 1 \Leftrightarrow |t| < 1$, so $R = 1$ for that series and also the series for $t/(1-t^8)$. By Theorem 2, the series for

$\displaystyle\int \frac{t}{1-t^8}\, dt$ also has $R = 1$.

26. $\dfrac{t}{1+t^3} = t \cdot \dfrac{1}{1-(-t^3)} = t\sum\limits_{n=0}^{\infty}(-t^3)^n = \sum\limits_{n=0}^{\infty}(-1)^n t^{3n+1} \quad\Rightarrow\quad \displaystyle\int \frac{t}{1+t^3}\,dt = C + \sum_{n=0}^{\infty}(-1)^n \frac{t^{3n+2}}{3n+2}$. The series for

$\dfrac{1}{1+t^3}$ converges when $\left|-t^3\right| < 1 \;\Leftrightarrow\; |t| < 1$, so $R = 1$ for that series and also for the series $\dfrac{t}{1+t^3}$. By Theorem 2, the

series for $\displaystyle\int \frac{t}{1+t^3}\,dt$ also has $R = 1$.

27. From Example 6, $\ln(1+x) = \sum\limits_{n=1}^{\infty}(-1)^{n-1}\dfrac{x^n}{n}$ for $|x| < 1$, so $x^2\ln(1+x) = \sum\limits_{n=1}^{\infty}(-1)^{n-1}\dfrac{x^{n+2}}{n}$ and

$\displaystyle\int x^2\ln(1+x)\,dx = C + \sum_{n=1}^{\infty}(-1)^{n-1}\frac{x^{n+3}}{n(n+3)}$. $R = 1$ for the series for $\ln(1+x)$, so $R = 1$ for the series representing

$x^2\ln(1+x)$ as well. By Theorem 2, the series for $\displaystyle\int x^2\ln(1+x)\,dx$ also has $R = 1$.

28. From Example 7, $\tan^{-1}x = \sum\limits_{n=0}^{\infty}(-1)^n\dfrac{x^{2n+1}}{2n+1}$ for $|x| < 1$, so $\dfrac{\tan^{-1}x}{x} = \sum\limits_{n=0}^{\infty}(-1)^n\dfrac{x^{2n}}{2n+1}$ and

$\displaystyle\int \frac{\tan^{-1}x}{x}\,dx = C + \sum_{n=0}^{\infty}(-1)^n\frac{x^{2n+1}}{(2n+1)^2}$. $R = 1$ for the series for $\tan^{-1}x$, so $R = 1$ for the series representing

$\dfrac{\tan^{-1}x}{x}$ as well. By Theorem 2, the series for $\displaystyle\int \frac{\tan^{-1}x}{x}\,dx$ also has $R = 1$.

29. $\dfrac{x}{1+x^3} = x\left[\dfrac{1}{1-(-x^3)}\right] = x\sum\limits_{n=0}^{\infty}(-x^3)^n = \sum\limits_{n=0}^{\infty}(-1)^n x^{3n+1} \quad\Rightarrow$

$\displaystyle\int \frac{x}{1+x^3}\,dx = \int \sum_{n=0}^{\infty}(-1)^n x^{3n+1}\,dx = C + \sum_{n=0}^{\infty}(-1)^n\frac{x^{3n+2}}{3n+2}$. Thus,

$$I = \int_0^{0.3}\frac{x}{1+x^3}\,dx = \left[\frac{x^2}{2} - \frac{x^5}{5} + \frac{x^8}{8} - \frac{x^{11}}{11} + \cdots\right]_0^{0.3} = \frac{(0.3)^2}{2} - \frac{(0.3)^5}{5} + \frac{(0.3)^8}{8} - \frac{(0.3)^{11}}{11} + \cdots.$$

The series is alternating, so if we use the first three terms, the error is at most $(0.3)^{11}/11 \approx 1.6\times 10^{-7}$. So

$I \approx (0.3)^2/2 - (0.3)^5/5 + (0.3)^8/8 \approx 0.044\,522$ to six decimal places.

30. We substitute $x/2$ for x in Example 7, and find that

$$\begin{aligned}\int \arctan(x/2)\,dx &= \int \sum_{n=0}^{\infty}(-1)^n\frac{(x/2)^{2n+1}}{2n+1}\,dx = \int \sum_{n=0}^{\infty}(-1)^n\frac{x^{2n+1}}{2^{2n+1}(2n+1)}\,dx\\ &= C + \sum_{n=0}^{\infty}(-1)^n\frac{x^{2n+2}}{2^{2n+1}(2n+1)(2n+2)}\end{aligned}$$

Thus,

$$\begin{aligned}I = \int_0^{1/2}\arctan(x/2)\,dx &= \left[\frac{x^2}{2(1)(2)} - \frac{x^4}{2^3(3)(4)} + \frac{x^6}{2^5(5)(6)} - \frac{x^8}{2^7(7)(8)} + \frac{x^{10}}{2^9(9)(10)} - \cdots\right]_0^{1/2}\\ &= \frac{1}{2^3(1)(2)} - \frac{1}{2^7(3)(4)} + \frac{1}{2^{11}(5)(6)} - \frac{1}{2^{15}(7)(8)} + \frac{1}{2^{19}(9)(10)} - \cdots\end{aligned}$$

[continued]

The series is alternating, so if we use four terms, the error is at most $1/(2^{19} \cdot 90) \approx 2.1 \times 10^{-8}$. So

$$I \approx \frac{1}{16} - \frac{1}{1536} + \frac{1}{61{,}440} - \frac{1}{1{,}835{,}008} \approx 0.061\,865 \text{ to six decimal places.}$$

Remark: The sum of the first three terms gives us the same answer to six decimal places, but the error is at most $1/1{,}835{,}008 \approx 5.5 \times 10^{-7}$, slightly too large to guarantee the desired accuracy.

31. We substitute x^2 for x in Example 6, and find that

$$\int x\ln(1+x^2)\,dx = \int x\sum_{n=1}^{\infty}(-1)^{n-1}\frac{(x^2)^n}{n}\,dx = \int \sum_{n=1}^{\infty}(-1)^{n-1}\frac{x^{2n+1}}{n}\,dx = C + \sum_{n=1}^{\infty}(-1)^{n-1}\frac{x^{2n+2}}{n(2n+2)}$$

Thus,

$$I \approx \int_0^{0.2} x\ln(1+x^2)\,dx = \left[\frac{x^4}{1(4)} - \frac{x^6}{2(6)} + \frac{x^8}{3(8)} - \frac{x^{10}}{4(10)} + \cdots\right]_0^{0.2} = \frac{(0.2)^4}{4} - \frac{(0.2)^6}{12} + \frac{(0.2)^8}{24} - \frac{(0.2)^{10}}{40} + \cdots$$

The series is alternating, so if we use two terms, the error is at most $(0.2)^8/24 \approx 1.1 \times 10^{-7}$. So

$$I \approx \frac{(0.2)^4}{4} - \frac{(0.2)^6}{12} \approx 0.000\,395 \text{ to six decimal places.}$$

32. $$\int_0^{0.3}\frac{x^2}{1+x^4}\,dx = \int_0^{0.3} x^2\sum_{n=0}^{\infty}(-1)^n x^{4n}\,dx = \sum_{n=0}^{\infty}\left[\frac{(-1)^n x^{4n+3}}{4n+3}\right]_0^{0.3} = \sum_{n=0}^{\infty}\frac{(-1)^n\,3^{4n+3}}{(4n+3)10^{4n+3}}$$
$$= \frac{3^3}{3\times 10^3} - \frac{3^7}{7\times 10^7} + \frac{3^{11}}{11\times 10^{11}} - \cdots$$

The series is alternating, so if we use only two terms, the error is at most $\dfrac{3^{11}}{11\times 10^{11}} \approx 0.000\,000\,16$. So, to six decimal places, $\displaystyle\int_0^{0.3}\frac{x^2}{1+x^4}\,dx \approx \frac{3^3}{3\times 10^3} - \frac{3^7}{7\times 10^7} \approx 0.008\,969$.

33. By Example 7, $\arctan x = x - \dfrac{x^3}{3} + \dfrac{x^5}{5} - \dfrac{x^7}{7} + \cdots$, so $\arctan 0.2 = 0.2 - \dfrac{(0.2)^3}{3} + \dfrac{(0.2)^5}{5} - \dfrac{(0.2)^7}{7} + \cdots$.

The series is alternating, so if we use three terms, the error is at most $\dfrac{(0.2)^7}{7} \approx 0.000\,002$.

Thus, to five decimal places, $\arctan 0.2 \approx 0.2 - \dfrac{(0.2)^3}{3} + \dfrac{(0.2)^5}{5} \approx 0.197\,40$.

34. $f(x) = \displaystyle\sum_{n=0}^{\infty}\frac{(-1)^n x^{2n}}{(2n)!} \quad\Rightarrow\quad f'(x) = \sum_{n=1}^{\infty}\frac{(-1)^n 2nx^{2n-1}}{(2n)!}$ [the first term disappears], so

$$f''(x) = \sum_{n=1}^{\infty}\frac{(-1)^n(2n)(2n-1)x^{2n-2}}{(2n)!} = \sum_{n=1}^{\infty}\frac{(-1)^n x^{2(n-1)}}{[2(n-1)]!} = \sum_{n=0}^{\infty}\frac{(-1)^{n+1}x^{2n}}{(2n)!} \quad \text{[substituting } n+1 \text{ for } n\text{]}$$
$$= -\sum_{n=0}^{\infty}\frac{(-1)^n x^{2n}}{(2n)!} = -f(x) \quad\Rightarrow\quad f''(x) + f(x) = 0.$$

35. (a) $J_0(x) = \sum_{n=0}^{\infty} \frac{(-1)^n x^{2n}}{2^{2n}(n!)^2}$, $J_0'(x) = \sum_{n=1}^{\infty} \frac{(-1)^n 2nx^{2n-1}}{2^{2n}(n!)^2}$, and $J_0''(x) = \sum_{n=1}^{\infty} \frac{(-1)^n 2n(2n-1)x^{2n-2}}{2^{2n}(n!)^2}$, so

$$\begin{aligned} x^2 J_0''(x) + xJ_0'(x) + x^2 J_0(x) &= \sum_{n=1}^{\infty} \frac{(-1)^n 2n(2n-1)x^{2n}}{2^{2n}(n!)^2} + \sum_{n=1}^{\infty} \frac{(-1)^n 2nx^{2n}}{2^{2n}(n!)^2} + \sum_{n=0}^{\infty} \frac{(-1)^n x^{2n+2}}{2^{2n}(n!)^2} \\ &= \sum_{n=1}^{\infty} \frac{(-1)^n 2n(2n-1)x^{2n}}{2^{2n}(n!)^2} + \sum_{n=1}^{\infty} \frac{(-1)^n 2nx^{2n}}{2^{2n}(n!)^2} + \sum_{n=1}^{\infty} \frac{(-1)^{n-1} x^{2n}}{2^{2n-2}\left[(n-1)!\right]^2} \\ &= \sum_{n=1}^{\infty} \frac{(-1)^n 2n(2n-1)x^{2n}}{2^{2n}(n!)^2} + \sum_{n=1}^{\infty} \frac{(-1)^n 2nx^{2n}}{2^{2n}(n!)^2} + \sum_{n=1}^{\infty} \frac{(-1)^n(-1)^{-1}2^2n^2x^{2n}}{2^{2n}(n!)^2} \\ &= \sum_{n=1}^{\infty} (-1)^n \left[\frac{2n(2n-1) + 2n - 2^2n^2}{2^{2n}(n!)^2}\right] x^{2n} \\ &= \sum_{n=1}^{\infty} (-1)^n \left[\frac{4n^2 - 2n + 2n - 4n^2}{2^{2n}(n!)^2}\right] x^{2n} = 0 \end{aligned}$$

(b) $$\begin{aligned} \int_0^1 J_0(x)\,dx &= \int_0^1 \left[\sum_{n=0}^{\infty} \frac{(-1)^n x^{2n}}{2^{2n}(n!)^2}\right] dx = \int_0^1 \left(1 - \frac{x^2}{4} + \frac{x^4}{64} - \frac{x^6}{2304} + \cdots\right) dx \\ &= \left[x - \frac{x^3}{3\cdot 4} + \frac{x^5}{5\cdot 64} - \frac{x^7}{7\cdot 2304} + \cdots\right]_0^1 = 1 - \frac{1}{12} + \frac{1}{320} - \frac{1}{16{,}128} + \cdots \end{aligned}$$

Since $\frac{1}{16{,}128} \approx 0.000062$, it follows from The Alternating Series Estimation Theorem that, correct to three decimal places, $\int_0^1 J_0(x)\,dx \approx 1 - \frac{1}{12} + \frac{1}{320} \approx 0.920$.

36. (a) $J_1(x) = \sum_{n=0}^{\infty} \frac{(-1)^n x^{2n+1}}{n!\,(n+1)!\,2^{2n+1}}$, $J_1'(x) = \sum_{n=0}^{\infty} \frac{(-1)^n (2n+1)\,x^{2n}}{n!\,(n+1)!\,2^{2n+1}}$, and $J_1''(x) = \sum_{n=1}^{\infty} \frac{(-1)^n (2n+1)(2n)\,x^{2n-1}}{n!\,(n+1)!\,2^{2n+1}}$.

$$\begin{aligned} &x^2 J_1''(x) + xJ_1'(x) + (x^2-1)J_1(x) \\ &\quad= \sum_{n=1}^{\infty} \frac{(-1)^n (2n+1)(2n)x^{2n+1}}{n!\,(n+1)!\,2^{2n+1}} + \sum_{n=0}^{\infty} \frac{(-1)^n (2n+1)x^{2n+1}}{n!\,(n+1)!\,2^{2n+1}} \\ &\qquad + \sum_{n=0}^{\infty} \frac{(-1)^n x^{2n+3}}{n!\,(n+1)!\,2^{2n+1}} - \sum_{n=0}^{\infty} \frac{(-1)^n x^{2n+1}}{n!\,(n+1)!\,2^{2n+1}} \\ &\quad= \sum_{n=1}^{\infty} \frac{(-1)^n (2n+1)(2n)x^{2n+1}}{n!\,(n+1)!\,2^{2n+1}} + \sum_{n=0}^{\infty} \frac{(-1)^n (2n+1)x^{2n+1}}{n!\,(n+1)!\,2^{2n+1}} \\ &\qquad - \sum_{n=1}^{\infty} \frac{(-1)^n x^{2n+1}}{(n-1)!\,n!\,2^{2n-1}} - \sum_{n=0}^{\infty} \frac{(-1)^n x^{2n+1}}{n!\,(n+1)!\,2^{2n+1}} \qquad \left[\begin{array}{c}\text{Replace } n \text{ with } n-1 \\ \text{in the third term}\end{array}\right] \\ &\quad= \frac{x}{2} - \frac{x}{2} + \sum_{n=1}^{\infty} (-1)^n \left[\frac{(2n+1)(2n) + (2n+1) - (n)(n+1)2^2 - 1}{n!\,(n+1)!\,2^{2n+1}}\right] x^{2n+1} = 0 \end{aligned}$$

(b) $J_0(x) = \sum_{n=0}^{\infty} \frac{(-1)^n x^{2n}}{2^{2n}(n!)^2} \quad\Rightarrow$

$$\begin{aligned} J_0'(x) &= \sum_{n=1}^{\infty} \frac{(-1)^n (2n)x^{2n-1}}{2^{2n}(n!)^2} = \sum_{n=0}^{\infty} \frac{(-1)^{n+1}\,2(n+1)x^{2n+1}}{2^{2n+2}\left[(n+1)!\right]^2} \qquad [\text{Replace } n \text{ with } n+1] \\ &= -\sum_{n=0}^{\infty} \frac{(-1)^n x^{2n+1}}{2^{2n+1}(n+1)!\,n!} \qquad [\text{cancel 2 and } n+1\text{; take } -1 \text{ outside sum}] \quad = -J_1(x) \end{aligned}$$

37. (a) $f(x) = \sum\limits_{n=0}^{\infty} \frac{x^n}{n!} \Rightarrow f'(x) = \sum\limits_{n=1}^{\infty} \frac{nx^{n-1}}{n!} = \sum\limits_{n=1}^{\infty} \frac{x^{n-1}}{(n-1)!} = \sum\limits_{n=0}^{\infty} \frac{x^n}{n!} = f(x)$

(b) By Theorem 9.4.2, the only solution to the differential equation $df(x)/dx = f(x)$ is $f(x) = Ke^x$, but $f(0) = 1$, so $K = 1$ and $f(x) = e^x$.

Or: We could solve the equation $df(x)/dx = f(x)$ as a separable differential equation.

38. $\frac{|\sin nx|}{n^2} \le \frac{1}{n^2}$, so $\sum\limits_{n=1}^{\infty} \frac{\sin nx}{n^2}$ converges by the Comparison Test. $\frac{d}{dx}\left(\frac{\sin nx}{n^2}\right) = \frac{\cos nx}{n}$, so when $x = 2k\pi$ [k an integer], $\sum\limits_{n=1}^{\infty} f_n'(x) = \sum\limits_{n=1}^{\infty} \frac{\cos(2kn\pi)}{n} = \sum\limits_{n=1}^{\infty} \frac{1}{n}$, which diverges [harmonic series]. $f_n''(x) = -\sin nx$, so $\sum\limits_{n=1}^{\infty} f_n''(x) = -\sum\limits_{n=1}^{\infty} \sin nx$, which converges only if $\sin nx = 0$, or $x = k\pi$ [k an integer].

39. If $a_n = \frac{x^n}{n^2}$, then by the Ratio Test, $\lim\limits_{n\to\infty} \left|\frac{a_{n+1}}{a_n}\right| = \lim\limits_{n\to\infty} \left|\frac{x^{n+1}}{(n+1)^2} \cdot \frac{n^2}{x^n}\right| = |x| \lim\limits_{n\to\infty} \left(\frac{n}{n+1}\right)^2 = |x| < 1$ for convergence, so $R = 1$. When $x = \pm 1$, $\sum\limits_{n=1}^{\infty} \left|\frac{x^n}{n^2}\right| = \sum\limits_{n=1}^{\infty} \frac{1}{n^2}$ which is a convergent p-series ($p = 2 > 1$), so the interval of convergence for f is $[-1, 1]$. By Theorem 2, the radii of convergence of f' and f'' are both 1, so we need only check the endpoints. $f(x) = \sum\limits_{n=1}^{\infty} \frac{x^n}{n^2} \Rightarrow f'(x) = \sum\limits_{n=1}^{\infty} \frac{nx^{n-1}}{n^2} = \sum\limits_{n=0}^{\infty} \frac{x^n}{n+1}$, and this series diverges for $x = 1$ (harmonic series) and converges for $x = -1$ (Alternating Series Test), so the interval of convergence is $[-1, 1)$. $f''(x) = \sum\limits_{n=1}^{\infty} \frac{nx^{n-1}}{n+1}$ diverges at both 1 and -1 (Test for Divergence) since $\lim\limits_{n\to\infty} \frac{n}{n+1} = 1 \ne 0$, so its interval of convergence is $(-1, 1)$.

40. (a) $\sum\limits_{n=1}^{\infty} nx^{n-1} = \sum\limits_{n=0}^{\infty} \frac{d}{dx} x^n = \frac{d}{dx}\left[\sum\limits_{n=0}^{\infty} x^n\right] = \frac{d}{dx}\left[\frac{1}{1-x}\right] = -\frac{1}{(1-x)^2}(-1) = \frac{1}{(1-x)^2}$, $|x| < 1$.

(b) (i) $\sum\limits_{n=1}^{\infty} nx^n = x \sum\limits_{n=1}^{\infty} nx^{n-1} = x\left[\frac{1}{(1-x)^2}\right]$ [from part (a)] $= \frac{x}{(1-x)^2}$ for $|x| < 1$.

(ii) Put $x = \frac{1}{2}$ in (i): $\sum\limits_{n=1}^{\infty} \frac{n}{2^n} = \sum\limits_{n=1}^{\infty} n\left(\frac{1}{2}\right)^n = \frac{1/2}{(1-1/2)^2} = 2$.

(c) (i) $\sum\limits_{n=2}^{\infty} n(n-1)x^n = x^2 \sum\limits_{n=2}^{\infty} n(n-1)x^{n-2} = x^2 \frac{d}{dx}\left[\sum\limits_{n=1}^{\infty} nx^{n-1}\right] = x^2 \frac{d}{dx} \frac{1}{(1-x)^2}$

$= x^2 \frac{2}{(1-x)^3} = \frac{2x^2}{(1-x)^3}$ for $|x| < 1$.

(ii) Put $x = \frac{1}{2}$ in (i): $\sum\limits_{n=2}^{\infty} \frac{n^2 - n}{2^n} = \sum\limits_{n=2}^{\infty} n(n-1)\left(\frac{1}{2}\right)^n = \frac{2(1/2)^2}{(1-1/2)^3} = 4$.

(iii) From (b)(ii) and (c)(ii), we have $\sum\limits_{n=1}^{\infty} \frac{n^2}{2^n} = \sum\limits_{n=1}^{\infty} \frac{n^2 - n}{2^n} + \sum\limits_{n=1}^{\infty} \frac{n}{2^n} = 4 + 2 = 6$.

41. By Example 7, $\tan^{-1} x = \sum_{n=0}^{\infty} (-1)^n \dfrac{x^{2n+1}}{2n+1}$ for $|x| < 1$. In particular, for $x = \dfrac{1}{\sqrt{3}}$, we

have $\dfrac{\pi}{6} = \tan^{-1}\left(\dfrac{1}{\sqrt{3}}\right) = \sum_{n=0}^{\infty} (-1)^n \dfrac{(1/\sqrt{3})^{2n+1}}{2n+1} = \sum_{n=0}^{\infty} (-1)^n \left(\dfrac{1}{3}\right)^n \dfrac{1}{\sqrt{3}} \dfrac{1}{2n+1}$, so

$$\pi = \frac{6}{\sqrt{3}} \sum_{n=0}^{\infty} \frac{(-1)^n}{(2n+1)3^n} = 2\sqrt{3} \sum_{n=0}^{\infty} \frac{(-1)^n}{(2n+1)3^n}.$$

42. (a) $\displaystyle\int_0^{1/2} \frac{dx}{x^2 - x + 1} = \int_0^{1/2} \frac{dx}{(x - 1/2)^2 + 3/4}$ $\quad\left[x - \frac{1}{2} = \frac{\sqrt{3}}{2} u,\ u = \frac{2}{\sqrt{3}}\left(x - \frac{1}{2}\right),\ dx = \frac{\sqrt{3}}{2}\, du\right]$

$$= \int_{-1/\sqrt{3}}^{0} \frac{(\sqrt{3}/2)\, du}{(3/4)(u^2 + 1)} = \frac{2\sqrt{3}}{3}\left[\tan^{-1} u\right]_{-1/\sqrt{3}}^{0} = \frac{2}{\sqrt{3}}\left[0 - \left(-\frac{\pi}{6}\right)\right] = \frac{\pi}{3\sqrt{3}}$$

(b) $\dfrac{1}{x^3 + 1} = \dfrac{1}{(x+1)(x^2 - x + 1)} \quad\Rightarrow$

$$\frac{1}{x^2 - x + 1} = (x+1)\left(\frac{1}{1 + x^3}\right) = (x+1)\frac{1}{1 - (-x^3)} = (x+1)\sum_{n=0}^{\infty} (-1)^n x^{3n}$$

$$= \sum_{n=0}^{\infty} (-1)^n x^{3n+1} + \sum_{n=0}^{\infty} (-1)^n x^{3n} \quad \text{for } |x| < 1 \quad\Rightarrow$$

$$\int \frac{dx}{x^2 - x + 1} = C + \sum_{n=0}^{\infty} (-1)^n \frac{x^{3n+2}}{3n+2} + \sum_{n=0}^{\infty} (-1)^n \frac{x^{3n+1}}{3n+1} \quad \text{for } |x| < 1 \quad\Rightarrow$$

$$\int_0^{1/2} \frac{dx}{x^2 - x + 1} = \sum_{n=0}^{\infty} (-1)^n \left[\frac{1}{4 \cdot 8^n (3n+2)} + \frac{1}{2 \cdot 8^n (3n+1)}\right] = \frac{1}{4} \sum_{n=0}^{\infty} \frac{(-1)^n}{8^n}\left(\frac{2}{3n+1} + \frac{1}{3n+2}\right).$$

By part (a), this equals $\dfrac{\pi}{3\sqrt{3}}$, so $\pi = \dfrac{3\sqrt{3}}{4} \sum_{n=0}^{\infty} \dfrac{(-1)^n}{8^n}\left(\dfrac{2}{3n+1} + \dfrac{1}{3n+2}\right)$.

11.10 Taylor and Maclaurin Series

1. Using Theorem 5 with $\sum_{n=0}^{\infty} b_n (x-5)^n$, $b_n = \dfrac{f^{(n)}(a)}{n!}$, so $b_8 = \dfrac{f^{(8)}(5)}{8!}$.

2. (a) Using Equation 6, a power series expansion of f at 1 must have the form $f(1) + f'(1)(x-1) + \cdots$. Comparing to the given series, $1.6 - 0.8(x-1) + \cdots$, we must have $f'(1) = -0.8$. But from the graph, $f'(1)$ is positive. Hence, the given series is *not* the Taylor series of f centered at 1.

(b) A power series expansion of f at 2 must have the form $f(2) + f'(2)(x-2) + \frac{1}{2}f''(2)(x-2)^2 + \cdots$. Comparing to the given series, $2.8 + 0.5(x-2) + 1.5(x-2)^2 - 0.1(x-2)^3 + \cdots$, we must have $\frac{1}{2}f''(2) = 1.5$; that is, $f''(2)$ is positive. But from the graph, f is concave downward near $x = 2$, so $f''(2)$ must be negative. Hence, the given series is *not* the Taylor series of f centered at 2.

3. Since $f^{(n)}(0) = (n+1)!$, Equation 7 gives the Maclaurin series

$\sum_{n=0}^{\infty} \dfrac{f^{(n)}(0)}{n!} x^n = \sum_{n=0}^{\infty} \dfrac{(n+1)!}{n!} x^n = \sum_{n=0}^{\infty} (n+1) x^n$. Applying the Ratio Test with $a_n = (n+1)x^n$ gives us

$\lim_{n\to\infty}\left|\dfrac{a_{n+1}}{a_n}\right| = \lim_{n\to\infty}\left|\dfrac{(n+2)x^{n+1}}{(n+1)x^n}\right| = |x|\lim_{n\to\infty}\dfrac{n+2}{n+1} = |x|\cdot 1 = |x|$. For convergence, we must have $|x| < 1$, so the radius of convergence $R = 1$.

4. Since $f^{(n)}(4) = \dfrac{(-1)^n\, n!}{3^n(n+1)}$, Equation 6 gives the Taylor series

$\sum\limits_{n=0}^{\infty}\dfrac{f^{(n)}(4)}{n!}(x-4)^n = \sum\limits_{n=0}^{\infty}\dfrac{(-1)^n\, n!}{3^n(n+1)\, n!}(x-4)^n = \sum\limits_{n=0}^{\infty}\dfrac{(-1)^n}{3^n(n+1)}(x-4)^n$, which is the Taylor series for f centered at 4. Apply the Ratio Test to find the radius of convergence R.

$$\begin{aligned}\lim_{n\to\infty}\left|\frac{a_{n+1}}{a_n}\right| &= \lim_{n\to\infty}\left|\frac{(-1)^{n+1}(x-4)^{n+1}}{3^{n+1}(n+2)}\cdot\frac{3^n(n+1)}{(-1)^n(x-4)^n}\right| = \lim_{n\to\infty}\left|\frac{(-1)(x-4)(n+1)}{3(n+2)}\right| \\ &= \frac{1}{3}|x-4|\lim_{n\to\infty}\frac{n+1}{n+2} = \frac{1}{3}|x-4|\end{aligned}$$

For convergence, $\frac{1}{3}|x-4| < 1 \;\Leftrightarrow\; |x-4| < 3$, so $R = 3$.

5.

n	$f^{(n)}(x)$	$f^{(n)}(0)$
0	xe^x	0
1	$(x+1)e^x$	1
2	$(x+2)e^x$	2
3	$(x+3)e^x$	3
4	$(x+4)e^x$	4

Using Equation 6 with $n = 0$ to 4 and $a = 0$, we get

$$\begin{aligned}\sum_{n=0}^{4}\frac{f^{(n)}(0)}{n!}(x-0)^n &= \frac{0}{0!}x^0 + \frac{1}{1!}x^1 + \frac{2}{2!}x^2 + \frac{3}{3!}x^3 + \frac{4}{4!}x^4 \\ &= x + x^2 + \tfrac{1}{2}x^3 + \tfrac{1}{6}x^4\end{aligned}$$

6.

n	$f^{(n)}(x)$	$f^{(n)}(2)$
0	$\dfrac{1}{1+x}$	$\frac{1}{3}$
1	$-\dfrac{1}{(1+x)^2}$	$-\frac{1}{9}$
2	$\dfrac{2}{(1+x)^3}$	$\frac{2}{27}$
3	$-\dfrac{6}{(1+x)^4}$	$-\frac{6}{81}$

$$\begin{aligned}\sum_{n=0}^{3}\frac{f^{(n)}(2)}{n!}(x-2)^n &= \frac{\frac{1}{3}}{0!}(x-2)^0 - \frac{\frac{1}{9}}{1!}(x-2)^1 \\ &\quad + \frac{\frac{2}{27}}{2!}(x-2)^2 - \frac{\frac{6}{81}}{3!}(x-2)^3 \\ &= \tfrac{1}{3} - \tfrac{1}{9}(x-2) + \tfrac{1}{27}(x-2)^2 - \tfrac{1}{81}(x-2)^3\end{aligned}$$

7.

n	$f^{(n)}(x)$	$f^{(n)}(8)$
0	$\sqrt[3]{x}$	2
1	$\dfrac{1}{3x^{2/3}}$	$\frac{1}{12}$
2	$-\dfrac{2}{9x^{5/3}}$	$-\frac{2}{288}$
3	$\dfrac{10}{27x^{8/3}}$	$\frac{10}{6912}$

$$\begin{aligned}\sum_{n=0}^{3}\frac{f^{(n)}(8)}{n!}(x-8)^n &= \frac{2}{0!}(x-8)^0 + \frac{\frac{1}{12}}{1!}(x-8)^1 \\ &\quad - \frac{\frac{2}{288}}{2!}(x-8)^2 + \frac{\frac{10}{6912}}{3!}(x-8)^3 \\ &= 2 + \tfrac{1}{12}(x-8) - \tfrac{1}{288}(x-8)^2 + \tfrac{5}{20{,}736}(x-8)^3\end{aligned}$$

8.

n	$f^{(n)}(x)$	$f^{(n)}(1)$
0	$\ln x$	0
1	$1/x$	1
2	$-1/x^2$	-1
3	$2/x^3$	2
4	$-6/x^4$	-6

$$\begin{aligned}\sum_{n=0}^{4}\frac{f^{(n)}(1)}{n!}(x-1)^n &= \frac{0}{0!}(x-1)^0+\frac{1}{1!}(x-1)^1-\frac{1}{2!}(x-1)^2\\&\quad+\frac{2}{3!}(x-1)^3-\frac{6}{4!}(x-1)^4\\&=(x-1)-\tfrac{1}{2}(x-1)^2+\tfrac{1}{3}(x-1)^3-\tfrac{1}{4}(x-1)^4\end{aligned}$$

9.

n	$f^{(n)}(x)$	$f^{(n)}(\pi/6)$
0	$\sin x$	$1/2$
1	$\cos x$	$\sqrt{3}/2$
2	$-\sin x$	$-1/2$
3	$-\cos x$	$-\sqrt{3}/2$

$$\begin{aligned}\sum_{n=0}^{3}\frac{f^{(n)}(\pi/6)}{n!}\left(x-\frac{\pi}{6}\right)^n &= \frac{1/2}{0!}\left(x-\frac{\pi}{6}\right)^0+\frac{\sqrt{3}/2}{1!}\left(x-\frac{\pi}{6}\right)^1-\frac{1/2}{2!}\left(x-\frac{\pi}{6}\right)^2-\frac{\sqrt{3}/2}{3!}\left(x-\frac{\pi}{6}\right)^3\\&=\frac{1}{2}+\frac{\sqrt{3}}{2}\left(x-\frac{\pi}{6}\right)-\frac{1}{4}\left(x-\frac{\pi}{6}\right)^2-\frac{\sqrt{3}}{12}\left(x-\frac{\pi}{6}\right)^3\end{aligned}$$

10.

n	$f^{(n)}(x)$	$f^{(n)}(0)$
0	$\cos^2 x$	1
1	$-2\cos x\,\sin x=-\sin 2x$	0
2	$-2\cos 2x$	-2
3	$4\sin 2x$	0
4	$8\cos 2x$	8
5	$-16\sin 2x$	0
6	$-32\cos 2x$	-32

$$\begin{aligned}\sum_{n=0}^{6}\frac{f^{(n)}(0)}{n!}(x-0)^n &= \frac{1}{0!}x^0-\frac{2}{2!}x^2+\frac{8}{4!}x^4-\frac{32}{6!}x^6\\&=1-x^2+\tfrac{1}{3}x^4-\tfrac{2}{45}x^6\end{aligned}$$

11.

n	$f^{(n)}(x)$	$f^{(n)}(0)$
0	$(1-x)^{-2}$	1
1	$2(1-x)^{-3}$	2
2	$6(1-x)^{-4}$	6
3	$24(1-x)^{-5}$	24
4	$120(1-x)^{-6}$	120
$\vdots$	$\vdots$	$\vdots$

$$\begin{aligned}(1-x)^{-2} &= f(0)+f'(0)x+\frac{f''(0)}{2!}x^2+\frac{f'''(0)}{3!}x^3+\frac{f^{(4)}(0)}{4!}x^4+\cdots\\&=1+2x+\tfrac{6}{2}x^2+\tfrac{24}{6}x^3+\tfrac{120}{24}x^4+\cdots\\&=1+2x+3x^2+4x^3+5x^4+\cdots=\sum_{n=0}^{\infty}(n+1)x^n\end{aligned}$$

$$\lim_{n\to\infty}\left|\frac{a_{n+1}}{a_n}\right|=\lim_{n\to\infty}\left|\frac{(n+2)x^{n+1}}{(n+1)x^n}\right|=|x|\lim_{n\to\infty}\frac{n+2}{n+1}=|x|\,(1)=|x|<1$$

for convergence, so $R=1$.

12.

n	$f^{(n)}(x)$	$f^{(n)}(0)$
0	$\ln(1+x)$	0
1	$(1+x)^{-1}$	1
2	$-(1+x)^{-2}$	-1
3	$2(1+x)^{-3}$	2
4	$-6(1+x)^{-4}$	-6
5	$24(1+x)^{-5}$	24
$\vdots$	$\vdots$	$\vdots$

$$\ln(1+x) = f(0) + f'(0)x + \frac{f''(0)}{2!}x^2 + \frac{f'''(0)}{3!}x^3 + \frac{f^{(4)}(0)}{4!}x^4 + \frac{f^{(5)}(0)}{5!}x^5 + \cdots$$

$$= 0 + x - \tfrac{1}{2}x^2 + \tfrac{2}{6}x^3 - \tfrac{6}{24}x^4 + \tfrac{24}{120}x^5 - \cdots$$

$$= x - \frac{x^2}{2} + \frac{x^3}{3} - \frac{x^4}{4} + \frac{x^5}{5} - \cdots = \sum_{n=1}^{\infty} \frac{(-1)^{n-1}}{n}x^n$$

$\lim\limits_{n\to\infty}\left|\dfrac{a_{n+1}}{a_n}\right| = \lim\limits_{n\to\infty}\left|\dfrac{x^{n+1}}{n+1}\cdot\dfrac{n}{x^n}\right| = \lim\limits_{n\to\infty}\dfrac{|x|}{1+1/n} = |x| < 1$ for convergence, so $R = 1$.

Notice that the answer agrees with the entry for $\ln(1+x)$ in Table 1, but we obtained it by a different method. (Compare with Example 11.9.6.)

13.

n	$f^{(n)}(x)$	$f^{(n)}(0)$
0	$\cos x$	1
1	$-\sin x$	0
2	$-\cos x$	-1
3	$\sin x$	0
4	$\cos x$	1
$\vdots$	$\vdots$	$\vdots$

$$\cos x = f(0) + f'(0)x + \frac{f''(0)}{2!}x^2 + \frac{f'''(0)}{3!}x^3 + \frac{f^{(4)}(0)}{4!}x^4 + \cdots$$

$$= 1 - \frac{1}{2!}x^2 + \frac{1}{4!}x^4 - \cdots$$

$$= \sum_{n=0}^{\infty}(-1)^n\frac{x^{2n}}{(2n)!} \qquad \text{[Equal to (16).]}$$

$\lim\limits_{n\to\infty}\left|\dfrac{a_{n+1}}{a_n}\right| = \lim\limits_{n\to\infty}\left|\dfrac{x^{2n+2}}{(2n+2)!}\cdot\dfrac{(2n)!}{x^{2n}}\right| = \lim\limits_{n\to\infty}\dfrac{x^2}{(2n+2)(2n+1)} = 0 < 1$ for all x, so $R = \infty$.

14.

n	$f^{(n)}(x)$	$f^{(n)}(0)$
0	e^{-2x}	1
1	$-2e^{-2x}$	-2
2	$4e^{-2x}$	4
3	$-8e^{-2x}$	-8
4	$16e^{-2x}$	16
$\vdots$	$\vdots$	$\vdots$

$$e^{-2x} = \sum_{n=0}^{\infty}\frac{f^{(n)}(0)}{n!}x^n = \sum_{n=0}^{\infty}\frac{(-2)^n}{n!}x^n.$$

$\lim\limits_{n\to\infty}\left|\dfrac{a_{n+1}}{a_n}\right| = \lim\limits_{n\to\infty}\left|\dfrac{(-2)^{n+1}x^{n+1}}{(n+1)!}\cdot\dfrac{n!}{(-2)^n x^n}\right| = \lim\limits_{n\to\infty}\dfrac{2|x|}{n+1} = 0 < 1$ for all x, so $R = \infty$.

15.

n	$f^{(n)}(x)$	$f^{(n)}(0)$
0	2^x	1
1	$2^x(\ln 2)$	$\ln 2$
2	$2^x(\ln 2)^2$	$(\ln 2)^2$
3	$2^x(\ln 2)^3$	$(\ln 2)^3$
4	$2^x(\ln 2)^4$	$(\ln 2)^4$
$\vdots$	$\vdots$	$\vdots$

$$2^x = \sum_{n=0}^{\infty}\frac{f^{(n)}(0)}{n!}x^n = \sum_{n=0}^{\infty}\frac{(\ln 2)^n}{n!}x^n.$$

$\lim\limits_{n\to\infty}\left|\dfrac{a_{n+1}}{a_n}\right| = \lim\limits_{n\to\infty}\left|\dfrac{(\ln 2)^{n+1}x^{n+1}}{(n+1)!}\cdot\dfrac{n!}{(\ln 2)^n x^n}\right| = \lim\limits_{n\to\infty}\dfrac{(\ln 2)|x|}{n+1} = 0 < 1$ for all x, so $R = \infty$.

16.

n	$f^{(n)}(x)$	$f^{(n)}(0)$
0	$x\cos x$	0
1	$-x\sin x+\cos x$	1
2	$-x\cos x-2\sin x$	0
3	$x\sin x-3\cos x$	-3
4	$x\cos x+4\sin x$	0
5	$-x\sin x+5\cos x$	5
6	$-x\cos x-6\sin x$	0
7	$x\sin x-7\cos x$	-7
$\vdots$	$\vdots$	$\vdots$

$$\begin{aligned} x\cos x &= f(0)+f'(0)x+\frac{f''(0)}{2!}x^2+\frac{f'''(0)}{3!}x^3+\frac{f^{(4)}(0)}{4!}x^4+\cdots \\ &= 0+1x+0-\frac{3}{3!}x^3+0+\frac{5}{5!}x^5+0-\frac{7}{7!}x^7+\cdots \\ &= x-\frac{1}{2!}x^3+\frac{1}{4!}x^5-\frac{1}{6!}x^7+\cdots \\ &= \sum_{n=0}^{\infty}(-1)^n\frac{1}{(2n)!}x^{2n+1} \end{aligned}$$

$$\begin{aligned} \lim_{n\to\infty}\left|\frac{a_{n+1}}{a_n}\right| &= \lim_{n\to\infty}\left|\frac{(-1)^{n+1}x^{2n+3}}{(2n+2)!}\cdot\frac{(2n)!}{(-1)^n x^{2n+1}}\right| \\ &= \lim_{n\to\infty}\frac{x^2}{(2n+2)(2n+1)}=0<1 \quad \text{for all } x\text{, so } R=\infty. \end{aligned}$$

17.

n	$f^{(n)}(x)$	$f^{(n)}(0)$
0	$\sinh x$	0
1	$\cosh x$	1
2	$\sinh x$	0
3	$\cosh x$	1
4	$\sinh x$	0
$\vdots$	$\vdots$	$\vdots$

$$f^{(n)}(0)=\begin{cases} 0 & \text{if } n \text{ is even} \\ 1 & \text{if } n \text{ is odd}\end{cases} \qquad \text{so } \sinh x=\sum_{n=0}^{\infty}\frac{x^{2n+1}}{(2n+1)!}.$$

Use the Ratio Test to find R. If $a_n=\dfrac{x^{2n+1}}{(2n+1)!}$, then

$$\lim_{n\to\infty}\left|\frac{a_{n+1}}{a_n}\right|=\lim_{n\to\infty}\left|\frac{x^{2n+3}}{(2n+3)!}\cdot\frac{(2n+1)!}{x^{2n+1}}\right|=x^2\cdot\lim_{n\to\infty}\frac{1}{(2n+3)(2n+2)}$$

$=0<1$ for all x, so $R=\infty$.

18.

n	$f^{(n)}(x)$	$f^{(n)}(0)$
0	$\cosh x$	1
1	$\sinh x$	0
2	$\cosh x$	1
3	$\sinh x$	0
$\vdots$	$\vdots$	$\vdots$

$$f^{(n)}(0)=\begin{cases} 1 & \text{if } n \text{ is even} \\ 0 & \text{if } n \text{ is odd}\end{cases} \qquad \text{so } \cosh x=\sum_{n=0}^{\infty}\frac{x^{2n}}{(2n)!}.$$

Use the Ratio Test to find R. If $a_n=\dfrac{x^{2n}}{(2n)!}$, then

$$\lim_{n\to\infty}\left|\frac{a_{n+1}}{a_n}\right|=\lim_{n\to\infty}\left|\frac{x^{2n+2}}{(2n+2)!}\cdot\frac{(2n)!}{x^{2n}}\right|=x^2\cdot\lim_{n\to\infty}\frac{1}{(2n+2)(2n+1)}$$

$=0<1$ for all x, so $R=\infty$.

19.

n	$f^{(n)}(x)$	$f^{(n)}(2)$
0	x^5+2x^3+x	50
1	$5x^4+6x^2+1$	105
2	$20x^3+12x$	184
3	$60x^2+12$	252
4	$120x$	240
5	120	120
6	0	0
7	0	0
$\vdots$	$\vdots$	$\vdots$

$f^{(n)}(x)=0$ for $n\geq 6$, so f has a finite expansion about $a=2$.

$$\begin{aligned} f(x)=x^5+2x^3+x &= \sum_{n=0}^{5}\frac{f^{(n)}(2)}{n!}(x-2)^n \\ &= \frac{50}{0!}(x-2)^0+\frac{105}{1!}(x-2)^1+\frac{184}{2!}(x-2)^2+\frac{252}{3!}(x-2)^3 \\ &\quad +\frac{240}{4!}(x-2)^4+\frac{120}{5!}(x-2)^5 \\ &= 50+105(x-2)+92(x-2)^2+42(x-2)^3 \\ &\quad +10(x-2)^4+(x-2)^5 \end{aligned}$$

A finite series converges for all x, so $R=\infty$.

20.

n	$f^{(n)}(x)$	$f^{(n)}(-2)$
0	$x^6 - x^4 + 2$	50
1	$6x^5 - 4x^3$	-160
2	$30x^4 - 12x^2$	432
3	$120x^3 - 24x$	-912
4	$360x^2 - 24$	1416
5	$720x$	-1440
6	720	720
7	0	0
8	0	0
$\vdots$	$\vdots$	$\vdots$

$f^{(n)}(x) = 0$ for $n \geq 7$, so f has a finite expansion about $a = -2$.

$$\begin{aligned} f(x) = x^6 - x^4 + 2 &= \sum_{n=0}^{6} \frac{f^{(n)}(-2)}{n!}(x+2)^n \\ &= \frac{50}{0!}(x+2)^0 - \frac{160}{1!}(x+2)^1 + \frac{432}{2!}(x+2)^2 - \frac{912}{3!}(x+2)^3 \\ &\quad + \frac{1416}{4!}(x+2)^4 - \frac{1440}{5!}(x+2)^5 + \frac{720}{6!}(x+2)^6 \\ &= 50 - 160(x+2) + 216(x+2)^2 - 152(x+2)^3 + 59(x+2)^4 - 12(x+2)^5 + (x+2)^6 \end{aligned}$$

A finite series converges for all x, so $R = \infty$.

21.

n	$f^{(n)}(x)$	$f^{(n)}(2)$
0	$\ln x$	$\ln 2$
1	$1/x$	$1/2$
2	$-1/x^2$	$-1/2^2$
3	$2/x^3$	$2/2^3$
4	$-6/x^4$	$-6/2^4$
5	$24/x^5$	$24/2^5$
$\vdots$	$\vdots$	$\vdots$

$$\begin{aligned} f(x) = \ln x &= \sum_{n=0}^{\infty} \frac{f^{(n)}(2)}{n!}(x-2)^n \\ &= \frac{\ln 2}{0!}(x-2)^0 + \frac{1}{1!\,2^1}(x-2)^1 + \frac{-1}{2!\,2^2}(x-2)^2 + \frac{2}{3!\,2^3}(x-2)^3 \\ &\quad + \frac{-6}{4!\,2^4}(x-2)^4 + \frac{24}{5!\,2^5}(x-2)^5 + \cdots \\ &= \ln 2 + \sum_{n=1}^{\infty} (-1)^{n+1} \frac{(n-1)!}{n!\,2^n}(x-2)^n \\ &= \ln 2 + \sum_{n=1}^{\infty} (-1)^{n+1} \frac{1}{n\,2^n}(x-2)^n \end{aligned}$$

$$\lim_{n\to\infty}\left|\frac{a_{n+1}}{a_n}\right| = \lim_{n\to\infty}\left|\frac{(-1)^{n+2}(x-2)^{n+1}}{(n+1)\,2^{n+1}} \cdot \frac{n\,2^n}{(-1)^{n+1}(x-2)^n}\right| = \lim_{n\to\infty}\left|\frac{(-1)(x-2)n}{(n+1)2}\right| = \lim_{n\to\infty}\left(\frac{n}{n+1}\right)\frac{|x-2|}{2}$$

$= \dfrac{|x-2|}{2} < 1$ for convergence, so $|x-2| < 2$ and $R = 2$.

22.

n	$f^{(n)}(x)$	$f^{(n)}(-3)$
0	$1/x$	$-1/3$
1	$-1/x^2$	$-1/3^2$
2	$2/x^3$	$-2/3^3$
3	$-6/x^4$	$-6/3^4$
4	$24/x^5$	$-24/3^5$
$\vdots$	$\vdots$	$\vdots$

$$\begin{aligned} f(x) = \frac{1}{x} &= \sum_{n=0}^{\infty} \frac{f^{(n)}(-3)}{n!}(x+3)^n \\ &= \frac{-1/3}{0!}(x+3)^0 + \frac{-1/3^2}{1!}(x+3)^1 + \frac{-2/3^3}{2!}(x+3)^2 \\ &\quad + \frac{-6/3^4}{3!}(x+3)^3 + \frac{-24/3^5}{4!}(x+3)^4 + \cdots \\ &= \sum_{n=0}^{\infty} \frac{-n!/3^{n+1}}{n!}(x+3)^n = -\sum_{n=0}^{\infty} \frac{(x+3)^n}{3^{n+1}} \end{aligned}$$

$$\lim_{n\to\infty}\left|\frac{a_{n+1}}{a_n}\right| = \lim_{n\to\infty}\left|\frac{(x+3)^{n+1}}{3^{n+2}} \cdot \frac{3^{n+1}}{(x+3)^n}\right| = \lim_{n\to\infty}\frac{|x+3|}{3} = \frac{|x+3|}{3} < 1 \quad \text{for convergence,}$$

so $|x+3| < 3$ and $R = 3$.

23.

n	$f^{(n)}(x)$	$f^{(n)}(3)$
0	e^{2x}	e^6
1	$2e^{2x}$	$2e^6$
2	2^2e^{2x}	$4e^6$
3	2^3e^{2x}	$8e^6$
4	2^4e^{2x}	$16e^6$
$\vdots$	$\vdots$	$\vdots$

$$\begin{aligned} f(x) = e^{2x} &= \sum_{n=0}^{\infty} \frac{f^{(n)}(3)}{n!}(x-3)^n \\ &= \frac{e^6}{0!}(x-3)^0 + \frac{2e^6}{1!}(x-3)^1 + \frac{4e^6}{2!}(x-3)^2 \\ &\quad + \frac{8e^6}{3!}(x-3)^3 + \frac{16e^6}{4!}(x-3)^4 + \cdots \\ &= \sum_{n=0}^{\infty} \frac{2^n e^6}{n!}(x-3)^n \end{aligned}$$

$$\lim_{n\to\infty}\left|\frac{a_{n+1}}{a_n}\right| = \lim_{n\to\infty}\left|\frac{2^{n+1}e^6(x-3)^{n+1}}{(n+1)!} \cdot \frac{n!}{2^n e^6 (x-3)^n}\right| = \lim_{n\to\infty}\frac{2\,|x-3|}{n+1} = 0 < 1 \quad \text{for all } x\text{, so } R = \infty.$$

24.

n	$f^{(n)}(x)$	$f^{(n)}(\pi/2)$
0	$\cos x$	0
1	$-\sin x$	-1
2	$-\cos x$	0
3	$\sin x$	1
4	$\cos x$	0
5	$-\sin x$	-1
6	$-\cos x$	0
7	$\sin x$	1
$\vdots$	$\vdots$	$\vdots$

$$\begin{aligned} f(x) = \cos x &= \sum_{n=0}^{\infty} \frac{f^{(n)}(\pi/2)}{n!}\left(x - \frac{\pi}{2}\right)^n \\ &= \frac{-1}{1!}\left(x-\frac{\pi}{2}\right)^1 + \frac{1}{3!}\left(x-\frac{\pi}{2}\right)^3 + \frac{-1}{5!}\left(x-\frac{\pi}{2}\right)^5 + \frac{1}{7!}\left(x-\frac{\pi}{2}\right)^7 + \cdots \\ &= \sum_{n=0}^{\infty} \frac{(-1)^{n+1}}{(2n+1)!}\left(x-\frac{\pi}{2}\right)^{2n+1} \end{aligned}$$

$$\begin{aligned} \lim_{n\to\infty}\left|\frac{a_{n+1}}{a_n}\right| &= \lim_{n\to\infty}\left|\frac{(-1)^{n+2}\left(x-\frac{\pi}{2}\right)^{2n+3}}{(2n+3)!} \cdot \frac{(2n+1)!}{(-1)^{n+1}\left(x-\frac{\pi}{2}\right)^{2n+1}}\right| \\ &= \lim_{n\to\infty}\frac{\left(x-\frac{\pi}{2}\right)^2}{(2n+3)(2n+2)} = 0 < 1 \quad \text{for all } x\text{, so } R = \infty. \end{aligned}$$

25.

n	$f^{(n)}(x)$	$f^{(n)}(\pi)$
0	$\sin x$	0
1	$\cos x$	-1
2	$-\sin x$	0
3	$-\cos x$	1
4	$\sin x$	0
5	$\cos x$	-1
6	$-\sin x$	0
7	$-\cos x$	1
$\vdots$	$\vdots$	$\vdots$

$$f(x) = \sin x = \sum_{n=0}^{\infty} \frac{f^{(n)}(\pi)}{n!}(x-\pi)^n$$

$$= \frac{-1}{1!}(x-\pi)^1 + \frac{1}{3!}(x-\pi)^3 + \frac{-1}{5!}(x-\pi)^5 + \frac{1}{7!}(x-\pi)^7 + \cdots$$

$$= \sum_{n=0}^{\infty} \frac{(-1)^{n+1}}{(2n+1)!}(x-\pi)^{2n+1}$$

$$\lim_{n\to\infty}\left|\frac{a_{n+1}}{a_n}\right| = \lim_{n\to\infty}\left|\frac{(-1)^{n+2}(x-\pi)^{2n+3}}{(2n+3)!}\cdot\frac{(2n+1)!}{(-1)^{n+1}(x-\pi)^{2n+1}}\right|$$

$$= \lim_{n\to\infty}\frac{(x-\pi)^2}{(2n+3)(2n+2)} = 0 < 1 \quad \text{for all } x \text{, so } R = \infty.$$

26.

n	$f^{(n)}(x)$	$f^{(n)}(16)$
0	$\sqrt{x}$	4
1	$\frac{1}{2}x^{-1/2}$	$\frac{1}{2}\cdot\frac{1}{4}$
2	$-\frac{1}{4}x^{-3/2}$	$-\frac{1}{4}\cdot\frac{1}{4^3}$
3	$\frac{3}{8}x^{-5/2}$	$\frac{3}{8}\cdot\frac{1}{4^5}$
4	$-\frac{15}{16}x^{-7/2}$	$-\frac{15}{16}\cdot\frac{1}{4^7}$
$\vdots$	$\vdots$	$\vdots$

$$f(x) = \sqrt{x} = \sum_{n=0}^{\infty}\frac{f^{(n)}(16)}{n!}(x-16)^n$$

$$= \frac{4}{0!}(x-16)^0 + \frac{1}{2}\cdot\frac{1}{4}\cdot\frac{1}{1!}(x-16)^1 - \frac{1}{4}\cdot\frac{1}{4^3}\cdot\frac{1}{2!}(x-16)^2$$

$$+ \frac{3}{8}\cdot\frac{1}{4^5}\cdot\frac{1}{3!}(x-16)^3 - \frac{15}{16}\cdot\frac{1}{4^7}\cdot\frac{1}{4!}(x-16)^4 + \cdots$$

$$= 4 + \frac{1}{8}(x-16) + \sum_{n=2}^{\infty}(-1)^{n-1}\frac{1\cdot3\cdot5\cdot\cdots\cdot(2n-3)}{2^n 4^{2n-1}\,n!}(x-16)^n$$

$$= 4 + \frac{1}{8}(x-16) + \sum_{n=2}^{\infty}(-1)^{n-1}\frac{1\cdot3\cdot5\cdot\cdots\cdot(2n-3)}{2^{5n-2}\,n!}(x-16)^n$$

$$\lim_{n\to\infty}\left|\frac{a_{n+1}}{a_n}\right| = \lim_{n\to\infty}\left|\frac{(-1)^n\,1\cdot3\cdot5\cdot\cdots\cdot(2n-1)(x-16)^{n+1}}{2^{5n+3}(n+1)!}\cdot\frac{2^{5n-2}n!}{(-1)^{n-1}\,1\cdot3\cdot5\cdot\cdots\cdot(2n-3)(x-16)^n}\right|$$

$$= \lim_{n\to\infty}\frac{(2n-1)\,|x-16|}{2^5(n+1)} = \frac{|x-16|}{32}\lim_{n\to\infty}\frac{2-1/n}{1+1/n} = \frac{|x-16|}{32}\cdot 2$$

$$= \frac{|x-16|}{16} < 1 \quad \text{for convergence, so } |x-16| < 16 \text{ and } R = 16.$$

27. If $f(x) = \cos x$, then $f^{(n+1)}(x) = \pm\sin x$ or $\pm\cos x$. In each case, $\left|f^{(n+1)}(x)\right| \le 1$, so by Formula 9 with $a = 0$ and $M = 1$, $|R_n(x)| \le \dfrac{1}{(n+1)!}|x|^{n+1}$. Thus, $|R_n(x)| \to 0$ as $n \to \infty$ by Equation 10. So $\lim_{n\to\infty} R_n(x) = 0$ and, by Theorem 8, the series in Exercise 13 represents $\cos x$ for all x.

28. If $f(x) = \sin x$, then $f^{(n+1)}(x) = \pm\sin x$ or $\pm\cos x$. In each case, $\left|f^{(n+1)}(x)\right| \le 1$, so by Formula 9 with $a = 0$ and $M = 1$, $|R_n(x)| \le \dfrac{1}{(n+1)!}|x-\pi|^{n+1}$. Thus, $|R_n(x)| \to 0$ as $n \to \infty$ by Equation 10. So $\lim_{n\to\infty} R_n(x) \to 0$ and, by Theorem 8, the series in Exercise 25 represents $\sin x$ for all x.

29. If $f(x) = \sinh x$, then for all n, $f^{(n+1)}(x) = \cosh x$ or $\sinh x$. Since $|\sinh x| < |\cosh x| = \cosh x$ for all x, we have $\left|f^{(n+1)}(x)\right| \le \cosh x$ for all n. If d is any positive number and $|x| \le d$, then $\left|f^{(n+1)}(x)\right| \le \cosh x \le \cosh d$, so by Formula 9 with $a = 0$ and $M = \cosh d$, we have $|R_n(x)| \le \dfrac{\cosh d}{(n+1)!}|x|^{n+1}$. It follows that $|R_n(x)| \to 0$ as $n \to \infty$ for $|x| \le d$ (by Equation 10). But d was an arbitrary positive number. So by Theorem 8, the series represents $\sinh x$ for all x.

30. If $f(x) = \cosh x$, then for all n, $f^{(n+1)}(x) = \cosh x$ or $\sinh x$. Since $|\sinh x| < |\cosh x| = \cosh x$ for all x, we have $\left|f^{(n+1)}(x)\right| \le \cosh x$ for all n. If d is any positive number and $|x| \le d$, then $\left|f^{(n+1)}(x)\right| \le \cosh x \le \cosh d$, so by Formula 9 with $a = 0$ and $M = \cosh d$, we have $|R_n(x)| \le \dfrac{\cosh d}{(n+1)!}|x|^{n+1}$. It follows that $|R_n(x)| \to 0$ as $n \to \infty$ for $|x| \le d$ (by Equation 10). But d was an arbitrary positive number. So by Theorem 8, the series represents $\cosh x$ for all x.

31. $$\sqrt[4]{1-x} = [1+(-x)]^{1/4} = \sum_{n=0}^{\infty}\binom{1/4}{n}(-x)^n = 1 + \tfrac{1}{4}(-x) + \frac{\frac{1}{4}\left(-\frac{3}{4}\right)}{2!}(-x)^2 + \frac{\frac{1}{4}\left(-\frac{3}{4}\right)\left(-\frac{7}{4}\right)}{3!}(-x)^3 + \cdots$$
$$= 1 - \frac{1}{4}x + \sum_{n=2}^{\infty}\frac{(-1)^{n-1}(-1)^n\cdot[3\cdot 7\cdot\cdots\cdot(4n-5)]}{4^n\cdot n!}x^n$$
$$= 1 - \frac{1}{4}x - \sum_{n=2}^{\infty}\frac{3\cdot 7\cdot\cdots\cdot(4n-5)}{4^n\cdot n!}x^n$$

and $|-x| < 1 \quad\Leftrightarrow\quad |x| < 1$, so $R = 1$.

32. $$\sqrt[3]{8+x} = \sqrt[3]{8\left(1+\frac{x}{8}\right)} = 2\left(1+\frac{x}{8}\right)^{1/3} = 2\sum_{n=0}^{\infty}\binom{1/3}{n}\left(\frac{x}{8}\right)^n$$
$$= 2\left[1 + \frac{1}{3}\left(\frac{x}{8}\right) + \frac{\frac{1}{3}\left(-\frac{2}{3}\right)}{2!}\left(\frac{x}{8}\right)^2 + \frac{\frac{1}{3}\left(-\frac{2}{3}\right)\left(-\frac{5}{3}\right)}{3!}\left(\frac{x}{8}\right)^3 + \cdots\right]$$
$$= 2\left[1 + \frac{1}{24}x + \sum_{n=2}^{\infty}\frac{(-1)^{n-1}\cdot[2\cdot 5\cdot\cdots\cdot(3n-4)]}{3^n\cdot 8^n\cdot n!}x^n\right]$$
$$= 2 + \frac{1}{12}x + 2\sum_{n=2}^{\infty}\frac{(-1)^{n-1}[2\cdot 5\cdot\cdots\cdot(3n-4)]}{24^n\cdot n!}x^n$$

and $\left|\dfrac{x}{8}\right| < 1 \quad\Leftrightarrow\quad |x| < 8$, so $R = 8$.

33. $\dfrac{1}{(2+x)^3} = \dfrac{1}{[2(1+x/2)]^3} = \dfrac{1}{8}\left(1+\dfrac{x}{2}\right)^{-3} = \dfrac{1}{8}\displaystyle\sum_{n=0}^{\infty}\binom{-3}{n}\left(\frac{x}{2}\right)^n$. The binomial coefficient is

$$\binom{-3}{n} = \frac{(-3)(-4)(-5)\cdot\cdots\cdot(-3-n+1)}{n!} = \frac{(-3)(-4)(-5)\cdot\cdots\cdot[-(n+2)]}{n!}$$
$$= \frac{(-1)^n\cdot 2\cdot 3\cdot 4\cdot 5\cdot\cdots\cdot(n+1)(n+2)}{2\cdot n!} = \frac{(-1)^n(n+1)(n+2)}{2}$$

Thus, $\dfrac{1}{(2+x)^3} = \dfrac{1}{8}\displaystyle\sum_{n=0}^{\infty}\frac{(-1)^n(n+1)(n+2)}{2}\frac{x^n}{2^n} = \sum_{n=0}^{\infty}\frac{(-1)^n(n+1)(n+2)x^n}{2^{n+4}}$ for $\left|\dfrac{x}{2}\right| < 1 \quad\Leftrightarrow\quad |x| < 2$, so $R = 2$.

34. $(1+x)^{3/4} = \sum\limits_{n=0}^{\infty} \binom{\frac{3}{4}}{n} x^n = 1 + \frac{3}{4}x + \frac{\frac{3}{4}\left(-\frac{1}{4}\right)}{2!}x^2 + \frac{\frac{3}{4}\left(-\frac{1}{4}\right)\left(-\frac{5}{4}\right)}{3!}x^3 + \cdots$

$$= 1 + \frac{3}{4}x + \sum_{n=2}^{\infty} \frac{(-1)^{n-1} \cdot 3 \cdot [1 \cdot 5 \cdot 9 \cdot \cdots \cdot (4n-7)]}{4^n \cdot n!} x^n$$

for $|x| < 1$, so $R = 1$.

35. $\arctan x = \sum\limits_{n=0}^{\infty} (-1)^n \frac{x^{2n+1}}{2n+1}$, so $f(x) = \arctan(x^2) = \sum\limits_{n=0}^{\infty} (-1)^n \frac{(x^2)^{2n+1}}{2n+1} = \sum\limits_{n=0}^{\infty} (-1)^n \frac{1}{2n+1} x^{4n+2}$, $R = 1$.

36. $\sin x = \sum\limits_{n=0}^{\infty} (-1)^n \frac{x^{2n+1}}{(2n+1)!}$, so $f(x) = \sin\left(\frac{\pi}{4}x\right) = \sum\limits_{n=0}^{\infty} (-1)^n \frac{\left(\frac{\pi}{4}x\right)^{2n+1}}{(2n+1)!} = \sum\limits_{n=0}^{\infty} (-1)^n \frac{\pi^{2n+1}}{4^{2n+1}(2n+1)!} x^{2n+1}$, $R = \infty$.

37. $\cos x = \sum\limits_{n=0}^{\infty} (-1)^n \frac{x^{2n}}{(2n)!} \quad \Rightarrow \quad \cos 2x = \sum\limits_{n=0}^{\infty} (-1)^n \frac{(2x)^{2n}}{(2n)!} = \sum\limits_{n=0}^{\infty} (-1)^n \frac{2^{2n}x^{2n}}{(2n)!}$, so

$f(x) = x\cos 2x = \sum\limits_{n=0}^{\infty} (-1)^n \frac{2^{2n}}{(2n)!} x^{2n+1}$, $R = \infty$.

38. $e^x = \sum\limits_{n=0}^{\infty} \frac{x^n}{n!}$, so $f(x) = e^{3x} - e^{2x} = \sum\limits_{n=0}^{\infty} \frac{(3x)^n}{n!} - \sum\limits_{n=0}^{\infty} \frac{(2x)^n}{n!} = \sum\limits_{n=0}^{\infty} \frac{3^n x^n}{n!} - \sum\limits_{n=0}^{\infty} \frac{2^n x^n}{n!} = \sum\limits_{n=0}^{\infty} \frac{3^n - 2^n}{n!} x^n$, $R = \infty$.

39. $\cos x = \sum\limits_{n=0}^{\infty} (-1)^n \frac{x^{2n}}{(2n)!} \quad \Rightarrow \quad \cos\left(\frac{1}{2}x^2\right) = \sum\limits_{n=0}^{\infty} (-1)^n \frac{\left(\frac{1}{2}x^2\right)^{2n}}{(2n)!} = \sum\limits_{n=0}^{\infty} (-1)^n \frac{x^{4n}}{2^{2n}(2n)!}$, so

$f(x) = x\cos\left(\frac{1}{2}x^2\right) = \sum\limits_{n=0}^{\infty} (-1)^n \frac{1}{2^{2n}(2n)!} x^{4n+1}$, $R = \infty$.

40. $\ln(1+x) = \sum\limits_{n=1}^{\infty} (-1)^{n-1} \frac{x^n}{n} \quad \Rightarrow \quad \ln(1+x^3) = \sum\limits_{n=1}^{\infty} (-1)^{n-1} \frac{x^{3n}}{n}$, so $f(x) = x^2 \ln(1+x^3) = \sum\limits_{n=1}^{\infty} (-1)^{n-1} \frac{x^{3n+2}}{n}$,

$R = 1$.

41. We must write the binomial in the form (1+ expression), so we'll factor out a 4.

$$\frac{x}{\sqrt{4+x^2}} = \frac{x}{\sqrt{4(1+x^2/4)}} = \frac{x}{2\sqrt{1+x^2/4}} = \frac{x}{2}\left(1+\frac{x^2}{4}\right)^{-1/2} = \frac{x}{2}\sum_{n=0}^{\infty}\binom{-\frac{1}{2}}{n}\left(\frac{x^2}{4}\right)^n$$

$$= \frac{x}{2}\left[1 + \left(-\tfrac{1}{2}\right)\frac{x^2}{4} + \frac{\left(-\frac{1}{2}\right)\left(-\frac{3}{2}\right)}{2!}\left(\frac{x^2}{4}\right)^2 + \frac{\left(-\frac{1}{2}\right)\left(-\frac{3}{2}\right)\left(-\frac{5}{2}\right)}{3!}\left(\frac{x^2}{4}\right)^3 + \cdots\right]$$

$$= \frac{x}{2} + \frac{x}{2}\sum_{n=1}^{\infty}(-1)^n \frac{1 \cdot 3 \cdot 5 \cdot \cdots \cdot (2n-1)}{2^n \cdot 4^n \cdot n!} x^{2n}$$

$$= \frac{x}{2} + \sum_{n=1}^{\infty}(-1)^n \frac{1 \cdot 3 \cdot 5 \cdot \cdots \cdot (2n-1)}{n!\, 2^{3n+1}} x^{2n+1} \text{ and } \frac{x^2}{4} < 1 \iff \frac{|x|}{2} < 1 \iff |x| < 2, \text{ so } R = 2.$$

42. $\dfrac{x^2}{\sqrt{2+x}} = \dfrac{x^2}{\sqrt{2(1+x/2)}} = \dfrac{x^2}{\sqrt{2}}\left(1+\dfrac{x}{2}\right)^{-1/2} = \dfrac{x^2}{\sqrt{2}}\sum\limits_{n=0}^{\infty}\binom{-\frac{1}{2}}{n}\left(\dfrac{x}{2}\right)^n$

$$= \frac{x^2}{\sqrt{2}}\left[1 + \left(-\tfrac{1}{2}\right)\left(\frac{x}{2}\right) + \frac{\left(-\frac{1}{2}\right)\left(-\frac{3}{2}\right)}{2!}\left(\frac{x}{2}\right)^2 + \frac{\left(-\frac{1}{2}\right)\left(-\frac{3}{2}\right)\left(-\frac{5}{2}\right)}{3!}\left(\frac{x}{2}\right)^3 + \cdots\right]$$

$$= \frac{x^2}{\sqrt{2}} + \frac{x^2}{\sqrt{2}}\sum_{n=1}^{\infty}(-1)^n\frac{1\cdot 3\cdot 5\cdot\cdots\cdot(2n-1)}{n!\,2^{2n}}x^n$$

$$= \frac{x^2}{\sqrt{2}} + \sum_{n=1}^{\infty}(-1)^n\frac{1\cdot 3\cdot 5\cdot\cdots\cdot(2n-1)}{n!\,2^{2n+1/2}}x^{n+2} \text{ and } \left|\frac{x}{2}\right| < 1 \iff |x| < 2, \text{ so } R = 2.$$

43. $\sin^2 x = \dfrac{1}{2}(1-\cos 2x) = \dfrac{1}{2}\left[1 - \sum\limits_{n=0}^{\infty}\dfrac{(-1)^n(2x)^{2n}}{(2n)!}\right] = \dfrac{1}{2}\left[1 - 1 - \sum\limits_{n=1}^{\infty}\dfrac{(-1)^n(2x)^{2n}}{(2n)!}\right] = \sum\limits_{n=1}^{\infty}\dfrac{(-1)^{n+1}2^{2n-1}x^{2n}}{(2n)!}$,

$R = \infty$

44. $\dfrac{x-\sin x}{x^3} = \dfrac{1}{x^3}\left[x - \sum\limits_{n=0}^{\infty}\dfrac{(-1)^n x^{2n+1}}{(2n+1)!}\right] = \dfrac{1}{x^3}\left[x - x - \sum\limits_{n=1}^{\infty}\dfrac{(-1)^n x^{2n+1}}{(2n+1)!}\right] = \dfrac{1}{x^3}\left[-\sum\limits_{n=0}^{\infty}\dfrac{(-1)^{n+1}x^{2n+3}}{(2n+3)!}\right]$

$$= \frac{1}{x^3}\sum_{n=0}^{\infty}\frac{(-1)^n x^{2n+3}}{(2n+3)!} = \sum_{n=0}^{\infty}\frac{(-1)^n x^{2n}}{(2n+3)!}$$

and this series also gives the required value at $x = 0$ (namely $1/6$); $R = \infty$.

45. $\cos x \overset{(16)}{=} \sum\limits_{n=0}^{\infty}(-1)^n\dfrac{x^{2n}}{(2n)!} \Rightarrow$

$$f(x) = \cos(x^2) = \sum_{n=0}^{\infty}\frac{(-1)^n(x^2)^{2n}}{(2n)!} = \sum_{n=0}^{\infty}\frac{(-1)^n x^{4n}}{(2n)!}$$

$$= 1 - \tfrac{1}{2}x^4 + \tfrac{1}{24}x^8 - \tfrac{1}{720}x^{12} + \cdots$$

The series for $\cos x$ converges for all x, so the same is true of the series for $f(x)$, that is, $R = \infty$. Notice that, as n increases, $T_n(x)$ becomes a better approximation to $f(x)$.

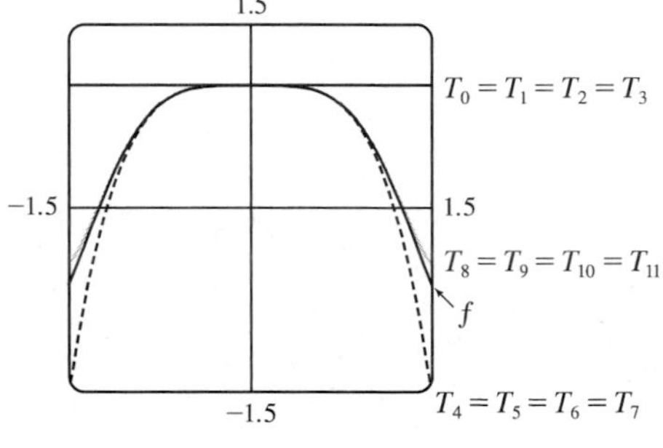

46. $\ln(1+x) = \sum\limits_{n=1}^{\infty}(-1)^{n-1}\dfrac{x^n}{n} \Rightarrow$

$$f(x) = \ln(1+x^2) = \sum_{n=1}^{\infty}\frac{(-1)^{n-1}(x^2)^n}{n} = \sum_{n=1}^{\infty}\frac{(-1)^{n-1}x^{2n}}{n}$$

$$= x^2 - \tfrac{1}{2}x^4 + \tfrac{1}{3}x^6 - \tfrac{1}{4}x^8 + \cdots$$

The series for $\ln(1+x)$ has $R = 1$ and $|x^2| < 1 \iff |x| < 1$, so the series for $f(x)$ also has $R = 1$. From the graphs of f and the first few Taylor polynomials, we see that $T_n(x)$ provides a closer fit to $f(x)$ near 0 as n increases.

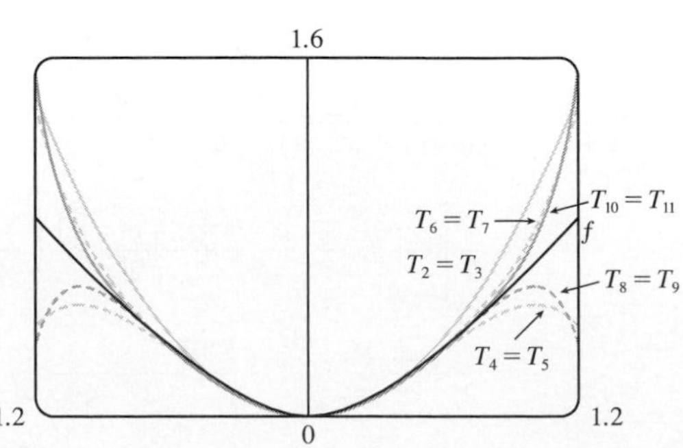

47. $e^x \overset{(11)}{=} \sum_{n=0}^{\infty} \frac{x^n}{n!}$, so $e^{-x} = \sum_{n=0}^{\infty} \frac{(-x)^n}{n!} = \sum_{n=0}^{\infty} (-1)^n \frac{x^n}{n!}$, so

$$f(x) = xe^{-x} = \sum_{n=0}^{\infty} (-1)^n \frac{1}{n!} x^{n+1}$$
$$= x - x^2 + \tfrac{1}{2}x^3 - \tfrac{1}{6}x^4 + \tfrac{1}{24}x^5 - \tfrac{1}{120}x^6 + \cdots$$
$$= \sum_{n=1}^{\infty} (-1)^{n-1} \frac{x^n}{(n-1)!}$$

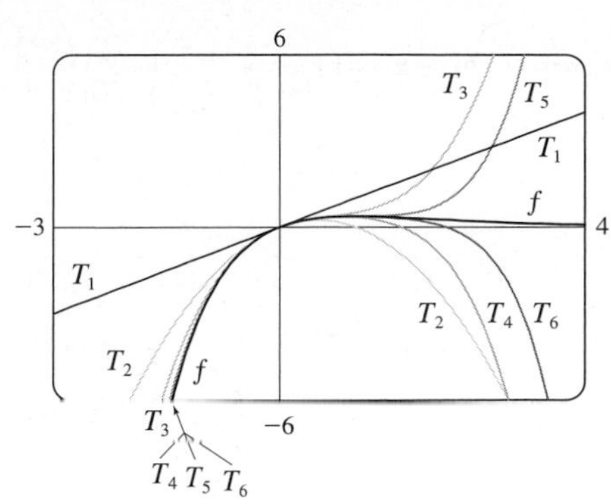

The series for e^x converges for all x, so the same is true of the series for $f(x)$; that is, $R = \infty$. From the graphs of f and the first few Taylor polynomials, we see that $T_n(x)$ provides a closer fit to $f(x)$ near 0 as n increases.

48. From Table 1, $\tan^{-1} x = \sum_{n=0}^{\infty} (-1)^n \frac{x^{2n+1}}{2n+1}$, so

$$f(x) = \tan^{-1}(x^3) = \sum_{n=0}^{\infty} (-1)^n \frac{(x^3)^{2n+1}}{2n+1} = \sum_{n=0}^{\infty} (-1)^n \frac{x^{6n+3}}{2n+1}$$
$$= x^3 - \tfrac{1}{3}x^9 + \tfrac{1}{5}x^{15} - \tfrac{1}{7}x^{21} + \cdots$$

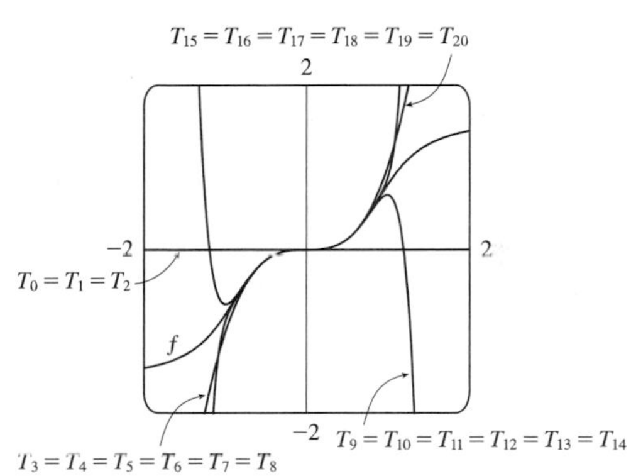

The series for $\tan^{-1} x$ has $R = 1$ and $|x^3| < 1 \;\Leftrightarrow\; |x| < 1$, so the series for $f(x)$ also has $R = 1$. From the graphs of f and the first few Taylor polynomials, we see that $T_n(x)$ provides a closer fit to $f(x)$ near 0 as n increases.

49. $5^\circ = 5^\circ\left(\frac{\pi}{180^\circ}\right) = \frac{\pi}{36}$ radians and $\cos x = \sum_{n=0}^{\infty} (-1)^n \frac{x^{2n}}{(2n)!} = 1 - \frac{x^2}{2!} + \frac{x^4}{4!} - \frac{x^6}{6!} + \cdots$, so

$\cos\frac{\pi}{36} = 1 - \frac{(\pi/36)^2}{2!} + \frac{(\pi/36)^4}{4!} - \frac{(\pi/36)^6}{6!} + \cdots$. Now $1 - \frac{(\pi/36)^2}{2!} \approx 0.99619$ and adding $\frac{(\pi/36)^4}{4!} \approx 2.4 \times 10^{-6}$ does not affect the fifth decimal place, so $\cos 5^\circ \approx 0.99619$ by the Alternating Series Estimation Theorem.

50. $1/\sqrt[10]{e} = e^{-1/10}$ and $e^x = \sum_{n=0}^{\infty} \frac{x^n}{n!} = 1 + x + \frac{x^2}{2!} + \frac{x^3}{3!} + \cdots$, so

$e^{-1/10} = 1 - \frac{1}{10} + \frac{(1/10)^2}{2!} - \frac{(1/10)^3}{3!} + \frac{(1/10)^4}{4!} - \frac{(1/10)^5}{5!} + \cdots$. Now

$1 - \frac{1}{10} + \frac{(1/10)^2}{2!} - \frac{(1/10)^3}{3!} + \frac{(1/10)^4}{4!} \approx 0.90484$ and subtracting $\frac{(1/10)^5}{5!} \approx 8.3 \times 10^{-8}$ does not affect the fifth decimal place, so $e^{-1/10} \approx 0.90484$ by the Alternating Series Estimation Theorem.

51. (a) $1/\sqrt{1-x^2} = \left[1 + (-x^2)\right]^{-1/2} = 1 + \left(-\tfrac{1}{2}\right)(-x^2) + \frac{\left(-\frac{1}{2}\right)\left(-\frac{3}{2}\right)}{2!}(-x^2)^2 + \frac{\left(-\frac{1}{2}\right)\left(-\frac{3}{2}\right)\left(-\frac{5}{2}\right)}{3!}(-x^2)^3 + \cdots$

$$= 1 + \sum_{n=1}^{\infty} \frac{1 \cdot 3 \cdot 5 \cdot \cdots \cdot (2n-1)}{2^n \cdot n!} x^{2n}$$

(b) $\sin^{-1} x = \int \frac{1}{\sqrt{1-x^2}}\,dx = C + x + \sum_{n=1}^{\infty} \frac{1 \cdot 3 \cdot 5 \cdot \cdots \cdot (2n-1)}{(2n+1)2^n \cdot n!} x^{2n+1}$

$$= x + \sum_{n=1}^{\infty} \frac{1 \cdot 3 \cdot 5 \cdot \cdots \cdot (2n-1)}{(2n+1)2^n \cdot n!} x^{2n+1} \quad \text{since } 0 = \sin^{-1} 0 = C.$$

52. (a) $1/\sqrt[4]{1+x} = (1+x)^{-1/4} = \sum_{n=0}^{\infty} \binom{-\frac{1}{4}}{n} x^n = 1 - \frac{1}{4}x + \frac{\left(-\frac{1}{4}\right)\left(-\frac{5}{4}\right)}{2!}x^2 + \frac{\left(-\frac{1}{4}\right)\left(-\frac{5}{4}\right)\left(-\frac{9}{4}\right)}{3!}x^3 + \cdots$

$$= 1 - \frac{1}{4}x + \sum_{n=2}^{\infty} (-1)^n \frac{1 \cdot 5 \cdot 9 \cdot \cdots \cdot (4n-3)}{4^n \cdot n!} x^n$$

(b) $1/\sqrt[4]{1+x} = 1 - \frac{1}{4}x + \frac{5}{32}x^2 - \frac{15}{128}x^3 + \frac{195}{2048}x^4 - \cdots$. $1/\sqrt[4]{1.1} = 1/\sqrt[4]{1+0.1}$, so let $x = 0.1$. The sum of the first four terms is then $1 - \frac{1}{4}(0.1) + \frac{5}{32}(0.1)^2 - \frac{15}{128}(0.1)^3 \approx 0.976$. The fifth term is $\frac{195}{2048}(0.1)^4 \approx 0.000\,009\,5$, which does not affect the third decimal place of the sum, so we have $1/\sqrt[4]{1.1} \approx 0.976$. (Note that the third decimal place of the sum of the first three terms is affected by the fourth term, so we need to use more than three terms for the sum.)

53. $\sqrt{1+x^3} = (1+x^3)^{1/2} = \sum_{n=0}^{\infty} \binom{\frac{1}{2}}{n} (x^3)^n = \sum_{n=0}^{\infty} \binom{\frac{1}{2}}{n} x^{3n} \Rightarrow \int \sqrt{1+x^3}\,dx = C + \sum_{n=0}^{\infty} \binom{\frac{1}{2}}{n} \frac{x^{3n+1}}{3n+1}$,

with $R = 1$.

54. $\sin x = \sum_{n=0}^{\infty} (-1)^n \frac{x^{2n+1}}{(2n+1)!} \Rightarrow \sin(x^2) = \sum_{n=0}^{\infty} (-1)^n \frac{(x^2)^{2n+1}}{(2n+1)!} = \sum_{n=0}^{\infty} (-1)^n \frac{x^{4n+2}}{(2n+1)!} \Rightarrow$

$x^2 \sin(x^2) = \sum_{n=0}^{\infty} (-1)^n \frac{x^{4n+4}}{(2n+1)!} \Rightarrow \int x^2 \sin(x^2)\,dx = C + \sum_{n=0}^{\infty} (-1)^n \frac{x^{4n+5}}{(2n+1)!(4n+5)}$, with $R = \infty$.

55. $\cos x \overset{(16)}{=} \sum_{n=0}^{\infty} (-1)^n \frac{x^{2n}}{(2n)!} \Rightarrow \cos x - 1 = \sum_{n=1}^{\infty} (-1)^n \frac{x^{2n}}{(2n)!} \Rightarrow \frac{\cos x - 1}{x} = \sum_{n=1}^{\infty} (-1)^n \frac{x^{2n-1}}{(2n)!} \Rightarrow$

$\int \frac{\cos x - 1}{x}\,dx = C + \sum_{n=1}^{\infty} (-1)^n \frac{x^{2n}}{2n \cdot (2n)!}$, with $R = \infty$.

56. $\arctan x = \sum_{n=0}^{\infty} (-1)^n \frac{x^{2n+1}}{2n+1} \Rightarrow \arctan(x^2) = \sum_{n=0}^{\infty} (-1)^n \frac{(x^2)^{2n+1}}{2n+1} = \sum_{n=0}^{\infty} (-1)^n \frac{x^{4n+2}}{2n+1} \Rightarrow$

$\int \arctan(x^2)\,dx = C + \sum_{n=0}^{\infty} (-1)^n \frac{x^{4n+3}}{(2n+1)(4n+3)}$, with $R = 1$.

57. $\arctan x = \sum_{n=0}^{\infty} (-1)^n \frac{x^{2n+1}}{2n+1}$ for $|x| < 1$, so $x^3 \arctan x = \sum_{n=0}^{\infty} (-1)^n \frac{x^{2n+4}}{2n+1}$ for $|x| < 1$ and

$\int x^3 \arctan x\,dx = C + \sum_{n=0}^{\infty} (-1)^n \frac{x^{2n+5}}{(2n+1)(2n+5)}$. Since $\frac{1}{2} < 1$, we have

$\int_0^{1/2} x^3 \arctan x\,dx = \sum_{n=0}^{\infty} (-1)^n \frac{(1/2)^{2n+5}}{(2n+1)(2n+5)} = \frac{(1/2)^5}{1 \cdot 5} - \frac{(1/2)^7}{3 \cdot 7} + \frac{(1/2)^9}{5 \cdot 9} - \frac{(1/2)^{11}}{7 \cdot 11} + \cdots$. Now

$\frac{(1/2)^5}{1 \cdot 5} - \frac{(1/2)^7}{3 \cdot 7} + \frac{(1/2)^9}{5 \cdot 9} \approx 0.0059$ and subtracting $\frac{(1/2)^{11}}{7 \cdot 11} \approx 6.3 \times 10^{-6}$ does not affect the fourth decimal place,

so $\int_0^{1/2} x^3 \arctan x\,dx \approx 0.0059$ by the Alternating Series Estimation Theorem.

58. $\sin x = \sum_{n=0}^{\infty} (-1)^n \frac{x^{2n+1}}{(2n+1)!}$ for all x, so $\sin(x^4) = \sum_{n=0}^{\infty} (-1)^n \frac{x^{8n+4}}{(2n+1)!}$ for all x and

$\int \sin(x^4)\,dx = C + \sum_{n=0}^{\infty} (-1)^n \frac{x^{8n+5}}{(2n+1)!\,(8n+5)}$. Thus,

$\displaystyle\int_0^1 \sin(x^4)\,dx = \sum_{n=0}^{\infty} (-1)^n \frac{1}{(2n+1)!\,(8n+5)} = \frac{1}{1!\cdot 5} - \frac{1}{3!\cdot 13} + \frac{1}{5!\cdot 21} - \frac{1}{7!\cdot 29} + \cdots$. Now

$\displaystyle\frac{1}{1!\cdot 5} - \frac{1}{3!\cdot 13} + \frac{1}{5!\cdot 21} \approx 0.1876$ and subtracting $\displaystyle\frac{1}{7!\cdot 29} \approx 6.84 \times 10^{-6}$ does not affect the fourth decimal place, so

$\int_0^1 \sin(x^4)\,dx \approx 0.1876$ by the Alternating Series Estimation Theorem.

59. $\displaystyle\sqrt{1+x^4} = (1+x^4)^{1/2} = \sum_{n=0}^{\infty} \binom{\frac{1}{2}}{n} (x^4)^n$, so $\displaystyle\int \sqrt{1+x^4}\,dx = C + \sum_{n=0}^{\infty} \binom{\frac{1}{2}}{n} \frac{x^{4n+1}}{4n+1}$ and hence, since $0.4 < 1$,

we have

$$I = \int_0^{0.4} \sqrt{1+x^4}\,dx = \sum_{n=0}^{\infty} \binom{\frac{1}{2}}{n} \frac{(0.4)^{4n+1}}{4n+1}$$

$$= (1)\frac{(0.4)^1}{0!} + \frac{\frac{1}{2}}{1!}\frac{(0.4)^5}{5} + \frac{\frac{1}{2}\left(-\frac{1}{2}\right)}{2!}\frac{(0.4)^9}{9} + \frac{\frac{1}{2}\left(-\frac{1}{2}\right)\left(-\frac{3}{2}\right)}{3!}\frac{(0.4)^{13}}{13} + \frac{\frac{1}{2}\left(-\frac{1}{2}\right)\left(-\frac{3}{2}\right)\left(-\frac{5}{2}\right)}{4!}\frac{(0.4)^{17}}{17} + \cdots$$

$$= 0.4 + \frac{(0.4)^5}{10} - \frac{(0.4)^9}{72} + \frac{(0.4)^{13}}{208} - \frac{5(0.4)^{17}}{2176} + \cdots$$

Now $\displaystyle\frac{(0.4)^9}{72} \approx 3.6 \times 10^{-6} < 5 \times 10^{-6}$, so by the Alternating Series Estimation Theorem, $\displaystyle I \approx 0.4 + \frac{(0.4)^5}{10} \approx 0.40102$

(correct to five decimal places).

60. $\displaystyle\int_0^{0.5} x^2 e^{-x^2}\,dx = \int_0^{0.5} \sum_{n=0}^{\infty} \frac{(-1)^n x^{2n+2}}{n!}\,dx = \sum_{n=0}^{\infty} \left[\frac{(-1)^n x^{2n+3}}{n!(2n+3)}\right]_0^{1/2} = \sum_{n=0}^{\infty} \frac{(-1)^n}{n!(2n+3)2^{2n+3}}$ and since the term

with $n = 2$ is $\displaystyle\frac{1}{1792} < 0.001$, we use $\displaystyle\sum_{n=0}^{1} \frac{(-1)^n}{n!(2n+3)2^{2n+3}} = \frac{1}{24} - \frac{1}{160} \approx 0.0354$.

61. $\displaystyle\lim_{x\to 0} \frac{x - \ln(1+x)}{x^2} = \lim_{x\to 0} \frac{x - \left(x - \frac{1}{2}x^2 + \frac{1}{3}x^3 - \frac{1}{4}x^4 + \frac{1}{5}x^5 - \cdots\right)}{x^2} = \lim_{x\to 0} \frac{\frac{1}{2}x^2 - \frac{1}{3}x^3 + \frac{1}{4}x^4 - \frac{1}{5}x^5 + \cdots}{x^2}$

$$= \lim_{x\to 0} \left(\tfrac{1}{2} - \tfrac{1}{3}x + \tfrac{1}{4}x^2 - \tfrac{1}{5}x^3 + \cdots\right) = \tfrac{1}{2}$$

since power series are continuous functions.

62. $\displaystyle\lim_{x\to 0} \frac{1-\cos x}{1+x-e^x} = \lim_{x\to 0} \frac{1 - \left(1 - \frac{1}{2!}x^2 + \frac{1}{4!}x^4 - \frac{1}{6!}x^6 + \cdots\right)}{1 + x - \left(1 + x + \frac{1}{2!}x^2 + \frac{1}{3!}x^3 + \frac{1}{4!}x^4 + \frac{1}{5!}x^5 + \frac{1}{6!}x^6 + \cdots\right)}$

$$= \lim_{x\to 0} \frac{\frac{1}{2!}x^2 - \frac{1}{4!}x^4 + \frac{1}{6!}x^6 - \cdots}{-\frac{1}{2!}x^2 - \frac{1}{3!}x^3 - \frac{1}{4!}x^4 - \frac{1}{5!}x^5 - \frac{1}{6!}x^6 - \cdots}$$

$$= \lim_{x\to 0} \frac{\frac{1}{2!} - \frac{1}{4!}x^2 + \frac{1}{6!}x^4 - \cdots}{-\frac{1}{2!} - \frac{1}{3!}x - \frac{1}{4!}x^2 - \frac{1}{5!}x^3 - \frac{1}{6!}x^4 - \cdots} = \frac{\frac{1}{2} - 0}{-\frac{1}{2} - 0} = -1$$

since power series are continuous functions.

63. $\displaystyle\lim_{x\to 0} \frac{\sin x - x + \frac{1}{6}x^3}{x^5} = \lim_{x\to 0} \frac{\left(x - \frac{1}{3!}x^3 + \frac{1}{5!}x^5 - \frac{1}{7!}x^7 + \cdots\right) - x + \frac{1}{6}x^3}{x^5}$

$$= \lim_{x\to 0} \frac{\frac{1}{5!}x^5 - \frac{1}{7!}x^7 + \cdots}{x^5} = \lim_{x\to 0} \left(\frac{1}{5!} - \frac{x^2}{7!} + \frac{x^4}{9!} - \cdots\right) = \frac{1}{5!} = \frac{1}{120}$$

since power series are continuous functions.

64. $\displaystyle \lim_{x\to 0}\frac{\sqrt{1+x}-1-\frac{1}{2}x}{x^2}=\lim_{x\to 0}\frac{\left(1+\frac{1}{2}x-\frac{1}{8}x^2+\frac{1}{16}x^3-\cdots\right)-1-\frac{1}{2}x}{x^2}=\lim_{x\to 0}\frac{-\frac{1}{8}x^2+\frac{1}{16}x^3-\cdots}{x^2}$

$\displaystyle =\lim_{x\to 0}\left(-\tfrac{1}{8}+\tfrac{1}{16}x-\cdots\right)=-\tfrac{1}{8}$ since power series are continuous functions.

65. $\displaystyle \lim_{x\to 0}\frac{x^3-3x+3\tan^{-1}x}{x^5}=\lim_{x\to 0}\frac{x^3-3x+3\left(x-\frac{1}{3}x^3+\frac{1}{5}x^5-\frac{1}{7}x^7+\cdots\right)}{x^5}$

$\displaystyle =\lim_{x\to 0}\frac{x^3-3x+3x-x^3+\frac{3}{5}x^5-\frac{3}{7}x^7+\cdots}{x^5}=\lim_{x\to 0}\frac{\frac{3}{5}x^5-\frac{3}{7}x^7+\cdots}{x^5}$

$\displaystyle =\lim_{x\to 0}\left(\tfrac{3}{5}-\tfrac{3}{7}x^2+\cdots\right)=\tfrac{3}{5}$ since power series are continuous functions.

66. $\displaystyle \lim_{x\to 0}\frac{\tan x-x}{x^3}=\lim_{x\to 0}\frac{\left(x+\frac{1}{3}x^3+\frac{2}{15}x^5+\cdots\right)-x}{x^3}=\lim_{x\to 0}\frac{\frac{1}{3}x^3+\frac{2}{15}x^5+\cdots}{x^3}=\lim_{x\to 0}\left(\tfrac{1}{3}+\tfrac{2}{15}x^2+\cdots\right)=\tfrac{1}{3}$

since power series are continuous functions.

67. From Equation 11, we have $e^{-x^2}=1-\dfrac{x^2}{1!}+\dfrac{x^4}{2!}-\dfrac{x^6}{3!}+\cdots$ and we know that $\cos x=1-\dfrac{x^2}{2!}+\dfrac{x^4}{4!}-\cdots$ from Equation 16. Therefore, $e^{-x^2}\cos x=\left(1-x^2+\frac{1}{2}x^4-\cdots\right)\left(1-\frac{1}{2}x^2+\frac{1}{24}x^4-\cdots\right)$. Writing only the terms with degree ≤ 4, we get $e^{-x^2}\cos x=1-\frac{1}{2}x^2+\frac{1}{24}x^4-x^2+\frac{1}{2}x^4+\frac{1}{2}x^4+\cdots=1-\frac{3}{2}x^2+\frac{25}{24}x^4+\cdots$.

68. $\displaystyle \sec x=\frac{1}{\cos x}\overset{(16)}{=}\frac{1}{1-\frac{1}{2}x^2+\frac{1}{24}x^4-\cdots}$.

$$
\begin{array}{r|l}
 & 1+\frac{1}{2}x^2+\frac{5}{24}x^4+\cdots \\ \hline
1-\frac{1}{2}x^2+\frac{1}{24}x^4-\cdots & 1 \\
 & 1-\frac{1}{2}x^2+\frac{1}{24}x^4-\cdots \\ \hline
 & \frac{1}{2}x^2-\frac{1}{24}x^4+\cdots \\
 & \frac{1}{2}x^2-\frac{1}{4}x^4+\cdots \\ \hline
 & \frac{5}{24}x^4+\cdots \\
 & \frac{5}{24}x^4+\cdots \\ \hline
 & \cdots
\end{array}
$$

From the long division above, $\sec x=1+\frac{1}{2}x^2+\frac{5}{24}x^4+\cdots$.

69. $\displaystyle \frac{x}{\sin x}\overset{(15)}{=}\frac{x}{x-\frac{1}{6}x^3+\frac{1}{120}x^5-\cdots}$.

$$
\begin{array}{r|l}
 & 1+\frac{1}{6}x^2+\frac{7}{360}x^4+\cdots \\ \hline
x-\frac{1}{6}x^3+\frac{1}{120}x^5-\cdots & x \\
 & x-\frac{1}{6}x^3+\frac{1}{120}x^5-\cdots \\ \hline
 & \frac{1}{6}x^3-\frac{1}{120}x^5+\cdots \\
 & \frac{1}{6}x^3-\frac{1}{36}x^5+\cdots \\ \hline
 & \frac{7}{360}x^5+\cdots \\
 & \frac{7}{360}x^5+\cdots \\ \hline
 & \cdots
\end{array}
$$

From the long division above, $\dfrac{x}{\sin x}=1+\frac{1}{6}x^2+\frac{7}{360}x^4+\cdots$.

70. From Table 1, we have $e^x = 1 + \frac{x}{1!} + \frac{x^2}{2!} + \frac{x^3}{3!} + \cdots$ and that $\ln(1+x) = x - \frac{x^2}{2} + \frac{x^3}{3} - \frac{x^4}{4} + \cdots$. Therefore,

$y = e^x \ln(1+x) = \left(1 + \frac{x}{1!} + \frac{x^2}{2!} + \frac{x^3}{3!} + \cdots\right)\left(x - \frac{x^2}{2} + \frac{x^3}{3} - \frac{x^4}{4} + \cdots\right)$. Writing only terms with degree ≤ 3,

we get $e^x \ln(1+x) = x - \frac{1}{2}x^2 + \frac{1}{3}x^3 + x^2 - \frac{1}{2}x^3 + \frac{1}{2}x^3 + \cdots = x + \frac{1}{2}x^2 + \frac{1}{3}x^3 + \cdots$.

71. $y = (\arctan x)^2 = \left(x - \frac{1}{3}x^3 + \frac{1}{5}x^5 - \frac{1}{7}x^7 + \cdots\right)\left(x - \frac{1}{3}x^3 + \frac{1}{5}x^5 - \frac{1}{7}x^7 + \cdots\right)$. Writing only the terms with degree ≤ 6, we get $(\arctan x)^2 = x^2 - \frac{1}{3}x^4 + \frac{1}{5}x^6 - \frac{1}{3}x^4 + \frac{1}{9}x^6 + \frac{1}{5}x^6 + \cdots = x^2 - \frac{2}{3}x^4 + \frac{23}{45}x^6 + \cdots$.

72. $y = e^x \sin^2 x = (e^x \sin x)\sin x = \left(x + x^2 + \frac{1}{3}x^3 + \cdots\right)\left(x - \frac{1}{6}x^3 + \cdots\right)$ [from Example 13]. Writing only the terms with degree ≤ 4, we get $e^x \sin^2 x = x^2 - \frac{1}{6}x^4 + x^3 + \frac{1}{3}x^4 + \cdots = x^2 + x^3 + \frac{1}{6}x^4 + \cdots$.

73. $\sum_{n=0}^{\infty} (-1)^n \frac{x^{4n}}{n!} = \sum_{n=0}^{\infty} \frac{(-x^4)^n}{n!} = e^{-x^4}$, by (11).

74. $\sum_{n=0}^{\infty} \frac{(-1)^n \pi^{2n}}{6^{2n}(2n)!} = \sum_{n=0}^{\infty} (-1)^n \frac{\left(\frac{\pi}{6}\right)^{2n}}{(2n)!} = \cos\frac{\pi}{6} = \frac{\sqrt{3}}{2}$, by (16).

75. $\sum_{n=1}^{\infty} (-1)^{n-1} \frac{3^n}{n5^n} = \sum_{n=1}^{\infty} (-1)^{n-1} \frac{(3/5)^n}{n} = \ln\left(1 + \frac{3}{5}\right)$ [from Table 1] $= \ln\frac{8}{5}$

76. $\sum_{n=0}^{\infty} \frac{3^n}{5^n\, n!} = \sum_{n=0}^{\infty} \frac{(3/5)^n}{n!} = e^{3/5}$, by (11).

77. $\sum_{n=0}^{\infty} \frac{(-1)^n \pi^{2n+1}}{4^{2n+1}(2n+1)!} = \sum_{n=0}^{\infty} \frac{(-1)^n \left(\frac{\pi}{4}\right)^{2n+1}}{(2n+1)!} = \sin\frac{\pi}{4} = \frac{1}{\sqrt{2}}$, by (15).

78. $1 - \ln 2 + \frac{(\ln 2)^2}{2!} - \frac{(\ln 2)^3}{3!} + \cdots = \sum_{n=0}^{\infty} \frac{(-\ln 2)^n}{n!} = e^{-\ln 2} = \left(e^{\ln 2}\right)^{-1} = 2^{-1} = \frac{1}{2}$, by (11).

79. $3 + \frac{9}{2!} + \frac{27}{3!} + \frac{81}{4!} + \cdots = \frac{3^1}{1!} + \frac{3^2}{2!} + \frac{3^3}{3!} + \frac{3^4}{4!} + \cdots = \sum_{n=1}^{\infty} \frac{3^n}{n!} = \sum_{n=0}^{\infty} \frac{3^n}{n!} - 1 = e^3 - 1$, by (11).

80. $\frac{1}{1 \cdot 2} - \frac{1}{3 \cdot 2^3} + \frac{1}{5 \cdot 2^5} - \frac{1}{7 \cdot 2^7} + \cdots = \sum_{n=0}^{\infty} (-1)^n \frac{1}{(2n+1)2^{2n+1}} = \sum_{n=0}^{\infty} (-1)^n \frac{(1/2)^{2n+1}}{2n+1} = \tan^{-1}\left(\frac{1}{2}\right)$ [from Table 1]

81. If p is an nth-degree polynomial, then $p^{(i)}(x) = 0$ for $i > n$, so its Taylor series at a is $p(x) = \sum_{i=0}^{n} \frac{p^{(i)}(a)}{i!}(x-a)^i$.

Put $x - a = 1$, so that $x = a + 1$. Then $p(a+1) = \sum_{i=0}^{n} \frac{p^{(i)}(a)}{i!}$.

This is true for any a, so replace a by x: $p(x+1) = \sum_{i=0}^{n} \frac{p^{(i)}(x)}{i!}$

82. The coefficient of x^{58} in the Maclaurin series of $f(x) = (1+x^3)^{30}$ is $\frac{f^{(58)}(0)}{58!}$. But the binomial series for $f(x)$ is

$(1+x^3)^{30} = \sum_{n=0}^{\infty} \binom{30}{n} x^{3n}$, so it involves only powers of x that are multiples of 3 and therefore the coefficient of x^{58} is 0.

So $f^{(58)}(0) = 0$.

83. Assume that $|f'''(x)| \le M$, so $f'''(x) \le M$ for $a \le x \le a + d$. Now $\int_a^x f'''(t)\,dt \le \int_a^x M\,dt \;\Rightarrow$

$f''(x) - f''(a) \le M(x-a) \;\Rightarrow\; f''(x) \le f''(a) + M(x-a)$. Thus, $\int_a^x f''(t)\,dt \le \int_a^x [f''(a) + M(t-a)]\,dt \;\Rightarrow$

$f'(x) - f'(a) \le f''(a)(x-a) + \frac{1}{2}M(x-a)^2 \;\Rightarrow\; f'(x) \le f'(a) + f''(a)(x-a) + \frac{1}{2}M(x-a)^2 \;\Rightarrow$

$\int_a^x f'(t)\,dt \le \int_a^x \left[f'(a) + f''(a)(t-a) + \frac{1}{2}M(t-a)^2\right] dt \;\Rightarrow$

$f(x) - f(a) \le f'(a)(x-a) + \frac{1}{2}f''(a)(x-a)^2 + \frac{1}{6}M(x-a)^3$. So

$f(x) - f(a) - f'(a)(x-a) - \frac{1}{2}f''(a)(x-a)^2 \le \frac{1}{6}M(x-a)^3$. But

$R_2(x) = f(x) - T_2(x) = f(x) - f(a) - f'(a)(x-a) - \frac{1}{2}f''(a)(x-a)^2$, so $R_2(x) \le \frac{1}{6}M(x-a)^3$.

A similar argument using $f'''(x) \ge -M$ shows that $R_2(x) \ge -\frac{1}{6}M(x-a)^3$. So $|R_2(x_2)| \le \frac{1}{6}M\,|x-a|^3$.

Although we have assumed that $x > a$, a similar calculation shows that this inequality is also true if $x < a$.

84. (a) $f(x) = \begin{cases} e^{-1/x^2} & \text{if } x \ne 0 \\ 0 & \text{if } x = 0 \end{cases}$ so $f'(0) = \lim\limits_{x\to 0} \dfrac{f(x) - f(0)}{x - 0} = \lim\limits_{x\to 0} \dfrac{e^{-1/x^2}}{x} = \lim\limits_{x\to 0} \dfrac{1/x}{e^{1/x^2}} = \lim\limits_{x\to 0} \dfrac{x}{2e^{1/x^2}} = 0$

(using l'Hospital's Rule and simplifying in the penultimate step). Similarly, we can use the definition of the derivative and l'Hospital's Rule to show that $f''(0) = 0$, $f^{(3)}(0) = 0, \ldots, f^{(n)}(0) = 0$, so that the Maclaurin series for f consists entirely of zero terms. But since $f(x) \ne 0$ except for $x = 0$, we see that f cannot equal its Maclaurin series except at $x = 0$.

(b)

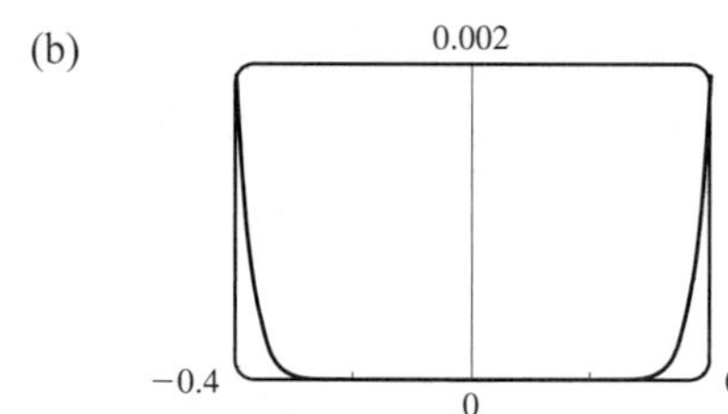

From the graph, it seems that the function is extremely flat at the origin. In fact, it could be said to be "infinitely flat" at $x = 0$, since all of its derivatives are 0 there.

85. (a) $g(x) = \sum\limits_{n=0}^{\infty} \binom{k}{n} x^n \;\Rightarrow\; g'(x) = \sum\limits_{n=1}^{\infty} \binom{k}{n} n x^{n-1}$, so

$$\begin{aligned}
(1+x)g'(x) &= (1+x)\sum_{n=1}^{\infty}\binom{k}{n} n x^{n-1} = \sum_{n=1}^{\infty}\binom{k}{n} n x^{n-1} + \sum_{n=1}^{\infty}\binom{k}{n} n x^{n} \\
&= \sum_{n=0}^{\infty}\binom{k}{n+1}(n+1)x^n + \sum_{n=0}^{\infty}\binom{k}{n} n x^n \qquad \left[\begin{array}{c}\text{Replace } n \text{ with } n+1 \\ \text{in the first series}\end{array}\right] \\
&= \sum_{n=0}^{\infty}(n+1)\frac{k(k-1)(k-2)\cdots(k-n+1)(k-n)}{(n+1)!}x^n + \sum_{n=0}^{\infty}\left[(n)\frac{k(k-1)(k-2)\cdots(k-n+1)}{n!}\right]x^n \\
&= \sum_{n=0}^{\infty}\frac{(n+1)k(k-1)(k-2)\cdots(k-n+1)}{(n+1)!}\,[(k-n)+n]\,x^n \\
&= k\sum_{n=0}^{\infty}\frac{k(k-1)(k-2)\cdots(k-n+1)}{n!}x^n = k\sum_{n=0}^{\infty}\binom{k}{n}x^n = kg(x)
\end{aligned}$$

Thus, $g'(x) = \dfrac{kg(x)}{1+x}$.

(b) $h(x) = (1+x)^{-k}\, g(x) \quad \Rightarrow$

$$h'(x) = -k(1+x)^{-k-1} g(x) + (1+x)^{-k}\, g'(x) \qquad \text{[Product Rule]}$$

$$= -k(1+x)^{-k-1} g(x) + (1+x)^{-k}\, \frac{kg(x)}{1+x} \qquad \text{[from part (a)]}$$

$$= -k(1+x)^{-k-1} g(x) + k(1+x)^{-k-1} g(x) = 0$$

(c) From part (b) we see that $h(x)$ must be constant for $x \in (-1, 1)$, so $h(x) = h(0) = 1$ for $x \in (-1, 1)$.

Thus, $h(x) = 1 = (1+x)^{-k}\, g(x) \quad \Leftrightarrow \quad g(x) = (1+x)^k$ for $x \in (-1, 1)$.

86. Using the binomial series to expand $\sqrt{1+x}$ as a power series as in Example 9, we get

$$\sqrt{1+x} = (1+x)^{1/2} = 1 + \frac{x}{2} + \sum_{n=2}^{\infty} \frac{(-1)^{n-1} 1 \cdot 3 \cdot 5 \cdot \cdots \cdot (2n-3)x^n}{2^n \cdot n!}, \text{ so}$$

$$(1-x^2)^{1/2} = 1 - \frac{1}{2}x^2 - \sum_{n=2}^{\infty} \frac{1 \cdot 3 \cdot 5 \cdot \cdots \cdot (2n-3)}{2^n \cdot n!} x^{2n} \text{ and}$$

$$\sqrt{1 - e^2 \sin^2 \theta} = 1 - \frac{1}{2} e^2 \sin^2 \theta - \sum_{n=2}^{\infty} \frac{1 \cdot 3 \cdot 5 \cdot \cdots \cdot (2n-3)}{2^n \cdot n!} e^{2n} \sin^{2n} \theta. \text{ Thus,}$$

$$L = 4a \int_0^{\pi/2} \sqrt{1 - e^2 \sin^2 \theta}\, d\theta = 4a \int_0^{\pi/2} \left(1 - \frac{1}{2} e^2 \sin^2 \theta - \sum_{n=2}^{\infty} \frac{1 \cdot 3 \cdot 5 \cdot \cdots \cdot (2n-3)}{2^n \cdot n!} e^{2n} \sin^{2n} \theta \right) d\theta$$

$$= 4a \left[\frac{\pi}{2} - \frac{e^2}{2} S_1 - \sum_{n=2}^{\infty} \frac{1 \cdot 3 \cdot 5 \cdot \cdots \cdot (2n-3)}{n!} \left(\frac{e^2}{2} \right)^n S_n \right]$$

where $S_n = \displaystyle\int_0^{\pi/2} \sin^{2n} \theta\, d\theta = \frac{1 \cdot 3 \cdot 5 \cdot \cdots \cdot (2n-1)}{2 \cdot 4 \cdot 6 \cdot \cdots \cdot 2n} \frac{\pi}{2}$ by Exercise 7.1.50.

$$L = 4a\left(\frac{\pi}{2}\right)\left[1 - \frac{e^2}{2} \cdot \frac{1}{2} - \sum_{n=2}^{\infty} \frac{1 \cdot 3 \cdot 5 \cdot \cdots \cdot (2n-3)}{n!} \left(\frac{e^2}{2}\right)^n \frac{1 \cdot 3 \cdot 5 \cdot \cdots \cdot (2n-1)}{2 \cdot 4 \cdot 6 \cdot \cdots \cdot 2n}\right]$$

$$= 2\pi a \left[1 - \frac{e^2}{4} - \sum_{n=2}^{\infty} \frac{e^{2n}}{2^n} \cdot \frac{1^2 \cdot 3^2 \cdot 5^2 \cdot \cdots \cdot (2n-3)^2 (2n-1)}{n! \cdot 2^n \cdot n!} \right]$$

$$= 2\pi a \left[1 - \frac{e^2}{4} - \sum_{n=2}^{\infty} \frac{e^{2n}}{4^n} \left(\frac{1 \cdot 3 \cdot \cdots \cdot (2n-3)}{n!} \right)^2 (2n-1) \right]$$

$$= 2\pi a \left[1 - \frac{e^2}{4} - \frac{3e^4}{64} - \frac{5e^6}{256} - \cdots \right] = \frac{\pi a}{128} (256 - 64e^2 - 12e^4 - 5e^6 - \cdots)$$

LABORATORY PROJECT An Elusive Limit

1. $f(x) = \dfrac{n(x)}{d(x)} = \dfrac{\sin(\tan x) - \tan(\sin x)}{\arcsin(\arctan x) - \arctan(\arcsin x)}$

The table of function values were obtained using Maple with 10 digits of precision. The results of this project will vary depending on the CAS and precision level. It appears that as $x \to 0^+$, $f(x) \to \frac{10}{3}$. Since f is an even function, we have $f(x) \to \frac{10}{3}$ as $x \to 0$.

x	$f(x)$
1	1.1838
0.1	0.9821
0.01	2.0000
0.001	3.3333
0.0001	3.3333

2. The graph is inconclusive about the limit of f as $x \to 0$.

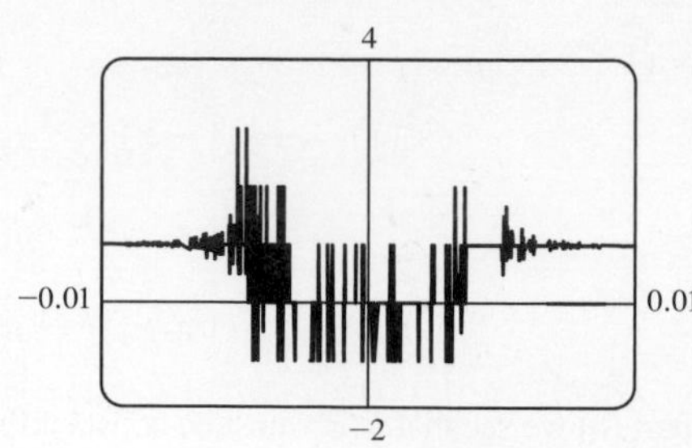

3. The limit has the indeterminate form $\frac{0}{0}$. Applying l'Hospital's Rule, we obtain the form $\frac{0}{0}$ six times. Finally, on the seventh application we obtain $\lim\limits_{x\to 0} \dfrac{n^{(7)}(x)}{d^{(7)}(x)} = \dfrac{-168}{-168} = 1$.

4. $\lim\limits_{x\to 0} f(x) = \lim\limits_{x\to 0} \dfrac{n(x)}{d(x)} \overset{\text{CAS}}{=} \lim\limits_{x\to 0} \dfrac{-\frac{1}{30}x^7 - \frac{29}{756}x^9 + \cdots}{-\frac{1}{30}x^7 + \frac{13}{756}x^9 + \cdots}$

$$= \lim_{x\to 0} \frac{\left(-\frac{1}{30}x^7 - \frac{29}{756}x^9 + \cdots\right)/x^7}{\left(-\frac{1}{30}x^7 + \frac{13}{756}x^9 + \cdots\right)/x^7} = \lim_{x\to 0} \frac{-\frac{1}{30} - \frac{29}{756}x^2 + \cdots}{-\frac{1}{30} + \frac{13}{756}x^2 + \cdots} = \frac{-\frac{1}{30}}{-\frac{1}{30}} = 1$$

Note that $n^{(7)}(x) = d^{(7)}(x) = -\frac{7!}{30} = -\frac{5040}{30} = -168$, which agrees with the result in Problem 3.

5. The limit command gives the result that $\lim\limits_{x\to 0} f(x) = 1$.

6. The strange results (with only 10 digits of precision) must be due to the fact that the terms being subtracted in the numerator and denominator are very close in value when $|x|$ is small. Thus, the differences are imprecise (have few correct digits).

11.11 Applications of Taylor Polynomials

1. (a)

n	$f^{(n)}(x)$	$f^{(n)}(0)$	$T_n(x)$
0	$\sin x$	0	0
1	$\cos x$	1	x
2	$-\sin x$	0	x
3	$-\cos x$	-1	$x - \frac{1}{6}x^3$
4	$\sin x$	0	$x - \frac{1}{6}x^3$
5	$\cos x$	1	$x - \frac{1}{6}x^3 + \frac{1}{120}x^5$

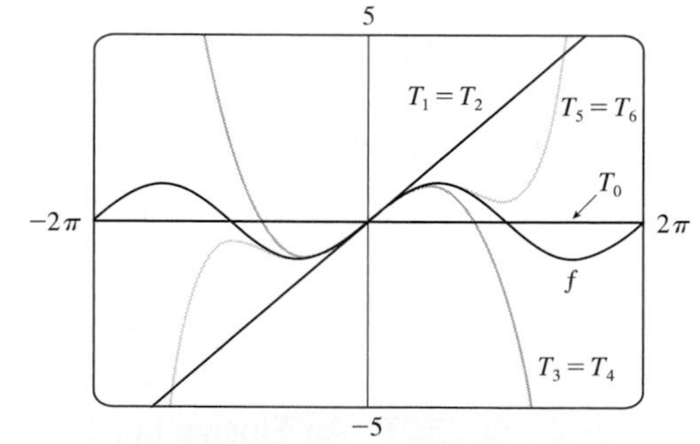

Note: $T_n(x) = \sum\limits_{k=0}^{n} \dfrac{f^{(k)}(0)}{k!}\, x^k$

(b)

x	f	$T_0(x)$	$T_1(x) = T_2(x)$	$T_3(x) = T_4(x)$	$T_5(x)$
$\frac{\pi}{4}$	0.7071	0	0.7854	0.7047	0.7071
$\frac{\pi}{2}$	1	0	1.5708	0.9248	1.0045
π	0	0	3.1416	−2.0261	0.5240

(c) As n increases, $T_n(x)$ is a good approximation to $f(x)$ on a larger and larger interval.

2. (a)

n	$f^{(n)}(x)$	$f^{(n)}(0)$	$T_n(x)$
0	$\tan x$	0	0
1	$\sec^2 x$	1	x
2	$2\sec^2 x\,\tan x$	0	x
3	$4\sec^2 x\,\tan^2 x + 2\sec^4 x$	2	$x + \frac{1}{3}x^3$

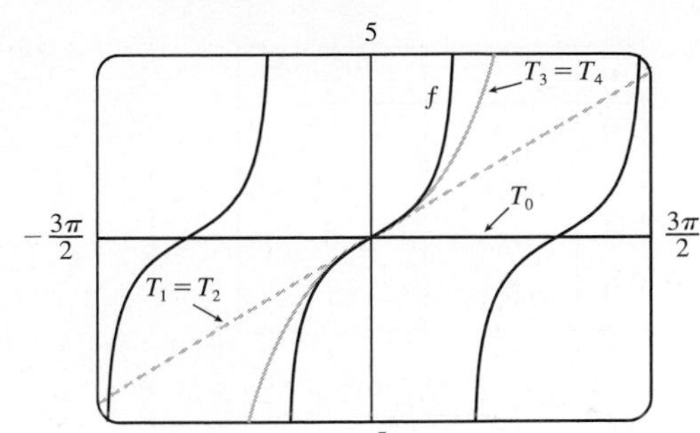

(b)

x	f	$T_0(x)$	$T_1(x) = T_2(x)$	$T_3(x)$
$\frac{\pi}{6}$	0.5774	0	0.5236	0.5714
$\frac{\pi}{4}$	1	0	0.7854	0.9469
$\frac{\pi}{3}$	1.7321	0	1.0472	1.4300

Note: $T_n(x) = \sum\limits_{k=0}^{n} \dfrac{f^{(k)}(0)}{k!}\,x^k$

(c) As n increases, $T_n(x)$ is a good approximation to $f(x)$ on a larger and larger interval. Because the Taylor polynomials are continuous, they cannot approximate the infinite discontinuities at $x = \pm\pi/2$. They can only approximate $\tan x$ on $(-\pi/2, \pi/2)$.

3.

n	$f^{(n)}(x)$	$f^{(n)}(1)$
0	e^x	e
1	e^x	e
2	c^x	c
3	e^x	e

$$\begin{aligned} T_3(x) &= \sum_{n=0}^{3} \frac{f^{(n)}(1)}{n!}\,(x-1)^n \\ &= \frac{e}{0!}\,(x-1)^0 + \frac{e}{1!}\,(x-1)^1 + \frac{e}{2!}\,(x-1)^2 + \frac{e}{3!}\,(x-1)^3 \\ &= e + e(x-1) + \tfrac{1}{2}e(x-1)^2 + \tfrac{1}{6}e(x-1)^3 \end{aligned}$$

4.

n	$f^{(n)}(x)$	$f^{(n)}(\pi/6)$
0	$\sin x$	$1/2$
1	$\cos x$	$\sqrt{3}/2$
2	$-\sin x$	$-1/2$
3	$-\cos x$	$-\sqrt{3}/2$

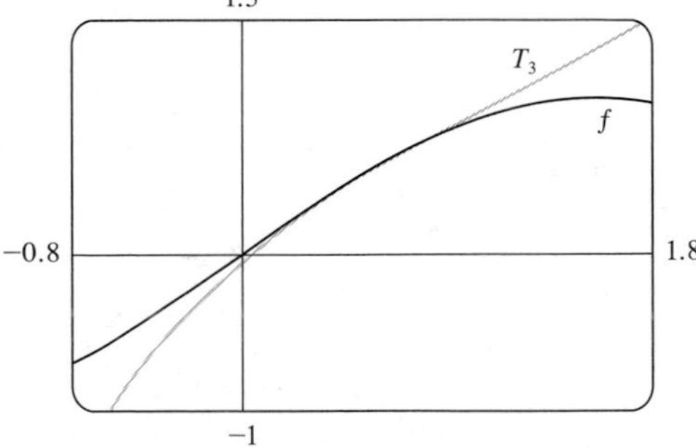

$$\begin{aligned} T_3(x) &= \sum_{n=0}^{3} \frac{f^{(n)}(\pi/6)}{n!}\left(x - \frac{\pi}{6}\right)^n \\ &= \frac{1/2}{0!}\left(x-\frac{\pi}{6}\right)^0 + \frac{\sqrt{3}/2}{1!}\left(x-\frac{\pi}{6}\right)^1 - \frac{1/2}{2!}\left(x-\frac{\pi}{6}\right)^2 + \frac{\sqrt{3}/2}{3!}\left(x-\frac{\pi}{6}\right)^3 \\ &= \frac{1}{2} + \frac{\sqrt{3}}{2}\left(x-\frac{\pi}{6}\right) - \frac{1}{4}\left(x-\frac{\pi}{6}\right)^2 - \frac{\sqrt{3}}{12}\left(x-\frac{\pi}{6}\right)^3 \end{aligned}$$

5.

n	$f^{(n)}(x)$	$f^{(n)}(\pi/2)$
0	$\cos x$	0
1	$-\sin x$	-1
2	$-\cos x$	0
3	$\sin x$	1

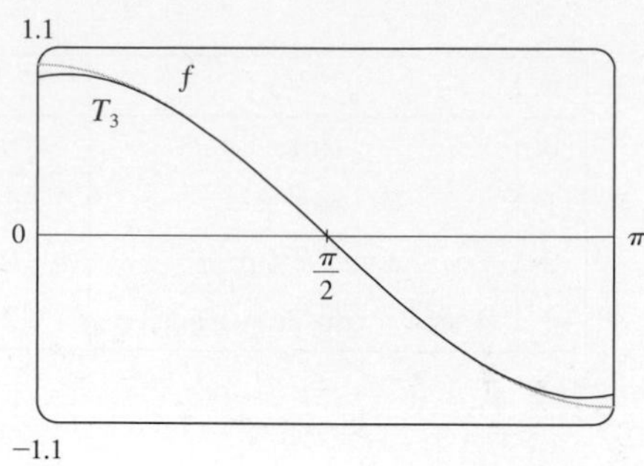

$$T_3(x) = \sum_{n=0}^{3} \frac{f^{(n)}(\pi/2)}{n!} \left(x - \tfrac{\pi}{2}\right)^n$$
$$= -\left(x - \tfrac{\pi}{2}\right) + \tfrac{1}{6}\left(x - \tfrac{\pi}{2}\right)^3$$

6.

n	$f^{(n)}(x)$	$f^{(n)}(0)$
0	$e^{-x} \sin x$	0
1	$e^{-x}(\cos x - \sin x)$	1
2	$-2e^{-x} \cos x$	-2
3	$2e^{-x}(\cos x + \sin x)$	2

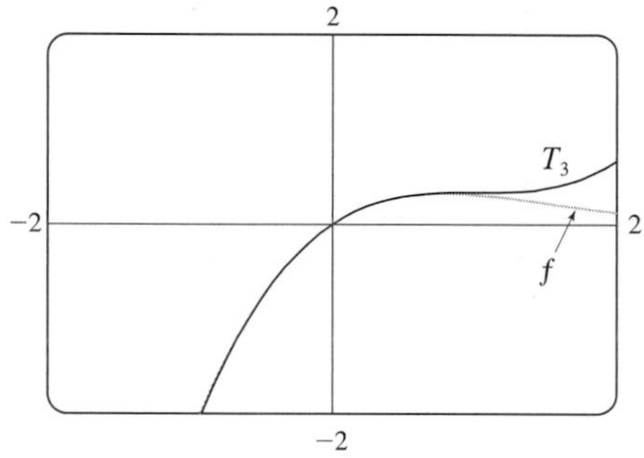

$$T_3(x) = \sum_{n=0}^{3} \frac{f^{(n)}(0)}{n!} x^n = x - x^2 + \tfrac{1}{3}x^3$$

7.

n	$f^{(n)}(x)$	$f^{(n)}(1)$
0	$\ln x$	0
1	$1/x$	1
2	$-1/x^2$	-1
3	$2/x^3$	2

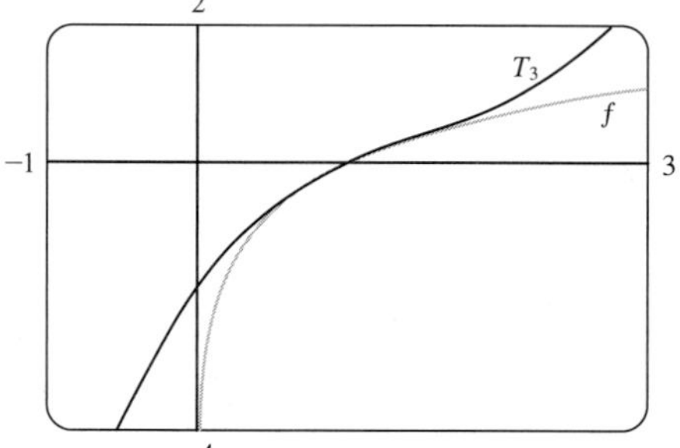

$$T_3(x) = \sum_{n=0}^{3} \frac{f^{(n)}(1)}{n!} (x-1)^n$$
$$= 0 + \frac{1}{1!}(x-1) + \frac{-1}{2!}(x-1)^2 + \frac{2}{3!}(x-1)^3$$
$$= (x-1) - \tfrac{1}{2}(x-1)^2 + \tfrac{1}{3}(x-1)^3$$

8.

n	$f^{(n)}(x)$	$f^{(n)}(0)$
0	$x \cos x$	0
1	$-x \sin x + \cos x$	1
2	$-x \cos x - 2 \sin x$	0
3	$x \sin x - 3 \cos x$	-3

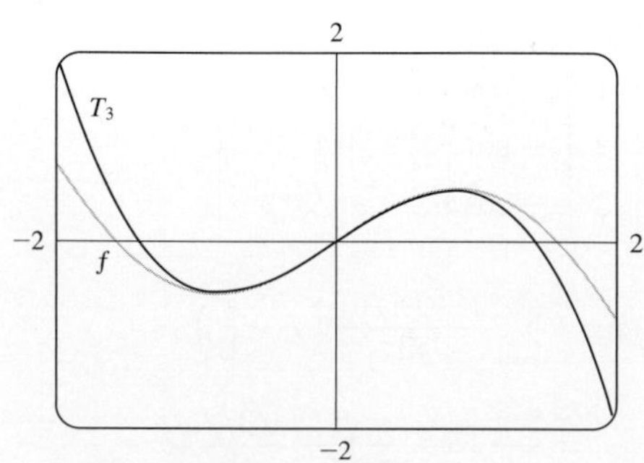

$$T_3(x) = \sum_{n=0}^{3} \frac{f^{(n)}(0)}{n!} x^n$$
$$= 0 + \frac{1}{1!}x + 0 + \frac{-3}{3!}x^3 = x - \frac{1}{2}x^3$$

9.

n	$f^{(n)}(x)$	$f^{(n)}(0)$
0	xe^{-2x}	0
1	$(1-2x)e^{-2x}$	1
2	$4(x-1)e^{-2x}$	-4
3	$4(3-2x)e^{-2x}$	12

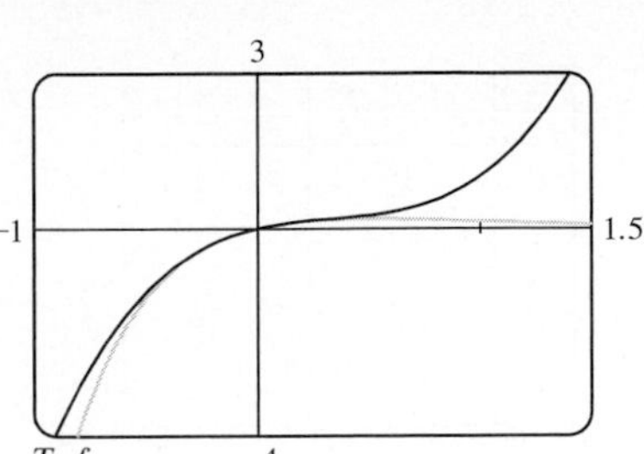

$$T_3(x) = \sum_{n=0}^{3} \frac{f^{(n)}(0)}{n!} x^n = \tfrac{0}{1} \cdot 1 + \tfrac{1}{1}x^1 + \tfrac{-4}{2}x^2 + \tfrac{12}{6}x^3 = x - 2x^2 + 2x^3$$

10.

n	$f^{(n)}(x)$	$f^{(n)}(1)$
0	$\tan^{-1} x$	$\frac{\pi}{4}$
1	$\dfrac{1}{1+x^2}$	$\frac{1}{2}$
2	$\dfrac{-2x}{(1+x^2)^2}$	$-\frac{1}{2}$
3	$\dfrac{6x^2-2}{(1+x^2)^3}$	$\frac{1}{2}$

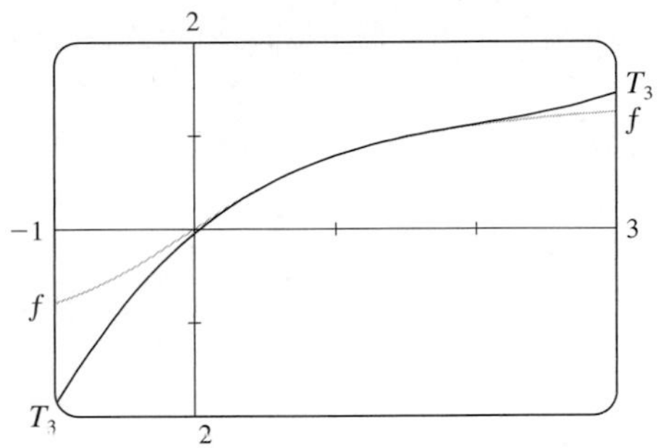

$$T_3(x) = \sum_{n=0}^{3} \frac{f^{(n)}(1)}{n!} (x-1)^n = \frac{\pi}{4} + \frac{1/2}{1}(x-1)^1 + \frac{-1/2}{2}(x-1)^2 + \frac{1/2}{6}(x-1)^3$$
$$= \tfrac{\pi}{4} + \tfrac{1}{2}(x-1) - \tfrac{1}{4}(x-1)^2 + \tfrac{1}{12}(x-1)^3$$

11. You may be able to simply find the Taylor polynomials for $f(x) = \cot x$ using your CAS. We will list the values of $f^{(n)}(\pi/4)$ for $n = 0$ to $n = 5$.

n	0	1	2	3	4	5
$f^{(n)}(\pi/4)$	1	-2	4	-16	80	-512

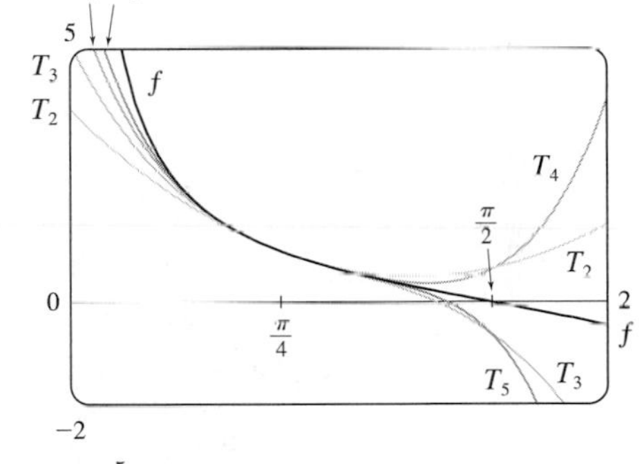

$$T_5(x) = \sum_{n=0}^{5} \frac{f^{(n)}(\pi/4)}{n!} \left(x - \tfrac{\pi}{4}\right)^n$$
$$= 1 - 2\left(x - \tfrac{\pi}{4}\right) + 2\left(x - \tfrac{\pi}{4}\right)^2 - \tfrac{8}{3}\left(x - \tfrac{\pi}{4}\right)^3 + \tfrac{10}{3}\left(x - \tfrac{\pi}{4}\right)^4 - \tfrac{64}{15}\left(x - \tfrac{\pi}{4}\right)^5$$

For $n = 2$ to $n = 5$, $T_n(x)$ is the polynomial consisting of all the terms up to and including the $\left(x - \frac{\pi}{4}\right)^n$ term.

12. You may be able to simply find the Taylor polynomials for $f(x) = \sqrt[3]{1+x^2}$ using your CAS. We will list the values of $f^{(n)}(0)$ for $n = 0$ to $n = 5$.

n	0	1	2	3	4	5
$f^{(n)}(0)$	1	0	$\frac{2}{3}$	0	$-\frac{8}{3}$	0

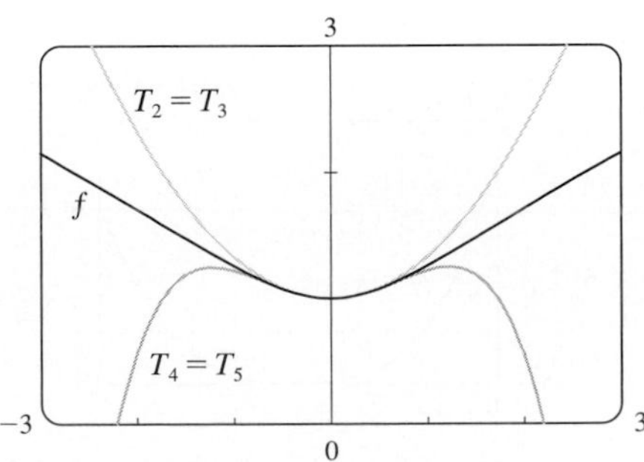

$$T_5(x) = \sum_{n=0}^{5} \frac{f^{(n)}(0)}{n!} x^n = 1 + \tfrac{1}{3}x^2 - \tfrac{1}{9}x^4$$

For $n = 2$ to $n = 5$, $T_n(x)$ is the polynomial consisting of all the terms up to and including the x^n term. Note that $T_2 = T_3$ and $T_4 = T_5$.

13. (a)

n	$f^{(n)}(x)$	$f^{(n)}(1)$
0	$1/x$	1
1	$-1/x^2$	-1
2	$2/x^3$	2
3	$-6/x^4$	

$$f(x) = 1/x \approx T_2(x)$$
$$= \frac{1}{0!}(x-1)^0 - \frac{1}{1!}(x-1)^1 + \frac{2}{2!}(x-1)^2$$
$$= 1 - (x-1) + (x-1)^2$$

(b) $|R_2(x)| \le \frac{M}{3!}|x-1|^3$, where $|f'''(x)| \le M$. Now $0.7 \le x \le 1.3 \;\Rightarrow\; |x-1| \le 0.3 \;\Rightarrow\; |x-1|^3 \le 0.027$.

Since $|f'''(x)|$ is decreasing on $[0.7, 1.3]$, we can take $M = |f'''(0.7)| = 6/(0.7)^4$, so

$|R_2(x)| \le \frac{6/(0.7)^4}{6}(0.027) = 0.112\,453\,1$.

(c)

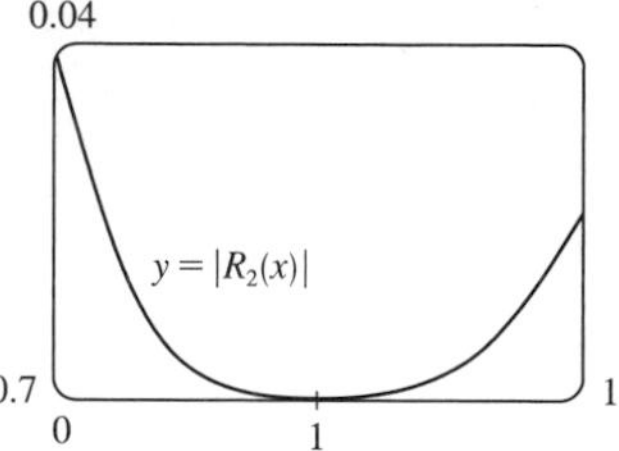

From the graph of $|R_2(x)| = \left|\frac{1}{x} - T_2(x)\right|$, it seems that the error is less than $0.038\,571$ on $[0.7, 1.3]$.

14. (a)

n	$f^{(n)}(x)$	$f^{(n)}(4)$
0	$x^{-1/2}$	$\frac{1}{2}$
1	$-\frac{1}{2}x^{-3/2}$	$-\frac{1}{16}$
2	$\frac{3}{4}x^{-5/2}$	$\frac{3}{128}$
3	$-\frac{15}{8}x^{-7/2}$	

$$f(x) = x^{-1/2} \approx T_2(x)$$
$$= \frac{1/2}{0!}(x-4)^0 - \frac{1/16}{1!}(x-4)^1 + \frac{3/128}{2!}(x-4)^2$$
$$= \tfrac{1}{2} - \tfrac{1}{16}(x-4) + \tfrac{3}{256}(x-4)^2$$

(b) $|R_2(x)| \le \frac{M}{3!}|x-4|^3$, where $|f'''(x)| \le M$. Now $3.5 \le x \le 4.5 \;\Rightarrow\; |x-4| \le 0.5 \;\Rightarrow\; |x-4|^3 \le 0.125$.

Since $|f'''(x)|$ is decreasing on $[3.5, 4.5]$, we can take $M = |f'''(3.5)| = \frac{15}{8(3.5)^{7/2}}$, so

$|R_2(x)| \le \frac{15}{6 \cdot 8(3.5)^{7/2}}(0.125) \approx 0.000\,487$.

(c)

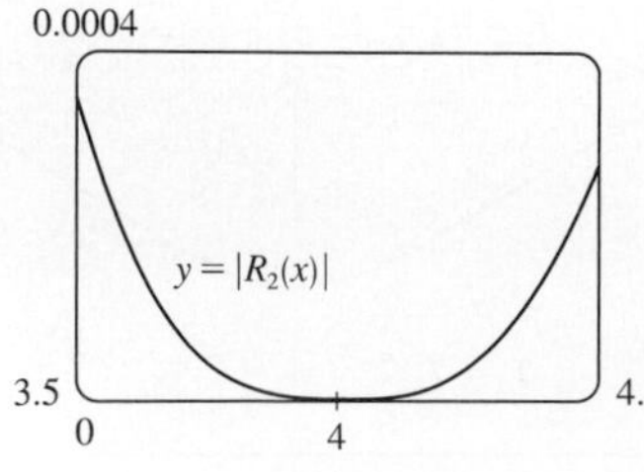

From the graph of $|R_2(x)| = \left|x^{-1/2} - T_2(x)\right|$, it seems that the error is less than $0.000\,343$ on $[3.5, 4.5]$.

15.

n	$f^{(n)}(x)$	$f^{(n)}(1)$
0	$x^{2/3}$	1
1	$\frac{2}{3}x^{-1/3}$	$\frac{2}{3}$
2	$-\frac{2}{9}x^{-4/3}$	$-\frac{2}{9}$
3	$\frac{8}{27}x^{-7/3}$	$\frac{8}{27}$
4	$-\frac{56}{81}x^{-10/3}$	

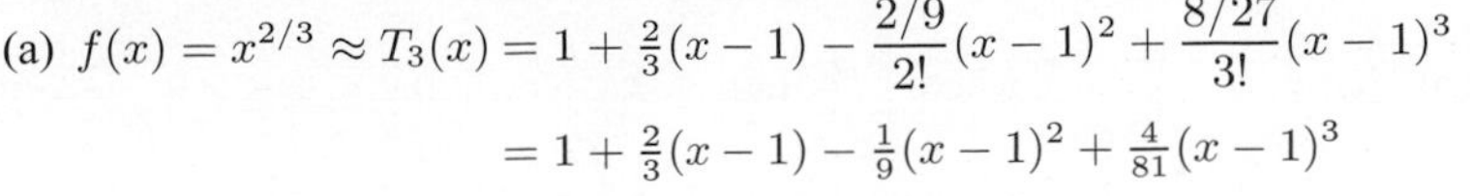

(a) $f(x) = x^{2/3} \approx T_3(x) = 1 + \frac{2}{3}(x-1) - \dfrac{2/9}{2!}(x-1)^2 + \dfrac{8/27}{3!}(x-1)^3$

$= 1 + \frac{2}{3}(x-1) - \frac{1}{9}(x-1)^2 + \frac{4}{81}(x-1)^3$

(b) $|R_3(x)| \le \dfrac{M}{4!}|x-1|^4$, where $\left|f^{(4)}(x)\right| \le M$. Now $0.8 \le x \le 1.2 \Rightarrow$ $|x-1| \le 0.2 \Rightarrow |x-1|^4 \le 0.0016$. Since $\left|f^{(4)}(x)\right|$ is decreasing on $[0.8, 1.2]$, we can take $M = \left|f^{(4)}(0.8)\right| = \frac{56}{81}(0.8)^{-10/3}$, so

$|R_3(x)| \le \dfrac{\frac{56}{81}(0.8)^{-10/3}}{24}(0.0016) \approx 0.000\,096\,97$.

(c)

From the graph of $|R_3(x)| = \left|x^{2/3} - T_3(x)\right|$, it seems that the error is less than $0.000\,053\,3$ on $[0.8, 1.2]$.

16.

n	$f^{(n)}(x)$	$f^{(n)}(\pi/6)$
0	$\sin x$	$1/2$
1	$\cos x$	$\sqrt{3}/2$
2	$-\sin x$	$-1/2$
3	$-\cos x$	$-\sqrt{3}/2$
4	$\sin x$	$1/2$
5	$\cos x$	

(a) $f(x) = \sin x \approx T_4(x)$

$= \frac{1}{2} + \frac{\sqrt{3}}{2}\left(x - \frac{\pi}{6}\right) - \frac{1}{4}\left(x - \frac{\pi}{6}\right)^2 - \frac{\sqrt{3}}{12}\left(x - \frac{\pi}{6}\right)^3 + \frac{1}{48}\left(x - \frac{\pi}{6}\right)^4$

(b) $|R_4(x)| \le \dfrac{M}{5!}\left|x - \frac{\pi}{6}\right|^5$, where $\left|f^{(5)}(x)\right| \le M$. Now $0 \le x \le \frac{\pi}{3} \Rightarrow -\frac{\pi}{6} \le x - \frac{\pi}{6} \le \frac{\pi}{6} \Rightarrow \left|x - \frac{\pi}{6}\right| \le \frac{\pi}{6} \Rightarrow$ $\left|x - \frac{\pi}{6}\right|^5 \le \left(\frac{\pi}{6}\right)^5$. Since $\left|f^{(5)}(x)\right|$ is decreasing on $\left[0, \frac{\pi}{3}\right]$, we can take $M = \left|f^{(5)}(0)\right| = \cos 0 = 1$, so $|R_4(x)| \le \dfrac{1}{5!}\left(\dfrac{\pi}{6}\right)^5 \approx 0.000\,328$.

(c)

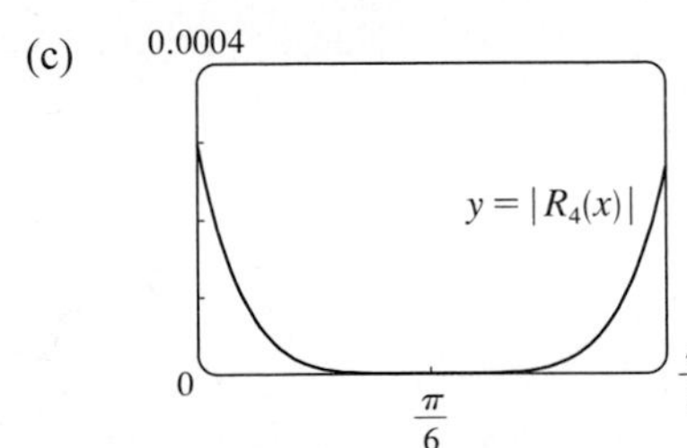

From the graph of $|R_4(x)| = |\sin x - T_4(x)|$, it seems that the error is less than $0.000\,297$ on $\left[0, \frac{\pi}{3}\right]$.

17.

n	$f^{(n)}(x)$	$f^{(n)}(0)$
0	$\sec x$	1
1	$\sec x \tan x$	0
2	$\sec x\,(2\sec^2 x - 1)$	1
3	$\sec x \tan x\,(6\sec^2 x - 1)$	

(a) $f(x) = \sec x \approx T_2(x) = 1 + \frac{1}{2}x^2$

(b) $|R_2(x)| \le \dfrac{M}{3!}|x|^3$, where $\left|f^{(3)}(x)\right| \le M$. Now $-0.2 \le x \le 0.2 \;\Rightarrow\; |x| \le 0.2 \;\Rightarrow\; |x|^3 \le (0.2)^3$.

$f^{(3)}(x)$ is an odd function and it is increasing on $[0, 0.2]$ since $\sec x$ and $\tan x$ are increasing on $[0, 0.2]$,

so $\left|f^{(3)}(x)\right| \le f^{(3)}(0.2) \approx 1.085\,158\,892$. Thus, $|R_2(x)| \le \dfrac{f^{(3)}(0.2)}{3!}(0.2)^3 \approx 0.001\,447$.

(c)

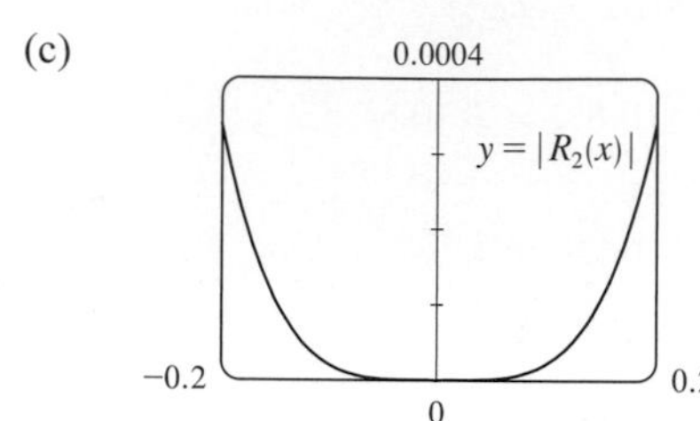

From the graph of $|R_2(x)| = |\sec x - T_2(x)|$, it seems that the error is less than $0.000\,339$ on $[-0.2, 0.2]$.

18.

n	$f^{(n)}(x)$	$f^{(n)}(1)$
0	$\ln(1+2x)$	$\ln 3$
1	$2/(1+2x)$	$\frac{2}{3}$
2	$-4/(1+2x)^2$	$-\frac{4}{9}$
3	$16/(1+2x)^3$	$\frac{16}{27}$
4	$-96/(1+2x)^4$	

(a) $f(x) = \ln(1+2x) \approx T_3(x)$

$$= \ln 3 + \tfrac{2}{3}(x-1) - \frac{4/9}{2!}(x-1)^2 + \frac{16/27}{3!}(x-1)^3$$

(b) $|R_3(x)| \le \dfrac{M}{4!}|x-1|^4$, where $\left|f^{(4)}(x)\right| \le M$. Now $0.5 \le x \le 1.5 \;\Rightarrow$

$-0.5 \le x - 1 \le 0.5 \;\Rightarrow\; |x-1| \le 0.5 \;\Rightarrow\; |x-1|^4 \le \frac{1}{16}$, and

letting $x = 0.5$ gives $M = 6$, so $|R_3(x)| \le \dfrac{6}{4!} \cdot \dfrac{1}{16} = \dfrac{1}{64} = 0.015\,625$.

(c)

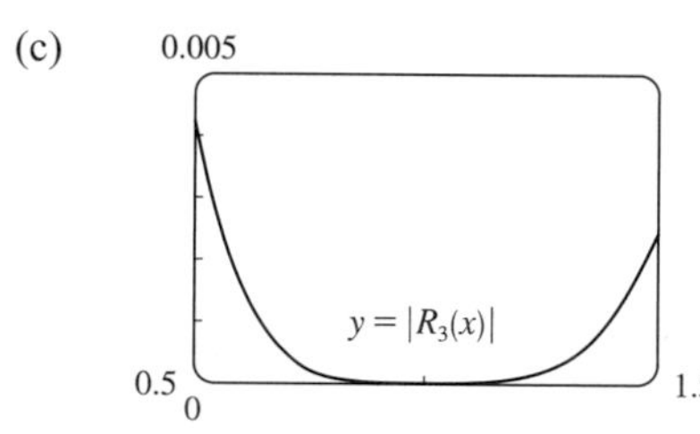

From the graph of $|R_3(x)| = |\ln(1+2x) - T_3(x)|$, it seems that the error is less than 0.005 on $[0.5, 1.5]$.

19.

n	$f^{(n)}(x)$	$f^{(n)}(0)$
0	e^{x^2}	1
1	$e^{x^2}(2x)$	0
2	$e^{x^2}(2+4x^2)$	2
3	$e^{x^2}(12x+8x^3)$	0
4	$e^{x^2}(12+48x^2+16x^4)$	

(a) $f(x) = e^{x^2} \approx T_3(x) = 1 + \dfrac{2}{2!}x^2 = 1 + x^2$

(b) $|R_3(x)| \le \dfrac{M}{4!}|x|^4$, where $\left|f^{(4)}(x)\right| \le M$. Now $0 \le x \le 0.1 \;\Rightarrow$

$x^4 \le (0.1)^4$, and letting $x = 0.1$ gives

$$|R_3(x)| \le \frac{e^{0.01}(12 + 0.48 + 0.0016)}{24}(0.1)^4 \approx 0.00006.$$

(c)

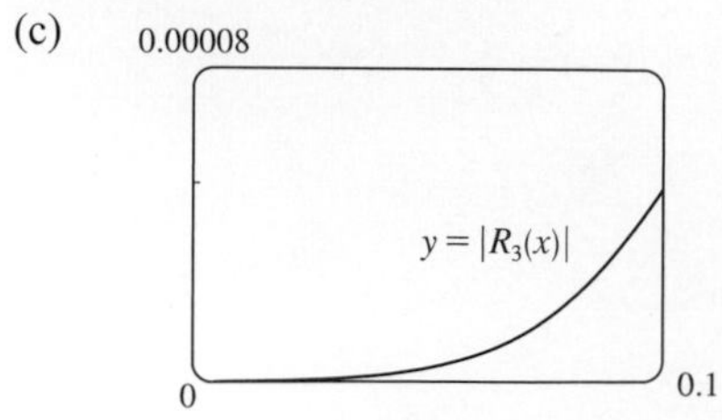

From the graph of $|R_3(x)| = \left|e^{x^2} - T_3(x)\right|$, it appears that the error is less than $0.000\,051$ on $[0, 0.1]$.

20.

n	$f^{(n)}(x)$	$f^{(n)}(1)$
0	$x \ln x$	0
1	$\ln x + 1$	1
2	$1/x$	1
3	$-1/x^2$	-1
4	$2/x^3$	

(a) $f(x) = x \ln x \approx T_3(x) = (x-1) + \frac{1}{2}(x-1)^2 - \frac{1}{6}(x-1)^3$

(b) $|R_3(x)| \le \dfrac{M}{4!}|x-1|^4$, where $\left|f^{(4)}(x)\right| \le M$. Now $0.5 \le x \le 1.5 \Rightarrow$ $|x-1| \le \frac{1}{2} \Rightarrow |x-1|^4 \le \frac{1}{16}$. Since $\left|f^{(4)}(x)\right|$ is decreasing on $[0.5, 1.5]$, we can take $M = \left|f^{(4)}(0.5)\right| = 2/(0.5)^3 = 16$, so $|R_3(x)| \le \frac{16}{24}(1/16) = \frac{1}{24} = 0.041\overline{6}$.

(c)

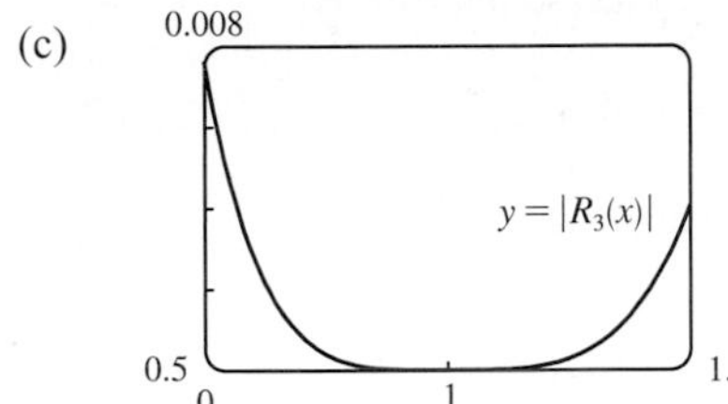

From the graph of $|R_3(x)| = |x \ln x - T_3(x)|$, it seems that the error is less than 0.0076 on $[0.5, 1.5]$.

21.

n	$f^{(n)}(x)$	$f^{(n)}(0)$
0	$x \sin x$	0
1	$\sin x + x \cos x$	0
2	$2\cos x - x \sin x$	2
3	$-3\sin x - x\cos x$	0
4	$-4\cos x + x \sin x$	-4
5	$5 \sin x + x \cos x$	

(a) $f(x) = x \sin x \approx T_4(x) = \dfrac{2}{2!}(x-0)^2 + \dfrac{-4}{4!}(x-0)^4 = x^2 - \dfrac{1}{6}x^4$

(b) $|R_4(x)| \le \dfrac{M}{5!}|x|^5$, where $\left|f^{(5)}(x)\right| \le M$. Now $-1 \le x \le 1 \Rightarrow$ $|x| \le 1$, and a graph of $f^{(5)}(x)$ shows that $\left|f^{(5)}(x)\right| \le 5$ for $-1 \le x \le 1$. Thus, we can take $M = 5$ and get $|R_4(x)| \le \dfrac{5}{5!} \cdot 1^5 = \dfrac{1}{24} = 0.041\overline{6}$.

(c)

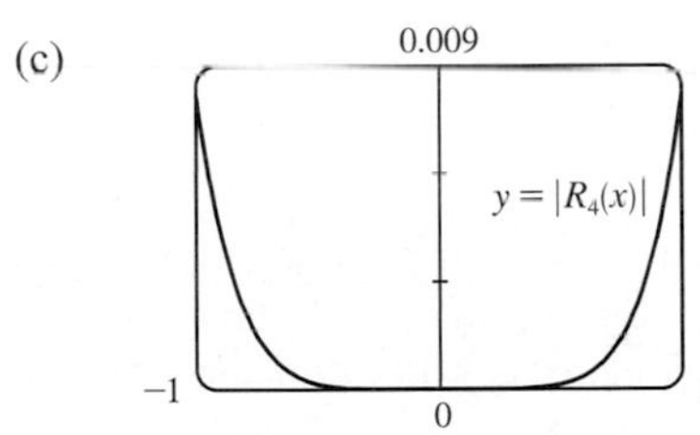

From the graph of $|R_4(x)| = |x \sin x - T_4(x)|$, it seems that the error is less than 0.0082 on $[-1, 1]$.

22.

n	$f^{(n)}(x)$	$f^{(n)}(0)$
0	$\sinh 2x$	0
1	$2\cosh 2x$	2
2	$4 \sinh 2x$	0
3	$8 \cosh 2x$	8
4	$16 \sinh 2x$	0
5	$32 \cosh 2x$	32
6	$64 \sinh 2x$	

(a) $f(x) = \sinh 2x \approx T_5(x) = 2x + \frac{8}{3!}x^3 + \frac{32}{5!}x^5 = 2x + \frac{4}{3}x^3 + \frac{4}{15}x^5$

(b) $|R_5(x)| \le \frac{M}{6!}|x|^6$, where $\left|f^{(6)}(x)\right| \le M$. For x in $[-1, 1]$, we have $|x| \le 1$. Since $f^{(6)}(x)$ is an increasing odd function on $[-1, 1]$, we see that $\left|f^{(6)}(x)\right| \le f^{(6)}(1) = 64 \sinh 2 = 32(e^2 - e^{-2}) \approx 232.119$, so we can take $M = 232.12$ and get $|R_5(x)| \le \frac{232.12}{720} \cdot 1^6 \approx 0.3224$.

(c)

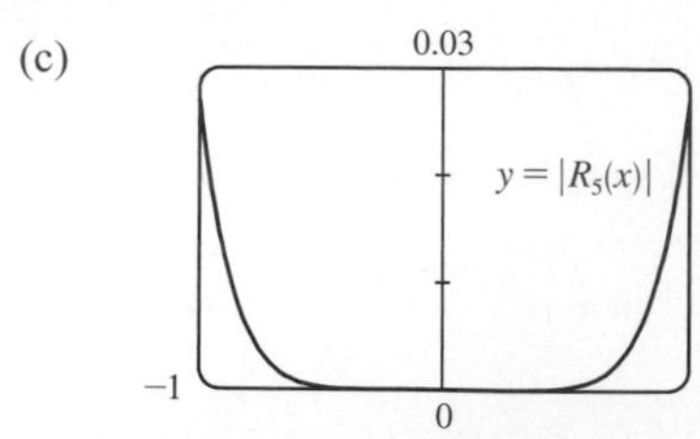

From the graph of $|R_5(x)| = |\sinh 2x - T_5(x)|$, it seems that the error is less than 0.027 on $[-1, 1]$.

23. From Exercise 5, $\cos x = -\left(x - \frac{\pi}{2}\right) + \frac{1}{6}\left(x - \frac{\pi}{2}\right)^3 + R_3(x)$, where $|R_3(x)| \le \dfrac{M}{4!}\left|x - \frac{\pi}{2}\right|^4$ with $\left|f^{(4)}(x)\right| = |\cos x| \le M = 1$. Now $x = 80^\circ = (90^\circ - 10^\circ) = \left(\frac{\pi}{2} - \frac{\pi}{18}\right) = \frac{4\pi}{9}$ radians, so the error is $\left|R_3\left(\frac{4\pi}{9}\right)\right| \le \frac{1}{24}\left(\frac{\pi}{18}\right)^4 \approx 0.000\,039$, which means our estimate would *not* be accurate to five decimal places. However, $T_3 = T_4$, so we can use $\left|R_4\left(\frac{4\pi}{9}\right)\right| \le \frac{1}{120}\left(\frac{\pi}{18}\right)^5 \approx 0.000\,001$. Therefore, to five decimal places, $\cos 80^\circ \approx -\left(-\frac{\pi}{18}\right) + \frac{1}{6}\left(-\frac{\pi}{18}\right)^3 \approx 0.17365$.

24. From Exercise 16, $\sin x = \frac{1}{2} + \frac{\sqrt{3}}{2}\left(x - \frac{\pi}{6}\right) - \frac{1}{4}\left(x - \frac{\pi}{6}\right)^2 - \frac{\sqrt{3}}{12}\left(x - \frac{\pi}{6}\right)^3 + \frac{1}{48}\left(x - \frac{\pi}{6}\right)^4 + R_4(x)$, where $|R_4(x)| \le \dfrac{M}{5!}\left|x - \frac{\pi}{6}\right|^5$ with $\left|f^{(5)}(x)\right| = |\cos x| \le M = 1$. Now $x = 38^\circ = (30^\circ + 8^\circ) = \left(\frac{\pi}{6} + \frac{2\pi}{45}\right)$ radians, so the error is $\left|R_4\left(\frac{38\pi}{180}\right)\right| \le \frac{1}{120}\left(\frac{2\pi}{45}\right)^5 \approx 0.000\,000\,44$, which means our estimate will be accurate to five decimal places. Therefore, to five decimal places, $\sin 38^\circ = \frac{1}{2} + \frac{\sqrt{3}}{2}\left(\frac{2\pi}{45}\right) - \frac{1}{4}\left(\frac{2\pi}{45}\right)^2 - \frac{\sqrt{3}}{12}\left(\frac{2\pi}{45}\right)^3 + \frac{1}{48}\left(\frac{2\pi}{45}\right)^4 \approx 0.61566$.

25. All derivatives of e^x are e^x, so $|R_n(x)| \le \dfrac{e^x}{(n+1)!}|x|^{n+1}$, where $0 < x < 0.1$. Letting $x = 0.1$, $R_n(0.1) \le \dfrac{e^{0.1}}{(n+1)!}(0.1)^{n+1} < 0.00001$, and by trial and error we find that $n = 3$ satisfies this inequality since $R_3(0.1) < 0.0000046$. Thus, by adding the four terms of the Maclaurin series for e^x corresponding to $n = 0, 1, 2$, and 3, we can estimate $e^{0.1}$ to within 0.00001. (In fact, this sum is $1.1051\overline{6}$ and $e^{0.1} \approx 1.10517$.)

26. From Table 1 in Section 11.10, $\ln(1 + x) = \sum\limits_{n=1}^{\infty}(-1)^{n-1}\dfrac{x^n}{n}$ for $|x| < 1$. Thus, $\ln 1.4 = \ln(1 + 0.4) = \sum\limits_{n=1}^{\infty}(-1)^{n-1}\dfrac{(0.4)^n}{n}$. Since this is an alternating series, the error is less than the first neglected term by the Alternating Series Estimation Theorem, and we find that $|a_6| = (0.4)^6/6 \approx 0.0007 < 0.001$. So we need the first five (nonzero) terms of the Maclaurin series for the desired accuracy. (In fact, this sum is approximately 0.33698 and $\ln 1.4 \approx 0.33647$.)

27. $\sin x = x - \dfrac{1}{3!}x^3 + \dfrac{1}{5!}x^5 - \cdots$. By the Alternating Series Estimation Theorem, the error in the approximation $\sin x = x - \dfrac{1}{3!}x^3$ is less than $\left|\dfrac{1}{5!}x^5\right| < 0.01 \iff$ $\left|x^5\right| < 120(0.01) \iff |x| < (1.2)^{1/5} \approx 1.037$. The curves $y = x - \frac{1}{6}x^3$ and $y = \sin x - 0.01$ intersect at $x \approx 1.043$, so the graph confirms our estimate. Since both the sine function

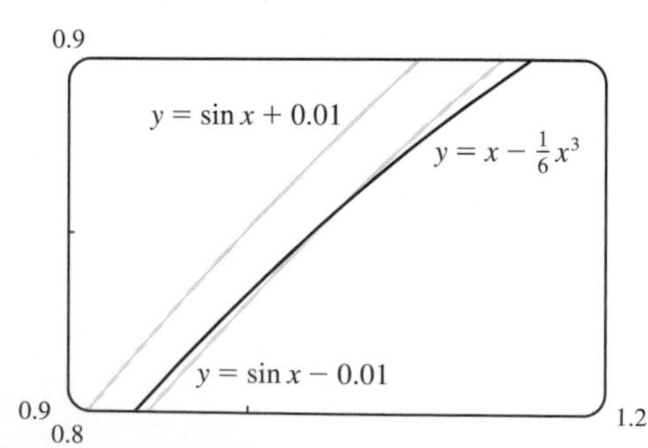

and the given approximation are odd functions, we need to check the estimate only for $x > 0$. Thus, the desired range of values for x is $-1.037 < x < 1.037$.

28. $\cos x = 1 - \frac{1}{2!}x^2 + \frac{1}{4!}x^4 - \frac{1}{6!}x^6 + \cdots$. By the Alternating Series Estimation Theorem, the error is less than $\left|-\frac{1}{6!}x^6\right| < 0.005 \Leftrightarrow$ $x^6 < 720(0.005) \Leftrightarrow |x| < (3.6)^{1/6} \approx 1.238$. The curves $y = 1 - \frac{1}{2}x^2 + \frac{1}{24}x^4$ and $y = \cos x + 0.005$ intersect at $x \approx 1.244$, so the graph confirms our estimate. Since both the cosine function and the given approximation are even functions, we need to check the estimate only for $x > 0$. Thus, the desired range of values for x is $-1.238 < x < 1.238$.

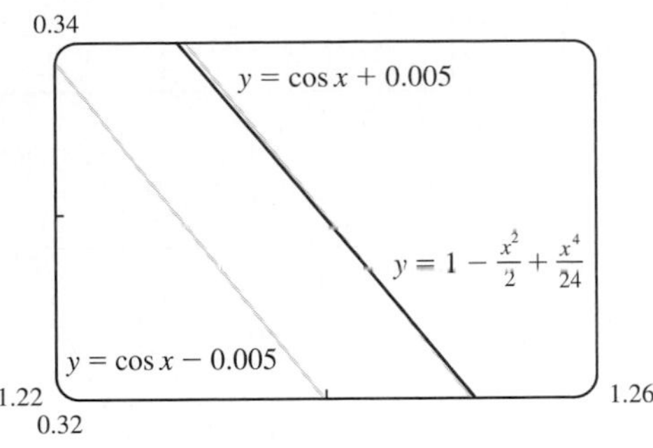

29. $\arctan x = x - \frac{x^3}{3} + \frac{x^5}{5} - \frac{x^7}{7} + \cdots$. By the Alternating Series Estimation Theorem, the error is less than $\left|-\frac{1}{7}x^7\right| < 0.05 \Leftrightarrow$ $\left|x^7\right| < 0.35 \Leftrightarrow |x| < (0.35)^{1/7} \approx 0.8607$. The curves $y = x - \frac{1}{3}x^3 + \frac{1}{5}x^5$ and $y = \arctan x + 0.05$ intersect at $x \approx 0.9245$, so the graph confirms our estimate. Since both the arctangent function and the given approximation are odd functions, we need to check the estimate only for $x > 0$. Thus, the desired range of values for x is $-0.86 < x < 0.86$.

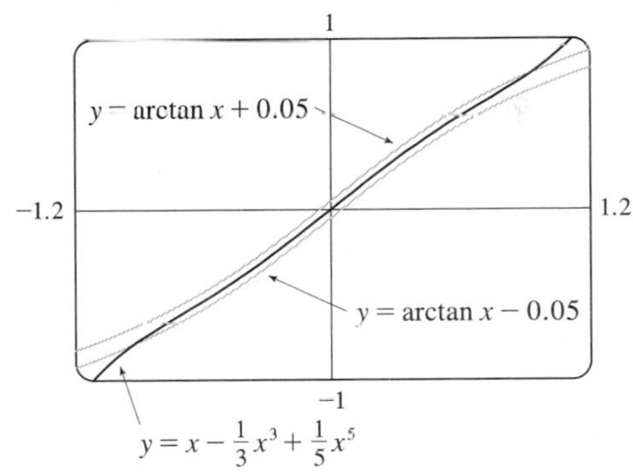

30. $f(x) = \sum_{n=0}^{\infty} \frac{f^{(n)}(4)}{n!}(x-4)^n = \sum_{n=0}^{\infty} \frac{(-1)^n\, n!}{3^n(n+1)\, n!}(x-4)^n = \sum_{n=0}^{\infty} \frac{(-1)^n}{3^n(n+1)}(x-4)^n$. Now

$f(5) = \sum_{n=0}^{\infty} \frac{(-1)^n}{3^n(n+1)} = \sum_{n=0}^{\infty}(-1)^n b_n$ is the sum of an alternating series that satisfies (i) $b_{n+1} \le b_n$ and (ii) $\lim_{n\to\infty} b_n = 0$, so by the Alternating Series Estimation Theorem, $|R_5(5)| = |f(5) - T_5(5)| \le b_6$, and

$b_6 = \frac{1}{3^6(7)} = \frac{1}{5103} \approx 0.000196 < 0.0002$; that is, the fifth-degree Taylor polynomial approximates $f(5)$ with error less than 0.0002.

31. Let $s(t)$ be the position function of the car, and for convenience set $s(0) = 0$. The velocity of the car is $v(t) = s'(t)$ and the acceleration is $a(t) = s''(t)$, so the second degree Taylor polynomial is $T_2(t) = s(0) + v(0)t + \frac{a(0)}{2}t^2 = 20t + t^2$. We estimate the distance traveled during the next second to be $s(1) \approx T_2(1) = 20 + 1 = 21$ m. The function $T_2(t)$ would not be accurate over a full minute, since the car could not possibly maintain an acceleration of 2 m/s^2 for that long (if it did, its final speed would be 140 m/s $\approx$ 313 mi/h!).

32. (a)

n	$\rho^{(n)}(t)$	$\rho^{(n)}(20)$
0	$\rho_{20}e^{\alpha(t-20)}$	ρ_{20}
1	$\alpha\rho_{20}e^{\alpha(t-20)}$	$\alpha\rho_{20}$
2	$\alpha^2\rho_{20}e^{\alpha(t-20)}$	$\alpha^2\rho_{20}$

The linear approximation is

$$T_1(t) = \rho(20) + \rho'(20)(t-20) = \rho_{20}[1+\alpha(t-20)]$$

The quadratic approximation is

$$T_2(t) = \rho(20) + \rho'(20)(t-20) + \frac{\rho''(20)}{2}(t-20)^2 = \rho_{20}\left[1+\alpha(t-20)+\tfrac{1}{2}\alpha^2(t-20)^2\right]$$

(b)

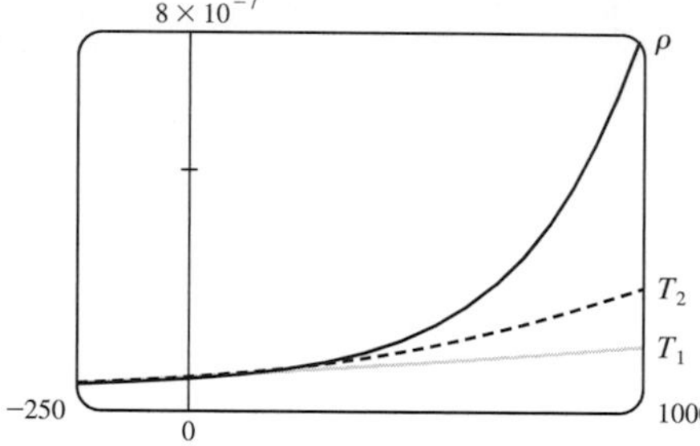

(c)

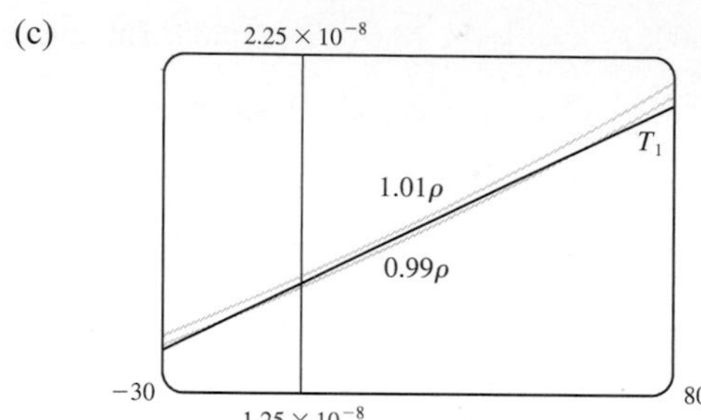

From the graph, it seems that $T_1(t)$ is within 1% of $\rho(t)$, that is, $0.99\rho(t) \le T_1(t) \le 1.01\rho(t)$, for $-14^\circ\text{C} \le t \le 58^\circ\text{C}$.

33. $E = \dfrac{q}{D^2} - \dfrac{q}{(D+d)^2} = \dfrac{q}{D^2} - \dfrac{q}{D^2(1+d/D)^2} = \dfrac{q}{D^2}\left[1-\left(1+\dfrac{d}{D}\right)^{-2}\right].$

We use the Binomial Series to expand $(1+d/D)^{-2}$:

$$E = \frac{q}{D^2}\left[1-\left(1-2\left(\frac{d}{D}\right)+\frac{2\cdot 3}{2!}\left(\frac{d}{D}\right)^2-\frac{2\cdot 3\cdot 4}{3!}\left(\frac{d}{D}\right)^3+\cdots\right)\right] = \frac{q}{D^2}\left[2\left(\frac{d}{D}\right)-3\left(\frac{d}{D}\right)^2+4\left(\frac{d}{D}\right)^3-\cdots\right]$$

$$\approx \frac{q}{D^2}\cdot 2\left(\frac{d}{D}\right) = 2qd\cdot\frac{1}{D^3}$$

when D is much larger than d; that is, when P is far away from the dipole.

34. (a) $\dfrac{n_1}{\ell_o} + \dfrac{n_2}{\ell_i} = \dfrac{1}{R}\left(\dfrac{n_2 s_i}{\ell_i} - \dfrac{n_1 s_o}{\ell_o}\right)$ [Equation 1] where

$$\ell_o = \sqrt{R^2+(s_o+R)^2-2R(s_o+R)\cos\phi} \quad\text{and}\quad \ell_i = \sqrt{R^2+(s_i-R)^2+2R(s_i-R)\cos\phi} \quad \textbf{(2)}$$

Using $\cos\phi \approx 1$ gives

$$\ell_o = \sqrt{R^2+(s_o+R)^2-2R(s_o+R)} = \sqrt{R^2+s_o^2+2Rs_o+R^2-2Rs_o-2R^2} = \sqrt{s_o^2} = s_o$$

and similarly, $\ell_i = s_i$. Thus, Equation 1 becomes $\dfrac{n_1}{s_o}+\dfrac{n_2}{s_i} = \dfrac{1}{R}\left(\dfrac{n_2 s_i}{s_i}-\dfrac{n_1 s_o}{s_o}\right) \Rightarrow \dfrac{n_1}{s_o}+\dfrac{n_2}{s_i} = \dfrac{n_2-n_1}{R}$.

(b) Using $\cos\phi \approx 1-\frac{1}{2}\phi^2$ in **(2)** gives us

$$\ell_o = \sqrt{R^2+(s_o+R)^2-2R(s_o+R)\left(1-\tfrac{1}{2}\phi^2\right)}$$

$$= \sqrt{R^2+s_o^2+2Rs_o+R^2-2Rs_o+Rs_o\phi^2-2R^2+R^2\phi^2} = \sqrt{s_o^2+Rs_o\phi^2+R^2\phi^2}$$

Anticipating that we will use the binomial series expansion $(1+x)^k \approx 1+kx$, we can write the last expression for ℓ_o as

$s_o\sqrt{1+\phi^2\left(\dfrac{R}{s_o}+\dfrac{R^2}{s_o^2}\right)}$ and similarly, $\ell_i = s_i\sqrt{1-\phi^2\left(\dfrac{R}{s_i}-\dfrac{R^2}{s_i^2}\right)}$. Thus, from Equation 1,

$$\frac{n_1}{\ell_o}+\frac{n_2}{\ell_i}=\frac{1}{R}\left(\frac{n_2 s_i}{\ell_i}-\frac{n_1 s_o}{\ell_o}\right) \Leftrightarrow n_1\ell_o^{-1}+n_2\ell_i^{-1}=\frac{n_2}{R}\cdot\frac{s_i}{\ell_i}-\frac{n_1}{R}\cdot\frac{s_o}{\ell_o} \Leftrightarrow$$

$$\frac{n_1}{s_o}\left[1+\phi^2\left(\frac{R}{s_o}+\frac{R^2}{s_o^2}\right)\right]^{-1/2}+\frac{n_2}{s_i}\left[1-\phi^2\left(\frac{R}{s_i}-\frac{R^2}{s_i^2}\right)\right]^{-1/2}$$
$$=\frac{n_2}{R}\left[1-\phi^2\left(\frac{R}{s_i}-\frac{R^2}{s_i^2}\right)\right]^{-1/2}-\frac{n_1}{R}\left[1+\phi^2\left(\frac{R}{s_o}+\frac{R^2}{s_o^2}\right)\right]^{-1/2}$$

Approximating the expressions for ℓ_o^{-1} and ℓ_i^{-1} by the first two terms in their binomial series, we get

$$\frac{n_1}{s_o}\left[1-\tfrac{1}{2}\phi^2\left(\frac{R}{s_o}+\frac{R^2}{s_o^2}\right)\right]+\frac{n_2}{s_i}\left[1+\tfrac{1}{2}\phi^2\left(\frac{R}{s_i}-\frac{R^2}{s_i^2}\right)\right]$$
$$=\frac{n_2}{R}\left[1+\tfrac{1}{2}\phi^2\left(\frac{R}{s_i}-\frac{R^2}{s_i^2}\right)\right]-\frac{n_1}{R}\left[1-\tfrac{1}{2}\phi^2\left(\frac{R}{s_o}+\frac{R^2}{s_o^2}\right)\right] \Leftrightarrow$$

$$\frac{n_1}{s_o}-\frac{n_1\phi^2}{2s_o}\left(\frac{R}{s_o}+\frac{R^2}{s_o^2}\right)+\frac{n_2}{s_i}+\frac{n_2\phi^2}{2s_i}\left(\frac{R}{s_i}-\frac{R^2}{s_i^2}\right)=\frac{n_2}{R}+\frac{n_2\phi^2}{2R}\left(\frac{R}{s_i}-\frac{R^2}{s_i^2}\right)-\frac{n_1}{R}+\frac{n_1\phi^2}{2R}\left(\frac{R}{s_o}+\frac{R^2}{s_o^2}\right) \Leftrightarrow$$

$$\frac{n_1}{s_o}+\frac{n_2}{s_i}=\frac{n_2}{R}-\frac{n_1}{R}+\frac{n_1\phi^2}{2s_o}\left(\frac{R}{s_o}+\frac{R^2}{s_o^2}\right)+\frac{n_1\phi^2}{2R}\left(\frac{R}{s_o}+\frac{R^2}{s_o^2}\right)+\frac{n_2\phi^2}{2R}\left(\frac{R}{s_i}-\frac{R^2}{s_i^2}\right)-\frac{n_2\phi^2}{2s_i}\left(\frac{R}{s_i}-\frac{R^2}{s_i^2}\right)$$
$$=\frac{n_2-n_1}{R}+\frac{n_1\phi^2}{2}\left(\frac{R}{s_o}+\frac{R^2}{s_o^2}\right)\left(\frac{1}{s_o}+\frac{1}{R}\right)+\frac{n_2\phi^2}{2}\left(\frac{R}{s_i}-\frac{R^2}{s_i^2}\right)\left(\frac{1}{R}-\frac{1}{s_i}\right)$$
$$=\frac{n_2-n_1}{R}+\frac{n_1\phi^2R^2}{2s_o}\left(\frac{1}{R}+\frac{1}{s_o}\right)\left(\frac{1}{R}+\frac{1}{s_o}\right)+\frac{n_2\phi^2R^2}{2s_i}\left(\frac{1}{R}-\frac{1}{s_i}\right)\left(\frac{1}{R}-\frac{1}{s_i}\right)$$
$$=\frac{n_2-n_1}{R}+\phi^2R^2\left[\frac{n_1}{2s_o}\left(\frac{1}{R}+\frac{1}{s_o}\right)^2+\frac{n_2}{2s_i}\left(\frac{1}{R}-\frac{1}{s_i}\right)^2\right]$$

From Figure 8, we see that $\sin\phi = h/R$. So if we approximate $\sin\phi$ with ϕ, we get $h = R\phi$ and $h^2 = \phi^2R^2$ and hence, Equation 4, as desired.

35. (a) If the water is deep, then $2\pi d/L$ is large, and we know that $\tanh x \to 1$ as $x \to \infty$. So we can approximate $\tanh(2\pi d/L) \approx 1$, and so $v^2 \approx gL/(2\pi) \Leftrightarrow v \approx \sqrt{gL/(2\pi)}$.

(b) From the table, the first term in the Maclaurin series of $\tanh x$ is x, so if the water is shallow, we can approximate $\tanh\dfrac{2\pi d}{L} \approx \dfrac{2\pi d}{L}$, and so $v^2 \approx \dfrac{gL}{2\pi}\cdot\dfrac{2\pi d}{L} \Leftrightarrow v \approx \sqrt{gd}$.

n	$f^{(n)}(x)$	$f^{(n)}(0)$
0	$\tanh x$	0
1	$\operatorname{sech}^2 x$	1
2	$-2\operatorname{sech}^2 x\tanh x$	0
3	$2\operatorname{sech}^2 x\,(3\tanh^2 x - 1)$	-2

(c) Since $\tanh x$ is an odd function, its Maclaurin series is alternating, so the error in the approximation

$\tanh \dfrac{2\pi d}{L} \approx \dfrac{2\pi d}{L}$ is less than the first neglected term, which is $\dfrac{|f'''(0)|}{3!}\left(\dfrac{2\pi d}{L}\right)^3 = \dfrac{1}{3}\left(\dfrac{2\pi d}{L}\right)^3$.

If $L > 10d$, then $\dfrac{1}{3}\left(\dfrac{2\pi d}{L}\right)^3 < \dfrac{1}{3}\left(2\pi \cdot \dfrac{1}{10}\right)^3 = \dfrac{\pi^3}{375}$, so the error in the approximation $v^2 = gd$ is less

than $\dfrac{gL}{2\pi} \cdot \dfrac{\pi^3}{375} \approx 0.0132gL$.

36. First note that

$$\begin{aligned} 2\left(\sqrt{d^2+R^2}-d\right) &= 2\left[\sqrt{d^2}\sqrt{1+\frac{R^2}{d^2}}-d\right] \\ &\approx 2\left[d\left(1+\frac{R^2}{d^2}\cdot\frac{1}{2}+\cdots\right)-d\right] \quad \left[\text{use the binomial series } 1+\tfrac{1}{2}x+\cdots \text{ for } \sqrt{1+x}\right] \\ &= 2\left[\left(d+\frac{R^2}{2d}+\cdots\right)-d\right] \approx \frac{R^2}{d} \end{aligned}$$

since for large d the other terms are comparatively small. Now $V = 2\pi k_e \sigma\left(\sqrt{d^2+R^2}-d\right) \approx \dfrac{\pi k_e R^2 \sigma}{d}$ by the preceding approximation.

37. (a) L is the length of the arc subtended by the angle θ, so $L = R\theta \;\Rightarrow$

$\theta = L/R$. Now $\sec\theta = (R+C)/R \;\Rightarrow\; R\sec\theta = R + C \;\Rightarrow$

$C = R\sec\theta - R = R\sec(L/R) - R$.

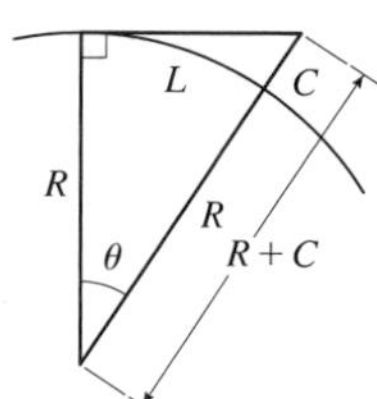

(b) First we'll find a Taylor polynomial $T_4(x)$ for $f(x) = \sec x$ at $x = 0$.

n	$f^{(n)}(x)$	$f^{(n)}(0)$
0	$\sec x$	1
1	$\sec x \tan x$	0
2	$\sec x(2\tan^2 x + 1)$	1
3	$\sec x \tan x(6\tan^2 x + 5)$	0
4	$\sec x(24\tan^4 x + 28\tan^2 x + 5)$	5

Thus, $f(x) = \sec x \approx T_4(x) = 1 + \frac{1}{2!}(x-0)^2 + \frac{5}{4!}(x-0)^4 = 1 + \frac{1}{2}x^2 + \frac{5}{24}x^4$. By part (a),

$$C \approx R\left[1+\frac{1}{2}\left(\frac{L}{R}\right)^2+\frac{5}{24}\left(\frac{L}{R}\right)^4\right]-R = R+\frac{1}{2}R\cdot\frac{L^2}{R^2}+\frac{5}{24}R\cdot\frac{L^4}{R^4}-R = \frac{L^2}{2R}+\frac{5L^4}{24R^3}.$$

(c) Taking $L = 100$ km and $R = 6370$ km, the formula in part (a) says that

$C = R\sec(L/R) - R = 6370\sec(100/6370) - 6370 \approx 0.785\,009\,965\,44$ km.

The formula in part (b) says that $C \approx \dfrac{L^2}{2R} + \dfrac{5L^4}{24R^3} = \dfrac{100^2}{2 \cdot 6370} + \dfrac{5 \cdot 100^4}{24 \cdot 6370^3} \approx 0.785\,009\,957\,36$ km.

The difference between these two results is only $0.000\,000\,008\,08$ km, or $0.000\,008\,08$ m!

38. (a) $4\sqrt{\dfrac{L}{g}}\displaystyle\int_0^{\pi/2}\frac{dx}{\sqrt{1-k^2\sin^2 x}} = 4\sqrt{\frac{L}{g}}\int_0^{\pi/2}\left[1+(-k^2\sin^2 x)\right]^{-1/2}dx$

$$= 4\sqrt{\frac{L}{g}}\int_0^{\pi/2}\left[1-\frac{1}{2}(-k^2\sin^2 x)+\frac{\frac{1}{2}\cdot\frac{3}{2}}{2!}(-k^2\sin^2 x)^2-\frac{\frac{1}{2}\cdot\frac{3}{2}\cdot\frac{5}{2}}{3!}(-k^2\sin^2 x)^3+\cdots\right]dx$$

$$= 4\sqrt{\frac{L}{g}}\int_0^{\pi/2}\left[1+\left(\frac{1}{2}\right)k^2\sin^2 x+\left(\frac{1\cdot 3}{2\cdot 4}\right)k^4\sin^4 x+\left(\frac{1\cdot 3\cdot 5}{2\cdot 4\cdot 6}\right)k^6\sin^6 x+\cdots\right]dx$$

$$= 4\sqrt{\frac{L}{g}}\left[\frac{\pi}{2}+\left(\frac{1}{2}\right)\left(\frac{1}{2}\cdot\frac{\pi}{2}\right)k^2+\left(\frac{1\cdot 3}{2\cdot 4}\right)\left(\frac{1\cdot 3}{2\cdot 4}\cdot\frac{\pi}{2}\right)k^4+\left(\frac{1\cdot 3\cdot 5}{2\cdot 4\cdot 6}\right)\left(\frac{1\cdot 3\cdot 5}{2\cdot 4\cdot 6}\cdot\frac{\pi}{2}\right)k^6+\cdots\right]$$

[split up the integral and use the result from Exercise 7.1.50]

$$= 2\pi\sqrt{\frac{L}{g}}\left[1+\frac{1^2}{2^2}k^2+\frac{1^2\cdot 3^2}{2^2\cdot 4^2}k^4+\frac{1^2\cdot 3^2\cdot 5^2}{2^2\cdot 4^2\cdot 6^2}k^6+\cdots\right]$$

(b) The first of the two inequalities is true because all of the terms in the series are positive. For the second,

$$T = 2\pi\sqrt{\frac{L}{g}}\left[1+\frac{1^2}{2^2}k^2+\frac{1^2\cdot 3^2}{2^2\cdot 4^2}k^4+\frac{1^2\cdot 3^2\cdot 5^2}{2^2\cdot 4^2\cdot 6^2}k^6+\frac{1^2\cdot 3^2\cdot 5^2\cdot 7^2}{2^2\cdot 4^2\cdot 6^2\cdot 8^2}k^8+\cdots\right]$$

$$\le 2\pi\sqrt{\frac{L}{g}}\left[1+\tfrac{1}{4}k^2+\tfrac{1}{4}k^4+\tfrac{1}{4}k^6+\tfrac{1}{4}k^8+\cdots\right]$$

The terms in brackets (after the first) form a geometric series with $a=\frac{1}{4}k^2$ and $r=k^2=\sin^2\left(\frac{1}{2}\theta_0\right)<1$.

So $T\le 2\pi\sqrt{\dfrac{L}{g}}\left[1+\dfrac{k^2/4}{1-k^2}\right]=2\pi\sqrt{\dfrac{L}{g}}\dfrac{4-3k^2}{4-4k^2}$.

(c) We substitute $L=1$, $g=9.8$, and $k=\sin(10°/2)\approx 0.08716$, and the inequality from part (b) becomes

$2.01090\le T\le 2.01093$, so $T\approx 2.0109$. The estimate $T\approx 2\pi\sqrt{L/g}\approx 2.0071$ differs by about 0.2%.

If $\theta_0=42°$, then $k\approx 0.35837$ and the inequality becomes $2.07153\le T\le 2.08103$, so $T\approx 2.0763$.

The one-term estimate is the same, and the discrepancy between the two estimates increases to about 3.4%.

39. Using $f(x)=T_n(x)+R_n(x)$ with $n=1$ and $x=r$, we have $f(r)=T_1(r)+R_1(r)$, where T_1 is the first-degree Taylor polynomial of f at a. Because $a=x_n$, $f(r)=f(x_n)+f'(x_n)(r-x_n)+R_1(r)$. But r is a root of f, so $f(r)=0$ and we have $0=f(x_n)+f'(x_n)(r-x_n)+R_1(r)$. Taking the first two terms to the left side gives us $f'(x_n)(x_n-r)-f(x_n)=R_1(r)$. Dividing by $f'(x_n)$, we get $x_n-r-\dfrac{f(x_n)}{f'(x_n)}=\dfrac{R_1(r)}{f'(x_n)}$. By the formula for Newton's method, the left side of the preceding equation is $x_{n+1}-r$, so $|x_{n+1}-r|=\left|\dfrac{R_1(r)}{f'(x_n)}\right|$. Taylor's Inequality gives us $|R_1(r)|\le\dfrac{|f''(r)|}{2!}|r-x_n|^2$. Combining this inequality with the facts $|f''(x)|\le M$ and $|f'(x)|\ge K$ gives us $|x_{n+1}-r|\le\dfrac{M}{2K}|x_n-r|^2$.

APPLIED PROJECT Radiation from the Stars

1. If we write $f(\lambda) = \dfrac{8\pi hc\lambda^{-5}}{e^{hc/(\lambda kT)} - 1} = \dfrac{a\lambda^{-5}}{e^{b/(\lambda T)} - 1}$, then as $\lambda \to 0^+$, it is of the form ∞/∞, and as $\lambda \to \infty$ it is of the form $0/0$, so in either case we can use l'Hospital's Rule. First of all,

$$\lim_{\lambda\to\infty} f(\lambda) \overset{\text{H}}{=} \lim_{\lambda\to\infty} \frac{a\left(-5\lambda^{-6}\right)}{-\dfrac{bT}{(\lambda T)^2} e^{b/(\lambda T)}} = 5\,\frac{aT}{b} \lim_{\lambda\to\infty} \frac{\lambda^2 \lambda^{-6}}{e^{b/(\lambda T)}} = 5\,\frac{aT}{b} \lim_{\lambda\to\infty} \frac{\lambda^{-4}}{e^{b/(\lambda T)}} = 0$$

Also,

$$\lim_{\lambda\to 0+} f(\lambda) \overset{\text{H}}{=} 5\,\frac{aT}{b} \lim_{\lambda\to 0+} \frac{\lambda^{-4}}{e^{b/(\lambda T)}} \overset{\text{H}}{=} 5\,\frac{aT}{b} \lim_{\lambda\to 0+} \frac{-4\lambda^{-5}}{-\dfrac{bT}{(\lambda T)^2} e^{b/(\lambda T)}} = 20\,\frac{aT^2}{b^2} \lim_{\lambda\to 0+} \frac{\lambda^{-3}}{e^{b/(\lambda T)}}$$

This is still indeterminate, but note that each time we use l'Hospital's Rule, we gain a factor of λ in the numerator, as well as a constant factor, and the denominator is unchanged. So if we use l'Hospital's Rule three more times, the exponent of λ in the numerator will become 0. That is, for some $\{k_i\}$, all constant,

$$\lim_{\lambda\to 0+} f(\lambda) \overset{\text{H}}{=} k_1 \lim_{\lambda\to 0+} \frac{\lambda^{-3}}{e^{b/(\lambda T)}} \overset{\text{H}}{=} k_2 \lim_{\lambda\to 0+} \frac{\lambda^{-2}}{e^{b/(\lambda T)}} \overset{\text{H}}{=} k_3 \lim_{\lambda\to 0+} \frac{\lambda^{-1}}{e^{b/(\lambda T)}} \overset{\text{H}}{=} k_4 \lim_{\lambda\to 0+} \frac{1}{e^{b/(\lambda T)}} = 0$$

2. We expand the denominator of Planck's Law using the Taylor series $e^x = 1 + x + \dfrac{x^2}{2!} + \dfrac{x^3}{3!} + \cdots$ with $x = \dfrac{hc}{\lambda kT}$, and use the fact that if λ is large, then all subsequent terms in the Taylor expansion are very small compared to the first one, so we can approximate using the Taylor polynomial T_1:

$$f(\lambda) = \frac{8\pi hc\lambda^{-5}}{e^{hc/(\lambda kT)} - 1} = \frac{8\pi hc\lambda^{-5}}{\left[1 + \dfrac{hc}{\lambda kT} + \dfrac{1}{2!}\left(\dfrac{hc}{\lambda kT}\right)^2 + \dfrac{1}{3!}\left(\dfrac{hc}{\lambda kT}\right)^3 + \cdots\right] - 1} \approx \frac{8\pi hc\lambda^{-5}}{\left(1 + \dfrac{hc}{\lambda kT}\right) - 1} = \frac{8\pi kT}{\lambda^4}$$

which is the Rayleigh-Jeans Law.

3. To convert to μm, we substitute $\lambda/10^6$ for λ in both laws. The first figure shows that the two laws are similar for large λ. The second figure shows that the two laws are very different for short wavelengths (Planck's Law gives a maximum at $\lambda \approx 0.51\ \mu$m; the Rayleigh-Jeans Law gives no minimum or maximum.).

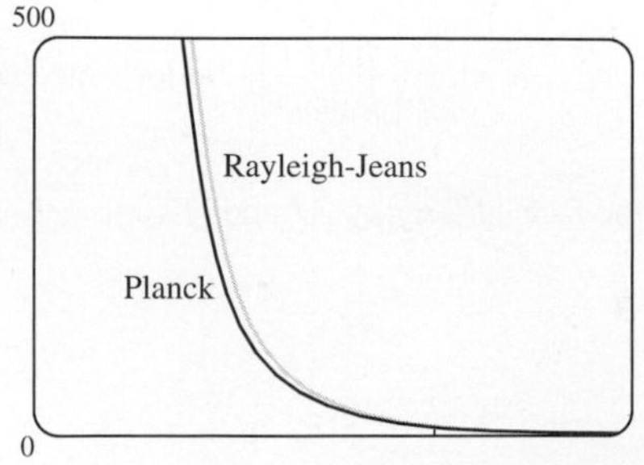

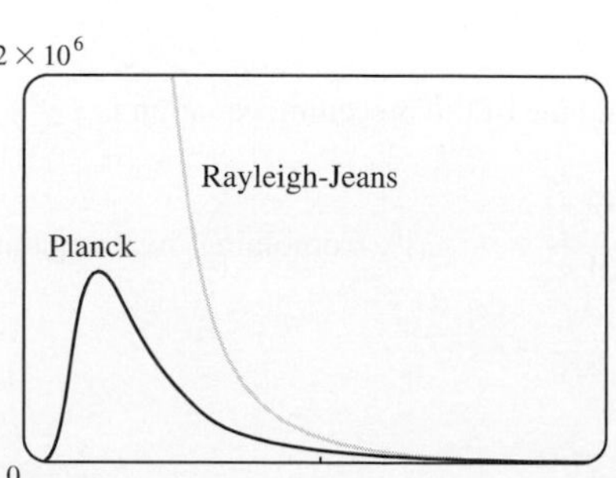

4. From the graph in Problem 3, $f(\lambda)$ has a maximum under Planck's Law at $\lambda \approx 0.51\ \mu$m.

5. 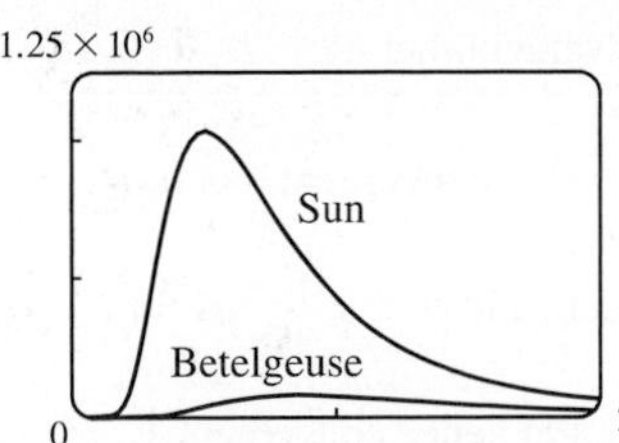

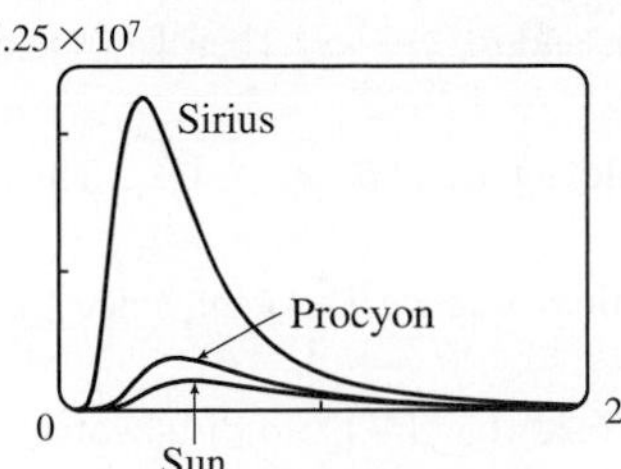

As T gets larger, the total area under the curve increases, as we would expect: the hotter the star, the more energy it emits. Also, as T increases, the λ-value of the maximum decreases, so the higher the temperature, the shorter the peak wavelength (and consequently the average wavelength) of light emitted. This is why Sirius is a blue star and Betelgeuse is a red star: most of Sirius's light is of a fairly short wavelength; that is, a higher frequency, toward the blue end of the spectrum, whereas most of Betelgeuse's light is of a lower frequency, toward the red end of the spectrum.

11 Review

TRUE-FALSE QUIZ

1. False. See Note 2 after Theorem 11.2.6.

2. False. The series $\sum_{n=1}^{\infty} n^{-\sin 1} = \sum_{n=1}^{\infty} \frac{1}{n^{\sin 1}}$ is a p-series with $p = \sin 1 \approx 0.84 \leq 1$, so the series diverges.

3. True. If $\lim_{n\to\infty} a_n = L$, then as $n \to \infty$, $2n+1 \to \infty$, so $a_{2n+1} \to L$.

4. True by Theorem 11.8.4.

Or: Use the Comparison Test to show that $\sum c_n(-2)^n$ converges absolutely.

5. False. For example, take $c_n = (-1)^n/(n6^n)$.

6. True by Theorem 11.8.4.

7. False, since $\lim_{n\to\infty} \left|\frac{a_{n+1}}{a_n}\right| = \lim_{n\to\infty} \left|\frac{1}{(n+1)^3} \cdot \frac{n^3}{1}\right| = \lim_{n\to\infty} \left|\frac{n^3}{(n+1)^3} \cdot \frac{1/n^3}{1/n^3}\right| = \lim_{n\to\infty} \frac{1}{(1+1/n)^3} = 1.$

8. True, since $\lim_{n\to\infty} \left|\frac{a_{n+1}}{a_n}\right| = \lim_{n\to\infty} \left|\frac{1}{(n+1)!} \cdot \frac{n!}{1}\right| = \lim_{n\to\infty} \frac{1}{n+1} = 0 < 1.$

9. False. See the note after Example 11.4.2.

10. True, since $\frac{1}{e} = e^{-1}$ and $e^x = \sum_{n=0}^{\infty} \frac{x^n}{n!}$, so $e^{-1} = \sum_{n=0}^{\infty} \frac{(-1)^n}{n!}$.

11. True. See (9) in Section 11.1.

12. True, because if $\sum |a_n|$ is convergent, then so is $\sum a_n$ by Theorem 11.6.3.

13. True. By Theorem 11.10.5 the coefficient of x^3 is $\frac{f'''(0)}{3!} = \frac{1}{3} \Rightarrow f'''(0) = 2$.

Or: Use Theorem 11.9.2 to differentiate f three times.

14. False. Let $a_n = n$ and $b_n = -n$. Then $\{a_n\}$ and $\{b_n\}$ are divergent, but $a_n + b_n = 0$, so $\{a_n + b_n\}$ is convergent.

15. False. For example, let $a_n = b_n = (-1)^n$. Then $\{a_n\}$ and $\{b_n\}$ are divergent, but $a_n b_n = 1$, so $\{a_n b_n\}$ is convergent.

16. True by the Monotonic Sequence Theorem, since $\{a_n\}$ is decreasing and $0 < a_n \leq a_1$ for all $n \Rightarrow \{a_n\}$ is bounded.

17. True by Theorem 11.6.3. $[\sum (-1)^n a_n$ is absolutely convergent and hence convergent.$]$

18. True. $\lim\limits_{n\to\infty} \dfrac{a_{n+1}}{a_n} < 1 \Rightarrow \sum a_n$ converges (Ratio Test) $\Rightarrow \lim\limits_{n\to\infty} a_n = 0$ [Theorem 11.2.6].

19. True. $0.99999\ldots = 0.9 + 0.9(0.1)^1 + 0.9(0.1)^2 + 0.9(0.1)^3 + \cdots = \sum\limits_{n=1}^{\infty} (0.9)(0.1)^{n-1} = \dfrac{0.9}{1 - 0.1} = 1$ by the formula for the sum of a geometric series $[S = a_1/(1 - r)]$ with ratio r satisfying $|r| < 1$.

20. True. Since $\lim\limits_{n\to\infty} a_n = 2$, we know that $\lim\limits_{n\to\infty} a_{n+3} = 2$. Thus, $\lim\limits_{n\to\infty} (a_{n+3} - a_n) = \lim\limits_{n\to\infty} a_{n+3} - \lim\limits_{n\to\infty} a_n = 2 - 2 = 0$.

21. True. A finite number of terms doesn't affect convergence or divergence of a series.

22. False. Let $a_n = (0.1)^n$ and $b_n = (0.2)^n$. Then $\sum\limits_{n=1}^{\infty} a_n = \sum\limits_{n=1}^{\infty} (0.1)^n = \dfrac{0.1}{1 - 0.1} = \dfrac{1}{9} = A$,

$\sum\limits_{n=1}^{\infty} b_n = \sum\limits_{n=1}^{\infty} (0.2)^n = \dfrac{0.2}{1 - 0.2} = \dfrac{1}{4} = B$, and $\sum\limits_{n=1}^{\infty} a_n b_n = \sum\limits_{n=1}^{\infty} (0.02)^n = \dfrac{0.02}{1 - 0.02} = \dfrac{1}{49}$, but

$AB = \frac{1}{9} \cdot \frac{1}{4} = \frac{1}{36}$.

EXERCISES

1. $\left\{\dfrac{2 + n^3}{1 + 2n^3}\right\}$ converges since $\lim\limits_{n\to\infty} \dfrac{2 + n^3}{1 + 2n^3} = \lim\limits_{n\to\infty} \dfrac{2/n^3 + 1}{1/n^3 + 2} = \dfrac{1}{2}$.

2. $a_n = \dfrac{9^{n+1}}{10^n} = 9 \cdot \left(\frac{9}{10}\right)^n$, so $\lim\limits_{n\to\infty} a_n = 9 \lim\limits_{n\to\infty} \left(\frac{9}{10}\right)^n = 9 \cdot 0 = 0$ by (11.1.9).

3. $\lim\limits_{n\to\infty} a_n = \lim\limits_{n\to\infty} \dfrac{n^3}{1 + n^2} = \lim\limits_{n\to\infty} \dfrac{n}{1/n^2 + 1} = \infty$, so the sequence diverges.

4. $a_n = \cos(n\pi/2)$, so $a_n = 0$ if n is odd and $a_n = \pm 1$ if n is even. As n increases, a_n keeps cycling through the values 0, 1, 0, -1, so the sequence $\{a_n\}$ is divergent.

5. $|a_n| = \left|\dfrac{n \sin n}{n^2 + 1}\right| \leq \dfrac{n}{n^2 + 1} < \dfrac{1}{n}$, so $|a_n| \to 0$ as $n \to \infty$. Thus, $\lim\limits_{n\to\infty} a_n = 0$. The sequence $\{a_n\}$ is convergent.

6. $a_n = \dfrac{\ln n}{\sqrt{n}}$. Let $f(x) = \dfrac{\ln x}{\sqrt{x}}$ for $x > 0$. Then $\lim\limits_{x\to\infty} f(x) = \lim\limits_{x\to\infty} \dfrac{\ln x}{\sqrt{x}} \overset{\text{H}}{=} \lim\limits_{x\to\infty} \dfrac{1/x}{1/(2\sqrt{x})} = \lim\limits_{x\to\infty} \dfrac{2}{\sqrt{x}} = 0$.

Thus, by Theorem 11.1.3, $\{a_n\}$ converges and $\lim\limits_{n\to\infty} a_n = 0$.

7. $\left\{\left(1+\frac{3}{n}\right)^{4n}\right\}$ is convergent. Let $y=\left(1+\frac{3}{x}\right)^{4x}$. Then

$$\lim_{x\to\infty}\ln y=\lim_{x\to\infty}4x\ln(1+3/x)=\lim_{x\to\infty}\frac{\ln(1+3/x)}{1/(4x)}\overset{\text{H}}{=}\lim_{x\to\infty}\frac{\dfrac{1}{1+3/x}\left(-\dfrac{3}{x^2}\right)}{-1/(4x^2)}=\lim_{x\to\infty}\frac{12}{1+3/x}=12,\text{ so}$$

$$\lim_{x\to\infty}y=\lim_{n\to\infty}\left(1+\frac{3}{n}\right)^{4n}=e^{12}.$$

8. $\left\{\dfrac{(-10)^n}{n!}\right\}$ converges, since $\dfrac{10^n}{n!}=\dfrac{10\cdot10\cdot10\cdot\cdots\cdot10}{1\cdot2\cdot3\cdot\cdots\cdot10}\cdot\dfrac{10\cdot10\cdot\cdots\cdot10}{11\cdot12\cdot\cdots\cdot n}\le10^{10}\left(\dfrac{10}{11}\right)^{n-10}\to0$ as $n\to\infty$, so

$\lim\limits_{n\to\infty}\dfrac{(-10)^n}{n!}=0$ [Squeeze Theorem]. *Or:* Use (11.10.10).

9. We use induction, hypothesizing that $a_{n-1}<a_n<2$. Note first that $1<a_2=\frac{1}{3}(1+4)=\frac{5}{3}<2$, so the hypothesis holds for $n=2$. Now assume that $a_{k-1}<a_k<2$. Then $a_k=\frac{1}{3}(a_{k-1}+4)<\frac{1}{3}(a_k+4)<\frac{1}{3}(2+4)=2$. So $a_k<a_{k+1}<2$, and the induction is complete. To find the limit of the sequence, we note that $L=\lim\limits_{n\to\infty}a_n=\lim\limits_{n\to\infty}a_{n+1}\Rightarrow$

$L=\frac{1}{3}(L+4)\quad\Rightarrow\quad L=2.$

10. $\lim\limits_{x\to\infty}\dfrac{x^4}{e^x}\overset{\text{H}}{=}\lim\limits_{x\to\infty}\dfrac{4x^3}{e^x}\overset{\text{H}}{=}\lim\limits_{x\to\infty}\dfrac{12x^2}{e^x}\overset{\text{H}}{=}\lim\limits_{x\to\infty}\dfrac{24x}{e^x}\overset{\text{H}}{=}\lim\limits_{x\to\infty}\dfrac{24}{e^x}=0$

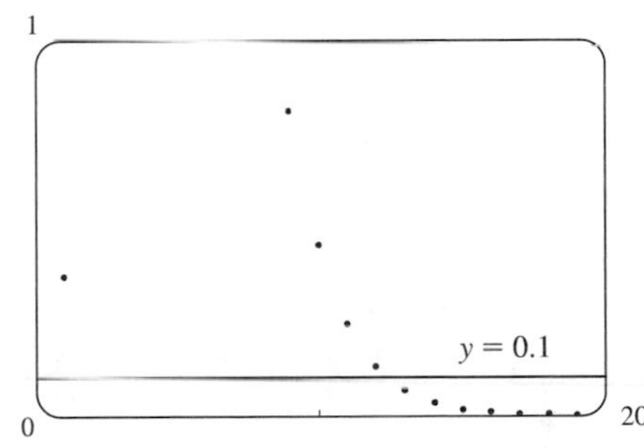

Then we conclude from Theorem 11.1.3 that $\lim\limits_{n\to\infty}n^4e^{-n}=0$.

From the graph, it seems that $12^4e^{-12}>0.1$, but $n^4e^{-n}<0.1$ whenever $n>12$. So the smallest value of N corresponding to $\varepsilon=0.1$ in the definition of the limit is $N=12$.

11. $\dfrac{n}{n^3+1}<\dfrac{n}{n^3}=\dfrac{1}{n^2}$, so $\sum\limits_{n=1}^{\infty}\dfrac{n}{n^3+1}$ converges by the Comparison Test with the convergent p-series $\sum\limits_{n=1}^{\infty}\dfrac{1}{n^2}$ $[p=2>1]$.

12. Let $a_n=\dfrac{n^2+1}{n^3+1}$ and $b_n=\dfrac{1}{n}$, so $\lim\limits_{n\to\infty}\dfrac{a_n}{b_n}=\lim\limits_{n\to\infty}\dfrac{n^3+n}{n^3+1}=\lim\limits_{n\to\infty}\dfrac{1+1/n^2}{1+1/n^3}=1>0$.

Since $\sum\limits_{n=1}^{\infty}b_n$ is the divergent harmonic series, $\sum\limits_{n=1}^{\infty}a_n$ also diverges by the Limit Comparison Test.

13. $\lim\limits_{n\to\infty}\left|\dfrac{a_{n+1}}{a_n}\right|=\lim\limits_{n\to\infty}\left[\dfrac{(n+1)^3}{5^{n+1}}\cdot\dfrac{5^n}{n^3}\right]=\lim\limits_{n\to\infty}\left(1+\dfrac{1}{n}\right)^3\cdot\dfrac{1}{5}=\dfrac{1}{5}<1$, so $\sum\limits_{n=1}^{\infty}\dfrac{n^3}{5^n}$ converges by the Ratio Test.

14. Let $b_n=\dfrac{1}{\sqrt{n+1}}$. Then b_n is positive for $n\ge1$, the sequence $\{b_n\}$ is decreasing, and $\lim\limits_{n\to\infty}b_n=0$, so the series $\sum\limits_{n=1}^{\infty}\dfrac{(-1)^n}{\sqrt{n+1}}$ converges by the Alternating Series Test.

15. Let $f(x) = \dfrac{1}{x\sqrt{\ln x}}$. Then f is continuous, positive, and decreasing on $[2, \infty)$, so the Integral Test applies.

$$\int_2^\infty f(x)\,dx = \lim_{t\to\infty}\int_2^t \frac{1}{x\sqrt{\ln x}}\,dx \quad \left[u = \ln x,\ du = \frac{1}{x}\,dx\right] = \lim_{t\to\infty}\int_{\ln 2}^{\ln t} u^{-1/2}\,du = \lim_{t\to\infty}\Big[2\sqrt{u}\Big]_{\ln 2}^{\ln t}$$
$$= \lim_{t\to\infty}\left(2\sqrt{\ln t} - 2\sqrt{\ln 2}\right) = \infty,$$

so the series $\sum\limits_{n=2}^{\infty} \dfrac{1}{n\sqrt{\ln n}}$ diverges.

16. $\lim\limits_{n\to\infty} \dfrac{n}{3n+1} = \dfrac{1}{3}$, so $\lim\limits_{n\to\infty} \ln\left(\dfrac{n}{3n+1}\right) = \ln\frac{1}{3} \neq 0$. Thus, the series $\sum\limits_{n=1}^{\infty} \ln\left(\dfrac{n}{3n+1}\right)$ diverges by the Test for Divergence.

17. $|a_n| = \left|\dfrac{\cos 3n}{1+(1.2)^n}\right| \le \dfrac{1}{1+(1.2)^n} < \dfrac{1}{(1.2)^n} = \left(\dfrac{5}{6}\right)^n$, so $\sum\limits_{n=1}^{\infty} |a_n|$ converges by comparison with the convergent geometric series $\sum\limits_{n=1}^{\infty} \left(\frac{5}{6}\right)^n$ $\left[r = \frac{5}{6} < 1\right]$. It follows that $\sum\limits_{n=1}^{\infty} a_n$ converges (by Theorem 11.6.3).

18. $\lim\limits_{n\to\infty} \sqrt[n]{|a_n|} = \lim\limits_{n\to\infty} \sqrt[n]{\left|\dfrac{n^{2n}}{(1+2n^2)^n}\right|} = \lim\limits_{n\to\infty} \dfrac{n^2}{1+2n^2} = \lim\limits_{n\to\infty} \dfrac{1}{1/n^2+2} = \dfrac{1}{2} < 1$, so $\sum\limits_{n=1}^{\infty} \dfrac{n^{2n}}{(1+2n^2)^n}$ converges by the Root Test.

19. $\lim\limits_{n\to\infty} \left|\dfrac{a_{n+1}}{a_n}\right| = \lim\limits_{n\to\infty} \dfrac{1\cdot 3\cdot 5\cdot\cdots\cdot(2n-1)(2n+1)}{5^{n+1}(n+1)!} \cdot \dfrac{5^n\, n!}{1\cdot 3\cdot 5\cdot\cdots\cdot(2n-1)} = \lim\limits_{n\to\infty} \dfrac{2n+1}{5(n+1)} = \dfrac{2}{5} < 1$, so the series converges by the Ratio Test.

20. $\sum\limits_{n=1}^{\infty} \dfrac{(-5)^{2n}}{n^2\, 9^n} = \sum\limits_{n=1}^{\infty} \dfrac{1}{n^2}\left(\dfrac{25}{9}\right)^n$. Now $\lim\limits_{n\to\infty} \left|\dfrac{a_{n+1}}{a_n}\right| = \lim\limits_{n\to\infty} \dfrac{25^{n+1}}{(n+1)^2\cdot 9^{n+1}} \cdot \dfrac{n^2\cdot 9^n}{25^n} = \lim\limits_{n\to\infty} \dfrac{25n^2}{9(n+1)^2} = \dfrac{25}{9} > 1$, so the series diverges by the Ratio Test.

21. $b_n = \dfrac{\sqrt{n}}{n+1} > 0$, $\{b_n\}$ is decreasing, and $\lim\limits_{n\to\infty} b_n = 0$, so the series $\sum\limits_{n=1}^{\infty} (-1)^{n-1} \dfrac{\sqrt{n}}{n+1}$ converges by the Alternating Series Test.

22. Use the Limit Comparison Test with $a_n = \dfrac{\sqrt{n+1}-\sqrt{n-1}}{n} = \dfrac{2}{n\left(\sqrt{n+1}+\sqrt{n-1}\right)}$ (rationalizing the numerator) and $b_n = \dfrac{1}{n^{3/2}}$. $\lim\limits_{n\to\infty} \dfrac{a_n}{b_n} = \lim\limits_{n\to\infty} \dfrac{2\sqrt{n}}{\sqrt{n+1}+\sqrt{n-1}} = 1$, so since $\sum\limits_{n=1}^{\infty} b_n$ converges $\left[p = \frac{3}{2} > 1\right]$, $\sum\limits_{n=1}^{\infty} a_n$ converges also.

23. Consider the series of absolute values: $\sum\limits_{n=1}^{\infty} n^{-1/3}$ is a p-series with $p = \frac{1}{3} \le 1$ and is therefore divergent. But if we apply the Alternating Series Test, we see that $b_n = \dfrac{1}{\sqrt[3]{n}} > 0$, $\{b_n\}$ is decreasing, and $\lim\limits_{n\to\infty} b_n = 0$, so the series $\sum\limits_{n=1}^{\infty} (-1)^{n-1}\, n^{-1/3}$ converges. Thus, $\sum\limits_{n=1}^{\infty} (-1)^{n-1}\, n^{-1/3}$ is conditionally convergent.

24. $\sum_{n=1}^{\infty} \left|(-1)^{n-1} n^{-3}\right| = \sum_{n=1}^{\infty} n^{-3}$ is a convergent p-series $[p = 3 > 1]$. Therefore, $\sum_{n=1}^{\infty} (-1)^{n-1} n^{-3}$ is absolutely convergent.

25. $\left|\dfrac{a_{n+1}}{a_n}\right| = \left|\dfrac{(-1)^{n+1}(n+2)3^{n+1}}{2^{2n+3}} \cdot \dfrac{2^{2n+1}}{(-1)^n(n+1)3^n}\right| = \dfrac{n+2}{n+1} \cdot \dfrac{3}{4} = \dfrac{1+(2/n)}{1+(1/n)} \cdot \dfrac{3}{4} \to \dfrac{3}{4} < 1$ as $n \to \infty$, so by the Ratio Test, $\sum_{n=1}^{\infty} \dfrac{(-1)^n(n+1)3^n}{2^{2n+1}}$ is absolutely convergent.

26. $\lim_{x\to\infty} \dfrac{\sqrt{x}}{\ln x} \overset{H}{=} \lim_{x\to\infty} \dfrac{1/(2\sqrt{x})}{1/x} = \lim_{x\to\infty} \dfrac{\sqrt{x}}{2} = \infty$. Therefore, $\lim_{n\to\infty} \dfrac{(-1)^n \sqrt{n}}{\ln n} \neq 0$, so the given series is divergent by the Test for Divergence.

27. $\sum_{n=1}^{\infty} \dfrac{(-3)^{n-1}}{2^{3n}} = \sum_{n=1}^{\infty} \dfrac{(-3)^{n-1}}{(2^3)^n} = \sum_{n=1}^{\infty} \dfrac{(-3)^{n-1}}{8^n} = \dfrac{1}{8} \sum_{n=1}^{\infty} \dfrac{(-3)^{n-1}}{8^{n-1}} = \dfrac{1}{8} \sum_{n=1}^{\infty} \left(-\dfrac{3}{8}\right)^{n-1} = \dfrac{1}{8}\left(\dfrac{1}{1-(-3/8)}\right)$

$= \dfrac{1}{8} \cdot \dfrac{8}{11} = \dfrac{1}{11}$

28. $\sum_{n=1}^{\infty} \dfrac{1}{n(n+3)} = \sum_{n=1}^{\infty} \left[\dfrac{1}{3n} - \dfrac{1}{3(n+3)}\right]$ [partial fractions].

$s_n = \sum_{i=1}^{n} \left[\dfrac{1}{3i} - \dfrac{1}{3(i+3)}\right] = \dfrac{1}{3} + \dfrac{1}{6} + \dfrac{1}{9} - \dfrac{1}{3(n+1)} - \dfrac{1}{3(n+2)} - \dfrac{1}{3(n+3)}$ (telescoping sum), so

$\sum_{n=1}^{\infty} \dfrac{1}{n(n+3)} = \lim_{n\to\infty} s_n = \dfrac{1}{3} + \dfrac{1}{6} + \dfrac{1}{9} = \dfrac{11}{18}.$

29. $\sum_{n=1}^{\infty} [\tan^{-1}(n+1) - \tan^{-1} n] = \lim_{n\to\infty} s_n$

$= \lim_{n\to\infty} [(\tan^{-1} 2 - \tan^{-1} 1) + (\tan^{-1} 3 - \tan^{-1} 2) + \cdots + (\tan^{-1}(n+1) - \tan^{-1} n)]$

$= \lim_{n\to\infty} [\tan^{-1}(n+1) - \tan^{-1} 1] = \frac{\pi}{2} - \frac{\pi}{4} = \frac{\pi}{4}$

30. $\sum_{n=0}^{\infty} \dfrac{(-1)^n \pi^n}{3^{2n}(2n)!} = \sum_{n=0}^{\infty} (-1)^n \dfrac{1}{(2n)!} \cdot \dfrac{\pi^n}{3^{2n}} = \sum_{n=0}^{\infty} (-1)^n \dfrac{1}{(2n)!} \cdot \left(\dfrac{\sqrt{\pi}}{3}\right)^{2n} = \cos\left(\dfrac{\sqrt{\pi}}{3}\right)$ since $\cos x = \sum_{n=0}^{\infty} (-1)^n \dfrac{x^{2n}}{(2n)!}$ for all x.

31. $1 - e + \dfrac{e^2}{2!} - \dfrac{e^3}{3!} + \dfrac{e^4}{4!} - \cdots = \sum_{n=0}^{\infty} (-1)^n \dfrac{e^n}{n!} = \sum_{n=0}^{\infty} \dfrac{(-e)^n}{n!} = e^{-e}$ since $e^x = \sum_{n=0}^{\infty} \dfrac{x^n}{n!}$ for all x.

32. $4.17\overline{326} = 4.17 + \dfrac{326}{10^5} + \dfrac{326}{10^8} + \cdots = 4.17 + \dfrac{326/10^5}{1 - 1/10^3} = \dfrac{417}{100} + \dfrac{326}{99{,}900} = \dfrac{416{,}909}{99{,}900}$

33. $\cosh x = \dfrac{1}{2}(e^x + e^{-x}) = \dfrac{1}{2}\left(\sum_{n=0}^{\infty} \dfrac{x^n}{n!} + \sum_{n=0}^{\infty} \dfrac{(-x)^n}{n!}\right)$

$= \dfrac{1}{2}\left[\left(1 + x + \dfrac{x^2}{2!} + \dfrac{x^3}{3!} + \dfrac{x^4}{4!} + \cdots\right) + \left(1 - x + \dfrac{x^2}{2!} - \dfrac{x^3}{3!} + \dfrac{x^4}{4!} - \cdots\right)\right]$

$= \dfrac{1}{2}\left(2 + 2 \cdot \dfrac{x^2}{2!} + 2 \cdot \dfrac{x^4}{4!} + \cdots\right) = 1 + \dfrac{1}{2}x^2 + \sum_{n=2}^{\infty} \dfrac{x^{2n}}{(2n)!} \geq 1 + \dfrac{1}{2}x^2$ for all x

34. $\sum\limits_{n=1}^{\infty} (\ln x)^n$ is a geometric series which converges whenever $|\ln x| < 1 \;\Rightarrow\; -1 < \ln x < 1 \;\Rightarrow\; e^{-1} < x < e$.

35. $\sum\limits_{n=1}^{\infty} \dfrac{(-1)^{n+1}}{n^5} = 1 - \dfrac{1}{32} + \dfrac{1}{243} - \dfrac{1}{1024} + \dfrac{1}{3125} - \dfrac{1}{7776} + \dfrac{1}{16{,}807} - \dfrac{1}{32{,}768} + \cdots$.

Since $b_8 = \dfrac{1}{8^5} = \dfrac{1}{32{,}768} < 0.000031$, $\sum\limits_{n=1}^{\infty} \dfrac{(-1)^{n+1}}{n^5} \approx \sum\limits_{n=1}^{7} \dfrac{(-1)^{n+1}}{n^5} \approx 0.9721$.

36. (a) $s_5 = \sum\limits_{n=1}^{5} \dfrac{1}{n^6} = 1 + \dfrac{1}{2^6} + \cdots + \dfrac{1}{5^6} \approx 1.017305$. The series $\sum\limits_{n=1}^{\infty} \dfrac{1}{n^6}$ converges by the Integral Test, so we estimate the remainder R_5 with (11.3.2): $R_5 \le \displaystyle\int_5^{\infty} \dfrac{dx}{x^6} = \left[-\dfrac{x^{-5}}{5}\right]_5^{\infty} = \dfrac{5^{-5}}{5} = 0.000064$. So the error is at most 0.000064.

(b) In general, $R_n \le \displaystyle\int_n^{\infty} \dfrac{dx}{x^6} = \dfrac{1}{5n^5}$. If we take $n = 9$, then $s_9 \approx 1.01734$ and $R_9 \le \dfrac{1}{5 \cdot 9^5} \approx 3.4 \times 10^{-6}$.

So to five decimal places, $\sum\limits_{n=1}^{\infty} \dfrac{1}{n^5} \approx \sum\limits_{n=1}^{9} \dfrac{1}{n^5} \approx 1.01734$.

Another method: Use (11.3.3) instead of (11.3.2).

37. $\sum\limits_{n=1}^{\infty} \dfrac{1}{2+5^n} \approx \sum\limits_{n=1}^{8} \dfrac{1}{2+5^n} \approx 0.18976224$. To estimate the error, note that $\dfrac{1}{2+5^n} < \dfrac{1}{5^n}$, so the remainder term is

$R_8 = \sum\limits_{n=9}^{\infty} \dfrac{1}{2+5^n} < \sum\limits_{n=9}^{\infty} \dfrac{1}{5^n} = \dfrac{1/5^9}{1-1/5} = 6.4 \times 10^{-7}$ [geometric series with $a = \frac{1}{5^9}$ and $r = \frac{1}{5}$].

38. (a) $\lim\limits_{n\to\infty} \left|\dfrac{a_{n+1}}{a_n}\right| = \lim\limits_{n\to\infty} \left|\dfrac{(n+1)^{n+1}}{[2(n+1)]!} \cdot \dfrac{(2n)!}{n^n}\right| = \lim\limits_{n\to\infty} \dfrac{(n+1)^n(n+1)^1}{(2n+2)(2n+1)n^n} = \lim\limits_{n\to\infty} \left(\dfrac{n+1}{n}\right)^n \dfrac{1}{2(2n+1)}$

$= \lim\limits_{n\to\infty} \left(1 + \dfrac{1}{n}\right)^n \dfrac{1}{2(2n+1)} = e \cdot 0 = 0 < 1$

so the series converges by the Ratio Test.

(b) The series in part (a) is convergent, so $\lim\limits_{n\to\infty} a_n = \lim\limits_{n\to\infty} \dfrac{n^n}{(2n)!} = 0$ by Theorem 11.2.6.

39. Use the Limit Comparison Test. $\lim\limits_{n\to\infty} \left|\dfrac{\left(\frac{n+1}{n}\right)a_n}{a_n}\right| = \lim\limits_{n\to\infty} \dfrac{n+1}{n} = \lim\limits_{n\to\infty} \left(1 + \dfrac{1}{n}\right) = 1 > 0$.

Since $\sum |a_n|$ is convergent, so is $\sum \left|\left(\dfrac{n+1}{n}\right)a_n\right|$, by the Limit Comparison Test.

40. $\lim\limits_{n\to\infty} \left|\dfrac{a_{n+1}}{a_n}\right| = \lim\limits_{n\to\infty} \left|\dfrac{x^{n+1}}{(n+1)^2\, 5^{n+1}} \cdot \dfrac{n^2 5^n}{x^n}\right| = \lim\limits_{n\to\infty} \dfrac{1}{(1+1/n)^2} \dfrac{|x|}{5} = \dfrac{|x|}{5}$, so by the Ratio Test, $\sum\limits_{n=1}^{\infty} (-1)^n \dfrac{x^n}{n^2\, 5^n}$ converges when $\dfrac{|x|}{5} < 1 \;\Leftrightarrow\; |x| < 5$, so $R = 5$. When $x = -5$, the series becomes the convergent p-series $\sum\limits_{n=1}^{\infty} \dfrac{1}{n^2}$ with $p = 2 > 1$. When $x = 5$, the series becomes $\sum\limits_{n=1}^{\infty} \dfrac{(-1)^n}{n^2}$, which converges by the Alternating Series Test. Thus, $I = [-5, 5]$.

41. $\lim_{n\to\infty}\left|\frac{a_{n+1}}{a_n}\right| = \lim_{n\to\infty}\left[\frac{|x+2|^{n+1}}{(n+1)\,4^{n+1}}\cdot\frac{n\,4^n}{|x+2|^n}\right] = \lim_{n\to\infty}\left[\frac{n}{n+1}\,\frac{|x+2|}{4}\right] = \frac{|x+2|}{4} < 1 \;\Leftrightarrow\; |x+2| < 4$, so $R = 4$.

$|x+2| < 4 \;\Leftrightarrow\; -4 < x+2 < 4 \;\Leftrightarrow\; -6 < x < 2$. If $x = -6$, then the series $\sum\limits_{n=1}^{\infty}\frac{(x+2)^n}{n\,4^n}$ becomes

$\sum\limits_{n=1}^{\infty}\frac{(-4)^n}{n4^n} = \sum\limits_{n=1}^{\infty}\frac{(-1)^n}{n}$, the alternating harmonic series, which converges by the Alternating Series Test. When $x = 2$, the

series becomes the harmonic series $\sum\limits_{n=1}^{\infty}\frac{1}{n}$, which diverges. Thus, $I = [-6, 2)$.

42. $\lim_{n\to\infty}\left|\frac{a_{n+1}}{a_n}\right| = \lim_{n\to\infty}\left|\frac{2^{n+1}\,(x-2)^{n+1}}{(n+3)!}\cdot\frac{(n+2)!}{2^n(x-2)^n}\right| = \lim_{n\to\infty}\frac{2}{n+3}\,|x-2| = 0 < 1$, so the series $\sum\limits_{n=1}^{\infty}\frac{2^n\,(x-2)^n}{(n+2)!}$

converges for all x. $R = \infty$ and $I = (-\infty, \infty)$.

43. $\lim_{n\to\infty}\left|\frac{a_{n+1}}{a_n}\right| = \lim_{n\to\infty}\left|\frac{2^{n+1}(x-3)^{n+1}}{\sqrt{n+4}}\cdot\frac{\sqrt{n+3}}{2^n(x-3)^n}\right| = 2\,|x-3|\lim_{n\to\infty}\sqrt{\frac{n+3}{n+4}} = 2\,|x-3| < 1 \;\Leftrightarrow\; |x-3| < \frac{1}{2}$,

so $R = \frac{1}{2}$. $|x-3| < \frac{1}{2} \;\Leftrightarrow\; -\frac{1}{2} < x-3 < \frac{1}{2} \;\Leftrightarrow\; \frac{5}{2} < x < \frac{7}{2}$. For $x = \frac{7}{2}$, the series $\sum\limits_{n=1}^{\infty}\frac{2^n\,(x-3)^n}{\sqrt{n+3}}$ becomes

$\sum\limits_{n=0}^{\infty}\frac{1}{\sqrt{n+3}} = \sum\limits_{n=3}^{\infty}\frac{1}{n^{1/2}}$, which diverges $\left[p = \frac{1}{2} \le 1\right]$, but for $x = \frac{5}{2}$, we get $\sum\limits_{n=0}^{\infty}\frac{(-1)^n}{\sqrt{n+3}}$, which is a convergent

alternating series, so $I = \left[\frac{5}{2}, \frac{7}{2}\right)$.

44. $\lim_{n\to\infty}\left|\frac{a_{n+1}}{a_n}\right| = \lim_{n\to\infty}\left|\frac{(2n+2)!\,x^{n+1}}{[(n+1)!]^2}\cdot\frac{(n!)^2}{(2n)!\,x^n}\right| = \lim_{n\to\infty}\frac{(2n+2)(2n+1)}{(n+1)(n+1)}\,|x| = 4\,|x|$.

To converge, we must have $4\,|x| < 1 \;\Leftrightarrow\; |x| < \frac{1}{4}$, so $R = \frac{1}{4}$.

45.

n	$f^{(n)}(x)$	$f^{(n)}\left(\frac{\pi}{6}\right)$
0	$\sin x$	$\frac{1}{2}$
1	$\cos x$	$\frac{\sqrt{3}}{2}$
2	$-\sin x$	$-\frac{1}{2}$
3	$-\cos x$	$-\frac{\sqrt{3}}{2}$
4	$\sin x$	$\frac{1}{2}$
$\vdots$	$\vdots$	$\vdots$

$$\begin{aligned}\sin x &= f\left(\frac{\pi}{6}\right) + f'\left(\frac{\pi}{6}\right)\left(x-\frac{\pi}{6}\right) + \frac{f''\left(\frac{\pi}{6}\right)}{2!}\left(x-\frac{\pi}{6}\right)^2 + \frac{f^{(3)}\left(\frac{\pi}{6}\right)}{3!}\left(x-\frac{\pi}{6}\right)^3 + \frac{f^{(4)}\left(\frac{\pi}{6}\right)}{4!}\left(x-\frac{\pi}{6}\right)^4 + \cdots \\ &= \frac{1}{2}\left[1 - \frac{1}{2!}\left(x-\frac{\pi}{6}\right)^2 + \frac{1}{4!}\left(x-\frac{\pi}{6}\right)^4 - \cdots\right] + \frac{\sqrt{3}}{2}\left[\left(x-\frac{\pi}{6}\right) - \frac{1}{3!}\left(x-\frac{\pi}{6}\right)^3 + \cdots\right] \\ &= \frac{1}{2}\sum_{n=0}^{\infty}(-1)^n\frac{1}{(2n)!}\left(x-\frac{\pi}{6}\right)^{2n} + \frac{\sqrt{3}}{2}\sum_{n=0}^{\infty}(-1)^n\frac{1}{(2n+1)!}\left(x-\frac{\pi}{6}\right)^{2n+1}\end{aligned}$$

46.

n	$f^{(n)}(x)$	$f^{(n)}\left(\frac{\pi}{3}\right)$
0	$\cos x$	$\frac{1}{2}$
1	$-\sin x$	$-\frac{\sqrt{3}}{2}$
2	$-\cos x$	$-\frac{1}{2}$
3	$\sin x$	$\frac{\sqrt{3}}{2}$
4	$\cos x$	$\frac{1}{2}$
$\vdots$	$\vdots$	$\vdots$

$$\cos x = f\left(\frac{\pi}{3}\right) + f'\left(\frac{\pi}{3}\right)\left(x - \frac{\pi}{3}\right) + \frac{f''\left(\frac{\pi}{3}\right)}{2!}\left(x - \frac{\pi}{3}\right)^2 + \frac{f^{(3)}\left(\frac{\pi}{3}\right)}{3!}\left(x - \frac{\pi}{3}\right)^3 + \frac{f^{(4)}\left(\frac{\pi}{3}\right)}{4!}\left(x - \frac{\pi}{3}\right)^4 + \cdots$$

$$= \frac{1}{2}\left[1 - \frac{1}{2!}\left(x - \frac{\pi}{3}\right)^2 + \frac{1}{4!}\left(x - \frac{\pi}{3}\right)^4 - \cdots\right] + \frac{\sqrt{3}}{2}\left[-\left(x - \frac{\pi}{3}\right) + \frac{1}{3!}\left(x - \frac{\pi}{3}\right)^3 - \cdots\right]$$

$$= \frac{1}{2}\sum_{n=0}^{\infty}(-1)^n \frac{1}{(2n)!}\left(x - \frac{\pi}{3}\right)^{2n} + \frac{\sqrt{3}}{2}\sum_{n=0}^{\infty}(-1)^{n+1}\frac{1}{(2n+1)!}\left(x - \frac{\pi}{3}\right)^{2n+1}$$

47. $\dfrac{1}{1+x} = \dfrac{1}{1-(-x)} = \displaystyle\sum_{n=0}^{\infty}(-x)^n = \sum_{n=0}^{\infty}(-1)^n x^n$ for $|x| < 1$ $\Rightarrow$ $\dfrac{x^2}{1+x} = \displaystyle\sum_{n=0}^{\infty}(-1)^n x^{n+2}$ with $R = 1$.

48. $\tan^{-1} x = \displaystyle\sum_{n=0}^{\infty}(-1)^n \frac{x^{2n+1}}{2n+1}$ with interval of convergence $[-1, 1]$, so

$\tan^{-1}(x^2) = \displaystyle\sum_{n=0}^{\infty}(-1)^n \frac{(x^2)^{2n+1}}{2n+1} = \sum_{n=0}^{\infty}(-1)^n \frac{x^{4n+2}}{2n+1}$, which converges when $x^2 \in [-1, 1]$ $\Leftrightarrow$ $x \in [-1, 1]$.

Therefore, $R = 1$.

49. $\displaystyle\int \frac{1}{4-x}\,dx = -\ln(4-x) + C$ and

$\displaystyle\int \frac{1}{4-x}\,dx = \frac{1}{4}\int \frac{1}{1-x/4}\,dx = \frac{1}{4}\int \sum_{n=0}^{\infty}\left(\frac{x}{4}\right)^n dx = \frac{1}{4}\int \sum_{n=0}^{\infty}\frac{x^n}{4^n}\,dx = \frac{1}{4}\sum_{n=0}^{\infty}\frac{x^{n+1}}{4^n(n+1)} + C$. So

$\ln(4-x) = -\dfrac{1}{4}\displaystyle\sum_{n=0}^{\infty}\frac{x^{n+1}}{4^n(n+1)} + C = -\sum_{n=0}^{\infty}\frac{x^{n+1}}{4^{n+1}(n+1)} + C = -\sum_{n=1}^{\infty}\frac{x^n}{n4^n} + C$. Putting $x = 0$, we get $C = \ln 4$.

Thus, $f(x) = \ln(4-x) = \ln 4 - \displaystyle\sum_{n=1}^{\infty}\frac{x^n}{n4^n}$. The series converges for $|x/4| < 1$ $\Leftrightarrow$ $|x| < 4$, so $R = 4$.

Another solution:

$$\ln(4-x) = \ln[4(1-x/4)] = \ln 4 + \ln(1-x/4) = \ln 4 + \ln[1+(-x/4)]$$

$$= \ln 4 + \sum_{n=1}^{\infty}(-1)^{n+1}\frac{(-x/4)^n}{n} \text{ [from Table 1] } = \ln 4 + \sum_{n=1}^{\infty}(-1)^{2n+1}\frac{x^n}{n4^n} = \ln 4 - \sum_{n=1}^{\infty}\frac{x^n}{n4^n}.$$

50. $e^x = \displaystyle\sum_{n=0}^{\infty}\frac{x^n}{n!}$ $\Rightarrow$ $e^{2x} = \displaystyle\sum_{n=0}^{\infty}\frac{(2x)^n}{n!}$ $\Rightarrow$ $xe^{2x} = x\displaystyle\sum_{n=0}^{\infty}\frac{2^n x^n}{n!} = \sum_{n=0}^{\infty}\frac{2^n x^{n+1}}{n!}$, $R = \infty$

51. $\sin x = \displaystyle\sum_{n=0}^{\infty}\frac{(-1)^n x^{2n+1}}{(2n+1)!}$ $\Rightarrow$ $\sin(x^4) = \displaystyle\sum_{n=0}^{\infty}\frac{(-1)^n (x^4)^{2n+1}}{(2n+1)!} = \sum_{n=0}^{\infty}\frac{(-1)^n x^{8n+4}}{(2n+1)!}$ for all x, so the radius of convergence is ∞.

52. $e^x = \displaystyle\sum_{n=0}^{\infty}\frac{x^n}{n!}$ $\Rightarrow$ $10^x = e^{(\ln 10)x} = \displaystyle\sum_{n=0}^{\infty}\frac{[(\ln 10)x]^n}{n!} = \sum_{n=0}^{\infty}\frac{(\ln 10)^n x^n}{n!}$, $R = \infty$

53. $f(x) = \dfrac{1}{\sqrt[4]{16-x}} = \dfrac{1}{\sqrt[4]{16(1-x/16)}} = \dfrac{1}{\sqrt[4]{16}\left(1-\frac{1}{16}x\right)^{1/4}} = \frac{1}{2}\left(1-\frac{1}{16}x\right)^{-1/4}$

$$= \frac{1}{2}\left[1 + \left(-\frac{1}{4}\right)\left(-\frac{x}{16}\right) + \frac{\left(-\frac{1}{4}\right)\left(-\frac{5}{4}\right)}{2!}\left(-\frac{x}{16}\right)^2 + \frac{\left(-\frac{1}{4}\right)\left(-\frac{5}{4}\right)\left(-\frac{9}{4}\right)}{3!}\left(-\frac{x}{16}\right)^3 + \cdots\right]$$

$$= \frac{1}{2} + \sum_{n=1}^{\infty} \frac{1\cdot 5\cdot 9\cdot\cdots\cdot(4n-3)}{2\cdot 4^n\cdot n!\cdot 16^n}\, x^n = \frac{1}{2} + \sum_{n=1}^{\infty} \frac{1\cdot 5\cdot 9\cdot\cdots\cdot(4n-3)}{2^{6n+1}\, n!}\, x^n$$

for $\left|-\dfrac{x}{16}\right| < 1 \Leftrightarrow |x| < 16$, so $R = 16$.

54. $(1-3x)^{-5} = \displaystyle\sum_{n=0}^{\infty}\binom{-5}{n}(-3x)^n = 1 + (-5)(-3x) + \frac{(-5)(-6)}{2!}(-3x)^2 + \frac{(-5)(-6)(-7)}{3!}(-3x)^3 + \cdots$

$$= 1 + \sum_{n=1}^{\infty}\frac{5\cdot 6\cdot 7\cdot\cdots\cdot(n+4)\cdot 3^n x^n}{n!} \quad \text{for } |-3x| < 1 \Leftrightarrow |x| < \tfrac{1}{3}, \text{ so } R = \tfrac{1}{3}.$$

55. $e^x = \displaystyle\sum_{n=0}^{\infty}\frac{x^n}{n!}$, so $\dfrac{e^x}{x} = \dfrac{1}{x}\displaystyle\sum_{n=0}^{\infty}\frac{x^n}{n!} = \sum_{n=0}^{\infty}\frac{x^{n-1}}{n!} = x^{-1} + \sum_{n=1}^{\infty}\frac{x^{n-1}}{n!} = \frac{1}{x} + \sum_{n=1}^{\infty}\frac{x^{n-1}}{n!}$ and

$$\int \frac{e^x}{x}\,dx = C + \ln|x| + \sum_{n=1}^{\infty}\frac{x^n}{n\cdot n!}.$$

56. $(1+x^4)^{1/2} = \displaystyle\sum_{n=0}^{\infty}\binom{\frac{1}{2}}{n}(x^4)^n = 1 + \left(\tfrac{1}{2}\right)x^4 + \frac{\left(\frac{1}{2}\right)\left(-\frac{1}{2}\right)}{2!}(x^4)^2 + \frac{\left(\frac{1}{2}\right)\left(-\frac{1}{2}\right)\left(-\frac{3}{2}\right)}{3!}(x^4)^3 + \cdots$

$$= 1 + \tfrac{1}{2}x^4 - \tfrac{1}{8}x^8 + \tfrac{1}{16}x^{12} - \cdots$$

so $\int_0^1 (1+x^4)^{1/2}\,dx = \left[x + \frac{1}{10}x^5 - \frac{1}{72}x^9 + \frac{1}{208}x^{13} - \cdots\right]_0^1 = 1 + \frac{1}{10} - \frac{1}{72} + \frac{1}{208} - \cdots$.

This is an alternating series, so by the Alternating Series Test, the error in the approximation $\int_0^1 (1+x^4)^{1/2}\,dx \approx 1 + \frac{1}{10} - \frac{1}{72} \approx 1.086$ is less than $\frac{1}{208}$, sufficient for the desired accuracy.

Thus, correct to two decimal places, $\int_0^1 (1+x^4)^{1/2}\,dx \approx 1.09$.

57. (a)

n	$f^{(n)}(x)$	$f^{(n)}(1)$
0	$x^{1/2}$	1
1	$\frac{1}{2}x^{-1/2}$	$\frac{1}{2}$
2	$-\frac{1}{4}x^{-3/2}$	$-\frac{1}{4}$
3	$\frac{3}{8}x^{-5/2}$	$\frac{3}{8}$
4	$-\frac{15}{16}x^{-7/2}$	$-\frac{15}{16}$
⋮	⋮	⋮

$$\sqrt{x} \approx T_3(x) = 1 + \frac{1/2}{1!}(x-1) - \frac{1/4}{2!}(x-1)^2 + \frac{3/8}{3!}(x-1)^3$$
$$= 1 + \tfrac{1}{2}(x-1) - \tfrac{1}{8}(x-1)^2 + \tfrac{1}{16}(x-1)^3$$

(b)

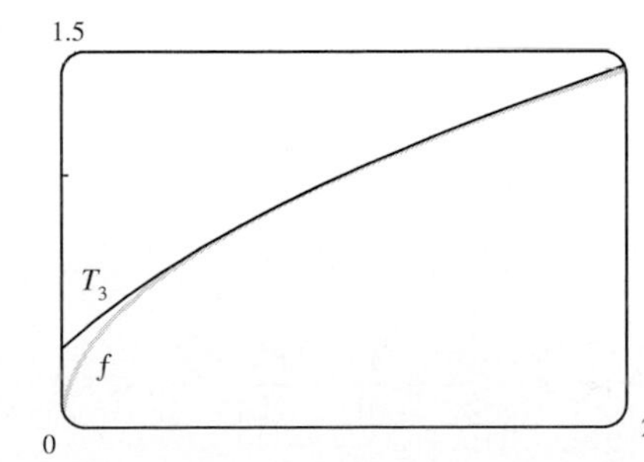

(c) $|R_3(x)| \le \dfrac{M}{4!}|x-1|^4$, where $\left|f^{(4)}(x)\right| \le M$ with $f^{(4)}(x) = -\frac{15}{16}x^{-7/2}$. Now $0.9 \le x \le 1.1 \Rightarrow$

$-0.1 \le x - 1 \le 0.1 \Rightarrow (x-1)^4 \le (0.1)^4$, and letting $x = 0.9$ gives $M = \dfrac{15}{16(0.9)^{7/2}}$, so

$$|R_3(x)| \le \frac{15}{16(0.9)^{7/2}\,4!}(0.1)^4 \approx 0.000\,005\,648 \approx 0.000\,006 = 6\times 10^{-6}.$$

(d)

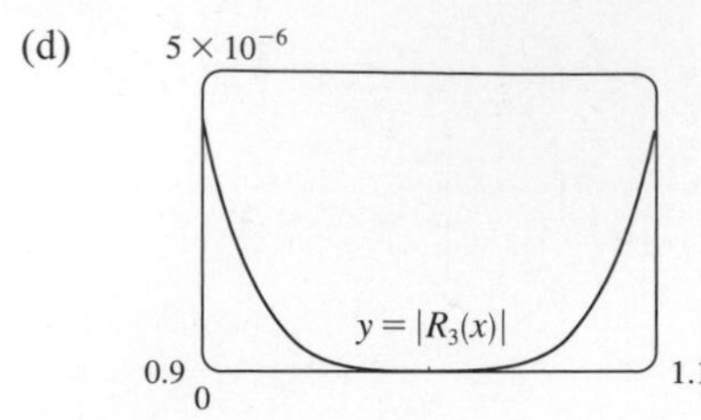

From the graph of $|R_3(x)| = |\sqrt{x} - T_3(x)|$, it appears that the error is less than 5×10^{-6} on $[0.9, 1.1]$.

58. (a)

n	$f^{(n)}(x)$	$f^{(n)}(0)$
0	$\sec x$	1
1	$\sec x \tan x$	0
2	$\sec x \tan^2 x + \sec^3 x$	1
3	$\sec x \tan^3 x + 5\sec^3 x \tan x$	0
$\vdots$	$\vdots$	$\vdots$

$\sec x \approx T_2(x) = 1 + \frac{1}{2}x^2$

(b)

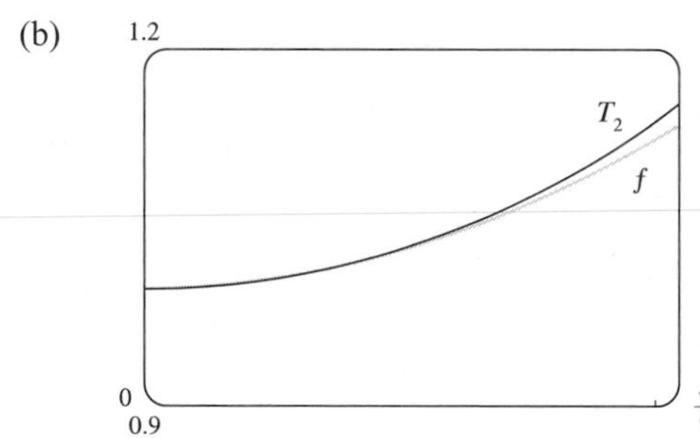

(c) $|R_2(x)| \leq \dfrac{M}{3!}|x|^3$, where $\left|f^{(3)}(x)\right| \leq M$ with $f^{(3)}(x) = \sec x \tan^3 x + 5\sec^3 x \tan x$.

Now $0 \leq x \leq \frac{\pi}{6} \;\Rightarrow\; x^3 \leq \left(\frac{\pi}{6}\right)^3$, and letting $x = \frac{\pi}{6}$ gives $M = \frac{14}{3}$, so $|R_2(x)| \leq \dfrac{14}{3 \cdot 6}\left(\dfrac{\pi}{6}\right)^3 \approx 0.111648$.

(d)

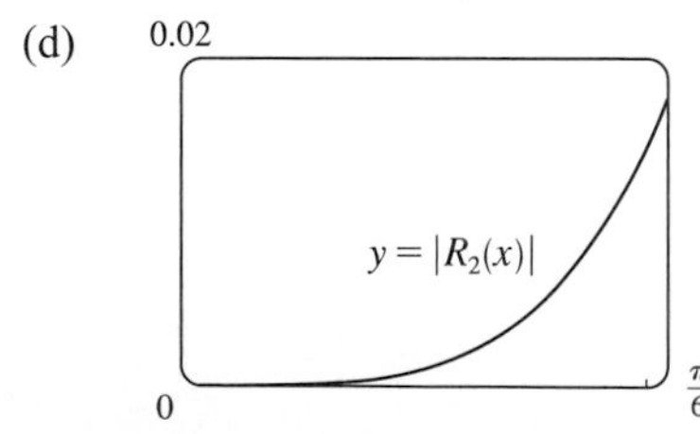

From the graph of $|R_2(x)| = |\sec x - T_2(x)|$, it appears that the error is less than 0.02 on $\left[0, \frac{\pi}{6}\right]$.

59. $\sin x = \sum\limits_{n=0}^{\infty} (-1)^n \dfrac{x^{2n+1}}{(2n+1)!} = x - \dfrac{x^3}{3!} + \dfrac{x^5}{5!} - \dfrac{x^7}{7!} + \cdots$, so $\sin x - x = -\dfrac{x^3}{3!} + \dfrac{x^5}{5!} - \dfrac{x^7}{7!} + \cdots$ and

$\dfrac{\sin x - x}{x^3} = -\dfrac{1}{3!} + \dfrac{x^2}{5!} - \dfrac{x^4}{7!} + \cdots$. Thus, $\lim\limits_{x\to 0} \dfrac{\sin x - x}{x^3} = \lim\limits_{x\to 0}\left(-\dfrac{1}{6} + \dfrac{x^2}{120} - \dfrac{x^4}{5040} + \cdots\right) = -\dfrac{1}{6}$.

60. (a) $F = \dfrac{mgR^2}{(R+h)^2} = \dfrac{mg}{(1+h/R)^2} = mg \sum\limits_{n=0}^{\infty} \dbinom{-2}{n}\left(\dfrac{h}{R}\right)^n$ [binomial series]

(b) We expand $F = mg\left[1 - 2(h/R) + 3(h/R)^2 - \cdots\right]$.

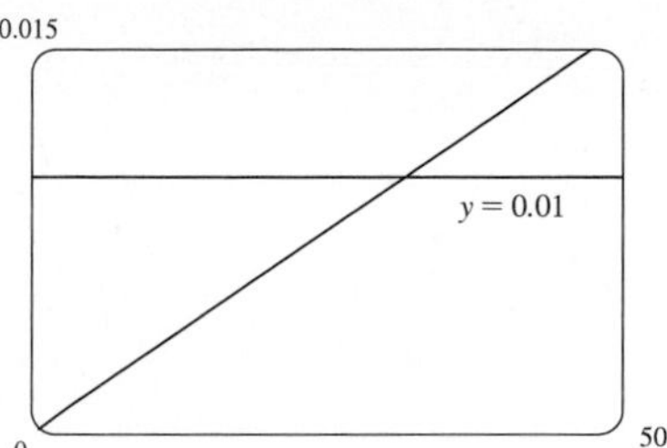

This is an alternating series, so by the Alternating Series Estimation Theorem, the error in the approximation $F = mg$ is less than $2mgh/R$, so for accuracy within 1% we want

$$\left|\frac{2mgh/R}{mgR^2/(R+h)^2}\right| < 0.01 \quad \Leftrightarrow \quad \frac{2h(R+h)^2}{R^3} < 0.01.$$

This inequality would be difficult to solve for h, so we substitute $R = 6{,}400$ km and plot both sides of the inequality. It appears that the approximation is accurate to within 1% for $h < 31$ km.

61. $f(x) = \sum\limits_{n=0}^{\infty} c_n x^n \;\Rightarrow\; f(-x) = \sum\limits_{n=0}^{\infty} c_n(-x)^n = \sum\limits_{n=0}^{\infty} (-1)^n c_n x^n$

(a) If f is an odd function, then $f(-x) = -f(x) \;\Rightarrow\; \sum\limits_{n=0}^{\infty} (-1)^n c_n x^n = \sum\limits_{n=0}^{\infty} -c_n x^n$. The coefficients of any power series are uniquely determined (by Theorem 11.10.5), so $(-1)^n c_n = -c_n$.

If n is even, then $(-1)^n = 1$, so $c_n = -c_n \;\Rightarrow\; 2c_n = 0 \;\Rightarrow\; c_n = 0$. Thus, all even coefficients are 0, that is, $c_0 = c_2 = c_4 = \cdots = 0$.

(b) If f is even, then $f(-x) = f(x) \;\Rightarrow\; \sum\limits_{n=0}^{\infty} (-1)^n c_n x^n = \sum\limits_{n=0}^{\infty} c_n x^n \;\Rightarrow\; (-1)^n c_n = c_n$.

If n is odd, then $(-1)^n = -1$, so $-c_n = c_n \;\Rightarrow\; 2c_n = 0 \;\Rightarrow\; c_n = 0$. Thus, all odd coefficients are 0, that is, $c_1 = c_3 = c_5 = \cdots = 0$.

62. $e^x = \sum\limits_{n=0}^{\infty} \dfrac{x^n}{n!} \;\Rightarrow\; f(x) = e^{x^2} = \sum\limits_{n=0}^{\infty} \dfrac{(x^2)^n}{n!} = \sum\limits_{n=0}^{\infty} \dfrac{x^{2n}}{n!} = \sum\limits_{n=0}^{\infty} \dfrac{1}{n!} x^{2n}$. By Theorem 11.10.6 with $a = 0$, we also have $f(x) = \sum\limits_{k=0}^{\infty} \dfrac{f^{(k)}(0)}{k!} x^k$. Comparing coefficients for $k = 2n$, we have $\dfrac{f^{(2n)}(0)}{(2n)!} = \dfrac{1}{n!} \;\Rightarrow\; f^{(2n)}(0) = \dfrac{(2n)!}{n!}$.

PROBLEMS PLUS

1. It would be far too much work to compute 15 derivatives of f. The key idea is to remember that $f^{(n)}(0)$ occurs in the coefficient of x^n in the Maclaurin series of f. We start with the Maclaurin series for sin: $\sin x = x - \dfrac{x^3}{3!} + \dfrac{x^5}{5!} - \cdots$.

Then $\sin(x^3) = x^3 - \dfrac{x^9}{3!} + \dfrac{x^{15}}{5!} - \cdots$, and so the coefficient of x^{15} is $\dfrac{f^{(15)}(0)}{15!} = \dfrac{1}{5!}$. Therefore,

$f^{(15)}(0) = \dfrac{15!}{5!} = 6 \cdot 7 \cdot 8 \cdot 9 \cdot 10 \cdot 11 \cdot 12 \cdot 13 \cdot 14 \cdot 15 = 10{,}897{,}286{,}400.$

2. We use the problem-solving strategy of taking cases:

Case (i): If $|x| < 1$, then $0 \le x^2 < 1$, so $\lim\limits_{n\to\infty} x^{2n} = 0$ [see Example 11.1.11]

and $f(x) = \lim\limits_{n\to\infty} \dfrac{x^{2n} - 1}{x^{2n} + 1} = \dfrac{0 - 1}{0 + 1} = -1.$[]

Case (ii): If $|x| = 1$, that is, $x = \pm 1$, then $x^2 = 1$, so $f(x) = \lim\limits_{n\to\infty} \dfrac{x^{2n} - 1}{x^{2n} + 1} = \lim\limits_{n\to\infty} \dfrac{1 - 1}{1 + 1} = 0.$

Case (iii): If $|x| > 1$, then $x^2 > 1$, so $\lim\limits_{n\to\infty} x^{2n} = \infty$ and $f(x) = \lim\limits_{n\to\infty} \dfrac{x^{2n} - 1}{x^{2n} + 1} = \lim\limits_{n\to\infty} \dfrac{1 - (1/x^{2n})}{1 + (1/x^{2n})} = \dfrac{1 - 0}{1 + 0} = 1.$

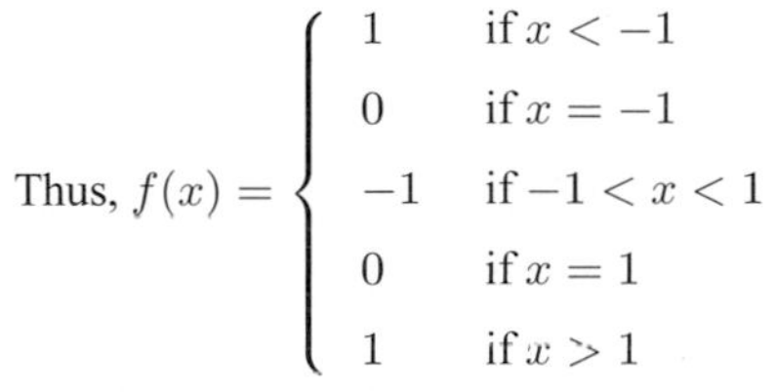

$$\text{Thus, } f(x) = \begin{cases} 1 & \text{if } x < -1 \\ 0 & \text{if } x = -1 \\ -1 & \text{if } -1 < x < 1 \\ 0 & \text{if } x = 1 \\ 1 & \text{if } x > 1 \end{cases}$$

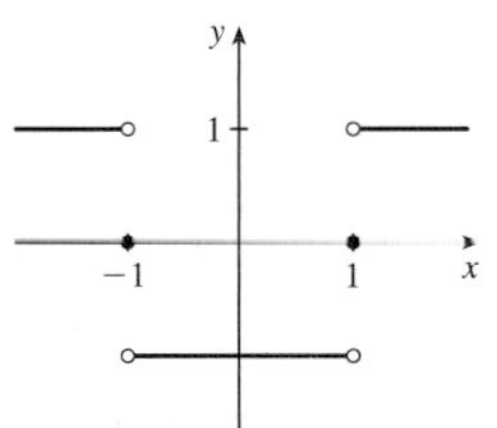

The graph shows that f is continuous everywhere except at $x = \pm 1$.

3. (a) From Formula 14a in Appendix D, with $x = y = \theta$, we get $\tan 2\theta = \dfrac{2\tan\theta}{1 - \tan^2\theta}$, so $\cot 2\theta = \dfrac{1 - \tan^2\theta}{2\tan\theta}$ $\Rightarrow$

$2\cot 2\theta = \dfrac{1 - \tan^2\theta}{\tan\theta} = \cot\theta - \tan\theta$. Replacing θ by $\frac{1}{2}x$, we get $2\cot x = \cot\frac{1}{2}x - \tan\frac{1}{2}x$, or

$\tan\frac{1}{2}x = \cot\frac{1}{2}x - 2\cot x$.

(b) From part (a) with $\dfrac{x}{2^{n-1}}$ in place of x, $\tan\dfrac{x}{2^n} = \cot\dfrac{x}{2^n} - 2\cot\dfrac{x}{2^{n-1}}$, so the nth partial sum of $\sum\limits_{n=1}^{\infty} \dfrac{1}{2^n}\tan\dfrac{x}{2^n}$ is

$$\begin{aligned} s_n &= \frac{\tan(x/2)}{2} + \frac{\tan(x/4)}{4} + \frac{\tan(x/8)}{8} + \cdots + \frac{\tan(x/2^n)}{2^n} \\ &= \left[\frac{\cot(x/2)}{2} - \cot x\right] + \left[\frac{\cot(x/4)}{4} - \frac{\cot(x/2)}{2}\right] + \left[\frac{\cot(x/8)}{8} - \frac{\cot(x/4)}{4}\right] + \cdots \\ &\qquad + \left[\frac{\cot(x/2^n)}{2^n} - \frac{\cot(x/2^{n-1})}{2^{n-1}}\right] = -\cot x + \frac{\cot(x/2^n)}{2^n} \quad \text{[telescoping sum]} \end{aligned}$$

Now $\dfrac{\cot(x/2^n)}{2^n} = \dfrac{\cos(x/2^n)}{2^n \sin(x/2^n)} = \dfrac{\cos(x/2^n)}{x} \cdot \dfrac{x/2^n}{\sin(x/2^n)} \to \dfrac{1}{x} \cdot 1 = \dfrac{1}{x}$ as $n \to \infty$ since $x/2^n \to 0$

for $x \neq 0$. Therefore, if $x \neq 0$ and $x \neq k\pi$ where k is any integer, then

$$\sum_{n=1}^{\infty} \frac{1}{2^n} \tan \frac{x}{2^n} = \lim_{n \to \infty} s_n = \lim_{n \to \infty} \left(-\cot x + \frac{1}{2^n} \cot \frac{x}{2^n} \right) = -\cot x + \frac{1}{x}$$

If $x = 0$, then all terms in the series are 0, so the sum is 0.

4. $|AP_2|^2 = 2$, $|AP_3|^2 = 2 + 2^2$, $|AP_4|^2 = 2 + 2^2 + (2^2)^2$, $|AP_5|^2 = 2 + 2^2 + (2^2)^2 + (2^3)^2, \ldots,$

$$|AP_n|^2 = 2 + 2^2 + (2^2)^2 + \cdots + (2^{n-2})^2 \quad [\text{for } n \geq 3] \quad = 2 + (4 + 4^2 + 4^3 + \cdots + 4^{n-2})$$

$$= 2 + \frac{4(4^{n-2} - 1)}{4 - 1} \quad [\text{finite geometric sum with } a = 4, r = 4] \quad = \frac{6}{3} + \frac{4^{n-1} - 4}{3} = \frac{2}{3} + \frac{4^{n-1}}{3}$$

So $\tan \angle P_n A P_{n+1} = \dfrac{|P_n P_{n+1}|}{|AP_n|} = \dfrac{2^{n-1}}{\sqrt{\dfrac{2}{3} + \dfrac{4^{n-1}}{3}}} = \dfrac{\sqrt{4^{n-1}}}{\sqrt{\dfrac{2}{3} + \dfrac{4^{n-1}}{3}}} = \dfrac{1}{\sqrt{\dfrac{2}{3 \cdot 4^{n-1}} + \dfrac{1}{3}}} \to \sqrt{3}$ as $n \to \infty$.

Thus, $\angle P_n A P_{n+1} \to \frac{\pi}{3}$ as $n \to \infty$.

5. (a) At each stage, each side is replaced by four shorter sides, each of length $\frac{1}{3}$ of the side length at the preceding stage. Writing s_0 and ℓ_0 for the number of sides and the length of the side of the initial triangle, we generate the table at right. In general, we have $s_n = 3 \cdot 4^n$ and $\ell_n = \left(\frac{1}{3}\right)^n$, so the length of the perimeter at the nth stage of construction is $p_n = s_n \ell_n = 3 \cdot 4^n \cdot \left(\frac{1}{3}\right)^n = 3 \cdot \left(\frac{4}{3}\right)^n$.

$s_0 = 3$	$\ell_0 = 1$
$s_1 = 3 \cdot 4$	$\ell_1 = 1/3$
$s_2 = 3 \cdot 4^2$	$\ell_2 = 1/3^2$
$s_3 = 3 \cdot 4^3$	$\ell_3 = 1/3^3$
$\vdots$	$\vdots$

(b) $p_n = \dfrac{4^n}{3^{n-1}} = 4\left(\dfrac{4}{3}\right)^{n-1}$. Since $\frac{4}{3} > 1$, $p_n \to \infty$ as $n \to \infty$.

(c) The area of each of the small triangles added at a given stage is one-ninth of the area of the triangle added at the preceding stage. Let a be the area of the original triangle. Then the area a_n of each of the small triangles added at stage n is $a_n = a \cdot \dfrac{1}{9^n} = \dfrac{a}{9^n}$. Since a small triangle is added to each side at every stage, it follows that the total area A_n added to the figure at the nth stage is $A_n = s_{n-1} \cdot a_n = 3 \cdot 4^{n-1} \cdot \dfrac{a}{9^n} = a \cdot \dfrac{4^{n-1}}{3^{2n-1}}$. Then the total area enclosed by the snowflake curve is $A = a + A_1 + A_2 + A_3 + \cdots = a + a \cdot \dfrac{1}{3} + a \cdot \dfrac{4}{3^3} + a \cdot \dfrac{4^2}{3^5} + a \cdot \dfrac{4^3}{3^7} + \cdots$. After the first term, this is a geometric series with common ratio $\dfrac{4}{9}$, so $A = a + \dfrac{a/3}{1 - \frac{4}{9}} = a + \dfrac{a}{3} \cdot \dfrac{9}{5} = \dfrac{8a}{5}$. But the area of the original equilateral triangle with side 1 is $a = \dfrac{1}{2} \cdot 1 \cdot \sin \dfrac{\pi}{3} = \dfrac{\sqrt{3}}{4}$. So the area enclosed by the snowflake curve is $\dfrac{8}{5} \cdot \dfrac{\sqrt{3}}{4} = \dfrac{2\sqrt{3}}{5}$.

6. Let the series $S = 1 + \frac{1}{2} + \frac{1}{3} + \frac{1}{4} + \frac{1}{6} + \frac{1}{8} + \frac{1}{9} + \frac{1}{12} + \cdots$. Then every term in S is of the form $\dfrac{1}{2^m 3^n}$, $m, n \geq 0$, and furthermore each term occurs only once. So we can write

$$S = \sum_{m=0}^{\infty} \sum_{n=0}^{\infty} \frac{1}{2^m 3^n} = \sum_{m=0}^{\infty} \sum_{n=0}^{\infty} \frac{1}{2^m} \frac{1}{3^n} = \sum_{m=0}^{\infty} \frac{1}{2^m} \sum_{n=0}^{\infty} \frac{1}{3^n} = \frac{1}{1 - \frac{1}{2}} \cdot \frac{1}{1 - \frac{1}{3}} = 2 \cdot \tfrac{3}{2} = 3$$

7. (a) Let $a = \arctan x$ and $b = \arctan y$. Then, from Formula 14b in Appendix D,

$$\tan(a-b) = \frac{\tan a - \tan b}{1 + \tan a \, \tan b} = \frac{\tan(\arctan x) - \tan(\arctan y)}{1 + \tan(\arctan x)\tan(\arctan y)} = \frac{x-y}{1+xy}$$

Now $\arctan x - \arctan y = a - b = \arctan(\tan(a-b)) = \arctan \dfrac{x-y}{1+xy}$ since $-\frac{\pi}{2} < a - b < \frac{\pi}{2}$.

(b) From part (a) we have

$$\arctan \tfrac{120}{119} - \arctan \tfrac{1}{239} = \arctan \frac{\frac{120}{119} - \frac{1}{239}}{1 + \frac{120}{119} \cdot \frac{1}{239}} = \arctan \frac{\frac{28{,}561}{28{,}441}}{\frac{28{,}561}{28{,}441}} = \arctan 1 = \tfrac{\pi}{4}$$

(c) Replacing y by $-y$ in the formula of part (a), we get $\arctan x + \arctan y = \arctan \dfrac{x+y}{1-xy}$. So

$$4\arctan \tfrac{1}{5} = 2\left(\arctan \tfrac{1}{5} + \arctan \tfrac{1}{5}\right) = 2\arctan \frac{\frac{1}{5} + \frac{1}{5}}{1 - \frac{1}{5} \cdot \frac{1}{5}} = 2\arctan \tfrac{5}{12} = \arctan \tfrac{5}{12} + \arctan \tfrac{5}{12}$$

$$= \arctan \frac{\frac{5}{12} + \frac{5}{12}}{1 - \frac{5}{12} \cdot \frac{5}{12}} = \arctan \tfrac{120}{119}$$

Thus, from part (b), we have $4\arctan \frac{1}{5} - \arctan \frac{1}{239} = \arctan \frac{120}{119} - \arctan \frac{1}{239} = \frac{\pi}{4}$.

(d) From Example 11.9.7 we have $\arctan x = x - \dfrac{x^3}{3} + \dfrac{x^5}{5} - \dfrac{x^7}{7} + \dfrac{x^9}{9} - \dfrac{x^{11}}{11} + \cdots$, so

$$\arctan \frac{1}{5} = \frac{1}{5} - \frac{1}{3 \cdot 5^3} + \frac{1}{5 \cdot 5^5} - \frac{1}{7 \cdot 5^7} + \frac{1}{9 \cdot 5^9} - \frac{1}{11 \cdot 5^{11}} + \cdots$$

This is an alternating series and the size of the terms decreases to 0, so by the Alternating Series Estimation Theorem, the sum lies between s_5 and s_6, that is, $0.197395560 < \arctan \frac{1}{5} < 0.197395562$.

(e) From the series in part (d) we get $\arctan \dfrac{1}{239} = \dfrac{1}{239} - \dfrac{1}{3 \cdot 239^3} + \dfrac{1}{5 \cdot 239^5} - \cdots$. The third term is less than 2.6×10^{-13}, so by the Alternating Series Estimation Theorem, we have, to nine decimal places, $\arctan \frac{1}{239} \approx s_2 \approx 0.004184076$. Thus, $0.004184075 < \arctan \frac{1}{239} < 0.004184077$.

(f) From part (c) we have $\pi = 16\arctan \frac{1}{5} - 4\arctan \frac{1}{239}$, so from parts (d) and (e) we have

$16(0.197395560) - 4(0.004184077) < \pi < 16(0.197395562) - 4(0.004184075) \quad \Rightarrow$

$3.141592652 < \pi < 3.141592692$. So, to 7 decimal places, $\pi \approx 3.1415927$.

8. (a) Let $a = \operatorname{arccot} x$ and $b = \operatorname{arccot} y$ where $0 < a - b < \pi$. Then

$$\cot(a-b) = \frac{1}{\tan(a-b)} = \frac{1 + \tan a \, \tan b}{\tan a - \tan b} = \frac{\frac{1}{\cot a} \cdot \frac{1}{\cot b} + 1}{\frac{1}{\cot a} - \frac{1}{\cot b}} \cdot \frac{\cot a \, \cot b}{\cot a \, \cot b}$$

$$= \frac{1 + \cot a \, \cot b}{\cot b - \cot a} = \frac{1 + \cot(\operatorname{arccot} x)\cot(\operatorname{arccot} y)}{\cot(\operatorname{arccot} y) - \cot(\operatorname{arccot} x)} = \frac{1+xy}{y-x}$$

Now $\operatorname{arccot} x - \operatorname{arccot} y = a - b = \operatorname{arccot}(\cot(a-b)) = \operatorname{arccot} \dfrac{1+xy}{y-x}$ since $0 < a - b < \pi$.

(b) From part (a), we want $\operatorname{arccot}(n^2+n+1)$ to equal $\operatorname{arccot}\dfrac{1+xy}{y-x}$. Note that $1+xy=n^2+n+1 \Leftrightarrow$

$xy=n^2+n=(n+1)n$, so if we let $x=n+1$ and $y=n$, then $y-x=1$. Therefore,

$$\operatorname{arccot}(n^2+n+1)=\operatorname{arccot}(1+n(n+1))=\operatorname{arccot}\frac{1+n(n+1)}{(n+1)-n}=\operatorname{arccot}n-\operatorname{arccot}(n+1)$$

Thus, we have a telescoping series with nth partial sum

$s_n=[\operatorname{arccot}0-\operatorname{arccot}1]+[\operatorname{arccot}1-\operatorname{arccot}2]+\cdots+[\operatorname{arccot}n-\operatorname{arccot}(n+1)]=\operatorname{arccot}0-\operatorname{arccot}(n+1)$.

Thus, $\sum\limits_{n=0}^{\infty}\operatorname{arccot}(n^2+n+1)=\lim\limits_{n\to\infty}s_n=\lim\limits_{n\to\infty}[\operatorname{arccot}0-\operatorname{arccot}(n+1)]=\frac{\pi}{2}-0=\frac{\pi}{2}$.

9. We want $\arctan\left(\dfrac{2}{n^2}\right)$ to equal $\arctan\dfrac{x-y}{1+xy}$. Note that $1+xy=n^2 \Leftrightarrow xy=n^2-1=(n+1)(n-1)$, so if we let $x=n+1$ and $y=n-1$, then $x-y=2$ and $xy\neq -1$. Thus, from Problem 7(a),

$\arctan\left(\dfrac{2}{n^2}\right)=\arctan\dfrac{x-y}{1+xy}=\arctan x-\arctan y=\arctan(n+1)-\arctan(n-1)$. Therefore,

$$\begin{aligned}\sum_{n=1}^{k}\arctan\left(\frac{2}{n^2}\right)&=\sum_{n=1}^{k}[\arctan(n+1)-\arctan(n-1)]\\&=\sum_{n=1}^{k}[\arctan(n+1)-\arctan n+\arctan n-\arctan(n-1)]\\&=\sum_{n=1}^{k}[\arctan(n+1)-\arctan n]+\sum_{n=1}^{k}[\arctan n-\arctan(n-1)]\\&=[\arctan(k+1)-\arctan 1]+[\arctan k-\arctan 0]\quad\text{[since both sums are telescoping]}\\&=\arctan(k+1)-\tfrac{\pi}{4}+\arctan k-0\end{aligned}$$

Now $\sum\limits_{n=1}^{k}\arctan\left(\dfrac{2}{n^2}\right)=\lim\limits_{k\to 0}\sum\limits_{n=1}^{k}\arctan\left(\dfrac{2}{n^2}\right)=\lim\limits_{k\to\infty}\left[\arctan(k+1)-\dfrac{\pi}{4}+\arctan k\right]=\dfrac{\pi}{2}-\dfrac{\pi}{4}+\dfrac{\pi}{2}=\dfrac{3\pi}{4}$.

Note: For all $n\geq 1$, $0\leq\arctan(n-1)<\arctan(n+1)<\frac{\pi}{2}$, so $-\frac{\pi}{2}<\arctan(n+1)-\arctan(n-1)<\frac{\pi}{2}$, and the identity in Problem 7(a) holds.

10. Let's first try the case $k=1$: $a_0+a_1=0 \Rightarrow a_1=-a_0 \Rightarrow$

$$\begin{aligned}\lim_{n\to\infty}\left(a_0\sqrt{n}+a_1\sqrt{n+1}\right)&=\lim_{n\to\infty}\left(a_0\sqrt{n}-a_0\sqrt{n+1}\right)=a_0\lim_{n\to\infty}\left(\sqrt{n}-\sqrt{n+1}\right)\frac{\sqrt{n}+\sqrt{n+1}}{\sqrt{n}+\sqrt{n+1}}\\&=a_0\lim_{n\to\infty}\frac{-1}{\sqrt{n}+\sqrt{n+1}}=0\end{aligned}$$

In general we have $a_0+a_1+\cdots+a_k=0 \Rightarrow a_k=-a_0-a_1-\cdots-a_{k-1} \Rightarrow$

$$\begin{aligned}&\lim_{n\to\infty}\left(a_0\sqrt{n}+a_1\sqrt{n+1}+a_2\sqrt{n+2}+\cdots+a_k\sqrt{n+k}\right)\\&\quad=\lim_{n\to\infty}\left(a_0\sqrt{n}+a_1\sqrt{n+1}+\cdots+a_{k-1}\sqrt{n+k-1}-a_0\sqrt{n+k}-a_1\sqrt{n+k}-\cdots-a_{k-1}\sqrt{n+k}\right)\\&\quad=a_0\lim_{n\to\infty}\left(\sqrt{n}-\sqrt{n+k}\right)+a_1\lim_{n\to\infty}\left(\sqrt{n+1}-\sqrt{n+k}\right)+\cdots+a_{k-1}\lim_{n\to\infty}\left(\sqrt{n+k-1}-\sqrt{n+k}\right)\end{aligned}$$

Each of these limits is 0 by the same type of simplification as in the case $k=1$. So we have

$$\lim_{n\to\infty}\left(a_0\sqrt{n}+a_1\sqrt{n+1}+a_2\sqrt{n+2}+\cdots+a_k\sqrt{n+k}\right)=a_0(0)+a_1(0)+\cdots+a_{k-1}(0)=0$$

11. We start with the geometric series $\sum_{n=0}^{\infty} x^n = \dfrac{1}{1-x}$, $|x| < 1$, and differentiate:

$$\sum_{n=1}^{\infty} nx^{n-1} = \frac{d}{dx}\left(\sum_{n=0}^{\infty} x^n\right) = \frac{d}{dx}\left(\frac{1}{1-x}\right) = \frac{1}{(1-x)^2} \text{ for } |x| < 1 \;\Rightarrow\; \sum_{n=1}^{\infty} nx^n = x\sum_{n=1}^{\infty} nx^{n-1} = \frac{x}{(1-x)^2}$$

for $|x| < 1$. Differentiate again:

$$\sum_{n=1}^{\infty} n^2x^{n-1} = \frac{d}{dx}\frac{x}{(1-x)^2} = \frac{(1-x)^2 - x\cdot 2(1-x)(-1)}{(1-x)^4} = \frac{x+1}{(1-x)^3} \;\Rightarrow\; \sum_{n=1}^{\infty} n^2x^n = \frac{x^2+x}{(1-x)^3} \;\Rightarrow$$

$$\sum_{n=1}^{\infty} n^3x^{n-1} = \frac{d}{dx}\frac{x^2+x}{(1-x)^3} = \frac{(1-x)^3(2x+1) - (x^2+x)3(1-x)^2(-1)}{(1-x)^6} = \frac{x^2+4x+1}{(1-x)^4} \;\Rightarrow$$

$\sum_{n=1}^{\infty} n^3x^n = \dfrac{x^3+4x^2+x}{(1-x)^4}$, $|x| < 1$. The radius of convergence is 1 because that is the radius of convergence for the geometric series we started with. If $x = \pm 1$, the series is $\sum n^3(\pm 1)^n$, which diverges by the Test For Divergence, so the interval of convergence is $(-1, 1)$.

12. Place the y-axis as shown and let the length of each book be L. We want to show that the center of mass of the system of n books lies above the table, that is, $\overline{x} < L$. The x-coordinates of the centers of mass of the books are

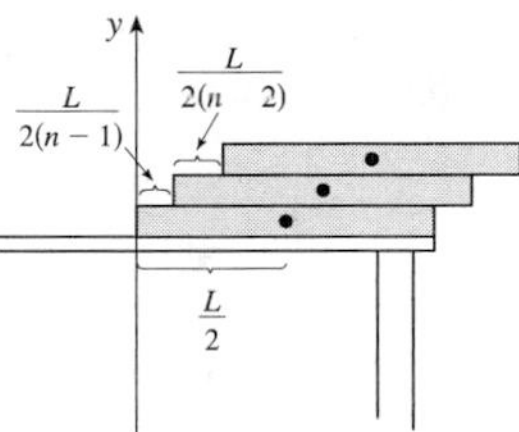

$x_1 = \dfrac{L}{2}$, $x_2 = \dfrac{L}{2(n-1)} + \dfrac{L}{2}$, $x_3 = \dfrac{L}{2(n-1)} + \dfrac{L}{2(n-2)} + \dfrac{L}{2}$, and so on.

Each book has the same mass m, so if there are n books, then

$$\begin{aligned}\overline{x} &= \frac{mx_1 + mx_2 + \cdots + mx_n}{mn} = \frac{x_1 + x_2 + \cdots + x_n}{n} \\ &= \frac{1}{n}\left[\frac{L}{2} + \left(\frac{L}{2(n-1)} + \frac{L}{2}\right) + \left(\frac{L}{2(n-1)} + \frac{L}{2(n-2)} + \frac{L}{2}\right) + \cdots \right. \\ &\qquad \left. + \left(\frac{L}{2(n-1)} + \frac{L}{2(n-2)} + \cdots + \frac{L}{4} + \frac{L}{2} + \frac{L}{2}\right)\right] \\ &= \frac{L}{n}\left[\frac{n-1}{2(n-1)} + \frac{n-2}{2(n-2)} + \cdots + \frac{2}{4} + \frac{1}{2} + \frac{n}{2}\right] = \frac{L}{n}\left[(n-1)\frac{1}{2} + \frac{n}{2}\right] = \frac{2n-1}{2n}L < L\end{aligned}$$

This shows that, no matter how many books are added according to the given scheme, the center of mass lies above the table. It remains to observe that the series $\frac{1}{2} + \frac{1}{4} + \frac{1}{6} + \frac{1}{8} + \cdots = \frac{1}{2}\sum(1/n)$ is divergent (harmonic series), so we can make the top book extend as far as we like beyond the edge of the table if we add enough books.

13.
$$\begin{aligned}\ln\left(1 - \frac{1}{n^2}\right) &= \ln\left(\frac{n^2-1}{n^2}\right) = \ln\frac{(n+1)(n-1)}{n^2} = \ln[(n+1)(n-1)] - \ln n^2 \\ &= \ln(n+1) + \ln(n-1) - 2\ln n = \ln(n-1) - \ln n - \ln n + \ln(n+1) \\ &= \ln\frac{n-1}{n} - [\ln n - \ln(n+1)] = \ln\frac{n-1}{n} - \ln\frac{n}{n+1}.\end{aligned}$$

Let $s_k = \sum_{n=2}^{k} \ln\left(1 - \dfrac{1}{n^2}\right) = \sum_{n=2}^{k}\left(\ln\dfrac{n-1}{n} - \ln\dfrac{n}{n+1}\right)$ for $k \geq 2$. Then

$$s_k = \left(\ln\frac{1}{2} - \ln\frac{2}{3}\right) + \left(\ln\frac{2}{3} - \ln\frac{3}{4}\right) + \cdots + \left(\ln\frac{k-1}{k} - \ln\frac{k}{k+1}\right) = \ln\frac{1}{2} - \ln\frac{k}{k+1}, \text{ so}$$

$$\sum_{n=2}^{\infty} \ln\left(1 - \frac{1}{n^2}\right) = \lim_{k\to\infty} s_k = \lim_{k\to\infty}\left(\ln\frac{1}{2} - \ln\frac{k}{k+1}\right) = \ln\frac{1}{2} - \ln 1 = \ln 1 - \ln 2 - \ln 1 = -\ln 2 \text{ (or } \ln\tfrac{1}{2}\text{)}.$$

14. First notice that both series are absolutely convergent (p-series with $p > 1$.) Let the given expression be called x. Then

$$x = \frac{1 + \frac{1}{2^p} + \frac{1}{3^p} + \frac{1}{4^p} + \cdots}{1 - \frac{1}{2^p} + \frac{1}{3^p} - \frac{1}{4^p} + \cdots} = \frac{1 + \left(2 \cdot \frac{1}{2^p} - \frac{1}{2^p}\right) + \frac{1}{3^p} + \left(2 \cdot \frac{1}{4^p} - \frac{1}{4^p}\right) + \cdots}{1 - \frac{1}{2^p} + \frac{1}{3^p} - \frac{1}{4^p} + \cdots}$$

$$= \frac{\left(1 - \frac{1}{2^p} + \frac{1}{3^p} - \frac{1}{4^p} + \cdots\right) + \left(2 \cdot \frac{1}{2^p} + 2 \cdot \frac{1}{4^p} + 2 \cdot \frac{1}{6^p} + \cdots\right)}{1 - \frac{1}{2^p} + \frac{1}{3^p} - \frac{1}{4^p} + \cdots}$$

$$= 1 + \frac{2\left(\frac{1}{2^p} + \frac{1}{4^p} + \frac{1}{6^p} + \frac{1}{8^p} + \cdots\right)}{1 - \frac{1}{2^p} + \frac{1}{3^p} - \frac{1}{4^p} + \cdots} = 1 + \frac{\frac{1}{2^{p-1}}\left(1 + \frac{1}{2^p} + \frac{1}{3^p} + \frac{1}{4^p} + \cdots\right)}{1 - \frac{1}{2^p} + \frac{1}{3^p} - \frac{1}{4^p} + \cdots} = 1 + 2^{1-p}x$$

Therefore, $x = 1 + 2^{1-p}x \;\Leftrightarrow\; x - 2^{1-p}x = 1 \;\Leftrightarrow\; x(1 - 2^{1-p}) = 1 \;\Leftrightarrow\; x = \dfrac{1}{1 - 2^{1-p}}$.

15. If L is the length of a side of the equilateral triangle, then the area is $A = \frac{1}{2}L \cdot \frac{\sqrt{3}}{2}L = \frac{\sqrt{3}}{4}L^2$ and so $L^2 = \frac{4}{\sqrt{3}}A$.

Let r be the radius of one of the circles. When there are n rows of circles, the figure shows that

$$L = \sqrt{3}\,r + r + (n-2)(2r) + r + \sqrt{3}\,r = r\left(2n - 2 + 2\sqrt{3}\right), \text{ so } r = \frac{L}{2\left(n + \sqrt{3} - 1\right)}.$$

The number of circles is $1 + 2 + \cdots + n = \dfrac{n(n+1)}{2}$, and so the total area of the circles is

$$\begin{aligned} A_n &= \frac{n(n+1)}{2}\pi r^2 = \frac{n(n+1)}{2}\pi \frac{L^2}{4\left(n + \sqrt{3} - 1\right)^2} \\ &= \frac{n(n+1)}{2}\pi \frac{4A/\sqrt{3}}{4\left(n + \sqrt{3} - 1\right)^2} = \frac{n(n+1)}{\left(n + \sqrt{3} - 1\right)^2}\frac{\pi A}{2\sqrt{3}} \quad \Rightarrow \\ \frac{A_n}{A} &= \frac{n(n+1)}{\left(n + \sqrt{3} - 1\right)^2}\frac{\pi}{2\sqrt{3}} \\ &= \frac{1 + 1/n}{\left[1 + \left(\sqrt{3} - 1\right)/n\right]^2}\frac{\pi}{2\sqrt{3}} \to \frac{\pi}{2\sqrt{3}} \text{ as } n \to \infty \end{aligned}$$

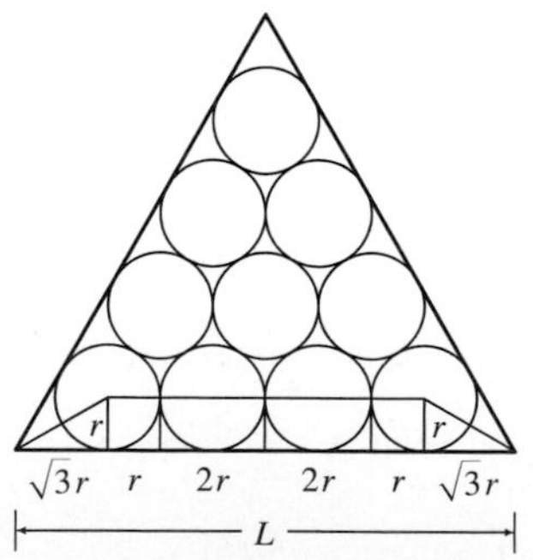

16. Given $a_0 = a_1 = 1$ and $a_n = \dfrac{(n-1)(n-2)a_{n-1} - (n-3)a_{n-2}}{n(n-1)}$, we calculate the next few terms of the sequence:

$a_2 = \dfrac{1 \cdot 0 \cdot a_1 - (-1)a_0}{2 \cdot 1} = \dfrac{1}{2}$, $a_3 = \dfrac{2 \cdot 1 \cdot a_2 - 0 \cdot a_1}{3 \cdot 2} = \dfrac{1}{6}$, $a_4 = \dfrac{3 \cdot 2 \cdot a_3 - 1 \cdot a_2}{4 \cdot 3} = \dfrac{1}{24}$. It seems that $a_n = \dfrac{1}{n!}$,

so we try to prove this by induction. The first step is done, so assume $a_k = \dfrac{1}{k!}$ and $a_{k-1} = \dfrac{1}{(k-1)!}$. Then

$$a_{k+1}=\frac{k(k-1)a_k-(k-2)a_{k-1}}{(k+1)k}=\frac{\dfrac{k(k-1)}{k!}-\dfrac{k-2}{(k-1)!}}{(k+1)k}=\frac{(k-1)-(k-2)}{[(k+1)(k)](k-1)!}=\frac{1}{(k+1)!}$$ and the induction is complete. Therefore, $\sum\limits_{n=0}^{\infty}a_n=\sum\limits_{n=0}^{\infty}\dfrac{1}{n!}=e.$

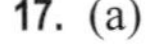

17. (a)

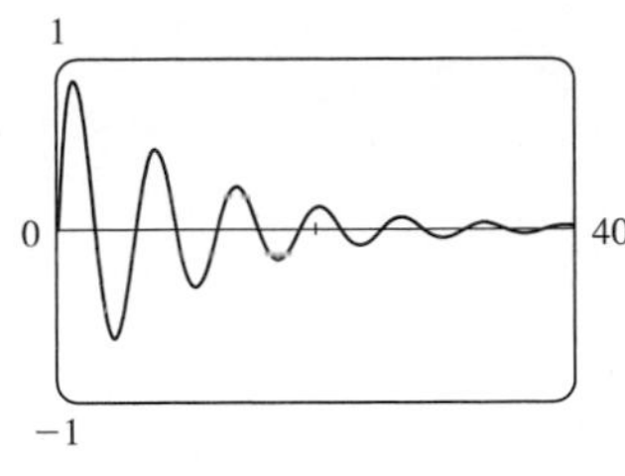

The x-intercepts of the curve occur where $\sin x=0 \Leftrightarrow x=n\pi$, n an integer. So using the formula for disks (and either a CAS or $\sin^2 x=\frac12(1-\cos 2x)$ and Formula 99 to evaluate the integral), the volume of the nth bead is

$$V_n=\pi\int_{(n-1)\pi}^{n\pi}(e^{-x/10}\sin x)^2\,dx=\pi\int_{(n-1)\pi}^{n\pi}e^{-x/5}\sin^2 x\,dx$$
$$=\tfrac{250\pi}{101}\left(e^{-(n-1)\pi/5}-e^{-n\pi/5}\right)$$

(b) The total volume is

$$\pi\int_0^{\infty}e^{-x/5}\sin^2 x\,dx=\sum_{n=1}^{\infty}V_n=\tfrac{250\pi}{101}\sum_{n=1}^{\infty}\left[e^{-(n-1)\pi/5}-e^{-n\pi/5}\right]=\tfrac{250\pi}{101}\quad\text{[telescoping sum]}.$$

Another method: If the volume in part (a) has been written as $V_n=\frac{250\pi}{101}e^{-n\pi/5}(e^{\pi/5}-1)$, then we recognize $\sum\limits_{n=1}^{\infty}V_n$ as a geometric series with $a=\frac{250\pi}{101}(1-e^{-\pi/5})$ and $r=e^{-\pi/5}$.

18. (a) Since P_n is defined as thc midpoint of $P_{n-4}P_{n-3}$, $x_n=\frac12(x_{n-4}+x_{n-3})$ for $n\geq 5$. So we prove by induction that $\frac12 x_n+x_{n+1}+x_{n+2}+x_{n+3}=2$. The case $n=1$ is immediate, since $\frac12\cdot 0+1+1+0=2$. Assume that the result holds for $n=k-1$, that is, $\frac12 x_{k-1}+x_k+x_{k+1}+x_{k+2}=2$. Then for $n=k$,

$$\tfrac12 x_k+x_{k+1}+x_{k+2}+x_{k+3}=\tfrac12 x_k+x_{k+1}+x_{k+2}+\tfrac12(x_{k+3-4}+x_{k+3-3})\quad\text{[by above]}$$
$$=\tfrac12 x_{k-1}+x_k+x_{k+1}+x_{k+2}=2\quad\text{[by the induction hypothesis]}$$

Similarly, for $n\geq 5$, $y_n=\frac12(y_{n-4}+y_{n-3})$, so the same argument as above holds for y, with 2 replaced by $\frac12 y_1+y_2+y_3+y_4=\frac12\cdot 1+1+0+0=\frac32$. So $\frac12 y_n+y_{n+1}+y_{n+2}+y_{n+3}=\frac32$ for all n.

(b) $\lim\limits_{n\to\infty}\left(\frac12 x_n+x_{n+1}+x_{n+2}+x_{n+3}\right)=\frac12\lim\limits_{n\to\infty}x_n+\lim\limits_{n\to\infty}x_{n+1}+\lim\limits_{n\to\infty}x_{n+2}+\lim\limits_{n\to\infty}x_{n+3}=2$. Since all the limits on the left hand side are the same, we get $\frac72\lim\limits_{n\to\infty}x_n=2 \Rightarrow \lim\limits_{n\to\infty}x_n=\frac47$. In the same way, $\frac72\lim\limits_{n\to\infty}y_n=\frac32 \Rightarrow \lim\limits_{n\to\infty}y_n=\frac37$, so $P=\left(\frac47,\frac37\right)$.

19. By Table 11.10.1, $\tan^{-1}x=\sum\limits_{n=0}^{\infty}(-1)^n\dfrac{x^{2n+1}}{2n+1}$ for $|x|<1$. In particular, for $x=\dfrac{1}{\sqrt3}$, we have $\dfrac{\pi}{6}=\tan^{-1}\left(\dfrac{1}{\sqrt3}\right)=\sum\limits_{n=0}^{\infty}(-1)^n\dfrac{(1/\sqrt3)^{2n+1}}{2n+1}=\sum\limits_{n=0}^{\infty}(-1)^n\left(\dfrac13\right)^n\dfrac{1}{\sqrt3}\dfrac{1}{2n+1}$, so

$$\pi=\frac{6}{\sqrt3}\sum_{n=0}^{\infty}\frac{(-1)^n}{(2n+1)3^n}=2\sqrt3\sum_{n=0}^{\infty}\frac{(-1)^n}{(2n+1)3^n}=2\sqrt3\left(1+\sum_{n=1}^{\infty}\frac{(-1)^n}{(2n+1)3^n}\right)\Rightarrow\sum_{n=1}^{\infty}\frac{(-1)^n}{(2n+1)3^n}=\frac{\pi}{2\sqrt3}-1.$$

20. (a) Using $s_n = a + ar + ar^2 + \cdots + ar^{n-1} = \dfrac{a(1-r^n)}{1-r}$,

$$1 - x + x^2 - x^3 + \cdots + x^{2n-2} - x^{2n-1} = \frac{1\left[1-(-x)^{2n}\right]}{1-(-x)} = \frac{1-x^{2n}}{1+x}.$$

(b) $\displaystyle\int_0^1 (1 - x + x^2 - x^3 + \cdots + x^{2n-2} - x^{2n-1})\,dx = \int_0^1 \frac{1-x^{2n}}{1+x}\,dx \Rightarrow$

$$\left[x - \frac{x^2}{2} + \frac{x^3}{3} - \frac{x^4}{4} + \cdots + \frac{x^{2n-1}}{2n-1} - \frac{x^{2n}}{2n}\right]_0^1 = \int_0^1 \frac{dx}{1+x} - \int_0^1 \frac{x^{2n}}{1+x}\,dx \Rightarrow$$

$$1 - \frac{1}{2} + \frac{1}{3} - \frac{1}{4} + \cdots + \frac{1}{2n-1} - \frac{1}{2n} = \int_0^1 \frac{dx}{1+x} - \int_0^1 \frac{x^{2n}}{1+x}\,dx$$

(c) Since $1 - \dfrac{1}{2} = \dfrac{1}{1\cdot 2}, \dfrac{1}{3} - \dfrac{1}{4} = \dfrac{1}{3\cdot 4}, \ldots, \dfrac{1}{2n-1} - \dfrac{1}{2n} = \dfrac{1}{(2n-1)(2n)}$, we see from part (b) that

$$\frac{1}{1\cdot 2} + \frac{1}{3\cdot 4} + \cdots + \frac{1}{(2n-1)(2n)} - \int_0^1 \frac{dx}{1+x} = -\int_0^1 \frac{x^{2n}}{1+x}\,dx. \text{ Thus,}$$

$$\left|\frac{1}{1\cdot 2} + \frac{1}{3\cdot 4} + \cdots + \frac{1}{(2n-1)(2n)} - \int_0^1 \frac{dx}{1+x}\right| = \int_0^1 \frac{x^{2n}}{1+x}\,dx < \int_0^1 x^{2n}\,dx$$

$\left[\text{since } \dfrac{x^{2n}}{1+x} < x^{2n} \text{ for } 0 < x \le 1\right]$.

(d) Note that $\displaystyle\int_0^1 \frac{dx}{1+x} = \Big[\ln(1+x)\Big]_0^1 = \ln 2$ and $\displaystyle\int_0^1 x^{2n}\,dx = \left[\frac{x^{2n+1}}{2n+1}\right]_0^1 = \frac{1}{2n+1}$. So part (c) becomes

$\left|\dfrac{1}{1\cdot 2} + \dfrac{1}{3\cdot 4} + \cdots + \dfrac{1}{(2n-1)(2n)} - \ln 2\right| < \dfrac{1}{2n+1}$. In other words, the nth partial sum s_n of the given series satisfies $|s_n - \ln 2| < \dfrac{1}{2n+1}$. Thus, $\displaystyle\lim_{n\to\infty} s_n = \ln 2$, that is, $\dfrac{1}{1\cdot 2} + \dfrac{1}{3\cdot 4} + \dfrac{1}{5\cdot 6} + \dfrac{1}{7\cdot 8} + \cdots = \ln 2$.

21. Let $f(x)$ denote the left-hand side of the equation $1 + \dfrac{x}{2!} + \dfrac{x^2}{4!} + \dfrac{x^3}{6!} + \dfrac{x^4}{8!} + \cdots = 0$. If $x \ge 0$, then $f(x) \ge 1$ and there are no solutions of the equation. Note that $f(-x^2) = 1 - \dfrac{x^2}{2!} + \dfrac{x^4}{4!} - \dfrac{x^6}{6!} + \dfrac{x^8}{8!} - \cdots = \cos x$. The solutions of $\cos x = 0$ for $x < 0$ are given by $x = \dfrac{\pi}{2} - \pi k$, where k is a positive integer. Thus, the solutions of $f(x) = 0$ are $x = -\left(\dfrac{\pi}{2} - \pi k\right)^2$, where k is a positive integer.

22. Suppose the base of the first right triangle has length a. Then by repeated use of the Pythagorean theorem, we find that the base of the second right triangle has length $\sqrt{1+a^2}$, the base of the third right triangle has length $\sqrt{2+a^2}$, and in general, the nth right triangle has base of length $\sqrt{n-1+a^2}$ and hypotenuse of length $\sqrt{n+a^2}$. Thus, $\theta_n = \tan^{-1}\left(1/\sqrt{n-1+a^2}\right)$ and $\displaystyle\sum_{n=1}^{\infty} \theta_n = \sum_{n=1}^{\infty} \tan^{-1}\left(\frac{1}{\sqrt{n-1+a^2}}\right) = \sum_{n=0}^{\infty} \tan^{-1}\left(\frac{1}{\sqrt{n+a^2}}\right)$. We wish to show that this series diverges.

First notice that the series $\displaystyle\sum_{n=1}^{\infty} \frac{1}{\sqrt{n+a^2}}$ diverges by the Limit Comparison Test with the divergent p-series $\displaystyle\sum_{n=1}^{\infty} \frac{1}{\sqrt{n}}$ $\left[p = \tfrac{1}{2} \le 1\right]$ since $\displaystyle\lim_{n\to\infty} \frac{1/\sqrt{n+a^2}}{1/\sqrt{n}} = \lim_{n\to\infty} \frac{\sqrt{n}}{\sqrt{n+a^2}} = \lim_{n\to\infty} \sqrt{\frac{n}{n+a^2}} = \lim_{n\to\infty} \sqrt{\frac{1}{1+a^2/n}} = 1 > 0$. Thus,

$\sum\limits_{n=0}^{\infty} \dfrac{1}{\sqrt{n+a^2}}$ also diverges. Now $\sum\limits_{n=0}^{\infty} \tan^{-1}\left(\dfrac{1}{\sqrt{n+a^2}}\right)$ diverges by the Limit Comparison Test with $\sum\limits_{n=0}^{\infty} \dfrac{1}{\sqrt{n+a^2}}$ since

$$\lim_{n\to\infty} \frac{\tan^{-1}(1/\sqrt{n+a^2}\,)}{1/\sqrt{n+a^2}} = \lim_{x\to\infty} \frac{\tan^{-1}(1/\sqrt{x+a^2}\,)}{1/\sqrt{x+a^2}} = \lim_{y\to\infty} \frac{\tan^{-1}(1/y)}{1/y} \qquad \left[y = \sqrt{x+a^2}\,\right]$$

$$= \lim_{z\to 0^+} \frac{\tan^{-1} z}{z} \quad [z = 1/y] \quad \overset{\text{H}}{=} \lim_{z\to 0^+} \frac{1/(1+z^2)}{1} = 1 > 0$$

Thus, $\sum\limits_{n=1}^{\infty} \theta_n$ is a divergent series.

23. Call the series S. We group the terms according to the number of digits in their denominators:

$$S = \underbrace{\left(\tfrac{1}{1} + \tfrac{1}{2} + \cdots + \tfrac{1}{8} + \tfrac{1}{9}\right)}_{g_1} + \underbrace{\left(\tfrac{1}{11} + \cdots + \tfrac{1}{99}\right)}_{g_2} + \underbrace{\left(\tfrac{1}{111} + \cdots + \tfrac{1}{999}\right)}_{g_3} + \cdots$$

Now in the group g_n, since we have 9 choices for each of the n digits in the denominator, there are 9^n terms.

Furthermore, each term in g_n is less than $\frac{1}{10^{n-1}}$ [except for the first term in g_1]. So $g_n < 9^n \cdot \frac{1}{10^{n-1}} = 9\left(\frac{9}{10}\right)^{n-1}$.

Now $\sum\limits_{n=1}^{\infty} 9\left(\frac{9}{10}\right)^{n-1}$ is a geometric series with $a = 9$ and $r = \frac{9}{10} < 1$. Therefore, by the Comparison Test,

$$S = \sum_{n=1}^{\infty} g_n < \sum_{n=1}^{\infty} 9\left(\tfrac{9}{10}\right)^{n-1} = \frac{9}{1 - 9/10} = 90.$$

24. (a) Let $f(x) = \dfrac{x}{1-x-x^2} = \sum\limits_{n=0}^{\infty} c_n x^n = c_0 + c_1 x + c_2 x^2 + c_3 x^3 + \cdots$. Then

$$x = (1 - x - x^2)(c_0 + c_1 x + c_2 x^2 + c_3 x^3 + \cdots)$$

$$\begin{aligned} x = c_0 + c_1 x + c_2 x^2 + c_3 x^3 &+ c_4 x^4 + c_5 x^5 + \cdots \\ - c_0 x - c_1 x^2 - c_2 x^3 &- c_3 x^4 - c_4 x^5 - \cdots \\ - c_0 x^2 - c_1 x^3 &- c_2 x^4 - c_3 x^5 - \cdots \end{aligned}$$

$$x = c_0 + (c_1 - c_0)x + (c_2 - c_1 - c_0)x^2 + (c_3 - c_2 - c_1)x^3 + \cdots$$

Comparing coefficients of powers of x gives us $c_0 = 0$ and

$$\begin{aligned} c_1 - c_0 &= 1 &&\Rightarrow\quad c_1 = c_0 + 1 = 1 \\ c_2 - c_1 - c_0 &= 0 &&\Rightarrow\quad c_2 = c_1 + c_0 = 1 + 0 = 1 \\ c_3 - c_2 - c_1 &= 0 &&\Rightarrow\quad c_3 = c_2 + c_1 = 1 + 1 = 2 \end{aligned}$$

In general, we have $c_n = c_{n-1} + c_{n-2}$ for $n \ge 3$. Each c_n is equal to the nth Fibonacci number, that is,

$$\sum_{n=0}^{\infty} c_n x^n = \sum_{n=1}^{\infty} c_n x^n = \sum_{n=1}^{\infty} f_n x^n$$

(b) Completing the square on $x^2 + x - 1$ gives us

$$\left(x^2 + x + \frac{1}{4}\right) - 1 - \frac{1}{4} = \left(x + \frac{1}{2}\right)^2 - \frac{5}{4} = \left(x + \frac{1}{2}\right)^2 - \left(\frac{\sqrt{5}}{2}\right)^2$$

$$= \left(x + \frac{1}{2} + \frac{\sqrt{5}}{2}\right)\left(x + \frac{1}{2} - \frac{\sqrt{5}}{2}\right) = \left(x + \frac{1+\sqrt{5}}{2}\right)\left(x + \frac{1-\sqrt{5}}{2}\right)$$

[continued]

So $\dfrac{x}{1-x-x^2} = \dfrac{-x}{x^2+x-1} = \dfrac{-x}{\left(x+\frac{1+\sqrt{5}}{2}\right)\left(x+\frac{1-\sqrt{5}}{2}\right)}$. The factors in the denominator are linear,

so the partial fraction decomposition is

$$\frac{-x}{\left(x+\frac{1+\sqrt{5}}{2}\right)\left(x+\frac{1-\sqrt{5}}{2}\right)} = \frac{A}{x+\frac{1+\sqrt{5}}{2}} + \frac{B}{x+\frac{1-\sqrt{5}}{2}} \quad -x = A\left(x+\tfrac{1-\sqrt{5}}{2}\right) + B\left(x+\tfrac{1+\sqrt{5}}{2}\right)$$

If $x = \frac{-1+\sqrt{5}}{2}$, then $-\frac{-1+\sqrt{5}}{2} = B\sqrt{5} \;\Rightarrow\; B = \frac{1-\sqrt{5}}{2\sqrt{5}}$.

If $x = \frac{-1-\sqrt{5}}{2}$, then $-\frac{-1-\sqrt{5}}{2} = A(-\sqrt{5}) \;\Rightarrow\; A = \frac{1+\sqrt{5}}{-2\sqrt{5}}$. Thus,

$$\frac{x}{1-x-x^2} = \frac{\dfrac{1+\sqrt{5}}{-2\sqrt{5}}}{x+\dfrac{1+\sqrt{5}}{2}} + \frac{\dfrac{1-\sqrt{5}}{2\sqrt{5}}}{x+\dfrac{1-\sqrt{5}}{2}} = \frac{\dfrac{1+\sqrt{5}}{-2\sqrt{5}}}{x+\dfrac{1+\sqrt{5}}{2}} \cdot \frac{\dfrac{2}{1+\sqrt{5}}}{\dfrac{2}{1+\sqrt{5}}} + \frac{\dfrac{1-\sqrt{5}}{2\sqrt{5}}}{x+\dfrac{1-\sqrt{5}}{2}} \cdot \frac{\dfrac{2}{1-\sqrt{5}}}{\dfrac{2}{1-\sqrt{5}}}$$

$$= \frac{-1/\sqrt{5}}{1+\dfrac{2}{1+\sqrt{5}}x} + \frac{1/\sqrt{5}}{1+\dfrac{2}{1-\sqrt{5}}x} = -\frac{1}{\sqrt{5}}\sum_{n=0}^{\infty}\left(-\frac{2}{1+\sqrt{5}}x\right)^n + \frac{1}{\sqrt{5}}\sum_{n=0}^{\infty}\left(-\frac{2}{1-\sqrt{5}}x\right)^n$$

$$= \frac{1}{\sqrt{5}}\sum_{n=0}^{\infty}\left[\left(\frac{-2}{1-\sqrt{5}}\right)^n - \left(\frac{-2}{1+\sqrt{5}}\right)^n\right]x^n$$

$$= \frac{1}{\sqrt{5}}\sum_{n=1}^{\infty}\left[\frac{(-2)^n\left(1+\sqrt{5}\right)^n - (-2)^n\left(1-\sqrt{5}\right)^n}{\left(1-\sqrt{5}\right)^n\left(1+\sqrt{5}\right)^n}\right]x^n \qquad \text{[the } n=0 \text{ term is 0]}$$

$$= \frac{1}{\sqrt{5}}\sum_{n=1}^{\infty}\left[\frac{(-2)^n\left(\left(1+\sqrt{5}\right)^n - \left(1-\sqrt{5}\right)^n\right)}{(1-5)^n}\right]x^n$$

$$= \frac{1}{\sqrt{5}}\sum_{n=1}^{\infty}\left[\frac{\left(1+\sqrt{5}\right)^n - \left(1-\sqrt{5}\right)^n}{2^n}\right]x^n \qquad [(-4)^n = (-2)^n \cdot 2^n]$$

From part (a), this series must equal $\sum\limits_{n=1}^{\infty} f_n x^n$, so $f_n = \dfrac{\left(1+\sqrt{5}\right)^n - \left(1-\sqrt{5}\right)^n}{2^n\sqrt{5}}$, which is an explicit formula for the nth Fibonacci number.

25. $u = 1 + \dfrac{x^3}{3!} + \dfrac{x^6}{6!} + \dfrac{x^9}{9!} + \cdots,\; v = x + \dfrac{x^4}{4!} + \dfrac{x^7}{7!} + \dfrac{x^{10}}{10!} + \cdots,\; w = \dfrac{x^2}{2!} + \dfrac{x^5}{5!} + \dfrac{x^8}{8!} + \cdots$.

Use the Ratio Test to show that the series for u, v, and w have positive radii of convergence (∞ in each case), so Theorem 11.9.2 applies, and hence, we may differentiate each of these series:

$$\frac{du}{dx} = \frac{3x^2}{3!} + \frac{6x^5}{6!} + \frac{9x^8}{9!} + \cdots = \frac{x^2}{2!} + \frac{x^5}{5!} + \frac{x^8}{8!} + \cdots = w$$

Similarly, $\dfrac{dv}{dx} = 1 + \dfrac{x^3}{3!} + \dfrac{x^6}{6!} + \dfrac{x^9}{9!} + \cdots = u$, and $\dfrac{dw}{dx} = x + \dfrac{x^4}{4!} + \dfrac{x^7}{7!} + \dfrac{x^{10}}{10!} + \cdots = v$.

So $u' = w$, $v' = u$, and $w' = v$. Now differentiate the left-hand side of the desired equation:

$$\frac{d}{dx}(u^3 + v^3 + w^3 - 3uvw) = 3u^2u' + 3v^2v' + 3w^2w' - 3(u'vw + uv'w + uvw')$$

$$= 3u^2w + 3v^2u + 3w^2v - 3(vw^2 + u^2w + uv^2) = 0 \quad \Rightarrow$$

$u^3 + v^3 + w^3 - 3uvw = C$. To find the value of the constant C, we put $x = 0$ in the last equation and get

$1^3 + 0^3 + 0^3 - 3(1 \cdot 0 \cdot 0) = C \Rightarrow C = 1$, so $u^3 + v^3 + w^3 - 3uvw = 1$.

26. **To prove:** If $n > 1$, then the nth partial sum $s_n = \sum\limits_{i=1}^{n} \dfrac{1}{i}$ of the harmonic series is not an integer.

Proof: Let 2^k be the largest power of 2 that is less than or equal to n and let M be the product of all the odd positive integers that are less than or equal to n. Suppose that $s_n = m$, an integer. Then $M2^k s_n = M2^k m$. Since $n \geq 2$, we have $k \geq 1$, and hence, $M2^k m$ is an even integer. We will show that $M2^k s_n$ is an odd integer, contradicting the equality $M2^k s_n = M2^k m$ and showing that the supposition that s_n is an integer must have been wrong.

$M2^k s_n = M2^k \sum\limits_{i=1}^{n} \dfrac{1}{i} = \sum\limits_{i=1}^{n} \dfrac{M2^k}{i}$. If $1 \leq i \leq n$ and i is odd, then $\dfrac{M}{i}$ is an odd integer since i is one of the odd integers that were multiplied together to form M. Thus, $\dfrac{M2^k}{i}$ is an even integer in this case. If $1 \leq i \leq n$ and i is even, then we can write $i = 2^r l$, where 2^r is the largest power of 2 dividing i and l is odd. If $r < k$, then $\dfrac{M2^k}{i} = \dfrac{2^k}{2^r} \cdot \dfrac{M}{l} = 2^{k-r} \dfrac{M}{l}$, which is an even integer, the product of the even integer 2^{k-r} and the odd integer $\dfrac{M}{l}$. If $r = k$, then $l = 1$, since $l > 1 = l \geq 2 \Rightarrow$ $i = 2^k l \geq 2^k \cdot 2 = 2^{k+1}$, contrary to the choice of 2^k as the largest power of 2 that is less than or equal to n. This shows that $r = k$ only when $i = 2^k$. In that case, $\dfrac{M2^k}{i} = M$, an *odd* integer. Since $\dfrac{M2^k}{i}$ is an even integer for every i except 2^k and $\dfrac{M2^k}{i}$ is an odd integer when $i = 2^k$, we see that $M2^k s_n$ is an odd integer. This concludes the proof.

APPENDIXES

A Numbers, Inequalities, and Absolute Values

1. $|5 - 23| = |-18| = 18$

2. $|5| - |-23| = 5 - 23 = -18$

3. $|-\pi| = \pi$ because $\pi > 0$.

4. $|\pi - 2| = \pi - 2$ because $\pi - 2 > 0$.

5. $\left|\sqrt{5} - 5\right| = -\left(\sqrt{5} - 5\right) = 5 - \sqrt{5}$ because $\sqrt{5} - 5 < 0$.

6. $\left||-2| - |-3|\right| = |2 - 3| = |-1| = 1$

7. If $x < 2$, $x - 2 < 0$, so $|x - 2| = -(x - 2) = 2 - x$.

8. If $x > 2$, $x - 2 > 0$, so $|x - 2| = x - 2$.

9. $|x + 1| = \begin{cases} x + 1 & \text{if } x + 1 \geq 0 \\ -(x + 1) & \text{if } x + 1 < 0 \end{cases} = \begin{cases} x + 1 & \text{if } x \geq -1 \\ -x - 1 & \text{if } x < -1 \end{cases}$

10. $|2x - 1| = \begin{cases} 2x - 1 & \text{if } 2x - 1 \geq 0 \\ -(2x - 1) & \text{if } 2x - 1 < 0 \end{cases} = \begin{cases} 2x - 1 & \text{if } x \geq \frac{1}{2} \\ 1 - 2x & \text{if } x < \frac{1}{2} \end{cases}$

11. $\left|x^2 + 1\right| = x^2 + 1$ [since $x^2 + 1 \geq 0$ for all x].

12. Determine when $1 - 2x^2 < 0 \quad \Leftrightarrow \quad 1 < 2x^2 \quad \Leftrightarrow \quad x^2 > \frac{1}{2} \quad \Leftrightarrow \quad \sqrt{x^2} > \sqrt{\frac{1}{2}} \quad \Leftrightarrow \quad |x| > \sqrt{\frac{1}{2}} \quad \Leftrightarrow$

$x < -\frac{1}{\sqrt{2}}$ or $x > \frac{1}{\sqrt{2}}$. Thus, $\left|1 - 2x^2\right| = \begin{cases} 1 - 2x^2 & \text{if } -\frac{1}{\sqrt{2}} \leq x \leq \frac{1}{\sqrt{2}} \\ 2x^2 - 1 & \text{if } x < -\frac{1}{\sqrt{2}} \text{ or } x > \frac{1}{\sqrt{2}} \end{cases}$

13. $2x + 7 > 3 \quad \Leftrightarrow \quad 2x > -4 \quad \Leftrightarrow \quad x > -2$, so $x \in (-2, \infty)$.

14. $3x - 11 < 4 \quad \Leftrightarrow \quad 3x < 15 \quad \Leftrightarrow \quad x < 5$, so $x \in (-\infty, 5)$.

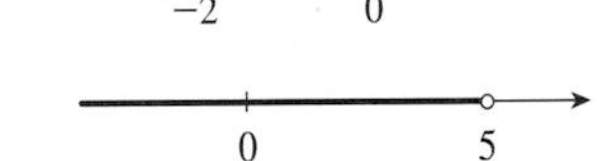

15. $1 - x \leq 2 \quad \Leftrightarrow \quad -x \leq 1 \quad \Leftrightarrow \quad x \geq -1$, so $x \in [-1, \infty)$.

16. $4 - 3x \geq 6 \quad \Leftrightarrow \quad -3x \geq 2 \quad \Leftrightarrow \quad x \leq -\frac{2}{3}$, so $x \in \left(-\infty, -\frac{2}{3}\right]$.

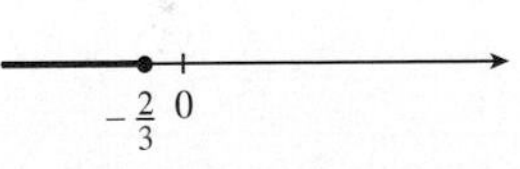

17. $2x + 1 < 5x - 8 \quad \Leftrightarrow \quad 9 < 3x \quad \Leftrightarrow \quad 3 < x$, so $x \in (3, \infty)$.

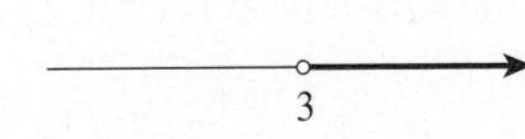

18. $1 + 5x > 5 - 3x \quad \Leftrightarrow \quad 8x > 4 \quad \Leftrightarrow \quad x > \frac{1}{2}$, so $x \in \left(\frac{1}{2}, \infty\right)$.

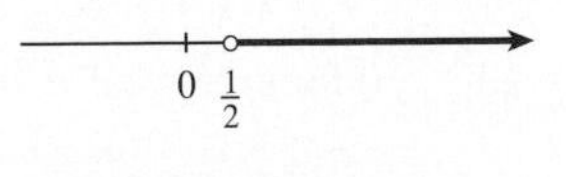

19. $-1 < 2x - 5 < 7 \quad \Leftrightarrow \quad 4 < 2x < 12 \quad \Leftrightarrow \quad 2 < x < 6$, so $x \in (2, 6)$.

20. $1 < 3x + 4 \le 16 \Leftrightarrow -3 < 3x \le 12 \Leftrightarrow -1 < x \le 4$, so $x \in (-1, 4]$.

21. $0 \le 1 - x < 1 \Leftrightarrow -1 \le -x < 0 \Leftrightarrow 1 \ge x > 0$, so $x \in (0, 1]$.

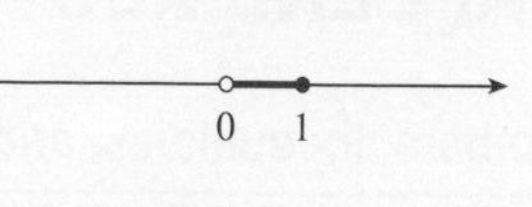

22. $-5 \le 3 - 2x \le 9 \Leftrightarrow -8 \le -2x \le 6 \Leftrightarrow 4 \ge x \ge -3$, so $x \in [-3, 4]$.

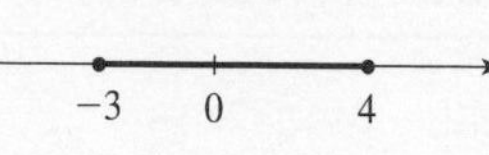

23. $4x < 2x + 1 \le 3x + 2$. So $4x < 2x + 1 \Leftrightarrow 2x < 1 \Leftrightarrow x < \frac{1}{2}$, and $2x + 1 \le 3x + 2 \Leftrightarrow -1 \le x$. Thus, $x \in \left[-1, \frac{1}{2}\right)$.

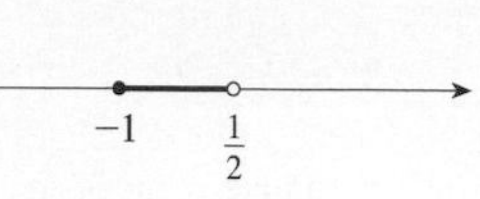

24. $2x - 3 < x + 4 < 3x - 2$. So $2x - 3 < x + 4 \Leftrightarrow x < 7$, and $x + 4 < 3x - 2 \Leftrightarrow 6 < 2x \Leftrightarrow 3 < x$, so $x \in (3, 7)$.

25. $(x - 1)(x - 2) > 0$.

Case 1: (both factors are positive, so their product is positive) $x - 1 > 0 \Leftrightarrow x > 1$, and $x - 2 > 0 \Leftrightarrow x > 2$, so $x \in (2, \infty)$.

Case 2: (both factors are negative, so their product is positive) $x - 1 < 0 \Leftrightarrow x < 1$, and $x - 2 < 0 \Leftrightarrow x < 2$, so $x \in (-\infty, 1)$.

Thus, the solution set is $(-\infty, 1) \cup (2, \infty)$.

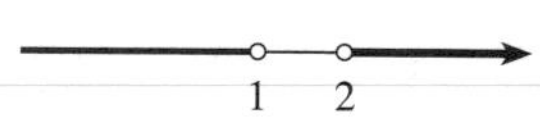

26. $(2x + 3)(x - 1) \ge 0$.

Case 1: $2x + 3 \ge 0 \Leftrightarrow x \ge -\frac{3}{2}$, and $x - 1 \ge 0 \Leftrightarrow x \ge 1$, so $x \in [1, \infty)$.

Case 2: $2x + 3 \le 0 \Leftrightarrow x \le -\frac{3}{2}$, and $x - 1 \le 0 \Leftrightarrow x \le 1$, so $x \in \left(-\infty, -\frac{3}{2}\right]$.

Thus, the solution set is $\left(-\infty, -\frac{3}{2}\right] \cup [1, \infty)$.

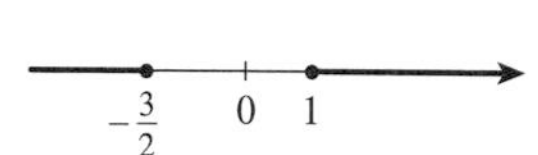

27. $2x^2 + x \le 1 \Leftrightarrow 2x^2 + x - 1 \le 0 \Leftrightarrow (2x - 1)(x + 1) \le 0$.

Case 1: $2x - 1 \ge 0 \Leftrightarrow x \ge \frac{1}{2}$, and $x + 1 \le 0 \Leftrightarrow x \le -1$, which is an impossible combination.

Case 2: $2x - 1 \le 0 \Leftrightarrow x \le \frac{1}{2}$, and $x + 1 \ge 0 \Leftrightarrow x \ge -1$, so $x \in \left[-1, \frac{1}{2}\right]$.

Thus, the solution set is $\left[-1, \frac{1}{2}\right]$.

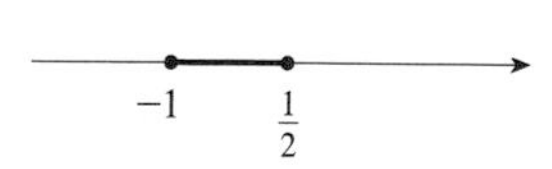

28. $x^2 < 2x + 8 \Leftrightarrow x^2 - 2x - 8 < 0 \Leftrightarrow (x - 4)(x + 2) < 0$.

Case 1: $x > 4$ and $x < -2$, which is impossible.

Case 2: $x < 4$ and $x > -2$.

Thus, the solution set is $(-2, 4)$.

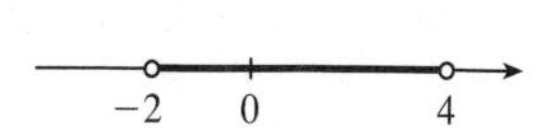

29. $x^2 + x + 1 > 0 \Leftrightarrow x^2 + x + \frac{1}{4} + \frac{3}{4} > 0 \Leftrightarrow \left(x + \frac{1}{2}\right)^2 + \frac{3}{4} > 0$. But since $\left(x + \frac{1}{2}\right)^2 \ge 0$ for every real x, the original inequality will be true for all real x as well.

Thus, the solution set is $(-\infty, \infty)$.

30. $x^2 + x > 1 \;\Leftrightarrow\; x^2 + x - 1 > 0$. Using the quadratic formula, we obtain

$x^2 + x - 1 = \left(x - \frac{-1-\sqrt{5}}{2}\right)\left(x - \frac{-1+\sqrt{5}}{2}\right) > 0$.

Case 1: $x - \frac{-1-\sqrt{5}}{2} > 0$ and $x - \frac{-1+\sqrt{5}}{2} > 0$, so that $x > \frac{-1+\sqrt{5}}{2}$.

Case 2: $x - \frac{-1-\sqrt{5}}{2} < 0$ and $x - \frac{-1+\sqrt{5}}{2} < 0$, so that $x < \frac{-1-\sqrt{5}}{2}$.

Thus, the solution set is $\left(-\infty, \frac{-1-\sqrt{5}}{2}\right) \cup \left(\frac{-1+\sqrt{5}}{2}, \infty\right)$.

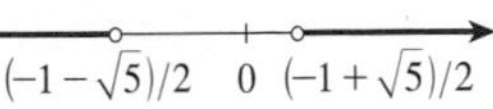

31. $x^2 < 3 \;\Leftrightarrow\; x^2 - 3 < 0 \;\Leftrightarrow\; (x - \sqrt{3})(x + \sqrt{3}) < 0$.

Case 1: $x > \sqrt{3}$ and $x < -\sqrt{3}$, which is impossible.

Case 2: $x < \sqrt{3}$ and $x > -\sqrt{3}$.

Thus, the solution set is $(-\sqrt{3}, \sqrt{3})$.

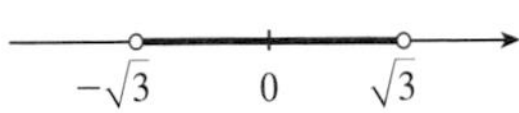

Another method: $x^2 < 3 \;\Leftrightarrow\; |x| < \sqrt{3} \;\Leftrightarrow\; -\sqrt{3} < x < \sqrt{3}$.

32. $x^2 \geq 5 \;\Leftrightarrow\; x^2 - 5 \geq 0 \;\Leftrightarrow\; (x - \sqrt{5})(x + \sqrt{5}) \geq 0$.

Case 1: $x \geq \sqrt{5}$ and $x \geq -\sqrt{5}$, so $x \in [\sqrt{5}, \infty)$.

Case 2: $x \leq \sqrt{5}$ and $x \leq -\sqrt{5}$, so $x \in (-\infty, -\sqrt{5}]$.

Thus, the solution set is $(-\infty, -\sqrt{5}] \cup [\sqrt{5}, \infty)$.

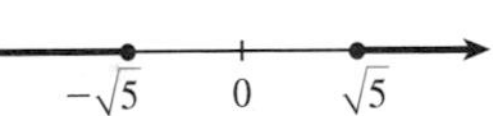

Another method: $x^2 \geq 5 \;\Leftrightarrow\; |x| \geq \sqrt{5} \;\Leftrightarrow\; x \geq \sqrt{5}$ or $x \leq -\sqrt{5}$.

33. $x^3 - x^2 \leq 0 \;\Leftrightarrow\; x^2(x - 1) \leq 0$. Since $x^2 \geq 0$ for all x, the inequality is satisfied when $x - 1 \leq 0 \;\Leftrightarrow\; x \leq 1$.

Thus, the solution set is $(-\infty, 1]$.

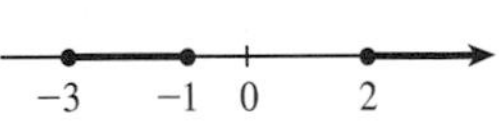

34. $(x + 1)(x - 2)(x + 3) = 0 \;\Leftrightarrow\; x = -1, 2,$ or 3. Construct a chart:

Interval	$x+1$	$x-2$	$x+3$	$(x+1)(x-2)(x+3)$
$x < -3$	$-$	$-$	$-$	$-$
$-3 < x < -1$	$-$	$-$	$+$	$+$
$-1 < x < 2$	$+$	$-$	$+$	$-$
$x > 2$	$+$	$+$	$+$	$+$

Thus, $(x + 1)(x - 2)(x + 3) \geq 0$ on $[-3, -1]$ and $[2, \infty)$, and the solution set is $[-3, -1] \cup [2, \infty)$.

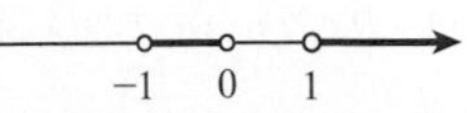

35. $x^3 > x \;\Leftrightarrow\; x^3 - x > 0 \;\Leftrightarrow\; x(x^2 - 1) > 0 \;\Leftrightarrow\; x(x - 1)(x + 1) > 0$. Construct a chart:

Interval	x	$x-1$	$x+1$	$x(x-1)(x+1)$
$x < -1$	$-$	$-$	$-$	$-$
$-1 < x < 0$	$-$	$-$	$+$	$+$
$0 < x < 1$	$+$	$-$	$+$	$-$
$x > 1$	$+$	$+$	$+$	$+$

Since $x^3 > x$ when the last column is positive, the solution set is $(-1, 0) \cup (1, \infty)$.

−1 0 1

36. $x^3 + 3x < 4x^2 \iff x^3 - 4x^2 + 3x < 0 \iff x(x^2 - 4x + 3) < 0 \iff x(x-1)(x-3) < 0.$

Interval	x	$x-1$	$x-3$	$x(x-1)(x-3)$
$x < 0$	$-$	$-$	$-$	$-$
$0 < x < 1$	$+$	$-$	$-$	$+$
$1 < x < 3$	$+$	$+$	$-$	$-$
$x > 3$	$+$	$+$	$+$	$+$

Thus, the solution set is $(-\infty, 0) \cup (1, 3)$.

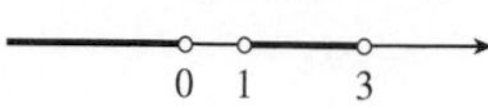

37. $1/x < 4$. This is clearly true for $x < 0$. So suppose $x > 0$. then $1/x < 4 \iff 1 < 4x \iff \frac{1}{4} < x$. Thus, the solution set is $(-\infty, 0) \cup \left(\frac{1}{4}, \infty\right)$.

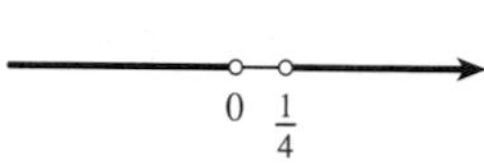

38. $-3 < 1/x \leq 1$. We solve the two inequalities separately and take the intersection of the solution sets. First, $-3 < 1/x$ is clearly true for $x > 0$. So suppose $x < 0$. Then $-3 < 1/x \iff -3x > 1 \iff x < -\frac{1}{3}$, so for this inequality, the solution set is $\left(-\infty, -\frac{1}{3}\right) \cup (0, \infty)$. Now $1/x \leq 1$ is clearly true if $x < 0$. So suppose $x > 0$. Then $1/x \leq 1 \iff 1 \leq x$, and the solution set here is $(-\infty, 0) \cup [1, \infty)$.

Taking the intersection of the two solution sets gives the final solution set: $\left(-\infty, -\frac{1}{3}\right) \cup [1, \infty)$.

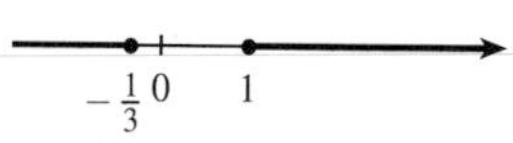

39. $C = \frac{5}{9}(F - 32) \Rightarrow F = \frac{9}{5}C + 32$. So $50 \leq F \leq 95 \Rightarrow 50 \leq \frac{9}{5}C + 32 \leq 95 \Rightarrow 18 \leq \frac{9}{5}C \leq 63 \Rightarrow 10 \leq C \leq 35$. So the interval is $[10, 35]$.

40. Since $20 \leq C \leq 30$ and $C = \frac{5}{9}(F - 32)$, we have $20 \leq \frac{5}{9}(F - 32) \leq 30 \Rightarrow 36 \leq F - 32 \leq 54 \Rightarrow 68 \leq F \leq 86$. So the interval is $[68, 86]$.

41. (a) Let T represent the temperature in degrees Celsius and h the height in km. $T = 20$ when $h = 0$ and T decreases by $10°$C for every km ($1°$C for each 100-m rise). Thus, $T = 20 - 10h$ when $0 \leq h \leq 12$.

(b) From part (a), $T = 20 - 10h \Rightarrow 10h = 20 - T \Rightarrow h = 2 - T/10$. So $0 \leq h \leq 5 \Rightarrow 0 \leq 2 - T/10 \leq 5 \Rightarrow -2 \leq -T/10 \leq 3 \Rightarrow -20 \leq -T \leq 30 \Rightarrow 20 \geq T \geq -30 \Rightarrow -30 \leq T \leq 20$. Thus, the range of temperatures (in $°$C) to be expected is $[-30, 20]$.

42. The ball will be at least 32 ft above the ground if $h \geq 32 \iff 128 + 16t - 16t^2 \geq 32 \iff 16t^2 - 16t - 96 \leq 0 \iff 16(t-3)(t+2) \leq 0$. $t = 3$ and $t = -2$ are endpoints of the interval we're looking for, and constructing a table gives $-2 \leq t \leq 3$. But $t \geq 0$, so the ball will be at least 32 ft above the ground in the time interval $[0, 3]$.

43. $|2x| = 3 \iff$ either $2x = 3$ or $2x = -3 \iff x = \frac{3}{2}$ or $x = -\frac{3}{2}$.

44. $|3x + 5| = 1 \iff$ either $3x + 5 = 1$ or -1. In the first case, $3x = -4 \iff x = -\frac{4}{3}$, and in the second case, $3x = -6 \iff x = -2$. So the solutions are -2 and $-\frac{4}{3}$.

45. $|x+3| = |2x+1| \quad \Leftrightarrow \quad$ either $x+3 = 2x+1$ or $x+3 = -(2x+1)$. In the first case, $x = 2$, and in the second case, $x+3 = -2x-1 \quad \Leftrightarrow \quad 3x = -4 \quad \Leftrightarrow \quad x = -\frac{4}{3}$. So the solutions are $-\frac{4}{3}$ and 2.

46. $\left|\dfrac{2x-1}{x+1}\right| = 3 \quad \Leftrightarrow \quad$ either $\dfrac{2x-1}{x+1} = 3$ or $\dfrac{2x-1}{x+1} = -3$. In the first case, $2x-1 = 3x+3 \quad \Leftrightarrow \quad x = -4$, and in the second case, $2x-1 = -3x-3 \quad \Leftrightarrow \quad x = -\frac{2}{5}$.

47. By Property 5 of absolute values, $|x| < 3 \quad \Leftrightarrow \quad -3 < x < 3$, so $x \in (-3, 3)$.

48. By Properties 4 and 6 of absolute values, $|x| \geq 3 \quad \Leftrightarrow \quad x \leq -3$ or $x \geq 3$, so $x \in (-\infty, -3] \cup [3, \infty)$.

49. $|x-4| < 1 \quad \Leftrightarrow \quad -1 < x-4 < 1 \quad \Leftrightarrow \quad 3 < x < 5$, so $x \in (3, 5)$.

50. $|x-6| < 0.1 \quad \Leftrightarrow \quad -0.1 < x-6 < 0.1 \quad \Leftrightarrow \quad 5.9 < x < 6.1$, so $x \in (5.9, 6.1)$.

51. $|x+5| \geq 2 \quad \Leftrightarrow \quad x+5 \geq 2$ or $x+5 \leq -2 \quad \Leftrightarrow \quad x \geq -3$ or $x \leq -7$, so $x \in (-\infty, -7] \cup [-3, \infty)$.

52. $|x+1| \geq 3 \quad \Leftrightarrow \quad x+1 \geq 3$ or $x+1 \leq -3 \quad \Leftrightarrow \quad x \geq 2$ or $x \leq -4$, so $x \in (-\infty, -4] \cup [2, \infty)$.

53. $|2x-3| \leq 0.4 \quad \Leftrightarrow \quad -0.4 \leq 2x-3 \leq 0.4 \quad \Leftrightarrow \quad 2.6 \leq 2x \leq 3.4 \quad \Leftrightarrow \quad 1.3 \leq x \leq 1.7$, so $x \in [1.3, 1.7]$.

54. $|5x-2| < 6 \quad \Leftrightarrow \quad -6 < 5x-2 < 6 \quad \Leftrightarrow \quad -4 < 5x < 8 \quad \Leftrightarrow \quad -\frac{4}{5} < x < \frac{8}{5}$, so $x \in \left(-\frac{4}{5}, \frac{8}{5}\right)$.

55. $1 \leq |x| \leq 4$. So either $1 \leq x \leq 4$ or $1 \leq -x \leq 4 \quad \Leftrightarrow \quad -1 \geq x \geq -4$. Thus, $x \in [-4, -1] \cup [1, 4]$.

56. $0 < |x-5| < \frac{1}{2}$. Clearly $0 < |x-5|$ for $x \neq 5$. Now $|x-5| < \frac{1}{2} \quad \Leftrightarrow \quad -\frac{1}{2} < x-5 < \frac{1}{2} \quad \Leftrightarrow \quad 4.5 < x < 5.5$. So the solution set is $(4.5, 5) \cup (5, 5.5)$.

57. $a(bx-c) \geq bc \quad \Leftrightarrow \quad bx - c \geq \dfrac{bc}{a} \quad \Leftrightarrow \quad bx \geq \dfrac{bc}{a} + c = \dfrac{bc+ac}{a} \quad \Leftrightarrow \quad x \geq \dfrac{bc+ac}{ab}$

58. $a \leq bx + c < 2a \quad \Leftrightarrow \quad a - c \leq bx < 2a - c \quad \Leftrightarrow \quad \dfrac{a-c}{b} \leq x < \dfrac{2a-c}{b}$ (since $b > 0$)

59. $ax + b < c \quad \Leftrightarrow \quad ax < c - b \quad \Leftrightarrow \quad x > \dfrac{c-b}{a}$ [since $a < 0$]

60. $\dfrac{ax+b}{c} \leq b \quad \Leftrightarrow \quad ax + b \geq bc$ [since $c < 0$] $\quad \Leftrightarrow \quad ax \geq bc - b \quad \Leftrightarrow \quad x \leq \dfrac{b(c-1)}{a}$ [since $a < 0$]

61. $|(x+y)-5| = |(x-2)+(y-3)| \leq |x-2| + |y-3| < 0.01 + 0.04 = 0.05$

62. Use the Triangle Inequality: $|x+3| < \frac{1}{2} \quad \Rightarrow$

$|4x+13| = |4(x+3)+1| \leq |4(x+3)| + |1| = 4|x+3| + 1 < 4\left(\frac{1}{2}\right) + 1 = 3$

Another method: $|x+3| < \frac{1}{2} \quad \Rightarrow \quad -\frac{1}{2} < x+3 < \frac{1}{2} \quad \Rightarrow \quad -2 < 4x+12 < 2 \quad \Rightarrow \quad -1 < 4x+13 < 3 \quad \Rightarrow$

$|4x+13| < 3$

63. If $a < b$ then $a + a < a + b$ and $a + b < b + b$. So $2a < a + b < 2b$. Dividing by 2, we get $a < \frac{1}{2}(a+b) < b$.

64. If $0 < a < b$, then $\dfrac{1}{ab} > 0$. So $a < b \;\Rightarrow\; \dfrac{1}{ab} \cdot a < \dfrac{1}{ab} \cdot b \;\Leftrightarrow\; \dfrac{1}{b} < \dfrac{1}{a}$.

65. $|ab| = \sqrt{(ab)^2} = \sqrt{a^2 b^2} = \sqrt{a^2}\,\sqrt{b^2} = |a|\,|b|$

66. $\left|\dfrac{a}{b}\right| |b| = \left|\dfrac{a}{b} \cdot b\right| = |a|$ [using the result of Exercise 65]. Dividing the equation through by $|b|$ gives $\left|\dfrac{a}{b}\right| = \dfrac{|a|}{|b|}$.

67. If $0 < a < b$, then $a \cdot a < a \cdot b$ and $a \cdot b < b \cdot b$ [using Rule 3 of Inequalities]. So $a^2 < ab < b^2$ and hence $a^2 < b^2$.

68. Following the hint, the Triangle Inequality becomes $|(x - y) + y| \le |x - y| + |y| \;\Leftrightarrow\; |x| \le |x - y| + |y| \;\Leftrightarrow$ $|x - y| \ge |x| - |y|$.

69. Observe that the sum, difference and product of two integers is always an integer. Let the rational numbers be represented by $r = m/n$ and $s = p/q$ (where m, n, p and q are integers with $n \neq 0$, $q \neq 0$). Now $r + s = \dfrac{m}{n} + \dfrac{p}{q} = \dfrac{mq + pn}{nq}$, but $mq + pn$ and nq are both integers, so $\dfrac{mq + pn}{nq} = r + s$ is a rational number by definition. Similarly, $r - s = \dfrac{m}{n} - \dfrac{p}{q} = \dfrac{mq - pn}{nq}$ is a rational number. Finally, $r \cdot s = \dfrac{m}{n} \cdot \dfrac{p}{q} = \dfrac{mp}{nq}$ but mp and nq are both integers, so $\dfrac{mp}{nq} = r \cdot s$ is a rational number by definition.

70. (a) No. Consider the case of $\sqrt{2}$ and $-\sqrt{2}$. Both are irrational numbers, yet $\sqrt{2} + \left(-\sqrt{2}\right) = 0$ and 0, being an integer, is not irrational.

(b) No. Consider the case of $\sqrt{2}$ and $\sqrt{2}$. Both are irrational numbers, yet $\sqrt{2} \cdot \sqrt{2} = 2$ is not irrational.

B Coordinate Geometry and Lines

1. Use the distance formula with $P_1(x_1, y_1) = (1, 1)$ and $P_2(x_2, y_2) = (4, 5)$ to get

$$|P_1 P_2| = \sqrt{(4 - 1)^2 + (5 - 1)^2} = \sqrt{3^2 + 4^2} = \sqrt{25} = 5$$

2. The distance from $(1, -3)$ to $(5, 7)$ is $\sqrt{(5 - 1)^2 + [7 - (-3)]^2} = \sqrt{4^2 + 10^2} = \sqrt{116} = 2\sqrt{29}$.

3. The distance from $(6, -2)$ to $(-1, 3)$ is $\sqrt{-1 - 6)^2 + [3 - (-2)]^2} = \sqrt{(-7)^2 + 5^2} = \sqrt{74}$.

4. The distance from $(1, -6)$ to $(-1, -3)$ is $\sqrt{(-1 - 1)^2 + [-3 - (-6)]^2} = \sqrt{(-2)^2 + 3^2} = \sqrt{13}$.

5. The distance from $(2, 5)$ to $(4, -7)$ is $\sqrt{(4 - 2)^2 + (-7 - 5)^2} = \sqrt{2^2 + (-12)^2} = \sqrt{148} = 2\sqrt{37}$.

6. The distance from (a, b) to (b, a) is $\sqrt{(b - a)^2 + (a - b)^2} = \sqrt{(a - b)^2 + (a - b)^2} = \sqrt{2(a - b)^2} = \sqrt{2}\,|a - b|$.

7. The slope m of the line through $P(1, 5)$ and $Q(4, 11)$ is $m = \dfrac{11 - 5}{4 - 1} = \dfrac{6}{3} = 2$.

8. The slope m of the line through $P(-1, 6)$ and $Q(4, -3)$ is $m = \dfrac{-3 - 6}{4 - (-1)} = -\dfrac{9}{5}$.

9. The slope m of the line through $P(-3,3)$ and $Q(-1,-6)$ is $m = \dfrac{-6-3}{-1-(-3)} = -\dfrac{9}{2}$.

10. The slope m of the line through $P(-1,-4)$ and $Q(6,0)$ is $m = \dfrac{0-(-4)}{6-(-1)} = \dfrac{4}{7}$.

11. Using $A(0,2)$, $B(-3,-1)$, and $C(-4,3)$, we have $|AC| = \sqrt{(-4-0)^2+(3-2)^2} = \sqrt{(-4)^2+1^2} = \sqrt{17}$ and $|BC| = \sqrt{[-4-(-3)]^2+[3-(-1)]^2} = \sqrt{(-1)^2+4^2} = \sqrt{17}$, so the triangle has two sides of equal length, and is isosceles.

12. (a) Using $A(6,-7)$, $B(11,-3)$, and $C(2,-2)$, we have

$|AB| = \sqrt{(11-6)^2+[-3-(-7)]^2} = \sqrt{5^2+4^2} = \sqrt{41}$,

$|AC| = \sqrt{(2-6)^2+[-2-(-7)]^2} = \sqrt{(-4)^2+5^2} = \sqrt{41}$, and

$|BC| = \sqrt{(2-11)^2+[-2-(-3)]^2} = \sqrt{(-9)^2+1^2} = \sqrt{82}$.

Thus, $|AB|^2+|AC|^2 = 41+41 = 82 = |BC|^2$ and so $\triangle ABC$ is a right triangle.

(b) $m_{AB} = \dfrac{-3-(-7)}{11-6} = \dfrac{4}{5}$ and $m_{AC} = \dfrac{2-(-7)}{2-6} = -\dfrac{5}{4}$. Thus $m_{AB}\cdot m_{AC} = -1$ and so AB is perpendicular to AC and $\triangle ABC$ must be a right triangle.

(c) Taking lengths from part (a), the base is $\sqrt{41}$ and the height is $\sqrt{41}$. Thus the area is $\frac{1}{2}bh = \frac{1}{2}\sqrt{41}\sqrt{41} = \frac{41}{2}$.

13. Using $A(-2,9)$, $B(4,6)$, $C(1,0)$, and $D(-5,3)$, we have

$|AB| = \sqrt{[4-(-2)]^2+(6-9)^2} = \sqrt{6^2+(-3)^2} = \sqrt{45} = \sqrt{9}\sqrt{5} = 3\sqrt{5}$,

$|BC| = \sqrt{(1-4)^2+(0-6)^2} = \sqrt{(-3)^2+(-6)^2} = \sqrt{45} = \sqrt{9}\sqrt{5} = 3\sqrt{5}$,

$|CD| = \sqrt{(-5-1)^2+(3-0)^2} = \sqrt{(-6)^2+3^2} = \sqrt{45} = \sqrt{9}\sqrt{5} = 3\sqrt{5}$, and

$|DA| = \sqrt{[-2-(-5)]^2+(9-3)^2} = \sqrt{3^2+6^2} = \sqrt{45} = \sqrt{9}\sqrt{5} = 3\sqrt{5}$. So all sides are of equal length and we have a rhombus. Moreover, $m_{AB} = \dfrac{6-9}{4-(-2)} = -\dfrac{1}{2}$, $m_{BC} = \dfrac{0-6}{1-4} = 2$, $m_{CD} = \dfrac{3-0}{-5-1} = -\dfrac{1}{2}$, and

$m_{DA} = \dfrac{9-3}{-2-(-5)} = 2$, so the sides are perpendicular. Thus, A, B, C, and D are vertices of a square.

14. (a) Using $A(-1,3)$, $B(3,11)$, and $C(5,15)$, we have

$|AB| = \sqrt{[3-(-1)]^2+(11-3)^2} = \sqrt{4^2+8^2} = \sqrt{80} = 4\sqrt{5}$,

$|BC| = \sqrt{(5-3)^2+(15-11)^2} = \sqrt{2^2+4^2} = \sqrt{20} = 2\sqrt{5}$, and

$|AC| = \sqrt{[5-(-1)]^2+(15-3)^2} = \sqrt{6^2+12^2} = \sqrt{180} = 6\sqrt{5}$. Thus, $|AC| = |AB|+|BC|$.

(b) $m_{AB} = \dfrac{11-3}{3-(-1)} = \dfrac{8}{4} = 2$ and $m_{AC} = \dfrac{15-3}{5-(-1)} = \dfrac{12}{6} = 2$. Since the segments AB and AC have the same slope, A, B and C must be collinear.

15. For the vertices $A(1,1)$, $B(7,4)$, $C(5,10)$, and $D(-1,7)$, the slope of the line segment AB is $\dfrac{4-1}{7-1}=\dfrac{1}{2}$, the slope of CD is $\dfrac{7-10}{-1-5}=\dfrac{1}{2}$, the slope of BC is $\dfrac{10-4}{5-7}=-3$, and the slope of DA is $\dfrac{1-7}{1-(-1)}=-3$. So AB is parallel to CD and BC is parallel to DA. Hence $ABCD$ is a parallelogram.

16. For the vertices $A(1,1)$, $B(11,3)$, $C(10,8)$, and $D(0,6)$, the slopes of the four sides are $m_{AB}=\dfrac{3-1}{11-1}=\dfrac{1}{5}$, $m_{BC}=\dfrac{8-3}{10-11}=-5$, $m_{CD}=\dfrac{6-8}{0-10}=\dfrac{1}{5}$, and $m_{DA}=\dfrac{1-6}{1-0}=-5$. Hence $AB \parallel CD$, $BC \parallel DA$, $AB \perp BC$, $BC \perp CD$, $CD \perp DA$, and $DA \perp AB$, and so $ABCD$ is a rectangle.

17. The graph of the equation $x=3$ is a vertical line with x-intercept 3. The line does not have a slope.

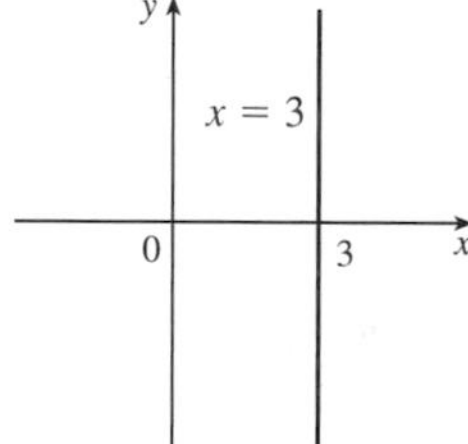

18. The graph of the equation $y=-2$ is a horizontal line with y-intercept -2. The line has slope 0.

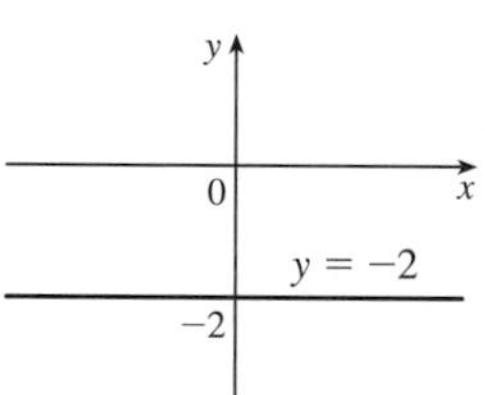

19. $xy=0 \;\Leftrightarrow\; x=0$ or $y=0$. The graph consists of the coordinate axes.

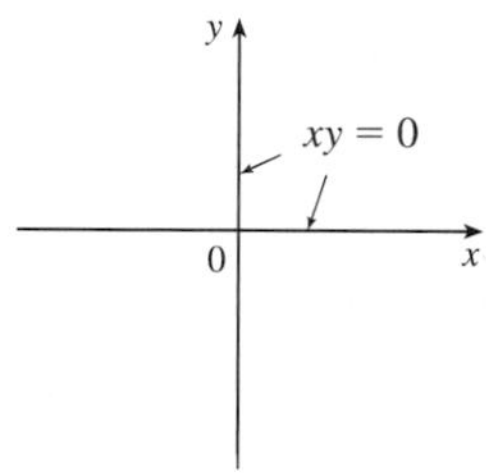

20. $|y|=1 \;\Leftrightarrow\; y=1$ or $y=-1$

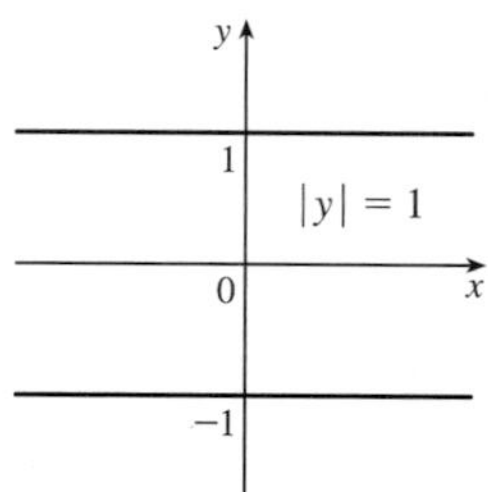

21. By the point-slope form of the equation of a line, an equation of the line through $(2,-3)$ with slope 6 is $y-(-3)=6(x-2)$ or $y=6x-15$.

22. $y-4=-3[x-(-1)]$ or $y=-3x+1$

23. $y-7=\frac{2}{3}(x-1)$ or $y=\frac{2}{3}x+\frac{19}{3}$

24. $y-(-5)=-\frac{7}{2}[x-(-3)]$ or $y=-\frac{7}{2}x-\frac{31}{2}$

25. The slope of the line through $(2,1)$ and $(1,6)$ is $m=\dfrac{6-1}{1-2}=-5$, so an equation of the line is $y-1=-5(x-2)$ or $y=-5x+11$.

26. For $(-1, -2)$ and $(4, 3)$, $m = \dfrac{3-(-2)}{4-(-1)} = 1$. An equation of the line is $y - 3 = 1(x - 4)$ or $y = x - 1$.

27. By the slope-intercept form of the equation of a line, an equation of the line is $y = 3x - 2$.

28. By the slope-intercept form of the equation of a line, an equation of the line is $y = \frac{2}{5}x + 4$.

29. Since the line passes through $(1, 0)$ and $(0, -3)$, its slope is $m = \dfrac{-3-0}{0-1} = 3$, so an equation is $y = 3x - 3$.

Another method: From Exercise 61, $\dfrac{x}{1} + \dfrac{y}{-3} = 1 \Rightarrow -3x + y = -3 \Rightarrow y = 3x - 3$.

30. For $(-8, 0)$ and $(0, 6)$, $m = \dfrac{6-0}{0-(-8)} = \dfrac{3}{4}$. So an equation is $y = \frac{3}{4}x + 6$.

Another method: From Exercise 61, $\dfrac{x}{-8} + \dfrac{y}{6} = 1 \Rightarrow -3x + 4y = 24 \Rightarrow y = \frac{3}{4}x + 6$.

31. The line is parallel to the x-axis, so it is horizontal and must have the form $y = k$. Since it goes through the point $(x, y) = (4, 5)$, the equation is $y = 5$.

32. The line is parallel to the y-axis, so it is vertical and must have the form $x = k$. Since it goes through the point $(x, y) = (4, 5)$, the equation is $x = 4$.

33. Putting the line $x + 2y = 6$ into its slope-intercept form gives us $y = -\frac{1}{2}x + 3$, so we see that this line has slope $-\frac{1}{2}$. Thus, we want the line of slope $-\frac{1}{2}$ that passes through the point $(1, -6)$: $y - (-6) = -\frac{1}{2}(x - 1) \Leftrightarrow y = -\frac{1}{2}x - \frac{11}{2}$.

34. $2x + 3y + 4 = 0 \Leftrightarrow y = \frac{2}{3}x - \frac{4}{3}$, so $m = -\frac{2}{3}$ and the required line is $y = -\frac{2}{3}x + 6$.

35. $2x + 5y + 8 = 0 \Leftrightarrow y = -\frac{2}{5}x - \frac{8}{5}$. Since this line has slope $-\frac{2}{5}$, a line perpendicular to it would have slope $\frac{5}{2}$, so the required line is $y - (-2) = \frac{5}{2}[x - (-1)] \Leftrightarrow y = \frac{5}{2}x + \frac{1}{2}$.

36. $4x - 8y = 1 \Leftrightarrow y = \frac{1}{2}x - \frac{1}{8}$. Since this line has slope $\frac{1}{2}$, a line perpendicular to it would have slope -2, so the required line is $y - (-\frac{2}{3}) = -2(x - \frac{1}{2}) \Leftrightarrow y = -2x + \frac{1}{3}$.

37. $x + 3y = 0 \Leftrightarrow y = -\frac{1}{3}x$, so the slope is $-\frac{1}{3}$ and the y-intercept is 0.

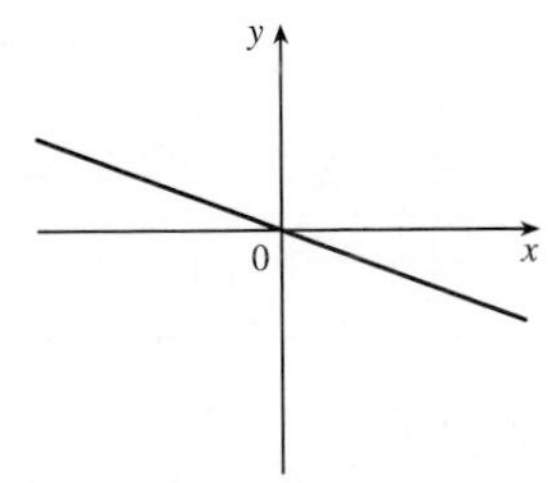

38. $2x - 5y = 0 \Leftrightarrow y = \frac{2}{5}x$, so the slope is $\frac{2}{5}$ and the y-intercept is 0.

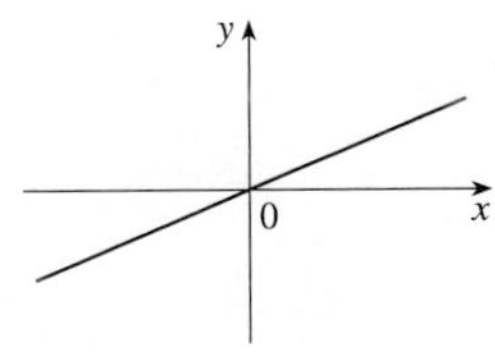

39. $y = -2$ is a horizontal line with slope 0 and y-intercept -2.

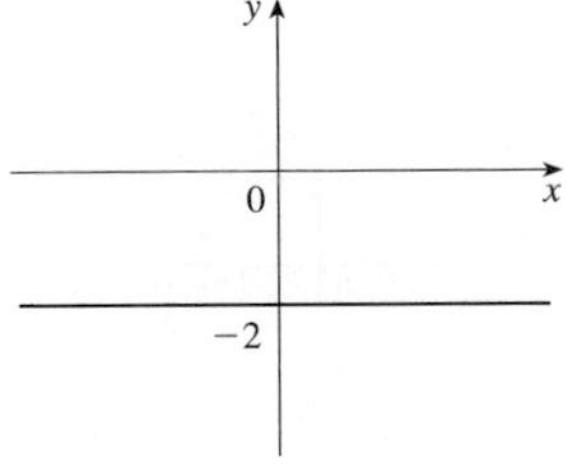

40. $2x - 3y + 6 = 0 \Leftrightarrow$
$y = \frac{2}{3}x + 2$, so the slope is $\frac{2}{3}$ and the y-intercept is 2.

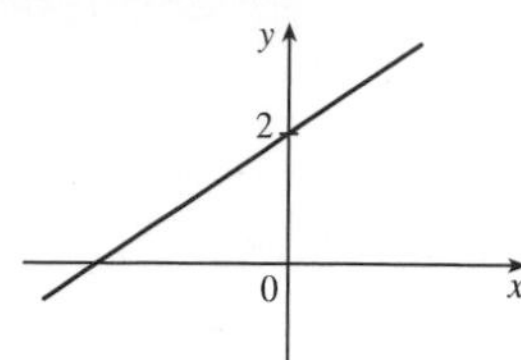

41. $3x - 4y = 12 \Leftrightarrow$
$y = \frac{3}{4}x - 3$, so the slope is $\frac{3}{4}$ and the y-intercept is -3.

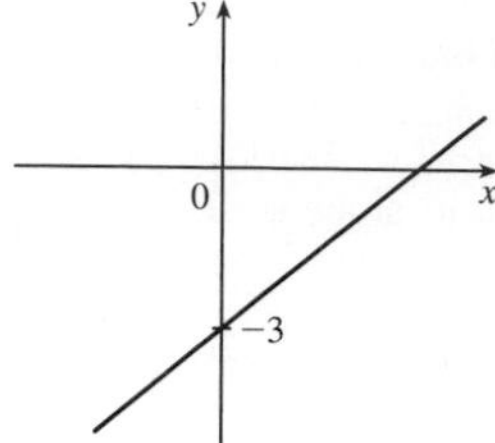

42. $4x + 5y = 10 \Leftrightarrow$
$y = -\frac{4}{5}x + 2$, so the slope is $-\frac{4}{5}$ and the y-intercept is 2.

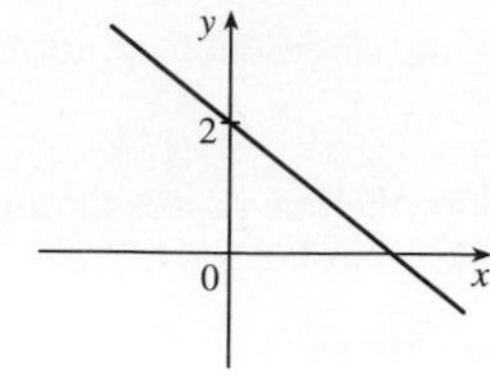

43. $\{(x, y) \mid x < 0\}$

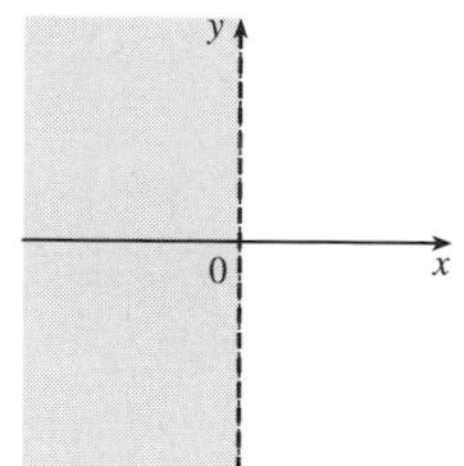

44. $\{(x, y) \mid y > 0\}$

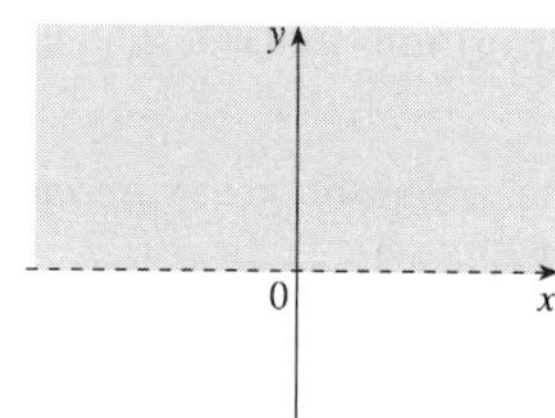

45. $\{(x, y) \mid xy < 0\} =$
$\{(x, y) \mid x < 0 \text{ and } y > 0\}$
$\cup \{(x, y) \mid x > 0 \text{ and } y < 0\}$

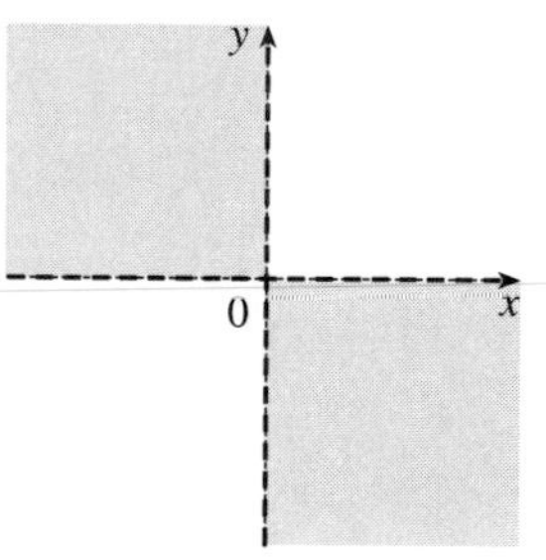

46. $\{(x, y) \mid x \geq 1 \text{ and } y < 3\}$

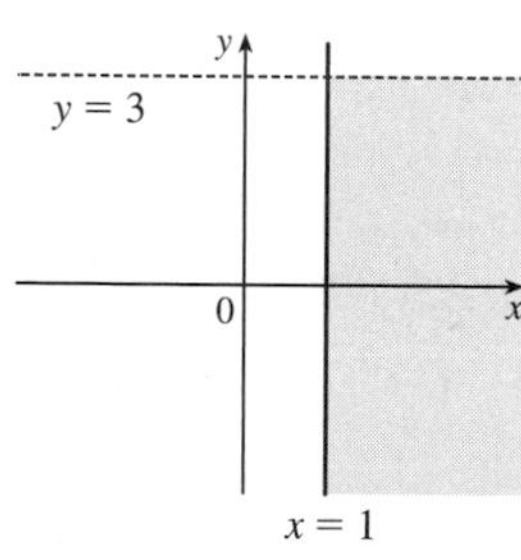

47. $\left\{(x, y) \,\middle|\, |x| \leq 2\right\} =$
$\{(x, y) \mid -2 \leq x \leq 2\}$

48. $\left\{(x, y) \,\middle|\, |x| < 3 \text{ and } |y| < 2\right\}$

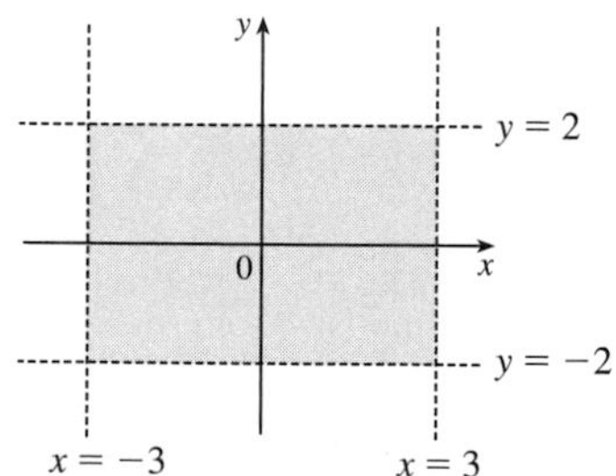

49. $\{(x, y) \mid 0 \leq y \leq 4, x \leq 2\}$

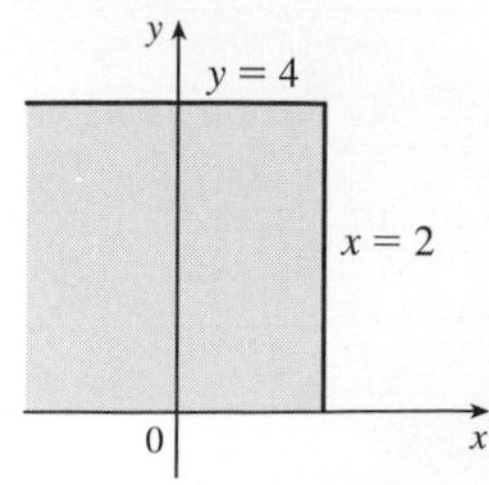

50. $\{(x, y) \mid y > 2x - 1\}$

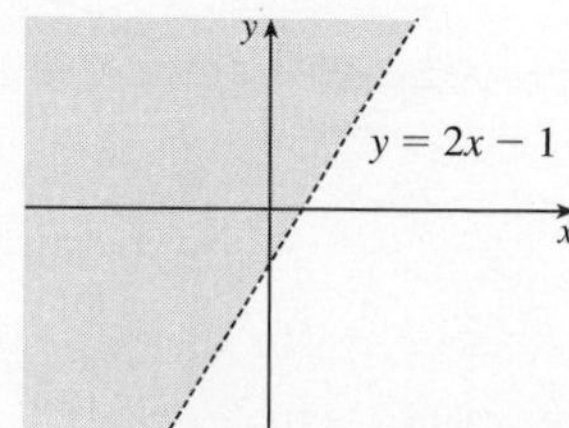

51. $\{(x, y) \mid 1 + x \leq y \leq 1 - 2x\}$

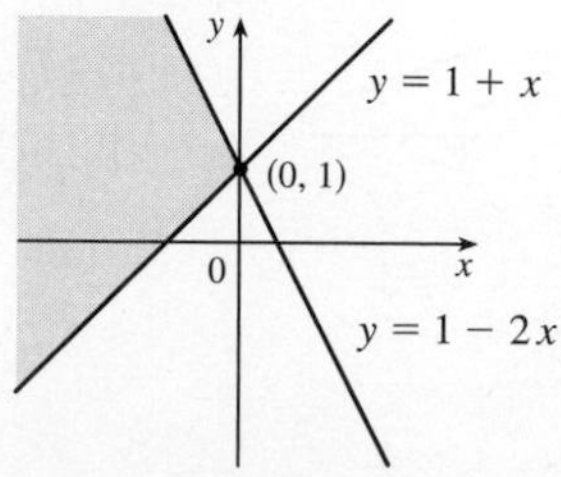

52. $\{(x,y) \mid -x \le y < \frac{1}{2}(x+3)\}$

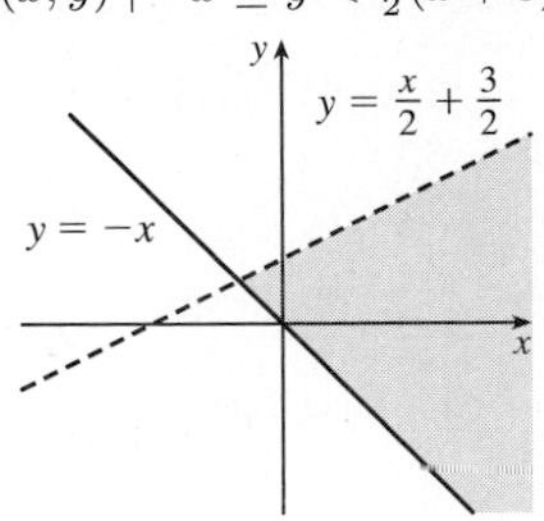

53. Let $P(0,y)$ be a point on the y-axis. The distance from P to $(5,-5)$ is $\sqrt{(5-0)^2+(-5-y)^2} = \sqrt{5^2+(y+5)^2}$. The distance from P to $(1,1)$ is $\sqrt{(1-0)^2+(1-y)^2} = \sqrt{1^2+(y-1)^2}$. We want these distances to be equal: $\sqrt{5^2+(y+5)^2} = \sqrt{1^2+(y-1)^2} \Leftrightarrow 5^2+(y+5)^2 = 1^2+(y-1)^2 \Leftrightarrow$ $25+(y^2+10y+25) = 1+(y^2-2y+1) \Leftrightarrow 12y=-48 \Leftrightarrow y=-4$. So the desired point is $(0,-4)$.

54. Let M be the point $\left(\dfrac{x_1+x_2}{2}, \dfrac{y_1+y_2}{2}\right)$. Then

$$|MP_1|^2 = \left(x_1 - \frac{x_1+x_2}{2}\right)^2 + \left(y_1 - \frac{y_1+y_2}{2}\right)^2 = \left(\frac{x_1-x_2}{2}\right)^2 + \left(\frac{y_1-y_2}{2}\right)^2$$

$$|MP_2|^2 = \left(x_2 - \frac{x_1+x_2}{2}\right)^2 + \left(y_2 - \frac{y_1+y_2}{2}\right)^2 = \left(\frac{x_2-x_1}{2}\right)^2 + \left(\frac{y_2-y_1}{2}\right)^2$$

Hence, $|MP_1| = |MP_2|$; that is, M is equidistant from P_1 and P_2.

55. (a) Using the midpoint formula from Exercise 54 with $(1,3)$ and $(7,15)$, we get $\left(\frac{1+7}{2}, \frac{3+15}{2}\right) = (4,9)$.

(b) Using the midpoint formula from Exercise 54 with $(-1,6)$ and $(8,-12)$, we get $\left(\frac{-1+8}{2}, \frac{6+(-12)}{2}\right) = \left(\frac{7}{2},-3\right)$.

56. With $A(1,0)$, $B(3,6)$, and $C(8,2)$, the midpoint M_1 of AB is $\left(\frac{1+3}{2}, \frac{0+6}{2}\right) = (2,3)$, the midpoint M_2 of BC is $\left(\frac{3+8}{2}, \frac{6+2}{2}\right) = \left(\frac{11}{2},4\right)$, and the midpoint M_3 of CA is $\left(\frac{8+1}{2}, \frac{2+0}{2}\right) = \left(\frac{9}{2},1\right)$. The lengths of the medians are

$$|AM_2| = \sqrt{\left(\tfrac{11}{2}-1\right)^2+(4-0)^2} = \sqrt{\left(\tfrac{9}{2}\right)^2+4^2} = \sqrt{\tfrac{145}{4}} = \tfrac{\sqrt{145}}{2}$$

$$|BM_3| = \sqrt{\left(\tfrac{9}{2}-3\right)^2+(1-6)^2} = \sqrt{\left(\tfrac{3}{2}\right)^2+(-5)^2} = \sqrt{\tfrac{109}{4}} = \tfrac{\sqrt{109}}{2}$$

$$|CM_1| = \sqrt{(2-8)^2+(3-2)^2} = \sqrt{(-6)^2+1^2} = \sqrt{37}$$

57. $2x-y=4 \Leftrightarrow y=2x-4 \Rightarrow m_1=2$ and $6x-2y=10 \Leftrightarrow 2y=6x-10 \Leftrightarrow y=3x-5 \Rightarrow m_2=3$. Since $m_1 \ne m_2$, the two lines are not parallel. To find the point of intersection: $2x-4=3x-5 \Leftrightarrow x=1 \Rightarrow$ $y=-2$. Thus, the point of intersection is $(1,-2)$.

58. $3x-5y+19=0 \Leftrightarrow 5y=3x+19 \Leftrightarrow y=\frac{3}{5}x+\frac{19}{5} \Rightarrow m_1=\frac{3}{5}$ and $10x+6y-50=0 \Leftrightarrow$ $6y=-10x+50 \Leftrightarrow y=-\frac{5}{3}x+\frac{25}{3} \Rightarrow m_2=-\frac{5}{3}$. Since $m_1m_2=\frac{3}{5}\left(-\frac{5}{3}\right)=-1$, the two lines are perpendicular. To find the point of intersection: $\frac{3}{5}x+\frac{19}{5}=-\frac{5}{3}x+\frac{25}{3} \Leftrightarrow 9x+57=-25x+125 \Leftrightarrow 34x=68 \Leftrightarrow x=2 \Rightarrow$ $y=\frac{3}{5}\cdot 2+\frac{19}{5}=\frac{25}{5}=5$. Thus, the point of intersection is $(2,5)$.

59. With $A(1,4)$ and $B(7,-2)$, the slope of segment AB is $\frac{-2-4}{7-1}=-1$, so its perpendicular bisector has slope 1. The midpoint of AB is $\left(\frac{1+7}{2}, \frac{4+(-2)}{2}\right) = (4,1)$, so an equation of the perpendicular bisector is $y-1=1(x-4)$ or $y=x-3$.

60. (a) Side PQ has slope $\frac{4-0}{3-1} = 2$, so its equation is $y - 0 = 2(x-1) \quad\Leftrightarrow\quad y = 2x - 2$. Side QR has slope $\frac{6-4}{-1-3} = -\frac{1}{2}$, so its equation is $y - 4 = -\frac{1}{2}(x-3) \quad\Leftrightarrow\quad y = -\frac{1}{2}x + \frac{11}{2}$. Side RP has slope $\frac{0-6}{1-(-1)} = -3$, so its equation is $y - 0 = -3(x-1) \quad\Leftrightarrow\quad y = -3x + 3$.

(b) M_1 (the midpoint of PQ) has coordinates $\left(\frac{1+3}{2}, \frac{0+4}{2}\right) = (2, 2)$. M_2 (the midpoint of QR) has coordinates $\left(\frac{3-1}{2}, \frac{4+6}{2}\right) = (1, 5)$. M_3 (the midpoint of RP) has coordinates $\left(\frac{1-1}{2}, \frac{0+6}{2}\right) = (0, 3)$. RM_1 has slope $\frac{2-6}{2-(-1)} = -\frac{4}{3}$ and hence equation $y - 2 = -\frac{4}{3}(x-2) \quad\Leftrightarrow\quad y = -\frac{4}{3}x + \frac{14}{3}$. PM_2 is a vertical line with equation $x = 1$. QM_3 has slope $\frac{3-4}{0-3} = \frac{1}{3}$ and hence equation $y - 3 = \frac{1}{3}(x-0) \quad\Leftrightarrow\quad y = \frac{1}{3}x + 3$. PM_2 and RM_1 intersect where $x = 1$ and $y = -\frac{4}{3}(1) + \frac{14}{3} = \frac{10}{3}$, or at $\left(1, \frac{10}{3}\right)$. PM_2 and QM_3 intersect where $x = 1$ and $y = \frac{1}{3}(1) + 3 = \frac{10}{3}$, or at $\left(1, \frac{10}{3}\right)$, so this is the point where all three medians intersect.

61. (a) Since the x-intercept is a, the point $(a, 0)$ is on the line, and similarly since the y-intercept is b, $(0, b)$ is on the line. Hence, the slope of the line is $m = \dfrac{b-0}{0-a} = -\dfrac{b}{a}$. Substituting into $y = mx + b$ gives $y = -\dfrac{b}{a}x + b \quad\Leftrightarrow\quad \dfrac{b}{a}x + y = b \quad\Leftrightarrow\quad \dfrac{x}{a} + \dfrac{y}{b} = 1$.

(b) Letting $a = 6$ and $b = -8$ gives $\dfrac{x}{6} + \dfrac{y}{-8} = 1 \quad\Leftrightarrow\quad -8x + 6y = -48$ [multiply by -48] $\quad\Leftrightarrow\quad 6y = 8x - 48 \quad\Leftrightarrow\quad 3y = 4x - 24 \quad\Leftrightarrow\quad y = \frac{4}{3}x - 8$.

62. (a) Let d = distance traveled (in miles) and t = time elapsed (in hours). At $t = 0$, $d = 0$ and at $t = 50$ minutes $= 50 \cdot \frac{1}{60} = \frac{5}{6}$ h, $d = 40$. Thus, we have two points: $(0, 0)$ and $\left(\frac{5}{6}, 40\right)$, so $m = \dfrac{40-0}{5/6-0} = 48$ and $d = 48t$.

(b)

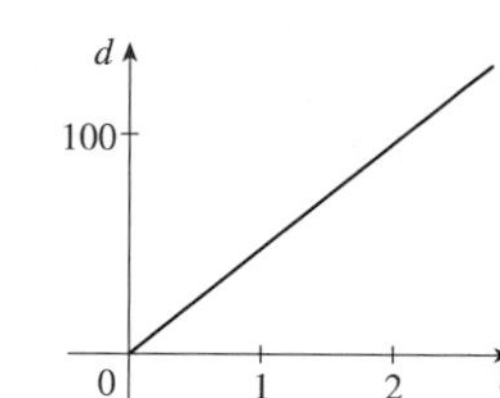

(c) The slope is 48 and represents the car's speed in mi/h.

C Graphs of Second-Degree Equations

1. An equation of the circle with center $(3, -1)$ and radius 5 is $(x-3)^2 + (y+1)^2 = 5^2 = 25$.

2. An equation of the circle with center $(-2, -8)$ and radius 10 is $(x+2)^2 + (y+8)^2 = 10^2 = 100$.

3. The equation has the form $x^2 + y^2 = r^2$. Since $(4, 7)$ lies on the circle, we have $4^2 + 7^2 = r^2 \quad\Rightarrow\quad r^2 = 65$. So the required equation is $x^2 + y^2 = 65$.

4. The equation has the form $(x+1)^2 + (y-5)^2 = r^2$. Since $(-4, -6)$ lies on the circle, we have $r^2 = (-4+1)^2 + (-6-5)^2 = 130$. So an equation is $(x+1)^2 + (y-5)^2 = 130$.

5. $x^2 + y^2 - 4x + 10y + 13 = 0 \quad\Leftrightarrow\quad x^2 - 4x + y^2 + 10y = -13 \quad\Leftrightarrow\quad (x^2 - 4x + 4) + (y^2 + 10y + 25) = -13 + 4 + 25 = 16 \quad\Leftrightarrow\quad (x-2)^2 + (y+5)^2 = 4^2$. Thus, we have a circle with center $(2, -5)$ and radius 4.

6. $x^2 + y^2 + 6y + 2 = 0 \quad\Leftrightarrow\quad x^2 + (y^2 + 6y + 9) = -2 + 9 \quad\Leftrightarrow\quad x^2 + (y+3)^2 = 7$. Thus, we have a circle with center $(0, -3)$ and radius $\sqrt{7}$.

7. $x^2 + y^2 + x = 0 \quad \Leftrightarrow \quad (x^2 + x + \frac{1}{4}) + y^2 = \frac{1}{4} \quad \Leftrightarrow \quad (x + \frac{1}{2})^2 + y^2 = (\frac{1}{2})^2$. Thus, we have a circle with center $(-\frac{1}{2}, 0)$ and radius $\frac{1}{2}$.

8. $16x^2 + 16y^2 + 8x + 32y + 1 = 0 \quad \Leftrightarrow \quad 16(x^2 + \frac{1}{2}x + \frac{1}{16}) + 16(y^2 + 2y + 1) = -1 + 1 + 16 \quad \Leftrightarrow$

$16(x + \frac{1}{4})^2 + 16(y + 1)^2 = 16 \quad \Leftrightarrow \quad (x + \frac{1}{4})^2 + (y + 1)^2 = 1$. Thus, we have a circle with center $(-\frac{1}{4}, -1)$ and radius 1.

9. $2x^2 + 2y^2 - x + y = 1 \quad \Leftrightarrow \quad 2(x^2 - \frac{1}{2}x + \frac{1}{16}) + 2(y^2 + \frac{1}{2}y + \frac{1}{16}) = 1 + \frac{1}{8} + \frac{1}{8} \quad \Leftrightarrow$

$2(x - \frac{1}{4})^2 + 2(y + \frac{1}{4})^2 = \frac{5}{4} \quad \Leftrightarrow \quad (x - \frac{1}{4})^2 + (y + \frac{1}{4})^2 = \frac{5}{8}$. Thus, we have a circle with center $(\frac{1}{4}, -\frac{1}{4})$ and radius $\frac{\sqrt{5}}{2\sqrt{2}} = \frac{\sqrt{10}}{4}$.

10. $x^2 + y^2 + ax + by + c = 0 \quad \Leftrightarrow \quad (x^2 + ax + \frac{1}{4}a^2) + (y^2 + by + \frac{1}{4}b^2) = -c + \frac{1}{4}a^2 + \frac{1}{4}b^2 \quad \Leftrightarrow$

$(x + \frac{1}{2}a)^2 + (y + \frac{1}{2}b)^2 = \frac{1}{4}(a^2 + b^2 - 4c)$. For this to represent a nondegenerate circle, $\frac{1}{4}(a^2 + b^2 - 4c) > 0$ or $a^2 + b^2 > 4c$. If this condition is satisfied, the circle has center $(-\frac{1}{2}a, -\frac{1}{2}b)$ and radius $\frac{1}{2}\sqrt{a^2 + b^2 - 4c}$.

11. $y = -x^2$. Parabola

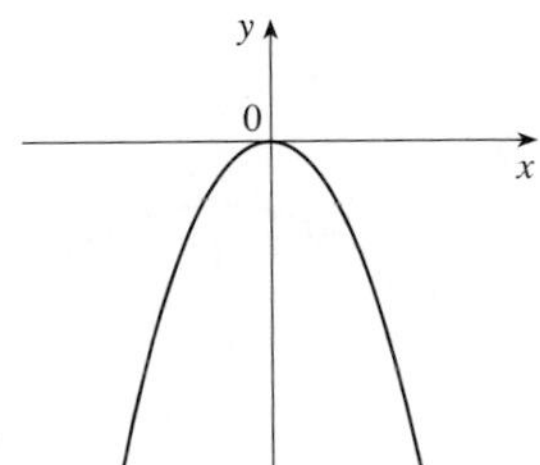

12. $y^2 - x^2 = 1$. Hyperbola

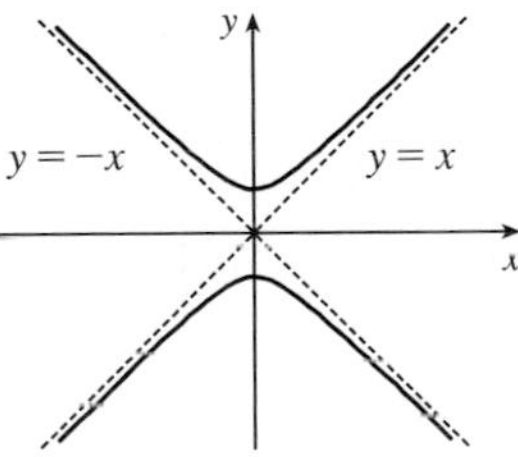

13. $x^2 + 4y^2 = 16 \quad \Leftrightarrow \quad \frac{x^2}{16} + \frac{y^2}{4} = 1$. Ellipse

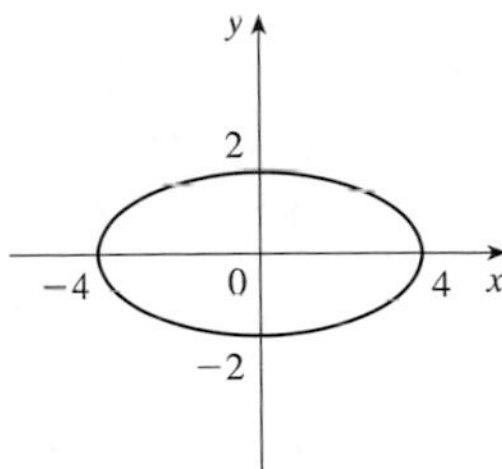

14. $x = -2y^2$. Parabola

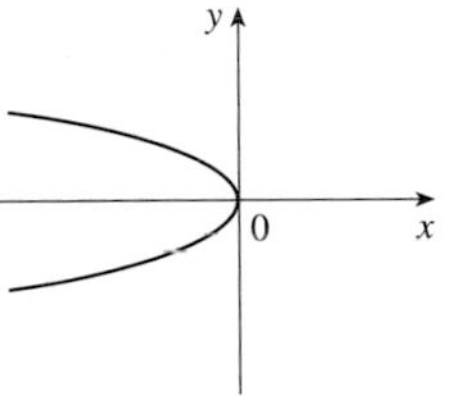

15. $16x^2 - 25y^2 = 400 \quad \Leftrightarrow \quad \frac{x^2}{25} - \frac{y^2}{16} = 1$. Hyperbola

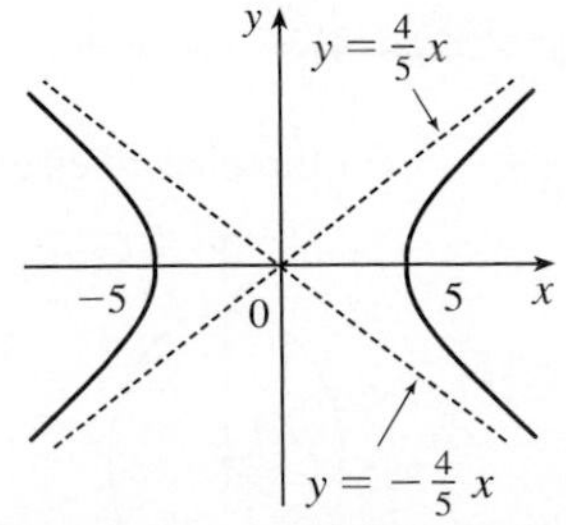

16. $25x^2 + 4y^2 = 100 \quad \Leftrightarrow \quad \frac{x^2}{4} + \frac{y^2}{25} = 1$. Ellipse

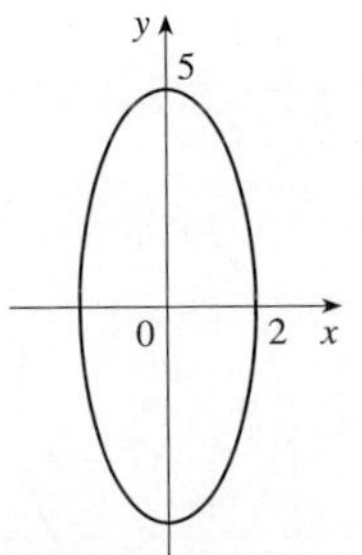

17. $4x^2 + y^2 = 1 \;\Leftrightarrow\; \dfrac{x^2}{1/4} + y^2 = 1$. Ellipse

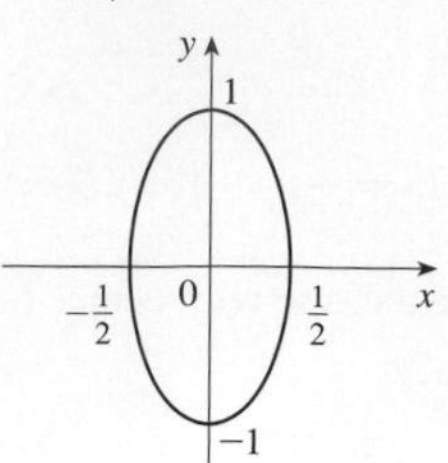

18. $y = x^2 + 2$. Parabola with vertex at $(0, 2)$

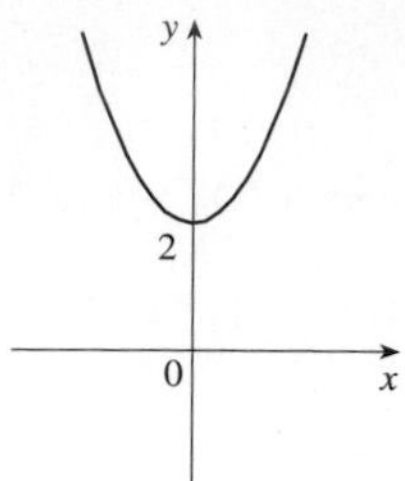

19. $x = y^2 - 1$. Parabola with vertex at $(-1, 0)$

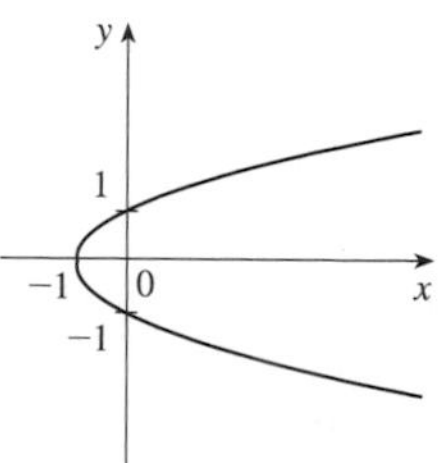

20. $9x^2 - 25y^2 = 225 \;\Leftrightarrow\; \dfrac{x^2}{25} - \dfrac{y^2}{9} = 1$. Hyperbola

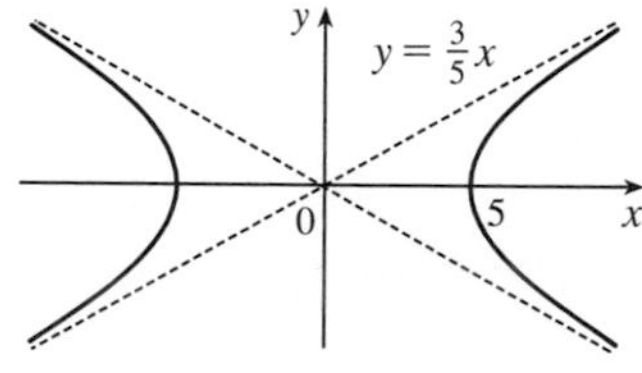

21. $9y^2 - x^2 = 9 \;\Leftrightarrow\; y^2 - \dfrac{x^2}{9} = 1$. Hyperbola

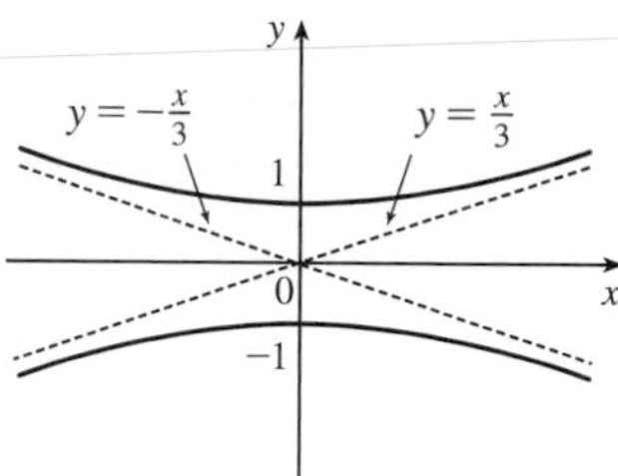

22. $2x^2 + 5y^2 = 10 \;\Leftrightarrow\; \dfrac{x^2}{5} + \dfrac{y^2}{2} = 1$. Ellipse

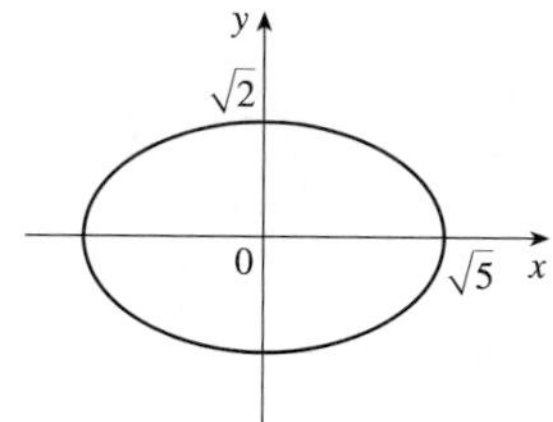

23. $xy = 4$. Hyperbola

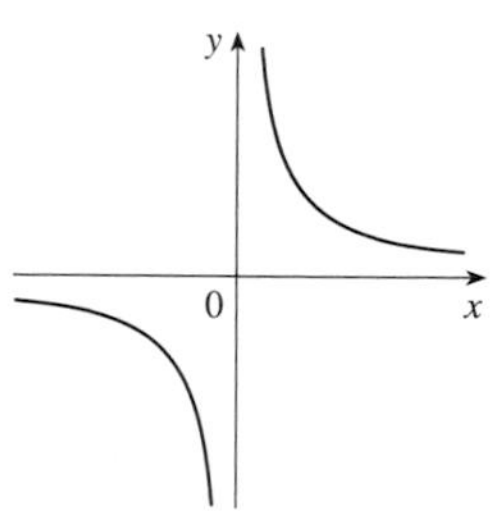

24. $y = x^2 + 2x = (x^2 + 2x + 1) - 1 = (x + 1)^2 - 1$.
Parabola with vertex at $(-1, -1)$

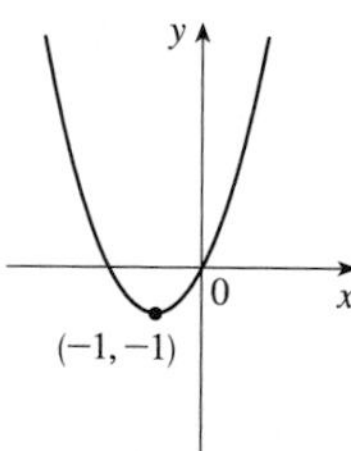

25. $9(x - 1)^2 + 4(y - 2)^2 = 36 \;\Leftrightarrow$
$\dfrac{(x - 1)^2}{4} + \dfrac{(y - 2)^2}{9} = 1$. Ellipse centered at $(1, 2)$

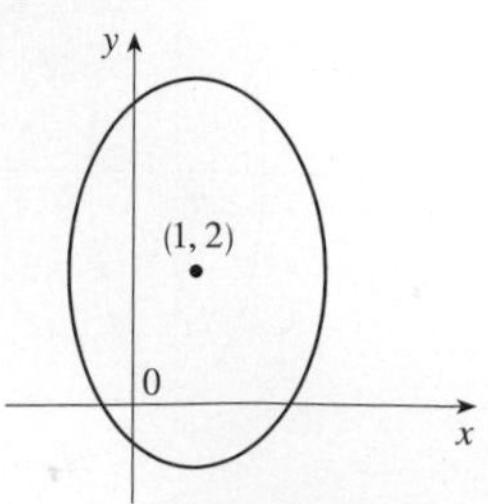

26. $16x^2 + 9y^2 - 36y = 108 \;\Leftrightarrow$
$16x^2 + 9(y^2 - 4y + 4) = 108 + 36 = 144 \;\Leftrightarrow$
$\dfrac{x^2}{9} + \dfrac{(y - 2)^2}{16} = 1$. Ellipse centered at $(0, 2)$

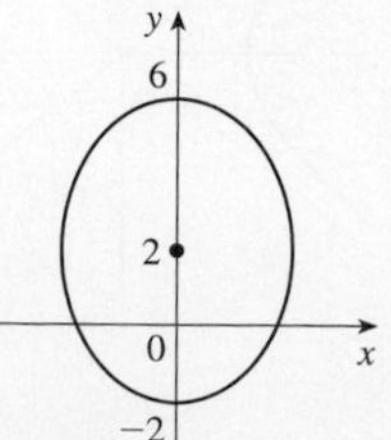

27. $y = x^2 - 6x + 13 = (x^2 - 6x + 9) + 4 = (x - 3)^2 + 4$.
Parabola with vertex at $(3, 4)$

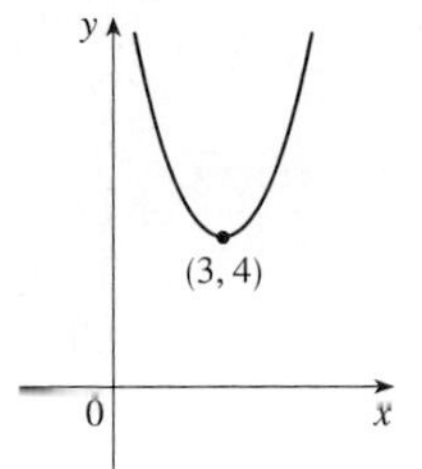

28. $x^2 - y^2 - 4x + 3 = 0 \quad \Leftrightarrow$
$(x^2 - 4x + 4) - y^2 = -3 + 4 = 1 \quad \Leftrightarrow$
$(x - 2)^2 - y^2 = 1$. Hyperbola centered at $(2, 0)$

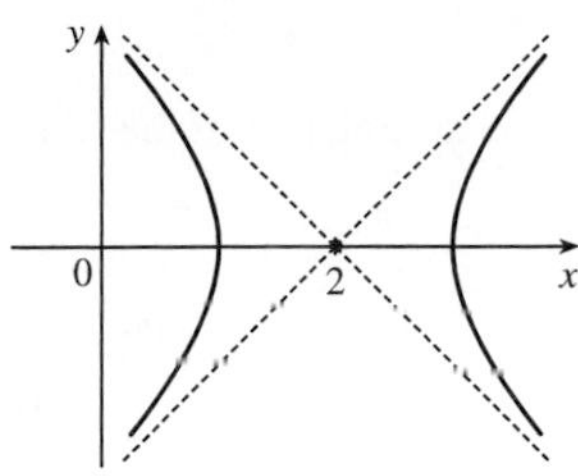

29. $x = 4 - y^2 = -y^2 + 4$. Parabola with vertex at $(4, 0)$

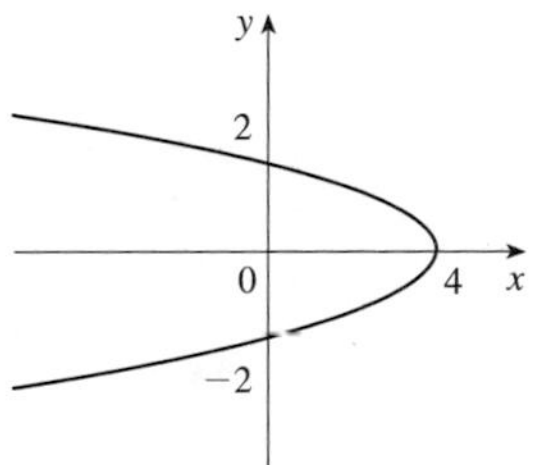

30. $y^2 - 2x + 6y + 5 = 0 \quad \Leftrightarrow \quad y^2 + 6y + 9 = 2x + 4 \quad \Leftrightarrow$
$(y + 3)^2 = 2(x + 2)$. Parabola with vertex $(-2, -3)$

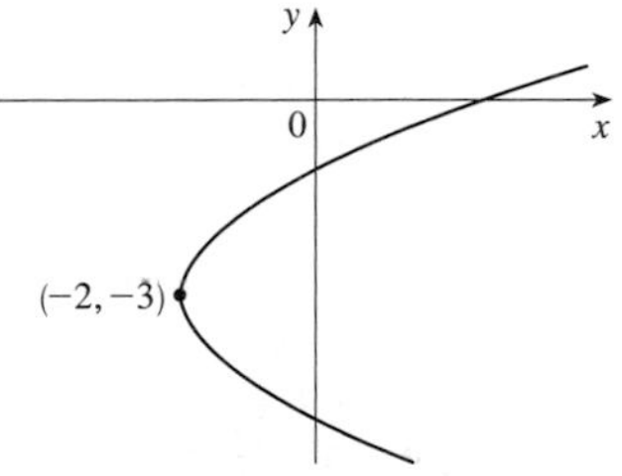

31. $x^2 + 4y^2 - 6x + 5 = 0 \quad \Leftrightarrow$
$(x^2 - 6x + 9) + 4y^2 = -5 + 9 = 4 \quad \Leftrightarrow$
$\dfrac{(x - 3)^2}{4} + y^2 = 1$. Ellipse centered at $(3, 0)$

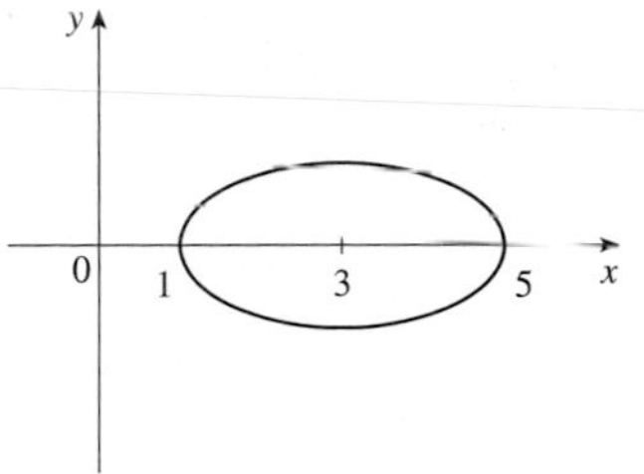

32. $4x^2 + 9y^2 - 16x + 54y + 61 = 0 \quad \Leftrightarrow$
$4(x^2 - 4x + 4) + 9(y^2 + 6y + 9) = -61 + 16 + 81 = 36$
$\Leftrightarrow \quad \dfrac{(x - 2)^2}{9} + \dfrac{(y + 3)^2}{4} = 1$. Ellipse centered at $(2, -3)$

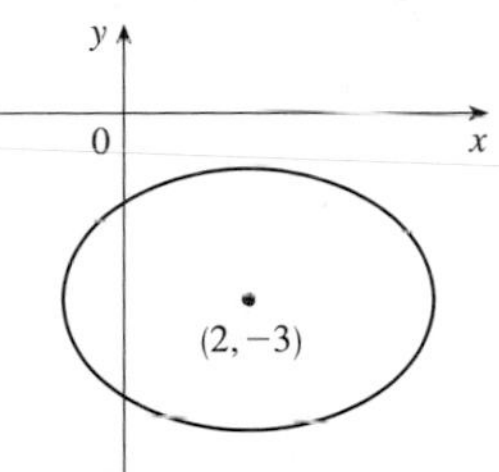

33. $y = 3x$ and $y = x^2$ intersect where $3x = x^2 \quad \Leftrightarrow$
$0 = x^2 - 3x = x(x - 3)$, that is, at $(0, 0)$ and $(3, 9)$.

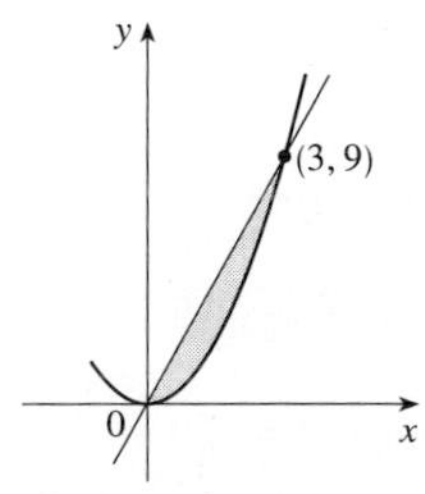

34. $y = 4 - x^2$, $x - 2y = 2$. Substitute y from the first equation into the second: $x - 2(4 - x^2) = 2 \quad \Leftrightarrow$
$2x^2 + x - 10 = 0 \quad \Leftrightarrow \quad (2x + 5)(x - 2) = 0 \quad \Leftrightarrow$
$x = -\frac{5}{2}$ or 2. So the points of intersection are $\left(-\frac{5}{2}, -\frac{9}{4}\right)$ and $(2, 0)$.

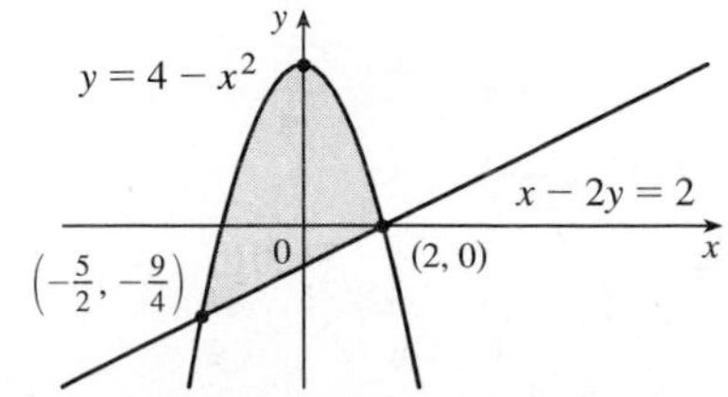

35. The parabola must have an equation of the form $y = a(x-1)^2 - 1$. Substituting $x = 3$ and $y = 3$ into the equation gives $3 = a(3-1)^2 - 1$, so $a = 1$, and the equation is $y = (x-1)^2 - 1 = x^2 - 2x$. Note that using the other point $(-1, 3)$ would have given the same value for a, and hence the same equation.

36. The ellipse has an equation of the form $\dfrac{x^2}{a^2} + \dfrac{y^2}{b^2} = 1$. Substituting $x = 1$ and $y = -\dfrac{10\sqrt{2}}{3}$ gives

$\dfrac{1^2}{a^2} + \dfrac{\left(-10\sqrt{2}/3\right)^2}{b^2} = \dfrac{1}{a^2} + \dfrac{200}{9b^2} = 1$. Substituting $x = -2$ and $y = \dfrac{5\sqrt{5}}{3}$ gives $\dfrac{(-2)^2}{a^2} + \dfrac{\left(5\sqrt{5}/3\right)^2}{b^2} = \dfrac{4}{a^2} + \dfrac{125}{9b^2} = 1$.

From the first equation, $\dfrac{1}{a^2} = 1 - \dfrac{200}{9b^2}$. Putting this into the second equation gives $4\left(1 - \dfrac{200}{9b^2}\right) + \dfrac{125}{9b^2} = 1 \;\Leftrightarrow$

$3 = \dfrac{675}{9b^2} \;\Leftrightarrow\; b^2 = \dfrac{675}{27} = 25$, so $b = 5$. Hence $\dfrac{1}{a^2} = 1 - \dfrac{200}{9(5)^2} = \dfrac{1}{9}$ and so $a = 3$. The equation of the ellipse is $\dfrac{x^2}{9} + \dfrac{y^2}{25} = 1$.

37. $\{(x, y) \mid x^2 + y^2 \le 1\}$

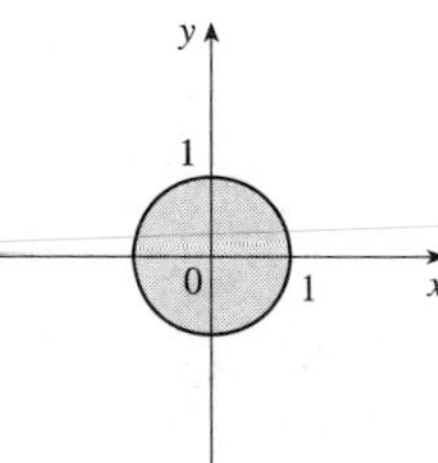

38. 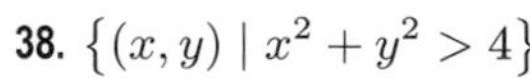 $\{(x, y) \mid x^2 + y^2 > 4\}$

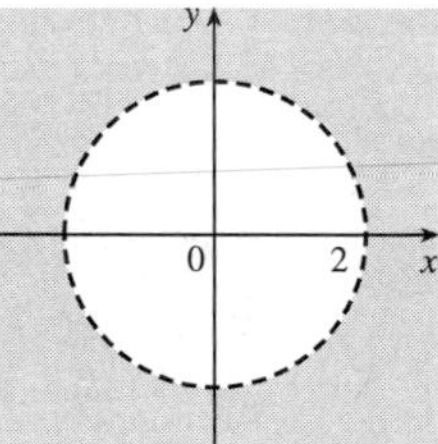

39. $\{(x, y) \mid y \ge x^2 - 1\}$

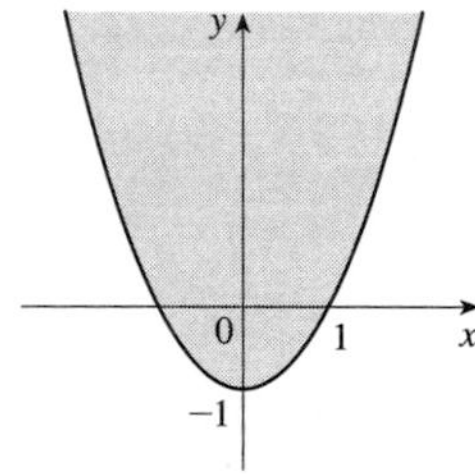

40. $\{(x, y) \mid x^2 + 4y^2 \le 4\}$

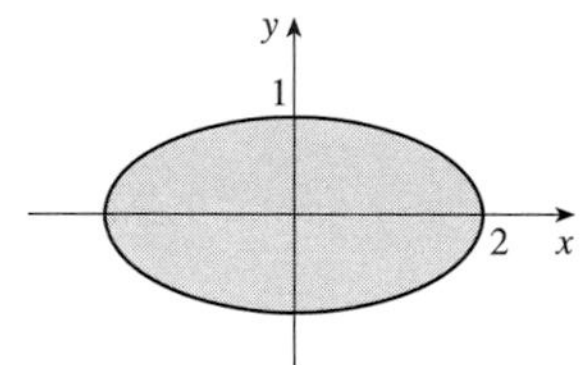

D Trigonometry

1. $210^\circ = 210^\circ\left(\frac{\pi}{180^\circ}\right) = \frac{7\pi}{6}$ rad

2. $300^\circ = 300^\circ\left(\frac{\pi}{180^\circ}\right) = \frac{5\pi}{3}$ rad

3. $9^\circ = 9^\circ\left(\frac{\pi}{180^\circ}\right) = \frac{\pi}{20}$ rad

4. $-315^\circ = -315^\circ\left(\frac{\pi}{180^\circ}\right) = -\frac{7\pi}{4}$ rad

5. $900^\circ = 900^\circ\left(\frac{\pi}{180^\circ}\right) = 5\pi$ rad

6. $36^\circ = 36^\circ\left(\frac{\pi}{180^\circ}\right) = \frac{\pi}{5}$ rad

7. 4π rad $= 4\pi\left(\frac{180^\circ}{\pi}\right) = 720^\circ$

8. $-\frac{7\pi}{2}$ rad $= -\frac{7\pi}{2}\left(\frac{180^\circ}{\pi}\right) = -630^\circ$

9. $\frac{5\pi}{12}$ rad $= \frac{5\pi}{12}\left(\frac{180^\circ}{\pi}\right) = 75^\circ$

10. $\frac{8\pi}{3}$ rad $= \frac{8\pi}{3}\left(\frac{180^\circ}{\pi}\right) = 480^\circ$

11. $-\frac{3\pi}{8}\text{ rad} = -\frac{3\pi}{8}\left(\frac{180^\circ}{\pi}\right) = -67.5^\circ$

12. $5\text{ rad} = 5\left(\frac{180^\circ}{\pi}\right) = \left(\frac{900}{\pi}\right)^\circ$

13. Using Formula 3, $a = r\theta = 36 \cdot \frac{\pi}{12} = 3\pi$ cm.

14. Using Formula 3, $a = r\theta = 10 \cdot 72^\circ\left(\frac{\pi}{180^\circ}\right) = 4\pi$ cm.

15. Using Formula 3, $\theta = a/r = \frac{1}{1.5} = \frac{2}{3}\text{ rad} = \frac{2}{3}\left(\frac{180^\circ}{\pi}\right) = \left(\frac{120}{\pi}\right)^\circ \approx 38.2^\circ$.

16. $a = r\theta \;\Rightarrow\; r = \frac{a}{\theta} = \frac{6}{3\pi/4} = \frac{8}{\pi}$ cm

17.

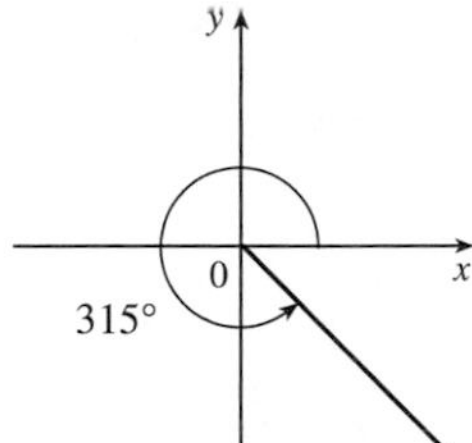

18.

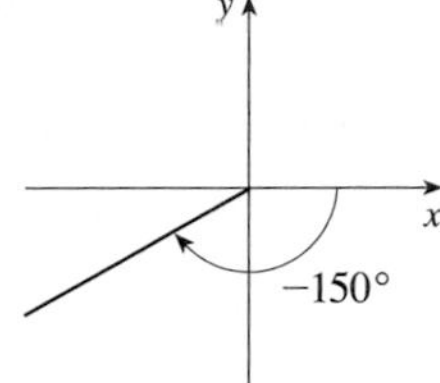

19.

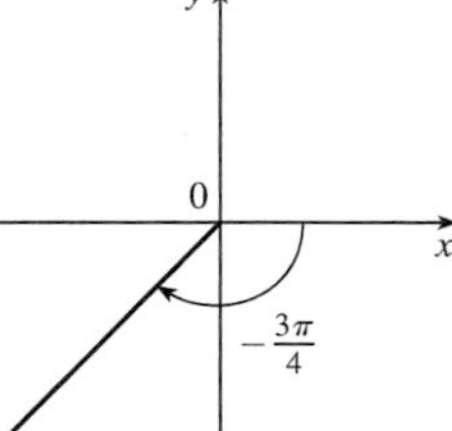

20.

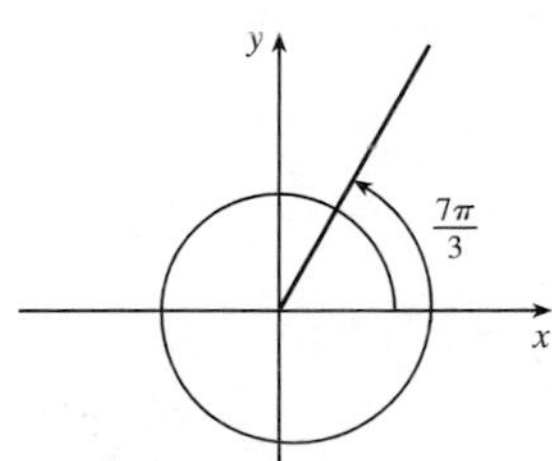

21.

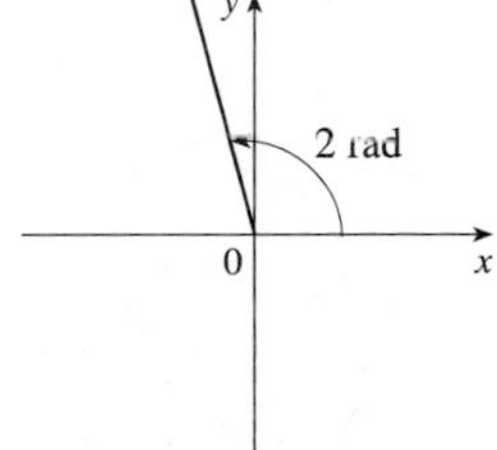

22.

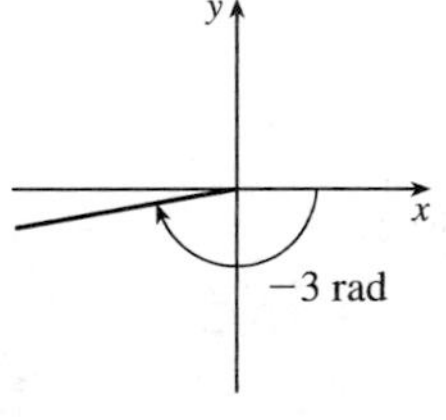

23.

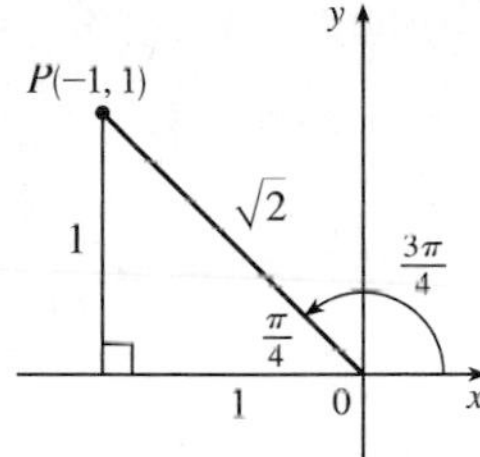

From the diagram we see that a point on the terminal side is $P(-1, 1)$. Therefore, taking $x = -1$, $y = 1$, $r = \sqrt{2}$ in the definitions of the trigonometric ratios, we have $\sin\frac{3\pi}{4} = \frac{1}{\sqrt{2}}$, $\cos\frac{3\pi}{4} = -\frac{1}{\sqrt{2}}$, $\tan\frac{3\pi}{4} = -1$, $\csc\frac{3\pi}{4} = \sqrt{2}$, $\sec\frac{3\pi}{4} = -\sqrt{2}$, and $\cot\frac{3\pi}{4} = -1$.

24.

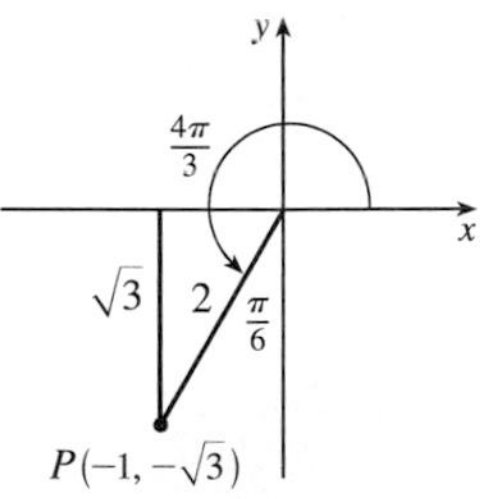

From the diagram and Figure 8, we see that a point on the terminal side is $P(-1, -\sqrt{3})$. Therefore, taking $x = -1$, $y = -\sqrt{3}$, $r = 2$ in the definitions of the trigonometric ratios, we have $\sin\frac{4\pi}{3} = -\frac{\sqrt{3}}{2}$, $\cos\frac{4\pi}{3} = -\frac{1}{2}$, $\tan\frac{4\pi}{3} = \sqrt{3}$, $\csc\frac{4\pi}{3} = -\frac{2}{\sqrt{3}}$, $\sec\frac{4\pi}{3} = -2$, and $\cot\frac{4\pi}{3} = \frac{1}{\sqrt{3}}$.

25.

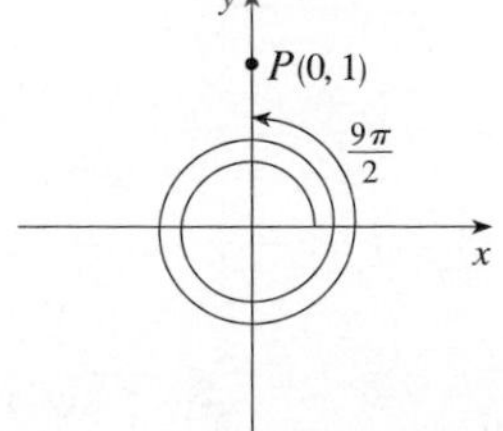

From the diagram we see that a point on the terminal side is $P(0, 1)$. Therefore taking $x = 0$, $y = 1$, $r = 1$ in the definitions of the trigonometric ratios, we have $\sin\frac{9\pi}{2} = 1$, $\cos\frac{9\pi}{2} = 0$, $\tan\frac{9\pi}{2} = y/x$ is undefined since $x = 0$, $\csc\frac{9\pi}{2} = 1$, $\sec\frac{9\pi}{2} = r/x$ is undefined since $x = 0$, and $\cot\frac{9\pi}{2} = 0$.

26.

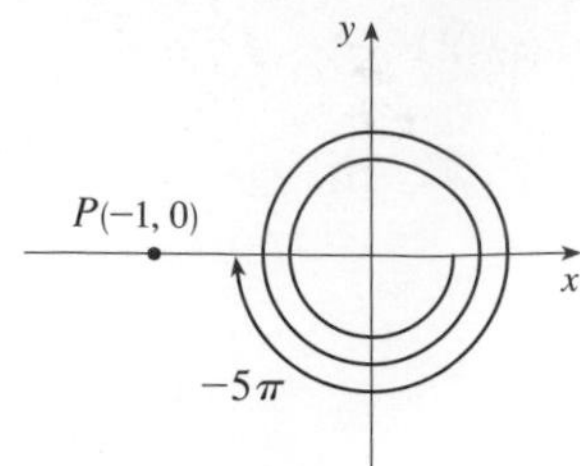

From the diagram, we see that a point on the terminal side is $P(-1, 0)$. Therefore taking $x = -1$, $y = 0$, $r = 1$ in the definitions of the trigonometric ratios we have $\sin(-5\pi) = 0$, $\cos(-5\pi) = -1$, $\tan(-5\pi) = 0$, $\csc(-5\pi)$ is undefined, $\sec(-5\pi) = -1$, and $\cot(-5\pi)$ is undefined.

27.

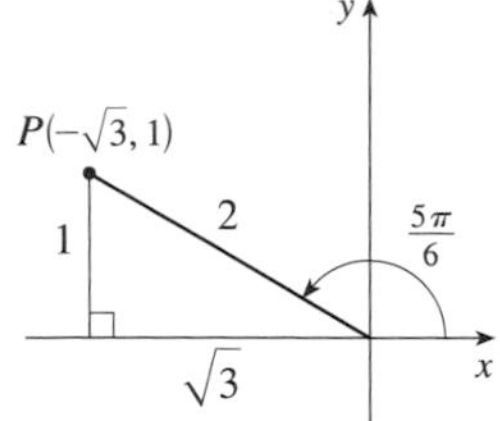

Using Figure 8 we see that a point on the terminal side is $P(-\sqrt{3}, 1)$. Therefore taking $x = -\sqrt{3}$, $y = 1$, $r = 2$ in the definitions of the trigonometric ratios, we have $\sin \frac{5\pi}{6} = \frac{1}{2}$, $\cos \frac{5\pi}{6} = -\frac{\sqrt{3}}{2}$, $\tan \frac{5\pi}{6} = -\frac{1}{\sqrt{3}}$, $\csc \frac{5\pi}{6} = 2$, $\sec \frac{5\pi}{6} = -\frac{2}{\sqrt{3}}$, and $\cot \frac{5\pi}{6} = -\sqrt{3}$.

28.

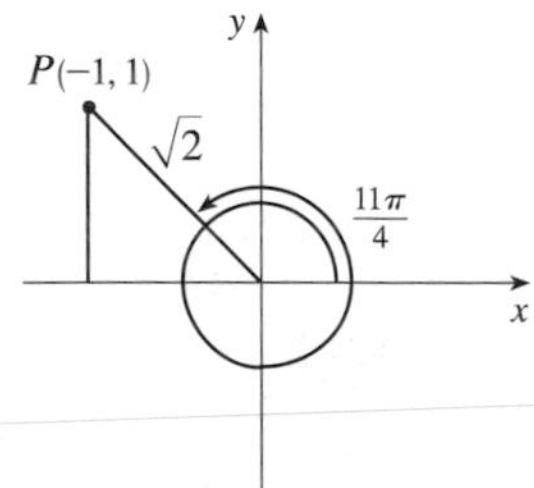

From the diagram, we see that a point on the terminal side is $P(-1, 1)$. Therefore taking $x = -1$, $y = 1$, $r = \sqrt{2}$ in the definitions of the trigonometric ratios we have $\sin \frac{11\pi}{4} = \frac{1}{\sqrt{2}}$, $\cos \frac{11\pi}{4} = -\frac{1}{\sqrt{2}}$, $\tan \frac{11\pi}{4} = -1$, $\csc \frac{11\pi}{4} = \sqrt{2}$, $\sec \frac{11\pi}{4} = -\sqrt{2}$, and $\cot \frac{11\pi}{4} = -1$.

29. $\sin\theta = y/r = \frac{3}{5} \Rightarrow y = 3$, $r = 5$, and $x = \sqrt{r^2 - y^2} = 4$ (since $0 < \theta < \frac{\pi}{2}$). Therefore taking $x = 4$, $y = 3$, $r = 5$ in the definitions of the trigonometric ratios, we have $\cos\theta = \frac{4}{5}$, $\tan\theta = \frac{3}{4}$, $\csc\theta = \frac{5}{3}$, $\sec\theta = \frac{5}{4}$, and $\cot\theta = \frac{4}{3}$.

30. Since $0 < \alpha < \frac{\pi}{2}$, α is in the first quadrant where x and y are both positive. Therefore, $\tan\alpha = y/x = \frac{2}{1} \Rightarrow y = 2$, $x = 1$, and $r = \sqrt{x^2 + y^2} = \sqrt{5}$. Taking $x = 1$, $y = 2$, $r = \sqrt{5}$ in the definitions of the trigonometric ratios, we have $\sin\alpha = \frac{2}{\sqrt{5}}$, $\cos\alpha = \frac{1}{\sqrt{5}}$, $\csc\alpha = \frac{\sqrt{5}}{2}$, $\sec\alpha = \sqrt{5}$, and $\cot\alpha = \frac{1}{2}$.

31. $\frac{\pi}{2} < \phi < \pi \Rightarrow \phi$ is in the second quadrant, where x is negative and y is positive. Therefore $\sec\phi = r/x = -1.5 = -\frac{3}{2} \Rightarrow r = 3$, $x = -2$, and $y = \sqrt{r^2 - x^2} = \sqrt{5}$. Taking $x = -2$, $y = \sqrt{5}$, and $r = 3$ in the definitions of the trigonometric ratios, we have $\sin\phi = \frac{\sqrt{5}}{3}$, $\cos\phi = -\frac{2}{3}$, $\tan\phi = -\frac{\sqrt{5}}{2}$, $\csc\phi = \frac{3}{\sqrt{5}}$, and $\cot\theta = -\frac{2}{\sqrt{5}}$.

32. Since $\pi < x < \frac{3\pi}{2}$, x is in the third quadrant where x and y are both negative. Therefore $\cos x = x/r = -\frac{1}{3} \Rightarrow x = -1$, $r = 3$, and $y = -\sqrt{r^2 - x^2} = -\sqrt{8} = -2\sqrt{2}$. Taking $x = -1$, $r = 3$, $y = -2\sqrt{2}$ in the definitions of the trigonometric ratios, we have $\sin x = -\frac{2\sqrt{2}}{3}$, $\tan x = 2\sqrt{2}$, $\csc x = -\frac{3}{2\sqrt{2}}$, $\sec x = -3$, and $\cot x = \frac{1}{2\sqrt{2}}$.

33. $\pi < \beta < 2\pi$ means that β is in the third or fourth quadrant where y is negative. Also since $\cot\beta = x/y = 3$ which is positive, x must also be negative. Therefore $\cot\beta = x/y = \frac{3}{1} \Rightarrow x = -3$, $y = -1$, and $r = \sqrt{x^2 + y^2} = \sqrt{10}$. Taking $x = -3$, $y = -1$ and $r = \sqrt{10}$ in the definitions of the trigonometric ratios, we have $\sin\beta = -\frac{1}{\sqrt{10}}$, $\cos\beta = -\frac{3}{\sqrt{10}}$, $\tan\beta = \frac{1}{3}$, $\csc\beta = -\sqrt{10}$, and $\sec\beta = -\frac{\sqrt{10}}{3}$.

34. Since $\frac{3\pi}{2} < \theta < 2\pi$, θ is in the fourth quadrant where x is positive and y is negative. Therefore $\csc\theta = r/y = -\frac{4}{3} \Rightarrow$ $r = 4$, $y = -3$, and $x = \sqrt{r^2 - y^2} = \sqrt{7}$. Taking $x = \sqrt{7}$, $y = -3$, and $r = 4$ in the definitions of the trigonometric ratios, we have $\sin\theta = -\frac{3}{4}$, $\cos\theta = \frac{\sqrt{7}}{4}$, $\tan\theta = -\frac{3}{\sqrt{7}}$, $\sec\theta = \frac{4}{\sqrt{7}}$, and $\cot\theta = -\frac{\sqrt{7}}{3}$.

35. $\sin 35^\circ = \dfrac{x}{10} \Rightarrow x = 10\sin 35^\circ \approx 5.73576$ cm

36. $\cos 40^\circ = \dfrac{x}{25} \Rightarrow x = 25\cos 40^\circ \approx 19.15111$ cm

37. $\tan\frac{2\pi}{5} = \dfrac{x}{8} \Rightarrow x = 8\tan\frac{2\pi}{5} \approx 24.62147$ cm

38. $\cos\dfrac{3\pi}{8} = \dfrac{22}{x} \Rightarrow x = \dfrac{22}{\cos\frac{3\pi}{8}} \approx 57.48877$ cm

39.

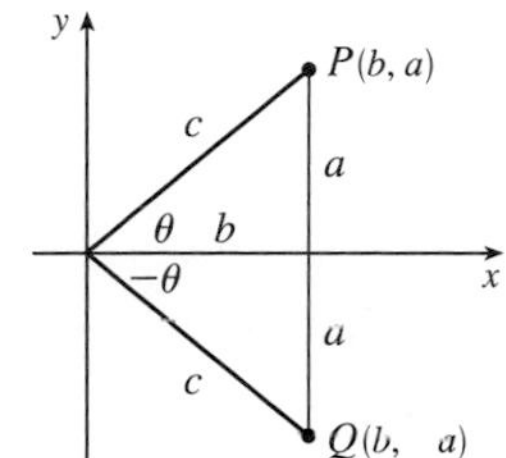

(a) From the diagram we see that $\sin\theta = \dfrac{y}{r} = \dfrac{a}{c}$, and $\sin(-\theta) = \dfrac{-a}{c} = -\dfrac{a}{c} = -\sin\theta$.

(b) Again from the diagram we see that $\cos\theta = \dfrac{x}{r} = \dfrac{b}{c} = \cos(-\theta)$.

40. (a) Using (12a) and (12b), we have

$$\tan(x+y) = \frac{\sin(x+y)}{\cos(x+y)} = \frac{\sin x\,\cos y + \cos x\,\sin y}{\cos x\,\cos y - \sin x\,\sin y} = \frac{\dfrac{\sin x\,\cos y}{\cos x\,\cos y} + \dfrac{\cos x\,\sin y}{\cos x\,\cos y}}{\dfrac{\cos x\,\cos y}{\cos x\,\cos y} - \dfrac{\sin x\,\sin y}{\cos x\,\cos y}} = \frac{\tan x + \tan y}{1 - \tan x\,\tan y}$$

(b) From (10a) and (10b), we have $\tan(-\theta) = -\tan\theta$, so (14a) implies that

$$\tan(x-y) = \tan(x + (-y)) = \frac{\tan x + \tan(-y)}{1 - \tan x\,\tan(-y)} = \frac{\tan x - \tan y}{1 + \tan x\,\tan y}$$

41. (a) Using (12a) and (13a), we have

$\frac{1}{2}[\sin(x+y) + \sin(x-y)] = \frac{1}{2}[\sin x\,\cos y + \cos x\,\sin y + \sin x\,\cos y - \cos x\,\sin y] = \frac{1}{2}(2\sin x\,\cos y) = \sin x\,\cos y$.

(b) This time, using (12b) and (13b), we have

$\frac{1}{2}[\cos(x+y) + \cos(x-y)] = \frac{1}{2}[\cos x\,\cos y - \sin x\,\sin y + \cos x\,\cos y + \sin x\,\sin y] = \frac{1}{2}(2\cos x\,\cos y) = \cos x\,\cos y$.

(c) Again using (12b) and (13b), we have

$$\begin{aligned}\tfrac{1}{2}[\cos(x-y) - \cos(x+y)] &= \tfrac{1}{2}[\cos x\,\cos y + \sin x\,\sin y - \cos x\,\cos y + \sin x\,\sin y] \\ &= \tfrac{1}{2}(2\sin x\,\sin y) = \sin x\,\sin y\end{aligned}$$

42. Using (13b), $\cos\left(\frac{\pi}{2} - x\right) = \cos\frac{\pi}{2}\cos x + \sin\frac{\pi}{2}\sin x = 0\cdot\cos x + 1\cdot\sin x = \sin x$.

43. Using (12a), we have $\sin\left(\frac{\pi}{2} + x\right) = \sin\frac{\pi}{2}\cos x + \cos\frac{\pi}{2}\sin x = 1\cdot\cos x + 0\cdot\sin x = \cos x$.

44. Using (13a), we have $\sin(\pi - x) = \sin\pi\cos x - \cos\pi\,\sin x = 0\cdot\cos x - (-1)\sin x = \sin x$.

45. Using (6), we have $\sin\theta\cot\theta = \sin\theta\cdot\dfrac{\cos\theta}{\sin\theta} = \cos\theta$.

46. $(\sin x + \cos x)^2 = \sin^2 x + 2\sin x\,\cos x + \cos^2 x = (\sin^2 x + \cos^2 x) + \sin 2x$ [by (15a)] $= 1 + \sin 2x$ [by (7)]

47. $\sec y - \cos y = \dfrac{1}{\cos y} - \cos y$ [by (6)] $= \dfrac{1 - \cos^2 y}{\cos y} = \dfrac{\sin^2 y}{\cos y}$ [by (7)] $= \dfrac{\sin y}{\cos y} \sin y = \tan y \sin y$ [by (6)]

48. $\tan^2 \alpha - \sin^2 \alpha = \dfrac{\sin^2 \alpha}{\cos^2 \alpha} - \sin^2 \alpha = \dfrac{\sin^2 \alpha - \sin^2 \alpha \cos^2 \alpha}{\cos^2 \alpha} = \dfrac{\sin^2 \alpha (1 - \cos^2 \alpha)}{\cos^2 \alpha} = \tan^2 \alpha \sin^2 \alpha$ [by (6), (7)]

49. $\cot^2 \theta + \sec^2 \theta = \dfrac{\cos^2 \theta}{\sin^2 \theta} + \dfrac{1}{\cos^2 \theta}$ [by (6)] $= \dfrac{\cos^2 \theta \cos^2 \theta + \sin^2 \theta}{\sin^2 \theta \cos^2 \theta}$

$= \dfrac{(1 - \sin^2 \theta)(1 - \sin^2 \theta) + \sin^2 \theta}{\sin^2 \theta \cos^2 \theta}$ [by (7)] $= \dfrac{1 - \sin^2 \theta + \sin^4 \theta}{\sin^2 \theta \cos^2 \theta}$

$= \dfrac{\cos^2 \theta + \sin^4 \theta}{\sin^2 \theta \cos^2 \theta}$ [by (7)] $= \dfrac{1}{\sin^2 \theta} + \dfrac{\sin^2 \theta}{\cos^2 \theta} = \csc^2 \theta + \tan^2 \theta$ [by (6)]

50. $2 \csc 2t = \dfrac{2}{\sin 2t} = \dfrac{2}{2 \sin t \cos t}$ [by (15a)] $= \dfrac{1}{\sin t \cos t} = \sec t \csc t$

51. Using (14a), we have $\tan 2\theta = \tan(\theta + \theta) = \dfrac{\tan \theta + \tan \theta}{1 - \tan \theta \tan \theta} = \dfrac{2 \tan \theta}{1 - \tan^2 \theta}$.

52. $\dfrac{1}{1 - \sin \theta} + \dfrac{1}{1 + \sin \theta} = \dfrac{1 + \sin \theta + 1 - \sin \theta}{(1 - \sin \theta)(1 + \sin \theta)} = \dfrac{2}{1 - \sin^2 \theta} = \dfrac{2}{\cos^2 \theta}$ [by (7)] $= 2 \sec^2 \theta$

53. Using (15a) and (16a),

$$\begin{aligned} \sin x \sin 2x + \cos x \cos 2x &= \sin x (2 \sin x \cos x) + \cos x (2 \cos^2 x - 1) = 2 \sin^2 x \cos x + 2 \cos^3 x - \cos x \\ &= 2(1 - \cos^2 x) \cos x + 2 \cos^3 x - \cos x \quad \text{[by (7)]} \\ &= 2 \cos x - 2 \cos^3 x + 2 \cos^3 x - \cos x = \cos x \end{aligned}$$

Or: $\sin x \sin 2x + \cos x \cos 2x = \cos (2x - x)$ [by 13(b)] $= \cos x$

54. We start with the right side using equations (12a) and (13a):

$$\begin{aligned} \sin(x + y) \sin(x - y) &= (\sin x \cos y + \cos x \sin y)(\sin x \cos y - \cos x \sin y) \\ &= \sin^2 x \cos^2 y - \sin x \cos y \cos x \sin y + \cos x \sin y \sin x \cos y - \cos^2 x \sin^2 y \\ &= \sin^2 x (1 - \sin^2 y) - (1 - \sin^2 x) \sin^2 y \quad \text{[by (7)]} \\ &= \sin^2 x - \sin^2 x \sin^2 y - \sin^2 y + \sin^2 x \sin^2 y = \sin^2 x - \sin^2 y \end{aligned}$$

55. $\dfrac{\sin \phi}{1 - \cos \phi} = \dfrac{\sin \phi}{1 - \cos \phi} \cdot \dfrac{1 + \cos \phi}{1 + \cos \phi} = \dfrac{\sin \phi (1 + \cos \phi)}{1 - \cos^2 \phi} = \dfrac{\sin \phi (1 + \cos \phi)}{\sin^2 \phi}$ [by (7)]

$= \dfrac{1 + \cos \phi}{\sin \phi} = \dfrac{1}{\sin \phi} + \dfrac{\cos \phi}{\sin \phi} = \csc \phi + \cot \phi$ [by (6)]

56. $\tan x + \tan y = \dfrac{\sin x}{\cos x} + \dfrac{\sin y}{\cos y} = \dfrac{\sin x \cos y + \cos x \sin y}{\cos x \cos y} = \dfrac{\sin(x + y)}{\cos x \cos y}$ [by (12a)]

57. Using (12a),

$$\begin{aligned} \sin 3\theta + \sin \theta &= \sin(2\theta + \theta) + \sin \theta = \sin 2\theta \cos \theta + \cos 2\theta \sin \theta + \sin \theta \\ &= \sin 2\theta \cos \theta + (2 \cos^2 \theta - 1) \sin \theta + \sin \theta \quad \text{[by (16a)]} \\ &= \sin 2\theta \cos \theta + 2 \cos^2 \theta \sin \theta - \sin \theta + \sin \theta = \sin 2\theta \cos \theta + \sin 2\theta \cos \theta \quad \text{[by (15a)]} \\ &= 2 \sin 2\theta \cos \theta \end{aligned}$$

58. We use (12b) with $x = 2\theta$, $y = \theta$ to get

$$\begin{aligned}\cos 3\theta &= \cos(2\theta + \theta) = \cos 2\theta \cos\theta - \sin 2\theta \sin\theta \\ &= (2\cos^2\theta - 1)\cos\theta - 2\sin^2\theta\cos\theta \quad \text{[by (16a) and (15a)]} \\ &= (2\cos^2\theta - 1)\cos\theta - 2(1 - \cos^2\theta)\cos\theta \quad \text{[by (7)]} \\ &= 2\cos^3\theta - \cos\theta - 2\cos\theta + 2\cos^3\theta = 4\cos^3\theta - 3\cos\theta\end{aligned}$$

59. Since $\sin x = \frac{1}{3}$ we can label the opposite side as having length 1, the hypotenuse as having length 3, and use the Pythagorean Theorem to get that the adjacent side has length $\sqrt{8}$. Then, from the diagram, $\cos x = \frac{\sqrt{8}}{3}$. Similarly we have that $\sin y = \frac{3}{5}$. Now use (12a):

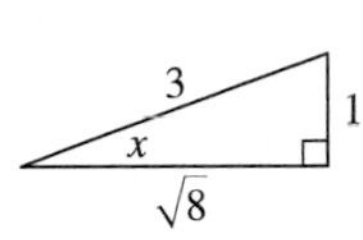

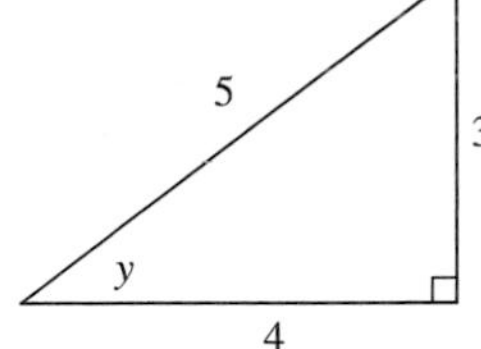

$\sin(x + y) = \sin x \cos y + \cos x \sin y = \frac{1}{3}\cdot\frac{4}{5} + \frac{\sqrt{8}}{3}\cdot\frac{3}{5} = \frac{4}{15} + \frac{3\sqrt{8}}{15} = \frac{4+6\sqrt{2}}{15}$.

60. Use (12b) and the values for $\sin y$ and $\cos x$ obtained in Exercise 59 to get

$$\cos(x + y) = \cos x \cos y - \sin x \sin y = \frac{\sqrt{8}}{3}\cdot\frac{4}{5} - \frac{1}{3}\cdot\frac{3}{5} = \frac{8\sqrt{2}-3}{15}$$

61. Using (13b) and the values for $\cos x$ and $\sin y$ obtained in Exercise 59, we have

$$\cos(x - y) = \cos x \cos y + \sin x \sin y = \frac{\sqrt{8}}{3}\cdot\frac{4}{5} + \frac{1}{3}\cdot\frac{3}{5} = \frac{8\sqrt{2}+3}{15}$$

62. Using (13a) and the values for $\sin y$ and $\cos x$ obtained in Exercise 59, we get

$$\sin(x - y) = \sin x \cos y - \cos x \sin y = \frac{1}{3}\cdot\frac{4}{5} - \frac{\sqrt{8}}{3}\cdot\frac{3}{5} = \frac{4-6\sqrt{2}}{15}$$

63. Using (15a) and the values for $\sin y$ and $\cos y$ obtained in Exercise 59, we have $\sin 2y = 2\sin y\cos y = 2\cdot\frac{3}{5}\cdot\frac{4}{5} = \frac{24}{25}$.

64. Using (16a) with $\cos y = \frac{4}{5}$, we have $\cos 2y = 2\cos^2 y - 1 = 2\left(\frac{4}{5}\right)^2 - 1 = \frac{32}{25} - 1 = \frac{7}{25}$.

65. $2\cos x - 1 = 0 \quad\Leftrightarrow\quad \cos x = \frac{1}{2} \quad\Rightarrow\quad x = \frac{\pi}{3}, \frac{5\pi}{3}$ for $x \in [0, 2\pi]$.

66. $3\cot^2 x = 1 \quad\Leftrightarrow\quad 3 = 1/\cot^2 x \quad\Leftrightarrow\quad \tan^2 x = 3 \quad\Leftrightarrow\quad \tan x = \pm\sqrt{3} \quad\Rightarrow\quad x = \frac{\pi}{3}, \frac{2\pi}{3}, \frac{4\pi}{3}$, and $\frac{5\pi}{3}$.

67. $2\sin^2 x = 1 \quad\Leftrightarrow\quad \sin^2 x = \frac{1}{2} \quad\Leftrightarrow\quad \sin x = \pm\frac{1}{\sqrt{2}} \quad\Rightarrow\quad x = \frac{\pi}{4}, \frac{3\pi}{4}, \frac{5\pi}{4}, \frac{7\pi}{4}$.

68. $|\tan x| = 1 \quad\Leftrightarrow\quad \tan x = -1$ or $\tan x = 1 \quad\Leftrightarrow\quad x = \frac{3\pi}{4}, \frac{7\pi}{4}$ or $x = \frac{\pi}{4}, \frac{5\pi}{4}$.

69. Using (15a), we have $\sin 2x = \cos x \quad\Leftrightarrow\quad 2\sin x\cos x - \cos x = 0 \quad\Leftrightarrow\quad \cos x(2\sin x - 1) = 0 \quad\Leftrightarrow\quad \cos x = 0$ or $2\sin x - 1 = 0 \quad\Rightarrow\quad x = \frac{\pi}{2}, \frac{3\pi}{2}$ or $\sin x = \frac{1}{2} \quad\Rightarrow\quad x = \frac{\pi}{6}$ or $\frac{5\pi}{6}$. Therefore, the solutions are $x = \frac{\pi}{6}, \frac{\pi}{2}, \frac{5\pi}{6}, \frac{3\pi}{2}$.

70. By (15a), $2\cos x + \sin 2x = 0 \quad\Leftrightarrow\quad 2\cos x + 2\sin x\cos x = 0 \quad\Leftrightarrow\quad 2\cos x(1 + \sin x) = 0 \quad\Leftrightarrow\quad \cos x = 0$ or $1 + \sin x = 0 \quad\Leftrightarrow\quad x = \frac{\pi}{2}, \frac{3\pi}{2}$ or $\sin x = -1 \quad\Rightarrow\quad x = \frac{3}{2}\pi$. So the solutions are $x = \frac{\pi}{2}, \frac{3\pi}{2}$.

71. $\sin x = \tan x \quad\Leftrightarrow\quad \sin x - \tan x = 0 \quad\Leftrightarrow\quad \sin x - \dfrac{\sin x}{\cos x} = 0 \quad\Leftrightarrow\quad \sin x\left(1 - \dfrac{1}{\cos x}\right) = 0 \quad\Leftrightarrow\quad \sin x = 0$ or $1 - \dfrac{1}{\cos x} = 0 \quad\Rightarrow\quad x = 0, \pi, 2\pi$ or $1 = \dfrac{1}{\cos x} \quad\Rightarrow\quad \cos x = 1 \quad\Rightarrow\quad x = 0, 2\pi$. Therefore the solutions are $x = 0, \pi, 2\pi$.

72. By (16a), $2 + \cos 2x = 3\cos x \iff 2 + 2\cos^2 x - 1 = 3\cos x \iff 2\cos^2 x - 3\cos x + 1 = 0 \iff$ $(2\cos x - 1)(\cos x - 1) = 0 \iff \cos x = 1$ or $\cos x = \frac{1}{2} \Rightarrow x = 0, 2\pi$ or $x = \frac{\pi}{3}, \frac{5\pi}{3}$.

73. We know that $\sin x = \frac{1}{2}$ when $x = \frac{\pi}{6}$ or $\frac{5\pi}{6}$, and from Figure 13(a), we see that $\sin x \leq \frac{1}{2} \Rightarrow 0 \leq x \leq \frac{\pi}{6}$ or $\frac{5\pi}{6} \leq x \leq 2\pi$ for $x \in [0, 2\pi]$.

74. $2\cos x + 1 > 0 \Rightarrow 2\cos x > -1 \Rightarrow \cos x > -\frac{1}{2}$. $\cos x = -\frac{1}{2}$ when $x = \frac{2\pi}{3}, \frac{4\pi}{3}$ and from Figure 13(b), we see that $\cos x > -\frac{1}{2}$ when $0 \leq x < \frac{2\pi}{3}, \frac{4\pi}{3} < x \leq 2\pi$.

75. $\tan x = -1$ when $x = \frac{3\pi}{4}, \frac{7\pi}{4}$, and $\tan x = 1$ when $x = \frac{\pi}{4}$ or $\frac{5\pi}{4}$. From Figure 14(a) we see that $-1 < \tan x < 1 \Rightarrow$ $0 \leq x < \frac{\pi}{4}, \frac{3\pi}{4} < x < \frac{5\pi}{4}$, and $\frac{7\pi}{4} < x \leq 2\pi$.

76. We know that $\sin x = \cos x$ when $x = \frac{\pi}{4}, \frac{5\pi}{4}$, and from the diagram we see that $\sin x > \cos x$ when $\frac{\pi}{4} < x < \frac{5\pi}{4}$.

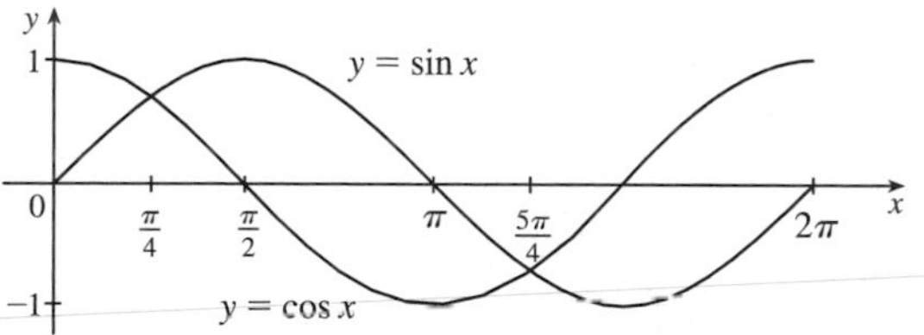

77. $y = \cos\left(x - \frac{\pi}{3}\right)$. We start with the graph of $y = \cos x$ and shift it $\frac{\pi}{3}$ units to the right.

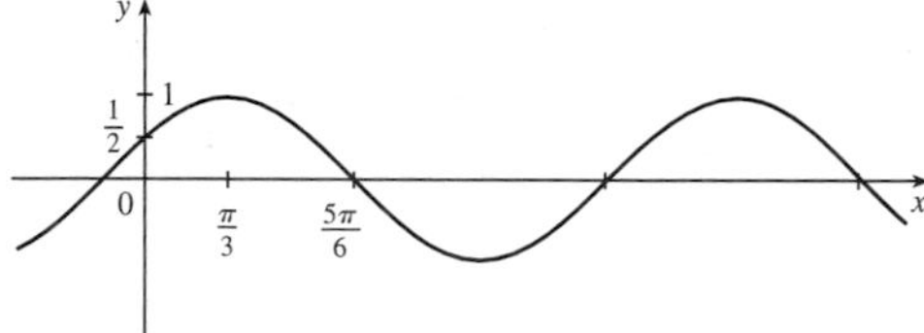

78. $y = \tan 2x$. Start with the graph of $y = \tan x$ with period π and compress it to a period of $\frac{\pi}{2}$.

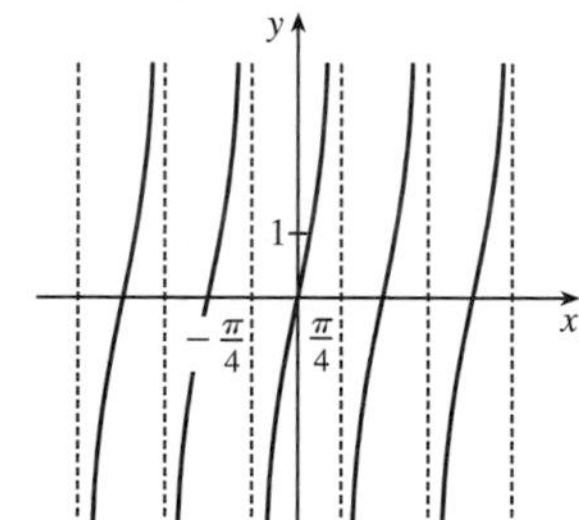

79. $y = \frac{1}{3}\tan\left(x - \frac{\pi}{2}\right)$. We start with the graph of $y = \tan x$, shift it $\frac{\pi}{2}$ units to the right and compress it to $\frac{1}{3}$ of its original vertical size.

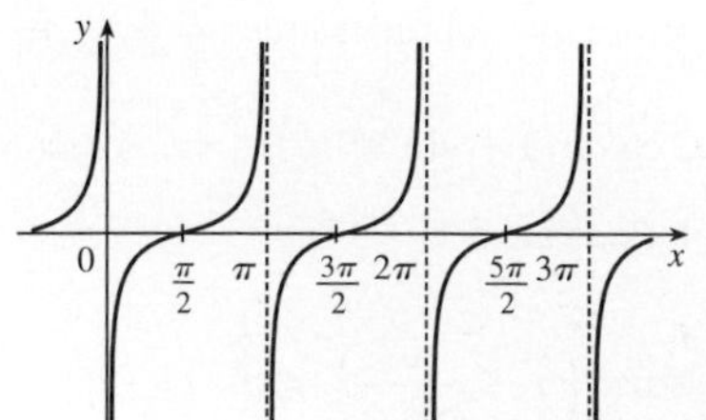

80. $y = 1 + \sec x$. Start with the graph of $y = \sec x$ and raise it by one unit.

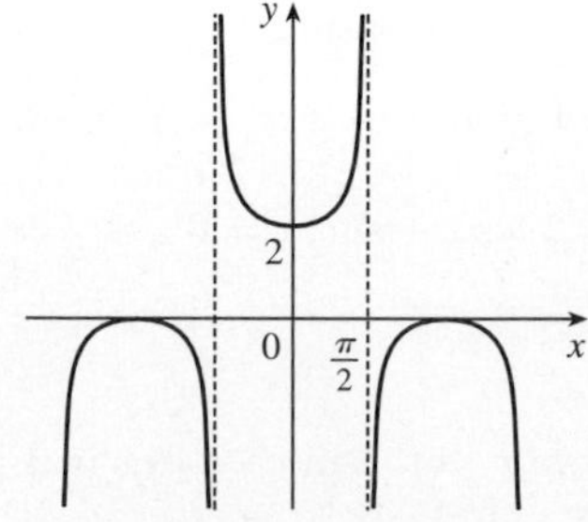

81. $y = |\sin x|$. We start with the graph of $y = \sin x$ and reflect the parts below the x-axis about the x-axis.

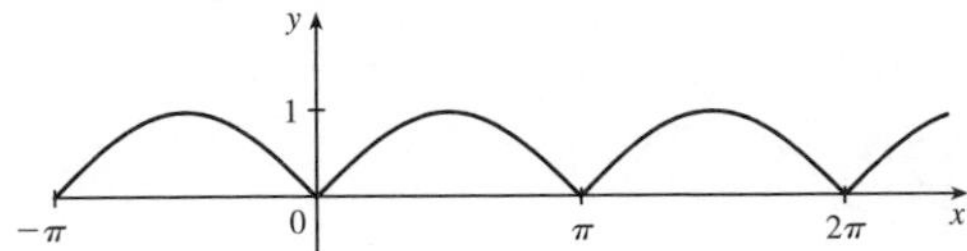

82. $y = 2 + \sin\left(x + \frac{\pi}{4}\right)$. Start with the graph of $y = \sin x$, and shift it $\frac{\pi}{4}$ units to the left and 2 units up.

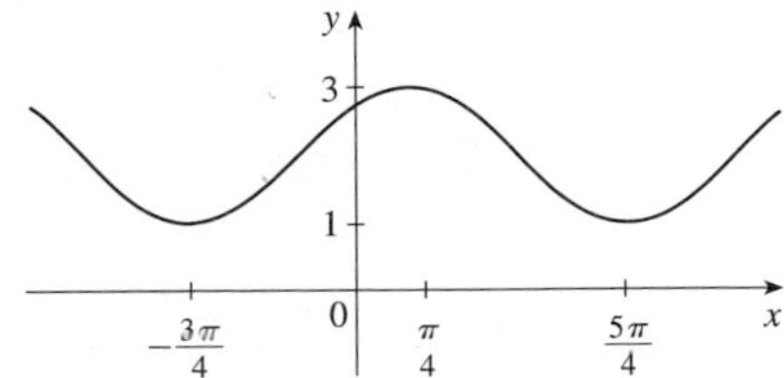

83. From the figure in the text, we see that $x = b\cos\theta$, $y = b\sin\theta$, and from the distance formula we have that the distance c from (x, y) to $(a, 0)$ is $c = \sqrt{(x-a)^2 + (y-0)^2} \;\;\Rightarrow$

$$c^2 = (b\cos\theta - a)^2 + (b\sin\theta)^2 = b^2\cos^2\theta - 2ab\cos\theta + a^2 + b^2\sin^2\theta$$
$$= a^2 + b^2(\cos^2\theta + \sin^2\theta) - 2ab\cos\theta = a^2 + b^2 - 2ab\cos\theta \quad \text{[by (7)]}$$

84. $|AB|^2 = |AC|^2 + |BC|^2 - 2\,|AC|\,|BC|\cos\angle C = (820)^2 + (910)^2 - 2(820)(910)\cos 103^\circ \approx 1{,}836{,}217 \;\;\Rightarrow$ $|AB| \approx 1355$ m

85. Using the Law of Cosines, we have $c^2 = 1^2 + 1^2 - 2(1)(1)\cos(\alpha - \beta) = 2\,[1 - \cos(\alpha - \beta)]$. Now, using the distance formula, $c^2 = |AB|^2 = (\cos\alpha - \cos\beta)^2 + (\sin\alpha - \sin\beta)^2$. Equating these two expressions for c^2, we get $2[1 - \cos(\alpha - \beta)] = \cos^2\alpha + \sin^2\alpha + \cos^2\beta + \sin^2\beta - 2\cos\alpha\,\cos\beta - 2\sin\alpha\,\sin\beta \;\;\Rightarrow$ $1 - \cos(\alpha - \beta) = 1 - \cos\alpha\,\cos\beta - \sin\alpha\,\sin\beta \;\;\Rightarrow\;\; \cos(\alpha - \beta) = \cos\alpha\,\cos\beta + \sin\alpha\,\sin\beta$.

86. $\cos(x + y) = \cos(x - (-y)) = \cos x\,\cos(-y) + \sin x\,\sin(-y)$
$= \cos x\,\cos y - \sin x\,\sin y$ [using Equations (10a) and (10b)]

87. In Exercise 86 we used the subtraction formula for cosine to prove the addition formula for cosine. Using that formula with $x = \frac{\pi}{2} - \alpha$, $y = \beta$, we get $\cos\left[\left(\frac{\pi}{2} - \alpha\right) + \beta\right] = \cos\left(\frac{\pi}{2} - \alpha\right)\cos\beta - \sin\left(\frac{\pi}{2} - \alpha\right)\sin\beta \;\;\Rightarrow$ $\cos\left[\frac{\pi}{2} - (\alpha - \beta)\right] = \cos\left(\frac{\pi}{2} - \alpha\right)\cos\beta - \sin\left(\frac{\pi}{2} - \alpha\right)\sin\beta$. Now we use the identities given in the problem, $\cos\left(\frac{\pi}{2} - \theta\right) = \sin\theta$ and $\sin\left(\frac{\pi}{2} - \theta\right) = \cos\theta$, to get $\sin(\alpha - \beta) = \sin\alpha\,\cos\beta - \cos\alpha\,\sin\beta$.

88. If $0 < \theta < \frac{\pi}{2}$, we have the case depicted in the first diagram. In this case, we see that the height of the triangle is $h = a\sin\theta$. If $\frac{\pi}{2} \le \theta < \pi$, we have the case depicted in the second diagram. In this case, the height of the triangle is $h = a\sin(\pi - \theta) = a\sin\theta$ (by the identity proved in Exercise 44). So in either case, the area of the triangle is $\frac{1}{2}bh = \frac{1}{2}ab\sin\theta$.

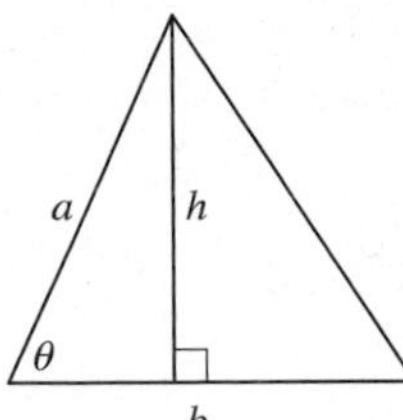

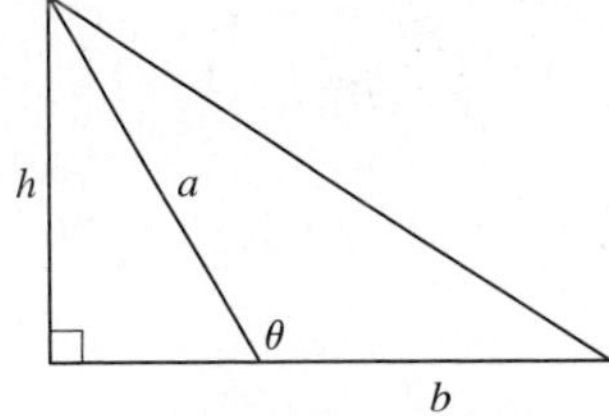

89. Using the formula from Exercise 88, the area of the triangle is $\frac{1}{2}(10)(3)\sin 107^\circ \approx 14.34457\ \text{cm}^2$.

E Sigma Notation

1. $\sum_{i=1}^{5} \sqrt{i} = \sqrt{1} + \sqrt{2} + \sqrt{3} + \sqrt{4} + \sqrt{5}$

2. $\sum_{i=1}^{6} \frac{1}{i+1} = \frac{1}{2} + \frac{1}{3} + \frac{1}{4} + \frac{1}{5} + \frac{1}{6} + \frac{1}{7}$

3. $\sum_{i=4}^{6} 3^i = 3^4 + 3^5 + 3^6$

4. $\sum_{i=4}^{6} i^3 = 4^3 + 5^3 + 6^3$

5. $\sum_{k=0}^{4} \frac{2k-1}{2k+1} = -1 + \frac{1}{3} + \frac{3}{5} + \frac{5}{7} + \frac{7}{9}$

6. $\sum_{k=5}^{8} x^k = x^5 + x^6 + x^7 + x^8$

7. $\sum_{i=1}^{n} i^{10} = 1^{10} + 2^{10} + 3^{10} + \cdots + n^{10}$

8. $\sum_{j=n}^{n+3} j^2 = n^2 + (n+1)^2 + (n+2)^2 + (n+3)^2$

9. $\sum_{j=0}^{n-1} (-1)^j = 1 - 1 + 1 - 1 + \cdots + (-1)^{n-1}$

10. $\sum_{i=1}^{n} f(x_i)\,\Delta x_i = f(x_1)\,\Delta x_1 + f(x_2)\,\Delta x_2 + f(x_3)\,\Delta x_3 + \cdots + f(x_n)\,\Delta x_n$

11. $1 + 2 + 3 + 4 + \cdots + 10 = \sum_{i=1}^{10} i$

12. $\sqrt{3} + \sqrt{4} + \sqrt{5} + \sqrt{6} + \sqrt{7} = \sum_{i=3}^{7} \sqrt{i}$

13. $\frac{1}{2} + \frac{2}{3} + \frac{3}{4} + \frac{4}{5} + \cdots + \frac{19}{20} = \sum_{i=1}^{19} \frac{i}{i+1}$

14. $\frac{3}{7} + \frac{4}{8} + \frac{5}{9} + \frac{6}{10} + \cdots + \frac{23}{27} = \sum_{i=3}^{23} \frac{i}{i+4}$

15. $2 + 4 + 6 + 8 + \cdots + 2n = \sum_{i=1}^{n} 2i$

16. $1 + 3 + 5 + 7 + \cdots + (2n-1) = \sum_{i=1}^{n} (2i-1)$

17. $1 + 2 + 4 + 8 + 16 + 32 = \sum_{i=0}^{5} 2^i$

18. $\frac{1}{1} + \frac{1}{4} + \frac{1}{9} + \frac{1}{16} + \frac{1}{25} + \frac{1}{36} = \sum_{i=1}^{6} \frac{1}{i^2}$

19. $x + x^2 + x^3 + \cdots + x^n = \sum_{i=1}^{n} x^i$

20. $1 - x + x^2 - x^3 + \cdots + (-1)^n x^n = \sum_{i=0}^{n} (-1)^i x^i$

21. $\sum_{i=4}^{8} (3i-2) = [3(4)-2] + [3(5)-2] + [3(6)-2] + [3(7)-2] + [3(8)-2] = 10 + 13 + 16 + 19 + 22 = 80$

22. $\sum_{i=3}^{6} i(i+2) = 3 \cdot 5 + 4 \cdot 6 + 5 \cdot 7 + 6 \cdot 8 = 15 + 24 + 35 + 48 = 122$

23. $\sum_{j=1}^{6} 3^{j+1} = 3^2 + 3^3 + 3^4 + 3^5 + 3^6 + 3^7 = 9 + 27 + 81 + 243 + 729 + 2187 = 3276$

(For a more general method, see Exercise 47.)

24. $\sum_{k=0}^{8} \cos k\pi = \cos 0 + \cos \pi + \cos 2\pi + \cos 3\pi + \cos 4\pi + \cos 5\pi + \cos 6\pi + \cos 7\pi + \cos 8\pi$

$$= 1 - 1 + 1 - 1 + 1 - 1 + 1 - 1 + 1 = 1$$

25. $\sum_{n=1}^{20} (-1)^n = -1 + 1 - 1 + 1 - 1 + 1 - 1 + 1 - 1 + 1 - 1 + 1 - 1 + 1 - 1 + 1 - 1 + 1 - 1 + 1 = 0$

26. $\sum_{i=1}^{100} 4 = \underbrace{4 + 4 + 4 + \cdots + 4}_{\text{(100 summands)}} = 100 \cdot 4 = 400$

27. $\sum_{i=0}^{4}(2^i + i^2) = (1+0) + (2+1) + (4+4) + (8+9) + (16+16) = 61$

28. $\sum_{i=-2}^{4} 2^{3-i} = 2^5 + 2^4 + 2^3 + 2^2 + 2^1 + 2^0 + 2^{-1} = 63.5$

29. $\sum_{i=1}^{n} 2i = 2\sum_{i=1}^{n} i = 2 \cdot \frac{n(n+1)}{2}$ [by Theorem 3(c)] $= n(n+1)$

30. $\sum_{i=1}^{n}(2-5i) = \sum_{i=1}^{n} 2 - \sum_{i=1}^{n} 5i = 2n - 5\sum_{i=1}^{n} i = 2n - \frac{5n(n+1)}{2} = \frac{4n}{2} - \frac{5n^2+5n}{2} = -\frac{n(5n+1)}{2}$

31. $\sum_{i=1}^{n}(i^2+3i+4) = \sum_{i=1}^{n} i^2 + 3\sum_{i=1}^{n} i + \sum_{i=1}^{n} 4 = \frac{n(n+1)(2n+1)}{6} + \frac{3n(n+1)}{2} + 4n$

$= \frac{1}{6}[(2n^3+3n^2+n) + (9n^2+9n) + 24n] = \frac{1}{6}(2n^3+12n^2+34n) = \frac{1}{3}n(n^2+6n+17)$

32. $\sum_{i=1}^{n}(3+2i)^2 = \sum_{i=1}^{n}(9+12i+4i^2) = \sum_{i=1}^{n} 9 + 12\sum_{i=1}^{n} i + 4\sum_{i=1}^{n} i^2 = 9n + 6n(n+1) + \frac{2n(n+1)(2n+1)}{3}$

$= \frac{27n + 18n^2 + 18n + 4n^3 + 6n^2 + 2n}{3} = \frac{1}{3}(4n^3+24n^2+47n) = \frac{1}{3}n(4n^2+24n+47)$

33. $\sum_{i=1}^{n}(i+1)(i+2) = \sum_{i=1}^{n}(i^2+3i+2) = \sum_{i=1}^{n} i^2 + 3\sum_{i=1}^{n} i + \sum_{i=1}^{n} 2 = \frac{n(n+1)(2n+1)}{6} + \frac{3n(n+1)}{2} + 2n$

$= \frac{n(n+1)}{6}[(2n+1)+9] + 2n = \frac{n(n+1)}{3}(n+5) + 2n$

$= \frac{n}{3}[(n+1)(n+5)+6] = \frac{n}{3}(n^2+6n+11)$

34. $\sum_{i=1}^{n} i(i+1)(i+2) = \sum_{i=1}^{n}(i^3+3i^2+2i) = \sum_{i=1}^{n} i^3 + 3\sum_{i=1}^{n} i^2 + 2\sum_{i=1}^{n} i$

$= \left[\frac{n(n+1)}{2}\right]^2 + \frac{3n(n+1)(2n+1)}{6} + \frac{2n(n+1)}{2}$

$= n(n+1)\left[\frac{n(n+1)}{4} + \frac{2n+1}{2} + 1\right] = \frac{n(n+1)}{4}(n^2+n+4n+2+4)$

$= \frac{n(n+1)}{4}(n^2+5n+6) = \frac{n(n+1)(n+2)(n+3)}{4}$

35. $\sum_{i=1}^{n}(i^3-i-2) = \sum_{i=1}^{n} i^3 - \sum_{i=1}^{n} i - \sum_{i=1}^{n} 2 = \left[\frac{n(n+1)}{2}\right]^2 - \frac{n(n+1)}{2} - 2n$

$= \frac{1}{4}n(n+1)[n(n+1)-2] - 2n = \frac{1}{4}n(n+1)(n+2)(n-1) - 2n$

$= \frac{1}{4}n[(n+1)(n-1)(n+2) - 8] = \frac{1}{4}n[(n^2-1)(n+2) - 8] = \frac{1}{4}n(n^3+2n^2-n-10)$

36. By Theorem 3(c) we have that $\sum_{i=1}^{n} i = \frac{n(n+1)}{2} = 78 \iff n(n+1) = 156 \iff n^2+n-156 = 0 \iff$ $(n+13)(n-12) = 0 \iff n = 12$ or -13. But $n = -13$ produces a negative answer for the sum, so $n = 12$.

37. By Theorem 2(a) and Example 3, $\sum_{i=1}^{n} c = c\sum_{i=1}^{n} 1 = cn$.

38. Let S_n be the statement that $\sum_{i=1}^{n} i^3 = \left[\frac{n(n+1)}{2}\right]^2$.

1. S_1 is true because $1^3 = \left(\frac{1 \cdot 2}{2}\right)^2$.

2. Assume S_k is true. Then $\sum_{i=1}^{k} i^3 = \left[\frac{k(k+1)}{2}\right]^2$, so

$$\sum_{i=1}^{k+1} i^3 = \left[\frac{k(k+1)}{2}\right]^2 + (k+1)^3 = \frac{(k+1)^2}{4}\left[k^2 + 4(k+1)\right] = \frac{(k+1)^2}{4}(k+2)^2 = \left(\frac{(k+1)[(k+1)+1]}{2}\right)^2$$

showing that S_{k+1} is true.

Therefore, S_n is true for all n by mathematical induction.

39. $$\sum_{i=1}^{n}\left[(i+1)^4 - i^4\right] = (2^4 - 1^4) + (3^4 - 2^4) + (4^4 - 3^4) + \cdots + \left[(n+1)^4 - n^4\right]$$
$$= (n+1)^4 - 1^4 = n^4 + 4n^3 + 6n^2 + 4n$$

On the other hand,

$$\sum_{i=1}^{n}\left[(i+1)^4 - i^4\right] = \sum_{i=1}^{n}(4i^3 + 6i^2 + 4i + 1) = 4\sum_{i=1}^{n} i^3 + 6\sum_{i=1}^{n} i^2 + 4\sum_{i=1}^{n} i + \sum_{i=1}^{n} 1$$
$$= 4S + n(n+1)(2n+1) + 2n(n+1) + n \qquad \left[\text{where } S = \sum_{i=1}^{n} i^3\right]$$
$$= 4S + 2n^3 + 3n^2 + n + 2n^2 + 2n + n = 4S + 2n^3 + 5n^2 + 4n$$

Thus, $n^4 + 4n^3 + 6n^2 + 4n = 4S + 2n^3 + 5n^2 + 4n$, from which it follows that

$4S = n^4 + 2n^3 + n^2 = n^2(n^2 + 2n + 1) = n^2(n+1)^2$ and $S = \left[\frac{n(n+1)}{2}\right]^2$.

40. The area of G_i is

$$\left(\sum_{k=1}^{i} k\right)^2 - \left(\sum_{k=1}^{i-1} k\right)^2 = \left[\frac{i(i+1)}{2}\right]^2 - \left[\frac{(i-1)i}{2}\right]^2 = \frac{i^2}{4}\left[(i+1)^2 - (i-1)^2\right]$$
$$= \frac{i^2}{4}\left[(i^2 + 2i + 1) - (i^2 - 2i + 1)\right] = \frac{i^2}{4}(4i) = i^3$$

Thus, the area of $ABCD$ is $\sum_{i=1}^{n} i^3 = \left[\frac{n(n+1)}{2}\right]^2$.

41. (a) $\sum_{i=1}^{n}\left[i^4 - (i-1)^4\right] = (1^4 - 0^4) + (2^4 - 1^4) + (3^4 - 2^4) + \cdots + \left[n^4 - (n-1)^4\right] = n^4 - 0 = n^4$

(b) $\sum_{i=1}^{100}(5^i - 5^{i-1}) = (5^1 - 5^0) + (5^2 - 5^1) + (5^3 - 5^2) + \cdots + (5^{100} - 5^{99}) = 5^{100} - 5^0 = 5^{100} - 1$

(c) $\sum_{i=3}^{99}\left(\frac{1}{i} - \frac{1}{i+1}\right) = \left(\frac{1}{3} - \frac{1}{4}\right) + \left(\frac{1}{4} - \frac{1}{5}\right) + \left(\frac{1}{5} - \frac{1}{6}\right) + \cdots + \left(\frac{1}{99} - \frac{1}{100}\right) = \frac{1}{3} - \frac{1}{100} = \frac{97}{300}$

(d) $\sum_{i=1}^{n}(a_i - a_{i-1}) = (a_1 - a_0) + (a_2 - a_1) + (a_3 - a_2) + \cdots + (a_n - a_{n-1}) = a_n - a_0$

42. Summing the inequalities $-|a_i| \le a_i \le |a_i|$ for $i = 1, 2, \ldots, n$, we get $-\sum_{i=1}^{n}|a_i| \le \sum_{i=1}^{n} a_i \le \sum_{i=1}^{n}|a_i|$. Since $|x| \le c \Leftrightarrow$ $-c \le x \le c$, we have $\left|\sum_{i=1}^{n} a_i\right| \le \sum_{i=1}^{n}|a_i|$. *Another method:* Use mathematical induction.

43. $\lim\limits_{n\to\infty}\sum\limits_{i=1}^{n}\frac{1}{n}\left(\frac{i}{n}\right)^2 = \lim\limits_{n\to\infty}\frac{1}{n^3}\sum\limits_{i=1}^{n}i^2 = \lim\limits_{n\to\infty}\frac{1}{n^3}\frac{n(n+1)(2n+1)}{6} = \lim\limits_{n\to\infty}\frac{1}{6}\left(1+\frac{1}{n}\right)\left(2+\frac{1}{n}\right) = \frac{1}{6}(1)(2) = \frac{1}{3}$

44. $\lim\limits_{n\to\infty}\sum\limits_{i=1}^{n}\frac{1}{n}\left[\left(\frac{i}{n}\right)^3+1\right] = \lim\limits_{n\to\infty}\sum\limits_{i=1}^{n}\left[\frac{i^3}{n^4}+\frac{1}{n}\right] = \lim\limits_{n\to\infty}\left[\frac{1}{n^4}\sum\limits_{i=1}^{n}i^3+\frac{1}{n}\sum\limits_{i=1}^{n}1\right] = \lim\limits_{n\to\infty}\left[\frac{1}{n^4}\left(\frac{n(n+1)}{2}\right)^2+\frac{1}{n}(n)\right]$

$$= \lim_{n\to\infty}\frac{1}{4}\left(1+\frac{1}{n}\right)^2+1 = \tfrac{1}{4}+1 = \tfrac{5}{4}$$

45. $\lim\limits_{n\to\infty}\sum\limits_{i=1}^{n}\frac{2}{n}\left[\left(\frac{2i}{n}\right)^3+5\left(\frac{2i}{n}\right)\right] = \lim\limits_{n\to\infty}\sum\limits_{i=1}^{n}\left[\frac{16}{n^4}i^3+\frac{20}{n^2}i\right] = \lim\limits_{n\to\infty}\left[\frac{16}{n^4}\sum\limits_{i=1}^{n}i^3+\frac{20}{n^2}\sum\limits_{i=1}^{n}i\right]$

$$= \lim_{n\to\infty}\left[\frac{16}{n^4}\frac{n^2(n+1)^2}{4}+\frac{20}{n^2}\frac{n(n+1)}{2}\right] = \lim_{n\to\infty}\left[\frac{4(n+1)^2}{n^2}+\frac{10n(n+1)}{n^2}\right]$$

$$= \lim_{n\to\infty}\left[4\left(1+\frac{1}{n}\right)^2+10\left(1+\frac{1}{n}\right)\right] = 4\cdot 1+10\cdot 1 = 14$$

46. $\lim\limits_{n\to\infty}\sum\limits_{i=1}^{n}\frac{3}{n}\left[\left(1+\frac{3i}{n}\right)^3-2\left(1+\frac{3i}{n}\right)\right] = \lim\limits_{n\to\infty}\sum\limits_{i=1}^{n}\frac{3}{n}\left[1+\frac{9i}{n}+\frac{27i^2}{n^2}+\frac{27i^3}{n^3}-2-\frac{6i}{n}\right]$

$$= \lim_{n\to\infty}\sum_{i=1}^{n}\left[\frac{81}{n^4}i^3+\frac{81}{n^3}i^2+\frac{9}{n^2}i-\frac{3}{n}\right]$$

$$= \lim_{n\to\infty}\left[\frac{81}{n^4}\frac{n^2(n+1)^2}{4}+\frac{81}{n^3}\frac{n(n+1)(2n+1)}{6}+\frac{9}{n^2}\frac{n(n+1)}{2}-\frac{3}{n}n\right]$$

$$= \lim_{n\to\infty}\left[\frac{81}{4}\left(1+\frac{1}{n}\right)^2+\frac{27}{2}\left(1+\frac{1}{n}\right)\left(2+\frac{1}{n}\right)+\frac{9}{2}\left(1+\frac{1}{n}\right)-3\right]$$

$$= \tfrac{81}{4}+\tfrac{54}{2}+\tfrac{9}{2}-3 = \tfrac{195}{4}$$

47. Let $S = \sum\limits_{i=1}^{n}ar^{i-1} = a+ar+ar^2+\cdots+ar^{n-1}$. Multiplying both sides by r gives us

$rS = ar+ar^2+\cdots+ar^{n-1}+ar^n$. Subtracting the first equation from the second, we find

$(r-1)S = ar^n-a = a(r^n-1)$, so $S = \dfrac{a(r^n-1)}{r-1}$ [since $r\neq 1$].

48. $\sum\limits_{i=1}^{n}\frac{3}{2^{i-1}} = 3\sum\limits_{i=1}^{n}\left(\frac{1}{2}\right)^{i-1} = \dfrac{3\left[\left(\frac{1}{2}\right)^n-1\right]}{\frac{1}{2}-1}$ [using Exercise 47 with $a=3$ and $r=\frac{1}{2}$] $= 6\left[1-\left(\frac{1}{2}\right)^n\right]$

49. $\sum\limits_{i=1}^{n}(2i+2^i) = 2\sum\limits_{i=1}^{n}i+\sum\limits_{i=1}^{n}2\cdot 2^{i-1} = 2\dfrac{n(n+1)}{2}+\dfrac{2(2^n-1)}{2-1} = 2^{n+1}+n^2+n-2.$

For the first sum we have used Theorems 2(a) and 3(c), and for the second, Exercise 47 with $a=r=2$.

50. $\sum\limits_{i=1}^{m}\left[\sum\limits_{j=1}^{n}(i+j)\right] = \sum\limits_{i=1}^{m}\left[\sum\limits_{j=1}^{n}i+\sum\limits_{j=1}^{n}j\right]$ [Theorem 2(b)] $= \sum\limits_{i=1}^{m}\left[ni+\dfrac{n(n+1)}{2}\right]$ [Theorem 3(b) and 3(c)]

$$= \sum_{i=1}^{m}ni+\sum_{i=1}^{m}\frac{n(n+1)}{2} = \frac{nm(m+1)}{2}+\frac{nm(n+1)}{2} = \frac{nm}{2}(m+n+2)$$

G The Logarithm Defined as an Integral

1. (a)

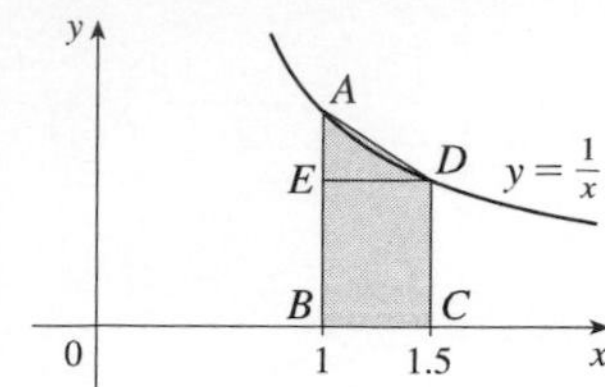

We interpret $\ln 1.5$ as the area under the curve $y = 1/x$ from $x = 1$ to $x = 1.5$. The area of the rectangle $BCDE$ is $\frac{1}{2} \cdot \frac{2}{3} = \frac{1}{3}$. The area of the trapezoid $ABCD$ is $\frac{1}{2} \cdot \frac{1}{2}\left(1 + \frac{2}{3}\right) = \frac{5}{12}$. Thus, by comparing areas, we observe that $\frac{1}{3} < \ln 1.5 < \frac{5}{12}$.

(b) $\ln x = \int_1^x (1/t)\,dt$, so $\ln 1.5 = \int_1^{1.5} (1/t)\,dt$. With $f(t) = 1/t$, $n = 10$, and $\Delta t = \frac{1.5 - 1}{10} = 0.05$, we have

$$\ln 1.5 = \int_1^{1.5} (1/t)\,dt \approx (0.05)[f(1.025) + f(1.075) + \cdots + f(1.475)] = (0.05)\left[\tfrac{1}{1.025} + \tfrac{1}{1.075} + \cdots + \tfrac{1}{1.475}\right]$$
$$\approx 0.4054$$

2. (a) $y = \dfrac{1}{t}$, $y' = -\dfrac{1}{t^2}$. The slope of the line through $A(1, 1)$ and $D\left(2, \frac{1}{2}\right)$ is $\dfrac{1/2 - 1}{2 - 1} = -\dfrac{1}{2}$. Let c be the t-coordinate of the point on $y = \dfrac{1}{t}$ with slope $-\dfrac{1}{2}$. Then $-\dfrac{1}{c^2} = -\dfrac{1}{2} \;\Rightarrow\; c^2 = 2 \;\Rightarrow\; c = \sqrt{2}$ since $c > 0$. Therefore, the tangent line is given by $y - \frac{1}{\sqrt{2}} = -\frac{1}{2}\left(t - \sqrt{2}\right)$, or $y = -\frac{1}{2}t + \sqrt{2}$.

(b) Since the graph of $y = 1/t$ is concave upward, the graph lies above the tangent line, that is, above the line segment BC. Now $|AB| = -\frac{1}{2} + \sqrt{2}$ and $|CD| = -1 + \sqrt{2}$. The area of the trapezoid $ABCD$ is $\frac{1}{2}\left[\left(-\frac{1}{2} + \sqrt{2}\right) + \left(-1 + \sqrt{2}\right)\right]1 = -\frac{3}{4} + \sqrt{2} \approx 0.6642$. So $\ln 2 >$ area of trapezoid $ABCD > 0.66$.

3. The area of R_i is $\dfrac{1}{i+1}$ and so $\dfrac{1}{2} + \dfrac{1}{3} + \cdots + \dfrac{1}{n} < \displaystyle\int_1^n \frac{1}{t}\,dt = \ln n$.

The area of S_i is $\dfrac{1}{i}$ and so $1 + \dfrac{1}{2} + \cdots + \dfrac{1}{n-1} > \displaystyle\int_1^n \frac{1}{t}\,dt = \ln n$.

Thus, $\dfrac{1}{2} + \dfrac{1}{3} + \cdots + \dfrac{1}{n} < \ln n < 1 + \dfrac{1}{2} + \cdots + \dfrac{1}{n-1}$.

4. (a) From the diagram, we see that the area under the graph of $y = 1/x$ between $x = 1$ and $x = 2$ is less than the area of the square, which is 1. So $\ln 2 = \int_1^2 (1/x)\,dx < 1$. To show the other side of the inequality, we must find an area larger than 1 which lies under the graph of $y = 1/x$ between $x = 1$ and $x = 3$. One way to do this is to partition the interval $[1, 3]$ into

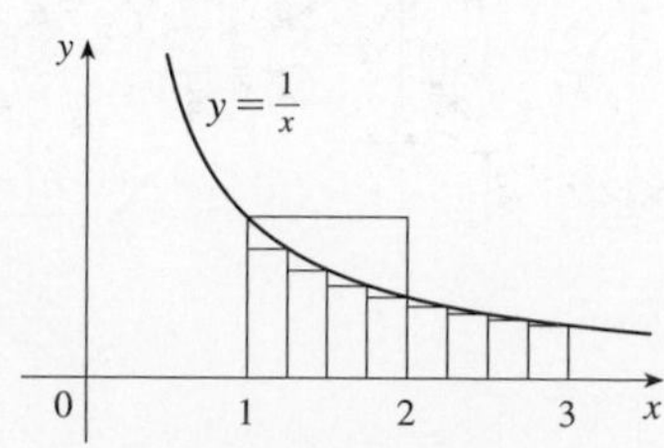

8 intervals of equal length and calculate the resulting Riemann sum, using the right endpoints:

$$\frac{1}{4}\left(\frac{1}{5/4}+\frac{1}{3/2}+\frac{1}{7/4}+\frac{1}{2}+\frac{1}{9/4}+\frac{1}{5/2}+\frac{1}{11/4}+\frac{1}{3}\right)=\frac{28{,}271}{27{,}720}>1$$

and therefore $1<\int_1^3 (1/x)\,dx=\ln 3$.

A slightly easier method uses the fact that since $y=1/x$ is concave upward, it lies above all its tangent lines. Drawing two such tangent lines at the points $\left(\frac{3}{2},\frac{2}{3}\right)$ and $\left(\frac{5}{2},\frac{2}{5}\right)$, we see that the area under the curve from $x=1$ to $x=3$ is more than the sum of the areas of the two trapezoids, that is, $\frac{2}{3}+\frac{2}{5}=\frac{16}{15}$. Thus, $1<\frac{16}{15}<\int_1^3 (1/x)\,dx=\ln 3$.

(b) By part (a), $\ln 2<1<\ln 3$. But e is defined such that $\ln e=1$, and because the natural logarithm function is increasing, we have $\ln 2<\ln e<\ln 3 \quad\Leftrightarrow\quad 2<e<3$.

5. If $f(x)=\ln(x^r)$, then $f'(x)=(1/x^r)\left(rx^{r-1}\right)=r/x$. But if $g(x)=r\ln x$, then $g'(x)=r/x$. So f and g must differ by a constant: $\ln(x^r)=r\ln x+C$. Put $x=1$: $\ln(1^r)=r\ln 1+C \;\Rightarrow\; C=0$, so $\ln(x^r)=r\ln x$.

6. Using the second law of logarithms and Equation 10, we have $\ln(e^x/e^y)=\ln e^x-\ln e^y=x-y=\ln(e^{x-y})$. Since ln is a one-to-one function, it follows that $e^x/e^y=e^{x-y}$.

7. Using the third law of logarithms and Equation 10, we have $\ln e^{rx}=rx=r\ln e^x=\ln(e^x)^r$. Since ln is a one-to-one function, it follows that $e^{rx}=(e^x)^r$.

8. Using Definition 13 and the second law of exponents for e^x, we have $a^{x-y}=e^{(x-y)\ln a}=e^{x\ln a-y\ln a}=\dfrac{e^{x\ln a}}{e^{y\ln a}}=\dfrac{a^x}{a^y}$.

9. Using Definition 13, the first law of logarithms, and the first law of exponents for e^x, we have

$(ab)^x=e^{x\ln(ab)}=e^{x(\ln a+\ln b)}=e^{x\ln a+x\ln b}=e^{x\ln a}e^{x\ln b}=a^xb^x$.

10. Let $\log_a x=r$ and $\log_a y=s$. Then $a^r=x$ and $a^s=y$.

(a) $xy=a^ra^s=a^{r+s} \;\Rightarrow\; \log_a(xy)=r+s=\log_a x+\log_a y$

(b) $\dfrac{x}{y}=\dfrac{a^r}{a^s}=a^{r-s} \;\Rightarrow\; \log_a\dfrac{x}{y}=r-s=\log_a x-\log_a y$

(c) $x^y=(a^r)^y=a^{ry} \;\Rightarrow\; \log_a(x^y)=ry=y\log_a x$

H Complex Numbers

1. $(5-6i)+(3+2i)=(5+3)+(-6+2)i=8+(-4)i=8-4i$

2. $\left(4-\frac{1}{2}i\right)-\left(9+\frac{5}{2}i\right)=(4-9)+\left(-\frac{1}{2}-\frac{5}{2}\right)i=-5+(-3)i=-5-3i$

3. $(2+5i)(4-i)=2(4)+2(-i)+(5i)(4)+(5i)(-i)=8-2i+20i-5i^2=8+18i-5(-1)$

$=8+18i+5=13+18i$

4. $(1-2i)(8-3i)=8-3i-16i+6(-1)=2-19i$

5. $\overline{12+7i}=12-7i$

6. $2i(\frac{1}{2}-i) = i - 2(-1) = 2 + i \;\Rightarrow\; \overline{2i(\frac{1}{2}-i)} = \overline{2+i} = 2 - i$

7. $\dfrac{1+4i}{3+2i} = \dfrac{1+4i}{3+2i}\cdot\dfrac{3-2i}{3-2i} = \dfrac{3-2i+12i-8(-1)}{3^2+2^2} = \dfrac{11+10i}{13} = \dfrac{11}{13} + \dfrac{10}{13}i$

8. $\dfrac{3+2i}{1-4i} = \dfrac{3+2i}{1-4i}\cdot\dfrac{1+4i}{1+4i} = \dfrac{3+12i+2i+8(-1)}{1^2+4^2} = \dfrac{-5+14i}{17} = -\dfrac{5}{17} + \dfrac{14}{17}i$

9. $\dfrac{1}{1+i} = \dfrac{1}{1+i}\cdot\dfrac{1-i}{1-i} = \dfrac{1-i}{1-(-1)} = \dfrac{1-i}{2} = \dfrac{1}{2} - \dfrac{1}{2}i$

10. $\dfrac{3}{4-3i} = \dfrac{3}{4-3i}\cdot\dfrac{4+3i}{4+3i} = \dfrac{12+9i}{16-9(-1)} = \dfrac{12}{25} + \dfrac{9}{25}i$

11. $i^3 = i^2 \cdot i = (-1)i = -i$

12. $i^{100} = (i^2)^{50} = (-1)^{50} = 1$

13. $\sqrt{-25} = \sqrt{25}\,i = 5i$

14. $\sqrt{-3}\,\sqrt{-12} = \sqrt{3}\,i\,\sqrt{12}\,i = \sqrt{3\cdot 12}\,i^2 = \sqrt{36}\,(-1) = -6$

15. $\overline{12-5i} = 12 + 15i$ and $|12 - 15i| = \sqrt{12^2 + (-5)^2} = \sqrt{144+25} = \sqrt{169} = 13$

16. $\overline{-1+2\sqrt{2}\,i} = -1 - 2\sqrt{2}\,i$ and $\left|-1+2\sqrt{2}\,i\right| = \sqrt{(-1)^2 + \left(2\sqrt{2}\right)^2} = \sqrt{1+8} = \sqrt{9} = 3$

17. $\overline{-4i} = \overline{0-4i} = 0 + 4i = 4i$ and $|-4i| = \sqrt{0^2+(-4)^2} = \sqrt{16} = 4$

18. Let $z = a + bi$ and $w = c + di$.

(a) $\overline{z+w} = \overline{(a+bi)+(c+di)} = \overline{(a+c)+(b+d)i} = (a+c)-(b+d)i = (a-bi)+(c-di) = \overline{z}+\overline{w}$

(b) $\overline{zw} = \overline{(a+bi)(c+di)} = \overline{(ac-bd)+(ad+bc)i} = (ac-bd)-(ad+bc)i.$

On the other hand, $\overline{z}\,\overline{w} = (a-bi)(c-di) = (ac-bd)-(ad+bc)i = \overline{zw}.$

(c) Use mathematical induction and part (b): Let S_n be the statement that $\overline{z^n} = \overline{z}^{\,n}$. S_1 is true because $\overline{z^1} = \overline{z} = \overline{z}^{\,1}$.

Assume S_k is true, that is $\overline{z^k} = \overline{z}^{\,k}$. Then $\overline{z^{k+1}} = \overline{z^{1+k}} = \overline{zz^k} = \overline{z}\,\overline{z^k}$ [part (b) with $w = z^k$] $= \overline{z}^1\overline{z}^k = \overline{z}^{1+k} = \overline{z}^{k+1}$,

which shows that S_{k+1} is true. Therefore, by mathematical induction, $\overline{z^n} = \overline{z}^n$ for every positive integer n.

Another proof: Use part (b) with $w = z$, and mathematical induction.

19. $4x^2 + 9 = 0 \;\Leftrightarrow\; 4x^2 = -9 \;\Leftrightarrow\; x^2 = -\frac{9}{4} \;\Leftrightarrow\; x = \pm\sqrt{-\frac{9}{4}} = \pm\sqrt{\frac{9}{4}}\,i = \pm\frac{3}{2}i.$

20. $x^4 = 1 \;\Leftrightarrow\; x^4 - 1 = 0 \;\Leftrightarrow\; (x^2-1)(x^2+1) = 0 \;\Leftrightarrow\; x^2 - 1 = 0$ or $x^2 + 1 = 0 \;\Leftrightarrow\; x = \pm 1$ or $x = \pm i$.

21. By the quadratic formula, $x^2 + 2x + 5 = 0 \;\Leftrightarrow\; x = \dfrac{-2 \pm \sqrt{2^2 - 4(1)(5)}}{2(1)} = \dfrac{-2\pm\sqrt{-16}}{2} = \dfrac{-2\pm 4i}{2} = -1 \pm 2i.$

22. $2x^2 - 2x + 1 = 0 \;\Leftrightarrow\; x = \dfrac{-(-2)\pm\sqrt{(-2)^2 - 4(2)(1)}}{2(2)} = \dfrac{2\pm\sqrt{-4}}{4} = \dfrac{2\pm 2i}{4} = \dfrac{1}{2} \pm \dfrac{1}{2}i$

23. By the quadratic formula, $z^2+z+2=0 \;\Leftrightarrow\; z=\dfrac{-1\pm\sqrt{1^2-4(1)(2)}}{2(1)}=\dfrac{-1\pm\sqrt{-7}}{2}=-\dfrac{1}{2}\pm\dfrac{\sqrt{7}}{2}i.$

24. $z^2+\frac{1}{2}z+\frac{1}{4}=0 \;\Leftrightarrow\; 4z^2+2z+1=0 \;\Leftrightarrow$

$$z=\frac{-2\pm\sqrt{2^2-4(4)(1)}}{2(4)}=\frac{-2\pm\sqrt{-12}}{8}=\frac{-2\pm2\sqrt{3}\,i}{8}=-\frac{1}{4}\pm\frac{\sqrt{3}}{4}i$$

25. For $z=-3+3i$, $r=\sqrt{(-3)^2+3^2}=3\sqrt{2}$ and $\tan\theta=\frac{3}{-3}=-1 \;\Rightarrow\; \theta=\frac{3\pi}{4}$ (since z lies in the second quadrant).

Therefore, $-3+3i=3\sqrt{2}\left(\cos\frac{3\pi}{4}+i\sin\frac{3\pi}{4}\right)$.

26. For $z=1-\sqrt{3}\,i$, $r=\sqrt{1^2+\left(-\sqrt{3}\right)^2}=2$ and $\tan\theta=\frac{-\sqrt{3}}{1}=-\sqrt{3} \;\Rightarrow\; \theta=\frac{5\pi}{3}$ (since z lies in the fourth quadrant).

Therefore, $1-\sqrt{3}\,i=2\left(\cos\frac{5\pi}{3}+i\sin\frac{5\pi}{3}\right)$.

27. For $z=3+4i$, $r=\sqrt{3^2+4^2}=5$ and $\tan\theta=\frac{4}{3} \;\Rightarrow\; \theta=\tan^{-1}\left(\frac{4}{3}\right)$ (since z lies in the first quadrant). Therefore,

$3+4i=5\left[\cos\left(\tan^{-1}\frac{4}{3}\right)+i\sin\left(\tan^{-1}\frac{4}{3}\right)\right]$.

28. For $z=8i$, $r=\sqrt{0^2+8^2}=8$ and $\tan\theta=\frac{8}{0}$ is undefined, so $\theta=\frac{\pi}{2}$ (since z lies on the positive imaginary axis). Therefore,

$8i=8\left(\cos\frac{\pi}{2}+i\sin\frac{\pi}{2}\right)$.

29. For $z=\sqrt{3}+i$, $r=\sqrt{\left(\sqrt{3}\right)^2+1^2}=2$ and $\tan\theta=\frac{1}{\sqrt{3}} \;\Rightarrow\; \theta=\frac{\pi}{6} \;\Rightarrow\; z=2\left(\cos\frac{\pi}{6}+i\sin\frac{\pi}{6}\right)$.

For $w=1+\sqrt{3}\,i$, $r=2$ and $\tan\theta=\sqrt{3} \;\Rightarrow\; \theta=\frac{\pi}{3} \;\Rightarrow\; w=2\left(\cos\frac{\pi}{3}+i\sin\frac{\pi}{3}\right)$.

Therefore, $zw=2\cdot2\left[\cos\left(\frac{\pi}{6}+\frac{\pi}{3}\right)+i\sin\left(\frac{\pi}{6}+\frac{\pi}{3}\right)\right]=4\left(\cos\frac{\pi}{2}+i\sin\frac{\pi}{2}\right)$,

$z/w=\frac{2}{2}\left[\cos\left(\frac{\pi}{6}-\frac{\pi}{3}\right)+i\sin\left(\frac{\pi}{6}-\frac{\pi}{3}\right)\right]=\cos\left(-\frac{\pi}{6}\right)+i\sin\left(-\frac{\pi}{6}\right)$, and $1=1+0i=1(\cos0+i\sin0) \;\Rightarrow$

$1/z=\frac{1}{2}\left[\cos\left(0-\frac{\pi}{6}\right)+i\sin\left(0-\frac{\pi}{6}\right)\right]=\frac{1}{2}\left[\cos\left(-\frac{\pi}{6}\right)+i\sin\left(-\frac{\pi}{6}\right)\right]$. For $1/z$, we could also use the formula that precedes Example 5 to obtain $1/z=\frac{1}{2}\left(\cos\frac{\pi}{6}-i\sin\frac{\pi}{6}\right)$.

30. For $z=4\sqrt{3}-4i$, $r=\sqrt{\left(4\sqrt{3}\right)^2+(-4)^2}=\sqrt{64}=8$ and $\tan\theta=\frac{-4}{4\sqrt{3}}=-\frac{1}{\sqrt{3}} \;\Rightarrow\; \theta=\frac{11\pi}{6} \;\Rightarrow$

$z=8\left(\cos\frac{11\pi}{6}+i\sin\frac{11\pi}{6}\right)$. For $w=8i$, $r=\sqrt{0^2+8^2}=8$ and $\tan\theta=\frac{8}{0}$ is undefined, so $\theta=\frac{\pi}{2} \;\Rightarrow$

$w=8\left(\cos\frac{\pi}{2}+i\sin\frac{\pi}{2}\right)$. Therefore, $zw=8\cdot8\left[\cos\left(\frac{11\pi}{6}+\frac{\pi}{2}\right)+i\sin\left(\frac{11\pi}{6}+\frac{\pi}{2}\right)\right]=64\left(\cos\frac{\pi}{3}+i\sin\frac{\pi}{3}\right)$,

$z/w=\frac{8}{8}\left[\cos\left(\frac{11\pi}{6}-\frac{\pi}{2}\right)+i\sin\left(\frac{11\pi}{6}-\frac{\pi}{2}\right)\right]=\cos\frac{4\pi}{3}+i\sin\frac{4\pi}{3}$, and

$1=1+0i=1(\cos0+i\sin0) \;\Rightarrow\; 1/z=\frac{1}{8}\left[\cos\left(0-\frac{11\pi}{6}\right)+i\sin\left(0-\frac{11\pi}{6}\right)\right]=\frac{1}{8}\left[\cos\left(\frac{\pi}{6}\right)+i\sin\left(\frac{\pi}{6}\right)\right]$.

For $1/z$, we could also use the formula that precedes Example 5 to obtain $1/z=\frac{1}{8}\left(\cos\frac{11\pi}{6}-i\sin\frac{11\pi}{6}\right)$.

31. For $z=2\sqrt{3}-2i$, $r=\sqrt{\left(2\sqrt{3}\right)^2+(-2)^2}=4$ and $\tan\theta=\frac{-2}{2\sqrt{3}}=-\frac{1}{\sqrt{3}} \;\Rightarrow\; \theta=-\frac{\pi}{6} \;\Rightarrow$

$z=4\left[\cos\left(-\frac{\pi}{6}\right)+i\sin\left(-\frac{\pi}{6}\right)\right]$. For $w=-1+i$, $r=\sqrt{2}$, $\tan\theta=\frac{1}{-1}=-1 \;\Rightarrow\; \theta=\frac{3\pi}{4} \;\Rightarrow$

$w=\sqrt{2}\left(\cos\frac{3\pi}{4}+i\sin\frac{3\pi}{4}\right)$. Therefore, $zw=4\sqrt{2}\left[\cos\left(-\frac{\pi}{6}+\frac{3\pi}{4}\right)+i\sin\left(-\frac{\pi}{6}+\frac{3\pi}{4}\right)\right]=4\sqrt{2}\left(\cos\frac{7\pi}{12}+i\sin\frac{7\pi}{12}\right)$,

$z/w=\frac{4}{\sqrt{2}}\left[\cos\left(-\frac{\pi}{6}-\frac{3\pi}{4}\right)+i\sin\left(-\frac{\pi}{6}-\frac{3\pi}{4}\right)\right]=\frac{4}{\sqrt{2}}\left[\cos\left(-\frac{11\pi}{12}\right)+i\sin\left(-\frac{11\pi}{12}\right)\right]=2\sqrt{2}\left(\cos\frac{13\pi}{12}+i\sin\frac{13\pi}{12}\right)$, and

$1/z=\frac{1}{4}\left[\cos\left(-\frac{\pi}{6}\right)-i\sin\left(-\frac{\pi}{6}\right)\right]=\frac{1}{4}\left(\cos\frac{\pi}{6}+i\sin\frac{\pi}{6}\right)$.

32. For $z = 4(\sqrt{3} + i) = 4\sqrt{3} + 4i$, $r = \sqrt{(4\sqrt{3})^2 + 4^2} = \sqrt{64} = 8$ and $\tan\theta = \frac{4}{4\sqrt{3}} = \frac{1}{\sqrt{3}} \Rightarrow \theta = \frac{\pi}{6} \Rightarrow$ $z = 8\left(\cos\frac{\pi}{6} + i\sin\frac{\pi}{6}\right)$. For $w = -3 - 3i$, $r = \sqrt{(-3)^2 + (-3)^2} = \sqrt{18} = 3\sqrt{2}$ and $\tan\theta = \frac{-3}{-3} = 1 \Rightarrow \theta = \frac{5\pi}{4} \Rightarrow$ $w = 3\sqrt{2}\left(\cos\frac{5\pi}{4} + i\sin\frac{5\pi}{4}\right)$. Therefore, $zw = 8 \cdot 3\sqrt{2}\left[\cos\left(\frac{\pi}{6} + \frac{5\pi}{4}\right) + i\sin\left(\frac{\pi}{6} + \frac{5\pi}{4}\right)\right] = 24\sqrt{2}\left(\cos\frac{17\pi}{12} + i\sin\frac{17\pi}{12}\right)$, $z/w = \frac{8}{3\sqrt{2}}\left[\cos\left(\frac{\pi}{6} - \frac{5\pi}{4}\right) + i\sin\left(\frac{\pi}{6} - \frac{5\pi}{4}\right)\right] = \frac{4\sqrt{2}}{3}\left[\cos\left(-\frac{13\pi}{12}\right) + i\sin\left(-\frac{13\pi}{12}\right)\right]$, and $1/z = \frac{1}{8}\left(\cos\frac{\pi}{6} - i\sin\frac{\pi}{6}\right)$.

33. For $z = 1 + i$, $r = \sqrt{2}$ and $\tan\theta = \frac{1}{1} = 1 \Rightarrow \theta = \frac{\pi}{4} \Rightarrow z = \sqrt{2}\left(\cos\frac{\pi}{4} + i\sin\frac{\pi}{4}\right)$. So by De Moivre's Theorem,

$$\begin{aligned}(1+i)^{20} &= \left[\sqrt{2}\left(\cos\tfrac{\pi}{4} + i\sin\tfrac{\pi}{4}\right)\right]^{20} = (2^{1/2})^{20}\left(\cos\tfrac{20\cdot\pi}{4} + i\sin\tfrac{20\cdot\pi}{4}\right) = 2^{10}(\cos 5\pi + i\sin 5\pi)\\ &= 2^{10}[-1 + i(0)] = -2^{10} = -1024\end{aligned}$$

34. For $z = 1 - \sqrt{3}\,i$, $r = \sqrt{1^2 + (-\sqrt{3})^2} = 2$ and $\tan\theta = \frac{-\sqrt{3}}{1} = -\sqrt{3} \Rightarrow \theta = \frac{5\pi}{3} \Rightarrow z = 2\left(\cos\frac{5\pi}{3} + i\sin\frac{5\pi}{3}\right)$. So by De Moivre's Theorem,

$$\begin{aligned}(1 - \sqrt{3}\,i)^5 &= \left[2\left(\cos\tfrac{5\pi}{3} + i\sin\tfrac{5\pi}{3}\right)\right]^5 = 2^5\left(\cos\tfrac{5\cdot5\pi}{3} + i\sin\tfrac{5\cdot5\pi}{3}\right) = 2^5\left(\cos\tfrac{\pi}{3} + i\sin\tfrac{\pi}{3}\right)\\ &= 32\left(\tfrac{1}{2} + \tfrac{\sqrt{3}}{2}\,i\right) = 16 + 16\sqrt{3}\,i\end{aligned}$$

35. For $z = 2\sqrt{3} + 2i$, $r = \sqrt{(2\sqrt{3})^2 + 2^2} = \sqrt{16} = 4$ and $\tan\theta = \frac{2}{2\sqrt{3}} = \frac{1}{\sqrt{3}} \Rightarrow \theta = \frac{\pi}{6} \Rightarrow z = 4\left(\cos\frac{\pi}{6} + i\sin\frac{\pi}{6}\right)$. So by De Moivre's Theorem,

$$(2\sqrt{3} + 2i)^5 = \left[4\left(\cos\tfrac{\pi}{6} + i\sin\tfrac{\pi}{6}\right)\right]^5 = 4^5\left(\cos\tfrac{5\pi}{6} + i\sin\tfrac{5\pi}{6}\right) = 1024\left[-\tfrac{\sqrt{3}}{2} + \tfrac{1}{2}i\right] = -512\sqrt{3} + 512i.$$

36. For $z = 1 - i$, $r = \sqrt{2}$ and $\tan\theta = \frac{-1}{1} = -1 \Rightarrow \theta = \frac{7\pi}{4} \Rightarrow z = \sqrt{2}\left(\cos\frac{7\pi}{4} + i\sin\frac{7\pi}{4}\right) \Rightarrow$

$$(1 - i)^8 = \left[\sqrt{2}\left(\cos\tfrac{7\pi}{4} + i\sin\tfrac{7\pi}{4}\right)\right]^8 = 2^4\left(\cos\tfrac{8\cdot7\pi}{4} + i\sin\tfrac{8\cdot7\pi}{4}\right) = 16(\cos 14\pi + i\sin 14\pi) = 16(1 + 0i) = 16.$$

37. $1 = 1 + 0i = 1(\cos 0 + i\sin 0)$. Using Equation 3 with $r = 1$, $n = 8$, and $\theta = 0$, we have

$$w_k = 1^{1/8}\left[\cos\left(\frac{0 + 2k\pi}{8}\right) + i\sin\left(\frac{0 + 2k\pi}{8}\right)\right] = \cos\frac{k\pi}{4} + i\sin\frac{k\pi}{4}, \text{ where } k = 0, 1, 2, \ldots, 7.$$

$w_0 = 1(\cos 0 + i\sin 0) = 1$, $w_1 = 1\left(\cos\frac{\pi}{4} + i\sin\frac{\pi}{4}\right) = \frac{1}{\sqrt{2}} + \frac{1}{\sqrt{2}}i$,

$w_2 = 1\left(\cos\frac{\pi}{2} + i\sin\frac{\pi}{2}\right) = i$, $w_3 = 1\left(\cos\frac{3\pi}{4} + i\sin\frac{3\pi}{4}\right) = -\frac{1}{\sqrt{2}} + \frac{1}{\sqrt{2}}i$,

$w_4 = 1(\cos\pi + i\sin\pi) = -1$, $w_5 = 1\left(\cos\frac{5\pi}{4} + i\sin\frac{5\pi}{4}\right) = -\frac{1}{\sqrt{2}} - \frac{1}{\sqrt{2}}i$,

$w_6 = 1\left(\cos\frac{3\pi}{2} + i\sin\frac{3\pi}{2}\right) = -i$, $w_7 = 1\left(\cos\frac{7\pi}{4} + i\sin\frac{7\pi}{4}\right) = \frac{1}{\sqrt{2}} - \frac{1}{\sqrt{2}}i$

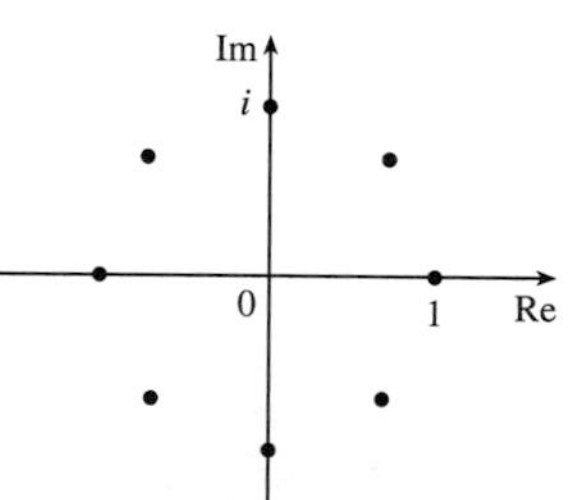

38. $32 = 32 + 0i = 32(\cos 0 + i\sin 0)$. Using Equation 3 with $r = 32$, $n = 5$, and $\theta = 0$, we have

$$w_k = 32^{1/5}\left[\cos\left(\frac{0 + 2k\pi}{5}\right) + i\sin\left(\frac{0 + 2k\pi}{5}\right)\right] = 2\left(\cos\tfrac{2}{5}\pi k + i\sin\tfrac{2}{5}\pi k\right), \text{ where } k = 0, 1, 2, 3, 4.$$

$w_0 = 2(\cos 0 + i\sin 0) = 2$

$w_1 = 2\left(\cos\frac{2\pi}{5} + i\sin\frac{2\pi}{5}\right)$

$w_2 = 2\left(\cos\frac{4\pi}{5} + i\sin\frac{4\pi}{5}\right)$

$w_3 = 2\left(\cos\frac{6\pi}{5} + i\sin\frac{6\pi}{5}\right)$

$w_4 = 2\left(\cos\frac{8\pi}{5} + i\sin\frac{8\pi}{5}\right)$

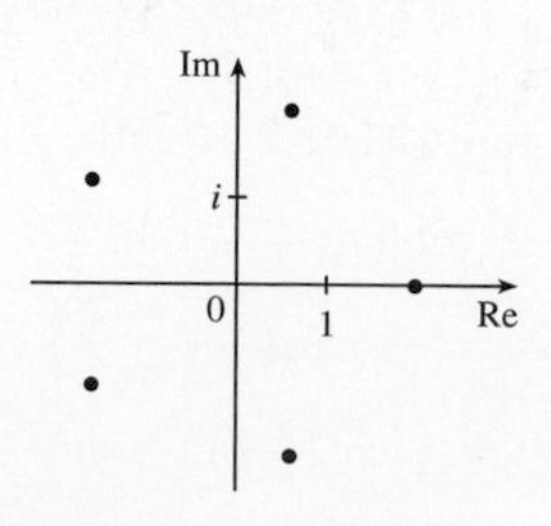

39. $i = 0 + i = 1\left(\cos\frac{\pi}{2} + i\sin\frac{\pi}{2}\right)$. Using Equation 3 with $r = 1$, $n = 3$, and $\theta = \frac{\pi}{2}$, we have

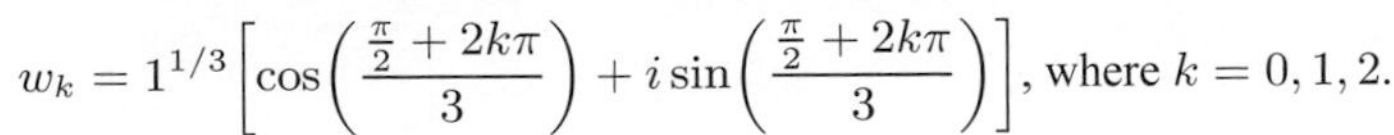

$$w_k = 1^{1/3}\left[\cos\left(\frac{\frac{\pi}{2} + 2k\pi}{3}\right) + i\sin\left(\frac{\frac{\pi}{2} + 2k\pi}{3}\right)\right], \text{ where } k = 0, 1, 2.$$

$$w_0 = \left(\cos\tfrac{\pi}{6} + i\sin\tfrac{\pi}{6}\right) = \tfrac{\sqrt{3}}{2} + \tfrac{1}{2}i$$

$$w_1 = \left(\cos\tfrac{5\pi}{6} + i\sin\tfrac{5\pi}{6}\right) = -\tfrac{\sqrt{3}}{2} + \tfrac{1}{2}i$$

$$w_2 = \left(\cos\tfrac{9\pi}{6} + i\sin\tfrac{9\pi}{6}\right) = -i$$

40. $1 + i = \sqrt{2}\left(\cos\frac{\pi}{4} + i\sin\frac{\pi}{4}\right)$. Using Equation 3 with $r = \sqrt{2}$, $n = 3$, and $\theta = \frac{\pi}{4}$, we have

$$w_k = \left(\sqrt{2}\right)^{1/3}\left[\cos\left(\frac{\frac{\pi}{4} + 2k\pi}{3}\right) + i\sin\left(\frac{\frac{\pi}{4} + 2k\pi}{3}\right)\right], \text{ where } k = 0, 1, 2.$$

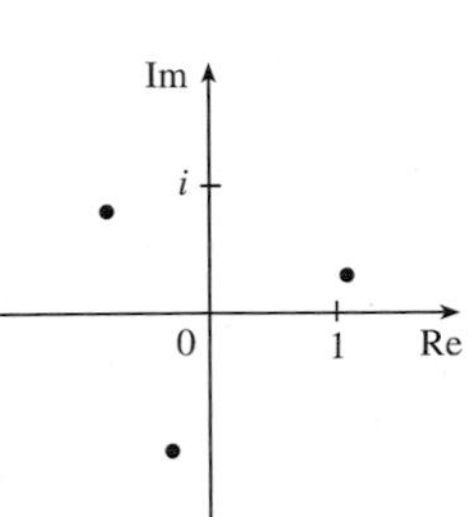

$$w_0 = 2^{1/6}\left(\cos\tfrac{\pi}{12} + i\sin\tfrac{\pi}{12}\right)$$

$$w_1 = 2^{1/6}\left(\cos\tfrac{3\pi}{4} + i\sin\tfrac{3\pi}{4}\right) = 2^{1/6}\left(-\tfrac{1}{\sqrt{2}} + \tfrac{1}{\sqrt{2}}i\right) = -2^{-1/3} + 2^{-1/3}i$$

$$w_2 = 2^{1/6}\left(\cos\tfrac{17\pi}{12} + i\sin\tfrac{17\pi}{12}\right)$$

41. Using Euler's formula (6) with $y = \frac{\pi}{2}$, we have $e^{i\pi/2} = \cos\frac{\pi}{2} + i\sin\frac{\pi}{2} = 0 + 1i = i$.

42. Using Euler's formula (6) with $y = 2\pi$, we have $e^{2\pi i} = \cos 2\pi + i\sin 2\pi = 1$.

43. Using Euler's formula (6) with $y = \dfrac{\pi}{3}$, we have $e^{i\pi/3} = \cos\dfrac{\pi}{3} + i\sin\dfrac{\pi}{3} = \dfrac{1}{2} + \dfrac{\sqrt{3}}{2}i$.

44. Using Euler's formula (6) with $y = -\pi$, we have $e^{-i\pi} = \cos(-\pi) + i\sin(-\pi) = -1$.

45. Using Equation 7 with $x = 2$ and $y = \pi$, we have $e^{2+i\pi} = e^2 e^{i\pi} = e^2(\cos\pi + i\sin\pi) = e^2(-1 + 0) = -e^2$.

46. Using Equation 7 with $x = \pi$ and $y = 1$, we have $e^{\pi+i} = e^\pi \cdot e^{1i} = e^\pi(\cos 1 + i\sin 1) = e^\pi\cos 1 + (e^\pi\sin 1)i$.

47. Take $r = 1$ and $n = 3$ in De Moivre's Theorem to get

$$\begin{aligned}
[1(\cos\theta + i\sin\theta)]^3 &= 1^3(\cos 3\theta + i\sin 3\theta)\\
(\cos\theta + i\sin\theta)^3 &= \cos 3\theta + i\sin 3\theta\\
\cos^3\theta + 3(\cos^2\theta)(i\sin\theta) + 3(\cos\theta)(i\sin\theta)^2 + (i\sin\theta)^3 &= \cos 3\theta + i\sin 3\theta\\
\cos^3\theta + (3\cos^2\theta\,\sin\theta)i - 3\cos\theta\,\sin^2\theta - (\sin^3\theta)i &= \cos 3\theta + i\sin 3\theta\\
(\cos^3\theta - 3\sin^2\theta\,\cos\theta) + (3\sin\theta\,\cos^2\theta - \sin^3\theta)i &= \cos 3\theta + i\sin 3\theta
\end{aligned}$$

Equating real and imaginary parts gives $\cos 3\theta = \cos^3\theta - 3\sin^2\theta\,\cos\theta$ and $\sin 3\theta = 3\sin\theta\,\cos^2\theta - \sin^3\theta$.

48. Using Formula 6,

$$e^{ix} + e^{-ix} = (\cos x + i\sin x) + [\cos(-x) + i\sin(-x)] = \cos x + i\sin x + \cos x - i\sin x = 2\cos x$$

Thus, $\cos x = \dfrac{e^{ix} + e^{-ix}}{2}$. Similarly,

$$e^{ix} - e^{-ix} = (\cos x + i\sin x) - [\cos(-x) + i\sin(-x)] = \cos x + i\sin x - \cos x - (-i\sin x) = 2i\sin x$$

Therefore, $\sin x = \dfrac{e^{ix} - e^{-ix}}{2i}$.

49. $F(x) = e^{rx} = e^{(a+bi)x} = e^{ax+bxi} = e^{ax}(\cos bx + i\sin bx) = e^{ax}\cos bx + i(e^{ax}\sin bx) \;\Rightarrow$

$$\begin{aligned} F'(x) &= (e^{ax}\cos bx)' + i(e^{ax}\sin bx)' \\ &= (ae^{ax}\cos bx - be^{ax}\sin bx) + i(ae^{ax}\sin bx + be^{ax}\cos bx) \\ &= a[e^{ax}(\cos bx + i\sin bx)] + b[e^{ax}(-\sin bx + i\cos bx)] \\ &= ae^{rx} + b[e^{ax}(i^2\sin bx + i\cos bx)] \\ &= ae^{rx} + bi[e^{ax}(\cos bx + i\sin bx)] = ae^{rx} + bie^{rx} = (a+bi)e^{rx} = re^{rx} \end{aligned}$$

50. (a) From Exercise 49, $F(x) = e^{(1+i)x} \;\Rightarrow\; F'(x) = (1+i)e^{(1+i)x}$. So

$$\int e^{(1+i)x}\,dx = \frac{1}{1+i}\int F'(x)\,dx = \frac{1}{1+i}F(x) + C = \frac{1-i}{2}F(x) + C = \frac{1-i}{2}e^{(1+i)x} + C$$

(b) $\int e^{(1+i)x}\,dx = \int e^x e^{ix}\,dx = \int e^x(\cos x + i\sin x)\,dx = \int e^x\cos x\,dx + i\int e^x\sin x$ **(1)**.

Also,

$$\begin{aligned} \frac{1-i}{2}e^{(1+i)x} &= \tfrac{1}{2}e^{(1+i)x} - \tfrac{1}{2}ie^{(1+i)x} = \tfrac{1}{2}e^{x+ix} - \tfrac{1}{2}ie^{x+ix} \\ &= \tfrac{1}{2}e^x(\cos x + i\sin x) - \tfrac{1}{2}ie^x(\cos x + i\sin x) \\ &= \tfrac{1}{2}e^x\cos x + \tfrac{1}{2}e^x\sin x + \tfrac{1}{2}ie^x\sin x - \tfrac{1}{2}ie^x\cos x \\ &= \tfrac{1}{2}e^x(\cos x + \sin x) + i\left[\tfrac{1}{2}e^x(\sin x - \cos x)\right] \quad \textbf{(2)} \end{aligned}$$

Equating the real and imaginary parts in **(1)** and **(2)**, we see that $\int e^x\cos x\,dx = \frac{1}{2}e^x(\cos x + \sin x) + C$ and $\int e^x\sin x\,dx = \frac{1}{2}e^x(\sin x - \cos x) + C$.